Physics
for Scientists and Engineers

TENTH
EDITION

Raymond A. Serway
Emeritus, James Madison University

John W. Jewett, Jr.
Emeritus, California State
Polytechnic University, Pomona

With contributions from Vahé Peroomian
University of Southern California

About the Cover
The cover shows a six-propeller drone carrying a pilot
cable almost 5 kilometers across the deep canyon through
which the Dadu River flows during the Xingkang Bridge
construction project in the Sichuan Province in China. This
method avoids the requirement to use boats on the fast-
flowing river or other methods such as manned helicopters
and small rockets. It also cuts the costs for laying the
cable to about 20% of that of traditional methods. Once
the pilot cable is laid, it can be used to pull heavier cables
across the gorge.

CENGAGE

Australia • Brazil • Mexico • Singapore • United Kingdom • United States

Physics for Scientists and Engineers,
Tenth Edition

Raymond A. Serway, John W. Jewett, Jr

Product Director: Dawn Giovanniello

Product Manager: Rebecca Berardy Schwartz

Content Developer: Ed Dodd

Product Assistant: Caitlyn Ghegan

Media Developer: Sheila Moran

Marketing Manager: Tom Ziolkowski

Content Project Manager: Tanya Nigh

Production Service: MPS Limited

Photo/Text Researcher: LDI

Art Director: Cate Barr

Cover/Text Designer: Shawn Girsberger

Cover and Title Page Image: Zhang Jian/Chengdu Economic Daily/VCG/Getty Images

Compositor: MPS Limited

For product information and technology assistance, contact us at
Cengage Customer & Sales Support, 1-800-354-9706.

For permission to use material from this text or product, submit all requests online at **www.cengage.com/permissions.**
Further permissions questions can be e-mailed to
permissionrequest@cengage.com.

Library of Congress Control Number: 2017953590

Student Edition:
ISBN: 978-1-337-55327-8

Loose-leaf Edition:
ISBN: 978-1-337-55344-5

Cengage
20 Channel Center Street
Boston, MA 02210
USA

Cengage is a leading provider of customized learning solutions with employees residing in nearly 40 different countries and sales in more than 125 countries around the world. Find your local representative at **www.cengage.com.**

Cengage products are represented in Canada by Nelson Education, Ltd.

To learn more about Cengage platforms and services, visit **www.cengage.com.** To register or access your online learning solution or purchase materials for your course, visit **www.cengagebrain.com.**

Printed in the United States of America
Print Number: 01   Print Year: 2017

We dedicate this book to our wives,

Elizabeth and Lisa,

and all our children and grandchildren

for their loving understanding

when we spent time on writing instead of being with them.

Brief Contents

Contents

About the Authors

Raymond A. Serway received his doctorate at Illinois Institute of Technology and is Professor Emeritus at James Madison University. In 2011, he was awarded with an honorary doctorate degree from his alma mater, Utica College. He received the 1990 Madison Scholar Award at James Madison University, where he taught for 17 years. Dr. Serway began his teaching career at Clarkson University, where he conducted research and taught from 1967 to 1980. He was the recipient of the Distinguished Teaching Award at Clarkson University in 1977 and the Alumni Achievement Award from Utica College in 1985. As Guest Scientist at the IBM Research Laboratory in Zurich, Switzerland, he worked with K. Alex Müller, 1987 Nobel Prize recipient. Dr. Serway also was a visiting scientist at Argonne National Laboratory, where he collaborated with his mentor and friend, the late Dr. Sam Marshall. Dr. Serway is the coauthor of *College Physics,* Eleventh Edition; *Principles of Physics,* Fifth Edition; *Essentials of College Physics*; *Modern Physics,* Third Edition; and the high school textbook *Physics,* published by Holt McDougal. In addition, Dr. Serway has published more than 40 research papers in the field of condensed matter physics and has given more than 60 presentations at professional meetings. Dr. Serway and his wife, Elizabeth, enjoy traveling, playing golf, fishing, gardening, singing in the church choir, and especially spending quality time with their four children, ten grandchildren, and a recent great grandson.

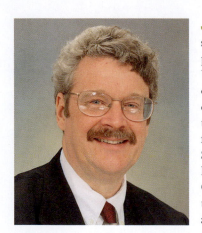

John W. Jewett, Jr. earned his undergraduate degree in physics at Drexel University and his doctorate at Ohio State University, specializing in optical and magnetic properties of condensed matter. Dr. Jewett began his academic career at Stockton University, where he taught from 1974 to 1984. He is currently Emeritus Professor of Physics at California State Polytechnic University, Pomona. Through his teaching career, Dr. Jewett has been active in promoting effective physics education. In addition to receiving four National Science Foundation grants in physics education, he helped found and direct the Southern California Area Modern Physics Institute (SCAMPI) and Science IMPACT (Institute for Modern Pedagogy and Creative Teaching). Dr. Jewett's honors include the Stockton Merit Award at Stockton University in 1980, selection as Outstanding Professor at California State Polytechnic University for 1991–1992, and the Excellence in Undergraduate Physics Teaching Award from the American Association of Physics Teachers (AAPT) in 1998. In 2010, he received an Alumni Lifetime Achievement Award from Drexel University in recognition of his contributions in physics education. He has given more than 100 presentations both domestically and abroad, including multiple presentations at national meetings of the AAPT. He has also published 25 research papers in condensed matter physics and physics education research. Dr. Jewett is the author of *The World of Physics: Mysteries, Magic, and Myth,* which provides many connections between physics and everyday experiences. In addition to his work as the coauthor for *Physics for Scientists and Engineers,* he is also the coauthor on *Principles of Physics,* Fifth Edition, as well as *Global Issues,* a four-volume set of instruction manuals in integrated science for high school. Dr. Jewett enjoys playing keyboard with his all-physicist band, traveling, underwater photography, learning foreign languages, and collecting antique quack medical devices that can be used as demonstration apparatus in physics lectures. Most importantly, he relishes spending time with his wife, Lisa, and their children and grandchildren.

Preface

n writing this Tenth Edition of *Physics for Scientists and Engineers,* we continue our ongoing efforts to improve the clarity of presentation and include new pedagogical features that help support the learning and teaching processes. Drawing on positive feedback from users of the Ninth Edition, data gathered from both professors and students who use WebAssign, as well as reviewers' suggestions, we have refined the text to better meet the needs of students and teachers.

This textbook is intended for a course in introductory physics for students majoring in science or engineering. The entire contents of the book in its extended version could be covered in a three-semester course, but it is possible to use the material in shorter sequences with the omission of selected chapters and sections. The mathematical background of the student taking this course should ideally include one semester of calculus. If that is not possible, the student should be enrolled in a concurrent course in introductory calculus.

Content

The material in this book covers fundamental topics in classical physics and provides an introduction to modern physics. The book is divided into six parts. Part 1 (Chapters 1 to 14) deals with the fundamentals of Newtonian mechanics and the physics of fluids; Part 2 (Chapters 15 to 17) covers oscillations, mechanical waves, and sound; Part 3 (Chapters 18 to 21) addresses heat and thermodynamics; Part 4 (Chapters 22 to 33) treats electricity and magnetism; Part 5 (Chapters 34 to 37) covers light and optics; and Part 6 (Chapters 38 to 44) deals with relativity and modern physics.

Objectives

This introductory physics textbook has three main objectives: to provide the student with a clear and logical presentation of the basic concepts and principles of physics, to strengthen an understanding of the concepts and principles through a broad range of interesting real-world applications, and to develop strong problem-solving skills through an effectively organized approach. To meet these objectives, we emphasize well-organized physical arguments and a focused problem-solving strategy. At the same time, we attempt to motivate the student through practical examples that demonstrate the role of physics in other disciplines, including engineering, chemistry, and medicine.

An Integrative Approach to Course Materials

This new edition takes an *integrative approach* to course material with an optimized, protected, online-only problem experience combined with rich textbook content designed to support an active classroom experience. This new optimized online homework set is built on contextual randomizations and answer-dependent student remediation for every problem. With this edition, you'll have an integrative approach that seamlessly matches curated content to the learning environment for which it was intended—from in-class group problem solving to online homework that utilizes targeted feedback. This approach engages and guides students where they are at—whether they are studying online or with the textbook.

Students often approach an online homework problem by googling to find the right equation or explanation of the relevant concept; however, this approach has eroded the value attributed to online homework as students leave the support of the program for unrelated help elsewhere and encounter imprecise information.

Students don't need to leave WebAssign to get help when they are stuck—each problem has feedback that addresses the misconception or error a student made to reach the wrong answer. Each optimized problem also features comprehensive written solutions, and many have supporting video solutions that go through one contextual variant of the problem one step at a time. Since the optimized problem set is not in print, the content is protected from "solution providers" and will be augmented every year with updates to the targeted feedback based on actual student answers.

Working in tandem with the optimized online homework, the printed textbook has been designed for an active learning experience that supports activities in the classroom as well as after-class practice and review. New content includes *Think–Pair–Share* activities, context-rich problems, and a greater emphasis on symbolic and conceptual problems. *All* of the printed textbook's problems will also be available to assign in WebAssign.

Changes in the Tenth Edition

A large number of changes and improvements were made for the Tenth Edition of this text. Some of the new features are based on our experiences and on current trends in science education. Other changes were incorporated in response to comments and suggestions offered by users of the Ninth Edition and by reviewers of the manuscript. The features listed here represent the major changes in the Tenth Edition.

WebAssign for *Physics for Scientists and Engineers*

WebAssign is a flexible and fully customizable online instructional solution that puts powerful tools in the hands of instructors, enabling you deploy assignments, instantly assess individual student and class performance, and help your students master the course concepts. With WebAssign's powerful digital platform and content specific to *Physics for Scientists and Engineers,* you can tailor your course with a wide range of assignment settings, add your own questions and content, and access student and course analytics and communication tools. WebAssign for *Physics for Scientists and Engineers* includes the following new features for this edition.

Optimized Problems. Only available online via WebAssign, this problem set combines new assessments with classic problems from *Physics for Scientists and Engineers* that have been optimized with just-in-time targeted feedback tailored to student responses and full student-focused solutions. Moving these problems so that they are only available online allows instructors to make full use of the capability of WebAssign to provide their students with dynamic assessment content, and reduces the opportunity for students to find online solutions through anti-search-engine optimizations. These problems reduce these opportunities both by making the text of the problem less searchable and by providing immediate assistance to students within the homework platform.

Interactive Video Vignettes (IVV) encourage students to address their alternate conceptions outside of the classroom and can be used for pre-lecture activities in traditional or even workshop physics classrooms. Interactive Video Vignettes include online video analysis and interactive individual tutorials to address learning difficulties identified by PER (Physics Education Research). Within the WebAssign platform there are additional conceptual questions immediately following each IVV in order to evaluate student engagement with the material and reinforce the message around these classic misconceptions. A screen shot from one of the Interactive Video Vignettes appears on the next page:

New MCAT-Style Passage Problem Modules. Available only in WebAssign, these 30 brand-new modules are modeled after the new MCAT exam's "passage problems." Each module starts with a text passage (often with accompanying photos/figures) followed by 5–6 multiple-choice questions. The passage and the questions are usually not confined to a single chapter, and feedback is available with each question.

New Life Science Problems. The online-only problems set for each chapter in WebAssign features two new life science problems that highlight the relevance of physics principles to those students taking the course who are majoring in one of the life sciences.

New What If? Problem Extensions. The online-only problems set for each chapter in WebAssign contains 6 new **What If? extensions** to existing problems. What If? extensions extend students' understanding of physics concepts beyond the simple act of arriving at a numerical result.

Pre-Lecture Explorations combine interactive simulations with conceptual and analytical questions that guide students to a deeper understanding and help promote a robust physical intuition.

An Expanded Offering of All-New Integrated Tutorials. These Integrated Tutorials strengthen students' problem-solving skills by guiding them through the steps in the book's problem-solving process, and include meaningful feedback at each step so students can practice the problem-solving process and improve their skills. The feedback also addresses student preconceptions and helps them to catch algebraic and other mathematical errors. Solutions are carried out symbolically as long as possible, with numerical values substituted at the end. This feature promotes conceptual understanding above memorization, helps students understand the effects of changing the values of each variable in the problem, avoids unnecessary repetitive substitution of the same numbers, and eliminates round-off errors.

Increased Number of Fully Worked-Out Problem Solutions. Hundreds of solutions have been newly added to online end-of-chapter problems. Solutions step through problem-solving strategies as they are applied to specific problems.

Objective and Conceptual Questions Now Exclusively Available in WebAssign. **Objective Questions** are multiple-choice, true/false, ranking, or other multiple-guess-type questions. Some require calculations designed to facilitate students' familiarity with the equations, the variables used, the concepts the variables represent, and the relationships between the concepts. Others are more conceptual in nature and designed to encourage conceptual thinking. Objective Questions are also written with the personal response system user in mind, and most of the questions could easily be used in these systems. **Conceptual Questions** are more traditional short-answer and essay-type questions that require students to think conceptually about a physical situation. More than 900 Objective and Conceptual Questions are available in WebAssign.

New Physics for Scientists and Engineers WebAssign Implementation Guide. The Implementation Guide provides instructors with occurrences of the different assignable problems, tutorials, questions, and activities that are available with each chapter of *Physics for Scientists and Engineers* in WebAssign. Instructors can use this manual when making decisions about which and how many assessment items to assign. To facilitate this, an overview of how the assignable items are integrated into the course is included.

New Assessment Items

New Context-Rich Problems. Context-rich problems (identified with a **CR** icon) always discuss "you" as the individual in the problem and have a real-world connection instead of discussing blocks on planes or balls on strings. They are structured like a short story and may not always explicitly identify the variable that needs to be evaluated. Context-rich problems may relate to the opening storyline of the chapter, might involve "expert witness" scenarios, which allow students to go beyond mathematical manipulation by designing an argument based on mathematical results, or ask for decisions to be made in real situations. Selected new context-rich problems will only appear online in WebAssign. An example of a new context-rich problem appears below:

> **20.** **CR** There is a 5K event coming up in your town. While talking to your grandmother, who uses an electric scooter for mobility, she says that she would like to accompany you on her scooter while you walk the 5.00-km distance. The manual that came with her scooter claims that the fully charged battery is capable of providing 120 Wh of energy before being depleted. In preparation for the race, you go for a "test drive": beginning with a fully charged battery, your grandmother rides beside you as you walk 5.00 km on flat ground. At the end of the walk, the battery usage indicator shows that 40.0% of the original energy in the battery remains. You also know that the combined weight of the scooter and your grandmother is 890 N. A few days later, filled with confidence that the battery has sufficient energy, you and your grandmother drive to the 5K event. Unbeknownst to you, the 5K route is not on flat ground, but is all uphill, ending at a point higher than the starting line. A race official tells you that the total amount of vertical displacement on the route is 150 m. Should your grandmother accompany you on the walk, or will she be stranded when her battery runs out of energy? Assume that the only difference between your test drive and the actual event is the vertical displacement.

New Think–Pair–Share Problems and Activities. Think–Pair–Share problems and activities are similar to context-rich problems, but tend to benefit more from group discussion because the solution is not as straightforward as for a single-concept problem. Some Think–Pair–Share problems require the group to discuss and make decisions; others are made more challenging by the fact that some information is not and cannot be known. All chapters in the text have at least one Think–Pair–Share problem or activity; several more per chapter will be available only in WebAssign. Examples of a Think–Pair–Share Problem and a Think–Pair–Share Activity appear on the next page:

1. You are working as a delivery person for a dairy store. In the back of your pickup truck is a crate of eggs. The dairy company has run out of bungee cords, so the crate is not tied down. You have been told to drive carefully because the coefficient of static friction between the crate and the bed of the truck is 0.600. You are not worried, because you are traveling on a road that appears perfectly straight. Due to your confidence and inattention, your speed has crept upward to 45.0 mi/h. Suddenly, you see a curve ahead with a warning sign saying, "Danger: unbanked curve with radius of curvature 35.0 m." You are 15.0 m from the beginning of the curve. What can you do to save the eggs: (i) take the curve at 45.0 mi/h, (ii) brake to a stop before entering the curve to think about it, or (iii) slow down to take the curve at a slower speed? Discuss these options in your group and determine if there is a best course of action.

3. **ACTIVITY** (a) Place ten pennies on a horizontal meterstick, with a penny at 10 cm, 20 cm, 30 cm, etc., out to 100 cm. Carefully pick up the meterstick, keeping it horizontal, and have a member of the group make a video recording of the following event, using a smartphone or other device. While the video recording is underway, release the 100-cm end of the meterstick while the 0-cm end rests on someone's finger or the edge of the desk. By stepping through the video images or watching the video in slow motion, determine which pennies first lose contact with the meterstick as it falls. (b) Make a theoretical determination of which pennies should first lose contact and compare to your experimental result.

Content Changes

Reorganized Chapter 16 (Wave Motion). This combination of Chapters 16 and 17 from the last edition brings all of the fundamental material on traveling mechanical waves on strings and sound waves through materials together in one chapter. This allows for more close comparisons between the features of the two types of waves that are similar, such as derivations of the speed of the wave. The section on reflection and transmission of waves, details of which are not necessary in a chapter on traveling waves, was moved into Chapter 17 (Superposition and Standing Waves) for this edition, where it fits more naturally in a discussion of the effects of boundary conditions on waves.

Reorganization of Chapters 22–24. Movement of the material on continuous distribution of charge out of Chapter 22 (Electric Fields) to Chapter 23 (Continuous Charge Distributions and Gauss's Law) results in a chapter that is a more gradual introduction for students into the new and challenging topic of electricity. The chapter now involves only electric fields due to point charges and uniform electric fields due to parallel plates.

Chapter 23 previously involved only the analysis of electric fields due to continuous charge distributions using Gauss's law. Movement of the material on continuous distribution of charge into Chapter 23 results in an entire chapter based on the analysis of fields from continuous charge distributions, using two techniques: integration and Gauss's law.

Chapter 23 previously contained a discussion of four properties of isolated charged conductors. Three of the properties were discussed and argued from basic principles, while the student was referred to necessary material in the next chapter (on Electric Potential) for a discussion of the fourth property. With the movement of this discussion into Chapter 24 for this edition, the student has learned all of the necessary basic material *before* the discussion of properties of isolated charged conductors, and all four properties can be argued from basic principles together.

Reorganized Chapter 43 (Nuclear Physics). Chapters 44 (Nuclear Structure) and 45 (Applications of Nuclear Physics) in the last edition have been combined in this edition. This new Chapter 43 allows all of the material on nuclear physics to be studied together. As a consequence, we now have a series of the final five chapters of the text that each cover in one chapter focused applications of the fundamental principles studied before: Chapter 40 (Quantum Mechanics), Chapter 41 (Atomic Physics), Chapter 42 (Molecules and Solids), Chapter 43 (Nuclear Physics), and Chapter 44 (Particle Physics).

New Storyline Approach to Chapter-Opening Text. Each chapter opens with a *Storyline* section. This feature provides a continuous storyline through the whole book of "you" as an inquisitive physics student observing and analyzing phenomena seen in everyday life. Many chapters' Storyline involves measurements made with a smartphone, observations of YouTube videos, or investigations on the Internet.

New Chapter-Opening **Connections.** The start of each chapter also features a *Connections* section that shows how the material in the chapter connects to previously studied material and to future material. The Connections section provides a "big picture" of the concepts, explains why this chapter is placed in this particular location relative to the other chapters, and shows how the structure of physics builds on previous material.

Text Features

Most instructors believe that the textbook selected for a course should be the student's primary guide for understanding and learning the subject matter. Furthermore, the textbook should be easily accessible and should be styled and written to facilitate instruction and learning. With these points in mind, we have included many pedagogical features, listed below, that are intended to enhance its usefulness to both students and instructors.

Problem Solving and Conceptual Understanding

Analysis Model Approach to Problem Solving. Students are faced with hundreds of problems during their physics courses. A relatively small number of fundamental principles form the basis of these problems. When faced with a new problem, a physicist forms a model of the problem that can be solved in a simple way by identifying the fundamental principle that is applicable in the problem. For example, many problems involve conservation of energy, Newton's second law, or kinematic equations. Because the physicist has studied these principles and their applications extensively, he or she can apply this knowledge as a model for solving a new problem. Although it would be ideal for students to follow this same process, most students have difficulty becoming familiar with the entire palette of fundamental principles that are available. It is easier for students to identify a situation rather than a fundamental principle.

The *Analysis Model approach* lays out a standard set of situations that appear in most physics problems. These situations are based on an entity in one of four simplification models: *particle, system, rigid object*, and *wave*. Once the simplification model is identified, the student thinks about what the entity is doing or how it interacts with its environment. This leads the student to identify a particular Analysis Model for the problem. For example, if an object is falling, the object is recognized as a particle experiencing an acceleration due to gravity that is constant. The student has learned that the Analysis Model of a *particle under constant acceleration* describes this situation. Furthermore, this model has a small number of equations associated with it for use in starting problems, the kinematic equations presented in Chapter 2. Therefore, an understanding of the situation has led to an Analysis Model, which then identifies a very small number of equations to start the problem, rather than the myriad equations that students see in the text. In this way, the use of Analysis Models leads the student to identify the fundamental principle. As the student gains more experience, he or she will lean less on the Analysis Model approach and begin to identify fundamental principles directly.

The Analysis Model Approach to Problem Solving is presented in full in Chapter 2 (Section 2.4, pages 30–32), and provides students with a structured process for solving problems. In all remaining chapters, the strategy is employed explicitly in every example so that students learn how it is applied. Students are encouraged to follow this strategy when working end-of-chapter problems.

Analysis Model descriptive boxes appear at the end of any section that introduces a new Analysis Model. This feature recaps the Analysis Model introduced in the section and provides examples of the types of problems that a student could solve using the Analysis Model. These boxes function as a "refresher" before students see the Analysis Models in use in the worked examples for a given section. The approach is further reinforced in the end-of-chapter summary under the heading *Analysis Models for Problem Solving*, and through the **Analysis Model Tutorials** that are based on selected end-of-chapter problems and appear in WebAssign.

Analysis Model Tutorials. John Jewett developed 165 tutorials (ones that appear in the printed text's problem sets are indicated by an AMT icon) that strengthen students' problem-solving skills by guiding them through the steps in the problem-solving process. Important first steps include making predictions and focusing on physics concepts before solving the problem quantitatively. A critical component of these tutorials is the selection of an appropriate Analysis Model to describe what is going on in the problem. This step allows students to make the important link between the situation in the problem and the mathematical representation of the situation. Analysis Model tutorials include meaningful feedback at each step to help students practice the problem-solving process and improve their skills. In addition, the feedback addresses student misconceptions and helps them to catch algebraic and other mathematical errors. Solutions are carried out symbolically as long as possible, with numerical values substituted at the end. This feature helps students understand the effects of changing the values of each variable in the problem, avoids unnecessary repetitive substitution of the same numbers, and eliminates round-off errors. Feedback at the end of the tutorial encourages students to compare the final answer with their original predictions.

Worked Examples. All in-text worked examples are presented in a two-column format to better reinforce physical concepts. The left column shows textual information that describes the steps for solving the problem. The right column shows the mathematical manipulations and results of taking these steps. This layout facilitates matching the concept with its mathematical execution and helps students organize their work. The examples closely follow the Analysis Model Approach to Problem Solving introduced in Section 2.4 to reinforce effective problem-solving habits. All worked examples in the text may be assigned for homework in WebAssign. A sample of a worked example can be found on the next page.

Examples consist of two types. The first (and most common) example type presents a problem and numerical answer. The second type of example is conceptual in nature. To accommodate increased emphasis on understanding physical concepts, the many conceptual examples are labeled as such and are designed to help students focus on the physical situation in the problem. Solutions in worked examples are presented symbolically as far as possible, with numerical values substituted at the end. This approach will help students think symbolically when they solve problems instead of unnecessarily inserting numbers into intermediate equations.

What If? Approximately one-third of the worked examples in the text contain a What If? feature. At the completion of the example solution, a What If? question offers a variation on the situation posed in the text of the example. This feature encourages students to think about the results of the example, and it also assists in conceptual understanding of the principles. What If? questions also prepare students to encounter novel problems that may be included on exams. Selected end-of-chapter problems also include this feature.

Quick Quizzes. Students are provided an opportunity to test their understanding of the physical concepts presented through Quick Quizzes. The questions require students to make decisions on the basis of sound reasoning, and some of the questions have been written to help students overcome common misconceptions. Quick Quizzes have been cast in an objective format, including multiple-choice, true–false, and ranking. Answers to all Quick Quiz questions are found at the end of the text. Many instructors choose to use such questions in a "peer instruction" teaching style or with the use of personal response system "clickers," but they can be used in standard quiz format as well. An example of a Quick Quiz follows below.

QUICK QUIZ 7.5 A dart is inserted into a spring-loaded dart gun by pushing the spring in by a distance x. For the next loading, the spring is compressed a distance $2x$. How much faster does the second dart leave the gun compared with the first? **(a)** four times as fast **(b)** two times as fast **(c)** the same **(d)** half as fast **(e)** one-fourth as fast

All worked examples are also available to be assigned as interactive examples in WebAssign.

Each solution has been written to closely follow the Analysis Model Approach to Problem Solving as outlined in Section 2.4 (pages 30–32), so as to reinforce good problem-solving habits.

Each step of the solution is detailed in a two-column format. The left column provides an explanation for each mathematical step in the right column, to better reinforce the physical concepts.

Example 3.2   A Vacation Trip

A car travels 20.0 km due north and then 35.0 km in a direction 60.0° west of north as shown in Figure 3.11a. Find the magnitude and direction of the car's resultant displacement.

SOLUTION

Conceptualize The two vectors $\vec{A}$ and $\vec{B}$ that appear in Figure 3.11a help us conceptualize the problem. The resultant vector $\vec{R}$ has also been drawn. We expect its magnitude to be a few tens of kilometers. The angle β that the resultant vector makes with the y axis is expected to be less than 60°, the angle that vector $\vec{B}$ makes with the y axis.

Figure 3.11 (Example 3.2) (a) Graphical method for finding the resultant displacement vector $\vec{R} = \vec{A} + \vec{B}$. (b) Adding the vectors in reverse order $(\vec{B} + \vec{A})$ gives the same result for $\vec{R}$.

Categorize We can categorize this example as a simple analysis problem in vector addition. The displacement $\vec{R}$ is the resultant when the two individual displacements $\vec{A}$ and $\vec{B}$ are added. We can further categorize it as a problem about the analysis of triangles, so we appeal to our expertise in geometry and trigonometry.

Analyze In this example, we show two ways to analyze the problem of finding the resultant of two vectors. The first way is to solve the problem geometrically, using graph paper and a protractor to measure the magnitude of $\vec{R}$ and its direction in Figure 3.11a. (In fact, even when you know you are going to be carrying out a calculation, you should sketch the vectors to check your results.) With an ordinary ruler and protractor, a large diagram typically gives answers to two-digit but not to three-digit precision. Try using these tools on $\vec{R}$ in Figure 3.11a and compare to the trigonometric analysis below!

The second way to solve the problem is to analyze it using algebra and trigonometry. The magnitude of $\vec{R}$ can be obtained from the law of cosines as applied to the triangle in Figure 3.11a (see Appendix B.4).

Use $R^2 = A^2 + B^2 - 2AB \cos \theta$ from the law of cosines to find R:	$R = \sqrt{A^2 + B^2 - 2AB \cos \theta}$
Substitute numerical values, noting that $\theta = 180° - 60° = 120°$:	$R = \sqrt{(20.0 \text{ km})^2 + (35.0 \text{ km})^2 - 2(20.0 \text{ km})(35.0 \text{ km}) \cos 120°}$ $= 48.2 \text{ km}$
Use the law of sines (Appendix B.4) to find the direction of $\vec{R}$ measured from the northerly direction:	$\dfrac{\sin \beta}{B} = \dfrac{\sin \theta}{R}$ $\sin \beta = \dfrac{B}{R} \sin \theta = \dfrac{35.0 \text{ km}}{48.2 \text{ km}} \sin 120° = 0.629$ $\beta = 38.9°$

The resultant displacement of the car is 48.2 km in a direction 38.9° west of north.

Finalize Does the angle β that we calculated agree with an estimate made by looking at Figure 3.11a or with an actual angle measured from the diagram using the graphical method? Is it reasonable that the magnitude of $\vec{R}$ is larger than that of both $\vec{A}$ and $\vec{B}$? Are the units of $\vec{R}$ correct?

Although the head to tail method of adding vectors works well, it suffers from two disadvantages. First, some people find using the laws of cosines and sines to be awkward. Second, a triangle only results if you are adding two vectors. If you are adding three or more vectors, the resulting geometric shape is usually not a triangle. In Section 3.4, we explore a new method of adding vectors that will address both of these disadvantages.

WHAT IF? Suppose the trip were taken with the two vectors in reverse order: 35.0 km at 60.0° west of north first and then 20.0 km due north. How would the magnitude and the direction of the resultant vector change?

Answer They would not change. The commutative law for vector addition tells us that the order of vectors in an addition is irrelevant. Graphically, Figure 3.11b shows that the vectors added in the reverse order give us the same resultant vector.

What If? statements appear in about one-third of the worked examples and offer a variation on the situation posed in the text of the example. For instance, this feature might explore the effects of changing the conditions of the situation, determine what happens when a quantity is taken to a particular limiting value, or question whether additional information can be determined about the problem situation. This feature encourages students to think about the results of the example and assists in conceptual understanding of the principles.

Pitfall Preventions. More than two hundred Pitfall Preventions (such as the one to the right) are provided to help students avoid common mistakes and misunderstandings. These features, which are placed in the margins of the text, address both common student misconceptions and situations in which students often follow unproductive paths.

Summaries. Each chapter contains a summary that reviews the important concepts and equations discussed in that chapter. The summary is divided into three sections: Definitions, Concepts and Principles, and Analysis Models for Problem Solving. In each section, flash card–type boxes focus on each separate definition, concept, principle, or analysis model.

Problems Sets. For the Tenth Edition, the authors reviewed each question and problem and incorporated revisions designed to improve both readability and assignability.

Problems. An extensive set of problems is included at the end of each chapter; in all, the printed textbook contains more than 2 000 problems, while another 1 500 optimized problems are available only in WebAssign. Answers for odd-numbered problems in the printed text are provided at the end of the book, and solutions for all printed text problems are found in the *Instructor's Solutions Manual.*

The end-of-chapter problems are organized by the sections in each chapter (about two-thirds of the problems are keyed to specific sections of the chapter). Within each section, the problems now "platform" students to higher-order thinking by presenting all the straightforward problems in the section first, followed by the intermediate problems. (The problem numbers for straightforward problems are printed in **black**; intermediate-level problems are in **blue.**) The *Additional Problems* section contains problems that are not keyed to specific sections. At the end of each chapter is the *Challenge Problems* section, which gathers the most difficult problems for a given chapter in one place. (Challenge Problems have problem numbers marked in **red.**)

There are several kinds of problems featured in this text:

V *Watch It* video solutions available in WebAssign explain fundamental problem-solving strategies to help students step through selected problems.

Q|C *Quantitative/Conceptual problems* contain parts that ask students to think both quantitatively and conceptually. An example of a Quantitative/Conceptual problem appears here:

> The problem is identified with a **Q|C** icon.

35. A horizontal spring attached to a wall has a force constant
Q|C of $k = 850$ N/m. A block of mass $m = 1.00$ kg is attached to the spring and rests on a frictionless, horizontal surface as in Figure P8.35. (a) The block is pulled to a position $x_i = 6.00$ cm from equilibrium and released. Find the elastic potential energy stored in the spring when the block is 6.00 cm from equilibrium and when the block passes through equilibrium. (b) Find the speed of the block as it passes through the equilibrium point. (c) What is the speed of the block when it is at a position $x_i/2 = 3.00$ cm? (d) Why isn't the answer to part (c) half the answer to part (b)?

> Parts (a)–(c) of the problem ask for quantitative calculations.

Figure P8.35

> Part (d) asks a conceptual question about the situation.

S *Symbolic problems* ask students to solve a problem using only symbolic manipulation. Reviewers of the Ninth Edition (as well as the majority of respondents to a large survey) asked specifically for an increase in the number of symbolic problems found in the text because it better reflects the way instructors want their students to think when solving physics problems. An example of a Symbolic problem appears on the next page:

PITFALL PREVENTION 16.2
Two Kinds of Speed/Velocity
Do not confuse v, the speed of the wave as it propagates along the string, with v_y, the transverse velocity of a point on the string. The speed v is constant for a uniform medium, whereas v_y varies sinusoidally.

The problem is identified with a **S** icon.

No numbers appear in the problem statement.

36. A truck is moving with constant acceleration *a* up a hill that makes an angle ϕ with the horizontal as in Figure P6.36. A small sphere of mass *m* is suspended from the ceiling of the truck by a light cord. If the pendulum makes a constant angle θ with the perpendicular to the ceiling, what is *a*?

Figure P6.36

The figure shows only symbolic quantities.

36. $g(\cos \phi \tan \theta - \sin \phi)$

The answer to the problem is purely symbolic.

GP *Guided Problems* help students break problems into steps. A physics problem typically asks for one physical quantity in a given context. Often, however, several concepts must be used and a number of calculations are required to obtain that final answer. Many students are not accustomed to this level of complexity and often don't know where to start. A Guided Problem breaks a standard problem into smaller steps, enabling students to grasp all the concepts and strategies required to arrive at a correct solution. Unlike standard physics problems, guidance is often built into the problem statement. Guided Problems are reminiscent of how a student might interact with a professor in an office visit. These problems (there is one in every chapter of the text) help train students to break down complex problems into a series of simpler problems, an essential problem-solving skill. An example of a Guided Problem appears here:

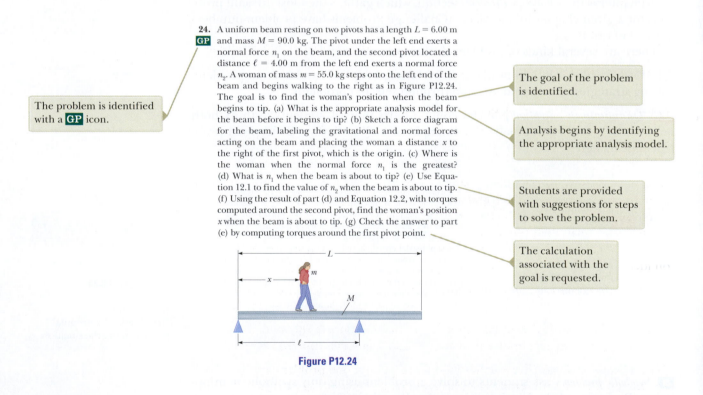

24. A uniform beam resting on two pivots has a length $L = 6.00$ m and mass $M = 90.0$ kg. The pivot under the left end exerts a normal force n_1 on the beam, and the second pivot located a distance $\ell = 4.00$ m from the left end exerts a normal force n_2. A woman of mass $m = 55.0$ kg steps onto the left end of the beam and begins walking to the right as in Figure P12.24. The goal is to find the woman's position when the beam begins to tip. (a) What is the appropriate analysis model for the beam before it begins to tip? (b) Sketch a force diagram for the beam, labeling the gravitational and normal forces acting on the beam and placing the woman a distance *x* to the right of the first pivot, which is the origin. (c) Where is the woman when the normal force n_1 is the greatest? (d) What is n_1 when the beam is about to tip? (e) Use Equation 12.1 to find the value of n_2 when the beam is about to tip. (f) Using the result of part (d) and Equation 12.2, with torques computed around the second pivot, find the woman's position *x* when the beam is about to tip. (g) Check the answer to part (e) by computing torques around the first pivot point.

The problem is identified with a **GP** icon.

The goal of the problem is identified.

Analysis begins by identifying the appropriate analysis model.

Students are provided with suggestions for steps to solve the problem.

The calculation associated with the goal is requested.

Figure P12.24

Biomedical problems. These problems (indicated with a **BIO** icon) highlight the relevance of physics principles to those students taking this course who are majoring in one of the life sciences.

T *Master It Tutorials* available in WebAssign help students solve problems by having them work through a stepped-out solution.

Impossibility problems. Physics education research has focused heavily on the problem-solving skills of students. Although most problems in this text are structured in the form of providing data and asking for a result of computation, two problems in each chapter, on average, are structured as impossibility problems. They begin with the phrase *Why is the following situation impossible?* That is followed by the description of a situation. The striking aspect of these problems is that no question is asked of the students, other than that in the initial italics. The student must determine what questions need to be asked and what calculations need to be performed. Based on the results of these calculations, the student must determine why the situation described is not possible. This determination may require information from personal experience, common sense, Internet or print research, measurement, mathematical skills, knowledge of human norms, or scientific thinking.

These problems can be assigned to build critical thinking skills in students. They are also fun, having the aspect of physics "mysteries" to be solved by students individually or in groups. An example of an impossibility problem appears here:

The initial phrase in italics signals an impossibility problem.

39. *Why is the following situation impossible?* Albert Pujols hits a home run so that the baseball just clears the top row of bleachers, 24.0 m high, located 130 m from home plate. The ball is hit at 41.7 m/s at an angle of 35.0° to the horizontal, and air resistance is negligible.

A situation is described.

No question is asked. The student must determine what needs to be calculated and why the situation is impossible.

Paired problems. These problems are otherwise identical, one asking for a numerical solution and one asking for a symbolic derivation. There is at least one pair of these problems in most chapters, indicated by cyan shading in the end-of-chapter problems set.

Review problems. Many chapters include review problems requiring the student to combine concepts covered in the chapter with those discussed in previous chapters. These problems (marked **Review**) reflect the cohesive nature of the principles in the text and verify that physics is not a scattered set of ideas. When facing a real-world issue such as global warming or nuclear weapons, it may be necessary to call on ideas in physics from several parts of a textbook such as this one.

"Fermi problems." One or more problems in most chapters ask the student to reason in order-of-magnitude terms.

Design problems. Several chapters contain problems that ask the student to determine design parameters for a practical device so that it can function as required.

Calculus-based problems. Every chapter contains at least one problem applying ideas and methods from differential calculus and one problem using integral calculus.

Artwork. Every piece of artwork in the Tenth Edition is in a modern style that helps express the physics principles at work in a clear and precise fashion. *Focus pointers* are included with many figures in the text; these either point out important aspects of a figure or guide students through a process illustrated by the artwork or photo. This format helps those students who are more visual learners. An example of a figure with a focus pointer appears on the next page.

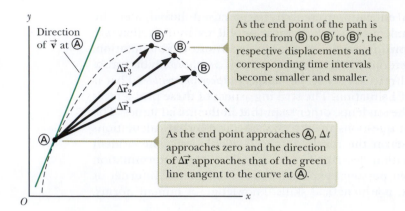

As the end point of the path is moved from Ⓑ to Ⓑ′ to Ⓑ″, the respective displacements and corresponding time intervals become smaller and smaller.

As the end point approaches Ⓐ, Δt approaches zero and the direction of $\Delta \vec{\mathbf{r}}$ approaches that of the green line tangent to the curve at Ⓐ.

Figure 4.2 As a particle moves between two points, its average velocity is in the direction of the displacement vector $\Delta \vec{\mathbf{r}}$. By definition, the instantaneous velocity at Ⓐ is directed along the line tangent to the curve at Ⓐ.

Math Appendix. The math appendix (Appendix B), a valuable tool for students, shows the math tools in a physics context. This resource is ideal for students who need a quick review on topics such as algebra, trigonometry, and calculus.

Helpful Features

Style. To facilitate rapid comprehension, we have written the book in a clear, logical, and engaging style. We have chosen a writing style that is somewhat informal and relaxed so that students will find the text appealing and enjoyable to read. New terms are carefully defined, and we have avoided the use of jargon.

Important Definitions and Equations. Most important definitions are set in **boldface** or are highlighted with a background screen for added emphasis and ease of review. Similarly, important equations are also highlighted with a background screen to facilitate location.

Marginal Notes. Comments and notes appearing in the margin with a ▶ icon can be used to locate important statements, equations, and concepts in the text.

Pedagogical Use of Color. Readers should consult the **pedagogical color chart** (inside the front cover) for a listing of the color-coded symbols used in the text diagrams. This system is followed consistently throughout the text.

Mathematical Level. We have introduced calculus gradually, keeping in mind that students often take introductory courses in calculus and physics concurrently. Most steps are shown when basic equations are developed, and reference is often made to mathematical appendices near the end of the textbook. Although vectors are discussed in detail in Chapter 3, vector products are introduced later in the text, where they are needed in physical applications. The dot product is introduced in Chapter 7, which addresses energy of a system; the cross product is introduced in Chapter 11, which deals with angular momentum.

Significant Figures. In both worked examples and end-of-chapter problems, significant figures have been handled with care. Most numerical examples are worked to either two or three significant figures, depending on the precision of the data provided. End-of-chapter problems regularly state data and answers to three-digit precision. When carrying out estimation calculations, we shall typically work with a single significant figure. (More discussion of significant figures can be found in Chapter 1, pages 13–15.)

Units. The international system of units (SI) is used throughout the text. The U.S. customary system of units is used only to a limited extent in the chapters on mechanics and thermodynamics.

Appendices and Endpapers. Several appendices are provided near the end of the textbook. Most of the appendix material represents a review of mathematical concepts and techniques used in the text, including scientific notation, algebra, geometry, trigonometry, differential calculus, and integral calculus. Reference to these appendices is made throughout the text. Most mathematical review sections in the appendices include worked examples and exercises with answers. In addition to the mathematical reviews, the appendices contain tables of physical data, conversion factors, and the SI units of physical quantities as well as a periodic table of the elements. Other useful information—fundamental constants and physical data, planetary data, a list of standard prefixes, mathematical symbols, the Greek alphabet, and standard abbreviations of units of measure—appears on the endpapers.

Course Solutions That Fit Your Teaching Goals and Your Students' Learning Needs

Recent advances in educational technology have made homework management systems and audience response systems powerful and affordable tools to enhance the way you teach your course. Whether you offer a more traditional text-based course, are interested in using or are currently using an online homework management system such as WebAssign, or are ready to turn your lecture into an interactive learning environment, you can be confident that the text's proven content provides the foundation for each and every component of our technology and ancillary package.

Lecture Presentation Resources

Cengage Learning Testing Powered by Cognero is a flexible, online system that allows you to author, edit, and manage test bank content from multiple Cengage Learning solutions, create multiple test versions in an instant, and deliver tests from your LMS, your classroom, or wherever you want.

Instructor Resource Website for **Serway/Jewett** *Physics for Scientists and Engineers, Tenth Edition.* The Instructor Resource Website contains a variety of resources to aid you in preparing and presenting text material in a manner that meets your personal preferences and course needs. The posted *Instructor's Solutions Manual* presents complete worked solutions for all of the printed textbook's end-of-chapter problems and answers for all even-numbered problems. Robust PowerPoint lecture outlines that have been designed for an active classroom are available, with reading check questions and Think–Pair–Share questions as well as the traditional section-by-section outline. Images from the textbook can be used to customize your own presentations. Available online via www.cengage.com/login.

CengageBrain.com

To register or access your online learning solution or purchase materials for your course, visit **www.cengagebrain.com**.

CENGAGE **brain**.com

Student Resources

Physics Laboratory Manual, Fourth Edition by David Loyd (Angelo State University) Ideal for use with any introductory physics text, Loyd's *Physics Laboratory Manual* is suitable for either calculus- or algebra/trigonometry-based physics courses. Designed to help students demonstrate a physical principle and teach techniques of careful measurement, Loyd's *Physics Laboratory Manual* also emphasizes conceptual understanding and includes a thorough discussion of physical theory to help students see the connection between the lab and the lecture. Many labs give students hands-on experience with statistical analysis, and now five computer-assisted data-entry labs are included in the printed manual. The fourth edition maintains the minimum equipment requirements to allow for maximum flexibility and to make the most of preexisting lab equipment. For instructors interested in using some of Loyd's experiments, a customized lab manual is another option available through the Cengage Learning Custom Solutions program. Now, you can select specific experiments from Loyd's *Physics Laboratory Manual*, include your own original lab experiments, and create one affordable bound book. Contact your Cengage Learning representative for more information on our Custom Solutions program. Available with InfoTrac® Student Collections http://gocengage.com/infotrac.

Physics Laboratory Experiments, Eighth Edition by Jerry D. Wilson (Lander College) and Cecilia A. Hernández (American River College). This market-leading manual for the first-year physics laboratory course offers a wide range of class-tested experiments designed specifically for use in small to midsize lab programs. A series of integrated experiments emphasizes the use of computerized instrumentation and includes a set of "computer-assisted experiments" to allow students and instructors to gain experience with modern equipment. It also lets instructors determine the appropriate balance of traditional versus computer-based experiments for their courses. By analyzing data through two different methods, students gain a greater understanding of the concepts behind the experiments. The Eighth Edition is updated with 4 new economical labs to accommodate shrinking department budgets and 30 new Pre-Lab Demonstrations, designed to capture students' interest prior to the lab and requiring only widely available materials and items.

Teaching Options

The topics in this textbook are presented in the following sequence: classical mechanics, oscillations and mechanical waves, and heat and thermodynamics, followed by electricity and magnetism, electromagnetic waves, optics, relativity, and modern physics. This presentation represents a traditional sequence, with the subject of mechanical waves being presented before electricity and magnetism. Some instructors may prefer to discuss both mechanical and electromagnetic waves together after completing electricity and magnetism. In this case, Chapters 16 and 17 could be covered along with Chapter 33. The chapter on relativity is placed near the end of the text because this topic often is treated as an introduction to the era of "modern physics." If time permits, instructors may choose to cover Chapter 38 after completing Chapter 13 as a conclusion to the material on Newtonian mechanics. For those instructors teaching a two-semester sequence, some sections and chapters could be deleted without any loss of continuity. The following sections can be considered optional for this purpose:

2.9	Kinematic Equations Derived from Calculus	**28.6**	The Hall Effect
4.6	Relative Velocity and Relative Acceleration	**29.6**	Magnetism in Matter
6.3	Motion in Accelerated Frames	**30.6**	Eddy Currents
6.4	Motion in the Presence of Resistive Forces	**33.6**	Production of Electromagnetic Waves by an Antenna
7.9	Energy Diagrams and Equilibrium of a System	**35.5**	Lens Aberrations
9.9	Rocket Propulsion	**35.6**	Optical Instruments
11.5	The Motion of Gyroscopes and Tops	**37.5**	Diffraction of X-Rays by Crystals
14.8	Other Applications of Fluid Dynamics	**38.9**	The General Theory of Relativity
15.6	Damped Oscillations	**40.6**	Applications of Tunneling
15.7	Forced Oscillations	**41.9**	Spontaneous and Stimulated Transitions
17.8	Nonsinusoidal Waveforms	**41.10**	Lasers
25.7	An Atomic Description of Dielectrics	**42.7**	Semiconductor Devices
26.5	Superconductors	**43.11**	Radiation Damage
27.5	Household Wiring and Electrical Safety	**43.12**	Uses of Radiation from the Nucleus
28.3	Applications Involving Charged Particles Moving in a Magnetic Field	**43.13**	Nuclear Magnetic Resonance and Magnetic Resonance Imaging

Acknowledgments

This Tenth Edition of *Physics for Scientists and Engineers* was prepared with the guidance and assistance of many professors who reviewed selections of the manuscript, the prerevision text, or both. We wish to acknowledge the following scholars and express our sincere appreciation for their suggestions, criticisms, and encouragement:

John F. DiTusa, *Louisiana State University;* Hani Dulli, *Texas Tech University;* Eric Hudson, *Pennsylvania State University;* David Joffe, *Kennesaw State University;* Yibin Pan, *University of Wisconsin-Madison;* Mark Rzchowski, *University of Wisconsin-Madison;* Joseph Scanio, *University of Cincinnati;* Brian Utter, *Bucknell University*

During our work on this revision, we worked with development partners to help us determine the course of the revision; we would like to thank them:

Tom Barrett, *The Ohio State University;* Ken Bolland, *The Ohio State University;* Colleen Countryman, *North Carolina State University;* Dawn Hollenbeck, *Rochester Institute of Technology;* Kathleen Koenig, *University of Cincinnati;* David Lamp, *Texas Tech University;* Rafael Lopez-Mobilia, *The University of Texas at San Antonio;* Yibin Pan, *University of Wisconsin-Madison;* Chandralekha Singh, *University of Pittsburgh;* Michael Thackston, *Kennesaw State University;* Michael Ziegler, *The Ohio State University*

Prior to our work on this revision, we conducted a survey of professors; their feedback and suggestions helped shape this revision, and so we would like to thank the survey participants:

Steve Alexander, *Southwestern University;* Sanjeev Arora, *Fort Valley State University;* Erik Aver, *Gonzaga University;* David Berube, *Loyola Marymount University;* Muhammad Bhatti, *The University of Texas Rio Grande Valley;* Jeffrey Bierman, *Gonzaga University;* Ken Bolland, *The Ohio State University;* John Bulman, *Loyola Marymount University;* Hani Dulli, *Texas Tech University;* Eric Hudson, *Pennsylvania State University;* Satyanaraya Kachiraju, *The University of Texas Rio Grande Valley;* Brent McDaniel, *Kennesaw State University;* Lisa Paulius, *Western Michigan University;* Linh Pham, *University of San Diego;* Charles Ruggiero, *The Ohio State University at Marion;* Mackay Salley, *Wofford College;* Jeff Sanny, *Loyola Marymount University;* Joseph Scanio, *University of Cincinnati;* Jeffrey Schwartz, *University of California at Los Angeles;* Amit Sharma, *Wright State University;* Mark Spraker, *University of North Georgia;* Anthony Teate, *James Madison University;* Lih-Sin The, *Clemson University*

This title was carefully checked for accuracy by Michael Faux, *SUNY College at Oneonta.* We thank him for his diligent efforts under schedule pressure.

Belal Abas, Zinoviy Akkerman, Eric Boyd, Hal Falk, Melanie Martin, Steve McCauley, and Glenn Stracher made corrections to problems taken from previous editions. Harvey Leff provided invaluable guidance on the restructuring of the discussion of entropy in Chapter 21. We are grateful to Vahé Peroomian for preparing an excellent *Instructor's Solutions Manual.* Matt Kohlmyer (Senior Instructional Content Developer at Cengage), Marllin L. Simon (Professor Emeritus, *Auburn University*), and Susan English all worked hard on the optimization of the online-only homework problems, and we thank them. Linnea Cookson provided an excellent accuracy check of the Analysis Model tutorials.

Special thanks and recognition go to the professional staff at Cengage—in particular, Rebecca Berardy Schwartz, Michael Jacobs, Ed Dodd, Tanya Nigh, Teresa Trego, Lorreen Towle, Tom Ziolkowski, Cate Barr, and Caitlin Ghegan—for their fine work during the development, production, and promotion of this textbook. We recognize the skilled production service and excellent artwork provided by Ed Dionne and the staff at MPS Limited and the dedicated image and text research of Cheryl Du Bois and Ragu Veeraragavan, respectively.

Finally, we are deeply indebted to our wives, children, and grandchildren for their love, support, and long-term sacrifices.

Raymond A. Serway
St. Petersburg, Florida

John W. Jewett, Jr.
Anaheim, California

To the Student

t is appropriate to offer some words of advice that should be of benefit to you, the student. Before doing so, we assume you have read the Preface, which describes the various features of the text and support materials that will help you through the course.

How to Study

Instructors are often asked, "How should I study physics and prepare for examinations?" There is no simple answer to this question, but we can offer some suggestions based on our own experiences in learning and teaching over the years.

First and foremost, maintain a positive attitude toward the subject matter, keeping in mind that physics is the most fundamental of all natural sciences. Other science courses that follow will use the same physical principles, so it is important that you understand and are able to apply the various concepts and theories discussed in the text.

Concepts and Principles

It is essential that you understand the basic concepts and principles before attempting to solve assigned problems. You can best accomplish this goal by carefully reading the textbook before you attend your lecture on the covered material. When reading the text, you should jot down those points that are not clear to you. Also be sure to make a diligent attempt at answering the questions in the Quick Quizzes as you come to them in your reading. We have worked hard to prepare questions that help you judge for yourself how well you understand the material. Study the **What If?** features that appear in many of the worked examples carefully. They will help you extend your understanding beyond the simple act of arriving at a numerical result. The Pitfall Preventions will also help guide you away from common misunderstandings about physics. During class, take careful notes and ask questions about those ideas that are unclear to you. Keep in mind that few people are able to absorb the full meaning of scientific material after only one reading; several readings of the text and your notes may be necessary. Your lectures and laboratory work supplement the textbook and should clarify some of the more difficult material. You should minimize your memorization of material. Successful memorization of passages from the text, equations, and derivations does not necessarily indicate that you understand the material. Your understanding of the material will be enhanced through a combination of efficient study habits, discussions with other students and with instructors, and your ability to solve the problems presented in the textbook. Ask questions whenever you believe that clarification of a concept is necessary.

Study Schedule

It is important that you set up a regular study schedule, preferably a daily one. Make sure that you read the syllabus for the course and adhere to the schedule set by your instructor. The lectures will make much more sense if you read the corresponding text material *before* attending them. As a general rule, you should devote about two hours of study time for each hour you are in class. If you are having trouble with the course, seek the advice of the instructor or other students who have taken the course. You may find it necessary to seek further instruction from experienced students. Very often, instructors offer review sessions in addition to regular class periods. Avoid the practice of delaying study until a day or two before an exam.

More often than not, this approach has disastrous results. Rather than undertake an all-night study session before a test, briefly review the basic concepts and equations, and then get a good night's rest.

You can purchase any Cengage Learning product at your local college store or at our preferred online store **CengageBrain.com.**

Use the Features

You should make full use of the various features of the text discussed in the Preface. For example, marginal notes are useful for locating and describing important equations and concepts, and **boldface** indicates important definitions. Many useful tables are contained in the appendices, but most are incorporated in the text where they are most often referenced. Appendix B is a convenient review of mathematical tools used in the text.

Answers to Quick Quizzes and odd-numbered problems are given at the end of the textbook. The table of contents provides an overview of the entire text, and the index enables you to locate specific material quickly. Footnotes are sometimes used to supplement the text or to cite other references on the subject discussed.

After reading a chapter, you should be able to define any new quantities introduced in that chapter and discuss the principles and assumptions that were used to arrive at certain key relations. In some cases, you may find it necessary to refer to the textbook's index to locate certain topics. You should be able to associate with each physical quantity the correct symbol used to represent that quantity and the unit in which the quantity is specified. Furthermore, you should be able to express each important equation in concise and accurate prose.

Problem Solving

R. P. Feynman, Nobel laureate in physics, once said, "You do not know anything until you have practiced." In keeping with this statement, we strongly advise you to develop the skills necessary to solve a wide range of problems. Your ability to solve problems will be one of the main tests of your knowledge of physics; therefore, you should try to solve as many problems as possible. It is essential that you understand basic concepts and principles before attempting to solve problems. It is good practice to try to find alternate solutions to the same problem. For example, you can solve problems in mechanics using Newton's laws, but very often an alternative method that draws on energy considerations is more direct. You should not deceive yourself into thinking that you understand a problem merely because you have seen it solved in class. You must be able to solve the problem and similar problems on your own.

The approach to solving problems should be carefully planned. A systematic plan is especially important when a problem involves several concepts. First, read the problem several times until you are confident you understand what is being asked. Look for any key words that will help you interpret the problem and perhaps allow you to make certain assumptions. Your ability to interpret a question properly is an integral part of problem solving. Second, you should acquire the habit of writing down the information given in a problem and those quantities that need to be found; for example, you might construct a table listing both the quantities given and the quantities to be found. This procedure is sometimes used in the worked examples of the textbook. Finally, after you have decided on the method you believe is appropriate for a given problem, proceed with your solution. The Analysis Model Approach to Problem Solving will guide you through complex problems. If you follow the steps of this procedure (*Conceptualize, Categorize, Analyze, Finalize*), you will find it easier to come up with a solution and gain more from your efforts. This strategy, located in Section 2.4 (pages 30–32), is used in all worked examples in the remaining chapters so that you can learn how to apply it. Specific problem-solving strategies for certain types of situations are included in the text and appear with a special heading. These specific strategies follow the outline of the Analysis Model Approach to Problem Solving.

Often, students fail to recognize the limitations of certain equations or physical laws in a particular situation. It is very important that you understand and remember the assumptions that underlie a particular theory or formalism. For example, certain equations in kinematics apply only to a particle moving with constant acceleration. These equations are not valid for describing motion whose acceleration is not constant, such as the motion of an object connected to a spring or the motion of an object through a fluid. Study the Analysis Models for Problem Solving in the chapter summaries carefully so that you know how each model can be applied to a specific situation. The analysis models provide you with a logical structure for solving problems and help you develop your thinking skills to become more like those of a physicist. Use the analysis model approach to save you hours of looking for the correct equation and to make you a faster and more efficient problem solver.

Experiments

Physics is a science based on experimental observations. Therefore, we recommend that you try to supplement the text by performing various types of "hands-on" experiments either at home or in the laboratory. These experiments can be used to test ideas and models discussed in class or in the textbook. For example, the common Slinky toy is excellent for studying traveling waves, a ball swinging on the end of a long string can be used to investigate pendulum motion, various masses attached to the end of a vertical spring or rubber band can be used to determine its elastic nature, an old pair of polarized sunglasses and some discarded lenses and a magnifying glass are the components of various experiments in optics, and an approximate measure of the free-fall acceleration can be determined simply by measuring with a stopwatch the time interval required for a ball to drop from a known height. The list of such experiments is endless. When physical models are not available, be imaginative and try to develop models of your own.

New Media

If available, we strongly encourage you to use the **WebAssign** product that is available with this textbook. It is far easier to understand physics if you see it in action, and the materials available in WebAssign will enable you to become a part of that action.

It is our sincere hope that you will find physics an exciting and enjoyable experience and that you will benefit from this experience, regardless of your chosen profession. Welcome to the exciting world of physics!

The scientist does not study nature because it is useful; he studies it because he delights in it, and he delights in it because it is beautiful. If nature were not beautiful, it would not be worth knowing, and if nature were not worth knowing, life would not be worth living.

—Henri Poincaré

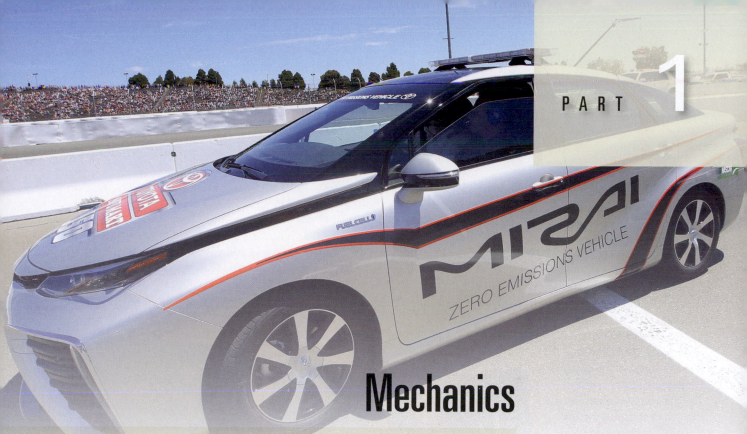

PART 1

Mechanics

Physics, the most fundamental physical science, is concerned with the fundamental principles of the Universe. It is the foundation upon which the other sciences—astronomy, biology, chemistry, and geology—are based. It is also the basis of a large number of engineering applications. The beauty of physics lies in the simplicity of its fundamental principles and in the manner in which just a small number of concepts and models can alter and expand our view of the world around us.

The study of physics can be divided into six main areas:

1. *classical mechanics,* concerning the motion of objects that are large relative to atoms and move at speeds much slower than the speed of light
2. *relativity,* a theory describing objects moving at any speed, even speeds approaching the speed of light
3. *thermodynamics,* dealing with heat, temperature, and the statistical behavior of systems with large numbers of particles
4. *electromagnetism,* concerning electricity, magnetism, and electromagnetic fields
5. *optics,* the study of the behavior of light and its interaction with materials
6. *quantum mechanics,* a collection of theories connecting the behavior of matter at the submicroscopic level to macroscopic observations

The disciplines of mechanics and electromagnetism are basic to all other branches of classical physics (developed before 1900) and modern physics (c. 1900–present). The first part of this textbook deals with classical mechanics, sometimes referred to as *Newtonian mechanics* or simply *mechanics.* Many principles and models used to understand mechanical systems retain their importance in the theories of other areas of physics and can later be used to describe many natural phenomena. Therefore, classical mechanics is of vital importance to students from all disciplines. ■

The Toyota Mirai, a fuel-cell-powered automobile available to the public, albeit in limited quantities. A fuel cell converts hydrogen fuel into electricity to drive the motor attached to the wheels of the car. Automobiles, whether powered by fuel cells, gasoline engines, or batteries, use many of the concepts and principles of mechanics that we will study in this first part of the book. Quantities that we can use to describe the operation of vehicles include position, velocity, acceleration, force, energy, and momentum. (*Chris Graythen/Getty Images Sport/ Getty Images*)

1 Physics and Measurement

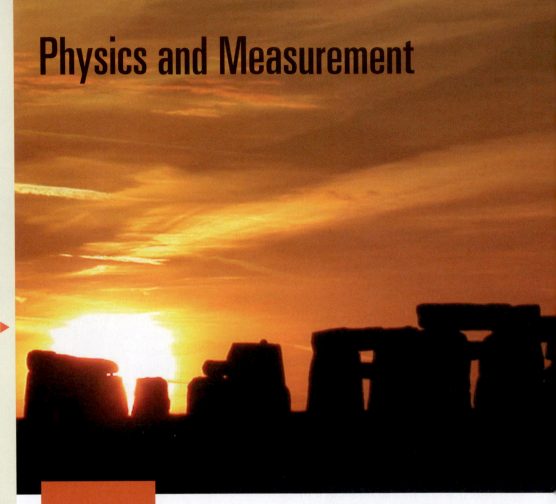

Stonehenge, in southern England, was built thousands of years ago. Various hypotheses have been proposed about its function, including a burial ground, a healing site, and a place for ancestor worship. One of the more intriguing ideas suggests that Stonehenge was an observatory, allowing measurements of some of the quantities discussed in this chapter, such as position of objects in space and time intervals between repeating celestial events. (*Image copyright Stephen Inglis. Used under license from Shutterstock.com*)

STORYLINE **Each chapter in this textbook will begin with a paragraph** related to a storyline that runs throughout the text. The storyline centers on *you*: an inquisitive physics student. You could live anywhere in the world, but let's say you live in southern California, where one of the authors lives. Most of your observations will occur there, although you will take trips to other locations. As you go through your everyday activities, you see physics in action all around you. In fact, you can't get away from physics! As you observe phenomena at the beginning of each chapter, you will ask yourself, "Why does that happen?" You might take measurements with your smartphone. You might look for related videos on YouTube or photographs on an image search site. You are lucky indeed because, in addition to those resources, you have this textbook and the expertise of your instructor to help you understand the exciting physics surrounding you. Let's look at your first observations as we begin your storyline. You have just bought this textbook and have flipped through some of its pages. You notice a page of conversions on the inside back cover. You notice in the entries under "Length" the unit of a *light-year*. You say, "Wait a minute! (You will say this often in the upcoming chapters.) How can a unit based on a *year* be a unit of *length*?" As you look farther down the page, you see 1 kg ≈ 2.2 lb (lb is the abbreviation for *pound*; lb is from Latin *libra pondo*) under the heading "Some Approximations Useful for Estimation Problems." Noticing the "approximately equal" sign (≈), you wonder what the *exact* conversion is and look upward on the page to the heading "Mass," since a kilogram is a unit of mass. The relation between kilograms and pounds is not there! Why not? Your physics adventure has begun!

CONNECTIONS The second paragraph in each chapter will explain how the material in the chapter connects to that in the previous chapter and/or future chapters. This feature will help you see that the textbook is not a collection of unrelated chapters, but rather is a structure of understanding that we are building, step by step. These paragraphs will provide a roadmap through the concepts and principles as they are introduced in the text. They will justify why the material in that chapter is presented at that time and help you to see the "big picture" of the study of physics. In this first chapter, of course, we cannot connect to a previous chapter. We will simply look ahead to the present chapter, in which we discuss some preliminary concepts of measurement, units, modeling, and estimation that we will need throughout *all* the chapters of the text.

1.1 Standards of Length, Mass, and Time

To describe natural phenomena, we must make measurements of various aspects of nature. Each measurement is associated with a physical quantity, such as the length of an object. The laws of physics are expressed as mathematical relationships among physical quantities that we will introduce and discuss throughout the book. In mechanics, the three fundamental quantities are *length, mass,* and *time*. All other quantities in mechanics can be expressed in terms of these three.

If we are to report the results of a measurement to someone who wishes to reproduce this measurement, a *standard* must be defined. For example, if someone reports that a wall is 2 meters high and our standard unit of length is defined to be 1 meter, we know that the height of the wall is twice our basic length unit. Whatever is chosen as a standard must be readily accessible and must possess some property that can be measured reliably. Measurement standards used by different people in different places—throughout the Universe—must yield the same result. In addition, standards used for measurements must not change with time.

In 1960, an international committee established a set of standards for the fundamental quantities of science. It is called the **SI** (Système International), and its fundamental units of length, mass, and time are the *meter, kilogram,* and *second,* respectively. Other standards for SI fundamental units established by the committee are those for temperature (the *kelvin*), electric current (the *ampere*), luminous intensity (the *candela*), and the amount of substance (the *mole*).

Length

We can identify **length** as the distance between two points in space. In 1120, the king of England decreed that the standard of length in his country would be named the *yard* and would be precisely equal to the distance from the tip of his nose to the end of his outstretched arm. Similarly, the original standard for the foot adopted by the French was the length of the royal foot of King Louis XIV. Neither of these standards is constant in time; when a new king took the throne, length measurements changed! The French standard prevailed until 1799, when the legal standard of length in France became the **meter** (m), defined as one ten-millionth of the distance from the equator to the North Pole along one particular longitudinal line that passes through Paris. Notice that this value is an Earth-based standard that does not satisfy the requirement that it can be used throughout the Universe.

Table 1.1 (page 4) lists approximate values of some measured lengths. You should study this table as well as the next two tables and begin to generate an intuition for what is meant by, for example, a length of 20 centimeters, a mass of 100 kilograms, or a time interval of 3.2×10^7 seconds.

As recently as 1960, the length of the meter was defined as the distance between two lines on a specific platinum–iridium bar stored under controlled conditions in France. Current requirements of science and technology, however, necessitate more accuracy than that with which the separation between the lines on the bar can be determined. In the 1960s and 1970s, the meter was defined to be equal to

PITFALL PREVENTION 1.1

Reasonable Values Generating intuition about typical values of quantities when solving problems is important because you must think about your end result and determine if it seems reasonable. For example, if you are calculating the mass of a housefly and arrive at a value of 100 kg, this answer is *unreasonable* and there is an error somewhere.

TABLE 1.1 Approximate Values of Some Measured Lengths

	Length (m)
Distance from the Earth to the most remote known quasar	2.7×10^{26}
Distance from the Earth to the most remote normal galaxies	3×10^{26}
Distance from the Earth to the nearest large galaxy (Andromeda)	2×10^{22}
Distance from the Sun to the nearest star (Proxima Centauri)	4×10^{16}
One light-year	9.46×10^{15}
Mean orbit radius of the Earth about the Sun	1.50×10^{11}
Mean distance from the Earth to the Moon	3.84×10^{8}
Distance from the equator to the North Pole	1.00×10^{7}
Mean radius of the Earth	6.37×10^{6}
Typical altitude (above the surface) of a satellite orbiting the Earth	2×10^{5}
Length of a football field	9.1×10^{1}
Length of a housefly	5×10^{-3}
Size of smallest dust particles	$\sim 10^{-4}$
Size of cells of most living organisms	$\sim 10^{-5}$
Diameter of a hydrogen atom	$\sim 10^{-10}$
Diameter of an atomic nucleus	$\sim 10^{-14}$
Diameter of a proton	$\sim 10^{-15}$

Jacques Brinon/AP Images

CSF1

Focke Strangmann/AP Images

b

Figure 1.1 (a) International Prototype of the Kilogram, an accurate copy of the International Standard Kilogram kept at Sèvres, France, is housed under a double bell jar in a vault at the National Institute of Standards and Technology. (b) A cesium fountain atomic clock. The clock will neither gain nor lose a second in 20 million years.

1 650 763.73 wavelengths[1] of orange-red light emitted from a krypton-86 lamp. In October 1983, however, the meter was redefined as **the distance traveled by light in vacuum during a time interval of 1/299 792 458 second.** In effect, this latest definition establishes that the speed of light in vacuum is precisely 299 792 458 meters per second. This definition of the meter is valid throughout the Universe based on our assumption that light is the same everywhere. The speed of light also allows us to define the **light-year**, as mentioned in the introductory storyline: the distance that light travels through empty space in one year. Use this definition and the speed of light to verify the length of a light-year in meters as given in Table 1.1.

Mass

We will find that the **mass** of an object is related to the amount of material that is present in the object, or to how much that object resists changes in its motion. Mass is an inherent property of an object and is independent of the object's surroundings and of the method used to measure it. The SI fundamental unit of mass, the **kilogram** (kg), is defined as **the mass of a specific platinum–iridium alloy cylinder kept at the International Bureau of Weights and Measures at Sèvres, France.** This mass standard was established in 1887 and has not been changed since that time because platinum–iridium is an unusually stable alloy. A duplicate of the Sèvres cylinder is kept at the National Institute of Standards and Technology (NIST) in Gaithersburg, Maryland (Fig. 1.1a). Table 1.2 lists approximate values of the masses of various objects.

In Chapter 5, we will discuss the difference between mass and weight. In anticipation of that discussion, let's look again at the approximate equivalence mentioned in the introductory storyline: 1 kg ≈ 2.2 lb. It would never be correct to claim that a number of kilograms *equals* a number of pounds, because these units represent different variables. A kilogram is a unit of *mass*, while a pound is a unit of *weight*. That's why an equality between kilograms and pounds is not given in the section of conversions for mass on the inside back cover of the textbook.

---

[1]We will use the standard international notation for numbers with more than three digits, in which groups of three digits are separated by spaces rather than commas. Therefore, 10 000 is the same as the common American notation of 10,000. Similarly, π = 3.14159265 is written as 3.141 592 65.

TABLE 1.2 Approximate Masses of Various Objects	
	Mass (kg)
Observable Universe	$\sim 10^{52}$
Milky Way galaxy	$\sim 10^{42}$
Sun	1.99×10^{30}
Earth	5.98×10^{24}
Moon	7.36×10^{22}
Shark	$\sim 10^{3}$
Human	$\sim 10^{2}$
Frog	$\sim 10^{-1}$
Mosquito	$\sim 10^{-5}$
Bacterium	$\sim 1 \times 10^{-15}$
Hydrogen atom	1.67×10^{-27}
Electron	9.11×10^{-31}

TABLE 1.3 Approximate Values of Some Time Intervals	
	Time Interval (s)
Age of the Universe	4×10^{17}
Age of the Earth	1.3×10^{17}
Average age of a college student	6.3×10^{8}
One year	3.2×10^{7}
One day	8.6×10^{4}
One class period	3.0×10^{3}
Time interval between normal heartbeats	8×10^{-1}
Period of audible sound waves	$\sim 10^{-3}$
Period of typical radio waves	$\sim 10^{-6}$
Period of vibration of an atom in a solid	$\sim 10^{-13}$
Period of visible light waves	$\sim 10^{-15}$
Duration of a nuclear collision	$\sim 10^{-22}$
Time interval for light to cross a proton	$\sim 10^{-24}$

Time

Before 1967, the standard of **time** was defined in terms of the *mean solar day.* (A solar day is the time interval between successive appearances of the Sun at the highest point it reaches in the sky each day.) The fundamental unit of a **second** (s) was defined as $\left(\frac{1}{60}\right)\left(\frac{1}{60}\right)\left(\frac{1}{24}\right)$ of a mean solar day. This definition is based on the rotation of one planet, the Earth. Therefore, this motion does not provide a time standard that is universal.

In 1967, the second was redefined to take advantage of the high precision attainable in a device known as an *atomic clock* (Fig. 1.1b), which measures vibrations of cesium atoms. One second is now defined as **9 192 631 770 times the period of vibration of radiation from the cesium-133 atom.**[2] Approximate values of time intervals are presented in Table 1.3.

You should note that we will use the notations *time* and *time interval* differently. A **time** is a description of an instant relative to a reference time. For example, $t = 10.0$ s refers to an instant 10.0 s after the instant we have identified as $t = 0$. As another example, a *time* of 11:30 a.m. means an instant 11.5 hours after our reference time of midnight. On the other hand, a **time interval** refers to *duration:* he required 30.0 minutes to finish the task. It is common to hear a "time of 30.0 minutes" in this latter example, but we will be careful to refer to measurements of duration as time intervals.

Units and Quantities In addition to SI, another system of units, the *U.S. customary system*, is still used in the United States despite acceptance of SI by the rest of the world. In this system, the units of length, mass, and time are the foot (ft), slug, and second, respectively. In this book, we shall use SI units because they are almost universally accepted in science and industry. We shall make some limited use of U.S. customary units in the study of classical mechanics.

In addition to the fundamental SI units of meter, kilogram, and second, we can also use other units, such as millimeters and nanoseconds, where the prefixes *milli-* and *nano-* denote multipliers of the basic units based on various powers of ten. Prefixes for the various powers of ten and their abbreviations are listed in Table 1.4 (page 6). For example, 10^{-3} m is equivalent to 1 millimeter (mm), and 10^{3} m corresponds to 1 kilometer (km). Likewise, 1 kilogram (kg) is 10^{3} grams (g), and 1 megavolt (MV) is 10^{6} volts (V).

---

[2]Period is defined as the time interval needed for one complete vibration.

TABLE 1.4   Prefixes for Powers of Ten

Power	Prefix	Abbreviation	Power	Prefix	Abbreviation
10^{-24}	yocto	y	10^{3}	kilo	k
10^{-21}	zepto	z	10^{6}	mega	M
10^{-18}	atto	a	10^{9}	giga	G
10^{-15}	femto	f	10^{12}	tera	T
10^{-12}	pico	p	10^{15}	peta	P
10^{-9}	nano	n	10^{18}	exa	E
10^{-6}	micro	μ	10^{21}	zetta	Z
10^{-3}	milli	m	10^{24}	yotta	Y
10^{-2}	centi	c			
10^{-1}	deci	d			

The variables length, mass, and time are examples of *fundamental quantities*. Most other variables are *derived quantities,* those that can be expressed as a mathematical combination of fundamental quantities. Common examples are *area* (a product of two lengths) and *speed* (a ratio of a length to a time interval).

Another example of a derived quantity is **density.** The density ρ (Greek letter rho) of any substance is defined as its *mass per unit volume:*

▶ A table of the letters in the Greek alphabet is provided on the back endpaper of this book.

$$\rho \equiv \frac{m}{V} \tag{1.1}$$

In terms of fundamental quantities, density is a ratio of a mass to a product of three lengths. Aluminum, for example, has a density of 2.70×10^{3} kg/m³, and iron has a density of 7.86×10^{3} kg/m³. An extreme difference in density can be imagined by thinking about holding a 10-centimeter (cm) cube of Styrofoam in one hand and a 10-cm cube of lead in the other. See Table 14.1 in Chapter 14 for densities of several materials.

QUICK QUIZ **1.1** In a machine shop, two cams are produced, one of aluminum and one of iron. Both cams have the same mass. Which cam is larger? **(a)** The aluminum cam is larger. **(b)** The iron cam is larger. **(c)** Both cams have the same size.

1.2 Modeling and Alternative Representations

Most courses in general physics require the student to learn the skills of problem solving, and examinations usually include problems that test such skills. This section describes some useful ideas that will enable you to enhance your understanding of physical concepts, increase your accuracy in solving problems, eliminate initial panic or lack of direction in approaching a problem, and organize your work.

One of the primary problem-solving methods in physics is to form an appropriate **model** of the problem. **A model is a simplified substitute for the real problem that allows us to solve the problem in a relatively simple way.** As long as the predictions of the model agree to our satisfaction with the actual behavior of the real system, the model is valid. If the predictions do not agree, the model must be refined or replaced with another model. The power of modeling is in its ability to reduce a wide variety of very complex problems to a limited number of classes of problems that can be approached in similar ways.

In science, a model is very different from, for example, an architect's scale model of a proposed building, which appears as a smaller version of what it represents.

A scientific model is a theoretical construct and may have no visual similarity to the physical problem. A simple application of modeling is presented in Example 1.1, and we shall encounter many more examples of models as the text progresses.

Models are needed because the actual operation of the Universe is extremely complicated. Suppose, for example, we are asked to solve a problem about the Earth's motion around the Sun. The Earth is very complicated, with many processes occurring simultaneously. These processes include weather, seismic activity, and ocean movements as well as the multitude of processes involving human activity. Trying to maintain knowledge and understanding of all these processes is an impossible task.

The modeling approach recognizes that none of these processes affects the motion of the Earth around the Sun to a measurable degree. Therefore, these details are all ignored. In addition, as we shall find in Chapter 13, the size of the Earth does not affect the gravitational force between the Earth and the Sun; only the masses of the Earth and Sun and the distance between their centers determine this force. In a simplified model, the Earth is imagined to be a particle, an object with mass but zero size. This replacement of an extended object by a particle is called the **particle model,** which is used extensively in physics. By analyzing the motion of a particle with the mass of the Earth in orbit around the Sun, we find that the predictions of the particle's motion are in excellent agreement with the actual motion of the Earth.

The two primary conditions for using the particle model are as follows:

- The size of the actual object is of no consequence in the analysis of its motion.
- Any internal processes occurring in the object are of no consequence in the analysis of its motion.

Both of these conditions are in action in modeling the Earth as a particle. Its radius is not a factor in determining its motion, and internal processes such as thunderstorms, earthquakes, and manufacturing processes can be ignored.

Four categories of models used in this book will help us understand and solve physics problems. The first category is the **geometric model.** In this model, we form a geometric construction that represents the real situation. We then set aside the real problem and perform an analysis of the geometric construction. Consider a popular problem in elementary trigonometry, as in the following example.

Example 1.1 **Finding the Height of a Tree**

You wish to find the height of a tree but cannot measure it directly. You stand 50.0 m from the tree and determine that a line of sight from the ground to the top of the tree makes an angle of 25.0° with the ground. How tall is the tree?

S O L U T I O N

Figure 1.2 shows the tree and a right triangle corresponding to the information in the problem superimposed over it. (We assume that the tree is exactly perpendicular to a perfectly flat ground.) In the triangle, we know the length of the horizontal leg and the angle between the hypotenuse and the horizontal leg. We can find the height of the tree by calculating the length of the vertical leg. We do so with the tangent function:

$$\tan \theta = \frac{\text{opposite side}}{\text{adjacent side}} = \frac{h}{50.0 \text{ m}}$$

$$h = (50.0 \text{ m}) \tan \theta = (50.0 \text{ m}) \tan 25.0° = \boxed{23.3 \text{ m}}$$

Figure 1.2 (Example 1.1) The height of a tree can be found by measuring the distance from the tree and the angle of sight to the top above the ground. This problem is a simple example of geometrically *modeling* the actual problem.

You may have solved a problem very similar to Example 1.1 but never thought about the notion of modeling. From the modeling approach, however, once we draw the triangle in Figure 1.2, the triangle is a geometric model of the real problem; it is a *substitute*. Until we reach the end of the problem, we do not imagine the problem to be about a *tree* but to be about a *triangle*. We use trigonometry to find the vertical leg of the triangle, leading to a value of 23.3 m. Because this leg *represents* the height of the tree, we can now return to the original problem and claim that the height of the tree is 23.3 m.

Other examples of geometric models include modeling the Earth as a perfect sphere, a pizza as a perfect disk, a meter stick as a long rod with no thickness, and an electric wire as a long, straight cylinder.

The particle model is an example of the second category of models, which we will call **simplification models.** In a simplification model, details that are not significant in determining the outcome of the problem are ignored. When we study rotation in Chapter 10, objects will be modeled as *rigid objects*. All the molecules in a rigid object maintain their exact positions with respect to one another. We adopt this simplification model because a spinning rock is much easier to analyze than a spinning block of gelatin, which is *not* a rigid object. Other simplification models will assume that quantities such as friction forces are negligible, remain constant, or are proportional to some power of the object's speed. We will assume *uniform* metal beams in Chapter 12, *laminar* flow of fluids in Chapter 14, *massless* springs in Chapter 15, *symmetric* distributions of electric charge in Chapter 23, *resistance-free* wires in Chapter 27, *thin* lenses in Chapter 34. These, and many more, are simplification models.

The third category is that of **analysis models,** which are general types of problems that we have solved before. An important technique in problem solving is to cast a new problem into a form similar to one we have already solved and which can be used as a model. As we shall see, there are about two dozen analysis models that can be used to solve most of the problems you will encounter. All of the analysis models in classical physics will be based on four simplification models: *particle, system, rigid object,* and *wave*. We will see our first analysis models in Chapter 2, where we will discuss them in more detail.

The fourth category of models is **structural models.** These models are generally used to understand the behavior of a system that is far different in scale from our macroscopic world—either much smaller or much larger—so that we cannot interact with it directly. As an example, the notion of a hydrogen atom as an electron in a circular orbit around a proton is a structural model of the atom. The ancient *geocentric* model of the Universe, in which the Earth is theorized to be at the center of the Universe, is an example of a structural model for something larger in scale than our macroscopic world.

Intimately related to the notion of modeling is that of forming **alternative representations** of the problem that you are solving. **A representation is a method of viewing or presenting the information related to the problem.** Scientists must be able to communicate complex ideas to individuals without scientific backgrounds. The best representation to use in conveying the information successfully will vary from one individual to the next. Some will be convinced by a well-drawn graph, and others will require a picture. Physicists are often persuaded to agree with a point of view by examining an equation, but non-physicists may not be convinced by this mathematical representation of the information.

A word problem, such as those at the ends of the chapters in this book, is one representation of a problem. In the "real world" that you will enter after graduation, the initial representation of a problem may be just an existing situation, such as the effects of climate change or a patient in danger of dying. You may have to identify the important data and information, and then cast the situation yourself into an equivalent word problem!

Considering alternative representations can help you think about the information in the problem in several different ways to help you understand and solve it. Several types of representations can be of assistance in this endeavor:

- **Mental representation.** From the description of the problem, imagine a scene that describes what is happening in the word problem, then let time progress so that you understand the situation and can predict what changes will occur in the situation. This step is critical in approaching *every* problem.
- **Pictorial representation.** Drawing a picture of the situation described in the word problem can be of great assistance in understanding the problem. In Example 1.1, the pictorial representation in Figure 1.2 allows us to identify the triangle as a geometric model of the problem. In architecture, a blueprint is a pictorial representation of a proposed building.

  Generally, a pictorial representation describes *what you would see* if you were observing the situation in the problem. For example, Figure 1.3 shows a pictorial representation of a baseball player hitting a short pop foul. Any coordinate axes included in your pictorial representation will be in two dimensions: *x* and *y* axes.

- **Simplified pictorial representation.** It is often useful to redraw the pictorial representation without complicating details by applying a simplification model. This process is similar to the discussion of the particle model described earlier. In a pictorial representation of the Earth in orbit around the Sun, you might draw the Earth and the Sun as spheres, with possibly some attempt to draw continents to identify which sphere is the Earth. In the simplified pictorial representation, the Earth and the Sun would be drawn simply as dots, representing particles, with appropriate labels. Figure 1.4 shows a simplified pictorial representation corresponding to the pictorial representation of the baseball trajectory in Figure 1.3. The notations v_x and v_y refer to the components of the velocity vector for the baseball. We will study vector components in Chapter 3. We shall use such simplified pictorial representations throughout the book.

- **Graphical representation.** In some problems, drawing a graph that describes the situation can be very helpful. In mechanics, for example, position–time graphs can be of great assistance. Similarly, in thermodynamics, pressure–volume graphs are essential to understanding. Figure 1.5 shows a graphical representation of the position as a function of time of a block on the end of a vertical spring as it oscillates up and down. Such a graph is helpful for understanding simple harmonic motion, which we study in Chapter 15.

  A graphical representation is different from a pictorial representation, which is also a two-dimensional display of information but whose axes, if any, represent *length* coordinates. In a graphical representation, the axes may represent *any* two related variables. For example, a graphical representation may have axes for temperature and time. The graph in Figure 1.5 has axes of vertical position *y* and time *t*. Therefore, in comparison to a pictorial representation, a graphical representation is generally *not* something you would see when observing the situation in the problem with your eyes.

- **Tabular representation.** It is sometimes helpful to organize the information in tabular form to help make it clearer. For example, some students find that making tables of known quantities and unknown quantities is helpful. The periodic table of the elements is an extremely useful tabular representation of information in chemistry and physics.

- **Mathematical representation.** The ultimate goal in solving a problem is often the mathematical representation. You want to move from the information contained in the word problem, through various representations of the problem that allow you to understand what is happening, to one or more equations that represent the situation in the problem and that can be solved mathematically for the desired result.

Figure 1.3 A pictorial representation of a pop foul being hit by a baseball player.

Figure 1.4 A simplified pictorial representation for the situation shown in Figure 1.3.

Figure 1.5 A graphical representation of the position as a function of time of a block hanging from a spring and oscillating.

1.3 Dimensional Analysis

In physics, the word *dimension* denotes the physical nature of a quantity. The distance between two points, for example, can be measured in feet, meters, or furlongs, which are all different units for expressing the dimension of length.

The symbols we use in this book to specify the dimensions of length, mass, and time are L, M, and T, respectively.[3] We shall often use brackets [ ] to denote the dimensions of a physical quantity. For example, the symbol we use for speed in this book is v, and in our notation, the dimensions of speed are written $[v] = \text{L}/\text{T}$. As another example, the dimensions of area A are $[A] = \text{L}^2$. The dimensions and units of area, volume, speed, and acceleration are listed in Table 1.5. The dimensions of other quantities, such as force and energy, will be described as they are introduced in the text.

In many situations, you may have to check a specific equation to see if it matches your expectations. A useful procedure for doing that, called **dimensional analysis,** can be used because dimensions can be treated as algebraic quantities. For example, quantities can be added or subtracted only if they have the same dimensions. Furthermore, the terms on both sides of an equation must have the same dimensions. By following these simple rules, you can use dimensional analysis to determine whether an expression has the correct form. Any relationship can be correct only if the dimensions on both sides of the equation are the same.

To illustrate this procedure, suppose you are interested in an equation for the position x of a car at a time t if the car starts from rest at $x = 0$ and moves with constant acceleration a. The correct expression for this situation is $x = \frac{1}{2}at^2$ as we show in Chapter 2. The quantity x on the left side has the dimension of length. For the equation to be dimensionally correct, the quantity on the right side must also have the dimension of length. We can perform a dimensional check by substituting the dimensions for acceleration, L/T^2 (Table 1.5), and time, T, into the equation. That is, the dimensional form of the equation $x = \frac{1}{2}at^2$ is

$$\text{L} = \frac{\text{L}}{\text{T}^2} \cdot \text{T}^2 = \text{L}$$

The dimensions of time cancel as shown, leaving the dimension of length on the right-hand side to match that on the left.

A more general procedure using dimensional analysis is to set up an expression of the form

$$x \propto a^n t^m$$

where n and m are exponents that must be determined and the symbol $\propto$ indicates a proportionality. This relationship is correct only if the dimensions of both sides are the same. Because the dimension of the left side is length, the dimension of the right side must also be length. That is,

$$[a^n t^m] = \text{L} = \text{L}^1\text{T}^0$$

PITFALL PREVENTION 1.2

Symbols for Quantities Some quantities have a small number of symbols that represent them. For example, the symbol for time is almost always t. Other quantities might have various symbols depending on the usage. Length may be described with symbols such as x, y, and z (for position); r (for radius); a, b, and c (for the legs of a right triangle); ℓ (for the length of an object); d (for a distance); h (for a height); and so forth.

TABLE 1.5 Dimensions and Units of Four Derived Quantities

Quantity	Area (A)	Volume (V)	Speed (v)	Acceleration (a)
Dimensions	L^2	L^3	L/T	L/T^2
SI units	m^2	m^3	m/s	m/s^2
U.S. customary units	ft^2	ft^3	ft/s	ft/s^2

---

[3]The *dimensions* of a quantity will be symbolized by a capitalized, nonitalic letter such as L or T. The *algebraic symbol* for the quantity itself will be an italicized letter such as L for the length of an object or t for time.

Because the dimensions of acceleration are L/T^2 and the dimension of time is T, we have

$$(L/T^2)^n T^m = L^1 T^0 \quad \rightarrow \quad (L^n T^{m-2n}) = L^1 T^0$$

The exponents of L and T must be the same on both sides of the equation. From the exponents of L, we see immediately that $n = 1$. From the exponents of T, we see that $m - 2n = 0$, which, once we substitute for n, gives us $m = 2$. Returning to our original expression $x \propto a^n t^m$, we conclude that $x \propto at^2$.

⊙ UICK QUIZ 1.2 True or False: Dimensional analysis can give you the numerical value of constants of proportionality that may appear in an algebraic expression.

Example **1.2** **Analysis of an Equation**

Show that the expression $v = at$, where v represents speed, a acceleration, and t an instant of time, is dimensionally correct.

SOLUTION

Identify the dimensions of v from Table 1.5:

$$[v] = \frac{L}{T}$$

Identify the dimensions of a from Table 1.5 and multiply by the dimensions of t:

$$[at] = \frac{L}{T^2} \, T = \frac{L}{T}$$

Therefore, $v = at$ is dimensionally correct because we have the same dimensions on both sides. (If the expression were given as $v = at^2$, it would be dimensionally *incorrect*. Try it and see!)

Example **1.3** **Analysis of a Power Law**

Suppose we are told that the acceleration a of a particle moving with uniform speed v in a circle of radius r is proportional to some power of r, say r^n, and some power of v, say v^m. Determine the values of n and m and write the simplest form of an equation for the acceleration.

SOLUTION

Write an expression for a with a dimensionless constant of proportionality k:

$$a = kr^n v^m$$

Substitute the dimensions of a, r, and v:

$$\frac{L}{T^2} = L^n \left(\frac{L}{T}\right)^m = \frac{L^{n+m}}{T^m}$$

Equate the exponents of L and T so that the dimensional equation is balanced:

$$n + m = 1 \text{ and } m = \boxed{2}$$

Solve the two equations for n:

$$n = \boxed{-1}$$

Write the acceleration expression:

$$a = kr^{-1} v^2 = \boxed{k \frac{v^2}{r}}$$

In Section 4.4 on uniform circular motion, we show that $k = 1$ if a consistent set of units is used. The constant k would not equal 1 if, for example, v were in km/h and you wanted a in m/s^2.

1.4 Conversion of Units

Sometimes it is necessary to convert units from one measurement system to another or convert within a system (for example, from kilometers to meters). Conversion factors between SI and U.S. customary units of length are as follows:

$$1 \text{ mile} = 1\ 609 \text{ m} = 1.609 \text{ km} \qquad 1 \text{ ft} = 0.304\ 8 \text{ m} = 30.48 \text{ cm}$$
$$1 \text{ m} = 39.37 \text{ in.} = 3.281 \text{ ft} \qquad 1 \text{ in.} = 0.025\ 4 \text{ m} = 2.54 \text{ cm (exactly)}$$

A more complete list of conversion factors can be found in Appendix A.

Like dimensions, units can be treated as algebraic quantities that can cancel each other. For example, suppose we wish to convert 15.0 in. to centimeters. Because 1 in. is defined as exactly 2.54 cm, we find that

$$15.0 \text{ in.} = (15.0 \text{ in.})\left(\frac{2.54 \text{ cm}}{1 \text{ in.}}\right) = 38.1 \text{ cm}$$

where the ratio in parentheses is equal to 1. We express 1 as 2.54 cm/1 in. (rather than 1 in./2.54 cm) so that the unit "inch" in the denominator cancels with the unit in the original quantity. The remaining unit is the centimeter, our desired result.

QUICK QUIZ 1.3 The distance between two cities is 100 mi. What is the number of kilometers between the two cities? **(a)** smaller than 100 **(b)** larger than 100 **(c)** equal to 100

Example 1.4 **Is He Speeding?**

On an interstate highway in a rural region of Wyoming, a car is traveling at a speed of 38.0 m/s. Is the driver exceeding the speed limit of 75.0 mi/h?

SOLUTION

Convert meters to miles and seconds to hours:

$$(38.0 \text{ m/s})\left(\frac{1 \text{ mi}}{1\ 609 \text{ m}}\right)\left(\frac{60 \text{ s}}{1 \text{ min}}\right)\left(\frac{60 \text{ min}}{1 \text{ h}}\right) = 85.0 \text{ mi/h}$$

The driver is indeed exceeding the speed limit and should slow down.

WHAT IF?   What if the driver were from outside the United States and is familiar with speeds measured in kilometers per hour? What is the speed of the car in km/h?

Answer   We can convert our final answer to the appropriate units:

$$(85.0 \text{ mi/h})\left(\frac{1.609 \text{ km}}{1 \text{ mi}}\right) = 137 \text{ km/h}$$

Figure 1.6 shows an automobile speedometer displaying speeds in both mi/h and km/h. Can you check the conversion we just performed using this photograph?

Figure 1.6 The speedometer of a vehicle that shows speeds in both miles per hour and kilometers per hour.

© Cengage

1.5 Estimates and Order-of-Magnitude Calculations

Suppose someone asks you the number of bits of data on a typical Blu-ray Disc. In response, it is not generally expected that you would provide the exact number but rather an estimate, which may be expressed in scientific notation. The estimate may be made even more approximate by expressing it as an **order of magnitude**, which is a power of 10 determined as follows:

1. Express the number in scientific notation, with the multiplier of the power of 10 between 1 and 10 and a unit.
2. If the multiplier is less than 3.162 (the square root of 10), the order of magnitude of the number is the power of 10 in the scientific notation. If the multiplier is greater than 3.162, the order of magnitude is one larger than the power of 10 in the scientific notation.

We use the symbol ~ for "is on the order of." Use the procedure above to verify the orders of magnitude for the following lengths:

$$0.008\,6\text{ m} \sim 10^{-2}\text{ m} \qquad 0.002\,1\text{ m} \sim 10^{-3}\text{ m} \qquad 720\text{ m} \sim 10^{3}\text{ m}$$

Usually, when an order-of-magnitude estimate is made, the results are reliable to within about a factor of 10.

Inaccuracies caused by guessing too low for one number are often canceled by other guesses that are too high. You will find that with practice your guessti-mates become better and better. Estimation problems can be fun to work because you freely drop digits, venture reasonable approximations for unknown numbers, make simplifying assumptions, and turn the question around into something you can answer in your head or with minimal mathematical manipulation on paper. Because of the simplicity of these types of calculations, they can be performed on a small scrap of paper and are often called *back-of-the-envelope calculations*.

Example **1.5**	**Breaths in a Lifetime**

Estimate the number of breaths taken during an average human lifetime.

SOLUTION

We start by guessing that the typical human lifetime is about 70 years. Think about the average number of breaths that a per-son takes in 1 min. This number varies depending on whether the person is exercising, sleeping, angry, serene, and so forth. To the nearest order of magnitude, we shall choose 10 breaths per minute as our estimate. (This estimate is certainly closer to the true average value than an estimate of 1 breath per minute or 100 breaths per minute.)

Find the approximate number of minutes in a year:

$$1\text{ yr}\left(\frac{400\text{ days}}{1\text{ yr}}\right)\left(\frac{25\text{ h}}{1\text{ day}}\right)\left(\frac{60\text{ min}}{1\text{ h}}\right) = 6 \times 10^{5}\text{ min}$$

Find the approximate number of minutes in a 70-year lifetime:

$$\text{number of minutes} = (70\text{ yr})(6 \times 10^{5}\text{ min/yr})$$
$$= 4 \times 10^{7}\text{ min}$$

Find the approximate number of breaths in a lifetime:

$$\text{number of breaths} = (10\text{ breaths/min})(4 \times 10^{7}\text{ min})$$
$$= 4 \times 10^{8}\text{ breaths}$$

Therefore, a person takes on the order of 10^{9} breaths in a lifetime. Notice how much simpler it is in the first calculation above to multiply 400×25 than it is to work with the more accurate 365×24.

WHAT IF? What if the average lifetime were estimated as 80 years instead of 70? Would that change our final estimate?

Answer We could claim that $(80\text{ yr})(6 \times 10^{5}\text{ min/yr}) = 5 \times 10^{7}\text{ min}$, so our final estimate should be 5×10^{8} breaths. This answer is still on the order of 10^{9} breaths, so an order-of-magnitude estimate would be unchanged.

1.6 Significant Figures

When certain quantities are measured, the measured values are known only to within the limits of the experimental uncertainty. The value of this uncertainty can depend on various factors, such as the quality of the apparatus, the skill of the experimenter, and the number of measurements performed. The number of

significant figures in a measurement can be used to express something about the uncertainty. The number of significant figures is related to the number of numerical digits used to express the measurement, as we discuss below.

As an example of significant figures, suppose we are asked to measure the radius of a Blu-ray Disc using a meterstick as a measuring instrument. Let us assume the accuracy to which we can measure the radius of the disc is ±0.1 cm. Because of the uncertainty of ±0.1 cm, if the radius is measured to be 6.0 cm, we can claim only that its radius lies somewhere between 5.9 cm and 6.1 cm. In this case, we say that the measured value of 6.0 cm has two significant figures. Note that *the significant figures include the first estimated digit*. Therefore, we could write the radius as (6.0 ± 0.1) cm.

Zeros may or may not be significant figures. Those used to position the decimal point in such numbers as 0.03 and 0.007 5 are not significant. Therefore, there are one and two significant figures, respectively, in these two values. When the zeros come after other digits, however, there is the possibility of misinterpretation. For example, suppose the mass of an object is given as 1 500 g. This value is ambiguous because we do not know whether the last two zeros are being used to locate the decimal point or whether they represent significant figures in the measurement. To remove this ambiguity, it is common to use scientific notation to indicate the number of significant figures. In this case, we would express the mass as 1.5×10^3 g if there are two significant figures in the measured value, 1.50×10^3 g if there are three significant figures, and 1.500×10^3 g if there are four. The same rule holds for numbers less than 1, so 2.3×10^{-4} has two significant figures (and therefore could be written 0.000 23) and 2.30×10^{-4} has three significant figures (and therefore written as 0.000 230).

In problem solving, we often combine quantities mathematically through multiplication, division, addition, subtraction, and so forth. When doing so, you must make sure that the result has the appropriate number of significant figures. A good rule of thumb to use in determining the number of significant figures that can be claimed in a multiplication or a division is as follows:

> When multiplying several quantities, the number of significant figures in the final answer is the same as the number of significant figures in the quantity having the smallest number of significant figures. The same rule applies to division.

Let's apply this rule to find the area of the Blu-ray Disc whose radius we measured above. Using the equation for the area of a circle,

$$A = \pi r^2 = \pi(6.0 \text{ cm})^2 = 1.1 \times 10^2 \text{ cm}^2$$

If you perform this calculation on your calculator, you will likely see 113.097 335 5. It should be clear that you don't want to keep all of these digits, but you might be tempted to report the result as 113 cm². This result is not justified because it has three significant figures, whereas the radius only has two. Therefore, we must report the result with only two significant figures as shown above.

For addition and subtraction, you must consider the number of decimal places when you are determining how many significant figures to report:

> When numbers are added or subtracted, the number of decimal places in the result should equal the smallest number of decimal places of any term in the sum or difference.

As an example of this rule, consider the sum

$$23.2 + 5.174 = 28.4$$

Notice that we do not report the answer as 28.374 because the lowest number of decimal places is one, for 23.2. Therefore, our answer must have only one decimal place.

PITFALL PREVENTION 1.4

Read Carefully Notice that the rule for addition and subtraction is different from that for multiplication and division. For addition and subtraction, the important consideration is the number of *decimal places*, not the number of *significant figures*.

The rule for addition and subtraction can often result in answers that have a different number of significant figures than the quantities with which you start. For example, consider these operations that satisfy the rule:

$$1.000\ 1 + 0.000\ 3 = 1.000\ 4$$
$$1.002 - 0.998 = 0.004$$

In the first example, the result has five significant figures even though one of the terms, 0.000 3, has only one significant figure. Similarly, in the second calculation, the result has only one significant figure even though the numbers being subtracted have four and three, respectively.

> In this book, most of the numerical examples and end-of-chapter problems will yield answers having three significant figures. When carrying out estimation calculations, we shall typically work with a single significant figure.

◀ Significant figure guidelines used in this book

If the number of significant figures in the result of a calculation must be reduced, there is a general rule for rounding numbers: the last digit retained is increased by 1 if the last digit dropped is greater than 5. (For example, 1.346 becomes 1.35.) If the last digit dropped is less than 5, the last digit retained remains as it is. (For example, 1.343 becomes 1.34.) If the last digit dropped is equal to 5, the remaining digit should be rounded to the nearest even number. (This rule helps avoid accumulation of errors in long arithmetic processes.)

In a long calculation involving multiple steps, it is very important to delay the rounding of numbers until you have the final result, in order to avoid error accumulation. Wait until you are ready to copy the final answer from your calculator before rounding to the correct number of significant figures. In this book, we display numerical values rounded off to two or three significant figures. This occasionally makes some mathematical manipulations look odd or incorrect. For instance, looking ahead to Example 3.5 on page 62, you will see the operation -17.7 km $+$ 34.6 km $= 17.0$ km. This looks like an incorrect subtraction, but that is only because we have rounded the numbers 17.7 km and 34.6 km for display. If all digits in these two intermediate numbers are retained and the rounding is only performed on the final number, the correct three-digit result of 17.0 km is obtained.

PITFALL PREVENTION 1.5
Symbolic Solutions When solving problems, it is very useful to perform the solution completely in algebraic form and wait until the very end to enter numerical values into the final symbolic expression. This method will save many calculator keystrokes, especially if some quantities cancel so that you never have to enter their values into your calculator! In addition, you will only need to round once, on the final result.

Example 1.6 Installing a Carpet

A carpet is to be installed in a rectangular room whose length is measured to be 12.71 m and whose width is measured to be 3.46 m. Find the area of the room.

SOLUTION

If you multiply 12.71 m by 3.46 m on your calculator, you will see an answer of 43.976 6 m². How many of these numbers should you claim? Our rule of thumb for multiplication tells us that you can claim only the number of significant figures in your answer as are present in the measured quantity having the lowest number of significant figures. In this example, the lowest number of significant figures is three in 3.46 m, so we should express our final answer as 44.0 m².

Summary

▶ Definitions

The three fundamental physical quantities of mechanics are **length, mass,** and **time,** which in the SI system have the units **meter** (m), **kilogram** (kg), and **second** (s), respectively. These fundamental quantities cannot be defined in terms of more basic quantities.

The **density** of a substance is defined as its *mass per unit volume:*

$$\rho \equiv \frac{m}{V} \qquad (1.1)$$

continued

Concepts and Principles

The method of **dimensional analysis** is very powerful in solving physics problems. Dimensions can be treated as algebraic quantities. By making estimates and performing order-of-magnitude calculations, you should be able to approximate the answer to a problem when there is not enough information available to specify an exact solution completely.

Problem-solving skills and physical understanding can be improved by **modeling** the problem and by constructing **alternative representations** of the problem. Models helpful in solving problems include **geometric, simplification, analysis,** and **structural models.** Helpful representations include the **mental, pictorial, simplified pictorial, graphical, tabular,** and **mathematical representations.**

When you compute a result from several measured numbers, each of which has a certain accuracy, you should give the result with the correct number of **significant figures.**

When **multiplying** several quantities, the number of significant figures in the final answer is the same as the number of significant figures in the quantity having the smallest number of significant figures. The same rule applies to **division.**

When numbers are **added** or **subtracted,** the number of decimal places in the result should equal the smallest number of decimal places of any term in the sum or difference.

Think–Pair–Share

See the Preface for an explanation of the icons used in this problems set. For additional assessment items for this section, go to ❖ **WEBASSIGN** From Cengage

1. A student is supplied with a stack of copy paper, ruler, compass, scissors, and a sensitive balance. He cuts out various shapes in various sizes, calculates their areas, measures their masses, and prepares the graph of Figure TP1.1. (a) Consider the fourth experimental point from the top. How far is it vertically from the best-fit straight line? Express your answer as a difference in vertical-axis coordinate. (b) Express your answer as a percentage. (c) Calculate the slope of the line. (d) State what the graph demonstrates, referring to the shape of the graph and the results of parts (b) and (c). (e) Describe whether this result should be expected theoretically. (f) Describe the physical meaning of the slope.

Figure TP1.1

2. **ACTIVITY** Have each person in the group measure the height of another person using a meter stick with metric distances on one side and U.S. customary distances, such as inches, on the other side. Record the height to the nearest centimeter and to the nearest half-inch. For each person, divide his or her height in centimeters by the height in inches. Compare the results of this division for everyone in your group. What can you say about the results?

3. **ACTIVITY** Gather together a number of U.S. pennies, either from your instructor or from the members of your group. Divide up the pennies into two samples: (1) those with dates of 1981 or earlier, and (2) those with dates of 1983 and later (exclude 1982 pennies from your sample). Find the total mass of all the pennies in each sample. Then divide each of these total masses by the number of pennies in its corresponding sample, to find the average penny mass in each sample. Discuss why the results are different for the two samples.

4. **ACTIVITY** Discuss in your group the process by which you can obtain the best measurement of the thickness of a single sheet of paper in Chapters 1–5 of this book. Perform that measurement and express it with an appropriate number of significant figures and uncertainty. From that measurement, predict the total thickness of the pages in Volume 1 of this book (Chapters 1–21). After making your prediction, measure the thickness of Volume 1. Is your measurement within the range of your prediction and its associated uncertainty?

Problems

See the Preface for an explanation of the icons used in this problems set.
For additional assessment items for this section, go to 🔴 **WEBASSIGN** From Cengage

> *Note:* Consult the endpapers, appendices, and tables in the text whenever necessary in solving problems. For this chapter, Table 14.1 and Appendix B.3 may be particularly useful. Answers to odd-numbered problems appear in the back of the book.

SECTION 1.1 Standards of Length, Mass, and Time

1. **Q|C** (a) Use information on the endpapers of this book to calculate the average density of the Earth. (b) Where does the value fit among those listed in Table 14.1 in Chapter 14? Look up the density of a typical surface rock like granite in another source and compare it with the density of the Earth.

2. **Q|C** A proton, which is the nucleus of a hydrogen atom, can be modeled as a sphere with a diameter of 2.4 fm and a mass of 1.67×10^{-27} kg. (a) Determine the density of the proton. (b) State how your answer to part (a) compares with the density of osmium, given in Table 14.1 in Chapter 14.

3. **V** Two spheres are cut from a certain uniform rock. One has radius 4.50 cm. The mass of the other is five times greater. Find its radius.

4. **S** What mass of a material with density ρ is required to make a hollow spherical shell having inner radius r_1 and outer radius r_2?

5. **CR** You have been hired by the defense attorney as an expert witness in a lawsuit. The plaintiff is someone who just returned from being a passenger on the first orbital space tourist flight. Based on a travel brochure offered by the space travel company, the plaintiff expected to be able to see the Great Wall of China from his orbital height of 200 km above the Earth's surface. He was unable to do so, and is now demanding that his fare be refunded and to receive additional financial compensation to cover his great disappointment. Construct the basis for an argument for the defense that shows that his expectation of seeing the Great Wall from orbit was unreasonable. The Wall is 7 m wide at its widest point and the normal visual acuity of the human eye is 3×10^{-4} rad. (Visual acuity is the smallest subtended angle that an object can make at the eye and still be recognized; the subtended angle in radians is the ratio of the width of an object to the distance of the object from your eyes.)

SECTION 1.2 Modeling and Alternative Representations

6. A surveyor measures the distance across a straight river by the following method (Fig. P1.6). Starting directly across from a tree on the opposite bank, she walks $d = 100$ m along the riverbank to establish a baseline. Then she sights across to the tree. The angle from her baseline to the tree is $\theta = 35.0°$. How wide is the river?

Figure P1.6

7. A crystalline solid consists of atoms stacked up in a repeating lattice structure. Consider a crystal as shown in Figure P1.7a. The atoms reside at the corners of cubes of side $L = 0.200$ nm. One piece of evidence for the regular arrangement of atoms comes from the flat surfaces along which a crystal separates, or cleaves, when it is broken. Suppose this crystal cleaves along a face diagonal as shown in Figure P1.7b. Calculate the spacing d between two adjacent atomic planes that separate when the crystal cleaves.

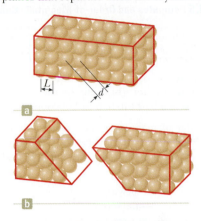

Figure P1.7

SECTION 1.3 Dimensional Analysis

8. The position of a particle moving under uniform acceleration is some function of time and the acceleration. Suppose we write this position as $x = ka^m t^n$, where k is a dimensionless constant. Show by dimensional analysis that this expression is satisfied if $m = 1$ and $n = 2$. Can this analysis give the value of k?

9. Which of the following equations are dimensionally correct? (a) $v_f = v_i + ax$ (b) $y = (2\text{ m}) \cos (kx)$, where $k = 2\text{ m}^{-1}$

10. (a) Assume the equation $x = At^3 + Bt$ describes the motion of a particular object, with x having the dimension of length and t having the dimension of time. Determine the dimensions of the constants A and B. (b) Determine the dimensions of the derivative $dx/dt = 3At^2 + B$.

SECTION 1.4 Conversion of Units

11. **V** A solid piece of lead has a mass of 23.94 g and a volume of 2.10 cm³. From these data, calculate the density of lead in SI units (kilograms per cubic meter).

12. *Why is the following situation impossible?* A student's dormitory room measures 3.8 m by 3.6 m, and its ceiling is 2.5 m high. After the student completes his physics course, he displays his dedication by completely wallpapering the walls of the room with the pages from his copy of volume 1 (Chapters 1–21) of this textbook. He even covers the door and window.

13. **T** One cubic meter (1.00 m³) of aluminum has a mass of 2.70×10^3 kg, and the same volume of iron has a mass of 7.86×10^3 kg. Find the radius of a solid aluminum sphere that will balance a solid iron sphere of radius 2.00 cm on an equal-arm balance.

14. Let ρ_{Al} represent the density of aluminum and ρ_{Fe} that of iron.
S Find the radius of a solid aluminum sphere that balances a solid iron sphere of radius r_{Fe} on an equal-arm balance.

15. One gallon of paint (volume $= 3.78 \times 10^{-3}$ m³) covers an
T area of 25.0 m². What is the thickness of the fresh paint on the wall?

16. An auditorium measures 40.0 m × 20.0 m × 12.0 m. The
V density of air is 1.20 kg/m³. What are (a) the volume of the room in cubic feet and (b) the weight of air in the room in pounds?

SECTION 1.5 Estimates and Order-of-Magnitude Calculations

Note: In your solutions to Problems 17 and 18, state the quantities you measure or estimate and the values you take for them.

17. (a) Compute the order of magnitude of the mass of a bathtub half full of water. (b) Compute the order of magnitude of the mass of a bathtub half full of copper coins.

18. To an order of magnitude, how many piano tuners reside in New York City? The physicist Enrico Fermi was famous for asking questions like this one on oral Ph.D. qualifying examinations.

19. Your roommate is playing a video game from the latest
CR *Star Wars* movie while you are studying physics. Distracted by the noise, you go to see what is on the screen. The game involves trying to fly a spacecraft through a crowded field of asteroids in the asteroid belt around the Sun. You say to him, "Do you know that the game you are playing is very unrealistic? The asteroid belt is not that crowded and you don't have to maneuver through it like that!" Distracted by your statement, he accidentally allows his spacecraft to strike an asteroid, just missing the high score. He turns to you in disgust and says, "Yeah, prove it." You say, "Okay, I've learned recently that the highest concentration of asteroids is in a doughnut-shaped region between the Kirkwood gaps at radii of 2.06 AU and 3.27 AU from the Sun. There are an estimated 10^9 asteroids of radius 100 m or larger, like those in your video game, in this region . . ." Finish your argument with a calculation to show that the number of asteroids in the space near a spacecraft is tiny. (An astronomical unit—AU—is the mean distance of the Earth from the Sun: 1 AU $= 1.496 \times 10^{11}$ m.)

SECTION 1.6 Significant Figures

Note: Appendix B.8 on propagation of uncertainty may be useful in solving some problems in this section.

20. How many significant figures are in the following numbers?
V (a) 78.9 ± 0.2 (b) 3.788×10^9 (c) 2.46×10^{-6} (d) 0.005 3

21. The *tropical year,* the time interval from one vernal equinox to the next vernal equinox, is the basis for our calendar. It contains 365.242 199 days. Find the number of seconds in a tropical year.

Note: The next seven problems call on mathematical skills from your prior education that will be useful throughout this course.

22. **Review.** The average density of the planet Uranus is 1.27×10^3 kg/m³. The ratio of the mass of Neptune to that of Uranus is 1.19. The ratio of the radius of Neptune to that of Uranus is 0.969. Find the average density of Neptune.

23. **Review.** In a community college parking lot, the number of ordinary cars is larger than the number of sport utility vehicles by 94.7%. The difference between the number of cars and the number of SUVs is 18. Find the number of SUVs in the lot.

24. **Review.** Find every angle θ between 0 and 360° for which the ratio of $\sin \theta$ to $\cos \theta$ is -3.00.

25. **Review.** The ratio of the number of sparrows visiting a bird feeder to the number of more interesting birds is 2.25. On a morning when altogether 91 birds visit the feeder, what is the number of sparrows?

26. **Review.** Prove that one solution of the equation

$$2.00x^4 - 3.00x^3 + 5.00x = 70.0$$

is $x = -2.22$.

27. **Review.** From the set of equations
S

$$p = 3q$$

$$pr = qs$$

$$\tfrac{1}{2}pr^2 + \tfrac{1}{2}qs^2 = \tfrac{1}{2}qt^2$$

involving the unknowns $p, q, r, s,$ and t, find the value of the ratio of t to r.

28. **Review.** Figure P1.28 shows students studying the ther-
Q|C mal conduction of energy into cylindrical blocks of ice. As
S we will see in Chapter 19, this process is described by the equation

$$\frac{Q}{\Delta t} = \frac{k\pi d^2(T_h - T_c)}{4L}$$

For experimental control, in one set of trials all quantities except d and Δt are constant. (a) If d is made three times larger, does the equation predict that Δt will get larger or get smaller? By what factor? (b) What pattern of proportionality of Δt to d does the equation predict? (c) To display this proportionality as a straight line on a graph, what quantities should you plot on the horizontal and vertical axes? (d) What expression represents the theoretical slope of this graph?

Figure P1.28

ADDITIONAL PROBLEMS

29. In a situation in which data are known to three significant digits, we write 6.379 m = 6.38 m and 6.374 m = 6.37 m. When a number ends in 5, we arbitrarily choose to write 6.375 m = 6.38 m. We could equally well write 6.375 m = 6.37 m, "rounding down" instead of "rounding up," because we would change the number 6.375 by equal increments in both cases. Now consider an order-of-magnitude estimate, in which factors of change rather than increments are important. We write 500 m ~ 10^3 m because 500 differs from 100 by a factor of 5 while it differs from 1 000 by only a factor of 2. We write 437 m ~ 10^3 m and 305 m ~ 10^2 m. What distance differs from 100 m and from 1 000 m by equal factors so that we could equally well choose to represent its order of magnitude as ~ 10^2 m or as ~ 10^3 m?

30. (a) What is the order of magnitude of the number of micro organisms in the human intestinal tract? A typical bacterial length scale is 10^{-6} m. Estimate the intestinal volume and assume 1% of it is occupied by bacteria. (b) Does the number of bacteria suggest whether the bacteria are beneficial, dangerous, or neutral for the human body? What functions could they serve?

31. The distance from the Sun to the nearest star is about 4×10^{16} m. The Milky Way galaxy (Fig. P1.31) is roughly a disk of diameter 10^{21} m and thickness ~ 10^{19} m. Find the order of magnitude of the number of stars in the Milky Way. Assume the distance between the Sun and our nearest neighbor is typical.

Figure P1.31 The Milky Way galaxy.

32. *Why is the following situation impossible?* In an effort to boost interest in a television game show, each weekly winner is offered an additional $1 million bonus prize if he or she can personally count out that exact amount from a supply of one-dollar bills. The winner must do this task under supervision by television show executives and within one 40-hour work week. To the dismay of the show's producers, most contestants succeed at the challenge.

33. Bacteria and other prokaryotes are found deep underground, in water, and in the air. One micron (10^{-6} m) is a typical length scale associated with these microbes. (a) Estimate the total number of bacteria and other prokaryotes on the Earth. (b) Estimate the total mass of all such microbes.

34. A spherical shell has an outside radius of 2.60 cm and an inside radius of a. The shell wall has uniform thickness and

is made of a material with density 4.70 g/cm^3. The space inside the shell is filled with a liquid having a density of 1.23 g/cm^3. (a) Find the mass m of the sphere, including its contents, as a function of a. (b) For what value of the variable a does m have its maximum possible value? (c) What is this maximum mass? (d) Explain whether the value from part (c) agrees with the result of a direct calculation of the mass of a solid sphere of uniform density made of the same material as the shell. (e) **What If?** Would the answer to part (a) change if the inner wall were not concentric with the outer wall?

35. Air is blown into a spherical balloon so that, when its radius is 6.50 cm, its radius is increasing at the rate 0.900 cm/s. (a) Find the rate at which the volume of the balloon is increasing. (b) If this volume flow rate of air entering the balloon is constant, at what rate will the radius be increasing when the radius is 13.0 cm? (c) Explain physically why the answer to part (b) is larger or smaller than 0.9 cm/s, if it is different.

36. In physics, it is important to use mathematical approximations. (a) Demonstrate that for small angles ($< 20°$)

$$\tan \alpha \approx \sin \alpha \approx \alpha = \frac{\pi \alpha'}{180°}$$

where α is in radians and α' is in degrees. (b) Use a calculator to find the largest angle for which $\tan \alpha$ may be approximated by α with an error less than 10.0%.

37. The consumption of natural gas by a company satisfies the empirical equation $V = 1.50t + 0.008\ 00t^2$, where V is the volume of gas in millions of cubic feet and t is the time in months. Express this equation in units of cubic feet and seconds. Assume a month is 30.0 days.

38. A woman wishing to know the height of a mountain measures the angle of elevation of the mountaintop as 12.0°. After walking 1.00 km closer to the mountain on level ground, she finds the angle to be 14.0°. (a) Draw a picture of the problem, neglecting the height of the woman's eyes above the ground. *Hint:* Use two triangles. (b) Using the symbol y to represent the mountain height and the symbol x to represent the woman's original distance from the mountain, label the picture. (c) Using the labeled picture, write two trigonometric equations relating the two selected variables. (d) Find the height y.

CHALLENGE PROBLEM

39. A woman stands at a horizontal distance x from a mountain and measures the angle of elevation of the mountaintop above the horizontal as θ. After walking a distance d closer to the mountain on level ground, she finds the angle to be ϕ. Find a general equation for the height y of the mountain in terms of d, ϕ, and θ, neglecting the height of her eyes above the ground.

2 Motion in One Dimension

The introductory storyline involves a long, straight road like this one, where the power poles are equally spaced. (*John Arehart/Shutterstock.com*)

STORYLINE **You are a passenger in a car being driven by a friend** down a straight road. You notice that the telephone poles, streetlight poles, or electric power poles on the side of the road are located at equal distances from each other. You pull out your smartphone and use it as a stopwatch to measure the time intervals required for you to pass between adjacent pairs of poles.[1] When your friend tells you that the car is moving at a fixed speed, you notice that all of these time intervals are the same. Now, the driver begins to slow down for a traffic light. You again measure the time intervals and find that each one is longer than the one before. After the car pulls away from the traffic light and speeds up, the time intervals between poles become shorter. Does this behavior make sense? When the car is moving at a constant speed again, you use the time interval between poles and the driving speed reported by your friend to calculate the distance between the poles. You excitedly tell your friend to pull over so you can pace out the distance between the poles. How accurate was your calculation?

CONNECTIONS We begin our study of physics with the topic of *kinematics*. In this broad topic, we generally investigate *motion*: the motion of objects without regard for interactions with the environment that influence the motion. Motion is what many of the early scientists studied. Early astronomers in Greece, China, the Middle East, and Central America observed the motion of objects in the night sky. Galileo Galilei studied the motion of objects rolling down inclined planes. Isaac Newton pondered the nature of falling objects. From everyday experience, we recognize that motion of an object represents a continuous change in the object's

---

[1]A number of specialized smartphone apps can be downloaded and used to make numerical measurements, such as speed and acceleration. In our storylines, however, we will restrict our smartphone use mostly to apps that are standard on the phone as purchased.

position. In this chapter, we will analyze the motion of an object along a straight line, like the car in the storyline. We will use measurements of length and time as described in Chapter 1 to quantify the motion. An object moving vertically and subject to gravity is an important application of one-dimensional motion, and will also be studied in this chapter. Remember our discussion of making models for physical situations in Section 1.2. In our study, we use the simplification model mentioned in that section and called the particle model, and describe the moving object as a particle regardless of its size. In general, a particle is a point-like object, that is, an object that has mass but is of infinitesimal size. In Section 1.2, we discussed the fact that the motion of the Earth around the Sun can be treated as if the Earth were a particle. We will return to this model for the Earth when we study planetary orbits in Chapter 13. As an example on a much smaller scale, it is possible to explain the pressure exerted by a gas on the walls of a container by treating the gas molecules as particles, without regard for the internal structure of the molecules; we will see this analysis in Chapter 20. For now, let us apply the particle model to a wide variety of moving objects in this chapter. An understanding of motion will be essential throughout the rest of this book: the motion of planets in Chapter 13 on gravity, the motion of electrons in electric circuits in Chapter 26, the motion of light waves in Chapter 34 on optics, the motion of quantum particles tunneling through barriers in Chapter 40.

2.1 Position, Velocity, and Speed of a Particle

A particle's **position** x is the location of the particle with respect to a chosen reference point that we can consider to be the origin of a coordinate system. The motion of a particle is completely known if the particle's position in space is known at all times.

◀ Position

Consider a car moving back and forth along the x axis as in Figure 2.1a (page 22). The numbers under the horizontal line are position markers for the car, similar to the equally spaced poles in the introductory storyline. When we begin collecting position data, the car is 30 m to the right of the reference position $x = 0$. We will use the particle model by identifying some point on the car, perhaps the front door handle, as a particle representing the entire car.

We start our clock, and once every 10 s we note the car's position. As you can see from Table 2.1, the car moves to the right (which we have defined as the positive direction) during the first 10 s of motion, from position Ⓐ to position Ⓑ. After Ⓑ, the position values begin to decrease, suggesting the car is backing up from position Ⓑ through position Ⓕ. In fact, at Ⓓ, 30 s after we start measuring, the car is at the origin of coordinates (see Fig. 2.1a). It continues moving to the left and is more than 50 m to the left of $x = 0$ when we stop recording information after our sixth data point. A graphical representation of this information is presented in Figure 2.1b. Such a plot is called a *position–time graph*.

Notice the alternative representations of information, as discussed in Section 1.2, that we have used for the motion of the car. Figure 2.1a is a pictorial representation, whereas Figure 2.1b is a graphical representation. Table 2.1 is a tabular representation of the same information. The ultimate goal, as mentioned in Section 1.2, is a mathematical representation, which can be analyzed to solve for some requested piece of information.

In the introductory storyline, you observed the change in the position of your car relative to the power poles. The **displacement** Δx of a particle is defined as its change in position in some time interval. As the particle moves from an initial position x_i to a final position x_f, its displacement is given by

TABLE 2.1 Position of the Car at Various Times

Position	t (s)	x (m)
Ⓐ	0	30
Ⓑ	10	52
Ⓒ	20	38
Ⓓ	30	0
Ⓔ	40	−37
Ⓕ	50	−53

$$\Delta x \equiv x_f - x_i \qquad (2.1)$$

◀ Displacement

Figure 2.1 A car moves back and forth along a straight line. Because we are interested only in the car's translational motion, we can model it as a particle. Several representations of the information about the motion of the car can be used. Table 2.1 is a tabular representation of the information. (a) A pictorial representation of the motion of the car. (b) A graphical representation (position–time graph) of the motion of the car.

Figure 2.2 On this basketball court, players run back and forth for the entire game. The distance that the players run over the duration of the game is nonzero. The displacement of the players over the duration of the game is approximately zero because they keep returning to the same point over and over again.

We use the capital Greek letter delta (Δ) to denote the *change* in a quantity. From this definition, we see that Δx is positive if x_f is greater than x_i and negative if x_f is less than x_i. Given the data in Table 2.1, we can easily determine the displacement of the car for various time intervals.

It is very important to recognize the difference between displacement and distance traveled. **Distance** is the length of a path followed by a particle. Consider, for example, the basketball players in Figure 2.2. If a player runs from his own team's basket down the court to the other team's basket and then returns to his own basket, the *displacement* of the player during this time interval is zero because he ended up at the same point as he started: $x_f = x_i$, so $\Delta x = 0$. During this time interval, however, he moved through a *distance* of twice the length of the basketball court. Distance is always represented as a positive number, whereas displacement can be either positive or negative.

Displacement is an example of a vector quantity. Many other physical quantities, including position, velocity, and acceleration, also are vectors. In general, a **vector quantity** requires the specification of both direction and magnitude. For example, in the case of the car in Figure 2.1, by how much did the position of the car change (*magnitude*) and in what *direction*—forward or backward? By contrast, a **scalar quantity** has a numerical value and no direction. Distance is a scalar: how far did the car move, as measured by its odometer, in a certain time interval? In this chapter, we use positive (+) and negative (−) signs to indicate vector direction. For example, for horizontal motion let us arbitrarily specify to the right as being the positive direction. It follows that any object always moving to the right undergoes a positive displacement $\Delta x > 0$, and any object moving to the left undergoes a negative displacement so that $\Delta x < 0$. We shall treat vector quantities in greater detail in Chapter 3.

One very important point has not yet been mentioned. Notice that the data in Table 2.1 result only in the six data points in the graph in Figure 2.1b. Therefore, the motion of the particle is not completely known because we don't know its position at *all* times. The smooth curve drawn through the six points in the graph is only a *possibility* of the actual motion of the car. We only have information about six instants of time; we have no idea what happened between the data points. The smooth curve is a *guess* as to what happened, but keep in mind that it is *only* a guess. If the smooth curve does represent the actual motion of the car, the graph contains complete information about the entire 50-s interval during which we watch the car move.

QUICK QUIZ 2.1 Which of the following choices best describes what can be determined exactly from Table 2.1 and Figure 2.1 for the entire 50-s interval? (a) The distance the car moved. (b) The displacement of the car. (c) Both (a) and (b). (d) Neither (a) nor (b).

It is much easier to see changes in position from the graph than from a verbal description or even a table of numbers. For example, it is clear that the car covers more ground during the middle of the 50-s interval than at the end. Between positions Ⓒ and Ⓓ, the car changes position by almost 40 m, but during the last 10 s, between positions Ⓔ and Ⓕ, it changes position by less than half that much. A common way of comparing these different motions is to divide the displacement Δx that occurs between two clock readings by the value of that particular time interval Δt. The result turns out to be a very useful ratio, one that we shall use many times. This ratio has been given a special name: the *average velocity*. The **average velocity** $v_{x,avg}$ of a particle is defined as the particle's displacement Δx divided by the time interval Δt during which that displacement occurs:

$$v_{x,avg} \equiv \frac{\Delta x}{\Delta t}$$

(2.2) ◀ Average velocity

where the subscript x indicates motion along the x axis. From this definition we see that average velocity has dimensions of length divided by time (L/T), or meters per second in SI units.

The average velocity of a particle moving in one dimension can be positive or negative, depending on the sign of the displacement. (The time interval Δt is always positive.) If the coordinate of the particle increases in time (that is, if $x_f > x_i$), Δx is positive and $v_{x,avg} = \Delta x/\Delta t$ is positive. This case corresponds to a particle moving in the positive x direction, that is, toward larger values of x. If the coordinate decreases in time (that is, if $x_f < x_i$), Δx is negative and hence $v_{x,avg}$ is negative. This case corresponds to a particle moving in the negative x direction.

We can interpret average velocity geometrically by drawing a straight line between any two points on the position–time graph in Figure 2.1b. This line forms the hypotenuse of a right triangle of height Δx and base Δt. The slope of this line is the ratio $\Delta x/\Delta t$, which is what we have defined as average velocity in Equation 2.2. For example, the line between positions Ⓐ and Ⓑ in Figure 2.1b has a slope equal to the average velocity of the car between those two times, (52 m − 30 m)/(10 s − 0) = 2.2 m/s.

In everyday usage, the terms *speed* and *velocity* are interchangeable. In physics, however, there is a clear distinction between these two quantities. Consider a marathon runner who runs a distance d of more than 40 km and yet ends up at her starting point. Her total displacement is zero, so her average velocity is zero! Nonetheless, we need to be able to quantify how fast she was running. A slightly different ratio accomplishes that for us. The **average speed** v_{avg} of a particle, a scalar quantity, is defined as the total distance d traveled divided by the total time interval required to travel that distance:

$$v_{avg} \equiv \frac{d}{\Delta t}$$

(2.3) ◀ Average speed

The SI unit of average speed is the same as the unit of average velocity: meters per second. Unlike average velocity, however, average speed has no direction and is always expressed as a positive number. Notice the clear distinction between the definitions of average velocity and average speed: average velocity (Eq. 2.2) is the *displacement* divided by the time interval, whereas average speed (Eq. 2.3) is the *distance* divided by the time interval.

Knowledge of the average velocity or average speed of a particle does not provide information about the details of the trip. For example, suppose it takes you 45.0 s to travel 100 m down a long, straight hallway toward your departure gate at an airport. At the 100-m mark, you realize you missed the restroom, and you return back 25.0 m along the same hallway, taking 10.0 s to make the return trip. The magnitude of your average *velocity* is +75.0 m/55.0 s = +1.36 m/s. The average *speed* for your trip is 125 m/55.0 s = 2.27 m/s. You may have traveled at various speeds during the walk and, of course, you changed direction. Neither average velocity nor average speed provides information about these details.

QUICK QUIZ **2.2** Under which of the following conditions is the magnitude of the average velocity of a particle moving in one dimension smaller than the average speed over some time interval? **(a)** A particle moves in the +x direction without reversing. **(b)** A particle moves in the −x direction without reversing. **(c)** A particle moves in the +x direction and then reverses the direction of its motion. **(d)** There are no conditions for which this is true.

Example 2.1 **Calculating the Average Velocity and Speed**

Find the displacement, average velocity, and average speed of the car in Figure 2.1a between positions Ⓐ and Ⓕ.

SOLUTION

Consult Figure 2.1 to form a mental image of the car and its motion. We model the car as a particle. From the position–time graph given in Figure 2.1b, notice that $x_Ⓐ = 30$ m at $t_Ⓐ = 0$ s and that $x_Ⓕ = -53$ m at $t_Ⓕ = 50$ s.

Use Equation 2.1 to find the displacement of the car:
$$\Delta x = x_Ⓕ - x_Ⓐ = -53 \text{ m} - 30 \text{ m} = \boxed{-83 \text{ m}}$$

This result means that the car ends up 83 m in the negative direction (to the left, in this case) from where it started. This number has the correct units and is of the same order of magnitude as the supplied data. A quick look at Figure 2.1a indicates that it is the correct answer.

Use Equation 2.2 to find the car's average velocity:
$$v_{x,\text{avg}} = \frac{x_Ⓕ - x_Ⓐ}{t_Ⓕ - t_Ⓐ}$$
$$= \frac{-53 \text{ m} - 30 \text{ m}}{50 \text{ s} - 0 \text{ s}} = \frac{-83 \text{ m}}{50 \text{ s}} = \boxed{-1.7 \text{ m/s}}$$

We cannot unambiguously find the average speed of the car from the data in Table 2.1 because we do not have information about the positions of the car between the data points. If we adopt the assumption that the details of the car's position are described by the curve in Figure 2.1b, the distance traveled is 22 m (from Ⓐ to Ⓑ) plus 105 m (from Ⓑ to Ⓕ), for a total of 127 m.

Use Equation 2.3 to find the car's average speed:
$$v_{\text{avg}} = \frac{127 \text{ m}}{50 \text{ s}} = \boxed{2.5 \text{ m/s}}$$

Notice that the average speed is positive, as it must be. Suppose the red-brown curve in Figure 2.1b were different so that between 0 s and 10 s it went from Ⓐ up to 100 m and then came back down to Ⓑ. The average speed of the car would change because the distance is different, but the average velocity would not change.

2.2 Instantaneous Velocity and Speed

Often we need to know the velocity of a particle at a particular instant in time t rather than the average velocity over a finite time interval Δt. In other words, you would like to be able to specify your velocity just as precisely as you can specify your position by noting what is happening at a specific clock reading, that is, at some

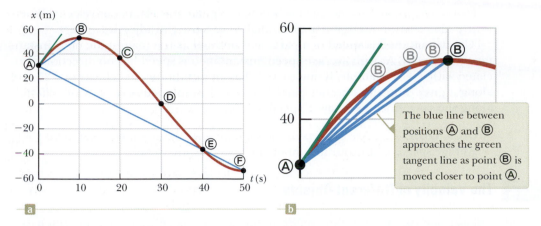

Figure 2.3 (a) Graph representing the motion of the car in Figure 2.1. (b) An enlargement of the upper-left-hand corner of the graph.

specific instant. What does it mean to talk about how quickly something is moving if we "freeze time" and talk only about an individual instant? If the time interval has a value of zero, the displacement of the object is also zero, so the average velocity from Equation 2.2 would seem to be 0/0. How do we evaluate that ratio? In the late 1600s, with the invention of calculus, scientists began to understand how to answer that question and describe an object's motion at any moment in time.

To see how that is done, consider Figure 2.3a, which is a reproduction of the graph in Figure 2.1b. What is the particle's velocity at $t = 0$? We have already discussed the average velocity for the interval during which the car moved from position Ⓐ to position Ⓑ (given by the slope of the blue line) and for the interval during which it moved from Ⓐ to Ⓕ (represented by the slope of the longer blue line and calculated in Example 2.1). The car starts out by moving to the right, which we defined to be the positive direction. Therefore, being positive, the value of the average velocity during the interval from Ⓐ to Ⓑ is more representative of the initial velocity than is the value of the average velocity during the interval from Ⓐ to Ⓕ, which we determined to be negative in Example 2.1. Now let us focus on the short blue line and imagine sliding point Ⓑ to the left along the curve, toward point Ⓐ, as in Figure 2.3b. The line between the points becomes steeper and steeper, and as the two points become extremely close together, the line becomes a tangent line to the curve, indicated by the green line in Figure 2.3b. The slope of this tangent line represents the velocity of the car at point Ⓐ. What we have done is determine the *instantaneous velocity* at that moment. In other words, the **instantaneous velocity** v_x equals the limiting value of the ratio $\Delta x/\Delta t$ as Δt approaches zero:[2]

$$v_x \equiv \lim_{\Delta t \to 0} \frac{\Delta x}{\Delta t} \tag{2.4}$$

In calculus notation, this limit is called the *derivative* of x with respect to t, written dx/dt:

$$v_x \equiv \lim_{\Delta t \to 0} \frac{\Delta x}{\Delta t} = \frac{dx}{dt} \tag{2.5}$$

◀ Instantaneous velocity

The instantaneous velocity can be positive, negative, or zero. When the slope of the position–time graph is positive, such as at any time during the first 10 s in Figure 2.3, v_x is positive and the car is moving toward larger values of x. After point Ⓑ, v_x is negative because the slope is negative and the car is moving toward smaller values of x. At point Ⓑ, the slope and the instantaneous velocity are zero and the car is momentarily at rest.

PITFALL PREVENTION 2.2

Slopes of Graphs In any graph of physical data, the *slope* represents the ratio of the change in the quantity represented on the vertical axis to the change in the quantity represented on the horizontal axis. Remember that *a slope has units* (unless both axes have the same units). The units of slope in Figures 2.1b and 2.3 are meters per second, the units of velocity.

PITFALL PREVENTION 2.3

Instantaneous Speed and Instantaneous Velocity In Pitfall Prevention 2.1, we argued that the magnitude of the average velocity is not the average speed. The magnitude of the instantaneous velocity, however, *is* the instantaneous speed. In an infinitesimal time interval, the magnitude of the displacement is equal to the distance traveled by the particle.

[2] As mentioned previously, the displacement Δx also approaches zero as Δt approaches zero, so the ratio $\Delta x/\Delta t$ looks like 0/0. The ratio can be evaluated in the limit in this situation, however. As Δx and Δt become smaller and smaller, the ratio $\Delta x/\Delta t$ approaches a value equal to the slope of the line tangent to the *x*-versus-*t* curve.

From here on, we use the word *velocity* to designate instantaneous velocity. When we are interested in *average velocity,* we shall always use the adjective *average.*

The **instantaneous speed** of a particle is defined as the magnitude of its instantaneous velocity. As with average speed, instantaneous speed has no direction associated with it. For example, if one particle has an instantaneous velocity of +25 m/s along a given line and another particle has an instantaneous velocity of −25 m/s along the same line, both have a speed[3] of 25 m/s.

QUICK QUIZ **2.3** Are officers in the highway patrol more interested in **(a)** your average speed or **(b)** your instantaneous speed as you drive?

Conceptual Example **2.2**   The Velocity of Different Objects

Consider the following one-dimensional motions: **(A)** a ball thrown directly upward rises to a highest point and falls back into the thrower's hand; **(B)** a race car starts from rest and speeds up to 100 m/s; and **(C)** a spacecraft drifts through space at constant velocity. Are there any points in the motion of these objects at which the instantaneous velocity has the same value as the average velocity over the entire motion? If so, identify the point(s).

SOLUTION

(A) The average velocity for the thrown ball is zero because the ball returns to the starting point; therefore, its displacement is zero. There is one point at which the instantaneous velocity is zero: at the top of the motion.

(B) The car's average velocity cannot be evaluated unambiguously with the information given, but it must have some value between 0 and 100 m/s. Because the car will have every instantaneous velocity between 0 and 100 m/s at some time during the interval, there must be some instant at which the instantaneous velocity is equal to the average velocity over the entire motion.

(C) Because the spacecraft's instantaneous velocity is constant, its instantaneous velocity at *any* time and its average velocity over *any* time interval are the same.

Example **2.3**   Average and Instantaneous Velocity

A particle moves along the x axis. Its position varies with time according to the expression $x = -4t + 2t^2$, where x is in meters and t is in seconds.[4] The position–time graph for this motion is shown in Figure 2.4a. Because the position of the particle is given by a mathematical function, the motion of the particle is known at all times, *unlike* that of the car in Figure 2.1, where data is only provided at six instants of time. Notice that the particle moves in the negative x direction for the first second of motion, is momentarily at rest at the moment $t = 1$ s, and moves in the positive x direction at times $t > 1$ s.

(A) Determine the displacement of the particle in the time intervals $t = 0$ to $t = 1$ s and $t = 1$ s to $t = 3$ s.

SOLUTION

From the graph in Figure 2.4a, form a mental representation of the particle's motion. Keep in mind that the particle does not move in a curved path in space such as that shown by the red-brown curve in the graphical representation. The particle moves only along the x axis in one dimension as shown in Figure 2.4b. At $t = 0$, is it moving to the right or to the left?

During the first time interval, the slope is negative and hence the average velocity is negative. Therefore, we know that the displacement between Ⓐ and Ⓑ must be a negative number having units of meters. Similarly, we expect the displacement between Ⓑ and Ⓓ to be positive.

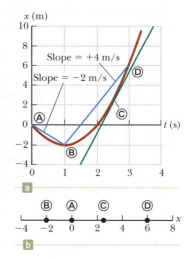

Figure 2.4 (Example 2.3) (a) Position–time graph for a particle having an x coordinate that varies in time according to the expression $x = -4t + 2t^2$. (b) The particle moves in one dimension along the x axis.

continued

---

[3]As with velocity, we drop the adjective for instantaneous speed. *Speed* means "instantaneous speed."

[4]Simply to make it easier to read, we write the expression as $x = -4t + 2t^2$ rather than as $x = (-4.00 \text{ m/s})t + (2.00 \text{ m/s}^2)t^{2.00}$. When an equation summarizes measurements, consider its coefficients and exponents to have as many significant figures as other data quoted in a problem. Consider its coefficients to have the units required for dimensional consistency. When we start our clocks at $t = 0$, we usually do not mean to limit the precision to a single digit. Consider any zero value in this book to have as many significant figures as you need.

2.3 continued

In the first time interval, set $t_i = t_{Ⓐ} = 0$ and $t_f = t_{Ⓑ} = 1$ s.
Substitute these values into $x = -4t + 2t^2$ and use
Equation 2.1 to find the displacement:

$$\Delta x_{Ⓐ \to Ⓑ} = x_f - x_i = x_{Ⓑ} - x_{Ⓐ}$$
$$= [-4(1) + 2(1)^2] - [-4(0) + 2(0)^2] = \boxed{-2 \text{ m}}$$

For the second time interval ($t = 1$ s to $t = 3$ s), set $t_i = t_{Ⓑ} = 1$ s and $t_f = t_{Ⓓ} = 3$ s:

$$\Delta x_{Ⓑ \to Ⓓ} = x_f - x_i = x_{Ⓓ} - x_{Ⓑ}$$
$$= [-4(3) + 2(3)^2] - [-4(1) + 2(1)^2] = \boxed{+8 \text{ m}}$$

These displacements can also be read directly from the position–time graph.

(B) Calculate the average velocity during these two time intervals.

SOLUTION

In the first time interval, use Equation 2.2 with $\Delta t = t_f - t_i = t_{Ⓑ} - t_{Ⓐ} = 1$ s:

$$v_{x,\text{avg}\,(Ⓐ \to Ⓑ)} = \frac{\Delta x_{Ⓐ \to Ⓑ}}{\Delta t} = \frac{-2 \text{ m}}{1 \text{ s}} = \boxed{-2 \text{ m/s}}$$

In the second time interval, $\Delta t = 2$ s:

$$v_{x,\text{avg}\,(Ⓑ \to Ⓓ)} = \frac{\Delta x_{Ⓑ \to Ⓓ}}{\Delta t} = \frac{8 \text{ m}}{2 \text{ s}} = \boxed{+4 \text{ m/s}}$$

These values are the same as the slopes of the blue lines joining these points in Figure 2.4a.

(C) Find the instantaneous velocity of the particle at $t = 2.5$ s.

SOLUTION

Calculate the slope of the green line at $t = 2.5$ s (point Ⓒ) in Figure 2.4a by reading position and time values for the ends of the green line from the graph:

$$v_x = \frac{10 \text{ m} - (-4 \text{ m})}{3.8 \text{ s} - 1.5 \text{ s}} = \boxed{+6 \text{ m/s}}$$

Notice that this instantaneous velocity is on the same order of magnitude as our previous results, that is, a few meters per second. Is that what you would have expected?

2.3 Analysis Model: Particle Under Constant Velocity

In Section 1.2 we discussed the importance of making models. As mentioned there, a particularly important model used in the solution to physics problems is an *analysis model*. **An analysis model is a common situation that occurs time and again when solving physics problems.** Because it represents a common situation, it also represents a common type of problem that we have solved before. When you identify an analysis model in a new problem, the solution to the new problem can be modeled after that of the previously solved problem. Analysis models help us to recognize those common situations and guide us toward a solution to the problem. The form that an analysis model takes is a description of either (1) the behavior of some physical entity or (2) the interaction between that entity and the environment. When you encounter a new problem, you should identify the fundamental details of the problem, ignore details that are not important, and attempt to recognize which of the situations you have already seen that might be used as a model for the new problem. For example, suppose an automobile is moving along a straight freeway at a constant speed. Is it important that it is an automobile? Is it important that it is a freeway? If the answers to both questions are no, but the car moves in a straight line at constant speed, we model the automobile as a *particle under constant velocity*, which we will discuss in this section. Once the problem has been modeled, it is no longer about an automobile. It is about a particle undergoing a certain type of motion, a motion that we have studied before.

◀ Analysis model

This method is somewhat similar to the common practice in the legal profession of finding "legal precedents." If a previously resolved case can be found that is very similar legally to the current one, it is used as a model and an argument is made in court to link them logically. The finding in the previous case can then be used to sway the finding in the current case. We will do something similar in physics. For a given problem, we search for a "physics precedent," a model with which we are already familiar and that can be applied to the current problem.

All of the analysis models that we will develop are based on four fundamental simplification models. The first of the four is the particle model discussed in the introduction to this chapter. We will look at a particle under various behaviors and environmental interactions. Further analysis models are introduced in later chapters based on simplification models of a *system*, a *rigid object*, and a *wave*. Once we have introduced these analysis models, we shall see that they appear again and again in different problem situations.

When solving a problem, you should avoid browsing through the chapter looking for an equation that contains the unknown variable that is requested in the problem. In many cases, the equation you find may have nothing to do with the problem you are attempting to solve. It is *much* better to take this first step: **Identify the analysis model that is appropriate for the problem.** To do so, think carefully about what is going on in the problem and match it to a situation you have seen before. Once the analysis model is identified, there are a small number of equations from which to choose that are appropriate for that model, sometimes only one equation. Therefore, **the model tells you which equation(s) to use for the mathematical representation.**

Let us use Equation 2.2 to build our first analysis model for solving problems. We imagine a particle moving with a constant velocity. The model of a **particle under constant velocity** can be applied in *any* situation in which an entity that can be modeled as a particle is moving with constant velocity. This situation occurs frequently, so this model is important.

If the velocity of a particle is constant, its instantaneous velocity at any instant during a time interval is the same as the average velocity over the interval. That is, $v_x = v_{x,\text{avg}}$. Therefore, substituting v_x for $v_{x,\text{avg}}$ in Equation 2.2 gives us an equation to be used in the mathematical representation of this situation:

$$v_x = \frac{\Delta x}{\Delta t} \qquad (2.6)$$

Remembering that $\Delta x = x_f - x_i$, we see that $v_x = (x_f - x_i)/\Delta t$, or

$$x_f = x_i + v_x \Delta t$$

This equation tells us that the position of the particle is given by the sum of its original position x_i at time $t = 0$ plus the displacement $v_x \Delta t$ that occurs during the time interval Δt. In practice, we usually choose the time at the beginning of the interval to be $t_i = 0$ and the time at the end of the interval to be $t_f = t$, so our equation becomes

Position as a function of ▶
time for the particle under
constant velocity model

$$x_f = x_i + v_x t \quad \text{(for constant } v_x) \qquad (2.7)$$

Equations 2.6 and 2.7 are the primary equations used in the model of a particle under constant velocity. Whenever you have identified the analysis model in a problem to be the particle under constant velocity, you can immediately turn to these equations.

Figure 2.5 is a graphical representation of the particle under constant velocity. On this position–time graph, the slope of the line representing the motion is constant and equal to the magnitude of the velocity. Equation 2.7, which is the equation of a straight line, is the mathematical representation of the particle under constant velocity model. The slope of the straight line is v_x and the y intercept is x_i in both representations.

In the opening storyline, the particle under constant velocity model was represented by the part of the motion taking place at "fixed speed." You found in the

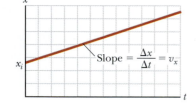

Figure 2.5 Position–time graph for a particle under constant velocity. The value of the constant velocity is the slope of the line.

storyline that the time intervals between poles were always the same in this case. Is this result consistent with Equation 2.7? Example 2.4 below shows a numerical application of the particle under constant velocity model.

Example 2.4 **Modeling a Runner as a Particle**

A kinesiologist is studying the biomechanics of the human body. (*Kinesiology* is the study of the movement of the human body. Notice the connection to the word *kinematics*.) She determines the velocity of an experimental subject while he runs along a straight line at a constant rate. The kinesiologist starts the stopwatch at the moment the runner passes a given point and stops it after the runner has passed another point 20 m away. The time interval indicated on the stopwatch is 4.0 s.

(A) What is the runner's velocity?

SOLUTION

We model the moving runner as a particle because the size of the runner and the movement of arms and legs are unnecessary details. Because the problem states that the subject runs "at a constant rate," we can model him as a *particle under constant velocity*.

Having identified the model, we can use Equation 2.6 to find the constant velocity of the runner:

$$v_x = \frac{\Delta x}{\Delta t} = \frac{x_f - x_i}{\Delta t} = \frac{20 \text{ m} - 0}{4.0 \text{ s}} = \boxed{5.0 \text{ m/s}}$$

(B) If the runner continues his motion after the stopwatch is stopped, what is his position after 10 s have passed?

SOLUTION

Use Equation 2.7 and the velocity found in part (A) to find the position of the particle at time $t = 10$ s:

$$x_f = x_i + v_x t = 0 + (5.0 \text{ m/s})(10 \text{ s}) = \boxed{50 \text{ m}}$$

Is the result for part (A) a reasonable speed for a human? How does it compare to world-record speeds in 100-m and 200-m sprints? Notice the value in part (B) is more than twice that of the 20-m position at which the stopwatch was stopped. Is this value consistent with the time of 10 s being more than twice the time of 4.0 s?

The mathematical manipulations for the particle under constant velocity stem from Equation 2.6 and its descendent, Equation 2.7. These equations can be used to solve for any variable in the equations that happens to be unknown if the other variables are known. For example, in part (B) of Example 2.4, we find the position when the velocity and the time are known. Similarly, if we know the velocity and the final position, we could use Equation 2.7 to find the time at which the runner is at this position.

A particle under constant velocity moves with a constant speed along a straight line. Now consider a particle moving with a constant speed through a distance d along a *curved* path. As we will see in Section 2.5 below, a change in the direction of motion of a particle signifies a change in the velocity of a particle even though its speed is constant; there is a change in the speed *vector*. Therefore, our particle moving along a curved path is not represented by the particle under constant velocity model. However, it can be represented with the model of a **particle under constant speed.** The primary equation for this model is Equation 2.3, with the average speed v_{avg} replaced by the constant speed v:

$$v = \frac{d}{\Delta t} \tag{2.8}$$

As an example, imagine a particle moving at a constant speed in a circular path. If the speed is 5.00 m/s and the radius of the path is 10.0 m, we can calculate the time interval required to complete one trip around the circle:

$$v = \frac{d}{\Delta t} \rightarrow \Delta t = \frac{d}{v} = \frac{2\pi r}{v} = \frac{2\pi(10.0 \text{ m})}{5.00 \text{ m/s}} = 12.6 \text{ s}$$

Particle Under Constant Velocity

Imagine a moving object that can be modeled as a particle. If it moves at a constant speed through a displacement Δx in a straight line in a time interval Δt, its constant velocity is

$$v_x = \frac{\Delta x}{\Delta t} \qquad (2.6)$$

The position of the particle as a function of time is given by

$$x_f = x_i + v_x t \qquad (2.7)$$

Examples:

- a meteoroid traveling through gravity-free space
- a car traveling at a constant speed on a straight highway
- a runner traveling at constant speed on a perfectly straight path
- an object moving at terminal speed through a viscous medium (Chapter 6)

Particle Under Constant Speed

Imagine a moving object that can be modeled as a particle. If it moves at a constant speed through a distance d along a straight line or a curved path in a time interval Δt, its constant speed is

$$v = \frac{d}{\Delta t} \qquad (2.8)$$

Examples:

- a planet traveling around a perfectly circular orbit
- a car traveling at a constant speed on a curved racetrack
- a runner traveling at constant speed on a curved path
- a charged particle moving through a uniform magnetic field (Chapter 28)

2.4 The Analysis Model Approach to Problem Solving

We have just seen our first analysis models: the particle under constant velocity and the particle under constant speed. Now, what do we do with these models? The analysis models fit into a general method of solving problems that we describe below. In particular, pay attention to the "Categorize" step in the discussion below. That is where you identify the analysis model to be applied to the problem. After that, the problem is solved using the equation or equations that you have already learned to be associated with that model. This is the way physicists approach complex situations and complicated problems, and break them into manageable pieces. It is an extremely useful skill for you to learn. It may look complicated at first, but it will become easier and of second nature as you practice it!

Conceptualize

- The first things to do when approaching a problem are to *think about* and *understand* the situation. Study carefully any representations of the information (for example, diagrams, graphs, tables, or photographs) that accompany the problem. Imagine a movie, running in your mind, of what happens in the problem: the mental representation.
- If a pictorial representation is not provided, you should almost always make a quick drawing of the situation. Indicate any known values, perhaps in a table or directly on your sketch.
- Now focus on what algebraic or numerical information is given in the problem. Carefully read the problem statement, looking for key phrases such as "starts from rest" ($v_i = 0$) or "stops" ($v_f = 0$).

- Now focus on the expected result of solving the problem. Exactly what is the question asking? Will the final result be numerical, algebraic, or verbal? Do you know what units to expect?
- Don't forget to incorporate information from your own experiences and common sense. What should a reasonable answer look like? For example, you wouldn't expect to calculate the speed of an automobile to be 5×10^6 m/s.

Categorize

- Once you have a good idea of what the problem is about, you need to *simplify* the problem. Use a simplification model to remove the details that are not important to the solution. For example, model a moving object as a particle. If appropriate, ignore air resistance or friction between a sliding object and a surface.
- Once the problem is simplified, it is important to *categorize* the problem in one of two ways. Is it a simple *substitution problem* such that numbers can be substituted into a simple equation or a definition? If so, the problem is likely to be finished when this substitution is done. If not, you face what we call an *analysis problem:* the situation must be analyzed more deeply to generate an appropriate equation and reach a solution.
- If it is an analysis problem, it needs to be categorized further. Have you seen this type of problem before? Does it fall into the growing list of types of problems that you have solved previously? If so, identify any *analysis model(s)* appropriate for the problem to prepare for the Analyze step below. Being able to classify a problem with an analysis model can make it much easier to lay out a plan to solve it.

Analyze

- Now you must analyze the problem and strive for a mathematical solution. Because you have already categorized the problem and identified an analysis model, it should not be too difficult to select relevant equations that apply to the type of situation in the problem. For example, if the problem involves a particle under constant velocity, Equation 2.7 is relevant.
- Use algebra (and calculus, if necessary) to solve symbolically for the unknown variable in terms of what is given. Finally, substitute in the appropriate numbers, calculate the result, and round it to the proper number of significant figures.

Finalize

- Examine your numerical answer. Does it have the correct units? Does it meet your expectations from your conceptualization of the problem? What about the algebraic form of the result? Does it make sense? Examine the variables in the problem to see whether the answer would change in a physically meaningful way if the variables were drastically increased or decreased or even became zero. Looking at limiting cases to see whether they yield expected values is a very useful way to make sure that you are obtaining reasonable results.
- Think about how this problem compared with others you have solved. How was it similar? In what critical ways did it differ? Why was this problem assigned? Can you figure out what you have learned by doing it? If it is a new category of problem, be sure you understand it so that you can use it as a model for solving similar problems in the future.

When solving complex problems, you may need to identify a series of subproblems and apply the Analysis Model Approach to each. For simple problems, you probably don't need this approach. When you are trying to solve a problem and you don't know what to do next, however, remember the steps in the approach and use them as a guide.

In the rest of this book, we will label the *Conceptualize, Categorize, Analyze,* and *Finalize* steps in the worked examples. If a worked example is identified as a substitution problem in the *Categorize* step, there will generally not be *Analyze* and *Finalize* sections labeled in the solution.

To show how to apply this approach, we reproduce Example 2.4 below, with the steps of the approach labeled.

Example 2.4   **Modeling a Runner as a Particle**

A kinesiologist is studying the biomechanics of the human body. (*Kinesiology* is the study of the movement of the human body. Notice the connection to the word *kinematics*.) She determines the velocity of an experimental subject while he runs along a straight line at a constant rate. The kinesiologist starts the stopwatch at the moment the runner passes a given point and stops it after the runner has passed another point 20 m away. The time interval indicated on the stopwatch is 4.0 s.

(A) What is the runner's velocity?

SOLUTION

Conceptualize   We model the moving runner as a particle because the size of the runner and the movement of arms and legs are unnecessary details.

Categorize   Because the problem states that the subject runs "at a constant rate," we can model him as a *particle under constant velocity.*

Analyze   Having identified the model, we can use Equation 2.6 to find the constant velocity of the runner:

$$v_x = \frac{\Delta x}{\Delta t} = \frac{x_f - x_i}{\Delta t} = \frac{20 \text{ m} - 0}{4.0 \text{ s}} = \boxed{5.0 \text{ m/s}}$$

(B) If the runner continues his motion after the stopwatch is stopped, what is his position after 10 s have passed?

SOLUTION

Use Equation 2.7 and the velocity found in part (A) to find the position of the particle at time $t = 10$ s:

$$x_f = x_i + v_x t = 0 + (5.0 \text{ m/s})(10 \text{ s}) = \boxed{50 \text{ m}}$$

Finalize   Is the result for part (A) a reasonable speed for a human? How does it compare to world-record speeds in 100-m and 200-m sprints? Notice the value in part (B) is more than twice that of the 20-m position at which the stopwatch was stopped. Is this value consistent with the time of 10 s being more than twice the time of 4.0 s?

2.5 **Acceleration**

In Example 2.3, we worked with a common situation in which the velocity of a particle changes while the particle is moving. When the velocity of a particle changes with time, the particle is said to be *accelerating.* For example, the magnitude of a car's velocity increases when you step on the gas and decreases when you apply the brakes. Both of these actions result in an acceleration of the car. Let us see how to quantify acceleration.

Suppose an object that can be modeled as a particle moving along the *x* axis has an initial velocity v_{xi} at time t_i at position Ⓐ and a final velocity v_{xf} at time t_f at position Ⓑ as in Figure 2.6a. The red-brown curve in Figure 2.6b shows how the velocity varies with time. The **average acceleration** $a_{x,\text{avg}}$ of the particle is defined as the *change in velocity* Δv_x divided by the time interval Δt during which that change occurs:

◀ Average acceleration

$$a_{x,\text{avg}} \equiv \frac{\Delta v_x}{\Delta t} = \frac{v_{xf} - v_{xi}}{t_f - t_i} \tag{2.9}$$

As with velocity, when the motion being analyzed is one dimensional, we can use positive and negative signs to indicate the direction of the acceleration. Because the dimensions of velocity are L/T and the dimension of time is T, acceleration has

The car moves with different velocities at points Ⓐ and Ⓑ.

The slope of the green line is the instantaneous acceleration of the car at point Ⓑ (Eq. 2.10).

The slope of the blue line connecting Ⓐ and Ⓑ is the average acceleration of the car during the time interval $\Delta t = t_f - t_i$ (Eq. 2.9).

Figure 2.6 (a) A car, modeled as a particle, moving along the x axis from Ⓐ to Ⓑ, has velocity v_{xi} at $t = t_i$ and velocity v_{xf} at $t = t_f$. (b) Velocity–time graph (red-brown) for the particle moving in a straight line.

dimensions of length divided by time squared, or L/T². The SI unit of acceleration is meters per second squared (m/s²). It might be easier to interpret these units if you think of them as meters per second per second. For example, suppose an object has an acceleration of +2 m/s². You can interpret this value by forming a mental image of the object having a velocity that is along a straight line and is increasing by 2 m/s during every time interval of 1 s. If the object starts from rest, you should be able to picture it moving at a velocity of +2 m/s after 1 s, at +4 m/s after 2 s, and so on.

When your friend sped up from the traffic light in the opening storyline, you found that the time intervals between poles on the side of the road decreased. Is that result consistent with your expectations? Each new displacement between poles is undertaken at a higher speed, so the time intervals between poles become smaller.

In some situations, the value of the average acceleration may be different over different time intervals. It is therefore useful to define the **instantaneous acceleration** as the limit of the average acceleration as Δt approaches zero. This concept is analogous to the definition of instantaneous velocity discussed in Section 2.2. If we imagine that point Ⓐ is brought closer and closer to point Ⓑ in Figure 2.6a and we take the limit of $\Delta v_x / \Delta t$ as Δt approaches zero, we obtain the instantaneous acceleration at point Ⓑ:

$$a_x \equiv \lim_{\Delta t \to 0} \frac{\Delta v_x}{\Delta t} = \frac{dv_x}{dt} \qquad (2.10)$$

That is, the instantaneous acceleration equals the derivative of the velocity with respect to time, which by definition is the slope of the velocity–time graph. The slope of the green line in Figure 2.6b is equal to the instantaneous acceleration at point Ⓑ. Notice that Figure 2.6b is a *velocity–time* graph, not a *position–time* graph like Figures 2.1b, 2.3, 2.4, and 2.5. Therefore, we see that just as the velocity of a moving particle is the slope at a point on the particle's x–t graph, the acceleration of a particle is the slope at a point on the particle's v_x–t graph. One can interpret the derivative of the velocity with respect to time as the time rate of change of velocity. If a_x is positive, the acceleration is in the positive x direction; if a_x is negative, the acceleration is in the negative x direction.

Figure 2.7 illustrates how an acceleration–time graph is related to a velocity–time graph. The acceleration at any time is the slope of the velocity–time graph at that time. Positive values of acceleration correspond to those points in Figure 2.7a where the velocity is increasing in the positive x direction. The acceleration reaches a maximum at time $t_Ⓐ$, when the slope of the velocity–time graph is a maximum. The acceleration then goes to zero at time $t_Ⓑ$, when the velocity is a maximum (that is, when the slope of the v_x–t graph is zero). The acceleration is negative when the velocity is decreasing in the positive x direction, and it reaches its most negative value at time $t_Ⓒ$.

◀ Instantaneous acceleration

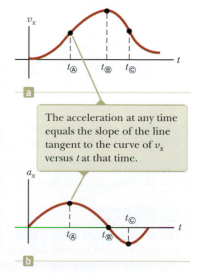

The acceleration at any time equals the slope of the line tangent to the curve of v_x versus t at that time.

Figure 2.7 (a) The velocity–time graph for a particle moving along the x axis. (b) The instantaneous acceleration can be obtained from the velocity–time graph.

QUICK QUIZ **2.4** Make a velocity–time graph for the car in Figure 2.1a. Suppose the speed limit for the road on which the car is driving is 30 km/h. True or False? The car exceeds the speed limit at some time within the time interval 0–50 s.

For the case of motion in a straight line, the direction of the velocity of an object and the direction of its acceleration are related as follows. When the object's velocity and acceleration are in the same direction, the object is speeding up. On the other hand, when the object's velocity and acceleration are in opposite directions, the object is slowing down.

To help with this discussion of the signs of velocity and acceleration, we can relate the acceleration of an object to the total *force* exerted on the object. In Chapter 5, we formally establish that **the force on an object is proportional to the acceleration of the object:**

$$F_x \propto a_x \qquad (2.11)$$

This proportionality indicates that acceleration is caused by force. Furthermore, force and acceleration are both vectors, and the vectors are in the same direction. Therefore, let us think about the signs of velocity and acceleration by imagining a force applied to an object and causing it to accelerate. Let us assume the velocity and acceleration are in the same direction. This situation corresponds to an object that experiences a force acting in the same direction as its velocity. In this case, the object speeds up! Now suppose the velocity and acceleration are in opposite directions. In this situation, the object moves in some direction and experiences a force acting in the opposite direction. Therefore, the object slows down! It is very useful to equate the direction of the acceleration to the direction of a force because it is easier from our everyday experience to think about what effect a force will have on an object than to think only in terms of the direction of the acceleration.

QUICK QUIZ **2.5** If a car is traveling eastward and slowing down, what is the direction of the force on the car that causes it to slow down? **(a)** eastward **(b)** westward **(c)** neither eastward nor westward

From now on, we shall use the term *acceleration* to mean instantaneous acceleration. When we mean average acceleration, we shall always use the adjective *average*. Because $v_x = dx/dt$, the acceleration can also be written as

$$a_x = \frac{dv_x}{dt} = \frac{d}{dt}\left(\frac{dx}{dt}\right) = \frac{d^2x}{dt^2} \qquad (2.12)$$

That is, in one-dimensional motion, the acceleration of a particle equals the *second derivative* of the particle's position x with respect to time.

PITFALL PREVENTION 2.4

Negative Acceleration Keep in mind that *negative acceleration does not necessarily mean that an object is slowing down*. If the acceleration is negative and the velocity is negative, the object is speeding up!

PITFALL PREVENTION 2.5

Deceleration The word *deceleration* has the common popular connotation of *slowing down*. We will not use this word in this book because it confuses the definition we have given for negative acceleration.

| Conceptual Example **2.5** | **Graphical Relationships Between x, v_x, and a_x** |

The position of an object moving along the x axis varies with time as in Figure 2.8a. Graph the velocity versus time and the acceleration versus time for the object.

SOLUTION

The velocity at any instant is the slope of the tangent to the x–t graph at that instant. Between $t = 0$ and $t = t_Ⓐ$, the slope of the x–t graph increases uniformly, so the velocity increases linearly as shown in Figure 2.8b. Between $t_Ⓐ$ and $t_Ⓑ$, the slope of the x–t graph is constant, so the velocity remains constant. Between $t_Ⓑ$ and $t_Ⓓ$, the slope of the x–t graph decreases, so the value of the velocity in the v_x–t graph decreases. At $t_Ⓓ$,

the slope of the x–t graph is zero, so the velocity is zero at that instant. Between $t_Ⓓ$ and $t_Ⓔ$, the slope of the x–t graph and therefore the velocity are negative and decrease uniformly in this interval. In the interval $t_Ⓔ$ to $t_Ⓕ$, the slope of the x–t graph is still negative, and at $t_Ⓕ$ it goes to zero. Finally, after $t_Ⓕ$, the slope of the x–t graph is zero, meaning that the object is at rest for $t > t_Ⓕ$.

continued

2.5 continued

The acceleration at any instant is the slope of the tangent to the v_x–t graph at that instant. The graph of acceleration versus time for this object is shown in Figure 2.8c. The acceleration is constant and positive between 0 and $t_Ⓐ$, where the slope of the v_x–t graph is positive. It is zero between $t_Ⓐ$ and $t_Ⓑ$ and for $t > t_Ⓕ$ because the slope of the v_x–t graph is zero at these times. It is negative between $t_Ⓑ$ and $t_Ⓔ$ because the slope of the v_x–t graph is negative during this interval. Between $t_Ⓔ$ and $t_Ⓕ$, the acceleration is positive like it is between 0 and $t_Ⓐ$, but higher in value because the slope of the v_x–t graph is steeper.

Notice that the sudden changes in acceleration shown in Figure 2.8c are unphysical. Such instantaneous changes cannot occur in reality.

Figure 2.8 (Conceptual Example 2.5) (a) Position–time graph for an object moving along the x axis. (b) The velocity–time graph for the object is obtained by measuring the slope of the position–time graph at each instant. (c) The acceleration–time graph for the object is obtained by measuring the slope of the velocity–time graph at each instant.

| Example **2.6** | **Average and Instantaneous Acceleration** |

The velocity of a particle moving along the x axis varies according to the expression $v_x = 40 - 5t^2$, where v_x is in meters per second and t is in seconds.

(A) Find the average acceleration in the time interval $t = 0$ to $t = 2.0$ s.

S O L U T I O N

Conceptualize Think about what the particle is doing from the mathematical representation. Is it moving at $t = 0$? In which direction? Does it speed up or slow down? Figure 2.9 is a v_x–t graph that was created from the velocity versus time expression given in the problem statement. Because the slope of the entire v_x–t curve is negative, we expect the acceleration to be negative.

Categorize The solution to this problem does not require either of the analysis models we have developed so far, and can be solved with simple mathematics. Therefore, we categorize the problem as a substitution problem.

The acceleration at Ⓑ is equal to the slope of the green tangent line at $t = 2$ s, which is -20 m/s^2.

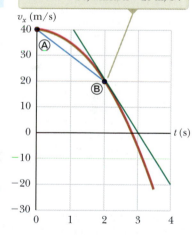

Figure 2.9 (Example 2.6) The velocity–time graph for a particle moving along the x axis according to the expression $v_x = 40 - 5t^2$.

Find the velocities at $t_i = t_Ⓐ = 0$ and $t_f = t_Ⓑ = 2.0$ s by substituting these values of t into the expression for the velocity:

$$v_{xⒶ} = 40 - 5t_Ⓐ^2 = 40 - 5(0)^2 = +40 \text{ m/s}$$

$$v_{xⒷ} = 40 - 5t_Ⓑ^2 = 40 - 5(2.0)^2 = +20 \text{ m/s}$$

Use Equation 2.9 to find the average acceleration in the specified time interval $\Delta t = t_Ⓑ - t_Ⓐ = 2.0$ s:

$$a_{x,\text{avg}} = \frac{v_{xf} - v_{xi}}{t_f - t_i} = \frac{v_{xⒷ} - v_{xⒶ}}{t_Ⓑ - t_Ⓐ} = \frac{20 \text{ m/s} - 40 \text{ m/s}}{2.0 \text{ s} - 0 \text{ s}}$$

$$= -10 \text{ m/s}^2$$

The negative sign is consistent with our expectations: the average acceleration, represented by the slope of the blue line joining the initial and final points on the velocity–time graph, is negative.

continued

2.6 continued

(B) Determine the acceleration at $t = 2.0$ s.

SOLUTION

Knowing that the initial velocity at any time t is $v_{xi} = 40 - 5t^2$, find the velocity at any later time $t + \Delta t$:

$$v_{xf} = 40 - 5(t + \Delta t)^2 = 40 - 5t^2 - 10t\,\Delta t - 5(\Delta t)^2$$

Find the change in velocity over the time interval Δt:

$$\Delta v_x = v_{xf} - v_{xi} = -10t\,\Delta t - 5(\Delta t)^2$$

To find the acceleration at any time t, divide this expression by Δt and take the limit of the result as Δt approaches zero:

$$a_x = \lim_{\Delta t \to 0} \frac{\Delta v_x}{\Delta t} = \lim_{\Delta t \to 0}(-10t - 5\,\Delta t) = -10t$$

Substitute $t = 2.0$ s:

$$a_x = (-10)(2.0) \text{ m/s}^2 = \boxed{-20 \text{ m/s}^2}$$

Because the velocity of the particle is positive and the acceleration is negative at this instant, the particle is slowing down.

Finalize   Notice that the answers to parts (A) and (B) are different. The average acceleration in part (A) is the slope of the blue line in Figure 2.9 connecting points Ⓐ and Ⓑ. The instantaneous acceleration in part (B) is the slope of the green line tangent to the curve at point Ⓑ. Notice also that the acceleration is *not* constant in this example. Situations involving constant acceleration are treated in Section 2.7.

So far, we have evaluated the derivatives of a function by starting with the definition of the function and then taking the limit of a specific ratio. If you are familiar with calculus, you should recognize that there are specific rules for taking derivatives. These rules, which are listed in Appendix B.6, enable us to evaluate derivatives quickly. For instance, one rule tells us that the derivative of any constant is zero. As another example, suppose x is proportional to some power of t such as in the expression

$$x = At^n$$

where A and n are constants. (This expression is a very common functional form.) The derivative of x with respect to t is

$$\frac{dx}{dt} = nAt^{n-1}$$

Applying these rules to Example 2.6, in which $v_x = 40 - 5t^2$, we quickly find that the acceleration is $a_x = dv_x/dt = -10t$, as we found in part (B) of the example.

2.6 Motion Diagrams

The concepts of velocity and acceleration are often confused with each other, but in fact they are quite different quantities. In forming a mental representation of a moving object, a pictorial representation called a *motion diagram* is sometimes useful to describe the velocity and acceleration while an object is in motion.

A motion diagram can be formed by imagining a *stroboscopic* photograph of a moving object, which shows several images of the object taken as the strobe light flashes at a constant rate. Figure 2.1a is a motion diagram for the car studied in Section 2.1. Figure 2.10 represents three sets of strobe photographs of cars moving along a straight roadway in a single direction, from left to right. The time intervals between flashes of the stroboscope are equal in each part of the diagram. So as to not confuse the two vector quantities, we use red arrows for velocity and purple arrows for acceleration in Figure 2.10. The arrows are shown at several instants during the motion of the object. Let us describe the motion of the car in each diagram.

In Figure 2.10a, the images of the car are equally spaced, showing us that the car moves through the same displacement in each time interval. This equal spacing is consistent with the car moving with *constant positive velocity* and *zero acceleration*. We

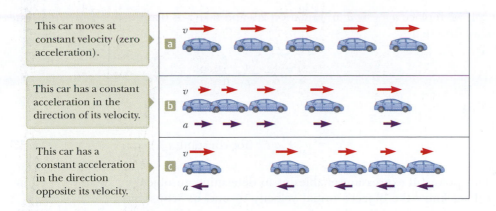

This car moves at constant velocity (zero acceleration).

This car has a constant acceleration in the direction of its velocity.

This car has a constant acceleration in the direction opposite its velocity.

Figure 2.10 Motion diagrams of a car moving along a straight roadway in a single direction. The velocity at each instant is indicated by a red arrow, and the constant acceleration is indicated by a purple arrow.

could model the car as a particle and describe it with the particle under constant velocity model. The red velocity arrows are all of equal length, and there is no purple acceleration arrow shown because it is of length zero.

In Figure 2.10b, the images become farther apart as time progresses. In this case, the red velocity arrows increase in length with time because the car's displacement between adjacent positions increases in time. These features suggest the car is moving with a *positive velocity* and a *positive acceleration*. The velocity and acceleration are in the same direction. In terms of our earlier force discussion, imagine a force pulling on the car in the same direction it is moving: it speeds up.

In Figure 2.10c, we can tell that the car slows as it moves to the right because its displacement between adjacent images decreases with time. This case suggests the car moves to the right with a negative acceleration. The lengths of the velocity arrows decrease in time and eventually reach zero. From this diagram, we see that the acceleration and velocity arrows are *not* in the same direction. The car is moving with a *positive velocity,* but with a *negative acceleration*. (This type of motion is exhibited by a car that skids to a stop after its brakes are applied.) The velocity and acceleration are in opposite directions. In terms of our earlier force discussion, imagine a force pulling on the car opposite to the direction it is moving: it slows down.

Each purple acceleration arrow in parts (b) and (c) of Figure 2.10 is the same length. Therefore, these diagrams represent motion of a *particle under constant acceleration*. This important analysis model will be discussed in the next section.

Q **UICK QUIZ 2.6** Which one of the following statements is true? **(a)** If a car is traveling eastward, its acceleration must be eastward. **(b)** If a car is slowing down, its acceleration must be negative. **(c)** A particle with constant acceleration can never stop and stay stopped.

2.7 Analysis Model: Particle Under Constant Acceleration

If the acceleration of a particle varies in time, its motion can be complex and difficult to analyze. A very common and simple type of one-dimensional motion, however, is that in which the acceleration is constant. In such a case, the average acceleration $a_{x,avg}$ over any time interval is numerically equal to the instantaneous acceleration a_x at any instant within the interval, and the velocity changes at the same rate throughout the motion. This situation occurs often enough that we identify it as an analysis model: the **particle under constant acceleration.** In the discussion that follows, we generate several equations that describe the motion of a particle for this model.

Figure 2.11 A particle under constant acceleration a_x moving along the x axis: (a) the position–time graph, (b) the velocity–time graph, and (c) the acceleration–time graph.

Position as a function of ▶ velocity and time for the particle under constant acceleration model

Position as a function of ▶ time for the particle under constant acceleration model

If we replace $a_{x,\text{avg}}$ by a_x in Equation 2.9 and take $t_i = 0$ and t_f to be any later time t, we find that

$$a_x = \frac{v_{xf} - v_{xi}}{t - 0}$$

or

$$v_{xf} = v_{xi} + a_x t \quad \text{(for constant } a_x\text{)} \tag{2.13}$$

This powerful expression enables us to determine an object's velocity at *any* time t if we know the object's initial velocity v_{xi} and its (constant) acceleration a_x. A velocity–time graph for this constant-acceleration motion is shown in Figure 2.11b. The graph is a straight line, the slope of which is the acceleration a_x; the (constant) slope is consistent with $a_x = dv_x/dt$ being a constant. Notice that the slope is positive, which indicates a positive acceleration. If the acceleration were negative, the slope of the line in Figure 2.11b would be negative. When the acceleration is constant, the graph of acceleration versus time (Fig. 2.11c) is a straight line having a slope of zero.

Because velocity at constant acceleration varies linearly in time according to Equation 2.13, we can express the average velocity in any time interval as the arithmetic mean of the initial velocity v_{xi} and the final velocity v_{xf}:

$$v_{x,\text{avg}} = \frac{v_{xi} + v_{xf}}{2} \quad \text{(for constant } a_x\text{)} \tag{2.14}$$

Notice that this expression for average velocity applies *only* in situations in which the acceleration is constant.

We can now use Equations 2.1, 2.2, and 2.14 to obtain the position of an object as a function of time. Recalling that Δx in Equation 2.2 represents $x_f - x_i$ and recognizing that $\Delta t = t_f - t_i = t - 0 = t$, we find that

$$x_f - x_i = v_{x,\text{avg}} t = \tfrac{1}{2}(v_{xi} + v_{xf})t$$

$$x_f = x_i + \tfrac{1}{2}(v_{xi} + v_{xf})t \quad \text{(for constant } a_x\text{)} \tag{2.15}$$

This equation provides the final position of the particle at time t in terms of the initial and final velocities.

We can obtain another useful expression for the position of a particle under constant acceleration by substituting Equation 2.13 into Equation 2.15:

$$x_f = x_i + \tfrac{1}{2}[v_{xi} + (v_{xi} + a_x t)]t$$

$$x_f = x_i + v_{xi}t + \tfrac{1}{2}a_x t^2 \quad \text{(for constant } a_x\text{)} \tag{2.16}$$

This equation provides the final position of the particle at time t in terms of the initial position, the initial velocity, and the constant acceleration.

The position–time graph for motion at constant (positive) acceleration shown in Figure 2.11a is obtained from Equation 2.16. Notice that the curve is a parabola. The slope of the tangent line to this curve at $t = 0$ equals the initial velocity v_{xi}, and the slope of the tangent line at any later time t equals the velocity v_{xf} at that time.

Finally, we can obtain an expression for the final velocity that does not contain time as a variable by substituting the value of t from Equation 2.13 into Equation 2.15:

$$x_f = x_i + \tfrac{1}{2}(v_{xi} + v_{xf})\left(\frac{v_{xf} - v_{xi}}{a_x}\right) = x_i + \frac{v_{xf}^2 - v_{xi}^2}{2a_x}$$

$$v_{xf}^2 = v_{xi}^2 + 2a_x(x_f - x_i) \quad \text{(for constant } a_x\text{)} \tag{2.17}$$

◀ Velocity as a function of position for the particle under constant acceleration model

This equation provides the final velocity in terms of the initial velocity, the constant acceleration, and the position of the particle.

For motion at *zero* acceleration, we see from Equations 2.13 and 2.16 that

$$\left.\begin{array}{c} v_{xf} = v_{xi} = v_x \\ x_f = x_i + v_x t \end{array}\right\} \quad \text{when } a_x = 0$$

That is, when the acceleration of a particle is zero, its velocity is constant and its position changes linearly with time. In terms of models, when the acceleration of a particle is zero, the particle under constant acceleration model reduces to the particle under constant velocity model (Section 2.3).

Equations 2.13 through 2.17 are **kinematic equations** that may be used to solve any problem involving a particle under constant acceleration in one dimension. These equations are listed together below for convenience. The choice of which equation you use in a given situation depends on what you know beforehand. Sometimes it is necessary to use two of these equations to solve for two unknowns. You should recognize that the quantities that vary during the motion are position x_f, velocity v_{xf}, and time t.

You will gain a great deal of experience in the use of these equations by solving a number of exercises and problems. Many times you will discover that more than one method can be used to obtain a solution. Remember that these equations of kinematics *cannot* be used in a situation in which the acceleration varies with time. They can be used only when the acceleration is constant.

Q **UICK QUIZ 2.7** In Figure 2.12, match each v_x–t graph on the top with the a_x–t graph on the bottom that best describes the motion.

a b c d e f

Figure 2.12 (Quick Quiz 2.7) Parts (a), (b), and (c) are v_x–t graphs of objects in one-dimensional motion. The possible accelerations of each object as a function of time are shown in scrambled order in (d), (e), and (f).

ANALYSIS MODEL Particle Under Constant Acceleration

Imagine a moving object that can be modeled as a particle. If it begins from position x_i and initial velocity v_{xi} and moves in a straight line with a constant acceleration a_x, its subsequent position and velocity are described by the following kinematic equations:

$$v_{xf} = v_{xi} + a_x t \tag{2.13}$$

$$v_{x,\text{avg}} = \frac{v_{xi} + v_{xf}}{2} \tag{2.14}$$

$$x_f = x_i + \tfrac{1}{2}(v_{xi} + v_{xf})t \tag{2.15}$$

$$x_f = x_i + v_{xi}t + \tfrac{1}{2}a_x t^2 \tag{2.16}$$

$$v_{xf}^2 = v_{xi}^2 + 2a_x(x_f - x_i) \tag{2.17}$$

Examples

- a car accelerating at a constant rate along a straight freeway
- a dropped object in the absence of air resistance (Section 2.8)
- an object on which a constant net force acts (Chapter 5)
- a charged particle in a uniform electric field (Chapter 22)

Example **2.7** | **Carrier Landing**

A jet lands on an aircraft carrier at a speed of 140 mi/h ($\approx$ 63 m/s).

(A) What is its acceleration (assumed constant) if it stops in 2.0 s due to an arresting cable that snags the jet and brings it to a stop?

SOLUTION

Conceptualize   You might have seen movies or television shows in which a jet lands on an aircraft carrier and is brought to rest surprisingly fast by an arresting cable. A careful reading of the problem reveals that in addition to being given the initial speed of 63 m/s, we also know that the final speed is zero. We define our x axis as the direction of motion of the jet. Notice that we have no information about the change in position of the jet while it is slowing down.

Categorize   Because the acceleration of the jet is assumed constant, we model it as a *particle under constant acceleration.*

Analyze   Equation 2.13 is the only equation in the particle under constant acceleration model that does not involve position, so we use it to find the acceleration of the jet, modeled as a particle:

$$a_x = \frac{v_{xf} - v_{xi}}{t} \approx \frac{0 - 63 \text{ m/s}}{2.0 \text{ s}}$$

$$= -32 \text{ m/s}^2$$

(B) If the jet touches down at position $x_i = 0$, what is its final position?

SOLUTION

Use Equation 2.15 to solve for the final position: $x_f = x_i + \frac{1}{2}(v_{xi} + v_{xf})t = 0 + \frac{1}{2}(63 \text{ m/s} + 0)(2.0 \text{ s}) = \boxed{63 \text{ m}}$

Finalize   Given the size of aircraft carriers, a length of 63 m seems reasonable for stopping the jet. The idea of using arresting cables to slow down landing aircraft and enable them to land safely on ships originated at about the time of World War I. The cables are still a vital part of the operation of modern aircraft carriers.

WHAT IF?   Suppose the jet lands on the deck of the aircraft carrier with a speed faster than 63 m/s but has the same acceleration due to the cable as that calculated in part (A). How will that change the answer to part (B)?

Answer   If the jet is traveling faster at the beginning, it will stop farther away from its starting point, so the answer to part (B) should be larger. Mathematically, we see in Equation 2.15 that if v_{xi} is larger, then x_f will be larger.

Example **2.8** | **Watch Out for the Speed Limit!**

You are driving at a constant speed of 45.0 m/s when you pass a trooper on a motorcycle hidden behind a billboard. One second after your car passes the billboard, the trooper sets out from the billboard to catch you, accelerating at a constant rate of 3.00 m/s². How long does it take the trooper to overtake your car?

Figure 2.13   (Example 2.8) You are in a speeding car that passes a hidden trooper.

SOLUTION

Conceptualize   This example represents a class of problems called *context-rich* problems. These problems involve real-world situations that one might encounter in one's daily life. These problems also involve "you" as opposed to an unspecified particle or object. With you as the character in the problem, *you* can make the connection between physics and everyday life!

Categorize   A pictorial representation (Fig. 2.13) helps clarify the sequence of events. Your car is modeled as a *particle under constant velocity*, and the trooper is modeled as a *particle under constant acceleration.*

Analyze   First, we write expressions for the position of each vehicle as a function of time. It is convenient to choose the position of the billboard as the origin and to set $t_\circledB = 0$ as the time the trooper begins moving. At that instant, your car has already

continued

2.8 continued

traveled a distance of 45.0 m from the billboard because it has traveled at a constant speed of $v_x = 45.0$ m/s for 1 s. Therefore, the initial position of your car is $x_\text{Ⓑ} = 45.0$ m.

Using the particle under constant velocity model, apply Equation 2.7 to give your car's position at any time t:

$$x_\text{car} = x_\text{Ⓑ} + v_{x\,\text{car}}t$$

A quick check shows that at $t = 0$, this expression gives your car's correct initial position when the trooper begins to move: $x_\text{car} = x_\text{Ⓑ} = 45.0$ m.

The trooper starts from rest at $t_\text{Ⓑ} = 0$ and accelerates at $a_x = 3.00$ m/s² away from the origin. Use Equation 2.16 to give her position at any time t:

$$x_f = x_i + v_{xi}t + \tfrac{1}{2}a_x t^2$$
$$x_\text{trooper} = 0 + (0)t + \tfrac{1}{2}a_x t^2 = \tfrac{1}{2}a_x t^2$$

Set the positions of your car and the trooper equal to represent the trooper overtaking your car at position Ⓒ:

$$x_\text{trooper} = x_\text{car}$$
$$\tfrac{1}{2}a_x t^2 = x_\text{Ⓑ} + v_{x\,\text{car}}t$$

Rearrange to give a quadratic equation:

$$\tfrac{1}{2}a_x t^2 - v_{x\,\text{car}}t - x_\text{Ⓑ} = 0$$

Solve the quadratic equation for the time at which the trooper catches your car (for help in solving quadratic equations, see Appendix B.2):

$$t = \frac{v_{x\,\text{car}} \pm \sqrt{v_{x\,\text{car}}^2 + 2a_x x_\text{Ⓑ}}}{a_x}$$

$$(1)\quad t = \frac{v_{x\,\text{car}}}{a_x} \pm \sqrt{\frac{v_{x\,\text{car}}^2}{a_x^2} + \frac{2x_\text{Ⓑ}}{a_x}}$$

Evaluate the solution, choosing the positive root because that is the only choice consistent with a time $t > 0$:

$$t = \frac{45.0 \text{ m/s}}{3.00 \text{ m/s}^2} + \sqrt{\frac{(45.0 \text{ m/s})^2}{(3.00 \text{ m/s}^2)^2} + \frac{2(45.0 \text{ m})}{3.00 \text{ m/s}^2}} = \boxed{31.0 \text{ s}}$$

Finalize   Why didn't we choose $t = 0$ as the time at which your car passes the trooper? If we did so, we would not be able to use the particle under constant acceleration model for the trooper. Her acceleration would be zero for the first second and then 3.00 m/s² for the remaining time. By defining the time $t = 0$ as when the trooper begins moving, we can use the particle under constant acceleration model for her movement for all positive times.

WHAT IF?   What if the trooper had a more powerful motorcycle with a larger acceleration? How would that change the time at which the trooper catches your car?

Answer   If the motorcycle has a larger acceleration, the trooper should catch up to your car sooner, so the answer for the time should be less than 31 s. Because all terms on the right side of Equation (1) have the acceleration a_x in the denominator, we see symbolically that increasing the acceleration will decrease the time at which the trooper catches your car.

2.8 Freely Falling Objects

It is well known that, in the absence of air resistance, all objects dropped near the Earth's surface fall toward the Earth with the same constant acceleration under the influence of the Earth's gravity, regardless of their mass. It was not until about 1600 that this conclusion was accepted. Before that time, the teachings of the Greek philosopher Aristotle (384–322 BC) had held that heavier objects fall faster than lighter ones.

The Italian Galileo Galilei (1564–1642) originated our present-day ideas concerning falling objects. There is a legend that he demonstrated the behavior of falling objects by observing that two different weights dropped simultaneously from the Leaning Tower of Pisa hit the ground at approximately the same time. Although there is some doubt that he carried out this particular experiment, it is well established that Galileo performed many experiments on objects moving on inclined planes. In his experiments, he rolled balls down a slight incline and measured the distances they covered in successive time intervals. The purpose of the incline was to reduce the acceleration, which made it possible for him to make accurate measurements of the time intervals. By gradually increasing the slope of the incline,

Galileo Galilei
Italian physicist and astronomer
(1564–1642)
Galileo formulated the laws that govern the motion of objects in free fall and made many other significant discoveries in physics and astronomy. Galileo publicly defended Nicolaus Copernicus's assertion that the Sun is at the center of the Universe (the heliocentric system).

he was finally able to draw conclusions about freely falling objects because a freely falling ball is equivalent to a ball moving down a vertical incline.

You might want to try the following experiment. Simultaneously drop a coin and a piece of paper from the same height. The coin will always reach the ground faster. Now, crumple the paper into a tight ball and repeat the experiment. Since you've minimized the effects of air resistance, the coin and the paper will have the same motion and will hit the floor at the same time. In the idealized case, in which air resistance is absent, such motion is referred to as *free-fall* motion. If this same experiment could be conducted in a vacuum, in which air resistance is truly negligible, the paper and the coin would fall with the same acceleration even when the paper is not crumpled. On August 2, 1971, astronaut David Scott conducted such a demonstration on the Moon. He simultaneously released a hammer and a feather, and the two objects fell together to the lunar surface. This simple demonstration surely would have pleased Galileo!

When we use the expression *freely falling object*, we do not necessarily refer to an object dropped from rest. A freely falling object is any object moving freely under the influence of gravity alone, regardless of its initial motion. Objects thrown upward or downward and those released from rest are all falling freely once they are released. Any freely falling object experiences an acceleration directed *downward*, regardless of its initial motion.

We shall denote the magnitude of the *free-fall acceleration*, also called the *acceleration due to gravity*, by the symbol *g*. The value of *g* decreases with increasing altitude above the Earth's surface. Furthermore, slight variations in *g* occur with changes in latitude. At the Earth's surface, the value of *g* is approximately 9.80 m/s^2. Unless stated otherwise, we shall use this value for *g* when performing calculations. For making quick estimates, use $g \sim 10$ m/s^2.

If we neglect air resistance and assume the free-fall acceleration does not vary with altitude over short vertical distances, the motion of a freely falling object moving vertically is equivalent to the motion of a particle under constant acceleration in one dimension. Therefore, the equations developed in Section 2.7 for the particle under constant acceleration model can be applied. The only modification for freely falling objects that we need to make in these equations is to note that the motion is in the vertical direction (the *y* direction) rather than in the horizontal direction (*x*) and that the acceleration is downward and has a magnitude of 9.80 m/s^2. Therefore, we choose $a_y = -g = -9.80$ m/s^2, where the negative sign means that the acceleration of a freely falling object is downward. In Chapter 13, we shall study how to deal with variations in *g* with altitude.

QUICK QUIZ **2.8** Consider the following choices: (a) increases, (b) decreases, (c) increases and then decreases, (d) decreases and then increases, (e) remains the same. From these choices, select what happens to **(i)** the acceleration and **(ii)** the speed of a ball after it is thrown upward into the air.

Conceptual Example 2.9   **The Daring Skydivers**

A skydiver jumps out of a hovering helicopter. A few seconds later, another skydiver jumps out, and they both fall along the same vertical line. Ignore air resistance so that both skydivers fall with the same acceleration. Does the difference in their speeds stay the same throughout the fall? Does the vertical distance between them stay the same throughout the fall?

SOLUTION

At any given instant, the speeds of the skydivers are different because one had a head start. In any time interval Δt after this instant, however, the two skydivers increase their speeds by the same amount because they have the same acceleration. Therefore, the difference in their speeds remains the same throughout the fall.

The first jumper always has a greater speed than the second. Therefore, in a given time interval, the first skydiver covers a greater distance than the second. Consequently, the separation distance between them increases.

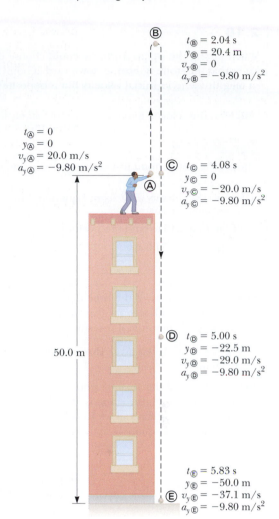

Example **2.10** **Not a Bad Throw for a Rookie!**

A stone thrown from the top of a building is given an initial velocity of 20.0 m/s straight upward. The stone is launched 50.0 m above the ground, and the stone just misses the edge of the roof on its way down as shown in Figure 2.14.

(A) Using $t_\text{Ⓐ} = 0$ as the time the stone leaves the thrower's hand at position Ⓐ, determine the time at which the stone reaches its maximum height.

SOLUTION

Conceptualize You most likely have experience with dropping objects or throwing them upward and watching them fall, so this problem should describe a familiar experience. To simulate this situation, toss a small object upward and notice the time interval required for it to fall to the floor. Now imagine throwing that object upward from the roof of a building.

Figure 2.14 (Example 2.10) Position, velocity, and acceleration values at various times for a freely falling stone thrown initially upward with a velocity $v_{yi} = 20.0$ m/s. Many of the quantities in the labels for points in the motion of the stone are calculated in the example. Can you verify the other values that are not?

$t_\text{Ⓐ} = 0$
$y_\text{Ⓐ} = 0$
$v_{y\text{Ⓐ}} = 20.0$ m/s
$a_{y\text{Ⓐ}} = -9.80$ m/s²

$t_\text{Ⓑ} = 2.04$ s
$y_\text{Ⓑ} = 20.4$ m
$v_{y\text{Ⓑ}} = 0$
$a_{y\text{Ⓑ}} = -9.80$ m/s²

$t_\text{Ⓒ} = 4.08$ s
$y_\text{Ⓒ} = 0$
$v_{y\text{Ⓒ}} = -20.0$ m/s
$a_{y\text{Ⓒ}} = -9.80$ m/s²

$t_\text{Ⓓ} = 5.00$ s
$y_\text{Ⓓ} = -22.5$ m
$v_{y\text{Ⓓ}} = -29.0$ m/s
$a_{y\text{Ⓓ}} = -9.80$ m/s²

50.0 m

$t_\text{Ⓔ} = 5.83$ s
$y_\text{Ⓔ} = -50.0$ m
$v_{y\text{Ⓔ}} = -37.1$ m/s
$a_{y\text{Ⓔ}} = -9.80$ m/s²

Categorize Because the stone is in free fall, it is modeled as a *particle under constant acceleration* due to gravity.

Analyze Recognize that the initial velocity is positive because the stone is launched upward. The velocity will change sign after the stone reaches its highest point, but the acceleration of the stone will *always* be downward so that it will always have a negative value. Choose an initial point just after the stone leaves the person's hand and a final point at the top of its flight.

Use Equation 2.13 to calculate the time at which the stone reaches its maximum height:

$$v_{yf} = v_{yi} + a_y t \rightarrow t = \frac{v_{yf} - v_{yi}}{a_y} = \frac{v_{y\text{Ⓑ}} - v_{y\text{Ⓐ}}}{-g}$$

Substitute numerical values, recognizing that $v = 0$ at point Ⓑ:

$$t = t_\text{Ⓑ} = \frac{0 - 20.0 \text{ m/s}}{-9.80 \text{ m/s}^2} = \boxed{2.04 \text{ s}}$$

(B) Find the maximum height of the stone.

SOLUTION

As in part (A), choose the initial and final points at the beginning and the end of the upward flight.

Set $y_\text{Ⓐ} = 0$ and substitute the time from part (A) into Equation 2.16 to find the maximum height:

$$y_\text{max} = y_\text{Ⓑ} = y_\text{Ⓐ} + v_{x\text{Ⓐ}} t + \tfrac{1}{2} a_y t^2$$

$$y_\text{Ⓑ} = 0 + (20.0 \text{ m/s})(2.04 \text{ s}) + \tfrac{1}{2}(-9.80 \text{ m/s}^2)(2.04 \text{ s})^2 = \boxed{20.4 \text{ m}}$$

(C) Determine the velocity of the stone when it returns to the height from which it was thrown.

SOLUTION

Choose the initial point where the stone is launched and the final point when it passes this position coming down.

Substitute known values into Equation 2.17:

$$v_{y\text{Ⓒ}}^2 = v_{y\text{Ⓐ}}^2 + 2a_y(y_\text{Ⓒ} - y_\text{Ⓐ})$$

$$v_{y\text{Ⓒ}}^2 = (20.0 \text{ m/s})^2 + 2(-9.80 \text{ m/s}^2)(0 - 0) = 400 \text{ m}^2/\text{s}^2$$

$$v_{y\text{Ⓒ}} = \boxed{-20.0 \text{ m/s}}$$

continued

2.10 continued

When taking the square root, we could choose either a positive or a negative root. We choose the negative root because we know that the stone is moving downward at point ©. The velocity of the stone when it arrives back at its original height is equal in magnitude to its initial velocity but is opposite in direction.

(D) Find the velocity and position of the stone at $t = 5.00$ s.

SOLUTION

Choose the initial point just after the throw and the final point 5.00 s later.

Calculate the velocity at Ⓓ from Equation 2.13:
$$v_{y\text{Ⓓ}} = v_{y\text{Ⓐ}} + a_y t = 20.0 \text{ m/s} + (-9.80 \text{ m/s}^2)(5.00 \text{ s}) = \boxed{-29.0 \text{ m/s}}$$

Use Equation 2.16 to find the position of the stone at $t_{\text{Ⓓ}} = 5.00$ s:
$$y_{\text{Ⓓ}} = y_{\text{Ⓐ}} + v_{y\text{Ⓐ}} t + \tfrac{1}{2} a_y t^2$$
$$= 0 + (20.0 \text{ m/s})(5.00 \text{ s}) + \tfrac{1}{2}(-9.80 \text{ m/s}^2)(5.00 \text{ s})^2$$
$$= \boxed{-22.5 \text{ m}}$$

Finalize The choice of the time defined as $t = 0$ is arbitrary and up to you to select as the problem solver. As an example of this arbitrariness, choose $t = 0$ as the time at which the stone is at the highest point in its motion. Then solve parts (C) and (D) again using this new initial instant and notice that your answers are the same as those above.

WHAT IF? What if the throw were from 30.0 m above the ground instead of 50.0 m? Which answers in parts (A) to (D) would change?

Answer None of the answers would change. All the motion takes place in the air during the first 5.00 s. (Notice that even for a throw from 30.0 m, the stone is above the ground at $t = 5.00$ s.) Therefore, the height from which the stone is thrown is not an issue. Mathematically, if we look back over our calculations, we see that we never entered the height from which the stone is thrown into any equation.

2.9 Kinematic Equations Derived from Calculus

The velocity of a particle moving in a straight line can be determined as the derivative of the position with respect to time. It is also possible to find the position of a particle if its velocity is known as a function of time. In calculus, the procedure used to perform this task is referred to either as *integration* or as finding the *antiderivative*.

Suppose the v_x–t graph for a particle moving along the x axis is as shown in Figure 2.15. Let us divide the time interval $t_f - t_i$ into many small intervals, each of duration Δt_n. From the definition of average velocity, we see that the displacement of the particle during any small interval, such as the one shaded in Figure 2.15, is given by $\Delta x_n = v_{xn,\text{avg}} \Delta t_n$, where $v_{xn,\text{avg}}$ is the average velocity in that interval. Therefore, the displacement during this small interval is simply the area of the shaded rectangle in Figure 2.15. The total displacement for the interval $t_f - t_i$ is the sum of the areas of all the rectangles from t_i to t_f:

$$\Delta x = \sum_n v_{xn,\text{avg}} \Delta t_n$$

where the symbol Σ (uppercase Greek sigma) signifies a sum over all terms, that is, over all values of n. Now, as the intervals are made smaller and smaller, the number of terms in the sum increases and the sum approaches a value equal to the area under the curve in the velocity–time graph. Therefore, in the limit $n \to \infty$, or $\Delta t_n \to 0$, the displacement is

$$\Delta x = \lim_{\Delta t_n \to 0} \sum_n v_{xn,\text{avg}} \Delta t_n \qquad (2.18)$$

The limit of the sum shown in Equation 2.18 is called a **definite integral** and so the displacement of the particle can be written as

Definite integral ▶

$$\Delta x = \int_{t_i}^{t_f} v_x(t)\, dt \qquad (2.19)$$

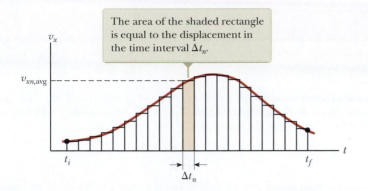

The area of the shaded rectangle is equal to the displacement in the time interval Δt_n.

Figure 2.15 Velocity versus time for a particle moving along the x axis. The total area under the curve is the total displacement of the particle.

where $v_x(t)$ denotes the velocity at any time t. If the explicit functional form of $v_x(t)$ is known and the limits are given, the integral can be evaluated.

Kinematic Equations

We now use the defining equations for acceleration and velocity to derive two of our kinematic equations, Equations 2.13 and 2.16.

The defining equation for acceleration (Eq. 2.10),

$$a_x = \frac{dv_x}{dt}$$

may be written as $dv_x = a_x\,dt$ or, in terms of an integral (or antiderivative), as

$$v_{xf} - v_{xi} = \int_0^t a_x\,dt$$

For the special case in which the acceleration is constant, a_x can be removed from the integral to give

$$v_{xf} - v_{xi} = a_x \int_0^t dt = a_x(t - 0) = a_x t \qquad (2.20)$$

which is Equation 2.13 in the particle under constant acceleration model.

Now let us consider the defining equation for velocity (Eq. 2.5):

$$v_x = \frac{dx}{dt}$$

We can write this equation as $dx = v_x\,dt$ or in integral form as

$$x_f - x_i = \int_0^t v_x\,dt$$

Because $v_x = v_{xf} = v_{xi} + a_x t$, this expression becomes

$$x_f - x_i = \int_0^t (v_{xi} + a_x t)\,dt = \int_0^t v_{xi}\,dt + a_x \int_0^t t\,dt = v_{xi}(t - 0) + a_x\left(\frac{t^2}{2} - 0\right)$$

$$x_f - x_i = v_{xi}t + \tfrac{1}{2}a_x t^2$$

which is Equation 2.16 in the particle under constant acceleration model.

PITFALL PREVENTION 2.10

Integration is an Area If this discussion of integration is confusing to you, just remember that the integral of a function is simply the area between the function and the x axis between the limits of integration. If the function has a simple shape, the area can be easily calculated without integration. For example, if the function is a constant, so that its graph is a horizontal line, the area is just that of the rectangle between the line and the x axis!

Summary

▶ Definitions

When a particle moves along the x axis from some initial position x_i to some final position x_f, its **displacement** is

$$\Delta x \equiv x_f - x_i \qquad (2.1)$$

The **average velocity** of a particle during some time interval is the displacement Δx divided by the time interval Δt during which that displacement occurs:

$$v_{x,\text{avg}} \equiv \frac{\Delta x}{\Delta t} \qquad (2.2)$$

The **average speed** of a particle is equal to the ratio of the total distance it travels to the total time interval during which it travels that distance:

$$v_{\text{avg}} \equiv \frac{d}{\Delta t} \qquad (2.3)$$

The **instantaneous velocity** of a particle is defined as the limit of the ratio $\Delta x/\Delta t$ as Δt approaches zero. By definition, this limit equals the derivative of x with respect to t, or the time rate of change of the position:

$$v_x \equiv \lim_{\Delta t \to 0} \frac{\Delta x}{\Delta t} = \frac{dx}{dt} \qquad (2.5)$$

The **instantaneous speed** of a particle is equal to the magnitude of its instantaneous velocity.

The **average acceleration** of a particle is defined as the ratio of the change in its velocity Δv_x divided by the time interval Δt during which that change occurs:

$$a_{x,\text{avg}} \equiv \frac{\Delta v_x}{\Delta t} = \frac{v_{xf} - v_{xi}}{t_f - t_i} \qquad (2.9)$$

The **instantaneous acceleration** is equal to the limit of the ratio $\Delta v_x/\Delta t$ as Δt approaches 0. By definition, this limit equals the derivative of v_x with respect to t, or the time rate of change of the velocity:

$$a_x \equiv \lim_{\Delta t \to 0} \frac{\Delta v_x}{\Delta t} = \frac{dv_x}{dt} \qquad (2.10)$$

▶ Concepts and Principles

When an object's velocity and acceleration are in the same direction, the object is speeding up. On the other hand, when the object's velocity and acceleration are in opposite directions, the object is slowing down. Remembering that $F_x \propto a_x$ is a useful way to identify the direction of the acceleration by associating it with a force.

An object falling freely in the presence of the Earth's gravity experiences free-fall acceleration directed toward the center of the Earth. If air resistance is neglected, if the motion occurs near the surface of the Earth, and if the range of the motion is small compared with the Earth's radius, the free-fall acceleration $a_y = -g$ is constant over the range of motion, where g is equal to 9.80 m/s².

Complicated problems are best approached in an organized manner. Recall and apply the *Conceptualize, Categorize, Analyze,* and *Finalize* steps of the **Analysis Model Approach to Problem Solving** when you need them.

An important aid to problem solving is the use of **analysis models.** Analysis models are situations that we have seen in previous problems. Each analysis model has one or more equations associated with it. When solving a new problem, identify the analysis model that corresponds to the problem. The model will tell you which equations to use. The first three analysis models introduced in this chapter are summarized below.

▶ Analysis Models for Problem Solving

Particle Under Constant Velocity. If a particle moves in a straight line with a constant speed v_x, its constant velocity is given by

$$v_x = \frac{\Delta x}{\Delta t} \qquad (2.6)$$

and its position is given by

$$x_f = x_i + v_x t \qquad (2.7)$$

Particle Under Constant Speed. If a particle moves a distance d along a curved or straight path with a constant speed, its constant speed is given by

$$v = \frac{d}{\Delta t} \qquad (2.8)$$

Particle Under Constant Acceleration. If a particle moves in a straight line with a constant acceleration a_x, its motion is described by the kinematic equations:

$$v_{xf} = v_{xi} + a_x t \tag{2.13}$$

$$v_{x,\text{avg}} = \frac{v_{xi} + v_{xf}}{2} \tag{2.14}$$

$$x_f = x_i + \tfrac{1}{2}(v_{xi} + v_{xf})t \tag{2.15}$$

$$x_f = x_i + v_{xi}t + \tfrac{1}{2}a_x t^2 \tag{2.16}$$

$$v_{xf}^{\,2} = v_{xi}^{\,2} + 2a_x(x_f - x_i) \tag{2.17}$$

Think–Pair–Share

See the Preface for an explanation of the icons used in this problems set. For additional assessment items for this section, go to **WEBASSIGN** From Cengage

1. You are at a carnival playing the "Strike-the-Bell" game, as shown in Figure TP2.1. The goal is to hit the end of the lever with a hammer, sending a hard object upward along the frictionless vertical track so as to strike a bell at the top. Showing off your control for the crowd, you hit the lever several times in a row in such a way that the hard object rises to a height $h = 4.50$ m and just touches the bell, which makes a gentle ringing sound. Now, to really impress the crowd, you swing the hammer with a mighty motion, hit the lever, and project the object upward with twice the initial speed of your previous demonstrations. Unbeknownst to you, on the previous demonstration, the bell came loose and slipped off to the side, so that, on this demonstration, the object bypasses the bell and is projected straight up into the air. What is the total time interval between when the object begins its upward motion and then later lands on the ground beside the apparatus?

Stephen Bjorck/Getty Images

Figure TP2.1

2. Your group is at the top of a cliff of height $h = 75.0$ m. At the bottom of the cliff is a pool of water. You split the group in two. The first half of the group volunteers a member to drop a rock from rest so that it falls straight downward and makes a splash in the water. The second half of the group volunteers a member to, after some time interval has passed since the first rock was dropped, throw a second rock straight downward so that both rocks arrive at the water at the same time. You test the performance by listening for a single splash made by the rocks simultaneously hitting the water. (a) If the second rock is thrown 1.00 s after the first rock is released, with what speed must the second rock be thrown? (b) If the fastest anyone in your group can throw the rock is 40.0 m/s, what is the longest time interval that can pass between the release of the rocks so that a single splash is heard? (c) If there is no limit as to how fast the rock can be thrown, what is the longest time interval that can pass between the release of the rocks so that a single splash is heard?

3. **ACTIVITY** Have your partner hold a ruler vertically with the zero end at the bottom. Place your open finger and thumb at the zero position. Without warning, your partner should release the ruler and you should catch it as soon as you see it moving. From the position of your finger on the ruler, determine your reaction time. Repeat the experiment a number of times to estimate the uncertainty in your reaction time. Have each member of your group catch the ruler and compare your reaction times.

4. **ACTIVITY** The Acela is an electric train on the Washington–New York–Boston run, carrying passengers at speeds as high as 170 mi/h. A velocity–time graph for the Acela is shown in Figure TP2.4. (a) Describe the train's motion in each successive time interval. (b) Find the train's peak positive acceleration in the motion graphed. (c) Find the train's displacement in miles between $t = 0$ and $t = 200$ s.

Figure TP2.4 Velocity–time graph for the Acela.

Problems

See the Preface for an explanation of the icons used in this problems set.
For additional assessment items for this section, go to WEB**ASSIGN**
From Cengage

SECTION 2.1 Position, Velocity, and Speed

1. The speed of a nerve impulse in the human body is about
BIO 100 m/s. If you accidentally stub your toe in the dark, estimate the time it takes the nerve impulse to travel to your brain.

2. A particle moves according to the equation $x = 10t^2$, where x
V is in meters and t is in seconds. (a) Find the average velocity for the time interval from 2.00 s to 3.00 s. (b) Find the average velocity for the time interval from 2.00 to 2.10 s.

3. The position of a pinewood derby car was observed at various times; the results are summarized in the following table. Find the average velocity of the car for (a) the first second, (b) the last 3 s, and (c) the entire period of observation.

t (s)	0	1.0	2.0	3.0	4.0	5.0
x (m)	0	2.3	9.2	20.7	36.8	57.5

SECTION 2.2 Instantaneous Velocity and Speed

4. An athlete leaves one end of a pool of length L at $t = 0$
S and arrives at the other end at time t_1. She swims back and arrives at the starting position at time t_2. If she is swimming initially in the positive x direction, determine her average velocities symbolically in (a) the first half of the swim, (b) the second half of the swim, and (c) the round trip. (d) What is her average speed for the round trip?

5. A position–time graph for a particle moving along the x axis is shown in Figure P2.5. (a) Find the average velocity in the time interval $t = 1.50$ s to $t = 4.00$ s. (b) Determine the instantaneous velocity at $t = 2.00$ s by measuring the slope of the tangent line shown in the graph. (c) At what value of t is the velocity zero?

Figure P2.5

SECTION 2.3 Analysis Model: Particle Under Constant Velocity

6. A car travels along a straight line at a constant speed of
AMT 60.0 mi/h for a distance d and then another distance d in the same direction at another constant speed. The average velocity for the entire trip is 30.0 mi/h. (a) What is the constant speed with which the car moved during the second distance d? (b) **What If?** Suppose the second distance d were traveled in the opposite direction; you forgot something and had to return home at the same constant speed as found in part (a). What is the average velocity for this trip? (c) What is the average speed for this new trip?

7. A person takes a trip, driving with a constant speed of 89.5
T km/h, except for a 22.0-min rest stop. If the person's average speed is 77.8 km/h, (a) how much time is spent on the trip and (b) how far does the person travel?

SECTION 2.5 Acceleration

8. A child rolls a marble on a bent track that is 100 cm long as shown in Figure P2.8. We use x to represent the position of the marble along the track. On the horizontal sections from $x = 0$ to $x = 20$ cm and from $x = 40$ cm to $x = 60$ cm, the marble rolls with constant speed. On the sloping sections, the marble's speed changes steadily. At the places where the slope changes, the marble stays on the track and does not undergo any sudden changes in speed. The child gives the marble some initial speed at $x = 0$ and $t = 0$ and then watches it roll to $x = 90$ cm, where it turns around, eventually returning to $x = 0$ with the same speed with which the child released it. Prepare graphs of x versus t, v_x versus t, and a_x versus t, vertically aligned with their time axes identical, to show the motion of the marble. You will not be able to place numbers other than zero on the horizontal axis or on the velocity or acceleration axes, but show the correct graph shapes.

Figure P2.8

9. Figure P2.9 shows a graph of v_x versus t for the motion of a motorcyclist as he starts from rest and moves along the road in a straight line. (a) Find the average acceleration for the time interval $t = 0$ to $t = 6.00$ s. (b) Estimate the time at which the acceleration has its greatest positive value and the value of the acceleration at that instant. (c) When is the acceleration zero? (d) Estimate the maximum negative value of the acceleration and the time at which it occurs.

Figure P2.9

10. (a) Use the data in Problem 3 to construct a smooth graph of position versus time. (b) By constructing tangents to the $x(t)$ curve, find the instantaneous velocity of the car at several instants. (c) Plot the instantaneous velocity versus time and, from this information, determine the average acceleration of the car. (d) What was the initial velocity of the car?

11. A particle starts from rest
and accelerates as shown
in Figure P2.11. Determine
(a) the particle's speed at t
= 10.0 s and at t = 20.0 s,
and (b) the distance traveled in the first 20.0 s.

Figure P2.11

SECTION 2.6 Motion Diagrams

12. Draw motion diagrams for
(a) an object moving to the
right at constant speed, (b) an object moving to the right
and speeding up at a constant rate, (c) an object moving
to the right and slowing down at a constant rate, (d) an
object moving to the left and speeding up at a constant rate,
and (e) an object moving to the left and slowing down at a
constant rate. (f) How would your drawings change if the
changes in speed were not uniform, that is, if the speed were
not changing at a constant rate?

13. Each of the strobe photographs (a), (b), and (c) in Figure P2.13
was taken of a single disk moving toward the right, which we
take as the positive direction. Within each photograph the time
interval between images is constant. For each photograph, prepare graphs of x versus t, v_x versus t, and a_x versus t, vertically
aligned with their time axes identical, to show the motion of
the disk. You will not be able to place numbers other than zero
on the axes, but show the correct shapes for the graph lines.

Charles D. Winters

Figure P2.13

SECTION 2.7 Analysis Model: Particle
Under Constant Acceleration

14. An electron in a cathode-ray tube accelerates uniformly
from 2.00×10^4 m/s to 6.00×10^6 m/s over 1.50 cm.
(a) In what time interval does the electron travel this 1.50 cm?
(b) What is its acceleration?

15. A parcel of air moving in a straight tube with a constant
acceleration of -4.00 m/s^2 has a velocity of 13.0 m/s at
10:05:00 a.m. (a) What is its velocity at 10:05:01 a.m.?
(b) At 10:05:04 a.m.? (c) At 10:04:59 a.m.? (d) Describe the
shape of a graph of velocity versus time for this parcel of air.
(e) Argue for or against the following statement: "Knowing

the single value of an object's constant acceleration is like
knowing a whole list of values for its velocity."

16. In Example 2.7, we investigated a jet landing on an aircraft
carrier. In a later maneuver, the jet comes in for a landing
on solid ground with a speed of 100 m/s, and its acceleration
can have a maximum magnitude of 5.00 m/s^2 as it comes to
rest. (a) From the instant the jet touches the runway, what
is the minimum time interval needed before it can come to
rest? (b) Can this jet land at a small tropical island airport
where the runway is 0.800 km long? (c) Explain your answer.

17. An object moving with uniform acceleration has a velocity
of 12.0 cm/s in the positive x direction when its x coordinate
is 3.00 cm. If its x coordinate 2.00 s later is -5.00 cm, what
is its acceleration?

18. Solve Example 2.8 by a graphical method. On the same
graph, plot position versus time for the car and the trooper.
From the intersection of the two curves, read the time at
which the trooper overtakes the car.

19. A glider of length ℓ moves through a stationary photogate
on an air track. A photogate (Fig. P2.19) is a device that measures the time interval Δt_d during which the glider blocks
a beam of infrared light passing across the photogate. The
ratio $v_d = \ell / \Delta t_d$ is the average velocity of the glider over this
part of its motion. Suppose the glider moves with constant
acceleration. (a) Argue for or against the idea that v_d is equal
to the instantaneous velocity of the glider when it is halfway
through the photogate in space. (b) Argue for or against
the idea that v_d is equal to the instantaneous velocity of the
glider when it is halfway through the photogate in time.

© Cengage

Figure P2.19 Problems 19 and 21.

20. *Why is the following situation impossible?* Starting from rest,
a charging rhinoceros moves 50.0 m in a straight line in
10.0 s. Her acceleration is constant during the entire motion,
and her final speed is 8.00 m/s.

21. A glider of length 12.4 cm moves on an air track with constant acceleration (Fig P2.19). A time interval of 0.628 s
elapses between the moment when its front end passes a
fixed point Ⓐ along the track and the moment when its
back end passes this point. Next, a time interval of 1.39 s
elapses between the moment when the back end of the
glider passes the point Ⓐ and the moment when the front
end of the glider passes a second point Ⓑ farther down the
track. After that, an additional 0.431 s elapses until the back
end of the glider passes point Ⓑ. (a) Find the average speed
of the glider as it passes point Ⓐ. (b) Find the acceleration
of the glider. (c) Explain how you can compute the acceleration without knowing the distance between points Ⓐ and Ⓑ.

22. In the particle under constant acceleration model, we
identify the variables and parameters v_{xi}, v_{xf}, a_x, t, and

$x_f - x_i$. Of the equations in the model, Equations 2.13–2.17, the first does not involve $x_f - x_i$, the second and third do not contain a_x, the fourth omits v_{xf}, and the last leaves out t. So, to complete the set, there should be an equation *not* involving v_{xi}. Derive it from the others.

23. At $t = 0$, one toy car is set rolling on a straight track with **QC** initial position 15.0 cm, initial velocity -3.50 cm/s, and constant acceleration 2.40 cm/s². At the same moment, another toy car is set rolling on an adjacent track with initial position 10.0 cm, initial velocity $+5.50$ cm/s, and constant acceleration zero. (a) At what time, if any, do the two cars have equal speeds? (b) What are their speeds at that time? (c) At what time(s), if any, do the cars pass each other? (d) What are their locations at that time? (e) Explain the difference between question (a) and question (c) as clearly as possible.

24. You are observing the poles along the side of the road as **CR** described in the opening storyline of the chapter. You have already stopped and measured the distance between adjacent poles as 40.0 m. You are now driving again and have activated your smartphone stopwatch. You start the stopwatch at $t = 0$ as you pass pole #1. At pole #2, the stopwatch reads 10.0 s. At pole #3, the stopwatch reads 25.0 s. Your friend tells you that he was pressing the brake and slowing down the car uniformly during the entire time interval from pole #1 to pole #3. (a) What was the acceleration of the car between poles #1 and #3? (b) What was the velocity of the car at pole #1? (c) If the motion of the car continues as described, what is the number of the *last* pole passed before the car comes to rest?

SECTION 2.8 Freely Falling Objects

Note: In all problems in this section, ignore the effects of air resistance.

25. *Why is the following situation impossible?* Emily challenges David to catch a $1 bill as follows. She holds the bill vertically as shown in Figure P2.25, with the center of the bill between but not touching David's index finger and thumb. Without warning, Emily releases the bill. David catches the bill without moving his hand downward. David's reaction time is equal to the average human reaction time.

Figure P2.25

© Cengage

26. An attacker at the base of a castle wall 3.65 m high throws a **QC** rock straight up with speed 7.40 m/s from a height of 1.55 m above the ground. (a) Will the rock reach the top of the wall? (b) If so, what is its speed at the top? If not, what initial speed must it have to reach the top? (c) Find the change in speed of a rock thrown straight down from the top of the wall at an initial speed of 7.40 m/s and moving between the same two points. (d) Does the change in speed of the downward-moving rock agree with the magnitude of the speed change of the rock moving upward between the same elevations? (e) Explain physically why it does or does not agree.

27. The height of a helicopter above the ground is given by $h = 3.00t^3$, where h is in meters and t is in seconds. At $t = 2.00$ s, the helicopter releases a small mailbag. How long after its release does the mailbag reach the ground?

28. A ball is thrown upward from the ground with an initial **T** speed of 25 m/s; at the same instant, another ball is dropped from a building 15 m high. After how long will the balls be at the same height above the ground?

29. A student throws a set of keys vertically upward to her soror- **T** ity sister, who is in a window 4.00 m above. The second student catches the keys 1.50 s later. (a) With what initial velocity were the keys thrown? (b) What was the velocity of the keys just before they were caught?

30. At time $t = 0$, a student throws a set of keys vertically upward **S** to her sorority sister, who is in a window at distance h above. The second student catches the keys at time t. (a) With what initial velocity were the keys thrown? (b) What was the velocity of the keys just before they were caught?

31. You have been hired by the prosecuting attorney as an expert **CR** witness in a robbery case. The defendant is accused of stealing an expensive and massive diamond ring in its box from a jewelry store. A witness to the alleged crime testified that she saw the defendant run from the store, stop next to an apartment building, and throw the box straight upward to an accomplice leaning out a fourth-floor window. When captured, the defendant did not have the stolen box with him and claimed innocence. When the witness testified in court about the defendant's throwing of the box to an accomplice, the defending attorney argued that it would be impossible to throw the box upward that high to reach the window in question. The bottom of the window is 19.0 m above the sidewalk. You have set up a demonstration in which the defendant was asked by the judge to throw a baseball horizontally as fast as he could and a radar device was used to determine that he can throw the ball at 20 m/s. (a) What testimony can you provide about the ability of the defendant to throw the box to the window in question? (b) What argument might the defense attorney make about the process used to develop your expert testimony? What might be your counter argument? Ignore any effects of air resistance on the box.

SECTION 2.9 Kinematic Equations Derived from Calculus

32. A student drives a moped along a straight road as described by the velocity–time graph in Figure P2.32. Sketch this graph in the middle of a sheet of graph paper. (a) Directly above your graph, sketch a graph of the position versus time, aligning the time coordinates of the two graphs. (b) Sketch a graph of the acceleration versus time directly below the velocity–time graph, again aligning the time coordinates. On each graph, show the numerical values of x and a_x for all points of inflection. (c) What is the acceleration at $t = 6.00$ s? (d) Find the position (relative to the starting point) at $t = 6.00$ s. (e) What is the moped's final position at $t = 9.00$ s?

Figure P2.32

33. Automotive engineers refer to the time rate of change of accel- **S** eration as the "jerk." Assume an object moves in one dimension such that its jerk J is constant. (a) Determine expressions for its acceleration $a_x(t)$, velocity $v_x(t)$, and position $x(t)$, given that its initial acceleration, velocity, and position are a_{xi}, v_{xi}, and x_i, respectively. (b) Show that $a_x^2 = a_{xi}^2 + 2J(v_x - v_{xi})$.

ADDITIONAL PROBLEMS

34. In Figure 2.11b, the area under the velocity–time graph
[QC] and between the vertical axis and time t (vertical dashed
[S] line) represents the displacement. As shown, this area consists of a rectangle and a triangle. (a) Compute their areas. (b) Explain how the sum of the two areas compares with the expression on the right-hand side of Equation 2.16.

35. The froghopper *Philaenus spumarius* is supposedly the best
[BIO] jumper in the animal kingdom. To start a jump, this insect can accelerate at 4.00 km/s^2 over a distance of 2.00 mm as it straightens its specially adapted "jumping legs." Assume the acceleration is constant. (a) Find the upward velocity with which the insect takes off. (b) In what time interval does it reach this velocity? (c) How high would the insect jump if air resistance were negligible? The actual height it reaches is about 70 cm, so air resistance must be a noticeable force on the leaping froghopper.

36. A woman is reported to have fallen 144 ft from the 17th floor of a building, landing on a metal ventilator box that she crushed to a depth of 18.0 in. She suffered only minor injuries. Ignoring air resistance, calculate (a) the speed of the woman just before she collided with the ventilator and (b) her average acceleration while in contact with the box. (c) Modeling her acceleration as constant, calculate the time interval it took to crush the box.

37. At $t = 0$, one athlete in a race running on a long, straight
[QC] track with a constant speed v_1 is a distance d_1 behind a sec-
[S] ond athlete running with a constant speed v_2. (a) Under what circumstances is the first athlete able to overtake the second athlete? (b) Find the time t at which the first athlete overtakes the second athlete, in terms of d_1, v_1, and v_2. (c) At what minimum distance d_2 from the leading athlete must the finish line be located so that the trailing athlete can at least tie for first place? Express d_2 in terms of d_1, v_1, and v_2 by using the result of part (b).

38. *Why is the following situation impossible?* A freight train is lumbering along at a constant speed of 16.0 m/s. Behind the freight train on the same track is a passenger train traveling in the same direction at 40.0 m/s. When the front of the passenger train is 58.5 m from the back of the freight train, the engineer on the passenger train recognizes the danger and hits the brakes of his train, causing the train to move with acceleration -3.00 m/s^2. Because of the engineer's action, the trains do not collide.

39. Hannah tests her new sports car by racing with Sam, an
[AMT] experienced racer. Both start from rest, but Hannah
[T] leaves the starting line 1.00 s after Sam does. Sam moves with a constant acceleration of 3.50 m/s^2, while Hannah maintains an acceleration of 4.90 m/s^2. Find (a) the time at which Hannah overtakes Sam, (b) the distance she travels before she catches him, and (c) the speeds of both cars at the instant Hannah overtakes Sam.

40. Two objects, A and B, are connected by hinges to a rigid
[QC] rod that has a length L. The objects slide along perpendic-
[S] ular guide rails as shown in Figure P2.40. Assume object A slides to the left with a constant speed v. (a) Find the velocity v_B of object B as a function of the angle θ. (b) Describe v_B

relative to v. Is v_B always smaller than v, larger than v, or the same as v, or does it have some other relationship?

Figure P2.40

41. Lisa rushes down onto a subway platform to find her train already departing. She stops and watches the cars go by. Each car is 8.60 m long. The first moves past her in 1.50 s and the second in 1.10 s. Find the constant acceleration of the train.

CHALLENGE PROBLEMS

42. Two thin rods are fastened to the inside of a circular ring as shown in Figure P2.42. One rod of length D is vertical, and the other of length L makes an angle θ with the horizontal. The two rods and the ring lie in a vertical plane. Two small beads are free to slide without friction along the rods. (a) If the two beads are released from rest simultaneously from the positions shown, use your intuition

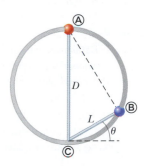

Figure P2.42

and guess which bead reaches the bottom first. (b) Find an expression for the time interval required for the red bead to fall from point Ⓐ to point Ⓒ in terms of g and D. (c) Find an expression for the time interval required for the blue bead to slide from point Ⓑ to point Ⓒ in terms of g, L, and θ. (d) Show that the two time intervals found in parts (b) and (c) are equal. *Hint:* What is the angle between the chords of the circle Ⓐ Ⓑ and Ⓑ Ⓒ? (e) Do these results surprise you? Was your intuitive guess in part (a) correct? This problem was inspired by an article by Thomas B. Greenslade, Jr., "Galileo's Paradox," *Phys. Teach.* **46**, 294 (May 2008).

43. In a women's 100-m race, accelerating uniformly, Laura takes 2.00 s and Healan 3.00 s to attain their maximum speeds, which they each maintain for the rest of the race. They cross the finish line simultaneously, both setting a world record of 10.4 s. (a) What is the acceleration of each sprinter? (b) What are their respective maximum speeds? (c) Which sprinter is ahead at the 6.00-s mark, and by how much? (d) What is the maximum distance by which Healan is behind Laura, and at what time does that occur?

44. **Review.** You are sitting in your car at rest at a traffic light with a bicyclist at rest next to you in the adjoining bicycle lane. As soon as the traffic light turns green, your car speeds up from rest to 50.0 mi/h with constant acceleration 9.00 mi/h/s and thereafter moves with a constant speed of 50.0 mi/h. At the same time, the cyclist speeds up from rest to 20.0 mi/h with constant acceleration 13.0 mi/h/s and thereafter moves with a constant speed of 20.0 mi/h. (a) For what time interval after the light turned green is the bicycle ahead of your car? (b) What is the maximum distance by which the bicycle leads your car during this time interval?

3 Vectors

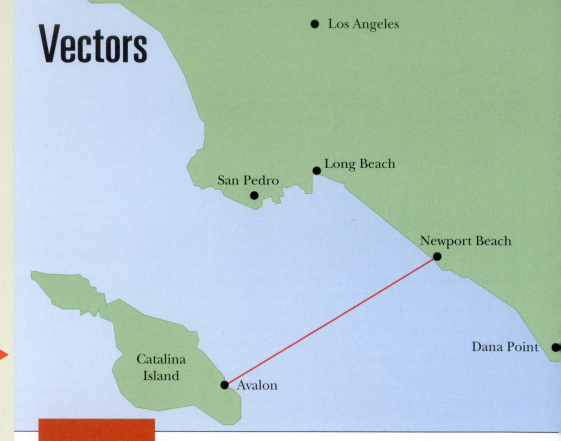

Catalina Island can be reached from different starting points along the Los Angeles–Orange County coast. The opening storyline refers to a trip to Avalon beginning in Newport Beach. ▶

STORYLINE **Your road trip in Chapter 2 takes you toward the ocean.** You end up in Newport Beach, California. Your friend who was driving the car in Chapter 2 owns a sailboat and asks you to pilot the boat from Newport Beach to Catalina Island, which is 26 miles off the coast. Your friend challenges you to pilot the boat along a perfectly straight line. Always up for a challenge, you agree, settle into the captain's chair, and then panic. You know you have to travel 26 miles in a straight line, but what should you set as the heading for the boat? The distance of 26 miles is not sufficient information to allow you to travel to Catalina Island in a straight line. You realize that your trip will require both the distance to Catalina Island and the *direction* in which you must travel. You ask your friend the appropriate direction to Catalina Island and he gives you a heading as an angle south of due west. You open the compass app on your smartphone, find the appropriate direction, and set sail!

CONNECTIONS If you move only along a straight line, as in the previous chapter, then a single number (with a positive or negative sign) can be used to specify your position with respect to the origin. In this chapter, we will study the positions of objects or points in two- or three-dimensional space that require two types of information: distance from a reference point and direction relative to a reference axis. Quantities that require these two types of information are called *vectors*. We will learn various properties of vectors and will see how to add and subtract vectors. Vector quantities are used throughout this text. In addition to the position vectors studied in this chapter, we will see other vector quantities in subsequent chapters, such as velocity, acceleration, force, and electric field. Therefore, it is imperative that you master the techniques discussed in this chapter.

3.1 Coordinate Systems

Many aspects of physics involve a description of a location in space. In Chapter 2, for example, we saw that the mathematical description of an object's motion requires a method for describing the object's position at various times. In two dimensions, this description is accomplished with the use of the Cartesian coordinate system, in which perpendicular axes intersect at a point defined as the origin O (Fig. 3.1). Cartesian coordinates of a point in space, representing the x and y values of the point, and expressed as (x, y), are also called *rectangular coordinates*.

Sometimes it is more convenient to represent a point in a plane by its *plane polar coordinates* (r, θ) as shown in Figure 3.2a. In this *polar coordinate system*, r is the distance from the origin to the point having Cartesian coordinates (x, y) and θ is the angle between a fixed axis and a line drawn from the origin to the point. The fixed axis is often the positive x axis, and θ is usually measured counterclockwise from it. From the right triangle in Figure 3.2b, we find that $\sin \theta = y/r$ and that $\cos \theta = x/r$. (A review of trigonometric functions is given in Appendix B.4.) Therefore, starting with the plane polar coordinates of any point, we can obtain the Cartesian coordinates by using the equations

$$x = r \cos \theta \qquad (3.1)$$

◀ Cartesian coordinates in terms of polar coordinates

$$y = r \sin \theta \qquad (3.2)$$

Conversely, if we know the Cartesian coordinates, the definitions of trigonometry tell us that the polar coordinates are given by

$$\tan \theta = \frac{y}{x} \qquad (3.3)$$

◀ Polar coordinates in terms of Cartesian coordinates

$$r = \sqrt{x^2 + y^2} \qquad (3.4)$$

Equation 3.4 is the familiar Pythagorean theorem.

These four expressions relating the coordinates (x, y) to the coordinates (r, θ) apply only when θ is defined as shown in Figure 3.2a—in other words, when positive θ is an angle measured counterclockwise from the positive x axis. (Some scientific calculators perform conversions between Cartesian and polar coordinates based on these standard conventions.) If the reference axis for the polar angle θ is chosen to be one other than the positive x axis or if the sense of increasing θ is chosen differently, the expressions relating the two sets of coordinates will be different from those above.

Figure 3.1 Designation of points in a Cartesian coordinate system. Every point is labeled with coordinates (x, y).

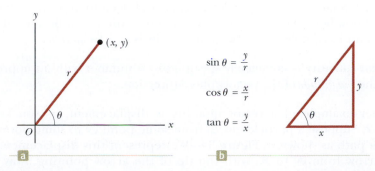

Figure 3.2 (a) The plane polar coordinates of a point are represented by the distance r and the angle θ, where θ is measured counterclockwise from the positive x axis. (b) The right triangle used to relate (x, y) to (r, θ).

Example **3.1**   **Polar Coordinates**

The Cartesian coordinates of a point in the xy plane are $(x, y) = (-3.50, -2.50)$ m as shown in Figure 3.3. Find the polar coordinates of this point.

S O L U T I O N

Conceptualize   The drawing in Figure 3.3 helps us conceptualize the problem. We wish to find r and θ. Based on the figure and the data given in the problem statement, we expect r to be a few meters and θ to be between 180° and 270°.

Categorize   Based on the statement of the problem and the Conceptualize step, we recognize that we are simply converting from Cartesian coordinates to polar coordinates. We therefore categorize this example as a substitution problem.

Figure 3.3   (Example 3.1)
Finding polar coordinates when
Cartesian coordinates are given.

As mentioned in Section 2.4, substitution problems generally do not have an extensive Analyze step other than the substitution of numbers into a given equation. Similarly, the Finalize step consists primarily of checking the units and making sure that the answer is reasonable and consistent with our expectations. Therefore, for substitution problems, we will not label Analyze or Finalize steps.

Use Equation 3.4 to find r:

$$r = \sqrt{x^2 + y^2} = \sqrt{(-3.50 \text{ m})^2 + (-2.50 \text{ m})^2} = \boxed{4.30 \text{ m}}$$

Use Equation 3.3 to find θ:

$$\tan \theta = \frac{y}{x} = \frac{-2.50 \text{ m}}{-3.50 \text{ m}} = 0.714$$

$$\theta = \boxed{216°}$$

Notice that you must use the signs of x and y to find that the point lies in the third quadrant of the coordinate system. That is, $\theta = 216°$, not 35.5°, whose tangent is also 0.714. Answers to both r and θ agree with our expectations in the Conceptualize step.

3.2   **Vector and Scalar Quantities**

We now formally describe the difference between scalar quantities and vector quantities. When you want to know the temperature outside so that you will know how to dress, the only information you need is a number and the unit "degrees C" or "degrees F." Temperature is therefore an example of a *scalar quantity*:

> A **scalar quantity** is completely specified by a single value with an appropriate unit and has no direction.

Other examples of scalar quantities are volume, mass, speed, time, and time intervals. Some scalars are always positive, such as mass and speed. Others, such as temperature, can have either positive or negative values. The rules of ordinary arithmetic are used to manipulate scalar quantities.

If you are preparing to pilot a small plane and need to know the wind velocity, you must know both the speed of the wind and its direction. Because direction is important for its complete specification, velocity is a *vector quantity*:

> A **vector quantity** is completely specified by a number with an appropriate unit (the *magnitude* of the vector) plus a direction.

Another example of a vector quantity is displacement, as you know from Chapter 2. Suppose a particle moves from some point Ⓐ to some point Ⓑ along a straight path as shown in Figure 3.4. We represent this displacement by drawing an arrow from Ⓐ to Ⓑ, with the tip of the arrow pointing away from the starting point. The direction of the arrowhead represents the direction of the displacement, and the length of the arrow represents the magnitude of the displacement. If the particle travels along some other path from Ⓐ to Ⓑ such as

Figure 3.4   As a particle moves from Ⓐ to Ⓑ either along the straight line *or* along an arbitrary path represented by the broken line, its displacement is a vector quantity shown by the arrow drawn from Ⓐ to Ⓑ.

shown by the broken line in Figure 3.4, its displacement is still the arrow drawn from Ⓐ to Ⓑ. Displacement depends only on the initial and final positions, so the displacement vector is independent of the path taken by the particle between these two points.

In this text, we use a boldface letter with an arrow over the letter, such as $\vec{A}$, to represent a vector. Another common notation for vectors with which you should be familiar is a simple boldface character: **A**. The magnitude of the vector $\vec{A}$ is written either A or $|\vec{A}|$. The magnitude of a vector has physical units, such as meters for displacement or meters per second for velocity. The magnitude of a vector is *always* a positive number.

What about the vector to follow in our opening storyline? What heading did your friend give you to Catalina Island? You can use a latitude and longitude finder online to find the coordinates for the opening of Newport Harbor and for Avalon Harbor. Then, putting these coordinates into a distance and azimuth calculator online, you find that the distance is 30.7 mi, with a heading of 236.2° relative to due east. (Note that Catalina is described as "26 miles across the sea" in a popular song from the 1950s, but we need to travel a bit farther to make this trip. An online calculation shows the distance between San Pedro and Avalon to be 27 miles, which might be the origin of the song.)

ⓆUICK QUIZ 3.1 Which of the following are vector quantities and which are scalar quantities? **(a)** your age **(b)** acceleration **(c)** velocity **(d)** speed **(e)** mass

3.3 Basic Vector Arithmetic

For many purposes, two vectors $\vec{A}$ and $\vec{B}$ may be defined to be *equal* if they have the same magnitude and if they point in the same direction. That is, $\vec{A} = \vec{B}$ only if $A = B$ and if $\vec{A}$ and $\vec{B}$ point in the same direction along parallel lines. For example, all the vectors in Figure 3.5 are equal even though they have different starting points. This property allows us to move a vector to a position parallel to itself in a diagram without affecting the vector.

The rules for **vector addition** are conveniently described by a graphical method. To add vector $\vec{B}$ to vector $\vec{A}$, first draw vector $\vec{A}$ on graph paper, with its magnitude represented by a convenient length scale, and then draw vector $\vec{B}$ to the same scale, with its tail starting from the tip of $\vec{A}$, as shown in Figure 3.6. The **resultant vector** $\vec{R} = \vec{A} + \vec{B}$ is the vector drawn from the tail of $\vec{A}$ to the tip of $\vec{B}$.

A geometric construction can also be used to add more than two vectors as shown in Figure 3.7 for the case of three vectors. The resultant vector $\vec{R} = \vec{A} + \vec{B} + \vec{C}$ is the vector that completes the polygon. In other words, $\vec{R}$ is the vector drawn from the tail of the first vector to the tip of the last vector. This technique for adding vectors is often called the "head to tail method."

PITFALL PREVENTION 3.1
Vector Addition Versus Scalar Addition Notice that $\vec{A} + \vec{B} = \vec{C}$ is very different from $A + B = C$. The first equation is a vector sum, which must be handled carefully, such as with the graphical method. The second equation is a simple algebraic addition of numbers that is handled with the normal rules of arithmetic.

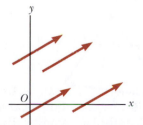

Figure 3.5 These four vectors are equal because they have equal lengths and point in the same direction.

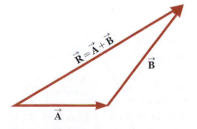

Figure 3.6 When vector $\vec{B}$ is added to vector $\vec{A}$ the resultant $\vec{R}$ is the vector that runs from the tail of $\vec{A}$ to the tip of $\vec{B}$.

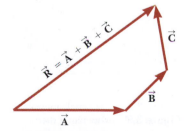

Figure 3.7 Geometric construction for summing three vectors. The resultant vector $\vec{R}$ is by definition the one that completes the polygon.

When two vectors are added, the sum is independent of the order of the addition. (This fact may seem trivial, but as you will see in Chapter 11, the order is important when vectors are multiplied. Procedures for multiplying vectors are discussed in Chapters 7 and 11.) This property, which can be seen from the geometric construction in Figure 3.8, is known as the **commutative law of addition:**

$$\vec{A} + \vec{B} = \vec{B} + \vec{A} \qquad (3.5)$$

When three or more vectors are added, their sum is independent of the way in which the individual vectors are grouped together. A geometric proof of this rule for three vectors is given in Figure 3.9, where two ways of adding the same three vectors are shown. This property is called the **associative law of addition:**

$$\vec{A} + (\vec{B} + \vec{C}) = (\vec{A} + \vec{B}) + \vec{C} \qquad (3.6)$$

We have described adding displacement vectors in this section because these types of vectors are easy to visualize. We can also add other types of vectors, such as velocity, force, and electric field vectors, which we will do in later chapters. When two or more vectors are added together, they must all have the same units and they must all be the same type of quantity. It would be meaningless to add a velocity vector (for example, 60 km/h to the east) to a displacement vector (for example, 200 km to the north) because these vectors represent different physical quantities. The same rule also applies to scalars. For example, it would be meaningless to add time intervals to temperatures.

The operation of **vector subtraction** makes use of the definition of the negative of a vector. The negative of the vector $\vec{A}$ is defined as the vector that when added to $\vec{A}$ gives zero for the vector sum. That is, $\vec{A} + (-\vec{A}) = 0$. The vectors $\vec{A}$ and $-\vec{A}$ have the same magnitude but point in opposite directions. We define the operation $\vec{A} - \vec{B}$ as vector $-\vec{B}$ added to vector $\vec{A}$:

$$\vec{A} - \vec{B} = \vec{A} + (-\vec{B}) \qquad (3.7)$$

The geometric construction for subtracting two vectors in this way is illustrated in Figure 3.10a.

Another way of looking at vector subtraction is to notice that the difference $\vec{A} - \vec{B}$ between two vectors $\vec{A}$ and $\vec{B}$ is what you have to add to the second vector to obtain the first. In this case, as Figure 3.10b shows, the vector $\vec{A} - \vec{B}$ points from the tip of the second vector to the tip of the first.

Scalar multiplication of vectors is straightforward. If vector $\vec{A}$ is multiplied by a positive scalar quantity m, the product $m\vec{A}$ is a vector that has the same direction as $\vec{A}$ and magnitude mA. If vector $\vec{A}$ is multiplied by a negative scalar quantity $-m$,

◀ Commutative law of addition

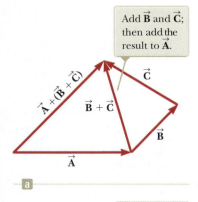

Figure 3.8 This construction shows that $\vec{A} + \vec{B} = \vec{B} + \vec{A}$ or, in other words, that vector addition is commutative.

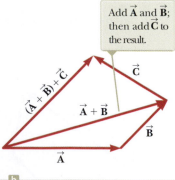

Figure 3.9 Geometric constructions for verifying the associative law of addition. (a) Vectors $\vec{B}$ and $\vec{C}$ are added first and added to $\vec{A}$. (b) Vectors $\vec{A}$ and $\vec{B}$ are added first, and then $\vec{C}$ is added.

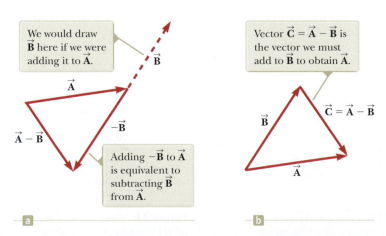

Figure 3.10 (a) Subtracting vector $\vec{B}$ from vector $\vec{A}$. The vector $-\vec{B}$ is equal in magnitude to vector $\vec{B}$ and points in the opposite direction. (b) A second way of looking at vector subtraction.

the product $-m\vec{\mathbf{A}}$ is directed opposite $\vec{\mathbf{A}}$. For example, the vector $5\vec{\mathbf{A}}$ is five times as long as $\vec{\mathbf{A}}$ and points in the same direction as $\vec{\mathbf{A}}$; the vector $-\frac{1}{3}\vec{\mathbf{A}}$ is one-third the length of $\vec{\mathbf{A}}$ and points in the direction opposite $\vec{\mathbf{A}}$.

QUICK QUIZ 3.2 The magnitudes of two vectors $\vec{\mathbf{A}}$ and $\vec{\mathbf{B}}$ are $A = 12$ units and $B = 8$ units. Which pair of numbers represents the *largest* and *smallest* possible values for the magnitude of the resultant vector $\vec{\mathbf{R}} = \vec{\mathbf{A}} + \vec{\mathbf{B}}$? **(a)** 14.4 units, 4 units **(b)** 12 units, 8 units **(c)** 20 units, 4 units **(d)** none of these answers

QUICK QUIZ 3.3 If vector $\vec{\mathbf{B}}$ is added to vector $\vec{\mathbf{A}}$, which *two* of the following choices must be true for the resultant vector to be equal to zero? **(a)** $\vec{\mathbf{A}}$ and $\vec{\mathbf{B}}$ are parallel and in the same direction. **(b)** $\vec{\mathbf{A}}$ and $\vec{\mathbf{B}}$ are parallel and in opposite directions. **(c)** $\vec{\mathbf{A}}$ and $\vec{\mathbf{B}}$ have the same magnitude. **(d)** $\vec{\mathbf{A}}$ and $\vec{\mathbf{B}}$ are perpendicular.

Example 3.2 | **A Vacation Trip**

A car travels 20.0 km due north and then 35.0 km in a direction 60.0° west of north as shown in Figure 3.11a. Find the magnitude and direction of the car's resultant displacement.

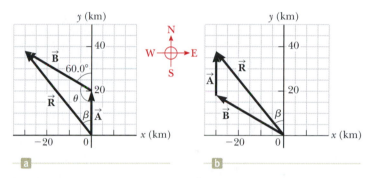

SOLUTION

Conceptualize The two vectors $\vec{\mathbf{A}}$ and $\vec{\mathbf{B}}$ that appear in Figure 3.11a help us conceptualize the problem. The resultant vector $\vec{\mathbf{R}}$ has also been drawn. We expect its magnitude to be a few tens of kilometers. The angle β that the resultant vector makes with the y axis is expected to be less than 60°, the angle that vector $\vec{\mathbf{B}}$ makes with the y axis.

Figure 3.11 (Example 3.2) (a) Graphical method for finding the resultant displacement vector $\vec{\mathbf{R}} = \vec{\mathbf{A}} + \vec{\mathbf{B}}$. (b) Adding the vectors in reverse order $(\vec{\mathbf{B}} + \vec{\mathbf{A}})$ gives the same result for $\vec{\mathbf{R}}$.

Categorize We can categorize this example as a simple analysis problem in vector addition. The displacement $\vec{\mathbf{R}}$ is the resultant when the two individual displacements $\vec{\mathbf{A}}$ and $\vec{\mathbf{B}}$ are added. We can further categorize it as a problem about the analysis of triangles, so we appeal to our expertise in geometry and trigonometry.

Analyze In this example, we show two ways to analyze the problem of finding the resultant of two vectors. The first way is to solve the problem geometrically, using graph paper and a protractor to measure the magnitude of $\vec{\mathbf{R}}$ and its direction in Figure 3.11a. (In fact, even when you know you are going to be carrying out a calculation, you should sketch the vectors to check your results.) With an ordinary ruler and protractor, a large diagram typically gives answers to two-digit but not to three-digit precision. Try using these tools on $\vec{\mathbf{R}}$ in Figure 3.11a and compare to the trigonometric analysis below!

The second way to solve the problem is to analyze it using algebra and trigonometry. The magnitude of $\vec{\mathbf{R}}$ can be obtained from the law of cosines as applied to the triangle in Figure 3.11a (see Appendix B.4).

Use $R^2 = A^2 + B^2 - 2AB\cos\theta$ from the law of cosines to find R:

$$R = \sqrt{A^2 + B^2 - 2AB\cos\theta}$$

Substitute numerical values, noting that $\theta = 180° - 60° = 120°$:

$$R = \sqrt{(20.0\text{ km})^2 + (35.0\text{ km})^2 - 2(20.0\text{ km})(35.0\text{ km})\cos 120°}$$
$$= \boxed{48.2\text{ km}}$$

Use the law of sines (Appendix B.4) to find the direction of $\vec{\mathbf{R}}$ measured from the northerly direction:

$$\frac{\sin\beta}{B} = \frac{\sin\theta}{R}$$

$$\sin\beta = \frac{B}{R}\sin\theta = \frac{35.0\text{ km}}{48.2\text{ km}}\sin 120° = 0.629$$

$$\beta = \boxed{38.9°}$$

The resultant displacement of the car is 48.2 km in a direction 38.9° west of north.

continued

3.2 continued

Finalize Does the angle β that we calculated agree with an estimate made by looking at Figure 3.11a or with an actual angle measured from the diagram using the graphical method? Is it reasonable that the magnitude of $\vec{R}$ is larger than that of both $\vec{A}$ and $\vec{B}$? Are the units of $\vec{R}$ correct?

Although the head to tail method of adding vectors works well, it suffers from two disadvantages. First, some people find using the laws of cosines and sines to be awkward. Second, a triangle only results if you are adding two vectors. If you are adding three or more vectors, the resulting geometric shape is usually not a triangle. In Section 3.4, we explore a new method of adding vectors that will address both of these disadvantages.

WHAT IF? Suppose the trip were taken with the two vectors in reverse order: 35.0 km at 60.0° west of north first and then 20.0 km due north. How would the magnitude and the direction of the resultant vector change?

Answer They would not change. The commutative law for vector addition tells us that the order of vectors in an addition is irrelevant. Graphically, Figure 3.11b shows that the vectors added in the reverse order give us the same resultant vector.

3.4 Components of a Vector and Unit Vectors

The graphical method of adding vectors is not recommended whenever high accuracy is required or in three-dimensional problems. In this section, we describe a method of adding vectors that makes use of the projections of vectors along coordinate axes. These projections are called the **components** of the vector or its **rectangular components**. Any vector can be completely described by its components.

Consider a vector $\vec{A}$ lying in the xy plane and making an arbitrary angle θ with the positive x axis as shown in Figure 3.12a. This vector can be expressed as the sum of two other *component vectors* $\vec{A}_x$, which is parallel to the x axis, and $\vec{A}_y$, which is parallel to the y axis. From the figure, we see that the three vectors form a right triangle and that $\vec{A} = \vec{A}_x + \vec{A}_y$. We shall often refer to the "components of a vector $\vec{A}$," written A_x and A_y (without the boldface notation). Figure 3.12b shows the component vector $\vec{A}_y$ moved to the left so that it lies along the y axis. We see that the component A_x represents the projection of $\vec{A}$ along the x axis, and the component A_y represents the projection of $\vec{A}$ along the y axis. These components can be positive or negative. The component A_x is positive if the component vector $\vec{A}_x$ points in the positive x direction and is negative if $\vec{A}_x$ points in the negative x direction. A similar statement is made for the component A_y.

From Figure 3.12 and the definition of sine and cosine, we see that $\cos \theta = A_x/A$ and that $\sin \theta = A_y/A$. Hence, the components of $\vec{A}$ are

$$A_x = A \cos \theta \qquad (3.8)$$

$$A_y = A \sin \theta \qquad (3.9)$$

> **PITFALL PREVENTION 3.2**
>
> **_x_ and _y_ Components** Equations 3.8 and 3.9 associate the cosine of the angle with the _x_ component and the sine of the angle with the _y_ component. This association is true *only* because we measured the angle θ with respect to the _x_ axis, so do not memorize these equations. If θ is measured with respect to the _y_ axis (as in some problems), these equations will be incorrect. Think about which side of the triangle containing the components is adjacent to the angle and which side is opposite and then assign the cosine and sine accordingly.

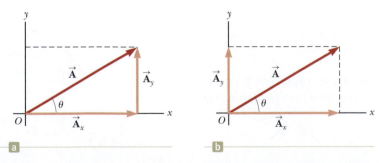

Figure 3.12 (a) A vector $\vec{A}$ lying in the xy plane can be represented as a vector sum of its component vectors $\vec{A}_x$ and $\vec{A}_y$. These three vectors form a right triangle. (b) The y component vector $\vec{A}_y$ can be moved to the left so that it lies along the y axis.

The magnitudes of these components are the lengths of the two sides of a right triangle with a hypotenuse of length A. Therefore, the magnitude and direction of $\vec{A}$ are related to its components through the expressions

$$A = \sqrt{A_x^2 + A_y^2} \tag{3.10}$$

$$\theta = \tan^{-1}\left(\frac{A_y}{A_x}\right) \tag{3.11}$$

Notice that the signs of the components A_x and A_y depend on the angle θ. For example, if $\theta = 120°$, A_x is negative and A_y is positive. If $\theta = 225°$, both A_x and A_y are negative. Figure 3.13 summarizes the directions of the component vectors and signs of the components when $\vec{A}$ lies in the various quadrants.

When solving problems in two dimensions, you can specify a vector $\vec{A}$ either with its components A_x and A_y or with its magnitude and direction A and θ.

In many applications, it is convenient to express the components in a coordinate system having axes that are not horizontal and vertical but that are still perpendicular to each other. For example, we will consider the motion of objects sliding down inclined planes. For these examples, it is often convenient to orient the x axis parallel to the plane and the y axis perpendicular to the plane.

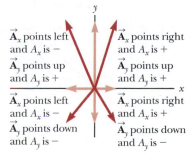

Figure 3.13 The signs of the components of a vector $\vec{A}$ depend on the quadrant in which the vector is located.

QUICK QUIZ 3.4 Choose the correct response to make the sentence true: A component of a vector is **(a)** always, **(b)** never, or **(c)** sometimes larger than the magnitude of the vector.

Vector quantities often are expressed in terms of unit vectors. A **unit vector** is a dimensionless vector having a magnitude of exactly 1. Unit vectors are used to specify a given direction and have no other physical significance. They are used solely as a bookkeeping convenience in describing a direction in space. We shall use the symbols $\hat{\mathbf{i}}$, $\hat{\mathbf{j}}$, and $\hat{\mathbf{k}}$ to represent unit vectors pointing in the positive x, y, and z directions, respectively. (The "hats," or circumflexes, on the symbols are a standard notation for unit vectors.) The unit vectors $\hat{\mathbf{i}}$, $\hat{\mathbf{j}}$, and $\hat{\mathbf{k}}$ form a set of mutually perpendicular vectors in a right-handed coordinate system as shown in Figure 3.14a. The magnitude of each unit vector equals 1; that is, $|\hat{\mathbf{i}}| = |\hat{\mathbf{j}}| = |\hat{\mathbf{k}}| = 1$.

Consider a vector $\vec{A}$ lying in the xy plane as shown in Figure 3.14b. The product of the component A_x and the unit vector $\hat{\mathbf{i}}$ is the component vector $\vec{A}_x = A_x\hat{\mathbf{i}}$, which lies on the x axis and has magnitude $|A_x|$. Likewise, $\vec{A}_y = A_y\hat{\mathbf{j}}$ is the component vector of magnitude $|A_y|$ lying on the y axis. Therefore, the unit-vector notation for the vector $\vec{A}$ is

$$\vec{A} = A_x\hat{\mathbf{i}} + A_y\hat{\mathbf{j}} \tag{3.12}$$

Consider now the polar coordinates shown for the point in Figure 3.2. The point in the first quadrant in that figure is reproduced in Figure 3.15. Notice that we can

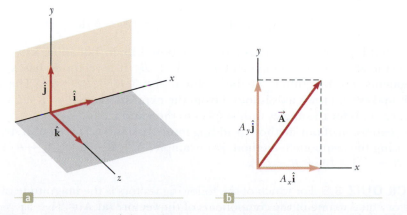

Figure 3.14 (a) The unit vectors $\hat{\mathbf{i}}$, $\hat{\mathbf{j}}$, and $\hat{\mathbf{k}}$ are directed along the x, y, and z axes, respectively. (b) Vector $\vec{A} = A_x\hat{\mathbf{i}} + A_y\hat{\mathbf{j}}$ lying in the xy plane has components A_x and A_y.

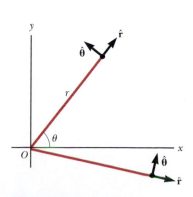

Figure 3.15 Unit vectors for a point described by polar coordinates.

identify radial and angular unit vectors $\hat{\mathbf{r}}$ and $\hat{\boldsymbol{\theta}}$. Just like for rectangular coordinates, these vectors are of unit length. Unlike rectangular coordinates, however, the directions of radial and angular unit vectors depend on the point, as shown by the point in the fourth quadrant in Figure 3.15.

Now let us see how to use components to add vectors when the graphical method is not sufficiently accurate. Suppose we wish to add vector $\vec{\mathbf{B}}$ to vector $\vec{\mathbf{A}}$ in Equation 3.12, where vector $\vec{\mathbf{B}}$ has components B_x and B_y. Because of the bookkeeping convenience of the unit vectors, all we do is add the x and y components separately. The resultant vector $\vec{\mathbf{R}}$ is

$$\vec{\mathbf{R}} = \vec{\mathbf{A}} + \vec{\mathbf{B}} = (A_x\hat{\mathbf{i}} + A_y\hat{\mathbf{j}}) + (B_x\hat{\mathbf{i}} + B_y\hat{\mathbf{j}})$$

or, rearranging terms,

$$\vec{\mathbf{R}} = (A_x + B_x)\hat{\mathbf{i}} + (A_y + B_y)\hat{\mathbf{j}} \tag{3.13}$$

Because $\vec{\mathbf{R}} = R_x\hat{\mathbf{i}} + R_y\hat{\mathbf{j}}$, we see that the components of the resultant vector are

$$R_x = A_x + B_x$$
$$R_y = A_y + B_y \tag{3.14}$$

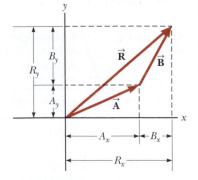

Figure 3.16 This geometric construction for the sum of two vectors shows the relationship between the components of the resultant $\vec{\mathbf{R}}$ and the components of the individual vectors.

Therefore, we see that in the component method of adding vectors, we add all the x components together to find the x component of the resultant vector and use the same process for the y components. We can check this addition by components with a geometric construction as shown in Figure 3.16.

The magnitude of $\vec{\mathbf{R}}$ and the angle it makes with the x axis are obtained from its components using the relationships

$$R = \sqrt{R_x^2 + R_y^2} = \sqrt{(A_x + B_x)^2 + (A_y + B_y)^2} \tag{3.15}$$

$$\tan\theta = \frac{R_y}{R_x} = \frac{A_y + B_y}{A_x + B_x} \tag{3.16}$$

PITFALL PREVENTION 3.3

Tangents on Calculators
Equation 3.16 involves the calculation of an angle by means of a tangent function. Generally, the inverse tangent function on calculators provides an angle between $-90°$ and $+90°$. As a consequence, if the vector you are studying lies in the second or third quadrant, the angle measured from the positive x axis will be the angle your calculator returns plus 180°.

At times, we need to consider situations involving motion in three component directions. The extension of our methods to three-dimensional vectors is straightforward. If $\vec{\mathbf{A}}$ and $\vec{\mathbf{B}}$ both have x, y, and z components, they can be expressed in the form

$$\vec{\mathbf{A}} = A_x\hat{\mathbf{i}} + A_y\hat{\mathbf{j}} + A_z\hat{\mathbf{k}} \tag{3.17}$$

$$\vec{\mathbf{B}} = B_x\hat{\mathbf{i}} + B_y\hat{\mathbf{j}} + B_z\hat{\mathbf{k}} \tag{3.18}$$

The sum of $\vec{\mathbf{A}}$ and $\vec{\mathbf{B}}$ is

$$\vec{\mathbf{R}} = (A_x + B_x)\hat{\mathbf{i}} + (A_y + B_y)\hat{\mathbf{j}} + (A_z + B_z)\hat{\mathbf{k}} \tag{3.19}$$

Notice that Equation 3.19 differs from Equation 3.13: in Equation 3.19, the resultant vector also has a z component $R_z = A_z + B_z$. If a vector $\vec{\mathbf{R}}$ has x, y, and z components, the magnitude of the vector is $R = \sqrt{R_x^2 + R_y^2 + R_z^2}$. The angle θ_x that $\vec{\mathbf{R}}$ makes with the x axis is found from the expression $\cos\theta_x = R_x/R$, with similar expressions for the angles with respect to the y and z axes.

The extension of our method to adding more than two vectors is also straightforward using the component method. For example, $\vec{\mathbf{A}} + \vec{\mathbf{B}} + \vec{\mathbf{C}} = (A_x + B_x + C_x)\hat{\mathbf{i}} + (A_y + B_y + C_y)\hat{\mathbf{j}} + (A_z + B_z + C_z)\hat{\mathbf{k}}$.

QUICK QUIZ 3.5 For which of the following vectors is the magnitude of the vector equal to one of the components of the vector? **(a)** $\vec{\mathbf{A}} = 2\hat{\mathbf{i}} + 5\hat{\mathbf{j}}$ **(b)** $\vec{\mathbf{B}} = -3\hat{\mathbf{j}}$ **(c)** $\vec{\mathbf{C}} = +5\hat{\mathbf{k}}$

Example 3.3 The Sum of Two Vectors

Find the sum of two vectors $\vec{A}$ and $\vec{B}$ lying in the xy plane and given by

$$\vec{A} = (2.0\,\hat{i} + 2.0\,\hat{j}) \quad \text{and} \quad \vec{B} = (2.0\,\hat{i} - 4.0\,\hat{j})$$

SOLUTION

Conceptualize You can conceptualize the situation by drawing the vectors on graph paper. Do this and then draw an approximation of the expected resultant vector.

Categorize We categorize this example as a simple substitution problem. Comparing this expression for $\vec{A}$ with the general expression $\vec{A} = A_x\hat{i} + A_y\hat{j} + A_z\hat{k}$, we see that $A_x = 2.0$, $A_y = 2.0$, and $A_z = 0$. Likewise, $B_x = 2.0$, $B_y = -4.0$, and $B_z = 0$. We can use a two-dimensional approach because there are no z components.

Use Equation 3.13 to obtain the resultant vector $\vec{R}$:

$$\vec{R} = (A_x + B_x)\hat{i} + (A_y + B_y)\hat{j} = (2.0 + 2.0)\hat{i} + (2.0 - 4.0)\hat{j}$$
$$= 4.0\,\hat{i} - 2.0\,\hat{j}$$

Use Equation 3.15 to find the magnitude of $\vec{R}$:

$$R = \sqrt{R_x^2 + R_y^2} = \sqrt{(4.0)^2 + (-2.0)^2} = \sqrt{20} = \boxed{4.5}$$

Find the direction of $\vec{R}$ from Equation 3.16:

$$\tan\theta = \frac{R_y}{R_x} = \frac{-2.0}{4.0} = -0.50$$

Your calculator likely gives the answer $-27°$ for $\theta = \tan^{-1}(-0.50)$. This answer is correct if we interpret it to mean 27° *clockwise* from the x axis. Our standard form has been to quote the angles measured *counterclockwise* from the $+x$ axis, and that angle for this vector is $\theta = \boxed{333°}$.

Example 3.4 The Resultant Displacement

A particle undergoes three consecutive displacements: $\Delta\vec{r}_1 = (15\,\hat{i} + 30\,\hat{j} + 12\,\hat{k})$ cm, $\Delta\vec{r}_2 = (23\,\hat{i} - 14\,\hat{j} - 5.0\,\hat{k})$ cm, and $\Delta\vec{r}_3 = (-13\,\hat{i} + 15\,\hat{j})$ cm. Find unit-vector notation for the resultant displacement and its magnitude.

SOLUTION

Conceptualize Although x is sufficient to locate a point in one dimension, we need a vector $\vec{r}$ to locate a point in two or three dimensions. The notation $\Delta\vec{r}$ is a generalization of the one-dimensional displacement Δx in Equation 2.1. Three-dimensional displacements are more difficult to conceptualize than those in two dimensions because they cannot be drawn on paper like the latter.

For this problem, let us imagine that you start with your pencil at the origin of a piece of graph paper on which you have drawn x and y axes. Move your pencil 15 cm to the right along the x axis, then 30 cm upward along the y axis, and then 12 cm *perpendicularly toward you away* from the graph paper. This procedure provides the displacement described by $\Delta\vec{r}_1$. From this point, move your pencil 23 cm to the right parallel to the x axis, then 14 cm parallel to the graph paper in the $-y$ direction, and then 5.0 cm perpendicularly away from you toward the graph paper. You are now at the displacement from the origin described by $\Delta\vec{r}_1 + \Delta\vec{r}_2$. From this point, move your pencil 13 cm to the left in the $-x$ direction, and (finally!) 15 cm parallel to the graph paper along the y axis. Your final position is at a displacement $\Delta\vec{r}_1 + \Delta\vec{r}_2 + \Delta\vec{r}_3$ from the origin.

Categorize Despite the difficulty in conceptualizing in three dimensions, we can categorize this problem as a substitution problem because of the careful bookkeeping methods that we have developed for vectors. The mathematical manipulation keeps track of this motion along the three perpendicular axes in an organized, compact way, as we see below.

To find the resultant displacement, add the three vectors:

$$\Delta\vec{r} = \Delta\vec{r}_1 + \Delta\vec{r}_2 + \Delta\vec{r}_3$$
$$= (15 + 23 - 13)\hat{i}\ \text{cm} + (30 - 14 + 15)\hat{j}\ \text{cm} + (12 - 5.0 + 0)\hat{k}\ \text{cm}$$
$$= \boxed{(25\,\hat{i} + 31\,\hat{j} + 7.0\,\hat{k})\ \text{cm}}$$

Find the magnitude of the resultant vector:

$$R = \sqrt{R_x^2 + R_y^2 + R_z^2}$$
$$= \sqrt{(25\ \text{cm})^2 + (31\ \text{cm})^2 + (7.0\ \text{cm})^2} = \boxed{40\ \text{cm}}$$

Example **3.5** | Taking a Hike

A hiker begins a trip by first walking 25.0 km southeast from her car. She stops and sets up her tent for the night. On the second day, she walks 40.0 km in a direction 60.0° north of east, at which point she discovers a forest ranger's tower.

(A) Determine the components of the hiker's displacement for each day.

SOLUTION

Conceptualize We conceptualize the problem by drawing a sketch as in Figure 3.17. If we denote the displacement vectors on the first and second days by $\vec{\mathbf{A}}$ and $\vec{\mathbf{B}}$, respectively, and use the car as the origin of coordinates, we obtain the vectors shown in Figure 3.17. The sketch allows us to estimate the resultant vector as shown.

Categorize Having drawn the resultant $\vec{\mathbf{R}}$, we can now categorize this problem as one we've solved before: an addition of two vectors. You should now have a hint of the power of categorization in that many new problems are very similar to problems we have already solved if we are careful to conceptualize them. Once we have drawn the displacement vectors and categorized the problem, this problem is no longer about a hiker, a walk, a car, a tent, or a tower. It is a problem about vector addition, one that we have already solved.

Figure 3.17 (Example 3.5) The total displacement of the hiker is the vector $\vec{\mathbf{R}} = \vec{\mathbf{A}} + \vec{\mathbf{B}}$.

Analyze Displacement $\vec{\mathbf{A}}$ has a magnitude of 25.0 km and is directed 45.0° below the positive x axis.

Find the components of $\vec{\mathbf{A}}$ using Equations 3.8 and 3.9:

$$A_x = A \cos (-45.0°) = (25.0 \text{ km})(0.707) = \boxed{17.7 \text{ km}}$$

$$A_y = A \sin (-45.0°) = (25.0 \text{ km})(-0.707) = \boxed{-17.7 \text{ km}}$$

The negative value of A_y indicates that the hiker ends up below the x axis on the first day. The signs of A_x and A_y also are evident from Figure 3.17.

Find the components of $\vec{\mathbf{B}}$ using Equations 3.8 and 3.9:

$$B_x = B \cos 60.0° = (40.0 \text{ km})(0.500) = \boxed{20.0 \text{ km}}$$

$$B_y = B \sin 60.0° = (40.0 \text{ km})(0.866) = \boxed{34.6 \text{ km}}$$

(B) Determine the components of the hiker's resultant displacement $\vec{\mathbf{R}}$ for the trip. Find an expression for $\vec{\mathbf{R}}$ in terms of unit vectors.

SOLUTION

Use Equation 3.14 to find the components of the resultant displacement $\vec{\mathbf{R}} = \vec{\mathbf{A}} + \vec{\mathbf{B}}$:

$$R_x = A_x + B_x = 17.7 \text{ km} + 20.0 \text{ km} = \boxed{37.7 \text{ km}}$$

$$R_y = A_y + B_y = -17.7 \text{ km} + 34.6 \text{ km} = \boxed{17.0 \text{ km}}$$

Write the total displacement in unit-vector form:

$$\vec{\mathbf{R}} = \boxed{(37.7\hat{\mathbf{i}} + 17.0\hat{\mathbf{j}}) \text{ km}}$$

Finalize Looking at the graphical representation in Figure 3.17, we estimate the position of the tower to be about (38 km, 17 km), which is consistent with the components of $\vec{\mathbf{R}}$ in our result for the final position of the hiker. Also, both components of $\vec{\mathbf{R}}$ are positive, putting the final position in the first quadrant of the coordinate system, which is also consistent with Figure 3.17.

WHAT IF? After reaching the tower, the hiker wishes to return to her car along a single straight line. What are the components of the vector representing this hike? What should the direction of the hike be?

Answer The desired vector $\vec{\mathbf{R}}_{\text{car}}$ is the negative of vector $\vec{\mathbf{R}}$:

$$\vec{\mathbf{R}}_{\text{car}} = -\vec{\mathbf{R}} = (-37.7\hat{\mathbf{i}} - 17.0\hat{\mathbf{j}}) \text{ km}$$

The direction is found by calculating the angle that the vector makes with the x axis:

$$\tan \theta = \frac{R_{\text{car},y}}{R_{\text{car},x}} = \frac{-17.0 \text{ km}}{-37.7 \text{ km}} = 0.450$$

which gives an angle of $\theta = 204.2°$, or 24.2° south of west.

Summary

> ## Definitions

Scalar quantities are those that have only a numerical value and no associated direction.

Vector quantities have both magnitude and direction and obey the laws of vector addition. The magnitude of a vector is *always* a positive number.

> ## Concepts and Principles

When two or more vectors are added together, they must all have the same units and they all must be the same type of quantity. We can add two vectors $\vec{A}$ and $\vec{B}$ graphically. In this method (Fig. 3.6), the resultant vector $\vec{R} = \vec{A} + \vec{B}$ runs from the tail of $\vec{A}$ to the tip of $\vec{B}$.

A second method of adding vectors involves **components** of the vectors. The x component A_x of the vector $\vec{A}$ is equal to the projection of $\vec{A}$ along the x axis of a coordinate system, where $A_x = A\cos\theta$. The y component A_y of $\vec{A}$ is the projection of $\vec{A}$ along the y axis, where $A_y = A\sin\theta$.

If a vector $\vec{A}$ has an x component A_x and a y component A_y, the vector can be expressed in unit-vector form as $\vec{A} = A_x\hat{i} + A_y\hat{j}$. In this notation, $\hat{i}$ is a unit vector pointing in the positive x direction and $\hat{j}$ is a unit vector pointing in the positive y direction. Because $\hat{i}$ and $\hat{j}$ are unit vectors, $|\hat{i}| = |\hat{j}| = 1$.

We can find the resultant of two or more vectors by resolving all vectors into their x and y components, adding their resultant x and y components, and then using the Pythagorean theorem to find the magnitude of the resultant vector. We can find the angle that the resultant vector makes with respect to the x axis by using a suitable trigonometric function.

Think–Pair–Share

See the Preface for an explanation of the icons used in this problems set. For additional assessment items for this section, go to **WEBASSIGN** *From Cengage*

1. You are working at a radar station for the Coast Guard. While everyone else is out to lunch, you hear a distress call from a sinking ship. The ship is located at a distance of 51.2 km from the station, at a bearing of 36° west of north. On your radar screen, you see the locations of four other ships as follows:

Ship #	Distance from Station (km)	Bearing	Maximum Speed (km/h)
1	36.1	42° W of N	30.0
2	37.3	61° W of N	38.0
3	10.2	36° W of N	32.0
4	51.2	79° W of N	45.0

Quick! Which ship do you contact to help the sinking ship? Which ship will get there in the shortest time interval? Assume that each ship would accelerate quickly to its maximum speed and then maintain that constant speed in a straight line for the entire trip to the sinking ship.

2. **ACTIVITY** On a paper map of the United States, locate Memphis, Albuquerque, and Chicago. Draw a vector from Albuquerque to Memphis and another vector from Memphis to Chicago. Using the scale on the map, determine the straight-line distances between Albuquerque and Memphis, and between Memphis and Chicago. Use a protractor to measure the angles of your two vectors with respect to latitude and longitude lines. From this information, determine the straight-line distance in miles between Albuquerque and Chicago.

Problems

See the Preface for an explanation of the icons used in this problems set. For additional assessment items for this section, go to **WEBASSIGN** *From Cengage*

SECTION 3.1 Coordinate Systems

1. Two points in the xy plane have Cartesian coordinates $(2.00, -4.00)$ m and $(-3.00, 3.00)$ m. Determine (a) the distance between these points and (b) their polar coordinates.

2. Two points in a plane have polar coordinates $(2.50\text{ m}, 30.0°)$ and $(3.80\text{ m}, 120.0°)$. Determine (a) the Cartesian coordinates of these points and (b) the distance between them.

3. The polar coordinates of a certain point are $(r = 4.30\text{ cm}, \theta = 214°)$. (a) Find its Cartesian coordinates x and y. Find the polar coordinates of the points with Cartesian coordinates (b) $(-x, y)$, (c) $(-2x, -2y)$, and (d) $(3x, -3y)$.

4. Let the polar coordinates of the point (x, y) be (r, θ). Determine the polar coordinates for the points (a) $(-x, y)$, (b) $(-2x, -2y)$, and (c) $(3x, -3y)$.

SECTION 3.2 Vector and Scalar Quantities

5. *Why is the following situation impossible?* A skater glides along a circular path. She defines a certain point on the circle as

her origin. Later on, she passes through a point at which the distance she has traveled along the path from the origin is smaller than the magnitude of her displacement vector from the origin.

SECTION 3.3 Basic Vector Arithmetic

6. Vector $\vec{\mathbf{A}}$ has a magnitude of 29 units and points in the positive y direction. When vector $\vec{\mathbf{B}}$ is added to $\vec{\mathbf{A}}$, the resultant vector $\vec{\mathbf{A}} + \vec{\mathbf{B}}$ points in the negative y direction with a magnitude of 14 units. Find the magnitude and direction of $\vec{\mathbf{B}}$.

7. A force $\vec{\mathbf{F}}_1$ of magnitude 6.00 units acts on an object at the origin in a direction $\theta = 30.0°$ above the positive x axis (Fig. P3.7). A second force $\vec{\mathbf{F}}_2$ of magnitude 5.00 units acts on the object in the direction of the positive y axis. Find graphically the magnitude and direction of the resultant force $\vec{\mathbf{F}}_1 + \vec{\mathbf{F}}_2$.

Figure P3.7

8. Three displacements are $\vec{\mathbf{A}} = 200$ m due south, $\vec{\mathbf{B}} = 250$ m due west, and $\vec{\mathbf{C}} = 150$ m at $30.0°$ east of north. (a) Construct a separate diagram for each of the following possible ways of adding these vectors: $\vec{\mathbf{R}}_1 = \vec{\mathbf{A}} + \vec{\mathbf{B}} + \vec{\mathbf{C}}$; $\vec{\mathbf{R}}_2 = \vec{\mathbf{B}} + \vec{\mathbf{C}} + \vec{\mathbf{A}}$; $\vec{\mathbf{R}}_3 = \vec{\mathbf{C}} + \vec{\mathbf{B}} + \vec{\mathbf{A}}$. (b) Explain what you can conclude from comparing the diagrams.

9. The displacement vectors $\vec{\mathbf{A}}$ and $\vec{\mathbf{B}}$ shown in Figure P3.9 both have magnitudes of 3.00 m. The direction of vector $\vec{\mathbf{A}}$ is $\theta = 30.0°$. Find graphically (a) $\vec{\mathbf{A}} + \vec{\mathbf{B}}$, (b) $\vec{\mathbf{A}} - \vec{\mathbf{B}}$, (c) $\vec{\mathbf{B}} - \vec{\mathbf{A}}$, and (d) $\vec{\mathbf{A}} - 2\vec{\mathbf{B}}$. (Report all angles counterclockwise from the positive x axis.)

Figure P3.9

Problems 9 and 25.

10. A roller-coaster car moves 200 ft horizontally and then rises 135 ft at an angle of $30.0°$ above the horizontal. It next travels 135 ft at an angle of $40.0°$ downward. What is its displacement from its starting point? Use graphical techniques.

SECTION 3.4 Components of a Vector and Unit Vectors

11. A minivan travels straight north in the right lane of a divided highway at 28.0 m/s. A camper passes the minivan and then changes from the left lane into the right lane. As it does so, the camper's path on the road is a straight displacement at $8.50°$ east of north. To avoid cutting off the minivan, the north–south distance between the camper's back bumper and the minivan's front bumper should not decrease. (a) Can the camper be driven to satisfy this requirement? (b) Explain your answer.

12. A person walks $25.0°$ north of east for 3.10 km. How far would she have to walk due north and due east to arrive at the same location?

13. Your dog is running around the grass in your back yard. He undergoes successive displacements 3.50 m south, 8.20 m northeast, and 15.0 m west. What is the resultant displacement?

14. Given the vectors $\vec{\mathbf{A}} = 2.00\hat{\mathbf{i}} + 6.00\hat{\mathbf{j}}$ and $\vec{\mathbf{B}} = 3.00\hat{\mathbf{i}} - 2.00\hat{\mathbf{j}}$, (a) draw the vector sum $\vec{\mathbf{C}} = \vec{\mathbf{A}} + \vec{\mathbf{B}}$ and the vector difference $\vec{\mathbf{D}} = \vec{\mathbf{A}} - \vec{\mathbf{B}}$. (b) Calculate

$\vec{\mathbf{C}}$ and $\vec{\mathbf{D}}$, in terms of unit vectors. (c) Calculate $\vec{\mathbf{C}}$ and $\vec{\mathbf{D}}$ in terms of polar coordinates, with angles measured with respect to the positive x axis.

15. The helicopter view in Fig. P3.15 shows two people pulling on a stubborn mule. The person on the right pulls with a force $\vec{\mathbf{F}}_1$ of magnitude 120 N and direction of $\theta_1 = 60.0°$. The person on the left pulls with a force $\vec{\mathbf{F}}_2$ of magnitude 80.0 N and direction of $\theta_2 = 75.0°$. Find (a) the single force that is equivalent to the two forces shown and (b) the force that a third person would have to exert on the mule to make the resultant force equal to zero. The forces are measured in units of newtons (symbolized N).

Figure P3.15

16. A snow-covered ski slope makes an angle of $35.0°$ with the horizontal. When a ski jumper plummets onto the hill, a parcel of splashed snow is thrown up to a maximum displacement of 1.50 m at $16.0°$ from the vertical in the uphill direction as shown in Figure P3.16. Find the components of its maximum displacement (a) parallel to the surface and (b) perpendicular to the surface.

Figure P3.16

17. Consider the three displacement vectors $\vec{\mathbf{A}} = (3\hat{\mathbf{i}} - 3\hat{\mathbf{j}})$ m, $\vec{\mathbf{B}} = (\hat{\mathbf{i}} - 4\hat{\mathbf{j}})$ m, and $\vec{\mathbf{C}} = (-2\hat{\mathbf{i}} + 5\hat{\mathbf{j}})$ m. Use the component method to determine (a) the magnitude and direction of $\vec{\mathbf{D}} = \vec{\mathbf{A}} + \vec{\mathbf{B}} + \vec{\mathbf{C}}$ and (b) the magnitude and direction of $\vec{\mathbf{E}} = -\vec{\mathbf{A}} - \vec{\mathbf{B}} + \vec{\mathbf{C}}$.

18. Vector $\vec{\mathbf{A}}$ has x and y components of -8.70 cm and 15.0 cm, respectively; vector $\vec{\mathbf{B}}$ has x and y components of 13.2 cm and -6.60 cm, respectively. If $\vec{\mathbf{A}} - \vec{\mathbf{B}} + 3\vec{\mathbf{C}} = 0$, what are the components of $\vec{\mathbf{C}}$?

19. The vector $\vec{\mathbf{A}}$ has x, y, and z components of 8.00, 12.0, and -4.00 units, respectively. (a) Write a vector expression for $\vec{\mathbf{A}}$ in unit-vector notation. (b) Obtain a unit-vector expression for a vector $\vec{\mathbf{B}}$ one-fourth the length of $\vec{\mathbf{A}}$ pointing in the same direction as $\vec{\mathbf{A}}$. (c) Obtain a unit-vector expression for a vector $\vec{\mathbf{C}}$ three times the length of $\vec{\mathbf{A}}$ pointing in the direction opposite the direction of $\vec{\mathbf{A}}$.

20. Given the displacement vectors $\vec{\mathbf{A}} = (3\hat{\mathbf{i}} - 4\hat{\mathbf{j}} + 4\hat{\mathbf{k}})$ m and $\vec{\mathbf{B}} = (2\hat{\mathbf{i}} + 3\hat{\mathbf{j}} - 7\hat{\mathbf{k}})$ m, find the magnitudes of the following vectors and express each in terms of

its rectangular components. (a) $\vec{C} = \vec{A} + \vec{B}$ (b) $\vec{D} = 2\vec{A} - \vec{B}$

21. Vector $\vec{A}$ has a negative x component 3.00 units in **T** length and a positive y component 2.00 units in length. (a) Determine an expression for $\vec{A}$ in unit-vector notation. (b) Determine the magnitude and direction of $\vec{A}$. (c) What vector $\vec{B}$ when added to $\vec{A}$ gives a resultant vector with no x component and a negative y component 4.00 units in length?

22. Three displacement vectors of a croquet ball are shown in Figure P3.22, where $|\vec{A}| = 20.0$ units, $|\vec{B}| = 40.0$ units, and $|\vec{C}| = 30.0$ units. Find (a) the resultant in unit-vector notation and (b) the magnitude and direction of the resultant displacement.

Figure P3.22

23. (a) Taking $\vec{A} = (6.00\,\hat{i} - 8.00\,\hat{j})$ units, **QC** $\vec{B} = (-8.00\,\hat{i} + 3.00\,\hat{j})$ units, and $\vec{C} = (26.0\,\hat{i} + 19.0\,\hat{j})$ units, determine a and b such that $a\vec{A} + b\vec{B} + \vec{C} = 0$. (b) A student has learned that a single equation cannot be solved to determine values for more than one unknown in it. How would you explain to him that both a and b can be determined from the single equation used in part (a)?

24. Vector $\vec{B}$ has x, y, and z components of 4.00, 6.00, and 3.00 units, respectively. Calculate (a) the magnitude of $\vec{B}$ and (b) the angle that $\vec{B}$ makes with each coordinate axis.

25. Use the component method to add the vectors $\vec{A}$ and $\vec{B}$ shown in Figure P3.9. Both vectors have magnitudes of 3.00 m and vector $\vec{A}$ makes an angle of $\theta = 30.0°$ with the x axis. Express the resultant $\vec{A} + \vec{B}$ in unit-vector notation.

26. A girl delivering newspapers covers her route by traveling 3.00 blocks west, 4.00 blocks north, and then 6.00 blocks east. (a) What is her resultant displacement? (b) What is the total distance she travels?

27. A man pushing a mop across a floor causes it to undergo **T** two displacements. The first has a magnitude of 150 cm and makes an angle of 120° with the positive x axis. The resultant displacement has a magnitude of 140 cm and is directed at an angle of 35.0° to the positive x axis. Find the magnitude and direction of the second displacement.

28. Figure P3.28 illustrates typical proportions of male (m) **BIO** and female (f) anatomies. The displacements $\vec{d}_{1m}$ and $\vec{d}_{1f}$ from the soles of the feet to the navel have magnitudes of

104 cm and 84.0 cm, respectively. The displacements $\vec{d}_{2m}$ and $\vec{d}_{2f}$ from the navel to outstretched fingertips have magnitudes of 100 cm and 86.0 cm, respectively. Find the vector sum of these displacements $\vec{d}_3 = \vec{d}_1 + \vec{d}_2$ for both people.

29. **Review.** As it passes over Grand Bahama Island, the **AMT** eye of a hurricane is moving in a direction 60.0° north **GP** of west with a speed of 41.0 km/h. (a) What is the unit-vector expression for the velocity of the hurricane? It maintains this velocity for 3.00 h, at which time the course of the hurricane suddenly shifts due north, and its speed slows to a constant 25.0 km/h. This new velocity is maintained for 1.50 h. (b) What is the unit-vector expression for the new velocity of the hurricane? (c) What is the unit-vector expression for the displacement of the hurricane during the first 3.00 h? (d) What is the unit-vector expression for the displacement of the hurricane during the latter 1.50 h? (e) How far from Grand Bahama is the eye 4.50 h after it passes over the island?

30. In an assembly operation illustrated in Figure P3.30, a robot moves an object first straight upward and then also to the east, around an arc forming one-quarter of a circle of radius 4.80 cm that lies in an east–west vertical plane. The robot then moves the object upward and to the north, through one-quarter of a circle of radius 3.70 cm that lies in a north–south vertical plane. Find (a) the magnitude of the total displacement of the object and (b) the angle the total displacement makes with the vertical.

Figure P3.30

31. **Review.** You are standing on the ground at the origin of a **AMT** coordinate system. An airplane flies over you with constant velocity parallel to the x axis and at a fixed height of 7.60×10^3 m. At time $t = 0$, the airplane is directly above you so that the vector leading from you to it is $\vec{P}_0 = 7.60 \times 10^3\,\hat{j}$ m. At $t = 30.0$ s, the position vector leading from you to the airplane is $\vec{P}_{30} = (8.04 \times 10^3\,\hat{i} + 7.60 \times 10^3\,\hat{j})$ m as suggested in Figure P3.31. Determine the magnitude and orientation of the airplane's position vector at $t = 45.0$ s.

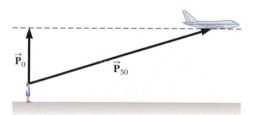

Figure P3.31

32. *Why is the following situation impossible?* A shopper pushing a cart through a market follows directions to the canned goods and moves through a displacement $8.00\,\hat{i}$ m down one aisle. He then makes a 90.0° turn and moves 3.00 m along the y axis. He then makes another 90.0° turn and moves 4.00 m along the x axis. *Every* shopper who follows these directions correctly ends up 5.00 m from the starting point.

Figure P3.28

33. In Figure P3.33, the line segment represents a path from the point with position vector $(5\hat{\mathbf{i}} + 3\hat{\mathbf{j}})$ m to the point with location $(16\hat{\mathbf{i}} + 12\hat{\mathbf{j}})$ m. Point Ⓐ is along this path, a fraction f of the way to the destination. (a) Find the position vector of point Ⓐ in terms of f. (b) Evaluate the expression from part (a) for $f = 0$. (c) Explain whether the result in part (b) is reasonable. (d) Evaluate the expression for $f = 1$. (e) Explain whether the result in part (d) is reasonable.

Figure P3.33 Point Ⓐ is a fraction f of the distance from the initial point (5, 3) to the final point (16, 12).

ADDITIONAL PROBLEMS

34. You are spending the summer as an assistant learning how to navigate on a large ship carrying freight across Lake Erie. One day, you and your ship are to travel across the lake a distance of 200 km traveling due north from your origin port to your destination port. Just as you leave your origin port, the navigation electronics go down. The captain continues sailing, claiming he can depend on his years of experience on the water as a guide. The engineers work on the navigation system while the ship continues to sail, and winds and waves push it off course. Eventually, enough of the navigation system comes back up to tell you your location. The system tells you that your current position is 50.0 km north of the origin port and 25.0 km east of the port. The captain is a little embarrassed that his ship is so far off course and barks an order to you to tell him immediately what heading he should set from your current position to the destination port. Give him an appropriate heading angle.

35. A person going for a walk follows the path shown in Figure P3.35. The total trip consists of four straight-line paths. At the end of the walk, what is the person's resultant displacement measured from the starting point?

Figure P3.35

36. A ferry transports tourists between three islands. It sails from the first island to the second island, 4.76 km away, in a direction 37.0° north of east. It then sails from the second island to the third island in a direction 69.0° west of north. Finally it returns to the first island, sailing in a direction 28.0° east of south. Calculate the distance between (a) the second and third islands and (b) the first and third islands.

37. Two vectors $\vec{\mathbf{A}}$ and $\vec{\mathbf{B}}$ have precisely equal magnitudes. For the magnitude of $\vec{\mathbf{A}} + \vec{\mathbf{B}}$ to be 100 times larger than the magnitude of $\vec{\mathbf{A}} - \vec{\mathbf{B}}$, what must be the angle between them?

38. Two vectors $\vec{\mathbf{A}}$ and $\vec{\mathbf{B}}$ have precisely equal magnitudes. For the magnitude of $\vec{\mathbf{A}} + \vec{\mathbf{B}}$ to be larger than the magnitude of $\vec{\mathbf{A}} - \vec{\mathbf{B}}$ by the factor n, what must be the angle between them?

39. Review. The biggest stuffed animal in the world is a snake 420 m long, constructed by Norwegian children. Suppose the snake is laid out in a park as shown in Figure P3.39, forming two straight sides of a 105° angle, with one side 240 m long. Olaf and Inge run a race they invent. Inge runs directly from the tail of the snake to its head, and Olaf starts from the same place at the same moment but runs along the snake. (a) If both children run steadily at 12.0 km/h, Inge reaches the head of the snake how much earlier than Olaf? (b) If Inge runs the race again at a constant speed of 12.0 km/h, at what constant speed must Olaf run to reach the end of the snake at the same time as Inge?

Figure P3.39

40. Ecotourists use their global positioning system indicator to determine their location inside a botanical garden as latitude 0.002 43 degree south of the equator, longitude 75.642 38 degrees west. They wish to visit a tree at latitude 0.001 62 degree north, longitude 75.644 26 degrees west. (a) Determine the straight-line distance and the direction in which they can walk to reach the tree as follows. First model the Earth as a sphere of radius 6.37×10^6 m to determine the westward and northward displacement components required, in meters. Then model the Earth as a flat surface to complete the calculation. (b) Explain why it is possible to use these two geometrical models together to solve the problem.

41. A vector is given by $\vec{\mathbf{R}} = 2\hat{\mathbf{i}} + \hat{\mathbf{j}} + 3\hat{\mathbf{k}}$. Find (a) the magnitudes of the x, y, and z components; (b) the magnitude of $\vec{\mathbf{R}}$; and (c) the angles between $\vec{\mathbf{R}}$ and the x, y, and z axes.

42. You are working as an assistant to an air-traffic controller at the local airport, from which small airplanes take off and land. Your job is to make sure that airplanes are not closer to each other than a minimum safe separation distance of 2.00 km. You observe two small aircraft on your radar screen, out over the ocean surface. The first is at altitude 800 m above the surface, horizontal distance 19.2 km, and 25.0° south of west. The second aircraft is at altitude 1 100 m, horizontal distance 17.6 km, and 20.0° south of west. Your supervisor is concerned that the two aircraft are too close together and asks for a separation distance for the two airplanes. (Place the x axis west, the y axis south, and the z axis vertical.)

43. Review. The instantaneous position of an object is specified by its position vector leading from a fixed origin to the location of the object, modeled as a particle. Suppose for a certain object the position vector is a function of time given by $\vec{\mathbf{r}} = 4\hat{\mathbf{i}} + 3\hat{\mathbf{j}} - 2t\,\hat{\mathbf{k}}$, where

$\vec{\mathbf{r}}$ is in meters and t is in seconds. (a) Evaluate $d\,\vec{\mathbf{r}}/dt$. (b) What physical quantity does $d\,\vec{\mathbf{r}}/dt$ represent about the object?

44. Vectors $\vec{\mathbf{A}}$ and $\vec{\mathbf{B}}$ have equal magnitudes of 5.00. The sum of $\vec{\mathbf{A}}$ and $\vec{\mathbf{B}}$ is the vector $6.00\,\hat{\mathbf{j}}$. Determine the angle between $\vec{\mathbf{A}}$ and $\vec{\mathbf{B}}$.

45. A rectangular parallelepiped has dimensions a, b, and c as **S** shown in Figure P3.45. (a) Obtain a vector expression for the face diagonal vector $\vec{\mathbf{R}}_1$. (b) What is the magnitude of this vector? (c) Notice that $\vec{\mathbf{R}}_1$, $c\hat{\mathbf{k}}$, and $\vec{\mathbf{R}}_2$ make a right triangle. Obtain a vector expression for the body diagonal vector $\vec{\mathbf{R}}_2$.

Figure P3.45

CHALLENGE PROBLEM

46. A pirate has buried his treasure on an island with five trees **QC** located at the points $(30.0\text{ m}, -20.0\text{ m})$, $(60.0\text{ m}, 80.0\text{ m})$, $(-10.0\text{ m}, -10.0\text{ m})$, $(40.0\text{ m}, -30.0\text{ m})$, and $(-70.0\text{ m},$ 60.0 m), all measured relative to some origin, as shown in Figure P3.46. His ship's log instructs you to start at tree A and move toward tree B, but to cover only one-half the distance between A and B. Then move toward tree C, covering one-third the distance between your current location and C. Next move toward tree D, covering one-fourth the distance between where you are and D. Finally move toward tree E, covering one-fifth the distance between you and E, stop, and dig. (a) Assume you have correctly determined the order in which the pirate labeled the trees as A, B, C, D, and E as shown in the figure. What are the coordinates of the point where his treasure is buried? (b) **What If?** What if you do not really know the way the pirate labeled the trees? What would happen to the answer if you rearranged the order of the trees, for instance, to B (30 m, -20 m), A (60 m, 80 m), E (-10 m, -10 m), C (40 m, -30 m), and D (-70 m, 60 m)? State reasoning to show that the answer does not depend on the order in which the trees are labeled.

Figure P3.46

Motion in Two Dimensions

4

STORYLINE   **In the preceding chapter, you are sailing for Catalina** Island in a sailboat. As you approach the island, you see teenagers doing flips off a cliff. You take out your smartphone, open the camera, and take photographs of a teenager as he travels along his trajectory. Using a special app on your phone, you combine several pictures into one so that you can see several images of the falling teenager. He seems to be following a path of a particular shape as he falls. What do you think that shape is? As you enter the harbor on Catalina Island, you see a welder performing repairs on a metal boat. A shower of sparks occurs. You look at the paths of the individual sparks and notice their shape. On shore, you see a fountain in a park in which streams of water are projected at an angle into the air and follow a certain shape as they come back down. You get a drink from a water fountain and notice the shape of the water projected from the fountain. What *is* that shape that you are seeing over and over again?

CONNECTIONS   In Chapter 2, we studied motion in one dimension. In Chapter 3, we learned about vector quantities in general, addition of vectors, and vector components. We focused there on position vectors. In this chapter, we will see how to use vectors from Chapter 3 to modify our mathematical expressions for position, velocity, and acceleration from Chapter 2 to account for motion in two dimensions. We will study two important types of two-dimensional motion: projectile motion, such as that of a thrown baseball or the diving teenager in the previous paragraph, and circular motion, such as the idealized motion of a planet around a star. We also discuss the concept of relative motion, which shows why observers in different frames of reference may measure different positions and velocities for a given particle. This chapter will complete our discussion of ways to describe the motion of a particle, and will set us up for Chapter 5, in which we study the cause of changes in the motion of a particle.

4.1 **The Position, Velocity, and Acceleration Vectors**

In one dimension, a single numerical value describes a particle's position, but in two dimensions, we indicate its position by its **position vector $\vec{r}$**, drawn from the origin of some coordinate system to the location of the particle in the xy plane as in Figure 4.1. At time t_i, the particle is at point Ⓐ, described by position vector $\vec{r}_i$. At some later time t_f, it is at point Ⓑ, described by position vector $\vec{r}_f$. The path followed by the particle from Ⓐ to Ⓑ is not necessarily a straight line. As the particle moves from Ⓐ to Ⓑ in the time interval $\Delta t = t_f - t_i$, its position vector changes from $\vec{r}_i$ to $\vec{r}_f$. As we learned in Chapter 2, displacement is a vector, and the displacement of the particle is the difference between its final position and its initial position. We now define the **displacement vector $\Delta\vec{r}$** for a particle such as the one in Figure 4.1 as the difference between its final position vector and its initial position vector:

$$\Delta\vec{r} \equiv \vec{r}_f - \vec{r}_i \qquad (4.1)$$

◀ Displacement vector (Compare to Equation 2.1)

The direction of $\Delta\vec{r}$ is indicated in Figure 4.1. As we see from the figure, the magnitude of $\Delta\vec{r}$ is *less* than the distance traveled along the curved path followed by the particle.

As we saw in Chapter 2, it is often useful to quantify motion by looking at the displacement divided by the time interval during which that displacement occurs, which gives the rate of change of position. Two-dimensional (or three-dimensional) kinematics is similar to one-dimensional kinematics, but we must now use full vector notation rather than positive and negative signs to indicate the direction of motion.

We define the **average velocity $\vec{v}_{avg}$** of a particle during the time interval Δt as the displacement of the particle divided by the time interval:

$$\vec{v}_{avg} \equiv \frac{\Delta\vec{r}}{\Delta t} \qquad (4.2)$$

◀ Average velocity (Compare to Equation 2.2)

Multiplying or dividing a vector quantity by a positive scalar quantity such as Δt changes only the magnitude of the vector, not its direction. Because displacement is a vector quantity and the time interval is a positive scalar quantity, we conclude that the average velocity is a vector quantity directed along $\Delta\vec{r}$.

The average velocity between points is *independent of the path* taken. That is because average velocity is proportional to displacement, which depends only on the initial and final position vectors and not on the path taken. As with one-dimensional motion, we conclude that if a particle starts its motion at some point and returns to this point via any path, its average velocity is zero for this trip because its displacement is zero. Consider again our basketball players on the court in Figure 2.2 (page 22). We previously considered only their one-dimensional motion back and forth between the baskets. In reality, however, they move over a two-dimensional surface, running back and forth between the baskets as well as left and right across the width of the court. Starting from one basket, a given player may follow a very complicated two-dimensional path. Upon returning to the original basket, however, a player's average velocity is zero because the player's displacement for the whole trip is zero.

Consider again the motion of a particle between two points in the xy plane as shown in Figure 4.2 (page 70). The dashed curve shows the path of the particle from point Ⓐ to point Ⓑ. As the time interval over which we observe the motion becomes smaller and smaller—that is, as Ⓑ is moved to Ⓑ′ and then to Ⓑ″ and so on—the direction of the displacement approaches that of the green line tangent to the path at Ⓐ. The **instantaneous velocity $\vec{v}$** is defined as the limit of the average velocity $\Delta\vec{r}/\Delta t$ as Δt approaches zero:

$$\vec{v} \equiv \lim_{\Delta t \to 0} \frac{\Delta\vec{r}}{\Delta t} = \frac{d\vec{r}}{dt} \qquad (4.3)$$

◀ Instantaneous velocity (Compare to Equation 2.5)

The displacement of the particle is the vector $\Delta\vec{r}$.

Figure 4.1 A particle moving in the xy plane is located with the position vector $\vec{r}$ drawn from the origin to the particle. The displacement of the particle as it moves from Ⓐ to Ⓑ in the time interval $\Delta t = t_f - t_i$ is equal to the vector $\Delta\vec{r} = \vec{r}_f - \vec{r}_i$.

Figure 4.2 As a particle moves between two points, its average velocity is in the direction of the displacement vector $\Delta \vec{\mathbf{r}}$. By definition, the instantaneous velocity at Ⓐ is directed along the line tangent to the curve at Ⓐ.

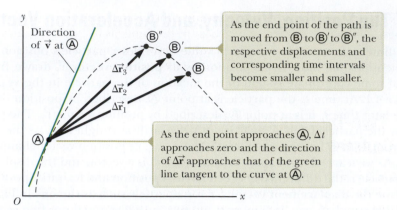

Direction of $\vec{\mathbf{v}}$ at Ⓐ

As the end point of the path is moved from Ⓑ to Ⓑ′ to Ⓑ″, the respective displacements and corresponding time intervals become smaller and smaller.

As the end point approaches Ⓐ, Δt approaches zero and the direction of $\Delta \vec{\mathbf{r}}$ approaches that of the green line tangent to the curve at Ⓐ.

That is, the instantaneous velocity at point Ⓐ equals the derivative of the position vector with respect to time, evaluated at point Ⓐ. The direction of the instantaneous velocity vector at any point in a particle's path is along a line tangent to the path at that point and in the direction of motion.

The magnitude of the instantaneous velocity vector $v = |\vec{\mathbf{v}}|$ of a particle is called the *speed* of the particle, which is a scalar quantity.

As a particle moves from one point to another along some path, its instantaneous velocity vector changes from $\vec{\mathbf{v}}_i$ at time t_i to $\vec{\mathbf{v}}_f$ at time t_f. Knowing the velocity at these points allows us to determine the average acceleration of the particle. The **average acceleration $\vec{\mathbf{a}}_{\text{avg}}$** of a particle is defined as the change in its instantaneous velocity vector $\Delta \vec{\mathbf{v}}$ divided by the time interval Δt during which that change occurs:

◀ Average acceleration
(Compare to Equation 2.9)

$$\vec{\mathbf{a}}_{\text{avg}} \equiv \frac{\Delta \vec{\mathbf{v}}}{\Delta t} = \frac{\vec{\mathbf{v}}_f - \vec{\mathbf{v}}_i}{t_f - t_i} \tag{4.4}$$

Because $\vec{\mathbf{a}}_{\text{avg}}$ is the ratio of a vector quantity $\Delta \vec{\mathbf{v}}$ and a positive scalar quantity Δt, we conclude that average acceleration is a vector quantity directed along $\Delta \vec{\mathbf{v}}$. As indicated in Figure 4.3, the vector $\Delta \vec{\mathbf{v}}$ is the difference between vectors $\vec{\mathbf{v}}_f$ and $\vec{\mathbf{v}}_i$: $\Delta \vec{\mathbf{v}} = \vec{\mathbf{v}}_f - \vec{\mathbf{v}}_i$.

When the average acceleration of a particle changes during different time intervals, it is useful to define its instantaneous acceleration. The **instantaneous acceleration $\vec{\mathbf{a}}$** is defined as the limiting value of the ratio $\Delta \vec{\mathbf{v}}/\Delta t$ as Δt approaches zero:

◀ Instantaneous acceleration
(Compare to Equation 2.10)

$$\vec{\mathbf{a}} \equiv \lim_{\Delta t \to 0} \frac{\Delta \vec{\mathbf{v}}}{\Delta t} = \frac{d\vec{\mathbf{v}}}{dt} \tag{4.5}$$

In other words, the instantaneous acceleration equals the derivative of the velocity vector with respect to time.

Various changes can occur when a particle accelerates in two dimensions. First, the magnitude of the velocity vector (the speed) may change with time as in one-dimensional motion. Second, the direction of the velocity vector may change with time even if its magnitude (speed) remains constant. Finally, both the magnitude and the direction of the velocity vector may change simultaneously.

PITFALL PREVENTION 4.1

Vector Addition As mentioned in Chapter 3, vector addition can be applied to *any* type of vector quantity. Figure 4.3, for example, shows the addition of *velocity* vectors using the graphical approach.

Figure 4.3 A particle moves from position Ⓐ to position Ⓑ. Its velocity vector changes from $\vec{\mathbf{v}}_i$ to $\vec{\mathbf{v}}_f$. The vector diagram at the right of the figure shows how to determine the vector $\Delta \vec{\mathbf{v}}$ from the initial and final velocities.

4.2 Two-Dimensional Motion with Constant Acceleration

In Section 2.7, we investigated one-dimensional motion of a particle under constant acceleration and developed the particle under constant acceleration model. Let us now consider two-dimensional motion during which the acceleration of a particle remains constant in both magnitude and direction. As we shall see, this approach is useful for analyzing some common types of motion.

In section 4.1, we considered position vectors for a particle and represented them as arrows. Now, let us recall our discussion of vector components in Section 3.4. Consider a particle located in the xy plane at a position having Cartesian coordinates (x, y) as in Figure 4.4. The point can be specified by the position vector $\vec{r}$, which in unit-vector form is given by

$$\vec{r} = x\hat{i} + y\hat{j} \qquad (4.6)$$

where x, y, and $\vec{r}$ change with time as the particle moves while the unit vectors $\hat{i}$ and $\hat{j}$ remain constant.

We need to emphasize an important point regarding two-dimensional motion. Imagine an air hockey puck moving in a straight line along a perfectly level, friction-free surface of an air hockey table. Figure 4.5a shows a motion diagram from an overhead point of view of this puck. Recall that in Section 2.4 we related the acceleration of an object to a force on the object. Because there are no forces on the puck in the horizontal plane, it moves with constant velocity in the x direction. Now suppose you blow a quick puff of air on the puck as it passes your position, with the force from your puff of air *exactly* in the y direction. Because the force from this puff of air has no component in the x direction, it causes no acceleration in the x direction. It only causes a momentary acceleration in the y direction, causing the puck to have a constant y component of velocity once the force from the puff of air is removed. After your puff of air on the puck, its velocity component in the x direction is unchanged as shown in Figure 4.5b. The y component of the puck in Equation 4.6 remained constant before the puff of air, but is increasing afterward. The generalization of this simple experiment is that **motion in two dimensions can be modeled as two *independent* motions in each of the two perpendicular directions associated with the x and y axes. That is, any influence in the y direction does not affect the motion in the x direction and vice versa.**

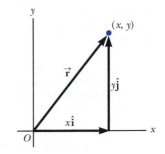

Figure 4.4 The point whose Cartesian coordinates are (x, y) can be represented by the position vector $\vec{r} = x\hat{i} + y\hat{j}$.

The horizontal red vectors, representing the x component of the velocity, are the same length in both parts of the figure, which demonstrates that motion in two dimensions can be modeled as two independent motions in perpendicular directions.

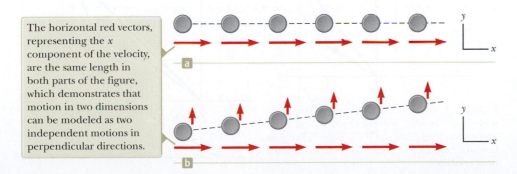

Figure 4.5 (a) A puck moves across a horizontal air hockey table at constant velocity in the x direction. (b) After a puff of air in the y direction is applied to the puck, the puck has gained a y component of velocity, but the x component is unaffected by the force in the perpendicular direction.

If the position vector of a particle is known, the velocity of the particle can be obtained from Equations 4.3 and 4.6, which give

$$\vec{\mathbf{v}} = \frac{d\vec{\mathbf{r}}}{dt} = \frac{dx}{dt}\hat{\mathbf{i}} + \frac{dy}{dt}\hat{\mathbf{j}} = v_x\hat{\mathbf{i}} + v_y\hat{\mathbf{j}} \tag{4.7}$$

Because the acceleration $\vec{\mathbf{a}}$ of the particle is assumed constant in this discussion, its components a_x and a_y also are constants. Therefore, we can model the particle as a particle under constant acceleration independently in each of the two directions and apply the equations of kinematics separately to the x and y components of the velocity vector. Substituting, from Equation 2.13, $v_{xf} = v_{xi} + a_x t$ and $v_{yf} = v_{yi} + a_y t$ into Equation 4.7 to determine the final velocity at any time t, we obtain

$$\vec{\mathbf{v}}_f = (v_{xi} + a_x t)\hat{\mathbf{i}} + (v_{yi} + a_y t)\hat{\mathbf{j}} = (v_{xi}\hat{\mathbf{i}} + v_{yi}\hat{\mathbf{j}}) + (a_x\hat{\mathbf{i}} + a_y\hat{\mathbf{j}})t$$

Velocity vector as ▶
a function of time for a
particle under constant
acceleration in two
dimensions (Compare to
Equation 2.13)

$$\boxed{\vec{\mathbf{v}}_f = \vec{\mathbf{v}}_i + \vec{\mathbf{a}}t \quad \text{(for constant } \vec{\mathbf{a}}\text{)}} \tag{4.8}$$

This result states that the velocity of a particle at some time t equals the vector sum of its initial velocity $\vec{\mathbf{v}}_i$ at time $t = 0$ and the additional velocity $\vec{\mathbf{a}}t$ acquired at time t as a result of constant acceleration.

Similarly, from Equation 2.16 we know that the x and y coordinates of a particle under constant acceleration are

$$x_f = x_i + v_{xi}t + \tfrac{1}{2}a_x t^2 \qquad y_f = y_i + v_{yi}t + \tfrac{1}{2}a_y t^2$$

Substituting these expressions into Equation 4.6 (and labeling the final position vector $\vec{\mathbf{r}}_f$) gives

$$\vec{\mathbf{r}}_f = (x_i + v_{xi}t + \tfrac{1}{2}a_x t^2)\hat{\mathbf{i}} + (y_i + v_{yi}t + \tfrac{1}{2}a_y t^2)\hat{\mathbf{j}}$$

$$= (x_i\hat{\mathbf{i}} + y_i\hat{\mathbf{j}}) + (v_{xi}\hat{\mathbf{i}} + v_{yi}\hat{\mathbf{j}})t + \tfrac{1}{2}(a_x\hat{\mathbf{i}} + a_y\hat{\mathbf{j}})t^2$$

Position vector as ▶
a function of time for a
particle under constant
acceleration in two
dimensions (Compare to
Equation 2.16)

$$\boxed{\vec{\mathbf{r}}_f = \vec{\mathbf{r}}_i + \vec{\mathbf{v}}_i t + \tfrac{1}{2}\vec{\mathbf{a}}t^2 \quad \text{(for constant } \vec{\mathbf{a}}\text{)}} \tag{4.9}$$

Equation 4.9 tells us that the position vector $\vec{\mathbf{r}}_f$ of a particle is the vector sum of the original position $\vec{\mathbf{r}}_i$, a displacement $\vec{\mathbf{v}}_i t$ arising from the initial velocity of the particle, and a displacement $\tfrac{1}{2}\vec{\mathbf{a}}t^2$ resulting from the constant acceleration of the particle.

We can consider Equations 4.8 and 4.9 to be the mathematical representation of a two-dimensional version of the particle under constant acceleration model. Graphical representations of Equations 4.8 and 4.9 are shown in Figure 4.6. The components of the position and velocity vectors are also illustrated in the figure.

Figure 4.6 Vector representations and components of (a) the velocity and (b) the position of a particle under constant acceleration in two dimensions.

Example **4.1** | **Motion in a Plane**

A particle moves in the xy plane, starting from the origin at $t = 0$ with an initial velocity having an x component of 20 m/s and a y component of −15 m/s. The particle experiences an acceleration in the x direction, given by $a_x = 4.0$ m/s².

(A) Determine the total velocity vector at any later time.

SOLUTION

Conceptualize The components of the initial velocity tell us that the particle starts by moving toward the right and downward. The x component of velocity starts at 20 m/s and increases by 4.0 m/s every second. The y component of velocity never changes from its initial value of −15 m/s. We sketch a motion diagram of the situation in Figure 4.7. Because the particle is accelerating in the $+x$ direction, its velocity component in this direction increases and the path curves as shown in the diagram. Notice that the spacing between successive images increases as time goes on because the speed is increasing. The placement of the acceleration (purple) and velocity (red) vectors in Figure 4.7 helps us further conceptualize the situation.

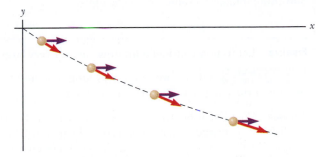

Figure 4.7 (Example 4.1) Motion diagram for the particle. Velocity vectors are shown in red and acceleration vectors in purple.

Categorize Because the initial velocity has components in both the x and y directions, we categorize this problem as one involving a particle moving in two dimensions. Because the particle only has an x component of acceleration, we model it as a *particle under constant acceleration* in the x direction and a *particle under constant velocity* in the y direction.

Analyze To begin the mathematical analysis, we set $v_{xi} = 20$ m/s, $v_{yi} = −15$ m/s, $a_x = 4.0$ m/s², and $a_y = 0$.

Use Equation 4.8 for the velocity vector:

$$\vec{v}_f = \vec{v}_i + \vec{a}t = (v_{xi} + a_x t)\hat{i} + (v_{yi} + a_y t)\hat{j}$$

Substitute numerical values in metric units:

$$\vec{v}_f = [20 + (4.0)t]\hat{i} + [−15 + (0)t]\hat{j}$$

$$(1) \quad \vec{v}_f = [(20 + 4.0t)\hat{i} − 15\hat{j}]$$

Finalize Notice from this expression that the x component of velocity increases in time while the y component remains constant; this result is consistent with our prediction.

(B) Calculate the velocity and speed of the particle at $t = 5.0$ s and the angle the velocity vector makes with the x axis.

SOLUTION

Analyze
Evaluate the result from Equation (1) at $t = 5.0$ s:

$$\vec{v}_f = \{[20 + 4.0(5.0)]\hat{i} − 15\hat{j}\} = (40\hat{i} − 15\hat{j}) \text{ m/s}$$

Determine the angle θ that $\vec{v}_f$ makes with the x axis at $t = 5.0$ s:

$$\theta = \tan^{-1}\left(\frac{v_{yf}}{v_{xf}}\right) = \tan^{-1}\left(\frac{−15 \text{ m/s}}{40 \text{ m/s}}\right) = −21°$$

Evaluate the speed of the particle as the magnitude of $\vec{v}_f$:

$$v_f = |\vec{v}_f| = \sqrt{v_{xf}^2 + v_{yf}^2} = \sqrt{(40)^2 + (−15)^2} \text{ m/s} = 43 \text{ m/s}$$

Finalize The negative sign for the angle θ indicates that the velocity vector is directed at an angle of 21° below the positive x axis. Notice that if we calculate v_i from the x and y components of $\vec{v}_i$, we find that $v_f > v_i$. Is that consistent with our prediction?

continued

4.1 continued

(C) Determine the x and y coordinates of the particle at any time t and its position vector at this time.

SOLUTION

Analyze

Use Equation 4.9 for the position vector:

$$\vec{r}_f = \vec{r}_i + \vec{v}_i t + \tfrac{1}{2}\vec{a}t^2 = \left(x_i + v_{xi}t + \tfrac{1}{2}a_x t^2\right)\hat{i} + \left(y_i + v_{yi}t + \tfrac{1}{2}a_y t^2\right)\hat{j}$$

Substitute numerical values in metric units:

$$\vec{r}_f = \left[0 + (20)t + \tfrac{1}{2}(4.0)t^2\right]\hat{i} + \left[0 + (-15)t + \tfrac{1}{2}(0)t^2\right]\hat{j}$$

$$\vec{r}_f = (20t + 2.0t^2)\hat{i} - 15t\,\hat{j}$$

Finalize   Let us now consider a limiting case for very large values of t.

WHAT IF?   What if we wait a very long time and then observe the motion of the particle? How would we describe the motion of the particle for large values of the time?

Answer   Looking at Figure 4.7, we see the path of the particle curving toward the x axis. There is no reason to assume this tendency will change, which suggests that the path will become more and more parallel to the x axis as time grows large. Mathematically, Equation (1) shows that the y component of the velocity remains constant while the x component grows linearly with t. Therefore, when t is very large, the x component of the velocity will be much larger than the y component, suggesting that the velocity vector becomes more and more parallel to the x axis. The magnitudes of both x_f and y_f continue to grow with time, although x_f grows much faster.

PITFALL PREVENTION 4.2
Acceleration at the Highest Point
As discussed in Pitfall Prevention 2.8, many people claim that the acceleration of a projectile at the topmost point of its trajectory is zero. This mistake arises from confusion between zero vertical velocity and zero acceleration. If the projectile were to experience zero acceleration at the highest point, its velocity at that point would not change; rather, the projectile would move horizontally at constant speed from then on! That does not happen, however, because the acceleration is *not* zero anywhere along the trajectory.

4.3 Projectile Motion

Anyone who has observed a baseball in motion has observed projectile motion. The ball moves in a curved path and returns to the ground. **Projectile motion** of an object is simple to analyze if we make two assumptions: (1) the free-fall acceleration is constant over the range of motion and is directed downward (i.e., $a_x = 0$, $a_y = -g$),[1] and (2) the effect of air resistance is negligible.[2] With these assumptions, we find that the path of a projectile, which we call its *trajectory*, is *always* a parabola as shown in Figure 4.8. **We use these assumptions throughout this chapter.** The parabola is the shape for *all* the trajectories described in the opening storyline for this chapter: the diving teenager, the sparks caused by the welder, the water in the park fountain, and the water in the drinking fountain.

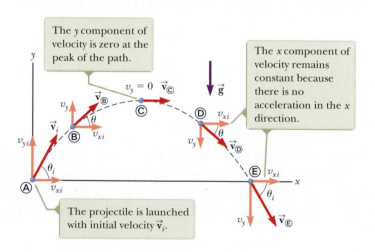

The y component of velocity is zero at the peak of the path.

The x component of velocity remains constant because there is no acceleration in the x direction.

$v_y = 0$

The projectile is launched with initial velocity $\vec{v}_i$.

Figure 4.8 The parabolic path of a projectile that leaves the origin with a velocity $\vec{v}_i$. The velocity vector $\vec{v}$ changes with time in both magnitude and direction. This change is the result of acceleration $\vec{a} = \vec{g}$ in the negative y direction.

---

[1]This assumption is reasonable as long as the range of motion is small compared with the radius of the Earth (6.4×10^6 m). In effect, this assumption is equivalent to assuming the Earth is flat over the range of motion considered.

[2]This assumption is often *not* justified, especially at high velocities. In addition, any spin imparted to a projectile, such as that applied when a pitcher throws a curve ball, can give rise to some very interesting effects associated with aerodynamic forces, which will be discussed in Chapter 14.

The expression for the position vector of the projectile as a function of time follows directly from Equation 4.9, with its acceleration being that due to gravity, $\vec{a} = \vec{g}$:

$$\vec{r}_f = \vec{r}_i + \vec{v}_i t + \tfrac{1}{2}\vec{g}t^2 \tag{4.10}$$

where the initial x and y components of the velocity of the projectile are

$$v_{xi} = v_i \cos\theta_i \qquad v_{yi} = v_i \sin\theta_i \tag{4.11}$$

A pictorial representation of the path of a particle described by the position function in Equation 4.10 is shown in Figure 4.9 for a projectile launched from the origin, so that $\vec{r}_i = 0$. The final position of a particle can be considered to be the superposition of its initial position $\vec{r}_i$; the term $\vec{v}_i t$, which is its displacement if no acceleration were present; and the term $\tfrac{1}{2}\vec{g}t^2$ that arises from its acceleration due to gravity. In other words, if there were no gravitational acceleration, the particle would continue to move along a straight path in the direction of $\vec{v}_i$. Therefore, the vertical distance $\tfrac{1}{2}\vec{g}t^2$ through which the particle "falls" off the straight-line path is the same distance that an object dropped from rest would fall during the same time interval.

In Section 4.2, we stated that two-dimensional motion with constant acceleration can be analyzed as a combination of two independent motions in the x and y directions, with accelerations a_x and a_y. Projectile motion can also be handled in this way, with acceleration $a_x = 0$ in the x direction and a constant acceleration $a_y = -g$ in the y direction. Therefore, when solving projectile motion problems, use two analysis models: (1) the particle under constant velocity in the horizontal direction (Eq. 2.7),

$$x_f = x_i + v_{xi}t \tag{4.12}$$

and (2) the particle under constant acceleration in the vertical direction (Eqs. 2.13–2.17 with x changed to y and $a_y = -g$),

$$v_{yf} = v_{yi} - gt \tag{4.13}$$

$$v_{y,\text{avg}} = \frac{v_{yi} + v_{yf}}{2} \tag{4.14}$$

$$y_f = y_i + \tfrac{1}{2}(v_{yi} + v_{yf})t \tag{4.15}$$

$$y_f = y_i + v_{yi}t - \tfrac{1}{2}gt^2 \tag{4.16}$$

$$v_{yf}^2 = v_{yi}^2 - 2g(y_f - y_i) \tag{4.17}$$

The horizontal and vertical components of a projectile's motion are completely independent of each other and can be handled separately, with time t as the common variable for both components.

QUICK QUIZ 4.2 (i) As a projectile thrown at an upward angle moves in its parabolic path (such as in Fig. 4.9), at what point along its path are the velocity and acceleration vectors for the projectile perpendicular to each other? (a) nowhere (b) the highest point (c) the launch point (ii) From the same choices, at what point are the velocity and acceleration vectors for the projectile parallel to each other?

Horizontal Range and Maximum Height of a Projectile

Before embarking on some examples, let us consider a special case of projectile motion that occurs often. Assume a projectile is launched from the origin at $t_i = 0$ with a positive v_{yi} component as shown in Figure 4.10 and returns to *the same horizontal level*. This situation is common in sports, where baseballs, footballs, and golf balls often land at the same level from which they were launched.

Figure 4.9 The position vector $\vec{r}_f$ of a projectile launched from the origin whose initial velocity at the origin is $\vec{v}_i$. The vector $\vec{v}_i t$ would be the displacement of the projectile if gravity were absent, and the vector $\tfrac{1}{2}\vec{g}t^2$ is its vertical displacement from a straight-line path due to its downward gravitational acceleration.

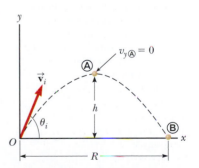

Figure 4.10 A projectile launched over a flat surface from the origin at $t_i = 0$ with an initial velocity $\vec{v}_i$. The maximum height of the projectile is h, and the horizontal range is R. At Ⓐ, the peak of the trajectory, the particle has coordinates $(R/2, h)$.

Two points in this motion are especially interesting to analyze: the peak point Ⓐ, which has Cartesian coordinates $(R/2, h)$, and the point Ⓑ, which has coordinates $(R, 0)$. The distance R is called the *horizontal range* of the projectile, and the distance h is its *maximum height*. Let us find h and R mathematically in terms of v_i, θ_i, and g.

We can determine h by noting that at the peak $v_{y\text{Ⓐ}} = 0$. Therefore, from the particle under constant acceleration model, we can use Equation 4.13 to determine the time $t_\text{Ⓐ}$ at which the projectile reaches the peak:

$$v_{yf} = v_{yi} - gt \quad \rightarrow \quad 0 = v_i \sin \theta_i - gt_\text{Ⓐ}$$

$$t_\text{Ⓐ} = \frac{v_i \sin \theta_i}{g} \tag{4.18}$$

Substituting this expression for $t_\text{Ⓐ}$ into Equation 4.16 and replacing $y_f = y_\text{Ⓐ}$ with h, we obtain an expression for h in terms of the magnitude and direction of the initial velocity vector:

$$y_f = y_i + v_{yi}t - \tfrac{1}{2}gt^2 \quad \rightarrow \quad h = (v_i \sin \theta_i)\frac{v_i \sin \theta_i}{g} - \tfrac{1}{2}g\left(\frac{v_i \sin \theta_i}{g}\right)^2$$

$$h = \frac{v_i^2 \sin^2 \theta_i}{2g} \tag{4.19}$$

Because of the symmetry of the trajectory, the projectile covers the upward part of the trajectory to the top in exactly the same time interval as it requires to come back to the ground from the topmost point. Therefore, the range R is the horizontal position of the projectile at a time that is twice the time at which it reaches its peak, that is, at time $t_\text{Ⓑ} = 2t_\text{Ⓐ}$. Using the particle under constant velocity model, noting that $v_{xi} = v_{x\text{Ⓑ}} = v_i \cos \theta_i$, and setting $x_\text{Ⓑ} = R$ at $t = 2t_\text{Ⓐ}$, we find from Equation 4.12 that

$$x_f = x_i + v_{xi}t \quad \rightarrow \quad R = v_{xi}t_\text{Ⓑ} = (v_i \cos \theta_i)2t_\text{Ⓐ}$$

$$= (v_i \cos \theta_i)\frac{2v_i \sin \theta_i}{g} = \frac{2v_i^2 \sin \theta_i \cos \theta_i}{g}$$

Using the identity $\sin 2\theta = 2 \sin \theta \cos \theta$ (see Appendix B.4), we can write R in the more compact form

$$R = \frac{v_i^2 \sin 2\theta_i}{g} \tag{4.20}$$

The maximum value of R from Equation 4.20 is $R_{max} = v_i^2/g$. This result makes sense because the maximum value of $\sin 2\theta_i$ is 1, which occurs when $2\theta_i = 90°$. Therefore, R is a maximum when $\theta_i = 45°$.

Figure 4.11 illustrates various trajectories for a projectile having a given initial speed but launched at different angles. As you can see, the range is a maximum for $\theta_i = 45°$. In addition, for any θ_i other than 45°, a point having Cartesian coordinates $(R, 0)$ can be reached by using either one of two complementary values of θ_i for

Figure 4.11 A projectile launched over a flat surface from the origin with an initial speed of 50 m/s at various angles of projection.

which $\sin 2\theta_i$ gives the same result, such as 75° and 15°. Of course, the maximum height and time of flight for one of these values of θ_i are different from the maximum height and time of flight for the complementary value. The time of flight depends only on v_{yi} and is independent of v_{xi}.

QUICK QUIZ 4.3 Rank the launch angles for the five paths in Figure 4.11 with respect to time of flight from the shortest time of flight to the longest.

PROBLEM-SOLVING STRATEGY **Projectile Motion**

We suggest you use the following approach when solving projectile motion problems.

1. Conceptualize. Think about what is going on physically in the problem. Establish the mental representation by imagining the projectile moving along its trajectory.

2. Categorize. Confirm that the problem involves a particle in free fall and that air resistance is neglected. Select a coordinate system with x in the horizontal direction and y in the vertical direction. Use the particle under constant velocity model for the x component of the motion. Use the particle under constant acceleration model for the y direction. In the special case of the projectile returning to the same level from which it was launched, use Equations 4.19 and 4.20.

3. Analyze. If the initial velocity vector is given, resolve it into x and y components. Select the appropriate equation(s) from the particle under constant acceleration model (4.13 through 4.17) for the vertical motion and use these along with Equation 4.12 for the horizontal motion to solve for the unknown(s).

4. Finalize. Once you have determined your result, check to see if your answers are consistent with the mental and pictorial representations and your results are realistic.

Example 4.2 **The Long Jump**

A long jumper (Fig. 4.12) leaves the ground at an angle of 20.0° above the horizontal and at a speed of 11.0 m/s.

(A) How far does he jump in the horizontal direction?

SOLUTION

Conceptualize The arms and legs of a long jumper move in a complicated way, but we will ignore this motion. We model the long jumper as a particle and conceptualize his motion as equivalent to that of a simple projectile.

Figure 4.12 (Example 4.2) Ashton Eaton of the United States competes in the men's decathlon long jump at the 2016 Rio de Janeiro Olympic Games.

Categorize We categorize this example as a projectile motion problem. Because the initial speed and launch angle are given and because the final height is the same as the initial height, we further categorize this problem as satisfying the conditions for which Equations 4.19 and 4.20 can be used. This approach is the most direct way to analyze this problem, although the general methods that have been described will always give the correct answer.

Analyze
Use Equation 4.20 to find the range of the jumper:

$$R = \frac{v_i^2 \sin 2\theta_i}{g} = \frac{(11.0 \text{ m/s})^2 \sin 2(20.0°)}{9.80 \text{ m/s}^2} = \boxed{7.94 \text{ m}}$$

(B) What is the maximum height reached?

SOLUTION

Analyze
Find the maximum height reached by using Equation 4.19:

$$h = \frac{v_i^2 \sin^2 \theta_i}{2g} = \frac{(11.0 \text{ m/s})^2 (\sin 20.0°)^2}{2(9.80 \text{ m/s}^2)} = \boxed{0.722 \text{ m}}$$

Finalize Find the answers to parts (A) and (B) using the general method. The results should agree. Treating the long jumper as a particle is an oversimplification. Nevertheless, the values obtained are consistent with experience in sports. We can model a complicated system such as a long jumper as a particle and still obtain reasonable results.

Example **4.3**   **A Bull's-Eye Every Time**

In a popular lecture demonstration, a projectile is aimed directly at a target and fired in such a way that the projectile leaves the gun at the same time the target is dropped from rest. Show that the projectile hits the falling target.

S O L U T I O N

Conceptualize   We conceptualize the problem by studying Figure 4.13a. Notice that the problem does not ask for numerical values. The expected result must involve an algebraic argument.

The velocity of the projectile (red arrows) changes in direction and magnitude, but its acceleration (purple arrows) remains constant.

Charles D. Winters

Figure 4.13   (Example 4.3) (a) Multiflash photograph of the projectile–target demonstration. If the gun is aimed directly at the target and is fired at the same instant the target begins to fall, the projectile will hit the target. (b) Schematic diagram of the projectile–target demonstration.

Categorize   Because both objects are subject only to gravity, we categorize this problem as one involving two objects in free fall, the target moving in one dimension and the projectile moving in two. The target T is modeled as a *particle under constant acceleration* in one dimension. The projectile P is modeled as a *particle under constant acceleration* in the *y* direction and a *particle under constant velocity* in the *x* direction.

Analyze   Figure 4.13b shows that the initial *y* coordinate y_{iT} of the target is $x_T \tan \theta_i$ and its initial velocity is zero. It falls with acceleration $a_y = -g$.

Write an expression for the *y* coordinate of the target at any moment after release, noting that its initial velocity is zero:

$$(1) \quad y_T = y_{iT} + (0)t - \tfrac{1}{2}gt^2 = x_T \tan \theta_i - \tfrac{1}{2}gt^2$$

Write an expression for the *y* coordinate of the projectile at any moment:

$$(2) \quad y_P = y_{iP} + v_{yiP}t - \tfrac{1}{2}gt^2 = 0 + (v_{iP}\sin\theta_i)t - \tfrac{1}{2}gt^2 = (v_{iP}\sin\theta_i)t - \tfrac{1}{2}gt^2$$

Write an expression for the *x* coordinate of the projectile at any moment:

$$x_P = x_{iP} + v_{xiP}t = 0 + (v_{iP}\cos\theta_i)t = (v_{iP}\cos\theta_i)t$$

Solve this expression for time as a function of the horizontal position of the projectile:

$$t = \frac{x_P}{v_{iP}\cos\theta_i}$$

Substitute this expression into Equation (2):

$$(3) \quad y_P = (v_{iP}\sin\theta_i)\left(\frac{x_P}{v_{iP}\cos\theta_i}\right) - \tfrac{1}{2}gt^2 = x_P \tan\theta_i - \tfrac{1}{2}gt^2$$

Finalize   Compare Equations (1) and (3). We see that when the *x* coordinates of the projectile and target are the same—that is, when $x_T = x_P$—their *y* coordinates given by Equations (1) and (3) are the same and a collision results.

Example **4.4**   **That's Quite an Arm!**

A stone is thrown from the top of a building upward at an angle of 30.0° to the horizontal with an initial speed of 20.0 m/s as shown in Figure 4.14. The height from which the stone is thrown is 45.0 m above the ground.

(A) How long does it take the stone to reach the ground?

SOLUTION

Conceptualize   Study Figure 4.14, in which we have indicated the trajectory and various parameters of the motion of the stone.

Categorize   We categorize this problem as a projectile motion problem. The stone is modeled as a *particle under constant acceleration* in the y direction and a *particle under constant velocity* in the x direction.

Analyze   We have the information $x_i = y_i = 0$, $y_f = -45.0$ m, $a_y = -g$, and $v_i = 20.0$ m/s (the numerical value of y_f is negative because we have chosen the point of the throw as the origin).

Figure 4.14   (Example 4.4) A stone is thrown from the top of a building.

Find the initial x and y components of the stone's velocity:

$$v_{xi} = v_i \cos\theta_i = (20.0 \text{ m/s}) \cos 30.0° = 17.3 \text{ m/s}$$

$$v_{yi} = v_i \sin\theta_i = (20.0 \text{ m/s}) \sin 30.0° = 10.0 \text{ m/s}$$

Express the vertical position of the stone from the particle under constant acceleration model:

$$y_f = y_i + v_{yi}t - \tfrac{1}{2}gt^2$$

Substitute numerical values:

$$-45.0 \text{ m} = 0 + (10.0 \text{ m/s})t + \tfrac{1}{2}(-9.80 \text{ m/s}^2)t^2$$

Solve the quadratic equation for t:

$$t = 4.22 \text{ s}$$

(B) What is the speed of the stone just before it strikes the ground?

SOLUTION

Analyze   Use the velocity equation in the particle under constant acceleration model to obtain the y component of the velocity of the stone just before it strikes the ground:

$$v_{yf} = v_{yi} - gt$$

Substitute numerical values, using $t = 4.22$ s:

$$v_{yf} = 10.0 \text{ m/s} + (-9.80 \text{ m/s}^2)(4.22 \text{ s}) = -31.3 \text{ m/s}$$

Use this component with the horizontal component $v_{xf} = v_{xi} = 17.3$ m/s to find the speed of the stone at $t = 4.22$ s:

$$v_f = \sqrt{v_{xf}^2 + v_{yf}^2} = \sqrt{(17.3 \text{ m/s})^2 + (-31.3 \text{ m/s})^2} = 35.8 \text{ m/s}$$

Finalize   Is it reasonable that the y component of the final velocity is negative? Is it reasonable that the final speed is larger than the initial speed of 20.0 m/s?

WHAT IF?   What if a horizontal wind is blowing in the same direction as the stone is thrown and it causes the stone to have a horizontal acceleration component $a_x = 0.500$ m/s²? Which part of this example, (A) or (B), will have a different answer?

Answer   Recall that the motions in the x and y directions are independent. Therefore, the horizontal wind cannot affect the vertical motion. The vertical motion determines the time of the projectile in the air, so the answer to part (A) does not change. The wind causes the horizontal velocity component to increase with time, so the final speed will be larger in part (B). Taking $a_x = 0.500$ m/s², we find $v_{xf} = 19.4$ m/s and $v_f = 36.9$ m/s.

Example **4.5**   **The End of the Ski Jump**

A ski jumper leaves the ski track moving in the horizontal direction with a speed of 25.0 m/s as shown in Figure 4.15. The landing incline below her falls off with a slope of 35.0°. Where does she land on the incline?

S O L U T I O N

Conceptualize   We can conceptualize this problem based on memories of observing winter ski jumping competitions. We estimate the skier to be airborne for perhaps 4 s and to travel a distance of about 100 m horizontally. We should expect the value of d, the distance traveled along the incline, to be of the same order of magnitude.

Categorize   We categorize the problem as one of a particle in projectile motion. As with other projectile motion problems, we use the *particle under constant velocity* model for the horizontal motion and the *particle under constant acceleration* model for the vertical motion.

Figure 4.15   (Example 4.5) A ski jumper leaves the track moving in a horizontal direction.

Analyze   It is convenient to select the beginning of the jump as the origin. The initial velocity components are v_{xi} = 25.0 m/s and v_{yi} = 0. From the right triangle in Figure 4.15, we see that the jumper's x and y coordinates at the landing point are given by $x_f = d \cos \phi$ and $y_f = -d \sin \phi$.

Express the coordinates of the jumper as a function of time, using the particle under constant velocity model for x and the position equation from the particle under constant acceleration model for y:

(1) $x_f = v_{xi}t$ → (2) $d \cos \phi = v_{xi}t$

(3) $y_f = v_{yi}t - \frac{1}{2}gt^2$ → (4) $-d \sin \phi = -\frac{1}{2}gt^2$

Solve Equation (2) for t and substitute the result into Equation (4):

$$-d \sin \phi = -\frac{1}{2}g\left(\frac{d \cos \phi}{v_{xi}}\right)^2$$

Solve for d and substitute numerical values:

$$d = \frac{2v_{xi}^2 \sin \phi}{g \cos^2 \phi} = \frac{2(25.0 \text{ m/s})^2 \sin 35.0°}{(9.80 \text{ m/s}^2) \cos^2 35.0°} = 109 \text{ m}$$

Evaluate the x and y coordinates of the point at which the skier lands:

$x_f = d \cos \phi = (109 \text{ m}) \cos 35.0° = \boxed{89.3 \text{ m}}$

$y_f = -d \sin \phi = -(109 \text{ m}) \sin 35.0° = \boxed{-62.5 \text{ m}}$

Finalize   Let us compare these results with our expectations. We expected the horizontal distance to be on the order of 100 m, and our result of 89.3 m is indeed on this order of magnitude. It might be useful to calculate the time interval that the jumper is in the air and compare it with our estimate of about 4 s.

W H A T   I F ?   Suppose everything in this example is the same except the ski jump is curved so that the jumper is projected upward at an angle from the end of the track. Is this design better in terms of maximizing the length of the jump?

Answer   If the initial velocity has an upward component, the skier will be in the air longer and should therefore travel farther. Tilting the initial velocity vector upward, however, will reduce the horizontal component of the initial velocity. Therefore, angling the end of the ski track upward at a *large* angle may actually *reduce* the distance. Consider the extreme case: the skier is projected at 90° to the horizontal and simply goes up and comes back down at the end of the ski track! This argument suggests that there must be an optimal angle between 0° and 90° that represents a balance between making the flight time longer and the horizontal velocity component smaller.

Let us find this optimal angle mathematically. We modify Equations (1) through (4) in the following way, assuming the skier is projected at an angle θ with respect to the horizontal over a landing incline sloped with an arbitrary angle ϕ:

(1) and (2)   → $x_f = (v_i \cos \theta)t = d \cos \phi$

(3) and (4)   → $y_f = (v_i \sin \theta)t - \frac{1}{2}gt^2 = -d \sin \phi$

By eliminating the time t between these equations and using differentiation to maximize d in terms of θ, we arrive (after several steps; see Problem 52) at the following equation for the angle θ that gives the maximum value of d:

$$\theta = 45° - \frac{\phi}{2}$$

For the slope angle in Figure 4.15, ϕ = 35.0°; this equation results in an optimal launch angle of θ = 27.5°. For a slope angle of ϕ = 0°, which represents a horizontal plane (no slope), this equation gives an optimal launch angle of θ = 45°, as we would expect (see Figure 4.11).

4.4 Analysis Model: Particle in Uniform Circular Motion

Figure 4.16a shows a car moving in a circular path; we describe this motion by calling it **circular motion.** If the car is moving on this path with *constant speed v,* we call it **uniform circular motion.** Because it occurs so often, this type of motion is recognized as an analysis model called the **particle in uniform circular motion.** We discuss this model in this section.

It is often surprising to students to find that even though an object moves at a constant speed in a circular path, *it still has an acceleration.* To see why, consider the defining equation for acceleration, $\vec{a} = d\vec{v}/dt$ (Eq. 4.5). Notice that the acceleration depends on the change in the *velocity.* Because velocity is a vector quantity, an acceleration can occur in two ways as mentioned in Section 4.1: by a change in the *magnitude* of the velocity and by a change in the *direction* of the velocity. The latter situation occurs for an object moving with constant speed in a circular path. The constant-magnitude velocity vector is always tangent to the path of the object and perpendicular to the radius of the circular path. Therefore, the direction of the velocity vector is always changing.

Let us first argue that the acceleration vector in uniform circular motion is always perpendicular to the path and, therefore, always points toward the center of the circle. If that were not true, there would be a component of the acceleration parallel to the path and therefore parallel to the velocity vector. Such an acceleration component would lead to a change in the speed of the particle along the path. This situation, however, is inconsistent with our setup of the situation: the particle moves with constant speed along the path. Therefore, for *uniform* circular motion, the acceleration vector can only have a component perpendicular to the path, which is toward the center of the circle.

Let us now find the magnitude of the acceleration of the particle. Consider the diagram of the position and velocity vectors in Figure 4.16b. The figure also shows the vector representing the change in position $\Delta\vec{r}$ for an arbitrary time interval. The particle follows a circular path of radius r, part of which is shown by the dashed curve. The particle is at Ⓐ at time t_i, and its velocity at that time is $\vec{v}_i$; it is at Ⓑ at some later time t_f, and its velocity at that time is $\vec{v}_f$. Let us also assume $\vec{v}_i$ and $\vec{v}_f$ differ only in direction; their magnitudes are the same (that is, $v_i = v_f = v$ because it is *uniform* circular motion).

In Figure 4.16c, the velocity vectors in Figure 4.16b have been redrawn tail to tail. The vector $\Delta\vec{v}$ connects the tips of the vectors, representing the vector addition $\vec{v}_f = \vec{v}_i + \Delta\vec{v}$. In both Figures 4.16b and 4.16c, we can identify triangles that help us analyze the motion. The angle $\Delta\theta$ between the two position vectors in Figure 4.16b is the same as the angle between the velocity vectors in Figure 4.16c because the velocity vector $\vec{v}$ is always perpendicular to the position vector $\vec{r}$. Therefore, the two triangles are *similar.* (Two triangles are similar if the angle between any two sides is the same for both triangles and if the ratio of the lengths of these sides is the same.) We can now write a relationship between the lengths of the sides for the two triangles in Figures 4.16b and 4.16c:

$$\frac{|\Delta\vec{v}|}{v} = \frac{|\Delta\vec{r}|}{r}$$

where $v = v_i = v_f$ and $r = r_i = r_f$. This equation can be solved for $|\Delta\vec{v}|$, and the expression obtained can be substituted into Equation 4.4, $\vec{a}_{avg} = \Delta\vec{v}/\Delta t$, to give

Figure 4.16 (a) A car moving along a circular path at constant speed experiences uniform circular motion. (b) As a particle moves along a portion of a circular path from Ⓐ to Ⓑ, its velocity vector changes from $\vec{v}_i$ to $\vec{v}_f$. (c) The construction for determining the direction of the change in velocity $\Delta\vec{v}$, which is toward the center of the circle for small $\Delta\vec{r}$.

the magnitude of the average acceleration over the time interval for the particle to move from Ⓐ to Ⓑ:

$$\left|\vec{\mathbf{a}}_{\text{avg}}\right| = \frac{|\Delta\vec{\mathbf{v}}|}{\Delta t} = \frac{v|\Delta\vec{\mathbf{r}}|}{r\Delta t}$$

Now imagine that points Ⓐ and Ⓑ in Figure 4.16b become extremely close together. As Ⓐ and Ⓑ approach each other, Δt approaches zero, $|\Delta\vec{\mathbf{r}}|$ approaches the distance traveled by the particle along the circular path, and the ratio $|\Delta\vec{\mathbf{r}}|/\Delta t$ approaches the speed v. In addition, the average acceleration becomes the instantaneous acceleration at point Ⓐ. Hence, in the limit $\Delta t \rightarrow 0$, the magnitude of the acceleration is

◀ Centripetal acceleration for a particle in uniform circular motion

$$a_c = \frac{v^2}{r} \tag{4.21}$$

An acceleration of this nature is called a **centripetal acceleration** (*centripetal* means *center-seeking*). The subscript on the acceleration symbol reminds us that the acceleration is centripetal.

In many situations, it is convenient to describe the motion of a particle moving with constant speed in a circle of radius r in terms of the **period** T, which is defined as the time interval required for one complete revolution of the particle. In the time interval T, the particle moves a distance of $2\pi r$, which is equal to the circumference of the particle's circular path. Therefore, because its speed is equal to the circumference of the circular path divided by the period, or $v = 2\pi r/T$, it follows that

◀ Period of circular motion for a particle in uniform circular motion

$$T = \frac{2\pi r}{v} \tag{4.22}$$

The period of a particle in uniform circular motion is a measure of the number of seconds for one revolution of the particle around the circle. The inverse of the period is the *rotation rate* and is measured in revolutions per second. Because one full revolution of the particle around the circle corresponds to an angle of 2π radians, the product of 2π and the rotation rate gives the **angular speed** ω of the particle, measured in radians/s or s^{-1}:

$$\omega = \frac{2\pi}{T} \tag{4.23}$$

Combining this equation with Equation 4.22, we find a relationship between angular speed and the translational speed with which the particle travels in the circular path:

$$\omega = 2\pi\left(\frac{v}{2\pi r}\right) = \frac{v}{r} \quad \rightarrow \quad v = r\omega \tag{4.24}$$

PITFALL PREVENTION 4.5

Centripetal Acceleration Is Not Constant We derived the magnitude of the centripetal acceleration vector and found it to be constant for uniform circular motion, but *the centripetal acceleration vector is not constant.* It always points toward the center of the circle, but it continuously changes direction as the object moves around the circular path.

Equation 4.24 demonstrates that, for a fixed angular speed, the translational speed becomes larger as the radial position becomes larger. Therefore, for example, if a merry-go-round rotates at a fixed angular speed ω, a rider at an outer position at large r will be traveling through space faster than a rider at an inner position at smaller r. We will investigate Equations 4.23 and 4.24 more deeply in Chapter 10.

We can express the centripetal acceleration of a particle in uniform circular motion in terms of angular speed by combining Equations 4.21 and 4.24:

$$a_c = \frac{(r\omega)^2}{r}$$

$$a_c = r\omega^2 \tag{4.25}$$

Equations 4.21–4.25 are to be used when the particle in uniform circular motion model is identified as appropriate for a given situation.

QUICK QUIZ 4.4 A particle moves in a circular path of radius r with speed v. It then increases its speed to $2v$ while traveling along the same circular path. **(i)** The centripetal acceleration of the particle has changed by what factor? Choose one: (a) 0.25 (b) 0.5 (c) 2 (d) 4 (e) impossible to determine **(ii)** From the same choices, by what factor has the period of the particle changed?

ANALYSIS MODEL **Particle in Uniform Circular Motion**

Imagine a moving object that can be modeled as a particle. If it moves in a circular path of radius r at a constant speed v, the magnitude of its centripetal acceleration is

$$a_c = \frac{v^2}{r} \qquad \text{(4.21)}$$

and the **period** of the particle's motion is given by

$$T = \frac{2\pi r}{v} \qquad \text{(4.22)}$$

The **angular speed** of the particle is

$$\omega = \frac{2\pi}{T} \qquad \text{(4.23)}$$

Examples:

- a rock twirled in a circle on a string of constant length
- a planet traveling around a perfectly circular orbit (Chapter 13)
- a charged particle moving in a uniform magnetic field (Chapter 28)
- an electron in orbit around a nucleus in the Bohr model of the hydrogen atom (Chapter 41)

Example 4.6 **The Centripetal Acceleration of the Earth**

(A) What is the centripetal acceleration of the Earth as it moves in its orbit around the Sun?

SOLUTION

Conceptualize We will model the Earth as a particle and approximate the Earth's orbit as circular (it's actually slightly elliptical, as we discuss in Chapter 13).

Categorize The Conceptualize step allows us to categorize this problem as one of a *particle in uniform circular motion*.

Analyze We do not know the orbital speed of the Earth to substitute into Equation 4.21. With the help of Equation 4.22, however, we can recast Equation 4.21 in terms of the period of the Earth's orbit, which we know is one year, and the radius of the Earth's orbit around the Sun, which is 1.496×10^{11} m.

Combine Equations 4.21 and 4.22:

$$a_c = \frac{v^2}{r} = \frac{\left(\dfrac{2\pi r}{T}\right)^2}{r} = \frac{4\pi^2 r}{T^2}$$

Substitute numerical values:

$$a_c = \frac{4\pi^2 (1.496 \times 10^{11} \text{ m})}{(1 \text{ yr})^2} \left(\frac{1 \text{ yr}}{3.156 \times 10^7 \text{ s}}\right)^2 = \boxed{5.93 \times 10^{-3} \text{ m/s}^2}$$

(B) What is the angular speed of the Earth in its orbit around the Sun?

SOLUTION

Analyze

Substitute numerical values into Equation 4.23:

$$\omega = \frac{2\pi}{1 \text{ yr}} \left(\frac{1 \text{ yr}}{3.156 \times 10^7 \text{ s}}\right) = \boxed{1.99 \times 10^{-7} \text{ s}^{-1}}$$

Finalize The acceleration in part (A) is much smaller than the free-fall acceleration on the surface of the Earth. An important technique we learned here is replacing the speed v in Equation 4.21 in terms of the period T of the motion. In many problems, it is more likely that T is known rather than v. In part (B), we see that the angular speed of the Earth is very small, which is to be expected because the Earth takes an entire year to go around the circular path once.

4.5 Tangential and Radial Acceleration

Let us consider a more general motion than that presented in Section 4.4. A particle moves to the right along a curved path, and its velocity changes *both* in direction and in magnitude as described in Figure 4.17. In this situation, the velocity vector is always tangent to the path; the acceleration vector $\vec{\mathbf{a}}$, however, is at some angle to the path. At each of three points Ⓐ, Ⓑ, and Ⓒ in Figure 4.17, the dashed blue circles represent the curvature of the actual path at each point. The radius of each circle is equal to the path's radius of curvature at each point.

As the particle moves along the curved path in Figure 4.17, the direction of the total acceleration vector $\vec{\mathbf{a}}$ changes from point to point. At any instant, this vector can be resolved into two components based on an origin at the center of the dashed circle corresponding to that instant: a radial component a_r along the radius of the circle and a tangential component a_t perpendicular to this radius. The *total* acceleration vector $\vec{\mathbf{a}}$ can be written as the vector sum of the component vectors:

Total acceleration ▶

$$\vec{\mathbf{a}} = \vec{\mathbf{a}}_r + \vec{\mathbf{a}}_t \tag{4.26}$$

The tangential acceleration component causes a change in the speed v of the particle. This component is parallel to the instantaneous velocity, and its magnitude is given by

Tangential acceleration ▶

$$a_t = \left| \frac{dv}{dt} \right| \tag{4.27}$$

The radial acceleration component arises from a change in direction of the velocity vector and is given by

Radial acceleration ▶

$$a_r = -a_c = -\frac{v^2}{r} \tag{4.28}$$

where r is the radius of curvature of the path at the point in question. We recognize the magnitude of the radial component of the acceleration as the centripetal acceleration discussed in Section 4.4 with regard to the particle in uniform circular motion model. Even in situations in which a particle moves along a curved path with a varying speed, however, Equation 4.21 can be used for the centripetal acceleration. In this situation, the equation gives the *instantaneous* centripetal acceleration at any time. The negative sign in Equation 4.28 indicates that the direction of the centripetal acceleration is toward the center of the circle representing the radius of curvature. The direction is opposite that of the radial unit vector $\hat{\mathbf{r}}$, which always points away from the origin at the center of the circle. (See Fig. 3.15.)

Because $\vec{\mathbf{a}}_r$ and $\vec{\mathbf{a}}_t$ are perpendicular component vectors of $\vec{\mathbf{a}}$, it follows that the magnitude of $\vec{\mathbf{a}}$ is $a = \sqrt{a_r^2 + a_t^2}$. At a given speed, a_r is large when the radius of curvature is small (as at points Ⓐ and Ⓑ in Fig. 4.17) and small when r is large (as at point Ⓒ). The direction of $\vec{\mathbf{a}}_t$ is either in the same direction as $\vec{\mathbf{v}}$ (if v is increasing) or opposite $\vec{\mathbf{v}}$ (if v is decreasing, as it must be at point Ⓑ).

In uniform circular motion, where v is constant, $a_t = 0$ and the acceleration is always completely radial as described in Section 4.4. In other words, uniform circular motion is a special case of motion along a general curved path. Furthermore, if the direction of $\vec{\mathbf{v}}$ does not change, there is no radial acceleration and the motion is one dimensional (in this case, $a_r = 0$, but a_t may not be zero).

Figure 4.17 The motion of a particle along an arbitrary curved path lying in the *xy* plane. If the velocity vector $\vec{\mathbf{v}}$ (always tangent to the path) changes in direction and magnitude, the components of the acceleration $\vec{\mathbf{a}}$ are a tangential component a_t and a radial component a_r.

A particle moves along a path, and its speed increases with time. **(i)** In which of the following cases are its acceleration and velocity vectors parallel? (a) when the path is circular (b) when the path is straight (c) when the path is a parabola (d) never **(ii)** From the same choices, in which case are its acceleration and velocity vectors perpendicular everywhere along the path?

Example 4.7 | **Over the Rise**

A car leaves a stop sign and exhibits a constant acceleration of 0.300 m/s^2 parallel to the roadway. The car passes over a rise in the roadway such that the top of the rise is shaped like an arc of a circle of radius 500 m. At the moment the car is at the top of the rise, its velocity vector is horizontal and has a magnitude of 6.00 m/s. What are the magnitude and direction of the total acceleration vector for the car at this instant?

a

SOLUTION

Conceptualize Conceptualize the situation using Figure 4.18a and any experiences you have had in driving over rises on a roadway.

Categorize Because the accelerating car is moving along a curved path, we categorize this problem as one involving a particle experiencing both tangential and radial acceleration. We recognize that it is a relatively simple substitution problem.

The tangential acceleration vector has magnitude 0.300 m/s^2 and is horizontal. The radial acceleration is given by Equation 4.28, with v = 6.00 m/s and r = 500 m. The radial acceleration vector is directed straight downward.

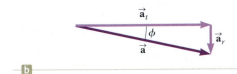

b

Figure 4.18 (Example 4.7) (a) A car passes over a rise that is shaped like an arc of a circle. (b) The total acceleration vector $\vec{a}$ is the sum of the tangential and radial acceleration vectors $\vec{a}_t$ and $\vec{a}_r$.

Evaluate the radial acceleration:

$$a_r = -\frac{v^2}{r} = -\frac{(6.00 \text{ m/s})^2}{500 \text{ m}} = -0.072\,0 \text{ m/s}^2$$

Find the magnitude of $\vec{a}$:

$$a = \sqrt{a_r^2 + a_t^2} = \sqrt{(-0.072\,0 \text{ m/s}^2)^2 + (0.300 \text{ m/s}^2)^2}$$
$$= 0.309 \text{ m/s}^2$$

Find the angle ϕ (see Fig. 4.18b) between $\vec{a}$ and the horizontal:

$$\phi = \tan^{-1}\frac{a_r}{a_t} = \tan^{-1}\left(\frac{-0.072\,0 \text{ m/s}^2}{0.300 \text{ m/s}^2}\right) = -13.5°$$

4.6 Relative Velocity and Relative Acceleration

a

In this section, we describe how observations made by different observers in different frames of reference are related to one another. A frame of reference can be described by a Cartesian coordinate system for which an observer is at rest with respect to the origin.

Let us conceptualize a sample situation in which there will be different observations for different observers. Consider the two observers A and B along the number line in Figure 4.19a. Observer A is located 5 units to the right of observer B. Both observers measure the position of point P, which is located 5 units to the right of observer A. Suppose each observer decides that he is located at the origin of an x axis as in Figure 4.19b. Notice that the two observers disagree on the value of the position of point P. Observer A claims point P is located at a position with a value of x_A = +5, whereas observer B claims it is located at a position with a value of x_B = +10. Both observers are correct, even though they make different measurements. Their measurements differ because they are making the measurement from different frames of reference.

Imagine now that observer B in Figure 4.19b is moving to the right along the x_B axis. Now the two measurements are even more different. Observer A claims point P

b

Figure 4.19 Different observers make different measurements. (a) Observer A is located 5 units to the right of Observer B. Both observers measure the position of a particle at P. (b) If both observers see themselves at the origin of their own coordinate system, they disagree on the value of the position of the particle at P.

The woman standing on the beltway sees the man moving with a slower speed than does the woman observing the man from the stationary floor.

Figure 4.20 Two observers measure the speed of a man walking on a moving beltway.

remains at rest at a position with a value of +5, whereas observer B claims the position of P continuously changes with time, even passing him and moving behind him! Again, both observers are correct, with the difference in their measurements arising from their different frames of reference.

We explore this phenomenon further by considering two observers watching a man walking on a moving beltway at an airport in Figure 4.20. The woman standing on the moving beltway sees the man moving at a normal walking speed. The woman observing from the stationary floor sees the man moving with a higher speed because the beltway speed combines with his walking speed. Both observers look at the same man and arrive at different values for his speed. Both are correct; the difference in their measurements results from the relative velocity of their frames of reference.

In a more general situation, consider a particle located at point P in Figure 4.21. Imagine that the motion of this particle is being described by two observers, observer A in a reference frame S_A fixed relative to the Earth and a second observer B in a reference frame S_B moving to the right relative to S_A (and therefore relative to the Earth) with a constant velocity $\vec{\mathbf{v}}_{BA}$. In this discussion of relative velocity, we use a double-subscript notation; the first subscript represents what is being observed, and the second represents who is doing the observing. Therefore, the notation $\vec{\mathbf{v}}_{BA}$ means the velocity of observer B (and the attached frame S_B) as measured by observer A. With this notation, observer B measures A to be moving to the left with a velocity $\vec{\mathbf{v}}_{AB} = -\vec{\mathbf{v}}_{BA}$. For purposes of this discussion, let us place each observer at her or his respective origin.

We define the time $t = 0$ as the instant at which the origins of the two reference frames coincide in space. Therefore, at time t, the origins of the reference frames will be separated by a distance $v_{BA}t$. We label the position P of the particle relative to observer A with the position vector $\vec{\mathbf{r}}_{PA}$ and that relative to observer B with the position vector $\vec{\mathbf{r}}_{PB}$, both at time t. From Figure 4.21, we see that the vectors $\vec{\mathbf{r}}_{PA}$ and $\vec{\mathbf{r}}_{PB}$ are related to each other through the expression

$$\vec{\mathbf{r}}_{PA} = \vec{\mathbf{r}}_{PB} + \vec{\mathbf{v}}_{BA}t \tag{4.29}$$

By differentiating Equation 4.29 with respect to time, noting that $\vec{\mathbf{v}}_{BA}$ is constant, we obtain

$$\frac{d\vec{\mathbf{r}}_{PA}}{dt} = \frac{d\vec{\mathbf{r}}_{PB}}{dt} + \vec{\mathbf{v}}_{BA}$$

$$\vec{\mathbf{u}}_{PA} = \vec{\mathbf{u}}_{PB} + \vec{\mathbf{v}}_{BA} \tag{4.30}$$

where $\vec{\mathbf{u}}_{PA}$ is the velocity of the particle at P measured by observer A and $\vec{\mathbf{u}}_{PB}$ is its velocity measured by B. (We use the symbol $\vec{\mathbf{u}}$ for particle velocity rather than $\vec{\mathbf{v}}$, which we have already used for the relative velocity of two reference frames.) Equation 4.30 is demonstrated by the red vectors at the top of Figure 4.21. Vector $\vec{\mathbf{u}}_{PB}$ is the velocity of the particle at time t as seen by observer B. When you add the relative velocity $\vec{\mathbf{v}}_{BA}$ of the frames, the sum is the velocity of the particle as measured by observer A.

Equations 4.29 and 4.30 are known as **Galilean transformation equations.** They relate the position and velocity of a particle as measured by observers in relative motion.

Although observers in two frames measure different velocities for the particle, they measure the *same acceleration* when $\vec{\mathbf{v}}_{BA}$ is constant. We can verify that by taking the time derivative of Equation 4.30:

$$\frac{d\vec{\mathbf{u}}_{PA}}{dt} = \frac{d\vec{\mathbf{u}}_{PB}}{dt} + \frac{d\vec{\mathbf{v}}_{BA}}{dt}$$

Because $\vec{\mathbf{v}}_{BA}$ is constant, $d\vec{\mathbf{v}}_{BA}/dt = 0$. Therefore, we conclude that $\vec{\mathbf{a}}_{PA} = \vec{\mathbf{a}}_{PB}$ because $\vec{\mathbf{a}}_{PA} = d\vec{\mathbf{u}}_{PA}/dt$ and $\vec{\mathbf{a}}_{PB} = d\vec{\mathbf{u}}_{PB}/dt$. That is, the acceleration of the particle measured by an observer in one frame of reference is the same as that measured by any other observer moving with constant velocity relative to the first frame.

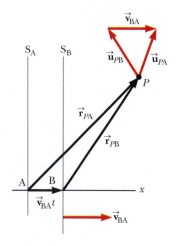

Figure 4.21 A particle located at P is described by two observers, one in the fixed frame of reference S_A and the other in the frame S_B, which moves to the right with a constant velocity $\vec{\mathbf{v}}_{BA}$. The vector $\vec{\mathbf{r}}_{PA}$ is the particle's position vector relative to S_A, and $\vec{\mathbf{r}}_{PB}$ is its position vector relative to S_B. The red vectors at the top of the figure show a vector addition for the velocities of the particle at time t, representing Equation 4.30.

Example 4.8 A Boat Crossing a River

A boat crossing a wide river moves with a speed of 10.0 km/h relative to the water. The water in the river has a uniform speed of 5.00 km/h due east relative to the Earth.

(A) If the boat heads due north, determine the velocity of the boat relative to an observer standing on either bank.

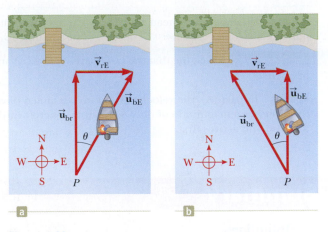

Figure 4.22 (Example 4.8) (a) A boat aims directly across a river and ends up downstream. (b) To move directly across the river, the boat must aim upstream.

SOLUTION

Conceptualize Imagine moving in a boat across a river while the current pushes you down the river. You will not be able to move directly across the river, but will end up downstream as suggested in Figure 4.22a. Imagine observer A on the shore, so that she is on the Earth, represented by letter E. Observer B is represented by letter r in the figure; this observer is on a cork floating in the river, at rest with respect to the water and carried along with the current. When the boat begins from point P and is aimed straight across the river, the velocities $\vec{u}_{br}$, the boat relative to the river, and $\vec{v}_{rE}$, the river relative to the Earth, add to give the velocity $\vec{v}_{bE}$, the velocity of the boat relative to observer A on the Earth. Compare the vector addition in Figure 4.22a to that in Figure 4.21. As the boat moves, it will follow along vector $\vec{v}_{bE}$, as suggested by its position after some time in Figure 4.22a.

Categorize Because of the combined velocities of you relative to the river and the river relative to the Earth, we can categorize this problem as one involving relative velocities.

Analyze We know $\vec{u}_{br}$, the velocity of the *boat* relative to the *river*, and $\vec{v}_{rE}$, the velocity of the *river* relative to the *Earth*. What we must find is $\vec{u}_{bE}$, the velocity of the *boat* relative to the *Earth*. The relationship between these three quantities is $\vec{u}_{bE} = \vec{u}_{br} + \vec{v}_{rE}$. The terms in the equation must be manipulated as vector quantities; the vectors are shown in Figure 4.22a. The quantity $\vec{u}_{br}$ is due north; $\vec{v}_{rE}$ is due east; and the vector sum of the two, $\vec{u}_{bE}$, is at an angle θ as defined in Figure 4.22a.

Find the speed u_{bE} of the boat relative to the Earth using the Pythagorean theorem:

$$u_{bE} = \sqrt{u_{br}^2 + v_{rE}^2} = \sqrt{(10.0 \text{ km/h})^2 + (5.00 \text{ km/h})^2}$$

$$= \boxed{11.2 \text{ km/h}}$$

Find the direction of $\vec{u}_{bE}$:

$$\theta = \tan^{-1}\left(\frac{v_{rE}}{u_{br}}\right) = \tan^{-1}\left(\frac{5.00}{10.0}\right) = \boxed{26.6°}$$

Finalize The boat is moving at a speed of 11.2 km/h in the direction 26.6° east of north relative to the Earth. Notice that the speed of 11.2 km/h is faster than your boat speed of 10.0 km/h. The current velocity adds to yours to give you a higher speed. Notice in Figure 4.22a that your resultant velocity is at an angle to the direction straight across the river, so you will end up downstream, as we predicted.

(B) If the boat travels with the same speed of 10.0 km/h relative to the river and is to travel due north as shown in Figure 4.22b, what should its heading be?

SOLUTION

Conceptualize/Categorize This question is an extension of part (A), so we have already conceptualized and categorized the problem. In this case, however, we must aim the boat upstream so as to go straight across the river.

Analyze The analysis now involves the new triangle shown in Figure 4.22b. As in part (A), we know $\vec{v}_{rE}$ and the magnitude of the vector $\vec{u}_{br}$, and we want $\vec{u}_{bE}$ to be directed across the river. Notice the difference between the triangle in Figure 4.22a and the one in Figure 4.22b: the hypotenuse in Figure 4.22b is no longer $\vec{u}_{bE}$.

Use the Pythagorean theorem to find u_{bE}:

$$u_{bE} = \sqrt{u_{br}^2 - v_{rE}^2} = \sqrt{(10.0 \text{ km/h})^2 - (5.00 \text{ km/h})^2} = 8.66 \text{ km/h}$$

Find the direction in which the boat is heading:

$$\theta = \tan^{-1}\left(\frac{v_{rE}}{u_{bE}}\right) = \tan^{-1}\left(\frac{5.00}{8.66}\right) = \boxed{30.0°}$$

continued

4.8 continued

Finalize   The boat must head upstream so as to travel directly northward across the river. For the given situation, the boat must steer a course 30.0° west of north. For faster currents, the boat must be aimed upstream at larger angles.

W H A T   I F ?   Imagine that the two boats in parts (A) and (B) are racing across the river. Which boat arrives at the opposite bank first?

Answer   In part (A), the velocity of 10 km/h is aimed directly across the river. In part (B), the velocity that is directed across the river has a magnitude of only 8.66 km/h. Therefore, the boat in part (A) has a larger velocity component directly across the river and arrives first.

Summary

▶ Definitions

The **displacement vector** $\Delta\vec{r}$ for a particle is the difference between its final position vector and its initial position vector:

$$\Delta\vec{r} \equiv \vec{r}_f - \vec{r}_i \tag{4.1}$$

The **average velocity** of a particle during the time interval Δt is defined as the displacement of the particle divided by the time interval:

$$\vec{v}_{avg} \equiv \frac{\Delta\vec{r}}{\Delta t} \tag{4.2}$$

The **instantaneous velocity** of a particle is defined as the limit of the average velocity as Δt approaches zero:

$$\vec{v} \equiv \lim_{\Delta t \to 0}\frac{\Delta\vec{r}}{\Delta t} = \frac{d\vec{r}}{dt} \tag{4.3}$$

The **average acceleration** of a particle is defined as the change in its instantaneous velocity vector divided by the time interval Δt during which that change occurs:

$$\vec{a}_{avg} \equiv \frac{\Delta\vec{v}}{\Delta t} = \frac{\vec{v}_f - \vec{v}_i}{t_f - t_i} \tag{4.4}$$

The **instantaneous acceleration** of a particle is defined as the limiting value of the average acceleration as Δt approaches zero:

$$\vec{a} \equiv \lim_{\Delta t \to 0}\frac{\Delta\vec{v}}{\Delta t} = \frac{d\vec{v}}{dt} \tag{4.5}$$

Projectile motion is one type of two-dimensional motion, exhibited by an object launched into the air near the Earth's surface and experiencing free fall. If the projectile is launched at an upward angle from the horizontal, it will follow a path described mathematically as a parabola.

A particle moving in a circular path with constant speed is exhibiting **uniform circular motion.**

▶ Concepts and Principles

If a particle moves with *constant* acceleration $\vec{a}$ and has velocity $\vec{v}_i$ and position $\vec{r}_i$ at $t = 0$, its velocity and position vectors at some later time t are

$$\vec{v}_f = \vec{v}_i + \vec{a}t \tag{4.8}$$

$$\vec{r}_f = \vec{r}_i + \vec{v}_i t + \tfrac{1}{2}\vec{a}t^2 \tag{4.9}$$

For two-dimensional motion in the xy plane under constant acceleration, each of these vector expressions is equivalent to two component expressions: one for the motion in the x direction and one for the motion in the y direction.

It is useful to think of projectile motion in terms of a combination of two analysis models: (1) the particle under constant velocity model in the x direction and (2) the particle under constant acceleration model in the vertical direction with a constant downward acceleration of magnitude $g = 9.80$ m/s².

If a particle moves along a curved path in such a way that both the magnitude and the direction of $\vec{v}$ change in time, the particle has an acceleration vector that can be described by two component vectors: (1) a radial component vector $\vec{a}_r$ that causes the change in direction of $\vec{v}$ and (2) a tangential component vector $\vec{a}_t$ that causes the change in magnitude of $\vec{v}$. The magnitude of $\vec{a}_r$ is v^2/r, and the magnitude of $\vec{a}_t$ is $|dv/dt|$.

A particle in uniform circular motion undergoes a radial acceleration $\vec{a}_r$ because the direction of $\vec{v}$ changes in time. This acceleration is called **centripetal acceleration,** and its direction is always toward the center of the circle.

The velocity $\vec{u}_{PA}$ of a particle measured in a fixed frame of reference S_A can be related to the velocity $\vec{u}_{PB}$ of the same particle measured in a moving frame of reference S_B by

$$\vec{u}_{PA} = \vec{u}_{PB} + \vec{v}_{BA} \tag{4.30}$$

where $\vec{v}_{BA}$ is the velocity of S_B relative to S_A.

> ## Analysis Model for Problem Solving

Particle in Uniform Circular Motion If a particle moves in a circular path of radius r with a constant speed v, the magnitude of its centripetal acceleration is given by

$$a_c = \frac{v^2}{r} \qquad (4.21)$$

and the **period** of the particle's motion is given by

$$T = \frac{2\pi r}{v} \qquad (4.22)$$

The **angular speed** of the particle is

$$\omega = \frac{2\pi}{T} \qquad (4.23)$$

Think–Pair–Share

See the Preface for an explanation of the icons used in this problems set. For additional assessment items for this section, go to **WEBASSIGN** *From Cengage*

1. You watch your toddler nephew rolling marbles toward the top of a staircase. There are 12 steps, each 30.0 cm deep horizontally, and separated by 20.0 cm vertically. The marbles leave the upper landing horizontally and are projected into the air, bouncing down the steps until they arrive at the lower floor. This gets you wondering the following: (a) How fast must the marble be rolled so that it misses bouncing off the *first* step below the upper landing? (b) How fast must the marble be rolled so that it misses bouncing off the *second* step below the upper landing? (c) Is it possible for your toddler nephew to roll the marble fast enough to miss *all* the steps? (d) Suppose the marble is projected with a speed such that it lands on the sixth step and bounces upward at the same angle at which it struck the step, with the same speed. Argue that the marble will not hit another step before striking the floor of the lower landing.

2. **ACTIVITY** Place a penny at the corner of a table as shown in the overhead view in Figure TP4.2. Place a ruler next to the penny and another penny on the top of the part of the ruler that hangs off the edge of the table. Hold the end of the ruler on the table with one hand and use your other hand to flick the end of the ruler with the penny parallel to the table surface. This will project the penny sitting on the corner of the table in a horizontal direction. At the same time, the ruler will slide out from under the second penny, which will fall straight down. Using your smartphone audio recorder, make an audio recording of the two falling pennies. From the recording, determine the time interval between the landing of the two pennies on the floor. What should the time interval be theoretically?

Flick the end of the ruler this way

Figure TP4.2

Problems

See the Preface for an explanation of the icons used in this problems set. For additional assessment items for this section, go to **WEBASSIGN** *From Cengage*

SECTION 4.1 The Position, Velocity, and Acceleration Vectors

1. Suppose the position vector for a particle is given as a function of time by $\vec{r}(t) = x(t)\hat{i} + y(t)\hat{j}$, with $x(t) = at + b$ and $y(t) = ct^2 + d$, where $a = 1.00$ m/s, $b = 1.00$ m, $c = 0.125$ m/s^2, and $d = 1.00$ m. (a) Calculate the average velocity during the time interval from $t = 2.00$ s to $t = 4.00$ s. (b) Determine the velocity and the speed at $t = 2.00$ s.

2. **Q|C** The coordinates of an object moving in the xy plane vary with time according to the equations $x = -5.00 \sin \omega t$ and $y = 4.00 - 5.00 \cos \omega t$, where ω is a constant, x and y are in meters, and t is in seconds. (a) Determine the components of velocity of the object at $t = 0$. (b) Determine the components of acceleration of the object at $t = 0$. (c) Write expressions for the position vector, the velocity vector, and the acceleration vector of the object at any time $t > 0$. (d) Describe the path of the object in an xy plot.

SECTION 4.2 Two-Dimensional Motion with Constant Acceleration

3. **V** The vector position of a particle varies in time according to the expression $\vec{r} = 3.00\hat{i} - 6.00t^2\hat{j}$, where $\vec{r}$ is in meters and t is in seconds. (a) Find an expression for the velocity of the particle as a function of time. (b) Determine the acceleration of the particle as a function of time. (c) Calculate the particle's position and velocity at $t = 1.00$ s.

4. It is not possible to see very small objects, such as viruses, using an ordinary light microscope. An electron microscope, however, can view such objects using an electron beam instead of a light beam. Electron microscopy has proved invaluable for investigations of viruses, cell membranes and subcellular structures, bacterial surfaces, visual receptors, chloroplasts, and the contractile properties of muscles. The "lenses" of an electron microscope consist of electric and magnetic fields that control the electron beam. As an example of the manipulation of an electron beam, consider an electron traveling away from the origin along the x axis in the xy plane with initial velocity $\vec{\mathbf{v}}_i = v_i\hat{\mathbf{i}}$. As it passes through the region $x = 0$ to $x = d$, the electron experiences acceleration $\vec{\mathbf{a}} = a_x\hat{\mathbf{i}} + a_y\hat{\mathbf{j}}$, where a_x and a_y are constants. For the case $v_i = 1.80 \times 10^7$ m/s, $a_x = 8.00 \times 10^{14}$ m/s^2, and $a_y = 1.60 \times 10^{15}$ m/s^2, determine at $x = d = 0.010\ 0$ m (a) the position of the electron, (b) the velocity of the electron, (c) the speed of the electron, and (d) the direction of travel of the electron (i.e., the angle between its velocity and the x axis).

5. Review. A snowmobile is originally at the point with position vector 29.0 m at 95.0° counterclockwise from the x axis, moving with velocity 4.50 m/s at 40.0°. It moves with constant acceleration 1.90 m/s^2 at 200°. After 5.00 s have elapsed, find (a) its velocity and (b) its position vector.

SECTION 4.3 Projectile Motion

Note: Ignore air resistance in all problems and take $g = 9.80$ m/s^2 at the Earth's surface.

6. In a local bar, a customer slides an empty beer mug down [S] the counter for a refill. The height of the counter is h. The mug slides off the counter and strikes the floor at distance d from the base of the counter. (a) With what velocity did the mug leave the counter? (b) What was the direction of the mug's velocity just before it hit the floor?

7. Mayan kings and many school sports teams are named [BIO] for the puma, cougar, or mountain lion—*Felis concolor*—the best jumper among animals. It can jump to a height of 12.0 ft when leaving the ground at an angle of 45.0°. With what speed, in SI units, does it leave the ground to make this leap?

8. A projectile is fired in such a way that its horizontal range is equal to three times its maximum height. What is the angle of projection?

9. The speed of a projectile when it reaches its maximum height is one-half its speed when it is at half its maximum height. What is the initial projection angle of the projectile?

10. A rock is thrown upward from level ground in such a way [QC] that the maximum height of its flight is equal to its hori-[S] zontal range R. (a) At what angle θ is the rock thrown? (b) In terms of its original range R, what is the range R_{max} the rock can attain if it is launched at the same speed but at the optimal angle for maximum range? (c) **What If?** Would your answer to part (a) be different if the rock is thrown with the same speed on a different planet? Explain.

11. A firefighter, a distance d from a burning building, directs [S] a stream of water from a fire hose at angle θ_i above the horizontal as shown in Figure P4.11. If the initial speed of the stream is v_i, at what height h does the water strike the building?

Figure P4.11

12. A basketball star covers 2.80 m horizontally in a jump to dunk the ball (Fig. P4.12a). His motion through space can be modeled precisely as that of a particle at his *center of mass*, which we will define in Chapter 9. His center of mass is at elevation 1.02 m when he leaves the floor. It reaches a maximum height of 1.85 m above the floor and is at elevation 0.900 m when he touches down again. Determine (a) his time of flight (his "hang time"), (b) his horizontal and (c) vertical velocity components at the instant of takeoff, and (d) his takeoff angle. (e) For comparison, determine the hang time of a whitetail deer making a jump (Fig. P4.12b) with center-of-mass elevations $y_i = 1.20$ m, $y_{max} = 2.50$ m, and $y_f = 0.700$ m.

a   b

Figure P4.12

13. A student stands at the edge of a [GP] cliff and throws a stone horizontally over the edge with a speed of $v_i = 18.0$ m/s. The cliff is $h = 50.0$ m above a body of water as shown in Figure P4.13. (a) What are the coordinates of the initial position of the stone? (b) What are the components of the initial velocity of the stone? (c) What is the appropriate analysis model for the vertical motion of the stone? (d) What is the appropriate analysis model for the horizontal motion of the stone? (e) Write symbolic equations for the x and y components of the velocity of the stone as a function of time. (f) Write symbolic equations for the position of the stone as a function of time. (g) How long after being released does the stone strike the water below the cliff? (h) With what speed and angle of impact does the stone land?

Figure P4.13

14. The record distance in the sport of throwing cowpats is QC 81.1 m. This record toss was set by Steve Urner of the United States in 1981. Assuming the initial launch angle was 45° and neglecting air resistance, determine (a) the initial speed of the projectile and (b) the total time interval the projectile was in flight. (c) How would the answers change if the range were the same but the launch angle were greater than 45°? Explain.

15. A home run is hit in such a way that the baseball just clears a T wall 21.0 m high, located 130 m from home plate. The ball is hit at an angle of 35.0° to the horizontal, and air resistance is negligible. Find (a) the initial speed of the ball, (b) the time it takes the ball to reach the wall, and (c) the velocity components and the speed of the ball when it reaches the wall. (Assume the ball is hit at a height of 1.00 m above the ground.)

16. A projectile is fired from the top of a cliff of height h above S the ocean below. The projectile is fired at an angle θ above the horizontal and with an initial speed v_i. (a) Find a symbolic expression in terms of the variables v_i, g, and θ for the time at which the projectile reaches its maximum height. (b) Using the result of part (a), find an expression for the maximum height h_{max} above the ocean attained by the projectile in terms of h, v_i, g, and θ.

17. A boy stands on a diving board and tosses a stone into a swimming pool. The stone is thrown from a height of 2.50 m above the water surface with a velocity of 4.00 m/s at an angle of 60.0° above the horizontal. As the stone strikes the water surface, it immediately slows down to exactly half the speed it had when it struck the water and maintains that speed while in the water. After the stone enters the water, it moves in a straight line in the direction of the velocity it had when it struck the water. If the pool is 3.00 m deep, how much time elapses between when the stone is thrown and when it strikes the bottom of the pool?

SECTION 4.4 Analysis Model: Particle in Uniform Circular Motion

Note: Problems 3 and 9 in Chapter 6 can also be assigned with this section.

18. In Example 4.6, we found the centripetal acceleration of the Earth as it revolves around the Sun. From information on the endpapers of this book, compute the centripetal acceleration of a point on the surface of the Earth at the equator caused by the rotation of the Earth about its axis.

19. The astronaut orbiting the Earth in Figure P4.19 is preparing to dock with a Westar VI satellite. The satellite is in a circular orbit 600 km above the Earth's surface, where the free-fall acceleration is 8.21 m/s². Take the radius of the Earth as 6 400 km. Determine the speed of the satellite and the time interval required to complete one orbit around the Earth, which is the period of the satellite.

Figure P4.19

20. An athlete swings a ball, connected to the end of a chain, in a horizontal circle. The athlete is able to rotate the ball at the rate of 8.00 rev/s when the length of the chain is 0.600 m. When he increases the length to 0.900 m, he is able to rotate the ball only 6.00 rev/s. (a) Which rate of rotation gives the greater speed for the ball? (b) What is the centripetal acceleration of the ball at 8.00 rev/s? (c) What is the centripetal acceleration at 6.00 rev/s?

21. The athlete shown in Figure P4.21 rotates a 1.00-kg discus along a circular path of radius 1.06 m. The maximum speed of the discus is 20.0 m/s. Determine the magnitude of the maximum radial acceleration of the discus.

Figure P4.21

22. A tire 0.500 m in radius rotates at a constant rate of 200 rev/min. Find the speed and acceleration of a small stone lodged in the tread of the tire (on its outer edge).

SECTION 4.5 Tangential and Radial Acceleration

23. (a) Can a particle moving with instantaneous speed 3.00 m/s on a path with radius of curvature 2.00 m have an acceleration of magnitude 6.00 m/s²? (b) Can it have an acceleration of magnitude 4.00 m/s²? In each case, if the answer is yes, explain how it can happen; if the answer is no, explain why not.

24. A ball swings counterclockwise in a vertical circle at the end of a rope 1.50 m long. When the ball is 36.9° past the lowest point on its way up, its total acceleration is $(-22.5\,\hat{\mathbf{i}} + 20.2\,\hat{\mathbf{j}})$ m/s². For that instant, (a) sketch a vector diagram showing the components of its acceleration, (b) determine the magnitude of its radial acceleration, and (c) determine the speed and velocity of the ball.

SECTION 4.6 Relative Velocity and Relative Acceleration

25. A bolt drops from the ceiling of a moving train car that is QC accelerating northward at a rate of 2.50 m/s². (a) What is the acceleration of the bolt relative to the train car? (b) What is the acceleration of the bolt relative to the Earth? (c) Describe the trajectory of the bolt as seen by an observer inside the train car. (d) Describe the trajectory of the bolt as seen by an observer fixed on the Earth.

26. The pilot of an airplane notes that the compass indicates a heading due west. The airplane's speed relative to the air is 150 km/h. The air is moving in a wind at 30.0 km/h toward the north. Find the velocity of the airplane relative to the ground.

27. You are taking flying lessons from an experienced pilot. CR You and the pilot are up in the plane, with you in the pilot seat. The control tower radios the plane, saying that, while you have been airborne, a 25-mi/h crosswind has arisen, with the direction of the wind perpendicular to the runway on which you plan to land. The pilot tells you that your normal airspeed as you land will be 80 mi/h relative to the ground. This speed is relative to the air, in the direction in which the nose of the airplane points. He asks you to determine the angle at which the aircraft must be "crabbed," that is, the angle between the centerline of the aircraft and the centerline of the runway that will allow the airplane's velocity relative to the ground to be parallel to the runway.

28. A car travels due east with a speed of 50.0 km/h. Raindrops are falling at a constant speed vertically with respect to the Earth. The traces of the rain on the side windows of the car make an angle of 60.0° with the vertical. Find the velocity of the rain with respect to (a) the car and (b) the Earth.

29. A science student is riding on a flatcar of a train traveling along a straight, horizontal track at a constant speed of 10.0 m/s. The student throws a ball into the air along a path that he judges to make an initial angle of 60.0° with the horizontal and to be in line with the track. The student's professor, who is standing on the ground nearby, observes the ball to rise vertically. How high does she see the ball rise?

30. A river has a steady speed of 0.500 m/s. A student swims upstream a distance of 1.00 km and swims back to the starting point. (a) If the student can swim at a speed of 1.20 m/s in still water, how long does the trip take? (b) How much time is required in still water for the same length swim? (c) Intuitively, why does the swim take longer when there is a current?

31. A river flows with a steady speed v. A student swims upstream a distance d and then back to the starting point. The student can swim at speed c in still water. (a) In terms of d, v, and c, what time interval is required for the round trip? (b) What time interval would be required if the water were still? (c) Which time interval is larger? Explain whether it is always larger.

32. You are participating in a summer internship with the Coast Guard. You have been assigned the duty of determining the direction in which a Coast Guard speedboat should travel to intercept unidentified vessels. One day, the radar operator detects an unidentified vessel at a distance of 20.0 km from the radar installation in the direction 15.0° east of north. The vessel is traveling at 26.0 km/h on a course at 40.0° east of north. The Coast Guard wishes to send a speedboat, which travels at 50.0 km/h, to travel in a straight line from the radar installation to intercept and investigate the vessel, and asks you for the heading for the speedboat to take. Express the direction as a compass bearing with respect to due north.

33. A farm truck moves due east with a constant velocity of 9.50 m/s on a limitless, horizontal stretch of road. A boy riding on the back of the truck throws a can of soda upward (Fig. P4.33) and catches the projectile at the same location on the truck bed, but 16.0 m farther down the road. (a) In the frame of reference of the truck, at what angle to the vertical does the boy throw the can? (b) What is the initial speed of the can relative to the truck? (c) What is the shape of the can's trajectory as seen by the boy?

Figure P4.33

An observer on the ground watches the boy throw the can and catch it. In this observer's frame of reference, (d) describe the shape of the can's path and (e) determine the initial velocity of the can.

ADDITIONAL PROBLEMS

34. A ball on the end of a string is whirled around in a horizontal circle of radius 0.300 m. The plane of the circle is 1.20 m above the ground. The string breaks and the ball lands 2.00 m (horizontally) away from the point on the ground directly beneath the ball's location when the string breaks. Find the radial acceleration of the ball during its circular motion.

35. *Why is the following situation impossible?* A normally proportioned adult walks briskly along a straight line in the $+x$ direction, standing straight up and holding his right arm vertical and next to his body so that the arm does not swing. His right hand holds a ball at his side a distance h above the floor. When the ball passes above a point marked as $x = 0$ on the horizontal floor, he opens his fingers to release the ball from rest relative to his hand. The ball strikes the ground for the first time at position $x = 7.00h$.

36. A particle starts from the origin with velocity $5\hat{\mathbf{i}}$ m/s at $t = 0$ and moves in the xy plane with a varying acceleration given by $\vec{\mathbf{a}} = (6\sqrt{t}\,\hat{\mathbf{j}})$, where $\vec{\mathbf{a}}$ is in meters per second squared and t is in seconds. (a) Determine the velocity of the particle as a function of time. (b) Determine the position of the particle as a function of time.

37. Lisa in her Lamborghini accelerates at $(3.00\hat{\mathbf{i}} - 2.00\hat{\mathbf{j}})$ m/s², while Jill in her Jaguar accelerates at $(1.00\hat{\mathbf{i}} + 3.00\hat{\mathbf{j}})$ m/s². They both start from rest at the origin. After 5.00 s, (a) what is Lisa's speed with respect to Jill, (b) how far apart are they, and (c) what is Lisa's acceleration relative to Jill?

38. A boy throws a stone horizontally from the top of a cliff of height h toward the ocean below. The stone strikes the ocean at distance d from the base of the cliff. In terms of h, d, and g, find expressions for (a) the time t at which the stone lands in the ocean, (b) the initial speed of the stone, (c) the speed of the stone immediately before it reaches the ocean, and (d) the direction of the stone's velocity immediately before it reaches the ocean.

39. *Why is the following situation impossible?* Albert Pujols hits a home run so that the baseball just clears the top row of bleachers, 24.0 m high, located 130 m from home plate. The ball is hit at 41.7 m/s at an angle of 35.0° to the horizontal, and air resistance is negligible.

40. As some molten metal splashes, one droplet flies off to the east with initial velocity v_i at angle θ_i above the horizontal, and another droplet flies off to the west with the same speed at the same angle above the horizontal as shown in Figure P4.40. In terms of v_i and θ_i, find the distance between the two droplets as a function of time.

Figure P4.40

41. An astronaut on the surface of the Moon fires a cannon to launch an experiment package, which leaves the barrel moving horizontally. Assume the free-fall acceleration on the Moon is one-sixth of that on the Earth. (a) What must the muzzle speed of the package be so that it travels completely around the Moon and returns to its original location? (b) What time interval does this trip around the Moon require?

42. A pendulum with a cord of length $r = 1.00$ m swings in a vertical plane (Fig. P4.42). When the pendulum is in the two horizontal positions $\theta = 90.0°$ and $\theta = 270°$, its speed is 5.00 m/s. Find the magnitude of (a) the radial acceleration and (b) the tangential acceleration for these positions. (c) Draw vector diagrams to determine the direction of the total acceleration for these two positions. (d) Calculate the magnitude and direction of the total acceleration at these two positions.

Figure P4.42

43. A spring cannon is located at the edge of a table that is
QC 1.20 m above the floor. A steel ball is launched from the cannon with speed v_i at 35.0° above the horizontal. (a) Find the horizontal position of the ball as a function of v_i at the instant it lands on the floor. We write this function as $x(v_i)$. Evaluate x for (b) $v_i = 0.100$ m/s and for (c) $v_i = 100$ m/s. (d) Assume v_i is close to but not equal to zero. Show that one term in the answer to part (a) dominates so that the function $x(v_i)$ reduces to a simpler form. (e) If v_i is very large, what is the approximate form of $x(v_i)$? (f) Describe the overall shape of the graph of the function $x(v_i)$.

44. A projectile is launched from the point $(x = 0, y = 0)$,
QC with velocity $(12.0\hat{\mathbf{i}} + 49.0\hat{\mathbf{j}})$ m/s, at $t = 0$. (a) Make a table listing the projectile's distance $|\vec{\mathbf{r}}|$ from the origin at the end of each second thereafter, for $0 \leq t \leq 10$ s. Tabulating the x and y coordinates and the components of velocity v_x and v_y will also be useful. (b) Notice that the projectile's distance from its starting point increases with time, goes through a maximum, and starts to decrease. Prove that the distance is a maximum when the position vector is perpendicular to the velocity. *Suggestion:* Argue that if $\vec{\mathbf{v}}$ is not perpendicular to $\vec{\mathbf{r}}$, then $|\vec{\mathbf{r}}|$ must be increasing or decreasing. (c) Determine the magnitude of the maximum displacement. (d) Explain your method for solving part (c).

45. A fisherman sets out upstream on a river. His small boat, powered by an outboard motor, travels at a constant speed v in still water. The water flows at a lower constant speed v_w. The fisherman has traveled upstream for 2.00 km when his ice chest falls out of the boat. He notices that the chest is missing only after he has gone upstream for another 15.0 min. At that point, he turns around and heads back downstream, all the time traveling at the same speed relative to the water. He catches up with the floating ice chest just as he returns to his starting point. How fast is the river flowing? Solve this problem in two ways. (a) First, use the Earth as a reference frame. With respect to the Earth, the boat travels upstream at speed $v - v_w$ and downstream at $v + v_w$. (b) A second much simpler

and more elegant solution is obtained by using the water as the reference frame. This approach has important applications in many more complicated problems; examples are calculating the motion of rockets and satellites and analyzing the scattering of subatomic particles from massive targets.

46. An outfielder throws a baseball to his catcher in an attempt to throw out a runner at home plate. The ball bounces once before reaching the catcher. Assume the angle at which the bounced ball leaves the ground is the same as the angle at which the outfielder threw it as shown in Figure P4.46, but that the ball's speed after the bounce is one-half of what it was before the bounce. (a) Assume the ball is always thrown with the same initial speed and ignore air resistance. At what angle θ should the fielder throw the ball to make it go the same distance D with one bounce (blue path) as a ball thrown upward at 45.0° with no bounce (green path)? (b) Determine the ratio of the time interval for the one-bounce throw to the flight time for the no-bounce throw.

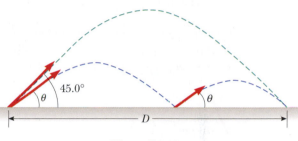

Figure P4.46

47. Do not hurt yourself; do not strike your hand against any-
QC thing. Within these limitations, describe what you do to give your hand a large acceleration. Compute an order-of-magnitude estimate of this acceleration, stating the quantities you measure or estimate and their values.

48. You are on the Pirates of the Caribbean attraction in the
CR Magic Kingdom at Disney World. Your boat rides through a pirate battle, in which cannons on a ship and in a fort are firing at each other. While you are aware that the splashes in the water do not represent actual cannonballs, you begin to wonder about such battles in the days of the pirates. Suppose the fort and the ship are separated by 75.0 m. You see that the cannons in the fort are aimed so that their cannonballs would be fired horizontally from a height of 7.00 m above the water. (a) You wonder at what speed they must be fired in order to hit the ship before falling in the water. (b) Then, you think about the sludge that must build up inside the barrel of a cannon. This sludge should slow down the cannonballs. A question occurs in your mind: if the cannonballs can be fired at only 50.0% of the speed found earlier, is it possible to fire them upward at some angle to the horizontal so that they would reach the ship?

CHALLENGE PROBLEMS

49. A skier leaves the ramp of a ski jump with a velocity of
QC $v = 10.0$ m/s at $\theta = 15.0°$ above the horizontal as shown in Figure P4.49 (page 94). The slope where she will land is inclined downward at $\phi = 50.0°$, and air resistance is negligible. Find (a) the distance from the end of the ramp to where the jumper lands and (b) her velocity components just before the landing. (c) Explain how you think the results might be affected if air resistance were included.

Figure P4.49

50. A projectile is fired up an incline (incline angle ϕ) with an initial speed v_i at an angle θ_i with respect to the horizontal ($\theta_i > \phi$) as shown in Figure P4.50. (a) Show that the projectile travels a distance d up the incline, where

$$d = \frac{2v_i^2 \cos\theta_i \sin(\theta_i - \phi)}{g \cos^2 \phi}$$

(b) For what value of θ_i is d a maximum, and what is that maximum value?

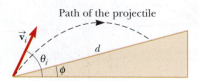

Figure P4.50

51. Two swimmers, Chris and Sarah, start together at the same point on the bank of a wide stream that flows with a speed v. Both move at the same speed c (where $c > v$) relative to the water. Chris swims downstream a distance L and then upstream the same distance. Sarah swims so that her motion relative to the Earth is perpendicular to the banks of the stream. She swims the distance L and then back the same distance, with both swimmers returning to the starting point. In terms of L, c, and v, find the time intervals required (a) for Chris's round trip and (b) for Sarah's round trip. (c) Explain which swimmer returns first.

52. In the What If? section of Example 4.5, it was claimed that the maximum range of a ski jumper occurs for a launch angle θ given by

$$\theta = 45° - \frac{\phi}{2}$$

where ϕ is the angle the hill makes with the horizontal in Figure 4.15. Prove this claim by deriving the equation above.

53. A fireworks rocket explodes at height h, the peak of its vertical trajectory. It throws out burning fragments in all directions, but all at the same speed v. Pellets of solidified metal fall to the ground without air resistance. Find the smallest angle that the final velocity of an impacting fragment makes with the horizontal.

The Laws of Motion

5

Your cousin prepares to catch a raw egg thrown to her at a birthday party. (*Sue McDonald/Shutterstock.com*)

STORYLINE **You have returned home from your trip to Catalina** Island in the previous two chapters. Your family is having a picnic to celebrate a birthday, so there are many people in your backyard on a beautiful day. Someone suggests an egg toss contest. You decide to offer some advice to your cousin and instruct her to move her hands backward just as she catches the egg. Your cousin looks you in the eye and says, "Why?" You are tempted to say, "Because that's just how you do it," but then consider the deeper implications of your cousin's question. Why is it that you move your hands backward? What happens if you hold your hands in a fixed position and catch the egg? Should you have your cousin try this? You take your cousin to the computer and have her search for YouTube videos involving catching an egg, and then you notice some videos showing the results of throwing an egg into a vertical sheet. As you and your cousin watch these videos, both of you begin to understand the physics of throwing and catching eggs.

CONNECTIONS In the previous chapters, we learned how to describe the motion of particles and objects that can be modeled as particles. We saw motion changing in various ways. The acceleration of a car is a change in its velocity. The direction of the velocity of a thrown baseball changes as it flies through the air. We can *describe* these changes with the material in the previous chapters, but what *causes* these changes? Such a question represents a transition from kinematics, the description of motion, to *dynamics*, the study of causes of changes in motion. We will see that *force* is the cause of changes in motion, and will study the effects of force through the laws of motion as handed down to us by Isaac Newton. The notion of force will be used again and again in future chapters: gravitational forces in Chapter 13, electric forces in Chapter 22, magnetic forces in Chapter 28, nuclear forces in Chapter 43, and more.

95

Isaac Newton
English physicist and mathematician
(1642–1727)
Isaac Newton was one of the most brilliant scientists in history. Before the age of 30, he formulated the basic concepts and laws of mechanics, discovered the law of universal gravitation, and invented the mathematical methods of calculus. As a consequence of his theories, Newton was able to explain the motions of the planets, the ebb and flow of the tides, and many special features of the motions of the Moon and the Earth. He also interpreted many fundamental observations concerning the nature of light. His contributions to physical theories dominated scientific thought for two centuries and remain important today.

5.1 The Concept of Force

Everyone has a basic understanding of the concept of force from everyday experience. When you push your empty dinner plate away, you exert a force on it. Similarly, you exert a force on a ball when you throw or kick it. In these examples, the word *force* refers to an interaction with an object by means of muscular activity and some change in the object's velocity. Forces do not always cause motion, however. For example, when you are sitting, a gravitational force acts on your body and yet you remain stationary. As a second example, you can push (in other words, exert a force) on a large boulder and not be able to move it.

What force (if any) causes the Moon to orbit the Earth? Newton answered this and related questions by stating that forces are what cause any change in the velocity of an object. The Moon's velocity changes in direction as it moves in a nearly circular orbit around the Earth. This change in velocity is caused by the gravitational force exerted by the Earth on the Moon.

When a coiled spring is pulled, as in Figure 5.1a, the spring stretches. When a stationary cart is pulled, as in Figure 5.1b, the cart moves. When a football is kicked, as in Figure 5.1c, it is both deformed and set in motion. These situations are all examples of a class of forces called *contact forces*. That is, they involve physical contact between two objects. Other examples of contact forces are the force exerted by gas molecules on the walls of a container and the force exerted by your feet on the floor.

Another class of forces, known as *field forces,* does not involve physical contact between two objects. These forces act through empty space. The gravitational force of attraction between two objects with mass, illustrated in Figure 5.1d, is an example of this class of force. The gravitational force keeps objects bound to the Earth and the planets in orbit around the Sun. Another common field force is the electric force that one electric charge exerts on another (Fig. 5.1e), such as the attractive electric force between an electron and a proton that form a hydrogen atom. A third example of a field force is the force a bar magnet exerts on a piece of iron (Fig. 5.1f).

The distinction between contact forces and field forces is not as sharp as you may have been led to believe by the previous discussion. When examined at the atomic level, all the forces we classify as contact forces turn out to be caused by electric (field) forces of the type illustrated in Figure 5.1e. Nevertheless, in developing models for macroscopic phenomena, it is convenient to use both classifications of forces. The only known *fundamental* forces in nature are all field forces: (1) *gravitational forces* between objects, (2) *electromagnetic forces* between electric charges, (3) *strong forces* between subatomic particles, and (4) *weak forces* that arise in certain radioactive decay processes. In classical physics, we are concerned only with gravitational and electromagnetic forces. We will discuss strong and weak forces in Chapter 44.

Figure 5.1 Some examples of applied forces. In each case, a force is exerted on the object within the boxed area. Some agent in the environment external to the boxed area exerts a force on the object.

A downward force $\vec{\mathbf{F}}_1$ elongates the spring 1.00 cm.

A downward force $\vec{\mathbf{F}}_2$ elongates the spring 2.00 cm.

When $\vec{\mathbf{F}}_1$ and $\vec{\mathbf{F}}_2$ are applied together in the same direction, the spring elongates by 3.00 cm.

When $\vec{\mathbf{F}}_1$ is downward and $\vec{\mathbf{F}}_2$ is horizontal, the combination of the two forces elongates the spring by 2.24 cm.

a b c d

Figure 5.2 The vector nature of a force is tested with a spring scale.

The Vector Nature of Force

It is possible to use the deformation of a spring to measure force. Suppose a vertical force is applied to a spring scale that has a fixed upper end as shown in Figure 5.2a. The spring elongates when the force is applied, and a pointer on the scale reads the extension of the spring. We can calibrate the spring by defining a reference force $\vec{\mathbf{F}}_1$ as the force that produces a pointer reading of 1.00 cm. If we now apply a different downward force $\vec{\mathbf{F}}_2$ whose magnitude is twice that of the reference force $\vec{\mathbf{F}}_1$ as seen in Figure 5.2b, the pointer moves to 2.00 cm. Figure 5.2c shows that the combined effect of the two collinear forces is the sum of the effects of the individual forces.

Now suppose the two forces are applied simultaneously with $\vec{\mathbf{F}}_1$ downward and $\vec{\mathbf{F}}_2$ horizontal as illustrated in Figure 5.2d. In this case, the pointer reads 2.24 cm. The single force $\vec{\mathbf{F}}$ that would produce this same reading is the sum of the two vectors $\vec{\mathbf{F}}_1$ and $\vec{\mathbf{F}}_2$ as described in Figure 5.2d. That is, $|\vec{\mathbf{F}}_1| = \sqrt{F_1^2 + F_2^2} = 2.24$ units, and its direction is $\theta = \tan^{-1}(-0.500) = -26.6°$. Because forces have been experimentally verified to behave as vectors, you *must* use the rules of vector addition to obtain the net force on an object.

5.2 Newton's First Law and Inertial Frames

We begin our study of forces by imagining some physical situations involving a puck on a perfectly level air hockey table (Fig. 5.3). You expect that the puck will remain stationary when it is placed gently at rest on the table. Now imagine your air hockey table is located on a train moving with constant velocity along a perfectly smooth track. If the puck is placed on the table, the puck again remains where it is placed. If the train were to accelerate, however, the puck would start moving along the table opposite the direction of the train's acceleration, just as a set of papers on your dashboard falls onto the floor of your car when you step on the accelerator.

As we saw in Section 4.6, a moving object can be observed from any number of reference frames. **Newton's first law of motion,** sometimes called the *law of inertia,* defines a special set of reference frames called *inertial frames.* This law can be stated

Figure 5.3 On an air hockey table, air blown through holes in the surface supports the puck and allows it to move almost without friction across the table. If the table is not accelerating, a puck placed on the table will remain at rest.

in a theoretical manner as follows:

▶ A theoretical statement
of Newton's first law

> If an object does not interact with other objects, it is possible to identify a reference frame in which the object has zero acceleration.

▶ Inertial frame of reference

Such a reference frame is called an **inertial frame of reference.** When the puck is on the air hockey table located on the ground, you are observing it from an inertial reference frame; there are no horizontal interactions of the puck with any other objects, and you observe it to have zero acceleration in that direction. When you are on the train moving at constant velocity, you are also observing the puck from an inertial reference frame. Any reference frame that moves with constant velocity relative to an inertial frame is itself an inertial frame. When you and the train accelerate, however, you are observing the puck from a **noninertial reference frame** because you and the train are accelerating relative to the inertial reference frame of the Earth's surface. While the puck appears to be accelerating according to your observations, a reference frame can be identified in which the puck has zero acceleration. For example, an observer standing outside the train on the ground sees the puck sliding relative to the table but always moving with the same velocity with respect to the ground as the train had before it started to accelerate (because there is almost no friction to "tie" the puck and the train together). Therefore, Newton's first law is still satisfied even though your observations as a rider on the train show an apparent acceleration relative to you.

A reference frame that moves with constant velocity relative to the distant stars is the best approximation of an inertial frame, and for our purposes we can consider the Earth as being such a frame. The Earth is not really an inertial frame because of its orbital motion around the Sun and its rotational motion about its own axis, both of which involve centripetal accelerations. These accelerations are small compared with g, however, and can often be neglected. For this reason, we model the Earth as an inertial frame, along with any other frame attached to it.

Let us assume we are observing an object from an inertial reference frame. (We will return to observations made in noninertial reference frames in Section 6.3.) Before about 1600, scientists believed that the natural state of matter was the state of rest. Observations showed that moving objects eventually stopped moving. Galileo was the first to take a different approach to motion and the natural state of matter. He devised thought experiments and concluded that it is not the nature of an object to stop once set in motion: rather, it is its nature to *resist changes in its motion.* In his words, "Any velocity once imparted to a moving body will be rigidly maintained as long as the external causes of retardation are removed." For example, a spacecraft drifting through empty space with its engine turned off will keep moving forever. It would *not* seek a "natural state" of rest.

Given our discussion of observations made from inertial reference frames, we can pose a more practical statement of Newton's first law of motion than that in the previous screened statement:

▶ A more practical statement
of Newton's first law

> In the absence of external forces and when viewed from an inertial reference frame, an object at rest remains at rest and an object in motion continues in motion with a constant velocity (that is, with a constant speed in a straight line).

In other words, **when no force acts on an object, the acceleration of the object is zero.** From the first law, we conclude that any *isolated object* (one that does not interact with its environment) is either at rest and stays at rest, or is moving with constant velocity. The tendency of an object to resist any attempt to change its velocity is called **inertia.** Given the statement of the first law above, we can conclude that an object that is accelerating must be experiencing a force. In turn, from the first law, we can define **force** as **that which causes a change in motion of an object.**

▶ Definition of force

PITFALL PREVENTION 5.1

Newton's First Law Newton's first law does *not* say what happens for an object with *zero net force,* that is, multiple forces that cancel; it says what happens *in the absence of external forces.* This subtle but important difference allows us to define force as that which causes a change in the motion. The description of an object under the effect of forces that balance is covered by Newton's second law.

QUICK QUIZ **5.1** Which of the following statements is correct? **(a)** It is possible for an object to have motion in the absence of forces on the object. **(b)** It is possible to have forces on an object in the absence of motion of the object. **(c)** Neither statement **(a)** nor statement **(b)** is correct. **(d)** Both statements **(a)** and **(b)** are correct.

5.3 Mass

Imagine playing catch with either a basketball or a bowling ball. Which ball is more likely to keep moving when you try to catch it? Which ball requires more effort to throw it? The bowling ball requires more effort. In the language of physics, we say that the bowling ball is more resistant to changes in its velocity than the basketball. How can we quantify this concept?

Mass is that property of an object that specifies how much resistance an object exhibits to changes in its velocity, and as we learned in Section 1.1, the SI unit of mass is the kilogram. Experiments show that the greater the mass of an object, the less that object accelerates under the action of a given applied force.

◀ Definition of mass

To describe mass quantitatively, we conduct experiments in which we compare the accelerations a given force produces on different objects. Suppose a force acting on an object of mass m_1 produces a change in motion of the object that we can quantify with the object's acceleration $\vec{a}_1$, and the *same force* acting on an object of mass m_2 produces an acceleration $\vec{a}_2$. The ratio of the two masses is defined as the *inverse* ratio of the magnitudes of the accelerations produced by the force:

$$\frac{m_1}{m_2} \equiv \frac{a_2}{a_1} \tag{5.1}$$

For example, if a given force acting on a 3-kg object produces an acceleration of 4 m/s^2, the same force applied to an object with twice the mass, 6 kg, produces an acceleration with half the magnitude, 2 m/s^2. According to a huge number of similar observations, we conclude that the magnitude of the acceleration of an object is inversely proportional to its mass when acted on by a given force. If one object has a known mass, the mass of the other object can be obtained from acceleration measurements.

As mentioned in Chapter 1, mass is an inherent property of an object and is independent of the object's surroundings and of the method used to measure it. Also, mass is a scalar quantity and thus obeys the rules of ordinary arithmetic. For example, if you combine a 3-kg mass with a 5-kg mass, the total mass is 8 kg. This result can be verified experimentally by comparing the acceleration that a known force gives to several objects separately with the acceleration that the same force gives to the same objects combined as one unit.

Mass should not be confused with weight. Mass and weight are two different quantities. The weight of an object is equal to the magnitude of the gravitational force exerted on the object and varies with location (see Section 5.5). For example, a person weighing 180 lb on the Earth weighs only about 30 lb on the Moon. On the other hand, the mass of an object is the same everywhere: an object having a mass of 2 kg on the Earth also has a mass of 2 kg on the Moon.

◀ Mass and weight are different quantities

5.4 Newton's Second Law

Newton's first law explains what happens to an object when *no* forces act on it: it maintains its original motion; it either remains at rest or moves in a straight line with constant speed. Newton's second law answers the question of what happens to an object when one or more forces act on it.

Imagine performing an experiment in which you push a block of mass m across a frictionless, horizontal surface. When you exert some horizontal force $\vec{F}$ on the block, it moves with some acceleration $\vec{a}$. If you apply a force twice as great on the same block, experimental results show that the acceleration of the block doubles;

PITFALL PREVENTION 5.2

Force Is the Cause of Changes in Motion An object can have motion in the absence of forces as described in Newton's first law. Therefore, don't interpret force as the cause of *motion*. Force is the cause of *changes in motion*.

if you increase the applied force to $3\vec{\mathbf{F}}$, the acceleration triples; and so on. From such observations, we conclude that the acceleration of an object is directly proportional to the force acting on it: $\vec{\mathbf{F}} \propto \vec{\mathbf{a}}$. This idea was first introduced in Section 2.4 when we discussed the direction of the acceleration of an object. We also know from Equation 5.1 that the magnitude of the acceleration of an object is inversely proportional to its mass: $|\vec{\mathbf{a}}| \propto 1/m$.

These experimental observations are summarized in **Newton's second law:**

> When viewed from an inertial reference frame, the acceleration of an object is directly proportional to the net force acting on it and inversely proportional to its mass:
>
> $$\vec{\mathbf{a}} \propto \frac{\sum \vec{\mathbf{F}}}{m}$$

If we choose a proportionality constant of 1, we can relate mass, acceleration, and force through the following mathematical statement of Newton's second law:[1]

◄ Newton's second law

$$\sum \vec{\mathbf{F}} = m\vec{\mathbf{a}} \tag{5.2}$$

In both the textual and mathematical statements of Newton's second law, we have indicated that the acceleration is due to the *net force* $\sum \vec{\mathbf{F}}$ acting on an object. The **net force** on an object is the vector sum of all forces acting on the object. (Other names used for the net force include the *total force*, the *resultant force*, and the *unbalanced force*.) In solving a problem using Newton's second law, it is imperative to determine the correct net force on an object. Many forces may be acting on an object, but there is only one acceleration of the object.

Equation 5.2 is a vector expression and hence is equivalent to three component equations:

◄ Newton's second law: component form

$$\sum F_x = ma_x \qquad \sum F_y = ma_y \qquad \sum F_z = ma_z \tag{5.3}$$

QUICK QUIZ 5.2 An object experiences no acceleration. Which of the following *cannot* be true for the object? **(a)** A single force acts on the object. **(b)** No forces act on the object. **(c)** Forces act on the object, but the forces cancel.

QUICK QUIZ 5.3 You push an object, initially at rest, across a frictionless floor with a constant force for a time interval Δt, resulting in a final speed of v for the object. You then repeat the experiment, but with a force that is twice as large. What time interval is now required to reach the same final speed v?
(a) $4\,\Delta t$ **(b)** $2\,\Delta t$ **(c)** Δt **(d)** $\Delta t/2$ **(e)** $\Delta t/4$

The SI unit of force is the **newton** (N). A force of 1 N is the force that, when acting on an object of mass 1 kg, produces an acceleration of 1 m/s^2. From this definition and Newton's second law, we see that the newton can be expressed in terms of the following fundamental units of mass, length, and time:

◄ Definition of the newton

$$1\text{ N} \equiv 1\text{ kg} \cdot \text{m/s}^2 \tag{5.4}$$

In the U.S. customary system, the unit of force is the **pound** (lb). A force of 1 lb is the force that, when acting on a 1-slug mass,[2] produces an acceleration of 1 ft/s^2:

$$1\text{ lb} \equiv 1\text{ slug} \cdot \text{ft/s}^2$$

A convenient approximation is $1\text{ N} \approx \frac{1}{4}\text{ lb}$.

PITFALL PREVENTION 5.3
$m\vec{\mathbf{a}}$ **Is Not a Force** Equation 5.2 does *not* say that the product $m\vec{\mathbf{a}}$ is a force. All forces on an object are added vectorially to generate the net force on the left side of the equation. This net force is then equated to the product of the mass of the object and the acceleration that results from the net force. Do *not* include an "$m\vec{\mathbf{a}}$ force" in your analysis of the forces on an object.

---

[1]Equation 5.2 is valid only when the speed of the object is much less than the speed of light. We treat the relativistic situation in Chapter 38.

[2]The *slug* is the unit of mass in the U.S. customary system and is that system's counterpart of the SI unit the *kilogram*. Because most of the calculations in our study of classical mechanics are in SI units, the slug is seldom used in this text.

Why do you move your hands backward when you catch the egg in the opening storyline? Imagine holding your hands stiffly and not moving them as you catch the egg. Then the egg will hit your hand and be brought to rest in a very short time interval. As a result, the magnitude of the acceleration of the egg will be large. According to Equation 5.2, this will require a large force from your hands. This large force is sufficient to break the shell of the egg. If you move your hands backward, however, and slowly bring the egg to rest, the acceleration is of a much smaller magnitude. This, in turn, requires a much smaller force, which can keep the shell of the egg intact.

Throwing the egg into the sheet is similar: when the egg strikes the sheet, the sheet moves in the same direction in response, bringing the egg to a lower velocity over a relatively long distance.

Example 5.1   An Accelerating Hockey Puck

A hockey puck having a mass of 0.30 kg slides on the frictionless, horizontal surface of an ice rink. Two hockey sticks strike the puck simultaneously, exerting the forces on the puck shown in Figure 5.4. The force $\vec{F}_1$ has a magnitude of 5.0 N, and is directed at $\theta = 20°$ below the x axis. The force $\vec{F}_2$ has a magnitude of 8.0 N and its direction is $\phi = 60°$ above the x axis. Determine both the magnitude and the direction of the puck's acceleration.

Figure 5.4 (Example 5.1) A hockey puck moving on a frictionless surface is subject to two forces $\vec{F}_1$ and $\vec{F}_2$.

$F_1 = 5.0$ N
$F_2 = 8.0$ N

SOLUTION

Conceptualize   Study Figure 5.4. Using your expertise in vector addition from Chapter 3, predict the approximate direction of the net force vector on the puck. The acceleration of the puck will be in the same direction.

Categorize   Because we can determine a net force and we want an acceleration, this problem is categorized as one that may be solved using Newton's second law. In Section 5.7, we will formally introduce the *particle under a net force* analysis model to describe a situation such as this one.

Analyze   Find the component of the net force acting on the puck in the x direction:

$$\sum F_x = F_{1x} + F_{2x} = F_1 \cos\theta + F_2 \cos\phi$$

Find the component of the net force acting on the puck in the y direction:

$$\sum F_y = F_{1y} + F_{2y} = F_1 \sin\theta + F_2 \sin\phi$$

Use Newton's second law in component form (Eq. 5.3) to find the x and y components of the puck's acceleration:

$$a_x = \frac{\sum F_x}{m} = \frac{F_1 \cos\theta + F_2 \cos\phi}{m}$$

$$a_y = \frac{\sum F_y}{m} = \frac{F_1 \sin\theta + F_2 \sin\phi}{m}$$

Substitute numerical values:

$$a_x = \frac{(5.0 \text{ N})\cos(-20°) + (8.0 \text{ N})\cos(60°)}{0.30 \text{ kg}} = 29 \text{ m/s}^2$$

$$a_y = \frac{(5.0 \text{ N})\sin(-20°) + (8.0 \text{ N})\sin(60°)}{0.30 \text{ kg}} = 17 \text{ m/s}^2$$

Find the magnitude of the acceleration:

$$a = \sqrt{(29 \text{ m/s}^2)^2 + (17 \text{ m/s}^2)^2} = \boxed{34 \text{ m/s}^2}$$

Find the direction of the acceleration relative to the positive x axis:

$$\theta = \tan^{-1}\left(\frac{a_y}{a_x}\right) = \tan^{-1}\left(\frac{17}{29}\right) = \boxed{31°}$$

Finalize   The vectors in Figure 5.4 can be added graphically to check the reasonableness of our answer. Because the acceleration vector is along the direction of the resultant force, a drawing showing the resultant force vector helps us check the validity of the answer. (Try it!)

continued

5.1 continued

WHAT IF?　Suppose three hockey sticks strike the puck simultaneously, with two of them exerting the forces shown in Figure 5.4. The result of the three forces is that the hockey puck shows *no* acceleration. What must be the components of the third force?

Answer　If there is zero acceleration, the net force acting on the puck must be zero. Therefore, the three forces must cancel. The components of the third force must be of equal magnitude and opposite sign compared to the components of the net force applied by the first two forces so that all the components add to zero. Therefore, $F_{3x} = -\sum F_x = -(0.30 \text{ kg})(29 \text{ m/s}^2) = -8.7 \text{ N}$ and $F_{3y} = -\sum F_y = -(0.30 \text{ kg})(17 \text{ m/s}^2) = -5.2 \text{ N}$.

5.5　The Gravitational Force and Weight

All objects are attracted to the Earth. The attractive force exerted by the Earth on an object is called the **gravitational force** $\vec{\mathbf{F}}_g$. This force is directed toward the center of the Earth,[3] and its magnitude is called the **weight** of the object.

We saw in Section 2.6 that a freely falling object experiences an acceleration $\vec{\mathbf{g}}$ acting toward the center of the Earth. Applying Newton's second law $\sum \vec{\mathbf{F}} = m\vec{\mathbf{a}}$ to a freely falling object of mass m, with $\vec{\mathbf{a}} = \vec{\mathbf{g}}$ and $\sum \vec{\mathbf{F}} = \vec{\mathbf{F}}_g$, gives

$$\vec{\mathbf{F}}_g = m\vec{\mathbf{g}} \qquad (5.5)$$

Therefore, the weight of an object, being defined as the magnitude of $\vec{\mathbf{F}}_g$, is given by

$$F_g = mg \qquad (5.6)$$

Because it depends on g, weight varies with geographic location. Because g decreases with increasing distance from the center of the Earth, objects weigh less at higher altitudes than at sea level. For example, a 1 000-kg pallet of bricks used in the construction of the Empire State Building in New York City weighed 9 800 N at street level, but weighed about 1 N less by the time it was lifted from sidewalk level to the top of the building. As another example, suppose a student has a mass of 70.0 kg. The student's weight in a location where $g = 9.80 \text{ m/s}^2$ is 686 N (about 150 lb). At the top of a mountain, however, where $g = 9.77 \text{ m/s}^2$, the student's weight is only 684 N. Therefore, if you want to lose weight without going on a diet, climb a mountain or weigh yourself at 30 000 ft during an airplane flight!

Equation 5.6 indicates that there is a clear difference between mass and weight. The life-support unit strapped to the back of astronaut Harrison Schmitt in Figure 5.5 weighed 300 lb on the Earth and had a mass of 136 kg. During his training, a 50-lb mockup with a mass of 23 kg was used. Although this strategy effectively simulated the reduced *weight* the unit would have on the Moon, it did not correctly mimic the unchanging *mass*. It was more difficult to accelerate the 136-kg unit (perhaps by jumping or twisting suddenly) on the Moon than it was to accelerate the 23-kg unit on the Earth.

Equation 5.6 quantifies the gravitational force on the object, but notice that this equation does not require the object to be moving. Even for a stationary object or for an object on which several forces act, Equation 5.6 can be used to calculate the magnitude of the gravitational force. The result is a subtle shift in the interpretation of m in the equation. The mass m in Equation 5.6 determines the strength of the gravitational attraction between the object and the Earth. This role is completely different from that previously described for mass, that of measuring the resistance to changes in motion in response to an external force. In that role, mass is also called **inertial mass.** We call m in Equation 5.6 the **gravitational mass.** Even though this quantity is different in behavior from inertial mass, it is one of the

PITFALL PREVENTION 5.4

"Weight of an Object"　We are familiar with the everyday phrase, the "weight of an object." Weight, however, is not an inherent property of an object; rather, it is a measure of the gravitational force between the object and the Earth (or other planet). Therefore, weight is a property of a *system* of items: the object and the Earth.

PITFALL PREVENTION 5.5

Kilogram Is Not a Unit of Weight　You may have seen the "conversion" 1 kg = 2.2 lb. Despite popular statements of weights expressed in kilograms, the kilogram is not a unit of *weight*, it is a unit of *mass*. The conversion statement is not an equality; it is an *equivalence* that is valid only on the Earth's surface. We first raised this issue in the Chapter 1 storyline.

Figure 5.5 Astronaut Harrison Schmitt carries a backpack on the Moon.

NASA/Eugene Cernan

---

[3]This statement ignores that the mass distribution of the Earth is not perfectly spherical.

experimental conclusions in Newtonian dynamics that gravitational mass and inertial mass have the same value.

Although this discussion has focused on the gravitational force on an object due to the Earth, the concept is generally valid on any planet. The value of g will vary from one planet to the next, but the magnitude of the gravitational force will always be given by the value of mg.

QUICK QUIZ 5.4 Suppose you are talking by interplanetary telephone to a friend who lives on the Moon. He tells you that he has just won a newton of gold in a contest. Excitedly, you tell him that you entered the Earth version of the same contest and also won a newton of gold! Who is richer? **(a)** You are. **(b)** Your friend is. **(c)** You are equally rich.

Conceptual Example 5.2 | **How Much Do You Weigh in an Elevator?**

You have most likely been in an elevator that accelerates upward as it moves toward a higher floor. In this case, you feel heavier. In fact, if you are standing on a bathroom scale at the time, the scale measures a force having a magnitude that is greater than your weight. Therefore, you have tactile and measured evidence that leads you to believe you are heavier in this situation. *Are* you heavier?

SOLUTION

No; your weight is unchanged. Your experiences are due to your being in a noninertial reference frame. To provide the acceleration upward, the floor or scale must exert on your feet an upward force that is greater in magnitude than your weight. It is this greater force you feel, which you interpret as feeling heavier. The scale reads the force with which it pushes up on you, not your weight (unless you are at rest), and so its reading increases. We will examine the effect of the acceleration of an elevator on apparent weight in Example 5.8.

5.6 Newton's Third Law

If you press against a corner of this textbook with your fingertip, the book pushes back and makes a small dent in your skin. If you push harder, the book does the same and the dent in your skin is a little larger. This simple activity illustrates that forces are *interactions* between two objects: when your finger pushes on the book, the book pushes back on your finger. This important principle is known as **Newton's third law:**

◀ Newton's third law

If two objects interact, the force $\vec{\mathbf{F}}_{12}$ exerted by object 1 on object 2 is equal in magnitude and opposite in direction to the force $\vec{\mathbf{F}}_{21}$ exerted by object 2 on object 1:

$$\vec{\mathbf{F}}_{12} = -\vec{\mathbf{F}}_{21} \tag{5.7}$$

When it is important to designate forces as interactions between two objects, we will use this subscript notation, where $\vec{\mathbf{F}}_{ab}$ means "the force exerted *by* a *on* b." The third law is illustrated in Figure 5.6. The force that object 1 exerts on object 2 is popularly called the *action force*, and the force of object 2 on object 1 is called the *reaction force*. These italicized terms are not scientific terms; furthermore, either force can be labeled the action or reaction force. We will use these terms for convenience. In all cases, the action and reaction forces act on *different* objects and must be of the same type (gravitational, electrical, etc.). For example, the force acting on a freely falling projectile is the gravitational force exerted by the Earth on the projectile $\vec{\mathbf{F}}_g = \vec{\mathbf{F}}_{Ep}$ (E = Earth, p = projectile), and the magnitude of this force is mg. The reaction to this force is the gravitational force exerted by the projectile on the Earth $\vec{\mathbf{F}}_{pE} = -\vec{\mathbf{F}}_{Ep}$. The reaction force $\vec{\mathbf{F}}_{pE}$ must accelerate the Earth toward the projectile just as the action force $\vec{\mathbf{F}}_{Ep}$ accelerates the projectile toward

Figure 5.6 Newton's third law. The force $\vec{\mathbf{F}}_{12}$ exerted by object 1 on object 2 is equal in magnitude and opposite in direction to the force $\vec{\mathbf{F}}_{21}$ exerted by object 2 on object 1.

Figure 5.7 (a) When a computer monitor is at rest on a table, the forces acting on the monitor are the normal force $\vec{\mathbf{n}}$ and the gravitational force $\vec{\mathbf{F}}_g$. The reaction to $\vec{\mathbf{n}}$ is the force $\vec{\mathbf{F}}_{mt}$ exerted by the monitor on the table. The reaction to $\vec{\mathbf{F}}_g$ is the force $\vec{\mathbf{F}}_{mE}$ exerted by the monitor on the Earth. (b) A *force diagram* shows the forces on the monitor. (c) A *free-body diagram* shows the monitor as a black dot with the forces acting on it.

PITFALL PREVENTION 5.6

n **Does Not Always Equal** *mg* In the situation shown in Figure 5.7 and in many others, we find that $n = mg$ (the normal force has the same magnitude as the gravitational force). This result, however, is *not* generally true. If an object is on an incline, if there are applied forces with vertical components, or if there is a vertical acceleration of the system, then $n \neq mg$. Always apply Newton's second law to find the relationship between *n* and *mg*.

PITFALL PREVENTION 5.7

Newton's Third Law Remember that Newton's third-law action and reaction forces act on *different* objects. For example, in Figure 5.7, $\vec{\mathbf{n}} = \vec{\mathbf{F}}_{tm} = -m\vec{\mathbf{g}} = -\vec{\mathbf{F}}_{Em}$. The forces $\vec{\mathbf{n}}$ and $m\vec{\mathbf{g}}$ are equal in magnitude and opposite in direction, but they do not represent an action–reaction pair because both forces act on the *same* object, the monitor.

PITFALL PREVENTION 5.8

Free-Body Diagrams The *most important* step in solving a problem using Newton's laws is to draw a proper sketch, the free-body diagram. Be sure to draw *only* those forces that act on the object you are isolating. Be sure to draw *all* forces acting on the object, including any field forces, such as the gravitational force.

the Earth. Because the Earth has such a large mass, however, its acceleration due to this reaction force is negligibly small.

Consider a computer monitor at rest on a table as in Figure 5.7a. The gravitational force on the monitor is $\vec{\mathbf{F}}_g = \vec{\mathbf{F}}_{Em}$. The reaction to this force is the force $\vec{\mathbf{F}}_{mE} = -\vec{\mathbf{F}}_{Em}$ exerted by the monitor on the Earth. The monitor does not accelerate because it is held up by the table. The table exerts on the monitor an upward force $\vec{\mathbf{n}} = \vec{\mathbf{F}}_{tm}$, called the **normal force.** (*Normal* in this context means *perpendicular*.) In general, whenever an object is in contact with a surface, the surface exerts a normal force on the object. The normal force on the monitor can have any value needed, up to the point of breaking the table. Because the monitor has zero acceleration, Newton's second law applied to the monitor gives us $\sum \vec{\mathbf{F}} = \vec{\mathbf{n}} + m\vec{\mathbf{g}} = 0$, so $n\hat{\mathbf{j}} - mg\hat{\mathbf{j}} = 0$, or $n = mg$. The normal force balances the gravitational force on the monitor, so the net force on the monitor is zero. The reaction force to $\vec{\mathbf{n}}$ is the force exerted by the monitor downward on the table, $\vec{\mathbf{F}}_{mt} = -\vec{\mathbf{F}}_{tm} = -\vec{\mathbf{n}}$.

Notice that the forces acting on the monitor are $\vec{\mathbf{F}}_g$ and $\vec{\mathbf{n}}$ as shown in Figure 5.7b. The two forces $\vec{\mathbf{F}}_{mE}$ and $\vec{\mathbf{F}}_{mt}$ are exerted on objects other than the monitor.

Figure 5.7 illustrates an extremely important step in solving problems involving forces. Figure 5.7a shows many of the forces in the situation: those acting on the monitor, one acting on the table, and one acting on the Earth. Figure 5.7b, by contrast, shows only the forces acting on *one object,* the monitor, and is called a **force diagram** or a *diagram showing the forces on the object.* The important pictorial representation in Figure 5.7c is called a **free-body diagram.** In a free-body diagram, the particle model is used by representing the object as a dot and showing the forces that act on the object as being applied to the dot. When analyzing an object subject to forces, we are interested in the net force acting on one object, which we will model as a particle. Therefore, a free-body diagram helps us isolate only those forces on the object and eliminate the other forces from our analysis.

QUICK QUIZ 5.5 (i) If a fly collides with the windshield of a fast-moving bus, which experiences an impact force with a larger magnitude? (a) The fly. (b) The bus. (c) The same force is experienced by both. **(ii)** Which experiences the greater acceleration? (a) The fly. (b) The bus. (c) The same acceleration is experienced by both.

| Conceptual Example **5.3** | **You Push Me and I'll Push You** |

A large man and a small boy stand facing each other on frictionless ice. They put their hands together and push against each other so that they move apart.

(A) Who moves away with the higher speed?

5.3 continued

SOLUTION

This situation is similar to what we saw in Quick Quiz 5.5. According to Newton's third law, the force exerted by the man on the boy and the force exerted by the boy on the man are a third-law pair of forces, so they must be equal in magnitude. (A bathroom scale placed between their hands would read the same, regardless of which way it faced.) Therefore, the boy, having the smaller mass, experiences the greater acceleration. Both individuals accelerate for the same amount of time, but the greater acceleration of the boy over this time interval results in his moving away from the interaction with the higher speed.

(B) Who moves farther while their hands are in contact?

SOLUTION

Because the boy has the greater acceleration and therefore the greater average velocity, he moves farther than the man during the time interval during which their hands are in contact.

5.7 Analysis Models Using Newton's Second Law

In this section, we discuss two analysis models for solving problems in which objects are either in equilibrium ($\vec{\mathbf{a}} = 0$) or accelerating under the action of constant external forces. Remember that when Newton's laws are applied to an object, we are interested only in external forces that act on the object. If the objects are modeled as particles, we need not worry about rotational motion such as spinning. For now, we also neglect the effects of friction in those problems involving motion, which is equivalent to stating that the surfaces are *frictionless*. (The friction force is discussed in Section 5.8.)

We usually neglect the mass of any ropes, strings, or cables involved. In this approximation, the magnitude of the force exerted by any element of the rope on the adjacent element is the same for all elements along the rope. In problem statements, the synonymous terms *light* and *of negligible mass* are used to indicate that a mass is to be ignored when you work the problems. When a rope attached to an object is pulling on the object, the rope exerts a force on the object in a direction away from the object, parallel to the rope. The magnitude T of that force is called the **tension** in the rope. Because it is the magnitude of a vector quantity, tension is a scalar quantity.

Analysis Model: The Particle in Equilibrium

If the acceleration of an object modeled as a particle is zero, the object is treated with the **particle in equilibrium** model. In this model, the net force on the object is zero:

$$\sum \vec{\mathbf{F}} = 0 \tag{5.8}$$

Consider a lamp suspended from a light chain fastened to the ceiling as in Figure 5.8a. The force diagram for the lamp (Fig. 5.8b) shows that the forces acting on the lamp are the downward gravitational force $\vec{\mathbf{F}}_g$ and the upward force $\vec{\mathbf{T}}$ exerted by the chain. Because there are no forces in the x direction, $\sum F_x = 0$ provides no helpful information. The condition $\sum F_y = 0$ gives

$$\sum F_y = T - F_g = 0 \text{ or } T = F_g$$

Again, notice that $\vec{\mathbf{T}}$ and $\vec{\mathbf{F}}_g$ are *not* an action–reaction pair because they act on the same object, the lamp. The reaction force to $\vec{\mathbf{T}}$ is a downward force exerted by the lamp on the chain.

Example 5.4 (page 107) shows an application of the particle in equilibrium model.

Figure 5.8 (a) A lamp suspended from a ceiling by a chain of negligible mass. (b) The forces acting on the lamp are the gravitational force $\vec{\mathbf{F}}_g$ and the force $\vec{\mathbf{T}}$ exerted by the chain.

Figure 5.9 (a) A crate being pulled to the right on a frictionless floor. (b) The free-body diagram representing the external forces acting on the crate.

Analysis Model: The Particle Under a Net Force

If an object experiences an acceleration, its motion can be analyzed with the **particle under a net force** model. The appropriate equation for this model is Newton's second law, Equation 5.2:

$$\sum \vec{\mathbf{F}} = m\vec{\mathbf{a}} \qquad (5.2)$$

Consider a crate being pulled to the right on a frictionless, horizontal floor as in Figure 5.9a. Of course, the floor directly under the boy must have friction; otherwise, his feet would simply slip when he tries to pull on the crate! Suppose you wish to find the acceleration of the crate and the force the floor exerts on it. The forces acting on the crate are illustrated in the free-body diagram in Figure 5.9b. Notice that the horizontal force $\vec{\mathbf{T}}$ being applied to the crate acts through the rope. The magnitude of $\vec{\mathbf{T}}$ is equal to the tension in the rope. In addition to the force $\vec{\mathbf{T}}$, the free-body diagram for the crate includes the gravitational force $\vec{\mathbf{F}}_g$ and the normal force $\vec{\mathbf{n}}$ exerted by the floor on the crate.

We can now apply Newton's second law in component form to the crate. The only force acting in the x direction is $\vec{\mathbf{T}}$. Applying $\sum F_x = ma_x$ to the horizontal motion gives

$$\sum F_x = T = ma_x \quad \text{or} \quad a_x = \frac{T}{m}$$

No acceleration occurs in the y direction because the crate moves only horizontally. Therefore, we use the particle in equilibrium model in the y direction. Applying the y component of Equation 5.8 yields

$$\sum F_y = n - F_g = 0 \quad \text{or} \quad n = F_g$$

That is, the normal force has the same magnitude as the gravitational force but acts in the opposite direction.

If $\vec{\mathbf{T}}$ is a constant force, the acceleration $a_x = T/m$ also is constant. Hence, the crate is also modeled as a particle under constant acceleration in the x direction, and the equations of kinematics from Chapter 2 can be used to obtain the crate's position x and velocity v_x as functions of time.

Notice from this discussion two concepts that will be important in future problem solving: (1) *In a given problem, it is possible to have different analysis models applied in different directions.* The crate in Figure 5.9 is a particle in equilibrium in the vertical direction and a particle under a net force in the horizontal direction. (2) *It is possible to describe an object by multiple analysis models.* The crate is a particle under a net force in the horizontal direction and is also a particle under constant acceleration in the same direction.

In the situation just described, the magnitude of the normal force $\vec{\mathbf{n}}$ is equal to the magnitude of $\vec{\mathbf{F}}_g$, but that is not always the case, as noted in Pitfall Prevention 5.6. For example, suppose a book is lying on a table and you push down on the book with a force $\vec{\mathbf{F}}$ as in Figure 5.10. Because the book is at rest and therefore not accelerating, $\sum F_y = 0$, which gives $n - F_g - F = 0$, or $n = F_g + F = mg + F$. In this situation, the normal force is *greater* than the gravitational force. Other examples in which $n \neq F_g$ are presented later.

Several examples below demonstrate the use of the particle in equilibrium model and the particle under a net force model.

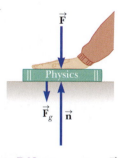

Figure 5.10 When a force $\vec{\mathbf{F}}$ pushes vertically downward on another object, the normal force $\vec{\mathbf{n}}$ on the object is greater than the gravitational force: $n = F_g + F$.

ANALYSIS MODEL   **Particle in Equilibrium**

Imagine an object that can be modeled as a particle. If it has several forces acting on it so that the forces all cancel, giving a net force of zero, the object will have an acceleration of zero. This condition is mathematically described as

$$\sum \vec{\mathbf{F}} = 0 \qquad (5.8)$$

$$\vec{\mathbf{a}} = 0$$
$$m$$
$$\sum \vec{\mathbf{F}} = 0$$

Particle in Equilibrium *continued*

Examples

- a chandelier hanging over a dining room table
- an object moving at terminal speed through a viscous medium (Chapter 6)
- a steel beam in the frame of a building (Chapter 12)
- a boat floating on a body of water (Chapter 14)

ANALYSIS MODEL **Particle Under a Net Force**

Imagine an object that can be modeled as a particle. If it has one or more forces acting on it so that there is a net force on the object, it will accelerate in the direction of the net force. The relationship between the net force and the acceleration is

$$\sum \vec{F} = m\vec{a} \qquad (5.2)$$

Examples:

- a crate pushed across a factory floor
- a falling object acted upon by a gravitational force
- a piston in an automobile engine pushed by hot gases (Chapter 21)
- a charged particle in an electric field (Chapter 22)

$$m \quad \vec{a}$$
$$\overrightarrow{\sum F}$$

Example **5.4** **A Traffic Light at Rest**

A traffic light weighing 122 N hangs from a cable tied to two other cables fastened to a support as in Figure 5.11a. The upper cables make angles of $\theta_1 = 37.0°$ and $\theta_2 = 53.0°$ with the horizontal. These upper cables are not as strong as the vertical cable and will break if the tension in them exceeds 100 N. Does the traffic light remain hanging in this situation, or will one of the cables break?

SOLUTION

Conceptualize Inspect the drawing in Figure 5.11a. Let us assume the cables do not break and nothing is moving.

Categorize If nothing is moving, no part of the system is accelerating. We can now model the light as a *particle in equilibrium* on which the net force is zero. Similarly, the net force on the knot (Fig. 5.11c) is zero, so it is also modeled as a *particle in equilibrium*.

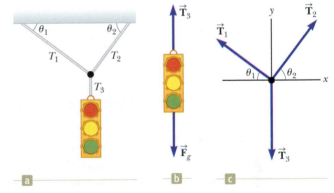

Figure 5.11 (Example 5.4) (a) A traffic light suspended by cables. (b) The forces acting on the traffic light. (c) The free-body diagram for the knot where the three cables are joined.

Analyze We construct a diagram of the forces acting on the traffic light, shown in Figure 5.11b, and a free-body diagram for the knot that holds the three cables together, shown in Figure 5.11c. This knot is a convenient object to choose because all the forces of interest act along lines passing through the knot.

From the particle in equilibrium model, apply Equation 5.8 for the traffic light in the *y* direction:

$$\sum F_y = 0 \rightarrow T_3 - F_g = 0$$
$$T_3 = F_g$$

Choose the coordinate axes as shown in Figure 5.11c and resolve the forces acting on the knot into their components:

Force	x Component	y Component
$\vec{T}_1$	$-T_1 \cos \theta_1$	$T_1 \sin \theta_1$
$\vec{T}_2$	$T_2 \cos \theta_2$	$T_2 \sin \theta_2$
$\vec{T}_3$	0	$-F_g$

Apply the particle in equilibrium model to the knot:

(1) $\sum F_x = -T_1 \cos \theta_1 + T_2 \cos \theta_2 = 0$

(2) $\sum F_y = T_1 \sin \theta_1 + T_2 \sin \theta_2 + (-F_g) = 0$

continued

5.4 continued

Equation (1) shows that the horizontal components of $\vec{T}_1$ and $\vec{T}_2$ must be equal in magnitude, and Equation (2) shows that the sum of the vertical components of $\vec{T}_1$ and $\vec{T}_2$ must balance the downward force $\vec{T}_3$, which is equal in magnitude to the weight of the light.

Solve Equation (1) for T_2 in terms of T_1:

$$(3) \quad T_2 = T_1\left(\frac{\cos\theta_1}{\cos\theta_2}\right)$$

Substitute this value for T_2 into Equation (2):

$$T_1\sin\theta_1 + T_1\left(\frac{\cos\theta_1}{\cos\theta_2}\right)(\sin\theta_2) - F_g = 0$$

Solve for T_1:

$$T_1 = \frac{F_g}{\sin\theta_1 + \cos\theta_1\tan\theta_2}$$

Substitute numerical values:

$$T_1 = \frac{122\ \text{N}}{\sin 37.0° + \cos 37.0°\tan 53.0°} = 73.4\ \text{N}$$

Using Equation (3), evaluate T_2:

$$T_2 = (73.4\ \text{N})\left(\frac{\cos 37.0°}{\cos 53.0°}\right) = 97.4\ \text{N}$$

Both values are less than 100 N (just barely for T_2), so the cables will not break.

Finalize   Let us finalize this problem by imagining a change in the system, as in the following What If?

WHAT IF?   Suppose the two angles in Figure 5.11a are equal. What would be the relationship between T_1 and T_2?

Answer   We can argue from the symmetry of the problem that the two tensions T_1 and T_2 would be equal to each other. Mathematically, if the equal angles are called θ, Equation (3) becomes

$$T_2 = T_1\left(\frac{\cos\theta}{\cos\theta}\right) = T_1$$

which also tells us that the tensions are equal. Without knowing the specific value of θ, we cannot find the values of T_1 and T_2. The tensions will be equal to each other, however, regardless of the value of θ.

Conceptual Example 5.5   **Forces Between Cars in a Train**

Train cars are connected by *couplers,* which are under tension as the locomotive pulls the train. Imagine you are on a train speeding up with a constant acceleration. As you move through the train from the locomotive to the last car, measuring the tension in each set of couplers, does the tension increase, decrease, or stay the same? When the engineer applies the brakes, the couplers are under compression. How does this compression force vary from the locomotive to the last car? (Assume only the brakes on the wheels of the engine are applied.)

SOLUTION

While the train is speeding up, tension decreases from the front of the train to the back. The coupler between the locomotive and the first car must apply enough force to accelerate the rest of the cars. As you move back along the train, each coupler is accelerating less mass behind it. The last coupler has to accelerate only the last car, and so it is under the least tension.

When the brakes are applied, the force again decreases from front to back. The coupler connecting the locomotive to the first car must apply a large force to slow down the rest of the cars, but the final coupler must apply a force large enough to slow down only the last car.

Example 5.6   **The Runaway Car**

A car of mass m is on an icy driveway inclined at an angle θ as in Figure 5.12a.

(A) Find the acceleration of the car, assuming the driveway is frictionless.

SOLUTION

Conceptualize   Use Figure 5.12a to conceptualize the situation. From everyday experience, we know that a car on an icy incline will accelerate down the incline. (The same thing happens to a car on a hill with its brakes not set.)

5.6 continued

Categorize We categorize the car as a *particle under a net force* because it accelerates. Furthermore, this example belongs to a very common category of problems in which an object moves under the influence of gravity on an inclined plane.

Analyze Figure 5.12b shows the free-body diagram for the car. The only forces acting on the car are the normal force $\vec{n}$ exerted by the inclined plane, which acts perpendicular to the plane, and the gravitational force $\vec{F}_g = m\vec{g}$, which acts vertically downward. For problems involving inclined planes, it is convenient to choose the coordinate axes with x along the incline and y perpendicular to it as in Figure 5.12b. Using similar triangles, we can show that the angle between the gravitational force $\vec{F}_g$ and the negative y axis in part b of Figure 5.12 is

Figure 5.12 (Example 5.6) (a) A car on a frictionless incline. (b) The free-body diagram for the car. The black dot represents the position of the center of mass of the car. We will learn about center of mass in Chapter 9.

equal to the angle θ that the incline makes with the horizontal in part a. With these axes, we represent the gravitational force by a component of magnitude $mg \sin \theta$ along the positive x axis and one of magnitude $mg \cos \theta$ along the negative y axis. Our choice of axes results in the car being modeled as a particle under a net force in the x direction and a particle in equilibrium in the y direction.

Apply these models to the car:

(1) $\sum F_x = mg \sin \theta = ma_x$

(2) $\sum F_y = n - mg \cos \theta = 0$

Solve Equation (1) for a_x:

(3) $a_x = g \sin \theta$

Finalize Note that the acceleration component a_x is independent of the mass of the car. It depends only on the angle of inclination and on g.

From Equation (2), we conclude that the component of $\vec{F}_g$ perpendicular to the incline is balanced by the normal force; that is, $n = mg \cos \theta$. This situation is a case in which the normal force is *not* equal in magnitude to the weight of the object (as discussed in Pitfall Prevention 5.6 on page 104).

It is possible, although inconvenient, to solve the problem with "standard" horizontal and vertical axes. You may want to try it, just for practice.

(B) Suppose the car is released from rest at the top of the incline and the distance from the car's front bumper to the bottom of the incline is d. How long does it take the front bumper to reach the bottom of the hill, and what is the car's speed as it arrives there?

SOLUTION

Conceptualize Imagine the car is sliding down the hill and you use a stopwatch to measure the entire time interval until it reaches the bottom.

Categorize This part of the problem belongs to kinematics rather than to dynamics, and Equation (3) shows that the acceleration a_x is constant. Therefore, you should categorize the car in this part of the problem as a *particle under constant acceleration*.

Analyze Defining the initial position of the front bumper as $x_i = 0$ and its final position as $x_f = d$, and recognizing that $v_{xi} = 0$, choose Equation 2.16 from the particle under constant acceleration model:

$x_f = x_i + v_{xi}t + \frac{1}{2}a_x t^2 \rightarrow d = \frac{1}{2}a_x t^2$

Solve for t:

(4) $t = \sqrt{\dfrac{2d}{a_x}} = \sqrt{\dfrac{2d}{g \sin \theta}}$

Use Equation 2.17, with $v_{xi} = 0$, to find the final velocity of the car:

$v_{xf}^2 = 2a_x d$

(5) $v_{xf} = \sqrt{2a_x d} = \sqrt{2gd \sin \theta}$

continued

5.6 continued

Finalize   We see from Equations (4) and (5) that the time t at which the car reaches the bottom and its final speed v_{xf} are independent of the car's mass, as was its acceleration. Notice that we have combined techniques from Chapter 2 with new techniques from this chapter in this example. As we learn more techniques in later chapters, this process of combining analysis models and information from several parts of the book will occur more often. In these cases, use the Analysis Model Approach to Problem Solving discussed in Chapter 2 to help you work your way through new problems.

W H A T   I F ?   What previously solved problem does this situation become if $\theta = 90°$?

Answer   Imagine θ going to 90° in Figure 5.12. The inclined plane becomes vertical, and the car is an object in free fall! Equation (3) becomes

$$a_x = g\sin\theta = g\sin 90° = g$$

which is indeed the free-fall acceleration. (We find $a_x = g$ rather than $a_x = -g$ because we have chosen positive x to be downward in Fig. 5.12.) Notice also that the condition $n = mg\cos\theta$ gives us $n = mg\cos 90° = 0$. That is consistent with the car falling downward *next to* the vertical plane, in which case there is no contact force between the car and the plane.

Example **5.7**   **One Block Pushes Another**

Two blocks of masses m_1 and m_2, with $m_1 > m_2$, are placed in contact with each other on a frictionless, horizontal surface as in Figure 5.13a. A constant horizontal force $\vec{F}$ is applied to m_1 as shown.

(A) Find the magnitude of the acceleration of the system.

S O L U T I O N

Conceptualize   Conceptualize the situation by using Figure 5.13a and realize that both blocks must experience the *same* acceleration because they are in contact with each other and remain in contact throughout the motion.

Categorize   We categorize this problem as one involving a *particle under a net force* because a force is applied to a system of blocks and we are looking for the acceleration of the system.

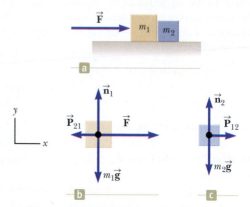

Figure 5.13   (Example 5.7) (a) A force is applied to a block of mass m_1, which pushes on a second block of mass m_2. (b) The forces acting on m_1. (c) The forces acting on m_2.

Analyze First model the combination of two blocks as a single particle under a net force. Apply Newton's second law to the combination in the x direction to find the acceleration:	$$\sum F_x = F = (m_1 + m_2)a_x$$ (1) $a_x = \dfrac{F}{m_1 + m_2}$

Finalize   The acceleration given by Equation (1) is the same as that of a single object of mass $m_1 + m_2$ and subject to the same force.

(B) Determine the magnitude of the contact force between the two blocks.

S O L U T I O N

Conceptualize   The contact force is internal to the system of two blocks. Therefore, we cannot find this force by modeling the whole system (the two blocks) as a single particle.

Categorize   Now consider each of the two blocks individually by categorizing each as a *particle under a net force*.

Analyze   We construct a diagram of forces acting on the object for each block as shown in Figures 5.13b and 5.13c, where the contact force is denoted by $\vec{P}$. From Figure 5.13c, we see that the only horizontal force acting on m_2 is the contact force $\vec{P}_{12}$ (the force exerted by m_1 on m_2), which is directed to the right.

Apply Newton's second law to m_2:

(2) $\sum F_x = P_{12} = m_2 a_x$

Substitute the value of the acceleration a_x given by Equation (1) into Equation (2):

(3) $P_{12} = m_2 a_x = \left(\dfrac{m_2}{m_1 + m_2}\right)F$

5.7 continued

Finalize This result shows that the contact force P_{12} is *less* than the applied force F. The force required to accelerate block 2 alone must be less than the force required to produce the same acceleration for the two-block system.

To finalize further, let us check this expression for P_{12} by considering the forces acting on m_1, shown in Figure 5.13b. The horizontal forces acting on m_1 are the applied force $\vec{F}$ to the right and the contact force $\vec{P}_{21}$ to the left (the force exerted by m_2 on m_1). From Newton's third law, $\vec{P}_{21}$ is the reaction force to $\vec{P}_{12}$, so $P_{21} = P_{12}$.

Apply Newton's second law to m_1:

$$(4) \quad \sum F_x = F - P_{21} = F - P_{12} = m_1 a_x$$

Solve for P_{12} and substitute the value of a_x from Equation (1):

$$P_{12} = F - m_1 a_x = F - m_1\left(\frac{F}{m_1 + m_2}\right) = \left(\frac{m_2}{m_1 + m_2}\right)F$$

This result agrees with Equation (3), as it must.

WHAT IF? Imagine that the force $\vec{F}$ in Figure 5.13 is applied toward the left on the right-hand block of mass m_2. Is the magnitude of the force $\vec{P}_{12}$ the same as it was when the force was applied toward the right on m_1?

Answer When the force is applied toward the left on m_2, the contact force must accelerate m_1. In the original situation, the contact force accelerates m_2. Because $m_1 > m_2$, more force is required, so the magnitude of $\vec{P}_{12}$ is greater than in the original situation. To see this mathematically, modify Equation (4) appropriately and solve for $\vec{P}_{12}$.

Example 5.8 Weighing a Fish in an Elevator

A person weighs a fish of mass m on a spring scale attached to the ceiling of an elevator as illustrated in Figure 5.14.

(A) Show that if the elevator accelerates either upward or downward, the spring scale gives a reading that is different from the weight of the fish.

SOLUTION

Conceptualize The reading on the scale is related to the extension of the spring in the scale, which is related to the force on the end of the spring as in Figure 5.2. Imagine that the fish is hanging on a string attached to the end of the spring. In this case, the magnitude of the force exerted on the spring is equal to the tension T in the string. Therefore, we are looking for T. The force $\vec{T}$ pulls down on the spring and pulls up on the fish.

Categorize We can categorize this problem by identifying the fish as a *particle in equilibrium* if the elevator is not accelerating or as a *particle under a net force* if the elevator is accelerating.

When the elevator accelerates upward, the spring scale reads a value greater than the weight of the fish.

When the elevator accelerates downward, the spring scale reads a value less than the weight of the fish.

Figure 5.14 (Example 5.8) A fish is weighed on a spring scale in an accelerating elevator car.

Analyze Inspect the diagrams of the forces acting on the fish in Figure 5.14 and notice that the external forces acting on the fish are the downward gravitational force $\vec{F}_g = m\vec{g}$ and the force $\vec{T}$ exerted by the string. If the elevator is either at rest or moving at constant velocity, the fish is a particle in equilibrium, so $\sum F_y = T - F_g = 0$ or $T = F_g = mg$. (Remember that the scalar mg is the weight of the fish.)

Now suppose the elevator is moving with an acceleration $\vec{a}$ relative to an observer standing outside the elevator in an inertial frame. The fish is now a particle under a net force.

Apply Newton's second law to the fish:

$$\sum F_y = T - mg = ma_y$$

Solve for T:

$$(1) \quad T = ma_y + mg = mg\left(\frac{a_y}{g} + 1\right) = F_g\left(\frac{a_y}{g} + 1\right)$$

where we have chosen upward as the positive y direction. We conclude from Equation (1) that the scale reading T is greater than the fish's weight mg if $\vec{a}$ is upward, so a_y is positive (Fig. 5.14a), and that the reading is less than mg if $\vec{a}$ is downward, so a_y is negative (Fig. 5.14b).

continued

5.8 continued

(B) Evaluate the scale readings for a 40.0-N fish if the elevator moves with an acceleration $a_y = \pm 2.00 \text{ m/s}^2$.

SOLUTION

Evaluate the scale reading from Equation (1) if $\vec{a}$ is upward: $T = (40.0 \text{ N})\left(\dfrac{2.00 \text{ m/s}^2}{9.80 \text{ m/s}^2} + 1\right) = \boxed{48.2 \text{ N}}$

Evaluate the scale reading from Equation (1) if $\vec{a}$ is downward: $T = (40.0 \text{ N})\left(\dfrac{-2.00 \text{ m/s}^2}{9.80 \text{ m/s}^2} + 1\right) = \boxed{31.8 \text{ N}}$

Finalize   Take this advice: if you buy a fish by weight in an elevator, make sure the fish is weighed while the elevator is either at rest or accelerating downward! Furthermore, notice that from the information given here, one cannot determine the direction of the velocity of the elevator.

WHAT IF?   Suppose the woman in Figure 5.14 tires of watching the scale and exits the elevator. Then the elevator cable breaks and the elevator and its remaining contents are in free fall. What happens to the reading on the scale?

Answer   If the elevator falls freely, the fish's acceleration is $a_y = -g$. We see from Equation (1) that the scale reading T is zero in this case; that is, the fish *appears* to be weightless.

Example 5.9 **The Atwood Machine**

When two objects of unequal mass are hung vertically over a frictionless pulley of negligible mass as in Figure 5.15a, the arrangement is called an *Atwood machine*. The device is sometimes used in the laboratory to determine the value of g by measuring the acceleration of the objects. Determine the magnitude of the acceleration of the two objects and the tension in the lightweight string.

SOLUTION

Conceptualize   Imagine the situation pictured in Figure 5.15a in action: as one object moves upward, the other object moves downward. Because the objects are connected by an inextensible string, the distance one object travels in a given time interval must be the same as the distance the other one travels, and their velocities and accelerations must be of equal magnitude.

Categorize   The objects in the Atwood machine are subject to the gravitational force as well as to the forces exerted by the strings connected to them. Therefore, we can categorize this problem as one involving two *particles under a net force*.

Figure 5.15 (Example 5.9) The Atwood machine. (a) Two objects connected by a massless inextensible string over a frictionless pulley. (b) The free-body diagrams for the two objects.

Analyze   The free-body diagrams for the two objects are shown in Figure 5.15b. Two forces act on each object: the upward force $\vec{T}$ exerted by the string and the downward gravitational force. In problems such as this one in which the pulley is modeled as massless and frictionless, the tension in the string on both sides of the pulley is the same. If the pulley has mass or is subject to friction, the tensions on either side are not the same and the situation requires techniques we will learn in Chapter 10.

We must be very careful with signs in problems such as this one. In Figure 5.15a, notice that if object 1 accelerates upward, object 2 accelerates downward. Therefore, for consistency with signs, if we define the upward direction as positive for object 1, we must define the downward direction as positive for object 2. With this sign convention, both objects accelerate in the same direction as defined by the choice of sign. Furthermore, according to this sign convention, the y component of the net force exerted on object 1 is $T - m_1 g$, and the y component of the net force exerted on object 2 is $m_2 g - T$.

From the particle under a net force model, apply Newton's second law to object 1:

$$(1) \quad \sum F_y = T - m_1 g = m_1 a_y$$

Apply Newton's second law to object 2:

$$(2) \quad \sum F_y = m_2 g - T = m_2 a_y$$

Add Equation (2) to Equation (1), noticing that T cancels:

$$-m_1 g + m_2 g = m_1 a_y + m_2 a_y$$

5.9 continued

Solve for the acceleration:

$$(3) \quad a_y = \left(\frac{m_2 - m_1}{m_1 + m_2} \right) g$$

Substitute Equation (3) into Equation (1) to find T:

$$(4) \quad T = m_1(g + a_y) = \left(\frac{2m_1 m_2}{m_1 + m_2} \right) g$$

Finalize The acceleration given by Equation (3) can be interpreted as the ratio of the magnitude of the unbalanced force on the system $(m_2 - m_1)g$ to the total mass of the system $(m_1 + m_2)$, as expected from Newton's second law. Notice that the sign of the acceleration depends on the relative masses of the two objects; if $m_2 > m_1$, the acceleration is positive, corresponding to downward motion for m_2 and upward for m_1. However, if $m_1 > m_2$, Equation (3) gives a negative acceleration, indicating that m_1 moves downward and m_2 moves upward.

WHAT IF? Describe the motion of the system if the objects have equal masses, that is, $m_1 = m_2$.

Answer If we have the same mass on both sides, the system is balanced and should not accelerate. Mathematically, we see that if $m_1 = m_2$, Equation (3) gives us $a_y = 0$.

WHAT IF? What if one of the masses is much larger than the other: $m_1 \gg m_2$?

Answer In the case in which one mass is infinitely larger than the other, we can ignore the effect of the smaller mass. Therefore, the larger mass should simply fall as if the smaller mass were not there. We see that if $m_1 \gg m_2$, Equation (3) gives us $a_y = -g$.

Example **5.10** Acceleration of Two Objects Connected by a Cord

A ball of mass m_1 and a block of mass m_2 are attached by a lightweight cord that passes over a frictionless pulley of negligible mass as in Figure 5.16a. The block lies on a frictionless incline of angle θ. Find the magnitude of the acceleration of the two objects and the tension in the cord.

SOLUTION

Conceptualize Imagine the objects in Figure 5.16 in motion. If m_2 moves down the incline, then m_1 moves upward. Because the objects are connected by a cord (which we assume does not stretch), their accelerations have the same magnitude. Notice the normal coordinate axes in Figure 5.16b for the ball and the "tilted" axes for the block in Figure 5.16c. Just as we chose the positive direction to be different for each of the objects in Example 5.9, we are free to choose entirely different coordinate axes for the two objects here.

Categorize We can identify forces on each of the two objects and we are looking for an acceleration, so we categorize the objects as *particles under a net force*. For the block, this model is only valid for the x' direction. In the y' direction, we apply the *particle in equilibrium* model because the block does not accelerate in that direction.

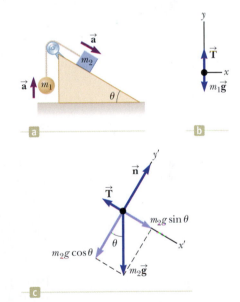

Figure 5.16 (Example 5.10) (a) Two objects connected by a lightweight cord strung over a frictionless pulley. (b) The free-body diagram for the ball. (c) The free-body diagram for the block. (The incline is frictionless.)

Analyze Consider the free-body diagrams shown in Figures 5.16b and 5.16c.

Apply Newton's second law in the y direction to the ball, choosing the upward direction as positive:

$$(1) \quad \sum F_y = T - m_1 g = m_1 a_y = m_1 a$$

For the ball to accelerate upward, it is necessary that $T > m_1 g$. In Equation (1), we replaced a_y with a because the acceleration has only a y component.

For the block, we have chosen the x' axis along the incline as in Figure 5.15c. For consistency with our choice for the ball, we choose the positive x' direction to be down the incline.

continued

5.10 continued

Apply the particle under a net force model to the block in the x' direction and the particle in equilibrium model in the y' direction:

(2) $\sum F_{x'} = m_2 g \sin\theta - T = m_2 a_{x'} = m_2 a$

(3) $\sum F_{y'} = n - m_2 g \cos\theta = 0$

In Equation (2), we replaced $a_{x'}$ with a because the two objects have accelerations of equal magnitude a.

Solve Equation (1) for T:

(4) $T = m_1(g + a)$

Substitute this expression for T into Equation (2):

$m_2 g \sin\theta - m_1(g + a) = m_2 a$

Solve for a:

(5) $a = \left(\dfrac{m_2 \sin\theta - m_1}{m_1 + m_2}\right)g$

Substitute this expression for a into Equation (4) to find T:

(6) $T = \left[\dfrac{m_1 m_2(\sin\theta + 1)}{m_1 + m_2}\right]g$

Finalize   The block accelerates down the incline only if $m_2 \sin\theta > m_1$. If $m_1 > m_2 \sin\theta$, the acceleration is up the incline for the block and downward for the ball. Also notice that the result for the acceleration, Equation (5), can be interpreted as the magnitude of the net external force acting on the ball–block system divided by the total mass of the system; this result is consistent with Newton's second law.

WHAT IF?   What happens in this situation if $\theta = 90°$?

Answer   If $\theta = 90°$, the inclined plane becomes vertical and there is no interaction between its surface and m_2. Therefore, this problem becomes the Atwood machine of Example 5.9. Letting $\theta \rightarrow 90°$ in Equations (5) and (6) causes them to reduce to Equations (3) and (4) of Example 5.9!

WHAT IF?   What if $m_1 = 0$?

Answer   If $m_1 = 0$, then m_2 is simply sliding down an inclined plane without interacting with m_1 through the string. Therefore, this problem becomes the sliding car problem in Example 5.6. Letting $m_1 \rightarrow 0$ in Equation (5) causes it to reduce to Equation (3) of Example 5.6!

5.8 Forces of Friction

When an object is in motion either on a surface or in a viscous medium such as air or water, there is resistance to the motion because the object interacts with its surroundings. We call such resistance a **force of friction.** Forces of friction are very important in our everyday lives. They allow us to walk or run and are necessary for the motion of wheeled vehicles.

Imagine that you are working in your garden and have filled a trash can with yard clippings. You then try to drag the trash can across the surface of your concrete patio as in Figure 5.17a. This surface is *real*, not an idealized, frictionless surface. If we apply an external horizontal force $\vec{F}$ to the trash can, acting to the right, the trash can remains stationary when $\vec{F}$ is small. The force on the trash can that counteracts $\vec{F}$ and keeps it from moving acts toward the left and is called the

Force of static friction ▶   **force of static friction** $\vec{f}_s$. As long as the trash can is not moving, $f_s = F$. Therefore, if $\vec{F}$ is increased, $\vec{f}_s$ also increases. Likewise, if $\vec{F}$ decreases, $\vec{f}_s$ also decreases.

Experiments show that the friction force arises from the nature of the two surfaces: because of their roughness, contact is made only at a few locations where peaks of the material touch. At these locations, the friction force arises in part because one peak physically blocks the motion of a peak from the opposing surface and in part from chemical bonding ("spot welds") of opposing peaks as they come into contact. Although the details of friction are quite complex at the atomic level, this force ultimately involves an electrical interaction between atoms or molecules.

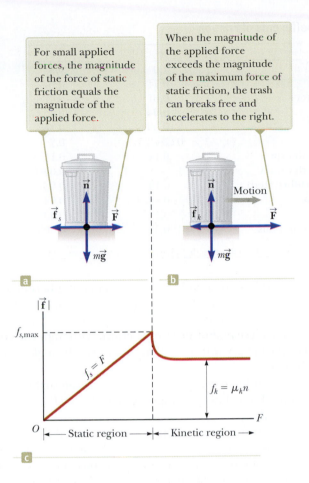

For small applied forces, the magnitude of the force of static friction equals the magnitude of the applied force.

When the magnitude of the applied force exceeds the magnitude of the maximum force of static friction, the trash can breaks free and accelerates to the right.

Figure 5.17 (a) and (b) When pulling on a trash can, the direction of the force of friction $\vec{\mathbf{f}}$ between the can and a rough surface is opposite the direction of the applied force $\vec{\mathbf{F}}$. (c) A graph of friction force versus applied force. Notice that $f_{s,\text{max}} > f_k$.

If we increase the magnitude of $\vec{\mathbf{F}}$ as in Figure 5.17b, the trash can eventually slips. When the trash can is on the verge of slipping, f_s has its maximum value $f_{s,\text{max}}$ as shown in Figure 5.17c. When F exceeds $f_{s,\text{max}}$, the trash can moves and accelerates to the right. We call the friction force for an object in motion the **force of kinetic friction** $\vec{\mathbf{f}}_k$. When the trash can is in motion, the force of kinetic friction on the can is less than $f_{s,\text{max}}$ (Fig. 5.17c). The net force $F - f_k$ in the x direction produces an acceleration to the right, according to Newton's second law. If $F = f_k$, the acceleration is zero and the trash can moves to the right with constant speed. If the applied force $\vec{\mathbf{F}}$ is removed from the moving can, the friction force $\vec{\mathbf{f}}_k$ acting to the left provides an acceleration of the trash can in the $-x$ direction and eventually brings it to rest, again consistent with Newton's second law.

◄ Force of kinetic friction

Experimentally, we find that, to a good approximation, both $f_{s,\text{max}}$ and f_k are proportional to the magnitude of the normal force exerted on an object by the surface. The following descriptions of the force of friction are based on experimental observations and serve as the simplification model we shall use for forces of friction in problem solving:

- The magnitude of the force of static friction between any two surfaces in contact can have the values

$$f_s \leq \mu_s n \tag{5.9}$$

where the dimensionless constant μ_s is called the **coefficient of static friction** and n is the magnitude of the normal force exerted by one surface on the other. The equality in Equation 5.9 holds when the surfaces are on the verge of slipping, that is, when $f_s = f_{s,\text{max}} = \mu_s n$. This situation is called *impending motion*. The inequality holds when the surfaces are not on the verge of slipping.

PITFALL PREVENTION 5.9
The Equal Sign Is Used in Limited Situations In Equation 5.9, the equal sign is used *only* in the case in which the surfaces are just about to break free and begin sliding. Do not fall into the common trap of using $f_s = \mu_s n$ in *any* static situation.

TABLE 5.1 Coefficients of Friction

	μ_s	μ_k
Rubber on concrete	1.0	0.8
Steel on steel	0.74	0.57
Aluminum on steel	0.61	0.47
Glass on glass	0.94	0.4
Copper on steel	0.53	0.36
Wood on wood	0.25–0.5	0.2
Waxed wood on wet snow	0.14	0.1
Waxed wood on dry snow	—	0.04
Metal on metal (lubricated)	0.15	0.06
Teflon on Teflon	0.04	0.04
Ice on ice	0.1	0.03
Synovial joints in humans	0.01	0.003

Note: All values are approximate. In some cases, the coefficient of friction can exceed 1.0.

- The magnitude of the force of kinetic friction acting between two surfaces is

$$f_k = \mu_k n \tag{5.10}$$

where μ_k is the **coefficient of kinetic friction.** Although the coefficient of kinetic friction can vary with speed, we shall usually neglect any such variations in this text.
- The values of μ_k and μ_s depend on the nature of the surfaces, but μ_k is generally less than μ_s. Typical values range from around 0.03 to 1.0. Table 5.1 lists some reported values.
- The direction of the friction force on an object is parallel to the surface with which the object is in contact and opposite to the actual motion (kinetic friction) or the impending motion (static friction) of the object relative to the surface.
- The coefficients of friction are nearly independent of the area of contact between the surfaces. We might expect that placing an object on the side having the most area might increase the friction force. Although this method provides more points in contact, the weight of the object is spread out over a larger area and the individual points are not pressed together as tightly. Because these effects approximately compensate for each other, the friction force is independent of the area.

Figure 5.18 (Quick Quiz 5.7) A father slides his daughter on a sled either by (a) pushing down on her shoulders or (b) pulling up on a rope.

QUICK QUIZ 5.6 You press your physics textbook flat against a vertical wall with your hand. What is the direction of the friction force exerted by the wall on the book? (a) downward (b) upward (c) out from the wall (d) into the wall

QUICK QUIZ 5.7 Charlie is playing with his daughter Torrey in the snow. She sits on a sled and asks him to slide her across a flat, horizontal field. Charlie has a choice of (a) pushing her from behind by applying a force downward on her shoulders at 30° below the horizontal (Fig. 5.18a) or (b) attaching a rope to the front of the sled and pulling with a force at 30° above the horizontal (Fig. 5.18b). Which would be easier for him and why?

Example 5.11   **Experimental Determination of μ_s and μ_k**

The following is a simple method of measuring coefficients of friction. Suppose a block is placed on a rough surface inclined relative to the horizontal as shown in Figure 5.19. The incline angle is increased until the block starts to move. Show that you can obtain μ_s by measuring the critical angle θ_c at which this slipping just occurs.

SOLUTION

Conceptualize Consider Figure 5.19 and imagine that the block tends to slide down the incline due to the gravitational force. To simulate the situation, place a coin on this book's cover and tilt the book until the coin begins to slide. Notice how this

5.11 continued

example differs from Example 5.6. When there is no friction on an incline, *any* angle of the incline will cause a stationary object to begin moving. When there is friction, however, there is no movement of the object for angles less than the critical angle.

Categorize The block is subject to various forces. Because we are raising the plane to the angle at which the block is just ready to begin to move but is not moving, we categorize the block as a *particle in equilibrium*.

Analyze The diagram in Figure 5.19 shows the forces on the block: the gravitational force $m\vec{g}$, the normal force $\vec{n}$, and the force of static friction $\vec{f}_s$. We choose x to be parallel to the plane and y perpendicular to it.

From the particle in equilibrium model, apply Equation 5.8 to the block in both the x and y directions:

$$(1) \quad \sum F_x = mg \sin \theta - f_s = 0$$

$$(2) \quad \sum F_y = n - mg \cos \theta = 0$$

Substitute $mg = n/\cos \theta$ from Equation (2) into Equation (1):

$$(3) \quad f_s = mg \sin \theta = \left(\frac{n}{\cos \theta}\right) \sin \theta = n \tan \theta$$

When the incline angle is increased until the block is on the verge of slipping, the force of static friction has reached its maximum value $\mu_s n$. The angle θ in this situation is the critical angle θ_c. Make these substitutions in Equation (3):

$$\mu_s n = n \tan \theta_c$$
$$\mu_s = \tan \theta_c$$

We have shown, as requested, that the coefficient of static friction is related only to the critical angle. For example, if the block just slips at $\theta_c = 20.0°$, we find that $\mu_s = \tan 20.0° = 0.364$.

Finalize Once the block starts to move at $\theta \geq \theta_c$, it accelerates down the incline and the force of friction is $f_k = \mu_k n$.

WHAT IF? How could you determine μ_k for the block and incline?

Answer If θ is reduced to a value less than θ_c, it may be possible to find an angle θ_c' such that the block moves down the incline with constant speed as a particle in equilibrium again ($a_x = 0$). In this case, use Equations (1) and (2) with f_s replaced by f_k to find μ_k: $\mu_k = \tan \theta_c'$, where $\theta_c' < \theta_c$.

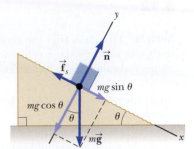

Figure 5.19 (Example 5.11) The external forces exerted on a block lying on a rough incline are the gravitational force $m\vec{g}$, the normal force $\vec{n}$, and the force of friction $\vec{f}_s$. For convenience, the gravitational force is resolved into a component $mg \sin \theta$ along the incline and a component $mg \cos \theta$ perpendicular to the incline.

Example 5.12 **The Sliding Hockey Puck**

A hockey puck on a frozen pond is given an initial speed of 20.0 m/s. If the puck always remains on the ice and slides 115 m before coming to rest, determine the coefficient of kinetic friction between the puck and ice.

SOLUTION

Conceptualize Imagine that the puck in Figure 5.20 slides to the right. The kinetic friction force acts to the left and slows the puck, which eventually comes to rest due to that force.

Categorize The forces acting on the puck are identified in Figure 5.20, but the text of the problem provides kinematic variables. Therefore, we categorize the problem in several ways. First, it involves modeling the puck as a *particle under a net force* in the horizontal direction: kinetic friction causes the puck to accelerate. There is no acceleration of the puck in the vertical direction, so we use the *particle in equilibrium* model for that direction. Furthermore, because we model the force of kinetic friction as independent of speed, the acceleration of the puck is constant. So, we can also categorize this problem by modeling the puck as a *particle under constant acceleration*.

Figure 5.20 (Example 5.12) After the puck is given an initial velocity to the right, the only external forces acting on it are the gravitational force $m\vec{g}$, the normal force $\vec{n}$, and the force of kinetic friction $\vec{f}_k$.

continued

5.12 continued

Analyze   First, let's find the acceleration algebraically in terms of the coefficient of kinetic friction, using Newton's second law. Once we know the acceleration of the puck and the distance it travels, the equations of kinematics can be used to find the numerical value of the coefficient of kinetic friction.

Apply the particle under a net force model in the x direction to the puck:

$$(1) \quad \sum F_x = -f_k = ma_x$$

Apply the particle in equilibrium model in the y direction to the puck:

$$(2) \quad \sum F_y = n - mg = 0$$

Substitute $n = mg$ from Equation (2) and $f_k = \mu_k n$ into Equation (1):

$$-\mu_k n = -\mu_k mg = ma_x$$
$$a_x = -\mu_k g$$

The negative sign means the acceleration is to the left in Figure 5.20. Because the velocity of the puck is to the right, the puck is slowing down. The acceleration is independent of the mass of the puck and is constant because we assume μ_k remains constant.

Apply the particle under constant acceleration model to the puck, choosing Equation 2.17 from the model, $v_{xf}^2 = v_{xi}^2 + 2a_x(x_f - x_i)$, with $x_i = 0$ and $v_{xf} = 0$:

$$0 = v_{xi}^2 + 2a_x x_f = v_{xi}^2 - 2\mu_k g x_f$$

Solve for the coefficient of kinetic friction:

$$\mu_k = \frac{v_{xi}^2}{2gx_f}$$

Substitute the numerical values:

$$\mu_k = \frac{(20.0 \text{ m/s})^2}{2(9.80 \text{ m/s}^2)(115 \text{ m})} = \boxed{0.177}$$

Finalize   Notice that μ_k is dimensionless, as it should be, and that it has a low value, consistent with an object sliding on ice.

Example 5.13   **Acceleration of Two Connected Objects When Friction Is Present**

A block of mass m_2 on a rough, horizontal surface is connected to a ball of mass m_1 by a lightweight cord over a lightweight, frictionless pulley as shown in Figure 5.21a. A force of magnitude F at an angle θ with the horizontal is applied to the block as shown, and the block slides to the right. The coefficient of kinetic friction between the block and surface is μ_k. Determine the magnitude of the acceleration of the two objects.

SOLUTION

Figure 5.21   (Example 5.13) (a) The external force $\vec{F}$ applied as shown can cause the block to accelerate to the right. (b, c) Diagrams showing the forces on the two objects, assuming the block accelerates to the right and the ball accelerates upward.

Conceptualize   Imagine what happens as $\vec{F}$ is applied to the block. Assuming $\vec{F}$ is large enough to break the block free from static friction but not large enough to lift the block, the block slides to the right and the ball rises.

Categorize   We can identify forces and we want an acceleration, so we categorize this problem as one involving two *particles under a net force*, the ball and the block. Because we assume that the block does not rise into the air due to the applied force, we model the block as a *particle in equilibrium* in the vertical direction.

Analyze   First draw force diagrams for the two objects as shown in Figures 5.21b and 5.21c. Notice that the string exerts a force of magnitude T on both objects. The applied force $\vec{F}$ has x and y components $F \cos \theta$ and $F \sin \theta$, respectively. Because the two objects are connected, we can equate the magnitudes of the x component of the acceleration of the block and the y component of the acceleration of the ball and call them both a. Let us assume the motion of the block is to the right.

Apply the particle under a net force model to the block in the horizontal direction:

$$(1) \quad \sum F_x = F\cos\theta - f_k - T = m_2 a_x = m_2 a$$

5.13 continued

Because the block moves only horizontally, apply the particle in equilibrium model to the block in the vertical direction:

$$(2) \quad \sum F_y = n + F \sin \theta - m_2 g = 0$$

Apply the particle under a net force model to the ball in the vertical direction:

$$(3) \quad \sum F_y = T - m_1 g = m_1 a_y = m_1 a$$

Solve Equation (2) for n:

$$n = m_2 g - F \sin \theta$$

Substitute n into $f_k = \mu_k n$ from Equation 5.10:

$$(4) \quad f_k = \mu_k (m_2 g - F \sin \theta)$$

Substitute Equation (4) and the value of T from Equation (3) into Equation (1):

$$F \cos \theta - \mu_k (m_2 g - F \sin \theta) - m_1 (a + g) = m_2 a$$

Solve for a:

$$(5) \quad a = \frac{F(\cos \theta + \mu_k \sin \theta) - (m_1 + \mu_k m_2) g}{m_1 + m_2}$$

Finalize   The acceleration of the block can be either to the right or to the left depending on the sign of the numerator in Equation (5). If the velocity is to the left, we must reverse the sign of f_k in Equation (1) because the force of kinetic friction must oppose the motion of the block relative to the surface. In this case, the value of a is the same as in Equation (5), with the two plus signs in the numerator changed to minus signs.

What does Equation (5) reduce to if the force $\vec{F}$ is removed and the surface becomes frictionless? Call this expression Equation (6). Does this algebraic expression match your intuition about the physical situation in this case? Now go back to Example 5.10 and let angle θ go to zero in Equation (5) of that example. How does the resulting equation compare with your Equation (6) here in Example 5.13? Should the algebraic expressions compare in this way based on the physical situations?

Summary

▶ Definitions

An **inertial frame of reference** is a frame in which an object that does not interact with other objects experiences zero acceleration. Any frame moving with constant velocity relative to an inertial frame is also an inertial frame.

We define **force** as **that which causes a change in motion of an object.**

▶ Concepts and Principles

Newton's first law states that it is possible to find an inertial frame in which an object that does not interact with other objects experiences zero acceleration, or, equivalently, in the absence of an external force, when viewed from an inertial frame, an object at rest remains at rest and an object in uniform motion in a straight line maintains that motion.

Newton's second law states that the acceleration of an object is directly proportional to the net force acting on it and inversely proportional to its mass.

Newton's third law states that if two objects interact, the force exerted by object 1 on object 2 is equal in magnitude and opposite in direction to the force exerted by object 2 on object 1.

The **gravitational force** exerted on an object is equal to the product of its mass (a scalar quantity) and the free-fall acceleration:

$$\vec{F}_g = m\vec{g} \quad (5.5)$$

The **weight** of an object is the magnitude of the gravitational force acting on the object:

$$F_g = mg \quad (5.6)$$

The maximum **force of static friction** $\vec{f}_{s,max}$ between an object and a surface is proportional to the normal force acting on the object. In general, $f_s \leq \mu_s n$, where μ_s is the **coefficient of static friction** and n is the magnitude of the normal force.

When an object slides over a surface, the magnitude of the **force of kinetic friction** $\vec{f}_k$ is given by $f_k = \mu_k n$, where μ_k is the **coefficient of kinetic friction.**

continued

Analysis Models for Problem Solving

Particle Under a Net Force If a particle of mass m experiences a nonzero net force, its acceleration is related to the net force by Newton's second law:

$$\sum \vec{F} = m\vec{a} \qquad (5.2)$$

Particle in Equilibrium If a particle maintains a constant velocity (so that $\vec{a} = 0$), which could include a velocity of zero, the forces on the particle balance and Newton's second law reduces to

$$\sum \vec{F} = 0 \qquad (5.8)$$

Think–Pair–Share

See the Preface for an explanation of the icons used in this problems set. For additional assessment items for this section, go to **WEBASSIGN** *From Cengage*

1. You are a member of an expert witness group that provides scientific services to the legal community. Your group has been asked by a defense attorney to argue at trial that a driver was not exceeding the speed limit. You are provided with the following data: The mass of the car is 1.50×10^3 kg. The mass of the driver is 95.0 kg. The coefficient of kinetic friction between the car's tires and the roadway is 0.580. The coefficient of static friction between the car's tires and the roadway is 0.820. The posted speed limit on the road is 25 mi/h. The roadway was dry and the weather was sunny at the time of the incident.

   You are also provided with the following description of the incident: The driver was driving up a hill that makes an angle of 17.5° with the horizontal. The driver saw a dog run into the street, slammed on the brakes and left a skid mark 17.0 m long. The car came to rest at the end of the skid mark. The driver did not hit the dog, but the sound of the screeching tires drew the attention of a nearby policeman, who ticketed the driver for speeding.

   (a) Should your group agree to offer testimony for the defense in this case? (b) Why or why not?

2. Consider the egg-catching activity discussed in the chapter-opening storyline. Discuss in your group and make estimates of the following. Identify a separation distance between the thrower and the catcher of the egg, and determine a typical speed with which the egg must be thrown so that it covers the distance without hitting the ground or passing over the head of the catcher. Estimate the mass and diameter of the egg (the shorter diameter, perpendicular to the longest dimension). From these data, estimate the force on the egg exerted by your hand if your hand is held stiffly and doesn't move when the egg hits it. Now, simulate moving your hands backward while you catch the egg. Have a group member estimate the distance over which your hands move in this process. From these data, estimate the force on the egg exerted by your hand in this catching process. Compare the forces of your hand on the egg between these two methods of catching the egg.

3. **ACTIVITY** A simple procedure can be followed to measure the coefficient of friction using the technique discussed in Example 5.11. Lay your book down on a table and place a coin on the cover far from the spine. Slowly open the book cover so that it forms an inclined plane down which the coin will slide. Watch carefully and stop opening the cover at the instant the coin begins to slide. (a) Measure this critical angle that the book cover makes with the horizontal with a protractor. From this angle, determine the coefficient of static friction between the coin and the cover. (b) Place a loop of tape between two coins and repeat the procedure above for the two-coin stack. How does the coefficient of static friction for the stack compare to that for the single coin?

Problems

See the Preface for an explanation of the icons used in this problems set. For additional assessment items for this section, go to **WEBASSIGN** *From Cengage*

SECTION 5.1 The Concept of Force

1. **BIO** A certain orthodontist uses a wire brace to align a patient's crooked tooth as in Figure P5.1. The tension in the wire is adjusted to have a magnitude of 18.0 N. Find the magnitude of the net force exerted by the wire on the crooked tooth.

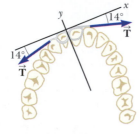

Figure P5.1

2. One or more external forces, large enough to be easily measured, are exerted on each object enclosed in a dashed box shown in Figure 5.1. Identify the reaction to each of these forces.

SECTION 5.4 Newton's Second Law

3. **V** A 3.00-kg object undergoes an acceleration given by $\vec{a} = (2.00\,\hat{i} + 5.00\,\hat{j})$ m/s². Find (a) the resultant force acting on the object and (b) the magnitude of the resultant force.

4. The average speed of a nitrogen molecule in air is about 6.70×10^2 m/s, and its mass is 4.68×10^{-26} kg. (a) If it takes 3.00×10^{-13} s for a nitrogen molecule to hit a wall and

rebound with the same speed but moving in the opposite direction, what is the average acceleration of the molecule during this time interval? (b) What average force does the molecule exert on the wall?

5. Two forces, $\vec{\mathbf{F}}_1 = (-6.00\hat{\mathbf{i}} - 4.00\hat{\mathbf{j}})$ N and $\vec{\mathbf{F}}_2 = (-3.00\hat{\mathbf{i}} + 7.00\hat{\mathbf{j}})$ N, act on a particle of mass 2.00 kg that is initially at rest at coordinates $(-2.00$ m, $+4.00$ m). (a) What are the components of the particle's velocity at $t = 10.0$ s? (b) In what direction is the particle moving at $t = 10.0$ s? (c) What displacement does the particle undergo during the first 10.0 s? (d) What are the coordinates of the particle at $t = 10.0$ s?

6. The force exerted by the wind on the sails of a sailboat is 390 N north. The water exerts a force of 180 N east. If the boat (including its crew) has a mass of 270 kg, what are the magnitude and direction of its acceleration?

7. Review. Three forces acting on an object are given by $\vec{\mathbf{F}}_1 = (-2.00\hat{\mathbf{i}} + 2.00\hat{\mathbf{j}})$ N, and $\vec{\mathbf{F}}_2 = (5.00\hat{\mathbf{i}} - 3.00\hat{\mathbf{j}})$ N, and $\vec{\mathbf{F}}_3 = (-45.0\hat{\mathbf{i}})$ N. The object experiences an acceleration of magnitude 3.75 m/s². (a) What is the direction of the acceleration? (b) What is the mass of the object? (c) If the object is initially at rest, what is its speed after 10.0 s? (d) What are the velocity components of the object after 10.0 s?

8. If a single constant force acts on an object that moves on a straight line, the object's velocity is a linear function of time. The equation $v = v_i + at$ gives its velocity v as a function of time, where a is its constant acceleration. What if velocity is instead a linear function of position? Assume that as a particular object moves through a resistive medium, its speed decreases as described by the equation $v = v_i - kx$, where k is a constant coefficient and x is the position of the object. Find the law describing the total force acting on this object.

SECTION 5.5 The Gravitational Force and Weight

9. Review. The gravitational force exerted on a baseball is 2.21 N down. A pitcher throws the ball horizontally with velocity 18.0 m/s by uniformly accelerating it along a straight horizontal line for a time interval of 170 ms. The ball starts from rest. (a) Through what distance does it move before its release? (b) What are the magnitude and direction of the force the pitcher exerts on the ball?

10. Review. The gravitational force exerted on a baseball is $-F_g\hat{\mathbf{j}}$. A pitcher throws the ball with velocity $v\hat{\mathbf{i}}$ by uniformly accelerating it along a straight horizontal line for a time interval of $\Delta t = t - 0 = t$. (a) Starting from rest, through what distance does the ball move before its release? (b) What force does the pitcher exert on the ball?

11. Review. An electron of mass 9.11×10^{-31} kg has an initial speed of 3.00×10^5 m/s. It travels in a straight line, and its speed increases to 7.00×10^5 m/s in a distance of 5.00 cm. Assuming its acceleration is constant, (a) determine the magnitude of the force exerted on the electron and (b) compare this force with the weight of the electron, which we ignored.

12. If a man weighs 900 N on the Earth, what would he weigh on Jupiter, where the free-fall acceleration is 25.9 m/s²?

13. You stand on the seat of a chair and then hop off. (a) During the time interval you are in flight down to the floor, the Earth moves toward you with an acceleration of what order of magnitude? In your solution, explain your logic. Model the Earth as a perfectly solid object. (b) The Earth moves toward you through a distance of what order of magnitude?

SECTION 5.6 Newton's Third Law

14. A brick of mass M has been placed on a rubber cushion of mass m. Together they are sliding to the right at constant velocity on an ice-covered parking lot. (a) Draw a free-body diagram of the brick and identify each force acting on it. (b) Draw a free-body diagram of the cushion and identify each force acting on it. (c) Identify all of the action–reaction pairs of forces in the brick–cushion–planet system.

SECTION 5.7 Analysis Models Using Newton's Second Law

15. Review. Figure P5.15 shows a worker poling a boat—a very efficient mode of transportation—across a shallow lake. He pushes parallel to the length of the light pole, exerting a force of magnitude 240 N on the bottom of the lake. Assume the pole lies in the vertical plane containing the keel of the boat. At one moment, the pole makes an angle of 35.0° with the vertical and the water exerts a horizontal drag force of 47.5 N on the boat, opposite to its forward velocity of magnitude 0.857 m/s. The mass of the boat including its cargo and the worker is 370 kg. (a) The water exerts a buoyant force vertically upward on the boat. Find the magnitude of this force. (b) Model the forces as constant over a short interval of time to find the velocity of the boat 0.450 s after the moment described.

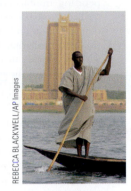

Figure P5.15

16. An iron bolt of mass 65.0 g hangs from a string 35.7 cm long. The top end of the string is fixed. Without touching it, a magnet attracts the bolt so that it remains stationary, but is displaced horizontally 28.0 cm to the right from the previously vertical line of the string. The magnet is located to the right of the bolt and on the same vertical level as the bolt in the final configuration. (a) Draw a free-body diagram of the bolt. (b) Find the tension in the string. (c) Find the magnetic force on the bolt.

17. A block slides down a frictionless plane having an inclination of $\theta = 15.0°$. The block starts from rest at the top, and the length of the incline is 2.00 m. (a) Draw a free-body diagram of the block. Find (b) the acceleration of the block and (c) its speed when it reaches the bottom of the incline.

18. A bag of cement whose weight is F_g hangs in equilibrium from three wires as shown in Figure P5.18. Two of the wires make angles θ_1 and θ_2 with the horizontal. Assuming the system is in equilibrium, show that the tension in the left-hand wire is

$$T_1 = \frac{F_g \cos\theta_2}{\sin(\theta_1 + \theta_2)}$$

Figure P5.18

19. The distance between two telephone poles is 50.0 m. When a 1.00-kg bird lands on the telephone wire midway between the poles, the wire sags 0.200 m. (a) Draw a free-body diagram of the bird. (b) How much tension does the bird produce in the wire? Ignore the weight of the wire.

20. An object of mass $m = 1.00$ kg is observed to have an acceleration $\vec{\mathbf{a}}$ with a magnitude of 10.0 m/s² in a direction 60.0°

east of north. Figure P5.20 shows a view of the object from above. The force $\vec{F}_2$ acting on the object has a magnitude of 5.00 N and is directed north. Determine the magnitude and direction of the one other horizontal force $\vec{F}_1$ acting on the object.

Figure P5.20

21. A simple accelerometer is constructed inside a car by suspending an object of mass m from a string of length L that is tied to the car's ceiling. As the car accelerates the string–object system makes a constant angle of θ with the vertical. (a) Assuming that the string mass is negligible compared with m, derive an expression for the car's acceleration in terms of θ and show that it is independent of the mass m and the length L. (b) Determine the acceleration of the car when $\theta = 23.0°$.

22. An object of mass $m_1 = 5.00$ kg placed on a frictionless, horizontal table is connected to a string that passes over a pulley and then is fastened to a hanging object of mass $m_2 = 9.00$ kg as shown in Figure P5.22. (a) Draw free-body diagrams of both objects. Find (b) the magnitude of the acceleration of the objects and (c) the tension in the string.

AMT
V

Figure P5.22
Problems 22 and 29.

23. In the system shown in Figure P5.23, a horizontal force $\vec{F}_x$ acts on an object of mass $m_2 = 8.00$ kg. The horizontal surface is frictionless. Consider the acceleration of the sliding object as a function of F_x. (a) For what values of F_x does the object of mass $m_1 = 2.00$ kg accelerate upward? (b) For what values of F_x is the tension in the cord zero? (c) Plot the acceleration of the m_2 object versus F_x. Include values of F_x from -100 N to $+100$ N.

T

Figure P5.23

24. A car is stuck in the mud. A tow truck pulls on the car with the arrangement shown in Figure P5.24. The tow cable is under a tension of 2 500 N and pulls downward and to the left on the pin at its upper end. The light pin is held in equilibrium by forces exerted by the two bars A and B. Each bar is a *strut*; that is, each is a bar whose weight is small compared to the forces it exerts and which exerts forces only through hinge pins at its ends. Each strut exerts a force directed parallel to its length. Determine the force of tension or compression in each strut. Proceed as follows. Make a guess as to which way (pushing or pulling) each force acts on the top pin. Draw a free-body diagram of the pin. Use the condition for equilibrium of the pin to translate the free-body diagram into equations. From the equations calculate the forces exerted by struts A and B. If you obtain a positive answer, you correctly guessed the direction of the force. A negative answer means that the direction should be reversed, but the absolute value correctly gives the

magnitude of the force. If a strut pulls on a pin, it is in tension. If it pushes, the strut is in compression. Identify whether each strut is in tension or in compression.

Figure P5.24

25. An object of mass m_1 hangs from a string that passes over a very light fixed pulley P_1 as shown in Figure P5.25. The string connects to a second very light pulley P_2. A second string passes around this pulley with one end attached to a wall and the other to an object of mass m_2 on a frictionless, horizontal table. (a) If a_1 and a_2 are the accelerations of m_1 and m_2, respectively, what is the relation between these accelerations? Find expressions for (b) the tensions in the strings and (c) the accelerations a_1 and a_2 in terms of the masses m_1 and m_2, and g.

S

Figure P5.25

SECTION 5.8 Forces of Friction

26. *Why is the following situation impossible?* Your 3.80-kg physics book is placed next to you on the horizontal seat of your car. The coefficient of static friction between the book and the seat is 0.650, and the coefficient of kinetic friction is 0.550. You are traveling forward at 72.0 km/h and brake to a stop with constant acceleration over a distance of 30.0 m. Your physics book remains on the seat rather than sliding forward onto the floor.

27. Consider a large truck carrying a heavy load, such as steel beams. A significant hazard for the driver is that the load may slide forward, crushing the cab, if the truck stops suddenly in an accident or even in braking. Assume, for example, that a 10 000-kg load sits on the flatbed of a 20 000-kg truck moving at 12.0 m/s. Assume that the load is not tied down to the truck, but has a coefficient of friction of 0.500 with the flatbed of the truck. (a) Calculate the minimum stopping distance for which the load will not slide forward relative to the truck. (b) Is any piece of data unnecessary for the solution?

28. Before 1960, people believed that the maximum attainable coefficient of static friction for an automobile tire on a roadway was $\mu_s = 1$. Around 1962, three companies independently developed racing tires with coefficients of 1.6. This problem shows that tires have improved further since then. The shortest time interval in which a piston-engine car initially at rest has covered a distance of one-quarter mile is

Q|C

about 4.43 s. (a) Assume the car's rear wheels lift the front wheels off the pavement as shown in Figure P5.28. What minimum value of μ_s is necessary to achieve the record time? (b) Suppose the driver were able to increase his or her engine power, keeping other things equal. How would this change affect the elapsed time?

Figure P5.28

29. A 9.00-kg hanging object is connected by a light, inextensible cord over a light, frictionless pulley to a 5.00-kg block that is sliding on a flat table (Fig. P5.22). Taking the coefficient of kinetic friction as 0.200, find the tension in the string.

30. The person in Figure P5.30 weighs 170 lb. As seen from the front, each light crutch makes an angle of 22.0° with the vertical. Half of the person's weight is supported by the crutches. The other half is supported by the vertical forces of the ground on the person's feet. Assuming that the person is moving with constant velocity and the force exerted by the ground on the crutches acts along the crutches, determine (a) the smallest possible coefficient of friction between crutches and ground and (b) the magnitude of the compression force in each crutch.

22.0°   22.0°

Figure P5.30

31. Three objects are connected on a table as shown in Figure P5.31. The coefficient of kinetic friction between the block of mass m_2 and the table is 0.350. The objects have masses of $m_1 = 4.00$ kg, $m_2 = 1.00$ kg, and $m_3 = 2.00$ kg, and the pulleys are frictionless. (a) Draw a free-body diagram of each object. (b) Determine the acceleration of each object, including its direction. (c) Determine the tensions in the two cords. **What If?** (d) If the tabletop were smooth, would the tensions increase, decrease, or remain the same? Explain.

m_1 m_2 m_3

Figure P5.31

32. You are working as a letter sorter in a U.S Post Office. Postal regulations require that employees' footwear must have a minimum coefficient of static friction of 0.5 on a specified tile surface. You are wearing athletic shoes for which you do not know the coefficient of static friction. In order to determine the coefficient, you imagine that there is an emergency and start running across the room. You have a coworker time you, and find that you can begin at rest and move 4.23 m in 1.20 s. If you try to move faster than this, your feet slip. Assuming your acceleration is constant, does your footwear qualify for the postal regulation?

33. You have been called as an expert witness for a trial in which a driver has been charged with speeding but is claiming innocence. He claims to have slammed on his brakes to avoid rear-ending another car, but tapped the back of the other car just as he came to rest. You have been hired by the prosecution to prove that the driver was indeed speeding. You have received data as follows from the police: Skid marks left by the driver are 56.0 m long and the roadway is level. Tires matching those on the car of the driver have been dragged over the same roadway to determine that the coefficient of kinetic friction between the tires and the roadway is 0.82 at all points along the skid mark. The speed limit on the road is 35 mi/h. Construct an argument to be used in court to show that the driver was indeed speeding.

34. A block of mass 3.00 kg is pushed up against a wall by a force $\vec{P}$ that makes an angle of $\theta = 50.0°$ with the horizontal as shown in Figure P5.34. The coefficient of static friction between the block and the wall is 0.250. (a) Determine the possible values for the magnitude of $\vec{P}$ that allow the block to remain stationary. (b) Describe what happens if $|\vec{P}|$ has a larger value and what happens if it is smaller. (c) Repeat parts (a) and (b), assuming the force makes an angle of $\theta = 13.0°$ with the horizontal.

θ

$\vec{P}$

Figure P5.34

35. **Review.** A Chinook salmon can swim underwater at 3.58 m/s, and it can also jump vertically upward, leaving the water with a speed of 6.26 m/s. A record salmon has length 1.50 m and mass 61.0 kg. Consider the fish swimming straight upward in the water below the surface of a lake. The gravitational force exerted on it is very nearly canceled out by a buoyant force exerted by the water as we will study in Chapter 14. The fish experiences an upward force P exerted by the water on its threshing tail fin and a downward fluid friction force that we model as acting on its front end. Assume the fluid friction force disappears as soon as the fish's head breaks the water surface and assume the force on its tail is constant. Model the gravitational force as suddenly switching full on when half the length of the fish is out of the water. Find the value of P.

36. A 5.00-kg block is placed on top of a 10.0-kg block (Fig. P5.36). A horizontal force of 45.0 N is applied to the 10-kg block, and the 5.00-kg block is tied to the wall. The coefficient of kinetic friction between all moving surfaces is 0.200. (a) Draw a free-body diagram for each block and identify the action–reaction forces between the blocks. (b) Determine the tension in the string and the magnitude of the acceleration of the 10.0-kg block.

5.00 kg

10.0 kg $F = 45.0$ N

Figure P5.36

ADDITIONAL PROBLEMS

37. A black aluminum glider floats on a film of air above a level aluminum air track. Aluminum feels essentially no force in a magnetic field, and air resistance is negligible. A strong magnet is attached to the top of the glider, forming a total mass of 240 g. A piece of scrap iron attached to one end stop on the track attracts the magnet with a force of 0.823 N when the iron and the magnet are separated by 2.50 cm. (a) Find the acceleration of the glider at this instant. (b) The scrap iron is now attached to another green glider, forming total mass 120 g. Find the acceleration of each glider when the gliders are simultaneously released at 2.50-cm separation.

38. *Why is the following situation impossible?* A book sits on an inclined plane on the surface of the Earth. The angle of the plane with the horizontal is 60.0°. The coefficient of kinetic friction between the book and the plane is 0.300. At time $t = 0$, the book is released from rest. The book then slides through a distance of 1.00 m, measured along the plane, in a time interval of 0.483 s.

39. Two blocks of masses m_1 and m_2 are placed on a table in con-
GP tact with each other as discussed in Example 5.7 and shown
S in Figure 5.13a. The coefficient of kinetic friction between the block of mass m_1 and the table is μ_1, and that between the block of mass m_2 and the table is μ_2. A horizontal force of magnitude F is applied to the block of mass m_1. We wish to find P, the magnitude of the contact force between the blocks. (a) Draw diagrams showing the forces for each block. (b) What is the net force on the system of two blocks? (c) What is the net force acting on m_1? (d) What is the net force acting on m_2? (e) Write Newton's second law in the x direction for each block. (f) Solve the two equations in two unknowns for the acceleration a of the blocks in terms of the masses, the applied force F, the coefficients of friction, and g. (g) Find the magnitude P of the contact force between the blocks in terms of the same quantities.

40. A 1.00-kg glider on a horizontal air track is pulled by a string at an angle θ. The taut string runs over a pulley and is attached to a hanging object of mass 0.500 kg as shown in Figure P5.40. (a) Show that the speed v_x of the glider and the speed v_y of the hanging object are related by $v_x = uv_y$, where $u = z(z^2 - h_0^2)^{-1/2}$. (b) The glider is released from rest. Show that at that instant the acceleration a_x of the glider and the acceleration a_y of the hanging object are related by $a_x = ua_y$. (c) Find the tension in the string at the instant the glider is released for $h_0 = 80.0$ cm and $\theta = 30.0°$.

Figure P5.40

41. An inventive child named Nick wants to reach an apple in a tree without climbing the tree. Sitting in a chair connected to a rope that passes over a frictionless pulley (Fig. P5.41), Nick pulls on the loose end of the rope with such a force that the spring scale reads 250 N. Nick's true weight is 320 N, and the chair weighs 160 N. Nick's feet are not touching the ground. (a) Draw one pair of diagrams showing the forces for Nick and the chair considered as separate systems and another diagram for Nick and the chair considered as one system. (b) Show that the acceleration of the system is *upward* and find its magnitude. (c) Find the force Nick exerts on the chair.

Figure P5.41 Problems 41 and 44.

42. A rope with mass m_r is attached
QC to a block with mass m_b as in Fig-
S ure P5.42. The block rests on a frictionless, horizontal surface. The rope does not stretch. The free end of the rope is pulled

Figure P5.42

to the right with a horizontal force $\vec{F}$. (a) Draw force diagrams for the rope and the block, noting that the tension in the rope is not uniform. (b) Find the acceleration of the system in terms of m_b, m_r, and F. (c) Find the magnitude of the force the rope exerts on the block. (d) What happens to the force on the block as the rope's mass approaches zero? What can you state about the tension in a *light* cord joining a pair of moving objects?

43. In Example 5.7, we pushed on two blocks on a table. Sup-
QC pose three blocks are in contact with one another on a frictionless, horizontal surface as shown in Figure P5.43. A horizontal force $\vec{F}$ is applied to m_1. Take $m_1 = 2.00$ kg, $m_2 = 3.00$ kg, $m_3 = 4.00$ kg, and $F = 18.0$ N. (a) Draw a separate free-body diagram for each block. (b) Determine the acceleration of the blocks. (c) Find the *resultant* force on each block. (d) Find the magnitudes of the contact forces between the blocks. (e) You are working on a construction project. A coworker is nailing up plasterboard on one side of a light partition, and you are on the opposite side, providing "backing" by leaning against the wall with your back pushing on it. Every hammer blow makes your back sting. The supervisor helps you put a heavy block of wood between the wall and your back. Using the situation analyzed in parts (a) through (d) as a model, explain how this change works to make your job more comfortable.

Figure P5.43

44. In the situation described in Problem 41 and Figure P5.41, **QC** the masses of the rope, spring balance, and pulley are negligible. Nick's feet are not touching the ground. (a) Assume Nick is momentarily at rest when he stops pulling down on the rope and passes the end of the rope to another child, of weight 440 N, who is standing on the ground next to him. The rope does not break. Describe the ensuing motion. (b) Instead, assume Nick is momentarily at rest when he ties the end of the rope to a strong hook projecting from the tree trunk. Explain why this action can make the rope break.

45. A crate of weight F_g is pushed by a force **S** $\vec{\mathbf{P}}$ on a horizontal floor as shown in Figure P5.45. The coefficient of static friction is μ_s, and $\vec{\mathbf{P}}$ is directed at angle θ below the horizontal. (a) Show that the minimum value of P that will move the crate is given by

Figure P5.45

$$P = \frac{\mu_s F_g \sec \theta}{1 - \mu_s \tan \theta}$$

(b) Find the condition on θ in terms of μ_s for which motion of the crate is impossible for any value of P.

46. In Figure P5.46, the pulleys and **QC** the cord are light, all surfaces are frictionless, and the cord does not stretch. (a) How does the acceleration of block 1 compare with the acceleration of block 2? Explain your reasoning. (b) The mass of block 2 is 1.30 kg. Find its acceleration as it depends on the mass m_1 of block 1. (c) **What If?** What does the result of part (b) predict if m_1 is very much less than 1.30 kg? (d) What does the result of part (b) predict if m_1 approaches infinity? (e) In this last case, what is the tension in the cord? (f) Could you anticipate the answers to parts (c), (d), and (e) without first doing part (b)? Explain.

Figure P5.46

47. You are working as an expert witness for the defense of a **CR** container ship captain whose ship ran into a reef surrounding a Caribbean island. The captain is being charged with intentionally running the ship into the reef. In discovery, the following information has been presented, and attorneys on both sides have stipulated that the information is correct: The ship was traveling at 2.50 m/s toward the reef when a mechanical failure caused the rudder to jam in the straight-ahead position. At that point in time, the ship was 900 m from the reef. The wind was blowing directly toward the reef, and exerting a constant force of 9.00×10^3 N on the boat in a direction toward the reef. The mass of the ship and its cargo was 5.50×10^7 kg. During the preparation for the trial, the captain claims that without control of the direction of travel, the only choice he had was to put the engines in reverse at maximum power, such that the total force exerted by the frictional drag force of the water and the force of the water on the propellers was 1.25×10^5 N in a direction away from the reef. From this information, construct a convincing argument that nothing the captain could do in this situation could have prevented the ship from striking the reef.

48. A flat cushion of mass m is **QC** released from rest at the corner of the roof of a building, at height h. A wind blowing along the side of the building exerts a constant horizontal force of magnitude F on the cushion as it drops as shown in Figure P5.48. The air exerts no vertical force. (a) Show that the path of the cushion is a straight line. (b) Does the cushion fall with constant velocity? Explain. (c) If $m = 1.20$ kg, $h = 8.00$ m, and $F = 2.40$ N, how far from the building will the cushion hit the level ground? **What If?** (d) If the cushion is thrown downward with a nonzero speed at the top of the building, what will be the shape of its trajectory? Explain.

Figure P5.48

49. What horizontal force must **S** be applied to a large block of mass M shown in Figure P5.49 so that the tan blocks remain stationary relative to M? Assume all surfaces and the pulley are frictionless. Notice that the force exerted by the string accelerates m_2.

Figure P5.49 Problems 49 and 53

50. An 8.40-kg object slides down a fixed, frictionless, inclined **QC** plane. Use a computer to determine and tabulate (a) the normal force exerted on the object and (b) its acceleration for a series of incline angles (measured from the horizontal) ranging from 0° to 90° in 5° increments. (c) Plot a graph of the normal force and the acceleration as functions of the incline angle. (d) In the limiting cases of 0° and 90°, are your results consistent with the known behavior?

CHALLENGE PROBLEMS

51. A block of mass 2.20 kg is accel- **QC** erated across a rough surface by a light cord passing over a small pulley as shown in Figure P5.51. The tension T in the cord is maintained at 10.0 N, and the pulley is 0.100 m above the top of the block. The coefficient of kinetic friction is 0.400. (a) Determine the acceleration of the block when $x = 0.400$ m. (b) Describe the general behavior of the acceleration as the block slides from a location where x is large to $x = 0$. (c) Find the maximum value of the acceleration and the position x for which it occurs. (d) Find the value of x for which the acceleration is zero.

Figure P5.51

52. *Why is the following situation impossible?* A 1.30-kg toaster is not plugged in. The coefficient of static friction between the toaster and a horizontal countertop is 0.350. To make the toaster start moving, you carelessly pull on its electric cord. Unfortunately, the cord has become frayed from your previous similar actions and will break if the tension in the cord exceeds 4.00 N. By pulling on the cord at a particular angle, you successfully start the toaster moving without breaking the cord.

53. Initially, the system of objects shown in Figure P5.49 is held
S motionless. The pulley and all surfaces and wheels are frictionless. Let the force $\vec{F}$ be zero and assume that m_1 can move only vertically. At the instant after the system of objects is released, find (a) the tension T in the string, (b) the acceleration of m_2, (c) the acceleration of M, and (d) the acceleration of m_1. (*Note:* The pulley accelerates along with the cart.)

54. A mobile is formed by supporting four metal butterflies
S of equal mass m from a string of length L. The points of support are evenly spaced a distance ℓ apart as shown in Figure P5.54. The string forms an angle θ_1 with the ceiling at each endpoint. The center section of string is horizontal. (a) Find the tension in each section of string in terms of θ_1, m, and g. (b) In terms of θ_1, find the angle θ_2 that the sections of string between the outside butterflies and the inside butterflies form with the horizontal. (c) Show that the distance D between the endpoints of the string is

$$D = \frac{L}{5}\left\{2\cos\theta_1 + 2\cos\left[\tan^{-1}\left(\tfrac{1}{2}\tan\theta_1\right)\right] + 1\right\}$$

Figure P5.54

55. In Figure P5.55, the incline has mass M and is fastened to
S the stationary horizontal tabletop. The block of mass m is placed near the bottom of the incline and is released with a quick push that sets it sliding upward. The block stops near the top of the incline as shown in the figure and then slides down again, always without friction. Find the force that the tabletop exerts on the incline throughout this motion in terms of m, M, g, and θ.

Figure P5.55

Circular Motion and Other Applications of Newton's Laws

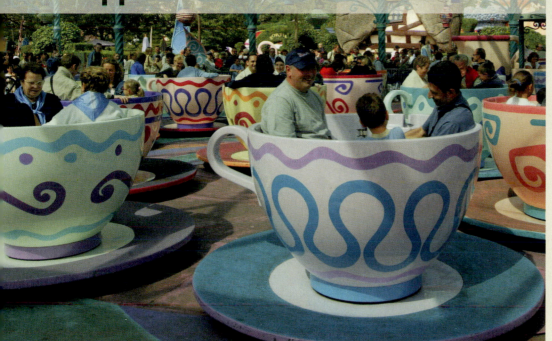

The Mad Tea Party at Disneyland is a ride with lots of circular motion. Each cup rotates around a central axis. In addition, six cups are mounted on a rotating turntable. Furthermore, three such turntables are mounted on a large turntable rotating in the opposite direction to the smaller turntables. (*Pascal Le Segretain/Getty Images News/Getty Images*)

STORYLINE **You have no classes today and decide to spend the day** at Disneyland with a friend. It's a weekday, so the lines are relatively short. In fact, even the line for the Mad Tea Party is short! You have read online that this ride is available at all of Disney's parks around the world, even at the newest Disneyland in Shanghai, China. For this ride, you and your friend sit in a large tea cup that spins rapidly. While your friend pulls on the wheel in the center of the teacup to make it spin quickly, you hang your smartphone from a string to form a pendulum. You dangle the pendulum from your hand at the rim of the tea cup. You notice that the pendulum does not hang straight down! You open a special app on the smartphone that gives you a readout of the angle of the phone with respect to the vertical and hang the phone again as a pendulum. Why does the pendulum deviate from the vertical? In what direction does the pendulum deviate from the vertical? What happens to the angle reading on the phone as you move your hand holding the pendulum toward the center of the tea cup? Why does the reading change in this way?

CONNECTIONS In this chapter, we expand on the circular motion we studied in Chapter 4, combining it with our new knowledge about force from Chapter 5. What forces act on an object when it is in circular motion? In addition, we consider some other cases in which Newton's laws help us to understand the motion. We will consider how the laws of physics appear when one is in an accelerated frame of reference, such as the spinning teacup in the storyline. We will also extend our discussions of friction from Chapter 5 by looking at resistive forces on an object, such as air resistance. Unlike in our model of kinetic friction, these forces vary in magnitude according to the speed of the object relative to the surrounding medium. In future chapters, we will see several examples of circular motion, such as planets in orbit in Chapter 13, charged particles moving in circular paths in magnetic fields in Chapter 28, and electrons in circular orbits in the Bohr theory of the hydrogen atom in Chapter 41. The action of particles undergoing resistive forces will appear in future chapters, and we will see electrical analogs to resistive forces in various kinds of electric circuits in Chapters 27 and 31. The material in this chapter on accelerated reference frames will be followed up with the discussion of general relativity in Chapter 38.

6.1 Extending the Particle in Uniform Circular Motion Model

In Section 4.4, we discussed the analysis model of a particle in uniform circular motion, in which a particle moves with constant speed v in a circular path having a radius r. The particle experiences an acceleration that has a magnitude

$$a_c = \frac{v^2}{r}$$

When the string breaks, the puck moves in the direction tangent to the circle.

Figure 6.2 The string holding the puck in its circular path breaks.

The acceleration is called *centripetal acceleration* because $\vec{a}_c$ is directed toward the center of the circle. Furthermore, $\vec{a}_c$ is *always* perpendicular to $\vec{v}$. (If there were a component of acceleration parallel to $\vec{v}$, the particle's speed would be changing.)

Let us now extend the particle in uniform circular motion model from Section 4.4 by incorporating the concept of force. Consider a puck of mass m that is tied to a string of length r and moves at constant speed in a horizontal, circular path as illustrated in the overhead view in Figure 6.1. Its weight is supported by the normal force from a frictionless table, and the string is anchored to a peg at the center of the circular path of the puck. Why does the puck move in a circle? According to Newton's first law, the puck would move in a straight line if there were no force on it; the string, however, prevents motion along a straight line by exerting on the puck a radial force $\vec{F}_r$ that makes it follow the circular path. This force is directed along the string toward the center of the circle as shown in Figure 6.1.

If Newton's second law is applied along the radial direction, the net force causing the centripetal acceleration can be related to the acceleration as follows:

◀ Force causing centripetal acceleration

$$\sum F = ma_c = m\frac{v^2}{r} \qquad (6.1)$$

A force causing a centripetal acceleration acts toward the center of the circular path and causes a change in the direction of the velocity vector. If that force should vanish, the object would no longer move in its circular path; instead, it would move along a straight-line path tangent to the circle. This idea is illustrated in Figure 6.2 for the puck moving in a circular path at the end of a string in a horizontal plane. If the string breaks at some instant, the puck moves along the straight-line path that is tangent to the circle at the position of the puck at this instant.

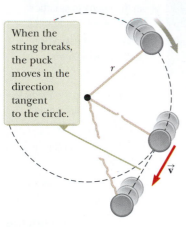

A force $\vec{F}_r$, directed toward the center of the circle, keeps the puck moving in its circular path.

Figure 6.1 An overhead view of a puck moving in a circular path in a horizontal plane.

PITFALL PREVENTION 6.1
Direction of Travel When the String Is Cut Study Figure 6.2 very carefully. Many students (wrongly) think that the puck will move *radially* away from the center of the circle when the string is cut. The velocity of the puck is *tangent* to the circle. By Newton's first law, the puck continues to move in the same direction in which it is moving just as the force from the string disappears.

QUICK QUIZ 6.1 You are riding on a Ferris wheel that is rotating with constant speed. The car in which you are riding always maintains its correct upward orientation; it does not invert. **(i)** What is the direction of the normal force on you from the seat when you are at the top of the wheel? (a) upward (b) downward (c) impossible to determine **(ii)** From the same choices, what is the direction of the net force on you when you are at the top of the wheel?

ANALYSIS MODEL **Particle in Uniform Circular Motion (Extension)**

Imagine a moving object that can be modeled as a particle. If it moves in a circular path of radius r at a constant speed v, it experiences a centripetal acceleration. Because the particle is accelerating, there must be a net force acting on the particle. That force is directed toward the center of the circular path and is given by

$$\sum F = ma_c = m\frac{v^2}{r} \qquad \textbf{(6.1)}$$

Examples

- the tension in a string of constant length acting on a rock twirled in a circle

- the gravitational force acting on a planet traveling around the Sun in a perfectly circular orbit (Chapter 13)
- the magnetic force acting on a charged particle moving in a uniform magnetic field (Chapter 28)
- the electric force acting on an electron in orbit around a nucleus in the Bohr model of the hydrogen atom (Chapter 41)

Example **6.1**　**The Conical Pendulum**

A small ball of mass m is suspended from a string of length L. The ball revolves with constant speed v in a horizontal circle of radius r as shown in Figure 6.3. (Because the string sweeps out the surface of a cone, the system is known as a *conical pendulum*.) Find an expression for v in terms of the length of the string and the angle it makes with the vertical in Figure 6.3.

SOLUTION

Conceptualize Imagine the motion of the ball in Figure 6.3a and convince yourself that the string sweeps out a cone and that the ball moves in a horizontal circle. What happens if the ball moves with a higher speed?

Categorize The ball in Figure 6.3 does not accelerate vertically. Therefore, we model it as a *particle in equilibrium* in the vertical direction. It experiences a centripetal acceleration in the horizontal direction, so it is modeled as a *particle in uniform circular motion* in this direction.

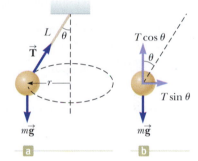

Figure 6.3 (Example 6.1) (a) A conical pendulum. The path of the ball is a horizontal circle. (b) The forces acting on the ball.

Analyze Let θ represent the angle between the string and the vertical. In the diagram of forces acting on the ball in Figure 6.3b, the force $\vec{\mathbf{T}}$ exerted by the string on the ball is resolved into a vertical component $T\cos\theta$ and a horizontal component $T\sin\theta$ acting toward the center of the circular path.

Apply the particle in equilibrium model in the vertical direction:

$$\sum F_y = T\cos\theta - mg = 0$$

$$(1) \quad T\cos\theta = mg$$

Use Equation 6.1 from the particle in uniform circular motion model in the horizontal direction:

$$(2) \quad \sum F_x = T\sin\theta = ma_c = \frac{mv^2}{r}$$

Divide Equation (2) by Equation (1) and use $\sin\theta/\cos\theta = \tan\theta$:

$$\tan\theta = \frac{v^2}{rg}$$

Solve for v:

$$v = \sqrt{rg\tan\theta}$$

Incorporate $r = L\sin\theta$ from the geometry in Figure 6.3a:

$$(3) \quad v = \sqrt{Lg\sin\theta\tan\theta}$$

Finalize Notice that the speed is independent of the mass of the ball. Consider what happens when θ goes to 90° so that the string is horizontal. Because the tangent of 90° is infinite, the speed v is infinite, which tells us the string cannot possibly be horizontal. If it were, there would be no vertical component of the force $\vec{\mathbf{T}}$ to balance the gravitational force on the ball. The situation in this problem is similar in some ways to your experience on the Mad Tea Party ride. As you move your hand holding the hanging smartphone toward the center of your spinning cup, the speed v of the phone changes, resulting in a different angle θ, as suggested by Equation (3).

Example **6.2**   How Fast Can It Spin?

A puck of mass 0.500 kg is attached to the end of a cord 1.50 m long. The puck moves in a horizontal circle as shown in Figure 6.1. If the cord can withstand a maximum tension of 50.0 N, what is the maximum speed at which the puck can move before the cord breaks? Assume the string remains horizontal during the motion.

SOLUTION

Conceptualize   It makes sense that the stronger the cord, the faster the puck can move before the cord breaks. Also, we expect a more massive puck to break the cord at a lower speed. (Imagine whirling a bowling ball on the cord!)

Categorize   Because the puck moves in a circular path, we model it as a *particle in uniform circular motion.*

Analyze   Incorporate the tension and the centripetal acceleration into Newton's second law as described by Equation 6.1:

$$T = m \frac{v^2}{r}$$

Solve for *v*:

$$(1) \quad v = \sqrt{\frac{Tr}{m}}$$

Find the maximum speed the puck can have, which corresponds to the maximum tension the string can withstand:

$$v_{\text{max}} = \sqrt{\frac{T_{\text{max}} r}{m}} = \sqrt{\frac{(50.0 \text{ N})(1.50 \text{ m})}{0.500 \text{ kg}}} = \boxed{12.2 \text{ m/s}}$$

Finalize   Equation (1) shows that *v* increases with *T* and decreases with larger *m*, as we expected from our conceptualization of the problem.

WHAT IF?   Suppose the puck moves in a circle of larger radius at the same speed *v*. Is the cord more likely or less likely to break?

Answer   The larger radius means that the change in the direction of the velocity vector will be smaller in a given time interval. Therefore, the acceleration is smaller and the required tension in the string is smaller. As a result, the string is less likely to break when the puck travels in a circle of larger radius.

Example **6.3**   What Is the Maximum Speed of the Car?

A 1 500-kg car moving on a flat, horizontal road negotiates a curve as shown in the overhead view in Figure 6.4a. If the radius of the curve is 35.0 m and the coefficient of static friction between the tires and dry pavement is 0.523, find the maximum speed the car can have and still make the turn successfully.

SOLUTION

Conceptualize   Imagine that the curved roadway is part of a large circle so that the car is moving in a circular path.

Categorize   Based on the Conceptualize step of the problem, we model the car as a *particle in uniform circular motion* in the horizontal direction. The car is not accelerating vertically, so it is modeled as a *particle in equilibrium* in the vertical direction.

Analyze   The back view in Figure 6.4b shows the forces on the car. The force that enables the car to remain in its circular path is the force of static friction. (It is *static* because no slipping occurs at the point of contact between road and tires. If this force of static friction were zero—for example, if the car were on an icy road—the car would continue in a straight line and slide off the curved road.) The maximum speed v_{max} the car can have around the curve is the speed at which it is on the verge of skidding outward. At this point, the friction force has its maximum value $f_{s,\text{max}} = \mu_s n$.

Figure 6.4 (Example 6.3) (a) The force of static friction directed toward the center of the curve keeps the car moving in a circular path. (b) The forces acting on the car.

6.3 continued

Apply Equation 6.1 from the particle in uniform circular motion model in the radial direction for the maximum speed condition:

$$(1) \quad f_{s,\text{max}} = \mu_s n = m \frac{v_{\text{max}}^2}{r}$$

Apply the particle in equilibrium model to the car in the vertical direction:

$$\sum F_y = 0 \quad \rightarrow \quad n - mg = 0 \quad \rightarrow \quad n = mg$$

Solve Equation (1) for the maximum speed and substitute for n:

$$(2) \quad v_{\text{max}} = \sqrt{\frac{\mu_s n r}{m}} = \sqrt{\frac{\mu_s m g r}{m}} = \sqrt{\mu_s g r}$$

Substitute numerical values:

$$v_{\text{max}} = \sqrt{(0.523)(9.80 \text{ m/s}^2)(35.0 \text{ m})} = \boxed{13.4 \text{ m/s}}$$

Finalize This speed is equivalent to 30.0 mi/h. Therefore, if the speed limit on this roadway is higher than 30 mi/h, this roadway could benefit greatly from some banking, as in the next example! Notice that the maximum speed does not depend on the mass of the car, which is why curved highways do not need multiple speed limits to cover the various masses of vehicles using the road.

WHAT IF? Suppose a car travels this curve on a wet day and begins to skid on the curve when its speed reaches only 8.00 m/s. What can we say about the coefficient of static friction in this case?

Answer The coefficient of static friction between the tires and a wet road should be smaller than that between the tires and a dry road. This expectation is consistent with experience with driving because a skid is more likely on a wet road than a dry road.

To check our suspicion, we can solve Equation (2) for the coefficient of static friction:

$$\mu_s = \frac{v_{\text{max}}^2}{gr}$$

Substituting the numerical values gives

$$\mu_s = \frac{v_{\text{max}}^2}{gr} = \frac{(8.00 \text{ m/s})^2}{(9.80 \text{ m/s}^2)(35.0 \text{ m})} = 0.187$$

which is indeed smaller than the coefficient of 0.523 for the dry road.

Example 6.4 | **The Banked Roadway**

You are a civil engineer who has been given the assignment to redesign the curved roadway in Example 6.3 in such a way that a car will not have to rely on friction to round the curve without skidding. In other words, a car moving at the designated speed can negotiate the curve even when the road is covered with ice. Such a road is usually *banked*, which means that the roadway is tilted toward the inside of the curve. Suppose the designated speed for the road is to be 13.4 m/s (30.0 mi/h) and the radius of the curve is 35.0 m. You need to determine the angle at which the roadway on the curve should be banked.

SOLUTION

Conceptualize The difference between this example and Example 6.3 is that the car is no longer moving on a flat roadway. Figure 6.5 shows the banked roadway, with the center of the circular path of the car far to the left of the figure. Notice that the horizontal component of the normal force participates in causing the car's centripetal acceleration. Note in Figure 6.5 that, unlike the motion on an inclined plane in Figure 5.12, we use a coordinate system with x horizontal and y upward. This choice is made so that the centripetal acceleration of the car is purely in the x direction.

Categorize As in Example 6.3, the car is modeled as a *particle in equilibrium* in the vertical direction and a *particle in uniform circular motion* in the horizontal direction.

Analyze On a level (unbanked) road, the force that causes the centripetal acceleration is the force of static friction between tires and the road as we saw in the preceding example.

Figure 6.5 (Example 6.4) A car moves into the page and is rounding a curve on a road banked at an angle θ to the horizontal. When friction is neglected, the force that causes the centripetal acceleration and keeps the car moving in its circular path is the horizontal component of the normal force.

continued

6.4 continued

If the road is banked at an angle θ as in Figure 6.5, however, the normal force $\vec{n}$ has a horizontal component toward the center of the curve. Because the road is to be designed so that the force of static friction is zero, the component $n_x = n \sin \theta$ is the only force that causes the centripetal acceleration.

Write Newton's second law for the car in the radial direction, which is the $-x$ direction:

$$(1) \quad \sum F_r = n \sin \theta = \frac{mv^2}{r}$$

Apply the particle in equilibrium model to the car in the vertical direction:

$$\sum F_y = n \cos \theta - mg = 0$$

$$(2) \quad n \cos \theta = mg$$

Divide Equation (1) by Equation (2):

$$(3) \quad \tan \theta = \frac{v^2}{rg}$$

Solve for the angle θ and substitute numerical values:

$$\theta = \tan^{-1}\left[\frac{(13.4 \text{ m/s})^2}{(35.0 \text{ m})(9.80 \text{ m/s}^2)}\right] = \boxed{27.6°}$$

Finalize   Equation (3) shows that the banking angle is independent of the mass of the vehicle negotiating the curve. If a car rounds the curve at a speed less than 13.4 m/s, the centripetal acceleration decreases. Therefore, the normal force, which is unchanged, is sufficient to cause *two* accelerations: the lower centripetal acceleration and an acceleration of the car down the inclined roadway. Consequently, an additional friction force parallel to the roadway and upward is needed to keep the car from sliding down the bank (to the left in Fig. 6.5). Similarly, a driver attempting to negotiate the curve at a speed greater than 13.4 m/s has to depend on friction to keep from sliding up the bank (to the right in Fig. 6.5). See problem 41 for an analysis of this situation.

WHAT IF?   Imagine that this same roadway were built on Mars in the future to connect different colony centers. Could it be traveled at the same speed?

Answer   The reduced gravitational force on Mars would mean that the car is not pressed as tightly to the roadway. The reduced normal force results in a smaller component of the normal force toward the center of the circle. This smaller component would not be sufficient to provide the centripetal acceleration associated with the original speed. The centripetal acceleration must be reduced, which can be done by reducing the speed v.

Mathematically, notice that Equation (3) shows that the speed v is proportional to the square root of g for a roadway of fixed radius r banked at a fixed angle θ. Therefore, if g is smaller, as it is on Mars, the speed v with which the roadway can be safely traveled is also smaller.

Example 6.5   **Riding the Ferris Wheel**

A child of mass m rides on a Ferris wheel as shown in Figure 6.6a. The child moves in a vertical circle of radius 10.0 m at a constant speed of 3.00 m/s.

(A) Determine the force exerted by the seat on the child at the bottom of the ride. Express your answer in terms of the weight of the child, mg.

SOLUTION

Conceptualize   Look carefully at Figure 6.6a. Based on experiences you may have had on a Ferris wheel or driving over small hills on a roadway, you would expect to feel lighter at the top of the path. Similarly, you would expect to feel heavier at the bottom of the path. At both the bottom of the path and the top, the normal and gravitational forces on the child act in *opposite* directions. The vector sum of these two forces gives a force of constant magnitude that keeps the child moving in a circular path at a constant speed. To yield net force vectors with the same magnitude, the normal force at the bottom must be greater than that at the top.

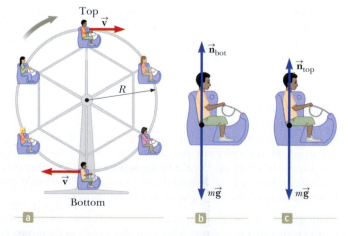

Figure 6.6   (Example 6.5) (a) A child rides on a Ferris wheel. (b) The forces acting on the child at the bottom of the path. (c) The forces acting on the child at the top of the path.

6.5 continued

Categorize Because the speed of the child is constant, we can categorize this problem as one involving a *particle* (the child) *in uniform circular motion*, complicated by the gravitational force acting at all times on the child.

Analyze We draw a diagram of forces acting on the child at the bottom of the ride as shown in Figure 6.6b. The only forces acting on him are the downward gravitational force $\vec{F}_g = m\vec{g}$ and the upward force $\vec{n}_{bot}$ exerted by the seat. The centripetal acceleration of the child at this point is upward and the net upward force on the child has a magnitude $n_{bot} - mg$.

Using the particle in uniform circular motion model, apply Newton's second law to the child in the radial direction when he is at the bottom of the ride:

$$\sum F = n_{bot} - mg = m\frac{v^2}{r}$$

Solve for the force exerted by the seat on the child:

$$n_{bot} = mg + m\frac{v^2}{r} = mg\left(1 + \frac{v^2}{rg}\right)$$

Substitute numerical values given for the speed and radius:

$$n_{bot} = mg\left[1 + \frac{(3.00 \text{ m/s})^2}{(10.0 \text{ m})(9.80 \text{ m/s}^2)}\right]$$

$$= \boxed{1.09 \; mg}$$

Hence, the magnitude of the force $\vec{n}_{bot}$ exerted by the seat on the child is *greater* than the weight of the child by a factor of 1.09. So, the child experiences an apparent weight that is greater than his true weight by a factor of 1.09.

(B) Determine the force exerted by the seat on the child at the top of the ride.

SOLUTION

Analyze The diagram of forces acting on the child at the top of the ride is shown in Figure 6.6c. The centripetal acceleration of the child at this point is downward and the net downward force has a magnitude $mg - n_{top}$.

Apply Newton's second law to the child at this position:

$$\sum F = mg - n_{top} = m\frac{v^2}{r}$$

Solve for the force exerted by the seat on the child:

$$n_{top} = mg - m\frac{v^2}{r} = mg\left(1 - \frac{v^2}{rg}\right)$$

Substitute numerical values:

$$n_{top} = mg\left[1 - \frac{(3.00 \text{ m/s})^2}{(10.0 \text{ m})(9.80 \text{ m/s}^2)}\right]$$

$$= \boxed{0.908 \; mg}$$

In this case, the magnitude of the force exerted by the seat on the child is *less* than his true weight by a factor of 0.908, and the child feels lighter.

Finalize The variations in the normal force are consistent with our prediction in the Conceptualize step of the problem.

WHAT IF? Suppose a defect in the Ferris wheel mechanism causes the speed of the child to increase to 10.0 m/s. What does the child experience at the top of the ride in this case?

Answer If the calculation above is performed with $v = 10.0$ m/s, the magnitude of the normal force at the top of the ride is negative, which is impossible. We interpret it to mean that the required downward centripetal acceleration of the child is larger than that due to gravity. As a result, the child will lose contact with the seat and will only stay in his circular path if there is a safety bar or a seat belt that provides a downward force on him to keep him in his seat. At the bottom of the ride, the normal force is 2.02 *mg*, which would be uncomfortable.

6.2 Nonuniform Circular Motion

In Chapter 4, we found that if a particle moves with varying speed in a circular path, there is, in addition to the radial component of acceleration, a tangential component having magnitude $|dv/dt|$. Therefore, the force acting on the particle must also have a tangential and a radial component. Because the total acceleration is $\vec{a} = \vec{a}_r + \vec{a}_t$, the total force exerted on the particle is $\sum \vec{F} = \sum \vec{F}_r + \sum \vec{F}_t$

The net force exerted on the particle is the vector sum of the radial force and the tangential force.

Figure 6.7 When the net force acting on a particle moving in a circular path has a tangential component ΣF_t, the particle's speed changes.

as shown in Figure 6.7. (We express the radial and tangential forces as net forces with the summation notation because each force could consist of multiple forces that combine.) The vector $\Sigma \vec{\mathbf{F}}_r$ is directed toward the center of the circle and is responsible for the centripetal acceleration. The vector $\Sigma \vec{\mathbf{F}}_t$ tangent to the circle is responsible for the tangential acceleration, which represents a change in the particle's speed with time.

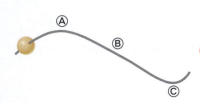

Figure 6.8 (Quick Quiz 6.2) A bead slides along a curved wire.

QUICK QUIZ 6.2 A bead slides at constant speed along a curved wire lying on a horizontal surface as shown in Figure 6.8. **(a)** Draw the vectors representing the force exerted by the wire on the bead at points Ⓐ, Ⓑ, and Ⓒ. **(b)** Suppose the bead in Figure 6.8 speeds up with constant tangential acceleration as it moves toward the right. Draw the vectors representing the forces on the bead at points Ⓐ, Ⓑ, and Ⓒ.

Example **6.6** **Keep Your Eye on the Ball**

A small sphere of mass m is attached to the end of a cord of length R and set into motion in a *vertical* circle about a fixed point O as illustrated in Figure 6.9. Determine the tangential acceleration of the sphere and the tension in the cord at any instant when the speed of the sphere is v and the cord makes an angle θ with the vertical.

S O L U T I O N

Conceptualize Compare the motion of the sphere in Figure 6.9 with that of the child in Figure 6.6a associated with Example 6.5. Both objects travel in a circular path. Unlike the child in Example 6.5, however, the speed of the sphere is *not* uniform in this example because, at most points along the path, a tangential component of acceleration arises from the gravitational force exerted on the sphere.

Categorize We model the sphere as a *particle under a net force* and moving in a circular path, but it is not a particle in *uniform* circular motion. We need to use the techniques discussed in this section on nonuniform circular motion.

Analyze From the force diagram in Figure 6.9, we see that the only forces acting on the sphere are the gravitational force $\vec{\mathbf{F}}_g = m\vec{\mathbf{g}}$ exerted by the Earth and the force $\vec{\mathbf{T}}$ exerted by the cord. We resolve $\vec{\mathbf{F}}_g$ into a tangential component $mg \sin \theta$ and a radial component $mg \cos \theta$.

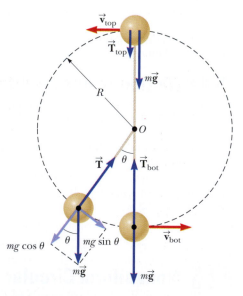

Figure 6.9 (Example 6.6) The forces acting on a sphere of mass m connected to a cord of length R and rotating in a vertical circle centered at O. Forces acting on the sphere are shown when the sphere is at the top and bottom of the circle and at an arbitrary location.

6.6 continued

From the particle under a net force model, apply Newton's second law to the sphere in the tangential direction:

$$\sum F_t = mg \sin \theta = ma_t$$

$$a_t = \boxed{g \sin \theta}$$

Apply Newton's second law to the forces acting on the sphere in the radial direction, noting that both $\vec{T}$ and $\vec{a}_r$ are directed toward O. As noted in Section 4.5, we can use Equation 4.21 for the instantaneous centripetal acceleration of a particle even when it moves in nonuniform circular motion:

$$\sum F_r = T - mg \cos \theta = \frac{mv^2}{R}$$

$$T = \boxed{mg\left(\frac{v^2}{Rg} + \cos \theta\right)}$$

Finalize Let us evaluate this result at the top and bottom of the circular path (Fig. 6.9):

$$T_{top} = mg\left(\frac{v_{top}^2}{Rg} - 1\right) \qquad T_{bot} = mg\left(\frac{v_{bot}^2}{Rg} + 1\right)$$

These results have similar mathematical forms as those for the normal forces n_{top} and n_{bot} on the child in Example 6.5, which is consistent with the normal force on the child playing a similar physical role in Example 6.5 as the tension in the string plays in this example. Keep in mind, however, that the normal force $\vec{n}$ on the child in Example 6.5 is always upward, whereas the force $\vec{T}$ in this example changes direction because it must always point inward along the string. Also note that v in the expressions above varies for different positions of the sphere, as indicated by the subscripts, whereas v in Example 6.5 is constant.

W H A T   I F ?   What if the sphere is set in motion with a slower speed?

(A) What speed would the sphere have as it passes over the top of the circle if the tension in the cord goes to zero instantaneously at this point?

Answer Let us set the tension equal to zero in the expression for T_{top}:

$$0 = mg\left(\frac{v_{top}^2}{Rg} - 1\right) \quad \rightarrow \quad v_{top} = \sqrt{gR}$$

(B) What if the sphere is set in motion such that the speed at the top is less than this value? What happens?

Answer In this case, the sphere never reaches the top of the circle. At some point on the way up, the tension in the string goes to zero and the sphere becomes a projectile. It follows a segment of a parabolic path, with its peak below the topmost position of the sphere shown in Figure 6.9, rejoining the circular path on the other side when the tension becomes nonzero again.

6.3 Motion in Accelerated Frames

Newton's laws of motion, which we introduced in Chapter 5, describe observations that are made in an inertial frame of reference. In this section, we analyze how Newton's laws are applied by an observer in a noninertial frame of reference, that is, one that is accelerating. For example, recall the discussion of the air hockey table on a train in Section 5.2. The train moving at constant velocity represents an inertial frame. An observer on the train sees the puck at rest remain at rest, and Newton's first law appears to be obeyed. The accelerating train is not an inertial frame. According to you as the observer on this train, there appears to be no force on the puck, yet it accelerates from rest toward the back of the train, appearing to violate Newton's first law. This property is a general property of observations made in noninertial frames: there appear to be unexplained accelerations of objects that are not "fastened" to the frame. Newton's first law is not violated, of course. It only appears to be violated because of observations made from a noninertial frame.

On the accelerating train, as you watch the puck accelerating toward the back of the train, you might conclude based on your belief in Newton's second law that a force has acted on the puck to cause it to accelerate. We call an apparent force such as this one a **fictitious force** because it is not a real force and is due only to

From the passenger's frame of reference, a force appears to push her toward the right door, but it is a fictitious force.

Fictitious force

Relative to the reference frame of the Earth, the car seat applies a real force (friction) toward the left on the passenger, causing her to change direction along with the rest of the car.

Real force

Figure 6.10 (a) A car approaching a curved exit ramp. What causes a passenger in the front seat to move toward the right-hand door? (b) Passenger's frame of reference. (c) Reference frame of the Earth.

observations made in an accelerated reference frame. A fictitious force appears to act on an object in the same way as a real force. Real forces are always interactions between two objects, however, and you cannot identify a second object for a fictitious force. (What second object is interacting with the puck to cause it to accelerate?) In general, simple fictitious forces appear to act in the direction *opposite* that of the acceleration of the noninertial frame. For example, the train accelerates forward and there appears to be a fictitious force causing the puck to slide toward the back of the train.

The train example describes a fictitious force due to a change in the train's speed. Another fictitious force is due to the change in the *direction* of the velocity vector. To understand the motion of a system that is noninertial because of a change in direction, consider a car traveling along a highway at a high speed and approaching a curved exit ramp on the left as shown in Figure 6.10a. As the car takes the sharp left turn on the ramp, a person sitting in the passenger seat leans or slides to the right and hits the door. At that point the force exerted by the door on the passenger keeps her from being ejected from the car. What causes her to move toward the door? A popular but incorrect explanation is that a force acting toward the right in Figure 6.10b pushes the passenger outward from the center of the circular path. Although often called the "centrifugal force," it is a fictitious force. The car represents a noninertial reference frame that has a centripetal acceleration toward the center of its circular path. As a result, the passenger feels an apparent force which is outward from the center of the circular path, or to the right in Figure 6.10b, in the direction opposite that of the acceleration.

Let us address this phenomenon in terms of Newton's laws. Before the car enters the ramp, the passenger is moving in a straight-line path. As the car enters the ramp and travels a curved path, the passenger tends to move along the original straight-line path, which is in accordance with Newton's first law: the natural tendency of an object is to continue moving in a straight line. If a sufficiently large force (toward the center of curvature) acts on the passenger as in Figure 6.10c, however, she moves in a curved path along with the car. This force is the force of friction between her and the car seat. If this friction force is not large enough, the seat follows a curved path while the passenger tends to continue in the straight-line path of the car before the car began the turn. Therefore, from the point of view of an observer in the car, the passenger leans or slides to the right relative to the seat. Eventually, she encounters the door, which provides a force large enough to enable her to follow the same curved path as the car.

Another interesting fictitious force is the "Coriolis force." It is an apparent force caused by changing the radial position of an object in a rotating coordinate system.

For example, suppose you and a friend are on opposite sides of a rotating circular platform and you decide to throw a baseball to your friend. Figure 6.11a represents what an observer would see if the ball is viewed while the observer is hovering at rest above the rotating platform. According to this observer, who is in an inertial frame, the ball follows a straight line as it must according to Newton's first law. At $t = 0$ you throw the ball toward your friend, but by the time t_f when the ball has crossed the platform, your friend has moved to a new position and can't catch the ball. Now, however, consider the situation from your friend's viewpoint. Your friend is in a noninertial reference frame because he is undergoing a centripetal acceleration relative to the inertial frame of the Earth's surface. He starts off seeing the baseball coming toward him, but as it crosses the platform, it veers to one side as shown in Figure 6.11b. Therefore, your friend on the rotating platform states that the ball does not obey Newton's first law and claims that a sideways force is causing the ball to follow a curved path. This fictitious force is called the Coriolis force.

Fictitious forces may not be real forces, but they can have real effects. An object on your dashboard *really* slides off if you press the accelerator of your car. As you ride on a merry-go-round, you feel pushed toward the outside as if due to the fictitious "centrifugal force." You are likely to fall over and injure yourself due to the

By the time t_f that the ball arrives at the other side of the platform, your friend is no longer there to catch it. According to this observer, the ball follows a straight-line path, consistent with Newton's laws.

From your friend's point of view, the ball veers to one side during its flight. Your friend introduces a fictitious force to explain this deviation from the expected path.

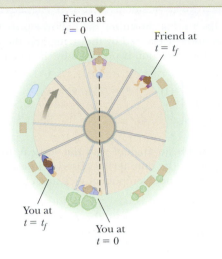

Friend at
$t = 0$

Friend at
$t = t_f$

You at
$t = t_f$

You at
$t = 0$

a

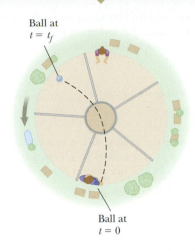

Ball at
$t = t_f$

Ball at
$t = 0$

b

Figure 6.11 You and your friend stand at the edge of a rotating circular platform. You throw the ball at $t = 0$ in the direction of your friend. (a) Overhead view observed by someone in an inertial reference frame attached to the Earth. The ground appears stationary, and the platform rotates clockwise. (b) Overhead view observed by someone in an inertial reference frame attached to the platform. The platform appears stationary, and the ground rotates counterclockwise.

Coriolis force if you walk along a radial line while a merry-go-round rotates. (One of the authors did so and suffered a separation of the ligaments from his ribs when he fell over.) The Coriolis force due to the rotation of the Earth is responsible for rotations of hurricanes and for large-scale ocean currents.

QUICK QUIZ **6.3** Consider the passenger in the car making a left turn in Figure 6.10. Which of the following is correct about forces in the horizontal direction if she is making contact with the right-hand door? **(a)** The passenger is in equilibrium between real forces acting to the right and real forces acting to the left. **(b)** The passenger is subject only to real forces acting to the right. **(c)** The passenger is subject only to real forces acting to the left. **(d)** None of those statements is true.

PITFALL PREVENTION 6.2
Centrifugal Force The commonly heard phrase "centrifugal force" is described as a force pulling *outward* on an object moving in a circular path. If you are feeling a "centrifugal force" on a rotating carnival ride, what is the other object with which you are interacting? You cannot identify another object because it is a fictitious force that occurs when you are in a noninertial reference frame.

Example **6.7** **Fictitious Forces in Circular Motion**

Consider the experiment described in the opening storyline: you are riding on the Mad Tea Party ride and holding your smartphone hanging from a string. Now suppose your friend stands on solid ground beside the ride watching you. You hold the upper end of the string above a point near the outer rim of the spinning tea cup. Both the inertial observer (your friend) and the noninertial observer (you) agree that the string makes an angle θ with respect to the vertical. You claim that a force, which we know to be fictitious, causes the observed deviation of the string from the vertical. How is the magnitude of this force related to the smartphone's centripetal acceleration measured by the inertial observer?

SOLUTION

Conceptualize Place yourself in the role of each of the two observers. The inertial observer on the ground knows that the smartphone has a centripetal acceleration and that the deviation of the string is related to this acceleration. As the noninertial observer on the teacup, imagine that you ignore any effects of the spinning of the teacup, so you have no knowledge of any centripetal acceleration. Because you are unaware of this acceleration, you claim that a force is pushing sideways on the smartphone to cause the deviation of the string from the vertical. To make the conceptualization more real, try running from rest while holding a hanging object on a string and notice that the string is at an angle to the vertical while you are accelerating, as if a force is pushing the object backward.

continued

6.7 continued

Categorize   For the inertial observer, we model the smartphone as a *particle under a net force* in the horizontal direction and a *particle in equilibrium* in the vertical direction. For the noninertial observer, the smartphone is modeled as a *particle in equilibrium* in both directions.

Analyze   The geometry for the spinning and hanging smartphone will be similar to that shown for the ball in Figure 6.3b. According to the inertial observer at rest, the forces on the smartphone are the force $\vec{T}$ exerted by the string and the gravitational force. The inertial observer concludes that the smartphone's centripetal acceleration is provided by the horizontal component of $\vec{T}$.

For this observer, apply the particle under a net force and particle in equilibrium models:

$$\text{Inertial observer} \begin{cases} (1) \quad \sum F_x = T \sin\theta = ma_c \\ (2) \quad \sum F_y = T \cos\theta - mg = 0 \end{cases}$$

According to the noninertial observer riding in the teacup, the string also makes an angle θ with the vertical; to that observer, however, the smartphone is at rest and so its acceleration is zero. Therefore, the noninertial observer introduces a force (which we know to be fictitious) in the horizontal direction to balance the horizontal component of $\vec{T}$ and claims that the net force on the smartphone is zero.

Apply the particle in equilibrium model for this observer in both directions:

$$\text{Noninertial observer} \begin{cases} \sum F'_x = T \sin\theta - F_{\text{fictitious}} = 0 \\ \sum F'_y = T \cos\theta - mg = 0 \end{cases}$$

These expressions are equivalent to Equations (1) and (2) if $F_{\text{fictitious}} = ma_c$, where a_c is the centripetal acceleration of the smartphone according to the inertial observer.

Finalize   The angle of the string will depend on where the upper end of the string is held relative to the center of the teacup. If the string is held directly over the center, for example, the smartphone is not moving in a circular path, it has no centripetal acceleration due to the motion of the teacup, and the string will not deviate from the vertical. (In practice, it may deviate slightly due to the rotation of the turntables on which the teacup is mounted.)

 W H A T   I F ?   Suppose you wish to measure the centripetal acceleration of the smartphone from your observations. How could you do so?

Answer   Our intuition tells us that the angle θ the string makes with the vertical should increase as the acceleration increases. By solving Equations (1) and (2) simultaneously for a_c, we find that $a_c = g \tan\theta$. Therefore, you can determine the magnitude of the centripetal acceleration of the smartphone by measuring the angle θ and using that relationship. Because the deflection of the string from the vertical serves as a measure of acceleration, *a simple pendulum can be used as an accelerometer.*

6.4   Motion in the Presence of Resistive Forces

In Chapter 5, we described the force of kinetic friction exerted on an object moving on some surface. We completely ignored any interaction between the object and the medium through which it moves. Now consider the effect of that medium, which can be either a liquid or a gas. The medium exerts a **resistive force** $\vec{R}$ on the object moving through it. Some examples are the air resistance associated with moving vehicles (sometimes called *air drag*) and the viscous forces that act on objects moving through a liquid. The magnitude of $\vec{R}$ depends on factors such as the speed of the object, and the direction of $\vec{R}$ is always opposite the direction of the object's motion relative to the medium. This direction may or may not be in the direction opposite the object's velocity according to the observer. For example, if a marble is dropped into a bottle of shampoo, the marble moves downward and the resistive force is upward, resisting the falling of the marble. In contrast, imagine the moment at which there is no wind and you are looking at a flag hanging limply on a flagpole. When a breeze begins to blow toward the right, the flag moves toward the right. In this case, the drag force on the flag from the moving air is to the right and the motion of the flag in response is also to the right, the *same* direction as the drag force. Because the air moves toward the right with respect to the flag, the flag moves to the left relative to the air. Therefore, the direction of the drag force is indeed opposite to the direction of the motion of the flag with respect to the air!

The magnitude of the resistive force can depend on speed in a complex way, and here we consider only two simplified models. In the first model, we assume the resistive force is proportional to the velocity of the moving object; this model is valid for objects falling slowly through a liquid and for very small objects, such as dust particles, moving through air. In the second model, we assume a resistive force that is proportional to the square of the speed of the moving object; large objects, such as skydivers moving through air in free fall, experience such a force.

Model 1: Resistive Force Proportional to Object Velocity

If we model the resistive force acting on an object moving through a liquid or gas as proportional to the object's velocity, the resistive force can be expressed as

$$\vec{\mathbf{R}} = -b\vec{\mathbf{v}} \tag{6.2}$$

where b is a constant whose value depends on the properties of the medium and on the shape and dimensions of the object and $\vec{\mathbf{v}}$ is the velocity of the object relative to the medium. The negative sign indicates that $\vec{\mathbf{R}}$ is in the opposite direction to $\vec{\mathbf{v}}$.

Consider a small sphere of mass m released from rest in a liquid as in Figure 6.12a. Assuming the only forces acting on the sphere are the resistive force $\vec{\mathbf{R}} = -b\vec{\mathbf{v}}$ and the gravitational force $\vec{\mathbf{F}}_g$, let us describe its motion.[1] We model the sphere as a particle under a net force. Applying Newton's second law to the vertical motion of the sphere and choosing the downward direction to be positive, we obtain

$$\sum F_y = ma \quad \rightarrow \quad mg - bv = ma \tag{6.3}$$

where the acceleration of the sphere is downward. Solving Equation 6.3 for a and noting that a is equal to dv/dt gives

$$\frac{dv}{dt} = g - \frac{b}{m}v \tag{6.4}$$

This equation is called a *differential equation*; it contains both v and the derivative of v. The methods of solving such an equation may not be familiar to you as yet. Notice, however, that initially when $v = 0$, the magnitude of the resistive force is

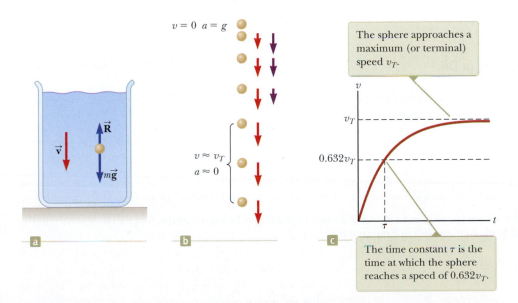

$v = 0 \ a = g$

$v \approx v_T$
$a \approx 0$

The sphere approaches a maximum (or terminal) speed v_T.

v_T

$0.632v_T$

τ

The time constant τ is the time at which the sphere reaches a speed of $0.632v_T$.

Figure 6.12 (a) A small sphere falling through a liquid. (b) A motion diagram of the sphere as it falls. Velocity vectors (red) and acceleration vectors (purple) are shown for each image after the first one. (c) A speed–time graph for the sphere.

[1] A *buoyant force* is also acting on the submerged object. This force is constant, and its magnitude is equal to the weight of the displaced liquid. This force can be modeled by changing the apparent weight of the sphere by a constant factor, so we will ignore the force here. We will discuss buoyant forces in Chapter 14.

also zero and the acceleration of the sphere is simply g. As t increases, the magnitude of the resistive force increases and the acceleration decreases. The acceleration approaches zero when the magnitude of the resistive force approaches the sphere's weight so that the net force on the sphere is zero. In this situation, the speed of the sphere approaches its **terminal speed** v_T.

◀ Terminal speed

The terminal speed is obtained from Equation 6.4 by setting $dv/dt = 0$, which gives

$$0 = g - \frac{b}{m}v_T \quad \text{or} \quad v_T = \frac{mg}{b} \tag{6.5}$$

Because you may not be familiar with differential equations yet, we won't show the explicit details of the process that gives the expression for v for all times t. If $v = 0$ at $t = 0$, this expression is

$$v = \frac{mg}{b}(1 - e^{-bt/m}) = v_T(1 - e^{-t/\tau}) \tag{6.6}$$

This function is plotted in Figure 6.12c. The symbol e represents the base of the natural logarithm and is also called *Euler's number:* $e = 2.718\,28$. The **time constant** $\tau = m/b$ (Greek letter tau) is the time at which the sphere released from rest at $t = 0$ reaches 63.2% of its terminal speed; when $t = \tau$, Equation 6.6 yields $v = 0.632v_T$. (The number 0.632 is $1 - e^{-1}$.)

We can check that Equation 6.6 is a solution to Equation 6.4 by direct differentiation:

$$\frac{dv}{dt} = \frac{d}{dt}\left[\frac{mg}{b}(1 - e^{-bt/m})\right] = \frac{mg}{b}\left(0 + \frac{b}{m}e^{-bt/m}\right) = ge^{-bt/m}$$

(See Appendix Table B.4 for the derivative of e raised to some power.) This is the left side of Equation 6.4. The right side is

$$g - \frac{b}{m}v = g - \frac{b}{m}\left[\frac{mg}{b}(1 - e^{-bt/m})\right]$$

$$= ge^{-bt/m}$$

Because the results for both sides of Equation 6.4 are the same, Equation 6.6 represents a solution to Equation 6.4.

Example 6.8 **Sphere Falling in Oil**

A small sphere of mass 2.00 g is released from rest in a large vessel filled with oil, where it experiences a resistive force proportional to its speed. The sphere reaches a terminal speed of 5.00 cm/s. Determine the time constant τ and the time at which the sphere reaches 90.0% of its terminal speed.

SOLUTION

Conceptualize With the help of Figure 6.12, imagine dropping the sphere into the oil and watching it sink to the bottom of the vessel. If you have some thick shampoo in a clear container, drop a marble in it and observe the motion of the marble.

Categorize We model the sphere as a *particle under a net force*, with one of the forces being a resistive force that depends on the speed of the sphere. This model leads to the result in Equation 6.5.

Analyze From Equation 6.5, evaluate the coefficient b:

$$b = \frac{mg}{v_T}$$

Evaluate the time constant τ:

$$\tau = \frac{m}{b} = m\left(\frac{v_T}{mg}\right) = \frac{v_T}{g}$$

Substitute numerical values:

$$\tau = \frac{5.00 \text{ cm/s}}{980 \text{ cm/s}^2} = 5.10 \times 10^{-3} \text{ s}$$

6.8 continued

Find the time t at which the sphere reaches a speed of $0.900v_T$ by setting $v = 0.900v_T$ in Equation 6.6 and solving for t:

$$0.900v_T = v_T(1 - e^{-t/\tau})$$

$$1 - e^{-t/\tau} = 0.900$$

$$e^{-t/\tau} = 0.100$$

$$-\frac{t}{\tau} = \ln(0.100) = -2.30$$

$$t = 2.30\tau = 2.30(5.10 \times 10^{-3} \text{ s}) = 11.7 \times 10^{-3} \text{ s}$$

$$= \boxed{11.7 \text{ ms}}$$

Finalize The sphere reaches 90.0% of its terminal speed in a very short time interval. You should have also seen this behavior if you performed the activity with the marble and the shampoo. Because of the short time interval required to reach terminal velocity, you may not have noticed the time interval at all. The marble may have appeared to immediately begin moving through the shampoo at a constant velocity.

Model 2: Resistive Force Proportional to Object Speed Squared

For objects moving at high speeds through air, such as airplanes, skydivers, cars, and baseballs, the resistive force is reasonably well modeled as proportional to the square of the speed. In these situations, the magnitude of the resistive force can be expressed as

$$R = \tfrac{1}{2}D\rho Av^2 \tag{6.7}$$

where D is a dimensionless empirical quantity called the *drag coefficient*, ρ is the density of air, and A is the cross-sectional area of the moving object measured in a plane perpendicular to its velocity. The drag coefficient has a value of about 0.5 for spherical objects but can have a value as great as 2 for irregularly shaped objects.

Let us analyze the motion of a falling object subject to an upward air resistive force of magnitude $R = \tfrac{1}{2}D\rho Av^2$. Suppose an object of mass m is released from rest. As Figure 6.13 shows, the object experiences two external forces:[2] the downward gravitational force $\vec{F}_g = m\vec{g}$ and the upward resistive force $\vec{R}$. Hence, the magnitude of the net force is

$$\sum F = mg - \tfrac{1}{2}D\rho Av^2 \tag{6.8}$$

where we have taken downward to be the positive vertical direction. Modeling the object as a particle under a net force, with the net force given by Equation 6.8, we find that the object has a downward acceleration of magnitude

$$a = g - \left(\frac{D\rho A}{2m}\right)v^2 \tag{6.9}$$

We can calculate the terminal speed v_T by noticing that when the gravitational force is balanced by the resistive force, the net force on the object is zero and therefore its acceleration is zero. Setting $a = 0$ in Equation 6.9 gives

$$0 = g - \left(\frac{D\rho A}{2m}\right)v_T^2$$

so, solving for v_T,

$$v_T = \sqrt{\frac{2mg}{D\rho A}} \tag{6.10}$$

Table 6.1 (page 142) lists the terminal speeds for several objects falling through air.

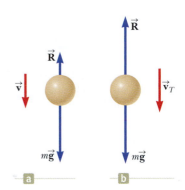

Figure 6.13 (a) An object falling through air experiences a resistive force $\vec{R}$ and a gravitational force $\vec{F}_g = m\vec{g}$. (b) The object reaches terminal speed when the net force acting on it is zero, that is, when $\vec{R} = -\vec{F}_g$ or $R = mg$.

---

[2]As with Model 1, there is also an upward buoyant force that we neglect.

TABLE 6.1 Terminal Speed for Various Objects Falling Through Air

Object	Mass (kg)	Cross-Sectional Area (m²)	v_T (m/s)
Skydiver	75	0.70	60
Baseball (radius 3.7 cm)	0.145	4.2×10^{-3}	43
Golf ball (radius 2.1 cm)	0.046	1.4×10^{-3}	44
Hailstone (radius 0.50 cm)	4.8×10^{-4}	7.9×10^{-5}	14
Raindrop (radius 0.20 cm)	3.4×10^{-5}	1.3×10^{-5}	9.0

QUICK QUIZ 6.4 A basketball and a 2-inch-diameter steel ball, having the same mass, are dropped through air from rest such that their bottoms are initially at the same height above the ground, on the order of 1 m or more. Which one strikes the ground first? **(a)** The steel ball strikes the ground first. **(b)** The basketball strikes the ground first. **(c)** Both strike the ground at the same time.

Conceptual Example **6.9** **The Skysurfer**

Consider a skysurfer (Fig. 6.14) who jumps from a plane with his feet attached firmly to his surfboard, does some tricks, and then opens his parachute. Describe the forces acting on him during these maneuvers.

SOLUTION

When the surfer first steps out of the plane, he has no vertical velocity. The downward gravitational force causes him and the board to accelerate toward the ground. As their downward speed increases, so does the upward resistive force exerted by the air on the surfer and the board. This upward force reduces their acceleration, and so their speed increases more slowly. Eventually, they are going so fast that the upward resistive force matches the downward gravitational force. Now the net force is zero and they no longer accelerate, but instead reach their terminal speed. At some point after reaching terminal speed, he opens his parachute, resulting in a drastic increase in the upward resistive force. The net force (and therefore the acceleration) is now upward, in the direction opposite the direction of the velocity. The downward velocity therefore decreases rapidly, and the resistive force on the parachute also decreases. Eventually, the upward resistive force and the downward gravitational force balance each other again and a much smaller terminal speed is reached, permitting a safe landing.

Figure 6.14 (Conceptual Example 6.9) A skysurfer.

(Contrary to popular belief, the velocity vector of a skydiver never points upward. You may have seen a video in which a skydiver appears to "rocket" upward once the parachute opens. In fact, what happens is that the skydiver slows down but the person holding the camera continues falling at high speed.)

Oliver Furrer/Jupiter Images

Example **6.10** **Resistive Force Exerted on a Baseball**

A pitcher hurls a 0.145-kg baseball past a batter at 40.2 m/s (= 90 mi/h). Find the resistive force acting on the ball at this speed.

SOLUTION

Conceptualize This example is different from the previous ones in that the object is now moving horizontally through the air instead of moving vertically under the influence of gravity and the resistive force. The resistive force causes the ball to slow down, and gravity causes its trajectory to curve downward. We simplify the situation by assuming the velocity vector is exactly horizontal at the instant it is traveling at 40.2 m/s.

Categorize In general, the ball is a *particle under a net force*. Because we are considering only one instant of time, however, we are not concerned about acceleration, so the problem involves only finding the value of one of the forces.

6.10 continued

Analyze To determine the drag coefficient D, imagine that we drop the baseball and allow it to reach terminal speed. Solve Equation 6.10 for D:

$$D = \frac{2mg}{v_T^2 \rho A}$$

Use this expression for D in Equation 6.7 to find an expression for the magnitude of the resistive force:

$$R = \tfrac{1}{2} D \rho A v^2 = \frac{1}{2}\left(\frac{2mg}{v_T^2 \rho A}\right)\rho A v^2 = mg\left(\frac{v}{v_T}\right)^2$$

Substitute numerical values, using the terminal speed from Table 6.1:

$$R = (0.145 \text{ kg})(9.80 \text{ m/s}^2)\left(\frac{40.2 \text{ m/s}}{43 \text{ m/s}}\right)^2 = 1.2 \text{ N}$$

Finalize The magnitude of the resistive force is similar in magnitude to the weight of the baseball, which is about 1.4 N. Therefore, air resistance plays a major role in the motion of the ball, as evidenced by the variety of curve balls, floaters, sinkers, and the like thrown by baseball pitchers.

Summary

> ## Concepts and Principles

A particle moving in uniform circular motion has a centripetal acceleration; this acceleration must be provided by a net force directed toward the center of the circular path.

An observer in a noninertial (accelerating) frame of reference introduces **fictitious forces** when applying Newton's second law in that frame.

An object moving through a liquid or gas experiences a speed-dependent **resistive force**. This resistive force is in a direction opposite that of the velocity of the object relative to the medium and generally increases with speed. The magnitude of the resistive force depends on the object's size and shape and on the properties of the medium through which the object is moving. In the limiting case for a falling object, when the magnitude of the resistive force equals the object's weight, the object reaches its **terminal speed.**

> ## Analysis Model for Problem Solving

Particle in Uniform Circular Motion (Extension) With our new knowledge of forces, we can extend the model of a particle in uniform circular motion, first introduced in Chapter 4. Newton's second law applied to a particle moving in uniform circular motion states that the net force causing the particle to undergo a centripetal acceleration (Eq. 4.21) is related to the acceleration according to

$$\sum F = ma_c = m\frac{v^2}{r} \tag{6.1}$$

Think–Pair–Share

See the Preface for an explanation of the icons used in this problems set. For additional assessment items for this section, go to **WEBASSIGN** From Cengage

1. You are working as a delivery person for a dairy store. In the back of your pickup truck is a crate of eggs. The dairy company has run out of bungee cords, so the crate is not tied down. You have been told to drive carefully because the coefficient of static friction between the crate and the bed of the truck is 0.600. You are not worried, because you are traveling on a road that appears perfectly straight. Due to your confidence and inattention, your speed has crept upward to 45.0 mi/h. Suddenly, you see a curve ahead with a warning sign saying, "Danger: unbanked curve with radius of curvature 35.0 m." You are 15.0 m from the beginning of the curve. What can you do to save the eggs: (i) take the curve at 45.0 mi/h, (ii) brake to a stop before entering the curve to think about it, or (iii) slow down to take the curve at a slower speed? Discuss these options in your group and determine if there is a best course of action.

2. **ACTIVITY** Find a YouTube video that shows the complete cycle for an amusement park ride called the "Roundup." In this ride, a rider stands against a wall at the edge of a disk

rotating around a vertical axis. When the disk reaches its operating speed, an arm raises the disk through an angle so that the disk rotates around an axis that is almost horizontal. As a result, the rider moves over the top of a vertical circle, seemingly unsupported, but does not fall downward. By using the height of a typical person on the ride, estimate the radius of the disk, using a stopped image of the disk at its highest angle. Begin the video again and use your smartphone stopwatch to measure the period of rotation of the disk. (a) From this information, calculate the centripetal acceleration of a rider at the top of the ride. (b) How does this acceleration compare to that due to gravity? (c) Why doesn't a rider at the top fall downward?

3. **ACTIVITY** Find a YouTube video that shows the complete cycle for an amusement park ride called the "Rotor." In this ride, a rider stands against a wall in a cylinder rotating around a vertical axis. When the cylinder reaches its operating speed, the floor drops away and riders remain suspended on the wall. By using the height of a typical person on the ride, estimate the radius of the cylinder, using a stopped image of the ride. Begin the video again and use your smartphone stopwatch to measure the period of rotation of the cylinder. From this information, determine the minimum coefficient of static friction necessary between the rider and the wall to keep the rider suspended.

Problems

See the Preface for an explanation of the icons used in this problems set. For additional assessment items for this section, go to ⚡ **WEBASSIGN**
From Cengage

SECTION 6.1 Extending the Particle in Uniform Circular Motion Model

1. In the Bohr model of the hydrogen atom, an electron moves in a circular path around a proton. The speed of the electron is approximately 2.20×10^6 m/s. Find (a) the force acting on the electron as it revolves in a circular orbit of radius 0.529×10^{-10} m and (b) the centripetal acceleration of the electron.

2. Whenever two *Apollo* astronauts were on the surface of the Moon, a third astronaut orbited the Moon. Assume the orbit to be circular and 100 km above the surface of the Moon, where the acceleration due to gravity is 1.52 m/s². The radius of the Moon is 1.70×10^6 m. Determine (a) the astronaut's orbital speed and (b) the period of the orbit.

3. A car initially traveling eastward turns north by traveling in a circular path at uniform speed as shown in Figure P6.3. The length of the arc *ABC* is 235 m, and the car completes the turn in 36.0 s. (a) What is the acceleration when the car is at *B* located at an angle of 35.0°? Express your answer in terms of the unit vectors $\hat{\mathbf{i}}$ and $\hat{\mathbf{j}}$. Determine (b) the car's average speed and (c) its average acceleration during the 36.0-s interval.

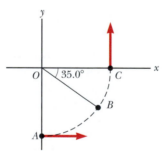

Figure P6.3

4. A curve in a road forms part of a horizontal circle. As a car goes around it at constant speed 14.0 m/s, the total horizontal force on the driver has magnitude 130 N. What is the total horizontal force on the driver if the speed on the same curve is 18.0 m/s instead?

5. In a cyclotron (one type of particle accelerator), a deuteron (of mass 2.00 u) reaches a final speed of 10.0% of the speed of light while moving in a circular path of radius 0.480 m. What magnitude of magnetic force is required to maintain the deuteron in a circular path?

6. *Why is the following situation impossible?* The object of mass $m = 4.00$ kg in Figure P6.6 is attached to a vertical rod by two strings of length $\ell = 2.00$ m. The strings are attached to the rod at points a distance $d = 3.00$ m apart. The object rotates in a horizontal circle at a constant speed of $v = 3.00$ m/s, and the strings remain taut. The rod rotates along with the object so that the strings do not wrap onto the rod. **What If?** Could this situation be possible on another planet?

Figure P6.6

7. You are working during your summer break as an amusement **CR** park ride operator. The ride you are controlling consists of a large vertical cylinder that spins about its axis fast enough that any person inside is held up against the wall when the floor drops away (Fig. P6.7). The coefficient of static friction between a person of mass m and the wall is μ_s, and the radius of the cylinder is R. You are rotating the ride with an angular speed ω suggested by your supervisor. (a) Suppose a very heavy person enters the ride. Do you need to increase the angular speed so that this person will not slide down the wall? (b) Suppose someone enters the ride wearing a very slippery satin workout outfit. In this case, do you need to increase the angular speed so that this person will not slide down the wall?

Figure P6.7

8. A driver is suing the state highway department after an **CR** accident on a curved freeway. The driver lost control and crashed into a tree located a short distance from the outside edge of the curved roadway. The driver is claiming that the radius of curvature of the unbanked roadway was too small for the speed limit, causing him to slide outward on the curve and hit the tree. You have been hired as an expert witness for the defense, and have been requested to use your knowledge of physics to testify that the radius of curvature of the roadway is appropriate for the speed limit. State regulations show that the radius of curvature of an unbanked roadway on which the speed limit is 65 mi/h must be at least 150 m. You build an accelerometer, which is a plumb bob

with a protractor that you attach to the roof of your car. An associate riding in your car with you observes that the plumb bob hangs at an angle of 15.0° from the vertical when the car is driven at a safer speed of 23.0 m/s on the curve in question. What is your testimony regarding the radius of the curve?

SECTION 6.2 Nonuniform Circular Motion

9. A hawk flies in a horizontal arc of radius 12.0 m at constant speed 4.00 m/s. (a) Find its centripetal acceleration. (b) It continues to fly along the same horizontal arc, but increases its speed at the rate of 1.20 m/s². Find the acceleration (magnitude and direction) in this situation at the moment the hawk's speed is 4.00 m/s.

10. **[T]** A 40.0-kg child swings in a swing supported by two chains, each 3.00 m long. The tension in each chain at the lowest point is 350 N. Find (a) the child's speed at the lowest point and (b) the force exerted by the seat on the child at the lowest point. (Ignore the mass of the seat.)

11. **[S]** A child of mass m swings in a swing supported by two chains, each of length R. If the tension in each chain at the lowest point is T, find (a) the child's speed at the lowest point and (b) the force exerted by the seat on the child at the lowest point. (Ignore the mass of the seat.)

12. **[QC]** One end of a cord is fixed and a small 0.500-kg object is attached to the other end, where it swings in a section of a vertical circle of radius 2.00 m as shown in Figure P6.12. When $\theta = 20.0°$, the speed of the object is 8.00 m/s. At this instant, find (a) the tension in the string, (b) the tangential and radial components of acceleration, and (c) the total acceleration. (d) Is your answer changed if the object is swinging down toward its lowest point instead of swinging up? (e) Explain your answer to part (d).

Figure P6.12

13. **[QC]** A roller coaster at the Six Flags Great America amusement park in Gurnee, Illinois, incorporates some clever design technology and some basic physics. Each vertical loop, instead of being circular, is shaped like a teardrop (Fig. P6.13). The cars ride on the inside of the loop at the top, and the speeds are fast enough to ensure the cars remain on the track. The biggest loop is 40.0 m high. Suppose the speed at the top of the loop is 13.0 m/s and the corresponding centripetal acceleration of the riders is 2g. (a) What is the radius of the arc of the teardrop at the top? (b) If the total mass of a car plus the riders is M, what force does the rail exert on the car at the top? (c) Suppose the roller coaster had a circular loop of radius 20.0 m. If the cars have the same speed, 13.0 m/s at the top, what is the centripetal acceleration of the riders at the top? (d) Comment on the normal force at the top in the situation described in part (c) and on the advantages of having teardrop-shaped loops.

Figure P6.13

Frank Cezus/Photographer's Choice/Getty Images

SECTION 6.3 Motion in Accelerated Frames

14. An object of mass $m = 5.00$ kg, attached to a spring scale, rests on a frictionless, horizontal surface as shown in Figure P6.14. The spring scale, attached to the front end of a boxcar, reads zero when the car is at rest. (a) Determine the acceleration of the car if the spring scale has a constant reading of 18.0 N when the car is in motion. (b) What constant reading will the spring scale show if the car moves with constant velocity? Describe the forces on the object as observed (c) by someone in the car and (d) by someone at rest outside the car.

Figure P6.14

15. **[T]** A person stands on a scale in an elevator. As the elevator starts, the scale has a constant reading of 591 N. As the elevator later stops, the scale reading is 391 N. Assuming the magnitude of the acceleration is the same during starting and stopping, determine (a) the weight of the person, (b) the person's mass, and (c) the acceleration of the elevator.

16. **[S]** **Review.** A student, along with her backpack on the floor next to her, is in an elevator that is accelerating upward with acceleration a. The student gives her backpack a quick kick at $t = 0$, imparting to it speed v and causing it to slide across the elevator floor. At time t, the backpack hits the opposite wall a distance L away from the student. Find the coefficient of kinetic friction μ_k between the backpack and the elevator floor.

17. A small container of water is placed on a turntable inside a microwave oven, at a radius of 12.0 cm from the center. The turntable rotates steadily, turning one revolution in each 7.25 s. What angle does the water surface make with the horizontal?

SECTION 6.4 Motion in the Presence of Resistive Forces

18. The mass of a sports car is 1 200 kg. The shape of the body is such that the aerodynamic drag coefficient is 0.250 and the frontal area is 2.20 m². Ignoring all other sources of friction, calculate the initial acceleration the car has if it has been traveling at 100 km/h and is now shifted into neutral and allowed to coast.

19. **[AMT]** **Review.** A window washer pulls a rubber squeegee down a very tall vertical window. The squeegee has mass 160 g and is mounted on the end of a light rod. The coefficient of kinetic friction between the squeegee and the dry glass is 0.900. The window washer presses it against the window with a force having a horizontal component of 4.00 N. (a) If she pulls the squeegee down the window at constant velocity, what vertical force component must she exert? (b) The window washer increases the downward force component by 25.0%, while all other forces remain the same. Find the squeegee's acceleration in this situation. (c) The squeegee is moved into a wet portion of the window, where its motion is resisted by a fluid drag force R proportional to its velocity according to $R = -20.0v$, where R is in newtons and v is in meters per second. Find the terminal velocity that the squeegee approaches, assuming the window washer exerts the same force described in part (b).

20. A small piece of Styrofoam packing material is dropped from a height of 2.00 m above the ground. Until it reaches terminal speed, the magnitude of its acceleration is given by $a = g - Bv$. After falling 0.500 m, the Styrofoam effectively reaches terminal speed and then takes 5.00 s more to reach the ground. (a) What is the value of the constant B? (b) What is the acceleration at $t = 0$? (c) What is the acceleration when the speed is 0.150 m/s?

21. A small, spherical bead of mass 3.00 g is released from rest at $t = 0$ from a point under the surface of a viscous liquid. The terminal speed is observed to be $v_T = 2.00$ cm/s. Find (a) the value of the constant b that appears in Equation 6.2, (b) the time t at which the bead reaches $0.632v_T$, and (c) the value of the resistive force when the bead reaches terminal speed.

22. Assume the resistive force acting on a speed skater is proportional to the square of the skater's speed v and is given by $f = -kmv^2$, where k is a constant and m is the skater's mass. The skater crosses the finish line of a straight-line race with speed v_i and then slows down by coasting on his skates. Show that the skater's speed at any time t after crossing the finish line is $v(t) = v_i/(1 + ktv_i)$.

23. You can feel a force of air drag on your hand if you stretch your arm out of the open window of a speeding car. *Note:* Do not endanger yourself. What is the order of magnitude of this force? In your solution, state the quantities you measure or estimate and their values.

ADDITIONAL PROBLEMS

24. A car travels clockwise at constant speed around a circular section of a horizontal road as shown in the aerial view of Figure P6.24. Find the directions of its velocity and acceleration at (a) position Ⓐ and (b) position Ⓑ.

Figure P6.24

25. A string under a tension of 50.0 N is used to whirl a rock in a horizontal circle of radius 2.50 m at a speed of 20.4 m/s on a frictionless surface as shown in Figure P6.25. As the string is pulled in, the speed of the rock increases. When the string on the table is 1.00 m long and the speed of the rock is 51.0 m/s, the string breaks. What is the breaking strength, in newtons, of the string?

Figure P6.25

26. Disturbed by speeding cars outside his workplace, Nobel laureate Arthur Holly Compton designed a speed bump (called the "Holly hump") and had it installed. Suppose a 1 800-kg car passes over a hump in a roadway that follows the arc of a circle of radius 20.4 m as shown in Figure P6.26. (a) If the car travels at

Figure P6.26

Problems 26 and 27.

30.0 km/h, what force does the road exert on the car as the car passes the highest point of the hump? (b) **What If?** What is the maximum speed the car can have without losing contact with the road as it passes this highest point?

27. A car of mass m passes over a hump in a road that follows the arc of a circle of radius R as shown in Figure P6.26. (a) If the car travels at a speed v, what force does the road exert on the car as the car passes the highest point of the hump? (b) **What If?** What is the maximum speed the car can have without losing contact with the road as it passes this highest point?

28. A child's toy consists of a small wedge that has an acute angle θ (Fig. P6.28). The sloping side of the wedge is frictionless, and an object of mass m on it remains at constant height if the wedge is spun at a certain constant speed. The wedge is spun by rotating, as an axis, a vertical rod that is firmly attached to the wedge at the bottom end. Show that, when the object sits at rest at a point at distance L up along the wedge, the speed of the object must be $v = (gL \sin \theta)^{1/2}$.

Figure P6.28

29. A seaplane of total mass m lands on a lake with initial speed $v_i\hat{\mathbf{i}}$. The only horizontal force on it is a resistive force on its pontoons from the water. The resistive force is proportional to the velocity of the seaplane: $\vec{\mathbf{R}} = -b\vec{\mathbf{v}}$. Newton's second law applied to the plane is $-bv\hat{\mathbf{i}} = m(dv/dt)\hat{\mathbf{i}}$. From the fundamental theorem of calculus, this differential equation implies that the speed changes according to

$$\int_{v_i}^{v} \frac{dv}{v} = -\frac{b}{m}\int_0^t dt$$

(a) Carry out the integration to determine the speed of the seaplane as a function of time. (b) Sketch a graph of the speed as a function of time. (c) Does the seaplane come to a complete stop after a finite interval of time? (d) Does the seaplane travel a finite distance in stopping?

30. An object of mass $m_1 = 4.00$ kg is tied to an object of mass $m_2 = 3.00$ kg with String 1 of length $\ell = 0.500$ m. The combination is swung in a vertical circular path on a second string, String 2, of length $\ell = 0.500$ m. During the motion, the two strings are collinear at all times as shown in Figure P6.30. At the top of its motion, m_2 is traveling at $v = 4.00$ m/s. (a) What is the tension in String 1 at this instant? (b) What is the tension in String 2 at this instant? (c) Which string will break first if the combination is rotated faster and faster?

Figure P6.30

31. A ball of mass $m = 0.275$ kg swings in a vertical circular path on a string $L = 0.850$ m long as in Figure P6.31. (a) What are the forces acting on the ball at any point on the path? (b) Draw force diagrams for the ball when it is at the bottom of the circle and when it is at the top. (c) If its speed is 5.20 m/s at the top of the circle, what is the tension in the string there? (d) If the string breaks when its tension exceeds 22.5 N, what is the maximum speed the ball can have at the bottom before that happens?

Figure P6.31

32. *Why is the following situation impossible?* A mischievous child goes to an amusement park with his family. On one ride, after a severe scolding from his mother, he slips out of his seat and climbs to the top of the ride's structure, which is shaped like a cone with its axis vertical and its sloped sides making an angle of $\theta = 20.0°$ with the horizontal as shown in Figure P6.32. This part of the structure rotates about the vertical central axis when the ride operates. The child sits on the sloped surface at a point $d = 5.32$ m down the sloped side from the center of the cone and pouts. The coefficient of static friction between the boy and the cone is 0.700. The ride operator does not notice that the child has slipped away from his seat and so continues to operate the ride. As a result, the sitting, pouting boy rotates in a circular path at a speed of 3.75 m/s.

Figure P6.32

33. The pilot of an airplane executes a loop-the-loop maneuver in a vertical circle. The speed of the airplane is 300 mi/h at the top of the loop and 450 mi/h at the bottom, and the radius of the circle is 1 200 ft. (a) What is the pilot's apparent weight at the lowest point if his true weight is 160 lb? (b) What is his apparent weight at the highest point? (c) **What If?** Describe how the pilot could experience weightlessness if both the radius and the speed can be varied. *Note:* His apparent weight is equal to the magnitude of the force exerted by the seat on his body.

34. A basin surrounding a drain has the shape of a circular cone opening upward, having everywhere an angle of 35.0° with the horizontal. A 25.0-g ice cube is set sliding around the cone without friction in a horizontal circle of radius R. (a) Find the speed the ice cube must have as a function of R. (b) Is any piece of data unnecessary for the solution? Suppose R is made two times larger. (c) Will the required speed increase, decrease, or stay constant? If it changes, by what factor? (d) Will the time interval required for each revolution increase, decrease, or stay constant? If it changes, by what factor? (e) Do the answers to parts (c) and (d) seem contradictory? Explain.

35. **Review.** While learning to drive, you are in a 1 200-kg car moving at 20.0 m/s across a large, vacant, level parking lot. Suddenly you realize you are heading straight toward the brick sidewall of a large supermarket and are in danger of running into it. The pavement can exert a maximum horizontal force of 7 000 N on the car. (a) Explain why you should expect the force to have a well-defined maximum value. (b) Suppose you apply the brakes and do not turn the steering wheel. Find the minimum distance you must be from the wall to avoid a collision. (c) If you do not brake but instead maintain constant speed and turn the steering wheel, what is the minimum distance you must be from the wall to avoid a collision? (d) Of the two methods in parts (b) and (c), which is better for avoiding a collision? Or should you use both the brakes and the steering wheel, or neither? Explain. (e) Does the conclusion in part (d) depend on the numerical values given in this problem, or is it true in general? Explain.

36. A truck is moving with constant acceleration a up a hill that makes an angle ϕ with the horizontal as in Figure P6.36. A small sphere of mass m is suspended from the ceiling of the truck by a light cord. If the pendulum makes a constant angle θ with the perpendicular to the ceiling, what is a?

Figure P6.36

37. Because the Earth rotates about its axis, a point on the equator experiences a centripetal acceleration of 0.033 7 m/s², whereas a point at the poles experiences no centripetal acceleration. If a person at the equator has a mass of 75.0 kg, calculate (a) the gravitational force (true weight) on the person and (b) the normal force (apparent weight) on the person. (c) Which force is greater? Assume the Earth is a uniform sphere and take $g = 9.800$ m/s².

38. A puck of mass m_1 is tied to a string and allowed to revolve in a circle of radius R on a frictionless, horizontal table. The other end of the string passes through a small hole in the center of the table, and an object of mass m_2 is tied to it (Fig. P6.38). The suspended object remains in equilibrium while the puck on the tabletop revolves. Find symbolic expressions for (a) the tension in the string, (b) the radial force acting on the puck, and (c) the speed of the puck. (d) Qualitatively describe what will happen in the motion of the puck if the value of m_2 is increased by placing a small additional load on the puck. (e) Qualitatively describe what will happen in the motion of the puck if the value of m_2 is instead decreased by removing a part of the hanging load.

Figure P6.38

39. Galileo thought about whether acceleration should be defined as the rate of change of velocity over time or as the rate of change in velocity over distance. He chose the former, so let's use the name "vroomosity" for the rate of change of velocity over distance. For motion of a particle on a straight line with constant acceleration, the equation

$v = v_i + at$ gives its velocity v as a function of time. Similarly, for a particle's linear motion with constant vroomosity k, the equation $v = v_i + kx$ gives the velocity as a function of the position x if the particle's speed is v_i at $x = 0$. (a) Find the law describing the total force acting on this object of mass m. Describe an example of such a motion or explain why it is unrealistic for (b) the possibility of k positive and (c) the possibility of k negative.

40. Members of a skydiving club were given the following data to use in planning their jumps. In the table, d is the distance fallen from rest by a skydiver in a "free-fall stable spread position" versus the time of fall t. (a) Convert the distances in feet into meters. (b) Graph d (in meters) versus t. (c) Determine the value of the terminal speed v_T by finding the slope of the straight portion of the curve. Use a least-squares fit to determine this slope.

t (s)	d (ft)	t (s)	d (ft)	t (s)	d (ft)
0	0	7	652	14	1 831
1	16	8	808	15	2 005
2	62	9	971	16	2 179
3	138	10	1 138	17	2 353
4	242	11	1 309	18	2 527
5	366	12	1 483	19	2 701
6	504	13	1 657	20	2 875

41. A car rounds a banked curve as discussed in Example [S] 6.4 and shown in Figure 6.5. The radius of curvature of the road is R, the banking angle is θ, and the coefficient of static friction is μ_s. (a) Determine the range of speeds the car can have without slipping up or down the road. (b) Find the minimum value for μ_s such that the minimum speed is zero.

42. In Example 6.5, we investigated the forces a child experi- [V] ences on a Ferris wheel. Assume the data in that example applies to this problem. What force (magnitude and direction) does the seat exert on a 40.0-kg child when the child is halfway between top and bottom?

43. **Review.** A piece of putty is initially located at point A on [GP] the rim of a grinding wheel rotating at constant angular [S] speed about a horizontal axis. The putty is dislodged from point A when the diameter through A is horizontal. It then rises vertically and returns to A at the instant the wheel completes one revolution. From this information, we wish to find the speed v of the putty when it leaves the wheel and the force holding it to the wheel. (a) What analysis model is appropriate for the motion of the putty as it rises and falls? (b) Use this model to find a symbolic expression for the time interval between when the putty leaves point A and when it arrives back at A, in terms of v and g. (c) What is the appropriate analysis model to describe point A on the wheel? (d) Find the period of the motion of point A in terms of the tangential speed v and the radius R of the wheel. (e) Set the time interval from part (b) equal to the period from part (d) and solve for the speed v of the putty as it leaves the wheel. (f) If the mass of the putty is m, what is the magnitude of the force that held it to the wheel before it was released?

44. A model airplane of mass 0.750 kg flies with a speed of [T] 35.0 m/s in a horizontal circle at the end of a 60.0-m-long control wire as shown in Figure P6.44a. The forces exerted

on the airplane are shown in Figure P6.44b: the tension in the control wire, the gravitational force, and aerodynamic lift that acts at $\theta = 20.0°$ inward from the vertical. Compute the tension in the wire, assuming it makes a constant angle of $\theta = 20.0°$ with the horizontal.

Figure P6.44

CHALLENGE PROBLEMS

45. A 9.00-kg object starting from rest falls through a viscous medium and experiences a resistive force given by Equation 6.2. The object reaches one half its terminal speed in 5.54 s. (a) Determine the terminal speed. (b) At what time is the speed of the object three-fourths the terminal speed? (c) How far has the object traveled in the first 5.54 s of motion?

46. For $t < 0$, an object of mass m experiences no force and [S] moves in the positive x direction with a constant speed v_i. Beginning at $t = 0$, when the object passes position $x = 0$, it experiences a net resistive force proportional to the square of its speed: $\vec{\mathbf{F}}_{net} = -mkv^2 \,\hat{\mathbf{i}}$, where k is a constant. The speed of the object after $t = 0$ is given by $v = v_i/(1 + kv_i t)$. (a) Find the position x of the object as a function of time. (b) Find the object's velocity as a function of position.

47. A golfer tees off from a location precisely at $\phi_i = 35.0°$ north latitude. He hits the ball due south, with range 285 m. The ball's initial velocity is at $48.0°$ above the horizontal. Suppose air resistance is negligible for the golf ball. (a) For how long is the ball in flight? The cup is due south of the golfer's location, and the golfer would have a hole-in-one if the Earth were not rotating. The Earth's rotation makes the tee move in a circle of radius $R_E \cos \phi_i = (6.37 \times 10^6 \text{ m}) \cos 35.0°$ as shown in Figure P6.47. The tee completes one revolution each day. (b) Find the eastward speed of the tee relative to the stars. The hole is also moving east, but it is 285 m farther south and thus at a slightly lower latitude ϕ_f. Because the hole moves in a slightly larger circle, its speed must be greater than that of the tee. (c) By how much does the hole's speed exceed that of the tee? During the time interval the ball is in flight, it moves upward and downward as well as southward with the projectile motion you studied in Chapter 4, but it also moves eastward with the speed you found in part (b). The hole moves to the east at a faster speed, however, pulling ahead of the ball with the relative speed you found in

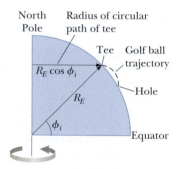

Figure P6.47

part (c). (d) How far to the west of the hole does the ball land?

48. A single bead can slide with negligible friction on a stiff wire that has been bent into a circular loop of radius 15.0 cm as shown in Figure P6.48. The circle is always in a vertical plane and rotates steadily about its vertical diameter with a period of 0.450 s. The position of the bead is described by the angle θ that the radial line, from the center of the loop to the bead, makes with the vertical. (a) At what angle up from the bottom of the circle can the bead stay motionless relative to the turning circle? (b) **What If?** Repeat the problem, this time taking the period of the circle's rotation as 0.850 s. (c) Describe how the solution to part (b) is different from the solution to part (a). (d) For any period or loop size, is there always an angle at which the bead can stand still relative to the loop? (e) Are there ever more than two angles? Arnold Arons suggested the idea for this problem.

Figure P6.48

49. Because of the Earth's rotation, a plumb bob does not hang exactly along a line directed to the center of the Earth. How much does the plumb bob deviate from a radial line at 35.0° north latitude? Assume the Earth is spherical.

50. You have a great job working at a major league baseball stadium for the summer! At this stadium, the speed of every pitch is measured using a radar gun aimed at the pitcher by an operator behind home plate. The operator has so much experience with this job that he has perfected a technique by which he can make each measurement at the exact instant at which the ball leaves the pitcher's hand. Your supervisor asks you to construct an algorithm that will provide the speed of the ball as it crosses home plate, 18.3 m from the pitcher, based on the measured speed v_i of the ball as it leaves the pitcher's hand. The speed at home plate will be lower due to the resistive force of the air on the baseball. The vertical motion of the ball is small, so, to a good approximation, we can consider only the horizontal motion of the ball. You begin to develop your algorithm by applying the particle under a net force to the baseball in the horizontal direction. A pitch is measured to have a speed of 40.2 m/s as it leaves the pitcher's hand. You need to tell your supervisor how fast it was traveling as it crossed home plate. (*Hint:* Use the chain rule to express acceleration in terms of a derivative with respect to x, and then solve a differential equation for v to find an expression for the speed of the baseball as a function of its position. The function will involve an exponential. Also make use of Table 6.1.)

7 Energy of a System

You use sandpaper to smooth the surface of a piece of wood. The sandpaper and wood both become warmer. How do we incorporate *warmth* into our growing list of physics concepts? (*DJTaylor/Shutterstock.com*)

STORYLINE **Your observations as an inquisitive physics student** have worn you out and you decide to spend a quiet day at home. You go into your garage to do further work on a carpentry project you started a while ago. You are thinking about how much you have learned about mechanics in your investigations as you find your sandpaper and a piece of wood that needs to be smoothed. You begin to sand the wood, still thinking that your studies of mechanics make a very complete description of nature and the Universe. Then you notice the sandpaper and the wood, along with your fingers, becoming *warmer* as you sand. "Wait. This is new!" you think. You are applying forces to the sandpaper, and it accelerates, therefore changing its velocity. There is friction between the sandpaper and the wood. This is all *mechanics*; you have thought about all of these concepts and have learned about them in previous chapters. But *warmth?* What's that all about? Maybe you have more thinking to do!

CONNECTIONS In this chapter, we are going to investigate a quantity that is very different from those studied in the previous chapters. Chapters 2 through 6 dealt with *change*. Velocity is a *change* in position, and acceleration is a *change* in velocity (Chapters 2 and 4). Force is the cause of *changes* in motion (Chapter 5). In this chapter and the next, we will study a quantity, energy, that is *conserved*. That is, the total energy in an isolated system *does not change* during any process that occurs in the system. Or if the total energy in a system does change, for example, if it increases, we find that the energy of the surroundings of the system decreases by the same amount! Therefore, the energy of the entire Universe is fixed; it has the same value at all times! Our analysis models presented in earlier chapters were based on the motion of a *particle*, or an object that could be modeled as a particle. We begin our new approach by focusing our attention on a new simplification model, a *system*, and analysis models based on the model of a system. These analysis models will be formally introduced in Chapter 8. In this chapter, we introduce systems and three ways to store energy in a system. We begin by making a connection between a familiar concept, force, and our new topic, *energy*. We will identify several forms in which energy can exist in a system. Even though this new quantity has a different nature from our previously studied quantities, it is very important and allows us to solve an entirely new class of problems. Furthermore, you might be happy to find out that energy is a scalar, so we don't have to perform complicated vector calculations! As we continue to study physics in the rest of the chapters in this book, we will see very often that we can take a force approach to a new area of study and we can also take an energy approach. The two approaches are complementary.

7.1 Systems and Environments

In the system model, we focus our attention on a small portion of the Universe—the **system**—and ignore details of the rest of the Universe outside of the system. A critical skill in applying the system model to problems is *identifying the system*.

A valid system

- may be a single object or particle
- may be a collection of objects or particles
- may be a region of space (such as the interior of an automobile engine combustion cylinder)
- may vary with time in size and shape (such as a rubber ball, which deforms upon striking a wall)

> **PITFALL PREVENTION 7.1**
> **Identify the System** The most important *first* step to take in solving a problem using the energy approach is to identify the appropriate system of interest.

Identifying the need for a system approach to solving a problem (as opposed to a particle approach) is part of the Categorize step in the Analysis Model Approach to Problem Solving outlined in Chapter 2. Identifying the particular system is a second part of this step.

No matter what the particular system is in a given problem, we identify a **system boundary,** an imaginary surface (not necessarily coinciding with a physical surface) that divides the Universe into the system and the **environment** surrounding the system.

As an example, imagine a force applied to an object in empty space. We can define the object as the system and its outer surface as the system boundary. The force applied to it is an influence on the system from the environment that acts across the system boundary. We will see how to analyze this situation from a system approach in a subsequent section of this chapter.

Another example was seen in Example 5.10, where the system can be defined as the combination of the ball, the block, and the cord. The influence from the environment includes the gravitational forces on the ball and the block, the normal and friction forces on the block, and the force exerted by the pulley on the cord. The forces exerted by the cord on the ball and the block are internal to the system and therefore are not included as an influence from the environment.

There are a number of mechanisms by which a system can be influenced by its environment. The first one we shall investigate is *work*.

7.2 Work Done by a Constant Force

Almost all the terms we have used thus far—velocity, acceleration, force, and so on—convey a similar meaning in physics as they do in everyday life. Now, however, we encounter a term whose meaning in physics is distinctly different from its everyday meaning: work.

a   b   c

Charles D. Winters

Figure 7.1 An eraser being pushed along a chalkboard tray by a force acting at different angles with respect to the horizontal direction.

PITFALL PREVENTION 7.2

Work Is Done by . . . on . . . Not only must you identify the system, you must also identify what agent in the environment is doing work on the system. When discussing work, always use the phrase, "the work done by . . . on . . . ." After "by," insert the part of the environment that is interacting directly with the system. After "on," insert the system. For example, "the work done by the hammer on the nail" identifies the nail as the system, and the force from the hammer represents the influence from the environment.

To understand what work as an influence on a system means to the physicist, consider the situation illustrated in Figure 7.1. A force $\vec{F}$ is applied to a chalkboard eraser, which we identify as the system, and the eraser slides along the tray. If we want to know how effective the force is in moving the eraser, we must consider not only the magnitude of the force but also its direction. Notice that the finger in Figure 7.1 applies forces in three different directions on the eraser. Assuming the magnitude of the applied force is the same in all three photographs, the push applied in Figure 7.1b is more effective in moving the eraser than the push in Figure 7.1a. On the other hand, Figure 7.1c shows a situation in which the applied force does not move the eraser at all, regardless of how hard it is pushed (unless, of course, we apply a force so great that we break the chalkboard tray!). These results suggest that when analyzing forces to determine the influence they have on the system, we must consider the vector nature of forces. We must also consider the magnitude of the force. Moving a force with a magnitude of $|\vec{F}| = 2$ N through a displacement represents a greater influence on the system than moving a force of magnitude 1 N through the same displacement. The magnitude of the displacement is also important. Moving the eraser 3 m along the tray represents a greater influence than moving it 2 cm if the same force is used in both cases.

Let us examine the situation in Figure 7.2, where the object (the system) undergoes a displacement along a straight line while acted on by a constant force of magnitude F that makes an angle θ with the direction of the displacement. We formally define the work done by the force on the system as follows:

> The **work** W done on a system by an agent exerting a constant force on the system is the product of the magnitude F of the force, the magnitude Δr of the displacement of the point of application of the force, and $\cos\theta$, where θ is the angle between the force and displacement vectors:

◄ Work done by a constant force

$$W \equiv F\,\Delta r\cos\theta \qquad (7.1)$$

Notice in Equation 7.1 that work is a scalar, even though it is defined in terms of two vectors in Figure 7.2, a force $\vec{F}$ and a displacement $\Delta\vec{r}$. In Section 7.3, we explore how to combine two vectors to generate a scalar quantity.

Notice also that the displacement in Equation 7.1 is that of *the point of application of the force*. If the force is applied to a particle or a rigid object that can be modeled as a particle, this displacement is the same as that of the particle. For a deformable system, however, these displacements are not the same. For example, imagine pressing in on the sides of a balloon with both hands. The center of the balloon moves through zero displacement. The points of application of the forces from your hands on the sides of the balloon, however, do indeed move through a displacement as the balloon is compressed, and that is the displacement to be used in Equation 7.1. We will see other examples of deformable systems, such as springs and samples of gas contained in a vessel.

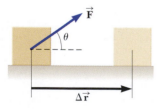

Figure 7.2 An object on a table undergoes a displacement $\Delta\vec{r}$ under the action of a constant force $\vec{F}$.

As an example of the distinction between the definition of work and our everyday understanding of the word, consider holding a heavy chair at arm's length for 3 min. At the end of this time interval, your tired arms may lead you to think you have done a considerable amount of work on the chair. According to our definition, however, you have done no work on it whatsoever. You exert a force to support the chair, but you do not move it. A force does no work on an object if the force does not move through a displacement. If $\Delta r = 0$, Equation 7.1 gives $W = 0$, which is the situation depicted in Figure 7.1c.

Also notice from Equation 7.1 that the work done by a force on a moving object is zero when the force applied is perpendicular to the displacement of its point of application. That is, if $\theta = 90°$, then $W = 0$ because $\cos 90° = 0$. For example, in Figure 7.3, the work done by the normal force on the object and the work done by the gravitational force on the object are both zero because both forces are perpendicular to the displacement and have zero components along an axis in the direction of $\Delta \vec{r}$.

The sign of the work also depends on the direction of $\vec{F}$ relative to $\Delta \vec{r}$. The work done by the applied force on a system is positive when the projection of $\vec{F}$ onto $\Delta \vec{r}$ is in the same direction as the displacement. For example, when an object is lifted, the work done by the applied force on the object is positive because the direction of that force is upward, in the same direction as the displacement of its point of application. When the projection of $\vec{F}$ onto $\Delta \vec{r}$ is in the direction opposite the displacement, W is negative. For example, as an object is lifted, the work done by the gravitational force on the object is negative. The factor $\cos \theta$ in the definition of W (Eq. 7.1) automatically takes care of the sign.

If an applied force $\vec{F}$ is in the same direction as the displacement $\Delta \vec{r}$, then $\theta = 0$ and $\cos 0 = 1$. In this case, Equation 7.1 gives

$$W = F\Delta r$$

The units of work are those of force multiplied by those of length. Therefore, the SI unit of work is the **newton · meter** (N · m = kg · m²/s²). This combination of units is used so frequently that it has been given a name of its own, the **joule** (J).

An important consideration for a system approach to problems is that **work is an energy transfer.** For now, **energy** sounds mysterious, because we have not studied it yet. It is difficult to define energy, other than to say that it is a physical quantity that is conserved. In that behavior, it is similar to *money*. When a financial transaction occurs in your checking account, money is transferred across the boundary of your account: for example, inward by deposits and outward by withdrawals. When a physical process occurs, energy is transferred across the boundary of a system. Our understanding of energy will improve as we investigate various examples in this chapter.

If W is the work done on a system and W is positive, energy is transferred *to* the system; if W is negative, energy is transferred *from* the system. Therefore, if a system interacts with its environment, this interaction can be described as a transfer of energy across the system boundary. The result is a change in the energy stored in the system. We will learn about the first type of energy storage in Section 7.5, after we investigate more aspects of work.

QUICK QUIZ 7.1 The gravitational force exerted by the Sun on the Earth holds the Earth in an orbit around the Sun. Let us assume that the orbit is perfectly circular. The work done by this gravitational force during a short time interval in which the Earth moves through a displacement in its orbital path is **(a)** zero **(b)** positive **(c)** negative **(d)** impossible to determine

QUICK QUIZ 7.2 Figure 7.4 shows four situations in which a force is applied to an object. In all four cases, the force has the same magnitude, and the displacement of the object is to the right and of the same magnitude. Rank the situations in order of the work done by the force on the object, from most positive to most negative.

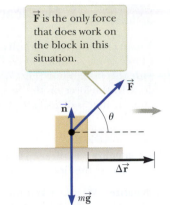

Figure 7.3 An object is displaced on a frictionless, horizontal surface. The normal force $\vec{n}$ and the gravitational force $m\vec{g}$ do no work on the object.

PITFALL PREVENTION 7.3

Cause of the Displacement We can calculate the work done by a force on an object, but that force is *not* necessarily the cause of the object's displacement. For example, if you lift an object, (negative) work is done on the object by the gravitational force, although gravity is not the cause of the object moving upward!

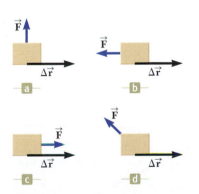

Figure 7.4 (Quick Quiz 7.2) A block is pulled by a force in four different directions. In each case, the displacement of the block is to the right and of the same magnitude.

Example 7.1 | Mr. Clean

A man cleaning a floor pulls a vacuum cleaner with a force of magnitude $F = 50.0$ N at an angle of 30.0° with the horizontal (Fig. 7.5). Calculate the work done by the force on the vacuum cleaner as the vacuum cleaner is displaced 3.00 m to the right.

Figure 7.5 (Example 7.1) A vacuum cleaner being pulled at an angle of 30.0° from the horizontal.

SOLUTION

Conceptualize Figure 7.5 helps conceptualize the situation. Think about an experience in your life in which you pulled an object across the floor with a rope or cord.

Categorize We are asked for the work done on an object by a force and are given the force on the object, the displacement of the object, and the angle between the two vectors, so we categorize this example as a substitution problem. We identify the vacuum cleaner as the system.

Use the definition of work (Eq. 7.1):

$$W = F\Delta r \cos\theta = (50.0 \text{ N})(3.00 \text{ m})(\cos 30.0°)$$
$$= 130 \text{ J}$$

Notice in this situation that the normal force $\vec{\mathbf{n}}$ and the gravitational $\vec{\mathbf{F}}_g = m\vec{\mathbf{g}}$ do no work on the vacuum cleaner because these forces are perpendicular to the displacements of their points of application. Furthermore, there was no mention of whether there was friction between the vacuum cleaner and the floor. The presence or absence of friction is not important when calculating the work done by the applied force. In addition, this work does not depend on whether the vacuum moved at constant velocity or if it accelerated.

7.3 The Scalar Product of Two Vectors

PITFALL PREVENTION 7.4
Work Is a Scalar Although Equation 7.3 defines the work in terms of two vectors, *work is a scalar;* there is no direction associated with it. *All* types of energy and energy transfer are scalars. This fact is a major advantage of the energy approach because we don't need vector calculations!

Because of the way the force and displacement vectors are combined in Equation 7.1, it is helpful to use a convenient mathematical tool called the **scalar product** of two vectors. We write this **scalar product** of vectors $\vec{\mathbf{A}}$ and $\vec{\mathbf{B}}$ as $\vec{\mathbf{A}} \cdot \vec{\mathbf{B}}$. (Because of the dot symbol, the scalar product is often called the **dot product.**)

The scalar product of any two vectors $\vec{\mathbf{A}}$ and $\vec{\mathbf{B}}$ is defined as a scalar quantity equal to the product of the magnitudes of the two vectors and the cosine of the angle θ between them:

$$\vec{\mathbf{A}} \cdot \vec{\mathbf{B}} \equiv AB \cos\theta \qquad (7.2)$$

◀ Scalar product of any two vectors $\vec{\mathbf{A}}$ and $\vec{\mathbf{B}}$

As is the case with any multiplication, $\vec{\mathbf{A}}$ and $\vec{\mathbf{B}}$ need not have the same units.

By comparing this definition with Equation 7.1, we can express Equation 7.1 as a scalar product:

$$W = F\Delta r \cos\theta = \vec{\mathbf{F}} \cdot \Delta\vec{\mathbf{r}} \qquad (7.3)$$

In other words, $\vec{\mathbf{F}} \cdot \Delta\vec{\mathbf{r}}$ is a shorthand notation for $F\Delta r \cos\theta$.

Before continuing with our discussion of work, let us investigate some properties of the dot product. Figure 7.6 shows two vectors $\vec{\mathbf{A}}$ and $\vec{\mathbf{B}}$ and the angle θ between them used in the definition of the dot product. In Figure 7.6, $B \cos\theta$ is the projection of $\vec{\mathbf{B}}$ onto $\vec{\mathbf{A}}$. Therefore, Equation 7.2 means that $\vec{\mathbf{A}} \cdot \vec{\mathbf{B}}$ is the product of the magnitude of $\vec{\mathbf{A}}$ and the projection of $\vec{\mathbf{B}}$ onto $\vec{\mathbf{A}}$.

From the right-hand side of Equation 7.2, we also see that the scalar product is **commutative.**[1] That is,

$$\vec{\mathbf{A}} \cdot \vec{\mathbf{B}} = \vec{\mathbf{B}} \cdot \vec{\mathbf{A}}$$

Figure 7.6 The scalar product $\vec{\mathbf{A}} \cdot \vec{\mathbf{B}}$ equals the magnitude of $\vec{\mathbf{A}}$ multiplied by $B \cos\theta$, which is the projection of $\vec{\mathbf{B}}$ onto $\vec{\mathbf{A}}$.

[1]The commutativity of the dot product means that $\vec{\mathbf{A}} \cdot \vec{\mathbf{B}}$ also equals the product of the magnitude of $\vec{\mathbf{B}}$ and the projection of $\vec{\mathbf{A}}$ onto $\vec{\mathbf{B}}$. In Chapter 11, you will see another way of combining vectors that proves useful in physics and is not commutative.

Finally, the scalar product obeys the distributive **law of multiplication,** so

$$\vec{A} \cdot (\vec{B} + \vec{C}) = \vec{A} \cdot \vec{B} + \vec{A} \cdot \vec{C}$$

The scalar product is simple to evaluate from Equation 7.2 when $\vec{A}$ is either perpendicular or parallel to $\vec{B}$. If $\vec{A}$ is perpendicular to $\vec{B}$ ($\theta = 90°$), then $\vec{A} \cdot \vec{B} = 0$. (The equality $\vec{A} \cdot \vec{B} = 0$ also holds in the more trivial case in which either $\vec{A}$ or $\vec{B}$ is zero.) If vector $\vec{A}$ is parallel to vector $\vec{B}$ and the two point in the same direction ($\theta = 0$), then $\vec{A} \cdot \vec{B} = AB$. If vector $\vec{A}$ is parallel to vector $\vec{B}$ but the two point in opposite directions ($\theta = 180°$), then $\vec{A} \cdot \vec{B} = -AB$. The scalar product is negative when $90° < \theta \le 180°$.

The unit vectors $\hat{i}$, $\hat{j}$, and $\hat{k}$, which were defined in Chapter 3, lie in the positive x, y, and z directions, respectively, of a right-handed coordinate system. Therefore, it follows from the definition of $\vec{A} \cdot \vec{B}$ that the scalar products of these unit vectors are

$$\hat{i} \cdot \hat{i} = \hat{j} \cdot \hat{j} = \hat{k} \cdot \hat{k} = 1 \tag{7.4}$$

$$\hat{i} \cdot \hat{j} = \hat{i} \cdot \hat{k} = \hat{j} \cdot \hat{k} = 0 \tag{7.5}$$

◀ Scalar products of unit vectors

Equations 3.17 and 3.18 state that two vectors $\vec{A}$ and $\vec{B}$ can be expressed in unit-vector form as

$$\vec{A} = A_x \hat{i} + A_y \hat{j} + A_z \hat{k}$$

$$\vec{B} = B_x \hat{i} + B_y \hat{j} + B_z \hat{k}$$

Using these expressions for the vectors and the information given in Equations 7.4 and 7.5 shows that the scalar product of $\vec{A}$ and $\vec{B}$ reduces to

$$\vec{A} \cdot \vec{B} = A_x B_x + A_y B_y + A_z B_z \tag{7.6}$$

(Details of the derivation are left for you in Problem 5 at the end of the chapter.) In the special case in which $\vec{A} = \vec{B}$, we see that

$$\vec{A} \cdot \vec{A} = A_x^2 + A_y^2 + A_z^2 = A^2$$

Ⓠ UICK QUIZ 7.3 Which of the following statements is true about the relationship between the dot product of two vectors and the product of the magnitudes of the vectors? **(a)** $\vec{A} \cdot \vec{B}$ is larger than AB. **(b)** $\vec{A} \cdot \vec{B}$ is smaller than AB. **(c)** $\vec{A} \cdot \vec{B}$ could be larger or smaller than AB, depending on the angle between the vectors. **(d)** $\vec{A} \cdot \vec{B}$ could be equal to AB.

Example 7.2 **The Scalar Product**

The vectors $\vec{A}$ and $\vec{B}$ are given by $\vec{A} = 2\hat{i} + 3\hat{j}$ and $\vec{B} = -\hat{i} + 2\hat{j}$.

(A) Determine the scalar product $\vec{A} \cdot \vec{B}$.

SOLUTION

Conceptualize   There is no physical system to imagine here. Rather, it is purely a mathematical exercise involving two vectors.

Categorize   Because we have a definition for the scalar product, we categorize this example as a substitution problem.

Substitute the specific vector expressions for $\vec{A}$ and $\vec{B}$:

$$\begin{aligned}
\vec{A} \cdot \vec{B} &= (2\hat{i} + 3\hat{j}) \cdot (-\hat{i} + 2\hat{j}) \\
&= -2\hat{i} \cdot \hat{i} + 2\hat{i} \cdot 2\hat{j} - 3\hat{j} \cdot \hat{i} + 3\hat{j} \cdot 2\hat{j} \\
&= -2(1) + 4(0) - 3(0) + 6(1) = -2 + 6 = \boxed{4}
\end{aligned}$$

The same result is obtained when we use Equation 7.6 directly, where $A_x = 2$, $A_y = 3$, $B_x = -1$, and $B_y = 2$.

continued

7.2 continued

(B) Find the angle θ between $\vec{\mathbf{A}}$ and $\vec{\mathbf{B}}$.

SOLUTION

Evaluate the magnitudes of $\vec{\mathbf{A}}$ and $\vec{\mathbf{B}}$ using the Pythagorean theorem:

$$A = \sqrt{A_x^2 + A_y^2} = \sqrt{(2)^2 + (3)^2} = \sqrt{13}$$

$$B = \sqrt{B_x^2 + B_y^2} = \sqrt{(-1)^2 + (2)^2} = \sqrt{5}$$

Use Equation 7.2 and the result from part (A) to find the angle:

$$\cos\theta = \frac{\vec{\mathbf{A}} \cdot \vec{\mathbf{B}}}{AB} = \frac{4}{\sqrt{13}\sqrt{5}} = \frac{4}{\sqrt{65}}$$

$$\theta = \cos^{-1}\frac{4}{\sqrt{65}} = \boxed{60.3°}$$

Example **7.3**   **Work Done by a Constant Force**

A particle moving in the xy plane undergoes a displacement given by $\Delta\vec{\mathbf{r}} = (2.0\hat{\mathbf{i}} + 3.0\hat{\mathbf{j}})$ m as a constant force $\vec{\mathbf{F}} = (5.0\hat{\mathbf{i}} + 2.0\hat{\mathbf{j}})$ N acts on the particle. Calculate the work done by $\vec{\mathbf{F}}$ on the particle.

SOLUTION

Conceptualize   Although this example is a little more physical than the previous one in that it identifies a force and a displacement, it is similar in terms of its mathematical structure.

Categorize   Because we are given force and displacement vectors and asked to find the work done by this force on the particle, we categorize this example as a substitution problem.

Substitute the expressions for $\vec{\mathbf{F}}$ and $\Delta\vec{\mathbf{r}}$ into Equation 7.3 and use Equations 7.4 and 7.5:

$$W = \vec{\mathbf{F}} \cdot \Delta\vec{\mathbf{r}} = [(5.0\hat{\mathbf{i}} + 2.0\hat{\mathbf{j}})\,\text{N}] \cdot [(2.0\hat{\mathbf{i}} + 3.0\hat{\mathbf{j}})\,\text{m}]$$

$$= (5.0\hat{\mathbf{i}} \cdot 2.0\hat{\mathbf{i}} + 5.0\hat{\mathbf{i}} \cdot 3.0\hat{\mathbf{j}} + 2.0\hat{\mathbf{j}} \cdot 2.0\hat{\mathbf{i}} + 2.0\hat{\mathbf{j}} \cdot 3.0\hat{\mathbf{j}})\,\text{N} \cdot \text{m}$$

$$= [10 + 0 + 0 + 6]\,\text{N} \cdot \text{m} = \boxed{16\,\text{J}}$$

7.4   Work Done by a Varying Force

Now consider a particle being displaced along the x axis under the action of a force that *varies* with position. In such a situation, we cannot use Equation 7.1 to calculate the work done by the force because this relationship applies only when $\vec{\mathbf{F}}$ is constant in magnitude and direction. The red-brown curve in Figure 7.7a shows a varying force applied on a particle that moves from initial position x_i to final position x_f. Imagine a particle undergoing a very small displacement Δx, shown in the figure. The x component F_x of the force is approximately constant over this small interval; for this small displacement, we can approximate the work done on the particle by the force using Equation 7.1 as

$$W \approx F_x \Delta x$$

which is the area of the shaded rectangle in Figure 7.7a. If the F_x versus x curve is divided into a large number of such intervals, the total work done for the displacement from x_i to x_f is approximately equal to the sum of a large number of such terms:

$$W \approx \sum_{x_i}^{x_f} F_x \Delta x$$

If the size of the small displacements is allowed to approach zero, the number of terms in the sum increases without limit but the value of the sum approaches a definite value equal to the area bounded by the F_x curve and the x axis, expressed as an integral:

$$\lim_{\Delta x \to 0} \sum_{x_i}^{x_f} F_x \Delta x = \int_{x_i}^{x_f} F_x dx$$

Therefore, we can express the work done by F_x on the system of the particle as it moves from x_i to x_f as

$$W = \int_{x_i}^{x_f} F_x \, dx \tag{7.7}$$

This equation reduces to Equation 7.1 when the component $F_x = F \cos \theta$ remains constant.

If more than one force acts on a system *and the system can be modeled as a particle,* the points of application of all forces move through the same displacement, and the total work done on the system is just the work done by the net force. If we express the net force in the x direction as $\sum F_x$, the total work, or *net work,* done as the particle moves from x_i to x_f is

$$\sum W = W_{\text{ext}} = \int_{x_i}^{x_f} \left(\sum F_x\right) dx \quad \text{(particle)}$$

For the general case of a net force $\sum \vec{\mathbf{F}}$ whose magnitude and direction may both vary, we use the scalar product,

$$\sum W = W_{\text{ext}} = \int \left(\sum \vec{\mathbf{F}}\right) \cdot d\vec{\mathbf{r}} \quad \text{(particle)} \tag{7.8}$$

where the integral is calculated over the path that the particle takes through space. The subscript "ext" on work reminds us that the net work is done by an *external* agent on the system. We will use this notation in this chapter as a reminder and to differentiate this work from an *internal* work to be described shortly.

If the system cannot be modeled as a particle (for example, if the system is deformable), we cannot use Equation 7.8 because different forces on the system may move through different displacements. In this case, we must evaluate the work done by each force separately and then add the works algebraically to find the net work done on the system:

$$\sum W = W_{\text{ext}} = \sum_{\text{forces}} \left(\int \vec{\mathbf{F}} \cdot d\vec{\mathbf{r}}\right) \quad \text{(deformable system)}$$

continued

The total work done for the displacement from x_i to x_f is approximately equal to the sum of the areas of all the rectangles.

a

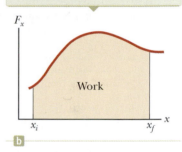

The work done by the component F_x of the varying force as the particle moves from x_i to x_f is *exactly* equal to the area under the curve.

b

Figure 7.7 (a) The work done on a particle by the force component F_x for the small displacement Δx is $F_x \Delta x$, which equals the area of the shaded rectangle. (b) The width Δx of each rectangle is shrunk to zero.

Example 7.4 Calculating Total Work Done from a Graph

A force acting on a particle varies with x as shown in Figure 7.8. Calculate the work done by the force on the particle as it moves from $x = 0$ to $x = 6.0$ m.

SOLUTION

Conceptualize Imagine a particle subject to the force in Figure 7.8. The force remains constant as the particle moves through the first 4.0 m and then decreases linearly to zero at 6.0 m. In terms of earlier discussions of motion, the particle could be modeled as a particle under constant acceleration for the first 4.0 m because the force is constant. Between 4.0 m and 6.0 m, however, the motion does not fit into one of our earlier analysis models because the acceleration of the particle is changing. If the particle starts from rest, its speed increases throughout the motion, and the particle is always moving in the positive x direction. These details about its speed and direction are not necessary for the calculation of the work done, however.

The net work done by this force is the area under the curve.

Figure 7.8 (Example 7.4) The force acting on a particle is constant for the first 4.0 m of motion and then decreases linearly with x from $x_{\text{Ⓑ}} = 4.0$ m to $x_{\text{Ⓒ}} = 6.0$ m.

7.4 continued

Categorize   Because the force varies during the motion of the particle, we must use the techniques for work done by varying forces. In this case, the graphical representation in Figure 7.8 can be used to evaluate the work done.

Analyze   The work done by the force is equal to the area under the curve from $x_{\circledA} = 0$ to $x_{\circledC} = 6.0$ m. This area is equal to the area of the rectangular section from $\circledA$ to $\circledB$ plus the area of the triangular section from $\circledB$ to $\circledC$.

Evaluate the area of the rectangle:

$$W_{\circledA \text{ to } \circledB} = (5.0\ \text{N})(4.0\ \text{m}) = 20\ \text{J}$$

Evaluate the area of the triangle:

$$W_{\circledB \text{ to } \circledC} = \tfrac{1}{2}(5.0\ \text{N})(2.0\ \text{m}) = 5.0\ \text{J}$$

Find the total work done by the force on the particle:

$$W_{\circledA \text{ to } \circledC} = W_{\circledA \text{ to } \circledB} + W_{\circledB \text{ to } \circledC} = 20\ \text{J} + 5.0\ \text{J} = \boxed{25\ \text{J}}$$

Finalize   Because the graph of the force consists of straight lines, we can use rules for finding the areas of simple geometric models to evaluate the total work done in this example. If a force does not vary linearly, as in Figure 7.7, such rules cannot be used and the force function must be integrated as in Equation 7.7 or 7.8.

Work Done by a Spring

A model of a common physical system on which the force varies with position is shown in Figure 7.9. The system is a block on a frictionless, horizontal surface and connected to a spring. For many springs, if the spring is either stretched or compressed a small distance from its unstretched (equilibrium) configuration, it exerts on the block a force component that can be mathematically modeled as

Spring force ▶

$$F_s = -kx \tag{7.9}$$

where x is the position of the block relative to its equilibrium ($x = 0$) position and k is a positive constant called the **force constant** or the **spring constant** of the spring. In other words, the force required to stretch or compress a spring is proportional

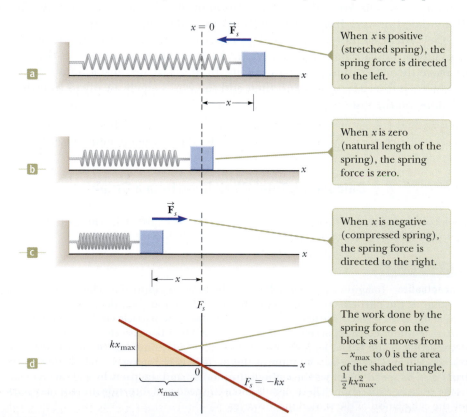

When x is positive (stretched spring), the spring force is directed to the left.

When x is zero (natural length of the spring), the spring force is zero.

When x is negative (compressed spring), the spring force is directed to the right.

The work done by the spring force on the block as it moves from $-x_{\text{max}}$ to 0 is the area of the shaded triangle, $\frac{1}{2}kx_{\text{max}}^2$.

Figure 7.9 The force exerted by a spring on a block varies with the block's position x relative to the equilibrium position $x = 0$. (a) x is positive. (b) x is zero. (c) x is negative. (d) Graph of F_s versus x for the block–spring system.

to the amount of stretch or compression x. This force law for springs is known as **Hooke's law.** The value of k is a measure of the *stiffness* of the spring. Stiff springs have large k values, and soft springs have small k values. As can be seen from Equation 7.9, the units of k are N/m.

The vector form of Equation 7.9 is

$$\vec{\mathbf{F}}_s = F_s \hat{\mathbf{i}} = -kx\hat{\mathbf{i}} \qquad (7.10)$$

where we have chosen the x axis to lie along the direction the spring extends or compresses.

The negative sign in Equations 7.9 and 7.10 signifies that the force exerted by the spring is always directed *opposite* the displacement from equilibrium. When $x > 0$ as in Figure 7.9a so that the block is to the right of the equilibrium position and the spring is stretched, the spring force is directed to the left, in the negative x direction. When $x < 0$ as in Figure 7.9c, the block is to the left of equilibrium, the spring is compressed, and the spring force is directed to the right, in the positive x direction. When $x = 0$ as in Figure 7.9b, the spring is unstretched and $F_s = 0$. Because the spring force always acts toward the equilibrium position ($x = 0$), it is sometimes called a *restoring force*.

If the spring is compressed until the block is at the point $-x_{max}$ and is then released, the block moves from $-x_{max}$ through zero to $+x_{max}$. It then reverses direction, returns to $-x_{max}$, and continues oscillating back and forth. We will study these oscillations in more detail in Chapter 15. For now, let's investigate the work done by the spring on the block over small portions of one oscillation.

Suppose the block has been pushed to the left to a position $-x_{max}$ and is then released as shown in Figure 7.10. We identify the block as our system and calculate the work W_s done by the spring force on the block as the block moves from $x_i = -x_{max}$ to $x_f = 0$. Applying Equation 7.8 and assuming the block may be modeled as a particle, we obtain

$$W_s = \int \vec{\mathbf{F}}_s \cdot d\vec{\mathbf{r}} = \int_{x_i}^{x_f} (-kx\hat{\mathbf{i}}) \cdot (dx\hat{\mathbf{i}}) = \int_{-x_{max}}^{0} (-kx)\,dx = \tfrac{1}{2}kx_{max}^2 \qquad (7.11)$$

where we have used the integral $\int x^n\,dx = x^{n+1}/(n+1)$ with $n = 1$. The work done by the spring force is positive because the force is in the same direction as its displacement (both are to the right in Figure 7.10 during the time interval considered). Because the block arrives at $x = 0$ with some speed, it will continue moving until it reaches a position $+x_{max}$. The work done by the spring force on the block as it moves from $x_i = 0$ to $x_f = x_{max}$ is $W_s = -\tfrac{1}{2}kx_{max}^2$. The work is negative because for this part of the motion the spring force is to the left and its displacement is to the right. Therefore, the *net* work done by the spring force on the block as it moves from $x_i = -x_{max}$ to $x_f = x_{max}$ is *zero*.

Figure 7.9d is a plot of F_s versus x. Equation 7.9 indicates that F_s is proportional to x, so the graph of F_s versus x is a straight line. The work calculated in Equation 7.11 is the area of the shaded triangle, corresponding to the displacement from $-x_{max}$ to 0. Because the triangle has base x_{max} and height kx_{max}, its area is $\tfrac{1}{2}kx_{max}^2$, agreeing with the work done by the spring calculated in Equation 7.11 by integration.

If the block undergoes an arbitrary displacement from $x = x_i$ to $x = x_f$, the work done by the spring force on the block is

$$W_s = \int_{x_i}^{x_f} (-kx)\,dx = \tfrac{1}{2}kx_i^2 - \tfrac{1}{2}kx_f^2 \qquad (7.12)$$

◀ Work done by a spring

From Equation 7.12, we see that the work done by the spring force is zero for any motion that ends where it began ($x_i = x_f$). We shall make use of this important result in Chapter 8 when we describe the motion of this system in greater detail.

Equations 7.11 and 7.12 describe the work done by the spring on the block. Now let us consider the work done on the block by an *external agent* as the agent applies

The force $\vec{\mathbf{F}}_s$ exerted by the spring performs work on the block as it moves to its final position.

$\vec{\mathbf{F}}_s$

$x_i = -x_{max}$ $x_f = 0$

Figure 7.10 A block is pushed to the initial position $x_i = -x_{max}$ and then released from rest. We identify the final position as the equilibrium position $x_f = 0$.

If the process of moving the block is carried out very slowly, then $\vec{\mathbf{F}}_{app}$ is equal in magnitude and opposite in direction to $\vec{\mathbf{F}}_s$ at all times.

$x_i = -x_{max}$ $x_f = 0$

Figure 7.11 A block moves from $x_i = -x_{max}$ to $x_f = 0$ on a frictionless surface as a force $\vec{\mathbf{F}}_{app}$ is applied to the block.

a force on the block and the block moves *very slowly* from $x_i = -x_{max}$ to $x_f = 0$ as in Figure 7.11. Compare the two figures carefully. In Figure 7.10, the spring expands freely. In Figure 7.11, however, the *applied force* $\vec{\mathbf{F}}_{app}$ pushes inward and prevents this free expansion. The magnitude of the applied force is adjusted so that the block moves to its final position very slowly. We can calculate the work done by the applied force by noting that at any value of the position, $\vec{\mathbf{F}}_{app}$ is equal in magnitude and opposite in direction to the spring force $\vec{\mathbf{F}}_s$, so $\vec{\mathbf{F}}_{app} = F_{app}\,\hat{\mathbf{i}} = -\vec{\mathbf{F}}_s = -(-kx\hat{\mathbf{i}}) = kx\hat{\mathbf{i}}$. Therefore, the work done by this applied force (the external agent) on the system of the block for the motion described is

$$W_{ext} = \int \vec{\mathbf{F}}_{app} \cdot d\vec{\mathbf{r}} = \int_{x_i}^{x_f} (kx\hat{\mathbf{i}}) \cdot (dx\hat{\mathbf{i}}) = \int_{-x_{max}}^{0} kx\,dx = -\tfrac{1}{2}kx_{max}^2$$

This work is equal to the negative of the work done by the spring force for this displacement (Eq. 7.11). The work is negative because the external agent must push inward on the spring to prevent it from expanding, and this direction is opposite the direction of the displacement of the point of application of the force as the block moves from $-x_{max}$ to 0.

For an arbitrary displacement of the block, the work done on the system by the external agent is

$$W_{ext} = \int_{x_i}^{x_f} kx\,dx = \tfrac{1}{2}kx_f^2 - \tfrac{1}{2}kx_i^2 \qquad (7.13)$$

Notice that this equation is the negative of Equation 7.12.

QUICK QUIZ 7.4 A dart is inserted into a spring-loaded dart gun by pushing the spring in by a distance x. For the next loading, the spring is compressed a distance $2x$. How much work is required to load the second dart compared with that required to load the first? **(a)** four times as much **(b)** two times as much **(c)** the same **(d)** half as much **(e)** one-fourth as much

---

Example **7.5**   **Measuring *k* for a Spring**

A common technique used to measure the force constant of a spring is demonstrated by the setup in Figure 7.12. The spring is hung vertically (Fig. 7.12a), and an object of mass m is attached to its lower end. Under the action of the "load" mg, the spring stretches a distance d from its equilibrium position (Fig. 7.12b).

(A) If a spring is stretched 2.0 cm by a suspended object having a mass of 0.55 kg, what is the force constant of the spring?

SOLUTION

Conceptualize   Figure 7.12b shows what happens to the spring when the object is attached to it. Simulate this situation by hanging an object on a rubber band.

Categorize   The object in Figure 7.12b is at rest and not accelerating, so it is modeled as a *particle in equilibrium*.

Analyze   Because the object is in equilibrium, the net force on it is zero and the upward spring force balances the downward gravitational force $m\vec{\mathbf{g}}$ (Fig. 7.12c).

Apply the particle in equilibrium model to the object: $\vec{\mathbf{F}}_s + m\vec{\mathbf{g}} = 0 \;\rightarrow\; F_s - mg = 0 \;\rightarrow\; F_s = mg$

Apply Hooke's law to give the magnitude $F_s = kd$ and solve for k: $k = \dfrac{mg}{d} = \dfrac{(0.55\text{ kg})(9.80\text{ m/s}^2)}{2.0 \times 10^{-2}\text{ m}} = 2.7 \times 10^2\text{ N/m}$

(B) How much work is done by the spring on the object as it stretches through this distance?

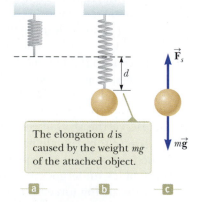

The elongation d is caused by the weight mg of the attached object.

a   b   c

Figure 7.12 (Example 7.5) Determining the force constant k of a spring.

7.5 continued

SOLUTION

Use Equation 7.12 to find the work done by the spring on the object:

$$W_s = 0 - \tfrac{1}{2}kd^2 = -\tfrac{1}{2}(2.7 \times 10^2 \text{ N/m})(2.0 \times 10^{-2} \text{ m})^2$$
$$= -5.4 \times 10^{-2} \text{ J}$$

Finalize This work is negative because the spring force acts upward on the object, but its point of application (where the spring attaches to the object) moves downward. As the object moves through the 2.0-cm distance, the gravitational force also does work on it. This work is positive because the gravitational force is downward and so is the displacement of the point of application of this force. Would we expect the work done by the gravitational force, as the applied force in a direction opposite to the spring force, to be the negative of the answer above? Let's find out.

Evaluate the work done by the gravitational force on the object:

$$W = \vec{\mathbf{F}} \cdot \Delta \vec{\mathbf{r}} = (mg)(d) \cos 0 = mgd$$
$$= (0.55 \text{ kg})(9.80 \text{ m/s}^2)(2.0 \times 10^{-2} \text{ m}) = 1.1 \times 10^{-1} \text{ J}$$

If you expected the work done by gravity simply to be that done by the spring with a positive sign, you may be surprised by this result! To understand why that is not the case, we need to explore further, as we do in the next section.

7.5 Kinetic Energy and the Work–Kinetic Energy Theorem

When energy transfers across the boundary of a system, the amount of energy stored in the system changes. We have investigated work in some depth and have identified it as a mechanism for transferring energy into a system. We have stated that work is an influence on a system from the environment, but we have not yet discussed the result of this influence on the system. One possible result of doing work on a system is that the system changes its speed: a common experience is to push on an object and observe it changing its state from rest to motion. In this section, we investigate this situation and introduce our first type of energy storage in a system, called *kinetic energy*.

Consider a system consisting of a single object. Figure 7.13 shows a block of mass m moving through a displacement directed to the right under the action of a net force $\sum \vec{\mathbf{F}}$, also directed to the right. We know from Newton's second law that the block moves with an acceleration $\vec{\mathbf{a}}$. If the block (and therefore the force) moves through a displacement $\Delta \vec{\mathbf{r}} = \Delta x \hat{\mathbf{i}} = (x_f - x_i)\hat{\mathbf{i}}$, the net work done on the block by the external net force $\sum \vec{\mathbf{F}}$ is given by Equation 7.7:

$$W_{\text{ext}} = \int_{x_i}^{x_f} \sum F \, dx \tag{7.14}$$

Using Newton's second law, we substitute for the magnitude of the net force $\sum F = ma$ and then perform the following chain-rule manipulations on the integrand:

$$W_{\text{ext}} = \int_{x_i}^{x_f} ma \, dx = \int_{x_i}^{x_f} m \frac{dv}{dt} \, dx = \int_{x_i}^{x_f} m \frac{dv}{dx} \frac{dx}{dt} \, dx = \int_{v_i}^{v_f} mv \, dv$$

$$W_{\text{ext}} = \tfrac{1}{2}mv_f^2 - \tfrac{1}{2}mv_i^2 \tag{7.15}$$

where v_i is the speed of the block at $x = x_i$ and v_f is its speed at x_f.

Equation 7.15 was generated for the specific situation of one-dimensional motion, but it is a general result. It tells us that the work done by the net force on a particle of mass m is equal to the difference between the initial and final values of a quantity $\tfrac{1}{2}mv^2$. This quantity is so important that it has been given a special name, **kinetic energy:**

$$K \equiv \tfrac{1}{2}mv^2 \tag{7.16}$$ ◀ Kinetic energy

Figure 7.13 An object undergoing a displacement $\Delta \vec{\mathbf{r}} = \Delta x \hat{\mathbf{i}}$ and a change in velocity under the action of a net force $\sum \vec{\mathbf{F}}$.

TABLE 7.1 Kinetic Energies for Various Objects

Object	Mass (kg)	Speed (m/s)	Kinetic Energy (J)
Earth orbiting the Sun	5.97×10^{24}	2.98×10^{4}	2.65×10^{33}
Moon orbiting the Earth	7.35×10^{22}	1.02×10^{3}	3.82×10^{28}
Rocket moving at escape speed[a]	500	1.12×10^{4}	3.14×10^{10}
Automobile at 65 mi/h	2 000	29	8.4×10^{5}
Running athlete	70	10	3 500
Stone dropped from 10 m	1.0	14	98
Golf ball at terminal speed	0.046	44	45
Raindrop at terminal speed	3.5×10^{-5}	9.0	1.4×10^{-3}
Oxygen molecule in air	5.3×10^{-26}	500	6.6×10^{-21}

[a]Escape speed is the minimum speed an object must reach near the Earth's surface to move infinitely far away from the Earth.

Kinetic energy represents the energy associated with the motion of the particle. Note that kinetic energy is a scalar quantity and has the same units as work. For example, a 2.0-kg object moving with a speed of 4.0 m/s has a kinetic energy of 16 J. Table 7.1 lists the kinetic energies for various objects.

Equation 7.15 states that the work done on a particle by a net force $\sum \vec{F}$ acting on it equals the change in kinetic energy of the particle. It is often convenient to write Equation 7.15 in the form

$$W_{\text{ext}} = K_f - K_i = \Delta K \qquad (7.17)$$

Another way to write it is $K_f = K_i + W_{\text{ext}}$, which tells us that the final kinetic energy of an object is equal to its initial kinetic energy plus the change in energy due to the net work done on it.

We have generated Equation 7.17 by imagining doing work on a particle. If we identify the particle as a system, we have increased the amount of energy stored in the system by doing work on it. We have stored the energy in the particular form of kinetic energy, represented by motion of the system through space. We could also do work on a deformable system, in which members of the system move with respect to one another. In this case, we also find that Equation 7.17 is valid as long as the net work is found by adding up the works done by each force and adding, as discussed earlier with regard to Equation 7.8. The kinetic energy K of the system is the sum of the kinetic energies of all members of the system.

Equation 7.17 is an important result known as the **work–kinetic energy theorem:**

◀ Work–kinetic energy theorem

> When work is done on a system and the only change in the system is in the speeds of its members, the net work done on the system equals the change in kinetic energy of the system, as expressed by Equation 7.17: $W = \Delta K$.

The work–kinetic energy theorem indicates that the kinetic energy of a system *increases* if the net work done on it is *positive*: energy is being transferred *into* the system. The kinetic energy *decreases* if the net work is *negative*: energy is being transferred *out of* the system.

Because we have so far only investigated translational motion through space, we arrived at the work–kinetic energy theorem by analyzing situations involving translational motion. Another type of motion is *rotational motion,* in which an object spins about an axis. We will study this type of motion in Chapter 10. The work–kinetic energy theorem is also valid for systems that undergo a change in the rotational speed due to work done on the system. A windmill serves as an example of work (done by the wind) causing rotational motion.

The work–kinetic energy theorem will clarify a result seen earlier in this chapter that may have seemed odd. In Section 7.4, we arrived at a result of zero net work done when we let a spring push a block from $x_i = -x_{max}$ to $x_f = x_{max}$. Notice that because the speed of the block is continually changing, it may seem complicated to analyze this process. The quantity ΔK in the work–kinetic energy theorem, however, only refers to the initial and final configurations of the system. It does not depend on the particular path followed by any members of the system. Therefore, because the speed of the block is zero at both the initial and final points of the motion, the net work done on the block is zero. We will often see this concept of path independence in similar approaches to problems.

Let us also return to the mystery in the Finalize step at the end of Example 7.5. Why was the work done by gravity not just the value of the work done by the spring with a positive sign? Notice that the work done by gravity is larger than the magnitude of the work done by the spring. Therefore, the total work done by all forces on the object is positive. Imagine now how to create the situation in which the *only* forces on the object are the spring force and the gravitational force. You must support the object at the highest point and then remove your hand and let the object fall. If you do so, you know that when the object reaches a position 2.0 cm below your hand, it will be *moving*, which is consistent with Equation 7.17. Positive net work is done on the object, and the result is that it has a kinetic energy as it passes through the 2.0-cm point.

The only way to prevent the object from having a kinetic energy after moving through 2.0 cm is to slowly lower it with your hand. Then, however, there is a third force doing work on the object, the normal force from your hand. If this work is calculated and added to that done by the spring force and the gravitational force, the net work done on the object is zero, which is consistent because it is not moving at the 2.0-cm point.

Earlier, we indicated that work can be considered as a mechanism for transferring energy into a system. Equation 7.17 is a mathematical statement of this concept. When work W_{ext} is done on a system, the result is a transfer of energy across the boundary of the system. The result on the system, in the case of Equation 7.17, is a change ΔK in kinetic energy. In the next section, we investigate another type of energy that can be stored in a system as a result of doing work on the system.

QUICK QUIZ 7.5 A dart is inserted into a spring-loaded dart gun by pushing the spring in by a distance x. For the next loading, the spring is compressed a distance $2x$. How much faster does the second dart leave the gun compared with the first? **(a)** four times as fast **(b)** two times as fast **(c)** the same **(d)** half as fast **(e)** one-fourth as fast

---

Example 7.6 | **A Block Pulled on a Frictionless Surface**

A 6.0-kg block initially at rest is pulled to the right along a frictionless, horizontal surface by a constant horizontal force of magnitude 12 N. Find the block's speed after it has moved through a horizontal distance of 3.0 m.

SOLUTION

Conceptualize Figure 7.14 illustrates this situation. Imagine pulling a toy car across a table with a horizontal rubber band attached to the front of the car. The force is maintained constant by ensuring that the stretched rubber band always has the same length.

Categorize We could apply the equations of kinematics to determine the answer, but let us practice the energy approach. The block is the system, and three external forces act on the system. The normal force balances the gravitational force on the block, and neither of these vertically acting forces does work on the block because their points of application are not vertically displaced.

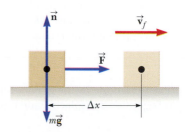

Figure 7.14 (Example 7.6) A block pulled to the right on a frictionless surface by a constant horizontal force.

continued

PITFALL PREVENTION 7.5

Conditions for the Work–Kinetic Energy Theorem The work–kinetic energy theorem is important but limited in its application; it is not a general principle. In many situations, other changes in the system occur besides its speed, and there are other interactions with the environment besides work. A more general principle involving energy is *conservation of energy* in Section 8.1.

PITFALL PREVENTION 7.6

The Work–Kinetic Energy Theorem: Speed, Not Velocity The work–kinetic energy theorem relates work to a change in the *speed* of a system, not a change in its velocity. For example, if an object is in uniform circular motion, its speed is constant. Even though its velocity is changing, no work is done on the object by the force causing the circular motion.

7.6 c o n t i n u e d

Analyze The net external force acting on the block is the horizontal 12-N force.

Use the work–kinetic energy theorem for the block, noting that its initial kinetic energy is zero:

$$W_{ext} = \Delta K = K_f - K_i = \tfrac{1}{2}mv_f^2 - 0 = \tfrac{1}{2}mv_f^2$$

Solve for v_f and use Equation 7.1 for the work done on the block by $\vec{F}$:

$$v_f = \sqrt{\frac{2W_{ext}}{m}} = \sqrt{\frac{2F\,\Delta x}{m}}$$

Substitute numerical values:

$$v_f = \sqrt{\frac{2(12\text{ N})(3.0\text{ m})}{6.0\text{ kg}}} = 3.5 \text{ m/s}$$

Finalize You should solve this problem again by modeling the block as a *particle under a net force* to find its acceleration and then as a *particle under constant acceleration* to find its final velocity. In Chapter 8, we will see that the energy procedure followed above is an example of the analysis model of the *nonisolated system*.

WHAT IF? Suppose the magnitude of the force in this example is doubled to $F' = 2F$. The 6.0-kg block accelerates to 3.5 m/s due to this applied force while moving through a displacement $\Delta x'$. How does the displacement $\Delta x'$ compare with the original displacement Δx?

Answer If we pull harder, the block should accelerate to a given speed in a shorter distance, so we expect that $\Delta x' < \Delta x$. In both cases, the block experiences the same change in kinetic energy ΔK. Therefore, the same work is done on the block in both cases. Mathematically, from the work–kinetic energy theorem, we find that

$$W_{ext} = F'\Delta x' = \Delta K = F\Delta x$$

$$\Delta x' = \frac{F}{F'}\,\Delta x = \frac{F}{2F}\,\Delta x = \tfrac{1}{2}\,\Delta x$$

and the distance is shorter as suggested by our conceptual argument.

Conceptual Example 7.7 **Does the Ramp Lessen the Work Required?**

A man wishes to load a refrigerator onto a truck using a ramp at angle θ as shown in Figure 7.15. He claims that less work would be required to load the truck if the length L of the ramp were increased so that the angle θ would be smaller. Is his claim valid?

SOLUTION

No. Suppose the refrigerator is wheeled on a hand truck up the ramp at constant speed. In this case, for the system of the refrigerator and the hand truck, $\Delta K = 0$. The normal force exerted by the ramp on the system is directed at 90° to the displacement of its point of application and so does no work on the system. Because $\Delta K = 0$, the work–kinetic energy theorem applied to the refrigerator gives

Figure 7.15 (Conceptual Example 7.7) A refrigerator attached to a frictionless, wheeled hand truck is moved up a ramp at constant speed.

$$W_{ext} = W_{by\;man} + W_{by\;gravity} = 0$$

The work done by the gravitational force equals the product of the weight mg of the system, the distance L through which the refrigerator is displaced, and $\cos (\theta + 90°)$. Therefore,

$$W_{by\;man} = -W_{by\;gravity} = -(mg)(L)[\cos (\theta + 90°)]$$
$$= mgL \sin \theta = mgh$$

where $h = L \sin \theta$ is the height of the ramp at the truck. Therefore, the man must do the same amount of work mgh on the system *regardless* of the length of the ramp. The work depends only on the height of the ramp. Although less force is required with a longer ramp, the point of application of that force moves through a greater displacement.

7.6 | Potential Energy of a System

So far in this chapter, we have defined a system in general, but have focused our attention primarily on single particles or objects under the influence of external forces. Let us now consider systems of two or more particles or objects interacting via a force that is *internal* to the system. The kinetic energy of such a system is the algebraic sum of the kinetic energies of all members of the system. There may be systems, however, in which one object is so massive that it can be modeled as stationary and its kinetic energy can be neglected. For example, if we consider a ball–Earth system as the ball falls to the Earth, the kinetic energy of the system can be considered as just the kinetic energy of the ball. The Earth moves so slowly in this process that we can ignore its kinetic energy. On the other hand, the kinetic energy of a system of two electrons must include the kinetic energies of both particles.

Let us imagine a system consisting of a book and the Earth, interacting via the gravitational force. We do some work on the system by lifting the book slowly from rest through a vertical displacement $\Delta \vec{r} = (y_f - y_i)\hat{j}$ as in Figure 7.16. According to our discussion of work as an energy transfer, this work done on the system must appear as an increase in energy of the system. The book is at rest before we perform the work and is at rest after we perform the work. Therefore, there is no change in the kinetic energy of the system.

Because the energy change of the system is not in the form of kinetic energy, the work-kinetic energy theorem does not apply here and the energy change must appear as some form of energy storage other than kinetic energy. After lifting the book, we could release it and let it fall back to the position y_i. Notice that the book (and therefore, the system) now has kinetic energy and that its source is in the work that was done in lifting the book. While the book was at the highest point, the system had the *potential* to possess kinetic energy, but it did not do so until the book was allowed to fall. Therefore, we call the energy storage mechanism before the book is released **potential energy.** We will find that the potential energy of a system can only be associated with specific types of forces acting between members of a system. The amount of potential energy in the system is determined by the *configuration* of the system. Moving members of the system to different positions or rotating them may change the configuration of the system and therefore its potential energy.

Let us now derive an expression for the potential energy associated with an object at a given location above the surface of the Earth. Consider an external agent lifting an object of mass m from an initial height y_i above the ground to a final height y_f as in Figure 7.16. We assume the lifting is done slowly, with no acceleration, so the applied force from the agent is equal in magnitude to the gravitational force on the object: the object is modeled as a particle in equilibrium moving at constant velocity. The work done by the external agent on the system (object and the Earth) as the object undergoes this upward displacement is given by the product of the upward applied force $\vec{F}_{app}$ and the upward displacement of this force, $\Delta \vec{r} = \Delta y \hat{j}$:

$$W_{ext} = (\vec{F}_{app}) \cdot \Delta \vec{r} = (mg\hat{j}) \cdot [(y_f - y_i)\hat{j}] = mgy_f - mgy_i \tag{7.18}$$

where this result is the net work done on the system because the applied force is the only force on the system from the environment. (Remember that the gravitational force is *internal* to the system.) Notice the similarity between Equation 7.18 and Equation 7.15. In each equation, the work done on a system equals a difference between the final and initial values of a quantity. In Equation 7.15, the work represents a transfer of energy into the system and the increase in energy of the system is kinetic in form. In Equation 7.18, the work represents a transfer of energy into the system and the system energy appears in a different form, which we have called potential energy.

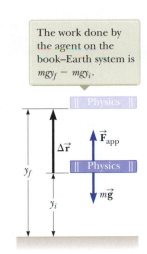

Figure 7.16 An external agent lifts a book slowly from a height y_i to a height y_f.

PITFALL PREVENTION 7.7

Potential Energy The phrase *potential energy* does not refer to something that has the potential to become energy. Potential energy *is* energy.

PITFALL PREVENTION 7.8

Potential Energy Belongs to a System Potential energy is always associated with a *system* of two or more interacting objects. When a small object moves near the surface of the Earth under the influence of gravity, we may sometimes refer to the potential energy "associated with the object" rather than the more proper "associated with the system" because the Earth does not move significantly. We will not, however, refer to the potential energy "of the object" because this wording ignores the role of the Earth.

Therefore, we can identify the quantity mgy as the **gravitational potential energy** U_g of the system of an object of mass m and the Earth:

◀ Gravitational potential energy

$$U_g \equiv mgy \qquad (7.19)$$

The units of gravitational potential energy are joules, the same as the units of work and kinetic energy. Potential energy, like work and kinetic energy, is a scalar quantity. Notice that Equation 7.19 is valid only for objects near the surface of the Earth, where g is approximately constant.[2]

Using our definition of gravitational potential energy, Equation 7.18 can now be rewritten as

$$W_{ext} = \Delta U_g \qquad (7.20)$$

which mathematically describes that the net external work done on the system in this situation appears as a change in the gravitational potential energy of the system.

Equation 7.20 is similar in form to the work–kinetic energy theorem, Equation 7.17. In Equation 7.17, work is done on a system and energy appears in the system as kinetic energy, representing *motion* of the members of the system. In Equation 7.20, work is done on the system and energy appears in the system as potential energy, representing a change in the *configuration* of the members of the system.

Gravitational potential energy depends only on the vertical height of the object above the surface of the Earth. The same amount of work must be done on an object–Earth system whether the object is lifted vertically from the Earth or is pushed starting from the same point up a frictionless incline, ending up at the same height. We verified this statement for a specific situation of rolling a refrigerator up a ramp in Conceptual Example 7.7. This statement can be shown to be true in general by calculating the work done on an object by an agent moving the object through a displacement having both vertical and horizontal components:

$$W_{ext} = (\vec{\mathbf{F}}_{app}) \cdot \Delta \vec{\mathbf{r}} = (mg\hat{\mathbf{j}}) \cdot [(x_f - x_i)\hat{\mathbf{i}} + (y_f - y_i)\hat{\mathbf{j}}] = mgy_f - mgy_i$$

where there is no term involving x in the final result because $\hat{\mathbf{j}} \cdot \hat{\mathbf{i}} = 0$.

In solving problems, you must choose a reference configuration for which the gravitational potential energy of the system is set equal to some reference value, which is normally zero. The choice of reference configuration is completely arbitrary because the important quantity is the *difference* in potential energy, and this difference is independent of the choice of reference configuration.

It is often convenient to choose as the reference configuration for zero gravitational potential energy the configuration in which an object is at the surface of the Earth, but this choice is not essential. Often, the statement of the problem suggests a convenient configuration to use.

QUICK QUIZ **7.6** Choose the correct answer. The gravitational potential energy of a system **(a)** is always positive **(b)** is always negative **(c)** can be negative or positive

---

Example **7.8** **The Proud Athlete and the Sore Toe**

A trophy being shown off by a careless athlete slips from the athlete's hands and drops on his foot. Choosing floor level as the $y = 0$ point of your coordinate system, estimate the change in gravitational potential energy of the trophy–Earth system as the trophy falls. Repeat the calculation, using the top of the athlete's head as the origin of coordinates.

SOLUTION

Conceptualize The trophy changes its vertical position with respect to the surface of the Earth. Associated with this change in position is a change in the gravitational potential energy of the trophy–Earth system.

---

[2]The assumption that g is constant is valid as long as the vertical displacement of the object is small compared with the Earth's radius.

7.8 continued

Categorize We evaluate a change in gravitational potential energy defined in this section, so we categorize this example as a substitution problem. Because there are no numbers provided in the problem statement, it is also an estimation problem.

The problem statement tells us that the reference configuration of the trophy–Earth system corresponding to zero potential energy is when the bottom of the trophy is at the floor. To find the change in potential energy for the system, we need to estimate a few values. Let's say the trophy has a mass of approximately 2 kg, and the top of a person's foot is about 0.05 m above the floor. Also, let's assume the trophy falls from a height of 1.4 m.

Calculate the gravitational potential energy of the trophy–Earth system just before the trophy is released:

$$U_i = mgy_i = (2 \text{ kg})(9.80 \text{ m/s}^2)(1.4 \text{ m}) = 27.4 \text{ J}$$

Calculate the gravitational potential energy of the trophy–Earth system when the trophy reaches the athlete's foot:

$$U_f = mgy_f = (2 \text{ kg})(9.80 \text{ m/s}^2)(0.05 \text{ m}) = 0.98 \text{ J}$$

Evaluate the change in gravitational potential energy of the trophy–Earth system:

$$\Delta U_g = 0.98 \text{ J} - 27.4 \text{ J} = -26.4 \text{ J}$$

We should probably keep only two digits because of the roughness of our estimates; therefore, we estimate that the change in gravitational potential energy is -26 J . The system had about 27 J of gravitational potential energy before the trophy began its fall and approximately 1 J of potential energy as the trophy reaches the top of the foot.

The second case presented indicates that the reference configuration of the system for zero potential energy is chosen to be when the trophy is on the athlete's head (even though the trophy is never at this position in its motion). We estimate this position to be 2.0 m above the floor).

Calculate the gravitational potential energy of the trophy–Earth system just before the trophy is released from its position 0.6 m below the athlete's head:

$$U_i = mgy_i = (2 \text{ kg})(9.80 \text{ m/s}^2)(-0.6 \text{ m}) = -11.8 \text{ J}$$

Calculate the gravitational potential energy of the trophy–Earth system when the trophy reaches the athlete's foot located 1.95 m below the athlete's head:

$$U_f = mgy_f = (2 \text{ kg})(9.80 \text{ m/s}^2)(-1.95 \text{ m}) = -38.2 \text{ J}$$

Evaluate the change in gravitational potential energy of the trophy–Earth system:

$$\Delta U_g = -38.2 \text{ J} - (-11.8 \text{ J}) = -26.4 \text{ J} \approx -26 \text{ J}$$

This value is the same as before, as it must be. The change in potential energy is independent of the choice of configuration of the system representing the zero of potential energy. If we wanted to keep only one digit in our estimates, we could write the final result as $3 \times 10^1 \text{ J}$.

Elastic Potential Energy

Because members of a system can interact with one another by means of different types of forces, it is possible that there are different types of potential energy in a system. We have just become familiar with gravitational potential energy of a system in which members interact via the gravitational force. Let us explore a second type of potential energy that a system can possess.

Consider a system consisting of a block and a spring as shown in Figure 7.17 (page 168). In Section 7.4, we identified *only* the block as the system. Now we include both the block and the spring in the system and recognize that the spring force is the interaction between the two members of the system. The force that the spring exerts on the block is given by $F_s = -kx$ (Eq. 7.9). The external work done by an applied force F_{app} on the block–spring system as the block moves from x_i to x_f is given by Equation 7.13:

$$W_{ext} = \tfrac{1}{2}kx_f^2 - \tfrac{1}{2}kx_i^2 \tag{7.21}$$

In this situation, the initial and final x coordinates of the block are measured from its equilibrium position, $x = 0$. Again (as in the gravitational case, Eq. 7.18) the

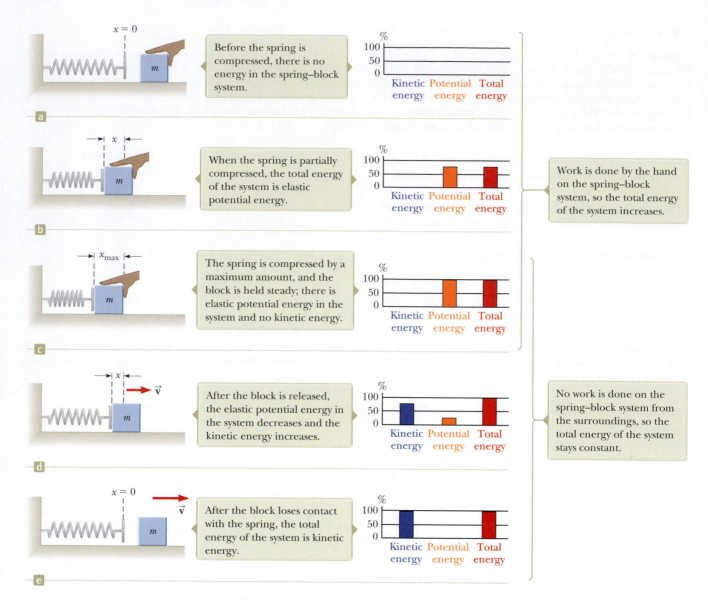

Figure 7.17 A spring on a frictionless, horizontal surface is compressed a distance x_{max} when a block of mass m is pushed against it. The block is then released and the spring pushes it to the right, where the block eventually loses contact with the spring. Parts (a) through (e) show various instants in the process. Energy bar charts on the right of each part of the figure help keep track of the energy in the system.

work done on the system is equal to the difference between the initial and final values of an expression related to the system's configuration. The **elastic potential energy** function associated with the block–spring system is defined by

Elastic potential energy ▶

$$U_s \equiv \tfrac{1}{2}kx^2 \tag{7.22}$$

Equation 7.21 can be expressed as

$$W_{ext} = \Delta U_s \tag{7.23}$$

Compare this equation to Equations 7.17 and 7.20. In all three situations, external work is done on a system and a form of energy storage in the system changes as a result.

The elastic potential energy of the system can be thought of as the energy stored in the deformed spring (one that is either compressed or stretched from its equilibrium position). The elastic potential energy stored in a spring is zero whenever the spring is undeformed ($x = 0$). Energy is stored in the spring only when the spring is either stretched or compressed. Because the elastic potential energy is proportional to x^2, we see that U_s is always positive in a deformed spring. Everyday examples of the storage of elastic potential energy can be found in old-style clocks or watches that operate from a wound-up spring and small wind-up toys for children.

Consider Figure 7.17 once again, which shows a spring on a frictionless, horizontal surface. When a block is pushed against the spring by an external agent, the elastic potential energy and the total energy of the system increase as indicated in Figure 7.17b. When the spring is compressed a distance x_{max} (Fig. 7.17c), the elastic potential energy stored in the spring is $\frac{1}{2}kx^2_{max}$. When the external force is removed, the only force on the block is that due to the spring, and the block moves to the right. The elastic potential energy of the system decreases, whereas the kinetic energy increases and the total energy remains fixed (Fig. 7.17d). When the spring returns to its original length, the stored elastic potential energy is completely transformed into kinetic energy of the block (Fig. 7.17e).

QUICK QUIZ 7.7 A ball is connected to a light spring suspended vertically as shown in Figure 7.18. When pulled downward from its equilibrium position and released, the ball oscillates up and down. **(i)** In the system of *the ball, the spring, and the Earth,* what forms of energy are there during the motion? (a) kinetic and elastic potential (b) kinetic and gravitational potential (c) kinetic, elastic potential, and gravitational potential (d) elastic potential and gravitational potential **(ii)** In the system of *the ball and the spring,* what forms of energy are there during the motion? Choose from the same possibilities (a) through (d).

Figure 7.18 (Quick Quiz 7.7) A ball connected to a massless spring suspended vertically. What forms of potential energy are associated with the system when the ball is displaced downward?

Energy Bar Charts

Figure 7.17 shows an important graphical representation of information related to energy of systems called an **energy bar chart.** The vertical axis represents the amount of energy of a given type in the system. The horizontal axis shows the types of energy in the system. The bar chart in Figure 7.17a shows that the system contains zero energy because the spring is relaxed and the block is not moving. Between Figure 7.17a and Figure 7.17c, the hand does work on the system, compressing the spring and storing elastic potential energy in the system. In Figure 7.17d, the block has been released and is moving to the right while still in contact with the spring. The height of the bar for the elastic potential energy of the system decreases, the kinetic energy bar increases, and the total energy bar remains fixed. In Figure 7.17e, the spring has returned to its relaxed length and the system now contains only kinetic energy associated with the moving block.

Energy bar charts can be a very useful representation for keeping track of the various types of energy in a system. For practice, try making energy bar charts for the book–Earth system in Figure 7.16 when the book is dropped from the higher position. Figure 7.18 associated with Quick Quiz 7.7 shows another system for which drawing an energy bar chart would be a good exercise. We will show energy bar charts in some figures in this chapter.

7.7 Conservative and Nonconservative Forces

We now introduce a third type of energy that a system can possess and store. Imagine that the book in Figure 7.19a (page 170) has been accelerated by your hand and is now sliding to the right on the surface of a heavy table and slowing down due to the friction force. Suppose the *surface* is the system. Then, from our discussion of work, we can argue that the friction force from the sliding book does work on the surface. The friction force on the surface is to the right and the displacement of the point of application of the force is to the right because the book has moved to the right. The work done on the surface is therefore positive, but the surface is not

Figure 7.19 (a) A book sliding to the right on a horizontal surface slows down in the presence of a force of kinetic friction acting to the left on the book. (b) An energy bar chart showing the energy in the system of the book and the surface at the initial instant of time. The energy of the system is all kinetic energy. (c) While the book is sliding, the kinetic energy of the system decreases as it is transformed to internal energy. (d) After the book has stopped, the energy of the system is all internal energy.

▶ Properties of conservative forces

moving after the book has stopped. Positive work has been done on the surface, yet there is no increase in the surface's kinetic energy. Nor is there any change in the potential energy of any system. So work has been done, but where is the energy?

From your everyday experience with sliding over surfaces with friction, you can probably guess that the surface will be *warmer* after the book slides over it. This is what you found when you sanded the wood in the opening storyline for this chapter. The work that was done on the surface has gone into warming the surface rather than increasing its speed or changing the configuration of a system. We call the energy associated with the temperature of a system its **internal energy,** symbolized E_{int}. (We will define internal energy more generally in Chapter 19.) In this case, the work done on the surface does indeed represent energy transferred into the system, but it appears in the system as internal energy rather than kinetic or potential energy.

Now consider the book and the surface in Figure 7.19a together as a system. After the book is released, and while it is slowing down, no work is done on this system. Initially, the system has kinetic energy because the book is moving. While the book is sliding, the internal energy of the system increases: the book and the surface are warmer than before. When the book stops, the kinetic energy has been completely transformed to internal energy. We can consider the friction force within the system—that is, between the book and the surface—as a *transformation mechanism* for energy. This force transforms the kinetic energy of the system into internal energy. Rub your hands together briskly to experience this effect!

Figures 7.19b through 7.19d show energy bar charts for the situation in Figure 7.19a. In Figure 7.19b, the bar chart shows that the system contains kinetic energy at the instant the book is released by your hand. We define the reference amount of internal energy in the system as zero at this instant. Figure 7.19c shows the kinetic energy transforming to internal energy as the book slows down due to the friction force. In Figure 7.19d, after the book has stopped sliding, the kinetic energy is zero, and the system now contains only internal energy E_{int}. Notice that the total energy bar in red has not changed during the process. The amount of internal energy in the system after the book has stopped is equal to the amount of kinetic energy in the system at the initial instant. This equality is described by an important principle called *conservation of energy.* We will explore this principle in Chapter 8.

Now consider in more detail an object moving downward near the surface of the Earth. The work done by the gravitational force on the object does not depend on whether it falls vertically or slides down a sloping incline with friction. All that matters is the change in the object's elevation. The energy transformation to internal energy due to friction on that incline, however, depends very much on the distance the object slides. The longer the incline, the more potential energy is transformed to internal energy. In other words, the path makes no difference when we consider the work done by the gravitational force, but it does make a difference when we consider the energy transformation due to friction forces. We can use this varying dependence on path to classify forces as either *conservative* or *nonconservative.* Of the two forces just mentioned, the gravitational force is conservative and the friction force is nonconservative.

Conservative Forces

Conservative forces have these two equivalent properties:

1. The work done by a conservative force on a particle moving between any two points is independent of the path taken by the particle.
2. The work done by a conservative force on a particle moving through any closed path is zero. (A closed path is one for which the beginning point and the endpoint are identical.)

The gravitational force is one example of a conservative force; the force that an ideal spring exerts on any object attached to the spring is another. The work done by the gravitational force on an object moving between any two points near the

Earth's surface is $W_g = -mg\hat{\mathbf{j}} \cdot [(y_f - y_i)\hat{\mathbf{j}}] = mgy_i - mgy_f$. From this equation, notice that W_g depends only on the initial and final y coordinates of the object and hence is independent of the path. Furthermore, W_g is zero when the object moves over any closed path (where $y_i = y_f$).

For the case of the object–spring system, the work W_s done by the spring force is given by $W_s = \frac{1}{2}kx_i^2 - \frac{1}{2}kx_f^2$ (Eq. 7.12). We see that the spring force is conservative because W_s depends only on the initial and final x coordinates of the object and is zero for any closed path (where $x_i = x_f$).

Nonconservative Forces

A force is **nonconservative** if it does not satisfy properties 1 and 2 above. The work done by a nonconservative force is path-dependent. We define the sum of the kinetic and potential energies of a system as the **mechanical energy** of the system:

$$E_{\text{mech}} \equiv K + U \tag{7.24}$$

where K includes the kinetic energy of all moving members of the system and U includes all types of potential energy in the system. For a book falling under the action of the gravitational force, the mechanical energy of the book–Earth system remains fixed; gravitational potential energy transforms to kinetic energy, and the total energy of the system remains constant. Nonconservative forces acting within a system, however, cause a *change* in the mechanical energy of the system. For example, for a book sent sliding on a horizontal surface that is not frictionless (Fig. 7.19a), the mechanical energy of the book–surface system is transformed to internal energy as we discussed earlier. Only part of the book's kinetic energy is transformed to internal energy in the book. The rest appears as internal energy in the surface. (When you trip and slide across a gymnasium floor, not only does the skin on your knees warm up, so does the floor!) Because the force of kinetic friction transforms the mechanical energy of a system into internal energy, it is a nonconservative force.

As an example of the path dependence of the work for a nonconservative force, consider Figure 7.20. Suppose you displace a book between two points on a table. If the book is displaced in a straight line along the blue path between points Ⓐ and Ⓑ in Figure 7.20, you do a certain amount of work against the kinetic friction force to keep the book moving at a constant speed. Now, imagine that you push the book along the brown semicircular path in Figure 7.20. You perform more work against friction along this curved path than along the straight path because the curved path is longer. The work done on the book depends on the path, so the friction force *cannot* be conservative.

The work done in moving the book is greater along the brown path than along the blue path.

Figure 7.20 The work done against the force of kinetic friction depends on the path taken as the book is moved from Ⓐ to Ⓑ.

7.8 Relationship Between Conservative Forces and Potential Energy

We can associate a **potential energy function** U for a system with a force acting between members of the system, but *we can do so only if the force is conservative*. In general, the work W_{int} done by a conservative force on an object that is a member of a system as the system changes from one configuration to another is equal to the initial value of the potential energy of the system minus the final value:

$$W_{\text{int}} = U_i - U_f = -\Delta U \tag{7.25}$$

The subscript "int" in Equation 7.25 reminds us that the work we are discussing is done by one member of the system on another member and is therefore *internal* to the system. It is different from the work W_{ext} done *on* the system as a whole by an external agent. As an example, compare Equation 7.25 with the equation for the work done by an external agent on a block–spring system (Eq. 7.23) as the extension of the spring changes.

PITFALL PREVENTION 7.9
Similar Equation Warning Compare Equation 7.25 with Equation 7.20. These equations are similar except for the negative sign, which is a common source of confusion. Equation 7.20 tells us that positive work done *by an outside agent* on a system causes an increase in the potential energy of the system (with no change in the kinetic or internal energy). Equation 7.25 states that positive work done on a component of a system by a conservative force *internal to the system* causes a decrease in the potential energy of the system.

Let us imagine a system of particles in which a conservative force $\vec{F}$ acts between the particles. Imagine also that the configuration of the system changes due to the motion of one particle along the x axis. Then we can evaluate the internal work done by this force as the particle moves along the x axis[3] using Equations 7.7 and 7.25:

$$W_{\text{int}} = \int_{x_i}^{x_f} F_x \, dx = -\Delta U \tag{7.26}$$

where F_x is the component of $\vec{F}$ in the direction of the displacement. We can also express Equation 7.26 as

$$\Delta U = U_f - U_i = -\int_{x_i}^{x_f} F_x \, dx \tag{7.27}$$

Therefore, ΔU is negative when F_x and dx are in the same direction, as when an object is lowered in a gravitational field or when a spring pushes an object toward equilibrium.

It is often convenient to establish some particular location x_i of one member of a system as representing a reference configuration and measure all potential energy differences with respect to it. We can then define the potential energy function as

$$U_f(x) = -\int_{x_i}^{x_f} F_x \, dx + U_i \tag{7.28}$$

The value of U_i is often taken to be zero for the reference configuration. It does not matter what value we assign to U_i because any nonzero value merely shifts $U_f(x)$ by a constant amount and only the *change* in potential energy is physically meaningful.

If the point of application of the force undergoes an infinitesimal displacement dx, we can express the infinitesimal change in the potential energy of the system dU as

$$dU = -F_x \, dx$$

Therefore, the conservative force is related to the potential energy function through the relationship[4]

◀ Relation of force between members of a system to the potential energy of the system

$$F_x = -\frac{dU}{dx} \tag{7.29}$$

That is, the x component of a conservative force acting on a member within a system equals the negative derivative of the potential energy of the system with respect to x.

We can easily check Equation 7.29 for the two examples already discussed. In the case of the deformed spring, $U_s = \frac{1}{2}kx^2$; therefore,

$$F_s = -\frac{dU_s}{dx} = -\frac{d}{dx}(\tfrac{1}{2}kx^2) = -kx$$

which corresponds to the restoring force in the spring (Hooke's law). Because the gravitational potential energy function is $U_g = mgy$, it follows from Equation 7.29 that $F_g = -mg$ when we differentiate U_g with respect to y instead of x.

We now see that U is an important function because a conservative force can be derived from it. Furthermore, Equation 7.29 should clarify that adding a constant to the potential energy is unimportant because the derivative of a constant is zero.

---

[3]For a general displacement, the work done in two or three dimensions also equals $-\Delta U$, where $U = U(x, y, z)$. We write this equation formally as $W_{\text{int}} = \int_i^f \vec{F} \cdot d\vec{r} = U_i - U_f$.

[4]In three dimensions, the expression is

$$\vec{F} = -\frac{\partial U}{\partial x}\hat{i} - \frac{\partial U}{\partial y}\hat{j} - \frac{\partial U}{\partial z}\hat{k}$$

where $(\partial U/\partial x)$ and so forth are partial derivatives. In the language of vector calculus, $\vec{F}$ equals the negative of the *gradient* of the scalar quantity $U(x, y, z)$.

QUICK QUIZ 7.8 What does the slope of a graph of $U(x)$ versus x represent? **(a)** the magnitude of the force on the object **(b)** the negative of the magnitude of the force on the object **(c)** the x component of the force on the object **(d)** the negative of the x component of the force on the object

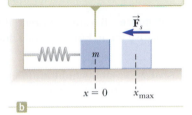

The restoring force exerted by the spring always acts toward $x = 0$, the position of stable equilibrium.

7.9 Energy Diagrams and Equilibrium of a System

The motion of a system can often be understood qualitatively through a graph of its potential energy versus the position of a member of the system. Consider the potential energy function for a block–spring system, given by $U_s = \frac{1}{2}kx^2$. This function is plotted versus x in Figure 7.21a, where x is the position of the block.

As we saw in Quick Quiz 7.8, the x component of the force is equal to the negative of the slope of the U-versus-x curve. When the block is placed at rest at the equilibrium position of the spring ($x = 0$), where $F_s = 0$, it will remain there unless some external force F_{ext} acts on it. If this external force stretches the spring from equilibrium, x is positive and the slope dU/dx is positive; therefore, the force F_s exerted by the spring is negative and the block accelerates back toward $x = 0$ when released. If the external force compresses the spring, x is negative and the slope is negative; therefore, F_s is positive and again the mass accelerates toward $x = 0$ upon release.

From this analysis, we conclude that the $x = 0$ position for a block–spring system is one of **stable equilibrium.** That is, any movement away from this position results in a force directed back toward $x = 0$. In general, configurations of a system in stable equilibrium correspond to those for which $U(x)$ for the system has a minimum.

Another simple mechanical system with a configuration of stable equilibrium is a ball rolling about in the bottom of a bowl. Anytime the ball is displaced from its lowest position, it tends to return to that position when released.

Now consider a particle moving along the x axis under the influence of a conservative force F_x, where the U-versus-x curve is as shown in Figure 7.22. Once again, $F_x = 0$ at $x = 0$, and so the particle is in equilibrium at this point. This position, however, is one of **unstable equilibrium** for the following reason. Suppose the particle is displaced to the right ($x > 0$). Because the slope is negative for $x > 0$, $F_x = -dU/dx$ is positive and the particle accelerates away from $x = 0$. If instead the particle is at $x = 0$ and is displaced to the left ($x < 0$), the force is negative because the slope is positive for $x < 0$ and the particle again accelerates away from the equilibrium position. The position $x = 0$ in this situation is one of unstable equilibrium because for any displacement from this point, the force pushes the particle farther away from equilibrium and toward a position of lower potential energy. A pencil balanced on its point is in a position of unstable equilibrium. If the pencil is displaced slightly from its absolutely vertical position and is then released, it will surely fall over. In general, configurations of a system in unstable equilibrium correspond to those for which $U(x)$ for the system has a maximum.

Finally, a configuration called **neutral equilibrium** arises when U is constant over some region. Small displacements of an object from a position in this region produce neither restoring nor disrupting forces. A ball lying on a flat, horizontal surface is an example of an object in neutral equilibrium.

Figure 7.21 (a) Potential energy as a function of x for the frictionless block–spring system shown in (b). For a given energy E of the system, the block oscillates between the turning points, which have the coordinates $x = \pm x_{max}$.

PITFALL PREVENTION 7.10

Energy Diagrams A common mistake is to think that potential energy on the graph in an energy diagram represents the height of some object. For example, that is not the case in Figure 7.21, where the block is only moving horizontally.

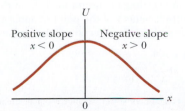

Figure 7.22 A plot of U versus x for a particle that has a position of unstable equilibrium located at $x = 0$. For any finite displacement of the particle, the force on the particle is directed away from $x = 0$.

Example **7.9** Force and Energy on an Atomic Scale

The potential energy associated with the force between two neutral atoms in a molecule can be modeled by the Lennard–Jones potential energy function:

$$U(r) = 4\epsilon\left[\left(\frac{\sigma}{r}\right)^{12} - \left(\frac{\sigma}{r}\right)^{6}\right]$$

where r is the separation of the atoms. The function $U(r)$ contains two parameters σ and ϵ that are determined from experiments. Sample values for the interaction between two atoms in a molecule are $\sigma = 0.263$ nm and $\epsilon = 1.51 \times 10^{-22}$ J. Using a spreadsheet or similar tool, graph this function and find the most likely distance between the two atoms.

SOLUTION

Conceptualize We identify the two atoms in the molecule as a system. Based on our understanding that stable molecules exist, we expect to find stable equilibrium when the two atoms are separated by some equilibrium distance.

Categorize Because a potential energy function exists, we categorize the force between the atoms as conservative. For a conservative force, Equation 7.29 describes the relationship between the force and the potential energy function.

Analyze Stable equilibrium exists for a separation distance at which the potential energy of the system of two atoms (the molecule) is a minimum.

Take the derivative of the function $U(r)$:

$$\frac{dU(r)}{dr} = 4\epsilon\frac{d}{dr}\left[\left(\frac{\sigma}{r}\right)^{12} - \left(\frac{\sigma}{r}\right)^{6}\right] = 4\epsilon\left[\frac{-12\sigma^{12}}{r^{13}} + \frac{6\sigma^{6}}{r^{7}}\right]$$

Minimize the function $U(r)$ by setting its derivative equal to zero:

$$4\epsilon\left[\frac{-12\sigma^{12}}{r_{eq}^{13}} + \frac{6\sigma^{6}}{r_{eq}^{7}}\right] = 0 \quad \rightarrow \quad r_{eq} = (2)^{1/6}\sigma$$

Evaluate r_{eq}, the equilibrium separation of the two atoms in the molecule:

$$r_{eq} = (2)^{1/6}(0.263 \text{ nm}) = \boxed{2.95 \times 10^{-10} \text{ m}}$$

We graph the Lennard–Jones function on both sides of this critical value to create our energy diagram as shown in Figure 7.23.

Finalize Notice that $U(r)$ is extremely large when the atoms are very close together, is a minimum when the atoms are at their critical separation, and then increases again as the atoms move apart. When $U(r)$ is a minimum, the atoms are in stable equilibrium, indicating that the most likely separation between them occurs at this point.

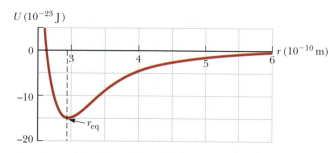

Figure 7.23 (Example 7.9) Potential energy curve associated with a molecule. The distance r is the separation between the two atoms making up the molecule.

Summary

▶ Definitions

A **system** is most often a single particle, a collection of particles, or a region of space, and may vary in size and shape. A **system boundary** separates the system from the **environment**.

The **work** W done on a system by an agent exerting a constant force $\vec{F}$ on the system is the product of the magnitude Δr of the displacement of the point of application of the force and the component $F\cos\theta$ of the force along the direction of the displacement $\Delta\vec{r}$:

$$W \equiv F\Delta r\cos\theta \tag{7.1}$$

If a varying force does work on a particle as the particle moves along the x axis from x_i to x_f, the work done by the force on the particle is given by

$$W = \int_{x_i}^{x_f} F_x \, dx \qquad (7.7)$$

where F_x is the component of force in the x direction.

The **scalar product** (dot product) of two vectors $\vec{A}$ and $\vec{B}$ is defined by the relationship

$$\vec{A} \cdot \vec{B} \equiv AB \cos \theta \qquad (7.2)$$

where the result is a scalar quantity and θ is the angle between the two vectors. The scalar product obeys the commutative and distributive laws.

The **kinetic energy** of a particle of mass m moving with a speed v is

$$K \equiv \tfrac{1}{2}mv^2 \qquad (7.16)$$

If a particle of mass m is at a distance y above the Earth's surface ($y = 0$), the **gravitational potential energy** of the particle–Earth system is

$$U_g \equiv mgy \qquad (7.19)$$

The **elastic potential energy** stored in a spring of force constant k is

$$U_s \equiv \tfrac{1}{2}kx^2 \qquad (7.22)$$

A force is **conservative** if the work it does on a particle that is a member of the system as the particle moves between two points is independent of the path the particle takes between the two points. Furthermore, a force is conservative if the work it does on a particle is zero when the particle moves through an arbitrary closed path and returns to its initial position. A force that does not meet these criteria is said to be **nonconservative.**

The **total mechanical energy of a system** is defined as the sum of the kinetic energy and the potential energy:

$$E_{\text{mech}} \equiv K + U \qquad (7.24)$$

▶ Concepts and Principles

The **work–kinetic energy theorem** states that if work is done on a system by external forces and the only change in the system is in the speeds of its members,

$$W_{\text{ext}} = K_f - K_i = \Delta K = \tfrac{1}{2}mv_f^2 - \tfrac{1}{2}mv_i^2 \qquad (7.15, 7.17)$$

If the only change is in the configuration of the system,

$$W_{\text{ext}} = \Delta U \qquad (7.20, 7.23)$$

A **potential energy function** U can be associated only with a conservative force. If a conservative force $\vec{F}$ acts between members of a system while one member moves along the x axis from x_i to x_f, the change in the potential energy of the system equals the negative of the work done by that force:

$$U_f - U_i = -\int_{x_i}^{x_f} F_x \, dx \qquad (7.27)$$

Systems can be in three types of equilibrium configurations when the net force on a member of the system is zero. Configurations of **stable equilibrium** correspond to those for which $U(x)$ has a minimum.

Configurations of **unstable equilibrium** correspond to those for which $U(x)$ has a maximum.

Neutral equilibrium arises when U is constant as a member of the system moves over some region.

Think–Pair–Share

See the Preface for an explanation of the icons used in this problems set. For additional assessment items for this section, go to **WEBASSIGN** From Cengage

1. You are working in a manufacturing plant. One of the machines in the plant uses a spring. For a new process, it would be desirable for springs with different force constants to be used during different portions of the process. Your boss asks you for advice on how to change out one spring for another quickly so that the entire process can take place in a reasonable amount of time. You suggest that, rather than changing out springs continuously, you can change the force constant of *one* long spring by clamping it at various locations to define a new fixed end of the spring. Then the effective spring consists only of those coils beyond the clamp. You design a system consisting of one long spring with N coils and a force constant k. You design a clamping system that will isolate part of the spring, leaving N' coils free beyond the fixed clamp. (a) Write an expression for the force constant k' of the free end of the spring in terms of k, N, and N'. (b) The end of the unclamped, relaxed spring is grasped and pulled outward by a distance x. In the process, the hand holding the end of the spring does work W on the spring. Now, the spring is returned to its relaxed state and then clamped at its center point. The free end of the clamped, relaxed spring is grasped and pulled outward by the same distance x. How much work does the hand do on the spring in this case? W? $2W$? $4W$? Another value?

2. **ACTIVITY** In the table of data, we see minimum stopping distances d as a function of the initial speed v of a car. Work in your group to answer the following. (a) If you double the initial speed, does it take twice the distance to stop the car? (b) Assume the stopping distance is proportional to the speed of the car raised to some power: $d \propto v^n$. Use graphing techniques to determine n. (c) Why does the stopping distance depend on the particular value of n that you found in (b)?

Speed (mi/h)	Stopping Distance (ft)
20	22.5
25	35.0
30	50.4
35	68.6
40	89.6
45	113.5
50	140.0
55	169.5
60	201.7
65	236.7
70	274.5

Problems

See the Preface for an explanation of the icons used in this problems set.
For additional assessment items for this section, go to **WEBASSIGN**
From Cengage

SECTION 7.2 Work Done by a Constant Force

1. **Q|C** A shopper in a supermarket pushes a cart with a force of 35.0 N directed at an angle of 25.0° below the horizontal. The force is just sufficient to balance various friction forces, so the cart moves at constant speed. (a) Find the work done by the shopper on the cart as she moves down a 50.0-m-long aisle. (b) The shopper goes down the next aisle, pushing horizontally and maintaining the same speed as before. If the friction force doesn't change, would the shopper's applied force be larger, smaller, or the same? (c) What about the work done on the cart by the shopper?

2. The record number of boat lifts, including the boat and its ten crew members, was achieved by Sami Heinonen and Juha Räsänen of Sweden in 2000. They lifted a total mass of 653.2 kg approximately 4 in. off the ground a total of 24 times. Estimate the total work done by the two men on the boat in this record lift, ignoring the negative work done by the men when they lowered the boat back to the ground.

3. In 1990, Walter Arfeuille of Belgium lifted a 281.5-kg object through a distance of 17.1 cm using only his teeth. (a) How much work was done on the object by Arfeuille in this lift, assuming the object was lifted at constant speed? (b) What total force was exerted on Arfeuille's teeth during the lift?

4. **T** Spiderman, whose mass is 80.0 kg, is dangling on the free end of a 12.0-m-long rope, the other end of which is fixed to a tree limb above. By repeatedly bending at the waist, he is able to get the rope in motion, eventually getting it to swing enough that he can reach a ledge when the rope makes a 60.0° angle with the vertical. How much work was done by the gravitational force on Spiderman in this maneuver?

SECTION 7.3 The Scalar Product of Two Vectors

5. **S** For any two vectors $\vec{A}$ and $\vec{B}$, show that $\vec{A} \cdot \vec{B} = A_x B_x + A_y B_y + A_z B_z$. *Suggestions:* Write $\vec{A}$ and $\vec{B}$ in unit-vector form and use Equations 7.4 and 7.5.

6. Vector $\vec{A}$ has a magnitude of 5.00 units, and vector $\vec{B}$ has a magnitude of 9.00 units. The two vectors make an angle of 50.0° with each other. Find $\vec{A} \cdot \vec{B}$.

Note: In Problems 7 and 8, calculate numerical answers to three significant figures as usual.

7. Find the scalar product of the vectors in Figure P7.7.

Figure P7.7

8. Using the definition of the scalar product, find the angles between (a) $\vec{A} = 3\hat{i} - 2\hat{j}$ and $\vec{B} = 4\hat{i} - 4\hat{j}$, (b) $\vec{A} = -2\hat{i} + 4\hat{j}$ and $\vec{B} = 3\hat{i} - 4\hat{j} + 2\hat{k}$, and (c) $\vec{A} = \hat{i} - 2\hat{j} + 2\hat{k}$ and $\vec{B} = 3\hat{j} + 4\hat{k}$.

SECTION 7.4 Work Done by a Varying Force

9. **V** A particle is subject to a force F_x that varies with position as shown in Figure P7.9. Find the work done by the force on the particle as it moves (a) from $x = 0$ to $x = 5.00$ m, (b) from $x = 5.00$ m to $x = 10.0$ m, and (c) from $x = 10.0$ m to $x = 15.0$ m. (d) What is the total work done by the force over the distance $x = 0$ to $x = 15.0$ m?

Figure P7.9 Problems 9 and 22.

10. In a control system, an accelerometer consists of a 4.70-g object sliding on a calibrated horizontal rail. A low-mass spring attaches the object to a flange at one end of the rail. Grease on the rail makes static friction negligible, but rapidly damps out vibrations of the sliding object. When subject to a steady acceleration of $0.800g$, the object should be at a location 0.500 cm away from its equilibrium position. Find the force constant of the spring required for the calibration to be correct.

11. When a 4.00-kg object is hung vertically on a certain light
AMT spring that obeys Hooke's law, the spring stretches 2.50 cm.
T If the 4.00-kg object is removed, (a) how far will the spring stretch if a 1.50-kg block is hung on it? (b) How much work must an external agent do to stretch the same spring 4.00 cm from its unstretched position?

12. Express the units of the force constant of a spring in SI fun-
S damental units.

13. The tray dispenser in your cafeteria has broken and is not
CR repairable. The custodian knows that you are good at designing things and asks you to help him build a new dispenser out of spare parts he has on his workbench. The tray dispenser supports a stack of trays on a shelf that is supported by four springs, one at each corner of the shelf. Each tray is rectangular, with dimensions 45.3 cm by 35.6 cm. Each tray is 0.450 cm thick and has a mass of 580 g. The custodian asks you to design a new four-spring dispenser such that when a tray is removed, the dispenser pushes up the remaining stack so that the top tray is at the same position as the just-removed tray was. He has a wide variety of springs that he can use to build the dispenser. Which springs should he use?

14. A light spring with force constant 3.85 N/m is compressed by 8.00 cm as it is held between a 0.250-kg block on the left and a 0.500-kg block on the right, both resting on a horizontal surface. The spring exerts a force on each block, tending to push the blocks apart. The blocks are simultaneously released from rest. Find the acceleration with which each block starts to move, given that the coefficient of kinetic friction between each block and the surface is (a) 0, (b) 0.100, and (c) 0.462.

15. A small particle of mass
S m is pulled to the top of a frictionless half-cylinder (of radius R) by a light cord that passes over the top of the cylinder as illustrated in Figure P7.15. (a) Assuming the particle moves at a constant speed,

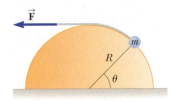

Figure P7.15

show that $F = mg \cos \theta$. *Note:* If the particle moves at constant speed, the component of its acceleration tangent to the cylinder must be zero at all times. (b) By directly integrating $W = \int \vec{F} \cdot d\vec{r}$, find the work done in moving the particle at constant speed from the bottom to the top of the half-cylinder.

16. The force acting on a particle is $F_x = (8x - 16)$, where F is in newtons and x is in meters. (a) Make a plot of this force versus x from $x = 0$ to $x = 3.00$ m. (b) From your graph, find the net work done by this force on the particle as it moves from $x = 0$ to $x = 3.00$ m.

17. When different loads hang on a spring, the spring
QC stretches to different lengths as shown in the following

table. (a) Make a graph of the applied force versus the extension of the spring. (b) By least-squares fitting, determine the straight line that best fits the data. (c) To complete part (b), do you want to use all the data points, or should you ignore some of them? Explain. (d) From the slope of the best-fit line, find the spring constant k. (e) If the spring is extended to 105 mm, what force does it exert on the suspended object?

F (N)	2.0	4.0	6.0	8.0	10	12	14	16	18	20	22
L (mm)	15	32	49	64	79	98	112	126	149	175	190

18. A 100-g bullet is fired from a rifle having a barrel 0.600 m long. Choose the origin to be at the location where the bullet begins to move. Then the force (in newtons) exerted by the expanding gas on the bullet is $15\,000 + 10\,000x - 25\,000x^2$, where x is in meters. (a) Determine the work done by the gas on the bullet as the bullet travels the length of the barrel. (b) **What If?** If the barrel is 1.00 m long, how much work is done, and (c) how does this value compare with the work calculated in part (a)?

19. (a) A force $\vec{F} = (4x\hat{i} + 3y\hat{j})$, where $\vec{F}$ is in newtons and x
V and y are in meters, acts on an object as the object moves in the x direction from the origin to $x = 5.00$ m. Find the work $W = \int \vec{F} \cdot d\vec{r}$ done by the force on the object. (b) **What If?** Find the work $W = \int \vec{F} \cdot d\vec{r}$ done by the force on the object if it moves from the origin to (5.00 m, 5.00 m) along a straight-line path making an angle of 45.0° with the positive x axis. Is the work done by this force dependent on the path taken between the initial and final points?

20. **Review.** The graph in Figure P7.20 specifies a functional relationship between the two variables u and v. (a) Find $\int_a^b u\,dv$. (b) Find $\int_b^a u\,dv$. (c) Find $\int_a^b v\,du$.

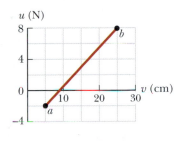

Figure P7.20

SECTION 7.5 Kinetic Energy and the Work–Kinetic Energy Theorem

21. A 0.600-kg particle has a speed of 2.00 m/s at point Ⓐ and
V kinetic energy of 7.50 J at point Ⓑ. What is (a) its kinetic energy at Ⓐ, (b) its speed at Ⓑ, and (c) the net work done on the particle by external forces as it moves from Ⓐ to Ⓑ?

22. A 4.00-kg particle is subject to a net force that varies with
V position as shown in Figure P7.9. The particle starts from rest at $x = 0$. What is its speed at (a) $x = 5.00$ m, (b) $x = 10.0$ m, and (c) $x = 15.0$ m?

23. A 2 100-kg pile driver is used to drive a steel I-beam into the
T ground. The pile driver falls 5.00 m before coming into contact with the top of the beam, and it drives the beam 12.0 cm farther into the ground before coming to rest. Using energy considerations, calculate the average force the beam exerts on the pile driver while the pile driver is brought to rest.

24. **Review.** In an electron microscope, there is an electron gun
AMT that contains two charged metallic plates 2.80 cm apart. An electric force accelerates each electron in the beam from rest to 9.60% of the speed of light over this distance.

(a) Determine the kinetic energy of the electron as it leaves the electron gun. Electrons carry this energy to a phosphorescent viewing screen where the microscope's image is formed, making it glow. For an electron passing between the plates in the electron gun, determine (b) the magnitude of the constant electric force acting on the electron, (c) the acceleration of the electron, and (d) the time interval the electron spends between the plates.

25. **Review.** You can think of the work–kinetic energy theorem as a second theory of motion, parallel to Newton's laws in describing how outside influences affect the motion of an object. In this problem, solve parts (a), (b), and (c) separately from parts (d) and (e) so you can compare the predictions of the two theories. A 15.0-g bullet is accelerated from rest to a speed of 780 m/s in a rifle barrel of length 72.0 cm. (a) Find the kinetic energy of the bullet as it leaves the barrel. (b) Use the work–kinetic energy theorem to find the net work that is done on the bullet. (c) Use your result to part (b) to find the magnitude of the average net force that acted on the bullet while it was in the barrel. (d) Now model the bullet as a particle under constant acceleration. Find the constant acceleration of a bullet that starts from rest and gains a speed of 780 m/s over a distance of 72.0 cm. (e) Modeling the bullet as a particle under a net force, find the net force that acted on it during its acceleration. (f) What conclusion can you draw from comparing your results of parts (c) and (e)?

26. You are lying in your bedroom, resting after doing your physics homework. As you stare at your ceiling, you come up with the idea for a new game. You grab a dart with a sticky nose and a mass of 19.0 g. You also grab a spring that has been lying on your desk from some previous project. You paint a target pattern on your ceiling. Your new game is to place the spring vertically on the floor, place the sticky-nose dart facing upward on the spring, and push the spring downward until the coils all press together, as on the right in Figure P7.26. You will then release the spring, firing the dart up toward the target on your ceiling, where its sticky nose will make it hang from the ceiling. The spring has an uncompressed end-to-end length of 5.00 cm, as shown on the left in Figure P7.26, and can be compressed to an end-to-end length of 1.00 cm when the coils are all pressed together. Before trying the game, you hold the upper end of the spring in one hand and hang a bundle of ten identical darts from the lower end of the spring. The spring extends by 1.00 cm due to the weight of the darts. You are so excited about the new game that, before doing a test of the game, you run out to gather your friends to show them. When your friends are in your room watching and you show them the first firing of your new game, why are you embarrassed?

Figure P7.26

27. **Review.** A 5.75-kg object passes through the origin at time $t = 0$ such that its x component of velocity is 5.00 m/s and its y component of velocity is -3.00 m/s. (a) What is the kinetic energy of the object at this time? (b) At a later time $t = 2.00$ s, the particle is located at $x = 8.50$ m and $y = 5.00$ m. What constant force acted on the object during this time interval? (c) What is the speed of the particle at $t = 2.00$ s?

28. **Review.** A 7.80-g bullet moving at 575 m/s strikes the hand of a superhero, causing the hand to move 5.50 cm in the direction of the bullet's velocity before stopping. (a) Use work and energy considerations to find the average force that stops the bullet. (b) Assuming the force is constant, determine how much time elapses between the moment the bullet strikes the hand and the moment it stops moving.

SECTION 7.6 Potential Energy of a System

29. A 0.20-kg stone is held 1.3 m above the top edge of a water well and then dropped into it. The well has a depth of 5.0 m. Relative to the configuration with the stone at the top edge of the well, what is the gravitational potential energy of the stone–Earth system (a) before the stone is released and (b) when it reaches the bottom of the well? (c) What is the change in gravitational potential energy of the system from release to reaching the bottom of the well?

30. A 1 000-kg roller coaster car is initially at the top of a rise, at point Ⓐ. It then moves 135 ft, at an angle of 40.0° below the horizontal, to a lower point Ⓑ. (a) Choose the car at point Ⓑ to be the zero configuration for gravitational potential energy of the roller coaster–Earth system. Find the potential energy of the system when the car is at points Ⓐ and Ⓑ, and the change in potential energy as the car moves between these points. (b) Repeat part (a), setting the zero configuration with the car at point Ⓐ.

SECTION 7.7 Conservative and Nonconservative Forces

31. A 4.00-kg particle moves from the origin to position Ⓒ, having coordinates $x = 5.00$ m and $y = 5.00$ m (Fig. P7.31). One force on the particle is the gravitational force acting in the negative y direction. Using Equation 7.3, calculate the work done by the gravitational force on the particle as it goes from O to Ⓒ along (a) the purple path, (b) the red path, and (c) the blue path. (d) Your results should all be identical. Why?

Figure P7.31
Problems 31 through 33

32. (a) Suppose a constant force acts on an object. The force does not vary with time or with the position or the velocity of the object. Start with the general definition for work done by a force

$$W = \int_i^f \vec{\mathbf{F}} \cdot d\vec{\mathbf{r}}$$

and show that the force is conservative. (b) As a special case, suppose the force $\vec{\mathbf{F}} = (3\hat{\mathbf{i}} + 4\hat{\mathbf{j}})$ N acts on a particle that moves from O to Ⓒ in Figure P7.31. Calculate the work done by $\vec{\mathbf{F}}$ on the particle as it moves along each one of the three paths shown in the figure and show that the work done along the three paths is identical. (c) **What If?** Is the work done also identical along the three paths for the force $\vec{F} = (4x\hat{i} + 3y\hat{j})$, where $\vec{F}$ is in newtons and x and y are in meters, from Problem 19? (d) **What If?** Suppose the force is given by $\vec{F} = (y\hat{i} - x\hat{j})$, where $\vec{F}$ is in newtons and x and y are in meters. Is the work done identical along the three paths for this force?

33. **Q|C** **T** A force acting on a particle moving in the xy plane is given by $\vec{F} = (2y\hat{i} + x^2\hat{j})$, where $\vec{F}$ is in newtons and x and y are in meters. The particle moves from the origin to a final position having coordinates $x = 5.00$ m and $y = 5.00$ m as shown in Figure P7.31. Calculate the work done by $\vec{F}$ on the particle as it moves along (a) the purple path, (b) the red path, and (c) the blue path. (d) Is $\vec{F}$ conservative or nonconservative? (e) Explain your answer to part (d).

SECTION 7.8 Relationship Between Conservative Forces and Potential Energy

34. *Why is the following situation impossible?* A librarian lifts a book from the ground to a high shelf, doing 20.0 J of work in the lifting process. As he turns his back, the book falls off the shelf back to the ground. The gravitational force from the Earth on the book does 20.0 J of work on the book while it falls. Because the work done was 20.0 J + 20.0 J = 40.0 J, the book hits the ground with 40.0 J of kinetic energy.

35. **T** A single conservative force acts on a 5.00-kg particle within a system due to its interaction with the rest of the system. The equation $F_x = 2x + 4$ describes the force, where F_x is in newtons and x is in meters. As the particle moves along the x axis from $x = 1.00$ m to $x = 5.00$ m, calculate (a) the work done by this force on the particle, (b) the change in the potential energy of the system, and (c) the kinetic energy the particle has at $x = 5.00$ m if its speed is 3.00 m/s at $x = 1.00$ m.

36. A potential energy function for a system in which a two-dimensional force acts is of the form $U = 3x^3y - 7x$. Find the force that acts at the point (x, y).

37. **S** The potential energy of a system of two particles separated by a distance r is given by $U(r) = A/r$, where A is a constant. Find the radial force $\vec{F}_r$ that each particle exerts on the other.

SECTION 7.9 Energy Diagrams and Equilibrium of a System

38. For the potential energy curve shown in Figure P7.38, (a) determine whether the force F_x is positive, negative, or zero at the five points indicated. (b) Indicate points of stable, unstable, and neutral equilibrium. (c) Sketch the curve for F_x versus x from $x = 0$ to $x = 9.5$ m.

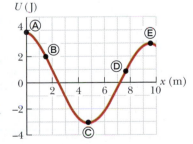

Figure P7.38

39. A right circular cone can theoretically be balanced on a horizontal surface in three different ways. Sketch these three equilibrium configurations and identify them as positions of stable, unstable, or neutral equilibrium.

ADDITIONAL PROBLEMS

40. The potential energy function for a system of particles is given by $U(x) = -x^3 + 2x^2 + 3x$, where x is the position of one particle in the system. (a) Determine the force F_x on the particle as a function of x. (b) For what values of x is the force equal to zero? (c) Plot $U(x)$ versus x and F_x versus x and indicate points of stable and unstable equilibrium.

41. **CR** You have a new internship, where you are helping to design a new freight yard for the train station in your city. There will be a number of dead-end sidings where single cars can be stored until they are needed. To keep the cars from running off the tracks at the end of the siding, you have designed a combination of two coiled springs as illustrated in Figure P7.41. When a car moves to the right in the figure and strikes the springs, they exert a force to the left on the car to slow it down.

Both springs are described by Hooke's law and have spring constants $k_1 = 1\,600$ N/m and $k_2 = 3\,400$ N/m. After the first spring compresses by a distance of $d = 30.0$ cm, the second spring acts with the first to increase the force to the left on the car in Figure P7.41. When the spring with spring constant k_2 compresses by 50.0 cm, the coils of both springs are pressed together, so that the springs can no longer compress. A typical car on the siding has a mass of 6 000 kg. When you present your design to your supervisor, he asks you for the maximum speed that a car can have and be stopped by your device.

Figure P7.41

42. **S** When an object is displaced by an amount x from stable equilibrium, a restoring force acts on it, tending to return the object to its equilibrium position. The magnitude of the restoring force can be a complicated function of x. In such cases, we can generally imagine the force function $F(x)$ to be expressed as a power series in x as $F(x) = -(k_1 x + k_2 x^2 + k_3 x^3 + \cdots)$. The first term here is just Hooke's law, which describes the force exerted by a simple spring for small displacements. For small excursions from equilibrium, we generally ignore the higher-order terms, but in some cases it may be desirable to keep the second term as well. If we model the restoring force as $F = -(k_1 x + k_2 x^2)$, how much work is done on an object in displacing it from $x = 0$ to $x = x_{max}$ by an applied force $-F$?

43. A particle moves along the x axis from $x = 12.8$ m to $x = 23.7$ m under the influence of a force

$$F = \frac{375}{x^3 + 3.75x}$$

where F is in newtons and x is in meters. Using numerical integration, determine the work done by this force on the particle during this displacement. Your result should be accurate to within 2%.

44. *Why is the following situation impossible?* In a new casino, a supersized pinball machine is introduced. Casino advertising boasts that a professional basketball player can lie on top of the machine and his head and feet will not hang off the edge! The ball launcher in the machine sends metal balls up one side of the machine and then into play. The spring in the launcher (Fig. P7.44) has a force constant of 1.20 N/cm. The surface on which the ball moves is inclined $\theta = 10.0°$ with respect to the horizontal. The spring is initially compressed its maximum distance $d = 5.00$ cm. A ball of mass 100 g is projected into play by releasing the plunger. Casino visitors find the play of the giant machine quite exciting.

Figure P7.44

45. **Review.** Two constant forces act on an object of mass $m = 5.00$ kg moving in the xy plane as shown in Figure P7.45. Force $\vec{F}_1$ is 25.0 N at 35.0°, and force $\vec{F}_2$ is 42.0 N at 150°. At time $t = 0$, the object is at the origin and has velocity $(4.00\hat{i} + 2.50\hat{j})$ m/s. (a) Express the two forces in unit-vector notation. Use unit-vector notation for your other answers. (b) Find the total force exerted on the object. (c) Find the object's acceleration. Now, considering the instant $t = 3.00$ s, find (d) the object's velocity, (e) its position, (f) its kinetic energy from $\frac{1}{2}mv_f^2$, and (g) its kinetic energy from $\frac{1}{2}mv_i^2 + \sum \vec{F} \cdot \Delta\vec{r}$. (h) What conclusion can you draw by comparing the answers to parts (f) and (g)?

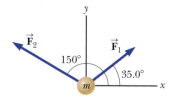

Figure P7.45

46. (a) Take $U = 5$ for a system with a particle at position $x = 0$ and calculate the potential energy of the system as a function of the particle position x. The force on the particle is given by $(8e^{-2x})\,\hat{i}$. (b) Explain whether the force is conservative or nonconservative and how you can tell.

47. An inclined plane of angle $\theta = 20.0°$ has a spring of force constant $k = 500$ N/m fastened securely at the bottom so that the spring is parallel to the surface as shown in Figure P7.47. A block of mass $m = 2.50$ kg is placed on the plane at a distance $d = 0.300$ m from

Figure P7.47
Problems 47 and 48.

the spring. From this position, the block is projected downward toward the spring with speed $v = 0.750$ m/s. By what distance is the spring compressed when the block momentarily comes to rest?

48. An inclined plane of angle θ has a spring of force constant k fastened securely at the bottom so that the spring is parallel to the surface. A block of mass m is placed on the plane at a distance d from the spring. From this position, the block is projected downward toward the spring with speed v as shown in Figure P7.47. By what distance is the spring compressed when the block momentarily comes to rest?

49. Over the Christmas break, you are making some extra money for buying presents by working in a factory, helping to move crates around. At one particular time, you find that all the handtrucks, dollies, and carts are in use, so you must move a crate across the room a straight-line distance of 35.0 m without the assistance of these devices. You notice that the crate has a rope attached to the middle of one of its vertical faces. You decide to move the crate by pulling on the rope. The crate has a mass of 130 kg, and the coefficient of kinetic friction between the crate and the concrete floor is 0.350. (a) Determine the angle relative to the horizontal at which you should pull upward on the rope so that you can move the crate over the desired distance with the force of the *smallest* magnitude. (b) At this angle of pulling on the rope, how much work do you do in dragging the crate over the desired distance?

CHALLENGE PROBLEM

50. A particle of mass $m = 1.18$ kg is attached between two identical springs on a frictionless, horizontal tabletop. Both springs have spring constant k and are initially unstressed, and the particle is at $x = 0$. (a) The particle is pulled a distance x along a direction perpendicular to the initial configuration of the springs as shown in Figure P7.50. Show that the force exerted by the springs on the particle is

$$\vec{F} = -2kx\left(1 - \frac{L}{\sqrt{x^2 + L^2}}\right)\hat{i}$$

(b) Show that the potential energy of the system is

$$U(x) = kx^2 + 2kL(L - \sqrt{x^2 + L^2})$$

(c) Make a plot of $U(x)$ versus x and identify all equilibrium points. Assume $L = 1.20$ m and $k = 40.0$ N/m. (d) If the particle is pulled 0.500 m to the right and then released, what is its speed when it reaches $x = 0$?

Overhead view

Figure P7.50

Conservation of Energy

◀ You use a table saw to make a cut in a piece of wood. How does the energy transfer to the saw to make the blade turn? (*George Rudy/Shutterstock.com*)

STORYLINE **In the previous chapter, you rubbed sandpaper on** wood and we associated the resulting warmth with internal energy. Now you look around the garage for more examples of energy. Your car is at rest in the garage now, but it has kinetic energy when in operation. How does it get that energy? From gasoline! But how did the gasoline get in the car? At the gasoline station! But where did the gasoline station get it? From the refinery! But where did the refinery get it? These questions go on and on! To get your mind off these questions, you start a long cut on a piece of wood with your table saw. Wait a minute! When in operation, the saw blade has rotational kinetic energy. Where does that energy come from? Ah-ha, you plugged it in, so energy is coming from the plug in the wall! But how did the energy get to the plug? It must come through the power lines from a power plant! But where does the power plant get the energy . . . ? As you continue to look around your garage, you see that energy must transfer into various devices for them to operate. And that energy must transfer out of something: a gasoline tank, a wall plug, batteries, and so on.

CONNECTIONS In the previous chapter, we found that energy can belong to a system in different forms. In this chapter, we will investigate ways that energy can *transfer* into or out of a system, or *transform* within a system. For example, in the system of the sandpaper and the wood for your carpentry project in Chapter 7, the kinetic energy of the sandpaper *transforms* to internal energy. On the other hand, for the system of your table saw in this chapter, energy *transfers* into the system by electricity to make it operate. We will see the full power of the energy approach in this chapter, embodied in the principle of *conservation of energy*. This approach will give us tools to solve problems that would be extremely difficult to solve with Newton's laws. In future chapters, we will see many cases in which the conservation of energy principle is applied in a variety of situations.

8.1 Analysis Model: Nonisolated System (Energy)

As we have seen, an object, modeled as a particle, can be acted on by various forces, resulting in a change in its kinetic energy according to the work–kinetic energy theorem from Chapter 7. If we choose the object as the system, this very simple situation is the first example of a *nonisolated system,* for which energy crosses the boundary of the system during some time interval due to an interaction with the environment. This scenario is common in physics problems. If a system does not interact with its environment, it is an *isolated system,* which we will study in Section 8.2.

The work–kinetic energy theorem (Eq. 7.17) is our first example of an energy equation appropriate for a nonisolated system. In the case of that theorem, the interaction of the system with its environment is the work done by the external force, and the quantity in the system that changes is the kinetic energy.

So far, we have seen only one way to transfer energy into a system: work. We mention below a few other ways to transfer energy into or out of a system. The details of these processes will be studied in other sections of the book, but they should be familiar to you from everyday experience. We illustrate mechanisms to transfer energy in Figure 8.1 and summarize them as follows.

Work, as we have learned in Chapter 7, is a method of transferring energy to a system by applying a force to the system such that the point of application of the force undergoes a displacement (Fig. 8.1a).

Mechanical waves (Chapters 16–17) are a means of transferring energy by allowing a disturbance to propagate through air or another medium. It is the method by which energy (which you detect as *sound*) leaves the system of your clock radio through the loudspeaker and enters your ears to stimulate the hearing process (Fig. 8.1b). Other examples of mechanical waves are seismic waves and ocean waves.

Energy is transferred to the block by *work.*

Energy leaves the radio from the speaker by *mechanical waves.*

Energy transfers to the handle of the spoon by *heat.*

Energy enters the automobile gas tank by *matter transfer.*

Energy enters the hair dryer by *electrical transmission.*

Energy leaves the light-bulb by *electromagnetic radiation.*

Figure 8.1 Energy transfer mechanisms. In each case, the system into which or from which energy is transferred is indicated.

Heat (Chapter 19) is a mechanism of energy transfer that is driven by a temperature difference between a system and its environment. For example, imagine dividing a metal spoon into two parts: the handle, which we identify as the system, and the portion submerged in a cup of coffee, which is part of the environment (Fig. 8.1c). The handle of the spoon becomes hot because fast-moving electrons and atoms in the submerged portion bump into slower ones in the nearby part of the handle. These particles move faster because of the collisions and bump into the next group of slow particles. Therefore, the internal energy of the spoon handle rises from energy transfer due to this collision process.

Matter transfer (Chapter 19) involves situations in which matter physically crosses the boundary of a system, carrying energy with it. Examples include filling the tank of your car in the opening storyline with gasoline (Fig. 8.1d) and carrying energy to the rooms of your home by circulating warm air from the furnace, a process called *convection*.

Electrical transmission (Chapters 26 and 27) involves energy transfer into or out of a system by means of electric currents. It is how energy transfers into your hair dryer (Fig. 8.1e), home theater system, or any other electrical device, such as the table saw in your garage in the opening storyline.

Electromagnetic radiation (Chapter 33) refers to electromagnetic waves such as visible light (Fig. 8.1f), microwaves, and radio waves crossing the boundary of a system. Examples of this method of transfer include cooking a baked potato in your microwave oven and energy traveling from the Sun to the Earth by light through space.[1]

A central feature of the energy approach is the notion that we can neither create nor destroy energy, that energy is always *conserved*. This feature has been tested in countless experiments, and no experiment has ever shown this statement to be incorrect. Therefore, **if the total amount of energy in a system changes, it can *only* be because energy has crossed the boundary of the system by a transfer mechanism such as one of the methods listed above.**

Energy is one of several quantities in physics that are conserved. We will see other conserved quantities in subsequent chapters. There are many physical quantities that do not obey a conservation principle. For example, there is no conservation of force principle or conservation of velocity principle. Similarly, in areas other than physical quantities, such as in everyday life, some quantities are conserved and some are not. For example, the money in the system of your bank account is a conserved quantity. The only way the account balance changes is if money crosses the boundary of the system by deposits or withdrawals. On the other hand, the number of people in the system of a country is not conserved. Although people indeed cross the boundary of the system, which changes the total population, the population can also change by people dying and by giving birth to new babies. Even if no people cross the system boundary, the births and deaths will change the number of people in the system. There is no equivalent in the concept of energy to dying or giving birth. The general statement of the principle of **conservation of energy** can be described mathematically with the **conservation of energy equation** as follows:

$$\Delta E_{system} = \sum T \tag{8.1}$$

◀ Conservation of energy

where E_{system} is the total energy of the system, including all methods of energy storage (kinetic, potential, and internal), T (for *transfer*) is the amount of energy transferred across the system boundary by some mechanism, and the sum is over all possible transfer mechanisms. Two of our transfer mechanisms have well-established symbolic notations. For work, $T_{work} = W$ as discussed in Chapter 7, and for heat, $T_{heat} = Q$ as defined in Chapter 19. (Now that we are familiar with work, we can simplify the appearance of equations by letting the simple symbol W represent the external work W_{ext} on a system. For internal work, we will *always* use

[1]Electromagnetic radiation and work done by field forces are the only energy transfer mechanisms that do not require molecules of the environment to be available at the system boundary. Therefore, systems surrounded by a vacuum (such as planets) can only exchange energy with the environment by means of these two possibilities.

W_{int} to differentiate it from W.) The other four members of our list do not have established symbols, so we will call them T_{MW} (mechanical waves), T_{MT} (matter transfer), T_{ET} (electrical transmission), and T_{ER} (electromagnetic radiation).

The full expansion of Equation 8.1 is

◀ The expanded conservation of energy equation

$$\Delta K + \Delta U + \Delta E_{\text{int}} = W + Q + T_{\text{MW}} + T_{\text{MT}} + T_{\text{ET}} + T_{\text{ER}} \tag{8.2}$$

which is the primary mathematical representation of the energy version of the analysis model of the **nonisolated system.** (We will see other versions of the nonisolated system model, involving linear momentum and angular momentum, in later chapters.) In most cases, Equation 8.2 reduces to a much simpler one because some of the terms are zero for the specific situation. If, for a given system, all terms on the right side of the conservation of energy equation are zero, the system is an *isolated system,* which we study in the next section.

The conservation of energy equation is no more complicated in theory than the process of balancing your checking account statement. If your account is the system, the change in the account balance for a given month is the sum of all the transfers: deposits, withdrawals, fees, interest, and checks written. You may find it useful to think of energy as the *currency of nature!*

Equation 8.2 represents a *general* situation; it covers all possibilities for situations in classical physics that we will find throughout this book. You don't need to memorize different energy equations for different situations. Equation 8.2 is the *only* equation you need to begin an energy approach to a problem solution. When using it to solve a problem, the procedure is to analyze the situation and set terms in Equation 8.2 that don't apply to the situation equal to zero. This will reduce Equation 8.2 to a smaller equation that is appropriate to the situation. For example, suppose a force is applied to a nonisolated system and the point of application of the force moves through a displacement. Now suppose the only change in the system is in the speed of one or more components of the system. Then Equation 8.2 reduces to

$$\Delta K = W \tag{8.3}$$

which is the work–kinetic energy theorem. This theorem is a special case of the more general principle of conservation of energy. We shall see several more special cases in future chapters.

QUICK QUIZ **8.1** Consider a block sliding over a horizontal surface with friction. Ignore any sound the sliding might make. **(i)** If the system is the *block,* this system is (a) isolated (b) nonisolated (c) impossible to determine **(ii)** If the system is the *surface,* describe the system from the same set of choices. **(iii)** If the system is the *block and the surface,* describe the system from the same set of choices.

ANALYSIS MODEL **Nonisolated System (Energy)**

Imagine you have identified a system to be analyzed and have defined a system boundary. Energy can exist in the system in three forms: kinetic, potential, and internal. The total of that energy can be changed when energy crosses the system boundary by any of six transfer methods shown in the diagram here. The total change in the energy in the system is equal to the total amount of energy that has crossed the system boundary. The mathematical statement of that concept is expressed in the **conservation of energy equation**:

$$\Delta E_{\text{system}} = \Sigma\, T \tag{8.1}$$

The full expansion of Equation 8.1 shows the specific types of energy storage and transfer:

$$\Delta K + \Delta U + \Delta E_{int} = W + Q + T_{MW} + T_{MT} + T_{ET} + T_{ER} \qquad (8.2)$$

For a specific problem, this equation is generally reduced to a smaller number of terms by eliminating the terms that are equal to zero because they are not appropriate to the situation. See Conceptual Example 8.1, below.

Examples:

- a force does work on a system of a single object, changing its speed: the work–kinetic energy theorem, $W = \Delta K$
- a gas contained in a vessel has work done on it and experiences a transfer of energy by heat, resulting in a change in its temperature: the first law of thermodynamics, $\Delta E_{int} = W + Q$ (Chapter 19)
- an incandescent light bulb is turned on, with energy entering the filament by electricity, causing its temperature to increase, and leaving by light: $\Delta E_{int} = T_{ET} + T_{ER}$ (Chapter 26)
- a photon enters a metal, causing an electron to be ejected from the metal: the photoelectric effect, $\Delta K + \Delta U = T_{ER}$ (Chapter 39)

Conceptual Example 8.1 **Reducing Equation 8.2 in Specific Situations**

When using Equation 8.2 to solve a problem, the following steps should be remembered: (1) define the system; (2) identify the beginning and end of the time interval of interest; (3) identify initial and final configurations of the system (positions of objects in gravitational situations, extensions of springs, etc.) and assign appropriate reference values of potential energy; (4) write Equation 8.2, eliminating or setting terms equal to zero that do not apply in the situation. Consider the following examples. For each example, the system is provided and the time interval is from before the device is turned on until it has been operating for a few moments.

(A) Your television set.

$$\cancel{\Delta K} + \cancel{\Delta U} + \Delta E_{int} = \cancel{W} + Q + T_{MW} + \cancel{T_{MT}} + T_{ET} + T_{ER} \quad \rightarrow \quad \Delta E_{int} = Q + T_{MW} + T_{ET} + T_{ER}$$

Your television set is a nonisolated system, warming up after it is turned on, taking in energy by electricity in order to operate, and emitting energy by sound from the speakers, light from the screen, and heat from warm surfaces.

(B) Your gasoline-powered lawn mower. The time interval includes the process of filling the tank with gasoline.

$$\Delta K + \Delta U + \Delta E_{int} = W + Q + T_{MW} + T_{MT} + \cancel{T_{ET}} + \cancel{T_{ER}} \quad \rightarrow \quad \Delta K + \Delta U + \Delta E_{int} = W + Q + T_{MW} + T_{MT}$$

Your lawn mower is a nonisolated system, with a moving blade, an increased potential energy in the fuel that was added, and an increasing temperature as it operates. Energy entered the lawn mower when it was filled with fuel and leaves by sound and heat from hot surfaces. The goal of the device is the work that it does on grass as the grass is cut. Notice that there may be electrical processes associated with the spark plug of the mower engine, but these processes are *internal* to the system, so they do not represent energy crossing the boundary.

(C) A cup of tea being warmed in a microwave oven.

$$\cancel{\Delta K} + \cancel{\Delta U} + \Delta E_{int} = \cancel{W} + Q + \cancel{T_{MW}} + \cancel{T_{MT}} + \cancel{T_{ET}} + T_{ER} \quad \rightarrow \quad \Delta E_{int} = Q + T_{ER}$$

Your cup of tea is a nonisolated system, with an increasing temperature. The input of energy is by electromagnetic radiation: the microwaves. The heat term represents some energy transferring out of the hot tea into the lower-temperature air surrounding the cup.

8.2 Analysis Model: Isolated System (Energy)

In this section, we study another very common scenario in physics problems: a system is chosen such that no energy crosses the system boundary by any method. We begin by considering a gravitational situation. Think about the book–Earth system in Figure 7.16 in the preceding chapter. After we have lifted the book, there is gravitational potential energy stored in the system, which can be calculated from the work done by the external agent on the system, using $W = \Delta U_g$.

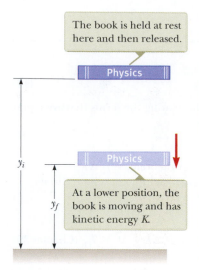

The book is held at rest here and then released.

Physics

y_i

Physics

y_f

At a lower position, the book is moving and has kinetic energy K.

Figure 8.2 A book is released from rest and falls due to work done by the gravitational force on the book.

PITFALL PREVENTION 8.2
Conditions on Equation 8.5 Equation 8.5 is only true for a system in which conservative forces act. We will see how to handle nonconservative forces in Sections 8.3 and 8.4.

The mechanical energy of ▶ an isolated system with no nonconservative forces acting is conserved.

The total energy of an ▶ isolated system is conserved.

(Check to see that this equation, which we've seen before as Eq. 7.20, is contained within Eq. 8.2 above.)

Now imagine dropping the book from the position to which you lifted it, as shown in Figure 8.2. The book–Earth system now does not interact with the environment, since your hand is no longer in contact with the book. As the book falls, the kinetic energy of the system, which is due to the motion of the book alone, increases, and the gravitational potential energy of the system decreases. From Equation 8.2, we see that

$$\Delta K + \Delta U_g = 0 \tag{8.4}$$

The left side of this equation represents a sum of changes of the energy stored in the system. There are no transfers of energy of any kind across the boundary of the system, so we set all terms on the right side of Equation 8.2 equal to zero; the book–Earth system is *isolated* from the environment. We developed this equation for a gravitational system, but it can be shown to be valid for a system with any type of potential energy. Therefore, for this isolated system,

$$\Delta K + \Delta U = 0 \tag{8.5}$$

(Check to see that this equation is contained within Eq. 8.2.) Notice what happens in this process. Energy is not *transferred* across the boundary of an isolated system. Rather, energy is *transformed* within the system, from one type to another. In the case of the falling book in Figure 8.2, the *transformation mechanism* is the internal work done on the book within the system by the gravitational force.

We defined in Chapter 7 the sum of the kinetic and potential energies of a system as its mechanical energy:

$$E_{\text{mech}} \equiv K + U \tag{8.6}$$

where U represents the total of *all* types of potential energy. Because the system under consideration is isolated, Equations 8.5 and 8.6 tell us that the mechanical energy of the system is conserved:

$$\Delta E_{\text{mech}} = 0 \tag{8.7}$$

Equation 8.7 is a statement of **conservation of mechanical energy** for an isolated system with no nonconservative forces acting. The mechanical energy in such a system is conserved: the sum of the kinetic and potential energies remains constant:

Let us now write the changes in energy in Equation 8.5 explicitly:

$$(K_f - K_i) + (U_f - U_i) = 0$$
$$K_f + U_f = K_i + U_i \tag{8.8}$$

For the gravitational situation of the falling book, Equation 8.8 can be written as

$$\tfrac{1}{2}mv_f^2 + mgy_f = \tfrac{1}{2}mv_i^2 + mgy_i \tag{8.9}$$

where $v_i = 0$ if the book in Figure 8.2 is dropped from rest. As the book falls to the Earth, the book–Earth system loses potential energy and gains kinetic energy such that the total of the two types of energy always remains constant: $E_{\text{total},i} = E_{\text{total},f}$.

If there are nonconservative forces acting within the system, mechanical energy is transformed to internal energy as discussed in Section 7.7. If nonconservative forces act in an isolated system, the total energy of the system is conserved, although the mechanical energy is not. In that case, we can express the conservation of energy of the system as

$$\Delta E_{\text{system}} = 0 \tag{8.10}$$

where E_{system} includes all kinetic, potential, and internal energies. This equation is the most general statement of the energy version of the **isolated system** model. It is equivalent to Equation 8.2 with all terms on the right-hand side equal to zero. When using the isolated or nonisolated system models as discussed here, we will add the qualifier *for energy* or *(energy)*. We will find in the next few chapters that there are isolated and nonisolated system models for other quantities as well.

Associated with this most general equation for the isolated system are a variety of new transformation mechanisms. Examples include nonconservative forces (the warming of the sandpaper in the Chapter 7 storyline), chemical reactions (an exploding firecracker), and nuclear reactions (operation of a nuclear reactor).

QUICK QUIZ 8.2 A rock of mass m is dropped to the ground from a height h. A second rock, with mass $2m$, is dropped from the same height. When the second rock strikes the ground, what is its kinetic energy? **(a)** twice that of the first rock **(b)** four times that of the first rock **(c)** the same as that of the first rock **(d)** half as much as that of the first rock **(e)** impossible to determine

QUICK QUIZ 8.3 Three identical balls are thrown from the top of a building, all with the same initial speed. As shown in Figure 8.3, the first is thrown horizontally, the second at some angle above the horizontal, and the third at some angle below the horizontal. Neglecting air resistance, rank the speeds of the balls at the instant each hits the ground.

Figure 8.3 (Quick Quiz 8.3) Three identical balls are thrown with the same initial speed from the top of a building.

ANALYSIS MODEL Isolated System (Energy)

Imagine you have identified a system to be analyzed and have defined a system boundary. Energy can exist in the system in three forms: kinetic, potential, and internal. Imagine also a situation in which no energy crosses the boundary of the system by any method. Then, the system is isolated; energy transforms from one form to another and Equation 8.2 becomes

System boundary

Kinetic energy
Potential energy
Internal energy

The total amount of energy in the system is constant. Energy transforms among the three possible types.

$$\Delta E_{system} = 0 \qquad (8.10)$$

If no nonconservative forces act within the isolated system, the mechanical energy of the system is conserved, so

$$\Delta E_{mech} = 0 \qquad (8.7)$$

Examples:

- an object is in free-fall; gravitational potential energy transforms to kinetic energy: $\Delta K + \Delta U = 0$
- a basketball rolling across a gym floor comes to rest; kinetic energy transforms to internal energy: $\Delta K + \Delta E_{int} = 0$
- a pendulum is raised and released with an initial speed; its motion eventually stops due to air resistance; gravitational potential energy and kinetic energy transform to internal energy, $\Delta K + \Delta U + \Delta E_{int} = 0$ (Chapter 15)
- a battery is connected to a resistor; chemical potential energy in the battery transforms to internal energy in both the battery and the resistor: $\Delta U + \Delta E_{int} = 0$ (Chapter 27)

Example 8.2 Ball in Free Fall

A ball of mass m is dropped from a height h above the ground as shown in Figure 8.4 (page 188).

(A) Neglecting air resistance, determine the speed of the ball when it is at a height y above the ground. Choose the system as the ball and the Earth.

SOLUTION

Conceptualize Figure 8.4 and our everyday experience with falling objects allow us to conceptualize the situation. Although we can readily solve this problem with the techniques of Chapter 2, let us practice an energy approach.

Categorize As suggested in the problem, we identify the system as the ball and the Earth. Because there is neither air resistance nor any other interaction between the system and the environment, the system is isolated and we use the *isolated system model* for energy. The only force between members of the system is the gravitational force, which is conservative.

continued

8.2 continued

Analyze   Because the system is isolated and there are no nonconservative forces acting within the system, we apply the principle of conservation of mechanical energy to the ball–Earth system. At the instant the ball is released, its kinetic energy is $K_i = 0$ and the gravitational potential energy of the system is $U_{gi} = mgh$. When the ball is at a position y above the ground, its kinetic energy is $K_f = \frac{1}{2}mv_f^2$ and the potential energy relative to the ground is $U_{gf} = mgy$.

Write the appropriate reduction of Equation 8.2, noting that the only types of energy in the system that change are kinetic energy and gravitational potential energy:

$$\Delta K + \Delta U_g = 0$$

Substitute for the energies:

$$(\tfrac{1}{2}mv_f^2 - 0) + (mgy - mgh) = 0$$

Solve for v_f:

$$v_f^2 = 2g(h - y) \;\rightarrow\; v_f = \boxed{\sqrt{2g(h - y)}}$$

The speed is always positive. If you had been asked to find the ball's velocity, you would use the negative value of the square root as the y component to indicate the downward motion.

(B) Find the speed of the ball again at height y by choosing the ball as the system.

SOLUTION

Categorize   In this case, the only type of energy in the system that changes is kinetic energy. A single object that can be modeled as a particle cannot possess potential energy. The effect of gravity is to do work on the ball across the boundary of the system. We use the *nonisolated system* model for energy.

Figure 8.4   (Example 8.2) A ball is dropped from a height h above the ground. Initially, the total energy of the ball–Earth system is gravitational potential energy, equal to mgh relative to the ground. At the position y, the total energy is the sum of the kinetic and potential energies.

*Figure labels:*
$y_i = h$, $U_{gi} = mgh$, $K_i = 0$
$y_f = y$, $U_{gf} = mgy$, $K_f = \frac{1}{2}mv_f^2$
$\vec{v}_f$
h, y
$y = 0$, $U_g = 0$

Analyze   Write the appropriate reduction of Equation 8.2:

$$\Delta K = W$$

Substitute for the initial and final kinetic energies and the work done by gravity:

$$(\tfrac{1}{2}mv_f^2 - 0) = \vec{\mathbf{F}}_g \cdot \Delta\vec{\mathbf{r}} = -mg\hat{\mathbf{j}} \cdot \Delta y\hat{\mathbf{j}}$$
$$= -mg\Delta y = -mg(y - h) = mg(h - y)$$

Solve for v_f:

$$v_f^2 = 2g(h - y) \;\rightarrow\; v_f = \boxed{\sqrt{2g(h - y)}}$$

Finalize   The final result is the same, regardless of the choice of system. In your future problem solving, keep in mind that the choice of system is yours to make. Sometimes the problem is much easier to solve if a judicious choice is made as to the system to analyze.

WHAT IF?   What if the ball were thrown downward from its highest position with a speed v_i? What would its speed be at height y?

Answer   If the ball is thrown downward initially, we would expect its speed at height y to be larger than if simply dropped. Make your choice of system, either the ball alone or the ball and the Earth. You should find that either choice gives you the following result:

$$v_f = \sqrt{v_i^2 + 2g(h - y)}$$

Example 8.3   **A Grand Entrance**

You are part of the stage crew for a theatrical company and are designing an apparatus to support an actor of mass 65.0 kg who is to "fly" down to the stage during the performance of a play. You attach the actor's harness to a 130-kg sandbag by means of a lightweight steel cable running smoothly over two frictionless pulleys as in Figure 8.5a. You need 3.00 m of cable between the harness and the nearest pulley so that the pulley can be hidden behind a curtain. For the apparatus to work successfully, the sandbag must never lift above the floor as the actor swings from above the stage to the floor. Let us call the initial angle that the actor's cable makes with the vertical θ. What is the maximum value θ can have before the sandbag lifts off the floor?

8.3 continued

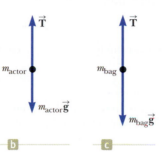

Figure 8.5 (Example 8.3) (a) An actor uses some clever staging to make his entrance. (b) The free-body diagram for the actor at the bottom of the circular path. (c) The free-body diagram for the sandbag if the normal force from the floor goes to zero.

SOLUTION

Conceptualize We must use several concepts to solve this problem. Imagine what happens as the actor approaches the bottom of the swing. At the bottom, the cable is vertical and must support his weight as well as provide centripetal acceleration of his body in the upward direction. At this point in his swing, the tension in the cable is the highest and the sandbag is most likely to lift off the floor.

Categorize Looking first at the swinging of the actor from the initial point to the lowest point, we model the actor and the Earth as an *isolated system* for energy. We ignore air resistance, so there are no nonconservative forces acting. You might initially be tempted to model the system as nonisolated because of the interaction of the system with the cable, which is in the environment. The force applied to the actor by the cable, however, is always perpendicular to each element of the displacement of the actor and hence does no work. Therefore, in terms of energy transfers across the boundary, the system is isolated.

Analyze We first find the actor's speed as he arrives at the floor as a function of the initial angle θ and the radius R of the circular path through which he swings. We use the particle model by choosing a particular point on the actor's body.

From the isolated system model, make the appropriate reduction of Equation 8.2 for the actor–Earth system:

$$(1) \quad \Delta K + \Delta U_g = 0$$

Let y_i be the initial height of the actor above the floor and v_f be his speed at the instant before he lands. (Notice that $K_i = 0$ because the actor starts from rest.) Insert the energies into Equation (1) and solve for the final speed of the actor.

$$\left(\tfrac{1}{2}m_{actor}v_f^2 - 0\right) + \left(m_{actor}gy_f - m_{actor}gy_i\right) = 0$$

$$(2) \quad v_f^2 = 2g(y_f - y_i)$$

From the geometry in Figure 8.5a, notice that $y_f - y_i = R - R\cos\theta = R(1 - \cos\theta)$. Use this relationship in Equation (2).

$$(3) \quad v_f^2 = 2gR(1 - \cos\theta)$$

Categorize Next, focus on the instant the actor is at the lowest point. Because the tension in the cable is transferred as a force applied to the sandbag, we model the actor at this instant as a *particle under a net force*. Because the actor moves along a circular arc, he experiences at the bottom of the swing a centripetal acceleration of v_f^2/R directed upward.

Analyze Apply Newton's second law from the particle under a net force model to the actor at the bottom of his path, using the free-body diagram in Figure 8.5b as a guide, and recognizing the acceleration as centripetal:

$$\sum F_y = T - m_{actor}g = m_{actor}\frac{v_f^2}{R}$$

$$(4) \quad T = m_{actor}g + m_{actor}\frac{v_f^2}{R}$$

Categorize Finally, notice that the sandbag lifts off the floor when the upward force exerted on it by the cable exceeds the gravitational force acting on it; the normal force from the floor is zero when that happens. We do *not*, however, want the sandbag to lift off the floor. The sandbag must remain at rest, so we model it as a *particle in equilibrium*.

continued

8.3 continued

Analyze   A force T of the magnitude given by Equation (4) is transmitted by the cable to the sandbag. If the sandbag remains at rest but is just ready to be lifted off the floor if any more force were applied by the cable, the normal force on it becomes zero and the particle in equilibrium model tells us that $T = m_{bag} g$ as in Figure 8.5c.

Substitute this condition and Equation (3) into Equation (4):

$$m_{bag}g = m_{actor}g + m_{actor}\frac{2gR(1 - \cos\theta)}{R}$$

Solve for $\cos\theta$ and substitute the given parameters:

$$\cos\theta = \frac{3m_{actor} - m_{bag}}{2m_{actor}} = \frac{3(65.0 \text{ kg}) - 130 \text{ kg}}{2(65.0 \text{ kg})} = 0.500$$

$$\theta = 60.0°$$

Finalize   Here we had to combine several analysis models from different areas of our study. Notice that the length R of the cable from the actor's harness to the leftmost pulley did not appear in the final algebraic equation for $\cos\theta$. Therefore, the final answer is independent of R.

Example **8.4**   **The Spring-Loaded Popgun**

The launching mechanism of a popgun consists of a trigger-released spring (Fig. 8.6a). The spring is compressed to a position $y_Ⓐ$ and the trigger is fired. The projectile of mass m rises to a maximum position $y_Ⓒ$ above the position at which it leaves the spring, indicated in Figure 8.6b as position $y_Ⓑ = 0$. Consider a firing of the gun for which $m = 35.0$ g, $y_Ⓐ = -0.120$ m, and $y_Ⓒ = 20.0$ m.

(A) Neglecting all resistive forces, determine the spring constant.

S O L U T I O N

Conceptualize   Imagine the process illustrated in parts (a) and (b) of Figure 8.6. The projectile starts from rest at Ⓐ, speeds up as the spring pushes upward on it, leaves the spring at Ⓑ, and then slows down as the gravitational force pulls downward on it, eventually coming to rest at point Ⓒ. Notice that there are two types of potential energy in this system: gravitational and elastic.

Figure 8.6 (Example 8.4) A spring-loaded popgun (a) before firing and (b) when the spring extends to its relaxed length. (c) An energy bar chart for the popgun–projectile–Earth system before the popgun is loaded. The energy in the system is zero. (d) The popgun is loaded by means of an external agent doing work on the system to push the spring downward. Therefore the system is nonisolated during this process. After the popgun is loaded, elastic potential energy is stored in the spring and the gravitational potential energy of the system is lower because the projectile is below point Ⓑ. (e) As the projectile passes through point Ⓑ, all of the energy of the isolated system is kinetic. (f) When the projectile reaches point Ⓒ, all of the energy of the isolated system is gravitational potential.

8.4 continued

Categorize We identify the system as the projectile, the spring, and the Earth. We ignore both air resistance on the projectile and friction in the gun, so we model the system as isolated for energy with no nonconservative forces acting.

Analyze Because the projectile starts from rest, its initial kinetic energy is zero. We choose the zero configuration for the gravitational potential energy of the system to be when the projectile leaves the spring at Ⓑ. For this configuration, the elastic potential energy is also zero.

After the gun is fired, the projectile rises to a maximum height $y_©$. The final kinetic energy of the projectile is zero.

From the isolated system model for energy, write a conservation of mechanical energy equation for the system between configurations when the projectile is at points Ⓐ and ©:

$$(1) \quad \Delta K + \Delta U_g + \Delta U_s = 0$$

Substitute for the initial and final energies:

$$(0 - 0) + (mgy_© - mgy_Ⓐ) + (0 - \tfrac{1}{2}kx^2) = 0$$

Solve for k:

$$k = \frac{2mg(y_© - y_Ⓐ)}{x^2}$$

Substitute numerical values:

$$k = \frac{2(0.035\ 0\ \text{kg})(9.80\ \text{m/s}^2)[20.0\ \text{m} - (-0.120\ \text{m})]}{(0.120\ \text{m})^2} = \boxed{958\ \text{N/m}}$$

(B) Find the speed of the projectile as it moves through the equilibrium position Ⓑ of the spring as shown in Figure 8.6b.

SOLUTION

Analyze The energy of the system as the projectile moves through the equilibrium position of the spring includes only the kinetic energy of the projectile $\tfrac{1}{2}mv_Ⓑ^2$. Both types of potential energy are equal to zero for this configuration of the system.

Write Equation (1) again for the system between configurations for which the projectile is at points Ⓐ and Ⓑ:

$$\Delta K + \Delta U_g + \Delta U_s = 0$$

Substitute for the initial and final energies:

$$(\tfrac{1}{2}mv_Ⓑ^2 - 0) + (0 - mgy_Ⓐ) + (0 - \tfrac{1}{2}kx^2) = 0$$

Solve for $v_Ⓑ$:

$$v_Ⓑ = \sqrt{\frac{kx^2}{m} + 2gy_Ⓐ}$$

Substitute numerical values:

$$v_Ⓑ = \sqrt{\frac{(958\ \text{N/m})(0.120\ \text{m})^2}{(0.035\ 0\ \text{kg})} + 2(9.80\ \text{m/s}^2)(-0.120\ \text{m})} = \boxed{19.8\ \text{m/s}}$$

Finalize This example is the first one we have seen in which we must include two different types of potential energy. Notice in part (A) that we never needed to consider anything about the speed of the ball between points Ⓐ and ©, which is part of the power of the energy approach: changes in kinetic and potential energy depend only on the initial and final values, not on what happens between the configurations corresponding to these values.

8.3 Situations Involving Kinetic Friction

Consider again the book in Figure 7.19a sliding to the right on the surface of a heavy table and slowing down due to the friction force. Work is done by the friction force on the book because there is a force and a displacement. Keep in mind, however, that our equations for work involve the displacement *of the point of application of the force*. A simple model of the friction force between the book and the surface is shown in Figure 8.7a (page 192). We have represented the entire friction force between the book and surface as being due to two identical teeth that have been spot-welded together.[2] One tooth projects upward from the surface, the other

---

[2]Figure 8.7 and its discussion are inspired by a classic article on friction: B. A. Sherwood and W. H. Bernard, "Work and heat transfer in the presence of sliding friction," *American Journal of Physics,* **52:**1001, 1984.

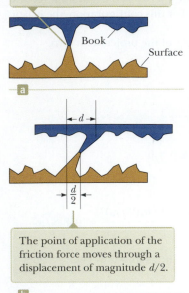

The entire friction force is modeled to be applied at the interface between two identical teeth projecting from the book and the surface.

Book

Surface

a

$\leftarrow d \rightarrow$

$\frac{d}{2}$

The point of application of the friction force moves through a displacement of magnitude $d/2$.

b

Figure 8.7 (a) A simplified model of friction between a book and a surface. (b) The book is moved to the right by a distance d.

downward from the book, and they are welded at the points where they touch. The friction force acts at the junction of the two teeth. Imagine that the book slides a small distance d to the right as in Figure 8.7b. Because the teeth are modeled as identical, the junction of the teeth moves to the right by a distance $d/2$. Therefore, the displacement of the point of application of the friction force is $d/2$, but the displacement of the book is d!

In reality, however, the friction force is spread out over the entire contact area of an object sliding on a surface, so the force is not localized at a point. In addition, because the magnitudes of the friction forces at various points are constantly changing as individual spot welds occur, the surface and the book deform locally, and so on, the displacement of the point of application of the friction force is not at all the same as the displacement of the book. In fact, the displacement of the point of application of the friction force is not calculable and so neither is the work done by the friction force.

The work–kinetic energy theorem is valid for a particle or an object that can be modeled as a particle. When a friction force acts, however, we cannot calculate the work done by friction. For such situations, Newton's second law is still valid for the system even though the work–kinetic energy theorem is not. The case of a nondeformable object like our book sliding on the surface[3] can be handled in a relatively straightforward way.

Starting from a situation in which forces, including friction, are applied to the book, we can follow a similar procedure to that done in developing Equation 7.17. Let us start by writing Equation 7.8 for all forces on an object other than friction:

$$\sum W_{\text{other forces}} = \int \left(\sum \vec{\mathbf{F}}_{\text{other forces}} \right) \cdot d\vec{\mathbf{r}} \tag{8.11}$$

The $d\vec{\mathbf{r}}$ in this equation is the displacement of the object because for forces other than friction, under the assumption that these forces do not deform the object, this displacement is the same as the displacement of the point of application of the forces. To each side of Equation 8.11 let us add the integral of the scalar product of the force of kinetic friction and $d\vec{\mathbf{r}}$. In doing so, we are not defining this quantity as work! We are simply saying that it is a quantity that can be calculated mathematically and will turn out to be useful to us in what follows.

$$\sum W_{\text{other forces}} + \int \vec{\mathbf{f}}_k \cdot d\vec{\mathbf{r}} = \int \left(\sum \vec{\mathbf{F}}_{\text{other forces}} \right) \cdot d\vec{\mathbf{r}} + \int \vec{\mathbf{f}}_k \cdot d\vec{\mathbf{r}}$$

$$= \int \left(\sum \vec{\mathbf{F}}_{\text{other forces}} + \vec{\mathbf{f}}_k \right) \cdot d\vec{\mathbf{r}}$$

The integrand on the right side of this equation is the net force $\sum \vec{\mathbf{F}}$ on the object, so

$$\sum W_{\text{other forces}} + \int \vec{\mathbf{f}}_k \cdot d\vec{\mathbf{r}} = \int \sum \vec{\mathbf{F}} \cdot d\vec{\mathbf{r}}$$

Incorporating Newton's second law $\sum \vec{\mathbf{F}} = m\vec{\mathbf{a}}$ gives

$$\sum W_{\text{other forces}} + \int \vec{\mathbf{f}}_k \cdot d\vec{\mathbf{r}} = \int m\vec{\mathbf{a}} \cdot d\vec{\mathbf{r}} = \int m\frac{d\vec{\mathbf{v}}}{dt} \cdot d\vec{\mathbf{r}} = \int_{t_i}^{t_f} m\frac{d\vec{\mathbf{v}}}{dt} \cdot \vec{\mathbf{v}}\, dt \tag{8.12}$$

where we have used Equation 4.3 to rewrite $d\vec{\mathbf{r}}$ as $\vec{\mathbf{v}}\, dt$. The scalar product obeys the product rule for differentiation (see Eq. B.30 in Appendix B.6), so the derivative of the scalar product of $\vec{\mathbf{v}}$ with itself can be written

$$\frac{d}{dt}(\vec{\mathbf{v}} \cdot \vec{\mathbf{v}}) = \frac{d\vec{\mathbf{v}}}{dt} \cdot \vec{\mathbf{v}} + \vec{\mathbf{v}} \cdot \frac{d\vec{\mathbf{v}}}{dt} = 2\frac{d\vec{\mathbf{v}}}{dt} \cdot \vec{\mathbf{v}}$$

---

[3]The overall shape of the book remains the same, which is why we say it is nondeformable. On a microscopic level, however, there is deformation of the book's face as it slides over the surface.

We used the commutative property of the scalar product to justify the final expression in this equation. Consequently,

$$\frac{d\vec{\mathbf{v}}}{dt}\cdot\vec{\mathbf{v}} = \tfrac{1}{2}\frac{d}{dt}(\vec{\mathbf{v}}\cdot\vec{\mathbf{v}}) = \tfrac{1}{2}\frac{dv^2}{dt}$$

Substituting this result into Equation 8.12 gives

$$\sum W_{\text{other forces}} + \int \vec{\mathbf{f}}_k\cdot d\vec{\mathbf{r}} = \int_{t_i}^{t_f} m\left(\tfrac{1}{2}\frac{dv^2}{dt}\right)dt = \tfrac{1}{2}m\int_{v_i}^{v_f} d(v^2) = \tfrac{1}{2}mv_f^2 - \tfrac{1}{2}mv_i^2 = \Delta K$$

Looking at the left side of this equation, notice that in the inertial frame of the surface, $\vec{\mathbf{f}}_k$ and $d\vec{\mathbf{r}}$ will be in opposite directions for every increment $d\vec{\mathbf{r}}$ of the path followed by the object. Therefore, $\vec{\mathbf{f}}_k\cdot d\vec{\mathbf{r}} = -f_k\,dr$. The previous expression now becomes

$$\sum W_{\text{other forces}} - \int f_k\,dr = \Delta K$$

In our model for friction, the magnitude of the kinetic friction force is constant, so f_k can be brought out of the integral. The remaining integral $\int dr$ is simply the sum of increments of length along the path, which is the total path length d. Therefore,

$$W - f_k d = \Delta K \tag{8.13}$$

where W represents the work done on the object by all forces other than friction. Equation 8.13 can be used when a friction force acts on an object. The change in kinetic energy is equal to the work done by all forces other than friction minus a term $f_k d$ associated with the friction force.

Considering the sliding book situation again, let's identify the larger system of the book *and* the surface as the book slows down under the influence of a friction force alone. There is no work done across the boundary of this system by other forces because the system does not interact with the environment. There are no other types of energy transfer occurring across the boundary of the system, assuming we ignore the inevitable sound the sliding book makes! In this case, Equation 8.2 becomes

$$\Delta K + \Delta E_{\text{int}} = 0$$

The change in kinetic energy of this book–surface system is the same as the change in kinetic energy of the book alone because the book is the only part of the system that is moving. Therefore, incorporating Equation 8.13 with no work done by other forces gives

$$-f_k d + \Delta E_{\text{int}} = 0$$

$$\Delta E_{\text{int}} = f_k d \tag{8.14}$$

◄ Change in internal energy due to a constant friction force within the system

Equation 8.14 tells us that the increase in internal energy of the system is equal to the product of the friction force and the path length through which the block moves. In summary, a friction force transforms kinetic energy in a system to internal energy. If work is done on the system by forces other than friction, Equation 8.13, with the help of Equation 8.14, can be written as

$$W = \Delta K + \Delta E_{\text{int}} \tag{8.15}$$

which is a reduced form of Equation 8.2 and represents the nonisolated system model for energy for a system within which a nonconservative force acts. For any system in which the force of kinetic friction acts between members of the system, we can write the full form of Equation 8.2, reduce it accordingly, and then use Equation 8.14 to substitute for the change in the internal energy.

QUICK QUIZ 8.4 You are traveling along a freeway at 65 mi/h. Your car has kinetic energy. You suddenly skid to a stop because of congestion in traffic. Where is the kinetic energy your car once had? **(a)** It is all in internal energy in the road. **(b)** It is all in internal energy in the tires. **(c)** Some of it has transformed to internal energy and some of it transferred away by mechanical waves. **(d)** It is all transferred away from your car by various mechanisms.

Example 8.5 **A Block Pulled on a Rough Surface**

A 6.0-kg block initially at rest is pulled to the right along a horizontal surface by a constant horizontal force of magnitude 12 N.

(A) Find the speed of the block after it has moved 3.0 m if the surfaces in contact have a coefficient of kinetic friction of 0.15.

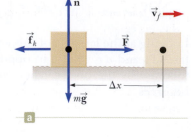

SOLUTION

Conceptualize   This example is similar to Example 7.6 (page 163), but modified so that the surface is no longer frictionless. The rough surface applies a friction force on the block opposite to the applied force. As a result, we expect the speed to be lower than that found in Example 7.6.

Categorize   The block is pulled by a force and the surface is rough, so the block and the surface are modeled as a *nonisolated system* for energy with a nonconservative force acting.

Analyze   Figure 8.8a illustrates this situation. Neither the normal force nor the gravitational force does work on the system because their points of application are displaced horizontally.

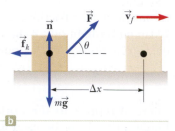

Figure 8.8 (Example 8.5) (a) A block pulled to the right on a rough surface by a constant horizontal force. (b) The applied force is at an angle θ to the horizontal.

Write the appropriate reduction of Equation 8.2:

(1) $\Delta K + \Delta E_{\text{int}} = W$

Find the work done on the system by the applied force just as in Example 7.6, noting that $\Delta x = d$ because the motion is in a straight line:

$W = F\Delta x = Fd$

Apply the *particle in equilibrium* model to the block in the vertical direction:

$\sum F_y = 0 \;\rightarrow\; n - mg = 0 \;\rightarrow\; n = mg$

Find the magnitude of the friction force:

$f_k = \mu_k n = \mu_k mg$

Substitute the energies into Equation (1), using Equation 8.14 for ΔE_{int}, and solve for the final speed of the block:

$(\tfrac{1}{2}mv_f{}^2 - 0) + (\mu_k mg)d = Fd$

$v_f = \sqrt{2d\left(\dfrac{F}{m} - \mu_k g\right)}$

Substitute numerical values:

$v_f = \sqrt{2(3.0\text{ m})\left[\dfrac{12\text{ N}}{6.0\text{ kg}} - (0.15)(9.80\text{ m/s}^2)\right]} = \boxed{1.8\text{ m/s}}$

Finalize   As expected, this value is less than the 3.5 m/s found in the case of the block sliding on a frictionless surface (see Example 7.6). The difference in kinetic energies between the block in Example 7.6 and the block in this example is equal to the increase in internal energy of the block–surface system in this example.

(B) Suppose the force $\vec{F}$ is applied at an angle θ as shown in Figure 8.8b. At what angle should the force be applied to achieve the largest possible speed after the block has moved 3.0 m to the right?

SOLUTION

Conceptualize   You might guess that $\theta = 0$ would give the largest speed because the force would have the largest component possible in the direction parallel to the surface. Think about $\vec{F}$ applied at an arbitrary nonzero angle, however. Although the horizontal component of the force would be reduced, the vertical component of the force would reduce the normal force, in turn reducing the force of friction, which suggests that the speed could be maximized by pulling at an angle other than $\theta = 0$.

8.5 continued

Categorize As in part (A), we model the block and the surface as a *nonisolated system* for energy with a nonconservative force acting.

Analyze Write the appropriate reduction of Equation 8.2:

(1) $\Delta K + \Delta E_{int} = W$

Find the work done by the applied force:

(2) $W = F\Delta x \cos\theta = Fd \cos\theta$

Apply the particle in equilibrium model to the block in the vertical direction:

$\sum F_y = n + F\sin\theta - mg = 0$

Solve for n:

(2) $n = mg - F\sin\theta$

Substitute for the energy changes in Equation (1) and solve for the final kinetic energy of the block:

$(K_f - 0) + f_k d = W \;\rightarrow\; K_f = W - f_k d$

Substitute the results found in Equations (1) and (2):

$K_f = Fd\cos\theta - \mu_k n d = Fd\cos\theta - \mu_k (mg - F\sin\theta)d$

Maximizing the speed is equivalent to maximizing the final kinetic energy. Consequently, differentiate K_f with respect to θ and set the result equal to zero:

$\dfrac{dK_f}{d\theta} = -Fd\sin\theta - \mu_k(0 - F\cos\theta)d = 0$

$-\sin\theta + \mu_k \cos\theta = 0$

$\tan\theta = \mu_k$

Evaluate θ for $\mu_k = 0.15$:

$\theta = \tan^{-1}(\mu_k) = \tan^{-1}(0.15) = \boxed{8.5°}$

Finalize Notice that the angle at which the speed of the block is a maximum is indeed not $\theta = 0$. When the angle exceeds 8.5°, the horizontal component of the applied force is too small to be compensated by the reduced friction force and the speed of the block begins to decrease from its maximum value.

Example 8.6   A Block–Spring System

A block of mass 1.6 kg is attached to a horizontal spring that has a force constant of 1 000 N/m as shown in Figure 8.9a. The spring is compressed 2.0 cm and is then released from rest as in Figure 8.9b.

(A) Calculate the speed of the block as it passes through the equilibrium position $x = 0$ if the surface is frictionless.

SOLUTION

Conceptualize This situation has been discussed before, and it is easy to visualize the block being pushed to the right by the spring and moving with some speed at $x = 0$.

Categorize We identify the system as the block and model the block as a *nonisolated system* for energy.

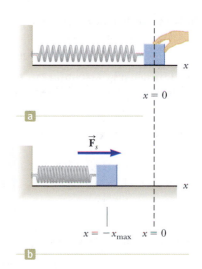

Figure 8.9 (Example 8.6) (a) A block attached to a spring is pushed inward from an initial position $x = 0$ by an external agent. (b) At position $x = -x_{max}$, the block is released from rest and the spring pushes it to the right.

Analyze Write the appropriate reduction of Equation 8.2 for the block being pushed by the spring:

(1) $\Delta K = W_s$

Use Equation 7.11 to find the work done by the spring on the system:

(2) $W_s = \frac{1}{2}kx_{max}^2$

Substitute the initial and final kinetic energies on the left of Equation (1) and the expression for the work in Equation (2) on the right:

$\left(\frac{1}{2}mv_f^2 - 0\right) = \frac{1}{2}kx_{max}^2 \;\rightarrow\; v_f = x_{max}\sqrt{\dfrac{k}{m}}$

Substitute numerical values:

$v_f = (0.020\text{ m})\sqrt{\dfrac{1\ 000\text{ N/m}}{1.6\text{ kg}}} = \boxed{0.50\text{ m/s}}$

continued

8.6 continued

Finalize   Although this problem could have been solved in Chapter 7, it is presented here to provide contrast with the following part (B), which requires the techniques of this chapter.

(B) Calculate the speed of the block as it passes through the equilibrium position if a constant friction force of 4.0 N retards its motion from the moment it is released.

SOLUTION

Conceptualize   The correct answer must be less than that found in part (A) because the friction force retards the motion.

Categorize   We identify the system as the block and the surface, a *nonisolated system* for energy because of the work done by the spring. There is a nonconservative force acting within the system: the friction between the block and the surface.

Analyze   Write the appropriate reduction of Equation 8.2 and substitute for the energy changes:

$$\Delta K + \Delta E_{\text{int}} = W_s \quad \rightarrow \quad (\tfrac{1}{2}mv_f^2 - 0) + f_k d = W_s$$

Solve for v_f:

$$v_f = \sqrt{\frac{2}{m}(W_s - f_k d)}$$

Substitute for the work done by the spring, found in part (A):

$$v_f = \sqrt{\frac{2}{m}(\tfrac{1}{2}kx_{\text{max}}^2 - f_k d)}$$

Substitute numerical values:

$$v_f = \sqrt{\frac{2}{1.6 \text{ kg}}[\tfrac{1}{2}(1\,000 \text{ N/m})(0.020 \text{ m})^2 - (4.0 \text{ N})(0.020 \text{ m})]} = 0.39 \text{ m/s}$$

Finalize   As expected, this value is less than the 0.50 m/s found in part (A).

WHAT IF?   What if the friction force were increased to 10.0 N? What is the block's speed at $x = 0$?

Answer   In this case, the value of $f_k d$ as the block moves to $x = 0$ is

$$f_k d = (10.0 \text{ N})(0.020 \text{ m}) = 0.20 \text{ J}$$

which is equal in magnitude to the kinetic energy at $x = 0$ for the frictionless case. (Verify it!). Therefore, all the kinetic energy has been transformed to internal energy by friction when the block arrives at $x = 0$, and its speed at this point is $v = 0$.

In this situation as well as that in part (B), the speed of the block reaches a maximum at some position other than $x = 0$. Problem 27 asks you to locate these positions.

8.4 Changes in Mechanical Energy for Nonconservative Forces

In the discussion leading to Equation 8.14, which identifies the change in internal energy of a system due to friction, we considered nonconservative forces that affected only the *kinetic* energy of the system. Now, however, suppose the book on the surface that we were discussing there is part of a system that also exhibits a change in potential energy. In this case, $f_k d$ is the change in internal energy due to a decrease in the *mechanical* energy of the system because of the force of kinetic friction. For example, if the book moves on an incline that is not frictionless, there is a change in both the kinetic energy and the gravitational potential energy of the book–incline–Earth system. Consequently, Equation 8.2 can be written as

$$\Delta K + \Delta U + \Delta E_{\text{int}} = 0 \tag{8.16}$$

where ΔE_{int} is given by Equation 8.14.

Example 8.7 | Crate Sliding Down a Ramp

A 3.00-kg crate slides down a ramp. The ramp is 1.00 m in length and inclined at an angle of 30.0° as shown in Figure 8.10. The crate starts from rest at the top and experiences a constant friction force of magnitude 5.00 N. The crate continues to move a short distance on the horizontal floor after it leaves the ramp, and then comes to rest.

(A) Use energy methods to determine the speed of the crate at the bottom of the ramp.

Figure 8.10 (Example 8.7) A crate slides down a ramp with friction under the influence of gravity. The potential energy of the system decreases, whereas the kinetic and internal energies increase.

SOLUTION

Conceptualize Imagine the crate sliding down the ramp in Figure 8.10. The larger the friction force, the more slowly the crate will slide.

Categorize We identify the crate, the surface, and the Earth as an *isolated system* for energy with a nonconservative force acting. We consider the time interval from when the crate leaves the top of the ramp until it reaches the bottom.

Analyze Because $v_i = 0$, the initial kinetic energy of the system when the crate is at the top of the ramp is zero. If the y coordinate is measured from the bottom of the ramp (the final position of the crate, for which we choose the gravitational potential energy of the system to be zero) with the upward direction being positive, then $y_i = 0.500$ m.

Write the conservation of energy equation (Eq. 8.2) for this system:

$$\Delta K + \Delta U + \Delta E_{int} = 0$$

Substitute for the energies:

$$(\tfrac{1}{2}mv_f^2 - 0) + (0 - mgy_i) + f_k d = 0$$

Solve for v_f:

$$(1) \quad v_f = \sqrt{\frac{2}{m}(mgy_i - f_k d)}$$

Substitute numerical values:

$$v_f = \sqrt{\frac{2}{3.00\ \text{kg}}\left[(3.00\ \text{kg})(9.80\ \text{m/s}^2)(0.500\ \text{m}) - (5.00\ \text{N})(1.00\ \text{m})\right]} = 2.54\ \text{m/s}$$

(B) How far does the crate slide on the horizontal floor if it continues to experience a friction force of magnitude 5.00 N?

SOLUTION

Analyze This part of the problem is handled in exactly the same way as part (A), but in this case we consider the time interval from the moment the crate begins to slide at the top of the ramp until it comes to rest on the floor.

Write the conservation of energy equation for this situation:

$$\Delta K + \Delta U + \Delta E_{int} = 0$$

Substitute for the energies noting that the crate slides over a total distance, ramp and floor, that we call d_{total}:

$$(0 - 0) + (0 - mgy_i) + f_k d_{total} = 0$$

Solve for the distance d_{total} and substitute numerical values:

$$d_{total} = \frac{mgy_i}{f_k} = \frac{(3.00\ \text{kg})(9.80\ \text{m/s}^2)(0.500\ \text{m})}{5.00\ \text{N}} = 2.94\ \text{m}$$

Subtracting the 1.00-m length of the ramp gives us 1.94 m that the crate slides across the floor.

Finalize For comparison, you may want to calculate the speed of the crate at the bottom of the ramp in the case in which the ramp is frictionless. Also notice that the increase in internal energy of the system for the entire motion of the crate is $f_k d_{total} = (5.00\ \text{N})(2.94\ \text{m}) = 14.7$ J. This energy is shared between the crate and the surface, each of which is a bit warmer than before.

Also notice that the distance d the object slides on the horizontal surface is infinite if the surface is frictionless. Is that consistent with your conceptualization of the situation?

WHAT IF? A cautious worker decides that the speed of the crate when it arrives at the bottom of the ramp may be so large that its contents may be damaged. Therefore, he replaces the ramp with a longer one such that the new ramp makes an angle of 25.0° with the ground. Does this new ramp reduce the speed of the crate as it reaches the ground?

continued

8.7 continued

Answer   Because the ramp is longer, the friction force acts over a longer distance and transforms more of the mechanical energy into internal energy. The result is a reduction in the kinetic energy of the crate, and we expect a lower speed as it reaches the ground.

Find the length d of the new ramp:
$$\sin 25.0° = \frac{0.500 \text{ m}}{d} \quad \rightarrow \quad d = \frac{0.500 \text{ m}}{\sin 25.0°} = 1.18 \text{ m}$$

Find v_f from Equation (1) in part (A):
$$v_f = \sqrt{\frac{2}{3.00 \text{ kg}}[(3.00 \text{ kg})(9.80 \text{ m/s}^2)(0.500 \text{ m}) - (5.00 \text{ N})(1.18 \text{ m})]} = 2.42 \text{ m/s}$$

The final speed is indeed lower than in the higher-angle case.

Example **8.8**   Block–Spring Collision

A block having a mass of 0.80 kg is given an initial velocity $v_{Ⓐ} = 1.2$ m/s to the right and collides with a spring whose mass is negligible and whose force constant is $k = 50$ N/m as shown in Figure 8.11.

(A) Assuming the surface to be frictionless, calculate the maximum compression of the spring after the collision.

SOLUTION

Conceptualize   The various parts of Figure 8.11 help us imagine what the block will do in this situation. All motion takes place in a horizontal plane, so we do not need to consider changes in gravitational potential energy. Before the collision, when the block is at Ⓐ, it has kinetic energy and the spring is uncompressed, so the elastic potential energy stored in the system is zero. Therefore, the total mechanical energy of the system before the collision is just $\frac{1}{2}mv_{Ⓐ}^2$. After the collision, when the block is at Ⓒ, the spring is fully compressed; now the block is at rest and so has zero kinetic energy. The elastic potential energy stored in the system, however, has its maximum value $\frac{1}{2}kx^2 = \frac{1}{2}kx_{max}^2$, where the

Figure 8.11   (Example 8.8) A block sliding on a frictionless, horizontal surface collides with a light spring. (a) Initially, the block slides to the right, approaching the spring. (b) The block strikes the spring and begins to compress it. (c) The block stops momentarily at the maximum compression of the spring. (d) The spring pushes the block to the left. As the spring returns to its equilibrium length, the block continues moving to the left. The energy equations at the right show the energies of the system in the frictionless case in part (A).

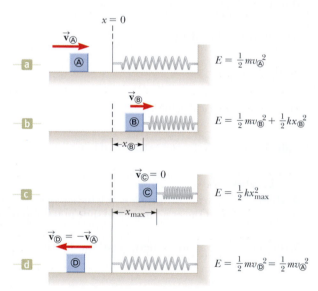

origin of coordinates $x = 0$ is chosen to be the equilibrium position of the spring and x_{max} is the maximum compression of the spring, which in this case happens to be $x_{Ⓒ}$. The total mechanical energy of the system is conserved because no nonconservative forces act on objects within the isolated system.

Categorize   We identify the system to be the block and the spring and model it as an *isolated system* for energy with no nonconservative forces acting.

Analyze   Write the appropriate reduction of Equation 8.2 between points Ⓐ and Ⓒ:
$$\Delta K + \Delta U = 0$$

Substitute for the energies:
$$(0 - \tfrac{1}{2}mv_{Ⓐ}^2) + (\tfrac{1}{2}kx_{max}^2 - 0) = 0$$

Solve for x_{max} and evaluate:
$$x_{max} = \sqrt{\frac{m}{k}}\,v_{Ⓐ} = \sqrt{\frac{0.80 \text{ kg}}{50 \text{ N/m}}}(1.2 \text{ m/s}) = \boxed{0.15 \text{ m}}$$

8.8 continued

(B) Suppose a constant force of kinetic friction acts between the block and the surface, with $\mu_k = 0.50$. If the speed of the block at the moment it collides with the spring is $v_\circledA = 1.2$ m/s, what is the maximum compression $x_\circledC$ in the spring?

SOLUTION

Conceptualize Because of the friction force, we expect the compression of the spring to be smaller than in part (A) because some of the block's kinetic energy is transformed to internal energy in the block and the surface.

Categorize We identify the system as the block, the surface, and the spring. This is an *isolated system* for energy but now involves a nonconservative force.

Analyze In this case, the mechanical energy $E_{\text{mech}} = K + U_s$ of the system is *not* conserved because a friction force acts on the block. From the *particle in equilibrium* model in the vertical direction, we see that $n = mg$.

Evaluate the magnitude of the friction force:

$$f_k = \mu_k n = \mu_k mg$$

Write the appropriate reduction of Equation 8.2 for this situation:

$$\Delta K + \Delta U + \Delta E_{\text{int}} = 0$$

Substitute the initial and final energies:

$$(0 - \tfrac{1}{2}mv_\circledA^2) + (\tfrac{1}{2}kx_\circledC^2 - 0) + \mu_k mgx_\circledC = 0$$

Rearrange the terms into a quadratic equation:

$$kx_\circledC^2 + 2\mu_k mgx_\circledC - mv_\circledA^2 = 0$$

Solve the quadratic equation:

$$x_\circledC = \frac{\mu_k mg}{k}\left(\pm\sqrt{1 + \frac{kv_\circledA^2}{\mu_k^2 m g^2}} - 1\right)$$

Substituting numerical values gives $x_\circledC = 0.092$ m and $x_\circledC = -0.25$ m. The physically meaningful root is $x_\circledC = 0.092$ m.

Finalize The negative root does not apply to this situation because the block must be to the right of the origin (positive value of x) when it comes to rest. Notice that the value of 0.092 m is less than the distance obtained in the frictionless case of part (A) as we expected.

Example 8.9 **Connected Blocks in Motion**

Two blocks are connected by a light string that passes over a frictionless pulley as shown in Figure 8.12. The block of mass m_1 lies on a horizontal surface and is connected to a spring of force constant k. The system is released from rest when the spring is unstretched. If the hanging block of mass m_2 falls a distance h before coming to rest, calculate the coefficient of kinetic friction between the block of mass m_1 and the surface.

SOLUTION

Conceptualize The key word *rest* appears twice in the problem statement. This word suggests that the configurations of the system associated with rest are good candidates for the initial and final configurations because the kinetic energy of the system is zero for these configurations.

Categorize In this situation, the system consists of the two blocks, the spring, the surface, and the Earth. This is an *isolated system* with a nonconservative force acting. We also model the sliding block as a *particle in equilibrium* in the vertical direction, leading to $n = m_1 g$.

Figure 8.12 (Example 8.9) As the hanging block moves from its highest elevation to its lowest, the system loses gravitational potential energy but gains elastic potential energy in the spring. Some mechanical energy is transformed to internal energy because of friction between the sliding block and the surface.

Analyze We need to consider two forms of potential energy for the system, gravitational and elastic: $\Delta U_g = U_{gf} - U_{gi}$ is the change in the system's gravitational potential energy, and $\Delta U_s = U_{sf} - U_{si}$ is the change in the system's elastic potential energy. The change in the gravitational potential energy of the system is associated with only the falling block because the vertical coordinate of the horizontally sliding block does not change.

continued

8.9 continued

Write the appropriate reduction of Equation 8.2:

(1) $\Delta K + \Delta U_g + \Delta U_s + \Delta E_{\text{int}} = 0$

Substitute for the energies for the time interval beginning upon release and ending when the system is again at rest, noting that as the hanging block falls a distance h, the horizontally moving block moves the same distance h to the right, and the spring stretches by a distance h:

$(0 - 0) + (0 - m_2gh) + (\frac{1}{2}kh^2 - 0) + f_kh = 0$

Substitute for the friction force:

$-m_2gh + \frac{1}{2}kh^2 + \mu_k m_1gh = 0$

Solve for μ_k:

$\mu_k = \dfrac{m_2g - \frac{1}{2}kh}{m_1g}$

Finalize   This setup represents a method of measuring the coefficient of kinetic friction between an object and some surface. Notice how we have solved the examples in this chapter using the energy approach. We begin with Equation 8.2 and then tailor it to the physical situation. This process may include deleting terms, such as all terms on the right-hand side of Equation 8.2 in this example. It can also include expanding terms, such as rewriting ΔU due to two types of potential energy in this example.

Conceptual Example 8.10   Interpreting the Energy Bars

The energy bar charts in Figure 8.13 show three instants in the motion of the system in Figure 8.12 and described in Example 8.9. For each bar chart, identify the configuration of the system that corresponds to the chart.

SOLUTION

In Figure 8.13a, there is no kinetic energy in the system. Therefore, nothing in the system is moving. The bar chart shows that the system contains only gravitational potential energy and no internal energy yet, which corresponds to the configuration with the darker blocks in Figure 8.12 and represents the instant just after the system is released.

In Figure 8.13b, the system contains four types of energy. The height of the gravitational potential energy bar is at 50%, which tells us that the hanging block has moved halfway between its position corresponding to Figure 8.13a and the position defined as $y = 0$. Therefore, in this configuration, the hanging block is between the dark and light images of the hanging block in Figure 8.12. The system has gained kinetic energy because the blocks are moving, elastic potential energy because the spring is stretching, and internal energy because of friction between the block of mass m_1 and the surface.

In Figure 8.13c, the height of the gravitational potential energy bar is zero, telling us that the hanging block is at $y = 0$. In addi-

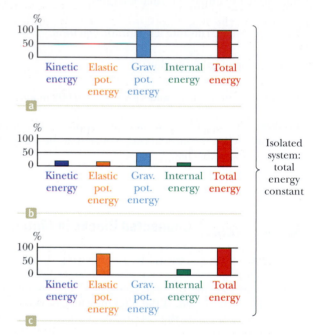

Figure 8.13 (Conceptual Example 8.10) Three energy bar charts are shown for the system in Figure 8.12.

tion, the height of the kinetic energy bar is zero, indicating that the blocks have stopped moving momentarily. Therefore, the configuration of the system is that shown by the light images of the blocks in Figure 8.12. The height of the elastic potential energy bar is high because the spring is stretched its maximum amount. The height of the internal energy bar is higher than in Figure 8.13b because the block of mass m_1 has continued to slide over the surface after the configuration shown in Figure 8.13b.

8.5 Power

Consider Conceptual Example 7.7 again, which involved rolling a refrigerator up a ramp into a truck. Suppose the man is not convinced the work is the same regardless of the ramp's length and sets up a long ramp with a gentle rise. Although he does the same amount of work as someone using a shorter ramp, he takes longer to do the work because he has to move the refrigerator over a greater distance. Although the work done on both ramps is the same, there is *something* different about the tasks: the *time interval* during which the work is done.

The time rate of energy transfer is called the **instantaneous power** P and is defined as

$$P \equiv \frac{dE}{dt} \tag{8.17}$$

◀ Definition of power

We will focus on work as the energy transfer method in this discussion, but keep in mind that the notion of power is valid for *any* means of energy transfer discussed in Section 8.1. If an external force is applied to an object (which we model as a particle) and if the work done by this force on the object in the time interval Δt is W, the **average power** during this interval is

$$P_{\text{avg}} = \frac{W}{\Delta t}$$

Therefore, in Conceptual Example 7.7, although the same work is done in rolling the refrigerator up both ramps, less power is required for the longer ramp.

In a manner similar to the way we approached the definition of velocity and acceleration, the instantaneous power is the limiting value of the average power as Δt approaches zero:

$$P = \lim_{\Delta t \to 0} \frac{W}{\Delta t} = \frac{dW}{dt}$$

where we have represented the infinitesimal value of the work done by dW. We find from Equation 7.3 that $dW = \vec{\mathbf{F}} \cdot d\vec{\mathbf{r}}$. Therefore, for a constant force, the instantaneous power can be written

$$P = \frac{dW}{dt} = \vec{\mathbf{F}} \cdot \frac{d\vec{\mathbf{r}}}{dt} = \vec{\mathbf{F}} \cdot \vec{\mathbf{v}} \tag{8.18}$$

where $\vec{\mathbf{v}} = d\vec{\mathbf{r}}/dt$.

The SI unit of power is joules per second (J/s), also called the **watt** (W) after James Watt:

$$1 \text{ W} = 1 \text{ J/s} = 1 \text{ kg} \cdot \text{m}^2/\text{s}^3$$

◀ The watt

A unit of power in the U.S. customary system is the **horsepower** (hp):

$$1 \text{ hp} = 746 \text{ W}$$

A unit of energy (or work) can now be defined in terms of the unit of power. One **kilowatt-hour** (kWh) is the energy transferred in 1 h at the constant rate of 1 kW = 1 000 J/s. The amount of energy represented by 1 kWh is

$$1 \text{ kWh} = (10^3 \text{ W})(3\,600 \text{ s}) = 3.60 \times 10^6 \text{ J}$$

A kilowatt-hour is a unit of energy, not power. When you pay your electric bill, you are buying energy, and the amount of energy transferred by electrical transmission into a home during the period represented by the electric bill is usually expressed in kilowatt-hours. For example, your bill may state that you used 900 kWh of energy during a month and that you are being charged at the rate of 11¢ per kilowatt-hour. Your obligation is then $99 for this amount of energy. As another example, suppose an electric bulb is rated at 100 W. In 1.00 h of operation, it would have energy transferred to it by electrical transmission in the amount of (0.100 kW)(1.00 h) = 0.100 kWh = 3.60 × 10^5 J.

PITFALL PREVENTION 8.3

W, *W*, and watts Do not confuse the symbol W for the watt with the italic symbol *W* for work. Also, remember that the watt already represents a rate of energy transfer, so "watts per second" does not make sense. The watt is *the same as* a joule per second.

Example **8.11**   **Power Delivered by an Elevator Motor**

An elevator car (Fig. 8.14a) has a mass of 1 600 kg and is carrying passengers having a combined mass of 200 kg. A constant friction force of 4 000 N retards its motion.

(A) How much power must a motor deliver to lift the elevator car and its passengers at a constant speed of 3.00 m/s?

SOLUTION

Conceptualize   The motor must supply the force of magnitude T that pulls the elevator car upward.

Categorize   The friction force increases the power necessary to lift the elevator. The problem states that the speed of the elevator is constant, which tells us that $a = 0$. We model the elevator as a *particle in equilibrium*.

Figure 8.14   (Example 8.11) (a) The motor exerts an upward force $\vec{\mathbf{T}}$ on the elevator car. The magnitude of this force is the total tension T in the cables connecting the car and motor. The downward forces acting on the car are a friction force $\vec{\mathbf{f}}$ and the gravitational force $\vec{\mathbf{F}}_g = M\vec{\mathbf{g}}$. (b) The free-body diagram for the elevator car.

Analyze   The free-body diagram in Figure 8.14b specifies the upward direction as positive. The *total* mass M of the system (car plus passengers) is equal to 1 800 kg.

Using the particle in equilibrium model, apply Newton's second law to the car:

$$\sum F_y = T - f - Mg = 0$$

Solve for T:

$$T = Mg + f$$

Use Equation 8.18 and that $\vec{\mathbf{T}}$ is in the same direction as $\vec{\mathbf{v}}$ to find the power:

$$P = \vec{\mathbf{T}} \cdot \vec{\mathbf{v}} = Tv = (Mg + f)v$$

Substitute numerical values:

$$P = [(1\,800\text{ kg})(9.80\text{ m/s}^2) + (4\,000\text{ N})](3.00\text{ m/s}) = \boxed{6.49 \times 10^4\text{ W}}$$

(B) What power must the motor deliver at the instant the speed of the elevator is v if the motor is designed to provide the elevator car with an upward acceleration of 1.00 m/s²?

SOLUTION

Conceptualize   In this case, the motor must supply the force of magnitude T that pulls the elevator car upward with an increasing speed. We expect that more power will be required to do that than in part (A) because the motor must now perform the additional task of accelerating the car.

Categorize   In this case, we model the elevator car as a *particle under a net force* because it is accelerating.

Analyze   Using the particle under a net force model, apply Newton's second law to the car:

$$\sum F_y = T - f - Mg = Ma$$

Solve for T:

$$T = M(a + g) + f$$

Use Equation 8.18 to obtain the required power:

$$P = Tv = [M(a + g) + f]v$$

Substitute numerical values:

$$P = [(1\,800\text{ kg})(1.00\text{ m/s}^2 + 9.80\text{ m/s}^2) + 4\,000\text{ N}]v$$

$$= \boxed{(2.34 \times 10^4)v}$$

where v is the instantaneous speed of the car in meters per second and P is in watts.

Finalize   To compare with part (A), let $v = 3.00$ m/s, giving a power of

$$P = (2.34 \times 10^4\text{ N})(3.00\text{ m/s}) = 7.02 \times 10^4\text{ W}$$

which is larger than the power found in part (A), as expected.

Summary

▶ Definitions

A **nonisolated system** is one for which energy crosses the boundary of the system. An **isolated system** is one for which no energy crosses the boundary of the system.

The **instantaneous power** P is defined as the time rate of energy transfer:

$$P \equiv \frac{dE}{dt} \qquad (8.17)$$

▶ Concepts and Principles

For a nonisolated system, we can equate the change in the total energy stored in the system to the sum of all the transfers of energy across the system boundary, which is a statement of **conservation of energy.** For an isolated system, the total energy is constant.

If a friction force of magnitude f_k acts over a distance d within a system, the change in internal energy of the system is

$$\Delta E_{int} = f_k d \qquad (8.14)$$

▶ Analysis Models for Problem Solving

Nonisolated System (Energy). The most general statement describing the behavior of a nonisolated system is the **conservation of energy equation:**

$$\Delta E_{system} = \Sigma\, T \qquad (8.1)$$

Including the types of energy storage and energy transfer that we have discussed gives

$$\Delta K + \Delta U + \Delta E_{int} = W + Q + T_{MW} + T_{MT} + T_{ET} + T_{ER} \qquad (8.2)$$

For a specific problem, this equation is generally reduced to a smaller number of terms by eliminating the terms that are not appropriate to the situation.

Isolated System (Energy). The total energy of an isolated system is conserved, so

$$\Delta E_{system} = 0 \qquad (8.10)$$

If no nonconservative forces act within the isolated system, the mechanical energy of the system is conserved, so

$$\Delta E_{mech} = 0 \qquad (8.7)$$

which can be written as

$$\Delta K + \Delta U = 0 \qquad (8.5)$$

If a nonconservative force such as friction acts within the system, there is a change in internal energy, so

$$\Delta K + \Delta U + \Delta E_{int} = 0 \qquad (8.16)$$

Think–Pair–Share

See the Preface for an explanation of the icons used in this problems set. For additional assessment items for this section, go to **WEBASSIGN** From Cengage

1. You are a member of an expert witness group that provides scientific services to the legal community. Your group has been asked by a defense attorney to prove at trial that a driver was not exceeding the speed limit. You are provided with the following data: The mass of the car is 1.50×10^3 kg. The mass of the driver is 95.0 kg. The coefficient of kinetic friction between the car's tires and the roadway is 0.580. The coefficient of static friction between the car's tires and the roadway is 0.820. The posted speed limit on the road is 25 mi/h. The roadway was dry and the weather was sunny at the time of the incident.

   You are also provided with the following description of the incident: The driver was driving up a hill that makes an angle of 17.5° with the horizontal. The driver saw a dog run into the street, slammed on the brakes and left a skid mark 17.0 m long. The car came to rest at the end of the skid mark. The driver did not hit the dog, but the sound of the screeching tires drew the attention of a nearby policeman, who ticketed the driver for speeding.

   Should your group agree to offer testimony for the defense in this case? (Notice that this problem is the same as Think–Pair–Share Problem 5.1 (see page 120), but we want to use an energy approach here for comparison.)

2. You are working on a team of expert witnesses for an automobile company. The company is being sued by a persistent inventor who is frustrated that the company will not adopt his idea of a car that is operated solely by solar power. Prepare an argument for your company to use at trial to show that there is simply not enough energy delivered to a normal-sized car by solar energy to operate the car on streets and highways. Use the fact that the maximum intensity of sunlight available, near the equator, is 1 000 W/m².

3. **ACTIVITY** (a) Draw a simple diagram of a house and indicate all major means of energy transfer between the house and the environment. (b) What means can be used to combat or take advantage of the energy transfers to keep the temperature of the house fixed at a lower monthly cost for utility bills? (c) The following are words used in architecture when discussing energy considerations for a building: Insolation, Infiltration. Assign these architectural words to specific corresponding terms in Eq. 8.2.

4. **ACTIVITY** Consider the popgun in Example 8.4. Suppose the projectile mass, compression distance, and spring constant remain the same as given or calculated in the example. Suppose, however, there is a friction force of magnitude 2.00 N acting on the projectile as it rubs against the interior of the barrel. The vertical length from point Ⓐ to the end of the barrel is 0.600 m. (a) After the spring is compressed and the popgun fired, to what height does the projectile rise above point Ⓑ? (b) Draw four energy bar charts for this situation, analogous to those in Figures 8.6c–d.

Problems

See the Preface for an explanation of the icons used in this problems set. For additional assessment items for this section, go to **WEBASSIGN** From Cengage

SECTION 8.1 Analysis Model: Nonisolated System (Energy)

1. A ball of mass m falls from a height h to the floor. (a) Write **S** the appropriate version of Equation 8.2 for the system of the ball and the Earth and use it to calculate the speed of the ball just before it strikes the Earth. (b) Write the appropriate version of Equation 8.2 for the system of the ball and use it to calculate the speed of the ball just before it strikes the Earth.

SECTION 8.2 Analysis Model: Isolated System (Energy)

2. A 20.0-kg cannonball is fired from a cannon with muzzle **V** speed of 1 000 m/s at an angle of 37.0° with the horizontal. A second ball is fired at an angle of 90.0°. Use the isolated system model to find (a) the maximum height reached by each ball and (b) the total mechanical energy of the ball–Earth system at the maximum height for each ball. Let $y = 0$ at the cannon.

3. A block of mass $m = 5.00$ kg is released from point Ⓐ and **V** slides on the frictionless track shown in Figure P8.3. Determine (a) the block's speed at points Ⓑ and Ⓒ and (b) the net work done by the gravitational force on the block as it moves from point Ⓐ to point Ⓒ.

Figure P8.3

4. At 11:00 a.m. on September 7, 2001, more than one million British schoolchildren jumped up and down for one minute to simulate an earthquake. (a) Find the energy stored in the children's bodies that was converted into internal energy in the ground and their bodies and propagated into the ground by seismic waves during the experiment. Assume 1 050 000 children of average mass 36.0 kg jumped 12 times each, raising their centers of mass by 25.0 cm each time and briefly resting between one jump and the next. (b) Of the energy that propagated into the ground, most produced high-frequency "microtremor" vibrations that were rapidly damped and did not travel far. Assume 0.01% of the total energy was carried away by long-range seismic

waves. The magnitude of an earthquake on the Richter scale is given by

$$M = \frac{\log E - 4.8}{1.5}$$

where E is the seismic wave energy in joules. According to this model, what was the magnitude of the demonstration quake?

5. A light, rigid rod is 77.0 cm long. Its top end is pivoted on a frictionless, horizontal axle. The rod hangs straight down at rest with a small, massive ball attached to its bottom end. You strike the ball, suddenly giving it a horizontal velocity so that it swings around in a full circle. What minimum speed at the bottom is required to make the ball go over the top of the circle?

6. **Review.** The system shown in Figure P8.6 consists of a light, inextensible cord, light, frictionless pulleys, and blocks of equal mass. Notice that block B is attached to one of the pulleys. The system is initially held at rest so that the blocks are at the same height above the ground. The blocks are then released. Find the speed of block A at the moment the vertical separation of the blocks is h.

Figure P8.6

SECTION 8.3 Situations Involving Kinetic Friction

7. A crate of mass 10.0 kg is pulled up a rough incline with an initial speed of 1.50 m/s. The pulling force is 100 N parallel to the incline, which makes an angle of 20.0° with the horizontal. The coefficient of kinetic friction is 0.400, and the crate is pulled 5.00 m. (a) How much work is done by the gravitational force on the crate? (b) Determine the increase in internal energy of the crate–incline system owing to friction. (c) How much work is done by the 100-N force on the crate? (d) What is the change in kinetic energy of the crate? (e) What is the speed of the crate after being pulled 5.00 m?

8. A 40.0-kg box initially at rest is pushed 5.00 m along a rough, horizontal floor with a constant applied horizontal force of 130 N. The coefficient of friction between box and floor is 0.300. Find (a) the work done by the applied force, (b) the increase in internal energy in the box–floor system as a result of friction, (c) the work done by the normal force, (d) the work done by the gravitational force, (e) the change in kinetic energy of the box, and (f) the final speed of the box.

9. A smooth circular hoop with a radius of 0.500 m is placed flat on the floor. A 0.400-kg particle slides around the inside edge of the hoop. The particle is given an initial speed of 8.00 m/s. After one revolution, its speed has dropped to 6.00 m/s because of friction with the floor. (a) Find the energy transformed from mechanical to internal in the particle–hoop–floor system as a result of friction in one revolution. (b) What is the total number of revolutions the particle makes before stopping? Assume the friction force remains constant during the entire motion.

SECTION 8.4 Changes in Mechanical Energy for Nonconservative Forces

10. **Q|C** As shown in Figure P8.10, a green bead of mass 25 g slides along a straight wire. The length of the wire from point Ⓐ to point Ⓑ is 0.600 m, and point Ⓐ is 0.200 m higher than point Ⓑ. A constant friction force of magnitude 0.025 0 N acts on the bead. (a) If the bead is released from rest at point Ⓐ, what is its speed at point Ⓑ? (b) A red bead of mass 25 g slides along a curved wire, subject to a friction force with the same constant magnitude as that on the green bead. If the green and red beads are released simultaneously from rest at point Ⓐ, which bead reaches point Ⓑ with a higher speed? Explain.

Figure P8.10

11. **Q|C** At time t_i, the kinetic energy of a particle is 30.0 J and the potential energy of the system to which it belongs is 10.0 J. At some later time t_f, the kinetic energy of the particle is 18.0 J. (a) If only conservative forces act on the particle, what are the potential energy and the total energy of the system at time t_f? (b) If the potential energy of the system at time t_f is 5.00 J, are any nonconservative forces acting on the particle? (c) Explain your answer to part (b).

12. A 1.50-kg object is held 1.20 m above a relaxed massless, vertical spring with a force constant of 320 N/m. The object is dropped onto the spring. (a) How far does the object compress the spring? (b) **What If?** Repeat part (a), but this time assume a constant air-resistance force of 0.700 N acts on the object during its motion. (c) **What If?** How far does the object compress the spring if the same experiment is performed on the Moon, where $g = 1.63$ m/s² and air resistance is neglected?

13. **GP Q|C S** A child of mass m starts from rest and slides without friction from a height h along a slide next to a pool (Fig. P8.13). She is launched from a height $h/5$ into the air over the pool. We wish to find the maximum height she reaches above the water in her projectile motion. (a) Is the child–Earth system isolated or nonisolated? Why? (b) Is there a nonconservative force acting within the system? (c) Define the configuration of the system when the child is at the water level as having zero gravitational potential energy. Express the total energy of the system when the child is at the top of the waterslide. (d) Express the total energy of the system when the child is at the launching point. (e) Express the total energy of the system when the child is at the highest point in her projectile motion. (f) From parts (c) and (d), determine her initial

Figure P8.13

speed v_i at the launch point in terms of g and h. (g) From parts (d), (e), and (f), determine her maximum airborne height y_{max} in terms of h and the launch angle θ. (h) Would your answers be the same if the waterslide were not frictionless? Explain.

14. **QC** An 80.0-kg skydiver jumps out of a balloon at an altitude of 1 000 m and opens his parachute at an altitude of 200 m. (a) Assuming the total retarding force on the skydiver is constant at 50.0 N with the parachute closed and constant at 3 600 N with the parachute open, find the speed of the skydiver when he lands on the ground. (b) Do you think the skydiver will be injured? Explain. (c) At what height should the parachute be opened so that the final speed of the skydiver when he hits the ground is 5.00 m/s? (d) How realistic is the assumption that the total retarding force is constant? Explain.

15. **CR** You have spent a long day skiing and are tired. You are standing at the top of a hill, looking at the lodge at the bottom of the hill. You are so tired that you want to simply start from rest and coast down the slope, without pushing with your poles or doing anything else to change your motion. You want to let gravity do all the work! You have a choice of two trails to reach the lodge. Both trails have the same coefficient of friction μ_k. In addition, both trails represent the same horizontal separation between the initial and final points. Trail A has a short, steep downslope and then a long, flat coast to the lodge. Trail B has a long, gentle downslope and then a short remaining flat coast to the lodge. Which trail will result in your arriving at the lodge with the highest final speed?

SECTION 8.5 Power

16. **QC** The electric motor of a model train accelerates the train from rest to 0.620 m/s in 21.0 ms. The total mass of the train is 875 g. (a) Find the minimum power delivered to the train by electrical transmission from the metal rails during the acceleration. (b) Why is it the minimum power?

17. An energy-efficient lightbulb, taking in 28.0 W of power, can produce the same level of brightness as a conventional lightbulb operating at power 100 W. The lifetime of the energy-efficient bulb is 10 000 h and its purchase price is $4.50, whereas the conventional bulb has a lifetime of 750 h and costs $0.42. Determine the total savings obtained by using one energy-efficient bulb over its lifetime as opposed to using conventional bulbs over the same time interval. Assume an energy cost of $0.200 per kilowatt-hour.

18. **S** An older-model car accelerates from 0 to speed v in a time interval of Δt. A newer, more powerful sports car accelerates from 0 to $2v$ in the same time period. Assuming the energy coming from the engine appears only as kinetic energy of the cars, compare the power of the two cars.

19. Make an order-of-magnitude estimate of the power a car engine contributes to speeding the car up to highway speed. In your solution, state the physical quantities you take as data and the values you measure or estimate for them. The mass of a vehicle is often given in the owner's manual.

20. **CR** There is a 5K event coming up in your town. While talking to your grandmother, who uses an electric scooter for mobility, she says that she would like to accompany you on her scooter while you walk the 5.00-km distance. The manual that came with her scooter claims that the fully charged battery is capable of providing 120 Wh of energy before being depleted. In preparation for the race, you go for a "test drive": beginning with a fully charged battery, your grandmother rides beside you as you walk 5.00 km on flat ground. At the end of the walk, the battery usage indicator shows that 40.0% of the original energy in the battery remains. You also know that the combined weight of the scooter and your grandmother is 890 N. A few days later, filled with confidence that the battery has sufficient energy, you and your grandmother drive to the 5K event. Unbeknownst to you, the 5K route is not on flat ground, but is all uphill, ending at a point higher than the starting line. A race official tells you that the total amount of vertical displacement on the route is 150 m. Should your grandmother accompany you on the walk, or will she be stranded when her battery runs out of energy? Assume that the only difference between your test drive and the actual event is the vertical displacement.

21. **BIO** For saving energy, bicycling and walking are far more efficient means of transportation than is travel by automobile. For example, when riding at 10.0 mi/h, a cyclist uses food energy at a rate of about 400 kcal/h above what he would use if merely sitting still. (In exercise physiology, power is often measured in kcal/h rather than in watts. Here 1 kcal = 1 nutritionist's Calorie = 4 186 J.) Walking at 3.00 mi/h requires about 220 kcal/h. It is interesting to compare these values with the energy consumption required for travel by car. Gasoline yields about 1.30×10^8 J/gal. Find the fuel economy in equivalent miles per gallon for a person (a) walking and (b) bicycling.

22. **BIO** Energy is conventionally measured in Calories as well as in joules. One Calorie in nutrition is one kilocalorie, defined as 1 kcal = 4 186 J. Metabolizing 1 g of fat can release 9.00 kcal. A student decides to try to lose weight by exercising. He plans to run up and down the stairs in a football stadium as fast as he can and as many times as necessary. To evaluate the program, suppose he runs up a flight of 80 steps, each 0.150 m high, in 65.0 s. For simplicity, ignore the energy he uses in coming down (which is small). Assume a typical efficiency for human muscles is 20.0%. This statement means that when your body converts 100 J from metabolizing fat, 20 J goes into doing mechanical work (here, climbing stairs). The remainder goes into extra internal energy. Assume the student's mass is 75.0 kg. (a) How many times must the student run the flight of stairs to lose 1.00 kg of fat? (b) What is his average power output, in watts and in horsepower, as he runs up the stairs? (c) Is this activity in itself a practical way to lose weight?

ADDITIONAL PROBLEMS

23. **QC** A block of mass $m = 200$ g is released from rest at point Ⓐ along the horizontal diameter on the inside of hemispherical bowl of radius $R = 30.0$ cm, and the surface of the bowl is rough (Fig. P8.23). The block's speed at point Ⓑ is 1.50 m/s.

Figure P8.23

(a) What is its kinetic energy at point Ⓑ? (b) How much mechanical energy is transformed into internal energy as the block moves from point Ⓐ to point Ⓑ? (c) Is it possible to determine the coefficient of friction from these results in any simple manner? (d) Explain your answer to part (c).

24. **BIO** Make an order-of-magnitude estimate of your power output as you climb stairs. In your solution, state the physical quantities you take as data and the values you measure or estimate for them. Do you consider your peak power or your sustainable power?

25. **CR** You are working with a team that is designing a new roller coaster–type amusement park ride for a major theme park. You are present for the testing of the ride, in which an empty 250-kg car is sent along the entire ride. Near the end of the ride, the car is at near rest at the top of a 110-m tall track. It then enters a final section, rolling down an undulating hill to ground level. The total length of track for this final section from the top to the ground is 250 m. For the first 230 m, a constant friction force of 50.0 N acts from computer-controlled brakes. For the last 20 m, which is horizontal at ground level, the computer increases the friction force to a value required for the speed to be reduced to zero just as the car arrives at the point on the track at which the passengers exit. (a) Determine the required constant friction force for the last 20 m for the empty test car. (b) Find the highest speed reached by the car during the final section of track length 250 m. (c) You are asked by your team supervisor to determine the answers to parts (a) and (b) for a fully loaded car with an upper limit of 450 kg of passenger mass. Find these new values. (d) The required friction force in part (c) is well within design limits. The fastest speed, however, is well below that of current leading rides, so you would like to increase the maximum speed. You can't make the tower taller above ground, so you decide to include a feature where part of the track goes *underground*. Determine the depth to which the underground part of the ride must go to increase the maximum speed to 55.0 m/s. Assume the overall length of the first part of the track remains at 230 m and the length of track from the top to the lowest underground point is 150 m. The same 50.0-N friction force acts on the entire 230-m section of track. (e) Is the construction in part (d) feasible?

26. **QC** **Review.** As shown in Figure P8.26, a light string that does not stretch changes from horizontal to vertical as it passes over the edge of a table. The string connects m_1, a 3.50-kg block originally at rest on the horizontal table at a height $h = 1.20$ m above the floor, to m_2, a hanging 1.90-kg block orig-

Figure P8.26

inally a distance $d = 0.900$ m above the floor. Neither the surface of the table nor its edge exerts a force of kinetic friction. The blocks start to move from rest. The sliding block m_1 is projected horizontally after reaching the edge of the table. The hanging block m_2 stops without bouncing when it strikes the floor. Consider the two blocks plus the Earth as the system. (a) Find the speed at which m_1 leaves the edge of the table. (b) Find the impact speed of m_1 on the floor. (c) What is the shortest length of the string so that

it does not go taut while m_1 is in flight? (d) Is the energy of the system when it is released from rest equal to the energy of the system just before m_1 strikes the ground? (e) Why or why not?

27. Consider the block–spring–surface system in part (B) of Example 8.6. (a) Using an energy approach, find the position x of the block at which its speed is a maximum. (b) In the **What If?** section of this example, we explored the effects of an increased friction force of 10.0 N. At what position of the block does its maximum speed occur in this situation?

28. *Why is the following situation impossible?* A softball pitcher has a strange technique: she begins with her hand at rest at the highest point she can reach and then quickly rotates her arm backward so that the ball moves through a half-circle path. She releases the ball when her hand reaches the bottom of the path. The pitcher maintains a component of force on the 0.180-kg ball of constant magnitude 12.0 N in the direction of motion around the complete path. As the ball arrives at the bottom of the path, it leaves her hand with a speed of 25.0 m/s.

29. **AMT** Jonathan is riding a bicycle and encounters a hill of height 7.30 m. At the base of the hill, he is traveling at 6.00 m/s. When he reaches the top of the hill, he is traveling at 1.00 m/s. Jonathan and his bicycle together have a mass of 85.0 kg. Ignore friction in the bicycle mechanism and between the bicycle tires and the road. (a) What is the total external work done on the system of Jonathan and the bicycle between the time he starts up the hill and the time he reaches the top? (b) What is the change in potential energy stored in Jonathan's body during this process? (c) How much work does Jonathan do on the bicycle pedals within the Jonathan–bicycle–Earth system during this process?

30. **S** Jonathan is riding a bicycle and encounters a hill of height h. At the base of the hill, he is traveling at a speed v_i. When he reaches the top of the hill, he is traveling at a speed v_f. Jonathan and his bicycle together have a mass m. Ignore friction in the bicycle mechanism and between the bicycle tires and the road. (a) What is the total external work done on the system of Jonathan and the bicycle between the time he starts up the hill and the time he reaches the top? (b) What is the change in potential energy stored in Jonathan's body during this process? (c) How much work does Jonathan do on the bicycle pedals within the Jonathan–bicycle–Earth system during this process?

31. **QC** As the driver steps on the gas pedal, a car of mass 1 160 kg accelerates from rest. During the first few seconds of motion, the car's acceleration increases with time according to the expression

$$a = 1.16t - 0.210t^2 + 0.240t^3$$

where t is in seconds and a is in m/s². (a) What is the change in kinetic energy of the car during the interval from $t = 0$ to $t = 2.50$ s? (b) What is the minimum average power output of the engine over this time interval? (c) Why is the value in part (b) described as the *minimum* value?

32. **QC** **S** As it plows a parking lot, a snowplow pushes an ever-growing pile of snow in front of it. Suppose a car moving through the air is similarly modeled as a cylinder of area A pushing a growing disk of air in front of it. The originally stationary air is set into motion at the constant speed v of the cylinder

as shown in Figure P8.32. In a time interval Δt, a new disk of air of mass Δm must be moved a distance $v\,\Delta t$ and hence must be given a kinetic energy $\frac{1}{2}(\Delta m)v^2$. Using this model, show that the car's power loss owing to air resistance is $\frac{1}{2}\rho A v^3$ and that the resistive force acting on the car is $\frac{1}{2}\rho A v^2$, where ρ is the density of air. Compare this result with the empirical expression $\frac{1}{2}D\rho A v^2$ for the resistive force.

Figure P8.32

33. Heedless of danger, a child leaps onto a pile of old mattresses to use them as a trampoline. His motion between two particular points is described by the energy conservation equation

$$\tfrac{1}{2}(46.0\text{ kg})(2.40\text{ m/s})^2 + (46.0\text{ kg})(9.80\text{ m/s}^2)(2.80\text{ m} + x) = \tfrac{1}{2}(1.94 \times 10^4\text{ N/m})x^2$$

(a) Solve the equation for x. (b) Compose the statement of a problem, including data, for which this equation gives the solution. (c) Add the two values of x obtained in part (a) and divide by 2. (d) What is the significance of the resulting value in part (c)?

34. **Review.** *Why is the following situation impossible?* A new high-speed roller coaster is claimed to be so safe that the passengers do not need to wear seat belts or any other restraining device. The coaster is designed with a vertical circular section over which the coaster travels on the inside of the circle so that the passengers are upside down for a short time interval. The radius of the circular section is 12.0 m, and the coaster enters the bottom of the circular section at a speed of 22.0 m/s. Assume the coaster moves without friction on the track and model the coaster as a particle.

35. A horizontal spring attached to a wall has a force constant of $k = 850$ N/m. A block of mass $m = 1.00$ kg is attached to the spring and rests on a frictionless, horizontal surface as in Figure P8.35. (a) The block is pulled to a position $x_i = 6.00$ cm from equilibrium and released. Find the elastic potential energy stored in the spring when the block is 6.00 cm from equilibrium and when the block passes through equilibrium. (b) Find the speed of the block as it passes through the equilibrium point. (c) What is the speed of the block when it is at a position $x_i/2 = 3.00$ cm? (d) Why isn't the answer to part (c) half the answer to part (b)?

$x = 0$ $x = x_i/2$ $x = x_i$

Figure P8.35

36. More than 2 300 years ago, the Greek teacher Aristotle wrote the first book called *Physics*. Put into more precise terminology, this passage is from the end of its Section Eta:

Let P be the power of an agent causing motion; w, the load moved; d, the distance covered; and Δt, the time interval required. Then (1) a power equal to P will in

an interval of time equal to Δt move $w/2$ a distance $2d$; or (2) it will move $w/2$ the given distance d in the time interval $\Delta t/2$. Also, if (3) the given power P moves the given load w a distance $d/2$ in time interval $\Delta t/2$, then (4) $P/2$ will move $w/2$ the given distance d in the given time interval Δt.

(a) Show that Aristotle's proportions are included in the equation $P\,\Delta t = bwd$, where b is a proportionality constant. (b) Show that our theory of motion includes this part of Aristotle's theory as one special case. In particular, describe a situation in which it is true, derive the equation representing Aristotle's proportions, and determine the proportionality constant.

37. **Review.** As a prank, someone has balanced a pumpkin at the highest point of a grain silo. The silo is topped with a hemispherical cap that is frictionless when wet. The line from the center of curvature of the cap to the pumpkin makes an angle $\theta_i = 0°$ with the vertical. While you happen to be standing nearby in the middle of a rainy night, a breath of wind makes the pumpkin start sliding downward from rest. It loses contact with the cap when the line from the center of the hemisphere to the pumpkin makes a certain angle with the vertical. What is this angle?

38. **Review.** *Why is the following situation impossible?* An athlete tests her hand strength by having an assistant hang weights from her belt as she hangs onto a horizontal bar with her hands. When the weights hanging on her belt have increased to 80% of her body weight, her hands can no longer support her and she drops to the floor. Frustrated at not meeting her hand-strength

Figure P8.38

goal, she decides to swing on a trapeze. The trapeze consists of a bar suspended by two parallel ropes, each of length ℓ, allowing performers to swing in a vertical circular arc (Fig. P8.38). The athlete holds the bar and steps off an elevated platform, starting from rest with the ropes at an angle $\theta_i = 60.0°$ with respect to the vertical. As she swings several times back and forth in a circular arc, she forgets her frustration related to the hand-strength test. Assume the size of the performer's body is small compared to the length ℓ and air resistance is negligible.

39. An airplane of mass 1.50×10^4 kg is in level flight, initially moving at 60.0 m/s. The resistive force exerted by air on the airplane has a magnitude of 4.0×10^4 N. By Newton's third law, if the engines exert a force on the exhaust gases to expel them out of the back of the engine, the exhaust gases exert a force on the engines in the direction of the airplane's travel. This force is called thrust, and the value of the thrust in this situation is 7.50×10^4 N. (a) Is the work done by the exhaust gases on the airplane during some time interval equal to the change in the airplane's kinetic energy? Explain. (b) Find the speed of the airplane after it has traveled 5.0×10^2 m.

40. A pendulum, comprising a light string of length L and a small sphere, swings in the vertical plane. The string hits a peg located a distance d below the point of suspension

(Fig. P8.40). (a) Show that if the sphere is released from a height below that of the peg, it will return to this height after the string strikes the peg. (b) Show that if the pendulum is released from rest at the horizontal position ($\theta = 90°$) and is to swing in a complete circle centered on the peg, the minimum value of d must be $3L/5$.

Figure P8.40

41. A ball whirls around in a *vertical* circle at the end of a string. The other end of the string is fixed at the center of the circle. Assuming the total energy of the ball–Earth system remains constant, show that the tension in the string at the bottom is greater than the tension at the top by six times the ball's weight.

42. You are working in the distribution center of a large online shopping site. Efforts are being made to increase the number of packages per unit time that are being loaded onto a conveyor belt to be carried to waiting trucks. But the motor driving the conveyor belt is having difficulty keeping up with the increased demands. Your supervisor has asked you to determine the requirements for a new motor that can provide enough power to keep the conveyor belt moving smoothly under the increased loading rate. You are given the following information: The design goal is to have 50.0-kg packages loaded onto the belt at several locations at an average rate of 5.00 packages per second. The belt moves at a horizontal speed of 1.35 m/s. Humans at the various locations along the belt place the package on the belt so that it is initially at rest relative to the floor of the building just before being dropped from negligible height onto the belt. Your task is to determine the minimum power the driving motor must have to accelerate these packages and keep the belt moving at constant speed.

43. Consider the block–spring collision discussed in Example 8.8. (a) For the situation in part (B), in which the surface exerts a friction force on the block, show that the block never arrives back at $x = 0$. (b) What is the maximum value of the coefficient of friction that would allow the block to return to $x = 0$?

CHALLENGE PROBLEMS

44. Starting from rest, a 64.0-kg person bungee jumps from a tethered hot-air balloon 65.0 m above the ground. The bungee cord has negligible mass and unstretched length 25.8 m. One end is tied to the basket of the balloon and the other end to a harness around the person's body. The cord is modeled as a spring that obeys Hooke's law with a spring constant of 81.0 N/m, and the person's body is modeled as a particle. The hot-air balloon does not move. (a) Express the gravitational potential energy of the person–Earth system as a function of the person's variable height y above the ground. (b) Express the elastic potential energy of the cord as a function of y. (c) Express the total potential energy of the person–cord–Earth system as a function of y. (d) Plot a graph of the gravitational, elastic, and total potential energies as functions of y. (e) Assume air resistance is negligible. Determine the minimum height of the person above the ground during his plunge. (f) Does the potential energy graph show any equilibrium position or positions? If so, at what elevations? Are they stable or unstable? (g) Determine the jumper's maximum speed.

45. **Review.** A uniform board of length L is sliding along a smooth, frictionless, horizontal plane as shown in Figure P8.45a. The board then slides across the boundary with a rough horizontal surface. The coefficient of kinetic friction between the board and the second surface is μ_k. (a) Find the acceleration of the board at the moment its front end has traveled a distance x beyond the boundary. (b) The board stops at the moment its back end reaches the boundary as shown in Figure P8.45b. Find the initial speed v of the board.

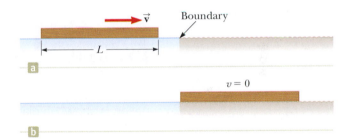

Figure P8.45

46. A uniform chain of length 8.00 m initially lies stretched out on a horizontal table. (a) Assuming the coefficient of static friction between chain and table is 0.600, show that the chain will begin to slide off the table if at least 3.00 m of it hangs over the edge of the table. (b) Determine the speed of the chain as its last link leaves the table, given that the coefficient of kinetic friction between the chain and the table is 0.400.

47. **What If?** Consider the roller coaster described in Problem 34. Because of some friction between the coaster and the track, the coaster enters the circular section at a speed of 15.0 m/s rather than the 22.0 m/s in Problem 34. Is this situation *more* or *less* dangerous for the passengers than that in Problem 34? Assume the circular section is still frictionless.

Linear Momentum and Collisions

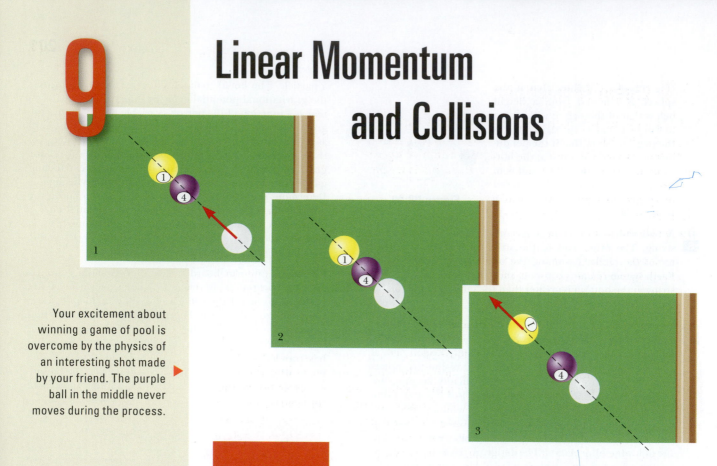

1

2

3

Your excitement about winning a game of pool is overcome by the physics of an interesting shot made by your friend. The purple ball in the middle never moves during the process.

STORYLINE **You decide to play pool at the student center at the** university. You and your friend are in the middle of a game when one shot made by your friend fascinates you and starts your mind thinking again. The initial situation is shown in diagram #1 above. Two balls, yellow and purple, are at rest and touching each other. You friend shoots the white cue ball along a line drawn through the centers of all three balls, and the cue ball makes a direct hit, so that the centers of all three balls are momentarily lined up, as in diagram #2. The cue ball stops and only the yellow ball moves away from the collision, as shown in diagram #3. The purple ball in the middle remains stationary during the entire interaction. You think, "Wait a minute! Why did that happen? The energy of the system of three balls must be conserved. So why couldn't *both* of the initially stationary balls move off after the collision at smaller speeds so that their kinetic energies add up to that of the cue ball?" Your friend pleads with you to continue the game, but your mind is elsewhere, analyzing this interesting situation.

CONNECTIONS While the energy approach studied in the previous chapters is powerful, there are still some problems we cannot solve in an easy way with the physics we've studied so far. In this chapter, we find that there is another conserved quantity besides energy. While this new quantity is a combination of mass and velocity, similar to kinetic energy, it is a vector, very different from energy. We find that the new conservation principle for this quantity, *momentum*, allows us to solve even more new types of problems, such as the one in the storyline. This conservation principle is particularly useful in analyzing collisions between two or more objects. As with energy, the analysis of systems is important; we will generate momentum principles for both isolated and nonisolated systems. Furthermore, our study of momentum in systems will lead to the important concept of the *center of mass* of a system of particles. The principles associated with momentum will join those associated with energy in several future chapters to allow us to understand many physical situations.

9.1 Linear Momentum

In Chapter 8, we studied situations that are difficult to analyze with Newton's laws. We were able to solve problems involving these situations by identifying a system and applying a conservation principle, conservation of energy. Let us consider another situation and see if we can solve it with the models we have developed so far:

A 60-kg archer stands at rest on frictionless ice and fires a 0.030-kg arrow horizontally at 85 m/s. With what velocity does the archer move across the ice after firing the arrow?

From Newton's third law, we know that the force that the bow exerts on the arrow is paired with a force in the opposite direction on the bow (and the archer). This force causes the archer to slide backward on the ice with the speed requested in the problem. We cannot determine this speed using motion models such as the particle under constant acceleration because we don't have any information about the acceleration of the archer. We cannot use force models such as the particle under a net force because we don't know anything about forces in this situation. Energy models are of no help because we know nothing about the work done in pulling the bowstring back or the elastic potential energy in the system related to the taut bowstring.

Despite our inability to solve the archer problem using models learned so far, this problem is very simple to solve if we introduce a new quantity that describes motion, *linear momentum*. To generate this new quantity, consider an isolated system of two particles (Fig. 9.1) with masses m_1 and m_2 moving with velocities $\vec{v}_1$ and $\vec{v}_2$ at an instant of time. Because the system is isolated, the only force on one particle is that from the other particle. If a force from particle 1 (for example, a gravitational force) acts on particle 2, there must be a second force—equal in magnitude but opposite in direction—that particle 2 exerts on particle 1. That is, the forces on the particles form a Newton's third law action–reaction pair, and $\vec{F}_{12} = -\vec{F}_{21}$. We can express this condition as

$$\vec{F}_{21} + \vec{F}_{12} = 0$$

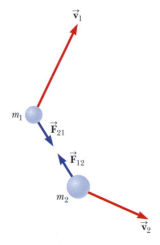

Figure 9.1 Two particles interact with each other. According to Newton's third law, we must have $\vec{F}_{12} = -\vec{F}_{21}$.

From a system point of view, this equation says that if we add up the forces on the particles in an isolated system, the sum is zero.

Let us further analyze this situation by incorporating Newton's second law, Equation 5.2. At the instant shown in Figure 9.1, the interacting particles in the system have accelerations corresponding to the forces on them. Therefore, replacing the force on each particle in the previous equation with $m\vec{a}$ for the particle gives

$$m_1\vec{a}_1 + m_2\vec{a}_2 = 0$$

Now we replace each acceleration with its definition from Equation 4.5:

$$m_1\frac{d\vec{v}_1}{dt} + m_2\frac{d\vec{v}_2}{dt} = 0$$

If the masses m_1 and m_2 are constant, we can bring them inside the derivative operation, which gives

$$\frac{d(m_1\vec{v}_1)}{dt} + \frac{d(m_2\vec{v}_2)}{dt} = 0$$

$$\frac{d}{dt}(m_1\vec{v}_1 + m_2\vec{v}_2) = 0 \tag{9.1}$$

Notice that the derivative of the sum $m_1\vec{v}_1 + m_2\vec{v}_2$ with respect to time is zero. Consequently, this sum must be constant over an arbitrary time interval. We saw in Chapter 8 that the total energy of an isolated system is constant over a time interval, because energy is conserved. We learn from this discussion that the quantity

$m\vec{\mathbf{v}}$ for a particle is important in that the sum of these quantities for an isolated system of particles is also conserved. We call this quantity *linear momentum:*

Definition of linear ▶
momentum of a particle

> The **linear momentum** $\vec{\mathbf{p}}$ of a particle or an object that can be modeled as a particle of mass m moving with a velocity $\vec{\mathbf{v}}$ is defined to be the product of the mass and velocity of the particle:
>
> $$\vec{\mathbf{p}} \equiv m\vec{\mathbf{v}} \tag{9.2}$$

Linear momentum is a vector quantity because it equals the product of a scalar quantity m and a vector quantity $\vec{\mathbf{v}}$. Its direction is along $\vec{\mathbf{v}}$, it has dimensions ML/T, and its SI unit is kg · m/s.

If a particle is moving in an arbitrary direction, $\vec{\mathbf{p}}$ has three components, and Equation 9.2 is equivalent to the component equations

$$p_x = mv_x \qquad p_y = mv_y \qquad p_z = mv_z$$

As you can see from its definition, the concept of momentum[1] provides a quantitative distinction between heavy and light particles moving at the same velocity. For example, the magnitude of the momentum of a bowling ball is much greater than that of a tennis ball moving at the same speed. Newton called the product $m\vec{\mathbf{v}}$ *quantity of motion;* this term is perhaps a more graphic description than our present-day word *momentum,* which comes from the Latin word for movement.

We have seen another quantity, kinetic energy, that is a combination of mass and speed. It would be a legitimate question to ask why we need a second quantity, momentum, based on mass and velocity. There are clear differences between kinetic energy and momentum. First, kinetic energy is a scalar, whereas momentum is a vector. Consider a system of two equal-mass particles heading toward each other along a line with equal speeds. There is kinetic energy associated with this system because members of the system are moving. Because of the vector nature of momentum, however, the momentum of this system is zero. A second major difference is that kinetic energy can transform to other types of energy, such as potential energy or internal energy. There is only one type of linear momentum, so we see no such transformations when using a momentum approach to a problem. These differences are sufficient to make models based on momentum separate from those based on energy, providing an independent tool to use in solving problems.

Using Newton's second law of motion, we can relate the linear momentum of a particle to the resultant force acting on the particle. We start with Newton's second law and substitute the definition of acceleration:

$$\sum \vec{\mathbf{F}} = m\vec{\mathbf{a}} = m\frac{d\vec{\mathbf{v}}}{dt}$$

In Newton's second law, the mass m is assumed to be constant. Therefore, we can bring m inside the derivative operation to give us

Newton's second law ▶
for a particle

$$\sum \vec{\mathbf{F}} = \frac{d(m\vec{\mathbf{v}})}{dt} = \frac{d\vec{\mathbf{p}}}{dt} \tag{9.3}$$

This equation shows that **the time rate of change of the linear momentum of a particle is equal to the net force acting on the particle.** In Chapter 5, we identified force as that which causes a change in the motion of an object (Section 5.2). In Newton's second law (Eq. 5.2), we used acceleration $\vec{\mathbf{a}}$ to represent the change in motion. We see now in Equation 9.3 that we can use the derivative of momentum $\vec{\mathbf{p}}$ with respect to time to represent the change in motion.

---

[1] In this chapter, the terms *momentum* and *linear momentum* have the same meaning. Later, in Chapter 11, we shall use the term *angular momentum* for a different quantity when dealing with rotational motion.

This alternative form of Newton's second law is the form in which Newton presented the law, and it is actually more general than the form introduced in Chapter 5. In addition to situations in which the velocity vector varies with time, we can use Equation 9.3 to study phenomena in which the mass changes. For example, the mass of a rocket changes as fuel is burned and ejected from the rocket. We cannot use $\sum \vec{F} = m\vec{a}$ to analyze rocket propulsion; we must use a momentum approach, as we will show in Section 9.9.

QUICK QUIZ 9.1 Two objects have equal kinetic energies. How do the magnitudes of their momenta compare? **(a)** $p_1 < p_2$ **(b)** $p_1 = p_2$ **(c)** $p_1 > p_2$ **(d)** not enough information to tell

QUICK QUIZ 9.2 Your physical education teacher throws a baseball to you at a certain speed and you catch it. The teacher is next going to throw you a medicine ball whose mass is ten times the mass of the baseball. You are given the following choices: You can have the medicine ball thrown with **(a)** the same speed as the baseball, **(b)** the same momentum, or **(c)** the same kinetic energy. Rank these choices from easiest to hardest to catch.

9.2 Analysis Model: Isolated System (Momentum)

Using the definition of momentum, Equation 9.1 can be written

$$\frac{d}{dt}(\vec{p}_1 + \vec{p}_2) = 0$$

Because the time derivative of the total momentum $\vec{p}_{tot} = \vec{p}_1 + \vec{p}_2$ is *zero*, we conclude that the *total* momentum of the isolated system of the two particles in Figure 9.1 must remain constant:

$$\vec{p}_{tot} = \text{constant} \tag{9.4}$$

or, equivalently, over some time interval,

$$\Delta \vec{p}_{tot} = 0 \tag{9.5}$$

Equation 9.5 can be written for a two-particle system as

$$\vec{p}_{1i} + \vec{p}_{2i} = \vec{p}_{1f} + \vec{p}_{2f}$$

where $\vec{p}_{1i}$ and $\vec{p}_{2i}$ are the initial values and $\vec{p}_{1f}$ and $\vec{p}_{2f}$ are the final values of the momenta for the two particles for the time interval during which the particles interact. This equation in component form demonstrates that the total momenta in the x, y, and z directions are all independently conserved:

$$p_{1ix} + p_{2ix} = p_{1fx} + p_{2fx} \qquad p_{1iy} + p_{2iy} = p_{1fy} + p_{2fy} \qquad p_{1iz} + p_{2iz} = p_{1fz} + p_{2fz} \tag{9.6}$$

Equation 9.5 is the mathematical statement of a new analysis model, the **isolated system (momentum)**. It can be extended to any number of particles in an isolated system, as we show in Section 9.7. For momentum, an isolated system is one on which no external forces act. We studied the energy version of the isolated system model in Chapter 8 ($\Delta E_{system} = 0$) and now we have a momentum version. In general, Equation 9.5 can be stated in words as follows:

Whenever two or more particles in an isolated system interact, the total momentum of the system does not change.

◀ The momentum version of the isolated system model

This statement tells us that the total momentum of an isolated system at all times equals its initial momentum.

PITFALL PREVENTION 9.1
Momentum of an Isolated *System* Is Conserved Although the momentum of an isolated *system* is conserved, the momentum of one *particle* within an isolated system is not necessarily conserved because other particles in the system may be interacting with it. Avoid applying conservation of momentum to a single particle.

Notice that we have made no statement concerning the type of forces acting on the particles of the system. Furthermore, we have not specified whether the forces are conservative or nonconservative. We have also not indicated whether or not the forces are constant. The only requirement is that the forces must be *internal* to the system. This single requirement should give you a hint about the power of this new model.

ANALYSIS MODEL   Isolated System (Momentum)

Imagine you have identified a system to be analyzed and have defined a system boundary. If there are no external forces on the system, the system is *isolated*. In that case, the total momentum of the system, which is the vector sum of the momenta of all members of the system, is conserved:

$$\Delta \vec{\mathbf{p}}_{\text{tot}} = 0 \qquad (9.5)$$

Examples:

- a cue ball strikes another ball on a pool table
- a spacecraft fires its rockets and moves faster through space (Section 9.9)
- molecules in a gas at a specific temperature move about and strike each other (Chapter 20)
- an incoming particle strikes a nucleus, creating a new nucleus and a different outgoing particle (Chapter 43)
- an electron and a positron annihilate to form two outgoing photons (Chapter 44)

System boundary

Momentum

If no external forces act on the system, the total momentum of the system is constant.

Example 9.1   The Archer

Let us consider the situation proposed at the beginning of Section 9.1. A 60-kg archer stands at rest on frictionless ice and fires a 0.030-kg arrow horizontally at 85 m/s (Fig. 9.2). With what velocity does the archer move across the ice after firing the arrow?

SOLUTION

Conceptualize   You may have conceptualized this problem already when it was introduced at the beginning of Section 9.1. Imagine the arrow being fired one way and the archer recoiling in the opposite direction.

Categorize   As discussed in Section 9.1, we cannot solve this problem with models based on motion, force, or energy. Nonetheless, we *can* solve this problem very easily with an approach involving momentum.

Let us take the system to consist of the archer (including the bow) and the arrow. The system is not isolated because the gravitational force and the normal force from the ice act on the system. These forces, however, are vertical and perpendicular to the motion of the system. There are no external forces in the horizontal direction, and we can apply the *isolated system (momentum)* model in terms of momentum components in this direction.

Figure 9.2 (Example 9.1) An archer fires an arrow horizontally to the right. Because he is standing on frictionless ice, he will begin to slide to the left across the ice.

Analyze   The total horizontal momentum of the system before the arrow is fired is zero because nothing in the system is moving. Therefore, the total horizontal momentum of the system after the arrow is fired must also be zero. We choose the direction of firing of the arrow as the positive x direction. Identifying the archer as particle 1 and the arrow as particle 2, we have $m_1 = 60$ kg, $m_2 = 0.030$ kg, and $\vec{\mathbf{v}}_{2f} = 85\hat{\mathbf{i}}$ m/s.

Using the isolated system (momentum) model, begin with Equation 9.5:

$$\Delta \vec{\mathbf{p}} = 0 \quad \rightarrow \quad \vec{\mathbf{p}}_f - \vec{\mathbf{p}}_i = 0 \quad \rightarrow \quad \vec{\mathbf{p}}_f = \vec{\mathbf{p}}_i \quad \rightarrow \quad m_1 \vec{\mathbf{v}}_{1f} + m_2 \vec{\mathbf{v}}_{2f} = 0$$

Solve this equation for $\vec{\mathbf{v}}_{1f}$ and substitute numerical values:

$$\vec{\mathbf{v}}_{1f} = -\frac{m_2}{m_1} \vec{\mathbf{v}}_{2f} = -\left(\frac{0.030 \text{ kg}}{60 \text{ kg}}\right)(85\hat{\mathbf{i}} \text{ m/s}) = -0.042\hat{\mathbf{i}} \text{ m/s}$$

9.1 continued

Finalize The negative sign for $\vec{v}_{1f}$ indicates that the archer is moving to the left in Figure 9.2 after the arrow is fired, in the direction opposite the direction of motion of the arrow, in accordance with Newton's third law. Because the archer is much more massive than the arrow, his acceleration and consequent velocity are much smaller than the acceleration and velocity of the arrow. Notice that this problem sounds very simple, but we could not solve it with models based on motion, force, or energy. Our new momentum model, however, shows us that it not only *sounds* simple, it *is* simple!

| WHAT IF? | What if the arrow were fired in a direction that makes an angle θ with the horizontal? How will that change the recoil velocity of the archer?

Answer The recoil velocity should decrease in magnitude because only a component of the velocity of the arrow is in the x direction. Conservation of momentum in the x direction gives

$$m_1 v_{1f} + m_2 v_{2f} \cos \theta = 0$$

leading to

$$v_{1f} = -\frac{m_2}{m_1} v_{2f} \cos \theta$$

For $\theta = 0$, $\cos \theta = 1$ and the final velocity of the archer reduces to the value when the arrow is fired horizontally. For nonzero values of θ, the cosine function is less than 1 and the recoil velocity is less than the value calculated for $\theta = 0$. If $\theta = 90°$, then $\cos \theta = 0$ and $v_{1f} = 0$, so there is no recoil velocity. In this case, the arrow is fired directly upward and the archer is simply pushed downward harder against the ice as the arrow is fired.

Example 9.2 | Can We Really Ignore the Kinetic Energy of the Earth?

In Section 7.6, we claimed that we can ignore the kinetic energy of the Earth when considering the energy of a system consisting of the Earth and a dropped ball. Verify this claim.

SOLUTION

Conceptualize Imagine dropping a ball at the surface of the Earth. From your point of view, the ball falls while the Earth remains stationary. By Newton's third law, however, the Earth experiences an upward force and therefore an upward acceleration while the ball falls. In the calculation below, we will show that this motion is extremely small and can be ignored.

Categorize We identify the system as the ball and the Earth. We assume there are no forces on the system from outer space, so the system is isolated. Let's use the *momentum* version of the *isolated system* model.

Analyze We begin by setting up a ratio of the kinetic energy of the Earth to that of the ball. We identify v_E and v_b as the speeds of the Earth and the ball, respectively, after the ball has fallen through some distance.

Use the definition of kinetic energy to set up this ratio:

$$(1) \quad \frac{K_E}{K_b} = \frac{\frac{1}{2}m_E v_E^2}{\frac{1}{2}m_b v_b^2} = \left(\frac{m_E}{m_b}\right)\left(\frac{v_E}{v_b}\right)^2$$

Apply the isolated system (momentum) model, recognizing that the initial momentum of the system is zero:

$$\Delta \vec{p} = 0 \quad \rightarrow \quad p_i = p_f \quad \rightarrow \quad 0 = m_b v_b + m_E v_E$$

Solve the equation for the ratio of velocity components:

$$\frac{v_E}{v_b} = -\frac{m_b}{m_E}$$

Take the absolute value of this ratio to make it a ratio of speeds and substitute for v_E/v_b in Equation (1):

$$\frac{K_E}{K_b} = \left(\frac{m_E}{m_b}\right)\left(\frac{m_b}{m_E}\right)^2 = \frac{m_b}{m_E}$$

Substitute order-of-magnitude numbers for the masses:

$$\frac{K_E}{K_b} = \frac{m_b}{m_E} = \frac{1 \text{ kg}}{10^{25} \text{ kg}} \sim 10^{-25}$$

Finalize The kinetic energy of the Earth is a very small fraction of the kinetic energy of the ball, so we are justified in ignoring it in the kinetic energy of the system.

9.3 Analysis Model: Nonisolated System (Momentum)

In the previous section, we found that the momentum of a system is conserved if there are no external forces on the system. What if there *is* an external force on the system? According to Equation 9.3, the momentum of a particle changes if a net force acts on the particle. The same can be said about a net force applied to a system as we will show explicitly in Section 9.7: the momentum of a system changes if a net force from the environment acts on the system. This may sound similar to our discussion of energy in Chapter 8: the energy of a system changes if energy crosses the boundary of the system to or from the environment. In this section, we consider a *nonisolated system*. For energy considerations, a system is nonisolated if energy transfers across the boundary of the system by any of the means listed in Section 8.1. For momentum considerations, a system is nonisolated if a net force acts on the system for a time interval. In this case, we can imagine momentum being transferred to the system from the environment by means of the net force. Knowing the change in momentum caused by a force is useful in solving some types of problems. To build a better understanding of this important concept, let us assume a net force $\sum \vec{\mathbf{F}}$ acts on a system consisting of a single particle and this force may vary with time. According to Newton's second law, in the form expressed in Equation 9.3, $\sum \vec{\mathbf{F}} = d\vec{\mathbf{p}}/dt$, we can write

$$d\vec{\mathbf{p}} = \sum \vec{\mathbf{F}} \, dt \tag{9.7}$$

We can integrate[2] this expression to find the change in the momentum of a particle when the force acts over some time interval. If the momentum of the particle changes from $\vec{\mathbf{p}}_i$ at time t_i to $\vec{\mathbf{p}}_f$ at time t_f, integrating Equation 9.7 gives

$$\Delta \vec{\mathbf{p}} = \vec{\mathbf{p}}_f - \vec{\mathbf{p}}_i = \int_{t_i}^{t_f} \sum \vec{\mathbf{F}} \, dt \tag{9.8}$$

To evaluate the integral, we need to know how the net force varies with time. The quantity on the right side of this equation is a vector called the **impulse** of the net force $\sum \vec{\mathbf{F}}$ acting on a particle over the time interval $\Delta t = t_f - t_i$:

Impulse of a force ▶

$$\vec{\mathbf{I}} \equiv \int_{t_i}^{t_f} \sum \vec{\mathbf{F}} \, dt \tag{9.9}$$

From its definition, we see that impulse $\vec{\mathbf{I}}$ is a vector quantity having a magnitude equal to the area under the force–time curve as described in Figure 9.3a. It is assumed the force varies in time in the general manner shown in the figure and is nonzero in the time interval $\Delta t = t_f - t_i$. The direction of the impulse vector is the same as the direction of the change in momentum. Impulse has the dimensions of momentum, that is, ML/T. Impulse is *not* a property of a particle; rather, it is a measure of the degree to which an external force changes the particle's momentum.

Because the net force imparting an impulse to a particle can generally vary in time, it is convenient to define a time-averaged net force:

$$\left(\sum \vec{\mathbf{F}} \right)_{\text{avg}} \equiv \frac{1}{\Delta t} \int_{t_i}^{t_f} \sum \vec{\mathbf{F}} \, dt \tag{9.10}$$

where $\Delta t = t_f - t_i$. (This equation is an application of the mean value theorem of calculus.) Therefore, we can express Equation 9.9 as

$$\vec{\mathbf{I}} = \left(\sum \vec{\mathbf{F}} \right)_{\text{avg}} \Delta t \tag{9.11}$$

This time-averaged force, shown in Figure 9.3b, can be interpreted as the constant force that would give to the particle in the time interval Δt the same impulse that the time-varying force gives over this same interval.

---

[2]Here we are integrating force with respect to time. Compare this strategy with our efforts in Chapter 7, where we integrated force with respect to position to find the work done by the force.

The impulse imparted to the particle by the force is the area under the curve.

a

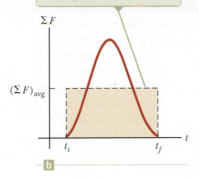

The time-averaged net force gives the same impulse to a particle as does the time-varying force in (a).

b

Figure 9.3 (a) A net force acting on a particle may vary in time. (b) The value of the constant force $(\Sigma F)_{avg}$ (horizontal dashed line) is chosen so that the area $(\Sigma F)_{avg} \Delta t$ of the rectangle is the same as the area under the curve in (a).

In principle, if $\Sigma \vec{F}$ is known as a function of time, the impulse can be calculated from Equation 9.9. The calculation becomes especially simple if the force acting on the particle is constant. In this case, $(\Sigma \vec{F})_{avg} = \Sigma \vec{F}$, where $\Sigma \vec{F}$ is the constant net force, and Equation 9.11 becomes

$$\vec{I} = \sum \vec{F} \, \Delta t \qquad \text{(constant net force)} \qquad (9.12)$$

Combining Equations 9.8 and 9.9 gives us an important statement known as the **impulse–momentum theorem:**

> The change in the momentum of a particle is equal to the impulse of the net force acting on the particle:
>
> $$\Delta \vec{p} = \vec{I} \qquad\qquad (9.13)$$

◀ Impulse–momentum theorem for a particle

This statement is equivalent to Newton's second law. When we say that an impulse is given to a particle, we mean that momentum is transferred from an external agent to that particle. Equation 9.13 is identical in form to the conservation of energy equation, Equation 8.1, and its full expansion, Equation 8.2. Equation 9.13 is the most general statement of the principle of **conservation of momentum** and is called the **conservation of momentum equation.** In the case of a momentum approach, isolated systems tend to appear in problems more often than nonisolated systems, so, in practice, the conservation of momentum equation is often identified as the special case of Equation 9.5.

The left side of Equation 9.13 represents the change in the momentum of the system, which in our discussion so far is a single particle. The right side is a measure of how much momentum crosses the boundary of the system due to the net force being applied to the system. Equation 9.13 is the mathematical statement of a new analysis model, the **nonisolated system (momentum)** model. Although this equation is similar in form to Equation 8.2, there are several differences in its application to problems. First, Equation 9.13 is a vector equation, whereas Equation 8.2 is a scalar equation. Therefore, directions are important for Equation 9.13. Second, there is only one type of momentum and therefore only one way to store momentum in a system. In contrast, as we see from Equation 8.2, there are three ways to store energy in a system: kinetic, potential, and internal. Third, there is only one way to transfer momentum into a system: by the application of a force on the system over a time interval. Equation 8.2 shows six ways we have identified as transferring energy into a system. Therefore, there is no expansion of Equation 9.13 analogous to Equation 8.2.

As a real-world example of Equation 9.13, consider the crash-test dummy in Figure 9.4 (page 218), representing a human driver in an accident. As the car

Figure 9.4 A crash-test dummy is brought to rest by an air bag in a test collision. The air bag increases the time interval during which the dummy is brought to rest, thereby decreasing the force on the dummy. Air bags in automobiles have saved countless human lives in accidents.

is brought to rest from its initial speed, the dummy experiences a given change in momentum. Now consider the impulse on the right side of Equation 9.13, expressed with Equation 9.11. The same impulse can occur with a large average force over a short time interval or a small average force over a long time interval. In the absence of an air bag, the dummy is brought to rest by the sudden collision of his head with the steering wheel or dashboard. This is an example of the former possibility, and the large average force could result in serious injury to a human driver. If an air bag is present, however, the dummy can be brought to rest gradually over a longer time interval, resulting in a smaller average force. As a result, there is a possibility of avoiding injury to a human driver.

In many physical situations, we shall use what is called the **impulse approximation,** in which we assume one of the forces exerted on a particle acts for a short time but is much greater than any other force present. In this case, the net force $\sum \vec{\mathbf{F}}$ in Equation 9.9 is replaced with a single force $\vec{\mathbf{F}}$ to find the impulse on the particle. This approximation is especially useful in treating collisions in which the duration of the collision is very short. When this approximation is made, the single force is referred to as an *impulsive force.* For example, when a baseball is struck with a bat, the time of the collision is about 0.01 s and the average force that the bat exerts on the ball in this time is typically several thousand newtons. Because this contact force is much greater than the magnitude of the gravitational force, the impulse approximation justifies our ignoring the gravitational forces exerted on the ball and bat during the collision. When we use this approximation, it is important to remember that $\vec{\mathbf{p}}_i$ and $\vec{\mathbf{p}}_f$ represent the momenta *immediately* before and after the collision, respectively. Therefore, in any situation in which it is proper to use the impulse approximation, the particle moves very little during the collision.

QUICK QUIZ **9.3** Two objects are at rest on a frictionless surface. Object 1 has a greater mass than object 2. **(i)** When a constant force is applied to object 1, it accelerates through a distance d in a straight line. The force is removed from object 1 and is applied to object 2. At the moment when object 2 has accelerated through the same distance d, which statements are true? (a) $p_1 < p_2$ (b) $p_1 = p_2$ (c) $p_1 > p_2$ (d) $K_1 < K_2$ (e) $K_1 = K_2$ (f) $K_1 > K_2$ **(ii)** When a force is applied to object 1, it accelerates for a time interval Δt. The force is removed from object 1 and is applied to object 2. From the same list of choices, which statements are true after object 2 has accelerated for the same time interval Δt?

QUICK QUIZ **9.4** Rank an automobile dashboard, seat belt, and air bag, each used alone in separate collisions from the same speed, in terms of (a) the impulse and (b) the average force each delivers to a front-seat passenger, from greatest to least.

ANALYSIS MODEL   **Nonisolated System (Momentum)**

Imagine you have identified a system to be analyzed and have defined a system boundary. If external forces are applied on the system, the system is *nonisolated.* In that case, the change in the total momentum of the system is equal to the impulse on the system, a statement known as the **impulse–momentum theorem:**

$$\Delta \vec{\mathbf{p}} = \vec{\mathbf{I}} \qquad (9.13)$$

Examples:

- a baseball is struck by a bat
- a spool sitting on a table is pulled by a string (Example 10.14 in Chapter 10)
- a gas molecule strikes the wall of the container holding the gas (Chapter 20)
- photons strike an absorbing surface and exert pressure on the surface (Chapter 33)

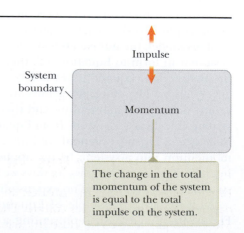

The change in the total momentum of the system is equal to the total impulse on the system.

| Example **9.3** | **How Good Are the Bumpers?** |

In a particular crash test, a car of mass 1 500 kg collides with a wall as shown in Figure 9.5. The initial and final velocities of the car are $\vec{\mathbf{v}}_i = -15.0\,\hat{\mathbf{i}}$ m/s and $\vec{\mathbf{v}}_f = 2.60\,\hat{\mathbf{i}}$ m/s, respectively. If the collision lasts 0.150 s, find the impulse on the car during the collision and the average net force exerted on the car.

Figure 9.5 (Example 9.3) (a) This car's momentum changes as a result of its collision with the wall. (b) In a crash test, much of the car's initial kinetic energy is transformed into energy associated with the damage to the car.

S O L U T I O N

Conceptualize The collision time is short, so we can imagine the car being brought to rest very rapidly and then moving back in the opposite direction with a reduced speed.

Categorize Let us assume the net force exerted on the car by the wall and friction from the ground is large compared with other forces on the car (such as air resistance). Furthermore, the gravitational force and the normal force exerted by the road on the car are perpendicular to the motion and therefore do not affect the horizontal momentum. Therefore, we categorize the problem as one in which we can apply the impulse approximation in the horizontal direction. We also see that the car's momentum changes due to an impulse from the environment. Therefore, we can apply the *nonisolated system (momentum)* model to the system of the car.

Analyze

Use Equation 9.13 to find the impulse on the car:

$$\vec{\mathbf{I}} = \Delta\vec{\mathbf{p}} = \vec{\mathbf{p}}_f - \vec{\mathbf{p}}_i = m\vec{\mathbf{v}}_f - m\vec{\mathbf{v}}_i = m(\vec{\mathbf{v}}_f - \vec{\mathbf{v}}_i)$$

$$= (1\,500 \text{ kg})[2.60\,\hat{\mathbf{i}} \text{ m/s} - (-15.0\,\hat{\mathbf{i}} \text{ m/s})] = \boxed{2.64 \times 10^4\,\hat{\mathbf{i}} \text{ kg} \cdot \text{m/s}}$$

Use Equation 9.11 to evaluate the average net force exerted on the car:

$$\left(\sum\vec{\mathbf{F}}\right)_{\text{avg}} = \frac{\vec{\mathbf{I}}}{\Delta t} = \frac{2.64 \times 10^4\,\hat{\mathbf{i}} \text{ kg} \cdot \text{m/s}}{0.150 \text{ s}} = \boxed{1.76 \times 10^5\,\hat{\mathbf{i}} \text{ N}}$$

Finalize The net force found above is a combination of the normal force on the car from the wall and any friction force between the tires and the ground as the front of the car crumples. If the brakes are not operating while the crash occurs and the crumpling metal does not interfere with the free rotation of the tires, this friction force could be relatively small due to the freely rotating wheels. Notice that the signs of the velocities in this example indicate the reversal of directions. What would the mathematics be describing if both the initial and final velocities had the same sign?

W H A T   I F ? What if the car did not rebound from the wall? Suppose the final velocity of the car is zero and the time interval of the collision remains at 0.150 s. Would that represent a larger or a smaller net force on the car?

Answer In the original situation in which the car rebounds, the net force on the car does two things during the time interval: (1) it stops the car, and (2) it causes the car to move away from the wall at 2.60 m/s after the collision. If the car does not rebound, the net force is only doing the first of these steps—stopping the car—which requires a *smaller* force.

Mathematically, in the case of the car that does not rebound, the impulse is

$$\vec{\mathbf{I}} = \Delta\vec{\mathbf{p}} = \vec{\mathbf{p}}_f - \vec{\mathbf{p}}_i = 0 - (1\,500 \text{ kg})(-15.0\,\hat{\mathbf{i}} \text{ m/s}) = 2.25 \times 10^4\,\hat{\mathbf{i}} \text{ kg} \cdot \text{m/s}$$

The average net force exerted on the car is

$$\left(\sum\vec{\mathbf{F}}\right)_{\text{avg}} = \frac{\vec{\mathbf{I}}}{\Delta t} = \frac{2.25 \times 10^4\,\hat{\mathbf{i}} \text{ kg} \cdot \text{m/s}}{0.150 \text{ s}} = 1.50 \times 10^5\,\hat{\mathbf{i}} \text{ N}$$

which is indeed smaller than the previously calculated value, as was argued conceptually.

9.4 Collisions in One Dimension

In this section, we use the isolated system (momentum) model to describe what happens when two particles collide. The term **collision** represents an event during which two particles come close to each other and interact by means of forces. The

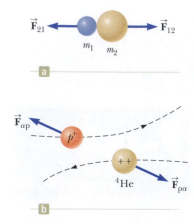

Figure 9.6 (a) The collision between two objects as the result of direct contact. (b) The "collision" between two charged particles.

interaction forces are assumed to be much greater than any external forces present, so we can use the impulse approximation.

A collision may involve physical contact between two macroscopic objects as described in Figure 9.6a, but the notion of what is meant by a collision must be generalized because "physical contact" on a submicroscopic scale is ill-defined and hence meaningless. To understand this concept, consider a collision on an atomic scale (Fig. 9.6b) such as the collision of a proton with an alpha particle (the nucleus of a helium atom). Because the particles are both positively charged, they repel each other due to the strong electrostatic force between them at close separations and never come into "physical contact."

When two particles of masses m_1 and m_2 collide as shown in Figure 9.6, the impulsive forces may vary in time in complicated ways, such as that shown in Figure 9.3. Regardless of the complexity of the time behavior of the impulsive force, however, this force is internal to the system of two particles. Therefore, the two particles form an isolated system and the momentum of the system must be conserved in *any* collision.

In contrast, the total kinetic energy of the system of particles may or may not be conserved, depending on the type of collision. In fact, collisions are categorized as being either *elastic* or *inelastic* depending on whether or not kinetic energy is conserved.

An **elastic collision** between two particles, or objects that can be modeled as particles, is one in which the total kinetic energy (as well as total momentum) of the system is the same before and after the collision. Collisions between certain objects in the macroscopic world, such as billiard balls, are only *approximately* elastic because some deformation and loss of kinetic energy take place. For example, you can hear a billiard ball collision, so you know that some of the energy is being transferred away from the system by sound. An elastic collision must be perfectly silent! *Truly* elastic collisions occur between atomic and subatomic particles. These collisions are described by the isolated system model for both energy and momentum.

An **inelastic collision** is one in which the total kinetic energy of the system is not the same before and after the collision (even though the momentum of the system is conserved). Inelastic collisions are further divided into two types. When the objects stick together after they collide, as happens when a meteorite collides with the Earth, the collision is called **perfectly inelastic.** When the colliding objects do not stick together but some kinetic energy is transformed or transferred away, the collision is called **inelastic** (with no modifying adverb). The collision of a rubber ball bouncing from a hard surface is inelastic, but not perfectly inelastic, because the ball does not stick to the surface. It is not elastic because some of the initial kinetic energy of the ball has been transformed to internal energy in the ball and the surface as the ball deformed during the time interval of contact.

In the remainder of this section, we investigate the mathematical details for collisions in one dimension and consider the two extreme cases, perfectly inelastic and elastic collisions.

Perfectly Inelastic Collisions

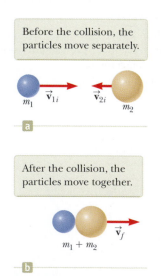

Figure 9.7 Schematic representation of a perfectly inelastic head-on collision between two particles.

Consider two particles of masses m_1 and m_2 moving with initial velocities $\vec{\mathbf{v}}_{1i}$ and $\vec{\mathbf{v}}_{2i}$ along the same straight line as shown in Figure 9.7. The two particles collide head-on, stick together, and then move with some common velocity $\vec{\mathbf{v}}_f$ after the collision. For example, two carts with Velcro on their bumpers colliding on an air track will behave in this way. Because the momentum of an isolated system is conserved in *any* collision, we can say that the total momentum before the collision equals the total momentum of the composite system after the collision:

$$\Delta \vec{\mathbf{p}} = 0 \quad \rightarrow \quad \vec{\mathbf{p}}_i = \vec{\mathbf{p}}_f \quad \rightarrow \quad m_1\vec{\mathbf{v}}_{1i} + m_2\vec{\mathbf{v}}_{2i} = (m_1 + m_2)\vec{\mathbf{v}}_f \quad (9.14)$$

Solving for the final velocity gives

$$\vec{\mathbf{v}}_f = \frac{m_1\vec{\mathbf{v}}_{1i} + m_2\vec{\mathbf{v}}_{2i}}{m_1 + m_2} \quad (9.15)$$

Elastic Collisions

Consider two particles of masses m_1 and m_2 moving with initial velocities $\vec{v}_{1i}$ and $\vec{v}_{2i}$ along the same straight line as shown in Figure 9.8. The two particles collide head-on and then leave the collision site with different velocities, $\vec{v}_{1f}$ and $\vec{v}_{2f}$. In an elastic collision, both the momentum and kinetic energy of the system are conserved. Therefore, considering velocities along the horizontal direction in Figure 9.8, we have

$$p_i = p_f \quad \rightarrow \quad m_1 v_{1i} + m_2 v_{2i} = m_1 v_{1f} + m_2 v_{2f} \tag{9.16}$$

$$K_i = K_f \quad \rightarrow \quad \tfrac{1}{2} m_1 v_{1i}{}^2 + \tfrac{1}{2} m_2 v_{2i}{}^2 = \tfrac{1}{2} m_1 v_{1f}{}^2 + \tfrac{1}{2} m_2 v_{2f}{}^2 \tag{9.17}$$

Because all velocities in Figure 9.8 are either to the left or the right, they can be represented by the corresponding speeds along with algebraic signs indicating direction. We shall indicate v as positive if a particle moves to the right and negative if it moves to the left.

In a typical problem involving elastic collisions, there are two unknown quantities, and Equations 9.16 and 9.17 can be solved simultaneously to find them. An alternative approach, however—one that involves a little mathematical manipulation of Equation 9.17—often simplifies this process. To see how, let us cancel the factor $\tfrac{1}{2}$ in Equation 9.17 and rewrite it by gathering terms with subscript 1 on the left and 2 on the right:

$$m_1(v_{1i}{}^2 - v_{1f}{}^2) = m_2(v_{2f}{}^2 - v_{2i}{}^2)$$

Factoring both sides of this equation gives

$$m_1(v_{1i} - v_{1f})(v_{1i} + v_{1f}) = m_2(v_{2f} - v_{2i})(v_{2f} + v_{2i}) \tag{9.18}$$

Next, let us separate the terms containing m_1 and m_2 in Equation 9.16 in a similar way to obtain

$$m_1(v_{1i} - v_{1f}) = m_2(v_{2f} - v_{2i}) \tag{9.19}$$

To obtain our final result, we divide Equation 9.18 by Equation 9.19 and obtain

$$v_{1i} + v_{1f} = v_{2f} + v_{2i}$$

Now rearrange terms once again so as to have initial quantities on the left and final quantities on the right:

$$v_{1i} - v_{2i} = -(v_{1f} - v_{2f}) \tag{9.20}$$

This equation, in combination with Equation 9.16, can be used to solve problems dealing with elastic collisions. This pair of equations (Eqs. 9.16 and 9.20) is easier to handle than the pair of Equations 9.16 and 9.17 because there are no quadratic terms like there are in Equation 9.17. According to Equation 9.20, the *relative* velocity of the two particles before the collision, $v_{1i} - v_{2i}$, equals the negative of their relative velocity after the collision, $-(v_{1f} - v_{2f})$.

Suppose the masses and initial velocities of both particles are known. Equations 9.16 and 9.20 can be solved for the final velocities in terms of the initial velocities because there are two equations and two unknowns:

$$v_{1f} = \left(\frac{m_1 - m_2}{m_1 + m_2}\right) v_{1i} + \left(\frac{2 m_2}{m_1 + m_2}\right) v_{2i} \tag{9.21}$$

$$v_{2f} = \left(\frac{2 m_1}{m_1 + m_2}\right) v_{1i} + \left(\frac{m_2 - m_1}{m_1 + m_2}\right) v_{2i} \tag{9.22}$$

It is important to use the appropriate signs for v_{1i} and v_{2i} in Equations 9.21 and 9.22.

Before the collision, the particles move separately.

a

After the collision, the particles continue to move separately with new velocities.

$\vec{v}_{1f}$ $\vec{v}_{2f}$

b

Figure 9.8 Schematic representation of an elastic head-on collision between two particles.

PITFALL PREVENTION 9.3
Not a General Equation Equation 9.20 can only be used in a very *specific* situation, a one-dimensional, elastic collision between two objects. The *general* concept is conservation of momentum (and conservation of kinetic energy if the collision is elastic) for an isolated system.

Let us consider some special cases. If $m_1 = m_2$, Equations 9.21 and 9.22 show that $v_{1f} = v_{2i}$ and $v_{2f} = v_{1i}$, which means that the particles exchange velocities if they have equal masses. That is approximately what one observes in head-on billiard ball collisions: the cue ball stops and the struck ball moves away from the collision with the same velocity the cue ball had.

If particle 2 is initially at rest, then $v_{2i} = 0$, and Equations 9.21 and 9.22 become

◄ Elastic collision: particle 2 initially at rest

$$v_{1f} = \left(\frac{m_1 - m_2}{m_1 + m_2} \right) v_{1i} \tag{9.23}$$

$$v_{2f} = \left(\frac{2m_1}{m_1 + m_2} \right) v_{1i} \tag{9.24}$$

If m_1 is much greater than m_2 and $v_{2i} = 0$, we see from Equations 9.23 and 9.24 that $v_{1f} \approx v_{1i}$ and $v_{2f} \approx 2v_{1i}$. That is, when a very heavy particle collides head-on with a very light one that is initially at rest, the heavy particle continues its motion unaltered after the collision and the light particle rebounds with a speed equal to about twice the initial speed of the heavy particle. An example of such a collision is that of a moving heavy atom, such as uranium, striking a light atom, such as hydrogen.

If m_2 is much greater than m_1 and particle 2 is initially at rest, then $v_{1f} \approx -v_{1i}$ and $v_{2f} \approx 0$. That is, when a very light particle collides head-on with a very heavy particle that is initially at rest, the light particle has its velocity reversed and the heavy one remains approximately at rest. For example, imagine what happens when you throw a table tennis ball at a bowling ball as in Quick Quiz 9.6 below.

QUICK QUIZ **9.5** In a perfectly inelastic one-dimensional collision between two moving objects, what condition alone is necessary so that the final kinetic energy of the system is zero after the collision? **(a)** The objects must have initial momenta with the same magnitude but opposite directions. **(b)** The objects must have the same mass. **(c)** The objects must have the same initial velocity. **(d)** The objects must have the same initial speed, with velocity vectors in opposite directions.

QUICK QUIZ **9.6** A table-tennis ball is thrown at a stationary bowling ball. The table-tennis ball makes a one-dimensional elastic collision and bounces back along the same line. Compared with the bowling ball after the collision, does the table-tennis ball have **(a)** a larger magnitude of momentum and more kinetic energy, **(b)** a smaller magnitude of momentum and more kinetic energy, **(c)** a larger magnitude of momentum and less kinetic energy, **(d)** a smaller magnitude of momentum and less kinetic energy, or **(e)** the same magnitude of momentum and the same kinetic energy?

PROBLEM-SOLVING STRATEGY   **One-Dimensional Collisions**

You should use the following approach when solving collision problems in one dimension:

1. Conceptualize. Imagine the collision occurring in your mind. Draw simple diagrams of the particles before and after the collision and include appropriate velocity vectors. At first, you may have to guess at the directions of the final velocity vectors.

2. Categorize. Is the system of particles isolated? If so, use the isolated system (momentum) model. Further categorize the collision as elastic, inelastic, or perfectly inelastic.

3. Analyze. Set up the appropriate mathematical representation for the problem. If the collision is perfectly inelastic, use Equation 9.15. If the collision is elastic, use Equations 9.16 and 9.20. If the collision is inelastic, use Equation 9.16. To find the final velocities in this case, you will need some additional information.

4. Finalize. Once you have determined your result, check to see if your answers are consistent with the mental and pictorial representations and that your results are realistic.

| Example **9.4** | **The Executive Stress Reliever** |

An ingenious device that illustrates conservation of momentum and kinetic energy is shown in Figure 9.9a. It consists of five identical hard balls supported by strings of equal lengths. When ball 1 is pulled out and released, after the almost-elastic collision between it and ball 2, ball 1 stops and ball 5 moves out as shown in Figure 9.9b. If balls 1 and 2 are pulled out and released, they stop after the collision and balls 4 and 5 swing out, and so forth. Even if four balls (1–4) are pulled out and released, four balls (2–5) swing out after the collision! Is it ever possible that when ball 1 is released, it stops after the collision and balls 4 and 5 will swing out on the opposite side and travel with half the speed of ball 1 as in Figure 9.9c?

Charles D. Winters

This can happen

This cannot happen

Figure 9.9 (Example 9.4) (a) An executive stress reliever. (b) If one ball swings down, we see one ball swing out at the other end. (c) Is it possible for one ball to swing down and two balls to leave the other end with half the speed of the first ball? In (b) and (c), the velocity vectors shown represent those of the balls immediately before and immediately after the collision.

SOLUTION

Conceptualize   With the help of Figure 9.9c, imagine one ball coming in from the left and two balls exiting the collision on the right. That is the phenomenon we want to test to see if it could ever happen.

Categorize   Because of the very short time interval between the arrival of the ball from the left and the departure of the ball(s) from the right, we can use the impulse approximation to ignore the gravitational forces on the balls and model the five balls as an *isolated system* in terms of both *momentum* and *energy*. Because the balls are hard, we can categorize the collisions between them as elastic for purposes of calculation.

Analyze   Let's consider the situation shown in Figure 9.9c. The momentum of the system just before the collision is mv, where m is the mass of ball 1 and v is its speed immediately before the collision. After the collision, we imagine that ball 1 stops and balls 4 and 5 swing out, each moving with speed $v/2$. The total momentum of the system after the collision would be $m(v/2) + m(v/2) = mv$. Therefore, the momentum of the system is conserved in the situation shown in Figure 9.9c!

The kinetic energy of the system immediately before the collision is $K_i = \frac{1}{2}mv^2$ and that after the collision is $K_f = \frac{1}{2}m(v/2)^2 + \frac{1}{2}m(v/2)^2 = \frac{1}{4}mv^2$. This calculation shows that the kinetic energy of the system is *not* conserved, which is inconsistent with our assumption that the collisions are elastic.

Finalize   Our analysis shows that it is *not* possible for balls 4 and 5 to swing out when only ball 1 is released. The only way to conserve both momentum and kinetic energy of the system is for one ball to move out when one ball is released, two balls to move out when two are released, and so on. A similar analysis can be applied to the billiard ball collision in the opening storyline. In that case, there are two billiard balls in contact rather than four steel balls as in Figure 9.9. When the cue ball strikes the pair of balls, the only way to conserve both momentum and kinetic energy for the system of three balls is for only one ball to leave the collision. Therefore, the purple ball remains stationary, just like balls 2 through 4 in Figure 9.9.

WHAT IF?   Consider what would happen if balls 4 and 5 are glued together. Now what happens when ball 1 is pulled out and released?

Answer   In this situation, balls 4 and 5 *must* move together as a single object after the collision. We have argued that both momentum and energy of the system cannot be conserved in this case. We assumed, however, ball 1 stopped after striking ball 2. What if we do not make this assumption? Consider the conservation equations with the assumption that ball 1 moves after the collision. For conservation of momentum,

$$p_i = p_f$$

$$mv_{1i} = mv_{1f} + 2mv_{4,5}$$

where $v_{4,5}$ refers to the final speed of the ball 4–ball 5 combination. Conservation of kinetic energy gives us

$$K_i = K_f$$

$$\tfrac{1}{2}mv_{1i}^2 = \tfrac{1}{2}mv_{1f}^2 + \tfrac{1}{2}(2m)v_{4,5}^2$$

continued

9.4 continued

Combining these equations gives

$$v_{4,5} = \tfrac{2}{3}v_{1i} \qquad v_{1f} = -\tfrac{1}{3}v_{1i}$$

Therefore, balls 4 and 5 move together as one object after the collision while ball 1 bounces back from the collision with one third of its original speed.

Example **9.5**   **Carry Collision Insurance!**

An 1 800-kg car stopped at a traffic light is struck from the rear by a 900-kg car. The two cars become entangled, moving along the same path as that of the originally moving car. If the smaller car were moving at 20.0 m/s before the collision, what is the velocity of the entangled cars after the collision?

SOLUTION

Conceptualize   This kind of collision is easily visualized, and one can predict that after the collision both cars will be moving in the same direction as that of the initially moving car. Because the initially moving car has only half the mass of the stationary car, we expect the final velocity of the cars to be relatively small.

Categorize   We identify the two cars as an *isolated system* in terms of *momentum* in the horizontal direction and apply the impulse approximation during the short time interval of the collision. The phrase "become entangled" tells us to categorize the collision as perfectly inelastic.

Analyze   The magnitude of the total momentum of the system before the collision is equal to that of the smaller car because the larger car is initially at rest.

Use the isolated system model for momentum:

$$\Delta \vec{\mathbf{p}} = 0 \quad \rightarrow \quad p_i = p_f \quad \rightarrow \quad m_1 v_i = (m_1 + m_2)v_f$$

Solve for v_f and substitute numerical values:

$$v_f = \frac{m_1 v_i}{m_1 + m_2} = \frac{(900 \text{ kg})(20.0 \text{ m/s})}{900 \text{ kg} + 1\ 800 \text{ kg}} = 6.67 \text{ m/s}$$

Finalize   Because the final velocity is positive, the direction of the final velocity of the combination is the same as the velocity of the initially moving car as predicted. The speed of the combination is also much lower than the initial speed of the moving car.

WHAT IF?   Suppose we reverse the masses of the cars. What if a stationary 900-kg car is struck by a moving 1 800-kg car? Is the final speed the same as before?

Answer   Intuitively, we can guess that the final speed of the combination is higher than 6.67 m/s if the initially moving car is the more massive car. Mathematically, that should be the case because the system has a larger momentum if the initially moving car is the more massive one. Solving for the new final velocity, we find

$$v_f = \frac{m_1 v_i}{m_1 + m_2} = \frac{(1\ 800 \text{ kg})(20.0 \text{ m/s})}{1\ 800 \text{ kg} + 900 \text{ kg}} = 13.3 \text{ m/s}$$

which is two times the previous final velocity.

Example **9.6**   **The Ballistic Pendulum**

The ballistic pendulum (Fig. 9.10) is an apparatus used to measure the speed of a fast-moving projectile such as a bullet. A projectile of mass m_1 is fired into a large block of wood of mass m_2 suspended from some light wires. The projectile embeds in the block, and the entire system swings through a height h. How can we determine the speed of the projectile from a measurement of h?

SOLUTION

Conceptualize   Figure 9.10a helps conceptualize the situation. Run the animation in your mind: the projectile enters the pendulum, which swings up to some height at which it momentarily comes to rest.

9.6 continued

Figure 9.10 (Example 9.6) (a) Diagram of a ballistic pendulum. Notice that $\vec{\mathbf{v}}_{1A}$ is the velocity of the projectile immediately before the collision and $\vec{\mathbf{v}}_B$ is the velocity of the projectile–block system immediately after the perfectly inelastic collision. (b) Multiflash photograph of a ballistic pendulum used in the laboratory.

Categorize   Let's focus first on the collision between the projectile and the block. The projectile and the block form an *isolated system* in terms of *momentum* in the horizontal direction if we identify configuration A as immediately before the collision and configuration B as immediately after the collision. Because the projectile imbeds in the block, we can categorize the collision between them as perfectly inelastic.

Analyze   To analyze the collision, we use Equation 9.15, which gives the speed of the system immediately after the collision when we assume the impulse approximation.

Noting that $v_{2A} = 0$, write Equation 9.15 for v_B:

$$(1) \quad v_B = \frac{m_1 v_{1A}}{m_1 + m_2}$$

Categorize   For the second process, during which the projectile–block combination swings upward to height h (ending at a configuration we'll call C), we focus on a *different* system, that of the projectile, the block, and the Earth. We categorize this part of the problem as one involving an *isolated system* for *energy* with no nonconservative forces acting.

Analyze   Write an expression for the total kinetic energy of the system immediately after the collision:

$$(2) \quad K_B = \tfrac{1}{2}(m_1 + m_2)v_B^{\,2}$$

Substitute the value of v_B from Equation (1) into Equation (2):

$$K_B = \frac{m_1^2 v_{1A}^{\,2}}{2(m_1 + m_2)}$$

This kinetic energy of the system immediately after the collision is *less* than the initial kinetic energy of the projectile as is expected in an inelastic collision.

   We define the gravitational potential energy of the system for configuration B to be zero. Therefore, $U_B = 0$, whereas $U_C = (m_1 + m_2)gh$.

Apply the isolated system model for energy (Eq. 8.2) to the system:

$$\Delta K + \Delta U = 0 \quad \rightarrow \quad (K_C - K_B) + (U_C - U_B) = 0$$

Substitute the energies:

$$\left[0 - \frac{m_1^2 v_{1A}^{\,2}}{2(m_1 + m_2)} \right] + [(m_1 + m_2)gh - 0] = 0$$

Solve for v_{1A}:

$$v_{1A} = \left(\frac{m_1 + m_2}{m_1} \right)\sqrt{2gh}$$

Finalize   We had to solve this problem in two steps. Each step involved a different system and a different analysis model: isolated system (momentum) for the first step and isolated system (energy) for the second. Because the collision was assumed to be perfectly inelastic, some mechanical energy was transformed to internal energy during the collision. Therefore, it would have been *incorrect* to apply the isolated system (energy) model to the entire process by equating the initial kinetic energy of the incoming projectile with the final gravitational potential energy of the projectile–block–Earth combination.

A Two-Body Collision with a Spring

A block of mass $m_1 = 1.60$ kg initially moving to the right with a speed of 4.00 m/s on a frictionless, horizontal track collides with a light spring attached to a second block of mass $m_2 = 2.10$ kg initially moving to the left with a speed of 2.50 m/s as shown in Figure 9.11a. The spring constant is 600 N/m.

(A) Find the velocities of the two blocks when they are again moving separately after the collision.

SOLUTION

Conceptualize With the help of Figure 9.11a, run an animation of the collision in your mind. Figure 9.11b shows an instant during the collision when the spring is compressed. Eventually, block 1 and the spring will again separate, so the system will look like Figure 9.11a again but with different velocity vectors for the two blocks.

Categorize Because the spring force is conservative, kinetic energy in the system of two blocks and the spring is not transformed to internal energy during the compression of the spring. Ignoring any sound made when the block hits the spring, we can categorize the collision as being elastic and categorize the two blocks and the spring as an *isolated system* for both *energy* and *momentum*.

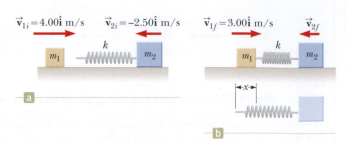

Figure 9.11 (Example 9.7) A moving block approaches a second moving block that is attached to a spring.

Analyze Because momentum of the system is conserved, apply Equation 9.16:

(1) $\quad m_1 v_{1i} + m_2 v_{2i} = m_1 v_{1f} + m_2 v_{2f}$

Because the collision is elastic, apply Equation 9.20:

(2) $\quad v_{1i} - v_{2i} = -(v_{1f} - v_{2f})$

Multiply Equation (2) by m_1:

(3) $\quad m_1 v_{1i} - m_1 v_{2i} = -m_1 v_{1f} + m_1 v_{2f}$

Add Equations (1) and (3):

$2m_1 v_{1i} + (m_2 - m_1)v_{2i} = (m_1 + m_2)v_{2f}$

Solve for v_{2f}:

$v_{2f} = \dfrac{2m_1 v_{1i} + (m_2 - m_1)v_{2i}}{m_1 + m_2}$

Substitute numerical values:

$v_{2f} = \dfrac{2(1.60 \text{ kg})(4.00 \text{ m/s}) + (2.10 \text{ kg} - 1.60 \text{ kg})(-2.50 \text{ m/s})}{1.60 \text{ kg} + 2.10 \text{ kg}} = \boxed{3.12 \text{ m/s}}$

Solve Equation (2) for v_{1f} and substitute numerical values:

$v_{1f} = v_{2f} - v_{1i} + v_{2i} = 3.12 \text{ m/s} - 4.00 \text{ m/s} + (-2.50 \text{ m/s}) = \boxed{-3.38 \text{ m/s}}$

Finalize Notice that both blocks have reversed direction due to the collision. Also notice that we did not need to know anything about the spring to find the answer in this part of the problem. The spring is just another mechanism for the two blocks to exert forces of equal magnitude and opposite direction on one another, just like those between the objects and particles shown in Figure 9.6.

(B) Determine the velocity of block 2 during the collision, at the instant block 1 is moving to the right with a velocity of +3.00 m/s as in Figure 9.11b.

SOLUTION

Conceptualize Focus your attention now on Figure 9.11b, which represents the final configuration of the system at the end of the time interval of interest.

Categorize Because the momentum of the *isolated system* of two blocks and the spring are conserved *throughout* the collision, the collision can be categorized as elastic for *any* final instant of time. Let us now choose the final instant to be when block 1 is moving with a velocity of +3.00 m/s.

9.7 continued

Analyze   Apply Equation 9.16:

$$m_1 v_{1i} + m_2 v_{2i} = m_1 v_{1f} + m_2 v_{2f}$$

Solve for v_{2f}:

$$v_{2f} = \frac{m_1 v_{1i} + m_2 v_{2i} - m_1 v_{1f}}{m_2}$$

Substitute numerical values:

$$v_{2f} = \frac{(1.60 \text{ kg})(4.00 \text{ m/s}) + (2.10 \text{ kg})(-2.50 \text{ m/s}) - (1.60 \text{ kg})(3.00 \text{ m/s})}{2.10 \text{ kg}}$$

$$= \boxed{-1.74 \text{ m/s}}$$

Finalize   The negative value for v_{2f} means that block 2 is still moving to the left at the instant we are considering.

(C) Determine the distance the spring is compressed at that instant.

SOLUTION

Conceptualize   Once again, focus on the configuration of the system shown in Figure 9.11b.

Categorize   For the system of the spring and two blocks, no friction or other nonconservative forces act within the system. Therefore, we categorize the system as an *isolated system* in terms of *energy* with no nonconservative forces acting. The system also remains an *isolated system* in terms of *momentum*.

Analyze   We choose the initial configuration of the system to be that existing immediately before block 1 strikes the spring and the final configuration to be that when block 1 is moving to the right at 3.00 m/s.

Write the appropriate reduction of Equation 8.2:

$$\Delta K + \Delta U = 0$$

Evaluate the energies, recognizing that two objects in the system have kinetic energy and that the potential energy is elastic:

$$[(\tfrac{1}{2}m_1 v_{1f}^2 + \tfrac{1}{2}m_2 v_{2f}^2) - (\tfrac{1}{2}m_1 v_{1i}^2 + \tfrac{1}{2}m_2 v_{2i}^2)] + (\tfrac{1}{2}kx^2 - 0) = 0$$

Solve for x^2:

$$x^2 = \frac{1}{k}[m_1(v_{1i}^2 - v_{1f}^2) + m_2(v_{2i}^2 - v_{2f}^2)]$$

Substitute numerical values:

$$x^2 = \left(\frac{1}{600 \text{ N/m}}\right)\{(1.60 \text{ kg})[(4.00 \text{ m/s})^2 - (3.00 \text{ m/s})^2] + (2.10 \text{ kg})[(2.50 \text{ m/s})^2 - (1.74 \text{ m/s})^2]\}$$

$$\rightarrow \quad x = \boxed{0.173 \text{ m}}$$

Finalize   This answer is not the maximum compression of the spring because the two blocks are still moving toward each other at the instant shown in Figure 9.11b. Can you determine the maximum compression of the spring?

9.5 Collisions in Two Dimensions

In Section 9.2, we showed that the momentum of a system of two particles is conserved when the system is isolated. For any collision of two particles, this result implies that the momentum in each of the directions x, y, and z is conserved. An important subset of collisions takes place in a plane. The game of billiards is a familiar example involving multiple collisions of objects moving on a two-dimensional surface. For such two-dimensional collisions between two particles, we obtain two component equations for conservation of momentum:

$$m_1 v_{1ix} + m_2 v_{2ix} = m_1 v_{1fx} + m_2 v_{2fx}$$

$$m_1 v_{1iy} + m_2 v_{2iy} = m_1 v_{1fy} + m_2 v_{2fy}$$

where the three subscripts on the velocity components in these equations represent, respectively, the identification of the object (1, 2), initial and final values (i, f), and the velocity component (x, y).

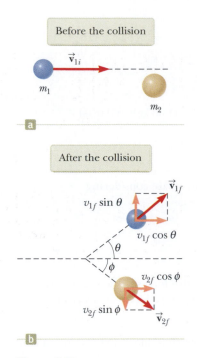

Before the collision

$\vec{\mathbf{v}}_{1i}$

m_1

m_2

a

After the collision

$\vec{\mathbf{v}}_{1f}$

$v_{1f} \sin \theta$

$v_{1f} \cos \theta$

θ

ϕ

$v_{2f} \cos \phi$

$v_{2f} \sin \phi$

$\vec{\mathbf{v}}_{2f}$

b

Figure 9.12 An elastic, glancing collision between two particles.

Let us consider a specific two-dimensional problem in which particle 1 of mass m_1 collides with particle 2 of mass m_2 initially at rest as in Figure 9.12a. After the collision (Fig. 9.12b), particle 1 moves at an angle θ with respect to the horizontal and particle 2 moves at an angle ϕ with respect to the horizontal. This event is called a *glancing* collision. Applying the law of conservation of momentum in component form and noting that the initial y component of the momentum of the two-particle system is zero gives

$$\Delta p_x = 0 \quad \rightarrow \quad p_{ix} = p_{fx} \quad \rightarrow \quad m_1 v_{1i} = m_1 v_{1f} \cos \theta + m_2 v_{2f} \cos \phi \quad (9.25)$$

$$\Delta p_y = 0 \quad \rightarrow \quad p_{iy} = p_{fy} \quad \rightarrow \quad 0 = m_1 v_{1f} \sin \theta - m_2 v_{2f} \sin \phi \quad (9.26)$$

where the minus sign in Equation 9.26 is included because after the collision particle 2 has a y component of velocity that is downward. (The symbols v in these particular equations are speeds, not velocity components. The direction of the component vector is indicated explicitly with plus or minus signs.) We now have two independent equations. As long as no more than two of the seven quantities in Equations 9.25 and 9.26 are unknown, we can solve the problem.

If the collision is elastic, we can also use Equation 9.17 (conservation of kinetic energy) with $v_{2i} = 0$:

$$K_i = K_f \quad \rightarrow \quad \tfrac{1}{2} m_1 v_{1i}^2 = \tfrac{1}{2} m_1 v_{1f}^2 + \tfrac{1}{2} m_2 v_{2f}^2 \quad (9.27)$$

Knowing the initial speed of particle 1 and both masses, we are left with four unknowns (v_{1f}, v_{2f}, θ, and ϕ). Because we have only three equations, one of the four remaining quantities must be given to determine the motion after the elastic collision from conservation principles alone.

If the collision is inelastic, kinetic energy is *not* conserved and Equation 9.27 does *not* apply. Then we have four unknowns and only two equations!

PITFALL PREVENTION 9.4

Don't Use Equation 9.20 Equation 9.20, relating the initial and final relative velocities of two colliding objects, is only valid for one-dimensional elastic collisions. Do not use this equation when analyzing two-dimensional collisions.

PROBLEM-SOLVING STRATEGY   **Two-Dimensional Collisions**

The following procedure is recommended when dealing with problems involving collisions between two particles in two dimensions.

1. Conceptualize. Imagine the collisions occurring and predict the approximate directions in which the particles will move after the collision. Set up a coordinate system and define your velocities in terms of that system. It is convenient to have the x axis coincide with one of the initial velocities. Sketch the coordinate system, draw and label all velocity vectors, and include all the given information.

2. Categorize. Is the system of particles truly isolated? If so, categorize the collision as elastic, inelastic, or perfectly inelastic.

3. Analyze. Write expressions for the x and y components of the momentum of each object before and after the collision. Remember to include the appropriate signs for the components of the velocity vectors and pay careful attention to signs throughout the calculation.

Apply the isolated system model for momentum $\Delta \vec{\mathbf{p}} = 0$. When applied in each direction, this equation will generally reduce to $p_{ix} = p_{fx}$ and $p_{iy} = p_{fy}$, where each of these terms refer to the sum of the momenta of all objects in the system. Write expressions for the *total* momentum in the x direction *before* and *after* the collision and equate the two. Repeat this procedure for the total momentum in the y direction.

Proceed to solve the momentum equations for the unknown quantities. If the collision is inelastic, kinetic energy is *not* conserved and additional information is probably required. If the collision is perfectly inelastic, the final velocities of the two objects are equal.

If the collision is elastic, kinetic energy is conserved and you can equate the total kinetic energy of the system before the collision to the total kinetic energy after the collision, providing an additional relationship between the velocity magnitudes.

4. Finalize. Once you have determined your result, check to see if your answers are consistent with the mental and pictorial representations and that your results are realistic.

Example **9.8** **Collision at an Intersection**

A 1 500-kg car traveling east with a speed of 25.0 m/s collides at an intersection with a 2 500-kg truck traveling north at a speed of 20.0 m/s as shown in Figure 9.13. Find the direction and magnitude of the velocity of the wreckage after the collision, assuming the vehicles stick together after the collision.

SOLUTION

Conceptualize Figure 9.13 should help you conceptualize the situation before and after the collision. Let us choose east to be along the positive x direction and north to be along the positive y direction.

Categorize Because we consider instants of time immediately before and immediately after the collision as defining our time interval, we ignore the small effect that friction would have on the wheels of the vehicles and model the two vehicles as an *isolated system* in terms of *momentum*. We also ignore the vehicles' sizes and model them as particles. The collision is perfectly inelastic because the car and the truck stick together after the collision.

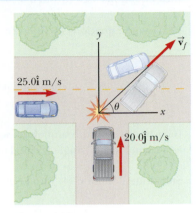

Figure 9.13 (Example 9.8) An eastbound car colliding with a northbound truck.

Analyze Before the collision, the only object having momentum in the x direction is the car. Therefore, the magnitude of the total initial momentum of the system (car plus truck) in the x direction is that of only the car. Similarly, the total initial momentum of the system in the y direction is that of the truck. Immediately after the collision, let us assume the wreckage moves at an angle θ with respect to the x axis with speed v_f.

Apply the isolated system model for momentum in the x direction:

$$\Delta p_x = 0 \quad \rightarrow \quad \sum p_{xi} = \sum p_{xf} \quad \rightarrow \quad (1) \quad m_1 v_{1i} = (m_1 + m_2) v_f \cos \theta$$

Apply the isolated system model for momentum in the y direction:

$$\Delta p_y = 0 \quad \rightarrow \quad \sum p_{yi} = \sum p_{yf} \quad \rightarrow \quad (2) \quad m_2 v_{2i} = (m_1 + m_2) v_f \sin \theta$$

Divide Equation (2) by Equation (1):

$$\frac{m_2 v_{2i}}{m_1 v_{1i}} = \frac{\sin \theta}{\cos \theta} = \tan \theta$$

Solve for θ and substitute numerical values:

$$\theta = \tan^{-1}\left(\frac{m_2 v_{2i}}{m_1 v_{1i}}\right) = \tan^{-1}\left[\frac{(2\,500 \text{ kg})(20.0 \text{ m/s})}{(1\,500 \text{ kg})(25.0 \text{ m/s})}\right] = \boxed{53.1°}$$

Use Equation (2) to find the value of v_f and substitute numerical values:

$$v_f = \frac{m_2 v_{2i}}{(m_1 + m_2) \sin \theta} = \frac{(2\,500 \text{ kg})(20.0 \text{ m/s})}{(1\,500 \text{ kg} + 2\,500 \text{ kg}) \sin 53.1°} = \boxed{15.6 \text{ m/s}}$$

Finalize Notice that the angle θ is qualitatively in agreement with Figure 9.13. Also notice that the final speed of the combination is less than the initial speeds of the two cars. This result is consistent with the kinetic energy of the system being reduced in an inelastic collision. It might help if you draw the momentum vectors of each vehicle before the collision and the two vehicles together after the collision.

Example **9.9** **Proton–Proton Collision**

A proton collides elastically with another proton that is initially at rest. The incoming proton has an initial speed of 3.50×10^5 m/s and makes a glancing collision with the second proton as in Figure 9.12. (At close separations, the protons exert a repulsive electrostatic force on each other.) After the collision, one proton moves off at an angle of 37.0° to the original direction of motion and the second deflects at an angle of ϕ to the same axis. Find the final speeds of the two protons and the angle ϕ.

SOLUTION

Conceptualize This collision is like that shown in Figure 9.12, which will help you conceptualize the behavior of the system. We define the x axis to be along the direction of the velocity vector of the initially moving proton.

Categorize The pair of protons form an *isolated system*. Both momentum and kinetic energy of the system are conserved in this glancing elastic collision.

continued

9.9 continued

Analyze   Using the isolated system model for both momentum and energy for a two-dimensional elastic collision, set up the mathematical representation with Equations 9.25 through 9.27:

(1) $v_{1i} = v_{1f}\cos\theta + v_{2f}\cos\phi$
(2) $0 = v_{1f}\sin\theta - v_{2f}\sin\phi$
(3) $v_{1i}^2 = v_{1f}^2 + v_{2f}^2$

Rearrange Equations (1) and (2):

$v_{2f}\cos\phi = v_{1i} - v_{1f}\cos\theta$

$v_{2f}\sin\phi = v_{1f}\sin\theta$

Square these two equations and add them:

$v_{2f}^2\cos^2\phi + v_{2f}^2\sin^2\phi =$
$\quad v_{1i}^2 - 2v_{1i}v_{1f}\cos\theta + v_{1f}^2\cos^2\theta + v_{1f}^2\sin^2\theta$

Incorporate that the sum of the squares of sine and cosine for *any* angle is equal to 1:

(4) $v_{2f}^2 = v_{1i}^2 - 2v_{1i}v_{1f}\cos\theta + v_{1f}^2$

Substitute Equation (4) into Equation (3):

$v_{1f}^2 + (v_{1i}^2 - 2v_{1i}v_{1f}\cos\theta + v_{1f}^2) = v_{1i}^2$

(5) $v_{1f}^2 - v_{1i}v_{1f}\cos\theta = 0$

One possible solution of Equation (5) is $v_{1f} = 0$, which corresponds to a head-on, one-dimensional collision in which the first proton stops and the second continues with the same speed in the same direction. That is not the solution we want.

Divide both sides of Equation (5) by v_{1f} and solve for the remaining factor of v_{1f}:

$v_{1f} = v_{1i}\cos\theta = (3.50 \times 10^5\,\text{m/s})\cos 37.0° = \boxed{2.80 \times 10^5\,\text{m/s}}$

Use Equation (3) to find v_{2f}:

$v_{2f} = \sqrt{v_{1i}^2 - v_{1f}^2} = \sqrt{(3.50 \times 10^5\,\text{m/s})^2 - (2.80 \times 10^5\,\text{m/s})^2}$

$= \boxed{2.11 \times 10^5\,\text{m/s}}$

Use Equation (2) to find ϕ:

(2) $\phi = \sin^{-1}\left(\dfrac{v_{1f}\sin\theta}{v_{2f}}\right) = \sin^{-1}\left[\dfrac{(2.80 \times 10^5\,\text{m/s})\sin 37.0°}{(2.11 \times 10^5\,\text{m/s})}\right]$

$= \boxed{53.0°}$

Finalize   It is interesting that $\theta + \phi = 90°$. This result is *not* accidental. Whenever two objects of equal mass collide elastically in a glancing collision and one of them is initially at rest, their final velocities are perpendicular to each other.

9.6 The Center of Mass

In this section, we describe the overall motion of a system in terms of a special point called the **center of mass** of the system. The system can be either a small number of distinct particles or an extended, continuous object, such as a gymnast leaping through the air. We shall see that the translational motion of the center of mass of the system is the same as if all the mass of the system were concentrated at that point. That is, the system moves as if the net external force were applied to a single particle located at the center of mass. This model, the *particle model,* was introduced in Chapter 2. This behavior is independent of other motion, such as rotation or vibration of the system or deformation of the system (for instance, when a gymnast folds her body).

Consider a system consisting of a pair of particles that have different masses and are connected by a light, rigid rod (Fig. 9.14). The position of the center of mass of a system can be described as being the *average position* of the system's mass. The center of mass of the system is located somewhere on the line joining the two particles and is closer to the particle having the larger mass. If a single force is applied at a point on the rod above the center of mass, the system rotates clockwise (see Fig. 9.14a). If the force is applied at a point on the rod below the center of mass, the system rotates counterclockwise (see Fig. 9.14b). If the force is applied at the center of mass, the system moves in the direction of the force without rotating (see Fig. 9.14c). The center of mass of an object can be located with this procedure.

The center of mass of the pair of particles described in Figure 9.15 is located on the x axis and lies somewhere between the particles. Its x coordinate is given by

$$x_{CM} \equiv \frac{m_1 x_1 + m_2 x_2}{m_1 + m_2} \tag{9.28}$$

For example, if $x_1 = 0$, $x_2 = d$, and $m_2 = 2m_1$, we find that $x_{CM} = \frac{2}{3}d$. That is, the center of mass lies closer to the more massive particle. If the two masses are equal, the center of mass lies midway between the particles.

We can extend this concept to a system of many particles with masses m_i in three dimensions. The x coordinate of the center of mass of n particles is defined to be

$$x_{CM} \equiv \frac{m_1 x_1 + m_2 x_2 + m_3 x_3 + \cdots + m_n x_n}{m_1 + m_2 + m_3 + \cdots + m_n} = \frac{\sum_i m_i x_i}{\sum_i m_i} = \frac{\sum_i m_i x_i}{M} = \frac{1}{M}\sum_i m_i x_i \tag{9.29}$$

where x_i is the x coordinate of the ith particle and the total mass is $M \equiv \sum_i m_i$ where the sum runs over all n particles. The y and z coordinates of the center of mass are similarly defined by the equations

$$y_{CM} \equiv \frac{1}{M}\sum_i m_i y_i \quad \text{and} \quad z_{CM} \equiv \frac{1}{M}\sum_i m_i z_i \tag{9.30}$$

The center of mass can be located in three dimensions by its position vector $\vec{r}_{CM}$. The components of this vector are x_{CM}, y_{CM}, and z_{CM}, defined in Equations 9.29 and 9.30. Therefore,

$$\vec{r}_{CM} = x_{CM}\hat{i} + y_{CM}\hat{j} + z_{CM}\hat{k} = \frac{1}{M}\sum_i m_i x_i \hat{i} + \frac{1}{M}\sum_i m_i y_i \hat{j} + \frac{1}{M}\sum_i m_i z_i \hat{k}$$

$$\vec{r}_{CM} \equiv \frac{1}{M}\sum_i m_i \vec{r}_i \tag{9.31}$$

where $\vec{r}_i$ is the position vector of the ith particle, defined by

$$\vec{r}_i \equiv x_i\hat{i} + y_i\hat{j} + z_i\hat{k}$$

Although locating the center of mass for an extended, continuous object is somewhat more cumbersome than locating the center of mass of a small number of particles, the basic ideas we have discussed still apply. Think of an extended object as a system containing a large number of small mass elements such as the cube in Figure 9.16 (page 232). Because the separation between elements is very small, the object can be considered to have a continuous mass distribution. By dividing the object into elements of mass Δm_i with coordinates x_i, y_i, z_i, we see that the x coordinate of the center of mass is approximately

$$x_{CM} \approx \frac{1}{M}\sum_i x_i \Delta m_i$$

with similar expressions for y_{CM} and z_{CM}. If we let the number of elements n approach infinity, the size of each element approaches zero and x_{CM} is given precisely. In this limit, we replace the sum by an integral and Δm_i by the differential element dm:

The system rotates clockwise when a force is applied above the center of mass.

a

The system rotates counterclockwise when a force is applied below the center of mass.

b

The system moves in the direction of the force without rotating when a force is applied at the center of mass.

c

Figure 9.14 A force is applied to a system of two particles of unequal mass connected by a light, rigid rod.

Figure 9.15 The center of mass of two particles of unequal mass on the x axis is located at x_{CM}, a point between the particles, closer to the one having the larger mass.

An extended object can be considered to be a distribution of small elements of mass Δm_i.

$$x_{CM} = \lim_{\Delta m_i \to 0} \frac{1}{M} \sum_i x_i \, \Delta m_i = \frac{1}{M} \int x \, dm \qquad (9.32)$$

Likewise, for y_{CM} and z_{CM} we obtain

$$y_{CM} = \frac{1}{M} \int y \, dm \quad \text{and} \quad z_{CM} = \frac{1}{M} \int z \, dm \qquad (9.33)$$

We can express the vector position of the center of mass of an extended object in the form

$$\vec{\mathbf{r}}_{CM} = \frac{1}{M} \int \vec{\mathbf{r}} \, dm \qquad (9.34)$$

Figure 9.16 The center of mass is located at the vector position $\vec{\mathbf{r}}_{CM}$, which has coordinates x_{CM}, y_{CM}, and z_{CM}.

which is equivalent to the three expressions given by Equations 9.32 and 9.33.

The center of mass of any symmetric object of uniform density lies on an axis of symmetry and on any plane of symmetry. For example, the center of mass of a uniform rod lies in the rod, midway between its ends. The center of mass of a sphere or a cube lies at its geometric center.

Because an extended object is a continuous distribution of mass, each small mass element is acted upon by the gravitational force. The net effect of all these forces is equivalent to the effect of a single force $M\vec{\mathbf{g}}$ acting through a special point, called the **center of gravity.** If $\vec{\mathbf{g}}$ is constant over the mass distribution, the center of gravity coincides with the center of mass. If an extended object is pivoted at its center of gravity, it balances in any orientation.

The center of gravity of an irregularly shaped object such as a wrench can be determined by suspending the object first from one point and then from another. In Figure 9.17, a wrench is hung from point A and a vertical line AB (which can be established with a plumb bob) is drawn when the wrench has stopped swinging. The wrench is then hung from point C, and a second vertical line CD is drawn. The center of gravity is halfway through the thickness of the wrench, under the intersection of these two lines. In general, if the wrench is hung freely from any point, the vertical line through this point must pass through the center of gravity.

The wrench is hung freely first from point A and then from point C.

The intersection of the two lines AB and CD locates the center of gravity.

Figure 9.17 An experimental technique for determining the center of gravity of a wrench.

QUICK QUIZ **9.7** A baseball bat of uniform density is cut at the location of its center of mass as shown in Figure 9.18. Which piece has the smaller mass? **(a)** the piece on the right **(b)** the piece on the left **(c)** both pieces have the same mass **(d)** impossible to determine

Figure 9.18 (Quick Quiz 9.7) A baseball bat cut at the location of its center of mass.

Example 9.10 | **The Center of Mass of Three Particles**

A system consists of three particles located as shown in Figure 9.19. Find the center of mass of the system. The masses of the particles are $m_1 = m_2 = 1.0$ kg and $m_3 = 2.0$ kg.

SOLUTION

Conceptualize Figure 9.19 shows the three masses. Your intuition should tell you that the center of mass is located somewhere in the region between the blue particle and the pair of tan particles as shown in the figure.

Figure 9.19 (Example 9.10) Two particles are located on the x axis, and a single particle is located on the y axis as shown. The vector indicates the location of the system's center of mass.

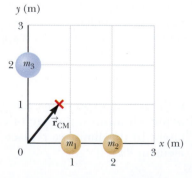

9.10 continued

Categorize We categorize this example as a substitution problem because we will be using the equations for the center of mass developed in this section.

Use the defining equations for the coordinates of the center of mass and notice that $z_{CM} = 0$:

$$x_{CM} = \frac{1}{M}\sum_i m_i x_i = \frac{m_1 x_1 + m_2 x_2 + m_3 x_3}{m_1 + m_2 + m_3}$$

$$= \frac{(1.0\text{ kg})(1.0\text{ m}) + (1.0\text{ kg})(2.0\text{ m}) + (2.0\text{ kg})(0)}{1.0\text{ kg} + 1.0\text{ kg} + 2.0\text{ kg}} = \frac{3.0\text{ kg}\cdot\text{m}}{4.0\text{ kg}} = 0.75\text{ m}$$

$$y_{CM} = \frac{1}{M}\sum_i m_i y_i = \frac{m_1 y_1 + m_2 y_2 + m_3 y_3}{m_1 + m_2 + m_3}$$

$$= \frac{(1.0\text{ kg})(0) + (1.0\text{ kg})(0) + (2.0\text{ kg})(2.0\text{ m})}{4.0\text{ kg}} = \frac{4.0\text{ kg}\cdot\text{m}}{4.0\text{ kg}} = 1.0\text{ m}$$

Write the position vector of the center of mass:

$$\vec{\mathbf{r}}_{CM} \equiv x_{CM}\hat{\mathbf{i}} + y_{CM}\hat{\mathbf{j}} = \boxed{(0.75\hat{\mathbf{i}} + 1.0\hat{\mathbf{j}})\text{ m}}$$

Example 9.11   **The Center of Mass of a Rod**

(A) Show that the center of mass of a rod of mass M and length L lies midway between its ends, assuming the rod has a uniform mass per unit length.

SOLUTION

Conceptualize The rod is shown aligned along the x axis in Figure 9.20, so $y_{CM} = z_{CM} = 0$. What is your prediction of the value of x_{CM}?

Categorize We categorize this example as an analysis problem because we need to divide the rod into small mass elements to perform the integration in Equation 9.32.

Figure 9.20 (Example 9.11) The geometry used to find the center of mass of a uniform rod.

Analyze The mass per unit length (this quantity is called the *linear mass density*) can be written as $\lambda = M/L$ for the uniform rod. If the rod is divided into elements of length dx, the mass of each element is $dm = \lambda\ dx$.

Use Equation 9.32 to find an expression for x_{CM}:

$$x_{CM} = \frac{1}{M}\int x\ dm = \frac{1}{M}\int_0^L x\lambda\ dx = \frac{\lambda}{M}\frac{x^2}{2}\Big|_0^L = \frac{\lambda L^2}{2M}$$

Substitute $\lambda = M/L$:

$$x_{CM} = \frac{L^2}{2M}\left(\frac{M}{L}\right) = \tfrac{1}{2}L$$

One can also use symmetry arguments to obtain the same result.

(B) Suppose a rod is *nonuniform* such that its mass per unit length varies linearly with x according to the expression $\lambda = \alpha x$, where α is a constant. Find the x coordinate of the center of mass as a fraction of L.

SOLUTION

Conceptualize Because the mass per unit length is not constant in this case but is proportional to x, elements of the rod to the right are more massive than elements near the left end of the rod.

Categorize This problem is categorized similarly to part (A), with the added twist that the linear mass density is not constant.

Analyze We replace dm in Equation 9.32 by $\lambda\ dx$, where, in this case, $\lambda = \alpha x$.

Use Equation 9.32 to find an expression for x_{CM}:

$$x_{CM} = \frac{1}{M}\int x\ dm = \frac{1}{M}\int_0^L x\lambda\ dx = \frac{1}{M}\int_0^L x\alpha x\ dx$$

$$= \frac{\alpha}{M}\int_0^L x^2\ dx = \frac{\alpha L^3}{3M}$$

continued

9.11 continued

Find the total mass of the rod:

$$M = \int dm = \int_0^L \lambda \, dx = \int_0^L \alpha x \, dx = \frac{\alpha L^2}{2}$$

Substitute M into the expression for x_{CM}:

$$x_{CM} = \frac{\alpha L^3}{3\alpha L^2/2} = \boxed{\tfrac{2}{3}L}$$

Finalize   Notice that the center of mass in part (B) is farther to the right than that in part (A). That result is reasonable because the elements of the rod become more massive as one moves to the right along the rod in part (B).

9.7 Systems of Many Particles

Consider a system of two or more particles for which we have identified the center of mass. We can begin to further understand the physical significance and utility of the center of mass concept by taking the time derivative of the position vector for the center of mass given by Equation 9.31. From Section 4.1, we know that the time derivative of a position vector is by definition the velocity vector. Assuming M remains constant for a system of particles—that is, no particles enter or leave the system—we obtain the following expression for the **velocity of the center of mass** of the system:

◀ Velocity of the center of mass of a system of particles

$$\vec{\mathbf{v}}_{CM} = \frac{d\vec{\mathbf{r}}_{CM}}{dt} = \frac{1}{M}\sum_i m_i \frac{d\vec{\mathbf{r}}_i}{dt} = \frac{1}{M}\sum_i m_i \vec{\mathbf{v}}_i \qquad (9.35)$$

where $\vec{\mathbf{v}}_i$ is the velocity of the ith particle. Rearranging Equation 9.35 gives

◀ Total momentum of a system of particles

$$M\vec{\mathbf{v}}_{CM} = \sum_i m_i \vec{\mathbf{v}}_i = \sum_i \vec{\mathbf{p}}_i = \vec{\mathbf{p}}_{tot} \qquad (9.36)$$

Therefore, the total linear momentum of the system equals the total mass multiplied by the velocity of the center of mass. In other words, the total linear momentum of the system is equal to that of a single particle of mass M moving with a velocity $\vec{\mathbf{v}}_{CM}$.

Differentiating Equation 9.35 with respect to time, we obtain the **acceleration of the center of mass** of the system:

◀ Acceleration of the center of mass of a system of particles

$$\vec{\mathbf{a}}_{CM} = \frac{d\vec{\mathbf{v}}_{CM}}{dt} = \frac{1}{M}\sum_i m_i \frac{d\vec{\mathbf{v}}_i}{dt} = \frac{1}{M}\sum_i m_i \vec{\mathbf{a}}_i \qquad (9.37)$$

Rearranging this expression and using Newton's second law gives

$$M\vec{\mathbf{a}}_{CM} = \sum_i m_i \vec{\mathbf{a}}_i = \sum_i \vec{\mathbf{F}}_i \qquad (9.38)$$

where $\vec{\mathbf{F}}_i$ is the net force on particle i.

The forces on any particle in the system may include both external forces (from outside the system) and internal forces (from within the system). By Newton's third law, however, the internal force exerted by particle 1 on particle 2, for example, is equal in magnitude and opposite in direction to the internal force exerted by particle 2 on particle 1. Therefore, when we sum over all internal force vectors in Equation 9.38, they cancel in pairs and we find that the net force on the system is caused *only* by external forces. We can then write Equation 9.38 in the form

◀ Newton's second law for a system of particles

$$\sum \vec{\mathbf{F}}_{ext} = M\vec{\mathbf{a}}_{CM} \qquad (9.39)$$

That is, the net external force on a system of particles equals the total mass of the system multiplied by the acceleration of the center of mass. Comparing Equation 9.39 with Newton's second law for a single particle, we see that the particle model we have used in several chapters can be described in terms of the center of mass:

> The center of mass of a system of particles having combined mass M moves like an equivalent single particle of mass M would move under the influence of the net external force on the system.

Let us integrate Equation 9.39 over a finite time interval:

$$\int \sum \vec{\mathbf{F}}_{ext}\, dt = \int M\vec{\mathbf{a}}_{CM}\, dt = \int M\frac{d\vec{\mathbf{v}}_{CM}}{dt}\, dt = M \int d\vec{\mathbf{v}}_{CM} = M\,\Delta\vec{\mathbf{v}}_{CM}$$

Notice that this equation can be written as

$$\Delta\vec{\mathbf{p}}_{tot} = \vec{\mathbf{I}} \qquad (9.40)$$

◀ Impulse–momentum theorem for a system of particles

where $\vec{\mathbf{I}}$ is the impulse imparted to the system by external forces and $\vec{\mathbf{p}}_{tot}$ is the momentum of the system. Equation 9.40 is the generalization of the impulse–momentum theorem for a particle (Eq. 9.13) to a system of many particles. It is also the mathematical representation of the nonisolated system (momentum) model for a system of many particles.

Finally, if the net external force on a system is zero so that the system is isolated, it follows from Equation 9.39 that

$$M\vec{\mathbf{a}}_{CM} = M\frac{d\vec{\mathbf{v}}_{CM}}{dt} = 0$$

Therefore, the isolated system model for momentum for a system of many particles is described by

$$\Delta\vec{\mathbf{p}}_{tot} = 0 \qquad (9.41)$$

which can be rewritten as

$$M\vec{\mathbf{v}}_{CM} = \vec{\mathbf{p}}_{tot} = \text{constant} \quad \left(\text{when } \sum \vec{\mathbf{F}}_{ext} = 0\right) \qquad (9.42)$$

That is, the total linear momentum of a system of particles is conserved if no net external force is acting on the system. It follows that for an isolated system of particles, both the total momentum and the velocity of the center of mass are constant in time. This statement is a generalization of the isolated system (momentum) model for a many-particle system.

Suppose the center of mass of an isolated system consisting of two or more members is at rest. The center of mass of the system remains at rest if there is no net force on the system. For example, consider a system of a swimmer standing on a raft, with the system initially at rest. When the swimmer dives horizontally off the raft, the raft moves in the direction opposite that of the swimmer and the center of mass of the system remains at rest (if we neglect friction between raft and water). Furthermore, the linear momentum of the diver is equal in magnitude to that of the raft, but opposite in direction.

ⓞUICK QUIZ 9.8 A cruise ship is moving at constant speed through the water. The vacationers on the ship are eager to arrive at their next destination. They decide to try to speed up the cruise ship by gathering at the bow (the front) and running together toward the stern (the back) of the ship. **(i)** While they are running toward the stern, is the speed of the ship (a) higher than it was before, (b) unchanged, (c) lower than it was before, or (d) impossible to determine? **(ii)** The vacationers stop running when they reach the stern of the ship. After they have all stopped running, is the speed of the ship (a) higher than it was before they started running, (b) unchanged from what it was before they started running, (c) lower than it was before they started running, or (d) impossible to determine?

Exploding Projectile

A projectile fired into the air suddenly explodes into several fragments (Fig. 9.21).

(A) What can be said about the motion of the center of mass of the system made up of all the fragments after the explosion?

SOLUTION

Neglecting air resistance, the only external force on the projectile is the gravitational force. Therefore, if the projectile did not explode, it would continue to move along the parabolic path indicated by the dashed line in Figure 9.21. Because the forces caused by the explosion are internal, they do not affect the motion of the center of mass of the system (the fragments). Therefore, after the explosion, the center of mass of the fragments follows the same parabolic path the projectile would have followed if no explosion had occurred.

Figure 9.21 (Conceptual Example 9.12) When a projectile explodes into several fragments, the center of mass of the system made up of all the fragments follows the same parabolic path the projectile would have taken had there been no explosion.

(B) If the projectile did not explode, it would land at a distance R from its launch point. Suppose the projectile explodes and splits into two pieces of equal mass. One piece lands at a distance $2R$ to the right of the launch point. Where does the other piece land?

SOLUTION

As discussed in part (A), the center of mass of the two-piece system lands at a distance R from the launch point. One of the pieces lands at a farther distance R from the landing point (or a distance $2R$ from the launch point), to the right in Figure 9.21. Because the two pieces have the same mass, the other piece must land a distance R to the left of the landing point in Figure 9.21, which places this piece right back at the launch point!

Example **9.13** **The Exploding Rocket**

A rocket is fired vertically upward. At the instant it reaches an altitude of 1 000 m and a speed of $v_i = 300$ m/s, it explodes into three fragments having equal mass. One fragment moves upward with a speed of $v_1 = 450$ m/s following the explosion. The second fragment has a speed of $v_2 = 240$ m/s and is moving east right after the explosion. What is the velocity of the third fragment immediately after the explosion?

SOLUTION

Conceptualize   Picture the explosion in your mind, with one piece going upward and a second piece moving horizontally toward the east. Do you have an intuitive feeling about the direction in which the third piece moves?

Categorize   This example is a two-dimensional problem because we have two fragments moving in perpendicular directions after the explosion as well as a third fragment moving in an unknown direction in the plane defined by the velocity vectors of the other two fragments. We assume the time interval of the explosion is very short, so we use the impulse approximation in which we ignore the gravitational force and air resistance. Because of the short time interval and the ignoring of external forces, the center of mass of the system remains fixed in space during the explosion. Therefore, the rocket is an *isolated system* in terms of *momentum*. Equation 9.41 describes the situation, and the total momentum $\vec{\mathbf{p}}_i$ of the rocket immediately before the explosion must equal the total momentum $\vec{\mathbf{p}}_f$ of the fragments immediately after the explosion.

Analyze   Because the three fragments have equal mass, the mass of each fragment is $M/3$, where M is the total mass of the rocket. We will let $\vec{\mathbf{v}}_3$ represent the unknown velocity of the third fragment.

Use the isolated system (momentum) model to equate the initial and final momenta of the system and express the momenta in terms of masses and velocities:

$$\Delta \vec{\mathbf{p}} = 0 \quad \rightarrow \quad \vec{\mathbf{p}}_i = \vec{\mathbf{p}}_f \quad \rightarrow \quad M\vec{\mathbf{v}}_i = \frac{M}{3}\vec{\mathbf{v}}_1 + \frac{M}{3}\vec{\mathbf{v}}_2 + \frac{M}{3}\vec{\mathbf{v}}_3$$

Solve for $\vec{\mathbf{v}}_3$:

$$\vec{\mathbf{v}}_3 = 3\vec{\mathbf{v}}_i - \vec{\mathbf{v}}_1 - \vec{\mathbf{v}}_2$$

Substitute the numerical values:

$$\vec{\mathbf{v}}_3 = 3(300\hat{\mathbf{j}} \text{ m/s}) - (450\hat{\mathbf{j}} \text{ m/s}) - (240\hat{\mathbf{i}} \text{ m/s}) = (-240\hat{\mathbf{i}} + 450\hat{\mathbf{j}}) \text{ m/s}$$

Finalize Notice that this event is the reverse of a perfectly inelastic collision. There is one object before the collision and three objects afterward. Imagine running a movie of the event backward: the three objects would come together and become a single object. In a perfectly inelastic collision, the kinetic energy of the system decreases. If you were to calculate the kinetic energy before and after the event in this example, you would find that the kinetic energy of the system increases. (Try it!) This increase in kinetic energy comes from the potential energy stored in whatever fuel exploded to cause the breakup of the rocket.

9.8 Deformable Systems

So far in our discussion of mechanics, we have analyzed the motion of particles or nondeformable objects that can be modeled as particles. The discussion in Section 9.7 can be applied to an analysis of the motion of deformable systems. For example, suppose you stand on a skateboard and push off a wall, setting yourself in motion away from the wall. Your body has deformed during this event: your arms were bent before the event, and they straightened out while you pushed off the wall. How would we describe this event?

The force from the wall on your hands moves through no displacement; the force is always located at the interface between the wall and your hands. Therefore, the force does no work on the system, which is you and your skateboard. Pushing off the wall, however, does indeed result in a change in the kinetic energy of the system. If you try to use the work–kinetic energy theorem, $W = \Delta K$, to describe this event, you will notice that the left side of the equation is zero but the right side is not zero. The work–kinetic energy theorem is not valid for this event and is often not valid for systems that are deformable.

To analyze the motion of deformable systems, we appeal to Equation 8.2, the conservation of energy equation, and Equation 9.40, the impulse–momentum theorem. For the example of you pushing off the wall on your skateboard, identifying the system as you and the skateboard, Equation 8.2 gives

$$\Delta K + \Delta U = 0$$

where ΔK is the change in kinetic energy, which is related to the increased speed of the system, and ΔU is the decrease in potential energy stored in your body from previous meals. This equation tells us that the system transformed potential energy in your body into kinetic energy by virtue of the muscular exertion necessary to push off the wall. Notice that the system is isolated in terms of energy but nonisolated in terms of momentum.

Applying Equation 9.40 to the system in this situation gives us

$$\Delta \vec{\mathbf{p}}_{tot} = \vec{\mathbf{I}} \quad \rightarrow \quad m\,\Delta\vec{\mathbf{v}} = \int \vec{\mathbf{F}}_{wall}\, dt$$

where $\vec{\mathbf{F}}_{wall}$ is the force exerted by the wall on your hands, m is the mass of you and the skateboard, and $\Delta\vec{\mathbf{v}}$ is the change in the velocity of the system during the event. To evaluate the right side of this equation, we would need to know how the force from the wall varies in time. In general, this process might be complicated. In the case of constant forces, or well-behaved forces, however, the integral on the right side of the equation can be evaluated.

Deformable systems occur often in common situations. Any time you run or jump, your body is a deformable system. A gymnast or a platform diver performing a routine is a deformable system. In Example 9.14 (page 238), we investigate a deformable system with two blocks and a spring. Beginning in Chapter 18, we will look at very important deformable systems: samples of gas changing in size as they undergo thermodynamic processes.

Example **9.14**   **Pushing on a Spring**[3]

As shown in Figure 9.22a, two blocks are at rest on a frictionless, level table. Both blocks have the same mass m, and they are connected by a spring of negligible mass. The separation distance of the blocks when the spring is relaxed is L. During a time interval Δt, a constant force of magnitude F is applied horizontally to the left block, moving it through a distance x_1 as shown in Figure 9.22b. During this time interval, the right block moves through a distance x_2. At the end of this time interval, the force F is removed.

(A) Find the resulting speed $\vec{\mathbf{v}}_{CM}$ of the center of mass of the system.

Figure 9.22 (Example 9.14) (a) Two blocks of equal mass are connected by a spring. (b) The left block is pushed with a constant force of magnitude F and moves a distance x_1 during some time interval. During this same time interval, the right block moves through a distance x_2.

SOLUTION

Conceptualize   Imagine what happens as you push on the left block. It begins to move to the right in Figure 9.22, and the spring begins to compress. As a result, the spring pushes to the right on the right block, which begins to move to the right. At any given time, the blocks are generally moving with different velocities. As the center of mass of the system moves to the right with a constant speed after the force is removed, the two blocks oscillate back and forth with respect to the center of mass.

Categorize   We apply three analysis models in this problem: the deformable system of two blocks and a spring is modeled as a *nonisolated system* in terms of *energy* because work is being done on it by the applied force. It is also modeled as a *nonisolated system* in terms of *momentum* because of the force acting on the system during a time interval. Because the applied force on the system is constant, the acceleration of its center of mass is constant and the center of mass is modeled as a *particle under constant acceleration*.

Analyze   Using the nonisolated system (momentum) model, we apply the impulse–momentum theorem to the system of two blocks, recognizing that the force F is constant during the time interval Δt while the force is applied.

Write Equation 9.40 for the system:

$$\Delta p_x = I_x \quad \rightarrow \quad (2m)(v_{CM} - 0) = F\Delta t$$

$$(1) \quad 2mv_{CM} = F\Delta t$$

During the time interval Δt, the center of mass of the system moves a distance $\frac{1}{2}(x_1 + x_2)$. Use this fact to express the time interval in terms of $v_{CM,avg}$:

$$\Delta t = \frac{\frac{1}{2}(x_1 + x_2)}{v_{CM,avg}}$$

Because the center of mass is modeled as a particle under constant acceleration, the average velocity of the center of mass is the average of the initial velocity, which is zero, and the final velocity v_{CM}:

$$\Delta t = \frac{\frac{1}{2}(x_1 + x_2)}{\frac{1}{2}(0 + v_{CM})} = \frac{(x_1 + x_2)}{v_{CM}}$$

Substitute this expression into Equation (1):

$$2mv_{CM} = F\frac{(x_1 + x_2)}{v_{CM}}$$

Solve for v_{CM}:

$$v_{CM} = \sqrt{F\frac{(x_1 + x_2)}{2m}}$$

(B) Find the total energy of the system associated with vibration relative to its center of mass after the force F is removed.

SOLUTION

Analyze   The vibrational energy is all the energy of the system other than the kinetic energy associated with translational motion of the center of mass. To find the vibrational energy, we apply the conservation of energy equation (Eq. 8.2). The kinetic energy of the system can be expressed as $K = K_{CM} + K_{vib}$, where K_{vib} is the kinetic energy of the blocks relative to the center of mass due to their vibration. The potential energy of the system is U_{vib}, which is the potential energy stored in the spring when the separation of the blocks is some value other than L.

---

[3]Example 9.14 was inspired in part by C. E. Mungan, "A primer on work–energy relationships for introductory physics," *The Physics Teacher* **43:**10, 2005.

9.14 c o n t i n u e d

From the nonisolated system (energy) model, express Equation 8.2 for this system:

$$(2) \quad \Delta K_{CM} + \Delta K_{vib} + \Delta U_{vib} = W$$

Express Equation (2) in an alternate form, noting that $K_{vib} + U_{vib} = E_{vib}$:

$$\Delta K_{CM} + \Delta E_{vib} = W$$

Substitute for each of the terms in this equation:

$$(K_{CM} - 0) + (E_{vib} - 0) = Fx_1 \quad \rightarrow \quad E_{vib} = Fx_1 - K_{CM}$$

Use the result from part (A):

$$E_{vib} = Fx_1 - \tfrac{1}{2}(2m)v_{CM}^2 = Fx_1 - \tfrac{1}{2}(2m)\left[F\frac{(x_1 + x_2)}{2m} \right] = F\frac{(x_1 - x_2)}{2}$$

Finalize Neither of the two answers in this example depends on the spring length, the spring constant, or the time interval. Notice also that the magnitude x_1 of the displacement of the point of application of the applied force is different from the magnitude $\tfrac{1}{2}(x_1 + x_2)$ of the displacement of the center of mass of the system. This difference reminds us that the displacement in the definition of work (Eq. 7.1) is that of the point of application of the force.

9.9 Rocket Propulsion

When ordinary vehicles such as cars are propelled, the driving force for the motion is friction. In the case of the car, the driving force is the force exerted by the road on the car. We can model the car as a nonisolated system in terms of momentum. An impulse is applied to the car from the roadway, and the result is a change in the momentum of the car as described by Equation 9.40.

A rocket moving in space, however, has no road to push against. The rocket is an isolated system in terms of momentum. Therefore, the source of the propulsion of a rocket must be something other than an external force. The operation of a rocket depends on the law of conservation of linear momentum as applied to an isolated system, where the system is the rocket plus its ejected fuel.

Rocket propulsion can be understood by first considering our archer standing on frictionless ice in Example 9.1. Imagine the archer fires several arrows horizontally. For each arrow fired, the archer receives a compensating momentum in the opposite direction. As more arrows are fired, the archer moves faster and faster across the ice. In addition to this analysis in terms of momentum, we can also understand this phenomenon in terms of Newton's second and third laws. Every time the bow pushes an arrow forward, the arrow pushes the bow (and the archer) backward, and these forces result in an acceleration of the archer. Figure 9.23a shows this mechanism used for maneuvering an astronaut in space. Instead of firing arrows like the archer, the astronaut fires short bursts of nitrogen gas.

In a similar manner, as a rocket moves in free space, its linear momentum changes when some of its mass is ejected in the form of exhaust gases. Because the gases are given momentum when they are ejected out of the engine, the rocket receives a compensating momentum in the opposite direction. Therefore, the rocket is accelerated as a result of the "push," or thrust, from the exhaust gases. In free space, the center of mass of the system (rocket plus expelled gases) moves uniformly, independent of the propulsion process.[4]

Suppose at some time t the magnitude of the momentum of a rocket plus its fuel is Mv, where v is the speed of the rocket relative to the Earth (Fig. 9.23b). Over a short time interval Δt, the rocket ejects fuel of mass Δm. At the end of this interval, the rocket's mass is $M - \Delta m$ and its speed is $v + \Delta v$, where Δv is the change in speed of

Figure 9.23 Rocket propulsion. (a) The force from a nitrogen-propelled, hand-controlled device allows an astronaut to move about freely in space without restrictive tethers, using the thrust force from the expelled nitrogen. (b) The initial mass of a rocket plus all its fuel is M at a time t, and its speed is v. (c) At a time $t + \Delta t$, the rocket's mass has been reduced to $M - \Delta m$ and an amount of fuel Δm has been ejected. The rocket's speed increases by an amount Δv.

---

[4]The rocket and the archer represent cases of the reverse of a perfectly inelastic collision: momentum is conserved, but the kinetic energy of the rocket–exhaust gas system increases (at the expense of chemical potential energy in the fuel), as does the kinetic energy of the archer–arrow system (at the expense of potential energy from the archer's previous meals when he pulls back on the bowstring and stretches it).

the rocket (Fig. 9.23c). If the fuel is ejected with a speed v_e relative to the rocket (the subscript e stands for *exhaust*, and v_e is usually called the *exhaust speed*), the velocity of the fuel relative to the Earth is $v - v_e$. Because the system of the rocket and the ejected fuel is isolated, we apply the isolated system model for momentum and obtain

$$\Delta p = 0 \quad \rightarrow \quad p_i = p_f \quad \rightarrow \quad Mv = (M - \Delta m)(v + \Delta v) + \Delta m(v - v_e)$$

Simplifying this expression gives

$$M\Delta v - \Delta m \Delta v = v_e \Delta m \tag{9.43}$$

Solving for the change in speed, we find

$$\Delta v = \frac{v_e \Delta m}{M - \Delta m} \tag{9.44}$$

This equation is valid for a one-time ejection of mass from the rocket. It is also valid for any situation in which an object ejects mass, causing the object to move in the opposite direction. The equation can be applied to the archer problem in Example 9.1, recognizing that the initial mass of the system was that of both the archer and the arrow, $M = 60.030$ kg.

If we now take the limit as Δt goes to zero, we let $\Delta v \rightarrow dv$ and $\Delta m \rightarrow dm$ in Equation 9.43. In addition, we ignore the term $dm\, dv$ because this product of two infinitesimal quantities is much smaller than the other terms in the equation. Furthermore, the increase in the exhaust mass dm corresponds to an equal decrease in the rocket mass, so $dm = -dM$. Using this fact gives

$$M\, dv = v_e\, dm = -v_e\, dM \tag{9.45}$$

Now divide the equation by M and integrate, taking the initial mass of the rocket plus fuel to be M_i and the final mass of the rocket plus its remaining fuel to be M_f. The result is

$$\int_{v_i}^{v_f} dv = -v_e \int_{M_i}^{M_f} \frac{dM}{M}$$

Expression for rocket ▶
propulsion

$$v_f - v_i = v_e \ln\left(\frac{M_i}{M_f}\right) \tag{9.46}$$

which is the basic expression for rocket propulsion. First, Equation 9.46 tells us that the increase in rocket speed is proportional to the exhaust speed v_e of the ejected gases. Therefore, the exhaust speed should be very high. Second, the increase in rocket speed is proportional to the natural logarithm of the ratio M_i/M_f. Therefore, this ratio should be as large as possible; that is, the mass of the rocket without its fuel should be as small as possible and the rocket should carry as much fuel as possible.

The **thrust** on the rocket is the force exerted on it by the ejected exhaust gases. We obtain the following expression for the thrust from Newton's second law and Equation 9.45:

$$\text{Thrust} = M\frac{dv}{dt} = \left| v_e \frac{dM}{dt} \right| \tag{9.47}$$

This expression shows that the thrust increases as the exhaust speed increases and as the rate of change of mass (called the *burn rate*) increases.

Example **9.15**   **A Rocket in Space**

A rocket moving in space, far from all other objects, has a speed of 3.0×10^3 m/s relative to the Earth. Its engines are turned on, and fuel is ejected in a direction opposite the rocket's motion at a speed of 5.0×10^3 m/s relative to the rocket.

(A) What is the speed of the rocket relative to the Earth once the rocket's mass is reduced to half its mass before ignition?

9.15 continued

SOLUTION

Conceptualize Figure 9.23 shows the situation in this problem. From the discussion in this section and scenes from science fiction movies, we can easily imagine the rocket accelerating to a higher speed as the engine operates.

Categorize This problem is a substitution problem in which we use given values in the equations derived in this section.

Solve Equation 9.46 for the final velocity and substitute the known values:

$$v_f = v_i + v_e \ln\left(\frac{M_i}{M_f}\right)$$

$$= 3.0 \times 10^3 \text{ m/s} + (5.0 \times 10^3 \text{ m/s}) \ln\left(\frac{M_i}{0.50 M_i}\right)$$

$$= \boxed{6.5 \times 10^3 \text{ m/s}}$$

(B) What is the thrust on the rocket if it burns fuel at the rate of 50 kg/s?

SOLUTION

Use Equation 9.47, noting that $dM/dt = 50$ kg/s:

$$\text{Thrust} = \left| v_e \frac{dM}{dt} \right| = (5.0 \times 10^3 \text{ m/s})(50 \text{ kg/s}) = \boxed{2.5 \times 10^5 \text{ N}}$$

Example 9.16 **Fighting a Fire**

Two firefighters must apply a total force of 600 N to steady a hose that is discharging water at the rate of 3 600 L/min. Estimate the speed of the water as it exits the nozzle.

SOLUTION

Conceptualize As the water leaves the hose, it acts in a way similar to the gases being ejected from a rocket engine. As a result, a force (thrust) acts on the firefighters in a direction opposite the direction of motion of the water. In this case, we want the end of the hose to be modeled as a particle in equilibrium rather than to accelerate as in the case of the rocket. Consequently, the firefighters must apply a force of magnitude equal to the thrust in the opposite direction to keep the end of the hose stationary.

Categorize This example is a substitution problem in which we use given values in an equation derived in this section. The water exits at 3 600 L/min, which is 60 L/s. Knowing that 1 L of water has a mass of 1 kg, we estimate that about 60 kg of water leaves the nozzle each second.

Use Equation 9.47 for the thrust:

$$\text{Thrust} = \left| v_e \frac{dM}{dt} \right|$$

Solve for the exhaust speed:

$$v_e = \frac{\text{Thrust}}{|dM/dt|}$$

Substitute numerical values:

$$v_e = \frac{600 \text{ N}}{60 \text{ kg/s}} = \boxed{10 \text{ m/s}}$$

Summary

▶ Definitions

The **linear momentum** $\vec{\mathbf{p}}$ of a particle of mass m moving with a velocity $\vec{\mathbf{v}}$ is

$$\vec{\mathbf{p}} \equiv m\vec{\mathbf{v}} \qquad (9.2)$$

The **impulse** imparted to a particle by a net force $\Sigma \vec{\mathbf{F}}$ is equal to the time integral of the force:

$$\vec{\mathbf{I}} \equiv \int_{t_i}^{t_f} \Sigma \vec{\mathbf{F}} \, dt \qquad (9.9)$$

continued

An **inelastic collision** is one for which the total kinetic energy of the system of colliding particles is not conserved. A **perfectly inelastic collision** is one in which the colliding particles stick together after the collision. An **elastic collision** is one in which the kinetic energy of the system is conserved.

The position vector of the **center of mass** of a system of particles is defined as

$$\vec{r}_{CM} \equiv \frac{1}{M} \sum_i m_i \vec{r}_i \qquad (9.31)$$

where $M = \sum_i m_i$ is the total mass of the system and $\vec{r}_i$ is the position vector of the ith particle.

Concepts and Principles

The position vector of the center of mass of an extended object can be obtained from the integral expression

$$\vec{r}_{CM} = \frac{1}{M} \int \vec{r} \, dm \qquad (9.34)$$

The velocity of the center of mass for a system of particles is

$$\vec{v}_{CM} = \frac{1}{M} \sum_i m_i \vec{v}_i \qquad (9.35)$$

The total momentum of a system of particles equals the total mass multiplied by the velocity of the center of mass.

Newton's second law applied to a system of particles is

$$\sum \vec{F}_{ext} = M\vec{a}_{CM} \qquad (9.39)$$

where $\vec{a}_{CM}$ is the acceleration of the center of mass and the sum is over all external forces. The center of mass moves like an imaginary particle of mass M under the influence of the resultant external force on the system.

Analysis Models for Problem Solving

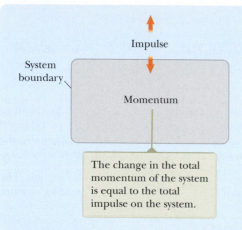

The change in the total momentum of the system is equal to the total impulse on the system.

Nonisolated System (Momentum). If a system interacts with its environment in the sense that there is an external force on the system, the behavior of the system is described by the **impulse–momentum theorem**:

$$\Delta \vec{p}_{tot} = \vec{I} \qquad (9.40)$$

If no external forces act on the system, the total momentum of the system is constant.

Isolated System (Momentum). The total momentum of an isolated system (no external forces) is conserved regardless of the nature of the forces between the members of the system:

$$\Delta \vec{p}_{tot} = 0 \qquad (9.41)$$

The system may be isolated in terms of momentum but nonisolated in terms of energy, as in the case of inelastic collisions.

Think–Pair–Share

See the Preface for an explanation of the icons used in this problems set.
For additional assessment items for this section, go to **WEBASSIGN** From Cengage

1. You are a football player on your school's team. In practice, you kick a 0.400-kg football from the ground at the 20-yard line of a football field laid out exactly along a north–south line. The ball's initial velocity vector is directed at 30.0° above the horizontal, toward the north, and has a magnitude of 27.0 m/s. You watch the ball as it rises in the air. Just as the ball reaches its highest point in its parabolic

trajectory, a 4.18-kg eagle flying due south and along a horizontal line collides with the ball. Assume that the highly inflated ball makes an elastic collision with the hard beak of the bird, and that the ball rebounds from the collision with a velocity vector that is horizontal and due south. The ball lands back at the exact point on the ground from which it was kicked. (a) How fast was the eagle flying? (Ignore air resistance.) (b) How fast and in what direction is the eagle moving just after the collision? (Assume no flapping of wings has occurred yet!) (c) In reality, the collision will not be elastic: some kinetic energy will be converted to other forms. With the assumption of some kinetic energy lost in an inelastic collision, will the required speed of the eagle be higher or lower than in part (a)?

2. **ACTIVITY** Carefully draw a right triangle on a piece of cardboard, such that one of its non-hypotenuse legs is 30–40 cm in length and the other leg is much shorter. Measure the exact midpoint of each of the three sides of the triangle and mark these three points. Draw a line across the triangle, from a corner of the triangle to the midpoint of the opposite side. Repeat for the other two corners. The three lines happen to intersect at the center of mass of the triangle. Draw a fourth line perpendicular to the longer non-hypotenuse leg, passing through the center of mass, and ending as it crosses the hypotenuse of the triangle. Punch a hole in the cardboard just inside the edge of the triangle where the fourth line crosses the hypotenuse. Carefully cut the triangle out of the cardboard. Tie a string through the hole and hang the triangle from the string. The longer side of the triangle should be parallel to the table. Why should this be true? Now measure the distance along the longer leg from the smaller angle to the fourth line. What fraction of the entire longer leg is this distance?

Problems

See the Preface for an explanation of the icons used in this problems set. For additional assessment items for this section, go to ⚡**WEBASSIGN** From Cengage

SECTION 9.1 Linear Momentum

1. A particle of mass m moves with momentum of magnitude p.
   S (a) Show that the kinetic energy of the particle is $K = p^2/2m$. (b) Express the magnitude of the particle's momentum in terms of its kinetic energy and mass.

2. A 3.00-kg particle has a velocity of $(3.00\hat{i} - 4.00\hat{j})$ m/s. (a) Find its x and y components of momentum. (b) Find the magnitude and direction of its momentum.

3. A baseball approaches home plate at a speed of 45.0 m/s, moving horizontally just before being hit by a bat. The batter hits a pop-up such that after hitting the bat, the baseball is moving at 55.0 m/s straight up. The ball has a mass of 145 g and is in contact with the bat for 2.00 ms. What is the average vector force the ball exerts on the bat during their interaction?

SECTION 9.2 Analysis Model: Isolated System (Momentum)

4. A 65.0-kg boy and his 40.0-kg sister, both wearing roller
   Q|C blades, face each other at rest. The girl pushes the boy hard, sending him backward with velocity 2.90 m/s toward the west. Ignore friction. (a) Describe the subsequent motion of the girl. (b) How much potential energy in the girl's body is converted into mechanical energy of the boy–girl system? (c) Is the momentum of the boy–girl system conserved in the pushing-apart process? If so, explain how that is possible considering (d) there are large forces acting and (e) there is no motion beforehand and plenty of motion afterward.

5. Two blocks of masses m and $3m$
   Q|C are placed on a frictionless, hor-
   V izontal surface. A light spring is attached to the more massive block, and the blocks are pushed together with the spring between them (Fig. P9.5). A cord initially holding the blocks together is burned; after that happens, the block of mass $3m$ moves to the right with a speed of 2.00 m/s. (a) What is the velocity of the block of mass m? (b) Find the system's original elastic potential energy, taking

Figure P9.5

$m = 0.350$ kg. (c) Is the original energy in the spring or in the cord? (d) Explain your answer to part (c). (e) Is the momentum of the system conserved in the bursting-apart process? Explain how that is possible considering (f) there are large forces acting and (g) there is no motion beforehand and plenty of motion afterward?

6. When you jump straight up as high as you can, what is the order of magnitude of the maximum recoil speed that you give to the Earth? Model the Earth as a perfectly solid object. In your solution, state the physical quantities you take as data and the values you measure or estimate for them.

SECTION 9.3 Analysis Model: Nonisolated System (Momentum)

7. A glider of mass m is free to slide along a horizontal air
   Q|C track. It is pushed against a launcher at one end of the
   S track. Model the launcher as a light spring of force constant k compressed by a distance x. The glider is released from rest. (a) Show that the glider attains a speed of $v = x(k/m)^{1/2}$. (b) Show that the magnitude of the impulse imparted to the glider is given by the expression $I = x(km)^{1/2}$. (c) Is more work done on a cart with a large or a small mass?

8. You and your brother argue often about how to safely secure
   CR a toddler in a moving car. You insist that special toddler seats are critical in improving the chances of a toddler surviving a crash. Your brother claims that, as long as his wife is buckled in next to him with a seat belt while he drives, she can hold onto their toddler on her lap in a crash. You decide to perform a calculation to try to convince your brother. Consider a hypothetical collision in which the 12-kg toddler and his parents are riding in a car traveling at 60 mi/h relative to the ground. The car strikes a wall, tree, or another car, and is brought to rest in 0.10 s. You wish to demonstrate to your brother the magnitude of the force necessary for his wife to hold onto their child during the collision.

9. The front 1.20 m of a 1 400-kg car is designed as a "crumple
   T zone" that collapses to absorb the shock of a collision. If a car traveling 25.0 m/s stops uniformly in 1.20 m, (a) how long does the collision last, (b) what is the magnitude of the average force on the car, and (c) what is the magnitude of the acceleration of the car? Express the acceleration as a multiple of the acceleration due to gravity.

10. The magnitude of the net force exerted in the x direction on a 2.50-kg particle varies in time as shown in Figure P9.10 (page 244). Find (a) the impulse of the force over the 5.00-s time interval, (b) the final velocity the particle attains if it is

originally at rest, (c) its final velocity if its original velocity is $-2.00\hat{\mathbf{i}}$ m/s, and (d) the average force exerted on the particle for the time interval between 0 and 5.00 s.

Figure P9.10

11. Water falls without splashing at a rate of 0.250 L/s from a height of 2.60 m into a bucket of mass 0.750 kg on a scale. If the bucket is originally empty, what does the scale read in newtons 3.00 s after water starts to accumulate in it?

SECTION 9.4 Collisions in One Dimension

12. **Q|C** A 1 200-kg car traveling initially at $v_{Ci} = 25.0$ m/s in an easterly direction crashes into the back of a 9 000-kg truck moving in the same direction at $v_{Ti} = 20.0$ m/s (Fig. P9.12). The velocity of the car immediately after the collision is $v_{Cf} = 18.0$ m/s to the east. (a) What is the velocity of the truck immediately after the collision? (b) What is the change in mechanical energy of the car–truck system in the collision? (c) Account for this change in mechanical energy.

Before   After

Figure P9.12

13. A railroad car of mass 2.50×10^4 kg is moving with a speed of 4.00 m/s. It collides and couples with three other coupled railroad cars, each of the same mass as the single car and moving in the same direction with an initial speed of 2.00 m/s. (a) What is the speed of the four cars after the collision? (b) What is the decrease in mechanical energy in the collision?

14. Four railroad cars, each of mass 2.50×10^4 kg, are coupled together and coasting along horizontal tracks at speed v_i toward the south. A very strong but foolish movie actor, riding on the second car, uncouples the front car and gives it a big push, increasing its speed to 4.00 m/s southward. The remaining three cars continue moving south, now at 2.00 m/s. (a) Find the initial speed of the four cars. (b) By how much did the potential energy within the body of the actor change? (c) State the relationship between the process described here and the process in Problem 13.

15. **S** A car of mass m moving at a speed v_1 collides and couples with the back of a truck of mass $2m$ moving initially in the same direction as the car at a lower speed v_2. (a) What is the speed v_f of the two vehicles immediately after the collision? (b) What is the change in kinetic energy of the car–truck system in the collision?

16. A 7.00-g bullet, when fired from a gun into a 1.00-kg block of wood held in a vise, penetrates the block to a depth of 8.00 cm. This block of wood is next placed on a frictionless horizontal surface, and a second 7.00-g bullet is fired from the gun into the block. To what depth will the bullet penetrate the block in this case?

17. **T** A tennis ball of mass 57.0 g is held just above a basketball of mass 590 g as shown in Figure P9.17. With their centers

vertically aligned, both balls are released from rest at the same time, to fall through a distance of 1.20 m. (a) Find the magnitude of the downward velocity with which the basketball reaches the ground. (b) Assume that an elastic collision with the ground instantaneously reverses the velocity of the basketball while the tennis ball is still moving down. Next, the two balls meet in an elastic collision. To what height does the tennis ball rebound?

Figure P9.17

18. **Q|C** (a) Three carts of masses $m_1 = 4.00$ kg, $m_2 = 10.0$ kg, and $m_3 = 3.00$ kg move on a frictionless, horizontal track with speeds of $v_1 = 5.00$ m/s to the right, $v_2 = 3.00$ m/s to the right, and $v_3 = 4.00$ m/s to the left as shown in Figure P9.18. Velcro couplers make the carts stick together after colliding. Find the final velocity of the train of three carts. (b) **What If?** Does your answer in part (a) require that all the carts collide and stick together at the same moment? What if they collide in a different order?

Figure P9.18

SECTION 9.5 Collisions in Two Dimensions

19. **CR** You have been hired as an expert witness by an attorney for a trial involving a traffic accident. The attorney's client, the plaintiff in this case, was traveling eastbound toward an intersection at 13.0 m/s as measured just before the accident by a roadside speed meter, and as seen by a trustworthy witness. As the plaintiff entered the intersection, his car was struck by a northbound driver, the defendant in this case, driving a car with identical mass to the plaintiff's. The vehicles stuck together after the collision and left parallel skid marks at an angle of $\theta = 55.0°$ north of east, as measured by accident investigators. The defendant is claiming that he was traveling within the 35-mi/h speed limit. What advice do you give to the attorney?

20. **V** Two shuffleboard disks of equal mass, one orange and the other yellow, are involved in an elastic, glancing collision. The yellow disk is initially at rest and is struck by the orange disk moving with a speed of 5.00 m/s. After the collision, the orange disk moves along a direction that makes an angle of 37.0° with its initial direction of motion. The velocities of the two disks are perpendicular after the collision. Determine the final speed of each disk.

21. **S** Two shuffleboard disks of equal mass, one orange and the other yellow, are involved in an elastic, glancing collision. The yellow disk is initially at rest and is struck by the orange disk moving with a speed v_i. After the collision, the orange disk moves along a direction that makes an angle θ with its initial direction of motion. The velocities of the two disks are perpendicular after the collision. Determine the final speed of each disk.

22. **Q|C** **V** A 90.0-kg fullback running east with a speed of 5.00 m/s is tackled by a 95.0-kg opponent running north with a speed of 3.00 m/s. (a) Explain why the successful tackle constitutes a perfectly inelastic collision. (b) Calculate the velocity of the players immediately after the tackle. (c) Determine

the decrease in mechanical energy as a result of the collision. Account for this decrease.

23. **S** A proton, moving with a velocity of $v_i\hat{\mathbf{i}}$, collides elastically with another proton that is initially at rest. Assuming that the two protons have equal speeds after the collision, find (a) the speed of each proton after the collision in terms of v_i and (b) the direction of the velocity vectors after the collision.

SECTION 9.6 The Center of Mass

24. **V** A uniform piece of sheet metal is shaped as shown in Figure P9.24. Compute the x and y coordinates of the center of mass of the piece.

Figure P9.24

25. Explorers in the jungle find an ancient monument in the shape of a large isosceles triangle as shown in Figure P9.25. The monument is made from tens of thousands of small stone blocks of density 3 800 kg/m³. The monument is 15.7 m high and 64.8 m wide at its base and is everywhere 3.60 m thick from front to back. Before the monument was built many years ago, all the stone blocks lay on the ground. How much work did laborers do on the blocks to put them in position while building the entire monument? *Note:* The gravitational potential energy of an object–Earth system is given by $U_g = Mgy_{CM}$, where M is the total mass of the object and y_{CM} is the elevation of its center of mass above the chosen reference level.

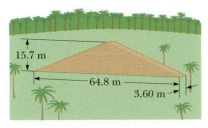

Figure P9.25

26. A rod of length 30.0 cm has linear density (mass per length) given by

$$\lambda = 50.0 + 20.0x$$

where x is the distance from one end, measured in meters, and λ is in grams/meter. (a) What is the mass of the rod? (b) How far from the $x = 0$ end is its center of mass?

SECTION 9.7 Systems of Many Particles

27. Consider a system of two particles in the xy plane: $m_1 = 2.00$ kg is at the location $\vec{\mathbf{r}}_1 = (1.00\hat{\mathbf{i}} + 2.00\hat{\mathbf{j}})$ m and has a velocity of $(3.00\hat{\mathbf{i}} + 0.500\hat{\mathbf{j}})$ m/s; $m_2 = 3.00$ kg is at $\vec{\mathbf{r}}_2 = (-4.00\hat{\mathbf{i}} - 3.00\hat{\mathbf{j}})$ m and has velocity $(3.00\hat{\mathbf{i}} - 2.00\hat{\mathbf{j}})$ m/s. (a) Plot these particles on a grid or graph paper. Draw their position vectors and show their velocities. (b) Find the position of the center of mass of the system and mark it on the grid. (c) Determine the velocity of the center of mass and also show it on the diagram. (d) What is the total linear momentum of the system?

28. The vector position of a 3.50-g particle moving in the xy plane varies in time according to $\vec{\mathbf{r}}_1 = (3\hat{\mathbf{i}} + 3\hat{\mathbf{j}})t + 2\hat{\mathbf{j}}t^2$, where t is in seconds and $\vec{\mathbf{r}}$ is in centimeters. At the same time, the vector position of a 5.50 g particle varies as $\vec{\mathbf{r}}_2 = 3\hat{\mathbf{i}} - 2\hat{\mathbf{i}}t^2 - 6\hat{\mathbf{j}}t$. At $t = 2.50$ s, determine (a) the vector position of the center of mass of the system, (b) the linear momentum of the system, (c) the velocity of the center of mass, (d) the acceleration of the center of mass, and (e) the net force exerted on the two-particle system.

29. **CR** You have been hired as an expert witness in an investigation of a quadcopter drone incident. The incident occurred during a very rare meteor shower during which several unusually massive chunks of meteoric material were passing through the atmosphere and striking the ground. The unmanned drone was hovering at rest over the center of a house on fire, having just dropped fire retardant, when it seemed to spontaneously explode into four large pieces. The locations of the four pieces on the ground were measured as follows, relative to the center of the house over which the drone was hovering:

Piece #	Mass (kg)	Distance from Center of House (m)	Direction from House
1	80.0	150	Due west
2	120	75.0	Due north
3	50.0	90.0	20.0° west of south
4	150	50.0	20.0° north of east

The fire department is suggesting that the drone was defective and exploded while in use. The drone manufacturer is suggesting that the drone was struck by a meteorite, causing the explosion. Perform a calculation that will show evidence suggesting agreement with one of these positions.

SECTION 9.8 Deformable Systems

30. **Q|C** For a technology project, a student has built a vehicle, of total mass 6.00 kg, that moves itself. As shown in Figure P9.30, it runs on four light wheels. A reel is attached to one of the axles, and a cord originally wound on the reel goes up over a pulley attached to the vehicle to support an elevated load. After the vehicle is released from rest, the load descends very slowly, unwinding the cord to turn the axle and make the vehicle move forward (to the left in Fig. P9.30). Friction is negligible in the pulley and axle bearings. The wheels do not slip on the floor. The reel has been constructed with a conical shape so that the load descends at a constant low speed while the vehicle moves horizontally across the floor with constant acceleration, reaching a final velocity of $3.00\hat{\mathbf{i}}$ m/s. (a) Does the floor impart impulse to the vehicle? If so, how much? (b) Does the floor do work on the vehicle? If so, how much? (c) Does it make sense to say that the final momentum of the vehicle came from the floor? If not, where did it come from? (d) Does it make sense to say that the final kinetic energy of the vehicle came from the floor? If not, where did it come from? (e) Can we say that one particular force causes the forward acceleration of the vehicle? What does cause it?

Figure P9.30

31. A 60.0-kg person bends his knees and then jumps straight up. After his feet leave the floor, his motion is unaffected by air resistance and his center of mass rises by a maximum of 15.0 cm. Model the floor as completely solid and motionless. (a) Does the floor impart impulse to the person? (b) Does the floor do work on the person? (c) With what momentum does the person leave the floor? (d) Does it make sense to say that this momentum came from the floor? Explain. (e) With what kinetic energy does the person leave the floor? (f) Does it make sense to say that this energy came from the floor? Explain.

SECTION 9.9 Rocket Propulsion

32. A garden hose is held as shown in Figure P9.32. The hose is originally full of motionless water. What additional force is necessary to hold the nozzle stationary after the water flow is turned on if the discharge rate is 0.600 kg/s with a speed of 25.0 m/s?

Figure P9.32

33. A rocket for use in deep space is to be capable of boosting a total load (payload plus rocket frame and engine) of 3.00 metric tons to a speed of 10 000 m/s. (a) It has an engine and fuel designed to produce an exhaust speed of 2 000 m/s. How much fuel plus oxidizer is required? (b) If a different fuel and engine design could give an exhaust speed of 5 000 m/s, what amount of fuel and oxidizer would be required for the same task? (c) Noting that the exhaust speed in part (b) is 2.50 times higher than that in part (a), explain why the required fuel mass is not simply smaller by a factor of 2.50.

34. A rocket has total mass $M_i = 360$ kg, including $M_{fuel} = 330$ kg of fuel and oxidizer. In interstellar space, it starts from rest at the position $x = 0$, turns on its engine at time $t = 0$, and puts out exhaust with relative speed $v_e = 1\,500$ m/s at the constant rate $k = 2.50$ kg/s. The fuel will last for a burn time of $T_b = M_{fuel}/k = 330$ kg/(2.5 kg/s) = 132 s. (a) Show that during the burn the velocity of the rocket as a function of time is given by

$$v(t) = -v_e \ln\left(1 - \frac{kt}{M_i}\right)$$

(b) Make a graph of the velocity of the rocket as a function of time for times running from 0 to 132 s. (c) Show that the acceleration of the rocket is

$$a(t) = \frac{kv_e}{M_i - kt}$$

(d) Graph the acceleration as a function of time. (e) Show that the position of the rocket is

$$x(t) = v_e\left(\frac{M_i}{k} - t\right)\ln\left(1 - \frac{kt}{M_i}\right) + v_e t$$

(f) Graph the position during the burn as a function of time.

ADDITIONAL PROBLEMS

35. An amateur skater of mass M is trapped in the middle of an ice rink and is unable to return to the side where there is no ice. Every motion she makes causes her to slip on the ice and remain in the same spot. She decides to try to return to safety by throwing her gloves of mass m in the direction opposite the safe side. (a) She throws the gloves as hard as she can, and they leave her hand with a horizontal velocity $\vec{v}_{gloves}$. Explain whether or not she moves. (b) If she does move, calculate her velocity $\vec{v}_{girl}$ relative to the Earth after she throws the gloves. (c) Discuss her motion from the point of view of the forces acting on her.

36. (a) Figure P9.36 shows three points in the operation of the ballistic pendulum discussed in Example 9.6 (and shown in Fig. 9.10b). The projectile approaches the pendulum in Figure P9.36a. Figure P9.36b shows the situation just after the projectile is captured in the pendulum. In Figure P9.36c, the pendulum arm has swung upward and come to rest momentarily at a height h above its initial position. Prove that the ratio of the kinetic energy of the projectile–pendulum system immediately after the collision to the kinetic energy immediately before is $m_1/(m_1 + m_2)$. (b) What is the ratio of the momentum of the system immediately after the collision to the momentum immediately before? (c) A student believes that such a large decrease in mechanical energy must be accompanied by at least a small decrease in momentum. How would you convince this student of the truth?

Figure P9.36 Problems 36 and 43. (a) A metal ball moves toward the pendulum. (b) The ball is captured by the pendulum. (c) The ball–pendulum combination swings up through a height h before coming to rest.

37. **Review.** A 60.0-kg person running at an initial speed of 4.00 m/s jumps onto a 120-kg cart initially at rest (Fig. P9.37). The person slides on the cart's top surface and finally comes to rest relative to the cart. The coefficient of kinetic friction between the person and the cart is 0.400. Friction between the cart and ground can be ignored. (a) Find the final velocity of the person and cart relative to the ground. (b) Find the friction force acting on the person while he is sliding across the top surface of the cart. (c) How long does the friction force act on the person? (d) Find the change in momentum of the person and the change in momentum of the cart. (e) Determine the displacement of the person relative to the ground while he is sliding on the cart. (f) Determine the displacement of the cart relative to the ground while the person is sliding. (g) Find the change in kinetic energy of

Figure P9.37

the person. (h) Find the change in kinetic energy of the cart. (i) Explain why the answers to (g) and (h) differ. (What kind of collision is this one, and what accounts for the loss of mechanical energy?)

38. A cannon is rigidly attached to a carriage, which can move along horizontal rails but is connected to a post by a large spring, initially unstretched and with force constant $k = 2.00 \times 10^4$ N/m, as shown in Figure P9.38. The cannon

Figure P9.38

fires a 200-kg projectile at a velocity of 125 m/s directed 45.0° above the horizontal. (a) Assuming that the mass of the cannon and its carriage is 5 000 kg, find the recoil speed of the cannon. (b) Determine the maximum extension of the spring. (c) Find the maximum force the spring exerts on the carriage. (d) Consider the system consisting of the cannon, carriage, and projectile. Is the momentum of this system conserved during the firing? Why or why not?

39. | QC | A 1.25-kg wooden block rests on a table over a large hole as in Figure P9.39. A 5.00-g bullet with an initial velocity v_i is fired upward into the bottom of the block and remains in the block after the collision. The block and bullet rise to a maximum height of 22.0 cm. (a) Describe how you would find the initial velocity of the bullet using ideas you have learned in this chapter. (b) Calculate the initial velocity of the bullet from the information provided.

Figure P9.39
Problems 39 and 40.

40. | QC | S | A wooden block of mass M rests on a table over a large hole as in Figure P9.39. A bullet of mass m with an initial velocity of v_i is fired upward into the bottom of the block and remains in the block after the collision. The block and bullet rise to a maximum height of h. (a) Describe how you would find the initial velocity of the bullet using ideas you have learned in this chapter. (b) Find an expression for the initial velocity of the bullet.

41. | S | Two gliders are set in motion on a horizontal air track. A light spring of force constant k is attached to the back end of the second glider. As shown in Figure P9.41, the first glider, of mass m_1, moves to the right with speed v_1, and the second glider, of mass m_2, moves more slowly to the right with speed v_2. When m_1 collides with the spring attached to m_2, the spring compresses by a distance x_{max}, and the gliders then move apart again. In terms of v_1, v_2, m_1, m_2, and k, find (a)

Figure P9.41

the speed v at maximum compression, (b) the maximum compression x_{max}, and (c) the velocity of each glider after m_1 has lost contact with the spring.

42. Pursued by ferocious wolves, you are in a sleigh with no horses, gliding without friction across an ice-covered lake. You take an action described by the equations

$$(270 \text{ kg})(7.50 \text{ m/s})\hat{\mathbf{i}} = (15.0 \text{ kg})(-v_{1f}\hat{\mathbf{i}}) + (255 \text{ kg})(v_{2f}\hat{\mathbf{i}})$$
$$v_{1f} + v_{2f} = 8.00 \text{ m/s}$$

(a) Complete the statement of the problem, giving the data and identifying the unknowns. (b) Find the values of v_{1f} and v_{2f}. (c) Find the amount of energy that has been transformed from potential energy stored in your body to kinetic energy of the system.

43. **Review.** | QC | A student performs a ballistic pendulum experiment using an apparatus similar to that discussed in Example 9.6 and shown in Figure P9.36. She obtains the following average data: $h = 8.68$ cm, projectile mass $m_1 = 68.8$ g, and pendulum mass $m_2 = 263$ g. (a) Determine the initial speed v_{1A} of the projectile. (b) The second part of her experiment is to obtain v_{1A} by firing the same projectile horizontally (with the pendulum removed from the path) and measuring its final horizontal position x and distance of fall y (Fig. P9.43). What numerical value does she obtain for v_{1A} based on her measured values of $x = 257$ cm and $y = 85.3$ cm? (c) What factors might account for the difference in this value compared with that obtained in part (a)?

Figure P9.43

44. *Why is the following situation impossible?* An astronaut, together with the equipment he carries, has a mass of 150 kg. He is taking a space walk outside his spacecraft, which is drifting through space with a constant velocity. The astronaut accidentally pushes against the spacecraft and begins moving away at 20.0 m/s, relative to the spacecraft, without a tether. To return, he takes equipment off his space suit and throws it in the direction away from the spacecraft. Because of his bulky space suit, he can throw equipment at a maximum speed of 5.00 m/s relative to himself. After throwing enough equipment, he starts moving back to the spacecraft and is able to grab onto it and climb inside.

45. **Review.** | T | A bullet of mass $m = 8.00$ g is fired into a block of mass $M = 250$ g that is initially at rest at the edge of a frictionless table of height $h = 1.00$ m (Fig. P9.45). The bullet

Figure P9.45 Problems 45 and 46.

remains in the block, and after the impact the block lands $d = 2.00$ m from the bottom of the table. Determine the initial speed of the bullet.

46. **Review.** A bullet of mass m is fired into a block of mass M initially at rest at the edge of a frictionless table of height h (Fig. P9.45). The bullet remains in the block, and after impact the block lands a distance d from the bottom of the table. Determine the initial speed of the bullet.

47. A 0.500-kg sphere moving with a velocity expressed as $(2.00\hat{\mathbf{i}} - 3.00\hat{\mathbf{j}} + 1.00\hat{\mathbf{k}})$ m/s strikes a second, lighter sphere of mass 1.50 kg moving with an initial velocity of $(-1.00\hat{\mathbf{i}} + 2.00\hat{\mathbf{j}} - 3.00\hat{\mathbf{k}})$ m/s. (a) The velocity of the 0.500-kg sphere after the collision is $(-1.00\hat{\mathbf{i}} + 3.00\hat{\mathbf{j}} - 8.00\hat{\mathbf{k}})$ m/s. Find the final velocity of the 1.50-kg sphere and identify the kind of collision (elastic, inelastic, or perfectly inelastic). (b) Now assume the velocity of the 0.500-kg sphere after the collision is $(-0.250\hat{\mathbf{i}} + 0.750\hat{\mathbf{j}} - 2.00\hat{\mathbf{k}})$ m/s. Find the final velocity of the 1.50-kg sphere and identify the kind of collision. (c) **What If?** Take the velocity of the 0.500-kg sphere after the collision as $(-1.00\hat{\mathbf{i}} + 3.00\hat{\mathbf{j}} + a\hat{\mathbf{k}})$ m/s. Find the value of a and the velocity of the 1.50-kg sphere after an elastic collision.

48. **Review.** A metal cannonball of mass m rests next to a tree at the very edge of a cliff 36.0 m above the surface of the ocean. In an effort to knock the cannonball off the cliff, some children tie one end of a rope around a stone of mass 80.0 kg and the other end to a tree limb just above the cannonball. They tighten the rope so that the stone just clears the ground and hangs next to the cannonball. The children manage to swing the stone back until it is held at rest 1.80 m above the ground. The children release the stone, which then swings down and makes a head-on, elastic collision with the cannonball, projecting it horizontally off the cliff. The cannonball lands in the ocean a horizontal distance R away from its initial position. (a) Find the horizontal component R of the cannonball's displacement as it depends on m. (b) What is the maximum possible value for R, and (c) to what value of m does it correspond? (d) For the stone–cannonball–Earth system, is mechanical energy conserved throughout the process? Is this principle sufficient to solve the entire problem? Explain. (e) **What if?** Show that R does not depend on the value of the gravitational acceleration. Is this result remarkable? State how one might make sense of it.

49. **Review.** A light spring of force constant 3.85 N/m is compressed by 8.00 cm and held between a 0.250-kg block on the left and a 0.500-kg block on the right. Both blocks are at rest on a horizontal surface. The blocks are released simultaneously so that the spring tends to push them apart. Find the maximum velocity each block attains if the coefficient of kinetic friction between each block and the surface is (a) 0, (b) 0.100, and (c) 0.462. Assume the coefficient of static friction is greater than the coefficient of kinetic friction in every case.

50. Consider as a system the Sun with the Earth in a circular orbit around it. Find the magnitude of the change in the velocity of the Sun relative to the center of mass of the system over a six-month period. Ignore the influence of other celestial objects. You may obtain the necessary astronomical data from the endpapers of the book.

51. **Review.** There are (one can say) three coequal theories of motion for a single particle: Newton's second law, stating that the total force on the particle causes its acceleration; the work–kinetic energy theorem, stating that the total work

on the particle causes its change in kinetic energy; and the impulse–momentum theorem, stating that the total impulse on the particle causes its change in momentum. In this problem, you compare predictions of the three theories in one particular case. A 3.00-kg object has velocity $7.00\hat{\mathbf{j}}$ m/s. Then, a constant net force $12.0\hat{\mathbf{i}}$ N acts on the object for 5.00 s. (a) Calculate the object's final velocity, using the impulse–momentum theorem. (b) Calculate its acceleration from $\vec{\mathbf{a}} = (\vec{\mathbf{v}}_f - \vec{\mathbf{v}}_i)/\Delta t$. (c) Calculate its acceleration from $\vec{\mathbf{a}} = \sum\vec{\mathbf{F}}/m$. (d) Find the object's vector displacement from $\Delta\vec{\mathbf{r}} = \vec{\mathbf{v}}_i t + \frac{1}{2}\vec{\mathbf{a}}t^2$. (e) Find the work done on the object from $W = \vec{\mathbf{F}}\cdot\Delta\vec{\mathbf{r}}$. (f) Find the final kinetic energy from $\frac{1}{2}mv_f^2 = \frac{1}{2}m\vec{\mathbf{v}}_f\cdot\vec{\mathbf{v}}_f$. (g) Find the final kinetic energy from $\frac{1}{2}mv_i^2 + W$. (h) State the result of comparing the answers to parts (b) and (c), and the answers to parts (f) and (g).

CHALLENGE PROBLEMS

52. Sand from a stationary hopper falls onto a moving conveyor belt at the rate of 5.00 kg/s as shown in Figure P9.52. The conveyor belt is supported by frictionless rollers and moves at a constant speed of $v = 0.750$ m/s under the action of a constant horizontal external force $\vec{\mathbf{F}}_{ext}$ supplied by the motor that drives the belt. Find (a) the sand's rate of change of momentum in the horizontal direction, (b) the force of friction exerted by the belt on the sand, (c) the external force $\vec{\mathbf{F}}_{ext}$, (d) the work done by $\vec{\mathbf{F}}_{ext}$ in 1 s, and (e) the kinetic energy acquired by the falling sand each second due to the change in its horizontal motion. (f) Why are the answers to parts (d) and (e) different?

Figure P9.52

53. Two particles with masses m and $3m$ are moving toward each other along the x axis with the same initial speeds v_i. Particle m is traveling to the left, and particle $3m$ is traveling to the right. They undergo an elastic glancing collision such that particle m is moving in the negative y direction after the collision at a right angle from its initial direction. (a) Find the final speeds of the two particles in terms of v_i. (b) What is the angle θ at which the particle $3m$ is scattered?

54. On a horizontal air track, a glider of mass m carries a Γ-shaped post. The post supports a small dense sphere, also of mass m, hanging just above the top of the glider on a cord of length L. The glider and sphere are initially at rest with the cord vertical. A constant horizontal force of magnitude F is applied to the glider, moving it through displacement x_1; then the force is removed. During the time interval when the force is applied, the sphere moves through a displacement with horizontal component x_2. (a) Find the horizontal component of the velocity of the center of mass of the glider–sphere system when the force is removed. (b) After the force is removed, the glider continues to move on the track and the sphere swings back and forth, both without friction. Find an expression for the largest angle the cord makes with the vertical.

A rusty bolt resists efforts to turn it with a wrench. How can you loosen the bolt? (*Scott Richardson/ Shutterstock*)

10

Rotation of a Rigid Object About a Fixed Axis

STORYLINE **You are back at home after your game of pool in the** previous chapter. You go back into your garage to work on another project. For this project, you need some pieces of metal from an older project. The pieces of metal have been joined together for years by nuts and bolts, which are now quite rusty. Using a wrench, you try to loosen a bolt. You are unable to do so because of the rust. You instinctively reach for a piece of hollow pipe that is longer than the handle of the wrench and slip it over the handle. Pushing on the far end of the pipe, you are now able to loosen the bolt. You say to yourself, "Wait a minute! How did I know to use a long piece of pipe? Why did the long pipe make it possible for me to loosen the rusted bolt?" Your project sits idle while you ponder this new development. Then your thoughts progress further. You applied a force on the pipe, like the forces studied in Chapter 5. But you didn't achieve an acceleration of something through space like the objects in Chapter 5. Something *rotated*: the bolt. This is new: force causes rotation. You have more thinking to do. Your project sits idle for the rest of the day.

CONNECTIONS We have focused our attention so far on particles in *translational* motion. When we analyzed the motion of objects with a size in previous chapters, we ignored any spinning motion of the object. It is now time to *not* ignore this spinning motion. In this chapter, we focus on the *rotational* motion of an object. We will be following the outline of earlier chapters for this new type of motion; we will find rotational analogs for position, speed, acceleration, mass, force, and energy. Many objects exhibit both translational and rotational motion at the same time. We will investigate how to reduce the apparently complicated motion of such an object to a combination of the two types of motion. In dealing

249

with a rotating object, analysis is greatly simplified by assuming the object is rigid. A **rigid object** is one that is nondeformable; that is, the relative locations of all particles of which the object is composed remain constant. All real objects are deformable to some extent; our rigid-object model, however, is useful in many situations in which deformation is negligible. We have developed analysis models based on particles and systems. In this chapter, we introduce another class of analysis models based on the simplification model of a rigid object. In future chapters, we will see rotating objects for which we will need these models: the spinning Earth in Chapter 13, the armature of a motor in Chapter 30, and an HCl molecule in Chapter 42, for example.

10.1 Angular Position, Velocity, and Acceleration

As mentioned in the introduction, we will develop our understanding of rotational motion in a manner parallel to that used for translational motion in previous chapters. We began in Chapter 2 by defining kinematic variables for translational motion: position, velocity, and acceleration. We do the same here for rotational motion.

Figure 10.1 illustrates an overhead view of a rotating Blu-ray Disc. The disc rotates about a fixed axis perpendicular to the plane of the figure and passing through the center of the disc at O. A small element of the disc modeled as a particle at P is at a fixed distance r from the origin and rotates about it in a circle of radius r. (In fact, *every* element of the disc undergoes circular motion about O.) It is convenient to represent the position of P with its polar coordinates (r, θ), where r is the distance from the origin to P and θ is measured *counterclockwise* from some reference line fixed in space as shown in Figure 10.1a. In this representation, the angle θ changes in time while r remains constant. As the particle moves along the circle from the reference line, which is at angle $\theta = 0$, it moves through an arc of length s as in Figure 10.1b. We can define the angle θ as the ratio of the arc length s to the radius r:

$$\theta = \frac{s}{r} \tag{10.1a}$$

Because θ is the ratio of an arc length and the radius of the circle, it is a pure number. Usually, however, we give θ the artificial unit **radian** (rad), where one radian is the angle subtended by an arc length equal to the radius of the arc. Because the circumference of a circle is $2\pi r$, it follows from Equation 10.1a that 360° corresponds to an angle of $(2\pi r/r)$ rad $= 2\pi$ rad. Hence, 1 rad $= 360°/2\pi \approx 57.3°$. To convert an angle in degrees to an angle in radians, we use that π rad $= 180°$, so

$$\theta(\text{rad}) = \frac{\pi}{180°}\,\theta(\text{deg})$$

For example, 60° equals $\pi/3$ rad and 45° equals $\pi/4$ rad.

Based on the definition of the angle θ in Equation 10.1a, we can express the arc length s through which the particle at P moves in Figure 10.1b as

$$s = r\theta \tag{10.1b}$$

Because the disc in Figure 10.1 is a rigid object, as the particle moves through an angle θ from the reference line, every other particle on the object rotates through the same angle θ. Therefore, we can associate the angle θ with the entire rigid object as well as with an individual particle, which allows us to define the *angular position* of a rigid object in its rotational motion. We choose a reference line on the object, such as a line connecting O and a chosen particle on the object. The **angular position** of the rigid object is the angle θ between this reference line on the object and the fixed reference line in space, which is often chosen as the x axis. Such identification is similar to the way we define the position of an object in one-dimensional translational

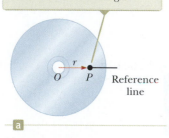

To define angular position for the disc, a reference line fixed in space is chosen. A particle at P is located at a distance r from the rotation axis through O.

a

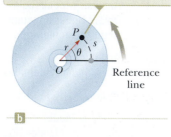

As the disc rotates, the particle at P moves through an arc length s on a circular path of radius r. The angular position of P is θ.

b

Figure 10.1 A Blu-ray Disc rotating about a fixed axis through O perpendicular to the plane of the figure.

PITFALL PREVENTION 10.1

Remember the Radian In rotational equations, you *must* use angles expressed in radians. Don't fall into the trap of using angles measured in degrees in rotational equations.

motion as the distance x between the object and the reference position, which is the origin, $x = 0$. Therefore, the angle θ plays the same role in rotational motion that the position x does in one-dimensional translational motion.

As the particle in question on our rigid object travels from position Ⓐ to position Ⓑ in a time interval Δt as in Figure 10.2, the reference line fixed to the object sweeps out an angle $\Delta\theta = \theta_f - \theta_i$. This quantity $\Delta\theta$ is defined as the **angular displacement** of the rigid object:

$$\Delta\theta \equiv \theta_f - \theta_i$$

◀ Angular displacement
(Compare to Equation 2.1)

The rate at which this angular displacement occurs can vary. If the rigid object spins rapidly, this displacement can occur in a short time interval. If it rotates slowly, this displacement occurs in a longer time interval. These different rotation rates can be quantified by defining the **average angular speed** ω_{avg} (Greek letter omega) as the ratio of the angular displacement of a rigid object to the time interval Δt during which the displacement occurs:

$$\omega_{avg} \equiv \frac{\Delta\theta}{\Delta t}$$ (10.2)

◀ Average angular speed
(Compare to Equation 2.2)

In analogy to translational speed, the **instantaneous angular speed** ω is defined as the limit of the average angular speed as Δt approaches zero:

$$\omega \equiv \lim_{\Delta t \to 0} \frac{\Delta\theta}{\Delta t} = \frac{d\theta}{dt}$$ (10.3)

◀ Instantaneous angular speed
(Compare to Equation 2.5)

Angular speed has units of radians per second (rad/s), which can be written as s^{-1} because radians are not dimensional. We take ω to be positive when θ is increasing (counterclockwise motion in Fig. 10.2) and negative when θ is decreasing (clockwise motion in Fig. 10.2).

QUICK QUIZ 10.1 A rigid object rotates in a counterclockwise sense around a fixed axis. Each of the following pairs of quantities represents an initial angular position and a final angular position of the rigid object. **(i)** Which of the sets can *only* occur if the rigid object rotates through more than 180°? (a) 3 rad, 6 rad (b) −1 rad, 1 rad (c) 1 rad, 5 rad **(ii)** Suppose the change in angular position for each of these pairs of values occurs in 1 s. Which choice represents the lowest average angular speed?

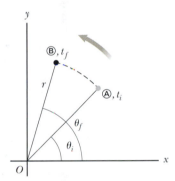

Figure 10.2 A particle on a rotating rigid object moves from Ⓐ to Ⓑ along the arc of a circle. In the time interval $\Delta t = t_f - t_i$, the radial line of length r moves through an angular displacement $\Delta\theta = \theta_f - \theta_i$.

If the instantaneous angular speed of an object changes from ω_i to ω_f in the time interval Δt, the object has an angular acceleration. The **average angular acceleration** α_{avg} (Greek letter alpha) of a rotating rigid object is defined as the ratio of the change in the angular speed to the time interval Δt during which the change in the angular speed occurs:

$$\alpha_{avg} \equiv \frac{\Delta\omega}{\Delta t} = \frac{\omega_f - \omega_i}{t_f - t_i}$$ (10.4)

◀ Average angular acceleration
(Compare to Equation 2.9)

In analogy to translational acceleration, the **instantaneous angular acceleration** is defined as the limit of the average angular acceleration as Δt approaches zero:

$$\alpha \equiv \lim_{\Delta t \to 0} \frac{\Delta\omega}{\Delta t} = \frac{d\omega}{dt}$$ (10.5)

◀ Instantaneous angular acceleration (Compare to Equation 2.10)

Angular acceleration has units of radians per second squared (rad/s²), or simply s^{-2}. Notice that α is positive when a rigid object rotating counterclockwise is speeding up or when a rigid object rotating clockwise is slowing down during some time interval.

When a rigid object is rotating about a *fixed* axis, every particle on the object rotates through the same angle in a given time interval and has the same angular speed and the same angular acceleration. Therefore, like the angular position θ,

PITFALL PREVENTION 10.2
Specify Your Axis In solving rotation problems, you must specify an axis of rotation. This new feature does not exist in our study of translational motion. The choice of axis is arbitrary, but once you make it, you must maintain that choice consistently throughout the problem. In some problems, the physical situation suggests a natural axis, such as one along the axle of an automobile wheel. In other problems, there may not be an obvious choice, and you must exercise judgment.

Figure 10.3 The right-hand rule for determining the direction of the angular velocity vector.

Rotational kinematic ▶
equations

the quantities ω and α characterize the rotational motion of the entire rigid object as well as individual particles in the object.

Angular position (θ), angular speed (ω), and angular acceleration (α) are analogous to translational position (x), translational speed (v), and translational acceleration (a). The variables θ, ω, and α differ dimensionally from the variables x, v, and a only by a factor having the unit of length. (See Section 10.3.)

We have not specified any direction in space for angular speed and angular acceleration. Strictly speaking, ω and α are the magnitudes of the angular velocity and the angular acceleration vectors[1] $\vec{\omega}$ and $\vec{\alpha}$, respectively, and they should always be positive. Because we are considering rotation about a fixed axis, however, we can use nonvector notation and indicate the vectors' directions by assigning a positive or negative sign to ω and α as discussed earlier with regard to Equations 10.3 and 10.5. For rotation about a fixed axis, the only direction that uniquely specifies the rotational motion is the direction along the axis of rotation. Therefore, the directions of $\vec{\omega}$ and $\vec{\alpha}$ are along this axis. If a particle rotates in the xy plane as in Figure 10.2, the direction of $\vec{\omega}$ for the particle is out of the plane of the diagram when the rotation is counterclockwise and into the plane of the diagram when the rotation is clockwise. To illustrate this convention, it is convenient to use the *right-hand rule* demonstrated in Figure 10.3. When the four fingers of the right hand are wrapped in the direction of rotation, the extended right thumb points in the direction of $\vec{\omega}$. The direction of $\vec{\alpha}$ follows from its definition $\vec{\alpha} \equiv d\vec{\omega}/dt$. It is in the same direction as $\vec{\omega}$ if the angular speed is increasing in time, and it is antiparallel to $\vec{\omega}$ if the angular speed is decreasing in time.

10.2 Analysis Model: Rigid Object Under Constant Angular Acceleration

In our study of translational motion, after introducing the kinematic variables, we considered the special case of a particle under constant acceleration. We follow the same procedure here for a rigid object under constant angular acceleration.

Imagine a rigid object such as the disc in Figure 10.1 rotates about a fixed axis and has a constant angular acceleration. In parallel with our analysis model of the particle under constant acceleration, we generate a new analysis model for rotational motion called the **rigid object under constant angular acceleration.** We develop kinematic relationships for this model in this section. Writing Equation 10.5 in the form $d\omega = \alpha\, dt$ and integrating from $t_i = 0$ to $t_f = t$ gives

$$\omega_f = \omega_i + \alpha t \quad \text{(for constant } \alpha\text{)} \tag{10.6}$$

where ω_i is the angular speed of the rigid object at time $t = 0$. Equation 10.6 allows us to find the angular speed ω_f of the object at any later time t. Substituting Equation 10.6 into Equation 10.3 and integrating once more, we obtain

$$\theta_f = \theta_i + \omega_i t + \tfrac{1}{2}\alpha t^2 \quad \text{(for constant } \alpha\text{)} \tag{10.7}$$

where θ_i is the angular position of the rigid object at time $t = 0$. Equation 10.7 allows us to find the angular position θ_f of the object at any later time t. Eliminating t from Equations 10.6 and 10.7 gives

$$\omega_f^2 = \omega_i^2 + 2\alpha(\theta_f - \theta_i) \quad \text{(for constant } \alpha\text{)} \tag{10.8}$$

---

[1]Although we do not verify it here, the instantaneous angular velocity and instantaneous angular acceleration are vector quantities, but the corresponding average values are not because angular displacements do not add as vector quantities for finite rotations.

This equation allows us to find the angular speed ω_f of the rigid object for any value of its angular position θ_f. If we eliminate α between Equations 10.6 and 10.7, we obtain

$$\theta_f = \theta_i + \tfrac{1}{2}(\omega_i + \omega_f)t \quad \text{(for constant } \alpha) \tag{10.9}$$

Notice that these kinematic expressions for the rigid object under constant angular acceleration are of the same mathematical form as those for a particle under constant acceleration (see Table 10.1). They can be generated from the equations for translational motion by making the substitutions $x \rightarrow \theta$, $v \rightarrow \omega$, and $a \rightarrow \alpha$. Table 10.1 compares the kinematic equations for the rigid object under constant angular acceleration and particle under constant acceleration models.

QUICK QUIZ 10.2 Consider again the pairs of angular positions for the rigid object in Quick Quiz 10.1. If the object starts from rest at the initial angular position, moves counterclockwise with constant angular acceleration, and arrives at the final angular position with the same angular speed in all three cases, for which choice is the angular acceleration the highest?

TABLE 10.1 Kinematic Equations for Rotational and Translational Motion

Rigid Object Under Constant Angular Acceleration		Particle Under Constant Acceleration	
$\omega_f = \omega_i + \alpha t$	(10.6)	$v_f = v_i + at$	(2.13)
$\theta_f = \theta_i + \omega_i t + \tfrac{1}{2}\alpha t^2$	(10.7)	$x_f = x_i + v_i t + \tfrac{1}{2}at^2$	(2.16)
$\omega_f^2 = \omega_i^2 + 2\alpha(\theta_f - \theta_i)$	(10.8)	$v_f^2 = v_i^2 + 2a(x_f - x_i)$	(2.17)
$\theta_f = \theta_i + \tfrac{1}{2}(\omega_i + \omega_f)t$	(10.9)	$x_f = x_i + \tfrac{1}{2}(v_i + v_f)t$	(2.15)

ANALYSIS MODEL **Rigid Object Under Constant Angular Acceleration**

Imagine an object that undergoes a spinning motion such that its angular acceleration is constant. The equations describing its angular position and angular speed are analogous to those for the particle under constant acceleration model:

α = constant

$$\omega_f = \omega_i + \alpha t \tag{10.6}$$

$$\theta_f = \theta_i + \omega_i t + \tfrac{1}{2}\alpha t^2 \tag{10.7}$$

$$\omega_f^2 = \omega_i^2 + 2\alpha(\theta_f - \theta_i) \tag{10.8}$$

$$\theta_f = \theta_i + \tfrac{1}{2}(\omega_i + \omega_f)t \tag{10.9}$$

Examples:

- during its spin cycle, the tub of a clothes washer begins from rest and accelerates up to its final spin speed
- a workshop grinding wheel is turned off and comes to rest under the action of a constant friction force in the bearings of the wheel
- a gyroscope is powered up and approaches its operating speed (Chapter 11)
- the crankshaft of a diesel engine changes to a higher angular speed (Chapter 21)

Example **10.1** **Rotating Wheel**

A wheel rotates with a constant angular acceleration of 3.50 rad/s².

(A) If the angular speed of the wheel is 2.00 rad/s at $t_i = 0$, through what angular displacement does the wheel rotate in 2.00 s?

SOLUTION

Conceptualize Look again at Figure 10.1. Imagine that the disc rotates with its angular speed increasing at a constant rate. You start your stopwatch when the disc is rotating at 2.00 rad/s. This mental image is a model for the motion of the wheel in this example.

continued

10.1 c o n t i n u e d

Categorize   The phrase "with a constant angular acceleration" tells us to apply the *rigid object under constant angular acceleration* model to the wheel.

Analyze   From the rigid object under constant angular acceleration model, choose Equation 10.7 and rearrange it so that it expresses the angular displacement of the wheel:

$$\Delta\theta = \theta_f - \theta_i = \omega_i t + \tfrac{1}{2}\alpha t^2$$

Substitute the known values to find the angular displacement at $t = 2.00$ s:

$$\Delta\theta = (2.00 \text{ rad/s})(2.00 \text{ s}) + \tfrac{1}{2}(3.50 \text{ rad/s}^2)(2.00 \text{ s})^2$$
$$= \boxed{11.0 \text{ rad}} = (11.0 \text{ rad})(180°/\pi \text{ rad}) = \boxed{630°}$$

(B) Through how many revolutions has the wheel turned during this time interval?

S O L U T I O N

Multiply the angular displacement found in part (A) by a conversion factor to find the number of revolutions:

$$\Delta\theta = 630°\left(\frac{1 \text{ rev}}{360°}\right) = \boxed{1.75 \text{ rev}}$$

(C) What is the angular speed of the wheel at $t = 2.00$ s?

S O L U T I O N

Use Equation 10.6 from the rigid object under constant angular acceleration model to find the angular speed at $t = 2.00$ s:

$$\omega_f = \omega_i + \alpha t = 2.00 \text{ rad/s} + (3.50 \text{ rad/s}^2)(2.00 \text{ s})$$
$$= \boxed{9.00 \text{ rad/s}}$$

Finalize   We could also obtain this result using Equation 10.8 and the results of part (A). (Try it!)

W H A T   I F ?   Suppose a particle moves along a straight line with a constant acceleration of 3.50 m/s². If the velocity of the particle is 2.00 m/s at $t_i = 0$, through what displacement does the particle move in 2.00 s? What is the velocity of the particle at $t = 2.00$ s?

Answer   Notice that these questions are translational analogs to parts (A) and (C) of the original problem. The mathematical solution follows exactly the same form. For the displacement, from the particle under constant acceleration model,

$$\Delta x = x_f - x_i = v_i t + \tfrac{1}{2}at^2$$
$$= (2.00 \text{ m/s})(2.00 \text{ s}) + \tfrac{1}{2}(3.50 \text{ m/s}^2)(2.00 \text{ s})^2 = 11.0 \text{ m}$$

and for the velocity,

$$v_f = v_i + at = 2.00 \text{ m/s} + (3.50 \text{ m/s}^2)(2.00 \text{ s}) = 9.00 \text{ m/s}$$

There is no translational analog to part (B) because translational motion under constant acceleration is not repetitive.

10.3   Angular and Translational Quantities

In this section, we derive some useful relationships between the angular speed and acceleration of a rotating rigid object and the translational speed and acceleration of a point in the object. To do so, we must keep in mind that when a rigid object rotates about a fixed axis, every particle of the object moves in a circle whose center is on the axis of rotation. We looked at a flat, circular object in Figure 10.1. Let us now generalize to an arbitrary, three-dimensional object, as in Figure 10.4. A reference axis fixed in space is chosen—the x axis in Figure 10.4—and we look at the motion of one point P contained within the object.

Because point P in Figure 10.4 moves in a circle, the translational velocity vector $\vec{\mathbf{v}}$ is always tangent to the circular path and hence is called *tangential velocity*. The magnitude of the tangential velocity of the point P is by definition the tangential speed $v = ds/dt$, where s is the distance traveled by this point measured along the circular path. Recalling that $s = r\theta$ (Eq. 10.1b) and noting that, for a given point on the object, r is constant, we obtain

$$v = \frac{ds}{dt} = r\frac{d\theta}{dt}$$

Because $d\theta/dt = \omega$ (see Eq. 10.3), it follows that

$$v = r\omega \tag{10.10}$$

As we saw in Equation 4.24, the tangential speed of a particle moving in a circle equals the distance of the particle from the center of the circle multiplied by the angular speed. We find the same relationship for particles at every point on a rigid object. Although every point on the rigid object has the same *angular* speed, not every point has the same *tangential* speed because r is not the same for all points on the object. Equation 10.10 shows that the tangential speed of a point on the rotating object increases as one moves outward from the center of rotation, as we would intuitively expect. For example, the outer end of a swinging golf club moves much faster than a point near the handle.

We can relate the angular acceleration of the rotating rigid object to the tangential acceleration of the point P by taking the time derivative of v in Equation 10.10:

$$a_t = \frac{dv}{dt} = r\frac{d\omega}{dt}$$

$$a_t = r\alpha \tag{10.11}$$

That is, the tangential component of the translational acceleration of a point on a rotating rigid object equals the point's perpendicular distance from the axis of rotation multiplied by the angular acceleration.

In Section 4.4, we found that a point moving in a circular path undergoes a radial acceleration a_r directed toward the center of rotation and whose magnitude is that of the centripetal acceleration v^2/r (Fig. 10.5). Because $v = r\omega$ for a point P on a rotating object, we can express the centripetal acceleration at that point in terms of angular speed as we did for a particle moving in a circular path in Equation 4.25:

$$a_c = \frac{v^2}{r} = r\omega^2 \tag{10.12}$$

The total acceleration vector at the point is $\vec{\mathbf{a}} = \vec{\mathbf{a}}_t + \vec{\mathbf{a}}_r$, where the magnitude of $\vec{\mathbf{a}}_r$ is the centripetal acceleration a_c. Because $\vec{\mathbf{a}}$ is a vector having a radial and a tangential component, the magnitude of $\vec{\mathbf{a}}$ at the point P on the rotating rigid object is

$$a = \sqrt{a_t^2 + a_r^2} = \sqrt{r^2\alpha^2 + r^2\omega^4} = r\sqrt{\alpha^2 + \omega^4} \tag{10.13}$$

QUICK QUIZ 10.3 Ethan and Rebecca are riding on a merry-go-round. Ethan rides on a horse at the outer rim of the circular platform, twice as far from the center of the circular platform as Rebecca, who rides on an inner horse. **(i)** When the merry-go-round is rotating at a constant angular speed, what is Ethan's angular speed? (a) twice Rebecca's (b) the same as Rebecca's (c) half of Rebecca's (d) impossible to determine **(ii)** When the merry-go-round is rotating at a constant angular speed, describe Ethan's tangential speed from the same list of choices.

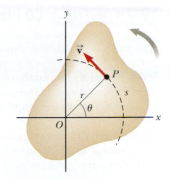

Figure 10.4 As a rigid object rotates about the fixed axis (the z axis) through O, the point P has a tangential velocity $\vec{\mathbf{v}}$ that is always tangent to the circular path of radius r.

◀ Relation between tangential acceleration and angular acceleration

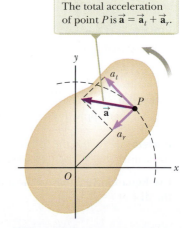

The total acceleration of point P is $\vec{\mathbf{a}} = \vec{\mathbf{a}}_t + \vec{\mathbf{a}}_r$.

Figure 10.5 As a rigid object rotates about a fixed axis (the z axis) through O, the point P experiences a tangential component of translational acceleration a_t and a radial component of translational acceleration a_r.

Example **10.2** CD Player

Despite the availability of music in digital form, the compact disc, or CD, remains a popular format for music and data. On a CD (Fig. 10.6), audio information is stored digitally in a series of pits and flat areas on the surface of the disc. The alternations between pits and flat areas on the surface represent binary ones and zeros to be read by the CD player and converted back to sound waves. The pits and flat areas are detected by a system consisting of a laser and lenses. The length of a string of ones and zeros representing one piece of information is the same everywhere on the disc, whether the information is near the center of the disc or near its outer edge. So that this length of ones and zeros always passes by the laser–lens system in the same time interval, the tangential speed of the disc surface at the location of the lens must be constant. According to Equation 10.10, the angular speed must therefore vary as the laser–lens system moves radially along the disc. In a typical CD player, the constant speed of the surface at the point of the laser–lens system is 1.3 m/s.

(A) Find the angular speed of the disc in revolutions per minute when information is being read from the innermost first track ($r = 23$ mm) and the outermost final track ($r = 58$ mm).

SOLUTION

Conceptualize Figure 10.6 shows a photograph of a compact disc. Trace your finger around the circle marked "23 mm" and mentally estimate the time interval to go around the circle once. Now trace your finger around the circle marked "58 mm," moving your finger across the surface of the page at the same speed as you did when tracing the smaller circle. Notice how much longer in time it takes your finger to go around the larger circle. If your finger represents the laser reading the disc, you can see that the disc rotates once in a longer time interval when the laser reads the information in the outer circle. Therefore, the disc must rotate more slowly when the laser is reading information from this part of the disc.

Figure 10.6 (Example 10.2) A compact disc.

Categorize This part of the example is categorized as a simple substitution problem. In later parts, we will need to identify analysis models.

Use Equation 10.10 to find the angular speed that gives the required tangential speed at the position of the inner track:

$$\omega_i = \frac{v}{r_i} = \frac{1.3 \text{ m/s}}{2.3 \times 10^{-2} \text{ m}} = 57 \text{ rad/s}$$

$$= (57 \text{ rad/s})\left(\frac{1 \text{ rev}}{2\pi \text{ rad}}\right)\left(\frac{60 \text{ s}}{1 \text{ min}}\right) = \boxed{5.4 \times 10^2 \text{ rev/min}}$$

Do the same for the outer track:

$$\omega_f = \frac{v}{r_f} = \frac{1.3 \text{ m/s}}{5.8 \times 10^{-2} \text{ m}} = 22 \text{ rad/s} = \boxed{2.1 \times 10^2 \text{ rev/min}}$$

The CD player adjusts the angular speed ω of the disc within this range so that information moves past the objective lens at a constant rate.

(B) The maximum playing time of a standard music disc is 74 min and 33 s. How many revolutions does the disc make during that time?

SOLUTION

Categorize From part (A), the angular speed decreases as the disc plays. Let us assume it decreases steadily, with α constant. We can then apply the *rigid object under constant angular acceleration* model to the disc.

Analyze If $t = 0$ is the instant the disc begins rotating, with angular speed of 57 rad/s, the final value of the time t is (74 min)(60 s/min) + 33 s = 4 473 s. We are looking for the angular displacement $\Delta\theta$ during this time interval.

Use Equation 10.9 to find the angular displacement of the disc at $t = 4\,473$ s:

$$\Delta\theta = \theta_f - \theta_i = \tfrac{1}{2}(\omega_i + \omega_f)t$$

$$= \tfrac{1}{2}(57 \text{ rad/s} + 22 \text{ rad/s})(4\,473 \text{ s}) = 1.8 \times 10^5 \text{ rad}$$

Convert this angular displacement to revolutions:

$$\Delta\theta = (1.8 \times 10^5 \text{ rad})\left(\frac{1 \text{ rev}}{2\pi \text{ rad}}\right) = \boxed{2.8 \times 10^4 \text{ rev}}$$

(c) What is the angular acceleration of the compact disc over the 4 473-s time interval?

SOLUTION

Categorize We again model the disc as a *rigid object under constant angular acceleration*. In this case, Equation 10.6 gives the value of the constant angular acceleration. Another approach is to use Equation 10.4 to find the average angular acceleration.

10.2 continued

In this case, we are not assuming the angular acceleration is constant. The answer is the same from both equations; only the interpretation of the result is different.

Analyze Use Equation 10.6 to find the angular acceleration:

$$\alpha = \frac{\omega_f - \omega_i}{t} = \frac{22 \text{ rad/s} - 57 \text{ rad/s}}{4\ 473 \text{ s}} = -7.6 \times 10^{-3} \text{ rad/s}^2$$

Finalize The disc experiences a very gradual decrease in its rotation rate, as expected from the long time interval required for the angular speed to change from the initial value to the final value. In reality, the angular acceleration of the disc is not constant. Problem 46 allows you to explore the actual time behavior of the angular acceleration.

10.4 Torque

In our study of translational motion, after investigating the description of motion in Chapters 2–4, we studied the cause of changes in motion: force, in Chapters 5–6. We follow the same plan here: What is the cause of changes in rotational motion?

When a force is exerted on a rigid object pivoted about an axis, the object tends to rotate about that axis. Imagine trying to rotate a door by applying a force of magnitude F perpendicular to the door surface near the hinges and then at various distances from the hinges. You will achieve a more rapid rate of rotation for the door by applying the force near the doorknob than by applying it near the hinges. Because the *same* force was applied at different positions on the door, this experiment indicates that the cause of changes in rotational motion must also depend on the *location* at which the force is applied.

The cause of changes in the rotational motion of an object about some axis is measured by a quantity called **torque** $\vec{\tau}$ (Greek letter tau). Torque is a vector, but we will consider only its magnitude here; we will explore its vector nature in Chapter 11.

Consider the wrench and bolt from the opening storyline for this chapter. We show these objects with some geometry added in Figure 10.7. We wish to rotate the wrench around an axis that is perpendicular to the page and passes through the center of the bolt. The applied force $\vec{F}$ acts at an angle ϕ to the horizontal. We define the magnitude of the torque associated with the force $\vec{F}$ around the axis passing through O by the expression

$$\tau \equiv rF \sin \phi = Fd \tag{10.14}$$

where r is the distance between the rotation axis and the point of application of $\vec{F}$, and d is the perpendicular distance from the rotation axis to the line of action of $\vec{F}$. (The *line of action* of a force is an imaginary line extending out both ends of the vector representing the force. The dashed line extending from the tail of $\vec{F}$ in Fig. 10.7 is part of the line of action of $\vec{F}$.) From the right triangle in Figure 10.7 that has the wrench as its hypotenuse, we see that $d = r \sin \phi$. The quantity d is called the **moment arm** (or *lever arm*) of $\vec{F}$.

In Figure 10.7, the only component of $\vec{F}$ that tends to cause rotation of the wrench around an axis through O is $F \sin \phi$, the component perpendicular to a line drawn from the rotation axis to the point of application of the force. The horizontal component $F \cos \phi$, because its line of action passes through O, has no tendency to produce rotation about an axis passing through O. From the definition of torque in Equation 10.14, the cause of changes in rotational motion increases as F increases and as d increases, which explains why it is easier to rotate a door if we push at the doorknob rather than at a point close to the hinges. We also want to apply our push as closely perpendicular to the door as we can so that ϕ is close to 90°, which maximizes the moment arm. Pushing sideways on the doorknob ($\phi = 0$) will not cause the door to rotate.

Equation 10.14 allows us to understand the use of the pipe to turn the wrench in the opening storyline. The maximum force you can apply to the wrench is not

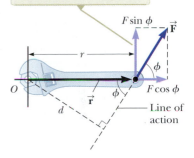

The component $F \sin \phi$ tends to rotate the wrench about an axis through O.

Figure 10.7 The force $\vec{F}$ has a greater rotating tendency about an axis through O as F increases and as the moment arm d increases.

PITFALL PREVENTION 10.4

Torque Depends on Your Choice of Axis There is no unique value of the torque on an object. Its value depends on your choice of rotation axis.

◀ Moment arm

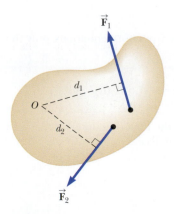

Figure 10.8 The force $\vec{\mathbf{F}}_1$ tends to rotate the object counterclockwise about an axis through O, and $\vec{\mathbf{F}}_2$ tends to rotate it clockwise.

enough to turn the bolt. You cannot apply more *force* $\vec{\mathbf{F}}$, but you can increase the *torque* on the bolt by putting the pipe over the wrench handle. This allows you to apply the same force at a larger distance d from the axis of rotation. You have increased the moment arm of the force and, therefore, increased the torque applied by the *same* force.

If two or more forces act on a rigid object as in Figure 10.8, each tends to produce rotation about the axis through O. In this example, $\vec{\mathbf{F}}_2$ tends to rotate the object clockwise and $\vec{\mathbf{F}}_1$ tends to rotate it counterclockwise. We use the convention that the sign of the torque resulting from a force is positive if the turning tendency of the force is counterclockwise and negative if the turning tendency is clockwise. For Example, in Figure 10.8, the torque resulting from $\vec{\mathbf{F}}_1$, which has a moment arm d_1, is positive and equal to $+F_1 d_1$; the torque from $\vec{\mathbf{F}}_2$ is negative and equal to $-F_2 d_2$. Hence, the *net* torque about an axis through O is

$$\sum \tau = \tau_1 + \tau_2 = F_1 d_1 - F_2 d_2$$

Torque should not be confused with force. Forces can cause a change in translational motion as described by Newton's second law. Forces can also cause a change in rotational motion, but the effectiveness of the forces in causing this change depends on both the magnitudes of the forces and the moment arms of the forces, in the combination we call *torque*. Torque has units of force times length—newton meters (N · m) in SI units—and should be reported in these units. Do not confuse torque and work (Chapter 7), which have the same units but are very different concepts.

QUICK QUIZ **10.4** If you are trying to loosen a stubborn screw from a piece of wood with a screwdriver and fail, should you find a screwdriver for which the handle is (a) longer or (b) fatter?

| Example **10.3** | **The Net Torque on a Cylinder** |

A one-piece cylinder is shaped as shown in Figure 10.9, with a core section protruding from the larger drum. The cylinder is free to rotate about the central z axis shown in the drawing. A rope wrapped around the drum, which has radius R_1, exerts a force $\vec{\mathbf{T}}_1$ to the right on the cylinder. A rope wrapped around the core, which has radius R_2, exerts a force $\vec{\mathbf{T}}_2$ downward on the cylinder.

(A) What is the net torque acting on the cylinder about the rotation axis (which is the z axis in Fig. 10.9)?

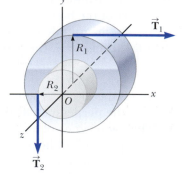

Figure 10.9 (Example 10.3) A solid cylinder pivoted about the z axis through O. The moment arm of $\vec{\mathbf{T}}_1$ is R_1, and the moment arm of $\vec{\mathbf{T}}_2$ is R_2.

S O L U T I O N

Conceptualize   Imagine that the cylinder in Figure 10.9 is a shaft in a machine. The force $\vec{\mathbf{T}}_1$ could be applied by a drive belt wrapped around the drum. The force $\vec{\mathbf{T}}_2$ could be applied by a friction brake at the surface of the core.

Categorize   This example is a substitution problem in which we evaluate the net torque using Equation 10.14.

The torque due to $\vec{\mathbf{T}}_1$ about the rotation axis is $-R_1 T_1$. (The sign is negative because the torque tends to produce clockwise rotation.) The torque due to $\vec{\mathbf{T}}_2$ is $+R_2 T_2$. (The sign is positive because the torque tends to produce counterclockwise rotation of the cylinder.)

Evaluate the net torque about the rotation axis:

$$\sum \tau = \tau_1 + \tau_2 = R_2 T_2 - R_1 T_1$$

As a quick check, notice that if the two forces are of equal magnitude, the net torque is negative because $R_1 > R_2$. Starting from rest with both forces of equal magnitude acting on it, the cylinder would rotate clockwise because $\vec{\mathbf{T}}_1$ would be more effective at turning it than would $\vec{\mathbf{T}}_2$.

10.3 continued

(B) Suppose $T_1 = 5.0$ N, $R_1 = 1.0$ m, $T_2 = 15$ N, and $R_2 = 0.50$ m. What is the net torque about the rotation axis, and which way does the cylinder rotate starting from rest?

S O L U T I O N

Substitute the given values:

$$\sum \tau = (0.50 \text{ m})(15 \text{ N}) - (1.0 \text{ m})(5.0 \text{ N}) = \boxed{2.5 \text{ N} \cdot \text{m}}$$

Because this net torque is positive, the cylinder begins to rotate in the counterclockwise direction.

10.5 Analysis Model: Rigid Object Under a Net Torque

In Chapter 5, we learned that a net force on an object causes an acceleration of the object and that the acceleration is proportional to the net force. These facts are the basis of the particle under a net force model whose mathematical representation is Newton's second law. In this section, we show the rotational analog of Newton's second law: the angular acceleration of a rigid object rotating about a fixed axis is proportional to the net torque acting about that axis. Before discussing the more complex case of rigid-object rotation, however, it is instructive first to discuss the case of a particle moving in a circular path about some fixed point under the influence of an external force.

Consider a particle of mass m rotating in a circle of radius r under the influence of a tangential net force $\sum \vec{\mathbf{F}}_t$ and a radial net force $\sum \vec{\mathbf{F}}_r$ as shown in Figure 10.10. The radial net force causes the particle to move in the circular path with a centripetal acceleration. The tangential force provides a tangential acceleration $\vec{\mathbf{a}}_t$, and

$$\sum F_t = ma_t$$

The magnitude of the net torque due to $\sum \vec{\mathbf{F}}_t$ on the particle about an axis perpendicular to the page through the center of the circle is

$$\sum \tau = \sum F_t r = (ma_t)r$$

Because the tangential acceleration is related to the angular acceleration through the relationship $a_t = r\alpha$ (Eq. 10.11), the net torque can be expressed as

$$\sum \tau = (mr\alpha)r = (mr^2)\alpha \tag{10.15}$$

Let us denote the quantity mr^2 with the symbol I for now. We will say more about this quantity below. Using this notation, Equation 10.15 can be written as

$$\sum \tau = I\alpha \tag{10.16}$$

That is, the net torque acting on the particle is proportional to its angular acceleration. Notice that $\sum \tau = I\alpha$ has the same mathematical form as Newton's second law of motion, $\sum F = ma$ (Eq. 5.2).

Now let us extend this discussion to a rigid object of arbitrary shape rotating about a fixed axis passing through a point O as in Figure 10.11. The object can be regarded as a collection of particles of mass m_i. If we impose a Cartesian coordinate system on the object, each particle rotates in a circle about the origin and each has a tangential acceleration a_i produced by an external tangential force of magnitude F_i. For any given particle, we know from Newton's second law that

$$F_i = m_i a_i$$

The external torque $\vec{\boldsymbol{\tau}}_i$ associated with the force $\vec{\mathbf{F}}_i$ acts about the origin and its magnitude is given by

$$\tau_i = r_i F_i = r_i m_i a_i$$

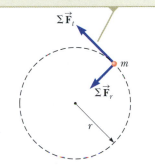

The tangential force on the particle results in a torque on the particle about an axis through the center of the circle.

Figure 10.10 A particle rotating in a circle under the influence of a tangential net force $\sum \vec{\mathbf{F}}_t$. A radial net force $\sum \vec{\mathbf{F}}_r$ also must be present to maintain the circular motion.

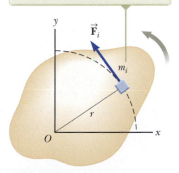

The particle of mass m_i of the rigid object experiences a torque in the same way that the particle in Figure 10.10 does.

Figure 10.11 A rigid object rotating about an axis through O. Each particle of mass m_i rotates about the axis with the same angular acceleration α.

Because $a_i = r_i\alpha$, the expression for τ_i becomes

$$\tau_i = m_i r_i^2 \alpha$$

Although different particles in the rigid object may have different translational accelerations a_i, they all have the *same* angular acceleration α. With that in mind, we can add the torques on all of the particles making up the rigid object to obtain the net torque on the object about an axis through O due to all external forces:

$$\sum \tau_{\text{ext}} = \sum_i \tau_i = \sum_i m_i r_i^2 \alpha = \left(\sum_i m_i r_i^2 \right)\alpha \tag{10.17}$$

where α can be taken outside the summation because it is common to all particles. Calling the quantity in parentheses I, the expression for $\sum \tau_{\text{ext}}$ becomes

◀ Torque on a rigid object is proportional to angular acceleration

$$\sum \tau_{\text{ext}} = I\alpha \tag{10.18}$$

This equation for a rigid object is the same as that found for a particle moving in a circular path (Eq. 10.16). The net torque about the rotation axis is proportional to the angular acceleration of the object, with the proportionality factor being I, a quantity that we have yet to describe fully. Equation 10.18 is the mathematical representation of the analysis model of a **rigid object under a net torque,** the rotational analog to the particle under a net force.

Let us now address the quantity I, defined as follows:

$$I = \sum_i m_i r_i^2 \tag{10.19}$$

PITFALL PREVENTION 10.5
No Single Moment of Inertia
There is one major difference between mass and moment of inertia. Mass is an inherent property of an object. The moment of inertia of an object depends on your choice of rotation axis. Therefore, there is no single value of the moment of inertia for an object. There is a *minimum* value of the moment of inertia, which is that calculated about an axis passing through the center of mass of the object.

This quantity is called the **moment of inertia** of the object, and depends on the masses of the particles making up the object and their distances from the rotation axis. Notice that Equation 10.19 reduces to $I = mr^2$ for a single particle, consistent with our use of the notation I in going from Equation 10.15 to Equation 10.16. Note that moment of inertia has units of $kg \cdot m^2$ in SI units.

Equation 10.18 has the same form as Newton's second law for a system of particles as expressed in Equation 9.39:

$$\sum \vec{\mathbf{F}}_{\text{ext}} = M\vec{\mathbf{a}}_{\text{CM}}$$

Consequently, the moment of inertia I must play the same role in rotational motion as the role that mass plays in translational motion: the moment of inertia is the resistance to changes in rotational motion. This resistance depends not only on the mass of the object, but also on how the mass is distributed around the rotation axis. Table 10.2 gives the moments of inertia[2] for a number of objects about specific axes. The moments of inertia of rigid objects with simple geometry (high symmetry) are relatively easy to calculate provided the rotation axis coincides with an axis of symmetry, as we show in the next section.

QUICK QUIZ 10.5 You turn off your electric drill and find that the time interval for the rotating bit to come to rest due to frictional torque in the drill is Δt. You replace the bit with a larger one that results in a doubling of the moment of inertia of the drill's entire rotating mechanism. When this larger bit is rotated at the same angular speed as the first and the drill is turned off, the frictional torque remains the same as that for the previous situation. What is the time interval for this second bit to come to rest? **(a)** $4\Delta t$ **(b)** $2\Delta t$ **(c)** Δt **(d)** $0.5\Delta t$ **(e)** $0.25\Delta t$ **(f)** impossible to determine

---

[2]Civil engineers use moment of inertia to characterize the elastic properties (rigidity) of such structures as loaded beams. Hence, it is often useful even in a nonrotational context.

TABLE 10.2 Moments of Inertia of Homogeneous Rigid Objects with Different Geometries

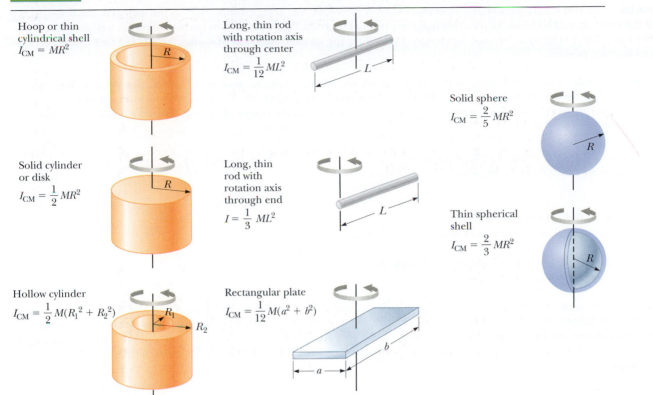

Hoop or thin cylindrical shell
$I_{CM} = MR^2$

Long, thin rod with rotation axis through center
$I_{CM} = \frac{1}{12}ML^2$

Solid sphere
$I_{CM} = \frac{2}{5}MR^2$

Solid cylinder or disk
$I_{CM} = \frac{1}{2}MR^2$

Long, thin rod with rotation axis through end
$I = \frac{1}{3}ML^2$

Thin spherical shell
$I_{CM} = \frac{2}{3}MR^2$

Hollow cylinder
$I_{CM} = \frac{1}{2}M(R_1{}^2 + R_2{}^2)$

Rectangular plate
$I_{CM} = \frac{1}{12}M(a^2 + b^2)$

ANALYSIS MODEL **Rigid Object Under a Net Torque**

Imagine you are analyzing the motion of an object that is free to rotate about a fixed axis. The cause of changes in rotational motion of this object is torque applied to the object and, in parallel to Newton's second law for translation motion, the torque is equal to the product of the moment of inertia of the object and the angular acceleration:

$$\sum \tau_{ext} = I\alpha \qquad \text{(10.18)}$$

The torque, the moment of inertia, and the angular acceleration must all be evaluated around the same rotation axis.

Examples:

- a bicycle chain around the sprocket of a bicycle causes the rear wheel of the bicycle to rotate
- an electric dipole moment in an electric field rotates due to the electric force from the field (Chapter 22)
- a magnetic dipole moment in a magnetic field rotates due to the magnetic force from the field (Chapter 28)
- the armature of a motor rotates due to the torque exerted by a surrounding magnetic field (Chapter 30)

Example **10.4** | **Rotating Rod**

A uniform rod of length L and mass M is attached at one end to a frictionless pivot and is free to rotate about the pivot in the vertical plane as in Figure 10.12. The rod is released from rest in the horizontal position. What are the initial angular acceleration of the rod and the initial translational acceleration of its right end?

SOLUTION

Conceptualize Imagine what happens to the rod in Figure 10.12 when it is released. It rotates clockwise around the pivot at the left end. When an object is pivoted at a point other than its center of mass, the gravitational force, assumed to be acting through the center of mass, provides a torque about the pivot.

Figure 10.12 (Example 10.4) A rod is free to rotate around a pivot at the left end. The gravitational force on the rod acts at its center of mass.

continued

10.4 c o n t i n u e d

Categorize   The rod is categorized as a *rigid object under a net torque*. The torque is due only to the gravitational force on the rod if the rotation axis is chosen to pass through the pivot in Figure 10.12. We *cannot* categorize the rod as a rigid object under constant angular acceleration because the torque exerted on the rod and therefore the angular acceleration of the rod vary with its angular position.

Analyze   The only force contributing to the torque about an axis through the pivot is the gravitational force $M\vec{g}$ exerted on the rod. (The force exerted by the pivot on the rod has zero torque about the pivot because its moment arm is zero.) To compute the torque on the rod, we assume the gravitational force acts at the center of mass of the rod a distance $L/2$ from the pivot as shown in Figure 10.12.

Write an expression for the magnitude of the net external torque due to the gravitational force about an axis through the pivot:

$$\sum \tau_{\text{ext}} = Mg\left(\frac{L}{2}\right)$$

Use Equation 10.18 to obtain the angular acceleration of the rod, using the moment of inertia for the rod from Table 10.2:

$$(1) \quad \alpha = \frac{\sum \tau_{\text{ext}}}{I} = \frac{Mg(L/2)}{\frac{1}{3}ML^2} = \frac{3g}{2L}$$

Use Equation 10.11 with $r = L$ to find the initial translational acceleration of the right end of the rod:

$$a_t = L\alpha = \tfrac{3}{2}g$$

Finalize   These values are the *initial* values of the angular and translational accelerations. Once the rod begins to rotate, the gravitational force is no longer perpendicular to the rod and the values of the two accelerations decrease, going to zero at the moment the rod passes through the vertical orientation. Also, because the value of a_t at a point on the rod depends on the distance of that point from the pivot, every point along the rod will have the *same* angular acceleration but a *different* tangential acceleration.

Conceptual Example 10.5   Falling Smokestacks and Tumbling Blocks

When a tall smokestack falls over, it often breaks somewhere along its length before it hits the ground as shown in Figure 10.13. Why?

S O L U T I O N

As the smokestack rotates around its base, each higher portion of the smokestack falls with a larger tangential acceleration than the portion below it according to Equation 10.11. The angular acceleration increases as the smokestack tips farther. Eventually, higher portions of the smokestack experience a tangential acceleration greater than the acceleration that could result from gravity alone; this situation is similar to that for the end of the rod in Example 10.4. That can happen only if these portions are being pulled downward by a force in addition to the gravitational force. The force that causes that to occur is the shear force from lower portions of the smokestack. Eventually, the shear force that provides this acceleration is greater than the smokestack can withstand, and the smokestack breaks. The same thing happens with a tall tower of children's toy blocks. Borrow some blocks from a child and build such a tower. Push it over and watch it come apart at some point before it strikes the floor.

Figure 10.13 (Conceptual Example 10.5) A falling smokestack breaks at some point along its length.

© Kevin Spreekmeester/AGE Fotostock

Example 10.6   Angular Acceleration of a Wheel

A wheel of radius R, mass M, and moment of inertia I is mounted on a frictionless, horizontal axle as in Figure 10.14. A light cord wrapped around the wheel supports an object of mass m. When the wheel is released, the object accelerates downward, the cord unwraps off the wheel, and the wheel rotates with an angular acceleration. Find expressions for the angular acceleration of the wheel, the translational acceleration of the object, and the tension in the cord.

S O L U T I O N

Conceptualize   Imagine that the object is a bucket in an old-fashioned water well. It is tied to a cord that passes around a cylinder equipped with a crank for raising the bucket. After the bucket has been raised, the system is released and the bucket accelerates downward while the cord unwinds off the cylinder.

10.6 continued

Categorize We apply two analysis models here. The object is modeled as a *particle under a net force*. The wheel is modeled as a *rigid object under a net torque*.

Analyze The magnitude of the torque acting on the wheel about its axis of rotation is $\tau = TR$, where T is the force exerted by the cord on the rim of the wheel. (The gravitational force exerted by the Earth on the wheel and the normal force exerted by the axle on the wheel both pass through the axis of rotation and therefore produce no torque about the axle.)

From the rigid object under a net torque model, write Equation 10.18:

$$\sum \tau_{ext} = I\alpha$$

Solve for α and substitute the net torque:

$$(1) \quad \alpha = \frac{\sum \tau_{ext}}{I} = \frac{TR}{I}$$

From the particle under a net force model, apply Newton's second law to the motion of the object, taking the downward direction to be positive:

$$\sum F_y = mg - T = ma$$

Solve for the acceleration a:

$$(2) \quad a = \frac{mg - T}{m}$$

Equations (1) and (2) have three unknowns: α, a, and T. Because the object and wheel are connected by a cord that does not slip, the translational acceleration of the suspended object is equal to the tangential acceleration of a point on the wheel's rim. Therefore, the angular acceleration α of the wheel and the translational acceleration of the object are related by $a = R\alpha$ (Eq. 10.11).

Use this fact together with Equations (1) and (2):

$$(3) \quad a = R\alpha = \frac{TR^2}{I} = \frac{mg - T}{m}$$

Solve for the tension T:

$$(4) \quad T = \frac{mg}{1 + (mR^2/I)}$$

Substitute Equation (4) into Equation (2) and solve for a:

$$(5) \quad a = \frac{g}{1 + (I/mR^2)}$$

Use $a = R\alpha$ and Equation (5) to solve for α:

$$\alpha = \frac{a}{R} = \frac{g}{R + (I/mR)}$$

Figure 10.14 (Example 10.6) An object hangs from a cord wrapped around a wheel.

Finalize We finalize this problem by imagining the behavior of the system in some extreme limits.

WHAT IF? What if the wheel were to become very massive so that I becomes very large? What happens to the acceleration a of the object and the tension T?

Answer If the wheel becomes infinitely massive, we can imagine that the object of mass m will simply hang from the cord without causing the wheel to rotate.

We can show that mathematically by taking the limit $I \rightarrow \infty$. Equation (5) then becomes

$$a = \lim_{I \rightarrow \infty} \frac{g}{1 + (I/mR^2)} = 0$$

which agrees with our conceptual conclusion that the object will hang at rest. Also, Equation (4) becomes

$$T = \lim_{I \rightarrow \infty} \frac{mg}{1 + (mR^2/I)} = mg$$

which is consistent because the object simply hangs at rest in equilibrium between the gravitational force and the tension in the string.

10.6 Calculation of Moments of Inertia

The moment of inertia of a system of discrete particles can be calculated in a straightforward way with Equation 10.19. On the other hand, suppose we consider a continuous rigid object. We can evaluate its moment of inertia by imagining the object to be divided into many small elements, each of which has mass Δm_i. We use the definition $I = \sum_i r_i^2\,\Delta m_i$ and take the limit of this sum as $\Delta m_i \to 0$. In this limit, the sum becomes an integral over the volume of the object:

◄ Moment of inertia of a rigid object

$$I = \lim_{\Delta m_i \to 0} \sum_i r_i^2\,\Delta m_i = \int r^2\,dm \qquad (10.20)$$

It is usually easier to calculate moments of inertia in terms of the volume of the elements rather than their mass, and we can easily make that change by using Equation 1.1, $\rho \equiv m/V$, where ρ is the density of the object and V is its volume. From this equation, the mass of a small element is $dm = \rho\,dV$. Substituting this result into Equation 10.20 gives

$$I = \int \rho r^2\,dV \qquad (10.21)$$

If the object is homogeneous, ρ is constant and the integral can be evaluated for a known geometry. If ρ is not constant, its variation with position must be known to complete the integration.

The density given by $\rho = m/V$ sometimes is referred to as *volumetric mass density* because it represents mass per unit volume. Often we use other ways of expressing density. For instance, when dealing with a sheet of uniform thickness t, we can define a *surface mass density* $\sigma = m/A = \rho t$, which represents *mass per unit area*. Finally, when mass is distributed along a rod of uniform cross-sectional area A, we sometimes use *linear mass density* $\lambda = m/L = \rho A$, which is the *mass per unit length*.

Example **10.7**   Uniform Rigid Rod

Calculate the moment of inertia of a uniform thin rod of length L and mass M (Fig. 10.15) about an axis perpendicular to the rod (the y axis) and passing through its center of mass.

SOLUTION

Conceptualize   Imagine twirling the rod in Figure 10.15 with your fingers around its midpoint. If you have a meterstick handy, use it to simulate the spinning of a thin rod and feel the resistance it offers to being spun.

Figure 10.15   (Example 10.7) A uniform rigid rod of length L. The moment of inertia about the y axis is less than that about the y' axis. The latter axis is examined in Example 10.9.

Categorize   This example is a substitution problem, using the definition of moment of inertia in Equation 10.20. As with any integration problem, the solution involves reducing the integrand to a single variable.

The shaded length element dx in Figure 10.15 has a mass dm equal to the mass per unit length λ multiplied by dx.

Express dm in terms of dx:

$$dm = \lambda\,dx = \frac{M}{L}\,dx$$

Substitute this expression into Equation 10.20, with $r^2 = x^2$:

$$I_y = \int r^2\,dm = \int_{-L/2}^{L/2} x^2\,\frac{M}{L}\,dx = \frac{M}{L}\int_{-L/2}^{L/2} x^2\,dx$$

$$= \frac{M}{L}\left[\frac{x^3}{3}\right]_{-L/2}^{L/2} = \tfrac{1}{12}ML^2$$

Check this result in Table 10.2.

Example **10.8**   Uniform Solid Cylinder

A uniform solid cylinder has a radius R, mass M, and length L. Calculate its moment of inertia about its central axis (the z axis in Fig. 10.16).

10.8 continued

SOLUTION

Conceptualize To simulate this situation, imagine twirling a can of frozen juice around its central axis. Don't twirl a nonfrozen can of vegetable soup; it is not a rigid object! The liquid is able to move relative to the metal can.

Categorize This example is a substitution problem, using the definition of moment of inertia. As with Example 10.7, we must reduce the integrand to a single variable.

It is convenient to divide the cylinder into many cylindrical shells, each having radius r, thickness dr, and length L as shown in Figure 10.16. The density of the cylinder is ρ. The volume dV of each shell is its cross-sectional area multiplied by its length: $dV = L\, dA = L(2\pi r)\, dr$.

Figure 10.16 (Example 10.8) Calculating I about the z axis for a uniform solid cylinder.

Express dm in terms of dr:

$$dm = \rho\, dV = \rho L(2\pi r)\, dr$$

Substitute this expression into Equation 10.20:

$$I_z = \int r^2\, dm = \int r^2[\rho L(2\pi r)\, dr] = 2\pi\rho L \int_0^R r^3\, dr = \tfrac{1}{2}\pi\rho L R^4$$

Use the total volume $\pi R^2 L$ of the cylinder to express its density:

$$\rho = \frac{M}{V} = \frac{M}{\pi R^2 L}$$

Substitute this value into the expression for I_z:

$$I_z = \tfrac{1}{2}\pi\left(\frac{M}{\pi R^2 L}\right)LR^4 = \tfrac{1}{2}MR^2$$

Check this result in Table 10.2.

WHAT IF?   What if the length of the cylinder in Figure 10.16 is increased to $2L$, while the mass M and radius R are held fixed? (The density becomes half as large.) How does that change the moment of inertia of the cylinder?

Answer Notice that the result for the moment of inertia of a cylinder does not depend on L, the length of the cylinder. It applies equally well to a long cylinder and a flat disk having the *same* mass M and radius R. Therefore, the moment of inertia of the cylinder around the central axis is not affected by how the mass is distributed along its length.

The calculation of moments of inertia of an object about an arbitrary axis can be cumbersome, even for a highly symmetric object. For example, imagine trying to find the moment of inertia of the cylinder in Figure 10.16 around an axis parallel to the z axis, but offset by the radius R of the cylinder, so that the axis just grazes along the outer surface of the cylinder. There is no symmetry around this axis! Fortunately, use of an important theorem, called the **parallel-axis theorem**, often simplifies the calculation.

To generate the parallel-axis theorem, suppose the object in Figure 10.17a (page 266) rotates about the z' axis. The moment of inertia does not depend on how the mass is distributed along the z' axis; as we found in Example 10.8, for example, the moment of inertia of a cylinder is independent of its length. Imagine collapsing the three-dimensional object in Figure 10.17a into a planar object of the same mass as in Figure 10.17b. In this imaginary process, all mass moves parallel to the z' axis until it lies in the $x'y'$ plane. The coordinates of the object's center of mass are now x'_{CM}, y'_{CM}, and $z'_{CM} = 0$. Let the mass element dm have coordinates $(x', y', 0)$ as shown in the view down the z' axis in Figure 10.17c. Because this element is a distance $r' = \sqrt{(x')^2 + (y')^2}$ from the z' axis, the moment of inertia of the entire object about the z' axis is

$$I = \int (r')^2\, dm = \int [(x')^2 + (y')^2]\, dm$$

We can relate the coordinates x', y' of the mass element dm to the coordinates of this same element located in a coordinate system having the object's center

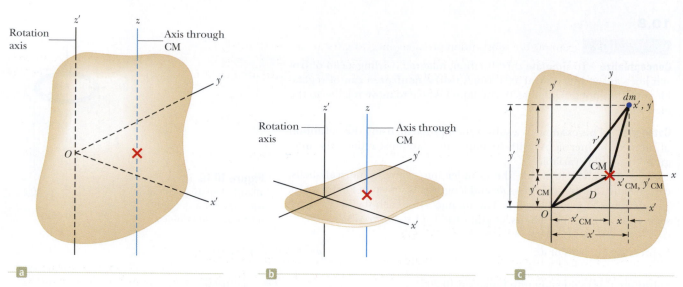

Figure 10.17 (a) An arbitrarily shaped rigid object. The origin of the coordinate system is not at the center of mass of the object. Imagine the object rotating about the z' axis. (b) All mass elements of the object are collapsed parallel to the z' axis to form a planar object. (c) An arbitrary mass element dm is indicated in blue in this view down the z' axis. The parallel axis theorem can be used with the geometry shown to determine the moment of inertia of the original object around the z' axis.

of mass as its origin. If the coordinates of the center of mass are x'_{CM}, and y'_{CM} in the original coordinate system centered on O, we see from Figure 10.17c that the relationships between the unprimed and primed coordinates are $x' = x + x'_{CM}$, $y' = y + y'_{CM}$, and $z' = z = 0$. Therefore,

$$I = \int [(x + x'_{CM})^2 + (y + y'_{CM})^2] \, dm$$

$$= \int (x^2 + y)^2 \, dm + 2x'_{CM} \int x \, dm + 2y'_{CM} \int y \, dm + (x'^2_{CM} + y'^2_{CM}) \int dm$$

The first integral is, by definition, the moment of inertia I_{CM} about an axis that is parallel to the z' axis and passes through the center of mass. The second two integrals are zero because, by definition of the center of mass, $\int x \, dm = \int y \, dm = 0$. The last integral is simply MD^2 because $\int dm = M$ and $D^2 = x'^2_{CM} + y'^2_{CM}$. Therefore, we conclude that

Parallel-axis theorem ▶

$$I = I_{CM} + MD^2 \tag{10.22}$$

The parallel axis theorem allows us to evaluate the moment of inertia of an object of mass M about any axis that is parallel to its central axis as the moment of inertia around the central axis plus the term MD^2, where D is the perpendicular distance between the axes.

Example **10.9** **Applying the Parallel-Axis Theorem**

Consider once again the uniform rigid rod of mass M and length L shown in Figure 10.15. Find the moment of inertia of the rod about an axis perpendicular to the rod through one end (the y' axis in Fig. 10.15).

S O L U T I O N

Conceptualize Imagine twirling the rod around an endpoint rather than the midpoint. If you have a meterstick handy, try it and notice the degree of difficulty in rotating it around the end compared with rotating it around the center.

10.9 continued

Categorize This example is a substitution problem, involving the parallel-axis theorem.

Intuitively, we expect the moment of inertia to be greater than the result $I_{CM} = \frac{1}{12}ML^2$ from Example 10.7 because there is mass up to a distance of L away from the rotation axis, whereas the farthest distance in Example 10.7 was only $L/2$. The distance between the center-of-mass axis and the y' axis is $D = L/2$.

Use the parallel-axis theorem:
$$I = I_{CM} + MD^2 = \tfrac{1}{12}ML^2 + M\left(\frac{L}{2}\right)^2 = \tfrac{1}{3}ML^2$$

Check this result in Table 10.2. As predicted in Example 10.7, it is more difficult to rotate the rod about one end than about the center of mass.

10.7 Rotational Kinetic Energy

After investigating the role of forces in our study of translational motion, we turned our attention to approaches involving energy in Chapters 7 and 8. We do the same thing in our current study of rotational motion.

In Chapter 7, we defined the kinetic energy of an object as the energy associated with its motion through space. An object rotating about a fixed axis remains stationary in space, so there is no kinetic energy associated with translational motion. The individual particles making up the rotating object, however, are moving through space; they follow circular paths. Consequently, there is kinetic energy associated with rotational motion.

Let us consider an object as a system of particles and assume it rotates about a fixed z axis with an angular speed ω. Figure 10.18 shows the rotating object and identifies one particle on the object located at a distance r_i from the rotation axis. If the mass of the ith particle is m_i and its tangential speed is v_i, its kinetic energy is

$$K_i = \tfrac{1}{2}m_i v_i^2$$

To proceed further, recall that although every particle in the rigid object has the same angular speed ω, the individual tangential speeds depend on the distance r_i from the axis of rotation according to Equation 10.10. The *total* kinetic energy of the rotating rigid object is the sum of the kinetic energies of the individual particles:

$$K_R = \sum_i K_i = \sum_i \tfrac{1}{2}m_i v_i^2 = \tfrac{1}{2}\sum_i m_i r_i^2 \omega^2$$

We can write this expression in the form

$$K_R = \tfrac{1}{2}\left(\sum_i m_i r_i^2\right)\omega^2 \qquad (10.23)$$

where we have factored ω^2 from the sum because it is common to every particle. We recognize the quantity in parentheses as the moment of inertia of the object, introduced in Section 10.5.

Therefore, Equation 10.23 can be written

$$K_R = \tfrac{1}{2}I\omega^2 \qquad (10.24)$$

◀ Rotational kinetic energy (Compare to Equation 7.16)

Compare Equation 10.24 to Equation 7.16 for the kinetic energy of an object in translational motion. Again, as in the discussion following Equation 10.19, we see that moment of inertia I plays the same role in rotational motion as mass m does in translational motion. Although we commonly refer to the quantity $\frac{1}{2}I\omega^2$ as **rotational kinetic energy,** it is not a new form of energy. It is ordinary kinetic energy because it is derived from a sum over individual kinetic energies of the particles contained in the rigid object. The mathematical form of the kinetic energy given by Equation 10.24 is convenient when we are dealing with rotational motion, provided we know how to calculate I.

The particle of mass m_i of the rigid object has the same kinetic energy as if it were moving through space with the same speed.

Figure 10.18 A rigid object rotating about the z axis with angular speed ω. The kinetic energy of the particle of mass m_i is $\frac{1}{2}m_i v_i^2$. The total kinetic energy of the object is called its rotational kinetic energy.

QUICK QUIZ **10.6** A section of hollow pipe and a solid cylinder have the same radius, mass, and length. They both rotate about their long central axes with the same angular speed. Which object has the higher rotational kinetic energy? **(a)** The hollow pipe does. **(b)** The solid cylinder does. **(c)** They have the same rotational kinetic energy. **(d)** It is impossible to determine.

Example 10.10 | **An Unusual Baton**

Four tiny spheres are fastened to the ends of two rods of negligible mass lying in the *xy* plane to form an unusual baton (Fig. 10.19). We shall assume the radii of the spheres are small compared with the dimensions of the rods.

(A) If the system rotates about the *y* axis (Fig. 10.19a) with an angular speed ω, find the moment of inertia and the rotational kinetic energy of the system about this axis.

SOLUTION

Conceptualize Figure 10.19 is a pictorial representation that helps conceptualize the system of spheres and how it spins. Model the spheres as particles. Notice that only the blue spheres contribute to the moment of inertia around the *y* axis.

Categorize This example is a substitution problem because it is a straightforward application of the definitions discussed in this section.

Figure 10.19 (Example 10.10) Four spheres form an unusual baton. (a) The baton is rotated about the *y* axis. (b) The baton is rotated about the *z* axis.

Apply Equation 10.19 to the system:

$$I_y = \sum_i m_i r_i^2 = Ma^2 + Ma^2 = \boxed{2Ma^2}$$

Evaluate the rotational kinetic energy using Equation 10.24:

$$K_R = \tfrac{1}{2}I_y\omega^2 = \tfrac{1}{2}(2Ma^2)\omega^2 = \boxed{Ma^2\omega^2}$$

That the two spheres of mass m do not enter into this result makes sense because they have no motion about the axis of rotation; hence, they have no rotational kinetic energy. By similar logic, we expect the moment of inertia about the x axis to be $I_x = 2mb^2$ with a rotational kinetic energy about that axis of $K_R = mb^2\omega^2$.

(B) Suppose the system rotates in the *xy* plane about an axis (the *z* axis) through the center of the baton (Fig. 10.19b). Calculate the moment of inertia and rotational kinetic energy about this axis.

SOLUTION

Apply Equation 10.19 for this new rotation axis:

$$I_z = \sum_i m_i r_i^2 = Ma^2 + Ma^2 + mb^2 + mb^2 = \boxed{2Ma^2 + 2mb^2}$$

Evaluate the rotational kinetic energy using Equation 10.24:

$$K_R = \tfrac{1}{2}I_z\omega^2 = \tfrac{1}{2}(2Ma^2 + 2mb^2)\omega^2 = \boxed{(Ma^2 + mb^2)\omega^2}$$

Comparing the results for parts (A) and (B), we conclude that the moment of inertia and therefore the rotational kinetic energy associated with a given angular speed depend on the axis of rotation. In part (B), we expect the result to include all four spheres and distances because all four spheres are rotating in the *xy* plane. Based on the work–kinetic energy theorem, the smaller rotational kinetic energy in part (A) than in part (B) indicates it would require less work to set the system into rotation about the *y* axis than about the *z* axis.

WHAT IF? What if the mass *M* is much larger than *m*? How do the answers to parts (A) and (B) compare?

Answer If $M \gg m$, then m can be neglected and the moment of inertia and the rotational kinetic energy in part (B) become

$$I_z = 2Ma^2 \quad \text{and} \quad K_R = Ma^2\omega^2$$

which are the same as the answers in part (A). If the masses *m* of the two tan spheres in Figure 10.19 are negligible, these spheres can be removed from the figure and rotations about the *y* and *z* axes are equivalent.

10.8 Energy Considerations in Rotational Motion

Having introduced rotational kinetic energy in Section 10.7, let us now see how an energy approach can be useful in solving rotational problems. We begin by considering the relationship between the torque acting on a rigid object and its resulting rotational motion so as to generate expressions for power and a rotational analog to the work–kinetic energy theorem. Consider the rigid object pivoted at O in Figure 10.20. Suppose a single external force $\vec{F}$ is applied at P, where $\vec{F}$ lies in the plane of the page. The work done on the object by $\vec{F}$ as its point of application rotates through an infinitesimal distance $ds = r\,d\theta$ is

$$dW = \vec{F} \cdot d\vec{s} = (F\sin\phi)r\,d\theta$$

where $F\sin\phi$ is the tangential component of $\vec{F}$, or, in other words, the component of the force along the displacement. Notice that the radial component vector of $\vec{F}$ does no work on the object because it is perpendicular to the displacement of the point of application of $\vec{F}$.

Because the magnitude of the torque due to $\vec{F}$ about an axis through O is defined as $rF\sin\phi$ by Equation 10.14, we can write the work done for the infinitesimal rotation as

$$dW = \tau\,d\theta \qquad (10.25)$$

The rate at which work is being done by $\vec{F}$ as the object rotates about the fixed axis through the angle $d\theta$ in a time interval dt is

$$\frac{dW}{dt} = \tau\frac{d\theta}{dt}$$

Because dW/dt is the instantaneous power P (see Section 8.5) delivered by the force and $d\theta/dt = \omega$, this expression reduces to

$$P = \frac{dW}{dt} = \tau\omega \qquad (10.26)$$

◀ Power delivered to a rotating rigid object

This equation is analogous to $P = Fv$ (Eq. 8.18) in the case of translational motion, and Equation 10.25 is analogous to $dW = F_x\,dx$.

In studying translational motion, we have seen that models based on an energy approach can be extremely useful in describing a system's behavior. From what we learned of translational motion, we expect that when a symmetric object rotates frictionlessly about a fixed axis, the work done by external forces equals the change in the rotational energy of the object.

To prove that fact, let us begin with the rigid object under a net torque model, whose mathematical representation is $\sum \tau_{\text{ext}} = I\alpha$. Using the chain rule from calculus, we can express the net torque as

$$\sum \tau_{\text{ext}} = I\alpha = I\frac{d\omega}{dt} = I\frac{d\omega}{d\theta}\frac{d\theta}{dt} = I\frac{d\omega}{d\theta}\omega$$

Rearranging this expression and noting that $\sum \tau_{\text{ext}}\,d\theta = dW$ from Equation 10.25 gives

$$\sum \tau_{\text{ext}}\,d\theta = dW = I\omega\,d\omega$$

Integrating this expression, we obtain for the work W done by the net external force acting on a rotating system

$$W = \int_{\omega_i}^{\omega_f} I\omega\,d\omega = \tfrac{1}{2}I\omega_f^2 - \tfrac{1}{2}I\omega_i^2 \qquad (10.27)$$

◀ Work–kinetic energy theorem for rotational motion

where the angular speed changes from ω_i to ω_f. Equation 10.27 is the **work–kinetic energy theorem for rotational motion.** Similar to the work–kinetic energy theorem in translational motion (Section 7.5), Equation 10.27 states that the net work done by external forces in rotating a symmetric rigid object about a fixed friction-free axis equals the change in the object's rotational energy.

Figure 10.20 A rigid object rotates about an axis through O under the action of an external force $\vec{F}$ applied at P.

TABLE 10.3 | Useful Equations in Rotational and Translational Motion

Rotational Motion About a Fixed Axis	Translational Motion
Angular speed $\omega = d\theta/dt$	Translational speed $v = dx/dt$
Angular acceleration $\alpha = d\omega/dt$	Translational acceleration $a = dv/dt$
Net torque $\Sigma\tau_{ext} = I\alpha$	Net force $\Sigma F = ma$
If $\alpha = $ constant $\begin{cases} \omega_f = \omega_i + \alpha t \\ \theta_f = \theta_i + \omega_i t + \frac{1}{2}\alpha t^2 \\ \omega_f^2 = \omega_i^2 + 2\alpha(\theta_f - \theta_i) \end{cases}$	If $a = $ constant $\begin{cases} v_f = v_i + at \\ x_f = x_i + v_i t + \frac{1}{2}at^2 \\ v_f^2 = v_i^2 + 2a(x_f - x_i) \end{cases}$
Work $W = \displaystyle\int_{\theta_i}^{\theta_f} \tau \, d\theta$	Work $W = \displaystyle\int_{x_i}^{x_f} F_x \, dx$
Rotational kinetic energy $K_R = \frac{1}{2}I\omega^2$	Kinetic energy $K = \frac{1}{2}mv^2$
Power $P = \tau\omega$	Power $P = Fv$
Angular momentum $L = I\omega$	Linear momentum $p = mv$
Net torque $\Sigma\tau = dL/dt$	Net force $\Sigma F = dp/dt$

Equation 10.27 is a form of the nonisolated system (energy) model discussed in Chapter 8. Work is done on the system of the rigid object, which represents a transfer of energy across the boundary of the system that appears as an increase in the object's rotational kinetic energy.

In general, we can combine Equation 10.27 with the translational form of the work–kinetic energy theorem from Chapter 7. Therefore, the net work done by external forces on an object is the change in its *total* kinetic energy, which is the sum of the translational and rotational kinetic energies. For example, when a pitcher throws a baseball, the work done by the pitcher's hands appears as kinetic energy associated with the ball moving through space as well as rotational kinetic energy associated with the spinning of the ball.

In addition to the work–kinetic energy theorem, other energy principles can also be applied to rotational situations. For example, if a system involving rotating objects is isolated and no nonconservative forces act within the system, the isolated system model and the principle of conservation of mechanical energy can be used to analyze the system as in Example 10.11 below. In general, Equation 8.2, the conservation of energy equation, applies to rotational situations, with the recognition that the change in kinetic energy ΔK will include changes in both translational and rotational kinetic energies.

Finally, in some situations an energy approach does not provide enough information to solve the problem and it must be combined with a momentum approach. Such a case is illustrated in Example 10.14 in Section 10.9.

Table 10.3 lists the various equations we have discussed pertaining to rotational motion together with the analogous expressions for translational motion. Notice the similar mathematical forms of the equations. The last two equations in the left-hand column of Table 10.3, involving angular momentum L, are discussed in Chapter 11 and are included here only for the sake of completeness.

Example **10.11**   **Rotating Rod Revisited**

A uniform rod of length L and mass M is free to rotate on a frictionless pin passing through one end (Fig 10.21). The rod is released from rest in the horizontal position.

(A) What is its angular speed when the rod reaches its lowest position?

S O L U T I O N

Conceptualize   Consider Figure 10.21 and imagine the rod rotating downward through a quarter turn about the pivot at the left end. Also look back at Example 10.4. This physical situation is the same.

10.11 continued

Categorize As mentioned in Example 10.4, the angular acceleration of the rod is not constant. Therefore, the kinematic equations for rotation (Section 10.2) cannot be used to solve this example. We categorize the system of the rod and the Earth as an *isolated system* in terms of *energy* with no nonconservative forces acting and use the principle of conservation of mechanical energy.

Figure 10.21 (Example 10.11) A uniform rigid rod pivoted at O rotates in a vertical plane under the action of the gravitational force.

Analyze We choose the configuration in which the rod is hanging straight down as the reference configuration for gravitational potential energy and assign a value of zero for this configuration. When the rod is in the horizontal position, it has no rotational kinetic energy. The potential energy of the system in this configuration relative to the reference configuration is $MgL/2$ because the center of mass of the rod is at a height $L/2$ higher than its position in the reference configuration. When the rod reaches its lowest position, the energy of the system is entirely rotational energy $\frac{1}{2}I\omega^2$, where I is the moment of inertia of the rod about an axis passing through the pivot.

Using the isolated system (energy) model, write an appropriate reduction of Equation 8.2:

$$\Delta K + \Delta U = 0$$

Substitute for each of the final and initial energies:

$$(\tfrac{1}{2}I\omega^2 - 0) + (0 - \tfrac{1}{2}MgL) = 0$$

Solve for ω and use $I = \frac{1}{3}ML^2$ (see Table 10.2) for the rod:

$$\omega = \sqrt{\frac{MgL}{I}} = \sqrt{\frac{MgL}{\frac{1}{3}ML^2}} = \boxed{\sqrt{\frac{3g}{L}}}$$

(B) Determine the tangential speed of the center of mass and the tangential speed of the lowest point on the rod when it is in the vertical position.

SOLUTION

Use Equation 10.10 and the result from part (A):

$$v_{CM} = r\omega = \frac{L}{2}\omega = \boxed{\tfrac{1}{2}\sqrt{3gL}}$$

Because r for the lowest point on the rod is twice what it is for the center of mass, the lowest point has a tangential speed twice that of the center of mass:

$$v = 2v_{CM} = \boxed{\sqrt{3gL}}$$

Finalize The initial configuration in this example is the same as that in Example 10.4. In Example 10.4, however, we could only find the initial angular acceleration of the rod. Applying an energy approach in the current example allows us to find additional information, the angular speed of the rod at the lowest point. Convince yourself that you could find the angular speed of the rod at any angular position by knowing the location of the center of mass at this position.

WHAT IF? What if we want to find the angular speed of the rod when the angle it makes with the horizontal is 45.0°? Because this angle is half of 90.0°, for which we solved the problem above, is the angular speed at this configuration half the answer in the calculation above, that is, $\frac{1}{2}\sqrt{3g/L}$?

Answer Imagine the rod in Figure 10.21 at the 45.0° position. Use a pencil or a ruler to represent the rod at this position. Notice that the center of mass has dropped through more than half of the distance $L/2$ in this configuration. Therefore, more than half of the initial gravitational potential energy has been transformed to rotational kinetic energy. So, we should not expect the value of the angular speed to be as simple as proposed above.

Note that the center of mass of the rod drops through a distance of $0.500L$ as the rod reaches the vertical configuration. When the rod is at 45.0° to the horizontal, we can show that the center of mass of the rod drops through a distance of $0.354L$. Continuing the calculation, we find that the angular speed of the rod at this configuration is $0.841\sqrt{3g/L}$, (not $\frac{1}{2}\sqrt{3g/L}$).

Example **10.12** **Energy and the Atwood Machine**

Two blocks having different masses m_1 and m_2 are connected by a string passing over a pulley as shown in Figure 10.22 (page 272). The pulley has a radius R and moment of inertia I about its axis of rotation. The string does not slip on the pulley, and the system is released from rest. Find the translational speeds of the blocks after block 2 descends through a distance h and find the angular speed of the pulley at this time.

continued

10.12 continued

Conceptualize   We have already seen the Atwood machine in Example 5.9, so the motion of the objects in Figure 10.22 should be easy to visualize.

Categorize   Because the string does not slip, the pulley rotates about the axle. We can neglect friction in the axle because the axle's radius is small relative to that of the pulley. Hence, the frictional torque is much smaller than the net torque applied by the two blocks provided that their masses are significantly different. Consequently, the system consisting of the two blocks, the pulley, and the Earth is an *isolated system* in terms of *energy* with no nonconservative forces acting.

Figure 10.22 (Example 10.12) An Atwood machine with a massive pulley.

Analyze   We define the zero configuration for gravitational potential energy as that which exists when the system is released. From Figure 10.22, we see that the descent of block 2 is associated with a decrease in system potential energy and that the rise of block 1 represents an increase in potential energy.

Using the isolated system (energy) model, write an appropriate reduction of the conservation of energy equation:

$$\Delta K + \Delta U = 0$$

Substitute for each of the energies:

$$[(\tfrac{1}{2}m_1 v_f^2 + \tfrac{1}{2}m_2 v_f^2 + \tfrac{1}{2}I\omega_f^2) - 0] + [(m_1 gh - m_2 gh) - 0] = 0$$

The two blocks, the string, and the outer rim of the pulley all move at the same speed. Therefore, use $v_f = R\omega_f$ to substitute for ω_f:

$$\tfrac{1}{2}m_1 v_f^2 + \tfrac{1}{2}m_2 v_f^2 + \tfrac{1}{2}I\frac{v_f^2}{R^2} = m_2 gh - m_1 gh$$

$$\tfrac{1}{2}\left(m_1 + m_2 + \frac{I}{R^2}\right)v_f^2 = (m_2 - m_1)gh$$

Solve for v_f:

$$(1) \quad v_f = \left[\frac{2(m_2 - m_1)gh}{m_1 + m_2 + I/R^2}\right]^{1/2}$$

Use $v_f = R\omega_f$ to solve for ω_f:

$$\omega_f = \frac{v_f}{R} = \frac{1}{R}\left[\frac{2(m_2 - m_1)gh}{m_1 + m_2 + I/R^2}\right]^{1/2}$$

Finalize   Each block can be modeled as a *particle under constant acceleration* because it experiences a constant net force. Think about what you would need to do to use Equation (1) to find the acceleration of one of the blocks. Then imagine the pulley becoming massless and determine the acceleration of a block. How does this result compare with the result of Example 5.9?

10.9 Rolling Motion of a Rigid Object

In this section, we treat the motion of a rigid object rolling along a flat surface. In general, such motion is complex. For example, suppose a cylinder is rolling on a straight path such that the axis of rotation remains parallel to its initial orientation in space. As Figure 10.23 shows, a point on the rim of the cylinder moves in a complex path called a *cycloid*. We can simplify matters, however, by focusing on the center of mass rather than on a point on the rim of the rolling object. As shown in

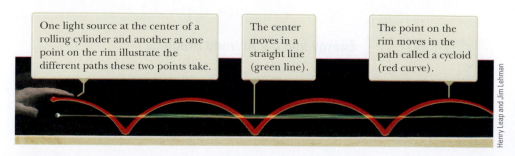

One light source at the center of a rolling cylinder and another at one point on the rim illustrate the different paths these two points take.

The center moves in a straight line (green line).

The point on the rim moves in the path called a cycloid (red curve).

Figure 10.23 Two points on a rolling object take different paths through space.

Henry Leap and Jim Lehman

Figure 10.23, the center of mass moves in *translational* motion in a straight line. If an object such as a cylinder rolls without slipping on the surface (called *pure rolling motion*), a simple relationship exists between its rotational and translational motions.

Consider a uniform cylinder of radius R rolling without slipping on a horizontal surface (Fig. 10.24). As the cylinder rotates through an angle θ, its center of mass moves a linear distance $s = R\theta$ (see Eq. 10.1b). Therefore, the translational speed of the center of mass for pure rolling motion is given by

$$v_{CM} = \frac{ds}{dt} = R\frac{d\theta}{dt} = R\omega \tag{10.28}$$

where ω is the angular speed of the cylinder. Equation 10.28 holds whenever a cylinder or sphere rolls without slipping and is the **condition for pure rolling motion.** The magnitude of the linear acceleration of the center of mass for pure rolling motion is

$$a_{CM} = \frac{dv_{CM}}{dt} = R\frac{d\omega}{dt} = R\alpha \tag{10.29}$$

where α is the angular acceleration of the cylinder.

Imagine that you are moving along with a rolling object at speed v_{CM}, staying in a frame of reference at rest with respect to the center of mass of the object. As you observe the object, you will see the object in pure rotation around its center of mass. Figure 10.25a shows the velocities of points at the top, center, and bottom of the object as observed by you. In addition to these velocities, every point on the object moves in the same direction with speed v_{CM} relative to the surface on which it rolls. Figure 10.25b shows these velocities for a nonrotating object. In the reference frame at rest with respect to the surface, the velocity of a given point on the object is the sum of the velocities shown in Figures 10.25a and 10.25b. Figure 10.25c shows the results of adding these velocities.

Notice that the contact point between the surface and object in Figure 10.25c has a translational speed of zero. At this instant, the rolling object is moving in exactly the same way as if the surface were removed and the object were pivoted at point P and spun about an axis passing through P. We can express the total kinetic energy of this imagined spinning object as

$$K = \tfrac{1}{2}I_P\omega^2 \tag{10.30}$$

where I_P is the moment of inertia about a rotation axis through P.

Because the motion of the imagined spinning object is the same at this instant as our actual rolling object, Equation 10.30 also gives the kinetic energy of the rolling object. Applying the parallel-axis theorem, we can substitute $I_P = I_{CM} + MR^2$ into Equation 10.30 to obtain

$$K = \tfrac{1}{2}I_{CM}\omega^2 + \tfrac{1}{2}MR^2\omega^2$$

Figure 10.24 For pure rolling motion, as the cylinder rotates through an angle θ its center moves a linear distance $s = R\theta$.

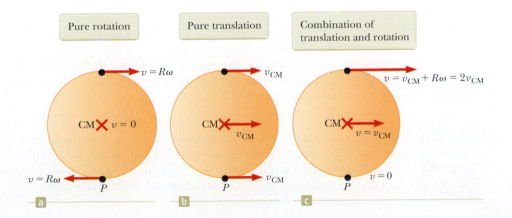

Figure 10.25 The motion of a rolling object can be modeled as a combination of pure translation and pure rotation. Equation 10.28 tells us that $v_{CM} = R\omega$.

Using $v_{CM} = R\omega$, this equation can be expressed as

◀ Total kinetic energy
of a rolling object

$$K = \tfrac{1}{2}I_{CM}\omega^2 + \tfrac{1}{2}Mv_{CM}^2 \qquad (10.31)$$

The term $\tfrac{1}{2}I_{CM}\omega^2$ represents the rotational kinetic energy of the object about its center of mass, and the term $\tfrac{1}{2}Mv_{CM}^2$ represents the kinetic energy the object would have if it were just translating through space without rotating. Therefore, the total kinetic energy of a rolling object is the sum of the rotational kinetic energy *about* the center of mass and the translational kinetic energy *of* the center of mass. This statement is consistent with the situation illustrated in Figure 10.25, which shows that the velocity of a point on the object is the sum of the velocity of the center of mass and the tangential velocity around the center of mass.

Energy methods can be used to treat a class of problems concerning the rolling motion of an object on a rough incline. For example, consider Figure 10.26, which shows a sphere rolling without slipping after being released from rest at the top of the incline. Accelerated rolling motion is possible only if a friction force is present between the sphere and the incline to produce a net torque about the center of mass. Despite the presence of friction, no loss of mechanical energy occurs because the contact point is at rest relative to the surface at any instant. (On the other hand, if the sphere were to slip, mechanical energy of the sphere–incline–Earth system would decrease due to the nonconservative force of kinetic friction.)

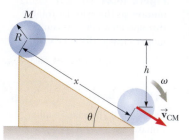

Figure 10.26 A sphere rolling down an incline. Mechanical energy of the sphere–Earth system is conserved if no slipping occurs.

In reality, *rolling friction* causes mechanical energy to transform to internal energy. Rolling friction is due to deformations of the surface and the rolling object. For example, automobile tires flex as they roll on a roadway, representing a transformation of mechanical energy to internal energy. The roadway also deforms a small amount, representing additional rolling friction. In our problem-solving models, we ignore rolling friction unless stated otherwise.

Using $v_{CM} = R\omega$ for pure rolling motion, we can express Equation 10.31 as

$$K = \tfrac{1}{2}I_{CM}\left(\frac{v_{CM}}{R}\right)^2 + \tfrac{1}{2}Mv_{CM}^2$$

$$K = \tfrac{1}{2}\left(\frac{I_{CM}}{R^2} + M\right)v_{CM}^2 \qquad (10.32)$$

For the sphere–Earth system in Figure 10.26, we define the zero configuration of gravitational potential energy to be when the sphere is at the bottom of the incline. Therefore, Equation 8.2 gives

$$\Delta K + \Delta U = 0$$

$$\left[\tfrac{1}{2}\left(\frac{I_{CM}}{R^2} + M\right)v_{CM}^2 - 0\right] + (0 - Mgh) = 0$$

$$v_{CM} = \left[\frac{2gh}{1 + (I_{CM}/MR^2)}\right]^{1/2} \qquad (10.33)$$

While this calculation was performed for the sphere in Figure 10.26, Equation 10.33 is general enough that it provides the speed of *any* object with a circular cross section that rolls from rest down an incline of height h.

QUICK QUIZ 10.7 A ball rolls without slipping down incline A, starting from rest. At the same time, a box starts from rest and slides down incline B, which is identical to incline A except that it is frictionless. Which arrives at the bottom first? **(a)** The ball arrives first. **(b)** The box arrives first. **(c)** Both arrive at the same time. **(d)** It is impossible to determine.

| Example **10.13** | **Sphere Rolling Down an Incline** |

Suppose the sphere shown in Figure 10.26 is solid and uniform. Calculate the translational speed of the center of mass at the bottom of the incline and the magnitude of the translational acceleration of the center of mass.

SOLUTION

Conceptualize Roll a golf ball or a marble down a ramp to visualize the motion of the sphere.

Categorize We model the sphere and the Earth as an *isolated system* in terms of *energy* with no nonconservative forces acting. This model is the one that led to Equation 10.33, so we can use that result. To find its acceleration, we will model the sphere as a *particle under constant acceleration.*

Analyze Evaluate the speed of the center of mass of the sphere from Equation 10.33, using the moment of inertia from Table 10.2:

$$(1) \quad v_{CM} = \left[\frac{2gh}{1 + (\frac{2}{5}MR^2/MR^2)} \right]^{1/2} = (\tfrac{10}{7}gh)^{1/2}$$

This result is less than $\sqrt{2gh}$, which is the speed an object would have if it simply slid down the incline without rotating. (Eliminate the rotation by setting $I_{CM} = 0$ in Eq. 10.33.)

To calculate the translational acceleration of the center of mass, notice that the vertical displacement of the sphere is related to the distance x it moves along the incline through the relationship $h = x \sin \theta$.

Use this relationship to rewrite Equation (1):

$$v_{CM}{}^2 = \tfrac{10}{7}gx \sin \theta$$

Write Equation 2.17 for an object starting from rest and moving through a distance x under constant acceleration:

$$v_{CM}{}^2 = 2a_{CM}x$$

Equate the preceding two expressions to find a_{CM}:

$$a_{CM} = \tfrac{5}{7}g \sin \theta$$

Finalize Both the speed and the acceleration of the center of mass are *independent* of the mass and the radius of the sphere. That is, all uniform solid spheres experience the same speed and acceleration on a given incline. Try to verify this statement experimentally with balls of different sizes, such as a marble and a croquet ball.

If we were to repeat the acceleration calculation for a hollow sphere, a solid cylinder, or a hoop, we would obtain similar results in which only the numerical factor in front of $g \sin \theta$ would differ. The numerical factors that appear in the expressions for v_{CM} and a_{CM} depend only on the moment of inertia about the center of mass for the specific object. In all cases, the acceleration of the center of mass is *less* than $g \sin \theta$, the value the acceleration would have if the incline were frictionless and no rolling occurred.

| Example **10.14** | **Pulling on a Spool**[3] |

A cylindrically symmetric spool of mass m and radius R sits at rest on a horizontal table with friction (Fig. 10.27). With your hand on a light string wrapped around the axle of radius r, you pull on the spool with a constant horizontal force of magnitude T to the right. As a result, the spool rolls without slipping a distance L along the table with no rolling friction.

(A) Find the final translational speed of the center of mass of the spool.

Figure 10.27 (Example 10.14)
A spool rests on a horizontal table. A string is wrapped around the axle and is pulled to the right by a hand.

SOLUTION

Conceptualize Use Figure 10.27 to visualize the motion of the spool when you pull the string. For the spool to roll through a distance L, notice that your hand on the string must pull through a distance *different* from L.

Categorize The spool is a *rigid object under a net torque*, but the net torque includes that due to the friction force at the bottom of the spool, about which we know nothing. Therefore, an approach based on the rigid object under a net torque model might be difficult. Work is done by your hand on the spool and string, which form a nonisolated system in terms of energy. Let's see if an approach based on the *nonisolated system (energy)* model is fruitful.

continued

---
[3]Example 10.14 was inspired in part by C. E. Mungan, "A primer on work–energy relationships for introductory physics," *The Physics Teacher,* **43**:10, 2005.

10.14 continued

Analyze The only type of energy that changes in the system is the kinetic energy of the spool. There is no rolling friction, so there is no change in internal energy. The only way that energy crosses the system's boundary is by the work done by your hand on the string. No work is done by the static force of friction on the bottom of the spool (to the left in Fig. 10.27) because the point of application of the force moves through no displacement.

Write the appropriate reduction of the conservation of energy equation, Equation 8.2:

$$(1) \quad W = \Delta K = \Delta K_{trans} + \Delta K_{rot}$$

In this expression, W is the work done on the string by your hand. To find this work, we need to find the displacement of your hand during the process. We first find the length of string that has unwound off the spool. If the spool rolls through a distance L, the total angle through which it rotates is $\theta = L/R$. The axle also rotates through this angle.

Use Equation 10.1b to find the total arc length through which the axle turns:

$$\ell = r\theta = \frac{r}{R}L$$

This result also gives the length of string pulled off the axle. Your hand will move through this distance *plus* the distance L through which the spool moves. Therefore, the magnitude of the displacement of the point of application of the force applied by your hand is $\ell + L = L(1 + r/R)$.

Evaluate the work done by your hand on the string:

$$(2) \quad W = TL\left(1 + \frac{r}{R}\right)$$

Substitute Equation (2) into Equation (1):

$$TL\left(1 + \frac{r}{R}\right) = \tfrac{1}{2}mv_{CM}^2 + \tfrac{1}{2}I\omega^2$$

where I is the moment of inertia of the spool about its center of mass and v_{CM} and ω are the final values after the wheel rolls through the distance L.

Apply the nonslip rolling condition $\omega = v_{CM}/R$:

$$TL\left(1 + \frac{r}{R}\right) = \tfrac{1}{2}mv_{CM}^2 + \tfrac{1}{2}I\frac{v_{CM}^2}{R^2}$$

Solve for v_{CM}:

$$(3) \quad v_{CM} = \sqrt{\frac{2TL(1 + r/R)}{m(1 + I/mR^2)}}$$

(B) Find the value of the friction force f.

SOLUTION

Categorize Because the friction force does no work, we cannot evaluate it from an energy approach. We model the spool as a *nonisolated system*, but this time in terms of *momentum*. The string applies a force across the boundary of the system, resulting in an impulse on the system. Because the forces on the spool are constant, we can model the spool's center of mass as a *particle under constant acceleration*.

Analyze Write the impulse–momentum theorem (Eq. 9.40) for the spool:

$$m(v_{CM} - 0) = (T - f)\Delta t$$
$$(4) \quad mv_{CM} = (T - f)\Delta t$$

For a particle under constant acceleration starting from rest, Equation 2.14 tells us that the average velocity of the center of mass is half the final velocity.

Use this fact and Equation 2.2 to find the time interval for the center of mass of the spool to move a distance L from rest to a final speed v_{CM}:

$$(5) \quad \Delta t = \frac{L}{v_{CM,avg}} = \frac{2L}{v_{CM}}$$

Substitute Equation (5) into Equation (4):

$$mv_{CM} = (T - f)\frac{2L}{v_{CM}}$$

Solve for the friction force f:

$$f = T - \frac{mv_{CM}^2}{2L}$$

10.14 continued

Substitute v_{CM} from Equation (3):

$$f = T - \frac{m}{2L}\left[\frac{2TL(1 + r/R)}{m(1 + I/mR^2)}\right]$$

$$= T - T\frac{(1 + r/R)}{(1 + I/mR^2)} = T\left[\frac{I - mrR}{I + mR^2}\right]$$

Finalize Notice that we could use the impulse–momentum theorem for the translational motion of the spool while ignoring that the spool is rotating! This fact demonstrates the power of our growing list of approaches to solving problems. To challenge yourself, solve part (A) again, using the rigid object under a net torque model for the spool and the particle under constant acceleration model for the center of mass of the spool, to derive Equation (3). Calculate torque and moment of inertia around the base of the spool to eliminate the unknown friction force from the torque equation.

Summary

▶ Definitions

The **angular position** of a rigid object is defined as the angle θ between a reference line attached to the object and a reference line fixed in space. The **angular displacement** of a particle moving in a circular path or a rigid object rotating about a fixed axis is $\Delta\theta \equiv \theta_f - \theta_i$.

The **instantaneous angular speed** of a particle moving in a circular path or of a rigid object rotating about a fixed axis is

$$\omega \equiv \frac{d\theta}{dt} \tag{10.3}$$

The **instantaneous angular acceleration** of a particle moving in a circular path or of a rigid object rotating about a fixed axis is

$$\alpha \equiv \frac{d\omega}{dt} \tag{10.5}$$

When a rigid object rotates about a fixed axis, every part of the object has the same angular speed and the same angular acceleration.

The magnitude of the **torque** associated with a force $\vec{F}$ acting on an object at a distance r from the rotation axis is

$$\tau = rF\sin\phi = Fd \tag{10.14}$$

where ϕ is the angle between the position vector of the point of application of the force and the force vector, and d is the moment arm of the force, which is the perpendicular distance from the rotation axis to the line of action of the force.

The **moment of inertia of a system of particles** is defined as

$$I \equiv \sum_i m_i r_i^2 \tag{10.19}$$

where m_i is the mass of the ith particle and r_i is its distance from the rotation axis.

▶ Concepts and Principles

When a rigid object rotates about a fixed axis, the angular position, angular speed, and angular acceleration are related to the translational position, translational speed, and translational acceleration through the relationships

$$s = r\theta \tag{10.1b}$$

$$v = r\omega \tag{10.10}$$

$$a_t = r\alpha \tag{10.11}$$

If a rigid object rotates about a fixed axis with angular speed ω, its **rotational kinetic energy** can be written

$$K_R = \tfrac{1}{2}I\omega^2 \tag{10.24}$$

where I is the moment of inertia of the object about the axis of rotation.

The **moment of inertia of a rigid object** is

$$I = \int r^2\, dm \tag{10.20}$$

where r is the distance from the mass element dm to the axis of rotation.

The rate at which work is done by an external force in rotating a rigid object about a fixed axis, or the **power** delivered, is

$$P = \tau\omega \tag{10.26}$$

If work is done on a rigid object and the only result of the work is rotation about a fixed axis, the net work done by external forces in rotating the object equals the change in the rotational kinetic energy of the object:

$$W = \tfrac{1}{2}I\omega_f^2 - \tfrac{1}{2}I\omega_i^2 \tag{10.27}$$

continued

The **total kinetic energy** of a rigid object rolling on a rough surface without slipping equals the rotational kinetic energy about its center of mass plus the translational kinetic energy of the center of mass:

$$K = \tfrac{1}{2} I_{CM} \omega^2 + \tfrac{1}{2} M v_{CM}^2 \qquad (10.31)$$

▶ Analysis Models for Problem Solving

Rigid Object Under Constant Angular Acceleration. If a rigid object rotates about a fixed axis under constant angular acceleration, one can apply equations of kinematics that are analogous to those for translational motion of a particle under constant acceleration:

α = constant

$$\omega_f = \omega_i + \alpha t \qquad (10.6)$$

$$\theta_f = \theta_i + \omega_i t + \tfrac{1}{2}\alpha t^2 \qquad (10.7)$$

$$\omega_f^2 = \omega_i^2 + 2\alpha(\theta_f - \theta_i) \qquad (10.8)$$

$$\theta_f = \theta_i + \tfrac{1}{2}(\omega_i + \omega_f)t \qquad (10.9)$$

Rigid Object Under a Net Torque. If a rigid object free to rotate about a fixed axis has a net external torque acting on it, the object undergoes an angular acceleration α, where

$$\sum \tau_{ext} = I\alpha \qquad (10.18)$$

This equation is the rotational analog to Newton's second law in the particle under a net force model.

Think–Pair–Share

See the Preface for an explanation of the icons used in this problems set. For additional assessment items for this section, go to **WEBASSIGN** From Cengage

1. You have a summer internship, working with other interns on an archeological dig. Your intern team has found a perfectly cylindrical object of an unknown material. Examination of the visible surface shows that the composition of the object seems to be uniform. The object has a mass of 15.7 kg and a radius of 5.00 cm. The lead archeologist wants to know if the artifact is hollow, but the x-ray machine and other scanning equipment have broken down, so there is no way to look inside. Your team comes up with the idea of building U-shaped supports from wood and laying the cylinder horizontally between the supports as shown in the end view in Figure TP10.1a. The wood can be sanded and oiled to almost eliminate friction. In this way, the cylindrical artifact is free to rotate around its long, horizontal axis. You wrap a long piece of twine several times around the cylinder

a

b

Figure TP10.1

and attach a 2.00-kg pickax to the free hanging end of the twine as shown in the side view in Figure TP10.1b. When the pickax is released from rest, it descends and causes the cylinder to rotate. (a) You measure the falling of the pickax and find that it falls 1.50 m in 1.45 s. Is the cylinder hollow? (b) Suppose you measure the falling of the pickax through the same distance and find it to take 1.13 s. What can you conclude about the cylinder now?

2. In order to save money on construction costs, a circular race track has been built with a flat roadway rather than a banked roadway, like that discussed in Example 6.4. During testing of the track, several race cars start, one at a time, at the beginning of the track and at the same radial distance from the center of the track, and undergo constant translational acceleration of magnitude a. All cars have identical tires. Show that all of the cars skid outward off the track at the same angular position around the track, regardless of their mass. To solve this problem, the stubborn owner still does not want to spend the money on banked roadways, so he simply has a circular track built with the same road material but a larger radius. What happens?

3. **ACTIVITY** (a) Place ten pennies on a horizontal meterstick, with a penny at 10 cm, 20 cm, 30 cm, etc., out to 100 cm. Carefully pick up the meterstick, keeping it horizontal, and have a member of the group make a video recording of the following event, using a smartphone or other device. While the video recording is underway, release the 100-cm end of the meterstick while the 0-cm end rests on someone's finger or the edge of the desk. By stepping through the video images or watching the video in slow motion, determine which pennies first lose contact with the meterstick as it falls. (b) Make a theoretical determination of which pennies should first lose contact and compare to your experimental result.

Problems

See the Preface for an explanation of the icons used in this problems set.
For additional assessment items for this section, go to ⚡ **WEBASSIGN**
From Cengage

SECTION 10.1 Angular Position, Velocity, and Acceleration

1. (a) Find the angular speed of the Earth's rotation about its
Q|C axis. (b) How does this rotation affect the shape of the Earth?

2. A bar on a hinge starts from rest and rotates with an
angular acceleration $\alpha = 10 + 6t$, where α is in rad/s^2
and t is in seconds. Determine the angle in radians through
which the bar turns in the first 4.00 s.

SECTION 10.2 Analysis Model: Rigid Object Under Constant Angular Acceleration

3. A wheel starts from rest and rotates with constant angular
V acceleration to reach an angular speed of 12.0 rad/s in 3.00 s.
Find (a) the magnitude of the angular acceleration of the
wheel and (b) the angle in radians through which it rotates
in this time interval.

4. A machine part rotates at an angular speed of 0.060 rad/s;
Q|C its speed is then increased to 2.2 rad/s at an angular accel-
eration of 0.70 rad/s^2. (a) Find the angle through which the
part rotates before reaching this final speed. (b) If both
the initial and final angular speeds are doubled and the
angular acceleration remains the same, by what factor is the
angular displacement changed? Why?

5. A dentist's drill starts from rest. After 3.20 s of constant angular acceleration, it turns at a rate of 2.51×10^4 rev/min. (a)
Find the drill's angular acceleration. (b) Determine the angle
(in radians) through which the drill rotates during this period.

6. *Why is the following situation impossible?* Starting from rest,
a disk rotates around a fixed axis through an angle of
50.0 rad in a time interval of 10.0 s. The angular acceleration of the disk is constant during the entire motion, and its
final angular speed is 8.00 rad/s.

7. **Review.** Consider a tall building located on the Earth's
Q|C equator. As the Earth rotates, a person on the top floor of
S the building moves faster than someone on the ground with
respect to an inertial reference frame because the person
on the ground is closer to the Earth's axis. Consequently,
if an object is dropped from the top floor to the ground a
distance h below, it lands east of the point vertically below
where it was dropped. (a) How far to the east will the object
land? Express your answer in terms of h, g, and the angular
speed ω of the Earth. Ignore air resistance and assume the
free-fall acceleration is constant over this range of heights.
(b) Evaluate the eastward displacement for $h = 50.0$ m.
(c) In your judgment, were we justified in ignoring this
aspect of the *Coriolis effect* in our previous study of free
fall? (d) Suppose the angular speed of the Earth were to
decrease with constant angular acceleration due to tidal
friction. Would the eastward displacement of the dropped
object increase or decrease compared with that in part (b)?

SECTION 10.3 Angular and Translational Quantities

8. Make an order-of-magnitude estimate of the number of revolutions through which a typical automobile tire turns in
one year. State the quantities you measure or estimate and
their values.

9. A discus thrower (Fig. P10.9)
V accelerates a discus from
rest to a speed of 25.0 m/s
by whirling it through
1.25 rev. Assume the dis-
cus moves on the arc of
a circle 1.00 m in radius.
(a) Calculate the final
angular speed of the discus.
(b) Determine the mag-
nitude of the angular
acceleration of the discus,
assuming it to be constant.
(c) Calculate the time interval required for the discus
to accelerate from rest to
25.0 m/s.

Herbert Kratky/Shutterstock.com

Figure P10.9

10. A straight ladder is leaning against the wall of a house. The
Q|C ladder has rails 4.90 m long, joined by rungs 0.410 m long.
Its bottom end is on solid but sloping ground so that the
top of the ladder is 0.690 m to the left of where it should
be, and the ladder is unsafe to climb. You want to put a
flat rock under one foot of the ladder to compensate for
the slope of the ground. (a) What should be the thickness
of the rock? (b) Does using ideas from this chapter make
it easier to explain the solution to part (a)? Explain your
answer.

11. A car accelerates uniformly from rest and reaches a speed
V of 22.0 m/s in 9.00 s. Assuming the diameter of a tire is
58.0 cm, (a) find the number of revolutions the tire makes
during this motion, assuming that no slipping occurs.
(b) What is the final angular speed of a tire in revolutions
per second?

12. **Review.** A small object with mass 4.00 kg moves counterclockwise with constant angular speed 1.50 rad/s in a circle
of radius 3.00 m centered at the origin. It starts at the point
with position vector $3.00\hat{\mathbf{i}}$ m. It then undergoes an angular
displacement of 9.00 rad. (a) What is its new position vector? Use unit-vector notation for all vector answers. (b) In
what quadrant is the particle located, and what angle does
its position vector make with the positive x axis? (c) What is
its velocity? (d) In what direction is it moving? (e) What is
its acceleration? (f) Make a sketch of its position, velocity,
and acceleration vectors. (g) What total force is exerted on
the object?

13. In a manufacturing process, a large, cylindrical roller is
used to flatten material fed beneath it. The diameter of
the roller is 1.00 m, and, while being driven into rotation
around a fixed axis, its angular position is expressed as

$$\theta = 2.50t^2 - 0.600t^3$$

where θ is in radians and t is in seconds. (a) Find the maximum
angular speed of the roller. (b) What is the maximum tangential speed of a point on the rim of the roller? (c) At what time t
should the driving force be removed from the roller so that the
roller does not reverse its direction of rotation? (d) Through
how many rotations has the roller turned between $t = 0$ and
the time found in part (c)?

SECTION 10.4 Torque

14. Find the net torque on the wheel in Figure P10.14 about the
T axle through O, taking $a = 10.0$ cm and $b = 25.0$ cm.

Figure P10.14

SECTION 10.5 Analysis Model: Rigid Object Under a Net Torque

15. A grinding wheel is in the form of a uniform solid disk of
AMT radius 7.00 cm and mass 2.00 kg. It starts from rest and
V accelerates uniformly under the action of the constant
torque of $0.600 \text{ N} \cdot \text{m}$ that the motor exerts on the wheel. (a)
How long does the wheel take to reach its final operating
speed of 1 200 rev/min? (b) Through how many revolutions
does it turn while accelerating?

16. **Review.** A block of mass $m_1 = 2.00$ kg and a block of mass
V $m_2 = 6.00$ kg are connected by a massless string over a pulley
in the shape of a solid disk having radius $R = 0.250$ m and
mass $M = 10.0$ kg. The fixed, wedge-shaped ramp makes an
angle of $\theta = 30.0°$ as shown in Figure P10.16. The coefficient
of kinetic friction is 0.360 for both blocks. (a) Draw force
diagrams of both blocks and of the pulley. Determine (b)
the acceleration of the two blocks and (c) the tensions in the
string on both sides of the pulley.

Figure P10.16

17. A model airplane with mass 0.750 kg is tethered to the
T ground by a wire so that it flies in a horizontal circle 30.0 m in
radius. The airplane engine provides a net thrust of 0.800 N
perpendicular to the tethering wire. (a) Find the torque the
net thrust produces about the center of the circle. (b) Find
the angular acceleration of the airplane. (c) Find the translational acceleration of the airplane tangent to its flight path.

18. A disk having moment of inertia $100 \text{ kg} \cdot \text{m}^2$ is free to
QC rotate without friction, starting from rest, about a fixed axis
through its center. A tangential force whose magnitude can
range from $F = 0$ to $F = 50.0$ N can be applied at any distance ranging from $R = 0$ to $R = 3.00$ m from the axis of
rotation. (a) Find a pair of values of F and R that cause the
disk to complete 2.00 rev in 10.0 s. (b) Is your answer for part
(a) a unique answer? How many answers exist?

19. Your grandmother enjoys creating pottery as a hobby. She
CR uses a potter's wheel, which is a stone disk of radius $R =$
0.500 m and mass $M = 100$ kg. In operation, the wheel

rotates at 50.0 rev/min. While the wheel is spinning, your
grandmother works clay at the center of the wheel with
her hands into a pot-shaped object with circular symmetry. When the correct shape is reached, she wants to stop
the wheel in as short a time interval as possible, so that the
shape of the pot is not further distorted by the rotation. She
pushes continuously with a wet rag as hard as she can radially inward on the edge of the wheel and the wheel stops in
6.00 s. (a) You would like to build a brake to stop the wheel
in a shorter time interval, but you must determine the coefficient of friction between the rag and the wheel in order to
design a better system. You determine that the maximum
pressing force your grandmother can sustain for 6.00 s is
70.0 N. (b) **What If?** If your grandmother instead chooses to
press down on the upper surface of the wheel a distance $r =$
0.300 m from the axis of rotation, what is the force needed
to stop the wheel in 6.00 s? Assume that the coefficient of
kinetic friction between the wet rag and the wheel remains
the same as before.

20. At a local mine, a cave-in has trapped a number of miners.
CR You and some classmates rush to the scene to see how you
can help. The trapped miners have been able to reach a
point in the mine at the bottom of a tall vertical shaft to
the surface, allowing them access to fresh air. But they are
in desperate need of fresh water and bandages for injuries.
Some rescue workers ask you to help pack a light plastic
cylindrical container with bottles of water and bandages.
Simply dropping the container into the shaft risks damaging
the container and contents and injuring the miners. Tying
a rope to the container and lowering it on the end of the
rope takes a long time. A quick and relatively safe method
is to wrap a lightweight rope around the container. One
end of the rope will be secured and the container will be
released into the vertical shaft. The container will unroll off
the rope like a falling yo-yo. (a) If immediate access to the
lightweight bandages is needed due to injuries, so that you
want the container to reach the bottom of the shaft in the
shortest possible time interval, should you pack the heavy
water bottles at the center of the container or near the outer
edges? (b) If the medical necessity is not so urgent and, for
safety considerations, you want the container to arrive at the
bottom of the shaft with the lowest possible speed, should
you pack the heavy water bottles at the center of the container or near the outer edges? Assume that the center of
mass of the container is at its center.

21. You have just bought a new bicycle. On your first riding trip,
CR it seems that the bike comes to rest relatively quickly after
you stop pedaling and let the bicycle coast on flat ground.
You call the bicycle shop from which you purchased the
vehicle and describe the problem. The technician says that
they will replace the bearings in the wheels or do whatever
else is necessary if you can prove that the frictional torque
in the axle of the wheels is worse than $-0.02 \text{ N} \cdot \text{m}$. At first,
you are discouraged by the technical sound of what you have
been told and by the absence of any tool to measure torque
in your garage. But then you remember that you are taking
a physics class! You take your bike into the garage, turn it
upside down and start spinning the wheel while you think
about how to determine the frictional torque. The driveway
outside the garage had a small puddle, so you notice that
droplets of water are flying off the edge of one point on the
tire tangentially, including drops that are projected straight

upward, as shown in Figure P10.21. Ah-ha! Here is your torque-measuring method! The upward-projected drops leave the rim of the wheel at the same level as the axle. You measure the height to which a drop rises from the level of the axle: $h_1 = 54.0$ cm. The wet spot on the tire makes one revolution and another drop is projected upward. You measure its highest point: $h_2 = 51.0$ cm. You measure the radius of the wheel: $r = 0.381$ m. Finally, you take the wheel off the bike and find its mass: $m = 0.850$ kg. Because most of the mass of the wheel is at the tire, you model the wheel as a hoop. What do you tell the technician when you call back?

Figure P10.21

SECTION 10.6 Calculation of Moments of Inertia

22. Imagine that you stand tall and turn about a vertical axis through the top of your head and the point halfway between your ankles. Compute an order-of-magnitude estimate for the moment of inertia of your body for this rotation. In your solution, state the quantities you measure or estimate and their values.

23. Following the procedure used in Example 10.7, prove that the moment of inertia about the y' axis of the rigid rod in Figure 10.15 is $\frac{1}{3}ML^2$.

24. Two balls with masses M and m are connected by a rigid rod of length L and negligible mass as shown in Figure P10.24. For an axis perpendicular to the rod, (a) show that the system has the minimum moment of inertia when the axis passes through the center of mass. (b) Show that this moment of inertia is $I = \mu L^2$, where $\mu = mM/(m + M)$.

Figure P10.24

SECTION 10.7 Rotational Kinetic Energy

25. Rigid rods of negligible mass lying along the y axis connect three particles (Fig. P10.25). The system rotates about the x axis with an angular speed of 2.00 rad/s. Find (a) the moment of inertia about the x axis, (b) the total rotational kinetic energy evaluated from $\frac{1}{2}I\omega^2$, (c) the tangential speed of each particle, and (d) the total kinetic energy evaluated from $\sum \frac{1}{2}m_i v_i^2$. (e) Compare the answers for kinetic energy in parts (a) and (b).

Figure P10.25

26. A *war-wolf* or *trebuchet* is a device used during the Middle Ages to throw rocks at castles and now sometimes used to fling large vegetables and pianos as a sport. A simple trebuchet is

shown in Figure P10.26. Model it as a stiff rod of negligible mass, 3.00 m long, joining particles of mass $m_1 = 0.120$ kg and $m_2 = 60.0$ kg at its ends. It can turn on a frictionless, horizontal axle perpendicular to the rod and 14.0 cm from the large-mass particle. The operator releases the trebuchet from rest in a horizontal orientation. (a) Find the maximum speed that the small-mass object attains. (b) While the small-mass object is gaining speed, does it move with constant acceleration? (c) Does it move with constant tangential acceleration? (d) Does the trebuchet move with constant angular acceleration? (e) Does it have constant momentum? (f) Does the trebuchet–Earth system have constant mechanical energy?

Figure P10.26

SECTION 10.8 Energy Considerations in Rotational Motion

27. Big Ben, the nickname for the clock in Elizabeth Tower (named after the Queen in 2012) in London, has an hour hand 2.70 m long with a mass of 60.0 kg and a minute hand 4.50 m long with a mass of 100 kg (Fig. P10.27). Calculate the total rotational kinetic energy of the two hands about the axis of rotation. (You may model the hands as long, thin rods rotated about one end. Assume the hour and minute hands are rotating at a constant rate of one revolution per 12 hours and 60 minutes, respectively.)

Peter Nadolski/Shutterstock

Figure P10.27 Problems 27 and 40.

28. Consider two objects with $m_1 > m_2$ connected by a light string that passes over a pulley having a moment of inertia of I about its axis of rotation as shown in Figure P10.28. The string does not slip on the pulley or stretch. The pulley turns without friction. The two objects are released from rest separated by a vertical distance $2h$. (a) Use the principle of conservation of energy to find the translational speeds of the objects as they pass each other. (b) Find the angular speed of the pulley at this time.

Figure P10.28

29. **Review.** An object with a mass of $m = 5.10$ kg is attached to the free end of a light string wrapped around a reel of radius $R = 0.250$ m and mass $M = 3.00$ kg. The reel is a solid disk, free

to rotate in a vertical plane about the horizontal axis passing through its center as shown in Figure P10.29. The suspended object is released from rest 6.00 m above the floor. Determine (a) the tension in the string, (b) the acceleration of the object, and (c) the speed with which the object hits the floor. (d) Verify your answer to part (c) by using the isolated system (energy) model.

Figure P10.29

30. *Why is the following situation impossible?* In a large city with an air-pollution problem, a bus has no combustion engine. It runs over its citywide route on energy drawn from a large, rapidly rotating flywheel under the floor of the bus. The flywheel is spun up to its maximum rotation rate of 3 000 rev/min by an electric motor at the bus terminal. Every time the bus speeds up, the flywheel slows down slightly. The bus is equipped with regenerative braking so that the flywheel can speed up when the bus slows down. The flywheel is a uniform solid cylinder with mass 1 200 kg and radius 0.500 m. The bus body does work against air resistance and rolling resistance at the average rate of 25.0 hp as it travels its route with an average speed of 35.0 km/h.

31. A uniform solid disk of radius R and mass M is free to rotate on a frictionless pivot through a point on its rim (Fig. P10.31). If the disk is released from rest in the position shown by the copper-colored circle, (a) what is the speed of its center of mass when the disk reaches the position indicated by the dashed circle? (b) What is the speed of the lowest point on the disk in the dashed position? (c) **What If?** Repeat part (a) using a uniform hoop.

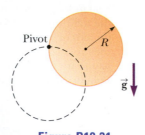

Figure P10.31

32. This problem describes one experimental method for determining the moment of inertia of an irregularly shaped object such as the payload for a satellite. Figure P10.32 shows a counterweight of mass m suspended by a cord wound around a spool of radius r, forming part of a turntable supporting the object. The turntable can rotate without friction. When the counterweight is released from rest, it descends through a distance h, acquiring a speed v. Show that the moment of inertia I of the rotating apparatus (including the turntable) is $mr^2(2gh/v^2 - 1)$.

Figure P10.32

SECTION 10.9 Rolling Motion of a Rigid Object

33. A tennis ball is a hollow sphere with a thin wall. It is set rolling without slipping at 4.03 m/s on a horizontal section of a track as shown in Figure P10.33. It rolls around the inside of a vertical circular loop of radius $r = 45.0$ cm. As the ball nears the bottom of the loop, the shape of the track deviates from a perfect circle so that the ball leaves the track at a point $h = 20.0$ cm below the horizontal section. (a) Find the ball's speed at the top of the loop. (b) Demonstrate that the ball will not fall from the track at the top of the loop. (c) Find the ball's speed as it leaves the track at the bottom. (d) **What If?** Suppose that static friction between ball and track were negligible so that the ball slid instead of rolling. Describe the speed of the ball at the top of the loop in this situation. (e) Explain your answer to part (d).

Figure P10.33

34. A smooth cube of mass m and edge length r slides with speed v on a horizontal surface with negligible friction. The cube then moves up a smooth incline that makes an angle θ with the horizontal. A cylinder of mass m and radius r rolls without slipping with its center of mass moving with speed v and encounters an incline of the same angle of inclination but with sufficient friction that the cylinder continues to roll without slipping. (a) Which object will go the greater distance up the incline? (b) Find the difference between the maximum distances the objects travel up the incline. (c) Explain what accounts for this difference in distances traveled.

35. A metal can containing condensed mushroom soup has mass 215 g, height 10.8 cm, and diameter 6.38 cm. It is placed at rest on its side at the top of a 3.00-m-long incline that is at 25.0° to the horizontal and is then released to roll straight down. It reaches the bottom of the incline after 1.50 s. (a) Assuming mechanical energy conservation, calculate the moment of inertia of the can. (b) Which pieces of data, if any, are unnecessary for calculating the solution? (c) Why can't the moment of inertia be calculated from $I = \frac{1}{2}mr^2$ for the cylindrical can?

ADDITIONAL PROBLEMS

36. You have been hired as an expert witness in the case of a factory owner suing a demolition company. The particular case involves a smokestack at a factory being demolished. In order to save money, the factory owner wanted to move the smokestack to a nearby factory that was being built. The demolition company guaranteed to deliver the undamaged smokestack to the new factory by toppling the smokestack freely onto a huge cushioned platform lying on the ground. The then-horizontal smokestack would have been loaded onto a long truck rig for transport to the new factory. However, as the smokestack toppled, it broke apart at a point along its length. The factory owner is blaming the demolition company for the destruction of his smokestack. The demolition company is claiming that there was a defect in the smokestack and that is the reason for its destruction. What advice do you give the attorney who is handling the case on the side of the factory owner?

37. A shaft is turning at 65.0 rad/s at time $t = 0$. Thereafter, its angular acceleration is given by

$$\alpha = -10.0 - 5.00t$$

where α is in rad/s^2 and t is in seconds. (a) Find the angular speed of the shaft at $t = 3.00$ s. (b) Through what angle does it turn between $t = 0$ and $t = 3.00$ s?

38. A shaft is turning at angular speed ω at time $t = 0$. There-
S after, its angular acceleration is given by

$$\alpha = A + Bt$$

(a) Find the angular speed of the shaft at time t. (b) Through what angle does it turn between $t = 0$ and t?

39. An elevator system in a tall building consists of a 800-kg car
QC and a 950-kg counterweight joined by a light cable of con-
stant length that passes over a pulley of mass 280 kg. The pulley, called a sheave, is a solid cylinder of radius 0.700 m turning on a horizontal axle. The cable does not slip on the sheave. A number n of people, each of mass 80.0 kg, are riding in the elevator car, moving upward at 3.00 m/s and approaching the floor where the car should stop. As an energy-conservation measure, a computer disconnects the elevator motor at just the right moment so that the sheave–car–counterweight system then coasts freely without friction and comes to rest at the floor desired. There it is caught by a sim-ple latch rather than by a massive brake. (a) Determine the distance d the car coasts upward as a function of n. Evaluate the distance for (b) $n = 2$, (c) $n = 12$, and (d) $n = 0$. (e) For what integer values of n does the expression in part (a) apply? (f) Explain your answer to part (e). (g) If an infinite number of people could fit on the elevator, what is the value of d?

40. The hour hand and the minute hand of Big Ben, the Eliza-beth Tower clock in London, are 2.70 m and 4.50 m long and have masses of 60.0 kg and 100 kg, respectively (see Fig. P10.27). (a) Determine the total torque due to the weight of these hands about the axis of rotation when the time reads (i) 3:00, (ii) 5:15, (iii) 6:00, (iv) 8:20, and (v) 9:45. (You may model the hands as long, thin, uniform rods.) (b) Determine all times when the total torque about the axis of rotation is zero. Determine the times to the nearest second, solving a transcendental equation numerically.

41. **Review.** A string is wound around a
S uniform disk of radius R and mass M. The disk is released from rest with the string vertical and its top end tied to a fixed bar (Fig. P10.41). Show that (a) the tension in the string is one third of the weight of the disk, (b) the magnitude of the acceleration of the center of mass is $2g/3$, and (c) the speed of the center of mass is $(4gh/3)^{1/2}$

Figure P10.41

after the disk has descended through distance h. (d) Verify your answer to part (c) using the energy approach.

42. **Review.** A spool of wire of mass M and radius R is unwound
S under a constant force $\vec{F}$ (Fig. P10.42). Assuming the spool is a uniform, solid cylinder that doesn't slip, show that (a) the acceleration of the center of mass is $4\vec{F}/3M$ and (b) the force of friction is to the *right* and equal in magnitude to $F/3$. (c) If the cylinder starts from rest and rolls without slipping, what is the speed of its center of mass after it has rolled through a distance d?

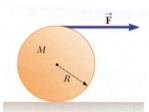

Figure P10.42

43. **Review.** A clown balances a small spherical grape at the top of his bald head, which also has the shape of a sphere. After drawing sufficient applause, the grape starts from rest and rolls down without slipping. It will leave contact with the clown's scalp when the radial line joining it to the center of curvature makes what angle with the vertical?

CHALLENGE PROBLEMS

44. As a gasoline engine operates, a flywheel turning with the crankshaft stores energy after each fuel explosion, provid-ing the energy required to compress the next charge of fuel and air. For the engine of a certain lawn tractor, suppose a flywheel must be no more than 18.0 cm in diameter. Its thickness, measured along its axis of rotation, must be no larger than 8.00 cm. The flywheel must release energy 60.0 J when its angular speed drops from 800 rev/min to 600 rev/min. Design a sturdy steel (density 7.85×10^3 kg/m^3) flywheel to meet these requirements with the smallest mass you can reasonably attain. Specify the shape and mass of the flywheel.

45. A spool of thread consists of a cylinder of radius R_1 with end
S caps of radius R_2 as depicted in the end view shown in Figure P10.45. The mass of the spool, including the thread, is m, and its moment of inertia about an axis through its center is I. The spool is placed on a rough, horizontal surface so that it rolls without slipping when a force $\vec{T}$ acting to the right is applied to the free end of the thread. (a) Show that the magnitude of the friction force exerted by the surface on the spool is given by

$$f = \left(\frac{I + mR_1R_2}{I + mR_2{}^2} \right) T$$

(b) Determine the direction of the force of friction.

Figure P10.45

46. To find the total angular displacement during the playing
S time of the compact disc in part (B) of Example 10.2, the disc was modeled as a rigid object under constant angular acceleration. In reality, the angular acceleration of a disc is not constant. In this problem, let us explore the actual time dependence of the angular acceleration. (a) Assume the track on the disc is a spiral such that adjacent loops of

the track are separated by a small distance h. Show that the radius r of a given portion of the track is given by

$$r = r_i + \frac{h\theta}{2\pi}$$

where r_i is the radius of the innermost portion of the track and θ is the angle through which the disc turns to arrive at the location of the track of radius r. (b) Show that the rate of change of the angle θ is given by

$$\frac{d\theta}{dt} = \frac{v}{r_i + (h\theta/2\pi)}$$

where v is the constant speed with which the disc surface passes the laser. (c) From the result in part (b), use integration to find an expression for the angle θ as a function of time. (d) From the result in part (c), use differentiation to find the angular acceleration of the disc as a function of time.

47. A uniform, hollow, cylindrical spool has inside radius $R/2$, outside radius R, and mass M (Fig. P10.47). It is mounted so that it rotates on a fixed, horizontal axle. A counterweight of mass m is connected to the end of a string wound around the spool. The counterweight falls from rest at $t = 0$ to a position y at time t. Show that the torque due to the friction forces between spool and axle is

Figure P10.47

$$\tau_f = R\left[m\left(g - \frac{2y}{t^2}\right) - M\frac{5y}{4t^2} \right]$$

48. A cord is wrapped around a pulley that is shaped like a disk of mass m and radius r. The cord's free end is connected to a block of mass M. The block starts from rest and then slides down an incline that makes an angle θ with the horizontal as shown in Figure P10.48. The coefficient of kinetic friction between block and incline is μ. (a) Use energy methods to show that the block's speed as a function of position d down the incline is

$$v = \sqrt{\frac{4Mgd(\sin\theta - \mu\cos\theta)}{m + 2M}}$$

(b) Find the magnitude of the acceleration of the block in terms of μ, m, M, g, and θ.

Figure P10.48

Angular Momentum

An Olympic diver does a fancy spinning dive. When she tucks into the position shown, she spins faster. Where does the extra rotational kinetic energy come from? (*Paolo Bona/Shutterstock.com*)

STORYLINE **While preparing to do your physics homework, you are** browsing among various YouTube videos and stumble on one of a spinning ice skater. You notice that he skates into a relatively slow spin, pulls his arms in, and then spins faster and faster. You wonder where the energy comes from to make him spin faster. Next, you find a slow-motion video of an Olympic high diver who is diving into a pool. You notice that after leaving the diving board, she spins slowly, but then tucks in and spins faster. Just like the ice skater, where did the additional rotational kinetic energy come from? In the suggested videos to the side of the web page, you see one about a falling cat. You watch that video and marvel about how a cat dropped upside down can always turn itself over and land on its feet. Just like the ice skater and the diver, there is rotational energy seemingly coming from nowhere. What's going on here? Spinning motion seems to have magical qualities associated with it!

CONNECTIONS The central topic of this chapter is *angular momentum,* a quantity that plays a key role in rotational dynamics. In analogy to the principle of conservation of linear momentum in Chapter 9, there is also a principle of conservation of angular momentum. The angular momentum of an isolated system is constant. For angular momentum, an isolated system is one for which no external torques act on the system. If a net external torque does act on a system, it is nonisolated, and the angular momentum of the system changes. Like the law of conservation of linear momentum, the law of conservation of angular momentum is a fundamental law of physics, equally valid for relativistic and quantum systems. This new fundamental principle allows us to understand more phenomena, such as the spinning skaters, divers, and cats in the opening

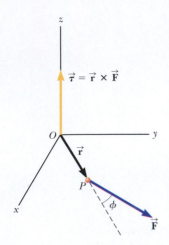

Figure 11.1 The torque vector $\vec{\boldsymbol{\tau}}$ on a particle lies in a direction perpendicular to the plane formed by the position vector $\vec{\mathbf{r}}$ of the particle and the applied force vector $\vec{\mathbf{F}}$. In the situation shown, $\vec{\mathbf{r}}$ and $\vec{\mathbf{F}}$ lie in the xy plane, so the torque is along the z axis.

The direction of $\vec{\mathbf{C}}$ is perpendicular to the plane formed by $\vec{\mathbf{A}}$ and $\vec{\mathbf{B}}$, and its direction is determined by the right-hand rule.

Figure 11.2 The vector product $\vec{\mathbf{A}} \times \vec{\mathbf{B}}$ is a third vector $\vec{\mathbf{C}}$ having a magnitude $AB \sin \theta$ equal to the area of the parallelogram shown.

Properties of the ▶
vector product

storyline. In addition, we will apply this new principle to the motion of planets in a solar system in Chapter 13, to atomic models in Chapter 41, and to molecular spectra in Chapter 42.

11.1 The Vector Product and Torque

An important consideration to address before defining angular momentum is the process of multiplying two vectors by means of the operation called the *vector product*. We will introduce the vector product by considering the vector nature of torque.

Consider a force $\vec{\mathbf{F}}$ acting on a particle located at point P and described by the vector position $\vec{\mathbf{r}}$ (Fig. 11.1). As we saw in Section 10.4, the *magnitude* of the torque on the particle due to this force about an axis through the origin is $rF \sin \phi$, where ϕ is the angle between $\vec{\mathbf{r}}$ and $\vec{\mathbf{F}}$. The axis about which $\vec{\mathbf{F}}$ tends to produce rotation is perpendicular to the plane formed by $\vec{\mathbf{r}}$ and $\vec{\mathbf{F}}$.

The torque vector $\vec{\boldsymbol{\tau}}$ is related to the two vectors $\vec{\mathbf{r}}$ and $\vec{\mathbf{F}}$. We can establish a mathematical relationship between $\vec{\boldsymbol{\tau}}$, $\vec{\mathbf{r}}$, and $\vec{\mathbf{F}}$ using a mathematical operation called the **vector product:**

$$\vec{\boldsymbol{\tau}} \equiv \vec{\mathbf{r}} \times \vec{\mathbf{F}} \tag{11.1}$$

We now give a formal definition of the vector product. Given any two vectors $\vec{\mathbf{A}}$ and $\vec{\mathbf{B}}$, the vector product $\vec{\mathbf{A}} \times \vec{\mathbf{B}}$ is defined as a third vector $\vec{\mathbf{C}}$, which has a magnitude of $AB \sin \theta$, where θ is the angle between $\vec{\mathbf{A}}$ and $\vec{\mathbf{B}}$. That is, if $\vec{\mathbf{C}}$ is given by

$$\vec{\mathbf{C}} = \vec{\mathbf{A}} \times \vec{\mathbf{B}} \tag{11.2}$$

its magnitude is

$$C = AB \sin \theta \tag{11.3}$$

The quantity $AB \sin \theta$ is equal to the area of the parallelogram formed by $\vec{\mathbf{A}}$ and $\vec{\mathbf{B}}$ as shown in Figure 11.2. The *direction* of $\vec{\mathbf{C}}$ is perpendicular to the plane formed by $\vec{\mathbf{A}}$ and $\vec{\mathbf{B}}$, and the best way to determine this direction is to use the right-hand rule illustrated in Figure 11.2. The four fingers of the right hand are pointed along $\vec{\mathbf{A}}$ and then "wrapped" in the direction that would rotate $\vec{\mathbf{A}}$ into $\vec{\mathbf{B}}$ through the angle θ. The direction of the upright thumb is the direction of $\vec{\mathbf{A}} \times \vec{\mathbf{B}} = \vec{\mathbf{C}}$. Because of the notation, $\vec{\mathbf{A}} \times \vec{\mathbf{B}}$ is often read "$\vec{\mathbf{A}}$ cross $\vec{\mathbf{B}}$," so the vector product is also called the **cross product.**

Some properties of the vector product that follow from its definition are as follows:

1. Unlike the scalar product, the vector product is *not* commutative. Instead, the order in which the two vectors are multiplied in a vector product is important:

$$\vec{\mathbf{A}} \times \vec{\mathbf{B}} = -\vec{\mathbf{B}} \times \vec{\mathbf{A}} \tag{11.4}$$

Therefore, if you change the order of the vectors in a vector product, you must change the sign. You can easily verify this relationship with the right-hand rule.

2. If $\vec{\mathbf{A}}$ is parallel to $\vec{\mathbf{B}}$ ($\theta = 0$ or $180°$), then $\vec{\mathbf{A}} \times \vec{\mathbf{B}} = 0$; therefore, it follows that $\vec{\mathbf{A}} \times \vec{\mathbf{A}} = 0$.

3. If $\vec{\mathbf{A}}$ is perpendicular to $\vec{\mathbf{B}}$, then $|\vec{\mathbf{A}} \times \vec{\mathbf{B}}| = AB$.

4. The vector product obeys the distributive law:

$$\vec{\mathbf{A}} \times (\vec{\mathbf{B}} + \vec{\mathbf{C}}) = \vec{\mathbf{A}} \times \vec{\mathbf{B}} + \vec{\mathbf{A}} \times \vec{\mathbf{C}} \tag{11.5}$$

5. The derivative of the vector product with respect to some variable such as t is

$$\frac{d}{dt}(\vec{A} \times \vec{B}) = \frac{d\vec{A}}{dt} \times \vec{B} + \vec{A} \times \frac{d\vec{B}}{dt} \tag{11.6}$$

where it is important to preserve the multiplicative order of the terms on the right side in view of Equation 11.4.

It is left as an exercise (Problem 4) to show from Equations 11.3 and 11.4 and from the definition of unit vectors that the cross products of the unit vectors $\hat{i}$, $\hat{j}$, and $\hat{k}$ obey the following rules:

$$\hat{i} \times \hat{i} = \hat{j} \times \hat{j} = \hat{k} \times \hat{k} = 0 \tag{11.7a}$$

$$\hat{i} \times \hat{j} = -\hat{j} \times \hat{i} = \hat{k} \tag{11.7b}$$

$$\hat{j} \times \hat{k} = -\hat{k} \times \hat{j} = \hat{i} \tag{11.7c}$$

$$\hat{k} \times \hat{i} = -\hat{i} \times \hat{k} = \hat{j} \tag{11.7d}$$

◀ Cross products of unit vectors

Signs are interchangeable in cross products. For example, $\vec{A} \times (-\vec{B}) = -\vec{A} \times \vec{B}$ and $\hat{i} \times (-\hat{j}) = -\hat{i} \times \hat{j}$.

The cross product of any two vectors $\vec{A}$ and $\vec{B}$ can be expressed in unit-vector notation by using the following determinant form:

$$\vec{A} \times \vec{B} = \begin{vmatrix} \hat{i} & \hat{j} & \hat{k} \\ A_x & A_y & A_z \\ B_x & B_y & B_z \end{vmatrix} = \begin{vmatrix} A_y & A_z \\ B_y & B_z \end{vmatrix} \hat{i} + \begin{vmatrix} A_z & A_x \\ B_z & B_x \end{vmatrix} \hat{j} + \begin{vmatrix} A_x & A_y \\ B_x & B_y \end{vmatrix} \hat{k}$$

Expanding these determinants gives the result

$$\vec{A} \times \vec{B} = (A_y B_z - A_z B_y)\hat{i} + (A_z B_x - A_x B_z)\hat{j} + (A_x B_y - A_y B_x)\hat{k} \tag{11.8}$$

Given the definition of the cross product, we can now assign a direction to the torque vector. If the force lies in the xy plane as in Figure 11.1, the torque $\vec{\tau}$ is represented by a vector parallel to the z axis. The force in Figure 11.1 creates a torque that tends to rotate the particle counterclockwise about the z axis; the direction of $\vec{\tau}$ is toward increasing z, and $\vec{\tau}$ is therefore in the positive z direction. If we reversed the direction of $\vec{F}$ in Figure 11.1, $\vec{\tau}$ would be in the negative z direction.

In Figure 11.1 and its discussion, we investigated the torque on a particle. Imagine that the particle is part of a rigid object free to rotate around the z axis. Then the torque that we found in Equation 11.1 is the torque applied to the entire rigid object due to the force $\vec{F}$.

PITFALL PREVENTION 11.1
The Vector Product Is a Vector
Remember that the result of taking a vector product between two vectors is *a third vector*. Equation 11.3 gives only the magnitude of this vector.

QUICK QUIZ 11.1 Which of the following statements about the relationship between the magnitude of the cross product of two vectors and the product of the magnitudes of the vectors is true? **(a)** $|\vec{A} \times \vec{B}|$ is larger than AB. **(b)** $|\vec{A} \times \vec{B}|$ is smaller than AB. **(c)** $|\vec{A} \times \vec{B}|$ could be larger or smaller than AB, depending on the angle between the vectors. **(d)** $|\vec{A} \times \vec{B}|$ could be equal to AB.

Example **11.1** **The Vector Product**

Two vectors lying in the xy plane are given by the equations $\vec{A} = 2\hat{i} + 3\hat{j}$ and $\vec{B} = -\hat{i} + 2\hat{j}$. Find $\vec{A} \times \vec{B}$ and verify that $\vec{A} \times \vec{B} = -\vec{B} \times \vec{A}$.

SOLUTION

Conceptualize Given the unit-vector notations of the vectors, think about the directions the vectors point in space. Draw them on graph paper and imagine the parallelogram shown in Figure 11.2 for these vectors.

continued

11.1 continued

Categorize   Because we use the definition of the cross product discussed in this section, we categorize this example as a substitution problem.

Write the cross product of the two vectors:

$$\vec{A} \times \vec{B} = (2\,\hat{i} + 3\,\hat{j}) \times (-\,\hat{i} + 2\,\hat{j})$$

Perform the multiplication using the distributive law:

$$\vec{A} \times \vec{B} = 2\,\hat{i} \times (-\,\hat{i}) + 2\,\hat{i} \times 2\,\hat{j} + 3\,\hat{j} \times (-\,\hat{i}) + 3\,\hat{j} \times 2\,\hat{j}$$

Use Equations 11.7a through 11.7d to evaluate the various terms:

$$\vec{A} \times \vec{B} = 0 + 4\hat{k} + 3\hat{k} + 0 = \boxed{7\hat{k}}$$

To verify that $\vec{A} \times \vec{B} = -\vec{B} \times \vec{A}$, evaluate $\vec{B} \times \vec{A}$:

$$\vec{B} \times \vec{A} = (-\,\hat{i} + 2\,\hat{j}) \times (2\,\hat{i} + 3\,\hat{j})$$

Perform the multiplication:

$$\vec{B} \times \vec{A} = (-\,\hat{i}) \times 2\,\hat{i} + (-\,\hat{i}) \times 3\,\hat{j} + 2\,\hat{j} \times 2\,\hat{i} + 2\,\hat{j} \times 3\,\hat{j}$$

Use Equations 11.7a through 11.7d to evaluate the various terms:

$$\vec{B} \times \vec{A} = 0 - 3\hat{k} - 4\hat{k} + 0 = \boxed{-7\hat{k}}$$

Therefore, $\vec{A} \times \vec{B} = -\vec{B} \times \vec{A}$. As an alternative method for finding $\vec{A} \times \vec{B}$, you could use Equation 11.8. Try it!

Example **11.2**   **The Torque Vector**

A force of $\vec{F} = (2.00\,\hat{i} + 3.00\,\hat{j})$ N is applied to a rigid object that is pivoted about a fixed axis aligned along the z coordinate axis. The force is applied at a point located at $\vec{r} = (4.00\,\hat{i} + 5.00\,\hat{j})$ m relative to the axis. Find the torque $\vec{\tau}$ applied to the object.

SOLUTION

Conceptualize   Given the unit-vector notations, think about the directions of the force and position vectors. If this force were applied at this position, in what direction would an object pivoted at the origin turn?

Categorize   Because we use the definition of the cross product discussed in this section, we categorize this example as a substitution problem.

Set up the torque vector using Equation 11.1:

$$\vec{\tau} = \vec{r} \times \vec{F} = [(4.00\,\hat{i} + 5.00\,\hat{j})\ \text{m}] \times [(2.00\,\hat{i} + 3.00\,\hat{j})\ \text{N}]$$

Perform the multiplication using the distributive law:

$$\vec{\tau} = [(4.00)(2.00)\,\hat{i} \times \hat{i} + (4.00)(3.00)\,\hat{i} \times \hat{j}$$
$$+ (5.00)(2.00)\hat{j} \times \hat{i} + (5.00)(3.00)\hat{j} \times \hat{j}]\,\text{N} \cdot \text{m}$$

Use Equations 11.7a through 11.7d to evaluate the various terms:

$$\vec{\tau} = [0 + 12.0\hat{k} - 10.0\hat{k} + 0]\,\text{N} \cdot \text{m} = \boxed{2.0\hat{k}\ \text{N} \cdot \text{m}}$$

Notice that both $\vec{r}$ and $\vec{F}$ are in the xy plane. As expected, the torque vector is perpendicular to this plane, having only a z component. We have followed the rules for significant figures discussed in Section 1.6, which lead to an answer with two significant figures. We have lost some precision because we ended up subtracting two numbers that are close.

11.2   **Analysis Model: Nonisolated System (Angular Momentum)**

Imagine a rigid pole sticking up through the ice on a frozen pond (Fig. 11.3). From the left in the figure, a skater glides rapidly along a straight line toward the pole, aiming a little to the side so that she does not hit it. As she passes the pole, she reaches out to her side and grabs it, an action that causes her to move in a circular path around the pole. Just as the idea of linear momentum helps us analyze translational motion, a rotational analog—*angular momentum*—helps us analyze the motion of this skater and other objects undergoing rotational motion.

In Chapter 9, we developed the mathematical form of linear momentum and then proceeded to show how this new quantity was valuable in problem solving. We will follow a similar procedure for angular momentum.

Consider a particle of mass m located at the vector position $\vec{\mathbf{r}}$ and moving with linear momentum $\vec{\mathbf{p}}$ as in Figure 11.4. In describing translational motion, we found that the net force on the particle equals the time rate of change of its linear momentum, $\sum \vec{\mathbf{F}} = d\vec{\mathbf{p}}/dt$ (see Eq. 9.3). Let us take the cross product of each side of Equation 9.3 with $\vec{\mathbf{r}}$, which gives the net torque on the particle on the left side of the equation:

$$\vec{\mathbf{r}} \times \sum \vec{\mathbf{F}} = \sum \vec{\boldsymbol{\tau}} = \vec{\mathbf{r}} \times \frac{d\vec{\mathbf{p}}}{dt} \qquad (11.9)$$

Now, let's write Equation 11.6 with $\vec{\mathbf{A}} = \vec{\mathbf{r}}$ and $\vec{\mathbf{B}} = \vec{\mathbf{p}}$:

$$\frac{d}{dt}(\vec{\mathbf{r}} \times \vec{\mathbf{p}}) = \frac{d\vec{\mathbf{r}}}{dt} \times \vec{\mathbf{p}} + \vec{\mathbf{r}} \times \frac{d\vec{\mathbf{p}}}{dt} = \vec{\mathbf{r}} \times \frac{d\vec{\mathbf{p}}}{dt} \qquad (11.10)$$

where we recognize that the cross product of $d\vec{\mathbf{r}}/dt = \vec{\mathbf{v}}$ with $\vec{\mathbf{p}} = m\vec{\mathbf{v}}$ is zero because $\vec{\mathbf{v}}$ and $\vec{\mathbf{p}}$ are parallel. Because the right sides of Equations 11.9 and 11.10 are the same, we equate the left sides:

$$\sum \vec{\boldsymbol{\tau}} = \frac{d(\vec{\mathbf{r}} \times \vec{\mathbf{p}})}{dt} \qquad (11.11)$$

which looks very similar in form to Equation 9.3, $\sum \vec{\mathbf{F}} = d\vec{\mathbf{p}}/dt$. Because torque plays the same role in rotational motion that force plays in translational motion, this result suggests that the combination $\vec{\mathbf{r}} \times \vec{\mathbf{p}}$ should play the same role in rotational motion that $\vec{\mathbf{p}}$ plays in translational motion. We call this combination the *angular momentum* of the particle:

> The instantaneous **angular momentum** $\vec{\mathbf{L}}$ of a particle relative to an axis through a chosen origin O is defined by the cross product of the particle's instantaneous position vector $\vec{\mathbf{r}}$ relative to that origin and its instantaneous linear momentum $\vec{\mathbf{p}}$:
>
> $$\vec{\mathbf{L}} \equiv \vec{\mathbf{r}} \times \vec{\mathbf{p}} \qquad (11.12)$$

We can now write Equation 11.11 as

$$\sum \vec{\boldsymbol{\tau}} = \frac{d\vec{\mathbf{L}}}{dt} \qquad (11.13)$$

which is the rotational analog of Newton's second law, $\sum \vec{\mathbf{F}} = d\vec{\mathbf{p}}/dt$. Torque causes the angular momentum $\vec{\mathbf{L}}$ to change just as force causes linear momentum $\vec{\mathbf{p}}$ to change.

Notice that Equation 11.13 is valid only if $\sum \vec{\boldsymbol{\tau}}$ and $\vec{\mathbf{L}}$ are measured about the same axis. Furthermore, the expression is valid for any axis fixed in an inertial frame.

The SI unit of angular momentum is kg · m²/s. Notice also that both the magnitude and the direction of $\vec{\mathbf{L}}$ depend on the choice of axis. Following the right-hand rule, we see that the direction of $\vec{\mathbf{L}}$ is perpendicular to the plane formed by

Figure 11.3 As the skater passes the pole, she grabs hold of it, which causes her to swing around the pole rapidly in a circular path.

The angular momentum $\vec{\mathbf{L}}$ of a particle about an axis is a vector perpendicular to both the particle's position $\vec{\mathbf{r}}$ relative to the axis and its momentum $\vec{\mathbf{p}}$.

Figure 11.4 The angular momentum $\vec{\mathbf{L}}$ of a particle is a vector given by $\vec{\mathbf{L}} = \vec{\mathbf{r}} \times \vec{\mathbf{p}}$.

◄ Angular momentum of a particle

$\vec{\mathbf{r}}$ and $\vec{\mathbf{p}}$. In Figure 11.4, $\vec{\mathbf{r}}$ and $\vec{\mathbf{p}}$ are in the xy plane, so $\vec{\mathbf{L}}$ points in the z direction. Because $\vec{\mathbf{p}} = m\vec{\mathbf{v}}$, the magnitude of $\vec{\mathbf{L}}$ is

$$L = mvr\sin\phi \qquad (11.14)$$

where ϕ is the angle between $\vec{\mathbf{r}}$ and $\vec{\mathbf{p}}$. It follows that L is zero when $\vec{\mathbf{r}}$ is parallel to $\vec{\mathbf{p}}$ ($\phi = 0$ or $180°$). In other words, when the translational velocity of the particle is along a line that passes through the axis, the particle has zero angular momentum with respect to the axis. On the other hand, if $\vec{\mathbf{r}}$ is perpendicular to $\vec{\mathbf{p}}$ ($\phi = 90°$), then $L = mvr$. At that instant, the particle moves exactly as if it were on the rim of a wheel rotating about the axis in a plane defined by $\vec{\mathbf{r}}$ and $\vec{\mathbf{p}}$.

> **PITFALL PREVENTION 11.2**
>
> **Is Rotation Necessary for Angular Momentum?** We can define angular momentum even if the particle is not moving in a circular path. A particle moving in a straight line such as the skater in Figure 11.3 has angular momentum about any axis displaced from the path of the particle, such as an axis through the pole in Figure 11.3. See the What If? in Example 11.3.

QUICK QUIZ **11.2** Recall the skater described at the beginning of this section. Let her mass be m. **(i)** What would be her angular momentum relative to the pole at the instant she is a distance d from the pole if she were skating directly toward it at speed v? (a) zero (b) mvd (c) impossible to determine **(ii)** What would be her angular momentum relative to the pole at the instant she is a distance d from the pole if she were skating at speed v along a straight path that is offset by a perpendicular distance a from the pole? (a) zero (b) mvd (c) mva (d) impossible to determine

---

Example 11.3   **Angular Momentum of a Particle in Uniform Circular Motion**

A particle moves at constant speed in the xy plane in a circular path of radius r as shown in Figure 11.5. Find the magnitude and direction of its angular momentum relative to an axis through O when its velocity is $\vec{\mathbf{v}}$.

SOLUTION

Conceptualize   The linear momentum of the particle is always changing in direction (but not in magnitude). You might therefore be tempted to conclude that the angular momentum of the particle is always changing. In this situation, however, that is not the case. Let's see why.

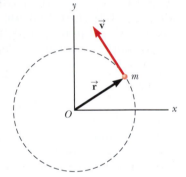

Figure 11.5 (Example 11.3) A particle moving in a circle of radius r has an angular momentum about an axis through O that has magnitude mvr. The vector $\vec{\mathbf{L}} = \vec{\mathbf{r}} \times \vec{\mathbf{p}}$ points *out* of the page.

Categorize   We use the definition of the angular momentum of a particle discussed in this section, so we categorize this example as a substitution problem.

Use Equation 11.14 to evaluate the magnitude of $\vec{\mathbf{L}}$: $L = mvr\sin 90° = \boxed{mvr}$

This value of L is constant because all three factors on the right are constant. The direction of $\vec{\mathbf{L}}$ also is constant, even though the direction of $\vec{\mathbf{p}} = m\vec{\mathbf{v}}$ keeps changing. To verify this statement, apply the right-hand rule to find the direction of $\vec{\mathbf{L}} = \vec{\mathbf{r}} \times \vec{\mathbf{p}} = m\vec{\mathbf{r}} \times \vec{\mathbf{v}}$ in Figure 11.5. Your thumb points out of the page, so that is the direction of $\vec{\mathbf{L}}$. Hence, we can write the vector expression $\vec{\mathbf{L}} = (mvr)\hat{\mathbf{k}}$. If the particle were to move clockwise, $\vec{\mathbf{L}}$ would point downward and into the page and $\vec{\mathbf{L}} = -(mvr)\hat{\mathbf{k}}$. A particle in *uniform* circular motion has a *constant* angular momentum about an axis through the center of its path.

WHAT IF?   The particle in Figure 11.4 moves in a straight line at constant speed along a path parallel to the linear momentum vector $\vec{\mathbf{p}}$. Is the angular momentum of the particle constant in this case?

Answer   Yes. In Equation 11.14, m and v are constant while r and ϕ vary in time. However, the product $r\sin\phi$ represents the perpendicular distance between the y axis and the path of the particle. This distance is constant. Therefore, L in Equation 11.14 has a fixed value even though the distance between the particle and the origin changes.

Angular Momentum of a System of Particles

Using the techniques of Section 9.7, we can show that Newton's second law for a system of particles is

$$\sum \vec{\mathbf{F}}_{\text{ext}} = \frac{d\vec{\mathbf{P}}_{\text{tot}}}{dt}$$

This equation states that the net external force on a system of particles is equal to the time rate of change of the total linear momentum of the system. Let's see if a similar statement can be made for rotational motion. The total angular momentum of a system of particles about some axis is defined as the vector sum of the angular momenta of the individual particles:

$$\vec{\mathbf{L}}_{tot} = \vec{\mathbf{L}}_1 + \vec{\mathbf{L}}_2 + \cdots + \vec{\mathbf{L}}_n = \sum_i \vec{\mathbf{L}}_i$$

where the vector sum is over all n particles in the system.

Differentiating this equation with respect to time gives

$$\frac{d\vec{\mathbf{L}}_{tot}}{dt} = \sum_i \frac{d\vec{\mathbf{L}}_i}{dt} = \sum_i \vec{\boldsymbol{\tau}}_i$$

where we have used Equation 11.13 to replace the time rate of change of the angular momentum of each particle with the net torque on the particle.

The torques acting on the individual particles of the system are those associated with internal forces between particles and those associated with external forces. The net torque associated with all internal forces, however, is zero. Recall that Newton's third law tells us that internal forces between particles of the system occur in pairs that are equal in magnitude and opposite in direction. If we assume these forces lie along the line of separation of each pair of particles, the total torque around some axis passing through an origin O due to each action–reaction force pair is zero (that is, the moment arm d from O to the line of action of the forces is equal for both particles, and the forces are in opposite directions). In the summation, therefore, the net internal torque is zero. We conclude that the total angular momentum of a system can vary with time only if a net external torque is acting on the system:

$$\sum \vec{\boldsymbol{\tau}}_{ext} = \frac{d\vec{\mathbf{L}}_{tot}}{dt} \qquad (11.15)$$

◀ The net external torque on a system equals the time rate of change of angular momentum of the system

This equation is indeed the rotational analog of $\sum \vec{\mathbf{F}}_{ext} = d\vec{\mathbf{p}}_{tot}/dt$ for a system of particles. Equation 11.15 is the mathematical representation of the **angular momentum version of the nonisolated system model.** If a system is nonisolated in the sense that there is a net external torque on it, the net external torque on the system is equal to the time rate of change of the angular momentum of the system.

Although we do not prove it here, this statement is true regardless of the motion of the center of mass. It applies even if the center of mass is accelerating, provided the torque and angular momentum are evaluated relative to an axis through the center of mass.

Equation 11.15 can be rearranged and integrated to give

$$\Delta \vec{\mathbf{L}}_{tot} = \int \left(\sum \vec{\boldsymbol{\tau}}_{ext} \right) dt$$

This equation represents the *angular impulse–angular momentum theorem.* Compare this equation to the translational version, Equation 9.40.

ANALYSIS MODEL **Nonisolated System (Angular Momentum)**

Imagine a system that rotates about an axis. If there is a net external torque acting on the system, the time rate of change of the angular momentum of the system is equal to the net external torque:

$$\sum \vec{\boldsymbol{\tau}}_{ext} = \frac{d\vec{\mathbf{L}}_{tot}}{dt} \qquad (11.15)$$

continued

ANALYSIS MODEL   **Nonisolated System (Angular Momentum)** *continued*

Examples:

- a flywheel in an automobile engine increases its angular momentum when the engine applies torque to it
- the tub of a washing machine decreases in angular momentum due to frictional torque after the machine is turned off
- the axis of the Earth undergoes a precessional motion due to the torque exerted on the Earth by the gravitational force from the Sun
- the armature of a motor increases its angular momentum due to the torque exerted by a surrounding magnetic field (Chapter 30)

The rate of change in the angular momentum of the nonisolated system is equal to the net external torque on the system.

Example 11.4   A System of Objects

A sphere of mass m_1 and a block of mass m_2 are connected by a light cord that passes over a pulley as shown in Figure 11.6. The radius of the pulley is R, and the mass of the thin rim is M. The spokes of the pulley have negligible mass. The block slides on a frictionless, horizontal surface. Find an expression for the linear acceleration of the two objects, using the concepts of angular momentum and torque.

SOLUTION

Conceptualize   When the system is released, the block slides to the left, the sphere drops downward, and the pulley rotates counterclockwise. This situation is similar to problems we have solved earlier except that now we want to use an angular momentum approach.

Categorize   We identify the block, pulley, and sphere as a *nonisolated system* for *angular momentum*, subject to the external torque due to the gravitational force on the sphere. We shall calculate the angular momentum about an axis that coincides with the axle of the pulley. The angular momentum of the system includes that of two objects moving translationally (the sphere and the block) and one object undergoing pure rotation (the pulley).

Figure 11.6   (Example 11.4) When the system is released, the sphere moves downward and the block moves to the left.

Analyze   At any instant of time, the sphere and the block have a common speed v, so the angular momentum of the sphere about the pulley axle is $m_1 vR$ and that of the block is $m_2 vR$. At the same instant, all points on the rim of the pulley also move with speed v, so the angular momentum of the pulley is MvR.

Now let's address the total external torque acting on the system about the pulley axle. Because it has a moment arm of zero, the force exerted by the axle on the pulley does not contribute to the torque. Furthermore, the normal force acting on the block is balanced by the gravitational force $m_2 \vec{\mathbf{g}}$, so these forces do not contribute to the torque. The gravitational force $m_1 \vec{\mathbf{g}}$ acting on the sphere produces a torque about the axle equal in magnitude to $m_1 gR$, where R is the moment arm of the force about the axle. This result is the total external torque about the pulley axle; that is, $\sum \tau_{\text{ext}} = m_1 gR$.

Write an expression for the total angular momentum of the system:	(1) $L = m_1 vR + m_2 vR + MvR = (m_1 + m_2 + M)vR$
Substitute this expression and the total external torque into Equation 11.15, the mathematical representation of the nonisolated system model for angular momentum:	$\sum \tau_{\text{ext}} = \dfrac{dL}{dt}$ $m_1 gR = \dfrac{d}{dt}[(m_1 + m_2 + M)vR]$ (2) $m_1 gR = (m_1 + m_2 + M)R\dfrac{dv}{dt}$
Recognizing that $dv/dt = a$, solve Equation (2) for a:	(3) $a = \dfrac{m_1 g}{m_1 + m_2 + M}$

11.4 c o n t i n u e d

Finalize When we evaluated the net torque about the axle, we did not include the forces that the cord exerts on the objects because these forces are internal to the system under consideration. Instead, we analyzed the system as a whole. Only *external* torques contribute to the change in the system's angular momentum. Let $M \to 0$ in Equation (3) and call the result Equation A. Now go back to Equation (5) in Example 5.10, let $\theta \to 0$, and call the result Equation B. Do Equations A and B match? Looking at Figures 5.16 and 11.6 in these limits, *should* the two equations match?

11.3 Angular Momentum of a Rotating Rigid Object

In Example 11.4, we considered the angular momentum of a deformable system of particles. Let us now restrict our attention to a nondeformable system, a rigid object. Consider a rigid object rotating about a fixed axis that coincides with the z axis of a coordinate system as shown in Figure 11.7. Let's determine the angular momentum of this object. Each *particle* of the object rotates in the xy plane about the z axis with an angular speed ω. The magnitude of the angular momentum of a particle of mass m_i about the z axis is $m_i v_i r_i$. Because $v_i = r_i \omega$ (Eq. 10.10), we can express the magnitude of the angular momentum of this particle as

$$L_i = m_i v_i r_i = m_i (r_i \omega) r_i = m_i r_i^2 \omega$$

The vector $\vec{\mathbf{L}}_i$ for this particle is directed along the z axis, as is the vector $\vec{\boldsymbol{\omega}}$.

We can now find the angular momentum (which in this situation has only a z component) of the whole object by taking the sum of L_i over all particles:

$$L_z = \sum_i L_i = \sum_i m_i r_i^2 \omega = \left(\sum_i m_i r_i^2 \right) \omega$$

$$L_z = I\omega \tag{11.16}$$

where we have recognized $\sum_i m_i r_i^2$ as the moment of inertia I of the object about the z axis (Eq. 10.19). Notice that Equation 11.16 is mathematically similar in form to Equation 9.2 for linear momentum: $\vec{\mathbf{p}} = m\vec{\mathbf{v}}$.

Now let's differentiate Equation 11.16 with respect to time, noting that I is constant for a rigid object:

$$\frac{dL_z}{dt} = I\frac{d\omega}{dt} = I\alpha \tag{11.17}$$

where α is the angular acceleration relative to the axis of rotation. Because dL_z/dt is equal to the net external torque (see Eq. 11.15), we can express Equation 11.17 as

$$\sum \tau_{\text{ext}} = I\alpha \tag{11.18}$$

◀ Rotational form of Newton's second law

That is, the net external torque acting on a rigid object rotating about a fixed axis equals the moment of inertia about the rotation axis multiplied by the object's angular acceleration relative to that axis. This result is the same as Equation 10.18, which was derived using a force approach, but we derived Equation 11.18 using the concept of angular momentum. As we saw in Section 10.5, Equation 11.18 is the mathematical representation of the rigid object under a net torque analysis model. This equation is also valid for a rigid object rotating about a moving axis, provided the moving axis (1) passes through the center of mass and (2) is a symmetry axis.

If a symmetrical object rotates about a fixed axis passing through its center of mass, you can write Equation 11.16 in vector form as $\vec{\mathbf{L}} = I\vec{\boldsymbol{\omega}}$, where $\vec{\mathbf{L}}$ is the total angular momentum of the object measured with respect to the axis of rotation.

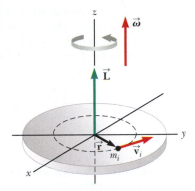

Figure 11.7 When a rigid object rotates about an axis, the angular momentum $\vec{\mathbf{L}}$ is in the same direction as the angular velocity $\vec{\boldsymbol{\omega}}$ according to the expression $\vec{\mathbf{L}} = I\vec{\boldsymbol{\omega}}$.

Furthermore, the expression is valid for any object, regardless of its symmetry, if $\vec{\mathbf{L}}$ stands for the component of angular momentum along the axis of rotation.[1]

QUICK QUIZ 11.3 A solid sphere and a hollow sphere have the same mass and radius. They are rotating with the same angular speed. Which one has the higher angular momentum? **(a)** the solid sphere **(b)** the hollow sphere **(c)** both have the same angular momentum **(d)** impossible to determine

---

Example **11.5**   The Seesaw

A father of mass m_f and his daughter of mass m_d sit on opposite ends of a seesaw at equal distances from the pivot at the center (Fig. 11.8). The seesaw is modeled as a rigid rod of mass M and length ℓ and is pivoted without friction. At a given moment, the combination rotates in a vertical plane with an angular speed ω.

(A) Find an expression for the magnitude of the system's angular momentum.

Figure 11.8 (Example 11.5) A father and daughter demonstrate angular momentum on a seesaw.

SOLUTION

Conceptualize   Identify the z axis through O as the axis of rotation in Figure 11.8. The rotating system has angular momentum about that axis.

Categorize   Ignore any movement of arms or legs of the father and daughter and model them both as particles. The system is therefore modeled as a rigid object. This first part of the example is categorized as a substitution problem.

The moment of inertia of the system equals the sum of the moments of inertia of the three components: the seesaw and the two individuals. We can refer to Table 10.2 to obtain the expression for the moment of inertia of the rod and use the particle expression $I = mr^2$ for each person.

Find the total moment of inertia of the system about the z axis through O:

$$I = \tfrac{1}{12}M\ell^2 + m_f\left(\frac{\ell}{2}\right)^2 + m_d\left(\frac{\ell}{2}\right)^2 = \frac{\ell^2}{4}\left(\frac{M}{3} + m_f + m_d\right)$$

Find the magnitude of the angular momentum of the system:

$$L = I\omega = \frac{\ell^2}{4}\left(\frac{M}{3} + m_f + m_d\right)\omega$$

(B) Find an expression for the magnitude of the angular acceleration of the system when the seesaw makes an angle θ with the horizontal.

SOLUTION

Conceptualize   Generally, fathers are more massive than daughters, so the system is not in equilibrium and has an angular acceleration. We expect the angular acceleration to be positive in Figure 11.8.

Categorize   The combination of the board, father, and daughter is a *rigid object under a net torque* because of the external torque associated with the gravitational forces on the father and daughter. We again identify the axis of rotation as the z axis in Figure 11.8.

Analyze   To find the angular acceleration of the system at any angle θ, we first calculate the net torque on the system and then use $\Sigma \tau_{\text{ext}} = I\alpha$ from the rigid object under a net torque model to obtain an expression for α.

Evaluate the torque due to the gravitational force on the father:

$$\tau_f = m_f g\,\frac{\ell}{2}\cos\theta \quad (\vec{\boldsymbol{\tau}}_f \text{ out of page})$$

Evaluate the torque due to the gravitational force on the daughter:

$$\tau_d = -m_d g\,\frac{\ell}{2}\cos\theta \quad (\vec{\boldsymbol{\tau}}_d \text{ into page})$$

---

[1]In general, the expression $\vec{\mathbf{L}} = I\vec{\boldsymbol{\omega}}$ is not always valid. If a rigid object rotates about an *arbitrary* axis, then $\vec{\mathbf{L}}$ and $\vec{\boldsymbol{\omega}}$ may point in different directions. In this case, the moment of inertia cannot be treated as a scalar. Strictly speaking, $\vec{\mathbf{L}} = I\vec{\boldsymbol{\omega}}$ applies only to rigid objects of any shape that rotate about one of three mutually perpendicular axes (called *principal axes*) through the center of mass. This concept is discussed in more advanced texts on mechanics.

11.5 continued

Evaluate the net external torque exerted on the system:

$$\sum \tau_{ext} = \tau_f + \tau_d = \tfrac{1}{2}(m_f - m_d)g\ell \cos \theta$$

Use Equation 11.18 and I from part (A) to find α:

$$\alpha = \frac{\sum \tau_{ext}}{I} = \frac{2(m_f - m_d)g \cos \theta}{\ell \left[(M/3) + m_f + m_d\right]}$$

Finalize For a father more massive than his daughter, the angular acceleration is positive as expected. If the seesaw begins in a horizontal orientation ($\theta = 0$) and is released, the rotation is counterclockwise in Figure 11.8 and the father's end of the seesaw drops, which is consistent with everyday experience.

WHAT IF? Imagine the father moves inward on the seesaw to a distance d from the pivot to try to balance the two sides. What is the angular acceleration of the system in this case when it is released from an arbitrary angle θ?

Answer The angular acceleration of the system should decrease if the system is more balanced.

Find the total moment of inertia about the z axis through O for the modified system:

$$I = \tfrac{1}{12}M\ell^2 + m_f d^2 + m_d \left(\frac{\ell}{2}\right)^2 = \frac{\ell^2}{4}\left(\frac{M}{3} + m_d\right) + m_f d^2$$

Find the net torque exerted on the modified system about an axis through O:

$$\sum \tau_{ext} = \tau_f + \tau_d = m_f g d \cos \theta - \tfrac{1}{2}m_d g\ell \cos \theta$$

Find the new angular acceleration of the system:

$$\alpha = \frac{\sum \tau_{ext}}{I} = \frac{(m_f d - \tfrac{1}{2}m_d \ell)g \cos \theta}{(\ell^2/4)\left[(M/3) + m_d\right] + m_f d^2}$$

WHAT IF? Where must the father sit for the seesaw to be balanced?

Answer The seesaw is balanced when the angular acceleration is zero. In this situation, both father and daughter can push off the ground and rise to the highest possible point.

For the seesaw to be balanced, the required position of the father is found by setting $\alpha = 0$:

$$\alpha = \frac{(m_f d - \tfrac{1}{2}m_d \ell)g \cos \theta}{(\ell^2/4)\left[(M/3) + m_d\right] + m_f d^2} = 0$$

$$m_f d - \tfrac{1}{2}m_d \ell = 0 \quad \rightarrow \quad d = \left(\frac{m_d}{m_f}\right)\frac{\ell}{2}$$

The heavier the father, the closer he must sit to the pivot to balance the seesaw. In the rare case that the father and daughter have the same mass, the father is located at the end of the seesaw, $d = \ell/2$.

11.4 Analysis Model: Isolated System (Angular Momentum)

In Chapter 9, we found that the total linear momentum of a system of particles remains constant if the system is isolated, that is, if the net external force acting on the system is zero. We have an analogous conservation law in rotational motion:

> The total angular momentum of a system is constant in both magnitude and direction if the net external torque acting on the system is zero, that is, if the system is isolated.

◄ Conservation of angular momentum

This statement is often called[2] the principle of **conservation of angular momentum** and is the basis of the **angular momentum version of the isolated system model**. This principle follows directly from Equation 11.15, which indicates that if

$$\sum \vec{\tau}_{ext} = \frac{d\vec{L}_{tot}}{dt} = 0$$

---

[2]The most general conservation of angular momentum equation is Equation 11.15, which describes how the system interacts with its environment.

then

$$\Delta \vec{\mathbf{L}}_{\text{tot}} = 0 \qquad (11.19)$$

Equation 11.19 can be written as

$$\vec{\mathbf{L}}_{\text{tot}} = \text{constant} \quad \text{or} \quad \vec{\mathbf{L}}_i = \vec{\mathbf{L}}_f \qquad (11.20)$$

For an isolated system consisting of a small number of particles, we write this conservation law as $\vec{\mathbf{L}}_{\text{tot}} = \sum \vec{\mathbf{L}}_n = \text{constant}$, where the index n denotes the nth particle in the system. If the system consists of a large number of particles, so that it is difficult to evaluate the individual L_n, then we can express the magnitude of the angular momentum of the system with Equation 11.16, $L = I\omega$.

If an isolated rotating system is deformable so that its mass undergoes redistribution in some way, the system's moment of inertia changes. Combining Equations 11.16 and 11.20, we see that conservation of angular momentum requires that the product of I and ω must remain constant. Therefore, a change in I for an isolated system requires a change in ω. In this case, we can express the principle of conservation of angular momentum as

$$I_i \omega_i = I_f \omega_f = \text{constant} \qquad (11.21)$$

This expression is valid both for rotation about a fixed axis and for rotation about an axis through the center of mass of a moving system as long as that axis remains fixed in direction. We require only that the net external torque be zero.

Many examples demonstrate conservation of angular momentum for a deformable system. You may have observed a figure skater spinning in the finale of a program (Fig. 11.9). The angular speed of the skater is large when his hands and feet are close to the trunk of his body. (Notice the skater's hair!) Ignoring friction between skater and ice, there are no external torques on the skater. The moment of inertia of his body increases as his hands and feet are moved away from his body, and therefore from the rotation axis, at the finish of the spin. According to the isolated system model for angular momentum, his angular speed must decrease, and he can perform his finishing flourish after coming to rest. In a similar way, when divers or acrobats wish to make several rotations, they pull their hands and feet close to their bodies to rotate at a higher rate, as shown in the photograph opening this chapter. In these cases, the external force due to gravity acts through the center of mass and hence exerts no torque about an axis through this point. Therefore, the angular momentum of the diver or acrobat about the center of mass must be conserved; that is, $I_i \omega_i = I_f \omega_f$. For example, when divers wish to double their angular speed, they must reduce their moment of inertia to half its initial value.

The introductory storyline to this chapter asked about the additional rotational kinetic energy possessed by spinning skaters and rotating divers when they pull their limbs inward. This energy comes from within the body. The muscles of the rotating athlete must do internal work to pull the limbs inward. This work is a transformation mechanism by which potential energy in the body from previous meals is transformed to rotational kinetic energy. The storyline also mentioned a falling cat, which is yet again an example of a deformable system, first introduced in Section 9.8. The cat is released with zero angular momentum, yet is able to rotate and right itself before landing. A number of theories have been proposed for this phenomenon, including a popular one modeling the cat as a pair of cylinders. Perform some online research to learn more about falling cats.

In Equation 11.20, we have a third version of the isolated system model. We can now state that the energy, linear momentum, and angular momentum of an

When his arms and legs are close to his body, the skater's moment of inertia is small and his angular speed is large.

To slow down for the finish of his spin, the skater moves his arms and legs outward, increasing his moment of inertia.

Figure 11.9 Angular momentum is conserved as Russian gold medalist Evgeni Plushenko performs during the Turin 2006 Winter Olympic Games.

isolated system are all constant:

$$\Delta E_{system} = 0 \quad \text{(if there are no energy transfers across the system boundary)}$$

$$\Delta \vec{p}_{tot} = 0 \quad \text{(if the net external force on the system is zero)}$$

$$\Delta \vec{L}_{tot} = 0 \quad \text{(if the net external torque on the system is zero)}$$

Notice that the definition of an isolated system varies for the three conserved quantities. A system may be isolated in terms of one of these quantities but not in terms of another. If a system is nonisolated in terms of momentum or angular momentum, it will often be nonisolated also in terms of energy because the system has a net force or torque on it and the net force or torque will do work on the system. We can, however, identify systems that are nonisolated in terms of energy but isolated in terms of momentum. For example, imagine pushing inward on a balloon (the system) between your hands. Work is done in compressing the balloon, so the system is nonisolated in terms of energy, but there is zero net force on the system, so the system is isolated in terms of momentum. A similar statement could be made about twisting the ends of a long, flat, springy piece of metal with both hands. Work is done on the metal (the system), so energy is stored in the nonisolated system as elastic potential energy, but the net torque on the system is zero. Therefore, the system is isolated in terms of angular momentum. Other examples are collisions of macroscopic objects, which represent isolated systems in terms of momentum but nonisolated systems in terms of energy because of the output of energy from the system by mechanical waves (sound).

QUICK QUIZ 11.4 A competitive diver leaves the diving board and falls toward the water with her body straight and rotating slowly. She pulls her arms and legs into a tight tuck position. What happens to her rotational kinetic energy? **(a)** It increases. **(b)** It decreases. **(c)** It stays the same. **(d)** It is impossible to determine.

ANALYSIS MODEL   **Isolated System (Angular Momentum)**

Imagine a system rotates about an axis. If there is no net external torque on the system, there is no change in the angular momentum of the system:

$$\Delta \vec{L}_{tot} = 0 \quad \textbf{(11.19)}$$

Applying this law of conservation of angular momentum to a system whose moment of inertia changes gives

$$I_i \omega_i = I_f \omega_f = \text{constant} \quad \textbf{(11.21)}$$

Examples:

- after a supernova explosion, the core of a star collapses to a small radius and spins at a much higher rate (Example 11.6, below).
- the square of the orbital period of a planet is proportional to the cube of its semimajor axis; Kepler's third law (Chapter 13)
- in atomic transitions, selection rules on the quantum numbers must be obeyed in order to conserve angular momentum (Chapter 41)
- in beta decay of a radioactive nucleus, a neutrino must be emitted in order to conserve angular momentum (Chapter 43)

Example 11.6   **Formation of a Neutron Star**

A star rotates with a period of 30 days about an axis through its center. The period is the time interval required for a point on the star's equator to make one complete revolution around the axis of rotation. After the star undergoes a supernova explosion, the stellar core, which had a radius of 1.0×10^4 km, collapses into a neutron star of radius 10.0 km. Determine the period of rotation of the neutron star.

SOLUTION

Conceptualize The change in the neutron star's motion is similar to that of the skater described earlier and illustrated in Figure 11.9, but in the reverse direction. As the mass of the star moves closer to the rotation axis, we expect the star to spin faster.

continued

11.6 continued

Categorize   Let us assume that during the collapse of the stellar core, (1) no external torque acts on it, (2) it remains spherical with the same relative mass distribution, and (3) its mass remains constant. We categorize the star as an *isolated system* in terms of *angular momentum*. We do not know the mass distribution of the star, but we have assumed the distribution is symmetric, so the moment of inertia can be expressed as kMR^2, where k is some numerical constant. (From Table 10.2, for example, we see that $k = \frac{2}{5}$ for a solid sphere and $k = \frac{2}{3}$ for a spherical shell.)

Analyze   Let's use the symbol T for the period, with T_i being the initial period of the star and T_f being the period of the neutron star. The star's angular speed is given by $\omega = 2\pi/T$.

From the isolated system model for angular momentum, write Equation 11.21 for the star:

$$I_i\omega_i = I_f\omega_f$$

Use $\omega = 2\pi/T$ to rewrite this equation in terms of the initial and final periods:

$$I_i\left(\frac{2\pi}{T_i}\right) = I_f\left(\frac{2\pi}{T_f}\right)$$

Substitute the moments of inertia in the preceding equation:

$$kMR_i^2\left(\frac{2\pi}{T_i}\right) = kMR_f^2\left(\frac{2\pi}{T_f}\right)$$

Solve for the final period of the star:

$$T_f = \left(\frac{R_f}{R_i}\right)^2 T_i$$

Substitute numerical values:

$$T_f = \left(\frac{10.0 \text{ km}}{1.0 \times 10^4 \text{ km}}\right)^2 (30 \text{ days}) = 3.0 \times 10^{-5} \text{ days} = \boxed{2.6 \text{ s}}$$

Finalize   The neutron star does indeed rotate faster after it collapses, as predicted. While it may seem difficult to believe that the core of a star could rotate as fast as once every 2.6 s, this is a relatively slow rotation rate. Some neutron stars rotate with a period of 1–2 milliseconds!

Example **11.7**   **The Merry-Go-Round**

A horizontal platform in the shape of a circular disk rotates freely in a horizontal plane about a frictionless, vertical axle (Fig. 11.10). The platform has a mass $M = 100$ kg and a radius $R = 2.0$ m. A student whose mass is $m = 60$ kg walks slowly from the rim of the disk toward its center. If the angular speed of the system is 2.0 rad/s when the student is at the rim, what is the angular speed when she reaches a point $r = 0.50$ m from the center?

S O L U T I O N

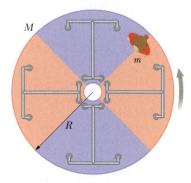

Figure 11.10 (Example 11.7) As the student walks toward the center of the rotating platform, the angular speed of the system increases because the angular momentum of the system remains constant.

Conceptualize   The speed change here is similar to those of the spinning skater and the neutron star in preceding discussions. This problem is different because part of the moment of inertia of the system changes (that of the student) while part remains fixed (that of the platform).

Categorize   Because the platform rotates on a frictionless axle, we identify the system of the student and the platform as an *isolated system* in terms of *angular momentum*.

Analyze   Let us denote the moment of inertia of the platform as I_p and that of the student as I_s. We model the student as a particle.

Write Equation 11.21 for the system:

$$I_i\omega_i = I_f\omega_f$$

Substitute the moments of inertia, using $r < R$ for the final position of the student:

$$(\tfrac{1}{2}MR^2 + mR^2)\omega_i = (\tfrac{1}{2}MR^2 + mr^2)\omega_f$$

11.7 continued

Solve for the final angular speed:

$$\omega_f = \left(\frac{\frac{1}{2}MR^2 + mR^2}{\frac{1}{2}MR^2 + mr^2} \right) \omega_i$$

Substitute numerical values:

$$\omega_f = \left[\frac{\frac{1}{2}(100 \text{ kg})(2.0 \text{ m})^2 + (60 \text{ kg})(2.0 \text{ m})^2}{\frac{1}{2}(100 \text{ kg})(2.0 \text{ m})^2 + (60 \text{ kg})(0.50 \text{ m})^2} \right](2.0 \text{ rad/s}) = \left[\frac{440 \text{ kg} \cdot \text{m}^2}{215 \text{ kg} \cdot \text{m}^2} \right](2.0 \text{ rad/s}) = \boxed{4.1 \text{ rad/s}}$$

Finalize As expected, the angular speed increases. The fastest that this system could spin would be when the student moves to the center of the platform. Do this calculation to show that this maximum angular speed is 4.4 rad/s. Notice that the activity described in this problem is dangerous as discussed with regard to the Coriolis force in Section 6.3.

WHAT IF? What if you measured the kinetic energy of the system before and after the student walks inward? Are the initial kinetic energy and the final kinetic energy the same?

Answer You may be tempted to say yes because the system is isolated. Remember, however, that energy can be transformed among several forms, so we have to handle an energy question carefully.

Find the initial kinetic energy:

$$K_i = \tfrac{1}{2}I_i\omega_i^2 = \tfrac{1}{2}(440 \text{ kg} \cdot \text{m}^2)(2.0 \text{ rad/s})^2 = 880 \text{ J}$$

Find the final kinetic energy:

$$K_f = \tfrac{1}{2}I_f\omega_f^2 = \tfrac{1}{2}(215 \text{ kg} \cdot \text{m}^2)(4.1 \text{ rad/s})^2 = 1.80 \times 10^3 \text{ J}$$

Therefore, the kinetic energy of the system *increases* by more than a factor of 2. The student must perform muscular activity to move herself closer to the center of rotation, so this extra kinetic energy comes from potential energy stored in the student's body from previous meals. The system is isolated in terms of energy, but a transformation process within the system changes potential energy to kinetic energy.

Example 11.8 Disk and Stick Collision

A 2.0-kg disk traveling at 3.0 m/s strikes a 1.0-kg stick of length 4.0 m that is lying flat on nearly frictionless ice as shown in the overhead view of Figure 11.11a. The disk strikes at the endpoint of the stick, at a distance $r = 2.0$ m from the stick's center. Assume the collision is elastic and the disk does not deviate from its original line of motion. Find the translational speed of the disk, the translational speed of the stick, and the angular speed of the stick after the collision. The moment of inertia of the stick about its center of mass is 1.33 kg · m².

SOLUTION

Conceptualize Examine Figure 11.11a and imagine what happens after the disk hits the stick. Figure 11.11b shows what you might expect: the disk continues to move at a slower speed, and the stick is in both translational and rotational motion. We assume the disk does not deviate from its original line of motion because the force exerted by the stick on the disk is parallel to the original path of the disk.

Categorize Because the ice is frictionless, the disk and stick form an *isolated system* in terms of *momentum* and *angular momentum*. Ignoring the sound made in the collision, we also model the system as an *isolated system* in terms of *energy*. In addition, because the collision is assumed to be elastic, the kinetic energy of the system is constant.

Figure 11.11 (Example 11.8) Overhead view of a disk striking a stick in an elastic collision. (a) Before the collision, the disk moves toward the stick. (b) The collision causes the stick to rotate and move to the right.

Analyze First notice that we have three unknowns, so we need three equations to solve simultaneously.

Apply the isolated system model for momentum to the system and then rearrange the result:

$$\Delta \vec{\mathbf{p}}_{tot} = 0 \quad \rightarrow \quad (m_d v_{df} + m_s v_s) - m_d v_{di} = 0$$

$$(1) \quad m_d(v_{di} - v_{df}) = m_s v_s$$

continued

11.8 continued

Apply the isolated system model for angular momentum to the system and rearrange the result. Use an axis passing through the center of the stick as the rotation axis so that the path of the disk is a distance r from the rotation axis:	$\Delta \vec{L}_{\text{tot}} = 0 \rightarrow (-rm_d v_{df} + I\omega) - (-rm_d v_{di}) = 0$ (2) $-rm_d(v_{di} - v_{df}) = I\omega$
Apply the isolated system model for energy to the system, rearrange the equation, and factor the combination of terms related to the disk:	$\Delta K = 0 \rightarrow (\frac{1}{2}m_d v_{df}^2 + \frac{1}{2}m_s v_s^2 + \frac{1}{2}I\omega^2) - \frac{1}{2}m_d v_{di}^2 = 0$ (3) $m_d(v_{di} - v_{df})(v_{di} + v_{df}) = m_s v_s^2 + I\omega^2$
Multiply Equation (1) by r and add to Equation (2):	$rm_d(v_{di} - v_{df}) = rm_s v_s$ $-rm_d(v_{di} - v_{df}) = I\omega$ $0 = rm_s v_s + I\omega$
Solve for ω:	(4) $\omega = -\dfrac{rm_s v_s}{I}$
Divide Equation (3) by Equation (1):	$\dfrac{m_d(v_{di} - v_{df})(v_{di} + v_{df})}{m_d(v_{di} - v_{df})} = \dfrac{m_s v_s^2 + I\omega^2}{m_s v_s}$ (5) $v_{di} + v_{df} = v_s + \dfrac{I\omega^2}{m_s v_s}$
Substitute Equation (4) into Equation (5):	(6) $v_{di} + v_{df} = v_s\left(1 + \dfrac{r^2 m_s}{I}\right)$
Substitute v_{df} from Equation (1) into Equation (6):	$v_{di} + \left(v_{di} - \dfrac{m_s}{m_d}v_s\right) = v_s\left(1 + \dfrac{r^2 m_s}{I}\right)$
Solve for v_s and substitute numerical values:	$v_s = \dfrac{2v_{di}}{1 + (m_s/m_d) + (r^2 m_s/I)}$ $= \dfrac{2(3.0 \text{ m/s})}{1 + (1.0 \text{ kg}/2.0 \text{ kg}) + [(2.0 \text{ m})^2(1.0 \text{ kg})/1.33 \text{ kg} \cdot \text{m}^2]} = \boxed{1.3 \text{ m/s}}$
Substitute numerical values into Equation (4):	$\omega = -\dfrac{(2.0 \text{ m})(1.0 \text{ kg})(1.3 \text{ m/s})}{1.33 \text{ kg} \cdot \text{m}^2} = \boxed{-2.0 \text{ rad/s}}$
Solve Equation (1) for v_{df} and substitute numerical values:	$v_{df} = v_{di} - \dfrac{m_s}{m_d}v_s = 3.0 \text{ m/s} - \dfrac{1.0 \text{ kg}}{2.0 \text{ kg}}(1.3 \text{ m/s}) = \boxed{2.3 \text{ m/s}}$

Finalize   These values seem reasonable. The disk is moving more slowly after the collision than it was before the collision. The stick has a small translational speed and is rotating clockwise. Table 11.1 summarizes the initial and final values of variables for the disk and the stick, and it verifies the conservation of linear momentum, angular momentum, and kinetic energy for the isolated system.

TABLE 11.1   Comparison of Values in Example 11.8 Before and After the Collision

	v (m/s)	ω (rad/s)	p (kg · m/s)	L (kg · m²/s)	K_{trans} (J)	K_{rot} (J)
Before						
Disk	3.0	—	6.0	−12	9.0	—
Stick	0	0	0	0	0	0
Total for system	—	—	6.0	−12	9.0	0
After						
Disk	2.3	—	4.7	−9.3	5.4	—
Stick	1.3	−2.0	1.3	−2.7	0.9	2.7
Total for system	—	—	6.0	−12	6.3	2.7

Note: Linear momentum, angular momentum, and total kinetic energy of the system are all conserved.

The Motion of Gyroscopes and Tops

An unusual and fascinating type of motion you have probably observed is that of a top spinning about its axis of symmetry as shown in Figure 11.12a. If the top spins rapidly, the symmetry axis rotates about the z axis, sweeping out a cone (see Fig. 11.12b). The motion of the symmetry axis about the vertical—known as **precessional motion**—is usually slow relative to the spinning motion of the top.

It is quite natural to wonder why the top does not fall over. Because the center of mass is not directly above the pivot point O, a net torque is acting on the top about an axis passing through O, a torque resulting from the gravitational force $M\vec{\mathbf{g}}$. The top would certainly fall over if it were not spinning. Because it is spinning, however, it has an angular momentum $\vec{\mathbf{L}}$ directed along its symmetry axis. We shall show that this symmetry axis moves about the z axis (precessional motion occurs) because the torque produces a change in the *direction* of the symmetry axis. This illustration is an excellent example of the importance of the vector nature of angular momentum.

The essential features of precessional motion can be illustrated by considering the top to act as a simple gyroscope. The two forces acting on the gyroscope are shown in Figure 11.12a: the downward gravitational force $M\vec{\mathbf{g}}$ and the normal force $\vec{\mathbf{n}}$ acting upward at the pivot point O. The normal force produces no torque about an axis passing through the pivot because its moment arm through that point is zero. The gravitational force, however, produces a torque $\vec{\boldsymbol{\tau}} = \vec{\mathbf{r}} \times M\vec{\mathbf{g}}$ about an axis passing through O, where the direction of $\vec{\boldsymbol{\tau}}$ is perpendicular to the plane formed by $\vec{\mathbf{r}}$ and $M\vec{\mathbf{g}}$. By necessity, the vector $\vec{\boldsymbol{\tau}}$ lies in a horizontal xy plane perpendicular to the angular momentum vector. The net torque and angular momentum of the gyroscope are related through Equation 11.15:

$$\sum \vec{\boldsymbol{\tau}}_{\text{ext}} = \frac{d\vec{\mathbf{L}}}{dt}$$

This expression shows that in the infinitesimal time interval dt, the nonzero torque produces a change in angular momentum $d\vec{\mathbf{L}}$, a change that is in the same direction as $\vec{\boldsymbol{\tau}}$. Therefore, like the torque vector, $d\vec{\mathbf{L}}$ must also be perpendicular to $\vec{\mathbf{L}}$. The overhead view in Figure 11.12c illustrates the resulting precessional motion of the symmetry axis of the gyroscope. In a time interval dt, the change in angular momentum is $d\vec{\mathbf{L}} = \vec{\mathbf{L}}_f - \vec{\mathbf{L}}_i = \vec{\boldsymbol{\tau}}\, dt$. Because $d\vec{\mathbf{L}}$ is perpendicular to $\vec{\mathbf{L}}$, the magnitude of $\vec{\mathbf{L}}$ does not change ($|\vec{\mathbf{L}}_i| = |\vec{\mathbf{L}}_f|$). Rather, what is changing is the *direction* of $\vec{\mathbf{L}}$. Because the change in angular momentum $d\vec{\mathbf{L}}$ is in the direction of $\vec{\boldsymbol{\tau}}$, which lies in the xy plane, the gyroscope undergoes precessional motion.

The vector diagram in Figure 11.12c shows that in the time interval dt, the angular momentum vector rotates through an angle $d\phi$, which is also the angle through which the gyroscope axle rotates. From the vector triangle formed by the vectors $\vec{\mathbf{L}}_i$, $\vec{\mathbf{L}}_f$, and $d\vec{\mathbf{L}}$, we see that

$$d\phi = \frac{dL}{L} = \frac{\sum \tau_{\text{ext}}\, dt}{L} = \frac{(Mgr_{\text{CM}})\, dt}{L}$$

Dividing through by dt and using the relationship $L = I\omega$, we find that the rate at which the axle rotates about the vertical axis is

$$\omega_p = \frac{d\phi}{dt} = \frac{Mgr_{\text{CM}}}{I\omega} \qquad (11.22)$$

The angular speed ω_p is called the **precessional frequency**. This result is valid only when $\omega_p \ll \omega$. Otherwise, a much more complicated motion is involved. As you can see from Equation 11.22, the condition $\omega_p \ll \omega$ is met when ω is large, that is, when the wheel spins rapidly. Furthermore, notice that the precessional frequency decreases as ω increases, that is, as the wheel spins faster about its axis of symmetry.

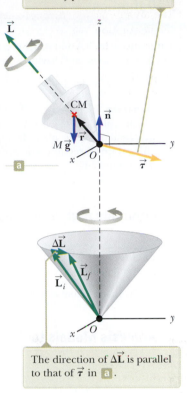

The right-hand rule indicates that $\vec{\boldsymbol{\tau}} = \vec{\mathbf{r}} \times \vec{\mathbf{F}} = \vec{\mathbf{r}} \times M\vec{\mathbf{g}}$ is in the xy plane.

a

The direction of $\Delta\vec{\mathbf{L}}$ is parallel to that of $\vec{\boldsymbol{\tau}}$ in **a**.

b

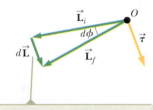

The torque results in a change in angular momentum $d\vec{\mathbf{L}}$ in a direction parallel to the torque vector. The gyroscope axle sweeps out an angle $d\phi$ in a time interval dt.

c

Figure 11.12 Precessional motion of a top spinning about its symmetry axis. (a) The only external forces acting on the top are the normal force $\vec{\mathbf{n}}$ and the gravitational force $M\vec{\mathbf{g}}$. The direction of the angular momentum $\vec{\mathbf{L}}$ is along the axis of symmetry. (b) Because $\vec{\mathbf{L}}_f = \Delta\vec{\mathbf{L}} + \vec{\mathbf{L}}_i$, the top precesses about the z axis. (c) Overhead view (looking down the z axis) of the gyroscope's initial and final angular momentum vectors for an infinitesimal time interval dt.

Summary

> ## Definitions

Given two vectors $\vec{A}$ and $\vec{B}$, the **vector product** $\vec{A} \times \vec{B}$ is a vector $\vec{C}$ having a magnitude

$$C = AB \sin \theta \qquad (11.3)$$

where θ is the angle between $\vec{A}$ and $\vec{B}$. The direction of the vector $\vec{C} = \vec{A} \times \vec{B}$ is perpendicular to the plane formed by $\vec{A}$ and $\vec{B}$, and this direction is determined by the right-hand rule.

The **torque** $\vec{\tau}$ on a particle due to a force $\vec{F}$ about an axis through the origin in an inertial frame is defined to be

$$\vec{\tau} \equiv \vec{r} \times \vec{F} \qquad (11.1)$$

The **angular momentum** $\vec{L}$ about an axis through the origin of a particle having linear momentum $\vec{p} = m\vec{v}$ is

$$\vec{L} \equiv \vec{r} \times \vec{p} \qquad (11.12)$$

where $\vec{r}$ is the vector position of the particle relative to the origin.

> ## Concepts and Principles

The z component of angular momentum of a rigid object rotating about a fixed z axis is

$$L_z = I\omega \qquad (11.16)$$

where I is the moment of inertia of the object about the axis of rotation and ω is its angular speed.

> ## Analysis Models for Problem Solving

The rate of change in the angular momentum of the nonisolated system is equal to the net external torque on the system.

Nonisolated System (Angular Momentum). If a system interacts with its environment in the sense that there is an external torque on the system, the net external torque acting on a system is equal to the time rate of change of its angular momentum:

$$\sum \vec{\tau}_{\text{ext}} = \frac{d\vec{L}_{\text{tot}}}{dt} \qquad (11.15)$$

The angular momentum of the isolated system is constant.

Isolated System (Angular Momentum). If a system experiences no external torque from the environment, the total angular momentum of the system is conserved:

$$\Delta \vec{L}_{\text{tot}} = 0 \qquad (11.20)$$

Applying this law of conservation of angular momentum to a system whose moment of inertia changes gives

$$I_i \omega_i = I_f \omega_f = \text{constant} \qquad (11.21)$$

Think–Pair–Share

See the Preface for an explanation of the icons used in this problems set. For additional assessment items for this section, go to **WEBASSIGN** From Cengage

1. **ACTIVITY** In your group, have someone donate temporarily a set of keys for the cause. Tie one end of a string about a meter long to the ring. To the other end attach a light weight, such as a paper clip or binder clip. If you hold the clip in your hand with the keys hanging on the string from your

hand and drop it, everything of course falls to the ground. Now try something different. Suspend the keys from a horizontal pencil held in your left hand so that they hang down from the pencil on a short segment of the string as shown in Figure TP11.1. The rest of the string lies over the pencil and extends horizontally to the binder clip, which you hold with your right hand. Now release the binder clip. What happens? Explain this result.

Figure TP11.1

2. A disk with moment of inertia I_1 rotates about a friction-less, vertical axle with angular speed ω_i. A second disk, this one having moment of inertia I_2 and initially not rotating, drops onto the first disk (Fig. TP11.2). Because of friction between the surfaces, the two eventually reach the same angular speed ω_f. Discuss in your group the following. (a) Calculate ω_f. (b) What fraction of the initial kinetic energy of the two-disk system remains after the disks rotate with the same angular speed? (c) Find the value of the answer in part (b) for the following limits: (i) $I_2 \to 0$, (ii) $I_1 = I_2$, (iii) $I_2 \to \infty$, and (iv) $I_1 \to \infty$. (d) Explain how each of the results in part (c) makes sense. (e) In the general case in which the kinetic energy of the system decreases in the process, where does that energy go? (f) **What If?** In Figure TP11.2, what is ω_f if the second disk is also rotating, but in the clockwise direction, opposite that of disk 1, with an angular speed of ω' before the collision?

Figure TP11.2

3. You are attending a county fair with your friend from your physics class. While walking around the fairgrounds, you discover a new game of skill. A thin rod of mass $M = 0.500$ kg and length $\ell = 2.00$ m hangs from a friction-free pivot at its upper end as shown in Figure TP11.3. The front surface of the rod is covered with Velcro. You are to throw a Velcro-covered ball of mass $m = 1.00$ kg at the rod in an attempt to make it swing backward and rotate all the way across the top. The ball must stick to the rod at all times after striking it. If you cause the rod to rotate over the top position, you win a stuffed animal. Your friend volunteers to try his luck. He feels that the most torque would be applied to the rod by striking it at its lowest end. After several tries, he fails to win the stuffed animal by throwing the ball so that it sticks at the end of the rod. He just couldn't throw the ball fast enough and accurately enough. (See Problem 43 to find out how fast he must throw the ball.) You analyze things differently from your friend. What if you were to throw the ball at a point *above* the end of the rod, a distance y below the pivot as shown in Figure TP11.3? This would reduce the torque on the rod, but torque is proportional to r, while moment of inertia is proportional to r^2. After the collision, the ball is part of the rotating system, so the moment of inertia of the system is reduced if the ball is stuck somewhere along the length of the rod, rather than at its end. (a) You pull out some sheets of paper and calculate an algebraic expression for the minimum required speed to spin the rod to the vertical position as a function of the point y along the rod at which the ball strikes and sticks to the rod. (b) Then, based on numerical values, you determine the point along the rod where you should strike it with the ball and make it go over the top by throwing the ball with the *lowest* speed. (c) Finally, you determine that lowest speed.

Figure TP11.3

Problems

See the Preface for an explanation of the icons used in this problems set. For additional assessment items for this section, go to *From Cengage*

SECTION 11.1 The Vector Product and Torque

1. Given $\vec{\mathbf{M}} = 2\hat{\mathbf{i}} - 3\hat{\mathbf{j}} + \hat{\mathbf{k}}$ and $\vec{\mathbf{N}} = 4\hat{\mathbf{i}} + 5\hat{\mathbf{j}} - 2\hat{\mathbf{k}}$, calculate the vector product $\vec{\mathbf{M}} \times \vec{\mathbf{N}}$.

2. The displacement vectors 42.0 cm at 15.0° and 23.0 cm at 65.0° both start from the origin and form two sides of a parallelogram. Both angles are measured counterclockwise from the *x* axis. (a) Find the area of the parallelogram. (b) Find the length of its longer diagonal.

3. If $|\vec{\mathbf{A}} \times \vec{\mathbf{B}}| = \vec{\mathbf{A}} \cdot \vec{\mathbf{B}}$, what is the angle between $\vec{\mathbf{A}}$ and $\vec{\mathbf{B}}$?

4. Use the definition of the vector product and the definitions of the unit vectors $\hat{\mathbf{i}}$, $\hat{\mathbf{j}}$, and $\hat{\mathbf{k}}$ to prove Equations 11.7. You may assume the *x* axis points to the right, the *y* axis up, and the *z* axis horizontally toward you (not away from you). This choice is said to make the coordinate system a *right-handed system*.

5. Two forces $\vec{\mathbf{F}}_1$ and $\vec{\mathbf{F}}_2$ act along the two sides of an equilateral triangle as shown in Figure P11.5. Point *O* is the intersection of the altitudes of the triangle. (a) Find the magnitude of a third force $\vec{\mathbf{F}}_3$ to be applied at *B* and along *BC* that will make the total torque zero about the point *O*. (b) **What If?** Will the total torque change if $\vec{\mathbf{F}}_3$ is applied not at *B* but at any other point along *BC*?

Figure P11.5

6. A student claims that he has found a vector $\vec{\mathbf{A}}$ such that $(2\hat{\mathbf{i}} - 3\hat{\mathbf{j}} + 4\hat{\mathbf{k}}) \times \vec{\mathbf{A}} = (4\hat{\mathbf{i}} + 3\hat{\mathbf{j}} - \hat{\mathbf{k}})$. (a) Do you believe this claim? (b) Explain why or why not.

7. A particle is located at a point described by the position vector $\vec{r} = (4.00\hat{i} + 6.00\hat{j})$ m, and a force exerted on it is given by $\vec{F} = (3.00\hat{i} + 2.00\hat{j})$ N. (a) What is the torque acting on the particle about the origin? (b) Can there be another point about which the torque caused by this force on this particle will be in the opposite direction and half as large in magnitude? (c) Can there be more than one such point? (d) Can such a point lie on the y axis? (e) Can more than one such point lie on the y axis? (f) Determine the position vector of one such point.

SECTION 11.2 Analysis Model: Nonisolated System (Angular Momentum)

8. A 1.50-kg particle moves in the xy plane with a velocity of $\vec{v} = (4.20\hat{i} - 3.60\hat{j})$ m/s. Determine the angular momentum of the particle about the origin when its position vector is $\vec{r} = (1.50\hat{i} + 2.20\hat{j})$ m.

9. A particle of mass m moves in the xy plane with a velocity of $\vec{v} = v_x\hat{i} + v_y\hat{j}$. Determine the angular momentum of the particle about the origin when its position vector is $\vec{r} = x\hat{i} + y\hat{j}$.

10. Heading straight toward the summit of Pike's Peak, an airplane of mass 12 000 kg flies over the plains of Kansas at nearly constant altitude 4.30 km with constant velocity 175 m/s west. (a) What is the airplane's vector angular momentum relative to a wheat farmer on the ground directly below the airplane? (b) Does this value change as the airplane continues its motion along a straight line? (c) **What If?** What is its angular momentum relative to the summit of Pike's Peak?

11. **Review.** A projectile of mass m is launched with an initial velocity $\vec{v}_i$ making an angle θ with the horizontal as shown in Figure P11.11. The projectile moves in the gravitational field of the Earth. Find the angular momentum of the projectile about the origin (a) when the projectile is at the origin, (b) when it is at the highest point of its trajectory, and (c) just before it hits the ground. (d) What torque causes its angular momentum to change?

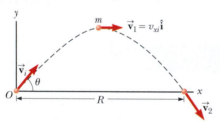

Figure P11.11

12. **Review.** A conical pendulum consists of a bob of mass m in motion in a circular path in a horizontal plane as shown in Figure P11.12. During the motion, the supporting wire of length ℓ maintains a constant angle θ with the vertical. Show that the magnitude of the angular momentum of the bob about the vertical dashed line is

$$L = \left(\frac{m^2 g \ell^3 \sin^4\theta}{\cos\theta}\right)^{1/2}$$

Figure P11.12

13. A particle of mass m moves in a circle of radius R at a constant speed v as shown in Figure P11.13. Time $t = 0$ is defined as

when the particle is at point Q. Determine the angular momentum of the particle about the axis perpendicular to the page through point P as a function of time.

Figure P11.13
Problems 13 and 24.

14. A 5.00-kg particle starts from the origin at time zero. Its velocity as a function of time is given by

$$\vec{v} = 6t^2\hat{i} + 2t\hat{j}$$

where $\vec{v}$ is in meters per second and t is in seconds. (a) Find its position as a function of time. (b) Describe its motion qualitatively. Find (c) its acceleration as a function of time, (d) the net force exerted on the particle as a function of time, (e) the net torque about the origin exerted on the particle as a function of time, (f) the angular momentum of the particle as a function of time, (g) the kinetic energy of the particle as a function of time, and (h) the power injected into the system of the particle as a function of time.

15. A ball having mass m is fastened at the end of a flagpole that is connected to the side of a tall building at point P as shown in Figure P11.15. The length of the flagpole is ℓ, and it makes an angle θ with the x axis. The ball becomes loose and starts to fall with acceleration $-g\hat{j}$. (a) Determine the angular momentum of the ball about point P as a function of time. (b) For what physical reason does the angular momentum change? (c) What is the rate of change of the angular momentum of the ball about point P?

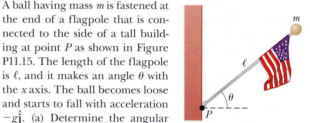

Figure P11.15

SECTION 11.3 Angular Momentum of a Rotating Rigid Object

16. A uniform solid sphere of radius $r = 0.500$ m and mass $m = 15.0$ kg turns counterclockwise about a vertical axis through its center. Find its vector angular momentum about this axis when its angular speed is 3.00 rad/s.

17. A uniform solid disk of mass $m = 3.00$ kg and radius $r = 0.200$ m rotates about a fixed axis perpendicular to its face with angular frequency 6.00 rad/s. Calculate the magnitude of the angular momentum of the disk when the axis of rotation (a) passes through its center of mass and (b) passes through a point midway between the center and the rim.

18. Show that the kinetic energy of an object rotating about a fixed axis with angular momentum $L = I\omega$ can be written as $K = L^2/2I$.

19. Big Ben (Fig. P10.27, page 281), the Parliament tower clock in London, has hour and minute hands with lengths of 2.70 m and 4.50 m and masses of 60.0 kg and 100 kg, respectively. Calculate the total angular momentum of these hands about the center point. (You may model the hands as long, thin rods rotating about one end. Assume the hour and minute hands are rotating at a constant rate of one revolution per 12 hours and 60 minutes, respectively.)

20. Model the Earth as a uniform sphere. (a) Calculate the angular momentum of the Earth due to its spinning motion about its axis. (b) Calculate the angular momentum of the Earth due to its orbital motion about the Sun. (c) Explain why the answer in part (b) is larger than that in part (a) even though it takes significantly longer for the Earth to go once around the Sun than to rotate once about its axis.

21. The distance between the centers of the wheels of a motorcycle is 155 cm. The center of mass of the motorcycle, including the rider, is 88.0 cm above the ground and halfway between the wheels. Assume the mass of each wheel is small compared with the body of the motorcycle. The engine drives the rear wheel only. What horizontal acceleration of the motorcycle will make the front wheel rise off the ground?

SECTION 11.4 Analysis Model: Isolated System (Angular Momentum)

22. You are working in an observatory, taking data on electromagnetic radiation from neutron stars. You happen to be analyzing results from the neutron star in Example 11.6, verifying that the period of the 10.0-km-radius neutron star is indeed 2.6 s. You go through weeks of data showing the same period. Suddenly, as you analyze the most recent data, you notice that the period has decreased to 2.3 s and remained at that level since that time. You ask your supervisor about this, who becomes excited and says that the neutron star must have undergone a *glitch*, which is a sudden shrinking of the radius of the star, resulting in a higher angular speed. As she runs to her computer to start writing a paper on the glitch, she calls back to you to calculate the new radius of the planet, assuming it has remained spherical. She is also talking about vortices and a superfluid core, but you don't understand those words.

23. A 60.0-kg woman stands at the western rim of a horizontal turntable having a moment of inertia of 500 kg · m² and a radius of 2.00 m. The turntable is initially at rest and is free to rotate about a frictionless, vertical axle through its center. The woman then starts walking around the rim clockwise (as viewed from above the system) at a constant speed of 1.50 m/s relative to the Earth. Consider the woman–turntable system as motion begins. (a) Is the mechanical energy of the system constant? (b) Is the momentum of the system constant? (c) Is the angular momentum of the system constant? (d) In what direction and with what angular speed does the turntable rotate? (e) How much potential energy in the woman's body is converted into mechanical energy of the woman–turntable system as the woman sets herself and the turntable into motion?

24. Figure P11.13 represents a small, flat puck with mass $m = 2.40$ kg sliding on a frictionless, horizontal surface. It is held in a circular orbit about a fixed axis by a rod with negligible mass and length $R = 1.50$ m, pivoted at one end. Initially, the puck has a speed of $v = 5.00$ m/s. A 1.30-kg ball of putty is dropped vertically onto the puck from a small distance above it and immediately sticks to the puck. (a) What is the new period of rotation? (b) Is the angular momentum of the puck–putty system about the axis of rotation constant in this process? (c) Is the momentum of the system constant in the process of the putty sticking to the

puck? (d) Is the mechanical energy of the system constant in the process?

25. A uniform cylindrical turntable of radius 1.90 m and mass 30.0 kg rotates counterclockwise in a horizontal plane with an initial angular speed of 4π rad/s. The fixed turntable bearing is frictionless. A lump of clay of mass 2.25 kg and negligible size is dropped onto the turntable from a small distance above it and immediately sticks to the turntable at a point 1.80 m to the east of the axis. (a) Find the final angular speed of the clay and turntable. (b) Is the mechanical energy of the turntable–clay system constant in this process? Explain and use numerical results to verify your answer. (c) Is the momentum of the system constant in this process? Explain your answer.

26. A puck of mass $m_1 = 80.0$ g and radius $r_1 = 4.00$ cm glides across an air table at a speed of $\vec{v} = 1.50$ m/s as shown in Figure P11.26a. It makes a glancing collision with a second puck of radius $r_2 = 6.00$ cm and mass $m_2 = 120$ g (initially at rest) such that their rims just touch. Because their rims are coated with instant-acting glue, the pucks stick together and rotate after the collision (Fig. P11.26b). (a) What is the angular momentum of the system relative to the center of mass? (b) What is the angular speed about the center of mass?

Figure P11.26

27. A wooden block of mass M resting on a frictionless, horizontal surface is attached to a rigid rod of length ℓ and of negligible mass (Fig. P11.27). The rod is pivoted at the other end. A bullet of mass m traveling parallel to the horizontal surface and perpendicular to the rod with speed v hits the block and becomes embedded in it. (a) What is the angular momentum of the bullet–block system about a vertical axis through the pivot? (b) What fraction of the original kinetic energy of the bullet is converted into internal energy in the system during the collision?

Figure P11.27

28. *Why is the following situation impossible?* A space station shaped like a giant wheel (Fig. P11.28, page 306) has a radius of $r = 100$ m and a moment of inertia of 5.00×10^8 kg · m². A crew of 150 people of average mass 65.0 kg is living on the rim, and the station's rotation causes the crew to experience an apparent free-fall acceleration of g. A research technician is assigned to perform an experiment in which a ball is dropped at the rim of the station every 15 minutes and the time interval for the ball to drop a given distance is measured as a test to make sure the apparent value of g is correctly maintained. One evening, 100 average people move to the center of the station for a union meeting. The research technician, who has already been performing his experiment for an hour before the meeting, is disappointed that he cannot attend the meeting, and his mood sours even further by his boring experiment in which every time interval for the dropped ball is identical for the entire evening.

Figure P11.28

29. A wad of sticky clay with mass m and velocity $\vec{v}_i$ is fired at a **Q|C** solid cylinder of mass M and radius R (Fig. P11.29). The cyl-**S** inder is initially at rest and is mounted on a fixed horizontal axle that runs through its center of mass. The line of motion of the projectile is perpendicular to the axle and at a distance $d < R$ from the center. (a) Find the angular speed of the system just after the clay strikes and sticks to the surface of the cylinder. (b) Is the mechanical energy of the clay–cylinder system constant in this process? Explain your answer. (c) Is the momentum of the clay–cylinder system constant in this process? Explain your answer.

Figure P11.29

30. A 0.005 00-kg bullet traveling horizontally with a speed of **Q|C** 1.00×10^3 m/s strikes an 18.0-kg door, embedding itself 10.0 cm from the side opposite the hinges as shown in Figure P11.30. The 1.00-m wide door is free to swing on its frictionless hinges. (a) Before it hits the door, does the bullet have angular momentum relative to the door's axis of rotation? (b) If so, evaluate this angular momentum. If not, explain why there is no angular momentum. (c) Is the mechanical energy of the bullet–door system constant during this collision? Answer without doing a calculation. (d) At what angular speed does the door swing open immediately after the collision? (e) Calculate the total energy of the bullet–door system and determine whether it is less than or equal to the kinetic energy of the bullet before the collision. (f) **What If?** Imagine now that the door is hanging vertically downward, hinged at the top, so that Figure P11.30 is a side view of the door and bullet during the collision. What is the maximum height that the bottom of the door will reach after the collision?

Figure P11.30 An overhead view of a bullet striking a door.

SECTION 11.5 The Motion of Gyroscopes and Tops

31. The angular momentum vector of a precessing gyroscope sweeps out a cone as shown in Figure P11.31. The angular speed of the tip of the angular momentum vector, called its precessional frequency, is given by $\omega_p = \tau/L$, where τ is the magnitude of the torque on the gyroscope and L is the magnitude of its angular momentum. In the motion called *precession of the equinoxes,* the Earth's axis of rotation precesses about the perpendicular to its orbital plane with a period of 2.58×10^4 yr. Model the Earth as a uniform sphere and calculate the torque on the Earth that is causing this precession.

Figure P11.31 A precessing angular momentum vector sweeps out a cone in space.

ADDITIONAL PROBLEMS

32. A light rope passes over a light, fric-**Q|C** tionless pulley. One end is fastened to a **S** bunch of bananas of mass M, and a monkey of mass M clings to the other end (Fig. P11.32). The monkey climbs the rope in an attempt to reach the bananas. (a) Treating the system as consisting of the monkey, bananas, rope, and pulley, find the net torque on the system about the pulley axis. (b) Using the result of part (a), determine the total angular momentum about the pulley axis and describe the motion of the system. (c) Will the monkey reach the bananas?

Figure P11.32

33. **Review.** A thin, uniform, rectangular signboard hangs vertically above the door of a shop. The sign is hinged to a stationary horizontal rod along its top edge. The mass of the sign is 2.40 kg, and its vertical dimension is 50.0 cm. The sign is swinging without friction, so it is a tempting target for children armed with snowballs. The maximum angular displacement of the sign is 25.0° on both sides of the vertical. At a moment when the sign is vertical and moving to the left, a snowball of mass 400 g, traveling horizontally with a velocity of 160 cm/s to the right, strikes perpendicularly at the lower edge of the sign and sticks there. (a) Calculate the angular speed of the sign immediately before the impact. (b) Calculate its angular speed immediately after the impact. (c) The spattered sign will swing up through what maximum angle?

34. You are advising a fellow student who wants to learn to **CR** perform multiple flips on the trampoline. You have him bounce vertically as high as he can, keeping his body perfectly straight and vertical. You determine that he can raise his center of mass by a distance of $h = 6.00$ m above its level when he initiates the jump. He can do a single flip by

bouncing gently, throwing his arms forward over his head, and tucking his body. You use your smartphone to make a video of him doing a single flip. Based on analysis of this video, you determine that his moment of inertia is $I_{straight} = 26.7$ kg $\cdot$ m^2 when his body is straight and $I_{tuck} = 5.62$ kg $\cdot$ m^2 in the tuck position. You suggest that he keep his body in the straight position for $\Delta t' = 0.400$ s after leaving the trampoline surface and then immediately go into a tuck position. As he lands, he should straighten his body out $\Delta t' = 0.400$ s before he lands. From analysis of the video recording, you determine that throwing his arms forward causes him to have an initial angular speed of $\omega_i = 2.88$ rad/s as he leaves the trampoline surface. If he tries to bounce as high as he can, do some flips, and land back on the same spot on the trampoline, predict how many flips he can safely do such that he lands on his feet on the trampoline.

35. We have all complained that there aren't enough hours in a day. In an attempt to fix that, suppose all the people in the world line up at the equator and all start running east at 2.50 m/s relative to the surface of the Earth. By how much does the length of a day increase? Assume the world population to be 7.00×10^9 people with an average mass of 55.0 kg each and the Earth to be a solid homogeneous sphere. In addition, depending on the details of your solution, you may need to use the approximation $1/(1 - x) \approx 1 + x$ for small x.

36. *Why is the following situation impossible?* A meteoroid strikes the Earth directly on the equator. At the time it lands, it is traveling exactly vertical and downward. Due to the impact, the time for the Earth to rotate once increases by 0.5 s, so the day is 0.5 s longer, undetectable to laypersons. After the impact, people on the Earth ignore the extra half-second each day and life goes on as normal. (Assume the density of the Earth is uniform.)

37. **S** A rigid, massless rod has three particles with equal masses attached to it as shown in Figure P11.37. The rod is free to rotate in a vertical plane about a frictionless axle perpendicular to the rod through the point *P* and is released from rest in the horizontal position at $t = 0$. Assuming m and d are known, find (a) the moment of inertia of the system of three particles about the pivot, (b) the torque acting on the system at $t = 0$, (c) the angular acceleration of the system at $t = 0$, (d) the linear acceleration of the particle labeled 3 at $t = 0$, (e) the maximum kinetic energy of the system, (f) the maximum angular speed reached by the rod, (g) the maximum angular momentum of the system, and (h) the maximum speed reached by the particle labeled 2.

Figure P11.37

38. **Review.** Two boys are sliding toward each other on a frictionless, ice-covered parking lot. Jacob, mass 45.0 kg, is gliding to the right at 8.00 m/s, and Ethan, mass 31.0 kg, is gliding to the left at 11.0 m/s along the same line. When they meet, they grab each other and hang on. (a) What is their velocity immediately thereafter? (b) What fraction of their original kinetic energy is still mechanical energy after their collision? That was so much fun that the boys repeat the collision with the same original velocities, this time moving along

parallel lines 1.20 m apart. At closest approach, they lock arms and start rotating about their common center of mass. Model the boys as particles and their arms as a cord that does not stretch. (c) Find the velocity of their center of mass. (d) Find their angular speed. (e) What fraction of their original kinetic energy is still mechanical energy after they link arms? (f) Why are the answers to parts (b) and (e) so different?

39. **T** Two astronauts (Fig. P11.39), each having a mass of 75.0 kg, are connected by a 10.0-m rope of negligible mass. They are isolated in space, orbiting their center of mass at speeds of 5.00 m/s. Treating the astronauts as particles, calculate (a) the magnitude of the angular momentum of the two-astronaut system and (b) the rotational energy of the system. By pulling on the rope, one astronaut shortens the distance between them to 5.00 m. (c) What is the new angular momentum of the system? (d) What are the astronauts' new speeds? (e) What is the new rotational energy of the system? (f) How much potential energy in the body of the astronaut was converted to mechanical energy in the system when he shortened the rope?

Figure P11.39 Problems 39 and 40.

40. **S** Two astronauts (Fig. P11.39), each having a mass M, are connected by a rope of length d having negligible mass. They are isolated in space, orbiting their center of mass at speeds v. Treating the astronauts as particles, calculate (a) the magnitude of the angular momentum of the two-astronaut system and (b) the rotational energy of the system. By pulling on the rope, one of the astronauts shortens the distance between them to $d/2$. (c) What is the new angular momentum of the system? (d) What are the astronauts' new speeds? (e) What is the new rotational energy of the system? (f) How much potential energy in the body of the astronaut was converted to mechanical energy in the system when he shortened the rope?

41. **Q|C** **S** Native people throughout North and South America used a *bola* to hunt for birds and animals. A bola can consist of three stones, each with mass m, at the ends of three light cords, each with length ℓ. The other ends of the cords are tied together to form a Y. The hunter holds one stone and swings the other two above his head (Figure P11.41a, page 308). Both these stones move together in a horizontal circle of radius 2ℓ with speed v_0. At a moment when the horizontal component of their velocity is directed toward the quarry, the hunter releases the stone in his hand. As the bola flies through the air, the cords quickly take a stable arrangement with constant 120-degree angles between them (Fig. P11.41b). In the vertical direction, the bola is in free fall. Gravitational forces exerted by the Earth make the junction of the cords move with the downward acceleration $\vec{g}$. You may ignore the vertical motion as you proceed to describe the horizontal motion of the bola. In terms of m, ℓ, and v_0, calculate (a) the magnitude of the momentum of

the bola at the moment of release and, after release, (b) the horizontal speed of the center of mass of the bola, and (c) the angular momentum of the bola about its center of mass. (d) Find the angular speed of the bola about its center of mass after it has settled into its Y shape. Calculate the kinetic energy of the bola (e) at the instant of release and (f) in its stable Y shape. (g) Explain how the conservation laws apply to the bola as its configuration changes. Robert Beichner suggested the idea for this problem.

Figure P11.41

42. Two children are playing on stools at a restaurant counter. **Q|C** Their feet do not reach the footrests, and the tops of the stools are free to rotate without friction on pedestals fixed to the floor. One of the children catches a tossed ball, in a process described by the equation

$$(0.730 \text{ kg} \cdot \text{m}^2)(2.40\hat{\mathbf{j}} \text{ rad/s})$$

$$+ (0.120 \text{ kg})(0.350\hat{\mathbf{i}} \text{ m}) \times (4.30\hat{\mathbf{k}} \text{ m/s})$$

$$= [0.730 \text{ kg} \cdot \text{m}^2 + (0.120 \text{ kg})(0.350 \text{ m})^2]\vec{\boldsymbol{\omega}}$$

(a) Solve the equation for the unknown $\vec{\boldsymbol{\omega}}$. (b) Complete the statement of the problem to which this equation applies. Your statement must include the given numerical information and specification of the unknown to be determined. (c) Could the equation equally well describe the other child throwing the ball? Explain your answer.

43. You are attending a county **CR** fair with your friend from your physics class. While walking around the fairgrounds, you discover a new game of skill. A thin rod of mass $M = 0.500$ kg and length $\ell = 2.00$ m hangs from a friction-free pivot at its upper end as shown in Figure P11.43. The front surface of the rod is covered with Velcro. You are to throw a Velcro-covered ball of mass $m = 1.00$ kg at the rod in an attempt to make it swing backward and rotate all the way across the top. The ball must stick to the rod at all times after striking it. If you cause the rod to rotate over the top position, you win a stuffed animal. Your friend volunteers to try his luck. He feels that the most torque would be applied to the rod by striking it at its lowest end. While he prepares to aim at the lowest point on the rod, you calculate how fast he must throw the ball to win the stuffed animal with this technique.

Pivot

M

Velcro

Incoming
Velcro-covered
ball

ℓ

m

Figure P11.43

44. A uniform rod of mass 300 g and length 50.0 cm rotates in **Q|C** a horizontal plane about a fixed, frictionless, vertical pin through its center. Two small, dense beads, each of mass m, are mounted on the rod so that they can slide without friction along its length. Initially, the beads are held by catches at positions 10.0 cm on each side of the center and the system is rotating at an angular speed of 36.0 rad/s. The catches are released simultaneously, and the beads slide outward along the rod. (a) Find an expression for the angular speed ω_f of the system at the instant the beads slide off the ends of the rod as it depends on m. (b) What are the maximum and the minimum possible values for ω_f and the values of m to which they correspond?

45. Global warming is a cause for concern because even small changes in the Earth's temperature can have significant consequences. For example, if the Earth's polar ice caps were to melt entirely, the resulting additional water in the oceans would flood many coastal areas. Model the polar ice as having mass 2.30×10^{19} kg and forming two flat disks of radius 6.00×10^5 m. Assume the water spreads into an unbroken thin, spherical shell after it melts. Calculate the resulting change in the duration of one day both in seconds and as a percentage.

46. The puck in Figure P11.46 has a mass of 0.120 kg. The distance of the puck from the center of rotation is originally 40.0 cm, and the puck is sliding with a speed of 80.0 cm/s. The string is pulled downward 15.0 cm through the hole in the frictionless table. Determine the work done on the puck. (*Suggestion:* Consider the change of kinetic energy.)

O R $\vec{\mathbf{v}}_i$

m

$\vec{\mathbf{F}}$

Figure P11.46

47. You operate a restaurant that has many large, circular **CR** tables. At the center of each table is a Lazy Susan that can turn to deliver salt, pepper, jam, hot sauce, bread, and other items to diners on the other side of the table. A fancy flower arrangement is located at the center of each Lazy Susan, and the turning of the flower arrangement is beautiful to you. Because of your interest in model trains, you decide to replace each Lazy Susan with a circular track on the table around which a model train will run. You can load the various condiments in the cars of the train and press a button to operate the train, causing the train to begin moving around the circle and deliver the load to your fellow diners! The train is of mass 1.96 kg and moves at a speed of 0.18 m/s relative to the track. After a few days, you realize that you miss the beautiful turning flower arrangements. So you come up with a new scheme. You return the Lazy Susan to the table and mount the circular track on the platform of the Lazy Susan, which has a friction-free axle at its center. The radius of the circular track is 40.0 cm (measured halfway between the rails) and the platform of the Lazy Susan is a uniform

disk of mass 3.00 kg and radius 48.0 cm. You finally equip all of your tables with the new apparatus and open your restaurant. As a demonstration to the diners, you mount one salt shaker and one pepper shaker, having a mass of 0.100 kg each, onto a flatcar and push the button to deliver the condiments to the other side of the table! How long does it take to deliver the condiments to the exact opposite side of the table? Ignore the moment of inertia of the flower arrangement, since its mass is all close to the rotation axis.

CHALLENGE PROBLEMS

48. A solid cube of wood of side $2a$ and mass M is resting on a horizontal surface. The cube is constrained to rotate about a fixed axis AB (Fig. P11.48). A bullet of mass m and speed v is shot at the face opposite $ABCD$ at a height of $4a/3$. The bullet becomes embedded in the cube. Find the minimum value of v required to tip the cube so that it falls on face $ABCD$. Assume $m \ll M$.

Figure P11.48

49. In Example 11.8, we investigated an elastic collision between a disk and a stick lying on a frictionless surface. Suppose everything is the same as in the example except that the collision is perfectly inelastic so that the disk adheres to the stick at the endpoint at which it strikes. Find (a) the speed of the center of mass of the system and (b) the angular speed of the system after the collision.

50. A solid cube of side $2a$ and mass M is sliding on a frictionless surface with uniform velocity $\vec{v}$ as shown in Figure P11.50a. It hits a small obstacle at the end of the table that causes the cube to tilt as shown in Figure P11.50b. Find the minimum value of the magnitude of $\vec{v}$ such that the cube tips over and falls off the table. *Note:* The cube undergoes an inelastic collision at the edge.

Figure P11.50

12

Static Equilibrium and Elasticity

STORYLINE **In the previous chapter, you were taking a break from** your physics homework and browsing videos online about rotational motion. While thinking about the spinning phenomena in those videos, you pick up a meterstick and start twirling it and sliding it back and forth in your hands. At one point, something about the meterstick's behavior makes you forget about the rotational motion videos. You say, "Wait a minute! What just happened there?" You reproduce the behavior as follows. You support the meterstick horizontally on one finger of each hand, the fingers pointing out horizontally in a forward direction from your body. One finger supports the meterstick near the 0-cm end and the other near the 100-cm end. Now you slowly start moving your hands toward each other. The meterstick slides on one finger while sticking to the other finger. Then, it switches to slide on the other finger! And then back to the first finger! This alternation continues until your fingers meet. No matter what efforts you make to move only one finger at a time, this sticking and sliding behavior always occurs, the meterstick always stays supported on your fingers, and your fingers always meet at the 50-cm mark!

CONNECTIONS In Chapters 10 and 11, we studied the dynamics of rigid objects in motion. This chapter addresses the conditions under which a rigid object is in equilibrium. The term *equilibrium* implies that the object moves with both constant velocity and constant angular velocity relative to an observer in an inertial reference frame. In this chapter, we consider only the special case in which both of these velocities are equal to zero. In this case, the object is in what is called *static equilibrium*. In the meterstick phenomenon described in the storyline, anytime you momentarily stop your fingers so that the meterstick is at rest relative to the ground, it is in static equilibrium. Static equilibrium represents a common situation in engineering practice, and the principles it involves are of special interest to civil engineers, architects, and mechanical engineers. If you are an engineering student, you will undoubtedly take an advanced course in statics in the near future. The last section of this chapter deals with how objects deform under load conditions. An elastic object returns to its original shape when the deforming forces are removed. Several elastic constants are defined, each corresponding to a different type of deformation. In future chapters, we will see examples of rigid objects in static equilibrium: for example, polarized molecules in an electric field and a loop of wire carrying a current in a magnetic field.

12.1 Analysis Model: Rigid Object in Equilibrium

In Chapter 5, we discussed the particle in equilibrium model, in which a particle moves with constant velocity because the net force acting on it is zero. The situation with real (extended) objects is more complex because these objects often cannot be modeled as particles. For an extended object to be in equilibrium, a second condition must be satisfied. This second condition involves the rotational motion of the extended object.

Consider a single force $\vec{F}$ acting on a rigid object at point P as shown in Figure 12.1. Recall that the torque associated with the force $\vec{F}$ about an axis through O is given by Equation 11.1:

$$\vec{\tau} = \vec{r} \times \vec{F}$$

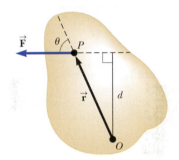

Figure 12.1 A single force $\vec{F}$ acts on a rigid object at the point P.

The magnitude of $\vec{\tau}$ is Fd (see Equation 10.14), where d is the moment arm shown in Figure 12.1. According to Equation 10.18, the net torque on a rigid object causes it to undergo an angular acceleration.

In this discussion, we investigate those rotational situations in which the angular acceleration of a rigid object is zero. Such an object is in **rotational equilibrium.** Because $\Sigma \tau_{\text{ext}} = I\alpha$ for rotation about a fixed axis, the necessary condition for rotational equilibrium is that the net torque about any axis must be zero. We now have two necessary conditions for equilibrium of a rigid object:

1. The net external force on the object must equal zero:

$$\sum \vec{F}_{\text{ext}} = 0 \tag{12.1}$$

2. The net external torque on the object about *any* axis must be zero:

$$\sum \vec{\tau}_{\text{ext}} = 0 \tag{12.2}$$

These conditions describe the **rigid object in equilibrium** analysis model. The first condition is a statement of translational equilibrium; it states that the translational acceleration of the object's center of mass must be zero when viewed from an inertial reference frame. The second condition is a statement of rotational equilibrium; it states that the angular acceleration about any axis must be zero. In the special case of **static equilibrium,** which is the main subject of this chapter, the object in equilibrium has a further requirement in addition to Equations 12.1 and 12.2: it is *at rest* relative to the observer and so has no translational or angular speed (that is, $v_{\text{CM}} = 0$ and $\omega = 0$).

PITFALL PREVENTION 12.1

Zero Torque Zero net torque does not mean an absence of rotational motion. An object that is rotating at a constant angular speed can be under the influence of a net torque of zero. This possibility is analogous to the translational situation: zero net force does not mean an absence of translational motion.

Figure 12.2 (Quick Quiz 12.1) Two forces of equal magnitude are applied at equal distances from the center of mass of a rigid object.

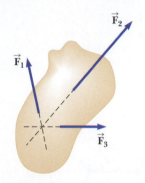

Figure 12.3 (Quick Quiz 12.2) Three forces act on an object. Notice that the lines of action of all three forces pass through a common point.

QUICK QUIZ 12.1 Consider the object subject to the two forces of equal magnitude in Figure 12.2. Choose the correct statement with regard to this situation. **(a)** The object is in force equilibrium but not torque equilibrium. **(b)** The object is in torque equilibrium but not force equilibrium. **(c)** The object is in both force equilibrium and torque equilibrium. **(d)** The object is in neither force equilibrium nor torque equilibrium.

QUICK QUIZ 12.2 Consider the object subject to the three forces in Figure 12.3. Choose the correct statement with regard to this situation from the choices **(a)–(d)** in Quick Quiz 12.1.

The two vector expressions given by Equations 12.1 and 12.2 are equivalent, in general, to six scalar equations: three from the first condition for equilibrium and three from the second (corresponding to x, y, and z components). Hence, in a complex system involving several forces acting in various directions, you could be faced with solving a set of equations with many unknowns. Here, we restrict our discussion to situations in which all the forces lie in the xy plane. (Forces whose vector representations are in the same plane are said to be *coplanar.*) With this restriction, we must deal with only three scalar equations. Two come from balancing the forces in the x and y directions. The third comes from the torque equation, namely that the net torque about a perpendicular axis through *any* point in the xy plane must be zero. This perpendicular axis will necessarily be parallel to the z axis, so the two conditions of the rigid object in equilibrium model provide the equations

$$\sum F_x = 0 \quad \sum F_y = 0 \quad \sum \tau_z = 0 \tag{12.3}$$

where the location of the axis of the torque equation is arbitrary.

ANALYSIS MODEL **Rigid Object in Equilibrium**

Imagine an object that can rotate, but is exhibiting no translational acceleration a and no rotational acceleration α. Such an object is in both translational *and* rotational equilibrium, so the net force *and* the net torque about any axis are both equal to zero:

$$\sum \vec{F}_{ext} = 0 \tag{12.1}$$

$$\sum \vec{\tau}_{ext} = 0 \tag{12.2}$$

Examples:

- a balcony juts out from a building and must support the weight of several humans without collapsing
- a gymnast performs the difficult *iron cross* maneuver in an Olympic event (Problem 37)
- a ship moves at constant speed through calm water and maintains a perfectly level orientation (Chapter 14)
- polarized molecules in a dielectric material in a constant electric field take on an average equilibrium orientation that remains fixed in time (Chapter 25)

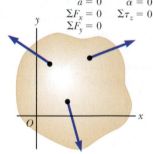

12.2 More on the Center of Gravity

Whenever we deal with a rigid object, one of the forces we must consider is the gravitational force acting on it, and we must know the point of application of this force. As we learned in Section 9.5, associated with every object is a special point called its center of gravity. The combination of the various gravitational forces acting on all the various mass elements of the object is equivalent to a single gravitational force acting through this point. Therefore, to compute the torque due to the gravitational force on an object of mass M, we need only consider the force $M\vec{g}$ acting at the object's center of gravity.

How do we find this special point? As mentioned in Section 9.5, if we assume $\vec{\mathbf{g}}$ is uniform over the object, the center of gravity of the object coincides with its center of mass. To see why, consider an object of arbitrary shape lying in the xy plane as illustrated in Figure 12.4. Suppose the object is divided into a large number of particles of masses m_1, m_2, m_3, ... having coordinates (x_1, y_1), (x_2, y_2), (x_3, y_3), ... In Equation 9.29, we defined the x coordinate of the center of mass of such an object to be

$$x_{\text{CM}} = \frac{m_1 x_1 + m_2 x_2 + m_3 x_3 + \cdots}{m_1 + m_2 + m_3 + \cdots} = \frac{\sum_i m_i x_i}{\sum_i m_i} \quad (9.29)$$

We use a similar equation to define the y coordinate of the center of mass, replacing each x with its y counterpart.

Let us now examine the situation from another point of view by considering the gravitational force exerted on each particle as shown in Figure 12.5. Each particle contributes a torque about an axis through the origin equal in magnitude to the particle's weight mg multiplied by its moment arm. For example, the magnitude of the torque due to the force $m_1 \vec{\mathbf{g}}_1$ is $m_1 g_1 x_1$, where g_1 is the value of the gravitational acceleration at the position of the particle of mass m_1. We wish to locate the center of gravity, the point at which application of the single gravitational force $M\vec{\mathbf{g}}_{\text{CG}}$ (where $M = m_1 + m_2 + m_3 + \cdots$ is the total mass of the object and $\vec{\mathbf{g}}_{\text{CG}}$ is the acceleration due to gravity at the location of the center of gravity) has the same effect on rotation as does the combined effect of all the individual gravitational forces $m_i \vec{\mathbf{g}}_i$. Equating the torque resulting from $M\vec{\mathbf{g}}_{\text{CG}}$ acting at the center of gravity to the sum of the torques acting on the individual particles gives

$$(m_1 + m_2 + m_3 + \cdots)g_{\text{CG}} \, x_{\text{CG}} = m_1 g_1 x_1 + m_2 g_2 x_2 + m_3 g_3 x_3 + \cdots$$

This expression accounts for the possibility that the value of g can in general vary over the object. If we assume uniform g over the object (as is usually the case), all the g factors are identical and cancel; we obtain

$$x_{\text{CG}} = \frac{m_1 x_1 + m_2 x_2 + m_3 x_3 + \cdots}{m_1 + m_2 + m_3 + \cdots} = \frac{\sum_i m_i x_i}{\sum_i m_i} \quad (12.4)$$

Comparing this result with Equation 9.29 shows that the center of gravity is located at the center of mass as long as $\vec{\mathbf{g}}$ is uniform over the entire object. Several examples in the next section deal with homogeneous, symmetric objects. The center of gravity for any such object coincides with its geometric center.

Ⓠ UICK QUIZ 12.3 A meterstick of uniform density is hung from a string tied at the 25-cm mark. A 0.50-kg object is hung from the zero end of the meterstick, and the meterstick is balanced horizontally. What is the mass of the meterstick? **(a)** 0.25 kg **(b)** 0.50 kg **(c)** 0.75 kg **(d)** 1.0 kg **(e)** 2.0 kg **(f)** impossible to determine

12.3 Examples of Rigid Objects in Static Equilibrium

The photograph of the one-bottle wine holder in Figure 12.6 shows one example of a balanced mechanical system that seems to defy gravity. For the system (wine holder plus bottle) to be in equilibrium, the net external force must be zero (see Eq. 12.1) and the net external torque around an axis passing through the support point must be zero (see Eq. 12.2). The second condition can be satisfied only when the center of gravity of the system in Figure 12.6 is directly over the support point.

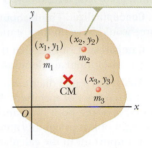

Each particle of the object has a specific mass and specific coordinates.

Figure 12.4 An object can be divided into many small particles. These particles can be used to locate the center of mass.

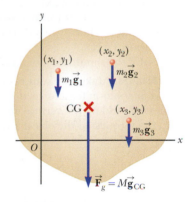

Figure 12.5 By dividing an object into many particles, we can find its center of gravity.

The center of gravity of the system (bottle plus holder) is directly over the support point.

Charles D. Winters

Figure 12.6 This one-bottle wine holder is a surprising display of static equilibrium.

| **PROBLEM-SOLVING STRATEGY** | **Rigid Object in Equilibrium** |

When analyzing a rigid object in equilibrium under the action of several external forces, use the following procedure.

1. Conceptualize. Think about the object that is in equilibrium and identify all the forces on it. Imagine what effect each force would have on the rotation of the object if it were the only force acting.

2. Categorize. Confirm that the object under consideration is indeed a rigid object in equilibrium. The object must have zero translational acceleration and zero angular acceleration.

3. Analyze. Draw a diagram and label all external forces acting on the object. Try to guess the correct direction for any forces that are not specified. When using the particle under a net force model, the object on which forces act can be represented in a free-body diagram with a dot because it does not matter where on the object the forces are applied. When using the rigid object in equilibrium model, however, we cannot use a dot to represent the object because the location where forces act is important in the calculation. Therefore, in a diagram showing the forces on an object, we must show the actual object or a simplified version of it.

Resolve all forces into rectangular components, choosing a convenient coordinate system. Then apply the first condition for equilibrium, Equation 12.1. Remember to keep track of the signs of the various force components.

Choose a convenient axis for calculating the net torque on the rigid object. Remember that the choice of the axis for the torque equation is arbitrary; therefore, choose an axis that simplifies your calculation as much as possible. Usually, the most convenient axis for calculating torques is one through a point through which the lines of action of several forces pass, so their torques around this axis are zero. If you don't know a force or don't need to know a force, it is often beneficial to choose an axis through the point at which this force acts. Apply the second condition for equilibrium, Equation 12.2.

Solve the simultaneous equations for the unknowns in terms of the known quantities.

4. Finalize. Make sure your results are consistent with your diagram. If you selected a direction that leads to a negative sign in your solution for a force, do not be alarmed; it merely means that the direction of the force is the opposite of what you guessed. Add up the vertical and horizontal forces on the object and confirm that each set of components adds to zero. Add up the torques on the object and confirm that the sum equals zero.

| Example **12.1** | **The Seesaw Revisited** |

A seesaw consisting of a uniform board of mass M and length ℓ supports at rest a father and daughter with masses m_f and m_d, respectively, as shown in Figure 12.7. The support (called the *fulcrum*) is under the center of gravity of the board, the father is a distance d from the center, and the daughter is a distance $\ell/2$ from the center.

(A) Determine the magnitude of the upward force $\vec{\mathbf{n}}$ exerted by the support on the board.

Figure 12.7 (Example 12.1) A balanced system.

> SOLUTION

Conceptualize Let us focus our attention on the board and consider the gravitational forces on the father and daughter as forces applied directly to the board. For force equilibrium, the location of the point of application of each force is not important.

Categorize Because the text of the problem states that the system is at rest, we model the board as a *rigid object in equilibrium*. Because we will only need the first condition of equilibrium to solve this part of the problem, however, we could also simply model the board as a *particle in equilibrium*.

Analyze Define upward as the positive y direction and substitute the forces on the board into Equation 12.1:

$$n - m_f g - m_d g - Mg = 0$$

Solve for the magnitude of the force $\vec{\mathbf{n}}$:

$$(1) \quad n = m_f g + m_d g + Mg = (m_f + m_d + M)g$$

(B) Determine where the father should sit to balance the system at rest.

12.1 continued

SOLUTION

Conceptualize For torque equilibrium, we need to pay attention to the location of the point of application of each force. The daughter would cause a clockwise rotation of the board around the support, whereas the father would cause a counterclockwise rotation.

Categorize This part of the problem requires the introduction of torque to find the position of the father, so we model the board as a *rigid object in equilibrium*.

Analyze The board's center of gravity is at its geometric center because we are told that the board is uniform. If we choose a rotation axis perpendicular to the page through the center of gravity of the board, the torques produced by $\vec{n}$ and the gravitational force on the board about this axis are zero.

Substitute expressions for the torques on the board due to the father and daughter into Equation 12.2:

$$(m_f g)(d) - (m_d g)\frac{\ell}{2} = 0$$

Solve for d:

$$d = \left(\frac{m_d}{m_f}\right)\frac{\ell}{2}$$

Finalize This result is the same one we obtained in Example 11.5 by evaluating the angular acceleration of the system and setting the angular acceleration equal to zero.

WHAT IF? Suppose we had chosen another point through which the rotation axis were to pass. For example, suppose the axis is perpendicular to the page and passes through the location of the father. Does that change the results to parts (A) and (B)?

Answer Part (A) is unaffected because the calculation of the net force does not involve a rotation axis. In part (B), we would conceptually expect there to be no change if a different rotation axis is chosen because the second condition of equilibrium claims that the torque is zero about *any* rotation axis.

Let's verify this answer mathematically. Recall that the sign of the torque associated with a force is positive if that force tends to rotate the system counterclockwise, whereas the sign of the torque is negative if the force tends to rotate the system clockwise. Let's choose a rotation axis perpendicular to the page and passing through the location of the father.

Substitute expressions for the torques on the board around this axis into Equation 12.2:

$$n(d) - (Mg)(d) - (m_d g)\left(d + \frac{\ell}{2}\right) = 0$$

Substitute from Equation (1) in part (A) and solve for d:

$$(m_f + m_d + M)g(d) - (Mg)(d) - (m_d g)\left(d + \frac{\ell}{2}\right) = 0$$

$$(m_f g)(d) - (m_d g)\left(\frac{\ell}{2}\right) = 0 \rightarrow d = \left(\frac{m_d}{m_f}\right)\frac{\ell}{2}$$

This result is in agreement with the one obtained in part (B).

Example 12.2 **Standing on a Horizontal Beam**

A uniform horizontal beam with a length of $\ell = 8.00$ m and a weight of $W_b = 200$ N is attached to a wall by a pin connection. Its far end is supported by a cable that makes an angle of $\phi = 53.0°$ with the beam (Fig. 12.8a, page 316). A person of weight $W_p = 600$ N stands a distance $d = 2.00$ m from the wall. Find the tension in the cable as well as the magnitude and direction of the force exerted by the wall on the beam.

SOLUTION

Conceptualize Imagine the person in Figure 12.8a moving outward on the beam. It seems reasonable that the farther he moves outward, the larger the torque he applies about the pivot and the larger the tension in the cable must be to balance this torque.

Categorize Because the system is at rest, we categorize the beam as a *rigid object in equilibrium*.

continued

12.2 continued

Figure 12.8 (Example 12.2) (a) A uniform beam supported by a cable. A person walks outward on the beam. (b) The force diagram for the beam. (c) The force diagram for the beam showing the components of $\vec{\mathbf{R}}$ and $\vec{\mathbf{T}}$.

Analyze   We identify all the external forces acting on the beam: the 200-N gravitational force, the force $\vec{\mathbf{T}}$ exerted by the cable, the force $\vec{\mathbf{R}}$ exerted by the wall at the pivot, and the 600-N force that the person exerts on the beam. These forces are all indicated in the force diagram for the beam shown in Figure 12.8b. When we assign directions for forces, it is sometimes helpful to imagine what would happen if a force were suddenly removed. For example, if the wall were to vanish suddenly, the left end of the beam would move to the left as it begins to fall. This scenario tells us that the wall is not only holding the beam up but is also pressing outward against it. Therefore, we draw the vector $\vec{\mathbf{R}}$ in the direction shown in Figure 12.8b. Figure 12.8c shows the horizontal and vertical components of $\vec{\mathbf{T}}$ and $\vec{\mathbf{R}}$.

Applying the first condition of equilibrium, substitute expressions for the forces on the beam into component equations from Equation 12.1:

(1) $\displaystyle\sum F_x = R\cos\theta - T\cos\phi = 0$

(2) $\displaystyle\sum F_y = R\sin\theta + T\sin\phi - W_p - W_b = 0$

where we have chosen rightward and upward as our positive directions. Because R, T, and θ are all unknown, we cannot obtain a solution from these expressions alone. (To solve for the unknowns, the number of simultaneous equations must generally equal the number of unknowns.)

Now let's invoke the condition for rotational equilibrium. A convenient axis to choose for our torque equation is the one that passes through the pin connection. The feature that makes this axis so convenient is that the force $\vec{\mathbf{R}}$ and the horizontal component of $\vec{\mathbf{T}}$ both have a moment arm of zero; hence, these forces produce no torque about this axis.

Substitute expressions for the torques on the beam into Equation 12.2:

$$\sum \tau_z = (T\sin\phi)(\ell) - W_p d - W_b\left(\frac{\ell}{2}\right) = 0$$

This equation contains only T as an unknown because of our choice of rotation axis. Solve for T and substitute numerical values:

$$T = \frac{W_p d + W_b(\ell/2)}{\ell\sin\phi} = \frac{(600\ \text{N})(2.00\ \text{m}) + (200\ \text{N})(4.00\ \text{m})}{(8.00\ \text{m})\sin 53.0°} = \boxed{313\ \text{N}}$$

Rearrange Equations (1) and (2) and then divide:

$$\frac{R\sin\theta}{R\cos\theta} = \tan\theta = \frac{W_p + W_b - T\sin\phi}{T\cos\phi}$$

Solve for θ and substitute numerical values:

$$\theta = \tan^{-1}\left(\frac{W_p + W_b - T\sin\phi}{T\cos\phi}\right)$$

$$= \tan^{-1}\left[\frac{600\ \text{N} + 200\ \text{N} - (313\ \text{N})\sin 53.0°}{(313\ \text{N})\cos 53.0°}\right] = \boxed{71.1°}$$

Solve Equation (1) for R and substitute numerical values:

$$R = \frac{T\cos\phi}{\cos\theta} = \frac{(313\ \text{N})\cos 53.0°}{\cos 71.1°} = \boxed{581\ \text{N}}$$

Finalize   The positive value for the angle θ indicates that our estimate of the direction of $\vec{\mathbf{R}}$ was accurate.

Had we selected some other axis for the torque equation, the solution might differ in the details but the answers would be the same. For example, had we chosen an axis through the center of gravity of the beam, the torque equation would involve both T and R. This equation, coupled with Equations (1) and (2), however, could still be solved for the unknowns. Try it!

12.2 continued

WHAT IF? What if the person walks farther out on the beam? Does *T* change? Does *R* change? Does θ change?

Answer *T* must increase because the gravitational force on the person exerts a larger torque about the pin connection, which must be countered by a larger torque in the opposite direction due to an increased value of *T*. If *T* increases, the vertical component of $\vec{\mathbf{R}}$ decreases to maintain force equilibrium in the vertical direction. Force equilibrium in the horizontal direction, however, requires an increased horizontal component of $\vec{\mathbf{R}}$ to balance the horizontal component of the increased $\vec{\mathbf{T}}$. This fact suggests that θ becomes smaller, but it is hard to predict what happens to *R*. Problem 50 asks you to explore the behavior of *R*.

Example **12.3** **The Leaning Ladder**

A uniform ladder of length ℓ rests against a smooth, vertical wall (Fig. 12.9a). The mass of the ladder is *m*, and the coefficient of static friction between the ladder and the ground is $\mu_s = 0.40$. Find the minimum angle θ_{min} at which the ladder does not slip.

SOLUTION

Conceptualize Think about any ladders you have climbed. Do you want a large friction force between the bottom of the ladder and the surface or a small one? If the friction force is zero, will the ladder stay up? Simulate a ladder with a ruler leaning against a vertical surface. Does the ruler slip at some angles and stay up at others?

Categorize We do not wish the ladder to slip, so we model it as a *rigid object in equilibrium*.

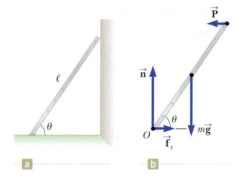

Figure 12.9 (Example 12.3) (a) A uniform ladder at rest, leaning against a smooth wall. The ground is rough. (b) The forces on the ladder.

Analyze A diagram showing all the external forces acting on the ladder is illustrated in Figure 12.9b. The force exerted by the ground on the ladder is the vector sum of a normal force $\vec{\mathbf{n}}$ and the force of static friction $\vec{\mathbf{f}}_s$. The wall exerts a normal force $\vec{\mathbf{P}}$ on the top of the ladder, but there is no friction force here because the wall is smooth. So the net force on the top of the ladder is perpendicular to the wall and of magnitude *P*.

Apply the first condition for equilibrium to the ladder in both the *x* and the *y* directions:

(1) $\sum F_x = f_s - P = 0$

(2) $\sum F_y = n - mg = 0$

Solve Equation (1) for *P*:

(3) $P = f_s$

Solve Equation (2) for *n*:

(4) $n = mg$

When the ladder is on the verge of slipping, the force of static friction must have its maximum value, which is given by $f_{s,max} = \mu_s n$. Combine this equation with Equations (3) and (4):

(5) $P_{max} = f_{s,max} = \mu_s n = \mu_s mg$

Apply the second condition for equilibrium to the ladder, evaluating torques about an axis perpendicular to the page through *O*:

$$\sum \tau_O = P\ell \sin \theta - mg\frac{\ell}{2} \cos \theta = 0$$

Solve for $\tan \theta$:

$$\frac{\sin \theta}{\cos \theta} = \tan \theta = \frac{mg}{2P} \rightarrow \theta = \tan^{-1}\left(\frac{mg}{2P}\right)$$

Under the conditions that the ladder is just ready to slip, θ becomes θ_{min} and P_{max} is given by Equation (5). Substitute:

$$\theta_{min} = \tan^{-1}\left(\frac{mg}{2P_{max}}\right) = \tan^{-1}\left(\frac{1}{2\mu_s}\right) = \tan^{-1}\left[\frac{1}{2(0.40)}\right] = \boxed{51°}$$

Finalize Notice that the angle depends only on the coefficient of friction, not on the mass or length of the ladder.

Example **12.4**   **Negotiating a Curb**

(A) Estimate the magnitude of the force $\vec{\mathbf{F}}$ a person must apply to a wheelchair's main wheel to roll up over a sidewalk curb (Fig. 12.10a). This main wheel that comes in contact with the curb has a radius r, and the height of the curb is h.

SOLUTION

Conceptualize Think about wheelchair access to buildings. Generally, there are ramps built for individuals in wheelchairs or scooters. Steplike structures such as curbs are serious barriers to a wheelchair.

Categorize Imagine the person exerts enough force so that the bottom of the main wheel just loses contact with the lower surface and hovers at rest. We model the wheel in this situation as a *rigid object in equilibrium*.

Analyze Usually, the person's hands supply the required force to a slightly smaller wheel that is concentric with the main wheel. For simplicity, let's assume the radius of this second wheel is the same as the radius of the main wheel. Let's estimate a combined gravitational force of magnitude $mg = 1\,400$ N for the person and the wheelchair, acting along a line of action passing through the axle of the main wheel, and choose a wheel radius of $r = 30$ cm. We also pick a curb height of $h = 10$ cm. Let's also assume the wheelchair and occupant are symmetric and each wheel supports a weight of 700 N. We then proceed to analyze only one of the main wheels. Figure 12.10b shows the geometry for a single wheel.

When the wheel is just about to be raised from the street, the normal force exerted by the ground on the wheel at point B goes to zero. Hence, at this time only three forces act on the wheel as shown in the force diagram in Figure 12.10c. The force $\vec{\mathbf{R}}$, which is the force exerted by the curb on the wheel, acts at point A, so if we choose to have our axis of rotation be perpendicular to the page and pass through point A, we do not need to include $\vec{\mathbf{R}}$ in our torque equation. The moment arm of $\vec{\mathbf{F}}$ relative to an axis through A is given by $2r - h$ (see Fig. 12.10c).

Figure 12.10 (Example 12.4) (a) A person in a wheelchair attempts to roll up over a curb. (b) Details of the wheel and curb. The person applies a force $\vec{\mathbf{F}}$ to the top of the wheel. (c) A force diagram for the wheel when it is just about to be raised. Three forces act on the wheel at this instant: $\vec{\mathbf{F}}$, which is exerted by the hand; $\vec{\mathbf{R}}$, which is exerted by the curb; and the gravitational force $m\vec{\mathbf{g}}$. (d) The vector sum of the three external forces acting on the wheel is zero.

Use the triangle OAC in Figure 12.10b to find the moment arm d of the gravitational force $m\vec{\mathbf{g}}$ acting on the wheel relative to an axis through point A:

$$(1) \quad d = \sqrt{r^2 - (r-h)^2} = \sqrt{2rh - h^2}$$

Apply the second condition for equilibrium to the wheel, taking torques about an axis through A:

$$(2) \quad \sum \tau_A = mgd - F(2r - h) = 0$$

Substitute for d from Equation (1):

$$mg\sqrt{2rh - h^2} - F(2r - h) = 0$$

Solve for F:

$$(3) \quad F = \frac{mg\sqrt{2rh - h^2}}{2r - h}$$

Simplify:

$$F = mg\,\frac{\sqrt{h}\,\sqrt{2r - h}}{2r - h} = mg\sqrt{\frac{h}{2r - h}}$$

Substitute numerical values:

$$F = (700\ \text{N})\sqrt{\frac{0.1\ \text{m}}{2(0.3\ \text{m}) - 0.1\ \text{m}}}$$

$$= 3 \times 10^2\ \text{N}$$

12.4 c o n t i n u e d

(B) Determine the magnitude and direction of $\vec{\mathbf{R}}$.

S O L U T I O N

Apply the first condition for equilibrium to the x and y components of the forces on the wheel:

$$(4) \quad \sum F_x = F - R\cos\theta = 0$$

$$(5) \quad \sum F_y = R\sin\theta - mg = 0$$

Divide Equation (5) by Equation (4):

$$\frac{R\sin\theta}{R\cos\theta} = \tan\theta = \frac{mg}{F}$$

Solve for the angle θ:

$$\theta = \tan^{-1}\left(\frac{mg}{F}\right) = \tan^{-1}\left(\frac{700\ \text{N}}{300\ \text{N}}\right) = \boxed{70°}$$

Solve Equation (5) for R and substitute numerical values:

$$R = \frac{mg}{\sin\theta} = \frac{700\ \text{N}}{\sin 70°} = \boxed{8 \times 10^2\ \text{N}}$$

Finalize Notice that we have kept only one digit as significant because we have guessed at some numbers and made some assumptions. (We have written the angle as 70° because $7 \times 10^{1°}$ is awkward!) For example, we assumed that the center of mass of the wheelchair–person system was directly over the axle of the wheel. How likely do you think that is to be true? The results indicate that the force that must be applied to each wheel is substantial. You may want to estimate the force required to roll a wheelchair up a typical sidewalk accessibility ramp for comparison.

W H A T   I F ? Would it be easier to negotiate the curb if the person grabbed the wheel at point D in Figure 12.10c and pulled *upward*?

Answer If the force $\vec{\mathbf{F}}$ in Figure 12.10c is rotated counterclockwise by 90° and applied at D, its moment arm about an axis through A is $d + r$. Let's call the magnitude of this new force F'.

Modify Equation (2) for this situation:

$$\sum \tau_A = mgd - F'(d + r) = 0$$

Solve this equation for F' and substitute for d:

$$F' = \frac{mgd}{d + r} = \frac{mg\sqrt{2rh - h^2}}{\sqrt{2rh - h^2} + r}$$

Take the ratio of this force to the original force from Equation (3) and express the result in terms of h/r, the ratio of the curb height to the wheel radius:

$$\frac{F'}{F} = \frac{\dfrac{mg\sqrt{2rh - h^2}}{\sqrt{2rh - h^2} + r}}{\dfrac{mg\sqrt{2rh - h^2}}{2r - h}} = \frac{2r - h}{\sqrt{2rh - h^2} + r} = \frac{2 - \left(\dfrac{h}{r}\right)}{\sqrt{2\left(\dfrac{h}{r}\right) - \left(\dfrac{h}{r}\right)^2} + 1}$$

Substitute the ratio $h/r = 0.33$ from the given values:

$$\frac{F'}{F} = \frac{2 - 0.33}{\sqrt{2(0.33) - (0.33)^2} + 1} = 0.96$$

This result tells us that, *for these values,* it is slightly easier to pull upward at D than horizontally at the top of the wheel. For very high curbs, so that h/r is close to 1, the ratio F'/F drops to about 0.5 because point A is located near the right edge of the wheel in Figure 12.10b. The force at D is applied at a distance of about $2r$ from A, whereas the force at the top of the wheel has a moment arm of only about r. For high curbs, then, it is best to pull upward at D, although a large value of the force is required. For small curbs, it is best to apply the force at the top of the wheel. The ratio F'/F becomes larger than 1 at about $h/r = 0.3$ because point A is now close to the bottom of the wheel and the force applied at the top of the wheel has a larger moment arm than when applied at D.

Finally, let's comment on the validity of these mathematical results. Consider Figure 12.10d and imagine that the vector $\vec{\mathbf{F}}$ is upward instead of to the right. There is no way the three vectors can add to equal zero as required by the first equilibrium condition. Therefore, our results above may be qualitatively valid, but not exact quantitatively. To cancel the horizontal component of $\vec{\mathbf{R}}$, the force at D must be applied at an angle to the vertical rather than straight upward. This feature makes the calculation more complicated and requires both conditions of equilibrium.

12.4 Elastic Properties of Solids

In Section 9.8, we explored deformable systems consisting of masses and springs. We continue and generalize that discussion in this section. We have assumed objects remain rigid when external forces act on them. In reality, all objects are deformable to some extent. That is, it is possible to change the shape or the size (or both) of an object by applying external forces. As these changes take place, however, internal forces in the object resist the deformation.

We shall discuss the deformation of solids in terms of the concepts of *stress* and *strain*. **Stress** is a quantity that is proportional to the force causing a deformation; more specifically, stress is the external force acting on an object per unit cross-sectional area. The result of a stress is **strain,** which is a measure of the degree of deformation. It is found that, for sufficiently small stresses, stress is proportional to strain; the constant of proportionality depends on the material being deformed and on the nature of the deformation. We call this proportionality constant the **elastic modulus.** The elastic modulus is therefore defined as the ratio of the stress to the resulting strain:

$$\text{Elastic modulus} \equiv \frac{\text{stress}}{\text{strain}} \qquad (12.5)$$

The elastic modulus in general relates what is done to a solid object (a force is applied) to how that object responds (it deforms to some extent). It is similar in nature to the spring constant k in Hooke's law (Eq. 7.9) that relates a force applied to a spring and the resultant deformation of the spring, measured by its extension or compression.

We consider three types of deformation and define an elastic modulus for each:

1. **Young's modulus** measures the resistance of a solid to a change in its length.
2. **Shear modulus** measures the resistance to motion of the planes within a solid parallel to each other.
3. **Bulk modulus** measures the resistance of solids or liquids to changes in their volume.

Young's Modulus: Elasticity in Length

Consider a long bar of cross-sectional area A and initial length L_i that is clamped at one end as in Figure 12.11. When an external force is applied perpendicular to the cross section, internal molecular forces in the bar resist distortion ("stretching"), but the bar reaches an equilibrium situation in which its final length L_f is greater than L_i and in which the external force is exactly balanced by the internal forces. In such a situation, the bar is said to be stressed. We define the **tensile stress** as the ratio of the magnitude of the external force F to the cross-sectional area A, where the cross section is perpendicular to the force vector. The **tensile strain** in this case is defined as the ratio of the change in length ΔL to the original length L_i. We define **Young's modulus** by a combination of these two ratios:

Young's modulus ▶

$$Y \equiv \frac{\text{tensile stress}}{\text{tensile strain}} = \frac{F/A}{\Delta L/L_i} \qquad (12.6)$$

Young's modulus is typically used to characterize a rod or wire stressed under either tension or compression. Because strain is a dimensionless quantity, Y has units of force per unit area. Typical values are given in Table 12.1.

The amount by which the length of the bar changes due to the applied force is ΔL.

Figure 12.11 A force $\vec{\mathbf{F}}$ is applied to the free end of a bar clamped at the other end.

TABLE 12.1 Typical Values for Elastic Moduli

Substance	Young's Modulus (N/m²)	Shear Modulus (N/m²)	Bulk Modulus (N/m²)
Tungsten	35×10^{10}	14×10^{10}	20×10^{10}
Steel	20×10^{10}	8.4×10^{10}	6×10^{10}
Copper	11×10^{10}	4.2×10^{10}	14×10^{10}
Brass	9.1×10^{10}	3.5×10^{10}	6.1×10^{10}
Aluminum	7.0×10^{10}	2.5×10^{10}	7.0×10^{10}
Glass	$6.5–7.8 \times 10^{10}$	$2.6–3.2 \times 10^{10}$	$5.0–5.5 \times 10^{10}$
Quartz	5.6×10^{10}	2.6×10^{10}	2.7×10^{10}
Water	—	—	0.21×10^{10}
Mercury	—	—	2.8×10^{10}

For relatively small stresses, the bar returns to its initial length when the force is removed. The **elastic limit** of a substance is defined as the maximum stress that can be applied to the substance before it becomes permanently deformed and does not return to its initial length. It is possible to exceed the elastic limit of a substance by applying a sufficiently large stress as seen in Figure 12.12. Initially, a stress-versus-strain curve is a straight line. As the stress increases, however, the curve is no longer a straight line. When the stress exceeds the elastic limit, the object is permanently distorted and does not return to its original shape after the stress is removed. As the stress is increased even further, the material ultimately breaks.

Shear Modulus: Elasticity of Shape

Another type of deformation occurs when an object is subjected to a force parallel to one of its faces while the opposite face is held fixed by another force (Fig. 12.13a). The stress in this case is called a *shear stress*. If the object is originally a rectangular block, a shear stress results in a shape whose cross section is a parallelogram. A book pushed sideways as shown in Figure 12.13b is an example of an object subjected to a shear stress. To a first approximation (for small distortions), no change in volume occurs with this deformation.

We define the **shear stress** as F/A, the ratio of the tangential force to the area A of the face being sheared. The **shear strain** is defined as the ratio $\Delta x/h$, where Δx is the horizontal distance that the sheared face moves and h is the height of the object. In terms of these quantities, the **shear modulus** is

$$S \equiv \frac{\text{shear stress}}{\text{shear strain}} = \frac{F/A}{\Delta x/h} \qquad (12.7)$$ ◀ Shear modulus

Values of the shear modulus for some representative materials are given in Table 12.1. Like Young's modulus, the unit of shear modulus is the ratio of that for force to that for area.

Bulk Modulus: Volume Elasticity

Bulk modulus characterizes the response of an object to changes in a force of uniform magnitude applied perpendicularly over the entire surface of the object as shown in Figure 12.14. (We assume here the object is made of a single substance.) As we shall see in Chapter 14, such a uniform distribution of forces occurs when an object is immersed in a fluid. An object subject to this type of stress undergoes a change in volume but no change in shape. The **volume stress** is defined as the ratio of the magnitude of the total force F exerted on a surface to the area A of the surface. The quantity $P = F/A$ is called **pressure,** which we shall study in more detail in Chapter 14. If the pressure on an object changes by an amount $\Delta P = \Delta F/A$, the object experiences a volume change ΔV. The **volume strain** is equal to the change in volume ΔV divided by the initial volume V_i. Therefore, from Equation 12.5,

Figure 12.12 Stress-versus-strain curve for an elastic solid.

The shear stress causes the top face of the block to move to the right relative to the bottom.

The shear stress causes the front cover of the book to move to the right relative to the back cover.

Figure 12.13 (a) A shear deformation in which a rectangular block is distorted by two forces of equal magnitude but opposite directions applied to two parallel faces. (b) A book is under shear stress when a hand placed on the cover applies a horizontal force away from the spine.

we can characterize a volume ("bulk") compression in terms of the **bulk modulus,** which is defined as

Bulk modulus ▶

$$B \equiv \frac{\text{volume stress}}{\text{volume strain}} = -\frac{\Delta F/A}{\Delta V/V_i} = -\frac{\Delta P}{\Delta V/V_i} \qquad (12.8)$$

A negative sign is inserted in this defining equation so that B is a positive number. This maneuver is necessary because an increase in pressure (positive ΔP) causes a decrease in volume (negative ΔV) and vice versa.

Table 12.1 lists bulk moduli for some materials. If you look up such values in a different source, you may find the reciprocal of the bulk modulus listed. The reciprocal of the bulk modulus is called the **compressibility** of the material.

Notice from Table 12.1 that both solids and liquids have a bulk modulus. No shear modulus and no Young's modulus are given for liquids, however, because a liquid does not sustain a shearing stress or a tensile stress. If a shearing force or a tensile force is applied to a liquid, the liquid simply flows in response.

The cube undergoes a change in volume but no change in shape.

Figure 12.14 A cube is under uniform pressure and is therefore compressed on all sides by forces normal to its six faces. The arrowheads of force vectors on the sides of the cube that are not visible are hidden by the cube.

QUICK QUIZ **12.4** For the three parts of this Quick Quiz, choose from the following choices the correct answer for the elastic modulus that describes the relationship between stress and strain for the system of interest, which is in italics: (a) Young's modulus (b) shear modulus (c) bulk modulus (d) none of those choices **(i)** A *block of iron* is sliding across a horizontal floor. The friction force between the sliding block and the floor causes the block to deform. **(ii)** A trapeze artist swings through a circular arc. At the bottom of the swing, the *wires* supporting the trapeze are longer than when the trapeze artist simply hangs from the trapeze due to the increased tension in them. **(iii)** A spacecraft carries a *steel sphere* to a planet on which atmospheric pressure is much higher than on the Earth. The higher pressure causes the radius of the sphere to decrease.

Prestressed Concrete

If the stress on a solid object exceeds a certain value, the object fractures. The maximum stress that can be applied before fracture occurs—called the *tensile strength, compressive strength,* or *shear strength*—depends on the nature of the material and on the type of applied stress. For example, concrete has a tensile strength of about 2×10^6 N/m², a compressive strength of 20×10^6 N/m², and a shear strength of 2×10^6 N/m². If the applied stress exceeds these values, the concrete fractures. It is common practice to use large safety factors to prevent failure in concrete structures.

Concrete is normally very brittle when it is cast in thin sections. Therefore, concrete slabs tend to sag and crack at unsupported areas as shown in Figure 12.15a. The slab can be strengthened by the use of steel rods to reinforce the concrete as illustrated in Figure 12.15b. Because concrete is much stronger under compression (squeezing) than under tension (stretching) or shear, vertical columns of concrete can support very heavy loads, whereas horizontal beams of concrete tend to

Figure 12.15 (a) A concrete slab with no reinforcement tends to crack under a heavy load. (b) The strength of the concrete is increased by using steel reinforcement rods. (c) The concrete is further strengthened by prestressing it with steel rods under tension.

sag and crack. A significant increase in shear strength is achieved, however, if the reinforced concrete is prestressed as shown in Figure 12.15c. As the concrete is being poured, the steel rods are held under tension by external forces. The external forces are released after the concrete cures; the result is a permanent tension in the steel and hence a compressive stress on the concrete. The concrete slab can now support a much heavier load.

Example 12.5 | **Stage Design**

In Example 8.3, we analyzed a cable used to support an actor as he swings onto the stage. Now suppose the tension in the cable is 940 N as the actor reaches the lowest point. What diameter should a 10-m-long steel cable have if we do not want it to stretch more than 0.50 cm under these conditions?

SOLUTION

Conceptualize Look back at Example 8.3 to recall what is happening in this situation. We ignored any stretching of the cable there, but we wish to address this phenomenon in this example.

Categorize We perform a simple calculation involving Equation 12.6, so we categorize this example as a substitution problem.

Solve Equation 12.6 for the cross-sectional area of the cable:

$$A = \frac{FL_i}{Y \Delta L}$$

Assuming the cross section is circular, find the diameter of the cable from $d = 2r$ and $A = \pi r^2$:

$$d = 2r = 2 \sqrt{\frac{A}{\pi}} = 2 \sqrt{\frac{FL_i}{\pi Y \Delta L}}$$

Substitute numerical values:

$$d = 2 \sqrt{\frac{(940 \text{ N})(10 \text{ m})}{\pi (20 \times 10^{10} \text{ N/m}^2)(0.005\,0 \text{ m})}} = 3.5 \times 10^{-3} \text{ m} = 3.5 \text{ mm}$$

To provide a large margin of safety, you would probably use a flexible cable made up of many smaller wires having a total cross-sectional area substantially greater than our calculated value.

Example 12.6 | **Squeezing a Brass Sphere**

A solid brass sphere is initially surrounded by air, and the air pressure exerted on it is $1.0 \times 10^5 \text{ N/m}^2$ (normal atmospheric pressure). The sphere is lowered into the ocean to a depth where the pressure is $2.0 \times 10^7 \text{ N/m}^2$. The volume of the sphere in air is 0.50 m^3. By how much does this volume change once the sphere is submerged?

SOLUTION

Conceptualize Think about movies or television shows you have seen in which divers go to great depths in the water in submersible vessels. These vessels must be very strong to withstand the large pressure under water. This pressure squeezes the vessel and reduces its volume.

Categorize We perform a simple calculation involving Equation 12.8, so we categorize this example as a substitution problem.

Solve Equation 12.8 for the volume change of the sphere:

$$\Delta V = -\frac{V_i \Delta P}{B}$$

Substitute numerical values:

$$\Delta V = -\frac{(0.50 \text{ m}^3)(2.0 \times 10^7 \text{ N/m}^2 - 1.0 \times 10^5 \text{ N/m}^2)}{6.1 \times 10^{10} \text{ N/m}^2}$$

$$= -1.6 \times 10^{-4} \text{ m}^3$$

The negative sign indicates that the volume of the sphere decreases.

Summary

Definitions

The gravitational force exerted on an object can be considered as acting at a single point called the **center of gravity.** An object's center of gravity coincides with its center of mass if the object is in a uniform gravitational field.

We can describe the elastic properties of a substance using the concepts of stress and strain. **Stress** is a quantity proportional to the force producing a deformation; **strain** is a measure of the degree of deformation. Stress is proportional to strain, and the constant of proportionality is the **elastic modulus:**

$$\text{Elastic modulus} \equiv \frac{\text{stress}}{\text{strain}} \qquad (12.5)$$

Concepts and Principles

Three common types of deformation are represented by (1) the resistance of a solid to elongation under a load, characterized by **Young's modulus** Y; (2) the resistance of a solid to the motion of internal planes sliding past each other, characterized by the **shear modulus** S; and (3) the resistance of a solid or fluid to a volume change, characterized by the **bulk modulus** B.

Analysis Model for Problem Solving

Rigid Object in Equilibrium A rigid object in equilibrium exhibits no translational or angular acceleration. The net external force acting on it is zero, and the net external torque on it is zero about any axis:

$$\sum \vec{\mathbf{F}}_{\text{ext}} = 0 \qquad (12.1)$$

$$\sum \vec{\boldsymbol{\tau}}_{\text{ext}} = 0 \qquad (12.2)$$

The first condition is the condition for translational equilibrium, and the second is the condition for rotational equilibrium.

Think–Pair–Share

See the Preface for an explanation of the icons used in this problems set. For additional assessment items for this section, go to WEB**ASSIGN** From Cengage

1. A father and his son are painting a wall. To reach the higher portions of the wall, they place a 20.0-kg plank of wood, 3.50 m long and of uniform consistency, on two sawhorses. A sawhorse is placed 1.00 m from each end of the plank. The son has a mass 50.0 kg. He and the father, of mass 85.0 kg, climb up and stand on the plank. Discuss in your group and respond to the following. (a) The father stands at rest on the plank, directly over one of the sawhorses. Can the son move to *any* position he wishes on the plank without the plank tipping? (b) After learning about the tipping possibilities in part (a), the father and son decide to use a more massive plank. What must the mass of the plank be so that both father and son are free to move anywhere on the plank that they wish? (c) They find a plank of just the mass found in part (b) and test it by standing on the right-hand end together. Will they be safe from tipping if they both stand on the *left-hand* end together? (d) After using the plank of the required mass determined in part (b) and being exhausted from moving it to new positions along the

wall, the father and son decide to set it aside and continue to use the 20.0-kg plank, while being careful to stay far apart on the plank. But the use of the massive plank in part (b) has damaged one of the sawhorses so that it can only support a force of 1.75×10^3 N. When the father and son get back on the 20.0-kg plank and move around, will the damaged sawhorse collapse?

2. **ACTIVITY** If an object is set on a table such that part of it extends off the edge of the table, the center of mass of the object must be over part of the table surface to avoid the object falling. If the center of mass is beyond the edge of the table, the gravitational force will exert a torque on the object and tip it off the table. Gather four metersticks together. (a) Determine a way that you can stack all four metersticks so that (1) all four metersticks are parallel; (2) each higher meterstick is further out over the edge of the table than the one below it, and (3) the topmost meterstick has *no* part of its length above the table surface. *Hint:* Begin from the top; put the topmost meterstick on the second one so that the top one does not tip off. Then put the stack of two on the third meterstick so that the top

two do not tip, and so on. (b) After successfully building the appropriate stack, calculate the position of the center mass of the stack and show that it is over the table surface, not over the air beyond the edge of the table. (c) What if you were to rotate the top meterstick by 90° around a vertical axis so that it sits on the end of the third meterstick, perpendicular to the other three metersticks? Would the system still be in equilibrium? (d) Now stack the metersticks as follows with the zero ends of the metersticks all to the right in the diagram at the top of the next column:

The right ends of the two metersticks at the upper left are above the 50-cm mark on the bottom meterstick. Place the system of metersticks on a table and move it off the edge to the right until a small additional outward movement would cause them to tip clockwise off the table. What reading on the bottom meterstick coincides with the edge of the table?

Problems

See the Preface for an explanation of the icons used in this problems set. For additional assessment items for this section, go to ✺ **WEB**ASSIGN
From Cengage

SECTION 12.1 Analysis Model: Rigid Object in Equilibrium

1. You are building additional storage space in your garage. You
CR decide to suspend a 10.0-kg sheet of plywood of dimensions 0.600 m wide by 2.25 m long from the ceiling. The plywood will be held in a horizontal orientation by four light vertical chains attached to the plywood at its corners and mounted to the ceiling. After you complete the job of suspending the plywood from the ceiling, you choose three cubic boxes to place on the shelf. Each box is 0.750 m on a side. Box 1 has a mass of 50.0 kg, box 2 has a mass of 100 kg, and box 3 has a mass of 125 kg. The mass of each box is uniformly distributed within the box and each box is centered on the front-to-back width of the shelf. Unbeknownst to you, one of the chains on the right-hand end of your shelf is defective and will break if subjected to a force of more than 700 N. There are six possible arrangements of the three boxes on the shelf, for example, from left to right, Box 1, Box 2, Box 3, and Box 1, Box 3, Box, 2, and four more. Which arrangements are safe (that is, the defective chain will not break if the boxes are arranged in this way), and which arrangements are dangerous?

2. *Why is the following situation impossible?* A uniform beam of mass $m_b = 3.00$ kg and length $\ell = 1.00$ m supports blocks with masses $m_1 = 5.00$ kg and $m_2 = 15.0$ kg at two positions as shown in Figure P12.2. The beam rests on two triangular blocks, with point P a distance $d = 0.300$ m to the right of the center of gravity of the beam. The position of the object of mass m_2 is adjusted along the length of the beam until the normal force on the beam at O is zero.

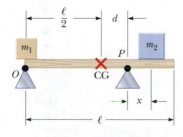

Figure P12.2

SECTION 12.2 More on the Center of Gravity

Problems 24 and 26 in Chapter 9 can also be assigned with this section.

3. A carpenter's square has the shape of an L as shown in
V Figure P12.3. Locate its center of gravity.

Figure P12.3

4. A circular pizza of radius R has a circular piece of radius
S $R/2$ removed from one side as shown in Figure P12.4. The center of gravity has moved from C to C' along the x axis. Show that the distance from C to C' is $R/6$. Assume the thickness and density of the pizza are uniform throughout.

Figure P12.4

5. Your brother is opening a skateboard shop. He has created
CR a sign for his shop made from a uniform material and in the shape shown in Figure P12.5. The shape of the sign represents one of the hills in the skateboard park he plans on building on land adjacent to the shop. The curve on the top of the sign is described by the function $y = (x - 3)^2/9$. When the sign arrives in his shop, your brother wants to hang it from a single wire outside the shop. But he doesn't know where on the sign to attach the wire so that the bottom edge of the sign will hang in a horizontal orientation. He asks for your help.

Figure P12.5

SECTION 12.3 Examples of Rigid Objects in Static Equilibrium

Problems 14, 16, 18, 19, 34, 45, and 52 in Chapter 5 can also be
assigned with this section.

6. A uniform beam of length 7.60 m and weight 4.50×10^2 N is
carried by two workers, Sam and Joe, as shown in Figure P12.6.
Determine the force that each person exerts on the beam.

Figure P12.6

7. Find the mass m of the counterweight needed to balance a
truck with mass $M = 1\,500$ kg on an incline of $\theta = 45°$
(Fig. P12.7). Assume both pulleys are frictionless and
massless.

Figure P12.7

8. A uniform beam of length L and
mass m shown in Figure P12.8
is inclined at an angle θ to the
horizontal. Its upper end is connected to a wall by a rope, and
its lower end rests on a rough,
horizontal surface. The coefficient of static friction between
the beam and surface is μ_s.
Assume the angle θ is such that

Figure P12.8

the static friction force is at its *maximum* value. (a) Draw a
force diagram for the beam. (b) Using the condition of rotational equilibrium, find an expression for the tension T in
the rope in terms of m, g, and θ. (c) Using the condition
of translational equilibrium, find a second expression for T
in terms of μ_s, m, and g. (d) Using the results from parts
(a) through (c), obtain an expression for μ_s involving only
the angle θ. (e) What happens if the ladder is lifted upward
and its base is placed back on the ground slightly to the left
of its position in Figure P12.8? Explain.

9. A flexible chain weighing 40.0 N hangs between two hooks
located at the same height (Fig. P12.9). At each hook, the
tangent to the chain makes an angle $\theta = 42.0°$ with the horizontal. Find (a) the magnitude of the force each hook exerts
on the chain and (b) the tension in the chain at its midpoint. *Suggestion:* For part (b), make a force diagram for half
of the chain.

Figure P12.9

10. A 20.0-kg floodlight in a park is supported
at the end of a horizontal beam of negligible mass that is hinged to a pole as shown
in Figure P12.10. A cable at an angle of
$\theta = 30.0°$ with the beam helps support
the light. (a) Draw a force diagram for
the beam. By computing torques about
an axis at the hinge at the left-hand end
of the beam, find (b) the tension in the
cable, (c) the horizontal component of the force exerted
by the pole on the beam, and (d) the vertical component
of this force. Now solve the same problem from the force
diagram from part (a) by computing torques around the
junction between the cable and the beam at the right-hand
end of the beam. Find (e) the vertical component of the
force exerted by the pole on the beam, (f) the tension in
the cable, and (g) the horizontal component of the force
exerted by the pole on the beam. (h) Compare the solution to parts (b) through (d) with the solution to parts
(e) through (g). Is either solution more accurate?

Figure P12.10

11. Sir Lost-a-Lot dons his armor and sets out from the castle on
his trusty steed (Fig. P12.11). Usually, the drawbridge is lowered to a horizontal position so that the end of the bridge
rests on the stone ledge. Unfortunately, Lost-a-Lot's squire
didn't lower the drawbridge far enough and stopped it at
$\theta = 20.0°$ above the horizontal. The knight and his horse
stop when their combined center of mass is $d = 1.00$ m from
the end of the bridge. The uniform bridge is $\ell = 8.00$ m
long and has mass 2 000 kg. The lift cable is attached to the
bridge 5.00 m from the hinge at the castle end and to a
point on the castle wall $h = 12.0$ m above the bridge. Lost-a-
Lot's mass combined with his armor and steed is 1 000 kg.
Determine (a) the tension in the cable and (b) the horizontal and (c) the vertical force components acting on the
bridge at the hinge.

Figure P12.11 Problems 11 and 12.

12. **Review.** While Lost-a-Lot ponders his next move in the situation described in Problem 11 and illustrated in Figure P12.11,

the enemy attacks! An incoming projectile breaks off the stone ledge so that the end of the drawbridge can be lowered past the wall where it usually rests. In addition, a fragment of the projectile bounces up and cuts the drawbridge cable! The hinge between the castle wall and the bridge is frictionless, and the bridge swings down freely until it is vertical and smacks into the vertical castle wall below the castle entrance. (a) How long does Lost-a-Lot stay in contact with the bridge while it swings downward? (b) Find the angular acceleration of the bridge just as it starts to move. (c) Find the angular speed of the bridge when it strikes the wall below the hinge. Find the force exerted by the hinge on the bridge (d) immediately after the cable breaks and (e) immediately before it strikes the castle wall.

13. Figure P12.13 shows a claw hammer being used to pull a nail out of a horizontal board. The mass of the hammer is 1.00 kg. A force of 150 N is exerted horizontally as shown, and the nail does not yet move relative to the board. Find (a) the force exerted by the hammer claws on the nail and (b) the force exerted by the surface on the point of contact with the hammer head. Assume the force the hammer exerts on the nail is parallel to the nail.

$\vec{F}$

30.0 cm

Single point of contact

30.0°

5.00 cm

Figure P12.13

14. A 10.0-kg monkey climbs a uniform ladder with weight 1.20×10^2 N and length $L = 3.00$ m as shown in Figure P12.14. The ladder rests against the wall and makes an angle of $\theta = 60.0°$ with the ground. The upper and lower ends of the ladder rest on frictionless surfaces. The lower end is connected to the wall by a horizontal rope that is frayed and can support a maximum tension of only 80.0 N. (a) Draw a force diagram for the ladder. (b) Find the normal force exerted on the bottom of the ladder. (c) Find the tension in the rope when the monkey is two-thirds of the way up the ladder. (d) Find the maximum distance d that the monkey can climb up the ladder before the rope breaks. (e) If the horizontal surface were rough and the rope were removed, how would your analysis of the problem change? What other information would you need to answer parts (c) and (d)?

L

θ Rope

Figure P12.14

15. John is pushing his daughter Rachel in a wheelbarrow when it is stopped by a brick 8.00 cm high (Fig. P12.15). The handles make an angle of $\theta = 15.0°$ with the ground. Due to

the weight of Rachel and the wheelbarrow, a downward force of 400 N is exerted at the center of the wheel, which has a radius of 20.0 cm. (a) What force must John apply along the handles to just start the wheel over the brick? (b) What is the force (magnitude and direction) that the brick exerts on the wheel just as the wheel begins to lift over the brick? In both parts, assume the brick remains fixed and does not slide along the ground. Also assume the force applied by John is directed exactly toward the center of the wheel.

θ

Figure P12.15 Problems 15 and 16.

16. John is pushing his daughter Rachel in a wheelbarrow when it is stopped by a brick of height h (Fig. P12.15). The handles make an angle of θ with the ground. Due to the weight of Rachel and the wheelbarrow, a downward force mg is exerted at the center of the wheel, which has a radius R. (a) What force F must John apply along the handles to just start the wheel over the brick? (b) What are the components of the force that the brick exerts on the wheel just as the wheel begins to lift over the brick? In both parts, assume the brick remains fixed and does not slide along the ground. Also assume the force applied by John is directed exactly toward the center of the wheel.

SECTION 12.4 Elastic Properties of Solids

17. The deepest point in the ocean is in the Mariana Trench, about 11 km deep, in the Pacific. The pressure at this depth is huge, about 1.13×10^8 N/m². (a) Calculate the change in volume of 1.00 m³ of seawater carried from the surface to this deepest point. (b) The density of seawater at the surface is 1.03×10^3 kg/m³. Find its density at the bottom. (c) Explain whether or when it is a good approximation to think of water as incompressible.

18. A steel wire of diameter 1 mm can support a tension of 0.2 kN. A steel cable to support a tension of 20 kN should have diameter of what order of magnitude?

19. A child slides across a floor in a pair of rubber-soled shoes. The friction force acting on each foot is 20.0 N. The footprint area of each shoe sole is 14.0 cm², and the thickness of each sole is 5.00 mm. Find the horizontal distance by which the upper and lower surfaces of each sole are offset. The shear modulus of the rubber is 3.00 MN/m².

20. Evaluate Young's modulus for the material whose stress–strain curve is shown in Figure 12.12.

21. Assume if the shear stress in steel exceeds about 4.00×10^8 N/m², the steel ruptures. Determine the shearing force necessary to (a) shear a steel bolt 1.00 cm in diameter and (b) punch a 1.00-cm-diameter hole in a steel plate 0.500 cm thick.

22. When water freezes, it expands by about 9.00%. What pressure increase would occur inside your automobile engine

block if the water in it froze? (The bulk modulus of ice is 2.00×10^9 N/m².)

23. **Review.** A 30.0-kg hammer, moving with speed 20.0 m/s, **AMT** strikes a steel spike 2.30 cm in diameter. The hammer rebounds with speed 10.0 m/s after 0.110 s. What is the average strain in the spike during the impact?

ADDITIONAL PROBLEMS

24. A uniform beam resting on two pivots has a length $L = 6.00$ m **GP** and mass $M = 90.0$ kg. The pivot under the left end exerts a normal force n_1 on the beam, and the second pivot located a distance $\ell = 4.00$ m from the left end exerts a normal force n_2. A woman of mass $m = 55.0$ kg steps onto the left end of the beam and begins walking to the right as in Figure P12.24. The goal is to find the woman's position when the beam begins to tip. (a) What is the appropriate analysis model for the beam before it begins to tip? (b) Sketch a force diagram for the beam, labeling the gravitational and normal forces acting on the beam and placing the woman a distance x to the right of the first pivot, which is the origin. (c) Where is the woman when the normal force n_1 is the greatest? (d) What is n_1 when the beam is about to tip? (e) Use Equation 12.1 to find the value of n_2 when the beam is about to tip. (f) Using the result of part (d) and Equation 12.2, with torques computed around the second pivot, find the woman's position x when the beam is about to tip. (g) Check the answer to part (e) by computing torques around the first pivot point.

Figure P12.24

25. A bridge of length 50.0 m and mass 8.00×10^4 kg is supported **T** on a smooth pier at each end as shown in Figure P12.25. A truck of mass 3.00×10^4 kg is located 15.0 m from one end. What are the forces on the bridge at the points of support?

Figure P12.25

26. In exercise physiology studies, it is sometimes important to **BIO** determine the location of a person's center of mass. This **V** determination can be done with the arrangement shown in Figure P12.26. A light plank rests on two scales, which read $F_{g1} = 380$ N and $F_{g2} = 320$ N. A distance of 1.65 m separates the scales. How far from the woman's feet is her center of mass?

Figure P12.26

27. The lintel of prestressed reinforced concrete in Figure P12.27 is 1.50 m long. The concrete encloses one steel reinforcing rod with cross-sectional area 1.50 cm². The rod joins two strong end plates. The cross-sectional area of the concrete perpendicular to the rod is 50.0 cm². Young's modulus for the concrete is 30.0×10^9 N/m². After the concrete cures and the original tension T_1 in the rod is released, the concrete is to be under compressive stress 8.00×10^6 N/m². (a) By what distance will the rod compress the concrete when the original tension in the rod is released? (b) What is the new tension T_2 in the rod? (c) The rod will then be how much longer than its unstressed length? (d) When the concrete was poured, the rod should have been stretched by what extension distance from its unstressed length? (e) Find the required original tension T_1 in the rod.

Figure P12.27

28. The following equations are obtained from a force diagram of a rectangular farm gate, supported by two hinges on the left-hand side. A bucket of grain is hanging from the latch.

$$-A + C = 0$$

$$+B - 392 \text{ N} - 50.0 \text{ N} = 0$$

$$A(0) + B(0) + C(1.80 \text{ m}) - 392 \text{ N}(1.50 \text{ m})$$
$$- 50.0 \text{ N}(3.00 \text{ m}) = 0$$

(a) Draw the force diagram and complete the statement of the problem, specifying the unknowns. (b) Determine the values of the unknowns and state the physical meaning of each.

29. A hungry bear weighing 700 N walks out on a beam in an **AMT** attempt to retrieve a basket of goodies hanging at the end of the beam (Fig. P12.29). The beam is uniform, weighs 200 N, and is 6.00 m long, and it is supported by a wire at an angle of $\theta = 60.0°$. The basket weighs 80.0 N. (a) Draw a force diagram for the beam. (b) When the bear is at $x = 1.00$ m, find the tension in the wire supporting the beam and the components of the force exerted by the wall on the left end of the

Figure P12.29

beam. (c) **What If?** If the wire can withstand a maximum tension of 900 N, what is the maximum distance the bear can walk before the wire breaks?

30. A 1 200-N uniform boom at $\phi = 65°$ to the vertical is supported by a cable at an angle $\theta = 25.0°$ to the horizontal as shown in Figure P12.30. The boom is pivoted at the bottom, and an object of weight $m = 2\,000$ N hangs from its top. Find (a) the tension in the support cable and (b) the components of the reaction force exerted by the floor on the boom.

Figure P12.30

31. A uniform sign of weight F_g and width $2L$ hangs from a **S** light, horizontal beam hinged at the wall and supported by a cable (Fig. P12.31). Determine (a) the tension in the cable and (b) the components of the reaction force exerted by the wall on the beam in terms of F_g, d, L, and θ.

Figure P12.31

32. When a person stands on tiptoe on one foot (a strenuous **BIO** position), the position of the foot is as shown in Figure P12.32a. The total gravitational force $\vec{F}_g$ on the body is supported by the normal force $\vec{n}$ exerted by the floor on the toes of one foot. A mechanical model of the situation is shown in Figure P12.32b, where $\vec{T}$ is the force exerted on the foot by the Achilles tendon and $\vec{R}$ is the force exerted on the foot by the tibia. Find the values of T, R, and θ when $F_g = 700$ N.

Figure P12.32

33. A 10 000-N shark is supported **T** by a rope attached to a 4.00-m rod that can pivot at the base. (a) Calculate the tension in the cable between the rod and the wall, assuming the cable is holding the system in the position shown in Figure P12.33. Find (b) the horizontal force and (c) the vertical force exerted on the base of the rod. Ignore the weight of the rod.

Figure P12.33

34. Assume a person bends for- **BIO** ward to lift a load "with his **QC** back" as shown in Figure P12.34a. The spine pivots mainly at the fifth lumbar vertebra, with the principal supporting force provided by the erector spinalis muscle in the back. To see the magnitude of the forces involved, consider the model shown in Figure P12.34b for a person bending forward to lift a 200-N object. The spine and upper body are represented as a uniform horizontal rod of weight 350 N, pivoted at the base of the spine. The erector spinalis muscle, attached at a point two-thirds of the way up the spine, maintains the position of the back. The angle between the spine and this muscle is $\theta = 12.0°$. Find (a) the tension T in the back muscle and (b) the compressional force in the spine. (c) Is this method a good way to lift a load? Explain your answer, using the results of parts (a) and (b). (d) Can you suggest a better method to lift a load?

Figure P12.34

35. A uniform beam of mass m is inclined at an angle θ to the **S** horizontal. Its upper end (point P) produces a 90° bend in a very rough rope tied to a wall, and its lower end rests on a rough floor (Fig. P12.35). Let μ_s represent the coefficient of static friction between beam and floor. Assume μ_s is less than the cotangent of θ. (a) Find an expression for the maximum mass M that can be suspended from the top before the beam slips. Determine (b) the magnitude of the reaction force at the floor and (c) the magnitude of the force exerted by the beam on the rope at P in terms of m, M, and μ_s.

Figure P12.35

36. *Why is the following situation impossible?* A worker in a factory pulls a cabinet across the floor using a rope as shown in Figure P12.36a. The rope make an angle $\theta = 37.0°$ with the floor and is tied $h_1 = 10.0$ cm from the bottom of the cabinet. The uniform rectangular cabinet has height $\ell = 100$ cm and width $w = 60.0$ cm, and it weighs 400 N. The cabinet slides with constant speed when a force $F = 300$ N is applied through the rope. The worker tires of walking backward. He fastens the rope to a point on the cabinet $h_2 = 65.0$ cm off the floor and lays the rope over his shoulder so that he can walk forward and pull as shown in Figure P12.36b. In this way, the rope again makes an angle of $\theta = 37.0°$ with the horizontal and again has a tension of 300 N. Using this technique, the worker is able to slide the cabinet over a long distance on the floor without tiring.

Figure P12.36 Problems 36 and 44.

37. When a circus performer performing on the rings executes the *iron cross*, he maintains the position at rest shown in Figure P12.37a. In this maneuver, the gymnast's feet (not shown) are off the floor. The primary muscles involved in supporting this position are the latissimus dorsi ("lats") and the pectoralis major ("pecs"). One of the rings exerts an upward force $\vec{\mathbf{F}}_h$ on a hand as shown in Figure P12.37b. The force $\vec{\mathbf{F}}_s$ is exerted by the shoulder joint on the arm. The latissimus dorsi and pectoralis major muscles exert a total force $\vec{\mathbf{F}}_m$ on the arm. (a) Using the information in the figure, find the

Figure P12.37

magnitude of the force $\vec{\mathbf{F}}_m$ for an athlete of weight 750 N. (b) Suppose a performer in training cannot perform the iron cross but can hold a position similar to the figure in which the arms make a 45° angle with the horizontal rather than being horizontal. Why is this position easier for the performer?

38. Figure P12.38 shows a light truss formed from three struts lying in a plane and joined by three smooth hinge pins at their ends. The truss supports a downward force of $\vec{\mathbf{F}} = 1\,000$ N applied at the point B. The truss has negligible weight. The piers at A and C are smooth. (a) Given $\theta_1 = 30.0°$ and $\theta_2 = 45.0°$, find n_A and n_C. (b) One can show that the force any strut exerts on a pin must be directed along the length of the strut as a force of tension or compression. Use that fact to identify the directions of the forces that the struts exert on the pins joining them. Find the force of tension or of compression in each of the three bars.

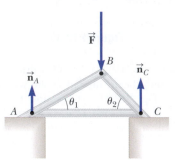

Figure P12.38

39. One side of a plant shelf is supported by a bracket mounted on a vertical wall by a single screw as shown in Figure P12.39. Ignore the weight of the bracket. (a) Find the horizontal component of the force that the screw exerts on the bracket when an 80.0 N vertical force is applied as shown. (b) As your grandfather waters his geraniums,

Figure P12.39

the 80.0-N load force is increasing at the rate 0.150 N/s. At what rate is the force exerted by the screw changing? *Suggestion:* Imagine that the bracket is slightly loose.

40. A stepladder of negligible weight is constructed as shown in Figure P12.40, with $AC = BC = \ell = 4.00$ m. A painter of mass $m = 70.0$ kg stands on the ladder $d = 3.00$ m from the bottom. Assuming the floor is frictionless, find (a) the tension in the horizontal bar DE connecting the two halves of the ladder, (b) the normal forces at A and B, and (c) the components of the reaction force at the single hinge C that the left half of the ladder exerts on the right half. *Suggestion:* Treat the

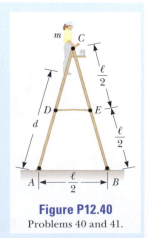

Figure P12.40
Problems 40 and 41.

ladder as a single object, but also treat each half of the ladder separately.

41. A stepladder of negligible weight is constructed as shown in Figure P12.40, with $AC = BC = \ell$. A painter of mass m stands on the ladder a distance d from the bottom. Assuming the floor is frictionless, find (a) the tension in the horizontal bar DE connecting the two halves of the ladder, (b) the normal forces at A and B, and (c) the components of the reaction force at the single hinge C that the left half of the ladder exerts on the right half. *Suggestion:* Treat the ladder as a single object, but also treat each half of the ladder separately.

42. Review. A wire of length L, Young's modulus Y, and cross-sectional area A is stretched elastically by an amount ΔL. By Hooke's law, the restoring force is $-k\,\Delta L$. (a) Show that $k = YA/L$. (b) Show that the work done in stretching the wire by an amount ΔL is $W = \frac{1}{2}YA(\Delta L)^2/L$.

43. Two racquetballs, each having a mass of 170 g, are placed in a glass jar as shown in Figure P12.43. Their centers lie on a straight line that makes a 45° angle with the horizontal. (a) Assume the walls are frictionless and determine P_1, P_2, and P_3. (b) Determine the magnitude of the force exerted by the left ball on the right ball.

Figure P12.43

44. Consider the rectangular cabinet of Problem 36 shown in Figure P12.36, but with a force $\vec{F}$ applied horizontally at the upper edge. (a) What is the minimum force required to start to tip the cabinet? (b) What is the minimum coefficient of static friction required for the cabinet not to slide with the application of a force of this magnitude? (c) Find the magnitude and direction of the minimum force required to tip the cabinet if the point of application can be chosen *anywhere* on the cabinet.

45. Review. An aluminum wire is 0.850 m long and has a circular cross section of diameter 0.780 mm. Fixed at the top end, the wire supports a 1.20-kg object that swings in a horizontal circle. Determine the angular speed of the object required to produce a strain of 1.00×10^{-3}.

46. You have been hired as an expert witness in a case involving an injury in a factory. The attorney who hired you represents the injured worker. The worker was told to lift one end of a long, heavy crate that was lying horizontally on the floor and tilt it up so that it is standing on end. He began lifting the end of the crate, always applying a force that was perpendicular to the top of the crate. As the end of the crate got higher, at a certain angle, the bottom of the crate slipped on the floor, and the worker, in trying to recover, stepped forward and the crate landed on his foot, injuring it badly. As part of your investigation, you go to the factory and measure the coefficient of static friction between a crate and the smooth concrete floor. You find it to be 0.340. Prepare an argument for the attorney showing that it was impossible to lift the crate in the manner described without it slipping on the floor.

47. A 500-N uniform rectangular sign 4.00 m wide and 3.00 m high is suspended from a horizontal, 6.00-m-long, uniform, 100-N rod as indicated in Figure P12.47. The left end of the rod is supported by a hinge, and the right end is supported by a thin cable making a 30.0° angle with the vertical. (a) Find the tension T in the cable. (b) Find the horizontal and vertical components of force exerted on the left end of the rod by the hinge.

Figure P12.47

48. A steel cable 3.00 cm² in cross-sectional area has a mass of 2.40 kg per meter of length. If 500 m of the cable is hung over a vertical cliff, how much does the cable stretch under its own weight? Take $Y_{\text{steel}} = 2.00 \times 10^{11}$ N/m².

CHALLENGE PROBLEMS

49. A uniform rod of weight F_g and length L is supported at its ends by a frictionless trough as shown in Figure P12.49. (a) Show that the center of gravity of the rod must be vertically over point O when the rod is in equilibrium. (b) Determine the equilibrium value of the angle θ. (c) Is the equilibrium of the rod stable or unstable?

Figure P12.49

50. In the What If? section of Example 12.2, let d represent the distance in meters between the person and the hinge at the left end of the beam. (a) Show that the cable tension is given by $T = 93.9d + 125$, with T in newtons. (b) Show that the direction angle θ of the hinge force is described by

$$\tan\theta = \left(\frac{32}{3d+4} - 1\right)\tan 53.0°$$

(c) Show that the magnitude of the hinge force is given by

$$R = \sqrt{8.82 \times 10^3 d^2 - 9.65 \times 10^4 d + 4.96 \times 10^5}$$

(d) Describe how the changes in T, θ, and R as d increases differ from one another.

13

Universal Gravitation

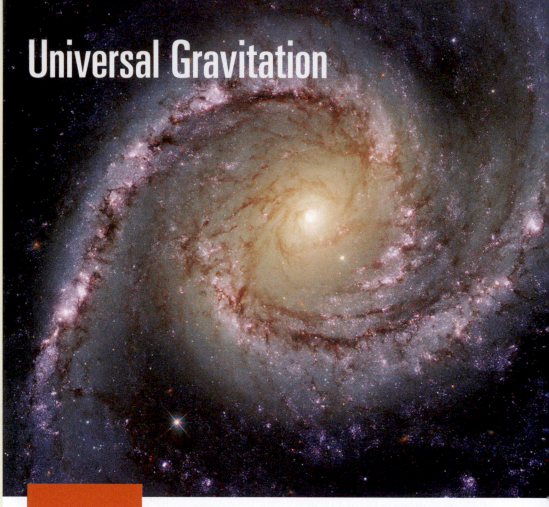

Hubble Space Telescope image of a spiral galaxy, NGC 1566, taken in 2014. In the spiral arms of the galaxy, hydrogen gas is compressed to create new stars. It is theorized that our own galaxy, the Milky Way, has a similar structure with spiral arms. (*ESA/Hubble & NASA*)

STORYLINE **In Chapter 11, you were trying to do your physics** homework and were distracted by spinning skaters and divers. In Chapter 12, you were distracted by a surprising phenomenon with a meterstick. Now, you finally begin to work on your physics homework and you open your physics textbook. You look again at the tables in the front, before the title page. In the table of Solar System Data, you notice an entry for the mass of the Sun. After remarking to yourself that that's a lot of mass, you say, "Wait a minute! How did they find the mass of the Sun? In fact, how did they find the mass of any of the planets?" That leads you to think about the mass of the entire Milky Way galaxy. Looking online, you find different estimates of the mass of the galaxy, some in the range of hundreds of billions of solar masses, and others in the trillions of solar masses. Why can't we come up with a single number for the mass of the galaxy? Your physics homework goes undone as you ponder these new questions.

CONNECTIONS We first studied gravity in Section 2.8, where we talked about freely falling objects. There, and in Section 4.3 on projectile motion, we considered the effects of gravity on objects near the surface of the Earth. In Section 5.5, we related the gravitational force on such objects to their weight. In Chapter 7, we related the gravitational force on an object near the surface of the Earth to gravitational potential energy of the object–Earth system. In this chapter, we remove the assumption that objects are near the surface of the Earth. How does the gravitational force on an object vary as we move the object far from the surface of the Earth? The answer to that question will allow us to understand the motion of planets around the Sun and has allowed scientists to place many objects in orbit around the Earth, the Moon, and Mars. The principle that allows this understanding is the *law of universal gravitation*. We emphasize a description of planetary motion because astronomical data provide an important test of this law's validity. After introducing this law, we will show connections between it and the angular momentum of Chapter 11 and the energy techniques in Chapters 7 and 8. As preparation for the remainder of the book, we recognize gravitation as one of four "fundamental forces" of nature. The others are the electromagnetic force (Chapters 22–33), the nuclear strong force (Chapters 43–44), and the weak force (Chapter 44).

13.1 Newton's Law of Universal Gravitation

You may have heard the legend that, while napping under a tree, Newton was struck on the head by a falling apple. This alleged accident supposedly prompted him to imagine that perhaps all objects in the Universe were attracted to each other in the same way the apple was attracted to the Earth. Newton analyzed astronomical data on the motion of the Moon around the Earth. From that analysis, he made the bold assertion that the force law governing the motion of planets was the *same* as the force law that attracted a falling apple to the Earth. This assertion was contradictory to earlier thought that had lasted for centuries, which claimed that the laws of physics on the Earth did not apply to the heavens.

In 1687, Newton published his work on the law of gravity in his treatise *Mathematical Principles of Natural Philosophy*. **Newton's law of universal gravitation** states that

> every particle in the Universe attracts every other particle with a force that is directly proportional to the product of their masses and inversely proportional to the square of the distance between them.

◄ The law of universal gravitation

If the particles have masses m_1 and m_2 and are separated by a distance r, the magnitude of this gravitational force is

$$F_g = G\frac{m_1 m_2}{r^2} \qquad (13.1)$$

where G is a constant, called the *universal gravitational constant*. Its value in SI units is

$$G = 6.674 \times 10^{-11} \ \text{N} \cdot \text{m}^2/\text{kg}^2 \qquad (13.2)$$

The universal gravitational constant G was first evaluated in the late nineteenth century, based on results of an important experiment by Sir Henry Cavendish (1731–1810) in 1798. The law of universal gravitation was not expressed by Newton in the form of Equation 13.1, and Newton did not mention a constant such as G. In fact, even by the time of Cavendish, a unit of force had not yet been included in the existing system of units. Cavendish's goal was to measure the density of the Earth. His results were then used by other scientists 100 years later to generate a value for G.

Cavendish's apparatus consists of two small spheres, each of mass m, fixed to the ends of a light, horizontal rod suspended by a fine fiber or thin metal wire as illustrated in Figure 13.1. When two large spheres, each of mass M, are placed near the smaller ones, the attractive gravitational force between smaller and larger spheres causes the rod to rotate and twist the wire suspension to a new equilibrium orientation. The angle of rotation is measured by the deflection of a light beam reflected from a mirror attached to the vertical suspension.

The form of the force law given by Equation 13.1 is often referred to as an **inverse-square law** because the magnitude of the force varies as the inverse square of the separation of the particles.[1] We shall see other examples of this type of force

The dashed line represents the original position of the rod.

Figure 13.1 Cavendish apparatus for measuring gravitational forces.

---

[1]An *inverse* proportionality between two quantities x and y is one in which $y = k/x$, where k is a constant. A *direct* proportion between x and y exists when $y = kx$.

Figure 13.2 The gravitational force between two particles is attractive. The unit vector $\hat{\mathbf{r}}_{12}$ is directed from particle 1 toward particle 2.

law in subsequent chapters. We can express this force in vector form by defining a unit vector $\hat{\mathbf{r}}_{12}$ (Fig. 13.2). Because this unit vector is directed from particle 1 toward particle 2, the force exerted by particle 1 on particle 2 is

$$\vec{\mathbf{F}}_{12} = -G\frac{m_1 m_2}{r^2}\hat{\mathbf{r}}_{12} \tag{13.3}$$

where the negative sign indicates that particle 2 is attracted to particle 1; hence, the force on particle 2 must be directed toward particle 1. By Newton's third law, the force exerted by particle 2 on particle 1, designated $\vec{\mathbf{F}}_{21}$, is equal in magnitude to $\vec{\mathbf{F}}_{12}$ and in the opposite direction. That is, these forces form an action–reaction pair, and $\vec{\mathbf{F}}_{21} = -\vec{\mathbf{F}}_{12}$.

Two features of Equation 13.3 deserve mention. First, the gravitational force is a field force that always exists between two particles, regardless of the medium that separates them. Second, because the force varies as the inverse square of the distance between the particles, it decreases rapidly with increasing separation.

Equation 13.3 can also be used to show that the gravitational force exerted by a finite-size, spherically symmetric mass distribution on a particle outside the distribution is the same as if the entire mass of the distribution were concentrated at the center. For example, the magnitude of the force exerted by the Earth on a particle of mass m near the Earth's surface is

$$F_g = G\frac{M_E m}{R_E^{\,2}} \tag{13.4}$$

where M_E is the Earth's mass and R_E its radius. This force is directed toward the center of the Earth.

PITFALL PREVENTION 13.1

Be Clear on g and G The symbol g represents the magnitude of the free-fall acceleration near a planet. At the surface of the Earth, g has an average value of 9.80 m/s². On the other hand, G is a universal constant that has the same value everywhere in the Universe.

QUICK QUIZ 13.1 A planet has two moons of equal mass. Moon 1 is in a circular orbit of radius r. Moon 2 is in a circular orbit of radius $2r$. What is the magnitude of the gravitational force exerted by the planet on Moon 2? **(a)** four times as large as that on Moon 1 **(b)** twice as large as that on Moon 1 **(c)** equal to that on Moon 1 **(d)** half as large as that on Moon 1 **(e)** one-fourth as large as that on Moon 1

Example 13.1 **Billiards, Anyone?**

Three 0.300-kg billiard balls are placed on a table at the corners of a right triangle as shown in Figure 13.3. The sides of the triangle are of lengths $a = 0.400$ m, $b = 0.300$ m, and $c = 0.500$ m. Calculate the gravitational force vector on the cue ball (designated m_1) resulting from the other two balls as well as the magnitude and direction of this force.

SOLUTION

Conceptualize Notice in Figure 13.3 that the cue ball is attracted to both other balls by the gravitational force. We can see graphically that the net force should point upward and toward the right. We locate our coordinate axes as shown in Figure 13.3, placing our origin at the position of the cue ball.

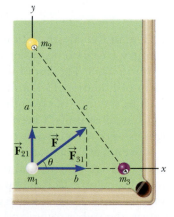

Figure 13.3 (Example 13.1) The resultant gravitational force acting on the cue ball is the vector sum $\vec{\mathbf{F}}_{21} + \vec{\mathbf{F}}_{31}$.

Categorize This problem involves evaluating the gravitational forces on the cue ball using Equation 13.3. Once these forces are evaluated, it becomes a vector addition problem to find the net force.

Analyze Find the force exerted by m_2 on the cue ball:

$$\vec{\mathbf{F}}_{21} = G\frac{m_2 m_1}{a^2}\hat{\mathbf{j}}$$

$$= (6.674 \times 10^{-11} \text{ N} \cdot \text{m}^2/\text{kg}^2)\frac{(0.300 \text{ kg})(0.300 \text{ kg})}{(0.400 \text{ m})^2}\hat{\mathbf{j}}$$

$$= 3.75 \times 10^{-11}\,\hat{\mathbf{j}} \text{ N}$$

13.1 continued

Find the force exerted by m_3 on the cue ball:

$$\vec{\mathbf{F}}_{31} = G\frac{m_3 m_1}{b^2}\hat{\mathbf{i}}$$

$$= (6.674 \times 10^{-11}\text{ N}\cdot\text{m}^2/\text{kg}^2)\frac{(0.300\text{ kg})(0.300\text{ kg})}{(0.300\text{ m})^2}\hat{\mathbf{i}}$$

$$= 6.67 \times 10^{-11}\,\hat{\mathbf{i}}\text{ N}$$

Find the net gravitational force on the cue ball by adding these force vectors:

$$\vec{\mathbf{F}} = \vec{\mathbf{F}}_{31} + \vec{\mathbf{F}}_{21} = (6.67\,\hat{\mathbf{i}} + 3.75\,\hat{\mathbf{j}}) \times 10^{-11}\text{ N}$$

Find the magnitude of this force:

$$F = \sqrt{F_{31}^2 + F_{21}^2} = \sqrt{(6.67)^2 + (3.75)^2} \times 10^{-11}\text{ N}$$

$$= 7.66 \times 10^{-11}\text{ N}$$

Find the tangent of the angle θ for the net force vector:

$$\tan\theta = \frac{F_y}{F_x} = \frac{F_{21}}{F_{31}} = \frac{3.75 \times 10^{-11}\text{ N}}{6.67 \times 10^{-11}\text{ N}} = 0.562$$

Evaluate the angle θ:

$$\theta = \tan^{-1}(0.562) = 29.4°$$

Finalize   The result for F shows that the gravitational forces between everyday objects have extremely small magnitudes.

13.2 Free-Fall Acceleration and the Gravitational Force

We have called the magnitude of the gravitational force on an object near the Earth's surface the *weight* of the object, where the weight is given by Equation 5.6, $F = mg$. Equation 13.4 is another expression for this force. Therefore, we can set Equations 5.6 and 13.4 equal to each other to obtain

$$mg = G\frac{M_E m}{R_E^2}$$

$$g = G\frac{M_E}{R_E^2} \tag{13.5}$$

Equation 13.5 relates the free-fall acceleration g to physical parameters of the Earth—its mass and radius—and explains the origin of the value of 9.80 m/s² that we have used in earlier chapters. Now consider an object of mass m located a distance h above the Earth's surface or a distance r from the Earth's center, where $r = R_E + h$. The magnitude of the gravitational force acting on this object is

$$F_g = G\frac{M_E m}{r^2} = G\frac{M_E m}{(R_E + h)^2}$$

The magnitude of the gravitational force acting on the object at this position is also $F_g = mg$, where g is the value of the free-fall acceleration at the altitude h. Substituting this expression for F_g into the last equation shows that g is given by

$$g = \frac{GM_E}{r^2} = \frac{GM_E}{(R_E + h)^2} \tag{13.6}$$

TABLE 13.1 Free-Fall Acceleration g at Various Altitudes Above the Earth's Surface

Altitude h (km)	g (m/s²)
0	9.80
1 000	7.33
2 000	5.68
3 000	4.53
4 000	3.70
5 000	3.08
6 000	2.60
7 000	2.23
8 000	1.93
9 000	1.69
10 000	1.49
50 000	0.13
∞	0

◀ Variation of g with altitude

Therefore, it follows that g *decreases* with *increasing altitude*. Values of g for the Earth at various altitudes are listed in Table 13.1. Because an object's weight is mg, we see that as $r \to \infty$, the weight of the object approaches zero.

⊙UICK QUIZ **13.2** Superman stands on top of a very tall mountain and throws a baseball horizontally with a speed such that the baseball goes into a circular orbit around the Earth. While the baseball is in orbit, what is the magnitude of the acceleration of the ball? **(a)** It depends on how fast the baseball is thrown. **(b)** It is zero because the ball does not fall to the ground. **(c)** It is slightly less than 9.80 m/s². **(d)** It is equal to 9.80 m/s².

Example **13.2**   **The Density of the Earth**

Using the known radius of the Earth and that $g = 9.80$ m/s² at the Earth's surface, find the average density of the Earth.

S O L U T I O N

Conceptualize   Assume the Earth is a perfect sphere. The density of material in the Earth varies, but let's adopt a simplified model in which we assume the density to be uniform throughout the Earth. The resulting density is the average density of the Earth.

Categorize   This example is a relatively simple substitution problem.

Using Equation 13.5, solve for the mass of the Earth:

$$M_E = \frac{gR_E^{\,2}}{G}$$

Substitute this mass and the volume of a sphere into the definition of density (Eq. 1.1):

$$\rho_E = \frac{M_E}{V_E} = \frac{gR_E^{\,2}/G}{\tfrac{4}{3}\pi R_E^{\,3}} = \tfrac{3}{4}\frac{g}{\pi G R_E}$$

$$= \tfrac{3}{4}\frac{9.80 \text{ m/s}^2}{\pi (6.674 \times 10^{-11} \text{ N} \cdot \text{m}^2/\text{kg}^2)(6.37 \times 10^6 \text{ m})} = 5.50 \times 10^3 \text{ kg/m}^3$$

W H A T   I F ?   What if you were told that a typical density of granite at the Earth's surface is 2.75×10^3 kg/m³? What would you conclude about the density of the material in the Earth's interior?

Answer   Because this value is about half the density we calculated as an average for the entire Earth, we would conclude that the inner core of the Earth has a density much higher than the average value. It is most amazing that the Cavendish experiment—which can be used to determine G and can be done today on a tabletop—combined with simple free-fall measurements of g provides information about the core of the Earth!

13.3 Analysis Model: Particle in a Field (Gravitational)

When Newton published his theory of universal gravitation, it was considered a success because it satisfactorily explained the motion of the planets. It represented strong evidence that the same laws that describe phenomena on the Earth can be used on large objects like planets and throughout the Universe. Since 1687, Newton's theory has been used to account for the motions of comets, the deflection of a Cavendish balance, the orbits of binary stars, and the rotation of galaxies. Nevertheless, both Newton's contemporaries and his successors found it difficult to accept the concept of a force that acts at a distance. They asked how it was possible for two objects such as the Sun and the Earth to interact when they were not in contact with each other. Newton himself could not answer that question.

An approach to describing interactions between objects that are not in contact came well after Newton's death. This approach enables us to look at the gravitational interaction in a different way, using the concept of a *gravitational field* that exists at every point in space. The concept of a *field* occurs often in physics. A **field** is a physical quantity that exists everywhere in space, is single-valued at all points, and is established by a source of some kind. For example, atmospheric pressure near the surface of the Earth is a field. At all points within the atmosphere, there is a value of the pressure. These values generally decrease with increasing altitude, and also change in time depending on current weather conditions. The source of atmospheric pressure is the air itself (see Chapter 14).

The source of a gravitational field is a *source particle* with mass M. Generally, this particle is planet- or star-sized, and can be modeled as a particle as long as we make observations outside of the planet or star. The source particle affects space about itself so that there is a quantity called the gravitational field everywhere in space.

This discussion of fields raises a couple of questions. First, how do we detect that a field exists at some point? And second, how do we define the value of the field at that point? To answer the first question, we must put a *test particle* at the point. A test particle is something that is sensitive to the altered space around the source. In the case of the atmosphere, imagine placing a helium-filled balloon at some point in the air. Because there is a pressure field at that point, and the pressure varies over the height of the balloon, the balloon will rise upward. (If a helium balloon were placed in empty space at a pressure of zero, it would remain stationary.) Therefore, the balloon detects the presence of the pressure field. In the case of the gravitational field, the test particle is a second particle, with mass m_0. If this particle is placed in the gravitational field, there is a gravitational force on the test particle. This force shows that a gravitational field exists at that point.

Now, how do we *define* the field so that we can assign a numerical value to it? For the balloon in the atmospheric pressure field, we could perhaps base the definition on the acceleration with which the balloon moves when released. In the case of gravity, we define the **gravitational field** $\vec{g}$ as

$$\vec{g} \equiv \frac{\vec{F}_g}{m_0} \tag{13.7}$$

That is, the gravitational field at a point in space equals the gravitational force $\vec{F}_g$ experienced by a test particle placed at that point divided by the mass m_0 of the test particle. Notice that the presence of the test particle is not necessary for the field to exist: the source particle creates the gravitational field. The gravitational field describes the effect that a source particle (for example, the Earth) has on the empty space around itself in terms of the force that *would* be present *if* a second object were somewhere in that space. It turns out to be useful to replace the direct gravitational force between two particles (Equation 13.1) with this "two-step" process: (1) one particle establishes a gravitational field, and (2) a second particle placed in the field experiences a force.[2]

The concept of a field is at the heart of the **particle in a field** analysis model. In Equation 13.7, the test particle of mass m_0 is placed in the field solely in order to determine the value of the gravitational field $\vec{g}$. Once the value is determined, any arbitrary particle of mass m can be placed in the field and will experience a force $m\vec{g}$. Therefore, the mathematical representation of the gravitational version of the particle in a field model is Equation 5.5:

$$\vec{F}_g = m\vec{g} \tag{5.5}$$

---

[2]We shall return to this idea of mass affecting the space around it when we discuss Einstein's theory of gravitation in Chapter 38.

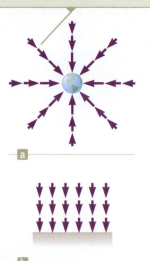

The field vectors point in the direction of the acceleration a particle would experience if it were placed in the field. The magnitude of the field vector at any location is the magnitude of the free-fall acceleration at that location.

a

b

Figure 13.4 (a) The gravitational field vectors in the vicinity of a uniform spherical mass such as the Earth vary in both direction and magnitude. (b) The gravitational field vectors in a small region near the Earth's surface are uniform in both direction and magnitude.

In the case of the pressure field in the atmosphere, we might recognize the force between the air and the balloon as a *contact force*, as discussed in Section 5.1. There is physical contact between the air and the balloon. What puzzled Newton and other scientists is that gravity is a *field force*: There is no physical contact between a star acting as a source particle and an orbiting planet placed in the resulting field.

In future chapters, we will see two other versions of the particle in a field model that turn out to be useful. In the electric version, the property of a source particle that results in an *electric field* is *electric charge*: when a second electrically-charged particle is placed in the electric field, it experiences a force. The magnitude of the force is the product of the electric charge and the field, in analogy with the gravitational force in Equation 5.5. In the magnetic version of the particle in a field model, a charged particle is placed in a *magnetic field*. One other property of this particle is required for the particle to experience a force: the particle must have a *velocity* at some nonzero angle to the magnetic field. The electric and magnetic versions of the particle in a field model are critical to the understanding of the principles of *electromagnetism*, which we will study in Chapters 22–33.

Because the gravitational force acting on a test particle of mass m_0 near the Earth has a magnitude $GM_E m_0/r^2$ (see Eq. 13.4), the gravitational field $\vec{\mathbf{g}}$ at a distance r from the center of the Earth is

$$\vec{\mathbf{g}} = \frac{\vec{\mathbf{F}}_g}{m_0} = -\frac{GM_E}{r^2}\hat{\mathbf{r}} \tag{13.8}$$

where $\hat{\mathbf{r}}$ is a unit vector pointing radially outward from the Earth (see Fig. 3.15) and the negative sign indicates that the field points toward the center of the Earth as illustrated in Figure 13.4a. The field vectors at different points surrounding the Earth vary in both direction and magnitude. In a small region near the Earth's surface, the downward field $\vec{\mathbf{g}}$ is approximately constant and uniform as indicated in Figure 13.4b. Equation 13.8 is valid at all points *outside* the Earth's surface, assuming the Earth is spherical. At the Earth's surface, where $r = R_E$, $\vec{\mathbf{g}}$ has a magnitude of 9.80 N/kg. (The unit N/kg is the same as m/s².)

ANALYSIS MODEL **Particle in a Field (Gravitational)**

Imagine an object with mass that we call a *source particle*. The source particle establishes a **gravitational field** $\vec{\mathbf{g}}$ throughout space. The gravitational field is evaluated by measuring the force on a test particle of mass m_0 and then using Equation 13.7. Now imagine a particle of mass m is placed in that field. The particle interacts with the gravitational field so that it experiences a gravitational force given by

$$\vec{\mathbf{F}}_g = m\vec{\mathbf{g}} \tag{5.5}$$

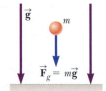

Examples:

- an object of mass m near the surface of the Earth has a *weight*, which is the result of the gravitational field established in space by the Earth
- a planet in the solar system is in orbit around the Sun, due to the gravitational force on the planet exerted by the gravitational field established by the Sun
- an object near a black hole is drawn into the black hole, never to escape, due to the tremendous gravitational field established by the black hole (Section 13.6)
- in the general theory of relativity, the gravitational field of a massive object is imagined to be described by a *curvature of spacetime* (Chapter 38)
- the gravitational field of a massive object is imagined to be mediated by particles called *gravitons*, which have never been detected (Chapter 44)

| Example **13.3** | **The Weight of the Space Station** |

The International Space Station operates at an altitude of 350 km. An online search for the station shows that a weight of 4.11×10^6 N, measured at the Earth's surface, has been lifted off the surface by various spacecraft during the construction process. What is the weight of the space station as it moves in its orbit?

SOLUTION

Conceptualize   The mass of the space station is fixed; it is independent of its location. Based on the discussions in this section and Section 13.2, we realize that the value of g is smaller at the height of the space station's orbit than at the surface of the Earth. Therefore, the weight of the Space Station is smaller than that at the surface of the Earth.

Categorize   We model the Space Station as a *particle in a gravitational field*.

Analyze   From the particle in a field model, find the mass of the space station from its weight at the surface of the Earth:

$$m = \frac{F_{g,\text{surface}}}{g_{\text{surface}}} = \frac{4.11 \times 10^6 \text{ N}}{9.80 \text{ m/s}^2} = 4.19 \times 10^5 \text{ kg}$$

Use Equation 13.6 with $h = 350$ km to find the magnitude of the gravitational field at the orbital location:

$$g_{\text{orbit}} = \frac{GM_E}{(R_E + h)^2}$$

$$= \frac{(6.674 \times 10^{-11} \text{ N} \cdot \text{m}^2/\text{kg}^2)(5.97 \times 10^{24} \text{ kg})}{(6.37 \times 10^6 \text{ m} + 0.350 \times 10^6 \text{ m})^2} = 8.82 \text{ m/s}^2$$

Use the particle in a field model again to find the space station's weight in orbit:

$$F_{g,\text{orbit}} = mg_{\text{orbit}} = (4.19 \times 10^5 \text{ kg})(8.82 \text{ m/s}^2) = \boxed{3.70 \times 10^6 \text{ N}}$$

Finalize   Notice that the weight of the Space Station is less when it is in orbit, as we expected. It has about 10% less weight than it has when on the Earth's surface, representing a 10% decrease in the magnitude of the gravitational field.

13.4 Kepler's Laws and the Motion of Planets

Humans have observed the movements of the planets, stars, and other celestial objects for thousands of years. In early history, these observations led scientists to design a structural model in which the Earth was regarded as the center of the Universe. This *geocentric model* was elaborated and formalized by the Greek astronomer Claudius Ptolemy (c. 100–c. 170) in the second century and was accepted for the next 1 400 years. In 1543, Polish astronomer Nicolaus Copernicus (1473–1543) offered another structural model for the solar system that suggested that the Earth and the other planets revolved in circular orbits around the Sun (the *heliocentric model*).[3]

Danish astronomer Tycho Brahe (1546–1601) performed more observations to determine how the heavens were constructed and pursued a project to measure the positions of both stars and planets. Those observations of the planets and 777 stars visible to the naked eye were carried out with only a large sextant and a compass. (The telescope had not yet been invented.)

German astronomer Johannes Kepler was Brahe's assistant for a short while before Brahe's death, whereupon he acquired his mentor's astronomical data and spent 16 years trying to deduce a mathematical model for the motion of the planets. Such data are difficult to sort out because the moving planets are observed from a moving Earth. After many laborious calculations, Kepler found that Brahe's data on the revolution of Mars around the Sun led to a successful model.

---

[3]The heliocentric model was proposed by Aristarchus of Samos (c. 310 BC–c. 230 BC) several centuries before Copernicus, but the theory was not widely accepted.

Kepler's structural model of planetary motion is summarized in three statements known as **Kepler's laws:**

1. All planets move in elliptical orbits with the Sun at one focus.
2. The radius vector drawn from the Sun to a planet sweeps out equal areas in equal time intervals.
3. The square of the orbital period of any planet is proportional to the cube of the semimajor axis of the elliptical orbit.

Kepler's First Law

Ptolemy's geocentric model and Copernicus's heliocentric model of the solar system both suggested circular orbits for heavenly bodies. Kepler's first law indicates that the circular orbit is a very special case and elliptical orbits are the general situation. This notion was difficult for scientists to accept because they believed, as had scientists for centuries before them, that perfect circular orbits of the planets reflected the perfection of heaven.

Figure 13.5 shows the geometry of an ellipse, which serves as our model for the elliptical orbit of a planet. An ellipse is mathematically defined by choosing two points F_1 and F_2, each of which is a called a **focus,** and then drawing a curve through points for which the sum of the distances r_1 and r_2 from F_1 and F_2, respectively, is a constant. The longest distance through the center between points on the ellipse (and passing through each focus) is called the **major axis,** and this distance is $2a$. In Figure 13.5, the major axis is drawn along the x direction. The distance a is called the **semimajor axis.** Similarly, the shortest distance through the center between points on the ellipse is called the **minor axis** of length $2b$, where the distance b is the **semiminor axis.** Either focus of the ellipse is located at a distance c from the center of the ellipse, where $a^2 = b^2 + c^2$. In the elliptical orbit of a planet around the Sun, the Sun is at one focus of the ellipse. There is nothing at the other focus.

The **eccentricity** of an ellipse is defined as $e = c/a$, and it describes the general shape of the ellipse. For a circle, $c = 0$, and the eccentricity is therefore zero. The smaller b is compared with a, the shorter the ellipse is along the y direction compared with its extent in the x direction in Figure 13.5. As b decreases, c increases and the eccentricity e increases. Therefore, higher values of eccentricity correspond to longer and thinner ellipses. The range of values of the eccentricity for an ellipse is $0 < e < 1$.

Eccentricities for planetary orbits vary widely in the solar system. The eccentricity of the Earth's orbit is 0.017, which makes it nearly circular. On the other hand, the eccentricity of Mercury's orbit is 0.21, the highest of the eight planets. Figure 13.6a shows an ellipse with an eccentricity equal to that of Mercury's orbit. Notice that even this highest-eccentricity orbit is difficult to distinguish from a circle, which is

Johannes Kepler
German astronomer (1571–1630)
Kepler is best known for developing the laws of planetary motion based on the careful observations of Tycho Brahe.

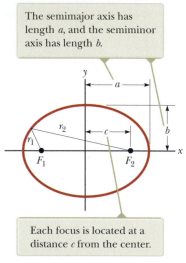

The semimajor axis has length a, and the semiminor axis has length b.

Each focus is located at a distance c from the center.

Figure 13.5 Plot of an ellipse.

Figure 13.6 (a) The shape of the orbit of Mercury, which has the highest eccentricity ($e = 0.21$) among the eight planets in the solar system. The broken line is *not* a circle. Measure the horizontal and vertical diameters. They differ by about 0.5 mm on the printed page. (Copy and enlarge to see the difference more easily!) (b) The shape of the orbit of Comet Halley. The shape of the orbit is correct; the comet and the Sun are shown larger than in reality for clarity.

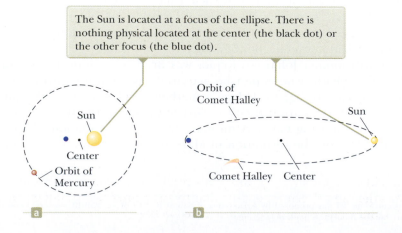

The Sun is located at a focus of the ellipse. There is nothing physical located at the center (the black dot) or the other focus (the blue dot).

one reason Kepler's first law is an admirable accomplishment. The eccentricity of the orbit of Comet Halley is 0.97, describing an orbit whose major axis is much longer than its minor axis, as shown in Figure 13.6b. As a result, Comet Halley spends much of its 76-year period far from the Sun and invisible from the Earth. It is only visible to the naked eye during a small part of its orbit when it is near the Sun.

Now imagine a planet in an elliptical orbit such as that shown in Figure 13.5, with the Sun at focus F_2. When the planet is at the far left in the diagram, the distance between the planet and the Sun is $a + c$. At this point, called the *aphelion*, the planet is at its maximum distance from the Sun. (For an object in orbit around the Earth, this point is called the *apogee*.) Conversely, when the planet is at the right end of the ellipse, the distance between the planet and the Sun is $a - c$. At this point, called the *perihelion* (for an Earth orbit, the *perigee*), the planet is at its minimum distance from the Sun.

Kepler's first law is a direct result of the inverse-square nature of the gravitational force. Circular and elliptical orbits correspond to objects that are *bound* to the gravitational force center. These objects include planets, asteroids, and comets that move repeatedly around the Sun as well as moons orbiting a planet. There are also *unbound* objects, such as a meteoroid from deep space that might pass by the Sun once and then never return. The gravitational force between the Sun and these objects also varies as the inverse square of the separation distance, and the allowed paths for these objects include parabolas ($e = 1$) and hyperbolas ($e > 1$).

Kepler's Second Law

Kepler's second law (page 340) can be shown to be a result of the isolated system model for angular momentum. Consider a planet of mass M_p moving about the Sun in an elliptical orbit (Fig. 13.7a). Let's consider the planet as a system. We model the Sun to be so much more massive than the planet that the Sun does not move. The gravitational force exerted by the Sun on the planet is a central force, always along the radius vector, directed toward the Sun (Fig. 13.7a). The torque on the planet due to this central force about an axis through the Sun is zero because $\vec{\mathbf{F}}_g$ is parallel to $\vec{\mathbf{r}}$.

Therefore, because the external torque on the planet is zero, it is modeled as an isolated system for angular momentum (Section 11.4), and the angular momentum $\vec{\mathbf{L}}$ of the planet is a constant of the motion:

$$\Delta\vec{\mathbf{L}} = 0 \quad \rightarrow \quad \vec{\mathbf{L}} = \text{constant}$$

Evaluating $\vec{\mathbf{L}}$ for the planet,

$$\vec{\mathbf{L}} = \vec{\mathbf{r}} \times \vec{\mathbf{p}} = M_p\vec{\mathbf{r}} \times \vec{\mathbf{v}} \quad \rightarrow \quad L = M_p|\vec{\mathbf{r}} \times \vec{\mathbf{v}}| \tag{13.9}$$

We can relate this result to the following geometric consideration. In a time interval dt, the radius vector $\vec{\mathbf{r}}$ in Figure 13.7b sweeps out the area dA, which equals half the area $|\vec{\mathbf{r}} \times d\vec{\mathbf{r}}|$ of the parallelogram formed by the vectors $\vec{\mathbf{r}}$ and $d\vec{\mathbf{r}}$. Because the displacement of the planet in the time interval dt is given by $d\vec{\mathbf{r}} = \vec{\mathbf{v}}\,dt$,

$$dA = \tfrac{1}{2}|\vec{\mathbf{r}} \times d\vec{\mathbf{r}}| = \tfrac{1}{2}|\vec{\mathbf{r}} \times \vec{\mathbf{v}}\,dt| = \tfrac{1}{2}|\vec{\mathbf{r}} \times \vec{\mathbf{v}}|\,dt$$

Substitute for the absolute value of the cross product from Equation 13.9:

$$dA = \tfrac{1}{2}\left(\frac{L}{M_p}\right)dt$$

Divide both sides by dt to obtain

$$\frac{dA}{dt} = \frac{L}{2M_p} \tag{13.10}$$

where L and M_p are both constants. This result shows that the derivative dA/dt is constant—the radius vector from the Sun to any planet sweeps out equal areas in equal time intervals as stated in Kepler's second law.

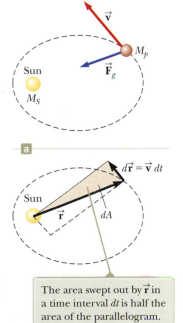

The area swept out by $\vec{\mathbf{r}}$ in a time interval dt is half the area of the parallelogram.

Figure 13.7 (a) The gravitational force acting on a planet is directed toward the Sun. (b) During a time interval dt, a parallelogram is formed by the vectors $\vec{\mathbf{r}}$ and $d\vec{\mathbf{r}} = \vec{\mathbf{v}}\,dt$.

This conclusion is a result of the gravitational force being a central force, which in turn implies that angular momentum of the planet is constant. Therefore, the law applies to *any* situation that involves a central force, whether inverse square or not.

Kepler's Third Law

Kepler's third law (page 340) can be predicted from the inverse-square law for circular orbits and our analysis models. Consider a planet of mass M_p that is assumed to be moving about the Sun (mass M_S) in a circular orbit as in Figure 13.8. Because the gravitational force provides the centripetal acceleration of the planet as it moves in a circle, we model the planet as a particle under a net force and as a particle in uniform circular motion and incorporate Newton's law of universal gravitation,

$$F_g = M_p a \quad \rightarrow \quad \frac{GM_S M_p}{r^2} = M_p \left(\frac{v^2}{r} \right)$$

Figure 13.8 A planet of mass M_p moving in a circular orbit around the Sun. The orbits of all planets except Mercury are nearly circular.

The orbital speed of the planet is $2\pi r/T$, where T is the period; therefore, the preceding expression becomes

$$\frac{GM_S}{r^2} = \frac{(2\pi r/T)^2}{r}$$

$$T^2 = \left(\frac{4\pi^2}{GM_S} \right) r^3 = K_S r^3$$

where K_S is a constant given by

$$K_S = \frac{4\pi^2}{GM_S} = 2.97 \times 10^{-19} \text{ s}^2/\text{m}^3$$

This equation is also valid for elliptical orbits if we replace r with the length a of the semimajor axis (Fig. 13.5):

◀ Kepler's third law

$$T^2 = \left(\frac{4\pi^2}{GM_S} \right) a^3 = K_S a^3 \tag{13.11}$$

Equation 13.11 is Kepler's third law: the square of the period is proportional to the cube of the semimajor axis. Because the semimajor axis of a circular orbit is its radius, this equation is valid for both circular and elliptical orbits. Notice that the constant of proportionality K_S is independent of the mass of the planet.[4] Equation 13.11 is therefore valid for *any* planet. If we were to consider the orbit of a satellite such as the Moon about the Earth, the constant would have a different value, with the Sun's mass replaced by the Earth's mass; that is, $K_E = 4\pi^2/GM_E$.

Table 13.2 is a collection of useful data for planets and other objects in the solar system. The far-right column verifies that the ratio T^2/r^3 is constant for all objects orbiting the Sun. The small variations in the values in this column are the result of uncertainties in the data measured for the periods and semimajor axes of the objects.

Recent astronomical work has revealed the existence of a large number of solar system objects beyond the orbit of Neptune. In general, these objects lie in the *Kuiper belt*, a region that extends from about 30 AU (the orbital radius of Neptune) to 50 AU. (An AU is an *astronomical unit*, equal to the radius of the Earth's orbit.) Current estimates identify at least 100 000 objects in this region with diameters larger than 100 km. The first Kuiper belt object (KBO) is Pluto, discovered in 1930 and formerly classified as a planet. Starting in 1992, many more KBOs have been detected. Several have diameters in the 1 000-km range, such as Varuna (discovered in 2000), Ixion (2001), Quaoar (2002), Sedna (2003), Haumea (2004), Orcus (2004), and Makemake (2005). One KBO, Eris, discovered in 2005, is believed to be

---

[4]Equation 13.11 is indeed a proportion because the ratio of the two quantities T^2 and a^3 is a constant. The variables in a proportion are not required to be limited to the first power only.

TABLE 13.2 Useful Planetary Data

Body	Mass (kg)	Mean Radius (m)	Period of Revolution (s)	Mean Distance from the Sun (m)	$\dfrac{T^2}{r^3}$ (s²/m³)
Mercury	3.30×10^{23}	2.44×10^6	7.60×10^6	5.79×10^{10}	2.98×10^{-19}
Venus	4.87×10^{24}	6.05×10^6	1.94×10^7	1.08×10^{11}	2.99×10^{-19}
Earth	5.97×10^{24}	6.37×10^6	3.156×10^7	1.496×10^{11}	2.97×10^{-19}
Mars	6.42×10^{23}	3.39×10^6	5.94×10^7	2.28×10^{11}	2.98×10^{-19}
Jupiter	1.90×10^{27}	6.99×10^7	3.74×10^8	7.78×10^{11}	2.97×10^{-19}
Saturn	5.68×10^{26}	5.82×10^7	9.29×10^8	1.43×10^{12}	2.95×10^{-19}
Uranus	8.68×10^{25}	2.54×10^7	2.65×10^9	2.87×10^{12}	2.97×10^{-19}
Neptune	1.02×10^{26}	2.46×10^7	5.18×10^9	4.50×10^{12}	2.94×10^{-19}
Pluto[a]	1.25×10^{22}	1.20×10^6	7.82×10^9	5.91×10^{12}	2.96×10^{-19}
Moon	7.35×10^{22}	1.74×10^6	—	—	—
Sun	1.989×10^{30}	6.96×10^8	—	—	—

[a]In August 2006, the International Astronomical Union adopted a definition of a planet that separates Pluto from the other eight planets. Pluto is now defined as a "dwarf planet" like the asteroid Ceres.

similar in size to Pluto and about 27% more massive. Other KBOs do not yet have names, but are currently indicated by their year of discovery and a code, such as 2010 EK139 and 2015 FG345.

A subset of about 1 400 KBOs are called "Plutinos" because, like Pluto, they exhibit a resonance phenomenon, orbiting the Sun two times in the same time interval as Neptune revolves three times. The contemporary application of Kepler's laws and such exotic proposals as planetary angular momentum exchange and migrating planets suggest the excitement of this active area of current research.

Ⓠ UICK QUIZ 13.3 An asteroid is in a highly eccentric elliptical orbit around the Sun. The period of the asteroid's orbit is 90 days. Which of the following statements is true about the possibility of a collision between this asteroid and the Earth? **(a)** There is no possible danger of a collision. **(b)** There is a possibility of a collision. **(c)** There is not enough information to determine whether there is danger of a collision.

Example **13.4** **The Mass of the Sun**

In the opening storyline, you were wondering how to determine the mass of the Sun. Now that we have discussed Kepler's third law, calculate the mass of the Sun.

SOLUTION

Conceptualize Based on the mathematical representation of Kepler's third law expressed in Equation 13.11, we realize that the mass of the central object in a gravitational system is related to the orbital size and period of objects in orbit around the central object.

Categorize This example is a relatively simple substitution problem.

Solve Equation 13.11 for the mass of the Sun:

$$M_S = \frac{4\pi^2 r^3}{GT^2}$$

Substitute numerical values, using data from Table 13.2:

$$M_S = \frac{4\pi^2(1.496 \times 10^{11}\text{ m})^3}{(6.674 \times 10^{-11}\text{ N} \cdot \text{m}^2/\text{kg}^2)(3.156 \times 10^7\text{ s})^2} = \boxed{1.99 \times 10^{30}\text{ kg}}$$

In Example 13.2, an understanding of gravitational forces enabled us to find out something about the density of the Earth's core, and now we have used this understanding to answer your question about the mass of the Sun! To answer your question about the masses of planets, we can perform the same calculation using the orbital size and period of a moon of a planet to find the planet mass. Neither Kepler's third law or Newton's law of universal gravitation can be used to determine the mass of the orbiting object, and the planets and KBOs for which we have precise mass data are the ones with moons or ones about which we have placed a spacecraft in orbit.

Example **13.5**   **A Geosynchronous Satellite**

Consider a satellite of mass m moving in a circular orbit around the Earth at a constant speed v and at an altitude h above the Earth's surface as illustrated in Figure 13.9.

(A) Determine the speed of the satellite in terms of G, h, R_E (the radius of the Earth), and M_E (the mass of the Earth).

S O L U T I O N

Conceptualize   Imagine the satellite moving around the Earth in a circular orbit under the influence of the gravitational force. Figure 13.9 is a polar view of this motion. This motion is similar to that of the International Space Station, the Hubble Space Telescope, and other objects in orbit around the Earth.

Categorize   The satellite moves in a circular orbit at a constant speed. Therefore, we categorize the satellite as a *particle in uniform circular motion* as well as a *particle under a net force*.

Figure 13.9   (Example 13.5) A satellite of mass m moving around the Earth in a circular orbit of radius r with constant speed v. The only force acting on the satellite is the gravitational force $\vec{F}_g$. (Not drawn to scale.)

Analyze   The only external force acting on the satellite is the gravitational force from the Earth, which acts toward the center of the Earth and keeps the satellite in its circular orbit.

Apply the particle under a net force and particle in uniform circular motion models to the satellite:

$$F_g = ma \;\rightarrow\; G\frac{M_E m}{r^2} = m\left(\frac{v^2}{r}\right)$$

Solve for v, noting that the distance r from the center of the Earth to the satellite is $r = R_E + h$:

$$(1) \quad v = \sqrt{\frac{GM_E}{r}} = \sqrt{\frac{GM_E}{R_E + h}}$$

(B) If the satellite is to be *geosynchronous* (that is, appearing to remain over a fixed position on the Earth), how fast is it moving through space?

S O L U T I O N

To appear to remain over a fixed position on the Earth, the period of the satellite must be 24 h = 86 400 s and the satellite must be in orbit directly over the equator.

Solve Kepler's third law (Equation 13.11, with $a = r$ and $M_S \rightarrow M_E$) for r:

$$r = \left(\frac{GM_E T^2}{4\pi^2}\right)^{1/3}$$

Substitute numerical values:

$$r = \left[\frac{(6.674 \times 10^{-11}\ \text{N} \cdot \text{m}^2/\text{kg}^2)(5.97 \times 10^{24}\ \text{kg})(86\ 400\ \text{s})^2}{4\pi^2}\right]^{1/3}$$

$$= 4.22 \times 10^7\ \text{m}$$

Use Equation (1) to find the speed of the satellite:

$$v = \sqrt{\frac{(6.674 \times 10^{-11}\ \text{N} \cdot \text{m}^2/\text{kg}^2)(5.97 \times 10^{24}\ \text{kg})}{4.22 \times 10^7\ \text{m}}}$$

$$= 3.07 \times 10^3\ \text{m/s}$$

Finalize   The value of r calculated here translates to a height of the satellite above the surface of the Earth of almost 36 000 km. Therefore, geosynchronous satellites have the advantage of allowing an earthbound antenna to be aimed in a fixed direction, but there is a disadvantage in that the signals between the Earth and the satellite must travel a long distance. It is difficult to use geosynchronous satellites for optical observation of the Earth's surface because of their high altitude.

W H A T   I F ?   What if the satellite motion in part (A) were taking place at height h above the surface of another planet more massive than the Earth but of the same radius? Would the satellite be moving at a higher speed or a lower speed than it does around the Earth?

Answer   If the planet exerts a larger gravitational force on the satellite due to its larger mass, the satellite must move with a higher speed to avoid moving toward the surface. This conclusion is consistent with the predictions of Equation (1), which shows that because the speed v is proportional to the square root of the mass of the planet, the speed increases as the mass of the planet increases.

Gravitational Potential Energy

13.5 Gravitational Potential Energy

In Chapter 7, we introduced the concept of gravitational potential energy, which is the energy associated with the configuration of a system of objects interacting via the gravitational force. The gravitational potential energy function $U_g = mgy$ (Eq. 7.19) for a particle–Earth system is restricted to situations where a very massive object (such as the Earth) establishes a gravitational field of magnitude g and a particle of much smaller mass m resides in that field. It is also restricted to positions of the object near the surface of the Earth, where g is independent of y. In reality, however, because the gravitational field varies as $1/r^2$ as shown in Equation 13.8, we expect that a more general potential energy function—one that is valid without the restrictions mentioned above—will be different from $U_g = mgy$.

Recall from Equation 7.27 that the change in the potential energy of a system associated with a given displacement of a member of the system is defined as the negative of the internal work done by the force on that member during the displacement:

$$\Delta U = U_f - U_i = -\int_{r_i}^{r_f} F(r)\, dr \tag{13.12}$$

We can use this result to evaluate the general gravitational potential energy function. Consider a particle of mass m moving between two points Ⓐ and Ⓑ above the Earth's surface (Fig. 13.10). The particle is subject to the gravitational force given by Equation 13.1. We can express this force as

$$F(r) = -\frac{GM_E m}{r^2}$$

where the negative sign indicates that the force is attractive. Substituting this expression for $F(r)$ into Equation 13.12, we can compute the change in the gravitational potential energy function for the particle–Earth system as the separation distance r changes:

$$U_f - U_i = GM_E m \int_{r_i}^{r_f} \frac{dr}{r^2} = GM_E m \left[-\frac{1}{r}\right]_{r_i}^{r_f}$$

$$U_f - U_i = -GM_E m \left(\frac{1}{r_f} - \frac{1}{r_i}\right) \tag{13.13}$$

As always, the choice of a reference configuration for the potential energy is completely arbitrary. It is customary to choose the reference configuration for zero potential energy to be the same as that for which the force is zero. Taking $U_i = 0$ at $r_i = \infty$, we obtain the important result

$$\boxed{U_g(r) = -\frac{GM_E m}{r}} \tag{13.14}$$

This expression applies when the particle is separated from the center of the Earth by a distance r, provided that $r \geq R_E$. The result is not valid for particles inside the Earth, where $r < R_E$. Because of our choice of U_i, the function U_g is always negative (Fig. 13.11).

Although Equation 13.14 was derived for the particle–Earth system, a similar form of the equation can be applied to any two particles. That is, the gravitational potential energy associated with a system of two particles of masses m_1 and m_2 separated by a distance r is

$$\boxed{U_g(r) = -\frac{Gm_1 m_2}{r}} \tag{13.15}$$

This expression shows that the gravitational potential energy for any pair of particles varies as $1/r$, whereas the force between them varies as $1/r^2$. Furthermore,

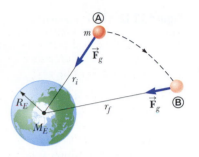

Figure 13.10 As a particle of mass m moves from Ⓐ to Ⓑ above the Earth's surface, the gravitational potential energy of the particle–Earth system changes according to Equation 13.12.

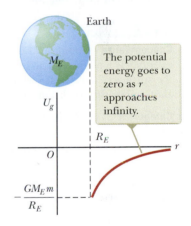

Figure 13.11 Graph of the gravitational potential energy U_g versus r for the system of an object above the Earth's surface.

◀ Gravitational potential energy of the Earth–particle system

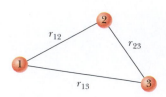

Figure 13.12 Three interacting particles.

the potential energy is negative because the force is attractive and we have chosen the potential energy as zero when the particle separation is infinite. Because the force between the particles is attractive, an external agent must do positive work to increase the separation between the particles. The work done by the external agent produces an increase in the potential energy as the two particles are separated. That is, U_g becomes less negative as r increases.

We can extend this concept to three or more particles. In this case, the total potential energy of the system is the sum over all pairs of particles. Each pair contributes a term of the form given by Equation 13.15. For example, if the system contains three particles as in Figure 13.12,

$$U_{\text{total}} = U_{12} + U_{13} + U_{23} = -G\left(\frac{m_1 m_2}{r_{12}} + \frac{m_1 m_3}{r_{13}} + \frac{m_2 m_3}{r_{23}}\right)$$

The absolute value of U_{total} represents the work needed to separate the three particles by an infinite distance.

Example 13.6 | The Change in Potential Energy

A particle of mass m is displaced through a small vertical distance Δy near the Earth's surface. Show that in this situation the general expression for the change in gravitational potential energy given by Equation 13.13 reduces to the familiar relationship $\Delta U_g = mg\,\Delta y$.

SOLUTION

Conceptualize Compare the two different situations for which we have developed expressions for gravitational potential energy: (1) a planet and an object that are far apart for which the energy expression is Equation 13.14 and (2) a small object at the surface of a planet for which the energy expression is Equation 7.19. We wish to show that these two expressions are equivalent under the conditions described in the problem.

Categorize This example is a substitution problem.

Combine the fractions in Equation 13.13:

$$(1) \quad \Delta U_g = -GM_E m\left(\frac{1}{r_f} - \frac{1}{r_i}\right) = GM_E m\left(\frac{r_f - r_i}{r_i r_f}\right)$$

Evaluate $r_f - r_i$ and $r_i r_f$ if both the initial and final positions of the particle are close to the Earth's surface:

$$r_f - r_i = \Delta y \quad r_i r_f \approx R_E^2$$

Substitute these expressions into Equation (1):

$$\Delta U_g \approx GM_E m\left(\frac{\Delta y}{R_E^2}\right) = m\left(\frac{GM_E}{R_E^2}\right)\Delta y = mg\,\Delta y$$

where $g = GM_E/R_E^2$ from Equation 13.5.

WHAT IF? Suppose you are performing upper-atmosphere studies and are asked by your supervisor to find the height in the Earth's atmosphere at which the "surface equation" $\Delta U_g = mg\,\Delta y$ gives a 1.0% error in the change in the potential energy. What is this height?

Answer Because the surface equation assumes a constant value for g, it will give a ΔU_g value that is larger than the value given by the general equation, Equation 13.13.

Set up a ratio reflecting a 1.0% error:

$$\frac{\Delta U_{\text{surface}}}{\Delta U_{\text{general}}} = 1.010$$

Substitute the expressions for each of these changes ΔU_g:

$$\frac{mg\,\Delta y}{GM_E m(\Delta y/r_i r_f)} = \frac{g r_i r_f}{GM_E} = 1.010$$

Substitute for r_i, r_f, and g from Equation 13.5:

$$\frac{(GM_E/R_E^2)R_E(R_E + \Delta y)}{GM_E} = \frac{R_E + \Delta y}{R_E} = 1 + \frac{\Delta y}{R_E} = 1.010$$

Solve for Δy:

$$\Delta y = 0.010 R_E = 0.010(6.37 \times 10^6 \text{ m}) = 6.37 \times 10^4 \text{ m} = 63.7 \text{ km}$$

13.6 Energy Considerations in Planetary and Satellite Motion

Given the general expression for gravitational potential energy developed in Section 13.5, we can now apply the analysis models that we have developed for energy to gravitational systems. Consider an object of mass m moving with a speed v in the vicinity of a massive object of mass M, where $M \gg m$. The system might be a planet moving around the Sun, a satellite in orbit around the Earth, or a comet making a one-time flyby of the Sun. If we assume the object of mass M is at rest in an inertial reference frame, the total mechanical energy E of the two-object system when the objects are separated by a distance r is the sum of the kinetic energy of the object of mass m and the gravitational potential energy of the system, given by Equation 13.15:

$$E = K + U_g$$

$$E = \tfrac{1}{2}mv^2 - \frac{GMm}{r} \tag{13.16}$$

If the system of objects of mass m and M is isolated, and there are no nonconservative forces acting within the system, the mechanical energy of the system given by Equation 13.16 is the total energy of the system and this energy is conserved:

$$\Delta K + \Delta U_g = 0 \;\;\rightarrow\;\; E_i = E_f$$

Therefore, as the object of mass m moves from Ⓐ to Ⓑ in Figure 13.10, the total energy remains constant and Equation 13.16 gives

$$\tfrac{1}{2}mv_i^2 - \frac{GMm}{r_i} = \tfrac{1}{2}mv_f^2 - \frac{GMm}{r_f} \tag{13.17}$$

Combining this statement of energy conservation with our earlier discussion of conservation of angular momentum, we see that both the total energy and the total angular momentum of a gravitationally bound, two-object system are constants of the motion.

Equation 13.16 shows that E may be positive, negative, or zero, depending on the value of v. For a bound system such as the Earth–Sun system, however, E is necessarily *less than zero* because we have chosen the convention that $U_g \rightarrow 0$ as $r \rightarrow \infty$.

We can easily establish that $E < 0$ for the system consisting of an object of mass m moving in a circular orbit about an object of mass $M \gg m$ (Fig. 13.13). Modeling the object of mass m as a particle under a net force and a particle in uniform circular motion gives

$$F_g = ma \;\;\rightarrow\;\; \frac{GMm}{r^2} = \frac{mv^2}{r}$$

Multiplying both sides by r and dividing by 2 gives

$$\tfrac{1}{2}mv^2 = \frac{GMm}{2r} \tag{13.18}$$

Substituting this equation into Equation 13.16, we obtain

$$E = \frac{GMm}{2r} - \frac{GMm}{r}$$

$$\boxed{E = -\frac{GMm}{2r}} \quad \text{(circular orbits)} \tag{13.19}$$

This result shows that the total mechanical energy is negative in the case of circular orbits. Notice that the kinetic energy is positive and equal to half the absolute value of the potential energy.

The total mechanical energy is also negative in the case of elliptical orbits. The expression for E for elliptical orbits is the same as Equation 13.19 with r replaced by

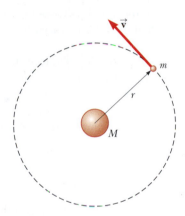

Figure 13.13 An object of mass m moving in a circular orbit about a much larger object of mass M.

◀ Total energy for circular orbits of an object of mass m around an object of mass $M \gg m$

the semimajor axis length a:

Total energy for elliptical ▶
orbits of an object of
mass m around an object of
mass $M \gg m$

$$E = -\frac{GMm}{2a} \quad \text{(elliptical orbits)}$$ (13.20)

QUICK QUIZ **13.4** A comet moves in an elliptical orbit around the Sun. Which point in its orbit (perihelion or aphelion) represents the highest value of **(a)** the speed of the comet, **(b)** the potential energy of the comet–Sun system, **(c)** the kinetic energy of the comet, and **(d)** the total energy of the comet–Sun system?

Example 13.7   **Changing the Orbit of a Satellite**

A space transportation vehicle releases a 470-kg communications satellite while in a circular orbit 280 km above the surface of the Earth. A rocket engine on the satellite boosts it into a geosynchronous orbit. How much energy does the engine have to provide for this boost?

SOLUTION

Conceptualize   Notice that the height of 280 km is much lower than that for a geosynchronous satellite, 36 000 km, as mentioned in Example 13.5. Therefore, energy must be expended to raise the satellite to this much higher position.

Categorize   This example is a substitution problem.

Find the initial radius of the satellite's orbit when it is still in the vehicle's cargo bay:

$$r_i = R_E + 280 \text{ km} = 6.65 \times 10^6 \text{ m}$$

Use Equation 13.19 to find the difference in energies for the satellite–Earth system with the satellite at the initial and final radii:

$$\Delta E = E_f - E_i = -\frac{GM_E m}{2r_f} - \left(-\frac{GM_E m}{2r_i}\right) = -\frac{GM_E m}{2}\left(\frac{1}{r_f} - \frac{1}{r_i}\right)$$

Substitute numerical values, using $r_f = 4.22 \times 10^7$ m from Example 13.5:

$$\Delta E = -\frac{(6.674 \times 10^{-11} \text{ N} \cdot \text{m}^2/\text{kg}^2)(5.97 \times 10^{24} \text{ kg})(470 \text{ kg})}{2}$$
$$\times \left(\frac{1}{4.22 \times 10^7 \text{ m}} - \frac{1}{6.65 \times 10^6 \text{ m}}\right) = \boxed{1.19 \times 10^{10} \text{ J}}$$

which is the energy equivalent of 89 gal of gasoline. NASA engineers must account for the changing mass of the spacecraft as it ejects burned fuel, something we have not done here. Would you expect the calculation that includes the effect of this changing mass to yield a greater or a lesser amount of energy required from the engine?

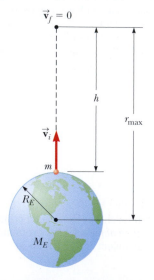

Figure 13.14 An object of mass m projected upward from the Earth's surface with an initial speed v_i reaches a maximum altitude h.

Escape Speed

Suppose an object of mass m is projected vertically upward from the Earth's surface with an initial speed v_i as illustrated in Figure 13.14. We can use energy considerations to find the value of the initial speed needed to allow the object to reach a certain distance away from the center of the Earth. Equation 13.16 gives the total energy of the system for any configuration. As the object is projected upward from the surface of the Earth, $v = v_i$ and $r = r_i = R_E$. When the object reaches its maximum altitude, $v = v_f = 0$ and $r = r_f = r_{max}$. Because the object–Earth system is isolated, we substitute these values into the isolated-system model expression given by Equation 13.17:

$$\tfrac{1}{2}mv_i^2 - \frac{GM_E m}{R_E} = -\frac{GM_E m}{r_{max}}$$

Solving for v_i^2 gives

$$v_i^2 = 2GM_E\left(\frac{1}{R_E} - \frac{1}{r_{max}}\right)$$ (13.21)

For a given maximum altitude $h = r_{max} - R_E$, we can use this equation to find the required initial speed.

We are now in a position to calculate the **escape speed,** which is the minimum speed the object must have at the Earth's surface to approach an infinite separation distance from the Earth. Traveling at this minimum speed, the object continues to move farther and farther away from the Earth as its speed asymptotically approaches zero. Letting $r_{max} \to \infty$ in Equation 13.21 and identifying v_i as v_{esc} gives

$$v_{esc} = \sqrt{\frac{2GM_E}{R_E}}$$

(13.22) ◀ Escape speed from the Earth

This expression for v_{esc} is independent of the mass of the object. In other words, a spacecraft has the same escape speed as a molecule. Furthermore, the result is independent of the direction of the velocity and ignores air resistance. Keep in mind that an object already has some initial speed at the surface of the Earth due to its rotation.

If the object is given an initial speed equal to v_{esc}, the total energy of the system is equal to zero. Notice that when $r \to \infty$, the object's kinetic energy and the potential energy of the system are both zero. If v_i is greater than v_{esc}, however, the total energy of the system is greater than zero and the object has some residual kinetic energy as $r \to \infty$.

> **PITFALL PREVENTION 13.3**
> **You Can't Really Escape** Although Equation 13.22 provides the "escape speed" from the Earth, *complete* escape from the Earth's gravitational influence is impossible because the gravitational force is of infinite range. In addition, escape from the Earth to an infinite distance also requires escape from the Sun, requiring additional energy.

Example 13.8 Escape Speed of a Rock

Superman picks up a 20-kg rock and hurls it into space. What minimum speed must it have at the Earth's surface to move infinitely far away from the Earth?

SOLUTION

Conceptualize Imagine Superman throwing the rock from the Earth's surface so that it moves farther and farther away, traveling more and more slowly, with its speed approaching zero. Its speed will never reach zero, however, so the rock will never turn around and come back.

Categorize This example is a substitution problem.

Use Equation 13.22 to find the escape speed:

$$v_{esc} = \sqrt{\frac{2GM_E}{R_E}} = \sqrt{\frac{2(6.674 \times 10^{-11}\ \text{N} \cdot \text{m}^2/\text{kg}^2)(5.97 \times 10^{24}\ \text{kg})}{6.37 \times 10^6\ \text{m}}}$$

$$= 1.12 \times 10^4\ \text{m/s}$$

The calculated escape speed corresponds to about 25 000 mi/h. The mass of the rock does not appear in the calculation. Therefore, this is also the escape speed for Superman throwing a 5 000-kg spacecraft from the surface of the Earth. Furthermore, if a spacecraft is in an orbit around the Earth, its orbital radius r is close to that of the Earth, R_E, so the escape speed we have calculated is also valid for the non-superhero situation of a spacecraft in orbit firing its engines to escape that orbit.

Equations 13.21 and 13.22 can be applied to objects projected from any planet. That is, in general, the escape speed from the surface of any planet of mass M and radius R is

$$v_{esc} = \sqrt{\frac{2GM}{R}}$$

(13.23) ◀ Escape speed from the surface of a planet of mass M and radius R

Escape speeds for the planets, the Moon, and the Sun are provided in Table 13.3 (page 350). The values vary from 2.3 km/s for the Moon to about 618 km/s for the Sun. These results, together with some ideas from the kinetic theory of gases (see Chapter 20), explain why some planets have atmospheres and others do not. As we shall see later, at a given temperature the average kinetic energy of a gas molecule depends only on the mass of the molecule. Lighter molecules, such as hydrogen and helium, have a higher average speed than heavier molecules at the same

TABLE 13.3 Escape Speeds from the Surfaces of the Planets, Moon, and Sun

Planet	v_{esc} (km/s)
Mercury	4.3
Venus	10.3
Earth	11.2
Mars	5.0
Jupiter	60
Saturn	36
Uranus	22
Neptune	24
Moon	2.3
Sun (from Earth orbit)	42
Sun (from Sun surface)	618

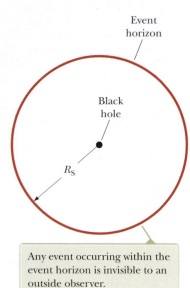

Any event occurring within the event horizon is invisible to an outside observer.

Figure 13.15 A black hole. The distance R_S equals the Schwarzschild radius.

Figure 13.16 The orbital speed v as a function of distance r from the Sun for the eight planets of the solar system. The theoretical curve is in red-brown, and the data points for the planets are in black.

temperature. When the *average* speed of the lighter molecules is not much less than the escape speed of a planet, a significant fraction of them are moving faster than the average speed and have a chance to escape the planet.

This mechanism also explains why the Earth does not retain hydrogen molecules and helium atoms in its atmosphere but does retain heavier molecules, such as oxygen and nitrogen. On the other hand, the very large escape speed for Jupiter enables that planet to retain hydrogen, the primary constituent of its atmosphere.

Black Holes

In Example 11.6, we briefly described a rare event called a supernova, the catastrophic explosion at the end of the life of a very massive star. The material that remains in the central core of such an object continues to collapse, and the core's ultimate fate depends on its mass. As the star collapses inward, electrons and protons fuse to form neutrons, and an object made purely of neutrons, a neutron star, is born, as discussed in Example 11.6. The inward collapse of such a star is halted by the quantum mechanical repulsion of neutrons called *neutron degeneracy pressure*, and the mass of the star is compressed to a radius of about 10 km. (On the Earth, a teaspoon of this material would weigh about 5 billion tons!). The escape speed from a neutron star is typically $> 0.5c$, where c is the speed of light.

An even more unusual death occurs when the core has a mass greater than about 2–3 solar masses. Neutron degeneracy pressure is unable to halt the collapse of the star, and the star becomes a very small object in space, commonly referred to as a **black hole.** In effect, black holes are remains of stars that have collapsed under their own gravitational force. If an object such as a spacecraft comes close to a black hole, the object experiences an extremely strong gravitational force and is trapped forever.

The escape speed for a black hole is very high because of the concentration of the star's mass into a sphere of very small radius (see Eq. 13.23). If the escape speed exceeds the speed of light c, radiation from the object (such as visible light) cannot escape and the object appears to be black (hence the origin of the terminology "black hole"). The critical radius R_S at which the escape speed is c is called the **Schwarzschild radius** (Fig. 13.15). The imaginary surface of a sphere of this radius surrounding the black hole is called the **event horizon,** which is the limit of how close you can approach the black hole and hope to escape.

There is evidence that supermassive black holes exist at the centers of galaxies, with masses very much larger than the Sun. (There is strong evidence of a supermassive black hole of mass 4.0–4.3 million solar masses at the center of our galaxy.)

Dark Matter

Equation (1) in Example 13.5 shows that the speed of an object in orbit around the Earth decreases as the object is moved farther away from the Earth:

$$v = \sqrt{\frac{GM_E}{r}} \tag{13.24}$$

Using data in Table 13.2 to find the speeds of planets in their orbits around the Sun, we find the same behavior for the planets. Figure 13.16 shows this behavior for the eight planets of our solar system. The theoretical prediction of the planet speed as a function of distance from the Sun is shown by the red-brown curve, using Equation 13.24 with the mass of the Earth replaced by the mass of the Sun. Data points for the individual planets lie right on this curve. This behavior results from the vast majority (99.9%) of the mass of the solar system being concentrated in a small space, i.e., the Sun.

Extending this concept further, we might expect the same behavior in a galaxy. Much of the visible galactic mass, including that of a supermassive black hole, is near the central core of a galaxy. Measurements of the speeds of faraway objects in the galaxy can be used in Kepler's third law to estimate the mass of the entire galaxy, a topic raised in the opening storyline. These estimates range from 0.8×10^{12} to 4.5×10^{12} solar masses. Measurements made on these faraway objects are difficult and the results depend on the method used to make the observations.

The opening photograph for this chapter shows the central core of the galaxy NGC 1566 as a very bright area surrounded by the "arms" of the galaxy, which contain material in orbit around the central core. Based on this distribution of matter in the galaxy, the speed of an object in the outer part of the galaxy would be smaller than that for objects closer to the center, just like for the planets of the solar system.

That is *not* what is observed, however. Figure 13.17 shows the results of measurements of the speeds of objects in the Andromeda galaxy as a function of distance from the galaxy's center.[5] The red-brown curve shows the expected speeds for these objects if they were traveling in circular orbits around the mass concentrated in the central core. The data for the individual objects in the galaxy shown by the black dots are all well above the theoretical curve. These data, as well as an extensive amount of data taken over the past half century, show that for objects outside the central core of the galaxy, the curve of speed versus distance from the center of the galaxy is approximately flat rather than decreasing at larger distances. Therefore, these objects (including our own Solar System in the Milky Way) are rotating faster than can be accounted for by gravity due to the visible galaxy! This surprising result means that there must be additional mass in a more extended distribution, causing these objects to orbit so fast, and has led scientists to propose the existence of **dark matter.** This matter is proposed to exist in a large halo around each galaxy (with a radius up to 10 times as large as the visible galaxy's radius). Because it is not luminous (i.e., does not emit electromagnetic radiation) it must be either very cold or electrically neutral. Therefore, we cannot "see" dark matter, except through its gravitational effects.

The proposed existence of dark matter is also implied by earlier observations made on larger gravitationally bound structures known as galaxy clusters.[6] These observations show that the orbital speeds of galaxies in a cluster are, on average, too large to be explained by the luminous matter in the cluster alone. The speeds of the individual galaxies are so high, they suggest that there is 50 times as much dark matter in galaxy clusters as in the galaxies themselves!

Why doesn't dark matter affect the orbital speeds of planets like it does those of a galaxy? It seems that a solar system is too small a structure to contain enough dark matter to affect the behavior of orbital speeds. A galaxy or galaxy cluster, on the other hand, contains huge amounts of dark matter, resulting in the surprising behavior.

What, though, *is* dark matter? At this time, no one knows. One hypothesis claims that dark matter is based on a particle called a weakly interacting massive particle, or WIMP. If this theory is correct, calculations show that about 200 WIMPs pass through a human body at any given time. The new Large Hadron Collider in Europe (see Chapter 44) is the first particle accelerator with enough energy to possibly generate and detect the existence of WIMPs, which has generated much current interest in dark matter. Keeping an eye on this research in the future should be exciting, and the creativity of physicists in generating whimsical names for newly proposed objects should be entertaining.

Figure 13.17 The orbital speed v of a galaxy object as a function of distance r from the center of the central core of the Andromeda galaxy. The theoretical curve is in red-brown, and the data points for the galaxy objects are in black. No data are provided on the left because the behavior inside the central core of the galaxy is more complicated.

[5] V. C. Rubin and W. K. Ford, "Rotation of the Andromeda Nebula from a Spectroscopic Survey of Emission Regions," *Astrophysical Journal* **159:** 379–403 (1970).

[6] F. Zwicky, "On the Masses of Nebulae and of Clusters of Nebulae," *Astrophysical Journal* **86:** 217–246 (1937).

Summary

> ## Definitions

The **gravitational field** at a point in space is defined as the gravitational force $\vec{\mathbf{F}}_g$ experienced by any test particle located at that point divided by the mass m_0 of the test particle:

$$\vec{\mathbf{g}} \equiv \frac{\vec{\mathbf{F}}_g}{m_0} \tag{13.7}$$

> ## Concepts and Principles

Newton's law of universal gravitation states that the gravitational force of attraction between any two particles of masses m_1 and m_2 separated by a distance r has the magnitude

$$F_g = G\frac{m_1 m_2}{r^2} \tag{13.1}$$

where $G = 6.674 \times 10^{-11}\ \text{N} \cdot \text{m}^2/\text{kg}^2$ is the **universal gravitational constant**. This equation enables us to calculate the force of attraction between masses under many circumstances.

An object at a distance h above the Earth's surface experiences a gravitational force of magnitude mg, where g is the free-fall acceleration at that elevation:

$$g = \frac{GM_E}{r^2} = \frac{GM_E}{(R_E + h)^2} \tag{13.6}$$

In this expression, M_E is the mass of the Earth and R_E is its radius. Therefore, the weight of an object decreases as the object moves away from the Earth's surface.

Kepler's laws of planetary motion state:

1. All planets move in elliptical orbits with the Sun at one focus.
2. The radius vector drawn from the Sun to a planet sweeps out equal areas in equal time intervals.
3. The square of the orbital period of any planet is proportional to the cube of the semimajor axis of the elliptical orbit.

Kepler's third law can be expressed as

$$T^2 = \left(\frac{4\pi^2}{GM_S}\right)a^3 \tag{13.11}$$

where M_S is the mass of the Sun and a is the semimajor axis. For a circular orbit, a can be replaced in Equation 13.11 by the radius r. Most planets have nearly circular orbits around the Sun.

The **gravitational potential energy** associated with a system of two particles of mass m_1 and m_2 separated by a distance r is

$$U_g(r) = -\frac{Gm_1 m_2}{r} \tag{13.15}$$

where U_g is taken to be zero as $r \to \infty$.

If an isolated system consists of an object of mass m moving with a speed v in the vicinity of a massive object of mass M, the total energy E of the system is the sum of the kinetic and potential energies:

$$E = \tfrac{1}{2}mv^2 - \frac{GMm}{r} \tag{13.16}$$

and the total energy of the system is a constant of the motion. If the object moves in an elliptical orbit of semimajor axis a around the massive object and $M \gg m$, the total energy of the system is

$$E = -\frac{GMm}{2a} \tag{13.20}$$

For a circular orbit, this same equation applies with $a = r$.

The **escape speed** for an object projected from the surface of a planet of mass M and radius R is

$$v_{\text{esc}} = \sqrt{\frac{2GM}{R}} \tag{13.23}$$

> ## Analysis Model for Problem Solving

Particle in a Field (Gravitational) A source particle with some mass establishes a **gravitational field** $\vec{\mathbf{g}}$ throughout space. When a particle of mass m is placed in that field, it experiences a gravitational force given by

$$\vec{\mathbf{F}}_g = m\vec{\mathbf{g}} \tag{5.5}$$

Think–Pair–Share

See the Preface for an explanation of the icons used in this problems set. For additional assessment items for this section, go to WEBASSIGN From Cengage

1. Kepler's third law should be obeyed for any group of objects orbiting a massive central object. Consider the five moons of Pluto: Charon, Styx, Nix, Kerberos, and Hydra. Data for these moons are given in the table below. If you evaluate the ratio T^2/a^3 for these moons, where T is the orbital period and a is the length of the semimajor axis, the results are all very close, *except* for Charon, which doesn't seem to follow Kepler's third law. Discuss with your group why the value of the ratio T^2/a^3 for Charon is different from those of the other moons.

Moon	Semimajor axis a (10^6 m)	Orbital period T (d)	Diameter (km)
Charon	17.54	6.387	1 208
Styx	42.66	20.16	~12
Nix	48.69	24.85	~40
Kerberos	57.78	32.17	~14
Hydra	64.74	38.20	~50

2. **ACTIVITY** Jupiter has over 60 moons, most of them discovered in the twenty-first century. The table shows astronomically measured data for the first fifteen moons, in order of semimajor axis of the orbit around Jupiter. (a) For these moons, show that Kepler's third law is satisfied within reasonable observational uncertainty for the data in the table. (b) Evaluate the ratio T^2/a^3, where T is the orbital period and a is the length of the semimajor axis, for these moons and compare the results to the value for the solar system (Table 13.2). Why is the value of this ratio larger for the moons of Jupiter than for the solar system? (c) Before calculating the value of T^2/a^3 from the data, could you have *predicted* what it would be?

Moon	Semimajor axis a (10^9 m)	Orbital period T (d)	Eccentricity	Inclination Angle
Moons discovered by Galileo:				
Io	0.421 7	1.769 1	0.004 1	0.05
Europa	0.671 0	3.551 2	0.009 4	0.47
Ganymede	1.070 4	7.154 6	0.001 1	0.20
Callisto	1.882 7	16.689	0.007 4	0.20
Inner moons:				
Metis	0.127 7	0.294 8	0.000 02	0.06
Adrastea	0.128 7	0.298 3	0.001 5	0.03
Amalthea	0.181 4	0.498 2	0.003 2	0.37
Thebe	0.221 9	0.674 5	0.017 5	1.08
Outer moons:				
Themisto	7.393 2	129.87	0.215 5	45.8
Leda	11.187 8	240.82	0.167 3	27.6
Himalia	11.452 0	250.23	0.151 3	30.5
Lysithea	11.740 6	259.89	0.132 2	27.0
Elara	11.778 0	257.62	0.194 8	29.7
Dia	12.570 4	287.93	0.205 8	27.6
Carpo	17.144 9	458.62	0.273 5	56.0

Problems

See the Preface for an explanation of the icons used in this problems set. For additional assessment items for this section, go to WEBASSIGN From Cengage

SECTION 13.1 Newton's Law of Universal Gravitation

1. In introductory physics laboratories, a typical Cavendish balance for measuring the gravitational constant G uses lead spheres with masses of 1.50 kg and 15.0 g whose centers are separated by about 4.50 cm. Calculate the gravitational force between these spheres, treating each as a particle located at the sphere's center.

2. During a solar eclipse, the Moon, the Earth, and the Sun all lie on the same line, with the Moon between the Earth and the Sun. (a) What force is exerted by the Sun on the Moon? (b) What force is exerted by the Earth on the Moon? (c) What force is exerted by the Sun on the Earth? (d) Compare the answers to parts (a) and (b). Why doesn't the Sun capture the Moon away from the Earth?

3. Determine the order of magnitude of the gravitational force that you exert on another person 2 m away. In your solution, state the quantities you measure or estimate and their values.

4. *Why is the following situation impossible?* The centers of two homogeneous spheres are 1.00 m apart. The spheres are each made of the same element from the periodic table. The gravitational force between the spheres is 1.00 N.

SECTION 13.2 Free-Fall Acceleration and the Gravitational Force

5. **Review.** Miranda, a satellite of Uranus, is shown in Figure P13.5a. It can be modeled as a sphere of radius 242 km and mass 6.68×10^{19} kg. (a) Find the free-fall acceleration on its surface. (b) A cliff on Miranda is 5.00 km high. It appears on the limb at the 11 o'clock position in Figure P13.5a and is magnified in Figure P13.5b. If a devotee of extreme sports

NASA/JPL

a   b

Figure P13.5

runs horizontally off the top of the cliff at 8.50 m/s, for what time interval is he in flight? (c) How far from the base of the vertical cliff does he strike the icy surface of Miranda? (d) What will be his vector impact velocity?

SECTION 13.3 Analysis Model: Particle in a Field (Gravitational)

6. (a) Compute the vector gravitational field at a point P on the perpendicular bisector of the line joining two objects of equal mass separated by a distance $2a$ as shown in Figure P13.6. (b) Explain physically why the field should approach zero as $r \to 0$. (c) Prove mathematically that the answer to part (a) behaves in this way. (d) Explain physically why the magnitude of the field should approach $2GM/r^2$ as $r \to \infty$. (e) Prove mathematically that the answer to part (a) behaves correctly in this limit.

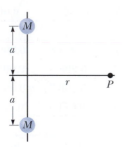

Figure P13.6

7. A spacecraft in the shape of a long cylinder has a length of 100 m, and its mass with occupants is 1 000 kg. It has strayed too close to a black hole having a mass 100 times that of the Sun (Fig. P13.7). The nose of the spacecraft points toward the black hole, and the distance between the nose and the center of the black hole is 10.0 km. (a) Determine the total force on the spacecraft. (b) What is the difference in the gravitational fields acting on the occupants in the nose of the ship and on those in the rear of the ship, farthest from the black hole? (This difference in accelerations grows rapidly as the ship approaches the black hole. It puts the body of the ship under extreme tension and eventually tears it apart.)

Black hole

Figure P13.7

SECTION 13.4 Kepler's Laws and the Motion of Planets

8. An artificial satellite circles the Earth in a circular orbit at a location where the acceleration due to gravity is 9.00 m/s². Determine the orbital period of the satellite.

9. You are out on a date, eating dinner in a restaurant that has several television screens. Most of the screens are showing sports events, but one near you and your date is showing a discussion of an upcoming voyage to Mars. (a) Your date says, "I wonder how long it takes to get to Mars?" Wanting to impress your date, you grab a napkin and draw Figure P13.9 on it. Even more impressively, you tell your date that the minimum-energy transfer orbit from Earth to Mars is an elliptical trajectory with the departure planet corresponding to the perihelion of the ellipse and the arrival planet at

the aphelion. You pull out your smartphone, activate the calculator feature, and perform a calculation on another napkin to answer the question above that your date asked about the transfer time interval to Mars on this particular trajectory. (b) **What If?** Your date is impressed, but then asks you to determine the transit time to an *inner* planet, like Venus.

Figure P13.9

10. A particle of mass m moves along a straight line with constant velocity $\vec{v}_0$ in the x direction, a distance b from the x axis (Fig. P13.10). (a) Does the particle possess any angular momentum about the origin? (b) Explain why the amount of its angular momentum should change or should stay constant. (c) Show that Kepler's second law is satisfied by showing that the two shaded triangles in the figure have the same area when $t_{\text{ⓓ}} - t_{\text{ⓒ}} = t_{\text{ⓑ}} - t_{\text{ⓐ}}$.

Figure P13.10

11. Use Kepler's third law to determine how many days it takes a spacecraft to travel in an elliptical orbit from a point 6 670 km from the Earth's center to the Moon, 385 000 km from the Earth's center.

12. The *Explorer VIII* satellite, placed into orbit November 3, 1960, to investigate the ionosphere, had the following orbit parameters: perigee, 459 km; apogee, 2 289 km (both distances above the Earth's surface); period, 112.7 min. Find the ratio v_p/v_a of the speed at perigee to that at apogee.

13. Suppose the Sun's gravity were switched off. The planets would leave their orbits and fly away in straight lines as described by Newton's first law. (a) Would Mercury ever be farther from the Sun than Pluto? (b) If so, find how long it would take Mercury to achieve this passage. If not, give a convincing argument that Pluto is always farther from the Sun than is Mercury.

14. (a) Given that the period of the Moon's orbit about the Earth is 27.32 days and the nearly constant distance between the center of the Earth and the center of the Moon is 3.84×10^8 m, use Equation 13.11 to calculate the mass of the Earth. (b) Why is the value you calculate a bit too large?

SECTION 13.5 Gravitational Potential Energy

Note: In Problems 15 through 23, assume $U_g = 0$ at $r = \infty$.

15. How much energy is required to move a 1 000-kg object from the Earth's surface to an altitude twice the Earth's radius?

16. An object is released from rest at an altitude h above the surface of the Earth. (a) Show that its speed at a distance r from the Earth's center, where $R_E \le r \le R_E + h$, is

$$v = \sqrt{2GM_E\left(\frac{1}{r} - \frac{1}{R_E + h}\right)}$$

(b) Assume the release altitude is 500 km. Perform the integral

$$\Delta t = \int_i^f dt = -\int_i^f \frac{dr}{v}$$

to find the time of fall as the object moves from the release point to the Earth's surface. The negative sign appears because the object is moving opposite to the radial direction, so its speed is $v = -dr/dt$. Perform the integral numerically.

17. **Q|C** A system consists of three particles, each of mass 5.00 g, located at the corners of an equilateral triangle with sides of 30.0 cm. (a) Calculate the gravitational potential energy of the system. (b) Assume the particles are released simultaneously. Describe the subsequent motion of each. Will any collisions take place? Explain.

SECTION 13.6 Energy Considerations in Planetary and Satellite Motion

18. **S** A "treetop satellite" moves in a circular orbit just above the surface of a planet, assumed to offer no air resistance. Show that its orbital speed v and the escape speed from the planet are related by the expression $v_{esc} = \sqrt{2}v$.

19. A 500-kg satellite is in a circular orbit at an altitude of 500 km above the Earth's surface. Because of air friction, the satellite eventually falls to the Earth's surface, where it hits the ground with a speed of 2.00 km/s. How much energy was transformed into internal energy by means of air friction?

20. **S** Derive an expression for the work required to move an Earth satellite of mass m from a circular orbit of radius $2R_E$ to one of radius $3R_E$.

21. An asteroid is on a collision course with Earth. An astronaut lands on the rock to bury explosive charges that will blow the asteroid apart. Most of the small fragments will miss the Earth, and those that fall into the atmosphere will produce only a beautiful meteor shower. The astronaut finds that the density of the spherical asteroid is equal to the average density of the Earth. To ensure its pulverization, she incorporates into the explosives the rocket fuel and oxidizer intended for her return journey. What maximum radius can the asteroid have for her to be able to leave it entirely simply by jumping straight up? On Earth she can jump to a height of 0.500 m.

22. (a) What is the minimum speed, relative to the Sun, necessary for a spacecraft to escape the solar system if it starts at the Earth's orbit? (b) *Voyager 1* achieved a maximum speed of 125 000 km/h on its way to photograph Jupiter. Beyond what distance from the Sun is this speed sufficient to escape the solar system?

23. **Q|C** Ganymede is the largest of Jupiter's moons. Consider a rocket on the surface of Ganymede, at the point farthest from the planet (Fig. P13.23). Model the rocket as a particle. (a) Does the presence of Ganymede make Jupiter exert a larger, smaller, or same size force on the rocket compared with the force it would exert if Ganymede were not interposed? (b) Determine the escape speed for the rocket from the planet–satellite system. The radius of Ganymede is 2.64×10^6 m, and its mass is 1.495×10^{23} kg. The distance between Jupiter and Ganymede is 1.071×10^9 m, and the mass of Jupiter is 1.90×10^{27} kg. Ignore the motion of Jupiter and Ganymede as they revolve about their center of mass.

Jupiter

Ganymede

Figure P13.23

ADDITIONAL PROBLEMS

24. **Q|C** A rocket is fired straight up through the atmosphere from the South Pole, burning out at an altitude of 250 km when traveling at 6.00 km/s. (a) What maximum distance from the Earth's surface does it travel before falling back to the Earth? (b) Would its maximum distance from the surface be larger if the same rocket were fired with the same fuel load from a launch site on the equator? Why or why not?

25. *Voyager 1* and *Voyager 2* surveyed the surface of Jupiter's moon Io and photographed active volcanoes spewing liquid sulfur to heights of 70 km above the surface of this moon. Find the speed with which the liquid sulfur left the volcano. Io's mass is 8.9×10^{22} kg, and its radius is 1 820 km.

26. **CR** After reading Sections 13.2 and 13.3, your classmate looks online to gather information about neutron stars. He finds that a typical radius of a neutron star is 10 km, a typical mass is two solar masses, and that a typical rotation period is as short as 1.4 ms. He suggests that if a spherical neutron star were spinning that fast, it seems that material at the equator of the sphere would be flung away because the gravity of the star could not supply the needed centripetal acceleration of the material. Prepare an argument that shows that the gravity at the surface of a neutron star is more than sufficient to provide the centripetal acceleration. (*Note:* Neutron stars typically have the same mass as our Sun.)

27. **CR** You are on a space station, in a circular orbit $h = 500$ km above the surface of the Earth. You complete your tasks several days early and must wait for the next mission from the surface to bring you home. After days of boredom, you decide to play some golf. Walking on the space station surface with magnetic shoes, you tee up a golf ball. You hit it with all of your might, sending it off with speed v_{rel}, relative to the space station, in a direction parallel to the velocity vector of the space station at the moment the ball is hit. You notice that you then orbit the Earth exactly $n = 2.00$ times and you reach up and catch the golf ball as it returns to the space station. With what speed v_{rel} was the golf ball hit? *Note:* Your result will be unrealistically high—much higher than it is possible for a human to hit a golf ball.

28. *Why is the following situation impossible?* A spacecraft is launched into a circular orbit around the Earth and circles the Earth once an hour.

29. Let Δg_M represent the difference in the gravitational fields produced by the Moon at the points on the Earth's surface nearest to and farthest from the Moon. Find the fraction $\Delta g_M/g$, where g is the Earth's gravitational field. (This difference is responsible for the occurrence of the *lunar tides* on the Earth.)

30. A sleeping area for a long space voyage consists of two cabins each connected by a cable to a central hub as shown in Figure P13.30. The cabins are set spinning around the hub axis, which is connected to the rest of the spacecraft to generate artificial gravity in the cabins. A space traveler lies in a bed parallel to the outer wall as shown in Figure P13.30. (a) With $r = 10.0$ m, what would the angular speed of the 60.0-kg traveler need to be if he is to experience half his normal Earth weight? (b) If the astronaut stands up perpendicular to the bed, without holding on to anything with his hands, will his head be moving at a faster, a slower, or the same tangential speed as his feet? Why? (c) Why is the action in part (b) dangerous?

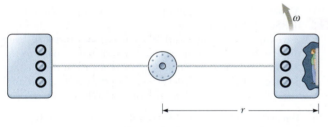

Figure P13.30

31. (a) A space vehicle is launched vertically upward from the Earth's surface with an initial speed of 8.76 km/s, which is less than the escape speed of 11.2 km/s. What maximum height does it attain? (b) A meteoroid falls toward the Earth. It is essentially at rest with respect to the Earth when it is at a height of 2.51×10^7 m above the Earth's surface. With what speed does the meteorite (a meteoroid that survives to impact the Earth's surface) strike the Earth?

32. (a) A space vehicle is launched vertically upward from the Earth's surface with an initial speed of v_i that is comparable to but less than the escape speed v_{esc}. What maximum height does it attain? (b) A meteoroid falls toward the Earth. It is essentially at rest with respect to the Earth when it is at a height h above the Earth's surface. With what speed does the meteorite (a meteoroid that survives to impact the Earth's surface) strike the Earth? (c) **What If?** Assume a baseball is tossed up with an initial speed that is very small compared to the escape speed. Show that the result from part (a) is consistent with Equation 4.19.

33. Assume you are agile enough to run across a horizontal surface at 8.50 m/s, independently of the value of the gravitational field. What would be (a) the radius and (b) the mass of an airless spherical asteroid of uniform density 1.10×10^3 kg/m^3 on which you could launch yourself into orbit by running? (c) What would be your period? (d) Would your running significantly affect the rotation of the asteroid? Explain.

34. Two spheres having masses M and $2M$ and radii R and $3R$, respectively, are simultaneously released from rest when the distance between their centers is $12R$. Assume the two spheres interact only with each other and we wish to find the speeds with which they collide. (a) What *two* isolated system models are appropriate for this system? (b) Write an equation from one of the models and solve it for $\vec{v}_1$, the velocity of the sphere of mass M at any time after release in terms of $\vec{v}_2$, the velocity of $2M$. (c) Write an equation from the other model and solve it for speed v_1 in terms of speed v_2 when the spheres collide. (d) Combine the two equations to find the two speeds v_1 and v_2 when the spheres collide.

35. (a) Show that the rate of change of the free-fall acceleration with vertical position near the Earth's surface is

$$\frac{dg}{dr} = -\frac{2GM_E}{R_E^{\,3}}$$

This rate of change with position is called a *gradient*. (b) Assuming h is small in comparison to the radius of the Earth, show that the difference in free-fall acceleration between two points separated by vertical distance h is

$$|\Delta g| = \frac{2GM_E h}{R_E^{\,3}}$$

(c) Evaluate this difference for $h = 6.00$ m, a typical height for a two-story building.

36. A certain quaternary star system consists of three stars, each of mass m, moving in the same circular orbit of radius r about a central star of mass M. The stars orbit in the same sense and are positioned one-third of a revolution apart from one another. Show that the period of each of the three stars is given by

$$T = 2\pi \sqrt{\frac{r^3}{G(M + m/\sqrt{3})}}$$

37. Studies of the relationship of the Sun to our galaxy—the Milky Way—have revealed that the Sun is located near the outer edge of the galactic disc, about 30 000 ly (1 ly $= 9.46 \times 10^{15}$ m) from the center. The Sun has an orbital speed of approximately 250 km/s around the galactic center. (a) What is the period of the Sun's galactic motion? (b) What is the order of magnitude of the mass of the Milky Way galaxy? (c) Suppose the galaxy is made mostly of stars of which the Sun is typical. What is the order of magnitude of the number of stars in the Milky Way?

38. **Review.** Two identical hard spheres, each of mass m and radius r, are released from rest in otherwise empty space with their centers separated by the distance R. They are allowed to collide under the influence of their gravitational attraction. (a) Show that the magnitude of the impulse received by each sphere before they make contact is given by $[Gm^3(1/2r - 1/R)]^{1/2}$. (b) **What If?** Find the magnitude of the impulse each receives during their contact if they collide elastically.

39. The maximum distance from the Earth to the Sun (at aphelion) is 1.521×10^{11} m, and the distance of closest approach (at perihelion) is 1.471×10^{11} m. The Earth's orbital speed at perihelion is 3.027×10^4 m/s. Determine (a) the Earth's orbital speed at aphelion and the kinetic and potential energies of the Earth–Sun system (b) at

perihelion and (c) at aphelion. (d) Is the total energy of the system constant? Explain. Ignore the effect of the Moon and other planets.

40. Many people assume air resistance acting on a moving
QC object will always make the object slow down. It can, however, actually be responsible for making the object speed up. Consider a 100-kg Earth satellite in a circular orbit at an altitude of 200 km. A small force of air resistance makes the satellite drop into a circular orbit with an altitude of 100 km. (a) Calculate the satellite's initial speed. (b) Calculate its final speed in this process. (c) Calculate the initial energy of the satellite–Earth system. (d) Calculate the final energy of the system. (e) Show that the mechanical energy of the system has decreased and find the amount of the decrease due to friction. (f) What force makes the satellite's speed increase? *Hint:* You will find a free-body diagram useful in explaining your answer.

41. Consider an object of mass m, not necessarily small compared with the mass of the Earth, released at a distance of 1.20×10^7 m from the center of the Earth. Assume the Earth and the object behave as a pair of particles, isolated from the rest of the Universe. (a) Find the magnitude of the acceleration a_{rel} with which each starts to move relative to the other as a function of m. Evaluate the acceleration (b) for $m = 5.00$ kg, (c) for $m = 2\,000$ kg, and (d) for $m = 2.00 \times 10^{24}$ kg. (e) Describe the pattern of variation of a_{rel} with m.

42. Show that the minimum period for a satellite in orbit
S around a spherical planet of uniform density ρ is

$$T_{min} = \sqrt{\frac{3\pi}{G\rho}}$$

independent of the planet's radius.

43. As thermonuclear fusion proceeds in its core, the Sun loses mass at a rate of 3.64×10^9 kg/s. During the 5 000-yr period of recorded history, by how much has the length of the year changed due to the loss of mass from the Sun? *Suggestions:* Assume the Earth's orbit is circular. No external torque acts on the Earth–Sun system, so the angular momentum of the Earth is constant.

44. Two stars of masses M and
S m, separated by a distance d, revolve in circular orbits about their center of mass (Fig. P13.44). Show that each star has a period given by

$$T^2 = \frac{4\pi^2 d^3}{G(M+m)}$$

Figure P13.44

CHALLENGE PROBLEM

45. The Solar and Heliospheric Observatory (SOHO) spacecraft has a special orbit, located between the Earth and the Sun along the line joining them, and it is always close enough to the Earth to transmit data easily. Both objects exert gravitational forces on the observatory. It moves around the Sun in a near-circular orbit that is smaller than the Earth's circular orbit. Its period, however, is not less than 1 yr but just equal to 1 yr. Show that its distance from the Earth must be 1.48×10^9 m. In 1772, Joseph Louis Lagrange determined theoretically the special location allowing this orbit. *Suggestions:* Use data that are precise to four digits. The mass of the Earth is 5.974×10^{24} kg. You will not be able to easily solve the equation you generate; instead, use a computer to verify that 1.48×10^9 m is the correct value.

14 Fluid Mechanics

An airplane takes off from an airport runway. How long does the runway have to be? (*F. JIMENEZ MECA/Shutterstock*)

STORYLINE **It is an academic holiday and you are spending some** time with your grandparents in Denver, Colorado. After visiting with them, your trip continues as you board an airplane to fly from Denver to see your other grandparents in Boston, Massachusetts. As the plane accelerates at the Denver airport, you notice that it is taking quite a while, compared to your previous flying experience, for the plane to leave the ground. You begin to worry that the plane will run out of runway length before it takes off. Finally, the plane lifts off and you breathe a sigh of relief. You think, "Why did it take so long for the plane to lift off? It didn't take that long when I took off for my flight from Los Angeles." Deciding that it would be worth it to pay for the Wi-Fi service on the plane, you look up runway lengths online. The longest runway at the airport at Los Angeles, at sea level, is 12 091 ft. The longest runway at Denver is 16 000 ft. The longest runway in the world is at Qamdo Bamda Airport in China: 18 045 ft. That airport is also at the second highest altitude in the world for an airport: 14 219 ft. (It was the highest until 2013.) Is there a relationship between airport altitude and runway length? Why?

CONNECTIONS In the previous chapters, we have considered the mechanics of particles, systems, and rigid objects. Forces on these particles and objects have been applied by hands, strings, inclined planes, gravity, etc. In this chapter, we consider the forces acting between an object and a *fluid*. A **fluid** is a collection of molecules that are randomly arranged and held together by weak cohesive forces between molecules and also by forces exerted by the container holding the fluid. Both liquids and gases are fluids. We discussed such a situation briefly in Section 6.4, when we considered the resistive forces on objects *moving* through fluids. Here we will discuss forces that fluids exert on objects that are *at rest* relative to the fluid. This discussion will lead to an important new quantity, *pressure,* and a force called the *buoyant force,* which is not a new type of force, but our familiar forces acting in a specific situation. We will also investigate the physics of moving fluids in the later sections of this chapter. Understanding the concepts of moving fluids is important for a wide range of applications, from plumbing systems to automobile aerodynamics to blood flow in veins and arteries.

14.1 Pressure

Fluids do not sustain shearing stresses or tensile stresses such as those discussed in Chapter 12; therefore, the only stress that can be exerted on an object submerged in a static fluid is one that tends to compress the object from all sides. In other words, the force exerted by a static fluid on an object is always perpendicular to the surfaces of the object as shown in Figure 14.1. We discussed this situation in Section 12.4.

The pressure in a fluid can be measured with the device pictured in Figure 14.2. The device consists of an evacuated cylinder that encloses a light piston connected to a spring. As the device is submerged in a fluid, the fluid presses on the piston and compresses the spring until the inward force exerted by the fluid is balanced by the outward force exerted by the spring. The fluid pressure can be measured directly if the spring is calibrated in advance. If F is the magnitude of the force exerted on the piston and A is the surface area of the piston, the **pressure** P of the fluid at the level to which the device has been submerged is defined as the ratio of the force exerted on the piston to its area:

$$P \equiv \frac{F}{A} \tag{14.1}$$

Pressure is a scalar quantity because it is proportional to the magnitude of the force on the piston.

If the pressure varies over an area, the infinitesimal force dF on an infinitesimal surface element of area dA is

$$dF = P \, dA \tag{14.2}$$

where P is the pressure at the location of the area dA. To calculate the total force exerted on a surface of a container, we must integrate Equation 14.2 over the surface.

The units of pressure are newtons per square meter (N/m^2) in the SI system. Another name for the SI unit of pressure is the **pascal** (Pa):

$$1 \text{ Pa} \equiv 1 \text{ N/m}^2 \tag{14.3}$$

For a tactile demonstration of the definition of pressure, hold a tack between your thumb and forefinger, with the point of the tack on your thumb and the head of the tack on your forefinger. Now *gently* press your thumb and forefinger together. Your thumb will begin to feel pain immediately while your forefinger will not. The tack is exerting the same force on both your thumb and forefinger, but the pressure on your thumb is much larger because of the small area over which the force is applied.

QUICK QUIZ 14.1 Suppose you are standing directly behind someone who steps back and accidentally stomps on your foot with the heel of one shoe. Would you be better off if that person were (a) a large, male professional basketball player wearing sneakers or (b) a petite woman wearing spike-heeled shoes?

At any point on the surface of the object, the force exerted by the fluid is perpendicular to the surface of the object.

Figure 14.1 The forces exerted by a fluid on the surfaces of a submerged object.

Figure 14.2 A simple device for measuring the pressure exerted by a fluid.

PITFALL PREVENTION 14.1
Force and Pressure Equations 14.1 and 14.2 make a clear distinction between force and pressure. Another important distinction is that *force is a vector* and *pressure is a scalar.* There is no direction associated with pressure, but the direction of the force associated with the pressure is perpendicular to the surface on which the pressure acts.

Example **14.1** | **The Water Bed**

The mattress of a water bed is 2.00 m long by 2.00 m wide and 30.0 cm deep.

(A) Find the weight of the water in the mattress.

SOLUTION

Conceptualize Think about carrying a jug of water and how heavy it is. Now imagine a sample of water the size of a water bed. We expect the weight to be relatively large.

Categorize This example is a substitution problem.

Find the volume of the water filling the mattress:

$$V = \ell w h$$

Use Equation 1.1 and the density of fresh water (see Table 14.1) to find the weight of the water bed:

$$Mg = (\rho V)g = \rho g \ell w h$$

Substitute numerical values:

$$Mg = (1\,000 \text{ kg/m}^3)(9.80 \text{ m/s}^2)(2.00 \text{ m})(2.00 \text{ m})(0.300 \text{ m})$$
$$= 1.18 \times 10^4 \text{ N}$$

which is approximately 2 650 lb. (A regular bed, including mattress, box spring, and metal frame, weighs approximately 300 lb.) Because this load is so great, it is best to place a water bed in the basement or on a sturdy, well-supported floor.

(B) Find the pressure exerted by the water bed on the floor when the bed rests in its normal position. Assume the entire lower surface of the bed makes contact with the floor.

SOLUTION

When the water bed is in its normal position, the area in contact with the floor is $A = \ell w$. Use Equation 14.1 to find the pressure:

$$P = \frac{Mg}{\ell w} = \frac{1.18 \times 10^4 \text{ N}}{(2.00 \text{ m})(2.00 \text{ m})} = 2.94 \times 10^3 \text{ Pa}$$

WHAT IF? What if the water bed is replaced by a 300-lb regular bed that is supported by four legs? Each leg has a circular cross section of radius 2.00 cm. What pressure does this bed exert on the floor?

Answer The weight of the regular bed is distributed over four circular cross sections at the bottom of the legs. Therefore, the pressure is

$$P = \frac{F}{A} = \frac{mg}{4(\pi r^2)} = \frac{300 \text{ lb}}{4\pi (0.020\,0 \text{ m})^2} \left(\frac{1 \text{ N}}{0.225 \text{ lb}} \right)$$
$$= 2.65 \times 10^5 \text{ Pa}$$

This result is almost 100 times larger than the pressure due to the water bed! The weight of the regular bed, even though it is much less than the weight of the water bed, is applied over the very small area of the four legs. The high pressure on the floor at the feet of a regular bed could cause dents in wood floors or permanently crush carpet pile.

14.2 Variation of Pressure with Depth

As divers well know, water pressure increases with depth, resulting in a feeling of discomfort in the ears of the diver. Likewise, atmospheric pressure decreases with increasing altitude; for this reason, aircraft flying at high altitudes must have pressurized cabins for the comfort of the passengers.

We now show details of how the pressure in a liquid increases with depth. As Equation 1.1 describes, the *density* of a substance is defined as its mass per unit volume; Table 14.1 lists the densities of various substances. These values vary slightly with temperature because the volume of a substance is dependent on temperature (as shown in Chapter 18). Under standard conditions (at 0°C and at atmospheric pressure), the densities of gases are about $\frac{1}{1\,000}$ the densities of solids

TABLE 14.1 Densities of Some Common Substances at Standard Temperature (0°C) and Pressure (Atmospheric)

Substance	ρ (kg/m³)	Substance	ρ (kg/m³)
Air	1.29	Iron	7.86×10^3
Air (at 20°C and		Lead	11.3×10^3
atmospheric pressure)	1.20	Mercury	13.6×10^3
Aluminum	2.70×10^3	Nitrogen gas	1.25
Benzene	0.879×10^3	Oak	0.710×10^3
Brass	8.4×10^3	Osmium	22.6×10^3
Copper	8.92×10^3	Oxygen gas	1.43
Ethyl alcohol	0.806×10^3	Pine	0.373×10^3
Fresh water	1.00×10^3	Platinum	21.4×10^3
Glycerin	1.26×10^3	Seawater	1.03×10^3
Gold	19.3×10^3	Silver	10.5×10^3
Helium gas	1.79×10^{-1}	Tin	7.30×10^3
Hydrogen gas	8.99×10^{-2}	Uranium	19.1×10^3
Ice	0.917×10^3		

and liquids. This difference in densities implies that the average molecular spacing in a gas under these conditions is about ten times greater than that in a solid or liquid.

Now consider a liquid of density ρ at rest as shown in Figure 14.3. We assume ρ is uniform throughout the liquid, which means the liquid is *incompressible*. Let us select a parcel of the liquid contained within an imaginary block of cross-sectional area A extending from depth d to depth $d + h$. The liquid external to our parcel exerts forces at all points on the surface of the parcel, perpendicular to the surface. The pressure exerted by the liquid on the bottom face of the parcel is P, and the pressure on the top face is P_0. Therefore, the upward force exerted by the outside fluid on the bottom of the parcel has a magnitude PA, and the downward force exerted on the top has a magnitude P_0A. The mass of liquid in the parcel is $M = \rho V = \rho Ah$; therefore, the weight of the liquid in the parcel is $Mg = \rho Ahg$. Because the parcel is at rest and remains at rest, it can be modeled as a particle in equilibrium, so that the net force acting on it must be zero. Choosing upward to be the positive y direction, we see that

$$\sum \vec{F} = PA\,\hat{j} - P_0A\,\hat{j} - Mg\,\hat{j} = 0$$

or

$$PA - P_0A - \rho Ahg = 0$$

$$P = P_0 + \rho gh \qquad (14.4)$$

◀ Variation of pressure with depth

That is, the pressure P at a depth h below a point in the liquid at which the pressure is P_0 is greater by an amount ρgh. If the liquid is open to the atmosphere and P_0 is the pressure at the surface of the liquid, then P_0 **is atmospheric pressure.** In our calculations and working of end-of-chapter problems, we usually take atmospheric pressure to be

$$P_0 = 1.00 \text{ atm} = 1.013 \times 10^5 \text{ Pa}$$

Equation 14.4 implies that the pressure is the same at all points having the same depth, independent of the shape of the container.

Because the pressure in a fluid depends on depth and on the value of P_0, any increase in pressure at the surface must be transmitted to every other point in the fluid. This concept was first recognized by French scientist Blaise Pascal (1623–1662) and is called **Pascal's law: a change in the pressure applied to a fluid is transmitted undiminished to every point of the fluid and to the walls of the container.**

◀ Pascal's law

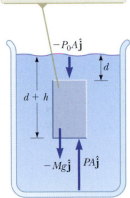

The parcel of fluid is in equilibrium, so the net force on it is zero.

$-P_0A\hat{j}$

d

$d + h$

$-Mg\hat{j}$ $PA\hat{j}$

Figure 14.3 A parcel of fluid in a larger volume of fluid is singled out.

Because the increase in pressure is the same on the two sides, a small force $\vec{\mathbf{F}}_1$ at the left produces a much greater force $\vec{\mathbf{F}}_2$ at the right.

Figure 14.4 (a) Diagram of a hydraulic lift. (b) A vehicle undergoing repair is supported by a hydraulic lift in a garage.

An important application of Pascal's law is the hydraulic lift illustrated in Figure 14.4a. A force of magnitude F_1 is applied to a small piston of surface area A_1. The pressure is transmitted through an incompressible liquid to a larger piston of surface area A_2. Because the pressure must be the same on both sides, $P = F_1/A_1 = F_2/A_2$. Therefore, the force F_2 is greater than the force F_1 by a factor of A_2/A_1. By designing a hydraulic lift with appropriate areas A_1 and A_2, a large output force can be applied by means of a small input force. Hydraulic brakes, car lifts, hydraulic jacks, and forklifts all make use of this principle (Fig. 14.4b).

Because liquid is neither added to nor removed from the system, the volume of liquid pushed down on the left in Figure 14.4a as the piston moves downward through a displacement Δx_1 equals the volume of liquid pushed up on the right as the right piston moves upward through a displacement Δx_2. That is, $A_1 \Delta x_1 = A_2 \Delta x_2$; therefore, $A_2/A_1 = \Delta x_1/\Delta x_2$. We have already shown that $A_2/A_1 = F_2/F_1$. Therefore, $F_2/F_1 = \Delta x_1/\Delta x_2$, so $F_1 \Delta x_1 = F_2 \Delta x_2$. Each side of this equation is the work done by the force on its respective piston. Therefore, the work done by $\vec{\mathbf{F}}_1$ on the input piston equals the work done by $\vec{\mathbf{F}}_2$ on the output piston, as it must to conserve energy. (The process can be modeled as a special case of the nonisolated system model: the *nonisolated system in steady state.* There is energy transfer into and out of the system, but these energy transfers balance, so that there is no net change in the energy of the system.) One could also consider the equation as indicating that you "trade" force for distance. Imagine jacking up a car. You can lift a heavy car with a relatively small force on the jack handle from your hand, but you have to move your hand through a very large total distance when you add up all the times you must move the end of the handle up and down!

Ⓠ UICK QUIZ 14.2 The pressure at the bottom of a filled glass of water ($\rho = 1\ 000\ \text{kg/m}^3$) is P. The water is poured out, and the glass is filled with ethyl alcohol ($\rho = 806\ \text{kg/m}^3$). What is the pressure at the bottom of the glass? **(a)** smaller than P **(b)** equal to P **(c)** larger than P **(d)** indeterminate

Example **14.2** **The Car Lift**

In a car lift used in a service station, compressed air exerts a force on a small piston that has a circular cross section of radius 5.00 cm. This pressure is transmitted by a liquid to a piston that has a radius of 15.0 cm.

(A) What force must the compressed air exert to lift a car weighing 13 300 N?

SOLUTION

Conceptualize Review the material just discussed about Pascal's law to understand the operation of a car lift.

14.2 continued

Categorize This example is a substitution problem.

Solve $F_1/A_1 = F_2/A_2$ for F_1:

$$F_1 = \left(\frac{A_1}{A_2}\right)F_2 = \frac{\pi(5.00 \times 10^{-2}\text{ m})^2}{\pi(15.0 \times 10^{-2}\text{ m})^2}(1.33 \times 10^4\text{ N})$$

$$= \boxed{1.48 \times 10^3\text{ N}}$$

(B) What air pressure produces this force?

SOLUTION

Use Equation 14.1 to find the air pressure that produces this force:

$$P = \frac{F_1}{A_1} = \frac{1.48 \times 10^3\text{ N}}{\pi(5.00 \times 10^{-2}\text{ m})^2}$$

$$= \boxed{1.88 \times 10^5\text{ Pa}}$$

This pressure is approximately twice atmospheric pressure.

Example 14.3 A Pain in Your Ear

Estimate the force exerted on your eardrum due to the water when you are swimming at the bottom of a pool that is 5.0 m deep.

SOLUTION

Conceptualize As you descend in the water, the pressure increases. You may have noticed this increased pressure in your ears while diving in a swimming pool, a lake, or the ocean. We can find the pressure difference exerted on the eardrum from the depth given in the problem; then, after estimating the ear drum's surface area, we can determine the net force the water exerts on it.

Categorize This example is a substitution problem.

The air inside the middle ear is normally at atmospheric pressure P_0. Therefore, to find the net force on the eardrum, we must consider the difference between the total pressure P_{bot} at the bottom of the pool and atmospheric pressure. Let's estimate the surface area of the eardrum to be approximately 1 cm² = 1×10^{-4} m².

Use Equation 14.4 to find this pressure difference:

$$P_{bot} - P_0 = \rho g h$$
$$= (1.00 \times 10^3\text{ kg/m}^3)(9.80\text{ m/s}^2)(5.0\text{ m}) = 4.9 \times 10^4\text{ Pa}$$

Use Equation 14.1 to find the magnitude of the net force on the ear:

$$F = (P_{bot} - P_0)A = (4.9 \times 10^4\text{ Pa})(1 \times 10^{-4}\text{ m}^2) \approx \boxed{5\text{ N}}$$

Because a force of this magnitude on the eardrum is extremely uncomfortable, swimmers often "pop their ears" while under water, an action that pushes air from the lungs into the middle ear. Using this technique equalizes the pressure on the two sides of the eardrum and relieves the discomfort.

Example 14.4 The Force on a Dam

Water is filled to a height H behind a dam of width w (Fig. 14.5, page 364). Determine the resultant force exerted by the water on the dam.

SOLUTION

Conceptualize Because pressure varies with depth, we cannot calculate the force simply by multiplying the area by the pressure. As the pressure in the water increases with depth, the force on the adjacent portion of the dam also increases.

Categorize Because of the variation of pressure with depth, we must use integration to solve this example, so we categorize it as an analysis problem.

continued

14.4 continued

Analyze   Let's imagine a vertical y axis, with $y = 0$ at the bottom of the dam. We divide the face of the dam into narrow horizontal strips at a distance y above the bottom, such as the red strip in Figure 14.5. The pressure on each such strip is due only to the water; atmospheric pressure acts on both sides of the dam.

Use Equation 14.4 to calculate the pressure due to the water at the depth h:

$$P = \rho g h = \rho g (H - y)$$

Use Equation 14.2 to find the force exerted on the shaded strip of area $dA = w\, dy$:

$$dF = P\, dA = \rho g (H - y) w\, dy$$

Integrate to find the total force on the dam:

$$F = \int P\, dA = \int_0^H \rho g (H - y) w\, dy = \tfrac{1}{2}\rho g w H^2$$

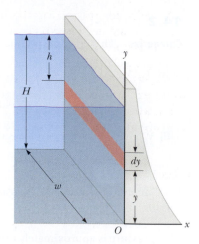

Figure 14.5 (Example 14.4) Water exerts a force on a dam.

Finalize   Notice that the thickness of the dam shown in Figure 14.5 increases with depth. This design accounts for the greater force the water exerts on the dam at greater depths.

WHAT IF?   What if you were asked to find this force without using calculus? How could you determine its value?

Answer   We know from Equation 14.4 that pressure varies linearly with depth. Therefore, the average pressure due to the water over the face of the dam is the average of the pressure at the top and the pressure at the bottom:

$$P_{avg} = \frac{P_{top} + P_{bottom}}{2} = \frac{0 + \rho g H}{2} = \tfrac{1}{2}\rho g H$$

The total force on the dam is equal to the product of the average pressure and the area of the face of the dam:

$$F = P_{avg} A = \left(\tfrac{1}{2}\rho g H\right)(Hw) = \tfrac{1}{2}\rho g w H^2$$

which is the same result we obtained using calculus.

Figure 14.6 Two devices for measuring pressure: (a) a mercury barometer and (b) an open-tube manometer.

14.3   Pressure Measurements

During the weather report on a television news program, the *barometric pressure* is often provided. This reading is the current local pressure of the atmosphere, which varies over a small range from the standard value for P_0 provided earlier. How is this pressure measured?

One instrument used to measure atmospheric pressure is the common barometer, invented by Evangelista Torricelli (1608–1647). A long tube closed at one end is filled with mercury and then inverted into a container of mercury (Fig. 14.6a). The closed end of the tube is nearly a vacuum, so the pressure at the top of the mercury column can be taken as zero. In Figure 14.6a, the pressure at point A, due to the column of mercury, must equal the pressure at point B, due to the atmosphere. If that were not the case, there would be a net force that would move mercury from one point to the other until equilibrium is established. Therefore, $P_0 = \rho_{Hg} g h$, where ρ_{Hg} is the density of the mercury and h is the height of the mercury column. As atmospheric pressure varies, the height of the mercury column varies, so the height can be calibrated to measure atmospheric pressure. Let us determine the height of a mercury column for one atmosphere of pressure, $P_0 = 1$ atm $= 1.013 \times 10^5$ Pa:

$$P_0 = \rho_{Hg} g h \rightarrow h = \frac{P_0}{\rho_{Hg} g} = \frac{1.013 \times 10^5 \text{ Pa}}{(13.6 \times 10^3 \text{ kg/m}^3)(9.80 \text{ m/s}^2)} = 0.760 \text{ m}$$

Based on such a calculation, one atmosphere of pressure is defined to be the pressure equivalent of a column of mercury that is exactly 0.760 0 m in height at 0°C.

A device for measuring the pressure of a gas contained in a vessel is the open-tube manometer illustrated in Figure 14.6b. One end of a U-shaped tube containing a liquid is open to the atmosphere, and the other end is connected to a container of gas at pressure P. In an equilibrium situation, the pressures at points A and B must be the same (otherwise, the curved portion of the liquid between points A and B would experience a net force and would accelerate), and the pressure at A is the unknown pressure of the gas. Therefore, equating the unknown pressure P to the pressure at point B, we see that $P = P_0 + \rho g h$. Again, we can calibrate the height h to the pressure P.

The difference in the pressures in each part of Figure 14.6 (that is, $P - P_0$) is equal to $\rho g h$. The pressure P is called the **absolute pressure,** and the difference $P - P_0$ is called the **gauge pressure.** For example, the pressure you measure in your bicycle tire is gauge pressure, the difference between the absolute pressure of the air inside the tire and the atmospheric pressure outside the tire.

QUICK QUIZ 14.3 Several common barometers are built, with a variety of fluids. For which of the following fluids will the column of fluid in the barometer be the highest? **(a)** mercury **(b)** water **(c)** ethyl alcohol **(d)** benzene

14.4 Buoyant Forces and Archimedes's Principle

Have you ever tried to push a beach ball down under water (Fig. 14.7a)? It is extremely difficult to do because of the large upward force exerted by the water on the ball. The upward force exerted by a fluid on any immersed object is called a **buoyant force.** The buoyant force is what allows huge ships made of steel to float on the surface of the ocean. We can determine the magnitude of a buoyant force by applying some logic. Imagine a beach ball–sized parcel of water beneath the water surface as in Figure 14.7b. Because this parcel can be modeled as a particle in equilibrium, there must be an upward force that balances the downward gravitational force on the parcel. This upward force is the buoyant force, and *its magnitude is equal to the weight of the water in the parcel.* The buoyant force is the resultant force on the parcel due to all forces applied on the parcel by the fluid surrounding the parcel.

Now imagine replacing the beach ball–sized parcel of water with an actual beach ball of the same size. The net force applied to the spherical volume indicated by the dashed line in Figure 14.7b is due to the surrounding fluid and is the same, regardless of whether it is applied to a beach ball or to a parcel of water. Consequently, **the magnitude of the buoyant force on an object always equals the weight of the fluid displaced by the object.** This statement is known as **Archimedes's principle.**

With the beach ball under water, the buoyant force, equal to the weight of a beach ball–sized parcel of water, is much larger than the weight of the beach ball.

Archimedes
Greek Mathematician, Physicist, and Engineer (c. 287–212 BC)
Archimedes was perhaps the greatest scientist of antiquity. He was the first to compute accurately the ratio of a circle's circumference to its diameter, and he also showed how to calculate the volume and surface area of spheres, cylinders, and other geometric shapes. He is well known for discovering the nature of the buoyant force and was also a gifted inventor. One of his practical inventions, still in use today, is Archimedes's screw, an inclined, rotating, coiled tube used originally to lift water from the holds of ships. He also invented the catapult and devised systems of levers, pulleys, and weights for raising heavy loads. Such inventions were successfully used to defend his native city, Syracuse, during a two-year siege by Romans.

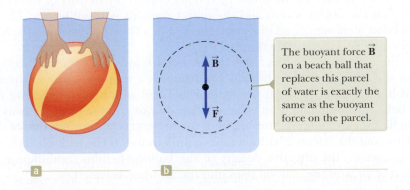

The buoyant force $\vec{\mathbf{B}}$ on a beach ball that replaces this parcel of water is exactly the same as the buoyant force on the parcel.

Figure 14.7 (a) A swimmer pushes a beach ball under water. (b) The forces on a beach ball–sized parcel of water.

The buoyant force on the cube is the resultant of the forces exerted on its top and bottom faces by the liquid.

Figure 14.8 The external forces acting on an immersed cube are the gravitational force $\vec{F}_g$ and the buoyant force $\vec{B}$.

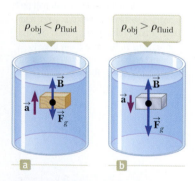

$\rho_{obj} < \rho_{fluid}$ $\rho_{obj} > \rho_{fluid}$

Figure 14.9 (a) A totally submerged object that is less dense than the fluid in which it is submerged experiences a net upward force and rises to the surface after it is released. (b) A totally submerged object that is denser than the fluid experiences a net downward force and sinks.

Therefore, there is a large net upward force on the ball, which explains why it is so hard to hold the beach ball under the water. Note that Archimedes's principle does not refer to the makeup of the object experiencing the buoyant force. The object's composition is not a factor in the buoyant force because the buoyant force is exerted by the surrounding fluid.

To better understand the origin of the buoyant force, consider a cube of solid material immersed in a liquid as in Figure 14.8. According to Equation 14.4, the pressure P_{bot} at the bottom of the cube is greater than the pressure P_{top} at the top by an amount $\rho_{fluid}gh$, where h is the height of the cube and ρ_{fluid} is the density of the fluid. The pressure at the bottom of the cube causes an *upward* force equal to $P_{bot}A$, where A is the area of the bottom face. The pressure at the top of the cube causes a *downward* force equal to $P_{top}A$. The resultant of these two forces is the buoyant force $\vec{B}$ with magnitude

$$B = (P_{bot} - P_{top})A = (\rho_{fluid}gh)A$$

$$B = \rho_{fluid}gV_{disp} \qquad (14.5)$$

where $V_{disp} = Ah$ is the volume of the fluid displaced by the cube. Because the product $\rho_{fluid}V_{disp}$ is equal to the mass of fluid displaced by the object,

$$B = M_{fluid}g$$

where $M_{fluid}g$ is the weight of the fluid displaced by the cube. This result is consistent with our initial statement about Archimedes's principle above, based on the discussion of the beach ball.

Buoyant forces are very important for the movement of fish through water. Under normal conditions, the weight of a fish is slightly greater than the buoyant force on the fish. Hence, the fish would sink if it did not have some mechanism for adjusting the buoyant force. The fish accomplishes that by internally regulating the size of its air-filled swim bladder to increase its volume and the magnitude of the buoyant force acting on it, according to Equation 14.5. In this manner, fish are able to swim to various depths.

Before we proceed with a few examples, it is instructive to discuss two common situations: a totally submerged object and a floating (partly submerged) object.

Case 1: Totally Submerged Object When an object is totally submerged in a fluid of density ρ_{fluid}, the volume V_{disp} of the displaced fluid is equal to the volume V_{obj} of the object; so, from Equation 14.5, the magnitude of the upward buoyant force is $B = \rho_{fluid}gV_{obj}$. If the object has a mass M and density ρ_{obj}, its weight is equal to $F_g = Mg = \rho_{obj}gV_{obj}$, and the net force on the object is $B - F_g = (\rho_{fluid} - \rho_{obj})gV_{obj}$. Hence, if the density of the object is *less* than the density of the fluid, the downward gravitational force is less than the buoyant force and the unsupported object accelerates upward (Fig. 14.9a). A block of wood held under water and released will rise to the surface. If the density of the object is *greater* than the density of the fluid, the upward buoyant force is less than the downward gravitational force and the unsupported object sinks (Fig. 14.9b). A rock will sink to the bottom when released in water. If the density of the submerged object *equals* the density of the fluid, the net force on the object is zero and the object remains in equilibrium. Therefore, the direction of motion of an object submerged in a fluid is determined *only* by the densities of the object and the fluid.

It is important to point out that gases exert buoyant forces also. Imagine a balloon surrounded by air. The balloon displaces a volume of air, so there is an upward buoyant force on it. If the balloon is filled with air, the effective density of the balloon–air combination is larger than that of air, due to the density of the balloon material. Therefore, the weight of the balloon is larger than that of the displaced air, and the released balloon falls to the ground. If the balloon is filled with helium, however, the effective density of the balloon-helium combination is less than that of air, and the balloon rises into the air when released.

Case 2: Floating Object   Now consider the object of volume V_{obj} and density $\rho_{obj} < \rho_{fluid}$ in Figure 14.9a after it reaches the surface. After bobbing a bit, it will settle into static equilibrium on the surface of the fluid: it will *float*, and will be only *partially* submerged (Fig. 14.10). In this case, the object is modeled as a particle in equilibrium: the upward buoyant force is balanced by the downward gravitational force acting on the object. We no longer have $V_{disp} = V_{obj}$ as in Case 1, because only a portion of the object's volume is below the surface of the fluid. If V_{disp} is the volume of the fluid displaced by the object (this volume is the same as the volume of that part of the object beneath the surface of the fluid), the buoyant force has a magnitude $B = \rho_{fluid}gV_{disp}$. Because the weight of the object is $F_g = Mg = \rho_{obj}gV_{obj}$ and because $F_g = B$, we see that $\rho_{fluid}gV_{disp} = \rho_{obj}gV_{obj}$, or

$$\frac{V_{disp}}{V_{obj}} = \frac{\rho_{obj}}{\rho_{fluid}} \tag{14.6}$$

This equation shows that the fraction of the volume of a floating object that is below the fluid surface is equal to the ratio of the density of the object to that of the fluid. For example, the density of ice is less than that of liquid water. Therefore, when an ice cube floats in your water glass or an iceberg floats on the surface of the ocean, part of the ice is below the water surface and part is above. We explore this situation in Example 14.6.

Because the object floats in equilibrium, $B = F_g$.

Figure 14.10 An object floating on the surface of a fluid experiences two forces, the gravitational force $\vec{F}_g$ and the buoyant force $\vec{B}$.

QUICK QUIZ 14.4 You are shipwrecked and floating in the middle of the ocean on a raft. Your cargo on the raft includes a treasure chest full of gold that you found before your ship sank, and the raft is just barely afloat. To keep you floating as high as possible in the water, should you **(a)** leave the treasure chest on top of the raft, **(b)** secure the treasure chest to the underside of the raft, or **(c)** hang the treasure chest in the water with a rope attached to the raft? (Assume throwing the treasure chest overboard is not an option you wish to consider.)

Example 14.5   **Eureka!**

Archimedes supposedly was asked to determine whether a crown made for the king consisted of pure gold. According to legend, he solved this problem by weighing the crown first in air and then in water as shown in Figure 14.11. Suppose the scale read 7.84 N when the crown was in air and 6.84 N when it was in water. What should Archimedes have told the king?

SOLUTION

Conceptualize   Figure 14.11 helps us imagine what is happening in this example. Because of the upward buoyant force on the crown, the scale reading is smaller in Figure 14.11b than in Figure 14.11a.

Categorize   This problem is an example of Case 1 discussed earlier because the crown is completely submerged. The scale reading is a measure of one of the forces on the crown, and the crown is stationary. Therefore, we can categorize the crown as a *particle in equilibrium*.

Analyze   When the crown is suspended in air, the scale reads the true weight $T_1 = F_g$ (neglecting the small buoyant force due to the surrounding air). When the crown is immersed in water, the buoyant force $\vec{B}$ due to the water reduces the scale reading to an *apparent* weight of $T_2 = F_g - B$.

Figure 14.11   (Example 14.5) (a) When the crown is suspended in air, the scale reads its true weight because $T_1 = F_g$ (the buoyancy of air is negligible). (b) When the crown is immersed in water, the buoyant force $\vec{B}$ changes the scale reading to a lower value $T_2 = F_g - B$.

continued

14.5 continued

Apply the particle in equilibrium model to the crown in water:

$$\sum F = B + T_2 - F_g = 0$$

Solve for B:

$$B = F_g - T_2 = m_c g - T_2$$

Because this buoyant force is equal in magnitude to the weight of the displaced water, $B = \rho_w g V_{disp}$, where V_{disp} is the volume of the displaced water and ρ_w is its density. Also, the volume of the crown V_c is equal to the volume of the displaced water because the crown is completely submerged, so $B = \rho_w g V_c$.

Find the density of the crown from Equation 1.1:

$$\rho_c = \frac{m_c}{V_c} = \frac{m_c g}{V_c g} = \frac{m_c g}{(B/\rho_w)} = \frac{m_c g \rho_w}{B} = \frac{m_c g \rho_w}{F_g - T_2} = \frac{m_c g \rho_w}{m_c g - T_2}$$

Substitute numerical values:

$$\rho_c = \frac{(7.84 \text{ N})(1\,000 \text{ kg/m}^3)}{7.84 \text{ N} - 6.84 \text{ N}} = 7.84 \times 10^3 \text{ kg/m}^3$$

Finalize From Table 14.1, we see that the density of gold is 19.3×10^3 kg/m^3. Therefore, Archimedes should have reported that the king had been cheated. Either the crown was hollow, or it was not made of pure gold.

WHAT IF? Suppose the crown has the same weight but is indeed pure gold and not hollow. What would the scale reading be when the crown is immersed in water?

Answer Find the buoyant force on the crown:

$$B = \rho_w g V_{disp} = \rho_w g V_c = \rho_w g \left(\frac{m_c}{\rho_c} \right) = \rho_w \left(\frac{m_c g}{\rho_c} \right)$$

Substitute numerical values:

$$B = (1.00 \times 10^3 \text{ kg/m}^3) \frac{7.84 \text{ N}}{19.3 \times 10^3 \text{ kg/m}^3} = 0.406 \text{ N}$$

Find the tension in the string hanging from the scale:

$$T_2 = m_c g - B = 7.84 \text{ N} - 0.406 \text{ N} = 7.43 \text{ N}$$

Example **14.6** A Titanic Surprise

An iceberg floating in seawater as shown in Figure 14.12a is extremely dangerous because most of the ice is below the surface. This hidden ice can damage a ship that is still a considerable distance from the visible ice. What fraction of the iceberg lies below the water level?

SOLUTION

Conceptualize You are likely familiar with the phrase, "That's only the tip of the iceberg." The origin of this popular saying is that most of the volume of a floating iceberg is beneath the surface of the water (Fig. 14.12b).

Categorize This example corresponds to Case 2 because only part of the iceberg is underneath the water. It is also a simple substitution problem involving Equation 14.6.

Figure 14.12 (Example 14.6) (a) Much of the volume of this iceberg is beneath the water. (b) A ship can be damaged even when it is not near the visible ice.

Evaluate Equation 14.6 using the densities of ice and seawater (Table 14.1):

$$f = \frac{V_{disp}}{V_{ice}} = \frac{\rho_{ice}}{\rho_{seawater}} = \frac{917 \text{ kg/m}^3}{1\,030 \text{ kg/m}^3} = \boxed{0.890 \text{ or } 89.0\%}$$

Therefore, the visible fraction of ice above the water's surface is about 11%. It is the unseen 89% below the water that represents the danger to a passing ship.

14.5 Fluid Dynamics

Thus far, our study of fluids has been restricted to fluids at rest. We now turn our attention to moving fluids. When fluid is in motion, its flow can be characterized as being one of two main types. The flow is said to be **steady,** or **laminar,** if each particle of the fluid follows a smooth path such that the paths of different particles never cross each other as shown in Figure 14.13. Laminar flow is predictable. If you determine the velocity vector of a fluid particle arriving at a certain position in space, every other particle arriving at that same position afterward will have the same velocity.

Above a certain critical speed, fluid flow becomes **turbulent.** Turbulent flow is irregular, unpredictable flow characterized by small whirlpool-like regions as shown in Figure 14.14.

The term *viscosity* is commonly used in the description of fluid flow to characterize the degree of internal friction in the fluid. This internal friction, or *viscous force,* is associated with the resistance that two adjacent layers of fluid have to moving relative to each other. Viscosity causes part of the fluid's kinetic energy to be transformed to internal energy. This mechanism is similar to the one by which the kinetic energy of an object sliding over a rough, horizontal surface decreases as discussed in Sections 8.3 and 8.4. We will address more details on viscosity in Section 14.7.

Because the motion of real fluids is very complex and not fully understood, we make some simplifying assumptions in our approach. In our simplification model of **ideal fluid flow,** we make the following four assumptions:

1. **The fluid is nonviscous.** In a nonviscous fluid, internal friction is neglected. An object moving through the fluid experiences no viscous force.
2. **The flow is laminar.** In laminar flow, all particles passing through a point have the same velocity and follow the same path.
3. **The fluid is incompressible.** The density of an incompressible fluid is the same throughout the fluid.
4. **The flow is irrotational.** In irrotational flow, the fluid has no angular momentum about any point. If a small paddle wheel placed anywhere in the fluid does not rotate about the wheel's center of mass, the flow is irrotational.

The path taken by a fluid particle under laminar flow is called a **streamline.** The velocity of the particle is always tangent to the streamline as shown in Figure 14.15. A set of streamlines like the ones shown in Figure 14.15 form a *tube of flow.* Fluid particles cannot flow into or out of the sides of this tube; if they could, the streamlines would cross one another.

Consider ideal fluid flow through a section of pipe of nonuniform size as illustrated in Figure 14.16. Let's focus our attention on a segment of fluid in the pipe. Figure 14.16a shows the segment at time $t = 0$ consisting of the gray portion between point 1 and point 2 and the short blue portion to the left of point 1. At this time, the fluid in the short blue portion is flowing through a cross section of area A_1 at speed v_1. During the time interval Δt, the small length Δx_1 of fluid in the blue portion moves into the section of pipe past point 1. During the same time interval, fluid at the right end of the segment moves out of the section of pipe past point 2.

Figure 14.13 Laminar flow of smoke over an automobile in a test wind tunnel.

Figure 14.14 Hot gases from a cigarette made visible by smoke particles. The smoke first moves in laminar flow at the bottom and then in turbulent flow above.

At $t = 0$, fluid in the blue portion is moving past point 1 at velocity $\vec{v}_1$.

After a time interval Δt, the fluid in the blue portion is moving past point 2 at velocity $\vec{v}_2$.

At each point along its path, the particle's velocity is tangent to the streamline.

Figure 14.15 A particle in laminar flow follows a streamline.

Figure 14.16 A fluid moving with steady flow through a pipe of varying cross-sectional area. (a) At $t = 0$, the small blue-colored portion of the fluid at the left is moving into the section of pipe through area A_1. (b) After a time interval Δt, the blue-colored portion shown here is that fluid that has moved out of the section of pipe through area A_2.

Figure 14.17 The speed of water spraying from the end of a garden hose increases as the size of the opening is decreased with the thumb.

© Cengage

Figure 14.16b shows the situation at the end of the time interval Δt. The blue portion at the right end represents the fluid that was originally in the pipe and has moved past point 2 through an area A_2 at a speed v_2.

The mass of fluid contained in the blue portion in Figure 14.16a is given by $m_1 = \rho A_1 \, \Delta x_1 = \rho A_1 v_1 \, \Delta t$, where ρ is the (unchanging) density of the ideal fluid. Similarly, the fluid in the blue portion in Figure 14.16b has a mass $m_2 = \rho A_2 \, \Delta x_2 = \rho A_2 v_2 \, \Delta t$. Because the fluid is incompressible and the flow is laminar, however, the mass of fluid that passes point 1 in a time interval Δt must equal the mass that passes point 2 in the same time interval. That is, $m_1 = m_2$ or $\rho A_1 v_1 \, \Delta t = \rho A_2 v_2 \, \Delta t$, which means that

◀ Equation of Continuity for Fluids

$$A_1 v_1 = A_2 v_2 = \text{constant} \tag{14.7}$$

This expression is called the **equation of continuity for fluids.** It states that the product of the area and the fluid speed at all points along a pipe is constant for an incompressible fluid. Equation 14.7 shows that the speed is high where the tube is constricted (small A) and low where the tube is wide (large A). The product Av, which has the dimensions of volume per unit time, is called either the *volume flux* or the *flow rate*. The condition $Av = $ constant is equivalent to the statement that the volume of fluid that enters one end of a tube in a given time interval equals the volume leaving the other end of the tube in the same time interval if no leaks are present.

You demonstrate the equation of continuity each time you water your garden with your thumb over the end of a garden hose as in Figure 14.17. By partially blocking the opening with your thumb, you reduce the cross-sectional area through which the water passes. As a result, the speed of the water increases as it exits the hose, and the water can be sprayed over a long distance.

Example 14.7 | **Watering a Garden**

A gardener uses a water hose to fill a 30.0-L bucket. The gardener notes that it takes 1.00 min to fill the bucket. A nozzle with an opening of cross-sectional area 0.500 cm^2 is then attached to the hose. The nozzle is held so that water is projected horizontally from a point 1.00 m above the ground. Over what horizontal distance can the water be projected?

SOLUTION

Conceptualize Imagine any past experience you have with projecting water from a horizontal hose or a pipe using either your thumb or a nozzle, which can be attached to the end of the hose. The faster the water is traveling as it leaves the hose, the farther it will land on the ground from the end of the hose.

Categorize Once the water leaves the hose, it is in free fall. Therefore, we categorize a given element of the water as a projectile. The element is modeled as a *particle under constant acceleration* (due to gravity) in the vertical direction and a *particle under constant velocity* in the horizontal direction. The horizontal distance over which the element is projected depends on the speed with which it is projected. This example involves a change in area for the pipe, so we also categorize it as one in which we use the continuity equation for fluids.

Analyze

Express the volume flow rate I_V in terms of area and speed of the water in the hose (we will discuss the origin for this notation in Section 14.7):

$$I_V = A_1 v_1$$

14.7 continued

Solve for the speed of the water in the hose:

$$v_1 = \frac{I_V}{A_1}$$

We have labeled this speed v_1 because we identify point 1 within the hose. We identify point 2 in the air just outside the nozzle. We must find the speed $v_2 = v_{xi}$ with which the water exits the nozzle (v_2) and begins its projectile motion (v_{xi}). The subscript i anticipates that it will be the *initial* velocity component of the water projected from the hose, and the subscript x indicates that the initial velocity vector of the projected water is horizontal.

Solve the continuity equation for fluids for v_2:

$$(1) \quad v_2 = v_{xi} = \frac{A_1}{A_2} v_1 = \frac{A_1}{A_2}\left(\frac{I_V}{A_1}\right) = \frac{I_V}{A_2}$$

We now shift our thinking away from fluids and to projectile motion. In the vertical direction, an element of the water starts from rest and falls through a vertical distance of 1.00 m.

Write Equation 2.16 for the vertical position of an element of water, modeled as a particle under constant acceleration:

$$y_f = y_i + v_{yi}t - \tfrac{1}{2}gt^2$$

Identify the origin as the initial position of the water as it leaves the hose, and recognize that the water begins with a vertical velocity component of zero. Solve for the time at which the water reaches the ground:

$$(2) \quad y_f = 0 + 0 - \tfrac{1}{2}gt^2 \quad \rightarrow \quad t = \sqrt{\frac{-2y_f}{g}}$$

Use Equation 2.7 to find the horizontal position of the element at this time, modeled as a particle under constant velocity:

$$x_f = x_i + v_{xi}t = 0 + v_2 t = v_2 t$$

Substitute from Equations (1) and (2):

$$x_f = \frac{I_V}{A_2}\sqrt{\frac{-2y_f}{g}}$$

Substitute numerical values:

$$x_f = \frac{30.0 \text{ L/min}}{0.500 \text{ cm}^2}\sqrt{\frac{-2(-1.00 \text{ m})}{9.80 \text{ m/s}^2}}\left(\frac{10^3 \text{ cm}^3}{1 \text{ L}}\right)\left(\frac{1 \text{ min}}{60 \text{ s}}\right) = 452 \text{ cm} = 4.52 \text{ m}$$

Finalize   The time interval for the element of water to fall to the ground is unchanged if the projection speed is changed because the projection is horizontal. Increasing the projection speed results in the water hitting the ground farther from the end of the hose, but requires the same time interval to strike the ground.

14.6   Bernoulli's Equation

You have probably experienced driving on a highway and having a large truck pass you at high speed. In this situation, you may have had the frightening feeling that your car was being pulled in toward the truck as it passed. We will investigate the origin of this effect in this section.

As a fluid moves through a region where its speed or elevation above the Earth's surface changes, the pressure in the fluid varies with these changes. The relationship between fluid speed, pressure, and elevation was first derived in 1738 by Swiss physicist Daniel Bernoulli. Consider the flow of a segment of an ideal fluid through a non-uniform section of pipe in a time interval Δt as illustrated in Figure 14.18 (page 372). This figure is very similar to Figure 14.16, which we used to develop the continuity equation. We have added two features: the forces on the outer ends of the blue portions of fluid and the heights of these portions above the reference position $y = 0$.

The force exerted on the segment by the fluid to the left of the blue portion in Figure 14.18a has a magnitude $P_1 A_1$. During a time interval Δt, the point of application of this force moves through a displacement of magnitude Δx_1, as the blue portion of fluid enters the section of pipe past point 1. The work done by this force on the segment in a time interval Δt is $W_1 = F_1 \Delta x_1 = P_1 A_1 \Delta x_1 = P_1 V$, where V is the volume of the blue portion of fluid passing point 1 in Figure 14.18a. In a similar manner, the work done on the segment by the fluid to the right of the segment in the same time interval Δt (Fig. 14.18b) is $W_2 = -P_2 A_2 \Delta x_2 = -P_2 V$, where V is the

© iStockphoto.com/ZU_09

Daniel Bernoulli
Swiss physicist (1700–1782)
Bernoulli made important discoveries in fluid dynamics. Bernoulli's most famous work, *Hydrodynamica*, was published in 1738; it is both a theoretical and a practical study of equilibrium, pressure, and speed in fluids. He showed that as the speed of a fluid increases, its pressure decreases. Referred to as "Bernoulli's principle," Bernoulli's work is used to produce a partial vacuum in chemical laboratories by connecting a vessel to a tube through which water is running rapidly.

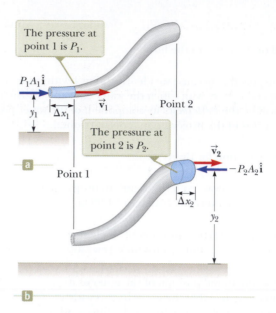

Figure 14.18 A fluid in laminar flow through a section of pipe. (a) A segment of the fluid at time $t = 0$. A small portion of the blue-colored fluid is at height y_1 above a reference position and is entering the section of pipe. (b) After a time interval Δt, the entire segment has moved to the right. The blue-colored portion of the fluid is that which has left the section of pipe at point 2 and is at height y_2.

volume of the blue portion of fluid passing point 2. (The volumes of the blue portions of fluid in Figures 14.18a and 14.18b are equal because the fluid is incompressible.) This work is negative because the force on the segment of fluid is to the left and the displacement of the point of application of the force is to the right. Therefore, the net work done on the segment by these forces in the time interval Δt is

$$W = (P_1 - P_2)V \tag{14.8}$$

Part of this work goes into changing the kinetic energy of the segment of fluid, and part goes into changing the gravitational potential energy of the segment–Earth system. The appropriate reduction of Equation 8.2 for the nonisolated system of the Earth and the segment of fluid is

$$\Delta K + \Delta U_g = W \tag{14.9}$$

Because we are assuming laminar flow, the kinetic energy K_{gray} of the gray portion of the segment is the same in both parts of Figure 14.18. Therefore, the change in the kinetic energy of the segment of fluid is

$$\Delta K = \left(\tfrac{1}{2}mv_2^2 + K_{gray}\right) - \left(\tfrac{1}{2}mv_1^2 + K_{gray}\right) = \tfrac{1}{2}mv_2^2 - \tfrac{1}{2}mv_1^2 \tag{14.10}$$

where m is the mass of the blue portions of fluid in both parts of Figure 14.18. (Because the volumes of both portions are the same, they also have the same mass.)

Considering the gravitational potential energy of the segment–Earth system, once again there is no change during the time interval for the gravitational potential energy U_{gray} associated with the gray portion of the fluid. Consequently, the change in gravitational potential energy of the system is

$$\Delta U_g = (mgy_2 + U_{gray}) - (mgy_1 + U_{gray}) = mgy_2 - mgy_1 \tag{14.11}$$

Substituting Equations 14.8, 14.10. and 14.11 into Equation 14.9 gives

$$\left(\tfrac{1}{2}mv_2^2 - \tfrac{1}{2}mv_1^2\right) - (mgy_2 - mgy_1) = (P_1 - P_2)V$$

If we divide each term by the blue portion volume V and recall that $\rho = m/V$, this expression reduces to

$$\tfrac{1}{2}\rho v_2^2 - \tfrac{1}{2}\rho v_1^2 + \rho gy_2 - \rho gy_1 = P_1 - P_2$$

Rearranging terms gives

$$P_1 + \tfrac{1}{2}\rho v_1^2 + \rho gy_1 = P_2 + \tfrac{1}{2}\rho v_2^2 + \rho gy_2 \tag{14.12}$$

which is **Bernoulli's equation** as applied to an ideal fluid. This equation is often expressed as

$$P + \tfrac{1}{2}\rho v^2 + \rho g y = \text{constant} \qquad (14.13)$$

◀ Bernoulli's equation

Bernoulli's equation shows that the pressure of a fluid decreases as the speed of the fluid increases. In addition, the pressure decreases as the elevation increases. This latter point explains why water pressure from faucets on the upper floors of a tall building is weak unless measures are taken to provide higher pressure for these upper floors.

When the fluid is at rest, $v_1 = v_2 = 0$ and Equation 14.12 becomes

$$P_1 - P_2 = \rho g(y_2 - y_1) = \rho g h$$

This result is in agreement with Equation 14.4.

Although Equation 14.13 was derived for an incompressible fluid, the general behavior of pressure with speed is true even for gases: as the speed increases, the pressure decreases. This *Bernoulli effect* explains the experience with the truck on the highway at the opening of this section. As air passes between you and the truck, it must pass through a relatively narrow channel. According to the continuity equation for fluids, the speed of the air in this channel is higher than that of the air on the other side of your car. According to the Bernoulli effect, this higher-speed air exerts less pressure on your car than the air on the other side. Therefore, there is a net force pushing you toward the truck!

Ⓠ UICK QUIZ 14.5 You observe two helium balloons floating next to each other at the ends of strings secured to a table. The facing surfaces of the balloons are separated by 1–2 cm. You blow through the small space between the balloons. What happens to the balloons? **(a)** They move toward each other. **(b)** They move away from each other. **(c)** They are unaffected.

Example 14.8 **The Venturi Tube**

The horizontal constricted pipe illustrated in Figure 14.19, known as a *Venturi tube,* can be used to measure the flow speed of an incompressible fluid. Determine the flow speed at point 2 of Figure 14.19a if the pressure difference $P_1 - P_2$ is known.

SOLUTION

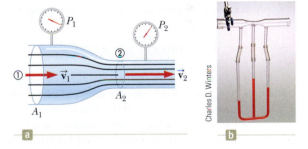

Figure 14.19 (Example 14.8) (a) Pressure P_1 is greater than pressure P_2 because $v_1 < v_2$. This device can be used to measure the speed of fluid flow. (b) A Venturi tube, located at the top of the photograph. Air is blown through the tube from the left. The higher level of fluid in the middle column shows that the pressure of the moving air at the top of that column, which is in the constricted region of the Venturi tube, is lower.

Conceptualize Bernoulli's equation shows how the pressure of an ideal fluid decreases as its speed increases. Therefore, we should be able to calibrate a device to give us the fluid speed if we can measure pressure.

Categorize Because the problem states that the fluid is incompressible, we can categorize it as one in which we can use the equation of continuity for fluids and Bernoulli's equation.

Analyze Apply Equation 14.12 to points 1 and 2, noting that $y_1 = y_2$ because the pipe is horizontal:

$$(1) \quad P_1 + \tfrac{1}{2}\rho v_1^2 = P_2 + \tfrac{1}{2}\rho v_2^2$$

Solve the equation of continuity for v_1:

$$v_1 = \frac{A_2}{A_1} v_2$$

Substitute this expression into Equation (1):

$$P_1 + \tfrac{1}{2}\rho \left(\frac{A_2}{A_1}\right)^2 v_2^2 = P_2 + \tfrac{1}{2}\rho v_2^2$$

Solve for v_2:

$$v_2 = A_1 \sqrt{\frac{2(P_1 - P_2)}{\rho(A_1^2 - A_2^2)}}$$

continued

14.8 continued

Finalize From the design of the tube (areas A_1 and A_2) and measurements of the pressure difference $P_1 - P_2$, we can calculate the speed of the fluid with this equation. To see the relationship between fluid speed and pressure difference, place two empty soda cans on their sides about 2 cm apart on a table. Gently blow a stream of air horizontally between the cans and watch them roll together slowly due to a modest pressure difference between the stagnant air on their outside edges and the moving air between them. Now blow more strongly and watch the increased pressure difference move the cans together more rapidly.

Example **14.9** Torricelli's Law

An enclosed tank containing a liquid of density ρ has a hole in its side at a distance y_1 from the tank's bottom (Fig. 14.20). The hole is open to the atmosphere, and its diameter is much smaller than the diameter of the tank. The air above the liquid is maintained at a pressure P. Determine the speed of the liquid as it leaves the hole when the liquid's level is a distance h above the hole.

Figure 14.20 (Example 14.9) A liquid leaves a hole in a tank at speed v_1.

SOLUTION

Conceptualize Imagine that the tank is a fire extinguisher. When the hole is opened, liquid leaves the hole with a certain speed. If the pressure P at the top of the liquid is increased, the liquid leaves with a higher speed. If the pressure P falls too low, the liquid leaves with a low speed and the extinguisher must be replaced. Because $A_2 \gg A_1$, the liquid is approximately at rest at the top of the tank, where the pressure is P, so $v_2 = 0$. At the hole, the liquid is open to the external atmosphere, so P_1 is equal to atmospheric pressure P_0.

Categorize Looking at Figure 14.20, we know the pressure at two points and the velocity at point 2. We wish to find the velocity at point 1. Therefore, we can categorize this example as one in which we can apply Bernoulli's equation.

Analyze

Apply Bernoulli's equation between points 1 and 2:
$$P_0 + \tfrac{1}{2}\rho v_1^2 + \rho g y_1 = P + \rho g y_2$$

Solve for v_1, noting that $y_2 - y_1 = h$:
$$v_1 = \sqrt{\frac{2(P - P_0)}{\rho} + 2gh}$$

Finalize

When P is much greater than P_0 (so that the term $2gh$ can be neglected), the exit speed of the water is mainly a function of P. If the tank is open at the top to the atmosphere, then $P = P_0$ and $v_1 = \sqrt{2gh}$. In other words, for an open tank, the speed of the liquid leaving a hole a distance h below the surface is equal to that acquired by an object falling freely through a vertical distance h. This phenomenon is known as *Torricelli's law*.

WHAT IF? What if the position of the hole in Figure 14.20 could be adjusted vertically? If the top of the tank is open to the atmosphere and sitting on a table, what position of the hole would cause the water to land on the table at the farthest distance from the tank?

Answer Model a parcel of water exiting the hole as a projectile. From the *particle under constant acceleration* model, find the time at which the parcel strikes the table from a hole at an arbitrary position y_1:
$$y_f = y_i + v_{yi}t - \tfrac{1}{2}gt^2$$
$$0 = y_1 + 0 - \tfrac{1}{2}gt^2$$
$$t = \sqrt{\frac{2y_1}{g}}$$

From the *particle under constant velocity* model, find the horizontal position of the parcel at the time it strikes the table:
$$x_f = x_i + v_{xi}t = 0 + \sqrt{2g(y_2 - y_1)}\ \sqrt{\frac{2y_1}{g}}$$
$$= 2\sqrt{(y_2 y_1 - y_1^2)}$$

Maximize the horizontal position by taking the derivative of x_f with respect to y_1 (because y_1, the height of the hole, is the variable that can be adjusted) and setting it equal to zero:
$$\frac{dx_f}{dy_1} = \tfrac{1}{2}(2)(y_2 y_1 - y_1^2)^{-1/2}(y_2 - 2y_1) = 0$$

Solve for y_1:
$$y_1 = \tfrac{1}{2}y_2$$

14.9 continued

Therefore, to maximize the horizontal distance, the hole should be halfway between the bottom of the tank and the upper surface of the water. Below this location, the water is projected at a higher speed but falls for a short time interval, reducing the horizontal range. Above this point, the water is in the air for a longer time interval but is projected with a smaller horizontal speed.

14.7 Flow of Viscous Fluids in Pipes

In Section 14.5, we discussed the flow of an ideal fluid. The results obtained there and in Section 14.6 are applicable to many situations. On the other hand, there are other situations in which we must investigate the flow of real, non-idealized fluids.

As an example, consider the flow of a fluid in a closed pipe, such as water in a plumbing system or blood in a human circulatory system. According to Bernoulli's principle, if the pipe were of uniform cross section, the pressure difference between two locations in a horizontal section of the pipe would be zero. Therefore, once set in motion, the fluid would flow without any external influence. This is not true in reality. If this were true, why would humans need hearts to continuously pump the blood?

In a real situation, the pressure differential to keep fluid moving at a fixed speed in a horizontal pipe is given by

$$\Delta P = I_V R \tag{14.14}$$

In this equation, I_V represents the volume rate of fluid flow in m^3/s. This quantity is equal to the product Av in Equation 14.7, as mentioned in Example 14.7. The parameter R is a measure of the resistance of the system to the movement of fluid in the pipe.

The notation I_V may appear odd, but it is chosen in order to make a comparison with a similar equation in electricity that we will see in Chapter 26:

$$\Delta V = IR \tag{14.15}$$

In this equation ΔV is an electric potential difference, which represents the external influence that attempts to move electrons through a wire. The quantity I is the current, representing the flow of electrons in the wire, and R is the resistance of the flow of those electrons through the wire. Compare this equation to Equation 14.14, in which ΔP represents a pressure difference that attempts to move fluid through a pipe. The quantity I_V represents the flow of fluid in the pipe, and R is the resistance of the flow of that fluid through the pipe. Equations 14.14 and 14.15 are both types of *transport equations,* in which an entity attempts to move something through space and encounters resistance to the effort. We will see a similar situation in Chapter 19, where a temperature difference attempts to move energy through a material by heat, and encounters resistance based on how good a thermal insulator the material is. We could even reverse the variables on the right side of Equation 5.2 to cast it in a form to compare with Equations 14.14 and 14.15:

$$\sum F = ma = am$$

Here, a net force on the left attempts to move an object through space, measured by its acceleration, and encounters resistance in the form of the mass of the object. Other transport situations also exist, such as a concentration difference driving a diffusion of molecules through another substance.

Now, what determines the resistance R for the fluid flow? One contribution to the physical origin of the resistance is the viscous resistive forces (Section 6.4) between the fluid and the inner wall of the pipe and between the layers of fluid that may be moving at different speeds relative to one another. To begin to evaluate the effect of these viscous forces, consider Figure 14.21 (page 376), which shows a layer of fluid of thickness h. Initially, the visible side of the layer forms a rectangle $ABCD$.

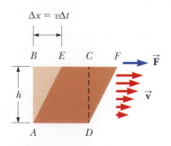

$\Delta x = v\Delta t$

Figure 14.21 The lower surface of a layer of liquid is held fixed while a force is applied to the upper surface. As a result, the layer deforms in shape.

Figure 14.22 Flow of a viscous fluid in a pipe. The red velocity vectors show the variation in speed of the fluid across a diameter of the pipe. The fluid flows fastest at the center and slowly at the walls of the pipe.

At $t = 0$, a force of magnitude F is applied to the right on the upper surface while the lower surface remains fixed. After a time interval Δt, the visible side of the layer now forms the parallelogram *AEFD*. The top surface of the liquid is moving with speed v, while lower portions of the layer move with progressively lower speeds.

Compare Figure 14.21 with Figure 12.13, in which we apply a shearing force to a solid material, and notice the similarities in the two situations. Based on this comparison, let us modify Equation 12.7 to fit the current circumstances. Solve Equation 12.7 for the force on the upper surface:

$$F = SA\frac{\Delta x}{h} \tag{14.16}$$

where S is the shear modulus of the material. This equation represents the deformation of a solid, but let us imagine it representing the deformation of the viscous liquid in Figure 14.21, with the top surface moving at speed v. The deformation behavior is the same, although we will ultimately replace the shear modulus with another parameter. Therefore, we can write Equation 14.16 as a proportionality:

$$F \propto A\frac{\Delta x}{h} \tag{14.17}$$

For a given time interval, Δx is proportional to the speed v of the upper surface, so we can write

$$F \propto A\frac{v}{h}$$

We can turn this inequality to an equality by introducing a proportionality constant η:

$$F = \eta A\frac{v}{h} \tag{14.18}$$

The constant η is called the **viscosity** of the fluid and has units of $N \cdot s/m^2 = Pa \cdot s$. Another common unit of viscosity is the *poise* (P), where $1\ Pa \cdot s = 10\ P$. Table 14.2 lists viscosities of some fluids. Notice that a "thick" fluid such as honey has a high viscosity, while fluids such as water and air have lower viscosity values.

Looking again at our representation of a viscous fluid in Figure 14.21, the speed of the bottom surface is zero and the speed of subsequent portions higher toward the top increases, with the highest surface having the highest speed. Applying this notion to the flow of fluid in a pipe, we find that, because of the viscous force between layers of fluid, the flow of a fluid in a pipe is not uniform across the area of the pipe. Figure 14.22 shows that the speed of the fluid is greatest at the center of the pipe and approaches zero at the pipe walls.

TABLE 14.2 Viscosities of Various Fluids[a]

Fluid	Viscosity (mPa · s)
Air	0.018
Helium	0.020
Liquid nitrogen (−196°C)	0.158
Acetone	0.306
Water	0.894
Ethanol	1.07
Blood (37.0°C)	2.70
Olive oil	81
Motor oil (SAE 40, 20°C)	319
Corn syrup	1 381
Glycerin	1 500
Honey[b]	2 000–10 000
Peanut butter	250 000

[a]All values at 25.0°C unless noted otherwise.
[b]Value depends on moisture content.

Let us return again to Equation 14.14. What is it that determines the resistance R to the flow of the fluid in the pipe? Clearly, the viscosity plays a role, but there are other factors. It can be shown that the resistance of the fluid in the segment of pipe in Figure 14.22 of length L and radius r is given by

$$R = \frac{8\eta L}{\pi r^4} \qquad (14.19)$$

Therefore, Equation 14.14 becomes

$$\Delta P = \frac{8\eta L}{\pi r^4} I_V \qquad (14.20)$$

◀ Poiseuille's Law (Hagen–Poiseuille equation)

This equation is known as **Poiseuille's law,** or the **Hagen–Poiseuille equation.**[1] Notice the important dependence of the pressure difference on r: the pressure difference is inversely proportional to r to the fourth power. Therefore, if the radius of the pipe drops by 50%, the pressure difference required to maintain the same flow through the pipe increases by a factor of 16.

This dependence is very important in blood flow in the human circulatory system. If a blood vessel becomes occluded by plaque so that the radius through which blood can flow is decreased, the pressure required to maintain the blood flow rises rapidly. Conversely, for a given pressure, the flow rate I_V of the blood is reduced.

14.8 Other Applications of Fluid Dynamics

Let's consider the opening storyline for this chapter. What forces are responsible for lifting an airplane into the air? Consider the streamlines that flow around an airplane wing as shown in Figure 14.23. Let's assume the airstream approaches the wing horizontally from the right with a velocity $\vec{\mathbf{v}}_1$, which is equivalent to the airplane moving to the right through still air. The tilt of the wing causes the airstream to be deflected downward with a velocity $\vec{\mathbf{v}}_2$. Because the airstream is deflected by the wing, the wing must exert a force on the airstream. According to Newton's third law, the airstream therefore exerts a force $\vec{\mathbf{F}}$ on the wing that is equal in magnitude and opposite in direction. This force has a vertical component called **lift** (or aerodynamic lift) and a horizontal component called **drag.** The lift depends on several factors, such as the speed of the airplane, the area of the wing, the wing's curvature, and the angle between the wing and the horizontal. The curvature of the wing surfaces causes the pressure above the wing to be lower than that below the wing due to the Bernoulli effect. This pressure difference assists with the lift on the wing. As the angle between the wing and the horizontal increases, turbulent flow can set in above the wing to reduce the lift.

The lift force exerted by the air on the wing according to Newton's law and the pressure difference between the top and bottom of the wing caused by the Bernoulli effect will both depend on the density of the air surrounding the wings. What do we know about the location of Denver in our opening storyline? Denver is often called the "Mile-High City." That is because it is in the Rocky Mountains, at an altitude of 1 610 m above sea level. Because of that altitude, the air is of lower density than that at the airport at Los Angeles, by an average of about 15%. Consequently, aircraft must attain a higher speed in order for the forces and pressure differences to be sufficient to lift the aircraft. This leads to a longer distance on the runway for the aircraft to move before reaching this higher speed.

In general, an object moving through a fluid experiences lift as the result of any effect that causes the fluid to change its direction as it flows past the object. Some factors that influence lift are the shape of the object, its orientation with respect to the fluid flow, any spinning motion it might have, and the texture of its surface.

The air approaching from the right is deflected downward by the wing.

Figure 14.23 Streamline flow around a moving airplane wing. By Newton's third law, the air deflected by the wing results in an upward force on the wing from the air: *lift*. Because of air resistance, there is also a force opposite the velocity of the wing: *drag*.

---

[1]Named after Jean Leonard Marie Poiseuille (1797–1869), a French physicist, and Gotthilf Heinrich Ludwig Hagen (1797–1884), a German civil engineer. The unit *poise* is named after Poiseuille.

For example, a golf ball struck with a club is given a rapid backspin due to the slant of the club. The dimples on the ball increase the friction force between the ball and the air so that air adheres to the ball's surface and is deflected downward as a result. Because the ball pushes the air down, the air must push up on the ball. Without the dimples, the friction force is lower and the golf ball does not travel as far. It may seem counterintuitive to increase the range by increasing the friction force, but the lift gained by spinning the ball more than compensates for the loss of range due to the effect of friction on the translational motion of the ball. For the same reason, a baseball's cover helps the spinning ball "grab" the air rushing by and helps deflect it when a "curve ball" is thrown.

Summary

> ## Definitions

The **pressure** P in a fluid is the force per unit area exerted by the fluid on a surface:

$$P \equiv \frac{F}{A} \tag{14.1}$$

In the SI system, pressure has units of newtons per square meter (N/m^2), and $1\ N/m^2 = 1$ **pascal** (Pa).

> ## Concepts and Principles

The pressure in a fluid at rest varies with depth h in the fluid according to the expression

$$P = P_0 + \rho g h \tag{14.4}$$

where P_0 is the pressure at $h = 0$ and ρ is the density of the fluid, assumed uniform.

Pascal's law states that when pressure is applied to an enclosed fluid, the pressure is transmitted undiminished to every point in the fluid and to every point on the walls of the container.

When an object is partially or fully submerged in a fluid, the fluid exerts on the object an upward force called the **buoyant force.** According to **Archimedes's principle,** the magnitude of the buoyant force is equal to the weight of the fluid displaced by the object:

$$B = \rho_{\text{fluid}} g V_{\text{disp}} \tag{14.5}$$

The flow rate (volume flux) through a pipe that varies in cross-sectional area is constant; that is equivalent to stating that the product of the cross-sectional area A and the speed v at any point is a constant. This result is expressed in the **equation of continuity for fluids:**

$$A_1 v_1 = A_2 v_2 = \text{constant} \tag{14.7}$$

The sum of the pressure, kinetic energy per unit volume, and gravitational potential energy per unit volume has the same value at all points along a streamline for an ideal fluid. This result is summarized in **Bernoulli's equation:**

$$P + \tfrac{1}{2}\rho v^2 + \rho g y = \text{constant} \tag{14.13}$$

Think–Pair–Share

See the Preface for an explanation of the icons used in this problems set. For additional assessment items for this section, go to **WEBASSIGN** From Cengage

1. You are a member of an expert witness group that provides scientific services to the legal community. Your group has been asked to defend an NFL football team, which has been caught in an embarrassing practice: they have been filling their teams' footballs with helium instead of air, believing that the helium would provide an increased buoyant force on the footballs, causing their passes and kicks to be longer. Despite your unhappiness with the efforts of the team to gain an unfair advantage, your legal group believes that

everyone deserves a defense, so you agree to take the case. (a) Develop an argument that filling the footballs with helium would *not* provide additional buoyant force on the footballs; therefore, the team was not trying to gain an advantage. (b) Develop a private argument to present to the team that filling the balls with helium will actually *decrease* the performance.

2. It is a warm day, and a student decides he would like to spend a few hours swimming in the pool. Using his snorkel equipment, he views the bottom of the pool. After a while, he notices a piece of PVC pipe leaning against the house and thinks about using it as a long snorkel. He takes the

mouthpiece off his snorkel and attaches it to the bottom of the PVC pipe. His plan is to have his mouth on the mouthpiece of the long snorkel and sink into the deep end of the pool, breathing all the way to the bottom! While he is preparing his long snorkel, his uncle, who is a pulmonologist, arrives for a visit. He asks the student what he is doing with the pipe and he explains. The uncle is horrified and says that it is very dangerous to attempt this activity. The student is disappointed, but then tells his uncle that he will hold his breath while he sinks to the bottom, keeping the mouthpiece closed with his thumb, put the mouthpiece on as he arrives at the bottom, and begin breathing. His uncle looks even more horrified, and tells him that that activity could be fatal! Discuss in your group: Why is deep snorkeling so *dangerous?*

3. **ACTIVITY** Perform the following activities and discuss the physics behind the results with your group members:

    (a) Place a can of diet soda and regular soda of the same brand in a large container of water. Do they both float? Explain the results.
    (b) Open a can of clear carbonated beverage and fill a transparent glass with the liquid. Now drop a few raisins into the beverage and record their behavior.

Problems

See the Preface for an explanation of the icons used in this problems set. For additional assessment items for this section, go to WEBASSIGN *From Cengage*

Note: In all problems, assume the density of air is the 20°C value from Table 14.1, 1.20 kg/m³, unless noted otherwise.

SECTION 14.1 Pressure

1. A large man sits on a four-legged chair with his feet off the floor. The combined mass of the man and chair is 95.0 kg. If the chair legs are circular and have a radius of 0.500 cm at the bottom, what pressure does each leg exert on the floor?

2. **QC** The nucleus of an atom can be modeled as several protons and neutrons closely packed together. Each particle has a mass of 1.67×10^{-27} kg and radius on the order of 10^{-15} m. (a) Use this model and the data provided to estimate the density of the nucleus of an atom. (b) Compare your result with the density of a material such as iron. What do your result and comparison suggest concerning the structure of matter?

3. Estimate the total mass of the Earth's atmosphere. (The radius of the Earth is 6.37×10^6 m, and atmospheric pressure at the surface is 1.013×10^5 Pa.)

SECTION 14.2 Variation of Pressure with Depth

4. *Why is the following situation impossible?* Figure P14.4 shows Superman attempting to drink cold water through a straw of length $\ell = 12.0$ m. The walls of the tubular straw are very strong and do not collapse. With his great strength, he achieves maximum possible suction and enjoys drinking the cold water.

Figure P14.4

5. **AMT** **T** What must be the contact area between a suction cup (completely evacuated) and a ceiling if the cup is to support the weight of an 80.0-kg student?

6. For the cellar of a new house, a hole is dug in the ground, with vertical sides going down 2.40 m. A concrete foundation wall is built all the way across the 9.60-m width of the excavation. This foundation wall is 0.183 m away from the front of the cellar hole. During a rainstorm, drainage from the street fills up the space in front of the concrete wall, but not the cellar behind the wall. The water does not soak into the clay soil. Find the force the water causes on the foundation wall. For comparison, the weight of the water is given by 2.40 m $\times$ 9.60 m $\times$ 0.183 m $\times$ 1 000 kg/m³ $\times$ 9.80 m/s² = 41.3 kN.

7. **Review.** A solid sphere of brass (bulk modulus of 14.0×10^{10} N/m²) with a diameter of 3.00 m is thrown into the ocean. By how much does the diameter of the sphere decrease as it sinks to a depth of 1.00 km?

SECTION 14.3 Pressure Measurements

8. **BIO** **QC** The human brain and spinal cord are immersed in the cerebrospinal fluid. The fluid is normally continuous between the cranial and spinal cavities and exerts a pressure of 100 to 200 mm of H_2O above the prevailing atmospheric pressure. In medical work, pressures are often measured in units of millimeters of H_2O because body fluids, including the cerebrospinal fluid, typically have the same density as water. The pressure of the cerebrospinal fluid can be measured by means of a *spinal tap* as illustrated in Figure P14.8. A hollow tube is inserted into the spinal column, and the height to which the fluid rises is observed. If the fluid rises to a height of 160 mm, we write its gauge pressure as 160 mm H_2O. (a) Express this pressure in pascals, in atmospheres, and in millimeters of mercury. (b) Some conditions that block or inhibit the flow of cerebrospinal fluid can be investigated by means of *Queckenstedt's test*. In this procedure, the veins in the patient's neck are compressed to make the blood pressure rise in the brain, which in turn should be transmitted

Figure P14.8

to the cerebrospinal fluid. Explain how the level of fluid in the spinal tap can be used as a diagnostic tool for the condition of the patient's spine.

9. Blaise Pascal duplicated Torricelli's barometer using a red Bordeaux wine, of density 984 kg/m^3, as the working liquid (Fig. P14.9). (a) What was the height h of the wine column for normal atmospheric pressure? (b) Would you expect the vacuum above the column to be as good as for mercury?

Figure P14.9

10. A tank with a flat bottom of area A and vertical sides is filled to a depth h with water. The pressure is P_0 at the top surface. (a) What is the absolute pressure at the bottom of the tank? (b) Suppose an object of mass M and density less than the density of water is placed into the tank and floats. No water overflows. What is the resulting increase in pressure at the bottom of the tank?

SECTION 14.4 Buoyant Forces and Archimedes's Principle

11. The gravitational force exerted on a solid object is 5.00 N. When the object is suspended from a spring scale and submerged in water, the scale reads 3.50 N (Fig. P14.11). Find the density of the object.

Figure P14.11 Problems 11 and 12.

12. A 10.0-kg block of metal measuring 12.0 cm by 10.0 cm by 10.0 cm is suspended from a scale and immersed in water as shown in Figure P14.11b. The 12.0-cm dimension is vertical, and the top of the block is 5.00 cm below the surface of the water. (a) What are the magnitudes of the forces acting on the top and on the bottom of the block due to the surrounding water? (b) What is the reading of the spring scale? (c) Show that the buoyant force equals the difference between the forces at the top and bottom of the block.

13. A plastic sphere floats in water with 50.0% of its volume submerged. This same sphere floats in glycerin with 40.0% of its volume submerged. Determine the densities of (a) the glycerin and (b) the sphere.

14. The weight of a rectangular block of low-density material is 15.0 N. With a thin string, the center of the horizontal bottom face of the block is tied to the bottom of a beaker partly filled with water. When 25.0% of the block's volume is submerged, the tension in the string is 10.0 N. (a) Find the buoyant force on the block. (b) Oil of density 800 kg/m^3 is now steadily added to the beaker, forming a layer above the water and surrounding the block. The oil exerts forces on each of the four sidewalls of the block that the oil touches. What are the directions of these forces? (c) What happens to the string tension as the oil is added? Explain how the oil has this effect on the string tension. (d) The string breaks when its tension reaches 60.0 N. At this moment, 25.0% of the block's volume is still below the water line. What additional fraction of the block's volume is below the top surface of the oil?

15. A wooden block of volume 5.24×10^{-4} m^3 floats in water, and a small steel object of mass m is placed on top of the block. When $m = 0.310$ kg, the system is in equilibrium and the top of the wooden block is at the level of the water. (a) What is the density of the wood? (b) What happens to the block when the steel object is replaced by an object whose mass is less than 0.310 kg? (c) What happens to the block when the steel object is replaced by an object whose mass is greater than 0.310 kg?

16. A *hydrometer* is an instrument used to determine liquid density. A simple one is sketched in Figure P14.16. The bulb of a syringe is squeezed and released to let the atmosphere lift a sample of the liquid of interest into a tube containing a calibrated rod of known density. The rod, of length L and average density ρ_0, floats partially immersed in the liquid of density ρ. A length h of the rod protrudes above the surface of the liquid. Show that the density of the liquid is given by

$$\rho = \frac{\rho_0 L}{L - h}$$

Figure P14.16 Problems 16 and 17

17. Refer to Problem 16 and Figure P14.16. A hydrometer is to be constructed with a cylindrical floating rod. Nine fiduciary marks are to be placed along the rod to indicate densities of 0.98 g/cm^3, 1.00 g/cm^3, 1.02 g/cm^3, 1.04 g/cm^3, ..., 1.14 g/cm^3. The row of marks is to start 0.200 cm from the top end of the rod and end 1.80 cm from the top end. (a) What is the required length of the rod? (b) What must be its average density? (c) Should the marks be equally spaced? Explain your answer.

18. On October 21, 2001, Ian Ashpole of the United Kingdom achieved a record altitude of 3.35 km (11 000 ft) powered by 600 toy balloons filled with helium. Each filled balloon had

a radius of about 0.50 m and an estimated mass of 0.30 kg. (a) Estimate the total buoyant force on the 600 balloons. (b) Estimate the net upward force on all 600 balloons. (c) Ashpole parachuted to the Earth after the balloons began to burst at the high altitude and the buoyant force decreased. Why did the balloons burst?

19. You have a job in a company that produces party supplies. **CR** You are designing helium-filled balloons to be sold as gifts. To save money on production costs, part of this design is to choose from an available selection the least massive token necessary to be tied to the lower end of the string hanging from the balloon in order to keep the balloon from rising off a table, hospital food tray, bedroom dresser, etc. In this way, the token remains stationary on the flat surface and the balloon is buoyed above the token at a fixed height, with the string straight. You are working on a balloon whose envelope is very thin and has a mass of 0.150 kg. The envelope is filled to a volume of 0.2300 m³ with helium at atmospheric pressure. The string has a mass of 0.070 0 kg. Among the selection of tokens are those with masses 10.0 g, 20.0 g, 30.0 g, 40.0 g, and 50.0 g. Chose the appropriate token for the balloon you are working on.

SECTION 14.5 Fluid Dynamics

20. Water flowing through a garden hose of diameter 2.74 cm fills a 25-L bucket in 1.50 min. (a) What is the speed of the water leaving the end of the hose? (b) A nozzle is now attached to the end of the hose. If the nozzle diameter is one-third the diameter of the hose, what is the speed of the water leaving the nozzle?

21. Water falls over a dam of height h with a mass flow rate of I_V, in units of kilograms per second. (a) Show that the power available from the water is

$$P = I_V gh$$

where g is the free-fall acceleration. (b) Each hydroelectric unit at the Grand Coulee Dam takes in water at a rate of 8.50×10^5 kg/s from a height of 87.0 m. The power developed by the falling water is converted to electric power with an efficiency of 85.0%. How much electric power does each hydroelectric unit produce?

SECTION 14.6 Bernoulli's Equation

22. A legendary Dutch boy saved Holland by plugging a hole of diameter 1.20 cm in a dike with his finger. If the hole was 2.00 m below the surface of the North Sea (density 1 030 kg/m³), (a) what was the force on his finger? (b) If he pulled his finger out of the hole, during what time interval would the released water fill 1 acre of land to a depth of 1 ft? Assume the hole remained constant in size.

23. Water is pumped up from the Colorado River to supply Grand Canyon Village, located on the rim of the canyon. The river is at an elevation of 564 m, and the village is at an elevation of 2 096 m. Imagine that the water is pumped through a single long pipe 15.0 cm in diameter, driven by a single pump at the bottom end. (a) What is the minimum pressure at which the water must be pumped if it is to arrive at the village? (b) If 4 500 m³ of water is pumped per day, what is the speed of the water in the pipe? *Note:* Assume the free-fall acceleration and the

density of air are constant over this range of elevations. The pressures you calculate are too high for an ordinary pipe. The water is actually lifted in stages by several pumps through shorter pipes.

24. In ideal flow, a liquid of density 850 kg/m³ moves from a **QC** horizontal tube of radius 1.00 cm into a second horizontal tube of radius 0.500 cm at the same elevation as the first tube. The pressure differs by ΔP between the liquid in one tube and the liquid in the second tube. (a) Find the volume flow rate as a function of ΔP. Evaluate the volume flow rate for (b) $\Delta P = 6.00$ kPa and (c) $\Delta P = 12.0$ kPa.

25. **Review.** Old Faithful Geyser in Yellowstone National Park erupts at approximately one-hour intervals, and the height of the water column reaches 40.0 m (Fig. P14.25). (a) Model the rising stream as a series of separate droplets. Analyze the free-fall motion of one of the droplets to determine the speed at which the water leaves the ground. (b) **What If?** Model the rising stream as an ideal fluid in streamline flow. Use Bernoulli's equation to determine the speed of the water as it leaves ground level. (c) How does the answer from part (a) compare with the answer from part (b)? (d) What is the pressure (above atmospheric) in the heated underground chamber if its depth is 175 m? Assume the chamber is large compared with the geyser's vent.

Videowokart, 2009/Used under license from Shutterstock.com

Figure P14.25

26. You are working as an expert witness for the owner of a sky- **CR** scraper complex in a downtown area. The owner is being sued by pedestrians on the streets below his buildings who were injured by falling glass when windows popped outward from the sides of the building. The Bernoulli effect can have important consequences for windows in such buildings. For example, wind can blow around a skyscraper at remarkably high speed, creating low pressure on the outside surface of the windows. The higher atmospheric pressure in the still air inside the buildings can cause windows to pop out. (a) In your research into the case, you find some overhead views of your client's project, as shown below. The project includes two tall skyscrapers and some park area on a square plot. Plan (i) (Fig. P14.26(i), page 382) was submitted by the original architects and planners. At the last minute, the owner decided he didn't want the park grounds to be divided into two areas and submitted Plan (ii) (Fig. P14.26(ii), which is the way the project was built. Explain to your client why Plan (ii) is a much more dangerous situation in terms of windows popping out than Plan (i). (b) Your client is not convinced by your conceptual argument in part (a), so you provide a numerical argument. Suppose a horizontal wind blows with a speed of 11.2 m/s outside a large pane of plate glass with dimensions 4.00 m × 1.50 m. Assume the density of the air to be constant at 1.20 kg/m³. The air inside the building is at atmospheric pressure. Calculate the total force exerted by air on the windowpane for your client. (c) **What If?** To further convince your client

Figure P14.26

of the problems with the building design, calculate the total force exerted by air on the windowpane if the wind speed between the buildings is 22.4 m/s, twice as high as in part (b).

SECTION 14.7 Flow of Viscous Fluids in Pipes

27. A thin 1.50-mm coating of glycerin has been placed between two microscope slides of width 1.00 cm and length 4.00 cm. Find the force required to pull one of the microscope slides at a constant speed of 0.300 m/s relative to the other slide.

28. **BIO** A hypodermic needle is 3.00 cm in length and 0.300 mm in diameter. What pressure difference between the input and output of the needle is required so that the flow rate of water through it will be 1.00 g/s? (Use 1.00×10^{-3} Pa · s as the viscosity of water.)

29. **BIO** What radius needle should be used to inject a volume of 500 cm^3 of a solution into a patient in 30.0 min? Assume the length of the needle is 2.50 cm and the solution is elevated 1.00 m above the point of injection. Further, assume the viscosity and density of the solution are those of pure water, and that the pressure inside the vein is atmospheric.

SECTION 14.8 Other Applications of Fluid Dynamics

30. **Q|C** An airplane has a mass of 1.60×10^4 kg, and each wing has an area of 40.0 m^2. During level flight, the pressure on the lower wing surface is 7.00×10^4 Pa. (a) Suppose the lift on the airplane were due to a pressure difference alone. Determine the pressure on the upper wing surface. (b) More realistically, a significant part of the lift is due to deflection of air downward by the wing. Does the inclusion of this force mean that the pressure in part (a) is higher or lower? Explain.

31. A siphon is used to drain water from a tank as illustrated in Figure P14.31. Assume steady flow without friction.

Figure P14.31

(a) If $h = 1.00$ m, find the speed of outflow at the end of the siphon. (b) **What If?** What is the limitation on the height of the top of the siphon above the end of the siphon? *Note:* For the flow of the liquid to be continuous, its pressure must not drop below its vapor pressure. Assume the water is at 20.0°C, at which the vapor pressure is 2.3 kPa.

ADDITIONAL PROBLEMS

32. **BIO** Decades ago, it was thought that huge herbivorous dinosaurs such as *Apatosaurus* and *Brachiosaurus* habitually walked on the bottom of lakes, extending their long necks up to the surface to breathe. *Brachiosaurus* had its nostrils on the top of its head. In 1977, Knut Schmidt-Nielsen pointed out that breathing would be too much work for such a creature. For a simple model, consider a sample consisting of 10.0 L of air at absolute pressure 2.00 atm, with density 2.40 kg/m^3, located at the surface of a freshwater lake. Find the work required to transport it to a depth of 10.3 m, with its temperature, volume, and pressure remaining constant. This energy investment is greater than the energy that can be obtained by metabolism of food with the oxygen in that quantity of air.

33. **GP** A helium-filled balloon (whose envelope has a mass of $m_b = 0.250$ kg) is tied to a uniform string of length $\ell = 2.00$ m and mass $m = 0.050\ 0$ kg. The balloon is spherical with a radius of $r = 0.400$ m. When released in air of temperature 20°C and density $\rho_{air} = 1.20$ kg/m^3, it lifts a length h of string and then remains stationary as shown in Figure P14.33. We wish to find the length of string lifted by the balloon. (a) When the balloon remains stationary, what is the appropriate analysis model to describe it? (b) Write a force equation for the balloon from this model in terms of the buoyant force B, the weight F_b of the balloon, the weight F_{He} of the helium, and the weight F_s of the segment of string of length h. (c) Make an appropriate substitution for each of these forces and solve symbolically for the mass m_s of the segment of string of length h in terms of m_b, r, ρ_{air}, and the density of helium ρ_{He}. (d) Find the numerical value of the mass m_s. (e) Find the length h numerically.

Figure P14.33

34. **S** The true weight of an object can be measured in a vacuum, where buoyant forces are absent. A measurement in air, however, is disturbed by buoyant forces. An object of volume V is weighed in air on an equal-arm balance with the use of counterweights of density ρ. Representing the density of air as ρ_{air} and the balance reading as F_g', show that the true weight F_g is

$$F_g = F_g' + \left(V - \frac{F_g'}{\rho g} \right) \rho_{air} g$$

35. To an order of magnitude, how many helium-filled toy balloons would be required to lift you? Because helium is an irreplaceable resource, develop a theoretical answer rather than an experimental answer. In your solution, state what physical quantities you take as data and the values you measure or estimate for them.

36. **Q|C** **Review.** Assume a certain liquid, with density 1 230 kg/m^3, exerts no friction force on spherical objects. A ball of mass 2.10 kg and radius 9.00 cm is dropped from rest into a deep tank of this liquid from a height of 3.30 m above the surface. (a) Find the speed at which the ball enters the

liquid. (b) Evaluate the magnitudes of the two forces that are exerted on the ball as it moves through the liquid. (c) Explain why the ball moves down only a limited distance into the liquid and calculate this distance. (d) With what speed will the ball pop up out of the liquid? (e) How does the time interval Δt_{down}, during which the ball moves from the surface down to its lowest point, compare with the time interval Δt_{up} for the return trip between the same two points? (f) **What If?** Now modify the model to suppose the liquid exerts a small friction force on the ball, opposite in direction to its motion. In this case, how do the time intervals Δt_{down} and Δt_{up} compare? Explain your answer with a conceptual argument rather than a numerical calculation.

37. Evangelista Torricelli was the first person to realize that we live at the bottom of an ocean of air. He correctly surmised that the pressure of our atmosphere is attributable to the weight of the air. The density of air at 0°C at the Earth's surface is 1.29 kg/m³. The density decreases with increasing altitude (as the atmosphere thins). On the other hand, if we assume the density is constant at 1.29 kg/m³ up to some altitude h and is zero above that altitude, then h would represent the depth of the ocean of air. (a) Use this model to determine the value of h that gives a pressure of 1.00 atm at the surface of the Earth. (b) Would the peak of Mount Everest rise above the surface of such an atmosphere?

38. A common parameter that can be used to predict turbu-
BIO lence in fluid flow is called the *Reynolds number*. The Rey-
QC nolds number for fluid flow in a pipe is a dimensionless quantity defined as

$$\text{Re} = \frac{\rho v d}{\eta}$$

where ρ is the density of the fluid, v is its speed, d is the inner diameter of the pipe, and η is the viscosity of the fluid. The criteria for the type of flow are as follows:

- If Re < 2 300, the flow is laminar.
- If 2 300 < Re < 4 000, the flow is in a transition region between laminar and turbulent.
- If Re > 4 000, the flow is turbulent.

(a) Let's model blood of density 1.06×10^3 kg/m³ and viscosity 3.00×10^{-3} Pa · s as a pure liquid, that is, ignore the fact that it contains red blood cells. Suppose it is flowing in a large artery of radius 1.50 cm with a speed of 0.067 0 m/s. Show that the flow is laminar. (b) Imagine that the artery ends in a *single* capillary so that the radius of the artery reduces to a much smaller value. What is the radius of the capillary that would cause the flow to become turbulent? (c) Actual capillaries have radii of about 5–10 micrometers, much smaller than the value in part (b). Why doesn't the flow in actual capillaries become turbulent?

39. In 1983, the United States began coining the one-cent piece out of copper-clad zinc rather than pure copper. The mass of the old copper penny is 3.083 g and that of the new cent is 2.517 g. The density of copper is 8.920 g/cm³ and that of zinc is 7.133 g/cm³. The new and old coins have the same volume. Calculate the percent of zinc (by volume) in the new cent.

40. **Review.** With reference to the dam studied in Example 14.4
S and shown in Figure 14.5, (a) show that the total torque exerted by the water behind the dam about a horizontal axis through O is $\frac{1}{6}\rho g w H^3$. (b) Show that the effective line of

action of the total force exerted by the water is at a distance $\frac{1}{3}H$ above O.

41. The *spirit-in-glass thermometer,* invented in Florence, Italy, around 1654, consists of a tube of liquid (the spirit) containing a number of submerged glass spheres with slightly different masses (Fig. P14.41). At sufficiently low temperatures, all the spheres float, but as the temperature rises, the spheres sink one after another. The device is a crude but interesting tool for measuring temperature. Suppose the tube is filled with ethyl alcohol, whose density is 0.789 45 g/cm³ at 20.0°C and decreases to 0.780 97 g/cm³ at 30.0°C. (a) Assuming that one of the spheres has a radius of 1.000 cm and is in equilibrium halfway up the tube at 20.0°C, determine its mass. (b) When the temperature increases to 30.0°C, what mass must a second sphere of the same radius have to be in equilibrium at the halfway point? (c) At 30.0°C, the first sphere has fallen to the bottom of the tube. What upward force does the bottom of the tube exert on this sphere?

Figure P14.41

42. A woman is draining her fish tank
S by siphoning the water into an outdoor drain as shown in Figure P14.42. The rectangular tank has footprint area A and depth h. The drain is located a distance d below the surface of the water in the tank, where $d \gg h$. The cross-sectional area of the siphon tube is A'. Model the water as flowing without friction. Show that the time interval required to empty the tank is given by

$$\Delta t = \frac{Ah}{A'\sqrt{2gd}}$$

Figure P14.42

43. **Review.** You and your father are designing a waterfall for
CR your backyard swimming pool. At the top of the waterfall is a tank containing water that is kept to a depth $d = 0.280$ m by a pump. As shown in Figure P14.43, there is a small hatch of height $h = 0.100$ m and width $w = 0.150$ m, hinged at the top, that can be used to turn on and turn off the supply of water to the waterfall. You want to attach a simple latch at the center of the bottom of the hatch and are trying to decide what type of latch to buy at your local hardware store. To make that decision, you need to determine the force that the latch must be able to withstand to keep the hatch closed.

Figure P14.43 Problems 43 and 44.

44. Review. You and your father are designing a waterfall for your backyard swimming pool. At the top of the waterfall is a tank containing water that is kept to a depth d by a pump. As shown in Figure P14.43, there is a small hatch of height h and width w, hinged at the top, that can be used to turn on and turn off the supply of water to the waterfall. You want to attach a simple latch at the center of the bottom of the hatch and are trying to decide what type of latch to buy at your local hardware store. To make that decision, you need to determine the force that the latch must be able to withstand to keep the hatch closed.

45. Review. A uniform disk of mass 10.0 kg and radius 0.250 m spins at 300 rev/min on a low-friction axle. It must be brought to a stop in 1.00 min by a brake pad that makes contact with the disk at an average distance 0.220 m from the axis. The coefficient of friction between pad and disk is 0.500. A piston in a cylinder of diameter 5.00 cm presses the brake pad against the disk. Find the pressure required for the brake fluid in the cylinder.

46. Review. In a water pistol, a piston drives water through a large tube of area A_1 into a smaller tube of area A_2 as shown in Figure P14.46. The radius of the large tube is 1.00 cm and that of the small tube is 1.00 mm. The smaller tube is 3.00 cm above the larger tube. (a) If the pistol is fired horizontally at a height of 1.50 m, determine the time interval required for the water to travel from the nozzle to the ground. Neglect air resistance and assume atmospheric pressure is 1.00 atm. (b) If the desired range of the stream is 8.00 m, with what speed v_2 must the stream leave the nozzle? (c) At what speed v_1 must the plunger be moved to achieve the desired range? (d) What is the pressure at the nozzle? (e) Find the pressure needed in the larger tube. (f) Calculate the force that must be exerted on the trigger to achieve the desired range. (The force that must be exerted is due to pressure over and above atmospheric pressure.)

Figure P14.46

47. An incompressible, nonviscous fluid is initially at rest in the vertical portion of the pipe shown in Figure P14.47a, where

Figure P14.47

$L = 2.00$ m. When the valve is opened, the fluid flows into the horizontal section of the pipe. What is the fluid's speed when all the fluid is in the horizontal section as shown in Figure P14.47b? Assume the cross-sectional area of the entire pipe is constant.

48. The hull of an experimental boat is to be lifted above the water by a hydrofoil mounted below its keel as shown in Figure P14.48. The hydrofoil has a shape like that of an airplane wing. Its area projected onto a horizontal surface is A. When the boat is towed at sufficiently high speed, water of density ρ moves in streamline flow so that its average speed at the top of the hydrofoil is n times larger than its speed v_b below the hydrofoil. (a) Ignoring the buoyant force, show that the upward lift force exerted by the water on the hydrofoil has a magnitude

$$F \approx \tfrac{1}{2}(n^2 - 1)\rho v_b^2 A$$

(b) The boat has mass M. Show that the liftoff speed is given by

$$v \approx \sqrt{\frac{2Mg}{(n^2 - 1)A\rho}}$$

Figure P14.48

CHALLENGE PROBLEMS

49. Show that the variation of atmospheric pressure with altitude is given by $P = P_0 e^{-\alpha y}$, where $\alpha = \rho_0 g / P_0$, P_0 is atmospheric pressure at some reference level $y = 0$, and ρ_0 is the atmospheric density at this level. Assume the decrease in atmospheric pressure over an infinitesimal change in altitude (so that the density is approximately uniform over the infinitesimal change) can be expressed from Equation 14.4 as $dP = -\rho g \, dy$. Also assume the density of air is proportional to the pressure, which, as we will see in Chapter 18, is equivalent to assuming the temperature of the air is the same at all altitudes.

50. *Why is the following situation impossible?* A barge is carrying a load of small pieces of iron along a river. The iron pile is in the shape of a cone for which the radius r of the base of the cone is equal to the central height h of the cone. The barge is square in shape, with vertical sides of length $2r$, so that the pile of iron comes just up to the edges of the barge. The barge approaches a low bridge, and the captain realizes that the top of the pile of iron is not going to make it under the bridge. The captain orders the crew to shovel iron pieces from the pile into the water to reduce the height of the pile. As iron is shoveled from the pile, the pile always has the shape of a cone whose diameter is equal to the side length of the barge. After a certain volume of iron is removed from the barge, it makes it under the bridge without the top of the pile striking the bridge.

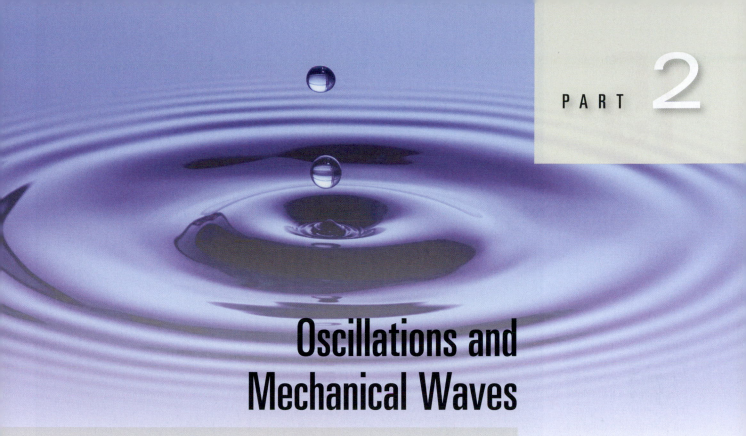

Oscillations and Mechanical Waves

In Part 1 of this text, we focused on one particular energy transfer term in Equation 8.2: work W. In Parts 2 through 5, we will focus our efforts in each part on a new term in Equation 8.2. Here in Part 2, we will investigate the term T_{MW}: transfer of energy by *mechanical waves*. We begin this new part of the text by studying a special type of motion called *periodic* motion, the repeating motion of an object in which it continues to return to a given position after a fixed time interval. The repetitive movements of such an object are called *oscillations*. We will focus our attention on a special case of periodic motion called *simple harmonic motion*. All periodic motions can be modeled as combinations of simple harmonic motions.

Simple harmonic motion also forms the basis for our understanding of mechanical waves. Sound waves, seismic waves, waves on stretched strings, and water waves are all produced by some source of oscillation. As a sound wave travels through the air, elements of the air oscillate back and forth; as a water wave travels across a pond, elements of the water oscillate up and down and backward and forward.

To explain many other phenomena in nature, we must understand the concepts of oscillations and waves. For instance, although skyscrapers and bridges appear to be rigid, they actually oscillate, something the architects and engineers who design and build them must take into account. To understand how radio and television work, we must understand the origin and nature of electromagnetic waves and how they propagate through space. Finally, much of what scientists have learned about atomic structure has come from information carried by waves. ■

Falling drops of water cause a water surface to oscillate. These oscillations are associated with circular waves moving away from the point at which the drops fall. In Part 2 of the text, we will explore the principles related to oscillations and waves. (*Ziga Camernik/Shutterstock*)

15

A grandfather clock keeps time in a room. The timing mechanism depends on the swinging of a pendulum. This repetitive swinging is an example of oscillatory motion. (*Antonio Gravante/ Shutterstock*)

Oscillatory Motion

STORYLINE **In the previous chapter, you were taking off in an** airplane from Denver, Colorado, to Boston, Massachusetts. You are now visiting your grandparents in Boston. There is an antique grandfather clock keeping time in one of the rooms. The gentle clicking of the swinging pendulum is relaxing to you. You recall that your grandparents used to live in Denver, like your other set of grandparents, and then moved from Denver to Boston, bringing the clock with them. Your grandmother enters the room and you mention your childhood memories of the clock. She tells you that she had it calibrated professionally in Denver and it kept perfect time for years. After they moved it here to their Massachusetts house, it has not been accurate. It runs too fast and has to be reset to the correct time every few days. You ask your grandmother what the clock shop in Denver did to calibrate the clock, but she doesn't know. You wonder—could *you* do something to calibrate the clock?

CONNECTIONS This is a bridging chapter. For the most part so far, we have considered motion that occurs once and does not repeat—a thrown ball, an accelerating car, a pushed crate. In Section 4.4, we saw our first example of *repeating* motion: a particle moving in a circular path returns to the starting point and performs the same motion over and over. In this chapter, we will be applying the principles of mechanics to the special case of an *oscillating* object. From this point of view, this chapter is based on understanding a new and important situation based on material we have studied in previous chapters. On the other hand, oscillations are the basis for understanding all types of *waves*. We mentioned mechanical waves and electromagnetic waves briefly in Section 8.1 and will study mechanical waves in the next two chapters and electromagnetic waves in Chapter 33. Therefore, this chapter, while based on principles from the *past*, is preparing us for our *future* study of waves.

15.1 Motion of an Object Attached to a Spring

As a model for oscillatory motion, consider a block of mass m attached to the end of a spring, with the block free to move on a frictionless, horizontal surface (Fig. 15.1). When the spring is neither stretched nor compressed, the block is at rest at the position called the **equilibrium position** of the system, which we identify as $x = 0$ (Fig. 15.1b). We know from experience that such a system oscillates back and forth if disturbed from its equilibrium position.

We can understand the oscillating motion of the block in Figure 15.1 qualitatively by first recalling that when the block is displaced to a position x, the spring exerts on the block a force that is proportional to the position and given by **Hooke's law** (see Section 7.4):

$$F_s = -kx \qquad (15.1)$$

◄ Hooke's law

We call F_s a **restoring force** because it is always directed toward the equilibrium position and therefore *opposite* the displacement of the block from equilibrium. That is, when the block is displaced to the right of $x = 0$ in Figure 15.1a, the position is positive and the restoring force is directed to the left. When the block is displaced to the left of $x = 0$ as in Figure 15.1c, the position is negative and the restoring force is directed to the right.

When the block is displaced from the equilibrium point and released, it is a particle under a net force and consequently undergoes an acceleration. Applying the particle under a net force model to the motion of the block, with Equation 15.1 providing the net force in the x direction, we obtain

$$\sum F_x = ma_x \;\to\; -kx = ma_x$$

$$a_x = -\frac{k}{m}x \qquad (15.2)$$

That is, the acceleration of the block is proportional to its position, and the direction of the acceleration is opposite the direction of the displacement of the block from equilibrium. Systems that behave in this way are said to exhibit **simple harmonic motion.** An object moves with simple harmonic motion whenever its acceleration is proportional to its position and is oppositely directed to the displacement from equilibrium.

PITFALL PREVENTION 15.1

The Orientation of the Spring Figure 15.1 shows a *horizontal* spring, with an attached block sliding on a frictionless surface. Another possibility is a block hanging from a *vertical* spring. All the results we discuss for the horizontal spring are the same for the vertical spring with one exception: when the block is placed on the vertical spring, its weight causes the spring to extend. If the resting position of the block on the extended spring is defined as $x = 0$, the results of this chapter also apply to this vertical system.

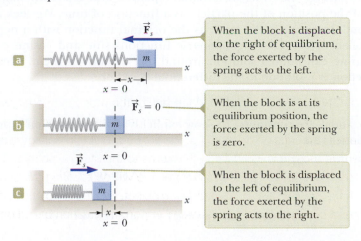

a

When the block is displaced to the right of equilibrium, the force exerted by the spring acts to the left.

b

When the block is at its equilibrium position, the force exerted by the spring is zero.

c

When the block is displaced to the left of equilibrium, the force exerted by the spring acts to the right.

Figure 15.1 A block attached to a spring and moving on a frictionless surface.

If the block in Figure 15.1 is displaced to a position $x = A$ and released from rest, its *initial* acceleration is $-kA/m$. When the block passes through the equilibrium position $x = 0$, its acceleration is zero. At this instant, its speed is a maximum because the acceleration changes sign. The block then continues to travel to the left of equilibrium with a positive acceleration and finally reaches $x = -A$, at which time its acceleration is $+kA/m$ and its speed is again zero as discussed in Sections 7.4 and 7.9. The block completes a full cycle of its motion by returning to the original position, again passing through $x = 0$ with maximum speed. Therefore, the block oscillates between the turning points $x = \pm A$. In the absence of friction, this idealized motion will continue forever because the force exerted by the spring is conservative. Real systems are generally subject to friction, so they do not oscillate forever. We shall explore the details of the situation with friction in Section 15.6.

QUICK QUIZ 15.1 A block on the end of a spring is pulled to position $x = A$ and released from rest. In one full cycle of its motion, through what total distance does it travel? **(a)** $A/2$ **(b)** A **(c)** $2A$ **(d)** $4A$

15.2 Analysis Model: Particle in Simple Harmonic Motion

The idealized motion described in the preceding section is the basis for so many real motions of objects that we identify the **particle in simple harmonic motion** model to represent such situations. To develop a mathematical representation for this model, we will generally choose x as the axis along which the oscillation of an object occurs; hence, we will drop the subscript-x notation in this discussion. Recall that, by definition, $a = dv/dt = d^2x/dt^2$, so we can express Equation 15.2 as

$$\frac{d^2x}{dt^2} = -\frac{k}{m}x \tag{15.3}$$

> **PITFALL PREVENTION 15.2**
>
> **A Nonconstant Acceleration** The acceleration of a particle in simple harmonic motion is not constant. Equation 15.3 shows that its acceleration varies with position x. Therefore, we *cannot* apply the kinematic equations of Chapter 2 in this situation. Those equations describe a particle under *constant acceleration*.

If we denote the ratio k/m with the symbol ω^2 (we choose ω^2 rather than ω so as to make the solution we develop below simpler in form), then

$$\omega^2 = \frac{k}{m} \tag{15.4}$$

and Equation 15.3 can be written in the form

$$\frac{d^2x}{dt^2} = -\omega^2 x \tag{15.5}$$

Let's now find a mathematical solution to Equation 15.5, that is, a function $x(t)$ that satisfies this second-order differential equation and is a mathematical representation of the position of the particle as a function of time. We seek a function whose second derivative is the same as the original function with a negative sign and multiplied by ω^2. The trigonometric functions sine and cosine exhibit this behavior, so we can build a solution around one or both of them. The following cosine function is a solution to the differential equation:

◀ Position versus time for a particle in simple harmonic motion

$$x(t) = A\cos(\omega t + \phi) \tag{15.6}$$

where A, ω, and ϕ are constants. To show explicitly that this solution satisfies Equation 15.5, notice that

$$\frac{dx}{dt} = A\frac{d}{dt}\cos(\omega t + \phi) = -\omega A\sin(\omega t + \phi) \tag{15.7}$$

$$\frac{d^2x}{dt^2} = -\omega A\frac{d}{dt}\sin(\omega t + \phi) = -\omega^2 A\cos(\omega t + \phi) \tag{15.8}$$

Comparing Equations 15.6 and 15.8, we see that $d^2x/dt^2 = -\omega^2 x$ and Equation 15.5 is satisfied.

The parameters A, ω, and ϕ are constants of the motion. To give physical significance to these constants, it is convenient to form a graphical representation of the motion by plotting x as a function of t as in Figure 15.2a. First, A, called the **amplitude** of the motion, is simply the maximum value of the position of the particle in either the positive or negative x direction. The constant ω is called the **angular frequency,** and it has units[1] of radians per second. It is a measure of how rapidly the oscillations are occurring; the more oscillations per unit time, the higher the value of ω. From Equation 15.4, the angular frequency is

$$\omega = \sqrt{\frac{k}{m}} \qquad (15.9)$$

The quantity $(\omega t + \phi)$ in Equation 15.6 is called the **phase** of the motion. The constant angle ϕ is called the **phase constant** (or initial phase angle) and, along with the amplitude A, is determined uniquely by the position and velocity of the particle at $t = 0$. Therefore, A and ϕ are two parameters that define the *initial conditions* of the motion of an oscillating object, just as x_i and v_i describe the initial conditions of an object undergoing constant acceleration in Equation 2.16. If the particle is at its maximum position $x = A$ at $t = 0$, the phase constant is $\phi = 0$ and the graphical representation of the motion is as shown in Figure 15.2b. Notice that the function $x(t)$ is periodic and its value is the same each time ωt increases by 2π radians.

Equations 15.1, 15.5, and 15.6 form the basis of the mathematical representation of the particle in simple harmonic motion model. If you are analyzing a situation and find that the force on an object modeled as a particle is of the mathematical form of Equation 15.1, you know the motion is that of a simple harmonic oscillator and the position of the particle is described by Equation 15.6. If you analyze a system and find that it is described by a differential equation of the form of Equation 15.5, the motion is that of a simple harmonic oscillator. If you analyze a situation and find that the position of a particle is described by Equation 15.6, you know the particle undergoes simple harmonic motion.

QUICK QUIZ 15.2 Consider a graphical representation (Fig. 15.3) of simple harmonic motion as described mathematically in Equation 15.6. When the particle is at point Ⓐ on the graph, what can you say about its position and velocity? **(a)** The position and velocity are both positive. **(b)** The position and velocity are both negative. **(c)** The position is positive, and the velocity is zero. **(d)** The position is negative, and the velocity is zero. **(e)** The position is positive, and the velocity is negative. **(f)** The position is negative, and the velocity is positive.

QUICK QUIZ 15.3 Figure 15.4 shows two curves representing particles undergoing simple harmonic motion. The correct description of these two motions is that the simple harmonic motion of particle B is **(a)** of larger angular frequency and larger amplitude than that of particle A, **(b)** of larger angular frequency and smaller amplitude than that of particle A, **(c)** of smaller angular frequency and larger amplitude than that of particle A, or **(d)** of smaller angular frequency and smaller amplitude than that of particle A.

Let us investigate further the mathematical description of simple harmonic motion. The **period** T of the motion is the time interval required for the particle to go through one full cycle of its motion (Fig. 15.2a). That is, the values of x and v

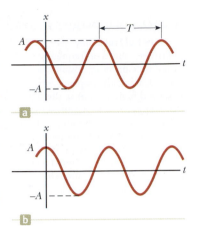

Figure 15.2 (a) An x–t graph for a particle undergoing simple harmonic motion. The amplitude of the motion is A, and the period is T. (b) The x–t graph for the special case in which $x = A$ at $t = 0$ and hence $\phi = 0$.

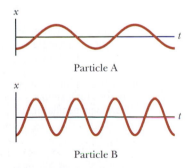

Figure 15.3 (Quick Quiz 15.2) An x–t graph for a particle undergoing simple harmonic motion. At a particular time, the particle's position is indicated by Ⓐ in the graph.

Particle A

Particle B

Figure 15.4 (Quick Quiz 15.3) Two x–t graphs for particles undergoing simple harmonic motion. The amplitudes and frequencies are different for the two particles.

---

[1] We have seen many examples in earlier chapters in which we evaluate a trigonometric function of an angle. The argument of a trigonometric function, such as the cosine function in Equation 15.6, *must* be a pure number with no units. The radian is a pure number because it is a ratio of lengths. Therefore, ω *must* be expressed in radians per second (and not, for example, in revolutions per second) if t is expressed in seconds.

for the particle at time t equal the values of x and v at time $t + T$. Because the phase increases by 2π radians in a time interval of T,

$$[\omega(t + T) + \phi] - (\omega t + \phi) = 2\pi$$

Simplifying this expression gives $\omega T = 2\pi$, or

$$T = \frac{2\pi}{\omega} \tag{15.10}$$

The inverse of the period is called the **frequency** f of the motion. Whereas the period is the time interval per oscillation, the frequency represents the number of oscillations the particle undergoes per unit time interval:

$$f = \frac{1}{T} = \frac{\omega}{2\pi} \tag{15.11}$$

The units of f are cycles per second, or **hertz** (Hz). Rearranging Equation 15.11 gives

$$\omega = 2\pi f = \frac{2\pi}{T} \tag{15.12}$$

Equations 15.9 through 15.11 can be used to express the period and frequency of the motion for the particle in simple harmonic motion in terms of the characteristics m and k of the system as

◀ Period of a simple harmonic oscillator

$$T = \frac{2\pi}{\omega} = 2\pi \sqrt{\frac{m}{k}} \tag{15.13}$$

◀ Frequency of a simple harmonic oscillator

$$f = \frac{1}{T} = \frac{1}{2\pi} \sqrt{\frac{k}{m}} \tag{15.14}$$

That is, the period and frequency depend *only* on the mass of the particle and the force constant of the spring and *not* on the parameters of the motion, such as A or ϕ. As we might expect, the frequency is larger for a stiffer spring (larger value of k) and decreases with increasing mass of the particle.

We can obtain the velocity and acceleration[2] of a particle undergoing simple harmonic motion from Equations 15.7 and 15.8:

◀ Velocity as a function of time for a simple harmonic oscillator

$$v = \frac{dx}{dt} = -\omega A \sin (\omega t + \phi) \tag{15.15}$$

◀ Acceleration as a function of time for a simple harmonic oscillator

$$a = \frac{d^2x}{dt^2} = -\omega^2 A \cos (\omega t + \phi) \tag{15.16}$$

From Equation 15.15, we see that because the sine and cosine functions oscillate between ± 1, the extreme values of the velocity v are $\pm \omega A$. Likewise, Equation 15.16 shows that the extreme values of the acceleration a are $\pm \omega^2 A$. Therefore, the *maximum* values of the magnitudes of the velocity and acceleration are

◀ Maximum magnitudes of velocity and acceleration in simple harmonic motion

$$v_{\text{max}} = \omega A = \sqrt{\frac{k}{m}} A \tag{15.17}$$

$$a_{\text{max}} = \omega^2 A = \frac{k}{m} A \tag{15.18}$$

---

[2]Because the motion of a simple harmonic oscillator takes place in one dimension, we denote velocity as v and acceleration as a, with the direction indicated by a positive or negative sign as in Chapter 2.

Figure 15.5a plots position versus time for an arbitrary value of the phase constant. The associated velocity–time and acceleration–time curves are illustrated in Figures 15.5b and 15.5c, respectively. It is evident that all three curves have the same general shape. The phase of the velocity, however, differs from the phase of the position by $\pi/2$ rad, or 90°. That is, when x is a maximum or a minimum, the velocity is zero. Likewise, when x is zero, the speed is a maximum. Furthermore, notice that the phase of the acceleration differs from the phase of the position by π radians, or 180°. For example, when x is a maximum, a has a maximum magnitude in the opposite direction.

QUICK QUIZ 15.4 An object of mass m is hung from a spring and set into oscillation. The period of the oscillation is measured and recorded as T. The object of mass m is removed and replaced with an object of mass $2m$. When this object is set into oscillation, what is the period of the motion? **(a)** $2T$ **(b)** $\sqrt{2}\,T$ **(c)** T **(d)** $T/\sqrt{2}$ **(e)** $T/2$

Equation 15.6 describes simple harmonic motion of a particle in terms of three constants of the motion. Let's now see how to evaluate these constants. The angular frequency ω is evaluated using Equation 15.9. The constants A and ϕ are evaluated from the initial conditions, that is, the state of the oscillator at $t = 0$.

Suppose a block is set into motion by pulling it from equilibrium by a distance A and releasing it from rest at $t = 0$ as in Figure 15.6. We must then require our solutions for $x(t)$ and $v(t)$ (Eqs. 15.6 and 15.15) to obey the initial conditions that $x(0) = A$ and $v(0) = 0$:

$$x(0) = A \cos \phi = A$$

$$v(0) = -\omega A \sin \phi = 0$$

These conditions are met if $\phi = 0$, giving $x = A \cos \omega t$ as our solution. To check this solution, notice that it satisfies the condition that $x(0) = A$ because $\cos 0 = 1$.

The position, velocity, and acceleration of the block versus time are plotted in Figure 15.7a for this special case. The acceleration reaches extreme values of $\mp\omega^2 A$ when the position has extreme values of $\pm A$. Furthermore, the velocity has extreme values of $\pm\omega A$, which both occur at $x = 0$. Hence, the quantitative solution agrees with our qualitative description of this system.

Let's consider another possibility. Suppose the system is oscillating and we define $t = 0$ as the instant the block passes through the unstretched position of the spring while moving to the right (Fig. 15.8). In this case, our solutions for $x(t)$ and $v(t)$ must obey the initial conditions that $x(0) = 0$ and $v(0) = v_i$:

$$x(0) = A \cos \phi = 0$$

$$v(0) = -\omega A \sin \phi = v_i$$

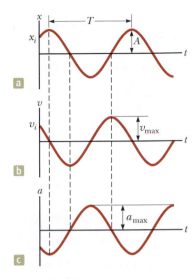

Figure 15.5 Graphical representation of simple harmonic motion. (a) Position versus time. (b) Velocity versus time. (c) Acceleration versus time. Notice that at any specified time the velocity is 90° out of phase with the position and the acceleration is 180° out of phase with the position.

Figure 15.6 A block–spring system that begins its motion from rest with the block at $x = A$ at $t = 0$.

Figure 15.8 The block–spring system is undergoing oscillation, and $t = 0$ is defined at an instant when the block passes through the equilibrium position $x = 0$ and is moving to the right with speed v_i.

Figure 15.7 (a) Position, velocity, and acceleration versus time for the block in Figure 15.6 under the initial conditions that at $t = 0$, $x(0) = A$, and $v(0) = 0$. (b) Position, velocity, and acceleration versus time for the block in Figure 15.8 under the initial conditions that at $t = 0$, $x(0) = 0$, and $v(0) = v_i$.

The first of these conditions tells us that $\phi = \pm\pi/2$. With these choices for ϕ, the second condition tells us that $A = \mp v_i/\omega$. Because the initial velocity is positive and the amplitude must be positive, we must have $\phi = -\pi/2$. Hence, the solution is

$$x = \frac{v_i}{\omega} \cos\left(\omega t - \frac{\pi}{2}\right)$$

The graphs of position, velocity, and acceleration versus time for this choice of $t = 0$ are shown in Figure 15.7b. Notice that these curves are the same as those in Figure 15.7a, but shifted to the right by one-fourth of a cycle. This shift is described mathematically by the phase constant $\phi = -\pi/2$, which is one-fourth of a full cycle of 2π.

ANALYSIS MODEL Particle in Simple Harmonic Motion

Imagine an object that is subject to a force that is proportional to the negative of the object's position, $F = -kx$ (Eq. 15.1). Such a force equation is known as Hooke's law, and it describes the force applied to an object attached to an ideal spring. The parameter k in Hooke's law is called the *spring constant* or the *force constant*. The position of an object acted on by a force described by Hooke's law is given by

$$x(t) = A \cos(\omega t + \phi) \tag{15.6}$$

where A is the **amplitude** of the motion, ω is the **angular frequency,** and ϕ is the **phase constant.** The values of A and ϕ depend on the initial position and initial velocity of the particle.

The **period** of the oscillation of the particle is

$$T = \frac{2\pi}{\omega} = 2\pi\sqrt{\frac{m}{k}} \tag{15.13}$$

and the inverse of the period is the **frequency.**

Examples:

- a bungee jumper hangs from a bungee cord and oscillates up and down
- a guitar string vibrates back and forth in a standing wave, with each element of the string moving in simple harmonic motion (Chapter 17)
- a piston in a gasoline engine oscillates up and down within the cylinder of the engine (Chapter 21)
- an atom in a diatomic molecule vibrates back and forth as if it is connected by a spring to the other atom in the molecule (Chapter 42)

Example 15.1 A Block–Spring System

A 200-g block connected to a light spring for which the force constant is 5.00 N/m is free to oscillate on a frictionless, horizontal surface. The block is displaced 5.00 cm from equilibrium and released from rest as in Figure 15.6.

(A) Find the period of its motion.

SOLUTION

Conceptualize   Study Figure 15.6 and imagine the block moving back and forth in simple harmonic motion once it is released. Set up an experimental model in the vertical direction by hanging a heavy object such as a stapler from a strong rubber band.

Categorize   The block is modeled as a *particle in simple harmonic motion.*

Analyze

Use Equation 15.9 to find the angular frequency of the block–spring system:

$$\omega = \sqrt{\frac{k}{m}} = \sqrt{\frac{5.00 \text{ N/m}}{200 \times 10^{-3} \text{ kg}}} = 5.00 \text{ rad/s}$$

Use Equation 15.13 to find the period of the system:

$$T = \frac{2\pi}{\omega} = \frac{2\pi}{5.00 \text{ rad/s}} = \boxed{1.26 \text{ s}}$$

15.1 continued

(B) Determine the maximum speed of the block.

SOLUTION

Use Equation 15.17 to find v_{max}:

$$v_{max} = \omega A = (5.00 \text{ rad/s})(5.00 \times 10^{-2} \text{ m}) = \boxed{0.250 \text{ m/s}}$$

(C) What is the maximum acceleration of the block?

SOLUTION

Use Equation 15.18 to find a_{max}:

$$a_{max} = \omega^2 A = (5.00 \text{ rad/s})^2(5.00 \times 10^{-2} \text{ m}) = \boxed{1.25 \text{ m/s}^2}$$

(D) Express the position, velocity, and acceleration as functions of time in SI units.

SOLUTION

Find the phase constant from the initial condition that $x = A$ at $t = 0$:

$$x(0) = A \cos \phi = A \rightarrow \phi = 0$$

Use Equation 15.6 to write an expression for $x(t)$:

$$x = A \cos (\omega t + \phi) = \boxed{0.050\ 0 \cos 5.00t}$$

Use Equation 15.15 to write an expression for $v(t)$:

$$v = -\omega A \sin (\omega t + \phi) = \boxed{-0.250 \sin 5.00t}$$

Use Equation 15.16 to write an expression for $a(t)$:

$$a = -\omega^2 A \cos (\omega t + \phi) = \boxed{-1.25 \cos 5.00t}$$

Finalize Consider part (a) of Figure 15.7, which shows the graphical representations of the motion of the block in this problem. Make sure that the mathematical representations found above in part (D) are consistent with these graphical representations.

WHAT IF? What if the block were released from the same initial position, $x_i = 5.00$ cm, but with an initial velocity of $v_i = -0.100$ m/s? Which parts of the solution change, and what are the new answers for those that do change?

Answers Part (A) does not change because the period is independent of how the oscillator is set into motion. Parts (B), (C), and (D) will change.

Write position and velocity expressions for the initial conditions:

$$(1) \quad x(0) = A \cos \phi = x_i$$

$$(2) \quad v(0) = -\omega A \sin \phi = v_i$$

Divide Equation (2) by Equation (1) to find the phase constant:

$$\frac{-\omega A \sin \phi}{A \cos \phi} = \frac{v_i}{x_i}$$

$$\tan \phi = -\frac{v_i}{\omega x_i} = -\frac{-0.100 \text{ m/s}}{(5.00 \text{ rad/s})(0.050\ 0 \text{ m})} = 0.400$$

$$\phi = \tan^{-1}(0.400) = 0.121\pi$$

Use Equation (1) to find A:

$$A = \frac{x_i}{\cos \phi} = \frac{0.050\ 0 \text{ m}}{\cos (0.121\pi)} = 0.053\ 9 \text{ m}$$

Find the new maximum speed:

$$v_{max} = \omega A = (5.00 \text{ rad/s})(5.39 \times 10^{-2} \text{ m}) = 0.269 \text{ m/s}$$

Find the new magnitude of the maximum acceleration:

$$a_{max} = \omega^2 A = (5.00 \text{ rad/s})^2(5.39 \times 10^{-2} \text{ m}) = 1.35 \text{ m/s}^2$$

Find new expressions for position, velocity, and acceleration in SI units:

$$x = 0.053\ 9 \cos (5.00t + 0.121\pi)$$

$$v = -0.269 \sin (5.00t + 0.121\pi)$$

$$a = -1.35 \cos (5.00t + 0.121\pi)$$

As we saw in Chapters 7 and 8, many problems are easier to solve using an energy approach rather than one based on variables of motion. This particular What If? is easier to solve from an energy approach. Therefore, we shall investigate the energy of the simple harmonic oscillator in the next section.

Example 15.2 | More Details of the Block–Spring System

Consider again the block–spring system in Example 15.1, whose position, velocity, and acceleration are given in part (D) of the problem. Find a general expression for all times at which the block is located at $x = +\frac{1}{2}A$.

SOLUTION

Conceptualize An important factor to keep in mind is that the block will be located at the requested position *twice* during each cycle. Our general expression should reflect that fact.

Categorize As in Example 15.1, the block is modeled as a *particle in simple harmonic motion*.

Analyze Write an expression for the position of the block knowing that the phase constant is equal to zero:

$$x = A \cos \omega t$$

Enter the condition that the position be half the amplitude and solve for t:

$$\tfrac{1}{2}A = A \cos \omega t \quad \rightarrow \quad t = \frac{1}{\omega} \cos^{-1}\left(\tfrac{1}{2}\right)$$

Recognize that there are two angles in the first cycle at which the inverse cosine is one-half, plus additional angles can be found by adding integral multiples of 2π:

$$\begin{cases} \cos^{-1}\left(\tfrac{1}{2}\right) = \dfrac{\pi}{3} + 2\pi n = \dfrac{\pi}{3}(1 + 6n) & n = 0, 1, 2, \ldots \\[2mm] \cos^{-1}\left(\tfrac{1}{2}\right) = \dfrac{5\pi}{3} + 2\pi n = \dfrac{\pi}{3}(5 + 6n) & n = 0, 1, 2, \ldots \end{cases}$$

Substitute these angles into the expression for t:

$$t = \frac{1}{5.00 \text{ s}^{-1}} \cos^{-1}\left(\tfrac{1}{2}\right) = \frac{\pi}{15.0 \text{ s}^{-1}}(1 + 6n) \quad \text{or} \quad \frac{\pi}{15.0 \text{ s}^{-1}}(5 + 6n) \quad n = 0, 1, 2, \ldots$$

Finalize Use these expressions to show that the first two times at which the block is at this position are 0.209 s and 1.05 s. These instants are shortly after the block is released and shortly before one full cycle has been completed at 1.26 s.

WHAT IF? Suppose we measure the speed of the block at the instants found in the problem. At these instants, will the speed of the block be half the maximum speed?

Answer The velocity of the block depends on the sine function. The angles at which the cosine function is equal to one-half will not be the same as the angles at which the sine function is equal to one-half. Therefore, we expect the answer to be *no*. Notice that we asked about the *speed* of the block, not the *velocity*. Perform the calculation and show that there are four expressions for the times at which the speed is one-half the maximum speed:

$$t = \frac{\pi}{30.0 \text{ s}^{-1}}(1 + 12n) \quad \text{or} \quad \frac{\pi}{30.0 \text{ s}^{-1}}(5 + 12n) \quad \text{or} \quad \frac{\pi}{30.0 \text{ s}^{-1}}(7 + 12n) \quad \text{or} \quad \frac{\pi}{30.0 \text{ s}^{-1}}(11 + 12n) \quad n = 0, 1, 2, \ldots$$

15.3 | Energy of the Simple Harmonic Oscillator

As we have done before, after studying the motion of an object modeled as a particle in a new situation (for example, as in Chapter 2) and investigating the forces involved in influencing that motion (for example, as in Chapter 5), we turn our attention to *energy* (for example, as in Chapter 7). Let us examine the mechanical energy of a system in which a particle undergoes simple harmonic motion, such as the block–spring system illustrated in Figure 15.1. Because the surface is frictionless and the normal and gravitational forces on the block cancel, the system can be modeled as isolated with no nonconservative forces acting, and we expect the total mechanical energy of the system to be constant. We assume a massless spring, so the kinetic energy of the system corresponds only to that of the block. We can use Equation 15.15 to express the kinetic energy of the block as

◀ Kinetic energy of a simple harmonic oscillator

$$K = \tfrac{1}{2}mv^2 = \tfrac{1}{2}m\omega^2 A^2 \sin^2 (\omega t + \phi) \tag{15.19}$$

The elastic potential energy stored in the spring for any elongation x is given by $\frac{1}{2}kx^2$ (see Eq. 7.22). Using Equation 15.6 gives

$$U_s = \tfrac{1}{2}kx^2 = \tfrac{1}{2}kA^2\cos^2(\omega t + \phi) \tag{15.20}$$

◀ Potential energy of a simple harmonic oscillator

We see that K and U_s are *always* positive quantities or zero. Because $\omega^2 = k/m$, we can express the total mechanical energy of the simple harmonic oscillator as

$$E = K + U_s = \tfrac{1}{2}kA^2[\sin^2(\omega t + \phi) + \cos^2(\omega t + \phi)]$$

From the identity $\sin^2\theta + \cos^2\theta = 1$, we see that the quantity in square brackets is unity. Therefore, this equation reduces to

$$E = \tfrac{1}{2}kA^2 \tag{15.21}$$

◀ Total energy of a simple harmonic oscillator

That is, the total mechanical energy of a simple harmonic oscillator is a constant of the motion and is proportional to the square of the amplitude. The total mechanical energy is equal to the maximum potential energy stored in the spring when $x = \pm A$ because $v = 0$ at these points and there is no kinetic energy. At the equilibrium position, where $U_s = 0$ because $x = 0$, the total energy, now all in the form of kinetic energy, still has the value $\frac{1}{2}kA^2$.

Plots of the kinetic and potential energies versus time appear in Figure 15.9a, where we have taken $\phi = 0$. At all times, the sum of the kinetic and potential energies is a constant equal to $\frac{1}{2}kA^2$, the total energy of the system.

The variations of K and U_s with the position x of the block are plotted in Figure 15.9b. Energy is continuously being transformed between potential energy stored in the spring and kinetic energy of the block.

Figure 15.10 (page 396) illustrates the position, velocity, acceleration, kinetic energy, and potential energy of the block–spring system for one full period of the motion. Most of the ideas discussed so far are incorporated in this important figure. Study it carefully.

Equation 15.15 gives the velocity of a particle in simple harmonic oscillation as function of time t. We can obtain the velocity of the block at an arbitrary *position* by expressing the total energy of the system at some arbitrary position x as

$$E = K + U_s = \tfrac{1}{2}mv^2 + \tfrac{1}{2}kx^2 = \tfrac{1}{2}kA^2$$

$$v = \pm\sqrt{\frac{k}{m}(A^2 - x^2)} = \pm\omega\sqrt{A^2 - x^2} \tag{15.22}$$

◀ Velocity as a function of position for a simple harmonic oscillator

When you check Equation 15.22 to see whether it agrees with known cases, you find that it verifies that the speed is a maximum at $x = 0$ and is zero at the turning points $x = \pm A$.

You may wonder why we are spending so much time studying simple harmonic oscillators. We do so because they are good models of a wide variety of physical phenomena.

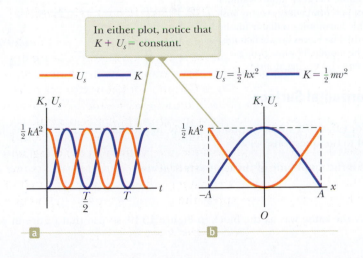

In either plot, notice that $K + U_s =$ constant.

— U_s — K — $U_s = \frac{1}{2}kx^2$ — $K = \frac{1}{2}mv^2$

a

b

Figure 15.9 (a) Kinetic energy and potential energy versus time for a simple harmonic oscillator with $\phi = 0$. (b) Kinetic energy and potential energy versus position for a simple harmonic oscillator.

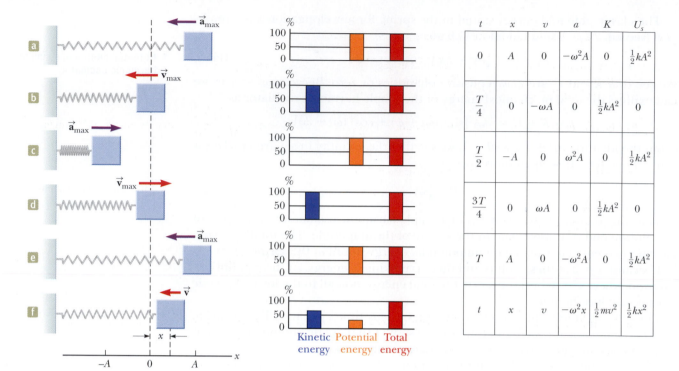

Figure 15.10 (a) through (e) Several instants in the simple harmonic motion for a block–spring system. Energy bar graphs show the distribution of the energy of the system at each instant. The parameters in the table at the right refer to the block–spring system, assuming at $t = 0$, $x = A$; hence, $x = A \cos \omega t$. For these five special instants, one of the types of energy is zero. (f) An arbitrary point in the motion of the oscillator. The system possesses both kinetic energy and potential energy at this instant as shown in the bar graph.

For example, recall the Lennard–Jones potential discussed in Example 7.9. This complicated function describes the forces holding atoms together. Figure 15.11a shows that for small displacements from the equilibrium position, the potential energy curve for this function approximates a parabola, which represents the potential energy function for a simple harmonic oscillator. Therefore, we can model the complex atomic binding forces as being due to tiny springs as depicted in Figure 15.11b.

The ideas presented in this chapter apply not only to block–spring systems and atoms, but also to a wide range of situations that include bungee jumping, playing a musical instrument, and viewing the light emitted by a laser. You will see more examples of simple harmonic oscillators as you work through this book.

Figure 15.11 (a) If the atoms in a molecule do not move too far from their equilibrium positions, a graph of potential energy versus separation distance between atoms is similar to the graph of potential energy versus position for a simple harmonic oscillator (dashed black curve). (b) The forces between atoms in a solid can be modeled by imagining springs between neighboring atoms.

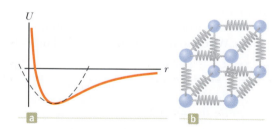

Example 15.3   **Oscillations on a Horizontal Surface**

A 0.500-kg cart connected to a light spring for which the force constant is 20.0 N/m oscillates on a frictionless, horizontal air track. Use an energy approach to respond to the questions below.

(A) Calculate the maximum speed of the cart if the amplitude of the motion is 3.00 cm.

SOLUTION

Conceptualize   The system oscillates in exactly the same way as the block in Figure 15.10, so use that figure in your mental image of the motion.

15.3 continued

Categorize The cart is modeled as a *particle in simple harmonic motion.*

Analyze Use Equation 15.21 to express the total energy of the oscillator system and equate it to the kinetic energy of the system when the cart is at $x = 0$:

$$E = \tfrac{1}{2}kA^2 = \tfrac{1}{2}mv_{max}^2$$

Solve for the maximum speed and substitute numerical values:

$$v_{max} = \sqrt{\frac{k}{m}}\, A = \sqrt{\frac{20.0\ \text{N/m}}{0.500\ \text{kg}}}(0.030\ 0\ \text{m}) = \boxed{0.190\ \text{m/s}}$$

(B) What is the velocity of the cart when the position is 2.00 cm?

SOLUTION

Use Equation 15.22 to evaluate the velocity:

$$v = \pm\sqrt{\frac{k}{m}(A^2 - x^2)}$$

$$= \pm\sqrt{\frac{20.0\ \text{N/m}}{0.500\ \text{kg}}[(0.030\ 0\ \text{m})^2 - (0.020\ 0\ \text{m})^2]}$$

$$= \boxed{\pm 0.141\ \text{m/s}}$$

The positive and negative signs indicate that the cart could be moving to either the right or the left at this instant.

(C) Compute the kinetic and potential energies of the system when the position of the cart is 2.00 cm.

SOLUTION

Use the result of part (B) to evaluate the kinetic energy at $x = 0.020\ 0$ m:

$$K = \tfrac{1}{2}mv^2 = \tfrac{1}{2}(0.500\ \text{kg})(0.141\ \text{m/s})^2 = \boxed{5.00 \times 10^{-3}\ \text{J}}$$

Evaluate the elastic potential energy at $x = 0.020\ 0$ m:

$$U_s = \tfrac{1}{2}kx^2 = \tfrac{1}{2}(20.0\ \text{N/m})(0.020\ 0\ \text{m})^2 = \boxed{4.00 \times 10^{-3}\ \text{J}}$$

Finalize The sum of the kinetic and potential energies in part (C) is equal to the total energy, which can be found from Equation 15.21. That must be true for *any* position of the cart.

WHAT IF? The cart in this example could have been set into motion by releasing the cart from rest at $x = 3.00$ cm. What if the cart were released from the same position, but with an initial velocity of $v = -0.100$ m/s? What are the new amplitude and maximum speed of the cart?

Answer This question is of the same type we asked at the end of Example 15.1, but here we apply an energy approach.

First calculate the total energy of the system at $t = 0$:

$$E = \tfrac{1}{2}mv^2 + \tfrac{1}{2}kx^2$$

$$= \tfrac{1}{2}(0.500\ \text{kg})(-0.100\ \text{m/s})^2 + \tfrac{1}{2}(20.0\ \text{N/m})(0.030\ 0\ \text{m})^2$$

$$= 1.15 \times 10^{-2}\ \text{J}$$

Equate this total energy to the potential energy of the system when the cart is at the endpoint of the motion:

$$E = \tfrac{1}{2}kA^2$$

Solve for the amplitude A:

$$A = \sqrt{\frac{2E}{k}} = \sqrt{\frac{2(1.15 \times 10^{-2}\ \text{J})}{20.0\ \text{N/m}}} = 0.033\ 9\ \text{m}$$

Equate the total energy to the kinetic energy of the system when the cart is at the equilibrium position:

$$E = \tfrac{1}{2}mv_{max}^2$$

Solve for the maximum speed:

$$v_{max} = \sqrt{\frac{2E}{m}} = \sqrt{\frac{2(1.15 \times 10^{-2}\ \text{J})}{0.500\ \text{kg}}} = 0.214\ \text{m/s}$$

The amplitude and maximum velocity are larger than the previous values because the cart was given an initial velocity at $t = 0$.

The back edge of the treadle goes up and down as one's feet rock the treadle.

The oscillation of the treadle causes circular motion of the drive wheel, eventually resulting in additional up and down motion—of the sewing needle.

John W. Jewett, Jr

Figure 15.12 The bottom of a treadle-style sewing machine from the early twentieth century. The treadle is the wide, flat foot pedal with the metal grillwork.

15.4 Comparing Simple Harmonic Motion with Uniform Circular Motion

Some common devices in everyday life exhibit a relationship between oscillatory motion and circular motion. For example, consider the drive mechanism for a non-electric sewing machine in Figure 15.12. The operator of the machine places her feet on the treadle and rocks them back and forth. This oscillatory motion causes the large wheel at the right to undergo circular motion. The red drive belt seen in the photograph transfers this circular motion to the sewing machine mechanism (above the photo) and eventually results in the oscillatory motion of the sewing needle. In this section, we explore this interesting relationship between these two types of motion.

Figure 15.13 is a view of an experimental arrangement that shows this relationship. A ball is attached to the rim of a turntable of radius A, which is illuminated from above by a lamp. The ball casts a shadow on a screen as the turntable rotates with constant angular speed. While the ball moves as a particle in uniform circular motion, the shadow of the ball moves back and forth on the screen as a particle in simple harmonic motion.

Consider a particle located at point P on the circumference of a circle of radius A as in Figure 15.14a, with the line OP making an angle ϕ with the x axis at $t = 0$. We call this circle a *reference circle* for comparing simple harmonic motion with uniform circular motion, and we choose the position of P at $t = 0$ as our reference position. If the particle moves counterclockwise along the circle with constant angular speed ω until OP makes an angle θ with the x axis as in Figure 15.14b, at some time $t > 0$ the angle between OP and the x axis is $\theta = \omega t + \phi$. As the particle moves along the circle, the projection of P on the x axis, labeled point Q, moves back and forth along the x axis between the limits $x = \pm A$.

Notice that points P and Q always have the same x coordinate. From the right triangle OPQ, we see that this x coordinate is

$$x(t) = A \cos(\omega t + \phi) \qquad (15.23)$$

This expression is the same as Equation 15.6 and shows that the point Q moves with simple harmonic motion along the x axis. Therefore, the motion of an object described by the analysis model of a particle in simple harmonic motion along a straight line can be represented by the projection of an object that can be modeled as a particle in uniform circular motion along a diameter of a reference circle.

This geometric interpretation shows that the time interval for one complete revolution of the point P on the reference circle is equal to the period of motion T for

The ball rotates like a particle in uniform circular motion.

Lamp

A

Turntable

A

Screen

The ball's shadow moves like a particle in simple harmonic motion.

Figure 15.13 An experimental setup for demonstrating the connection between a particle in simple harmonic motion and a corresponding particle in uniform circular motion.

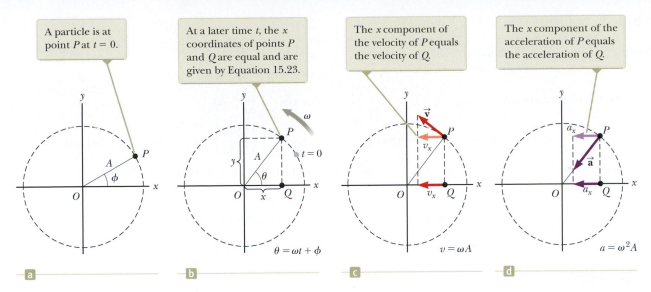

Figure 15.14 Relationship between the uniform circular motion of a point P and the simple harmonic motion of a point Q. A particle at P moves in a circle of radius A with constant angular speed ω.

simple harmonic motion between $x = \pm A$. Therefore, the angular speed ω of P is the same as the angular frequency ω of simple harmonic motion along the x axis (which is why we use the same symbol). The phase constant ϕ for simple harmonic motion corresponds to the initial angle OP makes with the x axis. The radius A of the reference circle equals the amplitude of the simple harmonic motion.

Because the relationship between linear and angular speed for circular motion is $v = r\omega$ (see Eq. 10.10), the particle moving on the reference circle of radius A has a velocity of magnitude ωA. From the geometry in Figure 15.14c, we see that the x component of this velocity is $-\omega A \sin(\omega t + \phi)$. By definition, point Q has a velocity given by dx/dt. Differentiating Equation 15.23 with respect to time, we find that the velocity of Q is the same as the x component of the velocity of P.

The acceleration of P on the reference circle is directed radially inward toward O and has a magnitude $v^2/A = \omega^2 A$. From the geometry in Figure 15.14d, we see that the x component of this acceleration is $-\omega^2 A \cos(\omega t + \phi)$. This value is also the acceleration of the projected point Q along the x axis, as you can verify by taking the second derivative of Equation 15.23.

ⓆUICK QUIZ 15.5 The ball in Figure 15.13 moves in a circle of radius 0.50 m. At $t = 0$, the ball is located on the left side of the turntable, exactly opposite its position in Figure 15.13. What are the correct values for the *amplitude* and *phase constant* (relative to an x axis to the right) of the simple harmonic motion of the shadow? **(a)** 0.50 m and 0 **(b)** 1.00 m and 0 **(c)** 0.50 m and π **(d)** 1.00 m and π

Example 15.4 **Circular Motion with Constant Angular Speed**

The ball in Figure 15.13 rotates counterclockwise in a circle of radius 3.00 m with a constant angular speed of 8.00 rad/s. At $t = 0$, its shadow has an x coordinate of 2.00 m and is moving to the right.

(A) Determine the x coordinate of the shadow as a function of time in SI units.

SOLUTION

Conceptualize Be sure you understand the relationship between circular motion of the ball and simple harmonic motion of its shadow as described in Figure 15.13. Notice that the shadow is *not* at is maximum position at $t = 0$.

Categorize The ball on the turntable is a *particle in uniform circular motion*. The shadow is modeled as a *particle in simple harmonic motion*.

continued

15.4 c o n t i n u e d

Analyze   Use Equation 15.23 to write an expression for the x coordinate of the rotating ball:

$$x = A \cos (\omega t + \phi)$$

Solve for the phase constant:

$$\phi = \cos^{-1} \left(\frac{x}{A} \right) - \omega t$$

Substitute numerical values for the initial conditions:

$$\phi = \cos^{-1} \left(\frac{2.00 \text{ m}}{3.00 \text{ m}} \right) - 0 = \pm 48.2° = \pm 0.841 \text{ rad}$$

If we were to take $\phi = +0.841$ rad as our answer, the shadow would be moving to the left at $t = 0$. Because the shadow is moving to the right at $t = 0$, we must choose $\phi = -0.841$ rad.

Write the x coordinate as a function of time:

$$x = 3.00 \cos (8.00t - 0.841)$$

(B) Find the x components of the shadow's velocity and acceleration at any time t.

S O L U T I O N

Differentiate the x coordinate with respect to time to find the velocity at any time in m/s:

$$v_x = \frac{dx}{dt} = (-3.00 \text{ m})(8.00 \text{ rad/s}) \sin (8.00t - 0.841)$$

$$= -24.0 \sin (8.00t - 0.841)$$

Differentiate the velocity with respect to time to find the acceleration at any time in m/s²:

$$a_x = \frac{dv_x}{dt} = (-24.0 \text{ m/s})(8.00 \text{ rad/s}) \cos (8.00t - 0.841)$$

$$= -192 \cos (8.00t - 0.841)$$

Finalize   Notice that the value of the phase constant puts the ball in the fourth quadrant of the xy coordinate system of Figure 15.14, which is consistent with the shadow having a positive value for x and moving toward the right.

15.5   The Pendulum

The **simple pendulum** is another mechanical system that exhibits periodic motion. It consists of a particle-like bob of mass m suspended by a light string of length L that is fixed at the upper end as shown in Figure 15.15. The motion occurs in the vertical plane and is driven by the gravitational force. We shall show that, provided the angle θ is small (less than about 10°), the motion is very close to that of a simple harmonic oscillator.

The forces acting on the bob are the force $\vec{\mathbf{T}}$ exerted by the string and the gravitational force $m\vec{\mathbf{g}}$. The tangential component $mg \sin \theta$ of the gravitational force always acts toward $\theta = 0$, opposite the displacement of the bob from the lowest position. Therefore, the tangential component is a restoring force, and we can apply Newton's second law for motion in the tangential direction:

$$F_t = ma_t \quad \rightarrow \quad -mg \sin \theta = m \frac{d^2s}{dt^2}$$

where the negative sign indicates that the tangential force acts toward the equilibrium (vertical) position and s is the bob's position measured along the arc. We have expressed the tangential acceleration as the second derivative of the position s. Because $s = L\theta$ (Eq. 10.1b with $r = L$) and L is constant, this equation reduces to

$$\frac{d^2\theta}{dt^2} = -\frac{g}{L} \sin \theta$$

Considering θ as the position, let us compare this equation with Equation 15.3. Does it have the same mathematical form? No! The right side is proportional to $\sin \theta$ rather than to θ; hence, we would not expect simple harmonic motion because this expression is not of the same mathematical form as Equation 15.3. If we

When θ is small, a simple pendulum's motion can be modeled as simple harmonic motion about the equilibrium position $\theta = 0$.

Figure 15.15   A simple pendulum.

assume θ is *small* (less than about 10° or 0.2 rad), however, we can use the **small angle approximation,** in which $\sin \theta \approx \theta$, where θ is measured in radians. Table 15.1 shows angles in degrees and radians and the sines of these angles. As long as θ is less than approximately 10°, the angle in radians and its sine are the same to within an accuracy of less than 1.0%. The table also shows the tangents of the angles, which we will use in the next chapter.

Therefore, for small angles, the equation of motion becomes

$$\frac{d^2\theta}{dt^2} = -\frac{g}{L}\theta \quad \text{(for small values of } \theta\text{)} \tag{15.24}$$

Equation 15.24 has the same mathematical form as Equation 15.3, so we conclude that the motion for small amplitudes of oscillation can be modeled as simple harmonic motion. Therefore, the solution of Equation 15.24 is modeled after Equation 15.6 and is given by $\theta = \theta_{max} \cos(\omega t + \phi)$, where θ_{max} is the *maximum angular position* and the angular frequency ω is

$$\omega = \sqrt{\frac{g}{L}} \tag{15.25}$$

◀ Angular frequency for a simple pendulum

The period of the motion is

$$T = \frac{2\pi}{\omega} = 2\pi\sqrt{\frac{L}{g}} \tag{15.26}$$

◀ Period of a simple pendulum

In other words, the period and frequency of a simple pendulum depend only on the length of the string and the acceleration due to gravity. Because the period is independent of the mass, we conclude that all simple pendula that are of equal length and are at the same location (so that g is the same) oscillate with the same period.

The simple pendulum can be used as a timekeeper because its period depends only on its length and the local value of g. It is also a convenient device for making precise measurements of the free-fall acceleration. Such measurements are important because variations in local values of g can provide information on the location of oil and other valuable underground resources.

QUICK QUIZ 15.6 The grandfather clock in the opening storyline depends on the period of a pendulum to keep correct time. **(i)** Suppose the clock is calibrated correctly and then a mischievous child slides the bob of the pendulum downward on the oscillating rod. Does the grandfather clock run (a) slow, (b) fast, or (c) correctly? **(ii)** Suppose a grandfather clock is calibrated correctly at sea level and is then taken to the top of a very tall mountain. Does the grandfather clock now run (a) slow, (b) fast, or (c) correctly?

Part (b) of Quick Quiz 15.6 relates to the grandfather clock at your grandparents' house in the opening storyline. The clock has been transferred from Denver,

TABLE 15.1 Sines and Tangents of Angles

Angle in Degrees	Angle in Radians	Sine of Angle	Percent Difference	Tangent of Angle	Percent Difference
0°	0.000 0	0.000 0	0.0%	0.000 0	0.0%
1°	0.017 5	0.017 5	0.0%	0.017 5	0.0%
2°	0.034 9	0.034 9	0.0%	0.034 9	0.0%
3°	0.052 4	0.052 3	0.0%	0.052 4	0.1%
5°	0.087 3	0.087 2	0.1%	0.087 5	0.3%
10°	0.174 5	0.173 6	0.5%	0.176 3	1.0%
15°	0.261 8	0.258 8	1.2%	0.267 9	2.3%
20°	0.349 1	0.342 0	2.1%	0.364 0	4.3%
30°	0.523 6	0.500 0	4.7%	0.577 4	10.3%

at an altitude of one mile, to Boston, essentially at sea level. As a result, the value of *g*, the acceleration due to gravity, has increased. As we can see from Equation 15.26, this decreases the period of the clock so that it runs fast. What can you do to adjust the clock? You can look at part (a) of Quick Quiz 15.6! The bob of the pendulum should have an adjustment mechanism that allows you to move the bob downward to increase the effective length of the pendulum and therefore increase the period.

Example 15.5   **A Connection Between Length and Time**

Christiaan Huygens (1629–1695), the greatest clockmaker in history, suggested that an international unit of length could be defined as the length of a simple pendulum having a period of exactly 1 s. How much shorter would our length unit be if his suggestion had been followed?

SOLUTION

Conceptualize   Imagine a pendulum that swings back and forth in exactly 1 second. Based on your experience in observing swinging objects, can you make an estimate of the required length? Hang a small object from a string and simulate the 1-s pendulum.

Categorize   This example involves a simple pendulum, so we categorize it as a substitution problem that applies the concepts introduced in this section.

Solve Equation 15.26 for the length and substitute numerical values:

$$L = \frac{T^2 g}{4\pi^2} = \frac{(1.00 \text{ s})^2 (9.80 \text{ m/s}^2)}{4\pi^2} = \boxed{0.248 \text{ m}}$$

The meter's length would be slightly less than one-fourth of its current length. Also, the number of significant digits depends only on how precisely we know *g* because the time has been defined to be exactly 1 s.

WHAT IF?   What if Huygens had been born on another planet? What would the value for *g* have to be on that planet such that the meter based on Huygens's pendulum would have the same value as our meter?

Answer   Solve Equation 15.26 for *g*:

$$g = \frac{4\pi^2 L}{T^2} = \frac{4\pi^2 (1.00 \text{ m})}{(1.00 \text{ s})^2} = 4\pi^2 \text{ m/s}^2 = 39.5 \text{ m/s}^2$$

No planet in our solar system has an acceleration due to gravity that large.

Physical Pendulum

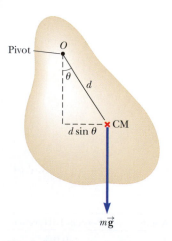

Pivot

d sin θ

Figure 15.16 A physical pendulum pivoted at *O*.

Suppose you balance a wire coat hanger so that the hook is supported by your extended index finger. When you give the hanger a small angular displacement with your other hand and then release it, it oscillates. If a hanging object oscillates about a fixed axis that does not pass through its center of mass and the object cannot be approximated as a point mass, we cannot treat the system as a simple pendulum. In this case, the system is called a **physical pendulum.**

Consider a rigid object pivoted at a point *O* that is a distance *d* from the center of mass (Fig. 15.16). The gravitational force provides a torque about an axis through *O*, and the magnitude of that torque is *mgd* sin *θ*, where *θ* is as shown in Figure 15.16. We apply the rigid object under a net torque analysis model to the object and use the rotational form of Newton's second law, $\sum \tau_{\text{ext}} = I\alpha$, where *I* is the moment of inertia of the object about the axis through *O*. The result is

$$-mgd \sin \theta = I \frac{d^2\theta}{dt^2}$$

The negative sign indicates that the torque about *O* tends to decrease *θ*. That is, the gravitational force produces a restoring torque. If we again assume *θ* is small, the approximation sin *θ* ≈ *θ* is valid and the equation of motion reduces to

$$\frac{d^2\theta}{dt^2} = -\left(\frac{mgd}{I}\right)\theta \qquad (15.27)$$

Because this equation is of the same mathematical form as Equation 15.3, its solution is modeled after that of the simple harmonic oscillator. That is, the solution of Equation 15.27 is given by $\theta = \theta_{max} \cos(\omega t + \phi)$, where θ_{max} is the maximum angular position and

$$\omega = \sqrt{\frac{mgd}{I}}$$

The period is

$$T = \frac{2\pi}{\omega} = 2\pi \sqrt{\frac{I}{mgd}}$$

(15.28)   ◀ Period of a physical pendulum

This result can be used to measure the moment of inertia of a flat, rigid object. If the location of the center of mass—and hence the value of d—is known, the moment of inertia can be obtained by measuring the period. Finally, notice that Equation 15.28 reduces to the period of a simple pendulum (Eq. 15.26) when $I = md^2$, that is, when all the mass is concentrated at the center of mass.

Example **15.6** **A Swinging Rod**

A uniform rod of mass M and length L is pivoted about one end and oscillates in a vertical plane (Fig. 15.17).

(A) Find the period of oscillation if the amplitude of the motion is small.

SOLUTION

Conceptualize Imagine a rod swinging back and forth when pivoted at one end. Try it with a meterstick or a scrap piece of wood.

Categorize Because the rod is not a point particle, we categorize it as a physical pendulum.

Figure 15.17 (Example 15.6) A rigid rod oscillating about a pivot through one end is a physical pendulum with $d = L/2$.

Analyze In Chapter 10, we found that the moment of inertia of a uniform rod about an axis through one end is $\frac{1}{3}ML^2$. The distance d from the pivot to the center of mass of the rod is $L/2$.

Substitute these quantities into Equation 15.28:

$$T = 2\pi \sqrt{\frac{\frac{1}{3}ML^2}{Mg(L/2)}} = 2\pi \sqrt{\frac{2L}{3g}}$$

(B) Suppose the pivot is moved to a small hole drilled in the rod at a distance $L/4$ from the upper end. What is the period of oscillation of the rod when it is hung from this pivot point and swings through small oscillations?

The moment of inertia in Equation 15.28 is now that about the new pivot point. Use the parallel axis theorem (Eq. 10.22):

$$I = I_{CM} + MD^2 = \tfrac{1}{12}ML^2 + M\left(\tfrac{1}{4}L\right)^2 = \tfrac{7}{48}ML^2$$

Substitute this moment of inertia and the new value of d into Equation 15.28:

$$T = 2\pi \sqrt{\frac{\frac{7}{48}ML^2}{Mg(L/4)}} = 2\pi \sqrt{\frac{7L}{12g}}$$

Finalize In one of the Moon landings, an astronaut walking on the Moon's surface had a belt hanging from his space suit, and the belt oscillated as a physical pendulum. A scientist on the Earth observed this motion on television and used it to estimate the free-fall acceleration on the Moon. How did the scientist make this calculation?

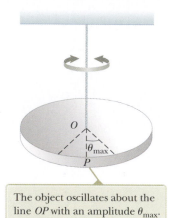

The object oscillates about the line *OP* with an amplitude θ_{max}.

Figure 15.18 A torsional pendulum.

Torsional Pendulum

Figure 15.18 shows a rigid object such as a disk suspended by a wire attached at the top to a fixed support. When the object is twisted through some angle θ, the twisted wire exerts on the object a restoring torque that is proportional to the angular position. That is,

$$\tau = -\kappa\theta$$

where κ (Greek letter kappa) is called the *torsion constant* of the support wire and is a rotational analog to the force constant k for a spring. The value of κ can be obtained by applying a known torque to twist the wire through a measurable angle θ. Applying Newton's second law for rotational motion, we find that

$$\sum \tau = I\alpha \quad \rightarrow \quad -\kappa\theta = I\frac{d^2\theta}{dt^2}$$

$$\frac{d^2\theta}{dt^2} = -\frac{\kappa}{I}\theta \tag{15.29}$$

Again, this result is the equation of motion for a simple harmonic oscillator, with $\omega = \sqrt{\kappa/I}$ and a period

$$T = 2\pi\sqrt{\frac{I}{\kappa}} \tag{15.30}$$

This system is called a *torsional pendulum*. There is no small-angle restriction in this situation as long as the elastic limit of the wire is not exceeded.

15.6 Damped Oscillations

Figure 15.19 One example of a damped oscillator is an object attached to a spring and submersed in a viscous liquid.

The oscillatory motions we have considered so far have been for ideal systems, that is, systems that oscillate indefinitely under the action of only one force, a linear restoring force. In many real systems, nonconservative forces such as friction or air resistance also act and retard the motion of the system. Consequently, the mechanical energy of the system diminishes in time, and the motion is said to be *damped*. The mechanical energy of the system is transformed into internal energy in the object and the retarding medium. Figure 15.19 depicts one such system: an object attached to a spring and submersed in a viscous liquid. Another example is a simple pendulum oscillating in air. After being set into motion, the pendulum eventually stops oscillating due to air resistance. Figure 15.20 depicts damped oscillations in practice. The spring-loaded devices mounted below the bridge are dampers that transform mechanical energy of the oscillating bridge into internal energy, reducing the swaying motion of the bridge.

One common type of retarding force is that discussed in Section 6.4, where the force is proportional to the speed of the moving object and acts in the direction opposite the velocity of the object with respect to the medium. This retarding force is often observed when an object moves through air, for instance. Because the retarding force can be expressed as $\vec{R} = -b\vec{v}$ (where b is a constant called the *damping coefficient*) and the restoring force of the system is $-kx$, we can write Newton's second law as

$$\sum F_x = -kx - bv_x = ma_x$$

which, by substituting derivatives for the velocity and acceleration, can be written as

$$m\frac{d^2x}{dt^2} + b\frac{dx}{dt} + kx = 0 \tag{15.31}$$

The solution to this equation requires mathematics that may be unfamiliar to you; we simply state it here without proof. When the retarding force is small compared

John W. Jewett, Jr.

Figure 15.20 The London Millennium Bridge over the River Thames in London. On opening day of the bridge, pedestrians noticed a swinging motion of the bridge, leading to its being named the "Wobbly Bridge." The bridge was closed after two days and remained closed for two years. Over 50 tuned mass dampers were added to the bridge: the pairs of spring-loaded structures on top of the cross members (arrow).

with the maximum restoring force—that is, when the damping coefficient b is small—the solution to Equation 15.31 is

$$x = Ae^{-(b/2m)t} \cos{(\omega t + \phi)} \tag{15.32}$$

where the angular frequency of oscillation is

$$\omega = \sqrt{\frac{k}{m} - \left(\frac{b}{2m}\right)^2} \tag{15.33}$$

This result can be verified by substituting Equation 15.32 into Equation 15.31. It is convenient to express the angular frequency of a damped oscillator in the form

$$\omega = \sqrt{\omega_0^2 - \left(\frac{b}{2m}\right)^2}$$

where $\omega_0 = \sqrt{k/m}$ represents the angular frequency in the absence of a retarding force (the undamped oscillator) and is called the **natural frequency** of the system.

Figure 15.21 shows the position as a function of time for an object oscillating in the presence of a retarding force. When the retarding force is small, the oscillatory character of the motion is preserved but the amplitude decreases exponentially in time, with the result that the motion ultimately becomes undetectable. Any system that behaves in this way is known as a **damped oscillator.** The dashed black lines in Figure 15.21, which define the *envelope* of the oscillatory curve, represent the exponential factor in Equation 15.32. This envelope shows that the amplitude decays exponentially with time. For motion with a given spring constant and object mass, the oscillations dampen more rapidly for larger values of the retarding force.

When the magnitude of the retarding force is small such that $b/2m < \omega_0$, the system is said to be **underdamped.** The resulting motion is represented by Figure 15.21 and the blue curve in Figure 15.22. As the value of b increases, the amplitude of the oscillations decreases more and more rapidly. When b reaches a critical value b_c such that $b_c/2m = \omega_0$, the system does not oscillate and is said to be **critically damped.** In this case, the system, once released from rest at some nonequilibrium position, approaches but does not pass through the equilibrium position. The graph of position versus time for this case is the red curve in Figure 15.22.

If the medium is so viscous that the retarding force is large compared with the restoring force—that is, if $b/2m > \omega_0$—the system is **overdamped.** Again, the displaced system, when free to move, does not oscillate but rather simply returns to its equilibrium position. As the damping increases, the time interval required for the system to approach equilibrium also increases as indicated by the black curve in Figure 15.22. For critically damped and overdamped systems, there is no angular frequency ω and the solution in Equation 15.32 is not valid.

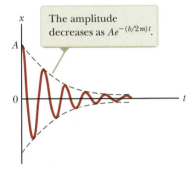

The amplitude decreases as $Ae^{-(b/2m)t}$.

Figure 15.21 Graph of position versus time for a damped oscillator.

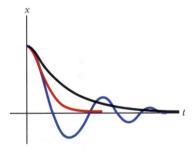

Figure 15.22 Graphs of position versus time for an underdamped oscillator (blue curve), a critically damped oscillator (red curve), and an overdamped oscillator (black curve).

15.7 Forced Oscillations

We have seen that the mechanical energy of a damped oscillator decreases in time as a result of the retarding force. It is possible to compensate for this energy decrease by applying a periodic external force that does positive work on the system. At any instant, energy can be transferred into the system by an applied force that acts in the direction of motion of the oscillator. For example, a child on a swing can be kept in motion by appropriately timed "pushes." The amplitude of motion remains constant if the energy input per cycle of motion exactly equals the decrease in mechanical energy in each cycle that results from retarding forces.

A common example of a forced oscillator is a damped oscillator driven by an external force that varies periodically, such as $F(t) = F_0 \sin{\omega t}$, where F_0 is a constant and ω is the angular frequency of the driving force. In general, the frequency ω of the driving force is variable, whereas the natural frequency ω_0 of the oscillator is fixed by the values of k and m. Modeling an oscillator with driving, retarding, and restoring forces as a particle under a net force, Newton's second law in this situation gives

$$\sum F_x = ma_x \quad \rightarrow \quad F_0 \sin \omega t - b\frac{dx}{dt} - kx = m\frac{d^2x}{dt^2} \tag{15.34}$$

Again, the solution of this equation is rather lengthy and will not be presented. After the driving force on an initially stationary object begins to act, the amplitude of the oscillation will increase. The system of the oscillator and the surrounding medium is a nonisolated system: work is done by the driving force, such that the vibrational energy of the system (kinetic energy of the object, elastic potential energy in the spring) and internal energy of the object and the medium increase. After a sufficiently long period of time, when the energy input per cycle from the driving force equals the amount of mechanical energy transformed to internal energy for each cycle, a steady-state condition is reached in which the oscillations proceed with constant amplitude. In this situation, the solution of Equation 15.34 is

$$x = A \cos(\omega t + \phi) \tag{15.35}$$

where

Amplitude of a ▶
driven oscillator

$$A = \frac{F_0/m}{\sqrt{(\omega^2 - \omega_0^2)^2 + \left(\dfrac{b\omega}{m}\right)^2}} \tag{15.36}$$

and where $\omega_0 = \sqrt{k/m}$ is the natural frequency of the undamped oscillator ($b = 0$).

Equations 15.35 and 15.36 show that the forced oscillator vibrates at the frequency of the driving force and that the amplitude of the oscillator is constant for a given driving force because it is being driven in steady-state by an external force. For small damping, the amplitude is large when the frequency of the driving force is near the natural frequency of oscillation, or when $\omega \approx \omega_0$. The dramatic increase in amplitude near the natural frequency is called **resonance**, and the natural frequency ω_0 is also called the **resonance frequency** of the system.

The reason for large-amplitude oscillations at the resonance frequency is that energy is being transferred to the system under the most favorable conditions. We can better understand this concept by taking the first time derivative of x in Equation 15.35, which gives an expression for the velocity of the oscillator. We find that v is proportional to $\sin(\omega t + \phi)$, which is the same trigonometric function as that describing the driving force. Therefore, the applied force $\vec{\mathbf{F}}$ is in phase with the velocity. The rate at which work is done on the oscillator by $\vec{\mathbf{F}}$ equals the dot product $\vec{\mathbf{F}} \cdot \vec{\mathbf{v}}$; this rate is the power delivered to the oscillator. Because the product $\vec{\mathbf{F}} \cdot \vec{\mathbf{v}}$ is a maximum when $\vec{\mathbf{F}}$ and $\vec{\mathbf{v}}$ are in phase, we conclude that at resonance, the applied force is in phase with the velocity and the power transferred to the oscillator is a maximum.

Figure 15.23 is a graph of amplitude as a function of driving frequency for a forced oscillator with and without damping. Notice that the amplitude increases with decreasing damping ($b \rightarrow 0$) and that the resonance curve broadens as the damping increases. In the absence of a damping force ($b = 0$), we see from Equation 15.36 that the steady-state amplitude approaches infinity as ω approaches ω_0. In other words, if there are no losses in the system and we continue to drive an initially motionless oscillator with a periodic force that is in phase with the velocity, the amplitude of motion builds without limit (see the red-brown curve in Fig. 15.23). This limitless building does not occur in practice because some damping is always present in reality.

Later in this book we shall see that resonance appears in other areas of physics. For example, certain electric circuits have natural frequencies and can be set into strong resonance by a varying voltage applied at a given frequency. A bridge has

When the frequency ω of the driving force equals the natural frequency ω_0 of the oscillator, resonance occurs.

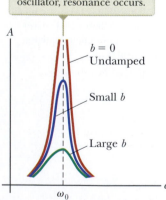

A

$b = 0$
Undamped

Small b

Large b

ω_0

ω

Figure 15.23 Graph of amplitude versus frequency for a damped oscillator when a periodic driving force is present. Notice that the shape of the resonance curve depends on the size of the damping coefficient b.

AP Images

Topham/The Image Works

a

b

Figure 15.24 (a) In 1940, turbulent winds set up torsional vibrations in the Tacoma Narrows Bridge, causing it to oscillate at a frequency near one of the natural frequencies of the bridge structure. (b) Once established, this resonance condition led to the bridge's collapse. (Mathematicians and physicists are currently challenging some aspects of this interpretation.)

natural frequencies that can be set into resonance by an appropriate driving force. A dramatic example of such resonance occurred in 1940 when the Tacoma Narrows Bridge in the state of Washington was destroyed by resonant vibrations. Although the winds were not particularly strong on that occasion, the "flapping" of the wind across the roadway (think of the "flapping" of a flag in a strong wind) provided a periodic driving force whose frequency matched that of the bridge. The resulting oscillations of the bridge caused it to ultimately collapse (Fig. 15.24) because the bridge design had inadequate built-in safety features.

Many other examples of resonant vibrations can be cited. A resonant vibration you may have experienced is the "singing" of telephone wires in the wind. Machines often break if one vibrating part is in resonance with some other moving part. Soldiers marching in cadence across a bridge have been known to set up resonant vibrations in the structure and thereby cause it to collapse. Whenever any real physical system is driven near its resonance frequency, you can expect oscillations of very large amplitudes.

Summary

> ## Concepts and Principles

The kinetic energy and potential energy for an object of mass m oscillating at the end of a spring of force constant k vary with time and are given by

$$K = \tfrac{1}{2}mv^2 = \tfrac{1}{2}m\omega^2 A^2 \sin^2(\omega t + \phi) \quad (15.19)$$

$$U_s = \tfrac{1}{2}kx^2 = \tfrac{1}{2}kA^2 \cos^2(\omega t + \phi) \quad (15.20)$$

The total energy of a simple harmonic oscillator is a constant of the motion and is given by

$$E = \tfrac{1}{2}kA^2 \quad (15.21)$$

A **simple pendulum** of length L can be modeled to move in simple harmonic motion for small angular displacements from the vertical. Its period is

$$T = 2\pi\sqrt{\frac{L}{g}} \quad (15.26)$$

A **physical pendulum** is an extended object that, for small angular displacements, can be modeled to move in simple harmonic motion about a pivot that does not go through the center of mass. The period of this motion is

$$T = 2\pi\sqrt{\frac{I}{mgd}} \quad (15.28)$$

where I is the moment of inertia of the object about an axis through the pivot and d is the distance from the pivot to the center of mass of the object.

If an oscillator experiences a damping force $\vec{R} = -b\vec{v}$, its position for small damping is described by

$$x = Ae^{-(b/2m)t} \cos(\omega t + \phi) \quad (15.32)$$

where

$$\omega = \sqrt{\frac{k}{m} - \left(\frac{b}{2m}\right)^2} \quad (15.33)$$

If an oscillator is subject to a sinusoidal driving force that is described by $F(t) = F_0 \sin \omega t$, it exhibits **resonance**, in which the amplitude is largest when the driving frequency ω matches the natural frequency $\omega_0 = \sqrt{k/m}$ of the oscillator.

continued

> ## Analysis Model for Problem Solving

Particle in Simple Harmonic Motion If a particle is subject to a force of the form of Hooke's law $F = -kx$, the particle exhibits **simple harmonic motion.** Its position is described by

$$x(t) = A \cos (\omega t + \phi) \tag{15.6}$$

where A is the **amplitude** of the motion, ω is the **angular frequency**, and ϕ is the **phase constant.** The value of ϕ depends on the initial position and initial velocity of the particle.

The **period** of the oscillation of the particle is

$$T = \frac{2\pi}{\omega} = 2\pi \sqrt{\frac{m}{k}} \tag{15.13}$$

and the inverse of the period is the **frequency.**

Think–Pair–Share

See the Preface for an explanation of the icons used in this problems set. For additional assessment items for this section, go to

1. Two identical steel balls, each of mass $m = 67.4$ g and diameter $d = 25.4$ mm, are moving in opposite directions, each at $v = 5.00$ m/s. They collide head-on and bounce apart elastically. (a) Split your group in two and have each half find the total time interval that the balls are in contact, using different models. Group (i): Model a given ball as having kinetic energy that is then completely transformed to elastic potential energy at the instant that the balls have momentarily come to rest. Assume the acceleration of the ball during this time interval is constant and use the particle under constant acceleration model to find the total time interval that the balls are in contact. Group (ii): By squeezing one of the balls in a vise while precise measurements are made of the resulting amount of compression, you have found that Hooke's law is a good model of the ball's elastic behavior. A force of $F = 16.0$ kN exerted by each jaw of the vise reduces the diameter by a distance $s = 0.200$ mm. The diameter returns to the original value when the force from the vise is removed. Model the motion of each ball, while the balls are in contact, as one-half of a cycle of simple harmonic motion. Find the total time interval that the balls are in contact. (b) Which result is more accurate?

2. **ACTIVITY** Divide your group in half. Each subgroup should work on one of the situations below:

   (i) A hanging spring stretches by 35.0 cm when an object of mass 450 g is hung on it at rest. In this situation, we define its position as $x = 0$, with positive x upward. The object is pulled down an additional 18.0 cm and released from rest to oscillate without friction.

   (ii) Another hanging spring stretches by 35.5 cm when an object of mass 440 g is hung on it at rest. We define this new position as $x = 0$. This object is pulled down an additional 18.0 cm and released from rest to oscillate without friction.

   (a) For each of these situations, answer the following two questions: (1) What is the position x of the object at a moment 84.4 s later? (2) What total distance has the vibrating object traveled in the 84.4-s time interval?

   When the calculations are finished, compare the results for the two situations. (b) Why are the answers to question 1 so different when the initial data in situations (i) and (ii) are so similar and the answers to question 2 are relatively close? (c) Does this circumstance reveal a fundamental difficulty in calculating the future?

3. **ACTIVITY** Online, you read about a group of physics students doing a simple pendulum lab. They used a small object attached to the end of a string to form a simple pendulum. The students measured the total time intervals for 50 oscillations of its harmonic motion for small angular displacements and three lengths. They posted their data online:

Length L (m)	Time interval for 50 oscillations (s)
1.000	99.8
0.750	86.6
0.500	71.1

Split your group in two and have each half find a value for g, the acceleration due to gravity, using different approaches. Group (i): Determine the period of motion T for each length of the pendulum. From that length, use Equation 15.26 to find a value of g for each length. Determine the mean value of g obtained from these three independent measurements and compare it with the accepted value. Group (ii): Determine the period of motion T for each length of the pendulum. Plot T^2 versus length L and obtain a value for g from the slope of your best-fit straight-line graph, using Equation 15.26. How do the values of g for the two groups compare?

Problems

Note: Ignore the mass of every spring.

SECTION 15.1 Motion of an Object Attached to a Spring

Problems 11, 12, 41 in Chapter 7 can also be assigned with this section.

1. A 0.60-kg block attached to a spring with force constant 130 N/m is free to move on a frictionless, horizontal surface as in Figure 15.1. The block is released from rest when the spring is stretched 0.13 m. At the instant the block is released, find (a) the force on the block and (b) its acceleration.

SECTION 15.2 Analysis Model: Particle in Simple Harmonic Motion

2. A piston in a gasoline engine is in simple harmonic motion. The engine is running at the rate of 3 600 rev/min. Taking the extremes of its position relative to its center point as ± 5.00 cm, find the magnitudes of the (a) maximum velocity and (b) maximum acceleration of the piston.

3. The position of a particle is given by the expression $x = 4.00 \cos (3.00\pi t + \pi)$, where x is in meters and t is in seconds. Determine (a) the frequency and (b) period of the motion, (c) the amplitude of the motion, (d) the phase constant, and (e) the position of the particle at $t = 0.250$ s.

4. A 7.00-kg object is hung from the bottom end of a vertical spring fastened to an overhead beam. The object is set into vertical oscillations having a period of 2.60 s. Find the force constant of the spring.

5. Review. A particle moves along the x axis. It is initially at the position 0.270 m, moving with velocity 0.140 m/s and acceleration -0.320 m/s^2. Suppose it moves as a particle under constant acceleration for 4.50 s. Find (a) its position and (b) its velocity at the end of this time interval. Next, assume it moves as a particle in simple harmonic motion for 4.50 s and $x = 0$ is its equilibrium position. Find (c) its position and (d) its velocity at the end of this time interval.

6. A ball dropped from a height of 4.00 m makes an elastic collision with the ground. Assuming no decrease in mechanical energy due to air resistance, (a) show that the ensuing motion is periodic and (b) determine the period of the motion. (c) Is the motion simple harmonic? Explain.

7. A particle moving along the x axis in simple harmonic motion starts from its equilibrium position, the origin, at $t = 0$ and moves to the right. The amplitude of its motion is 2.00 cm, and the frequency is 1.50 Hz. (a) Find an expression for the position of the particle as a function of time. Determine (b) the maximum speed of the particle and (c) the earliest time ($t > 0$) at which the particle has this speed. Find (d) the maximum positive acceleration of the particle and (e) the earliest time ($t > 0$) at which the particle has this acceleration. (f) Find the total distance traveled by the particle between $t = 0$ and $t = 1.00$ s.

8. The initial position, velocity, and acceleration of an object moving in simple harmonic motion are x_i, v_i, and a_i; the angular frequency of oscillation is ω. (a) Show that the position and velocity of the object for all time can be written as

$$x(t) = x_i \cos \omega t + \left(\frac{v_i}{\omega}\right) \sin \omega t$$

$$v(t) = -x_i \omega \sin \omega t + v_i \cos \omega t$$

(b) Using A to represent the amplitude of the motion, show that

$$v^2 - ax = v_i^2 - a_i x_i = \omega^2 A^2$$

9. You attach an object to the bottom end of a hanging vertical spring. It hangs at rest after extending the spring 18.3 cm. You then set the object vibrating. (a) Do you have enough information to find its period? (b) Explain your answer and state whatever you can about its period.

SECTION 15.3 Energy of the Simple Harmonic Oscillator

10. To test the resiliency of its bumper during low-speed collisions, a 1 000-kg automobile is driven into a brick wall. The car's bumper behaves like a spring with a force constant 5.00×10^6 N/m and compresses 3.16 cm as the car is brought to rest. What was the speed of the car before impact, assuming no mechanical energy is transformed or transferred away during impact with the wall?

11. A particle executes simple harmonic motion with an amplitude of 3.00 cm. At what position does its speed equal half of its maximum speed?

12. The amplitude of a system moving in simple harmonic motion is doubled. Determine the change in (a) the total energy, (b) the maximum speed, (c) the maximum acceleration, and (d) the period.

13. A simple harmonic oscillator of amplitude A has a total energy E. Determine (a) the kinetic energy and (b) the potential energy when the position is one-third the amplitude. (c) For what values of the position does the kinetic energy equal one-half the potential energy? (d) Are there any values of the position where the kinetic energy is greater than the maximum potential energy? Explain.

14. Review. A 65.0-kg bungee jumper steps off a bridge with a light bungee cord tied to her body and to the bridge. The unstretched length of the cord is 11.0 m. The jumper reaches the bottom of her motion 36.0 m below the bridge before bouncing back. We wish to find the time interval between her leaving the bridge and her arriving at the bottom of her motion. Her overall motion can be separated into an 11.0-m free fall and a 25.0-m section of simple harmonic oscillation. (a) For the free-fall part, what is the appropriate analysis model to describe her motion? (b) For what time interval is she in free fall? (c) For the simple harmonic oscillation part of the plunge, is the system of the bungee jumper, the spring, and the Earth isolated or nonisolated? (d) From your response in part (c) find the spring constant of the bungee cord. (e) What is the location of the equilibrium point where the spring force balances the gravitational force exerted on the jumper? (f) What is the angular frequency of the oscillation? (g) What time interval is required for the cord to stretch by 25.0 m? (h) What is the total time interval for the entire 36.0-m drop?

15. **Review.** A 0.250-kg block resting on a frictionless, horizon-
QC tal surface is attached to a spring whose force constant is
83.8 N/m as in Figure P15.15. A horizontal force $\vec{F}$ causes
the spring to stretch a distance of 5.46 cm from its equilibrium position. (a) Find the magnitude of $\vec{F}$. (b) What
is the total energy stored in the system when the spring is
stretched? (c) Find the magnitude of the acceleration of
the block just after the applied force is removed. (d) Find
the speed of the block when it first reaches the equilibrium
position. (e) If the surface is not frictionless but the block
still reaches the equilibrium position, would your answer to
part (d) be larger or smaller? (f) What other information
would you need to know to
find the actual answer to part
(d) in this case? (g) What is
the largest value of the coefficient of friction that would
allow the block to reach the
equilibrium position?

Figure P15.15

SECTION 15.4 Comparing Simple Harmonic Motion
with Uniform Circular Motion

16. While driving behind a car traveling at
QC 3.00 m/s, you notice that one of the car's
tires has a small hemispherical bump
on its rim as shown in Figure P15.16.
(a) Explain why the bump, from your
viewpoint behind the car, executes simple harmonic motion. (b) If the radii of
the car's tires are 0.300 m, what is the
bump's period of oscillation? (c) **What
If?** You hang a spring with spring con-
stant $k = 100$ N/m from the rear view mirror of your car.
What is the mass that needs to be hung from this spring to
produce simple harmonic motion with the same period as the
bump on the tire? (d) What would be the maximum speed of
the hanging mass in your car if you initially pulled the mass
down 8.00 cm beyond equilibrium before releasing it?

Bump

Figure P15.16

SECTION 15.5 The Pendulum

Problem 36 in Chapter 1 can also be assigned with this section.

17. A simple pendulum makes 120 complete oscillations in
3.00 min at a location where $g = 9.80$ m/s^2. Find (a) the
period of the pendulum and (b) its length.

18. A particle of mass m slides without friction inside a hemi-
S spherical bowl of radius R. Show that if the particle starts
from rest with a small displacement from equilibrium, it
moves in simple harmonic motion with an angular frequency equal to that of a simple pendulum of length R.
That is, $\omega = \sqrt{g/R}$.

19. A physical pendulum in the form of a planar object moves in
T simple harmonic motion with a frequency of 0.450 Hz. The
pendulum has a mass of 2.20 kg, and the pivot is located
0.350 m from the center of mass. Determine the moment of
inertia of the pendulum about the pivot point.

20. A physical pendulum in the form of a planar object moves in
S simple harmonic motion with a frequency f. The pendulum
has a mass m, and the pivot is located a distance d from the
center of mass. Determine the moment of inertia of the pendulum about the pivot point.

21. A simple pendulum has a mass of 0.250 kg and a length of
QC 1.00 m. It is displaced through an angle of 15.0° and then
released. Using the analysis model of a particle in simple
harmonic motion, what are (a) the maximum speed of
the bob, (b) its maximum angular acceleration, and (c) the
maximum restoring force on the bob? (d) **What If?** Solve
parts (a) through (c) again by using analysis models introduced in earlier chapters. (e) Compare the answers.

22. Consider the physical pendulum of Figure 15.16. (a) Rep-
S resent its moment of inertia about an axis passing through
its center of mass and parallel to the axis passing through its
pivot point as I_{CM}. Show that its period is

$$T = 2\pi \sqrt{\frac{I_{CM} + md^2}{mgd}}$$

where d is the distance
between the pivot point
and the center of mass.
(b) Show that the period
has a minimum value
when d satisfies $md^2 = I_{CM}$.

23. A watch balance wheel
(Fig. P15.23) has a period
of oscillation of 0.250 s.
The wheel is constructed
so that its mass of 20.0 g
is concentrated around
a rim of radius 0.500 cm.
What are (a) the wheel's
moment of inertia and
(b) the torsion constant of
the attached spring?

Balance wheel

© Cengage

Figure P15.23

SECTION 15.6 Damped Oscillations

24. Show that the time rate of change of mechanical
S energy for a damped, undriven oscillator is given by
$dE/dt = -bv^2$ and hence is always negative. To do so, differentiate the expression for the mechanical energy of an oscillator, $E = \frac{1}{2}mv^2 + \frac{1}{2}kx^2$, and use Equation 15.31.

25. Show that Equation 15.32 is a solution of Equation 15.31
S provided that $b^2 < 4mk$.

SECTION 15.7 Forced Oscillations

26. As you enter a fine restaurant, you realize that you have accidentally brought a small electronic timer from home instead
of your cell phone. In frustration, you drop the timer into a
side pocket of your suit coat, not realizing that the timer is
operating. The arm of your chair presses the light cloth of
your coat against your body at one spot. Fabric with a length
L hangs freely below that spot, with the timer at the bottom.
At one point during your dinner, the timer goes off and a
buzzer and a vibrator turn on and off with a frequency of
1.50 Hz. It makes the hanging part of your coat swing back
and forth with remarkably large amplitude, drawing everyone's attention. Find the value of L.

27. A 2.00-kg object attached to a spring moves without friction ($b = 0$) and is driven by an external force given by the
expression $F = 3.00 \sin(2\pi t)$, where F is in newtons and t is
in seconds. The force constant of the spring is 20.0 N/m.
Find (a) the resonance angular frequency of the system,

(b) the angular frequency of the driven system, and (c) the amplitude of the motion.

28. Considering an undamped, forced oscillator ($b = 0$), show
S that Equation 15.35 is a solution of Equation 15.34, with an amplitude given by Equation 15.36.

29. You have scored a part-
CR time job at a company that makes small probes to be released from satellites to study the very thin atmosphere at the location of satellite orbits. In order to keep the probes in a proper orientation in space, they will be spun about their axis before being released. It is important to know the moment of inertia of the odd-shaped probe. Your boss asks you to measure

Figure P15.29

its moment of inertia. You set up a system such as that in Figure 15.18, modifying it by adding a very light frame (Fig. P15.29) into which you can place objects, centering them on the disk. The frame is attached at the edges of the disk. The support wire is rigidly connected to the top of the frame so that it does not interfere with the objects you wish to place on the disk. The disk is of mass $M = 5.25$ kg and has a radius of $R = 25.8$ cm. You rotate the empty disk from its equilibrium position and let it operate as a torsional pendulum. You carefully measure its period of oscillation to be $T_{empty} = 10.8$ s. You then place the probe on the disk and adjust its position until the disk hangs exactly horizontal, so you know that the center of mass of the probe is directly over the center of the disk. You rotate the loaded disk from its equilibrium position and let it operate as a torsional pendulum. (a) You carefully measure its period of oscillation to be $T_{loaded} = 18.7$ s, and from this result you determine the moment of inertia of the probe about its center of mass. (b) When you present your results to your supervisor, she asks you about the moment of inertia of the frame you built. You go back to your desk and think about it. When you consider that the frame has some moment of inertia, is the value calculated in part (a) too high or too low?

30. You take on a research assistantship with a molecular
CR physicist. She is studying the vibrations of diatomic molecules. In these vibrations, the two atoms in the molecule move back and forth along the line connecting them (see Figure 20.5c). As an introduction to her research, she asks you to familiarize yourself with the Lennard–Jones potential (see Example 7.9), which describes the potential energy function for a diatomic molecule. She asks you to determine the effective spring constant, in terms of the parameters σ and ϵ, for the bond holding the atoms together in the molecule for small vibrations around the equilibrium separation r_{eq}. After being stumped for a while, you ask her for a hint. She responds, "Example 7.9 provides the derivative of the potential energy function. Compare that to Equation 7.29 to find the force between the atoms. You want to show that F is of the form $-kx$, and find k. Let the separation distance $r = r_{eq} + x$, where x is

small and take advantage of the series approximations in Appendix Section B.5." Wow, that's several hints! You sit down and get to work.

ADDITIONAL PROBLEMS

31. An object of mass m moves in simple harmonic motion with
QC amplitude 12.0 cm on a light spring. Its maximum acceleration is 108 cm/s². Regard m as a variable. (a) Find the period T of the object. (b) Find its frequency f. (c) Find the maximum speed v_{max} of the object. (d) Find the total energy E of the object–spring system. (e) Find the force constant k of the spring. (f) Describe the pattern of dependence of each of the quantities T, f, v_{max}, E, and k on m.

32. **Review.** This problem extends the reasoning of Problem 41
QC in Chapter 9. Two gliders are set in motion on an air track. Glider 1 has mass $m_1 = 0.240$ kg and moves to the right with speed 0.740 m/s. It will have a rear-end collision with glider 2, of mass $m_2 = 0.360$ kg, which initially moves to the right with speed 0.120 m/s. A light spring of force constant 45.0 N/m is attached to the back end of glider 2 as shown in Figure P9.41. When glider 1 touches the spring, superglue instantly and permanently makes it stick to its end of the spring. (a) Find the common speed the two gliders have when the spring is at maximum compression. (b) Find the maximum spring compression distance. The motion after the gliders become attached consists of a combination of (1) the constant-velocity motion of the center of mass of the two-glider system found in part (a) and (2) simple harmonic motion of the gliders relative to the center of mass. (c) Find the energy of the center-of-mass motion. (d) Find the energy of the oscillation.

33. An object attached to a spring vibrates with simple harmonic motion as described by Figure P15.33. For this motion, find (a) the amplitude, (b) the period, (c) the angular frequency, (d) the maximum speed, (e) the maximum acceleration, and (f) an equation for its position x as a function of time.

Figure P15.33

34. **Review.** A rock rests on a concrete sidewalk. An earthquake
QC strikes, making the ground move vertically in simple harmonic motion with a constant frequency of 2.40 Hz and with gradually increasing amplitude. (a) With what amplitude does the ground vibrate when the rock begins to lose contact with the sidewalk? Another rock is sitting on the concrete bottom of a swimming pool full of water. The earthquake produces only vertical motion, so the water does not slosh from side to side. (b) Present a convincing argument that when the ground vibrates with the amplitude found in part (a), the submerged rock also barely loses contact with the floor of the swimming pool.

35. A pendulum of length L and mass M has a spring of force constant k connected to it at a distance h below its point of suspension (Fig. P15.35). Find the frequency of vibration of the system for small values of the amplitude (small θ). Assume the vertical suspension rod of length L is rigid, but ignore its mass.

Figure P15.35

36. To account for the walking speed **BIO** of a bipedal or quadrupedal animal, model a leg that is not contacting the ground as a uniform rod of length ℓ, swinging as a physical pendulum through one-half of a cycle, in resonance. Let θ_{max} represent its amplitude. (a) Show that the animal's speed is given by the expression

$$v = \frac{\sqrt{6g\ell}\sin\theta_{max}}{\pi}$$

if θ_{max} is sufficiently small that the motion is nearly simple harmonic. An empirical relationship that is based on the same model and applies over a wider range of angles is

$$v = \frac{\sqrt{6g\ell}\cos(\theta_{max}/2)\sin\theta_{max}}{\pi}$$

(b) Evaluate the walking speed of a human with leg length 0.850 m and leg-swing amplitude 28.0°. (c) What leg length would give twice the speed for the same angular amplitude?

37. Review. A particle of mass 4.00 kg is attached to a spring **QC** with a force constant of 100 N/m. It is oscillating on a frictionless, horizontal surface with an amplitude of 2.00 m. A 6.00-kg object is dropped vertically on top of the 4.00-kg object as it passes through its equilibrium point. The two objects stick together. (a) What is the new amplitude of the vibrating system after the collision? (b) By what factor has the period of the system changed? (c) By how much does the energy of the system change as a result of the collision? (d) Account for the change in energy.

38. People who ride motorcycles and bicycles learn to look out **QC** for bumps in the road and especially for *washboarding,* a condition in which many equally spaced ridges are worn into the road. What is so bad about washboarding? A motorcycle has several springs and shock absorbers in its suspension, but you can model it as a single spring supporting a block. You can estimate the force constant by thinking about how far the spring compresses when a heavy rider sits on the seat. A motorcyclist traveling at highway speed must be particularly careful of washboard bumps that are a certain distance apart. What is the order of magnitude of their separation distance?

39. A ball of mass m is connected to two rubber bands of length **S** L, each under tension T as shown in Figure P15.39. The ball is displaced by a small distance y perpendicular to the length of the rubber bands. Assuming the tension does not change, show that (a) the restoring force is $-(2T/L)y$ and (b) the system exhibits simple harmonic motion with an angular frequency $\omega = \sqrt{2T/mL}$.

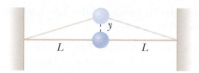

Figure P15.39

40. Consider the damped oscillator illustrated in Figure 15.19. The mass of the object is 375 g, the spring constant is 100 N/m, and $b = 0.100$ N · s/m. (a) Over what time interval does the amplitude drop to half its initial value? (b) **What If?** Over what time interval does the mechanical energy drop to half its initial value? (c) Show that, in general, the fractional rate at which the amplitude decreases in a damped harmonic oscillator is one-half the fractional rate at which the mechanical energy decreases.

41. Review. A lobsterman's buoy is a solid wooden cylinder of **S** radius r and mass M. It is weighted at one end so that it floats upright in calm seawater, having density ρ. A passing shark tugs on the slack rope mooring the buoy to a lobster trap, pulling the buoy down a distance x from its equilibrium position and releasing it. (a) Show that the buoy will execute simple harmonic motion if the resistive effects of the water are ignored. (b) Determine the period of the oscillations.

42. Your thumb squeaks on a plate you have just washed. Your **S** sneakers squeak on the gym floor. Car tires squeal when you start or stop abruptly. You can make a goblet sing by wiping your moistened finger around its rim. When chalk squeaks on a blackboard, you can see that it makes a row of regularly spaced dashes. As these examples suggest, vibration commonly results when friction acts on a moving elastic object. The oscillation is not simple harmonic motion, but is called *stick-and-slip.* This problem models stick-and-slip motion.

A block of mass m is attached to a fixed support by a horizontal spring with force constant k and negligible mass (Fig. P15.42). Hooke's law describes the spring both in extension and in compression. The block sits on a long horizontal board, with which it has coefficient of static friction μ_s and a smaller coefficient of kinetic friction μ_k. The board moves to the right at constant speed v. Assume the block spends most of its time sticking to the board and moving to the right with it, so the speed v is small in comparison to the average speed the block has as it slips back toward the left. (a) Show that the maximum extension of the spring from its unstressed position is very nearly given by $\mu_s mg/k$. (b) Show that the block oscillates around an equilibrium position at which the spring is stretched by $\mu_k mg/k$. (c) Graph the block's position versus time. (d) Show that the amplitude of the block's motion is

$$A = \frac{(\mu_s - \mu_k)mg}{k}$$

Figure P15.42

(e) Show that the period of the block's motion is

$$T = \frac{2(\mu_s - \mu_k)mg}{vk} + \pi\sqrt{\frac{m}{k}}$$

It is the excess of static over kinetic friction that is important for the vibration. "The squeaky wheel gets the grease" because even a viscous fluid cannot exert a force of static friction.

43. Your father is preparing the backyard for the installation
CR of new sod. He has finished cleaning the ground of roots and rocks, has raked it to the correct contours, and now must pull a heavy roller, shown in Figure P15.43a, over the ground several times to flatten and compact the dirt. He is tired after all of his work and asks you to do the rolling for him. He tells you that each section of the yard must be rolled over ten times with the roller. You are tired from your physics studying, but decide you can use your understanding of physics to make the job easier. You attach the roller to a spring as shown in Figure P15.43b, with the other end attached to a post pounded into the ground. You then just pull the roller out once and let it oscillate over each part of the yard for ten rolls while you sit back and relax. Before beginning, you wonder how much time you will have to relax at each location before you have to move the post and roller to a new location. The mass of the roller is $M = 400$ kg, and the spring constant is $k = 3\,500$ N/m. The flat, smooth ground supplies enough friction that the roller rolls instead of sliding, but the rolling friction is negligible.

Figure P15.43

44. *Why is the following situation impossible?* Your job involves building very small damped oscillators. One of your designs involves a spring–object oscillator with a spring of force constant $k = 10.0$ N/m and an object of mass $m = 1.00$ g. Your design objective is that the oscillator undergo many oscillations as its amplitude falls to 25.0% of its initial value in a certain time interval. Measurements on your latest design show that the amplitude falls to the 25.0% value in 23.1 ms. This time interval is too long for what is needed in your project. To shorten the time interval, you double the damping constant b for the oscillator. This doubling allows you to reach your design objective.

45. A block of mass m is connected to two springs of force con-
S stants k_1 and k_2 in two ways as shown in Figure P15.45. In both cases, the block moves on a frictionless table after it is displaced from equilibrium and released. Show that in the two cases the block exhibits simple harmonic motion with periods

(a) $T = 2\pi\sqrt{\dfrac{m(k_1 + k_2)}{k_1 k_2}}$ and (b) $T = 2\pi\sqrt{\dfrac{m}{k_1 + k_2}}$

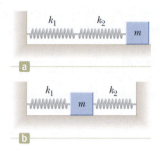

Figure P15.45

46. **Review.** A light balloon filled with helium of density 0.179 kg/m³ is tied to a light string of length $L = 3.00$ m. The string is tied to the ground forming an "inverted" simple pendulum (Fig. P15.46a). If the balloon is displaced slightly from equilibrium as in Figure P15.46b and released, (a) show that the motion is simple harmonic and (b) determine the period of the motion. Take the density of air to be 1.20 kg/m³. *Hint:* Use an analogy with the simple pendulum and see Chapter 14. Assume the air applies a buoyant force on the balloon but does not otherwise affect its motion.

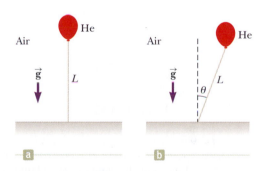

Figure P15.46

47. A particle with a mass of 0.500 kg is attached to a horizontal spring with a force constant of 50.0 N/m. At the moment $t = 0$, the particle has its maximum speed of 20.0 m/s and is moving to the left. (a) Determine the particle's equation of motion, specifying its position as a function of time. (b) Where in the motion is the potential energy three times the kinetic energy? (c) Find the minimum time interval required for the particle to move from $x = 0$ to $x = 1.00$ m. (d) Find the length of a simple pendulum with the same period.

CHALLENGE PROBLEMS

48. A smaller disk of radius r and
S mass m is attached rigidly to the face of a second larger disk of radius R and mass M as shown in Figure P15.48. The center of the small disk is located at the edge of the large disk. The large disk is mounted at its center on a frictionless axle. The assembly is rotated through a small angle θ from its equilibrium position and

Figure P15.48

released. (a) Show that the speed of the center of the small disk as it passes through the equilibrium position is

$$v = 2 \left[\frac{Rg(1 - \cos \theta)}{(M/m) + (r/R)^2 + 2} \right]^{1/2}$$

(b) Show that the period of the motion is

$$T = 2\pi \left[\frac{(M + 2m)R^2 + mr^2}{2mgR} \right]^{1/2}$$

49. **Review.** A system consists of a spring with force constant $k = 1\ 250$ N/m, length $L = 1.50$ m, and an object of mass $m = 5.00$ kg attached to the end (Fig. P15.49). The object is placed at the level of the point of attachment with the spring unstretched, at position $y_i = L$, and then it is released so that it swings like a pendulum. (a) Find the y position of the object at the lowest point. (b) Will the pendulum's period be greater or less than the period of a simple pendulum with the same mass m and length L? Explain.

Figure P15.49

50. **Review.** *Why is the following situation impossible?* You are in the high-speed package delivery business. Your competitor in the next building gains the right-of-way to build an evacuated tunnel just above the ground all the way around the Earth. By firing packages into this tunnel at just the right speed, your competitor is able to send the packages into orbit around the Earth in this tunnel so that they arrive on the exact opposite side of the Earth in a very short time interval. You come up with a competing idea. Figuring that the distance *through* the Earth is shorter than the distance *around*

the Earth, you obtain permits to build an evacuated tunnel through the center of the Earth (Fig. P15.50). By simply dropping packages into this tunnel, they fall downward and arrive at the other end of your tunnel, which is in a building right next to the other end of your competitor's tunnel. Because your packages arrive on the other side of the Earth in a shorter time interval, you win the competition and your business flourishes. *Note:* An object at a distance r from the center of the Earth is pulled toward the center of the Earth only by the mass within the sphere of radius r (the reddish region in Fig. P15.50). Assume the Earth has uniform density.

Figure P15.50

51. A light, cubical container of volume a^3 is initially filled with a liquid of mass density ρ as shown in Figure P15.51a. The cube is initially supported by a light string to form a simple pendulum of length L_i, measured from the center of mass of the filled container, where $L_i \gg a$. The liquid is allowed to flow from the bottom of the container at a constant rate (dM/dt). At any time t, the level of the liquid in the container is h and the length of the pendulum is L (measured relative to the instantaneous center of mass) as shown in Figure P15.51b. (a) Find the period of the pendulum as a function of time. (b) What is the period of the pendulum after the liquid completely runs out of the container?

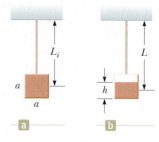

Figure P15.51

Wave Motion

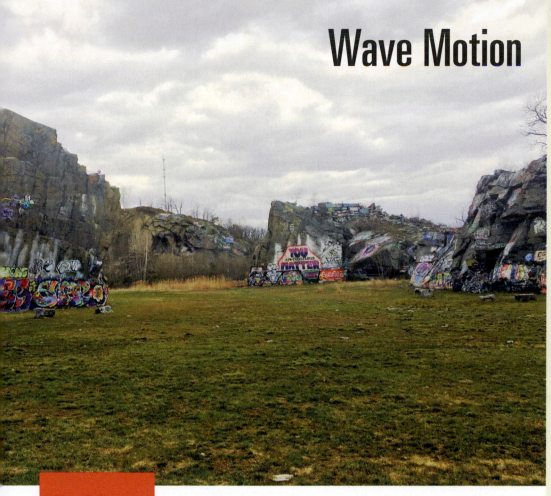

◄

At the Quincy Quarries Reservation in Quincy, Massachusetts, rainwater filled in an old granite quarry, so that the water was surrounded by rocks and cliffs. When Boston undertook the Big Dig, in which huge amounts of dirt were removed from beneath the city to make way for underground tunnels, the water at Quincy Quarry was filled in with that dirt. Consequently, there are now large flat areas between granite cliffs. (© *Cengage*)

STORYLINE **During your visit to your grandparents' house in Boston,** you and your grandfather take a day trip. You stop at Quincy Quarries Reservation, where there are large flat areas between the cliffs. While visiting, you notice that you can clap your hands and hear a distinct echo from a distant cliff. You say, "Watch this, Grandpa, let me show you something about cell phones." You pull out your smartphone and activate the digital recording app. You ask your grandfather to clap his hands once, just after you start recording. You stop the recording after the echo from the cliff arrives. Seeing the pulses on the app display representing both the clap and the echo, you determine the time interval for the sound of the clap to travel to the cliff and back. Then you use the GPS system on your phone to determine your latitude and longitude coordinates. At this point, you say, "Grandpa, let's go for a hike!" You hike across the former lake to the base of the cliff that provided the echo and determine your coordinates again. Based on the two sets of coordinates, you use a Web site to determine the distance between the cliff and your original position. From this distance and the time interval you measured for the echo to arrive, you make a reasonably accurate calculation of the speed of sound. Your grandfather is quite impressed with you.

CONNECTIONS In this chapter, we will continue crossing the bridge we mentioned at the beginning of Chapter 15. *Wave motion* represents phenomena in which a *disturbance* propagates through a medium. The disturbance carries energy from one point to another. But there is no matter that moves over that distance. For example, suppose you go bowling. You can knock the pins over by rolling the bowling ball at them. That is *not* wave motion. The energy is carried by the bowling ball—there is a transfer of matter. But suppose you could *shout* loud enough to knock the pins over. (Do *not* try this!) *That* would be energy transfer by waves. The energy is carried by the sound wave of your voice—no matter transfers from your mouth to the pins. In our discussions in this chapter and the next, we discuss *mechanical* waves. These waves require a *medium*. For example, we will study one-dimensional waves traveling on a string. The string is the medium. We will also consider mechanical waves in three dimensions: the waves can travel in any direction through a bulk medium. When the medium is air, we call such mechanical waves *sound*. We will relate phenomena associated with sound waves to our sense of hearing. We will use our information from this chapter to study waves under boundary conditions in Chapter 17, which will lead to an understanding of musical instruments. Furthermore, the material in this chapter will form the foundation of our study of electromagnetic waves in Chapters 33–37 and quantum physics in Chapters 39–44.

16.1 Propagation of a Disturbance

The introduction to this chapter alluded to the essence of wave motion: the transfer of energy through space without the accompanying transfer of matter. In the list of energy transfer mechanisms in Equation 8.2, two mechanisms—mechanical waves T_{MW} and electromagnetic radiation T_{ER}—depend on waves. By contrast, in another mechanism, matter transfer T_{MT}, the energy transfer is accompanied by a movement of matter through space with no wave character at all in the process.

All mechanical waves require (1) some source of disturbance, (2) a medium containing elements that can be disturbed, and (3) some physical mechanism through which elements of the medium can influence each other. One way to demonstrate wave motion is to flick one end of a long string that is under tension and has its opposite end fixed as shown in Figure 16.1. In this manner, a single bump (called a *pulse*) is formed and travels along the string with a constant speed. Figure 16.1 represents four consecutive "snapshots" of the creation and propagation of the traveling pulse. The hand is the source of the disturbance. The string is the medium through which the pulse travels—individual elements of the string are disturbed from their equilibrium position. Furthermore, the elements of the string are connected together so they influence each other: as one element goes up, it pulls the next one upward. The pulse has a definite height and a definite speed of propagation along the medium. The shape of the pulse changes very little as it travels along the string.[1]

As the pulse in Figure 16.1 travels to the right, each disturbed element of the string moves in the vertical direction, *perpendicular* to the direction of propagation. Figure 16.2 illustrates this point for one particular element, labeled *P*. Notice that no part of the string ever moves in the direction of the propagation. A traveling wave or pulse that causes the elements of the disturbed medium to move perpendicular to the direction of propagation is called **transverse.**

Compare the pulse in Figure 16.1 with another type of pulse, one moving down a long, stretched spring as shown in Figure 16.3. The left end of the spring is pushed briefly to the right and then pulled briefly to the left. This movement creates a sudden compression of a region of the coils. The compressed region travels along the spring (to the right in Fig. 16.3). Notice that the direction of the displacement of the coils is *parallel* to the direction of propagation of the compressed region. A traveling wave or pulse that causes the elements of the medium to move parallel to the direction of propagation is called **longitudinal.**

As the pulse moves along the string, new elements of the string are displaced from their equilibrium positions.

Figure 16.1 A hand moves the end of a stretched string up and down once (black, double-headed arrow), causing a pulse to travel along the string.

The direction of the displacement of any element at a point *P* on the string is perpendicular to the direction of propagation (red arrow).

Figure 16.2 The displacement of a particular string element for a transverse pulse traveling on a stretched string.

---

[1]In reality, the pulse changes shape and gradually spreads out during the motion. This effect, called *dispersion*, is common to many mechanical waves as well as to electromagnetic waves. We do not consider dispersion in this chapter.

The hand moves forward and back once to create a longitudinal pulse.

As the pulse passes by, the displacement of the coils is parallel to the direction of the propagation.

Figure 16.3 A longitudinal pulse along a stretched spring.

If the end of the string in Figure 16.1 were moved up and down continuously, the hand would generate a series of pulses called a **transverse wave.** We will study the details of waves such as this in Section 16.2. Sound waves, which we shall discuss later in this chapter, are an example of **longitudinal waves.** The disturbance in a sound wave is a series of high-pressure and low-pressure regions that travel through air, as we shall see in Section 16.6.

Some waves in nature exhibit a combination of transverse and longitudinal displacements. Surface-water waves such as those in the ocean are a good example. When a water wave travels on the surface of deep water, elements of water at the surface move in nearly circular paths as shown in Figure 16.4. The disturbance has both transverse and longitudinal components. The transverse displacements seen in Figure 16.4 represent the variations in vertical position of the water elements. The longitudinal displacements represent elements of water moving back and forth in a horizontal direction. A point in Figure 16.4 at which the displacement of the element from its normal position is highest is called the **crest** of the wave. The lowest point is called the **trough.**

An earthquake represents a disturbance that results in *seismic waves.* Two types of three-dimensional seismic waves travel out from a point under the Earth's surface at which an earthquake occurs: transverse and longitudinal. The longitudinal waves are the faster of the two, traveling at speeds in the range of 7 to 8 km/s near the surface. They are called **P waves,** with "P" standing for *primary,* because they travel faster than the transverse waves and arrive first at a seismograph (a device used to detect waves due to earthquakes). The slower transverse waves, called **S waves,** with "S" standing for *secondary,* travel through the Earth at 4 to 5 km/s near the surface. By recording the time interval between the arrivals of these two types of waves at a seismograph, the distance from the seismograph to the point of origin of the waves can be determined. This distance is the radius of an imaginary sphere centered on the seismograph. The origin of the waves is located somewhere on that sphere. The imaginary spheres from three or more monitoring stations located far apart from one another intersect at one region of the Earth, and this region is where the earthquake occurred.

QUICK QUIZ 16.1 **(i)** In a long line of people waiting to buy tickets, the first person leaves and a pulse of motion occurs as people step forward to fill the gap. As each person steps forward, the gap moves through the line. Is the propagation of this gap (a) transverse or (b) longitudinal? **(ii)** Consider "the wave" at a baseball game: people stand up and raise their arms as the pulse arrives at their location, and the resultant pulse moves around the stadium. Is this pulse (a) transverse or (b) longitudinal?

Consider a pulse traveling to the right on a long string as shown in Figure 16.5. At any time, the pulse can be represented by some mathematical function that we will write as $y(x, t)$. At $t = 0$, as in Figure 16.5a, let's write this as $y(x, 0) = f(x)$, where $f(x)$ describes the shape of the pulse in space.

The function $y(x, t)$, sometimes called the **wave function,** depends on the two variables x and t. For this reason, it is described as "y as a function of x and t."

It is important to understand the meaning of y. Consider an element of the string at point P in Figure 16.5, identified by a particular value of its x coordinate.

The elements at the surface move in nearly circular paths. Each element is displaced both horizontally and vertically from its equilibrium position.

Crest

Velocity of propagation

Trough

Figure 16.4 The motion of water elements on the surface of deep water in which a wave is propagating is a combination of transverse and longitudinal displacements.

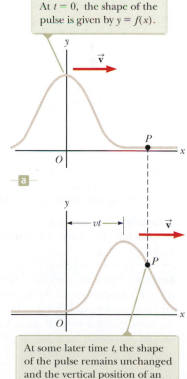

At $t = 0$, the shape of the pulse is given by $y = f(x)$.

At some later time t, the shape of the pulse remains unchanged and the vertical position of an element of the medium at any point P is given by $y = f(x - vt)$.

Figure 16.5 A one-dimensional pulse traveling to the right on a string with a speed v.

As the pulse passes through *P*, the *y* coordinate of this element increases, reaches a maximum, and then decreases to zero. The wave function $y(x, t)$ represents the *y* coordinate—the transverse position—of any element located at position *x* at any time *t*. If we were to view the pulse at a particular instant of time, such as in the case of taking a snapshot of the pulse, we would see something like Figure 16.5a or 16.5b. The geometric shape $f(x)$ of the pulse at a particular instant is called the **waveform.**

Because the speed of the pulse is *v*, the crest of the pulse has traveled to the right a distance *vt* at the time *t* (Fig. 16.5b). We assume the shape of the pulse does not change with time. Therefore, at time *t*, the shape of the pulse is the same as it was at time $t = 0$ as in Figure 16.5a. Consequently, an element of the string at *x* at this time has the same *y* position as an element located at $x - vt$ had at time $t = 0$:

$$y(x, t) = y(x - vt, 0)$$

In general, then, we can represent the transverse position *y* for all positions and times, measured in a stationary frame with the origin at *O*, as

$$y(x, t) = f(x - vt) \qquad (16.1)$$

Similarly, if the pulse travels to the left, the transverse positions of elements of the string are described by

$$y(x, t) = f(x + vt) \qquad (16.2)$$

Example 16.1 A Pulse Moving to the Right

A pulse moving to the right along the *x* axis is represented by the wave function

$$y(x, t) = \frac{2}{(x - 3.0t)^2 + 1}$$

where *x* and *y* are measured in centimeters and *t* is measured in seconds. Find expressions for the wave function at $t = 0$, $t = 1.0$ s, and $t = 2.0$ s.

SOLUTION

Conceptualize Figure 16.6a shows the pulse represented by this wave function at $t = 0$. Imagine this pulse moving to the right and maintaining its shape as suggested by Figures 16.6b and 16.6c.

Categorize We categorize this example as a relatively simple analysis problem in which we interpret the mathematical representation of a pulse.

Analyze The wave function is of the form $y = f(x - vt)$. Inspection of the expression for $y(x, t)$ and comparison to Equation 16.1 reveal that the wave speed is $v = 3.0$ cm/s. Furthermore, we can maximize the value of *y* by letting $x - 3.0t = 0$, and find that $y_{\max} = 2.0$ cm.

Figure 16.6
(Example 16.1) Graphs of the function $y(x, t) = 2/[(x - 3.0t)^2 + 1]$ at (a) $t = 0$, (b) $t = 1.0$ s, and (c) $t = 2.0$ s.

Write the wave function expression at $t = 0$:

$$y(x, 0) = \frac{2}{x^2 + 1}$$

Write the wave function expression at $t = 1.0$ s:

$$y(x, 1.0) = \frac{2}{(x - 3.0)^2 + 1}$$

Write the wave function expression at $t = 2.0$ s:

$$y(x, 2.0) = \frac{2}{(x - 6.0)^2 + 1}$$

16.1 continued

For each of these expressions, we can substitute various values of x and plot the wave function. This procedure yields the wave functions shown in the three parts of Figure 16.6.

Finalize These snapshots show that the pulse moves to the right without changing its shape and that it has a constant speed of 3.0 cm/s.

WHAT IF? What if the wave function were

$$y(x, t) = \frac{4}{(x + 3.0t)^2 + 1}$$

How would that change the situation?

Answer One new feature in this expression is the plus sign in the denominator rather than the minus sign. The new expression represents a pulse with a similar shape as that in Figure 16.6, but moving to the left as time progresses. Another new feature here is the numerator of 4 rather than 2. Therefore, the new expression represents a pulse with twice the height of that in Figure 16.6.

16.2 Analysis Model: Traveling Wave

To generate the pulse on the rope in Figure 16.1, we shook the end of the rope up and down *once*. In this section, we introduce an important wave function whose shape is shown in Figure 16.7 and is produced by shaking the end of the rope up and down *continuously* in simple harmonic motion. The wave represented by this curve is called a **sinusoidal wave** because the curve is the same as that of the function $\sin \theta$ plotted against θ. Because shaking the end of the rope in simple harmonic motion leads to a sinusoidal wave, we see that there is a close relationship between simple harmonic motion and sinusoidal waves.

The sinusoidal wave is the simplest example of a periodic continuous wave and can be used to build more complex waves (see Section 17.8). The brown curve in Figure 16.7 represents a snapshot of a traveling sinusoidal wave at $t = 0$, and the blue curve represents a snapshot of the wave at some later time t. Imagine two types of motion that can occur. First, the entire waveform in Figure 16.7 moves to the right so that the brown curve moves toward the right and eventually reaches the position of the blue curve. This movement is the motion of the *wave*. If we focus on one element of the medium, such as the element at $x = 0$, we see that each element moves up and down along the y axis in simple harmonic motion. This movement is the motion of the *elements of the medium*. It is important to differentiate between the motion of the wave and the motion of the elements of the medium. An element of the medium is described by the particle in simple harmonic motion model. A point on the wave, such as the crest, can be described with the particle under constant velocity model.

In the early chapters of this book, we developed several analysis models based on three simplification models: the particle, the system, and the rigid object. With our introduction to waves, we can develop a new simplification model, the **wave,** that will allow us to explore more analysis models for solving problems. In what follows, we will develop the principal features and mathematical representations of the analysis model of a **traveling wave.** This model is used in situations in which a wave moves through space without interacting with other waves or particles.

Figure 16.8a (page 420) shows a snapshot of a traveling wave moving through a medium. Figure 16.8b shows a graph of the position of one element of the medium as a function of time. Recall from Section 16.1 that the highest point on a wave is called the crest of the wave, and the lowest point is the trough. The distance from one crest to the next is called the **wavelength** λ (Greek letter lambda). More generally, the wavelength is the minimum distance between any two identical points on adjacent waves as shown in Figure 16.8a.

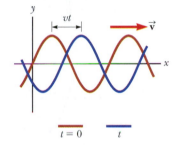

Figure 16.7 A one-dimensional sinusoidal wave traveling to the right with a speed v. The brown curve represents a snapshot of the wave at $t = 0$, and the blue curve represents a snapshot at some later time t.

The wavelength λ of a wave is the distance between adjacent crests or adjacent troughs.

a

The period T of a wave is the time interval required for the element to complete one cycle of its oscillation and for the wave to travel one wavelength.

b

Figure 16.8 (a) A snapshot of a sinusoidal wave. (b) The position of one element of the medium as a function of time.

PITFALL PREVENTION 16.1

What's the Difference Between Figures 16.8a and 16.8b? Notice the visual similarity between Figures 16.8a and 16.8b. The shapes are the same, but (a) is a graph of vertical position versus horizontal position, whereas (b) is vertical position versus time. Figure 16.8a is a pictorial representation of the wave *for a series of elements of the medium;* it is what you would see at an instant of time. Figure 16.8b is a graphical representation of the position of *one element of the medium* as a function of time. That both figures have the identical shape represents Equation 16.1: a wave is the *same* function of both *x* and *t*.

If you count the number of seconds between the arrivals of two adjacent crests at a given location in space, you measure the **period** T of the waves. In general, the period is the time interval required for an element of the medium to undergo a complete cycle and return to the same position as shown in Figure 16.8b. The period of the wave is the same as the period of the simple harmonic oscillation of one element of the medium.

The same information is more often given by the inverse of the period, which is called the **frequency** f. In general, the frequency of a periodic wave is the number of crests (or troughs, or any other point on the wave) that pass a given location in a unit time interval. The frequency of a sinusoidal wave is related to the period by the expression

$$f = \frac{1}{T} \tag{16.3}$$

The frequency of the wave is the same as the frequency of the simple harmonic oscillation of one element of the medium. The most common unit for frequency, as we learned in Chapter 15, is s^{-1}, or **hertz** (Hz). The corresponding unit for T is seconds.

An ideal particle has zero size. We can build physical objects with nonzero size as combinations of particles. Therefore, the particle can be considered a basic building block. An ideal wave has a single frequency and is infinitely long; that is, the wave exists throughout the Universe. (A wave of finite length must necessarily have a mixture of frequencies.) When this concept is explored in Section 17.8, we will find that ideal waves can be combined to build complex waves, just as we combined particles: the wave is a basic building block.

The maximum position of an element of the medium relative to its equilibrium position is called the **amplitude** A of the wave as indicated in Figure 16.8. Consider the sinusoidal wave in Figure 16.8a, which shows the position of the wave at $t = 0$. Because the wave is sinusoidal, we expect the wave function at this instant to be expressed as $y(x, 0) = A \sin ax$, where A is the amplitude and a is a constant to be determined. At $x = 0$, we see that $y(0, 0) = A \sin a(0) = 0$, consistent with Figure 16.8a. The next value of x for which y is zero is $x = \lambda/2$. Therefore,

$$y\left(\frac{\lambda}{2}, 0\right) = A \sin\left(a\frac{\lambda}{2}\right) = 0$$

For this equation to be true, we must have $a\lambda/2 = \pi$, or $a = 2\pi/\lambda$. Therefore, the function describing the positions of the elements of the medium through which the sinusoidal wave is traveling can be written

$$y(x, 0) = A \sin\left(\frac{2\pi}{\lambda} x\right) \tag{16.4}$$

where the constant A represents the wave amplitude and the constant λ is the wavelength. Notice that the vertical position of an element of the medium is the same whenever x is increased by an integral multiple of λ. Based on our discussion of Equation 16.1, if the wave moves to the right with a speed v, the wave function at some later time t is

$$y(x, t) = A \sin\left[\frac{2\pi}{\lambda}(x - vt)\right] \tag{16.5}$$

If the wave were traveling to the left, the quantity $x - vt$ would be replaced by $x + vt$ as we learned when we developed Equations 16.1 and 16.2.

By definition, the wave travels through a displacement Δx equal to one wavelength λ in a time interval Δt of one period T. Therefore, the wave speed, wavelength, and period are related by the expression

$$v = \frac{\Delta x}{\Delta t} = \frac{\lambda}{T} \tag{16.6}$$

Substituting this expression for v into Equation 16.5 gives

$$y(x, t) = A \sin \left[2\pi \left(\frac{x}{\lambda} - \frac{t}{T} \right) \right] \qquad (16.7)$$

This form of the wave function shows the *periodic* nature of $y(x, t)$. At any given time t, $y(x, t)$ has the *same* value at the positions x, $x + \lambda$, $x + 2\lambda$, and so on. Furthermore, at any given position x, the value of $y(x, t)$ is the same at times t, $t + T$, $t + 2T$, and so on.

We can express the wave function in a convenient form by defining two other quantities, the **angular wave number** k (usually called simply the **wave number**) and the **angular frequency** ω:

$$k \equiv \frac{2\pi}{\lambda} \qquad (16.8) \qquad \blacktriangleleft \text{ Angular wave number}$$

$$\omega \equiv \frac{2\pi}{T} = 2\pi f \qquad (16.9) \qquad \blacktriangleleft \text{ Angular frequency}$$

Using these definitions, Equation 16.7 can be written in the more compact form

$$y(x, t) = A \sin (kx - \omega t) \qquad (16.10) \qquad \blacktriangleleft \begin{array}{l} \text{Wave function for a} \\ \text{sinusoidal wave} \end{array}$$

Using Equations 16.3, 16.8, and 16.9, the wave speed v originally given in Equation 16.6 can be expressed in the following alternative forms:

$$v = \frac{\omega}{k} \qquad (16.11)$$

$$v = \lambda f \qquad (16.12) \qquad \blacktriangleleft \text{ Speed of a sinusoidal wave}$$

The wave function given by Equation 16.10 assumes the vertical position y of an element of the medium is zero at $x = 0$ and $t = 0$. That need not be the case. If it is not, we generally express the wave function in the form

$$y(x, t) = A \sin (kx - \omega t + \phi) \qquad (16.13) \qquad \blacktriangleleft \begin{array}{l} \text{General expression for a} \\ \text{sinusoidal wave} \end{array}$$

where ϕ is the **phase constant,** just as we learned in our study of periodic motion in Chapter 15. This constant can be determined from the initial conditions. The primary equations in the mathematical representation of the traveling wave analysis model are Equations 16.3, 16.10, and 16.12.

ⓆUICK QUIZ 16.2 A sinusoidal wave of frequency f is traveling along a stretched string. The string is brought to rest, and a second traveling wave of frequency $2f$ is established on the string. **(i)** What is the wave speed of the second wave? (a) twice that of the first wave (b) half that of the first wave (c) the same as that of the first wave (d) impossible to determine **(ii)** From the same choices, describe the wavelength of the second wave. **(iii)** From the same choices, describe the amplitude of the second wave.

Example **16.2** **A Traveling Sinusoidal Wave**

A sinusoidal wave traveling in the positive x direction has an amplitude of 15.0 cm, a wavelength of 40.0 cm, and a frequency of 8.00 Hz. The vertical position of an element of the medium at $t = 0$ and $x = 0$ is also 15.0 cm as shown in Figure 16.9 (page 422).

(A) Find the wave number k, period T, angular frequency ω, and speed v of the wave.

continued

16.2 continued

SOLUTION

Conceptualize   Figure 16.9 shows the wave at $t = 0$. Imagine this wave moving to the right and maintaining its shape.

Categorize   From the description in the problem statement, we see that we are analyzing a mechanical wave moving through a medium, so we categorize the problem with the *traveling wave* model.

Figure 16.9   (Example 16.2) A sinusoidal wave of wavelength $\lambda = 40.0$ cm and amplitude $A = 15.0$ cm.

Analyze

Evaluate the wave number from Equation 16.8:

$$k = \frac{2\pi}{\lambda} = \frac{2\pi \text{ rad}}{40.0 \text{ cm}} = 15.7 \text{ rad/m}$$

Evaluate the period of the wave from Equation 16.3:

$$T = \frac{1}{f} = \frac{1}{8.00 \text{ s}^{-1}} = 0.125 \text{ s}$$

Evaluate the angular frequency of the wave from Equation 16.9:

$$\omega = 2\pi f = 2\pi(8.00 \text{ s}^{-1}) = 50.3 \text{ rad/s}$$

Evaluate the wave speed from Equation 16.12:

$$v = \lambda f = (40.0 \text{ cm})(8.00 \text{ s}^{-1}) = 3.20 \text{ m/s}$$

(B) Determine the phase constant ϕ and write a general expression for the wave function.

SOLUTION

Substitute $A = 15.0$ cm, $y = 15.0$ cm, $x = 0$, and $t = 0$ into Equation 16.13:

$$15.0 = (15.0) \sin \phi \;\rightarrow\; \sin \phi = 1 \;\rightarrow\; \phi = \frac{\pi}{2} \text{ rad}$$

Write the wave function:

$$y(x, t) = A \sin \left(kx - \omega t + \frac{\pi}{2} \right) = A \cos (kx - \omega t)$$

Substitute the values for A, k, and ω in SI units into this expression:

$$y(x, t) = 0.150 \cos (15.7x - 50.3t)$$

Finalize   Review the results carefully and make sure you understand them. How would the graph in Figure 16.9 change if the phase angle were zero? How would the graph change if the amplitude were 30.0 cm? How would the graph change if the wavelength were 10.0 cm?

Sinusoidal Waves on Strings

In Figure 16.1, we demonstrated how to create a pulse by jerking a taut string up and down once. To create a series of such pulses—a wave—let's replace the hand with an oscillating blade whose end is vibrating in simple harmonic motion. Figure 16.10 represents snapshots of the wave created in this way at intervals of $T/4$. Because the end of the blade oscillates in simple harmonic motion, each element of the string, such as that at P, also oscillates vertically with simple harmonic motion. Therefore, every element of the string can be treated as a simple harmonic oscillator vibrating with a frequency equal to the frequency of oscillation of the blade.[2] Notice that while each element oscillates in the y direction, the wave travels to the right in the $+x$ direction with a speed v.

If we define $t = 0$ as the time for which the configuration of the string is as shown in Figure 16.10a, the wave function can be written from Equation 16.10 as

$$y = A \sin (kx - \omega t)$$

where we simplify $y(x, t)$ by writing it simply as y. We can use this expression to describe the motion of any element of the string. An element at point P (or any other element of the string) moves only vertically, and so its x coordinate remains

PITFALL PREVENTION 16.2

Two Kinds of Speed/Velocity
Do not confuse v, the speed of the wave as it propagates along the string, with v_y, the transverse velocity of a point on the string. The speed v is constant for a uniform medium, whereas v_y varies sinusoidally.

---

[2]In this arrangement, we are assuming that a string element always oscillates in a vertical line. The tension in the string would vary if an element were allowed to move sideways. Such motion would make the analysis very complex.

constant. Therefore, the **transverse speed** v_y (not to be confused with the wave speed v) and the **transverse acceleration** a_y of elements of the string are

$$v_y = \frac{dy}{dt}\bigg]_{x=\text{constant}} = \frac{\partial y}{\partial t} = -\omega A \cos(kx - \omega t) \qquad (16.14)$$

$$a_y = \frac{dv_y}{dt}\bigg]_{x=\text{constant}} = \frac{\partial v_y}{\partial t} = -\omega^2 A \sin(kx - \omega t) \qquad (16.15)$$

These expressions incorporate partial derivatives because y depends on both x and t. In the operation $\partial y / \partial t$, for example, we take a derivative with respect to t while holding x constant. The maximum magnitudes of the transverse speed and transverse acceleration are simply the absolute values of the coefficients of the cosine and sine functions:

$$v_{y,\text{max}} = \omega A \qquad (16.16)$$

$$a_{y,\text{max}} = \omega^2 A \qquad (16.17)$$

The transverse speed and transverse acceleration of elements of the string do not reach their maximum values simultaneously. The transverse speed reaches its maximum value (ωA) when $y = 0$, whereas the magnitude of the transverse acceleration reaches its maximum value ($\omega^2 A$) when $y = \pm A$. Finally, Equations 16.16 and 16.17 are identical in mathematical form to the corresponding equations for simple harmonic motion, Equations 15.17 and 15.18.

QUICK QUIZ 16.3 The amplitude of a wave is doubled, with no other changes made to the wave. As a result of this doubling, which of the following statements is correct? **(a)** The speed of the wave changes. **(b)** The frequency of the wave changes. **(c)** The maximum transverse speed of an element of the medium changes. **(d)** Statements **(a)** through **(c)** are all true. **(e)** None of statements **(a)** through **(c)** is true.

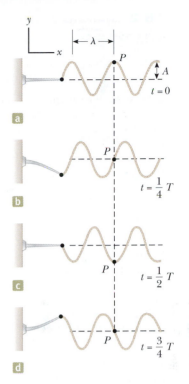

Figure 16.10 One method for producing a sinusoidal wave on a string. The left end of the string is connected to a blade that is set into oscillation. Every element of the string, such as that at point P, oscillates with simple harmonic motion in the vertical direction.

ANALYSIS MODEL **Traveling Wave**

Imagine a source vibrating such that it influences the medium that is in contact with the source. Such a source creates a disturbance that propagates through the medium. If the source vibrates in simple harmonic motion with period T, sinusoidal waves propagate through the medium at a speed given by

$$v = \frac{\lambda}{T} = \lambda f \qquad (16.6, 16.12)$$

where λ is the **wavelength** of the wave and f is its **frequency**. A sinusoidal wave can be expressed as

$$y(x, t) = A \sin(kx - \omega t) \qquad (16.10)$$

where A is the **amplitude** of the wave, k is its **wave number**, and ω is its **angular frequency**.

Examples:

- a vibrating blade sends a sinusoidal wave down a string attached to the blade
- a piston vibrates back and forth, emitting sound waves into a tube filled with gas (Section 16.6)
- a guitar body vibrates, emitting sound waves into the air (Chapter 17)
- a vibrating electric charge creates an electromagnetic wave that propagates into space at the speed of light (Chapter 33)

16.3 The Speed of Waves on Strings

One aspect of the behavior of *linear* mechanical waves is that the wave speed depends only on the properties of the medium through which the wave travels. Waves for which the amplitude A is small relative to the wavelength λ can be represented as linear waves. (See Section 16.5.) In this section, we determine the speed of a transverse wave traveling on a stretched string.

Figure 16.11 (a) In the reference frame of the Earth, a pulse moves to the right on a string with speed v. (b) In a frame of reference moving to the right with the pulse, the small element of length Δs moves to the left with speed v.

Let us use a mechanical analysis to derive the expression for the speed of a pulse traveling on a stretched string under tension T. Consider a pulse moving to the right with a uniform speed v, measured relative to a stationary (with respect to the Earth) inertial reference frame as shown in Figure 16.11a. Newton's laws are valid in any inertial reference frame. Therefore, let us view this pulse from a different inertial reference frame, one that moves along with the pulse at the same speed so that the pulse appears to be at rest in the frame as in Figure 16.11b. In this reference frame, the pulse remains fixed and each element of the string moves to the left through the pulse shape.

A short element of the string, of length Δs, forms an approximate arc of a circle of radius R as shown in the magnified view in Figure 16.11b. In our moving frame of reference, the element of the string moves to the left with speed v. As it travels through the arc, we can model the element as a particle in circular motion. This element has a centripetal (downward) acceleration of v^2/R, which is supplied by components of the force $\vec{T}$ whose magnitude is the tension in the string. The force $\vec{T}$ acts on each side of the element, tangent to the arc, as in Figure 16.11b. The horizontal components of $\vec{T}$ cancel, and each vertical component $T \sin \theta$ acts downward. Hence, the magnitude of the total radial force on the element is $2T \sin \theta$. Because the element is small, θ is small and we can use the small-angle approximation $\sin \theta \approx \theta$. Therefore, the magnitude of the total radial force is

$$F_r = 2T \sin \theta \approx 2T\theta$$

The element has mass $m = \mu \Delta s$, where μ is the mass per unit length of the string. Because the element forms part of a circle and subtends an angle of 2θ at the center, $\Delta s = R(2\theta)$, and

$$m = \mu \Delta s = 2\mu R\theta$$

The element of the string is modeled as a particle under a net force. Therefore, applying Newton's second law to this element in the radial direction gives

$$F_r = \frac{mv^2}{R} \quad \rightarrow \quad 2T\theta = \frac{2\mu R\theta v^2}{R} \quad \rightarrow \quad T = \mu v^2$$

Solving for v gives

Speed of a wave on ▶
a stretched string

$$v = \sqrt{\frac{T}{\mu}} \qquad (16.18)$$

PITFALL PREVENTION 16.3

Multiple T's Do not confuse the T in Equation 16.18 for the tension with the symbol T used in this chapter for the period of a wave. The context of the equation should help you identify which quantity is meant. There simply aren't enough letters in the alphabet to assign a unique letter to each variable!

Notice that this derivation is based on the assumption that the pulse height is small relative to the length of the pulse. Using this assumption, we were able to use the approximation $\sin \theta \approx \theta$. Furthermore, the model assumes that the tension T is not affected by the presence of the pulse, so T is the same at all points on the string. Finally, this proof does *not* assume any particular shape for the pulse. We therefore conclude that a pulse or a wave of *any shape* will travel on the string with speed $v = \sqrt{T/\mu}$, without any change in pulse shape.

Ⓠ UICK QUIZ 16.4 Suppose you create a pulse by moving the free end of a taut string up and down once with your hand beginning at $t = 0$. The string is attached at its other end to a distant wall. The pulse reaches the wall at time t. Which of the following actions, taken by itself, decreases the time interval required for the pulse to reach the wall? More than one choice may be correct. **(a)** moving your hand more quickly, but still only up and down once by the same amount **(b)** moving your hand more slowly, but still only up and down once by the same amount **(c)** moving your hand a greater distance up and down in the same amount of time **(d)** moving your hand a lesser distance up and down in the same amount of time **(e)** using a heavier string of the same length and under the same tension **(f)** using a lighter string of the same length and under the same tension **(g)** using a string of the same linear mass density but under decreased tension **(h)** using a string of the same linear mass density but under increased tension

Example 16.3 | The Speed of a Pulse on a Cord

A uniform string has a mass of 0.300 kg and a length of 6.00 m. The string passes over a pulley and supports a 2.00-kg object (Fig. 16.12). Find the speed of a pulse traveling along this string.

SOLUTION

Conceptualize In Figure 16.12, the hanging block establishes a tension in the horizontal string. This tension determines the speed with which waves move on the string.

Categorize To find the tension in the string, we model the hanging block as a *particle in equilibrium*. Then we use the tension to evaluate the wave speed on the string using Equation 16.18.

Figure 16.12 (Example 16.3) The tension T in the cord is maintained by the suspended object. The speed of any wave traveling along the cord is given by $v = \sqrt{T/\mu}$.

2.00 kg

Analyze Apply the particle in equilibrium model to the block:

$$\sum F_y = T - m_{block}g = 0$$

Solve for the tension in the string:

$$T = m_{block}g$$

Use Equation 16.18 to find the wave speed, using $\mu = m_{string}/\ell$ for the linear mass density of the string:

$$v = \sqrt{\frac{T}{\mu}} = \sqrt{\frac{m_{block}\,g\ell}{m_{string}}}$$

Substitute numerical values:

$$v = \sqrt{\frac{(2.00 \text{ kg})(9.80 \text{ m/s}^2)(6.00 \text{ m})}{0.300 \text{ kg}}} = 19.8 \text{ m/s}$$

Finalize The calculation of the tension neglects the small mass of the string. Strictly speaking, the string can never be exactly straight due to its weight; therefore, the tension is not uniform.

WHAT IF? What if the block were swinging back and forth with respect to the vertical like a pendulum? How would that affect the wave speed on the string?

Answer The swinging block is categorized as a *particle under a net force*. The magnitude of one of the forces on the block is the tension in the string, which determines the wave speed. As the block swings, the tension changes, so the wave speed changes.

When the block is at the bottom of the swing, the string is vertical and the tension is larger than the weight of the block because the net force must be upward to provide the centripetal acceleration of the block. Therefore, the wave speed must be greater than 19.8 m/s.

When the block is at its highest point at the end of a swing, it is momentarily at rest, so there is no centripetal acceleration at that instant. The block is a particle in equilibrium in the radial direction. The tension is balanced by a component of the gravitational force on the block. Therefore, the tension is smaller than the weight and the wave speed is less than 19.8 m/s. With what frequency does the speed of the wave vary? Is it the same frequency as the pendulum?

Example 16.4 | Rescuing the Hiker

An 80.0-kg hiker is trapped on a mountain ledge following a storm. A helicopter rescues the hiker by hovering above him and lowering a cable to him. The mass of the cable is 8.00 kg, and its length is 15.0 m. A sling of mass 70.0 kg is attached to the end of the cable. The hiker attaches himself to the sling, and the helicopter then accelerates upward. Terrified by hanging from the cable in midair, the hiker tries to signal the pilot by sending transverse pulses up the cable. A pulse takes 0.250 s to travel the length of the cable. What is the acceleration of the helicopter? Assume the tension in the cable is uniform.

SOLUTION

Conceptualize Imagine the effect of the acceleration of the helicopter on the cable. The greater the upward acceleration, the larger the tension in the cable. In turn, the larger the tension, the higher the speed of pulses on the cable.

Categorize This problem is a combination of one involving the speed of pulses on a string and one in which the hiker and sling are modeled as a *particle under a net force*.

continued

16.4 continued

Analyze   Solve Equation 16.18 for the tension in the cable:

$$(1) \quad v = \sqrt{\frac{T}{\mu}} \quad \rightarrow \quad T = \mu v^2$$

Model the hiker and sling as a particle under a net force, noting that the acceleration of this particle of mass m is the same as the acceleration of the helicopter:

$$\sum F = ma \quad \rightarrow \quad T - mg = ma$$

Solve for the acceleration and substitute the tension from Equation (1):

$$a = \frac{T}{m} - g = \frac{\mu v^2}{m} - g = \frac{m_{cable}}{\ell_{cable}} \frac{v^2}{m} - g = \frac{m_{cable}}{\ell_{cable} \, m} \left(\frac{\Delta x}{\Delta t}\right)^2 - g$$

Substitute numerical values:

$$a = \frac{(8.00 \text{ kg})}{(15.0 \text{ m})(150.0 \text{ kg})} \left(\frac{15.0 \text{ m}}{0.250 \text{ s}}\right)^2 - 9.80 \text{ m/s}^2 = 3.00 \text{ m/s}^2$$

Finalize   A real cable has stiffness in addition to tension. Stiffness tends to return a wire to its original straight-line shape even when it is not under tension. For example, a piano wire laid freely on a table straightens if released from a curved shape; package-wrapping string does not.

Stiffness represents a restoring force in addition to tension and increases the wave speed. Consequently, for a real cable, the acceleration of the helicopter is most likely smaller than what we calculated.

The pulse lifts the block, increasing the gravitational potential energy of the block–Earth system.

Figure 16.13 (a) A pulse travels to the right on a stretched string, carrying energy with it. (b) The energy of the pulse arrives at the hanging block.

Each element of the string is a simple harmonic oscillator and therefore has kinetic energy and potential energy associated with it.

Figure 16.14 A sinusoidal wave traveling along the x axis on a stretched string.

16.4   Rate of Energy Transfer by Sinusoidal Waves on Strings

Waves transport energy T_{MW} through a medium as they propagate. For example, suppose an object is hanging on a stretched string and a pulse is sent down the string as in Figure 16.13a. When the pulse meets the suspended object, the object is momentarily displaced upward as in Figure 16.13b. In the process, energy is transferred to the object and appears as an increase in the gravitational potential energy of the object–Earth system. This section examines the rate at which energy is transported along a string. We shall assume a one-dimensional sinusoidal wave in the calculation of the energy transferred.

Consider a sinusoidal wave traveling on a string (Fig. 16.14). The source of the energy is the vibrating blade at the left end of the string. We can consider the string to be a nonisolated system. As the blade performs work on the end of the string, moving it up and down, energy enters the system of the string and propagates along its length. Let's focus our attention on an infinitesimal element of the string of length dx and mass dm. We can model each element of the string as a particle in simple harmonic motion, with the oscillation in the y direction. All elements have the same angular frequency ω and the same amplitude A. The kinetic energy K associated with a moving particle is $K = \frac{1}{2}mv^2$. If we apply this equation to the infinitesimal element, the kinetic energy dK associated with the up and down motion of this element is

$$dK = \tfrac{1}{2}(dm)v_y^2$$

where v_y is the transverse speed of the element. If μ is the mass per unit length of the string, the mass dm of the element of length dx is equal to $\mu \, dx$. Hence, we can express the kinetic energy of an element of the string as

$$dK = \tfrac{1}{2}(\mu \, dx)v_y^2 \tag{16.19}$$

Substituting for the general transverse speed of an element of the medium using Equation 16.14 gives

$$dK = \tfrac{1}{2}\mu[-\omega A \cos (kx - \omega t)]^2 \, dx = \tfrac{1}{2}\mu\omega^2 A^2 \cos^2 (kx - \omega t) \, dx$$

If we take a snapshot of the wave at time $t = 0$, the kinetic energy of a given element is

$$dK = \tfrac{1}{2}\mu\omega^2 A^2 \cos^2 kx\, dx$$

Integrating this expression over all the string elements in a wavelength of the wave gives the total kinetic energy K_λ in one wavelength:

$$K_\lambda = \int dK = \int_0^\lambda \tfrac{1}{2}\mu\omega^2 A^2 \cos^2 kx\, dx = \tfrac{1}{2}\mu\omega^2 A^2 \int_0^\lambda \cos^2 kx\, dx$$

$$= \tfrac{1}{2}\mu\omega^2 A^2 \left[\tfrac{1}{2}x + \frac{1}{4k}\sin 2kx\right]_0^\lambda = \tfrac{1}{2}\mu\omega^2 A^2 \left[\tfrac{1}{2}\lambda\right] = \tfrac{1}{4}\mu\omega^2 A^2\lambda$$

In addition to kinetic energy, there is potential energy associated with each element of the string due to its displacement from the equilibrium position and the restoring forces from neighboring elements. A similar analysis to that above for the total potential energy U_λ in one wavelength gives exactly the same result:

$$U_\lambda = \tfrac{1}{4}\mu\omega^2 A^2\lambda$$

The total energy in one wavelength of the wave is the sum of the potential and kinetic energies:

$$E_\lambda = U_\lambda + K_\lambda = \tfrac{1}{2}\mu\omega^2 A^2\lambda \tag{16.20}$$

As the wave moves along the string, this amount of energy passes by a given point on the string during a time interval of one period of the oscillation. Therefore, the power P, or rate of energy transfer T_{MW} associated with the mechanical wave, is

$$P = \frac{T_{MW}}{\Delta t} = \frac{E_\lambda}{T} = \frac{\tfrac{1}{2}\mu\omega^2 A^2\lambda}{T} = \tfrac{1}{2}\mu\omega^2 A^2 \left(\frac{\lambda}{T}\right)$$

$$P = \tfrac{1}{2}\mu\omega^2 A^2 v \tag{16.21}$$

◀ Power of a wave

Equation 16.21 shows that the rate of energy transfer by a sinusoidal wave on a string is proportional to (a) the square of the frequency, (b) the square of the amplitude, and (c) the wave speed.

QUICK QUIZ 16.5 Which of the following, taken by itself, would be most effective in increasing the rate at which energy is transferred by a wave traveling along a string? **(a)** reducing the linear mass density of the string by one-half **(b)** doubling the wavelength of the wave **(c)** doubling the tension in the string **(d)** doubling the amplitude of the wave

| Example **16.5** | **Power Supplied to a Vibrating String** |

A taut string for which $\mu = 5.00 \times 10^{-2}$ kg/m is under a tension of 80.0 N. How much power must be supplied to the string to generate sinusoidal waves at a frequency of 60.0 Hz and an amplitude of 6.00 cm?

SOLUTION

Conceptualize Consider Figure 16.14 again and notice that the vibrating blade supplies energy to the string at a certain rate. This energy then propagates to the right along the string.

Categorize We evaluate quantities from equations developed in the chapter, so we categorize this example as a substitution problem.

continued

16.5 continued

Use Equation 16.21 to evaluate the power:

$$P = \tfrac{1}{2}\mu\omega^2 A^2 v$$

Use Equations 16.9 and 16.18 to substitute for ω and v:

$$P = \tfrac{1}{2}\mu(2\pi f)^2 A^2\left(\sqrt{\frac{T}{\mu}}\right) = 2\pi^2 f^2 A^2 \sqrt{\mu T}$$

Substitute numerical values:

$$P = 2\pi^2(60.0\text{ Hz})^2(0.060\ 0\text{ m})^2\sqrt{(0.050\ 0\text{ kg/m})(80.0\text{ N})} = \boxed{512\text{ W}}$$

WHAT IF?   What if the string is to transfer energy at a rate of 1 000 W? What must be the required amplitude if all other parameters remain the same?

Answer   Let us set up a ratio of the new and old power, reflecting only a change in the amplitude:

$$\frac{P_{\text{new}}}{P_{\text{old}}} = \frac{\tfrac{1}{2}\mu\omega^2 A^2_{\text{new}}\, v}{\tfrac{1}{2}\mu\omega^2 A^2_{\text{old}}\, v} = \frac{A^2_{\text{new}}}{A^2_{\text{old}}}$$

Solving for the new amplitude gives

$$A_{\text{new}} = A_{\text{old}}\sqrt{\frac{P_{\text{new}}}{P_{\text{old}}}} = (6.00\text{ cm})\sqrt{\frac{1\ 000\text{ W}}{512\text{ W}}} = 8.39\text{ cm}$$

16.5   The Linear Wave Equation

Figure 16.15 An element of a string under tension T.

In Section 16.1, we introduced the concept of the wave function to represent waves traveling on a string. All wave functions $y(x, t)$ represent solutions of an equation called the *linear wave equation*. This equation gives a complete description of the wave motion, and from it one can derive an expression for the wave speed. Furthermore, the linear wave equation is basic to many forms of wave motion. In this section, we derive this equation as applied to waves on strings.

Suppose a traveling wave is propagating along a string that is under a tension T. Let's consider one small string element of length Δx (Fig. 16.15). The ends of the element make small angles θ_A and θ_B with the x axis. Forces act on the string at its ends where it connects to neighboring elements. Therefore, the element is modeled as a particle under a net force. The net force acting on the element in the vertical direction is

$$\sum F_y = T\sin\theta_B - T\sin\theta_A = T(\sin\theta_B - \sin\theta_A)$$

Because the angles are small, we can use the approximation $\sin\theta \approx \tan\theta$ (see Table 15.1) to express the net force as

$$\sum F_y \approx T(\tan\theta_B - \tan\theta_A) \tag{16.22}$$

Imagine undergoing an infinitesimal displacement outward from the right end of the rope element in Figure 16.15 along the blue line representing the force $\vec{\mathbf{T}}$. This displacement has infinitesimal x and y components and can be represented by the vector $dx\,\hat{\mathbf{i}} + dy\,\hat{\mathbf{j}}$. The tangent of the angle with respect to the x axis for this displacement is dy/dx. Because we evaluate this tangent at a particular instant of time, we must express it in partial derivative form as $\partial y/\partial x$. Substituting for the tangents in Equation 16.22 gives

$$\sum F_y \approx T\left[\left(\frac{\partial y}{\partial x}\right)_B - \left(\frac{\partial y}{\partial x}\right)_A\right] \tag{16.23}$$

Now, from the particle under a net force model, let's apply Newton's second law to the element, with the mass of the element given by $m = \mu\,\Delta x$:

$$\sum F_y = ma_y = \mu\,\Delta x\left(\frac{\partial^2 y}{\partial t^2}\right) \tag{16.24}$$

Combining Equation 16.23 with Equation 16.24 gives

$$\mu \, \Delta x \left(\frac{\partial^2 y}{\partial t^2}\right) = T\left[\left(\frac{\partial y}{\partial x}\right)_B - \left(\frac{\partial y}{\partial x}\right)_A\right]$$

$$\frac{\mu}{T} \frac{\partial^2 y}{\partial t^2} = \frac{(\partial y/\partial x)_B - (\partial y/dx)_A}{\Delta x} \tag{16.25}$$

The right side of Equation 16.25 can be expressed in a different form if we note that the partial derivative of any function is defined as

$$\frac{\partial f}{\partial x} \equiv \lim_{\Delta x \to 0} \frac{f(x + \Delta x) - f(x)}{\Delta x}$$

Associating $f(x + \Delta x)$ with $(\partial y/\partial x)_B$ and $f(x)$ with $(\partial y/\partial x)_A$, we see that, in the limit $\Delta x \to 0$, Equation 16.25 becomes

$$\frac{\mu}{T} \frac{\partial^2 y}{\partial t^2} = \frac{\partial^2 y}{\partial x^2} \tag{16.26}$$

◀ Linear wave equation for a string

This expression is the linear wave equation as it applies to waves on a string.

The linear wave equation (Eq. 16.26) is often written in the form

$$\frac{\partial^2 y}{\partial x^2} = \frac{1}{v^2} \frac{\partial^2 y}{\partial t^2} \tag{16.27}$$

◀ Linear wave equation in general

Equation 16.27 applies in general to various types of traveling waves whose speed is v. For waves on strings, y represents the vertical position of elements of the string. For sound waves propagating through a gas, y corresponds to longitudinal position of elements of the gas from equilibrium or variations in either the pressure or the density of the gas. In the case of electromagnetic waves, y corresponds to electric or magnetic field components.

We have shown that the sinusoidal wave function (Eq. 16.10) is one solution of the linear wave equation (Eq. 16.27). Although we do not prove it here, the linear wave equation is satisfied by *any* wave function having the form $y = f(x \pm vt)$.

16.6 Sound Waves

We focus our attention now on **sound waves,** which travel through any material, but are most commonly experienced as the mechanical waves traveling through air that result in the human perception of hearing. As sound waves travel through air, elements of air are disturbed from their equilibrium positions. Accompanying these movements are changes in density and pressure of the air along the direction of wave motion. If the source of the sound waves vibrates sinusoidally, the density and pressure variations are also sinusoidal. The mathematical description of sinusoidal sound waves is very similar to that of sinusoidal waves on strings.

Sound waves are divided into three categories that cover different frequency ranges. (1) *Audible waves* lie within the range of sensitivity of the human ear. They can be generated in a variety of ways, such as by musical instruments, human voices, or loudspeakers. (2) *Infrasonic waves* have frequencies below the audible range. Elephants can use infrasonic waves to communicate with one another, even when separated by many kilometers. (3) *Ultrasonic waves* have frequencies above the audible range. You may have used a "silent" whistle to retrieve your dog. Dogs easily hear the ultrasonic sound this whistle emits, although humans cannot detect it at all. Ultrasonic waves are also used in medical imaging.

Earlier in the chapter, we began our investigation of waves by imagining the creation of a single pulse that traveled down a string (Figure 16.1) or a spring (Figure 16.3). Let's do something similar for sound. We describe pictorially the motion

Before the piston moves, the gas is undisturbed.

a

The gas is compressed by the motion of the piston.

b

When the piston stops, the compressed pulse continues through the gas.

$\vec{v}$

c

Figure 16.16 Motion of a longitudinal pulse through a compressible gas. The compression (darker region) is produced by the moving piston.

$\leftarrow\!\lambda\!\rightarrow$

Figure 16.17 A longitudinal wave propagating through a gas-filled tube. The source of the wave is an oscillating piston at the left.

of a one-dimensional longitudinal sound pulse moving through a long tube containing a compressible gas as shown in Figure 16.16. A piston at the left end can be quickly moved to the right to compress the gas and create the pulse. Before the piston is moved, the gas is undisturbed and of uniform density as represented by the uniformly shaded region in Figure 16.16a. When the piston is pushed to the right (Fig. 16.16b), the gas just in front of it is compressed (as represented by the more heavily shaded region); the pressure and density in this region are now higher than they were before the piston moved. When the piston comes to rest (Fig. 16.16c), the compressed region of the gas continues to move to the right, corresponding to a longitudinal pulse traveling through the tube with speed v.

One can produce a one-dimensional *periodic* sound wave in the tube of gas in Figure 16.16 by causing the piston to move in simple harmonic motion. The results are shown in Figure 16.17. The darker parts of the colored areas in this figure represent regions in which the gas is compressed and the density and pressure are above their equilibrium values. A compressed region is formed whenever the piston is pushed into the tube. This compressed region, called a **compression,** moves through the tube, continuously compressing the region just in front of itself. When the piston is pulled back, the gas in front of it expands and the pressure and density in this region fall below their equilibrium values (represented by the lighter parts of the colored areas in Fig. 16.17). These low-pressure regions, called **rarefactions,** also propagate along the tube, following the compressions. Both regions move at the speed of sound in the medium.

As the piston oscillates sinusoidally, regions of compression and rarefaction are continuously set up. The distance between two successive compressions (or two successive rarefactions) equals the wavelength λ of the sound wave. Because the sound wave is longitudinal, as the compressions and rarefactions travel through the tube, any small element of the gas moves with simple harmonic motion parallel to the direction of the wave. If $s(x, t)$ is the position of a small element relative to its equilibrium position,[3] we can express this harmonic position function as

$$s(x, t) = s_{max} \cos (kx - \omega t) \tag{16.28}$$

where s_{max} is the maximum position of the element relative to equilibrium. This parameter is often called the **displacement amplitude** of the wave. The parameter k is the wave number, and ω is the angular frequency of the wave as defined in Equations 16.8 and 16.9. Notice that the displacement of the element is along x, in the direction of propagation of the sound wave.

The variation in the gas pressure ΔP measured from the equilibrium value is also periodic with the same wave number and angular frequency as for the displacement in Equation 16.28. Therefore, we can write

$$\Delta P = \Delta P_{max} \sin (kx - \omega t) \tag{16.29}$$

where **the pressure amplitude ΔP_{max}** is the maximum change in pressure from the equilibrium value.

Notice that we have expressed the displacement by means of a cosine function and the pressure by means of a sine function. We will justify this choice in the procedure that follows and relate the pressure amplitude ΔP_{max} to the displacement amplitude s_{max}. Consider the piston–tube arrangement of Figure 16.16 once again. In Figure 16.18a, we focus our attention on a small cylindrical element of undisturbed gas of length Δx and area A. The volume of this element is $V_i = A \, \Delta x$.

Figure 16.18b shows this element of gas after a sound wave has moved it to a new position. The cylinder's two flat faces move through different distances s_1 and s_2. The change in volume ΔV of the element in the new position is equal to $A \, \Delta s$, where $\Delta s = s_1 - s_2$.

---

[3]We use $s(x, t)$ here instead of $y(x, t)$ because the displacement of elements of the medium is not perpendicular to the x direction.

From the definition of bulk modulus (see Eq. 12.8), we express the pressure variation in the element of gas as a function of its change in volume:

$$\Delta P = -B \frac{\Delta V}{V_i}$$

Let's substitute for the initial volume and the change in volume of the element:

$$\Delta P = -B \frac{A \, \Delta s}{A \, \Delta x}$$

Let the length Δx of the cylinder approach zero so that the ratio $\Delta s/\Delta x$ becomes a partial derivative:

$$\Delta P = -B \frac{\partial s}{\partial x} \qquad (16.30)$$

Substitute the position function given by Equation 16.28:

$$\Delta P = -B \frac{\partial}{\partial x}[s_{max} \cos (kx - \omega t)] = B s_{max} k \sin (kx - \omega t)$$

From this result, we see that a displacement described by a cosine function leads to a pressure described by a sine function. We also see that the displacement and pressure amplitudes are related by

$$\Delta P_{max} = B s_{max} k \qquad (16.31)$$

This relationship depends on the bulk modulus of the gas, which is not as readily available as is the density of the gas. Once we determine the speed of sound in a gas in Section 16.7, we will be able to provide an expression that relates ΔP_{max} and s_{max} in terms of the density of the gas.

This discussion shows that a sound wave in a gas may be described equally well in terms of either pressure or displacement. A comparison of Equations 16.28 and 16.29 shows that the pressure wave is 90° out of phase with the displacement wave. Graphs of these functions are shown in Figure 16.19. The pressure variation is a maximum when the displacement from equilibrium is zero, and the displacement from equilibrium is a maximum when the pressure variation is zero.

QUICK QUIZ **16.6** If you blow across the top of an empty soft-drink bottle, a pulse of sound travels down through the air in the bottle. At the moment the pulse reaches the bottom of the bottle, what is the correct description of the displacement of elements of air from their equilibrium positions and the pressure of the air at this point? **(a)** The displacement and pressure are both at a maximum. **(b)** The displacement and pressure are both at a minimum. **(c)** The displacement is zero, and the pressure is a maximum. **(d)** The displacement is zero, and the pressure is a minimum.

16.7 Speed of Sound Waves

We now extend the discussion begun in Section 16.6 to evaluate the speed of sound in a gas. In Figure 16.20a (page 432), consider the cylindrical element of gas between the piston and the dashed line. This element of gas is in equilibrium under the influence of forces of equal magnitude, from the piston on the left and from the rest of the gas on the right. The magnitude of each of these forces is PA, where P is the pressure in the gas and A is the cross-sectional area of the tube. The length of the undisturbed element of gas is chosen to be $v \, \Delta t$, where v is the speed of sound in the gas and Δt is the time interval between the configurations in Figures 16.20a and 16.20b.

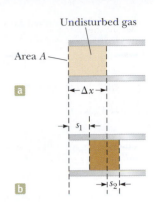

Figure 16.18 (a) An undisturbed cylindrical element of gas of length Δx in a tube of cross-sectional area A. (b) When a sound wave propagates through the gas, the element is moved to a new position and has a different length. The parameters s_1 and s_2 describe the displacements of the ends of the element from their equilibrium positions.

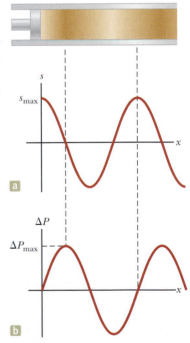

Figure 16.19 (a) Displacement amplitude and (b) pressure amplitude versus position for a sinusoidal longitudinal sound wave in a gas.

Figure 16.20 (a) An undisturbed element of gas of length $v \Delta t$ in a tube of cross-sectional area A. The element is in equilibrium between forces on either end. (b) When the piston moves inward at constant velocity v_x due to an increased force on the left, the element also moves with the same velocity.

Figure 16.20b shows the situation after this time interval Δt, during which the piston moves to the right at a constant speed v_x due to a force from the left on the piston that has increased in magnitude to $(P + \Delta P)A$. Because the speed of sound is v, the sound wave will just reach the right end of the cylindrical element of gas at the end of the time interval Δt. The gas to the right of the element is undisturbed because the sound wave has not reached it yet. At this moment every bit of gas in the element is moving with speed v_x. That will not be true in general for a macroscopic element of gas, but it will become true if we shrink the length of the element to an infinitesimal value.

The element of gas is modeled as a nonisolated system in terms of momentum. The force from the piston has provided an impulse to the element, which in turn exhibits a change in momentum. Therefore, we evaluate both sides of the impulse–momentum theorem, Equation 9.13:

$$\Delta \vec{\mathbf{p}} = \vec{\mathbf{I}} \tag{16.32}$$

On the right, the impulse is provided by the constant force due to the increased pressure on the piston:

$$\vec{\mathbf{I}} = \sum \vec{\mathbf{F}} \, \Delta t = (A \, \Delta P \, \Delta t) \hat{\mathbf{i}}$$

The pressure change ΔP can be related to the volume change and then to the speeds v and v_x through the bulk modulus:

$$\Delta P = -B \frac{\Delta V}{V_i} = -B \frac{(-v_x A \, \Delta t)}{vA \, \Delta t} = B \frac{v_x}{v}$$

Therefore, the impulse becomes

$$\vec{\mathbf{I}} = \left(AB \frac{v_x}{v} \Delta t \right) \hat{\mathbf{i}} \tag{16.33}$$

On the left-hand side of the impulse–momentum theorem, Equation 16.32, the change in momentum of the element of gas of mass m is as follows:

$$\Delta \vec{\mathbf{p}} = m \, \Delta \vec{\mathbf{v}} = (\rho V_i)(v_x \hat{\mathbf{i}} - 0) = (\rho v v_x A \, \Delta t) \hat{\mathbf{i}} \tag{16.34}$$

Substituting Equations 16.33 and 16.34 into Equation 16.32, we find

$$\rho v v_x A \, \Delta t = AB \frac{v_x}{v} \Delta t$$

which reduces to an expression for the speed of sound in a gas:

$$v = \sqrt{\frac{B}{\rho}} \tag{16.35}$$

It is interesting to compare this expression with Equation 16.18 for the speed of transverse waves on a string, $v = \sqrt{T/\mu}$. In both cases, the wave speed depends on an elastic property of the medium (bulk modulus B or string tension T) and on an inertial property of the medium (volume density ρ or linear density μ). In fact, the speed of all mechanical waves follows an expression of the general form

$$v = \sqrt{\frac{\text{elastic property}}{\text{inertial property}}}$$

For longitudinal sound waves in a solid rod of material, for example, the speed of sound depends on Young's modulus Y and the density ρ. Table 16.1 provides the speed of sound in several different materials.

TABLE 16.1 Speed of Sound in Various Media

Medium	v (m/s)	Medium	v (m/s)	Medium	v (m/s)
Gases		**Liquids at 25°C**		**Solids**[a]	
Hydrogen (0°C)	1 286	Glycerol	1 904	Pyrex glass	5 640
Helium (0°C)	972	Seawater	1 533	Iron	5 950
Air (20°C)	343	Water	1 493	Aluminum	6 420
Air (0°C)	331	Mercury	1 450	Brass	4 700
Oxygen (0°C)	317	Kerosene	1 324	Copper	5 010
		Methyl alcohol	1 143	Gold	3 240
		Carbon tetrachloride	926	Lucite	2 680
				Lead	1 960
				Rubber	1 600

[a]Values given are for propagation of longitudinal waves in bulk media. Speeds for longitudinal waves in thin rods are smaller, and speeds of transverse waves in bulk are smaller yet.

The speed of sound also depends on the temperature of the medium. For sound traveling through air, the relationship between wave speed and air temperature is

$$v = 331 \sqrt{1 + \frac{T_C}{273}} \qquad (16.36)$$

where v is in meters/second, 331 m/s is the speed of sound in air at 0°C, and T_C is the air temperature in degrees Celsius. Using this equation, one finds that at 20°C, the speed of sound in air is approximately 343 m/s.

Because the speed of sound is constant in a uniform medium, we can relate the speed to distance and time by modeling a pulse of sound as a particle under constant speed. For example, this model provides a convenient way to estimate the distance to a thunderstorm. First count the number of seconds between seeing the flash of lightning and hearing the thunder. Dividing this time interval by 3 gives the approximate distance to the lightning in kilometers because 343 m/s is approximately $\frac{1}{3}$ km/s. Dividing the time interval in seconds by 5 gives the approximate distance to the lightning in miles because the speed of sound is approximately $\frac{1}{5}$ mi/s.

Similarly, the particle under constant speed model allows the calculation described in the opening storyline. The GPS coordinates allow you to find the distance between you and the cliff. The sound of the clap echoes from the cliff and returns to you. So the distance traveled by the sound is twice the distance to the cliff. Dividing that distance by the time interval measured by the smartphone gives the speed of sound.

Having an expression (Eq. 16.35) for the speed of sound, we can now express the relationship between pressure amplitude and displacement amplitude for a sound wave (Eq. 16.31) as

$$\Delta P_{max} = B s_{max} k = (\rho v^2) s_{max} \left(\frac{\omega}{v}\right) = \rho v \omega s_{max} \qquad (16.37)$$

This expression is a bit more useful than Equation 16.31 because the density of a gas is more readily available than is the bulk modulus.

16.8 Intensity of Sound Waves

In Section 16.4, we showed that a wave traveling on a taut string transports energy, consistent with the notion of energy transfer T_{MW} by mechanical waves in Equation 8.2. Naturally, we would expect sound waves to also represent a transfer of energy. Consider the element of gas acted on by the piston in Figure 16.20. Imagine that the piston is moving back and forth in simple harmonic motion at angular frequency ω. Imagine also that the length of the element becomes very small so that the entire element moves with the same velocity as the piston. Then we can model the element as a particle on which the piston is doing work. The rate at which the piston is doing work on the element at any instant of time is given by Equation 8.18:

$$Power = \vec{\mathbf{F}} \cdot \vec{\mathbf{v}}_x$$

where we have used *Power* rather than *P* so that we don't confuse power *P* with pressure *P*! The force $\vec{\mathbf{F}}$ on the element of gas is related to the pressure and the velocity $\vec{\mathbf{v}}_x$ of the element is the derivative of the displacement function, so we find

$$Power = [\Delta P(x, t)A]\hat{\mathbf{i}} \cdot \frac{\partial}{\partial t}[s(x, t)\hat{\mathbf{i}}]$$

$$= [\rho v \omega A s_{max} \sin(kx - \omega t)]\left\{\frac{\partial}{\partial t}[s_{max} \cos(kx - \omega t)]\right\}$$

$$= [\rho v \omega A s_{max} \sin(kx - \omega t)][\omega s_{max} \sin(kx - \omega t)]$$

$$= \rho v \omega^2 A s_{max}^2 \sin^2(kx - \omega t)$$

We now find the time average power over one period of the oscillation. For any given value of *x*, which we can choose to be $x = 0$, the average value of $\sin^2(kx - \omega t)$ over one period *T* is

$$\frac{1}{T}\int_0^T \sin^2(0 - \omega t)\,dt = \frac{1}{T}\int_0^T \sin^2 \omega t\,dt = \frac{1}{T}\left(\frac{t}{2} + \frac{\sin 2\omega t}{2\omega}\right)\Bigg|_0^T = \tfrac{1}{2}$$

Therefore,

$$(Power)_{avg} = \tfrac{1}{2}\rho A \omega^2 s_{max}^2 v$$

Compare this equation to that for power transmitted on a string, Equation 16.21. The two equations have the same form! Be careful, though: *A* in Equation 16.21 is the amplitude of the string wave, while *A* here is the area of the piston in Figure 16.20.

We define the **intensity** *I* of a wave, or the power per unit area, as the rate at which the energy transported by the wave transfers through a unit area *A* perpendicular to the direction of travel of the wave:

Intensity of a sound wave ▶

$$I \equiv \frac{(Power)_{avg}}{A} \qquad (16.38)$$

In this case, the intensity is therefore

$$I = \tfrac{1}{2}\rho \omega^2 s_{max}^2 v$$

Hence, the intensity of a periodic sound wave is proportional to the square of the displacement amplitude and to the square of the angular frequency. This expression can also be written in terms of the pressure amplitude ΔP_{max}; in this case, we use Equation 16.37 to obtain

$$I = \frac{(\Delta P_{max})^2}{2\rho v} \qquad (16.39)$$

The sound waves we have studied with regard to Figures 16.16 through 16.18 and 16.20 are constrained to move in one dimension along the length of the tube. Sound waves, however, can move through three-dimensional bulk media, so let's place a sound source in the open air and study the results with regard to intensity.

Consider the special case of a point source emitting sound waves equally in all directions. If the air around the source is perfectly uniform, the sound power radiated in all directions is the same, and the speed of sound in all directions is the same. The result in this situation is called a **spherical wave.** Figure 16.21 shows these spherical waves as a series of circular arcs concentric with the source. Each arc represents a surface over which the phase of the wave is constant. For example, the arcs may connect corresponding crests on all the waves. We call such a surface of constant phase a **wave front.** The radial distance between adjacent wave fronts that have the same phase is the wavelength λ of the wave. The radial lines pointing outward from the source, representing the direction of propagation of the waves, are called **rays.**

The average power emitted by the source must be distributed uniformly over each spherical wave front of area $4\pi r^2$, where r is the distance from the point source to the wave front. Hence, the wave intensity at a distance r from the source is

$$I = \frac{(Power)_{avg}}{A} = \frac{(Power)_{avg}}{4\pi r^2} \tag{16.40}$$

The intensity decreases as the square of the distance from the source. This inverse-square law is reminiscent of the behavior of gravity in Chapter 13.

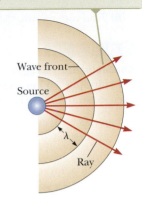

The rays are radial lines pointing outward from the source, perpendicular to the wave fronts.

Wave front—
Source
λ
Ray

Figure 16.21 Spherical waves emitted by a point source. The circular arcs represent the spherical wave fronts that are concentric with the source.

QUICK QUIZ **16.7** A vibrating guitar string makes very little sound if it is not mounted on the guitar body. Why does the sound have greater intensity if the string is attached to the guitar body? **(a)** The string vibrates with more energy. **(b)** The energy leaves the guitar more rapidly. **(c)** The sound power is spread over a larger area at the listener's position. **(d)** The sound power is concentrated over a smaller area at the listener's position. **(e)** The speed of sound is higher in the material of the guitar body. **(f)** None of these answers is correct.

Example **16.6** | **Hearing Limits**

The faintest sounds the human ear can detect at a frequency of 1 000 Hz correspond to an intensity of about 1.00×10^{-12} W/m², which is called *threshold of hearing*. The loudest sounds the ear can tolerate at this frequency correspond to an intensity of about 1.00 W/m², the *threshold of pain*. Determine the pressure amplitude and displacement amplitude associated with these two limits.

SOLUTION

Conceptualize Think about the quietest environment you have ever experienced. It is likely that the intensity of sound in even this quietest environment in your experience is higher than the threshold of hearing.

Categorize Because we are given intensities and asked to calculate pressure and displacement amplitudes, this problem is a substitution problem requiring the concepts discussed in this section.

To find the amplitude of the pressure variation at the threshold of hearing, use Equation 16.39, taking the speed of sound waves in air to be $v = 343$ m/s and the density of air to be $\rho = 1.20$ kg/m³:

$$\Delta P_{max} = \sqrt{2\rho vI}$$
$$= \sqrt{2(1.20 \text{ kg/m}^3)(343 \text{ m/s})(1.00 \times 10^{-12} \text{ W/m}^2)}$$
$$= 2.87 \times 10^{-5} \text{ N/m}^2$$

Calculate the corresponding displacement amplitude using Equation 16.37, recalling that $\omega = 2\pi f$ (Eq. 16.9):

$$s_{max} = \frac{\Delta P_{max}}{\rho v\omega} = \frac{2.87 \times 10^{-5} \text{ N/m}^2}{(1.20 \text{ kg/m}^3)(343 \text{ m/s})(2\pi \times 1\,000 \text{ Hz})}$$
$$= 1.11 \times 10^{-11} \text{ m}$$

In a similar manner, one finds that the loudest sounds the human ear can tolerate (the threshold of pain) correspond to a pressure amplitude of 28.7 N/m² and a displacement amplitude equal to 1.11×10^{-5} m .

Because atmospheric pressure is about 10^5 N/m², the result for the pressure amplitude tells us that the ear is sensitive to pressure fluctuations as small as 3 parts in 10^{10}! The displacement amplitude is also a remarkably small number! If we compare this result for s_{max} to the size of an atom (about 10^{-10} m), we see that the ear is an extremely sensitive detector of sound waves.

Example **16.7** **Intensity Variations of a Point Source**

A point source emits sound waves with an average power output of 80.0 W.

(A) Find the intensity 3.00 m from the source.

SOLUTION

Conceptualize Imagine a small loudspeaker sending sound out at an average rate of 80.0 W uniformly in all directions. You are standing 3.00 m away from the speakers. As the sound propagates, the energy of the sound waves is spread out over an ever-expanding sphere, so the intensity of the sound falls off with distance.

Categorize We evaluate the intensity from an equation generated in this section, so we categorize this example as a substitution problem.

Because a point source emits energy in the form of spherical waves, use Equation 16.40 to find the intensity:

$$I = \frac{(Power)_{avg}}{4\pi r^2} = \frac{80.0 \text{ W}}{4\pi (3.00 \text{ m})^2} = 0.707 \text{ W/m}^2$$

This intensity is close to the threshold of pain.

(B) Find the distance at which the intensity of the sound is 1.00×10^{-8} W/m^2.

SOLUTION

Solve for r in Equation 16.40 and use the given value for I:

$$r = \sqrt{\frac{(Power)_{avg}}{4\pi I}} = \sqrt{\frac{80.0 \text{ W}}{4\pi (1.00 \times 10^{-8} \text{ W/m}^2)}}$$

$$= 2.52 \times 10^4 \text{ m}$$

Sound Level in Decibels

Example 16.6 illustrates the wide range of intensities the human ear can detect. Because this range is so wide, it is convenient to use a logarithmic scale, where the **sound level** β (Greek letter beta) is defined by the equation

$$\beta \equiv 10 \log \left(\frac{I}{I_0} \right) \qquad (16.41)$$

TABLE 16.2 Sound Levels

Source of Sound	β (dB)
Nearby jet airplane	150
Jackhammer;	
machine gun	130
Siren; rock concert	120
Subway; power	
lawn mower	100
Busy traffic	80
Vacuum cleaner	70
Normal conversation	60
Mosquito buzzing	40
Whisper	30
Rustling leaves	10
Threshold of hearing	0

This process compresses the range of hearing into a narrower scale of numbers. The constant I_0 is the *reference intensity*, taken to be at the threshold of hearing ($I_0 = 1.00 \times 10^{-12}$ W/m^2), and I is the intensity in watts per square meter to which the sound level β corresponds, where β is measured[4] in **decibels** (dB). On this scale, the threshold of pain ($I = 1.00$ W/m^2) corresponds to a sound level of $\beta = 10 \log [(1 \text{ W/m}^2)/(10^{-12} \text{ W/m}^2)] = 10 \log (10^{12}) = 120$ dB, and the threshold of hearing corresponds to $\beta = 10 \log [(10^{-12} \text{ W/m}^2)/(10^{-12} \text{ W/m}^2)] = 0$ dB.

Prolonged exposure to high sound levels may seriously damage the human ear. Ear plugs are recommended whenever sound levels exceed 90 dB. Recent evidence suggests that "noise pollution" may be a contributing factor to high blood pressure, anxiety, and nervousness. Table 16.2 gives some typical sound levels.

QUICK QUIZ **16.8** Increasing the intensity of a sound by a factor of 100 causes the sound level to increase by what amount? **(a)** 100 dB **(b)** 20 dB **(c)** 10 dB **(d)** 2 dB

---

[4]The unit *bel* is named after the inventor of the telephone, Alexander Graham Bell (1847–1922). The prefix *deci-* is the SI prefix that stands for 10^{-1}.

| Example **16.8** | **Sound Levels** |

Two identical machines are positioned the same distance from a worker. The intensity of sound delivered by each operating machine at the worker's location is 2.0×10^{-7} W/m².

(A) Find the sound level heard by the worker when one machine is operating.

SOLUTION

Conceptualize Imagine a situation in which one source of sound is active and is then joined by a second identical source, such as one person speaking and then a second person speaking at the same time or one musical instrument playing and then being joined by a second instrument.

Categorize This example is a substitution problem requiring Equation 16.41.

Use Equation 16.41 to calculate the sound level at the worker's location with one machine operating:

$$\beta_1 = 10 \log \left(\frac{2.0 \times 10^{-7} \text{ W/m}^2}{1.00 \times 10^{-12} \text{ W/m}^2} \right) = 10 \log (2.0 \times 10^5) = \boxed{53 \text{ dB}}$$

(B) Find the sound level heard by the worker when two machines are operating.

SOLUTION

Use Equation 16.41 to calculate the sound level at the worker's location with double the intensity:

$$\beta_2 = 10 \log \left(\frac{4.0 \times 10^{-7} \text{ W/m}^2}{1.00 \times 10^{-12} \text{ W/m}^2} \right) = 10 \log (4.0 \times 10^5) = \boxed{56 \text{ dB}}$$

These results show that when the intensity is doubled, the sound level increases by only 3 dB. This 3-dB increase is independent of the original sound level. (Prove this to yourself!)

WHAT IF? *Loudness* is a psychological response to a sound. It depends on both the intensity and the frequency of the sound. As a rule of thumb, a doubling in loudness is approximately associated with an increase in sound level of 10 dB. (This rule of thumb is relatively inaccurate at very low or very high frequencies.) If the loudness of the machines in this example is to be doubled, how many machines at the same distance from the worker must be running?

Answer Using the rule of thumb, a doubling of loudness corresponds to a sound level increase of 10 dB. Therefore,

$$\beta_2 - \beta_1 = 10 \text{ dB} = 10 \log \left(\frac{I_2}{I_0} \right) - 10 \log \left(\frac{I_1}{I_0} \right) = 10 \log \left(\frac{I_2}{I_1} \right)$$

$$\log \left(\frac{I_2}{I_1} \right) = 1 \quad \rightarrow \quad I_2 = 10 I_1$$

Therefore, ten machines must be operating to double the loudness.

Loudness and Frequency

The discussion of sound level in decibels relates to a *physical* measurement of the strength of a sound. Let us now extend our discussion from the What If? section of Example 16.8 concerning the *psychological* "measurement" of the strength of a sound.

Of course, we don't have instruments in our bodies that can display numerical values of our reactions to stimuli. We have to "calibrate" our reactions somehow by comparing different sounds to a reference sound, but that is not easy to accomplish. For example, earlier we mentioned that the threshold intensity is 10^{-12} W/m², corresponding to an intensity level of 0 dB. In reality, this value is the threshold only for a sound of frequency 1 000 Hz, which is a standard reference frequency in acoustics. If we perform an experiment to measure the threshold intensity at other frequencies, we find a distinct variation of this threshold as a function of frequency. For example, at 100 Hz, a barely audible sound must have an intensity level of about

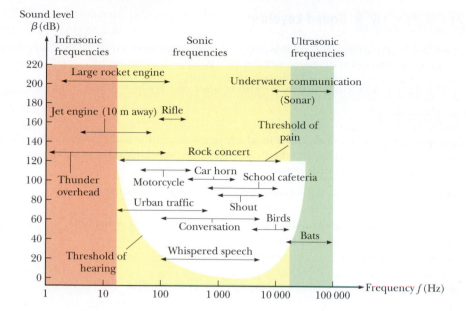

Figure 16.22 Approximate ranges of frequency and sound level of various sources and that of normal human hearing, shown by the white area. (From R. L. Reese, *University Physics*, Pacific Grove, Brooks/Cole, 2000.)

30 dB! Unfortunately, there is no simple relationship between physical measurements and psychological "measurements." The 100-Hz, 30-dB sound is psychologically "equal" in loudness to the 1 000-Hz, 0-dB sound (both are just barely audible), but they are not physically equal in sound level (30 dB ≠ 0 dB).

By using test subjects, the human response to sound has been studied, and the results are shown in the white area of Figure 16.22 along with the approximate frequency and sound-level ranges of other sound sources. The lower curve of the white area corresponds to the threshold of hearing. Its variation with frequency is clear from this diagram. Notice that humans are sensitive to frequencies ranging from about 20 Hz to about 20 000 Hz. The upper bound of the white area is the threshold of pain. Here the boundary of the white area appears straight because the psychological response is relatively independent of frequency at this high sound level.

> In all frames, the waves travel to the left, and their source is far to the right of the boat, out of the frame of the figure.

Figure 16.23 (a) Waves moving toward a stationary boat. (b) The boat moving toward the wave source. (c) The boat moving away from the wave source.

16.9 The Doppler Effect

Perhaps you have noticed how the sound of a vehicle's horn changes as the vehicle moves past you. The frequency of the sound you hear as the vehicle approaches you is higher than the frequency you hear as it moves away from you. This experience is one example of the **Doppler effect**.[5]

To see what causes this apparent frequency change, imagine you are in a boat that is lying at anchor on a gentle sea where the waves have a period of $T = 3.0$ s. Hence, every 3.0 s a crest hits your boat. Figure 16.23a shows this situation, with the water waves moving toward the left. If you set your watch to $t = 0$ just as one crest hits, the watch reads 3.0 s when the next crest hits, 6.0 s when the third crest hits, and so on. From these observations, you conclude that the wave frequency is $f = 1/T = 1/(3.0 \text{ s}) = 0.33$ Hz. Now suppose you pull up the anchor, start your motor, and head directly into the oncoming waves as in Figure 16.23b. Again you set your watch to $t = 0$ as a crest hits the front (the bow) of your boat. Now, however, because you are moving toward the next wave crest as it moves toward you, it hits you less than 3.0 s after the first hit. In other words, the period you observe is shorter than the 3.0-s period you observed when you were stationary. Because $f = 1/T$, you observe a higher wave frequency than when you were at rest.

---

[5] Named after Austrian physicist Christian Johann Doppler (1803–1853), who in 1842 predicted the effect for both sound waves and light waves.

If you turn around and move in the same direction as the waves (Fig. 16.23c), you observe the opposite effect. You set your watch to $t = 0$ as a crest hits the back (the stern) of the boat. Because you are now moving away from the next crest, more than 3.0 s has elapsed on your watch by the time that crest catches you. Therefore, you observe a lower frequency than when you were at rest.

These effects occur because the *relative* speed between your boat and the waves depends on the direction of travel and on the speed of your boat. (See Section 4.6.) When you are moving toward the right in Figure 16.23b, this relative speed is higher than that of the wave speed, which leads to the observation of an increased frequency. When you turn around and move to the left, the relative speed is lower, as is the observed frequency of the water waves.

Let's now examine an analogous situation with sound waves in which the water waves become sound waves, the water becomes the air, and the person on the boat becomes an observer listening to the sound. In this case, an observer O is moving with speed v_O and a sound source S is stationary with respect to the medium, air (Fig. 16.24).

If a point source emits sound waves and the medium is uniform, the waves move at the same speed in all directions radially away from the source; the result is a spherical wave as mentioned in Section 16.8. The distance between adjacent wave fronts equals the wavelength λ. In Figure 16.24, the circles are the intersections of these three-dimensional wave fronts with the two-dimensional paper.

We take the frequency of the source in Figure 16.24 to be f, the wavelength to be λ, and the speed of sound to be v. When the observer moves toward the source, the speed of the waves relative to the observer is $v' = v + v_O$, as in the case of the boat in Figure 16.23, but the wavelength λ is unchanged. Hence, using Equation 16.12, $v = \lambda f$, we can say that the frequency f' heard by the observer is *increased* and is given by

$$f' = \frac{v'}{\lambda} = \frac{v + v_O}{\lambda}$$

Because $\lambda = v/f$, we can express f' as

$$f' = \left(\frac{v + v_O}{v}\right) f \quad \text{(observer moving toward source)} \qquad (16.42)$$

If the observer is moving away from the source, the speed of the wave relative to the observer is $v' = v - v_O$. The frequency heard by the observer in this case is *decreased* and is given by

$$f' = \left(\frac{v - v_O}{v}\right) f \quad \text{(observer moving away from source)} \qquad (16.43)$$

These last two equations can be reduced to a single equation by adopting a sign convention. Whenever an observer moves with a speed v_O relative to a stationary source, the frequency heard by the observer is given by Equation 16.42, with v_O interpreted as follows: a positive value is substituted for v_O when the observer moves toward the source, and a negative value is substituted when the observer moves away from the source.

Now suppose the *source* is in motion and the observer is at rest. If the source moves directly toward observer A in Figure 16.25a (page 440), each new wave is emitted from a position to the right of the origin of the previous wave. As a result, the wave fronts heard by the observer are closer together than they would be if the source were not moving. (Fig. 16.25b shows this effect for waves moving on the surface of water.) As a result, the wavelength λ' measured by observer A is shorter than the wavelength λ of the source. During each vibration, which lasts for a time interval T (the period), the source moves a distance $v_S T = v_S/f$ and the wavelength is *shortened* by this amount. Therefore, the observed wavelength λ' is

$$\lambda' = \lambda - \Delta\lambda = \lambda - \frac{v_S}{f}$$

Figure 16.24 An observer O (the cyclist) moves with a speed v_O toward a stationary point source S, the horn of a parked truck. The observer hears a frequency f' that is greater than the source frequency f.

PITFALL PREVENTION 16.4
Doppler Effect Does Not Depend on Distance Some people think that the Doppler effect depends on the distance between the source and the observer. Although the *intensity* of a sound varies as the distance changes, the apparent *frequency* depends only on the relative speed of source and observer. As you listen to an approaching source, you will detect increasing intensity but constant frequency. As the source passes, you will hear the frequency suddenly drop to a new constant value and the intensity begin to decrease.

Figure 16.25 (a) A source *S* moving with a speed v_S toward a stationary observer A and away from a stationary observer B. Observer A hears an increased frequency, and observer B hears a decreased frequency. (b) The Doppler effect in water, observed in a ripple tank. Letters shown in the photo refer to Quick Quiz 16.9.

A point source is moving to the right with speed v_S.

Observer B

Observer A

S λ'

$\vec{v}_S$

a

b

Courtesy of the Educational Development Center, Newton, MA

Because $\lambda = v/f$, the frequency f' heard by observer A is

$$f' = \frac{v}{\lambda'} = \frac{v}{\lambda - (v_S/f)} = \frac{v}{(v/f) - (v_S/f)}$$

$$f' = \left(\frac{v}{v - v_S}\right)f \quad \text{(source moving toward observer)} \tag{16.44}$$

That is, the observed frequency is *increased* whenever the source is moving toward the observer.

When the source moves away from a stationary observer, as is the case for observer B in Figure 16.25a, the observer measures a wavelength λ' that is *greater* than λ and hears a *decreased* frequency:

$$f' = \left(\frac{v}{v + v_S}\right)f \quad \text{(source moving away from observer)} \tag{16.45}$$

We can express the general relationship for the observed frequency when a source is moving and an observer is at rest as Equation 16.44, with the same sign convention applied to v_S as was applied to v_O: a positive value is substituted for v_S when the source moves toward the observer, and a negative value is substituted when the source moves away from the observer.

Finally, combining Equations 16.42 and 16.44 gives the following general relationship for the observed frequency that includes all four conditions described by Equations 16.42 through 16.45:

General Doppler-shift ▶ expression

$$f' = \left(\frac{v + v_O}{v - v_S}\right)f \tag{16.46}$$

In this expression, the signs for the values substituted for v_O and v_S depend on the direction of the velocity. A positive value is used for motion of the observer or the source *toward* the other (associated with an *increase* in observed frequency), and a negative value is used for motion of one *away from* the other (associated with a *decrease* in observed frequency).

Although the Doppler effect is most typically experienced with sound waves, it is a phenomenon common to all waves. For example, the relative motion of source and observer produces a frequency shift in light waves. The Doppler effect is used in police radar systems to measure the speeds of motor vehicles. Likewise, astronomers use the effect to determine the speeds of stars, galaxies, and other celestial objects relative to the Earth.

Ⓠ UICK QUIZ 16.9 Consider detectors of water waves at three locations A, B, and C in Figure 16.25b. Which of the following statements is true? **(a)** The wave speed is highest at location A. **(b)** The wave speed is highest at location C. **(c)** The detected wavelength is largest at location B. **(d)** The detected wavelength is largest at location C. **(e)** The detected frequency is highest at location C. **(f)** The detected frequency is highest at location A.

Ⓠ UICK QUIZ 16.10 You stand on a platform at a train station and listen to a train approaching the station at a constant velocity. While the train approaches, but before it arrives, what do you hear? **(a)** the intensity and the frequency of the sound both increasing **(b)** the intensity and the frequency of the sound both decreasing **(c)** the intensity increasing and the frequency decreasing **(d)** the intensity decreasing and the frequency increasing **(e)** the intensity increasing and the frequency remaining the same **(f)** the intensity decreasing and the frequency remaining the same

---

Example 16.9 | The Broken Clock Radio

Your clock radio awakens you with a steady and irritating sound of frequency 600 Hz. One morning, it malfunctions and cannot be turned off. In frustration, you drop the clock radio from rest out of your fourth-story dorm window, 15.0 m from the ground. Assume the speed of sound is 343 m/s. As you listen to the falling clock radio, what frequency do you hear just before you hear it striking the ground?

SOLUTION

Conceptualize The speed of the clock radio increases as it falls. Therefore, it is a source of sound moving away from you with an increasing speed so the frequency you hear should be less than 600 Hz.

Categorize We categorize this problem as one in which we combine the *particle under constant acceleration* model for the falling radio with our understanding of the frequency shift of sound due to the Doppler effect.

Analyze Because the clock radio is modeled as a particle under constant acceleration due to gravity, use Equation 2.13 to express the speed of the source of sound:

$$(1) \quad v_S = v_{yi} + a_y t = 0 - gt = -gt$$

From Equation 2.16, find the time at which the clock radio strikes the ground:

$$y_f = y_i + v_{yi}t - \tfrac{1}{2}gt^2 = 0 + 0 - \tfrac{1}{2}gt^2 \;\rightarrow\; t = \sqrt{-\frac{2y_f}{g}}$$

Substitute into Equation (1):

$$v_S = (-g)\sqrt{-\frac{2y_f}{g}} = -\sqrt{-2gy_f}$$

Use Equation 16.46 to determine the Doppler-shifted frequency heard from the falling clock radio:

$$f' = \left[\frac{v + 0}{v - (-\sqrt{-2gy_f})}\right]f = \left(\frac{v}{v + \sqrt{-2gy_f}}\right)f$$

Substitute numerical values:

$$f' = \left[\frac{343 \text{ m/s}}{343 \text{ m/s} + \sqrt{-2(9.80 \text{ m/s}^2)(-15.0 \text{ m})}}\right](600 \text{ Hz})$$

$$= 571 \text{ Hz}$$

Finalize The frequency is lower than the actual frequency of 600 Hz because the clock radio is moving away from you. If it were to fall from a higher floor so that it passes below $y = -15.0$ m, the clock radio would continue to accelerate and the frequency you hear would continue to drop.

Example 16.10 | Doppler Submarines

A submarine (sub A) travels through water at a speed of 8.00 m/s, emitting a sonar wave at a frequency of 1 400 Hz. The speed of sound in the water is 1 533 m/s. A second submarine (sub B) is located such that both submarines are traveling directly toward each other. The second submarine is moving at 9.00 m/s.

(A) What frequency is detected by an observer riding on sub B as the subs approach each other?

continued

16.10 continued

SOLUTION

Conceptualize Even though the problem involves subs moving in water, there is a Doppler effect just like there is when you are in a moving car and listening to a sound moving through the air from another car.

Categorize Because both subs are moving, we categorize this problem as one involving the Doppler effect for both a moving source and a moving observer.

Analyze Use Equation 16.46 to find the Doppler-shifted frequency heard by the observer in sub B, being careful with the signs assigned to the source and observer speeds:

$$f' = \left(\frac{v + v_O}{v - v_S}\right)f$$

$$f' = \left[\frac{1\,533\ \text{m/s} + (+9.00\ \text{m/s})}{1\,533\ \text{m/s} - (+8.00\ \text{m/s})}\right](1\,400\ \text{Hz}) = \boxed{1\,416\ \text{Hz}}$$

(B) The subs barely miss each other and pass. What frequency is detected by an observer riding on sub B as the subs recede from each other?

SOLUTION

Use Equation 16.46 to find the Doppler-shifted frequency heard by the observer in sub B, again being careful with the signs assigned to the source and observer speeds:

$$f' = \left(\frac{v + v_O}{v - v_S}\right)f$$

$$f' = \left[\frac{1\,533\ \text{m/s} + (-9.00\ \text{m/s})}{1\,533\ \text{m/s} - (-8.00\ \text{m/s})}\right](1\,400\ \text{Hz}) = \boxed{1\,385\ \text{Hz}}$$

Notice that the frequency drops from 1 416 Hz to 1 385 Hz as the subs pass. This effect is similar to the drop in frequency you hear when a car passes by you while blowing its horn.

(C) While the subs are approaching each other, some of the sound from sub A reflects from sub B and returns to sub A. If this sound were to be detected by an observer on sub A, what is its frequency?

SOLUTION

The sound of apparent frequency 1 416 Hz found in part (A) is reflected from a moving source (sub B) and then detected by a moving observer (sub A). Find the frequency detected by sub A:

$$f'' = \left(\frac{v + v_O}{v - v_S}\right)f'$$

$$= \left[\frac{1\,533\ \text{m/s} + (+8.00\ \text{m/s})}{1\,533\ \text{m/s} - (+9.00\ \text{m/s})}\right](1\,416\ \text{Hz}) = \boxed{1\,432\ \text{Hz}}$$

Finalize This technique is used by police officers to measure the speed of a moving car. Microwaves are emitted from the police car and reflected by the moving car. By detecting the Doppler-shifted frequency of the reflected microwaves, the police officer can determine the speed of the moving car.

Shock Waves

Now consider what happens when the speed v_S of a source *exceeds* the wave speed v. This situation is depicted graphically in Figure 16.26a. The circles represent spherical wave fronts emitted by the source at various times during its motion. At $t = 0$,

Figure 16.26 (a) A representation of a shock wave produced when a source moves from S_0 to the right with a speed v_S that is greater than the wave speed v in the medium. (b) A stroboscopic photograph of a bullet moving at supersonic speed through the hot air above a candle.

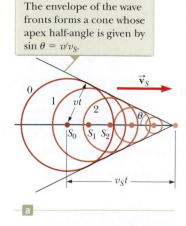

The envelope of the wave fronts forms a cone whose apex half-angle is given by $\sin \theta = v/v_S$.

Notice the shock wave in the vicinity of the bullet.

Omikron/Science Source/Getty Images

the source is at S_0 and moving toward the right. At later times, the source is at S_1, and then S_2, and so on. At the time t, the wave front centered at S_0 reaches a radius of vt. In this same time interval, the source travels a distance $v_S t$. Notice in Figure 16.26a that a straight line can be drawn tangent to all the wave fronts generated at various times. Therefore, the envelope of these wave fronts is a cone whose apex half-angle θ (the "Mach angle") is given by

$$\sin \theta = \frac{vt}{v_S t} = \frac{v}{v_S}$$

The inverse ratio v_S/v is referred to as the *Mach number*, and the conical wave front produced when $v_S > v$ (supersonic speeds) is known as a *shock wave*. An interesting analogy to shock waves is the V-shaped wave fronts produced by a boat when the boat's speed exceeds the speed of the surface-water waves (Fig. 16.27).

Jet airplanes traveling at supersonic speeds produce shock waves, which are responsible for the loud "sonic boom" one hears. The shock wave carries a great deal of energy concentrated on the surface of the cone, with correspondingly great pressure variations. Such shock waves are unpleasant to hear and can cause damage to buildings when aircraft fly supersonically at low altitudes. In fact, an airplane flying at supersonic speeds produces a double boom because two shock waves are formed, one from the nose of the plane and one from the tail.

Figure 16.27 The V-shaped bow wave of a boat is formed because the boat speed is greater than the speed of the water waves it generates. A bow wave is analogous to a shock wave formed by an airplane traveling faster than sound.

QUICK QUIZ **16.11** An airplane flying with a constant velocity moves from a cold air mass into a warm air mass. Does the Mach number **(a)** increase, **(b)** decrease, or **(c)** stay the same?

Summary

▶ Definitions

A **transverse wave** is one in which the elements of the medium move in a direction *perpendicular* to the direction of propagation.

A **longitudinal wave** is one in which the elements of the medium move in a direction *parallel* to the direction of propagation.

A one-dimensional **sinusoidal wave** is one for which the positions of the elements of the medium vary sinusoidally. A sinusoidal wave traveling to the right can be expressed with a **wave function**

$$y(x, t) = A \sin\left[\frac{2\pi}{\lambda}(x - vt)\right] \quad (16.5)$$

where A is the **amplitude**, λ is the **wavelength**, and v is the **wave speed**.

The **intensity** of a periodic sound wave, which is the power per unit area, is

$$I \equiv \frac{(Power)_{avg}}{A} = \frac{(\Delta P_{max})^2}{2\rho v} \quad (16.38, 16.39)$$

The **angular wave number** k and **angular frequency** ω of a wave are defined as follows:

$$k \equiv \frac{2\pi}{\lambda} \quad (16.8)$$

$$\omega \equiv \frac{2\pi}{T} = 2\pi f \quad (16.9)$$

where T is the **period** of the wave and f is its **frequency**.

The **sound level** of a sound wave in decibels is

$$\beta \equiv 10 \log\left(\frac{I}{I_0}\right) \quad (16.41)$$

The constant I_0 is a reference intensity, usually taken to be at the threshold of hearing (1.00×10^{-12} W/m²), and I is the intensity of the sound wave in watts per square meter.

continued

▶ Concepts and Principles

Any one-dimensional wave traveling with a speed v in the x direction can be represented by a wave function of the form

$$y(x, t) = f(x \pm vt) \qquad \text{(16.1, 16.2)}$$

where the positive sign applies to a wave traveling in the negative x direction and the negative sign applies to a wave traveling in the positive x direction. The shape of the wave at any instant in time (a snapshot of the wave) is obtained by holding t constant.

The speed of a wave traveling on a taut string of mass per unit length μ and tension T is

$$v = \sqrt{\frac{T}{\mu}} \qquad \text{(16.18)}$$

The **power** transmitted by a sinusoidal wave on a stretched string is

$$P = \tfrac{1}{2}\mu\omega^2 A^2 v \qquad \text{(16.21)}$$

Wave functions are solutions to a differential equation called the **linear wave equation:**

$$\frac{\partial^2 y}{\partial x^2} = \frac{1}{v^2}\frac{\partial^2 y}{\partial t^2} \qquad \text{(16.27)}$$

Sound waves are longitudinal and travel through a compressible medium with a speed that depends on the elastic and inertial properties of that medium. The speed of sound in a gas having a bulk modulus B and density ρ is

$$v = \sqrt{\frac{B}{\rho}} \qquad \text{(16.35)}$$

For sinusoidal sound waves, the variation in the position of an element of the medium is

$$s(x, t) = s_{\max} \cos(kx - \omega t) \qquad \text{(16.28)}$$

and the variation in pressure from the equilibrium value is

$$\Delta P = \Delta P_{\max} \sin(kx - \omega t) \qquad \text{(16.29)}$$

where $\Delta P_{\max}$ is the **pressure amplitude.** The pressure wave is 90° out of phase with the displacement wave. The relationship between $s_{\max}$ and $\Delta P_{\max}$ is

$$\Delta P_{\max} = \rho v \omega s_{\max} \qquad \text{(16.37)}$$

The change in frequency heard by an observer whenever there is relative motion between a source of sound waves and the observer is called the **Doppler effect.** The observed frequency is

$$f' = \left(\frac{v + v_O}{v - v_S}\right) f \qquad \text{(16.46)}$$

In this expression, the signs for the values substituted for v_O and v_S depend on the direction of the velocity. A positive value for the speed of the observer or source is substituted if the velocity of one is toward the other, whereas a negative value represents a velocity of one in a direction away from the other.

▶ Analysis Model for Problem Solving

Traveling Wave. The wave speed of a sinusoidal wave is

$$v = \frac{\lambda}{T} = \lambda f \qquad \text{(16.6, 16.12)}$$

A sinusoidal wave can be expressed as

$$y = A \sin(kx - \omega t) \qquad \text{(16.10)}$$

Think–Pair–Share

See the Preface for an explanation of the icons used in this problems set. For additional assessment items for this section, go to ⚡ **WEBASSIGN** From Cengage

1. You and your friends are working for the National Oceanic and Atmospheric Administration (NOAA) and are learning about tsunamis to prepare you to help at the Pacific Tsunami Warning Center in Hawaii. Your instructor tells you that a typical tsunami in the open ocean might have a speed of 800 km/h, a wavelength of 200 km, and an amplitude of 1.0 m. (a) If you were on a ship in the open ocean traveling to your position in Hawaii and a tsunami passed through the water around your ship, what effects would you experience? (b) If you were on the beach in Hawaii and saw the ocean water recede over a much larger distance than that due to

normal wave activity, representing the trough of a tsunami, how much time would you have to warn everyone to get off the beach? Note that the frequency of a water wave remains the same as it passes through different depths of water. The amplitude and speed of the wave do change, however. (c) Assume that the energy of the tsunami is conserved. That is, ignore any reflection of energy as the water wave transmits into the shallow water, where its speed changes. In that case, from Equation 16.21, let us express the power of the wave as

$$P \sim \omega^2 A^2 v$$

Based on our assumption of no reflection of energy, what would the amplitude of the wave described above be when it is passing through a shallow region where the wave speed is 75 km/h? (d) If half the energy of the wave is reflected as it enters the shallow water, what would be the amplitude of the wave in the shallow water described in part (c)?

2. An entrepreneur is designing a new outdoor concert venue. He wants to procure loudspeakers that will provide a sound level of 83 dB at a location 100 m from the speakers. (a) What sound power output is required for the speakers? Assume that the sound radiates as a spherical wave and that there are no reflections from the ground. (b) After having someone perform the calculation in part (a) for him, the entrepreneur goes to the electronics store and purchases a speaker that the salesman tells him is rated at 150 W, which he thinks should clearly do what he needs. When he tests the speaker at the venue and turns the sound up to the maximum that

the speaker can handle, however, the sound level at 100 m is only 70.8 dB. The angry entrepreneur runs to his attorney, who hires your team as expert witnesses to see if litigation is appropriate against the salesman. What is the power output of the speaker, based on the sound level data? (c) After performing some research on loudspeakers, argue that litigation should *not* be initiated against the salesman.

3. **ACTIVITY** The epicenter of an earthquake can be located by looking at the difference in arrival times between P and S waves on a seismograph. A single seismograph station can determine how far away the earthquake occurs. With data from three seismograph stations, triangulation can be used to determine the exact location. For near earthquakes, seismic waves travel through the crust of the Earth at typical speeds of 4.00 km/s for S waves and 8.00 km/s for P waves. The following table shows times of day for the *first* arrival of P and S waves at three different California seismograph stations. (a) Print out a map of California from the Internet and use the data below to determine what large California city represents the epicenter of the earthquake. (b) At what time did the earthquake occur?

Seismic Station	Arrival Time of P Wave (h:min:s)	Arrival Time of S Wave (h:min:s)
Sacramento	3:21:34.4 PM	3:22:04.8 PM
San Francisco	3:21:35.9 PM	3:22:07.8 PM
Los Angeles	3:21:52.1 PM	3:22:40.2 PM

Problems

See the Preface for an explanation of the icons used in this problems set.
For additional assessment items for this section, go to **WEBASSIGN** From Cengage

SECTION 16.1 Propagation of a Disturbance

1. A seismographic station receives S and P waves from an
V earthquake, separated in time by 17.3 s. Assume the waves have traveled over the same path at speeds of 4.50 km/s and 7.80 km/s. Find the distance from the seismograph to the focus of the quake.

2. Two points *A* and *B* on the surface of the Earth are at the same longitude and 60.0° apart in latitude as shown in Figure P16.2. Suppose an earthquake at point *A* creates a P wave that reaches point *B* by traveling straight through the body of the Earth at a constant speed of 7.80 km/s. The earthquake also radiates a Rayleigh wave that travels at 4.50 km/s. In addition to P and S waves, Rayleigh waves are a third type of seismic wave that travels along the *surface* of the Earth rather than through the *bulk* of the Earth. (a) Which of these two seismic waves arrives at *B* first? (b) What is the time difference between the arrivals of these two waves at *B*?

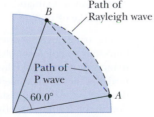

Figure P16.2

3. You are working for a plumber who is laying very long sec-
CR tions of copper pipe for a large building project. He spends a lot of time measuring the lengths of the sections with a

measuring tape. You suggest a faster way to measure the length. You know that the speed of a one-dimensional compressional wave traveling along a copper pipe is 3.56 km/s. You suggest that a worker give a sharp hammer blow at one end of the pipe. Using an oscilloscope app on your smartphone, you will measure the time interval Δt between the arrival of the two sound waves due to the blow: one through the 20.0°C air and the other through the pipe. (a) To measure the length, you must derive an equation that relates the length L of the pipe numerically to the time interval Δt. (b) You measure a time interval of $\Delta t = 127$ ms between the arrivals of the pulses and, from this value, determine the length of the pipe. (c) Your smartphone app claims an accuracy of 1.0% in measuring time intervals. So you calculate by how many centimeters your calculation of the length might be in error.

4. You are working on a senior project and are analyzing a
CR human "wave" at a sports stadium such as that shown in Figure P16.4 (page 446). You are trying to determine the effect of the wave on concession sales because people are standing up and sitting down while they participate in the wave, instead of buying food or drinks. You have made observations at a local stadium and have taken data on one particularly stable wave. This wave took 47.4 s to travel around a specific stadium row consisting of a circular ring of 974 seats. You also find that a typical time interval for spectators to stand and sit back down is 0.95 s. In this wave, how many people in the specific row were out of their seats at any given instant?

Figure P16.4 Problems 4 and 44.

SECTION 16.2 Analysis Model: Traveling Wave

5. When a particular wire is vibrating with a frequency of
   V 4.00 Hz, a transverse wave of wavelength 60.0 cm is produced. Determine the speed of waves along the wire.

6. (a) Plot y versus t at $x = 0$ for a sinusoidal wave of the form
   Q|C $y = 0.150 \cos(15.7x - 50.3t)$, where x and y are in meters
   and t is in seconds. (b) Determine the period of vibration.
   (c) State how your result compares with the value found in
   Example 16.2.

7. Consider the sinusoidal wave of Example 16.2 with the wave
   function

   $$y = 0.150 \cos(15.7x - 50.3t)$$

   where x and y are in meters and t is in seconds. At a certain
   instant, let point A be at the origin and point B be the closest point to A along the x axis where the wave is 60.0° out of
   phase with A. What is the coordinate of B?

8. A sinusoidal wave traveling in the negative x direction
   GP (to the left) has an amplitude of 20.0 cm, a wavelength of
   35.0 cm, and a frequency of 12.0 Hz. The transverse position
   of an element of the medium at $t = 0$, $x = 0$ is $y = -3.00$ cm,
   and the element has a positive velocity here. We wish to find
   an expression for the wave function describing this wave.
   (a) Sketch the wave at $t = 0$. (b) Find the angular wave number k from the wavelength. (c) Find the period T from the
   frequency. Find (d) the angular frequency ω and (e) the
   wave speed v. (f) From the information about $t = 0$, find
   the phase constant ϕ. (g) Write an expression for the wave
   function $y(x, t)$.

9. (a) Write the expression for y as a function of x and t in SI
   units for a sinusoidal wave traveling along a rope in the
   negative x direction with the following characteristics: $A =
   8.00$ cm, $\lambda = 80.0$ cm, $f = 3.00$ Hz, and $y(0, t) = 0$ at $t = 0$.
   (b) **What If?** Write the expression for y as a function of x
   and t for the wave in part (a) assuming $y(x, 0) = 0$ at the
   point $x = 10.0$ cm.

SECTION 16.3 The Speed of Waves on Strings

10. **Review.** The elastic limit of a steel wire is 2.70×10^8 Pa.
    What is the maximum speed at which transverse wave pulses
    can propagate along this wire without exceeding this stress?
    (The density of steel is 7.86×10^3 kg/m³.)

11. Transverse waves travel with a speed of 20.0 m/s on a string
    T under a tension of 6.00 N. What tension is required for a
    wave speed of 30.0 m/s on the same string?

12. *Why is the following situation impossible?* An astronaut on the
    Moon is studying wave motion using the apparatus dis-
    cussed in Example 16.3 and shown in Figure 16.12. He
    measures the time interval for pulses to travel along the
    horizontal wire. Assume the horizontal wire has a mass of
    4.00 g and a length of 1.60 m and assume a 3.00-kg object
    is suspended from its extension around the pulley. The
    astronaut finds that a pulse requires 26.1 ms to traverse the
    length of the wire.

13. Tension is maintained in a
    AMT string as in Figure P16.13. The
    observed wave speed is $v =
    24.0$ m/s when the suspended
    mass is $m = 3.00$ kg. (a) What is
    the mass per unit length of the
    string? (b) What is the wave
    speed when the suspended mass
    is $m = 2.00$ kg?

Figure P16.13
Problems 13 and 43.

14. Transverse pulses travel with a speed of 200 m/s along a taut
    V copper wire whose diameter is 1.50 mm. What is the tension
    in the wire? (The density of copper is 8.92 g/cm³.)

SECTION 16.4 Rate of Energy Transfer by Sinusoidal
Waves on Strings

15. Transverse waves are being generated on a rope under con-
    stant tension. By what factor is the required power increased
    or decreased if (a) the length of the rope is doubled and the
    angular frequency remains constant, (b) the amplitude is
    doubled and the angular frequency is halved, (c) both the
    wavelength and the amplitude are doubled, and (d) both
    the length of the rope and the wavelength are halved?

16. In a region far from the epicenter of an earthquake, a seis-
    Q|C mic wave can be modeled as transporting energy in a sin-
    gle direction without absorption, just as a string wave does.
    Suppose the seismic wave moves from granite into mudfill
    with similar density but with a much smaller bulk modulus.
    Assume the speed of the wave gradually drops by a factor
    of 25.0, with negligible reflection of the wave. (a) Explain
    whether the amplitude of the ground shaking will increase
    or decrease. (b) Does it change by a predictable factor?
    (This phenomenon led to the collapse of part of the Nim-
    itz Freeway in Oakland, California, during the Loma Prieta
    earthquake of 1989.)

17. A long string carries a wave; a 6.00-m segment of the string con-
    AMT tains four complete wavelengths and has a mass of 180 g. The
    string vibrates sinusoidally with a frequency of 50.0 Hz and a
    peak-to-valley displacement of 15.0 cm. (The "peak-to-valley"
    distance is the vertical distance from the farthest positive posi-
    tion to the farthest negative position.) (a) Write the function
    that describes this wave traveling in the positive x direction.
    (b) Determine the power being supplied to the string.

18. A two-dimensional water wave spreads in circular ripples.
    S Show that the amplitude A at a distance r from the initial
    disturbance is proportional to $1/\sqrt{r}$. *Suggestion:* Consider
    the energy carried by one outward-moving ripple.

19. A horizontal string can transmit a maximum power P_0
    S (without breaking) if a wave with amplitude A and angu-
    lar frequency ω is traveling along it. To increase this max-
    imum power, a student folds the string and uses this "dou-
    ble string" as a medium. Assuming the tension in the two

strands together is the same as the original tension in the single string and the angular frequency of the wave remains the same, determine the maximum power that can be transmitted along the "double string."

SECTION 16.5 The Linear Wave Equation

20. Show that the wave function $y = \ln [b(x - vt)]$ is a solution to
 S Equation 16.27, where b is a constant.

21. Show that the wave function $y = e^{b(x-vt)}$ is a solution of the
 S linear wave equation (Eq. 16.27), where b is a constant.

22. (a) Show that the function $y(x, t) = x^2 + v^2 t^2$ is a solution to
 S the wave equation. (b) Show that the function in part (a) can be written as $f(x + vt) + g(x - vt)$ and determine the functional forms for f and g. (c) **What If?** Repeat parts (a) and (b) for the function $y(x, t) = \sin (x) \cos (vt)$.

Note: In the rest of this chapter, for problems involving sound waves, pressure variations ΔP are measured relative to atmospheric pressure, 1.013×10^5 Pa.

SECTION 16.6 Sound Waves

23. A sinusoidal sound wave moves through a medium and is
 V described by the displacement wave function

$$s(x, t) = 2.00 \cos (15.7x - 858t)$$

where s is in micrometers, x is in meters, and t is in seconds. Find (a) the amplitude, (b) the wavelength, and (c) the speed of this wave. (d) Determine the instantaneous displacement from equilibrium of the elements of the medium at the position $x = 0.050$ 0 m at $t = 3.00$ ms. (e) Determine the maximum speed of the element's oscillatory motion.

SECTION 16.7 Speed of Sound Waves

Note: In the rest of this chapter, unless otherwise specified, the equilibrium density of air is $\rho = 1.20$ kg/m^3 and the speed of sound in air is $v = 343$ m/s. Use Table 16.1 to find speeds of sound in other media.

24. Earthquakes at fault lines in the Earth's crust create seismic waves, which are longitudinal (P waves) or transverse (S waves). The P waves have a speed of about 7 km/s. Estimate the average bulk modulus of the Earth's crust given that the density of rock is about 2 500 kg/m^3.

25. An experimenter wishes to generate in air a sound wave that
 T has a displacement amplitude of 5.50×10^{-6} m. The pressure amplitude is to be limited to 0.840 Pa. What is the minimum wavelength the sound wave can have?

26. A sound wave propagates in air at 27°C with frequency
 Q|C 4.00 kHz. It passes through a region where the temperature gradually changes and then moves through air at 0°C. Give numerical answers to the following questions to the extent possible and state your reasoning about what happens to the wave physically. (a) What happens to the speed of the wave? (b) What happens to its frequency? (c) What happens to its wavelength?

27. You are at Quincy Quarries Reservation with your grand-
 CR father, performing the activity described in the opening storyline. The coordinates of your position when your grandfather claps his hands are N 42.244 34°, W 71.033 78°. Your smartphone stopwatch tells you that the time interval

between the clap and the echo is 0.47 s. When you walk to the cliff, your coordinates are N 42.244 06°, W 71.034 66°. What speed of sound do you report to your grandfather? (*Hint:* Use an online resource to calculate the distance between the coordinates.)

28. A rescue plane flies horizontally at a constant speed searching
 V for a disabled boat. When the plane is directly above the boat, the boat's crew blows a loud horn. By the time the plane's sound detector receives the horn's sound, the plane has traveled a distance equal to half its altitude above the ocean. Assuming it takes the sound 2.00 s to reach the plane, determine (a) the speed of the plane and (b) its altitude.

29. The speed of sound in air (in meters per second) depends on temperature according to the approximate expression

$$v = 331.5 + 0.607 T_C$$

where T_C is the Celsius temperature. In dry air, the temperature decreases about 1°C for every 150-m rise in altitude. (a) Assume this change is constant up to an altitude of 9 000 m. What time interval is required for the sound from an airplane flying at 9 000 m to reach the ground on a day when the ground temperature is 30°C? (b) **What If?** Compare your answer with the time interval required if the air were uniformly at 30°C. Which time interval is longer?

30. A sound wave moves down a cylinder as in Figure 16.17.
 S Show that the pressure variation of the wave is described by $\Delta P = \pm \rho v \omega \sqrt{s_{max}^2 - s^2}$, where $s = s(x, t)$ is given by Equation 16.28.

SECTION 16.8 Intensity of Sound Waves

31. The intensity of a sound wave at a fixed distance from a speaker vibrating at 1.00 kHz is 0.600 W/m^2. (a) Determine the intensity that results if the frequency is increased to 2.50 kHz while a constant displacement amplitude is maintained. (b) Calculate the intensity if the frequency is reduced to 0.500 kHz and the displacement amplitude is doubled.

32. The intensity of a sound wave at a fixed distance from a
 S speaker vibrating at a frequency f is I. (a) Determine the intensity that results if the frequency is increased to f' while a constant displacement amplitude is maintained. (b) Calculate the intensity if the frequency is reduced to $f/2$ and the displacement amplitude is doubled.

33. The power output of a certain public-address speaker is
 V 6.00 W. Suppose it broadcasts equally in all directions. (a) Within what distance from the speaker would the sound be painful to the ear? (b) At what distance from the speaker would the sound be barely audible?

34. A fireworks rocket explodes at a height of 100 m above the ground. An observer on the ground directly under the explosion experiences an average sound intensity of 7.00×10^{-2} W/m^2 for 0.200 s. (a) What is the total amount of energy transferred away from the explosion by sound? (b) What is the sound level (in decibels) heard by the observer?

35. You are working at an open-air amphitheater, where rock
 CR concerts occur regularly. The venue has powerful loudspeakers mounted on 10.6-m-tall columns at various locations surrounding the audience. The loudspeakers emit sound uniformly in all directions. There are ladder steps sticking out from the columns, to help workers service the loudspeakers. Many times, audience members break

through the protective fencing around the columns and climb upward on the columns to get a better view of the performers. The upcoming concert is by a group that states that several very-high-volume pulses of sound occur in their concerts, and these sounds are part of their artistic expression. The amphitheater owners are worried about people climbing the columns and being too close to the loudspeakers when these peak sounds are emitted. They do not want to be held responsible for injuries to audience members' ears. Based on past performances of the group, you determine that the peak sound level is 150 dB measured 20.0 cm from the speakers on the columns. The owners ask you to determine the heights on the columns at which to mount impassable barricades to keep people from getting too close to the speakers and hearing sound above the threshold of pain.

36. *Why is the following situation impossible?* It is early on a Saturday morning, and much to your displeasure your next-door neighbor starts mowing his lawn. As you try to get back to sleep, your next-door neighbor on the other side of your house also begins to mow the lawn with an identical mower the same distance away. This situation annoys you greatly because the total sound now has twice the loudness it had when only one neighbor was mowing.

37. **S** Show that the difference between decibel levels β_1 and β_2 of a sound is related to the ratio of the distances r_1 and r_2 from the sound source by

$$\beta_2 - \beta_1 = 20 \log\left(\frac{r_1}{r_2}\right)$$

SECTION 16.9 The Doppler Effect

38. **GP** Submarine A travels horizontally at 11.0 m/s through ocean water. It emits a sonar signal of frequency $f = 5.27 \times 10^3$ Hz in the forward direction. Submarine B is in front of submarine A and traveling at 3.00 m/s relative to the water in the same direction as submarine A. A crewman in submarine B uses his equipment to detect the sound waves ("pings") from submarine A. We wish to determine what is heard by the crewman in submarine B. (a) An observer on which submarine detects a frequency f' as described by Equation 16.46? (b) In Equation 16.46, should the sign of v_S be positive or negative? (c) In Equation 16.46, should the sign of v_O be positive or negative? (d) In Equation 16.46, what speed of sound should be used? (e) Find the frequency of the sound detected by the crewman on submarine B.

39. When high-energy charged particles move through a transparent medium with a speed greater than the speed of light in that medium, a shock wave, or bow wave, of light is produced. This phenomenon is called the *Cerenkov effect*. When a nuclear reactor is shielded by a large pool of water, Cerenkov radiation can be seen as a blue glow in the vicinity of the reactor core due to high-speed electrons moving through the water (Fig. P16.39). In a particular

Figure P16.39

case, the Cerenkov radiation produces a wave front with an apex half-angle of 53.0°. Calculate the speed of the electrons in the water. The speed of light in water is 2.25×10^8 m/s.

40. *Why is the following situation impossible?* At the Summer Olympics, an athlete runs at a constant speed down a straight track while a spectator near the edge of the track blows a note on a horn with a fixed frequency. When the athlete passes the horn, she hears the frequency of the horn fall by the musical interval called a minor third. That is, the frequency she hears drops to five-sixths its original value.

41. **AMT** **Review.** A block with a speaker bolted to it is connected to a spring having spring constant $k = 20.0$ N/m and oscillates as shown in Figure P16.41. The total mass of the block and speaker is 5.00 kg, and the amplitude of this unit's motion is 0.500 m. The speaker emits sound waves of frequency 440 Hz. Determine (a) the highest and (b) the lowest frequencies heard by the person to the right of the speaker. (c) If the maximum sound level heard by the person is 60.0 dB when the speaker is at its closest distance $d = 1.00$ m from him, what is the minimum sound level heard by the observer?

Figure P16.41 Problems 41 and 42.

42. **S** **Review.** A block with a speaker bolted to it is connected to a spring having spring constant k and oscillates as shown in Figure P16.41. The total mass of the block and speaker is m, and the amplitude of this unit's motion is A. The speaker emits sound waves of frequency f. Determine (a) the highest and (b) the lowest frequencies heard by the person to the right of the speaker. (c) If the maximum sound level heard by the person is β when the speaker is at its closest distance d from him, what is the minimum sound level heard by the observer?

ADDITIONAL PROBLEMS

43. A sinusoidal wave in a rope is described by the wave function

$$y = 0.20 \sin (0.75\pi x + 18\pi t)$$

where x and y are in meters and t is in seconds. The rope has a linear mass density of 0.250 kg/m. The tension in the rope is provided by an arrangement like the one illustrated in Figure P16.13. What is the mass of the suspended object?

44. "The wave" is a particular type of pulse that can propagate through a large crowd gathered at a sports arena (Fig. P16.4). The elements of the medium are the spectators, with zero position corresponding to their being seated and maximum position corresponding to their standing and raising their arms. When a large fraction of the spectators

participates in the wave motion, a somewhat stable pulse shape can develop. The wave speed depends on people's reaction time, which is typically on the order of 0.1 s. Estimate the order of magnitude, in minutes, of the time interval required for such a pulse to make one circuit around a large sports stadium. State the quantities you measure or estimate and their values.

45. **BIO** Some studies suggest that the upper frequency limit of hearing is determined by the diameter of the eardrum. The diameter of the eardrum is approximately equal to half the wavelength of the sound wave at this upper limit. If the relationship holds exactly, what is the diameter of the eardrum of a person capable of hearing 20 000 Hz? (Assume a body temperature of 37.0°C.)

46. **QC** An undersea earthquake or a landslide can produce an ocean wave of short duration carrying great energy, called a tsunami. When its wavelength is large compared to the ocean depth d, the speed of a water wave is given approximately by $v = \sqrt{gd}$. Assume an earthquake occurs all along a tectonic plate boundary running north to south and produces a straight tsunami wave crest moving everywhere to the west. (a) What physical quantity can you consider to be constant in the motion of any one wave crest? (b) Explain why the amplitude of the wave increases as the wave approaches shore. (c) If the wave has amplitude 1.80 m when its speed is 200 m/s, what will be its amplitude where the water is 9.00 m deep? (d) Explain why the amplitude at the shore should be expected to be still greater, but cannot be meaningfully predicted by your model.

47. **QC** A sinusoidal wave in a string is described by the wave function

$$y = 0.150 \sin (0.800x - 50.0t)$$

where x and y are in meters and t is in seconds. The mass per length of the string is 12.0 g/m. (a) Find the maximum transverse acceleration of an element of this string. (b) Determine the maximum transverse force on a 1.00-cm segment of the string. (c) State how the force found in part (b) compares with the tension in the string.

48. **S** A rope of total mass m and length L is suspended vertically. Analysis shows that for short transverse pulses, the waves above a short distance from the free end of the rope can be represented to a good approximation by the linear wave equation discussed in Section 16.5. Show that a transverse pulse travels the length of the rope in a time interval that is given approximately by $\Delta t \approx 2\sqrt{L/g}$. Suggestion: First find an expression for the wave speed at any point a distance x from the lower end by considering the rope's tension as resulting from the weight of the segment below that point.

49. A wire of density ρ is tapered so that its cross-sectional area varies with x according to

$$A = 1.00 \times 10^{-5} x + 1.00 \times 10^{-6}$$

where A is in meters squared and x is in meters. The tension in the wire is T. (a) Derive a relationship for the speed of a wave as a function of position. (b) **What If?** Assume the wire is aluminum and is under a tension $T = 24.0$ N. Determine the wave speed at the origin and at $x = 10.0$ m.

50. *Why is the following situation impossible?* Tsunamis are ocean surface waves that have enormous wavelengths (100 to 200 km), and the propagation speed for these waves is $v \approx \sqrt{gd_{avg}}$, where d_{avg} is the average depth of the water. An earthquake on the ocean floor in the Gulf of Alaska produces a tsunami that reaches Hilo, Hawaii, 4 450 km away, in a time interval of 5.88 h. (This method was used in 1856 to estimate the average depth of the Pacific Ocean long before soundings were made to give a direct determination.)

51. **S** A pulse traveling along a string of linear mass density μ is described by the wave function

$$y = [A_0 e^{-bx}] \sin (kx - \omega t)$$

where the factor in brackets is said to be the amplitude. (a) What is the power $P(x)$ carried by this wave at a point x? (b) What is the power $P(0)$ carried by this wave at the origin? (c) Compute the ratio $P(x)/P(0)$.

52. A train whistle ($f = 400$ Hz) sounds higher or lower in frequency depending on whether it approaches or recedes. (a) Prove that the difference in frequency between the approaching and receding train whistle is

$$\Delta f = \frac{2u/v}{1 - u^2/v^2} f$$

where u is the speed of the train and v is the speed of sound. (b) Calculate this difference for a train moving at a speed of 130 km/h. Take the speed of sound in air to be 340 m/s.

53. **Review.** **AMT** **QC** A 150-g glider moves at $v_1 = 2.30$ m/s on an air track toward an originally stationary 200-g glider as shown in Figure P16.53. The gliders undergo a completely inelastic collision and latch together over a time interval of 7.00 ms. A student suggests roughly half the decrease in mechanical energy of the two-glider system is transferred to the environment by sound. Is this suggestion reasonable? To evaluate the idea, find the implied sound level at a position 0.800 m from the gliders. If the student's idea is unreasonable, suggest a better idea.

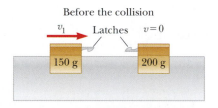

Figure P16.53

54. **QC** Consider the following wave function in SI units:

$$\Delta P(r, t) = \left(\frac{25.0}{r}\right) \sin (1.36r - 2\,030t)$$

Explain how this wave function can apply to a wave radiating from a small source, with r being the radial distance from the center of the source to any point outside the source. Give the most detailed description of the wave that you can. Include answers to such questions as the following and give representative values for any quantities that can be evaluated. (a) Does the wave move more toward the right or the left? (b) As it moves away from the source, what happens to its amplitude? (c) Its speed? (d) Its frequency? (e) Its wavelength? (f) Its power? (g) Its intensity?

55. With particular experimental methods, it is possible to produce and observe in a long, thin rod both a transverse wave whose speed depends primarily on tension in the rod and a longitudinal wave whose speed is determined by Young's modulus and the density of the material according to the expression $v = \sqrt{Y/\rho}$. The transverse wave can be modeled as a wave in a stretched string. A particular metal rod is 150 cm long and has a radius of 0.200 cm and a mass of 50.9 g. Young's modulus for the material is 6.80×10^{10} N/m^2. What must the tension in the rod be if the ratio of the speed of longitudinal waves to the speed of transverse waves is 8.00?

56. A large set of unoccupied football bleachers has solid seats and risers. You stand on the field in front of the bleachers and sharply clap two wooden boards together once. The sound pulse you produce has no definite frequency and no wavelength. The sound you hear reflected from the bleachers has an identifiable frequency and may remind you of a short toot on a trumpet, buzzer, or kazoo. (a) Explain what accounts for this sound. Compute order-of-magnitude estimates for (b) the frequency, (c) the wavelength, and (d) the duration of the sound on the basis of data you specify.

CHALLENGE PROBLEMS

57. A string on a musical instrument is held under tension T and extends from the point $x = 0$ to the point $x = L$. The string is overwound with wire in such a way that its mass per unit length $\mu(x)$ increases uniformly from μ_0 at $x = 0$ to μ_L at $x = L$. (a) Find an expression for $\mu(x)$ as a function of x over the range $0 \le x \le L$. (b) Find an expression for the time interval required for a transverse pulse to travel the length of the string.

58. Assume an object of mass M is suspended from the bottom of the rope of mass m and length L in Problem 48. (a) Show that the time interval for a transverse pulse to travel the length of the rope is

$$\Delta t = 2\sqrt{\frac{L}{mg}}(\sqrt{M + m} - \sqrt{M})$$

(b) **What If?** Show that the expression in part (a) reduces to the result of Problem 48 when $M = 0$. (c) Show that for $m \ll M$, the expression in part (a) reduces to

$$\Delta t = \sqrt{\frac{mL}{Mg}}$$

59. Equation 16.40 states that at distance r away from a point source with power $(Power)_{avg}$, the wave intensity is

$$I = \frac{(Power)_{avg}}{4\pi r^2}$$

Study Figure 16.25 and prove that at distance r straight in front of a point source with power $(Power)_{avg}$ moving with constant speed v_S the wave intensity is

$$I = \frac{(Power)_{avg}}{4\pi r^2}\left(\frac{v - v_S}{v}\right)$$

60. In Section 16.7, we derived the speed of sound in a gas using the impulse–momentum theorem applied to the cylinder of gas in Figure 16.20. Let us find the speed of sound in a gas using a different approach based on the element of gas in Figure 16.18. Proceed as follows. (a) Draw a force diagram for this element showing the forces exerted on the left and right surfaces due to the pressure of the gas on either side of the element. (b) By applying Newton's second law to the element, show that

$$-\frac{\partial(\Delta P)}{\partial x}A\,\Delta x = \rho A\,\Delta x\frac{\partial^2 s}{\partial t^2}$$

(c) By substituting $\Delta P = -(B\,\partial s/\partial x)$ (Eq. 16.30), derive the following wave equation for sound:

$$\frac{B}{\rho}\frac{\partial^2 s}{\partial x^2} = \frac{\partial^2 s}{\partial t^2}$$

(d) To a mathematical physicist, this equation demonstrates the existence of sound waves and determines their speed. As a physics student, you must take another step or two. Substitute into the wave equation the trial solution $s(x, t) = s_{max}\cos(kx - \omega t)$. Show that this function satisfies the wave equation, provided $\omega/k = v = \sqrt{B/\rho}$.

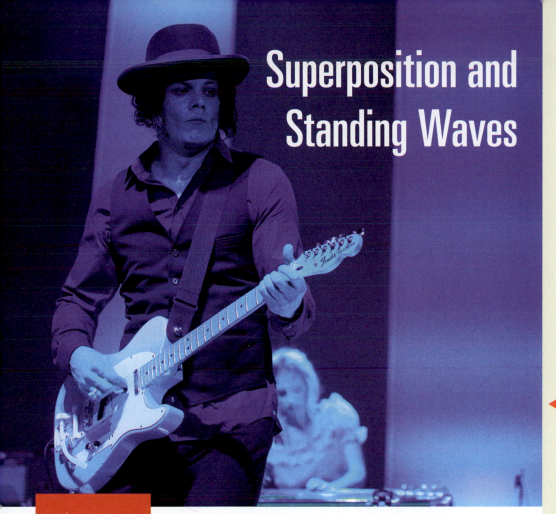

Superposition and Standing Waves

17

Guitarist Jack White takes advantage of standing waves on strings. He changes to higher notes on the guitar by pushing the strings against the frets on the fingerboard, shortening the lengths of the portions of the strings that vibrate. (*Mat Hayward/Shutterstock.com*)

STORYLINE **Your previous roommate moved out to live in an** apartment. You are just getting to know your new roommate. One evening, your roommate shows you her guitar that she uses when she performs in a musical group. You have no idea how to play a guitar, but you do know about music from your choir experiences in high school. You absentmindedly start plucking the strings while your roommate starts looking through your physics textbook. You've seen guitar players pressing their fingers on the frets, as Jack White is doing above, so you do the same. During your explorations, you notice the following. You pluck an open string and then you place your finger *lightly* on the midpoint of the string. When you pluck the string now, the note is an octave above that of the open string. And there is something different about the nature of the sound, aside from it being an octave higher. You continue experimenting and find that you can generate higher notes that have a musical relationship to the open string by lightly touching at other points, such as one-third the length of the string and one-fourth the length. You ask your roommate about this phenomenon. She mentions something about "harmonics" and tells you to read Chapter 17 in your textbook.

CONNECTIONS This chapter continues our studies of waves begun in Chapter 16. We have seen that waves are very different from particles. A particle is of zero size, whereas a wave has a characteristic size, its wavelength. Another important difference between waves and particles is that we can explore the possibility of two or more waves combining at one point in the same medium. Particles can be combined to form extended objects, but the particles must be at *different* locations. In contrast, two waves can both be present at the *same* location. The ramifications of this possibility are explored in this chapter. When waves are combined in systems with boundary conditions, only certain allowed frequencies can exist and we say the frequencies are *quantized*. Quantization is a notion that is at the heart of quantum mechanics, a subject introduced formally in Chapter 40. In Chapters 40–44, we show that analysis of waves under boundary conditions explains many of the quantum phenomena studied there. In this chapter, we use quantization to understand the behavior of the wide array of musical instruments that are based on strings and air columns.

17.1 Analysis Model: Waves in Interference

Many interesting wave phenomena in nature cannot be described by a single traveling wave. Instead, one must analyze these phenomena in terms of a combination of traveling waves. As noted in the introduction, waves have a remarkable difference from particles in that waves can be combined at the *same* location in space. To analyze such wave combinations, we make use of the **superposition principle:**

Superposition principle ▶

> If two or more traveling waves are moving through a medium, the resultant value of the wave function at any point where the waves both exist is the algebraic sum of the values of the wave functions of the individual waves at that point.

Waves that obey this principle are called *linear waves*. (See Section 16.5.) In the case of mechanical waves, linear waves are generally characterized by having amplitudes much smaller than their wavelengths. Waves that violate the superposition principle are called *nonlinear waves* and are often characterized by large amplitudes. In this book, we deal only with linear waves.

One consequence of the superposition principle is that two traveling waves can pass through each other without affecting one another. For instance, when two pebbles are thrown into a pond and hit the surface at different locations, the expanding circular surface waves from the two locations simply pass through each other with no permanent effect. The resulting complex pattern can be viewed as a combination of two independent sets of expanding circles.

Figure 17.1 is a pictorial representation of the superposition of two pulses moving on the same string. The wave function for the pulse moving to the right is y_1, and the wave function for the pulse moving to the left is y_2. The pulses have the same speed but different shapes, and the displacement of the elements of the medium is in the positive y direction for both pulses. When the waves overlap (Fig. 17.1b), the wave function for the resulting complex wave is given by $y_1 + y_2$. When the crests of the pulses coincide (Fig. 17.1c), the resulting wave given by $y_1 + y_2$ has a larger amplitude than that of the individual pulses. The two pulses finally separate and continue moving in their original directions (Fig. 17.1d). Notice that the pulse shapes remain unchanged after the interaction, as if the two pulses had never met!

The combination of separate waves in the same region of space to produce a resultant wave is called **interference.** When the displacements caused by the two pulses are in the same direction, as in Figure 17.1, we refer to their superposition as **constructive interference.**

Constructive interference ▶

Now consider two pulses traveling toward each other on a taut string where one pulse is inverted relative to the other as illustrated in Figure 17.2. When these pulses begin to overlap, the resultant pulse is given by $y_1 + y_2$, but the values of the function y_2 are negative. Therefore, at the instant shown in Figure 17.2c, the amplitude of the combined waves is *less* than that of the individual waves. Again, the two pulses pass through each other; because the displacements caused by the two pulses are in opposite directions, however, we refer to their superposition as **destructive interference.**

Destructive interference ▶

The superposition principle is the centerpiece of the analysis model called **waves in interference.** In many situations, both in acoustics and optics, waves

---

PITFALL PREVENTION 17.1

Do Waves Actually *Interfere?* In popular usage, the term *interfere* implies that an agent affects a situation in some way so as to preclude something from happening. For example, in American football, *pass interference* means that a defending player has affected the receiver so that the receiver is unable to catch the ball. This usage is very different from its use in physics, where waves pass through each other and interfere, but do not affect each other in any way. In physics, interference is similar to the notion of *combination* as described in this chapter.

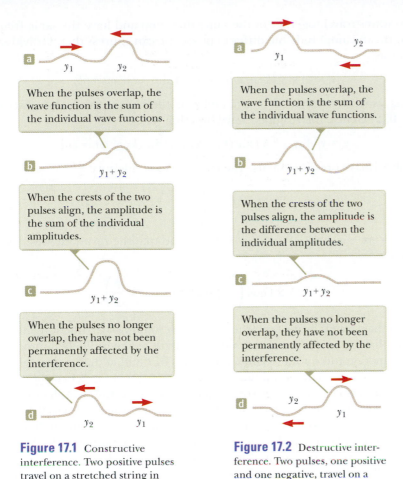

Figure 17.1 Constructive interference. Two positive pulses travel on a stretched string in opposite directions and overlap.

Figure 17.2 Destructive interference. Two pulses, one positive and one negative, travel on a stretched string in opposite directions and overlap.

combine according to this principle and exhibit interesting phenomena with practical applications.

QUICK QUIZ 17.1 Two pulses move in opposite directions on a string and are identical in shape and size except that one has positive displacements of the elements of the string and the other has negative displacements. At the moment the two pulses completely overlap on the string, what happens? **(a)** The energy associated with the pulses has disappeared. **(b)** The string is not moving. **(c)** The string forms a straight line. **(d)** The pulses have vanished and will not reappear.

Superposition of Sinusoidal Waves

Let us now apply the principle of superposition to two sinusoidal waves traveling in the *same* direction in a linear medium. Figure 17.3 shows a simple device that could create this situation for sound waves. Sound from a loudspeaker S is sent into a tube at point P, where there is a T-shaped junction. Half the sound energy travels in one direction, and half travels in the opposite direction. Therefore, the sound waves that reach the receiver R can travel along either of the two paths. The distance along any path from speaker to receiver is called the *path length r*. The lower path length r_1 is fixed, but the upper path length r_2 can be varied by sliding the U-shaped tube, which is similar to that on a slide trombone. This capability allows us to vary the phase difference between the waves arriving at Q. After the sound waves arrive at Q, they combine and travel together to the right from that point to the receiver R.

A sound wave from the speaker (S) propagates into the tube and splits into two parts at point P.

Path length r_2

Path length r_1

The two waves, which combine at the opposite side, are detected at the receiver (R).

Figure 17.3 An acoustical system for demonstrating interference of sound waves. The upper path length r_2 can be varied by sliding the upper section.

If two waves travel together in the same direction and have the same frequency, wavelength, and amplitude but differ in phase, we can express their individual wave functions as

$$y_1 = A \sin (kx - \omega t) \quad y_2 = A \sin (kx - \omega t + \phi)$$

where, as usual, $k = 2\pi/\lambda$, $\omega = 2\pi f$, and ϕ is the phase constant as discussed in Section 16.2. Hence, the resultant wave function y is

$$y = y_1 + y_2 = A [\sin (kx - \omega t) + \sin (kx - \omega t + \phi)]$$

To simplify this expression, we use the trigonometric identity

$$\sin a + \sin b = 2 \cos \left(\frac{a - b}{2} \right) \sin \left(\frac{a + b}{2} \right)$$

Letting $a = kx - \omega t$ and $b = kx - \omega t + \phi$, we find that the resultant wave function y reduces to

◀ **Resultant of two traveling sinusoidal waves**

$$y = 2A \cos \left(\frac{\phi}{2} \right) \sin \left(kx - \omega t + \frac{\phi}{2} \right)$$

This result has several important features. The resultant wave function y also is sinusoidal and has the same frequency and wavelength as the individual waves because the sine function incorporates the same values of k and ω that appear in the original wave functions. The amplitude of the resultant wave is $2A \cos (\phi/2)$, and its phase constant is $\phi/2$. Let's investigate the results for different values of ϕ. If the phase constant ϕ of the original wave equals 0, then $\cos (\phi/2) = \cos 0 = 1$ and the amplitude of the resultant wave is $2A$, twice the amplitude of either individual wave. This can occur in Figure 17.3 when the difference in the path lengths $\Delta r = |r_2 - r_1|$ is either zero or some integer multiple of the wavelength λ (that is, $\Delta r = n\lambda$, where $n = 0, 1, 2, 3, \ldots$). In this case, the crests of the two waves are at the same locations in space and the waves are said to be everywhere *in phase* and therefore interfere constructively. The individual waves y_1 and y_2 combine to form the red-brown curve y of amplitude $2A$ shown in Figure 17.4a. Because the individual waves are in phase, they are indistinguishable in Figure 17.4a, where they appear as a single blue curve. In general, constructive interference occurs when $\cos (\phi/2) = \pm 1$. That is true, for example, when $\phi = 0, 2\pi, 4\pi, \ldots$ rad, that is, when ϕ is an *even* multiple of π.

When ϕ is equal to π rad or to any *odd* multiple of π, then $\cos (\phi/2) = \cos (\pi/2) = 0$ and the crests of one wave occur at the same positions as the troughs of the second

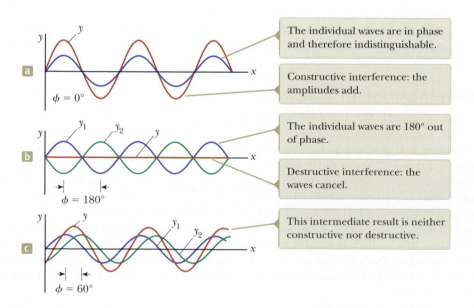

The individual waves are in phase and therefore indistinguishable.

Constructive interference: the amplitudes add.

The individual waves are 180° out of phase.

Destructive interference: the waves cancel.

This intermediate result is neither constructive nor destructive.

Figure 17.4 The superposition of two identical waves y_1 and y_2 (blue and green, respectively) to yield a resultant wave (red-brown).

wave, as shown by the blue and green curves in Figure 17.4b. This can be established in Figure 17.3 when the path length r_2 is adjusted so that the path difference $\Delta r = \lambda/2, 3\lambda/2, \ldots, n\lambda/2$ (for n odd). In this case, as a consequence of destructive interference, the resultant wave has *zero* amplitude everywhere as shown by the straight red-brown line in Figure 17.4b. Finally, when the phase constant has an arbitrary value other than 0 or an integer multiple of π rad (Fig. 17.4c), the resultant wave has an amplitude whose value is somewhere between 0 and 2A.

In the more general case in which the waves have the same wavelength but different amplitudes, the results are similar with the following exceptions. In the in-phase case, the amplitude of the resultant wave is not twice that of a single wave, but rather is the sum of the amplitudes of the two waves. (See the figure in the Analysis Model box below.) When the waves are π rad out of phase, they do not completely cancel as they do in Figure 17.4b. The result is a wave whose amplitude is the difference in the amplitudes of the individual waves.

ANALYSIS MODEL   Waves in Interference

Imagine two waves traveling in the same location through a medium. The displacement of elements of the medium is affected by both waves. According to the **principle of super-position**, the displacement of an element is the sum of the individual displacements that would be caused by each wave. When the waves are in phase, **constructive interference** occurs and the resultant displacement is larger than the individual displacements. **Destructive interference** occurs when the waves are out of phase.

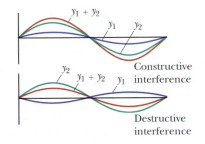

Constructive interference

Destructive interference

Examples:

- a piano tuner listens to a piano string and a tuning fork vibrating together and notices beats (Section 17.7)
- light waves from two coherent sources combine to form an interference pattern on a screen (Chapter 36)
- a thin film of oil on top of water shows swirls of color (Chapter 36)
- x-rays passing through a crystalline solid combine to form a Laue pattern (Chapter 37)

Example 17.1   Two Speakers Driven by the Same Source

Two identical loudspeakers placed 3.00 m apart are driven by the same oscillator (Fig. 17.5). A listener is originally at point O, located 8.00 m from the center of the line connecting the two speakers. The listener then moves to point P, which is a perpendicular distance 0.350 m from O, and she experiences the *first minimum* in sound intensity. What is the frequency of the oscillator?

SOLUTION

Conceptualize In Figure 17.3, a sound wave from one speaker enters a tube and is then *acoustically* split into two different paths before recombining at the other end. In this example, a signal representing the sound is *electrically* split and sent to two different loudspeakers. After leaving the speakers, the sound waves recombine at the position of the listener. Despite the difference in how the splitting occurs, the path difference discussion related to Figure 17.3 can be applied here.

Categorize Because the sound waves from two separate sources combine, we apply the *waves in interference* analysis model.

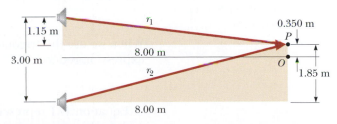

Figure 17.5 (Example 17.1) Two identical loudspeakers emit sound waves to a listener at P.

Analyze Figure 17.5 shows the physical arrangement of the speakers, along with two shaded right triangles that can be drawn on the basis of the lengths described in the problem. The first minimum occurs when the two waves reaching the listener at point P are 180° out of phase, in other words, when their path difference Δr equals $\lambda/2$.

continued

17.1 continued

From the shaded triangles, find the path lengths from the speakers to the listener:

$$r_1 = \sqrt{(8.00 \text{ m})^2 + (1.15 \text{ m})^2} = 8.08 \text{ m}$$

$$r_2 = \sqrt{(8.00 \text{ m})^2 + (1.85 \text{ m})^2} = 8.21 \text{ m}$$

Hence, the path difference is $r_2 - r_1 = 0.13$ m. Because this path difference must equal $\lambda/2$ for the first minimum, $\lambda = 0.26$ m.

To obtain the oscillator frequency, use Equation 16.12, $v = \lambda f$, where v is the speed of sound in air, 343 m/s:

$$f = \frac{v}{\lambda} = \frac{343 \text{ m/s}}{0.26 \text{ m}} = \boxed{1.3 \text{ kHz}}$$

Finalize   This example enables us to understand why the speaker wires in a stereo system should be connected properly. When connected the wrong way—that is, when the positive (or red) wire is connected to the negative (or black) terminal on one of the speakers and the other is correctly wired—the speakers are said to be "out of phase," with one speaker moving outward while the other moves inward. As a consequence, the sound wave coming from one speaker destructively interferes with the wave coming from the other at point O in Figure 17.5. A rarefaction region due to one speaker is superposed on a compression region from the other speaker. Although the two sounds probably do not completely cancel each other (because the left and right stereo signals are usually not identical), a substantial loss of sound quality occurs at point O.

WHAT IF?   What if the speakers were connected out of phase? What happens at point P in Figure 17.5?

Answer   In this situation, the path difference of $\lambda/2$ combines with a phase difference of $\lambda/2$ due to the incorrect wiring to give a full phase difference of λ. As a result, the waves are in phase and there is a *maximum* intensity at point P.

17.2   Standing Waves

Figure 17.6 Two identical loudspeakers emit sound waves toward each other. When they overlap, identical waves traveling in opposite directions will combine to form standing waves.

The sound waves from the pair of loudspeakers in Example 17.1 leave the speakers in the forward direction, and we considered interference at a point in front of the speakers. Suppose we turn each speaker by 90° so that they face each other as in Figure 17.6, and then have them emit sound of the same frequency and amplitude. In this situation, two identical waves travel in opposite directions in the same medium. These waves combine in accordance with the waves in interference model.

We can analyze such a situation by considering wave functions for two transverse sinusoidal waves having the same amplitude, frequency, and wavelength but traveling in opposite directions in the same medium:

$$y_1 = A \sin (kx - \omega t) \quad y_2 = A \sin (kx + \omega t)$$

where y_1 represents a wave traveling in the positive x direction and y_2 represents one traveling in the negative x direction. Adding these two functions according to the superposition principle gives the resultant wave function y:

$$y = y_1 + y_2 = A \sin (kx - \omega t) + A \sin (kx + \omega t)$$

When we use the trigonometric identity $\sin (a \pm b) = \sin a \cos b \pm \cos a \sin b$, this expression reduces to

$$y = (2A \sin kx) \cos \omega t \tag{17.1}$$

Equation 17.1 represents the wave function of a **standing wave.** A standing wave such as the one on a string shown in Figure 17.7 is an oscillation pattern *with a stationary outline* that results from the superposition of two identical waves traveling in opposite directions.

Notice that Equation 17.1 does not contain a function of $kx - \omega t$. Therefore, it is not an expression for a traveling wave. When you observe a standing wave, there is no sense of motion in the direction of propagation of either original wave. If you were to observe the motion of the string in Figure 17.7, you would not see any motion to the left or right. You would only see up and down motion of the elements of the string. Comparing Equation 17.1 with Equation 15.6, we see that it describes

The amplitude of the vertical oscillation of any element of the string depends on the horizontal position of the element. Each element vibrates within the confines of the envelope function $2A \sin kx$.

Antinode Antinode Node Node $2A \sin kx$

Richard Menga/Fundamental Photographs, NYC

Figure 17.7 Multiflash photograph of a standing wave on a string. The limits of motion of the string are seen as light blue and orange sine waves, while two intermediate positions of the string are seen as darker blue. The time behavior of the vertical displacement from equilibrium of an individual element of the string is given by $\cos \omega t$. That is, each element vibrates at an angular frequency ω.

a special kind of simple harmonic motion. Every element of the medium oscillates in simple harmonic motion with the same angular frequency ω (according to the $\cos \omega t$ factor in the equation). The amplitude of the simple harmonic motion of a given element (given by the factor $2A \sin kx$, the coefficient of the cosine function) depends on the location x of the element in the medium, however.

If you can find a noncordless telephone with a coiled cord connecting the handset to the base unit, you can see the difference between a standing wave and a traveling wave. Stretch the coiled cord out and flick it with a finger. You will see a pulse traveling along the cord. Now shake the handset up and down and adjust your shaking frequency until every coil on the cord is moving up at the same time and then down. That is a standing wave, formed from the combination of waves moving away from your hand and reflected from the base unit toward your hand. Notice that there is no sense of traveling along the cord like there was for the pulse. You only see up-and-down motion of the elements of the cord.

Equation 17.1 shows that the amplitude of the simple harmonic motion of an element of the medium has a minimum value of zero when x satisfies the condition $\sin kx = 0$, that is, when

$$kx = 0, \pi, 2\pi, 3\pi, \ldots$$

Because $k = 2\pi/\lambda$, these values for kx give

$$x = 0, \frac{\lambda}{2}, \lambda, \frac{3\lambda}{2}, \ldots = \frac{n\lambda}{2} \qquad n = 0, 1, 2, \ldots \qquad (17.2)$$

◀ Positions of nodes

These points of zero amplitude are called **nodes.** See if you can shake the coiled telephone cord at a higher frequency to generate a wave with a node in the middle, as shown in Figure 17.7.

The element of the medium with the *greatest* possible displacement from equilibrium has an amplitude of $2A$, which we define as the amplitude of the standing wave. The positions in the medium at which this maximum displacement occurs are called **antinodes.** The antinodes are located at positions for which the coordinate x satisfies the condition $\sin kx = \pm 1$, that is, when

$$kx = \frac{\pi}{2}, \frac{3\pi}{2}, \frac{5\pi}{2}, \ldots$$

Therefore, the positions of the antinodes are given by odd values of n:

$$x = \frac{\lambda}{4}, \frac{3\lambda}{4}, \frac{5\lambda}{4}, \ldots = \frac{n\lambda}{4} \qquad n = 1, 3, 5, \ldots \qquad (17.3)$$

◀ Positions of antinodes

PITFALL PREVENTION 17.2
Three Types of Amplitude We need to distinguish carefully here between the **amplitude of the individual waves,** which is A, and the **amplitude of the simple harmonic motion of the elements of the medium,** which is $2A \sin kx$. A given element in a standing wave vibrates within the constraints of the *envelope* function $2A \sin kx$, where x is that element's position in the medium. Such vibration is in contrast to traveling sinusoidal waves, in which elements at all positions oscillate with the same amplitude and the same frequency, and the amplitude A of the wave is the same as the amplitude A of the simple harmonic motion of the elements. Furthermore, we can identify the **amplitude of the standing wave** as $2A$.

Figure 17.8 Standing-wave patterns produced at various times by two waves of equal amplitude traveling in opposite directions. For the resultant wave y, the nodes (N) are points of zero displacement and the antinodes (A) are points of maximum displacement. Two wavelengths are shown for each traveling wave, so the standing wave patterns show twice as many antinodes as that in Figure 17.7.

Two nodes and two antinodes are labeled in the standing wave in Figure 17.7. The light blue curve labeled $2A \sin kx$ in Figure 17.7 represents one wavelength of the traveling waves that combine to form the standing wave. Figure 17.7 and Equations 17.2 and 17.3 provide the following important features of the locations of nodes and antinodes:

> The distance between adjacent antinodes is equal to $\lambda/2$.
> The distance between adjacent nodes is equal to $\lambda/2$.
> The distance between a node and an adjacent antinode is $\lambda/4$.

In the photograph in Figure 17.7, the frequency of the waves is so high that *several* oscillations of the elements of the string occur during the time interval during which the camera shutter is open. Let's slow things down a bit. Wave patterns of the elements of the medium produced at various times during half a cycle of oscillation for two transverse traveling waves moving in opposite directions are shown in Figures 17.8a–c. The blue and green curves are the wave patterns for the individual traveling waves, and the red-brown curves are the wave patterns for the resultant standing wave when they are combined. At $t = 0$ (Fig. 17.8a), the two traveling waves are in phase, giving a wave pattern in which each element of the medium is at rest and experiencing its maximum displacement from equilibrium. One-quarter of a period later, at $t = T/4$ (Fig. 17.8b), the traveling waves have moved one-fourth of a wavelength (one to the right and the other to the left). At this time, the traveling waves are out of phase, and each element of the medium is passing through the equilibrium position in its simple harmonic motion. The result is zero displacement for elements at all values of x; that is, the wave pattern is a straight line. At $t = T/2$ (Fig. 17.8c), the traveling waves are again in phase, producing a wave pattern that is inverted relative to the $t = 0$ pattern. In the standing wave, the elements of the medium alternate in time between the extremes shown in Figures 17.8a and 17.8c.

QUICK QUIZ 17.2 Consider the waves in Figure 17.8 to be waves on a stretched string. Define the velocity of elements of the string as positive if they are moving upward in the figure. **(i)** At the moment the string has the shape shown by the red-brown curve in Figure 17.8a, what is the instantaneous velocity of elements along the string? (a) zero for all elements (b) positive for all elements (c) negative for all elements (d) varies with the position of the element **(ii)** From the same choices, at the moment the string has the shape shown by the red-brown curve in Figure 17.8b, what is the instantaneous velocity of elements along the string?

Example 17.2 Formation of a Standing Wave

Two waves traveling in opposite directions produce a standing wave. The individual wave functions are

$$y_1 = 4.0 \sin (3.0x - 2.0t)$$
$$y_2 = 4.0 \sin (3.0x + 2.0t)$$

where x and y are measured in centimeters and t is in seconds.

(A) Find the amplitude of the simple harmonic motion of the element of the medium located at $x = 2.3$ cm.

SOLUTION

Conceptualize The waves described by the given equations are identical except for their directions of travel, so they indeed combine to form a standing wave as discussed in this section. We can represent the waves graphically by the blue and green curves in Figure 17.8.

Categorize We will substitute values into equations developed in this section, so we categorize this example as a substitution problem.

From the equations for the waves, we see that $A = 4.0$ cm, $k = 3.0$ rad/cm, and $\omega = 2.0$ rad/s. Use Equation 17.1 to write an expression for the standing wave:

$$y = (2A \sin kx) \cos \omega t = 8.0 \sin 3.0x \cos 2.0t$$

Find the amplitude of the simple harmonic motion of the element at the position $x = 2.3$ cm by evaluating the sine function at this position:

$$y_{max} = (8.0 \text{ cm}) \sin 3.0x \big|_{x = 2.3}$$
$$= (8.0 \text{ cm}) \sin (6.9 \text{ rad}) = \boxed{4.6 \text{ cm}}$$

(B) Find the positions of the nodes and antinodes if one end of the string is at $x = 0$.

SOLUTION

Find the wavelength of the traveling waves:

$$k = \frac{2\pi}{\lambda} = 3.0 \text{ rad/cm} \quad \rightarrow \quad \lambda = \frac{2\pi}{3.0} \text{ cm}$$

Use Equation 17.2 to find the locations of the nodes:

$$x = n\frac{\lambda}{2} = n\left(\frac{\pi}{3.0}\right) \text{ cm} \quad n = 0, 1, 2, 3, \dots$$

Use Equation 17.3 to find the locations of the antinodes:

$$x = n\frac{\lambda}{4} = n\left(\frac{\pi}{6.0}\right) \text{ cm} \quad n = 1, 3, 5, 7, \dots$$

17.3 Boundary Effects: Reflection and Transmission

So far in our discussion of waves, we have primarily considered waves traveling through a medium without interacting with any boundaries of the medium. The only exceptions have been references to reflections of waves, such as the echoes from the cliffs in the opening storyline for Chapter 16 and the reflection of waves on the coiled telephone cord from the base unit in Section 17.2. We now address the details of the interactions of waves with boundaries. For example, consider a pulse traveling on a string that is rigidly attached to a support at one end as in Figure 17.9. When the pulse reaches the support, the string ends. As a result, the pulse undergoes **reflection**; that is, the pulse moves back along the string in the opposite direction.

Notice that the reflected pulse is *inverted*. This inversion can be explained as follows. When the pulse reaches the fixed end of the string, the string produces an upward force on the support. By Newton's third law, the support must exert an equal-magnitude and oppositely directed (downward) reaction force on the string. This downward force causes a downward-oriented reflected pulse.

Figure 17.9 The reflection of a traveling pulse at the fixed end of a stretched string. The reflected pulse is inverted, but its shape is otherwise unchanged.

a

b

Reflected
pulse

c

Figure 17.10 The reflection of a
traveling pulse at the free end of
a stretched string. The reflected
pulse is not inverted.

Now consider another case. This time, the pulse arrives at the end of a string
that is free to move vertically as in Figure 17.10. The tension at the free end is
maintained because the string is tied to a ring of negligible mass that is free to
slide vertically on a smooth post without friction. Again, the pulse is reflected,
but this time it is not inverted. When it reaches the post, the pulse exerts a force
on the free end of the string, causing the ring to accelerate upward. The ring
rises as high as the incoming pulse, and then the downward component of the
tension force pulls the ring back down. This movement of the ring produces a
reflected pulse that is upward-oriented and that has the same amplitude as the
incoming pulse.

Finally, consider a situation in which the boundary is intermediate between
these two extremes. In this case, the medium does not end, but rather it changes in
some way and continues. When there is a change in the medium, part of the energy
in the incident pulse is reflected and part undergoes **transmission;** that is, some
of the energy passes through the boundary. For instance, suppose a light string is
attached to a heavier string as in Figure 17.11. When a pulse traveling on the light
string reaches the boundary between the two strings, part of the pulse is reflected
and inverted and part is transmitted to the heavier string. The reflected pulse is
inverted for the same reasons described earlier in the case of the string rigidly
attached to a support.

The reflected pulse has a smaller amplitude than the incident pulse. In Section 16.4, we showed that the energy carried by a wave is related to its amplitude.
According to the principle of conservation of energy, when the pulse breaks up
into a reflected pulse and a transmitted pulse at the boundary, the sum of the energies of these two pulses must equal the energy of the incident pulse. Because the
reflected pulse contains only part of the energy of the incident pulse, its amplitude
must be smaller.

When a pulse traveling on a heavy string strikes the boundary between the heavy
string and a lighter one as in Figure 17.12, again part is reflected and part is transmitted. In this case, the reflected pulse is not inverted.

According to Equation 16.18, the speed of a wave on a string increases as
the mass per unit length of the string decreases. In other words, a wave travels
more rapidly on a light string than on a heavy string if both are under the same
tension. The following general rules apply to reflected waves: When a wave or
pulse travels from medium A to medium B and $v_A > v_B$ (that is, when B is denser
than A), it is inverted upon reflection. When a wave or pulse travels from medium
A to medium B and $v_A < v_B$ (that is, when A is denser than B), it is not inverted
upon reflection.

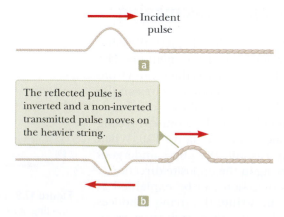

Figure 17.11 (a) A pulse traveling to the right on a
light string approaches the junction with a heavier
string. (b) The situation after the pulse reaches the
junction.

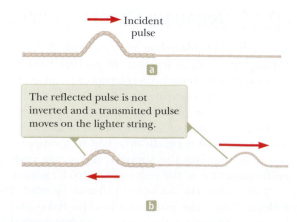

Figure 17.12 (a) A pulse traveling to the right on a
heavy string approaches the junction with a lighter
string. (b) The situation after the pulse reaches
the junction.

17.4 Analysis Model: Waves Under Boundary Conditions

Figure 17.13 A string of length L fixed at both ends.

In Section 17.2, we studied standing waves in a medium with no boundaries. In Section 17.3, we investigated the effect of a rigid boundary on waves in a medium: the waves reflect from the boundary. In this section, let us combine these ideas to see how the existence of boundaries affects the standing wave.

Consider a string of length L fixed at both ends as shown in Figure 17.13. We will use this system as a model for a guitar string or piano string. Waves can travel in both directions on the string due to reflections from the ends. Therefore, standing waves can be set up in the string by a continuous superposition of incident and reflected waves. Notice that there is a *boundary condition* for the waves on the string: because the ends of the string are fixed, they must necessarily have zero displacement and are therefore nodes by definition. The condition that both ends of the string must be nodes fixes the wavelength of the standing wave on the string: at the right end of the string, where $x = L$, Equation 17.2 gives us

$$L = \frac{n\lambda_n}{2} \tag{17.4}$$

where the subscript on λ indicates that different values of n will result in different values of the wavelength. The wavelength, in turn, determines the frequency of the wave according to Equation 16.12. The boundary condition results in the string having a number of discrete natural patterns of oscillation, called **normal modes,** each of which has a characteristic frequency that is easily calculated. This situation in which only certain frequencies of oscillation are allowed is called **quantization.** Quantization is a common occurrence when waves are subject to boundary conditions and is a central feature in our discussions of quantum physics in the extended version of this text. Notice in Figure 17.8 that there are no boundary conditions, so standing waves of *any* frequency can be established; there is no quantization without boundary conditions. Because boundary conditions occur so often for waves, we identify an analysis model called **waves under boundary conditions** for the discussion that follows.

The normal modes of oscillation for the string in Figure 17.13 can be described by imposing the boundary conditions that the ends be nodes and that the nodes be separated by one-half of a wavelength with antinodes halfway between the nodes. The first normal mode that is consistent with these requirements, shown in Figure 17.14a, has nodes at its ends and one antinode in the middle. This normal mode is the longest-wavelength mode that is consistent with our boundary conditions. The first normal mode occurs when the wavelength λ_1 is equal to twice the length of the string, or, from Equation 17.4, $\lambda_1 = 2L$. The section of a standing wave from one node to the next node is called a *loop*. In the first normal mode, the string is vibrating in one loop. In the second normal mode (see Fig. 17.14b), the string vibrates in two loops. When the left half of the string is moving upward, the right

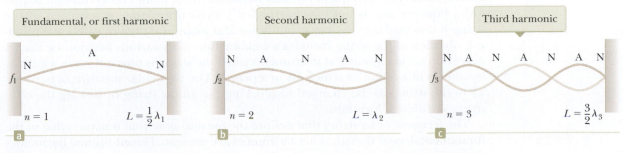

| Fundamental, or first harmonic | Second harmonic | Third harmonic |

$$f_1 \qquad n = 1 \qquad L = \tfrac{1}{2}\lambda_1$$

$$f_2 \qquad n = 2 \qquad L = \lambda_2$$

$$f_3 \qquad n = 3 \qquad L = \tfrac{3}{2}\lambda_3$$

a   b   c

Figure 17.14 The normal modes of vibration of the string in Figure 17.13 form a harmonic series. The string vibrates between the extremes shown.

half is moving downward. In this case, from Equation 17.4 with $n = 2$, the wavelength λ_2 is equal to the length of the string: $\lambda_2 = L$. The third normal mode (see Fig. 17.14c) corresponds to the case in which $\lambda_3 = 2L/3$, and the string vibrates in three loops. In general, the wavelengths of the various normal modes for a string of length L fixed at both ends are found by rearranging Equation 17.4:

Wavelengths of ▶
normal modes

$$\lambda_n = \frac{2L}{n} \quad n = 1, 2, 3, \ldots \tag{17.5}$$

where the index n refers to the nth normal mode of oscillation. These modes are *possible*. The *actual* modes that are excited on a string are discussed shortly.

The natural frequencies associated with the modes of oscillation are obtained from Equation 16.12, $f = v/\lambda$, where the wave speed v is the same for all frequencies. Using Equation 17.5, we find that the natural frequencies f_n of the normal modes are

Natural frequencies of ▶
normal modes as functions
of wave speed and length
of string

$$f_n = \frac{v}{\lambda_n} = n\frac{v}{2L} \quad n = 1, 2, 3, \ldots \tag{17.6}$$

These natural frequencies are also called the *quantized frequencies* associated with the vibrating string fixed at both ends.

Because $v = \sqrt{T/\mu}$ (see Eq. 16.18) for waves on a string, where T is the tension in the string and μ is its linear mass density, we can also express the natural frequencies of a taut string as

Natural frequencies of ▶
normal modes as functions
of string tension and
linear mass density

$$f_n = \frac{n}{2L}\sqrt{\frac{T}{\mu}} \quad n = 1, 2, 3, \ldots \tag{17.7}$$

The lowest frequency f_1, which corresponds to $n = 1$, is called either the **fundamental** or the **fundamental frequency** and is given by

Fundamental frequency ▶
of a taut string

$$f_1 = \frac{1}{2L}\sqrt{\frac{T}{\mu}} \tag{17.8}$$

The frequencies of the remaining normal modes are integer multiples of the fundamental frequency (Eq. 17.6). Frequencies of normal modes that exhibit such an integer-multiple relationship form a **harmonic series,** and the normal modes are called **harmonics.** The fundamental frequency f_1 is the frequency of the first harmonic, the frequency $f_2 = 2f_1$ is that of the second harmonic, and the frequency $f_n = nf_1$ is that of the nth harmonic. Other oscillating systems, such as a drumhead, exhibit normal modes, but the frequencies are not related as integer multiples of a fundamental. Therefore, we do not use the term *harmonic* in association with those types of systems.

Let us examine now how the various harmonics are actually excited in a string. To excite only a single harmonic, the string would have to be distorted into a shape that corresponds to that of the desired harmonic. After being released, the string would vibrate at the frequency of that harmonic. This maneuver is difficult to perform, however, and is not how a string of a musical instrument is excited. If the string is distorted into a general, nonsinusoidal shape and then released, the resulting vibration of the string includes a combination of its various harmonics. Such a distortion occurs in musical instruments when the string is plucked (as in a guitar), bowed (as in a cello), or struck (as in a piano). The particular mixture of harmonics in the string can be changed by plucking the guitar string or bowing the cello string at different locations.

The frequency of a string that defines the musical note that it plays is that of the fundamental, even though other harmonics are present. The additional harmonics determine the *quality*, or the *timbre*, of the sound without altering its frequency, as discussed further in Section 17.8. The quality of the sound is part of what allows you to identify instruments playing the same note. For example, you can differentiate between a guitar, banjo, or a sitar playing the same note.

The string's frequency can be varied by changing the string's tension or its length. For example, the tension in guitar and violin strings is varied by a screw adjustment mechanism or by tuning pegs located on the neck of the instrument. As the tension is increased, the frequency of the normal modes increases in accordance with Equation 17.7. Once the instrument is "tuned," players vary the frequency by moving their fingers along the neck, thereby changing the length L of the oscillating portion of the string. As the length is shortened, the frequency increases because, as Equation 17.7 specifies, the normal-mode frequencies are inversely proportional to string length.

In the opening storyline, when you pluck an open string on your roommate's guitar, the fundamental mode is that shown in Figure 17.4a. Then, you place your finger lightly at the midpoint of the string. Because your finger is pressing only lightly on the string, the entire string can still vibrate when you pluck it. But your finger imposes a node at the center of the string. Therefore, the fundamental mode of vibration now looks like Figure 17.4b. This is the $n = 2$ harmonic of the open string, so the frequency is twice as high: an octave.

QUICK QUIZ 17.3 When a standing wave is set up on a string fixed at both ends, which of the following statements is true? **(a)** The number of nodes is equal to the number of antinodes. **(b)** The wavelength is equal to the length of the string divided by an integer. **(c)** The frequency is equal to the number of nodes times the fundamental frequency. **(d)** The shape of the string at any instant shows a symmetry about the midpoint of the string.

ANALYSIS MODEL **Waves Under Boundary Conditions**

Imagine a wave that is not free to travel throughout all space as in the traveling wave model. If the wave is subject to boundary conditions, such that certain requirements must be met at specific locations in space, the wave is limited to a set of **normal modes** with quantized wavelengths and quantized natural frequencies.

For waves on a string fixed at both ends, the natural frequencies are

$$f_n = \frac{n}{2L}\sqrt{\frac{T}{\mu}} \quad n = 1, 2, 3, \ldots \qquad \text{(17.7)}$$

where T is the tension in the string and μ is its linear mass density.

Examples:

- waves traveling back and forth on a guitar string combine to form a standing wave
- sound waves traveling back and forth in a clarinet combine to form standing waves (Section 17.6)
- a microscopic particle confined to a small region of space is modeled as a wave and exhibits quantized energies (Chapter 40)
- the Fermi energy of a metal is determined by modeling electrons as wave-like particles in a box (Chapter 42)

Example 17.3 **Give Me a C Note!**

The middle C string on a piano has a fundamental frequency of 262 Hz, and the string for the first A above middle C has a fundamental frequency of 440 Hz.

(A) Calculate the frequencies of the next two harmonics of the C string.

SOLUTION

Conceptualize Remember that the harmonics of a vibrating string have frequencies that are related by integer multiples of the fundamental.

continued

17.3 continued

Categorize   This first part of the example is a simple substitution problem.

Knowing that the fundamental frequency is $f_1 = 262$ Hz, find the frequencies of the next harmonics by multiplying by integers:

$$f_2 = 2f_1 = \boxed{524 \text{ Hz}}$$

$$f_3 = 3f_1 = \boxed{786 \text{ Hz}}$$

(B) If the A and C strings have the same linear mass density μ and length L, determine the ratio of tensions in the two strings.

SOLUTION

Categorize   This part of the example is more of an analysis problem than is part (A) and uses the *waves under boundary conditions* model.

Analyze   Use Equation 17.8 to write expressions for the fundamental frequencies of the two strings:

$$f_{1A} = \frac{1}{2L}\sqrt{\frac{T_A}{\mu}} \quad \text{and} \quad f_{1C} = \frac{1}{2L}\sqrt{\frac{T_C}{\mu}}$$

Divide the first equation by the second and solve for the ratio of tensions:

$$\frac{f_{1A}}{f_{1C}} = \sqrt{\frac{T_A}{T_C}} \rightarrow \frac{T_A}{T_C} = \left(\frac{f_{1A}}{f_{1C}}\right)^2 = \left(\frac{440 \text{ Hz}}{262 \text{ Hz}}\right)^2 = \boxed{2.82}$$

Finalize   If the frequencies of piano strings were determined solely by tension, this result suggests that the ratio of tensions from the lowest string to the highest string on the piano would be enormous. Such large tensions would make it difficult to design a frame to support the strings. In reality, the frequencies of piano strings vary due to additional parameters, including the mass per unit length and the length of the string. The What If? below explores a variation in length.

WHAT IF?   If you look inside a real piano, you'll see that the assumption made in part (B) is only partially true. The strings are not likely to have the same length. The string densities for the given notes might be equal, but suppose the length of the A string is only 64% of the length of the C string. What is the ratio of their tensions?

Answer   Using Equation 17.8 again, we set up the ratio of frequencies:

$$\frac{f_{1A}}{f_{1C}} = \frac{L_C}{L_A}\sqrt{\frac{T_A}{T_C}} \rightarrow \frac{T_A}{T_C} = \left(\frac{L_A}{L_C}\right)^2\left(\frac{f_{1A}}{f_{1C}}\right)^2$$

$$\frac{T_A}{T_C} = (0.64)^2\left(\frac{440 \text{ Hz}}{262 \text{ Hz}}\right)^2 = 1.16$$

Notice that this result represents only a 16% increase in tension, compared with the 182% increase in part (B).

Example 17.4   Changing String Vibration with Water

One end of a horizontal string is attached to a vibrating blade, and the other end passes over a pulley as in Figure 17.15a. A sphere of mass 2.00 kg hangs on the end of the string. The string is vibrating in its second harmonic. A container of water is raised under the sphere so that the sphere is completely submerged. In this configuration, the string vibrates in its fifth harmonic as shown in Figure 17.15b. What is the radius of the sphere?

SOLUTION

Conceptualize   Imagine what happens when the sphere is immersed in the water. The buoyant force acts upward on the sphere, reducing the tension in the string. The change in tension causes a change in the speed of waves on the string, which in turn causes a change in the wavelength. This altered wavelength results in the string vibrating in its fifth normal mode rather than the second.

Categorize   The hanging sphere is modeled as a *particle in equilibrium*. One of the forces acting on it is the buoyant force from the water. We also apply the *waves under boundary conditions* model to the string.

Figure 17.15   (Example 17.4) (a) When the sphere hangs in air, the string vibrates in its second harmonic. (b) When the sphere is immersed in water, the string vibrates in its fifth harmonic.

17.4 continued

Analyze   Apply the particle in equilibrium model to the sphere in Figure 17.15a, identifying T_1 as the tension in the string as the sphere hangs in air:

$$\sum F = T_1 - mg = 0$$

$$T_1 = mg$$

Apply the particle in equilibrium model to the sphere in Figure 17.15b, where T_2 is the tension in the string as the sphere is immersed in water:

$$T_2 + B - mg = 0$$

$$(1) \quad B = mg - T_2$$

The desired quantity, the radius of the sphere, will appear in the expression for the buoyant force B. Before proceeding in this direction, however, we must evaluate T_2 from the information about the standing wave.

Write the equation for the frequency of a standing wave on a string (Eq. 17.7) twice, once before the sphere is immersed and once after. Notice that the frequency f is the same in both cases because it is determined by the vibrating blade. In addition, the linear mass density μ and the length L of the vibrating portion of the string are the same in both cases. Divide the equations:

$$f = \frac{n_1}{2L}\sqrt{\frac{T_1}{\mu}}$$
$$f = \frac{n_2}{2L}\sqrt{\frac{T_2}{\mu}} \quad \rightarrow \quad 1 = \frac{n_1}{n_2}\sqrt{\frac{T_1}{T_2}}$$

Solve for T_2:

$$T_2 = \left(\frac{n_1}{n_2}\right)^2 T_1 = \left(\frac{n_1}{n_2}\right)^2 mg$$

Substitute this result into Equation (1):

$$(2) \quad B = mg - \left(\frac{n_1}{n_2}\right)^2 mg = mg\left[1 - \left(\frac{n_1}{n_2}\right)^2\right]$$

Using Equation 14.5, express the buoyant force in terms of the radius of the sphere:

$$B = \rho_{water}gV_{sphere} = \rho_{water}g(\tfrac{4}{3}\pi r^3)$$

Solve for the radius of the sphere and substitute from Equation (2):

$$r = \left(\frac{3B}{4\pi\rho_{water}g}\right)^{1/3} = \left\{\frac{3m}{4\pi\rho_{water}}\left[1 - \left(\frac{n_1}{n_2}\right)^2\right]\right\}^{1/3}$$

Substitute numerical values:

$$r = \left\{\frac{3(2.00\ \text{kg})}{4\pi(1\ 000\ \text{kg/m}^3)}\left[1 - \left(\frac{2}{5}\right)^2\right]\right\}^{1/3}$$

$$= 0.073\ 7\ \text{m} = \boxed{7.37\ \text{cm}}$$

Finalize   Notice that only certain radii of the sphere will result in the string vibrating in a normal mode; the speed of waves on the string must be changed to a value such that the length of the string is an integer multiple of half wavelengths. This limitation is a feature of the *quantization* that was introduced earlier in this chapter: the sphere radii that cause the string to vibrate in a normal mode are *quantized*.

17.5   Resonance

We have seen that a system such as a taut string is capable of oscillating in one or more normal modes of oscillation. We find that if a periodic force is applied to such a system, the amplitude of the resulting motion of the string is greatest when the frequency of the applied force is equal to one of the natural frequencies of the system. This phenomenon, known as *resonance*, was discussed in Section 15.7 with regard to a simple harmonic oscillator. Although a block–spring system or a simple pendulum has only one natural frequency, standing-wave systems have a whole set of natural frequencies, such as that given by Equation 17.7 for a string. Because an oscillating system exhibits a large amplitude when driven at any of its natural frequencies, these frequencies are often referred to as **resonance frequencies.**

Consider Figure 17.16, which shows a string being driven by a vibrating blade. When the frequency of the blade equals one of the natural frequencies of the string,

When the blade vibrates at one of the natural frequencies of the string, large-amplitude standing waves are created.

Vibrating blade

Figure 17.16 Standing waves are set up in a string when one end is connected to a vibrating blade.

standing waves are produced and the string oscillates with a large amplitude. In this resonance case, the wave generated by the oscillating blade is in phase with the reflected wave and the string absorbs energy from the blade. If the string is driven at a frequency that is not one of its natural frequencies, the oscillations are of low amplitude and exhibit no stable pattern.

Resonance is very important in the excitation of musical instruments based on air columns. We shall discuss this application of resonance in Section 17.6.

17.6 Standing Waves in Air Columns

The waves under boundary conditions model can also be applied to sound waves in a column of air such as that inside an organ pipe or a clarinet. Standing waves in this case are the result of interference between longitudinal sound waves traveling in opposite directions.

In a pipe closed at one end, the closed end is a **displacement node** because the rigid barrier at this end does not allow longitudinal motion of the air. Because the pressure wave is 90° out of phase with the displacement wave (see Section 16.6), the closed end of an air column corresponds to a **pressure antinode** (that is, a point of maximum pressure variation).

The open end of an air column is approximately a **displacement antinode**[1] and a pressure node. We can understand why no pressure variation occurs at an open end by noting that the end of the air column is open to the atmosphere; therefore, the pressure at this end must remain constant at atmospheric pressure.

You may wonder how a sound wave can reflect from an open end because there may not appear to be a change in the medium at this point: the medium through which the sound wave moves is *air*, both inside and outside the pipe. Sound can be represented as a pressure wave, however, and a compression region of the sound wave is constrained by the sides of the pipe as long as the region is inside the pipe. As the compression region exits at the open end of the pipe, the constraint of the pipe is removed and the compressed air is free to expand into the atmosphere. Therefore, there is a change in the *character* of the medium between the inside of the pipe and the outside even though there is no change in the *material* of the medium. This change in character is sufficient to allow some reflection.

With the boundary conditions of nodes or antinodes at the ends of the air column, we have a set of normal modes of oscillation as is the case for the string fixed at both ends. Therefore, the air column has quantized frequencies.

The first three normal modes of oscillation of a pipe open at both ends are shown in Figure 17.17a. The diagrams in the left column show *graphical* representations of the *displacement* of elements of air from their equilibrium positions. The second column shows *pictorial* representations of the *pressure* in the air at various locations in the pipe, following the technique used in Figure 16.17. There is a lot of information in Figure 17.17. Study it carefully.

Notice that both ends of the pipe in Figure 17.17a are displacement antinodes (approximately) or pressure nodes. In the first normal mode, the standing wave extends between two adjacent displacement antinodes or two adjacent pressure nodes, which is a distance of half a wavelength. Therefore, the wavelength is twice the length of the pipe, and the fundamental frequency is $f_1 = v/2L$. As Figure 17.17a

[1]Strictly speaking, the open end of an air column is not exactly a displacement antinode. A compression reaching an open end does not reflect until it passes beyond the end. For a tube of circular cross section, an end correction equal to approximately 0.6R, where R is the tube's radius, must be added to the length of the air column. Hence, the effective length of the air column is longer than the true length L. We ignore this end correction in this discussion.

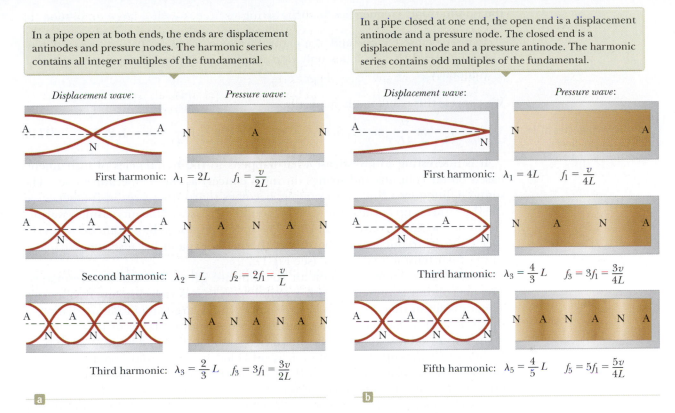

> In a pipe open at both ends, the ends are displacement antinodes and pressure nodes. The harmonic series contains all integer multiples of the fundamental.

> In a pipe closed at one end, the open end is a displacement antinode and a pressure node. The closed end is a displacement node and a pressure antinode. The harmonic series contains odd multiples of the fundamental.

Displacement wave: *Pressure wave:*

First harmonic: $\lambda_1 = 2L$ $f_1 = \dfrac{v}{2L}$

Second harmonic: $\lambda_2 = L$ $f_2 = 2f_1 = \dfrac{v}{L}$

Third harmonic: $\lambda_3 = \dfrac{2}{3}L$ $f_3 = 3f_1 = \dfrac{3v}{2L}$

Displacement wave: *Pressure wave:*

First harmonic: $\lambda_1 = 4L$ $f_1 = \dfrac{v}{4L}$

Third harmonic: $\lambda_3 = \dfrac{4}{3}L$ $f_3 = 3f_1 = \dfrac{3v}{4L}$

Fifth harmonic: $\lambda_5 = \dfrac{4}{5}L$ $f_5 = 5f_1 = \dfrac{5v}{4L}$

a   **b**

Figure 17.17 Standing longitudinal sound waves in air columns, showing the wave patterns for the three lowest frequencies. (a) In an open column, the standing waves are symmetric around the midpoint of the column. On the left are graphical representations of the displacement of elements of the air. On the right are pictorial representations of the pressure at various points in the wave. (b) In a column closed at one end, the standing waves are not symmetric. Again, the left-hand diagrams show graphical representations of the displacement of elements of the air, while the right-hand diagrams are pictorial representation of the pressure.

shows, the frequencies of the higher harmonics are $2f_1$, $3f_1$, ...

> In a pipe open at both ends, the natural frequencies of oscillation form a harmonic series that includes all integral multiples of the fundamental frequency.

Because the fundamental frequency is given by the same expression as that for a string (see Eq. 17.6), we can express the natural frequencies of oscillation as

$$f_n = n\frac{v}{2L} \quad n = 1, 2, 3, \ldots \qquad (17.9)$$

◀ Natural frequencies of a pipe open at both ends

Despite the similarity between Equations 17.6 and 17.9, you must remember that v in Equation 17.6 is the speed of waves on the string, whereas v in Equation 17.9 is the speed of sound in air.

If a pipe is closed at one end and open at the other, the closed end is a displacement node or a pressure antinode (see Fig. 17.17b). In this case, the standing wave for the fundamental mode extends from an antinode to the adjacent node, which is one-fourth of a wavelength. Hence, the wavelength for the first normal mode is $4L$, and the fundamental frequency is $f_1 = v/4L$. As Figure 17.17b shows, the higher-frequency waves that satisfy our conditions are those that have a node at the closed end and an antinode at the open end; hence, the higher harmonics do *not* include *all* integer multiples of the fundamental frequency, but rather have only the odd-multiple frequencies $3f_1$, $5f_1$, ...

> In a pipe closed at one end, the natural frequencies of oscillation form a harmonic series that includes only odd integral multiples of the fundamental frequency.

We express this result mathematically as

◀ Natural frequencies of a pipe closed at one end and open at the other

$$f_m = m\frac{v}{4L} \quad m = 1, 3, 5, \ldots \quad \text{or} \quad f_n = (2n-1)\frac{v}{4L} \quad n = 1, 2, 3, \ldots \quad (17.10)$$

It is interesting to investigate what happens to the frequencies of instruments based on air columns and strings during a concert as the temperature rises. The sound emitted by a flute, for example, becomes sharp (increases in frequency) as the flute warms up because the speed of sound increases in the increasingly warmer air inside the flute (consider Eq. 17.9). The sound produced by a violin becomes flat (decreases in frequency) as the strings thermally expand because the expansion causes their tension to decrease (see Eq. 17.7).

Musical instruments based on air columns are generally excited by resonance. The air column is presented with a sound wave that is rich in many frequencies. The air column then responds by resonance with a large-amplitude oscillation to the frequencies that match the quantized frequencies in its set of harmonics. In many woodwind instruments, the initial rich sound is provided by a vibrating reed. In brass instruments, this excitation is provided by the sound coming from the vibration of the player's lips. In a flute, the initial excitation comes from blowing over an edge at the mouthpiece of the instrument in a manner similar to blowing across the opening of a bottle with a narrow neck. The sound of the air rushing across the bottle opening has many frequencies, including one that sets the air cavity in the bottle into resonance.

QUICK QUIZ **17.4** A pipe open at both ends resonates at a fundamental frequency f_{open}. When one end is covered and the pipe is again made to resonate, the fundamental frequency is f_{closed}. Which of the following expressions describes how these two resonant frequencies compare? (a) $f_{\text{closed}} = f_{\text{open}}$ (b) $f_{\text{closed}} = \frac{1}{2}f_{\text{open}}$ (c) $f_{\text{closed}} = 2f_{\text{open}}$ (d) $f_{\text{closed}} = \frac{3}{2}f_{\text{open}}$

QUICK QUIZ **17.5** Balboa Park in San Diego has an outdoor organ. When the air temperature increases, the fundamental frequency of one of the organ pipes (a) stays the same, (b) goes down, (c) goes up, or (d) is impossible to determine.

Example 17.5   **Wind in a Culvert**

A section of drainage culvert 1.23 m in length makes a howling noise when the wind blows across its open ends.

Determine the frequencies of the first three harmonics of the culvert if it is cylindrical in shape and open at both ends. Take $v = 343$ m/s as the speed of sound in air.

SOLUTION

Conceptualize   The sound of the wind blowing across the end of the pipe contains many frequencies, and the culvert responds to the sound by resonance, vibrating at the natural frequencies of the air column.

Categorize   This example is a relatively simple substitution problem.

Find the frequency of the first harmonic of the culvert, modeling it as an air column open at both ends:

$$f_1 = \frac{v}{2L} = \frac{343 \text{ m/s}}{2(1.23 \text{ m})} = \boxed{139 \text{ Hz}}$$

Find the next harmonics by multiplying by integers:

$$f_2 = 2f_1 = \boxed{279 \text{ Hz}}$$

$$f_3 = 3f_1 = \boxed{418 \text{ Hz}}$$

| Example **17.6** | **Measuring the Frequency of a Tuning Fork** |

A simple apparatus for demonstrating resonance in an air column is depicted in Figure 17.18. A vertical pipe open at both ends is partially submerged in water, and a tuning fork vibrating at an unknown frequency is placed near the top of the pipe. The length L of the air column can be adjusted by moving the pipe vertically. The sound waves generated by the fork are reinforced when L corresponds to one of the resonance frequencies of the pipe. For a certain pipe, the smallest value of L for which a peak occurs in the sound intensity is 9.00 cm.

(A) What is the frequency of the tuning fork?

SOLUTION

Conceptualize Sound waves from the tuning fork enter the pipe at its upper end. Although the pipe is open at its lower end to allow the water to enter, the water's surface acts like a barrier, as if the end of the part of the pipe that is above the water were closed. The waves reflect from the water surface and combine with those moving downward to form a standing wave.

Categorize Because of the reflection of the sound waves from the water surface, we can model the part of the pipe that is above the water as open at the upper end and closed at the lower end. Therefore, we can apply the *waves under boundary conditions* model to this situation.

Figure 17.18 (Example 17.6) (a) Apparatus for demonstrating the resonance of sound waves in a pipe closed at one end. The length L of the air column is varied by moving the pipe vertically while it is partially submerged in water. (b) The first three normal modes of the system shown in (a).

Analyze

Use Equation 17.10 to find the fundamental frequency for $L = 0.090\ 0$ m:

$$f_1 = \frac{v}{4L} = \frac{343\ \text{m/s}}{4(0.090\ 0\ \text{m})} = 953\ \text{Hz}$$

Because the tuning fork causes the air column to resonate at this frequency, this frequency must also be that of the tuning fork.

(B) What are the values of L for the next two resonance conditions?

SOLUTION

Use Equation 16.12 to find the wavelength of the sound wave from the tuning fork:

$$\lambda = \frac{v}{f} = \frac{343\ \text{m/s}}{953\ \text{Hz}} = 0.360\ \text{m}$$

Notice from Figure 17.18b that the length of the air column above the water for the second resonance is $3\lambda/4$:

$$L = 3\lambda/4 = 0.270\ \text{m}$$

Notice from Figure 17.18b that the length of the air column above the water for the third resonance is $5\lambda/4$:

$$L = 5\lambda/4 = 0.450\ \text{m}$$

Finalize Consider how this problem differs from the preceding example. In the culvert, the length was fixed and the air column was presented with a mixture of many frequencies. The pipe in this example is presented with one single frequency from the tuning fork, and the length of the pipe above the water is varied until resonance is achieved.

17.7 Beats: Interference in Time

The interference phenomena we have studied so far involve the superposition of two or more waves having the same frequency. Because the amplitude of the oscillation of elements of the medium varies with the position in space of the element in such a wave, we refer to the phenomenon as *spatial interference*. Standing waves in strings and pipes are common examples of spatial interference.

Now let's consider another type of interference, one that results from the superposition of two waves having slightly *different* frequencies. In this case, when the two waves are observed at a point in space, they are periodically in and out of phase. That is, there is a *temporal* (time) alternation between constructive and destructive interference. As a consequence, we refer to this phenomenon as *interference in time* or *temporal interference*. For example, if two tuning forks of slightly different frequencies are struck, one hears a sound of periodically varying amplitude. This phenomenon is called **beating.**

> Beating is the periodic variation in amplitude at a given point due to the superposition of two waves having slightly different frequencies.

The number of amplitude maxima one hears per second, or the *beat frequency*, equals the difference in frequency between the two sources as we shall show below. The maximum beat frequency that the human ear can detect is about 20 beats/s. When the beat frequency exceeds this value, the beats blend indistinguishably with the sounds producing them.

Consider two sound waves of equal amplitude and slightly different frequencies f_1 and f_2 traveling through a medium. We use equations similar to Equation 16.13 to represent the wave functions for these two waves at a point that we identify as $x = 0$. We also choose the phase angle in Equation 16.13 as $\phi = \pi/2$:

$$y_1 = A \sin\left(\frac{\pi}{2} - \omega_1 t\right) = A \cos(2\pi f_1 t)$$

$$y_2 = A \sin\left(\frac{\pi}{2} - \omega_2 t\right) = A \cos(2\pi f_2 t)$$

Using the superposition principle, we find that the resultant wave function at this point is

$$y = y_1 + y_2 = A(\cos 2\pi f_1 t + \cos 2\pi f_2 t)$$

The trigonometric identity

$$\cos a + \cos b = 2 \cos\left(\frac{a - b}{2}\right) \cos\left(\frac{a + b}{2}\right)$$

allows us to write the expression for y as

$$y = \left[2A \cos 2\pi\left(\frac{f_1 - f_2}{2}\right)t\right] \cos 2\pi\left(\frac{f_1 + f_2}{2}\right)t \qquad (17.11)$$

Graphs of the individual waves and the resultant wave are shown in Figure 17.19. From the factors in Equation 17.11, we see that the resultant wave has an effective frequency equal to the average frequency $(f_1 + f_2)/2$. This wave is multiplied by an envelope wave given by the expression in the square brackets:

$$y_{\text{envelope}} = 2A \cos 2\pi\left(\frac{f_1 - f_2}{2}\right)t \qquad (17.12)$$

That is, the amplitude and therefore the intensity of the resultant sound vary in time. The dashed black line in Figure 17.19b is a graphical representation of the envelope wave in Equation 17.12 and is a sine wave varying with frequency $(f_1 - f_2)/2$.

A maximum in the amplitude of the resultant sound wave is detected whenever

$$\cos 2\pi\left(\frac{f_1 - f_2}{2}\right)t = \pm 1$$

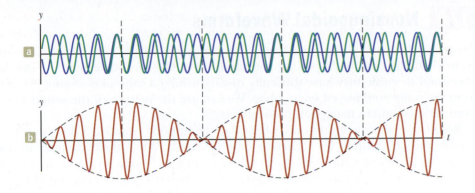

Figure 17.19 Beats are formed by the combination of two waves of slightly different frequencies. (a) The individual waves, shown in blue and green. (b) The combined wave. The envelope wave (dashed line) represents the beating of the combined sounds.

Hence, there are *two* maxima in each period of the envelope wave. Because the amplitude varies with frequency as $(f_1 - f_2)/2$, the number of beats per second, or the **beat frequency** f_{beat}, is twice this value. That is,

$$f_{beat} = |f_1 - f_2| \qquad (17.13)$$ ◀ Beat frequency

For instance, if one tuning fork vibrates at 438 Hz and a second one vibrates at 442 Hz, the resultant sound wave of the combination has a frequency of 440 Hz (the musical note A) and a beat frequency of 4 Hz. A listener would hear a 440-Hz sound wave go through an intensity maximum four times every second.

Example 17.7   **The Mistuned Piano Strings**

Two identical piano strings of length 0.750 m are each tuned exactly to 440 Hz. The tension in one of the strings is then increased by 1.0%. If they are now struck, what is the beat frequency between the fundamentals of the two strings?

SOLUTION

Conceptualize   As the tension in one of the strings is changed, its fundamental frequency changes. Therefore, when both strings are played, they will have different frequencies and beats will be heard.

Categorize   We must combine our understanding of the *waves under boundary conditions* model for strings with our new knowledge of beats.

Analyze Set up a ratio of the fundamental frequencies of the two strings using Equation 17.6 with $n = 1$:	$\dfrac{f_2}{f_1} = \dfrac{(v_2/2L)}{(v_1/2L)} = \dfrac{v_2}{v_1}$
Use Equation 16.18 to substitute for the wave speeds on the strings:	$\dfrac{f_2}{f_1} = \dfrac{\sqrt{T_2/\mu}}{\sqrt{T_1/\mu}} = \sqrt{\dfrac{T_2}{T_1}}$
Incorporate that the tension in one string is 1.0% larger than the other; that is, $T_2 = 1.010 T_1$:	$\dfrac{f_2}{f_1} = \sqrt{\dfrac{1.010 T_1}{T_1}} = 1.005$
Solve for the frequency of the tightened string:	$f_2 = 1.005 f_1 = 1.005(440 \text{ Hz}) = 442 \text{ Hz}$
Find the beat frequency using Equation 17.13:	$f_{beat} = 442 \text{ Hz} - 440 \text{ Hz} = \boxed{2 \text{ Hz}}$

Finalize   Notice that a 1.0% mistuning in tension leads to an easily audible beat frequency of 2 Hz. A piano tuner can use beats to tune a stringed instrument by "beating" a note against a reference tone of known frequency. The tuner can then adjust the string tension until the frequency of the sound it emits equals the frequency of the reference tone. The tuner does so by tightening or loosening the string until the beats produced by it and the reference source become too infrequent to notice.

17.8 Nonsinusoidal Waveforms

It is relatively easy to distinguish the sounds coming from a violin and a saxophone even when they are both playing the same note. On the other hand, a person untrained in music may have difficulty distinguishing a note played on a clarinet from the same note played on an oboe. We can use the pattern of the sound waves from various sources to explain these effects.

Recall that when a system under boundary conditions vibrates, it does so with a combination of frequencies occurring simultaneously. When those frequencies are integer multiples of a fundamental frequency, such as from a string or an air column, the result is a *musical* sound. A listener can assign a pitch to the sound based on the fundamental frequency. Pitch is a psychological reaction to a sound that allows the listener to place the sound on a scale from low to high (bass to treble). Combinations of frequencies that are not integer multiples of a fundamental, such as from a drumhead, result in a *noise* rather than a musical sound. It is much harder for a listener to assign a pitch to a noise than to a musical sound.

The wave patterns produced by a musical instrument are the result of the superposition of frequencies that are integer multiples of a fundamental. This superposition results in the corresponding richness of musical tones. The human perceptive response associated with various mixtures of harmonics is the *quality* or *timbre* of the sound. For instance, the sound of the trumpet is perceived to have a "brassy" quality (that is, we have learned to associate the adjective *brassy* with that sound); this quality enables us to distinguish the sound of the trumpet from that of the saxophone, whose quality is perceived as "reedy." The clarinet and oboe, however, both contain air columns excited by reeds; because of this similarity, they have similar mixtures of frequencies and it is more difficult for the human ear to distinguish them on the basis of their sound quality.

The sound waveforms produced by the majority of musical instruments are nonsinusoidal. Characteristic waveforms produced by a tuning fork, a flute, and a clarinet, each playing the same note, are shown in Figure 17.20. Each instrument has its own characteristic waveform. Notice, however, that despite the differences in the waveforms, each one is periodic. This point is important for our analysis of these waves. Notice that the frequencies at which the waveforms repeat are the same; the addition of higher harmonics does *not* affect the fundamental frequency of the sound.

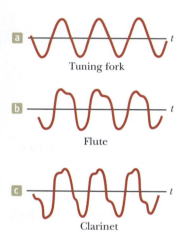

Figure 17.20 Sound waveforms produced by (a) a tuning fork, (b) a flute, and (c) a clarinet, each at approximately the same frequency.

The problem of analyzing nonsinusoidal waveforms appears at first sight to be a formidable task. If the waveform is periodic, however, it can be represented as closely as desired by the combination of a sufficiently large number of sinusoidal waves that form a harmonic series. In fact, we can represent any periodic function as a series of sine and cosine terms by using a mathematical technique based on **Fourier's theorem.**[2] The corresponding sum of terms that represents the periodic waveform is called a **Fourier series.** Let $y(t)$ be any function that is periodic in time with period T such that $y(t + T) = y(t)$. Fourier's theorem states that this function can be written as

◀ Fourier's theorem

$$y(t) = \sum (A_n \sin 2\pi f_n t + B_n \cos 2\pi f_n t) \qquad (17.14)$$

where the lowest frequency is $f_1 = 1/T$. The higher frequencies are integer multiples of the fundamental, $f_n = nf_1$, and the coefficients A_n and B_n represent the amplitudes of the various harmonics. Figure 17.21 represents a harmonic analysis of the waveforms shown in Figure 17.20. Each bar in the graph represents one of the terms in the series in Equation 17.14 up to $n = 9$. Notice that a struck tuning fork produces only one harmonic (the first), so that all coefficients except for A_1 are zero in Equation 17.14, and the waveform is a pure sine wave. On the other hand, the flute and clarinet produce the first harmonic and many higher ones.

---

[2]Developed by Jean Baptiste Joseph Fourier (1768–1830), a French physicist and mathematician.

Figure 17.21 Harmonics of the waveforms shown in Figure 17.20. Notice the variations in intensity of the various harmonics. Parts (a), (b), and (c) correspond to those in Figure 17.20.

Notice the variation in relative intensity of the various harmonics for the flute and the clarinet. In general, any musical sound consists of a fundamental frequency f plus other frequencies that are integer multiples of f, all having different intensities.

We have discussed the *analysis* of a waveform using Fourier's theorem. The analysis involves determining the coefficients of the harmonics in Equation 17.14 from a knowledge of the waveform. The reverse process, called *Fourier synthesis,* can also be performed. In this process, various harmonics are added together to form a resultant waveform. As an example of Fourier synthesis, consider the building of a square wave as shown in Figure 17.22. The symmetry of the square wave results in only odd multiples of the fundamental frequency combining in its synthesis. In Figure 17.22a, the blue curve shows the combination of f and $3f$, shown as black curves. In Figure 17.22b, we have added $5f$ to the combination and obtained the green curve. Notice how the general shape of the square wave is approximated, even though the upper and lower portions are not flat as they should be.

Figure 17.22c shows the result of adding odd frequencies up to $9f$. This approximation (red-brown curve) to the square wave is better than the approximations in Figures 17.22a and 17.22b. To approximate the square wave as closely as possible, we must add all odd multiples of the fundamental frequency, up to infinite frequency.

Using modern technology, musical sounds can be generated electronically by mixing different amplitudes of any number of harmonics. These widely used electronic music synthesizers are capable of producing an infinite variety of musical tones.

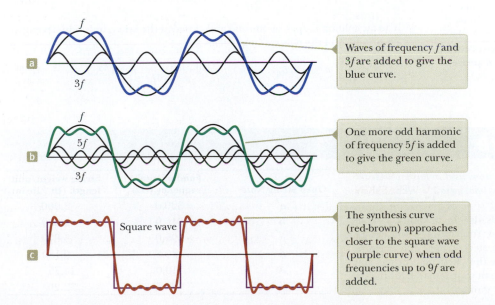

Waves of frequency f and $3f$ are added to give the blue curve.

One more odd harmonic of frequency $5f$ is added to give the green curve.

The synthesis curve (red-brown) approaches closer to the square wave (purple curve) when odd frequencies up to $9f$ are added.

Figure 17.22 Fourier synthesis of a square wave, represented by the sum of odd multiples of the first harmonic, which has frequency f.

Summary

▶ Concepts and Principles

The **superposition principle** specifies that when two or more waves move through a medium, the value of the resultant wave function equals the algebraic sum of the values of the individual wave functions.

The phenomenon of **beating** is the periodic variation in intensity at a given point due to the superposition of two waves having slightly different frequencies. The **beat frequency** is

$$f_{\text{beat}} = |f_1 - f_2| \tag{17.13}$$

where f_1 and f_2 are the frequencies of the individual waves.

Standing waves are formed from the combination of two sinusoidal waves having the same frequency, amplitude, and wavelength but traveling in opposite directions. The resultant standing wave is described by the wave function

$$y = (2A \sin kx) \cos \omega t \tag{17.1}$$

Hence, the amplitude of the standing wave is $2A$, and the amplitude of the simple harmonic motion of any element of the medium varies according to its position as $2A \sin kx$. The points of zero amplitude (called **nodes**) occur at $x = n\lambda/2$ ($n = 0, 1, 2, 3, \ldots$). The maximum amplitude points (called **antinodes**) occur at $x = n\lambda/4$ ($n = 1, 3, 5, \ldots$). Adjacent antinodes are separated by a distance $\lambda/2$. Adjacent nodes also are separated by a distance $\lambda/2$.

▶ Analysis Models for Problem Solving

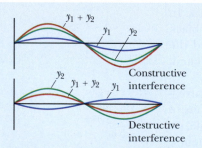

Waves in Interference. When two traveling waves having equal frequencies superimpose, the resultant wave is described by the **principle of superposition** and has an amplitude that depends on the phase angle ϕ between the two waves. **Constructive interference** occurs when the two waves are in phase, corresponding to $\phi = 0, 2\pi, 4\pi, \ldots$ rad. **Destructive interference** occurs when the two waves are 180° out of phase, corresponding to $\phi = \pi, 3\pi, 5\pi, \ldots$ rad.

Waves Under Boundary Conditions. When a wave is subject to boundary conditions, only certain natural frequencies are allowed; we say that the frequencies are *quantized*.

For waves on a string fixed at both ends, the natural frequencies are

$$f_n = \frac{n}{2L}\sqrt{\frac{T}{\mu}} \quad n = 1, 2, 3, \ldots \tag{17.7}$$

where T is the tension in the string and μ is its linear mass density.

For sound waves with speed v in an air column of length L open at both ends, the natural frequencies are

$$f_n = n\frac{v}{2L} \quad n = 1, 2, 3, \ldots \tag{17.9}$$

If an air column is open at one end and closed at the other, only odd harmonics are present and the natural frequencies are

$$f_m = m\frac{v}{4L} \quad m = 1, 3, 5, \ldots \quad \text{or} \quad f_n = (2n-1)\frac{v}{4L} \quad n = 1, 2, 3, \ldots \tag{17.10}$$

Think–Pair–Share

See the Preface for an explanation of the icons used in this problems set. For additional assessment items for this section, go to ⚡ **WEBASSIGN** From Cengage

1. **ACTIVITY** You and your friends decide to found a small start-up business in your garage. Your business will design and build acoustic guitars with steel strings. You perform research online and decide on the particular strings you will use. The table shows data from your research for the six strings that will be used on your guitars.

Open String Note	Fundamental Frequency (Hz)	String weight/unit length (10^{-5} lb/in)
e'	329.6	2.000
b	246.9	2.930
g	196.0	5.870
d	146.8	9.180
A	110.0	14.70
E	82.41	32.20

The system of naming notes used in the first column of the table is such that middle C (as played on a piano) is notated with lowercase c′. The notes in the octave above middle C use this lower case/primed notation. The C below middle C (and the notes in the octave starting with this C) is notated with a simple lower case c. The next C down (and the notes in the octave starting with this C) is notated with a capital letter C. Therefore, the E-note of the lowest string on the guitar is two octaves below the e′-note of the highest string.

You design your guitars so that the scale length of the string (the length of the vibrating portion of the string) is 25.50 in for all six strings. (a) After selecting the strings that will be used, your team needs to choose the particular wood that will be used in the guitar, and then design the thickness of the wood on the front face of the guitar. This choice and this design will depend on the total tension exerted by the strings on the front face. What is the total tension in all six strings? (b) Another part of your design relates to the wave speed in the strings. For your design, the wave speed in the e′-string should be about four times that in the E-string. Does the data above satisfy this design criterion? (c) Would any guitar designed like this one *not* satisfy the design criteria in part (b)?

2. **ACTIVITY** Set up four identical glass bottles filled with increasing levels of water from left to right. Have part of your group strike the bottles with a spoon from left to right and listen to how the frequencies change. Now have the other part blow into the top of each bottle from left to right and listen to how the frequencies change. Why do the frequencies change in opposite directions for these two experiments? What's vibrating in each case?

3. **ACTIVITY** Cost-free signal generator or function generator apps are available for download to your smartphone. Have two members of your group download an app that has the capability of performing a frequency sweep. Set one of the phones to play a continuous sine wave of frequency 4 000 Hz. Set the other to begin a downward sine wave sweep from 3 800 Hz to 3 000 Hz. Start both phones playing at the same time. Have each member of the group listen carefully to the combined sound. While you will clearly hear the frequency of the second phone going *down*, some members of the group may also hear a sound going *up* in frequency. It will be different in nature; it will sound as if it is coming from inside your ear rather than from the smartphones. Rotate your head back and forth, which may help you hear it. What is causing this sound?

Problems

See the Preface for an explanation of the icons used in this problems set. For additional assessment items for this section, go to ✸ **WEBASSIGN**
From Cengage

> *Note:* Unless otherwise specified, assume the speed of sound in air is 343 m/s, its value at an air temperature of 20.0°C. At any other Celsius temperature T_C, the speed of sound in air is described by
>
> $$v = 331 \sqrt{1 + \frac{T_C}{273}}$$
>
> where v is in m/s and T is in °C.

SECTION 17.1 Analysis Model: Waves in Interference

1. Two waves on one string are described by the wave functions

   V

   $$y_1 = 3.0 \cos (4.0x - 1.6t) \qquad y_2 = 4.0 \sin (5.0x - 2.0t)$$

   where x and y are in centimeters and t is in seconds. Find the values of $y_1 + y_2$ at the points (a) $x = 1.00$, $t = 1.00$; (b) $x = 1.00$, $t = 0.500$; and (c) $x = 0.500$, $t = 0$. *Note:* Remember that the arguments of the trigonometric functions are in radians.

2. Two pulses of different amplitudes approach each other, each having a speed of $v = 1.00$ m/s. Figure P17.2 shows the positions of the pulses at time $t = 0$. (a) Sketch the resultant wave at $t = 2.00$ s, 4.00 s, 5.00 s, and 6.00 s. (b) **What If?** If the pulse on the right is inverted so that it is upright, how would your sketches of the resultant wave change?

3. Two wave pulses A and B are moving in opposite directions, each with a speed $v = 2.00$ cm/s. The amplitude of A is twice the amplitude of B. The pulses are shown in Figure P17.3 at $t = 0$. Sketch the resultant wave at $t = 1.00$ s, 1.50 s, 2.00 s, 2.50 s, and 3.00 s.

Figure P17.3

4. *Why is the following situation impossible?* Two identical loudspeakers are driven by the same oscillator at frequency 200 Hz. They are located on the ground a distance $d = 4.00$ m from each other. Starting far from the speakers, a man walks straight toward the right-hand speaker as shown in Figure P17.4. After passing through three minima in sound intensity, he walks to the next maximum and stops. Ignore any sound reflection from the ground.

Figure P17.4

Figure P17.2

5. Two pulses traveling on the same string are described by

$$y_1 = \frac{5}{(3x - 4t)^2 + 2} \qquad y_2 = \frac{-5}{(3x + 4t - 6)^2 + 2}$$

(a) In which direction does each pulse travel? (b) At what instant do the two pulses cancel for all x? (c) At what point do the two pulses cancel at all times t?

6. Two identical loudspeakers 10.0 m apart are driven by the same oscillator with a frequency of $f = 21.5$ Hz (Fig. P17.6) in an area where the speed of sound is 344 m/s. (a) Show that a receiver at point A records a minimum in sound intensity from the two speakers. (b) If the receiver is moved in the plane of the speakers, show that the path it should take so that the intensity remains at a minimum is along the hyperbola $9x^2 - 16y^2 = 144$ (shown in red-brown in Fig. P17.6). (c) Can the receiver remain at a minimum and move very far away from the two sources? If so, determine the limiting form of the path it must take. If not, explain how far it can go.

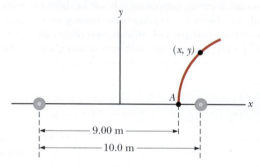

Figure P17.6

7. Two sinusoidal waves on a string are defined by the wave functions

$$y_1 = 2.00 \sin (20.0x - 32.0t) \qquad y_2 = 2.00 \sin (25.0x - 40.0t)$$

where x, y_1, and y_2 are in centimeters and t is in seconds. (a) What is the phase difference between these two waves at the point $x = 5.00$ cm at $t = 2.00$ s? (b) What is the positive x value closest to the origin for which the two phases differ by $\pm\pi$ at $t = 2.00$ s? (At that location, the two waves add to zero.)

SECTION 17.2 Standing Waves

8. Verify by direct substitution that the wave function for a standing wave given in Equation 17.1,

$$y = (2A \sin kx) \cos \omega t$$

is a solution of the general linear wave equation, Equation 16.27:

$$\frac{\partial^2 y}{\partial x^2} = \frac{1}{v^2} \frac{\partial^2 y}{\partial t^2}$$

9. Two waves simultaneously present on a long string have a phase difference ϕ between them so that a standing wave formed from their combination is described by

$$y(x, t) = 2A \sin \left(kx + \frac{\phi}{2}\right) \cos \left(\omega t - \frac{\phi}{2}\right)$$

(a) Despite the presence of the phase angle ϕ, is it still true that the nodes are one-half wavelength apart? Explain. (b) Are the nodes different in any way from the way they would be if ϕ were zero? Explain.

10. A standing wave is described by the wave function

$$y = 6 \sin \left(\frac{\pi}{2}x\right) \cos (100\pi t)$$

where x and y are in meters and t is in seconds. (a) Prepare graphs showing y as a function of x for five instants: $t = 0$, 5 ms, 10 ms, 15 ms, and 20 ms. (b) From the graph, identify the wavelength of the wave and explain how to do so. (c) From the graph, identify the frequency of the wave and explain how to do so. (d) From the equation, directly identify the wavelength of the wave and explain how to do so. (e) From the equation, directly identify the frequency and explain how to do so.

SECTION 17.4 Analysis Model: Waves Under Boundary Conditions

11. A standing wave is established in a 120-cm-long string fixed at both ends. The string vibrates in four segments when driven at 120 Hz. (a) Determine the wavelength. (b) What is the fundamental frequency of the string?

12. A taut string has a length of 2.60 m and is fixed at both ends. (a) Find the wavelength of the fundamental mode of vibration of the string. (b) Can you find the frequency of this mode? Explain why or why not.

13. A string that is 30.0 cm long and has a mass per unit length of 9.00×10^{-3} kg/m is stretched to a tension of 20.0 N. Find (a) the fundamental frequency and (b) the next three frequencies of possible standing-wave patterns on the string.

14. In the arrangement shown in Figure P17.14, an object of mass $m = 5.00$ kg hangs from a cord around a light pulley. The length of the cord between point P and the pulley is $L = 2.00$ m. (a) When the vibrator is set to a frequency of 150 Hz, a standing wave with six loops is formed. What must be the linear mass density of the cord? (b) How many loops (if any) will result if m is changed to 45.0 kg? (c) How many loops (if any) will result if m is changed to 10.0 kg?

Figure P17.14

15. **Review.** A sphere of mass $M = 1.00$ kg is supported by a string that passes over a pulley at the end of a horizontal rod of length $L = 0.300$ m (Fig. P17.15). The string makes an angle $\theta = 35.0°$ with the rod. The fundamental frequency of standing waves in the portion of the string above the rod is $f = 60.0$ Hz. Find the mass of the portion of the string above the rod.

Figure P17.15
Problems 15 and 16.

16. **Review.** A sphere of mass M is supported by a string that passes over a pulley at the end of a horizontal rod of length L (Fig. P17.15). The string makes an angle θ with the rod. The fundamental frequency of standing waves in the portion of the string above the rod is f. Find the mass of the portion of the string above the rod.

17. A violin string has a length of 0.350 m and is tuned to concert G, with $f_G = 392$ Hz. (a) How far from the end of the string must the violinist place her finger to play concert A, with $f_A = 440$ Hz? (b) If this position is to remain correct to one-half the width of a finger (that is, to within 0.600 cm), what is the maximum allowable percentage change in the string tension?

18. **Review.** A solid copper object hangs at the bottom of a steel wire of negligible mass. The top end of the wire is fixed. When the wire is struck, it emits sound with a fundamental frequency of 300 Hz. The copper object is then submerged in water so that half its volume is below the water line. Determine the new fundamental frequency.

SECTION 17.5 Resonance

19. The Bay of Fundy, Nova Scotia, has the highest tides in the world. Assume in midocean and at the mouth of the bay the Moon's gravity gradient and the Earth's rotation make the water surface oscillate with an amplitude of a few centimeters and a period of 12 h 24 min. At the head of the bay, the amplitude is several meters. Assume the bay has a length of 210 km and a uniform depth of 36.1 m. The speed of long-wavelength water waves is given by $v = \sqrt{gd}$, where d is the water's depth. Argue for or against the proposition that the tide is magnified by standing-wave resonance.

SECTION 17.6 Standing Waves in Air Columns

20. The windpipe of one typical whooping crane is 5.00 feet long. What is the fundamental resonant frequency of the bird's trachea, modeled as a narrow pipe closed at one end? Assume a temperature of 37°C.

21. The fundamental frequency of an open organ pipe corresponds to middle C (261.6 Hz on the chromatic musical scale). The third resonance of a closed organ pipe has the same frequency. What is the length of (a) the open pipe and (b) the closed pipe?

22. Ever since seeing Figure 16.22 in the previous chapter, you have been fascinated with the hearing response in humans. You have set up an apparatus that allows you to determine your own threshold of hearing as a function of frequency. After performing the experiment and recording the results, you graph the results, which look like Figure P17.22. You are

Figure P17.22

intrigued by the two dips in the curve at the right-hand side of the graph. You measure carefully and find that the minimum values of these dips occur at 3 800 Hz and 11 500 Hz. Performing some online research, you discover that the outer canal of the human ear can be modeled as an air column open at the outer end and closed at the inner end by the eardrum. You use this information to determine the length of the outer canal in your ear.

23. An air column in a glass tube is open at one end and closed at the other by a movable piston. The air in the tube is warmed above room temperature, and a 384-Hz tuning fork is held at the open end. Resonance is heard when the piston is at a distance $d_1 = 22.8$ cm from the open end and again when it is at a distance $d_2 = 68.3$ cm from the open end. (a) What speed of sound is implied by these data? (b) How far from the open end will the piston be when the next resonance is heard?

24. A shower stall has dimensions 86.0 cm × 86.0 cm × 210 cm. Assume the stall acts as a pipe closed at both ends, with nodes at opposite sides. Assume singing voices range from 130 Hz to 2 000 Hz and let the speed of sound in the hot air be 355 m/s. For someone singing in this shower, which frequencies would sound the richest (because of resonance)?

25. A glass tube (open at both ends) of length L is positioned near an audio speaker of frequency $f = 680$ Hz. For what values of L will the tube resonate with the speaker?

26. A tunnel under a river is 2.00 km long. (a) At what frequencies can the air in the tunnel resonate? (b) Explain whether it would be good to make a rule against blowing your car horn when you are in the tunnel.

27. As shown in Figure P17.27, water is pumped into a tall, vertical cylinder at a volume flow rate $R = 1.00$ L/min. The radius of the cylinder is $r = 5.00$ cm, and at the open top of the cylinder a tuning fork is vibrating with a frequency $f = 512$ Hz. As the water rises, what time interval elapses between successive resonances?

28. As shown in Figure P17.27, water is pumped into a tall, vertical cylinder at a volume flow rate R. The radius of the cylinder is r, and at the open top of the cylinder a tuning fork is vibrating with a frequency f. As the water rises, what time interval elapses between successive resonances?

Figure P17.27
Problems 27 and 28.

29. You are a flutist in a local orchestra. On a cold winter day, you are late to a performance. Arriving at the orchestra hall, you know that you have missed the group tune-up before the performance, so you tune your instrument in the cold air outside the stage door. After tuning, you run inside the auditorium, where the temperature is 22.2°C, take your seat, and begin playing the first song with the rest of the orchestra. You are quite embarrassed to notice that you are playing the song a half-step higher than your colleagues in the orchestra. Your excitement about physics overcomes your musical embarrassment as you realize that you can use this information to calculate the temperature outside. (Assume that the length of the instrument does not change with temperature. A half-step represents a frequency ratio of $2^{1/12}$.)

30. *Why is the following situation impossible?* A student is listening to the sounds from an air column that is 0.730 m long. He doesn't know if the column is open at both ends or open at only one end. He hears resonance from the air column at frequencies 235 Hz and 587 Hz.

SECTION 17.7 Beats: Interference in Time

31. **Review.** A student holds a tuning fork oscillating at 256 Hz. **T** He walks toward a wall at a constant speed of 1.33 m/s. (a) What beat frequency does he observe between the tuning fork and its echo? (b) How fast must he walk away from the wall to observe a beat frequency of 5.00 Hz?

32. While attempting to tune the note C at 523 Hz, a piano **V** tuner hears 2.00 beats/s between a reference oscillator and the string. (a) What are the possible frequencies of the string? (b) When she tightens the string slightly, she hears 3.00 beats/s. What is the frequency of the string now? (c) By what percentage should the piano tuner now change the tension in the string to bring it into tune?

SECTION 17.8 Nonsinusoidal Waveforms

33. Suppose a flutist plays a 523-Hz C note with first harmonic displacement amplitude $A_1 = 100$ nm. From Figure 17.21b read, by proportion, the displacement amplitudes of harmonics 2 through 7. Take these as the values A_2 through A_7 in the Fourier analysis of the sound and assume $B_1 = B_2 = \cdots = B_7 = 0$. Construct a graph of the waveform of the sound. Your waveform will not look exactly like the flute waveform in Figure 17.20b because you simplify by ignoring cosine terms; nevertheless, it produces the same sensation to human hearing.

ADDITIONAL PROBLEMS

34. Two strings are vibrating at the same frequency of 150 Hz. After the tension in one of the strings is decreased, an observer hears four beats each second when the strings vibrate together. Find the new frequency in the adjusted string.

35. The ship in Figure P17.35 travels along a straight line parallel to the shore and a distance $d = 600$ m from it. The ship's radio receives simultaneous signals of the same frequency from antennas A and B, separated by a distance $L = 800$ m. The signals interfere constructively at point C, which is equidistant from A and B. The signal goes through the first minimum at point D, which is directly outward from the shore from point B. Determine the wavelength of the radio waves.

Figure P17.35

36. A 2.00-m-long wire having a mass of 0.100 kg is fixed at both ends. The tension in the wire is maintained at 20.0 N. (a) What are the frequencies of the first three allowed modes of vibration? (b) If a node is observed at a point 0.400 m from one end, in what mode and with what frequency is it vibrating?

37. A string fixed at both ends and having a mass of 4.80 g, a **Q|C** length of 2.00 m, and a tension of 48.0 N vibrates in its second ($n = 2$) normal mode. (a) Is the wavelength in air of the sound emitted by this vibrating string larger or smaller than the wavelength of the wave on the string? (b) What is the ratio of the wavelength in air of the sound emitted by this vibrating string and the wavelength of the wave on the string?

38. You are working as an assistant to a landscape architect, **CR** who is designing the landscaping around a new commercial building. The architect plans to have a large rectangular water basin as part of his design. When you see this design, you mention to the architect that the project is located in an area prone to earthquakes. You point out that an earthquake could create a seiche in the basin by resonance, causing the water in the basin to spill out and enter nearby underground electrical transformers. A *seiche* is a standing wave in a body of water, in which the water sloshes back and forth with antinodes at the ends of the basin. (You may have created a seiche in a bathtub as a child by sliding your body back and forth along the length of the tub, leaving water on the floor for your parents to wipe up.) The architect dismisses your comments as unrealistic. While visiting your cousin the previous week in a non-earthquake-prone area, you had seen a water basin similar to the one planned by the architect. You call your cousin and find out that the water basin in his town has the same depth of water as that planned by the architect. You ask your cousin to create a pulse in the water by dropping a pebble, and determine how long the pulse takes to cross the basin. Based on this time interval and the length of your cousin's basin, you determine that a pulse will take 2.50 s to cross the basin planned by the architect. Show the architect that there will be several possible seiche resonances in the water basin for typical low frequencies of earthquakes in the range of 0–4 Hz.

39. **Review.** Consider the apparatus shown in Figure 17.15 **Q|C** and described in Example 17.4. Suppose the number of antinodes in Figure 17.15b is an arbitrary value n. (a) Find an expression for the radius of the sphere in the water as a function of only n. (b) What is the minimum allowed value of n for a sphere of nonzero size? (c) What is the radius of the largest sphere that will produce a standing wave on the string? (d) What happens if a larger sphere is used?

40. **Review.** For the arrangement shown in Figure P17.40, the **GP** inclined plane and the small pulley are frictionless; the string supports the object of mass M at the bottom of the plane; and the entire string has mass m. The system is in equilibrium, and the vertical part of the string has a length h. We wish to study standing waves set up in the vertical section

Figure P17.40

of the string. (a) What analysis model describes the object of mass M? (b) What analysis model describes the waves on the vertical part of the string? (c) Find the tension in the string. (d) Model the shape of the string as one leg and the hypotenuse of a right triangle. Find the whole length of the string. (e) Find the mass per unit length of the string. (f) Find the speed of waves on the string. (g) Find the lowest frequency for a standing wave on the vertical section of the string. (h) Evaluate this result for $M = 1.50$ kg, $m = 0.750$ g, $h = 0.500$ m, and $\theta = 30.0°$. (i) Find the numerical value for the lowest frequency for a standing wave on the sloped section of the string.

41. **Review.** A loudspeaker at the front of a room and an identical loudspeaker at the rear of the room are being driven by the same oscillator at 456 Hz. A student walks at a uniform rate of 1.50 m/s along the length of the room. She hears a single tone repeatedly becoming louder and softer. (a) Model these variations as beats between the Doppler-shifted sounds the student receives. Calculate the number of beats the student hears each second. (b) Model the two speakers as producing a standing wave in the room and the student as walking between antinodes. Calculate the number of intensity maxima the student hears each second.

42. Two speakers are driven by the same oscillator of frequency f. They are located a distance d from each other on a vertical pole. A man walks straight toward the lower speaker in a direction perpendicular to the pole as shown in Figure P17.42. (a) How many times will he hear a minimum in sound intensity? (b) How far is he from the pole at these moments? Let v represent the speed of sound and assume that the ground does not reflect sound. The man's ears are at the same level as the lower speaker.

Figure P17.42

43. A standing wave is set up in a string of variable length and
[S] tension by a vibrator of variable frequency. Both ends of the string are fixed. When the vibrator has a frequency f, in a string of length L and under tension T, n antinodes are set up in the string. (a) If the length of the string is doubled, by what factor should the frequency be changed so that the same number of antinodes is produced? (b) If the frequency and length are held constant, what tension will produce $n + 1$ antinodes? (c) If the frequency is tripled and the length of the string is halved, by what factor should the tension be changed so that twice as many antinodes are produced?

44. **Review.** The top end of a yo-yo string is held stationary.
[Q|C] The yo-yo itself is much more massive than the string. It starts from rest and moves down with constant acceleration 0.800 m/s² as it unwinds from the string. The

rubbing of the string against the edge of the yo-yo excites transverse standing-wave vibrations in the string. Both ends of the string are nodes even as the length of the string increases. Consider the instant 1.20 s after the motion begins from rest. (a) Show that the rate of change with time of the wavelength of the fundamental mode of oscillation is 1.92 m/s. (b) **What if?** Is the rate of change of the wavelength of the second harmonic also 1.92 m/s at this moment? Explain your answer. (c) **What if?** The experiment is repeated after more mass has been added to the yo-yo body. The mass distribution is kept the same so that the yo-yo still moves with downward acceleration 0.800 m/s². At the 1.20-s point in this case, is the rate of change of the fundamental wavelength of the string vibration still equal to 1.92 m/s? Explain. (d) Is the rate of change of the second harmonic wavelength the same as in part (b)? Explain.

45. **Review.** Consider the copper object hanging from the steel wire in Problem 18. The top end of the wire is fixed. When the wire is struck, it emits sound with a fundamental frequency of 300 Hz. The copper object is then submerged in water. If the object can be positioned with any desired fraction of its volume submerged, what is the lowest possible new fundamental frequency?

46. A string of linear density 1.60 g/m is stretched between clamps 48.0 cm apart. The string does not stretch appreciably as the tension in it is steadily raised from 15.0 N at $t = 0$ to 25.0 N at $t = 3.50$ s. Therefore, the tension as a function of time is given by the expression $T = 15.0 + 10.0t/3.50$, where T is in newtons and t is in seconds. The string is vibrating in its fundamental mode throughout this process. Find the number of oscillations it completes during the 3.50-s interval.

47. **Review.** A 12.0-kg object hangs in equilibrium from a string
[AMT] with a total length of $L = 5.00$ m and a linear mass density of $\mu = 0.001\ 00$ kg/m. The string is wrapped around two light, frictionless pulleys that are separated by a distance of $d = 2.00$ m (Fig. P17.47a). (a) Determine the tension in the string. (b) At what frequency must the string between the pulleys vibrate to form the standing-wave pattern shown in Figure P17.47b?

Figure P17.47 Problems 47 and 48.

48. **Review.** An object of mass m hangs in equilibrium from a
[S] string with a total length L and a linear mass density μ. The string is wrapped around two light, frictionless pulleys that are separated by a distance d (Fig. P17.47a). (a) Determine the tension in the string. (b) At what frequency must the string between the pulleys vibrate to form the standing-wave pattern shown in Figure P17.47b?

49. Two waves are described by the wave functions

$$y_1(x, t) = 5.00 \sin (2.00x - 10.0t)$$

$$y_2(x, t) = 10.0 \cos (2.00x - 10.0t)$$

where x, y_1, and y_2 are in meters and t is in seconds. (a) Show that the wave resulting from their superposition can be expressed as a single sine function. (b) Determine the amplitude and phase angle for this sinusoidal wave.

CHALLENGE PROBLEM

50. In Figures 17.22a and 17.22b, notice that the amplitude of the component wave for frequency f is large, that for $3f$ is smaller, and that for $5f$ smaller still. How do we know exactly how much amplitude to assign to each frequency component to build a square wave? This problem helps us find the answer to that question. Let the square wave in Figure 17.22c have an amplitude A and let $t = 0$ be at the extreme left of the figure. So, one period T of the square wave is described by

$$y(t) = \begin{cases} A & 0 < t < \dfrac{T}{2} \\ -A & \dfrac{T}{2} < t < T \end{cases}$$

Express Equation 17.14 with angular frequencies:

$$y(t) = \sum_n (A_n \sin n\omega t + B_n \cos n\omega t)$$

Now proceed as follows. (a) Multiply both sides of Equation 17.14 by $\sin m\omega t$ and integrate both sides over one period T. Show that the left-hand side of the resulting equation is equal to 0 if m is even and is equal to $4A/m\omega$ if m is odd. (b) Using trigonometric identities, show that all terms on the right-hand side involving B_n are equal to zero. (c) Using trigonometric identities, show that all terms on the right-hand side involving A_n are equal to zero *except* for the one case of $m = n$. (d) Show that the entire right-hand side of the equation reduces to $\frac{1}{2}A_m T$. (e) Show that the Fourier series expansion for a square wave is

$$y(t) = \sum_n \frac{4A}{n\pi} \sin n\omega t$$

PART 3

Thermodynamics

We now direct our attention to the study of thermodynamics, which involves situations in which the temperature or state (solid, liquid, gas) of a system changes due to energy transfers. In this part of the book, we will focus on the heat Q in Equation 8.2 and its effects on the thermal conditions of a system. We will also look at work W performed on deformable systems, such as an enclosed gas, as well as electromagnetic radiation T_{ER} across a system boundary. Each of these energy transfers can cause a change in the internal energy E_{int} of the system, which we can relate to *temperature*.

Historically, the development of thermodynamics paralleled the development of the atomic theory of matter. By the 1820s, chemical experiments had provided solid evidence for the existence of atoms. At that time, scientists recognized that a connection between thermodynamics and the structure of matter must exist. In 1827, botanist Robert Brown reported that grains of pollen suspended in a liquid move erratically from one place to another as if under constant agitation. In 1905, Albert Einstein used kinetic theory to explain the cause of this erratic motion, known today as *Brownian motion*. Einstein explained this phenomenon by assuming the grains are under constant bombardment by "invisible" molecules in the liquid, which themselves move erratically. The motion of the molecules is related to the temperature of the liquid. A connection was thus forged between the everyday world and the tiny, invisible building blocks that make up this world.

Thermodynamics also addresses more practical questions. Have you ever wondered how a refrigerator is able to cool its contents, or what types of transformations occur in a power plant or in the engine of your automobile, or what happens to the kinetic energy of a moving object when the object comes to rest? The laws of thermodynamics can be used to provide explanations for these and other phenomena. ■

A bubble in one of the many mud pots in Yellowstone National Park is caught just at the moment of popping. A mud pot is a pool of bubbling hot mud that demonstrates the existence of thermodynamic processes below the Earth's surface. (*Adambooth/Dreamstime.com*)

18

Temperature

A brick sidewalk exhibits buckling. In some cases, this is caused by a *mechanical* phenomenon: the growth of tree roots under the sidewalk. But we see no trees here that are close enough to the sidewalk to cause this effect. This buckling is caused by a *thermal* process, related to a high temperature. (*John W. Jewett, Jr.*)

STORYLINE **You have discovered that you are out of potato chips.** While driving to the store, you look at the high-voltage electric power transmission lines crossing the road ahead of you. You have seen these lines almost every day, but there is something different about them today. The power lines between the towers on either side of the road seem to be sagging lower today than they have in the past. Then you notice that a brick sidewalk on the side of the road has buckled, as shown above. What's causing these effects? Your hometown is experiencing severely high temperatures that have lasted for several days; it has been hotter than you can ever remember. Could this be related to these effects? Will the power lines rise back up when the weather cools? Will the sidewalk "unbuckle" when the temperature drops? What does the lizard checking out the situation from the curb think? Wait a minute! In all these questions, what exactly *is* temperature, what do *hot* and *cool* actually mean? This question haunts your visit to the store and you drive home without purchasing your potato chips.

CONNECTIONS Up to this point in the text, we have focused on *mechanical* situations, which generally involve macroscopic objects. For example, we looked at kinetic energies of cars, billiard balls, planets, and rolling wheels. We performed calculations using potential energies in systems of springs, a ball and the Earth, a planet and the Sun. In this chapter, we begin to investigate *thermal* phenomena. We introduced internal energy in Chapter 7, where we talked about something becoming warmer due to friction. That was our first hint of a thermal process. The hallmark of thermal processes is that they involve energy on a microscopic scale. As we shall see in Chapter 20, we can relate the temperature of an object to the kinetic energy of the molecules of the object. We introduced the energy transfer process of heat in Chapter 8 as a means of transferring energy into or out of a system; in Chapter 19, we will discuss this process in terms of microscopic collisions between molecules at the boundary of the system. To establish the basis for these discussions in the next chapters, we will first embark in this chapter on a macroscopic understanding of the concept of temperature and its effects. The chapter concludes with a study of ideal gases on the macroscopic scale. In this study, we connect the quantity of pressure from Chapter 14 to that of temperature from this chapter. Once we understand thermal phenomena, we will see, for example, thermal effects on electrical resistance in Chapter 26, on magnetic properties of materials in Chapter 29, on the radiation from a hot surface in Chapter 39, and so on.

18.1 Temperature and the Zeroth Law of Thermodynamics

We often associate the concept of temperature with how hot or cold an object feels when we touch it. In this way, our senses provide us with a qualitative indication of temperature. Our senses, however, are unreliable and often mislead us. For example, if you stand in bare feet with one foot on carpet and the other on an adjacent tile floor, the tile feels colder than the carpet *even though both are at the same temperature*. The two objects feel different because tile transfers energy by heat at a higher rate than carpet does. Your skin "measures" the rate of energy transfer by heat rather than the actual temperature. What we need is a reliable and reproducible method for measuring the relative hotness or coldness of objects rather than the rate of energy transfer. Scientists have developed a variety of thermometers for making such quantitative measurements.

Two objects at different initial temperatures eventually reach some intermediate temperature when placed in contact with each other. For example, when hot water and cold water are mixed in a bathtub, energy is transferred from the hot water to the cold water and the final temperature of the mixture is somewhere between the initial hot and cold temperatures.

The energy-transfer mechanisms from Chapter 8 that we will focus on in this current discussion are heat, Q in Eq. 8.2, and electromagnetic radiation, T_{ER}. For purposes of this discussion, let's assume two objects are in **thermal contact** with each other if energy can be exchanged between them by these processes due to a temperature difference. **Thermal equilibrium** is a situation in which two objects would not exchange energy by heat or electromagnetic radiation if they were placed in thermal contact.

Let's consider two objects A and B, which are not in thermal contact, and a third object C, which is our thermometer. We wish to determine whether A and B are in thermal equilibrium with each other. The thermometer (object C) is first placed in thermal contact with object A until thermal equilibrium is reached[1] as shown in Figure 18.1a (page 484). From that moment on, the thermometer's reading remains constant and we record this reading. The thermometer is then removed from object A and placed in thermal contact with object B as shown in Figure 18.1b. The reading is again recorded after thermal equilibrium is reached. If the two readings are the same, we can conclude that object A and object B are in thermal equilibrium with each other. If they are placed in contact with each other as in Figure 18.1c, there will be no exchange of energy between them.

---

[1] We assume a negligible amount of energy transfers between the thermometer and object A in the time interval during which they are in thermal contact. Without this assumption, which is also made for the thermometer and object B, the measurement of the temperature of an object disturbs the system so that the measured temperature is different from the initial temperature of the object. In practice, whenever you measure a temperature with a thermometer, you measure the disturbed system, not the original system.

The temperatures of A and B are measured to be the same by placing them in thermal contact with a thermometer (object C).

No energy will be exchanged between A and B when they are placed in thermal contact with each other.

Figure 18.1 The zeroth law of thermodynamics. In general, objects A and B can be of different sizes, different masses, and different materials. The zeroth law allows us to identify something that is the *same* for both objects: temperature.

We can summarize these results in a statement known as the **zeroth law of thermodynamics** (the law of equilibrium):

◀ Zeroth law of thermodynamics

> If objects A and B are separately in thermal equilibrium with a third object C, then A and B are in thermal equilibrium with each other.

This statement can easily be proved experimentally and is very important because *it enables us to define temperature.* We can think of **temperature** as the property that determines whether or not energy will transfer between two objects when they are in thermal contact. Two objects in thermal equilibrium with each other are at the same temperature. Conversely, if two objects have different temperatures, they are not in thermal equilibrium and energy will transfer between them when they are placed in thermal contact. In Figure 18.1, it is *only* the temperatures of A and B that determine whether energy will transfer from one to the other when they are placed in thermal contact—not *size, mass, material, density,* or anything else. For now, temperature is only defined for us in terms of the zeroth law. We will relate temperature to molecular motion in Chapter 20.

ⓆUICK QUIZ 18.1 Two objects, with different sizes, masses, and temperatures, are placed in thermal contact. In which direction does the energy travel? **(a)** Energy travels from the larger object to the smaller object. **(b)** Energy travels from the object with more mass to the one with less mass. **(c)** Energy travels from the object at higher temperature to the object at lower temperature.

18.2 Thermometers and the Celsius Temperature Scale

In Figure 18.1, we used a *thermometer* to measure the temperatures of A and B. All thermometers are based on the principle that some physical property of a system changes as the system's temperature changes. Some physical properties that change with temperature are (1) the volume of a liquid, (2) the dimensions of a solid, (3) the pressure of a gas at constant volume, (4) the volume of a gas at constant pressure, (5) the electric resistance of a conductor, and (6) the color of an object.

A common thermometer in everyday use consists of a mass of liquid—usually mercury or alcohol—that expands into a glass capillary tube when heated (Fig. 18.2). In this case, the physical property that changes is the volume of a liquid. Any temperature change in the range of the thermometer can be defined as being proportional to the change in length of the liquid column. The thermometer can be calibrated by placing it in thermal contact with a natural system that remains

The level of the mercury in the thermometer rises as the mercury is heated by water in the test tube.

20°C

30°C

Figure 18.2 A mercury thermometer before and after increasing its temperature.

at constant temperature. One such system is a mixture of water and ice in thermal equilibrium at atmospheric pressure. On the **Celsius temperature scale,** this mixture is defined to have a temperature of zero degrees Celsius, which is written as 0°C; this temperature is called the *ice point* of water. Another commonly used system is a mixture of water and steam in thermal equilibrium at atmospheric pressure; its temperature is defined as 100°C, which is the *steam point* of water. Once the liquid levels in the thermometer have been established at these two points, the length of the liquid column between the two points is divided into 100 equal segments to create the Celsius scale. Therefore, each segment denotes a change in temperature of one Celsius degree.

Thermometers calibrated in this way present problems when extremely accurate readings are needed. For instance, the readings given by an alcohol thermometer calibrated at the ice and steam points of water might agree with those given by a mercury thermometer only at the calibration points. Because mercury and alcohol have different thermal expansion properties, when one thermometer reads a temperature of, for example, 50°C, the other may indicate a slightly different value. The discrepancies between thermometers are especially large when the temperatures to be measured are far from the calibration points.[2]

An additional practical problem of any thermometer is the limited range of temperatures over which it can be used. A mercury thermometer, for example, cannot be used below the freezing point of mercury, which is −39°C, and an alcohol thermometer is not useful for measuring temperatures above 85°C, the boiling point of alcohol. To surmount this problem, we need a universal thermometer whose readings are independent of the substance used in it. The gas thermometer, discussed in the next section, approaches this requirement.

18.3 The Constant-Volume Gas Thermometer and the Absolute Temperature Scale

One version of a gas thermometer is the constant-volume apparatus shown in Figure 18.3. The physical change exploited in this device is the variation of pressure of a fixed volume of gas with temperature. The flask is immersed in an ice-water bath, and mercury reservoir *B* is raised or lowered. This will cause mercury to transfer

The volume of gas in the flask is kept constant by raising or lowering reservoir *B* to keep the mercury level in column *A* constant.

Scale

P Gas

Bath or environment to be measured

Flexible hose

Mercury reservoir

h

0

A

B

Figure 18.3 A constant-volume gas thermometer measures the pressure of the gas contained in the flask immersed in the bath.

[2]Two thermometers that use the same liquid may also give different readings, due in part to difficulties in constructing uniform-bore glass capillary tubes.

The two dots represent known reference temperatures (the ice and steam points of water).

Figure 18.4 A typical graph of pressure versus temperature taken with a constant-volume gas thermometer.

For all three trials, the pressure extrapolates to zero at the temperature −273.15°C.

Figure 18.5 Pressure versus temperature for experimental trials in which gases have different pressures in a constant-volume gas thermometer.

between reservoirs A and B through the flexible hose. Reservoir B is adjusted until the top of the mercury in column A is at the zero point on the scale. The height h, the difference between the mercury levels in reservoir B and column A, indicates the pressure in the flask at 0°C by means of Equation 14.4, $P = P_0 + \rho gh$, where P_0 is atmospheric pressure.

The flask is then immersed in water at the steam point. Reservoir B is readjusted until the top of the mercury in column A is again at zero on the scale, which ensures that the gas's volume is the same as it was when the flask was in the ice bath (hence the designation "constant-volume"). This adjustment of reservoir B gives a value for the gas pressure at 100°C. These two pressure and temperature values are then plotted as shown in Figure 18.4. The line connecting the two points serves as a calibration curve for unknown temperatures. (Other experiments show that a linear relationship between pressure and temperature is a very good assumption.) To measure the temperature of a substance, the gas flask of Figure 18.3 is placed in thermal contact with the substance and the height of reservoir B is adjusted until the top of the mercury column in A is at zero on the scale. The height of the mercury column in B indicates the pressure of the gas; knowing the pressure, the temperature of the substance is found using the graph in Figure 18.4.

Now suppose temperatures of different gases at different initial pressures are measured with gas thermometers. Experiments show that the thermometer readings are nearly independent of the type of gas used as long as the gas pressure is low and the temperature is well above the point at which the gas liquefies (Fig. 18.5). The agreement among thermometers using various gases improves as the pressure is reduced.

If we extend the solid-color straight lines in Figure 18.5 toward negative temperatures, we find a remarkable result: **in every case, the pressure is zero when the temperature is −273.15°C!** This finding suggests some special role that this particular temperature must play. It is used as the basis for the **absolute temperature scale,** which sets −273.15°C as its zero point. This temperature is often referred to as **absolute zero.** It is indicated as a zero because at a lower temperature, the pressure of the gas would become negative, which is meaningless. Therefore, absolute zero is a true, naturally defined zero of temperature. The size of one degree on the absolute temperature scale is chosen to be identical to the size of one degree on the Celsius scale. Therefore, the conversion between these temperatures is

$$T_C = T - 273.15 \tag{18.1}$$

where T_C is the Celsius temperature and T is the absolute temperature.

Because the ice and steam points are experimentally difficult to duplicate and depend on atmospheric pressure, an absolute temperature scale based on two new fixed points was adopted in 1954 by the International Committee on Weights and Measures. The first point is absolute zero, which does not depend on atmospheric pressure or on any particular material. The second reference temperature for this new scale was chosen as the **triple point of water,** which is the single combination of temperature and pressure at which liquid water, gaseous water, and ice (solid water) coexist in equilibrium. This triple point occurs at a temperature of 0.01°C and a pressure of 4.58 mm of mercury. The triple point of water is the same everywhere in the Universe. On the new scale, which uses the unit *kelvin,* the temperature of water at the triple point was set at 273.16 kelvins, abbreviated 273.16 K. This choice was made so that the old absolute temperature scale based on the ice and steam points would agree closely with the new scale based on the triple point. This new **absolute temperature scale** (also called the **Kelvin scale**) employs the SI unit of absolute temperature, the **kelvin,** which is defined to be 1/273.16 of the difference between absolute zero and the temperature of the triple point of water.

Figure 18.6 gives the absolute temperature for various physical processes and structures. The temperature of absolute zero (0 K) cannot be achieved, although laboratory experiments have come very close, reaching temperatures of less than one nanokelvin.

The Celsius, Fahrenheit, and Kelvin Temperature Scales[3]

Equation 18.1 shows that the Celsius temperature T_C is shifted from the absolute (Kelvin) temperature T by 273.15°. Because the size of one degree is the same on the two scales, a temperature difference of 5°C is equal to a temperature difference of 5 K. The two scales differ only in the choice of the zero point. Therefore, the ice-point temperature on the Kelvin scale, 273.15 K, corresponds to 0.00°C, and the Kelvin-scale steam point, 373.15 K, is equivalent to 100.00°C.

A common temperature scale in everyday use in the United States is the **Fahrenheit scale.** This scale sets the temperature of the ice point at 32°F and the temperature of the steam point at 212°F. The relationship between the Celsius and Fahrenheit temperature scales is

$$T_F = \tfrac{9}{5}T_C + 32°F \qquad (18.2)$$

We can use Equations 18.1 and 18.2 to find a relationship between changes in temperature on the Celsius, Kelvin, and Fahrenheit scales:

$$\Delta T_C = \Delta T = \tfrac{5}{9}\Delta T_F \qquad (18.3)$$

Of these three temperature scales, only the Kelvin scale is based on a true zero value of temperature. The Celsius and Fahrenheit scales are based on an arbitrary zero associated with one particular substance, water, on one particular planet, the Earth. Therefore, if you encounter an equation that calls for a temperature T or that involves a ratio of temperatures, you *must* convert all temperatures to kelvins. If the equation contains a change in temperature ΔT, using Celsius temperatures will give you the correct answer, in light of Equation 18.3, but it is always *safest* to convert temperatures to the Kelvin scale.

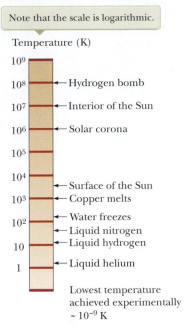

Note that the scale is logarithmic.

Temperature (K)

10^9
10^8 ← Hydrogen bomb
10^7 ← Interior of the Sun
10^6 ← Solar corona
10^5
10^4 ← Surface of the Sun
10^3 ← Copper melts
  ← Water freezes
10^2 ← Liquid nitrogen
  ← Liquid hydrogen
10
1 ← Liquid helium

Lowest temperature achieved experimentally ~ 10^{-9} K

Figure 18.6 Absolute temperatures at which various physical processes occur.

QUICK QUIZ 18.2 Consider the following pairs of materials. Which pair represents two materials, one of which is twice as hot as the other? **(a)** boiling water at 100°C, a glass of water at 50°C **(b)** boiling water at 100°C, frozen methane at −50°C **(c)** an ice cube at −20°C, flames from a circus fire-eater at 233°C **(d)** none of those pairs

---

Example 18.1 | **Converting Temperatures**

On a day when the temperature reaches 50°F, what is the temperature in degrees Celsius and in kelvins?

SOLUTION

Conceptualize In the United States, a temperature of 50°F is well understood. In many other parts of the world, however, this temperature might be meaningless because people are familiar with the Celsius temperature scale.

Categorize This example is a simple substitution problem.

Solve Equation 18.2 for the Celsius temperature and substitute numerical values:

$$T_C = \tfrac{5}{9}(T_F - 32) = \tfrac{5}{9}(50 - 32) = \boxed{10°C}$$

Use Equation 18.1 to find the Kelvin temperature:

$$T = T_C + 273.15 = 10°C + 273.15 = \boxed{283\ K}$$

A convenient set of weather-related temperature equivalents to keep in mind is that 0°C is (literally) freezing at 32°F, 10°C is cool at 50°F, 20°C is room temperature, 30°C is warm at 86°F, and 40°C is a hot day at 104°F.

---

[3]Named after Anders Celsius (1701–1744), Daniel Gabriel Fahrenheit (1686–1736), and William Thomson, Lord Kelvin (1824–1907), respectively.

Thermal Expansion of Solids and Liquids

Our discussion of the liquid thermometer makes use of one of the best-known thermal changes in a substance: as its temperature increases, its volume increases. This phenomenon, known as **thermal expansion,** plays an important role in numerous engineering applications. For example, thermal-expansion joints such as those shown in Figure 18.7 must be included in buildings, concrete highways, railroad tracks, brick walls, and bridges to compensate for dimensional changes that occur as the temperature changes.

Thermal expansion is responsible for the effects you saw in the opening storyline. On a hot day the power lines expand. The distance between the ends of the lines is fixed at the positions of the poles. Therefore, when the power line lengthens, it sags downward from its fixed ends. The brick sidewalk in the chapter opening photograph was likely installed with no expansion joints. As the temperature rises, the expansion of the sections of sidewalk causes them to buckle upward.

Thermal expansion is a consequence of the change in the *average* separation between the atoms in an object. To understand this concept, let's model the atoms as being connected by stiff springs as discussed in Section 15.3 and shown in Figure 15.11b. At ordinary temperatures, the atoms in a solid oscillate about their equilibrium positions with an amplitude of approximately 10^{-11} m and a frequency of approximately 10^{13} Hz. The average spacing between the atoms is about 10^{-10} m. As the temperature of the solid increases, the atoms oscillate with greater amplitudes; as a result, the average separation between them increases.[4] Consequently, the object expands.

If thermal expansion is sufficiently small relative to an object's initial dimensions, the change in any dimension is, to a good approximation, proportional to the first power of the temperature change. Suppose an object has an initial length L_i along some direction at some temperature and the length changes by an amount ΔL for a change in temperature ΔT. Because it is convenient to consider the fractional change in length per degree of temperature change, we define the **average coefficient of linear expansion** as

$$\alpha \equiv \frac{\Delta L/L_i}{\Delta T} \tag{18.4}$$

Without these joints to separate sections of roadway on bridges, the surface would buckle due to thermal expansion on very hot days or crack due to contraction on very cold days.

The long, vertical joint is filled with a soft material that allows the wall to expand and contract as the temperature of the bricks changes.

Figure 18.7 Thermal-expansion joints in (a) bridges and (b) walls.

---

[4] More precisely, thermal expansion arises from the *asymmetrical* nature of the potential energy curve for the atoms in a solid as shown in Figure 15.11a. If the oscillators were truly harmonic, the average atomic separations would not change regardless of the amplitude of vibration.

Experiments show that α is constant for small changes in temperature. For purposes of calculation, this equation is usually rewritten as

$$\Delta L = \alpha L_i\, \Delta T \qquad (18.5)$$

◀ Thermal expansion in one dimension

or as

$$L_f - L_i = \alpha L_i (T_f - T_i) \qquad (18.6)$$

where L_f is the final length, T_i and T_f are the initial and final temperatures, respectively, and the proportionality constant α is the average coefficient of linear expansion for a given material and has units of $(°C)^{-1}$. Equation 18.5 can be used for both thermal expansion, when the temperature of the material increases, and thermal contraction, when its temperature decreases.

It may be helpful to think of thermal expansion as an effective magnification or as a photographic enlargement of an object. For example, as a metal washer is heated (Fig. 18.8), all dimensions, including the radius of the hole, increase according to Equation 18.5. A cavity in a piece of material expands in the same way as if the cavity were filled with the material.

Table 18.1 lists the average coefficients of linear expansion for various materials. For these materials, α is positive, indicating an increase in length with increasing temperature. That is not always the case, however. Some substances—calcite ($CaCO_3$) is one example—expand along one dimension (positive α) and contract along another (negative α) as their temperatures are increased.

Because the linear dimensions of an object change with temperature, it follows that surface area and volume change as well. The change in volume is proportional to the initial volume V_i and to the change in temperature according to the relationship

$$\Delta V = \beta V_i\, \Delta T \qquad (18.7)$$

◀ Thermal expansion in three dimensions

where β is the **average coefficient of volume expansion.** To find the relationship between β and α, assume the average coefficient of linear expansion of the solid is the same in all directions; that is, assume the material is *isotropic*. Consider a solid box of dimensions ℓ, w, and h. Its volume at some temperature T_i is $V_i = \ell wh$. If the temperature changes to $T_i + \Delta T$, its volume changes to $V_i + \Delta V$, where each dimension changes according to Equation 18.5. Therefore,

$$V_i + \Delta V = (\ell + \Delta \ell)(w + \Delta w)(h + \Delta h)$$

$$= (\ell + \alpha \ell\, \Delta T)(w + \alpha w\, \Delta T)(h + \alpha h\, \Delta T)$$

$$= \ell wh (1 + \alpha\, \Delta T)^3$$

$$= V_i [1 + 3\alpha\, \Delta T + 3(\alpha\, \Delta T)^2 + (\alpha\, \Delta T)^3]$$

PITFALL PREVENTION 18.2
Do Holes Become Larger or Smaller? When an object's temperature is raised, every linear dimension increases in size. That includes any holes in the material, which expand in the same way as if the hole were filled with the material as shown in Figure 18.8.

TABLE 18.1 Average Expansion Coefficients for Some Materials Near Room Temperature

Material (Solids)	Average Linear Expansion Coefficient $(\alpha)(°C)^{-1}$	Material (Liquids and Gases)	Average Volume Expansion Coefficient $(\beta)(°C)^{-1}$
Aluminum	24×10^{-6}	Acetone	1.5×10^{-4}
Brass and bronze	19×10^{-6}	Alcohol, ethyl	1.12×10^{-4}
Concrete	12×10^{-6}	Benzene	1.24×10^{-4}
Copper	17×10^{-6}	Gasoline	9.6×10^{-4}
Glass (ordinary)	9×10^{-6}	Glycerin	4.85×10^{-4}
Glass (Pyrex)	3.2×10^{-6}	Mercury	1.82×10^{-4}
Invar (Ni–Fe alloy)	0.9×10^{-6}	Turpentine	9.0×10^{-4}
Lead	29×10^{-6}	Air[a] at 0°C	3.67×10^{-3}
Steel	11×10^{-6}	Helium[a]	3.665×10^{-3}

[a]Gases do not have a specific value for the volume expansion coefficient because the amount of expansion depends on the type of process through which the gas is taken. The values given here assume the gas undergoes an expansion at constant pressure.

As the washer is heated, all dimensions increase, including the radius of the hole.

Figure 18.8 Thermal expansion of a homogeneous metal washer. (The expansion is exaggerated in this figure.)

Dividing both sides by V_i and isolating the term $\Delta V/V_i$, we obtain the fractional change in volume:

$$\frac{\Delta V}{V_i} = 3\alpha\,\Delta T + 3(\alpha\,\Delta T)^2 + (\alpha\,\Delta T)^3$$

Because $\alpha\,\Delta T << 1$ for typical values of ΔT ($< \sim 100°C$), we can neglect the terms $3(\alpha\,\Delta T)^2$ and $(\alpha\,\Delta T)^3$. Upon making this approximation, we see that

$$\frac{\Delta V}{V_i} = 3\alpha\,\Delta T \quad \rightarrow \quad \Delta V = (3\alpha)V_i\,\Delta T$$

Comparing this expression to Equation 18.7 shows that

$$\beta = 3\alpha$$

In a similar way, you can show that the change in area of a rectangular plate is given by $\Delta A = 2\alpha A_i\,\Delta T$ (see Problem 37).

QUICK QUIZ 18.3 If you are asked to make a very sensitive glass thermometer, which of the following working liquids would you choose? **(a)** mercury **(b)** alcohol **(c)** gasoline **(d)** glycerin

QUICK QUIZ 18.4 Two spheres are made of the same metal and have the same radius, but one is hollow and the other is solid. The spheres are taken through the same temperature increase. Which sphere expands more? **(a)** The solid sphere expands more. **(b)** The hollow sphere expands more. **(c)** They expand by the same amount. **(d)** There is not enough information to say.

The Unusual Behavior of Water

Liquids generally increase in volume with increasing temperature and have average coefficients of volume expansion about ten times greater than those of solids. Water follows this general behavior *except* near 0°C, as you can see from its density-versus-temperature curve shown in Figure 18.9. As the temperature increases from 0°C to 4°C, water contracts and its density therefore increases. Above 4°C, water expands normally with increasing temperature and so its density decreases. Therefore, the density of water reaches a maximum value of 1.000 g/cm³ at 4°C.

We can use this unusual thermal-expansion behavior of water to explain why a pond begins freezing at the surface rather than at the bottom. When the air temperature drops from, for example, 7°C to 6°C, the surface water also cools and consequently decreases in volume. The surface water is denser than the water below it, which has not cooled and decreased in volume. As a result, the surface water sinks, and warmer water from below moves to the surface. When the air temperature is between 4°C and 0°C, however, the surface water expands as it cools, becoming less dense than the water below it. The mixing process stops, and eventually the surface water freezes. As the water freezes, the ice remains on the surface because ice is less dense than water. The ice continues to build up at the surface, while water near the bottom remains at 4°C. If that were not the case, fish and other forms of marine life would not survive.

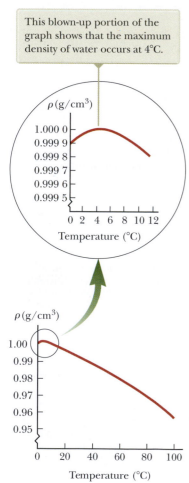

This blown-up portion of the graph shows that the maximum density of water occurs at 4°C.

Figure 18.9 The variation in the density of water at atmospheric pressure with temperature.

Example **18.2** **Expansion of a Railroad Track**

A segment of steel railroad track has a length of 30.000 m when the temperature is 0.0°C. What is its length when the temperature is 40.0°C?

18.2 continued

SOLUTION

Conceptualize Because the rail is relatively long, we expect to obtain a measurable increase in length for a 40°C temperature increase.

Categorize We will evaluate a length increase using the discussion of this section, so this example is a substitution problem.

Use Equation 18.5 and the value of the coefficient of linear expansion from Table 18.1:

$$\Delta L = \alpha L_i \Delta T = [11 \times 10^{-6} \ (\text{°C})^{-1}](30.000 \text{ m})(40.0\text{°C}) = 0.013 \text{ m}$$

Find the new length of the track:

$$L_f = 30.000 \text{ m} + 0.013 \text{ m} = \boxed{30.013 \text{ m}}$$

The expansion of 1.3 cm is indeed measurable as predicted in the Conceptualize step. If the section of rail is butted right against another rail, this expansion cannot occur and a *thermal stress* is developed in the rail. The thermal stress could bend the rail. This stress can be avoided by leaving small expansion gaps between the rails.

WHAT IF? What if the temperature drops to −40.0°C? What is the length of the unclamped segment?

Answer The expression for the change in length in Equation 18.5 is the same whether the temperature increases or decreases. Therefore, if there is an increase in length of 0.013 m when the temperature increases by 40°C, there is a decrease in length of 0.013 m when the temperature decreases by 40°C. (We assume α is constant over the entire range of temperatures.) The new length at the colder temperature is 30.000 m − 0.013 m = 29.987 m.

Example 18.3 The Thermal Electrical Short

A poorly designed electronic device has two bolts attached to different parts of the device that almost touch each other in its interior as in Figure 18.10. The steel and brass bolts are at different electric potentials, and if they touch, a short circuit will develop, damaging the device. (We will study electric potential in Chapter 24.) The initial gap between the ends of the bolts is $d = 5.0 \ \mu\text{m}$ at 27°C. At what temperature will the bolts touch? Assume the distance between the walls of the device is not affected by the temperature change.

Figure 18.10 (Example 18.3) Two bolts attached to different parts of an electrical device are almost touching when the temperature is 27°C. As the temperature increases, the ends of the bolts move toward each other.

SOLUTION

Conceptualize Imagine the ends of both bolts expanding into the gap between them as the temperature rises.

Categorize We categorize this example as a thermal expansion problem in which the *sum* of the changes in length of the two bolts must equal the length d of the initial gap between the ends.

Analyze Set the sum of the length changes equal to the width of the gap:

$$\Delta L_{br} + \Delta L_{st} = \alpha_{br} L_{i,br} \Delta T + \alpha_{st} L_{i,st} \Delta T = d$$

Solve for ΔT:

$$\Delta T = \frac{d}{\alpha_{br} L_{i,br} + \alpha_{st} L_{i,st}}$$

Substitute numerical values:

$$\Delta T = \frac{5.0 \times 10^{-6} \text{ m}}{[19 \times 10^{-6} \ (\text{°C})^{-1}](0.030 \text{ m}) + [11 \times 10^{-6} \ (\text{°C})^{-1}](0.010 \text{ m})} = 7.4\text{°C}$$

Find the temperature at which the bolts touch:

$$T = 27\text{°C} + 7.4\text{°C} = \boxed{34\text{°C}}$$

Finalize This temperature is possible if the air conditioning in the building housing the device fails for a long period on a very hot summer day.

18.5 Macroscopic Description of an Ideal Gas

The volume expansion equation $\Delta V = \beta V_i \Delta T$ (Eq. 18.7) is based on the assumption that the material has an initial volume V_i before the temperature change occurs. Such is the case for solids and liquids because a sample of solid or liquid has a fixed volume at a given temperature.

The case for gases is completely different. The interatomic forces within gases are very weak, and, in many cases, we can imagine these forces to be nonexistent and still make very good approximations. Therefore, *there is no equilibrium separation* for the atoms and no "standard" volume at a given temperature; the volume depends on the size of the container. As a result, we cannot express changes in volume ΔV in a process on a gas with Equation 18.7 because we have no defined volume V_i at the beginning of the process. Equations involving gases contain the volume V, rather than a *change* in the volume from an initial value, as a variable.

For a gas, it is useful to know how the quantities volume V, pressure P, and temperature T are related for a sample of gas of mass m. In general, the equation that interrelates these quantities, called the *equation of state,* is very complicated. If the gas is maintained at a very low pressure (or low density), however, the equation of state is quite simple and can be determined from experimental results. Such a low-density gas is commonly referred to as an **ideal gas.**[5] We can use a simplification model called the **ideal gas model** to make predictions that are adequate to describe the behavior of real gases at low pressures.

It is convenient to express the amount of gas in a given volume in terms of the number of moles n. One **mole** of any substance is that amount of the substance that contains **Avogadro's number** $N_A = 6.022 \times 10^{23}$ of constituent particles (atoms or molecules). The number of moles n of a substance is related to its mass m through the expression

$$n = \frac{m}{M} \tag{18.8}$$

Figure 18.11 An ideal gas confined to a cylinder whose volume can be varied by means of a movable piston.

where M is the molar mass of the substance. The molar mass of each chemical element is the atomic mass (from the periodic table; see Appendix C) expressed in grams per mole. For example, the mass of one helium (He) atom is 4.00 u (atomic mass units), so the molar mass of He is 4.00 g/mol.

Now suppose an ideal gas is confined to a cylindrical container whose volume can be varied by means of a movable piston as in Figure 18.11. If we assume the cylinder does not leak, the mass (or the number of moles) of the gas remains constant. For such a system, experiments measuring pressure, volume, and temperature provide the following information:

- When the gas is kept at a constant temperature, its pressure is inversely proportional to the volume. (This behavior is described historically as Boyle's law.)
- When the pressure of the gas is kept constant, the volume is directly proportional to the temperature. (This behavior is described historically as Charles's law.)
- When the volume of the gas is kept constant, the pressure is directly proportional to the temperature. (This behavior is described historically as Gay–Lussac's law, and justifies the straight lines we drew through the data points in the graph in Figure 18.5.)

These observations are summarized by the **equation of state for an ideal gas:**

Equation of state for ▶
an ideal gas

$$PV = nRT \tag{18.9}$$

[5] To be more specific, the assumptions here are that the temperature of the gas must not be too low (the gas must not condense into a liquid) or too high and that the pressure must be low. The concept of an ideal gas implies that the gas molecules do not interact except upon collision and that the molecular volume is negligible compared with the volume of the container. In reality, an ideal gas does not exist. The concept of an ideal gas is nonetheless very useful because real gases at low pressures are well-modeled as ideal gases.

In this expression, also known as the **ideal gas law,** n is the number of moles of gas in the sample and R is a constant. Experiments on numerous gases show that as the pressure approaches zero, the quantity PV/nT approaches the same value R for all gases. For this reason, R is called the **universal gas constant.** In SI units, in which pressure is expressed in pascals (1 Pa = 1 N/m^2) and volume in cubic meters, the product PV has units of newton · meters, or joules, and R has the value

$$R = 8.314 \text{ J/mol} \cdot \text{K} \tag{18.10}$$

If the pressure is expressed in atmospheres and the volume in liters (1 L = 10^3 cm^3 = 10^{-3} m^3), then R has the value

$$R = 0.082\,06 \text{ L} \cdot \text{atm/mol} \cdot \text{K}$$

Using this value of R and Equation 18.9 shows that the volume occupied by 1 mol of *any* gas at atmospheric pressure and at 0°C (273 K) is 22.4 L.

The ideal gas law states that if the volume and temperature of a fixed amount of gas do not change, the pressure also remains constant. Consider a bottle of champagne that is shaken and then spews liquid when opened as shown in Figure 18.12. A common misconception is that the pressure inside the bottle is increased when the bottle is shaken. On the contrary, because the temperature of the bottle and its contents remains constant as long as the bottle is sealed, so does the pressure, as can be shown by replacing the cork with a pressure gauge. The correct explanation is as follows. Carbon dioxide gas resides in the volume between the liquid surface and the cork. The pressure of the gas in this volume is set higher than atmospheric pressure in the bottling process. Shaking the bottle displaces some of the carbon dioxide gas into the liquid, where it forms bubbles, and these bubbles become attached to the inside of the bottle. (No new gas is generated by shaking.) When the bottle is opened, the pressure is reduced to atmospheric pressure, which causes the volume of the bubbles to increase suddenly. If the bubbles are attached to the bottle (beneath the liquid surface), their rapid expansion expels liquid from the bottle. If the sides and bottom of the bottle are first tapped until no bubbles remain beneath the surface, however, the drop in pressure does not force liquid from the bottle when the champagne is opened.

The ideal gas law is often expressed in terms of the total number of molecules N. Because the number of moles n equals the ratio of the total number of molecules and Avogadro's number N_A, we can write Equation 18.9 as

$$PV = nRT = \frac{N}{N_A} RT$$

$$PV = Nk_B T \tag{18.11}$$

where k_B is **Boltzmann's constant,** which has the value

$$k_B = \frac{R}{N_A} = 1.38 \times 10^{-23} \text{ J/K} \tag{18.12}$$

◀ Boltzmann's constant

It is common to call quantities such as P, V, and T the **thermodynamic variables** of an ideal gas. If the equation of state is known, one of the variables can always be expressed as some function of the other two.

QUICK QUIZ 18.5 A common material for cushioning objects in packages is made by trapping bubbles of air between sheets of plastic. Is this material more effective at keeping the contents of the package from moving around inside the package on **(a)** a hot day, **(b)** a cold day, or **(c)** either hot or cold days?

QUICK QUIZ 18.6 On a winter day, you turn on your furnace and the temperature of the air inside your home increases. Assume your home has the normal amount of leakage between inside air and outside air. Is the number of moles of air in your room at the higher temperature **(a)** larger than before, **(b)** smaller than before, or **(c)** the same as before?

Shutterstock.com

Figure 18.12 A bottle of champagne is shaken and opened. Liquid spews out of the opening. A common misconception is that the pressure inside the bottle is increased by the shaking.

PITFALL PREVENTION 18.3

So Many ks There are a variety of physical quantities for which the letter k is used. Two we have seen previously are the force constant for a spring (Chapter 15) and the wave number for a mechanical wave (Chapter 16). Boltzmann's constant is another k, and we will see k used for thermal conductivity in Chapter 19 and for an electrical constant in Chapter 22. To make some sense of this confusing state of affairs, we use a subscript B for Boltzmann's constant to help us recognize it. In this book, you will see Boltzmann's constant as k_B, but you may see Boltzmann's constant in other resources as simply k.

Example 18.4   Heating a Spray Can

A spray can containing a propellant gas at twice atmospheric pressure (202 kPa) and having a volume of 125.00 cm³ is at 22°C. It is then tossed into an open fire. (*Warning:* Do not do this experiment; it is very dangerous.) When the temperature of the gas in the can reaches 195°C, what is the pressure inside the can? Assume any change in the volume of the can is negligible.

SOLUTION

Conceptualize   Intuitively, you should expect that the pressure of the gas in the container increases because of the increasing temperature.

Categorize   We model the gas in the can as ideal and use the ideal gas law to calculate the new pressure.

Analyze   Rearrange Equation 18.9:

$$(1) \quad \frac{PV}{T} = nR$$

No air escapes during the compression, so n, and therefore nR, remains constant. Hence, set the initial value of the left side of Equation (1) equal to the final value:

$$(2) \quad \frac{P_i V_i}{T_i} = \frac{P_f V_f}{T_f}$$

Because the initial and final volumes of the gas are assumed to be equal, cancel the volumes:

$$(3) \quad \frac{P_i}{T_i} = \frac{P_f}{T_f}$$

Solve for P_f:

$$P_f = \left(\frac{T_f}{T_i}\right) P_i = \left(\frac{468 \text{ K}}{295 \text{ K}}\right)(202 \text{ kPa}) = \boxed{320 \text{ kPa}}$$

Finalize   The higher the temperature, the higher the pressure exerted by the trapped gas as expected. If the pressure increases sufficiently, the can may explode. Because of this possibility, you should never dispose of spray cans in a fire.

WHAT IF?   Suppose we include a volume change due to thermal expansion of the steel can as the temperature increases. Does that alter our answer for the final pressure significantly?

Answer   Because the thermal expansion coefficient of steel is very small, we do not expect much of an effect on our final answer.

Find the change in the volume of the can using Equation 18.7 and the value for α for steel from Table 18.1:

$$\Delta V = \beta V_i \, \Delta T = 3\alpha V_i \, \Delta T$$
$$= 3[11 \times 10^{-6} \, (°C)^{-1}](125.00 \text{ cm}^3)(173°C) = 0.71 \text{ cm}^3$$

Start from Equation (2) again and find an equation for the final pressure:

$$P_f = \left(\frac{T_f}{T_i}\right)\left(\frac{V_i}{V_f}\right) P_i$$

This result differs from Equation (3) only in the factor V_i/V_f. Evaluate this factor:

$$\frac{V_i}{V_f} = \frac{125.00 \text{ cm}^3}{(125.00 \text{ cm}^3 + 0.71 \text{ cm}^3)} = 0.994 = 99.4\%$$

Therefore, the final pressure will differ by only 0.6% from the value calculated without considering the thermal expansion of the can. Taking 99.4% of the previous final pressure, the final pressure including thermal expansion is 318 kPa.

Summary

▶ Definitions

Two objects are in **thermal equilibrium** with each other if they do not exchange energy when in thermal contact.

Temperature is the property that determines whether an object is in thermal equilibrium with other objects. Two objects in thermal equilibrium with each other are at the same temperature. The SI unit of absolute temperature is the **kelvin**, which is defined to be 1/273.16 of the difference between absolute zero and the temperature of the triple point of water.

Concepts and Principles

The **zeroth law of thermodynamics** states that if objects A and B are separately in thermal equilibrium with a third object C, then objects A and B are in thermal equilibrium with each other.

When the temperature of an object is changed by an amount ΔT, its length changes by an amount ΔL that is proportional to ΔT and to its initial length L_i:

$$\Delta L = \alpha L_i \, \Delta T \qquad (18.5)$$

where the constant α is the **average coefficient of linear expansion.** The **average coefficient of volume expansion** β for a solid is approximately equal to 3α.

An **ideal gas** is one for which PV/nT is constant. An ideal gas is described by the **equation of state,**

$$PV = nRT \qquad (18.9)$$

where n equals the number of moles of the gas, P is its pressure, V is its volume, R is the universal gas constant (8.314 J/mol $\cdot$ K), and T is the absolute temperature of the gas. A real gas behaves approximately as an ideal gas if it has a low density.

Think–Pair–Share

See the Preface for an explanation of the icons used in this problems set. For additional assessment items for this section, go to **WEBASSIGN** From Cengage

1. **Review.** In the storyline opening Chapter 15, we discussed the change in timing for a grandfather clock due to a change in the value of the acceleration g due to gravity. Have your group think about a change in the timing of the pendulum clock due to a change in temperature. Suppose the pendulum is made of brass and has a period of 1.000 s when the temperature is 20.0°C. During a warm summer week, the temperature remains in a very small range with an average of 30.0°C. (a) Does the clock lose time or gain time during that week? (b) By how much is the clock in error by the end of the week?

2. Your expert witness team has been hired by the city council in a lawsuit filed by a truck driver. While driving down a city street on a hot day, the top of the truck driven by the driver struck a power line hanging across the street, causing damage to his truck and injury to himself. The driver claims that the power line was sagging so much because of the temperature. As a result, it was hanging below the height clearance limit of 14 feet, 0 inches posted on a sign on the street. You go to the site of the accident and take the following measurements. The poles supporting the ends of the power line are 40.0 ft apart. Both ends of the line are supported at a height of 16.0 ft above the roadway surface. On the winter day you visit the site, the temperature is −25.0°C and the copper power line shows essentially no sag. On the day of the accident, the temperature was 38.0°C. Prepare an argument that the power line did not sag below the height clearance, and that the truck driver must have loaded his truck to a point higher than the posted clearance. (*Suggestion:* the shape of the sagging power line will be a curve, but assume an unrealistic shape for the line that will allow a simple calculation for the lowest possible point on the line.)

3. **ACTIVITY** For this activity, you'll need some containers, some straws, and some water. (a) Fill various containers to different levels with water and measure the depth h of the water in each cup carefully. Now immerse a straw vertically in each cup with the bottom of the straw resting on the bottom of the cup. Place your finger over the top end of the straw to seal it and lift the straw vertically out of the cup. Measure the length h' of the column of water trapped in the straw. You may need to take a photo with your smartphone of the straw and a ruler and carefully analyze an enlargement of the photo to make this measurement. Does the length of the column of water trapped in the straw agree with the depth of water in the cup? Should it agree? (b) In order for the column of water to be suspended in the straw, the pressure above the column (within the straw) must be less than atmospheric pressure. For that to happen, the column of water must move downward a bit when the straw is raised from the water, so that the air above the column expands in volume and the pressure decreases. Therefore, the two measured lengths should not be the same. But is the difference detectable? Show that the length h' of water suspended in the straw of length ℓ should be related to the depth h of the water in the cup by

$$h' = \tfrac{1}{2}\left(\ell + \frac{P_0}{\rho g}\right) - \tfrac{1}{2}\sqrt{\ell^2 + 2\left(\frac{P_0}{\rho g}\right)(\ell - 2h) + \left(\frac{P_0}{\rho g}\right)^2}$$

where P_0 is atmospheric pressure and ρ is the density of the water. (c) For a 30-cm straw, use the equation to determine the difference $h - h'$ for various levels of h from zero to 30 cm in increments of 1 cm. (d) For what values of h is the *percentage* drop in the water column the greatest when the straw is drawn out of the cup? (e) For what value of h is the value of $h - h'$ a maximum?

Problems

See the Preface for an explanation of the icons used in this problems set.
For additional assessment items for this section, go to ◆ **WEBASSIGN**
From Cengage

SECTION 18.2 Thermometers and the Celsius Temperature Scale

1. You are working as a research assistant for a professor
CR whose research area is thermodynamics. He points out to you that Daniel Fahrenheit used the best estimate of normal human body temperature as one of the points in defining the original Fahrenheit temperature scale. On the revised scale we now use, normal human body temperature is 98.6°F. Your professor proposes a new scale on which normal human body temperature would be exactly 100°N, where the unit °N is a degree on the *New* scale. The temperature of freezing water would be 0°N, as on the Celsius scale. Your professor asks you to determine the following temperatures on his new scale: (a) absolute zero, (b) the melting point of mercury (−37.9°F), (c) the boiling point of water, and, for publicity at his expected future press conference, (d) the highest recorded air temperature on the Earth's surface, 134.1°F on July 10, 1913, in Death Valley, California.

SECTION 18.3 The Constant-Volume Gas Thermometer and the Absolute Temperature Scale

2. A nurse measures the temperature of a patient to be 41.5°C.
BIO (a) What is this temperature on the Fahrenheit scale? (b) Do
QC you think the patient is seriously ill? Explain.

3. Convert the following temperatures to their values on the Fahrenheit and Kelvin scales: (a) the sublimation point of dry ice, −78.5°C; (b) human body temperature, 37.0°C.

4. Liquid nitrogen has a boiling point of −195.81°C at atmospheric pressure. Express this temperature (a) in degrees Fahrenheit and (b) in kelvins.

5. Death Valley holds the record for the highest recorded temperature in the United States. On July 10, 1913, at a place called Furnace Creek Ranch, the temperature rose to 134°F. The lowest U.S. temperature ever recorded occurred at Prospect Creek Camp in Alaska on January 23, 1971, when the temperature plummeted to −79.8°F. (a) Convert these temperatures to the Celsius scale. (b) Convert the Celsius temperatures to Kelvin.

SECTION 18.4 Thermal Expansion of Solids and Liquids

Note: Table 18.1 is available for use in solving problems in this section.

6. **Review.** Inside the wall of a house, an L-shaped section of hot-water pipe consists of three parts: a straight, horizontal piece $h = 28.0$ cm long; an elbow; and a straight, vertical piece $\ell = 134$ cm long (Fig. P18.6). A stud and a second-story floorboard hold the ends of this section of copper pipe stationary. Find the magnitude and direction of the displacement of the pipe elbow when the water flow is turned on, raising the temperature of the pipe from 18.0°C to 46.5°C.

Figure P18.6

7. A copper telephone wire has essentially no sag between
T poles 35.0 m apart on a winter day when the temperature is −20.0°C. How much longer is the wire on a summer day when the temperature is 35.0°C?

8. A pair of eyeglass frames is made of epoxy plastic. At room
BIO temperature (20.0°C), the frames have circular lens holes 2.20 cm in radius. To what temperature must the frames be heated if lenses 2.21 cm in radius are to be inserted in them? The average coefficient of linear expansion for epoxy is 1.30×10^{-4} (°C)$^{-1}$.

9. The Trans-Alaska pipeline is 1 300 km long, reaching from Prudhoe Bay to the port of Valdez. It experiences temperatures from −73°C to +35°C. How much does the steel pipeline expand because of the difference in temperature? How can this expansion be compensated for?

10. A square hole 8.00 cm along each side is cut in a sheet of copper. (a) Calculate the change in the area of this hole resulting when the temperature of the sheet is increased by 50.0 K. (b) Does this change represent an increase or a decrease in the area enclosed by the hole?

11. You are watching a new
CR bridge being built near your house. You notice during the construction that two concrete spans of the bridge of total length $L_i = 250$ m are placed end to end so that no room is allowed for expansion (Fig. P18.11a). In the opening story-line for this chapter, we talked about buckling sidewalks. The same thing will happen with spans on bridges if allowance is not made for expansion (Fig. P18.11b). You want to warn the construction crew about this dangerous situation, so you calculate the height y to which the spans will rise when they buckle in response to a temperature increase of $\Delta T = 20.0$°C.

Figure P18.11

12. You are watching a new bridge being built near your house.
CR You notice during the construction that two concrete spans
S are placed end to end to form a span of length L_i. However,

they are placed end to end so that no room is allowed for expansion (Fig. P18.11a). In the opening storyline for this chapter, we talked about buckling sidewalks. The same thing will happen with spans on bridges if allowance is not made for expansion (Fig. P18.11b). You want to warn the construction crew about this dangerous situation, so you calculate the height y to which the spans will rise when they buckle in response to a temperature increase of ΔT.

13. At 20.0°C, an aluminum ring has an inner diameter of 5.000 0 cm and a brass rod has a diameter of 5.050 0 cm. (a) If only the ring is warmed, what temperature must it reach so that it will just slip over the rod? (b) **What If?** If both the ring and the rod are warmed together, what temperature must they both reach so that the ring barely slips over the rod? (c) Would this latter process work? Explain. *Hint:* Consult Table 19.2 in the next chapter.

14. *Why is the following situation impossible?* A thin brass ring has an inner diameter 10.00 cm at 20.0°C. A solid aluminum cylinder has diameter 10.02 cm at 20.0°C. Assume the average coefficients of linear expansion of the two metals are constant. Both metals are cooled together to a temperature at which the ring can be slipped over the end of the cylinder.

15. A volumetric flask made of Pyrex is calibrated at 20.0°C. It is filled to the 100-mL mark with 35.0°C acetone. After the flask is filled, the acetone cools and the flask warms so that the combination of acetone and flask reaches a uniform temperature of 32.0°C. The combination is then cooled back to 20.0°C. (a) What is the volume of the acetone when it cools to 20.0°C? (b) At the temperature of 32.0°C, does the level of acetone lie above or below the 100-mL mark on the flask? Explain.

16. **Review.** On a day that the temperature is 20.0°C, a concrete walk is poured in such a way that the ends of the walk are unable to move. Take Young's modulus for concrete to be 7.00×10^9 N/m² and the compressive strength to be 2.00×10^9 N/m². (a) What is the stress in the cement on a hot day of 50.0°C? (b) Does the concrete fracture?

17. **Review.** The Golden Gate Bridge in San Francisco has a main span of length 1.28 km, one of the longest in the world. Imagine that a steel wire with this length and a cross-sectional area of 4.00×10^{-6} m² is laid in a straight line on the bridge deck with its ends attached to the towers of the bridge. On a summer day the temperature of the wire is 35.0°C. (a) When winter arrives, the towers stay the same distance apart and the bridge deck keeps the same shape as its expansion joints open. When the temperature drops to −10.0°C, what is the tension in the wire? Take Young's modulus for steel to be 20.0×10^{10} N/m². (b) Permanent deformation occurs if the stress in the steel exceeds its elastic limit of 3.00×10^8 N/m². At what temperature would the wire reach its elastic limit? (c) **What If?** Explain how your answers to parts (a) and (b) would change if the Golden Gate Bridge were twice as long.

SECTION 18.5 Macroscopic Description of an Ideal Gas

18. Your father and your younger brother are confronted with the same puzzle. Your father's garden sprayer and your brother's water cannon both have tanks with a capacity of 5.00 L (Fig. P18.18). Your father puts a negligible amount of concentrated fertilizer into his tank. They both pour in 4.00 L of water and seal up their tanks, so the tanks also contain air at atmospheric pressure. Next, each uses a hand-operated pump to inject more air until the absolute pressure in the tank reaches 2.40 atm. Now each uses his device to spray out water—not air—until the stream becomes feeble, which it does when the pressure in the tank reaches 1.20 atm. To accomplish spraying out all the water, each finds he must pump up the tank three times. Here is the puzzle: most of the water sprays out after the second pumping. The first and the third pumping-up processes seem just as difficult as the second but result in a much smaller amount of water coming out. Account for this phenomenon.

Figure P18.18

19. An auditorium has dimensions 10.0 m × 20.0 m × 30.0 m. How many molecules of air fill the auditorium at 20.0°C and a pressure of 101 kPa (1.00 atm)?

20. A container in the shape of a cube 10.0 cm on each edge contains air (with equivalent molar mass 28.9 g/mol) at atmospheric pressure and temperature 300 K. Find (a) the mass of the gas, (b) the gravitational force exerted on it, and (c) the force it exerts on each face of the cube. (d) Why does such a small sample exert such a great force?

21. (a) Find the number of moles in one cubic meter of an ideal gas at 20.0°C and atmospheric pressure. (b) For air, Avogadro's number of molecules has mass 28.9 g. Calculate the mass of one cubic meter of air. (c) State how this result compares with the tabulated density of air at 20.0°C.

22. Use the definition of Avogadro's number to find the mass of a helium atom.

23. In state-of-the-art vacuum systems, pressures as low as 1.00×10^{-9} Pa are being attained. Calculate the number of molecules in a 1.00-m³ vessel at this pressure and a temperature of 27.0°C.

24. You have scored a great internship with NASA, working on planning for an upcoming mission to Mars. The transfer orbit to Mars will last for several months and will require reclamation of the oxygen in the carbon dioxide exhaled by the crew. In one method of reclamation, 1.00 mol of carbon dioxide produces 1.00 mol of oxygen and 1.00 mol of methane as a byproduct. The methane is stored in a tank under pressure and is available to control the orientation of the spacecraft by controlled venting. A single astronaut exhales 1.09 kg of carbon dioxide each day. If the methane generated in the respiration recycling of three astronauts during one week of flight is stored in an originally empty 150-L tank at −45.0°C, what is the final pressure in the tank?

25. Review. The mass of a hot-air balloon and its cargo (not including the air inside) is 200 kg. The air outside is at 10.0°C and 101 kPa. The volume of the balloon is 400 m³. To what temperature must the air in the balloon be warmed before the balloon will lift off? (Air density at 10.0°C is 1.244 kg/m³.)

26. A room of volume V contains air having equivalent molar mass M (in g/mol). If the temperature of the room is raised from T_1 to T_2, what mass of air will leave the room? Assume that the air pressure in the room is maintained at P_0.

27. Estimate the mass of the air in your bedroom. State the quantities you take as data and the value you measure or estimate for each.

28. You are applying for a position with a sea rescue unit and are taking the qualifying exam. One question on the exam is about the use of a diving bell. The diving bell is in the shape of a cylinder with a vertical length of $L = 2.50$ m. It is closed at the upper circular end and open at the lower circular end. The bell is lowered from air into seawater ($\rho = 1.025$ g/cm³) and kept in its upright orientation as it is lowered. The air in the bell is initially at temperature $T_i = 20.0$°C. The bell, with two humans inside, is lowered to a depth (measured to the bottom of the bell) of 27.0 fathoms, or $h = 49.4$ m. At this depth the water temperature is $T_f = 4.0$°C, and the bell is in thermal equilibrium with the water. The exam question asks you to compare two situations: (i) No additional gas is added to the interior of the bell as it is submerged. Therefore, water enters the open bottom of the bell and the volume of the enclosed air decreases. (ii) The bell is fitted with pressurized air tanks, which deliver high-pressure air into the interior of the bell to keep the level of water at the bottom edge of the bell. This choice requires money and effort to attach the tanks. The exam question asks: Which scenario is better?

29. The pressure gauge on a cylinder of gas registers the gauge pressure, which is the difference between the interior pressure and the exterior pressure P_0. Let's call the gauge pressure P_g. When the cylinder is full, the mass of the gas in it is m_i at a gauge pressure of P_{gi}. Assuming the temperature of the cylinder remains constant, show that the mass of the gas *remaining* in the cylinder when the pressure reading is P_{gf} is given by

$$m_f = m_i \left(\frac{P_{gf} + P_0}{P_{gi} + P_0} \right)$$

ADDITIONAL PROBLEMS

30. A steel beam being used in the construction of a skyscraper has a length of 35.000 m when delivered on a cold day at a temperature of 15.000°F. What is the length of the beam when it is being installed later on a warm day when the temperature is 90.000°F?

31. Two metal bars are made of invar and a third bar is made of aluminum. At 0°C, each of the three bars is drilled with two holes 40.0 cm apart. Pins are put through the holes to assemble the bars into an equilateral triangle as in Figure P18.31. (a) First ignore the expansion of the invar. Find the angle between the invar bars as a function of Celsius temperature. (b) Is your answer accurate for negative as well as positive temperatures? (c) Is it accurate for 0°C? (d) Solve

the problem again, including the expansion of the invar. Aluminum melts at 660°C and invar at 1 427°C. Assume the tabulated expansion coefficients are constant. What are (e) the greatest and (f) the smallest attainable angles between the invar bars?

Figure P18.31

32. *Why is the following situation impossible?* An apparatus is designed so that steam initially at $T = 150$°C, $P = 1.00$ atm, and $V = 0.500$ m³ in a piston–cylinder apparatus undergoes a process in which (1) the volume remains constant and the pressure drops to 0.870 atm, followed by (2) an expansion in which the pressure remains constant and the volume increases to 1.00 m³, followed by (3) a return to the initial conditions. It is important that the pressure of the gas never fall below 0.850 atm so that the piston will support a delicate and very expensive part of the apparatus. Without such support, the delicate apparatus can be severely damaged and rendered useless. When the design is turned into a working prototype, it operates perfectly.

33. A student measures the length of a brass rod with a steel tape at 20.0°C. The reading is 95.00 cm. What will the tape indicate for the length of the rod when the rod and the tape are at (a) −15.0°C and (b) 55.0°C?

34. The density of gasoline is 730 kg/m³ at 0°C. Its average coefficient of volume expansion is 9.60×10^{-4} (°C)⁻¹. Assume 1.00 gal of gasoline occupies 0.003 80 m³. How many extra kilograms of gasoline would you receive if you bought 10.0 gal of gasoline at 0°C rather than at 20.0°C from a pump that is not temperature compensated?

35. A liquid has a density ρ. (a) Show that the fractional change in density for a change in temperature ΔT is $\Delta \rho / \rho = -\beta \, \Delta T$. (b) What does the negative sign signify? (c) Fresh water has a maximum density of 1.000 0 g/cm³ at 4.0°C. At 10.0°C, its density is 0.999 7 g/cm³. What is β for water over this temperature interval? (d) At 0°C, the density of water is 0.999 9 g/cm³. What is the value for β over the temperature range 0°C to 4.00°C?

36. (a) Take the definition of the coefficient of volume expansion to be

$$\beta = \frac{1}{V} \frac{dV}{dT} \bigg|_{P = \text{constant}} = \frac{1}{V} \frac{\partial V}{\partial T}$$

Use the equation of state for an ideal gas to show that the coefficient of volume expansion for an ideal gas at constant pressure is given by $\beta = 1/T$, where T is the absolute temperature. (b) What value does this expression predict for β at 0°C? State how this result compares with the experimental values for (c) helium and (d) air in Table 18.1. *Note:* These values are much larger than the coefficients of volume expansion for most liquids and solids.

37. The rectangular plate shown in Figure P18.37 has an area A_i equal to ℓw. If the temperature increases by ΔT, each dimension increases according to Equation 18.5, where α is the average coefficient of linear expansion. (a) Show that the increase in area is $\Delta A = 2\alpha A_i \Delta T$. (b) What approximation does this expression assume?

Figure P18.37

38. A bimetallic strip of length L is made of two ribbons of different metals bonded together. (a) First assume the strip is originally straight. As the strip is warmed, the metal with the greater average coefficient of expansion expands more than the other, forcing the strip into an arc with the outer radius having a greater circumference (Fig. P18.38). Derive an expression for the angle of bending θ as a function of the initial length of the strips, their average coefficients of linear expansion, the change in temperature, and the separation of the centers of the strips ($\Delta r = r_2 - r_1$). (b) Show that the angle of bending decreases to zero when ΔT decreases to zero and also when the two average coefficients of expansion become equal. (c) **What If?** What happens if the strip is cooled?

Figure P18.38

39. A copper rod and a steel rod are different in length by 5.00 cm at 0°C. The rods are warmed and cooled together. (a) Is it possible that the length difference remains constant at all temperatures? Explain. (b) If so, describe the lengths at 0°C as precisely as you can. Can you tell which rod is longer? Can you tell the lengths of the rods?

40. A vertical cylinder of cross-sectional area A is fitted with a tight-fitting, frictionless piston of mass m (Fig. P18.40). The piston is not restricted in its motion in any way and is supported by the gas at pressure P below it. Atmospheric pressure is P_0. We wish to find the height h in Figure P18.40. (a) What analysis model is appropriate to describe the piston? (b) Write an appropriate force equation for the piston from this analysis model in terms of P, P_0, m, A, and g. (c) Suppose n moles of an ideal gas are in the cylinder at a temperature of T. Substitute for P in your answer to part (b) to find the height h of the piston above the bottom of the cylinder.

Figure P18.40

41. **Review.** Consider an object with any one of the shapes displayed in Table 10.2. What is the percentage increase in the moment of inertia of the object when it is warmed from 0°C to 100°C if it is composed of (a) copper or (b) aluminum? Assume the average linear expansion coefficients shown in

Table 18.1 do not vary between 0°C and 100°C. (c) Why are the answers for parts (a) and (b) the same for all the shapes?

42. **Review.** Following a collision in outer space, a copper disk at 850°C is rotating about its axis with an angular speed of 25.0 rad/s. As the disk radiates infrared light, its temperature falls to 20.0°C. No external torque acts on the disk. (a) Does the angular speed change as the disk cools? Explain how it changes or why it does not. (b) What is its angular speed at the lower temperature?

43. Starting with Equation 18.11, show that the total pressure P in a container filled with a mixture of several ideal gases is $P = P_1 + P_2 + P_3 + \cdots$, where $P_1, P_2, \cdots$ are the pressures that each gas would exert if it alone filled the container. (These individual pressures are called the *partial pressures* of the respective gases.) This result is known as *Dalton's law of partial pressures.*

CHALLENGE PROBLEMS

44. **Review.** A house roof is a perfectly flat plane that makes an angle θ with the horizontal. When its temperature changes, between T_c before dawn each day and T_h in the middle of each afternoon, the roof expands and contracts uniformly with a coefficient of thermal expansion α_1. Resting on the roof is a flat, rectangular metal plate with expansion coefficient α_2, greater than α_1. The length of the plate is L, measured along the slope of the roof. The component of the plate's weight perpendicular to the roof is supported by a normal force uniformly distributed over the area of the plate. The coefficient of kinetic friction between the plate and the roof is μ_k. The plate is always at the same temperature as the roof, so we assume its temperature is continuously changing. Because of the difference in expansion coefficients, each bit of the plate is moving relative to the roof below it, except for points along a certain horizontal line running across the plate called the stationary line. If the temperature is rising, parts of the plate below the stationary line are moving down relative to the roof and feel a force of kinetic friction acting up the roof. Elements of area above the stationary line are sliding up the roof, and on them kinetic friction acts downward parallel to the roof. The stationary line occupies no area, so we assume no force of static friction acts on the plate while the temperature is changing. The plate as a whole is very nearly in equilibrium, so the net friction force on it must be equal to the component of its weight acting down the incline. (a) Prove that the stationary line is at a distance of

$$\frac{L}{2}\left(1 - \frac{\tan\theta}{\mu_k}\right)$$

below the top edge of the plate. (b) Analyze the forces that act on the plate when the temperature is falling and prove that the stationary line is at that same distance above the bottom edge of the plate. (c) Show that the plate steps down the roof like an inchworm, moving each day by the distance

$$\frac{L}{\mu_k}(\alpha_2 - \alpha_1)(T_h - T_c)\tan\theta$$

(d) Evaluate the distance an aluminum plate moves each day if its length is 1.20 m, the temperature cycles between

4.00°C and 36.0°C, and if the roof has slope 18.5°, coefficient of linear expansion 1.50×10^{-5} (°C)$^{-1}$, and coefficient of friction 0.420 with the plate. (e) **What If?** What if the expansion coefficient of the plate is less than that of the roof? Will the plate creep up the roof?

45. A 1.00-km steel railroad rail is fastened securely at both ends when the temperature is 20.0°C. As the temperature increases, the rail buckles, taking the shape of an arc of a vertical circle. Find the height h of the center of the rail when the temperature is 25.0°C. (You will need to solve a transcendental equation.)

46. Helium gas is sold in steel tanks that will rupture if subjected to tensile stress greater than its yield strength of 5×10^8 N/m². If the helium is used to inflate a balloon, could the balloon lift the spherical tank the helium came in? Justify your answer. *Suggestion:* You may consider a spherical steel shell of radius r and thickness t having the density of iron and on the verge of breaking apart into two hemispheres because it contains helium at high pressure.

The First Law of Thermodynamics

STORYLINE **It's a three-day weekend, and you decide to go RV** camping with other members of the Physics Club at Whitney Portal, California, the gateway to Mount Whitney, the tallest peak in the contiguous United States. This community is at an altitude of 2 393 m above sea level, so it should be a good place to do some astronomy observations. As your Club advisor's RV progresses toward Whitney Portal, gaining altitude with each minute, you notice a sign that says, "Caution: Bridge Freezes Before Road Surface." You wonder why that would happen. As you reach your destination, you climb out of your car and marvel at how cold it is. But then you think, "Wait a minute! I'm closer to the Sun than I was at sea level. Why isn't it warmer at the top of a mountain?" You set up camp and offer to make dinner for the group. You boil some eggs for three minutes, fry some hamburgers, and place a sheet of cookies in the oven. After the cookies are done, you put a homemade cake in the oven to bake. At the end of the meal, the results are mixed. The hamburgers are great, the eggs were not quite cooked enough, the cookies were too well done near the edges of the cookie sheet, and the cake fell. Why was your dinner so unsuccessful? You are unhappy with your cooking performance and go to bed. The next morning, you arise for a brisk walk and notice that there is frost on the cars, mailboxes, and the like, but only on the top surfaces of those items, not the sides. Why is the frost only on the top surfaces? There have been so many mysteries associated with this mountain trip and it's only the first morning! You hope there is cell phone service when you return to your RV from your walk so that you can spend some time investigating these mysteries online.

CONNECTIONS Equation 8.2, the conservation of energy equation, shows how the energy of a system can change due to mechanical transfers of energy, like work, and thermal transfers, like heat. It also shows that the energy of a system is divided between mechanical types (kinetic and potential energy) and a thermal type (internal energy). But this is our modern-day understanding of energy. Until about 1850, the fields of thermodynamics and mechanics were considered to be two distinct branches of science. The principle of conservation of energy seemed to describe only certain kinds of mechanical systems. Mid-19th-century experiments performed by Englishman James Joule and others, however, showed a strong connection between the transfer of energy by heat in thermal processes and the transfer of energy by work in mechanical processes. This connection led to what we know as Equation 8.2. This current chapter focuses on a reduced form of Equation 8.2, known as the *first law of thermodynamics*. The first law of thermodynamics describes systems in which the only energy change is that of internal energy and the transfers of energy are by heat and work. A major difference in our discussion of work in this chapter from that in most of the chapters on mechanics is that we will consider work done on *deformable* systems. We will see energy transfers associated with temperature and internal energy in a number of cases in the future, including, among others, the warming of electrical resistors in Chapter 26, cooking a potato in a microwave oven in Chapter 33, and thermal radiation from a black body in Chapter 39.

19.1 Heat and Internal Energy

In Chapter 7, we introduced *internal energy* E_{int}, which exhibits changes on the left side of Equation 8.2, and in Chapter 8, we introduced *heat* Q, which is a mechanism for energy transfer on the right hand side of the equation. These terms are often incorrectly used interchangeably in popular language. Therefore, let us define them carefully:

> **Internal energy** is all the energy of a system that is associated with its microscopic components—atoms and molecules—when viewed from a reference frame at rest with respect to the center of mass of the system.

The last part of this sentence ensures that any bulk kinetic energy of the system due to its motion through space is not included in internal energy. Internal energy includes kinetic energy of random translational, rotational, and vibrational motion of molecules; vibrational potential energy associated with forces between atoms in molecules; and electric potential energy associated with forces between molecules. In Chapter 7, we related internal energy to the temperature of an object, but this relationship is limited. We show in Section 19.3 that internal energy changes can also occur in the absence of temperature changes. In that discussion, we will investigate the internal energy of the system when there is a *physical change*, most often related to a phase change, such as melting or boiling.

We assign energy associated with *chemical changes*, related to chemical reactions, to the potential energy term in Equation 8.2, not to internal energy. Therefore, we discuss the *chemical potential energy* in, for example, a human body (due to previous meals), the gas tank of a car (due to an earlier transfer of fuel), and a battery of an electric circuit (stored in the battery during its construction in the manufacturing process).

Compare this description of internal energy with the following for heat:

> **Heat** is defined as a process of transferring energy across the boundary of a system because of a temperature difference between the system and its surroundings. It is also the amount of energy Q transferred by this process.

PITFALL PREVENTION 19.1

Internal Energy, Thermal Energy, and Bond Energy When reading other physics books, you may see terms such as *thermal energy* and *bond energy*. Thermal energy can be interpreted as that part of the internal energy associated with random motion of molecules and therefore related to temperature. Bond energy is the intermolecular potential energy. Therefore,

Internal energy =
   thermal energy + bond energy

Although this breakdown is presented here for clarification with regard to other books, we will not use these terms because there is no need for them.

PITFALL PREVENTION 19.2

Heat, Temperature, and Internal Energy Are Different As you read the newspaper or explore online, be alert for incorrectly used phrases including the word *heat* and think about the proper word to be used in place of *heat*. Incorrect examples include "As the truck braked to a stop, a large amount of heat was generated by friction" and "The heat of a hot summer day . . . ."

When you *heat* a substance, you are transferring energy into it by placing it in contact with surroundings that have a higher temperature. Such is the case, for example, when you place a pan of cold water on a stove burner. The burner is at a higher temperature than the water, and so the water gains energy by heat. For your cake in the oven in the opening storyline, energy transfers by heat from the hot air in the oven to the cake mixture.

Read this definition of heat (Q in Eq. 8.2) very carefully. In particular, notice what heat is *not* in the following common quotes. (1) Heat is *not* energy in a hot substance. For example, "The boiling water has a lot of heat" is incorrect; the boiling water has *internal energy E_{int}*. (2) Heat is *not* radiation. For example, "It was so hot during the bicycle race because the black roadway was radiating heat" is incorrect; energy is leaving the roadway by *electromagnetic radiation,* T_{ER} in Equation 8.2. (3) Heat is *not* warmth of an environment. For example, "The heat in the air was so oppressive" is incorrect; on a hot day, the air has a high *temperature T.*

As an analogy to the distinction between heat and internal energy, consider the distinction between work and mechanical energy discussed in Chapter 7. The work done on a system is a measure of the amount of energy transferred to the system from its surroundings, whereas the mechanical energy (kinetic energy plus potential energy) of a system is a consequence of the motion and configuration of the system. Therefore, when a person does work on a system, energy is transferred from the person to the system. It makes no sense to talk about the work *of* a system; one can refer only to the work done *on* or *by* a system when some process has occurred in which energy has been transferred to or from the system. Likewise, it makes no sense to talk about the heat *of* a system; one can refer to heat only when energy has been transferred to or from the system as a result of a temperature difference. Both heat and work are ways of transferring energy between a system and its surroundings, which is why they both appear on the right-hand side of Equation 8.2.

Units of Heat

Early studies of heat focused on the resultant increase in temperature of a substance, which was often water. Initial notions of heat were based on a fluid called *caloric* that flowed from one substance to another and caused changes in temperature. From the name of this mythical fluid came an energy unit related to thermal processes, the **calorie (cal),** which is defined as the amount of energy transfer necessary to raise the temperature of 1 g of water from 14.5°C to 15.5°C.[1] (The "Calorie," written with a capital "C" and used in describing the energy content of foods, is actually a kilocalorie.) The unit of energy in the U.S. customary system is the **British thermal unit (Btu),** which is defined as the amount of energy transfer required to raise the temperature of 1 lb of water from 63°F to 64°F.

Once the relationship between energy in thermal and mechanical processes became clear, there was no need for a separate unit related to thermal processes. We have already defined the *joule* as an energy unit based on mechanical processes. Scientists are increasingly turning away from the calorie and the Btu and are using the joule when describing thermal processes. In this textbook, heat, work, and internal energy are usually measured in joules.

The Mechanical Equivalent of Heat

In Chapters 7 and 8, we found that whenever friction is present in a mechanical system, the mechanical energy in the system decreases; in other words, mechanical energy is not conserved in the presence of nonconservative forces. Various experiments show that this mechanical energy does not simply disappear but is transformed into internal energy. You can perform such an experiment at home

James Prescott Joule
British physicist (1818–1889)
Joule received some formal education in mathematics, philosophy, and chemistry from John Dalton but was in large part self-educated. Joule's research led to the establishment of the principle of conservation of energy. His study of the quantitative relationship among electrical, mechanical, and chemical effects of heat culminated in his announcement in 1843 of the amount of work required to produce a unit of energy, called the mechanical equivalent of heat.

[1] Originally, the calorie was defined as the energy transfer necessary to raise the temperature of 1 g of water by 1°C. Careful measurements, however, showed that the amount of energy required to produce a 1°C change depends somewhat on the initial temperature; hence, a more precise definition evolved.

The falling blocks rotate the paddles, causing the temperature of the water to increase.

Thermal insulator

Figure 19.1 Joule's experiment for determining the mechanical equivalent of heat.

by hammering a nail into a scrap piece of wood. What happens to all the kinetic energy of the hammer once you have finished? For the nail and board as a nonisolated system, Equation 8.2 becomes $\Delta E_{int} = W + T_{MW}$, where W is the work done by the hammer on the nail, T_{MW} is the energy leaving the system by sound waves when the nail is struck, and ΔE_{int} represents the warmer nail and wood. Notice that there is *no* transfer of energy by heat in this process. Although this connection between mechanical and internal energy was first suggested by Benjamin Thompson, it was James Prescott Joule who established the equivalence of the decrease in mechanical energy and the increase in internal energy.

A schematic diagram of Joule's most famous experiment is shown in Figure 19.1. The system of interest is the Earth, the two blocks, and the water in a thermally insulated container. Work is done within the system on the water by a rotating paddle wheel, which is driven by heavy blocks falling at a constant speed. The energy transformed in the bearings and the energy passing through the walls by heat are neglected. After the blocks and paddle stop moving, the decrease in gravitational potential energy during the fall of the blocks equals the internal work done by the paddle wheel on the water and, in turn, the increase in internal energy of the water. If the two blocks fall through a distance h, the decrease in potential energy of the system is $2mgh$, where m is the mass of one block; this energy transforms to internal energy E_{int} of the water. By varying the conditions of the experiment, Joule found that the decrease in mechanical energy is proportional to the product of the mass of the water and the increase in water temperature. The proportionality constant was found to be approximately 4.18 J/g · °C. Hence, 4.18 J of mechanical energy raises the temperature of 1 g of water by 1°C. More precise measurements taken later demonstrated the proportionality to be 4.186 J/g · °C when the temperature of the water was raised from 14.5°C to 15.5°C. We adopt this "15-degree calorie" value:

$$1 \text{ cal} = 4.186 \text{ J} \tag{19.1}$$

This equality is known, for purely historical reasons, as the **mechanical equivalent of heat.** A more proper name would be the *conversion factor between calories and joules,* but the historical name is well entrenched in our language, despite the incorrect use of the word *heat.*

Example 19.1   **Losing Weight the Hard Way**

A student eats a dinner rated at 2 000 Calories. He wishes to do an equivalent amount of work in the gymnasium by lifting a 50.0-kg barbell. How many times must he raise the barbell to expend this much energy? Assume he raises the barbell 2.00 m each time he lifts it and he transfers no energy when he lowers the barbell.

SOLUTION

Conceptualize   Imagine the student raising the barbell. He is doing work on the system of the barbell and the Earth, so energy is leaving his body. The total amount of work that the student must do is 2 000 Calories.

Categorize   We model the system of the barbell and the Earth as a *nonisolated system* for *energy.*

Analyze   Reduce the conservation of energy equation, Equation 8.2, to the appropriate expression for the system of the barbell and the Earth:

(1) $\Delta U_{total} = W_{total}$

Express the change in gravitational potential energy of the system after the barbell is raised once:

$\Delta U = mgh$

Express the total amount of energy that must be transferred into the system by work for lifting the barbell n times, assuming energy is not regained when the barbell is lowered:

(2) $\Delta U_{total} = nmgh$

19.1 continued

Substitute Equation (2) into Equation (1):

$$nmgh = W_{total}$$

Solve for n:

$$n = \frac{W_{total}}{mgh}$$

Substitute numerical values:

$$n = \frac{(2\,000\ \text{Cal})}{(50.0\ \text{kg})(9.80\ \text{m/s}^2)(2.00\ \text{m})}\left(\frac{1.00 \times 10^3\ \text{cal}}{\text{Calorie}}\right)\left(\frac{4.186\ \text{J}}{1\ \text{cal}}\right)$$

$$= 8.54 \times 10^3\ \text{times}$$

Finalize If the student is in good shape and lifts the barbell once every 5 s, it will take him about 12 h to perform this feat. Clearly, it is much easier for this student to lose weight by dieting.

In reality, the human body is not 100% efficient. Therefore, not all the energy transformed within the body from the dinner transfers out of the body by work done on the barbell. Some of this energy is used to pump blood and perform other functions within the body. Therefore, the 2 000 Calories can be worked off in less time than 12 h when these other energy processes are included.

19.2 Specific Heat and Calorimetry

When energy is added to a system and there is no change in the kinetic or potential energy of the system, the temperature of the system usually rises. (An exception to this statement is the case in which a system undergoes a change of state—also called a *phase transition*—as discussed in the next section.) If the system consists of a sample of a substance, we find that the quantity of energy required to raise the temperature of a given mass of the substance by some amount varies from one substance to another. For example, the quantity of energy required to raise the temperature of 1 kg of water by 1°C is 4 186 J, but the quantity of energy required to raise the temperature of 1 kg of copper by 1°C is only 387 J. In the discussion that follows, we shall use heat as our example of energy transfer, but keep in mind that the temperature of the system could be changed by means of any method of energy transfer.

The **heat capacity** C of a particular sample is defined as the amount of energy needed to raise the temperature of that sample by 1°C. From this definition, we see that if energy Q produces a change ΔT in the temperature of a sample, then

$$Q = C\,\Delta T \tag{19.2}$$

The **specific heat** c of a substance is the heat capacity per unit mass. Therefore, if energy Q transfers to a sample of a substance with mass m and the temperature of the sample changes by ΔT, the specific heat of the substance is

$$c \equiv \frac{Q}{m\,\Delta T} \tag{19.3}$$

◀ Specific heat

Specific heat is essentially a measure of how thermally insensitive a substance is to the addition of energy. The greater a material's specific heat, the more energy must be added to a given mass of the material to cause a particular temperature change. Table 19.1 (page 506) lists representative specific heats.

From this definition, we can relate the energy Q transferred between a sample of mass m of a material and its surroundings to a temperature change ΔT as

$$Q = mc\,\Delta T \tag{19.4}$$

For example, the energy required to raise the temperature of 0.500 kg of water by 3.00°C is $Q = (0.500\ \text{kg})(4\,186\ \text{J/kg} \cdot \text{°C})(3.00\text{°C}) = 6.28 \times 10^3\ \text{J}$. Notice that when the temperature increases, Q and ΔT are taken to be positive and energy transfers into the system. When the temperature decreases, Q and ΔT are negative and energy transfers out of the system.

PITFALL PREVENTION 19.3

An Unfortunate Choice of Terminology The name *specific heat* is an unfortunate holdover from the days when thermodynamics and mechanics developed separately. A better name would be *specific energy transfer*, but the existing term is too entrenched to be replaced.

PITFALL PREVENTION 19.4

Energy Can Be Transferred by Any Method The symbol Q represents the amount of energy transferred, but keep in mind that the energy transfer in Equation 19.4 could be by *any* of the methods introduced in Chapter 8; it does not have to be heat. For example, repeatedly bending a wire coat hanger raises the temperature at the bending point by *work*.

TABLE 19.1 Specific Heats of Some Substances at 25°C and Atmospheric Pressure

Substance	Specific Heat (J/kg · °C)	Substance	Specific Heat (J/kg · °C)
Elemental solids		*Other solids*	
Aluminum	900	Brass	380
Beryllium	1 830	Glass	837
Cadmium	230	Ice (−5°C)	2 090
Copper	387	Marble	860
Germanium	322	Wood	1 700
Gold	129		
Iron	448	*Liquids*	
Lead	128	Alcohol (ethyl)	2 400
Silicon	703	Mercury	140
Silver	234	Water (15°C)	4 186
		Gas	
		Steam (100°C)	2 010

Note: To convert values to units of cal/g · °C, divide by 4 186.

We can identify $mc\,\Delta T$ as the change in internal energy of the system if we ignore any thermal expansion or contraction of the system, and if there are no phase changes. (Thermal expansion or contraction would result in a very small amount of work being done on the system by the surrounding air.) Then, Equation 19.4 is a reduced form of Equation 8.2: $\Delta E_{int} = Q$. The internal energy of the system can be changed by transferring energy into the system by any mechanism. For example, if the system is a baked potato in a microwave oven, Equation 8.2 reduces to the following analog to Equation 19.4: $\Delta E_{int} = T_{ER} = mc\,\Delta T$, where T_{ER} is the energy transferred to the potato from the microwave oven by electromagnetic radiation. If the system is the air in a bicycle pump, which becomes hot when the pump is operated, Equation 8.2 reduces to the following analog to Equation 19.4: $\Delta E_{int} = W = mc\,\Delta T$, where W is the work done on the pump by the operator. By identifying $mc\,\Delta T$ as ΔE_{int}, we have taken a step toward a better understanding of temperature: temperature is related to the energy of the molecules of a system. We will learn more details of this relationship in Chapter 20.

Specific heat varies with temperature. If, however, temperature intervals are not too great, the temperature variation can be ignored and c can be treated as a constant.[2] For example, the specific heat of water varies by only about 1% from 0°C to 100°C at atmospheric pressure. Unless stated otherwise, we shall neglect such variations.

QUICK QUIZ 19.1 Imagine you have 1 kg each of iron, glass, and water, and all three samples are at 10°C. **(a)** Rank the samples from highest to lowest temperature after 100 J of energy is added to each sample. **(b)** Rank the samples from greatest to least amount of energy transferred by heat if each sample increases in temperature by 20°C.

Notice from Table 19.1 that water has the highest specific heat of common materials. This high specific heat is in part responsible for the moderate climates found near large bodies of water. As the temperature of a body of water decreases during the winter, energy is transferred from the cooling water to the air by heat, increasing the internal energy of the air. Because of the high specific heat of water, a relatively large amount of energy is transferred to the air for even modest temperature changes of the water. The prevailing winds on the West Coast of

---

[2]The definition given by Equation 19.4 assumes the specific heat does not vary with temperature over the interval $\Delta T = T_f - T_i$. In general, if c varies with temperature over the interval, the correct expression for Q is $Q = m \int_{T_i}^{T_f} c(T)\, dT$.

the United States are toward the land (eastward). Hence, the energy liberated by the Pacific Ocean as it cools keeps coastal areas much warmer than they would otherwise be. As a result, West Coast states generally have more favorable winter weather than East Coast states, where the prevailing winds carry the energy away from land.

Calorimetry

One technique for measuring specific heat involves heating a sample to some known temperature T_x, placing it in a vessel containing water of known mass and temperature $T_w < T_x$, and measuring the temperature of the water after equilibrium has been reached. This technique is called **calorimetry,** and devices in which this energy transfer occurs are called **calorimeters.** Figure 19.2 shows the hot sample in the cold water and the resulting energy transfer by heat from the high-temperature part of the system to the low-temperature part. If the system of the sample and the water is isolated, the principle of conservation of energy requires that the amount of energy Q_{hot} that leaves the sample (of unknown specific heat) equal the amount of energy Q_{cold} that enters the water.[3] Conservation of energy allows us to write the mathematical representation of this energy statement as

$$Q_{cold} = -Q_{hot} \tag{19.5}$$

Suppose m_x is the mass of a sample of some substance whose specific heat we wish to determine. Let's call its specific heat c_x and its initial temperature T_x as shown in Figure 19.2. Likewise, let m_w, c_w, and T_w represent corresponding values for the water. If T_f is the final temperature after the system comes to equilibrium, Equation 19.4 shows that the energy transfer for the water is $m_w c_w (T_f - T_w)$, which is positive because $T_f > T_w$, and that the energy transfer for the sample of unknown specific heat is $m_x c_x (T_f - T_x)$, which is negative. Substituting these expressions into Equation 19.5 gives

$$m_w c_w (T_f - T_w) = -m_x c_x (T_f - T_x) \tag{19.6}$$

This equation can be solved for the unknown specific heat c_x.

Figure 19.2 In a calorimetry experiment, a hot sample whose specific heat is unknown is placed in cold water in a container that isolates the system from the environment.

PITFALL PREVENTION 19.5
Remember the Negative Sign It is *critical* to include the negative sign in Equation 19.5. The negative sign in the equation is necessary for consistency with our sign convention for energy transfer. The energy transfer Q_{hot} has a negative value because energy is leaving the hot substance. The negative sign in the equation ensures that the right side is a positive number, consistent with the left side, which is positive because energy is entering the cold water.

| Example **19.2** | **Fun Time for a Cowboy** |

A cowboy fires a silver bullet with a muzzle speed of 200 m/s into the pine wall of a saloon. Assume all the internal energy generated by the impact remains with the bullet. What is the temperature change of the bullet?

SOLUTION

Conceptualize Imagine similar experiences you may have had in which mechanical energy is transformed to internal energy when a moving object is stopped. For example, as mentioned in Section 19.1, a nail becomes warm after it is hit a few times with a hammer.

Categorize The bullet is modeled as an *isolated system*. No work is done on the system because the force from the wall moves through no displacement. This example is similar to the skateboarder pushing off a wall in Section 9.8. There, no work is done on the skateboarder by the wall, and potential energy stored in the body from previous meals is transformed to kinetic energy. Here, no work is done by the wall on the bullet, and kinetic energy of the bullet is transformed to internal energy of the silver comprising the bullet.

Analyze Reduce the conservation of energy equation, Equation 8.2, to the appropriate expression for the system of the bullet:

$$(1) \quad \Delta K + \Delta E_{int} = 0$$

continued

---

[3]For precise measurements, the water container should be included in our calculations because it also exchanges energy with the sample. Doing so would require that we know the container's mass and composition, however. If the mass of the water is much greater than that of the container, we can neglect the effects of the container.

19.2 continued

The change in the bullet's internal energy is related to its change in temperature:

$$(2) \quad \Delta E_{int} = mc\,\Delta T$$

Substitute Equation (2) into Equation (1):

$$(0 - \tfrac{1}{2}mv^2) + mc\,\Delta T = 0$$

Solve for ΔT, using 234 J/kg · °C as the specific heat of silver (see Table 19.1):

$$(3) \quad \Delta T = \frac{\tfrac{1}{2}mv^2}{mc} = \frac{v^2}{2c} = \frac{(200 \text{ m/s})^2}{2(234 \text{ J/kg} \cdot \text{°C})} = \boxed{85.5\text{°C}}$$

Finalize   Notice that the result does not depend on the mass of the bullet. (In reality, the wall also becomes warmer, so our analysis is simplified.)

W H A T I F ?   Suppose the cowboy runs out of silver bullets and fires a lead bullet at the same speed into the wall. Will the temperature change of the bullet be larger or smaller?

Answer   Table 19.1 shows that the specific heat of lead is 128 J/kg · °C, which is smaller than that for silver. Therefore, a given amount of energy input or transformation raises lead to a higher temperature than silver and the final temperature of the lead bullet will be larger. In Equation (3), let's substitute the new value for the specific heat:

$$\Delta T = \frac{v^2}{2c} = \frac{(200 \text{ m/s})^2}{2(128 \text{ J/kg} \cdot \text{°C})} = 156\text{°C}$$

There is no requirement that the silver and lead bullets have the same mass to determine this change in temperature. The only requirement is that they have the same speed.

Example **19.3**   Cooling a Hot Ingot

A 0.050 0-kg ingot of metal is heated to 200.0°C and then dropped into a calorimeter containing 0.400 kg of water initially at 20.0°C. The final equilibrium temperature of the mixed system is 22.4°C. Find the specific heat of the metal.

S O L U T I O N

Conceptualize   Imagine the process occurring in the isolated system of Figure 19.2. Energy leaves the hot ingot and goes into the cold water, so the ingot cools off and the water warms up. Once both are at the same temperature, the energy transfer stops.

Categorize   We use an equation developed in this section, so we categorize this example as a substitution problem.

Solve Equation 19.6 for c_x:

$$c_x = \frac{m_w c_w (T_f - T_w)}{m_x (T_x - T_f)}$$

Substitute numerical values:

$$c_x = \frac{(0.400 \text{ kg})(4\,186 \text{ J/kg} \cdot \text{°C})(22.4\text{°C} - 20.0\text{°C})}{(0.050\,0 \text{ kg})(200.0\text{°C} - 22.4\text{°C})}$$

$$= 453 \text{ J/kg} \cdot \text{°C}$$

The ingot is most likely iron as you can see by comparing this result with the data given in Table 19.1. The temperature of the ingot is initially above the steam point. Therefore, some of the water may vaporize when the ingot is dropped into the water. We assume the system is sealed and this steam cannot escape. Because the final equilibrium temperature is lower than the steam point, any steam that does result recondenses back into water.

W H A T I F ?   Suppose you are performing an experiment in the laboratory that uses this technique to determine the specific heat of a sample and you wish to decrease the overall uncertainty in your final result for c_x. Of the data given in this example, changing which value would be most effective in decreasing the uncertainty?

Answer   The largest experimental uncertainty is associated with the small difference in temperature of 2.4°C for the water. For example, using the rules for propagation of uncertainty in Appendix Section B.8, an uncertainty of 0.1°C in each of T_f and T_w leads to an 8% uncertainty in their difference. For this temperature difference to be larger experimentally, the most effective change is to *decrease the amount of water*.

As we have seen in the preceding section, a substance can undergo a change in temperature when energy is transferred between it and its surroundings. In some situations, however, the transfer of energy does not result in a change in temperature. That is the case whenever the physical characteristics of the substance change from one form to another; such a change is commonly referred to as a **phase change.** Two common phase changes are from solid to liquid (melting) and from liquid to gas (boiling); another is a change in the crystalline structure of a solid. All such phase changes involve a change in the system's internal energy but no change in its temperature. The increase in internal energy in boiling, for example, is represented by the breaking of bonds between molecules in the liquid state; this bond breaking allows the molecules to move farther apart in the gaseous state, with a corresponding increase in intermolecular potential energy.

As you might expect, different substances respond differently to the addition or removal of energy as they change phase because their internal molecular arrangements vary. Also, the amount of energy transferred during a phase change depends on the amount of substance involved. (It takes less energy to melt an ice cube than it does a frozen lake.) When discussing two phases of a material, we will use the term *higher-phase material* to mean the material existing at the higher temperature. So, for example, if we discuss water and ice, water is the higher-phase material, whereas steam is the higher-phase material in a discussion of steam and water. Consider a system containing a substance in two phases in equilibrium such as water and ice. The initial amount of the higher-phase material, water, in the system is m_i. Now imagine that energy Q enters the system. As a result, the final amount of water is m_f due to the melting of some of the ice. Therefore, the amount of ice that melted, equal to the amount of *new* water, is $\Delta m = m_f - m_i$. We define the **latent heat** for this phase change as

$$L \equiv \frac{Q}{\Delta m} \qquad (19.7)$$

This parameter is called latent heat (literally, the "hidden" heat) because this added or removed energy does not result in a temperature change. The value of L for a substance depends on the nature of the phase change as well as on the properties of the substance. If the entire amount of the lower-phase material undergoes a phase change, the change in mass Δm of the higher-phase material is equal to the initial mass of the lower-phase material. For example, if an ice cube of mass m on a plate melts completely, the change in mass of the water is $\Delta m = m_f - 0 = m$, which is the mass of new water and is also equal to the initial mass of the ice cube.

From the definition of latent heat, and again choosing heat as our energy transfer mechanism, the energy required to change the phase of a pure substance is

$$Q = L \, \Delta m \qquad (19.8)$$

where Δm is the change in mass of the higher-phase material.

Latent heat of fusion L_f is the term used when the phase change is from solid to liquid (*to fuse* means "to combine by melting"), and **latent heat of vaporization** L_v is the term used when the phase change is from liquid to gas (the liquid "vaporizes").[4] When energy enters a system, causing melting or vaporization, the amount of the higher-phase material increases, so Δm is positive and Q is positive, consistent with our sign convention. When energy is extracted from a system, causing freezing or condensation, the amount of the higher-phase material decreases, so Δm is negative and Q is negative, again consistent with our sign convention. Keep

PITFALL PREVENTION 19.6

Signs Are Critical Sign errors occur very often when students apply calorimetry equations. For phase changes, remember that Δm in Equation 19.8 is always the change in mass of the higher-phase material. In Equation 19.4, be sure your ΔT is *always* the final temperature minus the initial temperature. In addition, you must *always* include the negative sign on the right side of Equation 19.5.

◀ Energy transferred to a substance during a phase change

---

[4]When a gas cools, it eventually *condenses;* that is, it returns to the liquid phase. The energy given up per unit mass is called the *latent heat of condensation* and is numerically equal to the latent heat of vaporization. Likewise, when a liquid cools, it eventually solidifies, and the *latent heat of solidification* is numerically equal to the latent heat of fusion.

TABLE 19.2 Latent Heats of Fusion and Vaporization

Substance	Melting Point (°C)	Latent Heat of Fusion (J/kg)	Boiling Point (°C)	Latent Heat of Vaporization (J/kg)
Helium[a]	−272.2	5.23×10^3	−268.93	2.09×10^4
Oxygen	−218.79	1.38×10^4	−182.97	2.13×10^5
Nitrogen	−209.97	2.55×10^4	−195.81	2.01×10^5
Ethyl alcohol	−114	1.04×10^5	78	8.54×10^5
Water	0.00	3.33×10^5	100.00	2.26×10^6
Sulfur	119	3.81×10^4	444.60	3.26×10^5
Lead	327.3	2.45×10^4	1 750	8.70×10^5
Aluminum	660	3.97×10^5	2 450	1.14×10^7
Silver	960.80	8.82×10^4	2 193	2.33×10^6
Gold	1 063.00	6.44×10^4	2 660	1.58×10^6
Copper	1 083	1.34×10^5	1 187	5.06×10^6

[a]Helium does not solidify at atmospheric pressure. The melting point given here corresponds to a pressure of 2.5 MPa.

in mind that Δm in Equation 19.8 always refers to the higher-phase material. The latent heats of various substances vary considerably as data in Table 19.2 show.

To understand the role of latent heat in phase changes, consider the energy required to convert a system consisting of a 1.00-g cube of ice at −30.0°C to steam at 120.0°C. Figure 19.3 indicates the experimental results obtained when energy is gradually added to the ice. The results are presented as a graph of temperature of the system versus energy added to the system. Let's examine each portion of the red-brown curve, which is divided into parts A through E.

Part A. On this portion of the curve, the temperature of the system changes from −30.0°C to 0.0°C. Equation 19.4 indicates that the temperature varies linearly with the energy added, so the experimental result is a straight line on the graph. Because the specific heat of ice is 2 090 J/kg · °C, we can calculate the amount of energy added by using Equation 19.4:

$$Q = m_{ice}c_{ice}\,\Delta T = (1.00 \times 10^{-3}\ \text{kg})(2\,090\ \text{J/kg} \cdot {}^{\circ}\text{C})(30.0{}^{\circ}\text{C}) = 62.7\ \text{J}$$

Part B. When the temperature of the system reaches 0.0°C, the ice–water mixture remains at this temperature—even though energy is being added—until all the ice melts. The energy required to melt 1.00 g of ice at 0.0°C is, from Equation 19.8,

$$Q = L_f\Delta m_w = L_f m_{ice} = (3.33 \times 10^5\ \text{J/kg})(1.00 \times 10^{-3}\ \text{kg}) = 333\ \text{J}$$

At this point, we have moved to the 396 J (= 62.7 J + 333 J) mark on the energy axis in Figure 19.3.

Figure 19.3 A plot of temperature versus energy added when a system initially consisting of 1.00 g of ice at −30.0°C is converted to steam at 120.0°C.

Part C. Between 0.0°C and 100.0°C, nothing surprising happens. No phase change occurs, and so all energy added to the system, which is now water, is used to increase its temperature. The amount of energy necessary to increase the temperature from 0.0°C to 100.0°C is

$$Q = m_w c_w \Delta T = (1.00 \times 10^{-3} \text{ kg})(4.19 \times 10^3 \text{ J/kg} \cdot \text{°C})(100.0\text{°C}) = 419 \text{ J}$$

where m_w is the mass of the water in the system, which is the same as the mass m_{ice} of the original ice.

Part D. At 100.0°C, another phase change occurs as the system changes from water at 100.0°C to steam at 100.0°C. Similar to the ice–water mixture in part B, the water–steam mixture remains at a fixed temperature, this time 100.0°C—even though energy is being added—until all the liquid has been converted to steam. The energy required to convert 1.00 g of water to steam at 100.0°C is

$$Q = L_v \Delta m_s = L_v m_w = (2.26 \times 10^6 \text{ J/kg})(1.00 \times 10^{-3} \text{ kg}) = 2.26 \times 10^3 \text{ J}$$

Part E. On this portion of the curve, as in parts A and C, no phase change occurs; therefore, all energy added is used to increase the temperature of the system, which is now steam. The energy that must be added to raise the temperature of the steam from 100.0°C to 120.0°C is

$$Q = m_s c_s \Delta T = (1.00 \times 10^{-3} \text{ kg})(2.01 \times 10^3 \text{ J/kg} \cdot \text{°C})(20.0\text{°C}) = 40.2 \text{ J}$$

The total amount of energy that must be added to the system to change 1 g of ice at −30.0°C to steam at 120.0°C is the sum of the results from all five parts of the curve, which is 3.11×10^3 J. Conversely, to cool 1 g of steam at 120.0°C to ice at −30.0°C, we must remove 3.11×10^3 J of energy.

Notice in Figure 19.3 the relatively large amount of energy that is transferred into the water to vaporize it to steam. Imagine reversing this process, with a large amount of energy transferred out of steam to condense it into water. That is why a burn to your skin from steam at 100°C is much more damaging than exposure of your skin to water at 100°C. A very large amount of energy enters your skin from the steam, and the steam remains at 100°C for a long time while it condenses. Conversely, when your skin makes contact with water at 100°C, the water immediately begins to drop in temperature as energy transfers from the water to your skin.

Ⓠ UICK QUIZ 19.2 Suppose the same process of adding energy to the ice cube is performed as discussed above with regard to Figure 19.3, but instead we graph the internal energy of the system as a function of energy input. What would this graph look like?

If liquid water is held perfectly still in a very clean container, it is possible for the water to drop below 0°C without freezing into ice. This phenomenon, called **super-cooling,** arises because the water requires a disturbance of some sort for the molecules to move apart and start forming the large, open ice structure that makes the density of ice lower than that of water as discussed in Section 18.4. If supercooled water is disturbed, it suddenly freezes. The system drops into the lower-energy configuration of bound molecules of the ice structure, and the energy released raises the temperature back to 0°C.

Commercial hand warmers consist of liquid sodium acetate in a sealed plastic pouch. The solution in the pouch is in a stable supercooled state. When a disk in the pouch is clicked by your fingers, the liquid solidifies and the temperature increases, just like the supercooled water just mentioned. In this case, however, the freezing point of the liquid is higher than body temperature, so the pouch feels warm to the touch. To reuse the hand warmer, the pouch must be boiled until the solid liquefies. Then, as it cools, it passes below its freezing point into the super-cooled state.

It is also possible to create **superheating.** For example, clean water in a very clean cup placed in a microwave oven can sometimes rise in temperature beyond 100°C without boiling because the formation of a bubble of steam in the water requires scratches in the cup or some type of impurity in the water to serve as a nucleation site. When the cup is removed from the microwave oven, the superheated water can become explosive as bubbles form immediately and the hot water is forced upward out of the cup.

Example **19.4** Cooling the Steam

What mass of steam initially at 130°C is needed to warm 200 g of water in a 100-g glass container from 20.0°C to 50.0°C?

SOLUTION

Conceptualize   Imagine placing water and steam together in a closed insulated container. The steam cools and condenses into liquid water, and the system eventually reaches a uniform state of water with a final temperature of 50.0°C in equilibrium with the glass at the same temperature.

Categorize   Based on our conceptualization of this situation, we categorize this example as one involving calorimetry in which a phase change occurs. The calorimeter is an *isolated system* for *energy:* energy transfers between the components of the system but does not cross the boundary between the system and the environment.

Analyze   Write Equation 19.5 to describe the calorimetry process:

$$(1) \quad Q_{cold} = -Q_{hot}$$

The steam undergoes three processes: first a decrease in temperature to 100°C, then condensation into liquid water, and finally a decrease in temperature of the water to 50.0°C. Find the energy transfer in the first process using the unknown mass m_s of the steam:

$$Q_1 = m_s c_s \,\Delta T_s$$

Find the energy transfer in the second process:

$$Q_2 = L_v \,\Delta m_s = L_v(0 - m_s) = -m_s L_v$$

Find the energy transfer in the third process:

$$Q_3 = m_s c_w \,\Delta T_{hot\ water}$$

Add the energy transfers in these three stages:

$$(2) \quad Q_{hot} = Q_1 + Q_2 + Q_3 = m_s(c_s \,\Delta T_s - L_v + c_w \,\Delta T_{hot\ water})$$

The 20.0°C water and the glass undergo only one process, an increase in temperature to 50.0°C. Find the energy transfer in this process:

$$(3) \quad Q_{cold} = m_w c_w \,\Delta T_{cold\ water} + m_g c_g \,\Delta T_{glass}$$

Substitute Equations (2) and (3) into Equation (1):

$$m_w c_w \,\Delta T_{cold\ water} + m_g c_g \,\Delta T_{glass} = -m_s(c_s \,\Delta T_s - L_v + c_w \,\Delta T_{hot\ water})$$

Solve for m_s:

$$m_s = -\frac{m_w c_w \,\Delta T_{cold\ water} + m_g c_g \,\Delta T_{glass}}{c_s \,\Delta T_s - L_v + c_w \,\Delta T_{hot\ water}}$$

Substitute numerical values:

$$m_s = -\frac{(0.200 \text{ kg})(4\ 186 \text{ J/kg} \cdot \text{°C})(50.0\text{°C} - 20.0\text{°C}) + (0.100 \text{ kg})(837 \text{ J/kg} \cdot \text{°C})(50.0\text{°C} - 20.0\text{°C})}{(2\ 010 \text{ J/kg} \cdot \text{°C})(100\text{°C} - 130\text{°C}) - (2.26 \times 10^6 \text{ J/kg}) + (4\ 186 \text{ J/kg} \cdot \text{°C})(50.0\text{°C} - 100\text{°C})}$$

$$= 1.09 \times 10^{-2} \text{ kg} = \boxed{10.9 \text{ g}}$$

WHAT IF?   What if the final state of the system is water at 100°C? Would we need more steam or less steam? How would the analysis above change?

Answer   More steam would be needed to raise the temperature of the water and glass to 100°C instead of 50.0°C. There would be two major changes in the analysis. First, we would not have a term Q_3 for the steam because the water that condenses from the steam does not cool below 100°C. Second, in Q_{cold}, the temperature change would be 80.0°C instead of 30.0°C. For practice, show that the result is a required mass of steam of 31.8 g.

19.4 Work in Thermodynamic Processes

In thermodynamics, we describe the *state* of a system using such variables as pressure, volume, temperature, and internal energy. As a result, these quantities belong to a category called **state variables.** For any given configuration of the system, we can identify values of the state variables. (For mechanical systems, the state variables include kinetic energy K and potential energy U. For a single particle as a system, we could identify more state variables: its position x, its velocity v, and its acceleration a.) A state of a system can be specified only if the system is in thermal equilibrium internally. In the case of a gas in a container, internal thermal equilibrium requires that every part of the gas be at the same pressure and temperature.

A second category of variables in situations involving energy is **transfer variables.** These variables are those that appear on the right side of the conservation of energy equation, Equation 8.2. Such a variable has a nonzero value if a process occurs in which energy is transferred across the system's boundary. The transfer variable is positive or negative, depending on whether energy is entering or leaving the system. Because a transfer of energy across the boundary represents a change in the system, transfer variables are not associated with a given state of the system, but rather with a *change* in the state of the system.

In the previous sections, we discussed heat as a transfer variable. In this section, we study another important transfer variable for thermodynamic systems, work. Work performed on particles and nondeformable objects was studied extensively in Chapter 7, and here we investigate the work done on a *deformable* system, a gas. Consider a gas contained in a cylinder fitted with a movable piston (Fig. 19.4a). At equilibrium, the gas occupies a volume V and exerts a uniform pressure P on the cylinder's walls and on the piston. If the piston has a cross-sectional area A, the magnitude of the force exerted by the gas on the piston is $F = PA$. By Newton's third law, the magnitude of the force exerted by the piston on the gas is also PA. Now let's assume we push the piston inward and compress the gas **quasi-statically,** that is, slowly enough to allow the system to remain essentially in internal thermal equilibrium at all times. The point of application of the force on the gas is the bottom face of the piston. As the piston is pushed downward by an external force $\vec{F} = -F\hat{j}$ through a displacement of $d\vec{r} = dy\hat{j}$ (Fig. 19.4b), the work done on the gas is, according to our definition of work in Chapter 7,

$$dW = \vec{F} \cdot d\vec{r} = -F\hat{j} \cdot dy\,\hat{j} = -F\,dy = -PA\,dy$$

Because $A\,dy$ is the change in volume of the gas dV, we can express the work done on the gas as

$$dW = -P\,dV \tag{19.9}$$

If the gas is compressed, dV is negative and the work done on the gas is positive. If the gas expands, dV is positive and the work done on the gas is negative. If the volume remains constant, the work done on the gas is zero. The total work done on the gas as its volume changes from V_i to V_f is given by the integral of Equation 19.9:

$$W = -\int_{V_i}^{V_f} P\,dV \tag{19.10}$$

To evaluate this integral, you must know how the pressure varies with volume during the process.

In general, the pressure is not constant during a process followed by a gas, but depends on the volume and temperature. If the pressure and volume are known at each step of the process, the state of the gas at each step can be plotted on an important graphical representation called a ***PV* diagram** as in Figure 19.5. This type of diagram allows us to visualize a process through which a gas is progressing. The curve on a *PV* diagram is called the *path* taken between the initial and final states.

Figure 19.4 Work is done on a gas contained in a cylinder at a pressure P as the piston is pushed downward so that the gas is compressed.

The work done on a gas equals the negative of the area under the *PV* curve. The area is negative here because the volume is decreasing, resulting in positive work.

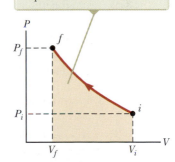

Figure 19.5 A gas is compressed quasi-statically (slowly) from state *i* to state *f*. An outside agent must do positive work on the gas to compress it.

◀ Work done on a gas

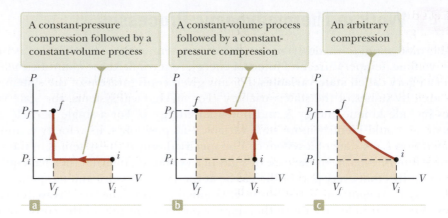

A constant-pressure compression followed by a constant-volume process

A constant-volume process followed by a constant-pressure compression

An arbitrary compression

Figure 19.6 The work done on a gas as it is taken from an initial state to a final state depends on the path between these states.

Notice that the integral in Equation 19.10 is equal to the area under a curve on a *PV* diagram. Therefore, we can identify an important use for *PV* diagrams:

> The work done on a gas in a quasi-static process that takes the gas from an initial state to a final state is the negative of the area under the curve on a *PV* diagram, evaluated between the initial and final states.

For the process of compressing a gas in a cylinder, the work done depends on the particular path taken between the initial and final states as Figure 19.5 suggests. To illustrate this important point, consider several different paths connecting *i* and *f* (Fig. 19.6). In the process depicted in Figure 19.6a, the volume of the gas is first reduced from V_i to V_f at constant pressure P_i and the pressure of the gas then increases from P_i to P_f by heating at constant volume V_f. The work done on the gas along this path is $-P_i(V_f - V_i)$. In Figure 19.6b, the pressure of the gas is increased from P_i to P_f at constant volume V_i and then the volume of the gas is reduced from V_i to V_f at constant pressure P_f. The work done on the gas is $-P_f(V_f - V_i)$. This value is greater than that for the process described in Figure 19.6a because the piston is moved through the same displacement by a larger force. Finally, for the process described in Figure 19.6c, where both *P* and *V* change continuously, the work done on the gas has some value between the values obtained in the first two processes. To evaluate the work in this case, the function $P(V)$ must be known so that we can evaluate the integral in Equation 19.10.

The energy transfer *Q* into or out of a system by heat also depends on the process. For example, in Chapter 20, we will show that a constant-volume process between two temperatures requires a different amount of heat than a constant-pressure process between the same temperatures.

19.5 The First Law of Thermodynamics

When we introduced the law of conservation of energy in Chapter 8, we stated that the change in the energy of a system is equal to the sum of all transfers of energy across the system's boundary (Eq. 8.2). The **first law of thermodynamics** is a special case of the law of conservation of energy that describes processes in which only the internal energy[5] changes and the only energy transfers are by heat and work:

First law of thermodynamics ▶

$$\Delta E_{\text{int}} = Q + W \tag{19.11}$$

Look back at Equation 8.2 to see that the first law of thermodynamics is contained within that more general equation.

---

[5]It is an unfortunate accident of history that the traditional symbol for internal energy is *U*, which is also the traditional symbol for potential energy as introduced in Chapter 7. To avoid confusion between potential energy and internal energy, we use the symbol E_{int} for internal energy in this book. If you take an advanced course in thermodynamics, however, be prepared to see *U* used as the symbol for internal energy in the first law.

Let's discuss each of the three terms in the first law for various processes through which a gas is taken. As a model, let's consider the sample of gas contained in the piston–cylinder apparatus in Figure 19.7. This figure shows work being done on the gas and energy transferring in by heat, so the internal energy of the gas is rising. In the following discussion of various processes, refer back to this figure and mentally alter the directions of the transfer of energy to reflect what is happening in the process.

First, consider an *isolated system,* that is, one that does not interact with its surroundings, as we have seen before. In this case, no energy transfer by heat takes place and the work done on the system is zero; hence, the internal energy remains constant. That is, because $Q = W = 0$, it follows that $\Delta E_{int} = 0$; therefore, $E_{int,i} = E_{int,f}$. We conclude that the internal energy E_{int} of an isolated system remains constant.

Next, consider the case of a system that can exchange energy with its surroundings and is taken through a **cyclic process,** that is, a process that starts and ends at the same state. On a *PV* diagram, a cyclic process appears as a closed curve, as shown in Figure 19.8. In this case, the change in the internal energy must again be zero because E_{int} is a state variable; therefore, the energy Q added to the system must equal the negative of the work W done on the system during the cycle. That is, in a cyclic process,

$$\Delta E_{int} = 0 \quad \text{and} \quad Q = -W \quad \text{(cyclic process)}$$

It can be shown that in a cyclic process for a gas, the net work done on the system per cycle equals the area enclosed by the path representing the process on a *PV* diagram.

A process that occurs at constant temperature is called an **isothermal process.** This process can be established by immersing the cylinder in Figure 19.7 in an ice–water bath or by putting the cylinder in contact with some other constant-temperature reservoir. A plot of *P* versus *V* at constant temperature for an ideal gas yields a hyperbolic curve called an *isotherm,* as shown in Figure 19.9. The ideal gas law (Eq. 18.9) with *T* constant indicates that the equation of this curve is $PV = nRT = $ constant. We show in Chapter 20 that the internal energy of an ideal gas is a function of temperature only. Hence, because the temperature does not change in an isothermal process involving an ideal gas, we must have $\Delta E_{int} = 0$. For an isothermal process, we conclude from the first law that the energy transfer Q must be equal to the negative of the work done on the gas; that is, $Q = -W$. Any energy that enters the system by heat is transferred out of the system by work.

Let's calculate the work done on the gas in the isothermal expansion from state *i* to state *f* in Figure 19.9. The work done on the gas is given by Equation 19.10.

Figure 19.7 The first law of thermodynamics equates the change in internal energy E_{int} in a system to the net energy transfer to the system by heat Q and work W. In the situation shown here, the internal energy of the gas increases.

PITFALL PREVENTION 19.7

Dual Sign Conventions Some physics and engineering books present the first law as $\Delta E_{int} = Q - W$, with a minus sign between the heat and work. The reason is that work is defined in these treatments as the work done *by* the gas rather than *on* the gas, as in our treatment. The equivalent equation to Equation 19.10 these treatments defines work as $W = \int_{V_i}^{V_f} P \, dV$. Therefore, if positive work is done by the gas, energy is leaving the system, leading to the negative sign in the first law.

In your studies in other chemistry or engineering courses, or in your reading of other physics books, be sure to note which sign convention is being used for the first law.

PITFALL PREVENTION 19.8

The First Law With our approach to energy in this book, the first law of thermodynamics is a special case of Equation 8.2. Some physicists argue that the first law is the general equation for energy conservation, equivalent to Equation 8.2. In this approach, the first law is applied to a closed system (so that there is no matter transfer), heat is interpreted so as to include electromagnetic radiation, and work is interpreted so as to include electrical transmission ("electrical work") and mechanical waves ("molecular work"). Keep that in mind if you run across the first law in your reading of other physics books.

Figure 19.8 A cyclic process on a gas forms a closed curve on a *PV* diagram.

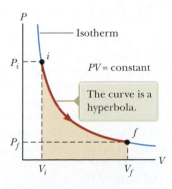

Figure 19.9 The *PV* diagram for an isothermal expansion of an ideal gas from an initial state to a final state.

Because the gas is ideal and the process is quasi-static, the ideal gas law is valid for each point on the path. Therefore,

$$W = -\int_{V_i}^{V_f} P \, dV = -\int_{V_i}^{V_f} \frac{nRT}{V} \, dV$$

Because T is constant in this case, it can be removed from the integral along with n and R:

$$W = -nRT \int_{V_i}^{V_f} \frac{dV}{V} = -nRT \ln V \Big|_{V_i}^{V_f}$$

To evaluate the integral, we used $\int (dx/x) = \ln x$. (See Appendix B.) Evaluating the result at the initial and final volumes gives

$$W = nRT \ln \left(\frac{V_i}{V_f} \right) \quad \text{(isothermal process)} \tag{19.12}$$

Numerically, this work W equals the negative of the shaded area under the PV curve shown in Figure 19.9. Because the gas expands, $V_f > V_i$ and the value for the work done on the gas is negative as we expect. If the gas is compressed, then $V_f < V_i$ and the work done on the gas is positive.

A process that occurs at constant pressure is called an **isobaric process.** In Figure 19.7, an isobaric process could be established by allowing the piston to move freely so that it is always in equilibrium between the net force from the gas pushing upward and the weight of the piston plus the force due to atmospheric pressure pushing downward. An isobaric process appears as a horizontal line on a PV diagram as shown in Figure 19.10. Find the isobaric processes in Figure 19.6.

In such a process, the values of the heat and the work are both usually nonzero. The work done on the gas in an isobaric process is simply

$$W = -P(V_f - V_i) \quad \text{(isobaric process)} \tag{19.13}$$

where P is the constant pressure of the gas during the process.

A process that takes place at constant volume is called an **isovolumetric process.** Another name for this type of process is *isochoric*. In Figure 19.7, clamping the piston at a fixed position would ensure an isovolumetric process. An isovolumetric process appears as a vertical line on a PV diagram as shown in Figure 19.11. Find the isovolumetric processes in Figure 19.6.

Because the volume of the gas does not change in such a process, the work given by Equation 19.10 is zero. Hence, from the first law we see that in an isovolumetric process, because $W = 0$,

$$\Delta E_{\text{int}} = Q \quad \text{(isovolumetric process)} \tag{19.14}$$

This expression specifies that if energy is added by heat to a system kept at constant volume, all the transferred energy remains in the system as an increase in its internal energy. For example, when a spray can is thrown in a fire, as in Example 18.4, energy enters the system (the gas in the can) by heat through the metal walls of the can. Consequently, the temperature, and therefore the pressure, in the can increases until the can possibly explodes.

An **adiabatic process** is one during which no energy enters or leaves the system by heat; that is, $Q = 0$. An adiabatic process can be achieved either by thermally insulating the walls of the system or by performing the process rapidly so that there is negligible time for energy to transfer by heat. Applying the first law of thermodynamics to an adiabatic process gives

$$\Delta E_{\text{int}} = W \quad \text{(adiabatic process)} \tag{19.15}$$

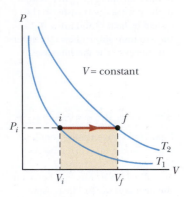

Figure 19.10 An isobaric process takes a gas between temperatures T_1 and T_2.

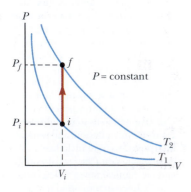

Figure 19.11 An isovolumetric process takes a gas between temperatures T_1 and T_2.

This result shows that if a gas is compressed adiabatically such that W is positive, then ΔE_{int} is positive and the temperature of the gas increases. Conversely, the temperature of a gas decreases when the gas expands adiabatically.

Adiabatic processes are very important in engineering practice. Some common examples are the expansion of hot gases in an internal combustion engine, the liquefaction of gases in a cooling system, and the compression stroke in a diesel engine. We will see PV diagrams for adiabatic processes and study them in more detail in Chapter 20.

QUICK QUIZ 19.3 In the last three columns of the following table, fill in the boxes with the correct signs ($-$, $+$, or 0) for Q, W, and ΔE_{int}. For each situation, the system to be considered is identified.

Situation	System	Q	W	ΔE_{int}
(a) Rapidly pumping up a bicycle tire	Air in the pump			
(b) Pan of room-temperature water sitting on a hot stove	Water in the pan			
(c) Air quickly leaking out of a balloon	Air originally in the balloon			

QUICK QUIZ 19.4 Characterize the paths in Figure 19.12 as isobaric, isovolumetric, isothermal, or adiabatic. For path B, $Q = 0$. The blue curves are isotherms.

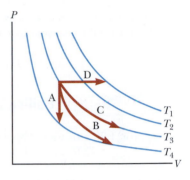

Figure 19.12 (Quick Quiz 19.4) Identify the nature of paths A, B, C, and D.

Example 19.5 **An Isothermal Expansion**

A 1.0-mol sample of an ideal gas is kept at 0.0°C during an expansion from 3.0 L to 10.0 L.

(A) How much work is done on the gas during the expansion?

SOLUTION

Conceptualize Run the process in your mind: the cylinder in Figure 19.7 is immersed in an ice-water bath, and the piston moves outward so that the volume of the gas increases. You can also use the graphical representation in Figure 19.9 to conceptualize the process.

Categorize We will evaluate parameters using equations developed in the preceding sections, so we categorize this example as a substitution problem. Because the temperature of the gas is fixed, the process is isothermal.

Substitute the given values into Equation 19.12:

$$W = nRT \ln\left(\frac{V_i}{V_f}\right)$$

$$= (1.0 \text{ mol})(8.31 \text{ J/mol} \cdot \text{K})(273 \text{ K}) \ln\left(\frac{3.0 \text{ L}}{10.0 \text{ L}}\right)$$

$$= -2.7 \times 10^3 \text{ J}$$

(B) How much energy transfer by heat occurs between the gas and its surroundings in this process?

SOLUTION

Find the heat from the first law:

$$\Delta E_{int} = Q + W$$

$$0 = Q + W$$

$$Q = -W = 2.7 \times 10^3 \text{ J}$$

continued

19.5 continued

(C) If the gas is returned to the original volume by means of an isobaric process, how much work is done on the gas?

SOLUTION

Use Equation 19.13. The pressure is not given, so incorporate the ideal gas law:

$$W = -P(V_f - V_i) = -\frac{nRT_i}{V_i}(V_f - V_i)$$

$$= -\frac{(1.0 \text{ mol})(8.31 \text{ J/mol} \cdot \text{K})(273 \text{ K})}{10.0 \times 10^{-3} \text{ m}^3}(3.0 \times 10^{-3} \text{ m}^3 - 10.0 \times 10^{-3} \text{ m}^3)$$

$$= \boxed{1.6 \times 10^3 \text{ J}}$$

We used the initial temperature and volume to calculate the work done because the final temperature was unknown. The work done on the gas is positive because the gas is being compressed.

Example **19.6**   Boiling Water

Suppose 1.00 g of water vaporizes isobarically at atmospheric pressure (1.013×10^5 Pa). Its volume in the liquid state is $V_i = V_{\text{liquid}} = 1.00 \text{ cm}^3$, and its volume in the vapor state is $V_f = V_{\text{vapor}} = 1\,671 \text{ cm}^3$. Find the work done in the expansion and the change in internal energy of the system. Ignore any mixing of the steam and the surrounding air; imagine that the steam simply pushes the surrounding air out of the way.

SOLUTION

Conceptualize   Notice that the temperature of the system does not change. There is a phase change occurring as the water evaporates to steam.

Categorize   Because the expansion takes place at constant pressure, we categorize the process as isobaric. We will use equations developed in the preceding sections, so we categorize this example as a substitution problem.

Use Equation 19.13 to find the work done on the system as the air is pushed out of the way:

$$W = -P(V_f - V_i)$$

$$= -(1.013 \times 10^5 \text{ Pa})(1\,671 \times 10^{-6} \text{ m}^3 - 1.00 \times 10^{-6} \text{ m}^3)$$

$$= \boxed{-169 \text{ J}}$$

Use Equation 19.8 and the latent heat of vaporization for water to find the energy transferred into the system by heat:

$$Q = L_v \Delta m_s = m_s L_v = (1.00 \times 10^{-3} \text{ kg})(2.26 \times 10^6 \text{ J/kg})$$

$$= 2\,260 \text{ J}$$

Use the first law to find the change in internal energy of the system:

$$\Delta E_{\text{int}} = Q + W = 2\,260 \text{ J} + (-169 \text{ J}) = \boxed{2.09 \text{ kJ}}$$

The positive value for ΔE_{int} indicates that the internal energy of the system increases. The largest fraction of the energy ($2\,090 \text{ J}/\,2\,260 \text{ J} = 93\%$) transferred to the liquid goes into increasing the internal energy of the system. The remaining 7% of the energy transferred leaves the system by work done by the steam on the surrounding atmosphere.

19.6   Energy Transfer Mechanisms in Thermal Processes

In Chapter 8, we introduced a global approach to the energy analysis of physical processes through Equation 8.2, where the energy transfer on the right hand side of the equation can occur by several mechanisms. Earlier in this chapter, we discussed two of the terms on the right side of this equation, work W and heat Q. In this section, we explore more details about heat as a means of energy transfer and two other energy transfer methods often related to temperature changes: convection (a form of matter transfer T_{MT}) and electromagnetic radiation T_{ER}.

Thermal Conduction

The process of energy transfer by heat (Q in Eq. 8.2) can also be called **conduction** or **thermal conduction.** In this process, the transfer can be represented on an atomic scale as an exchange of kinetic energy between microscopic particles—molecules, atoms, and free electrons—in which less-energetic particles gain energy in collisions with more-energetic particles. For example, if you hold one end of a long metal bar and insert the other end into a flame, you will find that the temperature of the metal in your hand soon increases. The energy reaches your hand by means of conduction. Initially, before the rod is inserted into the flame, the microscopic particles in the metal are vibrating about their equilibrium positions. As the flame raises the temperature of the rod, the particles near the flame begin to vibrate with greater and greater amplitudes. These particles, in turn, collide with their neighbors and transfer some of their energy in the collisions. Slowly, the amplitudes of vibration of metal atoms and electrons farther and farther from the flame increase until eventually those in the metal near your hand are affected. This increased vibration is detected by an increase in the temperature of the metal and of your potentially burned hand.

The rate of thermal conduction through a material depends on the properties of the material. For example, it is possible to hold a piece of asbestos in a flame indefinitely, which implies that very little energy is conducted through the asbestos. In general, metals are good thermal conductors and materials such as asbestos, cork, paper, and fiberglass are poor conductors. Gases also are poor conductors because the separation distance between the particles is so great. Metals are good thermal conductors because they contain large numbers of electrons that are relatively free to move through the metal and so can transport energy over large distances. Therefore, in a good conductor such as copper, conduction takes place by means of both the vibration of atoms and the motion of free electrons. The presence of free electrons in metals is also the reason that metals are good *electrical* conductors. We will study electrical conduction in metals in Chapter 26.

Conduction occurs only if there is a difference in temperature between two parts of the conducting medium. Consider a slab of material of thickness L and cross-sectional area A. One face of the slab is at a temperature T_c, and the other face is at a temperature $T_h > T_c$ (Fig. 19.13). Experimentally, it is found that energy Q transfers in a time interval Δt from the hotter face to the colder one. The energy Q that transfers is found to be proportional to the cross-sectional area, the temperature difference $\Delta T = T_h - T_c$, and the time interval, and inversely proportional to the thickness:

$$Q = kA\frac{\Delta T}{L}\Delta t \qquad (19.16)$$

where the proportionality constant k is the **thermal conductivity** of the material.

For a slab of infinitesimal thickness dx and temperature difference dT, we can write the **law of thermal conduction** as

$$P = kA\left|\frac{dT}{dx}\right| \qquad (19.17)$$

where $|dT/dx|$ is the **temperature gradient** (the rate at which temperature varies with position). Notice that P has units of watts when Q is in joules and Δt is in seconds. That is not surprising because P is power, the rate of energy transfer by heat.

Substances that are good thermal conductors have large thermal conductivity values, whereas good thermal insulators have low thermal conductivity values. Table 19.3 lists thermal conductivities for various substances. Notice that metals are generally better thermal conductors than nonmetals.

The opposite faces are at different temperatures where $T_h > T_c$.

Energy transfer for $T_h > T_c$

Figure 19.13 Energy transfer through a conducting slab with a cross-sectional area A and a thickness L.

TABLE 19.3 Thermal Conductivities

Substance	Thermal Conductivity (W/m · °C)
Metals (at 25°C)	
Aluminum	238
Copper	397
Gold	314
Iron	79.5
Lead	34.7
Silver	427
Nonmetals (approximate values)	
Asbestos	0.08
Concrete	0.8
Diamond	2 300
Glass	0.8
Ice	2
Rubber	0.2
Water	0.6
Wood	0.08
Gases (at 20°C)	
Air	0.023 4
Helium	0.138
Hydrogen	0.172
Nitrogen	0.023 4
Oxygen	0.023 8

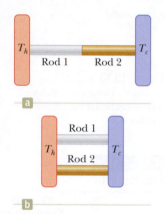

The opposite ends of the rod are in thermal contact with energy reservoirs at different temperatures.

Figure 19.14 Conduction of energy through a uniform, insulated rod of length L.

Suppose a long, uniform rod of length L is thermally insulated so that energy cannot escape by heat from its surface except at the ends as shown in Figure 19.14. One end is in thermal contact with an energy reservoir at temperature T_c, and the other end is in thermal contact with a reservoir at temperature $T_h > T_c$. When a steady state has been reached, the temperature at each point along the rod is constant in time. In this case, if we assume k is not a function of temperature, the temperature gradient is the same everywhere along the rod and is

$$\left| \frac{dT}{dx} \right| = \frac{T_h - T_c}{L}$$

Therefore, the rate of energy transfer by conduction through the rod is

$$P = kA\left(\frac{T_h - T_c}{L} \right) \tag{19.18}$$

For a compound slab containing several materials of thicknesses L_1, L_2, ... and thermal conductivities k_1, k_2, ..., the rate of energy transfer through the slab at steady state is

$$P = \frac{A(T_h - T_c)}{\sum_i (L_i/k_i)} \tag{19.19}$$

where T_h and T_c are the temperatures of the outer surfaces (which are held constant) and the summation is over all slabs. Example 19.7 shows how Equation 19.19 results from a consideration of two thicknesses of materials.

Figure 19.15 (Quick Quiz 19.5) In which case is the rate of energy transfer larger?

QUICK QUIZ 19.5 You have two rods of the same length and diameter, but they are formed from different materials. The rods are used to connect two regions at different temperatures so that energy transfers through the rods by heat. They can be connected in series as in Figure 19.15a or in parallel as in Figure 19.15b. In which case is the rate of energy transfer by heat larger? **(a)** The rate is larger when the rods are in series. **(b)** The rate is larger when the rods are in parallel. **(c)** The rate is the same in both cases.

Example 19.7 **Energy Transfer Through Two Slabs**

Two slabs of thickness L_1 and L_2 and thermal conductivities k_1 and k_2 are in thermal contact with each other as shown in Figure 19.16. The temperatures of their outer surfaces are T_c and T_h, respectively, and $T_h > T_c$. Determine the temperature at the interface and the rate of energy transfer by conduction through an area A of the slabs in the steady-state condition.

SOLUTION

Conceptualize Notice the phrase "in the steady-state condition." We interpret this phrase to mean that energy transfers through the compound slab at the same rate at all points. Otherwise, energy would be building up or disappearing at some point. Furthermore, the temperature varies with position in the two slabs, most likely at different rates in each part of the compound slab. When the system is in steady state, the interface is at some fixed temperature T.

Categorize We categorize this example as a thermal conduction problem and impose the condition that the power is the same in both slabs of material.

Figure 19.16 (Example 19.7) Energy transfer by conduction through two slabs in thermal contact with each other. At steady state, the rate of energy transfer through slab 1 equals the rate of energy transfer through slab 2.

19.7 continued

Analyze   Use Equation 19.18 to express the rate at which energy is transferred through an area A of slab 1:

$$(1) \quad P_1 = k_1 A \left(\frac{T - T_c}{L_1} \right)$$

Express the rate at which energy is transferred through the same area of slab 2:

$$(2) \quad P_2 = k_2 A \left(\frac{T_h - T}{L_2} \right)$$

Set these two rates equal to represent the steady-state situation:

$$k_1 A \left(\frac{T - T_c}{L_1} \right) = k_2 A \left(\frac{T_h - T}{L_2} \right)$$

Solve for T:

$$(3) \quad T = \frac{k_1 L_2 T_c + k_2 L_1 T_h}{k_1 L_2 + k_2 L_1}$$

Substitute Equation (3) into either Equation (1) or Equation (2):

$$(4) \quad P = \frac{A(T_h - T_c)}{(L_1/k_1) + (L_2/k_2)}$$

Finalize   Extension of this procedure to several slabs of materials leads to Equation 19.19. Equation (4) is Equation 19.19 with i ranging from 1 to 2.

WHAT IF?   Suppose you are building an insulated container with two layers of insulation and the rate of energy transfer determined by Equation (4) turns out to be too high. You have enough room to increase the thickness of one of the two layers by 20%. How would you decide which layer to choose?

Answer   To decrease the power as much as possible, you must increase the denominator in Equation (4) as much as possible. Whichever thickness you choose to increase, L_1 or L_2, you increase the corresponding term L/k in the denominator by 20%. For this percentage change to represent the largest absolute change, you want to take 20% of the larger term. Therefore, you should increase the thickness of the layer that has the larger value of L/k.

Home Insulation

In engineering practice, the term L/k for a particular substance is referred to as the **R-value** of the material. Therefore, Equation 19.19 reduces to

$$P = \frac{A(T_h - T_c)}{\sum_i R_i} \tag{19.20}$$

where $R_i = L_i/k_i$. The R-values for a few common building materials are given in Table 19.4. In the United States, the insulating properties of materials used in buildings are usually expressed in U.S. customary units, not SI units. Therefore,

TABLE 19.4   R-Values for Some Common Building Materials

Material	R-value (ft² · °F · h/Btu)
Hardwood siding (1 in. thick)	0.91
Wood shingles (lapped)	0.87
Brick (4 in. thick)	4.00
Concrete block (filled cores)	1.93
Fiberglass insulation (3.5 in. thick)	10.90
Fiberglass insulation (6 in. thick)	18.80
Fiberglass board (1 in. thick)	4.35
Cellulose fiber (1 in. thick)	3.70
Flat glass (0.125 in. thick)	0.89
Insulating glass (0.25-in. space)	1.54
Air space (3.5 in. thick)	1.01
Stagnant air layer	0.17
Drywall (0.5 in. thick)	0.45
Sheathing (0.5 in. thick)	1.32

in Table 19.4, *R*-values are given as a combination of British thermal units, feet, hours, and degrees Fahrenheit.

At any vertical surface open to the air, a very thin stagnant layer of air adheres to the surface. One must consider this layer when determining the *R*-value for a wall. The thickness of this stagnant layer on an outside wall depends on the speed of the wind. Energy transfer through the walls of a house on a windy day is greater than that on a day when the air is calm. A representative *R*-value for this stagnant layer of air is given in Table 19.4.

Example **19.8**   **The *R*-Value of a Typical Wall**

Calculate the total *R*-value for a wall constructed as shown in Figure 19.17a. Starting outside the house (toward the front in the figure) and moving inward, the wall consists of 4 in. of brick, 0.5 in. of sheathing, an air space 3.5 in. thick, and 0.5 in. of drywall.

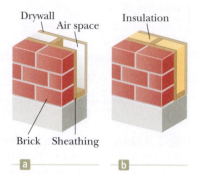

Figure 19.17 (Example 19.8) An exterior house wall containing (a) an air space and (b) insulation.

S O L U T I O N

Conceptualize Use Figure 19.17 to help conceptualize the structure of the wall. Do not forget the stagnant air layers inside and outside the house.

Categorize We will use specific equations developed in this section on home insulation, so we categorize this example as a substitution problem.

Use Table 19.4 to find the *R*-value of each layer:

$$R_1 \text{ (outside stagnant air layer)} = 0.17 \text{ ft}^2 \cdot {}^{\circ}\text{F} \cdot \text{h/Btu}$$

$$R_2 \text{ (brick)} = 4.00 \text{ ft}^2 \cdot {}^{\circ}\text{F} \cdot \text{h/Btu}$$

$$R_3 \text{ (sheathing)} = 1.32 \text{ ft}^2 \cdot {}^{\circ}\text{F} \cdot \text{h/Btu}$$

$$R_4 \text{ (air space)} = 1.01 \text{ ft}^2 \cdot {}^{\circ}\text{F} \cdot \text{h/Btu}$$

$$R_5 \text{ (drywall)} = 0.45 \text{ ft}^2 \cdot {}^{\circ}\text{F} \cdot \text{h/Btu}$$

$$R_6 \text{ (inside stagnant air layer)} = 0.17 \text{ ft}^2 \cdot {}^{\circ}\text{F} \cdot \text{h/Btu}$$

Add the *R*-values to obtain the total *R*-value for the wall:

$$R_{\text{total}} = R_1 + R_2 + R_3 + R_4 + R_5 + R_6 = 7.12 \text{ ft}^2 \cdot {}^{\circ}\text{F} \cdot \text{h/Btu}$$

W H AT I F ?   Suppose you are not happy with this total *R*-value for the wall. You cannot change the overall structure, but you can fill the air space as in Figure 19.17b. To *maximize* the total *R*-value, what material should you choose to fill the air space?

Answer   Looking at Table 19.4, we see that 3.5 in. of fiberglass insulation is more than ten times as effective as 3.5 in. of air. Therefore, we should fill the air space with fiberglass insulation. The result is that we add 10.90 ft^2 · °F · h/Btu of *R*-value, and we lose 1.01 ft^2 · °F · h/Btu due to the air space we have replaced. The new total *R*-value is equal to 7.12 ft^2 · °F · h/Btu + 9.89 ft^2 · °F · h/Btu = 17.01 ft^2 · °F · h/Btu.

Convection

At one time or another, you may have warmed your hands on a cold day by holding them over a toaster while it is operating. In this situation, the air in the toaster is warmed and expands. As a result, the density of this air decreases and the air rises. This hot air warms your hands as it flows by. Energy transferred by the movement of a warm substance is said to have been transferred by **convection,** which is a form of matter transfer, T_{MT} in Equation 8.2. When resulting from differences in density, as with air in the toaster, the process is referred to as *natural convection*. Airflow at an ocean coast (Section 19.2) is an example of natural convection, as is the mixing that occurs as surface water in a lake cools and sinks (see Section 18.4). When the heated substance is forced to move by a fan or pump, as in some hot-air and hot-water heating systems, the process is called *forced convection*.

If it were not for convection currents, it would be very difficult to boil water. As water is heated in a teakettle, the lower layers are warmed first. This water expands and rises to the top because its density is lowered. At the same time, the denser, cool water at the surface sinks to the bottom of the kettle and is heated.

Radiation

The third means of energy transfer we shall discuss is **electromagnetic radiation,** T_{ER} in Equation 8.2. All objects radiate energy continuously in the form of electromagnetic waves (see Chapter 33) produced by thermal vibrations of the molecules. You are likely familiar with electromagnetic radiation in the form of the orange glow from an electric stove burner or an electric space heater. In the toaster mentioned in the section on convection, energy reaches the bread by electromagnetic radiation from the glowing coils, which you can see if you look downward into the toaster.

The rate at which the surface of an object radiates energy is proportional to the fourth power of the absolute temperature of the surface. Known as **Stefan's law,** this behavior is expressed in equation form as

$$P = \sigma A e T^4 \qquad\qquad (19.21)$$

◀ Stefan's law

where P is the power in watts of electromagnetic waves radiated from the surface of the object, σ is a constant equal to $5.669\ 6 \times 10^{-8}$ W/m$^2 \cdot$ K^4, A is the surface area of the object in square meters, e is the **emissivity,** and T is the surface temperature in kelvins. The value of e can vary between zero and unity depending on the properties of the surface of the object. The emissivity is equal to the **absorptivity,** which is the fraction of the incoming radiation that the surface absorbs. A mirror has very low absorptivity because it reflects almost all incident light. Therefore, a mirror surface also has a very low emissivity. At the other extreme, a black surface has high absorptivity and high emissivity. An **ideal absorber** is defined as an object that absorbs all the energy incident on it, and for such an object, $e = 1$. An object for which $e = 1$ is often referred to as a **black body.** We shall investigate experimental and theoretical approaches to radiation from a black body in Chapter 39.

Every second, approximately 1 370 J of electromagnetic radiation from the Sun passes perpendicularly through each 1 m^2 at the top of the Earth's atmosphere. This radiation is primarily visible, infrared, and ultraviolet. We shall study these types of radiation in detail in Chapter 33. Enough energy arrives at the surface of the Earth each day to supply all our energy needs on this planet hundreds of times over, if only it could be captured and used efficiently. The growth in the number of solar energy–powered houses and solar energy "farms" in the world reflects the increasing efforts being made to use this abundant energy.

What happens to the atmospheric temperature at night is another example of the effects of energy transfer by radiation. If there is a cloud cover above the Earth, the water vapor in the clouds absorbs part of the infrared radiation emitted by the Earth and re-emits it back to the surface. Consequently, temperature levels at the surface remain moderate. In the absence of this cloud cover, there is less in the way to prevent this radiation from escaping into space; therefore, the temperature decreases more on a clear night than on a cloudy one.

As an object radiates energy at a rate given by Equation 19.21, it also absorbs electromagnetic radiation from the surroundings, which consist of other objects that radiate energy. If the latter process did not occur, an object would eventually radiate all its energy and its temperature would reach absolute zero. If an object is at a temperature T and its surroundings are at an average temperature T_0, the net rate of energy gained or lost by the object as a result of radiation is

$$P_{net} = \sigma Ae(T^4 - T_0^4) \tag{19.22}$$

When an object is in equilibrium with its surroundings, it radiates and absorbs energy at the same rate and its temperature remains constant. When an object is hotter than its surroundings, it radiates more energy than it absorbs and its temperature decreases.

Let's revisit your trip to the mountain in the opening storyline. Your first experience was seeing a sign, "Caution: Bridge Freezes Before Road Surface." A major contribution to this effect is that a roadway on the ground has energy transferring to it by heat Q from the warm ground underneath the roadway. A bridge has cold air underneath it, so it does not have this source of energy. Another factor is that the bridge roadway can radiate energy T_{ER} into the air from both upper and lower surfaces, losing internal energy E_{int} more rapidly than a roadway on the ground. Therefore, the bridge cools faster than the roadway, and water freezes on the bridge first. Your next thought was why mountain air is cold even though you are closer to the Sun. The change in distance to the Sun is miniscule compared to the distance to the Sun; that has no effect. Imagine air moving up to the mountain from sea level by convection T_{MT}. Because air is a poor thermal conductor, as a parcel of air moves from high pressure surroundings at sea level to lower-pressure surroundings on the mountain, it undergoes an adiabatic expansion, $Q = 0$. As mentioned in Section 19.5, an adiabatic expansion causes the temperature of the air to decrease: cold air at high altitudes.

Now, why was your meal on the mountain so unsuccessful? The eggs were undercooked. At lower atmospheric pressure on the mountain, the phase transition from water to steam takes place at a lower temperature. Therefore, you cooked your eggs at a temperature lower than 100°C, and they did not cook completely in the three-minute time interval. At higher altitudes, you need to boil food longer. The cookies near the edge of the baking sheet were too well done. The baking sheet is an object with a high temperature, so it radiates energy T_{ER} perpendicularly away from its surface. If the baking sheet has turned-up edges, this perpendicular direction from the edges is toward the cookies near the edge. Therefore, these cookies receive more energy by radiation than those near the center of the baking sheet, and they bake faster, even possibly burning while the ones near the center are perfect. Why did your cake fall? High-altitude baking is an art and requires careful adjustment of ingredients for successful cakes. One consideration in baking cakes is again the reduced boiling point of water. As a result, when the cake batter is placed in the oven, the water evaporates more rapidly than at sea level. As the too-dry batter rises, it cannot form "bubbles" of steam that build the regular cellular structure that supports the weight of the upper part of the cake.

After waking up the next morning and going for a walk, you noticed frost on cars and mailboxes, but only on the top surfaces. This effect is a demonstration of Equation 19.22. The side surfaces of cars and mailboxes are emitting energy T_{ER} horizontally. These surfaces are also absorbing radiation T_{ER} from other surrounding objects: houses, trees, other cars, and so on. As a result, the temperature of the side surfaces is relatively high, and the frost melts. On the other hand, upward-facing surfaces on top of the cars and mailboxes are radiating energy T_{ER} upward, but above them is open sky. There are no objects radiating energy downward into the top surfaces. As a result, the top surfaces are colder, and the frost doesn't melt as soon as that on the sides.

Notice that all of these effects involve transfers of energy like those discussed in this chapter and especially in this section. The only effect that does not depend on altitude is that of the well-done cookies near the edge of the baking sheet. There are many such thermal effects all around you: look for others!

The Dewar Flask

The *Dewar flask*[6] is a container designed to minimize energy transfers by conduction, convection, and radiation. Such a container is used to store cold or hot liquids for long periods of time. (An insulated bottle, such as a Thermos, is a common household equivalent of a Dewar flask.) The standard construction (Fig. 19.18) consists of a double-walled Pyrex glass vessel with silvered walls. The space between the walls is evacuated to minimize energy transfer by conduction and convection. The silvered surfaces minimize energy transfer by radiation because silver is a very good reflector and has very low emissivity. A further reduction in energy loss is obtained by reducing the size of the neck. Dewar flasks are commonly used to store liquid nitrogen (boiling point 77 K) and liquid oxygen (boiling point 90 K).

To confine liquid helium (boiling point 4.2 K), which has a very low heat of vaporization, it is often necessary to use a double Dewar system in which the Dewar flask containing the liquid is surrounded by a second Dewar flask. The space between the two flasks is filled with liquid nitrogen.

Newer designs of storage containers use "superinsulation" that consists of many layers of reflecting material separated by fiberglass. All this material is in a vacuum, and no liquid nitrogen is needed with this design.

---

[6]Invented by Sir James Dewar (1842–1923).

Figure 19.18 A cross-sectional view of a Dewar flask, which is used to store hot or cold substances.

Labels: Vacuum (white area); Hot or cold liquid; Silvered surfaces

Summary

▶ Definitions

Internal energy is a system's energy associated with its temperature and its physical state (solid, liquid, gas). Internal energy includes kinetic energy of random translation, rotation, and vibration of molecules; vibrational potential energy within molecules; and potential energy between molecules.

Heat is the process of energy transfer across the boundary of a system resulting from a temperature difference between the system and its surroundings. The symbol Q represents the amount of energy transferred by this process.

A **calorie** is the amount of energy necessary to raise the temperature of 1 g of water from 14.5°C to 15.5°C.

The **heat capacity** C of any sample is the amount of energy needed to raise the temperature of the sample by 1°C.

The **specific heat** c of a substance is the heat capacity per unit mass:

$$c \equiv \frac{Q}{m\,\Delta T} \qquad (19.3)$$

The **latent heat** of a substance is defined as the ratio of the energy input to a substance to the change in mass of the higher-phase material:

$$L \equiv \frac{Q}{\Delta m} \qquad (19.7)$$

▶ Concepts and Principles

The energy Q required to change the temperature of a mass m of a substance by an amount ΔT is

$$Q = mc\,\Delta T \qquad (19.4)$$

where c is the specific heat of the substance.

The energy required to change the phase of a pure substance is

$$Q = L\,\Delta m \qquad (19.8)$$

where L is the latent heat of the substance, which depends on the nature of the phase change and the substance, and Δm is the change in mass of the higher-phase material.

The **work** done on a gas as its volume changes from some initial value V_i to some final value V_f is

$$W = -\int_{V_i}^{V_f} P\,dV \qquad (19.10)$$

where P is the pressure of the gas, which may vary during the process. To evaluate W, the process must be fully specified; that is, P and V must be known during each step. The work done depends on the path taken between the initial and final states.

continued

The **first law of thermodynamics** is a specific reduction of the conservation of energy equation (Eq. 8.2) and states that when a system undergoes a change from one state to another, the change in its internal energy is

$$\Delta E_{int} = Q + W \tag{19.11}$$

where Q is the energy transferred into the system by heat and W is the work done on the system. Although Q and W both depend on the path taken from the initial state to the final state, the quantity ΔE_{int} does not depend on the path.

In a **cyclic process** (one that originates and terminates at the same state), $\Delta E_{int} = 0$ and therefore $Q = -W$. That is, the energy transferred into the system by heat equals the negative of the work done on the system during the process.

In an **adiabatic process,** no energy is transferred by heat between the system and its surroundings ($Q = 0$). In this case, the first law gives $\Delta E_{int} = W$.

An **isothermal process** is one that occurs at constant temperature. The work done on an ideal gas during an isothermal process is

$$W = nRT \ln \left(\frac{V_i}{V_f} \right) \tag{19.12}$$

An **isobaric process** is one that occurs at constant pressure. The work done on a gas in such a process is $W = -P(V_f - V_i)$.

An **isovolumetric process** is one that occurs at constant volume. No work is done in such a process, so $\Delta E_{int} = Q$.

Conduction can be viewed as an exchange of kinetic energy between colliding molecules or electrons. The rate of energy transfer by conduction through a slab of area A is

$$P = kA \left| \frac{dT}{dx} \right| \tag{19.17}$$

where k is the **thermal conductivity** of the material from which the slab is made and $|dT/dx|$ is the **temperature gradient.**

In **convection,** a warm substance transfers energy from one location to another.

All objects emit **electromagnetic radiation** in the form of electromagnetic waves at the rate

$$P = \sigma A e T^4 \tag{19.21}$$

Think–Pair–Share

See the Preface for an explanation of the icons used in this problems set. For additional assessment items for this section, go to **WEBASSIGN** From Cengage

1. Your team has been hired by a major builder who is designing simple homes for a new housing tract. He asks you to estimate the amount of natural gas that will be required to heat each house during the winter months. Figure TP19.1 shows the house you are currently working on. The average thermal conductivity of the walls (including the windows) and roof of the house depicted in the figure is 0.480 W/m · °C, and their average thickness is 21.0 cm. The heat of combustion (that is, the energy provided per cubic meter) of natural gas is 3.89×10^7 J/m³. (a) How many cubic meters of gas must be burned each day to maintain an inside temperature of 25.0°C in this house if the outside temperature is 0.0°C? Disregard radiation and the energy transferred by heat through the ground. (b) How will the answer to part (a) be affected (*increase* or *decrease* the gas requirements?) by the inclusion of (i) thermal conduction through the floor; (ii) radiation incident on the roof, walls, and windows during the daytime;

Figure TP19.1

(iii) operation of appliances, computers, entertainment systems; and (iv) leakage of air through cracks around doors and windows.

2. **ACTIVITY** Consider a spherical object of radius r with no atmosphere at a distance d from the Sun. Assume its emissivity is $e = 1$ for all kinds of electromagnetic waves and its temperature is uniform over its surface. At Earth's distance R from the Sun, the intensity of solar radiation is $I_S = 1\ 370$ W/m². This intensity varies as $1/d^2$ for distances other than R. A typical spherical object will *absorb* 70.0% of the solar radiation over its circular cross section πr^2. (The object will reflect about 30.0% of the incident radiation; the object appears circular when viewed from the Sun.) It will *emit* primarily infrared radiation from its entire surface area $4\pi r^2$. (a) Show that the equilibrium surface temperature of an object at a distance d from the Sun is

$$T = \left[\frac{(0.700) I_S}{4\sigma} \left(\frac{R}{d} \right)^2 \right]^{1/4} = (255 \text{ K}) \sqrt{\frac{R}{d}}$$

(b) Use the equation in part (a) to determine a theoretical surface temperature for the eight planets plus the dwarf planet Pluto, using the mean distance from the Sun given in Table 13.2. Also include the dwarf planet Ceres, at a distance of $d = 4.14 \times 10^{11}$ m from the Sun, for a total of ten objects in our solar system. (c) Make a bar graph of the temperatures found in part (c). (d) Add to your bar graph the measured and estimated surface temperatures, in kelvins, as provided by the Lunar and Planetary Institute, which are shown in the accompanying table. (e) Look first at our own planet, Earth. Is there a significance, in terms of life on this planet, to the fact that the theoretical temperature is below the freezing point of water, while the measured temperature is above it? (f) The actual temperature of Earth is raised by the atmospheric absorption of infrared radiation emitted from the surface. This effect is sometimes called the *greenhouse effect*. Consider the objects with the thinnest atmospheres: Mercury, Ceres, and Pluto. What do you notice about the comparison of theoretical and measured temperatures for these planets? (g) Consider the gas giants: Jupiter, Saturn, Uranus, and Neptune. These planets have no solid surface; the temperature data is provided for a point in the atmosphere where the pressure is the same as that at sea level on Earth. What do you notice about the comparison of theoretical and measured temperatures for these planets? (h) The clearest discrepancy between theoretical and measured temperatures in your graph is for Venus. Why is the measured temperature

so much higher than the theoretical temperature? (i) What can you conclude about the atmosphere of Mars from your graph?

Object	Surface Temperature (K) (from the Lunar and Planetary Institute)
Mercury	440
Venus	741
Earth	288
Mars	244
Ceres	173
Jupiter	165
Saturn	134
Uranus	77
Neptune	70
Pluto	40

Problems

See the Preface for an explanation of the icons used in this problems set.
For additional assessment items for this section, go to **WEBASSIGN**
From Cengage

SECTION 19.1 Heat and Internal Energy

1. A 55.0-kg woman eats a 540 Calorie (540 kcal) jelly doughnut for breakfast. (a) How many joules of energy are the equivalent of one jelly doughnut? (b) How many steps must the woman climb on a very tall stairway to change the gravitational potential energy of the woman–Earth system by a value equivalent to the food energy in one jelly doughnut? Assume the height of a single stair is 15.0 cm. (c) If the human body is only 25.0% efficient in converting chemical potential energy to mechanical energy, how many steps must the woman climb to work off her breakfast?

BIO

SECTION 19.2 Specific Heat and Calorimetry

2. The highest waterfall in the world is the Salto Angel in Venezuela. Its longest single falls has a height of 807 m. If water at the top of the falls is at 15.0°C, what is the maximum temperature of the water at the bottom of the falls? Assume all the kinetic energy of the water as it reaches the bottom goes into raising its temperature.

3. A combination of 0.250 kg of water at 20.0°C, 0.400 kg of aluminum at 26.0°C, and 0.100 kg of copper at 100°C is mixed in an insulated container and allowed to come to thermal equilibrium. Ignore any energy transfer to or from the container. What is the final temperature of the mixture?

4. The temperature of a silver bar rises by 10.0°C when it absorbs 1.23 kJ of energy by heat. The mass of the bar is 525 g. Determine the specific heat of silver from these data.

T

5. You are working in your kitchen preparing lunch for your family. You have decided to make egg salad sandwiches and are boiling six eggs, each of mass 55.5 g, in 0.750 L of water at 100°C. You wish to take all the eggs out of the boiling water and immediately place them in 23.0°C water to cool them down to a comfortable temperature to hold them

CR

and peel them. You decide that you wish the mixture of the water and the eggs to reach an equilibrium temperature of 40.0°C. Explaining this to a family member, she challenges you to determine *exactly* how much water at 23.0°C you need to achieve your desired equilibrium temperature. Take the average specific heat of an egg over the expected temperature range to be 3.27×10^3 J/kg · °C.

6. If water with a mass m_h at temperature T_h is poured into an aluminum cup of mass m_{Al} containing mass m_c of water at T_c, where $T_h > T_c$, what is the equilibrium temperature of the system?

S

7. An aluminum calorimeter with a mass of 100 g contains 250 g of water. The calorimeter and water are in thermal equilibrium at 10.0°C. Two metallic blocks are placed into the water. One is a 50.0-g piece of copper at 80.0°C. The other has a mass of 70.0 g and is originally at a temperature of 100°C. The entire system stabilizes at a final temperature of 20.0°C. (a) Determine the specific heat of the unknown sample. (b) Using the data in Table 19.1, can you make a positive identification of the unknown material? Can you identify a possible material? (c) Explain your answers for part (b).

Q|C
V

8. An electric drill with a steel drill bit of mass $m = 27.0$ g and diameter 0.635 cm is used to drill into a cubical steel block of mass $M = 240$ g. Assume steel has the same properties as iron. The cutting process can be modeled as happening at one point on the circumference of the bit. This point moves in a helix at constant tangential speed 40.0 m/s and exerts a force of constant magnitude 3.20 N on the block. As shown in Figure P19.8 (page 528), a groove in the bit carries the chips up to the top of the block, where they form a pile around the hole. The drill is turned on and drills into the block for a time interval of 15.0 s. Let's assume this time interval is long enough for conduction within the steel to bring it all to a uniform temperature. Furthermore, assume the steel objects lose a negligible amount of energy by conduction, convection, and radiation into their environment. (a) Suppose the

Q|C

drill bit cuts three-quarters of the way through the block during 15.0 s. Find the temperature change of the whole quantity of steel. (b) **What If?** Now suppose the drill bit is dull and cuts only one-eighth of the way through the block in 15.0 s. Identify the temperature change of the whole quantity of steel in this case. (c) What pieces of data, if any, are unnecessary for the solution? Explain.

Figure P19.8

9. A 3.00-g copper coin at 25.0°C drops 50.0 m to the ground.
[QC] (a) Assuming 60.0% of the change in gravitational potential energy of the coin–Earth system goes into increasing the internal energy of the coin, determine the coin's final temperature. (b) **What If?** Does the result depend on the mass of the coin? Explain.

SECTION 19.3 Latent Heat

10. How much energy is required to change a 40.0-g ice cube
[V] from ice at −10.0°C to steam at 110°C?

11. A 75.0-kg cross-country skier
[T] glides over snow as in Figure P19.11. The coefficient of friction between skis and snow is 0.200. Assume all the snow beneath his skis is at 0°C and that all the internal energy generated by friction is added to snow, which sticks to his skis until it melts. How far would he have to ski to melt 1.00 kg of snow?

Figure P19.11

12. A 3.00-g lead bullet at
[AMT] 30.0°C is fired at a speed of
[T] 240 m/s into a large block of ice at 0°C, in which it becomes embedded. What quantity of ice melts?

13. In an insulated vessel, 250 g of ice at 0°C is added to 600 g of water at 18.0°C. (a) What is the final temperature of the system? (b) How much ice remains when the system reaches equilibrium?

14. An automobile has a mass of 1 500 kg, and its aluminum
[QC] brakes have an overall mass of 6.00 kg. (a) Assume all the mechanical energy that transforms into internal energy when the car stops is deposited in the brakes and no energy is transferred out of the brakes by heat. The brakes are originally at 20.0°C. How many times can the car be stopped

from 25.0 m/s before the brakes start to melt? (b) Identify some effects ignored in part (a) that are important in a more realistic assessment of the warming of the brakes.

SECTION 19.4 Work in Thermodynamic Processes

15. One mole of an ideal gas is warmed slowly so that it goes
[QC] from the PV state (P_i, V_i) to $(3P_i, 3V_i)$ in such a way that
[S] the pressure of the gas is directly proportional to the volume. (a) How much work is done on the gas in the process? (b) How is the temperature of the gas related to its volume during this process?

16. (a) Determine the work done on a gas that expands from i
[V] to f as indicated in Figure P19.16. (b) **What If?** How much work is done on the gas if it is compressed from f to i along the same path?

Figure P19.16

SECTION 19.5 The First Law of Thermodynamics

17. A thermodynamic system undergoes a process in which its internal energy decreases by 500 J. Over the same time interval, 220 J of work is done on the system. Find the energy transferred from it by heat.

18. *Why is the following situation impossible?* An ideal gas undergoes a process with the following parameters: $Q = 10.0$ J, $W = 12.0$ J, and $\Delta T = -2.00$°C.

19. A 2.00-mol sample of helium gas initially at 300 K, and
[T] 0.400 atm is compressed isothermally to 1.20 atm. Noting that the helium behaves as an ideal gas, find (a) the final volume of the gas, (b) the work done on the gas, and (c) the energy transferred by heat.

20. (a) How much work is done on the steam when 1.00 mol of water at 100°C boils and becomes 1.00 mol of steam at 100°C at 1.00 atm pressure? Assume the steam to behave as an ideal gas. (b) Determine the change in internal energy of the system of the water and steam as the water vaporizes.

21. A 1.00-kg block of aluminum is warmed at atmospheric pressure so that its temperature increases from 22.0°C to 40.0°C. Find (a) the work done on the aluminum, (b) the energy added to it by heat, and (c) the change in its internal energy.

22. In Figure P19.22, the change in internal energy of a gas that is taken from A to C along the blue path is +800 J. The work done on the gas along the red path ABC is −500 J. (a) How much energy must be added to the system by heat as it goes from A through B to C? (b) If the pressure at point A is five times that of point C, what is the work done on the system in going from C to D?

Figure P19.22

(c) What is the energy exchanged with the surroundings by heat as the gas goes from *C* to *A* along the green path? (d) If the change in internal energy in going from point *D* to point *A* is +500 J, how much energy must be added to the system by heat as it goes from point *C* to point *D*?

SECTION 19.6 Energy Transfer Mechanisms in Thermal Processes

23. A student is trying to decide what to wear. His bedroom is **BIO** at 20.0°C. His skin temperature is 35.0°C. The area of his exposed skin is 1.50 m². People all over the world have skin that is dark in the infrared, with emissivity about 0.900. Find the net energy transfer from his body by radiation in 10.0 min.

24. A concrete slab is 12.0 cm thick and has an area of 5.00 m². Electric heating coils are installed under the slab to melt the ice on the surface in the winter months. What minimum power must be supplied to the coils to maintain a temperature difference of 20.0°C between the bottom of the slab and its surface? Assume all the energy transferred is through the slab.

25. Two lightbulbs have cylindrical filaments much greater in length than in diameter. The evacuated bulbs are identical except that one operates at a filament temperature of 2 100°C and the other operates at 2 000°C. (a) Find the ratio of the power emitted by the hotter lightbulb to that emitted by the cooler lightbulb. (b) With the bulbs operating at the same respective temperatures, the cooler lightbulb is to be altered by making its filament thicker so that it emits the same power as the hotter one. By what factor should the radius of this filament be increased?

26. The human body must maintain its core temperature inside **BIO** a rather narrow range around 37°C. Metabolic processes, notably muscular exertion, convert potential energy into internal energy deep in the interior. From the interior, energy must flow out to the skin or lungs to be expelled to the environment. During moderate exercise, an 80-kg man can metabolize food energy at the rate 300 kcal/h, do 60 kcal/h of mechanical work, and put out the remaining 240 kcal/h of energy by heat. Most of the energy is carried from the body interior out to the skin by forced convection, whereby blood is warmed in the interior and then cooled at the skin, which is a few degrees cooler than the body core. Without blood flow, living tissue is a good thermal insulator, with thermal conductivity about 0.210 W/m · °C. Show that blood flow is essential to cool the man's body by calculating the rate of energy conduction in kcal/h through the tissue layer under his skin. Assume that its area is 1.40 m², its thickness is 2.50 cm, and it is maintained at 37.0°C on one side and at 34.0°C on the other side.

27. (a) Calculate the *R*-value of a thermal window made of two single panes of glass each 0.125 in. thick and separated by a 0.250-in. air space. (b) By what factor is the transfer of energy by heat through the window reduced by using the thermal window instead of the single-pane window? Include the contributions of inside and outside stagnant air layers.

28. For bacteriological testing of water supplies and in medi-
BIO cal clinics, samples must routinely be incubated for 24 h at
QC 37°C. Peace Corps volunteer and MIT engineer Amy Smith invented a low-cost, low-maintenance incubator. The incubator consists of a foam-insulated box containing a waxy

material that melts at 37.0°C interspersed among tubes, dishes, or bottles containing the test samples and growth medium (bacteria food). Outside the box, the waxy material is first melted by a stove or solar energy collector. Then the waxy material is put into the box to keep the test samples warm as the material solidifies. The heat of fusion of the phase-change material is 205 kJ/kg. Model the insulation as a panel with surface area 0.490 m², thickness 4.50 cm, and conductivity 0.012 0 W/m · °C. Assume the exterior temperature is 23.0°C for 12.0 h and 16.0°C for 12.0 h. (a) What mass of the waxy material is required to conduct the bacteriological test? (b) Explain why your calculation can be done without knowing the mass of the test samples or of the insulation.

ADDITIONAL PROBLEMS

29. Gas in a container is at a pressure of 1.50 atm and a vol-
T ume of 4.00 m³. What is the work done on the gas (a) if it expands at constant pressure to twice its initial volume, and (b) if it is compressed at constant pressure to one-quarter its initial volume?

30. You are reading your textbook on Greek mythology. You
CR find a story about Daedalus and Icarus. Daedalus built two sets of wings out of feathers and wax, one set for him and one for his son Icarus. The father and son planned to use the wings to escape from their imprisonment on the island of Crete. The father warned Icarus not to fly too high because the proximity to the Sun might melt the wax in his wings. Of course, Icarus was overtaken by the thrill of flying and flew too close to the Sun. His wings melted and he fell into the sea. While reading this information, you think about your physics class, where your instructor has just discussed the equilibrium temperature of an object with no atmosphere at a given distance from the Sun. You look in your notes and find the following equation for this equilibrium temperature:

$$T = (255 \text{ K})\sqrt{\frac{R}{r}}$$

where *R* is the distance from the Sun to the Earth, *r* is the distance from the Sun to the object, and *T* is in kelvins. This raises a conundrum in your mind: If Icarus flew so close to the Sun that the wax in his wings melted, would there still be air at that location to allow him to fly to that location? Take the melting point of wax to be 65°C.

31. You have a particular interest in automobile engines, so you
CR have secured a co-op position at an automobile company while you attend school. Your supervisor is helping you to learn about the operation of an internal combustion engine. She gives you the following assignment, related to a simulation of a new engine she is designing. A gas, beginning at $P_A = 1.00$ atm, $V_A = 0.500$ L, and $T_A = 27.0°C$, is compressed from point *A* on the *PV* diagram in Figure P19.31 (page 530) to point *B*. This represents the compression stroke in a four-cycle gasoline engine. At that point, 132 J of energy is delivered to the gas at constant volume, taking the gas to point *C*. This represents the transformation of potential energy in the gasoline to internal energy when the spark plug fires. Your supervisor tells you that the internal energy of a gas is proportional to temperature (as we shall find in Chapter 20), the internal energy of the gas at point *A* is 200 J, and she wants to know what the temperature of the gas is at point *C*.

Figure P19.31

32. You are working in a condensed-matter laboratory for your senior project. Several of the ongoing projects use liquid helium, which is contained in a thermally insulated vessel that can hold up to a maximum of V_{max} = 240 L of the liquid at T_c = 4.20 K. Because some of the liquid helium has already been used, someone asks you to check to see if there is enough for the next day, on which four different experimental groups will need liquid helium. You are not sure how to measure the amount of liquid remaining, so you insert an aluminum rod of length L = 2.00 m and with a cross-sectional area A = 2.50 cm² into the vessel. By seeing how much of the lower end of the rod is frosted when you pull it out, you can estimate the depth of the liquid helium. After inserting the rod, however, one of the experimenters calls you over to perform a task and you forget about the rod, leaving it in the liquid helium until the next morning. How much liquid helium is available for the next day's experiments? (Aluminum has thermal conductivity of 3 100 W/m · K at 4.20 K; ignore its temperature variation. The density of liquid helium is 125 kg/m³.) Assume that gaseous helium can escape from the top of the vessel.

33. A *flow calorimeter* is an apparatus used to measure the specific heat of a liquid. The technique of flow calorimetry involves measuring the temperature difference between the input and output points of a flowing stream of the liquid while energy is added by heat at a known rate. A liquid of density 900 kg/m³ flows through the calorimeter with volume flow rate of 2.00 L/min. At steady state, a temperature difference 3.50°C is established between the input and output points when energy is supplied at the rate of 200 W. What is the specific heat of the liquid?

34. A *flow calorimeter* is an apparatus used to measure the specific heat of a liquid. The technique of flow calorimetry involves measuring the temperature difference between the input and output points of a flowing stream of the liquid while energy is added by heat at a known rate. A liquid of density ρ flows through the calorimeter with volume flow rate R. At steady state, a temperature difference ΔT is established between the input and output points when energy is supplied at the rate P. What is the specific heat of the liquid?

35. **Review.** Following a collision between a large spacecraft and an asteroid, a copper disk of radius 28.0 m and thickness 1.20 m at a temperature of 850°C is floating in space, rotating about its symmetry axis with an angular speed of 25.0 rad/s. As the disk radiates infrared light, its temperature falls to 20.0°C. No external torque acts on the disk.

(a) Find the change in kinetic energy of the disk. (b) Find the change in internal energy of the disk. (c) Find the amount of energy it radiates.

36. **Review.** Two speeding lead bullets, one of mass 12.0 g moving to the right at 300 m/s and one of mass 8.00 g moving to the left at 400 m/s, collide head-on, and all the material sticks together. Both bullets are originally at temperature 30.0°C. Assume the change in kinetic energy of the system appears entirely as increased internal energy. We would like to determine the temperature and phase of the bullets after the collision. (a) What two analysis models are appropriate for the system of two bullets for the time interval from before to after the collision? (b) From one of these models, what is the speed of the combined bullets after the collision? (c) How much of the initial kinetic energy has transformed to internal energy in the system after the collision? (d) Does all the lead melt due to the collision? (e) What is the temperature of the combined bullets after the collision? (f) What is the phase of the combined bullets after the collision?

37. An ice-cube tray is filled with 75.0 g of water. After the filled tray reaches an equilibrium temperature of 20.0°C, it is placed in a freezer set at −8.00°C to make ice cubes. (a) Describe the processes that occur as energy is being removed from the water to make ice. (b) Calculate the energy that must be removed from the water to make ice cubes at −8.00°C.

38. The rate at which a resting person converts food energy is called one's *basal metabolic rate* (BMR). Assume that the resulting internal energy leaves a person's body by radiation and convection of dry air. When you jog, most of the food energy you burn above your BMR becomes internal energy that would raise your body temperature if it were not eliminated. Assume that evaporation of perspiration is the mechanism for eliminating this energy. Suppose a person is jogging for "maximum fat burning," converting food energy at the rate 400 kcal/h above his BMR, and putting out energy by work at the rate 60.0 W. Assume that the heat of evaporation of water at body temperature is equal to its heat of vaporization at 100°C. (a) Determine the hourly rate at which water must evaporate from his skin. (b) When you metabolize fat, the hydrogen atoms in the fat molecule are transferred to oxygen to form water. Assume that metabolism of 1.00 g of fat generates 9.00 kcal of energy and produces 1.00 g of water. What fraction of the water the jogger needs is provided by fat metabolism?

39. An iron plate is held against an iron wheel so that a kinetic friction force of 50.0 N acts between the two pieces of metal. The relative speed at which the two surfaces slide over each other is 40.0 m/s. (a) Calculate the rate at which mechanical energy is converted to internal energy. (b) The plate and the wheel each have a mass of 5.00 kg, and each receives 50.0% of the internal energy. If the system is run as described for 10.0 s and each object is then allowed to reach a uniform internal temperature, what is the resultant temperature increase?

40. One mole of an ideal gas is contained in a cylinder with a movable piston. The initial pressure, volume, and temperature are P_i, V_i, and T_i, respectively. Find the work done on the gas in the following processes. In operational terms, describe how to carry out each process and show each

process on a *PV* diagram. (a) an isobaric compression in which the final volume is one-half the initial volume (b) an isothermal compression in which the final pressure is four times the initial pressure (c) an isovolumetric process in which the final pressure is three times the initial pressure

41. During periods of high activity, the Sun has more sunspots
Q|C than usual. Sunspots are cooler than the rest of the luminous layer of the Sun's atmosphere (the photosphere). Paradoxically, the total power output of the active Sun is not lower than average but is the same or slightly higher than average. Work out the details of the following crude model of this phenomenon. Consider a patch of the photosphere with an area of 5.10×10^{14} m². Its emissivity is 0.965. (a) Find the power it radiates if its temperature is uniformly 5 800 K, corresponding to the quiet Sun. (b) To represent a sunspot, assume 10.0% of the patch area is at 4 800 K and the other 90.0% is at 5 890 K. Find the power output of the patch. (c) State how the answer to part (b) compares with the answer to part (a). (d) Find the average temperature of the patch. Note that this cooler temperature results in a higher power output.

42. *Why is the following situation impossible?* A group of campers arises at 8:30 a.m. and uses a solar cooker, which consists of a curved, reflecting surface that concentrates sunlight onto the object to be warmed (Fig. P19.42). During the day, the maximum solar intensity reaching the Earth's surface at the cooker's location is $I = 600$ W/m². The cooker faces the Sun and has a face diameter of $d = 0.600$ m. Assume a fraction f of 40.0% of the incident energy is transferred to 1.50 L of water in an open container, initially at 20.0°C. The water comes to a boil, and the campers enjoy hot coffee for breakfast before hiking ten miles and returning by noon for lunch.

Figure P19.42

43. A cooking vessel on a slow burner contains 10.0 kg of water and an unknown mass of ice in equilibrium at 0°C at time $t = 0$. The temperature of the mixture is measured at various times, and the result is plotted in Figure P19.43.

Figure P19.43

During the first 50.0 min, the mixture remains at 0°C. From 50.0 min to 60.0 min, the temperature increases to 2.00°C. Ignoring the heat capacity of the vessel, determine the initial mass of the ice.

44. A student measures the following data in a calorimetry
Q|C experiment designed to determine the specific heat of aluminum:

Initial temperature of water and calorimeter:	70.0°C
Mass of water:	0.400 kg
Mass of calorimeter:	0.040 kg
Specific heat of calorimeter:	0.63 kJ/kg · °C
Initial temperature of aluminum:	27.0°C
Mass of aluminum:	0.200 kg
Final temperature of mixture:	66.3°C

(a) Use these data to determine the specific heat of aluminum. (b) Explain whether your result is within 15% of the value listed in Table 19.1.

CHALLENGE PROBLEMS

45. (a) The inside of a hollow cylinder is maintained at a temperature T_a, and the outside is at a lower temperature, T_b (Fig. P19.45). The wall of the cylinder has a thermal conductivity k. Ignoring end effects, show that the rate of energy conduction from the inner surface to the outer surface in the radial direction is

$$\frac{dQ}{dt} = 2\pi L k \left[\frac{T_a - T_b}{\ln(b/a)} \right]$$

Suggestions: The temperature gradient is dT/dr. A radial energy current passes through a concentric cylinder of area $2\pi rL$. (b) The passenger section of a jet airliner is in the shape of a cylindrical tube with a length of 35.0 m and an inner radius of 2.50 m. Its walls are lined with an insulating material 6.00 cm in thickness and having a thermal conductivity of 4.00×10^{-5} cal/s · cm · °C. A heater must maintain the interior temperature at 25.0°C while the outside temperature is −35.0°C. What power must be supplied to the heater?

Figure P19.45

46. A spherical shell has inner radius 3.00 cm and outer radius
Q|C 7.00 cm. It is made of material with thermal conductivity $k = 0.800$ W/m · °C. The interior is maintained at temperature 5°C and the exterior at 40°C. After an interval of time, the shell reaches a steady state with the temperature at each point within it remaining constant in time. (a) Explain why

the rate of energy transfer P must be the same through each spherical surface, of radius r, within the shell and must satisfy

$$\frac{dT}{dr} = \frac{P}{4\pi k r^2}$$

(b) Next, prove that

$$\int_5^{40} dT = \frac{P}{4\pi k} \int_{0.03}^{0.07} r^{-2}\, dr$$

where T is in degrees Celsius and r is in meters. (c) Find the rate of energy transfer through the shell. (d) Prove that

$$\int_5^T dT = 1.84 \int_{0.03}^r r^{-2}\, dr$$

where T is in degrees Celsius and r is in meters. (e) Find the temperature within the shell as a function of radius. (f) Find the temperature at $r = 5.00$ cm, halfway through the shell.

47. A pond of water at $0°C$ is covered with a layer of ice 4.00 cm thick. If the air temperature stays constant at $-10.0°C$, what time interval is required for the ice thickness to increase to 8.00 cm? *Suggestion:* Use Equation 19.18 in the form

$$\frac{dQ}{dt} = kA \frac{\Delta T}{x}$$

and note that the incremental energy dQ extracted from the water through the thickness x of ice is the amount required to freeze a thickness dx of ice. That is, $dQ = L_f \rho A\, dx$, where ρ is the density of the ice, A is the area, and L_f is the latent heat of fusion.

The Kinetic Theory of Gases

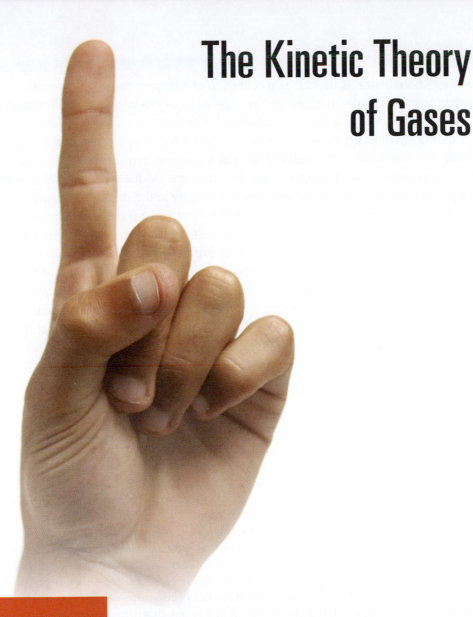

A wet finger is held upward to test the direction of the wind. Why is the finger cold on the side from which the wind is blowing? (*Joel Calheiros/Shutterstock*)

STORYLINE **You are still on the Physics Club camping trip to Whitney** Portal described at the beginning of Chapter 19. The evening plan is to build a campfire and huddle around it revisiting Physics Department stories. You are in charge of gathering the wood and setting up the fire. In order to locate the fire in the best place possible, you need to know the wind direction. As someone taught you years ago, you put your index finger in your mouth and then hold it vertically, knowing that the coldest side of your finger will be the direction from which the wind is coming. Your physics course kicks in again and you say, "Wait a minute! Why is that side of the finger cold in the wind?" While thinking about an answer, you reach down and grab a piece of wood. You scrape your finger painfully on it and go into the RV to receive medical treatment from your Club advisor. He puts some alcohol on the wound. Even though you know the alcohol is at the same temperature as the rest of the interior of the RV, the alcohol feels cold on your finger. Could the cold feeling of the alcohol be related to the cold feeling of your finger in the wind?

CONNECTIONS In Chapter 18, we discussed the properties of an ideal gas by using such *macroscopic* variables as pressure, volume, and temperature. Such large-scale properties can be related to a description of the gas on a *microscopic*

533

scale, where matter is treated as a collection of a huge number of molecules rather than as a single macroscopic sample. Applying Newton's laws of motion in a statistical manner to a collection of particles provides a reasonable description of thermodynamic processes. To keep the mathematics relatively simple, we shall consider primarily the behavior of gases because in gases the interactions between molecules are much weaker than they are in liquids or solids. We shall begin by relating pressure and temperature directly to the details of molecular motion in a sample of gas. Based on these results, we will make predictions of molar specific heats of gases. Some of these predictions will be correct and some will not. We will extend our model to explain those values that are not predicted correctly by the simpler model. Finally, we discuss the distribution of molecular speeds in a gas, and apply the results to a liquid. We shall find the concepts discussed in this chapter useful in the future when we analyze a situation on a microscopic scale, such as, for example, the analysis of the electrical characteristics of an *electron gas* in a conducting wire.

20.1   Molecular Model of an Ideal Gas

In Section 1.2, we introduced a number of types of models, one of which is the *structural model*. A structural model is a theoretical construct designed to represent a system that cannot be observed directly because it is too large or too small. For example, here on Earth we can only observe the solar system from the inside; we cannot travel outside the solar system and look back to see how it works. This restricted vantage point has led to the geocentric and heliocentric models of the solar system discussed in Section 13.4. An example of a system too small to observe directly is the hydrogen atom. Various structural models of this system have been developed, including the *Bohr model* (Section 41.3) and the *quantum model* (Section 41.4). Once a structural model is developed, its assumptions are used to make various predictions for experimental observations of the behavior of the system. For example, the geocentric model of the solar system makes predictions of how the movement of Mars should appear from the Earth. It turns out that those predictions do not match the actual observations. When this mismatch occurs with a structural model, the model must be modified or replaced with another model.

In this chapter, we will consider a structural model for an ideal gas, with the goal of relating *macroscopic* measurements of the gas (pressure, volume, temperature, etc.) to the behavior of its *microscopic* components—molecules. The structural model that we will develop is called **kinetic theory.** This model treats an ideal gas as a collection of molecules with the following assumptions:

1. *Physical components:*
   The gas consists of a number of identical molecules within a cubic container of side length d (Fig. 20.1). The number of molecules in the gas is large, and the average separation between them is large compared with their dimensions. Therefore, the molecules occupy a negligible volume in the container. This assumption is consistent with the ideal gas model, in which we imagine the molecules to be point-like.
2. *Behavior of the components:*
   (a) The molecules obey Newton's laws of motion, but as a whole their motion is isotropic: any molecule can move in any direction with any speed.
   (b) The molecules interact only by short-range forces during elastic collisions. This assumption is consistent with the ideal gas model, in which the molecules exert no long-range forces on one another.
   (c) The molecules make elastic collisions with the walls of the container.

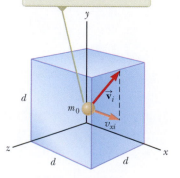

One molecule of the gas moves with velocity $\vec{\mathbf{v}}$ on its way toward a collision with the wall.

Figure 20.1 A cubical box with sides of length d containing an ideal gas.

Although we often picture an ideal gas as consisting of single atoms modeled as particles, the behavior of molecular gases approximates that of ideal gases rather well at low pressures. Usually, the internal structure of the molecule has no effect on the motions considered here.

For our first application of kinetic theory, let us relate the macroscope variable of pressure P to microscopic quantities. This will be a relatively long process, but each step will be based on a simple mathematical calculation, or on a principle or analysis model that we have studied in a previous chapter. Consider a collection of N molecules of an ideal gas in a container of volume V. As indicated in assumption 1, the container is a cube with edges of length d. We shall first focus our attention on one of these molecules of mass m_0 and assume it is moving so that its component of velocity in the x direction is v_{xi} as in Figure 20.1. (The subscript i here refers to the ith molecule in the collection, not to an initial value. We will combine the effects of all the molecules shortly.) Figure 20.2 shows the molecule making a collision with the wall of the container. As the molecule collides elastically with the wall (assumption 2(c) above), its velocity component perpendicular to the wall is reversed because the mass of the wall is far greater than the mass of the molecule. The molecule is modeled as a nonisolated system for which the impulse from the wall causes a change in the molecule's momentum. Because the momentum component p_{xi} of the molecule is $m_0 v_{xi}$ before the collision and $-m_0 v_{xi}$ after the collision, the change in the x component of the momentum of the molecule is

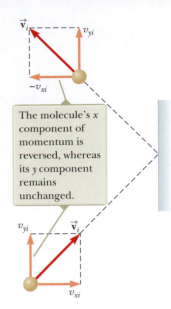

The molecule's x component of momentum is reversed, whereas its y component remains unchanged.

Figure 20.2 A molecule makes an elastic collision with the wall of the container. In this construction, we assume the molecule moves in the xy plane.

$$\Delta p_{xi} = -m_0 v_{xi} - (m_0 v_{xi}) = -2m_0 v_{xi} \tag{20.1}$$

From the nonisolated system model for momentum, we can apply the impulse-momentum theorem (Eqs. 9.11 and 9.13) to the molecule to give

$$\overline{F}_{i,\text{on molecule}} \Delta t_{\text{collision}} = \Delta p_{xi} = -2m_0 v_{xi} \tag{20.2}$$

where $\overline{F}_{i,\text{on molecule}}$ is the x component of the average force[1] the wall exerts on the molecule during the collision and $\Delta t_{\text{collision}}$ is the duration of the collision. For the molecule to make another collision with the same wall after this first collision, it must travel a distance of $2d$ in the x direction (across the cube and back). Therefore, from the particle under constant velocity model, the time interval between two collisions with the same wall is

$$\Delta t = \frac{2d}{v_{xi}} \tag{20.3}$$

The force that causes the change in momentum of the molecule in the collision with the wall occurs only during the collision. We can, however, find the long-term average force for many back-and-forth trips across the cube by averaging the force in Equation 20.2 over the time interval for the molecule to move across the cube and back, Equation 20.3. The average change in momentum per trip for the time interval for many trips is the same as that for the short duration of the collision. Therefore, we can rewrite Equation 20.2 as

$$\overline{F}_i \Delta t = -2m_0 v_{xi} \tag{20.4}$$

where $\overline{F}_i$ is the average force component over the time interval Δt for the molecule to move across the cube and back. Because exactly one collision occurs with the given wall for each such time interval, this result is also the long-term average force on the molecule over long time intervals containing any number of multiples of Δt.

---

[1]For this discussion, we use a bar over a variable to represent the average value of the variable, such as $\overline{F}$ for the average force, rather than the subscript "avg" that we have used before. This notation is to save confusion because we already have a number of subscripts on variables.

Equations 20.3 and 20.4 enable us to express the x component of the long-term average force exerted by the wall on the molecule as

$$\overline{F}_i = -\frac{2m_0 v_{xi}}{\Delta t} = -\frac{2m_0 v_{xi}^2}{2d} = -\frac{m_0 v_{xi}^2}{d} \qquad (20.5)$$

Now, by Newton's third law, the x component of the long-term average force exerted by the *molecule* on the *wall* is equal in magnitude and opposite in direction:

$$\overline{F}_{i,\text{on wall}} = -\overline{F}_i = -\left(-\frac{m_0 v_{xi}^2}{d} \right) = \frac{m_0 v_{xi}^2}{d} \qquad (20.6)$$

The total average force $\overline{F}$ exerted by the gas on the wall is found by adding the average forces exerted by *all* the individual molecules striking the wall. Adding terms such as those in Equation 20.6 for all molecules gives

$$\overline{F} = \sum_{i=1}^{N} \frac{m_0 v_{xi}^2}{d} = \frac{m_0}{d} \sum_{i=1}^{N} v_{xi}^2 \qquad (20.7)$$

where we have factored out the length of the box and the mass m_0 because assumption 1 tells us that all the molecules are the same. We now impose an additional feature from assumption 1, that the number of molecules is large. For a small number of molecules, the actual force on the wall would vary with time. It would be nonzero during the short interval of a collision of a molecule with the wall and zero when no molecule happens to be hitting the wall. For a very large number of molecules such as Avogadro's number, however, these variations in force are smoothed out so that the average force given above is the same over *any* time interval. Therefore, the *constant* force F on the wall due to the molecular collisions is

$$F = \frac{m_0}{d} \sum_{i=1}^{N} v_{xi}^2 \qquad (20.8)$$

To proceed further, let's consider part of the right-hand side of Equation 20.8: how do we express the average value of the square of the x component of the velocity for N molecules? The traditional average of a set of values is the sum of the values over the number of values:

$$\overline{v_x^2} = \frac{\sum_{i=1}^{N} v_{xi}^2}{N} \quad \rightarrow \quad \sum_{i=1}^{N} v_{xi}^2 = N\overline{v_x^2} \qquad (20.9)$$

Using Equation 20.9 to substitute for the sum in Equation 20.8 gives

$$F = \frac{m_0}{d} N\overline{v_x^2} \qquad (20.10)$$

Now let's focus again on one molecule with velocity components v_{xi}, v_{yi}, and v_{zi}. The Pythagorean theorem relates the square of the speed of the molecule to the squares of the velocity components:

$$v_i^2 = v_{xi}^2 + v_{yi}^2 + v_{zi}^2 \qquad (20.11)$$

Hence, the average value of v^2 for all the molecules in the container is related to the average values of v_x^2, v_y^2, and v_z^2 according to the expression

$$\overline{v^2} = \overline{v_x^2} + \overline{v_y^2} + \overline{v_z^2} \qquad (20.12)$$

Because the motion is isotropic (assumption 2(a) above), the average values $\overline{v_x^2}$, $\overline{v_y^2}$, and $\overline{v_z^2}$ are equal to one another. Using this fact and Equation 20.12, we find that

$$\overline{v^2} = 3\overline{v_x^2} \qquad (20.13)$$

Therefore, from Equation 20.10, the total force exerted on the wall is

$$F = \tfrac{1}{3}N\frac{m_0\overline{v^2}}{d} \tag{20.14}$$

Using this expression, we can find the total pressure exerted on the wall:

$$P = \frac{F}{A} = \frac{F}{d^2} = \tfrac{1}{3}N\frac{m_0\overline{v^2}}{d^3} = \tfrac{1}{3}\left(\frac{N}{V}\right)m_0\overline{v^2}$$

$$P = \tfrac{2}{3}\left(\frac{N}{V}\right)\left(\tfrac{1}{2}m_0\overline{v^2}\right) \tag{20.15}$$

◀ Relationship between pressure and molecular kinetic energy

where we have recognized the volume V of the cube as d^3.

We have finished the long process initiated at the beginning of this section. The reward for our patience and diligence is something profound: Equation 20.15 indicates that the pressure of a gas is proportional to (1) the number of molecules per unit volume and (2) the average translational kinetic energy of the molecules, $\tfrac{1}{2}m_0\overline{v^2}$. In analyzing this structural model of an ideal gas, we obtain an important result that relates the macroscopic quantity of pressure to a microscopic quantity, the average value of the square of the molecular speed. Therefore, a key link between the molecular world and the large-scale world has been established.

Notice that Equation 20.15 verifies some features of pressure with which you are probably familiar. One way to increase the pressure inside a container is to increase the number of molecules per unit volume N/V in the container. That is what you do when you add air to a tire. We will return to discuss the second set of parentheses in Equation 20.15 very shortly, after we discuss the macroscopic quantity of temperature.

We can gain some insight into the meaning of temperature by first writing Equation 20.15 in the form

$$PV = \tfrac{2}{3}N\left(\tfrac{1}{2}m_0\overline{v^2}\right) \tag{20.16}$$

Let's now compare this expression with the equation of state for an ideal gas (Eq. 18.11):

$$PV = Nk_B T \tag{20.17}$$

Equating the right sides of Equations 20.16 and 20.17 and solving for T gives

$$T = \frac{2}{3k_B}\left(\tfrac{1}{2}m_0\overline{v^2}\right) \tag{20.18}$$

◀ Relationship between temperature and molecular kinetic energy

This result tells us that *temperature is a direct measure of average molecular kinetic energy.* In Chapter 18, we could only define temperature macroscopically, in terms of the transfer of energy between two objects. In Equation 20.18, we have a deeper definition of temperature in terms of the microscopic motion of the molecules of a substance.

By rearranging Equation 20.18, we can relate the translational molecular kinetic energy to the temperature:

$$\tfrac{1}{2}m_0\overline{v^2} = \tfrac{3}{2}k_B T \tag{20.19}$$

◀ Average kinetic energy per molecule

Now look back at Equation 20.15. The quantity in the second set of parentheses in that equation is the same as the left-hand side of Equation 20.19. Therefore, we see that the pressure in Equation 20.15 depends on the temperature of the gas. With regard to the discussion of the air pressure in an automobile tire, the pressure can be raised by increasing the temperature of that air, which is why the pressure inside a tire increases as the tire warms up during long road trips. The continuous flexing of the tire as it moves along the road surface results in work done on the rubber as parts of the tire distort, causing an increase in internal energy of the rubber. The increased temperature of the rubber results in the transfer of energy by heat into the air inside the tire. This transfer increases the air's temperature, and this increase in temperature in turn produces an increase in pressure.

Equation 20.19 shows us that the average translational kinetic energy per molecule is $\frac{3}{2}k_B T$. Because $\overline{v_x^2} = \frac{1}{3}\overline{v^2}$ (Eq. 20.13), it follows that

$$\frac{1}{2}m_0 \overline{v_x^2} = \frac{1}{2}k_B T \tag{20.20}$$

In a similar manner, for the y and z directions,

$$\frac{1}{2}m_0 \overline{v_y^2} = \frac{1}{2}k_B T \quad \text{and} \quad \frac{1}{2}m_0 \overline{v_z^2} = \frac{1}{2}k_B T$$

Therefore, each translational degree of freedom contributes an equal amount of energy, $\frac{1}{2}k_B T$, to the gas. (In general, a "degree of freedom" refers to an independent means by which a molecule can possess energy.) A generalization of this result, known as the **theorem of equipartition of energy,** is as follows:

◀ Theorem of equipartition of energy

> Each degree of freedom contributes $\frac{1}{2}k_B T$ to the energy of a system, where possible degrees of freedom are those associated with translation, rotation, and vibration of molecules.

The total translational kinetic energy of N molecules of gas is simply N times the average energy per molecule, which is given by Equation 20.19:

◀ Total translational kinetic energy of N molecules

$$K_{\text{tot trans}} = N(\frac{1}{2}m_0 \overline{v^2}) = \frac{3}{2}Nk_B T = \frac{3}{2}nRT \tag{20.21}$$

where we have used $k_B = R/N_A$ for Boltzmann's constant and $N = nN_A$ for the number of molecules of gas. If the gas molecules possess only translational kinetic energy, *Equation 20.21 represents the internal energy of the gas.* This result implies that the internal energy of an ideal gas depends *only* on the temperature. We will follow up on this point in Section 20.2.

The square root of $\overline{v^2}$ is called the **root-mean-square (rms) speed** of the molecules. From Equation 20.19, we find that the rms speed is

◀ Root-mean-square speed

$$v_{\text{rms}} = \sqrt{\overline{v^2}} = \sqrt{\frac{3k_B T}{m_0}} = \sqrt{\frac{3RT}{M}} \tag{20.22}$$

where M is the molar mass in kilograms per mole and is equal to $m_0 N_A$. This expression shows that, at a given temperature, lighter molecules move faster, on the average, than do heavier molecules. For example, at a given temperature, hydrogen molecules, whose molar mass is 2.02×10^{-3} kg/mol, have an average speed approximately four times that of oxygen molecules, whose molar mass is 32.0×10^{-3} kg/mol. Table 20.1 lists the rms speeds for various molecules at 20°C.

PITFALL PREVENTION 20.1

The Square Root of the Square?
Taking the square root of $\overline{v^2}$ in Equation 20.22 does not "undo" the square because we have taken an average *between* squaring and taking the square root. Although the square root of $(\overline{v})^2$ is $\overline{v} = v_{\text{avg}}$ because the squaring is done after the averaging, the square root of $\overline{v^2}$ is *not* v_{avg}, but rather v_{rms}.

QUICK QUIZ **20.1** Two containers hold an ideal gas at the same temperature and pressure. Both containers hold the same type of gas, but container B has twice the volume of container A. **(i)** What is the average translational kinetic energy per molecule in container B? (a) twice that of container A (b) the same as that of container A (c) half that of container A (d) impossible to determine **(ii)** From the same choices, describe the internal energy of the gas in container B.

TABLE 20.1 Some Root-Mean-Square (rms) Speeds

Gas	Molar Mass (g/mol)	v_{rms} at 20°C (m/s)	Gas	Molar Mass (g/mol)	v_{rms} at 20°C (m/s)
H_2	2.02	1902	NO	30.0	494
He	4.00	1352	O_2	32.0	478
H_2O	18.0	637	CO_2	44.0	408
Ne	20.2	602	SO_2	64.1	338
N_2 or CO	28.0	511			

Example **20.1** A Tank of Helium

A tank used for filling helium balloons has a volume of 0.300 m³ and contains 2.00 mol of helium gas at 20.0°C. Assume the helium behaves like an ideal gas.

(A) What is the total translational kinetic energy of the gas molecules?

SOLUTION

Conceptualize   Imagine a microscopic model of a gas in which you can watch the molecules move about the container more rapidly as the temperature increases. Because the gas is monatomic, the only type of motion the particles of the gas can exhibit is translation, and the total translational kinetic energy of the molecules is the internal energy of the gas.

Categorize   We evaluate parameters with equations developed in the preceding discussion, so this example is a substitution problem.

Use Equation 20.21 with n = 2.00 mol and T = 293 K:

$$E_{int} = K_{tot\ trans} = \tfrac{3}{2}nRT = \tfrac{3}{2}(2.00\ mol)(8.31\ J/mol \cdot K)(293\ K)$$

$$= \boxed{7.30 \times 10^3\ J}$$

(B) What is the average kinetic energy per molecule?

SOLUTION

Use Equation 20.19:

$$K_{avg} = \tfrac{1}{2}m_0\overline{v^2} = \tfrac{3}{2}k_B T = \tfrac{3}{2}(1.38 \times 10^{-23}\ J/K)(293\ K)$$

$$= \boxed{6.07 \times 10^{-21}\ J}$$

WHAT IF?   What if the temperature is raised from 20.0°C to 40.0°C? Because 40.0 is twice as large as 20.0, is the total translational energy of the molecules of the gas twice as large at the higher temperature?

Answer   The expression for the total translational energy depends on the temperature, and the value for the temperature must be expressed in kelvins, not in degrees Celsius. Therefore, the ratio of 40.0 to 20.0 is *not* the appropriate ratio. Converting the Celsius temperatures to kelvins, 20.0°C is 293 K and 40.0°C is 313 K. Therefore, the total translational energy increases by a factor of only 313 K/293 K = 1.07.

20.2 Molar Specific Heat of an Ideal Gas

Let's use the results of Section 20.1 to investigate a macroscopic quantity associated with a gas: its specific heat. Consider an ideal gas undergoing several processes such that the change in temperature is $\Delta T = T_f - T_i$ for all processes. The temperature change can be achieved by taking a variety of paths from one isotherm to another as shown in Figure 20.3. Because ΔT is the same for all paths, the change in internal energy ΔE_{int} is the same for all paths. The work W done on the gas (the negative of the area under the curves), however, is different for each path, as we found in Section 19.4. Therefore, from the first law of thermodynamics, we can argue that the heat $Q = \Delta E_{int} - W$ associated with a given change in temperature does *not* have a unique value: the heat Q for a process taking place between two temperatures depends on the process. Therefore, in $Q = mc\Delta T$, the specific heat c does not have a unique value for a gas!

We can address this difficulty by defining specific heats for two special processes that we have studied: isovolumetric ($i \rightarrow f$ in Figure 20.4 on page 540) and isobaric ($i \rightarrow f'$ in Figure 20.4). Because the number of moles n is a convenient measure of the amount of gas, we define the **molar specific heats** associated with these processes as follows:

$$Q = nC_V \Delta T \quad \text{(constant volume)} \quad\quad (20.23)$$

$$Q = nC_P \Delta T \quad \text{(constant pressure)} \quad\quad (20.24)$$

where C_V is the **molar specific heat at constant volume** and C_P is the **molar specific heat at constant pressure.** At constant volume, no work is done on the gas; the

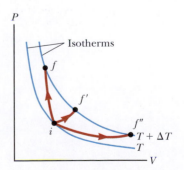

Figure 20.3 An ideal gas is taken from one isotherm at temperature T to another at temperature $T + \Delta T$ along three different paths.

For the constant-volume path, all the energy input goes into increasing the internal energy of the gas because no work is done.

Isotherms

$T + \Delta T$
T

Along the constant-pressure path, part of the energy transferred in by heat is transferred out by work.

Figure 20.4 Energy is transferred by heat to an ideal gas in two ways.

only energy change in the system is the change of internal energy. When energy is added to a gas by heat at constant pressure, however, not only does the internal energy of the gas increase, but (negative) work is done on the gas: the volume of the gas must increase to keep the pressure constant. Therefore, the heat Q in Equation 20.24 must account for both the increase in internal energy and the transfer of energy out of the system by work. For this reason, Q in Equation 20.24 is greater than that in Equation 20.23 for given values of n and ΔT. Therefore, C_P is greater than C_V.

Because no work is done on the gas in the constant-volume process in Figure 20.4, the first law of thermodynamics gives us

$$Q = \Delta E_{\text{int}} \qquad (20.25)$$

Substituting the expression for Q given by Equation 20.23 into Equation 20.25, we obtain

$$\Delta E_{\text{int}} = nC_V \Delta T \qquad (20.26)$$

where ΔT is the temperature difference between the two isotherms. This equation applies to all ideal gases, those gases having more than one atom per molecule as well as monatomic ideal gases. It also applies to *all processes* taking place between the same temperatures. All five processes in Figures 20.3 and 20.4 have the same change in internal energy because they all take place through the same temperature difference ΔT. This statement might seem surprising, because Equation 20.26 contains a molar specific heat for a specific process at constant volume. It is true, however, due to the fact that internal energy is a state variable, and its change depends only on the temperature change, so it is independent of the particular process. Equation 20.26 gives us the change in internal energy for all processes, and allows us to evaluate this change using the constant-volume specific heat.

In the limit of infinitesimal changes, we can use Equation 20.26 to express the molar specific heat at constant volume as

$$C_V = \frac{1}{n} \frac{dE_{\text{int}}}{dT} \qquad (20.27)$$

Let's now consider the simplest case of an ideal monatomic gas, that is, a gas containing one atom per molecule such as helium, neon, or argon. When energy is added to a monatomic gas in a container of fixed volume, all the added energy goes into increasing the translational kinetic energy of the atoms. There is no other way to store the energy in a monatomic gas. Therefore, from Equation 20.21, we see that the internal energy E_{int} of N molecules (or n mol) of an ideal monatomic gas is

$$E_{\text{int}} = K_{\text{tot trans}} = \tfrac{3}{2} N k_B T = \tfrac{3}{2} nRT \qquad (20.28)$$

Internal energy of an ideal ▶ monatomic gas

For a monatomic ideal gas, E_{int} is a function of T only and the functional relationship is given by Equation 20.28. In general, the internal energy of any ideal gas is a function of T only and the exact relationship depends on the type of gas.

Substituting the internal energy from Equation 20.28 into Equation 20.27 gives

$$C_V = \tfrac{3}{2} R = 12.5 \text{ J/mol} \cdot \text{K} \qquad (20.29)$$

This expression predicts a value of $C_V = \tfrac{3}{2} R$ for *all* monatomic gases. This prediction is in excellent agreement with measured values of molar specific heats for such gases as helium, neon, argon, and xenon over a wide range of temperatures (see the C_V column in Table 20.2). Small variations in Table 20.2 from the predicted values are because real gases are not ideal gases. In real gases, weak intermolecular interactions occur, which are not addressed in our ideal gas model.

Now suppose the gas is taken along the constant-pressure path $i \rightarrow f'$ shown in Figure 20.4. Along this path, the temperature again increases by ΔT. The energy that must be transferred by heat to the gas in this process is $Q = nC_P \Delta T$. Because the volume changes in this process, the work done on the gas is $W = -P\Delta V$, where

TABLE 20.2 Molar Specific Heats of Various Gases

Gas	Molar Specific Heat ($J/mol \cdot K$)[a]			
	C_P	C_V	$C_P - C_V$	$\gamma = C_P/C_V$
Monatomic gases				
He	20.8	12.5	8.33	1.67
Ar	20.8	12.5	8.33	1.67
Ne	20.8	12.7	8.12	1.64
Kr	20.8	12.3	8.49	1.69
Diatomic gases				
H_2	28.8	20.4	8.33	1.41
N_2	29.1	20.8	8.33	1.40
O_2	29.4	21.1	8.33	1.40
CO	29.3	21.0	8.33	1.40
Cl_2	34.7	25.7	8.96	1.35
Polyatomic gases				
CO_2	37.0	28.5	8.50	1.30
SO_2	40.4	31.4	9.00	1.29
H_2O	35.4	27.0	8.37	1.30
CH_4	35.5	27.1	8.41	1.31

[a] All values except that for water were obtained at 300 K.

P is the constant pressure at which the process occurs. Applying the first law of thermodynamics to this process, we have

$$\Delta E_{int} = Q + W = nC_P \Delta T + (-P \Delta V) \tag{20.30}$$

As discussed above, the change in internal energy for the process $i \rightarrow f'$ in Figure 20.4 is equal to that for process $i \rightarrow f$ because ΔT is the same for both processes. In addition, because $PV = nRT$, note that for a constant-pressure process, $P \Delta V = nR \Delta T$. Substituting this value for $P \Delta V$ into Equation 20.30 with $\Delta E_{int} = nC_V \Delta T$ (Eq. 20.26) gives

$$nC_V \Delta T = nC_P \Delta T - nR \Delta T$$

$$C_P - C_V = R \tag{20.31}$$

This expression applies to *any* ideal gas. It predicts that the molar specific heat of an ideal gas at constant pressure is greater than the molar specific heat at constant volume by an amount R, the universal gas constant (which has the value 8.31 $J/mol \cdot K$). This expression is applicable to real gases as the data in the $C_P - C_V$ column in Table 20.2 show.

Because $C_V = \frac{3}{2}R$ for a monatomic ideal gas, Equation 20.31 predicts a value $C_P = \frac{5}{2}R = 20.8$ $J/mol \cdot K$ for the molar specific heat of a monatomic gas at constant pressure. The ratio of these molar specific heats is a dimensionless quantity γ (Greek letter gamma):

$$\gamma = \frac{C_P}{C_V} = \frac{5R/2}{3R/2} = \frac{5}{3} = 1.67 \tag{20.32}$$

◀ Ratio of molar specific heats for a monatomic ideal gas

Theoretical values of C_V, C_P, and γ are in excellent agreement with experimental values obtained for monatomic gases, but they are in serious disagreement with the values for the more complex gases (see Table 20.2). That is not surprising; the value $C_V = \frac{3}{2}R$ was derived for a monatomic ideal gas, and we expect some additional contribution to the molar specific heat from the internal structure of the more complex molecules. In Section 20.3, we describe the effect of molecular structure on the molar specific heat of a gas. The internal energy—and hence the molar specific heat—of a complex gas must include contributions from the rotational and the vibrational motions of the molecule.

In the case of solids and liquids heated at constant pressure, very little work is done during such a process because the thermal expansion is small. Consequently, C_P and C_V are approximately equal for solids and liquids.

QUICK QUIZ **20.2**   **(i)** How does the internal energy of an ideal gas change as it follows path $i \rightarrow f$ in Figure 20.4? (a) E_{int} increases. (b) E_{int} decreases. (c) E_{int} stays the same. (d) There is not enough information to determine how E_{int} changes. **(ii)** From the same choices, how does the internal energy of an ideal gas change as it follows path $f \rightarrow f'$ along the isotherm labeled $T + \Delta T$ in Figure 20.4?

Example **20.2**   **Heating a Cylinder of Helium**

A cylinder contains 3.00 mol of helium gas at a temperature of 300 K.

(A) If the gas is heated at constant volume, how much energy must be transferred by heat to the gas for its temperature to increase to 500 K?

SOLUTION

Conceptualize   Run the process in your mind with the help of the piston–cylinder arrangement in Figure 19.7. Imagine that the piston is clamped in position to maintain the constant volume of the gas.

Categorize   We evaluate parameters with equations developed in the preceding discussion, so this example is a substitution problem.

Use Equation 20.23 to find the energy transfer: $Q_1 = nC_V \Delta T$

Substitute the given values: $Q_1 = (3.00 \text{ mol})(12.5 \text{ J/mol} \cdot \text{K})(500 \text{ K} - 300 \text{ K})$
          $= \boxed{7.50 \times 10^3 \text{ J}}$

(B) How much energy must be transferred by heat to the gas at constant pressure to raise the temperature to 500 K?

SOLUTION

Use Equation 20.24 to find the energy transfer: $Q_2 = nC_P \Delta T$

Substitute the given values: $Q_2 = (3.00 \text{ mol})(20.8 \text{ J/mol} \cdot \text{K})(500 \text{ K} - 300 \text{ K})$
          $= \boxed{12.5 \times 10^3 \text{ J}}$

This value is larger than Q_1 because of the transfer of energy out of the gas by work to raise the piston in the constant pressure process.

20.3   The Equipartition of Energy

In Section 20.1, we found that the temperature of a gas is a measure of the average translational kinetic energy of the gas molecules. This kinetic energy is associated with the motion of the center of mass of each molecule. It does not include the energy associated with the internal motion of the molecule, namely, vibrations and rotations about the center of mass. In this section, we introduce the contributions from rotation and vibration of the molecule to the specific heats of the gas.

Predictions based on our model for molar specific heat agree quite well with the behavior of monatomic gases, but not with the behavior of complex gases (see Table 20.2). The value predicted by the model for the quantity $C_P - C_V = R$, however, is the same for all gases. This similarity is not surprising because this difference is the result of the work done on the gas, which is independent of its molecular structure.

To clarify the variations in C_V and C_P in gases more complex than monatomic gases, let's explore further the origin of molar specific heat. So far, we have assumed the sole contribution to the internal energy of a gas is the translational kinetic energy of the molecules. The internal energy of a gas, however, includes contributions from the translational, vibrational, and rotational motion of the molecules. The rotational and vibrational motions of molecules can be activated by collisions and therefore are "coupled" to the translational motion of the molecules. The branch of physics known as *statistical mechanics* has shown that, for a large number of particles obeying the laws of Newtonian mechanics, the available energy is, on average, shared equally by each independent degree of freedom. Recall from Section 20.1 that the equipartition theorem states that, at equilibrium, each degree of freedom contributes $\frac{1}{2}k_BT$ of energy per molecule.

Let's consider a diatomic gas whose molecules have the shape of a dumbbell (Fig. 20.5). In this model, the center of mass of the molecule can translate in the x, y, and z directions. The gray arrow in Figure 20.5a shows a translation in the x direction. In addition, the molecule can rotate about three mutually perpendicular axes (Fig. 20.5b). The rotation about the y axis can be neglected because the molecule's moment of inertia I_y and its rotational energy $\frac{1}{2}I_y\omega^2$ about this axis are negligible compared with those associated with the x and z axes. (If the two atoms are modeled as particles, then I_y is identically zero.) Therefore, there are five degrees of freedom for translation and rotation: three associated with the translational motion (x, y, and z) and two associated with the rotational motion (x and z). Because each degree of freedom contributes, on average, $\frac{1}{2}k_BT$ of energy per molecule, the internal energy for a system of N molecules, ignoring vibration for now, is

$$E_{int} = 3N(\tfrac{1}{2}k_BT) + 2N(\tfrac{1}{2}k_BT) = \tfrac{5}{2}Nk_BT = \tfrac{5}{2}nRT \quad \textbf{(20.33)}$$

We can use this result and Equation 20.27 to find the molar specific heat at constant volume:

$$C_V = \frac{1}{n}\frac{dE_{int}}{dT} = \frac{1}{n}\frac{d}{dT}(\tfrac{5}{2}nRT) = \tfrac{5}{2}R = 20.8 \text{ J/mol} \cdot \text{K} \quad \textbf{(20.34)}$$

From Equations 20.31 and 20.32, we find that

$$C_P = C_V + R = \tfrac{7}{2}R = 29.1 \text{ J/mol} \cdot \text{K}$$

$$\gamma = \frac{C_P}{C_V} = \frac{\tfrac{7}{2}R}{\tfrac{5}{2}R} = \frac{7}{5} = 1.40$$

These results agree quite well with most of the data for diatomic molecules given in Table 20.2. That is rather surprising, however, because we have not yet accounted for the possible vibrations of the molecule.

In the model for vibration, the two atoms are joined by an imaginary spring (see Fig. 20.5c). The vibrational motion adds two more degrees of freedom, which correspond to the kinetic energy and the potential energy associated with vibrations along the length of the molecule. Hence, a model that includes all three types of motion predicts a total internal energy of

$$E_{int} = 3N(\tfrac{1}{2}k_BT) + 2N(\tfrac{1}{2}k_BT) + 2N(\tfrac{1}{2}k_BT) = \tfrac{7}{2}Nk_BT = \tfrac{7}{2}nRT$$

and a molar specific heat at constant volume of

$$C_V = \frac{1}{n}\frac{dE_{int}}{dT} = \frac{1}{n}\frac{d}{dT}(\tfrac{7}{2}nRT) = \tfrac{7}{2}R = 29.1 \text{ J/mol} \cdot \text{K} \quad \textbf{(20.35)}$$

This value is inconsistent with experimental data for molecules such as H_2 and N_2 (see Table 20.2) and suggests a breakdown of our model based on classical physics.

Translational motion of the center of mass

Rotational motion about the various axes

Vibrational motion along the molecular axis

Figure 20.5 Possible motions of a diatomic molecule.

The horizontal scale is logarithmic.

Hydrogen liquefies at 20 K.

Vibration

Rotation

Translation

$\frac{7}{2}R$

$\frac{5}{2}R$

$\frac{3}{2}R$

C_V (J/mol · K)

Temperature (K)

Figure 20.6 The molar specific heat of hydrogen as a function of temperature.

It might seem that our model is a failure for predicting molar specific heats for diatomic gases. We can claim some success for our model, however, if measurements of molar specific heat are made over a wide temperature range rather than at the single temperature that gives us the values in Table 20.2. Figure 20.6 shows the molar specific heat of hydrogen as a function of temperature. The remarkable feature about the three plateaus in the graph's curve is that they are at the values of the molar specific heat predicted by Equations 20.29, 20.34, and 20.35! For low temperatures, the diatomic hydrogen gas behaves like a monatomic gas. As the temperature rises to room temperature, its molar specific heat rises to a value for a diatomic gas, consistent with the inclusion of rotation but not vibration. For high temperatures, the molar specific heat is consistent with a model including all types of motion.

Before addressing the reason for this mysterious behavior, let's make some brief remarks about polyatomic gases. For nonlinear molecules with more than two atoms, three axes of rotation are available. The vibrations are more complex than for diatomic molecules. Therefore, the number of degrees of freedom is even larger. The result is an even higher predicted molar specific heat, which is in qualitative agreement with experiment. The molar specific heats for the polyatomic gases in Table 20.2 are higher than those for diatomic gases. The more degrees of freedom available to a molecule, the more "ways" there are to store energy, resulting in a higher molar specific heat.

A Hint of Energy Quantization

Our model for molar specific heats has been based so far on purely classical notions. It predicts a value of the specific heat for a diatomic gas that, according to Figure 20.6, only agrees with experimental measurements made at high temperatures. To explain why this value is only true at high temperatures and why the plateaus in Figure 20.6 exist, we must go beyond classical physics and introduce some quantum physics into the model. In Chapter 17, we discussed quantization of frequency for vibrating strings and air columns; only certain frequencies of standing waves can exist. That is a natural result whenever waves are subject to boundary conditions.

Quantum physics (Chapters 39 through 42) shows that atoms and molecules have a wavelike nature and can be analyzed with the waves under boundary conditions analysis model. Consequently, these waves have quantized frequencies. Furthermore, in quantum physics, the energy of a system is proportional to the frequency of the wave representing the system. Hence, **the energies of atoms and molecules are quantized:** only certain energies are allowed.

For a molecule, quantum physics tells us that the rotational and vibrational energies are quantized. Figure 20.7 shows an **energy-level diagram** for the rotational and vibrational quantum states of a diatomic molecule. The lowest allowed state is called the **ground state.** The three longer black lines represent allowed vibrational energies. These states are widely spaced in energy. Associated with each allowed vibrational energy is a set of more narrowly spaced rotational energies, represented by the shorter black lines.

If a molecule is in the ground state for rotation or vibration, then rotation or vibration does not contribute to the molar specific heat. These types of motion only contribute when there is a *transition* to an excited state. Section 20.1 tells us that molecular energies are proportional to temperature. Therefore, we can add labels on the right of Figure 20.7 that correspond roughly to temperatures at which the energy levels will be excited.

At low temperatures, the energy a molecule gains in collisions with its neighbors is generally not large enough to raise it to the first excited state of either rotation or vibration. Therefore, even though rotation and vibration are allowed according to classical physics, they do not occur in reality at low temperatures. All molecules are in the ground state for rotation and vibration. The only contribution to the molecules' average energy is from translation, and the specific heat is that predicted by Equation 20.29. The temperature T_A in Figure 20.7 might be 50 K for hydrogen: only the ground states for vibration or rotation are occupied; we are on the lowest plateau in Figure 20.6.

As the temperature is raised, the average energy of the molecules increases. In some collisions, a molecule may have enough energy transferred to it from another molecule to excite the first rotational state. As the temperature is raised further, more molecules can be excited to this state. The result is that rotation begins to contribute to the internal energy, and the molar specific heat rises. For hydrogen, the temperature T_B in Figure 20.7 might be 500 K: excited rotational levels are occupied, but for vibration, only the ground state is occupied; we are on the second plateau in Figure 20.6. The molar specific heat is now equal to the value predicted by Equation 20.34.

There is no contribution at room temperature from vibration because the molecules are still in the ground vibrational state. The temperature must be raised even further to excite the first vibrational state. For hydrogen, the temperature T_C in Figure 20.7 might be 5 000 K: excited rotational and vibrational levels are occupied; we are on the highest plateau in Figure 20.6 and the molar specific heat has the value predicted by Equation 20.35.

The predictions of this model are supportive of the theorem of equipartition of energy. In addition, the inclusion in the model of energy quantization from quantum physics allows a full understanding of Figure 20.6.

The rotational states lie closer together in energy than do the vibrational states.

Figure 20.7 An energy-level diagram for vibrational and rotational states of a diatomic molecule.

QUICK QUIZ **20.3** The molar specific heat of a diatomic gas is measured at constant volume and found to be 29.1 J/mol · K. What are the types of energy that are contributing to the molar specific heat? **(a)** translation only **(b)** translation and rotation only **(c)** translation and vibration only **(d)** translation, rotation, and vibration

QUICK QUIZ **20.4** The molar specific heat of a gas is measured at constant volume and found to be $11R/2$. Is the gas most likely to be **(a)** monatomic, **(b)** diatomic, or **(c)** polyatomic?

20.4 Adiabatic Processes for an Ideal Gas

Our study of molar specific heats allows us to complete our discussion of adiabatic processes begun in Section 19.5. As noted there, an **adiabatic process** is one in which no energy is transferred by heat between a system and its surroundings.

For example, if a gas is compressed (or expanded) rapidly, very little energy is transferred out of (or into) the system by heat, so the process is nearly adiabatic. Another example of an adiabatic process is the slow expansion of a gas that is thermally insulated from its surroundings. All three variables in the ideal gas law—P, V, and T—change during an adiabatic process.

Let's imagine an adiabatic gas process involving an infinitesimal change in volume dV and an accompanying infinitesimal change in temperature dT. The work done on the gas is $-P\,dV$. Because the internal energy of an ideal gas depends only on temperature, the change in the internal energy in an adiabatic process is the same as that for an isovolumetric process between the same temperatures, $dE_{\text{int}} = nC_V\,dT$ (Eq. 20.26). Hence, the first law of thermodynamics, $\Delta E_{\text{int}} = Q + W$, with $Q = 0$, becomes the infinitesimal form

$$dE_{\text{int}} = nC_V\,dT = -P\,dV \tag{20.36}$$

Taking the total differential of the equation of state of an ideal gas, $PV = nRT$, gives

$$P\,dV + V\,dP = nR\,dT \tag{20.37}$$

Eliminating dT between Equations 20.36 and 20.37, we find that

$$P\,dV + V\,dP = -\frac{R}{C_V}P\,dV$$

Substituting $R = C_P - C_V$ and dividing by PV gives

$$\frac{dV}{V} + \frac{dP}{P} = -\left(\frac{C_P - C_V}{C_V}\right)\frac{dV}{V} = (1 - \gamma)\frac{dV}{V}$$

$$\frac{dP}{P} + \gamma\frac{dV}{V} = 0$$

Integrating this expression, we have

$$\ln P + \gamma \ln V = \text{constant}$$

which is equivalent to

◀ **Relationship between P and V for an adiabatic process involving an ideal gas**

$$PV^\gamma = \text{constant} \tag{20.38}$$

The PV diagram for an adiabatic expansion is shown in Figure 20.8. Because $\gamma > 1$, the PV curve is steeper than it would be for an isothermal expansion, for which $PV = \text{constant}$. By the definition of an adiabatic process, no energy is transferred by heat into or out of the system. Hence, from the first law, we see that ΔE_{int} is negative (work is done *by* the gas, so its internal energy decreases) and so ΔT also is negative. Therefore, the temperature of the gas decreases ($T_f < T_i$) during an adiabatic expansion. Conversely, the temperature increases if the gas is compressed adiabatically. Applying Equation 20.38 to the initial and final states, we see that

$$P_iV_i^\gamma = P_fV_f^\gamma \tag{20.39}$$

Using the ideal gas law, we can express Equation 20.38 as

◀ **Relationship between T and V for an adiabatic process involving an ideal gas**

$$TV^{\gamma-1} = \text{constant} \tag{20.40}$$

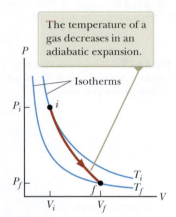

The temperature of a gas decreases in an adiabatic expansion.

Figure 20.8 The *PV* diagram for an adiabatic expansion of an ideal gas.

| Example **20.3** | **A Diesel Engine Cylinder** |

Air at 20.0°C in the cylinder of a diesel engine is compressed from an initial pressure of 1.00 atm and volume of 800.0 cm³ to a volume of 60.0 cm³. Assume air behaves as an ideal gas with $\gamma = 1.40$ and the compression is adiabatic. Find the final pressure and temperature of the air.

20.3 continued

SOLUTION

Conceptualize Imagine what happens if a gas is compressed into a smaller volume. Our discussion above and Figure 20.8 tell us that the pressure and temperature both increase.

Categorize We categorize this example as a problem involving an adiabatic process.

Analyze Use Equation 20.39 to find the final pressure:

$$P_f = P_i \left(\frac{V_i}{V_f}\right)^\gamma = (1.00 \text{ atm})\left(\frac{800.0 \text{ cm}^3}{60.0 \text{ cm}^3}\right)^{1.40}$$

$$= \boxed{37.6 \text{ atm}}$$

Use the ideal gas law to find the final temperature:

$$\frac{P_i V_i}{T_i} = \frac{P_f V_f}{T_f}$$

$$T_f = \frac{P_f V_f}{P_i V_i}T_i = \frac{(37.6 \text{ atm})(60.0 \text{ cm}^3)}{(1.00 \text{ atm})(800.0 \text{ cm}^3)}(293 \text{ K})$$

$$= 826 \text{ K} = \boxed{553°C}$$

Finalize The temperature of the gas increases by a factor of 826 K/293 K = 2.82. The high compression in a diesel engine raises the temperature of the gas enough to cause the combustion of fuel without the use of spark plugs.

20.5 Distribution of Molecular Speeds

Thus far, we have considered only average values of the energies of all the molecules in a gas and have not addressed the distribution of energies among individual molecules. The motion of the molecules is extremely chaotic. Any individual molecule collides with others at an enormous rate, typically a billion times per second. Each collision results in a change in the speed and direction of motion of each of the participant molecules. Equation 20.22 shows that rms molecular speeds increase with increasing temperature. At a given time, what is the relative number of molecules that possess some characteristic such as energy within a certain range?

We shall address this question by considering the **number density** $n_V(E)$. This quantity, called a *distribution function,* is defined so that $n_V(E)\,dE$ is the number of molecules per unit volume with energy between E and $E + dE$. In general, the number density is found from statistical mechanics to be

$$n_V(E) = n_0 e^{-E/k_B T} \tag{20.41}$$

where n_0 is defined such that $n_0\,dE$ is the number of molecules per unit volume having energy between $E = 0$ and $E = dE$. This equation, known as the **Boltzmann distribution law,** is important in describing the statistical mechanics of a large number of molecules. It states that the probability of finding the molecules in a particular energy state varies exponentially as the negative of the energy divided by $k_B T$. All the molecules would fall into the ground state if the thermal agitation at a temperature T did not excite the molecules to higher energy levels.

PITFALL PREVENTION 20.2
The Distribution Function
The distribution function $n_V(E)$ is defined in terms of the number of molecules with energy in the range E to $E + dE$ rather than in terms of the number of molecules with a specific energy E. Because the number of molecules is finite and the number of possible values of the energy is infinite, the number of molecules with an *exact* energy E may be zero.

◀ Boltzmann distribution law

Example **20.4** **Thermal Excitation of Atomic Energy Levels**

As discussed in Section 20.3, atoms can occupy only certain discrete energy levels. Consider a gas at a temperature of 2 500 K whose atoms can occupy only two energy levels separated by 1.50 eV, where 1 eV (electron volt) is an energy unit equal to 1.60×10^{-19} J. Determine the ratio of the number of atoms in the higher energy level to the number in the lower energy level.

continued

20.4 continued

SOLUTION

Conceptualize In your mental representation of this example, remember that only two possible states are allowed for the system of the atom. Figure 20.9 helps you visualize the two states on an energy-level diagram. In this case, the atom has two possible energies, E_1 and E_2, where $E_1 < E_2$.

Categorize We categorize this example as one in which we focus on particles in a two-state quantized system. We will apply the Boltzmann distribution law to this system.

Figure showing energy-level diagram with E_2 at top, E_1 at bottom, ENERGY arrow pointing up, and 1.50 eV between levels.

Figure 20.9 (Example 20.4) Energy-level diagram for a gas whose atoms can occupy two energy states.

Analyze Set up the ratio of the number density of atoms in the higher energy level to the number density in the lower energy level and use Equation 20.41 to express each number:

$$(1)\quad \frac{n_V(E_2)}{n_V(E_1)} = \frac{n_0 e^{-E_2/k_B T}}{n_0 e^{-E_1/k_B T}} = e^{-(E_2 - E_1)/k_B T}$$

Evaluate $k_B T$ in the exponent:

$$k_B T = (1.38 \times 10^{-23}\ \text{J/K})(2\,500\ \text{K})\left(\frac{1\ \text{eV}}{1.60 \times 10^{-19}\ \text{J}}\right) = 0.216\ \text{eV}$$

Substitute this value into Equation (1):

$$\frac{n_V(E_2)}{n_V(E_1)} = e^{-1.50\ \text{eV}/0.216\ \text{eV}} = e^{-6.96} = \boxed{9.52 \times 10^{-4}}$$

Finalize This result indicates that at $T = 2\,500$ K, only a small fraction of the atoms are in the higher energy level. In fact, for every atom in the higher energy level, there are about 1 000 atoms in the lower level. The number of atoms in the higher level increases at even higher temperatures, but the distribution law specifies that at equilibrium there are always more atoms in the lower level than in the higher level.

Ludwig Boltzmann
Austrian physicist (1844–1906)
Boltzmann made many important contributions to the development of the kinetic theory of gases, electromagnetism, and thermodynamics. His pioneering work in the field of kinetic theory led to the branch of physics known as statistical mechanics.

INTERFOTO/Alamy

Now that we have discussed the distribution of energies among molecules in a gas, let's think about the distribution of molecular speeds. In 1860, James Clerk Maxwell (1831–1879) derived an expression that describes the distribution of molecular speeds in a very definite manner. His work and subsequent developments by other scientists were highly controversial because direct detection of molecules could not be achieved experimentally at that time. About 60 years later, however, experiments were devised that confirmed Maxwell's predictions.

Let's consider a container of gas whose molecules have some distribution of speeds. Suppose we want to determine how many gas molecules have a speed in the range from, for example, 400 to 401 m/s. Intuitively, we expect the speed distribution to depend on temperature. Furthermore, we expect the distribution to peak in the vicinity of v_{rms}. That is, few molecules are expected to have speeds much less than or much greater than v_{rms} because these extreme speeds result only from an unlikely chain of collisions.

The observed speed distribution of gas molecules in thermal equilibrium is shown in Figure 20.10. The quantity N_v, called the **Maxwell–Boltzmann speed distribution function,** is defined as follows. If N is the total number of molecules, the number of molecules with speeds between v and $v + dv$ is $dN = N_v\,dv$, where the quantity N_v is dependent on temperature. This number is also equal to the area of the shaded rectangle in Figure 20.10. Furthermore, the fraction of molecules with speeds between v and $v + dv$ is $(N_v\,dv)/N$. This fraction is also equal to the probability that a molecule has a speed in the range v to $v + dv$.

The quantity N_v that describes the distribution of speeds of N gas molecules is

$$N_v = 4\pi N \left(\frac{m_0}{2\pi k_B T}\right)^{3/2} v^2 e^{-m_0 v^2/2k_B T} \tag{20.42}$$

where m_0 is the mass of a gas molecule, k_B is Boltzmann's constant, and T is the absolute temperature.[2] Observe the appearance of the Boltzmann factor $e^{-E/k_B T}$ with $E = \frac{1}{2}m_0 v^2$.

---

[2]For the derivation of this expression, see an advanced textbook on thermodynamics.

As indicated in Figure 20.10, the average speed is somewhat lower than the rms speed. The *most probable speed* v_{mp} is the speed at which the distribution curve reaches a peak. Using Equation 20.42, we find that

$$v_{rms} = \sqrt{\overline{v^2}} = \sqrt{\frac{3k_B T}{m_0}} = 1.73\sqrt{\frac{k_B T}{m_0}} \qquad (20.43)$$

$$v_{avg} = \sqrt{\frac{8k_B T}{\pi m_0}} = 1.60\sqrt{\frac{k_B T}{m_0}} \qquad (20.44)$$

$$v_{mp} = \sqrt{\frac{2k_B T}{m_0}} = 1.41\sqrt{\frac{k_B T}{m_0}} \qquad (20.45)$$

Equation 20.43 has previously appeared as Equation 20.22. The details of the derivations of these equations from Equation 20.42 are left for the end-of-chapter problems (see Problems 24 and 41). From these equations, we see that

$$v_{rms} > v_{avg} > v_{mp}$$

Figure 20.11 represents speed distribution curves for nitrogen, N_2, at two temperatures. Notice that the peak in the curve shifts to the right as T increases, indicating that the average speed increases with increasing temperature, as do the rms speed and the most probable speed, as expected from Equations 20.43–20.45. Because the lowest speed possible is zero and the upper classical limit of the speed is infinity, the curves are asymmetrical. (In Chapter 38, we show that the actual upper limit is the speed of light.)

Equation 20.42 shows that the distribution of molecular speeds in a gas depends on temperature as well as on the mass m_0 of the molecule. At a given temperature, the fraction of molecules with speeds exceeding a fixed value increases as the mass decreases. Hence, lighter molecules such as H_2 and He escape into space more readily from the Earth's atmosphere than do heavier molecules such as N_2 and O_2. (See the discussion of escape speed in Chapter 13. Gas molecules escape even more readily from the Moon's surface than from the Earth's because the escape speed on the Moon is lower than that on the Earth, leaving essentially no atmosphere.)

The speed distribution curves for molecules in a liquid are similar to those shown in Figure 20.11. We can understand the phenomenon of *evaporation* of a liquid from this distribution in speeds, given that some molecules in the liquid

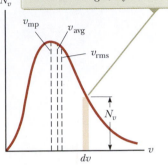

The number of molecules having speeds ranging from v to $v + dv$ equals the area of the tan rectangle, $N_v\,dv$.

Figure 20.10 The speed distribution of gas molecules at some temperature. The function N_v approaches zero as v approaches infinity.

The total area under either curve is equal to N, the total number of molecules. In this case, $N = 10^5$.

Note that $v_{rms} > v_{avg} > v_{mp}$.

Figure 20.11 The speed distribution function for 10^5 nitrogen molecules at 300 K and 900 K.

are more energetic than others. Some of the faster-moving molecules in the liquid penetrate the surface and even leave the liquid at temperatures well below the boiling point. The molecules that escape the liquid by evaporation are those that have sufficient energy to overcome the attractive forces of the molecules in the liquid phase. Consequently, the molecules left behind in the liquid phase have a lower average kinetic energy; as a result, the temperature of the liquid decreases. Hence, evaporation is a cooling process.

That evaporation is a cooling process explains the effects you noticed in the opening storyline. When the wind blows on one side of your upturned finger, the evaporation process is accelerated. Molecules of water vapor are blown away from the surface of your finger, reducing the water vapor pressure at the surface and making it easier for more molecules to leave the surface of the water. As a result, the cooling process is augmented, and that side of your finger feels cool. The opposite side of your finger is shielded from the wind, so that the evaporation is not as rapid. When your Club advisor put alcohol on your wound, it felt cold. Alcohol evaporates at a higher rate than water, so the evaporative cooling process makes your skin feel colder than the surrounding skin that is dry.

Example 20.5 | Molecular Speeds in a Hydrogen Gas

A 0.500-mol sample of hydrogen gas is at 300 K.

(A) Find the average speed, the rms speed, and the most probable speed of the H_2 molecules.

SOLUTION

Conceptualize Imagine the huge number of particles in a macroscopic sample of gas, all moving in random directions with different speeds.

Categorize We are dealing with a very large number of particles, so we can use the Maxwell–Boltzmann speed distribution function.

Analyze Use Equation 20.44 to find the average speed:

$$v_{avg} = 1.60 \sqrt{\frac{k_B T}{m_0}} = 1.60 \sqrt{\frac{(1.38 \times 10^{-23} \text{ J/K})(300 \text{ K})}{2(1.67 \times 10^{-27} \text{ kg})}}$$

$$= \boxed{1.78 \times 10^3 \text{ m/s}}$$

Use Equation 20.43 to find the rms speed:

$$v_{rms} = 1.73 \sqrt{\frac{k_B T}{m_0}} = 1.73 \sqrt{\frac{(1.38 \times 10^{-23} \text{ J/K})(300 \text{ K})}{2(1.67 \times 10^{-27} \text{ kg})}}$$

$$= \boxed{1.93 \times 10^3 \text{ m/s}}$$

Use Equation 20.45 to find the most probable speed:

$$v_{mp} = 1.41 \sqrt{\frac{k_B T}{m_0}} = 1.41 \sqrt{\frac{(1.38 \times 10^{-23} \text{ J/K})(300 \text{ K})}{2(1.67 \times 10^{-27} \text{ kg})}}$$

$$= \boxed{1.57 \times 10^3 \text{ m/s}}$$

(B) Find the number of molecules with speeds between 400 m/s and 401 m/s.

SOLUTION

Use Equation 20.42 to evaluate the number of molecules in a narrow speed range between v and $v + dv$:

$$(1) \quad N_v \, dv = 4\pi N \left(\frac{m_0}{2\pi k_B T} \right)^{3/2} v^2 e^{-m_0 v^2 / 2k_B T} \, dv$$

Evaluate the constant in front of v^2:

$$4\pi N \left(\frac{m_0}{2\pi k_B T} \right)^{3/2} = 4\pi n N_A \left(\frac{m_0}{2\pi k_B T} \right)^{3/2}$$

$$= 4\pi (0.500 \text{ mol})(6.02 \times 10^{23} \text{ mol}^{-1}) \left[\frac{2(1.67 \times 10^{-27} \text{ kg})}{2\pi (1.38 \times 10^{-23} \text{ J/K})(300 \text{ K})} \right]^{3/2}$$

$$= 1.74 \times 10^{14} \text{ s}^3/\text{m}^3$$

20.5 continued

Evaluate the exponent of e that appears in Equation (1):

$$-\frac{m_0 v^2}{2k_B T} = -\frac{2(1.67 \times 10^{-27}\ \text{kg})(400\ \text{m/s})^2}{2(1.38 \times 10^{-23}\ \text{J/K})(300\ \text{K})} = -0.064\ 5$$

Evaluate $N_v\, dv$ using these values in Equation (1) and evaluating v and dv:

$$N_v\, dv = (1.74 \times 10^{14}\ \text{s}^3/\text{m}^3)(400\ \text{m/s})^2 e^{-0.064\,5}(1\ \text{m/s})$$

$$= 2.61 \times 10^{19}\ \text{molecules}$$

Finalize   In this evaluation, we could calculate the result without integration because $dv = 1$ m/s is much smaller than $v = 400$ m/s. Had we sought the number of particles between, say, 400 m/s and 500 m/s, we would need to integrate Equation (1) between these speed limits.

Summary

> ## Concepts and Principles

The pressure of N molecules of an ideal gas contained in a volume V is

$$P = \frac{2}{3}\left(\frac{N}{V}\right)\left(\tfrac{1}{2}m_0\overline{v^2}\right) \tag{20.15}$$

The average translational kinetic energy per molecule of a gas, $\tfrac{1}{2}m_0\overline{v^2}$, is related to the temperature T of the gas through the expression

$$\tfrac{1}{2}m_0\overline{v^2} = \tfrac{3}{2}k_B T \tag{20.19}$$

where k_B is Boltzmann's constant. Each translational degree of freedom (x, y, or z) has $\tfrac{1}{2}k_B T$ of energy associated with it.

The change in internal energy for n mol of any ideal gas that undergoes a change in temperature ΔT is

$$\Delta E_{\text{int}} = nC_V\Delta T \tag{20.26}$$

where C_V is the **molar specific heat at constant volume.**

The internal energy of N molecules (or n mol) of an ideal monatomic gas is

$$E_{\text{int}} = \tfrac{3}{2}Nk_B T = \tfrac{3}{2}nRT \tag{20.28}$$

The molar specific heat of an ideal monatomic gas at constant volume is $C_V = \tfrac{3}{2}R$; the molar specific heat at constant pressure is $C_P = \tfrac{5}{2}R$. The ratio of specific heats is given by $\gamma = C_P/C_V = \tfrac{5}{3}$.

If an ideal gas undergoes an adiabatic expansion or compression, the first law of thermodynamics, together with the equation of state, shows that

$$PV^\gamma = \text{constant} \tag{20.38}$$

The **Boltzmann distribution law** describes the distribution of particles among available energy states. The relative number of particles having energy between E and $E + dE$ is $n_V(E)\, dE$, where

$$n_V(E) = n_0 e^{-E/k_B T} \tag{20.41}$$

The **Maxwell–Boltzmann speed distribution function** describes the distribution of speeds of molecules in a gas:

$$N_v = 4\pi N\left(\frac{m_0}{2\pi k_B T}\right)^{3/2} v^2 e^{-m_0 v^2/2k_B T} \tag{20.42}$$

Equation 20.42 enables us to calculate the **root-mean-square speed**, the **average speed**, and the **most probable speed** of molecules in a gas:

$$v_{\text{rms}} = \sqrt{\overline{v^2}} = \sqrt{\frac{3k_B T}{m_0}} = 1.73\sqrt{\frac{k_B T}{m_0}} \tag{20.43}$$

$$v_{\text{avg}} = \sqrt{\frac{8k_B T}{\pi m_0}} = 1.60\sqrt{\frac{k_B T}{m_0}} \tag{20.44}$$

$$v_{\text{mp}} = \sqrt{\frac{2k_B T}{m_0}} = 1.41\sqrt{\frac{k_B T}{m_0}} \tag{20.45}$$

Think–Pair–Share

See the Preface for an explanation of the icons used in this problems set. For additional assessment items for this section, go to **WEBASSIGN** From Cengage

1. Your group has been hired to do some consulting for a new baseball stadium that is being built in your city. It is going to be a totally enclosed stadium, and the architect is concerned about temperature regulation in the interior of the stadium. An eccentric architect on the development team has asked you to consider the following: every time a

baseball is hit or thrown, it eventually comes to rest because it is caught, it hits the fence, it lands in the stands, or it rolls on the ground to a stop. Whatever stops the ball, its initial kinetic energy is eventually transformed to internal energy in the interior of the stadium. The fear has been raised by this oddball character that an exciting game with lots of stopped baseballs will warm up the interior of the stadium too much for the air conditioning system to handle it. Perform a calculation to show the architect that even an exciting game will not challenge the air conditioning system for this reason.

2. **ACTIVITY** Consider the ten objects in our solar system listed below. Using the surface temperatures listed for these objects, determine the average speed of helium atoms located in the atmospheres of these objects. Using data in Table 13.2 and Equation 13.22, determine the escape speed for each of the objects. Finally, take the ratio of the escape speed to the average speed for helium atoms for each object. (a) What is a typical value for this ratio for an object with little to no atmosphere? (b) What is a typical value for this ratio for an object to have a robust atmosphere, but with little to no helium present? (c) What is a typical value for this ratio for an object to have a robust atmosphere with a significant amount of helium?

Object	Surface Temperature (K) (from the Lunar and Planetary Institute)	Atmosphere
Mercury	440	Little to none
Venus	741	Robust, little He
Earth	288	Robust, little He
Mars	244	Robust, little He
Ceres	173	Little to none
Jupiter	165	Robust, much He
Saturn	134	Robust, much He
Uranus	77	Robust, much He
Neptune	70	Robust, much He
Pluto	40	Little to none

3. **ACTIVITY** You are working as a teaching assistant to a physics professor. He has provided you with the following data on exam scores for two different sections of his physics class:

Section 1 Exam Scores	Section 2 Exam Scores
65	99
65	95
65	90
65	85
65	77
65	75
65	75
65	73
65	70
65	67
64	66
64	65
64	63
64	58
64	52
64	49
64	48
64	35
64	28
64	20

(a) Calculate the average score for each section and the rms average score for each section. (b) How the do the average scores for the two sections compare? (c) Compare the average and rms average scores for Section 1. Why do you think these two quantities have this relationship? (d) Compare the average and rms average scores for Section 2. (e) Will the rms average of a set of numbers always be larger than the straight average? Why or why not?

Problems

See the Preface for an explanation of the icons used in this problems set. For additional assessment items for this section, go to ⚡ **WEBASSIGN** From Cengage

SECTION 20.1 Molecular Model of an Ideal Gas

> Problem 20 in Chapter 18 can be assigned with this section.

1. **T** A spherical balloon of volume 4.00×10^3 cm^3 contains helium at a pressure of 1.20×10^5 Pa. How many moles of helium are in the balloon if the average kinetic energy of the helium atoms is 3.60×10^{-22} J?

2. **S** A spherical balloon of volume V contains helium at a pressure P. How many moles of helium are in the balloon if the average kinetic energy of the helium atoms is K_{avg}?

3. **V** A 2.00-mol sample of oxygen gas is confined to a 5.00-L vessel at a pressure of 8.00 atm. Find the average translational kinetic energy of the oxygen molecules under these conditions.

4. Oxygen, modeled as an ideal gas, is in a container and has a temperature of 77.0°C. What is the rms-average magnitude of the momentum of the gas molecules in the container?

5. A 5.00-L vessel contains nitrogen gas at 27.0°C and 3.00 atm. Find (a) the total translational kinetic energy of the gas molecules and (b) the average kinetic energy per molecule.

6. Calculate the mass of an atom of (a) helium, (b) iron, and (c) lead. Give your answers in kilograms. The atomic masses of these atoms are 4.00 u, 55.9 u, and 207 u, respectively.

7. **T** **V** In a period of 1.00 s, 5.00×10^{23} nitrogen molecules strike a wall with an area of 8.00 cm^2. Assume the molecules move with a speed of 300 m/s and strike the wall head-on in elastic collisions. What is the pressure exerted on the wall? *Note:* The mass of one N$_2$ molecule is 4.65×10^{-26} kg.

8. **Q** **C** A 7.00-L vessel contains 3.50 moles of gas at a pressure of 1.60×10^6 Pa. Find (a) the temperature of the gas and (b) the average kinetic energy of the gas molecules in the vessel. (c) What additional information would you need if you were asked to find the average speed of the gas molecules?

SECTION 20.2 Molar Specific Heat of an Ideal Gas

> *Note:* You may use data in Table 20.2 about particular gases. Here we define a "monatomic ideal gas" to have molar specific heats $C_V = \frac{3}{2}R$ and $C_P = \frac{5}{2}R$, and a "diatomic ideal gas" to have $C_V = \frac{5}{2}R$ and $C_P = \frac{7}{2}R$.

9. Calculate the change in internal energy of 3.00 mol of helium gas when its temperature is increased by 2.00 K.

10. Your sister, who is a realtor, is quite interested in your studies
CR in physics. Part of her job in selling properties involves being aware of the details of heating systems for the houses. She comes to you one day and says that a client told her the following: "If you measure the internal energy in the air in a house and then turn up the thermostat to a higher temperature, the internal energy of the air in the house is exactly the same as it was at the lower temperature." She finds this hard to believe, since you have added energy to the air by running the furnace. Help her to figure out if this statement is true or not.

11. In a constant-volume process, 209 J of energy is transferred
V by heat to 1.00 mol of an ideal monatomic gas initially at 300 K. Find (a) the work done on the gas, (b) the increase in internal energy of the gas, and (c) its final temperature.

12. A vertical cylinder with a heavy piston contains air at 300 K. The initial pressure is 2.00×10^5 Pa, and the initial volume is 0.350 m^3. Take the molar mass of air as 28.9 g/mol and assume $C_V = \frac{5}{2}R$. (a) Find the specific heat of air at constant volume in units of J/kg · °C. (b) Calculate the mass of the air in the cylinder. (c) Suppose the piston is held fixed. Find the energy input required to raise the temperature of the air to 700 K. (d) **What If?** Assume again the conditions of the initial state and assume the heavy piston is free to move. Find the energy input required to raise the temperature to 700 K.

13. A 1.00-L insulated bottle is full of tea at 90.0°C. You pour out one cup of tea and immediately screw the stopper back on the bottle. Make an order-of-magnitude estimate of the change in temperature of the tea remaining in the bottle that results from the admission of air at room temperature. State the quantities you take as data and the values you measure or estimate for them.

SECTION 20.3 The Equipartition of Energy

14. A certain molecule has f degrees of freedom. Show that an
S ideal gas consisting of such molecules has the following properties: (a) its total internal energy is $fnRT/2$, (b) its molar specific heat at constant volume is $fR/2$, (c) its molar specific heat at constant pressure is $(f + 2)R/2$, and (d) its specific heat ratio is $\gamma = C_P/C_V = (f + 2)/f$.

15. You are working for an automobile tire company. Your super-
CR visor is studying the effects of molecules striking the inner surface of the tire due to their thermal motion. He gives you the following data from a recent experiment. The air in a tire on a parked car was measured to have a gauge pressure of $P_i = 1.65$ atm on a day when the temperature was $T = 6.5°C$. The car was then driven for a while and then measurements were taken again. The gauge pressure in the tire was then $P_f = 1.95$ atm and the interior volume of the tire had increased by 5.00%. (a) Your supervisor asks you to determine by what factor the rms speed of the air molecules had increased from the first measurement to the second. (b) He also hints at a proposal he is going to make to replace air in tires with argon. Will this change the factor by which the average speed of the molecules changes in the conditions described?

16. *Why is the following situation impossible?* A team of researchers discovers a new gas, which has a value of $\gamma = C_P/C_V$ of 1.75.

17. You and your younger brother are designing an air rifle
CR that will shoot a lead pellet with mass $m = 1.10$ g and cross-sectional area $A = 0.030\ 0$ cm^2. The rifle works by allowing high-pressure air to expand, propelling the pellet down the rifle barrel. Because this process happens very quickly, no appreciable thermal conduction occurs and the expansion is essentially adiabatic. Your design is such that, once the pressure begins pushing on the pellet, it moves a distance of $L = 50.0$ cm before leaving the open end of the rifle at your desired speed of $v = 120$ m/s.

Your design also includes a chamber of volume $V = 12.0$ cm^3 in which the high-pressure air is stored until it is released. Your brother reminds you that you need to purchase a pump to pressurize the chamber. To determine what kind of pump to buy, you need to find what the pressure of the air must be in the chamber to achieve your desired muzzle speed. Ignore the effects of the air in front of the bullet and friction with the inside walls of the barrel.

SECTION 20.4 Adiabatic Processes for an Ideal Gas

18. During the compression stroke of a certain gasoline engine,
V the pressure increases from 1.00 atm to 20.0 atm. If the process is adiabatic and the air–fuel mixture behaves as a diatomic ideal gas, (a) by what factor does the volume change and (b) by what factor does the temperature change? Assuming the compression starts with 0.016 0 mol of gas at 27.0°C, find the values of (c) Q, (d) ΔE_{int}, and (e) W that characterize the process.

19. Air in a thundercloud expands as it rises. If its initial temper-
T ature is 300 K and no energy is lost by thermal conduction on expansion, what is its temperature when the initial volume has doubled?

20. *Why is the following situation impossible?* A new diesel engine that increases fuel economy over previous models is designed. Automobiles fitted with this design become incredible best sellers. Two design features are responsible for the increased fuel economy: (1) the engine is made entirely of aluminum to reduce the weight of the automobile, and (2) the exhaust of the engine is used to prewarm the air to 50°C before it enters the cylinder to increase the final temperature of the compressed gas. The engine has a *compression ratio*—that is, the ratio of the initial volume of the air to its final volume after compression—of 14.5. The compression process is adiabatic, and the air behaves as a diatomic ideal gas with $\gamma = 1.40$.

21. Air (a diatomic ideal gas) at 27.0°C and atmospheric pressure
GP is drawn into a bicycle pump that has a cylinder with an inner diameter of 2.50 cm and length 50.0 cm. The downstroke adiabatically compresses the air, which reaches a gauge pressure of 8.00×10^5 Pa before entering the tire. We wish to investigate the temperature increase of the pump. (a) What is the initial volume of the air in the pump? (b) What is the number of moles of air in the pump? (c) What is the absolute pressure of the compressed air? (d) What is the volume of the compressed air? (e) What is the temperature of the compressed air? (f) What is the increase in internal energy of the gas during the compression? **What If?** The pump is made of steel that is 2.00 mm thick. Assume 4.00 cm of the cylinder's length is allowed to come to thermal equilibrium with the air. (g) What is the volume of steel in this 4.00-cm length? (h) What is the mass of steel in this 4.00-cm length? (i) Assume the pump is compressed once. After the adiabatic expansion, conduction results in the energy increase in part (f) being shared between the gas and the 4.00-cm length of steel. What will be the increase in temperature of the steel after one compression?

SECTION 20.5 Distribution of Molecular Speeds

22. Two gases in a mixture diffuse through a filter at rates proportional to their rms speeds. (a) Find the ratio of speeds for the two isotopes of chlorine, ^{35}Cl and ^{37}Cl, as they diffuse through the air. (b) Which isotope moves faster?

23. **Review.** At what temperature would the average speed of helium atoms equal (a) the escape speed from the Earth, 1.12 $\times$ 10^4 m/s, and (b) the escape speed from the Moon, 2.37 $\times$ 10^3 m/s? *Note:* The mass of a helium atom is 6.64×10^{-27} kg.

24. From the Maxwell–Boltzmann speed distribution, show that the most probable speed of a gas molecule is given by Equation 20.45. *Note:* The most probable speed corresponds to the point at which the slope of the speed distribution curve dN_v/dv is zero.

25. Assume the Earth's atmosphere has a uniform temperature of 20.0°C and uniform composition, with an effective molar mass of 28.9 g/mol. (a) Show that the number density of molecules depends on height y above sea level according to

$$n_V(y) = n_0 e^{-m_0 gy/k_B T}$$

where n_0 is the number density at sea level (where $y = 0$). This result is called the *law of atmospheres*. (b) Commercial jetliners typically cruise at an altitude of 11.0 km. Find the ratio of the atmospheric density there to the density at sea level.

26. The law of atmospheres states that the number density of molecules in the atmosphere depends on height y above sea level according to

$$n_V(y) = n_0 e^{-m_0 gy/k_B T}$$

where n_0 is the number density at sea level (where $y = 0$). The average height of a molecule in the Earth's atmosphere is given by

$$y_{avg} = \frac{\int_0^\infty y n_V(y)\, dy}{\int_0^\infty n_V(y)\, dy} = \frac{\int_0^\infty y e^{-m_0 gy/k_B T}\, dy}{\int_0^\infty e^{-m_0 gy/k_B T}\, dy}$$

(a) Prove that this average height is equal to $k_B T/m_0 g$. (b) Evaluate the average height, assuming the temperature is 10.0°C and the molecular mass is 28.9 u, both uniform throughout the atmosphere.

ADDITIONAL PROBLEMS

27. Eight molecules have speeds of 3.00 km/s, 4.00 km/s, 5.80 km/s, 2.50 km/s, 3.60 km/s, 1.90 km/s, 3.80 km/s, and 6.60 km/s. Find (a) the average speed of the molecules and (b) the rms speed of the molecules.

28. In a sample of a solid metal, each atom is free to vibrate about some equilibrium position. The atom's energy consists of kinetic energy for motion in the x, y, and z directions plus elastic potential energy associated with the Hooke's law forces exerted by neighboring atoms on it in the x, y, and z directions. According to the theorem of equipartition of energy, assume the average energy of each atom is $\frac{1}{2}k_B T$ for each degree of freedom. (a) Prove that the molar specific heat of the solid is $3R$. The *Dulong–Petit law* states that this result generally describes pure solids at sufficiently high temperatures. (You may ignore the difference between the specific heat at constant pressure and the specific heat at constant volume.) (b) Evaluate the specific heat c of iron. Explain how it compares with the value listed in Table 19.1. (c) Repeat the evaluation and comparison for gold.

29. The dimensions of a classroom are 4.20 m $\times$ 3.00 m $\times$ 2.50 m. (a) Find the number of molecules of air in the classroom at atmospheric pressure and 20.0°C. (b) Find the mass of

this air, assuming the air consists of diatomic molecules with molar mass 28.9 g/mol. (c) Find the average kinetic energy of the molecules. (d) Find the rms molecular speed. (e) **What If?** Assume the molar specific heat of the air is independent of temperature. Find the change in internal energy of the air in the room as the temperature is raised to 25.0°C. (f) Explain how you could convince a fellow student that your answer to part (e) is correct, even though it sounds surprising.

30. The compressibility κ of a substance is defined as the fractional change in volume of that substance for a given change in pressure:

$$\kappa = -\frac{1}{V}\frac{dV}{dP}$$

(a) Explain why the negative sign in this expression ensures κ is always positive. (b) Show that if an ideal gas is compressed isothermally, its compressibility is given by $\kappa_1 = 1/P$. (c) **What If?** Show that if an ideal gas is compressed adiabatically, its compressibility is given by $\kappa_2 = 1/(\gamma P)$. Determine values for (d) κ_1 and (e) κ_2 for a monatomic ideal gas at a pressure of 2.00 atm.

31. The Earth's atmosphere consists primarily of oxygen (21%) and nitrogen (78%). The rms speed of oxygen molecules (O_2) in the atmosphere at a certain location is 535 m/s. (a) What is the temperature of the atmosphere at this location? (b) Would the rms speed of nitrogen molecules (N_2) at this location be higher, equal to, or lower than 535 m/s? Explain. (c) Determine the rms speed of N_2 at his location.

32. **Review.** As a sound wave passes through a gas, the compressions are either so rapid or so far apart that thermal conduction is prevented by a negligible time interval or by effective thickness of insulation. The compressions and rarefactions are adiabatic. (a) Show that the speed of sound in an ideal gas is

$$v = \sqrt{\frac{\gamma R T}{M}}$$

where M is the molar mass. The speed of sound in a gas is given by Equation 16.35; use that equation and the definition of the bulk modulus from Section 12.4. (b) Compute the theoretical speed of sound in air at 20.0°C and state how it compares with the value in Table 16.1. Take $M = 28.9$ g/mol. (c) Show that the speed of sound in an ideal gas is

$$v = \sqrt{\frac{\gamma k_B T}{m_0}}$$

where m_0 is the mass of one molecule. (d) State how the result in part (c) compares with the most probable, average, and rms molecular speeds.

33. Examine the data for polyatomic gases in Table 20.2 and give a reason why sulfur dioxide has a higher specific heat at constant volume than the other polyatomic gases at 300 K.

34. In a cylinder, a sample of an ideal gas with number of moles n undergoes an adiabatic process. (a) Starting with the expression $W = -\int P\, dV$ and using the condition $PV^\gamma =$ constant, show that the work done on the gas is

$$W = \left(\frac{1}{\gamma - 1}\right)(P_f V_f - P_i V_i)$$

(b) Starting with the first law of thermodynamics, show that the work done on the gas is equal to $nC_V(T_f - T_i)$. (c) Are these two results consistent with each other? Explain.

35. As a 1.00-mol sample of a monatomic ideal gas expands adiabatically, the work done on it is -2.50×10^3 J. The initial temperature and pressure of the gas are 500 K and 3.60 atm. Calculate (a) the final temperature and (b) the final pressure.

36. A sample consists of an amount n in moles of a monatomic ideal gas. The gas expands adiabatically, with work W done on it. (Work W is a negative number.) The initial temperature and pressure of the gas are T_i and P_i. Calculate (a) the final temperature and (b) the final pressure.

37. The latent heat of vaporization for water at room temperature is 2 430 J/g. Consider one particular molecule at the surface of a glass of liquid water, moving upward with sufficiently high speed that it will be the next molecule to join the vapor. (a) Find its translational kinetic energy. (b) Find its speed. Now consider a thin gas made only of molecules like that one. (c) What is its temperature? (d) Why are you not burned by water evaporating from a vessel at room temperature?

38. A vessel contains 1.00×10^4 oxygen molecules at 500 K. (a) Make an accurate graph of the Maxwell speed distribution function versus speed with points at speed intervals of 100 m/s. (b) Determine the most probable speed from this graph. (c) Calculate the average and rms speeds for the molecules and label these points on your graph. (d) From the graph, estimate the fraction of molecules with speeds in the range 300 m/s to 600 m/s.

39. For a Maxwellian gas, use a computer or programmable calculator to find the numerical value of the ratio $N_v(v)/N_v(v_{mp})$ for the following values of v: (a) $v = (v_{mp}/50.0)$, (b) $(v_{mp}/10.0)$, (c) $(v_{mp}/2.00)$, (d) v_{mp}, (e) $2.00v_{mp}$, (f) $10.0v_{mp}$, and (g) $50.0v_{mp}$. Give your results to three significant figures.

40. A triatomic molecule can have a linear configuration, as does CO_2 (Fig. P20.40a), or it can be nonlinear, like H_2O (Fig. P20.40b). Suppose the temperature of a gas of triatomic molecules is sufficiently low that vibrational motion is negligible. What is the molar specific heat at constant volume, expressed as a multiple of the universal gas constant, (a) if the molecules are linear and (b) if the molecules are nonlinear? At high temperatures, a triatomic molecule has two modes of vibration, and each contributes $\frac{1}{2}R$ to the molar specific heat for its kinetic energy and another $\frac{1}{2}R$ for its potential energy. Identify the high-temperature molar specific heat at constant volume for a triatomic ideal gas of (c) linear molecules and (d) nonlinear molecules. (e) Explain how specific heat data can be used to determine whether a triatomic molecule is linear or nonlinear. Are the data in Table 20.2 sufficient to make this determination?

Figure P20.40

41. Using the Maxwell–Boltzmann speed distribution function, verify Equations 20.43 and 20.44 for (a) the rms speed and (b) the average speed of the molecules of a gas at a temperature T. The average value of v^n is

$$\overline{v^n} = \frac{1}{N} \int_0^\infty v^n N_v \, dv$$

Use the table of integrals B.6 in Appendix B.

42. On the PV diagram for an ideal gas, one isothermal curve and one adiabatic curve pass through each point as shown in Figure P20.42. Prove that the slope of the adiabatic curve is steeper than the slope of the isotherm at that point by the factor γ.

Figure P20.42

43. Using multiple laser beams, physicists have been able to cool and trap sodium atoms in a small region. In one experiment, the temperature of the atoms was reduced to 0.240 mK. (a) Determine the rms speed of the sodium atoms at this temperature. The atoms can be trapped for about 1.00 s. The trap has a linear dimension of roughly 1.00 cm. (b) Over what approximate time interval would an atom wander out of the trap region if there were no trapping action?

44. Consider the particles in a gas centrifuge, a device used to separate particles of different mass by whirling them in a circular path of radius r at angular speed ω. The force acting on a gas molecule toward the center of the centrifuge is $m_0 \omega^2 r$. (a) Discuss how a gas centrifuge can be used to separate particles of different mass. (b) Suppose the centrifuge contains a gas of particles of identical mass. Show that the density of the particles as a function of r is

$$n(r) = n_0 e^{m_0 r^2 \omega^2 / 2k_B T}$$

CHALLENGE PROBLEM

45. Equations 20.43 and 20.44 show that $v_{rms} > v_{avg}$ for a collection of gas particles, which turns out to be true whenever the particles have a distribution of speeds. Let us explore this inequality for a two-particle gas. Let the speed of one particle be $v_1 = av_{avg}$ and the other particle have speed $v_2 = (2 - a)v_{avg}$. (a) Show that the average of these two speeds is v_{avg}. (b) Show that

$$v_{rms}^2 = v_{avg}^2 (2 - 2a + a^2)$$

(c) Argue that the equation in part (b) proves that, in general, $v_{rms} > v_{avg}$. (d) Under what special condition will $v_{rms} = v_{avg}$ for the two-particle gas?

21

Heat Engines, Entropy, and the Second Law of Thermodynamics

STORYLINE **You are still on your Physics Club camping trip. Uh-oh.**
The electric refrigerator in your Club advisor's RV has suddenly stopped working. You help your advisor take the refrigerator out from its mounting and inspect the workings. He looks around a bit, tries a few things, and then says that he suspects the compressor has gone bad. You ask him what the compressor does and he says that it compresses the gaseous refrigerant to a high temperature and pressure. This starts you thinking. If you are trying to keep food *cold* in a refrigerator, why would you want to make the refrigerant *hot*? Your advisor goes on to explain that air conditioners work the same way. Then he says that he thinks the refrigerator cannot be repaired and they should go to the camping supply store and buy an inexpensive propane-powered refrigerator to use for the rest of your trip. You say, "What!? You can *cool* food by *burning* propane? How can that possibly work?" You spend the next couple of hours online investigating refrigeration cycles.

CONNECTIONS Although the first law of thermodynamics, which we studied in Chapter 19, is very important, it makes no distinction between processes that occur spontaneously and those that do not. Only certain types of energy transformation and transfer processes actually take place in nature. The second law of thermodynamics, the major topic in this chapter, establishes which processes do and do not occur. In general, for example, it is common to see processes in which mechanical energy is transformed into internal energy. As a book sliding across a surface comes to rest, its kinetic energy has transformed to internal energy, which spreads out in the book and the surface. One would never expect this internal energy to somehow gather itself back into the book, so that the book begins to move again. Books at rest and in static equilibrium *always* remain at rest. It is also common to see energy transfer by heat from a hot object to a cold object with which it is in contact. One would never expect to add ice to room-temperature water and see the water become warmer and the ice colder. Energy *always* transfers from the warm water to the cold ice. The expected processes described here are *irreversible*; that is, they are processes that occur naturally in one direction only. No irreversible process has ever been observed to run backward. If it were to do so, it would violate the second law of thermodynamics.[1] The second law of thermodynamics is at work in all natural processes that we will study in future chapters. In this chapter, we study this law and a closely related quantity, *entropy*. We begin our quest to understand both the second law and entropy by investigating the thermodynamics of *heat engines*.

21.1 Heat Engines and the Second Law of Thermodynamics

A **heat engine** is a device that takes in energy by heat[2] and, operating in a cyclic process, expels a fraction of that energy by means of work. For instance, in a typical process by which a power plant produces electricity, a fuel such as natural gas is burned and the high-temperature gases produced are used to convert liquid water to steam. This steam is directed at the blades of a turbine. The steam does work on the blades of the turbine, setting it into rotation. The mechanical energy associated with this rotation is used to drive an electric generator. Another device that can be modeled as a heat engine is the internal combustion engine in an automobile. This device uses energy from a burning fuel to perform work on pistons that results in the motion of the automobile.

Let us consider the fundamental operation of a heat engine in more detail. A heat engine carries some working substance through a cyclic process during which (1) the working substance absorbs energy by heat from a high-temperature energy reservoir, (2) work is done by the engine, and (3) energy is expelled by heat to a lower-temperature reservoir. As an example, consider the operation of a steam engine (Fig. 21.1), which uses water as the working substance. The water in a boiler absorbs energy from burning fuel and evaporates to steam, which then does work by expanding against a piston. After the steam cools and condenses, the liquid water produced returns to the boiler and the cycle repeats.

It is useful to represent a heat engine schematically as in Figure 21.2 (page 558). The engine absorbs a quantity of energy $|Q_h|$ from the hot reservoir. For the mathematical discussion of heat engines, we use absolute values to make all energy transfers by heat positive, and the direction of transfer is indicated with an explicit positive or negative sign. The engine does work W_{eng} (so that *negative* work $W = -W_{eng}$ is done *on* the engine) and then gives up a quantity of energy $|Q_c|$ to the cold reservoir. Because the working substance in the engine goes through a cycle, its initial and final internal energies are equal: $\Delta E_{int} = 0$. Hence, from the first law of thermodynamics, for each cycle of the engine, $\Delta E_{int} = Q + W = Q_{net} - W_{eng} = 0$, and

Lord Kelvin
British physicist and mathematician (1824–1907)
Born William Thomson in Belfast, Kelvin was the first to propose the use of an absolute scale of temperature. The Kelvin temperature scale is named in his honor. Kelvin's work in thermodynamics led to the idea that energy cannot pass spontaneously from a colder object to a hotter object.

Figure 21.1 A steam-driven locomotive obtains its energy by burning wood or coal. The generated energy vaporizes water into steam, which powers the locomotive. Modern locomotives use diesel fuel instead of wood or coal. Whether old-fashioned or modern, such locomotives can be modeled as heat engines, which extract energy from a burning fuel and convert a fraction of it to mechanical energy.

---

[1]Although a process occurring in the time-reversed sense has never been *observed*, it is *possible* for it to occur. As we shall see later in this chapter, however, the probability of such a process occurring is infinitesimally small. From this viewpoint, processes occur with a vastly greater probability in one direction than in the opposite direction.

[2]We use heat as our model for energy transfer into a heat engine. Other methods of energy transfer are possible in the model of a heat engine, however. For example, the Earth's atmosphere can be modeled as a heat engine in which the input energy transfer is by means of electromagnetic radiation from the Sun. The output of the atmospheric heat engine causes the wind structure in the atmosphere.

the net work W_{eng} done by a heat engine is equal to the net energy Q_{net} transferred to it. As you can see from Figure 21.2, $Q_{net} = |Q_h| - |Q_c|$; therefore,

$$W_{eng} = |Q_h| - |Q_c| \tag{21.1}$$

The **thermal efficiency** e of a heat engine is defined as the ratio of the net work done by the engine during one cycle to the energy input at the higher temperature during the cycle:

◀ Thermal efficiency of a heat engine

$$e \equiv \frac{W_{eng}}{|Q_h|} = \frac{|Q_h| - |Q_c|}{|Q_h|} = 1 - \frac{|Q_c|}{|Q_h|} \tag{21.2}$$

You can think of the efficiency as the ratio of what you gain (work) to what you give (energy transfer at the higher temperature). In practice, all heat engines expel only a fraction of the input energy Q_h by mechanical work; consequently, their efficiency is always less than 100%. For example, a good automobile engine has an efficiency of about 20%, and diesel engines have efficiencies ranging from 35% to 40%.

Equation 21.2 shows that a heat engine has 100% efficiency ($e = 1$) only if $|Q_c| = 0$, that is, if no energy is expelled to the cold reservoir. In other words, a heat engine with perfect efficiency would have to expel all the input energy by work. Figure 21.3 is a schematic diagram of the "perfect" heat engine. The efficiencies of real engines are well below 100%, which is related to one form of the second law of thermodynamics. The **Kelvin–Planck form of the second law of thermodynamics** states the impossibility of an engine with 100% efficiency:

It is impossible to construct a heat engine that, operating in a cycle, produces no effect other than the input of energy by heat from a reservoir and the performance of an equal amount of work.

This statement of the second law means that during the operation of a heat engine, W_{eng} can never be equal to $|Q_h|$ or, alternatively, that some energy $|Q_c|$ *must* be rejected to the environment.

Figure 21.2 Schematic representation of a heat engine.

The engine does work W_{eng}.
Hot reservoir at T_h
Energy $|Q_h|$ enters the engine.
Q_h
Heat engine
W_{eng}
Energy $|Q_c|$ leaves the engine.
Q_c
Cold reservoir at T_c

PITFALL PREVENTION 21.1
The First and Second Laws Notice the distinction between the first and second laws of thermodynamics. If a gas undergoes a *one-time isothermal process*, then $\Delta E_{int} = Q + W = 0$ and $W = -Q$. Therefore, the first law allows *all* energy input by heat to be expelled by work. In a heat engine, however, in which a substance undergoes a *cyclic* process, only a *portion* of the energy input by heat can be expelled by work according to the second law.

QUICK QUIZ 21.1 The energy input to an engine is 4.00 times greater than the work it performs. **(i)** What is its thermal efficiency? (a) 4.00 (b) 1.00 (c) 0.250 (d) impossible to determine **(ii)** What fraction of the energy input is expelled to the cold reservoir? (a) 0.250 (b) 0.750 (c) 1.00 (d) impossible to determine

An impossible heat engine
Hot reservoir at T_h
Q_h
Heat engine
W_{eng}
Cold reservoir at T_c

Figure 21.3 Schematic diagram of a heat engine that takes in energy from a hot reservoir and does an equivalent amount of work. It is impossible to construct such a perfect engine.

| Example **21.1** | **The Efficiency of an Engine** |

An engine transfers 2.00×10^3 J of energy from a hot reservoir during a cycle and transfers 1.50×10^3 J as exhaust to a cold reservoir.

(A) Find the efficiency of the engine.

SOLUTION

Conceptualize   Review Figure 21.2; think about energy going into the engine from the hot reservoir and splitting, with part coming out by work and part by heat into the cold reservoir.

Categorize   This example involves evaluation of quantities from the equations introduced in this section, so we categorize it as a substitution problem.

Find the efficiency of the engine from Equation 21.2:

$$e = 1 - \frac{|Q_c|}{|Q_h|} = 1 - \frac{1.50 \times 10^3 \text{ J}}{2.00 \times 10^3 \text{ J}} = \boxed{0.250, \text{ or } 25.0\%}$$

(B) How much work does this engine do in one cycle?

SOLUTION

Find the work done by the engine by taking the difference between the input and output energies:

$$W_{\text{eng}} = |Q_h| - |Q_c| = 2.00 \times 10^3 \text{ J} - 1.50 \times 10^3 \text{ J}$$

$$= \boxed{5.0 \times 10^2 \text{ J}}$$

WHAT IF?   Suppose you were asked for the power output of this engine. Do you have sufficient information to answer this question?

Answer   No, you do not have enough information. The power of an engine is the *rate* at which work is done by the engine. You know how much work is done per cycle, but you have no information about the time interval associated with one cycle. If you were told that the engine operates at 2 000 rpm (revolutions per minute), however, you could relate this rate to the period of rotation T of the mechanism of the engine. Assuming there is one thermodynamic cycle per revolution, the power is

$$P = \frac{W_{\text{eng}}}{T} = \frac{5.0 \times 10^2 \text{ J}}{\left(\frac{1}{2\,000} \text{ min}\right)} \left(\frac{1 \text{ min}}{60 \text{ s}}\right) = 1.7 \times 10^4 \text{ W}$$

21.2 Heat Pumps and Refrigerators

In a heat engine, the direction of energy transfer is from the hot reservoir to the cold reservoir, which is the natural direction. The role of the heat engine is to process the energy from the hot reservoir so as to expel part of it by useful work. What if we wanted to transfer energy from the cold reservoir to the hot reservoir? Because that is not the natural direction of energy transfer, we must put some energy into a device to perform this task. Devices that transfer energy from a cold reservoir to a warm reservoir are called **heat pumps** and **refrigerators.** For example, homes in summer are cooled using heat pumps called *air conditioners*. The air conditioner transfers energy from the cool room in the home to the warm air outside.

In a refrigerator or a heat pump, the engine takes in energy $|Q_c|$ from a cold reservoir and expels energy $|Q_h|$ to a hot reservoir (Fig. 21.4), which can be accomplished only if work is done *on* the engine. From the first law, we know that the energy given up to the hot reservoir must equal the sum of the work done and the energy taken in from the cold reservoir. Therefore, the refrigerator or heat pump transfers energy from a colder body (for example, the contents of a kitchen refrigerator or the winter air outside a building) to a hotter body (the air in the kitchen or a room in the building). In practice, it is desirable to carry out this process with a minimum of work. If the process could be accomplished without doing

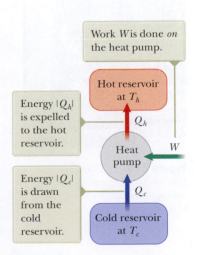

Figure 21.4 Schematic representation of a heat pump.

Work W is done *on* the heat pump.

Hot reservoir at T_h

Energy $|Q_h|$ is expelled to the hot reservoir.

Q_h

Heat pump

W

Energy $|Q_c|$ is drawn from the cold reservoir.

Q_c

Cold reservoir at T_c

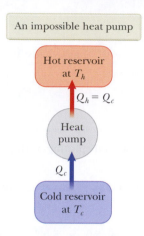

An impossible heat pump

Figure 21.5 Schematic diagram of an impossible heat pump or refrigerator, that is, one that takes in energy from a cold reservoir and expels an equivalent amount of energy to a hot reservoir without the input of energy by work.

The coils on the back of a refrigerator transfer energy by heat to the air.

Figure 21.6 The back of a household refrigerator. The air surrounding the coils is the hot reservoir.

any work, the refrigerator or heat pump would be "perfect" (Fig. 21.5). Again, the existence of such a device is impossible and is related to another form of the second law of thermodynamics. The **Clausius statement**[3] **of the second law of thermodynamics** states:

> It is impossible to construct a cyclical machine whose sole effect is to transfer energy continuously by heat from one object to another object at a higher temperature without the input of energy by work.

In simpler terms, energy does not transfer spontaneously by heat from a cold object to a hot object. Work input is required to run a refrigerator.

The Clausius and Kelvin–Planck statements of the second law of thermodynamics appear at first sight to be unrelated, but in fact they are equivalent in all respects. Although we do not prove so here, if either statement is false, so is the other.[4]

In practice, a heat pump includes a circulating fluid that passes through two sets of metal coils that can exchange energy with the surroundings. The fluid is cold and at low pressure when it is in the coils located in a cool environment, where it absorbs energy by heat. The resulting warm fluid is then compressed and enters the other coils as a hot, high-pressure fluid. There it releases its stored energy to the warm surroundings. In an air conditioner, energy is absorbed into the fluid in coils located in a building's interior; after the fluid is compressed, energy leaves the fluid through coils located outdoors. In a refrigerator, the external coils are behind the unit (Fig. 21.6) or underneath the unit. The internal coils are in the walls of the refrigerator and absorb energy from the food.

In the preceding paragraph, we see why we want the refrigerant to be hot, a question we asked in the opening storyline with regard to the refrigerator in your RV. But what about the propane refrigerator? This type of refrigerator also takes a substance through a cycle. In this case, the substance is ammonia, which combines at various parts of the cycle with water and hydrogen. The propane burner warms up the ammonia–water combination, which then travels to external coils to release energy into the air. The ammonia is separated from the water and then combines with hydrogen. It then evaporates. As we discussed in Chapter 20, evaporation is a cooling process. The cool ammonia is passed through the cooling coils, where it absorbs energy from the interior of the refrigerator, is mixed with water again, and then proceeds back to the beginning of the cycle.

The effectiveness of a heat pump is described in terms of a number called the **coefficient of performance** (COP). The COP is similar to the thermal efficiency for a heat engine in that it is a ratio of what you gain (energy transferred to or from a reservoir) to what you give (work input). For a heat pump operating in the cooling mode, "what you gain" is energy removed from the cold reservoir. The most effective refrigerator or air conditioner is one that removes the greatest amount of energy from the cold reservoir in exchange for the least amount of work. Therefore, for these devices operating in the cooling mode, we define the COP in terms of $|Q_c|$:

$$\text{COP (cooling mode)} = \frac{\text{energy transferred at low temperature}}{\text{work done on heat pump}} = \frac{|Q_c|}{W} \qquad (21.3)$$

A good refrigerator should have a high COP, typically 5 or 6.

In addition to cooling applications, heat pumps are becoming increasingly popular for heating purposes. In the heating mode, energy is absorbed from the cool air outside a building and warm air is released inside the building. The COP of a

[3]First expressed by Rudolf Clausius (1822–1888), a German physicist and mathematician.
[4]See an advanced textbook on thermodynamics for this proof.

heat pump is defined as the ratio of the energy transferred to the hot reservoir to the work required to transfer that energy:

$$\text{COP (heating mode)} = \frac{\text{energy transferred at high temperature}}{\text{work done on heat pump}} = \frac{|Q_h|}{W} \quad \text{(21.4)}$$

If the outside temperature is 25°F (−4°C) or higher, a typical value of the COP for a heat pump is about 4. That is, the amount of energy transferred to the building is about four times greater than the work done by the motor in the heat pump. As the outside temperature decreases, however, it becomes more difficult for the heat pump to extract sufficient energy from the air and so the COP decreases. Therefore, the use of heat pumps that extract energy from the air, although satisfactory in moderate climates, is not appropriate in areas where winter temperatures are very low. It is possible to use heat pumps in colder areas by burying the external coils deep in the ground. In that case, the energy is extracted from the ground, which tends to be warmer than the air in the winter.

QUICK QUIZ **21.2** The energy entering an electric heater by electrical transmission can be converted to internal energy with an efficiency of 100%. By what factor does the cost of heating your home change when you replace your electric heating system with an electric heat pump that has a COP of 4.00? Assume the motor running the heat pump is 100% efficient. **(a)** 4.00 **(b)** 2.00 **(c)** 0.500 **(d)** 0.250

Example **21.2**   **Freezing Water**

A certain refrigerator has a COP of 5.00. When the refrigerator is running, its power input is 500 W. A sample of water of mass 500 g and temperature 20.0°C is placed in the freezer compartment. How long does it take to freeze the water to ice at 0°C? Assume all other parts of the refrigerator stay at the same temperature and there is no leakage of energy from the exterior, so the operation of the refrigerator results only in energy being extracted from the water.

SOLUTION

Conceptualize   Energy leaves the water, reducing its temperature and then freezing it into ice. The time interval required for this entire process is related to the rate at which energy is withdrawn from the water, which, in turn, is related to the power input of the refrigerator.

Categorize   We categorize this example as one that combines our understanding of temperature changes and phase changes from Chapter 19 and our understanding of heat pumps from this chapter.

Analyze   Use the power rating of the refrigerator to find the time interval Δt required for the freezing process to occur:

$$P = \frac{W}{\Delta t} \;\rightarrow\; \Delta t = \frac{W}{P}$$

Use Equation 21.3 to relate the work W done on the heat pump to the energy $|Q_c|$ extracted from the water:

$$\Delta t = \frac{|Q_c|}{P(\text{COP})}$$

Use Equations 19.4 and 19.8 to substitute the amount of energy $|Q_c|$ that must be extracted from the water of mass m:

$$\Delta t = \frac{|mc\,\Delta T + L_f\,\Delta m|}{P(\text{COP})}$$

Recognize that the amount of water that freezes is $\Delta m = -m$ because all the water freezes:

$$\Delta t = \frac{|m(c\,\Delta T - L_f)|}{P(\text{COP})}$$

continued

21.2 continued

Substitute numerical values:

$$\Delta t = \frac{|(0.500 \text{ kg})[(4\ 186 \text{ J/kg} \cdot {}^\circ\text{C})(-20.0{}^\circ\text{C}) - 3.33 \times 10^5 \text{ J/kg}]|}{(500 \text{ W})(5.00)}$$

$$= \boxed{83.3 \text{ s}}$$

Finalize In reality, the time interval for the water to freeze in a refrigerator is much longer than 83.3 s, which suggests that the assumptions of our model are not valid. Only a small part of the energy extracted from the refrigerator interior in a given time interval comes from the water. Energy must also be extracted from the container in which the water is placed, and energy that continuously leaks into the interior from the exterior must be extracted.

21.3 Reversible and Irreversible Processes

Figure 21.7 Adiabatic free expansion of a gas.

In the next section, we will discuss a theoretical heat engine that is the most efficient possible. To understand its nature, we must first examine the meaning of reversible and irreversible processes. In a **reversible** process, the system undergoing the process can be returned to its initial conditions along the same path on a PV diagram, and every point along this path is an equilibrium state. A process that does not satisfy these requirements is **irreversible.**

All natural processes are known to be irreversible. Let's examine a unique process called the **adiabatic free expansion** of a gas, and show that it cannot be reversible. Consider a gas in a thermally insulated container as shown in Figure 21.7. A membrane separates the gas from a vacuum. When the membrane is punctured, the gas expands freely into the vacuum. As a result of the puncture, the system has changed because it occupies a greater volume after the expansion. Because the gas does not exert a force through a displacement, it does no work on the surroundings as it expands: $W = 0$. In addition, no energy is transferred to or from the gas by heat because the container is insulated from its surroundings: $Q = 0$. Therefore, from the first law of thermodynamics, the internal energy E_{int} of the gas does not change and, as a result, its temperature is the same after the expansion. In this process, the system has changed, but the surroundings have not.

For this process to be reversible, we must return the gas to its original volume and temperature without changing the surroundings. Imagine trying to reverse the process by compressing the gas to its original volume. To do so, we use an engine to force the piston shown in Figure 21.7 inward. During this process, the surroundings change because work is being done by an outside agent on the system. In addition, the system changes because the compression increases the temperature of the gas. The temperature of the gas can be lowered by allowing it to come into contact with an external energy reservoir. Although this step returns the gas to its original conditions, the surroundings are again affected because energy is being added to the surroundings from the gas. If this energy could be used to drive the engine that compressed the gas, the net energy transfer to the surroundings would be zero. In this way, the system and its surroundings could be returned to their initial conditions and we could identify the process as reversible. The Kelvin–Planck statement of the second law, however, specifies that the energy removed from the gas to return the temperature to its original value cannot be completely converted to mechanical energy by the process of work done by the engine in compressing the gas. Therefore, we must conclude that the process is irreversible.

We could also argue that the adiabatic free expansion is irreversible by relying on the portion of the definition of a reversible process that refers to equilibrium states. For example, during the sudden expansion, significant variations in pressure occur throughout the gas. Therefore, there is no single, well-defined value of the pressure for the entire system at any time between the initial and final states. In fact, the process cannot even be represented as a path on a *PV* diagram. The *PV* diagram for an adiabatic free expansion would show the initial and final conditions as points, but these points would not be connected by a path. Therefore, because the intermediate conditions between the initial and final states are not equilibrium states, the process is irreversible.

Although all real processes are irreversible, some are almost reversible. If a real process occurs very slowly such that the system is always very nearly in an equilibrium state, the process can be approximated as being reversible.

A general characteristic of a reversible process is that no nonconservative effects (such as turbulence or friction) that transform mechanical energy to internal energy can be present. Such effects can be impossible to eliminate completely. Hence, it is not surprising that real processes in nature are irreversible.

21.4 The Carnot Engine

In 1824, a French engineer named Sadi Carnot described a theoretical engine, now called a **Carnot engine,** that is of great importance from both practical and theoretical viewpoints. He showed that a heat engine operating in an ideal, reversible cycle—called a **Carnot cycle**—between two energy reservoirs is the most efficient engine possible. Such an ideal engine establishes an upper limit on the efficiencies of all other engines. That is, the net work done by a working substance taken through the Carnot cycle is the greatest amount of work possible for a given amount of energy supplied to the substance at the higher temperature. **Carnot's theorem** can be stated as follows:

> No real heat engine operating between two energy reservoirs can be more efficient than a Carnot engine operating between the same two reservoirs.

In this section, we will show that the efficiency of a Carnot engine depends only on the temperatures of the reservoirs. In turn, that efficiency represents the maximum possible efficiency for real engines. Let us confirm that the Carnot engine is the most efficient. We imagine a hypothetical engine with an efficiency greater than that of the Carnot engine. Consider Figure 21.8, which shows the hypothetical engine with $e > e_C$ on the left connected between hot and cold reservoirs. In addition, let us attach a Carnot engine between the same reservoirs. Because the Carnot cycle is reversible, the Carnot engine can be run in reverse as a Carnot heat pump as shown on the right in Figure 21.8. We match the output work of the engine to the input work of the heat pump, $W = W_C$, so there is no exchange of energy by work between the surroundings and the engine–heat pump combination.

Because of the proposed relation between the efficiencies of the heat engine and heat pump when both are operated as engines, we must have

$$e > e_C \;\rightarrow\; \frac{|W|}{|Q_h|} > \frac{|W_C|}{|Q_{hC}|}$$

The numerators of these two fractions cancel because the works have been matched in the configuration in Figure 21.8. Therefore, this expression becomes

$$|Q_{hC}| > |Q_h| \qquad (21.5)$$

From Equation 21.1, the equality of the works gives us

$$|W| = |W_C| \;\rightarrow\; |Q_h| - |Q_c| = |Q_{hC}| - |Q_{cC}|$$

Sadi Carnot
French engineer (1796–1832)
Carnot was the first to show the quantitative relationship between work and heat. In 1824, he published his only work, *Reflections on the Motive Power of Heat,* which reviewed the industrial, political, and economic importance of the steam engine. In it, he defined work as "weight lifted through a height."

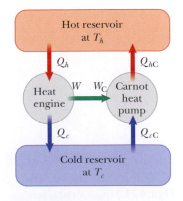

Figure 21.8 Two engines operate between two energy reservoirs: a Carnot engine operating as a heat pump and another engine with an efficiency that is proposed to be higher than that of the Carnot engine. The work output and input are matched.

which can be rewritten to put the energies exchanged with the cold reservoir on the left and those with the hot reservoir on the right:

$$|Q_{hC}| - |Q_h| = |Q_{cC}| - |Q_c| \qquad (21.6)$$

Note that, in light of Equation 21.5, the left side of Equation 21.6 is positive, so the right side must be positive also. We see that the net energy exchange with the hot reservoir is equal to the net energy exchange with the cold reservoir. As a result, for the combination of the heat engine and the heat pump, energy is transferring from the cold reservoir to the hot reservoir by heat with no input of energy by work from the surroundings.

This result is in violation of the Clausius statement of the second law. Therefore, our original assumption that $e > e_C$ must be incorrect, and we must conclude that the Carnot engine represents the highest possible efficiency for an engine. The key feature of the Carnot engine that makes it the most efficient is its *reversibility;* it can be run in reverse as a heat pump. All real engines are less efficient than the Carnot engine because they do not operate through a reversible cycle. The efficiency of a real engine is further reduced by such practical difficulties as friction and energy losses by conduction.

Let's now look at the details of the Carnot cycle for an engine operating between temperatures T_c and T_h. Assume the working substance is an ideal gas contained in a cylinder fitted with a movable piston at one end. The cylinder's walls and the piston are thermally nonconducting. Four stages of the Carnot cycle are shown in Figure 21.9,

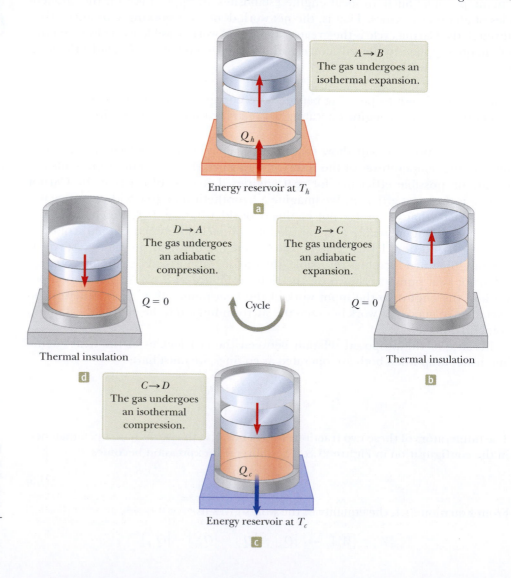

Figure 21.9 A pictorial representation of the Carnot cycle. The letters *A*, *B*, *C*, and *D* refer to the states of the gas shown in Figure 21.10. The arrows on the piston indicate the direction of its motion during each process. Compare to the graphical representation in Figure 21.10.

and the *PV* diagram for the cycle is shown in Figure 21.10. The Carnot cycle consists of two adiabatic processes and two isothermal processes, all reversible:

1. Process $A \rightarrow B$ (Fig. 21.9a) is an isothermal expansion at temperature T_h. The gas is placed in thermal contact with an energy reservoir at temperature T_h. During the expansion, the gas absorbs energy $|Q_h|$ from the reservoir through the base of the cylinder and does work W_{AB} in raising the piston.
2. In process $B \rightarrow C$ (Fig. 21.9b), the base of the cylinder is replaced by a thermally nonconducting wall and the gas expands adiabatically; that is, no energy enters or leaves the system by heat. During the expansion, the temperature of the gas decreases from T_h to T_c and the gas does work W_{BC} in raising the piston.
3. In process $C \rightarrow D$ (Fig. 21.9c), the gas is placed in thermal contact with an energy reservoir at temperature T_c and is compressed isothermally at temperature T_c. During this time, the gas expels energy $|Q_c|$ to the reservoir and the work done by the piston on the gas is W_{CD}.
4. In the final process $D \rightarrow A$ (Fig. 21.9d), the base of the cylinder is replaced by a nonconducting wall and the gas is compressed adiabatically. The temperature of the gas increases to T_h, and the work done by the piston on the gas is W_{DA}.

The thermal efficiency of the engine is given by Equation 21.2:

$$e = 1 - \frac{|Q_c|}{|Q_h|}$$

In Example 21.3, we show that for a Carnot cycle,

$$\frac{|Q_c|}{|Q_h|} = \frac{T_c}{T_h} \qquad (21.7)$$

Hence, the thermal efficiency of a Carnot engine is

$$e_C = 1 - \frac{T_c}{T_h} \qquad (21.8)$$

◀ Efficiency of a Carnot engine

The work done during the cycle equals the area enclosed by the path on the *PV* diagram.

Figure 21.10 *PV* diagram for the Carnot cycle represented pictorially in Figure 21.9. This is a graphical representation of the cycle. The net work done W_{eng} equals the net energy transferred into the Carnot engine in one cycle, $|Q_h| - |Q_c|$.

This result indicates that all Carnot engines operating between the same two temperatures have the same efficiency.[5]

Equation 21.8 can be applied to any working substance operating in a Carnot cycle between two energy reservoirs. According to this equation, the efficiency is zero if $T_c = T_h$, as one would expect. The efficiency increases as T_c is lowered and T_h is raised. The efficiency can be unity (100%), however, only if $T_c = 0$ K. A reservoir at absolute zero is not available; therefore, the maximum efficiency is always less than 100%. In most practical cases, T_c is near room temperature, which is about 300 K. Therefore, one usually strives to increase the efficiency by raising T_h.

Theoretically, a Carnot-cycle heat engine run in reverse constitutes the most effective heat pump possible, and it determines the maximum COP for a given

---

[5]For the processes in the Carnot cycle to be reversible, they must be carried out infinitesimally slowly. Therefore, although the Carnot engine is the most efficient engine possible, it has zero power output because it takes an infinite time interval to complete one cycle! For a real engine, the short time interval for each cycle results in the working substance reaching a high temperature lower than that of the hot reservoir and a low temperature higher than that of the cold reservoir. An engine undergoing a Carnot cycle between this narrower temperature range was analyzed by F. L. Curzon and B. Ahlborn ("Efficiency of a Carnot engine at maximum power output," *Am. J. Phys.* **43**(1), 22, 1975), who found that the efficiency at maximum power output depends only on the reservoir temperatures T_c and T_h and is given by $e_{C-A} = 1 - (T_c/T_h)^{1/2}$. The Curzon–Ahlborn efficiency e_{C-A} provides a closer approximation to the efficiencies of real engines than does the Carnot efficiency.

combination of hot and cold reservoir temperatures. Using Equations 21.1 and 21.4, we see that the maximum COP for a heat pump in its heating mode is

$$\text{COP}_C \text{ (heating mode)} = \frac{|Q_h|}{W}$$

$$= \frac{|Q_h|}{|Q_h| - |Q_c|} = \frac{1}{1 - \dfrac{|Q_c|}{|Q_h|}} = \frac{1}{1 - \dfrac{T_c}{T_h}} = \frac{T_h}{T_h - T_c}$$

The Carnot COP for a heat pump in the cooling mode is

$$\text{COP}_C \text{ (cooling mode)} = \frac{T_c}{T_h - T_c}$$

As the difference between the temperatures of the two reservoirs approaches zero in this expression, the theoretical COP approaches infinity. In practice, the low temperature of the cooling coils and the high temperature at the compressor limit the COP to values below 10.

QUICK QUIZ 21.3   Three engines operate between reservoirs separated in temperature by 300 K. The reservoir temperatures are as follows: Engine A: $T_h = 1\,000$ K, $T_c = 700$ K; Engine B: $T_h = 800$ K, $T_c = 500$ K; Engine C: $T_h = 600$ K, $T_c = 300$ K. Rank the engines in order of theoretically possible efficiency from highest to lowest.

Example **21.3**   **Efficiency of the Carnot Engine**

Show that the ratio of energy transfers by heat in a Carnot engine is equal to the ratio of reservoir temperatures, as given by Equation 21.7.

SOLUTION

Conceptualize   Make use of Figures 21.9 and 21.10 to help you visualize the processes in the Carnot cycle.

Categorize   Because of our understanding of the Carnot cycle, we can categorize the processes in the cycle as isothermal and adiabatic.

Analyze   For the isothermal expansion (process $A \rightarrow B$ in Fig. 21.9), find the energy transfer by heat from the hot reservoir using Equation 19.12 and the first law of thermodynamics:

$$|Q_h| = |\Delta E_{int} - W_{AB}| = |0 - W_{AB}| = nRT_h \ln \frac{V_B}{V_A}$$

In a similar manner, find the energy transfer to the cold reservoir during the isothermal compression $C \rightarrow D$:

$$|Q_c| = |\Delta E_{int} - W_{CD}| = |0 - W_{CD}| = nRT_c \ln \frac{V_C}{V_D}$$

Divide the second expression by the first:

$$(1) \quad \frac{|Q_c|}{|Q_h|} = \frac{T_c}{T_h} \frac{\ln (V_C/V_D)}{\ln (V_B/V_A)}$$

Apply Equation 20.40 to the adiabatic processes $B \rightarrow C$ and $D \rightarrow A$:

$$T_h V_B{}^{\gamma-1} = T_c V_C{}^{\gamma-1}$$
$$T_h V_A{}^{\gamma-1} = T_c V_D{}^{\gamma-1}$$

Divide the first equation by the second:

$$\left(\frac{V_B}{V_A}\right)^{\gamma-1} = \left(\frac{V_C}{V_D}\right)^{\gamma-1}$$

$$(2) \quad \frac{V_B}{V_A} = \frac{V_C}{V_D}$$

Substitute Equation (2) into Equation (1):

$$\frac{|Q_c|}{|Q_h|} = \frac{T_c}{T_h} \frac{\ln (V_C/V_D)}{\ln (V_B/V_A)} = \frac{T_c}{T_h} \frac{\ln (V_C/V_D)}{\ln (V_C/V_D)} = \frac{T_c}{T_h}$$

Finalize   This last equation is Equation 21.7, the one we set out to prove.

| Example **21.4** | **The Steam Engine** |

A steam engine has a boiler that operates at 500 K. The energy from the burning fuel changes water to steam, and this steam then drives a piston. The cold reservoir's temperature is that of the outside air, approximately 300 K. What is the maximum thermal efficiency of this steam engine?

SOLUTION

Conceptualize In a steam engine, the gas pushing on the piston in Figure 21.9 is steam. A real steam engine does not operate in a Carnot cycle, but, to find the maximum possible efficiency, imagine a Carnot steam engine.

Categorize We calculate an efficiency using Equation 21.8, so we categorize this example as a substitution problem.

Substitute the reservoir temperatures into Equation 21.8:
$$e_C = 1 - \frac{T_c}{T_h} = 1 - \frac{300 \text{ K}}{500 \text{ K}} = 0.400 \quad \text{or} \quad 40.0\%$$

This result is the highest *theoretical* efficiency of the engine. In practice, the efficiency is considerably lower.

WHAT IF? Suppose we wished to increase the theoretical efficiency of this engine. This increase can be achieved by raising T_h by ΔT or by decreasing T_c by the same ΔT. Which would be more effective?

Answer A given ΔT would have a larger fractional effect on a smaller temperature, so you would expect a larger change in efficiency if you alter T_c by ΔT. Let's test that numerically. Raising T_h by 50 K, corresponding to $T_h = 550$ K, would give a maximum efficiency of

$$e_C = 1 - \frac{T_c}{T_h} = 1 - \frac{300 \text{ K}}{550 \text{ K}} = 0.455$$

Decreasing T_c by 50 K, corresponding to $T_c = 250$ K, would give a maximum efficiency of

$$e_C = 1 - \frac{T_c}{T_h} = 1 - \frac{250 \text{ K}}{500 \text{ K}} = 0.500$$

Although changing T_c is *mathematically* more effective, often changing T_h is *practically* more feasible.

21.5 Gasoline and Diesel Engines

In a gasoline engine, four *strokes* occur in each cycle; in addition, two *events* have a significant effect on the state of the gas in the cylinder. These strokes and events are illustrated with a pictorial representation in Figure 21.11 (page 568). In this discussion, let's consider the interior of the cylinder above the piston to be the system that is taken through repeated cycles in the engine's operation. For a given cycle, the piston moves up and down twice, which represents a four-stroke cycle consisting of two upstrokes and two downstrokes. The processes in the cycle can be approximated by the **Otto cycle** shown in the *PV* diagram in Figure 21.12 (page 568), which is a graphical representation of the cycle. In the following discussion, note that the letter designations next to the piston in Figure 21.11 correspond to the states on the *PV* diagram in Figure 21.12.

1. During the *intake stroke* ($O \rightarrow A$ in Figures 21.11a and 21.12), the piston moves downward and a gaseous mixture of air and fuel is drawn into the cylinder at atmospheric pressure. That is the energy input part of the cycle: energy enters the system (the interior of the cylinder) by matter transfer as potential energy stored in the fuel. In this process, the volume increases from V_2 to V_1. This apparent backward numbering is based on the compression stroke (see 2 below), in which the air–fuel mixture is compressed from V_1 to V_2.

2. During the *compression stroke* ($A \rightarrow B$ in Figures 21.11b and 21.12), the piston moves upward, the air–fuel mixture is compressed adiabatically from volume V_1 to volume V_2, and the temperature increases from T_A to T_B. The work done on the gas is positive, and its value is equal to the negative of the area under the curve AB in Figure 21.12.

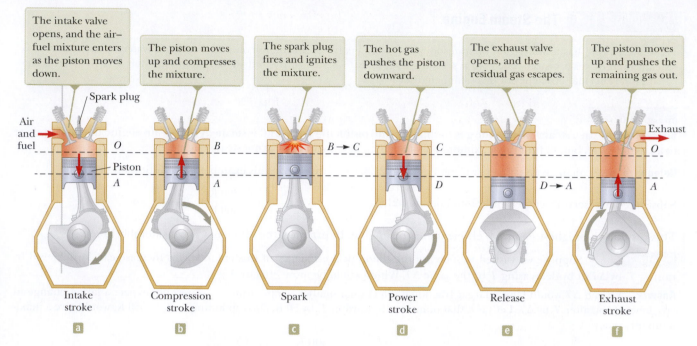

The intake valve opens, and the air–fuel mixture enters as the piston moves down.

The piston moves up and compresses the mixture.

The spark plug fires and ignites the mixture.

The hot gas pushes the piston downward.

The exhaust valve opens, and the residual gas escapes.

The piston moves up and pushes the remaining gas out.

Intake stroke **a** | Compression stroke **b** | Spark **c** | Power stroke **d** | Release **e** | Exhaust stroke **f**

Figure 21.11 The processes occurring during one cycle of a conventional gasoline engine. The broken lines show the extreme positions of the top of the piston and, therefore, represent the largest and smallest volumes of the gas in the cylinder. Parts a, b, d, and f represent *strokes* in the cycle, justifying the name of the device as a four-stroke engine. In a stroke, the piston moves up or down between its extreme positions. The red arrows show the direction of travel of the piston, and the letters next to the piston correspond to the states on the *PV* diagram in Figure 21.12. Parts c and e in the figure represent *events*, during which the piston does not move. In part c, the spark plug fires and the pressure and temperature of the gas shoot upward. In part e, the exhaust valve opens and the pressure and temperature of the gas plummet. The events in this figure correspond to the constant-volume processes in Figure 21.12. By comparing that figure with this one, convince yourself that the volumes at *O*, *B*, and *C* are all the same, as indicated by their positions on the upper broken line. Similarly, the volumes at *A* and *D* are the same.

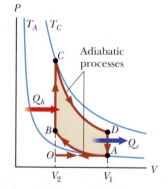

Figure 21.12 *PV* diagram for the Otto cycle, which approximately represents the processes occurring in an internal combustion engine.

3. Combustion occurs when the spark plug fires ($B \rightarrow C$ in Figures 21.11c and 21.12). This is the *combustion event* in the cycle; it occurs in a very short time interval while the piston is at its highest position. The combustion represents a rapid energy transformation from potential energy stored in chemical bonds in the fuel to internal energy associated with molecular motion, which is related to temperature. During this time interval, the mixture's pressure and temperature increase rapidly, with the temperature rising from T_B to T_C. The volume, however, remains approximately constant because of the short time interval. As a result, approximately no work is done on or by the gas. We can model this process in the *PV* diagram (Fig. 21.12) as that process in which the energy $|Q_h|$ enters the system. (In reality, however, this process is a *transformation* of energy already in the cylinder from process $O \rightarrow A$.)

4. In the *power stroke* ($C \rightarrow D$ in Figures 21.11d and 21.12), the gas expands adiabatically from V_2 to V_1. This expansion causes the temperature to drop from T_C to T_D. Work is done by the gas in pushing the piston downward, and the value of this work is equal to the area under the curve *CD*.

5. The *release event* in the cycle occurs when an exhaust valve is opened ($D \rightarrow A$ in Figures 21.11e and 21.12). The pressure suddenly drops for a short time interval. During this time interval, the piston is almost stationary and the volume is approximately constant. Energy is expelled from the interior of the cylinder and continues to be expelled during the next process.

6. In the final process, the *exhaust stroke* ($A \rightarrow O$ in Figures 21.11f and 21.12), the piston moves upward while the exhaust valve remains open. Residual gases are exhausted at atmospheric pressure, and the volume decreases from V_1 to V_2. The cycle then repeats.

If the air–fuel mixture is assumed to be an ideal gas, the efficiency of the Otto cycle is

$$e = 1 - \frac{1}{(V_1/V_2)^{\gamma-1}} \quad \text{(Otto cycle)} \tag{21.9}$$

where V_1/V_2 is the **compression ratio** and γ is the ratio of the molar specific heats C_P/C_V for the air–fuel mixture. Equation 21.9, which is derived in Example 21.5, shows that the efficiency increases as the compression ratio increases. For a typical compression ratio of 8 and with $\gamma = 1.4$, Equation 21.9 predicts a theoretical efficiency of 56% for an engine operating in the idealized Otto cycle. This value is much greater than that achieved in real engines (15% to 20%) because of such effects as friction, energy transfer by conduction through the cylinder walls, and incomplete combustion of the air–fuel mixture.

Diesel engines operate on a cycle similar to the Otto cycle, but they do not employ a spark plug. The compression ratio for a diesel engine is much greater than that for a gasoline engine. Air in the cylinder is compressed to a very small volume, and, as a consequence, the cylinder temperature at the end of the compression stroke is very high. At this point, fuel is injected into the cylinder. The temperature is high enough for the air–fuel mixture to ignite without the assistance of a spark plug. Diesel engines are more efficient than gasoline engines because of their greater compression ratios and resulting higher combustion temperatures.

Example 21.5 **Efficiency of the Otto Cycle**

Show that the thermal efficiency of an engine operating in an idealized Otto cycle (see Figs. 21.11 and 21.12) is given by Equation 21.9. Treat the working substance as an ideal gas.

SOLUTION

Conceptualize Study Figures 21.11 and 21.12 to make sure you understand the working of the Otto cycle.

Categorize As seen in Figure 21.12, we categorize the processes in the Otto cycle as isovolumetric and adiabatic.

Analyze Model the energy input and output as occurring by heat in processes $B \rightarrow C$ and $D \rightarrow A$. (In reality, most of the energy enters and leaves by matter transfer as the air–fuel mixture enters and leaves the cylinder.) Use Equation 20.23 to find the energy transfers by heat for these processes, which take place at constant volume:

$$B \rightarrow C \quad |Q_h| = nC_V(T_C - T_B)$$
$$D \rightarrow A \quad |Q_c| = nC_V(T_D - T_A)$$

Substitute these expressions into Equation 21.2:

$$(1) \quad e = 1 - \frac{|Q_c|}{|Q_h|} = 1 - \frac{T_D - T_A}{T_C - T_B}$$

Apply Equation 20.40 to the adiabatic processes $A \rightarrow B$ and $C \rightarrow D$:

$$A \rightarrow B \quad T_A V_A^{\gamma-1} = T_B V_B^{\gamma-1}$$
$$C \rightarrow D \quad T_C V_C^{\gamma-1} = T_D V_D^{\gamma-1}$$

Solve these equations for the temperatures T_A and T_D, noting that $V_A = V_D = V_1$ and $V_B = V_C = V_2$:

$$(2) \quad T_A = T_B \left(\frac{V_B}{V_A}\right)^{\gamma-1} = T_B \left(\frac{V_2}{V_1}\right)^{\gamma-1}$$

$$(3) \quad T_D = T_C \left(\frac{V_C}{V_D}\right)^{\gamma-1} = T_C \left(\frac{V_2}{V_1}\right)^{\gamma-1}$$

Subtract Equation (2) from Equation (3) and rearrange:

$$(4) \quad \frac{T_D - T_A}{T_C - T_B} = \left(\frac{V_2}{V_1}\right)^{\gamma-1}$$

Substitute Equation (4) into Equation (1):

$$e = 1 - \frac{1}{(V_1/V_2)^{\gamma-1}}$$

Finalize This final expression is Equation 21.9.

21.6 Entropy

The zeroth law of thermodynamics involves the concept of temperature, and the first law involves the concept of internal energy. Temperature and internal energy are both state variables; that is, the value of each depends only on the thermodynamic state of a system, not on the process that brought it to that state. Another state variable—this one related to the second law of thermodynamics—is *entropy*.

Entropy was originally formulated as a useful concept in thermodynamics. Its importance grew, however, as the field of statistical mechanics developed. In statistical mechanics, the behavior of a substance is described in terms of the statistical behavior of its atoms and molecules, as we did with our study of kinetic theory in Chapter 20. The analytical techniques of statistical mechanics provide an alternative means of interpreting entropy and a more global significance to the concept.

We will develop our understanding of entropy by first considering some non-thermodynamic systems, such as a pair of dice and poker hands. We will then expand on these ideas and use them to understand the concept of entropy as applied to thermodynamic systems.

We begin this process by distinguishing between *microstates* and *macrostates* of a system. A **microstate** is a particular configuration of the individual constituents of the system. A **macrostate** is a description of the system's conditions from a macroscopic point of view.

For any given macrostate of the system, a number of microstates are possible. For example, the macrostate of a 4 when a pair of six-sided dice are rolled can be formed from the possible microstates 1–3, 2–2, and 3–1. The macrostate of 2 has only one microstate, 1–1. It is assumed all microstates are equally probable. We can compare the two macrostates just mentioned in three ways. (1) *Uncertainty:* If we know that a macrostate of 4 exists, there is some uncertainty as to the microstate that exists, because there are multiple microstates that will result in a 4. In comparison, there is lower uncertainty (in fact, *zero* uncertainty) for a macrostate of 2 because only one microstate is possible. (2) *Choice:* There are more choices of microstates for a 4 than for a 2. (3) *Probability:* The macrostate of 4 has a higher probability than a macrostate of 2 because there are more ways (microstates) of achieving a 4. The notions of uncertainty, choice, and probability are central to the concept of entropy, as we discuss below.

Let's look at another example related to a five-card poker hand. There is only one microstate associated with the macrostate of a "royal flush" of five spades, laid out in order from ten to ace (Fig. 21.13a). Figure 21.13b shows another poker hand. The macrostate here is "worthless hand." The *particular* hand (the microstate) in Figure 21.13b and the hand in Figure 21.13a are equally probable. There are, however,

Figure 21.13 (a) A royal flush has low probability of occurring. (b) A worthless poker hand, one of many.

many other hands similar in value to that in Figure 21.13b; that is, there are many microstates that also qualify as worthless hands. If you, as a poker player, are told your opponent holds a macrostate of a royal flush in spades, there is *zero uncertainty* as to what five cards are in the hand, only *one choice* of what those cards are, and *low probability* that the hand actually occurred. In contrast, if you are told that your opponent has the macrostate of "worthless hand," there is *high uncertainty* as to what the five cards are, *many choices* of what they could be, and a *high probability* that a worthless hand occurred. Another variable in poker, of course, is the value of the hand, related to the probability: the higher the probability, the lower the value. The important point to take away from this discussion is that uncertainty, choice, and probability are related in these situations: if one is high, the others are high, and vice versa.[6]

For thermodynamic systems, the variable **entropy** S is used to represent the level of uncertainty, choice, and probability in the system. Consider Configuration 1 (a macrostate) in which all the oxygen molecules in the air in your room are located in the west half of the room and the nitrogen molecules in the east half. Compare that macrostate to the more common Configuration 2, in which the oxygen and nitrogen molecules are distributed uniformly throughout the room. Configuration 2 has the higher uncertainty as to where the molecules are located because they could be anywhere, not just in one-half of the room according to the type of molecule. Configuration 2 also represents more choices as to where to locate molecules. It also has a much higher probability of occurring; have you ever noticed your half of the room suddenly being empty of oxygen? Therefore, Configuration 2 represents a higher entropy.

For systems of dice and poker hands, the comparisons between probabilities for various macrostates involve relatively small numbers. For example, a macrostate of a 4 on a pair of dice is only three times as probable as a macrostate of 2. When we are talking about a macroscopic thermodynamic system containing on the order of Avogadro's number of molecules, however, the ratios of probabilities can be astronomical.

Let's explore this concept by considering 100 molecules in a container. Half of the molecules are oxygen and the other half are nitrogen. At any given moment, the probability of one molecule being in the left part of the container shown in Figure 21.14a as a result of random motion is $\frac{1}{2}$. If there are two molecules as shown in Figure 21.14b, the probability of both being in the left part is $\left(\frac{1}{2}\right)^2$, or 1 in 4. If there are three molecules (Fig. 21.14c), the probability of them all being in the left portion at the same moment is $\left(\frac{1}{2}\right)^3$, or 1 in 8. For 100 independently moving molecules, the probability that the 50 oxygen molecules will be found in the left part at any moment is $\left(\frac{1}{2}\right)^{50}$. Likewise, the probability that the remaining

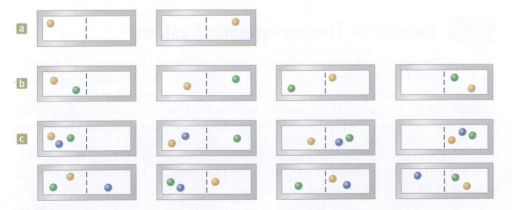

Figure 21.14 Possible distributions of identical molecules in a container. The colors used here exist only to allow us to distinguish among the molecules. (a) One molecule in a container has a 1-in-2 chance of being on the left side. (b) Two molecules have a 1-in-4 chance of being on the left side at the same time. (c) Three molecules have a 1-in-8 chance of being on the left side at the same time.

[6]Another way of describing macrostates is by means of "missing information." For high-probability macrostates with many microstates, there is a large amount of missing information, meaning we have very little information about what microstate actually exists.

50 nitrogen molecules will be found in the right part at any moment is $\left(\frac{1}{2}\right)^{50}$. Therefore, the probability of finding this oxygen–nitrogen separation as a result of random motion is the product $\left(\frac{1}{2}\right)^{50}\left(\frac{1}{2}\right)^{50} = \left(\frac{1}{2}\right)^{100}$, which corresponds to about 1 in 10^{30}. When this calculation is extrapolated from 100 molecules to the number in 1 mol of gas (6.02×10^{23}), the separated arrangement is found to be *extremely* improbable!

QUICK QUIZ **21.4** (a) Suppose you select four cards at random from a standard deck of playing cards and end up with a macrostate of four deuces. How many microstates are associated with this macrostate? (b) Suppose you pick up two cards and end up with a macrostate of two aces. How many microstates are associated with this macrostate?

Conceptual Example **21.6** **Let's Play Marbles!**

Suppose you have a bag of 100 marbles of which 50 are red and 50 are green. You are allowed to draw four marbles from the bag according to the following rules. Draw one marble, record its color, and return it to the bag. Shake the bag and then draw another marble. Continue this process until you have drawn and returned four marbles. What are the possible macrostates for this set of events? What is the most likely macrostate? What is the least likely macrostate?

SOLUTION

Because each marble is returned to the bag before the next one is drawn and the bag is then shaken, the probability of drawing a red marble is always the same as the probability of drawing a green one. All the possible microstates and macrostates are shown in Table 21.1. As this table indicates, there is only one way to draw a macrostate of four red marbles, so there is only one microstate for that macrostate. There are, however, four possible microstates that correspond to the macrostate of one green marble and three red marbles, six microstates that correspond to two green mar-

TABLE 21.1 Possible Results of Drawing Four Marbles from a Bag

Macrostate	Possible Microstates	Total Number of Microstates
All R	RRRR	1
1G, 3R	RRRG, RRGR, RGRR, GRRR	4
2G, 2R	RRGG, RGRG, GRRG, RGGR, GRGR, GGRR	6
3G, 1R	GGGR, GGRG, GRGG, RGGG	4
All G	GGGG	1

bles and two red marbles, four microstates that correspond to three green marbles and one red marble, and one microstate that corresponds to four green marbles. The most likely macrostate—two red marbles and two green marbles—corresponds to the largest number of choices of microstates, and, therefore, the most uncertainty as to what the exact microstate is. The least likely macrostates—four red marbles or four green marbles—correspond to only one choice of microstate and, therefore, zero uncertainty. There is no uncertainty for the least likely states: we know the colors of all four marbles.

21.7 Entropy in Thermodynamic Systems

We have investigated the notions of uncertainty, number of choices, and probability for some non-thermodynamic systems such as dice and cards, as well as for a small system of 100 oxygen and nitrogen molecules. We have argued that the concept of entropy can be related to these notions for macroscopic thermodynamic systems. In our discussion of entropy, there are two things we have *not* done yet: (1) indicate how to evaluate entropy numerically, and (2) discuss entropy for a macroscopic system with a huge number of particles. Both of these were performed through statistical means by Boltzmann in the 1870s and the numerical evaluation of entropy appears in its currently accepted form as

$$S = k_B \ln W \qquad (21.10)$$

where k_B is Boltzmann's constant. Boltzmann intended W, standing for *Wahrscheinlichkeit*, the German word for probability, to be proportional to the probability that a given macrostate exists. It is equivalent to let W be the number of microstates associated with the macrostate, so we can interpret W as representing the number of "ways" of achieving the macrostate. Therefore, macrostates with larger numbers of microstates have higher probability and, equivalently, higher entropy. Notice that the units of entropy are those of Boltzmann's constant, J/K.

In the kinetic theory of gases, gas molecules are represented as particles moving randomly. Suppose the gas is confined to a volume V. For a uniform distribution of gas in the volume, there are a large number of equivalent microstates, and the entropy of the gas can be related to the number of microstates corresponding to a given macrostate. Let us count the number of microstates by considering the variety of molecular locations available to the molecules. Let us assume each molecule occupies some microscopic volume V_m. The total number of possible locations of a single molecule in a macroscopic volume V is the ratio $w = V/V_m$, which is a huge number. We use lowercase w here to represent the number of ways a single molecule can be placed in the volume or the number of microstates for a single molecule, which is equivalent to the number of available locations. We assume the probabilities of a molecule occupying any of these locations are equal. As more molecules are added to the system, the number of possible ways the molecules can be positioned in the volume multiplies, as we saw in Figure 21.14. For example, if you consider two molecules, for every possible placement of the first, all possible placements of the second are available. Therefore, there are w ways of locating the first molecule, and for each way, there are w ways of locating the second molecule. The total number of ways of locating the two molecules is $W = w \times w = w^2 = (V/V_m)^2$. (Uppercase W represents the number of ways of putting multiple molecules into the volume and is not to be confused with work.)

Now consider placing N molecules of gas in the volume V. Neglecting the very small probability of having two molecules occupy the same location, each molecule may go into any of the V/V_m locations, and so the number of ways of locating N molecules in the volume becomes $W = w^N = (V/V_m)^N$. Therefore, the spatial part of the entropy of the gas, from Equation 21.10, is

$$S = k_B \ln W = k_B \ln \left(\frac{V}{V_m} \right)^N = N k_B \ln \left(\frac{V}{V_m} \right) = nR \ln \left(\frac{V}{V_m} \right) \qquad (21.11)$$

We will use this expression in the next section as we investigate changes in entropy for processes occurring in thermodynamic systems.

Notice that we have indicated Equation 21.11 as representing only the *spatial* portion of the entropy of the gas. There is also a temperature-dependent portion of the entropy that the discussion above does not address. For example, imagine an isovolumetric process in which the temperature of the gas increases. Equation 21.11 above shows no change in the spatial portion of the entropy for this situation. There *is* a change in entropy, however, associated with the increase in temperature. We can understand this by appealing again to a bit of quantum physics. Recall from Section 20.3 that the energies of the gas molecules are quantized. When the temperature of a gas changes, the distribution of energies of the gas molecules changes according to the Boltzmann distribution law, discussed in Section 20.5. Therefore, as the temperature of the gas increases, there is more uncertainty about the particular microstate that exists as gas molecules distribute themselves into higher available quantum states.

Thermodynamic systems are constantly in flux, changing continuously from one microstate to another. If the system is in equilibrium, a given macrostate exists, described by variables such as P, V, T, and E_{int} and the system fluctuates from one microstate associated with that macrostate to another. This change is unobservable because we are only able to detect the macrostate. Equilibrium states have tremendously higher probability than nonequilibrium states, so it is highly unlikely that an equilibrium state will spontaneously change to a nonequilibrium state. For example, we do not observe a spontaneous split into the oxygen–nitrogen separation discussed in Section 21.6.

What happens, however, if the system begins in a low-probability macrostate? What if the room *begins* with an oxygen–nitrogen separation? In this case, the system will progress from this low-probability macrostate to the much-higher probability state: the gases will disperse and mix throughout the room. Because entropy is related to probability, a spontaneous *increase* in entropy, such as in the latter situation, is natural. If the oxygen and nitrogen molecules were initially spread evenly throughout the room, the entropy of the mixture would *decrease* if the spontaneous splitting of molecules occurred.

One way of conceptualizing a change in entropy is to relate it to *energy spreading*. A natural tendency is for energy to undergo spatial spreading in time, representing an increase in entropy. If a basketball is dropped onto a floor, it bounces several times and eventually comes to rest. The initial gravitational potential energy in the basketball–Earth system has been transformed to internal energy in the ball and the floor. That energy is spreading outward by heat into the air and into regions of the floor farther from the drop point. In addition, some of the energy has spread throughout the room by sound. It would be unnatural for energy in the room and floor to reverse this motion and concentrate into the stationary ball so that it spontaneously begins to bounce again.

In the adiabatic free expansion discussed in Section 21.3, the spreading of energy accompanies the spreading of the molecules as the gas rushes into the evacuated half of the container. If a warm object is placed in thermal contact with a cool object, energy transfers from the warm object to the cool one by heat, representing a spread of internal energy until it is distributed more evenly between the two objects.

Now consider a mathematical representation of this spreading of energy or, equivalently, the change in entropy. The original formulation of entropy in thermodynamics involves the transfer of energy by heat during a reversible process. Consider any infinitesimal process in which a system changes from one equilibrium state to another. If dQ_r is the amount of energy transferred by heat when the system follows a reversible path between the states, the change in entropy dS can be shown to be equal to this amount of energy divided by the absolute temperature of the system:

Change in entropy for an ▶
infinitesimal process

$$dS = \frac{dQ_r}{T}$$

(21.12)

We have assumed the temperature is constant because the process is infinitesimal. Because entropy is a state variable, the change in entropy during a process depends only on the endpoints and therefore is independent of the actual path followed. Consequently, the entropy change for an irreversible process can be determined by calculating the entropy change for a *reversible* process that connects the same initial and final states.

Equation 21.10 defines entropy statistically. Evaluating W, however, is extremely difficult for a macroscopic system with a huge number of particles, on the order of Avogadro's number. On the other hand, Equation 21.12 defines changes in entropy in terms of macroscopic quantities, Q_r and T. Therefore, this equation is more practical than Equation 21.10.

The subscript r on the quantity dQ_r is a reminder that the transferred energy is to be measured along a reversible path even though the system may actually have followed some irreversible path. When energy is absorbed by the system, dQ_r is positive and the entropy of the system increases. When energy is expelled by the system, dQ_r is negative and the entropy of the system decreases. Notice that Equation 21.12 does not define entropy but rather the *change* in entropy. Hence, the meaningful quantity in describing a process is the *change* in entropy.

To calculate the change in entropy for a *finite* process, first recognize that T is generally not constant during the process. Therefore, we must integrate Equation 21.12:

$$\Delta S = \int_i^f dS = \int_i^f \frac{dQ_r}{T} \qquad (21.13)$$

◀ Change in entropy for a finite process

As with an infinitesimal process, the change in entropy ΔS of a system going from one state to another has the same value for *all* paths connecting the two states. That is, the finite change in entropy ΔS of a system depends only on the properties of the initial and final equilibrium states. Therefore, we are free to choose any convenient reversible path over which to evaluate the entropy in place of the actual path as long as the initial and final states are the same for both paths. This point is explored further on in this section.

From Equation 21.10, we see that a change in entropy is represented in the Boltzmann formulation as

$$\Delta S = k_B \ln\left(\frac{W_f}{W_i}\right) \qquad (21.14)$$

where W_i and W_f represent the initial and final numbers of microstates, respectively, for the initial and final configurations of the system. If $W_f > W_i$, the final state is more probable than the initial state (there are more choices of microstates), and the entropy increases. As mentioned above, however, evaluating W is extremely difficult for macroscopic systems.

QUICK QUIZ 21.5 An ideal gas is taken from an initial temperature T_i to a higher final temperature T_f along two different reversible paths as shown in Figure 21.15. Path A is at constant pressure, and path B is at constant volume. What is the relation between the entropy changes of the gas for these paths? **(a)** $\Delta S_A > \Delta S_B$ **(b)** $\Delta S_A = \Delta S_B$ **(c)** $\Delta S_A < \Delta S_B$

QUICK QUIZ 21.6 True or False: The entropy change in an adiabatic process must be zero because $Q = 0$.

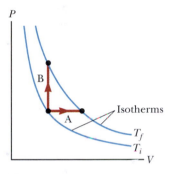

Figure 21.15 (Quick Quiz 21.5) An ideal gas is taken from temperature T_i to T_f via two different paths.

Example **21.7** **Change in Entropy: Melting**

A solid that has a latent heat of fusion L_f melts at a temperature T_m. Calculate the change in entropy of this substance when a mass m of the substance melts.

SOLUTION

Conceptualize We can choose any convenient reversible path to follow that connects the initial and final states. It is not necessary to identify the process or the path because, whatever it is, the effect is the same: energy enters the substance by heat and the substance melts. The mass m of the substance that melts is equal to Δm, the change in mass of the higher-phase (liquid) substance.

Categorize Because the melting takes place at a fixed temperature, we categorize the process as isothermal.

Analyze Use Equation 19.8 in Equation 21.13, noting that the temperature remains fixed:

$$\Delta S = \int \frac{dQ_r}{T} = \frac{1}{T_m}\int dQ_r = \frac{Q_r}{T_m} = \frac{L_f \Delta m}{T_m} = \boxed{\frac{L_f m}{T_m}}$$

Finalize Notice that Δm is positive so that ΔS is positive, representing that energy is added to the substance.

Entropy Change in a Carnot Cycle

Now that we have some understanding of entropy, let's consider the changes in entropy that occur in a Carnot heat engine that operates between the temperatures

T_c and T_h. In one cycle, the engine takes in energy $|Q_h|$ from the hot reservoir and expels energy $|Q_c|$ to the cold reservoir. These energy transfers occur only during the reversible, isothermal portions of the Carnot cycle; therefore, the constant temperature can be brought out in front of the integral sign in Equation 21.13. The integral then simply has the value of the total amount of energy transferred by heat. During the two adiabatic processes, for which $Q = 0$, the entropy changes are zero because these processes are reversible. Therefore, the total change in entropy for one cycle is

$$\Delta S = \frac{|Q_h|}{T_h} - \frac{|Q_c|}{T_c} \tag{21.15}$$

where the minus sign represents that energy is leaving the engine at temperature T_c. In Example 21.3, we showed that for a Carnot engine,

$$\frac{|Q_c|}{|Q_h|} = \frac{T_c}{T_h}$$

Using this result in Equation 21.15, we find that the total change in entropy for a Carnot engine operating in a cycle is *zero*:

$$\Delta S = 0$$

Now consider a system taken through an arbitrary (non-Carnot) reversible cycle. Because entropy is a state variable—and hence depends only on the properties of a given equilibrium state—we conclude that $\Delta S = 0$ for *any* reversible cycle. In general, we can write this condition as

$$\oint \frac{dQ_r}{T} = 0 \quad \text{(reversible cycle)} \tag{21.16}$$

where the symbol $\oint$ indicates that the integration is over a closed path.

Entropy Change in a Free Expansion

Let's again consider the adiabatic free expansion of a gas occupying an initial volume V_i (Fig. 21.16). In this situation, a membrane separating the gas from an evacuated region is broken and the gas expands to a volume V_f. This process is irreversible; the gas would not spontaneously crowd into half the volume after filling the entire volume. What is the change in entropy of the gas during this process? The process is neither reversible nor quasi-static. As argued in Section 21.3, the initial and final temperatures of the gas are the same.

To apply Equation 21.13, we cannot take $Q = 0$, the value for the irreversible process, but must instead find Q_r; that is, we must find an equivalent reversible path that shares the same initial and final states. A simple choice is an isothermal, reversible expansion in which the gas pushes slowly against a piston while energy enters the gas by heat from a reservoir to hold the temperature constant. Because T is constant in this process, Equation 21.13 gives

$$\Delta S = \int_i^f \frac{dQ_r}{T} = \frac{1}{T} \int_i^f dQ_r$$

For an isothermal process, the first law of thermodynamics specifies that $\int_i^f dQ_r$ is equal to the negative of the work done on the gas during the expansion from V_i to V_f, which is given by Equation 19.12. Using this result, we find that the entropy change for the gas is

$$\Delta S = nR \ln \left(\frac{V_f}{V_i} \right) \tag{21.17}$$

Because $V_f > V_i$, we conclude that ΔS is positive. This positive result indicates that the entropy of the gas *increases* as a result of the irreversible, adiabatic expansion.

When the membrane is ruptured, the gas will expand freely and irreversibly into the full volume.

Insulating wall

Vacuum

Membrane

Gas at T_i in volume V_i

Figure 21.16 Adiabatic free expansion of a gas. The container is thermally insulated from its surroundings; therefore, $Q = 0$.

It is easy to see that the energy has spread after the expansion. Instead of being concentrated in a relatively small space, the molecules and the energy associated with them are scattered over a larger region. In addition, there are more choices of the locations of the molecules, higher uncertainty as to their locations, and a higher probability for the molecules to be spread throughout the volume. The probability is indeed low for the molecules, in the absence of the membrane, to concentrate spontaneously in the lower half of the container.

Entropy Change in Thermal Conduction

Let us now consider a system consisting of a hot reservoir and a cold reservoir that are in thermal contact with each other and isolated from the rest of the Universe. A process occurs during which energy Q is transferred by heat from the hot reservoir at temperature T_h to the cold reservoir at temperature T_c. The process as described is irreversible (energy would not spontaneously flow from cold to hot), so we must find an equivalent reversible process. The overall process is a combination of two processes: energy leaving the hot reservoir and energy entering the cold reservoir. We will calculate the entropy change for the reservoir in each process and add to obtain the overall entropy change.

Consider first the process of energy entering the cold reservoir. Although the reservoir has absorbed some energy, the temperature of the reservoir has not changed. The energy that has entered the reservoir is the same as that which would enter by means of a reversible, isothermal process. The same is true for energy leaving the hot reservoir.

Because the cold reservoir absorbs energy Q, its entropy increases by Q/T_c. At the same time, the hot reservoir loses energy Q, so its entropy change is $-Q/T_h$. Therefore, the change in entropy of the system is

$$\Delta S = \frac{Q}{T_c} + \frac{-Q}{T_h} = Q\left(\frac{1}{T_c} - \frac{1}{T_h}\right) > 0 \qquad (21.18)$$

This increase is consistent with our interpretation of entropy changes as representing the spreading of energy. In the initial configuration, the hot reservoir has excess internal energy relative to the cold reservoir. The process that occurs spreads the energy into a more equitable distribution between the two reservoirs.

Example **21.8** | Adiabatic Free Expansion: Revisited

Let's verify that the macroscopic and microscopic approaches to the calculation of entropy lead to the same conclusion for the adiabatic free expansion of an ideal gas. Suppose the ideal gas in Figure 21.16 expands to four times its initial volume. As we have seen for this process, the initial and final temperatures are the same.

(A) Using a macroscopic approach, calculate the entropy change for the gas.

SOLUTION

Conceptualize Look back at Figure 21.16, which is a diagram of the system before the adiabatic free expansion. Imagine breaking the membrane so that the gas moves into the evacuated area. The expansion is irreversible.

Categorize We can replace the irreversible process with a reversible isothermal process between the same initial and final states. This approach is macroscopic, so we use a thermodynamic variable, in particular, the volume V.

Analyze Use Equation 21.17 to evaluate the entropy change: $\quad \Delta S = nR \ln\left(\frac{V_f}{V_i}\right) = nR \ln\left(\frac{4V_i}{V_i}\right) = nR \ln 4$

(B) Using statistical considerations, calculate the change in entropy for the gas and show that it agrees with the answer you obtained in part (A).

continued

21.8 continued

SOLUTION

Categorize This approach is microscopic, so we use variables related to the individual molecules.

Analyze As in the discussion leading to Equation 21.11, the number of microstates available to a single molecule in the initial volume V_i is $w_i = V_i/V_m$, where V_i is the initial volume of the gas and V_m is the microscopic volume occupied by the molecule. Use this number to find the number of available microstates for N molecules:

$$W_i = w_i^N = \left(\frac{V_i}{V_m}\right)^N$$

Find the number of available microstates for N molecules in the final volume $V_f = 4V_i$:

$$W_f = \left(\frac{V_f}{V_m}\right)^N = \left(\frac{4V_i}{V_m}\right)^N$$

Use Equation 21.14 to find the entropy change:

$$\Delta S = k_B \ln\left(\frac{W_f}{W_i}\right)$$

$$= k_B \ln\left(\frac{4V_i}{V_i}\right)^N = k_B \ln(4^N) = Nk_B \ln 4 = \boxed{nR \ln 4}$$

Finalize The answer is the same as that for part (A), which dealt with macroscopic parameters.

WHAT IF? In part (A), we used Equation 21.17, which was based on a reversible isothermal process connecting the initial and final states. Would you arrive at the same result if you chose a different reversible process?

Answer You *must* arrive at the same result because entropy is a state variable. For example, consider the two-step process in Figure 21.17: a reversible adiabatic expansion from V_i to $4V_i$ ($A \rightarrow B$) during which the temperature drops from T_1 to T_2 and a reversible isovolumetric process ($B \rightarrow C$) that takes the gas back to the initial temperature T_1. During the reversible adiabatic process, $\Delta S = 0$ because $Q_r = 0$.

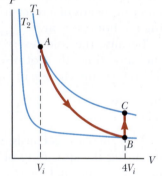

Figure 21.17 (Example 21.8) A gas expands to four times its initial volume and back to the initial temperature by means of a two-step process.

For the reversible isovolumetric process ($B \rightarrow C$), use Equation 21.13:

$$\Delta S = \int_i^f \frac{dQ_r}{T} = \int_{T_2}^{T_1} \frac{nC_V dT}{T} = nC_V \ln\left(\frac{T_1}{T_2}\right)$$

Find the ratio of temperature T_1 to T_2 from Equation 20.40 for the adiabatic process:

$$\frac{T_1}{T_2} = \left(\frac{4V_i}{V_i}\right)^{\gamma-1} = (4)^{\gamma-1}$$

Substitute to find ΔS:

$$\Delta S = nC_V \ln (4)^{\gamma-1} = nC_V(\gamma-1) \ln 4$$

$$= nC_V\left(\frac{C_P}{C_V} - 1\right) \ln 4 = n(C_P - C_V) \ln 4 = nR \ln 4$$

We do indeed obtain the exact same result for the entropy change.

21.8 Entropy and the Second Law

If we consider a system and its surroundings to include the entire Universe, the Universe is always moving toward a higher-probability macrostate, corresponding to the continuous spreading of energy. An alternative way of stating this behavior is yet another wording of the second law of thermodynamics:

◄ Entropy statement of the second law of thermodynamics

The entropy of the Universe increases in all real processes.

This statement can be shown to be equivalent to the Kelvin-Planck and Clausius statements.

Let us show this equivalence first for the Clausius statement. Looking at Figure 21.5, we see that, if the heat pump operates in the manner shown in the figure, energy is spontaneously flowing from the cold reservoir to the hot reservoir without an input of energy by work. As a result, the energy in the system is not spreading evenly between the two reservoirs, but is *concentrating* in the hot reservoir. Consequently, if the Clausius statement of the second law is not true, then the entropy statement is also not true, demonstrating their equivalence.

For the equivalence of the Kelvin–Planck statement, consider Figure 21.18, which shows the impossible engine of Figure 21.3 connected to a heat pump operating between the same reservoirs. The output work of the engine is used to drive the heat pump. The net effect of this combination is that energy leaves the cold reservoir and is delivered to the hot reservoir without the input of work. (The work done by the engine on the heat pump is *internal* to the system of both devices.) This is forbidden by the Clausius statement of the second law, which we have shown to be equivalent to the entropy statement. Therefore, the Kelvin–Planck statement of the second law is also equivalent to the entropy statement.

When dealing with a system that is not isolated from its surroundings, remember that the increase in entropy described in the second law is that of the system *and* its surroundings. When a system and its surroundings interact in an irreversible process, the increase in entropy of one is greater than the decrease in entropy of the other. Hence, the change in entropy of the Universe must be greater than zero for an irreversible process and equal to zero for a reversible process.

We can check this statement of the second law for the calculations of entropy change that we made in Section 21.7. Consider first the entropy change in a free expansion, described by Equation 21.17. Because the free expansion takes place in an insulated container, no energy is transferred by heat from the surroundings. Therefore, Equation 21.17 represents the entropy change of the entire Universe. Because $V_f > V_i$, the entropy change of the Universe is positive, consistent with the second law.

Now consider the entropy change in thermal conduction, described by Equation 21.18. Let each reservoir be half the Universe. (The larger the reservoir, the better is the assumption that its temperature remains constant!) Then the entropy change of the Universe is represented by Equation 21.18. Because $T_h > T_c$, this entropy change is positive, again consistent with the second law. The positive entropy change is also consistent with the notion of energy spreading. The warm portion of the Universe has excess internal energy relative to the cool portion. Thermal conduction represents a spreading of the energy more equitably throughout the Universe.

Finally, let us look at the entropy change in a Carnot cycle, given by Equation 21.15. The entropy change of the engine itself is zero. The entropy change of the reservoirs is

$$\Delta S = \frac{|Q_c|}{T_c} - \frac{|Q_h|}{T_h}$$

In light of Equation 21.7, this entropy change is also zero. Therefore, the entropy change of the Universe is only that associated with the work done by the engine. A portion of that work will be used to change the mechanical energy of a system external to the engine: speed up the shaft of a machine, raise a weight, and so on. There is no change in internal energy of the external system due to this portion of the work, or, equivalently, no energy spreading, so the entropy change is again zero. The other portion of the work will be used to overcome various friction forces or other nonconservative forces in the external system. This process will cause an increase in internal energy of that system. That same increase in internal energy could have happened via a reversible thermodynamic process in which energy Q_r is

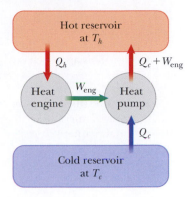

Figure 21.18 The impossible engine of Figure 21.3 transfers energy by work to a heat pump operating between two energy reservoirs. This situation is forbidden by the Clausius statement of the second law of thermodynamics.

transferred by heat, so the entropy change associated with that part of the work is positive. As a result, the overall entropy change of the Universe for the operation of the Carnot engine is positive, again consistent with the second law.

Ultimately, because real processes are irreversible, the entropy of the Universe should increase steadily and eventually reach a maximum value. At this value, assuming that the second law of thermodynamics, as formulated here on Earth, applies to the entire expanding Universe, the Universe will be in a state of uniform temperature and density. The total energy of the Universe will have spread more evenly throughout the Universe. All physical, chemical, and biological processes will have ceased at this time. This gloomy state of affairs is sometimes referred to as the *heat death* of the Universe.

Summary

▶ Definitions

The **thermal efficiency** e of a heat engine is

$$e \equiv \frac{W_{eng}}{|Q_h|} = \frac{|Q_h| - |Q_c|}{|Q_h|} = 1 - \frac{|Q_c|}{|Q_h|} \quad (21.2)$$

From a microscopic viewpoint, the **entropy** of a given macrostate is defined as

$$S \equiv k_B \ln W \quad (21.10)$$

where k_B is Boltzmann's constant and W is the number of microstates of the system corresponding to the macrostate.

The **microstate** of a system is the description of its individual components. The **macrostate** is a description of the system from a macroscopic point of view. A given macrostate can have many microstates.

In a **reversible** process, the system can be returned to its initial conditions along the same path on a PV diagram, and every point along this path is an equilibrium state. A process that does not satisfy these requirements is **irreversible.**

▶ Concepts and Principles

A **heat engine** is a device that takes in energy by heat and, operating in a cyclic process, expels a fraction of that energy by means of work. The net work done by a heat engine in carrying a working substance through a cyclic process ($\Delta E_{int} = 0$) is

$$W_{eng} = |Q_h| - |Q_c| \quad (21.1)$$

where $|Q_h|$ is the energy taken in from a hot reservoir and $|Q_c|$ is the energy expelled to a cold reservoir.

Two ways the **second law of thermodynamics** can be stated are as follows:

- It is impossible to construct a heat engine that, operating in a cycle, produces no effect other than the input of energy by heat from a reservoir and the performance of an equal amount of work (the Kelvin–Planck statement).
- It is impossible to construct a cyclical machine whose sole effect is to transfer energy continuously by heat from one object to another object at a higher temperature without the input of energy by work (the Clausius statement).

Carnot's theorem states that no real heat engine operating (irreversibly) between the temperatures T_c and T_h can be more efficient than an engine operating reversibly in a Carnot cycle between the same two temperatures.

The thermal efficiency of a heat engine operating in the Carnot cycle is

$$e_C = 1 - \frac{T_c}{T_h} \quad (21.8)$$

The macroscopic state of a system that has a large number of microstates has three qualities that are all related: (1) *uncertainty:* because of the large number of microstates, there is a large uncertainty as to which one actually exists; (2) *choice:* again because of the large number of microstates, there is a large number of choices from which to select as to which one exists; (3) *probability:* a macrostate with a large number of microstates is more likely to exist than one with a small number of microstates. For a thermodynamic system, all three of these can be related to the state variable of **entropy.**

The second law of thermodynamics states that when real (irreversible) processes occur, there is a spatial spreading of energy. This spreading of energy is related to a thermodynamic state variable called **entropy** S. Therefore, yet another way the second law can be stated is as follows:

- The entropy of the Universe increases in all real processes.

The **change in entropy** dS of a system during a process between two infinitesimally separated equilibrium states is

$$dS = \frac{dQ_r}{T} \qquad (21.12)$$

where dQ_r is the energy transfer by heat for the system for a reversible process that connects the initial and final states.

The change in entropy of a system during an arbitrary finite process between an initial state and a final state is

$$\Delta S = \int_i^f \frac{dQ_r}{T} \qquad (21.13)$$

The value of ΔS for the system is the same for all paths connecting the initial and final states. The change in entropy for a system undergoing any reversible, cyclic process is zero.

Think–Pair–Share

See the Preface for an explanation of the icons used in this problems set. For additional assessment items for this section, go to **WEBASSIGN** From Cengage

1. An engine operates in a Carnot cycle as follows. At point A in the cycle, 2.34 mol of a monatomic ideal gas has a pressure of 1 400 kPa, a volume of 10.0 L, and a temperature of 720 K. The gas expands isothermally to point B and then expands adiabatically to point C, where its volume is 24.0 L. An isothermal compression brings it to point D, where its volume is 15.0 L. An adiabatic process returns the gas to point A.
(a) Fill in the following table with the pressures, volumes, and temperatures at each of the four points in the cycle:

Point in Cycle	P(kPa)	V(L)	T(K)
A			
B			
C			
D			

(b) Fill in the following table with the energy transfer by heat, work done on the gas, and the change in internal energy of the gas for each of the four processes in the cycle:

Process in Cycle	Q(kJ)	W(kJ)	ΔE_{int} (kJ)
$A \rightarrow B$			
$B \rightarrow C$			
$C \rightarrow D$			
$D \rightarrow A$			

(c) Find the efficiency of the engine from the data in the table in part (b). (d) Find the efficiency of the engine from the data in the table in part (a).

2. **ACTIVITY** If two six-sided dice are rolled, possible results of adding up the number of dots on the upper faces range from 2 to 12. The probabilities of these results vary due to the number of ways a particular result can be achieved. For example, there is only one way that a 2 can be achieved: 1–1. But there are six ways to achieve a result of 7: 1–6, 2–5, 3–4, 4–3, 5–2, and 6–1. Therefore, a result of 7 is six times more probable than a 2. The bar graph in Figure TP21.2 shows the theoretical probability of all possible results for two dice. Experimentally, if the two dice were thrown 100 times and a histogram of the possible results were drawn, it would have a shape very similar to this probability graph. (a) What if we throw *three* dice? What will the histogram look like? Do this in your group. Throw three dice 100 times and make a histogram of the results. How does the shape of the resulting histogram differ from the probability curve shown in Figure TP21.2 for two dice? Make a theoretical probability graph for three dice like that

Figure TP21.2

in Figure TP21.2 and compare to your histogram. (b) What if you rolled Avogadro's number of dice? (Don't try this!) What would the histogram look like? (Make a prediction based on how the graph for three dice varies from that for two dice.) (c) Suppose you laboriously set up Avogadro's number of dice on a table, with all dice having a 1 on their upper face. Then you shook the table for a few seconds. When you added up all the numbers on the upper faces, what is the most likely result? (d) Now imagine shaking the table again. How likely is it that the dice could all return to having a 1 on their upper faces? (e) What does all this have to do with entropy?

3. **ACTIVITY** Let's consider the various liquids in the table below at their boiling points. The table provides the latent heat of vaporization of each liquid in kJ/mol (note the units), and the boiling point in °C. For each of the liquids, evaluate the entropy change of the liquid per mole when it vaporizes at the boiling point. What do you notice about the results?

	L_v (kJ/mol)	Boiling Point (°C)
Polar compounds		
HF	25.2	19.7
HCl	16.2	−84.8
HI	19.8	−35.6
H_2O	40.7	100
Nonpolar compounds		
C_3H_8	19.0	−42.1
C_4H_{10}	22.4	−0.50
Elements		
Hg	54.7	357
Pb	178	1 749
Cl_2	20.4	−34.0
Br_2	30.0	58.8

Problems

See the Preface for an explanation of the icons used in this problems set. For additional assessment items for this section, go to ⚙ WEBASSIGN From Cengage

SECTION 21.1 Heat Engines and the Second Law of Thermodynamics

1. A particular heat engine has a mechanical power output of 5.00 kW and an efficiency of 25.0%. The engine expels 8.00×10^3 J of exhaust energy in each cycle. Find (a) the energy taken in during each cycle and (b) the time interval for each cycle.

2. The work done by an engine equals one-fourth the energy it absorbs from a reservoir. (a) What is its thermal efficiency? (b) What fraction of the energy absorbed is expelled to the cold reservoir?

3. Suppose a heat engine is connected to two energy reservoirs, one a pool of molten aluminum (660°C) and the other a block of solid mercury ($-38.9°C$). The engine runs by freezing 1.00 g of aluminum and melting 15.0 g of mercury during each cycle. The heat of fusion of aluminum is 3.97×10^5 J/kg; the heat of fusion of mercury is 1.18×10^4 J/kg. What is the efficiency of this engine?

SECTION 21.2 Heat Pumps and Refrigerators

4. During each cycle, a refrigerator ejects 625 kJ of energy to a high-temperature reservoir and takes in 550 kJ of energy from a low-temperature reservoir. Determine (a) the work done on the refrigerant in each cycle and (b) the coefficient of performance of the refrigerator.

5. A freezer has a coefficient of performance of 6.30. It is advertised as using electricity at a rate of 457 kWh/yr. (a) On average, how much energy does it use in a single day? (b) On average, how much energy does it remove from the refrigerator in a single day? (c) What maximum mass of water at 20.0°C could the freezer freeze in a single day? *Note:* One kilowatt-hour (kWh) is an amount of energy equal to running a 1-kW appliance for one hour.

6. A heat pump has a coefficient of performance equal to 4.20 and requires a power of 1.75 kW to operate. (a) How much energy does the heat pump add to a home in one hour? (b) If the heat pump is reversed so that it acts as an air conditioner in the summer, what would be its coefficient of performance?

SECTION 21.4 The Carnot Engine

7. One of the most efficient heat engines ever built is a coal-fired steam turbine in the Ohio River valley, operating between 1 870°C and 430°C. (a) What is its maximum theoretical efficiency? (b) The actual efficiency of the engine is 42.0%. How much mechanical power does the engine deliver if it absorbs 1.40×10^5 J of energy each second from its hot reservoir?

8. *Why is the following situation impossible?* An inventor comes to a patent office with the claim that her heat engine, which employs water as a working substance, has a thermodynamic efficiency of 0.110. Although this efficiency is low compared with typical automobile engines, she explains that her engine operates between an energy reservoir at room temperature and a water–ice mixture at atmospheric pressure and therefore requires no fuel other than that to make the ice. The patent is approved, and working prototypes of the engine prove the inventor's efficiency claim.

9. If a 35.0%-efficient Carnot heat engine (Fig. 21.2) is run in reverse so as to form a refrigerator (Fig. 21.4), what would be this refrigerator's coefficient of performance?

10. An ideal refrigerator or ideal heat pump is equivalent to a Carnot engine running in reverse. That is, energy $|Q_c|$ is taken in from a cold reservoir and energy $|Q_h|$ is rejected to a hot reservoir. (a) Show that the work that must be supplied to run the refrigerator or heat pump is

$$W = \frac{T_h - T_c}{T_c} |Q_c|$$

(b) Show that the coefficient of performance (COP) of the ideal refrigerator is

$$\text{COP} = \frac{T_c}{T_h - T_c}$$

11. A heat engine is being designed to have a Carnot efficiency of 65.0% when operating between two energy reservoirs. (a) If the temperature of the cold reservoir is 20.0°C, what must be the temperature of the hot reservoir? (b) Can the actual efficiency of the engine be equal to 65.0%? Explain.

12. A power plant operates at a 32.0% efficiency during the summer when the seawater used for cooling is at 20.0°C. The plant uses 350°C steam to drive turbines. If the plant's efficiency changes in the same proportion as the ideal efficiency, what would be the plant's efficiency in the winter, when the seawater is at 10.0°C?

13. You are working on a summer job at a company that designs non-traditional energy systems. The company is working on a proposed electric power plant that would make use of the temperature gradient in the ocean. The system includes a heat engine that would operate between 20.0°C (surface-water temperature) and 5.00°C (water temperature at a depth of about 1 km). (a) Your supervisor asks you to determine the maximum efficiency of such a system. (b) In addition, if the electric power output of the plant is 75.0 MW and it operates at the maximum theoretically possible efficiency, you must determine the rate at which energy is taken in from the warm reservoir. (c) From this information, if an electric bill for a typical home shows a use of 950 kWh per month, your supervisor wants to know how many homes can be provided with power from this energy system operating at its maximum efficiency. (d) As energy is drawn from the warm surface water to operate the engine, it is replaced by energy absorbed from sunlight on the surface. If the average intensity absorbed from sunlight is 650 W/m² for 12 daylight hours on a clear day, you need to find the area of the ocean surface that is necessary for sunlight to replace the energy absorbed into the engine. (e) From this information, you need to determine if there is enough ocean surface on the Earth to use such engines to supply the electrical needs for all the homes associated with the Earth's population. Assume the energy use for a home in part (c) is an average over the entire planet. (f) In view of your results in this problem, your supervisor

has asked for your conclusion as to whether such a system is worthwhile to pursue. Note that the "fuel" (sunlight) is free.

14. **QC** A Carnot heat engine operates between temperatures T_h and T_c. (a) If $T_h = 500$ K and $T_c = 350$ K, what is the efficiency of the engine? (b) What is the change in its efficiency for each degree of increase in T_h above 500 K? (c) What is the change in its efficiency for each degree of change in T_c? (d) Does the answer to part (c) depend on T_c? Explain.

15. **QC** An electric generating station is designed to have an electric output power of 1.40 MW using a turbine with two-thirds the efficiency of a Carnot engine. The exhaust energy is transferred by heat into a cooling tower at 110°C. (a) Find the rate at which the station exhausts energy by heat as a function of the fuel combustion temperature T_h. (b) If the firebox is modified to run hotter by using more advanced combustion technology, how does the amount of energy exhaust change? (c) Find the exhaust power for $T_h = 800$°C. (d) Find the value of T_h for which the exhaust power would be only half as large as in part (c). (e) Find the value of T_h for which the exhaust power would be one-fourth as large as in part (c).

16. **QC S** Suppose you build a two-engine device with the exhaust energy output from one heat engine supplying the input energy for a second heat engine. We say that the two engines are running *in series*. Let e_1 and e_2 represent the efficiencies of the two engines. (a) The overall efficiency of the two-engine device is defined as the total work output divided by the energy put into the first engine by heat. Show that the overall efficiency e is given by

$$e = e_1 + e_2 - e_1 e_2$$

What If? For parts (b) through (e) that follow, assume the two engines are Carnot engines. Engine 1 operates between temperatures T_h and T_i. The gas in engine 2 varies in temperature between T_i and T_c. In terms of the temperatures, (b) what is the efficiency of the combination engine? (c) Does an improvement in net efficiency result from the use of two engines instead of one? (d) What value of the intermediate temperature T_i results in equal work being done by each of the two engines in series? (e) What value of T_i results in each of the two engines in series having the same efficiency?

17. A heat pump used for heating shown in Figure P21.17 is essentially an air conditioner installed backward. It extracts energy from colder air outside and deposits it in a warmer room. Suppose the ratio of the actual energy entering the room to the work done by the device's motor is 10.0% of the theoretical maximum ratio. Determine the energy entering the room per joule of work done by the motor given that the inside temperature is 20.0°C and the outside temperature is −5.00°C.

Figure P21.17

SECTION 21.5 Gasoline and Diesel Engines

Note: For problems in this section, assume the gas in the engine is diatomic with $\gamma = 1.40$.

18. A gasoline engine has a compression ratio of 6.00. (a) What is the efficiency of the engine if it operates in an idealized Otto cycle? (b) **What If?** If the actual efficiency is 15.0%, what fraction of the fuel is wasted as a result of friction and energy transfers by heat that could be avoided in a reversible engine? Assume complete combustion of the air–fuel mixture.

19. **S** An idealized diesel engine operates in a cycle known as the *air-standard diesel cycle* shown in Figure P21.19. Fuel is sprayed into the cylinder at the point of maximum compression, B. Combustion occurs during the expansion $B \rightarrow C$, which is modeled as an isobaric process. Show that the efficiency of an engine operating in this idealized diesel cycle is

$$e = 1 - \frac{1}{\gamma}\left(\frac{T_D - T_A}{T_C - T_B}\right)$$

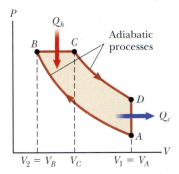

Figure P21.19

SECTION 21.6 Entropy

20. (a) Prepare a table like Table 21.1 for the following occurrence. You toss four coins into the air simultaneously and then record the results of your tosses in terms of the numbers of heads (H) and tails (T) that result. For example, HHTH and HTHH are two possible ways in which three heads and one tail can be achieved. (b) On the basis of your table, what is the most probable result recorded for a toss?

21. Prepare a table like Table 21.1 by using the same procedure (a) for the case in which you draw three marbles from your bag rather than four and (b) for the case in which you draw five marbles rather than four.

SECTION 21.7 Entropy in Thermodynamic Systems

22. A Styrofoam cup holding 125 g of hot water at 100°C cools to room temperature, 20.0°C. What is the change in entropy of the room? Neglect the specific heat of the cup and any change in temperature of the room.

23. **AMT** A 1 500-kg car is moving at 20.0 m/s. The driver brakes to a stop. The brakes cool off to the temperature of the surrounding air, which is nearly constant at 20.0°C. What is the total entropy change?

24. A 2.00-L container has a center partition that divides it into two equal parts as shown in Figure P21.24. The left side

Figure P21.24

contains 0.044 0 mol of H_2 gas, and the right side contains 0.044 0 mol of O_2 gas. Both gases are at room temperature and at atmospheric pressure. The partition is removed, and the gases are allowed to mix. What is the entropy increase of the system?

25. Calculate the change in entropy of 250 g of water warmed slowly from 20.0°C to 80.0°C.

26. What change in entropy occurs when a 27.9-g ice cube at −12°C is transformed into steam at 115°C?

SECTION 21.8 Entropy and the Second Law

27. When an aluminum bar is connected between a hot reservoir at 725 K and a cold reservoir at 310 K, 2.50 kJ of energy is transferred by heat from the hot reservoir to the cold reservoir. In this irreversible process, calculate the change in entropy of (a) the hot reservoir, (b) the cold reservoir, and (c) the Universe, neglecting any change in entropy of the aluminum rod.

28. When a metal bar is connected between a hot reservoir at T_h and a cold reservoir at T_c, the energy transferred by heat from the hot reservoir to the cold reservoir is Q. In this irreversible process, find expressions for the change in entropy of (a) the hot reservoir, (b) the cold reservoir, and (c) the Universe, neglecting any change in entropy of the metal rod.

29. How fast are you personally making the entropy of the Universe increase right now? Compute an order-of-magnitude estimate, stating what quantities you take as data and the values you measure or estimate for them.

ADDITIONAL PROBLEMS

30. Every second at Niagara Falls, some 5.00×10^3 m³ of water falls a distance of 50.0 m. What is the increase in entropy of the Universe per second due to the falling water? Assume the mass of the surroundings is so great that its temperature and that of the water stay nearly constant at 20.0°C. Also assume a negligible amount of water evaporates.

31. The energy absorbed by an engine is three times greater than the work it performs. (a) What is its thermal efficiency? (b) What fraction of the energy absorbed is expelled to the cold reservoir?

32. In 1993, the U.S. government instituted a requirement that all room air conditioners sold in the United States must have an energy efficiency ratio (EER) of 10 or higher. The EER is defined as the ratio of the cooling capacity of the air conditioner, measured in British thermal units per hour, or Btu/h, to its electrical power requirement in watts. (a) Convert the EER of 10.0 to dimensionless form, using the conversion 1 Btu = 1 055 J. (b) What is the appropriate name for this dimensionless quantity? (c) In the 1970s, it was common to find room air conditioners with EERs of 5 or lower. State how the operating costs compare for 10 000-Btu/h air conditioners with EERs of 5.00 and 10.0. Assume each air

conditioner operates for 1 500 h during the summer in a city where electricity costs 17.0¢ per kWh.

33. In 1816, Robert Stirling, a Scottish clergyman, patented the *Stirling engine*, which has found a wide variety of applications ever since, including current use in solar energy collectors to transform sunlight into electricity. Fuel is burned externally to warm one of the engine's two cylinders. A fixed quantity of inert gas moves cyclically between the cylinders, expanding in the hot one and contracting in the cold one. Figure P21.33 represents a model for its thermodynamic cycle. Consider n moles of an ideal monatomic gas being taken once through the cycle, consisting of two isothermal processes at temperatures $3T_i$ and T_i and two constant-volume processes. Let us find the efficiency of this engine. (a) Find the energy transferred by heat into the gas during the isovolumetric process AB. (b) Find the energy transferred by heat into the gas during the isothermal process BC. (c) Find the energy transferred by heat into the gas during the isovolumetric process CD. (d) Find the energy transferred by heat into the gas during the isothermal process DA. (e) Identify which of the results from parts (a) through (d) are positive and evaluate the energy input to the engine by heat. (f) From the first law of thermodynamics, find the work done by the engine. (g) From the results of parts (e) and (f), evaluate the efficiency of the engine. A Stirling engine is easier to manufacture than an internal combustion engine or a turbine. It can run on burning garbage. It can run on the energy transferred by sunlight and produce no material exhaust. Stirling engines are not currently used in automobiles due to long startup times and poor acceleration response.

Figure P21.33

34. Suppose an ideal (Carnot) heat pump could be constructed for use as an air conditioner. (a) Obtain an expression for the coefficient of performance (COP) for such an air conditioner in terms of T_h and T_c. (b) Would such an air conditioner operate on a smaller energy input if the difference in the operating temperatures were greater or smaller? (c) Compute the COP for such an air conditioner if the indoor temperature is 20.0°C and the outdoor temperature is 40.0°C.

35. **Review.** This problem complements Problem 44 in Chapter 10. In the operation of a single-cylinder internal combustion piston engine, one charge of fuel explodes to drive the piston outward in the *power stroke*. Part of its energy output is stored in a turning flywheel. This energy is then used to push the piston inward to compress the next charge of fuel and air. In this compression process, assume an original volume of 0.120 L of a diatomic ideal gas at atmospheric pressure

is compressed adiabatically to one-eighth of its original volume. (a) Find the work input required to compress the gas. (b) Assume the flywheel is a solid disk of mass 5.10 kg and radius 8.50 cm, turning freely without friction between the power stroke and the compression stroke. How fast must the flywheel turn immediately after the power stroke? This situation represents the minimum angular speed at which the engine can operate without stalling. (c) When the engine's operation is well above the point of stalling, assume the flywheel puts 5.00% of its maximum energy into compressing the next charge of fuel and air. Find its maximum angular speed in this case.

36. A firebox is at 750 K, and the ambient temperature is 300 K. **QC** The efficiency of a Carnot engine doing 150 J of work as it transports energy between these constant-temperature baths is 60.0%. The Carnot engine must take in energy 150 J/0.600 = 250 J from the hot reservoir and must put out 100 J of energy by heat into the environment. To follow Carnot's reasoning, suppose some other heat engine S could have an efficiency of 70.0%. (a) Find the energy input and exhaust energy output of engine S as it does 150 J of work. (b) Let engine S operate as in part (a) and run the Carnot engine in reverse between the same reservoirs. The output work of engine S is the input work for the Carnot refrigerator. Find the total energy transferred to or from the firebox and the total energy transferred to or from the environment as both engines operate together. (c) Explain how the results of parts (a) and (b) show that the Clausius statement of the second law of thermodynamics is violated. (d) Find the energy input and work output of engine S as it puts out exhaust energy of 100 J. Let engine S operate as in part (c) and contribute 150 J of its work output to running the Carnot engine in reverse. Find (e) the total energy the firebox puts out as both engines operate together, (f) the total work output, and (g) the total energy transferred to the environment. (h) Explain how the results show that the Kelvin–Planck statement of the second law is violated. Therefore, our assumption about the efficiency of engine S must be false. (i) Let the engines operate together through one cycle as in part (d). Find the change in entropy of the Universe. (j) Explain how the result of part (i) shows that the entropy statement of the second law is violated.

37. A 1.00-mol sample of an ideal monatomic gas is taken **QC** through the cycle shown in Figure P21.37. The process $A \rightarrow B$ is a reversible isothermal expansion. Calculate (a) the net work done by the gas, (b) the energy added to the gas by heat, (c) the energy exhausted from the gas by heat, and (d) the efficiency of the cycle. (e) Explain how the efficiency compares with that of a Carnot engine operating between the same temperature extremes.

Figure P21.37

38. A system consisting of n moles of an ideal gas with molar **QC** specific heat at constant pressure C_p undergoes two revers- **S** ible processes. It starts with pressure P_i and volume V_i, expands isothermally, and then contracts adiabatically to reach a final state with pressure P_i and volume $3V_i$. (a) Find its change in entropy in the isothermal process. (The entropy does not change in the adiabatic process.) (b) **What If?** Explain why the answer to part (a) must be the same as the answer to Problem 46. (You do not need to solve Problem 46 to answer this question.)

39. A heat engine operates between two reservoirs at $T_2 = 600$ K and $T_1 = 350$ K. It takes in 1.00×10^3 J of energy from the higher-temperature reservoir and performs 250 J of work. Find (a) the entropy change of the Universe ΔS_U for this process and (b) the work W that could have been done by an ideal Carnot engine operating between these two reservoirs. (c) Show that the difference between the amounts of work done in parts (a) and (b) is $T_1 \Delta S_U$.

40. You are working as an assistant to a physics professor. She **CR** has seen some presentations you have made to your classes and is aware of your expertise in preparing presentation slides. Her laptop has crashed and she cannot access the presentation slides she needs for her lecture coming up in one hour. Her lecture is on entropy in engine cycles. She asks you to quickly generate two slides on your laptop, both showing *TS* diagrams, (a) one for the Carnot cycle and (b) one for the Otto cycle. As she leaves, you think, "Uh-oh. What's a *TS diagram*?" Quick, you have no time to waste! Get to work!

41. You are working as an expert witness for an environmen- **CR** tal agency. A utility in a neighboring town has proposed a new power plant that produces 1.00 GW of electrical power from turbines. The utility claims that the plant will take in steam at 500 K and reject water at 300 K into a flowing cold-water river. The flow rate of the river is 6.00×10^4 kg/s. The agency supervisor is concerned about the effect of dumping warm water on the fish in the river. (a) The utility claims that the power plant operates with Carnot efficiency. With that assumption, you need to determine for a trial presentation by how much the temperature of the water downstream from the power plant will rise due to the rejected energy from the power plant. (b) If you abandon the utility's claim that the power plant operates at Carnot efficiency and assume a more realistic efficiency, you need to testify whether the increase in water temperature will be higher or lower than that found in part (a). (c) Finally, determine the increase in water temperature in the stream if the actual efficiency of the power plant were estimated by you and the agency physicist to be 15.0%.

42. You are working as an expert witness for an environmental **CR** agency. A utility in a neighboring town has proposed a new **S** power plant that produces electrical power P from turbines. The utility claims that the plant will take in steam at temperature T_h and reject water at temperature T_c into a flowing cold-water river. The flow rate of the river is $\Delta m/\Delta t$. The agency supervisor is concerned about the effect of dumping warm water on the fish in the river. (a) The utility claims that the power plant operates with Carnot efficiency. With that assumption, you need to determine for a trial presentation by how much the temperature of the water downstream from the power plant will rise due to the rejected energy from the power plant. (b) If you abandon the utility's claim that the power plant operates at Carnot efficiency and assume a more realistic efficiency e, you need to determine the increase in water temperature in the stream. (c) Finally, you need to testify whether the increase in water temperature in part (b) will be higher or lower than that found in part (a).

43. An athlete whose mass is 70.0 kg drinks 16.0 ounces (454 g)
BIO of refrigerated water. The water is at a temperature of 35.0°F.
QC (a) Ignoring the temperature change of the body that results from the water intake (so that the body is regarded as a reservoir always at 98.6°F), find the entropy increase of the entire system. (b) **What If?** Assume the entire body is cooled by the drink and the average specific heat of a person is equal to the specific heat of liquid water. Ignoring any other energy transfers by heat and any metabolic energy release, find the athlete's temperature after she drinks the cold water given an initial body temperature of 98.6°F. (c) Under these assumptions, what is the entropy increase of the entire system? (d) State how this result compares with the one you obtained in part (a).

44. *Why is the following situation impossible?* Two samples of water are mixed at constant pressure inside an insulated container: 1.00 kg of water at 10.0°C and 1.00 kg of water at 30.0°C. Because the container is insulated, there is no exchange of energy by heat between the water and the environment. Furthermore, the amount of energy that leaves the warm water by heat is equal to the amount that enters the cool water by heat. Therefore, the entropy change of the Universe is zero for this process.

45. A sample of an ideal gas expands isothermally, doubling in
QC volume. (a) Show that the work done on the gas in expand-
S ing is $W = -nRT \ln 2$. (b) Because the internal energy E_{int} of an ideal gas depends solely on its temperature, the change in internal energy is zero during the expansion. It follows from the first law that the energy input to the gas by heat during the expansion is equal to the energy output by work. Does this process have 100% efficiency in converting energy input by heat into work output? (c) Does this conversion violate the second law? Explain.

46. A sample consisting of n moles of an ideal gas undergoes a
S reversible isobaric expansion from volume V_i to volume $3V_i$. Find the change in entropy of the gas by calculating $\int_i^f dQ/T$, where $dQ = nC_P\,dT$.

CHALLENGE PROBLEM

47. The compression ratio of an Otto cycle as shown in Figure 21.12 is $V_A/V_B = 8.00$. At the beginning A of the compression process, 500 cm³ of gas is at 100 kPa and 20.0°C. At the beginning of the adiabatic expansion, the temperature is $T_C = 750$°C. Model the working fluid as an ideal gas with $\gamma = 1.40$. (a) Fill in this table to follow the states of the gas:

	T (K)	P (kPa)	V (cm³)
A	293	100	500
B			
C	1 023		
D			

(b) Fill in this table to follow the processes:

	Q	W	ΔE_{int}
$A \rightarrow B$			
$B \rightarrow C$			
$C \rightarrow D$			
$D \rightarrow A$			
$ABCDA$			

(c) Identify the energy input $|Q_h|$, (d) the energy exhaust $|Q_c|$, and (e) the net output work W_{eng}. (f) Calculate the thermal efficiency. (g) Find the number of crankshaft revolutions per minute required for a one-cylinder engine to have an output power of 1.00 kW = 1.34 hp. *Note:* The thermodynamic cycle involves four piston strokes.

PART 4

Electricity and Magnetism

We now study the branch of physics concerned with electric and magnetic phenomena. In this part of the book, we will focus on the term T_{ET}, representing energy transfer by electrical transmission in Equation 8.2. In the final chapter in this part, we will introduce the physics behind the term T_{ER} for electromagnetic radiation. The laws of electricity and magnetism play a central role in the operation of such devices as smartphones, televisions, electric motors, computers, high-energy accelerators, and other electronic devices. More fundamentally, the interatomic and intermolecular forces responsible for the formation of solids and liquids are electric in origin. In turn, electric forces are the basis for the science of chemistry, and are responsible for the development of biological organisms. Therefore, gravity plays a role in nature by allowing planets to exist, but life on that planet is due to electricity!

Not until the early part of the nineteenth century did scientists establish that electricity and magnetism are related phenomena. In 1819, Hans Oersted discovered that a compass needle is deflected when placed near a circuit carrying an electric current. In 1831, Michael Faraday and, almost simultaneously, Joseph Henry showed that when a wire is moved near a magnet (or, equivalently, when a magnet is moved near a wire), an electric current is established in the wire. In 1873, James Clerk Maxwell used these observations and other experimental facts as a basis for formulating the laws of electromagnetism as we know them today. (*Electromagnetism* is a name given to the combined study of electricity and magnetism.)

Maxwell's contributions to the field of electromagnetism were especially significant because the laws he formulated are basic to *all* forms of electromagnetic phenomena. His work is as important as Newton's work on the laws of motion and the theory of gravitation. ■

A Transrapid maglev train pulls into a station in Shanghai, China. The word *maglev* is an abbreviated form of *magnetic levitation.* This train makes no physical contact with its rails; its weight is totally supported by electromagnetic forces. In this part of the book, we will study these forces. (*Lee Prince/ Shutterstock*)

22

Electric Fields

An egg has code numbers printed on it. How do you print on an *egg*? (*Starstuff/Shutterstock*)

STORYLINE **Taking advantage of a weekend visit to your family** home, you have washed and dried your clothing and are removing the clothing from the dryer. You notice that your socks seem to be stuck to your shirts. Even shaking the shirt will not remove the socks. When you pull the socks off the shirt, they peel off with a crackling sound. As you carry your dry clothes into your bedroom, you wonder why those effects occurred. You are still wondering while you comb your hair in the bathroom. You turn on the faucet and unintentionally hold the comb you just used next to the stream of water. The stream of water bends to the side, toward the comb! You move the comb to different positions and notice that the stream of water deviates to the side by different amounts. Your father wanders by while you are doing this and says, "That's exactly the technique we use in designing our high-speed manufacturing printers. Go look at the cans of food in the kitchen. How do you think the expiration dates are printed on the cans? Even more fascinating, how do we print code numbers on eggs?" You never quite understood what your father does for a living, but are now quite intrigued. You know that he designs some kind of industrial printers. You ask him what he means about your bathroom experiment. He tells you to do some online research on continuous inkjet printing.

CONNECTIONS   In our earlier chapters on mechanics, we identified several types of forces: normal forces perpendicular to surfaces, friction forces parallel to surfaces, tension forces along strings, gravitational forces on planets, etc. Among these, the gravitational force, which we studied in detail in Chapter 13, is unique, as it is a fundamental force in nature. It turns out that the other forces in this list are all due to a second type of fundamental force, the electromagnetic force. In this chapter, we will begin our study of one manifestation of this force, the electric force. Our understanding of the gravitational force developed according to a conceptual structure that we built: we learned that the force exists between objects with mass. We then developed a mathematical law, Newton's law of universal gravitation, to describe the magnitude of the force. We then introduced the notion of a gravitational field. From there we discussed gravitational potential energy in a system of two or more massive objects. We will follow a similar conceptual development in our study of the electric force. We will learn that the force exists between objects with electric charge. We will develop a mathematical law, Coulomb's law, to describe the magnitude of the force. We will introduce the notion of an electric field, and we will discuss electric potential energy in a system of two or more charged objects. As we continue to study the electric force in the next few chapters, we will find that we have much more control over this force than we do over gravity. Sources of gravity are restricted to one shape: the spherical shape of planets and stars (with the exception of small asteroids and moons that might deviate slightly from spheres). On the other hand, we can formulate various shapes for electrical situations: spheres, plates, wires, and the like! Objects moving in gravitational fields are huge and massive; we can't control their motion. Objects moving in electric fields can be as small as electrons; we can easily change their motion! We have no control over gravity; it's always there. But we can turn electricity on and off! We can't adjust the strength of the gravitational field of the Earth. But we can easily turn a dial to change the strength of an electric field! Gravity is everywhere, inside and outside of everything. But some materials conduct electricity and others don't! And we can create electric field-free regions of space quite easily! This type of control that we have over electricity makes it the basis of our technological society. Phenomena associated with electrical charges will appear repeatedly in most of the remaining chapters in this book.

22.1 Properties of Electric Charges

A number of simple experiments demonstrate the existence of the electric force. For example, after rubbing a balloon on your hair on a dry day, you will find that the balloon attracts bits of paper. The *attractive* force is often strong enough to suspend the paper from the balloon. Figure 22.1a shows another effect of the electric force. The woman's body becomes charged, and, in this case, there is a *repulsive* force between all the hairs on her head. Figure 22.1b shows another *attractive* situation. A cat has rubbed its body against styrofoam peanuts while playing in a packing box. Upon exiting the box, the styrofoam peanuts are stuck to its body.

JENS SCHLUETER/Getty Images

Sean McGrath/Flickr

a

b

Figure 22.1 (a) This young woman is enjoying the effects of electrically charging her body. Each individual hair on her head becomes charged and exerts a repulsive force on the other hairs, resulting in the "stand-up" hairdo seen here. (b) An attractive electric force is demonstrated by a cat who got into a box of styrofoam peanuts.

A negatively charged rubber rod suspended by a string is attracted to a positively charged glass rod.

A negatively charged rubber rod is repelled by another negatively charged rubber rod.

Rubber

Rubber

$\vec{F}$

$\vec{F}$ $\vec{F}$ Glass

Rubber

$+ + + + + +$

$\vec{F}$

a

b

Figure 22.2   The electric force between (a) oppositely charged objects and (b) like-charged objects.

Because of conservation of charge, each electron adds negative charge to the silk and an equal positive charge is left on the glass rod.

Figure 22.3   When a glass rod is rubbed with silk, electrons are transferred from the glass to the silk.

Electric charge is conserved ▶

When materials behave in this way, they are said to be *electrified* or to have become **electrically charged.** The **electric force** is the force acting between electrically charged objects. You can easily electrify your body by vigorously rubbing your shoes on a wool rug. Evidence of the electric charge on your body can be detected by lightly touching (and startling) a friend. Under the right conditions, you will see a spark when you touch, and both of you will feel a slight tingle. (Experiments such as these work best on a dry day because an excessive amount of moisture in the air can cause any charge you build up to "leak" from your body to the Earth.)

In a series of simple experiments, it was found that there are two kinds of electric charges, which were given the names **positive** and **negative** by Benjamin Franklin (1706–1790). Electrons are identified as having negative charge, and protons are positively charged. To verify that there are two types of charge, suppose a hard rubber rod that has been rubbed on fur is suspended by a string as shown in Figure 22.2. When a glass rod that has been rubbed on silk is brought near the rubber rod, the two rods exhibit an *attractive* force on each other (Fig. 22.2a). On the other hand, if two charged rubber rods (or two charged glass rods) are brought near each other as shown in Figure 22.2b, the two rods exhibit a *repulsive* force on each other. This observation shows that the rubber and glass have two different types of charge on them. On the basis of these observations, we conclude that **charges of the same sign repel one another and charges with opposite signs attract one another.**

Using the convention suggested by Franklin, the electric charge on the glass rod is called positive and that on the rubber rod is called negative. Therefore, any charged object attracted to a charged rubber rod (or repelled by a charged glass rod) must have a positive charge, and any charged object repelled by a charged rubber rod (or attracted to a charged glass rod) must have a negative charge.

Another important aspect of electricity that arises from experimental observations is that **electric charge is always conserved** in an isolated system. That is, when one object is rubbed against another, charge is not created in the process. The electrified state is due to a *transfer* of charge from one object to the other. One object gains some amount of negative charge while the other gains an equal amount of positive charge. For example, when a glass rod is rubbed on silk as in Figure 22.3, the silk obtains a negative charge equal in magnitude to the positive charge on the glass rod. We now know from our understanding of atomic structure that electrons are transferred in the rubbing process from the glass to the silk. Similarly, when rubber is rubbed on fur, electrons are transferred from the fur to the rubber, giving the rubber a net negative charge and the fur a net positive charge. This process works because neutral, uncharged matter contains as many positive charges (protons within atomic nuclei) as negative charges (electrons). Conservation of electric charge for an isolated system is like conservation of energy, momentum, and angular momentum, but we don't identify an analysis model for this conservation principle because it is not used often enough in the mathematical solution to problems.

In 1909, Robert Millikan (1868–1953) discovered that electric charge always occurs as integral multiples of a fundamental amount of charge e. (We will provide a numerical value for e in Section 22.3.) In modern terms, the electric charge q is said to be **quantized,** where q is the standard symbol used for charge as a variable. That is, electric charge exists as discrete "packets," and we can write $q = \pm Ne$, where N is some integer. Other experiments in the same period showed that the electron has a charge $-e$ and the proton has a charge of equal magnitude but opposite sign $+e$. Some particles, such as the neutron, have no charge.

QUICK QUIZ 22.1 Three objects are brought close to each other, two at a time. When objects A and B are brought together, they repel. When objects B and C are brought together, they also repel. Which of the following are true? **(a)** Objects A and C possess charges of the same sign. **(b)** Objects A and C possess charges of opposite sign. **(c)** All three objects possess charges of the same sign. **(d)** One object is neutral. **(e)** Additional experiments must be performed to determine the signs of the charges.

22.2 Charging Objects by Induction

It is convenient to classify materials in terms of the ability of electrons to move through the material:

> Electrical **conductors** are materials in which some of the electrons are free electrons[1] that are not bound to atoms and can move relatively freely through the material; electrical **insulators** are materials in which all electrons are bound to atoms and cannot move freely through the material.

Materials such as glass, rubber, and dry wood fall into the category of electrical insulators. When such materials are charged by rubbing, only the area rubbed becomes charged and the charged particles are unable to move to other regions of the material.

In contrast, materials such as copper, aluminum, and silver are good electrical conductors. When such materials are charged in some small region, the charge readily distributes itself over the entire surface of the material.

Semiconductors are a third class of materials, and their electrical properties are somewhere between those of insulators and those of conductors. Silicon and germanium are well-known examples of semiconductors commonly used in the fabrication of a variety of electronic chips used in computers, cellular telephones, and home theater systems. The electrical properties of semiconductors can be changed over many orders of magnitude by the addition of controlled amounts of certain atoms to the materials.

To understand how to charge a conductor by a process known as **induction,** consider a neutral (uncharged) conducting sphere insulated from the ground as shown in Figure 22.4a (page 592). Electrons move freely within the conductor. These electrons originally belonged to the metal atoms before the atoms were combined into a macroscopic sample. Therefore, there is a lattice of atoms locked in place in the conductor, each missing an electron. The atoms are now called *ions* because they are charged, positively in this case due to the missing electron. We assume each atom releases one electron, so there are an equal number of free electrons and ions in the sphere if the charge on the sphere is exactly zero. When a negatively charged rubber

---

[1] A metal atom contains one or more outer electrons, which are weakly bound to the nucleus. When many atoms combine to form a metal, the *free electrons* are these outer electrons, which are not bound to any one atom. These electrons move about the metal in a manner similar to that of gas molecules moving in a container.

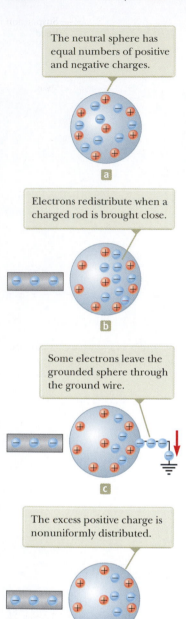

The neutral sphere has equal numbers of positive and negative charges.

a

Electrons redistribute when a charged rod is brought close.

b

Some electrons leave the grounded sphere through the ground wire.

c

The excess positive charge is nonuniformly distributed.

d

The remaining electrons redistribute uniformly, and there is a net uniform distribution of positive charge on the sphere.

e

Figure 22.4 Charging a metallic object by *induction*. (a) A neutral metallic sphere. (b) A charged rubber rod is placed near the sphere. (c) The sphere is grounded. (d) The ground connection is removed. (e) The rod is removed.

rod is brought near the sphere, electrons in the region nearest the rod experience a repulsive force and migrate to the opposite side of the sphere. This migration leaves the side of the sphere near the rod with an effective positive charge because of the diminished number of electrons as in Figure 22.4b. (The left side of the sphere in Figure 22.4b is positively charged *as if* positive charges moved into this region, but remember that only electrons are free to move.) This process occurs even though the rod never actually touches the sphere. If the same experiment is performed with a conducting wire connected from the sphere to the Earth (Fig. 22.4c), some of the electrons in the conductor are so strongly repelled by the presence of the negative charge in the rod that they move out of the sphere through the wire and into the Earth. The symbol ⏚ at the end of the wire in Figure 22.4c indicates that the wire is connected to **ground,** which means a reservoir, such as the Earth, that can accept or provide electrons freely with negligible effect on its electrical characteristics. If the wire to ground is then removed (Fig. 22.4d), the conducting sphere contains an excess of *induced* positive charge because it has fewer electrons than it needs to cancel out the positive charge of the ions. When the rubber rod is removed from the vicinity of the sphere (Fig. 22.4e), this induced positive charge remains on the ungrounded sphere. Notice that the rubber rod loses none of its negative charge during this process.

Charging an object by induction requires no contact with the object inducing the charge. That is in contrast to charging an object by rubbing (that is, by *conduction*), which does require contact between the two objects.

A process similar to induction in conductors takes place in insulators. In most neutral molecules, the center of positive charge coincides with the center of negative charge. In the presence of a charged object, however, these centers inside each molecule in an insulator may shift slightly, resulting in more positive charge on one side of the molecule than on the other. This realignment of charge within individual molecules produces a layer of charge on the surface of the insulator as shown in Figure 22.5. The proximity of the positive charges on the surface of the object and the negative charges on the surface of the insulator results in an attractive force between the object and the insulator. Your knowledge of induction in insulators should help you explain why the styrofoam peanuts stick to the cat in Figure 22.1b.

Ⓠ UICK QUIZ 22.2 Three objects are brought close to one another, two at a time. When objects A and B are brought together, they attract. When objects B and C are brought together, they repel. Which of the following are necessarily true? **(a)** Objects A and C possess charges of the same sign. **(b)** Objects A and C possess charges of opposite sign. **(c)** All three objects possess charges of the same sign. **(d)** One object is neutral. **(e)** Additional experiments must be performed to determine information about the charges on the objects.

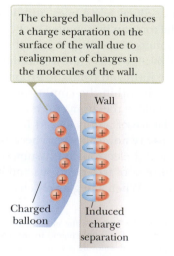

The charged balloon induces a charge separation on the surface of the wall due to realignment of charges in the molecules of the wall.

Wall

Charged balloon

Induced charge separation

Figure 22.5 A charged balloon is brought near an insulating wall.

22.3 Coulomb's Law

Charles Coulomb measured the magnitudes of the electric forces between charged objects using the torsion balance, which he invented (Fig. 22.6). The operating principle of the torsion balance is the same as that of the apparatus used by Cavendish to measure the density of the Earth (see Section 13.1), with the electrically neutral spheres replaced by charged ones. The electric force between charged spheres A and B in Figure 22.6 causes the spheres to either attract or repel each other, and the resulting motion causes the suspended fiber to twist. Because the restoring torque of the twisted fiber is proportional to the angle through which the fiber rotates, a measurement of this angle provides a quantitative measure of the electric force of attraction or repulsion. Once the spheres are charged by rubbing, the electric force between them is very large compared with the gravitational attraction, and so the gravitational force can be neglected.

From Coulomb's experiments, we can generalize the properties of the electric force (sometimes called the *electrostatic force*) between two stationary charged particles. We use the term **point charge** to refer to a charged particle of zero size. The electrical behavior of electrons and protons is very well described by modeling them as point charges. From experimental observations, we find that the magnitude of the electric force (sometimes called the *Coulomb force*) between two point charges is given by **Coulomb's law.**

Figure 22.6 Coulomb's balance, used to establish the inverse-square law for the electric force.

$$F_e = k_e \frac{|q_1||q_2|}{r^2} \qquad (22.1)$$

◀ Coulomb's law

where k_e is a constant called the **Coulomb constant.** In his experiments, Coulomb was able to show that the value of the exponent of r was 2 to within an uncertainty of a few percent. Modern experiments have shown that the exponent is 2 to within an uncertainty of a few parts in 10^{16}. Experiments also show that the electric force, like the gravitational force, is conservative.

The value of the Coulomb constant depends on the choice of units. The SI unit of charge is the **coulomb** (C). The Coulomb constant k_e in SI units has the value

$$k_e = 8.987\ 6 \times 10^9\ \mathrm{N \cdot m^2/C^2} \qquad (22.2)$$

◀ Coulomb constant

This constant is also written in the form

$$k_e = \frac{1}{4\pi\epsilon_0} \qquad (22.3)$$

where the constant ϵ_0 (Greek letter epsilon) is known as the **permittivity of free space** and has the value

$$\epsilon_0 = 8.854\ 2 \times 10^{-12}\ \mathrm{C^2/N \cdot m^2} \qquad (22.4)$$

The smallest unit of free charge e known in nature,[2] the charge on an electron $(-e)$ or a proton $(+e)$, has a magnitude

$$e = 1.602\ 18 \times 10^{-19}\ \mathrm{C} \qquad (22.5)$$

Therefore, 1 C of charge is approximately equal to the charge of 6.24×10^{18} electrons or protons. This number is very small when compared with the number of free electrons in 1 cm^3 of copper, which is on the order of 10^{23}. Nevertheless, 1 C is a substantial amount of charge. In typical experiments in which a rubber or glass

Charles Coulomb
French physicist (1736–1806)
Coulomb's major contributions to science were in the areas of electrostatics and magnetism. During his lifetime, he also investigated the strengths of materials, thereby contributing to the field of structural mechanics. In ergonomics, his research provided an understanding of the ways in which people and animals can best do work.

---

[2]No unit of charge smaller than e has been detected on a free particle; current theories, however, propose the existence of particles called *quarks* having charges $-e/3$ and $2e/3$. Although there is considerable experimental evidence for such particles inside nuclear matter, *free* quarks have never been detected. We discuss other properties of quarks in Chapter 44.

TABLE 22.1	Charge and Mass of the Electron, Proton, and Neutron	
Particle	Charge (C)	Mass (kg)
Electron (e)	$-1.602\ 176\ 5 \times 10^{-19}$	$9.109\ 4 \times 10^{-31}$
Proton (p)	$+1.602\ 176\ 5 \times 10^{-19}$	$1.672\ 62 \times 10^{-27}$
Neutron (n)	0	$1.674\ 93 \times 10^{-27}$

rod is charged by friction, a net charge on the order of 10^{-6} C is obtained. In other words, only a very small fraction of the total available charge is transferred between the rod and the rubbing material.

The charges and masses of the electron, proton, and neutron are given in Table 22.1. Notice that the electron and proton are identical in the magnitude of their charge but vastly different in mass. On the other hand, the proton and neutron are similar in mass but vastly different in charge. Chapter 44 will help us understand these interesting properties.

Example 22.1   The Hydrogen Atom

The electron and proton of a hydrogen atom are separated (on the average) by a distance of approximately 5.3×10^{-11} m. Find the magnitudes of the electric force and the gravitational force between the two particles.

SOLUTION

Conceptualize   Think about the two particles separated by the very small distance given in the problem statement. Because the particles have both electric charge and mass, there will be both an electric force and a gravitational force between them.

Categorize   The electric and gravitational forces will be evaluated from universal force laws, so we categorize this example as a substitution problem.

Use Coulomb's law to find the magnitude of the electric force:

$$F_e = k_e \frac{|e||-e|}{r^2} = (8.988 \times 10^9 \text{ N} \cdot \text{m}^2/\text{C}^2) \frac{(1.60 \times 10^{-19} \text{ C})^2}{(5.3 \times 10^{-11} \text{ m})^2}$$

$$= 8.2 \times 10^{-8} \text{ N}$$

Use Newton's law of universal gravitation and Table 22.1 (for the particle masses) to find the magnitude of the gravitational force:

$$F_g = G \frac{m_e m_p}{r^2}$$

$$= (6.674 \times 10^{-11} \text{ N} \cdot \text{m}^2/\text{kg}^2) \frac{(9.11 \times 10^{-31} \text{ kg})(1.67 \times 10^{-27} \text{ kg})}{(5.3 \times 10^{-11} \text{ m})^2}$$

$$= 3.6 \times 10^{-47} \text{ N}$$

The ratio $F_e/F_g \approx 2 \times 10^{39}$. Therefore, the gravitational force between charged atomic particles is negligible when compared with the electric force. Notice the similar mathematical forms of Newton's law of universal gravitation and Coulomb's law of electric forces. Other than the magnitude of the forces between elementary particles, what is a fundamental difference between the two forces?

When dealing with Coulomb's law, remember that force is a vector quantity and must be treated accordingly. Coulomb's law expressed in vector form for the electric force exerted by a charge q_1 on a second charge q_2, written $\vec{\mathbf{F}}_{12}$, is

◀ Vector form of Coulomb's law

$$\vec{\mathbf{F}}_{12} = k_e \frac{q_1 q_2}{r^2} \hat{\mathbf{r}}_{12} \qquad (22.6)$$

where $\hat{\mathbf{r}}_{12}$ is a unit vector directed from q_1 toward q_2 as shown in Figure 22.7a. Because the electric force obeys Newton's third law, the electric force exerted by q_2 on q_1 is equal in magnitude to the force exerted by q_1 on q_2 and in the opposite

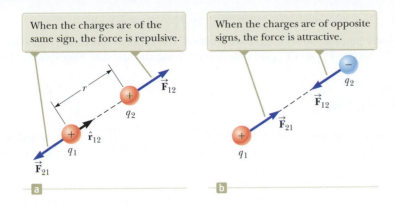

When the charges are of the same sign, the force is repulsive.

When the charges are of opposite signs, the force is attractive.

Figure 22.7 Two point charges separated by a distance r exert a force on each other that is given by Coulomb's law. The force $\vec{\mathbf{F}}_{21}$ exerted by q_2 on q_1 is equal in magnitude and opposite in direction to the force $\vec{\mathbf{F}}_{12}$, exerted by q_1 on q_2.

direction; that is, $\vec{\mathbf{F}}_{21} = -\vec{\mathbf{F}}_{12}$. Finally, Equation 22.6 shows that if q_1 and q_2 have the same sign as in Figure 22.7a, the product $q_1 q_2$ is positive and the electric force on one particle is directed away from the other particle. If q_1 and q_2 are of opposite sign as shown in Figure 22.7b, the product $q_1 q_2$ is negative and the electric force on one particle is directed toward the other particle. These signs describe the *relative* direction of the force but not the *absolute* direction. A negative product indicates an attractive force, and a positive product indicates a repulsive force. The *absolute* direction of the force on a charge depends on the location of the other charge. For example, if an x axis lies along the two charges in Figure 22.7a, the product $q_1 q_2$ is positive, but $\vec{\mathbf{F}}_{12}$ points in the positive x direction and $\vec{\mathbf{F}}_{21}$ points in the negative x direction.

When more than two charges are present, the force between any pair of them is given by Equation 22.6. The resultant force on any one of them is given by a superposition principle and equals the vector sum of the forces exerted by the other individual charges. For example, if four charges are present, the resultant force exerted by particles 2, 3, and 4 on particle 1 is

$$\sum \vec{\mathbf{F}}_1 = \vec{\mathbf{F}}_{21} + \vec{\mathbf{F}}_{31} + \vec{\mathbf{F}}_{41}$$

QUICK QUIZ **22.3** Object A has a charge of $+2\ \mu C$, and object B has a charge of $+6\ \mu C$. Which statement is true about the electric forces on the objects?
(a) $\vec{\mathbf{F}}_{AB} = -3\vec{\mathbf{F}}_{BA}$ (b) $\vec{\mathbf{F}}_{AB} = -\vec{\mathbf{F}}_{BA}$ (c) $3\vec{\mathbf{F}}_{AB} = -\vec{\mathbf{F}}_{BA}$ (d) $\vec{\mathbf{F}}_{AB} = 3\vec{\mathbf{F}}_{BA}$
(e) $\vec{\mathbf{F}}_{AB} = \vec{\mathbf{F}}_{BA}$ (f) $3\vec{\mathbf{F}}_{AB} = \vec{\mathbf{F}}_{BA}$

Example 22.2 Find the Resultant Force

Consider three point charges located at the corners of a right triangle as shown in Figure 22.8, where $q_1 = q_3 = 5.00\ \mu C$, $q_2 = -2.00\ \mu C$, and $a = 0.100$ m. Find the resultant force exerted on q_3.

SOLUTION

Conceptualize Think about the net force on q_3. Because charge q_3 is near two other charges, it will experience two electric forces. These forces are exerted in different directions as shown in Figure 22.8. Based on the forces shown in the figure, estimate the direction of the net force vector.

Categorize Because two forces are exerted on charge q_3, we categorize this example as a vector addition problem.

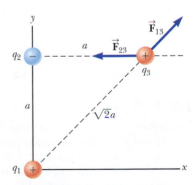

Figure 22.8 (Example 22.2) The force exerted by q_1 on q_3 is $\vec{\mathbf{F}}_{13}$. The force exerted by q_2 on q_3 is $\vec{\mathbf{F}}_{23}$. The resultant force $\vec{\mathbf{F}}_3$ exerted on q_3 is the vector sum $\vec{\mathbf{F}}_{13} + \vec{\mathbf{F}}_{23}$.

Analyze The individual forces exerted by q_1 and q_2 on q_3 have a direction determined by the pairs of charges; the forces are either attractive or repulsive. The vector forces on q_3 are shown in Figure 22.8. The force $\vec{\mathbf{F}}_{23}$ exerted by q_2 on q_3 is attractive because q_2 and q_3 have opposite signs. In the coordinate system shown in Figure 22.8, the attractive force $\vec{\mathbf{F}}_{23}$ is to the left (in the negative x direction).

continued

22.2 continued

The force $\vec{\mathbf{F}}_{13}$ exerted by q_1 on q_3 is repulsive because both charges are positive. The repulsive force $\vec{\mathbf{F}}_{13}$ makes an angle of 45.0° with the x axis. The magnitudes of the forces $\vec{\mathbf{F}}_{13}$ and $\vec{\mathbf{F}}_{23}$ are determined using the absolute magnitudes of the charges in Equation 22.1.

Use Equation 22.1 to find the magnitude of $\vec{\mathbf{F}}_{23}$:

$$F_{23} = k_e \frac{|q_2||q_3|}{a^2}$$

$$= (8.988 \times 10^9 \text{ N} \cdot \text{m}^2/\text{C}^2) \frac{(2.00 \times 10^{-6} \text{ C})(5.00 \times 10^{-6} \text{ C})}{(0.100 \text{ m})^2} = 8.99 \text{ N}$$

Find the magnitude of the force $\vec{\mathbf{F}}_{13}$:

$$F_{13} = k_e \frac{|q_1||q_3|}{(\sqrt{2}\,a)^2}$$

$$= (8.988 \times 10^9 \text{ N} \cdot \text{m}^2/\text{C}^2) \frac{(5.00 \times 10^{-6} \text{ C})(5.00 \times 10^{-6} \text{ C})}{2(0.100 \text{ m})^2} = 11.2 \text{ N}$$

Find the x and y components of the force $\vec{\mathbf{F}}_{13}$:

$$F_{13x} = (11.2 \text{ N}) \cos 45.0° = 7.94 \text{ N}$$
$$F_{13y} = (11.2 \text{ N}) \sin 45.0° = 7.94 \text{ N}$$

Find the components of the resultant force acting on q_3:

$$F_{3x} = F_{13x} + F_{23x} = 7.94 \text{ N} + (-8.99 \text{ N}) = -1.04 \text{ N}$$
$$F_{3y} = F_{13y} + F_{23y} = 7.94 \text{ N} + 0 = 7.94 \text{ N}$$

Express the resultant force acting on q_3 in unit-vector form:

$$\vec{\mathbf{F}}_3 = (-1.04\hat{\mathbf{i}} + 7.94\hat{\mathbf{j}}) \text{ N}$$

Finalize   The net force on q_3 is upward and toward the left in Figure 22.8. If q_3 moves in response to the net force, the distances between q_3 and the other charges change, so the net force changes. Therefore, if q_3 is free to move, it can be modeled as a particle under a net force as long as it is recognized that the force exerted on q_3 is *not* constant. As a reminder, we display most numerical values to three significant figures, which leads to operations such as 7.94 N + (−8.99 N) = −1.04 N above. If you carry all intermediate results to more significant figures, you will see that this operation is correct.

WHAT IF?   What if the signs of all three charges were changed to the opposite signs? How would that affect the result for $\vec{\mathbf{F}}_3$?

Answer   The charge q_3 would still be attracted toward q_2 and repelled from q_1 with forces of the same magnitude. Therefore, the final result for $\vec{\mathbf{F}}_3$ would be the same.

Example 22.3   **Where Is the Net Force Zero?**

Three point charges lie along the x axis as shown in Figure 22.9. The positive charge $q_1 = 15.0 \ \mu\text{C}$ is at $x = 2.00$ m, the positive charge $q_2 = 6.00 \ \mu\text{C}$ is at the origin, and the net force acting on q_3 is zero. What is the x coordinate of q_3?

SOLUTION

Conceptualize   Because q_3 is near two other charges, it experiences two electric forces. Unlike the preceding example, however, the forces lie along the same line in this problem as indicated in Figure 22.9. Because q_3 is negative and q_1 and q_2 are positive, the forces $\vec{\mathbf{F}}_{13}$ and $\vec{\mathbf{F}}_{23}$ are both attractive. Because q_2 is the smaller charge, the position of q_3 at which the force is zero should be closer to q_2 than to q_1.

Categorize   Because the net force on q_3 is zero, we model the point charge as a *particle in equilibrium*.

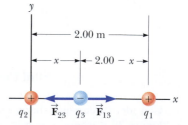

Figure 22.9 (Example 22.3) Three point charges are placed along the x axis. If the resultant force acting on q_3 is zero, the force $\vec{\mathbf{F}}_{13}$ exerted by q_1 on q_3 must be equal in magnitude and opposite in direction to the force $\vec{\mathbf{F}}_{23}$ exerted by q_2 on q_3.

Analyze   Write an expression for the net force on charge q_3 when it is in equilibrium:

$$\sum \vec{\mathbf{F}}_3 = \vec{\mathbf{F}}_{23} + \vec{\mathbf{F}}_{13} = -k_e \frac{|q_2||q_3|}{x^2} \hat{\mathbf{i}} + k_e \frac{|q_1||q_3|}{(2.00 - x)^2} \hat{\mathbf{i}} = 0$$

Move the second term to the right side of the equation and set the coefficients of the unit vector $\hat{\mathbf{i}}$ equal:

$$k_e \frac{|q_2||q_3|}{x^2} = k_e \frac{|q_1||q_3|}{(2.00 - x)^2}$$

22.3 continued

Eliminate k_e and $|q_3|$ and rearrange the equation:

$$(2.00 - x)^2|q_2| = x^2|q_1|$$

Take the square root of both sides of the equation:

$$(2.00 - x)\sqrt{|q_2|} = \pm x\sqrt{|q_1|}$$

Solve for x:

$$(1) \quad x = \frac{2.00\sqrt{|q_2|}}{\sqrt{|q_2|} \pm \sqrt{|q_1|}}$$

Substitute numerical values, choosing the plus sign:

$$x = \frac{2.00\sqrt{6.00 \times 10^{-6}\,\text{C}}}{\sqrt{6.00 \times 10^{-6}\,\text{C}} + \sqrt{15.0 \times 10^{-6}\,\text{C}}} = 0.775\,\text{m}$$

Finalize Notice that the movable charge is indeed closer to q_2 as we predicted in the Conceptualize step. Notice that the result in Equation (1) is independent of both the magnitude and the sign of charge q_3. If q_3 increases, both forces in Figure 22.9 increase in magnitude but still cancel. If q_3 changes sign, both forces reverse direction but still cancel. The second solution to the equation (if we choose the negative sign) is $x = -3.44$ m. That is another location where the *magnitudes* of the forces on q_3 are equal, but both forces are in the same direction, so they do not cancel.

W H A T   I F ? Suppose q_3 is constrained to move only along the x axis. From its initial position at $x = 0.775$ m, it is pulled a small distance along the x axis. When released, does it return to equilibrium, or is it pulled farther from equilibrium? That is, is the equilibrium stable or unstable?

Answer If q_3 is moved to the right, $\vec{\mathbf{F}}_{13}$ becomes larger and $\vec{\mathbf{F}}_{23}$ becomes smaller. The result is a net force to the right, in the same direction as the displacement. Therefore, the charge q_3 would continue to move to the right and the equilibrium is *unstable*. (See Section 7.9 for a review of stable and unstable equilibria.)

   If q_3 is constrained to stay at a *fixed x* coordinate but allowed to move up and down in Figure 22.9, the equilibrium is stable. In this case, if the charge is pulled upward (or downward) and released, it moves back toward the equilibrium position and oscillates about this point. Is the oscillation simple harmonic?

Example **22.4**   **Find the Charge on the Spheres**

Two identical small charged spheres, each having a mass of 3.00×10^{-2} kg, hang in equilibrium as shown in Figure 22.10a. The length L of each string is 0.150 m, and the angle θ is 5.00°. Find the magnitude of the charge on each sphere.

S O L U T I O N

Conceptualize Figure 22.10a helps us conceptualize this example. The two spheres exert repulsive forces on each other. If they are held close to each other and released, they move outward from the center and settle into the configuration in Figure 22.10a after the oscillations have vanished due to air resistance.

Categorize The key phrase "in equilibrium" helps us model each sphere as a *particle in equilibrium*. This example is similar to the particle in equilibrium problems in Chapter 5 with the added feature that one of the forces on a sphere is an electric force.

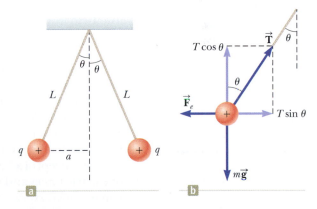

Figure 22.10 (Example 22.4) (a) Two identical spheres, each carrying the same charge q, suspended in equilibrium. (b) Diagram of the forces acting on the sphere on the left part of (a).

Analyze The force diagram for the left-hand sphere is shown in Figure 22.10b. The sphere is in equilibrium under the application of the force $\vec{\mathbf{T}}$ from the string, the electric force $\vec{\mathbf{F}}_e$ from the other sphere, and the gravitational force $m\vec{\mathbf{g}}$.

From the particle in equilibrium model, set the net force on the left-hand sphere equal to zero for each component:

$$(1) \quad \sum F_x = T\sin\theta - F_e = 0 \quad \rightarrow \quad T\sin\theta = F_e$$

$$(2) \quad \sum F_y = T\cos\theta - mg = 0 \quad \rightarrow \quad T\cos\theta = mg$$

Divide Equation (1) by Equation (2) to find F_e:

$$(3) \quad \tan\theta = \frac{F_e}{mg} \quad \rightarrow \quad F_e = mg\tan\theta$$

continued

22.4 continued

Use the geometry of the right triangle in Figure 22.10a to find a relationship between a, L, and θ:

(4) $\sin\theta = \dfrac{a}{L} \rightarrow a = L\sin\theta$

Solve Coulomb's law (Eq. 22.1) for the charge $|q|$ on each sphere and substitute from Equations (3) and (4):

$$|q| = \sqrt{\frac{F_e r^2}{k_e}} = \sqrt{\frac{F_e(2a)^2}{k_e}} = \sqrt{\frac{mg\tan\theta(2L\sin\theta)^2}{k_e}}$$

Substitute numerical values:

$$|q| = \sqrt{\frac{(3.00\times10^{-2}\text{ kg})(9.80\text{ m/s}^2)\tan(5.00°)[2(0.150\text{ m})\sin(5.00°)]^2}{8.988\times10^9\text{ N}\cdot\text{m}^2/\text{C}^2}}$$

$$= 4.42\times10^{-8}\text{ C}$$

Finalize   If the sign of the charges were not given in Figure 22.10, we could not determine them. In fact, the sign of the charge is not important. The situation is the same whether both spheres are positively charged or negatively charged.

WHAT IF?   Suppose your roommate proposes solving this problem without the assumption that the charges are of equal magnitude. She claims the symmetry of the problem is destroyed if the charges are not equal, so the strings would make two different angles with the vertical and the problem would be much more complicated. How would you respond?

Answer   The symmetry is not destroyed and the angles are not different. Newton's third law requires the magnitudes of the electric forces on the two spheres to be the same, regardless of the equality or nonequality of the charges. The solution to the example remains the same with one change: the value of $|q|$ in the solution is replaced by $\sqrt{|q_1 q_2|}$ in the new situation, where q_1 and q_2 are the values of the charges on the two spheres. The symmetry of the problem would be destroyed if the *masses* of the spheres were not the same. In this case, the strings would make different angles with the vertical and the problem would be more complicated.

22.4 Analysis Model: Particle in a Field (Electric)

In Section 5.1, we discussed the differences between contact forces and field forces. Two field forces—the gravitational force in Chapter 13 and the electric force here—have been introduced into our discussions so far. As pointed out earlier, field forces can act through space, producing an effect even when no physical contact occurs between interacting objects. Such an interaction can be modeled as a two-step process: a source particle establishes a field, and then a second particle interacts with the field and experiences a force. The gravitational field $\vec{g}$ at a point in space due to a source particle was defined in Section 13.3 to be equal to the gravitational force $\vec{F}_g$ acting on a test particle of mass m_0 divided by that mass: $\vec{g} \equiv \vec{F}_g/m_0$. Then the force exerted by the field on any particle of mass m is $\vec{F} = m\vec{g}$ (Eq. 5.5).

The concept of a field was developed by Michael Faraday (1791–1867) in the context of electric forces and is of such practical value that we shall devote much attention to it in the next several chapters. Figure 22.11 shows the two-step process for the electric force mentioned in the previous paragraph. An **electric field** is said to exist in the region of space around a charged object, the **source charge.** Figure 22.11a shows the source charge and the resulting electric field at one point P in the space external to the source charge. The presence of the electric field can be detected by placing a **test charge** in the field and noting the electric force on it, as is done in Figure 22.11b. We define the electric field due to the source charge at the location of the test charge to be the electric force on the test charge *per unit charge*, or, to be more specific, the **electric field vector** $\vec{E}$ at a point in space is defined as the electric force $\vec{F}_e$ acting on a positive test charge q_0 placed at that point divided by the test charge:[3]

Figure 22.11   An electric force between two particles is a two-step process: (a) A source charge q creates an electric field at a point P in space. (b) When another charge q_0 is placed at P, it feels the effect of that electric field as an electric force.

Definition of electric field ▶

$$\vec{E} \equiv \frac{\vec{F}_e}{q_0} \qquad (22.7)$$

[3]When using Equation 22.7, we must assume the test charge q_0 is small enough that it does not disturb the charge distribution responsible for the electric field. If the test charge is too large, the charge on the source might be redistributed and the electric field it sets up is different from the field it sets up in the absence of the test charge.

The vector $\vec{\mathbf{E}}$ has the SI units of newtons per coulomb (N/C). The direction of $\vec{\mathbf{E}}$ as shown in Figure 22.11a is the direction of the force a positive test charge experiences when placed in the field as shown in Figure 22.11b. Note that $\vec{\mathbf{E}}$ is the field produced by the source charge alone; the presence of the test charge is not necessary for the field to exist. The test charge serves only as a *detector* of the electric field: an electric field exists at a point if a test charge at that point experiences an electric force.

Now, imagine that we have established an electric field with a source charge and have evaluated its value at each point in space by using Equation 22.7. If an arbitrary charge q is placed in this electric field $\vec{\mathbf{E}}$, it experiences an electric force given by

$$\vec{\mathbf{F}}_e = q\vec{\mathbf{E}} \tag{22.8}$$

◀ Electric force on a charge in an electric field

This equation is the mathematical representation of the electric version of the **particle in a field** analysis model. Notice the similarity between Equation 22.8 and the corresponding equation from the gravitational version of the particle in a field model, $\vec{\mathbf{F}}_g = m\vec{\mathbf{g}}$ (Section 5.5). If q is positive, the force is in the same direction as the field. If q is negative, the force and the field are in opposite directions. Once the magnitude and direction of the electric field are known at some point, the electric force exerted on *any* charged particle placed at that point can be calculated from Equation 22.8.

To determine the vector form of an electric field, a test charge q_0 is placed at point P, a distance r from a source charge q, as in Figure 22.12a. We imagine using the test charge to determine the direction of the electric force and therefore that of the electric field. According to Coulomb's law, the force exerted by q on the test charge is

$$\vec{\mathbf{F}}_e = k_e \frac{q q_0}{r^2} \hat{\mathbf{r}}$$

where $\hat{\mathbf{r}}$ is a unit vector directed from q toward q_0. This force in Figure 22.12a is directed away from the source charge q. Because the electric field at P, the position of the test charge, is defined by Equation 22.7, $\vec{\mathbf{E}} = \vec{\mathbf{F}}_e / q_0$, the electric field at P created by q is

$$\vec{\mathbf{E}} = k_e \frac{q}{r^2} \hat{\mathbf{r}} \tag{22.9}$$

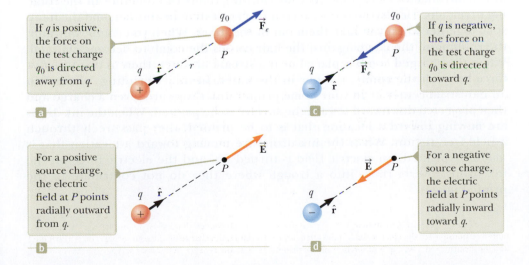

If q is positive, the force on the test charge q_0 is directed away from q.

If q is negative, the force on the test charge q_0 is directed toward q.

For a positive source charge, the electric field at P points radially outward from q.

For a negative source charge, the electric field at P points radially inward toward q.

Figure 22.12 (a), (c) When a test charge q_0 is placed near a source charge q, the test charge experiences a force. (b), (d) At a point P near a source charge q, there exists an electric field.

In this section, we have discussed the similarities between the electric field and the gravitational field introduced in Section 13.3. It is important to notice a subtle difference between the notations used for these two fields. The gravitational field, expressed in Equation 13.7 in terms of a source mass, is generally set up by an object whose mass is huge compared to that of an object placed in the field. Therefore, in Equation 13.8, we use the symbol M_E for the source, whereas in Equation 5.5, we use a separate symbol m for the mass of the object placed in the field. In an electric field, however, the charge of the source of the field is often similar in magnitude to the charge placed in the field. Therefore, we tend to use the same symbol q for both. In Figure 22.11 and Equation 22.9, q represents the source charge that sets up the electric field. In Equation 22.8, however, q represents the charge placed in the electric field. Whenever there is a possibility of confusion, we use subscripts to differentiate the charges, such as q_1 and q_2. If the source charge q is positive, Figure 22.12b shows the situation with the test charge removed: the source charge sets up an electric field at P, directed away from q. If q is negative as in Figure 22.12c, the force on the test charge is toward the source charge, so the electric field at P is directed toward the source charge as in Figure 22.12d.

To calculate the electric field at a point P due to a small number of point charges, we first calculate the electric field vectors at P individually using Equation 22.9 and then add them vectorially. In other words, at any point P, the total electric field due to a group of source charges equals the vector sum of the electric fields of all the charges. This superposition principle applied to fields follows directly from the vector addition of electric forces. Therefore, the electric field at point P due to a group of source charges can be expressed as the vector sum

◄ Electric field due to a finite number of point charges

$$\vec{\mathbf{E}} = k_e \sum_i \frac{q_i}{r_i^2} \hat{\mathbf{r}}_i \qquad (22.10)$$

where r_i is the distance from the ith source charge q_i to the point P and $\hat{\mathbf{r}}_i$ is a unit vector directed from q_i toward P.

In Example 22.6, we explore the electric field due to two charges using the superposition principle. Part (B) of the example focuses on an **electric dipole,** which is defined as a positive charge q and a negative charge $-q$ separated by a distance $2a$. The electric dipole is a good model of many molecules, such as hydrochloric acid (HCl). Neutral atoms and molecules behave as dipoles when placed in an external electric field. Furthermore, many molecules, such as HCl, are permanent dipoles. The effect of such dipoles on the behavior of materials subjected to electric fields is discussed in Chapter 25.

It is the electric force that is responsible for all three phenomena mentioned in the introductory storyline. As your clothing items rub together in the rotating dryer, electric charge is transferred from one item to another, and the items stick together when you take them out of the dryer. When you comb your hair, the rubbing of the comb against the hair causes the comb to become charged. When the charged comb is placed near a stream of water, there is an attractive force between the comb and ions in the water. In inkjet printing, whether in an industrial center or in your home printer, ink drops are given a charge and then projected downward toward the surface to be printed. When the ink drops are moving toward a location that is to be printed, they pass freely through a field-free region. When the ink drops are moving toward a location that is *not* to be printed, an electric field is turned on, and the electric force on the ink drops diverts them into a trough where they do not contribute to the printed image.

QUICK QUIZ 22.4 A test charge of $+3\ \mu C$ is at a point P where an external electric field is directed to the right and has a magnitude of 4×10^6 N/C. If the test charge is replaced with another test charge of $-3\ \mu C$, what happens to the external electric field at P? **(a)** It is unaffected. **(b)** It reverses direction. **(c)** It changes in a way that cannot be determined.

ANALYSIS MODEL | Particle in a Field (Electric)

Imagine an object with charge that we call a *source charge*. The source charge establishes an **electric field** $\vec{E}$ throughout space. Now imagine a particle with charge q is placed in that field. The particle interacts with the electric field so that the particle experiences an electric force given by

$$\vec{F}_e = q\vec{E} \qquad (22.8)$$

Examples:

- an electron moves between the deflection plates of a cathode ray oscilloscope and is deflected from its original path
- charged ions experience an electric force from the electric field in a velocity selector before entering a mass spectrometer (Chapter 28)
- an electron moves around the nucleus in the electric field established by the proton in a hydrogen atom as modeled by the Bohr theory (Chapter 41)
- a hole in a semiconducting material moves in response to the electric field established by applying a voltage to the material (Chapter 42)

Example 22.5 | A Suspended Water Droplet

A water droplet of mass 3.00×10^{-12} kg is located in the air near the ground during a stormy day. An atmospheric electric field of magnitude 6.00×10^3 N/C points vertically downward in the vicinity of the water droplet. The droplet remains suspended at rest in the air. What is the electric charge on the droplet?

SOLUTION

Conceptualize Imagine the water droplet hovering at rest in the air. This situation is not what is normally observed, so something must be holding the water droplet up.

Categorize The droplet can be modeled as a particle and is described by two analysis models associated with fields: the *particle in a field (gravitational)* and the *particle in a field (electric)*. Furthermore, because the droplet is subject to forces but remains at rest, it is also described by the *particle in equilibrium* model.

Analyze

Write Newton's second law from the particle in equilibrium model in the vertical direction:

$$(1) \quad \sum F_y = 0 \ \rightarrow \ F_e - F_g = 0$$

Using the two particle in a field models mentioned in the Categorize step, substitute for the forces in Equation (1), recognizing that the vertical component of the electric field is negative:

$$q(-E) - mg = 0$$

Solve for the charge on the water droplet:

$$q = -\frac{mg}{E}$$

Substitute numerical values:

$$q = -\frac{(3.00 \times 10^{-12}\ \text{kg})(9.80\ \text{m/s}^2)}{6.00 \times 10^3\ \text{N/C}} = -4.90 \times 10^{-15}\ \text{C}$$

Finalize Noting the smallest unit of free charge in Equation 22.5, the charge on the water droplet is a large number of these units. Notice that the electric *force* is upward to balance the downward gravitational force. The problem statement claims that the electric *field* is in the downward direction. Therefore, the charge found above is negative so that the electric force is in the direction opposite to the electric field.

Example 22.6 **Electric Field Due to Two Charges**

Charges q_1 and q_2 are located on the x axis, at distances a and b, respectively, from the origin as shown in Figure 22.13.

(A) Find the components of the net electric field at the point P, which is at position $(0, y)$.

SOLUTION

Conceptualize Compare this example with Example 22.2. There, we add vector forces to find the net force on a charged particle. Here, we add electric field vectors to find the net electric field at a point in space. If a charged particle were placed at P, we could use the particle in a field model to find the electric force on the particle.

Figure 22.13 (Example 22.6) The total electric field $\vec{E}$ at P equals the vector sum $\vec{E}_1 + \vec{E}_2$, where $\vec{E}_1$ is the field due to the positive charge q_1 and $\vec{E}_2$ is the field due to the negative charge q_2.

Categorize We have two source charges and wish to find the resultant electric field, so we categorize this example as one in which we can use the superposition principle represented by Equation 22.10.

Analyze Find the magnitude of the electric field at P due to charge q_1:

$$E_1 = k_e \frac{|q_1|}{r_1^2} = k_e \frac{|q_1|}{a^2 + y^2}$$

Find the magnitude of the electric field at P due to charge q_2:

$$E_2 = k_e \frac{|q_2|}{r_2^2} = k_e \frac{|q_2|}{b^2 + y^2}$$

Write the electric field vectors for each charge in unit-vector form:

$$\vec{E}_1 = k_e \frac{|q_1|}{a^2 + y^2} \cos\phi\,\hat{\mathbf{i}} + k_e \frac{|q_1|}{a^2 + y^2} \sin\phi\,\hat{\mathbf{j}}$$

$$\vec{E}_2 = k_e \frac{|q_2|}{b^2 + y^2} \cos\theta\,\hat{\mathbf{i}} - k_e \frac{|q_2|}{b^2 + y^2} \sin\theta\,\hat{\mathbf{j}}$$

Write the components of the net electric field vector:

$$(1) \quad E_x = E_{1x} + E_{2x} = k_e \frac{|q_1|}{a^2 + y^2} \cos\phi + k_e \frac{|q_2|}{b^2 + y^2} \cos\theta$$

$$(2) \quad E_y = E_{1y} + E_{2y} = k_e \frac{|q_1|}{a^2 + y^2} \sin\phi - k_e \frac{|q_2|}{b^2 + y^2} \sin\theta$$

(B) Evaluate the electric field at point P in the special case that $|q_1| = |q_2|$ and $a = b$.

SOLUTION

Conceptualize Figure 22.14 shows the situation in this special case. Notice the symmetry in the situation and that the charge distribution is now an electric dipole.

Categorize Because Figure 22.14 is a special case of the general case shown in Figure 22.13, we can categorize this example as one in which we can take the result of part (A) and substitute the appropriate values of the variables.

Analyze Based on the symmetry in Figure 22.14, evaluate Equations (1) and (2) from part (A) with $a = b$, $|q_1| = |q_2| = q$, and $\phi = \theta$:

$$(3) \quad E_x = k_e \frac{q}{a^2 + y^2} \cos\theta + k_e \frac{q}{a^2 + y^2} \cos\theta = 2k_e \frac{q}{a^2 + y^2} \cos\theta$$

$$E_y = k_e \frac{q}{a^2 + y^2} \sin\theta - k_e \frac{q}{a^2 + y^2} \sin\theta = 0$$

From the geometry in Figure 22.14, evaluate $\cos\theta$:

$$(4) \quad \cos\theta = \frac{a}{r} = \frac{a}{(a^2 + y^2)^{1/2}}$$

Substitute Equation (4) into Equation (3):

$$E_x = 2k_e \frac{q}{a^2 + y^2} \left[\frac{a}{(a^2 + y^2)^{1/2}}\right] = k_e \frac{2aq}{(a^2 + y^2)^{3/2}}$$

22.6 continued

(c) Find the electric field due to the electric dipole when point P is a distance $y \gg a$ from the origin.

SOLUTION

In the solution to part (B), because $y \gg a$, neglect a^2 compared with y^2 and write the expression for E in this case:

$$(5) \quad E \approx k_e \frac{2aq}{y^3}$$

Finalize From Equation (5), we see that at points far from a dipole but along the perpendicular bisector of the line joining the two charges, the magnitude of the electric field created by the dipole varies as $1/r^3$, whereas the more slowly varying field of a point charge varies as $1/r^2$ (see Eq. 22.9). That is because at distant points, the fields of the two charges of equal magnitude and opposite sign almost cancel each other. The $1/r^3$ variation in E for the dipole also is obtained for a distant point along the x axis and for any general distant point. In both parts (A) and (B), if a new charge q_1 is placed at point P, Equation 22.8 can be used to find the electric force on the charge: $\vec{F} = q_1\vec{E} = q_1 E_x \hat{i} + q_1 E_y \hat{j}$.

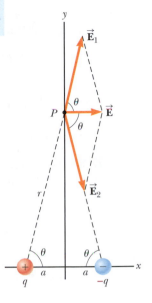

Figure 22.14 (Example 22.6) When the charges in Figure 22.13 are of equal magnitude and equidistant from the origin, the situation becomes symmetric as shown here.

22.5 Electric Field Lines

We have defined the electric field in the mathematical representation with Equation 22.7. Let's now explore a means of visualizing the electric field in a pictorial representation. A convenient way of visualizing electric field patterns is to draw lines, called **electric field lines** and first introduced by Faraday, that are related to the electric field in a region of space in the following manner:

- The electric field vector $\vec{E}$ is tangent to the electric field line at each point. The line has a direction, indicated by an arrowhead, that is the same as that of the electric field vector. The direction of the line is that of the force on a positive charge placed in the field according to the particle in a field model.
- The number of lines per unit area through a surface perpendicular to the lines is proportional to the magnitude of the electric field in that region. Therefore, the field lines are close together where the electric field is strong and far apart where the field is weak.

These properties are illustrated in Figure 22.15. The density of field lines through surface A is greater than the density of lines through surface B. Therefore, the magnitude of the electric field is larger on surface A than on surface B. Furthermore, because the lines at different locations point in different directions, the field is nonuniform.

Is this relationship between strength of the electric field and the density of field lines consistent with Equation 22.9, the expression we obtained for E using Coulomb's law? To answer this question, consider an imaginary spherical surface of radius r concentric with a point charge. From symmetry, we see that the magnitude of the electric field is the same everywhere on the surface of the sphere. The number of lines N that emerge from the charge is equal to the number that penetrate the spherical surface. Hence, the number of lines per unit area on the sphere is $N/4\pi r^2$ (where the surface area of the sphere is $4\pi r^2$). Because E is proportional to the number of lines per unit area, we see that E varies as $1/r^2$; this finding is consistent with Equation 22.9.

The magnitude of the field is greater on surface A than on surface B.

Figure 22.15 Electric field lines penetrating two surfaces.

PITFALL PREVENTION 22.2

Electric Field Lines Are Not Paths of Particles! Electric field lines represent the field at various locations. Except in very special cases, they *do not* represent the path of a charged particle moving in an electric field.

For a positive point charge, the field lines are directed radially outward.

For a negative point charge, the field lines are directed radially inward.

q

$-q$

a

b

Figure 22.16 The electric field lines for a point charge. Notice that the figures show only those field lines that lie in the plane of the page.

Representative electric field lines for the field due to a single positive point charge are shown in Figure 22.16a. This two-dimensional drawing shows only the field lines that lie in the plane containing the point charge. The lines are actually directed radially outward from the charge in all directions; therefore, instead of the flat "wheel" of lines shown, you should picture an entire spherical distribution of lines. Because a positive charge placed in this field would be repelled by the positive source charge, the lines are directed radially away from the source charge. The electric field lines representing the field due to a single negative point charge are directed toward the charge (Fig. 22.16b). In either case, the lines are along the radial direction and extend all the way to infinity. Notice that the lines become closer together as they approach the charge, indicating that the strength of the field increases as we move toward the source charge.

The rules for drawing electric field lines are as follows:

- The lines must begin on a positive charge and terminate on a negative charge. In the case of an excess of one type of charge, some lines will begin or end infinitely far away.
- The number of lines drawn leaving a positive charge or approaching a negative charge is proportional to the magnitude of the charge.
- No two field lines can cross.

We choose the number of field lines starting from any object with a positive charge q_+ to be Cq_+ and the number of lines ending on any object with a negative charge q_- to be $C|q_-|$, where C is an arbitrary proportionality constant. Once C is chosen, the number of lines is fixed. For example, in a two-charge system, if object 1 has charge Q_1 and object 2 has charge Q_2, the ratio of number of lines in contact with the charges is $N_2/N_1 = |Q_2/Q_1|$. The electric field lines for two point charges of equal magnitude but opposite signs (an electric dipole) are shown in Figure 22.17. Because the charges are of equal magnitude, the number of lines that begin at the positive charge must equal the number that terminate at the negative charge. At points very near the charges, the lines are nearly radial, as for a single isolated charge. The high density of lines between the charges indicates a region of strong electric field.

Figure 22.18 shows the electric field lines in the vicinity of two equal positive point charges. Again, the lines are nearly radial at points close to either charge, and the same number of lines emerges from each charge because the charges are equal in magnitude. Because there are no negative charges available, the electric field lines end infinitely far away. At great distances from the charges, the field is approximately equal to that of a single point charge of magnitude $2q$.

The number of field lines leaving the positive charge equals the number terminating at the negative charge.

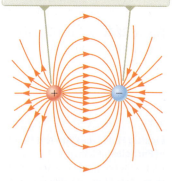

Figure 22.17 The electric field lines for two point charges of equal magnitude and opposite sign (an electric dipole).

Two field lines leave $+2q$ for every one that terminates on $-q$.

Figure 22.18 The electric field lines for two positive point charges. (The locations A, B, and C are discussed in Quick Quiz 22.5.)

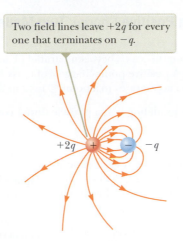

Figure 22.19 The electric field lines for a point charge $+2q$ and a second point charge $-q$.

Finally, in Figure 22.19, we sketch the electric field lines associated with a positive charge $+2q$ and a negative charge $-q$. In this case, the number of lines leaving $+2q$ is twice the number terminating at $-q$. Hence, only half the lines that leave the positive charge reach the negative charge. The remaining half terminate on a negative charge we assume to be at infinity. At distances much greater than the charge separation, the electric field lines are equivalent to those of a single charge $+q$.

QUICK QUIZ **22.5** Rank the magnitudes of the electric field at points A, B, and C shown in Figure 22.18 (greatest magnitude first).

22.6 Motion of a Charged Particle in a Uniform Electric Field

When a particle of charge q and mass m is placed in an electric field $\vec{E}$, the electric force exerted on the charge is $q\vec{E}$ according to Equation 22.8 in the particle in a field model. If that is the only force exerted on the particle, it must be the net force, and it causes the particle to accelerate according to the particle under a net force model. Therefore,

$$\sum \vec{F} = q\vec{E} = m\vec{a}$$

and the acceleration of the particle is

$$\vec{a} = \frac{q\vec{E}}{m} \qquad (22.11)$$

If $\vec{E}$ is uniform (that is, constant in magnitude and direction), and the particle is free to move, the electric force on the particle is constant and we can apply the particle under constant acceleration model to the motion of the particle. Therefore, the particle in this situation is described by *three* analysis models: particle in a field, particle under a net force, and particle under constant acceleration! If the particle has a positive charge, its acceleration is in the direction of the electric field. If the particle has a negative charge, its acceleration is in the direction opposite the electric field.

PITFALL PREVENTION 22.4
Just Another Force Electric forces and fields may seem abstract to you. Once $\vec{F}_e$ is evaluated, however, it causes a particle to move according to our well-established models of forces and motion from Chapters 2 through 6. Keeping this link with the past in mind should help you solve problems in this chapter.

| Example **22.7** | **An Accelerating Positive Charge: Two Models** |

A uniform electric field $\vec{\mathbf{E}}$ is directed along the x axis between parallel plates of charge separated by a distance d as shown in Figure 22.20. A positive point charge q of mass m is released from rest at a point Ⓐ next to the positive plate and accelerates to a point Ⓑ next to the negative plate.

(A) Find the speed of the particle at Ⓑ by modeling it as a particle under constant acceleration.

SOLUTION

Conceptualize When the positive charge is placed at Ⓐ, it experiences an electric force toward the right in Figure 22.20 due to the electric field directed toward the right. As a result, it will accelerate to the right and arrive at Ⓑ with some speed.

Figure 22.20 (Example 22.7) A positive point charge q in a uniform electric field $\vec{\mathbf{E}}$ undergoes constant acceleration in the direction of the field.

Categorize Because the electric field is uniform, a constant electric force acts on the charge. Therefore, as suggested in the discussion preceding the example and in the problem statement, the point charge can be modeled as a charged *particle under constant acceleration*.

Analyze Use Equation 2.17 to express the velocity of the particle as a function of position:

$$v_f^2 = v_i^2 + 2a(x_f - x_i) = 0 + 2a(d - 0) = 2ad$$

Solve for v_f and substitute for the magnitude of the acceleration from Equation 22.11:

$$v_f = \sqrt{2ad} = \sqrt{2\left(\frac{qE}{m}\right)d} = \sqrt{\frac{2qEd}{m}}$$

(B) Find the speed of the particle at Ⓑ by modeling it as a nonisolated system in terms of energy.

SOLUTION

Categorize The problem statement tells us that the charge is a *nonisolated system* for *energy*. The electric force, like any force, can do work on a system. Energy is transferred to the system of the charge by work done by the electric force exerted on the charge. The initial configuration of the system is when the particle is at rest at Ⓐ, and the final configuration is when it is moving with some speed at Ⓑ.

Analyze Write the appropriate reduction of the conservation of energy equation, Equation 8.2, for the system of the charged particle:

$$W = \Delta K$$

Replace the work and kinetic energies with values appropriate for this situation:

$$F_e \Delta x = K_Ⓑ - K_Ⓐ = \tfrac{1}{2}mv_f^2 - 0 \quad \rightarrow \quad v_f = \sqrt{\frac{2F_e \Delta x}{m}}$$

Substitute for the magnitudes of the electric force F_e from the particle in a field model and the displacement Δx:

$$v_f = \sqrt{\frac{2(qE)(d)}{m}} = \sqrt{\frac{2qEd}{m}}$$

Finalize The answer to part (B) is the same as that for part (A), as we expect. This problem can be solved with different approaches. We saw the same possibilities with mechanical problems.

| Example **22.8** | **An Accelerated Electron** |

An electron enters the region of a uniform electric field as shown in Figure 22.21, with $v_i = 3.00 \times 10^6$ m/s and $E = 200$ N/C. The horizontal length of the plates is $\ell = 0.100$ m.

(A) Find the acceleration of the electron while it is in the electric field.

22.8 continued

SOLUTION

Conceptualize This example differs from the preceding one because the velocity of the charged particle is initially perpendicular to the electric field lines. (In Example 22.7, the velocity of the charged particle is always parallel to the electric field lines.) As a result, the electron in this example follows a curved path as shown in Figure 22.21. The motion of the electron is the same as that of a massive particle projected horizontally in a gravitational field near the surface of the Earth.

Categorize The electron is a *particle in a field (electric)*. Because the electric field is uniform, a constant electric force is exerted on the electron. To find the acceleration of the electron, we can model it as a *particle under a net force*.

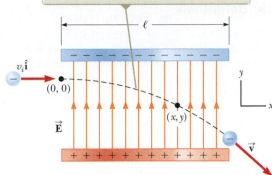

The electron undergoes a downward acceleration (opposite $\vec{\mathbf{E}}$), and its motion is parabolic while it is between the plates.

Figure 22.21 (Example 22.8) An electron is projected horizontally into a uniform electric field produced by two charged plates.

Analyze From the particle in a field model, we know that the direction of the electric force on the electron is downward in Figure 22.21, opposite the direction of the electric field lines. From the particle under a net force model, therefore, the acceleration of the electron is downward.

The particle under a net force model was used to develop Equation 22.11 in the case in which the electric force on a particle is the only force. Use this equation to evaluate the *y* component of the acceleration of the electron:

$$a_y = -\frac{eE}{m_e}$$

Substitute numerical values:

$$a_y = -\frac{(1.60 \times 10^{-19}\ \text{C})(200\ \text{N/C})}{9.11 \times 10^{-31}\ \text{kg}} = -3.51 \times 10^{13}\ \text{m/s}^2$$

(B) Assuming the electron enters the field at time $t = 0$, find the time at which it leaves the field.

SOLUTION

Categorize Because the electric force acts only in the vertical direction in Figure 22.21, the motion of the particle in the horizontal direction can be analyzed by modeling it as a *particle under constant velocity*.

Analyze Solve Equation 2.7 for the time at which the electron arrives at the right edges of the plates:

$$x_f = x_i + v_x t \ \rightarrow \ t = \frac{x_f - x_i}{v_x}$$

Substitute numerical values:

$$t = \frac{\ell - 0}{v_x} = \frac{0.100\ \text{m}}{3.00 \times 10^6\ \text{m/s}} = 3.33 \times 10^{-8}\ \text{s}$$

(C) Assuming the vertical position of the electron as it enters the field is $y_i = 0$, what is its vertical position when it leaves the field?

SOLUTION

Categorize Because the electric force is constant in Figure 22.21, the motion of the particle in the vertical direction can be analyzed by modeling it as a *particle under constant acceleration*.

Analyze Use Equation 2.16 to describe the position of the particle at any time t:

$$y_f = y_i + v_{yi} t + \tfrac{1}{2} a_y t^2$$

Substitute numerical values:

$$y_f = 0 + 0 + \tfrac{1}{2}(-3.51 \times 10^{13}\ \text{m/s}^2)(3.33 \times 10^{-8}\ \text{s})^2$$

$$= -0.019\,5\ \text{m} = -1.95\ \text{cm}$$

Finalize If the electron enters just below the negative plate in Figure 22.21 and the separation between the plates is less than the value just calculated, the electron will strike the positive plate.

Notice that we have used *four* analysis models to describe the electron in the various parts of this problem. We have neglected the gravitational force acting on the electron, which represents a good approximation when dealing with atomic particles. For an electric field of 200 N/C, the ratio of the magnitude of the electric force eE to the magnitude of the gravitational force mg is on the order of 10^{12} for an electron and on the order of 10^9 for a proton.

Summary

▶ Definitions

The **electric field** $\vec{\mathbf{E}}$ at some point in space is defined as the electric force $\vec{\mathbf{F}}_e$ that acts on a small positive test charge placed at that point divided by the magnitude q_0 of the test charge:

$$\vec{\mathbf{E}} \equiv \frac{\vec{\mathbf{F}}_e}{q_0} \qquad (22.7)$$

▶ Concepts and Principles

Electric charges have the following important properties:

- Charges of opposite sign attract one another, and charges of the same sign repel one another.
- The total charge in an isolated system is conserved.
- Charge is quantized.

Conductors are materials in which electrons move freely. **Insulators** are materials in which electrons do not move freely.

Coulomb's law states that the electric force exerted by a point charge q_1 on a second point charge q_2 is

$$\vec{\mathbf{F}}_{12} = k_e \frac{q_1 q_2}{r^2} \hat{\mathbf{r}}_{12} \qquad (22.6)$$

where r is the distance between the two charges and $\hat{\mathbf{r}}_{12}$ is a unit vector directed from q_1 toward q_2. The constant k_e, which is called the **Coulomb constant,** has the value $k_e = 8.988 \times 10^9 \text{ N} \cdot \text{m}^2/\text{C}^2$.

At a distance r from a point charge q, the electric field due to the charge is

$$\vec{\mathbf{E}} = k_e \frac{q}{r^2} \hat{\mathbf{r}} \qquad (22.9)$$

where $\hat{\mathbf{r}}$ is a unit vector directed from the charge toward the point in question. The electric field is directed radially outward from a positive charge and radially inward toward a negative charge.

The electric field due to a group of point charges can be obtained by using the superposition principle. That is, the total electric field at some point equals the vector sum of the electric fields of all the charges:

$$\vec{\mathbf{E}} = k_e \sum_i \frac{q_i}{r_i^2} \hat{\mathbf{r}}_i \qquad (22.10)$$

▶ Analysis Models for Problem Solving

Particle in a Field (Electric) A source particle with some electric charge establishes an **electric field** $\vec{\mathbf{E}}$ throughout space. When a particle with charge q is placed in that field, it experiences an electric force given by

$$\vec{\mathbf{F}}_e = q\vec{\mathbf{E}} \qquad (22.8)$$

Think–Pair–Share

See the Preface for an explanation of the icons used in this problems set. For additional assessment items for this section, go to **WEBASSIGN** From Cengage

1. You and your fellow students form an intern group at a nanotechnology company. The company is having trouble manufacturing nanoparticles of uniform mass and your group has been asked to devise a system that will determine the mass of a charged nanoparticle by having it pass through a uniform electric field between parallel plates in a manner identical to that of the electron in Figure 22.21. The entire system is located in an evacuated glass chamber so that air resistance is not a factor. The mass can be determined by the amount of deflection of the nanoparticle as it passes through the field after having been projected into the field just under the left edge of the top plate with an initial velocity in a direction parallel to the plates. Your group works well together and designs a pair of parallel plates of length $\ell = 1.00$ m, with the negative plate situated a distance $d = 8.00$ mm vertically above the positive plate, with a uniform

electric field of magnitude $E = 2.00 \times 10^4$ N/C between them directed perpendicular to the plates. You arrange to have nanoparticles with mass $m = 6.50 \times 10^{-13}$ g, carrying a charge of $-e$, to be projected into the field at a speed of $v = 30.0$ m/s. Your deadline is approaching as you finish the construction of the device and you don't have time to test it before you are called in to demonstrate the device to the research group leaders. When you demonstrate the device, why is your group embarrassed?

2. **ACTIVITY** Figure TP22.2 shows the *triboelectric series*. It is used with regard to the rubbing experiments described in Section 22.1. If a material from high on the list is rubbed against a material from a lower portion, each material will become electrically charged according to the signs at the top and bottom of the list. The farther apart two materials are on the list, the greater will be the amount of electric charge when the materials are rubbed together. As an example, consider the glass rod rubbed with silk in Figure 22.3. Based on the triboelectric series shown, do we expect the silk to become negatively charged and the glass to become positively charged? Although both are on the positive side of the series, the glass is much higher, so it takes on a positive charge and the silk becomes negatively charged.

For the objects rubbed together below, identify the sign of the charge on the rod, pipe, fork, or balloon after it is rubbed with the other material:

(a) A glass rod is rubbed with a wool cloth.
(b) A glass rod is rubbed with cat fur.
(c) A PVC pipe is rubbed with a paper towel.
(d) A Sterling silver fork is rubbed with a nylon cloth.
(e) A silicone rubber rod is rubbed with a cotton cloth.
(f) A hard rubber rod is rubbed with a paper towel.
(g) A copper pipe is rubbed with cat fur.
(h) An aluminum rod is rubbed with a polyester shirt.
(i) A lead pipe is rubbed with a paper towel.
(j) A rubber balloon is rubbed on your hair.

(k) From the ten choices in Parts (a)–(j), which do you think represents the greatest transfer of charge?

Figure TP22.2

Problems

See the Preface for an explanation of the icons used in this problems set. For additional assessment items for this section, go to **WEBASSIGN** From Cengage

SECTION 22.1 Properties of Electric Charges

1. Find to three significant digits the charge and the mass of the following particles. *Suggestion:* Begin by looking up the mass of a neutral atom on the periodic table of the elements in Appendix C. (a) an ionized hydrogen atom, represented as H^+ (b) a singly ionized sodium atom, Na^+ (c) a chloride ion Cl^- (d) a doubly ionized calcium atom, $Ca^{++} = Ca^{2+}$ (e) the center of an ammonia molecule, modeled as an N^{3-} ion (f) quadruply ionized nitrogen atoms, N^{4+}, found in plasma in a hot star (g) the nucleus of a nitrogen atom (h) the molecular ion H_2O^-

SECTION 22.3 Coulomb's Law

2. (a) Find the magnitude of the electric force between a Na^+ ion and a Cl^- ion separated by 0.50 nm. (b) Would the answer change if the sodium ion were replaced by Li^+ and the chloride ion by Br^-? Explain.

3. In a thundercloud, there may be electric charges of +40.0 C near the top of the cloud and −40.0 C near the bottom of the cloud. These charges are separated by 2.00 km. What is the electric force on the top charge?

4. Nobel laureate Richard Feynman (1918–1988) once said that if two persons stood at arm's length from each other and each person had 1% more electrons than protons, the force of repulsion between them would be enough to lift a "weight" equal to that of the entire Earth. Carry out an order-of-magnitude calculation to substantiate this assertion.

5. A 7.50-nC point charge is located 1.80 m from a 4.20-nC point charge. (a) Find the magnitude of the electric force that one particle exerts on the other. (b) Is the force attractive or repulsive?

6. This afternoon, you have a physics symposium class, and you are the presenter. You will be presenting a topic to physics majors and faculty. You have been so busy that you have not had time to prepare and you don't even have an idea for a topic. You are frantically reading your physics textbook looking for an idea. In your reading, you have learned that the Earth carries a charge on its surface of about 10^5 C, which results in electric fields in the atmosphere. This gets you very excited about a new theory. Suppose the Moon also carries a

charge on the order of 10^5 C, with the opposite sign! Maybe the orbit of the Moon around the Earth is due to electrical attraction between the Moon and the Earth! There's an idea for your symposium presentation! You quickly jot down a few notes and run off to your symposium. While you are speaking, you notice one of the professors doing some calculations on a scrap of paper. Uh-oh! He has just raised his hand with a question. Why are you embarrassed?

7. Two small beads having positive charges $q_1 = 3q$ and $q_2 = q$ are fixed at the opposite ends of a horizontal insulating rod of length $d = 1.50$ m. The bead with charge q_1 is at the origin. As shown in Figure P22.7, a third small, charged bead is free to slide on the rod. (a) At what position x is the third bead in equilibrium? (b) Can the equilibrium be stable?

Figure P22.7 Problems 7 and 8.

8. Two small beads having charges q_1 and q_2 of the same sign are fixed at the opposite ends of a horizontal insulating rod of length d. The bead with charge q_1 is at the origin. As shown in Figure P22.8, a third small, charged bead is free to slide on the rod. (a) At what position x is the third bead in equilibrium? (b) Can the equilibrium be stable?

9. Review. In the Bohr theory of the hydrogen atom, an electron moves in a circular orbit about a proton, where the radius of the orbit is 5.29×10^{-11} m. (a) Find the magnitude of the electric force exerted on each particle. (b) If this force causes the centripetal acceleration of the electron, what is the speed of the electron?

10. Three point charges lie along a straight line as shown in Figure P22.10, where $q_1 = 6.00$ μC, $q_2 = 1.50$ μC, and $q_3 = -2.00$ μC. The separation distances are $d_1 = 3.00$ cm and $d_2 = 2.00$ cm. Calculate the magnitude and direction of the net electric force on (a) q_1, (b) q_2, and (c) q_3.

Figure P22.10

11. A point charge $+2Q$ is at the origin and a point charge $-Q$ is located along the x axis at $x = d$ as in Figure P22.11. Find a symbolic expression for the net force on a third point charge $+Q$ located along the y axis at $y = d$.

Figure P22.11

12. Particle A of charge 3.00×10^{-4} C is at the origin, particle B of charge -6.00×10^{-4} C is at (4.00 m, 0), and particle C of charge 1.00×10^{-4} C is at (0, 3.00 m). We wish to find

the net electric force on C. (a) What is the x component of the electric force exerted by A on C? (b) What is the y component of the force exerted by A on C? (c) Find the magnitude of the force exerted by B on C. (d) Calculate the x component of the force exerted by B on C. (e) Calculate the y component of the force exerted by B on C. (f) Sum the two x components from parts (a) and (d) to obtain the resultant x component of the electric force acting on C. (g) Similarly, find the y component of the resultant force vector acting on C. (h) Find the magnitude and direction of the resultant electric force acting on C.

13. Review. Two identical particles, each having charge $+q$, are fixed in space and separated by a distance d. A third particle with charge $-Q$ is free to move and lies initially at rest on the perpendicular bisector of the two fixed charges a distance x from the midpoint between those charges (Fig. P22.13). (a) Show that if x is small compared with d, the motion of $-Q$ is simple harmonic along the perpendicular bisector. (b) Determine the period of that motion. (c) How fast will the charge $-Q$ be moving when it is at the midpoint between the two fixed charges if initially it is released at a distance $a \ll d$ from the midpoint?

Figure P22.13

14. *Why is the following situation impossible?* Two identical dust particles of mass 1.00 μg are floating in empty space, far from any external sources of large gravitational or electric fields, and at rest with respect to each other. Both particles carry electric charges that are identical in magnitude and sign. The gravitational and electric forces between the particles happen to have the same magnitude, so each particle experiences zero net force and the distance between the particles remains constant.

SECTION 22.4 Analysis Model: Particle in a Field (Electric)

15. What are the magnitude and direction of the electric field that will balance the weight of (a) an electron and (b) a proton? (You may use the data in Table 22.1.)

16. Consider n equal positively charged particles each of magnitude Q/n placed symmetrically around a circle of radius a. Calculate the magnitude of the electric field at a point a distance x from the center of the circle and on the line passing through the center and perpendicular to the plane of the circle.

17. Two equal positively charged particles are at opposite corners of a trapezoid as shown in Figure P22.17. Find symbolic expressions for the total electric field at (a) the point P and (b) the point P'.

Figure P22.17

18. Two charged particles are located on the x axis. The first is a charge $+Q$ at $x = -a$. The second is an unknown charge located at $x = +3a$. The net electric field these charges produce at the origin has a magnitude of $2k_eQ/a^2$. Explain how many values are possible for the unknown charge and find the possible values.

19. Three point charges are located on a circular arc as shown in Figure P22.19. (a) What is the total electric field at P, the center of the arc? (b) Find the electric force that would be exerted on a -5.00-nC point charge placed at P.

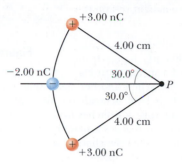

Figure P22.19

20. Two 2.00-μC point charges are located on the x axis. One is at $x = 1.00$ m, and the other is at $x = -1.00$ m. (a) Determine the electric field on the y axis at $y = 0.500$ m. (b) Calculate the electric force on a -3.00-μC charge placed on the y axis at $y = 0.500$ m.

21. Three point charges are arranged as shown in Figure P22.21. (a) Find the vector electric field that the 6.00-nC and -3.00-nC charges together create at the origin. (b) Find the vector force on the 5.00-nC charge.

Figure P22.21

22. Consider the electric dipole shown in Figure P22.22.
    S Show that the electric field at a *distant* point on the $+x$ axis is $E_x \approx 4k_e qa/x^3$.

Figure P22.22

SECTION 22.5 Electric Field Lines

23. Three equal positive charges q
    S are at the corners of an equilateral triangle of side a as shown in Figure P22.23. Assume the three charges together create an electric field. (a) Sketch the field lines in the plane of the charges. (b) Find the location of one point (other than ∞) where the electric field is zero. What

Figure P22.23

are (c) the magnitude and (d) the direction of the electric field at P due to the two charges at the base?

SECTION 22.6 Motion of a Charged Particle in a Uniform Electric Field

24. A proton accelerates from rest in a uniform electric field of
    AMT 640 N/C. At one later moment, its speed is 1.20 Mm/s (non-
    T relativistic because v is much less than the speed of light). (a) Find the acceleration of the proton. (b) Over what time interval does the proton reach this speed? (c) How far does it move in this time interval? (d) What is its kinetic energy at the end of this interval?

25. A proton moves at 4.50×10^5 m/s in the horizontal direc-
    T tion. It enters a uniform vertical electric field with a magnitude of 9.60×10^3 N/C. Ignoring any gravitational effects, find (a) the time interval required for the proton to travel 5.00 cm horizontally, (b) its vertical displacement during the time interval in which it travels 5.00 cm horizontally, and (c) the horizontal and vertical components of its velocity after it has traveled 5.00 cm horizontally.

26. Protons are projected with an initial speed $v_i = $
    GP 9.55 km/s from a field-free region through a plane and into a region where a uniform electric field $\vec{\mathbf{E}} = -720\hat{\mathbf{j}}$ N/C is present above the plane as shown in Figure P22.26. The initial velocity vector of the protons makes an angle θ with the plane. The protons are to hit a target that lies at a horizontal distance of $R = 1.27$ mm from the point where the protons cross the plane and enter the electric field. We wish to find the angle θ at which the protons must pass through the plane to strike the target. (a) What analysis model describes the horizontal motion of the protons above the plane? (b) What analysis model describes the vertical motion of the protons above the plane? (c) Argue that Equation 4.20 would be applicable to the protons in this situation. (d) Use Equation 4.20 to write an expression for R in terms of v_i, E, the charge and mass of the proton, and the angle θ. (e) Find the two possible values of the angle θ. (f) Find the time interval during which the proton is above the plane in Figure P22.26 for each of the two possible values of θ.

Figure P22.26

27. You are still fascinated by the process of inkjet printing,
    CR as described in the opening storyline for this chapter. You convince your father to take you to his manufacturing facility to see the machines that print expiration dates on eggs. You strike up a conversation with the technician operating the machine. He tells you that the ink drops are created using a piezoelectric crystal, acoustic waves, and the

Plateau–Rayleigh instability, which creates uniform drops of mass $m = 1.25 \times 10^{-8}$ g. While you don't understand the fancy words, you do recognize mass! The technician also tells you that the drops are charged to a controllable value of q and then projected vertically downward between parallel deflecting plates at a constant terminal speed of 18.5 m/s. The plates are $\ell = 2.25$ cm long and have a uniform electric field of magnitude $E = 6.35 \times 10^4$ N/C between them. Noting your interest in the process, the technician asks you, "If the position on the egg at which the drop is to be deposited requires that its deflection at the bottom end of the plates be 0.17 mm, what is the required charge on the drop?" You quickly get to work to find the answer.

28. You are working on a research project in which you must **CR** control the direction of travel of electrons using deflection plates. You have devised the apparatus shown in Figure P22.28. The plates are of length $\ell = 0.500$ m and are separated by a distance $d = 3.00$ cm. Electrons are fired at $v_i = 5.00 \times 10^6$ m/s into a uniform electric field from the left edge of the lower, positive plate, aimed directly at the right edge of the upper, negative plate. Therefore, if there is no electric field between the plates, the electrons will follow the broken line in the figure. With an electric field existing between the plates, the electrons will follow a curved path, bending downward. You need to determine (a) the range of angles over which the electron can leave the apparatus and (b) the electric field required to give the maximum possible deviation angle.

Figure P22.28

ADDITIONAL PROBLEMS

29. Consider an infinite number of identical particles, each **S** with charge q, placed along the x axis at distances a, $2a$, $3a$, $4a$, ... from the origin. What is the electric field at the origin due to this distribution? *Suggestion:* Use

$$1 + \frac{1}{2^2} + \frac{1}{3^2} + \frac{1}{4^2} + \cdots = \frac{\pi^2}{6}$$

30. A particle with charge -3.00 nC is at the origin, and a particle with negative charge of magnitude Q is at $x = 50.0$ cm. A third particle with a positive charge is in equilibrium at $x = 20.9$ cm. What is Q?

31. A small block of mass m and charge Q is placed on an insu- **AMT** lated, frictionless, inclined plane of angle θ as in Figure P22.31. An electric field is applied parallel to the incline. (a) Find an expression for the magnitude of the electric field that enables the block to remain at rest. (b) If $m = 5.40$ g,

$Q = -7.00$ μC, and $\theta = 25.0°$, determine the magnitude and the direction of the electric field that enables the block to remain at rest on the incline.

Figure P22.31

32. A small sphere of charge $q_1 = 0.800$ μC hangs from the end of a spring as in Figure P22.32a. When another small sphere of charge $q_2 = -0.600$ μC is held beneath the first sphere as in Figure P22.32b, the spring stretches by $d = 3.50$ cm from its original length and reaches a new equilibrium position with a separation between the charges of $r = 5.00$ cm. What is the force constant of the spring?

Figure P22.32

33. A charged cork ball of mass **T** 1.00 g is suspended on a light string in the presence of a uniform electric field as shown in Figure P22.33. When $\vec{\mathbf{E}} = (3.00\hat{\mathbf{i}} + 5.00\hat{\mathbf{j}}) \times 10^5$ N/C, the ball is in equilibrium at $\theta = 37.0°$. Find (a) the charge on the ball and (b) the tension in the string.

Figure P22.33
Problems 33 and 34

34. A charged cork ball of mass m **S** is suspended on a light string in the presence of a uniform electric field as shown in Figure P22.33. When $\vec{\mathbf{E}} = A\hat{\mathbf{i}} + B\hat{\mathbf{j}}$, where A and B are positive quantities, the ball is in equilibrium at the angle θ. Find (a) the charge on the ball and (b) the tension in the string.

35. Three charged particles are aligned along the x axis as shown in Figure P22.35. Find the electric field at (a) the position $(2.00 \text{ m}, 0)$ and (b) the position $(0, 2.00 \text{ m})$.

Figure P22.35

36. Two point charges $q_A = -12.0\ \mu C$ and $q_B = 45.0\ \mu C$ and a
Q|C third particle with unknown charge q_C are located on the x axis. The particle q_A is at the origin, and q_B is at $x = 15.0$ cm. The third particle is to be placed so that each particle is in equilibrium under the action of the electric forces exerted by the other two particles. (a) Is this situation possible? If so, is it possible in more than one way? Explain. Find (b) the required location and (c) the magnitude and the sign of the charge of the third particle.

37. Two small spheres hang in equilibrium at the bottom ends of threads, 40.0 cm long, that have their top ends tied to the same fixed point. One sphere has mass 2.40 g and charge +300 nC. The other sphere has the same mass and charge +200 nC. Find the distance between the centers of the spheres.

38. Four identical charged particles ($q = +10.0\ \mu C$) are located on the corners of a rectangle as shown in Figure P22.38. The dimensions of the rectangle are $L = 60.0$ cm and $W = 15.0$ cm. Calculate (a) the magnitude and (b) the direction of the total electric force exerted on the charge at the lower left corner by the other three charges.

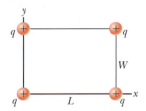

Figure P22.38

39. Review. Two identical blocks resting on a frictionless, horizontal surface are connected by a light spring having a spring constant $k = 100$ N/m and an unstretched length $L_i = 0.400$ m as shown in Figure P22.39a. A charge Q is slowly placed on each block, causing the spring to stretch to an equilibrium length $L = 0.500$ m as shown in Figure P22.39b. Determine the value of Q, modeling the blocks as charged particles.

Figure P22.39 Problems 39 and 40

40. Review. Two identical blocks resting on a frictionless, horizontal surface are connected by a light spring having a
S spring constant k and an unstretched length L_i as shown in Figure P22.39a. A charge Q is slowly placed on each block, causing the spring to stretch to an equilibrium length L as shown in Figure P22.39b. Determine the value of Q, modeling the blocks as charged particles.

41. Three identical point charges, each of mass $m = 0.100$ kg, hang from three strings as shown in Figure P22.41. If the lengths of the left and right strings are each $L = 30.0$ cm and the angle θ is 45.0°, determine the value of q.

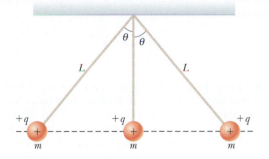

Figure P22.41

42. *Why is the following situation impossible?* An electron enters a region of uniform electric field between two parallel plates. The plates are used in a cathode-ray tube to adjust the position of an electron beam on a distant fluorescent screen. The magnitude of the electric field between the plates is 200 N/C. The plates are 0.200 m in length and are separated by 1.50 cm. The electron enters the region at a speed of 3.00×10^6 m/s, traveling parallel to the plane of the plates in the direction of their length. It leaves the plates heading toward its correct location on the fluorescent screen.

43. Two hard rubber spheres, each of mass $m = 15.0$ g, are rubbed with fur on a dry day and are then suspended with two insulating strings of length $L = 5.00$ cm whose support points are a distance $d = 3.00$ cm from each other as shown in Figure P22.43. During the rubbing process, one sphere receives exactly twice the charge of the other. They are observed to hang at equilibrium, each at an angle of $\theta = 10.0°$ with the vertical. Find the amount of charge on each sphere.

Figure P22.43

44. Two identical beads each have a mass m and charge q. When
S placed in a hemispherical bowl of radius R with frictionless, nonconducting walls, the beads move, and at equilibrium, they are a distance d apart (Fig. P22.44). (a) Determine the charge q on each bead. (b) Determine the charge required for d to become equal to $2R$.

Figure P22.44

45. Two small spheres of mass m are suspended from strings of length ℓ that are connected at a common point. One sphere has charge Q and the other charge $2Q$. The strings make angles θ_1 and θ_2 with the vertical. (a) Explain how θ_1 and θ_2 are related. (b) Assume θ_1 and θ_2 are small. Show that the distance r between the spheres is approximately

$$r \approx \left(\frac{4k_eQ^2\ell}{mg}\right)^{1/3}$$

46. You are working as an expert witness for an inventor. The inventor devised a system that allows an 85.0-kg human to hover above the ground at the surface of the Earth due to the repulsive force between a charge q applied to his body and the normal electric charge on the Earth. The normal charge on the Earth is such that the electric field is uniform near the Earth's surface, directed downward toward the surface, and is of magnitude 130 N/C at the location of the engineer's experiments. Everything went well until the engineer tried a new experiment. He attempted to transfer the same amount of charge q to each of *two* experimental subjects standing next to each other, so they could hover and work close together on a task. The charged, hovering experimental subjects repelled each other and were injured as they flew away in opposite directions. Both experimental subjects are now suing the inventor for their injuries. The inventor is claiming that it is not his fault if the subjects find each other repulsive. To find out whether the inventor has a good defense, determine the initial acceleration of each subject if they are working 1.00 m apart.

47. **Review.** A 1.00-g cork ball with charge 2.00 μC is suspended vertically on a 0.500-m-long light string in the presence of a uniform, downward-directed electric field of magnitude $E = 1.00 \times 10^5$ N/C. If the ball is displaced slightly from the vertical, it oscillates like a simple pendulum. (a) Determine the period of this oscillation. (b) Should the effect of gravitation be included in the calculation for part (a)? Explain.

CHALLENGE PROBLEMS

48. Eight charged particles, each of magnitude q, are located on the corners of a cube of edge s as shown in Figure P22.48. (a) Determine the x, y, and z components of the total force exerted by the other charges on the charge located at point A. What are (b) the magnitude and (c) the direction of this total force?

Figure P22.48

49. Two particles, each with charge 52.0 nC, are located on the y axis at $y = 25.0$ cm and $y = -25.0$ cm. (a) Find the vector electric field at a point on the x axis as a function of x. (b) Find the field at $x = 36.0$ cm. (c) At what location is the field $1.00\hat{\mathbf{i}}$ kN/C? You may need a computer to solve this equation. (d) At what location is the field $16.0\hat{\mathbf{i}}$ kN/C?

50. **Review.** An electric dipole in a uniform horizontal electric field is displaced slightly from its equilibrium position as shown in Figure P22.50, where θ is small. The separation of the charges is $2a$, and each of the two particles has mass m. (a) Assuming the dipole is released from this position, show that its angular orientation exhibits simple harmonic motion with a frequency

$$f = \frac{1}{2\pi}\sqrt{\frac{qE}{ma}}$$

What If? (b) Suppose the masses of the two charged particles in the dipole are not the same even though each particle continues to have charge q. Let the masses of the particles be m_1 and m_2. Show that the frequency of the oscillation in this case is

$$f = \frac{1}{2\pi}\sqrt{\frac{qE(m_1 + m_2)}{2am_1m_2}}$$

Figure P22.50

23

◄ Lightning is a dramatic example of electricity in nature. If you are in a lightning storm, should you walk through the circular metal gate? (*Courtesy of straysparks.com*)

Continuous Charge Distributions and Gauss's Law

STORYLINE **It's spring break! You have arranged to travel to Florida** with some fellow students for a spring break getaway. As your plane lands, you notice some storm clouds in the distance and then see some lightning. Because lightning is relatively rare in southern California, you are fascinated to see the bright flashes, and maybe a little worried. After settling into your accommodations, you are walking around a large park when the lightning storm approaches your location. You want to run back to the park entrance to avoid the lightning, but you must pass through a circular "moongate" as shown in the photograph. You have read that the ground obtains a large charge during a lightning storm and you are concerned that the metal moongate will also have a charge, because it is connected to the ground. If the moongate has a charge on it, that charge will create an electric field. Could that be dangerous? What would be the safest thing for you to do? Should you run through the moongate or try to avoid it?

CONNECTIONS In Chapter 22, we showed how to evaluate the electric field due to a point charge or to a collection of a relatively small number of point charges. In this chapter, we imagine a *continuous* charge distribution: a number of charges so large that the distribution can be considered to be continuous, such as the distribution of charges on the moongate in the opening photograph. We discover in this chapter two ways to evaluate the electric field due to a continuous distribution of charge. One way is to use the superposition principle from Equation 22.10. The sum in that equation will become an integral over the distribution. The second means of finding the electric field for certain types of continuous distributions of charge is to use Gauss's law. Gauss's law is based on the inverse square behavior of the electric force between point charges. Although Gauss's law is a direct consequence of Coulomb's law, it is more convenient for calculating the electric fields of highly symmetric charge distributions and makes it possible to deal with complicated problems using qualitative reasoning. This new tool will be added to our kit of techniques to be used in evaluating electric fields and can be used in future chapters whenever we encounter the electric field due to a continuous, symmetric charge distribution.

23.1 Electric Field of a Continuous Charge Distribution

In Chapter 22, we investigated the electric fields due to point charges and the effects of external electric fields on point charges. Equation 22.10 is useful for calculating the electric field due to a small number of charges. In many cases, we have a *continuous* distribution of charge rather than a collection of discrete charges. The charge in these situations can be described as continuously distributed along some line, over some surface, or throughout some volume.

To set up the process for evaluating the electric field created by a continuous charge distribution, let's use the following procedure. First, divide the charge distribution into small elements, each of which contains a small charge Δq as shown in Figure 23.1. Next, use Equation 22.9 to calculate the electric field due to one of these elements at a point P. Finally, evaluate the total electric field at P due to the charge distribution by summing the contributions of all the charge elements (that is, by applying the superposition principle).

The electric field at P due to one charge element carrying charge Δq is

$$\Delta \vec{\mathbf{E}} = k_e \frac{\Delta q}{r^2} \hat{\mathbf{r}}$$

where r is the distance from the charge element to point P and $\hat{\mathbf{r}}$ is a unit vector directed from the element toward P. The total electric field at P due to all elements in the charge distribution is approximately

$$\vec{\mathbf{E}} \approx k_e \sum_i \frac{\Delta q_i}{r_i^2} \hat{\mathbf{r}}_i$$

where the index i refers to the ith element in the distribution. Because the number of elements is very large and the charge distribution is modeled as continuous, the total field at P in the limit $\Delta q_i \to 0$ is

$$\vec{\mathbf{E}} = k_e \lim_{\Delta q_i \to 0} \sum_i \frac{\Delta q_i}{r_i^2} \hat{\mathbf{r}}_i = k_e \int \frac{dq}{r^2} \hat{\mathbf{r}} \qquad (23.1)$$

◀ Electric field due to a continuous charge distribution

where the integration is over the entire charge distribution. The integration in Equation 23.1 is a vector operation and must be treated appropriately.

Let's illustrate this type of calculation with several examples in which the charge is distributed on a line, on a surface, or throughout a volume. When performing such calculations, it is convenient to use the concept of a *charge density* along with the following notations:

- If a charge Q is uniformly distributed throughout a volume V, the **volume charge density** ρ is defined by

◀ Volume charge density

$$\rho \equiv \frac{Q}{V}$$

where ρ has units of coulombs per cubic meter (C/m^3).

Figure 23.1 The electric field at P due to a continuous charge distribution is the vector sum of the fields $\Delta \vec{\mathbf{E}}_i$ due to all the elements Δq_i of the charge distribution. Three sample elements are shown.

- If a charge Q is uniformly distributed on a surface of area A, the **surface charge density** σ (Greek letter sigma) is defined by

$$\sigma \equiv \frac{Q}{A}$$

◀ Surface charge density

where σ has units of coulombs per square meter (C/m^2).

- If a charge Q is uniformly distributed along a line of length ℓ, the **linear charge density** λ is defined by

$$\lambda \equiv \frac{Q}{\ell}$$

◀ Linear charge density

where λ has units of coulombs per meter (C/m).

- If the charge is nonuniformly distributed over a volume, surface, or line, the amounts of charge dq in a small volume, surface, or length element are

$$dq = \rho \, dV \qquad dq = \sigma \, dA \qquad dq = \lambda \, d\ell$$

PROBLEM-SOLVING STRATEGY   Calculating the Electric Field

The following procedure is recommended for solving problems that involve the determination of an electric field due to individual charges or a charge distribution.

1. Conceptualize. Establish a mental representation of the problem: think carefully about the individual charges or the charge distribution and imagine what type of electric field it would create. Appeal to any symmetry in the arrangement of charges to help you visualize the electric field.

2. Categorize. Are you analyzing a group of individual charges or a continuous charge distribution? The answer to this question tells you how to proceed in the Analyze step.

3. Analyze.

(a) If you are analyzing a group of individual charges, use the superposition principle: when several point charges are present, the resultant field at a point in space is the *vector sum* of the individual fields due to the individual charges (Eq. 22.10). Be very careful in the manipulation of vector quantities. It may be useful to review the material on vector addition in Chapter 3. Example 22.6 in the previous chapter demonstrated this procedure.

(b) If you are analyzing a continuous charge distribution, the superposition principle is applied by replacing the vector sums for evaluating the total electric field from individual charges by vector integrals. The charge distribution is divided into infinitesimal pieces, and the vector sum is carried out by integrating over the entire charge distribution (Eq. 23.1). Examples 23.1 through 23.3 demonstrate such procedures.

Consider symmetry when dealing with either a distribution of point charges or a continuous charge distribution. Take advantage of any symmetry in the system you observed in the Conceptualize step to simplify your calculations. The cancellation of field components perpendicular to the axis in Example 23.2 is an example of the application of symmetry.

4. Finalize. Check to see if your electric field expression is consistent with the mental representation and if it reflects any symmetry that you noted previously. Imagine varying parameters such as the distance of the observation point from the charges or the radius of any circular objects to see if the mathematical result changes in a reasonable way.

Example **23.1**   **The Electric Field Due to a Charged Rod**

A rod of length ℓ has a uniform positive charge per unit length λ and a total charge Q. Calculate the electric field at a point P that is located along the long axis of the rod and a distance a from one end (Fig. 23.2).

Figure 23.2 (Example 23.1) The electric field at P due to a uniformly charged rod lying along the x axis. We choose the location of point P to be the origin.

S O L U T I O N

Conceptualize   The field $d\vec{E}$ at P due to each segment of charge on the rod is in the negative x direction because every segment carries a positive charge. Figure 23.2 shows the appropriate geometry. In our result, we expect the electric field to become smaller as the distance a becomes larger because point P is farther from the charge distribution.

Categorize   Because the rod is continuous, we are evaluating the field due to a continuous charge distribution rather than a group of individual charges. Because every segment of the rod produces an electric field in the negative x direction, the vector sum of their contributions is easy to determine.

Analyze   Let's assume the rod is lying along the x axis, dx is the length of one small segment, and dq is the charge on that segment. Because the rod has a charge per unit length λ, the charge dq on the small segment is $dq = \lambda \, dx$.

Find the magnitude of the electric field at P due to one segment of the rod having a charge dq:

$$dE = k_e \frac{dq}{x^2} = k_e \frac{\lambda \, dx}{x^2}$$

Find the total field at P using[1] Equation 23.1:

$$E = \int_a^{\ell+a} k_e \lambda \frac{dx}{x^2}$$

Noting that k_e and $\lambda = Q/\ell$ are constants and can be removed from the integral, evaluate the integral:

$$E = k_e \lambda \int_a^{\ell+a} \frac{dx}{x^2} = k_e \lambda \left[-\frac{1}{x} \right]_a^{\ell+a}$$

$$(1) \quad E = k_e \frac{Q}{\ell} \left(\frac{1}{a} - \frac{1}{\ell+a} \right) = \frac{k_e Q}{a(\ell+a)}$$

Finalize   We see that our prediction is correct; if a becomes larger, the denominator of the fraction grows larger, and E becomes smaller. On the other hand, if $a \to 0$, which corresponds to sliding the bar to the left until its left end is at the origin, then $E \to \infty$. That represents the condition in which the observation point P is at zero distance from the charge at the end of the rod, so the field becomes infinite. We explore large values of a below.

W H A T   I F ?   Suppose point P is very far away from the rod. What is the nature of the electric field at such a point?

Answer   If P is far from the rod ($a \gg \ell$), then ℓ in the denominator of Equation (1) can be neglected and $E \approx k_e Q/a^2$. That is exactly the form you would expect for a point charge. Therefore, at large values of a/ℓ, the charge distribution appears to be a point charge of magnitude Q; the point P is so far away from the rod we cannot distinguish that it has a size. The use of the limiting technique ($a/\ell \to \infty$) is often a good method for checking a mathematical expression.

Example **23.2**   **The Electric Field of a Uniform Ring of Charge**

A ring of radius a carries a uniformly distributed positive total charge Q. Calculate the electric field due to the ring at a point P lying a distance x from its center along the central axis perpendicular to the plane of the ring (Fig. 23.3a).

S O L U T I O N

Conceptualize   Figure 23.3a shows the electric field contribution $d\vec{E}$ at P due to a single segment of charge at the top of the ring. This field vector can be resolved into components dE_x parallel to the axis of the ring and $dE_\perp$ perpendicular to the axis.

---

[1]To carry out integrations such as this one, first express the charge element dq in terms of the other variables in the integral. (In this example, there is one variable, x, so we made the change $dq = \lambda \, dx$.) The integral must be over scalar quantities; therefore, express the electric field in terms of components, if necessary. (In this example, the field has only an x component, so this detail is of no concern.) Then, reduce your expression to an integral over a single variable (or to multiple integrals, each over a single variable). In examples that have spherical or cylindrical symmetry, the single variable is a radial coordinate.

23.2 continued

Figure 23.3b shows the electric field contributions from two segments on opposite sides of the ring. Because of the symmetry of the situation, the perpendicular components $dE_\perp$ of the field cancel. That is true for all pairs of segments around the ring, so we can ignore the perpendicular component of the field and focus solely on the parallel components dE_x, which simply add.

Categorize Because the ring is continuous, we are evaluating the field due to a continuous charge distribution rather than a group of individual charges.

Figure 23.3 (Example 23.2) A uniformly charged ring of radius a. (a) The field at P on the x axis due to an element of charge dq. (b) The perpendicular component of the field at P due to segment 1 is canceled by the perpendicular component due to segment 2.

Analyze Evaluate the parallel component of an electric field contribution from a segment of charge dq on the ring:

$$(1) \quad dE_x = k_e \frac{dq}{r^2} \cos\theta = k_e \frac{dq}{a^2 + x^2} \cos\theta$$

From the geometry in Figure 23.3a, evaluate $\cos\theta$:

$$(2) \quad \cos\theta = \frac{x}{r} = \frac{x}{(a^2 + x^2)^{1/2}}$$

Substitute Equation (2) into Equation (1):

$$dE_x = k_e \frac{dq}{a^2 + x^2}\left[\frac{x}{(a^2 + x^2)^{1/2}}\right] = \frac{k_e x}{(a^2 + x^2)^{3/2}}\, dq$$

All segments of the ring make the same contribution to the field at P because they are all equidistant from this point. Integrate over the circumference of the ring to obtain the total field at P:

$$E_x = \int \frac{k_e x}{(a^2 + x^2)^{3/2}}\, dq = \frac{k_e x}{(a^2 + x^2)^{3/2}} \int dq$$

$$(3) \quad E = \frac{k_e x}{(a^2 + x^2)^{3/2}}\, Q$$

Finalize The electric field at P is of this magnitude and directed along the x axis, away from the ring. This result shows that the field is zero at $x = 0$. Is that consistent with the symmetry in the problem? Furthermore, notice that Equation (3) reduces to $k_e Q/x^2$ if $x \gg a$, so the ring acts like a point charge for locations far away from the ring. From a faraway point, we cannot distinguish the ring shape of the charge.

W H A T   I F ? Suppose a negative charge is placed at the center of the ring in Figure 23.3 and displaced slightly by a distance $x \ll a$ along the x axis. When the charge is released, what type of motion does it exhibit?

Answer In the expression for the field due to a ring of charge, let $x \ll a$, which results in

$$E_x = \frac{k_e Q}{a^3}\, x$$

Therefore, from Equation 22.8, the force on a charge $-q$ placed near the center of the ring is

$$F_x = -\frac{k_e qQ}{a^3}\, x$$

Because this force has the form of Hooke's law (Eq. 15.1), the motion of the negative charge is described with the *particle in simple harmonic motion model!*

Example 23.2 relates to our opening storyline. We wanted to know if the region inside the circular moongate would be safe if the metal of the moongate were charged. Equation (3) in Example 23.2 shows that the electric field is zero at the exact center of one ring of the moongate. The opening photograph shows two rings in the moongate, so you may experience zero field at the center of one but will experience a nonzero field from the other! You will also experience a radial field at points other than the exact center of the moongate, but these fields will be small in magnitude. As you approach the moongate, you will experience a maximum value of the electric field on the axis of the ring at a point a distance $a/2^{1/2}$ from the center of the ring, as shown in Problem 40. Therefore, to get to the zero field inside the ring, you have to pass through a maximum value. Maybe the best choice is to take a little extra time and run *around* the moongate!

Example 23.3 The Electric Field of a Uniformly Charged Disk

A disk of radius R has a uniform surface charge density σ. Calculate the electric field at a point P that lies along the central perpendicular axis of the disk and a distance x from the center of the disk (Fig. 23.4).

SOLUTION

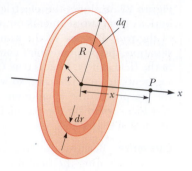

Conceptualize If the disk is considered to be a set of concentric rings, we can use our result from Example 23.2—which gives the field created by a single ring of radius a—and sum the contributions of all rings making up the disk. By symmetry, the field at an axial point must be along the central axis.

Figure 23.4 (Example 23.3) A uniformly charged disk of radius R. The electric field at an axial point P is directed along the central axis, perpendicular to the plane of the disk.

Categorize Because the disk is continuous, we are evaluating the field due to a continuous charge distribution rather than a group of individual charges.

Analyze Find the amount of charge dq on the surface area of a ring of radius r and width dr as shown in Figure 23.4:

$$dq = \sigma \, dA = \sigma\,(2\pi r \, dr) = 2\pi\sigma r \, dr$$

Use this result in Equation (3) in Example 23.2 (with a replaced by r and Q replaced by dq) to find the field due to the ring:

$$dE_x = \frac{k_e x}{(r^2 + x^2)^{3/2}}\,(2\pi\sigma r \, dr)$$

To obtain the total field at P, integrate this expression over the limits $r = 0$ to $r = R$, noting that x is a constant in this situation:

$$E_x = k_e x \pi \sigma \int_0^R \frac{2r \, dr}{(r^2 + x^2)^{3/2}}$$

$$= k_e x \pi \sigma \int_0^R (r^2 + x^2)^{-3/2}\,d(r^2)$$

$$= k_e x \pi \sigma \left[\frac{(r^2 + x^2)^{-1/2}}{-1/2}\right]_0^R = 2\pi k_e \sigma \left[1 - \frac{x}{(R^2 + x^2)^{1/2}}\right]$$

Finalize This result is valid for all values of $x > 0$. For large values of x, the result above can be evaluated by a series expansion and shown to be equivalent to the electric field of a point charge Q. We can calculate the field close to the disk along the axis by assuming $x \ll R$; in this case, the expression in brackets reduces to unity to give us the near-field approximation

$$E = 2\pi k_e \sigma = \frac{\sigma}{2\epsilon_0}$$

where ϵ_0 is the permittivity of free space.

WHAT IF? What if we let the radius of the disk grow so that the disk becomes an infinite plane of charge?

Answer The result of letting $R \to \infty$ in the final result of the example is that the magnitude of the electric field becomes

$$E = 2\pi k_e \sigma = \frac{\sigma}{2\epsilon_0}$$

This is the same expression that we obtained for $x \ll R$. If $R \to \infty$, *everywhere* is near-field—the result is independent of the position at which you measure the electric field. Therefore, the electric field due to an infinite plane of charge is uniform throughout space.

An infinite plane of charge is impossible in practice. If two planes of charge are placed close to each other, however, with one plane positively charged, and the other negatively, the electric field between the plates is very close to uniform at points far from the edges. Such a configuration will be investigated in Chapter 25.

23.2 Electric Flux

The concept of electric field lines was described qualitatively in Chapter 22. We now treat electric field lines in a more quantitative way.

Consider an electric field that is uniform in both magnitude and direction as shown in Figure 23.5. The field lines penetrate a rectangular surface of area A, whose plane is oriented perpendicular to the field. Recall from Section 22.5 that the number of lines per unit area (in other words, the *line density*) is proportional to the magnitude of the electric field. Therefore, the total number of lines penetrating the surface is proportional to the product *EA*. This product of the magnitude of the electric field *E* and surface area *A* perpendicular to the field is called the **electric flux** Φ_E (uppercase Greek letter phi):

$$\Phi_E = EA \tag{23.2}$$

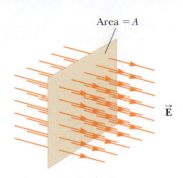

Figure 23.5 Field lines representing a uniform electric field penetrating a plane of area *A* perpendicular to the field.

From the SI units of *E* and *A*, we see that Φ_E has units of newton meters squared per coulomb ($N \cdot m^2/C$).

If the surface under consideration is not perpendicular to the field, the flux through it must be less than that given by Equation 23.2. Consider Figure 23.6, where the surface of area *A* is at an angle θ to a plane of area $A_\perp$ that is perpendicular to the uniform electric field. Notice that the number of lines that cross this area *A* is equal to the number of lines that cross the area $A_\perp$. The area *A* is the product of the length and the width of the surface: $A = \ell w$. At the left edge of the figure, we see that the widths of the surfaces are related by $w_\perp = w \cos \theta$. The area $A_\perp$ is given by $A_\perp = \ell w_\perp = \ell w \cos \theta$ and we see that the two areas are related by $A_\perp = A \cos \theta$. Because the flux through *A* equals the flux through $A_\perp$, the flux through *A* is

$$\Phi_E = EA_\perp = EA \cos \theta \tag{23.3}$$

The number of field lines that go through the area $A_\perp$ is the same as the number that go through area *A*.

From this result, we see that the flux through a surface of fixed area *A* has a maximum value *EA* when the surface is perpendicular to the field (when the normal to the surface is parallel to the field, that is, when $\theta = 0°$ in Fig. 23.6); the flux is zero when the surface is parallel to the field (when the normal to the surface is perpendicular to the field, that is, when $\theta = 90°$).

In this discussion, the angle θ is used to describe the orientation of the surface of area *A*. We can also interpret the angle as that between the electric field vector and the normal to the surface as shown in Figure 23.6. In this case, the product $E \cos \theta$ in Equation 23.3 is the component of the electric field perpendicular to the surface. The flux through the surface can then be written $\Phi_E = (E \cos \theta)A = E_n A$, where we use E_n as the component of the electric field normal to the surface.

Figure 23.6 Field lines representing a uniform electric field penetrating an area *A* whose normal is at an angle θ to the field.

We assumed a uniform electric field in the preceding discussion. In more general situations, the electric field may vary over a large surface. Therefore, the definition of flux given by Equation 23.3 has meaning only for a small element of area over which the field is approximately constant. Consider a general surface divided into a large number of small elements, each of area ΔA_i. It is convenient to define a vector $\Delta \vec{A}_i$ whose magnitude represents the area of the *i*th element of the large surface and whose direction is defined to be *perpendicular* to the surface element as shown in Figure 23.7. The electric field $\vec{E}_i$ at the location of this element makes an angle θ_i with the vector $\Delta \vec{A}_i$. The electric flux $\Phi_{E,i}$ through this element is

$$\Phi_{E,i} = E_i \, \Delta A_i \cos \theta_i = \vec{E}_i \cdot \Delta \vec{A}_i$$

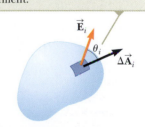

The electric field makes an angle θ_i with the vector $\Delta \vec{A}_i$, defined as being normal to the surface element.

where we have used the definition of the scalar product of two vectors ($\vec{A} \cdot \vec{B} \equiv AB \cos \theta$; see Section 7.3). Summing the contributions of all elements gives an approximation to the total flux through the surface:

$$\Phi_E \approx \sum \vec{E}_i \cdot \Delta \vec{A}_i$$

If the area of each element approaches zero, the number of elements approaches infinity and the sum is replaced by an integral. Therefore, the general definition of electric flux is

$$\Phi_E \equiv \int_{\text{surface}} \vec{E} \cdot d\vec{A} \tag{23.4}$$

Figure 23.7 A small element of surface area ΔA_i in an electric field.

◄ Definition of electric flux

Equation 23.4 is a *surface integral,* which means it must be evaluated over the surface in question. In general, the value of Φ_E depends both on the field pattern and on the surface.

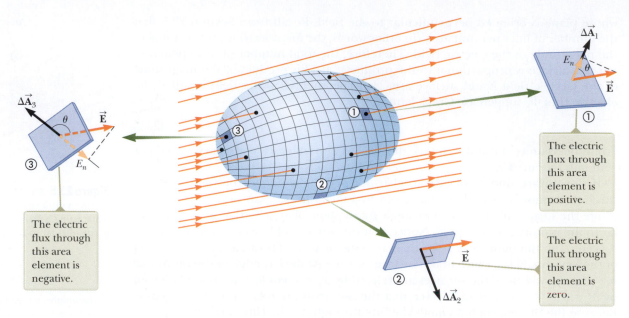

The electric flux through this area element is negative.

The electric flux through this area element is positive.

The electric flux through this area element is zero.

Figure 23.8 A closed surface in an electric field. The area vectors are, by convention, normal to the surface and point outward.

We are often interested in evaluating the flux through a *closed surface*, defined as a surface that divides space into an inside and an outside region so that one cannot move from one region to the other without crossing the surface. The surface of a sphere, for example, is a closed surface. By convention, if the area element in Equation 23.4 is part of a closed surface, the direction of the area vector is chosen so that the vector points outward from the surface. If the area element is not part of a closed surface, the direction of the area vector is chosen so that the angle between the area vector and the electric field vector is less than or equal to 90°.

Consider the closed surface in Figure 23.8. The vectors $\Delta\vec{A}_i$ point in different directions for the various surface elements, but for each element they are normal to the surface and point outward. At the element labeled ①, the field lines are crossing the surface from the inside to the outside and $\theta < 90°$; hence, the flux $\Phi_{E,1} = \vec{E} \cdot \Delta\vec{A}_1$ through this element is positive. For element ②, the field lines graze the surface (perpendicular to $\Delta\vec{A}_2$); therefore, $\theta = 90°$ and the flux is zero. For elements such as ③, where the field lines are crossing the surface from outside to inside, $180° > \theta > 90°$ and the flux is negative because $\cos\theta$ is negative. The *net* flux through the surface is proportional to the net number of lines leaving the surface, where the net number means *the number of lines leaving the surface minus the number of lines entering the surface.* If more lines are leaving than entering, the net flux is positive. If more lines are entering than leaving, the net flux is negative. Using the symbol $\oint$ to represent an integral over a closed surface, we can write the net flux Φ_E through a closed surface as

$$\Phi_E = \oint \vec{E} \cdot d\vec{A} = \oint E_n \, dA \qquad (23.5)$$

where E_n represents the component of the electric field normal to the surface.

QUICK QUIZ 23.1 Suppose a point charge is located at the center of a spherical surface. The electric field at the surface of the sphere and the total flux through the sphere are determined. Now the radius of the sphere is halved. What happens to the flux through the sphere and the magnitude of the electric field at the surface of the sphere? **(a)** The flux and field both increase. **(b)** The flux and field both decrease. **(c)** The flux increases, and the field decreases. **(d)** The flux decreases, and the field increases. **(e)** The flux remains the same, and the field increases. **(f)** The flux decreases, and the field remains the same.

Example **23.4** Flux Through a Cube

Consider a uniform electric field $\vec{E}$ oriented in the x direction in empty space. A cube of edge length ℓ is placed in the field, oriented as shown in Figure 23.9. Find the net electric flux through the surface of the cube.

SOLUTION

Conceptualize Examine Figure 23.9 carefully. Notice that the electric field lines pass through two faces perpendicularly and are parallel to four other faces of the cube.

Categorize We evaluate the flux from its definition, so we categorize this example as a substitution problem.

The flux through four of the faces (③, ④, and the unnumbered faces) is zero because $\vec{E}$ is parallel to the four faces and therefore perpendicular to $d\vec{A}$ on these faces.

Figure 23.9 (Example 23.4) A closed surface in the shape of a cube in a uniform electric field oriented parallel to the x axis. Side ④ is the bottom of the cube, and side ① is opposite side ②.

Write the integrals for the net flux through faces ① and ②:

$$\Phi_E = \int_1 \vec{E} \cdot d\vec{A} + \int_2 \vec{E} \cdot d\vec{A}$$

For face ①, $\vec{E}$ is constant and directed inward but $d\vec{A}_1$ is directed outward ($\theta = 180°$). Find the flux through this face:

$$\int_1 \vec{E} \cdot d\vec{A} = \int_1 E(\cos 180°)\, dA = -E\int_1 dA = -EA = -E\ell^2$$

For face ②, $\vec{E}$ is constant and outward and in the same direction as $d\vec{A}_2$ ($\theta = 0°$). Find the flux through this face:

$$\int_2 \vec{E} \cdot d\vec{A} = \int_2 E(\cos 0°)\, dA = E\int_2 dA = +EA = E\ell^2$$

Find the net flux by adding the flux over all six faces:

$$\Phi_E = -E\ell^2 + E\ell^2 + 0 + 0 + 0 + 0 = \boxed{0}$$

In the next section, we generate a fundamental principle that explains this zero value.

23.3 Gauss's Law

In this section, we describe a general relationship between the net electric flux through a closed surface (often called a *gaussian surface*) and the charge enclosed by the surface. This relationship, known as *Gauss's law,* is of fundamental importance in the study of electric fields.

Consider a positive point charge q located at the center of a sphere of radius r as shown in Figure 23.10. From Equation 22.9, we know that the magnitude of the electric field everywhere on the surface of the sphere is $E = k_e q/r^2$. The field lines are directed radially outward and hence are perpendicular to the surface at every point on the surface. That is, at each surface point, $\vec{E}$ is parallel to the vector $\Delta\vec{A}_i$ representing a local element of area ΔA_i surrounding the surface point. Therefore,

$$\vec{E} \cdot \Delta\vec{A}_i = E\,\Delta A_i$$

and, from Equation 23.5, we find that the net flux through the gaussian surface is

$$\Phi_E = \oint \vec{E} \cdot d\vec{A} = \oint E\, dA = E\oint dA$$

where we have moved E outside of the integral because, by symmetry, E is constant over the surface. The value of E is given by $E = k_e q/r^2$. Furthermore, because the surface is spherical, $\oint dA = A = 4\pi r^2$. Hence, the net flux through the gaussian surface is

$$\Phi_E = k_e \frac{q}{r^2}(4\pi r^2) = 4\pi k_e q$$

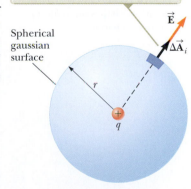

When the charge is at the center of the sphere, the electric field is everywhere normal to the surface and constant in magnitude.

Figure 23.10 A spherical gaussian surface of radius r surrounding a positive point charge q.

Karl Friedrich Gauss
German mathematician and astronomer (1777–1855)
Gauss received a doctoral degree in mathematics from the University of Helmstedt in 1799. In addition to his work in electromagnetism, he made contributions to mathematics and science in number theory, statistics, non-Euclidean geometry, and cometary orbital mechanics. He was a founder of the German Magnetic Union, which studies the Earth's magnetic field on a continual basis.

Recalling from Equation 22.3 that $k_e = 1/4\pi\epsilon_0$, we can write this equation in the form

$$\Phi_E = \frac{q}{\epsilon_0} \tag{23.6}$$

Equation 23.6 shows that the net flux through the spherical surface is proportional to the charge inside the surface. The flux is independent of the radius r because the area of the spherical surface is proportional to r^2, whereas the electric field is proportional to $1/r^2$. Therefore, in the product of area and electric field, the dependence on r cancels.

Now consider several closed surfaces surrounding a charge q as shown in Figure 23.11. Surface S_1 is spherical, but surfaces S_2 and S_3 are not. From Equation 23.6, the flux that passes through S_1 has the value q/ϵ_0. As discussed in the preceding section, flux is proportional to the number of electric field lines passing through a surface. In Figure 23.11, every field line that passes through S_1 also passes through the nonspherical surfaces S_2 and S_3. Therefore,

> the net flux through *any* closed surface surrounding a point charge q is given by q/ϵ_0 and is independent of the shape of that surface.

Now consider a point charge located *outside* a closed surface of arbitrary shape as shown in Figure 23.12. As can be seen from this construction, any electric field line entering the surface leaves the surface at another point. The number of electric field lines entering the surface equals the number leaving the surface. Therefore, the net electric flux through a closed surface that surrounds no charge is zero. Applying this result to Example 23.4, we see that the net flux through the cube is zero because there is no charge inside the cube.

Let's extend these arguments to two generalized cases: (1) that of many point charges and (2) that of a continuous distribution of charge. We once again use the superposition principle, which states that the electric field due to many charges is the vector sum of the electric fields produced by the individual charges. Therefore, the flux through any closed surface can be expressed as

$$\oint \vec{\mathbf{E}} \cdot d\vec{\mathbf{A}} = \oint (\vec{\mathbf{E}}_1 + \vec{\mathbf{E}}_2 + \cdots) \cdot d\vec{\mathbf{A}}$$

where $\vec{\mathbf{E}}$ is the total electric field at any point on the surface produced by the vector addition of the electric fields at that point due to the individual charges. Consider the system of charges shown in Figure 23.13. The surface S surrounds only one

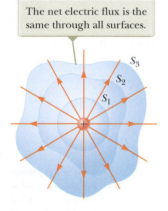

The net electric flux is the same through all surfaces.

Figure 23.11 Closed surfaces of various shapes surrounding a positive charge.

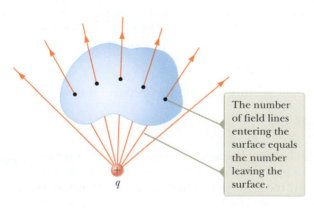

The number of field lines entering the surface equals the number leaving the surface.

Figure 23.12 A point charge located *outside* a closed surface.

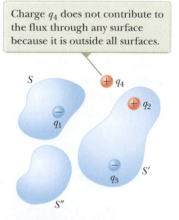

Charge q_4 does not contribute to the flux through any surface because it is outside all surfaces.

Figure 23.13 The net electric flux through any closed surface depends only on the charge *inside* that surface. The net flux through surface S is q_1/ϵ_0, the net flux through surface S' is $(q_2 + q_3)/\epsilon_0$, and the net flux through surface S'' is zero.

charge, q_1; hence, the net flux through S is q_1/ϵ_0. The flux through S due to charges q_2, q_3, and q_4 outside it is zero because each electric field line from these charges that enters S at one point leaves it at another. The surface S' surrounds charges q_2 and q_3; hence, the net flux through it is $(q_2 + q_3)/\epsilon_0$. Finally, the net flux through surface S'' is zero because there is no charge inside this surface. That is, *all* the electric field lines that enter S'' at one point leave at another. Charge q_4 does not contribute to the net flux through any of the surfaces.

The mathematical form of **Gauss's law** is a generalization of what we have just described and states that the net flux through *any* closed surface is

$$\Phi_E = \oint \vec{\mathbf{E}} \cdot d\vec{\mathbf{A}} = \frac{q_{in}}{\epsilon_0} \tag{23.7}$$

where $\vec{\mathbf{E}}$ represents the electric field at any point on the surface and q_{in} represents the net charge inside the surface.

When using Equation 23.7, you should note that although the charge q_{in} is the net charge inside the gaussian surface, $\vec{\mathbf{E}}$ represents the *total electric field,* which includes contributions from charges both inside and outside the surface.

In principle, Gauss's law can be solved for $\vec{\mathbf{E}}$ to determine the electric field due to a system of charges or a continuous distribution of charge. In practice, however, this type of solution is applicable only in a limited number of highly symmetric situations. In the next section, we use Gauss's law to evaluate the electric field for charge distributions that have spherical, cylindrical, or planar symmetry. If one chooses the gaussian surface surrounding the charge distribution carefully, the integral in Equation 23.7 can be simplified and the electric field determined.

QUICK QUIZ 23.2 If the net flux through a gaussian surface is *zero,* the following four statements *could be true.* Which of the statements *must be true?* **(a)** There are no charges inside the surface. **(b)** The net charge inside the surface is zero. **(c)** The electric field is zero everywhere on the surface. **(d)** The number of electric field lines entering the surface equals the number leaving the surface.

> **PITFALL PREVENTION 23.1**
> **Zero Flux Is Not Zero Field** In two situations, there is zero flux through a closed surface: either (1) there are no charged particles enclosed by the surface or (2) there are charged particles enclosed, but the net charge inside the surface is zero. For either situation, it is *incorrect* to conclude that the electric field on the surface is zero. Gauss's law states that the electric *flux* is proportional to the enclosed charge, not the electric *field.*

Conceptual Example 23.5 **Flux Due to a Point Charge**

A spherical gaussian surface surrounds a point charge q. Describe what happens to the total flux through the surface if **(A)** the charge is tripled, **(B)** the radius of the sphere is doubled, **(C)** the surface is changed to a cube, and **(D)** the charge is moved to another location inside the surface.

SOLUTION

(A) The flux through the surface is tripled because flux is proportional to the amount of charge inside the surface.

(B) The flux does not change because all electric field lines from the charge pass through the sphere, regardless of its radius.

(C) The flux does not change when the shape of the gaussian surface changes because all electric field lines from the charge pass through the surface, regardless of its shape.

(D) The flux does not change when the charge is moved to another location inside that surface because Gauss's law refers to the total charge enclosed, regardless of where the charge is located inside the surface.

23.4 Application of Gauss's Law to Various Charge Distributions

As mentioned earlier, Gauss's law is useful for determining electric fields when the charge distribution is highly symmetric. The following examples demonstrate ways of choosing the gaussian surface over which the surface integral given by

Equation 23.7 can be simplified and the electric field determined. In choosing the surface, always take advantage of the symmetry of the charge distribution so that E can be removed from the integral. The goal in this type of calculation is to determine a surface for which each portion of the surface satisfies one or more of the following conditions:

1. The value of the electric field can be argued by symmetry to be constant over the portion of the surface.
2. The dot product in Equation 23.7 can be expressed as a simple algebraic product $E \, dA$ because $\vec{\mathbf{E}}$ and $d\vec{\mathbf{A}}$ are parallel.
3. The dot product in Equation 23.7 is zero because $\vec{\mathbf{E}}$ and $d\vec{\mathbf{A}}$ are perpendicular.
4. The electric field is zero over the portion of the surface.

Different portions of the gaussian surface can satisfy different conditions as long as every portion satisfies at least one condition. All four conditions are used in examples throughout the remainder of this chapter and the next, and will be identified by number. If the charge distribution does not have sufficient symmetry such that a gaussian surface that satisfies these conditions can be found, Gauss's law is still true, but is not useful for determining the electric field for that charge distribution.

Example 23.6 | A Spherically Symmetric Charge Distribution

An insulating solid sphere of radius a has a uniform volume charge density ρ and carries a total positive charge Q (Fig. 23.14).

(A) Calculate the magnitude of the electric field at a point outside the sphere.

SOLUTION

Conceptualize The electric field due to point charges was discussed in Section 22.4. Now we are considering the electric field due to a distribution of charge. We found the field for various distributions of charge in Section 23.1 by integrating over the distribution. This example demonstrates a difference from our discussions in Section 23.1. In this section, we find the electric field using Gauss's law.

Categorize Because the charge is distributed uniformly throughout the sphere, the charge distribution has spherical symmetry and we can apply Gauss's law to find the electric field.

For points outside the sphere, a large, spherical gaussian surface is drawn concentric with the sphere.

For points inside the sphere, a spherical gaussian surface smaller than the sphere is drawn.

Figure 23.14 (Example 23.6) A uniformly charged insulating sphere of radius a and total charge Q. In diagrams such as this one, the dotted line represents the intersection of the gaussian surface with the plane of the page.

Analyze To reflect the spherical symmetry, let's choose a spherical gaussian surface of radius r, concentric with the sphere, as shown in Figure 23.14a. For this choice, condition (2) is satisfied everywhere on the surface and $\vec{\mathbf{E}} \cdot d\vec{\mathbf{A}} = E \, dA$.

Replace $\vec{\mathbf{E}} \cdot d\vec{\mathbf{A}}$ in Gauss's law with $E \, dA$:

$$\Phi_E = \oint \vec{\mathbf{E}} \cdot d\vec{\mathbf{A}} = \oint E \, dA = \frac{Q}{\epsilon_0}$$

By symmetry, E has the same value everywhere on the surface, which satisfies condition (1), so we can remove E from the integral:

$$\oint E \, dA = E \oint dA = E(4\pi r^2) = \frac{Q}{\epsilon_0}$$

Solve for E:

$$(1) \quad E = \frac{Q}{4\pi\epsilon_0 r^2} = k_e \frac{Q}{r^2} \quad \text{(for } r > a\text{)}$$

23.6 continued

Finalize This field is identical to that for a point charge. Therefore, **the electric field due to a uniformly charged sphere in the region external to the sphere is *equivalent* to that of a point charge located at the center of the sphere.**

(B) Find the magnitude of the electric field at a point inside the sphere.

SOLUTION

Analyze In this case, let's choose a spherical gaussian surface having radius $r < a$, concentric with the insulating sphere (Fig. 23.14b). Let V' be the volume of this smaller sphere. To apply Gauss's law in this situation, recognize that the charge q_{in} within the gaussian surface of volume V' is less than Q.

Calculate q_{in} by using $q_{in} = \rho V'$:

$$q_{in} = \rho V' = \rho\left(\tfrac{4}{3}\pi r^3\right)$$

Notice that conditions (1) and (2) are satisfied everywhere on the gaussian surface in Figure 23.14b. Apply Gauss's law in the region $r < a$:

$$\oint E\,dA = E\oint dA = E(4\pi r^2) = \frac{q_{in}}{\epsilon_0}$$

Solve for E and substitute for q_{in}:

$$E = \frac{q_{in}}{4\pi\epsilon_0 r^2} = \frac{\rho\left(\tfrac{4}{3}\pi r^3\right)}{4\pi\epsilon_0 r^2} = \frac{\rho}{3\epsilon_0}r$$

Substitute $\rho = Q/\tfrac{4}{3}\pi a^3$ and $\epsilon_0 = 1/4\pi k_e$:

$$(2)\quad E = \frac{Q/\tfrac{4}{3}\pi a^3}{3(1/4\pi k_e)}r = k_e\frac{Q}{a^3}r \quad (\text{for } r < a)$$

Finalize This result for E differs from the one obtained in part (A). It shows that $E \to 0$ as $r \to 0$. Therefore, the result eliminates the problem that would exist at $r = 0$ if E varied as $1/r^2$ inside the sphere as it does outside the sphere. That is, if $E \propto 1/r^2$ for $r < a$, the field would be infinite at $r = 0$, which is physically impossible.

WHAT IF? Suppose the radial position $r = a$ is approached from inside the sphere and from outside. Do we obtain the same value of the electric field from both directions?

Answer Equation (1) shows that the electric field approaches a value from the outside given by

$$E = \lim_{r \to a}\left(k_e\frac{Q}{r^2}\right) = k_e\frac{Q}{a^2}$$

From the inside, Equation (2) gives

$$E = \lim_{r \to a}\left(k_e\frac{Q}{a^3}r\right) = k_e\frac{Q}{a^3}a = k_e\frac{Q}{a^2}$$

Therefore, the value of the field is the same as the surface is approached from both directions. A plot of E versus r is shown in Figure 23.15. Notice that the magnitude of the field is continuous.

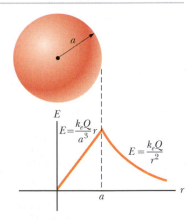

Figure 23.15 (Example 23.6) A plot of E versus r for a uniformly charged insulating sphere. The electric field inside the sphere ($r < a$) varies linearly with r. The field outside the sphere ($r > a$) is the same as that of a point charge Q located at $r = 0$.

Example 23.7 **A Cylindrically Symmetric Charge Distribution**

Find the electric field a distance r from a line of positive charge of infinite length and constant charge per unit length λ (Fig. 23.16a, page 628).

SOLUTION

Conceptualize The line of charge is *infinitely* long. Therefore, the field is the same at all points equidistant from the line, regardless of the vertical position of the point in Figure 23.16a. We expect the field to become weaker as we move farther away radially from the line of charge.

continued

23.7 continued

Categorize Because the charge is distributed uniformly along the line, the charge distribution has cylindrical symmetry and we can apply Gauss's law to find the electric field.

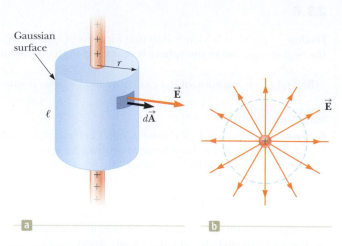

Analyze The symmetry of the charge distribution requires that $\vec{E}$ be perpendicular to the line charge and directed outward as shown in Figure 23.16b. To reflect the symmetry of the charge distribution, let's choose a cylindrical gaussian surface of radius r and length ℓ that is coaxial with the line charge. For the curved part of this surface, $\vec{E}$ is constant in magnitude and perpendicular to the surface at each point, satisfying conditions (1) and (2). Furthermore, the flux through the ends of the gaussian cylinder is zero because $\vec{E}$ is parallel to these surfaces. That is the first application we have seen of condition (3).

Figure 23.16 (Example 23.7) (a) An infinite line of charge surrounded by a cylindrical gaussian surface concentric with the line. (b) An end view shows that the electric field at the cylindrical surface is constant in magnitude and perpendicular to the surface.

We must take the surface integral in Gauss's law over the entire gaussian surface. Because $\vec{E} \cdot d\vec{A}$ is zero for the flat ends of the cylinder, however, we restrict our attention to only the curved surface of the cylinder.

Apply Gauss's law and conditions (1) and (2) for the curved surface, noting that the total charge inside our gaussian surface is $\lambda\ell$:

$$\Phi_E = \oint \vec{E} \cdot d\vec{A} = E \oint dA = EA = \frac{q_{in}}{\epsilon_0} = \frac{\lambda\ell}{\epsilon_0}$$

Substitute the area $A = 2\pi r\ell$ of the curved surface:

$$E(2\pi r\ell) = \frac{\lambda\ell}{\epsilon_0}$$

Solve for the magnitude of the electric field:

$$E = \frac{\lambda}{2\pi\epsilon_0 r} = 2k_e\frac{\lambda}{r} \qquad (23.8)$$

Finalize This result shows that the electric field due to a cylindrically symmetric charge distribution varies as $1/r$, whereas the field external to a spherically symmetric charge distribution varies as $1/r^2$. Equation 23.8 can also be derived by direct integration over the charge distribution. (See Problem 8.)

WHAT IF?   What if the line segment in this example were not infinitely long?

Answer If the line charge in this example were of finite length, the electric field would not be given by Equation 23.8. A finite line charge does not possess sufficient symmetry to make use of Gauss's law because the magnitude of the electric field is no longer constant over the surface of the gaussian cylinder: the field near the ends of the line would be different from that far from the ends. Therefore, condition (1) would not be satisfied in this situation. Furthermore, $\vec{E}$ is not perpendicular to the cylindrical surface at all points: the field vectors near the ends would have a component parallel to the line. Therefore, condition (2) would not be satisfied. For points close to a finite line charge and far from the ends, Equation 23.8 gives a good approximation of the value of the field.

It is left for you to show (see Problem 31) that the electric field inside a uniformly charged rod of finite radius and infinite length is proportional to r.

Example **23.8**   **A Plane of Charge**

Find the electric field due to an infinite plane of positive charge with uniform surface charge density σ.

SOLUTION

Conceptualize Notice that the plane of charge is *infinitely* large. Therefore, the electric field should be the same at all points equidistant from the plane. How would you expect the electric field to depend on the distance from the plane?

23.8 continued

Categorize Because the charge is distributed uniformly on the plane, the charge distribution is symmetric; hence, we can use Gauss's law to find the electric field.

Analyze By symmetry, $\vec{E}$ must be perpendicular to the plane at all points. The direction of $\vec{E}$ is away from positive charges, indicating that the direction of $\vec{E}$ on one side of the plane must be opposite its direction on the other side as shown in Figure 23.17. A gaussian surface that reflects the symmetry is a small cylinder whose axis is perpendicular to the plane and whose ends each have an area A and are equidistant from the plane. Because $\vec{E}$ is parallel to the curved surface of the cylinder—and therefore perpendicular to $d\vec{A}$ at all points on this surface—condition (3) is satisfied and there is no contribution to the surface integral from this surface. For the flat ends of the cylinder, conditions (1) and (2) are satisfied. The flux through each end of the cylinder is EA; hence, the total flux through the entire gaussian surface is just that through the ends, $\Phi_E = 2EA$.

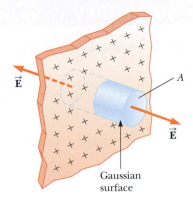

Figure 23.17 (Example 23.8) A cylindrical gaussian surface penetrating an infinite plane of charge. The flux is EA through each end of the gaussian surface and zero through its curved surface.

Write Gauss's law for this surface, noting that the enclosed charge is $q_{in} = \sigma A$:

$$\Phi_E = 2EA = \frac{q_{in}}{\epsilon_0} = \frac{\sigma A}{\epsilon_0}$$

Solve for E:

$$E = \frac{\sigma}{2\epsilon_0} \qquad (23.9)$$

Finalize Because the distance from each flat end of the cylinder to the plane does not appear in Equation 23.9, we conclude that $E = \sigma/2\epsilon_0$ at *any* distance from the plane. That is, the field is uniform everywhere. Notice that this is the same result as that obtained in Example 23.3, where we let the radius of a disk of charge become infinite. Figure 23.18 shows this uniform field due to an infinite plane of charge, seen edge-on.

WHAT IF? Suppose two infinite planes of charge are parallel to each other, one positively charged and the other negatively charged. The surface charge densities of both planes are of the same magnitude. What does the electric field look like in this situation?

Answer We first addressed this configuration in the **What If?** section of Example 23.3. The electric fields due to the two planes add in the region between the planes, resulting in a uniform field of magnitude σ/ϵ_0, and cancel elsewhere to give a field of zero. Figure 23.19 shows the field lines for such a configuration. This method is a practical way to achieve uniform electric fields with finite-sized planes placed close to each other.

Figure 23.18 (Example 23.8) The electric field lines due to an infinite plane of positive charge.

Figure 23.19 (Example 23.8) The electric field lines between two infinite planes of charge, one positive and one negative. In practice, the field lines near the edges of finite-sized sheets of charge will curve outward.

Conceptual Example 23.9 **Don't Use Gauss's Law Here!**

Explain why Gauss's law cannot be used to calculate the electric field near an electric dipole, a charged disk, or a triangle with a point charge at each corner.

SOLUTION

The charge distributions of all these configurations do not have sufficient symmetry to make the use of Gauss's law practical. We cannot find a closed surface surrounding any of these distributions for which all portions of the surface satisfy one or more of conditions (1) through (4) listed at the beginning of this section.

Summary

> ## Definitions

Electric flux is proportional to the number of electric field lines that penetrate a surface. If the electric field is uniform and makes an angle θ with the normal to a surface of area A, the electric flux through the surface is

$$\Phi_E = EA \cos \theta \tag{23.3}$$

In general, the electric flux through a surface is

$$\Phi_E \equiv \int_{\text{surface}} \vec{\mathbf{E}} \cdot d\vec{\mathbf{A}} \tag{23.4}$$

> ## Concepts and Principles

Gauss's law says that the net electric flux Φ_E through any closed gaussian surface is equal to the *net* charge q_{in} inside the surface divided by ϵ_0:

$$\Phi_E = \oint \vec{\mathbf{E}} \cdot d\vec{\mathbf{A}} = \frac{q_{\text{in}}}{\epsilon_0} \tag{23.7}$$

Using Gauss's law, you can calculate the electric field due to various symmetric charge distributions.

The electric field at some point due to a continuous charge distribution is

$$\vec{\mathbf{E}} = k_e \int \frac{dq}{r^2} \hat{\mathbf{r}} \tag{23.1}$$

where dq is the charge on one element of the charge distribution and r is the distance from the element to the point in question.

Think–Pair–Share

See the Preface for an explanation of the icons used in this problems set. For additional assessment items for this section, go to

1. An early (incorrect) model of the hydrogen atom, suggested by J. J. Thomson, proposed that a positive cloud of charge $+e$ was uniformly distributed throughout the volume of a sphere of radius a, with the electron (a particle of zero size and charge $-e$) at the center. Imagine the electron is displaced a small distance r from the center of the cloud of positive charge. Part (b) of Example 23.6 gives the magnitude of the electric field of the cloud at a distance r from the center. (a) Discuss in your group a method to show mathematically that the displaced electron would exhibit simple harmonic motion through the center of the cloud if released and carry out your method. (b) Find an expression for the frequency f of simple harmonic oscillations that an electron of mass m_e would make. (c) Calculate a numerical value for a that would result in a frequency of 2.47×10^{15} Hz, the frequency of the light radiated in the most intense line in the hydrogen spectrum. (d) Is this value consistent with estimated size of a hydrogen atom?

2. **ACTIVITY** Suppose you are in orbit around the Earth on the International Space Station. You have finished reading the books you brought and are looking for something to help you pass the time. You attach a tube to the water supply in your cabin and mount the open end of the tube in a fixed position in the air in the middle of your cabin. In the open end of the tube, you mount a small, spherical sponge that will cause the water coming out of the end of the tube to spread out with spherical symmetry in all directions. You turn the water on at a low volume flow rate I_V (see Section 14.7), so that the water exits the sponge at the open end and joins the water that has already left the end of the tube. Because you are in free-fall, you are in a reference frame in which there is no effective gravity, so the water collects at the end of the tube in an expanding sphere centered on the end of the tube. All of the water in the sphere is moving radially outward. The velocity of the water has a value at every point within the sphere, so the velocity can be represented as a vector field. (a) Show that the magnitude of the water velocity field falls off as $1/r^2$. (b) Imagine a nonspherical closed surface in the water and surrounding the end of the tube. Draw a diagram showing a two-dimensional version of the tube delivering the water, the outer surface of the sphere, the spherical surface in part (a), the nonspherical closed surface suggested here, and vectors $\vec{\mathbf{v}}$ and $d\vec{\mathbf{A}}$ at some point on the nonspherical surface, where is the velocity vector field, and $d\vec{\mathbf{A}}$ is a small area element on the nonspherical closed surface. (c) Show that

$$I_V = \oint \vec{\mathbf{v}} \cdot d\vec{\mathbf{A}}$$

where I_V is the flow rate of water coming from the end of the tube. (d) Discuss the similarities between this equation and Gauss's law. What is the analog to the electric field? What is the analog to the enclosed charge?

Problems

See the Preface for an explanation of the icons used in this problems set.
For additional assessment items for this section, go to **WEBASSIGN**
From Cengage

SECTION 23.1 Electric Field of a Continuous Charge Distribution

1. A negatively charged rod of finite length carries charge with a uniform charge per unit length. Sketch the electric field lines in a plane containing the rod.

2. A positively charged disk has a uniform charge per unit area σ as described in Example 23.3. Sketch the electric field lines in a plane perpendicular to the plane of the disk passing through its center.

3. **T** A uniformly charged ring of radius 10.0 cm has a total charge of 75.0 μC. Find the electric field on the axis of the ring at (a) 1.00 cm, (b) 5.00 cm, (c) 30.0 cm, and (d) 100 cm from the center of the ring.

4. **S** The electric field along the axis of a uniformly charged disk of radius R and total charge Q was calculated in Example 23.3. Show that the electric field at distances x that are large compared with R approaches that of a particle with charge $Q = \sigma\pi R^2$. *Suggestion:* First show that $x/(x^2 + R^2)^{1/2} = (1 + R^2/x^2)^{-1/2}$ and use the binomial expansion $(1 + \delta)^n \approx 1 + n\delta$, when $\delta \ll 1$.

5. **Q|C** Example 23.3 derives the exact expression for the electric field at a point on the axis of a uniformly charged disk. Consider a disk of radius $R = 3.00$ cm having a uniformly distributed charge of +5.20 μC. (a) Using the result of Example 23.3, compute the electric field at a point on the axis and 3.00 mm from the center. (b) **What If?** Explain how the answer to part (a) compares with the field computed from the near-field approximation $E = \sigma/2\epsilon_0$. (We derived this expression in Example 23.3.) (c) Using the result of Example 23.3, compute the electric field at a point on the axis and 30.0 cm from the center of the disk. (d) **What If?** Explain how the answer to part (c) compares with the electric field obtained by treating the disk as a +5.20-μC charged particle at a distance of 30.0 cm.

6. **Q|C S** A uniformly charged rod of length L and total charge Q lies along the x axis as shown in Figure P23.6. (a) Find the components of the electric field at the point P on the y axis a distance d from the origin. (b) What are the approximate values of the field components when $d \gg L$? Explain why you would expect these results.

Figure P23.6

7. **S V** A continuous line of charge lies along the x axis, extending from $x = +x_0$ to positive infinity. The line carries positive charge with a uniform linear charge density λ_0. What are (a) the magnitude and (b) the direction of the electric field at the origin?

8. **S** A thin rod of length ℓ and uniform charge per unit length λ lies along the x axis as shown in Figure P23.8. (a) Show that the electric field at P, a distance d from the rod along its perpendicular bisector, has no x component and is given by

$E = 2k_e\lambda \sin\theta_0/d$. (b) **What If?** Using your result to part (a), show that the field of a rod of infinite length is $E = 2k_e\lambda/d$.

Figure P23.8

9. **S** (a) Consider a uniformly charged, thin-walled, right circular cylindrical shell having total charge Q, radius R, and length ℓ. Determine the electric field at a point a distance d from the right side of the cylinder as shown in Figure P23.9. *Suggestion:* Use the result of Example 23.2 and treat the cylinder as a collection of ring charges. (b) **What If?** Consider now a solid cylinder with the same dimensions and carrying the same charge, uniformly distributed through its volume. Use the result of Example 23.3 to find the field it creates at the same point.

Figure P23.9

SECTION 23.2 Electric Flux

10. **V** A vertical electric field of magnitude 2.00×10^4 N/C exists above the Earth's surface on a day when a thunderstorm is brewing. A car with a rectangular size of 6.00 m by 3.00 m is traveling along a dry gravel roadway sloping downward at 10.0°. Determine the electric flux through the bottom of the car.

11. A flat surface of area 3.20 m² is rotated in a uniform electric field of magnitude $E = 6.20 \times 10^5$ N/C. Determine the electric flux through this area (a) when the electric field is perpendicular to the surface and (b) when the electric field is parallel to the surface.

12. **S** A nonuniform electric field is given by the expression

$$\vec{E} = ay\,\hat{i} + bz\,\hat{j} + cx\,\hat{k}$$

where a, b, and c are constants. Determine the electric flux through a rectangular surface in the xy plane, extending from $x = 0$ to $x = w$ and from $y = 0$ to $y = h$.

SECTION 23.3 Gauss's Law

13. An uncharged, nonconducting, hollow sphere of radius 10.0 cm surrounds a 10.0-μC charge located at the origin of a Cartesian coordinate system. A drill with a radius of 1.00 mm is aligned along the z axis, and a hole is drilled in the sphere. Calculate the electric flux through the hole.

14. Find the net electric flux through the spherical closed surface shown in Figure P23.14. The two charges on the right are inside the spherical surface.

+2.00 nC

+1.00 nC

−3.00 nC

Figure P23.14

15. Four closed surfaces, S_1 through S_4, together with the charges −2Q, Q, and −Q are sketched in Figure P23.15. (The colored lines are the intersections of the surfaces with the page.) Find the electric flux through each surface.

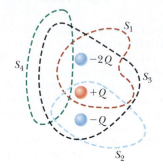

S_4

S_1

−2Q

S_3

+Q

−Q

S_2

Figure P23.15

16. A charge of 170 μC is at the center of a cube of edge 80.0 cm. No other charges are nearby. (a) Find the flux through each face of the cube. (b) Find the flux through the whole surface of the cube. (c) **What If?** Would your answers to either part (a) or part (b) change if the charge were not at the center? Explain.

17. (a) Find the net electric flux through the cube shown in Figure P23.17.(b) Can you use Gauss's law to find the electric field on the surface of this cube? Explain.

+3.00 nC

−8.00 nC

Figure P23.17

18. A particle with charge of 12.0 μC is placed at the center of a spherical shell of radius 22.0 cm. What is the total electric flux through (a) the surface of the shell and (b) any hemispherical surface of the shell? (c) Do the results depend on the radius? Explain.

19. A particle with charge Q = 5.00 μC is located at the center of a cube of edge L = 0.100 m. In addition, six other identical charged particles having q = −1.00 μC are positioned

symmetrically around Q as shown in Figure P23.19. Determine the electric flux through one face of the cube.

L

L

q

q

q

Q

q

q

q

L

Figure P23.19
Problems 19 and 20.

20. A particle with charge Q is located at the center of a cube of edge L. In addition, six other identical charged particles q are positioned symmetrically around Q as shown in Figure P23.19. For each of these particles, q is a negative number. Determine the electric flux through one face of the cube.

21. (a) A particle with charge q is located a distance d from an infinite plane. Determine the electric flux through the plane due to the charged particle. (b) **What If?** A particle with charge q is located a *very small* distance from the center of a *very large* square on the line perpendicular to the square and going through its center. Determine the approximate electric flux through the square due to the charged particle. (c) How do the answers to parts (a) and (b) compare? Explain.

22. Find the net electric flux through (a) the closed spherical surface in a uniform electric field shown in Figure P23.22a and (b) the closed cylindrical surface shown in Figure P23.22b. (c) What can you conclude about the charges, if any, inside the cylindrical surface?

$\vec{E}$

R

$\vec{E}$

2R

a

b

Figure P23.22

23. Figure P23.23 represents the top view of a cubic gaussian surface in a uniform electric field $\vec{E}$ oriented parallel to the top and bottom faces of the cube. The field makes an angle θ with side ①, and the area of each face is A. In symbolic form, find the electric flux through (a) face ①,

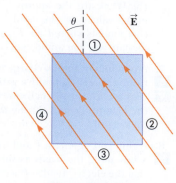

θ

$\vec{E}$

①

④

②

③

Figure P23.23

(b) face ②, (c) face ③, (d) face ④, and (e) the top and bottom faces of the cube. (f) What is the net electric flux through the cube? (g) How much charge is enclosed within the gaussian surface?

SECTION 23.4 Application of Gauss's Law to Various Charge Distributions

24. Determine the magnitude of the electric field at the surface of a lead-208 nucleus, which contains 82 protons and 126 neutrons. Assume the lead nucleus has a volume 208 times that of one proton and consider a proton to be a sphere of radius 1.20×10^{-15} m.

25. In nuclear fission, a nucleus of uranium-238, which contains 92 protons, can divide into two smaller spheres, each having 46 protons and a radius of 5.90×10^{-15} m. What is the magnitude of the repulsive electric force pushing the two spheres apart?

26. Suppose you fill two rubber balloons with air, suspend both of them from the same point, and let them hang down on strings of equal length. You then rub each with wool or on your hair so that the balloons hang apart with a noticeable separation between them. Make order-of-magnitude estimates of (a) the force on each, (b) the charge on each, (c) the field each creates at the center of the other, and (d) the total flux of electric field created by each balloon. In your solution, state the quantities you take as data and the values you measure or estimate for them.

27. A large, flat, horizontal sheet of charge has a charge per unit area of $9.00 \ \mu C/m^2$. Find the electric field just above the middle of the sheet.

28. A nonconducting wall carries charge with a uniform density of $8.60 \ \mu C/cm^2$. (a) What is the electric field 7.00 cm in front of the wall if 7.00 cm is small compared with the dimensions of the wall? (b) Does your result change as the distance from the wall varies? Explain.

29. A uniformly charged, straight filament 7.00 m in length has a total positive charge of $2.00 \ \mu C$. An uncharged cardboard cylinder 2.00 cm in length and 10.0 cm in radius surrounds the filament at its center, with the filament as the axis of the cylinder. Using reasonable approximations, find (a) the electric field at the surface of the cylinder and (b) the total electric flux through the cylinder.

30. You are working on a laboratory device that includes a small sphere with a large electric charge Q. Because of this charged sphere, there is a strong electric field surrounding your device. Other researchers in your laboratory are complaining that your electric field is affecting their equipment. You think about how you can obtain the large electric field that you need close to the sphere but prohibit the field from reaching your colleagues. You decide to surround your device with a spherical transparent plastic shell. The nonconducting shell is given a uniform charge distribution. (a) The shell is placed so that the small sphere is at the exact center of the shell. Determine the charge that must be placed on the shell to completely eliminate the electric field outside of the shell. (b) What if the shell moves? Does the small sphere have to be at the center of the shell for this scheme to work?

31. Consider a long, cylindrical charge distribution of radius R with a uniform charge density ρ. Find the electric field at distance r from the axis, where $r < R$.

32. Assume the magnitude of the electric field on each face of the cube of edge $L = 1.00$ m in Figure P23.32 is uniform and the directions of the fields on each face are as indicated. Find (a) the net electric flux through the cube and (b) the net charge inside the cube. (c) Could the net charge be a single point charge?

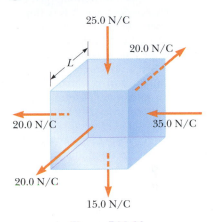

Figure P23.32

33. A solid sphere of radius 40.0 cm has a total positive charge of $26.0 \ \mu C$ uniformly distributed throughout its volume. Calculate the magnitude of the electric field (a) 0 cm, (b) 10.0 cm, (c) 40.0 cm, and (d) 60.0 cm from the center of the sphere.

34. A cylindrical shell of radius 7.00 cm and length 2.40 m has its charge uniformly distributed on its curved surface. The magnitude of the electric field at a point 19.0 cm radially outward from its axis (measured from the midpoint of the shell) is 36.0 kN/C. Find (a) the net charge on the shell and (b) the electric field at a point 4.00 cm from the axis, measured radially outward from the midpoint of the shell.

35. You are working for the summer at a research laboratory. Your research director has devised a scheme for holding small charged particles at fixed positions. The scheme is shown in Figure P23.35. A large insulating sphere of radius a carries a total positive charge Q with a uniform volume charge density. A very thin tunnel is drilled through a diameter of the sphere and two small spheres with charge q are placed in the tunnel. These spheres are represented by the blue dots in the figure. They find equilibrium positions at a distance of r on either side of the center of the sphere. Your research director has had great success with this scheme. (a) Determine the specific value of r at which equilibrium exists. (b) Your research director asks

Figure P23.35

you to see if he can extend the system as follows. Determine if it is possible to add transparent plastic tubes as extensions of the tunnel and have the small spheres be in equilibrium at a position for which $r > a$.

36. You are working for the summer at a research laboratory. [CR] Your research director has devised a scheme for holding small charged particles at fixed positions. The scheme is shown in Figure P23.36. An insulating cylinder of radius a and length $L \gg a$ is positively charged and carries a uniform volume charge density ρ. A very thin tunnel is drilled through a diameter of the cylinder and two small spheres with charge q are placed in the tunnel. These spheres are represented by the blue dots in the figure. They find equilibrium positions at a distance of r on opposite sides of the axis of the cylinder. Your research director has had great success with this scheme. (a) Determine the specific value of r at which equilibrium exists. (b) Your research director asks you see if he can extend the system as follows. Determine if it is possible to add transparent plastic tubes as extensions of the tunnel and have the small spheres be in equilibrium at a position for which $r > a$.

Figure P23.36

ADDITIONAL PROBLEMS

37. Find the electric flux through the plane surface shown in Figure P23.37 if $\theta = 60.0°$, $E = 350$ N/C, and $d = 5.00$ cm. The electric field is uniform over the entire area of the surface.

Figure P23.37

38. Three solid plastic cylinders all have radius 2.50 cm and length 6.00 cm. Find the charge of each cylinder given the following additional information about each one. Cylinder (a) carries charge with uniform density 15.0 nC/m² everywhere on its surface. Cylinder (b) carries charge with uniform density 15.0 nC/m² on its curved lateral surface only. Cylinder (c) carries charge with uniform density 500 nC/m³ throughout the plastic.

39. A line of charge starts at $x = +x_0$ and extends to positive [S] infinity. The linear charge density is $\lambda = \lambda_0 x_0/x$, where λ_0 is a constant. Determine the electric field at the origin.

40. Show that the maximum magnitude E_{max} of the electric field [S] along the axis of a uniformly charged ring occurs at $x = a/\sqrt{2}$ (see Fig. 23.3) and has the value $Q/(6\sqrt{3}\pi\epsilon_0 a^2)$.

41. A line of positive charge is formed into a semicircle of radius $R = 60.0$ cm as shown in Figure P23.41. The charge per unit length along the semicircle is given by the expression $\lambda = \lambda_0 \cos\theta$. The total charge on the semicircle is 12.0 μC. Calculate the total force on a charge of 3.00 μC placed at the center of curvature P.

Figure P23.41

42. A very large conducting plate lying in the xy plane carries a [S] charge per unit area of σ. A second such plate located above the first plate at $z = z_0$ and oriented parallel to the xy plane carries a charge per unit area of -2σ. Find the electric field for (a) $z < 0$, (b) $0 < z < z_0$, and (c) $z > z_0$.

43. A sphere of radius $R = 1.00$ m surrounds a particle with charge $Q = 50.0$ μC located at its center as shown in Figure P23.43. Find the electric flux through a circular cap of half-angle $\theta = 45.0°$.

44. A sphere of radius R surrounds a [S] particle with charge Q located at its center as shown in Figure P23.43. Find the electric flux through a circular cap of half-angle θ.

Figure P23.43
Problems 43 and 44.

CHALLENGE PROBLEMS

45. A slab of insulating material has a [S] nonuniform positive charge density $\rho = Cx^2$, where x is measured from the center of the slab as shown in Figure P23.45 and C is a constant. The slab is infinite in the y and z directions. Derive expressions for the electric field in (a) the exterior regions ($|x| > d/2$) and (b) the interior region of the slab ($-d/2 < x < d/2$).

Figure P23.45
Problems 45 and 49.

46. A sphere of radius $2a$ is made of [S] a nonconducting material that has a uniform volume charge density ρ. Assume the material does not affect the electric field. A spherical cavity of radius a is now removed from the sphere as shown in Figure P23.46. Show that the electric field within the cavity is uniform and is given by $E_x = 0$ and $E_y = \rho a/3\epsilon_0$.

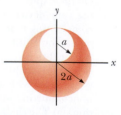

Figure P23.46

47. An infinitely long insulating cylinder of radius R has [S] a volume charge density that varies with the radius as

$$\rho = \rho_0\left(a - \frac{r}{b}\right)$$

where ρ_0, a, and b are positive constants and r is the distance from the axis of the cylinder. Use Gauss's law to determine the magnitude of the electric field at radial distances (a) $r < R$ and (b) $r > R$.

48. A particle with charge Q is located
S on the axis of a circle of radius R at a distance b from the plane of the circle (Fig. P23.48). Show that if one-fourth of the electric flux from the charge passes through the circle, then $R = \sqrt{3}b$.

Figure P23.48

49. Review. A slab of insulating material
S (infinite in the y and z directions) has a thickness d and a uniform positive charge density ρ. An edge view of the slab is shown in Figure P23.45. (a) Show that the magnitude of the electric field a distance x from its center and inside the slab is $E = \rho x/\epsilon_0$. (b) **What If?** Suppose an electron of charge $-e$ and mass m_e can move freely within the slab. It is released from rest at a distance x from the center. Show that the electron exhibits simple harmonic motion with a frequency

$$f = \frac{1}{2\pi} \sqrt{\frac{\rho e}{m_e \epsilon_0}}$$

50. Identical thin rods of length $2a$ carry equal charges $+Q$
S uniformly distributed along their lengths. The rods lie along the x axis with their centers separated by a distance $b > 2a$ (Fig. P23.50). Show that the magnitude of the force exerted by the left rod on the right one is

$$F = \left(\frac{k_e Q^2}{4a^2}\right) \ln \left(\frac{b^2}{b^2 - 4a^2}\right)$$

Figure P23.50

51. A solid insulating sphere of radius R has a nonuniform
S charge density that varies with r according to the expression $\rho = Ar^2$, where A is a constant and $r < R$ is measured from the center of the sphere. (a) Show that the magnitude of the electric field outside $(r > R)$ the sphere is $E = AR^5/5\epsilon_0 r^2$. (b) Show that the magnitude of the electric field inside $(r < R)$ the sphere is $E = Ar^3/5\epsilon_0$. *Note:* The volume element dV for a spherical shell of radius r and thickness dr is equal to $4\pi r^2 dr$.

24 Electric Potential

Processes occurring during thunderstorms cause large differences in electric potential between a thundercloud and the ground. The result of this potential difference is an electrical discharge that we call lightning, such as this display. Notice at the left that a downward channel of lightning (a *stepped leader*) is about to make contact with a channel coming up from the ground (a *return stroke*). (*Costazzurra/Shutterstock*)

STORYLINE **You are still in Florida during your spring break. You** have visited some open areas and have observed several lightning flashes, as described in the storyline in the previous chapter. You captured a couple of flashes on your smartphone video and are watching them with fascination. That night, in your hotel room, you do online research on your smartphone and find that when lightning occurs, there is a potential difference of hundreds of thousands of volts between a cloud and the ground. You are not familiar with the phrase *potential difference* and wonder what that means. You have heard about *volts* because you know your electrical devices at home are rated at 120 volts and your electric clothes dryer is rated at 240 volts. Your smartphone charger plug will operate at either 120 volts or 240 volts. But what exactly *is* a volt?

CONNECTIONS In Chapter 22, we linked our new study of electromagnetism to our earlier studies of *force*. In this chapter, we make a link between electromagnetism and our earlier investigations into *energy*. The concept of potential energy was introduced in Chapter 7 in connection with such conservative forces as the gravitational force and the elastic force exerted by a spring. By using the law of conservation of energy, we could solve various problems in mechanics that were not solvable with an approach using forces. Because the electrostatic force is conservative, electrostatic phenomena can be conveniently described in terms of an *electric potential energy*, which is of great value in the study of electricity. This idea enables us to define a related quantity known as *electric potential*. Because the electric potential at any point in an electric field is a scalar quantity, we can use it to describe electrostatic phenomena more simply than if we were to rely only on vector quantities such as the electric field and electric forces. The concept of electric potential is of great practical value in the operation of electric circuits and devices that we will study in later chapters.

24.1 Electric Potential and Potential Difference

When a charge q is placed in an electric field $\vec{\mathbf{E}}$ created by some source charge distribution, the particle in a field model tells us that there is an electric force $q\vec{\mathbf{E}}$ acting on the charge. This force is conservative because the force between charges described by Coulomb's law is conservative. Let us identify the charge and the field as a system. If the charge is free to move, it will do so in response to the electric force. Therefore, the electric field will be doing work on the charge. This work is *internal* to the system. This situation is similar to that in a gravitational system: When an object is released near the surface of the Earth, the gravitational force does work on the object. This work is internal to the object–Earth system as discussed in Section 7.8.

When analyzing electric and magnetic fields, it is common practice to use the notation $d\vec{\mathbf{s}}$ to represent an infinitesimal displacement vector that is oriented tangent to a path through space. This path may be straight or curved, and an integral performed along this path is called either a *path integral* or a *line integral* (the two terms are synonymous).

For an infinitesimal displacement $d\vec{\mathbf{s}}$ of a point charge q immersed in an electric field, the work done within the charge–field system by the electric field on the charge is $W_{\text{int}} = \vec{\mathbf{F}}_e \cdot d\vec{\mathbf{s}} = q\vec{\mathbf{E}} \cdot d\vec{\mathbf{s}}$ Recall from Equation 7.26 that internal work done in a gravitational system is equal to the negative of the change in the gravitational potential energy of the system: $W_{\text{int}} = -\Delta U_g$. Because internal work is done when a charge is moved in an electric field, we can identify an **electric potential energy** U_E for the charge–field system, where $W_{\text{int}} = -\Delta U_E$. From Equation 7.26, we see that, as the charge q is displaced, the electric potential energy of the charge–field system is changed by an amount $dU_E = -W_{\text{int}} = -q\vec{\mathbf{E}} \cdot d\vec{\mathbf{s}}$. For a finite displacement of the charge from some point Ⓐ in space to some other point Ⓑ, the change in electric potential energy of the system is

$$\Delta U_E = -q \int_{Ⓐ}^{Ⓑ} \vec{\mathbf{E}} \cdot d\vec{\mathbf{s}} \qquad (24.1)$$

◀ Change in electric potential energy of a system

The integration is performed along the path that q follows as it moves from Ⓐ to Ⓑ. Because the force $q\vec{\mathbf{E}}$ is conservative, this line integral does not depend on the path taken from Ⓐ to Ⓑ.

For a given position of the charge in the field, the charge–field system has a potential energy U_E relative to the configuration of the system that is defined as $U_E = 0$. Dividing the potential energy by the charge gives a physical quantity that depends only on the source charge distribution and has a value at every point in an electric field. This quantity is called the **electric potential** (or simply the **potential**) V:

$$V = \frac{U_E}{q} \qquad (24.2)$$

Because potential energy is a scalar quantity, electric potential also is a scalar quantity.

The **potential difference** $\Delta V = V_Ⓑ - V_Ⓐ$ between two points Ⓐ and Ⓑ in an electric field is defined as the change in electric potential energy of the system when a charge q is moved between the points (Eq. 24.1) divided by the charge:

$$\Delta V \equiv \frac{\Delta U_E}{q} = -\int_{Ⓐ}^{Ⓑ} \vec{\mathbf{E}} \cdot d\vec{\mathbf{s}} \qquad (24.3)$$

◀ Potential difference between two points

PITFALL PREVENTION 24.1
Potential and Potential Energy
The *potential is characteristic of the field only*, independent of a charged particle that may be placed in the field. *Potential energy is characteristic of the charge-field system* due to an interaction between the field and a charged particle placed in the field.

In this definition, the infinitesimal displacement $d\vec{\mathbf{s}}$ is interpreted as the displacement between two points in space rather than the displacement of a point charge as in Equation 24.1.

Just as with potential energy, only *differences* in electric potential are meaningful. We often take the value of the electric potential to be zero at some convenient point in an electric field.

Potential difference should not be confused with difference in potential energy. The potential *difference* between Ⓐ and Ⓑ exists solely because of a source charge and depends on the source charge distribution (consider points Ⓐ and Ⓑ in the discussion above *without* the presence of the charge q). For a potential *energy* to exist, we must have a system of two or more charges. The potential energy belongs to the system and changes only if a charge is moved relative to the rest of the system. This situation is similar to that for the electric field. An electric *field* exists solely because of a source charge. An electric *force* requires two charges: the source charge to set up the field and another charge placed within that field.

Let's now consider the situation in which an external agent moves the charge in the field. If the agent moves the charge from Ⓐ to Ⓑ without changing the kinetic energy of the charge, the agent performs work that changes the potential energy of the system: $W = \Delta U_E$. From Equation 24.3, the work done by an external agent in moving a charge q through an electric field at constant velocity is

$$W = q\,\Delta V \qquad (24.4)$$

Because electric potential is a measure of potential energy per unit charge, the SI unit of both electric potential and potential difference is joules per coulomb, which is defined as a **volt** (V):

$$1\text{ V} \equiv 1\text{ J/C}$$

That is, as we can see from Equation 24.4, 1 J of work must be done to move a 1-C charge through a potential difference of 1 V.

Equation 24.3 shows that potential difference also has units of electric field times distance. It follows that the SI unit of electric field (N/C) can also be expressed in volts per meter:

$$1\text{ N/C} = 1\text{ V/m}$$

Therefore, we can state a new interpretation of the electric field:

> The electric field is a measure of the rate of change of the electric potential with respect to position.

A unit of energy commonly used in atomic and nuclear physics is the **electron volt** (eV), which is defined as the energy a charge–field system gains or loses when a charge of magnitude e (that is, an electron or a proton) is moved through a potential difference of 1 V. We can find the relation between electron volts and joules by imagining that 1 eV of work is done in Equation 24.4 and using Equation 22.5 for e:

$$1\text{ eV} = (1.602\,18 \times 10^{-19}\text{ C})(1\text{ V}) = 1.602\,18 \times 10^{-19}\text{ J} \qquad (24.5)$$

For instance, an electron in the beam of a typical dental x-ray machine may have a speed of 1.4×10^8 m/s. This speed corresponds to a kinetic energy 1.1×10^{-14} J (using relativistic calculations as discussed in Chapter 38), which is equivalent to 6.7×10^4 eV. Such an electron has to be accelerated from rest through a potential difference of 67 kV to reach this speed.

QUICK QUIZ **24.1** In Figure 24.1, two points Ⓐ and Ⓑ are located within a region in which there is an electric field. **(i)** How would you describe the potential difference $\Delta V = V_{\text{Ⓑ}} - V_{\text{Ⓐ}}$? (a) It is positive. (b) It is negative. (c) It is zero. **(ii)** A negative charge is placed at Ⓐ and then moved to Ⓑ. How would you describe the change in potential energy of the charge–field system for this process? Choose from the same possibilities.

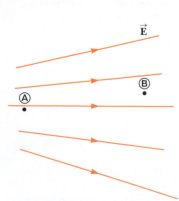

Figure 24.1 (Quick Quiz 24.1) Two points in an electric field.

The definition of the volt above answers in some ways our question in the opening storyline. But you may still not be completely comfortable with the concept of electric potential. One reason for this discomfort is that, despite the many *similarities* between the gravitational force and the electric force, we do not define a gravitational potential: $V_{grav} = U_g/m$, with a unit of J/kg. We do not do that because there is no benefit to it. Defining a gravitational potential does not allow us to solve any more gravitational problems. One major *difference* between gravity and electricity makes the definition of electric potential very beneficial: we can change the shape of electric situations and even build electric circuits with different kinds of circuit elements. Nothing like that is possible for gravity. We will find ourselves using electric potential continuously in our discussions of electric circuits.

24.2 Potential Difference in a Uniform Electric Field

Equations 24.1 and 24.3 hold in all electric fields, whether uniform or varying, but they can be simplified for the special case of a uniform field. First, consider a uniform electric field directed along the negative y axis as shown in Figure 24.2a. Let's calculate the potential difference between two points Ⓐ and Ⓑ separated by a distance d, where the displacement $\vec{s}$ points from Ⓐ toward Ⓑ and is parallel to the field lines. Equation 24.3 gives

$$V_{Ⓑ} - V_{Ⓐ} = \Delta V = -\int_{Ⓐ}^{Ⓑ} \vec{E} \cdot d\vec{s} = -\int_{Ⓐ}^{Ⓑ} E\, ds(\cos 0°) = -\int_{Ⓐ}^{Ⓑ} E\, ds$$

Because E is constant, it can be removed from the integral sign, which gives

$$\Delta V = -E \int_{Ⓐ}^{Ⓑ} ds$$

$$\Delta V = -Ed \qquad (24.6)$$

◄ Potential difference between two points in a uniform electric field

The negative sign indicates that the electric potential at point Ⓑ is lower than at point Ⓐ; that is, $V_{Ⓑ} < V_{Ⓐ}$. Electric field lines *always* point in the direction of decreasing electric potential as shown in Figure 24.2a.

Now suppose a charge q moves from Ⓐ to Ⓑ. We can calculate the change in the potential energy of the charge–field system from Equations 24.3 and 24.6:

$$\Delta U_E = q\Delta V = -qEd \qquad (24.7)$$

This result shows that if q is positive, then ΔU_E is negative. Therefore, in a system consisting of a positive charge and an electric field, the electric potential energy

PITFALL PREVENTION 24.4

The Sign of ΔV The negative sign in Equation 24.6 is due to the fact that we started at point Ⓐ and moved to a new point in the *same* direction as the electric field lines. If we started from Ⓑ and moved to Ⓐ, the potential difference would be $+Ed$. In a uniform electric field, the magnitude of the potential difference is Ed and the sign can be determined by the direction of travel.

When a positive charge moves from point Ⓐ to point Ⓑ, the electric potential energy of the charge–field system decreases.

When an object with mass moves from point Ⓐ to point Ⓑ, the gravitational potential energy of the object–field system decreases.

$\vec{E}$

a

$\vec{g}$

b

Figure 24.2 (a) When the electric field $\vec{E}$ is directed downward, point Ⓑ is at a lower electric potential than point Ⓐ. (b) A gravitational analog to the situation in (a).

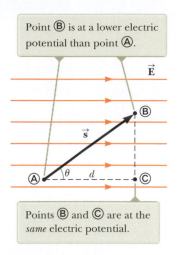

Point Ⓑ is at a lower electric potential than point Ⓐ.

Points Ⓑ and Ⓒ are at the *same* electric potential.

Figure 24.3 A uniform electric field directed along the positive *x* axis. Three points in the electric field are labeled.

of the system decreases when the charge moves in the direction of the field. If a positive charge is released from rest in this electric field, it experiences an electric force $q\vec{\mathbf{E}}$ in the direction of $\vec{\mathbf{E}}$ (downward in Fig. 24.2a). Therefore, it accelerates downward, gaining kinetic energy. As the charged particle gains kinetic energy, the electric potential energy of the charge–field system decreases by an equal amount. This equivalence should not be surprising; it is simply conservation of mechanical energy in an isolated system as introduced in Chapter 8.

Figure 24.2b shows an analogous situation with a gravitational field. When a particle with mass *m* is released in a gravitational field, it accelerates downward, gaining kinetic energy. At the same time, the gravitational potential energy of the object–field system decreases.

The comparison between a system of a positive charge residing in an electrical field and an object with mass residing in a gravitational field in Figure 24.2 is useful for conceptualizing electrical behavior. The electrical situation, however, has one feature that the gravitational situation does not: the charge can be negative. If *q* is negative, then ΔU_E in Equation 24.7 is positive and the situation is reversed. A system consisting of a negative charge and an electric field *gains* electric potential energy when the charge moves in the direction of the field. If a negative charge is released from rest in an electric field, it accelerates in a direction *opposite* the direction of the field. For the negative charge to move in the direction of the field, an external agent must apply a force and do positive work on the charge.

Now consider the more general case of a charged particle that moves between Ⓐ and Ⓑ in a uniform electric field such that the vector $\vec{\mathbf{s}}$ is *not* parallel to the field lines as shown in Figure 24.3. In this case, Equation 24.3 gives

◄ Change in potential between two points in a uniform electric field

$$\Delta V = -\int_{Ⓐ}^{Ⓑ} \vec{\mathbf{E}} \cdot d\vec{\mathbf{s}} = -\vec{\mathbf{E}} \cdot \int_{Ⓐ}^{Ⓑ} d\vec{\mathbf{s}} = -\vec{\mathbf{E}} \cdot \vec{\mathbf{s}} \qquad (24.8)$$

where again $\vec{\mathbf{E}}$ was removed from the integral because it is constant. The change in potential energy of the charge–field system is

$$\Delta U_E = q\Delta V = -q\vec{\mathbf{E}} \cdot \vec{\mathbf{s}} \qquad (24.9)$$

Finally, we conclude from Equation 24.8 that all points in a plane perpendicular to a uniform electric field are at the same electric potential. We can see that in Figure 24.3, where the potential difference $V_Ⓑ - V_Ⓐ$ is equal to the potential difference $V_Ⓒ - V_Ⓐ$. (Prove this fact to yourself by working out two dot products for $\vec{\mathbf{E}} \cdot \vec{\mathbf{s}}$: one for $\vec{\mathbf{s}}_{Ⓐ \to Ⓑ}$, where the angle θ between $\vec{\mathbf{E}}$ and $\vec{\mathbf{s}}$ is arbitrary as shown in Figure 24.3, and one for $\vec{\mathbf{s}}_{Ⓐ \to Ⓒ}$, where $\theta = 0$.) Therefore, $V_Ⓑ = V_Ⓒ$. The name **equipotential surface** is given to any surface consisting of a continuous distribution of points having the same electric potential.

The equipotential surfaces associated with a uniform electric field consist of a family of parallel planes that are all perpendicular to the field. Equipotential surfaces associated with fields having other symmetries are described in later sections.

Figure 24.4 (Quick Quiz 24.2) Four equipotential surfaces.

ⓆUICK QUIZ **24.2** The labeled points in Figure 24.4 are on a series of equipotential surfaces associated with an electric field. Rank (from greatest to least) the work done by the electric field on a positively charged particle that moves from Ⓐ to Ⓑ, from Ⓑ to Ⓒ, from Ⓒ to Ⓓ, and from Ⓓ to Ⓔ.

Example 24.1   **The Electric Field Between Two Parallel Plates of Opposite Charge**

A battery has a specified potential difference ΔV between its terminals and establishes that potential difference between conductors attached to the terminals. A 12-V battery is connected between two parallel plates as shown in Figure 24.5. The separation between the plates is $d = 0.30$ cm, and we assume the electric field between the plates to be uniform. (This assumption is reasonable if the plate separation is small relative to the plate dimensions and we do not consider locations near the plate edges.) Find the magnitude of the electric field between the plates.

24.1 continued

Figure 24.5 (Example 24.1) A 12-V battery connected to two parallel plates. The electric field between the plates has a magnitude given by the potential difference ΔV divided by the plate separation d.

$\Delta V = 12$ V

SOLUTION

Conceptualize In Example 23.8, we illustrated the uniform electric field between parallel plates. The new feature to this problem is that the electric field is related to the new concept of electric potential.

Categorize The electric field is evaluated from a relationship between field and potential given in this section, so we categorize this example as a substitution problem.

Use Equation 24.6 to evaluate the magnitude of the electric field between the plates:

$$E = \frac{|V_B - V_A|}{d} = \frac{12 \text{ V}}{0.30 \times 10^{-2} \text{ m}} = 4.0 \times 10^3 \text{ V/m}$$

The configuration of plates in Figure 24.5 is called a *parallel-plate capacitor* and is examined in greater detail in Chapter 25.

Example 24.2 **Motion of a Proton in a Uniform Electric Field**

A proton is released from rest at point Ⓐ in a uniform electric field that has a magnitude of 8.0×10^4 V/m (Fig. 24.6). The proton undergoes a displacement of magnitude $d = 0.50$ m to point Ⓑ in the direction of $\vec{E}$. Find the speed of the proton after completing the displacement.

SOLUTION

Conceptualize Visualize the proton in Figure 24.6 moving downward through the potential difference. The situation is analogous to an object falling through a gravitational field. Also compare this example to Example 22.7 where a positive charge was moving in a uniform electric field. In that example, we applied the particle under constant acceleration and nonisolated system models. Now that we have investigated electric potential energy, what model can we use here?

Categorize The system of the proton and the two plates in Figure 24.6 does not interact with the environment, so we model it as an *isolated system* for *energy*.

Figure 24.6 (Example 24.2) A proton accelerates from Ⓐ to Ⓑ in the direction of the electric field.

Analyze

Write the appropriate reduction of Equation 8.2, the conservation of energy equation, for the isolated system of the charge and the electric field:

$$\Delta K + \Delta U_E = 0$$

Substitute the changes in energy for both terms:

$$\left(\tfrac{1}{2}mv^2 - 0\right) + e\,\Delta V = 0$$

Solve for the final speed of the proton and substitute for ΔV from Equation 24.6:

$$v = \sqrt{\frac{-2e\,\Delta V}{m}} = \sqrt{\frac{-2e(-Ed)}{m}} = \sqrt{\frac{2eEd}{m}}$$

Substitute numerical values:

$$v = \sqrt{\frac{2(1.6 \times 10^{-19} \text{ C})(8.0 \times 10^4 \text{ V})(0.50 \text{ m})}{1.67 \times 10^{-27} \text{ kg}}}$$

$$= 2.8 \times 10^6 \text{ m/s}$$

Finalize Because ΔV is negative for the field, ΔU_E is also negative for the proton–field system. The negative value of ΔU_E means the potential energy of the system decreases as the proton moves in the direction of the electric field. As the proton accelerates in the direction of the field, it gains kinetic energy while the electric potential energy of the system decreases at the same time.

continued

24.2 continued

Figure 24.6 is oriented so that the proton moves downward. The proton's motion is analogous to that of an object falling in a gravitational field. Although the gravitational field is always downward at the surface of the Earth, an electric field can be in any direction, depending on the orientation of the plates creating the field. Therefore, Figure 24.6 could be rotated 90° or 180° and the proton could move horizontally or upward in the electric field!

24.3 Electric Potential and Potential Energy Due to Point Charges

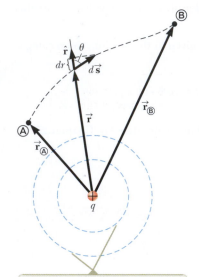

Figure 24.7 The potential difference between points Ⓐ and Ⓑ due to a point charge q depends *only* on the initial and final radial coordinates $r_{Ⓐ}$ and $r_{Ⓑ}$.

The two dashed circles represent intersections of spherical equipotential surfaces with the page.

As discussed in Section 22.4, an isolated positive point charge q produces an electric field directed radially outward from the charge. To find the electric potential at a point located a distance r from the charge, let's begin with the general expression for potential difference, Equation 24.3,

$$V_{Ⓑ} - V_{Ⓐ} = -\int_{Ⓐ}^{Ⓑ} \vec{E} \cdot d\vec{s}$$

where Ⓐ and Ⓑ are the two arbitrary points shown in Figure 24.7. At any point in space, the electric field due to the point charge is $\vec{E} = (k_e q/r^2)\hat{r}$ (Eq. 22.9), where $\hat{r}$ is a unit vector directed radially outward from the charge. Therefore, the quantity $\vec{E} \cdot d\vec{s}$ can be expressed as

$$\vec{E} \cdot d\vec{s} = k_e \frac{q}{r^2} \hat{r} \cdot d\vec{s}$$

Because the magnitude of $\hat{r}$ is 1, the dot product $\hat{r} \cdot d\vec{s} = ds \cos\theta$, where θ is the angle between $\hat{r}$ and $d\vec{s}$. Furthermore, $ds \cos\theta$ is the projection of $d\vec{s}$ onto $\hat{r}$; therefore, $ds \cos\theta = dr$. That is, any displacement $d\vec{s}$ along the path from point Ⓐ to point Ⓑ produces a change dr in the magnitude of $\vec{r}$, the position vector of the point relative to the charge creating the field. Making these substitutions, we find that $\vec{E} \cdot d\vec{s} = (k_e q/r^2) dr$; hence, the expression for the potential difference becomes

$$V_{Ⓑ} - V_{Ⓐ} = -k_e q \int_{r_{Ⓐ}}^{r_{Ⓑ}} \frac{dr}{r^2} = k_e \frac{q}{r} \Big|_{r_{Ⓐ}}^{r_{Ⓑ}}$$

$$V_{Ⓑ} - V_{Ⓐ} = k_e q \left[\frac{1}{r_{Ⓑ}} - \frac{1}{r_{Ⓐ}} \right] \tag{24.10}$$

PITFALL PREVENTION 24.5

Similar Equation Warning Do not confuse Equation 24.11 for the electric potential of a point charge with Equation 22.9 for the electric field of a point charge. Potential is proportional to $1/r$, whereas the magnitude of the field is proportional to $1/r^2$. The effect of a charge on the space surrounding it can be described in two ways. The charge sets up a vector electric field $\vec{E}$, which is related to the force experienced by a charge placed in the field. It also sets up a scalar potential V, which is related to the potential energy of the two-charge system when a charge is placed in the field.

Equation 24.10 shows us that the integral of $\vec{E} \cdot d\vec{s}$ is *independent* of the path between points Ⓐ and Ⓑ. Multiplying by a charge q_0 that moves between points Ⓐ and Ⓑ, we see that the integral of $q_0\vec{E} \cdot d\vec{s}$ is also independent of path. This latter integral, which is the work done by the electric force on the charge q_0, shows that the electric force is conservative (see Section 7.7). We define a field that is related to a conservative force as a **conservative field.** Therefore, Equation 24.10 tells us that the electric field of a fixed point charge q is conservative. Furthermore, Equation 24.10 expresses the important result that the potential difference between any two points Ⓐ and Ⓑ in a field created by a point charge depends only on the radial coordinates $r_{Ⓐ}$ and $r_{Ⓑ}$. It is customary to choose the reference of electric potential for a point charge to be $V_{Ⓐ} = 0$ at $r_{Ⓐ} = \infty$. With this reference choice, the electric potential due to a point charge at any distance r from the charge is

$$V = k_e \frac{q}{r} \tag{24.11}$$

We obtain the electric potential resulting from two or more point charges by applying the superposition principle. That is, the total electric potential at some

Figure 24.8 (a) Charge q_1 establishes an electric potential V_1 at point P. (b) Charge q_2 is brought from infinity to point P.

point P due to several point charges is the sum of the potentials due to the individual charges. For a group of point charges, we can write the total electric potential at P as

$$V = k_e \sum_i \frac{q_i}{r_i} \qquad (24.12)$$

◄ Electric potential due to several point charges

Figure 24.8a shows a charge q_1, which sets up an electric field throughout space. The charge also establishes an electric potential at all points, including point P, where the electric potential is V_1. Now imagine that an external agent brings a charge q_2 from infinity to point P. The work that must be done to do this is given by Equation 24.4, $W = q_2 \Delta V$. This work represents a transfer of energy across the boundary of the two-charge system, and the energy appears in the system as potential energy U_E when the particles are separated by a distance r_{12} as in Figure 24.8b. From Equation 8.2, we have $W = \Delta U_E$. Therefore, the electric potential energy of a pair of point charges[1] can be found as follows:

$$\Delta U_E = W = q_2 \Delta V \quad \rightarrow \quad U_E - 0 = q_2 \left(k_e \frac{q_1}{r_{12}} - 0 \right)$$

$$U_E = k_e \frac{q_1 q_2}{r_{12}} \qquad (24.13)$$

If the charges are of the same sign, then U_E is positive. Positive work must be done by an external agent on the system to bring the two charges near each other (because charges of the same sign repel). If the charges are of opposite sign, as in Figure 24.8b, then U_E is negative. Negative work is done by an external agent against the attractive force between the charges of opposite sign as they are brought near each other; a force must be applied opposite the displacement to prevent q_2 from accelerating toward q_1.

If the system consists of more than two charged particles, we can obtain the total potential energy of the system by calculating U_E for every *pair* of charges and summing the terms algebraically. For example, the total potential energy of the system of three charges shown in Figure 24.9 is

$$U_E = k_e \left(\frac{q_1 q_2}{r_{12}} + \frac{q_1 q_3}{r_{13}} + \frac{q_2 q_3}{r_{23}} \right) \qquad (24.14)$$

Physically, this result can be interpreted as follows. Begin with all three charges infinitely far apart. Now bring charge q_1 to its position in Figure 24.9. No work is required for an external agent to do this because $V = 0$ due to other charges and there are no other charges in the vicinity. Now, the work that the agent must do

Figure 24.9 Three point charges are fixed at the positions shown.

---

[1] The expression for the electric potential energy of a system made up of two point charges, Equation 24.13, is of the *same* form as the equation for the gravitational potential energy of a system made up of two point masses, $-Gm_1 m_2/r$ (see Chapter 13). The similarity is not surprising considering that both expressions are derived from an inverse-square force law.

to bring q_2 from infinity to its position near q_1 is $k_e q_1 q_2 / r_{12}$, which is the first term in Equation 24.14. The last two terms represent the work required to bring q_3 from infinity to its position near q_1 and q_2. (The result is independent of the order in which the charges are transported.)

QUICK QUIZ 24.3 In Figure 24.8b, take q_2 to be a negative source charge and q_1 to be a second charge whose sign can be changed. **(i)** If q_1 is initially positive and is changed to a charge of the same magnitude but negative, what happens to the potential at the position of q_1 due to q_2? (a) It increases. (b) It decreases. (c) It remains the same. **(ii)** When q_1 is changed from positive to negative, what happens to the potential energy of the two-charge system? Choose from the same possibilities.

Example 24.3   **The Electric Potential Due to Two Point Charges**

As shown in Figure 24.10a, a charge $q_1 = 2.00\ \mu C$ is located at the origin and a charge $q_2 = -6.00\ \mu C$ is located at $(0, 3.00)$ m.

(A) Find the total electric potential due to these charges at the point P, whose coordinates are $(4.00, 0)$ m.

SOLUTION

Conceptualize   Recognize first that the 2.00-μC and -6.00-μC charges are source charges and set up an electric field as well as a potential at all points in space, including point P.

Figure 24.10 (Example 24.3) (a) The electric potential at P due to the two charges q_1 and q_2 is the algebraic sum of the potentials due to the individual charges. (b) A third charge $q_3 = 3.00\ \mu C$ is brought from infinity to point P.

Categorize   The potential is evaluated using an equation developed in this chapter, so we categorize this example as a substitution problem.

Use Equation 24.12 for the system of two source charges:

$$V_P = k_e \left(\frac{q_1}{r_1} + \frac{q_2}{r_2} \right)$$

Substitute numerical values:

$$V_P = (8.988 \times 10^9\ \text{N} \cdot \text{m}^2/\text{C}^2) \left(\frac{2.00 \times 10^{-6}\ \text{C}}{4.00\ \text{m}} + \frac{-6.00 \times 10^{-6}\ \text{C}}{5.00\ \text{m}} \right)$$

$$= \boxed{-6.29 \times 10^3\ \text{V}}$$

(B) Find the change in potential energy of the system of two charges plus a third charge $q_3 = 3.00\ \mu C$ as the latter charge moves from infinity to point P (Fig. 24.10b).

SOLUTION

Assign $U_i = 0$ for the system to the initial configuration in which the charge q_3 is at infinity. Use Equation 24.2 to evaluate the potential energy for the configuration in which the charge is at P:

$$U_f = q_3 V_P$$

Substitute numerical values to evaluate ΔU_E:

$$\Delta U_E = U_f - U_i = q_3 V_P - 0 = (3.00 \times 10^{-6}\ \text{C})(-6.29 \times 10^3\ \text{V})$$

$$= \boxed{-1.89 \times 10^{-2}\ \text{J}}$$

Therefore, because the potential energy of the system has decreased, an external agent has to do positive work to remove the charge q_3 from point P back to infinity.

WHAT IF?   You are working through this example with a classmate and she says, "Wait a minute! In part (B), we ignored the potential energy associated with the pair of charges q_1 and q_2!" How would you respond?

24.3 c o n t i n u e d

Answer Given the statement of the problem, it is not necessary to include this potential energy because part (B) asks for the *change* in potential energy of the system as q_3 is brought in from infinity. Because the configuration of charges q_1 and q_2 does not change in the process, there is no ΔU_E associated with these charges. Had part (B) asked to find the change in potential energy when *all three* charges start out infinitely far apart and are then brought to the positions in Figure 24.10b, however, you would have to calculate the change using Equation 24.14.

24.4 Obtaining the Value of the Electric Field from the Electric Potential

The electric field $\vec{\mathbf{E}}$ and the electric potential V are related as shown in Equation 24.3, which tells us how to find ΔV if the electric field $\vec{\mathbf{E}}$ is known. What if the situation is reversed? How do we calculate the value of the electric field if the electric potential is known in a certain region?

From Equation 24.3, the potential difference dV between two points a distance ds apart can be expressed as

$$dV = -\vec{\mathbf{E}} \cdot d\vec{\mathbf{s}} \qquad (24.15)$$

If the electric field has only one component E_x, then $\vec{\mathbf{E}} \cdot d\vec{\mathbf{s}} = E_x dx$. Therefore, Equation 24.15 becomes $dV = -E_x\, dx$, or

$$E_x = -\frac{dV}{dx} \qquad (24.16)$$

That is, the x component of the electric field is equal to the negative of the derivative of the electric potential with respect to x. Similar statements can be made about the y and z components. Equation 24.16 is the mathematical statement of the electric field being a measure of the rate of change with position of the electric potential as mentioned in Section 24.1.

Experimentally, electric potential and position can be measured easily with a voltmeter (a device for measuring potential difference) and a meterstick. Consequently, an electric field can be determined by measuring the electric potential at several positions in the field and making a graph of the results. According to Equation 24.16, the slope of a graph of V versus x at a given point provides the magnitude of the electric field at that point.

Imagine starting at a point and then moving through a displacement $d\vec{\mathbf{s}}$ along an equipotential surface. For this motion, $dV = 0$ because the potential is constant along an equipotential surface. From Equation 24.15, we see that $dV = -\vec{\mathbf{E}} \cdot d\vec{\mathbf{s}} = 0$; therefore, because the dot product is zero, $\vec{\mathbf{E}}$ must be perpendicular to the displacement along the equipotential surface. This result shows that the equipotential surfaces must always be perpendicular to the electric field lines passing through them.

As mentioned at the end of Section 24.2, the equipotential surfaces associated with a uniform electric field consist of a family of planes perpendicular to the field lines. Figure 24.11a (page 646) shows some representative equipotential surfaces for this situation.

If the charge distribution creating an electric field has spherical symmetry such that the volume charge density depends only on the radial distance r, the electric field is radial. In this case, $\vec{\mathbf{E}} \cdot d\vec{\mathbf{s}} = E_r\, dr$, and we can express dV as $dV = -E_r\, dr$. Therefore,

$$E_r = -\frac{dV}{dr} \qquad (24.17)$$

For example, the electric potential of a point charge is $V = k_e q/r$. Because V is a function of r only, the potential function has spherical symmetry. Applying Equation 24.17, we find that the magnitude of the electric field due to the point

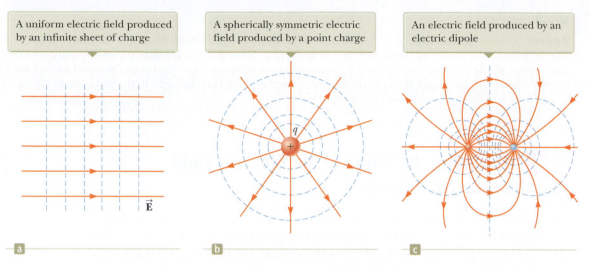

A uniform electric field produced by an infinite sheet of charge

A spherically symmetric electric field produced by a point charge

An electric field produced by an electric dipole

a

b

c

Figure 24.11 Equipotential surfaces (the dashed blue lines are intersections of these surfaces with the page) and electric field lines. In all cases, the equipotential surfaces are *perpendicular* to the electric field lines at every point.

charge is $E_r = k_e q/r^2$, a familiar result. Notice that the potential changes only in the radial direction, not in any direction perpendicular to r. Therefore, V (like E_r) is a function only of r, which is again consistent with the idea that equipotential surfaces are perpendicular to field lines. In this case, the equipotential surfaces are a family of spheres concentric with the spherically symmetric charge distribution (Fig. 24.11b). The equipotential surfaces for an electric dipole are sketched in Figure 24.11c.

In general, the electric potential is a function of all three spatial coordinates. If $V(r)$ is given in terms of the Cartesian coordinates, the electric field components E_x, E_y, and E_z can readily be found from $V(x, y, z)$ as the partial derivatives[2]

◄ Finding the electric field from the potential

$$E_x = -\frac{\partial V}{\partial x} \qquad E_y = -\frac{\partial V}{\partial y} \qquad E_z = -\frac{\partial V}{\partial z} \qquad \text{(24.18)}$$

QUICK QUIZ **24.4** In a certain region of space, the electric potential is zero everywhere along the x axis. **(i)** From this information, you can conclude that the x component of the electric field in this region is (a) zero, (b) in the positive x direction, or (c) in the negative x direction. **(ii)** Suppose the electric potential is +2 V everywhere along the x axis. From the same choices, what can you conclude about the x component of the electric field now?

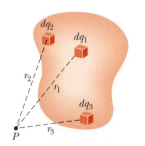

Figure 24.12 The electric potential at point P due to a continuous charge distribution can be calculated by dividing the charge distribution into elements of charge dq and summing the electric potential contributions over all elements. Three sample elements of charge are shown.

24.5 Electric Potential Due to Continuous Charge Distributions

In Section 24.3, we found how to determine the electric potential due to a small number of charges. What if we wish to find the potential due to a continuous distribution of charge? The electric potential in this situation can be calculated using two different methods. The first method is as follows. If the charge distribution is known, we consider the potential due to a small charge element dq, treating this element as a point charge (Fig. 24.12). From Equation 24.11, the electric potential

---

[2]In vector notation, $\vec{\mathbf{E}}$ is often written in Cartesian coordinate systems as

$$\vec{\mathbf{E}} = -\nabla V = -\left(\hat{\mathbf{i}}\frac{\partial}{\partial x} + \hat{\mathbf{j}}\frac{\partial}{\partial y} + \hat{\mathbf{k}}\frac{\partial}{\partial z}\right)V$$

where ∇ is called the *gradient operator*.

dV at some point P due to the charge element dq is

$$dV = k_e \frac{dq}{r} \qquad (24.19)$$

where r is the distance from the charge element to point P. To obtain the total potential at point P, we integrate Equation 24.19 to include contributions from all elements of the charge distribution. Because each element is, in general, a different distance from point P and k_e is constant, we can express V as

$$V = k_e \int \frac{dq}{r} \qquad (24.20)$$

◀ Electric potential due to a continuous charge distribution

In effect, we have replaced the sum in Equation 24.12 with an integral. In this expression for V, the electric potential is taken to be zero when point P is infinitely far from the charge distribution.

The second method for calculating the electric potential is used if the electric field is already known from other considerations such as Gauss's law. If the charge distribution has sufficient symmetry, we first evaluate $\vec{E}$ using Gauss's law and then substitute the value obtained into Equation 24.3 to determine the potential difference ΔV between any two points. We then choose the electric potential V to be zero at some convenient point.

PROBLEM-SOLVING STRATEGY **Calculating Electric Potential**

The following procedure is recommended for solving problems that involve the determination of an electric potential due to a charge distribution.

1. Conceptualize. Think carefully about the individual charges or the charge distribution you have in the problem and imagine what type of potential would be created. Appeal to any symmetry in the arrangement of charges to help you visualize the potential.

2. Categorize. Are you analyzing a group of individual charges or a continuous charge distribution? The answer to this question will tell you how to proceed in the *Analyze* step.

3. Analyze. When working problems involving electric potential, remember that it is a *scalar quantity*, so there are no vector components to consider. Therefore, when using the superposition principle to evaluate the electric potential at a point, simply take the algebraic sum of the potentials due to each charge. You must keep track of signs, however.

As with potential energy in mechanics, only *changes* in electric potential are significant; hence, the point where the potential is set at zero is arbitrary. When dealing with point charges or a finite-sized charge distribution, we usually define $V = 0$ to be at a point infinitely far from the charges. If the charge distribution itself extends to infinity, however, some other nearby point must be selected as the reference point.

(a) *If you are analyzing a group of individual charges:* Use the superposition principle, which states that when several point charges are present, the resultant potential at a point P in space is the *algebraic sum* of the individual potentials at P due to the individual charges (Eq. 24.12). Example 24.4 demonstrates this procedure.

(b) *If you are analyzing a continuous charge distribution:* Replace the sums for evaluating the total potential at some point P from individual charges by integrals (Eq. 24.20). The total potential at P is obtained by integrating over the entire charge distribution. For many problems, it is possible in performing the integration to express dq and r in terms of a single variable. To simplify the integration, give careful consideration to the geometry involved in the problem. Examples 24.5 through 24.7 demonstrate such a procedure.

To obtain the potential from the electric field: Another method used to obtain the potential is to start with the definition of the potential difference given by Equation 24.3. If $\vec{E}$ is known or can be obtained easily (such as from Gauss's law), the line integral of $\vec{E} \cdot d\vec{s}$ can be evaluated.

4. Finalize. Check to see if your expression for the potential is consistent with your mental representation and reflects any symmetry you noted previously. Imagine varying parameters such as the distance of the observation point from the charges or the radius of any circular objects to see if the mathematical result changes in a reasonable way.

Example 24.4 The Electric Potential Due to a Dipole

An electric dipole consists of two charges of equal magnitude and opposite sign separated by a distance $2a$ as shown in Figure 24.13. The dipole is along the x axis and is centered at the origin.

(A) Calculate the electric potential at point P on the y axis.

SOLUTION

Conceptualize Compare this situation to that in part (B) of Example 22.6. It is the same situation, but here we are seeking the electric potential rather than the electric field.

Categorize We categorize the problem as one in which we have a small number of particles rather than a continuous distribution of charge. The electric potential can be evaluated by summing the potentials due to the individual charges.

Figure 24.13 (Example 24.4) An electric dipole located on the x axis.

Analyze Use Equation 24.12 to find the electric potential at P due to the two charges:

$$V_P = k_e \sum_i \frac{q_i}{r_i} = k_e \left(\frac{q}{\sqrt{a^2 + y^2}} + \frac{-q}{\sqrt{a^2 + y^2}} \right) = 0$$

(B) Calculate the electric potential at point R on the positive x axis.

SOLUTION

Use Equation 24.12 to find the electric potential at R due to the two charges:

$$V_R = k_e \sum_i \frac{q_i}{r_i} = k_e \left(\frac{-q}{x - a} + \frac{q}{x + a} \right) = -\frac{2k_e qa}{x^2 - a^2}$$

(C) Calculate V and E_x at a point on the x axis far from the dipole.

SOLUTION

For point R far from the dipole such that $x \gg a$, neglect a^2 in the denominator of the answer to part (B) and write V in this limit:

$$V_R = \lim_{x \gg a} \left(-\frac{2k_e qa}{x^2 - a^2} \right) \approx -\frac{2k_e qa}{x^2} \quad (x \gg a)$$

Use Equation 24.16 and this result to calculate the x component of the electric field at a point on the x axis far from the dipole:

$$E_x = -\frac{dV}{dx} = -\frac{d}{dx} \left(-\frac{2k_e qa}{x^2} \right)$$

$$= 2k_e qa \frac{d}{dx} \left(\frac{1}{x^2} \right) = -\frac{4k_e qa}{x^3} \quad (x \gg a)$$

Finalize The potentials in parts (B) and (C) are negative because points on the positive x axis are closer to the negative charge than to the positive charge. For the same reason, the x component of the electric field is negative. Notice that we have a $1/r^3$ falloff of the electric field with distance far from the dipole, similar to the behavior of the electric field on the y axis in Example 22.6.

WHAT IF? Suppose you want to find the electric field at a point P on the y axis. In part (A), the electric potential was found to be zero for all values of y. Is the electric field zero at all points on the y axis?

Answer No. That there is no change in the potential along the y axis tells us only that the y component of the electric field is zero. Look back at Figure 22.13 in Example 22.6. We showed there that the electric field of a dipole on the y axis has only an x component. We could not find the x component in the current example because we have only a single value of the potential: $V_P = 0$. We do not have an expression for the potential near the y axis as a function of x.

Example 24.5 Electric Potential Due to a Uniformly Charged Ring

(A) Find an expression for the electric potential at a point P located on the perpendicular central axis of a uniformly charged ring of radius a and total charge Q.

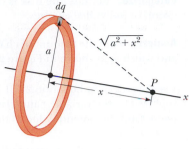

SOLUTION

Conceptualize Study Figure 24.14, in which the ring is oriented so that its plane is perpendicular to the x axis and its center is at the origin. Notice that the symmetry of the situation means that all the charges on the ring are the same distance from point P. Compare this example to Example 23.2. Notice that no vector considerations are necessary here because electric potential is a scalar.

Categorize Because the ring consists of a continuous distribution of charge rather than a set of discrete charges, we must use the integration technique represented by Equation 24.20 in this example.

Figure 24.14 (Example 24.5) A uniformly charged ring of radius a lies in a plane perpendicular to the x axis. All elements dq of the ring are the same distance from a point P lying on the x axis.

Analyze We take point P to be at a distance x from the center of the ring as shown in Figure 24.14.

Use Equation 24.20 to express V in terms of the geometry:

$$V = k_e \int \frac{dq}{r} = k_e \int \frac{dq}{\sqrt{a^2 + x^2}}$$

Noting that a and x do not vary for an integration over the ring, bring $\sqrt{a^2 + x^2}$ in front of the integral sign and integrate over the ring:

$$V = \frac{k_e}{\sqrt{a^2 + x^2}} \int dq = \frac{k_e Q}{\sqrt{a^2 + x^2}} \qquad (24.21)$$

(B) Find an expression for the magnitude of the electric field at point P.

SOLUTION

From symmetry, notice that along the x axis $\vec{E}$ can have only an x component. Therefore, apply Equation 24.16 to Equation 24.21:

$$E_x = -\frac{dV}{dx} = -k_e Q \frac{d}{dx}(a^2 + x^2)^{-1/2}$$

$$= -k_e Q(-\tfrac{1}{2})(a^2 + x^2)^{-3/2}(2x)$$

$$E_x = \frac{k_e x}{(a^2 + x^2)^{3/2}} Q \qquad (24.22)$$

Finalize The only variable in the expressions for V and E_x is x. That is not surprising because our calculation is valid only for points along the x axis, where y and z are both zero. This result for the electric field agrees with that obtained by direct integration (see Example 23.2). For practice, use the result of part (B) in Equation 24.3 to verify that the potential is given by the expression in part (A).

Example 24.6 Electric Potential Due to a Uniformly Charged Disk

A uniformly charged disk has radius R and surface charge density σ.

(A) Find the electric potential at a point P along the perpendicular central axis of the disk.

SOLUTION

Figure 24.15 (Example 24.6) A uniformly charged disk of radius R lies in a plane perpendicular to the x axis. The calculation of the electric potential at any point P on the x axis is simplified by dividing the disk into many rings of radius r and width dr, with area $2\pi r\, dr$.

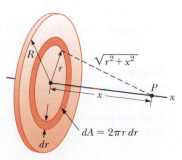

Conceptualize If we consider the disk to be a set of concentric rings, we can use our result from Example 24.5—which gives the potential due to a ring of radius a—and sum the contributions of all rings making up the disk. Figure 24.15 shows one such ring. Because point P is on the central axis of the disk, symmetry again tells us that all points in a given ring are the same distance from P.

continued

24.6 continued

Categorize Because the disk is continuous, we evaluate the potential due to a continuous charge distribution rather than a group of individual charges.

Analyze Find the amount of charge dq on a ring of radius r and width dr as shown in Figure 24.15:

$$dq = \sigma \, dA = \sigma(2\pi r \, dr) = 2\pi\sigma r \, dr$$

Use this result in Equation 24.21 in Example 24.5 (with a replaced by the variable r and Q replaced by the differential dq) to find the potential due to the ring:

$$dV = \frac{k_e \, dq}{\sqrt{r^2 + x^2}} = \frac{k_e(2\pi\sigma r \, dr)}{\sqrt{r^2 + x^2}}$$

To obtain the total potential at P, integrate this expression over the limits $r = 0$ to $r = R$, noting that x is a constant:

$$V = \pi k_e \sigma \int_0^R \frac{2r \, dr}{\sqrt{r^2 + x^2}} = \pi k_e \sigma \int_0^R (r^2 + x^2)^{-1/2} \, d(r^2)$$

This integral is of the common form $\int u^n \, du$, where $n = -\frac{1}{2}$ and $u = r^2 + x^2$, and has the value $u^{n+1}/(n + 1)$. Use this result to evaluate the integral:

$$V = 2\pi k_e \sigma[(R^2 + x^2)^{1/2} - x] \quad (24.23)$$

(B) Find the x component of the electric field at a point P along the perpendicular central axis of the disk.

SOLUTION

As in Example 24.5, use Equation 24.16 to find the electric field at any axial point:

$$E_x = -\frac{dV}{dx} = 2\pi k_e \sigma\left[1 - \frac{x}{(R^2 + x^2)^{1/2}}\right] \quad (24.24)$$

Finalize Compare Equation 24.24 with the result of Example 23.3. They are the same. The calculation of V and $\vec{\mathbf{E}}$ for an arbitrary point off the x axis is more difficult to perform because of the absence of symmetry and we do not treat that situation in this book.

Example 24.7 Electric Potential Due to a Finite Line of Charge

A rod of length ℓ located along the x axis has a total charge Q and a uniform linear charge density λ. Find the electric potential at a point P located on the y axis a distance a from the origin (Fig. 24.16).

SOLUTION

Conceptualize The potential at P due to every segment of charge on the rod is positive because every segment carries a positive charge. Notice that we have no symmetry to appeal to here, but the simple geometry should make the problem solvable.

Categorize Because the rod is continuous, we evaluate the potential due to a continuous charge distribution rather than a group of individual charges.

Figure 24.16 (Example 24.7) A uniform line charge of length ℓ located along the x axis. To calculate the electric potential at P, the line charge is divided into segments each of length dx and each carrying a charge $dq = \lambda \, dx$.

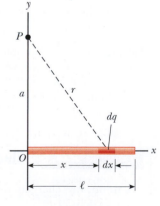

Analyze In Figure 24.16, the rod lies along the x axis, dx is the length of one small segment, and dq is the charge on that segment. Because the rod has a charge per unit length λ, the charge dq on the small segment is $dq = \lambda \, dx$.

Find the potential at P due to one segment of the rod at an arbitrary position x:

$$dV = k_e \frac{dq}{r} = k_e \frac{\lambda \, dx}{\sqrt{a^2 + x^2}}$$

Find the total potential at P by integrating this expression over the limits $x = 0$ to $x = \ell$:

$$V = \int_0^\ell k_e \frac{\lambda \, dx}{\sqrt{a^2 + x^2}}$$

24.7 continued

Noting that k_e and $\lambda = Q/\ell$ are constants and can be removed from the integral, evaluate the integral with the help of Appendix B:

$$V = k_e \lambda \int_0^\ell \frac{dx}{\sqrt{a^2 + x^2}} = k_e \frac{Q}{\ell} \ln (x + \sqrt{a^2 + x^2}) \Big|_0^\ell$$

Evaluate the result between the limits:

$$V = k_e \frac{Q}{\ell} [\ln(\ell + \sqrt{a^2 + \ell^2}) - \ln a] = k_e \frac{Q}{\ell} \ln \left(\frac{\ell + \sqrt{a^2 + \ell^2}}{a} \right) \quad (24.25)$$

Finalize If $\ell \ll a$, the potential at P should approach that of a point charge because the rod is very short compared to the distance from the rod to P. By using a series expansion for the natural logarithm from Appendix B.5, it is easy to show that Equation 24.25 becomes $V = k_e Q/a$.

WHAT IF? What if you were asked to find the electric field at point P? Would that be a simple calculation?

Answer Calculating the electric field by means of Equation 23.1 would be a little messy. There is no symmetry to appeal to, and the integration over the line of charge would represent a vector addition of electric fields at point P. Using Equation 24.18, you could find E_y by replacing a with y in Equation 24.25 and performing the differentiation with

respect to y. Because the charged rod in Figure 24.16 lies entirely to the right of $x = 0$, the electric field at point P would have an x component to the left if the rod is charged positively. You cannot use Equation 24.18 to find the x component of the field, however, because the potential due to the rod was evaluated at a specific value of x ($x = 0$) rather than a general value of x. You would have to find the potential as a function of both x and y to be able to find the x and y components of the electric field using Equation 24.18.

24.6 Conductors in Electrostatic Equilibrium

As we learned in Section 22.2, a good electrical conductor contains charges (electrons) that are not bound to any atom and therefore are free to move about within the material. When there is no net motion of charge within a conductor, the conductor is in **electrostatic equilibrium**. A conductor in electrostatic equilibrium has the following properties:

1. The electric field is zero everywhere inside the conductor, whether the conductor is solid or hollow.
2. If the conductor is isolated and carries a charge, the charge resides on its surface.
3. The electric field at a point just outside a charged conductor is perpendicular to the surface of the conductor and has a magnitude σ/ϵ_0, where σ is the surface charge density at that point.
4. On an irregularly shaped conductor, the surface charge density is greatest at locations where the radius of curvature of the surface is smallest.

◀ Properties of a conductor in electrostatic equilibrium

We can understand the first property by considering a conducting slab placed in an external field $\vec{E}$ (Fig. 24.17). The electric field inside the conductor *must* be zero, assuming electrostatic equilibrium exists. If the field were not zero, free electrons in the conductor would experience an electric force ($\vec{F} = q\vec{E}$) and would accelerate due to this force. This motion of electrons, however, would mean that the conductor is not in electrostatic equilibrium. Therefore, the existence of electrostatic equilibrium is consistent only with a zero field in the conductor.

Let's investigate how this zero field is accomplished. Before the external field is applied, free electrons are uniformly distributed throughout the conductor. When the external field is applied, the free electrons accelerate to the left in Figure 24.17, causing a plane of negative charge to accumulate on the left surface. The movement of electrons to the left results in a plane of positive charge on the right surface. These planes of charge create an additional electric field inside the conductor that opposes the external field. As the electrons move, the surface charge densities on the left and right surfaces increase until the magnitude of

Figure 24.17 A conducting slab in an external electric field $\vec{E}$. The charges induced on the two surfaces of the slab produce an electric field that opposes the external field, giving a resultant field of zero inside the slab.

Figure 24.18 A conductor of arbitrary shape. The broken line represents a gaussian surface that can be just inside the conductor's surface.

The flux through the gaussian surface is *EA*.

Figure 24.19 A gaussian surface in the shape of a small cylinder is used to calculate the electric field immediately outside a charged conductor.

the internal field equals that of the external field, resulting in a net field of zero inside the conductor. The time it takes a good conductor to reach equilibrium is on the order of 10^{-16} s, which for most purposes can be considered instantaneous.

Gauss's law can be used to verify the second property of a conductor in electrostatic equilibrium. Figure 24.18 shows an arbitrarily shaped conductor. A gaussian surface is drawn inside the conductor and can be very close to the conductor's surface. As we have just shown, the electric field everywhere inside the conductor is zero when it is in electrostatic equilibrium. Therefore, the electric field must be zero at every point on the gaussian surface, in accordance with condition (4) in Section 23.4, and the net flux through this gaussian surface is zero. From this result and Gauss's law, we conclude that the net charge inside the gaussian surface is zero. Because there can be no net charge inside the gaussian surface (which is arbitrarily close to the conductor's surface), any net charge on the conductor must reside on its surface. Gauss's law does not indicate how this excess charge is distributed on the conductor's surface, only that it resides exclusively on the surface.

To verify the third property, let's begin with the perpendicularity of the field to the surface. If the field vector $\vec{\mathbf{E}}$ had a component parallel to the conductor's surface, free electrons would experience an electric force and move along the surface; in such a case, the conductor would not be in equilibrium. Therefore, the field vector must be perpendicular to the surface.

To determine the magnitude of the electric field, we use Gauss's law and draw a gaussian surface in the shape of a small cylinder whose end faces are parallel to the conductor's surface (Fig. 24.19). Part of the cylinder is just outside the conductor, and part is inside. The field is perpendicular to the conductor's surface from the condition of electrostatic equilibrium. Therefore, condition (3) in Section 23.4 is satisfied for the curved part of the cylindrical gaussian surface: there is no flux through this part of the gaussian surface because $\vec{\mathbf{E}}$ is parallel to the surface. There is no flux through the flat face of the cylinder inside the conductor because here $\vec{\mathbf{E}} = 0$, which satisfies condition (4). Hence, the net flux through the gaussian surface is equal to that through only the flat face outside the conductor, where the field is perpendicular to the gaussian surface. Using conditions (1) and (2) for this face, the flux is EA, where E is the electric field just outside the conductor and A is the area of the cylinder's face. Applying Gauss's law to this surface gives

$$\Phi_E = \oint E\, dA = EA = \frac{q_{\text{in}}}{\epsilon_0} = \frac{\sigma A}{\epsilon_0}$$

where we have used $q_{\text{in}} = \sigma A$. Solving for E gives for the electric field immediately outside a charged conductor:

$$E = \frac{\sigma}{\epsilon_0} \qquad (24.26)$$

Let's now verify property 4 listed at the beginning of this section for a charged conductor in electrostatic equilibrium. Consider two points Ⓐ and Ⓑ on the surface of a charged conductor as shown in Figure 24.20. Along a surface path connecting these points, $\vec{\mathbf{E}}$ is always perpendicular to the displacement $d\vec{\mathbf{s}}$; therefore, $\vec{\mathbf{E}} \cdot d\vec{\mathbf{s}} = 0$. Using this result and Equation 24.3, we conclude that the potential difference between Ⓐ and Ⓑ is necessarily zero:

$$V_Ⓑ - V_Ⓐ = -\int_Ⓐ^Ⓑ \vec{\mathbf{E}} \cdot d\vec{\mathbf{s}} = 0$$

This result applies to any two points on the surface. Therefore, V is constant

PITFALL PREVENTION 24.6

Potential May Not Be Zero
The electric potential inside the conductor is not necessarily zero in Figure 24.20, even though the electric field is zero. Equation 24.15 shows that a zero value of the field results in no *change* in the potential from one point to another inside the conductor. Therefore, the potential everywhere inside the conductor, including the surface, has the same value, which may or may not be zero, depending on where the zero of potential is defined.

Notice from the spacing of the positive signs that the surface charge density is nonuniform.

Figure 24.20 An arbitrarily shaped conductor carrying a positive charge. When the conductor is in electrostatic equilibrium, all the charge resides at the surface, $\vec{\mathbf{E}} = 0$ inside the conductor, and the direction of $\vec{\mathbf{E}}$ immediately outside the conductor is perpendicular to the surface. The electric potential is constant inside the conductor and is equal to the potential at the surface.

Figure 24.21 (a) The excess charge on a conducting sphere of radius R is uniformly distributed on its surface. (b) Electric potential versus distance r from the center of the charged conducting sphere. (c) Electric field magnitude versus distance r from the center of the charged conducting sphere.

everywhere on the surface of a charged conductor in equilibrium. That is,

the surface of any charged conductor in electrostatic equilibrium is an equipotential surface: every point on the surface of a charged conductor in equilibrium is at the same electric potential. Furthermore, because the electric field is zero inside the conductor, the electric potential is constant everywhere inside the conductor and equal to its value at the surface.

Because of the constant value of the potential, no work is required to move a charge from the interior of a charged conductor to its surface.

Consider a solid metal conducting sphere of radius R and total positive charge Q as shown in Figure 24.21a. As determined in part (A) of Example 23.6, the electric field outside the sphere is $k_e Q/r^2$ and points radially outward. Because the field outside a spherically symmetric charge distribution is identical to that of a point charge, we expect the potential to also be that of a point charge, $k_e Q/r$. At the surface of the conducting sphere in Figure 24.21a, the potential must be $k_e Q/R$. Because the entire sphere must be at the same potential, the potential at any point within the sphere must also be $k_e Q/R$. Figure 24.21b is a plot of the electric potential as a function of r, and Figure 25.18c shows how the electric field varies with r.

When a net charge is placed on a spherical conductor, the surface charge density is uniform as indicated in Figure 24.21a. If the conductor is nonspherical as in Figure 24.20, however, we find that the surface charge density is high where the radius of curvature is small and low where the radius of curvature is large. Let us show theoretically why this is true. Consider the two spheres in Figure 24.22, connected by a wire. If we imagine the spheres to be very far apart, one sphere will not affect the charge distribution of the other, and we can express the potential at the surface of each sphere using Equation 24.11:

$$V = k_e \frac{q_1}{r_1} = k_e \frac{q_2}{r_2}$$

where we have set the potentials equal because the connecting wire assures that the whole system is a single conductor. Now set up the ratio of the electric fields at the surfaces of the two spheres:

$$\frac{E_1}{E_2} = \frac{k_e \dfrac{q_1}{r_1^2}}{k_e \dfrac{q_2}{r_2^2}} = \frac{\dfrac{1}{r_1}V}{\dfrac{1}{r_2}V} = \frac{r_2}{r_1}$$

Figure 24.22 Two charged spherical conductors connected by a conducting wire. The spheres are at the *same* electric potential V.

The electric field in the cavity is zero regardless of the charge on the conductor.

Figure 24.23 A conductor in electrostatic equilibrium containing a cavity.

The ratio of the magnitudes of the electric field is equal to the inverse ratio of the radii of the spheres. Therefore, the field is strong when the radius is small, and the field is weaker when the radius is larger. The electric field reaches very high values at sharp points. In turn, Equation 24.26 tells us that the surface charge density is high when the radius is small.

A Cavity Within a Conductor

Suppose a conductor of arbitrary shape contains a cavity as shown in Figure 24.23. Let's assume no charges are inside the cavity. In this case, the electric field inside the cavity must be *zero* regardless of the charge distribution on the outside surface of the conductor. Furthermore, the field in the cavity is zero even if an electric field exists outside the conductor.

To prove this point, remember that every point on the conductor is at the same electric potential; therefore, any two points Ⓐ and Ⓑ on the cavity's surface must be at the same potential. Now imagine a field $\vec{E}$ exists in the cavity and evaluate the potential difference $V_{Ⓑ} - V_{Ⓐ}$ defined by Equation 24.3:

$$V_{Ⓑ} - V_{Ⓐ} = -\int_{Ⓐ}^{Ⓑ} \vec{E} \cdot d\vec{s}$$

Because $V_{Ⓑ} - V_{Ⓐ} = 0$, the integral of $\vec{E} \cdot d\vec{s}$ must be zero for all paths between any two points Ⓐ and Ⓑ on the conductor. The only way that can be true for *all* paths is if $\vec{E}$ is zero *everywhere* in the cavity. Therefore, a cavity surrounded by conducting walls is a field-free region as long as no charges are inside the cavity.

This phenomenon is used in a *Faraday cage*, which is a conducting material, either solid or mesh, surrounding an interior space. A Faraday cage is used to protect sensitive electronic equipment, and it protects you if you are inside a car during a lightning storm. The metal body of the car acts as a Faraday cage; any charge on the car due to the strong electric fields in the car are on the outer surface of the car, and the electric field inside the car must be zero. Faraday cages often have a negative effect, such as the loss of cellphone service inside a metal elevator car.

QUICK QUIZ 24.5 Your younger brother likes to rub his feet on the carpet and then touch you to give you a shock. While you are trying to escape the shock treatment, you discover a hollow metal cylinder in your basement, large enough to climb inside. In which of the following cases will you *not* be shocked? **(a)** You climb inside the cylinder, making contact with the inner surface, and your charged brother touches the outer metal surface. **(b)** Your charged brother is inside touching the inner metal surface and you are outside, touching the outer metal surface. **(c)** Both of you are outside the cylinder, touching its outer metal surface but not touching each other directly.

Example 24.8   A Sphere Inside a Spherical Shell

A solid insulating sphere of radius a carries a net positive charge Q uniformly distributed throughout its volume. A conducting spherical shell of inner radius b and outer radius c is concentric with the solid sphere and carries a net charge $-2Q$. Using Gauss's law, find the electric field in the regions labeled ①, ②, ③, and ④ in Figure 24.24 and the charge distribution on the shell when the entire system is in electrostatic equilibrium.

Figure 24.24 (Example 24.8) An insulating sphere of radius a and carrying a charge Q surrounded by a conducting spherical shell carrying a charge $-2Q$.

SOLUTION

Conceptualize Notice how this problem differs from Example 23.6 in the previous chapter. The charged sphere in Figure 23.14 appears in Figure 24.24, but it is now surrounded by a shell carrying a charge $-2Q$. Think about how the presence of the shell will affect the electric field of the sphere.

24.8 continued

Categorize The charge is distributed uniformly throughout the sphere, and we know that the charge on the conducting shell distributes itself uniformly on the surfaces. Therefore, the system has spherical symmetry and we can apply Gauss's law to find the electric field in the various regions.

Analyze In region ②—between the surface of the solid sphere and the inner surface of the shell—we construct a spherical gaussian surface of radius r, where $a < r < b$, noting that the charge inside this surface is $+Q$ (the charge on the solid sphere). Because of the spherical symmetry, the electric field lines must be directed radially outward and be constant in magnitude on the gaussian surface.

The charge on the conducting shell creates zero electric field in the region $r < b$, so the shell has no effect on the field in region ② due to the sphere. Therefore, write an expression for the field in region ② as that due to the sphere from part (A) of Example 23.6:

$$E_2 = k_e \frac{Q}{r^2} \quad \text{(for } a < r < b\text{)}$$

Because the conducting shell creates zero field inside itself, it also has no effect on the field inside the sphere. Therefore, write an expression for the field in region ① as that due to the sphere from part (B) of Example 23.6:

$$E_1 = k_e \frac{Q}{a^3} r \quad \text{(for } r < a\text{)}$$

In region ④, where $r > c$, construct a spherical gaussian surface; this surface surrounds a total charge $q_{in} = Q + (-2Q) = -Q$. Therefore, model the charge distribution as a sphere with charge $-Q$ and write an expression for the field in region ④ from part (A) of Example 23.6:

$$E_4 = -k_e \frac{Q}{r^2} \quad \text{(for } r > c\text{)}$$

In region ③, we need property 1 from this section: the electric field must be zero because the spherical shell is a conductor in equilibrium:

$$E_3 = 0 \quad \text{(for } b < r < c\text{)}$$

Construct a gaussian surface of radius r in region ③, where $b < r < c$, and note that q_{in} must be zero because $E_3 = 0$. Find the amount of charge q_{inner} on the inner surface of the shell:

$$q_{in} = q_{sphere} + q_{inner}$$
$$q_{inner} = q_{in} - q_{sphere} = 0 - Q = -Q$$

Finalize The charge on the inner surface of the spherical shell must be $-Q$ to cancel the charge $+Q$ on the solid sphere and give zero electric field in the material of the shell. Because the net charge on the shell is $-2Q$, its outer surface must carry a charge $-Q$.

WHAT IF? How would the results of this problem differ if the sphere were conducting instead of insulating?

Answer The only change would be in region ①, where $r < a$. Because there can be no charge inside a conductor in electrostatic equilibrium, $q_{in} = 0$ for a gaussian surface of radius $r < a$; therefore, on the basis of Gauss's law and symmetry, $E_1 = 0$. In regions ②, ③, and ④, there would be no way to determine from observations of the electric field whether the sphere is conducting or insulating.

Summary

▶ Definitions

The **potential difference** ΔV between points Ⓐ and Ⓑ in an electric field $\vec{E}$ is defined as

$$\Delta V \equiv \frac{\Delta U_E}{q} = -\int_Ⓐ^Ⓑ \vec{E} \cdot d\vec{s} \quad (24.3)$$

where ΔU_E is given by Equation 24.1 on page 637. The **electric potential** $V = U_E/q$ is a scalar quantity and has the units of joules per coulomb, where $1 \text{ J/C} \equiv 1 \text{ V}$.

An **equipotential surface** is one on which all points are at the same electric potential. Equipotential surfaces are perpendicular to electric field lines.

continued

❯ **Concepts and Principles**

When a positive charge q is moved between points Ⓐ and Ⓑ in an electric field $\vec{\mathbf{E}}$, the change in the potential energy of the charge–field system is

$$\Delta U_E = -q \int_{Ⓐ}^{Ⓑ} \vec{\mathbf{E}} \cdot d\vec{\mathbf{s}} \quad \text{(24.1)}$$

A conductor in **electrostatic equilibrium** has the following properties:

1. The electric field is zero everywhere inside the conductor, whether the conductor is solid or hollow.

2. If the conductor is isolated and carries a charge, the charge resides on its surface.

3. The electric field at a point just outside a charged conductor is perpendicular to the surface of the conductor and has a magnitude σ/ϵ_0, where σ is the surface charge density at that point.

4. On an irregularly shaped conductor, the surface charge density is greatest at locations where the radius of curvature of the surface is smallest.

The potential difference between two points separated by a distance d in a uniform electric field $\vec{\mathbf{E}}$ is

$$\Delta V = -Ed \quad \text{(24.6)}$$

if the direction of travel between the points is in the same direction as the electric field.

If we define $V = 0$ at $r = \infty$, the electric potential due to a point charge at any distance r from the charge is

$$V = k_e \frac{q}{r} \quad \text{(24.11)}$$

The electric potential associated with a group of point charges is obtained by summing the potentials due to the individual charges.

The **electric potential energy** associated with a pair of point charges separated by a distance r_{12} is

$$U_E = k_e \frac{q_1 q_2}{r_{12}} \quad \text{(24.13)}$$

We obtain the potential energy of a distribution of point charges by summing terms like Equation 24.13 over all pairs of particles.

If the electric potential is known as a function of coordinates x, y, and z, we can obtain the components of the electric field by taking the negative derivative of the electric potential with respect to the coordinates. For example, the x component of the electric field is

$$E_x = -\frac{dV}{dx} \quad \text{(24.16)}$$

The electric potential due to a continuous charge distribution is

$$V = k_e \int \frac{dq}{r} \quad \text{(24.20)}$$

Every point on the surface of a charged conductor in electrostatic equilibrium is at the same electric potential. The potential is constant everywhere inside the conductor and equal to its value at the surface.

Think–Pair–Share

See the Preface for an explanation of the icons used in this problems set.
For additional assessment items for this section, go to ❖ WEB**ASSIGN** From Cengage

1. Robert Millikan performed a brilliant set of experiments from 1909 to 1913 in which he measured e, the magnitude of the elementary charge on an electron, and demonstrated the quantized nature of this charge. Because the charge of an electron is negative, we express its charge as $-e$. His apparatus, diagrammed in Figure TP24.1, contains two parallel metallic plates separated by a distance d. Oil droplets are sprayed from an atomizer above the upper plate. Some of these oil droplets pass through a small hole in the upper plate. Millikan used x-rays to ionize the air in the chamber so that freed electrons would adhere to the oil drops, giving them a negative charge $-q$. A horizontally directed light beam is used to illuminate the oil droplets, which are viewed through a telescope whose long axis is also horizontal but perpendicular to the light beam. Also visible through the

Oil droplets

Pinhole

Telescope with scale in eyepiece

Figure TP24.1

telescope is a calibrated scale that shows the vertical position of the oil drop at any time. Discuss this experiment in your group and perform the following tasks. (a) If a downward electric field is established between the plates, a given oil drop can be suspended at rest between the upward electric force and the downward gravitational force on the drop. Show that the required potential difference between the plates to keep an oil drop suspended at rest is given by

$$\Delta V_{rest} = \frac{4\pi\rho_{oil}gd}{3q}\, r^3$$

where ρ_{oil} is the density of the oil, g is the acceleration due to gravity, and r is the radius of the oil drop. (b) To determine q from the equation in part (a), we need to know all the other quantities, which we do, except for r. The radius of the oil drops is too small to measure directly. Therefore, Millikan performed a second measurement. With the electric field removed, the oil drops drift downward at terminal speed because of the resistive force on them, given by Equation 6.2. For a sphere moving slowly through a viscous fluid, Equation 6.2 can be modified to become what is known as *Stokes's law*:

$$\vec{R} = -6\pi\eta r\vec{v}$$

where η is the *viscosity* of the fluid, given in units of N · s/m². In addition, the oil drops experience an upward buoyant force due to the surrounding air, given by Equation 14.5:

$$B = \rho_{air}gV_{disp}$$

where V_{disp} is the volume of air displaced by an oil drop or, equivalently, the volume of the oil drop. The falling oil drop is acted on by gravity, the resistive force of the air, and the buoyant force. Show from the particle in equilibrium model applied to the oil drop that the radius of the drop is

$$r = 3\sqrt{\frac{\eta v_T}{2g(\rho_{oil} - \rho_{air})}}$$

(c) Combine the first and last equations to find an expression for the charge q on an oil drop in terms of quantities, all of which can be measured by the two experiments described.

2. **ACTIVITY** An electric field in a region of space is parallel to the x axis. The electric potential varies with position as shown in Figure TP24.2. Discuss in your group how the electric field would vary with position x and then graph the x component of the electric field versus position in this region of space.

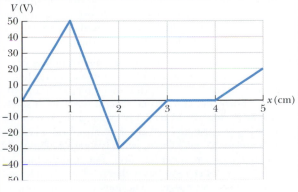

Figure TP24.2

3. **ACTIVITY** Robert Millikan performed a brilliant set of experiments from 1909 to 1913 in which he measured e, the magnitude of the elementary charge on an electron, and demonstrated the quantized nature of this charge. Because the charge of an electron is negative, we express its charge as $-e$. His apparatus, diagrammed in Figure TP24.1, contains two parallel metallic plates separated by a distance d. Oil droplets are sprayed from an atomizer above the upper plate. Some of these oil droplets pass through a small hole in the upper plate. Millikan used x-rays to ionize the air in the chamber so that freed electrons would adhere to the oil drops, giving them a negative charge $-q$. A horizontally directed light beam is used to illuminate the oil droplets, which are viewed through a telescope whose long axis is also horizontal but perpendicular to the light beam. Also visible through the telescope is a calibrated scale that shows the vertical position of the oil drop at any time.

In his experiment, two types of measurements were made. First, the oil drops were allowed to fall freely while the time interval for them to fall through a distance of $\Delta y = 1.00$ mm against a calibrated scale was measured. From these data, the terminal velocity v_T of a drop can be measured. Knowing the terminal velocity allows us to find the radius r of the oil drop from the following equation:

$$r = 3\sqrt{\frac{\eta v_T}{2g\rho_{oil}}}$$

where η is the viscosity of air, $\eta = 1.81 \times 10^{-5}$ N · s/m², g is the acceleration due to gravity, and ρ_{oil} is the density of the oil used in the experiment, $\rho_{oil} = 824$ kg/m³.

In another experiment, a voltage is applied between the plates separated by $d = 1.00$ mm in Figure TP24.1 and adjusted to a value ΔV_{rest} at which the same drop from the first experiment is suspended at rest. This measurement allows a calculation of the magnitude q of the charge on the oil drop:

$$q = \frac{4\pi\rho_{oil}gd}{3\Delta V_{rest}}\, r^3$$

Below are ten sets of data for different drops measured in this experiment. (a) Find the charge on each drop. Make a histogram of the values of the ten charges found and determine the value e of the elementary charge from these data. (b) The manufacturer of the atomizer claims that it should provide oil drops with a relatively consistent radius, ranging between 0.1 and 1 μm. Are the ten drops below consistent with this claim?

Drop	Δt to fall 1.00 mm (s)	ΔV_{rest} (V)
1	40.0	8.95
2	31.8	9.44
3	22.7	12.55
4	40.3	26.65
5	64.1	13.32
6	38.3	14.25
7	31.6	7.65
8	49.2	19.62
9	112	5.70
10	27.3	23.7

Problems

See the Preface for an explanation of the icons used in this problems set.
For additional assessment items for this section, go to 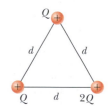 WEB**ASSIGN** From Cengage

SECTION 24.1 Electric Potential and Potential Difference

1. **V** How much work is done (by a battery, generator, or some other source of potential difference) in moving Avogadro's number of electrons from an initial point where the electric potential is 9.00 V to a point where the electric potential is −5.00 V? (The potential in each case is measured relative to a common reference point.)

2. **Q|C** (a) Find the electric potential difference ΔV_e required to stop an electron (called a "stopping potential") moving with an initial speed of 2.85×10^7 m/s. (b) Would a proton traveling at the same speed require a greater or lesser magnitude of electric potential difference? Explain. (c) Find a symbolic expression for the ratio of the proton stopping potential and the electron stopping potential, $\Delta V_p/\Delta V_e$.

SECTION 24.2 Potential Difference in a Uniform Electric Field

3. **T** Oppositely charged parallel plates are separated by 5.33 mm. A potential difference of 600 V exists between the plates. (a) What is the magnitude of the electric field between the plates? (b) What is the magnitude of the force on an electron between the plates? (c) How much work must be done on the electron to move it to the negative plate if it is initially positioned 2.90 mm from the positive plate?

4. **Q|C** **S** Starting with the definition of work, prove that at every point on an equipotential surface, the surface must be perpendicular to the electric field there.

5. **Q|C** An insulating rod having linear charge density $\lambda = 40.0$ μC/m and linear mass density $\mu = 0.100$ kg/m is released from rest in a uniform electric field $E = 100$ V/m directed perpendicular to the rod (Fig. P24.5). (a) Determine the speed of the rod after it has traveled 2.00 m. (b) **What If?** How does your answer to part (a) change if the electric field is not perpendicular to the rod? Explain.

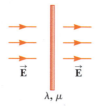

Figure P24.5

6. **GP** **Q|C** **Review.** A block having mass m and charge $+Q$ is connected to an insulating spring having a force constant k. The block lies on a frictionless, insulating, horizontal track, and the system is immersed in a uniform electric field of magnitude E directed as shown in Figure P24.6. The block is released from rest when the spring is unstretched (at $x = 0$). We wish to show that the ensuing motion of the block is simple harmonic. (a) Consider the system of the block, the spring, and the electric field. Is this system isolated or nonisolated? (b) What kinds of potential energy exist within this system? (c) Call the initial configuration of the system that existing just as the block is released from rest. The final configuration is when the block momentarily comes to rest again. What is the value of x when the block comes to rest momentarily? (d) At some value of x we will call $x = x_0$, the block has zero net force on it. What analysis model describes the particle in this situation? (e) What is the value of x_0? (f) Define a new coordinate system x' such that $x' = x - x_0$. Show that x' satisfies a differential equation for simple harmonic motion. (g) Find the period of the simple harmonic motion. (h) How does the period depend on the electric field magnitude?

SECTION 24.3 Electric Potential and Potential Energy Due to Point Charges

Note: Unless stated otherwise, assume the reference level of potential is $V = 0$ at $r = \infty$.

7. **S** Three positive charges are located at the corners of an equilateral triangle as in Figure P24.7. Find an expression for the electric potential at the center of the triangle.

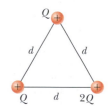

Figure P24.7

8. **Q|C** **T** Two point charges $Q_1 = +5.00$ nC and $Q_2 = -3.00$ nC are separated by 35.0 cm. (a) What is the electric potential at a point midway between the charges? (b) What is the potential energy of the pair of charges? What is the significance of the algebraic sign of your answer?

9. **CR** You are working on a laboratory device that includes a small sphere with a large electric charge Q. Because of this charged sphere, there is a strong electric field surrounding your device. Other researchers in your laboratory are complaining that your electric field is affecting their equipment. You think about how you can obtain the large electric field that you need close to the sphere but prohibit the field from reaching your colleagues. You decide to surround your device with a spherical transparent plastic shell of radius R. The plastic has a very thin coating of conducting material on the outside that only minimally reduces the transparency of the material. The shell is placed so that the small sphere is at the exact center of the shell. Determine to what electric potential the outer shell must be raised to completely eliminate the electric field outside of the shell.

10. **CR** Your roommate is having trouble understanding why solids form. He asks, "Why would atoms bond into solids rather than just floating freely with respect to each other?" To help him understand at least one type of bonding in solids, you decide to embark on an energy explanation. You show him a drawing of a primitive cell of a sodium chloride crystal, NaCl, or simple table salt. The drawing is shown in Figure P24.10, where the orange spheres are Na$^+$ ions and the blue spheres are Cl$^-$ ions. Each ion has a charge of magnitude equal to the elementary charge e. The ions lie on the corners of a cube of side d. You explain to your roommate that the electrical potential energy is defined as zero when all eight charges

Figure P24.6

Figure P24.10

are infinitely far apart from each other. Then you bring them together to form the crystal structure shown. (a) Evaluate the electric potential energy of the crystal as shown and (b) show that it is energetically favorable for such crystals to form.

11. [S] Four point charges each having charge Q are located at the corners of a square having sides of length a. Find expressions for (a) the total electric potential at the center of the square due to the four charges and (b) the work required to bring a fifth charge q from infinity to the center of the square.

12. The two charges in Figure P24.12 are separated by a distance $d = 2.00$ cm, and $Q = +5.00$ nC. Find (a) the electric potential at A, (b) the electric potential at B, and (c) the electric potential difference between B and A.

Figure P24.12

13. [S] Show that the amount of work required to assemble four identical charged particles of magnitude Q at the corners of a square of side s is $5.41 k_e Q^2/s$.

14. [S] Two charged particles of equal magnitude are located along the y axis equal distances above and below the x axis as shown in Figure P24.14. (a) Plot a graph of the electric potential at points along the x axis over the interval $-3a < x < 3a$. You should plot the potential in units of k_eQ/a. (b) Let the charge of the particle located at $y = -a$ be negative. Plot the potential along the y axis over the interval $-4a < y < 4a$.

Figure P24.14

15. [S] Three particles with equal positive charges q are at the corners of an equilateral triangle of side a as shown in Figure P24.15. (a) At what point, if any, in the plane of the particles is the electric potential zero? (b) What is the electric potential at the position of one of the particles due to the other two particles in the triangle?

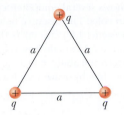

Figure P24.15

16. [S] **Review.** A light, unstressed spring has length d. Two identical particles, each with charge q, are connected to the opposite ends of the spring. The particles are held stationary a distance d apart and then released at the same moment. The system then oscillates on a frictionless, horizontal table. The spring has a bit of internal kinetic friction, so the oscillation is damped. The particles eventually stop vibrating when the distance between them is $3d$. Assume the system of the spring and two charged particles is isolated. Find the increase in internal energy that appears in the spring during the oscillations.

17. [AMT] [Q|C] **Review.** Two insulating spheres have radii 0.300 cm and 0.500 cm, masses 0.100 kg and 0.700 kg, and uniformly distributed charges -2.00 μC and 3.00 μC. They are released from rest when their centers are separated by 1.00 m. (a) How fast will each be moving when they collide? (b) **What If?** If the spheres were conductors, would the speeds be greater or less than those calculated in part (a)? Explain.

18. [Q|C] [S] **Review.** Two insulating spheres have radii r_1 and r_2, masses m_1 and m_2, and uniformly distributed charges $-q_1$ and q_2. They are released from rest when their centers are separated by a distance d. (a) How fast is each moving when they collide? (b) **What If?** If the spheres were conductors, would their speeds be greater or less than those calculated in part (a)? Explain.

19. [S] How much work is required to assemble eight identical charged particles, each of magnitude q, at the corners of a cube of side s?

20. [S] Four identical particles, each having charge q and mass m, are released from rest at the vertices of a square of side L. How fast is each particle moving when their distance from the center of the square doubles?

SECTION 24.4 Obtaining the Value of the Electric Field from the Electric Potential

21. [S] It is shown in Example 24.7 that the potential at a point P a distance a above one end of a uniformly charged rod of length ℓ lying along the x axis is

$$V = k_e \frac{Q}{\ell} \ln \left(\frac{\ell + \sqrt{a^2 + \ell^2}}{a} \right)$$

Use this result to derive an expression for the y component of the electric field at P.

22. Figure P24.22 represents a graph of the electric potential in a region of space versus position x, where the electric field is parallel to the x axis. Draw a graph of the x component of the electric field versus x in this region.

Figure P24.22

23. Figure P24.23 shows several equipotential lines, each labeled by its potential in volts. The distance between the lines of the square grid represents 1.00 cm. (a) Is the magnitude of the field larger at A or at B? Explain how you can tell. (b) Explain what you can determine about $\vec{E}$ at B. (c) Represent what the electric field looks like by drawing at least eight field lines.

Numerical values are in volts.

Figure P24.23

24. An electric field in a region of space is parallel to the x axis. The electric potential varies with position as shown in Figure P24.24. Graph the x component of the electric field versus position in this region of space.

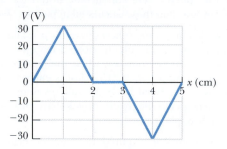

Figure P24.24

SECTION 24.5 Electric Potential Due to Continuous Charge Distributions

25. A rod of length L (Fig. P24.25) lies along the x axis with its
S left end at the origin. It has a nonuniform charge density $\lambda = \alpha x$, where α is a positive constant. (a) What are the units of α? (b) Calculate the electric potential at A.

Figure P24.25 Problems 25 and 26.

26. For the arrangement described in Problem 25, calculate the
S electric potential at point B, which lies on the perpendicular bisector of the rod a distance b above the x axis.

27. A wire having a uniform linear charge density λ is bent into
S the shape shown in Figure P24.27. Find the electric poten-
V tial at point O.

Figure P24.27

28. You are a coach for the Physics Olympics team participating
CR in a competition overseas. You are given the following sample problem to present to your team of students, which you need to solve very quickly: A person is standing on the midline of a soccer field. At one end of the field, as shown in Figure P24.28, is a letter D, consisting of a semicircular metallic ring of radius R and a long straight metal rod of length $2R$, the diameter of the ring. The plane of the ring is perpendicular to the ground and perpendicular to the midline of the field shown by the broken line in Figure P24.28. Because of an approaching lightning storm, the semicircular ring and the rod become charged. The ring and the rod each attain a charge Q. What is the electric potential at point P, which is at a position x along the midline of the field, measured from the center of the rod, due to the letter D? Think quickly and use all resources available to you, which include your physics textbook: you are in competition!

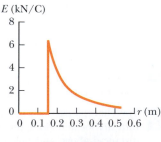

Figure P24.28

SECTION 24.6 Conductors in Electrostatic Equilibrium

29. The electric field magnitude on the surface of an irregularly shaped conductor varies from 56.0 kN/C to 28.0 kN/C. Can you evaluate the electric potential on the conductor? If so, find its value. If not, explain why not.

30. *Why is the following situation impossible?* A solid copper sphere of radius 15.0 cm is in electrostatic equilibrium and carries a charge of 40.0 nC. Figure P24.30 shows the magnitude of the electric field as a function of radial position r measured from the center of the sphere.

Figure P24.30

31. A solid metallic sphere of radius a carries total charge Q.
Q|C No other charges are nearby. The electric field just outside
S its surface is $k_e Q/a^2$ radially outward. At this close point, the
V uniformly charged surface of the sphere looks exactly like a uniform flat sheet of charge. Is the electric field here given by σ/ϵ_0 or by $\sigma/2\epsilon_0$?

32. A positively charged particle is at a distance $R/2$ from the center of an uncharged thin, conducting, spherical shell of radius R. Sketch the electric field lines set up by this arrangement both inside and outside the shell.

33. A very large, thin, flat plate of aluminum of area A has a
S total charge Q uniformly distributed over its surfaces.

Assuming the same charge is spread uniformly over the *upper* surface of an otherwise identical glass plate, compare the electric fields just above the center of the upper surface of each plate.

34. A solid conducting sphere of radius 2.00 cm has a charge of 8.00 μC. A conducting spherical shell of inner radius 4.00 cm and outer radius 5.00 cm is concentric with the solid sphere and has a charge of −4.00 μC. Find the electric field at (a) $r = 1.00$ cm, (b) $r = 3.00$ cm, (c) $r = 4.50$ cm, and (d) $r = 7.00$ cm from the center of this charge configuration.

35. A spherical conductor has a radius of 14.0 cm and a charge of 26.0 μC. Calculate the electric field and the electric potential at (a) $r = 10.0$ cm, (b) $r = 20.0$ cm, and (c) $r = 14.0$ cm from the center.

36. A long, straight wire is surrounded by a hollow metal cylinder whose axis coincides with that of the wire. The wire has a charge per unit length of λ, and the cylinder has a net charge per unit length of 2λ. From this information, use Gauss's law to find (a) the charge per unit length on the inner surface of the cylinder, (b) the charge per unit length on the outer surface of the cylinder, and (c) the electric field outside the cylinder a distance r from the axis.

ADDITIONAL PROBLEMS

37. *Why is the following situation impossible?* In the Bohr model of the hydrogen atom, an electron moves in a circular orbit about a proton. The model states that the electron can exist only in certain allowed orbits around the proton: those whose radius r satisfies $r = n^2(0.052\ 9$ nm$)$, where $n = 1, 2, 3, \ldots$. For one of the possible allowed states of the atom, the electric potential energy of the system is −13.6 eV.

38. On a dry winter day, you scuff your leather-soled shoes across a carpet and get a shock when you extend the tip of one finger toward a metal doorknob. In a dark room, you see a spark perhaps 5 mm long. Make order-of-magnitude estimates of (a) your electric potential and (b) the charge on your body before you touch the doorknob. Explain your reasoning.

39. (a) Use the exact result from Example 24.4 to find the electric potential created by the dipole described in the example at the point $(3a, 0)$. (b) Explain how this answer compares with the result of the approximate expression that is valid when x is much greater than a.

40. *Why is the following situation impossible?* You set up an apparatus in your laboratory as follows. The x axis is the symmetry axis of a stationary, uniformly charged ring of radius $R = 0.500$ m and charge $Q = 50.0$ μC (Fig. P24.40). You place a particle with charge $Q = 50.0$ μC and mass $m = 0.100$ kg at the center of the ring and arrange for it to be constrained to move only along the x axis. When it is displaced slightly, the particle is repelled by the ring and accelerates along the x axis. The particle moves faster than you expected and strikes the opposite wall of your laboratory at 40.0 m/s.

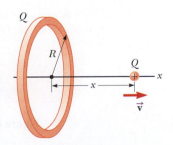

Figure P24.40

41. The thin, uniformly charged rod shown in Figure P24.41 has a linear charge density λ. Find an expression for the electric potential at *P*.

Figure P24.41

42. A Geiger–Mueller tube is a radiation detector that consists of a closed, hollow, metal cylinder (the cathode) of inner radius r_a and a coaxial cylindrical wire (the anode) of radius r_b (Fig. P24.42a). The charge per unit length on the anode is λ, and the charge per unit length on the cathode is −λ. A gas fills the space between the electrodes. When the tube is in use (for example, in measuring radioactivity from fruit in Fig. P24.42b) and a high-energy elementary particle passes through this space, it can ionize an atom of the gas. The strong electric field makes the resulting ion and electron accelerate in opposite directions. They strike other molecules of the gas to ionize them, producing an avalanche of electrical discharge. The pulse of electric current between the wire and the cylinder is counted by an external circuit. (a) Show that the magnitude of the electric potential difference between the wire and the cylinder is

$$\Delta V = 2k_e\lambda \ln\left(\frac{r_a}{r_b}\right)$$

(b) Show that the magnitude of the electric field in the space between cathode and anode is

$$E = \frac{\Delta V}{\ln(r_a/r_b)}\left(\frac{1}{r}\right)$$

where r is the distance from the axis of the anode to the point where the field is to be calculated.

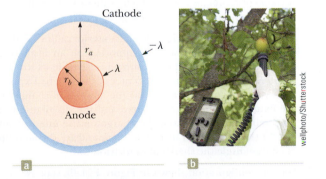

Figure P24.42

43. **Review.** Two parallel plates having charges of equal magnitude but opposite sign are separated by 12.0 cm. Each plate has a surface charge density of 36.0 nC/m². A proton is released from rest at the positive plate. Determine (a) the magnitude of the electric field between the plates from the charge density, (b) the potential difference between the plates, (c) the kinetic energy of the proton when it reaches the negative plate, (d) the speed of the proton just before it strikes the negative plate, (e) the acceleration of the proton, and (f) the force on the proton. (g) From the force, find the magnitude of the electric field. (h) How does your value of the electric field compare with that found in part (a)?

44. When an uncharged conducting sphere of radius a is placed at the origin of an *xyz* coordinate system that lies in an initially uniform electric field $\vec{E} = E_0\hat{k}$, the resulting electric potential is $V(x, y, z) = V_0$ for points inside the sphere and

$$V(x, y, z) = V_0 - E_0 z + \frac{E_0 a^3 z}{(x^2 + y^2 + z^2)^{3/2}}$$

for points outside the sphere, where V_0 is the (constant) electric potential on the conductor. Use this equation to determine the x, y, and z components of the resulting electric field (a) inside the sphere and (b) outside the sphere.

45. A solid, insulating sphere of radius a has a uniform charge density throughout its volume and a total charge Q. Concentric with this sphere is an uncharged, conducting, hollow sphere whose inner and outer radii are b and c as shown in Figure P24.45. We wish to understand completely

Insulator

Conductor

Figure P24.45

Problems 45 and 47.

the charges and electric fields at all locations. (a) Find the charge contained within a sphere of radius $r < a$. (b) From this value, find the magnitude of the electric field for $r < a$. (c) What charge is contained within a sphere of radius r when $a < r < b$? (d) From this value, find the magnitude of the electric field for r when $a < r < b$. (e) Now consider r when $b < r < c$. What is the magnitude of the electric field for this range of values of r? (f) From this value, what must be the charge on the inner surface of the hollow sphere? (g) From part (f), what must be the charge on the outer surface of the hollow sphere? (h) Consider the three spherical surfaces of radii a, b, and c. Which of these surfaces has the largest magnitude of surface charge density?

46. A hollow, metallic, spherical shell has exterior radius 0.750 m, carries no net charge, and is supported on an insulating stand. The electric field everywhere just outside its surface is 890 N/C radially toward the center of the sphere. Explain what you can conclude about (a) the amount of charge on the exterior surface of the sphere and the distribution of this charge, (b) the amount of charge on the interior surface of the sphere and its distribution, and (c) the amount of charge inside the shell and its distribution.

47. For the configuration shown in Figure P24.45, suppose $a = 5.00$ cm, $b = 20.0$ cm, and $c = 25.0$ cm. Furthermore, suppose the electric field at a point 10.0 cm from the center is measured to be 3.60×10^3 N/C radially inward and the electric field at a point 50.0 cm from the center is of magnitude 200 N/C and points radially outward. From this information, find (a) the charge on the insulating sphere, (b) the net charge on the hollow conducting sphere, (c) the charge on the inner surface of the hollow conducting sphere, and (d) the charge on the outer surface of the hollow conducting sphere.

CHALLENGE PROBLEMS

48. An electric dipole is located along the y axis as shown in Figure P24.48. The magnitude of its electric dipole moment is defined as $p = 2aq$. (a) At a point P, which is far from the dipole ($r \gg a$), show that the electric potential is

$$V = \frac{k_e p \cos \theta}{r^2}$$

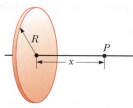

Figure P24.48

(b) Calculate the radial component E_r and the perpendicular component E_θ of the associated electric field. Note that $E_\theta = -(1/r)(\partial V/\partial \theta)$. Do these results seem reasonable for (c) $\theta = 90°$ and $0°$? (d) For $r = 0$? (e) For the dipole arrangement shown in Figure P24.48, express V in terms of Cartesian coordinates using $r = (x^2 + y^2)^{1/2}$ and

$$\cos \theta = \frac{y}{(x^2 + y^2)^{1/2}}$$

(f) Using these results and again taking $r \gg a$, calculate the field components E_x and E_y.

49. A disk of radius R (Fig. P24.49) has a nonuniform surface charge density $\sigma = Cr$, where C is a constant and r is measured from the center of the disk to a point on the surface of the disk. Find (by direct integration) the electric potential at P.

Figure P24.49

50. A particle with charge q is located at $x = -R$, and a particle with charge $-2q$ is located at the origin. Prove that the equipotential surface that has zero potential is a sphere centered at $(-4R/3, 0, 0)$ and having a radius $r = \frac{2}{3}R$.

51. (a) A uniformly charged cylindrical shell with no end caps has total charge Q, radius R, and length h. Determine the electric potential at a point a distance d from the right end of the cylinder as shown in Figure P24.51. *Suggestion:* Use the result of Example 24.5 by treating the cylinder as a collection of ring charges. (b) **What If?** Use the result of Example 24.6 to solve the same problem for a solid cylinder.

Figure P24.51

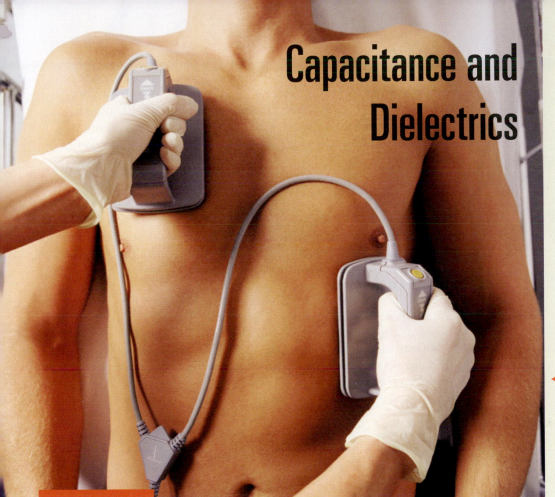

Capacitance and Dielectrics

When a patient receives a shock from a defibrillator, the energy delivered to the patient is initially stored in a *capacitor*. We will study capacitors and capacitance in this chapter. (*Andrew Olney/Getty Images*)

STORYLINE **Your spring break trip to Florida with your fellow** students continues. You have been there several days and all of you have noticed that there is a *lot* of lightning in Florida! One of your friends suggests calling your electrical engineer uncle and asking him about all the lightning. You call your uncle and mention your observation about the amount of lightning in Florida. He agrees with you and mentions that he did atmospheric modeling when he was younger. He says to you, "Did you know that lightning can be analyzed by modeling the entire atmosphere of the Earth as a giant capacitor with a capacitance of about 1 *farad*? That's a huge capacitance! For example, a defibrillator used in an emergency situation to send a powerful jolt into a patient might have a maximum capacitance of only a couple hundred *microfarads*!" After an embarrassing moment of silence, you make what you think is an appropriate response and excuse yourself. Ignoring the questioning looks of your friends, you run into another room, snatch your smartphone from your pocket, and, before you forget the words, start doing online research for the words *capacitor, capacitance, defibrillator,* and *farad*.

CONNECTIONS In the introduction to the previous chapter, we mentioned electric circuits. Electric circuits consist of various *circuit elements* that are connected by wires. In this chapter, we introduce the first of three simple circuit elements that we will discuss. Electric circuits are the basis for the vast majority of the devices used in our society. Here we shall discuss *capacitors*, devices that store electric charge. This discussion is followed by the study of resistors in Chapter 26 and *inductors* in Chapter 31. In later chapters, we will study more sophisticated circuit elements such as *transformers* and *transistors*. Capacitors are commonly used in a variety of electric circuits. For instance, they are used to tune the frequency of radio receivers, as filters in power supplies, as energy storage devices for cardiac defibrillators, and as accelerometers in smartphones. We will be combining capacitors with other circuit elements in future chapters.

25.1 Definition of Capacitance

Any combination of two separated conductors can act as an electrical circuit element and is called a **capacitor.** The conductors are called *plates.* If the conductors carry charges of equal magnitude and opposite sign as shown in Figure 25.1, a potential difference ΔV exists between them.

It seems reasonable that increasing the amount of charge on the conductors would increase the potential difference between them, but what is the exact relationship between charge and potential difference for a capacitor? Experiments show that the quantity of charge Q on a capacitor[1] is linearly proportional to the potential difference between the conductors; that is, $Q \propto \Delta V$. The proportionality constant depends on the shape and separation of the conductors.[2] Because of this proportionality, the ratio of charge to potential difference is a constant defined as follows:

◀ Definition of capacitance

> The **capacitance** C of a capacitor is defined as the ratio of the magnitude of the charge on either conductor to the magnitude of the potential difference between the conductors:
>
> $$C \equiv \frac{Q}{\Delta V} \tag{25.1}$$

By definition *capacitance is always a positive quantity.* Furthermore, the charge Q and the potential difference ΔV are always expressed in Equation 25.1 as positive quantities.

From Equation 25.1, we see that capacitance has SI units of coulombs per volt. Named in honor of Michael Faraday, the SI unit of capacitance is the **farad** (F):

$$1 \text{ F} = 1 \text{ C/V}$$

The farad is a very large unit of capacitance. In practice, typical devices have capacitances ranging from microfarads (10^{-6} F) to picofarads (10^{-12} F). We shall use the symbol μF to represent microfarads. In practice, to avoid the use of Greek letters,

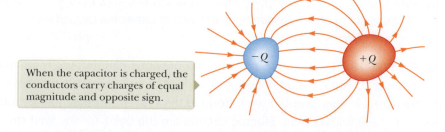

When the capacitor is charged, the conductors carry charges of equal magnitude and opposite sign.

Figure 25.1 A capacitor consists of two conductors. If the capacitor is charged as shown here, a potential difference ΔV exists between the conductors.

---

[1]Although the total charge on the capacitor is zero (because there is as much excess positive charge on one conductor as there is excess negative charge on the other), it is common practice to refer to the magnitude of the charge on either conductor as "the charge on the capacitor."

[2]The proportionality between Q and ΔV can be proven from Coulomb's law or by experiment.

physical capacitors are often labeled "mF" for microfarads and "mmF" for micromicrofarads or, equivalently, "pF" for picofarads.

Let's consider a capacitor formed from a pair of parallel plates as shown in Figure 25.2. Each plate is connected to one terminal of a battery, which acts as a source of potential difference. If the capacitor is initially uncharged, the battery establishes an electric field in the connecting wires when the connections are made. Let's focus on the plate connected to the negative terminal of the battery. The electric field in the wire applies a force on electrons in the wire immediately outside this plate; this force causes the electrons to move onto the plate. The movement continues until the plate, the wire, and the terminal are all at the same electric potential. Once this equilibrium situation is attained, a potential difference no longer exists between the terminal and the plate; as a result, no electric field is present in the wire and the electrons stop moving. The plate now carries a negative charge. A similar process occurs at the other capacitor plate, where electrons move from the plate to the wire, leaving the plate positively charged. In this final configuration, the potential difference across the capacitor plates is the same as that between the terminals of the battery.

Now suppose we disconnect the battery from the plates. The plates are not connected with a wire to anything now, so the plates remain charged. The capacitor has *stored* the charge. It has also stored energy, associated with the separation of charges. We will explore these ideas and the uses of capacitors after performing a bit more mathematical analysis.

QUICK QUIZ 25.1 A capacitor stores charge Q at a potential difference ΔV. What happens if the voltage applied to the capacitor by a battery is doubled to $2\,\Delta V$? **(a)** The capacitance falls to half its initial value, and the charge remains the same. **(b)** The capacitance and the charge both fall to half their initial values. **(c)** The capacitance and the charge both double. **(d)** The capacitance remains the same, and the charge doubles.

When the capacitor is connected to the terminals of a battery, electrons transfer between the plates and the wires so that the plates become charged.

Figure 25.2 A parallel-plate capacitor consists of two parallel conducting plates, each of area A, separated by a distance d.

25.2 Calculating Capacitance

We can derive an expression for the capacitance of a pair of conductors in the following manner. First we calculate the potential difference between the conductors, assuming that they have a charge Q, using the techniques described in Chapter 24. We then use Equation 25.1 to evaluate the capacitance. The calculation is relatively easy if the geometry of the capacitor is simple. Let's look at some examples.

Although the most common situation is that of two conductors, a single conductor also has a capacitance. For example, imagine a single spherical, charged conductor. The electric field lines around this conductor are exactly the same as if there were a conducting, spherical shell of infinite radius, concentric with the sphere and carrying a charge of the same magnitude but opposite sign. Therefore, we can identify the imaginary shell as the second conductor of a two-conductor capacitor. The electric potential of the sphere of radius a is simply $k_e Q/a$ (see Section 24.6), and setting $V = 0$ for the infinitely large shell gives

$$C = \frac{Q}{\Delta V} = \frac{Q}{k_e Q/a} = \frac{a}{k_e} = 4\pi\epsilon_0 a \qquad (25.2)$$

◀ Capacitance of an isolated charged sphere

This expression shows that the capacitance of an isolated, charged sphere is proportional to its radius and is independent of both the charge on the sphere and its potential, as is the case with all capacitors. Equation 25.1 is the general definition of the capacitance of an arbitrary capacitor in terms of electrical parameters, but the capacitance of a given capacitor will depend only on the geometry of the plates.

PITFALL PREVENTION 25.3

Too Many Cs Do not confuse an italic C for capacitance with a non-italic C for the unit coulomb.

Let's now imagine two parallel, metallic plates of equal area A, separated by a distance d as shown in Figure 25.2. One plate carries a charge $+Q$, and the other carries a charge $-Q$. The surface charge density on each plate is $\sigma = Q/A$. If the plates are very close together (in comparison with their length and width), we can assume the electric field is uniform between the plates and zero elsewhere. According to the What If? feature of Example 23.8, the value of the electric field between the plates is

$$E = \frac{\sigma}{\epsilon_0} = \frac{Q}{\epsilon_0 A}$$

Because the field between the plates is uniform, the magnitude of the potential difference between the plates equals Ed (see Eq. 24.6); therefore,

$$\Delta V = Ed = \frac{Qd}{\epsilon_0 A}$$

Substituting this result into Equation 25.1, we find that the capacitance is

$$C = \frac{Q}{\Delta V} = \frac{Q}{Qd/\epsilon_0 A}$$

Capacitance of parallel plates ▶

$$C = \frac{\epsilon_0 A}{d} \qquad (25.3)$$

That is, the capacitance of a parallel-plate capacitor is proportional to the area of its plates and inversely proportional to the plate separation.

Let's consider how the geometry of these conductors influences the capacity of the pair of plates to store charge. As a capacitor is being charged by a battery, electrons flow into the negative plate and out of the positive plate. If the capacitor plates are large, the accumulated charges are able to distribute themselves over a substantial area and the amount of charge that can be stored on a plate for a given potential difference increases as the plate area is increased. Therefore, it is reasonable that the capacitance is proportional to the plate area A as in Equation 25.3.

Now consider the region that separates the plates. Imagine moving the plates closer together. Consider the situation before any charges have had a chance to move in response to this change. Because no charges have moved, the electric field between the plates has the same value but extends over a shorter distance. Therefore, the magnitude of the potential difference between the plates $\Delta V = Ed$ (Eq. 24.6) is smaller. The difference between this new capacitor voltage and the terminal voltage of the battery appears as a potential difference across the wires connecting the battery to the capacitor, resulting in an electric field in the wires that drives more charge onto the plates and increases the potential difference between the plates. When the potential difference between the plates again matches that of the battery, the flow of charge stops. Therefore, moving the plates closer together causes the charge on the capacitor to increase. If d is increased, the charge decreases. As a result, the inverse relationship between C and d in Equation 25.3 is reasonable.

Key

Movable plate

Insulator

Fixed plate

Figure 25.3 (Quick Quiz 25.2) One type of computer keyboard button.

QUICK QUIZ **25.2** Many computer keyboard buttons are constructed of capacitors as shown in Figure 25.3. When a key is pushed down, the soft insulator between the movable plate and the fixed plate is compressed. When the key is pressed, what happens to the capacitance? **(a)** It increases. **(b)** It decreases. **(c)** It changes in a way you cannot determine because the electric circuit connected to the keyboard button may cause a change in ΔV.

Example 25.1 The Cylindrical Capacitor

A solid cylindrical conductor of radius a is coaxial with a cylindrical shell of negligible thickness and radius $b > a$ (Fig. 25.4a). Find the capacitance of this cylindrical capacitor if its length is $\ell \gg b$.

SOLUTION

Conceptualize Recall that any pair of conductors qualifies as a capacitor, so the system described in this example therefore qualifies. Figure 25.4b helps visualize the electric field between the conductors if the capacitor carries a charge Q. We expect the capacitance to depend only on geometric factors, which, in this case, are a, b, and ℓ.

Categorize Because of the cylindrical symmetry of the system, we can use results from previous studies of cylindrical systems to find the capacitance.

Analyze Assuming the capacitor carries a charge Q and ℓ is much greater than a and b, we can neglect end effects. In this case, the electric field is perpendicular to the long axis of the cylinders and is confined to the region between them (Fig. 25.4b).

Figure 25.4 (Example 25.1) (a) A cylindrical capacitor consists of a solid cylindrical conductor of radius a and length ℓ surrounded by a coaxial cylindrical shell of radius b. (b) End view of the capacitor if it is charged. The electric field lines are radial.

Write an expression for the potential difference between the two charged cylinders from Equation 24.3:

$$V_b - V_a = -\int_a^b \vec{\mathbf{E}} \cdot d\vec{\mathbf{s}}$$

Notice from Figure 25.4b that $\vec{\mathbf{E}}$ is parallel to $d\vec{\mathbf{s}}$ along a radial line and apply Equation 23.8 for the electric field outside a cylindrically symmetric charge distribution:

$$V_b - V_a = -\int_a^b E_r\, dr = -2k_e\lambda\int_a^b \frac{dr}{r} = -2k_e\lambda \ln\left(\frac{b}{a}\right)$$

Substitute the absolute value of ΔV into Equation 25.1 and use $\lambda = Q/\ell$:

$$C = \frac{Q}{\Delta V} = \frac{Q}{(2k_e Q/\ell)\ln(b/a)} = \frac{\ell}{2k_e \ln(b/a)} \qquad (25.4)$$

Finalize The capacitance depends on the radii a and b and is proportional to the length of the cylinders. Equation 25.4 shows that the capacitance per unit length of a combination of concentric cylindrical conductors is

$$\frac{C}{\ell} = \frac{1}{2k_e \ln(b/a)} \qquad (25.5)$$

An example of this type of geometric arrangement is a *coaxial cable*, which consists of two concentric cylindrical conductors separated by an insulator. You probably have a coaxial cable attached to your television set if you are a subscriber to cable television. The coaxial cable is especially useful for shielding electrical signals from any possible external influences.

WHAT IF? Suppose $b = 2.00a$ for the cylindrical capacitor. You would like to increase the capacitance, and you can do so by choosing to increase either ℓ by 10% or a by 10%. Which choice is more effective at increasing the capacitance?

Answer According to Equation 25.4, C is proportional to ℓ, so increasing ℓ by 10% results in a 10% increase in C. For the result of the change in a, let's use Equation 25.4 to set up a ratio of the capacitance C' for the enlarged cylinder radius a' to the original capacitance:

$$\frac{C'}{C} = \frac{\ell/2k_e \ln(b/a')}{\ell/2k_e \ln(b/a)} = \frac{\ln(b/a)}{\ln(b/a')}$$

We now substitute $b = 2.00a$ and $a' = 1.10a$, representing a 10% increase in a:

$$\frac{C'}{C} = \frac{\ln(2.00a/a)}{\ln(2.00a/1.10a)} = \frac{\ln 2.00}{\ln 1.82} = 1.16$$

which corresponds to a 16% increase in capacitance. Therefore, it is more effective to increase a than to increase ℓ.

Note two more extensions of this problem. First, it is advantageous to increase a only for a range of relationships between a and b. If $b > 2.85a$, increasing ℓ by 10% is more effective than increasing a (see Problem 44). Second, if b decreases, the capacitance increases. Increasing a or decreasing b has the effect of bringing the plates closer together, which increases the capacitance.

Example **25.2**   **The Spherical Capacitor**

A spherical capacitor consists of a spherical conducting shell of radius b concentric with a smaller conducting sphere of radius a (Fig. 25.5). Find the capacitance of this device.

Figure 25.5 (Example 25.2) A spherical capacitor consists of an inner sphere of radius a surrounded by a concentric spherical shell of radius b. The diagram shows the capacitor carrying a charge Q. The electric field between the spheres is directed radially outward when the inner sphere is positively charged.

SOLUTION

Conceptualize   As with Example 25.1, this system involves a pair of conductors and qualifies as a capacitor. We expect the capacitance to depend on the spherical radii a and b.

Categorize   Because of the spherical symmetry of the system, we can use results from previous studies of spherical systems to find the capacitance.

Analyze   Imagine the capacitor carries a charge Q with the inner sphere positive, as shown in Figure 25.5. As shown in Chapter 23, the direction of the electric field outside a spherically symmetric charge distribution is radial and its magnitude is given by the expression $E = k_e Q / r^2$. In this case, this result applies to the field *between* the spheres ($a < r < b$).

Write an expression for the potential difference between the two charged conductors from Equation 24.3:

$$V_b - V_a = -\int_a^b \vec{\mathbf{E}} \cdot d\vec{\mathbf{s}}$$

Notice that $\vec{\mathbf{E}}$ is parallel to $d\vec{\mathbf{s}}$ along a radial line and apply the result of Example 23.6 for the electric field outside a spherically symmetric charge distribution:

$$V_b - V_a = -\int_a^b E_r \, dr = -k_e Q \int_a^b \frac{dr}{r^2} = k_e Q \left[\frac{1}{r} \right]_a^b$$

$$(1) \quad V_b - V_a = k_e Q \left(\frac{1}{b} - \frac{1}{a} \right) = k_e Q \frac{a-b}{ab}$$

Substitute the absolute value of ΔV into Equation 25.1:

$$C = \frac{Q}{\Delta V} = \frac{Q}{|V_b - V_a|} = \frac{ab}{k_e(b-a)} \qquad (25.6)$$

Finalize   The capacitance depends on a and b as expected. The potential difference between the spheres in Equation (1) is negative because Q is positive and $b > a$. Therefore, in Equation 25.6, when we take the absolute value, we change $a - b$ to $b - a$. The result is a positive number.

WHAT IF?   If the radius b of the outer sphere approaches infinity, what does the capacitance become?

Answer   In Equation 25.6, we let $b \to \infty$:

$$C = \lim_{b \to \infty} \frac{ab}{k_e(b-a)} = \frac{ab}{k_e(b)} = \frac{a}{k_e} = 4\pi\epsilon_0 a$$

Notice that this expression is the same as Equation 25.2, the capacitance of an isolated spherical conductor.

Capacitor symbol

Battery symbol

Switch symbol   Open

Closed

Figure 25.6 Circuit symbols for capacitors, batteries, and switches. Notice that capacitors are in blue, batteries are in green, and switches are in red. The closed switch represents an electrical connection between the red circles, while the open switch represents no connection.

25.3 Combinations of Capacitors

Two or more capacitors often are combined in electric circuits. We can calculate the equivalent capacitance of certain combinations using methods described in this section. Throughout this section, we assume the capacitors to be combined are initially uncharged.

In studying electric circuits, we use a simplified pictorial representation called a **circuit diagram.** Such a diagram uses **circuit symbols** to represent various circuit elements. The circuit symbols are connected by straight lines that represent the wires between the circuit elements. The circuit symbols for capacitors, batteries, and switches as well as the color codes used for them in this text are given in Figure 25.6. The symbol for the capacitor reflects the geometry of the most common model for a capacitor, a pair of parallel plates. The positive terminal of the battery is at the higher potential and is represented in the circuit symbol by the longer line.

Parallel Combination

Two capacitors connected as shown in Figure 25.7a are known as a **parallel combination** of capacitors. Figure 25.7b shows a circuit diagram for this combination of capacitors. The left plates of the capacitors are connected to the positive terminal of the battery by a conducting wire and are therefore both at the same electric potential as the positive terminal. Likewise, the right plates are connected to the negative terminal and so are both at the same potential as the negative terminal. Therefore, the individual potential differences across capacitors connected in parallel are the same and are equal to the potential difference applied across the combination. That is,

$$\Delta V_1 = \Delta V_2 = \Delta V \qquad (25.7)$$

where ΔV is the battery terminal voltage.

After the battery is attached to the circuit, the capacitors quickly reach their maximum charge. Let's call the maximum charges on the two capacitors Q_1 and Q_2, where $Q_1 = C_1\Delta V_1$ and $Q_2 = C_2\Delta V_2$. The *total charge* Q_{tot} stored by the combination of two capacitors is the sum of the charges on the individual capacitors:

$$Q_{tot} = Q_1 + Q_2 = C_1\Delta V_1 + C_2\Delta V_2 \qquad (25.8)$$

Suppose you wish to replace these two capacitors by one *equivalent capacitor* having a capacitance C_{eq} as in Figure 25.7c. The effect this equivalent capacitor has on the circuit must be exactly the same as the effect of the combination of the two individual capacitors. That is, the equivalent capacitor must store charge Q_{tot} when connected to the battery. Figure 25.7c shows that the voltage across the equivalent capacitor is ΔV because the equivalent capacitor is connected directly across the battery terminals. Therefore, for the equivalent capacitor,

$$Q_{tot} = C_{eq}\,\Delta V$$

Substituting this result into Equation 25.8 gives

$$C_{eq}\,\Delta V = C_1\,\Delta V_1 + C_2\,\Delta V_2$$

$$C_{eq} = C_1 + C_2 \quad \text{(parallel combination)}$$

where we have canceled the voltages because they are all the same (Eq. 25.7). If this

A pictorial representation of two capacitors connected in parallel to a battery

A circuit diagram showing the two capacitors connected in parallel to a battery

A circuit diagram showing the equivalent capacitance of the capacitors in parallel

Figure 25.7 Two capacitors connected in parallel. All three diagrams are equivalent.

treatment is extended to three or more capacitors connected in parallel, the **equivalent capacitance** is found to be

◀ Equivalent capacitance for capacitors in parallel

$$C_{eq} = C_1 + C_2 + C_3 + \cdots \quad \text{(parallel combination)} \quad (25.9)$$

Therefore, the equivalent capacitance of a parallel combination of capacitors is (1) the algebraic sum of the individual capacitances and (2) greater than any of the individual capacitances. Statement (2) makes sense in light of Equation 25.3 because we are essentially combining the areas of all the capacitor plates when they are connected with conducting wire, and capacitance of parallel plates is proportional to area.

Series Combination

Two capacitors connected as shown in Figure 25.8a and the equivalent circuit diagram in Figure 25.8b are known as a **series combination** of capacitors. The left plate of capacitor 1 and the right plate of capacitor 2 are connected to the terminals of a battery. The other two plates are connected to each other and to nothing else; hence, they form an isolated system that is initially uncharged and must continue to have zero net charge. To analyze this combination, let's first consider the uncharged capacitors and then follow what happens immediately after a battery is connected to the circuit. When the battery is connected, electrons are transferred out of the left plate of C_1 and into the right plate of C_2. As this negative charge accumulates on the right plate of C_2, an equivalent amount of negative charge is forced off the left plate of C_2, and this left plate therefore has an excess positive charge. The negative charge leaving the left plate of C_2 causes negative charges to accumulate on the right plate of C_1. As a result, both right plates end up with a charge $-Q$ and both left plates end up with a charge $+Q$. Therefore, the charges on capacitors connected in series are the same:

$$Q_1 = Q_2 = Q \quad (25.10)$$

where Q is the charge that moved between a wire and the connected outside plate of one of the capacitors.

Figure 25.8a shows the individual voltages ΔV_1 and ΔV_2 across the capacitors. These voltages add to give the total voltage ΔV_{tot} across the combination:

$$\Delta V_{tot} = \Delta V_1 + \Delta V_2 = \frac{Q_1}{C_1} + \frac{Q_2}{C_2} \quad (25.11)$$

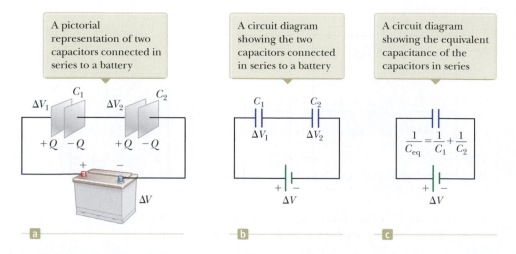

A pictorial representation of two capacitors connected in series to a battery

A circuit diagram showing the two capacitors connected in series to a battery

A circuit diagram showing the equivalent capacitance of the capacitors in series

$$\frac{1}{C_{eq}} = \frac{1}{C_1} + \frac{1}{C_2}$$

Figure 25.8 Two capacitors connected in series. All three diagrams are equivalent.

In general, the total potential difference across any number of capacitors connected in series is the sum of the potential differences across the individual capacitors.

Suppose the equivalent single capacitor in Figure 25.8c has the same effect on the circuit as the series combination when it is connected to the battery. After it is fully charged, the equivalent capacitor must have a charge of $-Q$ on its right plate and a charge of $+Q$ on its left plate. Applying the definition of capacitance to the circuit in Figure 25.8c gives

$$\Delta V_{\text{tot}} = \frac{Q}{C_{\text{eq}}}$$

Substituting this result into Equation 25.11, we have

$$\frac{Q}{C_{\text{eq}}} = \frac{Q_1}{C_1} + \frac{Q_2}{C_2}$$

Canceling the charges because they are all the same (Eq. 25.10) gives

$$\frac{1}{C_{\text{eq}}} = \frac{1}{C_1} + \frac{1}{C_2} \quad \text{(series combination)}$$

When this analysis is applied to three or more capacitors connected in series, the relationship for the **equivalent capacitance** is

$$\frac{1}{C_{\text{eq}}} = \frac{1}{C_1} + \frac{1}{C_2} + \frac{1}{C_3} + \cdots \quad \text{(series combination)} \qquad (25.12)$$

◀ Equivalent capacitance for capacitors in series

This expression shows that (1) the inverse of the equivalent capacitance is the algebraic sum of the inverses of the individual capacitances and (2) the equivalent capacitance of a series combination is always less than any individual capacitance in the combination.

QUICK QUIZ **25.3** Two capacitors are identical. They can be connected in series or in parallel. If you want the *smallest* equivalent capacitance for the combination, how should you connect them? **(a)** in series **(b)** in parallel **(c)** either way because both combinations have the same capacitance

---

Example 25.3 **Equivalent Capacitance**

Find the equivalent capacitance between a and b for the combination of capacitors shown in Figure 25.9a. All capacitances are in microfarads.

SOLUTION

Conceptualize Study Figure 25.9a carefully and make sure you understand how the capacitors are connected. Verify that there are only series and parallel connections between capacitors.

Categorize Figure 25.9a shows that the circuit contains both series and parallel connections, so we use the rules for series and parallel combinations discussed in this section.

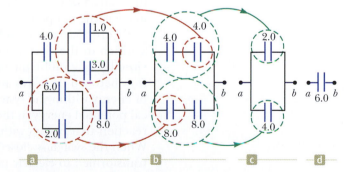

Figure 25.9 (Example 25.3) To find the equivalent capacitance of the capacitors in (a), we reduce the various combinations in steps as indicated in (b), (c), and (d), using the series and parallel rules described in the text. All capacitances are in microfarads.

Analyze Using Equations 25.9 and 25.12, we reduce the combination step by step as indicated in the figure. As you follow along below, notice that in each step we replace the combination of two capacitors in the circuit diagram with a single capacitor having the equivalent capacitance.

The 1.0-μF and 3.0-μF capacitors (upper red-brown circle in Fig. 25.9a) are in parallel. Find the equivalent capacitance from Equation 25.9:

$$C_{\text{eq}} = C_1 + C_2 = 4.0 \ \mu\text{F}$$

continued

25.3 continued

The 2.0-μF and 6.0-μF capacitors (lower red-brown circle in Fig. 25.9a) are also in parallel:

$$C_{eq} = C_1 + C_2 = 8.0 \ \mu F$$

The circuit now looks like Figure 25.9b. The two 4.0-μF capacitors (upper green circle in Fig. 25.9b) are in series. Find the equivalent capacitance from Equation 25.12:

$$\frac{1}{C_{eq}} = \frac{1}{C_1} + \frac{1}{C_2} = \frac{1}{4.0 \ \mu F} + \frac{1}{4.0 \ \mu F} = \frac{1}{2.0 \ \mu F}$$

$$C_{eq} = 2.0 \ \mu F$$

The two 8.0-μF capacitors (lower green circle in Fig. 25.9b) are also in series. Find the equivalent capacitance from Equation 25.12:

$$\frac{1}{C_{eq}} = \frac{1}{C_1} + \frac{1}{C_2} = \frac{1}{8.0 \ \mu F} + \frac{1}{8.0 \ \mu F} = \frac{1}{4.0 \ \mu F}$$

$$C_{eq} = 4.0 \ \mu F$$

The circuit now looks like Figure 25.9c. The 2.0-μF and 4.0-μF capacitors are in parallel:

$$C_{eq} = C_1 + C_2 = \boxed{6.0 \ \mu F}$$

Finalize   This final value is that of the single equivalent capacitor shown in Figure 25.9d. For further practice in treating circuits with combinations of capacitors, imagine a battery is connected between points a and b in Figure 25.9a so that a potential difference ΔV is established across the combination. Can you find the voltage across and the charge on each capacitor?

25.4 Energy Stored in a Charged Capacitor

Because positive and negative charges are separated in the system of two conductors in a charged capacitor, electric potential energy is stored in the system. Many of those who work with electronic equipment have at some time verified that a capacitor can store energy. If the plates of a charged capacitor are connected by a conductor such as a wire, charge moves between each plate and its connecting wire until the capacitor is uncharged. The discharge can often be observed as a visible spark. If you accidentally touch the opposite plates of a charged capacitor, your fingers act as a pathway for discharge and the result is an electric shock. The degree of shock you receive depends on the capacitance and the voltage applied to the capacitor. Such a shock could be dangerous if high voltages are present as in the power supply of a home theater system, for example. Because the charges can be stored in a capacitor even when the system is turned off, unplugging the system does not make it safe to open the case and touch the components inside.

Figure 25.10a shows a battery connected to a single parallel-plate capacitor with a switch in the circuit. Let us identify the circuit as a system. When the switch is closed (Fig. 25.10b), the battery establishes an electric field in the wires and charges flow between the wires and the capacitor. As that occurs, there is a transformation of energy within the system. Before the switch is closed, energy is stored as chemical potential energy in the battery. This energy is transformed during the chemical reaction that occurs within the battery when it is operating in an electric circuit. When the switch is closed, some of the chemical potential energy in the battery is transformed to electric potential energy associated with the separation of positive and negative charges on the plates.

To calculate the energy stored in the capacitor, we shall assume a charging process that is different from the actual process described in Section 25.1 but that gives the same final result. This assumption is justified because the energy in the final configuration does not depend on the actual charge-transfer process.[3] Imagine the plates are disconnected from the battery and you transfer the charge mechanically through the space between the plates as follows. You grab a small amount of

---

[3]This discussion is similar to that of state variables in thermodynamics. The change in a state variable such as temperature is independent of the path followed between the initial and final states. The potential energy of a capacitor (or any system) is also a state variable, so its change does not depend on the process followed to charge the capacitor.

Figure 25.10 (a) A circuit consisting of a capacitor, a battery, and a switch. (b) When the switch is closed, the battery establishes an electric field in the wire and the capacitor becomes charged.

positive charge on one plate and apply a force that causes this positive charge to move over to the other plate. Therefore, you do work on the charge as it is transferred from one plate to the other. At first, no work is required to transfer a small amount of charge dq from one plate to the other,[4] but once this charge has been transferred, a small potential difference exists between the plates. Therefore, work must be done to move additional charge through this potential difference. As more and more charge is transferred from one plate to the other, the potential difference increases in proportion and more work is required. The overall process is described by the nonisolated system model for energy. Equation 8.2 reduces to $W = \Delta U_E$; the work done on the system by the external agent appears as an increase in electric potential energy in the system.

Suppose q is the charge on the capacitor at some instant during the charging process. At the same instant, the potential difference across the capacitor is $\Delta V = q/C$. This relationship is graphed in Figure 25.11. From Section 24.1, we know that the work necessary to transfer an increment of charge dq from the plate carrying charge $-q$ to the plate carrying charge q (which is at the higher electric potential) is

$$dW = \Delta V \, dq = \frac{q}{C} \, dq$$

The work required to transfer the charge dq is the area of the tan rectangle in Figure 25.11. Because 1 V = 1 J/C, the unit for the area is the joule. The total work required to charge the capacitor from $q = 0$ to some final charge $q = Q$ is

$$W = \int_0^Q \frac{q}{C} \, dq = \frac{1}{C} \int_0^Q q \, dq = \frac{Q^2}{2C}$$

The work done in charging the capacitor appears as electric potential energy U_E stored in the capacitor. Using Equation 25.1, we can express the potential energy stored in a charged capacitor as

$$U_E = \frac{Q^2}{2C} = \tfrac{1}{2} Q \, \Delta V = \tfrac{1}{2} C (\Delta V)^2 \qquad (25.13)$$

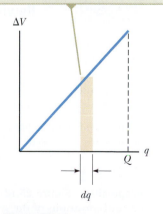

Figure 25.11 A plot of potential difference versus charge for a capacitor is a straight line having slope $1/C$.

◄ Energy stored in a charged capacitor

---

[4]We shall use lowercase q for the time-varying charge on the capacitor while it is charging to distinguish it from uppercase Q, which is the total charge on the capacitor after it is completely charged.

Because the curve in Figure 25.11 is a straight line, the total area under the curve is that of a triangle of base Q and height ΔV.

Equation 25.13 applies to any capacitor, regardless of its geometry. For a given capacitance, the stored energy increases as the charge and the potential difference increase. In practice, there is a limit to the maximum energy (or charge) that can be stored because, at a sufficiently large value of ΔV, discharge ultimately occurs between the plates. For this reason, capacitors are usually labeled with a maximum operating voltage.

We can consider the energy in a capacitor to be stored in the electric field created between the plates as the capacitor is charged. This description is reasonable because the electric field is proportional to the charge on the capacitor. For a parallel-plate capacitor, the potential difference is related to the electric field through the relationship $\Delta V = Ed$. Furthermore, its capacitance is $C = \epsilon_0 A/d$ (Eq. 25.3). Substituting these expressions into Equation 25.13 gives

$$U_E = \tfrac{1}{2}\left(\frac{\epsilon_0 A}{d}\right)(Ed)^2 = \tfrac{1}{2}(\epsilon_0 Ad)E^2 \qquad (25.14)$$

Because the volume occupied by the electric field is Ad, the *energy per unit volume* $u_E = U_E/Ad$, known as the *energy density,* is

◀ Energy density in an electric field

$$u_E = \tfrac{1}{2}\epsilon_0 E^2 \qquad (25.15)$$

Although Equation 25.15 was derived for a parallel-plate capacitor, the expression is generally valid regardless of the source of the electric field. That is, the energy density in *any* electric field is proportional to the square of the magnitude of the electric field at a given point.

QUICK QUIZ **25.4** You have three capacitors and a battery. In which of the following combinations of the three capacitors is the maximum possible energy stored when the combination is attached to the battery? **(a)** series **(b)** parallel **(c)** no difference because both combinations store the same amount of energy

---

Example 25.4 | **Rewiring Two Charged Capacitors**

Two capacitors C_1 and C_2 (where $C_1 > C_2$) are charged to the same initial potential difference ΔV_i. The charged capacitors are removed from the battery, and their plates are connected with opposite polarity as in Figure 25.12a. The switches S_1 and S_2 are then closed as in Figure 25.12b.

(A) Find the final potential difference ΔV_f between a and b after the switches are closed.

Figure 25.12 (Example 25.4) (a) Two capacitors are charged to the same initial potential difference and connected together with plates of opposite sign to be in contact when the switches are closed. (b) When the switches are closed, the charges redistribute.

SOLUTION

Conceptualize Figure 25.12 helps us understand the initial and final configurations of the system. When the switches are closed, the charge on the system will redistribute between the capacitors until both capacitors have the same final potential difference. Because $C_1 > C_2$, more charge exists on C_1 than on C_2, so the final configuration will have positive charge on the left plates as shown in Figure 25.12b.

Categorize In Figure 25.12b, it might appear as if the capacitors are connected in parallel, but there is no battery in this circuit to apply a voltage across the combination. Therefore, we *cannot* categorize this problem as one in which capacitors are connected in parallel. We *can* categorize it as a problem involving an isolated system for electric charge. The left-hand plates of the capacitors form an isolated system because they are not connected to the right-hand plates by conductors.

25.4 continued

Analyze Write an expression for the total charge on the left-hand plates of the system before the switches are closed, noting that a negative sign for Q_{2i} is necessary because the charge on the left plate of capacitor C_2 is negative:

$$(1) \quad Q_i = Q_{1i} + Q_{2i} = C_1 \, \Delta V_i - C_2 \, \Delta V_i = (C_1 - C_2) \Delta V_i$$

After the switches are closed, the charges on the individual capacitors change to new values Q_{1f} and Q_{2f} such that the potential difference is again the same across both capacitors, with a value of ΔV_f. Write an expression for the total charge on the left-hand plates of the system after the switches are closed:

$$(2) \quad Q_f = Q_{1f} + Q_{2f} = C_1 \, \Delta V_f + C_2 \, \Delta V_f = (C_1 + C_2) \Delta V_f$$

Because the system is isolated, the initial and final charges on the system must be the same. Use this condition and Equations (1) and (2) to solve for ΔV_f:

$$Q_f = Q_i \quad \rightarrow \quad (C_1 + C_2) \, \Delta V_f = (C_1 - C_2) \, \Delta V_i$$

$$(3) \quad \Delta V_f = \left(\frac{C_1 - C_2}{C_1 + C_2} \right) \Delta V_i$$

(B) Find the total energy stored in the capacitors before and after the switches are closed and determine the ratio of the final energy to the initial energy.

SOLUTION

Use Equation 25.13 to find an expression for the total energy stored in the capacitors before the switches are closed:

$$(4) \quad U_i = \tfrac{1}{2} C_1 (\Delta V_i)^2 + \tfrac{1}{2} C_2 (\Delta V_i)^2 = \tfrac{1}{2} (C_1 + C_2)(\Delta V_i)^2$$

Write an expression for the total energy stored in the capacitors after the switches are closed:

$$U_f = \tfrac{1}{2} C_1 (\Delta V_f)^2 + \tfrac{1}{2} C_2 (\Delta V_f)^2 = \tfrac{1}{2} (C_1 + C_2)(\Delta V_f)^2$$

Use the results of part (A) to rewrite this expression in terms of ΔV_i:

$$(5) \quad U_f = \tfrac{1}{2}(C_1 + C_2)\left[\left(\frac{C_1 - C_2}{C_1 + C_2} \right) \Delta V_i \right]^2 = \tfrac{1}{2} \frac{(C_1 - C_2)^2 (\Delta V_i)^2}{C_1 + C_2}$$

Divide Equation (5) by Equation (4) to obtain the ratio of the energies stored in the system:

$$\frac{U_f}{U_i} = \frac{\tfrac{1}{2}(C_1 - C_2)^2 (\Delta V_i)^2 / (C_1 + C_2)}{\tfrac{1}{2}(C_1 + C_2)(\Delta V_i)^2}$$

$$(6) \quad \frac{U_f}{U_i} = \left(\frac{C_1 - C_2}{C_1 + C_2} \right)^2$$

Finalize The ratio of energies is *less* than unity, indicating that the final energy is *less* than the initial energy. At first, you might think the law of energy conservation has been violated, but that is not the case. The "missing" energy is transferred out of the system by the mechanism of electromagnetic waves (T_{ER} in Eq. 8.2), as we shall see in Chapter 33. Therefore, this system is isolated for electric charge, but nonisolated for energy.

WHAT IF? What if the two capacitors have the same capacitance? What would you expect to happen when the switches are closed?

Answer Because both capacitors have the same initial potential difference applied to them, the charges on the identical capacitors have the same magnitude. When the capacitors with opposite polarities are connected together, the equal-magnitude charges should cancel each other, leaving the capacitors uncharged.

Let's test our results to see if that is the case mathematically. In Equation (1), because the capacitances are equal, the initial charge Q_i on the system of left-hand plates is zero. Equation (3) shows that $\Delta V_f = 0$, which is consistent with uncharged capacitors. Finally, Equation (5) shows that $U_f = 0$, which is also consistent with uncharged capacitors.

One device in which capacitors have an important role is the portable *defibrillator* discussed in the opening storyline. When cardiac fibrillation (random contractions) occurs, the heart produces a rapid, irregular pattern of beats. A fast discharge of energy through the heart can return the organ to its normal beat pattern. Emergency medical teams use portable defibrillators that contain batteries capable of charging a capacitor to a high voltage. (The circuitry actually permits the capacitor to be charged to a much higher voltage than that of the battery.) Up to 360 J is stored in the electric field of a large capacitor in a defibrillator when it is fully charged. The stored energy is released through the heart by conducting electrodes, called paddles, which are placed on both sides of the victim's chest. The defibrillator can deliver the energy to a patient in about 2 ms (roughly equivalent to 3 000 times the power delivered to a 60-W lightbulb!). The paramedics must wait between applications of the energy because of the time interval necessary for the capacitors to become fully charged. In this application and others (e.g., camera flash units and lasers used for fusion experiments), capacitors serve as energy reservoirs that can be slowly charged and then quickly discharged to provide large amounts of energy in a short pulse.

In the opening storyline, your uncle also mentioned modeling the atmosphere of the Earth as a huge capacitor. The surface of the Earth is one plate, negatively charged, and the other plate is a spherical shell representing the average position of positive charges located in the air. Because there are freely moving charged particles in the air between the plates of this capacitor, there is an electrical leakage between the plates, tending to continuously reduce the charge on the capacitor. But the process of lightning delivers negative charge to the ground and recharges the capacitor. An equilibrium situation is reached in which the rate of leakage through the air is balanced by the rate of lightning strikes over the surface of the globe.

25.5 Capacitors with Dielectrics

A **dielectric** is a nonconducting material such as rubber, glass, or waxed paper. We can perform the following experiment to illustrate the effect of a dielectric in a capacitor. Consider a parallel-plate capacitor that without a dielectric has a charge Q_0 and a capacitance C_0, where, for this discussion, we will use the subscript 0 to represent parameters related to a capacitor with nothing but air between the plates. The potential difference across the capacitor is $\Delta V_0 = Q_0/C_0$. Figure 25.13a illustrates this situation. The potential difference is measured by a device called a *voltmeter*. Notice that no battery is shown in the figure; the battery was used to charge the capacitor and was then removed. Also, we must assume no charge can flow through an ideal voltmeter. Hence, there is no path by which charge can flow and alter the charge on the capacitor. If a dielectric is now inserted between the plates as in Figure 25.13b, we find that the voltmeter indicates that the voltage between the plates decreases to a value ΔV. The voltages with and without the dielectric are related by a factor κ as follows:

$$\Delta V = \frac{\Delta V_0}{\kappa}$$

Because $\Delta V < \Delta V_0$, we see that $\kappa > 1$. The dimensionless factor κ is called the **dielectric constant** of the material. The dielectric constant varies from one material to another. In this section, we analyze the change in capacitance due to a dielectric in terms of electrical parameters such as electric charge, electric field, and potential difference; Section 25.7 describes the microscopic origin of these changes.

Because the charge Q_0 on the capacitor in Figure 25.13 does not change, the capacitance must change to the value

$$C = \frac{Q_0}{\Delta V} = \frac{Q_0}{\Delta V_0/\kappa} = \kappa \frac{Q_0}{\Delta V_0}$$

◀ Capacitance of a capacitor filled with a material of dielectric constant κ

$$C = \kappa C_0 \qquad (25.16)$$

> **PITFALL PREVENTION 25.5**
>
> **Is the Capacitor Connected to a Battery?** For problems in which a capacitor is modified (by insertion of a dielectric, for example), you must note whether modifications to the capacitor are being made while the capacitor is connected to a battery or after it is disconnected. If the capacitor remains connected to the battery, the voltage across the capacitor necessarily remains the same. If you disconnect the capacitor from the battery before making any modifications to the capacitor, the capacitor is an isolated system for electric charge and its charge remains the same.

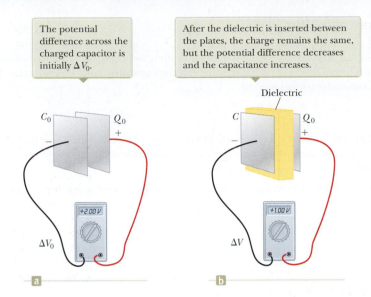

The potential difference across the charged capacitor is initially ΔV_0.

After the dielectric is inserted between the plates, the charge remains the same, but the potential difference decreases and the capacitance increases.

Figure 25.13 A charged capacitor (a) before and (b) after insertion of a dielectric with $\kappa = 2.00$ between the plates.

That is, the capacitance *increases* by the factor κ when the dielectric completely fills the region between the plates. Because $C_0 = \epsilon_0 A/d$ (Eq. 25.3) for a parallel-plate capacitor, we can express the capacitance of a parallel-plate capacitor filled with a dielectric as

$$C = \kappa \frac{\epsilon_0 A}{d} \qquad (25.17)$$

In Figure 25.13a, the plates were charged with a battery to voltage ΔV_0. Then the battery was removed and replaced with a voltmeter. Suppose the battery remains connected to the plates as we insert the dielectric. In this case, the voltage between the plates is fixed by the battery and cannot change. What we find is that charges flow between the battery and the plates in order to hold the voltage constant. We find that the charge on the capacitor changes to $Q = \kappa Q_0$ after the dielectric is inserted. Evaluating the capacitance in this situation gives the same result as in Equation 25.16.

From Equation 25.17, it would appear that the capacitance could be made very large by inserting a dielectric between the plates and decreasing d. In practice, the lowest value of d is limited by the electric discharge that could occur through the dielectric medium separating the plates. For any given separation d, the maximum voltage that can be applied to a capacitor without causing a discharge depends on the **dielectric strength** (maximum electric field) of the dielectric. If the magnitude of the electric field in the dielectric exceeds the dielectric strength, the insulating properties break down and the dielectric begins to conduct.

Physical capacitors have a specification called by a variety of names, including *working voltage*, *breakdown voltage*, and *rated voltage*. This parameter represents the largest voltage that can be applied to the capacitor without exceeding the dielectric strength of the dielectric material in the capacitor. Consequently, when selecting a capacitor for a given application, you must consider its capacitance as well as the expected voltage across the capacitor in the circuit, making sure the expected voltage is smaller than the rated voltage of the capacitor.

Insulating materials have values of κ greater than unity and dielectric strengths greater than that of air as Table 25.1 indicates. Therefore, a dielectric provides the following advantages:

- An increase in capacitance
- An increase in maximum operating voltage
- Possible mechanical support between the plates, which allows the plates to be close together without touching, thereby decreasing d and increasing C

TABLE 25.1 Approximate Dielectric Constants and Dielectric Strengths of Various Materials at Room Temperature

Material	Dielectric Constant κ	Dielectric Strength[a] $(10^6$ V/m)
Air (dry)	1.000 59	3
Bakelite	4.9	24
Fused quartz	3.78	8
Mylar	3.2	7
Neoprene rubber	6.7	12
Nylon	3.4	14
Paper	3.7	16
Paraffin-impregnated paper	3.5	11
Polyethylene	2.30	18
Polystyrene	2.56	24
Polyvinyl chloride	3.4	40
Porcelain	6	12
Pyrex glass	5.6	14
Silicone oil	2.5	15
Strontium titanate	233	8
Teflon	2.1	60
Vacuum	1.000 00	—

[a]The dielectric strength equals the maximum electric field that can exist in a dielectric without electrical breakdown. These values depend strongly on the presence of impurities and flaws in the materials.

| Example **25.5** | **Energy Stored Before and After** |

A parallel-plate capacitor is charged with a battery to a charge Q_0. The battery is then removed, and a slab of material that has a dielectric constant κ is inserted between the plates. Identify the system as the capacitor and the dielectric. Find the energy stored in the system before and after the dielectric is inserted.

SOLUTION

Conceptualize Think about what happens when the dielectric is inserted between the plates. Because the battery has been removed, the charge on the capacitor must remain the same. We know from our earlier discussion, however, that the capacitance must change. Therefore, we expect a change in the energy of the system.

Categorize Because we expect the energy of the system to change, we model it as a *nonisolated system* for *energy* involving a capacitor and a dielectric.

Analyze From Equation 25.13, find the energy stored in the absence of the dielectric:

$$U_0 = \frac{Q_0^{\,2}}{2C_0}$$

Find the energy stored in the capacitor after the dielectric is inserted between the plates:

$$U_E = \frac{Q_0^{\,2}}{2C}$$

Use Equation 25.16 to replace the capacitance C:

$$U_E = \frac{Q_0^{\,2}}{2\kappa C_0} = \frac{U_0}{\kappa}$$

Finalize Because $\kappa > 1$, the final energy is less than the initial energy. We can account for the decrease in energy of the system by performing an experiment and noting that the dielectric, when inserted, is pulled into the device. To keep the dielectric from accelerating, an external agent must do negative work on the dielectric. Equation 8.2 becomes $\Delta U_E = W$, where both sides of the equation are negative.

The materials between the plates of the capacitor are the wallboard and air.

Capacitor plates

Stud finder

Stud

Wallboard

When the capacitor moves across a stud in the wall, the materials between the plates are the wallboard and the wood stud. The change in the dielectric constant causes a signal light to illuminate.

a   b

Figure 25.14 (Quick Quiz 25.5) A stud finder.

QUICK QUIZ 25.5 If you have ever tried to hang a picture or a mirror, you know it can be difficult to locate a wooden stud in which to anchor your nail or screw. A carpenter's stud finder is a capacitor with its plates arranged side by side instead of facing each other as shown in Figure 25.14. When the device is moved over a stud, does the capacitance **(a)** increase or **(b)** decrease?

25.6 Electric Dipole in an Electric Field

We have discussed the effect on the capacitance of placing a dielectric between the plates of a capacitor. In Section 25.7, we shall describe the microscopic origin of this effect. Before we can do so, however, let's expand the discussion of the electric dipole introduced in Section 22.4 (see Example 22.6). The electric dipole consists of two charges of equal magnitude and opposite sign separated by a distance $2a$ as shown in Figure 25.15. The **electric dipole moment** of this configuration is defined as the vector $\vec{p}$ directed from $-q$ toward $+q$ along the line joining the charges and having magnitude

$$p \equiv 2aq \tag{25.18}$$

Now suppose an electric dipole is placed in a uniform electric field $\vec{E}$ and makes an angle θ with the field as shown in Figure 25.16. We identify $\vec{E}$ as the field *external* to the dipole, established by some other charge distribution, to distinguish it from the field *due to* the dipole, which we discussed in Section 22.4.

Each of the charges is modeled as a particle in an electric field. The electric forces acting on the two charges are equal in magnitude ($F = qE$) and opposite in direction as shown in Figure 25.16. Therefore, the net force on the dipole is zero. The two forces produce a net torque on the dipole, however; the dipole is therefore described by the rigid object under a net torque model. As a result, the dipole rotates in the direction that brings the dipole moment vector into greater alignment with the field. The torque due to the force on the positive charge about an axis through O in Figure 25.16 has magnitude $Fa \sin \theta$, where $a \sin \theta$ is the moment arm of F about O. This force tends to produce a clockwise rotation. The torque about O on the negative charge is also of magnitude $Fa \sin \theta$; here again, the force tends to produce a clockwise rotation. Therefore, the magnitude of the net torque about O is

$$\tau = 2Fa \sin \theta$$

Because $F = qE$ and $p = 2aq$, we can express τ as

$$\tau = 2aqE \sin \theta = pE \sin \theta \tag{25.19}$$

Based on this expression, it is convenient to express the torque in vector form as the cross product of the vectors $\vec{p}$ and $\vec{E}$:

$$\vec{\tau} = \vec{p} \times \vec{E} \tag{25.20}$$

We can also model the system of the dipole and the external electric field as an isolated system for energy. Let's determine the potential energy of the system as a function of the dipole's orientation with respect to the field. To do so, recognize that work must be done by an external agent to rotate the dipole through an angle so as to cause the dipole moment vector to become less aligned with the field. The work done is then stored as electric potential energy in the system. Notice that this potential energy is associated with a *rotational* configuration of the system. Previously, we have seen potential energies associated with *translational* configurations: an object with mass was moved in a gravitational field, a charge was moved in an electric field, or a spring was extended. The work dW required to rotate the dipole through an angle $d\theta$ is $dW = \tau \, d\theta$ (see Eq. 10.25). Because $\tau = pE \sin \theta$ and the work results in an increase in the electric potential energy U_E, we find that for a rotation from θ_i to θ_f, the change in potential energy of the system is

$$U_f - U_i = \int_{\theta_i}^{\theta_f} \tau \, d\theta = \int_{\theta_i}^{\theta_f} pE \sin \theta \, d\theta = pE \int_{\theta_i}^{\theta_f} \sin \theta \, d\theta$$

$$= pE[-\cos \theta]_{\theta_i}^{\theta_f} = pE(\cos \theta_i - \cos \theta_f)$$

The term that contains $\cos \theta_i$ is a constant that depends on the initial orientation of the dipole. It is convenient to choose a reference angle of $\theta_i = 90°$ so that $\cos \theta_i = \cos 90° = 0$. Furthermore, let's choose $U_i = 0$ at $\theta_i = 90°$ as our reference value of potential energy. Hence, we can express a general value of $U_E = U_f$ as

$$U_E = -pE \cos \theta \tag{25.21}$$

We can write this expression for the potential energy of a dipole in an electric field as the dot product of the vectors $\vec{p}$ and $\vec{E}$:

$$\boxed{U_E = -\vec{p} \cdot \vec{E}} \tag{25.22}$$

To develop a conceptual understanding of Equation 25.21, compare it with the expression for the potential energy of the system of an object in the Earth's gravitational field, $U_g = mgy$ (Eq. 7.19). First, both expressions contain a parameter of the entity placed in the field: mass for the object, dipole moment for the dipole. Second, both expressions contain the field, g for the object, E for the dipole. Finally, both expressions contain a configuration description: translational position y for the object, rotational position θ for the dipole. In both cases, once the configuration is changed, the system tends to return to the original configuration when the object is released: the object of mass m falls toward the ground, and the dipole begins to rotate back toward the configuration in which it is aligned with the field.

The electric dipole moment $\vec{p}$ is directed from $-q$ toward $+q$.

Figure 25.15 An electric dipole consists of two charges of equal magnitude and opposite sign separated by a distance of $2a$.

◄ Torque on an electric dipole in an external electric field

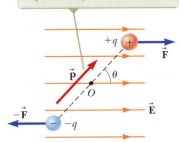

The dipole moment $\vec{p}$ is at an angle θ to the field, causing the dipole to experience a torque.

Figure 25.16 An electric dipole in a uniform external electric field.

◄ Potential energy of the system of an electric dipole in an external electric field

The center of the positive charge distribution is at the point **×**.

Figure 25.17 The water molecule, H_2O, has a permanent polarization resulting from its nonlinear geometry.

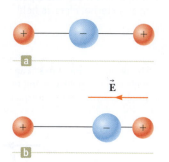

Figure 25.18 (a) A linear symmetric molecule has no permanent polarization. (b) An external electric field induces a polarization in the molecule.

Molecules are said to be *polarized* when a separation exists between the average position of the negative charges and the average position of the positive charges in the molecule. In some molecules such as water, this condition is always present; such molecules are called **polar molecules.** Molecules that do not possess a permanent polarization are called **nonpolar molecules.**

We can understand the permanent polarization of water by inspecting the geometry of the water molecule. The oxygen atom in the water molecule is bonded to the hydrogen atoms such that an angle of 105° is formed between the two bonds (Fig. 25.17). The center of the negative charge distribution is near the oxygen atom, and the center of the positive charge distribution lies at a point midway along the line joining the hydrogen atoms (the point labeled **×** in Fig. 25.17). We can model the water molecule and other polar molecules as dipoles because the average positions of the positive and negative charges act as point charges. As a result, we can apply our discussion of dipoles to the behavior of polar molecules.

Washing with soap and water is a household scenario in which the dipole structure of water is exploited. Grease and oil are made up of nonpolar molecules, which are generally not attracted to water. Plain water is not very useful for removing this type of grime. Soap contains long molecules called *surfactants*. In a long molecule, the polarity characteristics of one end of the molecule can be different from those at the other end. In a surfactant molecule, one end acts like a nonpolar molecule and the other acts like a polar molecule. The nonpolar end can attach to a grease or oil molecule, and the polar end can attach to a water molecule. Therefore, the soap serves as a chain, linking the dirt and water molecules together. When the water is rinsed away, the grease and oil go with it.

A symmetric molecule (Fig. 25.18a) has no permanent polarization, but polarization can be induced by placing the molecule in an electric field. A field directed to the left as in Figure 25.18b causes the center of the negative charge distribution to shift to the right relative to the positive charges. This *induced polarization* is the effect that predominates in most materials used as dielectrics in capacitors.

Example 25.6   **The H_2O Molecule**

The water (H_2O) molecule has an electric dipole moment of 6.3×10^{-30} C · m. A sample contains 10^{21} water molecules, with the dipole moments all oriented in the direction of an electric field of magnitude 2.5×10^5 N/C. How much work is required to rotate the dipoles from this orientation ($\theta = 0°$) to one in which all the moments are perpendicular to the field ($\theta = 90°$)?

SOLUTION

Conceptualize   When all the dipoles are aligned with the electric field, the dipoles–electric field system has the minimum potential energy. This energy has a negative value given by the product of the right side of Equation 25.21, evaluated at 0°, and the number N of dipoles.

Categorize   The combination of the dipoles and the electric field is identified as a system. We use the *nonisolated system* model because an external agent performs work on the system to change its potential energy.

Analyze   Write the appropriate reduction of the conservation of energy equation, Equation 8.2, for this situation:

$$(1) \quad \Delta U_E = W$$

Use Equation 25.21 to evaluate the initial and final potential energies of the system and Equation (1) to calculate the work required to rotate the dipoles:

$$W = U_{90°} - U_{0°} = (-NpE \cos 90°) - (-NpE \cos 0°)$$
$$= NpE = (10^{21})(6.3 \times 10^{-30} \text{ C} \cdot \text{m})(2.5 \times 10^5 \text{ N/C})$$
$$= 1.6 \times 10^{-3} \text{ J}$$

Finalize   Notice that the work done on the system is positive because the potential energy of the system has been raised from a negative value to a value of zero.

25.7 An Atomic Description of Dielectrics

In Section 25.5, we found that the potential difference ΔV_0 between the plates of an empty capacitor is reduced to $\Delta V_0/\kappa$ when a dielectric is introduced between the plates. The potential difference is reduced because the magnitude of the electric field decreases between the plates. In particular, if $\vec{\mathbf{E}}_0$ is the electric field without the dielectric, the field in the presence of a dielectric is

$$\vec{\mathbf{E}} = \frac{\vec{\mathbf{E}}_0}{\kappa} \qquad (25.23)$$

First consider a dielectric made up of polar molecules placed in the electric field between the plates of a capacitor. The dipoles (that is, the polar molecules making up the dielectric) are randomly oriented in the absence of an electric field as shown in Figure 25.19a. When an external field $\vec{\mathbf{E}}_0$ due to charges on the capacitor plates is applied, a torque is exerted on the dipoles, causing them to partially align with the field as shown in Figure 25.19b. The dielectric is now polarized. The degree of alignment of the molecules with the electric field depends on temperature and the magnitude of the field. In general, the alignment increases with decreasing temperature and with increasing electric field.

If the molecules of the dielectric are nonpolar, the electric field due to the plates produces an induced polarization in the molecule. These induced dipole moments tend to align with the external field, and the dielectric is polarized. Therefore, a dielectric can be polarized by an external field regardless of whether the molecules in the dielectric are polar or nonpolar.

With these ideas in mind, consider a slab of dielectric material placed between the plates of a capacitor with each plate carrying a charge density with a magnitude σ. This will result in a uniform electric field $\vec{\mathbf{E}}_0$ as shown in Figure 25.19c. The electric field due to the plates is directed to the right and polarizes the dielectric. The net effect on the dielectric is the formation of an *induced* positive surface charge density σ_{ind} on the right face and an equal-magnitude negative surface charge density $-\sigma_{ind}$ on the left face as shown in Figure 25.19c. Because we can model these surface charge distributions as being due to charged parallel plates, the induced surface charges on the dielectric give rise to an induced electric field $\vec{\mathbf{E}}_{ind}$ in the direction opposite the external field $\vec{\mathbf{E}}_0$. Therefore, the net electric field $\vec{\mathbf{E}}$ in the dielectric has a magnitude

$$E = E_0 - E_{ind} \qquad (25.24)$$

In the parallel-plate capacitor shown in Figure 25.19c, the external field E_0 is related to the charge density σ on the plates through the relationship $E_0 = \sigma/\epsilon_0$.

Polar molecules are randomly oriented in the absence of an external electric field.

When an external electric field is applied, the molecules partially align with the field.

The charged edges of the dielectric can be modeled as an additional pair of parallel plates establishing an electric field $\vec{\mathbf{E}}_{ind}$ in the direction opposite that of $\vec{\mathbf{E}}_0$.

Figure 25.19 (a) Polar molecules in a dielectric. (b) An electric field is applied to the dielectric. (c) Details of the electric field inside the dielectric.

The induced electric field in the dielectric is related to the induced charge density σ_{ind} through the relationship $E_{\text{ind}} = \sigma_{\text{ind}}/\epsilon_0$. Because $E = E_0/\kappa = \sigma/\kappa\epsilon_0$, substitution into Equation 25.24 gives

$$\frac{\sigma}{\kappa\epsilon_0} = \frac{\sigma}{\epsilon_0} - \frac{\sigma_{\text{ind}}}{\epsilon_0}$$

$$\sigma_{\text{ind}} = \left(\frac{\kappa - 1}{\kappa}\right)\sigma \qquad (25.25)$$

Because $\kappa > 1$, this expression shows that the charge density σ_{ind} induced on the dielectric is less than the charge density σ on the plates. For instance, if $\kappa = 3$, the induced charge density is two-thirds the charge density on the plates. If no dielectric is present, then $\kappa = 1$ and $\sigma_{\text{ind}} = 0$ as expected. If the dielectric is replaced by an electrical conductor for which $E = 0$, however, Equation 25.24 indicates that $E_0 = E_{\text{ind}}$, which corresponds to $\sigma_{\text{ind}} = \sigma$. That is, the surface charge induced on the conductor is equal in magnitude but opposite in sign to that on the plates, resulting in a net electric field of zero in the conductor (see Fig. 24.17).

Example **25.7**   Effect of a Metallic Slab

A parallel-plate capacitor has a plate separation d and plate area A. An uncharged metallic slab of thickness a is inserted midway between the plates.

(A) Find the capacitance of the device.

S O L U T I O N

Conceptualize   Figure 25.20a shows the metallic slab between the plates of the capacitor. Any charge that appears on one plate of the capacitor must induce a charge of equal magnitude and opposite sign on the near side of the slab as shown in Figure 25.20a. Consequently, the net charge on the slab remains zero and the electric field inside the slab is zero.

Categorize   The planes of charge on the metallic slab's upper and lower edges are identical to the distribution of charges on the plates of a capacitor. The metal between the slab's edges serves only to make an electrical connection between the edges. Therefore, we can model the edges of the slab as conducting planes and the bulk of the slab as a wire. As a result, the capacitor in Figure 25.20a is equivalent to two capacitors in series, each having a plate separation $(d - a)/2$ as shown in Figure 25.20b.

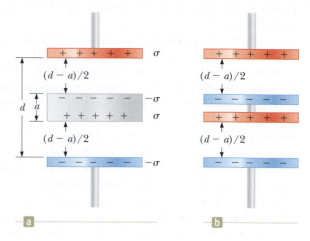

Figure 25.20   (Example 25.7) (a) A parallel-plate capacitor of plate separation d partially filled with a metallic slab of thickness a. (b) The equivalent circuit of the device in (a) consists of two capacitors in series, each having a plate separation $(d - a)/2$.

Analyze   Use Equation 25.3 and the rule for adding two capacitors in series (Eq. 25.12) to find the equivalent capacitance in Figure 25.20b:

$$\frac{1}{C} = \frac{1}{C_1} + \frac{1}{C_2} = \frac{1}{\dfrac{\epsilon_0 A}{(d - a)/2}} + \frac{1}{\dfrac{\epsilon_0 A}{(d - a)/2}}$$

$$C = \frac{\epsilon_0 A}{d - a}$$

(B) Show that the capacitance of the original capacitor is unaffected by the insertion of the metallic slab if the slab is infinitesimally thin.

S O L U T I O N

In the result for part (A), let $a \to 0$:

$$C = \lim_{a \to 0}\left(\frac{\epsilon_0 A}{d - a}\right) = \frac{\epsilon_0 A}{d}$$

25.7 continued

Finalize The result of part (B) is the original capacitance before the slab is inserted, which tells us that we can insert an infinitesimally thin metallic sheet between the plates of a capacitor without affecting the capacitance. We use this fact in the next example.

WHAT IF? What if the metallic slab in part (A) is not midway between the plates? How would that affect the capacitance?

Answer Let's imagine moving the slab in Figure 25.20a upward so that the distance between the upper edge of the slab and the upper plate is b. Then, the distance between the lower edge of the slab and the lower plate is $d - b - a$. As in part (A), we find the total capacitance of the series combination:

$$\frac{1}{C} = \frac{1}{C_1} + \frac{1}{C_2} = \frac{1}{\epsilon_0 A/b} + \frac{1}{\epsilon_0 A/(d - b - a)}$$

$$= \frac{b}{\epsilon_0 A} + \frac{d - b - a}{\epsilon_0 A} = \frac{d - a}{\epsilon_0 A} \rightarrow C = \frac{\epsilon_0 A}{d - a}$$

which is the same result as found in part (A). The capacitance is independent of the value of b, so it does not matter where the slab is located. In Figure 25.20b, when the central structure is moved up or down, the decrease in plate separation of one capacitor is compensated by the increase in plate separation for the other.

Example **25.8** **A Partially Filled Capacitor**

A parallel-plate capacitor with a plate separation d has a capacitance C_0 in the absence of a dielectric. What is the capacitance when a slab of dielectric material of dielectric constant κ and thickness fd is inserted between the plates (Fig. 25.21a), where f is a fraction between 0 and 1?

SOLUTION

Conceptualize In our previous discussions of dielectrics between the plates of a capacitor, the dielectric filled the volume between the plates. In this example, only part of the volume between the plates contains the dielectric material.

Categorize In Example 25.7, we found that an infinitesimally thin metallic sheet inserted between the plates of a capacitor does not affect the capacitance. Imagine sliding an infinitesimally thin metallic slab along the bottom face of the dielectric shown in Figure 25.21a. We can model this system as a series combination of two capacitors as shown in Figure 25.21b. One capacitor has a plate separation fd and is filled with a dielectric; the other has a plate separation $(1 - f)d$ and has air between its plates.

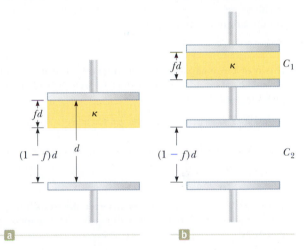

Figure 25.21 (Example 25.8) (a) A parallel-plate capacitor of plate separation d partially filled with a dielectric of thickness fd. (b) The equivalent circuit of the capacitor consists of two capacitors connected in series.

Analyze Evaluate the two capacitances in Figure 25.21b from Equation 25.17:

$$C_1 = \frac{\kappa \epsilon_0 A}{fd} \quad \text{and} \quad C_2 = \frac{\epsilon_0 A}{(1 - f)d}$$

Find the equivalent capacitance C from Equation 25.12 for two capacitors combined in series:

$$\frac{1}{C} = \frac{1}{C_1} + \frac{1}{C_2} = \frac{fd}{\kappa \epsilon_0 A} + \frac{(1 - f)d}{\epsilon_0 A}$$

$$\frac{1}{C} = \frac{fd}{\kappa \epsilon_0 A} + \frac{\kappa(1 - f)d}{\kappa \epsilon_0 A} = \frac{f + \kappa(1 - f)}{\kappa} \frac{d}{\epsilon_0 A}$$

Invert and substitute for the capacitance without the dielectric, $C_0 = \epsilon_0 A/d$:

$$C = \frac{\kappa}{f + \kappa(1 - f)} \frac{\epsilon_0 A}{d} = \frac{\kappa}{f + \kappa(1 - f)} C_0$$

Finalize Let's test this result for some known limits. If $f \rightarrow 0$, the dielectric should disappear. In this limit, $C \rightarrow C_0$, which is consistent with a capacitor with air between the plates. If $f \rightarrow 1$, the dielectric fills the volume between the plates. In this limit, $C \rightarrow \kappa C_0$, which is consistent with Equation 25.16.

Summary

> ### Definitions

A **capacitor** consists of two separated conductors called *plates*. If the capacitor is charged, the plates carry charges of equal magnitude and opposite sign. The **capacitance** C of any capacitor is the ratio of the charge Q on either conductor to the potential difference ΔV between them:

$$C \equiv \frac{Q}{\Delta V} \qquad (25.1)$$

The capacitance depends only on the geometry of the conductors and not on an external source of charge or potential difference. The SI unit of capacitance is coulombs per volt, or the **farad** (F): 1 F = 1 C/V.

The **electric dipole moment** $\vec{\mathbf{p}}$ of an electric dipole has a magnitude

$$p \equiv 2aq \qquad (25.18)$$

where $2a$ is the distance between the charges q and $-q$. The direction of the electric dipole moment vector is from the negative charge toward the positive charge.

> ### Concepts and Principles

If two or more capacitors are connected in parallel, the potential difference is the same across all capacitors. The equivalent capacitance of a **parallel combination** of capacitors is

$$C_{eq} = C_1 + C_2 + C_3 + \cdots \qquad (25.9)$$

If two or more capacitors are connected in series, the charge is the same on all capacitors, and the equivalent capacitance of the **series combination** is given by

$$\frac{1}{C_{eq}} = \frac{1}{C_1} + \frac{1}{C_2} + \frac{1}{C_3} + \cdots \qquad (25.12)$$

These two equations enable you to simplify many electric circuits by replacing multiple capacitors with a single equivalent capacitance.

Energy is stored in a charged capacitor because the charging process is equivalent to the transfer of charges from one conductor at a lower electric potential to another conductor at a higher potential. The energy stored in a capacitor of capacitance C with charge Q and potential difference ΔV is

$$U_E = \frac{Q^2}{2C} = \tfrac{1}{2}Q\,\Delta V = \tfrac{1}{2}C\,(\Delta V)^2 \qquad (25.13)$$

When a dielectric material is inserted between the plates of a capacitor, the capacitance increases by a dimensionless factor κ, called the **dielectric constant**:

$$C = \kappa C_0 \qquad (25.16)$$

where C_0 is the capacitance in the absence of the dielectric.

The torque acting on an electric dipole in a uniform electric field $\vec{\mathbf{E}}$ is

$$\vec{\boldsymbol{\tau}} = \vec{\mathbf{p}} \times \vec{\mathbf{E}} \qquad (25.20)$$

The potential energy of the system of an electric dipole in a uniform external electric field $\vec{\mathbf{E}}$ is

$$U_E = -\vec{\mathbf{p}} \cdot \vec{\mathbf{E}} \qquad (25.22)$$

Think–Pair–Share

See the Preface for an explanation of the icons used in this problems set. For additional assessment items for this section, go to **WEBASSIGN** From Cengage

1. Your group is a team of teaching assistants for a physics professor. The professor asks your group to set up and test the apparatus shown in Figure TP25.1 for a classroom demonstration showing the force between the plates in a capacitor. The negative plate of the capacitor at the right of the figure is clamped in place. The upper, positive plate is free to move up and down. Both plates are of area A. A potential difference ΔV is applied across the capacitor. (a) What is the separation distance d between the capacitor plates when the gravitational force on the hanging ball on the left is balanced by the electrical force

Figure TP25.1

on the upper plate on the right? (b) Is this equilibrium stable or unstable?

2. **ACTIVITY** Your group is performing electrical experiments in your physics laboratory. Your supply of capacitors in the stockroom has run low and you have only two different values of capacitance available: 20 μF and 50 μF. You have a large number of each of these capacitors. (a) Your experiments require a 45-μF capacitor and a 35-μF capacitor. Split your group into two halves. Group (i) will determine how to form a 45-μF capacitor from your supply, while group (ii) will do the same for a 35-μF capacitor. (b) After working on the experiment for a while, you realize you now need a 105-μF capacitor. Have your whole group work together to find at least three ways to combine the capacitors in your stockroom to generate 105 μF.

Problems

See the Preface for an explanation of the icons used in this problems set.
For additional assessment items for this section, go to **WEBASSIGN** From Cengage

SECTION 25.1 Definition of Capacitance

1. (a) When a battery is connected to the plates of a 3.00-μF capacitor, it stores a charge of 27.0 μC. What is the voltage of the battery? (b) If the same capacitor is connected to another battery and 36.0 μC of charge is stored on the capacitor, what is the voltage of the battery?

2. Two conductors having net charges of +10.0 μC and -10.0 μC have a potential difference of 10.0 V between them. (a) Determine the capacitance of the system. (b) What is the potential difference between the two conductors if the charges on each are increased to +100 μC and -100 μC?

SECTION 25.2 Calculating Capacitance

3. When a potential difference of 150 V is applied to the plates of a parallel-plate capacitor, the plates carry a surface charge density of 30.0 nC/cm². What is the spacing between the plates?

4. An air-filled parallel-plate capacitor has plates of area 2.30 cm² separated by 1.50 mm. (a) Find the value of its capacitance. The capacitor is connected to a 12.0-V battery. (b) What is the charge on the capacitor? (c) What is the magnitude of the uniform electric field between the plates?

5. A variable air capacitor used in a radio tuning circuit is made of N semicircular plates, each of radius R and positioned a distance d from its neighbors, to which it is electrically connected. As shown in Figure P25.5, a second identical set of plates is enmeshed with the first set. Each plate in the second set is halfway between two plates of the first set. The second set can rotate as a unit. Determine the capacitance as a function of the angle of rotation θ, where $\theta = 0$ corresponds to the maximum capacitance.

Figure P25.5

6. **Review.** A small object of mass m carries a charge q and is suspended by a thread between the vertical plates of a parallel-plate capacitor. The plate separation is d. If the thread makes an angle θ with the vertical, what is the potential difference between the plates?

SECTION 25.3 Combinations of Capacitors

7. Find the equivalent capacitance of a 4.20-μF capacitor and an 8.50-μF capacitor when they are connected (a) in series and (b) in parallel.

8. *Why is the following situation impossible?* A technician is testing a circuit that contains a capacitance C. He realizes that a better design for the circuit would include a capacitance $\frac{7}{3}C$ rather than C. He has three additional capacitors, each with capacitance C. By combining these additional capacitors in a certain combination that is then placed in parallel with the original capacitor, he achieves the desired capacitance.

9. A group of identical capacitors is connected first in series and then in parallel. The combined capacitance in parallel is 100 times larger than for the series connection. How many capacitors are in the group?

10. Three capacitors are connected to a battery as shown in Figure P25.10. Their capacitances are $C_1 = 3C$, $C_2 = C$, and $C_3 = 5C$. (a) What is the equivalent capacitance of this set of capacitors? (b) State the ranking of the capacitors according to the charge they store from largest to smallest. (c) Rank the capacitors according to the potential differences across them from largest to smallest. (d) **What If?** Assume C_3 is increased. Explain what happens to the charge stored by each capacitor.

Figure P25.10

11. Four capacitors are connected as shown in Figure P25.11. (a) Find the equivalent capacitance between points a and b. (b) Calculate the charge on each capacitor, taking $\Delta V_{ab} = 15.0$ V.

Figure P25.11

12. (a) Find the equivalent capacitance between points a and b for the group of capacitors connected as shown in Figure P25.12 (page 686). Take $C_1 = 5.00$ μF, $C_2 = 10.0$ μF, and

$C_3 = 2.00 \ \mu F$. (b) What charge is stored on C_3 if the potential difference between points a and b is 60.0 V?

Figure P25.12

13. Find the equivalent capacitance between points a and b in the combination of capacitors shown in Figure P25.13.

Figure P25.13

14. **CR** You are working at an electronics fabrication shop. Your current project is on the team producing capacitors for the timer circuit that delays the closing of an elevator door. According to its design specification, the timer circuit is to have a capacitance of 32.0 μF between two points A and B. As your capacitors come off the assembly line, you find that they have a variation of ±5.00% from this value. After a team meeting to evaluate this situation, the team decides that capacitances in the range 32.0 ± 0.5 μF are acceptable and do not need modification. For capacitances outside this range, the director does not wish to discard the capacitors, but rather to add extra capacitors in series or parallel with the main capacitor to bring the total equivalent capacitance to the exact design value of 32.0 μF. You are put in charge of procuring the extra capacitors. What range of capacitances for these extra capacitors do you need to cover the entire range of variation of ±5.00%? All capacitances can be measured to three significant figures.

15. Two capacitors give an equivalent capacitance of 9.00 pF when connected in parallel and an equivalent capacitance of 2.00 pF when connected in series. What is the capacitance of each capacitor?

16. **S** Two capacitors give an equivalent capacitance of C_p when connected in parallel and an equivalent capacitance of C_s when connected in series. What is the capacitance of each capacitor?

SECTION 25.4 Energy Stored in a Charged Capacitor

17. **V** A 3.00-μF capacitor is connected to a 12.0-V battery. How much energy is stored in the capacitor? (b) Had the capacitor been connected to a 6.00-V battery, how much energy would have been stored?

18. **QC** Two capacitors, $C_1 = 18.0 \ \mu F$ and $C_2 = 36.0 \ \mu F$, are connected in series, and a 12.0-V battery is connected across

the two capacitors. Find (a) the equivalent capacitance and (b) the energy stored in this equivalent capacitance. (c) Find the energy stored in each individual capacitor. (d) Show that the sum of these two energies is the same as the energy found in part (b). (e) Will this equality always be true, or does it depend on the number of capacitors and their capacitances? (f) If the same capacitors were connected in parallel, what potential difference would be required across them so that the combination stores the same energy as in part (a)? (g) Which capacitor stores more energy in this situation, C_1 or C_2?

19. **QC** Two identical parallel-plate capacitors, each with capacitance 10.0 μF, are charged to potential difference 50.0 V and then disconnected from the battery. They are then connected to each other in parallel with plates of like sign connected. Finally, the plate separation in one of the capacitors is doubled. (a) Find the total energy of the system of two capacitors *before* the plate separation is doubled. (b) Find the potential difference across each capacitor *after* the plate separation is doubled. (c) Find the total energy of the system *after* the plate separation is doubled. (d) Reconcile the difference in the answers to parts (a) and (c) with the law of conservation of energy.

20. **QC** **S** Two identical parallel-plate capacitors, each with capacitance C, are charged to potential difference ΔV and then disconnected from the battery. They are then connected to each other in parallel with plates of like sign connected. Finally, the plate separation in one of the capacitors is doubled. (a) Find the total energy of the system of two capacitors *before* the plate separation is doubled. (b) Find the potential difference across each capacitor *after* the plate separation is doubled. (c) Find the total energy of the system *after* the plate separation is doubled. (d) Reconcile the difference in the answers to parts (a) and (c) with the law of conservation of energy.

21. Two capacitors, $C_1 = 25.0 \ \mu F$ and $C_2 = 5.00 \ \mu F$, are connected in parallel and charged with a 100-V power supply. (a) Draw a circuit diagram and (b) calculate the total energy stored in the two capacitors. (c) **What If?** What potential difference would be required across the same two capacitors connected in series for the combination to store the same amount of energy as in part (b)? (d) Draw a circuit diagram of the circuit described in part (c).

22. **S** A parallel-plate capacitor has a charge Q and plates of area A. What force acts on one plate to attract it toward the other plate? Because the electric field between the plates is $E = Q/A\epsilon_0$, you might think the force is $F = QE = Q^2/A\epsilon_0$. This conclusion is wrong because the field E includes contributions from both plates, and the field created by the positive plate cannot exert any force on the positive plate. Show that the force exerted on each plate is actually $F = Q^2/2A\epsilon_0$. *Suggestion:* Let $C = \epsilon_0 A/x$ for an arbitrary plate separation x and note that the work done in separating the two charged plates is $W = \int F \, dx$.

23. **GP** **S** Consider two conducting spheres with radii R_1 and R_2 separated by a distance much greater than either radius. A total charge Q is shared between the spheres. We wish to show that when the electric potential energy of the system has a minimum value, the potential difference between the spheres is zero. The total charge Q is equal to $q_1 + q_2$, where q_1 represents the charge on the first sphere and q_2 the charge on the second. Because the spheres are very far

apart, you can assume the charge of each is uniformly distributed over its surface. (a) Show that the energy associated with a single conducting sphere of radius R and charge q surrounded by a vacuum is $U_E = k_e q^2/2R$. (b) Find the total energy of the system of two spheres in terms of q_1, the total charge Q, and the radii R_1 and R_2. (c) To minimize the energy, differentiate the result to part (b) with respect to q_1 and set the derivative equal to zero. Solve for q_1 in terms of Q and the radii. (d) From the result to part (c), find the charge q_2. (e) Find the potential of each sphere. (f) What is the potential difference between the spheres?

SECTION 25.5 Capacitors with Dielectrics

24. A supermarket sells rolls of aluminum foil, plastic wrap, and waxed paper. (a) Describe a capacitor made from such materials. Compute order-of-magnitude estimates for (b) its capacitance and (c) its breakdown voltage.

25. **[V]** Determine (a) the capacitance and (b) the maximum potential difference that can be applied to a Teflon-filled parallel-plate capacitor having a plate area of 1.75 cm² and a plate separation of 0.040 0 mm.

26. **[QC]** The voltage across an air-filled parallel-plate capacitor is measured to be 85.0 V as shown in Figure P25.26a. When a dielectric is inserted and completely fills the space between the plates as in Figure P25.26b, the voltage drops to 25.0 V. (a) What is the dielectric constant of the inserted material? (b) Can you identify the dielectric? If so, what is it? (c) If the dielectric does not completely fill the space between the plates, what could you conclude about the voltage across the plates?

Figure P25.26

27. A commercial capacitor is to be constructed as shown in Figure P25.27. This particular capacitor is made from two strips of aluminum foil separated by a strip of paraffin-coated paper. Each strip of foil and paper is 7.00 cm wide. The foil is 0.004 00 mm thick, and the paper is 0.025 0 mm thick and has a dielectric constant of 3.70. What length should the strips have if a

Figure P25.27

capacitance of 9.50×10^{-8} F is desired before the capacitor is rolled up? (Adding a second strip of paper and rolling the capacitor would effectively double its capacitance by allowing charge storage on both sides of each strip of foil.)

28. Each capacitor in the combination shown in Figure P25.28 has a breakdown voltage of 15.0 V. What is the breakdown voltage of the combination between points a and b?

Figure P25.28

29. **[AMT]** A 2.00-nF parallel-plate capacitor is charged to an initial potential difference $\Delta V_i = 100$ V and is then isolated. The dielectric material between the plates is mica, with a dielectric constant of 5.00. (a) How much work is required to withdraw the mica sheet? (b) What is the potential difference across the capacitor after the mica is withdrawn?

SECTION 25.6 Electric Dipole in an Electric Field

30. **[AMT]** An infinite line of positive charge lies along the y axis, with charge density $\lambda = 2.00 \ \mu C/m$. A dipole is placed with its center along the x axis at $x = 25.0$ cm. The dipole consists of two charges $\pm 10.0 \ \mu C$ separated by 2.00 cm. The axis of the dipole makes an angle of 35.0° with the x axis, and the positive charge is farther from the line of charge than the negative charge. Find the net force exerted on the dipole.

31. **[S]** A small object with electric dipole moment $\vec{\mathbf{p}}$ is placed in a nonuniform electric field $\vec{\mathbf{E}} = E(x)\hat{\mathbf{i}}$. That is, the field is in the x direction, and its magnitude depends only on the coordinate x. Let θ represent the angle between the dipole moment and the x direction. Prove that the net force on the dipole is

$$F = p\left(\frac{dE}{dx}\right)\cos\theta$$

acting in the direction of increasing field.

SECTION 25.7 An Atomic Description of Dielectrics

32. **[S]** The general form of Gauss's law describes how a charge creates an electric field in a material, as well as in vacuum:

$$\int \vec{\mathbf{E}} \cdot d\vec{\mathbf{A}} = \frac{q_{in}}{\epsilon}$$

where $\epsilon = \kappa \epsilon_0$ is the permittivity of the material. (a) A sheet with charge Q uniformly distributed over its area A is surrounded by a dielectric. Show that the sheet creates a uniform electric field at nearby points with magnitude $E = Q/2A\epsilon$. (b) Two large sheets of area A, carrying opposite charges of equal magnitude Q, are a small distance d apart. Show that they create uniform electric field in the space between them with magnitude $E = Q/A\epsilon$. (c) Assume the negative plate is at zero potential. Show that the positive plate is at potential $Qd/A\epsilon$. (d) Show that the capacitance of the pair of plates is given by $C = A\epsilon/d = \kappa A\epsilon_0/d$.

ADDITIONAL PROBLEMS

33. You are working in a laboratory, using very sensitive mea-
BIO surement equipment. Your supervisor has explained that
CR the equipment is also very sensitive to electrical discharge
from human operators. Specification tables for the equipment indicate that an electrical discharge providing even a
very small amount of energy of 250 μJ is enough to damage
the equipment. Your supervisor wants to install an apparatus that will be used to remove the electrical charge from
individuals' bodies before they touch the equipment. To
do this, she asks you to estimate (a) the capacitance of the
human body and determine (b) the charge on the body
and (c) the electric potential of the body, relative to a point
infinitely far away, corresponding to the energy transfer
that will damage the equipment.

34. Four parallel metal plates P_1, P_2, P_3, and P_4, each of area
7.50 cm², are separated successively by a distance $d =$
1.19 mm as shown in Figure P25.34. Plate P_1 is connected
to the negative terminal of a battery, and P_2 is connected
to the positive terminal. The battery maintains a potential
difference of 12.0 V. (a) If P_3 is connected to the negative
terminal, what is the capacitance of the three-plate system
$P_1P_2P_3$? (b) What is the charge on P_2? (c) If P_4 is now connected to the positive terminal, what is the capacitance of
the four-plate system $P_1P_2P_3P_4$? (d) What is the charge on P_4?

Figure P25.34

35. A uniform electric field $E = 3\,000$ V/m exists within a certain region. What volume of space contains an energy equal
to 1.00×10^{-7} J? Express your answer in cubic meters and
in liters.

36. Two large, parallel metal plates, each of area A, are oriented
S horizontally and separated by a distance $3d$. A grounded
conducting wire joins them, and initially each plate carries no charge. Now a third identical plate carrying charge
Q is inserted between the two plates, parallel to them and
located a distance d from the upper plate as shown in Figure P25.36. (a) What induced charge appears on each of the
two original plates? (b) What potential difference appears
between the middle plate and each of the other plates?

Figure P25.36

37. A parallel-plate capacitor with vacuum between its horizon-
QC tal plates has a capacitance of 25.0 μF. A nonconducting

liquid with dielectric constant 6.50 is poured into the space
between the plates, filling up a fraction f of its volume.
(a) Find the new capacitance as a function of f. (b) What
should you expect the capacitance to be when $f = 0$? Does
your expression from part (a) agree with your answer? (c)
What capacitance should you expect when $f = 1$? Does the
expression from part (a) agree with your answer?

38. *Why is the following situation impossible?* A 10.0-μF capacitor has plates with vacuum between them. The capac-
itor is charged so that it stores 0.050 0 J of energy. A
particle with charge -3.00 μC is fired from the positive
plate toward the negative plate with an initial kinetic energy
equal to 1.00×10^{-4} J. The particle arrives at the negative
plate with a reduced kinetic energy.

39. Two square plates of sides ℓ are placed parallel to each
QC other with separation d as suggested in Figure P25.39. You
S may assume d is much less than ℓ. The plates carry uniformly distributed static charges $+Q_0$ and $-Q_0$. A block
of metal has width ℓ, length ℓ, and thickness slightly less
than d. It is inserted a distance x into the space between
the plates. The charges on the plates remain uniformly distributed as the block slides in. In a static situation, a metal
prevents an electric field from penetrating inside it. The
metal can be thought of as a perfect dielectric, with $\kappa \to \infty$.
(a) Calculate the stored energy in the system as a function
of x. (b) Find the direction and magnitude of the force
that acts on the metallic block. (c) The area of the advanc-
ing front face of the block is essentially equal to ℓd. Considering the force on the block as acting on this face, find the
stress (force per area) on it. (d) Express the energy density
in the electric field between the charged plates in terms
of Q_0, ℓ, d, and ϵ_0. (e) Explain how the answers to parts
(c) and (d) compare with each other.

Figure P25.39

40. (a) Two spheres have radii a and b, and their centers
S are a distance d apart. Show that the capacitance of this system is

$$C = \frac{4\pi\epsilon_0}{\dfrac{1}{a} + \dfrac{1}{b} - \dfrac{2}{d}}$$

provided d is large compared with a and b. *Suggestion:*
Because the spheres are far apart, assume the potential of
each equals the sum of the potentials due to each sphere.
(b) Show that as d approaches infinity, the above result
reduces to that of two spherical capacitors in series.

41. Assume that the internal diameter of the Geiger–
Mueller tube described in Problem 42 in Chapter 24 is
2.50 cm and that the wire along the axis has a diameter
of 0.200 mm. The dielectric strength of the gas between
the central wire and the cylinder is 1.20×10^6 V/m.
Use the result of that problem to calculate the maximum
potential difference that can be applied between the wire
and the cylinder before breakdown occurs in the gas.

42. A parallel-plate capacitor of plate separation d is charged
to a potential difference ΔV_0. A dielectric slab of thickness d and dielectric constant κ is introduced between the plates while the battery remains connected to the plates. (a) Show that the ratio of energy stored after the dielectric is introduced to the energy stored in the empty capacitor is $U_E/U_0 = \kappa$. (b) Give a physical explanation for this increase in stored energy. (c) What happens to the charge on the capacitor? *Note:* This situation is not the same as in Example 25.5, in which the battery was removed from the circuit before the dielectric was introduced.

43. To repair a power supply for a stereo amplifier, an electronics technician needs a 100-μF capacitor capable of withstanding a potential difference of 90 V between the plates. The immediately available supply is a box of five 100-μF capacitors, each having a maximum voltage capability of 50 V. (a) What combination of these capacitors has the proper electrical characteristics? Will the technician use all the capacitors in the box? Explain your answers. (b) In the combination of capacitors obtained in part (a), what will be the maximum voltage across each of the capacitors used?

44. Example 25.1 explored a cylindrical capacitor of length ℓ with radii a and b for the two conductors. In the What If? section of that example, it was claimed that increasing ℓ by 10% is more effective in terms of increasing the capacitance than increasing a by 10% if $b > 2.85a$. Verify this claim mathematically.

45. You are part of a team working in a machine parts mechanic's shop. An important customer has asked your company to provide springs with a very precise force constant k. You devise the electrical circuit shown in Figure P25.45 to measure the spring constant of each of the springs to be provided to the customer. The circuit consists of two identical, parallel metal plates free to move, other than being connected to identical metal springs, a switch, and a battery with terminal voltage ΔV. With the switch open, the plates are uncharged, are separated by a distance d, and have a capacitance C. When the switch is closed, the plates become charged and attract each other. The distance between the plates changes by a factor f, after which the plates are in equilibrium between the spring forces and the attractive electric force between the plates. To keep the plates from going into oscillations, you hold each plate with insulating gloves as the switch is closed and apply a force on the plates that allows them to move together at a slow constant speed until they are at the equilibrium separation, at which point you can release the plates. You determine an expression for the spring constant in terms of C, d, ΔV, and f.

Figure P25.45 Problems 45 and 50.

CHALLENGE PROBLEMS

46. Consider two long, parallel, and oppositely charged wires of radius r with their centers separated by a distance D that is much larger than r. Assuming the charge is distributed uniformly on the surface of each wire, show that the capacitance per unit length of this pair of wires is

$$\frac{C}{\ell} = \frac{\pi\epsilon_0}{\ln(D/r)}$$

47. Some physical systems possessing capacitance continuously distributed over space can be modeled as an infinite array of discrete circuit elements. Examples are a microwave waveguide and the axon of a nerve cell. To practice analysis of an infinite array, determine the equivalent capacitance C between terminals X and Y of the infinite set of capacitors represented in Figure P25.47. Each capacitor has capacitance C_0. *Suggestions:* Imagine that the ladder is cut at the line AB and note that the equivalent capacitance of the infinite section to the right of AB is also C.

Figure P25.47

48. A parallel-plate capacitor with plates of area LW and plate separation t has the region between its plates filled with wedges of two dielectric materials as shown in Figure P25.48. Assume t is much less than both L and W. (a) Determine its capacitance. (b) Should the capacitance be the same if the labels κ_1 and κ_2 are interchanged? Demonstrate that your expression does or does not have this property. (c) Show that if κ_1 and κ_2 approach equality to a common value κ, your result becomes the same as the capacitance of a capacitor containing a single dielectric: $C = \kappa\epsilon_0 LW/t$.

Figure P25.48

49. A capacitor is constructed from two square, metallic plates of sides ℓ and separation d. Charges $+Q$ and $-Q$ are placed on the plates, and the power supply is then removed. A material of dielectric constant κ is inserted a distance x into the capacitor as shown in Figure P25.49 (page 690). Assume d is much smaller than x. (a) Find the equivalent capacitance of the device. (b) Calculate the energy stored in the capacitor. (c) Find the direction and magnitude of the force exerted by the plates on the dielectric. (d) Obtain a numerical value for the force when $x = \ell/2$, assuming $\ell = 5.00$ cm, $d = 2.00$ mm, the dielectric is glass ($\kappa = 4.50$), and the capacitor was charged to 2.00×10^3 V before the dielectric was inserted. *Suggestion:* The system can be considered as two capacitors connected in parallel.

Figure P25.49

50. This problem is a continuation of Problem 45. You are
CR part of a team working in a machine parts mechanic's shop. An important customer has asked your company to provide springs with a very precise force constant k. You devise the electrical circuit shown in Figure P25.45 to measure the spring constant of each of the springs to be provided to the customer.

The circuit consists of two identical, parallel metal plates connected to identical metal springs, a switch, and a battery with emf ΔV. With the switch open, the plates are uncharged, are separated by a distance d, and have a capacitance C.

To provide a comparison value for the spring constant that you found in Problem 45, you slide a slab of material with dielectric constant κ and thickness t between the plates, so that it is in contact with one of the plates as shown in Figure P25.50. When the switch is closed, the plates become charged and attract each other. The distance between the plates changes by a factor f, after which the plates are in equilibrium between the spring forces and the attractive electric force between the plates. To keep the plates from going into oscillations, you hold each plate with insulating gloves as the switch is closed and apply a force on the plates that allows them to move together at a slow constant speed until they are at the equilibrium separation, at which point you can release the plates. Find an expression for the spring constant in terms of C, d, ΔV, k, t, and f.

Figure P25.50

Current and Resistance

STORYLINE **You have returned from your spring break vacation in** Florida and are back in classes. While driving to campus one day, you pass an electrical substation and begin to wonder what goes on there. You park and walk up to the fence surrounding the substation. A sign says, "Danger! High Voltage! Keep out!" This starts you wondering about high voltage. The voltage in your home, you know, is delivered at 240 volts. Why would the voltage at the substation be any higher? You pull out your smartphone and do a little online searching. You find that substations are designed to change the voltage from high values to low values. You also find that energy is transferred by electric power lines from the source at potential differences as high as 765 *kilovolts*! You say, "Wait a minute! Why is that necessary? Why don't we just transfer electricity at 240 volts, and we won't need all this substation stuff?"

CONNECTIONS So far in our study of electricity, we have considered primarily situations involving *stationary* electric charge distributions, which result in electric fields in a region of space and electric forces on charged particles (Chapter 22), potential differences between points in space (Chapter 24), and capacitance associated with a pair of conductors (Chapter 25). In this chapter, we consider situations involving electric charges that are in motion through some region of space. We use the term *electric current*, or simply *current*, to describe the rate of flow of charge. Most practical applications of electricity deal with electric currents, including a variety of home appliances. For example, the voltage from a wall plug produces a current in the coils of a toaster when it is turned on. In these common situations, current exists in a conductor such as a copper wire. Currents can also exist outside a conductor. For instance, a beam of electrons in a particle accelerator constitutes

691

a current. Our study of current will allow us to define electrical *resistance* and introduce a new circuit element, the *resistor*. We conclude the chapter by returning to the concept of energy and discussing the rate at which energy is transferred to a device in an electric circuit. The energy transfer mechanism in Equation 8.2 that corresponds to this process is electrical transmission T_{ET}. Our discussions in this chapter will prepare us to design electric circuits in Chapter 27 and beyond.

26.1 Electric Current

In this first section, we study the basic behavior of a flow of electric charges through a region of space. The flow of charges between two points in space is driven by a potential difference between the points. Whenever there is a net flow of charge through some region, an electric *current* is said to exist. The amount of current depends both on the potential difference and any material that may fill the space through which the charges flow.

It is instructive to compare electric current to the flow of viscous fluids in pipes discussed in Section 14.7. For example, the flow of water in a plumbing pipe is driven by a pressure difference and can be quantified as the volume flow rate, often measured in liters per minute. A river current can be characterized by describing the rate at which the water flows past a particular location. For example, the flow over the brink at Niagara Falls is maintained at rates between 1 400 m³/s and 2 800 m³/s.

There is also an analogy between thermal conduction and current. In Section 19.6, we discussed the flow of energy by heat through a sample of material. The rate of energy flow is determined by the material as well as the temperature difference across the material as described by Equation 19.17. Another analogy is *diffusion*, for example, when a drop of food coloring is placed in a cup of water and spreads to eventually fill the cup. The flow of atoms or molecules in diffusion is driven by a concentration difference.

To define current quantitatively, suppose charges are moving perpendicular to a surface of area A as shown in Figure 26.1. (This area could be the cross-sectional area of a wire, for example.) The **current** is defined as the rate at which charge flows through this surface. If ΔQ is the amount of charge that passes through this surface in a time interval Δt, the **average current** I_{avg} is equal to the charge that passes through A per unit time:

$$I_{avg} = \frac{\Delta Q}{\Delta t} \tag{26.1}$$

If the rate at which charge flows varies in time, the current varies in time; we define the **instantaneous current** I as the limit of the average current as $\Delta t \rightarrow 0$:

◀ Electric current

$$I \equiv \frac{dQ}{dt} \tag{26.2}$$

The SI unit of current is the **ampere** (A):

$$1\text{ A} = 1\text{ C/s} \tag{26.3}$$

That is, 1 A of current is equivalent to 1 C of charge passing through a surface in 1 s.

The charged particles passing through the surface in Figure 26.1 can be positive, negative, or both. It is conventional to assign to the current a direction that is the same as that of the flow of positive charge. In electrical conductors such as copper or aluminum, the current results from the motion of negatively charged electrons. Therefore, in an ordinary conductor, the direction of the current is opposite the direction of flow of electrons. For a beam of positively charged protons in an accelerator, however, the current is in the direction of motion of the protons. In some cases—such as those involving gases and electrolytes, for instance—the

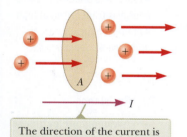

The direction of the current is the direction in which positive charges flow when free to do so.

Figure 26.1 Charges in motion through an area A. The time rate at which charge flows through the area is defined as the current I.

current is the result of the flow of both positive and negative charges. It is common to refer to a moving charge (positive or negative) as a mobile **charge carrier.**

Microscopic Model of Current

We can relate current to the motion of the charge carriers by describing a microscopic model of conduction in a metal. Consider the current in a cylindrical conductor of cross-sectional area A (Fig. 26.2). The volume of a segment of the conductor of length Δx (between the two circular cross sections shown in Fig. 26.2) is $A \Delta x$. If n represents the number of mobile charge carriers per unit volume (in other words, the charge carrier density), the number of carriers in the segment is $nA \Delta x$. Therefore, the total charge ΔQ in this segment is

$$\Delta Q = (nA \Delta x)q$$

Figure 26.2 A segment of a uniform conductor of cross-sectional area A.

where q is the charge on each carrier. If the carriers move with a velocity $\vec{v}_d$ parallel to the axis of the cylinder, the magnitude of the displacement they experience in the x direction in a time interval Δt is $\Delta x = v_d \Delta t$. Let Δt be the time interval required for the charge carriers in the segment to move through a displacement whose magnitude is equal to the length of the segment. This time interval is also the same as that required for all the charge carriers in the segment to pass through the circular area at one end. With this choice, we can write ΔQ as

$$\Delta Q = (nAv_d \Delta t)q$$

Dividing both sides of this equation by Δt, we find that the average current in the conductor is

$$I_{\text{avg}} = \frac{\Delta Q}{\Delta t} = nqv_d A \tag{26.4}$$

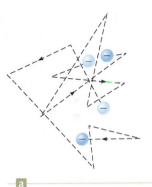

In reality, the speed of the charge carriers v_d is an average speed called the **drift speed.** To understand the meaning of drift speed, consider an isolated conductor in which the charge carriers are free electrons, as discussed in Section 22.2. These electrons undergo random thermal motion that is analogous to the motion of gas molecules. The electrons collide repeatedly with the metal atoms, and their resultant motion is complicated and zigzagged as in Figure 26.3a. When a potential difference is applied across the conductor (for example, by means of a battery), an electric field is set up in the conductor; this field exerts an electric force on the electrons, producing a current. In addition to the zigzag motion due to the collisions with the metal atoms, the electrons move slowly along the conductor (in a direction opposite that of $\vec{E}$) at the **drift velocity** $\vec{v}_d$ as shown in Figure 26.3b.

You can think of the atom–electron collisions in a conductor as an effective internal friction (or drag force) similar to that experienced by a liquid's molecules flowing through a pipe stuffed with steel wool. The energy transferred from the electrons to the metal atoms during collisions causes an increase in the atom's vibrational energy and a corresponding increase in the conductor's temperature.

The random motion of the charge carriers is modified by the field, and they have a drift velocity opposite the direction of the electric field.

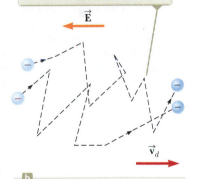

Figure 26.3 (a) A schematic diagram of the random motion of two charge carriers in a conductor in the absence of an electric field. The drift velocity is zero. (b) The motion of the charge carriers in a conductor in the presence of an electric field. Because of the acceleration of the charge carriers due to the electric force, the paths are actually parabolic. The drift speed, however, is much smaller than the average speed, so the parabolic shape is not visible on this scale.

QUICK QUIZ 26.1 Consider positive and negative charges of equal magnitude moving horizontally through the four regions shown in Figure 26.4. Rank the current in these four regions from highest to lowest.

Figure 26.4 (Quick Quiz (26.1)) Charges move through four regions.

| Example **26.1** | **Drift Speed in a Copper Wire** |

The 12-gauge copper wire in a typical residential building has a cross-sectional area of 3.31×10^{-6} m². It carries a constant current of 10.0 A. What is the drift speed of the electrons in the wire? Assume each copper atom contributes one free electron to the current. The density of copper is 8.92 g/cm³.

SOLUTION

Conceptualize Imagine electrons following a zigzag motion such as that in Figure 26.3a, with a drift velocity parallel to the wire superimposed on the motion as in Figure 26.3b. As mentioned earlier, the drift speed is small, and this example helps us quantify the speed.

Categorize We evaluate the drift speed using Equation 26.4. Because the current is constant, the average current during any time interval is the same as the constant current: $I_{avg} = I$.

Analyze The periodic table of the elements in Appendix C shows that the molar mass of copper is $M = 63.5$ g/mol. Recall that 1 mol of any substance contains Avogadro's number of atoms ($N_A = 6.02 \times 10^{23}$ mol^{-1}).

Use the molar mass and the density of copper to find the volume of 1 mole of copper:

$$V = \frac{M}{\rho}$$

From the assumption that each copper atom contributes one free electron to the current, find the electron density in copper:

$$n = \frac{N_A}{V} = \frac{N_A \rho}{M}$$

Solve Equation 26.4 for the drift speed and substitute for the electron density:

$$v_d = \frac{I_{avg}}{nqA} = \frac{I}{nqA} = \frac{IM}{qAN_A \rho}$$

Substitute numerical values:

$$v_d = \frac{(10.0 \text{ A})(0.063\ 5 \text{ kg/mol})}{(1.60 \times 10^{-19} \text{ C})(3.31 \times 10^{-6} \text{ m}^2)(6.02 \times 10^{23} \text{ mol}^{-1})(8\ 920 \text{ kg/m}^3)}$$

$$= 2.23 \times 10^{-4} \text{ m/s}$$

Finalize This result shows that typical drift speeds are very small. For instance, electrons traveling with a speed of 2.23×10^{-4} m/s would take about 75 min to travel 1 m! You might therefore wonder why a light turns on almost instantaneously when its switch is thrown. In a conductor, changes in the electric field that drives the free electrons according to the particle in a field model travel through the conductor with a speed close to that of light. So, when you flip on a light switch, electrons *already in* the lightbulb experience electric forces and begin moving after a time interval on the order of nanoseconds.

26.2 Resistance

In Section 24.6, we argued that the electric field inside a conductor is zero. This statement is true, however, *only* if the conductor is in static equilibrium as stated in that discussion. If a wire is connected across the terminals of a battery, the conductor is *not* in static equilibrium. In this case, there is a nonzero electric field in the conductor, and a current exists in the wire.

Consider a conductor of cross-sectional area A carrying a current I. The **current density** J in the conductor is defined as the current per unit area. Because the current $I = nqv_d A$, the current density is

Current density ▶

$$J \equiv \frac{I}{A} = nqv_d \tag{26.5}$$

where J has SI units of amperes per meter squared. This expression is valid only if the current density is uniform and only if the surface of cross-sectional area A is perpendicular to the direction of the current.

A current density and an electric field are established in a conductor whenever a potential difference is maintained across the conductor. In some materials, the current density is proportional to the electric field:

$$J = \sigma E \tag{26.6}$$

where the constant of proportionality σ is called the **conductivity** of the conductor.[1] Materials that obey Equation 26.6 are said to follow **Ohm's law,** named after Georg Simon Ohm. More specifically, Ohm's law states the following:

> For many materials (including most metals), the ratio of the current density to the electric field is a constant σ that is independent of the electric field producing the current.

Materials and devices that obey Ohm's law and hence demonstrate this simple relationship between E and J are said to be *ohmic*. Experimentally, however, it is found that not all materials and devices have this property. Those that do not obey Ohm's law are said to be *nonohmic*. Ohm's law is not a fundamental law of nature; rather, it is an empirical relationship valid only for certain situations.

We can obtain an equation useful in practical applications by considering a segment of straight wire of uniform cross-sectional area A and length ℓ as shown in Figure 26.5. A potential difference $\Delta V = V_b - V_a$ is maintained across the wire, creating in the wire an electric field and a current. If the field is assumed to be uniform, the magnitude of the potential difference across the wire is related to the field within the wire through Equation 24.6,

$$\Delta V = E\ell$$

Therefore, using Equation 26.6, we can express the potential difference across the wire as

$$\Delta V = \frac{\ell J}{\sigma}$$

Because $J = I/A$, the potential difference across the wire is

$$\Delta V = \left(\frac{\ell}{\sigma A}\right) I = R I$$

The quantity $R = \ell/\sigma A$ is called the **resistance** of the conductor. We define the resistance in terms of dynamic variables as the ratio of the potential difference across a conductor to the current in the conductor:

$$R \equiv \frac{\Delta V}{I} \tag{26.7}$$

We will use this equation again and again when studying electric circuits. This result shows that resistance has SI units of volts per ampere. One volt per ampere is defined to be one **ohm** (Ω):

$$1\ \Omega \equiv 1\ \text{V/A} \tag{26.8}$$

Equation 26.7 shows that if a potential difference of 1 V across a conductor causes a current of 1 A, the resistance of the conductor is 1 Ω. For example, if an electrical appliance connected to a 120-V source of potential difference carries a current of 6 A, its resistance is 20 Ω.

Most electric circuits use circuit elements called **resistors** to control the current in the various parts of the circuit. As with capacitors in Chapter 25, many resistors are built into integrated circuit chips (Section 42.7), but stand-alone resistors are still available and widely used. Two common types are the *composition resistor*, which

Georg Simon Ohm
German physicist (1789–1854)
Ohm, a high school teacher and later a professor at the University of Munich, formulated the concept of resistance and discovered the proportionalities expressed in Equations 26.6 and 26.7.

A potential difference $\Delta V = V_b - V_a$ maintained across the conductor sets up an electric field $\vec{\mathbf{E}}$, and this field produces a current I that is proportional to the potential difference.

Figure 26.5 A uniform conductor of length ℓ and cross-sectional area A.

PITFALL PREVENTION 26.2
Equation 26.7 Is Not Ohm's Law
Many individuals call Equation 26.7 Ohm's law, but that is incorrect. This equation is simply the definition of resistance, and it provides an important relationship between voltage, current, and resistance. Ohm's law is related to a proportionality of J to E (Eq. 26.6) or, equivalently, of I to ΔV, which, from Equation 26.7, indicates that the resistance is constant, independent of the applied voltage. We will see some devices for which Equation 26.7 correctly describes their resistance, but that do *not* obey Ohm's law. Equation 26.7 can be written as $\Delta V = IR$. Compare to Equation 14.14.

---

[1] Do not confuse conductivity σ with surface charge density, for which the same symbol is used.

The colored bands on this resistor are yellow, violet, black, and gold.

dexns/Shutterstock

Figure 26.6 A close-up view of a circuit board shows the color coding on a resistor. The gold band on the left tells us that the resistor is oriented "backward" in this view and we need to read the colors from right to left.

TABLE 26.1   Color Coding for Resistors

Color	Number	Multiplier	Tolerance
Black	0	1	
Brown	1	10^1	
Red	2	10^2	
Orange	3	10^3	
Yellow	4	10^4	
Green	5	10^5	
Blue	6	10^6	
Violet	7	10^7	
Gray	8	10^8	
White	9	10^9	
Gold		10^{-1}	5%
Silver		10^{-2}	10%
Colorless			20%

contains carbon, and the *wire-wound resistor*, which consists of a coil of wire. Values of resistors in ohms are normally indicated by color coding as shown in Figure 26.6 and Table 26.1. The first two colors on a resistor give the first two digits in the resistance value, with the decimal place to the right of the second digit. The third color represents the power of 10 for the multiplier of the resistance value. The last color is the tolerance of the resistance value. As an example, the four colors on the resistor at the bottom of Figure 26.6 are yellow (= 4), violet (= 7), black (= 10^0), and gold (= 5%), and so the resistance value is $47 \times 10^0 = 47\ \Omega$ with a tolerance value of 5% = 2 Ω.

The inverse of conductivity[2] is **resistivity** ρ:

◀ Resistivity is the inverse of conductivity

$$\rho = \frac{1}{\sigma}$$ (26.9)

where ρ has the units ohm · meters (Ω · m). Because $R = \ell/\sigma A$, we can express the resistance of a uniform block of material along the length ℓ as

◀ Resistance of a uniform material along the length ℓ

$$R = \rho\,\frac{\ell}{A}$$ (26.10)

Every ohmic material has a characteristic resistivity that depends on the properties of the material and on temperature. In addition, as you can see from Equation 26.10, the resistance of a sample of the material depends on the geometry of the sample as well as on the resistivity of the material. Table 26.2 gives the resistivities of a variety of materials at 20°C. Notice the enormous range, from very low values for good conductors such as copper and silver to very high values for good insulators such as glass and rubber. An ideal conductor would have zero resistivity, and an ideal insulator would have infinite resistivity.

Equation 26.10 shows that the resistance of a given cylindrical conductor such as a wire is proportional to its length and inversely proportional to its cross-sectional area. If the length of a wire is doubled, its resistance doubles. If its cross-sectional area is doubled, its resistance decreases by one-half. The situation is analogous to the flow of a liquid through a pipe. As the pipe's length is increased, the resistance to flow increases. As the pipe's cross-sectional area is increased, more liquid crosses a given cross section of the pipe per unit time interval. Therefore, more liquid flows for the same pressure differential applied to the pipe, and the resistance to flow decreases.

Ohmic materials and devices have a linear current–potential difference relationship over a broad range of applied potential differences (Fig. 26.7a). The slope of

PITFALL PREVENTION 26.3

Resistance and Resistivity
Resistivity is a property of a *substance*, whereas resistance is a property of an *object*. We have seen similar pairs of variables before. For example, density is a property of a substance, whereas mass is a property of an object. Equation 26.10 relates resistance to resistivity, and Equation 1.1 relates mass to density. Similarly, specific heat, which we studied in Chapter 19, is a property of a substance, while heat capacity is a property of an object.

---

[2] Do not confuse resistivity ρ with mass density or charge density, for which the same symbol is used.

TABLE 26.2 Resistivities and Temperature Coefficients of Resistivity for Various Materials

Material	Resistivity[a] ($\Omega \cdot m$)	Temperature Coefficient[b] α [(°C)$^{-1}$]
Silver	1.59×10^{-8}	3.8×10^{-3}
Copper	1.7×10^{-8}	3.9×10^{-3}
Gold	2.44×10^{-8}	3.4×10^{-3}
Aluminum	2.82×10^{-8}	3.9×10^{-3}
Tungsten	5.6×10^{-8}	4.5×10^{-3}
Iron	10×10^{-8}	5.0×10^{-3}
Platinum	11×10^{-8}	3.92×10^{-3}
Lead	22×10^{-8}	3.9×10^{-3}
Nichrome[c]	1.00×10^{-6}	0.4×10^{-3}
Carbon	3.5×10^{-5}	-0.5×10^{-3}
Germanium	0.46	-48×10^{-3}
Silicon[d]	2.3×10^{3}	-75×10^{-3}
Glass	10^{10} to 10^{14}	
Hard rubber	$\sim 10^{13}$	
Sulfur	10^{15}	
Quartz (fused)	75×10^{16}	

[a] All values at 20°C. All elements in this table are assumed to be free of impurities.
[b] See Section 26.4.
[c] A nickel–chromium alloy commonly used in heating elements. The resistivity of Nichrome varies with composition and ranges between 1.00×10^{-6} and 1.50×10^{-6} $\Omega \cdot m$.
[d] The resistivity of silicon is very sensitive to purity. The value can be changed by several orders of magnitude when it is doped with other atoms.

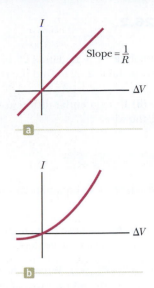

Figure 26.7 (a) The current–potential difference curve for an ohmic material. The curve is linear, and the slope is equal to the inverse of the resistance of the conductor. (b) A nonlinear current–potential difference curve for a junction diode. This device does not obey Ohm's law.

the I-versus-ΔV curve in the linear region yields a value for $1/R$. Nonohmic materials have a nonlinear current–potential difference relationship. One common semiconducting device with nonlinear I-versus-ΔV characteristics is the *junction diode* (Fig. 26.7b), which we discuss in Chapter 42 of the extended version of this text. The resistance of this device is low for currents in one direction (positive ΔV) and high for currents in the reverse direction (negative ΔV). In fact, most modern electronic devices, such as transistors, have nonlinear current–potential difference relationships; their proper operation depends on the particular way they violate Ohm's law.

QUICK QUIZ 26.2 A cylindrical wire has a radius r and length ℓ. If both r and ℓ are doubled, does the resistance of the wire (a) increase, (b) decrease, or (c) remain the same?

QUICK QUIZ 26.3 In Figure 26.7b, as the applied voltage increases, does the resistance of the diode (a) increase, (b) decrease, or (c) remain the same?

Example 26.2 **The Resistance of Nichrome Wire**

The radius of 22-gauge Nichrome wire is 0.32 mm.

(A) Calculate the resistance per unit length of this wire.

SOLUTION

Conceptualize Table 26.2 shows that Nichrome has a resistivity two orders of magnitude larger than the best conductors in the table. Therefore, we expect it to have some special practical applications that the best conductors may not have.

Categorize We model the wire as a cylinder so that a simple geometric analysis can be applied to find the resistance.

continued

26.2 continued

Analyze   Use Equation 26.10 and the resistivity of Nichrome from Table 26.2 to find the resistance per unit length:

$$\frac{R}{\ell} = \frac{\rho}{A} = \frac{\rho}{\pi r^2} = \frac{1.0 \times 10^{-6}\,\Omega \cdot \text{m}}{\pi(0.32 \times 10^{-3}\,\text{m})^2} = \boxed{3.1\,\Omega/\text{m}}$$

(B) If a potential difference of 10 V is maintained across a 1.0-m length of the Nichrome wire, what is the current in the wire?

SOLUTION

Analyze   Use Equation 26.7 to find the current:

$$I = \frac{\Delta V}{R} = \frac{\Delta V}{(R/\ell)\ell} = \frac{10\,\text{V}}{(3.1\,\Omega/\text{m})(1.0\,\text{m})} = \boxed{3.2\,\text{A}}$$

Finalize   Because of its high resistivity and resistance to oxidation, Nichrome is often used for heating elements in toasters, irons, and electric heaters.

WHAT IF?   What if the wire were composed of copper instead of Nichrome? How would the values of the resistance per unit length and the current change?

Answer   Table 26.2 shows us that copper has a resistivity two orders of magnitude smaller than that for Nichrome. Therefore, we expect the answer to part (A) to be smaller and the answer to part (B) to be larger. Calculations show that a copper wire of the same radius would have a resistance per unit length of only 0.053 Ω/m. A 1.0-m length of copper wire of the same radius would carry a current of 190 A with an applied potential difference of 10 V.

Example 26.3   **The Radial Resistance of a Coaxial Cable**

Coaxial cables are used extensively for cable television and other electronic applications. A coaxial cable consists of two concentric cylindrical conductors. The region between the conductors is completely filled with polyethylene plastic as shown in Figure 26.8a. Current leakage through the plastic, in the *radial* direction, is unwanted. (The cable is designed to conduct current along its length, but that is *not* the current being considered here.) The radius of the inner conductor is $a = 0.500$ cm, the radius of the outer conductor is $b = 1.75$ cm, and the length is $L = 15.0$ cm. The resistivity of the plastic is $1.0 \times 10^{13}\,\Omega \cdot \text{m}$. Calculate the radial resistance of the plastic between the two conductors.

SOLUTION

Conceptualize   Imagine two currents as suggested in the text of the problem. The desired current is along the cable, carried within the conductors. The undesired current corresponds to leakage through the plastic, and its direction is radial.

Categorize   Because the resistivity and the geometry of the plastic are known, we categorize this problem as one in which we find the resistance of the plastic from these parameters. Equation 26.10, however, represents the resistance of a block of material. We have a more complicated geometry in this situation. Because the area through which the charges pass depends on the radial position, we must use integral calculus to determine the answer.

Analyze   We divide the plastic into concentric cylindrical shells of infinitesimal thickness dr (Fig. 26.8b). Any charge passing from the inner to the outer conductor must move radially through this shell. Use a differential form of Equation 26.10, replacing ℓ with dr for the length variable: $dR = \rho\,dr/A$, where dR is the resistance of a shell of plastic of thickness dr and surface area A.

Figure 26.8   (Example 26.3) A coaxial cable. (a) Polyethylene plastic fills the gap between the two conductors. (b) End view, showing current leakage.

Write an expression for the resistance of our hollow cylindrical shell of plastic representing the area as the surface area of the shell:

$$dR = \frac{\rho\,dr}{A} = \frac{\rho}{2\pi r L}\,dr$$

26.3 continued

Integrate this expression from $r = a$ to $r = b$:

$$(1) \quad R = \int dR = \frac{\rho}{2\pi L} \int_a^b \frac{dr}{r} = \frac{\rho}{2\pi L} \ln\left(\frac{b}{a}\right)$$

Substitute the values given:

$$R = \frac{1.0 \times 10^{13}\ \Omega \cdot \text{m}}{2\pi(0.150\ \text{m})} \ln\left(\frac{1.75\ \text{cm}}{0.500\ \text{cm}}\right) = 1.33 \times 10^{13}\ \Omega$$

Finalize Let's compare this resistance to that of the inner copper conductor of the cable along the 15.0-cm length.

Use Equation 26.10 to find the resistance of the copper cylinder:

$$R_{\text{Cu}} = \rho \frac{\ell}{A} = (1.7 \times 10^{-8}\ \Omega \cdot \text{m})\left[\frac{0.150\ \text{m}}{\pi(5.00 \times 10^{-3}\ \text{m})^2}\right]$$

$$= 3.2 \times 10^{-5}\ \Omega$$

This resistance is 18 orders of magnitude smaller than the radial resistance. Therefore, almost all the current corresponds to charge moving along the length of the cable, with a very small fraction leaking in the radial direction.

W H A T I F ? Suppose the coaxial cable is enlarged to twice the overall diameter with two possible choices: (1) the ratio b/a is held fixed, or (2) the difference $b - a$ is held fixed. For which choice does the leakage current between the inner and outer conductors increase when the voltage is applied between them?

Answer For the current to increase, the resistance must decrease. For choice (1), in which b/a is held fixed,

Equation (1) shows that the resistance is unaffected. For choice (2), we do not have an equation involving the difference $b - a$ to inspect. Looking at Figure 26.8b, however, we see that increasing b and a while holding the difference constant results in charge flowing through the same thickness of plastic but through a larger area perpendicular to the flow. This larger area results in lower resistance and a higher current.

26.3 A Model for Electrical Conduction

In this section, we describe a structural model of electrical conduction in metals that was first proposed by Paul Drude (1863–1906) in 1900. (See Sections 1.2 and 20.1 for a review of structural models.) This model leads to Ohm's law and shows that resistivity can be related to the motion of electrons in metals. Although the Drude model described here has limitations, it introduces concepts that are applied in more elaborate treatments.

Following the outline of structural models from Section 20.1, the Drude model for electrical conduction has the following properties:

1. *Physical components:*
   Consider a conductor as a regular array of atoms plus a collection of free electrons, which are sometimes called *conduction* electrons. We identify the system as the combination of the atoms and the conduction electrons. The conduction electrons, although bound to their respective atoms when the atoms are not part of a solid, become free when the atoms condense into a solid.
2. *Behavior of the components:*
   (a) In the absence of an electric field, the conduction electrons move in random directions through the conductor (Fig. 26.3a). The situation is similar to the motion of gas molecules confined in a vessel. In fact, some scientists refer to conduction electrons in a metal as an *electron gas*.
   (b) When an electric field is applied to the system, the free electrons drift slowly in a direction opposite that of the electric field (Fig. 26.3b), with an average drift speed v_d that is much smaller (typically 10^{-4} m/s) than their average speed v_{avg} between collisions (typically 10^6 m/s).
   (c) The electron's motion after a collision is independent of its motion before the collision. The excess energy acquired by the electrons due to the work done on them by the electric field is transferred to the atoms of the conductor when the electrons and atoms collide.

With regard to property 2(c) above, the energy transferred to the atoms causes the internal energy of the system and, therefore, the temperature of the conductor to increase.

We are now in a position to derive an expression for the drift velocity, using several of our analysis models. When a free electron of mass m_e and charge q ($= -e$) is subjected to an electric field $\vec{\mathbf{E}}$, it is described by the particle in a field model and experiences a force $\vec{\mathbf{F}} = q\vec{\mathbf{E}}$. The electron is a particle under a net force, and its acceleration can be found from Newton's second law, $\sum \vec{\mathbf{F}} = m\vec{\mathbf{a}}$:

$$\vec{\mathbf{a}} = \frac{\sum \vec{\mathbf{F}}}{m} = \frac{q\vec{\mathbf{E}}}{m_e} \tag{26.11}$$

Because the electric field is uniform, the electron's acceleration is constant, so the electron can be modeled as a particle under constant acceleration. If $\vec{\mathbf{v}}_i$ is the electron's initial velocity the instant after a collision (which occurs at a time defined as $t = 0$), the velocity of the electron at a very short time t later (immediately before the next collision occurs) is, from Equation 4.8,

$$\vec{\mathbf{v}}_f = \vec{\mathbf{v}}_i + \vec{\mathbf{a}}t = \vec{\mathbf{v}}_i + \frac{q\vec{\mathbf{E}}}{m_e} t \tag{26.12}$$

Let's now take the average value of $\vec{\mathbf{v}}_f$ for all the electrons in the wire over all possible collision times t and all possible values of $\vec{\mathbf{v}}_i$. Assuming the initial velocities are randomly distributed over all possible directions (property 2(a) above), the average value of $\vec{\mathbf{v}}_i$ is zero. The average value of the second term of Equation 26.12 is $(q\vec{\mathbf{E}}/m_e)\tau$, where τ is the *average time interval between successive collisions*. Because the average value of $\vec{\mathbf{v}}_f$ is equal to the drift velocity,

◀ Drift velocity in terms of microscopic quantities

$$\vec{\mathbf{v}}_{f,\text{avg}} = \vec{\mathbf{v}}_d = \frac{q\vec{\mathbf{E}}}{m_e} \tau \tag{26.13}$$

The value of τ depends on the size of the metal atoms and the number of electrons per unit volume. We can relate this expression for drift velocity in Equation 26.13 to the current in the conductor. Substituting the magnitude of the velocity from Equation 26.13 into Equation 26.4, the average current in the conductor is given by

$$I_{\text{avg}} = nq\left(\frac{qE}{m_e}\tau\right)A = \frac{nq^2\tau A}{m_e} E \tag{26.14}$$

Because the current density J is the current divided by the area A,

◀ Current density in terms of microscopic quantities

$$J = \frac{nq^2\tau}{m_e} E$$

where n is the number of electrons per unit volume. Comparing this expression with Ohm's law, $J = \sigma E$, we obtain the following relationships for conductivity and resistivity of a conductor:

◀ Conductivity in terms of microscopic quantities

$$\sigma = \frac{nq^2\tau}{m_e} \tag{26.15}$$

◀ Resistivity in terms of microscopic quantities

$$\rho = \frac{1}{\sigma} = \frac{m_e}{nq^2\tau} \tag{26.16}$$

According to this classical model, conductivity and resistivity do not depend on the strength of the electric field. This feature is characteristic of a conductor obeying Ohm's law.

The model shows that the resistivity can be calculated from a knowledge of the density of the electrons, their charge and mass, and the average time interval τ between collisions. This time interval is related to the average distance between collisions ℓ_{avg} (the *mean free path*) and the average speed v_{avg} through the expression[3]

$$\tau = \frac{\ell_{avg}}{v_{avg}} \tag{26.17}$$

Although this structural model of conduction is consistent with Ohm's law, it does not correctly predict the values of resistivity or the behavior of the resistivity with temperature. For example, the results of classical calculations for v_{avg} using the ideal gas model for the electrons are about a factor of ten smaller than the actual values, which results in incorrect predictions of values of resistivity from Equation 26.16. Furthermore, according to Equations 26.16 and 26.17, the resistivity is predicted to vary with temperature as does v_{avg}, which, according to an ideal-gas model (Chapter 20, Eq. 20.44), is proportional to $\sqrt{T}$. This behavior is in disagreement with the experimentally observed linear dependence of resistivity with temperature for pure metals. (See Section 26.4.) Because of these incorrect predictions, we must modify our structural model. We shall call the model that we have developed so far the *classical* model for electrical conduction. To account for the incorrect predictions of the classical model, we develop it further into a *quantum mechanical* model, which can be found in advanced textbooks.

26.4 Resistance and Temperature

At the end of the preceding section, we discussed the temperature variation of resistivity. Over a limited temperature range, the resistivity of a conductor varies approximately linearly with temperature according to the expression

$$\rho = \rho_0[1 + \alpha(T - T_0)] \tag{26.18}$$

◀ Variation of ρ with temperature

where ρ is the resistivity at some temperature T (in degrees Celsius), ρ_0 is the resistivity at some reference temperature T_0 (usually taken to be 20°C), and α is the **temperature coefficient of resistivity.** From Equation 26.18, the temperature coefficient of resistivity can be expressed as

$$\alpha = \frac{\Delta\rho/\rho_0}{\Delta T} \tag{26.19}$$

where $\Delta\rho = \rho - \rho_0$ is the change in resistivity in the temperature interval $\Delta T = T - T_0$. Compare the form of Equation 26.19 with Equation 18.4 for the coefficient of thermal expansion.

The temperature coefficients of resistivity for various materials are given in Table 26.2. Notice that the unit for α is degrees Celsius^{-1} [(°C)$^{-1}$]. Because resistance is proportional to resistivity (Eq. 26.10), the variation of resistance of a sample is

$$R = R_0[1 + \alpha(T - T_0)] \tag{26.20}$$

where R_0 is the resistance at temperature T_0. Use of this property enables precise temperature measurements through careful monitoring of the resistance of a probe made from a particular material.

For some metals such as copper, resistivity is nearly proportional to temperature as shown in Figure 26.9. A nonlinear region always exists at very low temperatures, however, and the resistivity usually reaches some finite value as the temperature approaches absolute zero. This residual resistivity near absolute zero is caused primarily by the collision of electrons with impurities and imperfections in the metal. In contrast, high-temperature resistivity (the linear region) is predominantly characterized by collisions between electrons and metal atoms.

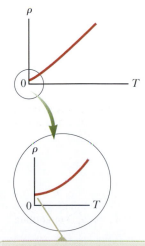

As T approaches absolute zero, the resistivity approaches a nonzero value.

Figure 26.9 Resistivity versus temperature for a metal such as copper. The curve is linear over a wide range of temperatures, and ρ increases with increasing temperature.

---

[3]Recall that the average speed of a group of particles depends on the temperature of the group (Chapter 20) and is not the same as the drift speed v_d.

Notice that three of the α values in Table 26.2 are negative, indicating that the resistivity of these materials decreases with increasing temperature. This behavior is indicative of a class of materials called *semiconductors,* first introduced in Section 22.2, and is due to an increase in the density of charge carriers at higher temperatures.

QUICK QUIZ 26.4 When does an incandescent lightbulb carry more current, **(a)** immediately after it is turned on and the glow of the metal filament is increasing or **(b)** after it has been on for a few milliseconds and the glow is steady?

26.5 Superconductors

There is a class of metals and compounds whose resistance decreases to zero when they are below a certain temperature T_c, known as the **critical temperature.** These materials are known as **superconductors.** The resistance–temperature graph for a superconductor follows that of a normal metal at temperatures above T_c (Fig. 26.10). When the temperature is at or below T_c, the resistivity drops suddenly to zero. This phenomenon was discovered in 1911 by Dutch physicist Heike Kamerlingh-Onnes (1853–1926) as he worked with mercury, which is a superconductor below 4.2 K. Measurements have shown that the resistivities of superconductors below their T_c values are less than 4×10^{-25} $\Omega \cdot$ m, or approximately 10^{17} times smaller than the resistivity of copper. In practice, these resistivities are considered to be zero.

Today, thousands of superconductors are known, and as Table 26.3 illustrates, the critical temperatures of recently discovered superconductors are substantially higher than initially thought possible. Two kinds of superconductors are recognized. The more recently identified ones are essentially ceramics with high critical temperatures, whereas superconducting materials such as those observed by Kamerlingh-Onnes are metals. If a room-temperature superconductor is ever identified, its effect on technology could be tremendous.

The value of T_c is sensitive to chemical composition, pressure, and molecular structure. Copper, silver, and gold, which are excellent room-temperature conductors, do not exhibit superconductivity.

One truly remarkable feature of superconductors is that once a current is set up in them, it persists *without any applied potential difference* (because $R = 0$). Steady currents have been observed to persist in superconducting loops for several years with no apparent decay!

An important and useful application of superconductivity is in the development of superconducting magnets, in which the magnitudes of the magnetic field are approximately ten times greater than those produced by the best normal

The resistance drops discontinuously to zero at T_c, which is 4.15 K for mercury.

Figure 26.10 Resistance versus temperature for a sample of mercury (Hg). The graph follows that of a normal metal above the critical temperature T_c.

TABLE 26.3 Critical Temperatures for Various Superconductors

Material	T_c (K)
$HgBa_2Ca_2Cu_3O_8$	134
Tl—Ba—Ca—Cu—O	125
Bi—Sr—Ca—Cu—O	105
$YBa_2Cu_3O_7$	92
Nb_3Ge	23.2
Nb_3Sn	18.05
Nb	9.46
Pb	7.18
Hg	4.15
Sn	3.72
Al	1.19
Zn	0.88

electromagnets. Such superconducting magnets are being considered as a means of storing energy. Superconducting magnets are currently used in medical magnetic resonance imaging, or MRI, units, which produce high-quality images of internal organs without the need for excessive exposure of patients to x-rays or other harmful radiation.

A successful theory for superconductivity in metals was published in 1957 by John Bardeen (1908–1991), L. N. Cooper (b. 1930), and J. R. Schrieffer (b. 1931); it is generally called BCS theory, based on the first letters of their last names. This theory led to a Nobel Prize in Physics for the three scientists in 1972.

An important development in physics that elicited much excitement in the scientific community was the discovery of high-temperature copper oxide–based superconductors. The excitement began with a 1986 publication by J. Georg Bednorz (b. 1950) and K. Alex Müller (b. 1927), scientists at the IBM Zurich Research Laboratory in Switzerland. In their seminal paper,[4] Bednorz and Müller reported strong evidence for superconductivity at 30 K in an oxide of barium, lanthanum, and copper. They were awarded the Nobel Prize in Physics in 1987 for their remarkable discovery. Later that year, teams of scientists from Japan and the United States reported superconductivity at 105 K in an oxide of bismuth, strontium, calcium, and copper. Superconductivity at temperatures as high as 150 K have been reported in an oxide containing mercury. The search for novel superconducting materials continues both for scientific reasons and because practical applications become more probable and widespread as the critical temperature is raised.

26.6 Electrical Power

In typical electric circuits, energy T_{ET} is transferred by electrical transmission from a source such as a battery to some device such as a lightbulb or a radio receiver. Let's determine an expression that will allow us to calculate the rate of this energy transfer. First, consider the simple circuit in Figure 26.11, where energy is delivered to a resistor. (Resistors are designated by the circuit symbol —⋀⋀⋀—.) Because the connecting wires also have resistance, some energy is delivered to the wires and some to the resistor. Unless noted otherwise, we shall assume the resistance of the wires is small compared with the resistance of the circuit element so that the energy delivered to the wires is negligible.

Imagine following a positive quantity of charge Q moving clockwise around the circuit in Figure 26.11 from point a through the battery and resistor back to point a. We identify the entire circuit as our system. As the charge moves from a to b through the battery, the electric potential energy of the system *increases* by an amount $Q \, \Delta V$ while the chemical potential energy in the battery *decreases* by the same amount. (Recall from Eq. 24.3 that $\Delta U_E = q \, \Delta V$.) As the charge moves from c to d through the resistor, however, the electric potential energy of the system decreases due to collisions of electrons with atoms in the resistor. In this process, the electric potential energy is transformed to internal energy corresponding to increased vibrational motion of the atoms in the resistor. Because the resistance of the interconnecting wires is neglected, no energy transformation occurs for paths bc and da. When the charge returns to point a, the net result is that some of the chemical potential energy in the battery has been delivered to the resistor and resides in the resistor as internal energy E_{int} associated with molecular vibration.

Let us analyze the energy situation in Figure 26.11 in terms of Equation 8.2. If we choose the resistor as the system, Equation 8.2 becomes, during a time interval after the switch is closed,

$$\Delta E_{int} = Q + T_{ET} + T_{ER}$$

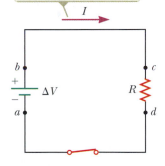

The direction of the effective flow of positive charge is clockwise.

Figure 26.11 A circuit consisting of a resistor of resistance R and a battery having a potential difference ΔV across its terminals.

PITFALL PREVENTION 26.4

Charges Do Not Move All the Way Around a Circuit in a Short Time In terms of understanding the energy transfer in a circuit, it is useful to *imagine* a charge moving all the way around the circuit even though it would take hours to do so.

[4]J. G. Bednorz and K. A. Müller, *Z. Phys. B* **64**:189, 1986.

The left side represents the increasing temperature of the resistor as it receives energy T_{ET} from the battery and radiates energy T_{ER} into the surroundings. The heat Q is another process by which the resistor can transfer energy because it is in thermal contact with the air and it is warmer than the air. Once the resistor reaches a steady-state temperature, the left side of the equation becomes zero, and the input electrical energy is balanced by the output heat and radiation.

If we choose the entire circuit as the system, Equation 8.2 for this same time interval becomes

$$\Delta U_c + \Delta E_{int} = Q + T_{ER}$$

The term ΔU_c on the left side represents the decreasing chemical potential energy of the battery as it delivers energy to the resistor. The right side represents thermal conduction of energy Q into the air and energy T_{ER} radiating into the surroundings from the resistor. Once the resistor reaches a steady-state temperature, the term ΔE_{int} becomes zero. The equation then represents the continuous draining of energy from the battery, with the energy leaving the circuit by heat and radiation, eventually becoming internal energy in the surroundings. Notice that T_{ET} does not appear in this equation, because the transfer of energy by electrical transmission occurs *within* this system.

Some electrical devices include *heat sinks*[5] connected to parts of the circuit to prevent these parts from reaching dangerously high temperatures. Heat sinks are pieces of metal with many fins. Because the metal's high thermal conductivity provides a rapid transfer of energy by heat away from the hot component and the large number of fins provides a large surface area in contact with the air, energy can transfer by radiation and into the air by heat at a high rate.

Let's now investigate the rate at which the electric potential energy of the system decreases as the charge Q passes through the resistor, using the first part of Equation 24.3:

$$\frac{dU_E}{dt} = \frac{d}{dt}(Q\,\Delta V) = \frac{dQ}{dt}\Delta V = I\,\Delta V$$

where I is the current in the circuit. The system regains this potential energy when the charge passes through the battery, at the expense of chemical energy in the battery. The rate at which the potential energy of the system decreases as the charge passes through the resistor is equal to the rate at which the system gains internal energy in the resistor. Therefore, the power P, representing the rate at which energy is delivered to the resistor, is

$$P = I\,\Delta V \tag{26.21}$$

We derived this result by considering a battery delivering energy to a resistor. Equation 26.21, however, can be used to calculate the power delivered by a voltage source to *any* device carrying a current I and having a potential difference ΔV between its terminals.

Using Equation 26.21 and $\Delta V = IR$ for a resistor, we can express the power delivered to the resistor in the alternative forms

$$P = I^2 R = \frac{(\Delta V)^2}{R} \tag{26.22}$$

When I is expressed in amperes, ΔV in volts, and R in ohms, the SI unit of power is the watt, as it was in Chapter 8 in our discussion of mechanical power. The process by which energy is transformed to internal energy in a conductor of resistance R is often called *joule heating*;[6] this transformation is also often referred to as

---

[5]This usage is another misuse of the word *heat* that is ingrained in our common language.

[6]It is commonly called *joule heating* even though the process of heat does not occur when energy delivered to a resistor appears as internal energy. It is another example of incorrect usage of the word *heat* that has become entrenched in our language.

an I^2R loss. When evaluating the electrical power delivered to a device in a circuit, keep in mind that Equation 26.21 can be used *generally*, but Equation 26.22 is only to be used when evaluating the power delivered to a *resistor*.

Let us now address the question raised in the opening storyline: Why is energy transported through electrical wires at very high voltages? When transporting energy by electricity through power lines (Fig. 26.12), we cannot assume the lines have zero resistance. Real power lines do indeed have resistance, and power is delivered to the resistance of these wires. Utility companies seek to minimize the energy transformed to internal energy in the lines and maximize the energy delivered to the consumer. Because $P = I\,\Delta V$, the same amount of energy can be transported either at high currents and low potential differences or at low currents and high potential differences. Utility companies choose to transport energy at low currents and high potential differences primarily for economic reasons. Copper wire is very expensive, so it is cheaper to use high-resistance wire (that is, wire having a small cross-sectional area; see Eq. 26.10). Therefore, in the expression for the power delivered to a resistor, $P = I^2R$, the resistance of the wire is fixed at a relatively high value for economic considerations. The I^2R loss can be reduced by keeping the current I as low as possible, which means transferring the energy at a high voltage. At the substation, the potential difference is usually reduced by a device called a *transformer*. Of course, when the potential difference decreases, the current increases by the same factor and the power remains the same. We shall discuss transformers in greater detail in Chapter 32.

Naufal MQ/Shutterstock

Figure 26.12 These power lines transfer energy from the electric company to homes and businesses. The energy is transferred at a very high voltage, possibly hundreds of thousands of volts in some cases. Even though it makes power lines very dangerous, the high voltage results in less loss of energy due to resistance in the wires.

Example 26.4 **Power in an Electric Heater**

An electric heater is constructed by applying a potential difference of 120 V across a Nichrome wire that has a total resistance of 8.00 Ω. Find the current carried by the wire and the power rating of the heater.

SOLUTION

Conceptualize As discussed in Example 26.2, Nichrome wire has high resistivity and is often used for heating elements in toasters, irons, and electric heaters. Therefore, we expect the power delivered to the wire to be relatively high.

Categorize We evaluate the power from Equation 26.22, so we categorize this example as a substitution problem.

Use Equation 26.7 to find the current in the wire:

$$I = \frac{\Delta V}{R} = \frac{120\ \text{V}}{8.00\ \Omega} = \boxed{15.0\ \text{A}}$$

Find the power rating using the expression $P = I^2R$ from Equation 26.22:

$$P = I^2R = (15.0\ \text{A})^2(8.00\ \Omega) = 1.80 \times 10^3\ \text{W} = \boxed{1.80\ \text{kW}}$$

WHAT IF? What if the heater were accidentally connected to a 240-V supply? (That is difficult to do because the shape and orientation of the metal contacts in 240-V plugs are different from those in 120-V plugs.) How would that affect the current carried by the heater and the power rating of the heater, assuming the resistance remains constant?

Answer If the applied potential difference were doubled, Equation 26.7 shows that the current would double. According to Equation 26.22, $P = (\Delta V)^2/R$, the power would be four times larger.

Example 26.5 **Linking Electricity and Thermodynamics**

An immersion heater must increase the temperature of 1.50 kg of water from 10.0°C to 50.0°C in 10.0 min while operating at 110 V.

(A) What is the required resistance of the heater?

continued

26.5 continued

SOLUTION

Conceptualize   An immersion heater is a resistor that is inserted into a container of water. As energy is delivered to the immersion heater, raising its temperature, energy leaves the surface of the resistor by heat, going into the water. When the immersion heater reaches a constant temperature, the rate of energy delivered to the resistance by electrical transmission (T_{ET}) is equal to the rate of energy delivered by heat (Q) to the water.

Categorize   This example allows us to link our new understanding of power in electricity with our experience with specific heat in thermodynamics (Chapter 19). The water is a *nonisolated system*. Its internal energy is rising because of energy transferred into the water by heat from the resistor, so Equation 8.2 reduces to $\Delta E_{int} = Q$. In our model, we assume the energy that enters the water from the heater remains in the water.

Analyze   To simplify the analysis, let's ignore the initial period during which the temperature of the resistor increases and also ignore any variation of resistance with temperature. Therefore, we imagine a constant rate of energy transfer for the entire 10.0 min.

Set the rate of energy delivered to the resistor equal to the rate of energy Q entering the water by heat:

$$P = \frac{(\Delta V)^2}{R} = \frac{Q}{\Delta t}$$

Use Equation 19.4, $Q = mc\,\Delta T$, to relate the energy input by heat to the resulting temperature change of the water and solve for the resistance:

$$\frac{(\Delta V)^2}{R} = \frac{mc\,\Delta T}{\Delta t} \quad \rightarrow \quad R = \frac{(\Delta V)^2\,\Delta t}{mc\,\Delta T}$$

Substitute the values given in the statement of the problem:

$$R = \frac{(110\text{ V})^2(600\text{ s})}{(1.50\text{ kg})(4\,186\text{ J/kg}\cdot{}^\circ\text{C})(50.0{}^\circ\text{C} - 10.0{}^\circ\text{C})} = \boxed{28.9\ \Omega}$$

(B) Estimate the cost of heating the water.

SOLUTION

Multiply the power by the time interval to find the amount of energy transferred to the resistor:

$$T_{ET} = P\,\Delta t = \frac{(\Delta V)^2}{R}\,\Delta t = \frac{(110\text{ V})^2}{28.9\ \Omega}(10.0\text{ min})\left(\frac{1\text{ h}}{60.0\text{ min}}\right)$$

$$= 69.8\text{ Wh} = 0.069\,8\text{ kWh}$$

Find the cost knowing that energy is purchased at an estimated price of 11¢ per kilowatt-hour:

$$\text{Cost} = (0.069\,8\text{ kWh})(\$0.11/\text{kWh}) = \$0.008 = \boxed{0.8¢}$$

Finalize   The cost to heat the water is very low, less than one cent. In reality, the cost is higher because some energy is transferred from the water into the surroundings by heat and electromagnetic radiation while its temperature is increasing. If you have electrical devices in your home with power ratings on them, use this power rating and an approximate time interval of use to estimate the cost for one use of the device.

Summary

> ## Definitions

The electric **current** I in a conductor is defined as

$$I \equiv \frac{dQ}{dt} \qquad (26.2)$$

where dQ is the charge that passes through a cross section of the conductor in a time interval dt. The SI unit of current is the **ampere** (A), where 1 A = 1 C/s.

The **current density** J in a conductor is the current per unit area:

$$J \equiv \frac{I}{A} \qquad (26.5)$$

The **resistance** R of a conductor is defined as

$$R \equiv \frac{\Delta V}{I} \qquad (26.7)$$

where ΔV is the potential difference across the conductor and I is the current it carries. The SI unit of resistance is volts per ampere, which is defined to be 1 **ohm** (Ω); that is, 1 Ω = 1 V/A.

❯ Concepts and Principles

The average current in a conductor is related to the motion of the charge carriers through the relationship

$$I_{avg} = nqv_d A \qquad (26.4)$$

where n is the density of charge carriers, q is the charge on each carrier, v_d is the drift speed, and A is the cross-sectional area of the conductor.

The current density in an ohmic conductor is proportional to the electric field according to the expression

$$J = \sigma E \qquad (26.6)$$

The proportionality constant σ is called the **conductivity** of the material of which the conductor is made. The inverse of σ is known as **resistivity** ρ (that is, $\rho = 1/\sigma$). Equation 26.6 is known as **Ohm's law,** and a material is said to obey this law if the ratio of its current density to its applied electric field is a constant that is independent of the applied field.

For a uniform block of material of cross-sectional area A and length ℓ, the resistance over the length ℓ is

$$R = \rho \frac{\ell}{A} \qquad (26.10)$$

where ρ is the resistivity of the material.

In a classical model of electrical conduction in metals, the electrons are treated as molecules of a gas. In the absence of an electric field, the average velocity of the electrons is zero. When an electric field is applied, the electrons move (on average) with a **drift velocity** $\vec{v}_d$ that is opposite the electric field. The drift velocity is given by

$$\vec{v}_d = \frac{q\vec{E}}{m_e}\tau \qquad (26.13)$$

where q is the electron's charge, m_e is the mass of the electron, and τ is the average time interval between electron–atom collisions. According to this model, the resistivity of the metal is

$$\rho = \frac{m_e}{nq^2\tau} \qquad (26.16)$$

where n is the number of free electrons per unit volume.

The resistivity of a conductor varies approximately linearly with temperature according to the expression

$$\rho = \rho_0[1 + \alpha(T - T_0)] \qquad (26.18)$$

where ρ_0 is the resistivity at some reference temperature T_0 and α is the **temperature coefficient of resistivity.**

If a potential difference ΔV is maintained across a circuit element, the **power,** or rate at which energy is supplied to the element, is

$$P = I\,\Delta V \qquad (26.21)$$

Because the potential difference across a resistor is given by $\Delta V = IR$, we can express the power delivered to a resistor as

$$P = I^2 R = \frac{(\Delta V)^2}{R} \qquad (26.22)$$

The energy delivered to a resistor by electrical transmission T_{ET} appears in the form of internal energy E_{int} in the resistor.

Think–Pair–Share

See the Preface for an explanation of the icons used in this problems set. For additional assessment items for this section, go to **WEBASSIGN** From Cengage

1. Imagine a system run by a gasoline engine that operates a conveyor belt. The belt carries basketballs to the top of a hill. At the top, the basketballs are dumped out and roll down the hill. They roll at an approximately constant speed because of the continuous collisions the balls make with trees, grass, and shrubs on the way down. The average speed of the balls downhill is the same as that with which they move on the conveyor belt. At the bottom of the hill, the balls arrive at the conveyor belt and are carried again to the top. Imagine that this system is a mechanical analogy to the electrical circuit in Figure 26.11. Discuss in your group the following: In the mechanical system, what is the analog to (a) the resistance, (b) the battery, (c) the electrons in the wires, (d) the electric field in the wires, and (e) the terminal voltage of the battery. (f) What is the analog to the situation in which the battery eventually runs out of energy? (g) Write the appropriate reduction of Equation 8.2 for the system of the conveyor belt, the basketballs, and the Earth (excluding the atmosphere) and a time interval from before the system is turned on (and all the basketballs are at the bottom of the hill) until a few minutes after it is turned on. Ignore air resistance on the basketballs. (h) Write the equation in part (g) for a time interval well after the system is turned on and the temperature of the system has stabilized. Compare the equation to the analogous energy equation for the circuit in Figure 26.11 and comment on the comparison. (i) In part (g), it is mentioned that the basketballs all start at the bottom of the hill when the system is first turned on. How is this *not* analogous to the circuit in Figure 26.11?

2. **ACTIVITY** With your group, consider the table below that shows voltage and current data for two different electrical devices. Which of the devices obeys Ohm's law?

Voltage Applied to Device (V)	Current in Device 1 (A)	Current in Device 2 (A)
1.00	0.123	0.123
2.00	0.249	0.250
3.00	0.365	0.389
4.00	0.486	0.545
5.00	0.621	0.701
6.00	0.745	0.909
7.00	0.854	1.230
8.00	0.984	1.550
9.00	1.102	1.719
10.0	1.241	1.747

3. **S** The ratio of the electric charge to the mass of an electron, e/m_e, was measured in 1897 by J.J. Thomson using a cathode-ray tube. (See Section 28.3.) Another way to find this ratio is by means of the *Tolman–Stewart experiment*. In this experiment, a coil consisting of a number N of circular turns of wire is rotated at a constant angular speed ω_0 about an axis perpendicular to the plane of the coil and passing through the common center of the turns. Because the electrons in the coil are carried around a circular path, there is a current I_0 associated with the movement of the electrons. The coil is then quickly brought to a stop and the amount of charge that passes through a slice of the coil containing all N turns during the stopping of the coil is measured by specialized apparatus. Because electrons are not attached to atoms, they will keep rotating for a very short time interval after the coil is stopped until they make enough collisions so that their motion is random. With your group, follow the steps outlined here to find an expression for the charge to mass ratio of the electron. (a) Assume that the current falls off exponentially with time as the coil is brought to rest: $I = I_0e^{-bt}$, where b is some unknown parameter. Find the total charge Q, in terms of I_0 and b, that passes through a measuring device located at some position around the circular ring from $t = 0$ when the rotation first begins to decrease until $t = \infty$. (b) Find the change ΔE_{int} in internal energy of the coil of N turns, due to the current and the resistance R in the ring during the slowing-down process from $t = 0$ to $t = \infty$. Express the change in terms of N, I_0, R, and Q. (c) Find the change in kinetic energy of all the electrons in the coil as it slows down, in terms of m, r, and ω_0, where m is the mass of all of the electrons in the coil and r is the radius of the coil. (d) From Equation 8.2, the energy equation for the system of the coil from $t = 0$ to $t = \infty$ is $\Delta K + \Delta E_{int} = 0$, where we imagine that the coil is stopped essentially immediately by a brake located at some point on the circumference of the coil. Even though the coil itself stops, we are interested only in the electrons in the coil. By stopping the coil immediately, we do no work on the coil that might increase its internal energy beyond that due to the collisions of the electrons with the lattice atoms. Substitute your results from parts (b) and (c) into this reduced form of Equation 8.2. With the help of Equation 26.4, solve the equation in part (d) to find the ratio e/m_e in terms of r, ω_0, R, and Q. All of these quantities can be measured in the experiment.

4. **ACTIVITY** Wire to be used in buildings and homes in the United States is standardized according to the AWG system (American Wire Gauge). Wires are assigned a gauge number according to their diameter. The scale ranges from gauge 0000 to gauge 36. The table below shows a portion of the data table for wire gauges. In a home, a 15-A circuit usually is constructed from 14-gauge copper wire, while a 20-A circuit uses 12-gauge copper wire. These numbers are indicated by the column headed "Ampacity," which is a shortened version of "ampere capacity," or the highest safe current that can be carried by the wire.

Wire Gauge	Diameter (mm)	(Copper Wire) Ampacity (A)	(Copper Wire) Resistance per Unit Length (mΩ/m)
1	7.348	110	0.406 6
2	6.544	95	0.512 7
3	5.827	85	0.646 5
4	5.189	70	0.815 2
5	4.621		1.028
6	4.115	55	1.296
7	3.665		1.634
8	3.264	40	2.061
9	2.906		2.599
10	2.588	30	3.277
11	2.305		4.132
12	2.053	20	5.211
13	1.828		6.571
14	1.628	15	8.286
15	1.450		10.45

In your group, perform the following activities related to these data. (a) From the data, determine the resistivity of copper. Compare to the value in Table 26.2. (b) There is not a linear relationship between the wire gauge and the diameter of the wire. Find an equation that relates the wire gauge to the logarithm (base 10) of the diameter. (c) A graph of ampacity versus diameter of the wire is approximately a straight line. But a combination of Equations 26.7 and 26.10 shows that current is proportional to the *square* of the diameter of the wire. What argument can you make that this equation could be modified so that the maximum safe current is proportional to the diameter? Remember that the ampacity is the largest *safe* current and the major worry about high currents in wires in an increase in their temperature, which could cause a fire.

Problems

See the Preface for an explanation of the icons used in this problems set. For additional assessment items for this section, go to WEB**ASSIGN** From Cengage

SECTION 26.1 Electric Current

1. A 200-km-long high-voltage transmission line 2.00 cm in diameter carries a steady current of 1 000 A. If the conductor is copper with a free charge density of 8.50×10^{28} electrons per cubic meter, how many years does it take one electron to travel the full length of the cable?

2. A small sphere that carries a charge q is whirled in a circle at the end of an insulating string. The angular frequency of revolution is ω. What average current does this revolving charge represent?

3. In the Bohr model of the hydrogen atom (which will be covered in detail in Chapter 41), an electron in the lowest energy state moves at a speed of 2.19×10^6 m/s in a circular path of radius 5.29×10^{-11} m. What is the effective current associated with this orbiting electron?

4. A copper wire has a circular cross section with a radius of 1.25 mm. (a) If the wire carries a current of 3.70 A, find the drift speed of the electrons in this wire. (b) All other things being equal, what happens to the drift speed in wires made of metal having a larger number of conduction electrons per atom than copper? Explain.

5. Suppose the current in a conductor decreases exponentially with time according to the equation $I(t) = I_0 e^{-t/\tau}$, where I_0 is the initial current (at $t = 0$) and τ is a constant having dimensions of time. Consider a fixed observation point within the conductor. (a) How much charge passes this point between $t = 0$ and $t = \tau$? (b) How much charge passes this point between $t = 0$ and $t = 10\tau$? (c) **What If?** How much charge passes this point between $t = 0$ and $t = \infty$?

6. Figure P26.6 represents a section of a conductor of nonuniform diameter carrying a current of $I = 5.00$ A. The radius of cross-section A_1 is $r_1 = 0.400$ cm. (a) What is the magnitude of the current density across A_1? The radius r_2 at A_2 is larger than the radius r_1 at A_1. (b) Is the current at A_2 larger, smaller, or the same? (c) Is the current density at A_2 larger, smaller, or the same? Assume $A_2 = 4A_1$. Specify the (d) radius, (e) current, and (f) current density at A_2.

Figure P26.6

7. The quantity of charge q (in coulombs) that has passed through a surface of area 2.00 cm² varies with time according to the equation $q = 4t^3 + 5t + 6$, where t is in seconds. (a) What is the instantaneous current through the surface at $t = 1.00$ s? (b) What is the value of the current density?

8. A Van de Graaff generator (see Problem 24) produces a beam of 2.00-MeV deuterons, which are heavy hydrogen nuclei containing a proton and a neutron. (a) If the beam current is 10.0 μA, what is the average separation of the deuterons? (b) Is the electrical force of repulsion among them a significant factor in beam stability? Explain.

9. An electric current in a conductor varies with time according to the expression $I(t) = 100 \sin (120\pi t)$, where I is in amperes and t is in seconds. What is the total charge passing a given point in the conductor from $t = 0$ to $t = \frac{1}{240}$ s?

SECTION 26.2 Resistance

10. A wire 50.0 m long and 2.00 mm in diameter is connected to a source with a potential difference of 9.11 V, and the current is found to be 36.0 A. Assume a temperature of 20.0°C and, using Table 26.2, identify the metal out of which the wire is made.

11. An electric heater carries a current of 13.5 A when operating at a voltage of 120 V. What is the resistance of the heater?

12. You are working at a company that manufactures electrical wire. Gold is the most ductile of all metals: it can be stretched into incredibly long, thin wires. The company has developed a new technique that will stretch 1.00 g of gold into a wire of length $L = 2.40$ km and uniform diameter. Your supervisor gives you the task of determining the resistance of such a wire at 20.0°C.

13. Suppose you wish to fabricate a uniform wire from 1.00 g of copper. If the wire is to have a resistance of $R = 0.500 \ \Omega$ and all the copper is to be used, what must be (a) the length and (b) the diameter of this wire?

14. Suppose you wish to fabricate a uniform wire from a mass m of a metal with density ρ_m and resistivity ρ. If the wire is to have a resistance of R and all the metal is to be used, what must be (a) the length and (b) the diameter of this wire?

SECTION 26.3 A Model for Electrical Conduction

15. A current density of 6.00×10^{-13} A/m² exists in the atmosphere at a location where the electric field is 100 V/m. Calculate the electrical conductivity of the Earth's atmosphere in this region.

16. An iron wire has a cross-sectional area of 5.00×10^{-6} m². Carry out the following steps to determine the drift speed of the conduction electrons in the wire if it carries a current of 30.0 A. (a) How many kilograms are there in 1.00 mole of iron? (b) Starting with the density of iron and the result of part (a), compute the molar density of iron (the number of moles of iron per cubic meter). (c) Calculate the number density of iron atoms using Avogadro's number. (d) Obtain the number density of conduction electrons given that there are two conduction electrons per iron atom. (e) Calculate the drift speed of conduction electrons in this wire.

SECTION 26.4 Resistance and Temperature

17. What is the fractional change in the resistance of an iron filament when its temperature changes from 25.0°C to 50.0°C?

18. A certain lightbulb has a tungsten filament with a resistance of 19.0 Ω when at 20.0°C and 140 Ω when hot. Assume the resistivity of tungsten varies linearly with temperature even over the large temperature range involved here. Find the temperature of the hot filament.

19. **T** An aluminum wire with a diameter of 0.100 mm has a uniform electric field of 0.200 V/m imposed along its entire length. The temperature of the wire is 50.0°C. Assume one free electron per atom. (a) Use the information in Table 26.2 to determine the resistivity of aluminum at this temperature. (b) What is the current density in the wire? (c) What is the total current in the wire? (d) What is the drift speed of the conduction electrons? (e) What potential difference must exist between the ends of a 2.00-m length of the wire to produce the stated electric field?

20. **BIO** Plethysmographs are devices used for measuring changes in the volume of internal organs or limbs. In one form of this device, a rubber capillary tube with an inside diameter of 1.00 mm is filled with mercury at 20.0°C. The resistance of the mercury is measured with the aid of electrodes sealed into the ends of the tube. If 100 cm of the tube is wound in a helix around a patient's upper arm, the blood flow during a heartbeat causes the arm to expand, stretching the length of the tube by 0.040 0 cm. From this observation and assuming cylindrical symmetry, you can find the change in volume of the arm, which gives an indication of blood flow. Taking the resistivity of mercury to be 9.58×10^{-7} Ω · m, calculate (a) the resistance of the mercury and (b) the fractional change in resistance during the heartbeat. *Hint:* The fraction by which the cross-sectional area of the mercury column decreases is the fraction by which the length increases because the volume of mercury is constant.

21. At what temperature will aluminum have a resistivity that is three times the resistivity copper has at room temperature?

22. **CR** You are working in a laboratory that studies the effects of currents in various crystals. One of the experiments involves a requirement for a steady current of $I = 0.500$ A in a wire that delivers the current to the crystal. Both the wire and the crystal are in a chamber whose interior temperature T will vary from −40.0°C to 150°C. The wire is made of tungsten and is of length $L = 25.0$ cm and radius $r = 1.00$ mm. A test run is being made before the crystal is added to the circuit. Your supervisor asks you to determine the range of voltages that must be supplied to the wire in the test run to maintain its current at 0.500 A.

SECTION 26.6 Electrical Power

23. Assume that global lightning on the Earth constitutes a constant current of 1.00 kA between the ground and an atmospheric layer at potential 300 kV. (a) Find the power of terrestrial lightning. (b) For comparison, find the power of sunlight falling on the Earth. Sunlight has an intensity of 1 370 W/m² above the atmosphere. Sunlight falls perpendicularly on the circular projected area that the Earth presents to the Sun.

24. The Van de Graaff generator, diagrammed in Figure P26.24, is an electrostatic device that can raise the metal dome to a high voltage. The dome of such a generator is seen on the left in Figure 22.1a. In the device, charge is delivered continuously to the high-potential dome by means of a moving belt of insulating material. The belt is charged at

point Ⓐ by means of a discharge between comb-like metallic needles and a grounded grid. The needles are maintained at a positive electric potential of typically 10^4 V. The positive charge on the moving belt is transferred to the dome by a second comb of needles at point Ⓑ. Because the electric field inside the dome is negligible, the positive charge on the belt is easily transferred to the dome from its interior regardless of its potential. Suppose the generator is operating so that the potential difference between the high potential dome Ⓑ and the charging needles at Ⓐ is 15.0 kV. Calculate the power required to drive the belt against electrical forces at an instant when the effective current delivered to the dome is 500 μA.

Figure P26.24

25. A 100-W lightbulb connected to a 120-V source experiences a voltage surge that produces 140 V for a moment. By what percentage does its power output increase? Assume its resistance does not change.

26. **BIO** The potential difference across a resting neuron in the human body is about 75.0 mV and carries a current of about 0.200 mA. How much power does the neuron release?

27. The cost of energy delivered to residences by electrical transmission varies from $0.070/kWh to $0.258/kWh throughout the United States; $0.110/kWh is the average value. At this average price, calculate the cost of (a) leaving a 40.0-W porch light on for two weeks while you are on vacation, (b) making a piece of dark toast in 3.00 min with a 970-W toaster, and (c) drying a load of clothes in 40.0 min in a 5.20×10^3-W dryer.

28. **Q|C** **V** Residential building codes typically require the use of 12-gauge copper wire (diameter 0.205 cm) for wiring receptacles. Such circuits carry currents as large as 20.0 A. If a wire of smaller diameter (with a higher gauge number) carried that much current, the wire could rise to a high temperature and cause a fire. (a) Calculate the rate at which internal energy is produced in 1.00 m of 12-gauge copper wire carrying 20.0 A. (b) **What If?** Repeat the calculation for a 12-gauge aluminum wire. (c) Explain whether a 12-gauge aluminum wire would be as safe as a copper wire.

29. **T** Assuming the cost of energy from the electric company is $0.110/kWh, compute the cost per day of operating a lamp that draws a current of 1.70 A from a 110-V line.

30. An 11.0-W energy-efficient fluorescent lightbulb is designed to produce the same illumination as a conventional 40.0-W incandescent lightbulb. Assuming a cost of $0.110/kWh for energy from the electric company, how much money does the user of the energy-efficient bulb save during 100 h of use?

31. A 500-W heating coil designed to operate from 110 V is made of Nichrome wire 0.500 mm in diameter. (a) Assuming the resistivity of the Nichrome remains constant at its

20.0°C value, find the length of wire used. (b) **What If?** Now consider the variation of resistivity with temperature. What power is delivered to the coil of part (a) when it is warmed to 1 200°C?

32. *Why is the following situation impossible?* A politician is decrying wasteful uses of energy and decides to focus on energy used to operate plug-in electric clocks in the United States. While many people use their smartphone as an alarm clock, he estimates that there are still 270 million plug-in alarm clocks in continuous use. The clocks transform energy taken in by electrical transmission at the average rate 2.50 W. The politician gives a speech in which he complains that, at today's electrical rates, the nation is losing $100 million every year to operate these clocks.

33. Make an order-of-magnitude estimate of the cost of one person's routine use of a handheld hair dryer for 1 year. If you do not use a hair dryer yourself, observe or interview someone who does. State the quantities you estimate and their values.

ADDITIONAL PROBLEMS

34. **Q|C** Lightbulb A is marked "25 W 120 V," and lightbulb B is marked "100 W 120 V." These labels mean that each lightbulb has its respective power delivered to it when it is connected to a constant 120-V source. (a) Find the resistance of each lightbulb. (b) During what time interval does 1.00 C pass into lightbulb A? (c) Is this charge different upon its exit versus its entry into the lightbulb? Explain. (d) In what time interval does 1.00 J pass into lightbulb A? (e) By what mechanisms does this energy enter and exit the lightbulb? Explain. (f) Find the cost of running lightbulb A continuously for 30.0 days, assuming the electric company sells its product at $0.110 per kWh.

35. **V** One wire in a high-voltage transmission line carries 1 000 A starting at 700 kV for a distance of 100 mi. If the resistance in the wire is 0.500 Ω/mi, what is the power loss due to the resistance of the wire?

36. **CR** You are working with an oceanographer who is studying how the ion concentration in seawater depends on depth. She shows you the device that she uses to measure the resistivity of water from a boat. It consists of a pair of concentric metallic cylinders at the end of a cable (Fig. P26.36). Seawater flows freely between the two cylindrical shells. She makes a measurement by lowering the device into the water and applying a potential difference ΔV between the inner and outer cylinders. This produces an outward radial current I in the seawater between the shells. She shows you the current and voltage data for the water at a particular depth and is then called away to answer a long call on her cellphone about a laboratory issue back on the mainland. As she leaves, she says, "Have the resistivity of the water calculated when I get back." She forgot to show you any tables or formulas to use to determine the resistivity, so you are on your own. Quick! Find an expression for the resistivity in terms of I and ΔV before she finishes her phone call!

Figure P26.36

37. **S** A charge Q is placed on a capacitor of capacitance C. The capacitor is connected into the circuit shown in Figure P26.37, with an open switch, a resistor, and an initially

Figure P26.37

uncharged capacitor of capacitance $3C$. The switch is then closed, and the circuit comes to equilibrium. In terms of Q and C, find (a) the final potential difference between the plates of each capacitor, (b) the charge on each capacitor, and (c) the final energy stored in each capacitor. (d) Find the internal energy appearing in the resistor.

38. **Q|C** An experiment is conducted to measure the electrical resistivity of Nichrome in the form of wires with different lengths and cross-sectional areas. For one set of measurements, a student uses 30-gauge wire, which has a cross-sectional area of 7.30×10^{-8} m². The student measures the potential difference across the wire and the current in the wire with a voltmeter and an ammeter, respectively. (a) For each set of measurements given in the table taken on wires of three different lengths, calculate the resistance of the wires and the corresponding values of the resistivity. (b) What is the average value of the resistivity? (c) Explain how this value compares with the value given in Table 26.2.

L (m)	ΔV (V)	I (A)	R (Ω)	ρ (Ω · m)
0.540	5.22	0.72		
1.028	5.82	0.414		
1.543	5.94	0.281		

39. A straight, cylindrical wire lying along the x axis has a length of 0.500 m and a diameter of 0.200 mm. It is made of a material described by Ohm's law with a resistivity of $\rho = 4.00 \times 10^{-8}$ Ω · m. Assume a potential of 4.00 V is maintained at the left end of the wire at $x = 0$. Also assume $V = 0$ at $x = 0.500$ m. Find (a) the magnitude and direction of the electric field in the wire, (b) the resistance of the wire, (c) the magnitude and direction of the electric current in the wire, and (d) the current density in the wire. (e) Show that $E = \rho J$.

40. **S** A straight, cylindrical wire lying along the x axis has a length L and a diameter d. It is made of a material described by Ohm's law with a resistivity ρ. Assume potential V is maintained at the left end of the wire at $x = 0$. Also assume the potential is zero at $x = L$. In terms of L, d, V, ρ, and physical constants, derive expressions for (a) the magnitude and direction of the electric field in the wire, (b) the resistance of the wire, (c) the magnitude and direction of the electric current in the wire, and (d) the current density in the wire. (e) Show that $E = \rho J$.

41. **Review.** An office worker uses an immersion heater to warm 250 g of water in a light, covered, insulated cup from 20.0°C to 100°C in 4.00 min. The heater is a Nichrome resistance wire connected to a 120-V power supply. Assume the wire is at 100°C throughout the 4.00-min time interval. (a) Specify a relationship between a diameter and a length that the wire can have. (b) Can it be made from less than 0.500 cm³ of Nichrome?

42. The strain in a wire can be monitored and computed by mea-
QC suring the resistance of the wire. Let L_i represent the orig-
S   inal length of the wire, A_i its original cross-sectional area,
    $R_i = \rho L_i/A_i$ the original resistance between its ends, and
    $\delta = \Delta L/L_i = (L - L_i)/L_i$ the strain resulting from the applica-
    tion of tension. Assume the resistivity and the volume of the
    wire do not change as the wire stretches. (a) Show that the
    resistance between the ends of the wire under strain is given
    by $R = R_i(1 + 2\delta + \delta^2)$. (b) If the assumptions are precisely
    true, is this result exact or approximate? Explain your answer.

43. A close analogy exists between the flow of energy by heat
QC because of a temperature difference (see Section 19.6) and
S   the flow of electric charge because of a potential difference.
    In a metal, energy dQ and electrical charge dq are both
    transported by free electrons. Consequently, a good electri-
    cal conductor is usually a good thermal conductor as well.
    Consider a thin conducting slab of thickness dx, area A,
    and electrical conductivity σ, with a potential difference dV
    between opposite faces. (a) Show that the current $I = dq/dt$
    is given by the equation on the left:

Charge conduction	Thermal conduction
$\dfrac{dq}{dt} = \sigma A \left\| \dfrac{dV}{dx} \right\|$	$\dfrac{dQ}{dt} = kA \left\| \dfrac{dT}{dx} \right\|$

   In the analogous thermal conduction equation on the right
   (Eq. 19.17), the rate dQ/dt of energy flow by heat (in SI units
   of joules per second) is due to a temperature gradient
   dT/dx in a material of thermal conductivity k. (b) State analogous rules relating the direction of the electric current to
   the change in potential and relating the direction of energy
   flow to the change in temperature.

44. The dielectric material between the plates of a parallel-
   plate capacitor always has some nonzero conductivity σ.
   Let A represent the area of each plate and d the distance
   between them. Let κ represent the dielectric constant of the
   material. (a) Show that the resistance R and the capacitance
   C of the capacitor are related by

$$RC = \frac{\kappa \epsilon_0}{\sigma}$$

   (b) Find the resistance between the plates of a 14.0-nF
   capacitor with a fused quartz dielectric.

45. Review. A parallel-plate capacitor consists of square plates
S   of edge length ℓ that are separated by a distance d, where
    $d \ll \ell$. A potential difference ΔV is maintained between the
    plates. A material of dielectric constant κ fills half the space
    between the plates. The dielectric slab is withdrawn from
    the capacitor as shown in Figure P26.45. (a) Find the capac-
    itance when the left edge of the dielectric is at a distance
    x from the center of the capacitor. (b) If the dielectric is
    removed at a constant speed v, what is the current in the
    circuit as the dielectric is being withdrawn?

Figure P26.45

46. The current–voltage characteristic curve for a semiconductor diode as a function of temperature T is given by

$$I = I_0(e^{e\,\Delta V/k_B T} - 1)$$

   Here the first symbol e represents Euler's number, the base
   of natural logarithms. The second e is the magnitude of the
   electron charge, the k_B stands for Boltzmann's constant,
   and T is the absolute temperature. (a) Set up a spreadsheet
   to calculate I and $R = \Delta V/I$ for $\Delta V = 0.400$ V to 0.600 V
   in increments of 0.005 V. Assume $I_0 = 1.00$ nA. (b) Plot R
   versus ΔV for $T = 280$ K, 300 K, and 320 K.

47. *Why is the following situation impossible?* An inquisitive physics student takes a 100-W incandescent lightbulb out of its
   socket and measures its resistance with an ohmmeter. He
   measures a value of $10.5\ \Omega$. He is able to connect an ammeter to the lightbulb socket to correctly measure the current
   drawn by the bulb while operating. Inserting the bulb back
   into the socket and operating the bulb from a 120-V source,
   he measures the current to be 11.4 A.

CHALLENGE PROBLEMS

48. A more general definition of the temperature coefficient of
S   resistivity is

$$\alpha = \frac{1}{\rho}\frac{d\rho}{dT}$$

   where ρ is the resistivity at temperature T. (a) Assuming α is
   constant, show that

$$\rho = \rho_0 e^{\alpha(T - T_0)}$$

   where ρ_0 is the resistivity at temperature T_0. (b) Using the
   series expansion $e^x \approx 1 + x$ for $x \ll 1$, show that the resistivity is given approximately by the expression

$$\rho = \rho_0[1 + \alpha(T - T_0)] \quad \text{for } \alpha(T - T_0) \ll 1$$

49. A spherical shell with inner radius r_a and outer radius r_b is
S   formed from a material of resistivity ρ. It carries current
   radially, with uniform density in all directions. Show that
   its resistance is

$$R = \frac{\rho}{4\pi}\left(\frac{1}{r_a} - \frac{1}{r_b}\right)$$

50. Material with uniform resistivity ρ is formed into a wedge as
S   shown in Figure P26.50. Show that the resistance between
   face A and face B of this wedge is

$$R = \rho\,\frac{L}{w(y_2 - y_1)}\ln\frac{y_2}{y_1}$$

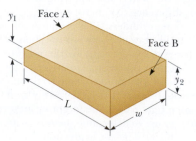

Figure P26.50

Direct-Current Circuits

A bird rests on a powerline. Why is the bird safe from electrocution? (*Hydromet/Shutterstock*)

STORYLINE **You are doing your homework on a Saturday and** notice that the sky is darkening. You look out the window and are surprised to see a rare display of lightning in California. You excitedly call your electrical engineer uncle in Florida, with whom you had earlier discussions. You tell him that you are going to go out in an open field and observe the lightning. He tells you that it is dangerous to be the highest point in an open field because you might be struck by the lightning. You explain that if the lightning gets close, you will lie down on the ground to be safe. He says, "Oh, no, don't do that! After all, cows are killed by lightning more often than chickens!" Puzzled, you begin to ask what he means, but then he is called away from the phone by your aunt and he must hang up. What did his cryptic comment mean? While contemplating this question, you look out the window and see a bird sitting on a high-voltage power line. Now, why isn't that bird electrocuted? Should you think about that, or go out to the field and watch the lightning?

CONNECTIONS In previous chapters, we have introduced two types of circuit elements: capacitors and resistors. We can combine these elements with batteries to form a variety of electric circuits, which we analyze in this chapter. We were first introduced to circuit diagrams in Section 25.3. In this chapter, we will make frequent use of circuit diagrams to help us understand the behavior of more complex circuits. Some circuits containing multiple resistors can be combined using simple rules. The analysis of more complicated circuits is simplified using *Kirchhoff's rules*, which follow from the laws of conservation of energy and conservation of electric charge for isolated systems. Most of the circuits analyzed are assumed to be in steady state, which means that currents in the circuit

713

a

b

Figure 27.1 (a) Circuit diagram of a source of emf ε (in this case, a battery), of internal resistance r, connected to an external resistor of resistance R. (b) Graphical representation showing how the electric potential changes as the circuit in (a) is traversed clockwise.

are constant in magnitude and direction. A current that is constant in direction is called a *direct current* (DC). We will study *alternating current* (AC), in which the current changes direction periodically, in Chapter 32.

27.1 Electromotive Force

Consider the simple circuit shown in Figure 26.11 (page 703). This circuit is easy to analyze. Suppose we expand this circuit by adding another resistor, another battery, and, possibly, one or more capacitors. Now it may not seem so easy for you to analyze. Let's embark on our journey to see how to analyze these kinds of circuits.

We will generally use a battery as a source of energy for circuits in our discussion. A battery is called either a *source of electromotive force* or, more commonly, a *source of emf*. (The phrase *electromotive force* is an unfortunate historical term, describing not a force, but rather a potential difference in volts.) The **emf ε of a battery is the maximum possible voltage the battery can provide between its terminals.** You can think of a source of emf as a "charge pump." When an electric potential difference exists between two points, the source moves charges "uphill" from the lower potential to the higher.

Because the potential difference at the battery terminals is constant in a particular circuit, the current in the circuit is constant in magnitude and direction and is called **direct current.** We shall generally assume the connecting wires in a circuit have no resistance. The positive terminal of a battery is at a higher potential than the negative terminal. Because a real battery is made of matter, there is resistance to the flow of charge within the battery. This resistance is called **internal resistance** r. For an idealized battery with zero internal resistance, the potential difference across the battery (called its *terminal voltage*) equals its emf. For a real battery, however, the terminal voltage is *not* equal to the emf for a battery in a circuit in which there is a current. To understand why, consider the circuit diagram in Figure 27.1a. We model the battery as shown in the diagram; it is represented by the dashed rectangle containing an ideal, resistance-free emf ε in series with an internal resistance r. A resistor of resistance R is connected across the terminals of the battery. Now imagine moving through the battery from a to d and measuring the electric potential at various locations. Passing from the negative terminal to the positive terminal, the potential increases by an amount ε. As we move through the resistance r, however, the potential *decreases* by an amount Ir, where I is the current in the circuit. Therefore, the terminal voltage of the battery $\Delta V = V_d - V_a$ is

$$\Delta V = \varepsilon - Ir \tag{27.1}$$

Figure 27.1b is a graphical representation of the changes in electric potential as the circuit is traversed in the clockwise direction. From Equation 27.1, notice that ε is equivalent to the **open-circuit voltage,** that is, the terminal voltage when the current is zero. The emf is the voltage labeled on a battery; for example, the emf of a D cell is 1.5 V. The actual potential difference between a battery's terminals depends on the current in the battery as described by Equation 27.1.

Figure 27.1a shows that the terminal voltage ΔV must equal the potential difference across the external resistance R, often called the **load resistance.** The load resistor might be a simple resistive circuit element as in Figure 27.1a, or it could be the resistance of some electrical device (such as a toaster, electric heater, or lightbulb) connected to the battery (or, in the case of household devices, to the wall outlet). The resistor represents a *load* on the battery because the battery must supply energy to operate the device containing the resistance. The potential difference across the load resistance is $\Delta V = IR$. Combining this expression with Equation 27.1, we see that

$$\varepsilon = IR + Ir \tag{27.2}$$

Solving for the current gives

$$I = \frac{\mathcal{E}}{R + r} \quad (27.3)$$

Equation 27.3 shows that the current in this simple circuit depends on both the load resistance R external to the battery and the internal resistance r. If R is much greater than r, as it is in many real-world circuits, we can neglect r.

Multiplying Equation 27.2 by the current I in the circuit gives

$$I\mathcal{E} = I^2R + I^2r \quad (27.4)$$

Equation 27.4 indicates that because power $P = I\,\Delta V$ (see Eq. 26.21), the total power output $I\mathcal{E}$ associated with the emf of the battery is delivered to the external load resistance in the amount I^2R and to the internal resistance in the amount I^2r.

QUICK QUIZ 27.1 To maximize the percentage of the power from the emf of a battery that is delivered to a device external to the battery, what should the internal resistance of the battery be? **(a)** It should be as low as possible. **(b)** It should be as high as possible. **(c)** The percentage does not depend on the internal resistance.

PITFALL PREVENTION 27.1

Batteries Do Not Supply Electrons
A battery does not supply electrons to the circuit. It establishes the electric field that exerts a force on electrons already in the wires and elements of the circuit.

PITFALL PREVENTION 27.2

What Is Constant in a Battery?
It is a common misconception that a battery is a source of constant current. Equation 27.3 shows that is not true. The current in the circuit depends on the resistance R connected to the battery. It is also not true that a battery is a source of constant terminal voltage as shown by Equation 27.1. **A battery is a source of constant emf.**

Example 27.1 Terminal Voltage of a Battery

A battery has an emf of 12.0 V and an internal resistance of 0.050 0 Ω. Its terminals are connected to a load resistance of 3.00 Ω.

(A) Find the current in the circuit and the terminal voltage of the battery.

SOLUTION

Conceptualize Study Figure 27.1a, which shows a circuit consistent with the problem statement. The battery delivers energy to the load resistor.

Categorize This example involves simple calculations from this section, so we categorize it as a substitution problem.

Use Equation 27.3 to find the current in the circuit:

$$I = \frac{\mathcal{E}}{R + r} = \frac{12.0\ \text{V}}{3.00\ \Omega + 0.050\ 0\ \Omega} = \boxed{3.93\ \text{A}}$$

Use Equation 27.1 to find the terminal voltage:

$$\Delta V = \mathcal{E} - Ir = 12.0\ \text{V} - (3.93\ \text{A})(0.050\ 0\ \Omega) = \boxed{11.8\ \text{V}}$$

To check this result, calculate the voltage across the load resistance R:

$$\Delta V = IR = (3.93\ \text{A})(3.00\ \Omega) = 11.8\ \text{V}$$

(B) Calculate the power delivered to the load resistor, the power delivered to the internal resistance of the battery, and the power delivered by the battery.

SOLUTION

Use Equation 26.22 to find the power delivered to the load resistor:

$$P_R = I^2R = (3.93\ \text{A})^2(3.00\ \Omega) = \boxed{46.3\ \text{W}}$$

Find the power delivered to the internal resistance:

$$P_r = I^2r = (3.93\ \text{A})^2(0.050\ 0\ \Omega) = \boxed{0.772\ \text{W}}$$

Find the power delivered by the battery by adding these quantities:

$$P = P_R + P_r = 46.3\ \text{W} + 0.772\ \text{W} = \boxed{47.1\ \text{W}}$$

WHAT IF? As a battery ages, its internal resistance increases. Suppose the internal resistance of this battery rises to 2.00 Ω toward the end of its useful life. How does that alter the battery's ability to deliver energy?

Answer Let's connect the same 3.00-Ω load resistor to the battery.

continued

27.1 continued

Find the new current in the battery:

$$I = \frac{\mathcal{E}}{R + r} = \frac{12.0 \text{ V}}{3.00 \text{ } \Omega + 2.00 \text{ } \Omega} = 2.40 \text{ A}$$

Find the new terminal voltage:

$$\Delta V = \mathcal{E} - Ir = 12.0 \text{ V} - (2.40 \text{ A})(2.00 \text{ } \Omega) = 7.2 \text{ V}$$

Find the new powers delivered to the load resistor and internal resistance:

$$P_R = I^2R = (2.40 \text{ A})^2(3.00 \text{ } \Omega) = 17.3 \text{ W}$$
$$P_r = I^2r = (2.40 \text{ A})^2(2.00 \text{ } \Omega) = 11.5 \text{ W}$$

In this situation, the terminal voltage is only 60% of the emf. Notice that 40% of the power from the battery is delivered to the internal resistance when r is 2.00 Ω. When r is 0.050 0 Ω as in part (B), this percentage is only 1.6%. Consequently, even though the emf remains fixed, the increasing internal resistance of the battery significantly reduces the battery's ability to deliver energy to an external load.

Example **27.2**   Matching the Load

Find the load resistance R for which the maximum power is delivered to the load resistance in Figure 27.1a.

SOLUTION

Conceptualize Think about varying the load resistance in Figure 27.1a and the effect on the power delivered to the load resistance. When R is large, there is very little current, so the power I^2R delivered to the load resistor is small. When R is small, let's say $R \ll r$, the current is large and the power delivered to the internal resistance is $I^2r \gg I^2R$. Therefore, the power delivered to the load resistor is small compared to that delivered to the internal resistance. For some intermediate value of the resistance R, the power must maximize.

Categorize We categorize this example as an analysis problem because we must undertake a procedure to maximize the power. The circuit is the same as that in Example 27.1. The load resistance R in this case, however, is a variable.

Figure 27.2 (Example 27.2) Graph of the power P delivered by a battery to a load resistor of resistance R as a function of R.

Analyze Find the power delivered to the load resistance using Equation 26.22, with I given by Equation 27.3:

$$(1) \quad P = I^2R = \frac{\mathcal{E}^2R}{(R + r)^2}$$

Differentiate the power with respect to the load resistance R and set the derivative equal to zero to maximize the power:

$$\frac{dP}{dR} = \frac{d}{dR}\left[\frac{\mathcal{E}^2R}{(R + r)^2}\right] = \frac{d}{dR}[\mathcal{E}^2R(R + r)^{-2}] = 0$$

$$[\mathcal{E}^2(R + r)^{-2}] + [\mathcal{E}^2R(-2)(R + r)^{-3}] = 0$$

$$\frac{\mathcal{E}^2(R + r)}{(R + r)^3} - \frac{2\mathcal{E}^2R}{(R + r)^3} = \frac{\mathcal{E}^2(r - R)}{(R + r)^3} = 0$$

Solve for R:

$$R = r$$

Finalize To check this result, let's plot P versus R as in Figure 27.2. The graph shows that P reaches a maximum value at $R = r$. Equation (1) shows that this maximum value is $P_{max} = \mathcal{E}^2/4r$.

27.2   Resistors in Series and Parallel

In Section 25.3, we studied capacitors in series and parallel connections. In this section, we connect resistors in series and parallel and analyze the results. We will make multiple uses of Equation 26.7 in this process.

When two or more resistors are connected together as are the incandescent lightbulbs in Figure 27.3a, they are said to be in a **series combination**. Figure 27.3b

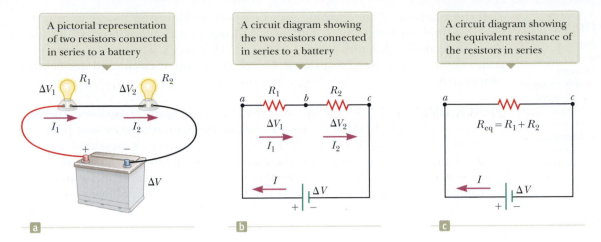

A pictorial representation of two resistors connected in series to a battery

A circuit diagram showing the two resistors connected in series to a battery

A circuit diagram showing the equivalent resistance of the resistors in series

Figure 27.3 Two incandescent lightbulbs with resistances R_1 and R_2 connected in series. All three diagrams are equivalent.

is the circuit diagram for the lightbulbs, shown as resistors, and the battery. What if you wanted to replace the series combination with a single resistor that would draw the same current from the battery? What would be its value? In a series connection, if an amount of charge Q exits resistor R_1, charge Q must also enter the second resistor R_2. Otherwise, charge would accumulate on the wire between the resistors. Therefore, the same amount of charge passes through both resistors in a given time interval and the currents are the same in both resistors:

$$I = I_1 = I_2 \tag{27.5}$$

where I is the current leaving the battery, I_1 is the current in resistor R_1, and I_2 is the current in resistor R_2.

The potential difference applied across the series combination of resistors divides between the resistors. In Figure 27.3b, because the voltage drop[1] from a to b equals I_1R_1 and the voltage drop from b to c equals I_2R_2, the voltage drop from a to c is

$$\Delta V = \Delta V_1 + \Delta V_2 = I_1R_1 + I_2R_2 \tag{27.6}$$

The potential difference across the battery is also applied to the **equivalent resistance** R_{eq} in Figure 27.3c:

$$\Delta V = IR_{eq}$$

where the equivalent resistance has the same effect on the circuit as the series combination because it results in the same current I in the battery. Substituting this expression into Equation 27.6 gives

$$IR_{eq} = I_1R_1 + I_2R_2 \quad \rightarrow \quad R_{eq} = R_1 + R_2 \tag{27.7}$$

where we have canceled the currents I, I_1, and I_2 because they are all the same (Eq. 27.5). We see that we can replace the two resistors in series with a single equivalent resistance whose value is the *sum* of the individual resistances.

The equivalent resistance of three or more resistors connected in series is

$$R_{eq} = R_1 + R_2 + R_3 + \cdots \tag{27.8}$$

◀ The equivalent resistance of a series combination of resistors

This relationship indicates that the equivalent resistance of a series combination of resistors is the numerical sum of the individual resistances and is always greater than any individual resistance.

Looking back at Equation 27.3, we see that the denominator of the right-hand side is the simple algebraic sum of the external and internal resistances. That is consistent with the internal and external resistances being in series in Figure 27.1a.

---

[1]The term *voltage drop* is synonymous with a decrease in electric potential across a resistor. It is often used by individuals working with electric circuits.

If the filament of one incandescent lightbulb in Figure 27.3 were to fail, the circuit would no longer be complete (resulting in an open-circuit condition) and the second lightbulb would also go out. This fact is a general feature of a series circuit: if one device in the series creates an open circuit, all devices are inoperative.

QUICK QUIZ 27.2 With the switch in the circuit of Figure 27.4a closed, there is no current in R_2 because the current has an alternate zero-resistance path through the switch. There is current in R_1, and this current is measured with the ammeter (a device for measuring current) at the bottom of the circuit. If the switch is opened (Fig. 27.4b), there is current in R_2. What happens to the reading on the ammeter when the switch is opened? **(a)** The reading goes up. **(b)** The reading goes down. **(c)** The reading does not change.

Figure 27.4 (Quick Quiz 27.2) What happens when the switch is opened?

Now consider two resistors in a **parallel combination** as shown in Figure 27.5. As with the series combination, what is the value of the single resistor that could replace the combination and draw the same current from the battery? Notice that both resistors are connected directly across the terminals of the battery. Therefore, the potential differences across the resistors are the same:

$$\Delta V = \Delta V_1 = \Delta V_2 \qquad (27.9)$$

where ΔV is the terminal voltage of the battery.

When charges reach point a in Figure 27.5b, they split into two parts, with some going toward R_1 and the rest going toward R_2. A **junction** is any such point

Figure 27.5 Two incandescent lightbulbs with resistances R_1 and R_2 connected in parallel. All three diagrams are equivalent.

A pictorial representation of two resistors connected in parallel to a battery

A circuit diagram showing the two resistors connected in parallel to a battery

A circuit diagram showing the equivalent resistance of the resistors in parallel in a circuit where a current can split. This split results in less current in each individual resistor than the current leaving the battery. Because electric charge is conserved, the current I that enters point a must equal the total current leaving that point:

$$I = I_1 + I_2 = \frac{\Delta V_1}{R_1} + \frac{\Delta V_2}{R_2} \qquad (27.10)$$

where I_1 is the current in R_1 and I_2 is the current in R_2.

The current in the equivalent resistance R_{eq} in Figure 27.5c is

$$I = \frac{\Delta V}{R_{eq}}$$

where the equivalent resistance has the same effect on the circuit as the two resistors in parallel; that is, the equivalent resistance draws the same current I from the battery. Substituting this last equation into Equation 27.10, we see that the equivalent resistance of two resistors in parallel is given by

$$\frac{\Delta V}{R_{eq}} = \frac{\Delta V_1}{R_1} + \frac{\Delta V_2}{R_2} \quad \rightarrow \quad \frac{1}{R_{eq}} = \frac{1}{R_1} + \frac{1}{R_2} \qquad (27.11)$$

where we have canceled ΔV, ΔV_1, and ΔV_2 because they are all the same (Eq. 27.9).

An extension of this analysis to three or more resistors in parallel gives

$$\frac{1}{R_{eq}} = \frac{1}{R_1} + \frac{1}{R_2} + \frac{1}{R_3} + \cdots \qquad (27.12)$$

◀ The equivalent resistance of a parallel combination of resistors

This expression shows that the inverse of the equivalent resistance of two or more resistors in a parallel combination is equal to the sum of the inverses of the individual resistances. Furthermore, the equivalent resistance is always less than the smallest resistance in the group.

Household circuits are always wired such that the appliances are connected in parallel. Each device operates independently of the others so that if one is switched off, the others remain on. In addition, in this type of connection, all the devices operate on the same voltage.

Let's use the concepts in this section to address our opening storyline. First, why don't you want to stand in an open field during a lightning storm? A stroke of lightning begins with a *stepped leader* moving downward from a charged cloud. This is a column of negative charge moving toward the ground along a zig-zag path at high speed. The stepped leader is *not* the flash of light that you associate with lightning. From the ground, a column of positive charge, called a *return stroke*, begins to move upward from a point of large electric field. When a stepped leader and a return stroke meet in the air, as is about to happen at the left side of the photo on page 636, a conducting channel opens up between the cloud and the ground, a large current suddenly exists, and a bright flash of light is emitted.

If you are standing in an open field, your head represents a sharp point relative to the flat field. Therefore, because you and the ground are both charged, and as we found out in Section 24.6, your head as a sharp point has a very strong electric field at its surface. That increases the probability that a return stroke will begin from your head rather than from the flat ground, endangering your safety.

So why not lie on the ground to remove your head as a sharp point? When lightning strikes, the current in the air also exists in the surface of the ground, spreading out radially from the point from which the return stroke began. If you lie on the ground with your body along a radial line to the return stroke, your body is placed *in parallel* with the current from the lightning stroke. Therefore, some current could take a path through your body, from the contact point at the upper part of the your body to the contact point at your feet. This is why cows are killed by lightning. They have contact points at their front feet and also at their hind feet.

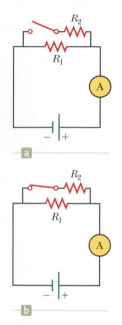

Figure 27.6 (Quick Quiz 27.3) What happens when the switch is closed?

If their body is aimed at the return stroke, a significant amount of current can exist in their bodies. Chickens have two contact points also, but their feet are close together. Therefore, the resistance of the ground is smaller between their feet than that for the cow. As a result, the potential difference between the two contact points on the ground is smaller for the chicken, so less current exists in their bodies than for the cows.

Finally, what about the bird on the wire? It may be that the wire is insulated, keeping the bird safe. Even if the wire is not insulated, however, for the types of birds that land on wires, their feet are even closer together than chicken feet. Furthermore, the wire most likely has a smaller resistivity than the ground. Both factors lead to a very small potential difference between the feet of the bird when it is connected in parallel with the wire. In turn, there is very little current in the body of the bird.

QUICK QUIZ 27.3 With the switch in the circuit of Figure 27.6a open, there is no current in R_2. There is current in R_1, however, and it is measured with the ammeter at the right side of the circuit. If the switch is closed (Fig. 27.6b), there is current in R_2. What happens to the reading on the ammeter when the switch is closed? **(a)** The reading increases. **(b)** The reading decreases. **(c)** The reading does not change.

QUICK QUIZ 27.4 Consider the following choices: (a) increases, (b) decreases, (c) remains the same. From these choices, choose the best answer for the following situations. **(i)** In Figure 27.3, a third resistor is added in series with the first two. What happens to the current in the battery? **(ii)** What happens to the terminal voltage of the battery? **(iii)** In Figure 27.5, a third resistor is added in parallel with the first two. What happens to the current in the battery? **(iv)** What happens to the terminal voltage of the battery?

Conceptual Example 27.3   **Landscape Lights**

A homeowner wishes to install low-voltage landscape lighting in his back yard. To save money, he purchases inexpensive 18-gauge cable, which has a relatively high resistance per unit length. This cable consists of two side-by-side wires separated by insulation, like the cord on an appliance. He runs a 200-foot length of this cable from the power supply to the farthest point at which he plans to position a light fixture. He attaches light fixtures across the two wires on the cable at 10-foot intervals so that the light fixtures are in parallel. Because of the cable's resistance, the brightness of the lightbulbs in the fixtures is not as desired. Which of the following problems does the homeowner have? (a) All the lightbulbs glow equally less brightly than they would if lower-resistance cable had been used. (b) The brightness of the lightbulbs decreases as you move farther from the power supply.

SOLUTION

A circuit diagram for the system appears in Figure 27.7. The horizontal resistors with letter subscripts (such as R_A) represent the resistance of the wires in the cable between the light fixtures, and the vertical resistors with number subscripts (such as R_1) represent the resistance of the light fixtures themselves. Part of the terminal voltage of the power supply is dropped across resistors R_A and R_B. Therefore, the voltage across light fixture R_1 is less than the terminal voltage. There is a further voltage drop across resistors R_C and R_D. Consequently, the voltage across light fixture R_2 is smaller than that across R_1. This pattern continues down the line of light fixtures, so the correct choice is (b). Each successive light fixture has a smaller voltage across it and glows less brightly than the one before.

Figure 27.7 (Conceptual Example 27.3) The circuit diagram for a set of landscape light fixtures connected in parallel across the two wires of a two-wire cable.

Example 27.4 Find the Equivalent Resistance

Four resistors are connected as shown in Figure 27.8a.

(A) Find the equivalent resistance between points a and c.

SOLUTION

Conceptualize Imagine charges flowing into and through this combination from the left. All charges must pass from a to b through the first two resistors, but the charges split at b into two different paths when encountering the combination of the 6.0-Ω and the 3.0-Ω resistors.

Categorize Because of the simple nature of the combination of resistors in Figure 27.8, we categorize this example as one for which we can use the rules for series and parallel combinations of resistors.

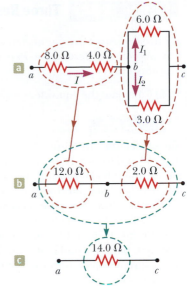

Figure 27.8 (Example 27.4) The original network of resistors is reduced to a single equivalent resistance.

Analyze The combination of resistors can be reduced in steps as shown in Figure 27.8.

Find the equivalent resistance between a and b of the 8.0-Ω and 4.0-Ω resistors, which are in series (left-hand red-brown circles):

$$R_{eq} = 8.0\ \Omega + 4.0\ \Omega = 12.0\ \Omega$$

Find the equivalent resistance between b and c of the 6.0-Ω and 3.0-Ω resistors, which are in parallel (right-hand red-brown circles):

$$\frac{1}{R_{eq}} = \frac{1}{6.0\ \Omega} + \frac{1}{3.0\ \Omega} = \frac{3}{6.0\ \Omega}$$

$$R_{eq} = \frac{6.0\ \Omega}{3} = 2.0\ \Omega$$

The circuit of equivalent resistances now looks like Figure 27.8b. The 12.0-Ω and 2.0-Ω resistors are in series (green circles). Find the equivalent resistance from a to c:

$$R_{eq} = 12.0\ \Omega + 2.0\ \Omega = \boxed{14.0\ \Omega}$$

This resistance is that of the single equivalent resistor in Figure 27.8c.

(B) What is the current in each resistor if a potential difference of 42 V is maintained between a and c?

SOLUTION

The currents in the 8.0-Ω and 4.0-Ω resistors are the same because they are in series. In addition, they carry the same current that would exist in the 14.0-Ω equivalent resistor subject to the 42-V potential difference.

Use Equation 26.7 ($R = \Delta V/I$) and the result from part (A) to find the current in the 8.0-Ω and 4.0-Ω resistors:

$$I = \frac{\Delta V_{ac}}{R_{eq}} = \frac{42\ \text{V}}{14.0\ \Omega} = \boxed{3.0\ \text{A}}$$

Set the voltages across the resistors in parallel in Figure 27.8a equal to find a relationship between the currents:

$$\Delta V_1 = \Delta V_2 \rightarrow (6.0\ \Omega)I_1 = (3.0\ \Omega)I_2 \rightarrow I_2 = 2I_1$$

Use $I_1 + I_2 = 3.0$ A to find I_1:

$$I_1 + I_2 = 3.0\ \text{A} \rightarrow I_1 + 2I_1 = 3.0\ \text{A} \rightarrow I_1 = \boxed{1.0\ \text{A}}$$

Find I_2:

$$I_2 = 2I_1 = 2(1.0\ \text{A}) = \boxed{2.0\ \text{A}}$$

Finalize As a final check of our results, note that $\Delta V_{bc} = (6.0\ \Omega)I_1 = (3.0\ \Omega)I_2 = 6.0$ V and $\Delta V_{ab} = (12.0\ \Omega)I = 36$ V; therefore, $\Delta V_{ac} = \Delta V_{ab} + \Delta V_{bc} = 42$ V, as it must.

Example 27.5 Three Resistors in Parallel

Three resistors are connected as shown in Figure 27.9a. A potential difference of 18.0 V is maintained between points *a* and *b*.

(A) Calculate the equivalent resistance of the circuit.

SOLUTION

Conceptualize In Figure 27.9a, it should be clear that the 6.00-W and 9.00-W resistors are connected in parallel. What about the 3.00-W resistor? Imagine sliding that resistor to the left, without altering the connections, around the corner and halfway down the vertical wire marked with the current I_1. This does not change the electrical characteristics of the circuit. It should be clear now that we are dealing with a simple parallel combination of three resistors. Notice that the current I splits into three currents I_1, I_2, and I_3 in the three resistors.

Figure 27.9 (Example 27.5) (a) Three resistors connected in parallel. The voltage across each resistor is 18.0 V. (b) Another circuit with three resistors and a battery. Is it equivalent to the circuit in (a)?

Categorize This problem can be solved with rules developed in this section, so we categorize it as a substitution problem. Because the three resistors are connected in parallel, we can use the rule for resistors in parallel, Equation 27.12, to evaluate the equivalent resistance.

Use Equation 27.12 to find R_{eq}:

$$\frac{1}{R_{eq}} = \frac{1}{3.00\ \Omega} + \frac{1}{6.00\ \Omega} + \frac{1}{9.00\ \Omega} = \frac{11}{18.0\ \Omega}$$

$$R_{eq} = \frac{18.0\ \Omega}{11} = \boxed{1.64\ \Omega}$$

(B) Find the current in each resistor.

SOLUTION

The potential difference across each resistor is 18.0 V. Apply the relationship $\Delta V = IR$ to find the currents:

$$I_1 = \frac{\Delta V}{R_1} = \frac{18.0\ \text{V}}{3.00\ \Omega} = \boxed{6.00\ \text{A}}$$

$$I_2 = \frac{\Delta V}{R_2} = \frac{18.0\ \text{V}}{6.00\ \Omega} = \boxed{3.00\ \text{A}}$$

$$I_3 = \frac{\Delta V}{R_3} = \frac{18.0\ \text{V}}{9.00\ \Omega} = \boxed{2.00\ \text{A}}$$

(C) Calculate the power delivered to each resistor and the total power delivered to the combination of resistors.

SOLUTION

Apply the relationship $P = I^2R$ to each resistor using the currents calculated in part (B):

$3.00\text{-}\Omega$: $P_1 = I_1^2 R_1 = (6.00\ \text{A})^2(3.00\ \Omega) = \boxed{108\ \text{W}}$

$6.00\text{-}\Omega$: $P_2 = I_2^2 R_2 = (3.00\ \text{A})^2(6.00\ \Omega) = \boxed{54\ \text{W}}$

$9.00\text{-}\Omega$: $P_3 = I_3^2 R_3 = (2.00\ \text{A})^2(9.00\ \Omega) = \boxed{36\ \text{W}}$

These results show that the smallest resistor receives the most power. Summing the three quantities gives a total power of 198 W. We could have calculated this final result from part (A) by considering the equivalent resistance as follows: $P = (\Delta V)^2/R_{eq} = (18.0\ \text{V})^2/1.64\ \Omega = 198\ \text{W}$.

WHAT IF? What if the circuit were as shown in Figure 27.9b instead of as in Figure 27.9a? How would that affect the calculation?

Answer There would be no effect on the calculation. The physical placement of a circuit element is not important, as we saw when we moved the 3.00-Ω resistor in Figure 27.9a. Only the electrical arrangement is important. In Figure 27.9b, the battery still maintains a potential difference of 18.0 V between points *a* and *b*, so the two circuits in the figure are electrically identical.

27.3 Kirchhoff's Rules

As we saw in the preceding section, combinations of resistors can be simplified and analyzed using the rules for series and parallel combinations of resistors. Very often, however, it is not possible to reduce a circuit to a single loop using these rules. For example, consider the circuit in Figure 27.10, which is the same as that in Figure 27.9b, but with the addition of one battery. This circuit cannot be reduced to a simple combination of resistors in series and parallel. The procedure for analyzing more complex circuits is made possible by using the following two principles, called **Kirchhoff's rules.**

1. **Junction rule.** At any junction, the sum of the currents must equal zero:

$$\sum_{\text{junction}} I = 0 \qquad (27.13)$$

2. **Loop rule.** The sum of the potential differences across all elements around any closed circuit loop must be zero:

$$\sum_{\text{closed loop}} \Delta V = 0 \qquad (27.14)$$

Kirchhoff's first rule is a statement of conservation of electric charge. All charges that enter a given point in a circuit must leave that point because charge cannot build up or disappear at a point. Currents directed into the junction are entered into the sum in the junction rule as $+I$, whereas currents directed out of a junction are entered as $-I$. Applying this rule to the junction in Figure 27.11a gives

$$I_1 - I_2 - I_3 = 0$$

Figure 27.11b represents a mechanical analog of this situation, in which water flows through a branched pipe having no leaks. Because water does not build up anywhere in the pipe, the flow rate into the pipe on the left equals the total flow rate out of the two branches on the right.

Kirchhoff's loop rule arises because the electric force is conservative. The electric potential is similar to a state variable, such as internal energy, in thermodynamics. For a given state of a thermodynamic system, the internal energy has a definite value. At any point in a circuit, the electric potential has a definite value. Now imagine starting at a given point in a circuit and moving around the circuit, measuring the potential. The potential will rise as you pass through some circuit elements and fall as you pass through others. When you arrive back at the starting point, you *must* measure the same potential as when you started. The thermodynamic analog is that when you return to the initial point on a PV diagram, the internal energy of the system must have the same value as when you started.

The only way you can arrive at the same potential when you return to the starting point is if the sum of the increases in potential in some circuit elements equals the sum of the decreases as you pass through others. This is the loop rule. Look again at Figure 27.1b as an example. The potential was defined as zero at point a, and returned to zero at point f, which is connected to point a with a resistance-free wire.

Figure 27.10 The circuit of Figure 27.9b with one battery added in the left branch.

The total amount of charge flowing in the branches on the right must equal the amount flowing in the single branch on the left.

The total amount of water flowing out of the branches on the right must equal the amount flowing into the single branch on the left.

Figure 27.11 (a) Kirchhoff's junction rule. (b) A mechanical analog of the junction rule.

In each diagram, $\Delta V = V_b - V_a$ and the circuit element is traversed from a to b, left to right.

a

a $\Delta V = -IR$ b

b

a $\Delta V = +IR$ b

c

a $\Delta V = +\mathcal{E}$ b

d

a $\Delta V = -\mathcal{E}$ b

Figure 27.12 Rules for determining the signs of the potential differences across a resistor and a battery. (The battery is assumed to have no internal resistance.)

Figure 27.12 shows the sign conventions for the changes in electric potential as you travel through batteries and resistors in multiloop circuits:

- Charges move from the high-potential end of a resistor toward the low-potential end, so if a resistor is traversed in the direction of the current, the potential difference ΔV across the resistor is $-IR$ (Fig. 27.12a).
- If a resistor is traversed in the direction *opposite* the current, the potential difference ΔV across the resistor is $+IR$ (Fig. 27.12b).
- If a source of emf (assumed to have zero internal resistance) is traversed in the direction of the emf (from negative to positive), the potential difference ΔV is $+\mathcal{E}$ (Fig. 27.12c).
- If a source of emf (assumed to have zero internal resistance) is traversed in the direction opposite the emf (from positive to negative), the potential difference ΔV is $-\mathcal{E}$ (Fig. 27.12d).

There are limits on the number of times you can usefully apply Kirchhoff's rules in analyzing a circuit. You can use the junction rule as often as you need as long as you include in it a current that has not been used in a preceding junction-rule equation. In general, the number of times you can use the junction rule is one fewer than the number of junction points in the circuit. You can apply the loop rule as often as needed as long as a new circuit element (resistor or battery) or a new current appears in each new equation. In general, to solve a particular circuit problem, the number of independent equations you need to obtain from the two rules equals the number of unknown currents.

The following examples illustrate how to use Kirchhoff's rules. In all cases, it is assumed the circuits have reached steady-state conditions; in other words, the currents in the various branches are constant. Any capacitor acts as an open branch in a circuit; that is, the current in the branch containing the capacitor is zero under steady-state conditions.

Gustav Kirchhoff
German Physicist (1824–1887)
Kirchhoff, a professor at Heidelberg, and Robert Bunsen invented the spectroscope and founded the science of spectroscopy, which we shall study in Chapter 41. They discovered the elements cesium and rubidium and invented astronomical spectroscopy.

PROBLEM-SOLVING STRATEGY   Kirchhoff's Rules

The following procedure is recommended for solving problems that involve circuits that cannot be reduced by the rules for combining resistors in series or parallel.

1. Conceptualize. Study the circuit diagram and make sure you recognize all elements in the circuit. Identify the polarity of each battery and try to imagine the directions in which the current would exist in the batteries.

2. Categorize. Determine whether the circuit can be reduced by means of combining series and parallel resistors. If so, use the techniques of Section 27.2. If not, apply Kirchhoff's rules according to the *Analyze* step below.

3. Analyze. Assign labels to all known quantities and symbols to all unknown quantities. You must assign *directions* to the currents in each part of the circuit. Although the assignment of current directions is arbitrary, you must adhere *rigorously* to the directions you assign when you apply Kirchhoff's rules.

Apply the junction rule (Kirchhoff's first rule) to all junctions in the circuit except one. Now apply the loop rule (Kirchhoff's second rule) to as many loops in the circuit as are needed to obtain, in combination with the equations from the junction rule, as many equations as there are unknowns. To apply this rule, you must choose a direction in which to travel around the loop (either clockwise or counterclockwise) and correctly identify the change in potential as you cross each element. Be careful with signs! Follow the rules in Figure 27.12 carefully.

Solve the equations simultaneously for the unknown quantities.

4. Finalize. Check your numerical answers for consistency. Do not be alarmed if any of the resulting currents have a negative value. That only means you have guessed the direction of that current incorrectly, but *its magnitude will be correct*.

Example 27.6 A Single-Loop Circuit

A single-loop circuit contains two resistors and two batteries as shown in Figure 27.13. (Neglect the internal resistances of the batteries.) Find the current in the circuit.

Figure 27.13 (Example 27.6) A series circuit containing two batteries and two resistors, where the polarities of the batteries are in opposition.

SOLUTION

Conceptualize Figure 27.13 shows the polarities of the batteries and a guess at the direction of the current. The 12-V battery is the stronger of the two, so the current should be counterclockwise. Therefore, we expect our guess for the direction of the current to be wrong, but we will continue and see how this incorrect guess is represented by our final answer.

Categorize We do not need Kirchhoff's rules to analyze this simple circuit, but let's use them anyway simply to see how they are applied. There are no junctions in this single-loop circuit; therefore, the current is the same in all elements.

Analyze Let's assume the current is clockwise as shown in Figure 27.13. Traversing the circuit in the clockwise direction, starting at a, we see that $a \rightarrow b$ represents a potential difference of $+\mathcal{E}_1$, $b \rightarrow c$ represents a potential difference of $-IR_1$, $c \rightarrow d$ represents a potential difference of $-\mathcal{E}_2$, and $d \rightarrow a$ represents a potential difference of $-IR_2$.

Apply Kirchhoff's loop rule to the single loop in the circuit:

$$\sum \Delta V = 0 \quad \rightarrow \quad \mathcal{E}_1 - IR_1 - \mathcal{E}_2 - IR_2 = 0$$

Solve for I and use the values given in Figure 27.13:

$$(1) \quad I = \frac{\mathcal{E}_1 - \mathcal{E}_2}{R_1 + R_2} = \frac{6.0 \text{ V} - 12 \text{ V}}{8.0 \text{ }\Omega + 10 \text{ }\Omega} = \boxed{-0.33 \text{ A}}$$

Finalize The negative sign for I indicates that the direction of the current is opposite the assumed direction. The emfs in the numerator subtract because the batteries in Figure 27.13 have opposite polarities. The resistances in the denominator add because the two resistors are in series.

WHAT IF? What if the polarity of the 12.0-V battery were reversed? How would that affect the circuit?

Answer Although we could repeat the Kirchhoff's rules calculation, let's instead examine Equation (1) and modify it accordingly. Because the polarities of the two batteries are now in the same direction, the signs of $\mathcal{E}_1$ and $\mathcal{E}_2$ are the same and Equation (1) becomes

$$I = \frac{\mathcal{E}_1 + \mathcal{E}_2}{R_1 + R_2} = \frac{6.0 \text{ V} + 12 \text{ V}}{8.0 \text{ }\Omega + 10 \text{ }\Omega} = 1.0 \text{ A}$$

Example 27.7 A Multiloop Circuit

Find the currents I_1, I_2, and I_3 in the circuit shown in Figure 27.14.

SOLUTION

Conceptualize Imagine physically rearranging the circuit while keeping it electrically the same. Can you rearrange it so that it consists of simple series or parallel combinations of resistors? You should find that you cannot. (If the 10.0-V battery were removed and replaced by a wire from b to the 6.0-Ω resistor, the circuit would consist of only series and parallel combinations.)

Figure 27.14 (Example 27.7) A circuit containing different branches.

Categorize We cannot simplify the circuit by the rules associated with combining resistances in series and in parallel. Therefore, this problem is one in which we must use Kirchhoff's rules.

Analyze We identify three different currents and arbitrarily choose their directions as labeled in Figure 27.14.

continued

27.7 continued

Apply Kirchhoff's junction rule to junction c:

(1) $I_1 + I_2 - I_3 = 0$

We now have one equation with three unknowns: I_1, I_2, and I_3. There are three loops in the circuit: $abcda$, $befcb$, and $aefda$. We need only two loop equations to determine the unknown currents. (The third equation would give no new information.) Let's choose to traverse these loops in the clockwise direction. Apply Kirchhoff's loop rule to loops $abcda$ and $befcb$:

$abcda$: (2) $10.0\text{ V} - (6.0\ \Omega)I_1 - (2.0\ \Omega)I_3 = 0$

$befcb$: $-(4.0\ \Omega)I_2 - 14.0\text{ V} + (6.0\ \Omega)I_1 - 10.0\text{ V} = 0$

(3) $-24.0\text{ V} + (6.0\ \Omega)I_1 - (4.0\ \Omega)I_2 = 0$

Solve Equation (1) for I_3 and substitute into Equation (2):

$10.0\text{ V} - (6.0\ \Omega)I_1 - (2.0\ \Omega)(I_1 + I_2) = 0$

(4) $10.0\text{ V} - (8.0\ \Omega)I_1 - (2.0\ \Omega)I_2 = 0$

Multiply each term in Equation (3) by 4 and each term in Equation (4) by 3:

(5) $-96.0\text{ V} + (24.0\ \Omega)I_1 - (16.0\ \Omega)I_2 = 0$

(6) $30.0\text{ V} - (24.0\ \Omega)I_1 - (6.0\ \Omega)I_2 = 0$

Add Equation (6) to Equation (5) to eliminate I_1 and find I_2:

$-66.0\text{ V} - (22.0\ \Omega)I_2 = 0$

$I_2 = -3.0\text{ A}$

Use this value of I_2 in Equation (3) to find I_1:

$-24.0\text{ V} + (6.0\ \Omega)I_1 - (4.0\ \Omega)(-3.0\text{ A}) = 0$

$-24.0\text{ V} + (6.0\ \Omega)I_1 + 12.0\text{ V} = 0$

$I_1 = 2.0\text{ A}$

Use Equation (1) to find I_3:

$I_3 = I_1 + I_2 = 2.0\text{ A} - 3.0\text{ A} = -1.0\text{ A}$

Finalize Because our values for I_2 and I_3 are negative, the directions of these currents are opposite those indicated in Figure 27.14. The numerical values for the currents are correct. Despite the incorrect direction, we *must* continue to use these negative values in subsequent calculations because our equations were established with our original choice of direction. What would have happened had we left the current directions as labeled in Figure 27.14 but traversed the loops in the opposite direction?

27.4 *RC* Circuits

So far, we have analyzed direct-current circuits in which the current is constant. In DC circuits containing capacitors, the current is always in the same direction but may vary in magnitude at different times. A circuit containing a series combination of a resistor and a capacitor is called an *RC* **circuit.**

Charging a Capacitor

Figure 27.15 shows a simple series *RC* circuit. Let's assume the capacitor in this circuit is initially uncharged. There is no current while the switch is open (Fig. 27.15a). If the switch is thrown to position a at $t = 0$ (Fig. 27.15b), however, charge begins to flow, setting up a current in the circuit, and the capacitor begins to charge.[2] Notice that during charging, charges do not jump across the capacitor plates because the gap between the plates represents an open circuit. Instead, charge is transferred between each plate and its connecting wires due to the electric field established in the wires by the battery until the capacitor is fully charged. As the plates are being charged, the potential difference across the capacitor increases. The value of the maximum charge on the plates depends on the voltage of the battery. Once the maximum charge is reached, the current in the circuit is zero because the potential difference across the capacitor matches that supplied by the battery.

---

[2]In previous discussions of capacitors, we assumed a steady-state situation, in which no current was present in any branch of the circuit containing a capacitor. Now we are considering the case *before* the steady-state condition is realized; in this situation, charges are moving and a current exists in the wires connected to the capacitor.

To analyze this circuit quantitatively, let's apply Kirchhoff's loop rule to the circuit after the switch is thrown to position *a*. Traversing the loop in Figure 27.15b clockwise gives

$$\mathcal{E} - \frac{q}{C} - iR = 0 \qquad (27.15)$$

where q/C is the potential difference across the capacitor and iR is the potential difference across the resistor. When we study electric circuits in which charge or current is changing in time, we will use lower case q and i for the time-varying values. We will reserve uppercase letters for initial, final, or steady-state values. We have used the sign conventions discussed earlier for the signs on $\mathcal{E}$ and iR in Equation 27.15. The capacitor is traversed in the direction from the positive plate to the negative plate, which represents a decrease in potential. Therefore, we use a negative sign for this potential difference in Equation 27.15.

We can use Equation 27.15 to find the initial current I_i in the circuit and the maximum charge Q_{max} on the capacitor. At the instant the switch is thrown to position *a* ($t = 0$), the charge on the capacitor is zero. Equation 27.15 shows that the initial current I_i in the circuit is a maximum and is given by

$$I_i = \frac{\mathcal{E}}{R} \quad \text{(current at } t = 0) \qquad (27.16)$$

At this time, the potential difference from the battery terminals appears entirely across the resistor. Later, when the capacitor is charged to its maximum value Q_{max}, charges cease to flow, the current in the circuit is zero, and the potential difference from the battery terminals appears entirely across the capacitor. Substituting $i = 0$ into Equation 27.15 gives the maximum charge on the capacitor:

$$Q_{max} = C\mathcal{E} \quad \text{(maximum charge)} \qquad (27.17)$$

To determine analytical expressions for the time dependence of the charge and current, we must solve Equation 27.15, a single equation containing two variables q and i. The current in all parts of the series circuit must be the same. Therefore, the current in the resistance R must be the same as the current between each capacitor plate and the wire connected to it. This current is equal to the time rate of change of the charge on the capacitor plates. Therefore, we substitute $i = dq/dt$ into Equation 27.15 and rearrange the equation:

$$\frac{dq}{dt} = \frac{\mathcal{E}}{R} - \frac{q}{RC}$$

To find an expression for q, we solve this separable differential equation as follows. First combine the terms on the right-hand side:

$$\frac{dq}{dt} = \frac{C\mathcal{E}}{RC} - \frac{q}{RC} = -\frac{q - C\mathcal{E}}{RC}$$

Multiply this equation by dt and divide by $q - C\mathcal{E}$:

$$\frac{dq}{q - C\mathcal{E}} = -\frac{1}{RC}\,dt$$

Integrate this expression, using $q = 0$ at $t = 0$:

$$\int_0^q \frac{dq}{q - C\mathcal{E}} = -\frac{1}{RC}\int_0^t dt$$

$$\ln\left(\frac{q - C\mathcal{E}}{-C\mathcal{E}}\right) = -\frac{t}{RC}$$

When the switch is thrown to position *a*, the capacitor begins to charge up.

When the switch is thrown to position *b*, the capacitor discharges.

Figure 27.15 A capacitor in series with a resistor, switch, and battery.

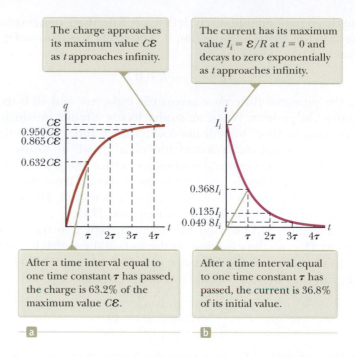

The charge approaches its maximum value $C\mathcal{E}$ as t approaches infinity.

The current has its maximum value $I_i = \mathcal{E}/R$ at $t = 0$ and decays to zero exponentially as t approaches infinity.

After a time interval equal to one time constant τ has passed, the charge is 63.2% of the maximum value $C\mathcal{E}$.

After a time interval equal to one time constant τ has passed, the current is 36.8% of its initial value.

Figure 27.16 (a) Plot of capacitor charge versus time for the circuit shown in Figure 27.15b. (b) Plot of current versus time for the circuit shown in Figure 27.15b.

From the definition of the natural logarithm, we can write this expression as

◄ Charge as a function of time for a capacitor being charged

$$q(t) = C\mathcal{E}(1 - e^{-t/RC}) = Q_{max}(1 - e^{-t/RC}) \tag{27.18}$$

where e is the base of the natural logarithm and we have made the substitution from Equation 27.17.

We can find an expression for the current in the circuit as a function of time by differentiating Equation 27.18 with respect to time. Using $i = dq/dt$, we find that

◄ Current as a function of time for a capacitor being charged

$$i(t) = \frac{\mathcal{E}}{R}e^{-t/RC} \tag{27.19}$$

Plots of capacitor charge and circuit current versus time are shown in Figure 27.16. Notice that the charge is zero at $t = 0$ and approaches the maximum value $C\mathcal{E}$ as $t \to \infty$. The current has its maximum value $I_i = \mathcal{E}/R$ at $t = 0$ and decays exponentially to zero as $t \to \infty$. The quantity RC, which appears in the exponents of Equations 27.18 and 27.19, is called the **time constant** τ of the circuit:

$$\tau = RC \tag{27.20}$$

The time constant represents the time interval during which the current decreases to $1/e$ of its initial value; that is, after a time interval τ, the current decreases to $i = e^{-1}I_i = 0.368I_i$. After a time interval 2τ, the current decreases to $i = e^{-2}I_i = 0.135I_i$, and so forth. Likewise, in a time interval τ, the charge increases from zero to $C\mathcal{E}[1 - e^{-1}] = 0.632C\mathcal{E}$.

The energy supplied by the battery during the time interval required to fully charge the capacitor is $Q_{max}\mathcal{E} = C\mathcal{E}^2$. After the capacitor is fully charged, the energy stored in the capacitor is $\frac{1}{2}Q_{max}\mathcal{E} = \frac{1}{2}C\mathcal{E}^2$, which is only half the energy output of the battery. It is left as a problem (Problem 44) to show that the remaining half of the energy supplied by the battery appears as internal energy in the resistor.

Discharging a Capacitor

Imagine that the capacitor in Figure 27.15b is completely charged. An initial potential difference Q_i/C exists across the capacitor, and there is zero potential difference across the resistor because $i = 0$. If the switch is now thrown to position b at $t = 0$ (Fig. 27.15c), the capacitor begins to discharge through the resistor. At some time t during the discharge, the current in the circuit is i and the charge on the capacitor is q. The circuit in Figure 27.15c is the same as the circuit in Figure 27.15b except for the absence of the battery. Therefore, we eliminate the emf $\mathcal{E}$ from Equation 27.15 to obtain the appropriate loop equation for the circuit in Figure 27.15c:

$$-\frac{q}{C} - iR = 0 \qquad (27.21)$$

When we substitute $i = dq/dt$ into this expression, it becomes

$$-R\frac{dq}{dt} = \frac{q}{C}$$

$$\frac{dq}{q} = -\frac{1}{RC}\,dt$$

Integrating this expression using $q = Q_i$ at $t = 0$ gives

$$\int_{Q_i}^{q} \frac{dq}{q} = -\frac{1}{RC}\int_{0}^{t} dt$$

$$\ln\left(\frac{q}{Q_i}\right) = -\frac{t}{RC}$$

$$q(t) = Q_i e^{-t/RC} \qquad (27.22)$$

◀ Charge as a function of time for a discharging capacitor

Differentiating Equation 27.22 with respect to time gives the instantaneous current as a function of time:

$$i(t) = -\frac{Q_i}{RC} e^{-t/RC} \qquad (27.23)$$

◀ Current as a function of time for a discharging capacitor

where $Q_i/RC = I_i$ is the initial current. Figure 27.15b shows a downward current in the resistor that we guessed in order to apply Kirchhoff's rules and generate Equations 27.15 and 27.21. Equation 27.23 shows that the current in the discharging capacitor is negative, indicating that the current is *upward* in the resistor in Figure 27.15c. Both the charge on the capacitor and the current decay exponentially at a rate characterized by the time constant $\tau = RC$.

QUICK QUIZ 27.5 Consider the circuit in Figure 27.17 and assume the battery has no internal resistance. **(i)** Just after the switch is closed, what is the current in the battery? (a) 0 (b) $\mathcal{E}/2R$ (c) $2\mathcal{E}/R$ (d) $\mathcal{E}/R$ (e) impossible to determine **(ii)** After a very long time, what is the current in the battery? Choose from the same choices.

Figure 27.17 (Quick Quiz 27.5) How does the current vary after the switch is closed?

Conceptual Example 27.8 | Intermittent Windshield Wipers

Many automobiles are equipped with windshield wipers that can operate intermittently during a light rainfall. How does the operation of such wipers depend on the charging and discharging of a capacitor?

SOLUTION

The wipers are part of an *RC* circuit whose time constant can be varied by selecting different values of R through a multiposition switch. As the voltage across the capacitor increases, the capacitor reaches a point at which it discharges and triggers the wipers. The circuit then begins another charging cycle. The time interval between the individual sweeps of the wipers is determined by the value of the time constant.

Example 27.9 | Charging a Capacitor in an *RC* Circuit

An uncharged capacitor and a resistor are connected in series to a battery as shown in Figure 27.15, where $\mathcal{E} = 12.0$ V, $C = 5.00\ \mu$F, and $R = 8.00 \times 10^5\ \Omega$. The switch is thrown to position a. Find the time constant of the circuit, the maximum charge on the capacitor, the maximum current in the circuit, and the charge and current as functions of time.

continued

27.9 continued

SOLUTION

Conceptualize Study Figure 27.15 and imagine throwing the switch to position a as shown in Figure 27.15b. Upon doing so, the capacitor begins to charge.

Categorize We evaluate our results using equations developed in this section, so we categorize this example as a substitution problem.

Evaluate the time constant of the circuit from Equation 27.20:

$$\tau = RC = (8.00 \times 10^5 \ \Omega)(5.00 \times 10^{-6} \ \text{F}) = \boxed{4.00 \ \text{s}}$$

Evaluate the maximum charge on the capacitor, which occurs as $t \rightarrow \infty$, from Equation 27.17:

$$Q_{max} = C\mathcal{E} = (5.00 \ \mu\text{F})(12.0 \ \text{V}) = \boxed{60.0 \ \mu\text{C}}$$

Evaluate the maximum current in the circuit, which occurs at $t = 0$, from Equation 27.16:

$$I_i = \frac{\mathcal{E}}{R} = \frac{12.0 \ \text{V}}{8.00 \times 10^5 \ \Omega} = \boxed{15.0 \ \mu\text{A}}$$

Use these values in Equations 27.18 and 27.19 to find the charge and current as functions of time:

(1) $\ q(t) = \boxed{60.0(1 - e^{-t/4.00})}$

(2) $\ i(t) = \boxed{15.0e^{-t/4.00}}$

In Equations (1) and (2), q is in microcoulombs, i is in microamperes, and t is in seconds.

Example **27.10** **Discharging a Capacitor in an *RC* Circuit**

Consider a capacitor of capacitance C that is being discharged through a resistor of resistance R as shown in Figure 27.15c.

(A) After how many time constants is the charge on the capacitor one-fourth its initial value?

SOLUTION

Conceptualize Study Figure 27.15 and imagine throwing the switch to position b as shown in Figure 27.15c. Upon doing so, the capacitor begins to discharge.

Categorize We categorize the example as one involving a discharging capacitor and use the appropriate equations.

Analyze Substitute $q(t) = Q_i/4$ into Equation 27.22:

$$\frac{Q_i}{4} = Q_i e^{-t/RC}$$
$$\tfrac{1}{4} = e^{-t/RC}$$

Take the logarithm of both sides of the equation and solve for t:

$$-\ln 4 = -\frac{t}{RC}$$
$$t = RC \ln 4 = 1.39RC = \boxed{1.39\tau}$$

(B) The energy stored in the capacitor decreases with time as the capacitor discharges. After how many time constants is this stored energy one-fourth its initial value?

SOLUTION

Use Equations 25.13 and 27.22 to express the energy stored in the capacitor at any time t:

(1) $\ U(t) = \dfrac{q^2}{2C} = \dfrac{Q_i^2}{2C} e^{-2t/RC}$

Substitute $U(t) = \tfrac{1}{4}(Q_i^2/2C)$ into Equation (1):

$$\frac{1}{4}\frac{Q_i^2}{2C} = \frac{Q_i^2}{2C} e^{-2t/RC}$$
$$\tfrac{1}{4} = e^{-2t/RC}$$

Take the logarithm of both sides of the equation and solve for t:

$$-\ln 4 = -\frac{2t}{RC}$$
$$t = \tfrac{1}{2}RC \ln 4 = 0.693RC = \boxed{0.693\tau}$$

27.10 continued

Finalize Notice that because the energy depends on the square of the charge, the energy in the capacitor drops more rapidly than the charge on the capacitor.

> **WHAT IF?** What if you want to describe the circuit in terms of the time interval required for the charge to fall to one-half its original value rather than by the time constant τ? That would give a parameter for the circuit called its *half-life* $t_{1/2}$. How is the half-life related to the time constant?
>
> **Answer** In one half-life, the charge falls from Q_i to $Q_i/2$. Therefore, from Equation 27.22,
>
> $$\frac{Q_i}{2} = Q_i e^{-t_{1/2}/RC} \rightarrow \tfrac{1}{2} = e^{-t_{1/2}/RC}$$
>
> which leads to
>
> $$t_{1/2} = 0.693\tau$$
>
> The concept of half-life will be important to us when we study nuclear decay in Chapter 43. The radioactive decay of an unstable sample behaves in a mathematically similar manner to a discharging capacitor in an *RC* circuit.

Example **27.11** **Energy Delivered to a Resistor**

A 5.00-μF capacitor is charged to a potential difference of 800 V and then discharged through a resistor. How much energy is delivered to the resistor in the time interval required to fully discharge the capacitor?

SOLUTION

Conceptualize In part (B) of Example 27.10, we considered the energy decrease in a discharging capacitor to a value of one-fourth the initial energy. In this example, the capacitor fully discharges.

Categorize We solve this example using two approaches. The first approach is to model the circuit as an *isolated system* for *energy*. Because energy in an isolated system is conserved, the initial electric potential energy U_E stored in the capacitor is transformed into internal energy $E_{int} = E_R$ in the resistor. The second approach is to model the resistor as a *nonisolated system* for *energy*. Energy enters the resistor by electrical transmission from the capacitor, causing an increase in the resistor's internal energy.

Analyze We begin with the isolated system approach.

Write the appropriate reduction of the conservation of energy equation, Equation 8.2:

$$\Delta U + \Delta E_{int} = 0$$

Substitute the initial and final values of the energies:

$$(0 - U_E) + (E_{int} - 0) = 0 \rightarrow E_R = U_E$$

Use Equation 25.13 for the electric potential energy in the capacitor:

$$E_R = \tfrac{1}{2}C\mathcal{E}^2$$

Substitute numerical values:

$$E_R = \tfrac{1}{2}(5.00 \times 10^{-6}\text{ F})(800\text{ V})^2 = \boxed{1.60\text{ J}}$$

The second approach, which is more difficult but perhaps more instructive, is to note that as the capacitor discharges through the resistor, the rate at which energy is delivered to the resistor by electrical transmission is i^2R, where i is the instantaneous current given by Equation 27.23.

Evaluate the energy delivered to the resistor by integrating the power over all time because it takes an infinite time interval for the capacitor to completely discharge:

$$P = \frac{dE}{dt} \rightarrow E_R = \int_0^\infty P\,dt$$

Substitute for the instantaneous power delivered to the resistor from Equation 26.22:

$$E_R = \int_0^\infty i^2R\,dt$$

continued

27.11 continued

Substitute for the current from Equation 27.23:

$$E_R = \int_0^\infty \left(-\frac{Q_i}{RC} e^{-t/RC} \right)^2 R\, dt = \frac{Q_i^2}{RC^2} \int_0^\infty e^{-2t/RC}\, dt = \frac{\mathcal{E}^2}{R} \int_0^\infty e^{-2t/RC}\, dt$$

Substitute the value of the integral, which is $RC/2$ (see Problem 28):

$$E_R = \frac{\mathcal{E}^2}{R} \left(\frac{RC}{2} \right) = \tfrac{1}{2}C\mathcal{E}^2$$

Finalize This result agrees with that obtained using the isolated system approach, as it must. We can use this second approach to find the total energy delivered to the resistor at *any* time after the switch is closed by simply replacing the upper limit in the integral with that specific value of *t*.

27.5 Household Wiring and Electrical Safety

Many considerations are important in the design of an electrical system of a home that will provide adequate electrical service for the occupants while maximizing their safety. We discuss some aspects of a home electrical system in this section.

Household Wiring

Household circuits represent a practical application of some of the ideas presented in this chapter. In our world of electrical appliances, it is useful to understand the power requirements and limitations of conventional electrical systems and the safety measures that prevent accidents.

In a conventional installation, the utility company distributes electric power to individual homes by means of a pair of wires, with each home connected in parallel to these wires. One wire is called the *live wire*[3] as illustrated in Figure 27.18, and the other is called the *neutral wire*. The neutral wire is grounded; that is, its electric potential is taken to be zero. The potential difference between the live and neutral wires is approximately 120 V. This voltage alternates in time, and the potential of the live wire oscillates relative to ground. Much of what we have learned so far in this chapter for the constant-emf situation (direct current) can also be applied to the alternating current that power companies supply to businesses and households. (Alternating voltage and current are discussed in detail in Chapter 32.)

To record a household's energy consumption, a meter is connected in series with the live wire entering the house. After the meter, the wire splits so that there are several separate circuits in parallel distributed throughout the house. Each circuit contains a circuit breaker (or, in older installations, a fuse). A circuit breaker is a special switch that opens if the current exceeds the rated value for the circuit breaker. The wire and circuit breaker for each circuit are carefully selected to meet the current requirements for that circuit. If a circuit is to carry currents as large as 30 A, a heavy wire and an appropriate circuit breaker must be selected to handle this current. A circuit used to power only lamps and small appliances often requires only 20 A. Each circuit has its own circuit breaker to provide protection for that part of the entire electrical system of the house.

As an example, consider a circuit in which a toaster oven, a microwave oven, and a coffee maker are connected (corresponding to R_1, R_2, and R_3 in Fig. 27.18). We can calculate the current in each appliance by using the expression $P = I\,\Delta V$. The toaster oven, rated at 1 000 W, draws a current of 1 000 W/120 V = 8.33 A. The microwave oven, rated at 1 300 W, draws 10.8 A, and the coffee maker, rated at 800 W, draws 6.67 A. When the three appliances are operated simultaneously, they draw a total current of 25.8 A. Therefore, the circuit must be wired to handle at

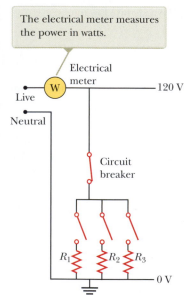

The electrical meter measures the power in watts.

Live

Neutral

Electrical meter

120 V

Circuit breaker

R_1 R_2 R_3

0 V

Figure 27.18 Wiring diagram for one parallel circuit in a household electrical system. The resistances represent appliances or other electrical devices that operate with an applied voltage of 120 V.

---

[3]*Live wire* is a common expression for a conductor whose electric potential is above or below ground potential.

least this much current. If the rating of the circuit breaker protecting the circuit is too small—say, 20 A—the breaker will be tripped when the third appliance is turned on, preventing all three appliances from operating. To avoid this situation, the toaster oven and coffee maker can be operated on one 20-A circuit and the microwave oven on a separate 20-A circuit.

Many heavy-duty appliances such as electric ranges and clothes dryers require 240 V for their operation. The power company supplies this voltage by providing a third wire that is 120 V below ground potential (Fig. 27.19). The potential difference between this live wire and the other live wire (which is 120 V above ground potential) is 240 V. An appliance that operates from a 240-V line requires half as much current compared with operating it at 120 V; therefore, smaller wires can be used in the higher-voltage circuit without overheating.

Electrical Safety

When the live wire of an electrical outlet is connected directly to ground, the circuit is completed and a *short-circuit condition* exists. A short circuit occurs when almost zero resistance exists between two points at different potentials, and the result is a very large current. When that happens accidentally, a properly operating circuit breaker opens the circuit and no damage is done. A person in contact with ground, however, can be electrocuted by touching the live wire of a frayed cord or other exposed conductor. An exceptionally effective (and dangerous!) ground contact is made when the person either touches a water pipe (normally at ground potential) or stands on the ground with wet feet. The latter situation represents effective ground contact because normal, nondistilled water is a conductor due to the large number of ions associated with impurities. This situation should be avoided at all cost.

Many 120-V outlets are designed to accept a three-pronged power cord. (This feature is required in all new electrical installations.) One of these prongs is the live wire at a nominal potential of 120 V. The second is the neutral wire, nominally at 0 V, which carries current to ground. Figure 27.20a shows a connection to an electric drill with only these two wires. If the live wire accidentally makes contact with the casing of the electric drill (which can occur if the wire insulation wears off), current can be carried to ground by way of the person, resulting in an electric shock. The third wire in a three-pronged power cord, the round prong, is a safety ground wire that normally carries no current. It is both grounded and connected directly to the casing of the appliance. If the live wire is accidentally shorted to the

Figure 27.19 (a) An outlet for connection to a 240-V supply. (b) The connections for each of the openings in a 240-V outlet.

In the situation shown, the live wire has come into contact with the drill case. As a result, the person holding the drill acts as a current path to ground and receives an electric shock.

In this situation, the drill case remains at ground potential and no current exists in the person.

Figure 27.20 (a) A diagram of the circuit for an electric drill with only two connecting wires. The normal current path is from the live wire through the motor connections and back to ground through the neutral wire. (b) This shock can be avoided by connecting the drill case to ground through a third ground wire. The wire colors represent electrical standards in the United States: the "hot" wire is black, the ground wire is green, and the neutral wire is white (shown as gray in the figure).

casing in this situation, most of the current takes the low-resistance path through the appliance to ground as shown in Figure 27.20b.

Special power outlets called *ground-fault circuit interrupters*, or GFCIs, are used in kitchens, bathrooms, basements, exterior outlets, and other hazardous areas of homes. These devices are designed to protect persons from electric shock by sensing small currents (< 5 mA) leaking to ground. (The principle of their operation is described in Chapter 30.) When an excessive leakage current is detected, the current is shut off in less than 1 ms.

Electric shock can result in fatal burns or can cause the muscles of vital organs such as the heart to malfunction. The degree of damage to the body depends on the magnitude of the current, the length of time it acts, the part of the body touched by the live wire, and the part of the body in which the current exists. Currents of 5 mA or less cause a sensation of shock, but ordinarily do little or no damage. If the current is larger than about 10 mA, the muscles contract and the person may be unable to release the live wire. If the body carries a current of about 100 mA for only a few seconds, the result can be fatal. Such a large current paralyzes the respiratory muscles and prevents breathing. In some cases, currents of approximately 1 A can produce serious (and sometimes fatal) burns. In practice, no contact with live wires is regarded as safe whenever the voltage is greater than 24 V.

Summary

▶ Definitions

The **emf** of a battery is equal to the voltage across its terminals when the current is zero. That is, the emf is equivalent to the **open-circuit voltage** of the battery.

▶ Concepts and Principles

The **equivalent resistance** of a set of resistors connected in a **series combination** is

$$R_{eq} = R_1 + R_2 + R_3 + \cdots \quad (27.8)$$

The **equivalent resistance** of a set of resistors connected in a **parallel combination** is found from the relationship

$$\frac{1}{R_{eq}} = \frac{1}{R_1} + \frac{1}{R_2} + \frac{1}{R_3} + \cdots \quad (27.12)$$

Circuits involving more than one loop are conveniently analyzed with the use of **Kirchhoff's rules:**

1. **Junction rule.** At any junction, the sum of the currents must equal zero:

$$\sum_{junction} I = 0 \quad (27.13)$$

2. **Loop rule.** The sum of the potential differences across all elements around any circuit loop must be zero:

$$\sum_{closed\ loop} \Delta V = 0 \quad (27.14)$$

When a resistor is traversed in the direction of the current, the potential difference ΔV across the resistor is $-IR$. When a resistor is traversed in the direction opposite the current, $\Delta V = +IR$. When a source of emf is traversed in the direction of the emf (negative terminal to positive terminal), the potential difference is $+\mathcal{E}$. When a source of emf is traversed opposite the emf (positive to negative), the potential difference is $-\mathcal{E}$.

If a capacitor is charged with a battery through a resistor of resistance R, the charge on the capacitor and the current in the circuit vary in time according to the expressions

$$q(t) = Q_{max}(1 - e^{-t/RC}) \quad (27.18)$$

$$i(t) = \frac{\mathcal{E}}{R} e^{-t/RC} \quad (27.19)$$

where $Q_{max} = C\mathcal{E}$ is the maximum charge on the capacitor. The product RC is called the **time constant** τ of the circuit.

If a charged capacitor of capacitance C is discharged through a resistor of resistance R, the charge and current decrease exponentially in time according to the expressions

$$q(t) = Q_i e^{-t/RC} \quad (27.22)$$

$$i(t) = -\frac{Q_i}{RC} e^{-t/RC} \quad (27.23)$$

where Q_i is the initial charge on the capacitor and Q_i/RC is the initial current in the circuit.

Think–Pair–Share

See the Preface for an explanation of the icons used in this problems set. For additional assessment items for this section, go to ❖ WEBASSIGN From Cengage

1. You and your cousins are playing with your niece, showing her some simple electric circuits. You connect a 9.0-V battery to an electric buzzer with two wires equipped with alligator clips on each end. Both wires are copper and of diameter 0.500 mm. Wire 1 is 1.00 m long, while wire 2 is 0.250 m long. You expect your niece to be thrilled with the sound from the buzzer, but something is wrong. The sound is much too weak. You happen to have a multimeter, so you can take voltage and current measurements. Half of your group takes measurements, while the circuit is operating: voltage across battery, 8.60 V; voltage across buzzer, 6.40 V; voltage across wire 2, 0.010 V current in circuit, 0.500 A. The other half of your group looks online and finds that the battery you are using should have an internal resistance less than $0.900 \, \Omega$, and that the buzzer you are using should have no more than $15.0 \, \Omega$ of resistance while in operation. The halves of your group now reconvene and determine which component is defective: the battery, the buzzer, wire 1, or wire 2.

2. **ACTIVITY** Your group is performing electrical experiments in your physics laboratory. Your supply of resistors in the stockroom has run low and you only have two different values of resistance: $20 \, \Omega$ and $50 \, \Omega$. (a) Your experiments require a 45-Ω resistor and a 35-Ω resistor. Split your group into two halves. Group (i) will determine how to form a 45-Ω resistor from your supply, while group (ii) will do the same for a 35-Ω resistor. (b) After working on the experiment for a while, you realize you now need a 105-Ω resistor. Have your whole group work together to find at least three ways to combine the resistors in your stockroom to generate $105 \, \Omega$.

3. Consider the table below, which shows a typical amount of power utilized for a number of household appliances. In the third column, discuss in your group and enter the number of such appliances in a typical home. Many of these will be 1, some will be 0 if the item is not used, and others will be greater than one, such as the number of lightbulbs in the home. In the time column, estimate the amount of time in hours that each appliance would be used in a day. Multiply columns 2, 3, and 4 to obtain the energy use in one day for each appliance in the last column. From these results, estimate (a) the monthly energy usage of this household in kWh and (b) the monthly electric bill if electricity costs 11¢ per kWh.

Appliance	Power (W)	Number of items	Time of use in one day (h)	Energy in one day (kWh)
Household:				
Central air conditioner	5 000			
Vacuum cleaner	500			
Electric water heater	475			

Appliance	Power (W)	Number of items	Time of use in one day (h)	Energy in one day (kWh)
LCD television	215			
100-watt incandescent bulb	100			
CFL bulb	25			
Ceiling fan	100			
LED bulb	10			
Table fan	20			
Garage door opener	350			
Satellite dish	30			
Plasma television	340			
Kitchen:				
Oven	3 000			
Dishwasher	1 200			
Coffee machine	1 500			
Microwave oven	1 500			
Toaster	1 100			
Refrigerator	400			
Bathroom:				
Hair dryer	1 500			
Electric shaver	15			
Curling iron	90			
Laundry:				
Electric clothes dryer	3 400			
Iron	1 100			
Washing machine	500			
Computer:				
Laptop computer	100			
Desktop computer	120			
Cell phone recharger	4			
Internet router	15			
Inkjet printer	25			
Bedroom:				
Electric blanket	200			
Clock radio	2			

Problems

See the Preface for an explanation of the icons used in this problems set.
For additional assessment items for this section, go to ⚡ **WEBASSIGN**
From Cengage

SECTION 27.1 Electromotive Force

1. Two 1.50-V batteries—with their positive terminals in the
AMT same direction—are inserted in series into a flashlight. One
battery has an internal resistance of 0.255 Ω, and the other
has an internal resistance of 0.153 Ω. When the switch is
closed, the bulb carries a current of 600 mA. (a) What is the
bulb's resistance? (b) What fraction of the chemical energy
transformed appears as internal energy in the batteries?

2. As in Example 27.2, consider a power supply with fixed emf $\mathcal{E}$
Q|C and internal resistance r causing current in a load resistance
R. In this problem, R is fixed and r is a variable. The efficiency is defined as the energy delivered to the load divided
by the energy delivered by the emf. (a) When the internal
resistance is adjusted for maximum power transfer, what is
the efficiency? (b) What should be the internal resistance for
maximum possible efficiency? (c) When the electric company
sells energy to a customer, does it have a goal of high efficiency or of maximum power transfer? Explain. (d) When
a student connects a loudspeaker to an amplifier, does she
most want high efficiency or high power transfer? Explain.

SECTION 27.2 Resistors in Series and Parallel

3. Figure P27.3 shows the interior of a *three-way* incandescent
lightbulb, which provides three levels of light intensity. The
socket of the lamp is equipped with a *four-position* switch for
selecting different light intensities, with the positions
described as follows: (1) off (switches S_1 and S_2 both open),
(2) switch S_1 closed, (3) switch S_2 closed, and (4) switches
S_1 and S_2 both closed. The lightbulb contains two filaments.
When the lamp is connected to a 120-V source, one filament
receives 100 W of power and the other receives 75 W. What is
the total power input to the light bulb when (a) only switch
S_1 is closed, (b) only switch S_2 is closed, and (c) both switches
are closed? (d) **What If?** Suppose the 75-W filament breaks
and no longer is able to carry a current. How many switch
positions will result in light leaving the bulb and what will
be the power input to the bulb in those positions?

100-W filament

75-W filament

S_1

S_2

120 V

Figure P27.3

4. A lightbulb marked "75 W [at] 120 V" is screwed into a socket
Q|C at one end of a long extension cord, in which each of the
two conductors has resistance 0.800 Ω. The other end of the
extension cord is plugged into a 120-V outlet. (a) Explain

why the actual power delivered to the lightbulb cannot be 75
W in this situation. (b) Draw a circuit diagram. (c) Find the
actual power delivered to the lightbulb in this circuit.

5. Consider the two circuits shown in Figure P27.5 in which
Q|C the batteries are identical. The resistance of each lightbulb
S is R. Neglect the internal resistances of the batteries. (a)
Find expressions for the currents in each lightbulb. (b) How
does the brightness of B compare with that of C? Explain.
(c) How does the brightness of A compare with that of B and
of C? Explain.

A

B   C

$+$ $-$ $+$ $-$
$\mathcal{E}$ $\mathcal{E}$

Figure P27.5

6. Consider strings of
incandescent lights
that are used for
many ornamental
purposes, such as
decorating Christ-
mas trees. Over the
years, both parallel
and series connec-
tions have been used
for strings of lights,
and the bulbs have
varied in design.
Because series-wired
lightbulbs operate
with less energy per
bulb and at a lower
temperature, they are
safer than parallel-

I   I

Filament

Jumper

Glass insulator

a   b

Figure P27.6

wired lightbulbs, where each bulb operates at 120 V. To prevent
the failure of one lightbulb from causing the entire string to go
out for the bulbs wired in series, a new design was developed.
Figure P27.6a shows one of these types of miniature lightbulb
designed to operate in a series connection. When the filament
breaks in one of these lightbulbs, the break in the filament represents the largest resistance in the series, much larger than
that of the intact filaments. As a result, most of the applied
120 V appears across the lightbulb with the broken filament.
Inside the lightbulb, a small jumper loop covered by an insulating material is wrapped around the filament leads. When
the filament fails and 120 V appears across the lightbulb, an
arc burns the insulation on the jumper and connects the filament leads, as shown in Figure P27.6b. This connection now
provides a low-resistance path through the lightbulb, even
though its filament is no longer active, and the voltage across
the bulb drops to zero. All the other lightbulbs not only stay
on, but they glow more brightly because the total resistance
of the string is reduced and consequently the current in each
remaining lightbulb increases. Suppose you have a string of
48 bulbs, each one with a resistance of 8.00 Ω. Assume the

resistance of a bulb with its filament broken drops to zero. Suppose that a bulb becomes dangerously warm, so that it could set something on fire, if it receives a power of 1.75 W. How many bulbs can fail before the string of lights becomes dangerous?

7. You are working at an electronics fabrication shop. Your current project is on the team producing resistors for the timer circuit that delays the closing of an elevator door. According to its design specification, the timer circuit is to have a resistance of 32.0 Ω between two points A and B. As your resistors come off the assembly line, you find that they have a variation of ±5.00% from this value. After a team meeting to evaluate this situation, the team decides that resistances in the range 32.0 ± 0.5 Ω are acceptable and do not need modification. For resistances outside this range, the director does not wish to discard the resistors, but rather to add extra resistors in series or parallel with the main resistor to bring the total equivalent resistance to the exact design value of 32.0 Ω. You are put in charge of procuring the extra resistors. What range of resistances for these extra resistors do you need to cover the entire range of variation of ±5.00%? All resistances can be measured to three significant figures.

8. In your new job at an engineering company, your supervisor asks you to fabricate a resistor that has a resistance of $R = 0.100$ Ω and *no* change in resistance with temperature. She suggests making the resistor from lengths of cylindrical carbon and Nichrome wires of equal radius, placed end-to-end. She wants the combination to fit into a machine that allows for a radius of the resistor to be $r = 1.50$ mm. What are the lengths of the two segments of the resistor?

9. A battery with $\mathcal{E} = 6.00$ V and no internal resistance supplies current to the circuit shown in Figure P27.9. When the double-throw switch S is open as shown in the figure, the current in the battery is 1.00 mA. When the switch is closed in position *a*, the current in the battery is 1.20 mA. When the switch is closed in position *b*, the current in the battery is 2.00 mA. Find the resistances (a) R_1, (b) R_2, and (c) R_3.

Figure P27.9
Problems 9 and 10.

10. A battery with emf $\mathcal{E}$ and no internal resistance supplies current to the circuit shown in Figure P27.9. When the double-throw switch S is open as shown in the figure, the current in the battery is I_0. When the switch is closed in position *a*, the current in the battery is I_a. When the switch is closed in position *b*, the current in the battery is I_b. Find the resistances (a) R_1, (b) R_2, and (c) R_3.

11. Today's class on current and resistance is about to begin and you await your professor, who is known for unorthodox demonstrations. He walks in just at the beginning time for the class, and is carrying hot dogs! He then proceeds to set up a demonstration using an older style of hot dog cooker in which the hot dogs are directly connected across 120 V from the wall socket. He has modified the cooker so it simultaneously applies the 120 V to three combinations: across the ends of a single hot dog, across the ends of two hot dogs in parallel, and across the outer ends of two hot dogs in series. He explains that he has measured the resistance of a hot dog to be 11.0 Ω, and that a hot dog requires 75.0 kJ of energy to cook it. He says he will give extra credit to anyone who, before any hot dog begins smoking, can determine (a) which hot dog(s) will cook first, and (b) the time interval for each hot dog to cook. Quick! Get to work!

12. *Why is the following situation impossible?* A technician is testing a circuit that contains a resistance R. He realizes that a better design for the circuit would include a resistance $\frac{7}{3}R$ rather than R. He has three additional resistors, each with resistance R. By combining these additional resistors in a certain combination that is then placed in series with the original resistor, he achieves the desired resistance.

13. Calculate the power delivered to each resistor in the circuit shown in Figure P27.13.

Figure P27.13

14. For the purpose of measuring the electric resistance of shoes through the body of the wearer standing on a metal ground plate, the American National Standards Institute (ANSI) specifies the circuit shown in Figure P27.14. The potential difference ΔV across the 1.00-MΩ resistor is measured with an ideal voltmeter. (a) Show that the resistance of the footwear is

$$R_{\text{shoes}} = \frac{50.0\ \text{V} - \Delta V}{\Delta V}$$

(b) In a medical test, a current through the human body should not exceed 150 μA. Can the current delivered by the ANSI-specified circuit exceed 150 μA? To decide, consider a person standing barefoot on the ground plate.

Figure P27.14

15. Four resistors are connected to a battery as shown in Figure P27.15. (a) Determine the potential difference across each resistor in terms of $\mathcal{E}$. (b) Determine the current in each resistor in terms of I. (c) **What If?** If R_3 is increased, explain what happens to the current in each of the resistors. (d) In the limit that $R_3 \to \infty$, what are the new values of the current in each resistor in terms of I, the original current in the battery?

Figure P27.15

16. You have a faculty position at a community college and are
CR teaching a class in automotive technology. You are deep in a
discussion of using jumper cables to start a car with a dead
battery from a car with a fresh battery. You have drawn the
circuit diagram in Figure P27.16 to explain the process. The
battery on the left is the live battery in the correctly functioning car, with emf $\mathcal{E}$ and internal resistance R_L, where
the L subscript refers to "live." Its terminals are connected
directly across those of the dead battery, in the middle of
the diagram, with emf $\mathcal{E}$ and internal resistance R_D, where
the D subscript refers to "dead." Then, the starter in the car
with the dead battery is activated by closing the ignition
switch, allowing the car to start. The resistance of the starter
is R_S. A student raises his hand and asks, "So is the dead
battery being charged while the starter is operating?" How
do you respond?

Figure P27.16

SECTION 27.3 Kirchhoff's Rules

17. The circuit shown in Figure P27.17 is connected for
Q|C 2.00 min. (a) Determine the current in each branch of
T the circuit. (b) Find the energy delivered by each
battery. (c) Find the energy delivered to each resistor.
(d) Identify the type of energy storage transformation that
occurs in the operation of the circuit. (e) Find the total
amount of energy transformed into internal energy in the
resistors.

Figure P27.17

18. The following equations describe an electric circuit:

$$-I_1 (220\ \Omega) + 5.80\ \text{V} - I_2 (370\ \Omega) = 0$$

$$+I_2 (370\ \Omega) + I_3 (150\ \Omega) - 3.10\ \text{V} = 0$$

$$I_1 + I_3 - I_2 = 0$$

(a) Draw a diagram of the circuit. (b) Calculate the unknowns
and identify the physical meaning of each unknown.

19. Taking $R = 1.00\ \text{k}\Omega$ and $\mathcal{E} = 250\ \text{V}$ in Figure P27.19, determine the direction and magnitude of the current in the
horizontal wire between a and e.

Figure P27.19

20. In the circuit of Figure P27.20, the current $I_1 = 3.00\ \text{A}$
Q|C and the values of $\mathcal{E}$ for the ideal battery and R are unknown.
What are the currents (a) I_2 and (b) I_3? (c) Can you find the
values of $\mathcal{E}$ and R? If so, find their values. If not, explain.

Figure P27.20

21. (a) Can the circuit shown in Figure P27.21 be reduced to a
Q|C single resistor connected to a battery? Explain. Calculate
the currents (b) I_1, (c) I_2, and (d) I_3.

Figure P27.21

22. For the circuit shown in Figure P27.22, we wish to find the
GP currents I_1, I_2, and I_3. Use Kirchhoff's rules to obtain equa-
Q|C tions for (a) the upper loop, (b) the lower loop, and (c) the
junction on the left side. In each case, suppress units for
clarity and simplify, combining the terms. (d) Solve the
junction equation for I_3. (e) Using the equation found in
part (d), eliminate I_3 from the equation found in part (b).
(f) Solve the equa-
tions found in parts
(a) and (e) simulta-
neously for the two
unknowns I_1 and I_2.
(g) Substitute the
answers found in
part (f) into the junc-
tion equation found
in part (d), solving
for I_3. (h) What is
the significance of
the negative answer
for I_2?

Figure P27.22

SECTION 27.4 *RC* Circuits

23. An uncharged capacitor and a resistor are connected in series to a source of emf. If $\mathcal{E}$ = 9.00 V, C = 20.0 μF, and R = 100 Ω, find (a) the time constant of the circuit, (b) the maximum charge on the capacitor, and (c) the charge on the capacitor at a time equal to one time constant after the battery is connected.

24. Show that the time constant in Equation 27.20 has units of time.

25. In the circuit of Figure P27.25, the switch S has been open for a long time. It is then suddenly closed. Take $\mathcal{E}$ = 10.0 V, R_1 = 50.0 kΩ, R_2 = 100 kΩ, and C = 10.0 μF. Determine the time constant (a) before the switch is closed and (b) after the switch is closed. (c) Let the switch be closed at $t = 0$. Determine the current in the switch as a function of time.

Figure P27.25 Problems 25 and 26.

26. In the circuit of Figure P27.25, the switch S has been open for a long time. It is then suddenly closed. Determine the time constant (a) before the switch is closed and (b) after the switch is closed. (c) Let the switch be closed at $t = 0$. Determine the current in the switch as a function of time.

27. A 10.0-μF capacitor is charged by a 10.0-V battery through a resistance R. The capacitor reaches a potential difference of 4.00 V in a time interval of 3.00 s after charging begins. Find R.

28. Show that the integral $\int_0^{\infty} e^{-2t/RC}\, dt$ in Example 27.11 has the value $\frac{1}{2}RC$.

SECTION 27.5 Household Wiring and Electrical Safety

29. You and your roommates are studying hard for your physics exam. You study late into the night and then fall into your bed for some sleep. You all wake early before the exam and scramble groggily around making breakfast. You can't agree on what to have, so one of you cooks waffles on a 990-watt waffle iron while another toasts bread in a 900-watt toaster. You want to make coffee with a 650-watt coffeemaker, and you plug it into the same power strip into which the waffle iron and toaster are plugged. Will the 20-A circuit breaker remain operational?

30. An electric heater is rated at 1.50×10^3 W, a toaster at 750 W, and an electric grill at 1.00×10^3 W. The three appliances are connected to a common 120-V household circuit. (a) How much current does each draw? (b) If the circuit is protected with a 25.0-A circuit breaker, will the circuit breaker be tripped in this situation? Explain your answer.

31. Turn on your desk lamp. Pick up the cord, with your thumb and index finger spanning the width of the cord. (a) Compute an order-of-magnitude estimate for the current in your hand. Assume the conductor inside the lamp cord next to your thumb is at potential $\sim 10^2$ V at a typical instant and the conductor next to your index finger is at ground potential (0 V). The resistance of your hand depends strongly on the thickness and the moisture content of the outer layers of your skin. Assume the resistance of your hand between fingertip and thumb tip is $\sim 10^4\ \Omega$. You may model the cord as having rubber insulation. State the other quantities you measure or estimate and their values. Explain your reasoning. (b) Suppose your body is isolated from any other charges or currents. In order-of-magnitude terms, estimate the potential difference between your thumb where it contacts the cord and your finger where it touches the cord.

ADDITIONAL PROBLEMS

32. Four resistors are connected in parallel across a 9.20-V battery. They carry currents of 150 mA, 45.0 mA, 14.0 mA, and 4.00 mA. If the resistor with the largest resistance is replaced with one having twice the resistance, (a) what is the ratio of the new current in the battery to the original current? (b) **What If?** If instead the resistor with the smallest resistance is replaced with one having twice the resistance, what is the ratio of the new total current to the original current? (c) On a February night, energy leaves a house by several energy leaks, including 1.50×10^3 W by conduction through the ceiling, 450 W by infiltration (airflow) around the windows, 140 W by conduction through the basement wall above the foundation sill, and 40.0 W by conduction through the plywood door to the attic. To produce the biggest saving in heating bills, which one of these energy transfers should be reduced first? Explain how you decide. Clifford Swartz suggested the idea for this problem.

33. Find the equivalent resistance between points a and b in Figure P27.33.

Figure P27.33

34. The circuit in Figure P27.34a consists of three resistors and one battery with no internal resistance. (a) Find the current in the 5.00-Ω resistor. (b) Find the power delivered to the 5.00-Ω resistor. (c) In each of the circuits in Figures P27.34b, P27.34c, and P27.34d, an additional 15.0-V battery has

Figure P27.34

been inserted into the circuit. Which diagram or diagrams represent a circuit that requires the use of Kirchhoff's rules to find the currents? Explain why. (d) In which of these three new circuits is the smallest amount of power delivered to the 10.0-Ω resistor? (You need not calculate the power in each circuit if you explain your answer.)

35. The circuit in Figure P27.35 has been connected for several seconds. Find the current (a) in the 4.00-V battery, (b) in the 3.00-Ω resistor, (c) in the 8.00-V battery, and (d) in the 3.00-V battery. (e) Find the charge on the capacitor.

Figure P27.35

36. The resistance between terminals a and b in Figure P27.36 is 75.0 Ω. If the resistors labeled R have the same value, determine R.

Figure P27.36

37. (a) Calculate the potential difference between points a and b in Figure P27.37 and (b) identify which point is at the higher potential.

Figure P27.37

38. *Why is the following situation impossible?* A battery has an emf of $\mathcal{E} = 9.20$ V and an internal resistance of $r = 1.20$ Ω. A resistance R is connected across the battery and extracts from it a power of $P = 21.2$ W.

39. When two unknown resistors are connected in series with a battery, the battery delivers 225 W and carries a total current of 5.00 A. For the same total current, 50.0 W is delivered when the resistors are connected in parallel. Determine the value of each resistor.

40. When two unknown resistors are connected in series with [S] a battery, the battery delivers total power P_s and carries a total current of I. For the same total current, a total power

P_p is delivered when the resistors are connected in parallel. Determine the value of each resistor.

41. The circuit in Figure P27.41 contains two resistors, $R_1 = 2.00$ kΩ and $R_2 = 3.00$ kΩ, and two capacitors, $C_1 = 2.00$ μF and $C_2 = 3.00$ μF, connected to a battery with emf $\mathcal{E} = 120$ V. If there are no charges on the capacitors before switch S is closed, determine the charges on capacitors (a) C_1 and (b) C_2 as functions of time, after the switch is closed.

Figure P27.41

42. Two resistors R_1 and R_2 are in parallel with each other. [S] Together they carry total current I. (a) Determine the current in each resistor. (b) Prove that this division of the total current I between the two resistors results in less power delivered to the combination than any other division. It is a general principle that *current in a direct current circuit distributes itself so that the total power delivered to the circuit is a minimum.*

43. A power supply has an open-circuit voltage of 40.0 V and an internal resistance of 2.00 Ω. It is used to charge two storage batteries connected in series, each having an emf of 6.00 V and internal resistance of 0.300 Ω. If the charging current is to be 4.00 A, (a) what additional resistance should be added in series? At what rate does the internal energy increase in (b) the supply, (c) in the batteries, and (d) in the added series resistance? (e) At what rate does the chemical energy increase in the batteries?

44. A battery is used to charge a capacitor through a resistor as [S] shown in Figure P27.44. Show that half the energy supplied by the battery appears as internal energy in the resistor and half is stored in the capacitor.

Figure P27.44

45. An ideal voltmeter connected across a cer-[QC] tain fresh 9-V battery reads 9.30 V, and an ideal ammeter briefly connected across the same battery reads 3.70 A. We say the battery has an open-circuit voltage of 9.30 V and a short-circuit current of 3.70 A. Model the battery as a source of emf $\mathcal{E}$ in series with an internal resistance r as in Figure 27.1a. Determine both (a) $\mathcal{E}$ and (b) r. An experimenter connects two of these identical batteries together as shown in Figure P27.45. Find (c) the open-circuit

Figure P27.45

voltage and (d) the short-circuit current of the pair of connected batteries. (e) The experimenter connects a 12.0-Ω
resistor between the exposed terminals of the connected
batteries. Find the current in the resistor. (f) Find the power
delivered to the resistor. (g) The experimenter connects a
second identical resistor in parallel with the first. Find the
power delivered to each resistor. (h) Because the same pair
of batteries is connected across both resistors as was connected across the single resistor, why is the power in part
(g) not the same as that in part (f)?

46. (a) Determine the equilibrium charge on the capacitor in
Q|C the circuit of Figure P27.46 as a function of R. (b) Evaluate
the charge when $R = 10.0 \ \Omega$. (c) Can the charge on the
capacitor be zero? If so, for what value of R? (d) What is the
maximum possible magnitude of the charge on the capacitor? For what value of R is it achieved? (e) Is it experimentally meaningful to take $R = \infty$? Explain your answer. If so,
what charge magnitude does it imply?

Figure P27.46

47. In Figure P27.47, suppose the switch has been closed for a
time interval sufficiently long for the capacitor to become
fully charged. Find (a) the steady-state current in each resistor and (b) the charge Q_{max} on the capacitor. (c) The switch
is now opened at $t = 0$. Write an equation for the current in
R_2 as a function of time and (d) find the time interval
required for the charge on the capacitor to fall to one-fifth
its initial value.

Figure P27.47

48. Figure P27.48 shows a circuit model for the transmission of an
S electrical signal such as cable TV to a large number of subscribers. Each subscriber connects a load resistance R_L between the
transmission line and the ground. The ground is assumed to
be at zero potential and able to carry any current between any
ground connections with negligible resistance. The resistance
of the transmission line between the connection points of different subscribers is modeled as the constant resistance R_T.
Show that the equivalent resistance across the signal source is

$$R_{eq} = \tfrac{1}{2}[(4R_T R_L + R_T^2)^{1/2} + R_T]$$

Suggestion: Because the number of subscribers is large,
the equivalent resistance would not change noticeably
if the first subscriber canceled the service. Consequently,
the equivalent resistance of the section of the circuit to the
right of the first load resistor is nearly equal to R_{eq}.

Figure P27.48

49. The student engineer of a
campus radio station wishes
to verify the effectiveness
of the lightning rod on the
antenna mast (Fig. P27.49).
The unknown resistance R_x
is between points C and E.
Point E is a true ground, but
it is inaccessible for direct
measurement because this

Figure P27.49

stratum is several meters below the Earth's surface. Two
identical rods are driven into the ground at A and B, introducing an unknown resistance R_y. The procedure is as follows. Measure resistance R_1 between points A and B, then
connect A and B with a heavy conducting wire and measure
resistance R_2 between points A and C. (a) Derive an equation for R_x in terms of the observable resistances, R_1 and R_2.
(b) A satisfactory ground resistance would be $R_x < 2.00 \ \Omega$. Is
the grounding of the station adequate if measurements give
$R_1 = 13.0 \ \Omega$ and $R_2 = 6.00 \ \Omega$? Explain.

50. A voltage ΔV is applied to a series configuration of n resistors,
S each of resistance R. The circuit components are reconnected
in a parallel configuration, and voltage ΔV is again applied.
Show that the power delivered to the series configuration is
$1/n^2$ times the power delivered to the parallel configuration.

CHALLENGE PROBLEM

51. The switch in Figure P27.51a closes when $\Delta V_c > \tfrac{2}{3}\Delta V$ and
S opens when $\Delta V_c < \tfrac{1}{3}\Delta V$. The ideal voltmeter reads a potential difference as plotted in Figure P27.51b. What is the
period T of the waveform in terms of R_1, R_2, and C?

Figure P27.51

28

Magnetic Fields

STORYLINE **Your family is taking a trip to British Columbia by air.** After landing in Vancouver, you take a small private plane to a local airport. You are amazed that you can look right into the cockpit and see what the pilot sees out the front window. Your view is especially exciting as you are landing. You notice the runway ahead of you has the characters "35R" painted on it. You wonder how that runway number is determined. After landing at the local airport, your family's plan is to go into the wilderness and try to find their way around using a smartphone compass. You pull out your smartphone, open the compass app, and point in a direction, saying, "That's North!" Another member of your family tells you that that's not true North, because the west coast of Canada has a relatively large magnetic declination. Not having heard of magnetic declination, you quietly put your phone away as the other family member uses some math to determine the actual direction of true North. You vow to spend some time in your hotel room tonight looking for *magnetic declination* online.

CONNECTIONS At the beginning of Chapter 22, we investigated some interesting phenomena related to *electricity*: a balloon rubbed on your hair attracts bits of paper, rubbing your shoes on a wool rug and touching a friend creates a spark. Many of us have experienced effects of *magnetism* in our lives, also. As children, we might have played with magnets or explored with a compass. In this chapter, we explore a new type of field, the *magnetic field*. This investigation will connect to the last few chapters in that we find the magnetic field exerts forces on electrically charged particles. These forces occur only if the charged particles are moving, however. This fact will form the basis of all that we study in this chapter. In the subsequent chapter, we will investigate another strong connection between electricity and magnetism by showing that the *source* of a magnetic field is moving electric charges. As we move forward, we shall find that these connections between electricity and magnetism lead to the existence of electromagnetic waves, which we study in Chapter 33. The existence of these waves, in turn, leads to the entire topic of optics, about which we learn in Chapters 34 to 37.

28.1 Analysis Model: Particle in a Field (Magnetic)

In our study of electricity in Chapter 22, we first discussed electric forces between charged particles. We then made great progress in our understanding by introducing the notion of an electric field. We studied in detail the effects on a charged particle placed in an electric field. We will follow a similar process here, with some differences. Because magnetic effects, such as that of a bar magnet picking up a paper clip, occur at a distance without need for physical contact, we will simply assume the existence of a *magnetic field* for now. We will explore the behavior of charged particles residing in that field. Because the source of a magnetic field is more complicated than that of an electric field, we will postpone a detailed discussion of the source of a magnetic field until Chapter 29, and devote that entire chapter to that discussion. Let us just say for now that the region of space surrounding any *moving* electric charge contains a **magnetic field.** A magnetic field also surrounds a magnetic substance making up a permanent magnet.

The source of any magnetic field possesses two poles, a north pole and a south pole. The poles received their names because of the way a magnet, such as that in a compass, behaves in the presence of the Earth's magnetic field. If a bar magnet is suspended from its midpoint and can swing freely in a horizontal plane, it will rotate until its magnetic north pole points toward the Earth's geographic North Pole and its magnetic south pole points toward the Earth's geographic South Pole. The poles of a magnet have some similarities to electric charges: experiments show that magnetic poles exert attractive or repulsive forces on each other and that these forces vary as the inverse square of the distance between the interacting poles. There are major differences between electric charges and magnetic poles, however. For example, electric charges can be isolated (witness the electron and proton), whereas a single magnetic pole has never been isolated; magnetic poles are always found in pairs. All attempts thus far to detect an isolated magnetic pole have been unsuccessful. No matter how many times a permanent magnet is cut in two, each piece always has a north and a south pole.

Historically, the symbol $\vec{\mathbf{B}}$ has been used to represent a magnetic field, and we use this notation in this book. The direction of the magnetic field $\vec{\mathbf{B}}$ at any location is the direction in which the north pole of a compass needle points at that location. As with the electric field, we can represent the magnetic field by means of drawings with *magnetic field lines*.

Figure 28.1 shows how the magnetic field lines of a bar magnet can be traced with the aid of a compass. Notice that the magnetic field lines outside the magnet point away from the north pole and toward the south pole. One can display magnetic field patterns of a bar magnet using small iron filings as shown in Figure 28.2 (page 744).

The configuration of the Earth's magnetic field, pictured in Figure 28.3 (page 744), is very much like the one that would be achieved by burying a gigantic bar magnet deep in the Earth's interior. The reason that the north pole of a magnetic is attracted toward the north geographic pole of the Earth is that the south pole of the model bar magnet is presently located near the north geographic pole. If a compass needle is supported by bearings that allow it to rotate in the vertical plane as well as in the horizontal plane, the needle is horizontal with respect to

Hans Christian Oersted
Danish Physicist and Chemist
(1777–1851)
Oersted is best known for observing that a compass needle deflects when placed near a wire carrying a current. This important discovery was the first evidence of the connection between electric and magnetic phenomena. Oersted was also the first to prepare pure aluminum.

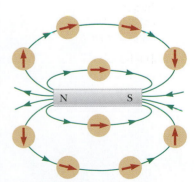

Figure 28.1 Compass needles can be used to trace the magnetic field lines in the region outside a bar magnet.

Magnetic field pattern surrounding a bar magnet

Magnetic field pattern between *opposite* poles (N–S) of two bar magnets

Magnetic field pattern between *like* poles (N–N) of two bar magnets

a

b

c

Courtesy of Henry Leap and Jim Lehman

Figure 28.2 Magnetic field patterns can be displayed with iron filings sprinkled on paper near magnets.

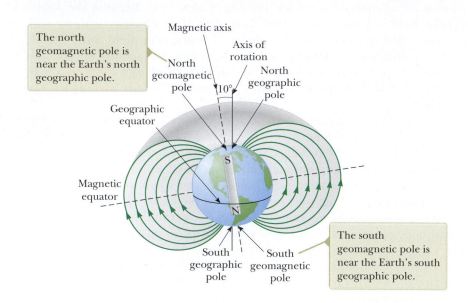

The north geomagnetic pole is near the Earth's north geographic pole.

Magnetic axis

Axis of rotation

North geomagnetic pole

North geographic pole

10°

Geographic equator

Magnetic equator

S

N

The south geomagnetic pole is near the Earth's south geographic pole.

South geographic pole

South geomagnetic pole

Figure 28.3 The Earth's magnetic field lines can be modeled as originating from a bar magnet. Notice that the *north* geomagnetic pole is associated with the *south* pole of the model magnet.

the Earth's surface only near the equator. As the compass is moved northward, the needle rotates so that it points more and more toward the Earth's surface. Finally, at some point, the north pole of the needle points directly downward.

The bar magnet model in Figure 28.3 is simplified. Because the magnetic field of the Earth is not exactly the same as that of a simple bar magnet, and because there are magnetic anomalies in the Earth's crust, there are actually two types of magnetic poles. The point on the surface at which a compass needle points straight down is the North **magnetic pole.** The position of the magnetic pole has moved over hundreds of miles since the year 1900, from latitude 70° N to its present latitude of 86° N. The North **geomagnetic pole** is the point on the Earth's surface at which the magnetic axis of the model bar magnet intersects the surface. By contrast, this point has moved only over a relatively small distance at about latitude 80° N since 1900. While the north magnetic pole has been moving northward, closer to the north geographic pole, the south magnetic pole has also been moving northward, *away* from the south geographic pole. It is currently near latitude 65° S, while the south geomagnetic pole is near 80° S.

It is this difference between the geographic and magnetic north poles that causes the difficulty in determining North with a compass mentioned in the opening storyline. Your compass will point toward the magnetic north pole, but in a region like British Columbia, that direction is quite different from the direction along a longitude line toward the geographic North Pole. For example, in Vancouver, your compass needle will point about 17° to the east of true north. The number in the runway marker in the chapter opening photograph refers to the direction of the runway relative to magnetic north, divided by 10. Therefore, runway 35 is oriented 350°, measured clockwise, from magnetic north. The letter R tells us that there are at least two parallel runways in this direction, and this one is on the *right*. The other runway is marked 35L, for *left*, and there may even be a 35C for *center*. At Vancouver International Airport, there are two runways at an orientation of 100° relative to true north. Because of magnetic declination, however, they are not marked as runway 10, but rather as runways 8L and 8R, because they are oriented 83° from magnetic north.

The direction of the Earth's magnetic field has reversed several times during the last million years. Evidence for this reversal is provided by basalt, a type of rock that contains iron. Basalt forms from material spewed forth by volcanic activity on the ocean floor. As the lava cools, it solidifies and retains a picture of the Earth's magnetic field direction. The rocks are dated by other means to provide a time line for these periodic reversals of the magnetic field.

We can quantify the magnetic field $\vec{\mathbf{B}}$ by using our model of a particle in a field, like the model discussed for gravity in Chapter 13 and for electricity in Chapter 22. The existence of a magnetic field at some point in space can be determined by measuring the **magnetic force** $\vec{\mathbf{F}}_B$ exerted on an appropriate test particle placed at that point. This process is the same one we followed in defining the electric field in Chapter 22. If we perform such an experiment by placing a particle with charge q in the magnetic field, we find the following results that are similar to those for experiments on electric forces, where $\vec{\mathbf{F}}_e = q\vec{\mathbf{E}}$ (Eq. 22.8):

- The magnetic force is proportional to the charge q of the particle.
- The magnetic force on a negative charge is directed opposite to the force on a positive charge moving in the same direction.
- The magnetic force is proportional to the magnitude of the magnetic field vector $\vec{\mathbf{B}}$.

We also find the following results, which are *totally different* from those for experiments on electric forces:

- The magnetic force is proportional to the speed v of the particle.
- If the velocity vector makes an angle θ with the magnetic field, the magnitude of the magnetic force is proportional to $\sin\theta$.
- When a charged particle moves *parallel* to the magnetic field vector, the magnetic force on the charge is zero.
- When a charged particle moves in a direction *not* parallel to the magnetic field vector, the magnetic force acts in a direction perpendicular to both $\vec{\mathbf{v}}$ and $\vec{\mathbf{B}}$; that is, the magnetic force is perpendicular to the plane formed by $\vec{\mathbf{v}}$ and $\vec{\mathbf{B}}$.

These results show that the magnetic force on a particle is more complicated than the electric force. The magnetic force is distinctive because it depends on the velocity of the particle and because its direction is perpendicular to both $\vec{\mathbf{v}}$ and $\vec{\mathbf{B}}$. Figure 28.4 (page 746) shows the details of the direction of the magnetic force on a charged particle. Despite this complicated behavior, these observations can be summarized in a compact way by writing the magnetic force in the form

$$\vec{\mathbf{F}}_B = q\vec{\mathbf{v}} \times \vec{\mathbf{B}} \tag{28.1}$$

◀ Vector expression for the magnetic force on a charged particle moving in a magnetic field

which by definition of the cross product (see Section 11.1) is perpendicular to both $\vec{\mathbf{v}}$ and $\vec{\mathbf{B}}$. We can regard this equation as an operational definition of the

Figure 28.4 (a) The direction of the magnetic force $\vec{\mathbf{F}}_B$ acting on a charged particle moving with a velocity $\vec{\mathbf{v}}$ in the presence of a magnetic field $\vec{\mathbf{B}}$. (b) Magnetic forces on positive and negative charges. A uniform magnetic field is represented by uniformly spaced magnetic field lines. At any point, the magnetic field vector $\vec{\mathbf{B}}$ is parallel to the field line. The dashed lines show the paths of the particles, which are investigated in Section 28.2.

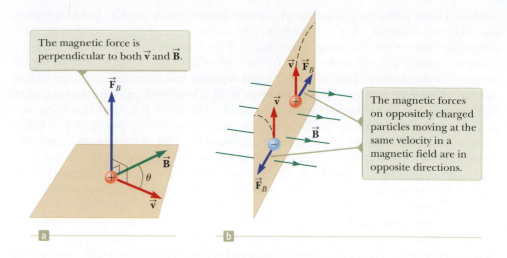

The magnetic force is perpendicular to both $\vec{\mathbf{v}}$ and $\vec{\mathbf{B}}$.

The magnetic forces on oppositely charged particles moving at the same velocity in a magnetic field are in opposite directions.

magnetic field at some point in space. That is, the magnetic field is defined in terms of the force acting on a moving charged particle. Equation 28.1 is the mathematical representation of the magnetic version of the **particle in a field** analysis model and is the magnetic analog to Equation 22.8.

Figure 28.5 reviews two right-hand rules for determining the direction of the cross product $\vec{\mathbf{v}} \times \vec{\mathbf{B}}$ and determining the direction of $\vec{\mathbf{F}}_B$. The rule in Figure 28.5a depends on our right-hand rule for the cross product in Figure 11.2. Point the four fingers of your right hand along the direction of $\vec{\mathbf{v}}$ with the palm facing $\vec{\mathbf{B}}$ and curl them toward $\vec{\mathbf{B}}$. Your extended thumb, which is at a right angle to your fingers, points in the direction of $\vec{\mathbf{v}} \times \vec{\mathbf{B}}$. Because $\vec{\mathbf{F}}_B = q\vec{\mathbf{v}} \times \vec{\mathbf{B}}$, $\vec{\mathbf{F}}_B$ is in the direction of your thumb if q is positive and is opposite the direction of your thumb if q is negative. (If you need more help understanding the cross product, you should review Section 11.1, including Fig. 11.2.)

An alternative rule is shown in Figure 28.5b. Here the thumb points in the direction of $\vec{\mathbf{v}}$ and the extended fingers in the direction of $\vec{\mathbf{B}}$. Now, the force $\vec{\mathbf{F}}_B$ on a positive charge extends outward from the palm. The advantage of this rule is that the force on the charge is in the direction you would push on something with your hand: outward from your palm. The force on a negative charge is in the opposite direction. You can use either of these two right-hand rules.

Figure 28.5 Two right-hand rules for determining the direction of the magnetic force $\vec{\mathbf{F}}_B = q\vec{\mathbf{v}} \times \vec{\mathbf{B}}$ acting on a particle with charge q moving with a velocity $\vec{\mathbf{v}}$ in a magnetic field $\vec{\mathbf{B}}$. (a) In this rule, the magnetic force is in the direction in which your thumb points. (b) In this rule, the magnetic force is in the direction of your palm, as if you are pushing the particle with your hand.

(2) Your upright thumb shows the direction of the magnetic force on a positive particle.

(1) Point your fingers in the direction of $\vec{\mathbf{v}}$ and then curl them toward the direction of $\vec{\mathbf{B}}$.

(1) Point your fingers in the direction of $\vec{\mathbf{B}}$, with $\vec{\mathbf{v}}$ coming out of your thumb.

(2) The magnetic force on a positive particle is in the direction you would push with your palm.

Based on Equation 11.3, the magnitude of the magnetic force on a charged particle is

$$F_B = |q|vB\sin\theta \tag{28.2}$$

◀ Magnitude of the magnetic force on a charged particle moving in a magnetic field

where θ is the smaller angle between $\vec{v}$ and $\vec{B}$. From this expression, we see that F_B is zero when $\vec{v}$ is parallel or antiparallel to $\vec{B}$ ($\theta = 0$ or $180°$) and maximum when $\vec{v}$ is perpendicular to $\vec{B}$ ($\theta = 90°$).

Ⓠ UICK QUIZ 28.1 An electron moves in the plane of this paper toward the top of the page. A magnetic field is also in the plane of the page and directed toward the right. What is the direction of the magnetic force on the electron? **(a)** toward the top of the page **(b)** toward the bottom of the page **(c)** toward the left edge of the page **(d)** toward the right edge of the page **(e)** upward out of the page **(f)** downward into the page

Let's compare the important differences between the electric and magnetic versions of the particle in a field model:

- The electric force vector is along the direction of the electric field, whereas the magnetic force vector is perpendicular to the magnetic field.
- The electric force acts on a charged particle regardless of whether the particle is moving, whereas the magnetic force acts on a charged particle only when the particle is in motion.
- The electric force does work in displacing a charged particle, whereas the magnetic force associated with a steady magnetic field does no work when a particle is displaced because the force is perpendicular to the displacement of its point of application.

From the last statement and on the basis of the work–kinetic energy theorem, we conclude that the kinetic energy of a charged particle moving through a magnetic field cannot be altered by the magnetic field alone. The field can alter the direction of the velocity vector, but it cannot change the speed or kinetic energy of the particle.

From Equation 28.2, we see that the SI unit of magnetic field is the newton per coulomb-meter per second, which is called the **tesla** (T):

$$1\text{ T} = 1\,\frac{\text{N}}{\text{C}\cdot\text{m/s}}$$

◀ The tesla

Because a coulomb per second is defined to be an ampere,

$$1\text{ T} = 1\,\frac{\text{N}}{\text{A}\cdot\text{m}}$$

A non-SI magnetic-field unit in common use, called the *gauss* (G), is related to the tesla through the conversion $1\text{ T} = 10^4$ G. Table 28.1 shows some typical values of magnetic fields.

TABLE 28.1 Some Approximate Magnetic Field Magnitudes

Source of Field	Field Magnitude (T)
Strong superconducting laboratory magnet	30
Strong conventional laboratory magnet	2
Medical MRI unit	1.5
Bar magnet	10^{-2}
Surface of the Sun	10^{-2}
Surface of the Earth	5×10^{-5}
Inside human brain (due to nerve impulses)	10^{-13}

Particle in a Field (Magnetic)

Imagine some source (which we will investigate later) establishes a **magnetic field** $\vec{B}$ throughout space. Now imagine a particle with charge q is placed in that field. The particle interacts with the magnetic field so that the particle experiences a magnetic force given by

$$\vec{F}_B = q\vec{v} \times \vec{B} \qquad \textbf{(28.1)}$$

Examples:

- an ion moves in a circular path in the magnetic field of a mass spectrometer (Section 28.3)
- a current exists in a conducting bar when it is moved in a magnetic field (Chapter 30)
- a coil in a motor rotates in response to the magnetic field in the motor (Chapter 30)
- in a bubble chamber, particles created in collisions follow curved paths in a magnetic field, allowing the particles to be identified (Chapter 44)

Example **28.1**   **An Electron Moving in a Magnetic Field**

An electron moves through space as a cosmic ray (see page 752) with a speed of 8.0×10^6 m/s along the x axis (Fig. 28.6). At its location, the magnetic field of the Earth has a magnitude of 0.050 mT, and is directed at an angle of 60° to the x axis, lying in the xy plane. Calculate the magnetic force on the electron.

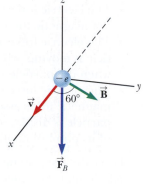

Figure 28.6 (Example 28.1) The magnetic force $\vec{F}_B$ acting on the electron is in the negative z direction when $\vec{v}$ and $\vec{B}$ lie in the xy plane.

SOLUTION

Conceptualize   Recall that the magnetic force on a charged particle is perpendicular to the plane formed by the velocity and magnetic field vectors. Use one of the right-hand rules in Figure 28.5 to convince yourself that the direction of the force on the electron is downward, in the $-z$ direction in Figure 28.6.

Categorize   We evaluate the magnetic force using the *magnetic* version of the *particle in a field* model.

Analyze   Use Equation 28.2 to find the magnitude of the magnetic force:

$$F_B = |q|vB \sin\theta$$

$$= (1.6 \times 10^{-19}\text{ C})(8.0 \times 10^6\text{ m/s})(5.0 \times 10^{-5}\text{ T})(\sin 60°)$$

$$= 5.5 \times 10^{-17}\text{ N}$$

Finalize   For practice using the vector product, evaluate this force in vector notation using Equation 28.1. The magnitude of the magnetic force may seem small to you, but remember that it is acting on a very small particle, the electron. To convince yourself that this is a substantial force for an electron, calculate the initial acceleration of the electron due to this force.

28.2 Motion of a Charged Particle in a Uniform Magnetic Field

In Figure 28.4b, we show two charged particles in a uniform magnetic field at an instant of time. The dashed lines suggest the subsequent motion of the two particles in response to the magnetic force on them. In this section, we investigate more details about this motion and the paths followed by the particles.

Before we continue our discussion, some explanation of the notation used in this book is in order. To indicate the direction of $\vec{B}$ in illustrations, we sometimes present perspective views such as those in Figure 28.6. If $\vec{B}$ lies in the plane of the page or is present in a perspective drawing, we use green vectors or green field lines with arrowheads. In nonperspective illustrations, we depict a magnetic field perpendicular to and directed out of the page with a series of green dots, which represent the tips of arrows coming toward you (see Fig. 28.7a). In this case, the field is labeled $\vec{\mathbf{B}}_{out}$. If $\vec{\mathbf{B}}$ is directed perpendicularly into the page, we use green crosses, which represent the feathered tails of archery-type arrows fired away from you, as in Figure 28.7b. In this case, the field is labeled $\vec{\mathbf{B}}_{in}$, where the subscript "in" indicates "into the page." The same notation with crosses and dots is also used for other quantities that might be perpendicular to the page such as forces, velocities, and current directions.

Now consider the special case of a positively charged particle moving in a uniform magnetic field with the initial velocity vector of the particle perpendicular to the field. Let's assume the direction of the magnetic field is into the page as in Figure 28.8. The particle in a field model tells us that the magnetic force on the particle is perpendicular to both the magnetic field lines and the velocity of the particle. The fact that there is a force on the particle tells us to apply the particle under a net force model to the particle. As the particle changes the direction of its velocity in response to the magnetic force, the magnetic force remains perpendicular to the velocity. As we found in Section 6.1, if the force is always perpendicular to the velocity, the path of the particle is a circle! Figure 28.8 shows the particle moving in a circle in a plane perpendicular to the magnetic field. Although magnetism and magnetic forces may be new and unfamiliar to you now, we see a magnetic effect that results in something with which we are familiar: the particle in uniform circular motion model!

The particle moves in a circle because the magnetic force $\vec{\mathbf{F}}_B$ is perpendicular to $\vec{\mathbf{v}}$ and $\vec{\mathbf{B}}$ and has a constant magnitude qvB. As Figure 28.8 illustrates, the rotation is counterclockwise for a positive charge in a magnetic field directed into the page. If q were negative, the rotation would be clockwise. We use the particle under a net force model to write Newton's second law for the particle:

$$\sum F = F_B = ma$$

Because the particle moves in a circle, we also model it as a particle in uniform circular motion and we replace the acceleration with centripetal acceleration:

$$F_B = qvB = \frac{mv^2}{r}$$

This expression leads to the following equation for the radius of the circular path:

$$r = \frac{mv}{qB} \qquad (28.3)$$

That is, the radius of the path is proportional to the linear momentum mv of the particle and inversely proportional to the magnitude of the charge on the particle and to the magnitude of the magnetic field. The angular speed of the particle (from Eq. 10.10) is

$$\omega = \frac{v}{r} = \frac{qB}{m} \qquad (28.4)$$

The period of the motion (the time interval the particle requires to complete one revolution) is equal to the circumference of the circle divided by the speed of the particle:

$$T = \frac{2\pi r}{v} = \frac{2\pi}{\omega} = \frac{2\pi m}{qB} \qquad (28.5)$$

These results show that the angular speed of the particle and the period of the circular motion do not depend on the speed of the particle or on the radius of the

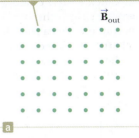

Magnetic field lines coming out of the paper are indicated by dots, representing the tips of arrows coming outward.

$\vec{\mathbf{B}}_{out}$

a

Magnetic field lines going into the paper are indicated by crosses, representing the feathers of arrows going inward.

$\vec{\mathbf{B}}_{in}$

b

Figure 28.7 Representations of magnetic field lines perpendicular to the page.

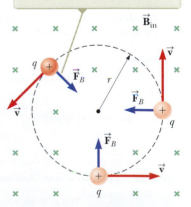

The magnetic force $\vec{\mathbf{F}}_B$ acting on the charge is always directed toward the center of the circle.

Figure 28.8 When the velocity of a charged particle is perpendicular to a uniform magnetic field, the particle moves in a circular path in a plane perpendicular to $\vec{\mathbf{B}}$.

orbit. The angular speed ω is often referred to as the **cyclotron frequency** because charged particles circulate at this angular frequency in the type of accelerator called a *cyclotron*, which is discussed in Section 28.3.

QUICK QUIZ **28.2** A charged particle is moving perpendicular to a magnetic field in a circle with a radius *r*. **(i)** An identical particle enters the field, with $\vec{\mathbf{v}}$ perpendicular to $\vec{\mathbf{B}}$, but with a higher speed than the first particle. Compared with the radius of the circle for the first particle, is the radius of the circular path for the second particle (a) smaller, (b) larger, or (c) equal in size? **(ii)** The magnitude of the magnetic field is increased. From the same choices, compare the radius of the new circular path of the first particle with the radius of its initial path.

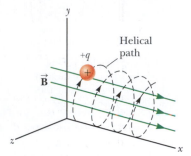

Figure 28.9 A charged particle having a velocity vector that has a component parallel to a uniform magnetic field moves in a helical path.

In Figure 28.8, the velocity vector $\vec{\mathbf{v}}$ is perpendicular to the magnetic field $\vec{\mathbf{B}}$. If a charged particle moves in a uniform magnetic field with its velocity at some *arbitrary* angle with respect to $\vec{\mathbf{B}}$, its path is a helix. For example, if the field is directed in the *x* direction as shown in Figure 28.9, there is no component of magnetic force in the *x* direction. As a result, $a_x = 0$, and the *x* component of the velocity of the particle remains constant; the charged particle is a particle in equilibrium in this direction. The magnetic force $q\vec{\mathbf{v}} \times \vec{\mathbf{B}}$ causes the components v_y and v_z to change in time, however, and the resulting motion is a helix whose axis is parallel to the magnetic field. The projection of the path onto the *yz* plane (viewed along the *x* axis) is a circle. (The projections of the path onto the *xy* and *xz* planes are sinusoids!) Equations 28.3 to 28.5 still apply provided *v* is replaced by $v_\perp = \sqrt{v_y^2 + v_z^2}$.

Example 28.2   **A Proton Moving Perpendicular to a Uniform Magnetic Field**

A proton is moving in a circular orbit of radius 14 cm in a uniform 0.35-T magnetic field perpendicular to the velocity of the proton. Find the speed of the proton.

SOLUTION

Conceptualize From our discussion in this section, we know the proton follows a circular path when moving perpendicular to a uniform magnetic field. In Chapter 38, we will learn that the highest possible speed for a particle is the speed of light, 3.00×10^8 m/s, so the speed of the particle in this problem must come out to be smaller than that value.

Categorize The proton is described by both the *particle in a field* model and the *particle in uniform circular motion* model. These models led to Equation 28.3.

Analyze

Solve Equation 28.3 for the speed of the particle:

$$v = \frac{qBr}{m_p}$$

Substitute numerical values:

$$v = \frac{(1.60 \times 10^{-19}\ \text{C})(0.35\ \text{T})(0.14\ \text{m})}{1.67 \times 10^{-27}\ \text{kg}}$$

$$= 4.7 \times 10^6\ \text{m/s}$$

Finalize The speed is indeed smaller than the speed of light, as required.

WHAT IF? What if an electron, rather than a proton, moves in a direction perpendicular to the same magnetic field with this same speed? Will the radius of its orbit be different?

Answer An electron has a much smaller mass than a proton, so the magnetic force should be able to change its velocity much more easily than that for the proton. Therefore, we expect the radius to be smaller. Equation 28.3 shows that *r* is proportional to *m* with *q*, *B*, and *v* the same for the electron as for the proton. Consequently, the radius will be smaller by the same factor as the ratio of masses m_e/m_p.

Example 28.3 Bending an Electron Beam

In an experiment designed to measure the magnitude of a uniform magnetic field, electrons are accelerated from rest through a potential difference of 350 V and then enter a uniform magnetic field that is perpendicular to the velocity vector of the electrons. The electrons travel along a curved path because of the magnetic force exerted on them, and the radius of the path is measured to be 7.5 cm. (Such a curved beam of electrons is shown in Fig. 28.10.)

(A) What is the magnitude of the magnetic field?

SOLUTION

Conceptualize This example involves electrons accelerating from rest due to an electric force and then moving in a circular path due to a magnetic force. With the help of Figures 28.8 and 28.10, visualize the circular motion of the electrons.

Figure 28.10 (Example 28.3) The bending of an electron beam in a magnetic field.

Categorize Equation 28.3 shows that we need the speed v of the electron to find the magnetic field magnitude, and v is not given. Consequently, we must find the speed of the electron based on the potential difference through which it is accelerated. To do so, we categorize the first part of the problem by modeling an electron and the electric field as an *isolated system* in terms of *energy*. Once the electron enters the magnetic field, we categorize the second part of the problem as one involving a *particle in a field* and a *particle in uniform circular motion*, as we have done in this section.

Analyze Write the appropriate reduction of the conservation of energy equation, Equation 8.2, for the electron–electric field system:

$$\Delta K + \Delta U_E = 0$$

Substitute the appropriate initial and final energies for the time interval during which the electron accelerates from rest:

$$(\tfrac{1}{2}m_e v^2 - 0) + (q\,\Delta V) = 0$$

Solve for the final speed of the electron:

$$v = \sqrt{\frac{-2q\,\Delta V}{m_e}}$$

Substitute numerical values:

$$v = \sqrt{\frac{-2(-1.60 \times 10^{-19}\ \text{C})(350\ \text{V})}{9.11 \times 10^{-31}\ \text{kg}}} = 1.11 \times 10^7\ \text{m/s}$$

Now imagine the electron entering the magnetic field with this speed. Solve Equation 28.3 for the magnitude of the magnetic field:

$$B = \frac{m_e v}{er}$$

Substitute numerical values:

$$B = \frac{(9.11 \times 10^{-31}\ \text{kg})(1.11 \times 10^7\ \text{m/s})}{(1.60 \times 10^{-19}\ \text{C})(0.075\ \text{m})} = 8.4 \times 10^{-4}\ \text{T}$$

(B) What is the angular speed of the electrons?

SOLUTION

Use Equation 10.10:

$$\omega = \frac{v}{r} = \frac{1.11 \times 10^7\ \text{m/s}}{0.075\ \text{m}} = 1.5 \times 10^8\ \text{rad/s}$$

Finalize The angular speed can be represented as $\omega = (1.5 \times 10^8\ \text{rad/s})(1\ \text{rev}/2\pi\ \text{rad}) = 2.4 \times 10^7\ \text{rev/s}$. The electrons travel around the circle 24 million times per second! This answer is consistent with the very high speed found in part (A).

WHAT IF? What if a sudden voltage surge causes the accelerating voltage to increase to 400 V? How does that affect the angular speed of the electrons, assuming the magnetic field remains constant?

Answer The increase in accelerating voltage ΔV causes the electrons to enter the magnetic field with a higher speed v. This higher speed causes them to travel in a circle with a larger radius r. The angular speed is the ratio of v to r. Both v and r increase by the same factor, so the effects cancel and the angular speed remains the same. Equation 28.4 is an expression for the cyclotron frequency, which is the same as the angular speed of the electrons. The cyclotron frequency depends only on the charge q, the magnetic field B, and the mass m_e, none of which have changed. Therefore, the voltage surge has no effect on the angular speed. (In reality, however, the voltage surge may also increase the magnetic field if the magnetic field is powered by the same source as the accelerating voltage. In that case, the angular speed increases according to Eq. 28.4.)

The magnetic force exerted on the particle near either end of the bottle has a component that causes the particle to spiral back toward the center.

Path of particle

Figure 28.11 A charged particle moving in a nonuniform magnetic field (a magnetic bottle) spirals about the field and oscillates between the endpoints.

Figure 28.12 The Van Allen belts are made up of charged particles trapped by the Earth's nonuniform magnetic field. The magnetic field lines are in green, and the particle paths are dashed black lines.

When charged particles move in a nonuniform magnetic field, the motion is complex. For example, in a magnetic field that is strong at the ends and weak in the middle such as that shown in Figure 28.11, the particles can oscillate between two positions. A charged particle starting at one end spirals along the field lines until it reaches the other end, where it reverses its path and spirals back. This configuration is known as a *magnetic bottle* because charged particles can be trapped within it.

The Van Allen radiation belts consist of charged particles (mostly electrons and protons) surrounding the Earth in doughnut-shaped regions (Fig. 28.12). The particles, trapped by the Earth's nonuniform magnetic field, spiral around the field lines from pole to pole, covering the distance in only a few seconds. These particles originate mainly from the Sun, but some come from stars and other heavenly objects. For this reason, the particles are called *cosmic rays*. Most cosmic rays are deflected by the Earth's magnetic field and never reach the atmosphere. Some of the particles become trapped, however, and it is these particles that make up the Van Allen belts. When the particles are located over the poles, they sometimes collide with atoms in the atmosphere, causing the atoms to emit visible light. Such collisions are the origin of the beautiful aurora borealis, or northern lights, in the northern hemisphere and the aurora australis in the southern hemisphere. Auroras are usually confined to the polar regions because the Van Allen belts are nearest the Earth's surface there. Occasionally, though, solar activity causes larger numbers of charged particles to enter the belts and significantly distort the normal magnetic field lines associated with the Earth. In these situations, an aurora can sometimes be seen at lower latitudes.

28.3 Applications Involving Charged Particles Moving in a Magnetic Field

A charge moving with a velocity $\vec{v}$ in the presence of both an electric field $\vec{E}$ and a magnetic field $\vec{B}$ is described by two particle in a field models. It experiences both an electric force $q\vec{E}$ and a magnetic force $q\vec{v} \times \vec{B}$. The total force (called the Lorentz force) acting on the charge is

$$\vec{F} = q\vec{E} + q\vec{v} \times \vec{B} \qquad (28.6)$$

Let us investigate some devices that take advantage of the Lorentz force.

Figure 28.13 A velocity selector. When a positively charged particle is moving with velocity $\vec{v}$ in the presence of a magnetic field directed into the page and an electric field directed to the right, it experiences an electric force $q\vec{E}$ to the right and a magnetic force $q\vec{v} \times \vec{B}$ to the left.

Velocity Selector

In many experiments involving moving charged particles, it is important that all particles move with essentially the same velocity, which can be achieved by applying a combination of an electric field, created by parallel plates, and a magnetic field oriented as shown in Figure 28.13. A uniform electric field is directed to the right (in the plane of the page in Fig. 28.13), and a uniform magnetic field is applied in the direction perpendicular to the electric field (into the page in Fig. 28.13). If q is positive and the velocity $\vec{v}$ is upward, the magnetic force $q\vec{v} \times \vec{B}$ is to the left and the electric force $q\vec{E}$ is to the right. When the magnitudes of the two fields are chosen so that $qE = qvB$, the forces cancel. The charged particle is modeled as a particle in equilibrium and moves in a straight vertical line through the region of the fields. From the expression $qE = qvB$, we find that

$$v = \frac{E}{B} \qquad (28.7)$$

Only those particles having this speed pass undeflected through the mutually perpendicular electric and magnetic fields. The magnetic force exerted on particles moving at speeds greater than that is stronger than the electric force, and the particles are deflected to the left. Those moving at slower speeds are deflected to the right.

The Mass Spectrometer

A **mass spectrometer** separates ions according to their mass-to-charge ratio. In one version of this device, known as the *Bainbridge mass spectrometer,* a beam of ions first passes through a velocity selector and then enters a second uniform magnetic field $\vec{B}_0$ that has the same direction as the magnetic field in the selector (Fig. 28.14). Upon entering the second magnetic field, the ions are described by the particle in uniform circular motion model. They move in a semicircle of radius r before striking a detector array at P. If the ions are positively charged, the beam deflects to the left as Figure 28.14 shows. If the ions are negatively charged, the beam deflects to the right. From Equation 28.3, we can express the ratio m/q as

$$\frac{m}{q} = \frac{rB_0}{v}$$

Using Equation 28.7 gives

$$\frac{m}{q} = \frac{rB_0 B}{E} \tag{28.8}$$

Therefore, we can determine m/q by measuring the radius of curvature and knowing the field magnitudes B, B_0, and E. In practice, one usually measures the masses of various isotopes of a given ion, with the ions all carrying the same charge q. In this way, the mass ratios can be determined even if q is unknown.

A variation of this technique was used by J. J. Thomson (1856–1940) in 1897 to measure the ratio e/m_e for electrons. Figure 28.15a shows the basic apparatus he used. Electrons are accelerated from the cathode and pass through two slits. They then drift into a region of perpendicular electric and magnetic fields. The magnitudes of the two fields are first adjusted to produce an undeflected beam. When the magnetic field is turned off, the electric field produces a measurable beam deflection that is recorded on the fluorescent screen. From the size of the deflection and the measured values of E and B, the charge-to-mass ratio can be determined. The results of this crucial experiment represent the discovery of the electron as a fundamental particle of nature.

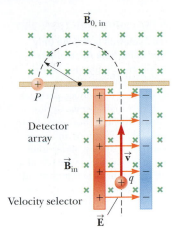

Figure 28.14 A mass spectrometer. Positively charged particles are sent first through a velocity selector and then into a region where the magnetic field $\vec{B}_0$ causes the particles to move in a semicircular path and strike a detector array at P.

Electrons are accelerated from the cathode, pass through two slits, and are deflected by both an electric field (formed by the charged deflection plates) and a magnetic field (directed perpendicular to the electric field). The beam of electrons then strikes a fluorescent screen.

Figure 28.15 (a) Thomson's apparatus for measuring e/m_e. (b) J. J. Thomson (*left*) in the Cavendish Laboratory, University of Cambridge. The man on the right, Frank Baldwin Jewett, is a distant relative of John W. Jewett, Jr., coauthor of this text.

The Cyclotron

A **cyclotron** is a device that can accelerate charged particles to very high speeds. The energetic particles produced are used to bombard atomic nuclei and thereby produce nuclear reactions of interest to researchers. A number of hospitals use cyclotron facilities to produce radioactive substances for diagnosis and treatment.

Both electric and magnetic forces play key roles in the operation of a cyclotron, a schematic drawing of which is shown in Figure 28.16a. The charges move inside two semicircular containers D_1 and D_2, referred to as *dees* because of their shape like the letter D. A high-frequency alternating potential difference is applied to the dees, and a uniform magnetic field is directed perpendicular to them. A positive ion released at P near the center of the magnet in one dee moves in a semicircular path (indicated by the dashed black line in the drawing) and arrives back at the gap in a time interval $T/2$, where T is the time interval needed to make one complete trip around the two dees, given by Equation 28.5. The frequency of the applied potential difference is adjusted so that the polarity of the dees is reversed in the same time interval during which the ion travels around one dee. If the applied potential difference is adjusted such that D_1 is at a lower electric potential than D_2 by an amount ΔV, the ion accelerates across the gap to D_1 and its kinetic energy increases by an amount $q\,\Delta V$. It then moves around D_1 in a semicircular path of greater radius (because its speed has increased). After a time interval $T/2$, it again arrives at the gap between the dees. By this time, the polarity across the dees has again been reversed and the ion is given another "kick" across the gap. The motion continues so that for each half-circle trip around one dee, the ion gains additional kinetic energy equal to $q\,\Delta V$. When the radius of its path is nearly that of the dees, the energetic ion leaves the system through the exit slit. The cyclotron's operation depends on T being independent of the speed of the ion and of the radius of the circular path (Eq. 28.5).

We can obtain an expression for the kinetic energy of the ion when it exits the cyclotron in terms of the radius R of the dees. From Equation 28.3, we know that $v = qBR/m$. Hence, the kinetic energy is

$$K = \tfrac{1}{2}mv^2 = \frac{q^2 B^2 R^2}{2m} \tag{28.9}$$

When the energy of the ions in a cyclotron exceeds about 20 MeV, relativistic effects come into play. (Such effects are discussed in Chapter 38.) Observations

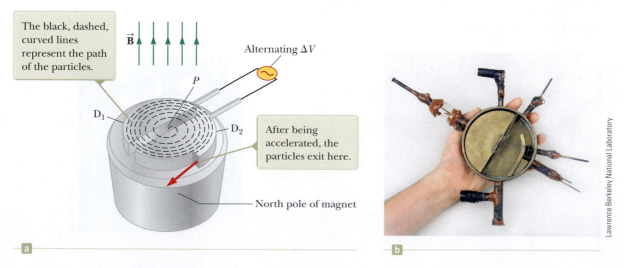

The black, dashed, curved lines represent the path of the particles.

$\vec{B}$

Alternating ΔV

P

D_1

D_2

After being accelerated, the particles exit here.

North pole of magnet

Lawrence Berkeley National Laboratory

a

b

Figure 28.16 (a) A cyclotron consists of an ion source at P, two dees D_1 and D_2 across which an alternating potential difference is applied, and a uniform magnetic field. (The south pole of the magnet is not shown.) (b) The first cyclotron, invented by E. O. Lawrence and M. S. Livingston in 1934.

show that T increases and the moving ions do not remain in phase with the applied potential difference. Some accelerators overcome this problem by modifying the period of the applied potential difference so that it remains in phase with the moving ions.

28.4 Magnetic Force Acting on a Current-Carrying Conductor

If a magnetic force is exerted on a single charged particle when the particle moves through a magnetic field, it should not surprise you that a current-carrying wire also experiences a force when placed in a magnetic field. The current is a collection of many charged particles in motion; hence, the resultant force exerted by the field on the wire is the vector sum of the individual forces exerted on all the charged particles making up the current. The force exerted on the particles is transmitted to the wire when the particles collide with the atoms making up the wire.

One can demonstrate the magnetic force acting on a current-carrying conductor by hanging a wire between the poles of a magnet as shown in Figure 28.17a. For ease in visualization, part of the horseshoe magnet in part (a) is removed to show the end face of the south pole in parts (b) through (d) of Figure 28.17. The magnetic field is directed into the page and covers the region within the shaded squares. When the current in the wire is zero, the wire remains vertical as in Figure 28.17b. When the wire carries a current directed upward as in Figure 28.17c, however, the wire deflects to the left. If the current is reversed as in Figure 28.17d, the wire deflects to the right.

Let's quantify this discussion by considering a straight segment of wire of length L and cross-sectional area A carrying a current I in a uniform magnetic field $\vec{\mathbf{B}}$ as in Figure 28.18. According to the magnetic version of the particle in a field model, the magnetic force exerted on a charge q moving with a drift velocity $\vec{\mathbf{v}}_d$ is $q\vec{\mathbf{v}}_d \times \vec{\mathbf{B}}$. To find the total force acting on the wire, we multiply the force $q\vec{\mathbf{v}}_d \times \vec{\mathbf{B}}$ exerted on one charge by the number of charges in the segment. Because the volume of the segment is AL, the number of charges in the segment is nAL, where n is

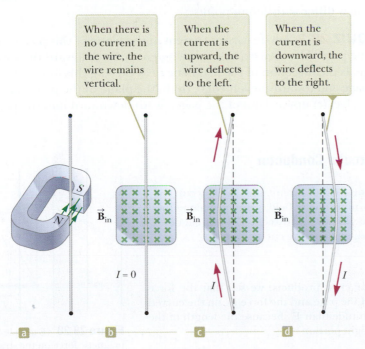

When there is no current in the wire, the wire remains vertical.

When the current is upward, the wire deflects to the left.

When the current is downward, the wire deflects to the right.

$\vec{\mathbf{B}}_{in}$ $\vec{\mathbf{B}}_{in}$ $\vec{\mathbf{B}}_{in}$

$I = 0$ I I

a   b   c   d

Figure 28.17 (a) A wire suspended vertically between the poles of a magnet. (b)–(d) The setup shown in (a) as seen looking at the south pole of the magnet so that the magnetic field (green crosses) is directed into the page.

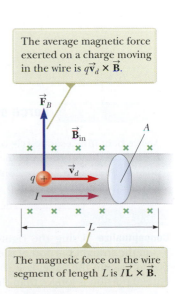

The average magnetic force exerted on a charge moving in the wire is $q\vec{\mathbf{v}}_d \times \vec{\mathbf{B}}$.

$\vec{\mathbf{F}}_B$

$\vec{\mathbf{B}}_{in}$ A

q $\oplus$ $\vec{\mathbf{v}}_d$

I

L

The magnetic force on the wire segment of length L is $I\vec{\mathbf{L}} \times \vec{\mathbf{B}}$.

Figure 28.18 A segment of a current-carrying wire in a magnetic field $\vec{\mathbf{B}}$.

the number of mobile charge carriers per unit volume. Hence, the total magnetic force on the segment of wire of length L is

$$\vec{F}_B = (q\vec{v}_d \times \vec{B})nAL$$

We can write this expression in a more convenient form by noting that, from Equation 26.4, the current in the wire is $I = nqv_dA$. Therefore,

◀ Force on a segment of current-carrying wire in a uniform magnetic field

$$\vec{F}_B = I\vec{L} \times \vec{B} \tag{28.10}$$

where $\vec{L}$ is a vector that points in the direction of the current I and has a magnitude equal to the length L of the segment. This expression applies only to a straight segment of wire in a uniform magnetic field.

Now consider an arbitrarily shaped wire segment of uniform cross section in a magnetic field as shown in Figure 28.19. It follows from Equation 28.10 that the magnetic force exerted on a small segment of vector length $d\vec{s}$ in the presence of a field $\vec{B}$ is

The magnetic force on any segment $d\vec{s}$ is $I\,d\vec{s} \times \vec{B}$ and is directed out of the page.

$$d\vec{F}_B = I\,d\vec{s} \times \vec{B} \tag{28.11}$$

where $d\vec{F}_B$ is directed out of the page for the directions of $\vec{B}$ and $d\vec{s}$ in Figure 28.19. Equation 28.11 can be considered as an alternative definition of $\vec{B}$. That is, we can define the magnetic field $\vec{B}$ in terms of a measurable force exerted on a current element, where the force is a maximum when $\vec{B}$ is perpendicular to the element and zero when $\vec{B}$ is parallel to the element.

To calculate the total force $\vec{F}_B$ acting on the wire shown in Figure 28.19, we integrate Equation 28.11 over the length of the wire:

Figure 28.19 A wire segment of arbitrary shape carrying a current I in a magnetic field $\vec{B}$ experiences a magnetic force.

$$\vec{F}_B = I\int_a^b d\vec{s} \times \vec{B} \tag{28.12}$$

where a and b represent the endpoints of the wire. When this integration is carried out, the magnitude of the magnetic field and the direction the field makes with the vector $d\vec{s}$ may differ at different points.

QUICK QUIZ 28.3 A wire carries current in the plane of this paper toward the top of the page. The wire experiences a magnetic force toward the right edge of the page. Is the direction of the magnetic field causing this force **(a)** in the plane of the page and toward the left edge, **(b)** in the plane of the page and toward the bottom edge, **(c)** upward out of the page, or **(d)** downward into the page?

Example 28.4 **Force on a Semicircular Conductor**

A wire bent into a semicircle of radius R forms a closed circuit and carries a current I. The wire lies in the xy plane, and a uniform magnetic field is directed along the positive y axis as in Figure 28.20. Find the magnitude and direction of the magnetic force acting on the straight portion of the wire and on the curved portion.

SOLUTION

Conceptualize Using the right-hand rule for cross products, we see that the force $\vec{F}_1$ on the straight portion of the wire is out of the page and the force $\vec{F}_2$ on the curved portion is into the page. Is $\vec{F}_2$ larger in magnitude than $\vec{F}_1$ because the length of the curved portion is longer than that of the straight portion?

Categorize Because we are dealing with a current-carrying wire in a magnetic field rather than a single charged particle, we must use Equation 28.12 to find the total force on each portion of the wire.

Figure 28.20 (Example 28.4) The magnetic force on the straight portion of the loop is directed out of the page, and the magnetic force on the curved portion is directed into the page.

28.4 continued

Analyze Notice that $d\vec{\mathbf{s}}$ is perpendicular to $\vec{\mathbf{B}}$ everywhere on the straight portion of the wire. Use Equation 28.12 to find the force on this portion:

$$\vec{\mathbf{F}}_1 = I \int_a^b d\vec{\mathbf{s}} \times \vec{\mathbf{B}} = I \int_{-R}^{R} B \, dx \, \hat{\mathbf{k}} = \boxed{2IRB \, \hat{\mathbf{k}}}$$

To find the magnetic force on the curved part, first write an expression for the magnetic force $d\vec{\mathbf{F}}_2$ on the element $d\vec{\mathbf{s}}$ in Figure 28.20:

(1) $d\vec{\mathbf{F}}_2 = I \, d\vec{\mathbf{s}} \times \vec{\mathbf{B}} = -IB \sin\theta \, ds \, \hat{\mathbf{k}}$

From the geometry in Figure 28.20, write an expression for ds:

(2) $ds = R \, d\theta$

Substitute Equation (2) into Equation (1) and integrate over the angle θ from 0 to π:

$$\vec{\mathbf{F}}_2 = -\int_0^{\pi} IRB \sin\theta \, d\theta \, \hat{\mathbf{k}} = -IRB \int_0^{\pi} \sin\theta \, d\theta \, \hat{\mathbf{k}} = -IRB[-\cos\theta]_0^{\pi} \, \hat{\mathbf{k}}$$

$$= IRB(\cos\pi - \cos 0)\hat{\mathbf{k}} = IRB(-1-1)\hat{\mathbf{k}} = \boxed{-2IRB \, \hat{\mathbf{k}}}$$

Finalize Two very important general statements follow from this example. First, the force on the curved portion is the same in magnitude as the force on a straight wire between the same two points. In general, the magnetic force on a curved current-carrying wire in a uniform magnetic field is equal to that on a straight wire connecting the endpoints and carrying the same current. Furthermore, $\vec{\mathbf{F}}_1 + \vec{\mathbf{F}}_2 = 0$ is also a general result: the net magnetic force acting on any closed current loop in a uniform magnetic field is zero.

28.5 Torque on a Current Loop in a Uniform Magnetic Field

In Example 28.4, we found the net force on the loop in Figure 28.20 to be zero. What if the loop were mounted on pivots at its lowest corners so that it is free to rotate around the x axis? There would be zero net force on the loop, but would it remain at rest when released? Keep in mind that zero net force does not necessarily mean zero net *torque*!

Consider a rectangular loop carrying a current I in the presence of a uniform magnetic field directed parallel to the plane of the loop as shown in Figure 28.21a.

No magnetic forces act on sides ① and ③ because these sides are parallel to $\vec{\mathbf{B}}$.

Sides ② and ④ are perpendicular to the magnetic field and experience forces.

The magnetic forces $\vec{\mathbf{F}}_2$ and $\vec{\mathbf{F}}_4$ exerted on sides ② and ④ create a torque that tends to rotate the loop clockwise.

Figure 28.21 (a) Overhead view of a rectangular current loop in a uniform magnetic field. (b) Edge view of the loop sighting down sides ② and ④. The side you see in this view is side ③. The purple dot in the left circle represents current in wire ② directed toward you; the purple cross in the right circle represents current in wire ④ directed away from you.

No magnetic forces act on sides ① and ③ because these wires are parallel to the field; hence, $\vec{\mathbf{L}} \times \vec{\mathbf{B}} = 0$ for these sides. Magnetic forces do, however, act on sides ② and ④ because these sides are oriented perpendicular to the field. The magnitude of these forces is, from Equation 28.10,

$$F_2 = F_4 = IaB$$

The direction of $\vec{\mathbf{F}}_2$, the magnetic force exerted on wire ②, is out of the page in the view shown in Figure 28.20a and that of $\vec{\mathbf{F}}_4$, the magnetic force exerted on wire ④, is into the page in the same view. If we view the loop from side ③ and sight along sides ② and ④, we see the view shown in Figure 28.21b, and the two magnetic forces $\vec{\mathbf{F}}_2$ and $\vec{\mathbf{F}}_4$ are directed as shown. Notice that the two forces point in opposite directions but are *not* directed along the same line of action. If the loop is pivoted so that it can rotate about point O, these two forces produce about O a torque that rotates the loop clockwise. The magnitude of this torque τ_{max} is

$$\tau_{max} = F_2 \frac{b}{2} + F_4 \frac{b}{2} = (IaB) \frac{b}{2} + (IaB) \frac{b}{2} = IabB$$

where the moment arm about O is $b/2$ for each force. Because the area enclosed by the loop is $A = ab$, we can express the maximum torque as

$$\tau_{max} = IAB \tag{28.13}$$

Imagine that the loop is released from rest. The loop is modeled as a rigid object under a net torque (Chapter 10), and will begin to rotate in response to the net torque. The sense of the rotation is clockwise when viewed from side ③ as indicated in Figure 28.21b. If the current direction were reversed, the force directions would also reverse and the rotational tendency would be counterclockwise. This behavior is exploited practically in a device called a *motor*, which we discuss in Chapter 30.

Equation 28.13 uses the subscript "max" because the torque has its maximum value when the magnetic field is parallel to the loop. Now suppose the uniform magnetic field makes an angle $\theta < 90°$ with a line perpendicular to the plane of the loop as in Figure 28.22. For convenience, let's assume $\vec{\mathbf{B}}$ is perpendicular to sides ② and ④. In this case, the magnetic forces $\vec{\mathbf{F}}_1$ and $\vec{\mathbf{F}}_3$ exerted on sides ① and ③ cancel each other and produce no torque because they act along the same line. The magnetic forces $\vec{\mathbf{F}}_2$ and $\vec{\mathbf{F}}_4$ acting on sides ② and ④, however, produce a torque about *any point*. Referring to the edge view shown in Figure 28.22, we see that the moment arm of $\vec{\mathbf{F}}_2$ about the point O is equal to $(b/2) \sin \theta$. Likewise, the moment arm of $\vec{\mathbf{F}}_4$ about O is also equal to $(b/2) \sin \theta$. Because $F_2 = F_4 = IaB$, the magnitude of the net torque about O is

$$\begin{aligned} \tau &= F_2 \frac{b}{2} \sin \theta + F_4 \frac{b}{2} \sin \theta \\ &= IaB \left(\frac{b}{2} \sin \theta \right) + IaB \left(\frac{b}{2} \sin \theta \right) = IabB \sin \theta \\ &= IAB \sin \theta \end{aligned} \tag{28.14}$$

where $A = ab$ is the area of the loop. This result shows that the torque has its maximum value IAB when the field is perpendicular to the normal to the plane of the loop ($\theta = 90°$) as discussed with regard to Figure 28.21 and is zero when the field is parallel to the normal to the plane of the loop ($\theta = 0$).

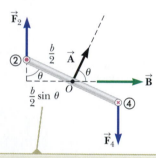

When the normal to the loop makes an angle θ with the magnetic field, the moment arm for the torque is $(b/2) \sin \theta$.

Figure 28.22 An edge view of the loop in Figure 28.21 with the normal to the loop at an angle θ with respect to the magnetic field.

Comparing Equation 28.14 with Equation 11.3, we see that a convenient vector expression for the torque exerted on a loop about an axis passing through its center when it is placed in a uniform magnetic field $\vec{B}$ is

$$\vec{\tau} = I\vec{A} \times \vec{B} \qquad (28.15)$$

where $\vec{A}$, the vector shown in Figure 28.22, is perpendicular to the plane of the loop and has a magnitude equal to the area of the loop. To determine the direction of $\vec{A}$, use the right-hand rule described in Figure 28.23. When you curl the fingers of your right hand in the direction of the current in the loop, your thumb points in the direction of $\vec{A}$. Figure 28.22 shows that the loop tends to rotate in the direction of decreasing values of θ (that is, such that the area vector $\vec{A}$ rotates toward the direction of the magnetic field).

The product $I\vec{A}$ is defined to be the **magnetic dipole moment** $\vec{\mu}$ (often simply called the "magnetic moment") of the loop:

$$\vec{\mu} \equiv I\vec{A} \qquad (28.16)$$

◀ Magnetic dipole moment of a current loop

The SI unit of magnetic dipole moment is the ampere-meter^2 (A · m^2). If a coil of wire contains N loops of the same area, the magnetic moment of the coil is

$$\vec{\mu}_{\text{coil}} = NI\vec{A} \qquad (28.17)$$

Using Equation 28.16, we can express the torque exerted on a current-carrying loop in a magnetic field $\vec{B}$ as

$$\vec{\tau} = \vec{\mu} \times \vec{B} \qquad (28.18)$$

◀ Torque on a magnetic moment in a magnetic field

This result is analogous to Equation 25.20, $\vec{\tau} = \vec{p} \times \vec{E}$, for the torque exerted on an electric dipole in the presence of an electric field $\vec{E}$, where $\vec{p}$ is the electric dipole moment.

Although we obtained the torque for a particular orientation of $\vec{B}$ with respect to the loop, the equation $\vec{\tau} = \vec{\mu} \times \vec{B}$ is valid for any orientation. Furthermore, although we derived the torque expression for a rectangular loop, the result is valid for a loop of any shape.

In Section 25.6, we found that the potential energy of a system of an electric dipole in an electric field is given by $U_E = -\vec{p} \cdot \vec{E}$. This energy depends on the orientation of the dipole in the electric field. Likewise, the potential energy of a system of a magnetic dipole in a magnetic field depends on the orientation of the dipole in the magnetic field and is given by

$$U_B = -\vec{\mu} \cdot \vec{B} \qquad (28.19)$$

◀ Potential energy of a system of a magnetic moment in a magnetic field

This expression shows that the system has its lowest energy $U_{\text{min}} = -\mu B$ when $\vec{\mu}$ points in the same direction as $\vec{B}$. The system has its highest energy $U_{\text{max}} = +\mu B$ when $\vec{\mu}$ points in the direction opposite $\vec{B}$.

(1) Curl your fingers in the direction of the current around the loop.

(2) Your thumb points in the direction of $\vec{A}$ and $\vec{\mu}$.

Figure 28.23 Right-hand rule for determining the direction of the vector $\vec{A}$ for a current loop. The direction of the magnetic moment $\vec{\mu}$ is the same as the direction of $\vec{A}$.

Ⓠ UICK QUIZ 28.4 **(i)** Rank the magnitudes of the torques acting on the rectangular loops (a), (b), and (c) shown edge-on in Figure 28.24 (page 760) from highest to lowest. All loops are identical and carry the same current. **(ii)** Rank the magnitudes of the net forces acting on the rectangular loops shown in Figure 28.24 from highest to lowest.

Figure 28.24 (Quick Quiz 28.4) Which current loop (seen edge-on) experiences the greatest torque, (a), (b), or (c)? Which experiences the greatest net force?

Example 28.5 Rotating a Coil

Consider the loop of wire in Figure 28.25a. Imagine it is pivoted along side ④, which is parallel to the z axis and fastened so that side ④ remains fixed and the rest of the loop hangs vertically in the gravitational field of the Earth but can rotate around side ④ (Fig. 28.25b). The mass of the loop is 50.0 g, and the sides are of lengths $a = 0.200$ m and $b = 0.100$ m. The loop carries a current of 3.50 A and is immersed in a vertical uniform magnetic field of magnitude 0.010 0 T in the positive y direction (Fig. 28.25c). What angle does the plane of the loop make with the vertical?

SOLUTION

Conceptualize In the edge view of Figure 28.25b, notice that the magnetic moment of the loop is to the left. Therefore, when the loop is in the magnetic field, the magnetic torque on the loop causes it to rotate in a clockwise direction around side ④, which we choose as the rotation axis. Imagine the loop making this clockwise rotation so that the plane of the loop is at some angle θ to the vertical as in Figure 28.25c. The gravitational force on the loop exerts a torque that would cause a rotation in the counterclockwise direction if the magnetic field were turned off.

The loop hangs vertically and is pivoted so that it can rotate around side ④.

The magnetic torque causes the loop to rotate in a clockwise direction around side ④, whereas the gravitational torque is in the opposite direction.

Figure 28.25 (Example 28.5) (a) The dimensions of a rectangular current loop. (b) Edge view of the loop sighting down sides ② and ④. (c) An edge view of the loop in (b) rotated through an angle with respect to the horizontal when it is placed in a magnetic field.

Categorize At some angle of the loop, the two torques described in the Conceptualize step are equal in magnitude and the loop is at rest. We therefore model the loop as a *rigid object in equilibrium*.

Analyze Evaluate the magnetic torque on the loop about side ④ from Equation 28.18:

$$\vec{\tau}_B = \vec{\mu} \times \vec{B} = -\mu B \sin(90° - \theta)\,\hat{\mathbf{k}} = -IAB\cos\theta\,\hat{\mathbf{k}} = -IabB\cos\theta\,\hat{\mathbf{k}}$$

Evaluate the gravitational torque on the loop about side ④, noting that the gravitational force can be modeled to act at the center of the loop:

$$\vec{\tau}_g = \vec{r} \times m\vec{g} = mg\frac{b}{2}\sin\theta\,\hat{\mathbf{k}}$$

From the rigid body in equilibrium model, add the torques and set the net torque equal to zero:

$$\sum \vec{\tau} = -IabB\cos\theta\,\hat{\mathbf{k}} + mg\frac{b}{2}\sin\theta\,\hat{\mathbf{k}} = 0$$

Solve for θ:

$$IabB\cos\theta = mg\frac{b}{2}\sin\theta \quad \rightarrow \quad \tan\theta = \frac{2IaB}{mg}$$

$$\theta = \tan^{-1}\left(\frac{2IaB}{mg}\right)$$

Substitute numerical values:

$$\theta = \tan^{-1}\left[\frac{2(3.50\text{ A})(0.200\text{ m})(0.010\ 0\text{ T})}{(0.050\ 0\text{ kg})(9.80\text{ m/s}^2)}\right] = \boxed{1.64°}$$

Finalize The angle is relatively small, so the loop still hangs almost vertically. If the current I or the magnetic field B is increased, however, the angle increases as the magnetic torque becomes stronger.

28.6 The Hall Effect

When a current-carrying conductor is placed in a magnetic field, a potential difference is generated between two points lying along a direction perpendicular to both the current and the magnetic field. This phenomenon, first observed by Edwin Hall (1855–1938) in 1879, is known as the *Hall effect*. The arrangement for observing the Hall effect consists of a flat conductor carrying a current I in the x direction as shown in Figure 28.26. A uniform magnetic field $\vec{B}$ is applied in the y direction. If the charge carriers are electrons moving in the negative x direction with a drift velocity $\vec{v}_d$, they experience an upward magnetic force $\vec{F}_B = q\vec{v}_d \times \vec{B}$, are deflected upward, and accumulate at the upper edge of the flat conductor, leaving an excess of positive charge at the lower edge (Fig. 28.27a). This accumulation of charge at the edges establishes an electric field in the conductor and increases until the electric force on carriers remaining in the bulk of the conductor balances the magnetic force acting on the carriers. The moving electrons can now be described by the particle in equilibrium model, and they are no longer deflected upward. A sensitive voltmeter connected across the sample as shown in Figure 28.27 can measure the potential difference, known as the **Hall voltage** ΔV_{H}, generated across the conductor.

If the charge carriers are positive and hence move in the positive x direction (for rightward current) as shown in Figures 28.26 and 28.27b, they also experience an upward magnetic force $q\vec{v}_d \times \vec{B}$, which produces a buildup of positive charge on the upper edge and leaves an excess of negative charge on the lower edge. Hence, the sign of the Hall voltage generated in the sample is opposite the sign of the Hall voltage resulting from the deflection of electrons. The sign of the charge carriers can therefore be determined from measuring the polarity of the Hall voltage.

In deriving an expression for the Hall voltage, first note that the magnetic force exerted on the carriers has magnitude qv_dB. In equilibrium, this force is balanced by the electric force qE_{H}, where E_{H} is the magnitude of the electric field due to the charge separation (sometimes referred to as the *Hall field*). Therefore,

$$qv_dB = qE_{\text{H}}$$

$$E_{\text{H}} = v_dB$$

If d is the width of the conductor, the Hall voltage is

$$\Delta V_{\text{H}} = E_{\text{H}}d = v_dBd \qquad (28.20)$$

Therefore, the measured Hall voltage gives a value for the drift speed of the charge carriers if d and B are known.

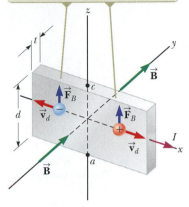

When I is in the x direction and $\vec{B}$ in the y direction, both positive and negative charge carriers are deflected upward in the magnetic field.

Figure 28.26 To observe the Hall effect, a magnetic field is applied to a current-carrying conductor. The Hall voltage is measured between points a and c.

When the charge carriers are negative, the upper edge of the conductor becomes negatively charged and c is at a lower electric potential than a.

The charge carriers are no longer deflected when the edges become sufficiently charged that there is a balance between the electric force and the magnetic force.

When the charge carriers are positive, the upper edge of the conductor becomes positively charged and c is at a higher potential than a.

Figure 28.27 The sign of the Hall voltage depends on the sign of the charge carriers.

We can obtain the charge-carrier density n by measuring the current in the sample. From Equation 26.4, we can express the drift speed as

$$v_d = \frac{I}{nqA} \tag{28.21}$$

where A is the cross-sectional area of the conductor. Substituting Equation 28.21 into Equation 28.20 and solving for B gives

$$B = \frac{nqA}{Id}\,\Delta V_{\mathrm{H}} \tag{28.22}$$

Because $A = td$, where t is the thickness of the conductor, we can also express Equation 28.22 as

The Hall voltage ▶

$$B = \frac{nqt}{I}\,\Delta V_{\mathrm{H}} \tag{28.23}$$

This relationship shows that a properly calibrated conductor can be used to measure the magnitude of an unknown magnetic field.

Example 28.6 The Hall Effect for Copper

A rectangular copper strip 1.5 cm wide and 0.10 cm thick carries a current of 5.0 A. Find the Hall voltage for a 1.2-T magnetic field applied in a direction perpendicular to the strip.

SOLUTION

Conceptualize Study Figures 28.26 and 28.27 carefully and make sure you understand that a Hall voltage is developed between the top and bottom edges of the strip.

Categorize We evaluate the Hall voltage using an equation developed in this section, so we categorize this example as a substitution problem.

Assuming one electron per atom is available for conduction, find the charge-carrier density in terms of the molar mass M and density ρ of copper:

$$(1) \quad n = \frac{N_{\mathrm{A}}}{V} = \frac{N_{\mathrm{A}}\rho}{M}$$

Solve Equation 28.23 for the Hall voltage and substitute Equation (1):

$$\Delta V_{\mathrm{H}} = \frac{IB}{nqt} = \frac{MIB}{N_{\mathrm{A}}\rho qt}$$

Substitute numerical values:

$$\Delta V_{\mathrm{H}} = \frac{(0.063\,5\ \mathrm{kg/mol})(5.0\ \mathrm{A})(1.2\ \mathrm{T})}{(6.02 \times 10^{23}\ \mathrm{mol^{-1}})(8\,920\ \mathrm{kg/m^3})(1.60 \times 10^{-19}\ \mathrm{C})(0.001\,0\ \mathrm{m})}$$

$$= 0.44\ \mu\mathrm{V}$$

Such an extremely small Hall voltage is expected in good conductors. (Notice that the width of the conductor is not needed in this calculation.)

WHAT IF? What if the strip has the same dimensions but is made of a semiconductor? Will the Hall voltage be smaller or larger?

Answer In semiconductors, n is much smaller than it is in metals that contribute one electron per atom to the current; hence, the Hall voltage is usually larger because it varies as the inverse of n. Currents on the order of 0.1 mA are generally used for such materials. Consider a piece of silicon that has the same dimensions as the copper strip in this example and whose value for n is 1.0×10^{20} electrons/m³. Taking $B = 1.2$ T and $I = 0.10$ mA, we find that $\Delta V_{\mathrm{H}} = 7.5$ mV. A potential difference of this magnitude is readily measured.

Summary

▶ Definition

The **magnetic dipole moment** $\vec{\mu}$ of a loop carrying a current I is

$$\vec{\mu} \equiv I\vec{A} \tag{28.16}$$

where the area vector $\vec{A}$ is perpendicular to the plane of the loop and $|\vec{A}|$ is equal to the area of the loop. The SI unit of $\vec{\mu}$ is A · m^2.

▶ Concepts and Principles

If a charged particle moves in a uniform magnetic field so that its initial velocity is perpendicular to the field, the particle moves in a circle, the plane of which is perpendicular to the magnetic field. The radius of the circular path is

$$r = \frac{mv}{qB} \tag{28.3}$$

where m is the mass of the particle and q is its charge. The angular speed of the charged particle is

$$\omega = \frac{qB}{m} \tag{28.4}$$

If a straight conductor of length L carries a current I, the force exerted on that conductor when it is placed in a uniform magnetic field $\vec{B}$ is

$$\vec{F}_B = I\vec{L} \times \vec{B} \tag{28.10}$$

where the direction of $\vec{L}$ is in the direction of the current and $|\vec{L}| = L$.

If an arbitrarily shaped wire carrying a current I is placed in a magnetic field, the magnetic force exerted on a very small segment $d\vec{s}$ is

$$d\vec{F}_B = I\, d\vec{s} \times \vec{B} \tag{28.11}$$

To determine the total magnetic force on the wire, one must integrate Equation 28.11 over the wire, keeping in mind that both $\vec{B}$ and $d\vec{s}$ may vary at each point.

The torque $\vec{\tau}$ on a current loop placed in a uniform magnetic field $\vec{B}$ is

$$\vec{\tau} = \vec{\mu} \times \vec{B} \tag{28.18}$$

The potential energy of the system of a magnetic dipole in a magnetic field is

$$U_B = -\vec{\mu} \cdot \vec{B} \tag{28.19}$$

▶ Analysis Models for Problem Solving

Particle in a Field (Magnetic) A source (to be discussed in Chapter 29) establishes a **magnetic field** $\vec{B}$ throughout space. When a particle with charge q and moving with velocity $\vec{v}$ is placed in that field, it experiences a magnetic force given by

$$\vec{F}_B = q\vec{v} \times \vec{B} \tag{28.1}$$

The direction of this magnetic force is perpendicular both to the velocity of the particle and to the magnetic field. The magnitude of this force is

$$F_B = |q|vB \sin\theta \tag{28.2}$$

where θ is the smaller angle between $\vec{v}$ and $\vec{B}$. The SI unit of $\vec{B}$ is the **tesla** (T), where 1 T = 1 N/A · m.

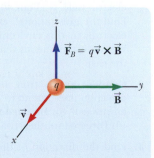

Think–Pair–Share

See the Preface for an explanation of the icons used in this problems set. For additional assessment items for this section, go to **WEBASSIGN** From Cengage

1. Your team is working for a spring manufacturing company. The team has been asked to design a new method of measuring the spring constant of a spring. Together, you have come up with the design shown in Figure TP28.1. Part (a) of the figure shows a side view of a conducting loop of area $A = ab$. The loop is pivoted at O. The center point of the left side of the loop is attached to the upper end of a spring of force constant k. The lower end of the spring is attached to a rigid support. The loop is in a magnetic field of magnitude B directed toward the right. When a current I exists in the loop, the loop rotates to a new equilibrium position at angle θ, as shown in part (b) of Figure TP28.1. Find an expression for the spring constant k of the spring. Assume the rotation of the loop is small enough that the spring remains essentially vertical.

2. **ACTIVITY** Your group has just completed taking data for a Hall voltage experiment in your physics laboratory, using the setup shown in Figure 28.27. The following table shows the measurements. If the measurements were taken with a current of 0.200 A, and the sample is made from a material having a charge-carrier density of 1.00×10^{26} carriers/m³, what is the thickness of the sample?

B (T)	ΔV_H (μV)
0.00	0
0.10	11.0
0.20	19.0
0.30	28.0
0.40	42.0
0.50	50.0
0.60	61.0
0.70	68.0
0.80	79.0
0.90	90.0
1.00	102

Figure TP28.1

Problems

See the Preface for an explanation of the icons used in this problems set. For additional assessment items for this section, go to **WEBASSIGN** From Cengage

SECTION 28.1 Analysis Model: Particle in a Field (Magnetic)

Problems 1–2 and 4 in Chapter 11 can be assigned with this section as review for the vector product.

1. At the equator, near the surface of the Earth, the magnetic field is approximately 50.0 μT northward, and the electric field is about 100 N/C downward in fair weather. Find the gravitational, electric, and magnetic forces on an electron in this environment, assuming that the electron has an instantaneous velocity of 6.00×10^6 m/s directed to the east.

2. Consider an electron near the Earth's equator. In which direction does it tend to deflect if its velocity is (a) directed downward? (b) Directed northward? (c) Directed westward? (d) Directed southeastward?

3. Find the direction of the magnetic field acting on a positively charged particle moving in the various situations shown in Figure P28.3 if the direction of the magnetic force acting on it is as indicated.

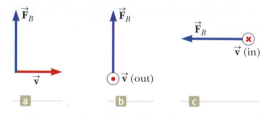

Figure P28.3

4. A proton moving at 4.00×10^6 m/s through a magnetic field of magnitude 1.70 T experiences a magnetic force of magnitude 8.20×10^{-13} N. What is the angle between the proton's velocity and the field?

5. A proton travels with a speed of 5.02×10^6 m/s in a direction that makes an angle of 60.0° with the direction of a magnetic field of magnitude 0.180 T in the positive x direction. What are the magnitudes of (a) the magnetic force on the proton and (b) the proton's acceleration?

6. A laboratory electromagnet produces a magnetic field of magnitude 1.50 T. A proton moves through this field with a speed of 6.00×10^6 m/s. (a) Find the magnitude of the

maximum magnetic force that could be exerted on the proton. (b) What is the magnitude of the maximum acceleration of the proton? (c) Would the field exert the same magnetic force on an electron moving through the field with the same speed? (d) Would the electron experience the same acceleration? Explain.

7. A proton moves perpendicular to a uniform magnetic field **T** $\vec{B}$ at a speed of 1.00×10^7 m/s and experiences an acceleration of 2.00×10^{13} m/s² in the positive x direction when its velocity is in the positive z direction. Determine the magnitude and direction of the field.

SECTION 28.2 Motion of a Charged Particle in a Uniform Magnetic Field

8. An accelerating voltage of 2.50×10^3 V is applied to an elec-
Q|C tron gun, producing a beam of electrons originally traveling horizontally north in vacuum toward the center of a viewing screen 35.0 cm away. What are (a) the magnitude and (b) the direction of the deflection on the screen caused by the Earth's gravitational field? What are (c) the magnitude and (d) the direction of the deflection on the screen caused by the vertical component of the Earth's magnetic field, taken as 20.0 μT down? (e) Does an electron in this vertical magnetic field move as a projectile, with constant vector acceleration perpendicular to a constant northward component of velocity? (f) Is it a good approximation to assume it has this projectile motion? Explain.

9. A proton (charge $+e$, mass m_p), a deuteron (charge $+e$, mass
S $2m_p$), and an alpha particle (charge $+2e$, mass $4m_p$) are accelerated from rest through a common potential difference ΔV. Each of the particles enters a uniform magnetic field $\vec{B}$, with its velocity in a direction perpendicular to $\vec{B}$. The proton moves in a circular path of radius r_p. In terms of r_p, determine (a) the radius r_d of the circular orbit for the deuteron and (b) the radius r_α for the alpha particle.

10. **Review.** A 30.0-g metal ball having net charge $Q = 5.00$ μC
Q|C is thrown out of a window horizontally north at a speed $v = 20.0$ m/s. The window is at a height $h = 20.0$ m above the ground. A uniform, horizontal magnetic field of magnitude $B = 0.010\ 0$ T is perpendicular to the plane of the ball's trajectory and directed toward the west. (a) Assuming the ball follows the same trajectory as it would in the absence of the magnetic field, find the magnetic force acting on the ball just before it hits the ground. (b) Based on the result of part (a), is it justified for three-significant-digit precision to assume the trajectory is unaffected by the magnetic field? Explain.

11. **Review.** One electron collides elastically with a second elec-
AMT tron initially at rest. After the collision, the radii of their trajectories are 1.00 cm and 2.40 cm. The trajectories are perpendicular to a uniform magnetic field of magnitude 0.044 0 T. Determine the energy (in keV) of the incident electron.

12. **Review.** One electron collides elastically with a second elec-
S tron initially at rest. After the collision, the radii of their trajectories are r_1 and r_2. The trajectories are perpendicular to a uniform magnetic field of magnitude B. Determine the energy of the incident electron.

13. **Review.** An electron moves in a circular path perpendicular to a constant magnetic field of magnitude 1.00 mT. The

angular momentum of the electron about the center of the circle is 4.00×10^{-25} kg $\cdot$ m²/s. Determine (a) the radius of the circular path and (b) the speed of the electron.

SECTION 28.3 Applications Involving Charged Particles Moving in a Magnetic Field

14. A cyclotron designed to accelerate protons has a magnetic
T field of magnitude 0.450 T over a region of radius 1.20 m. What are (a) the cyclotron frequency and (b) the maximum speed acquired by the protons?

15. You are working as a medical assistant at a proton beam
CR facility, where high-speed protons are used to bombard cancer cells. The protons are accelerated with a cyclotron, which you find very interesting because of your background in physics. You are explaining this to a patient who has some familiarity with cyclotrons. She asks, "How many revolutions does a proton make in the cyclotron before it reaches its exit kinetic energy?" You are taken aback, both by the high quality of her question and the fact that you never thought of such a question before. You tell her you will try to get her an answer before she finishes her treatment today. When you are finished preparing her for treatment, you go into the cyclotron room and look at the machine. Only three numbers are available on the machine labeling: the exit energy $K = 250$ MeV, the radius at which the protons exit, $r = 0.850$ m, and the accelerating potential difference between the dees, $\Delta V = 800$ V. You go back to the patient prepared to give her a total number of times the protons go around the cyclotron before exiting.

16. Singly charged uranium-238 ions are accelerated through
Q|C a potential difference of 2.00 kV and enter a uniform magnetic field of magnitude 1.20 T directed perpendicular to their velocities. (a) Determine the radius of their circular path. (b) Repeat this calculation for uranium-235 ions. (c) **What If?** How does the ratio of these path radii depend on the accelerating voltage? (d) On the magnitude of the magnetic field?

17. A cyclotron (Fig. 28.16) designed to accelerate protons has an outer radius of 0.350 m. The protons are emitted nearly at rest from a source at the center and are accelerated through 600 V each time they cross the gap between the dees. The dees are between the poles of an electromagnet where the field is 0.800 T. (a) Find the cyclotron frequency for the protons in this cyclotron. Find (b) the speed at which protons exit the cyclotron and (c) their maximum kinetic energy. (d) How many revolutions does a proton make in the cyclotron? (e) For what time interval does the proton accelerate?

18. A particle in the cyclotron shown in Figure 28.16a gains
Q|C energy $q\,\Delta V$ from the alternating power supply each time it passes from one dee to the other. The time interval for each full orbit is

$$T = \frac{2\pi}{\omega} = \frac{2\pi m}{qB}$$

so the particle's average rate of increase in energy is

$$\frac{2q\,\Delta V}{T} = \frac{q^2 B\,\Delta V}{\pi m}$$

Notice that this power input is constant in time. On the other hand, the rate of increase in the radius r of its path is *not* constant. (a) Show that the rate of increase in the radius *r* of the particle's path is given by

$$\frac{dr}{dt} = \frac{1}{r}\frac{\Delta V}{\pi B}$$

(b) Describe how the path of the particles in Figure 28.16a is consistent with the result of part (a). (c) At what rate is the radial position of the protons in a cyclotron increasing immediately before the protons leave the cyclotron? Assume the cyclotron has an outer radius of 0.350 m, an accelerating voltage of $\Delta V = 600$ V, and a magnetic field of magnitude 0.800 T. (d) By how much does the radius of the protons' path increase during their last full revolution?

19. In his experiments on "cathode rays" during which he discovered the electron, J. J. Thomson showed that the same beam deflections resulted with tubes having cathodes made of *different* materials and containing *various* gases before evacuation. (a) Are these observations important? Explain your answer. (b) When he applied various potential differences to the deflection plates and turned on the magnetic coils, alone or in combination with the deflection plates, Thomson observed that the fluorescent screen continued to show a *single small* glowing patch. Argue whether his observation is important. (c) Do calculations to show that the charge-to-mass ratio Thomson obtained was huge compared with that of any macroscopic object or of any ionized atom or molecule. How can one make sense of this comparison? (d) Could Thomson observe any deflection of the beam due to gravitation? Do a calculation to argue for your answer. *Note:* To obtain a visibly glowing patch on the fluorescent screen, the potential difference between the slits and the cathode must be 100 V or more.

SECTION 28.4 Magnetic Force Acting on a Current-Carrying Conductor

20. A straight wire carrying a 3.00-A current is placed in a uniform magnetic field of magnitude 0.280 T directed perpendicular to the wire. (a) Find the magnitude of the magnetic force on a section of the wire having a length of 14.0 cm. (b) Explain why you can't determine the direction of the magnetic force from the information given in the problem.

21. A wire carries a steady current of 2.40 A. A straight section of the wire is 0.750 m long and lies along the *x* axis within a uniform magnetic field, $\vec{B} = 1.60\hat{k}$ T. If the current is in the positive *x* direction, what is the magnetic force on the section of wire?

22. *Why is the following situation impossible?* Imagine a copper wire with radius 1.00 mm encircling the Earth at its magnetic equator, where the field direction is horizontal. A power supply delivers 100 MW to the wire to maintain a current in it, in a direction such that the magnetic force from the Earth's magnetic field is upward. Due to this force, the wire is levitated immediately above the ground.

23. Review. A rod of mass 0.720 kg and radius 6.00 cm rests on two parallel rails (Fig. P28.23) that are *d* = 12.0 cm apart and *L* = 45.0 cm long. The rod carries a current of *I* = 48.0 A in the direction shown and rolls along the rails without slipping. A uniform magnetic field of magnitude 0.240 T is directed perpendicular to the rod and the rails. If it starts from rest, what is the speed of the rod as it leaves the rails?

Figure P28.23 Problems 23 and 24.

24. Review. A rod of mass *m* and radius *R* rests on two parallel rails (Fig. P28.23) that are a distance *d* apart and have a length *L*. The rod carries a current *I* in the direction shown and rolls along the rails without slipping. A uniform magnetic field *B* is directed perpendicular to the rod and the rails. If it starts from rest, what is the speed of the rod as it leaves the rails?

25. A wire having a mass per unit length of 0.500 g/cm carries a 2.00-A current horizontally to the south. What are (a) the direction and (b) the magnitude of the minimum magnetic field needed to lift this wire vertically upward?

26. Consider the system pictured in Figure P28.26. A 15.0-cm horizontal wire of mass 15.0 g is placed between two thin, vertical conductors, and a uniform magnetic field acts perpendicular to the page. The wire is free to move vertically without friction on the two vertical conductors. When a 5.00-A current is directed as shown in the figure, the horizontal wire moves upward at constant velocity in the presence of gravity. (a) What forces act on the horizontal wire, and (b) under what condition is the wire able to move upward at constant velocity? (c) Find the magnitude and direction of the minimum magnetic field required to move the wire at constant speed. (d) What happens if the magnetic field exceeds this minimum value?

Figure P28.26

27. A strong magnet is placed under a horizontal conducting ring of radius *r* that carries current *I* as shown in Figure P28.27. If the magnetic field $\vec{B}$ makes an angle θ with the vertical at the ring's location, what are (a) the magnitude and (b) the direction of the resultant magnetic force on the ring?

Figure P28.27

28. In Figure P28.28, the cube is 40.0 cm on each edge. Four straight segments of wire—*ab*, *bc*, *cd*, and *da*—form a closed loop that carries a current *I* = 5.00 A in the direction shown. A uniform magnetic field of magnitude *B* = 0.020 0 T is in the positive *y* direction. Determine the magnetic force vector on (a) *ab*, (b) *bc*, (c) *cd*, and (d) *da*.

Figure P28.28

(e) Explain how you could find the force exerted on the fourth of these segments from the forces on the other three, without further calculation involving the magnetic field.

SECTION 28.5 Torque on a Current Loop in a Uniform Magnetic Field

29. A magnetized sewing needle has a magnetic moment of **QC** 9.70 mA · m². At its location, the Earth's magnetic field is 55.0 μT northward at 48.0° below the horizontal. Identify the orientations of the needle that represent (a) the minimum potential energy and (b) the maximum potential energy of the needle–field system. (c) How much work must be done on the system to move the needle from the minimum to the maximum potential energy orientation?

30. A 50.0-turn circular coil of radius 5.00 cm can be oriented in any direction in a uniform magnetic field having a magnitude of 0.500 T. If the coil carries a current of 25.0 mA, find the magnitude of the maximum possible torque exerted on the coil.

31. You are in charge of planning a physics magic show for an **CR** open house on your campus. You come up with the following plan for one trick. You will place a sphere on a rough inclined plane of angle θ, as shown in Figure P28.31, and it will *not* roll down the incline. Here is the secret that only you know: The sphere is nonconducting, has a mass of 80.0 g, and a radius 20.0 cm. A flat, compact coil of wire with five turns is wrapped tightly around it, with each turn concentric with the sphere. The sphere is placed on the incline so that the coil is parallel to the plane. You establish a uniform magnetic field of 0.350 T vertically upward in the region of the sphere. (a) What current in the coil do you need to make this trick work? (b) You explain the trick to a friend in confidence and he suggests lowering the angle θ of the plane to make the required current lower. How do you respond?

Figure P28.31

32. You are working in your dream job: an assistant for the **CR** special effects department of a movie studio. You have just been given this assignment: the star of a horror movie is walking down a spooky hallway when suddenly, due to some unknown and strange supernatural forces, all the pictures hanging on the wall start rotating about their upper edges until they are sticking straight out from the wall! To set up this effect, you attach the pictures to the wall with hinges along their upper end and wrap 20 turns of wire around the outside frame of the picture, as shown in Figure P28.32a. You set up a uniform magnetic field in the hallway that is directed upward and oriented at an angle of $\gamma = 5.00°$ to the vertical, with its horizontal component directed perpendicularly into the wall. When you send a current of $I = 10.0$ A through the wire around each picture, the frame swings up perpendicular to the wall as shown in Figure P28.32b. Consider a particular picture of width $w = 40.6$ cm, height $h = 50.8$ cm, and mass $m = 0.750$ kg. (a) Your supervisor asks you to determine the magnetic field magnitude that is necessary for this picture to rotate so that its face is parallel to the floor and perpendicular to the wall, as in Figure P28.32b. (b) She also asks about any dangers associated with this magnetic field.

Figure P28.32

33. A rectangular coil consists of $N = 100$ closely wrapped **T** turns and has dimensions $a = 0.400$ m and $b = 0.300$ m. The coil is hinged along the y axis, and its plane makes an angle $\theta = 30.0°$ with the x axis (Fig. P28.33). (a) What is the magnitude of the torque exerted on the coil by a uniform magnetic field $B = 0.800$ T directed in the positive x direction when the current is $I = 1.20$ A in the direction shown? (b) What is the expected direction of rotation of the coil?

Figure P28.33

34. A rectangular loop of wire has dimensions 0.500 m by
0.300 m. The loop is pivoted at the x axis and lies in the
xy plane as shown in Figure P28.34. A uniform magnetic
field of magnitude 1.50 T is directed at an angle of 40.0°
with respect to the y axis with field lines parallel to the yz
plane. The loop carries a current of 0.900 A in the direction shown. (Ignore gravitation.) We wish to evaluate the
torque on the current loop. (a) What is the direction of
the magnetic force exerted on wire segment ab? (b) What
is the direction of the torque associated with this force
about an axis through the origin? (c) What is the direction
of the magnetic force exerted on segment cd? (d) What is
the direction of the torque associated with this force about
an axis through the origin? (e) Can the forces examined in
parts (a) and (c) combine to cause the loop to rotate around
the x axis? (f) Can they affect the motion of the loop in any
way? Explain. (g) What is the direction of the magnetic
force exerted on segment bc? (h) What is the direction of
the torque associated with this force about an axis through
the origin? (i) What is the torque on segment ad about an
axis through the origin? (j) From the point of view of Figure
P28.34, once the loop is released from rest at the position
shown, will it rotate clockwise or counterclockwise around
the x axis? (k) Compute the magnitude of the magnetic
moment of the loop. (l) What is the angle between the magnetic moment vector and the magnetic field? (m) Compute
the torque on the loop using the results to parts (k) and (l).

Figure P28.34

35. A wire is formed into a circle having a diameter of 10.0 cm
and is placed in a uniform magnetic field of 3.00 mT. The
wire carries a current of 5.00 A. Find (a) the maximum
torque on the wire and (b) the range of potential energies of
the wire–field system for different orientations of the circle.

SECTION 28.6 The Hall Effect

36. A Hall-effect probe operates with a 120-mA current. When
the probe is placed in a uniform magnetic field of magnitude 0.080 0 T, it produces a Hall voltage of 0.700 μV.
(a) When it is used to measure an unknown magnetic field,
the Hall voltage is 0.330 μV. What is the magnitude of the
unknown field? (b) The thickness of the probe in the direction of $\vec{B}$ is 2.00 mm. Find the density of the charge carriers,
each of which has charge of magnitude e.

ADDITIONAL PROBLEMS

37. Carbon-14 and carbon-12 ions (each with charge of magnitude e) are accelerated in a cyclotron. If the cyclotron has a
magnetic field of magnitude 2.40 T, what is the difference
in cyclotron frequencies for the two ions?

38. Figure 28.11 shows a charged particle traveling in a nonuniform magnetic field forming a magnetic bottle. (a) Explain
why the positively charged particle in the figure must be

moving clockwise when viewed from the right of the figure.
The particle travels along a helix whose radius decreases
and whose pitch decreases as the particle moves into a stronger magnetic field. If the particle is moving to the right
along the x axis, its velocity in this direction will be reduced
to zero and it will be reflected from the right-hand side of
the bottle, acting as a "magnetic mirror." The particle ends
up bouncing back and forth between the ends of the bottle.
(b) Explain qualitatively why the axial velocity is reduced to
zero as the particle moves into the region of strong magnetic field at the end of the bottle. (c) Explain why the tangential velocity increases as the particle approaches the end
of the bottle. (d) Explain why the orbiting particle has a
magnetic dipole moment.

39. Within a cylindrical region of space of radius 100 Mm, a
magnetic field is uniform with a magnitude 25.0 μT and oriented parallel to the axis of the cylinder. The magnetic field
is zero outside this cylinder. A cosmic-ray proton traveling
at one-tenth the speed of light is heading directly toward
the center of the cylinder, moving perpendicular to the
cylinder's axis. (a) Find the radius of curvature of the path
the proton follows when it enters the region of the field. (b)
Explain whether the proton will arrive at the center of the
cylinder.

40. Heart–lung machines and artificial kidney machines
employ electromagnetic blood pumps. The blood is confined to an electrically insulating tube, cylindrical in practice but represented here for simplicity as a rectangle of
interior width w and height h. Figure P28.40 shows a rectangular section of blood within the tube. Two electrodes
fit into the top and the bottom of the tube. The potential
difference between them establishes an electric current
through the blood, with current density J over the section of
length L shown in Figure P28.40. A perpendicular magnetic
field exists in the same region. (a) Explain why this arrangement produces on the liquid a force that is directed along
the length of the pipe. (b) Show that the section of liquid
in the magnetic field experiences a pressure increase JLB.
(c) After the blood leaves the pump, is it charged? (d) Is it
carrying current? (e) Is it magnetized? (The same electromagnetic pump can be used for any fluid that conducts electricity, such as liquid sodium in a nuclear reactor.)

Figure P28.40

41. **Review.** A proton is at rest at the plane boundary of a region
containing a uniform magnetic field B (Fig. P28.41). An
alpha particle moving horizontally makes a head-on elastic
collision with the proton. Immediately after the collision,
both particles enter the magnetic field, moving perpendicular to the direction of the field. The radius of the proton's
trajectory is R. The mass of the alpha particle is four times
that of the proton, and its charge is twice that of the proton.
Find the radius of the alpha particle's trajectory.

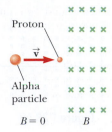

Figure P28.41

42. (a) A proton moving with velocity $\vec{\mathbf{v}} = v_i \hat{\mathbf{i}}$ experiences a magnetic force $\vec{\mathbf{F}} = F_i \hat{\mathbf{j}}$. Explain what you can and cannot infer about $\vec{\mathbf{B}}$ from this information. (b) **What If?** In terms of F_i, what would be the force on a proton in the same field moving with velocity $\vec{\mathbf{v}} = -v_i \hat{\mathbf{i}}$? (c) What would be the force on an electron in the same field moving with velocity $\vec{\mathbf{v}} = -v_i \hat{\mathbf{i}}$?

43. A proton having an initial velocity of $20.0 \hat{\mathbf{i}}$ Mm/s enters a uniform magnetic field of magnitude 0.300 T with a direction perpendicular to the proton's velocity. It leaves the field-filled region with velocity $-20.0 \hat{\mathbf{j}}$ Mm/s. Determine (a) the direction of the magnetic field, (b) the radius of curvature of the proton's path while in the field, (c) the distance the proton traveled in the field, and (d) the time interval during which the proton is in the field.

44. You have been called in as an expert witness in a civil case. The case involves a dispute between neighbors. The plaintiff neighbor is complaining about a buzzing noise during the night that prevents the plaintiff from sleeping. He claims that the buzzing is coming from a light fixture on the defendant's porch ceiling. The defendant likes to do installations and repairs himself and has done a sloppy job of installing the light fixture. The fixture hangs vertically from a single wire that is attached through the porch ceiling and down the wall to one connector in a nearby electrical outlet. The second wire is hung horizontally with strings at the level of the light and then runs down the wall to the other connector in the outlet. The defendant leaves the light on all night long for security. Recalling his high school physics, the plaintiff states that the combination of the 60-Hz household voltage and the magnetic field of the Earth results in an oscillating driving force on the single wire from which the light fixture hangs vertically. This, in turn, sets up a standing wave in the wire, and that is the cause of the buzz. You have been hired by the defense attorney. Upon hearing the details of the case, you obtain permission from the defendant and make measurements. The mass of the light fixture is 17.5 kg. The vertical wire from which it hangs is 0.150 m long and has a mass of 0.030 kg. Is the plaintiff correct that the magnetic field of the Earth is causing the buzzing of the wire? Ignore any effect of the second wire.

45. Model the electric motor in a handheld electric mixer as a single flat, compact, circular coil carrying electric current in a region where a magnetic field is produced by an external permanent magnet. You need consider only one instant in the operation of the motor. (We will consider motors again in Chapter 30.) Make order-of-magnitude estimates of (a) the magnetic field, (b) the torque on the coil, (c) the current in the coil, (d) the coil's area, and (e) the number of turns in the coil. The input power to the motor is electric, given by $P = I \Delta V$, and the useful output power is mechanical, $P = \tau \omega$.

46. *Why is the following situation impossible?* Figure P28.46 shows an experimental technique for altering the direction of travel for a charged particle. A particle of charge $q = 1.00 \ \mu\text{C}$ and mass $m = 2.00 \times 10^{-13}$ kg enters the bottom of the region of uniform magnetic field at speed $v = 2.00 \times 10^5$ m/s, with a velocity vector perpendicular to the field lines. The magnetic force on the particle causes its direction of travel to change so that it leaves the region of the magnetic field at the top traveling at an angle from its original direction. The magnetic field has magnitude $B = 0.400$ T and is directed out of the page. The length h of the magnetic field region is 0.110 m. An experimenter performs the technique and measures the angle θ at which the particles exit the top of the field. She finds that the angles of deviation are exactly as predicted.

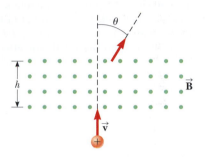

Figure P28.46

47. A heart surgeon monitors the flow rate of blood through an artery using an electromagnetic flowmeter (Fig. P28.47). Electrodes A and B make contact with the outer surface of the blood vessel, which has a diameter of 3.00 mm. (a) For a magnetic field magnitude of 0.040 0 T, an emf of 160 μV appears between the electrodes. Calculate the speed of the blood. (b) Explain why electrode A has to be positive as shown. (c) Does the sign of the emf depend on whether the mobile ions in the blood are predominantly positively or negatively charged? Explain.

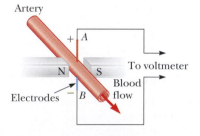

Figure P28.47

48. **Review.** (a) Show that a magnetic dipole in a uniform magnetic field, displaced from its equilibrium orientation and released, can oscillate as a torsional pendulum (Section 15.5) in simple harmonic motion. (b) Is this statement true for all angular displacements, for all displacements less than 180°, or only for small angular displacements? Explain. (c) Assume the dipole is a compass needle—a light bar magnet—with a magnetic moment of magnitude μ. It has moment of inertia I about its center, where it is mounted on a frictionless, vertical axle, and it is placed in a horizontal magnetic field of magnitude B. Determine its frequency of oscillation. (d) Explain how the compass needle can be conveniently used as an indicator of the magnitude of the external magnetic field. (e) If its frequency is 0.680 Hz in the Earth's local field, with a horizontal component of 39.2 μT, what is the magnitude of a field parallel to the needle in which its frequency of oscillation is 4.90 Hz?

CHALLENGE PROBLEMS

49. Consider an electron orbiting a proton and maintained in a fixed circular path of radius $R = 5.29 \times 10^{-11}$ m by the Coulomb force. Treat the orbiting particle as a current loop. Calculate the resulting torque when the electron–proton system is placed in a magnetic field of 0.400 T directed perpendicular to the magnetic moment of the loop.

50. Protons having a kinetic energy of 5.00 MeV (1 eV = 1.60 $\times$ 10^{-19} J) are moving in the positive x direction and enter a magnetic field $\vec{\mathbf{B}} = 0.050\ 0\hat{\mathbf{k}}$ T directed out of the plane of the page and extending from $x = 0$ to $x = 1.00$ m as shown in Figure P28.50. (a) Ignoring relativistic effects, find the angle α between the initial velocity vector of the proton

Figure P28.50

beam and the velocity vector after the beam emerges from the field. (b) Calculate the y component of the protons' momenta as they leave the magnetic field.

51. **Review.** A wire having a linear mass density of 1.00 g/cm is placed on a horizontal surface that has a coefficient of kinetic friction of 0.200. The wire carries a current of 1.50 A toward the east and slides horizontally to the north at constant velocity. What are (a) the magnitude and (b) the direction of the smallest magnetic field that enables the wire to move in this fashion?

A technician prepares a patient to receive a scan from a magnetic resonance imaging (MRI) machine in a hospital. Superconducting wires (Section 26.5) are used to create a very strong magnetic field in the interior of the machine, as well as around the machine. (*James Steidl/Shutterstock*)

Sources of the Magnetic Field

STORYLINE **You begin working part-time as a janitor at a local** hospital to earn some extra spending money. Your supervisor is going through your orientation. She is showing you the janitorial equipment and supplies. She then tells you to listen very carefully while she tells you about cleaning the room in which an MRI (Magnetic Resonance Imaging) machine is located. She explains that the MRI magnet is always on, even when the room is not being used during the night. Metal objects can be attracted to the very strong magnetic field of the MRI machine. She stresses that there have been accidents, even fatal accidents, involving MRI machines, including many involving janitorial equipment. There was even a case where a gun was pulled away from a police officer by an MRI and discharged as it struck the machine. As a result, there are special types of cleaning equipment and supplies that must be used in the area of the MRI machine. She makes you promise that you will use only the special janitorial equipment and supplies in the MRI room. After the orientation, you wonder how the MRI machine can create such a strong magnetic field. You also carefully study the special equipment for use in the MRI room to find out what makes it safe to use near the MRI machine.

CONNECTIONS In Chapter 28, we discussed the magnetic force exerted on a charged particle moving in a magnetic field. To complete the description of the magnetic interaction, this chapter explores where the magnetic field comes from in the first place: moving charges. We begin by showing how to use the law of Biot and Savart to calculate the magnetic field produced at some point in space by a small current element. This formalism is used to investigate several magnetic configurations. We eventually generate *Ampère's law*, which is useful in calculating the magnetic field of a highly symmetric configuration carrying a steady current, and is reminiscent of Gauss's law in Chapter 23. After completing this chapter, we will be prepared for the subsequent chapters, which address the combination of electric and magnetic effects that is called *electromagnetism*. Electromagnetism is the basis of many physical phenomena that we will study for the rest of the book.

29.1 The Biot–Savart Law

The relationship between magnetism and electricity was discovered in 1819 when, during a lecture demonstration, Hans Christian Oersted found that an electric current in a wire deflected a nearby compass needle. In the 1820s, further connections between electricity and magnetism were demonstrated independently by Faraday and Joseph Henry (1797–1878). They showed that an electric current can be produced in a circuit either by moving a magnet near the circuit or by changing the current in a nearby circuit. These observations demonstrate that a changing magnetic field creates an electric field. Years later, theoretical work by Maxwell showed that the reverse is also true: a changing electric field creates a magnetic field. In general, then, the source of a magnetic field is a *moving electric charge*.

Shortly after Oersted's discovery, Jean-Baptiste Biot (1774–1862) and Félix Savart (1791–1841) performed quantitative experiments on the force exerted by an electric current on a nearby magnet. From their experimental results, Biot and Savart arrived at a mathematical expression that gives the magnetic field at some point in space in terms of the current that produces the field. In Chapter 22, we showed that the mathematical expression for the electric field due to a single charge (Eq. 22.9) is relatively simple. We will find that the mathematical expression for the magnetic field is not quite so simple. That expression is based on the following experimental observations for the magnetic field $d\vec{\mathbf{B}}$ at a point P associated with a length element $d\vec{\mathbf{s}}$ of a wire carrying a steady current I (Fig. 29.1):

- The vector $d\vec{\mathbf{B}}$ is perpendicular both to $d\vec{\mathbf{s}}$ (which points in the direction of the current) and to the unit vector $\hat{\mathbf{r}}$ directed from $d\vec{\mathbf{s}}$ toward P.
- The magnitude of $d\vec{\mathbf{B}}$ is inversely proportional to r^2, where r is the distance from $d\vec{\mathbf{s}}$ to P.
- The magnitude of $d\vec{\mathbf{B}}$ is proportional to the current I and to the magnitude ds of the length element $d\vec{\mathbf{s}}$.
- The magnitude of $d\vec{\mathbf{B}}$ is proportional to $\sin\theta$, where θ is the angle between the vectors $d\vec{\mathbf{s}}$ and $\hat{\mathbf{r}}$.

These observations are summarized in the mathematical expression known today as the **Biot–Savart law:**

Biot–Savart law ▶

$$d\vec{\mathbf{B}} = \frac{\mu_0}{4\pi}\frac{I\,d\vec{\mathbf{s}} \times \hat{\mathbf{r}}}{r^2}$$ (29.1)

where μ_0 is a constant called the **permeability of free space:**

Permeability of free space ▶

$$\mu_0 = 4\pi \times 10^{-7} \text{ T} \cdot \text{m/A}$$ (29.2)

Interesting similarities and differences exist between Equation 29.1 for the magnetic field due to a current element and Equation 22.9 for the electric field due to a point charge. The magnitude of the magnetic field varies as the inverse square of the distance from the source, as does the electric field due to a point charge. The directions of the two fields are quite different, however. The electric field created by a point charge is radial, but the magnetic field created by a current element is perpendicular to both the length element $d\vec{\mathbf{s}}$ and the unit vector $\hat{\mathbf{r}}$ as described by the cross product in Equation 29.1. Hence, if the conductor lies in

The direction of the field is out of the page at P.

$d\vec{\mathbf{B}}_{\text{out}}$ P

$\hat{\mathbf{r}}$ θ
$\hat{\mathbf{r}}$ $d\vec{\mathbf{s}}$ I
P'
$d\vec{\mathbf{B}}_{\text{in}}$

The direction of the field is into the page at P'.

Figure 29.1 The magnetic field $d\vec{\mathbf{B}}$ at a point due to the current I through a length element $d\vec{\mathbf{s}}$ is given by the Biot–Savart law.

PITFALL PREVENTION 29.1

The Biot–Savart Law The magnetic field described by the Biot–Savart law is the field *due to* a given current-carrying conductor. Do not confuse this field with any *external* field that may be applied to the conductor from some other source.

the plane of the page as shown in Figure 29.1, $d\vec{\mathbf{B}}$ points out of the page at P and into the page at P'.

Notice that the field $d\vec{\mathbf{B}}$ in Equation 29.1 is the field created at a point by the current in only a small length element $d\vec{\mathbf{s}}$ of the conductor. To find the *total* magnetic field $\vec{\mathbf{B}}$ created at some point by a current of finite size, we must sum up contributions from all current elements $I\,d\vec{\mathbf{s}}$ that make up the current. That is, we must evaluate $\vec{\mathbf{B}}$ by integrating Equation 29.1:

$$\vec{\mathbf{B}} = \frac{\mu_0 I}{4\pi} \int \frac{d\vec{\mathbf{s}} \times \hat{\mathbf{r}}}{r^2} \qquad (29.3)$$

where the integral is taken over the entire current distribution. This expression must be handled with special care because the integrand is a cross product and therefore a vector quantity. We shall see one case of such an integration in Example 29.1.

Although the Biot–Savart law was discussed for a current-carrying wire, it is also valid for a current consisting of charges flowing through space such as the particle beam in an accelerator. In that case, $d\vec{\mathbf{s}}$ represents the length of a small segment of space in which the charges flow.

The Biot–Savart law gives the magnetic field of an isolated current element at some point, but such an isolated current element cannot exist the way an isolated electric charge can. A current element *must* be part of an extended current distribution because a complete circuit is needed for charges to flow. Therefore, the Biot–Savart law (Eq. 29.1) is only the first step in a calculation of a magnetic field; it must be followed by an integration over the current distribution as in Equation 29.3.

QUICK QUIZ 29.1 Consider the magnetic field due to the current in the wire shown in Figure 29.2. Rank the points A, B, and C in terms of magnitude of the magnetic field that is due to the current in just the length element $d\vec{\mathbf{s}}$ shown from greatest to least.

Figure 29.2 (Quick Quiz 29.1) Where is the magnetic field due to the current element the greatest?

Example 29.1 below investigates the magnetic field due to a long, straight wire. This geometry is important because it occurs often. Figure 29.3 is a perspective view of the magnetic field surrounding a long, straight, current-carrying wire. Because of the wire's symmetry, the magnetic field lines are circles concentric with the wire and lie in planes perpendicular to the wire. The magnitude of $\vec{\mathbf{B}}$ is constant on any circle of radius a and will be found in Example 29.1. A convenient rule for determining the direction of $\vec{\mathbf{B}}$ is to grasp the wire with the right hand, positioning the thumb along the direction of the current. The four fingers wrap in the direction of the magnetic field.

Figure 29.3 also shows that the magnetic field line has no beginning and no end. Rather, it forms a closed loop. That is a major difference between magnetic field lines and electric field lines, which begin on positive charges and end on negative charges. We will explore this feature of magnetic field lines further in Section 29.5. Test the right-hand rule for the field vectors in Figure 29.1.

Figure 29.3 The right-hand rule for determining the direction of the magnetic field surrounding a long, straight wire carrying a current. Notice that the magnetic field lines form circles around the wire.

Example 29.1 **Magnetic Field Surrounding a Thin, Straight Conductor**

Consider a thin, straight wire of finite length carrying a constant current I and placed along the x axis as shown in Figure 29.4 (page 774). Determine the magnitude and direction of the magnetic field at point P due to this current.

SOLUTION

Conceptualize From the Biot–Savart law, we expect that the magnitude of the field is proportional to the current in the wire and decreases as the distance a from the wire to point P increases. We also expect the field to depend on the angles θ_1 and θ_2 in Figure 29.4b. We place the origin at O and let point P be along the positive y axis.

continued

29.1 continued

Categorize   We are asked to find the magnetic field due to a simple current distribution, so this example is a typical problem for which the Biot–Savart law is appropriate. We must find the field contribution from a small element of current and then integrate over the current distribution. When finding the electric field due to continuous linear charge distribution in Section 23.1, we expressed the element of charge dq in Equation 23.1 as $dq = \lambda\,dx$. When using the Biot–Savart law, we have a similar element in $I\,d\vec{s}$. In this element, however, I is constant and $d\vec{s}$ is chosen to be parallel to the direction of I.

Figure 29.4   (Example 29.1) (a) A thin, straight wire carrying a current I. (b) The angles θ_1 and θ_2, from point P to the ends of the wire, are used for determining the net field.

Analyze   Let's start by considering a length element $d\vec{s}$ located a distance r from P as shown in Figure 29.4a. The direction of the magnetic field at point P due to the current in this element is out of the page because $d\vec{s} \times \hat{r}$ is out of the page. (Check this direction using the right-hand rule in Figure 29.3.) In fact, because *all* the current elements $I\,d\vec{s}$ lie in the plane of the page, they all produce a magnetic field directed out of the page at point P. Therefore, the direction of the magnetic field at point P is out of the page and we need only find the magnitude of the field.

Evaluate the cross product in the Biot–Savart law:

$$d\vec{s} \times \hat{r} = |d\vec{s} \times \hat{r}|\hat{k} = \left[dx\sin\left(\frac{\pi}{2} - \theta\right)\right]\hat{k} = (dx\cos\theta)\hat{k}$$

Substitute into Equation 29.1:

$$(1) \quad d\vec{B} = (dB)\hat{k} = \frac{\mu_0 I}{4\pi}\frac{dx\cos\theta}{r^2}\hat{k}$$

From the geometry in Figure 29.4a, express r in terms of θ:

$$(2) \quad r = \frac{a}{\cos\theta}$$

Notice that $\tan\theta = -x/a$ from the right triangle in Figure 29.4a (the negative sign is necessary because $d\vec{s}$ is located at a negative value of x) and solve for x:

$$x = -a\tan\theta$$

Find the differential dx:

$$(3) \quad dx = -a\sec^2\theta\,d\theta = -\frac{a\,d\theta}{\cos^2\theta}$$

Substitute Equations (2) and (3) into the expression for the z component of the field from Equation (1):

$$(4) \quad dB = -\frac{\mu_0 I}{4\pi}\left(\frac{a\,d\theta}{\cos^2\theta}\right)\left(\frac{\cos^2\theta}{a^2}\right)\cos\theta = -\frac{\mu_0 I}{4\pi a}\cos\theta\,d\theta$$

Integrate Equation (4) over all length elements on the wire, where the subtending angles range from θ_1 to θ_2 as defined in Figure 29.4b:

$$B = -\frac{\mu_0 I}{4\pi a}\int_{\theta_1}^{\theta_2}\cos\theta\,d\theta = \boxed{\frac{\mu_0 I}{4\pi a}(\sin\theta_1 - \sin\theta_2)} \qquad (29.4)$$

Finalize   We can use this result to find the magnitude of the magnetic field of *any* straight current-carrying wire if we know the geometry and hence the angles θ_1 and θ_2. Consider the special case of an infinitely long, straight wire. If the wire in Figure 29.4b becomes infinitely long, we see that $\theta_1 = \pi/2$ and $\theta_2 = -\pi/2$ for length elements ranging between positions $x = -\infty$ and $x = +\infty$. Because $(\sin\theta_1 - \sin\theta_2) = [\sin\pi/2 - \sin(-\pi/2)] = 2$, Equation 29.4 becomes

$$B = \frac{\mu_0 I}{2\pi a} \qquad (29.5)$$

Equations 29.4 and 29.5 both show that the magnitude of the magnetic field is proportional to the current and decreases with increasing distance from the wire, as expected. Equation 29.5 has the same mathematical form as the expression for the magnitude of the electric field due to a long charged wire (see Eq. 23.8).

Example 29.2   **Magnetic Field Due to a Curved Wire Segment**

Calculate the magnetic field at point O for the current-carrying wire segment shown in Figure 29.5. The wire consists of two straight portions and a circular arc of radius a, which subtends an angle θ.

29.2 continued

SOLUTION

Conceptualize The magnetic field at O due to the current in the straight segments AA' and CC' is zero because $d\vec{s}$ is parallel to $\hat{r}$ along these paths, which means that $d\vec{s} \times \hat{r} = 0$ for these paths. Therefore, we expect the magnetic field at O to be due only to the current in the curved portion of the wire.

Categorize Because we can ignore segments AA' and CC', this example is categorized as an application of the Biot–Savart law to the curved wire segment AC.

Analyze Each length element $d\vec{s}$ along path AC is at the same distance a from O, and the current in each contributes a field element $d\vec{B}$ directed into the page at O. Furthermore, at every point on AC, $d\vec{s}$ is perpendicular to $\hat{r}$; hence, $|d\vec{s} \times \hat{r}| = ds$.

Figure 29.5 (Example 29.2) A portion of a wire forms an arc of a circle. The length of the curved segment AC is s.

From Equation 29.1, find the magnitude of the field at O due to the current in an element of length ds:

$$dB = \frac{\mu_0}{4\pi} \frac{I\, ds}{a^2}$$

Integrate this expression over the curved path AC, noting that I and a are constants:

$$B = \frac{\mu_0 I}{4\pi a^2} \int ds = \frac{\mu_0 I}{4\pi a^2} s$$

From the geometry, note that $s = a\theta$ and substitute:

$$B = \frac{\mu_0 I}{4\pi a^2}(a\theta) = \frac{\mu_0 I}{4\pi a}\theta \qquad (29.6)$$

Finalize Equation 29.6 gives the magnitude of the magnetic field at O. The direction of $\vec{B}$ is into the page at O because $d\vec{s} \times \hat{r}$ is into the page for every length element.

WHAT IF? What if you were asked to find the magnetic field at the center of a circular wire loop of radius R that carries a current I? Can this question be answered at this point in our understanding of the source of magnetic fields?

Answer Yes, it can. The straight wires in Figure 29.5 do not contribute to the magnetic field. The only contribution is from the curved segment. As the angle θ increases, the curved segment becomes a full circle when $\theta = 2\pi$. Therefore, you can find the magnetic field at the center of a wire loop by letting $\theta = 2\pi$ in Equation 29.6:

$$B = \frac{\mu_0 I}{4\pi a} 2\pi = \frac{\mu_0 I}{2a}$$

This result is a limiting case of a more general result discussed in Example 29.3.

Example **29.3** **Magnetic Field on the Axis of a Circular Current Loop**

Consider a circular wire loop of radius a located in the yz plane and carrying a steady current I as in Figure 29.6. Calculate the magnetic field at an axial point P a distance x from the center of the loop.

SOLUTION

Conceptualize Compare this problem to Example 23.2 for the electric field due to a ring of charge. Figure 29.6 shows the magnetic field contribution $d\vec{B}$ at P due to a single current element at the top of the ring. This field vector can be resolved into components dB_x parallel to the axis of the ring and $dB_\perp$ perpendicular to the axis. Think about the magnetic field contribution from a current element at the bottom of the loop. Because of the symmetry of the situation, the perpendicular components of the field due to elements at the top and bottom of the ring cancel. This cancellation occurs for all pairs of segments around the ring, so we can ignore the perpendicular component of the field and focus solely on the parallel components, which simply add.

Figure 29.6 (Example 29.3) Geometry for calculating the magnetic field at a point P lying on the axis of a current loop. By symmetry, the total field $\vec{B}$ is along this axis.

continued

29.3 continued

Categorize   We are asked to find the magnetic field due to a simple current distribution, so this example is a typical problem for which the Biot–Savart law is appropriate.

Analyze   In this situation, every length element $d\vec{\mathbf{s}}$ is perpendicular to the vector $\hat{\mathbf{r}}$ at the location of the element. Therefore, for any element, $|d\vec{\mathbf{s}} \times \hat{\mathbf{r}}| = (ds)(1) \sin 90° = ds$. Furthermore, all length elements around the loop are at the same distance r from P, where $r^2 = a^2 + x^2$.

Use Equation 29.1 to find the magnitude of $d\vec{\mathbf{B}}$ due to the current in any length element $d\vec{\mathbf{s}}$:

$$dB = \frac{\mu_0 I}{4\pi} \frac{|d\vec{\mathbf{s}} \times \hat{\mathbf{r}}|}{r^2} = \frac{\mu_0 I}{4\pi} \frac{ds}{(a^2 + x^2)}$$

Find the x component of the field element:

$$(1) \quad dB_x = \frac{\mu_0 I}{4\pi} \frac{ds}{(a^2 + x^2)} \cos\theta$$

From the geometry, evaluate $\cos\theta$:

$$\cos\theta = \frac{a}{(a^2 + x^2)^{1/2}}$$

Substitute into Equation (1) and integrate over the entire loop, noting that x and a are both constant:

$$B_x = \frac{\mu_0 I}{4\pi} \oint \frac{ds}{a^2 + x^2} \left[\frac{a}{(a^2 + x^2)^{1/2}} \right] = \frac{\mu_0 I}{4\pi} \frac{a}{(a^2 + x^2)^{3/2}} \oint ds$$

The remaining integral is the circumference of the loop:

$$B_x = \frac{\mu_0 I}{4\pi} \frac{a}{(a^2 + x^2)^{3/2}} (2\pi a) = \boxed{\frac{\mu_0 I a^2}{2(a^2 + x^2)^{3/2}}} \qquad (29.7)$$

Finalize   To find the magnetic field at the center of the loop, set $x = 0$ in Equation 29.7. At this special point,

$$B = \frac{\mu_0 I}{2a} \quad \text{(at } x = 0) \qquad (29.8)$$

which is consistent with the result of the **What If?** feature of Example 29.2.

The pattern of magnetic field lines for a circular current loop is shown in Figure 29.7a. For clarity, the lines are drawn for only the plane that contains the axis of the loop. The field-line pattern is axially symmetric and is similar to the pattern around a bar magnet, which is shown in Figure 29.7b.

WHAT IF?   What if we consider points on the x axis very far from the loop? How does the magnetic field behave at these distant points?

Answer   In this case, in which $x \gg a$, we can neglect the term a^2 in the denominator of Equation 29.7 and obtain

$$B \approx \frac{\mu_0 I a^2}{2x^3} \quad \text{(for } x \gg a) \qquad (29.9)$$

Figure 29.7 (Example 29.3) (a) Magnetic field lines surrounding a current loop. (b) Magnetic field lines surrounding a bar magnet. Notice the similarity between this line pattern and that of a current loop.

The magnitude of the magnetic moment μ of the loop is defined as the product of current and loop area (see Eq. 28.16): $\mu = I(\pi a^2)$ for our circular loop. Therefore, we can express Equation 29.9 as

$$B \approx \frac{\mu_0}{2\pi} \frac{\mu}{x^3} \qquad (29.10)$$

This result is similar in form to the expression for the electric field due to an electric dipole, $E = k_e(p/y^3)$ (see Example 22.6), where $p = 2aq$ is the electric dipole moment as defined in Equation 25.18.

Although the Earth's magnetic field pattern (Fig. 28.3) is similar to the one that would be set up by a bar magnet (Fig. 29.7b) deep within the Earth, the source of this magnetic field cannot be large masses of permanently magnetized material. The Earth does indeed have large deposits of iron ore deep beneath its surface, but the high temperatures in the Earth's core prevent the iron from retaining any permanent magnetization. (See Section 29.6.) Scientists consider it more likely that the source of the Earth's magnetic field is convection currents in the Earth's outer core. Charged ions or electrons circulating in the liquid outer core could produce a magnetic field just like a current loop does, as in Example 29.3. There is also strong evidence that the magnitude of a planet's magnetic field is related to the planet's rate of rotation. For example, Jupiter rotates faster than the Earth, and space probes indicate that Jupiter's magnetic field is stronger than the Earth's. Venus, on the other hand, rotates more slowly than the Earth, and its magnetic field is found to be weaker. Investigation into the cause of the Earth's magnetism is ongoing.

29.2 The Magnetic Force Between Two Parallel Conductors

In Chapter 28, we described the magnetic force that acts on a current-carrying conductor placed in an external magnetic field. Because a current in a conductor acts as a source of a magnetic field, it is easy to understand that two current-carrying conductors exert magnetic forces on each other. One wire establishes the magnetic field and the other wire is modeled as a collection of particles in a magnetic field. Such forces between wires can be used as the basis for defining the ampere and the coulomb.

Consider two long, straight, parallel wires separated by a distance a and carrying currents I_1 and I_2 in the same direction as in Figure 29.8. Let's determine the force exerted on one wire due to the magnetic field set up by the other wire. Wire 2, which carries a current I_2 and is identified arbitrarily as the source wire, creates a magnetic field $\vec{B}_2$ at the location of wire 1, the test wire. The magnitude of this magnetic field is the same at all points on wire 1. The direction of $\vec{B}_2$ can be found using the right-hand rule in Figure 29.3, and is perpendicular to wire 1 as shown in Figure 29.8. According to Equation 28.10, the magnetic force on a length ℓ of wire 1 is $\vec{F}_1 = I_1 \vec{\ell} \times \vec{B}_2$. Because $\vec{\ell}$ is perpendicular to $\vec{B}_2$ in this situation, the magnitude of $\vec{F}_1$ is $F_1 = I_1 \ell B_2$. Because the magnitude of $\vec{B}_2$ is given by Equation 29.5,

$$F_1 = I_1 \ell B_2 = I_1 \ell \left(\frac{\mu_0 I_2}{2\pi a} \right) = \frac{\mu_0 I_1 I_2}{2\pi a} \ell \tag{29.11}$$

The direction of $\vec{F}_1$ is toward wire 2 because $\vec{\ell} \times \vec{B}_2$ is in that direction. When the field set up at wire 2 by wire 1 is calculated, the force $\vec{F}_2$ acting on wire 2 is found to be equal in magnitude and opposite in direction to $\vec{F}_1$, which is what we expect because Newton's third law must be obeyed. When the currents are in opposite directions (that is, when one of the currents is reversed in Fig. 29.8), the forces are reversed and the wires repel each other. Hence, parallel conductors carrying currents in the *same* direction *attract* each other, and parallel conductors carrying currents in *opposite* directions *repel* each other.

Because the magnitudes of the forces are the same on both wires, we denote the magnitude of the magnetic force between the wires as simply F_B. We can rewrite this magnitude in terms of the force per unit length:

$$\frac{F_B}{\ell} = \frac{\mu_0 I_1 I_2}{2\pi a} \tag{29.12}$$

The force between two parallel wires is used to define the **ampere** as follows:

> When the magnitude of the force per unit length between two long, parallel wires that carry identical currents and are separated by 1 m is 2×10^{-7} N/m, the current in each wire is defined to be 1 A.

◀ Definition of the ampere

The value 2×10^{-7} N/m is obtained from Equation 29.12 with $I_1 = I_2 = 1$ A and $a = 1$ m. Because this definition is based on a force, a mechanical measurement can be used to standardize the ampere. For instance, the National Institute of

The field $\vec{B}_2$ due to the current in wire 2 exerts a magnetic force of magnitude $F_1 = I_1 \ell B_2$ on wire 1.

Figure 29.8 Two parallel wires that each carry a steady current exert a magnetic force on each other. The force is attractive if the currents are parallel (as shown) and repulsive if the currents are antiparallel.

Standards and Technology uses an instrument called a *current balance* for primary current measurements. The results are then used to standardize other, more conventional electrical instruments such as ammeters.

The SI unit of charge, the **coulomb,** is defined in terms of the ampere: When a conductor carries a steady current of 1 A, the quantity of charge that flows through a cross section of the conductor in 1 s is 1 C.

In deriving Equations 29.11 and 29.12, we assumed both wires are long compared with their separation distance. In fact, only one wire needs to be long. The equations accurately describe the forces exerted on each other by a long wire and a straight, parallel wire of limited length ℓ.

QUICK QUIZ 29.2 A loose spiral spring carrying no current is hung from a ceiling. When a switch is thrown so that a current exists in the spring, do the coils **(a)** move closer together, **(b)** move farther apart, or **(c)** not move at all?

Example 29.4 **Suspending a Wire**

Two infinitely long, parallel wires are lying on the ground a distance $a = 1.00$ cm apart as shown in Figure 29.9a. A third wire, of length $L = 10.0$ m and mass 400 g, carries a current of $I_1 = 100$ A and is levitated above the first two wires, at a horizontal position midway between them. The infinitely long wires carry equal currents I_2 in the same direction, but in the direction opposite that in the levitated wire. What current must the infinitely long wires carry so that the three wires form an equilateral triangle?

Figure 29.9 (Example 29.4) (a) Two current-carrying wires lie on the ground and suspend a third wire in the air by magnetic forces. (b) End view. In the situation described in the example, the three wires form an equilateral triangle. The two magnetic forces on the levitated wire are $\vec{\mathbf{F}}_{B,L}$, the force due to the left-hand lower wire, and $\vec{\mathbf{F}}_{B,R}$, the force due to the right-hand wire. The gravitational force $\vec{\mathbf{F}}_g$ on the levitated wire is also shown.

SOLUTION

Conceptualize Because the current in the short wire is opposite those in the long wires, the short wire is repelled from both of the others. Imagine the currents in the long wires in Figure 29.9a are increased. The repulsive force becomes stronger, and the levitated wire rises to the point at which the wire is once again levitated in equilibrium at a higher position. Figure 29.9b shows the desired situation with the three wires forming an equilateral triangle.

Categorize Because the levitated wire is subject to forces but does not accelerate, it is modeled as a *particle in equilibrium.*

Analyze The horizontal components of the magnetic forces on the levitated wire cancel. The vertical components are both positive and add together. Choose the z axis to be upward through the top wire in Figure 29.9b and in the plane of the page.

Find the total magnetic force in the upward direction on the levitated wire:

$$\vec{\mathbf{F}}_B = 2\left(\frac{\mu_0 I_1 I_2}{2\pi a}\ell\right)\cos\theta\,\hat{\mathbf{k}} = \frac{\mu_0 I_1 I_2}{\pi a}\ell\cos\theta\,\hat{\mathbf{k}}$$

Find the gravitational force on the levitated wire:

$$\vec{\mathbf{F}}_g = -mg\hat{\mathbf{k}}$$

Apply the particle in equilibrium model by adding the forces and setting the net force equal to zero:

$$\sum\vec{\mathbf{F}} = \vec{\mathbf{F}}_B + \vec{\mathbf{F}}_g = \frac{\mu_0 I_1 I_2}{\pi a}\ell\cos\theta\,\hat{\mathbf{k}} - mg\hat{\mathbf{k}} = 0$$

Solve for the current in the wires on the ground:

$$I_2 = \frac{mg\pi a}{\mu_0 I_1 \ell\cos\theta}$$

Substitute numerical values:

$$I_2 = \frac{(0.400\text{ kg})(9.80\text{ m/s}^2)\pi(0.010\,0\text{ m})}{(4\pi\times 10^{-7}\text{ T}\cdot\text{m/A})(100\text{ A})(10.0\text{ m})\cos 30.0°}$$

$$= 113\text{ A}$$

Finalize The currents in all wires are on the order of 10^2 A. Such large currents would require specialized equipment. Therefore, this situation would be difficult to establish in practice. Is the equilibrium of wire 1 stable or unstable?

29.3 Ampère's Law

In Figure 28.1, we show how compasses can be used to map out the magnetic field lines around a bar magnet. Figure 29.10 shows how compasses can be used to demonstrate the magnetic field lines around a long, vertical wire. When no current is present in the wire (Fig. 20.10a), all the compass needles point in the same direction (that of the horizontal component of the Earth's magnetic field) as expected. When the wire carries a strong, steady current, the needles all deflect in a direction tangent to the circle as in Figure 29.10b. These observations demonstrate that the direction of the magnetic field produced by the current in the wire is consistent with the right-hand rule described in Figure 29.3. When the current is reversed, the needles in Figure 29.10b also reverse.

Now let's evaluate the product $\vec{\mathbf{B}} \cdot d\vec{\mathbf{s}}$ for a small length element $d\vec{\mathbf{s}}$ on the circular path defined by the compass needles and sum the products for all elements over the closed circular path.[1] Along this path, the vectors $d\vec{\mathbf{s}}$ and $\vec{\mathbf{B}}$ are parallel at each point (see Fig. 29.10b), so $\vec{\mathbf{B}} \cdot d\vec{\mathbf{s}} = B\,ds$. Furthermore, the magnitude of $\vec{\mathbf{B}}$ is constant on this circle and is given by Equation 29.5. Therefore, the sum of the products $B\,ds$ over the closed path, which is equivalent to the line integral of $\vec{\mathbf{B}} \cdot d\vec{\mathbf{s}}$, is

$$\oint \vec{\mathbf{B}} \cdot d\vec{\mathbf{s}} = B \oint ds = \frac{\mu_0 I}{2\pi r}(2\pi r) = \mu_0 I$$

where $\oint ds = 2\pi r$ is the circumference of the circular path of radius r. Although this result was calculated for the special case of a circular path surrounding a wire of infinite length, it holds for a closed path of *any* shape (an *amperian loop*) surrounding a current that exists in an unbroken circuit. The general case, known as **Ampère's law,** can be stated as follows:

> The line integral of $\vec{\mathbf{B}} \cdot d\vec{\mathbf{s}}$ around *any* closed path equals $\mu_0 I$, where I is the total steady current passing through *any* surface bounded by the closed path:
>
> $$\oint \vec{\mathbf{B}} \cdot d\vec{\mathbf{s}} = \mu_0 I \qquad (29.13)$$

◄ Ampère's law

When no current is present in the wire, all compass needles point in the same direction (toward the Earth's north pole).

When the wire carries a strong current, the compass needles deflect in a direction tangent to the circle, which is the direction of the magnetic field created by the current.

$I = 0$

I

$\vec{\mathbf{B}}$

$d\vec{\mathbf{s}}$

a

b

c

Richard Megna/Fundamental Photographs

iStockphoto.com/HultonArchive

Andre-Marie Ampère
French Physicist (1775–1836)
Ampère is credited with the discovery of electromagnetism, which is the relationship between electric currents and magnetic fields. Ampère's genius, particularly in mathematics, became evident by the time he was 12 years old; his personal life, however, was filled with tragedy. His father, a wealthy city official, was guillotined during the French Revolution, and his wife died young, in 1803. Ampère died at the age of 61 of pneumonia.

Figure 29.10 (a) and (b) Compasses show the effects of the current in a nearby wire. (c) Circular magnetic field lines surrounding a current-carrying conductor, displayed with iron filings.

[1]You may wonder why we would choose to evaluate this scalar product. The origin of Ampère's law is in 19th-century science, in which a "magnetic charge" (the supposed analog to an isolated electric charge) was imagined to be moved around a circular field line. The work done on the charge was related to $\vec{\mathbf{B}} \cdot d\vec{\mathbf{s}}$, just as the work done moving an electric charge in an electric field is related to $\vec{\mathbf{E}} \cdot d\vec{\mathbf{s}}$. Therefore, Ampère's law, a valid and useful principle, arose from an erroneous and abandoned work calculation!

Notice the italics in the shaded box. We can choose *any* path and *any* surface bounded by that path. In most cases, we choose paths that are simple, such as circles or rectangles. Also, in most cases, we choose the flat surface bounded by the path. Imagine a drumhead: the circular rim is the path and the flat drumhead membrane is the surface. If the membrane vibrates, however, there are instants of time when the path remains fixed, but the surface is not flat; the membrane bows upward or downward from its equilibrium position. We will see in Chapter 33 a situation in which we need to consider a surface other than the flat one bounded by the path.

Ampère's law describes the creation of magnetic fields by all continuous current configurations, but at our mathematical level it is useful only for calculating the magnetic field of current configurations having a high degree of symmetry. Its use is similar to that of Gauss's law in calculating electric fields for highly symmetric charge distributions.

QUICK QUIZ 29.3 Rank the magnitudes of $\oint \vec{\mathbf{B}} \cdot d\vec{\mathbf{s}}$ for the closed paths *a* through *d* in Figure 29.11 from greatest to least.

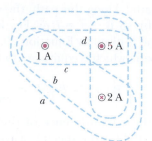

Figure 29.11 (Quick Quiz 29.3) Four closed paths around three current-carrying wires.

QUICK QUIZ 29.4 Rank the magnitudes of $\oint \vec{\mathbf{B}} \cdot d\vec{\mathbf{s}}$ for the closed paths *a* through *d* in Figure 29.12 from greatest to least.

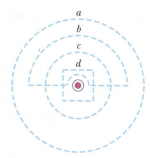

Figure 29.12 (Quick Quiz 29.4) Several closed paths near a single current-carrying wire.

Example 29.5 The Magnetic Field Created by a Long Current-Carrying Wire

A long, straight wire of radius R carries a steady current I that is uniformly distributed through the cross section of the wire (Fig. 29.13). Calculate the magnetic field a distance r from the center of the wire in the regions $r \geq R$ and $r < R$.

SOLUTION

Conceptualize Study Figure 29.13 to understand the structure of the wire and the current in the wire. The current creates magnetic fields everywhere, both inside and outside the wire. Based on our discussions about long, straight wires, we expect the magnetic field lines to be circles centered on the central axis of the wire. In Example 29.1, we used *a* for the distance from a wire of negligible radius. In this example, the wire has a radius R. We will use r for the distance from the center of the wire and compare regions both inside and outside the wire.

Categorize Because the wire has a high degree of symmetry, we categorize this example as an Ampère's law problem. For the $r \geq R$ case, we should arrive at the same result as was obtained in Example 29.1, where we applied the Biot–Savart law to the same situation.

Analyze For the magnetic field exterior to the wire, let us choose for our path of integration circle 1 in Figure 29.13. From symmetry, $\vec{\mathbf{B}}$ must be constant in magnitude and parallel to $d\vec{\mathbf{s}}$ at every point on this circle.

Figure 29.13 (Example 29.5) A long, straight wire of radius R carrying a steady current I uniformly distributed across the cross section of the wire. The magnetic field at any point can be calculated from Ampère's law using a circular path of radius r, concentric with the wire.

29.5 continued

Note that the total current passing through the plane of the circle is I and apply Ampère's law:

$$\oint \vec{B} \cdot d\vec{s} = B\oint ds = B(2\pi r) = \mu_0 I$$

Solve for B:

$$B = \frac{\mu_0 I}{2\pi r} \quad \text{(for } r \geq R) \tag{29.14}$$

Now consider the interior of the wire, where $r < R$. Here the current I' passing through the plane of circle 2 is less than the total current I. Because the current is *uniformly distributed across the cross section of the wire*, the current density J (Eq. 26.5) is constant in the interior of the wire. Therefore, for any area A of the interior perpendicular to the length of the wire, the current passing through that area is $I' = JA$.

Set the ratio of the current I' enclosed by circle 2 to the entire current I equal to the ratio of the area πr^2 enclosed by circle 2 to the cross-sectional area πR^2 of the wire:

$$\frac{I'}{I} = \frac{JA'}{JA} = \frac{\pi r^2}{\pi R^2}$$

Solve for I':

$$I' = \frac{r^2}{R^2} I$$

Apply Ampère's law to circle 2:

$$\oint \vec{B} \cdot d\vec{s} = B(2\pi r) = \mu_0 I' = \mu_0 \left(\frac{r^2}{R^2} I \right)$$

Solve for B:

$$B = \left(\frac{\mu_0 I}{2\pi R^2} \right) r \quad \text{(for } r < R) \tag{29.15}$$

Finalize The magnetic field exterior to the wire (Eq. 29.14) is identical in form to Equation 29.5, except for the substitution of r for a. As is often the case in highly symmetric situations, it is much easier to use Ampère's law than the Biot–Savart law (Example 29.1). The magnetic field interior to the wire is similar in form to the expression for the electric field inside a uniformly charged sphere (see Example 23.6). The magnitude of the magnetic field versus r for this configuration is plotted in Figure 29.14. Inside the wire, $B \to 0$ as $r \to 0$. Furthermore, Equations 29.14 and 29.15 give the same value of the magnetic field at $r = R$, demonstrating that the magnetic field is continuous at the surface of the wire.

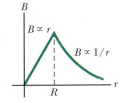

Figure 29.14 (Example 29.5) Magnitude of the magnetic field versus r for the wire shown in Figure 29.13. The field is proportional to r inside the wire and varies as $1/r$ outside the wire.

Example 29.6 **The Magnetic Field Created by a Toroid**

A device called a *toroid* (Fig. 29.15) is often used to create an almost uniform magnetic field in some enclosed area. The device consists of a conducting wire wrapped around a ring (a *torus*, which is shaped like a doughnut) made of a nonconducting material. For a toroid having N closely spaced turns of wire, calculate the magnetic field in the region occupied by the torus, a distance r from the center.

SOLUTION

Conceptualize Study Figure 29.15 carefully to understand how the wire is wrapped around the torus. The torus could be a solid material or it could be air, with a stiff wire wrapped into the shape shown in Figure 29.15 to form an empty toroid. Imagine each turn of the wire to be a circular loop as in Example 29.3. The magnetic field at the center of the loop is perpendicular to the plane of the loop. Therefore, the magnetic field lines of the collection of loops will form circles within the toroid such as suggested by loop 1 in Figure 29.15.

Categorize Because the toroid has a high degree of symmetry, we categorize this example as an Ampère's law problem.

Analyze Consider the circular amperian loop (loop 1) of radius r in the plane of Figure 29.15. By symmetry, the magnitude of the field is constant

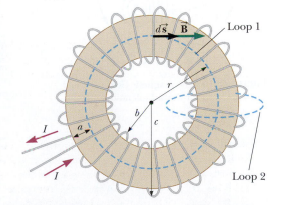

Figure 29.15 (Example 29.6) A toroid consisting of many turns of wire. If the turns are closely spaced, the magnetic field in the interior of the toroid is tangent to the dashed circle (loop 1) and varies as $1/r$. The dimension a is the cross-sectional radius of the torus. The field outside the toroid is very small and can be described by using the amperian loop (loop 2) at the right side, perpendicular to the page.

continued

29.6 continued

on this circle and tangent to it, so $\vec{\mathbf{B}} \cdot d\vec{\mathbf{s}} = B\,ds$. Furthermore, the wire passes through the loop N times, so the total current through the loop is NI.

Apply Ampère's law to loop 1:

$$\oint \vec{\mathbf{B}} \cdot d\vec{\mathbf{s}} = B \oint ds = B(2\pi r) = \mu_0 NI$$

Solve for B:

$$B = \frac{\mu_0 NI}{2\pi r} \tag{29.16}$$

Finalize This result shows that B varies as $1/r$ and hence is *nonuniform* in the region occupied by the torus. If, however, r is very large compared with the cross-sectional radius a of the torus, the field is approximately uniform inside the torus.

For an ideal toroid, in which the turns are closely spaced, the external magnetic field is close to zero, but it is not exactly zero. In Figure 29.15, imagine the radius r of amperian loop 1 to be either smaller than b or larger than c. In either case, the loop encloses zero net current, so $\oint \vec{\mathbf{B}} \cdot d\vec{\mathbf{s}} = 0$. You might think this result proves that $\vec{\mathbf{B}} = 0$, but it does not. Consider the amperian loop (loop 2) on the right side of the toroid in

Figure 29.15. The plane of this loop is perpendicular to the page, and the toroid passes through the loop. As charges enter the toroid as indicated by the current directions in Figure 29.15, they work their way counterclockwise around the toroid. Therefore, there is a counterclockwise current around the toroid, so that a current passes through amperian loop 2! This current is small, but not zero. As a result, the toroid acts as a current loop and produces a weak external field of the form shown in Figure 29.7. The reason $\oint \vec{\mathbf{B}} \cdot d\vec{\mathbf{s}} = 0$ for amperian loop 1 of radius $r < b$ or $r > c$ is that the field lines are perpendicular to $d\vec{\mathbf{s}}$, *not* because $\vec{\mathbf{B}} = 0$.

29.4 The Magnetic Field of a Solenoid

A **solenoid** is a long wire wound in the form of a helix. With this configuration, a reasonably uniform magnetic field can be produced in the space surrounded by the turns of wire—which we shall call the *interior* of the solenoid—when the solenoid carries a current. When the turns are closely spaced, each can be approximated as a circular loop; the net magnetic field is the vector sum of the fields resulting from all the turns.

Figure 29.16 shows the magnetic field lines surrounding a loosely wound solenoid. The field lines in the interior are nearly parallel to one another, are uniformly distributed, and are close together, indicating that the field in this space is strong and almost uniform.

If the turns are closely spaced and the solenoid is of finite length, the external magnetic field lines are as shown in Figure 29.17a. This field line distribution is similar to that surrounding a bar magnet (Fig. 29.17b). Hence, one end of the

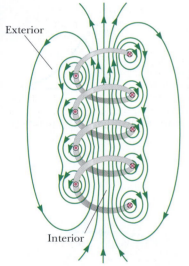

Exterior

Interior

Figure 29.16 The magnetic field lines for a loosely wound solenoid.

The magnetic field lines resemble those of a bar magnet, meaning that the solenoid effectively has north and south poles.

N

S

a

b

Henry Leap and Jim Lehman

Figure 29.17 (a) Magnetic field lines for a tightly wound solenoid of finite length, carrying a steady current. The field in the interior space is strong and nearly uniform. (b) The magnetic field pattern of a bar magnet, displayed with small iron filings on a sheet of paper.

solenoid behaves like the north pole of a magnet and the opposite end behaves like the south pole. As the length of the solenoid increases, the interior field becomes more uniform and the exterior field becomes weaker. An *ideal solenoid* is approached when the turns are closely spaced and the length is much greater than the radius of the turns. Figure 29.18 shows a longitudinal cross section of part of such a solenoid carrying a current I. In this case, the external field is close to zero and the interior field is uniform over a great volume.

Consider the amperian loop (loop 1) perpendicular to the page in Figure 29.18, surrounding the ideal solenoid. This loop encloses a small current as the charges in the wire move coil by coil along the length of the solenoid. Therefore, there is a nonzero magnetic field outside the solenoid. It is a weak field, with circular field lines, like those due to a line of current as in Figure 29.3. For an ideal solenoid, this weak field is the only field external to the solenoid.

We can use Ampère's law to obtain a quantitative expression for the interior magnetic field in an ideal solenoid. Because the solenoid is ideal, $\vec{\mathbf{B}}$ in the interior space is uniform and parallel to the axis and the magnetic field lines in the exterior space form circles around the solenoid. The planes of these circles are perpendicular to the page. Consider the rectangular path (loop 2) of length ℓ and width w shown in Figure 29.18. Let's apply Ampère's law to this path by evaluating the integral of $\vec{\mathbf{B}} \cdot d\vec{\mathbf{s}}$ over each side of the rectangle. The contribution along side 3 is zero because the external magnetic field lines are perpendicular to the path in this region. The contributions from sides 2 and 4 are both zero, again because $\vec{\mathbf{B}}$ is perpendicular to $d\vec{\mathbf{s}}$ along these paths, both inside and outside the solenoid. Side 1 gives a contribution to the integral because along this path $\vec{\mathbf{B}}$ is uniform and parallel to $d\vec{\mathbf{s}}$. The integral over the closed rectangular path is therefore

$$\oint \vec{\mathbf{B}} \cdot d\vec{\mathbf{s}} = \int_{\text{path 1}} \vec{\mathbf{B}} \cdot d\vec{\mathbf{s}} = B \int_{\text{path 1}} ds = B\ell$$

The right side of Ampère's law involves the total current I through the area bounded by the path of integration. In this case, the total current through the rectangular path equals the current through each turn multiplied by the number of turns. If N is the number of turns in the length ℓ, the total current through the rectangle is NI. Therefore, Ampère's law applied to this path gives

$$\oint \vec{\mathbf{B}} \cdot d\vec{\mathbf{s}} = B\ell = \mu_0 NI$$

$$B = \mu_0 \frac{N}{\ell} I = \mu_0 nI \qquad (29.17)$$

◀ Magnetic field inside a solenoid

where $n = N/\ell$ is the number of turns per unit length.

We also could obtain this result by reconsidering the magnetic field of a toroid (see Example 29.6). If the radius r of the torus in Figure 29.15 containing N turns is much greater than the toroid's cross-sectional radius a, a short section of the toroid approximates a solenoid for which $n = N/2\pi r$. In this limit, Equation 29.16 agrees with Equation 29.17.

Equation 29.17 is valid only for points near the center of the length (that is, far from the ends) of a very long solenoid. As you might expect, the field near each end is smaller than the value given by Equation 29.17. As the length of a solenoid increases, the magnitude of the field at the end approaches half the magnitude at the center (see Problem 45).

QUICK QUIZ 29.5 Consider a solenoid that is very long compared with its radius. Of the following choices, what is the most effective way to increase the magnetic field in the interior of the solenoid? **(a)** double its length, keeping the number of turns per unit length constant **(b)** reduce its radius by half, keeping the number of turns per unit length constant **(c)** overwrap the entire solenoid with an additional layer of current-carrying wire

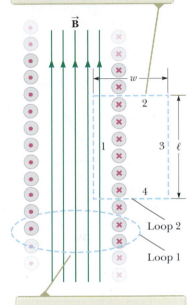

Ampère's law applied to the rectangular dashed path can be used to calculate the magnitude of the interior field.

Ampère's law applied to the circular path whose plane is perpendicular to the page can be used to show that there is a weak field outside the solenoid.

Figure 29.18 Cross-sectional view of an ideal solenoid, where the interior magnetic field is uniform and the exterior field is close to zero.

29.5 Gauss's Law in Magnetism

The flux associated with a magnetic field is defined in a manner similar to that used to define electric flux (see Eq. 23.4). Consider an element of area dA on an arbitrarily shaped surface as shown in Figure 29.19. If the magnetic field at this element is $\vec{B}$, the magnetic flux through the element is $\vec{B} \cdot d\vec{A}$, where $d\vec{A}$ is a vector that is perpendicular to the surface and has a magnitude equal to the area dA. Therefore, the total magnetic flux Φ_B through the surface is

◀ Definition of magnetic flux

$$\Phi_B \equiv \int \vec{B} \cdot d\vec{A} \qquad (29.18)$$

Consider the special case of a plane of area A in a uniform field $\vec{B}$ that makes an angle θ with $d\vec{A}$. The magnetic flux through the plane in this case is

$$\Phi_B = BA \cos \theta \qquad (29.19)$$

If the magnetic field is parallel to the plane as in Figure 29.20a, then $\theta = 90°$ and the flux through the plane is zero. If the field is perpendicular to the plane as in Figure 29.20b, then $\theta = 0$ and the flux through the plane is BA (the maximum value).

The unit of magnetic flux is T · m², which is defined as a *weber* (Wb); 1 Wb = 1 T · m².

Figure 29.19 The magnetic flux through an area element dA is $\vec{B} \cdot d\vec{A} = B\,dA \cos \theta$, where $d\vec{A}$ is a vector perpendicular to the surface.

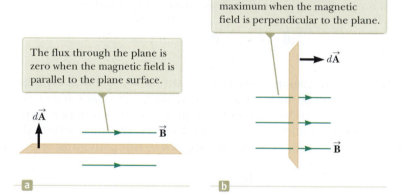

The flux through the plane is zero when the magnetic field is parallel to the plane surface.

The flux through the plane is a maximum when the magnetic field is perpendicular to the plane.

Figure 29.20 Magnetic flux through a plane lying in a magnetic field.

Example 29.7   **Magnetic Flux Through a Rectangular Loop**

A rectangular loop of width a and length b is located near a long wire carrying a current I (Fig. 29.21). The distance between the wire and the closest side of the loop is c. The wire is parallel to the long side of the loop. Find the total magnetic flux through the loop due to the current in the wire.

SOLUTION

Conceptualize   As we saw in Figure 29.3, the magnetic field lines due to the wire will be circles, many of which will pass through the rectangular loop. We know that the magnitude of the magnetic field is a function of distance r from a long wire. Therefore, the magnetic field varies over the area of the rectangular loop.

Categorize   Because the magnetic field varies over the area of the loop, we must integrate over this area to find the total flux. That identifies this as an analysis problem.

Figure 29.21 (Example 29.7) The magnetic field due to the wire carrying a current I is not uniform over the rectangular loop.

29.7 continued

Analyze Noting that $\vec{\mathbf{B}}$ is parallel to $d\vec{\mathbf{A}}$ at any point within the loop, find the magnetic flux through the rectangular area using Equation 29.18 and incorporate Equation 29.14 for the magnetic field:

$$\Phi_B = \int \vec{\mathbf{B}} \cdot d\vec{\mathbf{A}} = \int B\, dA = \int \frac{\mu_0 I}{2\pi r}\, dA$$

Express the area element (the tan strip in Fig. 29.21) as $dA = b\, dr$ and substitute:

$$\Phi_B = \int \frac{\mu_0 I}{2\pi r}\, b\, dr = \frac{\mu_0 Ib}{2\pi} \int \frac{dr}{r}$$

Integrate from $r = c$ to $r = a + c$:

$$\Phi_B = \frac{\mu_0 Ib}{2\pi} \int_c^{a+c} \frac{dr}{r} = \frac{\mu_0 Ib}{2\pi} \ln r \Big|_c^{a+c}$$

$$= \frac{\mu_0 Ib}{2\pi} \ln\left(\frac{a+c}{c}\right) = \frac{\mu_0 Ib}{2\pi} \ln\left(1 + \frac{a}{c}\right)$$

Finalize Notice how the flux depends on the size of the loop. Increasing either a or b increases the flux as expected. If c becomes large such that the loop is very far from the wire, the flux approaches zero, also as expected. If c goes to zero, the flux becomes infinite. In principle, this infinite value occurs because the field becomes infinite at $r = 0$ (assuming an infinitesimally thin wire). That will not happen in reality because the thickness of the wire prevents the left edge of the loop from reaching $r = 0$.

In Chapter 23, we found that the electric flux through a closed surface surrounding a net charge is proportional to that charge (Gauss's law). In other words, the number of electric field lines leaving the surface depends only on the net charge within it. This behavior exists because electric field lines originate and terminate on electric charges.

The situation is quite different for magnetic fields, which are continuous and form closed loops. In other words, as illustrated by the magnetic field lines of a current in Figure 29.3 and of a bar magnet in Figure 29.22, magnetic field lines do not begin or end at any point. For any closed surface such as the one outlined by the dashed line in Figure 29.22, the number of lines entering the surface equals the number leaving the surface; therefore, the net magnetic flux is zero. In contrast, for a closed surface surrounding one charge of an electric dipole (Fig. 29.23), the net electric flux is not zero.

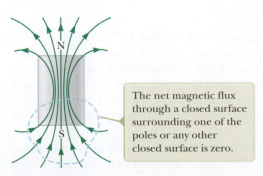

The net magnetic flux through a closed surface surrounding one of the poles or any other closed surface is zero.

Figure 29.22 The magnetic field lines of a bar magnet form closed loops. (The dashed line represents the intersection of a closed surface with the page.)

The electric flux through a closed surface surrounding one of the charges is not zero.

Figure 29.23 The electric field lines surrounding an electric dipole begin on the positive charge and terminate on the negative charge.

Gauss's law in magnetism states that

the net magnetic flux through any closed surface is always zero:

$$\oint \vec{\mathbf{B}} \cdot d\vec{\mathbf{A}} = 0 \qquad \text{(29.20)}$$

This statement represents that isolated magnetic poles (monopoles) have never been detected and perhaps do not exist. Nonetheless, scientists continue the search because certain theories that are otherwise successful in explaining fundamental physical behavior suggest the possible existence of magnetic monopoles.

29.6 Magnetism in Matter

The magnetic field produced by a current in a coil of wire gives us a hint as to what causes certain materials to exhibit strong magnetic properties. Earlier we found that a solenoid like the one shown in Figure 29.17a has a north pole and a south pole. In general, *any* current loop has a magnetic field and therefore has a magnetic dipole moment, including the atomic-level current loops described in some models of the atom.

The Magnetic Moments of Atoms

The electron has an angular momentum $\vec{\mathbf{L}}$ in one direction and a magnetic moment $\vec{\boldsymbol{\mu}}$ in the opposite direction.

Figure 29.24 An electron moving in the direction of the gray arrow in a circular orbit of radius *r*. Because the electron carries a negative charge, the direction of the current due to its motion about the nucleus is opposite the direction of that motion.

Let's begin our discussion with a classical model of the atom in which electrons move in circular orbits around the much more massive nucleus as shown in Figure 29.24. In this model, an orbiting electron constitutes a tiny current loop (because it is a moving charge). Although this model has many deficiencies, some of its predictions are in good agreement with the correct theory, which is expressed in terms of quantum physics.

Let's call the current of the orbiting electron *I*. Because the orbit has an area *A*, there is a magnetic moment associated with the orbiting electron of magnitude $\mu = IA$. In addition, the electron has an angular momentum about the nucleus of magnitude $L = m_e v r$, where m_e is the mass of the electron and v is its orbital speed. Because the electron is negatively charged, the vectors $\vec{\boldsymbol{\mu}}$ and $\vec{\mathbf{L}}$ point in *opposite* directions. Both vectors are perpendicular to the plane of the orbit as indicated in Figure 29.24.

A fundamental outcome of quantum physics is that orbital angular momentum is quantized and is equal to multiples of $\hbar = h/2\pi = 1.05 \times 10^{-34}\,\text{J} \cdot \text{s}$, where *h* is Planck's constant (see Chapter 39). The smallest nonzero value of the electron's magnetic moment resulting from its orbital motion is

$$\mu = \sqrt{2}\,\frac{e}{2m_e}\,\hbar \qquad \text{(29.21)}$$

We shall see in Chapter 41 how expressions such as Equation 29.21 arise.

Because all substances contain electrons, you may wonder why most substances are not magnetic. The main reason is that, in most substances, the magnetic moment of one electron in an atom is canceled by that of another electron orbiting in the opposite direction. The net result is that, for most materials, the magnetic effect produced by the orbital motion of the electrons is either zero or very small.

In addition to its orbital magnetic moment, an electron (as well as protons, neutrons, and other particles) has an intrinsic property called **spin** that also contributes to its magnetic moment. Classically, the electron might be viewed as spinning about its axis as shown in Figure 29.25, but you should be very careful with the classical interpretation. The magnitude of the angular momentum $\vec{\mathbf{S}}$ associated with spin

PITFALL PREVENTION 29.3

The Electron Does Not Spin The electron is *not* physically spinning. It has an intrinsic angular momentum *as if it were spinning*, but the notion of rotation for a point particle is meaningless. Rotation applies only to a *rigid object*, with an extent in space, as in Chapter 10. Spin angular momentum is actually a relativistic effect.

is on the same order of magnitude as the magnitude of the angular momentum $\vec{\mathbf{L}}$ due to the orbital motion. The magnitude of the spin angular momentum of an electron predicted by quantum theory is

$$S = \frac{\sqrt{3}}{2}\hbar$$

The magnetic moment characteristically associated with the spin of an electron has the value

$$\mu_{spin} = \frac{e\hbar}{2m_e} \tag{29.22}$$

This combination of constants is called the **Bohr magneton** μ_B:

$$\mu_B = \frac{e\hbar}{2m_e} = 9.27 \times 10^{-24} \, \text{J/T} \tag{29.23}$$

Therefore, atomic magnetic moments can be expressed as multiples of the Bohr magneton. (Note that $1 \, \text{J/T} = 1 \, \text{A} \cdot \text{m}^2$.)

In atoms containing many electrons, the electrons usually pair up with their spins opposite each other; therefore, the spin magnetic moments cancel. Atoms containing an odd number of electrons, however, must have at least one unpaired electron and therefore some spin magnetic moment. The total magnetic moment of an atom is the vector sum of the orbital and spin magnetic moments, and a few examples are given in Table 29.1. Notice that helium and neon have zero moments because their individual spin and orbital moments cancel.

The nucleus of an atom also has a magnetic moment associated with its constituent protons and neutrons. The magnetic moment of a proton or neutron, however, is much smaller than that of an electron and can usually be neglected. We can understand this smaller value by inspecting Equation 29.23 and replacing the mass of the electron with the mass of a proton or a neutron. Because the masses of the proton and neutron are much greater than that of the electron, their magnetic moments are on the order of 10^3 times smaller than that of the electron.

Ferromagnetism

A small number of substances exhibit strong magnetic effects called **ferromagnetism.** Some examples of ferromagnetic substances are iron, cobalt, nickel, gadolinium, and dysprosium. These substances contain permanent atomic magnetic moments that tend to align parallel to each other even in a weak external magnetic field. The substance remains magnetized after the external field is removed, leading to a permanent magnet. This permanent alignment is due to a strong coupling between neighboring moments, a coupling that can be understood only in quantum-mechanical terms.

All ferromagnetic materials are made up of microscopic regions called **domains,** regions within which all magnetic moments are aligned. These domains have volumes of about 10^{-12} to 10^{-8} m^3 and contain 10^{17} to 10^{21} atoms. The boundaries between the various domains having different orientations are called **domain walls.** In an unmagnetized sample, the magnetic moments in the domains are randomly oriented so that the net magnetic moment is zero as in Figure 29.26a (page 788). When the sample is placed in an external magnetic field $\vec{\mathbf{B}}$, the size of those domains with magnetic moments aligned with the field grows, which results in a magnetized sample as in Figure 29.26b. As the external field becomes very strong as in Figure 29.26c, the domains in which the magnetic moments are not aligned with the field become very small. When the external field is removed, the sample may retain a net magnetization in the direction of the original field. At ordinary temperatures, thermal agitation is not sufficient to disrupt this preferred orientation of magnetic moments.

Figure 29.25 Classical model of a spinning electron. We can adopt this model to remind ourselves that electrons have an intrinsic angular momentum. The model should not be pushed too far, however; it gives an incorrect magnitude for the magnetic moment, incorrect quantum numbers, and too many degrees of freedom.

TABLE 29.1 Magnetic Moments of Some Atoms and Ions

Atom or Ion	Magnetic Moment (10^{-24} J/T)
H	9.27
He	0
Ne	0
Ce^{3+}	19.8
Yb^{3+}	37.1

In an unmagnetized substance, the atomic magnetic dipoles are randomly oriented.

a

When an external field $\vec{B}$ is applied, the domains with components of magnetic moment in the same direction as $\vec{B}$ grow larger, giving the sample a net magnetization.

$\longrightarrow \vec{B}$

b

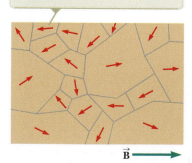

As the field is made even stronger, the domains with magnetic moment vectors not aligned with the external field become very small.

$\vec{B} \longrightarrow$

c

Figure 29.26 Orientation of magnetic dipoles before and after a magnetic field is applied to a ferromagnetic substance.

TABLE 29.2 Curie Temperatures for Several Ferromagnetic Substances	
Substance	T_{Curie} (K)
Iron	1 043
Cobalt	1 394
Nickel	631
Gadolinium	317
Fe_2O_3	893

When the temperature of a ferromagnetic substance reaches or exceeds a critical temperature called the **Curie temperature,** the substance loses its residual magnetization. Below the Curie temperature, the magnetic moments are aligned and the substance is ferromagnetic. Above the Curie temperature, the thermal agitation is great enough to cause a random orientation of the moments and the substance becomes paramagnetic. Curie temperatures for several ferromagnetic substances are given in Table 29.2.

Paramagnetism

Paramagnetic substances have a weak magnetism resulting from the presence of atoms (or ions) that have permanent magnetic moments. These moments interact only weakly with one another and are randomly oriented in the absence of an external magnetic field. When a paramagnetic substance is placed in an external magnetic field, its atomic moments tend to line up with the field. This alignment process, however, must compete with thermal motion, which tends to randomize the magnetic moment orientations.

Diamagnetism

When an external magnetic field is applied to a diamagnetic substance, a weak magnetic moment is induced in the direction opposite the applied field, causing diamagnetic substances to be weakly repelled by a magnet. Although diamagnetism is present in all matter, its effects are much smaller than those of paramagnetism or ferromagnetism and are evident only when those other effects do not exist.

We can attain some understanding of diamagnetism by considering a classical model of two atomic electrons orbiting the nucleus in opposite directions but with the same speed. The electrons remain in their circular orbits because of the attractive electrostatic force exerted by the positively charged nucleus. Because the magnetic moments of the two electrons are equal in magnitude and opposite in direction, they cancel each other and the magnetic moment of the atom is zero. When an external magnetic field is applied, the electrons experience an additional magnetic force $q\vec{v} \times \vec{B}$. This added magnetic force combines with the electrostatic force to increase the orbital speed of the electron whose magnetic moment is antiparallel to the field and to decrease the speed of the electron whose magnetic moment is parallel to the field. As a result, the two magnetic moments of the electrons no longer cancel and the substance acquires a net magnetic moment that is opposite the applied field.

As you recall from Chapter 26, a superconductor is a substance in which the electrical resistance is zero below some critical temperature. Certain types of superconductors also exhibit perfect diamagnetism in the superconducting state. As a result, an applied magnetic field is expelled by the superconductor so that the field is zero in its interior. This phenomenon is known as the **Meissner effect.** If a permanent magnet is brought near a superconductor, the two objects repel each other. This

In the Meissner effect, the small magnet at the top induces currents in the superconducting disk below, which is cooled to −321°F (77 K). The currents create a repulsive magnetic force on the magnet causing it to levitate above the superconducting disk.

Courtesy of Argonne National Laboratory

Figure 29.27 An illustration of the Meissner effect, shown by this magnet suspended above a cooled ceramic superconductor disk, has become our most visual image of high-temperature superconductivity. Superconductivity is the loss of all resistance to electrical current and is a key to more-efficient energy use.

repulsion is illustrated in Figure 29.27, which shows a small permanent magnet levitated above a superconductor maintained at 77 K.

Superconducting wires in the MRI's solenoid are what allow the MRI machine in the opening storyline to provide such a large magnetic field. Because the resistance of the wires is zero, a very high current is possible, creating a very strong magnetic field, whose magnitude is given approximately by Equation 29.17. As shown in Table 28.1, a typical field magnitude in an MRI machine is 1.5 T. Because the solenoid is not infinite, there is an external magnetic field as discussed in Section 29.4. As a result, if a ferromagnetic material is present near the MRI, it can be attracted strongly to the machine, leading to the possibility of a violent event. The special equipment that you use in the MRI room must have no ferromagnetic material, and, ideally, no paramagnetic material.

Summary

▶ Definition

The **magnetic flux** Φ_B through a surface is defined by the surface integral

$$\Phi_B \equiv \int \vec{\mathbf{B}} \cdot d\vec{\mathbf{A}} \tag{29.18}$$

▶ Concepts and Principles

The **Biot–Savart law** says that the magnetic field $d\vec{\mathbf{B}}$ at a point P due to a length element $d\vec{\mathbf{s}}$ that carries a steady current I is

$$d\vec{\mathbf{B}} = \frac{\mu_0}{4\pi} \frac{I\, d\vec{\mathbf{s}} \times \hat{\mathbf{r}}}{r^2} \tag{29.1}$$

where μ_0 is the **permeability of free space,** r is the distance from the element to the point P, and $\hat{\mathbf{r}}$ is a unit vector pointing from $d\vec{\mathbf{s}}$ toward point P. We find the total field at P by integrating this expression over the entire current distribution.

The magnetic force per unit length between two parallel wires separated by a distance a and carrying currents I_1 and I_2 has a magnitude

$$\frac{F_B}{\ell} = \frac{\mu_0 I_1 I_2}{2\pi a} \tag{29.12}$$

The force is attractive if the currents are in the same direction and repulsive if they are in opposite directions.

continued

Ampère's law says that the line integral of $\vec{\mathbf{B}} \cdot d\vec{\mathbf{s}}$ around any closed path equals $\mu_0 I$, where I is the total steady current through any surface bounded by the closed path:

$$\oint \vec{\mathbf{B}} \cdot d\vec{\mathbf{s}} = \mu_0 I \qquad (29.13)$$

The magnitude of the magnetic field at a distance r from a long, straight wire carrying an electric current I is

$$B = \frac{\mu_0 I}{2\pi r} \qquad (29.14)$$

The field lines are circles concentric with the wire.

The magnitudes of the fields inside a toroid and solenoid are

$$B = \frac{\mu_0 N I}{2\pi r} \quad \text{(toroid)} \qquad (29.16)$$

$$B = \mu_0 \frac{N}{\ell} I = \mu_0 n I \quad \text{(solenoid)} \qquad (29.17)$$

where N is the total number of turns.

Gauss's law of magnetism states that the net magnetic flux through any closed surface is zero:

$$\oint \vec{\mathbf{B}} \cdot d\vec{\mathbf{A}} = 0 \qquad (29.20)$$

Substances can be classified into one of three categories that describe their magnetic behavior. **Diamagnetic** substances are those in which the magnetic moment is weak and opposite the applied magnetic field. **Paramagnetic** substances are those in which the magnetic moment is weak and in the same direction as the applied magnetic field. In **ferromagnetic** substances, interactions between atoms cause magnetic moments to align and create a strong magnetization that remains after the external field is removed.

Think–Pair–Share

See the Preface for an explanation of the icons used in this problems set. For additional assessment items for this section, go to WEBASSIGN From Cengage

1. **S** Your professor sometimes employs group exams in his class, in which students work together and everyone receives the same grade. Your group is working on one of these exams and is faced with the following situation. Figure TP29.1 shows a side view of a tightly wound solenoid of length ℓ, radius a, and n turns per unit length. The solenoid, which carries a current I, is located with its center at $x = 0$. Notice that the solenoid is *not* infinitely long. Your group is tasked with determining the magnetic field at arbitrary points along the axis of the solenoid.

Toward this end, carry out the steps in the following plan: (a) By modeling each short length dx of the solenoid as a circular current loop, find the magnetic field due to the entire solenoid at point P on the x axis, a distance d from its center. (b) Let $d \to 0$ to find the magnetic field on the x axis at the midpoint of the solenoid. (c) In the expression in part (b), let $\ell \to \infty$ to find the field inside an infinitely long solenoid. (d) How does the result in part (c) compare to Equation 29.17?

2. **ACTIVITY** For this activity, your group will need a large nail, a length of wire, a battery, and some paper clips. (a) One member of the group wraps the wire four times around the nail, laying each new turn of wire next to the previous one, and leaving enough length of both ends of the wire to connect the ends to a battery. Another connects the ends of the wire to a battery. The apparatus is now an electromagnet. See how many paper clips the electromagnet will pick up off the table so they are suspended from the tip of the nail. Record your group's results. (b) Now continue the experiment by wrapping two more turns of wire next to the previous ones, for a total of 6 turns, and measuring the number of paper clips picked up. Repeat the experiment for 8 turns, 10 turns, 12 turns, and 14 turns. From your group's data, predict the number of paper clips the nail will pick up for 20 turns of wire. (c) Wrap 20 turns and measure the number of paper clips. How did your group's prediction match the results? (d) Now wrap 20 turns of wire, but with the final 10 turns wrapped on top of the first layer of 10 turns. How many paper clips does this magnet pick up?

Figure TP29.1

Problems

See the Preface for an explanation of the icons used in this problems set. For additional assessment items for this section, go to ⚡ **WEBASSIGN** From Cengage

SECTION 29.1 The Biot–Savart Law

1. Calculate the magnitude of the magnetic field at a point 25.0 cm from a long, thin conductor carrying a current of 2.00 A.

2. You are working as an expert witness in a civil case. You have been hired by the attorney for a company that manufactures compasses. The company is being sued by a novice hiker who used one of the company's top-level compasses. The hiker claims that the compass was defective, sending him off in a different direction from his desired direction. After taking off in the erroneous direction, he dropped and lost his compass so that he could not take subsequent measurements. As a result, he became lost for days, with the subsequent ill effects on his health and lost wages from missed days at work. The hiker has provided the exact location at which he took the erroneous compass reading. You take a trip to this location and look around. You notice that there is an electric power transmission line directly above your location, running in a north–south direction. Using trigonometry, you determine that the power line is a vertical distance of 6.65 m above the ground. Upon returning to your office, you contact employees of the electric power company, who tell you that that particular rural power line actually carries DC current with a typical magnitude during the day of 135 A. (a) In order to provide advice in this case, you calculate the magnetic field caused by the power line at the location of the hiker. (b) What advice do you give to the attorney?

3. In Niels Bohr's 1913 model of the hydrogen atom, an electron circles the proton at a distance of 5.29×10^{-11} m with a speed of 2.19×10^6 m/s. Compute the magnitude of the magnetic field this motion produces at the location of the proton.

4. An infinitely long wire carrying a current I is bent at a right angle as shown in Figure P29.4. Determine the magnetic field at point P, located a distance x from the corner of the wire.

Figure P29.4

5. A long, straight wire carries a current I. A right-angle bend is made in the middle of the wire. The bend forms an arc of a circle of radius r as shown in Figure P29.5. Determine the magnetic field at point P, the center of the arc.

Figure P29.5

6. Consider a flat, circular current loop of radius R carrying a current I. Choose the x axis to be along the axis of the loop, with the origin at the loop's center. Plot a graph of the ratio of the magnitude of the magnetic field at coordinate x to that at the origin for $x = 0$ to $x = 5R$. It may be helpful to use a programmable calculator or a computer to solve this problem.

7. Three long, parallel conductors each carry a current of $I = 2.00$ A. Figure P29.7 is an end view of the conductors, with each current coming out of the page. Taking $a = 1.00$ cm, determine the magnitude and direction of the magnetic field at (a) point A, (b) point B, and (c) point C.

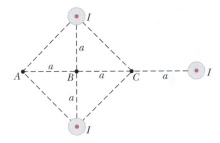

Figure P29.7

8. One long wire carries current 30.0 A to the left along the x axis. A second long wire carries current 50.0 A to the right along the line ($y = 0.280$ m, $z = 0$). (a) Where in the plane of the two wires is the total magnetic field equal to zero? (b) A particle with a charge of $-2.00\ \mu$C is moving with a velocity of $150\hat{i}$ Mm/s along the line ($y = 0.100$ m, $z = 0$). Calculate the vector magnetic force acting on the particle. (c) **What If?** A uniform electric field is applied to allow this particle to pass through this region undeflected. Calculate the required vector electric field.

9. Determine the magnetic field (in terms of I, a, and d) at the origin due to the current loop in Figure P29.9. The loop extends to infinity above the figure.

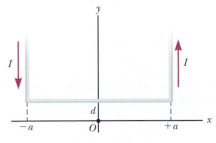

Figure P29.9

10. A wire carrying a current I is bent into the shape of an equilateral triangle of side L. (a) Find the magnitude of the magnetic field at the center of the triangle. (b) At a point halfway between the center and any vertex, is the field stronger or weaker than at the center? Give a qualitative argument for your answer.

11. Two long, parallel wires carry currents of $I_1 = 3.00$ A and $I_2 = 5.00$ A in the directions indicated in Figure P29.11 (page 792). (a) Find the magnitude and direction of the magnetic field at a point midway between the wires. (b) Find the

magnitude and direction of the magnetic field at point *P*, located *d* = 20.0 cm above the wire carrying the 5.00-A current.

Figure P29.11

SECTION 29.2 The Magnetic Force Between Two Parallel Conductors

12. Two parallel wires separated by 4.00 cm repel each other with a force per unit length of 2.00×10^{-4} N/m. The current in one wire is 5.00 A. (a) Find the current in the other wire. (b) Are the currents in the same direction or in opposite directions? (c) What would happen if the direction of one current were reversed and doubled?

13. Two parallel wires are separated by 6.00 cm, each carrying 3.00 A of current in the same direction. (a) What is the magnitude of the force per unit length between the wires? (b) Is the force attractive or repulsive?

14. Two long wires hang vertically. Wire 1 carries an upward current of 1.50 A. Wire 2, 20.0 cm to the right of wire 1, carries a downward current of 4.00 A. A third wire, wire 3, is to be hung vertically and located such that when it carries a certain current, each wire experiences no net force. (a) Is this situation possible? Is it possible in more than one way? Describe (b) the position of wire 3 and (c) the magnitude and direction of the current in wire 3.

15. You are part of a team working in a machine parts mechanic's shop. An important customer has asked your company to provide springs with a very precise force constant *k*. To measure the spring constant, you fasten two of the springs between the ends of two very long wires of length *L*, separated by the unstretched length ℓ of the springs as shown in Figure P29.15. The specific attachment method that you use insulates the springs from the wires so that no current passes through the springs. You lay the apparatus flat on a table and then pass a current of magnitude *I* through the wires, in opposite directions. As a result the springs stretch by a distance *d* and come to equilibrium. You determine an expression for the spring constant in terms of *L*, *I*, ℓ, and *d*.

Figure P29.15

16. *Why is the following situation impossible?* Two parallel copper conductors each have length ℓ = 0.500 m and radius *r* = 250 µm. They carry currents *I* = 10.0 A in opposite directions and repel each other with a magnetic force F_B = 1.00 N.

17. The unit of magnetic flux is named for Wilhelm Weber. A practical-size unit of magnetic field is named for Johann Karl Friedrich Gauss. Along with their individual accomplishments, Weber and Gauss built a telegraph in 1833 that consisted of a battery and switch, at one end of a transmission line 3 km long, operating an electromagnet at the other end. Suppose their transmission line was as diagrammed in Figure P29.17. Two long, parallel wires, each having a mass per unit length of 40.0 g/m, are supported in a horizontal plane by strings ℓ = 6.00 cm long. When both wires carry the same current *I*, the wires repel each other so that the angle between the supporting strings is θ = 16.0°. (a) Are the currents in the same direction or in opposite directions? (b) Find the magnitude of the current. (c) If this transmission line were taken to Mars, would the current required to separate the wires by the same angle be larger or smaller than that required on the Earth? Why?

Figure P29.17

SECTION 29.3 Ampère's Law

18. Niobium metal becomes a superconductor when cooled below 9 K. Its superconductivity is destroyed when the surface magnetic field exceeds 0.100 T. In the absence of any external magnetic field, determine the maximum current a 2.00-mm-diameter niobium wire can carry and remain superconducting.

19. The magnetic coils of a tokamak fusion reactor are in the shape of a toroid having an inner radius of 0.700 m and an outer radius of 1.30 m. The toroid has 900 turns of large-diameter wire, each of which carries a current of 14.0 kA. Find the magnitude of the magnetic field inside the toroid along (a) the inner radius and (b) the outer radius.

20. A packed bundle of 100 long, straight, insulated wires forms a cylinder of radius *R* = 0.500 cm. If each wire carries 2.00 A, what are (a) the magnitude and (b) the direction of the magnetic force per unit length acting on a wire located 0.200 cm from the center of the bundle? (c) **What If?** Would a wire on the outer edge of the bundle experience a force greater or smaller than the value calculated in parts (a) and (b)? Give a qualitative argument for your answer.

21. The magnetic field 40.0 cm away from a long, straight wire carrying current 2.00 A is 1.00 µT. (a) At what distance is it 0.100 µT? (b) **What If?** At one instant, the two conductors in a long household extension cord carry equal 2.00-A currents in opposite directions. The two wires are 3.00 mm apart. Find the magnetic field 40.0 cm away from the middle of the straight cord, in the plane of the two wires. (c) At what distance is it one-tenth as large? (d) The center wire in a coaxial cable carries current 2.00 A in one direction, and the sheath around it carries current 2.00 A in the opposite direction. What magnetic field does the cable create at points outside the cable?

22. A long, cylindrical conductor of radius *R* carries a current *I* as shown in Figure P29.22. The current density *J*, however, is not uniform over the cross section of the conductor but rather is a function of the radius according to $J = br$, where b is a constant. Find an expression for the magnetic field magnitude B (a) at a distance $r_1 < R$ and (b) at a distance $r_2 > R$, measured from the center of the conductor.

Figure P29.22

SECTION 29.4 The Magnetic Field of a Solenoid

23. A long solenoid that has 1 000 turns uniformly distributed
T over a length of 0.400 m produces a magnetic field of magnitude 1.00×10^{-4} T at its center. What current is required in the windings for that to occur?

24. A certain superconducting magnet in the form of a solenoid of length 0.500 m can generate a magnetic field of 9.00 T in its core when its coils carry a current of 75.0 A. Find the number of turns in the solenoid.

25. You are working at a company that manufactures solenoids
CR for industrial and research use. A client has ordered a solenoid that will be operated by a 1 000-V power supply and must be of length $\ell = 25.0$ cm. A cylindrical experimental package of radius $r_s = 1.00$ cm must fit inside the solenoid. The client wants the largest possible magnetic field inside the solenoid. The thinnest copper wires allowed by your company are AWG 36, which corresponds to a wire diameter of $d_w = 0.127$ mm. You determine the maximum magnitude of magnetic field that can be created in the solenoid to report to the client.

26. You are given a certain volume of copper from which you
Q|C can make copper wire. To insulate the wire, you can have as much enamel as you like. You will use the wire to make a tightly wound solenoid 20 cm long having the greatest possible magnetic field at the center and using a power supply that can deliver a current of 5 A. The solenoid can be wrapped with wire in one or more layers. (a) Should you make the wire long and thin or shorter and thick? Explain. (b) Should you make the radius of the solenoid small or large? Explain.

SECTION 29.5 Gauss's Law in Magnetism

27. Consider the hemispherical closed surface in Figure P29.27.
S The hemisphere is in a uniform magnetic field that makes an angle θ with the vertical. Calculate the magnetic flux through (a) the flat surface S_1 and (b) the hemispherical surface S_2.

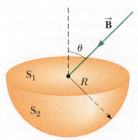

Figure P29.27

28. You are working for a company that creates special magnetic
CR environments. Your new supervisor has come from the financial side of the organization rather than the technical side. He has promised a client that the company can provide a device that will create a magnetic field inside a cylindrical chamber that is directed along the cylinder axis at all points in the chamber and increases in the axial direction as the square of the value of y, where y is the in the axial direction and $y = 0$ is at the bottom end of the cylinder. Prepare a calculation to show that the field requested by your supervisor and promised to a client is impossible.

29. A solenoid of radius $r = 1.25$ cm and length $\ell = 30.0$ cm has
V 300 turns and carries 12.0 A. (a) Calculate the flux through the surface of a disk-shaped area of radius $R = 5.00$ cm that is positioned perpendicular to and centered on the axis of the solenoid as shown in Figure P29.29a. (b) Figure P29.29b shows an enlarged end view of the same solenoid. Calculate the flux through the tan area, which is an annulus with an inner radius of $a = 0.400$ cm and an outer radius of $b = 0.800$ cm.

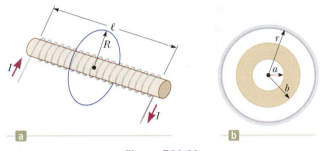

Figure P29.29

SECTION 29.6 Magnetism in Matter

30. The magnetic moment of the Earth is approximately
T 8.00×10^{22} A·m². Imagine that the planetary magnetic field were caused by the complete magnetization of a huge iron deposit with density 7 900 kg/m³ and approximately 8.50×10^{28} iron atoms/m³. (a) How many unpaired electrons, each with a magnetic moment of 9.27×10^{-24} A·m², would participate? (b) At two unpaired electrons per iron atom, how many kilograms of iron would be present in the deposit?

ADDITIONAL PROBLEMS

31. A 30.0-turn solenoid of length 6.00 cm produces a magnetic field of magnitude 2.00 mT at its center. Find the current in the solenoid.

32. *Why is the following situation impossible?* The magnitude of the Earth's magnetic field at either pole is approximately 7.00×10^{-5} T. Suppose the field fades away to zero before its next reversal. Several scientists propose plans for artificially generating a replacement magnetic field to assist with devices that depend on the presence of the field. The plan that is selected is to lay a copper wire around the equator and supply it with a current that would generate a magnetic field of magnitude 7.00×10^{-5} T at the poles. (Ignore magnetization of any materials inside the Earth.) The plan is implemented and is highly successful.

33. Suppose you install a compass on the center of a car's dashboard. (a) Assuming the dashboard is made mostly of plastic, compute an order-of-magnitude estimate for the magnetic field at this location produced by the current when you switch on the car's headlights. (b) How does this estimate compare with the Earth's magnetic field?

34. A very long, thin strip of metal of width w carries a current I
S along its length as shown in Figure P29.34. The current is distributed uniformly across the width of the strip. Find the magnetic field at point P in the diagram. Point P is in the plane of the strip at distance b away from its edge.

Figure P29.34

35. A nonconducting ring of radius 10.0 cm is uniformly
T charged with a total positive charge 10.0 μC. The ring rotates at a constant angular speed 20.0 rad/s about an axis through its center, perpendicular to the plane of the ring. What is the magnitude of the magnetic field on the axis of the ring 5.00 cm from its center?

36. A nonconducting ring of radius R is uniformly charged with a
S total positive charge q. The ring rotates at a constant angular speed ω about an axis through its center, perpendicular to the plane of the ring. What is the magnitude of the magnetic field on the axis of the ring a distance $\frac{1}{2}R$ from its center?

37. A very large parallel-plate capacitor has uniform charge per
S unit area $+\sigma$ on the upper plate and $-\sigma$ on the lower plate. The plates are horizontal, and both move horizontally with speed v to the right. (a) What is the magnetic field between the plates? (b) What is the magnetic field just above or just below the plates? (c) What are the magnitude and direction of the magnetic force per unit area on the upper plate? (d) At what extrapolated speed v will the magnetic force on a plate balance the electric force on the plate? *Suggestion:* Use Ampere's law and choose a path that closes between the plates of the capacitor.

38. Two circular coils of radius R, each with N turns, are perpen-
S dicular to a common axis. The coil centers are a distance R apart. Each coil carries a steady current I in the same direction as shown in Figure P29.38. (a) Show that the magnetic field on the axis at a distance x from the center of one coil is

$$B = \frac{N\mu_0 I R^2}{2}\left[\frac{1}{(R^2 + x^2)^{3/2}} + \frac{1}{(2R^2 + x^2 - 2Rx)^{3/2}}\right]$$

Figure P29.38 Problems 38 and 39.

(b) Show that dB/dx and d^2B/dx^2 are both zero at the point midway between the coils. We may then conclude that the magnetic field in the region midway between the coils is uniform. Coils in this configuration are called *Helmholtz coils*.

39. Two identical, flat, circular coils of wire each have 100 turns and radius $R = 0.500$ m. The coils are arranged as a set of Helmholtz coils so that the separation distance between the coils is equal to the radius of the coils (see Fig. P29.38). Each coil carries current $I = 10.0$ A. Determine the magnitude of the magnetic field at a point on the common axis of the coils and halfway between them.

40. Two circular loops are parallel, coaxial, and almost in con-
AMT tact, with their centers 1.00 mm apart (Fig. P29.40). Each
QC loop is 10.0 cm in radius. The top loop carries a clockwise current of $I = 140$ A. The bottom loop carries a counterclockwise current of $I = 140$ A. (a) Calculate the magnetic force exerted by the bottom loop on the top loop. (b) Suppose a student thinks the first step in solving part (a) is to use Equation 29.7 to find the magnetic field created by one of the loops. How would you argue for or against this idea? (c) The upper loop has a mass of 0.021 0 kg. Calculate its acceleration, assuming the only forces acting on it are the force in part (a) and the gravitational force.

Figure P29.40

41. As seen in previous chapters, any object with electric charge, stationary or moving, other than the charged object that created the field, experiences a force in an electric field. Also, any object with electric charge, stationary or moving, can create an electric field (Chapter 22). Similarly, an electric current or a moving electric charge, other than the current or charge that created the field, experiences a force in a magnetic field (Chapter 28), and an electric current creates a magnetic field (Section 29.1). (a) To understand how a moving charge can also create a magnetic field, consider a particle with charge q moving with velocity $\vec{v}$. Define the position vector $\vec{r} = r\hat{r}$ leading from the particle to some location. Show that the magnetic field at that location is

$$\vec{B} = \frac{\mu_0}{4\pi}\frac{q\vec{v}\times\hat{r}}{r^2}$$

(b) Find the magnitude of the magnetic field 1.00 mm to the side of a proton moving at 2.00×10^7 m/s. (c) Find the magnetic force on a second proton at this point, moving with the same speed in the opposite direction. (d) Find the electric force on the second proton.

42. Review. Rail guns have been suggested for launching projec-
AMT tiles into space without chemical rockets. A tabletop model
GP rail gun (Fig. P29.42) consists of two long, parallel, horizontal rails $\ell = 3.50$ cm apart, bridged by a bar of mass $m = 3.00$ g that is free to slide without friction. The rails and bar have low electric resistance, and the current is limited to a constant $I = 24.0$ A by a power supply that is far to the left of the figure, so it has no magnetic effect on the bar. Figure P29.42 shows the bar at rest at the midpoint of the rails at

Figure P29.42

the moment the current is established. We wish to find the speed with which the bar leaves the rails after being released from the midpoint of the rails. (a) Find the magnitude of the magnetic field at a distance of 1.75 cm from a single long wire carrying a current of 2.40 A. (b) For purposes of evaluating the magnetic field, model the rails as infinitely long. Using the result of part (a), find the magnitude and direction of the magnetic field at the midpoint of the bar. (c) Argue that this value of the field will be the same at all positions of the bar to the right of the midpoint of the rails. At other points along the bar, the field is in the same direction as at the midpoint, but is larger in magnitude. Assume the average effective magnetic field along the bar is five times larger than the field at the midpoint. With this assumption, find (d) the magnitude and (e) the direction of the force on the bar. (f) Is the bar properly modeled as a particle under constant acceleration? (g) Find the velocity of the bar after it has traveled a distance $d = 130$ cm to the end of the rails.

43. Fifty turns of insulated wire 0.100 cm in diameter are tightly wound to form a flat spiral. The spiral fills a disk surrounding a circle of radius 5.00 cm and extending to a radius 10.00 cm at the outer edge. Assume the wire carries a current I at the center of its cross section. Approximate each turn of wire as a circle. Then a loop of current exists at radius 5.05 cm, another at 5.15 cm, and so on. Numerically calculate the magnetic field at the center of the coil.

44. An infinitely long, straight wire carrying a current I_1 is partially surrounded by a loop as shown in Figure P29.44. The loop has a length L and radius R, and it carries a current I_2. The axis of the loop coincides with the wire. Calculate the magnetic force exerted on the loop.

Figure P29.44

CHALLENGE PROBLEMS

45. Consider a solenoid of length ℓ and radius a containing N closely spaced turns and carrying a steady current I. (a) In terms of these parameters, find the magnetic field at a point along the axis as a function of position x from the end of the solenoid. (b) Show that as ℓ becomes very long, B approaches $\mu_0 NI/2\ell$ at each end of the solenoid.

46. We have seen that a long solenoid produces a uniform magnetic field directed along the axis of a cylindrical region. To produce a uniform magnetic field directed parallel to a *diameter* of a cylindrical region, however, one can use the *saddle coils* illustrated in Figure P29.46. The loops are wrapped over a long, somewhat flattened tube. Figure P29.46a shows one wrapping of wire around the tube. This wrapping is continued in this manner until the visible side has many long sections of wire carrying current to the left in Figure P29.46a and the back side has many lengths carrying current to the right. The end view of the tube in Figure P29.46b shows these wires and the currents they carry. By wrapping the wires carefully, the distribution of wires can take the shape suggested in the end view such that the overall current distribution is approximately the superposition of two overlapping, circular cylinders of radius R (shown by the dashed lines) with uniformly distributed current, one toward you and one away from you. The current density J is the same for each cylinder. The center of one cylinder is described by a position vector $\vec{d}$ relative to the center of the other cylinder. Prove that the magnetic field inside the hollow tube is $\mu_0 Jd/2$ downward. *Suggestion:* The use of vector methods simplifies the calculation.

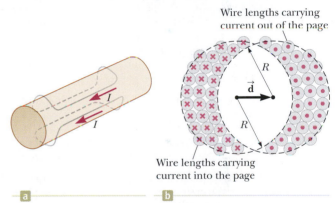

Wire lengths carrying current out of the page

Wire lengths carrying current into the page

Figure P29.46

47. A wire carrying a current I is bent into the shape of an exponential spiral, $r = e^\theta$, from $\theta = 0$ to $\theta = 2\pi$ as suggested in Figure P29.47. To complete a loop, the ends of the spiral are connected by a straight wire along the x axis. (a) The angle β between a radial line and its tangent line at any point on a curve $r = f(\theta)$ is related to the function by

$$\tan \beta = \frac{r}{dr/d\theta}$$

Use this fact to show that $\beta = \pi/4$. (b) Find the magnetic field at the origin.

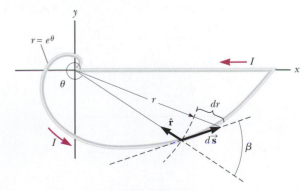

Figure P29.47

48. A sphere of radius R has a uniform volume charge density ρ. When the sphere rotates as a rigid object with angular speed ω about an axis through its center (Fig. P29.48), determine (a) the magnetic field at the center of the sphere and (b) the magnetic moment of the sphere.

Figure P29.48

49. A long, cylindrical conductor of radius a has two cylindrical cavities each of diameter a through its entire length as shown in the end view of Figure P29.49. A current I is directed out of the page and is uniform through a cross section of the conducting material. Find the magnitude and direction of the magnetic field in terms of μ_0, I, r, and a at (a) point P_1 and (b) point P_2.

Figure P29.49

50. A wire is formed into the shape of a square of edge length L (Fig. P29.50). Show that when the current in the loop is I, the magnetic field at point P a distance x from the center of the square along its axis is

$$B = \frac{\mu_0 I L^2}{2\pi(x^2 + L^2/4)\sqrt{x^2 + L^2/2}}$$

Figure P29.50

51. The magnitude of the force on a magnetic dipole $\vec{\mu}$ aligned with a nonuniform magnetic field in the positive x direction is $F_x = |\vec{\mu}| \, dB/dx$. Suppose two flat loops of wire each have radius R and carry a current I. (a) The loops are parallel to each other and share the same axis. They are separated by a variable distance $x \gg R$. Show that the magnetic force between them varies as $1/x^4$. (b) Find the magnitude of this force, taking $I = 10.0$ A, $R = 0.500$ cm, and $x = 5.00$ cm.

Faraday's Law

A windmill takes energy from the wind and converts it to electricity. How does it do that? What's inside the box behind the blades? (*Lukasz Janyst/Shutterstock*)

STORYLINE **You are taking a weekend drive outside of town and** enjoying the scenery. At one point, you pass by a wind farm and are impressed with the number and size of the windmills that are generating electricity. While you know that the energy to drive the windmill comes from the wind, you are not sure how the energy is converted to electricity. You look carefully at the windmills and notice that each one has a box or an enclosure behind the blades. Is the enclosure just a support mechanism for the blades, or is there something going on inside that enclosure? Is that where the electricity comes from?

CONNECTIONS So far, our studies in electricity and magnetism have treated electric fields and magnetic fields as separate entities. Electric fields are caused by stationary charges and magnetic fields are caused by moving charges. Experiments conducted by Michael Faraday in England in 1831 and independently by Joseph Henry in the United States that same year showed interesting effects when a *changing* magnetic field exists in a region of space. One effect occurs when a battery-free circuit is placed in the region of changing magnetic field. We find that a current exists in the circuit! As we study this type of phenomenon further, we find that, even if the circuit is not present, there is an *electric* field in the region of changing *magnetic* field! These results suggest a deep relationship between electric and magnetic fields. We use the term *induced* to describe the effects: there is an induced current in the circuit and an induced electric field in the region of changing magnetic field. The mathematical relationship between electric and magnetic fields that we generate in this chapter is called *Faraday's law of induction*. This will be our first introduction to *electromagnetism*, a topic that has revolutionized research in physics, and has allowed the development of

Michael Faraday
British Physicist and Chemist
(1791–1867)
Faraday is often regarded as the greatest experimental scientist of the 1800s. His many contributions to the study of electricity include the invention of the electric motor, electric generator, and transformer as well as the discovery of electromagnetic induction and the laws of electrolysis. Greatly influenced by religion, he refused to work on the development of poison gas for the British military.

myriad electronic devices, such as your smartphone. Our study of electromagnetism will lead to electromagnetic waves, providing the foundation for our study of optics in Chapters 34–37.

30.1 Faraday's Law of Induction

Let's begin our investigation into the induced currents mentioned in the introduction by considering the experimental results obtained when a loop of wire is connected to a sensitive ammeter as illustrated in Figure 30.1. Notice first that there is no battery providing energy to the loop of wire. When a magnet is held stationary near the loop, as in Figure 30.1a, no current is measured in the loop. The passage of the static magnetic field lines through the loop have no electrical effect. But now move the magnet toward the loop, as in Figure 30.1b. A quite remarkable thing happens: a current is induced in the loop, as measured by the ammeter! When the magnet stops moving, the current falls to zero again. Now pull the magnet away from the loop, as in Figure 30.1c. Again, an induced current is registered, in the opposite direction to that in Figure 30.1b.

This simple experiment suggests a fundamental connection between electric and magnetic fields. A stationary charge establishes an electric field, as discussed in Chapter 22. If the charge moves, the electric field at a point in space near the charge must change with time. A moving charge, however, is a current. And, as we found in Chapter 29, a current establishes a magnetic field. Therefore, a changing electric field results in a magnetic field. With the experiment described in the previous paragraph, we see the reverse: a changing magnetic field induces a current, which is due to an electric field in the wire!

Now let's describe an experiment conducted by Faraday and illustrated in Figure 30.2. A primary coil is wrapped around an iron ring and connected to a switch and a battery. A secondary coil also is wrapped around the ring and is connected to a sensitive ammeter. No battery is present in the secondary circuit, and the secondary coil is not electrically connected to the primary coil.

When the switch is open as shown in Figure 30.2, no current is detected in the secondary circuit, as indicated by the reading on the ammeter. But what happens if we close the switch? We find that the current reading jumps momentarily as the switch is thrown closed and then falls back to zero. As the switch remains closed,

When a magnet is held stationary near a loop of wire connected to a sensitive ammeter, there is no induced current in the loop, even when the magnet is inside the loop.

When the magnet is moved toward the loop of wire, the ammeter shows that a current is induced in the loop.

When the magnet is moved away from the loop, the ammeter shows that the induced current is opposite that shown in part **b**.

Figure 30.1 A simple experiment showing that a current is induced in a loop when a magnet is moved toward or away from the loop.

The current induced in the secondary circuit is caused by the changing magnetic field through the secondary coil.

When the switch in the primary circuit is closed, the ammeter reading in the secondary circuit changes momentarily.

Battery   Iron

Primary   Secondary
coil   coil

Figure 30.2 Faraday's experiment.

the current reading in the secondary circuit remains at zero even though there is current in the primary circuit. Now, when the switch is opened, the current reading again jumps momentarily, with the opposite sign as that when the switch was opened, and then falls back to zero.

The experiments shown in Figures 30.1 and 30.2 have one thing in common: in each case, a current is induced in a loop when the magnetic flux through the loop *changes* with time. In Figure 30.1, the magnetic field changes because the magnet is moved relative to the loop. In Figure 30.2, the closing of the switch allows a current to exist in the primary coil. This current establishes a magnetic field in the iron ring that changes from zero to its equilibrium value after the switch closes. The changing magnetic field through the secondary coil induces a current. When the switch is reopened, the magnetic field drops back to zero and there is momentary induced current in the secondary coil in the opposite direction. We have discussed that a current is due to an emf (Section 27.1), so we say that an emf is induced by a changing magnetic field. Experiments show that the induced emf in a loop of wire is related to the time rate of change of the magnetic flux through the loop. This statement can be written mathematically as **Faraday's law of induction:**

$$\mathcal{E} = -\frac{d\Phi_B}{dt}$$ (30.1)

◀ Faraday's law of induction

where $\Phi_B = \int \vec{B} \cdot d\vec{A}$ is the magnetic flux through the loop. (See Section 29.5.)

If a coil consists of N loops with the same area, like the secondary loop in Figure 30.2, and Φ_B is the magnetic flux through one loop, an emf is induced in every loop. The loops are in series, so their emfs add; therefore, the total induced emf in the coil is given by

$$\mathcal{E} = -N\frac{d\Phi_B}{dt}$$ (30.2)

The negative sign in Equations 30.1 and 30.2 is of important physical significance and will be discussed in Section 30.3.

Suppose a loop enclosing an area A lies in a uniform magnetic field $\vec{B}$ as in Figure 30.3. The magnetic flux through the loop is equal to $BA\cos\theta$, where θ is the angle between the magnetic field and the normal to the loop; hence, the induced emf can be expressed as

$$\mathcal{E} = -\frac{d}{dt}(BA\cos\theta)$$ (30.3)

Figure 30.3 A conducting loop that encloses an area A in the presence of a uniform magnetic field $\vec{B}$. The angle between $\vec{B}$ and the normal to the loop is θ.

Figure 30.4 Essential components of a ground fault circuit interrupter.

Figure 30.5 (a) In an electric guitar, a vibrating magnetized string induces an emf in a pickup coil. (b) The pickups (the circles beneath the metallic strings) of this electric guitar detect the vibrations of the strings and send this information through an amplifier and into speakers. (A switch on the guitar allows the musician to select which set of six pickups is used.)

From this expression, we see that an emf can be induced in the circuit in several ways:

- The magnitude of $\vec{B}$ can change with time.
- The area enclosed by the loop can change with time.
- The angle θ between $\vec{B}$ and the normal to the loop can change with time.
- Any combination of the above can occur.

QUICK QUIZ 30.1 A circular loop of wire is held in a uniform magnetic field, with the plane of the loop perpendicular to the field lines. Which of the following will *not* cause a current to be induced in the loop? **(a)** crushing the loop **(b)** rotating the loop about an axis perpendicular to the field lines **(c)** keeping the orientation of the loop fixed and moving it along the field lines **(d)** pulling the loop out of the field

Some Applications of Faraday's Law

The ground fault circuit interrupter (GFCI), mentioned in Section 27.5, is an interesting safety device that protects users of electrical appliances in the home against electric shock. Its operation makes use of Faraday's law. In the GFCI shown in Figure 30.4, wire 1 leads from the wall outlet to the appliance to be protected and wire 2 leads from the appliance back to the wall outlet. An iron ring surrounds the two wires, and a sensing coil is wrapped around part of the ring. Because the currents in the wires are in opposite directions and of equal magnitude, there is zero net current flowing through the ring and the net magnetic flux through the sensing coil is zero. Now suppose the return current in wire 2 changes so that the two currents are not equal in magnitude. (That can happen if, for example, the appliance becomes wet, enabling current to leak to ground.) Then the net current through the ring is not zero and the magnetic flux through the sensing coil is no longer zero. Because household current is alternating (meaning that its direction keeps reversing), the magnetic flux through the sensing coil changes with time, inducing an emf in the coil. This induced emf is used to trigger a circuit breaker, which stops the current before it is able to reach a harmful level.

Another interesting application of Faraday's law is the production of sound in an electric guitar. The coil in this case, called the *pickup coil*, is placed near the vibrating guitar string, which is made of a metal that can be magnetized. A permanent magnet inside the coil magnetizes the portion of the string nearest the coil (Fig. 30.5a). When the string vibrates at some frequency, its magnetized segment produces a changing magnetic flux through the coil. The changing flux induces an emf in the coil that is fed to an amplifier. The output of the amplifier is sent to the loudspeakers, which produce the sound waves we hear.

| Example **30.1** | **Inducing an emf in a Coil** |

A coil consists of 200 turns of wire. Each turn is a square of side $d = 18$ cm, and a uniform magnetic field directed perpendicular to the plane of the coil is turned on. Figure 30.6 shows the behavior of the magnitude of the magnetic field with time. From $t = 0$ to $t = 0.80$ s, the field changes linearly from 0 to 0.50 T. After $t = 0.80$ s, the magnitude of the field decays in time according to the expression $B = B_{max}e^{-at}$, where a is some constant and $B_{max} = 0.50$ T.

(A) What is the magnitude of the induced emf in the coil between $t = 0$ and $t = 0.80$ s?

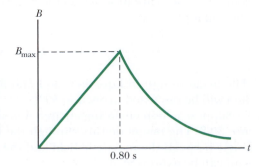

Figure 30.6 (Example 30.1) The magnitude of the uniform magnetic field through a loop of wire increases linearly and then decreases exponentially.

30.1 continued

SOLUTION

Conceptualize From the description in the problem, imagine magnetic field lines passing through the coil. Because the magnetic field is changing in magnitude, an emf is induced in the coil.

Categorize We will evaluate the emf using Faraday's law from this section, so we categorize this example as a substitution problem.

Evaluate Equation 30.2 for the situation described here, noting that the magnetic field changes linearly with time:

$$|\mathcal{E}| = N\frac{\Delta\Phi_B}{\Delta t} = N\frac{\Delta(BA)}{\Delta t} = NA\frac{\Delta B}{\Delta t} = Nd^2\frac{B_f - B_i}{\Delta t}$$

Substitute numerical values:

$$|\mathcal{E}| = (200)(0.18 \text{ m})^2\frac{(0.50 \text{ T} - 0)}{0.80 \text{ s}} = 4.0 \text{ V}$$

(B) What is the magnitude of the induced emf in the coil after $t = 0.80$ s?

Evaluate Equation 30.2 for the situation described here:

$$\mathcal{E} = -N\frac{d\Phi_B}{dt} = -N\frac{d}{dt}(AB_{max}\,e^{-at}) = -NAB_{max}\frac{d}{dt}e^{-at} = aNd^2B_{max}\,e^{-at}$$

Substitute numerical values:

$$\mathcal{E} = a(200)(0.18 \text{ m})^2(0.50 \text{ T})e^{-at} = 3.2ae^{-at}$$

This expression indicates that the emf in the loop decays exponentially after $t = 0.80$ s. The initial magnitude of the emf depends on the unknown parameter a.

WHAT IF? What if you were asked to find the magnitude of the induced current in the coil while the field is changing during the first 0.80 s? Can you answer that question?

Answer If the ends of the coil are not connected to a circuit, the answer to this question is easy: the current is zero! (Charges move within the wire of the coil, but they cannot move into or out of the ends of the coil.) For a steady current to exist, the ends of the coil must be connected to each other or to an external circuit. Let's assume the coil is connected to a circuit and the total resistance of the coil and the circuit is 2.0 Ω. Then, the magnitude of the induced current in the coil is

$$I = \frac{|\mathcal{E}|}{R} = \frac{4.0 \text{ V}}{2.0 \text{ }\Omega} = 2.0 \text{ A}$$

30.2 Motional emf

In Section 30.1 and Example 30.1, we analyzed a situation in which a coil of wire was stationary and the magnetic field changed in time. Let's now look at something different. Suppose that a magnetic field is uniform and constant, and we move a conductor in the field. We find that there is an emf induced in the conductor. We call such an emf a **motional emf.**

The straight, isolated conductor of length ℓ shown in Figure 30.7 is moving through a uniform magnetic field directed into the page. For simplicity, let's assume the conductor is moving in a direction perpendicular to the field with constant velocity under the influence of some external agent. From the magnetic version of the particle in a field model, the electrons in the conductor experience a force $\vec{\mathbf{F}}_B = q\vec{\mathbf{v}} \times \vec{\mathbf{B}}$ (Eq. 28.1) that is directed along the length ℓ, perpendicular to both $\vec{\mathbf{v}}$ and $\vec{\mathbf{B}}$. Under the influence of this force, the electrons move to the lower end of the conductor and accumulate there, leaving a net positive charge at the upper end. As a result of this charge separation, an electric field $\vec{\mathbf{E}}$ is produced inside the conductor. Therefore, the electrons in the wire are also described by the electric version of the particle in a field model. The charges accumulate at both ends until the downward magnetic force qvB on electrons between the ends of the conductor is balanced by the upward electric force qE. The electrons are then described by the particle in equilibrium model:

$$\sum F = 0 \quad \rightarrow \quad qE - qvB = 0 \quad \rightarrow \quad E = vB$$

In steady state, the electric and magnetic forces on an electron in the conductor are balanced.

Due to the magnetic force on electrons, the ends of the conductor become oppositely charged, which establishes an electric field in the conductor.

Figure 30.7 A straight electrical conductor of length ℓ moving with a velocity $\vec{\mathbf{v}}$ through a uniform magnetic field $\vec{\mathbf{B}}$ directed perpendicular to $\vec{\mathbf{v}}$.

The magnitude of the electric field produced in the conductor is related to the potential difference across the ends of the conductor according to the relationship $\Delta V = E\ell$ (Eq. 24.6). Therefore, for the equilibrium condition,

$$\Delta V = E\ell = B\ell v \tag{30.4}$$

where the upper end of the conductor in Figure 30.7 is at a higher electric potential than the lower end. Therefore, a potential difference is maintained between the ends of the conductor as long as the conductor continues to move through the uniform magnetic field. If the direction of the motion is reversed, the polarity of the potential difference is also reversed.

A more interesting situation occurs when the moving conductor is part of a closed conducting path. Consider a circuit consisting of a conducting bar of length ℓ sliding along two fixed, parallel conducting rails as shown in Figure 30.8a. For simplicity, let's assume the bar has zero resistance and the stationary part of the circuit has a resistance R. A uniform and constant magnetic field $\vec{\mathbf{B}}$ is applied perpendicular to the plane of the circuit. As the bar is pulled to the right with a velocity $\vec{\mathbf{v}}$ under the influence of an applied force $\vec{\mathbf{F}}_{\text{app}}$, electrons in the bar are moving particles in a magnetic field, so they experience a magnetic force directed downward in the bar, as in Figure 30.7. As a result, a potential difference is established between the ends of the moving bar. Because of the closed conducting path in this situation, the bar is part of a complete circuit, as shown in Figure 30.8b. The moving bar acts as a source of emf for the circuit. Electrons everywhere in the circuit move clockwise around the circuit, constituting a current I in the counterclockwise direction.

Let's address this situation by noting that the flux through the circuit in Figure 30.8a changes because the area of the circuit changes as the bar moves. Because the area enclosed by the circuit at any instant is ℓx, where x is the position of the bar, the magnetic flux through that area is

$$\Phi_B = B\ell x$$

Using Faraday's law and noting that x changes with time at a rate $dx/dt = v$, the speed of the bar, we find that the induced motional emf is

$$\varepsilon = -\frac{d\Phi_B}{dt} = -\frac{d}{dt}(B\ell x) = -B\ell \frac{dx}{dt}$$

◀ Motional emf

$$\varepsilon = -B\ell v \tag{30.5}$$

A counterclockwise current I is induced in the loop. The magnetic force $\vec{\mathbf{F}}_B$ on the bar carrying this current opposes the motion.

Figure 30.8 (a) A conducting bar sliding with a velocity $\vec{\mathbf{v}}$ along two conducting rails under the action of an applied force $\vec{\mathbf{F}}_{\text{app}}$. Because of the current in the bar, there is a magnetic force $\vec{\mathbf{F}}_B$ on the bar in the direction opposite to the applied force. (b) The equivalent circuit diagram for the setup shown in (a).

The magnitude of the emf is the same result that we obtained in Equation 30.4 using a force model! Because the resistance of the circuit is R, the magnitude of the induced current is

$$I = \frac{|\mathcal{E}|}{R} = \frac{B\ell v}{R} \qquad (30.6)$$

Using Figure 30.7, we analyzed the motional emf generated in a moving bar using a force model. With the help of Figure 30.8, we used Faraday's law to generate the same expression for the motional emf. Now let's consider an energy approach. You may wonder about the source of the energy delivered to the resistor in Figure 30.8 because there is no battery in the circuit. Assume that the bar is modeled as a particle in equilibrium, moving at constant speed under the influence of two forces of equal magnitude: $F_{app} = F_B$. The applied force must do work on the bar to keep it moving at this constant speed against the magnetic force $\vec{\mathbf{F}}_B$ on the moving electrons. The transfer of energy represented by this work results in the warming up of the resistor!

Let's verify this statement mathematically. Identifying the bar and magnetic field as a nonisolated system for energy, the appropriate reduction of Equation 8.2 is $0 = W_{app} + T_{ET}$, where W_{app} is the work done by the agent moving the bar and T_{ET} is the energy transferred out of the bar and into the resistor by electrical transmission. Taking a time derivative of this equation gives $dW_{app}/dt = -dT_{ET}/dt$ or $P_{app} = -P_{elec}$. In this expression, P_{app} is the power input from the agent moving the bar and P_{elec} is the rate of energy transferring from the bar to the resistor by electricity. The power P_{elec} is a negative number because energy is leaving the bar by this method. Let us verify this equation by using, respectively, Equations 8.18, 5.8, 28.10, 30.5, 26.7, and 26.22:

$$P_{app} = F_{app}v = F_B v = (I\ell B)v = I(B\ell v) = I\mathcal{E} = I(IR) = I^2 R = -P_{elec} \qquad (30.7)$$

In the final step, we recognize that $P = I^2 R$ is the rate of energy delivered to the resistor, so $-I^2 R$ is the rate at which energy leaves the bar.

QUICK QUIZ **30.2** In Figure 30.8a, a given applied force of magnitude F results in a constant speed v and a power input P. Imagine that the force is increased so that the constant speed of the bar is doubled to $2v$. Under these conditions, what are the new force and the new power input? **(a)** $2F$ and $2P$ **(b)** $4F$ and $2P$ **(c)** $2F$ and $4P$ **(d)** $4F$ and $4P$

| Example **30.2** | **Magnetic Force Acting on a Sliding Bar** |

The conducting bar illustrated in Figure 30.9 moves on two frictionless, parallel rails in the presence of a uniform magnetic field directed into the page. The bar has mass m, and its length is ℓ. The bar is given an initial velocity $\vec{\mathbf{v}}_i$ to the right and is released at $t = 0$.

(A) Using Newton's laws, find the speed of the bar as a function of time after it is released.

SOLUTION

Conceptualize As the bar slides to the right in Figure 30.9, a counterclockwise current is established in the circuit consisting of the bar, the rails, and the resistor. The upward current in the bar results in a magnetic force to the left on the bar as shown in the figure. Therefore, the bar must slow down, so our mathematical solution should demonstrate that.

Figure 30.9 (Example 30.2) A conducting bar of length ℓ on two fixed conducting rails is given an initial velocity $\vec{\mathbf{v}}_i$ to the right.

Categorize The text already categorizes this problem as one that uses Newton's laws. We model the bar as a *particle under a net force.*

continued

30.2 continued

Analyze   From Equation 28.10, the magnetic force is $F_B = -I\ell B$, where the negative sign indicates that the force is to the left. The magnetic force is the *only* horizontal force acting on the bar.

Using the particle under a net force model, apply Newton's second law to the bar in the horizontal direction:

$$F_x = ma \quad \rightarrow \quad -I\ell B = m\frac{dv}{dt}$$

Substitute $I = B\ell v/R$ from Equation 30.6:

$$m\frac{dv}{dt} = -\left(\frac{B\ell v}{R}\right)\ell B = -\frac{B^2\ell^2}{R}v$$

Rearrange the equation so that all occurrences of the variable v are on the left and those of t are on the right:

$$\frac{dv}{v} = -\left(\frac{B^2\ell^2}{mR}\right)dt$$

Integrate this equation using the initial condition that $v = v_i$ at $t = 0$ and noting that $(B^2\ell^2/mR)$ is a constant:

$$\int_{v_i}^{v}\frac{dv}{v} = -\frac{B^2\ell^2}{mR}\int_0^t dt$$

$$\ln\left(\frac{v}{v_i}\right) = -\left(\frac{B^2\ell^2}{mR}\right)t$$

Define the constant $\tau = mR/B^2\ell^2$ and solve for the speed:

$$(1) \quad v = v_i e^{-t/\tau}$$

Finalize   This expression for v indicates that the speed of the bar decreases with time under the action of the magnetic force as expected from our conceptualization of the problem. The mathematical form of the decrease is exponential.

(B) Show that the same result is found by using an energy approach.

SOLUTION

Categorize   The text of this part of the problem tells us to use an energy approach for the same situation. We model the bar in Figure 30.9 as a *nonisolated system for energy*.

Analyze   The appropriate reduction of Equation 8.2 is $\Delta K = T_{ET}$. The term on the left represents the change in the speed of the bar, while the right-hand term represents energy transferred out of the bar by electricity.

Differentiate the reduction of Equation 8.2 with respect to time:

$$\frac{dK}{dt} = \frac{dT_{ET}}{dt} = P_{elec} = -I^2R$$

Substitute for the kinetic energy of the bar from Equation 7.16 and the current from Equation 30.6:

$$\frac{d}{dt}\left(\tfrac{1}{2}mv^2\right) = -\left(\frac{B\ell v}{R}\right)^2 R \quad \rightarrow \quad mv\frac{dv}{dt} = -\frac{(B\ell v)^2}{R}$$

Rearrange terms:

$$\frac{dv}{v} = -\left(\frac{B^2\ell^2}{mR}\right)dt$$

Finalize   This result is the same expression to be integrated that we found in part (A).

WHAT IF?   Suppose you wished to increase the distance through which the bar moves between the time it is initially projected and the time it essentially comes to rest. You can do so by changing one of three variables—v_i, R, or B—by a factor of 2 or $\frac{1}{2}$. Which variable should you change to maximize the distance, and would you double it or halve it?

Answer   Increasing v_i would make the bar move farther. Increasing R would decrease the current and therefore the magnetic force, making the bar move farther. Decreasing B would decrease the magnetic force and make the bar move farther. Which method is most effective, though?

Use Equation (1) to find the distance the bar moves by integration:

$$v = \frac{dx}{dt} = v_i e^{-t/\tau}$$

$$x = \int_0^{\infty} v_i e^{-t/\tau}\, dt = -v_i \tau e^{-t/\tau}\Big|_0^{\infty}$$

$$= -v_i\tau(0-1) = v_i\tau = v_i\left(\frac{mR}{B^2\ell^2}\right)$$

This expression shows that doubling v_i or R will double the distance. Changing B by a factor of $\frac{1}{2}$, however, causes the distance to be four times as great!

Example 30.3　Motional emf Induced in a Rotating Bar

A conducting bar of length ℓ rotates with a constant angular speed ω about a pivot at one end. A uniform magnetic field $\vec{\mathbf{B}}$ is directed perpendicular to the plane of rotation as shown in Figure 30.10. Find the motional emf induced between the ends of the bar.

Figure 30.10　(Example 30.3) A conducting bar rotating around a pivot at one end in a uniform magnetic field that is perpendicular to the plane of rotation. A motional emf is induced between the ends of the bar.

SOLUTION

Conceptualize　The rotating bar is different in nature from the bar moving translationally in Figure 30.7. Consider a small segment of the bar, however. It is a short length of conductor moving in a magnetic field and has an emf generated in it like the moving bar in Figure 30.7. By thinking of each small segment as a source of emf, we see that all segments are in series and the emfs add over the length of the bar.

Categorize　Based on the conceptualization of the problem, we approach this example as we did in the discussion leading to Equation 30.5, with the added feature that the short segments of the bar are traveling in circular paths.

Analyze　Evaluate the magnitude of the emf induced in a segment of the bar of length dr having a velocity $\vec{\mathbf{v}}$ from Equation 30.5:

$$d\mathcal{E} = Bv\,dr$$

Find the total emf between the ends of the bar by adding the emfs induced across all segments:

$$\mathcal{E} = \int Bv\,dr$$

The tangential speed v of an element is related to the angular speed ω through the relationship $v = r\omega$ (Eq. 10.10); use that fact and integrate:

$$\mathcal{E} = B\int v\,dr = B\omega\int_0^\ell r\,dr = \tfrac{1}{2}B\omega\ell^2$$

Finalize　In Equation 30.5 for a sliding bar, we can increase $\mathcal{E}$ by increasing B, ℓ, or v. Increasing any one of these variables by a given factor increases $\mathcal{E}$ by the same factor. Therefore, you would choose whichever of these three variables is most convenient to increase. For the rotating rod, however, there is an advantage to increasing the length of the rod to raise the emf because ℓ is squared. Doubling the length gives four times the emf, whereas doubling the angular speed only doubles the emf.

WHAT IF?　Suppose, after reading through this example, you come up with a brilliant idea. A Ferris wheel has radial metallic spokes between the hub and the circular rim. These spokes move in the magnetic field of the Earth, so each spoke acts like the bar in Figure 30.10. You plan to use the emf generated by the rotation of the Ferris wheel to power the incandescent lightbulbs on the wheel. Will this idea work?

Answer　If you calculate a typical emf generated by a spoke, you find that it is about 1 mV, too small to operate an incandescent lightbulb. (Try this calculation!)

An additional difficulty is related to energy. Even assuming you could find lightbulbs that operate using a potential difference on the order of millivolts, a spoke must be part of a circuit to provide a voltage to the lightbulbs. Consequently, the spoke must carry a current. Because this current-carrying spoke is in a magnetic field, a magnetic force is exerted on the spoke in the direction opposite its direction of motion. As a result, the motor of the Ferris wheel must supply more energy to perform work against this magnetic drag force. The motor must ultimately provide the energy that is operating the lightbulbs, and you have not gained anything for free!

30.3　Lenz's Law

Faraday's law (Eq. 30.1) indicates that the induced emf and the change in flux have opposite algebraic signs. This feature has a very real physical interpretation that has come to be known as **Lenz's law:**[1]

> The induced current in a loop is in the direction that creates a magnetic field that opposes the change in magnetic flux through the area enclosed by the loop.

◀ Lenz's law

---

[1]Developed by German physicist Heinrich Lenz (1804–1865).

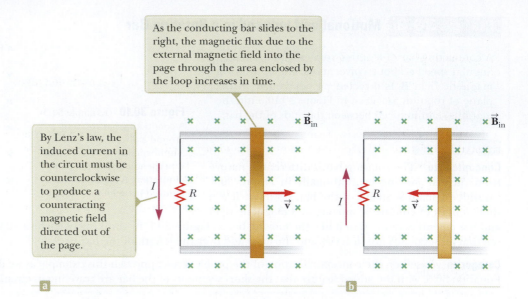

As the conducting bar slides to the right, the magnetic flux due to the external magnetic field into the page through the area enclosed by the loop increases in time.

By Lenz's law, the induced current in the circuit must be counterclockwise to produce a counteracting magnetic field directed out of the page.

Figure 30.11 (a) Lenz's law can be used to determine the direction of the induced current. (b) When the bar moves to the left, the induced current must be clockwise. Why?

That is, the induced current tends to keep the original magnetic flux through the loop from changing. We shall show that this law is a consequence of the law of conservation of energy.

To understand Lenz's law, let's return to the example of a bar moving to the right on two parallel rails in the presence of a uniform magnetic field (the *external* magnetic field, shown by the green crosses in Fig. 30.11a). As the bar moves to the right, the magnetic flux through the area enclosed by the circuit increases with time because the area increases. Lenz's law states that the induced current in the circuit must be directed so that the magnetic field it produces opposes the change in the external magnetic flux. Because the magnetic flux due to an external field directed into the page is increasing, the induced current in the circuit—if it is to oppose this change—must produce a field directed out of the page in the area enclosed by the circuit. Hence, the induced current in the loop of the circuit must be directed counterclockwise when the bar moves to the right. (Use the right-hand rule to verify this direction.) If the bar is moving to the left as in Figure 30.11b, the external magnetic flux through the area enclosed by the loop decreases with time. Because the field is directed into the page, the direction of the induced current must be clockwise if it is to produce a field that also is directed into the page. In either case, the induced current attempts to maintain the original flux through the area enclosed by the current loop.

Let's examine this situation using energy considerations. Suppose the bar is given a slight push to the right. In the preceding analysis, we found that this motion sets up a counterclockwise current in the loop. What happens if we assume the current is clockwise? In this case, the current is downward in the bar. According to Equation 28.10, the direction of the magnetic force exerted on the bar is to the right. This force would accelerate the rod and increase its speed, which in turn would cause the area enclosed by the loop to increase more rapidly. The result would be an increase in the induced current, which would cause an increase in the force, which would produce an increase in the current, and so on—a runaway situation. In effect, the system would acquire energy with no input of energy. This behavior is clearly inconsistent with all experience and violates the law of conservation of energy. Therefore, the current must be counterclockwise.

QUICK QUIZ 30.3 Figure 30.12 shows a circular loop of wire falling toward a wire carrying a current to the left. What is the direction of the induced current in the loop of wire? **(a)** clockwise **(b)** counterclockwise **(c)** zero **(d)** impossible to determine

Figure 30.12 (Quick Quiz 30.3)

Conceptual Example **30.4**   **Application of Lenz's Law**

A magnet is placed near a metal loop as shown in Figure 30.13a.

(A) Find the direction of the induced current in the loop when the magnet is pushed toward the loop.

SOLUTION

This is the same situation as in Figure 30.1b. In the discussion of that figure, we said that the direction of the current was opposite to that in Figure 30.1c, but at that point we could not specify the direction further than that. Now, we can. As the magnet moves to the right toward the loop in Figure 30.13a, the external magnetic flux through the loop increases with time. To counteract this increase in flux due to a field toward the right, the induced current produces its own magnetic field to the left as illustrated in Figure 30.13b; hence, the induced current is in the direction shown. Knowing that like magnetic poles repel each other, we conclude that the left face of the current loop acts like a north pole and the right face acts like a south pole.

(B) Find the direction of the induced current in the loop when the magnet is pulled away from the loop.

SOLUTION

If the magnet moves to the left as in Figure 30.13c, its flux through the area enclosed by the loop decreases in time. Now the induced current in the loop is in the direction shown in Figure 30.13d because this current direction produces a magnetic field in the same direction as the external field. In this case, the left face of the loop is a south pole and the right face is a north pole.

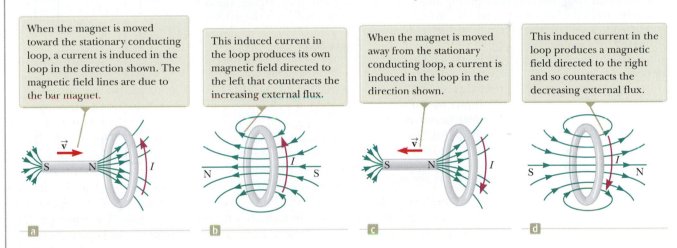

When the magnet is moved toward the stationary conducting loop, a current is induced in the loop in the direction shown. The magnetic field lines are due to the bar magnet.

This induced current in the loop produces its own magnetic field directed to the left that counteracts the increasing external flux.

When the magnet is moved away from the stationary conducting loop, a current is induced in the loop in the direction shown.

This induced current in the loop produces a magnetic field directed to the right and so counteracts the decreasing external flux.

Figure 30.13 (Conceptual Example 30.4) A moving bar magnet induces a current in a conducting loop.

Conceptual Example **30.5**   **A Loop Moving Through a Magnetic Field**

A rectangular metallic loop of dimensions ℓ and w and resistance R moves with constant speed v to the right as in Figure 30.14a (page 808). The loop passes through a uniform magnetic field $\vec{B}$ directed into the page and extending a distance $3w$ along the x axis. Define x as the position of the right side of the loop along the x axis.

(A) Plot the magnetic flux through the area enclosed by the loop as a function of x.

SOLUTION

Figure 30.14b shows the flux through the area enclosed by the loop as a function of x. Before the loop enters the field, the flux through the loop is zero. As the loop enters the field, the flux increases linearly with position until the left edge of the loop is just inside the field. As the loop moves through the uniform field, the flux through the loop remains constant. Finally, the flux through the loop decreases linearly to zero as the loop leaves the field.

(B) Plot the induced motional emf in the loop as a function of x.

SOLUTION

Before the loop enters the field, no motional emf is induced in it because no field is present (Fig. 30.14c). As the right side of the loop enters the field, the magnetic flux directed into the page increases. Hence, according to Lenz's law, the induced current is

continued

30.5 continued

counterclockwise because it must produce its own magnetic field directed out of the page. The motional emf $-B\ell v$ (from Eq. 30.5) arises from the magnetic force experienced by charges in the right side of the loop. When the loop is entirely in the field, the change in magnetic flux through the loop is zero; hence, the motional emf vanishes. That happens because once the left side of the loop enters the field, the motional emf induced in it cancels the motional emf present in the right side of the loop. As the right side of the loop leaves the field, the flux through the loop begins to decrease, a clockwise current is induced, and the induced emf is $B\ell v$. As soon as the left side leaves the field, the emf decreases to zero.

(C) Plot the external applied force necessary to counter the magnetic force and keep v constant as a function of x.

S O L U T I O N

The external force that must be applied to the loop to maintain this motion is plotted in Figure 31.14d. Before the loop enters the field, no magnetic force acts on it; hence, the applied force must be zero if v is constant. When the right side of the loop enters the field, the applied force on the loop must increase to maintain constant speed of the loop. The applied force must be equal in magnitude and opposite in direction to the magnetic force exerted on that side, so that the loop is a particle in equilibrium. When the loop is entirely in the field, the flux through the loop is not changing with time. Hence, the net emf induced in the loop is zero and the current also is zero. Therefore, no external force is needed to maintain the motion: the applied force drops to zero. Finally, as the right side leaves the field, the applied force must be equal in magnitude and opposite in direction to the magnetic force acting on the left side of the loop, which is still in the field.

From this analysis, we conclude that power is supplied only when the loop is either entering or leaving the field. Furthermore, this example shows that the motional emf induced in the loop can be zero even when there is motion through the field! A motional emf is induced *only* when the magnetic flux through the loop *changes in time*.

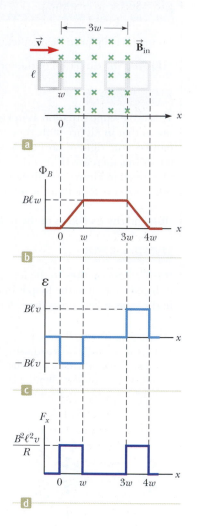

Figure 30.14 (Conceptual Example 30.5) (a) A conducting rectangular loop of width w and length ℓ moving with a velocity $\vec{v}$ through a uniform magnetic field extending a distance $3w$. (b) Magnetic flux through the area enclosed by the loop as a function of loop position. (c) Induced emf as a function of loop position. (d) Applied force required for constant velocity as a function of loop position.

30.4 The General Form of Faraday's Law

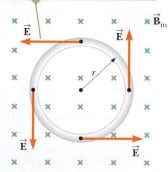

If $\vec{B}$ changes in time, an electric field is induced in a direction tangent to the circumference of the loop.

Figure 30.15 A conducting loop of radius r in a uniform magnetic field perpendicular to the plane of the loop.

Let's look once again at a conducting loop in a magnetic field, as in Figure 30.15. Imagine that the magnetic field changes in time. According to Equation 29.18, the changing field results in a changing magnetic flux through the loop. According to Equation 30.1, the changing flux causes an emf in the loop. According to Equation 26.7, the emf causes a current in the loop. According to Equation 26.6, current is driven by an electric field in the loop. The electric field that is driving the current in the loop is shown at several points in Figure 30.15.

We have found, then, that a changing magnetic field has generated an electric field. We alluded to this phenomenon at the beginning of Section 30.1, and now we are prepared to discuss more details. Here's the question that will lead to a remarkable feature of this discussion: *what if we take the conducting loop away?* This, of course, would take away the charges moving around the loop, but we find the following: the electric field is still there! The moving charges in the loop simply demonstrated that the electric field was there, but the loop is not necessary for the existence of the electric field. Its existence is due solely to the changes in the magnetic field.

Let's try to quantify this new type of electric field. In Equation 24.3, we found that a potential difference between two points in space was equal to the line integral of the dot product of the electric field and an infinitesimal displacement along a path between the points. Let's apply this to a trip around our conducting loop in

Figure 30.15, where the potential difference around the loop will be represented by the emf induced in the loop by the changing magnetic field. Using the integral for the emf in Equation 30.1, we find

$$\oint \vec{E} \cdot d\vec{s} = -\frac{d\Phi_B}{dt} \qquad (30.8)$$

◀ General form of Faraday's law

Equation 30.8 is the general form of Faraday's law. It represents all situations in which a changing magnetic field generates an electric field. In Chapter 33, we will discuss a generalization of Ampère's law (Eq. 29.13) that represents magnetic fields generated by changing electric fields. In that chapter, we will gather together all of these and other important equations into a set called *Maxwell's equations*, which will form the basis of all electromagnetic phenomena.

Let's use Equation 30.8 to find the electric field generated by the changing magnetic field in Figure 30.15. The electric field is everywhere parallel to the displacement vectors on the loop, so the dot product becomes simply $E\,ds$. Because the magnetic field is uniform, the symmetry of the loop tells us that E is the same everywhere on the loop. Therefore, Equation 30.8 becomes

$$E\oint ds = -\frac{d}{dt}(BA) \rightarrow E(2\pi r) = -\frac{dB}{dt}(\pi r^2) \rightarrow E = -\frac{r}{2}\frac{dB}{dt} \qquad (30.9)$$

If the time variation of the magnetic field is specified, the induced electric field can be calculated from Equation 30.9.

Equation 24.3 evaluates the potential difference between two points in space as an integral between those two points of the electric field created by some source charges. Suppose you integrate around a circular path in space in a region containing such an electric field, returning to the same point. If the two points in Equation 24.3 are the same, the integral reduces to zero, which makes sense: the potential difference between two points in space that are the same has to be zero.

But the integral in Equation 30.8 is the same integral and we discussed taking a trip around the circular loop in Figure 30.15. The value of the integral is *not* zero in this case. What's going on? This is evidence that the electric field we are discussing here is different in nature from that formed by the stationary charges in Chapter 23. We describe an induced electric field as *nonconservative*, because the integral around a closed path is not zero. Despite this difference in nature, the induced electric field has many of the same properties as electric fields due to source charges. For example, an induced electric field can apply forces on charged particles according to Equation 22.8.

Example 30.6   Electric Field Induced by a Changing Magnetic Field in a Solenoid

A long solenoid of radius R has n turns of wire per unit length and carries a time-varying current that varies sinusoidally as $I = I_{max}\cos\omega t$, where I_{max} is the maximum current and ω is the angular frequency of the alternating current source (Fig. 30.16).

(A) Determine the magnitude of the induced electric field outside the solenoid at a distance $r > R$ from its long central axis.

SOLUTION

Conceptualize Figure 30.16 shows the physical situation. As the current in the coil changes, imagine a changing magnetic field at all points in space as well as an induced electric field.

Categorize In this analysis problem, because the current varies in time, the magnetic field is changing, leading to an induced electric field as opposed to the electrostatic electric fields due to stationary electric charges.

Analyze First consider an external point and take the path for the line integral to be a circle of radius r centered on the solenoid as illustrated in Figure 30.16.

Figure 30.16 (Example 30.6) A long solenoid carrying a time-varying current given by $I = I_{max}\cos\omega t$. An electric field is induced both inside and outside the solenoid.

continued

30.6 continued

Evaluate the right side of Equation 30.8, noting that the magnetic field $\vec{B}$ inside the solenoid is perpendicular to the circle bounded by the path of integration:

$$(1) \quad -\frac{d\Phi_B}{dt} = -\frac{d}{dt}(B\pi R^2) = -\pi R^2 \frac{dB}{dt}$$

Evaluate the magnetic field inside the solenoid from Equation 29.17:

$$(2) \quad B = \mu_0 nI = \mu_0 nI_{max} \cos \omega t$$

Substitute Equation (2) into Equation (1):

$$(3) \quad -\frac{d\Phi_B}{dt} = -\pi R^2 \mu_0 nI_{max} \frac{d}{dt}(\cos \omega t) = \pi R^2 \mu_0 nI_{max} \omega \sin \omega t$$

Evaluate the left side of Equation 30.8, noting that the magnitude of $\vec{E}$ is constant on the path of integration and $\vec{E}$ is tangent to it:

$$(4) \quad \oint \vec{E} \cdot d\vec{s} = E(2\pi r)$$

Substitute Equations (3) and (4) into Equation 30.8:

$$E(2\pi r) = \pi R^2 \mu_0 nI_{max} \omega \sin \omega t$$

Solve for the magnitude of the electric field:

$$E = \frac{\mu_0 nI_{max} \omega R^2}{2r} \sin \omega t \quad \text{(for } r > R)$$

Finalize   This result shows that the amplitude of the electric field outside the solenoid falls off as $1/r$ and varies sinusoidally with time. It is proportional to the current I as well as to the frequency ω, consistent with the fact that a larger value of ω means more change in magnetic flux per unit time. As we will learn in Chapter 33, the time-varying electric field creates an additional contribution to the magnetic field. The magnetic field can be somewhat stronger than we first stated, both inside and outside the solenoid. The correction to the magnetic field is small if the angular frequency ω is small. At high frequencies, however, a new phenomenon can dominate: The electric and magnetic fields, each re-creating the other, constitute an electromagnetic wave radiated by the solenoid as we will study in Chapter 33.

(B) What is the magnitude of the induced electric field inside the solenoid, a distance r from its axis?

SOLUTION

Analyze   For an interior point ($r < R$), the magnetic flux through an integration loop is given by $\Phi_B = B\pi r^2$.

Evaluate the right side of Equation 30.8:

$$(5) \quad -\frac{d\Phi_B}{dt} = -\frac{d}{dt}(B\pi r^2) = -\pi r^2 \frac{dB}{dt}$$

Substitute Equation (2) into Equation (5):

$$(6) \quad -\frac{d\Phi_B}{dt} = -\pi r^2 \mu_0 nI_{max} \frac{d}{dt}(\cos \omega t) = \pi r^2 \mu_0 nI_{max} \omega \sin \omega t$$

Substitute Equations (4) and (6) into Equation 30.8:

$$E(2\pi r) = \pi r^2 \mu_0 nI_{max} \omega \sin \omega t$$

Solve for the magnitude of the electric field:

$$E = \frac{\mu_0 nI_{max} \omega}{2} r \sin \omega t \quad \text{(for } r < R)$$

Finalize   This result shows that the amplitude of the electric field induced inside the solenoid by the changing magnetic flux through the solenoid increases linearly with r and varies sinusoidally with time. As with the field outside the solenoid, the field inside is proportional to the current I and the frequency ω.

30.5   Generators and Motors

You may own an emergency flashlight or radio that operates by turning a crank with your hand. Inside this device is a *generator* that converts the energy from your hand into electrical potential energy. Let us look first at the **direct-current (DC) generator,** which is illustrated in Figure 30.17a. In its simplest form, it consists of a loop of wire rotated by some external means in a magnetic field. As the loop

Commutator

N

S

ε

t

Brush

a

b

Figure 30.17 (a) Schematic diagram of a DC generator. (b) The magnitude of the emf varies in time, but the polarity never changes.

rotates, the magnetic flux through the area enclosed by the loop changes with time, and this change induces an emf and a current in the loop according to Faraday's law. The ends of the loop are connected to a split ring device, called a *commutator*, that rotates with the loop. Connections from the commutator, which act as output terminals of the generator, to the external circuit are made by stationary metallic brushes in contact with the commutator.

In this configuration, the output voltage always has the same polarity and pulsates with time as shown in Figure 30.17b. We can understand why by noting that the contacts to the split ring reverse their roles every half cycle. At the same time, the polarity of the induced emf reverses; hence, the polarity of the split ring (which is the same as the polarity of the output voltage) remains the same.

A pulsating DC current is not suitable for most applications. To obtain a steadier DC current, commercial DC generators use many coils and commutators distributed so that the sinusoidal pulses from the various coils are out of phase. When these pulses are superimposed, the DC output is almost free of fluctuations.

Let us now consider the **alternating-current (AC) generator.** As with the DC generator, it consists of a loop of wire rotated by some external means in a magnetic field (Fig. 30.18a). The ends of the loop are connected to two *slip rings* that rotate with the loop. Connections from these slip rings, which act as output terminals of the generator, to the external circuit are made by stationary metallic brushes in contact with the slip rings.

In commercial power plants, the energy required to rotate the loop can be derived from a variety of sources. For example, in a hydroelectric plant, falling water directed against the blades of a turbine produces the rotary motion. For the windmills in the opening storyline, the rotation of the blades from the wind

An emf is induced in a loop that rotates in a magnetic field.

Slip rings

N

S

ε

ε_{max}

t

External circuit

Brushes

a

b

Figure 30.18 (a) Schematic diagram of an AC generator. (b) The alternating emf induced in the loop plotted as a function of time.

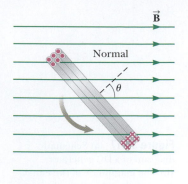

Figure 30.19 A cutaway view of a loop enclosing an area A and containing N turns, rotating with constant angular speed ω in a magnetic field. The emf induced in the loop varies sinusoidally in time.

causes the rotation of a loop in a generator, which is what is in the box behind the blades.

Instead of a single turn, suppose a coil with N turns (a more practical situation), with the same area A, rotates in a magnetic field with a constant angular speed ω. If θ is the angle between the magnetic field and the normal to the plane of the coil as in Figure 30.19, the magnetic flux through the coil at any time t is

$$\Phi_B = BA \cos \theta = BA \cos \omega t$$

where we have used the relationship $\theta = \omega t$ between angular position and angular speed (see Eq. 10.3). (We have set the clock so that $t = 0$ when $\theta = 0$.) Hence, the induced emf in the coil is

$$\mathcal{E} = -N \frac{d\Phi_B}{dt} = -NBA \frac{d}{dt}(\cos \omega t) = NBA\omega \sin \omega t \tag{30.10}$$

This result shows that the emf varies sinusoidally with time as plotted in Figure 30.18b. Equation 30.10 shows that the maximum emf has the value

$$\mathcal{E}_{max} = NBA\omega \tag{30.11}$$

which occurs when $\omega t = 90°$ or $270°$. In other words, $\mathcal{E} = \mathcal{E}_{max}$ when the magnetic field is in the plane of the coil and the time rate of change of flux is a maximum. Furthermore, the emf is zero when $\omega t = 0$ or $180°$, that is, when $\vec{B}$ is perpendicular to the plane of the coil and the time rate of change of flux is zero.

The frequency for commercial generators in the United States and Canada is 60 Hz, whereas in some European countries it is 50 Hz. (Recall that $\omega = 2\pi f$, where f is the frequency in hertz.)

QUICK QUIZ 30.4 In an AC generator, a coil with N turns of wire spins in a magnetic field. Of the following choices, which does *not* cause an increase in the emf generated in the coil? **(a)** replacing the coil wire with one of lower resistance **(b)** spinning the coil faster **(c)** increasing the magnetic field **(d)** increasing the number of turns of wire on the coil

---

Example 30.7   **emf Induced in a Generator**

The coil in an AC generator consists of 8 turns of wire, each of area $A = 0.090\ 0\ \text{m}^2$, and the total resistance of the wire is $12.0\ \Omega$. The coil rotates in a 0.500-T magnetic field at a constant frequency of 60.0 Hz.

(A) Find the maximum induced emf in the coil.

SOLUTION

Conceptualize   Study Figure 30.18 to make sure you understand the operation of an AC generator.

Categorize   We evaluate parameters using equations developed in this section, so we categorize this example as a substitution problem.

Use Equation 30.11 to find the maximum induced emf:

$$\mathcal{E}_{max} = NBA\omega = NBA(2\pi f)$$

Substitute numerical values:

$$\mathcal{E}_{max} = 8(0.500\ \text{T})(0.090\ 0\ \text{m}^2)(2\pi)(60.0\ \text{Hz}) = \boxed{136\ \text{V}}$$

(B) What is the maximum induced current in the coil when the output terminals are connected to a low-resistance conductor?

SOLUTION

Use Equation 26.7 and the result to part (A):

$$I_{max} = \frac{\mathcal{E}_{max}}{R} = \frac{136\ \text{V}}{12.0\ \Omega} = \boxed{11.3\ \text{A}}$$

A **motor** is a device into which energy is transferred by electrical transmission while energy is transferred out by work. A motor is essentially a generator operating in reverse. Instead of generating a current by rotating a coil, a current is supplied to the coil by a battery, and the torque acting on the current-carrying coil (Section 28.5) causes it to rotate.

Useful mechanical work can be done by attaching the rotating coil of a motor to some external device. As the coil rotates in a magnetic field, however, the changing magnetic flux induces an emf in the coil; consistent with Lenz's law, this induced emf always acts to reduce the current in the coil. The phrase *back emf* is used to indicate an emf that tends to reduce the supplied current. Because the voltage available to supply current equals the difference between the supply voltage and the back emf, the current in the rotating coil is limited by the back emf.

When a motor is first turned on, there is initially no back emf, and the current is very large because it is limited only by the resistance of the coil. As the coil begins to rotate, the induced back emf increases with the speed of the coil and opposes the applied voltage, and the current in the coil decreases. If the motor is allowed to run under no mechanical load, the back emf reduces the current to a value just large enough to overcome energy losses due to internal energy and friction. If a very heavy load jams the motor so that it cannot rotate, the lack of a back emf can lead to dangerously high current in the motor's wire. This dangerous situation is explored in the What If? section of Example 30.8.

Example 30.8 | **The Induced Current in a Motor**

A motor contains a coil with a total resistance of 10 Ω and is supplied by a voltage of 120 V. When the motor is running at its maximum speed, the back emf is 70 V.

(A) Find the current in the coil at the instant the motor is turned on.

SOLUTION

Conceptualize Think about the motor just after it is turned on. It has not yet moved, so there is no back emf generated. As a result, the current in the motor is high. After the motor begins to turn, a back emf is generated and the current decreases.

Categorize We need to combine our new understanding about motors with the relationship between current, voltage, and resistance in this substitution problem.

Evaluate the current in the coil from Equation 26.7 with no back emf generated:

$$I = \frac{\varepsilon}{R} = \frac{120 \text{ V}}{10 \text{ } \Omega} = 12 \text{ A}$$

(B) Find the current in the coil when the motor has reached maximum speed.

SOLUTION

Evaluate the current in the coil with the maximum back emf generated:

$$I = \frac{\varepsilon - \varepsilon_{\text{back}}}{R} = \frac{120 \text{ V} - 70 \text{ V}}{10 \text{ } \Omega} = \frac{50 \text{ V}}{10 \text{ } \Omega} = 5.0 \text{ A}$$

The current drawn by the motor when operating at its maximum speed is significantly less than that drawn before it begins to turn.

WHAT IF? Suppose this motor is in a circular saw. When you are operating the saw, the blade becomes jammed in a piece of wood and the motor cannot turn. By what percentage does the power input to the motor increase when it is jammed?

Answer You may have everyday experiences with motors becoming warm when they are prevented from turning. That is due to the increased power input to the motor. The higher rate of energy transfer results in an increase in the internal energy of the coil, an undesirable effect.

continued

30.8 continued

Set up the ratio of power input to the motor when jammed, using the current calculated in part (A), to that when it is not jammed, part (B):

$$\frac{P_{\text{jammed}}}{P_{\text{not jammed}}} = \frac{I_A^2 R}{I_B^2 R} = \frac{I_A^2}{I_B^2}$$

Substitute numerical values:

$$\frac{P_{\text{jammed}}}{P_{\text{not jammed}}} = \frac{(12\ \text{A})^2}{(5.0\ \text{A})^2} = 5.76$$

That represents a 476% increase in the input power! Such a high power input can cause the coil to become so hot that it is damaged.

30.6 Eddy Currents

As we have seen, an emf and a current are induced in a loop of wire by a changing magnetic flux. Imagine now a plate of metal, such as the one hanging from a rod in Figure 30.20. A metal plate can be considered to be a combination of many concentric circular conducting loops of various radii. Therefore, circulating currents called **eddy currents** are induced in bulk pieces of metal moving through a magnetic field. This phenomenon can be demonstrated by allowing the plate in Figure 30.20 to swing back and forth through a magnetic field. As the plate enters the field, the changing magnetic flux induces an emf in the plate, which in turn causes the free electrons in the plate to move, producing the swirling eddy currents. According to Lenz's law, the direction of the eddy currents is such that they create magnetic fields that oppose the change that causes the currents. For this reason, the eddy currents must produce effective magnetic poles on the plate, which are repelled by the poles of the magnet; this situation gives rise to a repulsive force that opposes the motion of the plate. (If the opposite were true, the plate would accelerate and its energy would increase after each swing, in violation of the law of conservation of energy.)

As indicated in Figure 30.21a, with $\vec{\mathbf{B}}$ directed into the page, the induced eddy current is counterclockwise as the swinging plate enters the field at position 1 because the flux due to the external magnetic field into the page through the plate is increasing. Hence, by Lenz's law, the induced current must provide its own

As the plate enters or leaves the field, the changing magnetic flux induces an emf, which causes eddy currents in the plate.

Figure 30.20 Formation of eddy currents in a conducting plate moving through a magnetic field.

Figure 30.21 When a conducting plate swings through a magnetic field, eddy currents are induced and the magnetic force $\vec{\mathbf{F}}_B$ on the plate opposes its velocity, causing it to eventually come to rest.

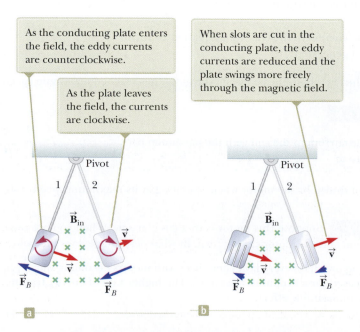

As the conducting plate enters the field, the eddy currents are counterclockwise.

As the plate leaves the field, the currents are clockwise.

When slots are cut in the conducting plate, the eddy currents are reduced and the plate swings more freely through the magnetic field.

magnetic field out of the page. The opposite is true as the plate leaves the field at position 2, where the current is clockwise. Because the induced eddy current always produces a magnetic retarding force $\vec{\mathbf{F}}_B$ when the plate enters or leaves the field, the swinging plate eventually comes to rest. If slots are cut in the plate as shown in Figure 30.21b, many of the conducting loops in the plate are cut and the eddy currents and, therefore, the corresponding retarding force, are greatly reduced.

The braking systems on many subway and rapid-transit cars make use of electromagnetic induction and eddy currents. An electromagnet attached to the train is positioned near the steel rails. (An electromagnet is essentially a solenoid with an iron core.) The braking action occurs when a large current is passed through the electromagnet. The relative motion of the magnet and rails induces eddy currents in the rails, and the direction of these currents produces a drag force on the moving train. Because the eddy currents decrease steadily in magnitude as the train slows down, the braking effect is quite smooth. As a safety measure, some power tools use eddy currents to stop rapidly spinning blades once the device is turned off.

Eddy currents are often undesirable because they represent a transformation of mechanical energy to internal energy in the resistance of the metal. To reduce this energy loss, conducting parts are often laminated; that is, they are built up in thin layers separated by a nonconducting material such as lacquer or a metal oxide. This layered structure prevents large current loops and effectively confines the currents to small loops in individual layers. Such a laminated structure is used in transformer cores (see Section 32.8) and motors to minimize eddy currents and thereby increase the efficiency of these devices.

Summary

▶ Concepts and Principles

Faraday's law of induction states that the emf induced in a loop is directly proportional to the time rate of change of magnetic flux through the loop, or

$$\mathcal{E} = -\frac{d\Phi_B}{dt} \tag{30.1}$$

where $\Phi_B = \int \vec{\mathbf{B}} \cdot d\vec{\mathbf{A}}$ is the magnetic flux through the loop.

When a conducting bar of length ℓ moves at a velocity $\vec{\mathbf{v}}$ through a magnetic field $\vec{\mathbf{B}}$, where $\vec{\mathbf{B}}$ is perpendicular to the bar and to $\vec{\mathbf{v}}$, the **motional emf** induced in the bar is

$$\mathcal{E} = -B\ell v \tag{30.5}$$

Lenz's law states that the induced current and induced emf in a conductor are in such a direction as to set up a magnetic field that opposes the change that produced them.

A general form of **Faraday's law of induction** is

$$\oint \vec{\mathbf{E}} \cdot d\vec{\mathbf{s}} = -\frac{d\Phi_B}{dt} \tag{30.8}$$

where $\vec{\mathbf{E}}$ is the nonconservative electric field that is produced by the changing magnetic flux.

Think–Pair–Share

See the Preface for an explanation of the icons used in this problems set. For additional assessment items for this section, go to **WEBASSIGN** From Cengage

1. **ACTIVITY** In this activity, your group will build a simple *homopolar motor*. You will need the following components:

   AA battery
   Wood screw
   Disk magnet
   Insulated wire, about 10 cm long, with ends stripped

Place the disk magnet on the head of the screw, where it should be attracted and stay attached. Hang the point of the screw from the flat bottom of the battery, where it will be magnetically attracted and hang. Figure TP30.1 (page 816) shows the structure of the motor. Put one end of the wire on the button at the top of the battery and hold it with a finger or tape. Touch the other end of the wire to the edge of the disk magnet, and watch the screw spin! The tip of the screw may wander as it spins. If it arrives at the edge of the flat part of the bottom of the battery, it may find a spot where the friction at the tip prevents its rotation. Record the sense of rotation of the screw and report what happens if (a) the disk magnet is turned over to reverse its poles, and (b) the battery is turned over so that the screw hangs from the button.

Figure TP30.1

2. Your group has an idea for the design of a new fitness apparatus that will make use of induced emf. You plan to construct a cylindrical cage, with a horizontal axis, using 32 metal rods of length $\ell = 0.800$ m mounted between two circular metal endpieces of radius $r = 25.0$ cm as shown in part (a) of Figure TP30.2. Between the endpieces you will connect a resistor. This cage will be mounted on a pivot so that it can rotate around the centers of the metal endpieces. In addition, an electromagnet is mounted with one pole inside the cage and the other outside, as shown in part (b) of Figure TP30.2. The figure shows an endview of the poles of the magnet, which extend into the page a distance ℓ, so that the magnetic poles are as long as the metal rods. The magnet produces a magnetic field of magnitude $B = 0.250$ T between its poles. In use, the operator stands facing the cage and continuously pulls downward on the metal rods at the left of the cage as shown in Figure TP30.2b. This sets the cage in rotation, causing the metal rods to pass between the poles of the magnet. The emf generated in the metal rods creates a current, and the magnetic force on this current provides a resistive force to the rotation of the cage. (a) Discuss in your group the following and perform the requested calculation: Your design goal is that the operator should do work at the rate of 100 watts when the cage is rotating at a constant angular speed of 5.00 rad/s. Before building the apparatus, see if your design goal is feasible by calculating the resistance required between the endpieces. (b) Discuss in your group the following questions: How safe is this device? Can you see a safety flaw?

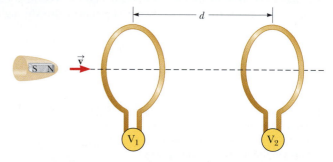

Figure TP30.2

Problems

See the Preface for an explanation of the icons used in this problems set. For additional assessment items for this section, go to **WEBASSIGN** From Cengage

SECTION 30.1 Faraday's Law of Induction

1. A circular loop of wire of radius 12.0 cm is placed in a magnetic field directed perpendicular to the plane of the loop as in Figure P30.1. If the field decreases at the rate of 0.050 0 T/s in some time interval, find the magnitude of the emf induced in the loop during this interval.

Figure P30.1

2. An instrument based on induced emf has been used to measure projectile speeds up to 6 km/s. A small magnet is imbedded in the projectile as shown in Figure P30.2. The projectile passes through two coils separated by a distance d. As the projectile passes through each coil, a pulse of emf is induced in the coil. The time interval between pulses

Figure P30.2

can be measured accurately with an oscilloscope, and thus the speed can be determined. (a) Sketch a graph of ΔV versus t for the arrangement shown. Consider a current that flows counterclockwise as viewed from the starting point of the projectile as positive. On your graph, indicate which pulse is from coil 1 and which is from coil 2. (b) If the pulse separation is 2.40 ms and $d = 1.50$ m, what is the projectile speed?

3. Scientific work is currently under way to determine whether **BIO** weak oscillating magnetic fields can affect human health. For example, one study found that drivers of trains had a higher incidence of blood cancer than other railway workers, possibly due to long exposure to mechanical devices in the train engine cab. Consider a magnetic field of magnitude 1.00×10^{-3} T, oscillating sinusoidally at 60.0 Hz.

If the diameter of a red blood cell is 8.00 μm, determine the maximum emf that can be generated around the perimeter of a cell in this field.

4. A long solenoid has $n = 400$ turns per meter and carries a current given by $I = 30.0(1 - e^{-1.60t})$, where I is in amperes and t is in seconds. Inside the solenoid and coaxial with it is a coil that has a radius of $R = 6.00$ cm and consists of a total of $N = 250$ turns of fine wire (Fig. P30.4). What emf is induced in the coil by the changing current?

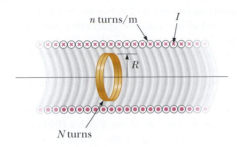

Figure P30.4

5. An aluminum ring of radius $r_1 = 5.00$ cm and resistance 3.00×10^{-4} Ω is placed around one end of a long air-core solenoid with 1 000 turns per meter and radius $r_2 = 3.00$ cm as shown in Figure P30.5. Assume the axial component of the field produced by the solenoid is one-half as strong over the area of the end of the solenoid as at the center of the solenoid. Also assume the solenoid produces negligible field outside its cross-sectional area. The current in the solenoid is increasing at a rate of 270 A/s. (a) What is the induced current in the ring? At the center of the ring, what are (b) the magnitude and (c) the direction of the magnetic field produced by the induced current in the ring?

Figure P30.5 Problems 5 and 6.

6. An aluminum ring of radius r_1 and resistance R is placed around one end of a long air-core solenoid with n turns per meter and smaller radius r_2 as shown in Figure P30.5. Assume the axial component of the field produced by the solenoid over the area of the end of the solenoid is one-half as strong as at the center of the solenoid. Also assume the solenoid produces negligible field outside its cross-sectional area. The current in the solenoid is increasing at a rate of $\Delta I/\Delta t$. (a) What is the induced current in the ring? (b) At the center of the ring, what is the magnetic field produced by the induced current in the ring? (c) What is the direction of this field?

7. A coil formed by wrapping 50 turns of wire in the shape of a square is positioned in a magnetic field so that the normal to the plane of the coil makes an angle of 30.0° with the direction of the field. When the magnetic field is increased

uniformly from 200 μT to 600 μT in 0.400 s, an emf of magnitude 80.0 mV is induced in the coil. What is the total length of the wire in the coil?

8. When a wire carries an AC current with a known frequency, you can use a *Rogowski coil* to determine the amplitude I_{max} of the current without disconnecting the wire to shunt the current through a meter. The Rogowski coil, shown in Figure P30.8, simply clips around the wire. It consists of a toroidal conductor wrapped around a circular return cord. Let n represent the number of turns in the toroid per unit distance along it. Let A represent the cross-sectional area of the toroid. Let $I(t) = I_{max} \sin \omega t$ represent the current to be measured. (a) Show that the amplitude of the emf induced in the Rogowski coil is $\mathcal{E}_{max} = \mu_0 nA\omega I_{max}$. (b) Explain why the wire carrying the unknown current need not be at the center of the Rogowski coil and why the coil will not respond to nearby currents that it does not enclose.

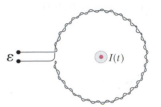

Figure P30.8

9. A toroid having a rectangular cross section ($a = 2.00$ cm by $b = 3.00$ cm) and inner radius $R = 4.00$ cm consists of $N = 500$ turns of wire that carry a sinusoidal current $I = I_{max} \sin \omega t$, with $I_{max} = 50.0$ A and a frequency $f = \omega/2\pi = 60.0$ Hz. A coil that consists of $N' = 20$ turns of wire is wrapped around one section of the toroid as shown in Figure P30.9. Determine the emf induced in the coil as a function of time.

Figure P30.9

SECTION 30.2 Motional emf

Problem 47 in Chapter 28 can be assigned with this section.

10. A small airplane with a wingspan of 14.0 m is flying due north at a speed of 70.0 m/s over a region where the vertical component of the Earth's magnetic field is 1.20 μT downward. (a) What potential difference is developed between the airplane's wingtips? (b) Which wingtip is at higher potential? (c) **What If?** How would the answers to parts (a) and (b) change if the plane turned to fly due east? (d) Can this emf be used to power a lightbulb in the passenger compartment? Explain your answer.

11. A helicopter (Fig. P30.11) has blades of length 3.00 m, extending out from a central hub and rotating at 2.00 rev/s. If the vertical component of the Earth's magnetic field is 50.0 μT, what is the emf induced between the blade tip and the center hub?

Figure P30.11

12. A 2.00-m length of wire is held in an east–west direction and moves horizontally to the north with a speed of 0.500 m/s. The Earth's magnetic field in this region is of magnitude 50.0 μT and is directed northward and 53.0° below the horizontal. (a) Calculate the magnitude of the induced emf between the ends of the wire and (b) determine which end is positive.

13. A metal rod of mass m slides without friction along two **S** parallel horizontal rails, separated by a distance ℓ and connected by a resistor R, as shown in Figure P30.13. A uniform vertical magnetic field of magnitude B is applied perpendicular to the plane of the paper. The applied force shown in the figure acts only for a moment, to give the rod a speed v. In terms of m, ℓ, R, B, and v, find the distance the rod will then slide as it coasts to a stop.

Figure P30.13

14. *Why is the following situation impossible?* An automobile has a vertical radio antenna of length $\ell = 1.20$ m. The automobile travels on a curvy, horizontal road where the Earth's magnetic field has a magnitude of $B = 50.0$ μT and is directed toward the north and downward at an angle of $\theta = 65.0°$ below the horizontal. The motional emf developed between the top and bottom of the antenna varies with the speed and direction of the automobile's travel and has a maximum value of 4.50 mV.

15. A conducting bar of length ℓ moves to the right on two fric- **Q|C** tionless rails as shown in Figure P30.15. A uniform magnetic field directed into the page has a magnitude of 0.300 T. Assume $R = 9.00$ Ω and $\ell = 0.350$ m. (a) At what constant speed should the bar move to produce an 8.50-mA current in the resistor? (b) What is the direction of the induced current? (c) At what rate is energy delivered to the resistor? (d) Explain the origin of the energy being delivered to the resistor.

Figure P30.15

16. An astronaut is connected to her spacecraft by a 25.0-m-long **Q|C** tether cord as she and the spacecraft orbit the Earth in a circular path at a speed of 7.80×10^3 m/s. At one instant, the emf between the ends of a wire embedded in the cord is measured to be 1.17 V. Assume the long dimension of the cord is perpendicular to the Earth's magnetic field at that instant. Assume also the tether's center of mass moves with a velocity perpendicular to the Earth's magnetic field. (a) What is the magnitude of the Earth's field at this location? (b) Does the emf change as the system moves from one location to another? Explain. (c) Provide two conditions under which the emf would be zero even though the magnetic field is not zero.

17. You are working for a company that manufactures motors **CR** and generators. At the end of your first day of work, your supervisor explains to you that you will be assigned to a team that is designing a new *homopolar generator*. You have no idea what that is, but agree wholeheartedly to the assignment. At home that evening, you go online to learn about the homopolar generator and find the following. The homopolar generator, also called the *Faraday disk,* is a low-voltage, high-current electric generator. It consists of a rotating conducting disk with one stationary brush (a sliding electrical contact) at its axle and another at a point on its circumference as shown in Figure P30.17. A uniform magnetic field is applied perpendicular to the plane of the disk. When superconducting coils are used to produce a large magnetic field, a homopolar generator can have a power output of several megawatts. Such a generator is useful, for example, in purifying metals by electrolysis. If a voltage is applied to the output terminals of the generator, it runs in reverse as a *homopolar motor* capable of providing great torque, useful in ship propulsion. At work the next morning, your supervisor tells you that the homopolar generator under consideration will have a magnetic field of magnitude $B = 0.900$ T and the radius of the disk is $r = 0.400$ m. The desired emf to be generated with the device is $\varepsilon = 25.0$ V. Your supervisor asks you to determine the required angular speed of the disk to achieve this result.

Figure P30.17

18. You are working in a laboratory that uses motional emf to make magnetic measurements. You have found that it is difficult to create a uniform magnetic field across the entire sliding-bar apparatus shown in Figure 30.8a, with a resistance R connected between the rails. You decide to investigate creating the magnetic field with a long, straight, current-carrying conductor lying next to and parallel to one of the rails, as shown in Figure P30.18. This will create a nonuniform field across the plane of the bar and rails. You set up the apparatus in this way, with the current-carrying wire a distance a from the upper rail. You wish to find an expression for the force necessary to slide the bar at a constant speed of v to the right in Figure P30.18 if the wire carries a current I. (*Hint:* Two separate integrations will be required.)

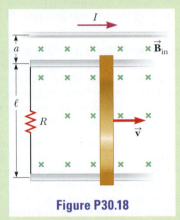

Figure P30.18

19. You are working in a factory that produces long bars of copper with a square cross section. In one section of the production process, the bars must slide down a plane inclined at an angle $\theta = 21.0°$ to the horizontal. It has been found that the bars travel with too high a speed and become dented or bent when they arrive at the bottom of the plane and must be discarded. In order to prevent this waste, you devise a way to deliver the bars at the bottom of the plane at a lower speed. You replace the inclined plane with a pair of parallel metal rails, shown in Figure P30.19, separated by a distance $\ell = 2.00$ m. The smooth bars of mass $m = 1.00$ kg will slide down the smooth rails, with the length of the bar always perpendicular to the rails. The rails are immersed in a magnetic field of magnitude B, and a resistor of resistance $R = 1.00\ \Omega$ is connected between the upper ends of the rails. Determine the magnetic field necessary in your device so that the bars will arrive at the bottom of the plane with a maximum speed $v = 1.00$ m/s.

Figure P30.19
Problems 19 and 20.

20. You are working in a factory that produces long bars of copper with a square cross section. In one section of the production process, the bars must slide down an inclined plane of angle θ. It has been found that the bars travel with too high a speed and become dented or bent when they arrive at the bottom of the plane and must be discarded. In order to prevent this waste, you devise a way to deliver the bars at the bottom of the plane at a lower speed. You replace the inclined plane with a pair of parallel metal rails, shown in Figure P30.19, separated by a distance ℓ. The smooth bars of mass m will slide down the smooth rails, with the length of the bar always perpendicular to the rails. The rails are immersed in a magnetic field of magnitude B, and a resistor of resistance R is connected between the upper ends of the rails. Determine the magnetic field necessary in your device so that the bars will arrive at the bottom of the plane with a maximum speed v_{max}.

SECTION 30.4 The General Form of Faraday's Law

21. Within the green dashed circle shown in Figure P30.21, the magnetic field changes with time according to the expression $B = 2.00t^3 - 4.00t^2 + 0.800$, where B is in teslas, t is in seconds, and $R = 2.50$ cm. When $t = 2.00$ s, calculate (a) the magnitude and (b) the direction of the force exerted on an electron located at point P, which is at a distance $r = 5.00$ cm from the center of the circular field region. (c) At what instant is this force equal to zero?

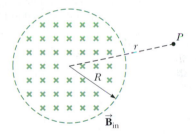

Figure P30.21

22. A long solenoid with 1.00×10^3 turns per meter and radius 2.00 cm carries an oscillating current $I = 5.00 \sin 100\pi t$, where I is in amperes and t is in seconds. (a) What is the electric field induced at a radius $r = 1.00$ cm from the axis of the solenoid? (b) What is the direction of this electric field when the current is increasing counterclockwise in the solenoid?

SECTION 30.5 Generators and Motors

Problem 45 in Chapter 28 can be assigned with this section.

23. A generator produces 24.0 V when turning at 900 rev/min. What emf does it produce when turning at 500 rev/min?

24. Figure P30.24 (page 820) is a graph of the induced emf versus time for a coil of N turns rotating with angular speed ω in a uniform magnetic field directed perpendicular to the coil's axis of rotation. **What If?** Copy this sketch (on a larger scale) and on the same set of axes show the graph of emf versus t (a) if the number of turns in the coil is doubled, (b) if instead the angular speed is doubled, and (c) if the angular speed is doubled while the number of turns in the coil is halved.

Figure P30.24

25. The rotating loop in an AC generator is a square 10.0 cm on each side. It is rotated at 60.0 Hz in a uniform magnetic field of 0.800 T. Calculate (a) the flux through the loop as a function of time, (b) the emf induced in the loop, (c) the current induced in the loop for a loop resistance of 1.00 Ω, (d) the power delivered to the loop, and (e) the torque that must be exerted to rotate the loop.

26. **QC** In Figure P30.26, a semicircular conductor of radius $R = 0.250$ m is rotated about the axis AC at a constant rate of 120 rev/min. A uniform magnetic field of magnitude 1.30 T fills the entire region below the axis and is directed out of the page. (a) Calculate the maximum value of the emf induced between the ends of the conductor. (b) What is the value of the average induced emf for each complete rotation? (c) **What If?** How would your answers to parts (a) and (b) change if the magnetic field were allowed to extend a distance R above the axis of rotation? Sketch the emf versus time (d) when the field is as drawn in Figure P30.26 and (e) when the field is extended as described in part (c).

Figure P30.26

SECTION 30.6 Eddy Currents

27. **QC** Figure P30.27 represents an electromagnetic brake that uses eddy currents. An electromagnet hangs from a railroad car near one rail. To stop the car, a large current is sent through

Figure P30.27

the coils of the electromagnet. The moving electromagnet induces eddy currents in the rails, whose fields oppose the change in the electromagnet's field. The magnetic fields of the eddy currents exert force on the current in the electromagnet, thereby slowing the car. The direction of the car's motion and the direction of the current in the electromagnet are shown correctly in the picture. Determine which of the eddy currents shown on the rails is correct. Explain your answer.

ADDITIONAL PROBLEMS

28. Suppose you wrap wire onto the core from a roll of Scotch tape to make a coil. Describe how you can use a bar magnet to produce an induced voltage in the coil. What is the order of magnitude of the emf you generate? State the quantities you take as data and their values.

29. A rectangular loop of area $A = 0.160$ m² is placed in a region where the magnetic field is perpendicular to the plane of the loop. The magnitude of the field is allowed to vary in time according to $B = 0.350 \, e^{-t/2.00}$, where B is in teslas and t is in seconds. The field has the constant value 0.350 T for $t < 0$. What is the value for $\mathcal{E}$ at $t = 4.00$ s?

30. **S** A rectangular loop of area A is placed in a region where the magnetic field is perpendicular to the plane of the loop. The magnitude of the field is allowed to vary in time according to $B = B_{max} e^{-t/\tau}$, where B_{max} and τ are constants. The field has the constant value B_{max} for $t < 0$. Find the emf induced in the loop as a function of time.

31. **T** A circular coil enclosing an area of 100 cm² is made of 200 turns of copper wire (Figure P30.31). The wire making up the coil has no resistance; the ends of the wire are connected across a 5.00-Ω resistor to form a closed circuit. Initially, a 1.10-T uniform magnetic field points perpendicularly upward through the plane of the coil. The direction of the field then reverses so that the final magnetic field has a magnitude of 1.10 T and points downward through the coil. If the time interval required for the field to reverse directions is 0.100 s, what is the average current in the coil during that interval?

Figure P30.31

32. **GP** Consider the apparatus shown in Figure P30.32: a conducting bar is moved along two rails connected to an incandescent lightbulb. The whole system is immersed in a magnetic field of magnitude $B = 0.400$ T perpendicular and into the page. The distance between the horizontal rails is $\ell = 0.800$ m. The resistance of the lightbulb is $R = 48.0$ Ω, assumed to be constant. The bar and rails have negligible resistance. The bar is moved toward the right by a constant

Figure P30.32

force of magnitude $F = 0.600$ N. We wish to find the maximum power delivered to the lightbulb. (a) Find an expression for the current in the lightbulb as a function of B, ℓ, R, and v, the speed of the bar. (b) When the maximum power is delivered to the lightbulb, what analysis model properly describes the moving bar? (c) Use the analysis model in part (b) to find a numerical value for the speed v of the bar when the maximum power is being delivered to the lightbulb. (d) Find the current in the lightbulb when maximum power is being delivered to it. (e) Using $P = I^2R$, what is the maximum power delivered to the lightbulb? (f) What is the maximum mechanical input power delivered to the bar by the force F? (g) We have assumed the resistance of the lightbulb is constant. In reality, as the power delivered to the lightbulb increases, the filament temperature increases and the resistance increases. Does the speed found in part (c) change if the resistance increases and all other quantities are held constant? (h) If so, does the speed found in part (c) increase or decrease? If not, explain. (i) With the assumption that the resistance of the lightbulb increases as the current increases, does the power found in part (f) change? (j) If so, is the power found in part (f) larger or smaller? If not, explain.

33. A guitar's steel string vibrates (see Fig. 30.5). The component of magnetic field perpendicular to the area of a pickup coil nearby is given by

$$B = 50.0 + 3.20 \sin 1\,046\pi t$$

where B is in milliteslas and t is in seconds. The circular pickup coil has 30 turns and radius 2.70 mm. Find the emf induced in the coil as a function of time.

34. *Why is the following situation impossible?* A conducting rectangular loop of mass $M = 0.100$ kg, resistance $R = 1.00\ \Omega$, and dimensions $w = 50.0$ cm by $\ell = 90.0$ cm is held with its lower edge just above a region with a uniform magnetic field of magnitude $B = 1.00$ T as shown in Figure P30.34. The loop is released from rest. Just as the top edge of the loop reaches the region containing the field, the loop moves with a speed 4.00 m/s.

Figure P30.34

35. A conducting rod of length $\ell = 35.0$ cm is free to slide on two parallel conducting bars as shown in Figure P30.35. Two resistors $R_1 = 2.00\ \Omega$ and $R_2 = 5.00\ \Omega$ are connected across the ends of the bars to form a loop. A constant magnetic field $B = 2.50$ T is directed perpendicularly into the page. An external agent pulls the rod to the left with a constant speed of $v = 8.00$ m/s. Find (a) the currents in both resistors, (b) the total power delivered to the resistance of the circuit, and (c) the magnitude of the applied force that is needed to move the rod with this constant velocity.

Figure P30.35

36. Magnetic field values are often determined by using a device known as a *search coil*. This technique depends on the measurement of the total charge passing through a coil in a time interval during which the magnetic flux linking the windings changes either because of the coil's motion or because of a change in the value of B. (a) Show that as the flux through the coil changes from Φ_1 to Φ_2, the charge transferred through the coil is given by $Q = N(\Phi_2 - \Phi_1)/R$, where R is the resistance of the coil and N is the number of turns. (b) As a specific example, calculate B when a total charge of 5.00×10^{-4} C passes through a 100-turn coil of resistance 200 Ω and cross-sectional area 40.0 cm² as it is rotated in a uniform field from a position where the plane of the coil is perpendicular to the field to a position where it is parallel to the field.

37. The plane of a square loop of wire with edge length $a = 0.200$ m is oriented vertically and along an east–west axis. The Earth's magnetic field at this point is of magnitude $B = 35.0\ \mu$T and is directed northward at 35.0° below the horizontal. The total resistance of the loop and the wires connecting it to a sensitive ammeter is 0.500 Ω. If the loop is suddenly collapsed by horizontal forces as shown in Figure P30.37, what total charge enters one terminal of the ammeter?

Figure P30.37

38. In Figure P30.38, the rolling axle, 1.50 m long, is pushed along horizontal rails at a constant speed $v = 3.00$ m/s. A resistor $R = 0.400\ \Omega$ is connected to the rails at points a and b, directly opposite each other. The wheels make good electrical contact with the rails, so the axle, rails, and R

Figure P30.38

form a closed-loop circuit. The only significant resistance in the circuit is R. A uniform magnetic field $B = 0.080\ 0$ T is vertically downward. (a) Find the induced current I in the resistor. (b) What horizontal force F is required to keep the axle rolling at constant speed? (c) Which end of the resistor, a or b, is at the higher electric potential? (d) **What If?** After the axle rolls past the resistor, does the current in R reverse direction? Explain your answer.

39. Figure P30.39 shows a stationary conductor whose shape is similar to the letter e. The radius of its circular portion is $a = 50.0$ cm. It is placed in a constant magnetic field of 0.500 T directed out of the page. A straight conducting rod, 50.0 cm long, is pivoted about point O and rotates with a constant angular speed of 2.00 rad/s. (a) Determine the induced emf in the loop POQ. Note that the area of the loop is $\theta a^2/2$. (b) If all the conducting material has a resistance per length of 5.00 Ω/m, what is the induced current in the loop POQ at the instant 0.250 s after point P passes point Q?

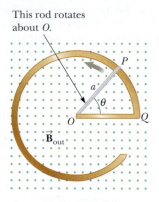

This rod rotates about O.

Figure P30.39

40. A conducting rod moves with a constant velocity in a direction perpendicular to a long, straight wire carrying a current I as shown in Figure P30.40. Show that the magnitude of the emf generated between the ends of the rod is

$$|\mathcal{E}| = \frac{\mu_0 vI\ell}{2\pi r}$$

In this case, note that the emf decreases with increasing r as you might expect.

Figure P30.40

41. Figure P30.41 shows a compact, circular coil with 220 turns and radius 12.0 cm immersed in a uniform magnetic field parallel to the axis of the coil. The rate of change of the field has the constant magnitude 20.0 mT/s. (a) What

Figure P30.41

additional information is necessary to determine whether the coil is carrying clockwise or counterclockwise current? (b) The coil overheats if more than 160 W of power is delivered to it. What resistance would the coil have at this critical point? (c) To run cooler, should it have lower resistance or higher resistance?

42. **Review.** In Figure P30.42, a uniform magnetic field decreases at a constant rate $dB/dt = -K$, where K is a positive constant. A circular loop of wire of radius a containing a resistance R and a capacitance C is placed with its plane normal to the field. (a) Find the charge Q on the capacitor when it is fully charged. (b) Which plate, upper or lower, is at the higher potential? (c) Discuss the force that causes the separation of charges.

Figure P30.42

43. An N-turn square coil with side ℓ and resistance R is pulled to the right at constant speed v in the presence of a uniform magnetic field B acting perpendicular to the coil as shown in Figure P30.43. At $t = 0$, the right side of the coil has just departed the right edge of the field. At time t, the left side of the coil enters the region where $B = 0$. In terms of the quantities N, B, ℓ, v, and R, find symbolic expressions for (a) the magnitude of the induced emf in the loop during the time interval from $t = 0$ to t, (b) the magnitude of the induced current in the coil, (c) the power delivered to the coil, and (d) the force required to remove the coil from the field. (e) What is the direction of the induced current in the loop? (f) What is the direction of the magnetic force on the loop while it is being pulled out of the field?

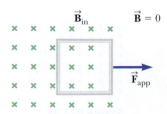

Figure P30.43

44. A conducting rod of length ℓ moves with velocity $\vec{v}$ parallel to a long wire carrying a steady current I. The axis of the rod is maintained perpendicular to the wire with the near end a distance r away (Fig. P30.44). Show that the magnitude of the emf induced in the rod is

$$|\mathcal{E}| = \frac{\mu_0 Iv}{2\pi} \ln \left(1 + \frac{\ell}{r}\right)$$

Figure P30.44

45. A long, straight wire carries a current given by $I = I_{max} \sin(\omega t + \phi)$. The wire lies in the plane of a rectangular coil of N turns of wire as shown in Figure P30.45. The quantities I_{max}, ω, and ϕ are all constants. Assume $I_{max} = 50.0$ A, $\omega = 200\pi$ s^{-1}, $N = 100$, $h = w = 5.00$ cm, and $L = 20.0$ cm. Determine the emf induced in the coil by the magnetic field created by the current in the straight wire.

Figure P30.45

46. A rectangular loop of dimensions ℓ and w moves with a constant velocity $\vec{v}$ away from a long wire that carries a current I in the plane of the loop (Fig. P30.46). The total resistance of the loop is R. Derive an expression that gives the current in the loop at the instant the near side is a distance r from the wire.

Figure P30.46

47. A thin wire $\ell = 30.0$ cm long is held parallel to and $d = 80.0$ cm above a long, thin wire carrying $I = 200$ A and fixed in position (Fig. P30.47). The 30.0-cm wire is released at the instant $t = 0$ and falls, remaining parallel to the current-carrying wire as it falls. Assume the falling wire accelerates at 9.80 m/s^2. (a) Derive an equation for the emf induced in it as a function of time. (b) What is the minimum value of the emf? (c) What is the maximum value? (d) What is the induced emf 0.300 s after the wire is released?

Figure P30.47

CHALLENGE PROBLEMS

48. An *induction furnace* uses electromagnetic induction to produce eddy currents in a conductor, thereby raising the conductor's temperature. Commercial units operate at frequencies ranging from 60 Hz to about 1 MHz and deliver powers from a few watts to several megawatts. Induction heating can be used for warming a metal pan on a kitchen stove. It can be used to avoid oxidation and contamination of the metal when welding in a vacuum enclosure. To explore induction heating, consider a flat conducting disk of radius R, thickness b, and resistivity ρ. A sinusoidal magnetic field $B_{max} \cos \omega t$ is applied perpendicular to the disk. Assume the eddy currents occur in circles concentric with the disk. (a) Calculate the average power delivered to the disk. (b) **What If?** By what factor does the power change when the amplitude of the field doubles? (c) When the frequency doubles? (d) When the radius of the disk doubles?

49. A bar of mass m and resistance R slides without friction in a horizontal plane, moving on parallel rails as shown in Figure P30.49. The rails are separated by a distance d. A battery that maintains a constant emf $\mathcal{E}$ is connected between the rails, and a constant magnetic field $\vec{B}$ is directed perpendicularly out of the page. Assuming the bar starts from rest at time $t = 0$, show that at time t it moves with a speed

$$v = \frac{\mathcal{E}}{Bd}(1 - e^{-B^2 d^2 t/mR})$$

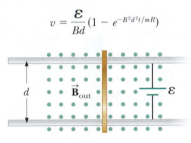

Figure P30.49

50. A *betatron* is a device that accelerates electrons to energies in the MeV range by means of electromagnetic induction. Electrons in a vacuum chamber are held in a circular orbit by a magnetic field perpendicular to the orbital plane. The magnetic field is gradually increased to induce an electric field around the orbit. (a) Show that the electric field is in the correct direction to make the electrons speed up. (b) Assume the radius of the orbit remains constant. Show that the average magnetic field over the area enclosed by the orbit must be twice as large as the magnetic field at the circle's circumference.

51. **Review.** The bar of mass m in Figure P30.51 is pulled horizontally across parallel, frictionless rails by a massless string that passes over a light, frictionless pulley and is attached to a suspended object of mass M. The uniform upward magnetic field has a magnitude B, and the distance between the rails is ℓ. The only significant electrical resistance is the load resistor R shown connecting the rails at one end. Assuming the suspended object is released with the bar at rest at $t = 0$, derive an expression that gives the bar's horizontal speed as a function of time.

Figure P30.51

31

Inductance

At an intersection controlled by a traffic light, one can often see circles cut into the pavement. The photograph shows a roadway with several such circles, all ▶ connected by dark straight lines to a point near the bottom of the image. What is the purpose of these circles? (*John W. Jewett, Jr.*)

STORYLINE **You are still on your weekend trip that you began in** the previous chapter. You pull up to a stop at a traffic light where there is very little traffic. The light immediately turns green for you. You have noticed this phenomenon before, but now, after having studied physics, you say, "Wait a minute! How exactly does the traffic light know that my car is here?" You try it again, pulling into a left-turn lane. The left-turn green arrow illuminates! As you approach more traffic lights, you look around for some type of structure that might contain something that detects the presence of your car. You don't see anything mounted on poles or overhead that might serve as a detector. Then you notice that there are circular grooves that appear to be sawn into the roadway near each intersection. Could that have something to do with it?

CONNECTIONS In Chapter 30, we saw that an emf and a current are induced in a loop of wire when the magnetic flux through the area enclosed by the loop changes with time. This phenomenon of induction has some practical consequences. In this chapter, we first describe an effect known as *self-induction*, in which a time-varying current in a circuit produces an induced emf opposing the emf that initially set up the time-varying current. Self-induction is the basis of the *inductor*, a new circuit element. We can combine the inductor into electric circuits with our previously introduced circuit elements, the capacitor and the resistor. We discuss the energy stored in the magnetic field of an inductor and the energy density associated with the magnetic field. We will find that circuits including inductors can have behavior similar to the simple harmonic oscillator that we studied back in Chapter 15. In addition, our understanding of inductors will allow us to move forward and understand the operation of AC circuits in Chapter 32.

31.1 Self-Induction and Inductance

Now that we have studied Faraday's law, we need to distinguish carefully between emfs and currents that are caused by physical sources such as batteries and those that are induced by changing magnetic fields. When we use a term (such as *emf* or *current*) without an adjective, we are describing the parameters associated with a physical source. We use the adjective *induced* to describe those emfs and currents caused by a changing magnetic field.

Consider a circuit consisting of a switch, a resistor, and a source of emf as shown in Figure 31.1. The circuit diagram is represented in perspective to show the orientations of some of the magnetic field lines due to the current in the circuit. When the switch is thrown to its closed position, we observe that the current does not immediately jump from zero to its maximum value $\mathcal{E}/R$. Faraday's law of electromagnetic induction (Eq. 30.1) can be used to describe this effect as follows. The circuit is a current loop. Therefore, it is a source of a magnetic field. The magnetic field lines from this field pass through the loop of the circuit itself. As the current increases with time after the switch is thrown closed, there is an increasing magnetic flux through the loop of the circuit. This increasing flux creates an induced emf in the circuit. The direction of the induced emf is such that it would cause an induced current in the loop (if the loop did not already carry a current), which would establish a magnetic field opposing the change in the original magnetic field. Therefore, the direction of the induced emf is opposite the direction of the emf of the battery, which results in a gradual rather than instantaneous increase in the current to its final equilibrium value. Because of the direction of the induced emf, it is also called a *back emf,* similar to that in a motor as discussed in Chapter 30. This effect is called **self-induction** because the changing flux through the circuit and the resultant induced emf arise from the circuit itself. The emf $\mathcal{E}_L$ set up in this case is called a **self-induced emf.**

To obtain a quantitative description of self-induction, recall from Faraday's law that the induced emf is equal to the negative of the time rate of change of the magnetic flux. The magnetic flux is proportional to the magnetic field, which in turn is proportional to the current in the circuit. Therefore, a self-induced emf in a circuit is always proportional to the time rate of change of the current in the circuit. For any loop of wire, we can write this proportionality as

$$\mathcal{E}_L = -L \frac{di}{dt} \tag{31.1}$$

where L is a proportionality constant—called the **inductance** of the loop—that depends on the geometry of the loop and other physical characteristics. If we consider a closely spaced coil of N turns (a toroid or an ideal solenoid) carrying a current i and containing N turns, Faraday's law tells us that $\mathcal{E}_L = -N \, d\Phi_B/dt$ (Eq. 30.2). Comparing this expression with Equation 31.1, we see that

$$L = \frac{N\Phi_B}{i} \tag{31.2}$$

where it is assumed the same magnetic flux passes through each turn and L is the inductance of the entire coil.

From Equation 31.1, we can also write the inductance as the ratio

$$L = -\frac{\mathcal{E}_L}{di/dt} \tag{31.3}$$

The SI unit of inductance is the **henry** (H), which as we can see from Equation 31.3 is 1 volt-second per ampere: 1 H = 1 V · s/A. Recall that resistance is a measure of the opposition to current as given by Equation 26.7, $R = \Delta V/I$; in comparison,

Joseph Henry
American Physicist (1797–1878)
Henry became the first director of the Smithsonian Institution and first president of the Academy of Natural Science. He improved the design of the electromagnet and constructed one of the first motors. He also discovered the phenomenon of self-induction, but he failed to publish his findings. The unit of inductance, the henry, is named in his honor.

After the switch is closed, the current produces a magnetic flux through the area enclosed by the loop. As the current increases toward its equilibrium value, this magnetic flux changes in time and induces an emf in the loop.

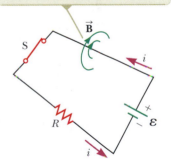

Figure 31.1 Self-induction in a simple circuit.

◀ Inductance of an *N*-turn coil

Equation 31.3, being of the same mathematical form as Equation 26.7, shows us that inductance is a measure of the opposition to a *change* in current.

As shown in Example 31.1, the inductance of a coil depends on its geometry. This dependence is analogous to the capacitance of a capacitor depending on the geometry of its plates as we found in Equation 25.3 and the resistance of a resistor depending on the length and area of the conducting material in Equation 26.10. Inductance calculations can be quite difficult to perform for complicated geometries, but the examples below involve simple situations for which inductances are easily evaluated.

QUICK QUIZ **31.1** A coil with zero resistance has its ends labeled a and b. The potential at a is higher than at b. Which of the following could be consistent with this situation? **(a)** The current is constant and is directed from a to b. **(b)** The current is constant and is directed from b to a. **(c)** The current is increasing and is directed from a to b. **(d)** The current is decreasing and is directed from a to b. **(e)** The current is increasing and is directed from b to a. **(f)** The current is decreasing and is directed from b to a.

Example **31.1**   **Inductance of a Solenoid**

Consider a uniformly wound solenoid having N turns and length ℓ. Assume ℓ is much longer than the radius of the windings and the core of the solenoid is air.

(A) Find the inductance of the solenoid.

SOLUTION

Conceptualize The magnetic field lines from each turn of the solenoid pass through all the turns, so an induced emf in each coil opposes changes in the current.

Categorize We categorize this example as a substitution problem. Because the solenoid is long, we can use the results for an ideal solenoid obtained in Chapter 29.

Find the magnetic flux through each turn of area A in the solenoid, using the expression for the magnetic field from Equation 29.17:

$$\Phi_B = BA = \mu_0 niA = \mu_0 \frac{N}{\ell} iA$$

Substitute this expression into Equation 31.2:

$$L = \frac{N\Phi_B}{i} = \mu_0 \frac{N^2}{\ell} A \qquad (31.4)$$

(B) Calculate the inductance of the solenoid if it contains 300 turns, its length is 25.0 cm, and its cross-sectional area is 4.00 cm².

SOLUTION

Substitute numerical values into Equation 31.4:

$$L = (4\pi \times 10^{-7}\,\text{T}\cdot\text{m/A}) \frac{300^2}{25.0 \times 10^{-2}\,\text{m}} (4.00 \times 10^{-4}\,\text{m}^2)$$

$$= 1.81 \times 10^{-4}\,\text{T}\cdot\text{m}^2/\text{A} = \boxed{0.181\,\text{mH}}$$

(C) Calculate the self-induced emf in the solenoid if the current it carries decreases at the rate of 50.0 A/s.

SOLUTION

Substitute $di/dt = -50.0$ A/s and the answer to part (B) into Equation 31.1:

$$\boldsymbol{\mathcal{E}}_L = -L\frac{di}{dt} = -(1.81 \times 10^{-4}\,\text{H})(-50.0\,\text{A/s})$$

$$= \boxed{9.05\,\text{mV}}$$

The result for part (A) shows that L depends on geometry and is proportional to the square of the number of turns. Because $N = n\ell$, we can also express the result in the form

$$L = \mu_0 \frac{(n\ell)^2}{\ell} A = \mu_0 n^2 A\ell = \mu_0 n^2 V \qquad (31.5)$$

where $V = A\ell$ is the interior volume of the solenoid.

31.2 *RL* Circuits

If a circuit contains a coil such as a solenoid, the inductance of the coil prevents the current in the circuit from increasing or decreasing instantaneously. A circuit element that has a large inductance is called an **inductor** and has the circuit symbol —⟋⟋⟋—. We always assume the inductance of the remainder of a circuit is negligible compared with that of the inductor. Keep in mind, however, that even a circuit without a coil has some inductance that can affect the circuit's behavior.

Because the inductance of an inductor results in a back emf, an inductor in a circuit opposes changes in the current in that circuit. If the battery voltage in the circuit is increased so that the current rises, the inductor opposes this change and the rise is not instantaneous. If the battery voltage is decreased, the inductor causes a slow drop in the current rather than an immediate drop. Therefore, the inductor causes the circuit to be "sluggish" as it reacts to changes in the voltage.

Consider the circuit shown in Figure 31.2a, which contains a battery of negligible internal resistance. This circuit is an *RL circuit* because the elements connected to the battery are a resistor and an inductor. The curved lines on switch S_2 suggest this switch can never be open; it is always set to either a or b. (If the switch is connected to neither a nor b, any current in the circuit suddenly stops.) Suppose S_2 is set to a and switch S_1 open for $t < 0$ and then thrown closed at $t = 0$ as shown in Figure 31.2b. The current in the circuit begins to increase, and a back emf (Eq. 31.1) that opposes the increasing current is induced in the inductor.

At an instant when the current is changing, let's apply Kirchhoff's loop rule to this circuit, traversing the outer loop in Figure 31.2b in the clockwise direction:

$$\mathcal{E} - iR - L\frac{di}{dt} = 0 \tag{31.6}$$

where iR is the voltage drop across the resistor. (Kirchhoff's rules were developed for circuits with steady currents, but they can also be applied to a circuit in which the current is changing if we imagine them to represent the circuit at one *instant* of time.) Now let's find a solution to this differential equation, which is similar to that for the *RC* circuit (see Section 27.4).

A mathematical solution of Equation 31.6 represents the current in the circuit as a function of time. To find this solution, we change variables for convenience, letting $x = (\mathcal{E}/R) - i$, so $dx = -di$. With these substitutions, Equation 31.6 becomes

$$x + \frac{L}{R}\frac{dx}{dt} = 0$$

Rearranging and integrating this last expression gives

$$\int_{x_0}^{x} \frac{dx}{x} = -\frac{R}{L}\int_0^t dt \;\rightarrow\; \ln\frac{x}{x_0} = -\frac{R}{L}t$$

where x_0 is the value of x at time $t = 0$. Taking the antilogarithm of this result gives

$$x = x_0 e^{-Rt/L}$$

Because $i = 0$ at $t = 0$, note from the definition of x that $x_0 = \mathcal{E}/R$. Hence, this last expression is equivalent to

$$\frac{\mathcal{E}}{R} - i = \frac{\mathcal{E}}{R}e^{-Rt/L} \;\rightarrow\; i = \frac{\mathcal{E}}{R}(1 - e^{-Rt/L})$$

This expression shows how the inductor affects the current. The current does not increase instantly to its final equilibrium value when the switch is closed, but instead increases according to an exponential function. If the inductance is removed from the circuit, which corresponds to letting L approach zero, the exponential term becomes zero and there is no time dependence of the current in this case; the current increases instantaneously to its final equilibrium value in the absence of the inductance.

When switch S_1 is open, there is no current anywhere in the circuit.

a

When switch S_1 is thrown closed, the current increases and an emf that opposes the increasing current is induced in the inductor.

b

Figure 31.2 An *RL* circuit. (a) We begin with switch S_1 open, so that the battery is not connected to the other elements in the circuit. (b) When switch S_1 is thrown closed, the battery is connected and a current begins to build in the circuit.

After switch S_1 is thrown closed at $t = 0$, the current increases toward its maximum value $\mathcal{E}/R$.

Figure 31.3 Plot of the current versus time for the RL circuit shown in Figure 31.2b. The time constant τ is the time interval required for i to reach 63.2% of its maximum value.

The time rate of change of current is a maximum at $t = 0$, which is the instant at which switch S_1 is thrown closed.

Figure 31.4 Plot of di/dt versus time for the RL circuit shown in Figure 31.2b. The rate of change of the current decreases exponentially with time as i increases toward its maximum value.

When switch S_1 has been closed for a long time, the current in the circuit reaches its steady state value I.

a

When the switch S_2 is thrown to position b, the battery is no longer part of the circuit and the current decreases.

b

Figure 31.5 (a) The condition shown in Figure 31.2b has existed for a long time and the current has its maximum value. (b) When switch S_2 is thrown to position b, the current begins to fall.

We can also write this expression as

$$i = \frac{\mathcal{E}}{R}(1 - e^{-t/\tau}) \tag{31.7}$$

where the constant τ is the **time constant** of the RL circuit:

$$\tau = \frac{L}{R} \tag{31.8}$$

Physically, τ is the time interval required for the current in the circuit to reach $(1 - e^{-1}) = 0.632 = 63.2\%$ of its final value $\mathcal{E}/R$. The time constant is a useful parameter for comparing the time responses of various circuits.

Figure 31.3 shows a graph of the current versus time in the RL circuit. Notice that the equilibrium value of the current, which occurs as t approaches infinity, is $\mathcal{E}/R$. That can be seen by setting di/dt equal to zero in Equation 31.6 and solving for the current i. (At equilibrium, the change in the current is zero.) Therefore, the current initially increases very rapidly and then gradually approaches the equilibrium value $\mathcal{E}/R$ as t approaches infinity.

Let's also investigate the time rate of change of the current. Taking the first time derivative of Equation 31.7 gives

$$\frac{di}{dt} = \frac{\mathcal{E}}{L} e^{-t/\tau} \tag{31.9}$$

This result shows that the time rate of change of the current is a maximum (equal to $\mathcal{E}/L$) at $t = 0$ and falls off exponentially to zero as t approaches infinity (Fig. 31.4).

Suppose switch S_2 in the circuit in Figure 31.2b has been set at position a long enough (and switch S_1 remains closed) to allow the current to reach its equilibrium value $I = \mathcal{E}/R$, as shown in Figure 31.5a. If S_2 is thrown from a to b, the circuit is now described by only the right-hand loop as seen in Figure 31.5b. Therefore, the battery has been eliminated from the circuit. Setting $\mathcal{E} = 0$ in Equation 31.6 gives

$$iR + L\frac{di}{dt} = 0$$

It is left as a problem (Problem 12) to show that the solution of this differential equation is

$$i = \frac{\mathcal{E}}{R} e^{-t/\tau} = I_i e^{-t/\tau} \qquad (31.10)$$

where $\mathcal{E}$ is the emf of the battery and $I_i = \mathcal{E}/R$ is the initial current at the instant the switch is thrown to b.

If the circuit did not contain an inductor, the current would immediately decrease to zero when the battery is removed. When the inductor is present, it opposes the decrease in the current and causes the current to decrease exponentially. A graph of the current in the circuit versus time (Fig. 31.6) shows that the current is continuously decreasing with time.

At $t = 0$, switch S_2 is thrown to position b and the current has its maximum value $\mathcal{E}/R$.

Figure 31.6 Current versus time for the circuit shown in Figure 31.5b. For $t < 0$, switch S_2 is at position a.

Q**UICK QUIZ 31.2** Consider the circuit in Figure 31.2a with S_1 open and S_2 at position a. Switch S_1 is now thrown closed as in Figure 31.2b. **(i)** At the instant it is closed, across which circuit element is the voltage equal to the emf of the battery? (a) the resistor (b) the inductor (c) both the inductor and resistor **(ii)** After a very long time, across which circuit element is the voltage equal to the emf of the battery? Choose from among the same answers.

Example **31.2** **Time Constant of an *RL* Circuit**

Consider the circuit in Figure 31.2 again. Suppose the circuit elements have the following values: $\mathcal{E} = 12.0$ V, $R = 6.00$ Ω, and $L = 30.0$ mH.

(A) Find the time constant of the circuit.

SOLUTION

Conceptualize You should understand the operation and behavior of the circuit in Figure 31.2 from the discussion in this section.

Categorize We evaluate the results using equations developed in this section, so this example is a substitution problem.

Evaluate the time constant from Equation 31.8:

$$\tau = \frac{L}{R} = \frac{30.0 \times 10^{-3}\ \text{H}}{6.00\ \Omega} = 5.00\ \text{ms}$$

(B) Switch S_2 is at position a, and switch S_1 is thrown closed at $t = 0$. The circuit now appears as in Figure 31.2b. Calculate the current in the circuit at $t = 2.00$ ms.

SOLUTION

Evaluate the current at $t = 2.00$ ms from Equation 31.7:

$$i = \frac{\mathcal{E}}{R}(1 - e^{-t/\tau}) = \frac{12.0\ \text{V}}{6.00\ \Omega}(1 - e^{-2.00\ \text{ms}/5.00\ \text{ms}}) = 2.00\ \text{A}\ (1 - e^{-0.400})$$

$$= 0.659\ \text{A}$$

(C) Compare the potential difference across the resistor with that across the inductor.

SOLUTION

At the instant Switch S_1 is closed, there is no current and therefore no potential difference across the resistor. At this instant, the battery voltage appears entirely across the inductor in the form of a back emf of 12.0 V as the inductor tries to maintain the zero-current condition. (The top end of the inductor in Fig. 31.2b is at a higher electric potential than the bottom end.) As time passes, the emf across the inductor decreases and the current in the resistor (and hence the voltage across it) increases as shown in Figure 31.7 (page 830). The sum of the two voltages at all times is 12.0 V.

WHAT IF? In Figure 31.7, the voltages across the resistor and inductor are equal at 3.4 ms. What if you wanted to delay the condition in which the voltages are equal to some later instant, such as $t = 10.0$ ms? Which parameter, L or R, would require the least adjustment, in terms of a percentage change, to achieve that?

continued

31.2 continued

Answer Figure 31.7 shows that the voltages are equal when the voltage across the inductor has fallen to half its original value. Therefore, the time interval required for the voltages to become equal is the *half-life* $t_{1/2}$ of the decay. We introduced the half-life in the What If? section of Example 27.10 to describe the exponential decay in *RC* circuits, where $t_{1/2} = 0.693\tau$.

Figure 31.7 (Example 31.2) The time behavior of the voltages across the resistor and inductor in Figure 31.2b given the values provided in this example.

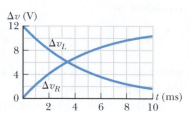

From the desired half-life of 10.0 ms, use the result from Example 27.10 to find the time constant of the circuit:

$$\tau = \frac{t_{1/2}}{0.693} = \frac{10.0 \text{ ms}}{0.693} = 14.4 \text{ ms}$$

Hold *L* fixed and find the value of *R* that gives this time constant:

$$\tau = \frac{L}{R} \rightarrow R = \frac{L}{\tau} = \frac{30.0 \times 10^{-3} \text{ H}}{14.4 \text{ ms}} = 2.08 \ \Omega$$

Now hold *R* fixed and find the appropriate value of *L*:

$$\tau = \frac{L}{R} \rightarrow L = \tau R = (14.4 \text{ ms})(6.00 \ \Omega) = 86.4 \times 10^{-3} \text{ H}$$

The change in *R* corresponds to a 65% decrease compared with the initial resistance. The change in *L* represents a 188% increase in inductance! Therefore, a much smaller percentage adjustment in *R* can achieve the desired effect than would an adjustment in *L*.

31.3 Energy in a Magnetic Field

As we move forward, we will find that we will need to address energy considerations in circuits with inductors. In general, a battery in a circuit containing an inductor must provide more energy than one in a circuit without the inductor. When switch S_1 in Figure 31.2b is thrown closed, part of the energy supplied by the battery appears as internal energy in the resistance in the circuit, and the remaining energy is stored in the magnetic field of the inductor. Multiplying each term in Equation 31.6 by i and rearranging the expression gives

$$i\mathcal{E} = i^2 R + Li\frac{di}{dt} \tag{31.11}$$

Recognizing $i\mathcal{E}$ as the rate at which energy is supplied by the battery and i^2R as the rate at which energy is delivered to the resistor, we see that $Li(di/dt)$ must represent the rate at which energy is being stored in the inductor, associated with the magnetic field of the inductor. If U_B is the energy stored in the inductor at any time, we can write the rate dU_B/dt at which energy is stored as

$$\frac{dU_B}{dt} = Li\frac{di}{dt}$$

To find the total energy stored in the inductor at any instant, let's rewrite this expression as $dU_B = Li\,di$ and integrate:

$$U_B = \int dU_B = \int_0^i Li\,di = L\int_0^i i\,di$$

Energy stored in an inductor ▶

$$U_B = \tfrac{1}{2}Li^2 \tag{31.12}$$

where *L* has been removed from the integral because it is constant. Equation 31.12 represents the energy stored in the magnetic field of the inductor when the current is *i*. It is similar in form to Equation 25.13 for the energy stored in the electric field of a capacitor, $U_E = \tfrac{1}{2}C(\Delta V)^2$. In either case, energy is required to establish a field.

We can also determine the energy density of a magnetic field and compare it to the energy density of an electric field, found in Section 25.4. For simplicity, consider a solenoid whose inductance is given by Equation 31.5:

$$L = \mu_0 n^2 V$$

The magnetic field of a solenoid is given by Equation 29.17:

$$B = \mu_0 n i$$

Substituting the expression for L and $i = B/\mu_0 n$ into Equation 31.12 gives

$$U_B = \tfrac{1}{2} L i^2 = \tfrac{1}{2} \mu_0 n^2 V \left(\frac{B}{\mu_0 n} \right)^2 = \frac{B^2}{2\mu_0} V \qquad (31.13)$$

The magnetic energy density, or the energy stored per unit volume in the magnetic field of the inductor, is $u_B = U_B / V$, or

$$u_B = \frac{B^2}{2\mu_0} \qquad (31.14)$$ ◀ Magnetic energy density

Although this expression was derived for the special case of a solenoid, it is valid for any region of space in which a magnetic field exists. Equation 31.14 is similar in form to Equation 25.15 for the energy per unit volume stored in an electric field, $u_E = \tfrac{1}{2}\epsilon_0 E^2$. In both cases, the energy density is proportional to the square of the field magnitude.

Ⓠ UICK QUIZ 31.3 You are performing an experiment that requires the highest-possible magnetic energy density in the interior of a very long current-carrying solenoid. Which of the following adjustments increases the energy density? (More than one choice may be correct.) **(a)** increasing the number of turns per unit length on the solenoid **(b)** increasing the cross-sectional area of the solenoid **(c)** increasing only the length of the solenoid while keeping the number of turns per unit length fixed **(d)** increasing the current in the solenoid

---

Example 31.3 | **What Happens to the Energy in the Inductor?**

Consider once again the RL circuit shown in Figure 31.5a, with switch S_2 at position a and the current having reached its steady-state value. When S_2 is thrown to position b, as in Figure 31.5b, the current in the right-hand loop decays exponentially with time according to Equation 31.10. Show that all the energy initially stored in the magnetic field of the inductor appears as internal energy in the resistor as the current decays to zero.

SOLUTION

Conceptualize Before S_2 is thrown to b, the current is constant at its maximum value, and energy is being delivered at a constant rate to the resistor from the battery. A constant amount of energy is stored in the magnetic field of the inductor. After $t = 0$, when S_2 is thrown to b, the battery can no longer provide energy and energy is delivered to the resistor only from the inductor.

Categorize We model the right-hand loop of the circuit as an *isolated system* so that energy is transferred between components of the system but does not leave the system.

Analyze We begin by evaluating the energy delivered to the resistor, which appears as internal energy in the resistor.

Begin with Equation 26.22 and recognize that the rate of change of internal energy in the resistor is the power delivered to the resistor:

$$\frac{dE_{int}}{dt} = P = i^2 R$$

Substitute the current given by Equation 31.10 into this equation:

$$\frac{dE_{int}}{dt} = i^2 R = (I_i e^{-Rt/L})^2 R = I_i^2 R e^{-2Rt/L}$$

Solve for dE_{int} and integrate this expression over the limits $t = 0$ to $t \rightarrow \infty$:

$$E_{int} = \int_0^\infty I_i^2 R e^{-2Rt/L}\, dt = I_i^2 R \int_0^\infty e^{-2Rt/L}\, dt$$

The value of the definite integral can be shown to be $L/2R$ (see Problem 22). Use this result to evaluate E_{int}:

$$E_{int} = I_i^2 R \left(\frac{L}{2R} \right) = \tfrac{1}{2} L I_i^2$$

Finalize This result is equal to the initial energy stored in the magnetic field of the inductor, given by Equation 31.12, as we set out to prove.

Example **31.4**   The Coaxial Cable

Coaxial cables are often used to connect electrical devices, such as your video system, and in receiving signals in television cable systems. Model a long coaxial cable as a thin, cylindrical conducting shell of radius b concentric with a solid cylinder of radius a as in Figure 31.8. The conductors carry the same current i in opposite directions. Calculate the inductance L of a length ℓ of this cable.

SOLUTION

Conceptualize   Consider Figure 31.8. Although we do not have a visible coil in this geometry, imagine a thin, radial slice of the coaxial cable such as the light gold rectangle in Figure 31.8. If the inner and outer conductors are connected at the ends of the cable (above and below the figure), this slice represents one large conducting loop. The current in the loop sets up a magnetic field between the inner and outer conductors that passes through this loop. If the current changes, the magnetic field changes and the induced emf opposes the original change in the current in the conductors.

Figure 31.8 (Example 31.4) Section of a long coaxial cable. The inner and outer conductors carry equal currents in opposite directions.

Categorize   We categorize this situation as one in which we must return to the fundamental definition of inductance, Equation 31.2.

Analyze   We must find the magnetic flux through the light gold rectangle in Figure 31.8. Ampère's law (see Section 29.3) tells us that the magnetic field in the region between the conductors is due to the inner conductor alone and that its magnitude is $B = \mu_0 i / 2\pi r$, where r is measured from the common center of the cylinders. A sample circular field line is shown in Figure 31.8, along with a field vector tangent to the field line.

The magnetic field is perpendicular to the light gold rectangle of length ℓ and width $b - a$, the cross section of interest. Because the magnetic field varies with radial position across this rectangle, we must use calculus to find the total magnetic flux.

Divide the light gold rectangle into strips of width dr such as the darker strip in Figure 31.8. Evaluate the magnetic flux through such a strip:

$$d\Phi_B = B\, dA = B\ell\, dr$$

Substitute for the magnetic field and integrate over the entire light gold rectangle:

$$\Phi_B = \int_a^b \frac{\mu_0 i}{2\pi r} \ell\, dr = \frac{\mu_0 i\ell}{2\pi} \int_a^b \frac{dr}{r} = \frac{\mu_0 i\ell}{2\pi} \ln\left(\frac{b}{a}\right)$$

Use Equation 31.2 to find the inductance of the cable:

$$L = \frac{\Phi_B}{i} = \frac{\mu_0 \ell}{2\pi} \ln\left(\frac{b}{a}\right)$$

Finalize   The inductance depends only on geometric factors related to the cable. It increases if ℓ increases, if b increases, or if a decreases. This result is consistent with our conceptualization: any of these changes increases the size of the loop represented by our radial slice and through which the magnetic field passes, increasing the inductance.

31.4   Mutual Inductance

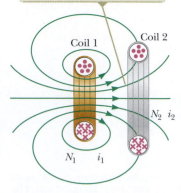

A current in coil 1 sets up a magnetic field, and some of the magnetic field lines pass through coil 2.

Very often, the magnetic flux through the area enclosed by a circuit varies with time because of time-varying currents in nearby circuits. This condition induces an emf through a process known as *mutual induction,* so named because it depends on the interaction of two circuits.

Consider the two closely wound coils of wire shown in cross-sectional view in Figure 31.9. The current i_1 in coil 1, which has N_1 turns, creates a magnetic field. Some of the magnetic field lines pass through coil 2, which has N_2 turns. The magnetic flux caused by the current in coil 1 and passing through coil 2 is represented by Φ_{12}. In analogy to Equation 31.2, we can identify the **mutual inductance** M_{12} of coil 2 with respect to coil 1:

$$M_{12} = \frac{N_2 \Phi_{12}}{i_1} \tag{31.15}$$

Figure 31.9 A cross-sectional view of two adjacent coils.

Mutual inductance depends on the geometry of both circuits and on their orientation with respect to each other. As the circuit separation distance increases, the mutual inductance decreases because the flux linking the circuits decreases.

If the current i_1 varies with time, we see from Faraday's law and Equation 31.15 that the emf induced by coil 1 in coil 2 is

$$\mathcal{E}_2 = -N_2 \frac{d\Phi_{12}}{dt} = -N_2 \frac{d}{dt}\left(\frac{M_{12}i_1}{N_2}\right) = -M_{12}\frac{di_1}{dt} \qquad (31.16)$$

In the preceding discussion, it was assumed the current is in coil 1. Let's also imagine a current i_2 in coil 2. The preceding discussion can be repeated to show that there is a mutual inductance M_{21}. If the current i_2 varies with time, the emf induced by coil 2 in coil 1 is

$$\mathcal{E}_1 = -M_{21}\frac{di_2}{dt} \qquad (31.17)$$

In mutual induction, the emf induced in one coil is always proportional to the rate at which the current in the other coil is changing. Although the proportionality constants M_{12} and M_{21} have been treated separately, it can be shown that they are equal. Therefore, with $M_{12} = M_{21} = M$, Equations 31.16 and 31.17 become

$$\mathcal{E}_2 = -M\frac{di_1}{dt} \quad \text{and} \quad \mathcal{E}_1 = -M\frac{di_2}{dt}$$

These two equations are similar in form to Equation 31.1 for the self-induced emf $\mathcal{E} = -L\,(di/dt)$. The unit of mutual inductance is the henry.

QUICK QUIZ 31.4 In Figure 31.9, coil 1 is moved closer to coil 2, with the orientation of both coils remaining fixed. Because of this movement, the mutual induction of the two coils **(a)** increases, **(b)** decreases, or **(c)** is unaffected.

Example 31.5 | **Wireless Battery Charger**

An electric toothbrush has a base designed to hold the toothbrush handle when not in use. As shown in Figure 31.10a, the handle has a cylindrical hole that fits loosely over a matching cylinder on the base. When the handle is placed on the base, a changing current in a solenoid inside the base cylinder induces a current in a coil inside the handle. This induced current charges the battery in the handle.

We can model the base as a solenoid of length ℓ with N_B turns (Fig. 31.10b), carrying a current i, and having a cross-sectional area A. The handle coil contains N_H turns and completely surrounds the base coil. Find the mutual inductance of the system.

© by Braun GmbH, Kronberg

Figure 31.10 (Example 31.5) (a) This electric toothbrush uses the mutual induction of solenoids as part of its battery-charging system. (b) A coil of N_H turns wrapped around the center of a solenoid of N_B turns.

SOLUTION

Conceptualize Be sure you can identify the two coils in the situation and understand that a changing current in one coil induces a current in the second coil.

Categorize We will determine the result using concepts discussed in this section, so we categorize this example as a substitution problem.

Use Equation 29.17 to express the magnetic field in the interior of the base solenoid:

$$B = \mu_0 \frac{N_B}{\ell} i$$

Find the mutual inductance, noting that the magnetic flux Φ_{BH} through the handle's coil caused by the magnetic field of the base coil is BA:

$$M = \frac{N_H \Phi_{BH}}{i} = \frac{N_H BA}{i} = \mu_0 \frac{N_B N_H}{\ell} A$$

Wireless charging is used in a number of other "cordless" devices. One significant example is the inductive charging used by some manufacturers of electric cars that avoids direct metal-to-metal contact between the car and the charging apparatus.

Figure 31.11 A simple *LC* circuit. The capacitor has an initial charge Q_{max}. The switch is at position *a* for $t < 0$ and then thrown to position *b* at $t = 0$.

31.5 Oscillations in an *LC* Circuit

In Section 31.2, we connected the new circuit element we are studying, the inductor, with a resistor and studied the behavior of the circuit. Let's now connect an inductor with a capacitor, as shown in Figure 31.11. This combination is an ***LC* circuit.** When the switch is in position *a* as in Figure 31.11, the battery is charging the capacitor. Because we assume no resistance in the circuit, this charging process is essentially instantaneous. In addition, the absence of resistance means that no energy in the circuit is transformed to internal energy. We also assume an idealized situation in which energy is not radiated away from the circuit (T_{ER} in Equation 8.2). In reality, this radiation will occur, and is discussed in Chapter 33.

When the capacitor is fully charged, energy is stored in the capacitor's electric field and is equal to $U_E = Q^2_{max}/2C$ (Eq. 25.13). With the switch at position *a*, the inductor is not in the circuit, so no energy is stored in the inductor. Now imagine the switch in Figure 31.11 is thrown to position *b*. The capacitor begins to discharge; the rate at which charges leave the capacitor plates (which is also the rate at which the charge on the capacitor changes) is equal to the current in the circuit. The energy stored in the electric field of the capacitor decreases. Because there is a current in the circuit, some energy is now stored in the magnetic field of the inductor. Therefore, energy is transferred from the electric field of the capacitor to the magnetic field of the inductor. When the capacitor is fully discharged, it stores no energy. At this time, the current reaches its maximum value and all the energy in the circuit is stored in the inductor. The current continues in the same direction, decreasing in magnitude, with the capacitor eventually becoming fully charged again but with the polarity of its plates now opposite the initial polarity. At that point, the current stops and there is no energy stored in the inductor. This process is followed by another discharge until the circuit returns to its original state of maximum charge Q_{max} and the plate polarity shown in Figure 31.11. The energy continues to transfer back and forth between inductor and capacitor.

Consider some arbitrary time *t* after the switch in Figure 31.11 has been thrown to position *b*. At such a time, the capacitor has a charge $q < Q_{max}$ and the current is $i < I_{max}$. Because we have assumed the circuit resistance to be zero, there is no change in the internal energy in the circuit. In addition, we have assumed no electromagnetic radiation. With these assumptions, the circuit is an isolated system for energy, and Equation 8.2 becomes

Total energy stored in ▶
an *LC* circuit

$$\Delta U_E + \Delta U_B = 0 \tag{31.18}$$

Differentiate Equation 31.18 with respect to time, using Equations 25.13 and 31.12 for the energies stored in the capacitor and inductor, respectively:

$$\frac{d}{dt}\left(\frac{q^2}{2C} + \tfrac{1}{2}Li^2\right) = \frac{q}{C}\frac{dq}{dt} + Li\frac{di}{dt} = 0 \tag{31.19}$$

We can reduce this result to a differential equation in one variable by remembering that the current in the circuit is equal to the rate at which the charge on the capacitor changes: $i = dq/dt$. It then follows that $di/dt = d^2q/dt^2$. Substitution of these relationships into Equation 31.19 and rearranging gives us

$$\frac{d^2q}{dt^2} = -\frac{1}{LC}q \tag{31.20}$$

Let's solve for *q* by noting that this expression is of the same form as the analogous Equation 15.3 for a particle in simple harmonic motion:

$$\frac{d^2x}{dt^2} = -\frac{k}{m}x$$

where k is the spring constant, m is the mass of the particle, and the charge q plays the same mathematical role in Equation 31.20 as does x in Equation 15.3. The solution of this mechanical equation has the general form (Eq. 15.6):

$$x = A \cos (\omega t + \phi)$$

where A is the amplitude of the simple harmonic motion (the maximum value of x), ω is the angular frequency of this motion, given by Equation 15.9, and ϕ is the phase constant; the values of A and ϕ depend on the initial conditions. Because Equation 31.20 is of the same mathematical form as the differential equation of the simple harmonic oscillator, it has a solution of the same form:

$$q = Q_{max} \cos (\omega t + \phi) \tag{31.21}$$

◀ Charge as a function of time for an ideal *LC* circuit

where Q_{max} is the maximum charge on the capacitor and the angular frequency ω is the square root of the coefficient of q in Equation 31.20 (see Equations 15.3 and 15.5):

$$\omega = \frac{1}{\sqrt{LC}} \tag{31.22}$$

◀ Angular frequency of oscillation in an *LC* circuit

The charge on the capacitor undergoes a simple harmonic oscillation, alternating between polarities. Note that the angular frequency of the oscillations depends solely on the inductance and capacitance of the circuit. Equation 31.22 gives the *natural frequency* of oscillation of the *LC* circuit.

Because q varies sinusoidally with time, the current in the circuit also varies sinusoidally. We can show that by differentiating Equation 31.21 with respect to time:

$$i = \frac{dq}{dt} = -\omega Q_{max} \sin (\omega t + \phi) = -I_{max} \sin(\omega t + \phi) \tag{31.23}$$

◀ Current as a function of time for an ideal *LC* current

A representation of the energy oscillations in an *LC* circuit is shown in Figure 31.12 (page 836). As mentioned, the behavior of the circuit is analogous to that of the particle in simple harmonic motion studied in Chapter 15. For example, consider the block–spring system shown in Figure 15.10. The oscillations of this mechanical system are shown at the right of Figure 31.12. The potential energy $\frac{1}{2}kx^2$ stored in the stretched spring is analogous to the potential energy $q^2/2C$ stored in the capacitor in Figure 31.12. The kinetic energy $\frac{1}{2}mv^2$ of the moving block is analogous to the magnetic energy $\frac{1}{2}Li^2$ stored in the inductor, which requires the presence of moving charges. In Figure 31.12a, all the energy is stored as electric potential energy in the capacitor at $t = 0$ (because $i = 0$), just as all the energy in a block–spring system is initially stored as potential energy in the spring if it is stretched and released at $t = 0$. In Figure 31.12b, all the energy is stored as magnetic energy $\frac{1}{2}LI_{max}^2$ in the inductor, where I_{max} is the maximum current. Figures 31.12c and 31.12d show subsequent quarter-cycle situations in which the energy is all electric or all magnetic. At intermediate points, part of the energy is electric and part is magnetic (Fig. 31.12e).

Graphs of q versus t and i versus t are shown in Figure 31.13 (page 836). The charge on the capacitor oscillates between the extreme values Q_{max} and $-Q_{max}$, and the current oscillates between I_{max} and $-I_{max}$. Furthermore, the current is 90° out of phase with the charge. That is, when the charge is a maximum, the current is zero, and when the charge is zero, the current has its maximum value.

Plots of the time variations of U_E and U_B are shown in Figure 31.14 (page 836). The sum $U_E + U_B$ is a constant and is equal to the total energy $Q_{max}^2/2C$, or $\frac{1}{2}LI_{max}^2$. Analytical verification is straightforward. The amplitudes of the two graphs in Figure 31.14 must be equal because the maximum energy stored in the capacitor (when $I = 0$) must equal the maximum energy stored in the inductor (when $q = 0$). This equality is expressed mathematically as

$$\frac{Q_{max}^2}{2C} = \frac{LI_{max}^2}{2}$$

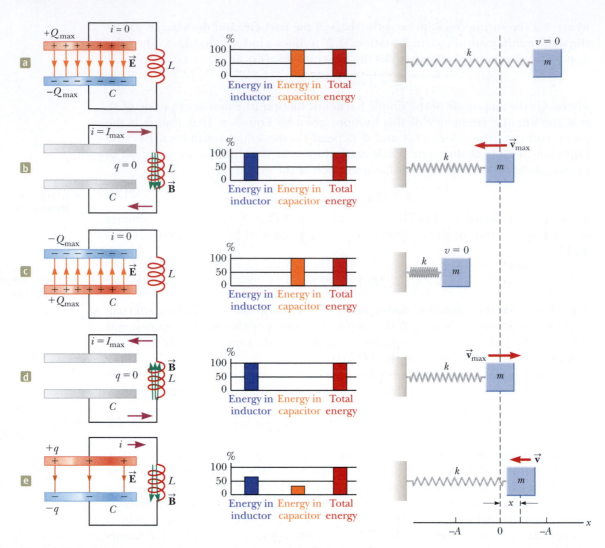

Figure 31.12 Energy transfer in a resistanceless, nonradiating *LC* circuit. The capacitor has a charge Q_{max} at $t = 0$, the instant at which the switch in Figure 31.11 is thrown to position *b*. The mechanical analog of this circuit is the particle in simple harmonic motion, represented by the block–spring system at the right of the figure. (a)–(d) At these special instants, all of the energy in the circuit resides in one of the circuit elements. (e) At an arbitrary instant, the energy is split between the capacitor and the inductor.

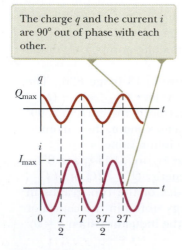

Figure 31.13 Graphs of charge versus time and current versus time for a resistanceless, nonradiating *LC* circuit.

Figure 31.14 Plots of U_E versus t and U_B versus t for a resistanceless, nonradiating *LC* circuit.

In our idealized situation, the oscillations in the circuit persist indefinitely; the total energy U of the circuit, however, remains constant only if energy transfers and transformations are neglected. In actual circuits, there is always some resistance and some energy is therefore transformed to internal energy. We mentioned at the beginning of this section that we are also ignoring radiation from the circuit. In reality, radiation is inevitable in this type of circuit, and the total energy in the circuit continuously decreases as a result of this process.

QUICK QUIZ 31.5 **(i)** At an instant of time during the oscillations of an *LC* circuit, the current is at its maximum value. At this instant, what happens to the voltage across the capacitor? (a) It is different from that across the inductor. (b) It is zero. (c) It has its maximum value. (d) It is impossible to determine. **(ii)** Now consider an instant when the current is momentarily zero. From the same choices, describe the magnitude of the voltage across the capacitor at this instant.

Example 31.6 **Oscillations in an *LC* Circuit**

In Figure 31.11, the battery has an emf of 12.0 V, the inductance is 2.81 mH, and the capacitance is 9.00 pF. The switch has been set to position a for a long time so that the capacitor is charged. The switch is then thrown to position b, removing the battery from the circuit and connecting the capacitor directly across the inductor.

(A) Find the frequency of oscillation of the circuit.

SOLUTION

Conceptualize When the switch is thrown to position b, the active part of the circuit is the right-hand loop, which is an *LC* circuit.

Categorize We use equations developed in this section, so we categorize this example as a substitution problem.

Use Equation 31.22 to find the frequency:

$$f = \frac{\omega}{2\pi} = \frac{1}{2\pi\sqrt{LC}}$$

Substitute numerical values:

$$f = \frac{1}{2\pi[(2.81 \times 10^{-3}\text{ H})(9.00 \times 10^{-12}\text{ F})]^{1/2}} = \boxed{1.00 \times 10^6 \text{ Hz}}$$

(B) What are the maximum values of charge on the capacitor and current in the circuit?

SOLUTION

Find the initial charge on the capacitor, which equals the maximum charge:

$$Q_{\text{max}} = C\,\Delta V = (9.00 \times 10^{-12}\text{ F})(12.0\text{ V}) = \boxed{1.08 \times 10^{-10}\text{ C}}$$

Use Equation 31.23 to express the maximum current in terms of the maximum charge:

$$I_{\text{max}} = \omega Q_{\text{max}} = 2\pi f Q_{\text{max}} = (2\pi \times 10^6 \text{ s}^{-1})(1.08 \times 10^{-10}\text{ C})$$

$$= \boxed{6.79 \times 10^{-4}\text{ A}}$$

31.6 The *RLC* Circuit

The *LC* circuit studied in Section 31.5 was idealized: the resistance of the circuit was zero. Let's now turn our attention to a more realistic circuit consisting of a resistor, an inductor, and a capacitor connected as shown in Figure 31.15a (page 838). We assume the resistance of the resistor represents all the resistance in the circuit. Suppose the switch has been at position a for a long time so that the capacitor has an initial charge Q_{max}. The switch is now thrown to position b as shown in Figure 31.15b. The three circuit elements are now connected in series. Continuing to ignore electromagnetic radiation from the circuit, we can write the appropriate reduction of Equation 8.2 for the circuit as

$$\Delta U_E + \Delta U_B + \Delta E_{\text{int}} = 0$$

The switch is set first to position *a*, and the capacitor is charged.

The switch is thrown to position *b* and oscillations begin.

Figure 31.15 A series *RLC* circuit. (a) With the switch in position *a*, the capacitor is charged by the battery. (b) When the switch is thrown to position *b*, the battery is removed from the circuit and the current in the *RLC* circuit oscillates.

where the internal energy represents the warming up of the resistor. Now differentiate this equation with respect to time:

$$\frac{dU_E}{dt} + \frac{dU_B}{dt} + \frac{dE_{\text{int}}}{dt} = 0$$

Use Equations 25.13 and 31.12 to evaluate the first two derivatives, and recognize that the third term is the rate at which energy is delivered to the resistor:

$$\frac{q}{C}\frac{dq}{dt} + Li\frac{di}{dt} + i^2 R = 0$$

Recognizing that the current in the circuit is equal to the time rate of change of charge on the capacitor, substitute $i = dq/dt$, and rearrange:

$$L\frac{d^2q}{dt^2} + R\frac{dq}{dt} + \frac{q}{C} = 0 \qquad (31.24)$$

The *RLC* circuit is analogous to the damped harmonic oscillator discussed in Section 15.6 and illustrated in Figure 15.19. The equation of motion for a particle undergoing damped harmonic oscillation is, from Equation 15.31,

$$m\frac{d^2x}{dt^2} + b\frac{dx}{dt} + kx = 0 \qquad (31.25)$$

Comparing Equations 31.24 and 31.25, we see that q corresponds to the position x of the particle at any instant, L to the mass m of the particle, R to the damping coefficient b, and C to $1/k$, where k is the force constant of the spring. These and other relationships are listed in Table 31.1.

Because the analytical solution of Equation 31.24 is cumbersome, we give only a qualitative description of the circuit behavior. In the simplest case, when $R = 0$, Equation 31.24 reduces to that of a simple *LC* circuit as expected, and the charge

TABLE 31.1 Analogies Between the *RLC* Circuit and the Particle in Damped Harmonic Motion

RLC Circuit		One-Dimensional Particle in Damped Harmonic Motion
Charge	$q \leftrightarrow x$	Position
Current	$i \leftrightarrow v_x$	Velocity
Potential difference	$\Delta V \leftrightarrow F_x$	Force
Resistance	$R \leftrightarrow b$	Viscous damping coefficient
Capacitance	$C \leftrightarrow 1/k$	(k = spring constant)
Inductance	$L \leftrightarrow m$	Mass
Current = time derivative of charge	$i = \dfrac{dq}{dt} \leftrightarrow v_x = \dfrac{dx}{dt}$	Velocity = time derivative of position
Rate of change of current = second time derivative of charge	$\dfrac{di}{dt} = \dfrac{d^2q}{dt^2} \leftrightarrow a_x = \dfrac{dv_x}{dt} = \dfrac{d^2x}{dt^2}$	Acceleration = second time derivative of position
Energy in inductor	$U_B = \frac{1}{2}Li^2 \leftrightarrow K = \frac{1}{2}mv^2$	Kinetic energy of moving particle
Energy in capacitor	$U_E = \frac{1}{2}\dfrac{q^2}{C} \leftrightarrow U = \frac{1}{2}kx^2$	Potential energy stored in a spring
Rate of energy loss due to resistance	$i^2R \leftrightarrow bv^2$	Rate of energy loss due to friction
RLC circuit	$L\dfrac{d^2q}{dt^2} + R\dfrac{dq}{dt} + \dfrac{q}{C} = 0 \leftrightarrow m\dfrac{d^2x}{dt^2} + b\dfrac{dx}{dt} + kx = 0$	Particle in damped harmonic motion

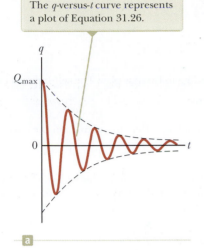

The *q*-versus-*t* curve represents a plot of Equation 31.26.

iStockphoto.com/A_Carina

Figure 31.16 (a) Charge versus time for a damped *RLC* circuit. The charge decays in this way when $R < \sqrt{4L/C}$. (b) Oscilloscope pattern showing the decay in the oscillations of an *RLC* circuit.

and the current oscillate sinusoidally in time. This situation is equivalent to removing all damping in the mechanical oscillator.

When *R* is small, a situation that is analogous to light damping in the mechanical oscillator, the solution of Equation 31.24 is

$$q = Q_{max}e^{-Rt/2L}\cos\omega_d t \qquad (31.26)$$

where ω_d, the angular frequency at which the circuit oscillates, is given by

$$\omega_d = \left[\frac{1}{LC} - \left(\frac{R}{2L}\right)^2\right]^{1/2} \qquad (31.27)$$

That is, the value of the charge on the capacitor undergoes a damped harmonic oscillation in analogy with a particle–spring system moving in a viscous medium. Equation 31.27 shows that when $R \ll \sqrt{4L/C}$ (so that the second term in the brackets is much smaller than the first), the frequency ω_d of the damped oscillator is close to that of the undamped oscillator, $1/\sqrt{LC}$. Because $i = dq/dt$, it follows that the current also undergoes damped harmonic oscillation. A plot of the charge versus time for the damped oscillator is shown in Figure 31.16a, and an oscilloscope trace for a real *RLC* circuit is shown in Figure 31.16b. The maximum value of *q* decreases after each oscillation, just as the amplitude of a damped particle–spring system decreases in time.

For larger values of *R*, the oscillations damp out more rapidly; in fact, there exists a critical resistance value $R_c = \sqrt{4L/C}$ above which no oscillations occur. A system with $R = R_c$ is said to be *critically damped*. When *R* exceeds R_c, the system is said to be *overdamped*.

Now, what's going on with the circles cut into the street in the opening storyline? In those circles, there is a loop of wire buried under the street surface. As we know from this chapter, a loop of wire acts as an inductor. The loop under the street is connected to a circuit, by wires buried in the straight-line cuts we see in the photograph. The circuit is an *RLC* circuit, with the loop in the street acting as the primary induction *L* in the circuit. The control electronics drives the *RLC* circuit with an oscillating voltage so that the circuit oscillates at its natural frequency, given by Equation 31.27. When a vehicle drives up and stops over the loop, the inductance of the loop changes for two reasons: (1) The metal of the car represents a magnetic material (Section 29.6), which alters the magnetic field passing through the loop; and (2) eddy currents (Section 30.6) are induced in the metal of the car, producing additional magnetic field lines that pass through the inductance loop. The change in the inductance *L* of the loop changes its natural frequency of oscillation. The control electronics detects this change, and sends a signal to change the light from red to green.

Summary

> ## Concepts and Principles

When the current in a loop of wire changes with time, an emf is induced in the loop according to Faraday's law. The **self-induced emf** is

$$\varepsilon_L = -L\frac{di}{dt} \tag{31.1}$$

where L is the **inductance** of the loop. Inductance is a measure of how much opposition a loop offers to a change in the current in the loop. Inductance has the SI unit of **henry** (H), where $1\ \text{H} = 1\ \text{V}\cdot\text{s/A}$.

The inductance of any coil is

$$L = \frac{N\Phi_B}{i} \tag{31.2}$$

where N is the total number of turns and Φ_B is the magnetic flux through the coil. The inductance of a device depends on its geometry. For example, the inductance of an air-core solenoid is

$$L = \mu_0\frac{N^2}{\ell}A \tag{31.4}$$

where ℓ is the length of the solenoid and A is the cross-sectional area.

If a resistor and inductor are connected in series to a battery of emf ε at time $t = 0$, the current in the circuit varies in time according to the expression

$$i = \frac{\varepsilon}{R}(1 - e^{-t/\tau}) \tag{31.7}$$

where $\tau = L/R$ is the **time constant** of the RL circuit. If we replace the battery in the circuit by a resistanceless wire, the current decays exponentially with time according to the expression

$$i = \frac{\varepsilon}{R}e^{-t/\tau} \tag{31.10}$$

where ε/R is the initial current in the circuit.

The energy stored in the magnetic field of an inductor carrying a current i is

$$U_B = \tfrac{1}{2}Li^2 \tag{31.12}$$

This energy is the magnetic counterpart to the energy stored in the electric field of a charged capacitor.

The energy density at a point where the magnetic field is B is

$$u_B = \frac{B^2}{2\mu_0} \tag{31.14}$$

The **mutual inductance** of a system of two coils is

$$M_{12} = \frac{N_2\Phi_{12}}{i_1} = M_{21} = \frac{N_1\Phi_{21}}{i_2} = M \tag{31.15}$$

This mutual inductance allows us to relate the induced emf in a coil to the changing source current in a nearby coil using the relationships

$$\varepsilon_2 = -M_{12}\frac{di_1}{dt} \quad\text{and}\quad \varepsilon_1 = -M_{21}\frac{di_2}{dt} \tag{31.16, 31.17}$$

In an LC circuit that has zero resistance and does not radiate electromagnetically (an idealization), the values of the charge on the capacitor and the current in the circuit vary sinusoidally in time at an angular frequency given by

$$\omega = \frac{1}{\sqrt{LC}} \tag{31.22}$$

The energy in an LC circuit continuously transfers between energy stored in the capacitor and energy stored in the inductor.

In an RLC circuit with small resistance, the charge on the capacitor varies with time according to

$$q = Q_{max}e^{-Rt/2L}\cos\omega_d t \tag{31.26}$$

where

$$\omega_d = \left[\frac{1}{LC} - \left(\frac{R}{2L}\right)^2\right]^{1/2} \tag{31.27}$$

Think–Pair–Share

See the Preface for an explanation of the icons used in this problems set. For additional assessment items for this section, go to **WEBASSIGN** From Cengage

1. One member of your group suggests that the RL circuit in Figure 31.5a be modified so that a third switch S_3 is added as shown in Figure TP31.1. Switch S_1 has been closed for a long time and switch S_2 has been at position a for a long time, so that the current in the inductor has reached its maximum value. The inductor has a resistance R_L in its windings. Now, switch S_3 is thrown closed at $t = 0$. The group member who suggested the circuit challenges the rest of the group to discuss and solve the following problem: Find the subsequent

Figure TP31.1

current as a function of time in (a) the resistor R, (b) the inductor L, and (c) the switch S_3. As a bonus question, the group is given the following challenge: (d) What is the current in switch S_3 at time $t = 0$?

2. **ACTIVITY** Your team has used a data logger with graphing software to take data on the voltage across the capacitor in the series RLC circuit in Figure 31.15. Switch S has been at position a for a very long time and is then thrown to position b. You know that the capacitance C is 15.0 μF. Your data logger makes a graph of the voltage across the capacitor as a function of time, with $t = 0$ being the instant the switch was thrown to position b. The graph is shown in Figure TP31.2. Using the graph, discuss this situation in your group and determine (a) the emf of the battery, (b) the inductance L, and (c) the resistance R.

Figure TP31.2

Problems

See the Preface for an explanation of the icons used in this problems set. For additional assessment items for this section, go to **WEBASSIGN** From Cengage

SECTION 31.1 Self-Induction and Inductance

1. A 2.00-H inductor carries a steady current of 0.500 A. When the switch in the circuit is opened, the current is effectively zero after 10.0 ms. What is the average induced emf in the inductor during this time interval?

2. A coiled telephone cord forms a spiral with 70.0 turns, a diameter of 1.30 cm, and an unstretched length of 60.0 cm. Determine the inductance of one conductor in the unstretched cord.

3. An emf of 24.0 mV is induced in a 500-turn coil when the current is changing at the rate of 10.0 A/s. What is the magnetic flux through each turn of the coil at an instant when the current is 4.00 A?

4. **QC** A 40.0-mA current is carried by a uniformly wound air-core solenoid with 450 turns, a 15.0-mm diameter, and 12.0-cm length. Compute (a) the magnetic field inside the solenoid, (b) the magnetic flux through each turn, and (c) the inductance of the solenoid. (d) **What If?** If the current were different, which of these quantities would change?

5. **S** A self-induced emf in a solenoid of inductance L changes in time as $\mathcal{E} = \mathcal{E}_0 e^{-kt}$. Assuming the charge is finite, find the total charge that passes a point in the wire of the solenoid.

6. **S** A toroid has a major radius R and a minor radius r and is tightly wound with N turns of wire on a hollow cardboard torus. Figure P31.6 shows half of this toroid, allowing us to see its cross section. If $R \gg r$, the magnetic field in the region enclosed by the wire is essentially the same as the magnetic field of a solenoid that has been bent into a large circle of radius R.

Figure P31.6

Modeling the field as the uniform field of a long solenoid, show that the inductance of such a toroid is approximately

$$L \approx \tfrac{1}{2}\mu_0 N^2 \frac{r^2}{R}$$

7. **T** A 10.0-mH inductor carries a current $i = I_{max} \sin \omega t$, with $I_{max} = 5.00$ A and $f = \omega/2\pi = 60.0$ Hz. What is the self-induced emf as a function of time?

8. The current in a 4.00 mH-inductor varies in time as shown in Figure P31.8. Construct a graph of the self-induced emf across the inductor over the time interval $t = 0$ to $t = 12.0$ ms.

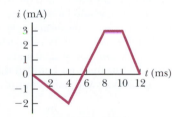

Figure P31.8

9. **CR** You are working as an electrical technician. One day, out in the field, you need an inductor but cannot find one. Looking in your wire supply cabinet, you find a cardboard tube with single-conductor wire wrapped uniformly around to it form a solenoid. You carefully count the turns of wire and find that there are 580 turns. The diameter of the tube is 8.00 cm, and the length of the wire-wrapped portion is 36.0 cm. You pull out your calculator to determine (a) the inductance of the coil and (b) the emf generated in it if the current in the wire increases at the rate of 4.00 A/s.

SECTION 31.2 *RL* Circuits

10. A 510-turn solenoid has a radius of 8.00 mm and an overall length of 14.0 cm. (a) What is its inductance? (b) If the solenoid is connected in series with a 2.50-Ω resistor and a battery, what is the time constant of the circuit?

11. A series RL circuit with $L = 3.00$ H and a series RC circuit with $C = 3.00$ μF have equal time constants. If the two circuits contain the same resistance R, (a) what is the value of R? (b) What is the time constant?

12. Show that $i = I_i e^{-t/\tau}$ is a solution of the differential equation

$$iR + L\frac{di}{dt} = 0$$

where I_i is the current at $t = 0$ and $\tau = L/R$.

13. A circuit consists of a coil, a switch, and a battery, all in series. The internal resistance of the battery is negligible compared with that of the coil. The switch is originally open. It is thrown closed, and after a time interval Δt, the current in the circuit reaches 80.0% of its final value. The switch then remains closed for a time interval much longer than Δt. The wires connected to the terminals of the battery are then short-circuited with another wire and removed from the battery, so that the current is uninterrupted. (a) At an instant that is a time interval Δt after the short circuit, the current is what percentage of its maximum value? (b) At the moment $2\Delta t$ after the coil is short-circuited, the current in the coil is what percentage of its maximum value?

14. **[CR]** You are working as a demonstration assistant for a physics professor. He shows you the circuit in Figure P31.14, which he wants you to build for an upcoming class. The lightbulb is a household incandescent bulb that receives energy at the rate of 40.0 W when operating at 120 V. It has a resistance R_1, which, for simplicity, we will assume is constant at all operating voltages. The battery in the circuit has an emf of 12.0 V. When the switch has been closed for a long time, the bulb glows dimly, since it is powered by only 12.0 V. When the switch is opened, however, the bulb flashes brightly and then gradually dims to darkness. Your professor wants you to determine two values: (a) the resistance R_2 that is necessary for the bulb to initially flash, when the switch is opened, at the same brightness it would have if plugged into a 120-V socket; (b) the inductance L necessary to keep the current in the lightbulb above 50.0% of its value when the switch is opened, for a time interval of 2.00 s after it is opened. Assume a resistance-free inductor and that the resistance of the lightbulb does not vary with temperature.

Figure P31.14

15. The switch in Figure P31.15 is open for $t < 0$ and is then thrown closed at time $t = 0$. Assume $R = 4.00$ Ω, $L = 1.00$ H,

Figure P31.15 Problems 15, 16, and 38.

and $\mathcal{E} = 10.0$ V. Find (a) the current in the inductor and (b) the current in the switch as functions of time thereafter.

16. **[S]** The switch in Figure P31.15 is open for $t < 0$ and is then thrown closed at time $t = 0$. Find (a) the current in the inductor and (b) the current in the switch as functions of time thereafter.

17. An inductor that has an inductance of 15.0 H and a resistance of 30.0 Ω is connected across a 100-V battery. What is the rate of increase of the current (a) at $t = 0$ and (b) at $t = 1.50$ s?

18. **[QC][S]** Two ideal inductors, L_1 and L_2, have *zero* internal resistance and are far apart, so their magnetic fields do not influence each other. (a) Assuming these inductors are connected in series, show that they are equivalent to a single ideal inductor having $L_{eq} = L_1 + L_2$. (b) Assuming these same two inductors are connected in parallel, show that they are equivalent to a single ideal inductor having $1/L_{eq} = 1/L_1 + 1/L_2$. (c) **What If?** Now consider two inductors L_1 and L_2 that have *nonzero* internal resistances R_1 and R_2, respectively. Assume they are still far apart, so their mutual inductance is zero, and assume they are connected in series. Show that they are equivalent to a single inductor having $L_{eq} = L_1 + L_2$ and $R_{eq} = R_1 + R_2$. (d) If these same inductors are now connected in parallel, is it necessarily true that they are equivalent to a single ideal inductor having $1/L_{eq} = 1/L_1 + 1/L_2$ and $1/R_{eq} = 1/R_1 + 1/R_2$? Explain your answer.

19. Consider the current pulse $i(t)$ shown in Figure P31.19a. The current begins at zero, becomes 10.0 A between $t = 0$ and $t = 200$ μs, and then is zero once again. This pulse is applied to the input of the partial circuit shown in Figure P31.19b. Determine the current in the inductor as a function of time.

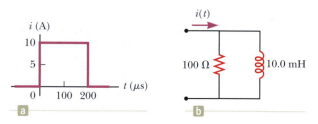

Figure P31.19

SECTION 31.3 Energy in a Magnetic Field

20. Calculate the energy associated with the magnetic field of a 200-turn solenoid in which a current of 1.75 A produces a magnetic flux of 3.70×10^{-4} T $\cdot$ m^2 in each turn.

21. **[T]** An air-core solenoid with 68 turns is 8.00 cm long and has a diameter of 1.20 cm. When the solenoid carries a current of 0.770 A, how much energy is stored in its magnetic field?

22. **[S]** Complete the calculation in Example 31.3 by proving that

$$\int_0^\infty e^{-2Rt/L}\, dt = \frac{L}{2R}$$

23. **[T]** A 24.0-V battery is connected in series with a resistor and an inductor, with $R = 8.00$ Ω and $L = 4.00$ H, respectively. Find the energy stored in the inductor (a) when the current reaches its maximum value and (b) at an instant that is a time interval of one time constant after the switch is closed.

24. **[QC]** A flat coil of wire has an inductance of 40.0 mH and a resistance of 5.00 Ω. It is connected to a 22.0-V battery at the instant $t = 0$. Consider the moment when the current is 3.00 A.

(a) At what rate is energy being delivered by the battery? (b) What is the power being delivered to the resistance of the coil? (c) At what rate is energy being stored in the magnetic field of the coil? (d) What is the relationship among these three power values? (e) Is the relationship described in part (d) true at other instants as well? (f) Explain the relationship at the moment immediately after $t = 0$ and at a moment several seconds later.

SECTION 31.4 Mutual Inductance

25. An emf of 96.0 mV is induced in the windings of a coil when the current in a nearby coil is increasing at the rate of 1.20 A/s. What is the mutual inductance of the two coils?

26. **T** Two solenoids A and B, spaced close to each other and sharing the same cylindrical axis, have 400 and 700 turns, respectively. A current of 3.50 A in solenoid A produces an average flux of 300 μWb through each turn of A and a flux of 90.0 μWb through each turn of B. (a) Calculate the mutual inductance of the two solenoids. (b) What is the inductance of A? (c) What emf is induced in B when the current in A changes at the rate of 0.500 A/s?

27. **Q|C S** Solenoid S_1 has N_1 turns, radius R_1, and length ℓ. It is so long that its magnetic field is uniform nearly everywhere inside it and is nearly zero outside. Solenoid S_2 has N_2 turns, radius $R_2 < R_1$, and the same length as S_1. It lies inside S_1, with their axes parallel. (a) Assume S_1 carries variable current i. Compute the mutual inductance characterizing the emf induced in S_2. (b) Now assume S_2 carries current i. Compute the mutual inductance to which the emf in S_1 is proportional. (c) State how the results of parts (a) and (b) compare with each other.

28. Two single-turn circular loops of wire have radii R and r, with $R \gg r$. The loops lie in the same plane and are concentric. (a) Show that the mutual inductance of the pair is approximately $M = \mu_0 \pi r^2 / 2R$. (b) Evaluate M for $r = 2.00$ cm and $R = 20.0$ cm.

SECTION 31.5 Oscillations in an *LC* Circuit

29. In the circuit of Figure P31.29, the battery emf is 50.0 V, the resistance is 250 Ω, and the capacitance is 0.500 μF. The switch S is closed for a long time interval, and zero potential difference is measured across the capacitor. After the switch is opened, the potential difference across the capacitor reaches a maximum value of 150 V. What is the value of the inductance?

Figure P31.29

30. *Why is the following situation impossible?* The *LC* circuit shown in Figure P31.30 has $L = 30.0$ mH and $C = 50.0$ μF. The capacitor has an initial charge of 200 μC. The switch is closed, and the circuit undergoes undamped *LC* oscillations. At periodic instants, the energies stored by the capacitor and the inductor are equal, with each of the two components storing 250 μJ.

Figure P31.30 Problems 30 and 32.

31. An *LC* circuit consists of a 20.0-mH inductor and a 0.500-μF capacitor. If the maximum instantaneous current in the circuit is 0.100 A, what is the greatest potential difference across the capacitor?

32. **T** An *LC* circuit like that in Figure P31.30 consists of a 3.30-H inductor and an 840-pF capacitor that initially carries a 105-μC charge. The switch is open for $t < 0$ and is then thrown closed at $t = 0$. Compute the following quantities at $t = 2.00$ ms: (a) the energy stored in the capacitor, (b) the energy stored in the inductor, and (c) the total energy in the circuit.

SECTION 31.6 The *RLC* Circuit

33. In Figure 31.15, let $R = 7.60$ Ω, $L = 2.20$ mH, and $C = 1.80$ μF. (a) Calculate the frequency of the damped oscillation of the circuit when the switch is thrown to position *b*. (b) What is the critical resistance for damped oscillations?

34. **S** Show that Equation 31.24 in the text is Kirchhoff's loop rule as applied to the circuit in Figure 31.15b.

35. **S** Electrical oscillations are initiated in a series circuit containing a capacitance C, inductance L, and resistance R. (a) If $R \ll \sqrt{4L/C}$ (weak damping), what time interval elapses before the amplitude of the current oscillation falls to 50.0% of its initial value? (b) Over what time interval does the energy decrease to 50.0% of its initial value?

ADDITIONAL PROBLEMS

36. **Review.** Consider a capacitor with vacuum between its large, closely spaced, oppositely charged parallel plates. (a) Show that the force on one plate can be accounted for by thinking of the electric field between the plates as exerting a "negative pressure" equal to the energy density of the electric field. (b) Consider two infinite plane sheets carrying electric currents in opposite directions with equal linear current densities J_s. Calculate the force per area acting on one sheet due to the magnetic field, of magnitude $\mu_0 J_s / 2$, created by the other sheet. (c) Calculate the net magnetic field between the sheets and the field outside of the volume between them. (d) Calculate the energy density in the magnetic field between the sheets. (e) Show that the force on one sheet can be accounted for by thinking of the magnetic field between the sheets as exerting a positive pressure equal to its energy density. This result for magnetic pressure applies to all current configurations, not only to sheets of current.

37. **S** A capacitor in a series *LC* circuit has an initial charge Q and is being discharged. When the charge on the capacitor is $Q/2$, find the flux through each of the N turns in the coil of the inductor in terms of Q, N, L, and C.

38. **AMT Q|C** In the circuit diagrammed in Figure P31.15, assume the switch has been closed for a long time interval and is opened at $t = 0$. Also assume $R = 4.00$ Ω, $L = 1.00$ H, and $\mathcal{E} = 10.0$ V.

(a) Before the switch is opened, does the inductor behave as an open circuit, a short circuit, a resistor of some particular resistance, or none of those choices? (b) What current does the inductor carry? (c) How much energy is stored in the inductor for $t < 0$? (d) After the switch is opened, what happens to the energy previously stored in the inductor? (e) Sketch a graph of the current in the inductor for $t \geq 0$. Label the initial and final values and the time constant.

39. (a) A flat, circular coil does not actually produce a uniform magnetic field in the area it encloses. Nevertheless, estimate the inductance of a flat, compact, circular coil with radius R and N turns by assuming the field at its center is uniform over its area. (b) A circuit on a laboratory table consists of a 1.50-volt battery, a 270-Ω resistor, a switch, and three 30.0-cm-long patch cords connecting them. Suppose the circuit is arranged to be circular. Think of it as a flat coil with one turn. Compute the order of magnitude of its inductance and (c) of the time constant describing how fast the current increases when you close the switch.

40. At the moment $t = 0$, a 24.0-V battery is connected to a 5.00-mH coil and a 6.00-Ω resistor. (a) Immediately thereafter, how does the potential difference across the resistor compare to the emf across the coil? (b) Answer the same question about the circuit several seconds later. (c) Is there an instant at which these two voltages are equal in magnitude? If so, when? Is there more than one such instant? (d) After a 4.00-A current is established in the resistor and coil, the battery is suddenly replaced by a short circuit. Answer parts (a), (b), and (c) again with reference to this new circuit.

41. You are working on an LC circuit for an experiment you are performing in your basement. You have an appropriate capacitor, but you need to build your own inductor. You wish to cut a wooden ring with a rectangular cross section, as shown in Figure P31.41, from wood with thickness $h = 1.00$ cm. You want to wrap 500 turns of wire around it to form a toroidal inductor. For your experiment, you need to have 1.82×10^{-4} J of energy stored in the inductor when it carries a current of 2.00 A. In order to cut the appropriate wooden ring, you need to determine the ratio b/a. Ignore any effect of the wood core on the magnetic field.

Figure P31.41 Problems 41 and 42.

42. You are working on an LC circuit for an experiment you are performing in your basement. You have an appropriate capacitor, but you need to build your own inductor. You wish to cut a wooden ring with a rectangular cross section, as shown in Figure P31.41, from wood with thickness h. You want to wrap N turns of wire around it to form a toroidal inductor. For your experiment, you need to have energy U_B stored in the inductor when it carries a current i. In order to cut the appropriate wooden ring, you need to determine the ratio b/a. Ignore any effect of the wood core on the magnetic field.

43. You are trying out to represent your campus in the Physics Olympics. You have just been given a problem involving the circuit shown in Figure P31.43. The values of the circuit elements are $\mathcal{E} = 12.0$ V, $R = 10.0\ \Omega$, $C = 5.00\ \mu$F, and $L = 2.00$ mH. The inductor is resistance-free and the capacitor begins with zero charge. Switch S has been set to position a for a long time. At $t = 0$, switch S is thrown to position b. What is the charge on the capacitor at $t = 1.00$ s? To qualify for the team, you must be the first contestant to determine the answer! Ready? Go!

Figure P31.43

44. *Why is the following situation impossible?* You are working on an experiment involving a series circuit consisting of a charged 500-μF capacitor, a 32.0-mH inductor, and a resistor R. You discharge the capacitor through the inductor and resistor and observe the decaying oscillations of the current in the circuit. When the resistance R is 8.00 Ω, the decay in the oscillations is too slow for your experimental design. To make the decay faster, you double the resistance. As a result, you generate decaying oscillations of the current that are perfect for your needs.

45. A time-varying current i is sent through a 50.0-mH inductor from a source as shown in Figure P31.45a. The current is constant at $i = -1.00$ mA until $t = 0$ and then varies with time afterward as shown in Figure P31.45b. Make a graph of the emf across the inductor as a function of time.

Figure P31.45

46. At $t = 0$, the open switch in Figure P31.46 is thrown closed. We wish to find a symbolic expression for the current in the inductor for time $t > 0$. Let this current be called i and choose it to be downward in the inductor in Figure P31.46. Identify i_1 as the current to the right through R_1 and i_2 as the current downward through R_2. (a) Use Kirchhoff's junction rule to find a relation among the three currents. (b) Use Kirchhoff's loop rule around the left loop to find another relationship. (c) Use Kirchhoff's loop rule around the outer loop to find a third relationship. (d) Eliminate i_1 and i_2 among the three equations to find an equation involving only the current i. (e) Compare the equation in part (d) with Equation 31.6 in the text. Use this comparison

Figure P31.46

to rewrite Equation 31.7 in the text for the situation in this problem and show that

$$i(t) = \frac{\varepsilon}{R_1}[1 - e^{-(R'/L)t}]$$

where $R' = R_1R_2/(R_1 + R_2)$.

Problems 47 and 48 apply ideas from this and earlier chapters to some properties of superconductors, which were introduced in Section 26.5.

47. **Review.** The use of superconductors has been proposed for power transmission lines. A single coaxial cable (Fig. P31.47) could carry a power of 1.00×10^3 MW (the output of a large power plant) at 200 kV, DC, over a distance of 1.00×10^3 km without loss. An inner wire of radius $a = 2.00$ cm, made from the superconductor Nb_3Sn, carries the current I in one direction. A surrounding superconducting cylinder of radius $b = 5.00$ cm would carry the return current I. In such a system, what is the magnetic field (a) at the surface of the inner conductor and (b) at the inner surface of the outer conductor? (c) How much energy would be stored in the magnetic field in the space between the conductors in a 1.00×10^3 km superconducting line? (d) What is the pressure exerted on the outer conductor due to the current in the inner conductor?

Figure P31.47

48. **Review.** A fundamental property of a type I superconducting material is *perfect diamagnetism,* or demonstration of the *Meissner effect,* illustrated in Figure 29.27 in Section 29.6 and described as follows. If a sample of superconducting material is placed into an externally produced magnetic field or is cooled to become superconducting while it is in a magnetic field, electric currents appear on the surface of the sample. The currents have precisely the strength and orientation required to make the total magnetic field be zero throughout the interior of the sample. This problem will help you understand the magnetic force that can then act on the sample. Compare this problem with Problem 39 in Chapter 25, pertaining to the force attracting a perfect dielectric into a strong electric field.

A vertical solenoid with a length of 120 cm and a diameter of 2.50 cm consists of 1 400 turns of copper wire carrying a counterclockwise current (when viewed from above) of 2.00 A as shown in Figure P31.48a. (a) Find the magnetic

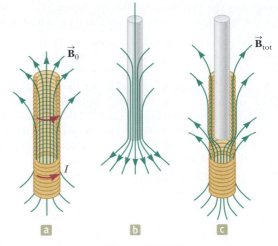

Figure P31.48

field in the vacuum inside the solenoid. (b) Find the energy density of the magnetic field. Now a superconducting bar 2.20 cm in diameter is inserted partway into the solenoid. Its upper end is far outside the solenoid, where the magnetic field is negligible. The lower end of the bar is deep inside the solenoid. (c) Explain how you identify the direction required for the current on the curved surface of the bar so that the total magnetic field is zero within the bar. The field created by the supercurrents is sketched in Figure P31.48b, and the total field is sketched in Figure P31.48c. (d) The field of the solenoid exerts a force on the current in the superconductor. Explain how you determine the direction of the force on the bar. (e) Noting that the units J/m^3 of energy density are the as the units N/m^2 of pressure, calculate the magnitude of the force by multiplying the energy density of the solenoid field times the area of the bottom end of the superconducting bar.

49. A wire of nonmagnetic material, with radius R, carries current uniformly distributed over its cross section. The total current carried by the wire is I. Show that the magnetic energy per unit length inside the wire is $\mu_0 I^2/16\pi$.

CHALLENGE PROBLEMS

50. In earlier times when many households received nondigital television signals from an antenna, the lead-in wires from the antenna were often constructed in the form of two parallel wires (Fig. P31.50). The two wires carry currents of equal magnitude in opposite directions. The center-to-center separation of the wires is w, and a is their radius. Assume w is large enough compared with a that the wires carry the current uniformly distributed over their surfaces and negligible magnetic field exists inside the wires. (a) Why does this configuration of conductors have an inductance? (b) What

Figure P31.50

constitutes the flux loop for this configuration? (c) Show that the inductance of a length x of this type of lead-in is

$$L = \frac{\mu_0 x}{\pi} \ln\left(\frac{w - a}{a}\right)$$

51. Assume the magnitude of the magnetic field outside a sphere of radius R is $B = B_0(R/r)^2$, where B_0 is a constant. (a) Determine the total energy stored in the magnetic field outside the sphere. (b) Evaluate your result from part (a) for $B_0 = 5.00 \times 10^{-5}$ T and $R = 6.00 \times 10^6$ m, values appropriate for the Earth's magnetic field.

52. In Figure P31.52, the battery has emf $\mathcal{E} = 18.0$ V and the other circuit elements have values $L = 0.400$ H, $R_1 = 2.00$ kΩ, and $R_2 = 6.00$ kΩ. The switch is closed for $t < 0$, and steady-state conditions are established. The switch is then opened at $t = 0$. (a) Find the emf across L immediately after $t = 0$. (b) Which end of

Figure P31.52

the coil, a or b, is at the higher potential? (c) Make graphs of the currents in R_1 and in R_2 as a function of time, treating the steady-state directions as positive. Show values before and after $t = 0$. (d) At what moment after $t = 0$ does the current in R_2 have the value 2.00 mA?

53. Two inductors having inductances L_1 and L_2 are connected in parallel as shown in Figure P31.53a. The mutual inductance between the two inductors is M. Determine the equivalent inductance L_{eq} for the system (Fig. P31.53b).

Figure P31.53

Alternating-Current Circuits

32

STORYLINE **You are in your garage gathering some tools to** investigate an electrical receptacle in your house that seems to be giving you some trouble. You grab an extension cord and decide to test it to make sure its ground wire is operating properly. To do so, you use a *receptacle tester*. You plug this device into an electrical receptacle and the combination of lights indicates whether the receptacle is properly grounded and operating correctly. You plug this device into the end of your extension cord. You then accidentally drop the end of the cord and notice something interesting about the appearance of the light from the device. To investigate further, you pick up the tester, turn off the lights in the garage, and swing the end of the cord with the tester in a circle. Whoa! The lights on the receptacle tester appears as a circular series of bright and dark segments. What causes this effect? You spend the next few minutes trying out different radii of the circle, different angular speeds, etc.

CONNECTIONS In earlier chapters, we have studied a number of circuit elements: capacitors, resistors, and inductors. Beginning in Chapter 25, we connected these elements to batteries, forming *direct-current (DC) circuits*, in which the current always travels in the same direction. In subsequent chapters, we found a number of interesting effects when we combined elements in *RC*, *LC*, *RL*, and *RLC* circuits. But, so far, we have only used batteries as the power source. Every time you turn on a television set, a computer, or any of a multitude of other electrical appliances in a home, you are calling on an *alternating-current (AC) circuit* to provide the power to operate them. In this type of circuit, the power source does not provide a fixed voltage like a battery, but rather supplies an alternating voltage, usually sinusoidal. An understanding of AC circuits will allow investigations well beyond the scope of this text, from home electrical systems up to studies of the power grid of a public utility that is providing energy on a large scale to homes and businesses.

32.1   AC Sources

An AC circuit consists of circuit elements and a power source that provides an alternating voltage Δv. This time-varying voltage from the source is described by

$$\Delta v = \Delta V_{max} \sin \omega t \tag{32.1}$$

where ΔV_{max} is the maximum output voltage of the source, or the **voltage amplitude**, and ω is the angular frequency of the source. There are various possibilities for AC sources, including generators as discussed in Section 30.5 and electrical oscillators. In a home, each electrical outlet serves as an AC source. Because the output voltage of an AC source varies sinusoidally with time, the voltage is positive during one half of the cycle and negative during the other half as in Figure 32.1. Likewise, the current in any circuit driven by an AC source is an alternating current that also varies sinusoidally with time.

From Equation 15.12, the angular frequency of the AC voltage is

$$\omega = 2\pi f = \frac{2\pi}{T}$$

where f is the frequency of the source and T is the period. The source determines the frequency of the current in any circuit connected to it. Commercial electric-power plants in the United States use a frequency of 60.0 Hz, which corresponds to an angular frequency of 377 rad/s.

The receptacle into which you plug your receptacle tester in the opening storyline is an AC source. Therefore, the lights on the device are actually flashing on and off many times per second. The flashing is too fast for you to detect if you hold the tester still. But when you rotate it in a circle, the on-and-off flashing of the lights on the tester is apparent.

Figure 32.1 The voltage supplied by an AC source is sinusoidal with a period T.

32.2   Resistors in an AC Circuit

With this introduction to AC sources, let's apply an AC source to our familiar circuit elements individually, and then apply it to a combination of all the elements. We begin with a resistor.

Consider a simple AC circuit consisting of a resistor and an AC source as shown in Figure 32.2. At any instant, the algebraic sum of the voltages around a closed loop in a circuit must be zero (Kirchhoff's loop rule). Therefore, $\Delta v + \Delta v_R = 0$. At the instant in the figure, the current is clockwise around the circuit; let's travel around the circuit in the same direction. Using the sign conventions in Figure 27.12, we see that the voltage across the resistor is negative. For the current to be in a clockwise direction, the left side of the AC source must be momentarily positive, so, again from Figure 27.12, the voltage across the AC source is positive. Therefore,

$$\Delta v - i_R R = 0 \tag{32.2}$$

If we rearrange Equation 32.2 and substitute $\Delta V_{max} \sin \omega t$ for Δv, the instantaneous current in the resistor is

$$i_R = \frac{\Delta v}{R} = \frac{\Delta V_{max}}{R} \sin \omega t = I_{max} \sin \omega t \tag{32.3}$$

$$\Delta v = \Delta V_{max} \sin \omega t$$

Figure 32.2 A circuit consisting of a resistor of resistance R connected to an AC source, designated by the symbol

. At the moment depicted in the figure, the current is to the right in the resistor.

where I_{max} is the maximum current:

$$I_{max} = \frac{\Delta V_{max}}{R} \qquad (32.4)$$

◀ Maximum current in a resistor

Equation 26.7 shows that the instantaneous voltage across the resistor is

$$\Delta v_R = i_R R = I_{max} R \sin \omega t \qquad (32.5)$$

◀ Voltage across a resistor

A plot of voltage and current versus time for this circuit is shown in Figure 32.3a. At point *a*, the current has a maximum value in one direction, arbitrarily called the positive direction. Between points *a* and *b*, the current is decreasing in magnitude but is still in the positive direction. At point *b*, the current is momentarily zero; it then begins to increase in the negative direction between points *b* and *c*. At point *c*, the current has reached its maximum value in the negative direction.

The current and voltage are in step with each other because they both vary as sin ωt and reach their maximum values at the same time as shown in Figure 32.3a. They are said to be **in phase,** similar to the way two waves can be in phase as discussed in our study of wave motion in Chapter 17. For resistors in AC circuits, there are no new concepts to learn. Resistors behave essentially the same way in both DC and AC circuits. That, however, is not the case for capacitors and inductors, as we shall see.

To simplify our analysis of circuits containing two or more elements, we use a graphical representation called a *phasor diagram*. A **phasor** is a vector whose length is proportional to the maximum value of the variable it represents (ΔV_{max} for voltage and I_{max} for current in this discussion). The phasor rotates counterclockwise at an angular speed equal to the angular frequency associated with the variable. The projection of the phasor onto the vertical axis represents the instantaneous value of the quantity it represents.

Figure 32.3b shows voltage and current phasors for the circuit of Figure 32.2 at some instant of time. Notice that Figure 32.3a shows the current and voltage at *all* times along the *t* axis. Figure 32.3b shows the current and voltage phasors at *only one* time. As time progresses, the phasors rotate counterclockwise. The projections of the phasor arrows onto the vertical axis are determined by a sine function of the angle of the phasor with respect to the horizontal axis. For example, the projection of the current phasor in Figure 32.3b is $i_R = I_{max} \sin \omega t$. Notice that this expression is the same as Equation 32.3. Therefore, the projection of a current phasor represents a current that varies sinusoidally in time. We can do the same with time-varying voltages. The advantage of this approach is

Figure 32.3 (a) Plots of the instantaneous current i_R and instantaneous voltage Δv_R across a resistor as functions of time. At time $t = T$, one cycle of the time-varying voltage and current has been completed. (b) Phasor diagram for the resistive circuit showing that the current is in phase with the voltage.

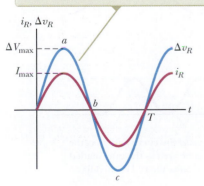

The current and the voltage are in phase: they simultaneously reach their maximum values, their minimum values, and their zero values.

The current and the voltage phasors are in the same direction because the current is in phase with the voltage.

Figure 32.4 (Quick Quiz 32.1) A voltage phasor is shown at three instants of time, (a), (b), and (c).

that the phase relationships among currents and voltages can be represented as vector additions of phasors using the vector addition techniques discussed in Chapter 3.

In the case of the single-loop resistive circuit of Figure 32.2, the current and voltage phasors are in the same direction in Figure 32.3b because i_R and Δv_R are in phase. The current and voltage in circuits containing capacitors and inductors have different phase relationships.

QUICK QUIZ 32.1 Consider the voltage phasor in Figure 32.4, shown at three instants of time. **(i)** Choose the part of the figure, (a), (b), or (c), that represents the instant of time at which the instantaneous value of the voltage has the largest magnitude. **(ii)** Choose the part of the figure that represents the instant of time at which the instantaneous value of the voltage has the smallest magnitude.

For the simple resistive circuit in Figure 32.2, notice that the average value of the current over one cycle is zero. That is, the current is maintained in the positive direction for the same amount of time and at the same magnitude as it is maintained in the negative direction. The direction of the current, however, has no effect on the behavior of the resistor. We can understand this concept by realizing that collisions between electrons and the fixed atoms of the resistor result in an increase in the resistor's temperature at all times; it doesn't matter which way the electrons are going.

We can make this discussion quantitative by recalling from Equation 26.22 that the rate at which energy is delivered to a resistor is the power $P = i^2R$, where i is the instantaneous current in the resistor. Because this rate is proportional to the square of the current, it makes no difference whether the current is direct or alternating, that is, whether the sign associated with the current is positive or negative. The temperature increase produced by an alternating current having a maximum value I_{max}, however, is not the same as that produced by a direct current equal to I_{max} because the alternating current has this maximum value for only an instant during each cycle (Fig. 32.5a). What is of importance in an AC circuit is an average value of current, referred to as the **rms current**. As we learned in Section 20.1, the notation *rms* stands for *root-mean-square*, which in this case means the square root of the mean (average) value of the square of the current: $I_{rms} = \sqrt{(i^2)_{avg}}$. Because i^2 varies as $\sin^2 \omega t$ and because the average value of i^2 is $\frac{1}{2}I^2_{max}$ (see Fig. 32.5b), the rms current is

rms current ▶

$$I_{rms} = \frac{I_{max}}{\sqrt{2}} = 0.707 I_{max} \qquad (32.6)$$

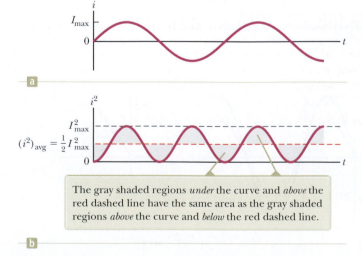

Figure 32.5 (a) Graph of the current in a resistor as a function of time. (b) Graph of the current squared in a resistor as a function of time, showing that the red dashed line is the average of $I^2_{max} \sin^2 \omega t$. In general, the average value of $\sin^2 \omega t$ or $\cos^2 \omega t$ over one cycle is $\frac{1}{2}$.

The gray shaded regions *under* the curve and *above* the red dashed line have the same area as the gray shaded regions *above* the curve and *below* the red dashed line.

This equation states that an alternating current whose maximum value is 2.00 A delivers to a resistor the same power as a direct current that has a value of $(0.707)(2.00 \text{ A}) = 1.41 \text{ A}$. The average power delivered to a resistor that carries an alternating current is

$$P_{avg} = I_{rms}^2 R \qquad (32.7)$$

◀ Average power delivered to a resistor

Alternating voltage is also best discussed in terms of rms voltage, and the relationship is identical to that for current:

$$\Delta V_{rms} = \frac{\Delta V_{max}}{\sqrt{2}} = 0.707 \,\Delta V_{max} \qquad (32.8)$$

◀ rms voltage

When we speak of measuring a 120-V alternating voltage from an electrical outlet in the home, we are referring to an rms voltage of 120 V. A calculation using Equation 32.8 shows that such an alternating voltage has a maximum value of about 170 V. One reason rms values are often used when discussing alternating currents and voltages is that AC ammeters and voltmeters are designed to read rms values. Furthermore, with rms values, many of the equations we use have the same form as their direct-current counterparts.

Example 32.1 **What Is the rms Current?**

The voltage output of an AC source is given by the expression $\Delta v = 200 \sin \omega t$, where Δv is in volts. Find the rms current in the circuit when this source is connected to a 47.0-Ω resistor.

SOLUTION

Conceptualize Figure 32.2 shows the physical situation for this problem.

Categorize We evaluate the current with an equation developed in this section, so we categorize this example as a substitution problem.

Combine Equations 32.4 and 32.6 to find the rms current:

$$I_{rms} = \frac{I_{max}}{\sqrt{2}} = \frac{\Delta V_{max}}{\sqrt{2}\,R}$$

Comparing the expression for voltage output with the general form $\Delta v = \Delta V_{max} \sin \omega t$ shows that $\Delta V_{max} = 200$ V. Substitute numerical values:

$$I_{rms} = \frac{200 \text{ V}}{\sqrt{2}\,(47.0 \ \Omega)} = \boxed{3.01 \text{ A}}$$

32.3 Inductors in an AC Circuit

Now let's move on to another circuit element to which we will apply an alternating voltage. Consider an AC circuit consisting only of an inductor connected to the terminals of an AC source as shown in Figure 32.6. Because $\Delta v_L = -L(di_L/dt)$ is the self-induced instantaneous voltage across the inductor (see Eq. 31.1), Kirchhoff's loop rule applied to this circuit gives $\Delta v + \Delta v_L = 0$, or

$$\Delta v - L\frac{di_L}{dt} = 0$$

Substituting $\Delta V_{max} \sin \omega t$ for Δv and rearranging gives

$$\Delta v = L\frac{di_L}{dt} = \Delta V_{max} \sin \omega t \qquad (32.9)$$

Figure 32.6 A circuit consisting of an inductor of inductance L connected to an AC source.

Solving this equation for di_L gives

$$di_L = \frac{\Delta V_{max}}{L} \sin \omega t \, dt$$

Integrating this expression[1] gives the instantaneous current i_L in the inductor as a function of time:

$$i_L = \frac{\Delta V_{max}}{L} \int \sin \omega t \, dt = -\frac{\Delta V_{max}}{\omega L} \cos \omega t \qquad (32.10)$$

Using the trigonometric identity $\cos \omega t = -\sin(\omega t - \pi/2)$, we can express Equation 32.10 as

Current in an inductor ▶
in an AC circuit

$$i_L = \frac{\Delta V_{max}}{\omega L} \sin \left(\omega t - \frac{\pi}{2} \right) \qquad (32.11)$$

Comparing this result with Equation 32.9 shows that the instantaneous current i_L in the inductor and the instantaneous voltage Δv_L across the inductor are *out* of phase by $\pi/2$ rad = 90°.

A plot of voltage and current versus time is shown in Figure 32.7a. When the voltage Δv_L across the inductor is a maximum (point *a* in Fig. 32.7a), the current in the inductor has a value of zero (point *d*), but is changing at its highest rate. When the voltage is zero (point *b*), the current has its maximum value (point *e*). Notice that the voltage reaches its maximum value one-quarter of a period before the current reaches its maximum value. Therefore, for a sinusoidal applied voltage, the current in an inductor always *lags* behind the voltage across the inductor by 90° (one-quarter cycle in time).

As with the relationship between current and voltage for a resistor, we can represent this relationship for an inductor with a phasor diagram as in Figure 32.7b. The phasors are at 90° to each other, representing the 90° phase difference between current and voltage. (See Eqs. 32.9 and 32.11.)

Equation 32.10 shows that the current in an inductive circuit reaches its maximum value when $\cos \omega t = \pm 1$:

Maximum current in ▶
an inductor

$$I_{max} = \frac{\Delta V_{max}}{\omega L} \qquad (32.12)$$

Notice that this expression is similar in form to the relationship between current, voltage, and resistance in a DC circuit, $I = \Delta V/R$ (Eq. 26.7). Because I_{max} has units of amperes and ΔV_{max} has units of volts, ωL must have units of ohms. Therefore, ωL

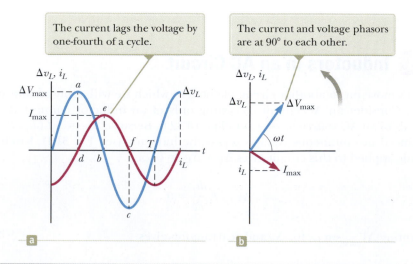

Figure 32.7 (a) Plots of the instantaneous current i_L and instantaneous voltage Δv_L across an inductor as functions of time. (b) Phasor diagram for the inductive circuit.

The current lags the voltage by one-fourth of a cycle.

The current and voltage phasors are at 90° to each other.

[1] We neglect the constant of integration here because it depends on the initial conditions, which are not important for this situation.

has the same units as resistance and is related to current and voltage in the same way as resistance. It must behave in a manner similar to resistance in the sense that it represents opposition to the flow of charge. Because ωL depends on the applied frequency ω, the inductor *reacts* differently, in terms of offering opposition to current, for different frequencies. For this reason, we define ωL as the **inductive reactance** X_L:

$$X_L \equiv \omega L \tag{32.13}$$

◀ Inductive reactance

Therefore, we can write Equation 32.12 as

$$I_{max} = \frac{\Delta V_{max}}{X_L} \tag{32.14}$$

The expression for the rms current in an inductor is similar to Equation 32.14, with I_{max} replaced by I_{rms} and ΔV_{max} replaced by ΔV_{rms}.

Equation 32.13 indicates that, for a given applied voltage, the inductive reactance increases as the frequency increases. This conclusion is consistent with Faraday's law: the greater the rate of change of current in the inductor, the larger the back emf. The larger back emf translates to an increase in the reactance and a decrease in the current.

Using Equations 32.9 and 32.14, we find that the instantaneous voltage across the inductor is

$$\Delta v_L = -L\frac{di_L}{dt} = -\Delta V_{max}\sin\omega t = -I_{max}X_L\sin\omega t \tag{32.15}$$

◀ Voltage across an inductor

Figure 32.8 (Quick Quiz 32.2) At what frequencies does the lightbulb glow the brightest?

QUICK QUIZ 32.2 Consider the AC circuit in Figure 32.8. The frequency of the AC source is adjusted while its voltage amplitude is held constant. When does the lightbulb glow the brightest? **(a)** It glows brightest at high frequencies. **(b)** It glows brightest at low frequencies. **(c)** The brightness is the same at all frequencies.

Example 32.2 **A Purely Inductive AC Circuit**

In a purely inductive AC circuit, $L = 25.0$ mH and the rms voltage is 150 V. Calculate the inductive reactance and rms current in the circuit if the frequency is 60.0 Hz.

SOLUTION

Conceptualize Figure 32.6 shows the physical situation for this problem. Keep in mind that inductive reactance increases with increasing frequency of the applied voltage.

Categorize We determine the reactance and the current from equations developed in this section, so we categorize this example as a substitution problem.

Use Equation 32.13 to find the inductive reactance:

$$X_L = \omega L = 2\pi f L = 2\pi (60.0 \text{ Hz})(25.0 \times 10^{-3} \text{ H})$$

$$= 9.42 \ \Omega$$

Use an rms version of Equation 32.14 to find the rms current:

$$I_{rms} = \frac{\Delta V_{rms}}{X_L} = \frac{150 \text{ V}}{9.42 \ \Omega} = 15.9 \text{ A}$$

WHAT IF? If the frequency increases to 6.00 kHz, what happens to the rms current in the circuit?

Answer If the frequency increases, the inductive reactance also increases because the current is changing at a higher rate. The increase in inductive reactance results in a lower current.

Let's calculate the new inductive reactance and the new rms current:

$$X_L = 2\pi (6.00 \times 10^3 \text{ Hz})(25.0 \times 10^{-3} \text{ H}) = 942 \ \Omega$$

$$I_{rms} = \frac{150 \text{ V}}{942 \ \Omega} = 0.159 \text{ A}$$

$\Delta v = \Delta V_{max} \sin \omega t$

Figure 32.9 A circuit consisting of a capacitor of capacitance C connected to an AC source.

32.4 Capacitors in an AC Circuit

Figure 32.9 shows an AC circuit consisting of a capacitor connected across the terminals of an AC source. Kirchhoff's loop rule applied to this circuit gives $\Delta v + \Delta v_C = 0$, or

$$\Delta v - \frac{q}{C} = 0 \tag{32.16}$$

where q is the charge on the capacitor and the negative sign is due to the fact that the sign of the potential difference across the capacitor is opposite that of the source, as we discussed with regard to Figure 27.15. Substituting $\Delta V_{max} \sin \omega t$ for Δv and rearranging gives

$$q = C \Delta V_{max} \sin \omega t \tag{32.17}$$

where q is the instantaneous charge on the capacitor. Differentiating Equation 32.17 with respect to time gives the instantaneous current in the circuit:

$$i_C = \frac{dq}{dt} = \omega C \Delta V_{max} \cos \omega t \tag{32.18}$$

Using the trigonometric identity $\cos \omega t = \sin(\omega t + \pi/2)$, we can express Equation 32.18 in the alternative form

◀ **Current in a capacitor in an AC circuit**

$$i_C = \omega C \Delta V_{max} \sin\left(\omega t + \frac{\pi}{2}\right) \tag{32.19}$$

Comparing this expression with $\Delta v = \Delta V_{max} \sin \omega t$ shows that the current is $\pi/2$ rad = 90° out of phase with the voltage across the capacitor. A plot of current and voltage versus time for the capacitor appears in Figure 32.10.

Consider a point such as a in Figure 32.10a where the voltage across the capacitor is a maximum. That occurs when the capacitor reaches its maximum charge. At this instant, the current is zero (point d). At points such as e, the current has a maximum magnitude, which occurs at those instants when the charge on the capacitor reaches zero and the capacitor begins to recharge with the opposite polarity. When the charge is zero, the voltage across the capacitor is zero (point b).

As with inductors, we can represent the current and voltage for a capacitor on a phasor diagram. The phasor diagram in Figure 32.10b shows that for a sinusoidally applied voltage, the current always *leads* the voltage across a capacitor by 90°.

Equation 32.18 shows that the current in the circuit reaches its maximum value when $\cos \omega t = \pm 1$:

$$I_{max} = \omega C \Delta V_{max} = \frac{\Delta V_{max}}{(1/\omega C)} \tag{32.20}$$

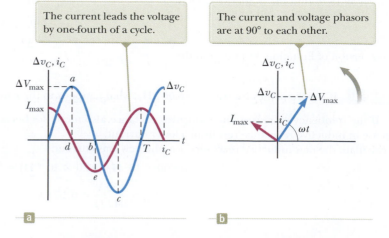

Figure 32.10 (a) Plots of the instantaneous current i_C and instantaneous voltage Δv_C across a capacitor as functions of time. (b) Phasor diagram for the capacitive circuit.

As in the case with inductors, this looks like Equation 26.7, so the denominator plays the role of resistance, with units of ohms. We give the combination $1/\omega C$ the symbol X_C, and because this function varies with frequency, we define it as the **capacitive reactance:**

$$X_C \equiv \frac{1}{\omega C} \qquad (32.21)$$

◀ Capacitive reactance

We can now write Equation 32.20 as

$$I_{max} = \frac{\Delta V_{max}}{X_C} \qquad (32.22)$$

◀ Maximum current in a capacitor

The rms current is given by an expression similar to Equation 32.22, with I_{max} replaced by I_{rms} and ΔV_{max} replaced by ΔV_{rms}.

Using Equation 32.22, we can express the instantaneous voltage across the capacitor as

$$\Delta v_C = \Delta V_{max} \sin \omega t = I_{max} X_C \sin \omega t \qquad (32.23)$$

◀ Voltage across a capacitor

Equations 32.21 and 32.22 indicate that as the frequency of the voltage source increases, the capacitive reactance decreases and the maximum current therefore increases. The frequency of the current is determined by the frequency of the voltage source driving the circuit. As the frequency approaches zero, the capacitive reactance approaches infinity and the current therefore approaches zero. This conclusion makes sense because the circuit approaches direct current conditions as ω approaches zero and the capacitor represents an open circuit.

QUICK QUIZ 32.3 Consider the AC circuit in Figure 32.11. The frequency of the AC source is adjusted while its voltage amplitude is held constant. When does the lightbulb glow the brightest? **(a)** It glows brightest at high frequencies. **(b)** It glows brightest at low frequencies. **(c)** The brightness is the same at all frequencies.

Figure 32.11 (Quick Quiz 32.3)

QUICK QUIZ 32.4 Consider the AC circuit in Figure 32.12. The frequency of the AC source is adjusted while its voltage amplitude is held constant. When does the lightbulb glow the brightest? **(a)** It glows brightest at high frequencies. **(b)** It glows brightest at low frequencies. **(c)** The brightness is the same at all frequencies.

Figure 32.12 (Quick Quiz 32.4)

Example **32.3** **A Purely Capacitive AC Circuit**

An 8.00-μF capacitor is connected to the terminals of a 60.0-Hz AC source whose rms voltage is 150 V. Find the capacitive reactance and the rms current in the circuit.

SOLUTION

Conceptualize Figure 32.9 shows the physical situation for this problem. Keep in mind that capacitive reactance decreases with increasing frequency of the applied voltage.

Categorize We determine the reactance and the current from equations developed in this section, so we categorize this example as a substitution problem.

Use Equation 32.21 to find the capacitive reactance:

$$X_C = \frac{1}{\omega C} = \frac{1}{2\pi f C} = \frac{1}{2\pi (60.0 \text{ Hz})(8.00 \times 10^{-6} \text{ F})} = 332 \ \Omega$$

Use an rms version of Equation 32.22 to find the rms current:

$$I_{rms} = \frac{\Delta V_{rms}}{X_C} = \frac{150 \text{ V}}{332 \ \Omega} = 0.452 \text{ A}$$

continued

32.3 continued

WHAT IF?   What if the frequency is doubled? What happens to the rms current in the circuit?

Answer   If the frequency increases, the capacitive reactance decreases, which is just the opposite from the case of an inductor. The decrease in capacitive reactance results in an increase in the current.

Let's calculate the new capacitive reactance and the new rms current:

$$X_C = \frac{1}{\omega C} = \frac{1}{2\pi(120 \text{ Hz})(8.00 \times 10^{-6} \text{ F})} = 166 \ \Omega$$

$$I_{rms} = \frac{150 \text{ V}}{166 \ \Omega} = 0.904 \text{ A}$$

Figure 32.13 A series circuit consisting of a resistor, an inductor, and a capacitor connected to an AC source.

32.5 The *RLC* Series Circuit

In the previous sections, we considered individual circuit elements connected to an AC source. Figure 32.13 shows a circuit that contains a combination of all three of our circuit elements: a resistor, an inductor, and a capacitor connected in series. We studied this circuit in Section 31.6, where we charged the capacitor and then closed a switch to connect the other two circuit elements. That led to oscillations of the circuit that were damped due to the resistance. We compared the circuit to a mechanical oscillator (Figure 15.1), where we pull a block outward, stretching a spring, and let it go, watching the damped oscillations due to friction. In Section 15.7, we applied a sinusoidal driving force to a mechanical oscillator. Let us look at the electrical analog to this situation here, where we connect an alternating-voltage source across the series connection of circuit elements. If the applied voltage varies sinusoidally with time, the instantaneous applied voltage is

$$\Delta v = \Delta V_{max} \sin \omega t$$

The current in the circuit is given by

$$i = I_{max} \sin(\omega t - \phi)$$

where ϕ is some **phase angle** between the current and the applied voltage. Based on our discussions of phase in Sections 32.3 and 32.4, we expect that the current will generally not be in phase with the voltage in an *RLC* circuit.

Because the circuit elements in Figure 32.13 are in series, the current everywhere in the circuit must be the same at any instant. That is, the current at all points in a series AC circuit has the same amplitude and phase. Based on the preceding sections, we know that the voltage across each element has a different amplitude and phase. In particular, the voltage across the resistor is in phase with the current, the voltage across the inductor leads the current by 90°, and the voltage across the capacitor lags behind the current by 90°. Using these phase relationships, we can express the instantaneous voltages across the three circuit elements as

$$\Delta v_R = I_{max} R \sin \omega t = \Delta V_R \sin \omega t \tag{32.24}$$

$$\Delta v_L = I_{max} X_L \sin\left(\omega t + \frac{\pi}{2}\right) = \Delta V_L \cos \omega t \tag{32.25}$$

$$\Delta v_C = I_{max} X_C \sin\left(\omega t - \frac{\pi}{2}\right) = -\Delta V_C \cos \omega t \tag{32.26}$$

The sum of these three voltages must equal the instantaneous voltage Δv from the AC source, but it is important to recognize that because the three voltages have different phase relationships with the current, they cannot be added directly. Figure 32.14 represents the phasors at an instant at which the current in all three elements is momentarily zero. The zero current is represented by the current phasor along the horizontal axis in each part of the figure. Next the voltage phasor is drawn at the appropriate phase angle to the current for each element.

Because phasors are rotating vectors, the voltage phasors in Figure 32.14 can be combined using vector addition as in Figure 32.15. In Figure 32.15a, the voltage phasors in Figure 32.14 are combined on the same coordinate axes. Figure 32.15b shows the vector addition of the voltage phasors. The voltage phasors ΔV_L and ΔV_C are in *opposite* directions along the same line, so we can construct the difference phasor $\Delta V_L - \Delta V_C$, which is perpendicular to the phasor ΔV_R. This diagram shows that the vector sum of the voltage amplitudes ΔV_R, ΔV_L, and ΔV_C equals a phasor whose length is the maximum applied voltage ΔV_{max} and that makes an angle ϕ with the current phasor I_{max}. From the right triangle in Figure 32.15b, we see that

$$\Delta V_{max} = \sqrt{\Delta V_R^2 + (\Delta V_L - \Delta V_C)^2} = \sqrt{(I_{max}R)^2 + (I_{max}X_L - I_{max}X_C)^2}$$

$$\Delta V_{max} = I_{max}\sqrt{R^2 + (X_L - X_C)^2}$$

Therefore, we can express the maximum current as

$$I_{max} = \frac{\Delta V_{max}}{\sqrt{R^2 + (X_L - X_C)^2}} \qquad (32.27)$$

Once again, this expression has the same mathematical form as Equation 26.7. The denominator of the fraction plays the role of resistance and is called the **impedance** Z of the circuit:

$$Z \equiv \sqrt{R^2 + (X_L - X_C)^2} \qquad (32.28)$$

where impedance also has units of ohms. Therefore, Equation 32.27 can be written in the form

$$I_{max} = \frac{\Delta V_{max}}{Z} \qquad (32.29)$$

Equation 32.29 is the AC equivalent of Equation 26.7. Note that the impedance and therefore the current in an AC circuit depend on the resistance, the inductance, the capacitance, and the frequency (because the reactances are frequency dependent).

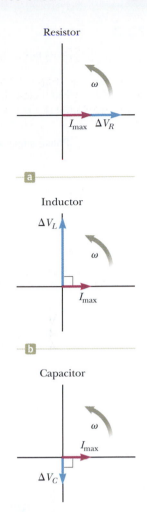

Figure 32.14 Phase relationships between the voltage and current phasors for (a) a resistor, (b) an inductor, and (c) a capacitor connected in series.

The phasors of Figure 32.14 are combined on a single set of axes.

The total voltage ΔV_{max} makes an angle ϕ with I_{max}.

Figure 32.15 (a) The phasors in Figure 32.14 for the elements in a series *RLC* circuit are combined on a single diagram. (b) The inductance and capacitance phasors are added together and then added vectorially to the resistance phasor.

From the right triangle in the phasor diagram in Figure 32.15b, the phase angle ϕ between the current and the voltage is found as follows:

$$\phi = \tan^{-1}\left(\frac{\Delta V_L - \Delta V_C}{\Delta V_R}\right) = \tan^{-1}\left(\frac{I_{max}X_L - I_{max}X_C}{I_{max}R}\right)$$

Phase angle ▶

$$\phi = \tan^{-1}\left(\frac{X_L - X_C}{R}\right) \tag{32.30}$$

When $X_L > X_C$ (which occurs at high frequencies), the phase angle is positive, signifying that the current lags the applied voltage as in Figure 32.15b. We describe this situation by saying that the circuit is *more inductive than capacitive*. When $X_L < X_C$, the phase angle is negative, signifying that the current leads the applied voltage, and the circuit is *more capacitive than inductive*. When $X_L = X_C$, the phase angle is zero and the circuit is *purely resistive*.

QUICK QUIZ 32.5 Label each part of Figure 32.16, (a), (b), and (c), as representing $X_L > X_C$, $X_L = X_C$, or $X_L < X_C$.

Figure 32.16 (Quick Quiz 32.5) Match the phasor diagrams to the relationships between the reactances.

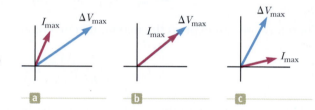

Example 32.4 **Analyzing a Series *RLC* Circuit**

A series *RLC* circuit has $R = 425\ \Omega$, $L = 1.25$ H, and $C = 3.50\ \mu$F. It is connected to an AC source with $f = 60.0$ Hz and $\Delta V_{max} = 150$ V.

(A) Determine the inductive reactance, the capacitive reactance, and the impedance of the circuit.

SOLUTION

Conceptualize   The circuit of interest in this example is shown in Figure 32.13. The current in the combination of the resistor, inductor, and capacitor oscillates at a particular phase angle with respect to the applied voltage.

Categorize   The circuit is a simple series *RLC* circuit, so we can use the approach discussed in this section.

Analyze   Find the angular frequency:

$$\omega = 2\pi f = 2\pi(60.0\text{ Hz}) = 377\text{ s}^{-1}$$

Use Equation 32.13 to find the inductive reactance:

$$X_L = \omega L = (377\text{ s}^{-1})(1.25\text{ H}) = \boxed{471\ \Omega}$$

Use Equation 32.21 to find the capacitive reactance:

$$X_C = \frac{1}{\omega C} = \frac{1}{(377\text{ s}^{-1})(3.50 \times 10^{-6}\text{ F})} = \boxed{758\ \Omega}$$

Use Equation 32.28 to find the impedance:

$$Z = \sqrt{R^2 + (X_L - X_C)^2}$$

$$= \sqrt{(425\ \Omega)^2 + (471\ \Omega - 758\ \Omega)^2} = \boxed{513\ \Omega}$$

(B) Find the maximum current in the circuit.

SOLUTION

Use Equation 32.29 to find the maximum current:

$$I_{max} = \frac{\Delta V_{max}}{Z} = \frac{150\text{ V}}{513\ \Omega} = \boxed{0.293\text{ A}}$$

32.4 continued

(C) Find the phase angle between the current and voltage.

SOLUTION

Use Equation 32.30 to calculate the phase angle:

$$\phi = \tan^{-1}\left(\frac{X_L - X_C}{R}\right) = \tan^{-1}\left(\frac{471\ \Omega - 758\ \Omega}{425\ \Omega}\right) = -34.0°$$

(D) Find the maximum voltage across each element.

SOLUTION

Use Equations 32.4, 32.14, and 32.22 to calculate the maximum voltages:

$$\Delta V_R = I_{max}R = (0.293\ \text{A})(425\ \Omega) = \boxed{124\ \text{V}}$$

$$\Delta V_L = I_{max}X_L = (0.293\ \text{A})(471\ \Omega) = \boxed{138\ \text{V}}$$

$$\Delta V_C = I_{max}X_C = (0.293\ \text{A})(758\ \Omega) = \boxed{222\ \text{V}}$$

(E) What replacement value of L should an engineer analyzing the circuit choose such that the current leads the applied voltage by 30.0° rather than 34.0°? All other values in the circuit stay the same.

SOLUTION

Solve Equation 32.30 for the inductive reactance:

$$X_L = X_C + R \tan \phi$$

Substitute Equations 32.13 and 32.21 into this expression:

$$\omega L = \frac{1}{\omega C} + R \tan \phi$$

Solve for L:

$$L = \frac{1}{\omega}\left(\frac{1}{\omega C} + R \tan \phi\right)$$

Substitute the given values:

$$L = \frac{1}{(377\ \text{s}^{-1})}\left[\frac{1}{(377\ \text{s}^{-1})(3.50 \times 10^{-6}\ \text{F})} + (425\ \Omega)\tan(-30.0°)\right]$$

$$L = \boxed{1.36\ \text{H}}$$

Finalize Because the capacitive reactance is larger than the inductive reactance, the circuit is more capacitive than inductive. In this case, the phase angle ϕ is negative, so the current leads the applied voltage.

Using Equations 32.24, 32.25, and 32.26, the instantaneous voltages across the three elements are

$$\Delta v_R = (124\ \text{V})\sin 377t$$

$$\Delta v_L = (138\ \text{V})\cos 377t$$

$$\Delta v_C = (-222\ \text{V})\cos 377t$$

WHAT IF? What if you added up the maximum voltages across the three circuit elements? Is that a physically meaningful quantity?

Answer The sum of the maximum voltages across the elements is $\Delta V_R + \Delta V_L + \Delta V_C = 484\ \text{V}$. This sum is much greater than the maximum voltage of the source, 150 V. The sum of the maximum voltages is a meaningless quantity because when sinusoidally varying quantities are added, *both their amplitudes and their phases* must be taken into account. The maximum voltages across the various elements occur at different times. Therefore, the voltages must be added in a way that takes account of the different phases as shown in Figure 32.15.

32.6 Power in an AC Circuit

Now let's take an energy approach to analyzing AC circuits and consider the transfer of energy from the AC source to the circuit. The power delivered by a battery to an external DC circuit is equal to the product of the current and the terminal voltage of the battery. Likewise, the instantaneous power delivered by an AC source to a circuit is the product of the current and the applied voltage. For the *RLC* circuit shown in Figure 32.13, we can express the instantaneous power p as

$$p = i\,\Delta v = [I_{max}\sin(\omega t - \phi)][\Delta V_{max}\sin\omega t]$$

$$p = I_{max}\,\Delta V_{max}\sin\omega t\sin(\omega t - \phi) \tag{32.31}$$

This result is a complicated function of time and is therefore not very useful from a practical viewpoint. What is generally of interest is the average power over one or more cycles. Such an average can be computed by first using the trigonometric identity $\sin(\omega t - \phi) = \sin\omega t\cos\phi - \cos\omega t\sin\phi$. Substituting this identity into Equation 32.31 gives

$$p = I_{max}\,\Delta V_{max}\sin^2\omega t\cos\phi - I_{max}\,\Delta V_{max}\sin\omega t\cos\omega t\sin\phi \tag{32.32}$$

Let's now take the time average of p over one or more cycles, noting that I_{max}, ΔV_{max}, ϕ, and ω are all constants. The time average of the first term on the right of the equal sign in Equation 32.32 involves the average value of $\sin^2\omega t$, which is $\frac{1}{2}$. The time average of the second term on the right of the equal sign is identically zero because $\sin\omega t\cos\omega t = \frac{1}{2}\sin 2\omega t$, and the average value of $\sin 2\omega t$ is zero. Therefore, we can express the **average power** P_{avg} as

$$P_{avg} = \tfrac{1}{2}I_{max}\,\Delta V_{max}\cos\phi \tag{32.33}$$

It is convenient to express the average power in terms of the rms current and rms voltage defined by Equations 32.6 and 32.8:

◀ Average power delivered to an *RLC* circuit

$$P_{avg} = I_{rms}\,\Delta V_{rms}\cos\phi \tag{32.34}$$

where the quantity $\cos\phi$ is called the **power factor.** Figure 32.15b shows that the maximum voltage across the resistor is given by $\Delta V_R = \Delta V_{max}\cos\phi = I_{max}R$. Therefore,

$$\cos\phi = I_{max}\frac{R}{\Delta V_{max}} = \frac{R}{Z} \tag{32.35}$$

and we can express P_{avg} as

$$P_{avg} = I_{rms}\,\Delta V_{rms}\cos\phi = I_{rms}\,\Delta V_{rms}\left(\frac{R}{Z}\right) = I_{rms}\left(\frac{\Delta V_{rms}}{Z}\right)R$$

Recognizing that $\Delta V_{rms}/Z = I_{rms}$ gives

$$P_{avg} = I_{rms}^2 R \tag{32.36}$$

The average power delivered by the source is converted to internal energy in the resistor, just as in the case of a DC circuit. When the load is purely resistive, $\phi = 0$, $\cos\phi = 1$, and, from Equation 32.34, we see that

$$P_{avg} = I_{rms}\,\Delta V_{rms}$$

Note that no power losses are associated with ideal capacitors and ideal inductors in an AC circuit. Energy is temporarily stored as U_E in a capacitor and U_B in an inductor, but no energy is transformed to E_{int} in these circuit elements.

Equation 32.34 shows that the power delivered by an AC source to any circuit depends on the phase, a result that has many interesting applications. For example, a factory that uses large motors in machines, generators, or transformers has a large inductive load (because of all the windings). To deliver greater power to such devices in the factory without using excessively high voltages, technicians introduce capacitance in the circuits to shift the phase.

○**UICK QUIZ 32.6** An AC source drives an *RLC* circuit with a fixed voltage amplitude. If the driving frequency is ω_1, the circuit is more capacitive than inductive and the phase angle is $-10°$. If the driving frequency is ω_2, the circuit is more inductive than capacitive and the phase angle is $+10°$. At what frequency is the largest amount of power delivered to the circuit? **(a)** It is largest at ω_1. **(b)** It is largest at ω_2. **(c)** The same amount of power is delivered at both frequencies.

Example **32.5** Average Power in an *RLC* Series Circuit

Calculate the average power delivered to the series *RLC* circuit described in Example 32.4.

SOLUTION

Conceptualize Consider the circuit in Figure 32.13 and imagine energy being delivered to the circuit by the AC source. Review Example 32.4 for other details about this circuit.

Categorize We find the result by using equations developed in this section, so we categorize this example as a substitution problem.

Use Equation 32.8 and the maximum voltage from Example 32.4 to find the rms voltage from the source:

$$\Delta V_{rms} = \frac{\Delta V_{max}}{\sqrt{2}} = \frac{150 \text{ V}}{\sqrt{2}} = 106 \text{ V}$$

Similarly, find the rms current in the circuit:

$$I_{rms} = \frac{I_{max}}{\sqrt{2}} = \frac{0.293 \text{ A}}{\sqrt{2}} = 0.207 \text{ A}$$

Use Equation 32.34 to find the power delivered by the source:

$$P_{avg} = I_{rms} V_{rms} \cos \phi = (0.207 \text{ A})(106 \text{ V}) \cos (-34.0°)$$

$$= 18.2 \text{ W}$$

32.7 Resonance in a Series *RLC* Circuit

We investigated mechanical oscillating systems in Chapter 15. In Section 15.7, we considered the situation in which an oscillating system is driven by an external force that varies sinusoidally in time. This led to the concept of **resonance,** where the system exhibits its maximum response when driven at its natural frequency. As shown in Section 31.6, a series *RLC* circuit is an electrical oscillating system with a natural frequency. Imagine now driving the circuit with a sinusoidal voltage like that in Equation 32.1. We expect a resonance phenomenon. Such a circuit is said to be in resonance when the driving frequency is such that the rms current has its maximum value. In general, the rms current in the circuit can be written

$$I_{rms} = \frac{\Delta V_{rms}}{Z} \tag{32.37}$$

where Z is the impedance of the circuit. Substituting the expression for Z from Equation 32.28 into Equation 32.37 gives

$$I_{rms} = \frac{\Delta V_{rms}}{\sqrt{R^2 + (X_L - X_C)^2}} \tag{32.38}$$

Because the impedance depends on the frequency of the source, the current in the *RLC* circuit also depends on the frequency. The angular frequency ω_0 at which $X_L - X_C = 0$ is called the **resonance frequency** of the circuit. At this frequency, the circuit will exhibit its maximum response: the rms current in Equation 32.38 will have its largest value. To find ω_0, we set $X_L = X_C$, which gives $\omega_0 L = 1/\omega_0 C$, or

$$\omega_0 = \frac{1}{\sqrt{LC}} \tag{32.39}$$ ◀ Resonance frequency

This frequency matches the natural frequency of oscillation of an *LC* circuit (see Section 31.5). Therefore, as we expect, the rms current in a series *RLC* circuit has its maximum value when the frequency of the applied voltage matches the natural oscillator frequency, which depends only on *L* and *C*. Furthermore, at the resonance frequency, the current is in phase with the applied voltage.

QUICK QUIZ **32.7** What is the impedance of a series *RLC* circuit at resonance? **(a)** larger than *R* **(b)** less than *R* **(c)** equal to *R* **(d)** impossible to determine

A plot of rms current versus angular frequency for a series *RLC* circuit is shown in Figure 32.17a. The data assume a constant $\Delta V_{rms} = 5.0$ mV, $L = 5.0\ \mu$H, and $C = 2.0$ nF. The three curves correspond to three values of *R*. In each case, the rms current has its maximum value at the resonance frequency ω_0. Furthermore, the curves become narrower and taller as the resistance decreases.

Equation 32.38 shows that when $R = 0$, the current becomes infinite at resonance. Real circuits, however, always have some resistance, which limits the value of the current to some finite value. The analog in a mechanical oscillating system is that the amplitude of the oscillation cannot become infinite, because there is always some friction in the system.

We can also calculate the average power as a function of frequency for a series *RLC* circuit. Using Equations 32.36, 32.37, and 32.28 gives

$$P_{avg} = I_{rms}^2\, R = \frac{(\Delta V_{rms})^2}{Z^2}\, R = \frac{(\Delta V_{rms})^2 R}{R^2 + (X_L - X_C)^2} \tag{32.40}$$

Because $X_L = \omega L$, $X_C = 1/\omega C$, and $\omega_0^2 = 1/LC$, the term $(X_L - X_C)^2$ can be expressed as

$$(X_L - X_C)^2 = \left(\omega L - \frac{1}{\omega C}\right)^2 = \frac{L^2}{\omega^2}(\omega^2 - \omega_0^2)^2$$

Using this result in Equation 32.40 gives

◀ Average power as a function of frequency in an *RLC* circuit

$$P_{avg} = \frac{(\Delta V_{rms})^2\, R\omega^2}{R^2\,\omega^2 + L^2(\omega^2 - \omega_0^2)^2} \tag{32.41}$$

Equation 32.41 shows that at resonance, when $\omega = \omega_0$, the average power is a maximum and has the value $(\Delta V_{rms})^2/R$. Figure 32.17b is a plot of average power versus

Figure 32.17 (a) The rms current versus frequency for a series *RLC* circuit for three values of *R*. (b) Average power delivered to the circuit versus frequency for the series *RLC* circuit for three values of *R*.

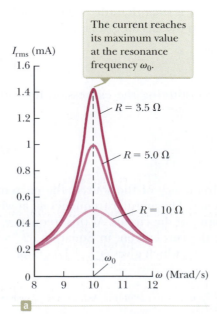

The current reaches its maximum value at the resonance frequency ω_0.

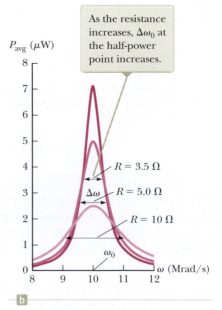

As the resistance increases, $\Delta\omega_0$ at the half-power point increases.

frequency for three values of R in a series RLC circuit. As the resistance is made smaller, the curve becomes sharper in the vicinity of the resonance frequency. This curve sharpness is usually described by a dimensionless parameter known as the **quality factor,**[2] denoted by Q:

$$Q = \frac{\omega_0}{\Delta\omega} \qquad (32.42)$$

◀ Quality factor

where $\Delta\omega$ is the width of the curve measured between the two values of ω for which P_{avg} has one-half its maximum value, called the *half-power points* (see Fig. 32.17b). It is left as a problem (Problem 48) to show that the width at the half-power points has the value $\Delta\omega = R/L$ so that

$$Q = \frac{\omega_0 L}{R} \qquad (32.43)$$

The receiving circuit of a radio is an important application of a resonance RLC circuit. The driving voltages on the circuit come from a large number of radio signals from nearby transmitting stations causing electromagnetic oscillations in the antenna of the radio. Despite being driven by many voltages simultaneously from different radio stations, the circuit will respond only to one: the one whose frequency matches that of the resonance frequency of the radio. The resonance frequency can be varied by adjusting the capacitance of the circuit, which you do when you turn the tuning knob. The one signal to which the circuit responds is then passed on to the amplifier and loudspeakers. Because many signals over a range of frequencies drive the tuning circuit, it is important to design a high-Q circuit to eliminate unwanted signals. In this manner, stations whose frequencies are near but not equal to the resonance frequency have a response at the receiver that is negligibly small relative to the signal that matches the resonance frequency.

Example 32.6 A Resonating Series *RLC* Circuit

Consider a series RLC circuit for which $R = 150\ \Omega$, $L = 20.0$ mH, $\Delta V_{rms} = 20.0$ V, and $\omega = 5\,000$ s^{-1}. Determine the value of the capacitance for which the current is a maximum.

SOLUTION

Conceptualize In this problem, the driving frequency is fixed. We wish to design the circuit in Figure 32.13 so that its resonance frequency matches the driving frequency.

Categorize We find the result by using equations developed in this section, so we categorize this example as a substitution problem.

Use Equation 32.39 to solve for the required capacitance in terms of the resonance frequency:

$$\omega_0 = \frac{1}{\sqrt{LC}} \quad \rightarrow \quad C = \frac{1}{\omega_0^2 L}$$

Note that we only need two of the quantities provided. Substitute numerical values:

$$C = \frac{1}{(5.00 \times 10^3\ \text{s}^{-1})^2 (20.0 \times 10^{-3}\ \text{H})} = 2.00\ \mu\text{F}$$

32.8 The Transformer and Power Transmission

As discussed in Section 26.6, it is economical to use a high voltage and a low current to minimize the I^2R loss in transmission lines when electric power is transmitted over great distances. Consequently, 350-kV lines are common, and in many areas, even higher-voltage (765-kV) lines are used. At the receiving end of such lines, the

---

[2]The quality factor is also defined as the ratio $2\pi E/\Delta E$, where E is the energy stored in the oscillating system and ΔE is the energy decrease per cycle of oscillation due to the resistance.

© Cengage

Figure 32.18 The transformer on this power pole steps down AC voltage from 4 000 V to 240 V for distribution to individual residences.

An alternating voltage Δv_1 is applied to the primary coil, and the output voltage Δv_2 is across the resistor of resistance R_L.

Soft iron

Δv_1 N_1 N_2 Δv_2 R_L

Primary (input)   Secondary (output)

Figure 32.19 An ideal transformer consists of two coils wound on the same iron core.

I_1 I_2

Δv_1 R_L Δv_2

N_1 N_2

Figure 32.20 Circuit diagram for a transformer.

consumer requires power at a low voltage (for safety and for efficiency in design). In practice, the voltage is decreased to approximately 20 000 V at a distribution substation, then to 4 000 V for delivery to residential areas, and finally to 120 V and 240 V at the customer's site. Therefore, a device is needed that can change the alternating voltage and current without causing appreciable changes in the power delivered. The AC transformer is that device. Figure 32.18 shows a typical transformer in a residential area.

In its simplest form, the **AC transformer** consists of two coils of wire wound around a core of iron as illustrated in Figure 32.19. (Compare this arrangement to Faraday's experiment in Figure 30.2.) The coil on the left, which is connected to the input alternating-voltage source and has N_1 turns, is called the *primary winding* (or the *primary*). The coil on the right, consisting of N_2 turns and connected to a load resistor R_L, is called the *secondary winding* (or the *secondary*). The purposes of the iron core are to increase the magnetic flux through the coil and to provide a medium in which nearly all the magnetic field lines through one coil pass through the other coil. In this way, the iron core increases the mutual induction of the coils.

Faraday's law (Eq. 30.1) gives the relationship between the voltage Δv_1 across the primary and the flux Φ_B through each turn of the primary:

$$\Delta v_1 = -N_1 \frac{d\Phi_B}{dt} \tag{32.44}$$

If we assume all magnetic field lines remain within the iron core, the flux through each turn of the primary equals the flux through each turn of the secondary. Hence, the voltage across the secondary is

$$\Delta v_2 = -N_2 \frac{d\Phi_B}{dt} \tag{32.45}$$

Solving Equation 32.44 for $d\Phi_B/dt$ and substituting the result into Equation 32.45 gives

$$\Delta v_2 = \frac{N_2}{N_1} \Delta v_1 \tag{32.46}$$

When $N_2 > N_1$, the output voltage Δv_2 exceeds the input voltage Δv_1. This configuration is referred to as a *step-up transformer*. When $N_2 < N_1$, the output voltage is less than the input voltage, and we have a *step-down transformer*. A circuit diagram for a transformer connected to a load resistance is shown in Figure 32.20.

When a current I_1 exists in the primary circuit, a current I_2 is induced in the secondary. (In this discussion, uppercase I and ΔV refer to rms values.) If the load in the secondary circuit is a pure resistance, the induced current is in phase with the induced voltage. The power supplied to the secondary circuit must be provided by the AC source connected to the primary circuit. In an ideal transformer where there are no losses, the power $I_1 \Delta V_1$ supplied by the source is equal to the power $I_2 \Delta V_2$ in the secondary circuit. That is,

$$I_1 \Delta V_1 = I_2 \Delta V_2 \tag{32.47}$$

The value of the load resistance R_L determines the value of the secondary current because $I_2 = \Delta V_2 / R_L$. Furthermore, the current in the primary is $I_1 = \Delta V_1 / R_{eq}$, where

$$R_{eq} = \left(\frac{N_1}{N_2} \right)^2 R_L \tag{32.48}$$

is the equivalent resistance of the load resistance when viewed from the primary side. We see from this analysis that a transformer may be used to match resistances between the primary circuit and the load. In this manner, maximum power transfer can be achieved between a given power source and the load resistance.

The primary winding in this transformer is attached to the prongs of the plug, whereas the secondary winding is connected to the power cord on the right.

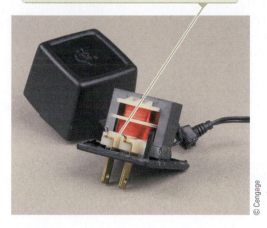

© Cengage

Figure 32.21 Electronic devices are often powered by AC adaptors containing transformers such as this one. These adaptors alter the AC voltage. In many applications, the adaptors also convert alternating current to direct current.

Bettmann/Getty Images

Nikola Tesla
American Physicist (1856–1943)
Tesla was born in Croatia, but he spent most of his professional life as an inventor in the United States. He was a key figure in the development of alternating-current electricity, high-voltage transformers, and the transport of electrical power using AC transmission lines. Tesla's viewpoint was at odds with the ideas of Thomas Edison, who committed himself to the use of direct current in power transmission. Tesla's AC approach won out.

For example, a transformer connected between the 1-kΩ output of an audio amplifier and an 8-Ω speaker ensures that as much of the audio signal as possible is transferred into the speaker. In stereo terminology, this process is called *impedance matching*.

In the discussion above, we have assumed an ideal transformer in which the energy losses in the windings and core are zero. In reality, there are possibilities for energy loss, and there are ways of minimizing them. Eddy currents can be generated in the iron core, leading to an increase in internal energy due to the resistance of the iron. These losses can be reduced by using a laminated core as discussed in Section 30.6. Transformation of energy to internal energy in the finite resistance of the coil wires is usually quite small. Typical transformers have power efficiencies from 90% to 99%.

To operate properly, many common household electronic devices require low voltages. A small transformer that plugs directly into the wall like the one illustrated in Figure 32.21 can provide the proper voltage. The photograph shows the two windings wrapped around a common iron core that is found inside all these little "black boxes." This particular transformer converts the 120-V AC in the wall socket to 12.5-V AC. (Can you determine the ratio of the numbers of turns in the two coils?)

Example 32.7 | The Economics of AC Power

An electricity-generating station needs to deliver energy at a rate of 20 MW to a city 1.0 km away. A common voltage for commercial power generators is 22 kV, but a step-up transformer is used to boost the voltage to 230 kV before transmission.

(A) If the resistance of the wires is 2.0 Ω and the energy costs are about 11¢/kWh, estimate the cost of the energy converted to internal energy in the wires during one day.

SOLUTION

Conceptualize The resistance of the wires is in series with the resistance representing the load (homes and businesses). Therefore, there is a voltage drop in the wires, which means that some of the transmitted energy is converted to internal energy in the wires and never reaches the load.

Categorize This problem involves finding the power delivered to a resistive load in an AC circuit. Let's ignore any capacitive or inductive characteristics of the load and set the power factor equal to 1.

continued

32.7 continued

Analyze Calculate the energy T_{ET} delivered to the wires over a time interval Δt:

$$T_{ET} = P_{wires}\,\Delta t$$

Use Equation 32.36 to evaluate the power delivered to the wires:

$$T_{ET} = I_{rms}^2 R_{wires}\,\Delta t$$

Use Equation 32.36 to evaluate the rms current:

$$(1)\quad T_{ET} = \frac{P_{avg}^2}{\Delta V_{rms}^2} R_{wires}\,\Delta t$$

Substitute numerical values:

$$T_{ET} = \frac{(20\times10^6\,\text{W})^2}{(230\times10^3\,\text{V})^2}(2.0\,\Omega)(24\,\text{h}) = 3.6\times10^5\,\text{Wh} = 360\,\text{kWh}$$

Find the cost of this energy at a rate of 11¢/kWh:

$$\text{Cost} = (360\,\text{kWh})(\$0.11/\text{kWh}) = \boxed{\$40}$$

(B) Repeat the calculation for the hypothetical situation in which the power plant delivers the energy at its original voltage of 22 kV.

SOLUTION

Use the new voltage in Equation (1):

$$T_{ET} = \frac{(20\times10^6\,\text{W})^2}{(22\times10^3\,\text{V})^2}(2.0\,\Omega)(24\,\text{h}) = 4.0\times10^7\,\text{Wh} = 4.0\times10^4\,\text{kWh}$$

Find the cost of this energy at a rate of 11¢/kWh:

$$\text{Cost} = (4.0\times10^4\,\text{kWh})(\$0.11/\text{kWh}) = \boxed{\$4.4\times10^3}$$

Finalize Notice the tremendous savings that are possible through the use of transformers and high-voltage transmission lines. Such savings in combination with the efficiency of using alternating current to operate motors led to the universal adoption of alternating current instead of direct current for commercial power grids.

Summary

▶ Definitions

In AC circuits that contain inductors and capacitors, it is useful to define the **inductive reactance** X_L and the **capacitive reactance** X_C as

$$X_L \equiv \omega L \tag{32.13}$$

$$X_C \equiv \frac{1}{\omega C} \tag{32.21}$$

where ω is the angular frequency of the AC source. The SI unit of reactance is the ohm.

The **impedance** Z of an RLC series AC circuit is

$$Z \equiv \sqrt{R^2 + (X_L - X_C)^2} \tag{32.28}$$

This expression illustrates that we cannot simply add the resistance and reactances in a circuit. We must account for the applied voltage and current being out of phase, with the **phase angle** ϕ between the current and voltage being

$$\phi = \tan^{-1}\left(\frac{X_L - X_C}{R}\right) \tag{32.30}$$

The sign of ϕ can be positive or negative, depending on whether X_L is greater or less than X_C. The phase angle is zero when $X_L = X_C$.

▶ Concepts and Principles

The **rms current** and **rms voltage** in an AC circuit in which the voltages and current vary sinusoidally are given by

$$I_{rms} = \frac{I_{max}}{\sqrt{2}} = 0.707 I_{max} \tag{32.6}$$

$$\Delta V_{rms} = \frac{\Delta V_{max}}{\sqrt{2}} = 0.707\,\Delta V_{max} \tag{32.8}$$

where I_{max} and ΔV_{max} are the maximum values.

If an AC circuit consists of a source and a resistor, the current is in phase with the voltage. That is, the current and voltage reach their maximum values at the same time.

If an AC circuit consists of a source and an inductor, the current lags the voltage by 90°. That is, the voltage reaches its maximum value one-quarter of a period before the current reaches its maximum value.

If an AC circuit consists of a source and a capacitor, the current leads the voltage by 90°. That is, the current reaches its maximum value one-quarter of a period before the voltage reaches its maximum value.

The rms current in a series *RLC* circuit is

$$I_{rms} = \frac{\Delta V_{rms}}{\sqrt{R^2 + (X_L - X_C)^2}} \qquad (32.38)$$

A series *RLC* circuit is in resonance when the inductive reactance equals the capacitive reactance. When this condition is met, the rms current given by Equation 32.38 has its maximum value. The **resonance frequency** ω_0 of the circuit is

$$\omega_0 = \frac{1}{\sqrt{LC}} \qquad (32.39)$$

The rms current in a series *RLC* circuit has its maximum value when the frequency of the source equals ω_0, that is, when the "driving" frequency matches the resonance frequency.

The **average power** delivered by the source in an *RLC* circuit is

$$P_{avg} = I_{rms} \Delta V_{rms} \cos \phi \qquad (32.34)$$

An equivalent expression for the average power is

$$P_{avg} = I_{rms}^2 R \qquad (32.36)$$

The average power delivered by the source results in increasing internal energy in the resistor. No power loss occurs in an ideal inductor or capacitor.

AC transformers allow for easy changes in alternating voltage according to

$$\Delta v_2 = \frac{N_2}{N_1} \Delta v_1 \qquad (32.46)$$

where N_1 and N_2 are the numbers of windings on the primary and secondary coils, respectively, and Δv_1 and Δv_2 are the voltages on these coils.

Think–Pair–Share

See the Preface for an explanation of the icons used in this problems set.
For additional assessment items for this section, go to **WEBASSIGN** From Cengage

1. Figure TP32.1 shows an *RLC* circuit with three possibilities for measuring an output voltage Δv_{out}. The curves

Figure TP32.1

in Figure 32.17a show the rms current measured in the resistance as the source frequency is varied. The output voltage $\Delta V_R = I_{rms}R$ will exhibit a similar behavior as the frequency varies. The output voltages across the inductor and the capacitor will show different behavior. (a) Show that the magnitudes of the output voltages across the capacitor and the inductor have the same value at resonance. (b) Show that the magnitude of the output voltage across the resistor at resonance can be either smaller or larger than those across the inductor and capacitor, and find the value of *R* at which all three output voltages are the same at resonance. (c) The magnitudes of the output voltages do not all reach the maximum value at the same frequency. Show that, for the resistance from part (b), the frequency at which the magnitude of the inductor voltage maximizes is twice that at which the capacitor voltage maximizes. *Suggestion:* For part (c), split your group in two and have one group work on the inductor and the other one work on the capacitor.

2. **ACTIVITY** In Section 27.4, we studied *RC* circuits, with a DC voltage applied. We learned about both charging and discharging the capacitor in that discussion. Let's now apply an AC voltage to an *RC* circuit. Figure TP32.2 (page 868) shows an *RC* circuit with two possibilities for measuring an output voltage Δv_{out}. Discuss the following questions in your group. (a) For which possibility will the output voltage be small at lower frequencies and equal to the input voltage at high frequencies? (b) For which possibility will the output voltage be small at high frequencies and equal to the input voltage at low frequencies?

Figure TP32.2

3. In Sections 32.5 and 32.7, we investigated the *series RLC* circuit. Figure TP32.3a shows a *parallel RLC* circuit. Discuss in your group how this circuit would be different from the series circuit. As with any parallel circuit, the instantaneous voltages (and rms voltages) across each of the three circuit elements are the same. Furthermore, each voltage is in phase with the current in the resistor. The currents in C

Figure TP32.3

and L lead or lag the current in the resistor as shown in the current phasor diagram, Figure TP32.3b. Work with your group to perform the following for this circuit: (a) Show that the rms current delivered by the source is

$$I_{rms} = \Delta V_{rms}\left[\frac{1}{R^2} + \left(\omega C - \frac{1}{\omega L}\right)^2\right]^{1/2}$$

(b) Show that the phase angle ϕ between ΔV_{rms} and I_{rms} is given by

$$\tan\phi = R\left(\omega C - \frac{1}{\omega L}\right)$$

(c) Show that the current delivered by the source reaches a *minimum* when the circuit is driven at its resonance frequency. (d) Find an expression for the impedance of the parallel *RLC* circuit.

4. **ACTIVITY** As a group, consider and discuss the circuit shown in Figure 32.6, with an inductor and an AC source. The inductor has an inductance of 3.50 mH. Suppose that the AC source does *not* provide a sinusoidal voltage, although the voltage is still periodic. The result of the AC source is that the current in the circuit can be expressed during the first second after the AC source is turned on as

$$i(t) = \begin{cases} (6.00 \times 10^{-3})t & 0 < t < 0.600 \text{ s} \\ 9.00 \times 10^{-3} - (9.00 \times 10^{-3})t & 0.600 \text{ s} < t < 1.00 \text{ s} \end{cases}$$

where the current i is in amperes and the time t is in seconds. This pattern in the current repeats for every subsequent second of time.
 (a) Draw a graph of the current in the circuit as a function of time. (b) Draw a graph of the voltage across the inductor as a function of time.

Problems

See the Preface for an explanation of the icons used in this problems set. For additional assessment items for this section, go to ⚡ **WEBASSIGN** From Cengage

SECTION 32.2 Resistors in an AC Circuit

1. (a) What is the resistance of a lightbulb that uses an average power of 75.0 W when connected to a 60.0-Hz power source having a maximum voltage of 170 V? (b) **What If?** What is the resistance of a 100-W lightbulb?

2. **QC** A certain lightbulb is rated at 60.0 W when operating at an rms voltage of 120 V. (a) What is the peak voltage applied across the bulb? (b) What is the resistance of the bulb? (c) Does a 100-W bulb have greater or less resistance than a 60.0-W bulb? Explain.

3. The current in the circuit shown in Figure P32.3 equals 60.0% of the peak current at $t = 7.00$ ms. What is the lowest source frequency that gives this current?

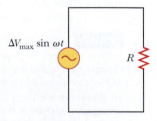

$\Delta V_{max} \sin \omega t$ R

Figure P32.3
Problems 3 and 5.

4. Figure P32.4 shows three lightbulbs connected to a 120-V AC (rms) household supply voltage. Bulbs 1 and 2 have a

120 V   1   2   3

Figure P32.4

power rating of 150 W, and bulb 3 has a 100-W rating. Find (a) the rms current in each bulb and (b) the resistance of each bulb. (c) What is the total resistance of the combination of the three lightbulbs?

5. In the AC circuit shown in Figure P32.3, $R = 70.0\ \Omega$ and the output voltage of the AC source is $\Delta V_{max} \sin \omega t$. (a) If $\Delta V_R = 0.250\ \Delta V_{max}$ for the first time at $t = 0.010\ 0$ s, what is the angular frequency of the source? (b) What is the next value of t for which $\Delta V_R = 0.250\ \Delta V_{max}$?

SECTION 32.3 Inductors in an AC Circuit

6. **T** In a purely inductive AC circuit as shown in Figure P32.6, $\Delta V_{max} = 100$ V. (a) The maximum current is 7.50 A at 50.0 Hz. Calculate the inductance L. (b) **What If?** At what angular frequency ω is the maximum current 2.50 A?

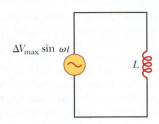

Figure P32.6 Problems 6 and 7.

7. For the circuit shown in Figure P32.6, ΔV_{max} = 80.0 V, ω = 65.0π rad/s, and L = 70.0 mH. Calculate the current in the inductor at t = 15.5 ms.

8. A 20.0-mH inductor is connected to a North American electrical outlet (ΔV_{rms} = 120 V, f = 60.0 Hz). Assuming the energy stored in the inductor is zero at t = 0, determine the energy stored at $t = \frac{1}{180}$ s.

9. An AC source has an output rms voltage of 78.0 V at a frequency of 80.0 Hz. If the source is connected across a 25.0-mH inductor, what are (a) the inductive reactance of the circuit, (b) the rms current in the circuit, and (c) the maximum current in the circuit?

10. **Review.** Determine the maximum magnetic flux through an inductor connected to a North American electrical outlet (ΔV_{rms} = 120 V, f = 60.0 Hz).

SECTION 32.4 Capacitors in an AC Circuit

11. A 1.00-mF capacitor is connected to a North American electrical outlet (ΔV_{rms} = 120 V, f = 60.0 Hz). Assuming the energy stored in the capacitor is zero at t = 0, determine the magnitude of the current in the wires at $t = \frac{1}{180}$ s.

12. An AC source with an output rms voltage of 36.0 V at a frequency of 60.0 Hz is connected across a 12.0-μF capacitor. Find (a) the capacitive reactance, (b) the rms current, and (c) the maximum current in the circuit. (d) Does the capacitor have its maximum charge when the current has its maximum value? Explain.

13. What is the maximum current in a 2.20-μF capacitor when it is connected across (a) a North American electrical outlet having ΔV_{rms} = 120 V and f = 60.0 Hz and (b) a European electrical outlet having ΔV_{rms} = 240 V and f = 50.0 Hz?

14. A capacitor C is connected to a power supply that operates at a frequency f and produces an rms voltage ΔV. What is the maximum charge that appears on either capacitor plate?

SECTION 32.5 The RLC Series Circuit

15. In addition to phasor diagrams showing voltages such as in Figure 32.15, we can draw phasor diagrams with resistance and reactances. The resultant of adding the phasors is the impedance. Draw to scale a phasor diagram showing Z, X_L, X_C, and ϕ for an AC series circuit for which R = 300 Ω, C = 11.0 μF, L = 0.200 H, and f = 500/π Hz.

16. An AC source with ΔV_{max} = 150 V and f = 50.0 Hz is connected between points a and d in Figure P32.16. Calculate

Figure P32.16 Problems 16 and 51.

the maximum voltages between (a) points a and b, (b) points b and c, (c) points c and d, and (d) points b and d.

17. You are working in a factory and have been tasked with **CR** determining the electrical needs for a new motor that will be installed on an assembly line. The motor has been tested under load conditions and found to have a resistance of 35.0 Ω and an inductive reactance of 50.0 Ω. We can model the motor as a series RL circuit. The motor will have its own dedicated circuit with an rms voltage of 480 V. You need to determine the peak current drawn by the motor to determine the size of the circuit breaker needed to protect the circuit.

18. Draw phasors to scale for the following voltages in SI units: (a) 25.0 sin ωt at ωt = 90.0°, (b) 30.0 sin ωt at ωt = 60.0°, and (c) 18.0 sin ωt at ωt = 300°.

19. An RLC circuit consists of a 150-Ω resistor, a 21.0-μF capacitor, and a 460-mH inductor connected in series with a 120-V, 60.0-Hz power supply. (a) What is the phase angle between the current and the applied voltage? (b) Which reaches its maximum earlier, the current or the voltage?

20. A 60.0-Ω resistor is connected in series with a 30.0-μF capacitor and a source whose maximum voltage is 120 V, operating at 60.0 Hz. Find (a) the capacitive reactance of the circuit, (b) the impedance of the circuit, and (c) the maximum current in the circuit. (d) Does the voltage lead or lag the current? (e) How will adding an inductor in series with the existing resistor and capacitor affect the current? Explain.

SECTION 32.6 Power in an AC Circuit

21. A series RLC circuit has a resistance of 45.0 Ω and an impedance of 75.0 Ω. What average power is delivered to this circuit when ΔV_{rms} = 210 V?

22. *Why is the following situation impossible?* A series circuit consists of an ideal AC source (no inductance or capacitance in the source itself) with an rms voltage of ΔV at a frequency f and a magnetic buzzer with a resistance R and an inductance L. By carefully adjusting the inductance L of the circuit, a power factor of exactly 1.00 is attained.

23. A series RLC circuit has a resistance of 22.0 Ω and an impedance of 80.0 Ω. If the rms voltage applied to the circuit is 160 V, what average power is delivered to the circuit?

24. An AC voltage of the form Δv = 90.0 sin 350t, where Δv is in volts and t is in seconds, is applied to a series RLC circuit. If R = 50.0 Ω, C = 25.0 μF, and L = 0.200 H, find (a) the impedance of the circuit, (b) the rms current in the circuit, and (c) the average power delivered to the circuit.

SECTION 32.7 Resonance in a Series RLC Circuit

25. The LC circuit of a radar transmitter oscillates at 9.00 GHz. (a) What inductance is required for the circuit to resonate at this frequency if its capacitance is 2.00 pF? (b) What is the inductive reactance of the circuit at this frequency?

26. A series RLC circuit has components with the following values: L = 20.0 mH, C = 100 nF, R = 20.0 Ω, and ΔV_{max} = 100 V, with Δv = ΔV_{max} sin ωt. Find (a) the resonant frequency of the circuit, (b) the amplitude of the current at the resonant frequency, (c) the Q of the circuit, and (d) the amplitude of the voltage across the inductor at resonance.

27. You wish to build a series *RLC* circuit for a project you are
CR working on. Looking in your electronics parts box, you are
disappointed to find that you have only two resistors, each
of resistance 47.0 Ω, two capacitors, each of capacitance
5.00 nF, and one inductor of inductance 5.00 mH. You need
to determine the lowest possible angular frequency at resonance that you can obtain from all five components by connecting the inductor in series with a combination of the two
resistors and a combination of the two capacitors.

28. A 10.0-Ω resistor, 10.0-mH inductor, and 100-μF capacitor
are connected in series to a 50.0-V (rms) source having variable frequency. If the operating frequency is twice the resonance frequency, find the energy delivered to the circuit
during one period.

29. A resistor *R*, inductor *L*, and capacitor *C* are connected in
S series to an AC source of rms voltage Δ*V* and variable frequency. If the operating frequency is twice the resonance
frequency, find the energy delivered to the circuit during
one period.

SECTION 32.8 The Transformer and Power Transmission

30. The primary coil of a transformer has $N_1 = 350$ turns, and
T the secondary coil has $N_2 = 2\,000$ turns. If the input voltage across the primary coil is Δ*v* = 170 cos ω*t*, where Δ*v* is
in volts and *t* is in seconds, what rms voltage is developed
across the secondary coil?

31. A person is working near the secondary of a transformer
BIO as shown in Figure P32.31. The primary voltage is 120 V at
60.0 Hz. The secondary voltage is 5 000 V. The capacitance
C_s, which is the stray capacitance between the hand and the
secondary winding, is 20.0 pF. Assuming the person has a
body resistance to ground of $R_b = 50.0$ kΩ, determine the
rms voltage across the body. *Suggestion:* Model the secondary
of the transformer as an AC source.

Figure P32.31

32. A transmission line that has a resistance per unit length of
AMT 4.50×10^{-4} Ω/m is to be used to transmit 5.00 MW across
QC 400 mi (6.44 × 10⁵ m). The output voltage of the source is
4.50 kV. (a) What is the line loss if a transformer is used to
step up the voltage to 500 kV? (b) What fraction of the input
power is lost to the line under these circumstances? (c) **What
If?** What difficulties would be encountered in attempting to
transmit the 5.00 MW at the source voltage of 4.50 kV?

ADDITIONAL PROBLEMS

33. *Why is the following situation impossible?* An *RLC* circuit is used
in a radio to tune into a North American AM commercial
radio station. The values of the circuit components are *R* =
15.0 Ω, *L* = 2.80 μH, and *C* = 0.910 pF.

34. A 400-Ω resistor, an inductor, and a capacitor are in series
QC with an AC source. The reactance of the inductor is 700 Ω,
and the circuit impedance is 760 Ω. (a) What are the possible values of the reactance of the capacitor? (b) If you find
that the power delivered to the circuit decreases as you raise
the frequency, what is the capacitive reactance in the original circuit? (c) Repeat part (a) assuming the resistance is
200 Ω instead of 400 Ω and the circuit impedance continues
to be 760 Ω.

35. Energy is to be transmitted over a pair of copper wires in a
transmission line at the rate of 20.0 kW with only a 1.00%
loss over a distance of 18.0 km at potential difference
$\Delta V_{rms} = 1.50 \times 10^3$ V between the wires. Assuming the current density is uniform in the conductors, what is the diameter required for each of the two wires?

36. Energy is to be transmitted over a pair of copper wires in
S a transmission line at a rate *P* with only a fractional loss
f over a distance ℓ at potential difference ΔV_{rms} between
the wires. Assuming the current density is uniform in the
conductors, what is the diameter required for each of the
two wires?

37. A transformer may be used to provide maximum power
transfer between two AC circuits that have different
impedances Z_1 and Z_2. This process is called *impedance
matching*. (a) Show that the ratio of turns N_1/N_2 for this
transformer is

$$\frac{N_1}{N_2} = \sqrt{\frac{Z_1}{Z_2}}$$

(b) Suppose you want to use a transformer as an
impedance-matching device between an audio amplifier
that has an output impedance of 8.00 kΩ and a speaker that
has an input impedance of 8.00 Ω. What should your N_1/N_2
ratio be?

38. Show that the rms value for the sawtooth voltage shown in
S Figure P33.38 is $\Delta V_{max}/\sqrt{3}$.

Figure P33.38

39. Marie Cornu, a physicist at the Polytechnic Institute in
QC Paris, invented phasors in about 1880. This problem
V helps you see their general utility in representing oscillations. Two mechanical vibrations are represented by the
expressions

$$y_1 = 12.0 \sin 4.50t$$

and

$$y_2 = 12.0 \sin (4.50t + 70.0°)$$

where y_1 and y_2 are in centimeters and *t* is in seconds. Find
the amplitude and phase constant of the sum of these
functions (a) by using a trigonometric identity (as from
Appendix B) and (b) by representing the oscillations as
phasors. (c) State the result of comparing the answers to
parts (a) and (b). (d) Phasors make it equally easy to add

traveling waves. Find the amplitude and phase constant of the sum of the three waves represented by

$$y_1 = 12.0 \sin (15.0x - 4.50t + 70.0°)$$

$$y_2 = 15.5 \sin (15.0x - 4.50t - 80.0°)$$

$$y_3 = 17.0 \sin (15.0x - 4.50t + 160°)$$

where x, y_1, y_2, and y_3 are in centimeters and t is in seconds.

40. A series RLC circuit has resonance angular frequency 2.00×10^3 rad/s. When it is operating at some input frequency, $X_L = 12.0$ Ω and $X_C = 8.00$ Ω. (a) Is this input frequency higher than, lower than, or the same as the resonance frequency? Explain how you can tell. (b) Explain whether it is possible to determine the values of both L and C. (c) If it is possible, find L and C. If it is not possible, give a compact expression for the condition that L and C must satisfy.

41. Review. One insulated conductor from a household extension cord has a mass per length of 19.0 g/m. A section of this conductor is held under tension between two clamps. A subsection is located in a magnetic field of magnitude 15.3 mT directed perpendicular to the length of the cord. When the cord carries an AC current of 9.00 A at a frequency of 60.0 Hz, it vibrates in resonance in its simplest standing-wave vibration mode. (a) Determine the relationship that must be satisfied between the separation d of the clamps and the tension T in the cord. (b) Determine one possible combination of values for these variables.

42. (a) Sketch a graph of the phase angle for an RLC series circuit as a function of angular frequency from zero to a frequency much higher than the resonance frequency. (b) Identify the value of ϕ at the resonance angular frequency ω_0. (c) Prove that the slope of the graph of ϕ versus ω at the resonance point is $2Q/\omega_0$.

43. A series RLC circuit contains the following components: $R = 150$ Ω, $L = 0.250$ H, $C = 2.00$ μF, and a source with $\Delta V_{max} = 210$ V operating at 50.0 Hz. Our goal is to find the phase angle, the power factor, and the power input for this circuit. (a) Find the inductive reactance in the circuit. (b) Find the capacitive reactance in the circuit. (c) Find the impedance in the circuit. (d) Calculate the maximum current in the circuit. (e) Determine the phase angle between the current and source voltage. (f) Find the power factor for the circuit. (g) Find the power input to the circuit.

44. Review. In the circuit shown in Figure P32.44, assume all parameters except C are given. Find (a) the current in the circuit as a function of time and (b) the power delivered to the circuit. (c) Find the current as a function of time after *only* switch 1 is opened. (d) After switch 2 is *also* opened, the

Figure P32.44

current and voltage are in phase. Find the capacitance C. Find (e) the impedance of the circuit when both switches are open, (f) the maximum energy stored in the capacitor during oscillations, and (g) the maximum energy stored in the inductor during oscillations. (h) Now the frequency of the voltage source is doubled. Find the phase difference between the current and the voltage. (i) Find the frequency that makes the inductive reactance one-half the capacitive reactance.

45. You have decided to build your own speaker system for your home entertainment system. The system will consist of two loudspeakers: a large "woofer," to which you want to send low audio frequencies (*bass*), and a small "tweeter," which should receive high audio frequencies (*treble*). To separate the high and low frequencies of the audio signal, you build the "crossover network" shown in Figure P32.45. The input voltage is the audio output of the amplifier in your system, shown in the figure as an AC source. You have two outputs as shown: one across the resistor and one across the capacitor. (a) Across which element should you connect the woofer? (b) Across which element should you connect the tweeter? (c) To choose the appropriate values of R and C, you need to determine an expression for the ratio of the output voltage to the input voltage as a function of angular frequency ω for the resistor as an output. (d) You need to determine a similar expression for the ratio of the output voltage to the input voltage as a function of angular frequency ω for the capacitor as an output.

Figure P32.45

46. A series RLC circuit is operating at 2.00×10^3 Hz. At this frequency, $X_L = X_C = 1884$ Ω. The resistance of the circuit is 40.0 Ω. (a) Prepare a table showing the values of X_L, X_C, and Z for $f = 300$, 600, 800, 1.00×10^3, 1.50×10^3, 2.00×10^3, 3.00×10^3, 4.00×10^3, 6.00×10^3, and 1.00×10^4 Hz. (b) Plot on the same set of axes X_L, X_C, and Z as a function of $\ln f$.

47. You are trying to become a member of the Physics Olympics team. Your physics professor is training you and some other students by having you compete with each other to solve problems as accurately and quickly as you can. During one session, he springs the RLC circuit shown in Figure P32.47 (page 872) on you. Figure P32.47a shows the circuit with a battery as the energy source. The battery has an emf ε and internal resistance r. He tells you the following, assuming switch S has been at position a for a long time:

- If only switch S_L is closed, and switches S_C and S_R are open, and then switch S is thrown to position b, the time constant of the circuit is $\tau_1 = 0.200$ ms. Switch S is returned to position a for a long time.

Figure P32.47

- If only switch S_C is closed, and switches S_L and S_R are open, and then switch S is thrown to position b, the time constant of the circuit is $\tau_2 = 0.050\ 0$ ms. Switch S is returned to position a for a long time.

In Figure P32.47b, an AC source with a variable frequency has been added to the same circuit, and switch S is thrown to position b. Switches S_C, S_L, and S_R are all open. At what angular frequency ω should the AC source be set so that the circuit exhibits resonance? Quick! Get to work!

48. A series *RLC* circuit in which $R = 1.00\ \Omega$, $L = 1.00$ mH, and $C = 1.00$ nF is connected to an AC source delivering 1.00 V (rms). (a) Make a precise graph of the power delivered to the circuit as a function of the frequency and (b) verify that the full width of the resonance peak at half-maximum is $R/2\pi L$.

CHALLENGE PROBLEMS

49. The resistor in Figure P32.49 represents the midrange speaker in a three-speaker system. Assume its resistance to be constant at $8.00\ \Omega$. The source represents an audio amplifier producing signals of uniform amplitude $\Delta V_{max} = 10.0$ V at all audio frequencies. The inductor and capacitor are to function as a *band-pass filter* with $\Delta V_{out}/\Delta V_{in} = \frac{1}{2}$ at 200 Hz and at 4.00×10^3 Hz. Determine the required values of (a) L and (b) C. Find (c) the maximum value of the ratio $\Delta V_{out}/\Delta V_{in}$; (d) the frequency f_0 at which the ratio has its maximum value; (e) the phase shift between Δv_{in} and Δv_{out} at 200 Hz, at f_0, and at 4.00×10^3 Hz; and (f) the average power transferred to the speaker at 200 Hz, at f_0, and at 4.00×10^3 Hz. (g) Recognizing that the diagram represents an *RLC* circuit driven by an AC source, find its quality factor.

Figure P32.49

50. An $80.0\text{-}\Omega$ resistor and a 200-mH inductor are connected in *parallel* across a 100-V (rms), 60.0-Hz source. (a) What is the rms current in the resistor? (b) By what angle does the total current lead or lag behind the voltage?

51. An AC source with $\Delta V_{rms} = 120$ V is connected between points a and d in Figure P32.16. At what frequency will it deliver a power of 250 W? Explain your answer.

33

Electromagnetic Waves

STORYLINE **You are performing online research on your smartphone,** using the Wi-Fi signal from your home network. The signal suddenly cuts out, and you go to the Wi-Fi settings on your smartphone to investigate. In the list of available networks, you see that your network has come back up, but you also notice that you are receiving a signal from your next-door neighbor's house. This starts you wondering if your neighbors are receiving *your* Wi-Fi signal. You take your smartphone outside and start walking away from your house, monitoring the Wi-Fi signal strength indicator on your smartphone. You are surprised to see that the signal is available outside. How does the signal go through the walls of your house? In fact, what exactly *is* a Wi-Fi signal? As you walk away from your house, the signal strength drops off. Why does that happen? What's going on here?

CONNECTIONS This chapter represents a very strong connection between *three* parts of the book. In *Part 2*, we discussed mechanical waves, such as sound, ocean waves, and waves on strings. Here in *Part 4*, we are investigating the principles of electromagnetism. In this chapter, we connect these seemingly disparate sets of chapters. We find that the principles of electromagnetism predict the possibility of electromagnetic waves. This theoretical prediction is clearly borne out in practice, of course, by our experience with a wide variety of electromagnetic waves: light, radio, microwaves, x-rays, etc. The behavior of these waves has clear similarities with mechanical waves, with one strong difference:

873

James Clerk Maxwell
Scottish Theoretical Physicist
(1831–1879)
Maxwell developed the electromagnetic theory of light and the kinetic theory of gases, and explained the nature of Saturn's rings and color vision. Maxwell's successful interpretation of the electromagnetic field resulted in the field equations that bear his name. Formidable mathematical ability combined with great insight enabled him to lead the way in the study of electromagnetism and kinetic theory. He died of cancer before he was 50.

The conduction current I in the wire passes only through S_1, which leads to a contradiction in Ampère's law that is resolved only if one postulates a displacement current through S_2.

Figure 33.1 Two surfaces S_1 and S_2 near the plate of a capacitor are bounded by the same path P.

electromagnetic waves do not require a medium; they can propagate through empty space! The understanding of electromagnetic waves has led to many practical communication systems, including radio, television, cell phone systems, wireless Internet connectivity, and optoelectronics. Furthermore, the study of electromagnetic waves prepares us for *Part 5* of the book, in which we study *optics*, which describes the detailed behavior of electromagnetic waves.

33.1 Displacement Current and the General Form of Ampère's Law

In Chapter 29, we discussed using Ampère's law (Eq. 29.13) to analyze the magnetic fields created by currents:

$$\oint \vec{\mathbf{B}} \cdot d\vec{\mathbf{s}} = \mu_0 I$$

In this equation, the line integral is over any closed path through which conduction current passes, where conduction current is defined by the expression $I = dq/dt$. (In this section, we use the term *conduction current* to refer to the current carried by charge carriers in the wire to distinguish it from a new type of current we shall introduce shortly.) We have accepted Ampère's law as a fundamental equation in electromagnetism. But suppose we found a situation where it doesn't apply? James Clerk Maxwell recognized such a situation and modified Ampère's law accordingly.

Consider a capacitor being charged as illustrated in Figure 33.1. When a conduction current is present, the charge on the positive plate changes, but no conduction current exists in the gap between the plates because there are no charge carriers in the gap. Now consider the two surfaces S_1 and S_2 in Figure 33.1, bounded by the same path P. Surface S_1 is a flat circular area, through which the wire passes. Surface S_2 is a hemisphere, sharing the same path P with surface S_1. The surface of the hemisphere passes through the space between the capacitor plates. Ampère's law states that $\oint \vec{\mathbf{B}} \cdot d\vec{\mathbf{s}}$ around this path must equal $\mu_0 I$, where I is the total current through *any* surface bounded by the path P.

When the path P is considered to be the boundary of S_1, $\oint \vec{\mathbf{B}} \cdot d\vec{\mathbf{s}} = \mu_0 I$ because the conduction current I passes through S_1. When the path is considered to be the boundary of S_2, however, $\oint \vec{\mathbf{B}} \cdot d\vec{\mathbf{s}} = 0$ because no conduction current passes through S_2. Therefore, we have a contradictory situation that arises from the discontinuity of the current! Ampère's law gives two different answers for the two surfaces!

Maxwell solved this problem by postulating an additional term on the right side of Ampère's law, which includes a factor called the **displacement current** I_d defined as[1]

$$I_d \equiv \epsilon_0 \frac{d\Phi_E}{dt} \tag{33.1}$$

where ϵ_0 is the permittivity of free space (see Section 22.3) and $\Phi_E \equiv \int \vec{\mathbf{E}} \cdot d\vec{\mathbf{A}}$ is the electric flux (see Eq. 23.4) for the electric field between the plates of the capacitor. This flux passes through S_2 but not S_1.

As the capacitor is being charged (or discharged), the changing electric field between the plates may be considered equivalent to a current given by Equation 33.1 that acts as a continuation of the conduction current in the wire. When

---

[1]*Displacement* in this context does not have the meaning it does in Chapter 2. Despite the inaccurate implications, the word is historically entrenched in the language of physics, so we continue to use it.

the expression for the displacement current is added to the conduction current on the right side of Ampère's law, the difficulty represented in Figure 33.1 is resolved. No matter which surface bounded by the path *P* is chosen, either a conduction current or a displacement current passes through it. With this new term I_d, we can express the general form of Ampère's law (sometimes called the **Ampère–Maxwell law**) as

$$\oint \vec{\mathbf{B}} \cdot d\vec{\mathbf{s}} = \mu_0(I + I_d) = \mu_0 I + \mu_0 \epsilon_0 \frac{d\Phi_E}{dt} \qquad (33.2)$$

◀ Ampère–Maxwell law

This result was a remarkable example of theoretical work by Maxwell, and it contributed to major advances in the understanding of electromagnetism.

We can understand the meaning of Equation 33.2 by considering Figure 33.2, which is similar to Figure 33.1, but we have now identified a flat plane S passing between the plates of the capacitor. The electric flux through surface S is $\Phi_E = \int \vec{\mathbf{E}} \cdot d\vec{\mathbf{A}} = EA$, where *A* is the area of the capacitor plates and *E* is the magnitude of the uniform electric field between the plates. If *q* is the charge on the plates at any instant, then $E = q/(\epsilon_0 A)$ (see the What If? in Example 23.8). Therefore, the electric flux through S is

$$\Phi_E = EA = \frac{q}{\epsilon_0}$$

The electric field lines between the plates create an electric flux through surface S.

Hence, the displacement current through S is

$$I_d = \epsilon_0 \frac{d\Phi_E}{dt} = \frac{dq}{dt} \qquad (33.3)$$

That is, the displacement current I_d through S is precisely equal to the conduction current *I* in the wires connected to the capacitor!

QUICK QUIZ 33.1 In an *RC* circuit, the capacitor begins to discharge. **(i)** During the discharge, in the region of space between the plates of the capacitor, is there (a) conduction current but no displacement current, (b) displacement current but no conduction current, (c) both conduction and displacement current, or (d) no current of any type? **(ii)** In the same region of space, is there (a) an electric field but no magnetic field, (b) a magnetic field but no electric field, (c) both electric and magnetic fields, or (d) no fields of any type?

Figure 33.2 When a conduction current exists in the wires, a changing electric field $\vec{\mathbf{E}}$ exists between the plates of the capacitor.

| Example **33.1** | **Displacement Current in a Capacitor** |

A sinusoidally varying voltage is applied across a capacitor as shown in Figure 33.3. The capacitance is $C = 8.00\ \mu\text{F}$, the frequency of the applied voltage is $f = 3.00$ kHz, and the voltage amplitude is $\Delta V_{max} = 30.0$ V. Find the displacement current in the capacitor.

SOLUTION

Conceptualize Figure 33.3 represents the circuit diagram for this situation. Figure 33.2 shows a close-up of the capacitor and the electric field between the plates.

Categorize We determine results using equations discussed in this section, so we categorize this example as a substitution problem.

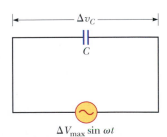

Figure 33.3 (Example 33.1)

Use Equation 33.3 to find the displacement current as a function of time. Note that the charge on the capacitor is $q = C \Delta v_C$:

$$i_d = \frac{dq}{dt} = \frac{d}{dt}(C \Delta v_C) = C \frac{d}{dt}(\Delta V_{max} \sin \omega t)$$
$$= \omega C \Delta V_{max} \cos \omega t = 2\pi f C \Delta V_{max} \cos(2\pi f t)$$

Substitute numerical values to obtain the current in amperes:

$$i_d = 2\pi(3.00 \times 10^3\ \text{Hz})(8.00 \times 10^{-6}\ \text{F})(30.0\ \text{V}) \cos[2\pi(3.00 \times 10^3\ \text{Hz})t]$$
$$= 4.52 \cos(1.88 \times 10^4\ t)$$

33.2   Maxwell's Equations and Hertz's Discoveries

We now present four equations that are regarded as the basis of all electrical and magnetic phenomena. We've seen all four equations before. These equations, developed by Maxwell, are as fundamental to electromagnetic phenomena as Newton's laws are to mechanical phenomena. In fact, the theory that Maxwell developed was more far-reaching than even he imagined because it turned out to be in agreement with the special theory of relativity, as Einstein showed in 1905.

Maxwell's equations represent the laws of electricity and magnetism that we have already discussed, but they have additional important consequences. For simplicity, we present **Maxwell's equations** as applied to free space, that is, in the absence of any dielectric or magnetic material. The four equations are

Gauss's law ▶   Equation 23.7   → $$\oint \vec{\mathbf{E}} \cdot d\vec{\mathbf{A}} = \frac{q}{\epsilon_0}$$ (33.4)

Gauss's law in magnetism ▶   Equation 29.20   → $$\oint \vec{\mathbf{B}} \cdot d\vec{\mathbf{A}} = 0$$ (33.5)

Faraday's law ▶   Equation 30.8   → $$\oint \vec{\mathbf{E}} \cdot d\vec{\mathbf{s}} = -\frac{d\Phi_B}{dt}$$ (33.6)

Ampère–Maxwell law ▶   Equation 33.2   → $$\oint \vec{\mathbf{B}} \cdot d\vec{\mathbf{s}} = \mu_0 I + \epsilon_0 \mu_0 \frac{d\Phi_E}{dt}$$ (33.7)

Equation 33.4 is Gauss's law: the total electric flux through any closed surface equals the net charge inside that surface divided by ϵ_0. This law relates an electric field to the charge distribution that creates it.

Equation 33.5 is Gauss's law in magnetism, and it states that the net magnetic flux through a closed surface is zero. That is, the number of magnetic field lines that enter a closed volume must equal the number that leave that volume, which implies that magnetic field lines cannot begin or end at any point. If they did, it would mean that isolated magnetic monopoles existed at those points. That isolated magnetic monopoles have not been observed in nature can be taken as a confirmation of Equation 33.5.

Equation 33.6 is Faraday's law of induction, which describes the creation of an electric field by a changing magnetic flux. This law states that the emf, which is the line integral of the electric field around any closed path, equals the rate of change of magnetic flux through any surface bounded by that path.

Equation 33.7 is the Ampère–Maxwell law, discussed in Section 33.1, and it describes the creation of a magnetic field by a changing electric field and by electric current: the line integral of the magnetic field around any closed path is the sum of μ_0 multiplied by the net current through that path and $\epsilon_0 \mu_0$ multiplied by the rate of change of electric flux through any surface bounded by that path.

Once the electric and magnetic fields are known at some point in space, the force acting on a particle of charge q can be calculated from the electric and magnetic versions of the particle in a field model:

Lorentz force law ▶ $$\vec{\mathbf{F}} = q\vec{\mathbf{E}} + q\vec{\mathbf{v}} \times \vec{\mathbf{B}}$$ (33.8)

This relationship is called the **Lorentz force law.** (We saw this relationship earlier as Eq. 28.6.) Maxwell's equations, together with this force law, completely describe all classical electromagnetic interactions in a vacuum.

In earlier chapters we looked at charge q and current I as sources of electric and magnetic fields. Now, let's imagine a region of space that contains no charges and no currents. Under these conditions, Maxwell's equations become

$$\oint \vec{\mathbf{E}} \cdot d\vec{\mathbf{A}} = 0 \tag{33.9}$$

$$\oint \vec{\mathbf{B}} \cdot d\vec{\mathbf{A}} = 0 \tag{33.10}$$

$$\oint \vec{\mathbf{E}} \cdot d\vec{\mathbf{s}} = -\frac{d\Phi_B}{dt} \tag{33.11}$$

$$\oint \vec{\mathbf{B}} \cdot d\vec{\mathbf{s}} = \epsilon_0 \mu_0 \frac{d\Phi_E}{dt} \tag{33.12}$$

Notice the symmetry of Maxwell's equations in charge-free, current-free space. Equations 33.9 and 33.10 are of the same form. Furthermore, Equations 33.11 and 33.12 are symmetric in that the line integrals of $\vec{\mathbf{E}}$ and $\vec{\mathbf{B}}$ around a closed path are related to the rate of change of magnetic flux and electric flux, respectively. These equations suggest that electric and magnetic fields can exist in charge-free, current-free space! They do that by regenerating each other as described by the last two equations. Equation 33.11 tells that a time variation in a *B*-field generates an *E*-field. And Equation 33.12 tells us the reverse also happens.

In the next section, we show that Equations 33.11 and 33.12 can be combined to obtain a wave equation for both the electric field and the magnetic field. In empty space, where $q = 0$ and $I = 0$, the solution to these two equations shows that the speed at which electromagnetic waves travel equals the measured speed of light. This result led Maxwell to predict that light waves are a form of electromagnetic radiation.

Hertz performed experiments that verified Maxwell's prediction. The experimental apparatus Hertz used to generate and detect electromagnetic waves is shown schematically in Figure 33.4. An induction coil is connected to a transmitter made up of two spherical electrodes separated by a narrow gap. The coil provides short voltage surges to the electrodes, making one positive and the other negative. A spark is generated between the spheres when the electric field near either electrode surpasses the dielectric strength for air (3×10^6 V/m; see Table 25.1). Free electrons in a strong electric field are accelerated and gain enough energy to ionize any molecules they strike. This ionization provides more electrons, which can accelerate and cause further ionizations. As the air in the gap is ionized, it becomes a much better conductor and the discharge between the electrodes exhibits an oscillatory behavior at a very high frequency. From an electric-circuit viewpoint, this experimental apparatus is equivalent to an *LC* circuit in which the inductance is that of the coil and the capacitance is due to the spherical electrodes.

Because *L* and *C* are small in Hertz's apparatus, the frequency of oscillation is high, on the order of 100 MHz. (Recall from Eq. 31.22 that $\omega = 1/\sqrt{LC}$ for an *LC* circuit.) Electromagnetic waves are radiated at this frequency as a result of the oscillation of free charges in the transmitter circuit. Hertz was able to detect these waves by resonance using a single loop of wire with its own spark gap (the receiver). Such a receiver loop, placed several meters from the transmitter, has its own effective inductance, capacitance, and natural frequency of oscillation. In Hertz's experiment, sparks were induced across the gap of the receiving electrodes when the receiver's frequency was adjusted to match that of the transmitter. In this way, Hertz demonstrated that the oscillating current induced in the receiver was produced by electromagnetic waves radiated by the transmitter. His experiment is analogous to the mechanical phenomenon in which a tuning fork responds to acoustic vibrations from an identical tuning fork that is oscillating nearby.

The transmitter consists of two spherical electrodes connected to an induction coil, which provides short voltage surges to the spheres, setting up oscillations in the discharge between the electrodes.

The receiver is a nearby loop of wire containing a second spark gap.

Figure 33.4 Schematic diagram of Hertz's apparatus for generating and detecting electromagnetic waves.

Heinrich Rudolf Hertz
German Physicist (1857–1894)
Hertz made his most important discovery of electromagnetic waves in 1887. After finding that the speed of an electromagnetic wave was the same as that of light, Hertz showed that electromagnetic waves, like light waves, could be reflected, refracted, and diffracted. The hertz, equal to one complete vibration or cycle per second, is named after him.

In addition, Hertz showed in a series of experiments that the radiation generated by his spark-gap device exhibited the wave properties of interference, diffraction, reflection, refraction, and polarization, which are all properties exhibited by light as we shall see in Part 5. Therefore, it became evident that the radio-frequency waves Hertz was generating had properties similar to those of light waves and that they differed only in frequency and wavelength. Perhaps his most convincing experiment was the measurement of the speed of this radiation. Waves of known frequency were reflected from a metal sheet and created a standing-wave interference pattern whose nodal points could be detected. The measured distance between the nodal points enabled determination of the wavelength λ. Using the relationship $v = \lambda f$ (Eq. 16.12) from the traveling wave model, Hertz found that v was close to 3×10^8 m/s, the known speed c of visible light.

We've argued that electric and magnetic fields can support each other in free space, and Hertz's experiments verified the existence of electromagnetic waves. But where does the electromagnetic wave come from in the first place? Stationary charges and steady currents cannot produce electromagnetic waves. If the current in a wire changes with time, however, the wire emits electromagnetic waves. The fundamental mechanism responsible for this radiation is the acceleration of a charged particle. **Whenever a charged particle accelerates, energy is transferred away from the particle by electromagnetic radiation.** Let's now investigate the properties of those waves.

33.3 Plane Electromagnetic Waves

The properties of electromagnetic waves can be deduced from Maxwell's equations. One approach to deriving these properties is to solve the second-order differential equation obtained from Equations 33.11 and 33.12. A rigorous mathematical treatment of that sort is beyond the scope of this text. To circumvent this problem, let's assume the vectors for the electric field and magnetic field in an electromagnetic wave have a specific space–time behavior that is simple but consistent with Maxwell equations.

To understand the prediction of electromagnetic waves more fully, let's focus our attention on an electromagnetic wave that travels in the x direction (the *direction of propagation*). Figure 30.15 shows us that the electric field generated by a changing magnetic field is perpendicular to the magnetic field. Figure 33.2 shows a changing electric field as an effective current, which would generate circular magnetic field lines around the electric field lines. Therefore, the magnetic field generated by a changing electric field is perpendicular to the electric field. Following on these suggestions of perpendicularity, let us design a simple electromagnetic wave for which the electric field $\vec{\mathbf{E}}$ is in the y direction and the magnetic field $\vec{\mathbf{B}}$ is in the z direction as shown in Figure 33.5. Furthermore, let's assume the field magnitudes E and B depend on x and t only, not on the y or z coordinate.

Figure 33.5 shows field vectors for a wave propagating along the x axis, as suggested by the vector $\vec{\mathbf{c}}$, with magnitude c, the speed of light. Imagine a source in the yz plane that emits a large number of such waves from *all* positions in the plane, not just the origin, with all waves traveling in the x direction. If we define a **ray** as in Section 16.8 as the line along which the wave travels, all rays for these waves are parallel. This entire collection of waves is often called a **plane wave.**

To generate the prediction of plane electromagnetic waves, we start with Faraday's law, Equation 33.11, which describes an electric field generated by a changing magnetic field. To apply this equation to the wave in Figure 33.5, consider a rectangle of width dx and height ℓ lying in the xy plane as shown in Figure 33.6. Let's first evaluate the line integral of $\vec{\mathbf{E}} \cdot d\vec{\mathbf{s}}$ around this rectangle in the counterclockwise direction at an instant of time when the wave is passing through the rectangle. The contributions from the top and bottom of the rectangle are zero

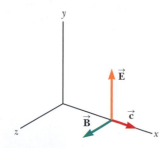

Figure 33.5 Electric and magnetic fields of an electromagnetic wave traveling at velocity $\vec{\mathbf{c}}$ in the positive x direction. The field vectors are shown at one instant of time and at one position in space. These fields depend on x and t.

According to Equation 33.15, this spatial variation in $\vec{\mathbf{E}}$ gives rise to a time-varying magnetic field along the z direction.

According to Equation 33.18, this spatial variation in $\vec{\mathbf{B}}$ gives rise to a time-varying electric field along the y direction.

Figure 33.6 At an instant when a plane wave moving in the positive x direction passes through a rectangular path of width dx lying in the xy plane, the electric field in the y direction varies from $\vec{\mathbf{E}}(x)$ to $\vec{\mathbf{E}}(x + dx)$.

Figure 33.7 At an instant when a plane wave passes through a rectangular path of width dx lying in the xz plane, the magnetic field in the z direction varies from $\vec{\mathbf{B}}(x)$ to $\vec{\mathbf{B}}(x + dx)$.

because $\vec{\mathbf{E}}$ is perpendicular to $d\vec{\mathbf{s}}$ for these paths. We can express the electric field on the right side of the rectangle as

$$E(x + dx) \approx E(x) + \frac{dE}{dx}\bigg|_{t\,constant} dx = E(x) + \frac{\partial E}{\partial x}\,dx$$

where $E(x)$ is the field on the left side of the rectangle at this instant.[2] Therefore, the line integral over this rectangle is approximately

$$\oint \vec{\mathbf{E}} \cdot d\vec{\mathbf{s}} = \left[E(x) + \frac{\partial E}{\partial x}dx\right]\ell - [E(x)]\ell \approx \ell\left(\frac{\partial E}{\partial x}\right)dx \qquad (33.13)$$

Because the magnetic field is in the z direction, the magnetic flux through the rectangle of area $\ell\,dx$ is approximately $\Phi_B = B\ell\,dx$ (assuming dx is very small compared with the wavelength of the wave). Taking the time derivative of the magnetic flux gives

$$\frac{d\Phi_B}{dt} = \ell\,dx\,\frac{dB}{dt}\bigg|_{x\,constant} = \ell\,dx\,\frac{\partial B}{\partial t} \qquad (33.14)$$

Substituting Equations 33.13 and 33.14 into Equation 33.11 gives

$$\ell\left(\frac{\partial E}{\partial x}\right)dx = -\ell\,dx\,\frac{\partial B}{\partial t}$$

$$\frac{\partial E}{\partial x} = -\frac{\partial B}{\partial t} \qquad (33.15)$$

In a similar manner, we can derive a second equation by starting with Maxwell's fourth equation in empty space (Eq. 33.12), which describes a magnetic field generated by a changing electric field. In this case, the line integral of $\vec{\mathbf{B}} \cdot d\vec{\mathbf{s}}$ is evaluated around a rectangle lying in the xz plane and having width dx and length ℓ as in Figure 33.7. Noting that the magnitude of the magnetic field changes from $B(x)$ to $B(x + dx)$ over the width dx and that the direction for taking the line integral is

---

[2]Because dE/dx in this equation is expressed as the change in E with x at a given instant t, dE/dx is equivalent to the partial derivative $\partial E/\partial x$. Likewise, dB/dt means the change in B with time at a particular position x; therefore, in Equation 33.14, we can replace dB/dt with $\partial B/\partial t$.

counterclockwise when viewed from above in Figure 33.7, the line integral over this rectangle is found to be approximately

$$\oint \vec{\mathbf{B}} \cdot d\vec{\mathbf{s}} = [B(x)]\ell - \left[B(x) + \frac{\partial B}{\partial x} dx \right]\ell \approx -\ell \left(\frac{\partial B}{\partial x} \right) dx \tag{33.16}$$

The electric flux through the rectangle is $\Phi_E = E\ell\, dx$, which, when differentiated with respect to time, gives

$$\frac{\partial \Phi_E}{\partial t} = \ell\, dx \frac{\partial E}{\partial t} \tag{33.17}$$

Substituting Equations 33.16 and 33.17 into Equation 33.12 gives

$$-\ell \left(\frac{\partial B}{\partial x} \right) dx = \mu_0 \epsilon_0 \ell\, dx \left(\frac{\partial E}{\partial t} \right)$$

$$\frac{\partial B}{\partial x} = -\mu_0 \epsilon_0 \frac{\partial E}{\partial t} \tag{33.18}$$

Taking the derivative of Equation 33.15 with respect to x and combining the result with Equation 33.18 gives

$$\frac{\partial^2 E}{\partial x^2} = -\frac{\partial}{\partial x} \left(\frac{\partial B}{\partial t} \right) = -\frac{\partial}{\partial t} \left(\frac{\partial B}{\partial x} \right) = -\frac{\partial}{\partial t} \left(-\mu_0 \epsilon_0 \frac{\partial E}{\partial t} \right)$$

$$\frac{\partial^2 E}{\partial x^2} = \mu_0 \epsilon_0 \frac{\partial^2 E}{\partial t^2} \tag{33.19}$$

In the same manner, taking the derivative of Equation 33.18 with respect to x and combining it with Equation 33.15 gives

$$\frac{\partial^2 B}{\partial x^2} = \mu_0 \epsilon_0 \frac{\partial^2 B}{\partial t^2} \tag{33.20}$$

Equations 33.19 and 33.20 both have the form of the linear wave equation, Equation 16.27 from Section 16.5. In that equation, the coefficient of the time derivative is the inverse of the wave speed. Calling this speed c for light, we see that

◀ **Speed of electromagnetic waves**

$$c = \frac{1}{\sqrt{\mu_0 \epsilon_0}} \tag{33.21}$$

Let's evaluate this speed numerically:

$$c = \frac{1}{\sqrt{(4\pi \times 10^{-7}\ \mathrm{T \cdot m/A})(8.854\,19 \times 10^{-12}\ \mathrm{C^2/N \cdot m^2})}}$$

$$= 2.997\,92 \times 10^8\ \mathrm{m/s}$$

This speed is precisely the same as the experimentally measured speed of light in empty space! We are led to believe (correctly) that light is an electromagnetic wave.

The simplest solution to Equations 33.19 and 33.20 is a sinusoidal wave for which the field magnitudes E and B vary with x and t according to the expressions

◀ **Sinusoidal electric and magnetic fields**

$$E = E_{\max} \cos(kx - \omega t) \tag{33.22}$$

$$B = B_{\max} \cos(kx - \omega t) \tag{33.23}$$

where $E_{\max}$ and $B_{\max}$ are the maximum values of the fields. The angular wave number is $k = 2\pi/\lambda$, where λ is the wavelength. The angular frequency is $\omega = 2\pi f$, where f is the wave frequency. According to the traveling wave model of Section 16.2, the ratio ω/k equals the speed of an electromagnetic wave, c:

$$\frac{\omega}{k} = \frac{2\pi f}{2\pi/\lambda} = \lambda f = c$$

where we have used Equation 16.12, $v = c = \lambda f$, which relates the speed, frequency, and wavelength of a sinusoidal wave. Figure 33.8 is a pictorial representation, at one instant, of a sinusoidal electromagnetic wave moving in the positive x direction based on Equations 33.22 and 33.23. Such a wave, in which the electric and magnetic fields are restricted to being parallel to a pair of perpendicular axes, is said to be a **linearly polarized wave.**

We can generate other mathematical representations of the traveling wave model for electromagnetic waves. Taking partial derivatives of Equations 33.22 (with respect to x) and 33.23 (with respect to t) gives

$$\frac{\partial E}{\partial x} = -kE_{max} \sin (kx - \omega t)$$

$$\frac{\partial B}{\partial t} = \omega B_{max} \sin (kx - \omega t)$$

Substituting these results into Equation 33.15 shows that, at any instant,

$$kE_{max} = \omega B_{max}$$

$$\frac{E_{max}}{B_{max}} = \frac{\omega}{k} = c$$

Using these results together with Equations 33.22 and 33.23 gives

$$\frac{E_{max}}{B_{max}} = \frac{E}{B} = c \qquad (33.24)$$

That is, at every instant, the ratio of the magnitude of the electric field to the magnitude of the magnetic field in an electromagnetic wave equals the speed of light.

QUICK QUIZ 33.2 What is the phase difference between the sinusoidal oscillations of the electric and magnetic fields in Figure 33.8? **(a)** 180° **(b)** 90° **(c)** 0 **(d)** impossible to determine

QUICK QUIZ 33.3 An electromagnetic wave propagates in the negative y direction. The electric field at a point in space is momentarily oriented in the positive x direction. In which direction is the magnetic field at that point momentarily oriented? **(a)** the negative x direction **(b)** the positive y direction **(c)** the positive z direction **(d)** the negative z direction

Figure 33.8 A sinusoidal electromagnetic wave moves in the positive x direction with a speed c.

PITFALL PREVENTION 33.2

$\vec{E}$ **Stronger than** $\vec{B}$**?** Because the value of c is so large, some students incorrectly interpret Equation 33.24 as meaning that the electric field is much stronger than the magnetic field. Electric and magnetic fields are measured in different units, however, so they cannot be directly compared. In Section 33.4, we find that the electric and magnetic fields contribute equally to the wave's energy.

Example 33.2 **An Electromagnetic Wave**

A sinusoidal electromagnetic wave of frequency 40.0 MHz travels in free space in the x direction as in Figure 33.9.

(A) Determine the wavelength and period of the wave.

SOLUTION

Conceptualize Imagine the wave in Figure 33.9 moving to the right along the x axis, with the electric and magnetic fields oscillating in phase.

Categorize We use the mathematical representation of the *traveling wave* model for electromagnetic waves.

Analyze

Solve Equation 16.12 to find the wavelength of the wave:

$$\lambda = \frac{c}{f} = \frac{3.00 \times 10^8 \text{ m/s}}{40.0 \times 10^6 \text{ Hz}} = \boxed{7.50 \text{ m}}$$

Find the period T of the wave as the inverse of the frequency:

$$T = \frac{1}{f} = \frac{1}{40.0 \times 10^6 \text{ Hz}} = \boxed{2.50 \times 10^{-8} \text{ s}}$$

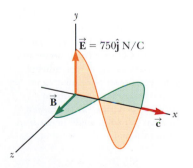

Figure 33.9 (Example 33.2) At some instant, a plane electromagnetic wave moving in the x direction has a maximum electric field of 750 N/C in the positive y direction.

$\vec{E} = 750\hat{j}$ N/C

continued

33.2 continued

(B) At some point and at some instant, the electric field has its maximum value of 750 N/C and is directed along the y axis. Calculate the magnitude and direction of the magnetic field at this position and time.

SOLUTION

Use Equation 33.24 to find the magnitude of the magnetic field:

$$B_{max} = \frac{E_{max}}{c} = \frac{750 \text{ N/C}}{3.00 \times 10^8 \text{ m/s}} = 2.50 \times 10^{-6} \text{ T}$$

Because $\vec{E}$ and $\vec{B}$ must be perpendicular to each other and perpendicular to the direction of wave propagation (x in this case), we conclude that $\vec{B}$ is in the z direction.

Finalize Notice that the wavelength is several meters. This is relatively long for an electromagnetic wave. As we will see in Section 33.7, this wave belongs to the radio range of frequencies.

33.4 Energy Carried by Electromagnetic Waves

In our discussion of the nonisolated system model for energy in Section 8.1, we identified electromagnetic radiation as one method of energy transfer across the boundary of a system. The amount of energy transferred by electromagnetic waves is symbolized as T_{ER} in Equation 8.2. The rate of transfer of energy by an electromagnetic wave is described by a vector $\vec{S}$, called the **Poynting vector,** which is defined by the expression

◄ Poynting vector

$$\vec{S} \equiv \frac{1}{\mu_0} \vec{E} \times \vec{B} \tag{33.25}$$

From the definition of the vector product (Section 11.1), we see that $\vec{S}$ is in the direction of the propagation of the wave (Fig. 33.10). The units for $\vec{S}$ can be found by dimensional analysis (Section 1.3):

$$[\vec{S}] = \frac{[\vec{E}][\vec{B}]}{[\mu_0]} = \frac{(\text{N/C})(\text{T})}{\text{T} \cdot \text{m/A}} = \frac{\text{N} \cdot \text{m}}{\text{m}^2 \cdot \text{s}} = \frac{\text{J}}{\text{m}^2 \cdot \text{s}} = \frac{\text{W}}{\text{m}^2}$$

The magnitude of the Poynting vector represents the *intensity*, the rate at which energy passes through a unit surface area perpendicular to the direction of wave propagation. Therefore, the magnitude of $\vec{S}$ represents *power per unit area*.

As an example, let's evaluate the magnitude of $\vec{S}$ for a plane electromagnetic wave where $|\vec{E} \times \vec{B}| = EB$. In this case,

$$S = \frac{EB}{\mu_0} \tag{33.26}$$

Because $B = E/c$, we can also express this result as

$$S = \frac{E^2}{\mu_0 c} = \frac{cB^2}{\mu_0}$$

These equations for S apply at any instant of time and represent the *instantaneous* rate at which energy is passing through a unit area in terms of the instantaneous values of E and B.

What is of greater interest for a sinusoidal plane electromagnetic wave is the time average of S over one or more cycles, which is called the *wave intensity I*. (We discussed the intensity of sound waves in Chapter 16.) When this average is taken, we obtain an expression involving the time average of $\cos^2 (kx - \omega t)$, which equals $\frac{1}{2}$. Hence, the average value of S (in other words, the intensity of the wave) is

◄ Wave intensity

$$I = S_{avg} = \frac{E_{max} B_{max}}{2\mu_0} = \frac{E_{max}^2}{2\mu_0 c} = \frac{cB_{max}^2}{2\mu_0} \tag{33.27}$$

PITFALL PREVENTION 33.3

An Instantaneous Value The Poynting vector given by Equation 33.25 is time dependent. Its magnitude varies in time, reaching a maximum value at the same instant the magnitudes of $\vec{E}$ and $\vec{B}$ do. The *average* rate of energy transfer is given by Equation 33.27.

PITFALL PREVENTION 33.4

Irradiance In this discussion, intensity is defined in the same way as in Chapter 16 (as power per unit area). In the optics industry, however, power per unit area is called the *irradiance*. Radiant intensity is defined as the power in watts per solid angle (measured in steradians).

Recall that the energy per unit volume associated with an electric field, which is the instantaneous energy density u_E, is given by Equation 25.15:

$$u_E = \tfrac{1}{2}\epsilon_0 E^2 \qquad (33.28)$$

Also recall that the instantaneous energy density u_B associated with a magnetic field is given by Equation 31.14:

$$u_B = \frac{B^2}{2\mu_0} \qquad (33.29)$$

Figure 33.10 The Poynting vector $\vec{S}$ for a plane electromagnetic wave is along the direction of wave propagation.

Because E and B vary with time for an electromagnetic wave, the energy densities also vary with time. Using the relationships $B = E/c$ and $c = 1/\sqrt{\mu_0\epsilon_0}$, the expression for u_B becomes

$$u_B = \frac{(E/c)^2}{2\mu_0} = \frac{\mu_0\epsilon_0}{2\mu_0}E^2 = \tfrac{1}{2}\epsilon_0 E^2 = u_E \qquad (33.30)$$

That is, the instantaneous energy density associated with the magnetic field of an electromagnetic wave equals the instantaneous energy density associated with the electric field. Hence, in a given volume, the energy is equally shared by the two fields.

The **total instantaneous energy density** u is equal to the sum of the energy densities associated with the electric and magnetic fields:

$$u = u_E + u_B = 2u_E = 2u_B = \epsilon_0 E^2 = \frac{B^2}{\mu_0} \qquad (33.31)$$

◀ Total instantaneous energy density of an electromagnetic wave

When this total instantaneous energy density is averaged over one or more cycles of an electromagnetic wave, we again obtain a factor of $\tfrac{1}{2}$. Hence, from Equation 33.31, for any electromagnetic wave, the total average energy per unit volume is

$$u_{avg} = \epsilon_0 (E^2)_{avg} = \tfrac{1}{2}\epsilon_0 E_{max}^2 = \frac{B_{max}^2}{2\mu_0} \qquad (33.32)$$

◀ Average energy density of an electromagnetic wave

Comparing this result with Equation 33.27 for the average value of S, we see that

$$I = S_{avg} = cu_{avg} \qquad (33.33)$$

Therefore, the intensity of an electromagnetic wave equals the average energy density multiplied by the speed of light.

The Sun delivers about 10^3 W/m^2 of energy to the Earth's surface via electromagnetic radiation which represents the intensity, or the average magnitude of the Poynting vector, for solar radiation. Let's calculate the total power that is incident on the roof of a home. The roof's dimensions are 8.00 m × 20.0 m and we assume the radiation is incident normal to the roof. Because intensity represents power per unit area, we obtain

$$P_{avg} = S_{avg}A = (1\,000 \text{ W/m}^2)(8.00 \text{ m} \times 20.0 \text{ m}) = 1.60 \times 10^5 \text{ W}$$

This power is large compared with the power requirements of a typical home. If this power could be absorbed and made available to electrical devices, it would provide more than enough energy for the average home. Solar energy is not easily harnessed, however, and the prospects for large-scale conversion are not as bright as may appear from this calculation. For example, the efficiency of conversion from solar energy is typically 12–18% for photovoltaic cells, reducing the available power by an order of magnitude. Other considerations reduce the power even further. Depending on location, the radiation is most likely not incident normal to the roof and, even if it is, this situation exists for only a short time near the middle of the day. No energy is available for about half of each day during the nighttime hours, and cloudy days further reduce the available energy. Finally, while energy is arriving at a large rate during the middle of the day, some of it must be stored for later use, requiring batteries or other storage devices. Despite these difficulties, conversion of homes to solar operation can be cost-effective, and many homeowners are making the conversion.

Example 33.3   Fields on the Page

Estimate the maximum magnitudes of the electric and magnetic fields of the light that is incident on this page because of the visible light coming from your incandescent desk lamp. Treat the lightbulb as a point source of electromagnetic radiation that is 5% efficient at transforming energy coming in by electrical transmission to energy leaving by visible light.

SOLUTION

Conceptualize The filament in your incandescent lightbulb emits electromagnetic radiation. The brighter the light, the larger the magnitudes of the electric and magnetic fields.

Categorize Because the lightbulb is to be treated as a point source, it emits equally in all directions, so the outgoing electromagnetic radiation can be modeled as a spherical wave.

Analyze Recall from Equation 16.40 that the intensity I of a sound wave a distance r from a point source is $I = P_{avg}/4\pi r^2$, where P_{avg} is the average power output of the source and $4\pi r^2$ is the area of a sphere of radius r centered on the source. This expression is also valid for electromagnetic waves.

Set this expression for I equal to the intensity of an electromagnetic wave given by Equation 33.27:

$$I = \frac{P_{avg}}{4\pi r^2} = \frac{E_{max}^2}{2\mu_0 c}$$

Solve for the electric field magnitude:

$$E_{max} = \sqrt{\frac{\mu_0 c P_{avg}}{2\pi r^2}}$$

Let's make some assumptions about numbers to enter in this equation. The visible light output of a 60-W lightbulb operating at 5% efficiency is approximately 3.0 W by visible light. (The remaining energy transfers out of the lightbulb by thermal conduction and invisible radiation.) A reasonable distance from the lightbulb to the page might be 0.30 m.

Substitute these values:

$$E_{max} = \sqrt{\frac{(4\pi \times 10^{-7}\ \text{T} \cdot \text{m/A})(3.00 \times 10^8\ \text{m/s})(3.0\ \text{W})}{2\pi(0.30\ \text{m})^2}}$$

$$= \boxed{45\ \text{V/m}}$$

Use Equation 33.24 to find the magnetic field magnitude:

$$B_{max} = \frac{E_{max}}{c} = \frac{45\ \text{V/m}}{3.00 \times 10^8\ \text{m/s}} = \boxed{1.5 \times 10^{-7}\ \text{T}}$$

Finalize This value of the magnetic field magnitude is two orders of magnitude smaller than the Earth's magnetic field.

33.5 Momentum and Radiation Pressure

Electromagnetic waves transport linear momentum as well as energy. As this momentum is absorbed by some surface, pressure is exerted on the surface. Therefore, the surface is a nonisolated system for momentum. In this discussion, let's assume the electromagnetic wave strikes the surface at normal incidence and transports a total energy T_{ER} to the surface in a time interval Δt. Maxwell showed that if the surface absorbs all the incident energy T_{ER} in this time interval (as does a black body, introduced in Section 19.6), the total momentum $\vec{\mathbf{p}}$ transported to the surface has a magnitude

◀ Momentum transported to a perfectly absorbing surface

$$p = \frac{T_{ER}}{c} \qquad \text{(complete absorption)} \tag{33.34}$$

The pressure P exerted on the surface is defined as force per unit area F/A, which when combined with Newton's second law gives

$$P = \frac{F}{A} = \frac{1}{A}\frac{dp}{dt}$$

Substituting Equation 33.34 into this expression for pressure P gives

$$P = \frac{1}{A}\frac{dp}{dt} = \frac{1}{A}\frac{d}{dt}\left(\frac{T_{ER}}{c}\right) = \frac{1}{c}\frac{(dT_{ER}/dt)}{A}$$

We recognize $(dT_{ER}/dt)/A$ as the rate at which energy is arriving at the surface per unit area, which is the magnitude of the Poynting vector. Therefore, the radiation pressure P exerted on the perfectly absorbing surface is

$$P = \frac{S}{c} \quad \text{(complete absorption)} \tag{33.35}$$

◀ Radiation pressure exerted on a perfectly absorbing surface

If the surface is a perfect reflector (such as a mirror) and incidence is normal, the momentum transported to the surface in a time interval Δt is twice that given by Equation 33.34. That is, the momentum transferred to the surface by the incoming light is $p = T_{ER}/c$ and that transferred by the reflected light is also $p = T_{ER}/c$. Therefore,

$$p = \frac{2T_{ER}}{c} \quad \text{(complete reflection)} \tag{33.36}$$

> **PITFALL PREVENTION 33.5**
> **So Many p's** We have p for momentum and P for pressure, and they are both related to P for power! Be sure to keep all these symbols straight.

The radiation pressure exerted on a perfectly reflecting surface for normal incidence of the wave is

$$P = \frac{2S}{c} \quad \text{(complete reflection)} \tag{33.37}$$

◀ Radiation pressure exerted on a perfectly reflecting surface

For a surface that is neither a perfect absorber nor a perfect reflector, we can write the pressure as

$$P = (1 + f)\frac{S}{c} \tag{33.38}$$

where f is the fraction of the incident light that is reflected from the surface.

Although radiation pressures are very small (about 5×10^{-6} N/m^2 for direct sunlight), *solar sailing* is a low-cost means of sending spacecraft to the planets. Large sheets experience radiation pressure from sunlight and are used in much the way canvas sheets are used on earthbound sailboats. In 2010, the Japan Aerospace Exploration Agency (JAXA) launched the first spacecraft to use solar sailing as its primary propulsion, *IKAROS* (Interplanetary Kite-craft Accelerated by Radiation of the Sun). This spacecraft completed its planned mission and is now in orbit around the Sun, sending back data when it is close enough to the Sun for the solar panels to provide power.

QUICK QUIZ 33.4 To maximize the radiation pressure on the sails of a spacecraft using solar sailing, should the sheets be **(a)** very black to absorb as much sunlight as possible or **(b)** very shiny to reflect as much sunlight as possible?

Conceptual Example 33.4 **Sweeping the Solar System**

A great amount of dust exists in interplanetary space. Although in theory these dust particles can vary in size from molecular size to a much larger size, very little of the dust in our solar system is smaller than about 0.2 μm. Why?

SOLUTION

The dust particles are subject to two significant forces: the gravitational force that draws them toward the Sun and the radiation-pressure force that pushes them away from the Sun. The gravitational force is proportional to the cube of the radius of a spherical dust particle because it is proportional to the mass and therefore to the volume $4\pi r^3/3$ of the particle. The radiation pressure is proportional to the square of the radius because it depends on the planar cross section of the particle. For large particles, the gravitational force is greater than the force from radiation pressure. For particles having radii less than about 0.2 μm, the radiation-pressure force is greater than the gravitational force. As a result, these particles are swept out of our solar system by sunlight.

Example 33.5   **Pressure from a Laser Pointer**

When giving presentations, many people use a laser pointer to direct the attention of the audience to information on a screen. If a 3.0-mW pointer creates a spot on a screen that is 2.0 mm in diameter, determine the radiation pressure on a screen that reflects 70% of the light that strikes it. The power 3.0 mW is a time-averaged value.

SOLUTION

Conceptualize   Imagine the waves striking the screen and exerting a radiation pressure on it. The pressure should not be very large. Note that the radiation from a laser is very different from that due to a point source. The point source sends out radiation uniformly in all directions, while a laser concentrates the radiation into a narrow beam in a single direction.

Categorize   This problem involves a calculation of radiation pressure for a surface that is neither a perfect absorber nor a perfect reflector.

Analyze   We begin by determining the magnitude of the beam's Poynting vector.

Divide the time-averaged power delivered via the electromagnetic wave by the cross-sectional area of the beam:

$$S_{avg} = \frac{(Power)_{avg}}{A} = \frac{(Power)_{avg}}{\pi r^2} = \frac{3.0 \times 10^{-3} \text{ W}}{\pi \left(\dfrac{2.0 \times 10^{-3} \text{ m}}{2}\right)^2} = 955 \text{ W/m}^2$$

Use Equation 33.38 to find the pressure on the surface:

$$P_{avg} = (1 + f)\frac{S_{avg}}{c}$$

$$= (1 + 0.70)\frac{955 \text{ W/m}^2}{3.0 \times 10^8 \text{ m/s}} = \boxed{5.4 \times 10^{-6} \text{ N/m}^2}$$

Finalize   The pressure has an extremely small value, as expected. (Recall from Section 14.2 that atmospheric pressure is approximately 10^5 N/m^2.) Consider the magnitude of the Poynting vector, $S_{avg} = 955$ W/m^2. It is about the same as the intensity of sunlight at the Earth's surface. For this reason, it is not safe to shine the beam of a laser pointer into a person's eyes, which may be more dangerous than looking directly at the Sun.

WHAT IF?   What if the laser pointer is moved twice as far away from the screen? Does that affect the radiation pressure on the screen?

Answer   Because a laser beam is popularly represented as a beam of light with constant cross section, you might think that the intensity of radiation, and therefore the radiation pressure, is independent of distance from the screen. A laser beam, however, does not have a constant cross section at all distances from the source; rather, there is a small but measurable divergence of the beam. If the laser is moved farther away from the screen, the area of illumination on the screen increases, decreasing the intensity. In turn, the radiation pressure is reduced.

In addition, the doubled distance from the screen results in more loss of energy from the beam due to scattering from air molecules and dust particles as the light travels from the laser to the screen. This energy loss further reduces the radiation pressure on the screen.

33.6   Production of Electromagnetic Waves by an Antenna

We mentioned in Section 33.2 that the source of electromagnetic waves is accelerated charges. Let's investigate details of that process in the emission of radiation from charges in an antenna. Consider first a *half-wave antenna*. In this arrangement, two conducting rods are connected to a source of alternating voltage (such as an *LC* oscillator) as shown in Figure 33.11. The length of each rod is equal to one-quarter the wavelength of the radiation emitted when the oscillator operates at frequency *f*. The oscillator forces charges to accelerate back and forth between the two rods. Figure 33.11 shows the configuration of the electric and magnetic fields at some instant when the current is upward. The separation of charges in the upper and lower portions of the antenna make the electric field lines resemble those of an electric dipole. (As a result, this type of antenna is sometimes called a *dipole antenna*.) Because these charges are continuously oscillating between the two rods, the antenna can be approximated by an oscillating electric dipole. The current representing the movement of charges between the ends of the antenna produces magnetic field lines forming concentric circles around the antenna that are perpendicular to the electric field lines at all points. The magnetic field is zero at all points along the axis of the antenna. Furthermore, $\vec{E}$ and $\vec{B}$ are 90° out of phase in time; for example, the current is zero when the charges at the outer ends of the rods are at a maximum.

At the two points where the magnetic field is shown in Figure 33.11, the Poynting vector $\vec{S}$ is directed radially outward, indicating that energy is flowing away from the antenna at this instant. At later times, the fields and the Poynting vector reverse direction as the current alternates. Because $\vec{E}$ and $\vec{B}$ are 90° out of phase at points near the dipole, the net energy flow is zero. From this fact, you might conclude (incorrectly) that no energy is radiated by the dipole.

Energy is indeed radiated, however. Because the dipole fields fall off as $1/r^3$ (as shown in Example 22.6 for the electric field of a static dipole), they are negligible at great distances from the antenna. At these great distances, something else causes a type of radiation different from that close to the antenna. The source of this radiation is the continuous induction of an electric field by the time-varying magnetic field and the induction of a magnetic field by the time-varying electric field, predicted by Equations 33.11 and 33.12. The electric and magnetic fields produced in this manner are in phase with each other and vary as $1/r$. The result is an outward flow of energy at all times.

The angular dependence of the radiation intensity produced by a dipole antenna is shown in Figure 33.12. Notice that the intensity and the power radiated are a maximum in a plane that is perpendicular to the antenna and passing through its midpoint. Furthermore, the power radiated is zero along the antenna's axis. A mathematical solution to Maxwell's equations for the dipole antenna shows that the intensity of the radiation varies as $(\sin^2 \theta)/r^2$, where θ is measured from the axis of the antenna.

Electromagnetic waves can also induce currents in a receiving antenna. The response of a dipole receiving antenna at a given position is a maximum when the antenna axis is parallel to the electric field at that point and zero when the axis is perpendicular to the electric field.

QUICK QUIZ 33.5 If the antenna in Figure 33.11 represents the source of a distant radio station, what would be the best orientation for your portable radio antenna located to the right of the figure? (a) up-down along the page (b) left-right along the page (c) perpendicular to the page

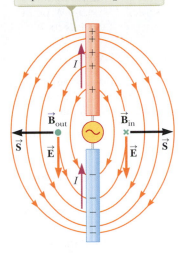

The electric field lines resemble those of an electric dipole (shown in Fig. 22.17).

Figure 33.11 A half-wave antenna consists of two metal rods connected to an alternating voltage source. This diagram shows $\vec{E}$ and $\vec{B}$ at an arbitrary instant when the current is upward.

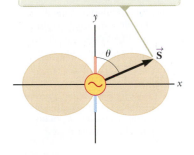

The distance from the origin to a point on the edge of the tan shape is proportional to the magnitude of the Poynting vector and the intensity of radiation in that direction.

Figure 33.12 Angular dependence of the intensity of radiation produced by an oscillating electric dipole.

33.7 The Spectrum of Electromagnetic Waves

The various types of electromagnetic waves are listed in Figure 33.13 (page 888), which shows the **electromagnetic spectrum.** Notice the wide ranges of frequencies and wavelengths. No sharp dividing point exists between one type of wave and the next. Remember that all forms of the various types of radiation are produced by the same phenomenon: acceleration of electric charges. The names given to the types of waves are simply a convenient way to describe the region of the spectrum in which they lie.

As noted in the discussion of Equation 8.2, energy can be transferred by electromagnetic waves; this transfer is represented by the term T_{ER} in the equation. We will see in future chapters that the energy carried by electromagnetic waves is proportional to the frequency. In Figure 33.13, therefore, the axis of increasing frequency can also be considered as an axis of increasing energy.

Radio waves, whose wavelengths range from more than 10^4 m to about 0.1 m, are the result of charges accelerating through conducting wires or in antennas. They are generated by such electronic devices as LC oscillators and are used in radio and television communication systems.

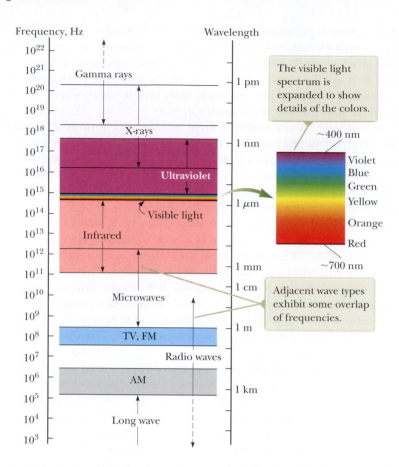

Figure 33.13 The electromagnetic spectrum.

TABLE 33.1 Approximate Correspondence Between Wavelengths of Visible Light and Color

Wavelength Range (nm)	Color Description
400–430	Violet
430–485	Blue
485–560	Green
560–590	Yellow
590–625	Orange
625–700	Red

Note: The wavelength ranges here are approximate. Different people will describe colors differently.

Microwaves have wavelengths ranging from approximately 0.3 m to 10^{-4} m and are also generated by electronic devices. Because of their short wavelengths, they are well suited for radar systems and for studying the atomic and molecular properties of matter. Microwave ovens are an interesting domestic application of these waves. It has been suggested that solar energy could be harnessed by beaming microwaves to the Earth from a solar collector in space.

Infrared waves have wavelengths ranging from approximately 10^{-3} m to the longest wavelength of visible light, 7×10^{-7} m. These waves, produced by molecules and room-temperature objects, are readily absorbed by most materials. The infrared (IR) energy absorbed by a substance appears as internal energy because the energy agitates the object's atoms, increasing their vibrational or translational motion, which results in a temperature increase. Infrared radiation has practical and scientific applications in many areas, including physical therapy, IR photography, and vibrational spectroscopy.

Visible light, the most familiar form of electromagnetic waves, is the part of the electromagnetic spectrum the human eye can detect. Light is produced by the rearrangement of electrons in atoms and molecules. The various wavelengths of visible light, which correspond to different colors, range from red ($\lambda \approx 7 \times 10^{-7}$ m) to violet ($\lambda \approx 4 \times 10^{-7}$ m). The sensitivity of the human eye is a function of wavelength, being a maximum at a wavelength of about 5.5×10^{-7} m. With that in mind, why do you suppose tennis balls often have a yellow-green color? Table 33.1 provides approximate correspondences between the wavelength of visible light and the color assigned to it by humans. Light is the basis of the science of optics and optical instruments, to be discussed in Chapters 34 through 37.

Ultraviolet waves cover wavelengths ranging from approximately 4×10^{-7} m to 6×10^{-10} m. The Sun is an important source of ultraviolet (UV) light, which is the main cause of sunburn. Recall from the beginning of this section that increasing frequency correlates with increasing energy. For UV light and the next two categories, x-rays and gamma rays, the energy is high enough for the radiation to penetrate into the skin. Sunscreen lotions are transparent to visible light but absorb most UV light. The higher a sunscreen's solar protection factor, or SPF, the greater the percentage of UV light absorbed. Ultraviolet rays have also been implicated in the formation of cataracts, a clouding of the lens inside the eye. Sunglasses that block UV light are critical, as noted in Figure 33.14.

Most of the UV light from the Sun is absorbed by ozone (O_3) molecules in the Earth's upper atmosphere, in a layer called the stratosphere. This ozone shield converts lethal high-energy UV radiation to IR radiation, which in turn warms the stratosphere.

X-rays have wavelengths in the range from approximately 10^{-8} m to 10^{-12} m. The most common source of x-rays is the stopping of high-energy electrons upon bombarding a metal target. X-rays are used as a diagnostic tool in medicine and as a treatment for certain forms of cancer. Because x-rays can damage or destroy living tissues and organisms, care must be taken to avoid unnecessary exposure or overexposure. X-rays are also used in the study of crystal structure because x-ray wavelengths are comparable to the atomic separation distances in solids (about 0.1 nm).

Gamma rays are electromagnetic waves emitted by radioactive nuclei and during certain nuclear reactions. High-energy gamma rays are a component of cosmic rays that enter the Earth's atmosphere from space. They have wavelengths ranging from approximately 10^{-10} m to less than 10^{-14} m. Gamma rays are highly penetrating and produce serious damage when absorbed by living tissues. Consequently, those working near such dangerous radiation must be protected with heavily absorbing materials such as thick layers of lead.

So what's going on with your Wi-Fi signal in the opening storyline? A Wi-Fi signal is a *radio* signal, often at 2.4 or 5 GHz. Therefore, in Figure 33.13, we see that Wi-Fi signals lie in the microwave region, near the upper end of the long vertical arrow representing the range of radio waves. Is it a surprise that your Wi-Fi signal goes through the walls of your home? No, not if you think about radio signals coming from outside your home to old-time portable radios, or television signals arriving at old-time indoor antennas. Why does the Wi-Fi signal strength fall off as your walk away from your house? This is just an example of Equation 16.40; the intensity falls off as the square of the distance from the source. In fact, the Wi-Fi symbol on the phone in the chapter-opening photograph suggests that effect: a point source is sending out spherical waves that spread out in space.

Figure 33.14 Wearing sunglasses that do not block ultraviolet (UV) light is worse for your eyes than wearing no sunglasses at all. The lenses of any sunglasses absorb some visible light, thereby causing the wearer's pupils to dilate. If the glasses do not also block UV light, more damage may be done to the lens of the eye because of the dilated pupils. If you wear no sunglasses at all, your pupils are contracted, you squint, and much less UV light enters your eyes. High-quality sunglasses block nearly all the eye-damaging UV light.

QUICK QUIZ 33.6 In many kitchens, a microwave oven is used to cook food. The frequency of the microwaves is on the order of 10^{10} Hz. Are the wavelengths of these microwaves on the order of **(a)** kilometers, **(b)** meters, **(c)** centimeters, or **(d)** micrometers?

QUICK QUIZ 33.7 A radio wave of frequency on the order of 10^5 Hz is used to carry a sound wave with a frequency on the order of 10^3 Hz. Is the wavelength of this radio wave on the order of **(a)** kilometers, **(b)** meters, **(c)** centimeters, or **(d)** micrometers?

Summary

> ## Definitions

In a region of space in which there is a changing electric field, there is a **displacement current** defined as

$$I_d \equiv \epsilon_0 \frac{d\Phi_E}{dt} \qquad (33.1)$$

where ϵ_0 is the permittivity of free space (see Section 22.3) and $\Phi_E = \int \vec{\mathbf{E}} \cdot d\vec{\mathbf{A}}$ is the electric flux.

The rate at which energy passes through a unit area by electromagnetic radiation is described by the **Poynting vector $\vec{\mathbf{S}}$**, where

$$\vec{\mathbf{S}} \equiv \frac{1}{\mu_0} \vec{\mathbf{E}} \times \vec{\mathbf{B}} \qquad (33.25)$$

continued

❯ Concepts and Principles

When used with the **Lorentz force law**, $\vec{F} = q\vec{E} + q\vec{v} \times \vec{B}$, **Maxwell's equations** describe all electromagnetic phenomena:

$$\oint \vec{E} \cdot d\vec{A} = \frac{q}{\epsilon_0} \qquad (33.4)$$

$$\oint \vec{E} \cdot d\vec{s} = -\frac{d\Phi_B}{dt} \qquad (33.6)$$

$$\oint \vec{B} \cdot d\vec{A} = 0 \qquad (33.5)$$

$$\oint \vec{B} \cdot d\vec{s} = \mu_0 I + \epsilon_0 \mu_0 \frac{d\Phi_E}{dt} \qquad (33.7)$$

Electromagnetic waves, which are predicted by Maxwell's equations, have the following properties and are described by the following mathematical representations of the traveling wave model for electromagnetic waves.

- The electric field and the magnetic field each satisfy a wave equation. These two wave equations, which can be obtained from Maxwell's third and fourth equations, are

$$\frac{\partial^2 E}{\partial x^2} = \mu_0 \epsilon_0 \frac{\partial^2 E}{\partial t^2} \qquad (33.19)$$

$$\frac{\partial^2 B}{\partial x^2} = \mu_0 \epsilon_0 \frac{\partial^2 B}{\partial t^2} \qquad (33.20)$$

- The waves travel through a vacuum with the speed of light c, where

$$c = \frac{1}{\sqrt{\mu_0 \epsilon_0}} \qquad (33.21)$$

- Numerically, the speed of electromagnetic waves in a vacuum is 3.00×10^8 m/s.
- The electric and magnetic fields are perpendicular to each other and perpendicular to the direction of wave propagation.
- The instantaneous magnitudes of $\vec{E}$ and $\vec{B}$ in an electromagnetic wave are related by the expression

$$\frac{E}{B} = c \qquad (33.24)$$

- Electromagnetic waves carry energy.
- Electromagnetic waves carry momentum.

Because electromagnetic waves carry momentum, they exert pressure on surfaces. If an electromagnetic wave whose Poynting vector is $\vec{S}$ is completely absorbed by a surface upon which it is normally incident, the radiation pressure on that surface is

$$P = \frac{S}{c} \quad \text{(complete absorption)} \qquad (33.35)$$

If the surface totally reflects a normally incident wave, the pressure is doubled.

The electric and magnetic fields of a sinusoidal plane electromagnetic wave propagating in the positive x direction can be written as

$$E = E_{max} \cos (kx - \omega t) \qquad (33.22)$$

$$B = B_{max} \cos (kx - \omega t) \qquad (33.23)$$

where k is the angular wave number and ω is the angular frequency of the wave. These equations represent special solutions to the wave equations for E and B.

The average value of the Poynting vector for a plane electromagnetic wave has a magnitude

$$S_{avg} = \frac{E_{max} B_{max}}{2\mu_0} = \frac{E_{max}^2}{2\mu_0 c} = \frac{cB_{max}^2}{2\mu_0} \qquad (33.27)$$

The intensity of a sinusoidal plane electromagnetic wave equals the average value of the Poynting vector taken over one or more cycles.

The electromagnetic spectrum includes waves covering a broad range of wavelengths, from long radio waves at more than 10^4 m to gamma rays at less than 10^{-14} m.

Think–Pair–Share

See the Preface for an explanation of the icons used in this problems set. For additional assessment items for this section, go to ⚡ **WEBASSIGN** From Cengage

1. You are working for NASA and have joined the team investigating the concept of solar sailing. (a) Your supervisor has asked your team to determine the minimum area of a perfectly reflecting solar sail that will be necessary to keep a 15 000-kg solar spacecraft moving away from the gravitational attraction of the Sun, which has a power output of 3.85×10^{26} W, and then make a presentation on your results. (b) After your

team presents your results and your co-workers are shocked at the size of the sail needed, your supervisor asks you, "But the needed area of the sail can be smaller as we move farther from the Sun, right, because the gravitational force will decrease?" How do you respond?

2. **ACTIVITY** Planets orbiting stars other than the Sun are called *exoplanets*. Your group has found the table below, which lists data on some exoplanets that have been discovered, in order of the year of discovery.

Planet	Mass M (M_{Earth})	Radius R (R_{Earth})	Orbital Radius a (AU)	Luminosity L_{star} of Star (P_{Sun})	Year of Discovery
GJ 436b	22.3	4.17	0.028 9	0.025	2004
GJ 674b	12.7	12.4	0.039	0.016	2007
Gliese 581c	5.40	1.50	0.073	0.013	2007
HAT-P-11b	26.2	4.63	0.053	0.26	2009
GJ 3470b	13.9	3.14	0.035 6	0.029	2012
Kepler-42b	2.86	0.768	0.011 6	0.002 4	2012
Kepler-42c	1.91	0.713	0.006 0	0.002 4	2012
Kepler-42d	0.955	0.209	0.015 4	0.002 4	2012
Kepler-138b	21.3	0.571	0.074 6	0.060	2014
HD 219134b	3.82	1.57	0.038 5	0.28	2015
Kepler-452b	2.86	1.50	1.05	1.2	2015

Suppose that the evolution of life like ours on Earth will occur for a surface temperature (without accounting for an atmosphere) between 250 K and 320 K and for accelerations due to gravity at the surface of the planet between 5.00 m/s² and 15.0 m/s². How many of the planets above are candidates for the evolution of life like ours? Assume that all planets reflect 30.0% of the light incident on them. The power output P_{Sun} of the Sun is 3.85×10^{26} W.

3. Your group has been hired as expert witnesses for an astronaut team that, after returning to Earth, is being reprimanded by the military for losing a very expensive, high-powered laser weapon that was to be tested from an orbital position. Here's the story: The team was watching their fellow astronaut take a space walk to move the laser to an orbital position away from the International Space Station (ISS) when an accident occurred. While everyone else was safely enclosed within the ISS, she was on her space walk, 10.0 m away from the entrance to the ISS, at rest with respect to the station. Her maneuvering unit had failed and she was not attached with a tether. Her total mass, including spacesuit and all gear, is 250 kg. This total includes the laser of mass 103 kg, which puts out a parallel beam of light of power 9.50×10^4 W. She was told by radio to aim the laser away from the ISS and turn it on, so it would act as a photon rocket, propelling her toward the station. She answered back, suggesting an alternative plan: she could just throw the laser in a direction away from the station, which would cause her to move toward the station. She estimated that she could throw the laser at a speed of 0.200 m/s relative to the station. The team discussed the options and decided to have her throw the laser. Prepare an argument that can hopefully support the astronaut team, arguing that using the laser as a photon rocket would have required too much time because she only had 1.00 h of oxygen left in her life support system, and that throwing the laser was the only option available to keep her alive. Divide up the work between two halves of your group and have one half work on the laser-throwing possibility and the other on the photon rocket possibility.

Problems

See the Preface for an explanation of the icons used in this problems set. For additional assessment items for this section, go to **WEBASSIGN** From Cengage

SECTION 33.1 Displacement Current and the General Form of Ampère's Law

1. **V** A 0.200-A current is charging a capacitor that has circular plates 10.0 cm in radius. If the plate separation is 4.00 mm, (a) what is the time rate of increase of electric field between the plates? (b) What is the magnetic field between the plates 5.00 cm from the center?

SECTION 33.2 Maxwell's Equations and Hertz's Discoveries

2. A very long, thin rod carries electric charge with the linear density 35.0 nC/m. It lies along the x axis and moves in the x direction at a speed of 1.50×10^7 m/s. (a) Find the electric field the rod creates at the point ($x = 0$, $y = 20.0$ cm, $z = 0$). (b) Find the magnetic field it creates at the same point. (c) Find the force exerted on an electron at this point, moving with a velocity of $(2.40 \times 10^8)\hat{\mathbf{i}}$ m/s.

3. **T** A proton moves through a region containing a uniform electric field given by $\vec{\mathbf{E}} = 50.0\hat{\mathbf{j}}$ V/m and a uniform magnetic field $\vec{\mathbf{B}} = (0.200\hat{\mathbf{i}} + 0.300\hat{\mathbf{j}} + 0.400\hat{\mathbf{k}})$ T. Determine the acceleration of the proton when it has a velocity $\vec{\mathbf{v}} = 200\hat{\mathbf{i}}$ m/s.

SECTION 33.3 Plane Electromagnetic Waves

Note: Assume the medium is vacuum unless specified otherwise.

4. **BIO** A diathermy machine, used in physiotherapy, generates electromagnetic radiation that gives the effect of "deep heat" when absorbed in tissue. One assigned frequency for diathermy is 27.33 MHz. What is the wavelength of this radiation?

5. The distance to the North Star, Polaris, is approximately 6.44×10^{18} m. (a) If Polaris were to burn out today, how many years from now would we see it disappear? (b) What time interval is required for sunlight to reach the Earth? (c) What time interval is required for a microwave signal to travel from the Earth to the Moon and back?

6. A radar pulse returns to the transmitter–receiver after a total travel time of 4.00×10^{-4} s. How far away is the object that reflected the wave?

7. The speed of an electromagnetic wave traveling in a transparent nonmagnetic substance is $v = 1/\sqrt{\kappa \mu_0 \epsilon_0}$, where κ is the dielectric constant of the substance. Determine the speed of light in water, which has a dielectric constant of 1.78 at optical frequencies.

8. **CR** You are working for SETI, the Search for Extraterrestrial Intelligence. One day, you receive a radio communication from an alien intelligence. Although you cannot understand their language, they have included some photos from an *I Love Lucy* episode. The photos allow you to determine that it is the episode in which Lucy makes a television commercial on Vitameatavegamin. This episode first aired on CBS on May 5, 1952. Before running to your supervisor to tell him the news, you quickly determine how far away in light-years the alien civilization is.

9. **AMT** **Review.** A microwave oven is powered by a magnetron, an electronic device that generates electromagnetic waves of frequency 2.45 GHz. The microwaves enter the oven and are reflected by the walls. The standing-wave pattern produced in the oven can cook food unevenly, with hot spots in the food at antinodes and cool spots at nodes, so a turntable is often used to rotate the food and distribute the energy. If a microwave oven intended for use with a turntable is instead used with a cooking dish in a fixed position, the antinodes can appear as burn marks on foods such as carrot strips or cheese. The separation distance between the burns is measured to be 6 cm ± 5%. From these data, calculate the speed of the microwaves.

10. Verify by substitution that the following equations are solutions to Equations 33.19 and 33.20, respectively:

$$E = E_{\max} \cos (kx - \omega t)$$
$$B = B_{\max} \cos (kx - \omega t)$$

11. *Why is the following situation impossible?* An electromagnetic wave travels through empty space with electric and magnetic fields described by

$$E = 9.00 \times 10^3 \cos [(9.00 \times 10^6)x - (3.00 \times 10^{15})t]$$
$$B = 3.00 \times 10^{-5} \cos [(9.00 \times 10^6)x - (3.00 \times 10^{15})t]$$

where all numerical values and variables are in SI units.

SECTION 33.4 Energy Carried by Electromagnetic Waves

12. At what distance from the Sun is the intensity of sunlight three times the value at the Earth? (The average Earth–Sun separation is 1.496×10^{11} m.)

13. **V** If the intensity of sunlight at the Earth's surface under a fairly clear sky is 1 000 W/m^2, how much electromagnetic energy per cubic meter is contained in sunlight?

14. **QC** The power of sunlight reaching each square meter of the Earth's surface on a clear day in the tropics is close to 1 000 W. On a winter day in Manitoba, the power concentration of sunlight can be 100 W/m^2. Many human activities are described by a power per unit area on the order of 10^2 W/m^2 or less. (a) Consider, for example, a family of four paying $66 to the electric company every 30 days for 600 kWh of energy carried by electrical transmission to their house, which has floor dimensions of 13.0 m by 9.50 m. Compute the power per unit area used by the family. (b) Consider a car 2.10 m wide and 4.90 m long traveling at 55.0 mi/h using gasoline having "heat of combustion" 44.0 MJ/kg with fuel economy 25.0 mi/gal. One gallon of gasoline has a mass of 2.54 kg. Find the power per unit area used by the car. (c) Explain why direct use of solar energy is not practical for running a conventional automobile. (d) What are some uses of solar energy that are more practical?

15. **V** High-power lasers in factories are used to cut through cloth and metal (Fig. P33.15). One such laser has a beam diameter of 1.00 mm and generates an electric field having an amplitude of 0.700 MV/m at the target. Find (a) the amplitude of the magnetic field produced, (b) the intensity of the laser, and (c) the power delivered by the laser.

Philippe Plailly/SPL/Science Source

Figure P33.15

16. **Review.** Model the electromagnetic wave in a microwave oven as a plane traveling wave moving to the left, with an intensity of 25.0 kW/m^2. An oven contains two cubical containers of small mass, each full of water. One has an edge length of 6.00 cm, and the other, 12.0 cm. Energy falls perpendicularly on one face of each container. The water in the smaller container absorbs 70.0% of the energy that falls on it. The water in the larger container absorbs 91.0%. That is, the fraction 0.300 of the incoming microwave energy passes through a 6.00-cm thickness of water, and the fraction $(0.300)(0.300) = 0.090$ passes through a 12.0-cm thickness. Assume a negligible amount of energy leaves either container by heat. Find the temperature change of the water in each container over a time interval of 480 s.

17. **CR** You are serving as an expert witness for the city council of a community. The council is exploring the concept of providing the electrical needs of the community by building a facility with photovoltaic cells to convert sunlight to electric potential energy. But they are facing resistance from members of the community, who claim that there is not enough open land in the community to build such a facility. The opposition is building toward a lawsuit, which the city council wants to avoid. The community requires 1.00 MW of power, and the best photovoltaic cells on the market at the time have an efficiency of 30.0%. In your community, an average intensity of sunlight during the day is 1 000 W/m^2. The council members have no idea how much land is needed, so they have asked you to estimate the area of land that must be found to construct this facility.

18. **QC** Assuming the antenna of a 10.0-kW radio station radiates spherical electromagnetic waves, (a) compute the maximum

value of the magnetic field 5.00 km from the antenna and (b) state how this value compares with the surface magnetic field of the Earth.

19. At what distance from a 100-W electromagnetic wave point source does $E_{max} = 15.0$ V/m?

20. At one location on the Earth, the rms value of the magnetic field caused by solar radiation is 1.80 μT. From this value, calculate (a) the rms electric field due to solar radiation, (b) the average energy density of the solar component of electromagnetic radiation at this location, and (c) the average magnitude of the Poynting vector for the Sun's radiation.

SECTION 33.5 Momentum and Radiation Pressure

21. A 25.0-mW laser beam of diameter 2.00 mm is reflected at normal incidence by a perfectly reflecting mirror. Calculate the radiation pressure on the mirror.

22. **Q|C** The intensity of sunlight at the Earth's distance from the Sun is 1 370 W/m^2. Assume the Earth absorbs all the sunlight incident upon it. (a) Find the total force the Sun exerts on the Earth due to radiation pressure. (b) Explain how this force compares with the Sun's gravitational attraction.

23. **T** A 15.0-mW helium–neon laser emits a beam of circular cross section with a diameter of 2.00 mm. (a) Find the maximum electric field in the beam. (b) What total energy is contained in a 1.00-m length of the beam? (c) Find the momentum carried by a 1.00-m length of the beam.

24. **S** A helium–neon laser emits a beam of circular cross section with a radius r and a power P. (a) Find the maximum electric field in the beam. (b) What total energy is contained in a length ℓ of the beam? (c) Find the momentum carried by a length ℓ of the beam.

25. **Q|C** A plane electromagnetic wave of intensity 6.00 W/m^2, moving in the x direction, strikes a small perfectly reflecting pocket mirror, of area 40.0 cm^2, held in the yz plane. (a) What momentum does the wave transfer to the mirror each second? (b) Find the force the wave exerts on the mirror. (c) Explain the relationship between the answers to parts (a) and (b).

26. **Q|C** Assume the intensity of solar radiation incident on the upper atmosphere of the Earth is 1 370 W/m^2 and use data from Table 13.2 as necessary. Determine (a) the intensity of solar radiation incident on Mars, (b) the total power incident on Mars, and (c) the radiation force that acts on that planet if it absorbs nearly all the light. (d) State how this force compares with the gravitational attraction exerted by the Sun on Mars. (e) Compare the ratio of the gravitational force to the light-pressure force exerted on the Earth and the ratio of these forces exerted on Mars, found in part (d).

SECTION 33.6 Production of Electromagnetic Waves by an Antenna

27. **Q|C** Extremely low-frequency (ELF) waves that can penetrate the oceans are the only practical means of communicating with distant submarines. (a) Calculate the length of a quarter-wavelength antenna for a transmitter generating ELF waves of frequency 75.0 Hz into air. (b) How practical is this means of communication?

28. A large, flat sheet carries a uniformly distributed electric current with current per unit width J_s. This current creates a magnetic field on both sides of the sheet, parallel to the sheet and perpendicular to the current, with magnitude $B = \frac{1}{2}\mu_0 J_s$. If the current is in the y direction and oscillates in time according to

$$J_{max}(\cos \omega t)\hat{\mathbf{j}} = J_{max}[\cos(-\omega t)]\hat{\mathbf{j}}$$

the sheet radiates an electromagnetic wave. Figure P33.28 shows such a wave emitted from one point on the sheet chosen to be the origin. Such electromagnetic waves are emitted from all points on the sheet. The magnetic field of the wave to the right of the sheet is described by the wave function

$$\vec{\mathbf{B}} = \frac{1}{2}\mu_0 J_{max}[\cos(kx - \omega t)]\hat{\mathbf{k}}$$

(a) Find the wave function for the electric field of the wave to the right of the sheet. (b) Find the Poynting vector as a function of x and t. (c) Find the intensity of the wave. (d) **What If?** If the sheet is to emit radiation in each direction (normal to the plane of the sheet) with intensity 570 W/m^2, what maximum value of sinusoidal current density is required?

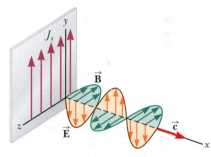

Figure P33.28

29. **Review.** Accelerating charges radiate electromagnetic waves. Calculate the wavelength of radiation produced by a proton in a cyclotron with a magnetic field of 0.350 T.

30. **S** **Review.** Accelerating charges radiate electromagnetic waves. Calculate the wavelength of radiation produced by a proton of mass m_p moving in a circular path perpendicular to a magnetic field of magnitude B.

SECTION 33.7 The Spectrum of Electromagnetic Waves

31. **V** Compute an order-of-magnitude estimate for the frequency of an electromagnetic wave with wavelength equal to (a) your height and (b) the thickness of a sheet of paper. How is each wave classified on the electromagnetic spectrum?

32. **Q|C** An important news announcement is transmitted by radio waves to people sitting next to their radios 100 km from the station and by sound waves to people sitting across the newsroom 3.00 m from the newscaster. Taking the speed of sound in air to be 343 m/s, who receives the news first? Explain.

ADDITIONAL PROBLEMS

33. Assume the intensity of solar radiation incident on the cloud tops of the Earth is 1 370 W/m^2. (a) Taking the average Earth–Sun separation to be 1.496×10^{11} m, calculate the total power radiated by the Sun. Determine the maximum values of (b) the electric field and (c) the magnetic field in the sunlight at the Earth's location.

34. Classify waves with frequencies of 2 Hz, 2 kHz, 2 MHz, 2 GHz, 2 THz, 2 PHz, 2 EHz, 2 ZHz, and 2 YHz on the

electromagnetic spectrum. Classify waves with wavelengths of 2 km, 2 m, 2 mm, 2 μm, 2 nm, 2 pm, 2 fm, and 2 am.

35. The eye is most sensitive to light having a frequency of 5.45×10^{14} Hz, which is in the green-yellow region of the visible electromagnetic spectrum. What is the wavelength of this light?

36. Write expressions for the electric and magnetic fields of a sinusoidal plane electromagnetic wave having an electric field amplitude of 300 V/m and a frequency of 3.00 GHz and traveling in the positive x direction.

37. You are working as a radio technician. One day, you set **CR** up a standing wave pattern with radio waves between two metal sheets 2.00 m apart. You cannot achieve a standing wave pattern with any smaller distances between the sheets. From this information, you determine the frequency of the radio waves.

38. One goal of the Russian space program is to illuminate dark northern cities with sunlight reflected to the Earth from a 200-m diameter mirrored surface in orbit. Several smaller prototypes have already been constructed and put into orbit. (a) Assume that sunlight with intensity 1 370 W/m² falls on the mirror nearly perpendicularly and that the atmosphere of the Earth allows 74.6% of the energy of sunlight to pass though it in clear weather. What is the power received by a city when the space mirror is reflecting light to it? (b) The plan is for the reflected sunlight to cover a circle of diameter 8.00 km. What is the intensity of light (the average magnitude of the Poynting vector) received by the city? (c) This intensity is what percentage of the vertical component of sunlight at St. Petersburg in January, when the sun reaches an angle of 7.00° above the horizon at noon?

39. The intensity of solar radiation at the top of the Earth's atmosphere is 1 370 W/m². Assuming 60% of the incoming solar energy reaches the Earth's surface and you absorb 50% of the incident energy, make an order-of-magnitude estimate of the amount of solar energy you absorb if you sunbathe for 60 minutes.

40. The Earth reflects approximately 38.0% of the incident sun-**Q|C** light from its clouds and surface. (a) Given that the intensity of solar radiation at the top of the atmosphere is 1 370 W/m², find the radiation pressure on the Earth, in pascals, at the location where the Sun is straight overhead. (b) State how this quantity compares with normal atmospheric pressure at the Earth's surface, which is 101 kPa.

41. Consider a small, spherical particle of radius r located in **AMT** space a distance $R = 3.75 \times 10^{11}$ m from the Sun. Assume the particle has a perfectly absorbing surface and a mass density of $\rho = 1.50$ g/cm³. Use $S = 214$ W/m² as the value of the solar intensity at the location of the particle. Calculate the value of r for which the particle is in equilibrium between the gravitational force and the force exerted by solar radiation.

42. Consider a small, spherical particle of radius r located in **S** space a distance R from the Sun, of mass M_S. Assume the particle has a perfectly absorbing surface and a mass density ρ. The value of the solar intensity at the particle's location is S. Calculate the value of r for which the particle is in equilibrium between the gravitational force and the force exerted by solar radiation. Your answer should be in terms of S, R, ρ, and other constants.

43. **Review.** A 1.00-m-diameter circular mirror focuses the Sun's rays onto a circular absorbing plate 2.00 cm in radius, which holds a can containing 1.00 L of water at 20.0°C. (a) If the solar intensity is 1.00 kW/m², what is the intensity on the absorbing plate? At the plate, what are the maximum magnitudes of the fields (b) $\vec{\mathbf{E}}$ and (c) $\vec{\mathbf{B}}$? (d) If 40.0% of the energy is absorbed, what time interval is required to bring the water to its boiling point?

44. (a) A stationary charged particle at the origin creates an **Q|C** electric flux of 487 N · m²/C through any closed surface surrounding the charge. Find the electric field it creates in the empty space around it as a function of radial distance r away from the particle. (b) A small source at the origin emits an electromagnetic wave with a single frequency into vacuum, equally in all directions, with power 25.0 W. Find the electric field amplitude as a function of radial distance away from the source. (c) At what distance is the amplitude of the electric field in the wave equal to 3.00 MV/m, representing the dielectric strength of air? (d) As the distance from the source doubles, what happens to the field amplitude? (e) State how the behavior shown in part (d) compares with the behavior of the field in part (a).

45. **Review.** (a) A homeowner has a solar water heater installed on the roof of his house (Fig. P33.45). The heater is a flat, closed box with excellent thermal insulation. Its interior is painted black, and its front face is made of insulating glass. Its emissivity for visible light is 0.900, and its emissivity for infrared light is 0.700. Light from the noontime Sun is incident perpendicular to the glass with an intensity of 1 000 W/m², and no water enters or leaves the box. Find the steady-state temperature of the box's interior. (b) **What If?** The homeowner builds an identical box with no water tubes. It lies flat on the ground in front of the house. He uses it as a cold frame, where he plants seeds in early spring. Assuming the same noontime Sun is at an elevation angle of 50.0°, find the steady-state temperature of the interior of the box when its ventilation slots are tightly closed.

Figure P33.45

46. You may wish to review Sections 16.4 and 16.8 on the trans-**GP** port of energy by string waves and sound. Figure P33.46 is a **Q|C** graphical representation of an electromagnetic wave moving in the x direction. We wish to find an expression for the intensity of this wave by means of a different process from that by which Equation 33.27 was generated. (a) Sketch a graph of the electric field in this wave at the instant $t = 0$, letting your flat paper represent the xy plane. (b) Compute

Figure P33.46

the energy density u_E in the electric field as a function of x at the instant $t = 0$. (c) Compute the energy density in the magnetic field u_B as a function of x at that instant. (d) Find the total energy density u as a function of x, expressed in terms of only the electric field amplitude. (e) The energy in a "shoebox" of length λ and frontal area A is $E_\lambda = \int_0^\lambda uA\,dx$. (The symbol E_λ for energy in a wavelength imitates the notation of Section 16.4.) Perform the integration to compute the amount of this energy in terms of A, λ, E_{max}, and universal constants. (f) We may think of the energy transport by the whole wave as a series of these shoeboxes going past as if carried on a conveyor belt. Each shoebox passes by a point in a time interval defined as the period $T = 1/f$ of the wave. Find the power the wave carries through area A. (g) The intensity of the wave is the power per unit area through which the wave passes. Compute this intensity in terms of E_{max} and universal constants. (h) Explain how your result compares with that given in Equation 33.27.

47. CR You are working at NASA, in a division that is studying the possibility of rotating small spacecraft using radiation pressure from the Sun. You have built a scale model of a spacecraft as shown in Figure P33.47. The central body is a spherical shell with mass $m = 0.500$ kg and radius $R = 15.0$ cm. The thin rod extending from each side of the sphere is of mass $m_r = 50.0$ g and of total length $\ell = 1.00$ m. At each end of the rod are circular plates of mass $m_p = 10.0$ g and radius $r_p = 2.00$ cm, with the center of each plate located at the end of the rod. One plate is perfectly reflecting and the other is perfectly absorbing. The initial configuration of this model is that it is at rest, mounted on a vertical axle with very low friction. To begin the simulation, you expose the model to sunlight of intensity $I_s = 1\,000$ W/m², directed perpendicularly to the plates, for a time interval of $\Delta t = 2.00$ min. The sunlight is then removed from the model. Determine the angular velocity ω with which the model now rotates about the axle.

Figure P33.47

48. The electromagnetic power radiated by a nonrelativistic particle with charge q moving with acceleration a is

$$P = \frac{q^2 a^2}{6\pi\epsilon_0 c^3}$$

where ϵ_0 is the permittivity of free space (also called the permittivity of vacuum) and c is the speed of light in vacuum. (a) Show that the right side of this equation has units of watts. An electron is placed in a constant electric field of magnitude 100 N/C. Determine (b) the acceleration of the electron and (c) the electromagnetic power radiated by this electron. (d) **What If?** If a proton is placed in a cyclotron with a radius of 0.500 m and a magnetic field of magnitude 0.350 T, what electromagnetic power does this proton radiate just before leaving the cyclotron?

49. Review. A 5.50-kg black cat and her four black kittens, each with mass 0.800 kg, sleep snuggled together on a mat on a cool night, with their bodies forming a hemisphere. Assume the hemisphere has a surface temperature of 31.0°C, an emissivity of 0.970, and a uniform density of 990 kg/m³. Find (a) the radius of the hemisphere, (b) the area of its curved surface, (c) the radiated power emitted by the cats at their curved surface, and (d) the intensity of radiation at this surface. You may think of the emitted electromagnetic wave as having a single predominant frequency. Find (e) the amplitude of the electric field in the electromagnetic wave just outside the surface of the cozy pile and (f) the amplitude of the magnetic field. (g) **What If?** The next night, the kittens all sleep alone, curling up into separate hemispheres like their mother. Find the total radiated power of the family. (For simplicity, ignore the cats' absorption of radiation from the environment.)

CHALLENGE PROBLEMS

50. Review. In the absence of cable input or a satellite dish, a television set can use a dipole-receiving antenna for VHF channels and a loop antenna for UHF channels. The VHF antenna consisted of two straight metal rods that were often called "rabbit ears." The UHF antenna produces an emf from the changing magnetic flux through the loop. The television station broadcasts a signal with a frequency f, and the signal has an electric field amplitude E_{max} and a magnetic field amplitude B_{max} at the location of the receiving antenna. (a) Using Faraday's law, derive an expression for the amplitude of the emf that appears in a single-turn, circular loop antenna with a radius r that is small compared with the wavelength of the wave. (b) If the electric field in the signal points vertically, what orientation of the loop gives the best reception?

51. A plane electromagnetic wave varies sinusoidally at 90.0 MHz as it travels through vacuum along the positive x direction. The peak value of the electric field is 2.00 mV/m, and it is directed along the positive y direction. Find (a) the wavelength, (b) the period, and (c) the maximum value of the magnetic field. (d) Write expressions in SI units for the space and time variations of the electric field and of the magnetic field. Include both numerical values and unit vectors to indicate directions. (e) Find the average power per unit area this wave carries through space. (f) Find the average energy density in the radiation (in joules per cubic meter). (g) What radiation pressure would this wave exert upon a perfectly reflecting surface at normal incidence?

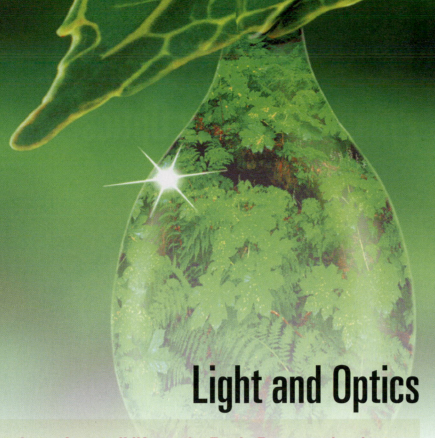

PART 5

Light and Optics

Light is basic to almost all life on the Earth. For example, plants convert the energy transferred by sunlight to chemical energy through photosynthesis. In addition, light is the principal means by which we are able to transmit and receive information to and from objects around us and throughout the Universe. Light is a form of electromagnetic radiation and represents energy transfer from the source to the observer. It is represented by T_{ER} in Equation 8.2.

Many phenomena in our everyday life depend on the properties of light. When you watch a television or view photos on a computer monitor, you are seeing millions of colors formed from combinations of only three colors that are physically on the screen: red, blue, and green. The blue color of the daytime sky is a result of the optical phenomenon of *scattering* of light by air molecules, as are the red and orange colors of sunrises and sunsets. You see your image in your bathroom mirror in the morning or the images of other cars in your rearview mirror when you are driving. These images result from *reflection* of light. If you wear glasses or contact lenses, you are depending on *refraction* of light for clear vision. The colors of a rainbow result from *dispersion* of light as it passes through raindrops hovering in the sky after a rainstorm. If you have ever seen the colored circles of the glory surrounding the shadow of your airplane on clouds as you fly above them, you are seeing an effect that results from *interference* of light. The phenomena mentioned here have been studied by scientists and are well understood.

In the introduction to Chapter 34, we briefly discuss the dual nature of light. In some cases, it is best to model light as a stream of particles; in others, a wave model works better. Chapters 34 through 37 concentrate on those aspects of light that are best understood through the wave model of light. In Part 6, we will investigate the particle nature of light. ■

The light rays coming from the leaves in the background of this scene did not form a focused image in the camera that took this photograph. Consequently, the background appears very blurry. Light rays passing though the raindrop, however, have been altered so as to form a focused image of the background leaves for the camera. The optical principles we study in this part of the book will explain phenomena such as this one. (*Don Hammond Photography*)

34

The Nature of Light and the Principles of Ray Optics

This photograph of a rainbow shows the range of colors from red on the top to violet on the bottom. The appearance of the rainbow depends on three optical phenomena discussed in this chapter: reflection, refraction, and dispersion. The faint pastel-colored bows beneath the main rainbow are called supernumerary bows. They are formed by interference between rays of light leaving raindrops below those causing the main rainbow. (*John W. Jewett Jr*)

STORYLINE **In the previous chapter, you took a walk outside your** home to investigate the signal strength of your home Wi-Fi system. You are now standing on the sidewalk contemplating your results. You glance over at your shadow on the dewdrop-encrusted grass. You see a bright glow around the shadow of your head. Startled by this effect, you look upward and see a rainbow in the sky. And there are some faint pastel-colored bands below the main rainbow, as in the photo above. You think that all these effects must have something to do with the Sun behind you, so you turn to look up at the Sun. You are startled to see two bright areas in the sky far off to either side of the Sun. You then look down the street and see what appears to be a puddle of water in the street. But the street is dry where you are standing. You walk down to where the puddle was seen. The roadway is dry there, too! What's going on? What's causing all these effects?

CONNECTIONS   In the previous chapter, we introduced the notion of electromagnetic waves. In these next few chapters on optics, we will focus on light as our representative electromagnetic wave, because we have everyday experience with light. This first chapter on optics begins by discussing the nature of light and early methods for measuring the speed of light. Next, we study the fundamental phenomena of geometric optics: reflection of light from a surface and refraction as the light crosses the boundary between two media. We also study the dispersion of light as it refracts into materials, resulting in visual displays such as the rainbow. Finally, we investigate the phenomenon of total internal reflection, which is the basis for the operation of optical fibers and the technology of fiber optics. These investigations will set up what we need to form optical images using mirrors and lenses in Chapter 35. As we continue through our investigations into the behavior of light in these next few chapters, we will be setting ourselves up for our studies in Part 6, where much of quantum physics deals with the interaction between light and matter.

34.1 The Nature of Light

Before the beginning of the 19th century, light was considered to be a stream of particles that either was emitted by the object being viewed or emanated from the eyes of the viewer. Newton, the chief architect of the particle model of light, held that particles were emitted from a light source and that these particles stimulated the sense of sight upon entering the eye. Using this idea, he was able to explain reflection and refraction.

Most scientists accepted Newton's particle model. During Newton's lifetime, however, another model was proposed, one that argued that light might be some sort of wave motion. In 1678, Dutch physicist and astronomer Christiaan Huygens showed that a wave model of light could also explain reflection and refraction.

In 1801, Thomas Young (1773–1829) provided the first clear *experimental* demonstration of the wave nature of light. Young showed that under appropriate conditions light rays interfere with one another according to the waves in interference model, just like mechanical waves (Chapter 17). Such behavior could not be explained at that time by a particle model because there was no conceivable way in which two or more particles could come together and cancel one another. Additional developments during the 19th century led to the general acceptance of the wave model of light, the most important resulting from the work of Maxwell, who in 1873 asserted that light was a form of high-frequency electromagnetic wave. As discussed in Chapter 33, Hertz provided experimental confirmation of Maxwell's theory in 1887 by producing and detecting electromagnetic waves.

These results represented convincing information that light has a wave nature, and scientists accepted the wave nature of light. Surprisingly, in the early twentieth century, new experiments indicated that light *also* has a particle nature! The particles of light are called *photons*.

We will explore the wave nature of light in these next few chapters on optics and delay the study of the particle nature of light until Chapter 39. Let's begin by looking at how the speed of light was measured historically.

Light travels at such a high speed (to three digits, $c = 3.00 \times 10^8$ m/s) that early attempts to measure its speed were unsuccessful. Galileo attempted to measure the speed of light by positioning two observers in towers separated by approximately 10 km. Each observer carried a shuttered lantern. One observer would open his shutter first, and then the other would open his shutter at the moment he saw the light from the first lantern. Galileo reasoned that by knowing the transit time of the light beams from one lantern to the other and the distance between the two lanterns, he could obtain the speed. His results were inconclusive. Today, we realize (as Galileo concluded) that it is impossible to measure the speed of light in this manner because the transit time for the light is so much less than the reaction time of the observers in opening the shutters. Let's look at two later methods that were more successful.

Historical/Getty Images

Christiaan Huygens
Dutch Physicist and Astronomer
(1629–1695)
Huygens is best known for his contributions to the fields of optics and dynamics. To Huygens, light was a type of vibratory motion, spreading out and producing the sensation of light when impinging on the eye. On the basis of this theory, he deduced the laws of reflection and refraction and explained the phenomenon of double refraction.

Roemer's Method

In 1675, Danish astronomer Ole Roemer (1644–1710) made observations that led to the first successful estimate of the speed of light. Roemer's technique involved astronomical observations of Io, one of the moons of Jupiter. Io has a period of revolution around Jupiter of approximately 42.5 h. The period is measured by observing eclipses of Io as it revolves about Jupiter.

When Earth is far from Jupiter like at position E_1, the eclipses of Io appear late. When Earth is at positions like E_2, the eclipses appear early.

Figure 34.1 Roemer's method for measuring the speed of light (drawing not to scale).

Figure 34.2 Fizeau's method for measuring the speed of light using a rotating toothed wheel. The light source is considered to be at the location of the wheel; therefore, the distance d is known.

An observer using the orbital motion of Io as a clock would expect the orbit to have a constant period. After collecting data for more than a year, however, Roemer observed a systematic variation in Io's period. He found that the eclipses were later than average when the Earth was in a position in its orbit like E_1 in Figure 34.1 on the opposite side of the Sun from Jupiter and far from it, and earlier than average when the Earth was on the same side of the Sun as Jupiter and closer to it as at position E_2. Roemer attributed this variation in the observed period to the extra time interval required for the light representing the eclipse to travel across the diameter of the Earth's orbit.

Using Roemer's data, Huygens estimated the lower limit for the speed of light to be approximately 2.3×10^8 m/s. This experiment is important historically because it demonstrated that light does have a finite speed and gave an estimate of this speed.

Fizeau's Method

The first successful method for measuring the speed of light by means of purely terrestrial techniques was developed in 1849 by French physicist Armand H. L. Fizeau (1819–1896). Figure 34.2 represents a simplified diagram of Fizeau's apparatus. The basic procedure is to measure the total time interval during which light travels from some point to a distant mirror and back. If d is the distance between the light source and the mirror and if the time interval for one round trip is Δt, the speed of light is $c = 2d/\Delta t$.

To measure the transit time, Fizeau used a rotating toothed wheel, which converts a continuous beam of light into a series of light pulses. Therefore, the wheel acts as the light source and defines one end of the distance d. The observer looks through the teeth and determines whether or not the reflected light is observable. For example, if the pulse traveling toward the mirror and passing the opening at point A in Figure 34.2 should reflect from the mirror and return to the wheel at the instant tooth B had rotated into position to cover the return path, the pulse would not reach the observer. At a greater rate of rotation, the opening at point C could move into position to allow the reflected pulse to reach the observer. Knowing the distance d, the number of teeth in the wheel, and the angular speed of the wheel, Fizeau arrived at a value of 3.1×10^8 m/s. Similar measurements made by subsequent investigators yielded more precise values for c, which led to the currently accepted value of $2.997\ 924\ 58 \times 10^8$ m/s.

| Example **34.1** | **Measuring the Speed of Light with Fizeau's Wheel** |

Assume Fizeau's wheel has 360 teeth and rotates at 55.0 rev/s when a pulse of light passing through opening A in Figure 34.2 passes through opening C on its return. If the distance to the mirror is 7 500 m, what is the speed of light?

SOLUTION

Conceptualize   Imagine a pulse of light passing through opening A in Figure 34.2 and reflecting from the mirror. By the time the pulse arrives back at the wheel, tooth B has passed by, and opening C has rotated into the position previously occupied by opening A.

Categorize   The wheel is a rigid object rotating at constant angular speed. We model the pulse of light as a *particle under constant speed*.

Analyze   The wheel has 360 teeth, so it must have 360 openings. Therefore, because the light passes through opening A and reflects back through the opening immediately adjacent to A, the wheel must rotate through an angular displacement of $\frac{1}{360}$ rev in the time interval during which the light pulse makes its round trip.

From the particle under constant speed model, find the speed of the pulse of light, and use Equation 10.2 to substitute for the time interval for the pulse's round trip:

$$c = \frac{2d}{\Delta t} = \frac{2d\omega}{\Delta\theta}$$

34.1 continued

Substitute numerical values:

$$c = \frac{2(7\,500 \text{ m})(55.0 \text{ rev/s})}{\frac{1}{360} \text{ rev}} = 2.97 \times 10^8 \text{ m/s}$$

Finalize   This result is very close to the actual value of the speed of light.

34.2 The Ray Approximation in Ray Optics

The field of **ray optics** (sometimes called *geometric optics*) involves the study of the propagation of light. Ray optics assumes light travels in a fixed direction in a straight line as it passes through a uniform medium and changes its direction when it meets the surface of a different medium or if the optical properties of the medium are nonuniform in either space or time. In our study of ray optics here and in Chapter 35, we use what is called the **ray approximation.** To understand this approximation, first recall that the rays of a given wave are straight lines perpendicular to the wave fronts as illustrated in Figure 34.3 for a plane wave. In the ray approximation, a wave moving through a uniform medium travels in a straight line in the direction of its rays.

If the wave meets a barrier in which there is a circular opening whose diameter is much larger than the wavelength as in Figure 34.4a, the wave emerging from the opening continues to move in a straight line (apart from some small edge effects); hence, the ray approximation is valid. If the diameter of the opening is on the order of the wavelength as in Figure 34.4b, the waves spread out from the opening in all directions. This effect, called *diffraction,* will be studied in Chapter 37. Finally, if the opening is much smaller than the wavelength, the waves to the right of the barrier can be approximated as if there is a point source at the opening as shown in Fig. 34.4c.

Similar effects are seen when waves encounter an opaque object of dimension d. In that case, when $\lambda << d$, the object casts a sharp shadow.

The ray approximation and the assumption that $\lambda << d$ are used in this chapter and in Chapter 35, both of which deal with ray optics. This approximation is very good for the study of mirrors, lenses, prisms, and associated optical instruments such as telescopes, cameras, and eyeglasses. When we study interference, diffraction, and polarization in Chapters 36 and 37, we will need to look more closely at the wave characteristics of light.

The rays, which always point in the direction of the wave propagation, are straight lines perpendicular to the wave fronts.

Rays

Wave fronts

Figure 34.3 A plane wave propagating to the right.

When $\lambda << d$, the rays continue in a straight-line path and the ray approximation remains valid.

When $\lambda \approx d$, the rays spread out after passing through the opening.

When $\lambda >> d$, the opening behaves as a point source emitting spherical waves.

$\lambda << d$

$\lambda \approx d$

$\lambda >> d$

a   **b**   **c**

Figure 34.4 A plane wave of wavelength λ is incident on a barrier in which there is an opening of diameter d.

Figure 34.5 Schematic representation of (a) specular reflection, where the reflected rays are all parallel to one another, and (b) diffuse reflection, where the reflected rays travel in random directions. (c) and (d) Photographs of specular and diffuse reflection using laser light.

The incident ray, the reflected ray, and the normal all lie in the same plane, and $\theta_1' = \theta_1$.

Normal

Incident ray

Reflected ray

θ_1 θ_1'

Figure 34.6 The wave under reflection model.

PITFALL PREVENTION 34.1

Subscript Notation The subscript 1 refers to parameters for the light in the initial medium. When light travels from one medium to another, we use the subscript 2 for the parameters associated with the light in the new medium. In this discussion, the light stays in the same medium, so we only have to use the subscript 1.

Law of reflection ▶

34.3 Analysis Model: Wave Under Reflection

We introduced the concept of reflection of waves in a discussion of waves on strings in Section 17.3. As with waves on strings, when a light ray traveling in one medium encounters a boundary with another medium, part of the incident light is reflected. For waves on a one-dimensional string, the reflected wave must necessarily be restricted to a direction along the string. For light waves free to travel in three-dimensional space, no such restriction applies and the reflected light waves can be in directions different from the direction of the incident waves. Figure 34.5a shows several rays of a beam of light incident on a smooth, mirror-like, reflecting surface. The reflected rays are parallel to one another as indicated in the figure. The direction of a reflected ray is in the plane perpendicular to the reflecting surface that contains the incident ray. Reflection of light from such a smooth surface is called **specular reflection.** If the reflecting surface is rough as in Figure 34.5b, the surface reflects the rays not as a parallel set but in various directions. Reflection from any rough surface is known as **diffuse reflection.** A surface behaves as a smooth surface as long as the surface variations are much smaller than the wavelength of the incident light. Your bathroom mirror exhibits specular reflection, whereas light reflecting from this page experiences diffuse reflection.

The difference between these two kinds of reflection explains why it is more difficult to see while driving on a rainy night than on a dry night. If the road is wet, the smooth surface of the water specularly reflects most of your headlight beams away from your car (and perhaps into the eyes of oncoming drivers). When the road is dry, its rough surface diffusely reflects part of your headlight beam back toward you, allowing you to see the road more clearly. In this book, we restrict our study to specular reflection and use the term *reflection* to mean specular reflection.

Consider a light ray traveling in air and incident at an angle on a flat, smooth surface as shown in Figure 34.6. The incident and reflected rays make angles θ_1 and θ_1', respectively, where the angles are measured between the normal and the rays. (The normal is a line drawn perpendicular to the surface at the point where the incident ray strikes the surface.) Experiments and theory show that **the angle of reflection equals the angle of incidence:**

$$\theta_1' = \theta_1 \tag{34.1}$$

This relationship is called the **law of reflection.** Because reflection of waves from an interface between two media is a common phenomenon, we identify an analysis model for this situation: the **wave under reflection.** Equation 34.1 is the mathematical representation of this model.

QUICK QUIZ 34.1 In the movies, you sometimes see an actor looking in a mirror and you can see his face in the mirror. It can be said with certainty that during the filming of such a scene, the actor sees in the mirror: **(a)** his face **(b)** your face **(c)** the director's face **(d)** the movie camera **(e)** impossible to determine

---

Example 34.2   The Double-Reflected Light Ray

Two mirrors make an angle of 120° with each other as illustrated in Figure 34.7a. A ray is incident on mirror M_1 at an angle of 65° to the normal. Find the direction of the ray after it is reflected from mirror M_2.

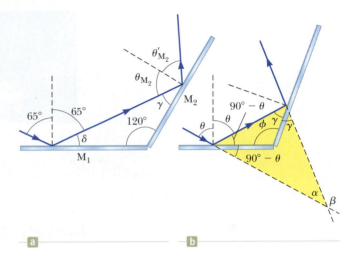

SOLUTION

Conceptualize Figure 34.7a helps conceptualize this situation. The incoming ray reflects from the first mirror, and the reflected ray is directed toward the second mirror. Therefore, there is a second reflection from the second mirror.

Categorize Because the interactions with both mirrors are simple reflections, we apply the *wave under reflection* model and some geometry.

Figure 34.7 (Example 34.2) (a) Mirrors M_1 and M_2 make an angle of 120° with each other. (b) The geometry for an arbitrary mirror angle.

Analyze From the law of reflection, the first reflected ray makes an angle of 65° with the normal.

Find the angle the first reflected ray makes with the horizontal:

$$\delta = 90° - 65° = 25°$$

From the triangle made by the first reflected ray and the two mirrors, find the angle the reflected ray makes with M_2:

$$\gamma = 180° - 25° - 120° = 35°$$

Find the angle the first reflected ray makes with the normal to M_2:

$$\theta_{M_2} = 90° - 35° = 55°$$

From the law of reflection, find the angle the second reflected ray makes with the normal to M_2:

$$\theta'_{M_2} = \theta_{M_2} = \boxed{55°}$$

Finalize Let's explore variations in the angle between the mirrors as follows.

WHAT IF? Notice that the angle between the incident and reflected rays at Mirror M_1 is 65° + 65° = 130°. Therefore, the angle by which the direction of the light ray changes from its original direction is 180° − 130° = 50°. Similarly, for the reflection at mirror M_2, the change of direction is 70°. Therefore, the *overall* change in direction of the light ray for two reflections is 50° + 70° = 120°. Interesting! This angle is the same as that between the mirrors! What if the angle between the mirrors is changed? Is the overall change in the direction of the light ray always equal to the angle between the mirrors?

Answer Making a general statement based on one data point or one observation is always a dangerous practice! Let's investigate the change in direction for a general situation. Figure 34.7b shows the mirrors at an arbitrary angle ϕ and the incoming light ray striking the mirror at an arbitrary angle θ with respect to the normal to the mirror surface. In accordance with the law of reflection and the sum of the interior angles of a triangle, the angle γ is given by $\gamma = 180° - (90° - \theta) - \phi = 90° + \theta - \phi$.

Consider the triangle highlighted in yellow in Figure 34.7b and determine α:

$$\alpha + 2\gamma + 2(90° - \theta) = 180° \quad \rightarrow \quad \alpha = 2(\theta - \gamma)$$

Notice from Figure 34.7b that the overall change in direction of the light ray from its original direction is angle β. Use the geometry in the figure to solve for β:

$$\beta = 180° - \alpha = 180° - 2(\theta - \gamma)$$
$$= 180° - 2[\theta - (90° + \theta - \phi)] = 360° - 2\phi$$

Notice that β is not equal to ϕ. For $\phi = 120°$, we obtain $\beta = 120°$, which happens to be the same as the mirror angle; that is true only for this special angle between the mirrors, however. For example, if $\phi = 90°$, we obtain $\beta = 180°$. In that case, the light is reflected straight back to its origin.

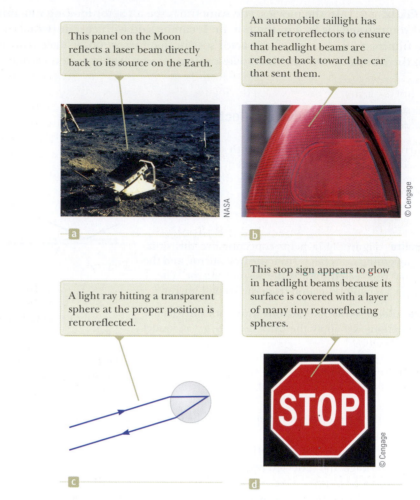

This panel on the Moon reflects a laser beam directly back to its source on the Earth.

An automobile taillight has small retroreflectors to ensure that headlight beams are reflected back toward the car that sent them.

A light ray hitting a transparent sphere at the proper position is retroreflected.

This stop sign appears to glow in headlight beams because its surface is covered with a layer of many tiny retroreflecting spheres.

Figure 34.8 Applications of retroreflection.

This leg of an ant gives a scale for the size of the mirrors.

The mirror on the left is "on," and the one on the right is "off."

Figure 34.9 (a) An array of mirrors on the surface of a digital micromirror device. Each mirror has an area of approximately 16 μm². (b) A close-up view of two single micromirrors.

If the angle between two mirrors is 90°, the reflected beam returns to the source parallel to its original path as discussed in the What If? section of the preceding example. This phenomenon, called *retroreflection,* has many practical applications. If a third mirror is placed perpendicular to the first two so that the three form the corner of a cube, the retroreflection phenomenon works in three dimensions. In 1969, a panel of many small reflectors was placed on the Moon by the *Apollo 11* astronauts (Fig. 34.8a). A laser beam from the Earth is reflected directly back on itself, and its transit time is measured. This information is used to determine the distance to the Moon with an uncertainty of 15 cm. (Imagine how difficult it would be to align a regular flat mirror on the Moon so that the reflected laser beam would hit a particular location on the Earth!) A more everyday application is found in automobile taillights. Part of the plastic making up the taillight is formed into many tiny cube corners (Fig. 34.8b) so that headlight beams from cars approaching from the rear are reflected back to the drivers. Instead of cube corners, small spherical bumps are sometimes used (Fig. 34.8c). Tiny clear spheres are used in a coating material found on many road signs. Due to retroreflection from these spheres, the stop sign in Figure 34.8d appears much brighter than it would if it were simply a flat, shiny surface. Retroreflectors are also used for reflective panels on running shoes and running clothing to allow joggers to be seen at night.

Another practical application of the law of reflection is the digital projection of movies, television shows, and computer presentations. A digital projector uses an optical semiconductor chip called a *digital micromirror device.* This device contains an array of tiny mirrors (Fig. 34.9a) that can be individually tilted by means of signals to an address electrode underneath the edge of the mirror. Each mirror corresponds to a pixel in the projected image. When the pixel corresponding to a given mirror is to be bright, the mirror is in the "on" position and is oriented so as to reflect light from a source illuminating the array to the screen (Fig. 34.9b). When the pixel for this mirror is to be dark, the mirror is "off" and is tilted so that the light is reflected away from the screen. The brightness of the pixel is determined by the total time interval during which the mirror is in the "on" position during the display of one image.

Digital movie projectors use three micromirror devices, one for each of the primary colors red, blue, and green, so that movies can be displayed with up to 35 trillion colors. Because information is stored as binary data, a digital movie does not degrade with time as does film. Furthermore, because the movie is entirely in the form of computer software, it can be delivered to theaters by means of satellites, optical discs, or optical fiber networks.

 ANALYSIS MODEL **Wave Under Reflection**

Imagine a wave (electromagnetic or mechanical) traveling through space and striking a flat surface at an angle θ_1 with respect to the normal to the surface. The wave will reflect from the surface in a direction described by the **law of reflection**—the angle of reflection θ_1' equals the angle of incidence θ_1:

$$\theta_1' = \theta_1 \qquad \text{(34.1)}$$

Examples:

- Sound waves from an orchestra reflect from a bandshell out to the audience.
- A mirror is used to deflect a laser beam in a laser light show.
- Your bathroom mirror reflects light from your face back to you to form an image of your face (Chapter 35).
- X-rays reflected from a crystalline material create an optical pattern that can be used to understand the structure of the solid (Chapter 37).

34.4 Analysis Model: Wave Under Refraction

In addition to the phenomenon of reflection discussed for waves on strings in Section 17.3, we also found that some of the energy of the incident wave transmits into the new medium. For example, consider Figures 17.11 and 17.12, in which a pulse on a string approaching a junction with another string both reflects from and transmits past the junction and into the second string. Similarly, when a ray of light traveling through a transparent medium encounters a boundary leading into another transparent medium as shown in Figure 34.10, part of the energy is reflected and part enters the second medium. As with reflection, the direction of the transmitted wave exhibits an interesting behavior because of the three-dimensional nature of the light waves. The ray that enters the second medium changes its direction of propagation at the boundary, bending toward or away from the normal, and is said to be **refracted.** The incident ray, the reflected ray, and the refracted ray all lie in the same plane. The **angle of refraction,** θ_2 in Figure 34.10a, depends on the properties of the two media and on the angle of incidence θ_1 through the relationship

$$\frac{\sin \theta_2}{\sin \theta_1} = \frac{v_2}{v_1} \qquad \text{(34.2)}$$

where v_1 is the speed of light in the first medium and v_2 is the speed of light in the second medium. We have stated this equation without proof, but it will be derived in Section 34.5.

Ⓠ**UICK QUIZ 34.2** If beam ① is the incoming beam in Figure 34.10b, which of the other four red lines are reflected beams and which are refracted beams?

All rays and the normal lie in the same plane, and the refracted ray is bent toward the normal because $v_2 < v_1$.

Figure 34.10 (a) The wave under refraction model. (b) Light incident on the Lucite block refracts both when it enters the block and when it leaves the block.

When the light beam moves from air into glass, the light slows down upon entering the glass and its path is bent toward the normal.

θ_1

Normal

Air
Glass

$\theta_2 < \theta_1$ v_1

v_2

θ_2 $v_2 < v_1$

Normal

θ_1

Glass
Air

$\theta_2 > \theta_1$ v_1

v_2

θ_2 $v_2 > v_1$

When the beam moves from glass into air, the light speeds up upon entering the air and its path is bent away from the normal.

a

b

Figure 34.11 The refraction of light as it (a) moves from air into glass and (b) moves from glass into air.

The path of a light ray through a refracting surface is reversible. For example, the ray shown in Figure 34.10a travels from point A to point B. If the ray originated at B, it would travel upward to the point of incidence at the surface, bend away from the normal, and then reach point A. The reflected ray would point downward and to the left in the glass.

From Equation 34.2, we can infer that when light moves from a material in which its speed is high to a material in which its speed is lower as shown in Figure 34.11a, the angle of refraction θ_2 is less than the angle of incidence θ_1 and the ray is bent toward the normal. If the ray moves from a material in which light moves slowly to a material in which it moves more rapidly as illustrated in Figure 34.11b, then θ_2 is greater than θ_1 and the ray is bent away from the normal.

A mechanical analog of refraction is shown in Figure 34.12. When the left end of the rolling barrel reaches the grass, it slows down, whereas the right end remains on the concrete and moves at its original speed. This difference in speeds causes the barrel to pivot, which changes the direction of travel.

The behavior of light as it passes from air into another substance and then re-emerges into air is often a source of confusion to students. When light travels in air, its speed is 3.00×10^8 m/s, but this speed is reduced to approximately 2×10^8 m/s when the light enters a block of glass. When the light re-emerges into air, its speed instantaneously increases to its original value of 3.00×10^8 m/s. This effect is far different from what happens, for example, when a bullet is fired through a block of wood. In that case, the speed of the bullet decreases as it moves through the wood because some of its original energy is used to tear apart the wood fibers. When the bullet enters the air once again, it emerges at a speed lower than it had when it entered the wood, consistent with its reduced kinetic energy.

But light is a wave. Its speed in air is always the same. Therefore, when the light leaves the block and enters the air, it *must* travel at the speed with which it entered. Similar to the reduced energy of the bullet, the light also has less energy: it is less *intense*. The outgoing beam of light will appear dimmer than the incident beam. Some of the energy has been absorbed within the glass.

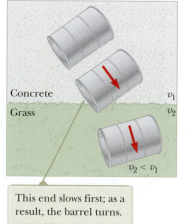

Concrete v_1
Grass v_2

$v_2 < v_1$

This end slows first; as a result, the barrel turns.

Figure 34.12 Overhead view of a barrel rolling from concrete onto grass.

Index of Refraction

In general, the speed of light in any material is *less* than its speed in vacuum. In fact, *light travels at its maximum speed c in vacuum*. It is convenient to define the **index of refraction** n of a medium to be the ratio

Index of refraction ▶

$$n \equiv \frac{\text{speed of light in vacuum}}{\text{speed of light in a medium}} \equiv \frac{c}{v} \qquad (34.3)$$

This definition shows that the index of refraction is a dimensionless number greater than unity because v is always less than c. Furthermore, n is equal to unity for vacuum. The indices of refraction for various substances are listed in Table 34.1. Because the value of n for air is so close to 1, we will use $n = 1$ for air in this chapter.

TABLE 34.1 Indices of Refraction

Substance	Index of Refraction	Substance	Index of Refraction
Solids at 20°C		*Liquids at 20°C*	
Cubic zirconia	2.20	Benzene	1.501
Diamond (C)	2.419	Carbon disulfide	1.628
Fluorite (CaF_2)	1.434	Carbon tetrachloride	1.461
Fused quartz (SiO_2)	1.458	Ethyl alcohol	1.361
Gallium phosphide	3.50	Glycerin	1.473
Glass, crown	1.52	Water	1.333
Glass, flint	1.66		
Ice (H_2O)	1.309	*Gases at 0°C, 1 atm*	
Polystyrene	1.49	Air	1.000 293
Sodium chloride (NaCl)	1.544	Carbon dioxide	1.000 45

Note: All values are for light having a wavelength of 589 nm in vacuum.

As light travels from one medium to another, its frequency does not change but its wavelength does. To see why that is true, consider Figure 34.13. Waves pass an observer at point A in medium 1 with a certain frequency and are incident on the boundary between medium 1 and medium 2. The frequency with which the waves pass an observer at point B in medium 2 must equal the frequency at which they pass point A. If that were not the case, energy would be piling up or disappearing at the boundary. Because there is no mechanism for that to occur, the frequency must be a constant as a light ray passes from one medium into another. Therefore, because the relationship $v = \lambda f$ (Eq. 16.12) from the traveling wave model must be valid in both media and because $f_1 = f_2 = f$, we see that

$$v_1 = \lambda_1 f \quad \text{and} \quad v_2 = \lambda_2 f \tag{34.4}$$

Because $v_1 \neq v_2$, it follows that $\lambda_1 \neq \lambda_2$ as shown in Figure 34.13.

We can obtain a relationship between index of refraction and wavelength by dividing the first Equation 34.4 by the second and then using Equation 34.3:

$$\frac{\lambda_1}{\lambda_2} = \frac{v_1}{v_2} = \frac{c/n_1}{c/n_2} = \frac{n_2}{n_1} \tag{34.5}$$

This expression gives

$$\lambda_1 n_1 = \lambda_2 n_2$$

If medium 1 is vacuum or air, then $n_1 = 1$. Hence, it follows from Equation 34.5 that the index of refraction of any medium can be expressed as the ratio

$$n = \frac{\lambda}{\lambda_n} \tag{34.6}$$

where λ is the wavelength of light in vacuum and λ_n is the wavelength of light in the medium whose index of refraction is n. From Equation 34.6, we see that because $n > 1$, $\lambda_n < \lambda$. We see the shortening of the wavelength in Figure 34.13.

We are now in a position to express Equation 34.2 in an alternative form. Replacing the v_2/v_1 term in Equation 34.2 with n_1/n_2 from Equation 34.5 gives

$$n_1 \sin \theta_1 = n_2 \sin \theta_2 \tag{34.7}$$

The experimental discovery of this relationship is usually credited to Willebrord Snell (1591–1626) and it is therefore known as **Snell's law of refraction.** (There is some evidence that the law was developed centuries earlier in the Middle East.)

As a wave moves between the media, its wavelength changes but its frequency remains constant.

Figure 34.13 A wave travels from medium 1 to medium 2, in which it moves with lower speed.

PITFALL PREVENTION 34.2

An Inverse Relationship The index of refraction is *inversely* proportional to the wave speed. As the wave speed v decreases, the index of refraction n increases. Therefore, the higher the index of refraction of a material, the more it *slows down* light from its speed in vacuum. The more the light slows down, the more θ_2 differs from θ_1 in Equation 34.7.

◀ Snell's law of refraction

We shall examine this equation further in Section 34.5. Refraction of waves at an interface between two media is a common phenomenon, so we identify an analysis model for this situation: the **wave under refraction.** Equation 34.7 is the mathematical representation of this model for electromagnetic radiation. Other waves, such as seismic waves and sound waves, also exhibit refraction according to this model, and the mathematical representation of the model for these waves is Equation 34.2.

QUICK QUIZ 34.3   Light passes from a material with index of refraction 1.3 into one with index of refraction 1.2. Compared to the incident ray, what happens to the refracted ray? **(a)** It bends toward the normal. **(b)** It is undeflected. **(c)** It bends away from the normal.

ANALYSIS MODEL   **Wave Under Refraction**

Imagine a wave (electromagnetic or mechanical) traveling through space and striking a flat surface at an angle θ_1 with respect to the normal to the surface. Some of the energy of the wave refracts into the medium below the surface in a direction θ_2 described by the **law of refraction**—

$$\frac{\sin \theta_2}{\sin \theta_1} = \frac{v_2}{v_1} \qquad (34.2)$$

where v_1 and v_2 are the speeds of the wave in medium 1 and medium 2, respectively.

For light waves, **Snell's law of refraction** states that

$$n_1 \sin \theta_1 = n_2 \sin \theta_2 \qquad (34.7)$$

where n_1 and n_2 are the indices of refraction in the two media.

Examples:

- Sound waves moving upward from the shore of a lake refract in warmer layers of air higher above the lake and travel downward to a listener in a boat, making sounds from the shore louder than expected.
- Light from the sky approaches a hot roadway at a grazing angle and refracts upward so as to miss the roadway and enter a driver's eye, giving the illusion of a pool of water on the distant roadway.
- Light is sent over long distances in an optical fiber because of a difference in index of refraction between the fiber and the surrounding material (Section 34.7).
- A magnifying glass forms an enlarged image of a postage stamp due to refraction of light through the lens (Chapter 35).

Example 34.3   **Angle of Refraction for Glass**

A light ray of wavelength 589 nm traveling through air is incident on a smooth, flat slab of crown glass at an angle of 30.0° to the normal.

(A) Find the angle of refraction.

SOLUTION

Conceptualize   Study Figure 34.11a, which illustrates the refraction process occurring in this problem. We expect that $\theta_2 < \theta_1$ because the speed of light is lower in the glass.

Categorize   This is a typical problem in which we apply the *wave under refraction* model.

Analyze   Rearrange Snell's law of refraction to find $\sin \theta_2$:

$$\sin \theta_2 = \frac{n_1}{n_2} \sin \theta_1$$

Solve for θ_2:

$$\theta_2 = \sin^{-1}\left(\frac{n_1}{n_2} \sin \theta_1\right)$$

Substitute indices of refraction from Table 34.1 and the incident angle:

$$\theta_2 = \sin^{-1}\left(\frac{1.00}{1.52} \sin 30.0°\right) = \boxed{19.2°}$$

34.3 continued

(B) Find the speed of this light once it enters the glass.

SOLUTION

Solve Equation 34.3 for the speed of light in the glass:

$$v = \frac{c}{n}$$

Substitute numerical values:

$$v = \frac{3.00 \times 10^8 \text{ m/s}}{1.52} = 1.97 \times 10^8 \text{ m/s}$$

(C) What is the wavelength of this light in the glass?

SOLUTION

Use Equation 34.6 to find the wavelength in the glass:

$$\lambda_n = \frac{\lambda}{n} = \frac{589 \text{ nm}}{1.52} = 388 \text{ nm}$$

Finalize In part (A), note that $\theta_2 < \theta_1$, consistent with the slower speed of the light found in part (B). In part (C), we see that the wavelength of the light is shorter in the glass than in the air.

Example **34.4** **Light Passing Through a Slab**

A light beam passes from medium 1 to medium 2, with the latter medium being a thick slab of material whose index of refraction is n_2 (Fig. 34.14). Show that the beam emerging into medium 1 from the other side is parallel to the incident beam.

SOLUTION

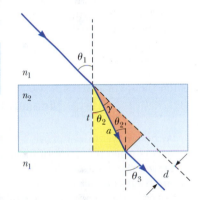

Conceptualize Follow the path of the light beam as it enters and exits the slab of material in Figure 34.14, where we have assumed that $n_2 > n_1$. The ray bends toward the normal upon entering and away from the normal upon leaving.

Figure 34.14 (Example 34.4) The dashed line drawn parallel to the ray coming out the bottom of the slab represents the path the light would take were the slab not there.

Categorize Like Example 34.3, this is another typical problem in which we apply the *wave under refraction* model.

Analyze Apply Snell's law of refraction to the upper surface:

$$(1) \quad \sin \theta_2 = \frac{n_1}{n_2} \sin \theta_1$$

Apply Snell's law to the lower surface:

$$(2) \quad \sin \theta_3 = \frac{n_2}{n_1} \sin \theta_2$$

Substitute Equation (1) into Equation (2):

$$\sin \theta_3 = \frac{n_2}{n_1} \left(\frac{n_1}{n_2} \sin \theta_1 \right) = \sin \theta_1$$

Finalize Therefore, $\theta_3 = \theta_1$ and the slab does not alter the direction of the beam. It does, however, offset the beam parallel to itself by the distance d shown in Figure 34.14.

WHAT IF? What if the thickness t of the slab is doubled? Does the offset distance d also double?

Answer Consider the region of the light path within the slab in Figure 34.14. The distance a is the common hypotenuse of the red and yellow right triangles.

Find an expression for a from the yellow triangle:

$$a = \frac{t}{\cos \theta_2}$$

Find an expression for d from the red triangle:

$$d = a \sin \gamma = a \sin (\theta_1 - \theta_2)$$

Combine these equations:

$$d = \frac{t}{\cos \theta_2} \sin (\theta_1 - \theta_2)$$

For a given incident angle θ_1, the refracted angle θ_2 is determined solely by the index of refraction, so the offset distance d is proportional to t. If the thickness doubles, so does the offset distance.

The apex angle Φ is the angle between the sides of the prism through which the light enters and leaves.

Figure 34.15 A prism refracts a single-wavelength light ray through an angle of deviation δ.

In Example 34.4, the light passes through a slab of material with parallel sides. What happens when light strikes a prism with nonparallel sides as shown in Figure 34.15? In this case, the outgoing ray does not propagate in the same direction as the incoming ray. A ray of single-wavelength light incident on the prism from the left emerges at angle δ from its original direction of travel. This angle δ is called the **angle of deviation.** The **apex angle** Φ of the prism, shown in the figure, is defined as the angle between the surface at which the light enters the prism and the second surface that the light encounters.

Example 34.5   **Measuring *n* Using a Prism**

Although we do not prove it here, the minimum angle of deviation δ_{min} for a prism occurs when the angle of incidence θ_1 is such that the refracted ray inside the prism makes the same angle with the normal to the two prism faces[1] as shown in Figure 34.16. Obtain an expression for the index of refraction of the prism material in terms of the minimum angle of deviation and the apex angle Φ.

SOLUTION

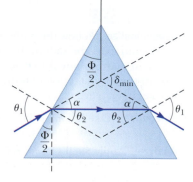

Figure 34.16 (Example 34.5) A light ray passing through a prism at the minimum angle of deviation δ_{min}.

Conceptualize Study Figure 34.16 carefully and be sure you understand why the light ray comes out of the prism traveling in a different direction.

Categorize In this example, light enters a material through one surface and leaves the material at another surface. Let's apply the *wave under refraction* model to the light passing through the prism.

Analyze Consider the geometry in Figure 34.16, where we have used symmetry to label several angles. The reproduction of the angle Φ/2 at the location of the incoming light ray shows that $\theta_2 = \Phi/2$. The theorem that an exterior angle of any triangle equals the sum of the two opposite interior angles shows that $\delta_{min} = 2\alpha$. The geometry also shows that $\theta_1 = \theta_2 + \alpha$.

Combine these three geometric results:

$$\theta_1 = \theta_2 + \alpha = \frac{\Phi}{2} + \frac{\delta_{min}}{2} = \frac{\Phi + \delta_{min}}{2}$$

Apply the wave under refraction model at the left surface and solve for *n*:

$$(1.00)\sin\theta_1 = n\sin\theta_2 \rightarrow n = \frac{\sin\theta_1}{\sin\theta_2}$$

Substitute for the incident and refracted angles:

$$n = \frac{\sin\left(\dfrac{\Phi + \delta_{min}}{2}\right)}{\sin(\Phi/2)} \tag{34.8}$$

Finalize Knowing the apex angle Φ of the prism and measuring δ_{min}, you can calculate the index of refraction of the prism material. Furthermore, a hollow prism can be used to determine the values of *n* for various liquids filling the prism.

---

[1] The details of this proof are available in texts on optics.

34.5 Huygens's Principle

The laws of reflection and refraction were stated earlier in this chapter without proof. In this section, we develop these laws by using a geometric method proposed by Huygens in 1678. **Huygens's principle** is a geometric construction for using knowledge of an earlier wave front to determine the position of a new wave front at some instant:

> All points on a given wave front are taken as point sources for the production of spherical secondary waves, called wavelets, that propagate outward through a medium with speeds characteristic of waves in that medium. After some time interval has passed, the new position of the wave front is the surface tangent to the wavelets.

First, consider a plane wave moving through free space as shown in Figure 34.17a. At $t = 0$, the wave front is indicated by the plane labeled AA' and oriented perpendicular to the page. The wave moves to the right in the plane of the page. In Huygens's construction, each point on this wave front is considered a point source for spherical wavelets. For clarity, only three point sources on AA' are shown as black dots. With these sources for the wavelets, we draw circular arcs, each of radius $c \, \Delta t$, where c is the speed of light in vacuum and Δt is some time interval during which the wave propagates. The surface drawn tangent to these wavelets is the plane BB', which is the wave front at a later time, and is parallel to AA'. In a similar manner, Figure 34.17b shows Huygens's construction for a spherical wave using four sources on the original wavefront.

PITFALL PREVENTION 34.4
Of What Use Is Huygens's Principle? At this point, the importance of Huygens's principle may not be evident. Predicting the position of a future wave front may not seem to be very critical. We will use Huygens's principle here to generate the laws of reflection and refraction and in later chapters to explain additional wave phenomena for light.

Huygens's Principle Applied to Reflection and Refraction

We now derive the laws of reflection and refraction, using Huygens's principle.

For the law of reflection, refer to Figure 34.18. The line AB represents a plane wave front of the incident light just as ray 1 strikes the surface. At this instant, the wave at A sends out a Huygens wavelet (appearing at a later time as the light brown circular arc passing through D); the reflected light makes an angle γ' with the surface. At the same time, the wave at B emits a Huygens wavelet (the light brown circular arc passing through C) with the incident light making an angle γ with the surface. Figure 34.18 shows these wavelets after a time interval Δt, after which ray 2 strikes the surface. Because both rays 1 and 2 move with the same speed, we must have $AD = BC = c \, \Delta t$.

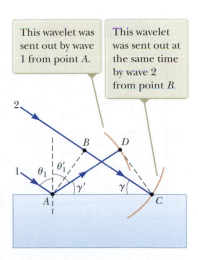

Figure 34.17 Huygens's construction for (a) a plane wave propagating to the right and (b) a spherical wave propagating from a point source.

Figure 34.18 Huygens's construction for proving the law of reflection.

The remainder of our analysis depends on geometry. Notice that the two triangles ABC and ADC are congruent because they have the same hypotenuse AC and because $AD = BC$. Figure 34.18 shows that

$$\cos \gamma = \frac{BC}{AC} \quad \text{and} \quad \cos \gamma' = \frac{AD}{AC}$$

where $\gamma = 90° - \theta_1$ and $\gamma' = 90° - \theta_1'$. Because $AD = BC$,

$$\cos \gamma = \cos \gamma'$$

Therefore,

$$\gamma = \gamma'$$
$$90° - \theta_1 = 90° - \theta_1'$$

and

$$\theta_1 = \theta_1'$$

which is the law of reflection.

Now let's use Huygens's principle to derive Snell's law of refraction. We focus our attention on the instant ray 1 strikes the surface and the subsequent time interval until ray 2 strikes the surface as in Figure 34.19. During this time interval, the wave at A sends out a Huygens wavelet (the light brown arc passing through D) and the light refracts into the material, making an angle θ_2 with the normal to the surface. In the same time interval, the wave at B sends out a Huygens wavelet (the light brown arc passing through C) and the light continues to propagate in the same direction. Because these two wavelets travel through different media, the radii of the wavelets are different. The radius of the wavelet from A is $AD = v_2 \Delta t$, where v_2 is the wave speed in the second medium. The radius of the wavelet from B is $BC = v_1 \Delta t$, where v_1 is the wave speed in the original medium.

From triangles ABC and ADC, we find that

$$\sin \theta_1 = \frac{BC}{AC} = \frac{v_1 \Delta t}{AC} \quad \text{and} \quad \sin \theta_2 = \frac{AD}{AC} = \frac{v_2 \Delta t}{AC}$$

Dividing the first equation by the second gives

$$\frac{\sin \theta_1}{\sin \theta_2} = \frac{v_1}{v_2}$$

which is Equation 34.2. From Equation 34.3, we know that $v_1 = c/n_1$ and $v_2 = c/n_2$. Therefore,

$$\frac{\sin \theta_1}{\sin \theta_2} = \frac{c/n_1}{c/n_2} = \frac{n_2}{n_1}$$

and

$$n_1 \sin \theta_1 = n_2 \sin \theta_2$$

which is Snell's law of refraction, Equation 34.7.

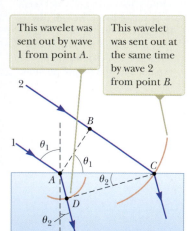

Figure 34.19 Huygens's construction for proving Snell's law of refraction.

Figure 34.20 Variation of index of refraction with vacuum wavelength for three materials.

34.6 Dispersion

An important property of the index of refraction n is that, for a given material, the index varies with the wavelength of the light passing through the material as Figure 34.20 shows. This behavior is called **dispersion**. Because n is a function of wavelength, Snell's law of refraction indicates that light of different wavelengths is refracted at different angles when incident on a material.

The colors in the refracted beam are separated because dispersion in the prism causes different wavelengths of light to be refracted through different angles.

David Parker/Science Source

Figure 34.21 White light enters a glass prism at the upper left.

The violet light refracts through larger angles than the red light.

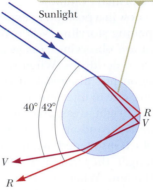

Sunlight

40° 42°

R
V

V

R

Figure 34.22 Path of sunlight through a spherical raindrop. Light following this path contributes to the visible rainbow.

PITFALL PREVENTION 34.5
A Rainbow of Many Light Rays
Pictorial representations such as Figure 34.22 are subject to misinterpretation. The figure shows one ray of light entering the raindrop and undergoing reflection and refraction, exiting the raindrop in a range of 40° to 42° from the entering ray. This illustration might be interpreted incorrectly as meaning that *all* light entering the raindrop exits in this small range of angles. In reality, light enters the raindrop at all positions on its surface and exits the raindrop over a much larger range of angles, from 0° to 42°. A careful analysis of the reflection and refraction from the spherical raindrop shows that the range of 40° to 42° is where the *highest-intensity light* exits the raindrop.

Figure 34.20 shows that the index of refraction generally decreases with increasing wavelength. For example, violet light refracts more than red light does when passing into a material.

Now suppose a beam of *white light* (a combination of all visible wavelengths) is incident on a prism as illustrated in Figure 34.21. The angle of deviation δ (Fig. 34.15) depends on the index of refraction n, so, in turn, the angle depends on wavelength. The rays that emerge spread out in a series of colors known as the **visible spectrum.** These colors, in order of decreasing wavelength, are red, orange, yellow, green, blue, and violet. Newton showed that each color has a particular angle of deviation and that the colors can be recombined to form the original white light.

The dispersion of light into a spectrum is demonstrated most vividly in nature by the formation of a rainbow, which is often seen by an observer positioned between the Sun and a rain shower. To understand how a rainbow is formed, consider Figure 34.22. We will need to apply both the wave under reflection and wave under refraction models. A ray of sunlight (which is white light) strikes a drop of water in the atmosphere and is refracted and reflected as follows. It is first refracted at the front surface of the drop, with the violet light deviating the most and the red light the least. At the back surface of the drop, the light is reflected and returns to the front surface, where it again undergoes refraction as it moves from water into air. The rays leave the drop such that the angle between the incident white light and the most intense returning violet ray is 40° and the angle between the incident white light and the most intense returning red ray is 42°. This small angular difference between the returning rays causes us to see a colored bow.

Now suppose an observer is viewing a rainbow as shown in Figure 34.23. If a raindrop high in the sky is being observed, the most intense red light returning from the drop reaches the observer because it is deviated the least; the most intense violet light from this drop, however, passes over the observer because it is deviated the most. Hence, the observer sees red light coming from this drop. Similarly, a drop lower in the sky directs the most intense violet light toward the observer and appears violet to the observer. (The most intense red light from this drop passes below the observer's eye and is not seen.) The most intense light from other colors of the spectrum reaches the observer from raindrops lying between these two extreme positions.

Figure 34.24 (page 914) shows a *double rainbow*. The secondary rainbow is fainter than the primary rainbow, and the colors are reversed. The secondary rainbow arises from light that makes two reflections from the interior surface before exiting the raindrop. In the laboratory, rainbows have been observed in which the light

The highest-intensity light traveling from higher raindrops toward the eyes of the observer is red, whereas the most intense light from lower drops is violet.

White

White

42°
40°

42°
40°
R

V

V

R

Figure 34.23 The formation of a rainbow seen by an observer standing with the Sun behind his back.

Figure 34.24 This photograph of a rainbow shows a distinct secondary rainbow with the colors reversed.

makes more than 30 reflections before exiting the water drop. Because each reflection involves some loss of light due to refraction of part of the incident light out of the water drop, the intensity of these higher-order rainbows is small compared with that of the primary rainbow.

We are now in a position to explain all of the optical phenomena that you observe in the opening storyline. When you glance at your shadow on the dew-encrusted grass, you are observing *Heiligenschein*. This effect is due to retroreflection from the spherical dewdrops. When you turn and view the rainbow, you are seeing the phenomenon just discussed in this section. The pastel-colored bands below the rainbow are due to interference, to be discussed in Chapter 36. Turning to the Sun, you see the bright areas, at about a 22° angle on either side of the Sun. These are called *Sun dogs*, and are due to refraction of sunlight through hexagonal-shaped crystals of ice suspended in a horizontal orientation in the atmosphere. If the crystals are longer, they will be in a variety of orientations, and you will see a full *halo* around the Sun. When you look at the hot, black roadway, you see puddles in the distance. The absorbed sunlight warms the air above the surface and changes its index of refraction. Light coming from the sky at a shallow angle toward you along the roadway experiences continuous refraction, and actually *misses* the roadway, curving back upward to enter your eye. As a result, you see an image of the sky on the roadway, making it look wet. This is a common *mirage*.

QUICK QUIZ 34.4 In photography, lenses in a camera use refraction to form an image on a light-sensitive surface. Ideally, you want all the colors in the light from the object being photographed to be refracted by the same amount. Of the materials shown in Figure 34.20, which would you choose for a single-element camera lens? **(a)** crown glass **(b)** acrylic **(c)** fused quartz **(d)** impossible to determine

34.7 Total Internal Reflection

An interesting effect called **total internal reflection** can occur when light is directed from a medium having a given index of refraction toward one having a lower index of refraction. Consider Figure 34.25a, in which a light ray travels in

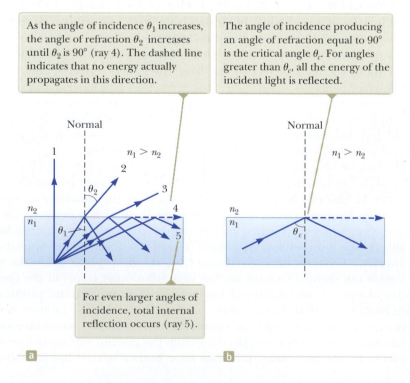

As the angle of incidence θ_1 increases, the angle of refraction θ_2 increases until θ_2 is 90° (ray 4). The dashed line indicates that no energy actually propagates in this direction.

The angle of incidence producing an angle of refraction equal to 90° is the critical angle θ_c. For angles greater than θ_c, all the energy of the incident light is reflected.

For even larger angles of incidence, total internal reflection occurs (ray 5).

Figure 34.25 (a) Rays travel from a medium of index of refraction n_1 into a medium of index of refraction n_2, where $n_2 < n_1$. (b) Ray 4 is singled out.

medium 1 and meets the boundary between medium 1 and medium 2, where n_1 is greater than n_2. In the figure, labels 1 through 5 indicate various possible directions of the ray consistent with the wave under refraction model. The refracted rays are bent away from the normal because n_1 is greater than n_2. As θ_1 is increased, θ_2 also become larger and the refracted ray bends away from the normal so much that it approaches a direction parallel to the interface. At some particular angle of incidence θ_c, called the **critical angle,** the refracted light ray reaches this condition and is indeed parallel to the boundary so that $\theta_2 = 90°$ (see ray 4 in Fig. 34.25a and the dashed ray in Fig. 34.25b). For angles of incidence greater than θ_c, the ray cannot escape from the material and is entirely reflected at the boundary as shown by ray 5 in Figure 34.25a.

We can use Snell's law of refraction to find the critical angle. When $\theta_1 = \theta_c$, $\theta_2 = 90°$ and Equation 34.7 gives

$$n_1 \sin \theta_c = n_2 \sin 90° = n_2$$

$$\sin \theta_c = \frac{n_2}{n_1} \quad \text{(for } n_1 > n_2\text{)} \qquad (34.9)$$

◄ Critical angle for total internal reflection

This equation can be used only when n_1 is greater than n_2. That is, total internal reflection occurs only when light is directed from a medium of a given index of refraction toward a medium of lower index of refraction. If n_1 were less than n_2, Equation 34.9 would give $\sin \theta_c > 1$, which is a meaningless result because the sine of an angle can never be greater than unity.

The critical angle for total internal reflection is small when n_1 is considerably greater than n_2. For example, the critical angle for a diamond in air is 24°. Any ray inside the diamond that approaches the surface at an angle greater than 24° is completely reflected back into the crystal. This property, combined with proper faceting, causes diamonds to sparkle. The angles of the facets are cut so that light is "caught" inside the crystal through multiple internal reflections. These multiple reflections give the light a long path through the medium, and substantial dispersion of colors occurs. By the time the light exits through the top surface of the crystal, the rays associated with different colors have been fairly widely separated from one another.

Cubic zirconia also has a high index of refraction and can be made to sparkle very much like a diamond. If a suspect jewel is immersed in corn syrup, the difference in n for the cubic zirconia and that for the corn syrup is small and the critical angle is therefore great. Hence, more rays escape sooner; as a result, the sparkle completely disappears. A real diamond does not lose all its sparkle when placed in corn syrup.

QUICK QUIZ 34.5 In Figure 34.26, five light rays enter a glass prism from the left. **(i)** How many of these rays undergo total internal reflection at the slanted surface of the prism? (a) one (b) two (c) three (d) four (e) five **(ii)** Suppose the prism in Figure 34.26 can be rotated in the plane of the paper. For *all five* rays to experience total internal reflection from the slanted surface, should the prism be rotated (a) clockwise or (b) counterclockwise?

Courtesy of Henry Leap and Jim Lehman

Figure 34.26 (Quick Quiz 34.5) Five nonparallel light rays enter a glass prism from the left.

Example **34.6** **A View from the Fish's Eye**

Find the critical angle for an air–water boundary. (Assume the index of refraction of water is 1.33.)

SOLUTION

Conceptualize Study Figure 34.25 to understand the concept of total internal reflection and the significance of the critical angle.

continued

34.6 continued

Categorize We use concepts developed in this section, so we categorize this example as a substitution problem.

Apply Equation 34.9 to the air–water interface:

$$\sin \theta_c = \frac{n_2}{n_1} = \frac{1.00}{1.33} = 0.752$$

$$\theta_c = \boxed{48.8°}$$

WHAT IF? What if a fish in a still pond looks upward toward the water's surface at different angles relative to the surface as in Figure 34.27? What does it see?

Answer Because the path of a light ray is reversible, light traveling from medium 2 into medium 1 in Figure 34.25a follows the paths shown, but in the *opposite* direction. A fish looking upward toward the water surface as in Figure 34.27 can see out of the water if it looks toward the surface at an angle less than the critical angle. Therefore, when the fish's line of vision makes an angle of $\theta = 40°$ with the normal to the surface, for example, light from above the water reaches the fish's eye. At $\theta = 48.8°$, the critical angle for water, the light has to skim along the water's surface before being refracted to the fish's eye; at this angle, the fish can, in principle, see the entire shore of the pond. At angles greater than the critical angle, the light reaching the fish comes by means of total internal reflection at the surface of light originating in the water. Therefore, at $\theta = 60°$, the fish sees a reflection of the bottom of the pond.

Figure 34.27 (Example 34.6) **What If?** A fish looks upward toward the water surface.

Optical Fibers

Another interesting application of total internal reflection is the use of glass or transparent plastic rods to "pipe" light from one place to another. As indicated in Figure 34.28, light is confined to traveling within a rod, even around curves, as the result of successive total internal reflections. Such a light pipe is flexible if thin fibers are used rather than thick rods. A flexible light pipe is called an **optical fiber.** If a bundle of parallel fibers is used to construct an optical transmission line, images can be transferred from one point to another. Part of the 2009 Nobel Prize in Physics was awarded to Charles K. Kao (b. 1933) for his discovery of how to transmit light signals over long distances through thin glass fibers. This discovery has led to the development of a sizable industry known as *fiber optics.*

Figure 34.28 Light travels in a curved transparent rod by multiple internal reflections.

A practical optical fiber consists of a transparent core surrounded by a *cladding,* a material that has a lower index of refraction than the core. The combination may be surrounded by a plastic *jacket* to prevent mechanical damage. Figure 34.29 shows a cutaway view of this construction. Because the index of refraction of the cladding is less than that of the core, light traveling in the core experiences total internal reflection if it arrives at the interface between the core and the cladding at an angle of incidence that exceeds the critical angle. In this case, light "bounces" along the core of the optical fiber, losing very little of its intensity as it travels.

Any loss in intensity in an optical fiber is essentially due to reflections from the two ends and absorption by the fiber material. Optical fiber devices are particularly useful for viewing an object at an inaccessible location. For example, physicians often use such devices to examine internal organs of the body or to perform surgery without making large incisions. Optical fiber cables are replacing copper wiring and coaxial cables for telecommunications because the fibers can carry a much greater volume of telephone calls or other forms of communication than electrical wires can.

Glass or plastic core

Jacket

Cladding

Figure 34.29 The construction of an optical fiber. Light travels in the core, which is surrounded by a cladding and a protective jacket.

Figure 34.30a shows a bundle of optical fibers gathered into an optical cable that can be used to carry communication signals. Many computers and other electronic equipment now have optical ports as well as electrical ports for transferring information (Figure 34.30b).

Figure 34.30 (a) Strands of glass optical fibers are used to carry voice, video, and data signals in telecommunication networks. (b) A technician works on optical fiber connections in a computer networking system.

Summary

▶ Definition

The **index of refraction** n of a medium is defined by the ratio

$$n \equiv \frac{c}{v} \tag{34.3}$$

where c is the speed of light in vacuum and v is the speed of light in the medium.

▶ Concepts and Principles

In geometric optics, we use the **ray approximation,** in which a wave travels through a uniform medium in straight lines in the direction of the rays.

Total internal reflection occurs when light travels from a medium of high index of refraction to one of lower index of refraction. The **critical angle** θ_c for which total internal reflection occurs at an interface is given by

$$\sin \theta_c = \frac{n_2}{n_1} \quad \text{(for } n_1 > n_2\text{)} \tag{34.9}$$

▶ Analysis Models for Problem Solving

Wave Under Reflection. The **law of reflection** states that for a light ray (or other type of wave) incident on a smooth surface, the angle of reflection θ_1' equals the angle of incidence θ_1:

$$\theta_1' = \theta_1 \tag{34.1}$$

Wave Under Refraction. A wave crossing a boundary as it travels from medium 1 to medium 2 is **refracted.** The angle of refraction θ_2 is related to the incident angle θ_1 by the relationship

$$\frac{\sin \theta_2}{\sin \theta_1} = \frac{v_2}{v_1} \tag{34.2}$$

where v_1 and v_2 are the speeds of the wave in medium 1 and medium 2, respectively. The incident ray, the reflected ray, the refracted ray, and the normal to the surface all lie in the same plane.

For light waves, **Snell's law of refraction** states that

$$n_1 \sin \theta_1 = n_2 \sin \theta_2 \tag{34.7}$$

where n_1 and n_2 are the indices of refraction in the two media.

Think–Pair–Share

See the Preface for an explanation of the icons used in this problems set. For additional assessment items for this section, go to **WEBASSIGN** From Cengage

1. Discuss the following situation in your group: Three sheets of plastic have unknown indices of refraction. They are placed on top of each other two at a time and a laser beam is directed onto the sheets from above. In all three cases, the laser beam is adjusted so that it strikes the interface between the top sheet and the bottom sheet (*not* between the air and the top sheet) at an angle of 26.5° with the normal. (i) Sheet 1 is placed on top of sheet 2. The refracted beam inside sheet 2 makes an angle of 31.7° with the normal. (ii) Sheet 3 is placed on top of sheet 2. The refracted beam inside sheet 2 makes an angle of 36.7° with the normal. (iii) Sheet 1 is placed on top of sheet 3. The refracted beam inside sheet 3 makes an angle of 23.1° with the normal. Now generate a consensus answer in your group: Do you now have enough information to find the index of refraction for each sheet?

2. **ACTIVITY** Your group has been assigned to learn about Claudius Ptolemy (100–170) and analyze his early results on the refraction of light. In his book, *Optics,* which survives in an Arabic translation and a Latin translation of the Arabic, Ptolemy includes a table of incident angles and refracted angles for light entering water. This is the earliest surviving set of data of this type. There is some controversy about whether the data truly is experimental or if it is generated mathematically, but either way, having refractive data from the second century is remarkable. Ptolemy's data are shown in the following table.

Angle of Incidence (°)	Angle of Refraction (°)
10.0	8
20.0	15.5
30.0	22.5
40.0	29
50.0	35
60.0	40.5
70.0	45.5
80.0	50

What is the index of refraction of water according to Ptolemy's data?

Problems

See the Preface for an explanation of the icons used in this problems set. For additional assessment items for this section, go to **WEBASSIGN** From Cengage

SECTION 34.1 The Nature of Light

1. In an experiment to measure the speed of light using the apparatus of Armand H. L. Fizeau (see Fig. 34.2), the distance between light source and mirror was 11.45 km and the wheel had 720 notches. The experimentally determined value of c was 2.998×10^8 m/s when the outgoing light passed through one notch and then returned through the next notch. Calculate the minimum angular speed of the wheel for this experiment.

2. The *Apollo 11* astronauts set up a panel of efficient corner-cube retroreflectors on the Moon's surface (Fig. 34.8a). The speed of light can be found by measuring the time interval required for a laser beam to travel from the Earth, reflect from the panel, and return to the Earth. Assume this interval is measured to be 2.51 s at a station where the Moon is at the zenith and take the center-to-center distance from the Earth to the Moon to be equal to 3.84×10^8 m. (a) What is the measured speed of light? (b) Explain whether it is necessary to consider the sizes of the Earth and the Moon in your calculation.

3. As a result of his observations, Ole Roemer concluded that eclipses of Io by Jupiter were delayed by 22 min during a six-month period as the Earth moved from the point in its orbit where it is closest to Jupiter to the diametrically opposite point where it is farthest from Jupiter. Using the value 1.50×10^8 km as the average radius of the Earth's orbit around the Sun, calculate the speed of light from these data.

SECTION 34.3 Analysis Model: Wave Under Reflection

4. A dance hall is built without pillars and with a horizontal ceiling 7.20 m above the floor. A mirror is fastened flat against one section of the ceiling. Following an earthquake, the mirror is in place and unbroken. An engineer makes a quick check of whether the ceiling is sagging by directing a vertical beam of laser light up at the mirror and observing its reflection on the floor. (a) Show that if the mirror has rotated to make an angle ϕ with the horizontal, the normal to the mirror makes an angle ϕ with the vertical. (b) Show that the reflected laser light makes an angle 2ϕ with the vertical. (c) Assume the reflected laser light makes a spot on the floor 1.40 cm away from the point vertically below the laser. Find the angle ϕ.

5. You are working for an optical research company during a summer break. Part of the apparatus in one particular experiment is shown in Figure 34.7b. In fact, the experimenter used this textbook to set up this part of the experiment, and also used the result in the What If? section to determine the angular change in direction of the light beam:

$$\beta = 360° - 2\phi \quad (1)$$

The experimenter is constantly grumbling that the measuring device to determine the angle ϕ on the inside of the mirrors is constantly getting in the way of the light beam and making his life difficult. You quickly draw Figure P34.5 and then say, "Then why don't you use the measuring device to measure the angle δ *outside* the mirror, and then your device won't get in the way of the light?"

The experimenter, who has never thought of this, tries to save face and says to you, "Well, Smarty, then tell me how angle β depends on angle δ!" You provide him the answer quickly.

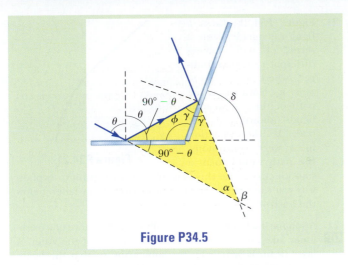

Figure P34.5

6. The reflecting surfaces of two intersecting flat mirrors are at an angle θ ($0° < \theta < 90°$) as shown in Figure P34.6. For a light ray that strikes the horizontal mirror, show that the emerging ray will intersect the incident ray at an angle $\beta = 180° - 2\theta$.

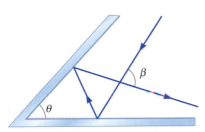

Figure P34.6

7. The two mirrors illustrated in Figure P34.7 meet at a right angle. The beam of light in the vertical plane indicated by the dashed lines strikes mirror 1 as shown. (a) Determine the distance the reflected light beam travels before striking mirror 2. (b) In what direction does the light beam travel after being reflected from mirror 2?

Figure P34.7

8. Two flat, rectangular mirrors, both perpendicular to a horizontal sheet of paper, are set edge to edge with their reflecting surfaces perpendicular to each other. (a) A light ray in the plane of the paper strikes one of the mirrors at an arbitrary angle of incidence θ_1. Prove that the final direction of the ray, after reflection from both mirrors, is opposite its initial direction. (b) **What If?** Now assume the paper is replaced with a third flat mirror, touching edges with the other two and perpendicular to both, creating a corner-cube retroreflector (Fig. 34.8a). A ray of light is incident from any direction within the octant of space bounded by the reflecting surfaces. Argue that the ray will reflect once from each mirror and that its final direction will be opposite its original direction. The *Apollo 11* astronauts placed a panel of corner-cube retroreflectors on the Moon. Analysis of timing data taken with it reveals that the radius of the Moon's orbit is increasing at the rate of 3.8 cm/yr as it loses kinetic energy because of tidal friction.

SECTION 34.4 Analysis Model: Wave Under Refraction

Notes: You may look up indices of refraction in Table 34.1. Unless indicated otherwise, assume the medium surrounding a piece of material is air with $n = 1.000\,293$.

9. Find the speed of light in (a) flint glass, (b) water, and (c) cubic zirconia.

10. A ray of light strikes a flat block of glass ($n = 1.50$) of thickness 2.00 cm at an angle of 30.0° with the normal. Trace the light beam through the glass and find the angles of incidence and refraction at each surface.

11. A ray of light travels from air into another medium, making an angle of $\theta_1 = 45.0°$ with the normal as in Figure P34.11. Find the angle of refraction θ_2 if the second medium is (a) fused quartz, (b) carbon disulfide, and (c) water.

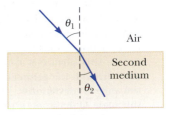

Figure P34.11

12. A plane sound wave in air at 20°C, with wavelength 589 mm, is incident on a smooth surface of water at 25°C at an angle of incidence of 13.0°. Determine (a) the angle of refraction for the sound wave and (b) the wavelength of the sound in water. A narrow beam of sodium yellow light, with wavelength 589 nm in vacuum, is incident from air onto a smooth water surface at an angle of incidence of 13.0°. Determine (c) the angle of refraction and (d) the wavelength of the light in water. (e) Compare and contrast the behavior of the sound and light waves in this problem.

13. A laser beam is incident at an angle of 30.0° from the vertical onto a solution of corn syrup in water. The beam is refracted to 19.24° from the vertical. (a) What is the index of refraction of the corn syrup solution? Assume that the light is red, with vacuum wavelength 632.8 nm. Find its (b) wavelength, (c) frequency, and (d) speed in the solution.

14. A ray of light strikes the midpoint of one face of an equiangular (60°–60°–60°) glass prism ($n = 1.5$) at an angle of incidence of 30°. (a) Trace the path of the light ray through the glass and find the angles of incidence and refraction at each surface. (b) If a small fraction of light is also reflected at each surface, what are the angles of reflection at the surfaces?

15. When you look through a window, by what time interval is the light you see delayed by having to go through glass instead of air? Make an order-of-magnitude estimate on the basis of data you specify. By how many wavelengths is it delayed?

16. Light passes from air into flint glass at a nonzero angle of incidence. (a) Is it possible for the component of its velocity perpendicular to the interface to remain constant? Explain your answer. (b) **What If?** Can the component of velocity parallel to the interface remain constant during refraction? Explain your answer.

17. You have just installed a new bathroom in your home. Your shower doors have frosted glass to provide privacy for the person using the shower. The frosted surface is on the outside of the shower door, facing the rest of the bathroom.

The frosting is done by acid etching the surface so that light incident on the rough surface is scattered in all directions. Proud of your new bathroom, you take a photo of it with your smartphone. You notice in the photograph that you can see a reflection of the flash in the shower doors and the reflection is surrounded by a halo of light. Curious, you turn on a laser pointer and aim it at the shower door. Looking closely at the reflection, you again see a halo that consists of a dark area surrounding the reflection of the pointer and then an area of brightness outside this dark ring. You grab a micrometer and a ruler and measure the thickness of the glass to be 6.35 mm and the inner radius of the bright halo to be 10.7 mm. From these measurements, you determine the index of refraction of the glass.

18. A triangular glass prism with apex angle 60.0° has an index of refraction of 1.50. (a) Show that if its angle of incidence on the first surface is $\theta_1 = 48.6°$, light will pass symmetrically through the prism as shown in Figure 34.16. (b) Find the angle of deviation δ_{min} for $\theta_1 = 48.6°$. (c) **What If?** Find the angle of deviation if the angle of incidence on the first surface is 45.6°. (d) Find the angle of deviation if $\theta_1 = 51.6°$.

19. You are working at your university swimming center. The
CR athletic department decides that it would like to install a flag pole of height 10.0 m at the south end of one of the outdoor pools, which lies along a north–south axis. The pool is 3.00 m deep and the flag pole is to be installed 4.00 m from the south edge of the pool, midway along the length of the south edge. (a) Your supervisor knows of your expertise in physics and asks you to determine the distance of the shadow of the tip of the flag pole on the bottom of the pool from the south wall of the pool on a summer day when the Sun appears directly south and at an angle of 65.0° above the horizon. (b) Your supervisor also asks if there is any time during the year that the flag pole will *not* cast a shadow along the bottom of the pool when the Sun is due south. The highest the Sun reaches in the sky at this location is 68.5° at the summer solstice.

20. A person looking into an empty container is able to see the far edge of the container's bottom as shown in Figure P34.20a. The height of the container is h, and its width is d. When the container is completely filled with a fluid of index of refraction n and viewed from the same angle, the person can see the center of a coin at the middle of the container's bottom as shown in Figure P34.20b. (a) Show that the ratio h/d is given by

$$\frac{h}{d} = \sqrt{\frac{n^2 - 1}{4 - n^2}}$$

(b) Assuming the container has a width of 8.00 cm and is filled with water, use the expression above to find the height of the container. (c) For what range of values of n will the center of the coin not be visible for any values of h and d?

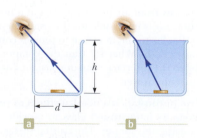

Figure P34.20

21. Figure P34.21 shows a light ray incident on a series of slabs having different refractive indices, where $n_1 < n_2 < n_3 < n_4$. Notice that the path of the ray steadily bends toward the normal. If the variation in n were continuous, the path would form a smooth curve. Use this idea and a ray diagram to explain why you can see the Sun at sunset after it has fallen below the horizon.

Figure P34.21

22. A submarine is 300 m horizontally from the shore of a fresh-
GP water lake and 100 m beneath the surface of the water. A laser beam is sent from the submarine so that the beam strikes the surface of the water 210 m from the shore. A building stands on the shore, and the laser beam hits a target at the top of the building. The goal is to find the height of the target above sea level. (a) Draw a diagram of the situation, identifying the two triangles that are important in finding the solution. (b) Find the angle of incidence of the beam striking the water–air interface. (c) Find the angle of refraction. (d) What angle does the refracted beam make with the horizontal? (e) Find the height of the target above sea level.

23. A beam of light both reflects and
S refracts at the surface between air and glass as shown in Figure P34.23. If the refractive index of the glass is n_g, find the angle of incidence θ_1 in the air that would result in the reflected ray and the refracted ray being perpendicular to each other.

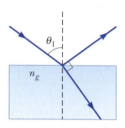

Figure P34.23

SECTION 34.6 Dispersion

24. A light beam containing red and violet wavelengths is incident on a slab of quartz at an angle of incidence of 50.0°. The index of refraction of quartz is 1.455 at 600 nm (red light), and its index of refraction is 1.468 at 410 nm (violet light). Find the dispersion of the slab, which is defined as the difference in the angles of refraction for the two wavelengths.

25. The index of refraction for violet light in silica flint glass is
S n_V, and that for red light is n_R. What is the angular spread of visible light passing through a prism of apex angle Φ if the angle of incidence is θ? See Figure P34.25.

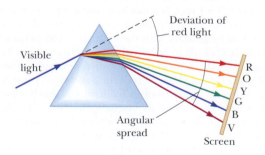

Figure P34.25

26. The speed of a water wave is described by $v = \sqrt{gd}$, where d is the water depth, assumed to be small compared to the wavelength. Because their speed changes, water waves refract when moving into a region of different depth. (a) Sketch a map of an ocean beach on the eastern side of a landmass. Show contour lines of constant depth under water, assuming a reasonably uniform slope. (b) Suppose waves approach the coast from a storm far away to the north–northeast. Demonstrate that the waves move nearly perpendicular to the shoreline when they reach the beach. (c) Sketch a map of a coastline with alternating bays and headlands as suggested in Figure P34.26. Again make a reasonable guess about the shape of contour lines of constant depth. (d) Suppose waves approach the coast, carrying energy with uniform density along originally straight wave fronts. Show that the energy reaching the coast is concentrated at the headlands and has lower intensity in the bays.

Figure P34.26

Kevin Miller/Getty Images

SECTION 34.7 Total Internal Reflection

27. For 589-nm light, calculate the critical angle for the following materials surrounded by air: (a) cubic zirconia, (b) flint glass, and (c) ice.

28. Consider a light ray traveling between air and a diamond cut in the shape shown in Figure P34.28. (a) Find the critical angle for total internal reflection for light in the diamond incident on the interface between the diamond and the outside air. (b) Consider the light ray incident normally on the top surface of the diamond as shown in Figure P34.28. Show that the light traveling toward point P in the diamond is totally reflected. **What If?** Suppose the diamond is immersed in water. (c) What is the critical angle at the diamond–water interface? (d) When the diamond is immersed in water, does the light ray entering the top surface in Figure P34.28 undergo total internal reflection at P? Explain. (e) If the light ray entering the diamond remains vertical as shown in Figure P34.28, which way should the diamond

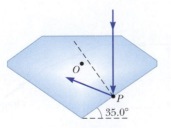

Figure P34.28

in the water be rotated about an axis perpendicular to the page through O so that light will exit the diamond at P? (f) At what angle of rotation in part (e) will light first exit the diamond at point P?

29. A room contains air in which the speed of sound is 343 m/s. The walls of the room are made of concrete in which the speed of sound is 1 850 m/s. (a) Find the critical angle for total internal reflection of sound at the concrete–air boundary. (b) In which medium must the sound be initially traveling if it is to undergo total internal reflection? (c) "A bare concrete wall is a highly efficient mirror for sound." Give evidence for or against this statement.

30. Around 1968, Richard A. Thorud, an engineer at The Toro Company, invented a gasoline gauge for small engines diagrammed in Figure P34.30. The gauge has no moving parts. It consists of a flat slab of transparent plastic fitting vertically into a slot in the cap on the gas tank. None of the plastic has a reflective coating. The plastic projects from the horizontal top down nearly to the bottom of the opaque tank. Its lower edge is cut with facets making angles of

Figure P34.30

45° with the horizontal. A lawn mower operator looks down from above and sees a boundary between bright and dark on the gauge. The location of the boundary, across the width of the plastic, indicates the quantity of gasoline in the tank. (a) Explain how the gauge works. (b) Explain the design requirements, if any, for the index of refraction of the plastic.

31. An optical fiber has an index of refraction n and diameter d. It is surrounded by vacuum. Light is sent into the fiber along its axis as shown in Figure P34.31. (a) Find the smallest outside radius R_{min} permitted for a bend in the fiber if no light is to escape. (b) **What If?** What result does part (a) predict as d approaches zero? Is this behavior reasonable? Explain. (c) As n

Figure P34.31

increases? (d) As n approaches 1? (e) Evaluate R_{min} assuming the fiber diameter is 100 μm and its index of refraction is 1.40.

ADDITIONAL PROBLEMS

32. Consider a horizontal interface between air above and glass of index of refraction 1.55 below. (a) Draw a light ray incident from the air at angle of incidence 30.0°. Determine the angles of the reflected and refracted rays and show them on the diagram. (b) **What If?** Now suppose the light ray is incident from the glass at an angle of 30.0°. Determine the angles of the reflected and refracted rays and show all three rays on a new diagram. (c) For rays incident from the air onto the air–glass surface, determine and tabulate the angles of reflection and refraction for all the angles of incidence at 10.0° intervals from 0° to 90.0°. (d) Do the same for light rays coming up to the interface through the glass.

33. How many times will the incident beam in Figure P34.33 (page 922) be reflected by each of the parallel mirrors?

Figure P34.33

34. Consider a beam of light from the left entering a prism of apex angle Φ as shown in Figure P34.34. Two angles of incidence, θ_1 and θ_3, are shown as well as two angles of refraction, θ_2 and θ_4. Show that $\Phi = \theta_2 + \theta_3$.

Figure P34.34

35. *Why is the following situation impossible?* While at the bottom of a calm freshwater lake, a scuba diver sees the Sun at an apparent angle of 38.0° above the horizontal.

36. *Why is the following situation impossible?* A laser beam strikes one end of a slab of material of length $L = 42.0$ cm and thickness $t = 3.10$ mm as shown in Figure P34.36 (not to scale). It enters the material at the center of the left end, striking it at an angle of incidence of $\theta = 50.0°$. The index of refraction of the slab is $n = 1.48$. The light makes 85 internal reflections from the top and bottom of the slab before exiting at the other end.

Figure P34.36

37. When light is incident normally on the interface between two transparent optical media, the intensity of the reflected light is given by the expression

$$S_1' = \left(\frac{n_2 - n_1}{n_2 + n_1}\right)^2 S_1$$

In this equation, S_1 represents the average magnitude of the Poynting vector in the incident light (the incident intensity), S_1' is the reflected intensity, and n_1 and n_2 are the refractive indices of the two media. (a) What fraction of the incident intensity is reflected for 589-nm light normally incident on an interface between air and crown glass? (b) Does it matter in part (a) whether the light is in the air or in the glass as it strikes the interface?

38. Refer to Problem 37 for its description of the reflected intensity of light normally incident on an interface between two transparent media. (a) For light normally incident on an interface between vacuum and a transparent medium of index n, show that the intensity S_2 of the transmitted light is given by $S_2/S_1 = 4n/(n + 1)^2$. (b) Light travels perpendicularly through a diamond slab, surrounded by air, with

parallel surfaces of entry and exit. Apply the transmission fraction in part (a) to find the approximate overall transmission through the slab of diamond, as a percentage. Ignore light reflected back and forth within the slab.

39. **T** A light ray enters the atmosphere of the Earth and descends vertically to the surface a distance $h = 100$ km below. The index of refraction where the light enters the atmosphere is 1.00, and it increases linearly with distance to have the value $n = 1.000\ 293$ at the Earth's surface. (a) Over what time interval does the light traverse this path? (b) By what percentage is the time interval larger than that required in the absence of the Earth's atmosphere?

40. **S** A light ray enters the atmosphere of a planet and descends vertically to the surface a distance h below. The index of refraction where the light enters the atmosphere is 1.00, and it increases linearly with distance to have the value n at the planet surface. (a) Over what time interval does the light traverse this path? (b) By what fraction is the time interval larger than that required in the absence of an atmosphere?

41. **Q|C** A light ray of wavelength 589 nm is incident at an angle θ on the top surface of a block of polystyrene as shown in Figure P34.41. (a) Find the maximum value of θ for which the refracted ray undergoes total internal reflection at the point P located at the left vertical face of the block. **What If?** Repeat the calculation for the case in which the polystyrene block is immersed in (b) water and (c) carbon disulfide. Explain your answers.

Figure P34.41

42. One technique for measuring the apex angle of a prism is shown in Figure P34.42. Two parallel rays of light are directed onto the apex of the prism so that the rays reflect from opposite faces of the prism. The angular separation γ of the two reflected rays can be measured. Show that $\Phi = \frac{1}{2}\gamma$.

Figure P34.42

43. **S** A material having an index of refraction n is surrounded by vacuum and is in the shape of a quarter circle of radius R (Fig. P34.43). A light ray parallel to the base of the material is incident from the left at a distance L above the base and emerges from the material at the angle θ. Determine an expression for θ in terms of n, R, and L.

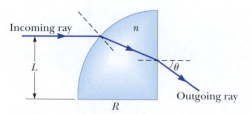

Figure P34.43

44. Review. A mirror is often "silvered" with aluminum. By adjusting the thickness of the metallic film, one can make a sheet of glass into a mirror that reflects anything between 3% and 98% of the incident light, transmitting the rest. Prove that it is impossible to construct a "one-way mirror" that would reflect 90% of the electromagnetic waves incident from one side and reflect 10% of those incident from the other side. *Suggestion:* Use Clausius's statement of the second law of thermodynamics.

45. Figure P34.45 shows the path of a light beam through several slabs with different indices of refraction. (a) If $\theta_1 = 30.0°$, what is the angle θ_2 of the emerging beam? (b) What must the incident angle θ_1 be to have total internal reflection at the surface between the medium with $n = 1.20$ and the medium with $n = 1.00$?

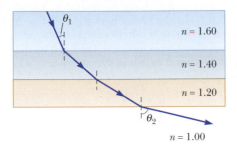

Figure P34.45

46. As sunlight enters the Earth's atmosphere, it changes direction due to the small difference between the speeds of light in vacuum and in air. The duration of an *optical* day is defined as the time interval between the instant when the top of the rising Sun is just visible above the horizon and the instant when the top of the Sun just disappears below the horizontal plane. The duration of the *geometric* day is defined as the time interval between the instant a mathematically straight line between an observer and the top of the Sun just clears the horizon and the instant this line just dips below the horizon. (a) Explain which is longer, an optical day or a geometric day. (b) Find the difference between these two time intervals. Model the Earth's atmosphere as uniform, with index of refraction 1.000 293, a sharply defined upper surface, and depth 8 614 m. Assume the observer is at the Earth's equator so that the apparent path of the rising and setting Sun is perpendicular to the horizon.

47. A ray of light passes from air into water. For its deviation angle $\delta = |\theta_1 - \theta_2|$ to be 10.0°, what must its angle of incidence be?

48. In your work for an optical research company, you are asked to consider the triangular shaped prism shown in Figure P34.48. Light enters the left slanted side of the prism from air at normal incidence, reflects from the top surface by total internal reflection, and then refracts out of the right slanted surface. (a) Your supervisor asks you to determine the range of angles over which visible light exits the right slanted surface due to dispersion in the material. (b) An actual physical prism of the shape in Figure P34.48 is then made from cubic zirconia with $\alpha = 60°$ and $\gamma = 30°$, and it doesn't work as planned. Explain to your supervisor why not.

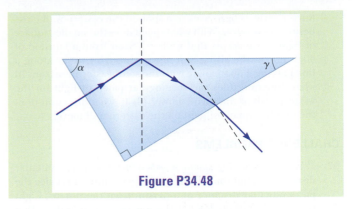

Figure P34.48

49. A. H. Pfund's method for measuring the index of refraction of glass is illustrated in Figure P34.49. One face of a slab of thickness t is painted white, and a small hole scraped clear at point P serves as a source of diverging rays when the slab is illuminated from below. Ray PBB' strikes the clear surface at the critical angle and is totally reflected, as are rays such as PCC'. Rays such as PAA' emerge from the clear surface. On the painted surface, there appears a dark circle of diameter d surrounded by an illuminated region, or halo. (a) Derive an equation for n in terms of the measured quantities d and t. (b) What is the diameter of the dark circle if $n = 1.52$ for a slab 0.600 cm thick? (c) If white light is used, dispersion causes the critical angle to depend on color. Is the inner edge of the white halo tinged with red light or with violet light? Explain.

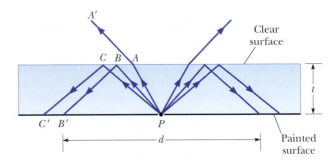

Figure P34.49

50. Figure P34.50 shows a top view of a square enclosure. The inner surfaces are plane mirrors. A ray of light enters a small hole in the center of one mirror. (a) At what angle θ must the ray enter if it exits through the hole after being reflected once by each of the other three mirrors? (b) **What If?** Are there other values of θ for which the ray can exit after multiple reflections? If so, sketch one of the ray's paths.

Figure P34.50

51. The walls of an ancient shrine are perpendicular to the four cardinal compass directions. On the first day of spring, light from the rising Sun enters a rectangular window in the eastern wall. The light traverses 2.37 m horizontally to shine perpendicularly on the wall opposite the window. A tourist observes the patch of light moving across this western wall. (a) With what speed does the illuminated rectangle move? (b) The tourist holds a small, square mirror flat against the western wall at one corner of the rectangle of light. The mirror reflects light back to a spot on the eastern wall close beside the window. With what speed does the smaller square of light move across that wall? (c) Seen from a latitude of 40.0° north, the rising Sun moves through the sky along a line making a 50.0° angle with the southeastern horizon. In what direction does the rectangular patch of light on the western wall of the shrine move? (d) In what direction does the smaller square of light on the eastern wall move?

CHALLENGE PROBLEMS

52. *Why is the following situation impossible?* The perpendicular distance of a lightbulb from a large plane mirror is twice the perpendicular distance of a person from the mirror. Light from the lightbulb reaches the person by two paths: (1) it travels to the mirror and reflects from the mirror to the person, and (2) it travels directly to the person without reflecting off the mirror. The total distance traveled by the light in the first case is 3.10 times the distance traveled by the light in the second case.

53. Figure P34.53 shows an overhead view of a room of square floor area and side L. At the center of the room is a mirror set in a vertical plane and rotating on a vertical shaft at angular speed ω about an axis coming out of the page. A bright red laser beam enters from the center point on one wall of the room and strikes the mirror. As the mirror rotates, the reflected laser beam creates a red spot sweeping across the walls of the room. (a) When the spot of light on the wall is at distance x from point O, what is its speed? (b) What value of x corresponds to the minimum value for the speed? (c) What is the minimum value for the speed? (d) What is the maximum speed of the spot on the wall? (e) In what time interval does the spot change from its minimum to its maximum speed?

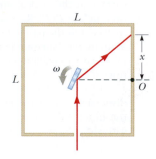

Figure P34.53

54. Pierre de Fermat (1601–1665) showed that whenever light travels from one point to another, its actual path is the path that requires the smallest time interval. This statement is known as *Fermat's principle*. The simplest example is for light propagating in a homogeneous medium. It moves in a straight line because a straight line is the shortest distance between two points. Derive Snell's law of refraction from Fermat's principle. Proceed as follows. In Figure P34.54, a light ray travels from point P in medium 1 to point Q in

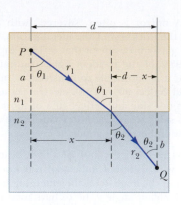

Figure P34.54 Problems 54 and 55.

medium 2. The two points are, respectively, at perpendicular distances a and b from the interface. The displacement from P to Q has the component d parallel to the interface, and we let x represent the coordinate of the point where the ray enters the second medium. Let $t = 0$ be the instant the light starts from P. (a) Show that the time at which the light arrives at Q is

$$t = \frac{r_1}{v_1} + \frac{r_2}{v_2} = \frac{n_1\sqrt{a^2 + x^2}}{c} + \frac{n_2\sqrt{b^2 + (d - x)^2}}{c}$$

(b) To obtain the value of x for which t has its minimum value, differentiate t with respect to x and set the derivative equal to zero. Show that the result implies

$$\frac{n_1 x}{\sqrt{a^2 + x^2}} = \frac{n_2(d - x)}{\sqrt{b^2 + (d - x)^2}}$$

(c) Show that this expression in turn gives Snell's law,

$$n_1 \sin \theta_1 = n_2 \sin \theta_2$$

55. Refer to Problem 54 for the statement of Fermat's principle of least time. Derive the law of reflection (Eq. 34.1) from Fermat's principle.

56. Suppose a luminous sphere of radius R_1 (such as the Sun) is surrounded by a uniform atmosphere of radius $R_2 > R_1$ and index of refraction n. When the sphere is viewed from a location far away in vacuum, what is its apparent radius (a) when $R_2 > nR_1$ and (b) when $R_2 < nR_1$?

57. This problem builds upon the results of Problems 37 and 38. Light travels perpendicularly through a diamond slab, surrounded by air, with parallel surfaces of entry and exit. The intensity of the transmitted light is what fraction of the incident intensity? Include the effects of light reflected back and forth inside the slab.

Image Formation

A smartphone display and a magnifying glass can be used to perform simple optics experiments. *(iStockphoto.com/alexsl)*

STORYLINE **While thinking about what might be coming up in this** new chapter, you notice a magnifying glass in your desk drawer. You take it out of the drawer; this device might be useful for your study of image formation! In a pitch-dark room, you turn on your smartphone and lay it down on a table, counter, or desk so that the display is upward. You hold the magnifying glass lens horizontally a few inches above the display. You notice when the lens is at a certain position above the display a clear image of the display forms on the ceiling. You measure the distance between the display and the lens. You then move to a room with a window and hold the lens of the magnifying glass vertically near the wall of the room opposite the window. As you move the lens back and forth, you notice that a clear image of the window and the building across the street forms when the lens is at a certain distance from the wall. You measure the distance between the wall and the lens. You notice that this horizontal distance from the wall is very similar to the vertical distance between your smartphone and the lens in the first experiment! You find that the distances are *similar* but not quite *equal*. Should they be the same?

CONNECTIONS In this chapter, we apply the laws of reflection and refraction from Chapter 34 in order to investigate the images that result when light rays encounter flat or curved surfaces between two media. We can design mirrors and lenses to form images with desired characteristics. In this study, we continue to use the ray approximation and assume light travels in straight lines. We first study the geometry of images formed by mirrors and lenses, and then determine techniques for locating an image and predicting its size. Then we investigate how to combine these elements into several useful optical instruments such as microscopes and telescopes. The control of light with reflecting and refracting optical instruments is critical to research that allows us to understand the material in upcoming chapters.

35.1 Images Formed by Flat Mirrors

Image formation by mirrors can be understood through the behavior of light rays as described by the wave under reflection analysis model. Let's begin with an image you see every day: your face in the bathroom mirror. This image is formed by the simplest possible mirror, the flat mirror. Consider a point source of light placed at O in Figure 35.1, a distance p in front of a flat mirror. In the figure, the mirror surface is the dark blue edge. The lighter blue band represents the structural support for the mirrored surface, such as a piece of glass on which a silvered reflecting surface is deposited. The mirror is perpendicular to the page, so we see the intersection of the mirror with the page. The distance p is called the **object distance**, the name anticipating that we will place *objects* in front of mirrors and study their images. Diverging light rays leave the source and are reflected from the mirror, obeying the law of reflection. Upon reflection, the rays continue to diverge. The dashed lines in Figure 35.1 are backward extensions of the diverging rays back to a point of intersection at I. The diverging rays appear to the viewer to originate at the point I behind the mirror. Point I, which is a distance q behind the mirror, is called the **image** of the object at O. The distance q is called the **image distance.** Regardless of the system under study, images can always be located by extending diverging rays back to a point at which they intersect.

Images are located either at a point from which rays of light *actually* diverge, or at a point from which they *appear* to diverge, as in Figure 35.1. This difference allows us to classify images as *real* or *virtual*. A **real image** is formed when all light rays pass through and diverge from the image point; a **virtual image** is formed when most if not all of the light rays do *not* pass through the image point but only appear to diverge from that point. The image formed by the mirror in Figure 35.1 is virtual. No light rays from the object exist behind the mirror, at the location of the image, so the light rays in front of the mirror only seem to be diverging from I. The image of an object seen in a flat mirror is *always* virtual. Real images can be displayed on a screen (as at a movie theater), but virtual images cannot be displayed on a screen. We shall see an example of a real image in Section 35.2.

Figure 35.1 shows the image of a *point object*. We can use the simple geometry in Figure 35.2 to examine the properties of the images of *extended objects* formed by flat mirrors. The gray arrow is the object. The image of the bottom of the arrow is located behind the mirror at distance q, just like the point object in Figure 35.1. Even though there are an infinite number of choices of other points on the object and directions in which light rays could leave those points, we need to choose only two rays leaving the top of the arrow at point P to determine where the image is formed. The red ray starts at P, follows a path perpendicular to the mirror to Q, and reflects back on itself. The blue ray follows the oblique path PR and reflects as shown in Figure 35.2 according to the law of reflection. An observer in front of the mirror would extend the two reflected rays back to the point at which they appear to have originated, which is point P' behind the mirror. A continuation of this

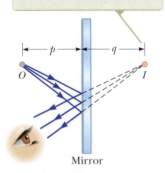

The image point I is located behind the mirror a distance q from the mirror. The image is virtual.

Figure 35.1 An image formed by reflection from a flat mirror.

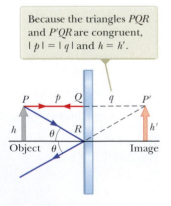

Because the triangles PQR and $P'QR$ are congruent, $|p| = |q|$ and $h = h'$.

Figure 35.2 A geometric construction that is used to locate the image of an object placed in front of a flat mirror.

process for points other than P on the object would result in a virtual image (represented by a pink arrow) of the entire object behind the mirror. Because triangles PQR and $P'QR$ are congruent, $PQ = P'Q$, so $|p| = |q|$. (We use absolute value signs here because we will find that q is negative for this type of image.) Therefore, the image formed of an object placed in front of a flat mirror is as far behind the mirror as the object is in front of the mirror.

The geometry in Figure 35.2 also reveals that the object height h equals the image height h'. Let us define **lateral magnification** M of an image as follows:

$$M = \frac{\text{image height}}{\text{object height}} = \frac{h'}{h} \tag{35.1}$$

This general definition of the lateral magnification for an image from any type of mirror is also valid for images formed by lenses, which we study in Section 35.4. For a flat mirror, $M = +1$ for any image because $h' = h$. The positive value of the magnification signifies that the image is upright. (By upright we mean that if the object arrow points upward as in Figure 35.2, so does the image arrow.)

A flat mirror produces an image that has an *apparent* left–right reversal. You can see this reversal by standing in front of a mirror and raising your right hand as shown in Figure 35.3. The image you see raises its left hand. Likewise, your hair appears to be parted on the side opposite your real part, and a mole on your right cheek appears to be on your left cheek.

This reversal is not *actually* a left–right reversal. Imagine, for example, lying on your left side on the floor with your body parallel to the mirror surface. Now your head is on the left and your feet are on the right. If you shake your feet, the image does not shake its head! If you raise your right hand, however, the image again raises its left hand. Therefore, the mirror again appears to produce a left–right reversal but in the up–down direction!

The reversal is actually a *front–back reversal*, caused by the light rays going forward toward the mirror and then reflecting back from it. An interesting exercise is to stand in front of a mirror while holding an overhead transparency in front of you so that you can read the writing on the transparency. You will also be able to read the writing on the image of the transparency. You may have had a similar experience if you have attached a transparent decal with words on it to the rear window of your car. If the decal can be read from outside the car, you can also read it when looking into your rearview mirror from inside the car.

The thumb is on the left side of both real hands and on the left side of the image. That the thumb is not on the right side of the image indicates that there is no left-to-right reversal.

© Cengage

Figure 35.3 The image in the mirror of a person's right hand is reversed front to back, which makes the right hand appear to be a left hand.

QUICK QUIZ 35.1 You are standing approximately 2 m away from a mirror. The mirror has water spots on its surface. True or False: It is possible for you to see the water spots and your image both in focus at the same time.

Conceptual Example 35.1   **Multiple Images Formed by Two Mirrors**

Two flat mirrors are perpendicular to each other as in Figure 35.4, and an object is placed at point O. In this situation, multiple images are formed. Locate the positions of these images.

SOLUTION

Figure 35.4 (Conceptual Example 35.1) When an object is placed in front of two mutually perpendicular mirrors as shown, three images are formed. Follow the different-colored light rays to understand the formation of each image.

The image of the object is at I_1 in mirror 1 (green rays) and at I_2 in mirror 2 (red rays). In addition, a third image is formed at I_3 (blue rays). To form this image at I_3, the rays reflect twice, once from each mirror, after leaving the object at O. This third image can be considered as the image of I_1 in mirror 2, as indicated by the dashed lines leaving I_1 toward the upper right. That is, the image at I_1 serves as the object for I_3. It is also possible to show that I_3 is the image of I_2 in mirror 1.

Many rearview mirrors in cars have a day setting and a night setting. The night setting greatly diminishes the intensity of the image so that lights from trailing vehicles do not temporarily blind the driver. How does such a mirror work?

SOLUTION

Figure 35.5 shows a cross-sectional view of a rearview mirror for each setting. The unit consists of a reflective coating on the back of a wedge of glass. In the day setting (Fig. 35.5a), the light from an object behind the car strikes the glass wedge at point 1. Most of the light enters the wedge, refracting as it crosses the front surface, and reflects from the back surface at 2 to return to the front surface at 3, where it is refracted again as it re-enters the air as ray *B* (for *bright*). In addition, a small portion of the light is reflected at the front surface of the glass at 1 as indicated by ray *D* (for *dim*).

This dim reflected light is responsible for the image observed when the mirror is in the night setting (Fig. 35.5b). In that case, the wedge is rotated so that the path followed by the bright light (ray *B*) does not lead to the eye. Instead, the dim light reflected from the front surface of the wedge travels to the eye, and the brightness of trailing headlights does not become a hazard.

Figure 35.5 (Conceptual Example 35.2) Cross-sectional views of a rearview mirror.

35.2 Images Formed by Spherical Mirrors

In the preceding section, we considered images formed by flat mirrors. Now we study images formed by curved mirrors. Although a variety of curvatures are possible, we will restrict our investigation to spherical mirrors. As its name implies, a **spherical mirror** has the shape of a section of a sphere.

Concave Mirrors

We first consider reflection of light from the inner, concave surface of a spherical mirror as shown in Figure 35.6. The curved line in the figure represents the intersection of the bowl-shaped section of a sphere with the page. We can determine all of the properties of images formed by spherical mirrors by considering only rays in the plane of the page. The solid, curved dark blue line is the reflecting surface of the mirror. This type of reflecting surface is called a **concave mirror.** Figure 35.6a shows that the mirror has a radius of curvature *R*, and its center of curvature is point *C*. Point *V* is the center of the spherical section, and a line through *C* and *V* is called the **principal axis** of the mirror.

Figure 35.6 (a) A concave mirror of radius *R*. The center of curvature *C* is located on the principal axis. (b) A point object placed at *O* in front of a concave spherical mirror of radius *R*, where *O* is any point on the principal axis farther than *R* from the mirror surface, forms a real image at *I*.

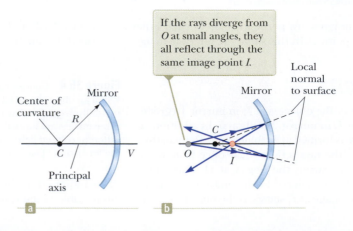

Now consider a point source of light placed at point O in Figure 35.6b, where O is any point on the principal axis to the left of C. Two diverging light rays that originate at O are shown. After reflecting from the mirror, obeying the law of reflection from the wave under reflection analysis model, these rays converge and cross at the image point I. They then continue to diverge from I as if an object were there. An observer to the left of O would see the light rays diverging from I. As a result, the image at point I is real.

In this section, we shall consider only rays that diverge from the object and make a small angle with the principal axis. Such rays are called **paraxial rays.** All paraxial rays reflect through the image point. Rays that are far from the principal axis such as those shown in Figure 35.7 converge to other points on the principal axis, producing a blurred image. This effect, called *spherical aberration,* is present to some extent for any spherical mirror and is discussed in Section 35.5.

If the object distance p and radius of curvature R are known, we can use Figure 35.8 to calculate the image distance q. By convention, these distances are measured from point V. Figure 35.8 shows two rays leaving the tip of the object. The red ray passes through the center of curvature C of the mirror, hitting the mirror perpendicular to the mirror surface and reflecting back on itself. The blue ray strikes the mirror at its center (point V) and reflects as shown, obeying the law of reflection. The image of the tip of the arrow is located at the point where these two rays intersect. From the large, red right triangle in Figure 35.8, we see that $\tan \theta = h/p$, and from the yellow right triangle, we see that $\tan \theta = -h'/q$. The negative sign is introduced because the image is inverted, so h' is taken to be negative. Therefore, from Equation 35.1 and these results, we find that the magnification of the image is

$$M = \frac{h'}{h} = -\frac{q}{p} \tag{35.2}$$

Also notice from the green right triangle in Figure 35.8 and the smaller red right triangle that

$$\tan \alpha = \frac{-h'}{R - q} \quad \text{and} \quad \tan \alpha = \frac{h}{p - R}$$

from which it follows that

$$\frac{h'}{h} = -\frac{R - q}{p - R} \tag{35.3}$$

Comparing Equations 35.2 and 35.3 gives

$$\frac{R - q}{p - R} = \frac{q}{p}$$

The reflected rays intersect at different points on the principal axis.

Figure 35.7 A spherical concave mirror exhibits spherical aberration when light rays make large angles with the principal axis.

PITFALL PREVENTION 35.1

Magnification Does Not Necessarily Imply Enlargement For optical elements other than flat mirrors, the magnification defined in Equation 35.2 can result in a number with a magnitude larger *or* smaller than 1. Therefore, despite the cultural usage of the word *magnification* to mean *enlargement,* the image could be smaller than the object.

The real image lies at the location at which the reflected rays cross.

h

Principal axis

O α C α I θ V h' θ

q

R

p

Figure 35.8 The image formed by a spherical concave mirror when the object O lies outside the center of curvature C. This geometric construction is used to derive Equation 35.6.

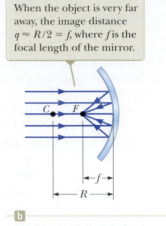

When the object is very far away, the image distance $q \approx R/2 = f$, where f is the focal length of the mirror.

Courtesy of Henry Leap and Jim Lehman

Figure 35.9 (a) Light rays from an object far to the left are parallel as they arrive at the mirror. Upon reflection, they all pass through the *focal point*. (b) The *focal length f* is half the radius of curvature of the mirror.

iStockPhoto.com/Maria Barski

Figure 35.10 A satellite-dish antenna is a concave reflector for television signals from a satellite in orbit around the Earth. Because the satellite is so far away, the signals are carried by microwaves that are parallel when they arrive at the dish. These waves reflect from the dish and are focused on the receiver.

Simple algebra reduces this expression to

$$\frac{1}{p} + \frac{1}{q} = \frac{2}{R} \tag{35.4}$$

which is called the *mirror equation*. We present a modified version of this equation shortly.

If the object is very far from the mirror—that is, if p is so much greater than R that p can be said to approach infinity—then the light rays reaching the mirror from the object are parallel. Figure 35.9a shows the results of parallel rays in the laboratory. The rays reflect from the mirror and pass through a single point called the **focal point** F. Figure 35.9b shows a geometric construction for parallel rays striking the mirror. From the law of reflection, we see that the focal point must lie between the mirror surface and the center of curvature C of the mirror. If we let p approach infinity in Equation 35.4, we see that $q \approx R/2$. For parallel rays, the image point must be the focal point. Therefore, we see that the focal point is located a distance from the mirror called the **focal length** f, and

◀ Focal length for a mirror

$$f = \frac{R}{2} \tag{35.5}$$

Figure 35.10 shows a practical application of the situation in Figure 35.9. Parallel microwaves carrying television signals from a satellite far above the Earth's surface strike the curved surface of the antenna and are focused at a receiver placed at the focal point of the surface.

Combining Equations 35.4 and 35.5, the **mirror equation** can now be expressed in terms of the focal length:

◀ Mirror equation

$$\frac{1}{p} + \frac{1}{q} = \frac{1}{f} \tag{35.6}$$

Notice from Equation 35.5 that the focal length of a mirror depends only on the curvature of the mirror and not on the material from which the mirror is made because the formation of the image results from rays reflected from the surface of the material. The situation is different for lenses; in that case, the light actually passes through the material and the focal length depends on the type of material from which the lens is made. (See Section 35.4.)

Convex Mirrors

Figure 35.11 shows the formation of an image by a **convex mirror,** that is, one silvered so that light is reflected from the outer, convex surface. It is sometimes

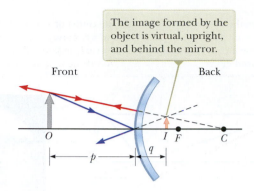

The image formed by the object is virtual, upright, and behind the mirror.

Figure 35.11 Formation of an image by a spherical convex mirror.

called a **diverging mirror** because the rays from any point on an object diverge after reflection as though they were coming from some point behind the mirror. The image in Figure 35.11 is virtual because the reflected rays only appear to originate at the image point as indicated by the dashed lines. Furthermore, the image is always upright and smaller than the object. This type of mirror is often used in stores to foil shoplifters. A single mirror can be used to survey a large field of view because it forms a smaller, upright image of the interior of the store.

We do not derive any equations for convex spherical mirrors because Equations 35.2, 35.4, and 35.6 can be used for either concave or convex mirrors if we adhere to a strict sign convention. We will refer to the region in which light rays originate and move toward the mirror as the *front side* of the mirror and the other side as the *back side*. For example, in Figures 35.8 and 35.11, the side to the left of the mirrors is the front side and the side to the right of the mirrors is the back side. Figure 35.12 states the sign conventions for object and image distances for any type of mirror, and Table 35.1 summarizes the sign conventions for all quantities. One entry in the table, a *virtual object,* is formally introduced in Section 35.4.

Ray Diagrams for Mirrors

In Figures 35.2, 35.8, and 35.11, we have located an image using two rays: a red one hitting the mirror perpendicular to the surface and reflecting straight back, and a blue ray striking the mirror at the principal axis and obeying the law of reflection. If we wanted to draw a precision diagram to locate the image position carefully, we would need a protractor to make the sure the incident and reflected angles at the principal axis are the same for the blue ray. Our new knowledge of the focal point, however, makes things easier. Let's investigate the building of *ray diagrams* that are accurate but do not require a protractor. These pictorial representations reveal the nature of the image and can be used to check results calculated from the mathematical representation using the mirror and magnification equations. To draw a ray diagram, you must know the position of the object and the locations of the mirror's focal point and center of curvature. You then draw three rays from the top of the object to locate the image as shown by the examples in

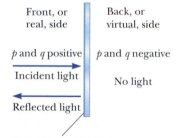

Front, or real, side   Back, or virtual, side

p and q positive p and q negative

Incident light   No light

Reflected light

Flat, convex, or concave mirrored surface

Figure 35.12 Signs of p and q for all types of mirrors.

TABLE 35.1 Sign Conventions for Mirrors

Quantity	Positive When ...	Negative When ...
Object location (p)	object is in front of mirror (real object).	object is in back of mirror (virtual object).
Image location (q)	image is in front of mirror (real image).	image is in back of mirror (virtual image).
Image height (h')	image is upright.	image is inverted.
Focal length (f) and radius (R)	mirror is concave.	mirror is convex.
Magnification (M)	image is upright.	image is inverted.

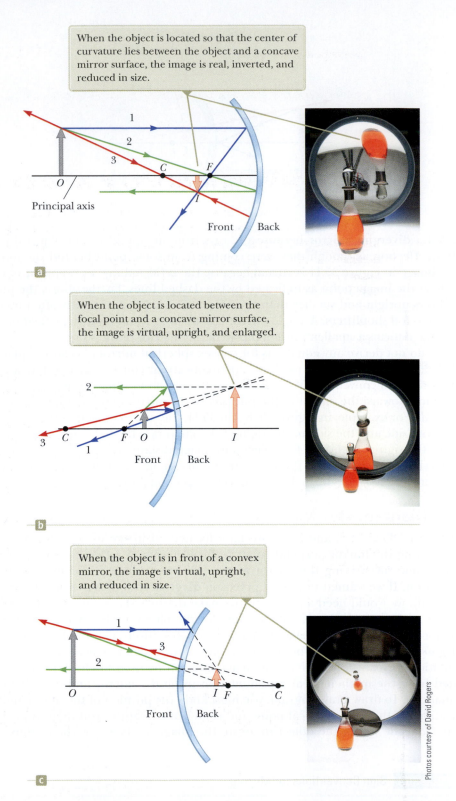

When the object is located so that the center of curvature lies between the object and a concave mirror surface, the image is real, inverted, and reduced in size.

Principal axis

Front   Back

a

When the object is located between the focal point and a concave mirror surface, the image is virtual, upright, and enlarged.

Front   Back

b

When the object is in front of a convex mirror, the image is virtual, upright, and reduced in size.

Front   Back

c

Figure 35.13 Ray diagrams for spherical mirrors along with corresponding photographs of the images of bottles. (a) In this photograph, both the object and the image are in front of the mirror, so you would need to focus your eyes on points in front of the mirror to see the object and image clearly. (b), (c) In these photographs, the object is in front of the mirror, but you would need to focus your eyes on a point behind the mirror to see the image clearly.

Figure 35.13. We will keep the red ray from our earlier diagrams and add two new rays. For concave mirrors (see Figs. 35.13a and 35.13b), draw the following three rays, noting the colors indicated in Figure 35.13:

- Ray 1 (blue) is drawn from the top of the object parallel to the principal axis and is reflected through the focal point *F*.

- Ray 2 (green) is drawn from the top of the object through the focal point (or as if coming from the focal point if $p < f$) and is reflected parallel to the principal axis.
- Ray 3 (red) is drawn from the top of the object through the center of curvature C (or as if coming from the center C if $p < 2f$) and is reflected back on itself.

PITFALL PREVENTION 35.4
Choose a Small Number of Rays
A *huge* number of light rays leave each point on an object (and pass through each point on an image). In a ray diagram, which displays the characteristics of the image, we choose only a few rays that follow simply stated rules. Locating the image by calculation complements the diagram.

The intersection of any two of these rays locates the image. The third ray serves as a check of the construction. The image point obtained in this fashion must always agree with the value of q calculated from the mirror equation. With concave mirrors, notice what happens as the object is moved closer to the mirror. The real, inverted image in Figure 35.13a moves to the left and becomes larger as the object approaches the center of curvature C. When the object is at C, a distance $p = 2f$ from the mirror, Equation 35.6 shows that $q = 2f$: the image is located at the location of the object, with a magnification of −1. As the object continues to move from the center of curvature toward the focal point, the image grows larger ($|M| > 1$) and moves to the left. When the object is at the focal point, the image is infinitely far to the left. When the object lies between the focal point and the mirror surface as shown in Figure 35.13b, however, the image is to the right, behind the object, and virtual, upright, and enlarged. This latter situation applies when you use a shaving mirror or a makeup mirror, both of which are concave. Your face is closer to the mirror than the focal point, and you see an upright, enlarged image of your face.

For convex mirrors (see Fig. 35.13c), draw the following three rays:

- Ray 1 (blue) is drawn from the top of the object parallel to the principal axis and is reflected *away from* the focal point F.
- Ray 2 (green) is drawn from the top of the object toward the focal point on the back side of the mirror and is reflected parallel to the principal axis.
- Ray 3 (red) is drawn from the top of the object toward the center of curvature C on the back side of the mirror and is reflected back on itself.

In a convex mirror, the image of an object is always virtual, upright, and reduced in size as shown in Figure 35.13c. In this case, as the object distance decreases, the virtual image increases in size and moves away from the focal point toward the mirror as the object approaches the mirror. You should construct other diagrams to verify how image position varies with object position.

QUICK QUIZ 35.2 You wish to start a fire by reflecting sunlight from a mirror onto some paper under a pile of wood. Which would be the best choice for the type of mirror? **(a)** flat **(b)** concave **(c)** convex

QUICK QUIZ 35.3 Consider the image in the mirror in Figure 35.14. Based on the appearance of this image, would you conclude that **(a)** the mirror is concave and the image is real, **(b)** the mirror is concave and the image is virtual, **(c)** the mirror is convex and the image is real, or **(d)** the mirror is convex and the image is virtual?

Figure 35.14 (Quick Quiz 35.3) What type of mirror is shown here?

Example **35.3** **The Image Formed by a Concave Mirror**

A spherical mirror has a focal length of +10.0 cm.

(A) Locate and describe the image for an object distance of 25.0 cm.

SOLUTION

Conceptualize Because the focal length of the mirror is positive, it is a concave mirror (see Table 35.1). We expect the possibilities of both real and virtual images.

continued

35.3 continued

Categorize   Because the object distance in this part of the problem is larger than the focal length, we expect the image to be real. This situation is analogous to that in Figure 35.13a.

Analyze   Find the image distance by using Equation 35.6:

$$\frac{1}{q} = \frac{1}{f} - \frac{1}{p} \rightarrow q = \frac{fp}{p-f} \tag{35.7}$$

Substitute numerical values:

$$q = \frac{(10.0 \text{ cm})(25.0 \text{ cm})}{25.0 \text{ cm} - 10.0 \text{ cm}} = \boxed{16.7 \text{ cm}}$$

Find the magnification of the image from Equation 35.2:

$$M = -\frac{q}{p} = -\frac{16.7 \text{ cm}}{25.0 \text{ cm}} = \boxed{-0.667}$$

Finalize   The absolute value of M is less than unity, so the image is smaller than the object, and the negative sign for M tells us that the image is inverted. Because q is positive, the image is located on the front side of the mirror and is real. Look into the bowl of a shiny spoon or stand far away from a shaving mirror to see this image.

(B) Locate and describe the image for an object distance of 5.00 cm.

SOLUTION

Categorize   Because the object distance is smaller than the focal length, we expect the image to be virtual. This situation is analogous to that in Figure 35.13b.

Analyze   Find the image distance by using Equation 35.7:

$$q = \frac{fp}{p-f}$$

Substitute numerical values:

$$q = \frac{(10.0 \text{ cm})(5.0 \text{ cm})}{5.0 \text{ cm} - 10.0 \text{ cm}} = \boxed{-10.0 \text{ cm}}$$

Find the magnification of the image from Equation 35.2:

$$M = -\frac{q}{p} = -\left(\frac{-10.0 \text{ cm}}{5.00 \text{ cm}}\right) = \boxed{+2.00}$$

Finalize   The image is twice as large as the object, and the positive sign for M indicates that the image is upright (see Fig. 35.13b). The negative value of the image distance tells us that the image is virtual, as expected. Put your face close to a shaving mirror to see this type of image.

WHAT IF?   Suppose you set up the bottle and mirror apparatus illustrated in Figure 35.13a and described here in part (A). While adjusting the apparatus, you accidentally bump the bottle and it begins to slide toward the mirror at speed v_p. How fast does the image of the bottle move?

Answer   Begin with Equation 35.7:

$$q = \frac{fp}{p-f}$$

Differentiate this equation with respect to time to find the velocity of the image:

$$(1) \quad v_q = \frac{dq}{dt} = \frac{d}{dt}\left(\frac{fp}{p-f}\right) = -\frac{f^2}{(p-f)^2}\frac{dp}{dt} = -\frac{f^2 v_p}{(p-f)^2}$$

Substitute numerical values from part (A):

$$v_q = -\frac{(10.0 \text{ cm})^2 v_p}{(25.0 \text{ cm} - 10.0 \text{ cm})^2} = -0.444 v_p$$

Therefore, the speed of the image is less than that of the object in this case.

We can see two interesting behaviors of the function for v_q in Equation (1). First, the velocity is negative regardless of the value of p or f. Therefore, if the object moves toward the mirror, the image moves toward the left in Figure 35.13 without regard for the side of the focal point at which the object is located or whether the mirror is concave or convex. Second, in the limit of $p \rightarrow 0$, the velocity v_q approaches $-v_p$. As the object moves very close to the mirror, the mirror looks like a plane mirror, the image is as far behind the mirror as the object is in front, and both the object and the image move with the same speed.

Example 35.4   The Image Formed by a Convex Mirror

An automobile sideview mirror on the passenger side as shown in Figure 35.15 shows an image of a truck located 50.0 m from the mirror. The focal length of the mirror is −0.60 m.

(A) Find the position of the image of the truck.

Figure 35.15 (Example 35.4) An approaching truck is seen in a convex mirror on the passenger side of an automobile. Notice that the image of the truck is in focus, but the frame of the mirror is not, which demonstrates that the image is not at the same location as the mirror surface.

Brian A Jackson/Shutterstock

SOLUTION

Conceptualize   This situation is depicted in Figure 35.13c.

Categorize   Because the mirror is convex, we expect it to form an upright, reduced, virtual image for any object position.

Analyze   Find the image distance by using Equation 35.7: $q = \dfrac{fp}{p-f} = \dfrac{(-0.60\text{ cm})(50.0\text{ cm})}{50.0\text{ cm} - (-0.60\text{ cm})} = -0.59\text{ cm}$

(B) Find the magnification of the image.

SOLUTION

Analyze   Use Equation 35.2: $M = -\dfrac{q}{p} = -\left(\dfrac{-0.59\text{ m}}{50.0\text{ m}}\right) = +0.012$

Finalize   The negative value of q in part (A) indicates that the image is virtual, or behind the mirror, as shown in Figure 35.13c. The magnification in part (B) indicates that the image is much smaller than the truck and is upright because M is positive. The image is reduced in size, so the truck appears to be farther away than it actually is. Because of the image's small size, these mirrors often carry the inscription, "Objects in this mirror are closer than they appear." Look into your passenger-side rearview mirror or the back side of a shiny spoon to see an image of this type.

35.3   Images Formed by Refraction

In this section, we describe how images are formed when light rays follow the wave under refraction model at the boundary between two materials. Consider two transparent media having indices of refraction n_1 and n_2, where the boundary between the two media is a spherical surface of radius R as in Figure 35.16. The object at O is in the medium for which the index of refraction is n_1. Let's consider the paraxial rays leaving O. Figure 35.16 is the refraction analog to the reflections from the mirror in Figure 35.6b.

QUICK QUIZ 35.4   In Figure 35.16, what happens to the image point I as the object point O is moved from very far to the left to very close to the refracting surface? **(a)** It is always to the right of the surface. **(b)** It is always to the left of the surface. **(c)** It starts off to the left, and at some position of O, I moves to the right of the surface. **(d)** It starts off to the right, and at some position of O, I moves to the left of the surface.

Figure 35.17 (page 936) shows a single ray leaving point O and refracting to point I. Snell's law of refraction in the wave under refraction model applied to this ray gives

$$n_1 \sin \theta_1 = n_2 \sin \theta_2$$

Because θ_1 and θ_2 are assumed to be small, we can use the small-angle approximation $\sin \theta \approx \theta$ (with angles in radians) and write Snell's law as

$$n_1 \theta_1 = n_2 \theta_2$$

Rays making small angles with the principal axis diverge from a point object at O and are refracted through the image point I.

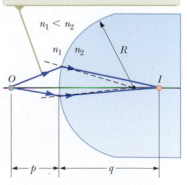

Figure 35.16 An image formed by refraction at a spherical surface.

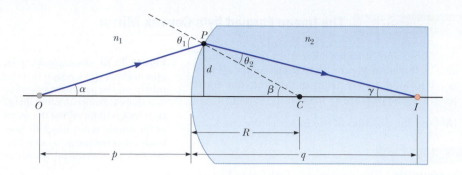

Figure 35.17 Geometry used to derive Equation 35.9, assuming $n_1 < n_2$.

We know that an exterior angle of any triangle equals the sum of the two opposite interior angles, so applying this rule to triangles *OPC* and *PIC* in Figure 35.17 gives

$$\theta_1 = \alpha + \beta$$

$$\beta = \theta_2 + \gamma$$

Combining all three expressions and eliminating θ_1 and θ_2 gives

$$n_1\alpha + n_2\gamma = (n_2 - n_1)\beta \tag{35.8}$$

Figure 35.17 shows three right triangles that have a common vertical leg of length d. For paraxial rays (unlike the relatively large-angle ray shown in Fig. 35.17), the horizontal legs of these triangles are approximately p for the triangle containing angle α, R for the triangle containing angle β, and q for the triangle containing angle γ. In the small-angle approximation, $\tan \theta \approx \theta$, so we can write the approximate relationships from these triangles as follows:

$$\tan \alpha \approx \alpha \approx \frac{d}{p} \quad \tan \beta \approx \beta \approx \frac{d}{R} \quad \tan \gamma \approx \gamma \approx \frac{d}{q}$$

Substituting these expressions into Equation 35.8 and dividing through by d gives

$$\frac{n_1}{p} + \frac{n_2}{q} = \frac{n_2 - n_1}{R} \tag{35.9}$$

◀ Relation between object and image distance for a refracting surface

For a fixed object distance p, the image distance q is independent of the angle the ray makes with the axis. This result tells us that all paraxial rays focus at the same point I. The magnification of an image due to a refracting surface is given by (see Problem 20)

$$M = -\frac{n_1 q}{n_2 p} \tag{35.10}$$

As with mirrors, we must use a sign convention to apply Equation 35.9 to a variety of cases. We define the side of the surface in which light rays originate as the front side. The other side is called the back side. In contrast with mirrors, where real images are formed in front of the reflecting surface, real images are formed by refraction of light rays to the back of the surface. This is consistent with the facts that mirrors reflect light back to the same side, while transparent materials allow light to pass through to the other side. Because of the difference in location of real images, the refraction sign conventions for q and R are opposite the reflection sign conventions. For example, q and R are both positive in Figure 35.17. The sign conventions for spherical refracting surfaces are summarized in Table 35.2.

We derived Equation 35.9 from an assumption that $n_1 < n_2$ in Figure 35.17. This assumption is not necessary, however. Equation 35.9 is valid regardless of which index of refraction is greater.

For a concave mirror, we found that both real and virtual images can be formed, depending on the location of the object relative to the focal point. Figures 35.13a

TABLE 35.2 Sign Conventions for Refracting Surfaces

Quantity	Positive When . . .	Negative When . . .
Object location (p)	object is in front of surface (real object).	object is in back of surface (virtual object).
Image location (q)	image is in back of surface (real image).	image is in front of surface (virtual image).
Image height (h')	image is upright.	image is inverted.
Radius (R)	center of curvature is in back of surface.	center of curvature is in front of surface.

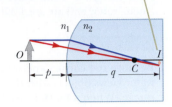

The image due to the surface is real, so I is to the right of the surface.

and 35.13b show these possibilities. We find that we can create both real and virtual images with a refracting surface, also. Figure 35.18 shows these two possibilities for a refracting surface surrounded by a medium of index n_1. These two types of images can be created by placing the object at different positions relative to the surface. The value for p that determines whether an image is real or virtual can be found by letting $q \to \infty$ in Equation 35.9. Solving for p, we find

$$p = \frac{n_1}{n_2 - n_1} R$$

When the object is at this distance from the surface, the image is infinitely far away. If the object is moved farther from the mirror than this position, the image is real as in Figure 35.18a. If the object is moved closer to the surface than this distance, the image is virtual as in Figure 35.18b.

Now consider placing an object *inside* a material of index n_1 and looking at the image from a material of index n_2, where $n_1 > n_2$. Equation 35.9 gives us

$$q = \frac{pR}{\left(1 - \dfrac{n_1}{n_2}\right)p - \dfrac{n_1}{n_2}R}$$

For the condition $n_1 > n_2$, the value of q is *always* negative. Therefore, the image is always virtual. This is analogous to all images being virtual for a convex mirror, as in Figure 35.13c. See Figure 35.19 and Examples 35.6 and 35.7 for situations in which the object is inside a material.

The image due to the surface is virtual, so I is to the left of the surface.

Figure 35.18 A refracting surface can create a (a) real image or a (b) virtual image.

Flat Refracting Surfaces

If a refracting surface is flat, then R is infinite and Equation 35.9 reduces to

$$\frac{n_1}{p} = -\frac{n_2}{q}$$

$$q = -\frac{n_2}{n_1} p \qquad (35.11)$$

From this expression, we see that the sign of q is opposite that of p. Therefore, according to Table 35.2, the image formed by a flat refracting surface is on the same side of the surface as the object as illustrated in Figure 35.19 for the situation in which the object is in the medium of index n_1 and n_1 is greater than n_2. In this case, a virtual image is formed between the object and the surface. If n_1 is less than n_2, the rays on the back side diverge from one another at smaller angles than those in Figure 35.19. As a result, the virtual image is formed to the left of the object.

QUICK QUIZ 35.5 In Figure 35.19, what happens to the image point I as the object point O moves toward the right-hand surface of the material of index of refraction n_1? **(a)** It always remains between O and the surface, arriving at the surface just as O does. **(b)** It moves toward the surface more slowly than O so that eventually O passes I. **(c)** It approaches the surface and then moves to the right of the surface.

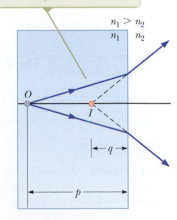

The image is virtual and on the same side of the surface as the object.

Figure 35.19 The image formed by a flat refracting surface. All rays are assumed to be paraxial.

Conceptual Example 35.5 | **Let's Go Scuba Diving!**

Objects viewed under water with the naked eye appear blurred and out of focus. A scuba diver using a mask, however, has a clear view of underwater objects. Explain how that works, using the information that the indices of refraction of the cornea, water, and air are 1.376, 1.333, and 1.000 29, respectively.

S O L U T I O N

Because the cornea and water have almost identical indices of refraction, very little refraction occurs when a person under water views objects with the naked eye. In this case, light rays from an object focus behind the retina, resulting in a blurred image. When a mask is used, however, the air space between the eye and the mask surface provides the normal amount of refraction at the eye–air interface; consequently, the light from the object focuses on the retina.

Example 35.6 | **Gaze into the Crystal Ball**

A set of coins is embedded in a spherical plastic paperweight having a radius of 3.0 cm. The index of refraction of the plastic is $n_1 = 1.50$. One coin is located 2.0 cm from the edge of the sphere (Fig. 35.20). Find the position of the image of the coin.

Figure 35.20 (Example 35.6) Light rays from a coin embedded in a plastic sphere form a virtual image between the surface of the object and the sphere surface. Because the object is inside the sphere, the front of the refracting surface is the *interior* of the sphere.

S O L U T I O N

Conceptualize Because $n_1 > n_2$, where $n_2 = 1.00$ is the index of refraction for air, the rays originating from the coin in Figure 35.20 are refracted away from the normal at the surface and diverge outward. Extending the outgoing rays backward shows an image point within the sphere.

Categorize Because the light rays originate in one material and then pass through a curved surface into another material, this example involves an image formed by refraction.

Analyze Apply Equation 35.9, noting from Table 35.2 that R is negative:

$$\frac{n_2}{q} = \frac{n_2 - n_1}{R} - \frac{n_1}{p}$$

Substitute numerical values and solve for q:

$$\frac{1}{q} = \frac{1.00 - 1.50}{-3.0 \text{ cm}} - \frac{1.50}{2.0 \text{ cm}}$$

$$q = \boxed{-1.7 \text{ cm}}$$

Finalize The negative sign for q indicates that the image is in front of the surface; in other words, it is in the same medium as the object as shown in Figure 35.20. Therefore, the image must be virtual. (See Table 35.2.) The coin appears to be closer to the paperweight surface than it actually is.

Example 35.7 | **The One That Got Away**

A small fish is swimming at a depth d below the surface of a pond (Fig. 35.21).

(A) What is the apparent depth of the fish as viewed from directly overhead?

S O L U T I O N

Conceptualize Because $n_1 > n_2$, where $n_2 = 1.00$ is the index of refraction for air, the rays originating from the fish in Figure 35.21 are refracted away from the normal at the surface and diverge outward. Extending the outgoing rays backward shows an image point under the water.

Categorize Because the refracting surface is flat, R is infinite. Hence, we can use Equation 35.11 to determine the location of the image with $p = d$.

Figure 35.21 (Example 35.7) The apparent depth q of the fish is less than the true depth d. All rays are assumed to be paraxial.

35.7 continued

Analyze Use the indices of refraction given in Figure 35.21 in Equation 35.11:

$$q = -\frac{n_2}{n_1}p = -\frac{1.00}{1.33}d = \boxed{-0.752d}$$

Finalize Because q is negative, the image is virtual as indicated by the dashed lines in Figure 35.21. The apparent depth is approximately three-fourths the actual depth. Any source of light underwater appears to be closer to the surface. Therefore, considering the light leaving the bottom of a stream, all streams appear to have a depth that is about three-fourths their actual depth.

(B) If your face is a distance d above the water surface, at what apparent distance above the surface does the fish see your face?

S O L U T I O N

Conceptualize Imagine light rays leaving your face and moving downward toward the water. Upon entering the water, they will refract toward the normal. Draw a ray diagram and show that your face appears to the fish to be higher than it actually is.

Categorize Because the refracting surface is flat, R is infinite. Hence, we can use Equation 35.11 to determine the location of the image with $p = d$.

Analyze Use Equation 35.11 to find the image distance:

$$q = -\frac{n_2}{n_1}p = -\frac{1.33}{1.00}d = \boxed{-1.33d}$$

Finalize The negative sign for q indicates that the image is in the medium from which the light originated, which is the air above the water.

W H A T I F ? What if you look more carefully at the fish and measure its apparent *height* from its upper fin to its lower fin? Is the apparent height h' of the fish different from the actual height h?

Answer Because all points on the fish appear to be fractionally closer to the observer, we expect the height to be smaller. Let the distance d in Figure 35.21 be measured to the top fin and let the distance to the bottom fin be $d + h$. Then the images of the top and bottom of the fish are located at

$$q_{\text{top}} = -0.752d$$

$$q_{\text{bottom}} = -0.752(d + h)$$

The apparent height h' of the fish is

$$h' = q_{\text{top}} - q_{\text{bottom}} = -0.752d - [-0.752(d + h)] = 0.752h$$

Hence, the fish appears to be approximately three-fourths its actual height.

35.4 Images Formed by Thin Lenses

Lenses are commonly used to form images by refraction in optical instruments such as cameras, telescopes, and microscopes. Let's use what we just learned about images formed by refracting surfaces to help locate the image formed by a lens. Light passing through a lens experiences refraction at two surfaces. The development we shall follow is based on the notion that the image formed by one refracting surface serves as the object for the second surface. We shall analyze a thick lens first and then let the thickness of the lens shrink to approximately zero.

Consider Figure 35.18 and imagine that the right side of the material does not continue indefinitely, but ends in another curved surface. Then we have a refracting material with two spherical surfaces with radii of curvature R_1 and R_2, separated by a distance t as in Figure 35.22 (page 940). (Notice that R_1 is the radius of curvature of the lens surface the light from the object reaches first and R_2 is the radius of curvature of the other surface of the lens.)

Figure 35.22 shows the real and virtual images formed by the first surface, as we saw previously in Figure 35.18. Using Equation 35.9 and assuming $n_1 = 1$ because

Figure 35.22 To locate the image formed by a lens, we use the virtual image at I_1 formed by surface 1 as the object for the image formed by surface 2. The point C_1 is the center of curvature of surface 1.

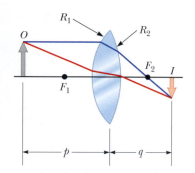

Figure 35.23 Simplified geometry for a thin lens. The dots labeled F_1 and F_2 are focal points.

Lens-makers' equation ▶

the lens is surrounded by air, we find that the image I_1 formed by surface 1 satisfies the equation

$$\frac{1}{p_1} + \frac{n}{q_1} = \frac{n-1}{R_1} \qquad (35.12)$$

where q_1 is the position of the image formed by surface 1. If the image formed by surface 1 is real (Fig. 35.22a), then q_1 is positive; it is negative if the image is virtual (Fig. 35.22b).

Now let's apply Equation 35.9 to surface 2, taking $n_1 = n$ and $n_2 = 1$. (We make this switch in index because the light rays approaching surface 2 are *in the material of the lens,* and this material has index n.) Taking p_2 as the object distance for surface 2 and q_2 as the image distance gives

$$\frac{n}{p_2} + \frac{1}{q_2} = \frac{1-n}{R_2} \qquad (35.13)$$

We now introduce mathematically that the image formed by the first surface acts as the object for the second surface. If the image from surface 1 is real as in Figure 35.22a, we see that the physical distance p_2 is $q_1 - t$. But I_1 serves as a **virtual object** for the second surface, because it is to the right of the surface. To express p_2 as an object distance for optical purposes, it must be negative (Table 35.2). Therefore, for optical purposes, we must have $p_2 = -q_1 + t$. If the image from surface 1 is virtual as in Figure 35.22b, we see that $p_2 = -q_1 + t$, where we have used $-q_1$ because q_1 is negative for a virtual object. Therefore, regardless of the type of image from surface 1, the same equation describes the location of the object for surface 2 based on our sign convention. For a *thin* lens (one whose thickness is small compared with the radii of curvature), we can neglect t. In this approximation, $p_2 = -q_1$ for either type of image from surface 1. Hence, Equation 35.13 becomes

$$-\frac{n}{q_1} + \frac{1}{q_2} = \frac{1-n}{R_2} \qquad (35.14)$$

Adding Equations 35.12 and 35.14 gives

$$\frac{1}{p_1} + \frac{1}{q_2} = (n-1)\left(\frac{1}{R_1} - \frac{1}{R_2}\right) \qquad (35.15)$$

For a thin lens, p_1 is the position p of the object and q_2 is the position q of the final image as in Figure 35.23. Hence, we can write Equation 35.15 as

$$\frac{1}{p} + \frac{1}{q} = (n-1)\left(\frac{1}{R_1} - \frac{1}{R_2}\right) \qquad (35.16)$$

This expression relates the image distance q of the image formed by a thin lens to the object distance p and to the lens properties (index of refraction and radii of curvature). It is valid only for paraxial rays and only when the lens thickness is much less than R_1 and R_2.

The **focal length** f of a thin lens is the image distance that corresponds to an infinite object distance, just as with mirrors. Letting p approach ∞ and q approach f in Equation 35.16, we see that the inverse of the focal length for a thin lens is

$$\frac{1}{f} = (n-1)\left(\frac{1}{R_1} - \frac{1}{R_2}\right) \qquad (35.17)$$

This relationship is called the **lens-makers' equation** because it can be used to determine the values of R_1 and R_2 needed for a given index of refraction and a desired focal length f. Conversely, if the index of refraction and the radii of curvature of a lens are given, this equation can be used to find the focal length. If the lens is immersed in something other than air, this same equation can be used, with

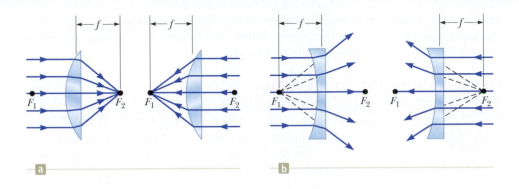

Figure 35.24 Parallel light rays
pass through (a) a converging lens
and (b) a diverging lens. The focal
length is the same for light rays
passing through a given lens in
either direction. Both focal points
F_1 and F_2 are the same distance
from the lens.

n interpreted as the *ratio* of the index of refraction of the lens material to that of
the surrounding fluid.

Using Equation 35.17, we can write Equation 35.16 in a form identical to Equation 35.6 for mirrors:

$$\frac{1}{p} + \frac{1}{q} = \frac{1}{f} \qquad (35.18)$$

◀ Thin lens equation

This equation, called the **thin lens equation,** can be used to relate the image distance and object distance for a thin lens.

Because light can travel in either direction through a lens, each lens has two focal
points, one for light rays passing through in one direction and one for rays passing
through in the other direction. These two focal points are illustrated in Figure 35.24
for a plano-convex lens (a converging lens) and a plano-concave lens (a diverging lens).

Figure 35.25 is useful for obtaining the signs of p and q, and Table 35.3 gives the
sign conventions for thin lenses. These sign conventions are the *same* as those for
refracting surfaces (see Table 35.2).

Various lens shapes are shown in Figure 35.26 (page 942). Notice that a converging lens is thicker at the center than at the edge, whereas a diverging lens is thinner
at the center than at the edge.

The magnifying glass used in the experiments described in the opening storyline is acting as a thin lens to form real images. In Think–Pair–Share Problem 35.3,
you will have a chance to perform these experiments and answer the question at
the end of the storyline.

Magnification of Images

Consider a thin lens through which light rays from an object pass. As with mirrors
(Eq. 35.2), a geometric construction shows that the lateral magnification of the image is

$$M = \frac{h'}{h} = -\frac{q}{p} \qquad (35.19)$$

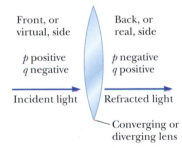

Front, or
virtual, side

Back, or
real, side

p positive
q negative

p negative
q positive

Incident light

Refracted light

Converging or
diverging lens

Figure 35.25 A diagram for
obtaining the signs of p and q for
a thin lens. (This diagram also
applies to a refracting surface.)

TABLE 35.3   Sign Conventions for Thin Lenses

Quantity	Positive When . . .	Negative When . . .
Object location (p)	object is in front of lens (real object).	object is in back of lens (virtual object).
Image location (q)	image is in back of lens (real image).	image is in front of lens (virtual image).
Image height (h')	image is upright.	image is inverted.
R_1 and R_2	center of curvature is in back of lens.	center of curvature is in front of lens.
Focal length (f)	a converging lens.	a diverging lens.

PITFALL PREVENTION 35.5

**A Lens Has Two Focal Points
but Only One Focal Length** A lens
has a focal point on each side,
front and back. There is only
one focal length, however; each
of the two focal points is located
the same distance from the lens
(Fig. 35.24). As a result, the lens
forms an image of an object at
the same point if it is turned
around. In practice, that might
not happen because real lenses
are not infinitesimally thin.

Biconvex   Convex-   Plano-
            concave   convex

Biconcave   Convex-   Plano-
             concave   concave

Figure 35.26 Various lens shapes. (a) Converging lenses have a positive focal length and are thickest at the middle. (b) Diverging lenses have a negative focal length and are thickest at the edges.

From this expression, it follows that when M is positive, the image is upright and on the same side of the lens as the object. When M is negative, the image is inverted and on the side of the lens opposite the object.

Ray Diagrams for Thin Lenses

As for mirrors, ray diagrams are convenient for locating the images formed by thin lenses or systems of lenses. They also help clarify our sign conventions. Figure 35.27 shows such diagrams for three single-lens situations. In Figures 35.23 and 35.24, we showed the path of light rays refracting at both surfaces of the lens. Because we model lenses to have zero thickness, and because we know the importance of the focal point of a lens, in future diagrams we will show the light ray simply refracting at the center of the lens as shown in Figure 35.27.

To locate the image of a *converging* lens (Figs. 35.27a and 35.27b), the following three rays are drawn from the top of the object, noting the colors indicated in Figure 35.27:

- Ray 1 (blue) is drawn parallel to the principal axis. After being refracted by the lens, this ray passes through the focal point on the back side of the lens.
- Ray 2 (green) is drawn through the focal point on the front side of the lens (or as if coming from the focal point if $p < f$) and emerges from the lens parallel to the principal axis.
- Ray 3 (red) is drawn through the center of the lens and continues in a straight line.

To locate the image of a *diverging* lens (Fig. 35.27c), the following three rays are drawn from the top of the object:

- Ray 1 (blue) is drawn parallel to the principal axis. After being refracted by the lens, this ray emerges directed *away from* the focal point on the front side of the lens.
- Ray 2 (green) is drawn in the direction toward the focal point on the back side of the lens and emerges from the lens parallel to the principal axis.
- Ray 3 (red) is drawn through the center of the lens and continues in a straight line.

For the converging lens in Figure 35.27a, where the object is to the left of the focal point ($p > f$), the image is real and inverted and the lens acts like a video projector. When the object is between the focal point and the lens ($p < f$) as in Figure 35.27b, the image is virtual and upright. In that case, the lens acts as a magnifying glass, which we study in more detail in Section 35.6. For a diverging lens

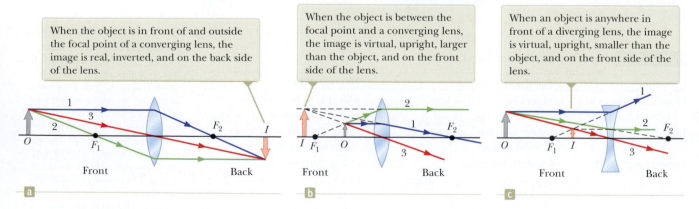

When the object is in front of and outside the focal point of a converging lens, the image is real, inverted, and on the back side of the lens.

When the object is between the focal point and a converging lens, the image is virtual, upright, larger than the object, and on the front side of the lens.

When an object is anywhere in front of a diverging lens, the image is virtual, upright, smaller than the object, and on the front side of the lens.

Figure 35.27 Ray diagrams for locating the image formed by a thin lens.

Figure 35.28 A side view of the construction of a Fresnel lens. (a) The thick lens refracts a light ray as shown. (b) Lens material in the bulk of the lens is cut away, leaving only the material close to the curved surface. (c) The small pieces of remaining material are moved to the left to form a flat surface on the left of the Fresnel lens with ridges on the right surface. From a front view, these ridges would be circular in shape. This new lens refracts light in the same way as the lens in (a). (d) A Fresnel lens used in a lighthouse shows several segments with the ridges discussed in (c).

(Fig. 35.27c), the image is always virtual and upright, regardless of where the object is placed, like the image in a door peephole. These geometric constructions are reasonably accurate only if the distance between the rays and the principal axis is much less than the radii of the lens surfaces.

Refraction occurs only at the surfaces of the lens. The uniform material inside the lens simply propagates the light, but does not affect the direction in which it travels. A certain lens design takes advantage of this behavior to produce the *Fresnel lens*, a powerful lens without great thickness. Because only the surface curvature is important in the refracting qualities of the lens, material in the middle of a Fresnel lens is removed as shown in the cross sections of lenses in Figure 35.28. Because the edges of the curved segments cause some distortion, Fresnel lenses are generally used only in situations in which image quality is less important than reduction of weight. A classroom overhead projector often uses a Fresnel lens; the circular edges between segments of the lens can be seen by looking closely at the light projected onto a screen.

QUICK QUIZ 35.6 What is the focal length of a pane of window glass? **(a)** zero **(b)** infinity **(c)** the thickness of the glass **(d)** impossible to determine

---

Example 35.8 | **Images Formed by a Converging Lens**

A converging lens has a focal length of 10.0 cm.

(A) An object is placed 30.0 cm from the lens. Construct a ray diagram, find the image distance, and describe the image.

SOLUTION

Conceptualize Because the lens is converging, the focal length is positive (see Table 35.3). We expect the possibilities of both real and virtual images.

Categorize Because the object distance is larger than the focal length, we expect the image to be real. The ray diagram for this situation is shown in Figure 35.29a.

The object is farther from the lens than the focal point.

The object is closer to the lens than the focal point.

Figure 35.29 (Example 35.8) An image is formed by a converging lens.

Analyze Because Equation 35.18 for lenses is identical to Equation 35.6 for mirrors, we can use Equation 35.7 for lenses:

$$q = \frac{fp}{p - f} = \frac{(10.0 \text{ cm})(30.0 \text{ cm})}{30.0 \text{ cm} - 10.0 \text{ cm}} = +15.0 \text{ cm}$$

continued

35.8 continued

Find the magnification of the image from Equation 35.19:

$$M = -\frac{q}{p} = -\frac{15.0 \text{ cm}}{30.0 \text{ cm}} = \boxed{-0.500}$$

Finalize   The positive sign for the image distance tells us that the image is indeed real and on the back side of the lens. The magnification of the image tells us that the image is reduced in height by one half, and the negative sign for M tells us that the image is inverted.

(B) An object is placed 5.00 cm from the lens. Construct a ray diagram, find the image distance, and describe the image.

SOLUTION

Categorize   Because the object distance is smaller than the focal length, we expect the image to be virtual. The ray diagram for this situation is shown in Figure 35.29b.

Analyze   Find the image distance by using Equation 35.7:

$$q = \frac{fp}{p-f} = \frac{(10.0 \text{ cm})(5.00 \text{ cm})}{5.00 \text{ cm} - 10.0 \text{ cm}} = \boxed{-10.0 \text{ cm}}$$

Find the magnification of the image from Equation 35.19:

$$M = -\frac{q}{p} = -\left(\frac{-10.0 \text{ cm}}{5.00 \text{ cm}}\right) = \boxed{+2.00}$$

Finalize   The negative image distance tells us that the image is virtual and formed on the side of the lens from which the light is incident, the front side. The image is enlarged, and the positive sign for M tells us that the image is upright.

WHAT IF?   What if the object moves right up to the lens surface so that $p \rightarrow 0$? Where is the image?

Answer   In this case, because $p \ll R$, where R is either of the radii of the surfaces of the lens, the curvature of the lens can be ignored. The lens should appear to have the same effect as a flat piece of material, which suggests that the image is just on the front side of the lens, at $q = 0$. This conclusion can be verified mathematically by rearranging the thin lens equation:

$$\frac{1}{q} = \frac{1}{f} - \frac{1}{p}$$

If we let $p \rightarrow 0$, the second term on the right becomes very large compared with the first and we can neglect $1/f$. The equation becomes

$$\frac{1}{q} = -\frac{1}{p} \quad \rightarrow \quad q = -p = 0$$

Therefore, q is on the front side of the lens (because it has the opposite sign as p) and right at the lens surface.

Example 35.9   **Images Formed by a Diverging Lens**

A diverging lens has a focal length of 10.0 cm.

(A) An object is placed 15.0 cm from the lens. Construct a ray diagram, find the image distance, and describe the image.

SOLUTION

Conceptualize   Because the lens is diverging, the focal length is negative (see Table 35.3). The ray diagram for this situation is shown in Figure 35.30a.

Categorize   Because the lens is diverging, we expect it to form an upright, reduced, virtual image for any object position.

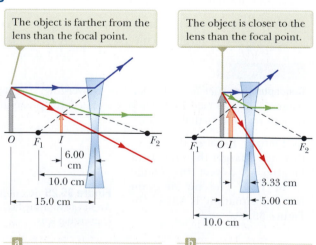

Figure 35.30   (Example 35.9) An image is formed by a diverging lens.

Analyze   Find the image distance by using Equation 35.7:

$$q = \frac{fp}{p-f} = \frac{(-10.0 \text{ cm})(15.0 \text{ cm})}{15.0 \text{ cm} - (-10.0 \text{ cm})} = \boxed{-6.00 \text{ cm}}$$

35.9 continued

Find the magnification of the image from Equation 35.19:

$$M = -\frac{q}{p} = -\left(\frac{-6.00 \text{ cm}}{15.0 \text{ cm}}\right) = \boxed{+0.400}$$

Finalize This result confirms that the image is virtual, smaller than the object, and upright.

(B) An object is placed 5.00 cm from the lens. Construct a ray diagram, find the image distance, and describe the image.

SOLUTION

The ray diagram for this situation is shown in Figure 35.30b.

Analyze Find the image distance by using Equation 35.7:

$$q = \frac{fp}{p-f} = \frac{(-10.0 \text{ cm})(5.00 \text{ cm})}{5.00 \text{ cm} - (-10.0 \text{ cm})} = \boxed{-3.33 \text{ cm}}$$

Find the magnification of the image from Equation 35.19:

$$M = -\left(\frac{-3.33 \text{ cm}}{5.00 \text{ cm}}\right) = \boxed{+0.667}$$

Finalize For both object positions, the image position is negative and the magnification is a positive number smaller than 1, which confirms that the image is virtual, smaller than the object, and upright.

Combinations of Thin Lenses

If two thin lenses separated by a distance d are used to form an image, the system can be treated in the following manner. First, the image formed by the first lens is located as if the second lens were not present. Then a ray diagram is drawn for the second lens, with the image formed by the first lens now serving as the object for the second lens. The second image formed is the final image of the system. If the image formed by the first lens lies on the back side of the second lens, that image is treated as a virtual object for the second lens (that is, in the thin lens equation, p is negative). The same procedure can be extended to a system of three or more lenses. Because the magnification due to the second lens is performed on the magnified image due to the first lens, the overall magnification of the image due to the combination of lenses is the product of the individual magnifications:

$$M = M_1 M_2 \tag{35.20}$$

This equation can be used for combinations of any optical elements such as a lens and a mirror. For more than two optical elements, the magnifications due to all elements are multiplied together.

Example 35.10 **Where Is the Final Image?**

Two thin converging lenses of focal lengths $f_1 = 10.0$ cm and $f_2 = 20.0$ cm are separated by $d = 20.0$ cm as illustrated in Figure 35.31. An object is placed 30.0 cm to the left of lens 1. Find the position and the magnification of the final image.

SOLUTION

Conceptualize Imagine light rays passing through the first lens and forming a real image (because $p > f$) in the absence of a second lens. Figure 35.31 shows these light rays forming the inverted image I_1. Once the light rays converge to the image point, they do not stop. They continue through the image point and interact with the second lens. The rays leaving the image point behave in the same way as the rays leaving an object. Therefore, the image of the first lens serves as the object of the second lens.

Figure 35.31 (Example 35.10) A combination of two converging lenses. The ray diagram shows the location of the final image (I_2) due to the combination of lenses. The black dots are the focal points of lens 1, and the red dots are the focal points of lens 2. Notice that the green ray for lens 1 becomes the blue ray for lens 2. Also, because the focal point for lens 2 is at the center of lens 1, the red ray for lens 1 becomes the green ray for lens 2.

continued

35.10 continued

Categorize   We categorize this problem as one in which the thin lens equation is applied in a stepwise fashion to the two lenses.

Analyze   Find the location of the image formed by lens 1 from the thin lens equation:

$$q_1 = \frac{f_1 p_1}{p_1 - f_1} = \frac{(10.0 \text{ cm})(30.0 \text{ cm})}{30.0 \text{ cm} - 10.0 \text{ cm}} = \boxed{+15.0 \text{ cm}}$$

Find the magnification of the image from Equation 35.19:

$$M_1 = -\frac{q_1}{p_1} = -\frac{15.0 \text{ cm}}{30.0 \text{ cm}} = -0.500$$

The image formed by this lens acts as the object for the second lens. Therefore, the object distance for the second lens is $p_2 = d - q_1 = 20.0 \text{ cm} - 15.0 \text{ cm} = 5.00 \text{ cm}$.

Find the location of the image formed by lens 2 from the thin lens equation:

$$q_2 = \frac{f_2 p_2}{p_2 - f_2} = \frac{(20.0 \text{ cm})(5.00 \text{ cm})}{5.00 \text{ cm} - 20.0 \text{ cm}} = \boxed{-6.67 \text{ cm}}$$

Find the magnification of the image from Equation 35.19:

$$M_2 = -\frac{q_2}{p_2} = -\frac{(-6.67 \text{ cm})}{5.00 \text{ cm}} = +1.33$$

Find the overall magnification of the system from Equation 35.20:

$$M = M_1 M_2 = (-0.500)(1.33) = \boxed{-0.667}$$

Finalize   The negative sign on the overall magnification indicates that the final image is inverted with respect to the initial object. Because the absolute value of the magnification is less than 1, the final image is smaller than the object.

Because q_2 is negative, the final image is on the front, or left, side of lens 2. These conclusions are consistent with the ray diagram in Figure 35.31.

WHAT IF?   Suppose you want to create an upright image with this system of two lenses. How must the second lens be moved?

Answer   Because the object is farther from the first lens than the focal length of that lens, the first image is inverted. Consequently, the second lens must invert the image once again so that the final image is upright. An inverted image is only formed by a converging lens if the object is outside the focal point. Therefore, the image formed by the first lens must be to the left of the focal point of the second lens in Figure 35.31. To make that happen, you must move the second lens at least as far away from the first lens as the sum $q_1 + f_2 = 15.0 \text{ cm} + 20.0 \text{ cm} = 35.0 \text{ cm}$.

Let's consider the special case of a system of two lenses of focal lengths f_1 and f_2 in contact with each other, so that d goes to zero. If $p_1 = p$ is the object distance for the combination, application of the thin lens equation (Eq. 35.18) to the first lens gives

$$\frac{1}{p} + \frac{1}{q_1} = \frac{1}{f_1}$$

where q_1 is the image distance for the first lens. Treating this image as the object for the second lens, we see that the object distance for the second lens must be $p_2 = -q_1$. (The distances are the same because the lenses are in contact and assumed to be infinitesimally thin. The object distance is negative because the object is virtual if the image from the first lens is real.) Therefore, for the second lens,

$$\frac{1}{p_2} + \frac{1}{q_2} = \frac{1}{f_2} \quad \rightarrow \quad -\frac{1}{q_1} + \frac{1}{q} = \frac{1}{f_2}$$

where $q = q_2$ is the final image distance from the second lens, which is the image distance for the combination. Adding the equations for the two lenses eliminates q_1 and gives

$$\frac{1}{p} + \frac{1}{q} = \frac{1}{f_1} + \frac{1}{f_2}$$

If the combination is replaced with a single lens that forms an image at the same location, its focal length must be related to the individual focal lengths by the expression

◀ Focal length for a combination of two thin lenses in contact

$$\frac{1}{f} = \frac{1}{f_1} + \frac{1}{f_2} \tag{35.21}$$

Therefore, two thin lenses in contact with each other are equivalent to a single thin lens having a focal length given by Equation 35.21.

35.5 Lens Aberrations

Our analysis of mirrors and lenses assumes rays make small angles with the principal axis and the lenses are thin. In this simple model, all rays leaving a point source focus at a single point, producing a sharp image. Clearly, that is not always true. When the approximations used in this analysis do not hold, imperfect images are formed.

A precise analysis of image formation requires tracing each ray, using Snell's law at each refracting surface and the law of reflection at each reflecting surface. This procedure shows that the rays from a point object do not focus at a single point, with the result that the image is blurred. The departures of actual images from the ideal predicted by our simplified model are called **aberrations.**

Spherical Aberration

Spherical aberration occurs because the focal points of rays far from the principal axis of a spherical lens (or mirror) are different from the focal points of rays of the same wavelength passing near the axis. Figure 35.32 illustrates spherical aberration for parallel rays passing through a converging lens. Rays passing through points near the center of the lens are imaged farther from the lens than rays passing through points near the edges. Figure 35.7 earlier in the chapter shows spherical aberration for light rays leaving a point object and striking a spherical mirror.

Many cameras have an adjustable aperture to control light intensity and reduce spherical aberration. (An aperture is an opening that controls the amount of light passing through the lens.) Sharper images are produced as the aperture size is reduced; with a small aperture, only the central portion of the lens is exposed to the light and therefore a greater percentage of the rays are paraxial. At the same time, however, less light passes through the lens. To compensate for this lower light intensity, a longer exposure time is used.

In the case of mirrors, spherical aberration can be minimized through the use of a parabolic reflecting surface rather than a spherical surface. Parabolic surfaces are not used often, however, because those with high-quality optics are very expensive to make. Parallel light rays incident on a parabolic surface focus at a common point, regardless of their distance from the principal axis. Parabolic reflecting surfaces are used in many astronomical telescopes to enhance image quality.

Chromatic Aberration

In Chapter 34, we described dispersion, whereby a material's index of refraction varies with wavelength. Because of this phenomenon, violet rays are refracted more than red rays when white light passes through a lens (Fig. 35.33). The figure shows that the focal length of a lens is greater for red light than for violet light. Other wavelengths (not shown in Fig. 35.33) have focal points intermediate between those of red and violet, which causes a blurred image and is called **chromatic aberration.**

Chromatic aberration for a diverging lens also results in a shorter focal length for violet light than for red light, but on the front side of the lens. Chromatic aberration can be greatly reduced by combining a converging lens made of one type of glass and a diverging lens made of another type of glass.

35.6 Optical Instruments

The Camera

The photographic **camera** is a simple optical instrument whose essential features are shown in Figure 35.34. It consists of a light-tight chamber, a converging lens that produces a real image, and a light-sensitive component behind the lens on which the image is formed.

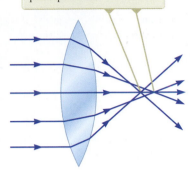

Figure 35.32 Spherical aberration caused by a converging lens. Does a diverging lens cause spherical aberration?

The refracted rays intersect at different points on the principal axis.

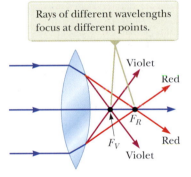

Figure 35.33 Chromatic aberration caused by a converging lens.

Rays of different wavelengths focus at different points.

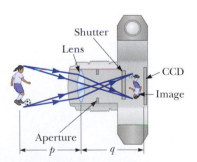

Figure 35.34 Cross-sectional view of a simple digital camera. The CCD is the light-sensitive component of the camera. In a nondigital camera, the light from the lens falls onto photographic film. In reality, $p \gg q$.

The image in a digital camera is formed on a *charge-coupled device* (CCD), which digitizes the image, turning it into binary code. (A CCD is described in Section 39.2.) The digital information is then stored on a memory chip for playback on the camera's display screen, or it can be downloaded to a computer. Film cameras are similar to digital cameras except that the light forms an image on light-sensitive film rather than on a CCD. The film must then be chemically processed to produce the image on paper. In the discussion that follows, we assume the camera is digital.

A camera is focused by varying the distance between the lens and the CCD. For proper focusing—which is necessary for the formation of sharp images—the lens-to-CCD distance depends on the object distance as well as the focal length of the lens.

The shutter, positioned behind the lens, is a mechanical device that is opened for selected time intervals, called *exposure times*. You can photograph moving objects by using short exposure times or photograph dark scenes (with low light levels) by using long exposure times. If this adjustment were not available, it would be impossible to take stop-action photographs. For example, a rapidly moving vehicle could move enough in the time interval during which the shutter is open to produce a blurred image. Another major cause of blurred images is the movement of the camera while the shutter is open. To prevent such movement, either short exposure times or a tripod should be used, even for stationary objects. Typical shutter speeds (that is, exposure times) are $\frac{1}{30}$ s, $\frac{1}{60}$ s, $\frac{1}{125}$ s, and $\frac{1}{250}$ s. In practice, stationary objects are normally shot with an intermediate shutter speed of $\frac{1}{60}$ s.

The Eye

Like a camera, a normal eye focuses light and produces a sharp image. The mechanisms by which the eye controls the amount of light admitted and adjusts to produce correctly focused images, however, are far more complex, intricate, and effective than those in even the most sophisticated camera. In all respects, the eye is a physiological wonder.

Figure 35.35 shows the basic parts of the human eye. Light entering the eye passes through a transparent structure called the *cornea* (Fig. 35.36), behind which are a clear liquid (the *aqueous humor*), a variable aperture (the *pupil*, which is an opening in the *iris*), and the *crystalline lens*. Most of the refraction occurs at the outer surface of the eye, where the cornea is covered with a film of tears. Relatively little refraction occurs in the crystalline lens because the aqueous humor in contact with the lens has an average index of refraction close to that of the lens. The iris, which is the colored portion of the eye, is a muscular diaphragm that controls pupil size. The iris regulates the amount of light entering the eye by dilating, or opening, the pupil in low-light conditions and contracting, or closing, the pupil in high-light conditions.

Figure 35.35 Important parts of the eye.

Figure 35.36 The cornea is the curved outer surface of the eye.

The cornea–lens system focuses light onto the back surface of the eye, the *retina*, which consists of millions of sensitive receptors called *rods* and *cones*. When stimulated by light, these receptors send impulses via the optic nerve to the brain, where an image is perceived. By this process, a distinct image of an object is observed when the image falls on the retina.

The eye focuses on an object by varying the shape of the pliable crystalline lens through a process called **accommodation.** The lens adjustments take place so swiftly that we are not even aware of the change. Accommodation is limited in that objects very close to the eye produce blurred images. The **near point** is the closest distance for which the lens can accommodate to focus light on the retina. This distance usually increases with age and has an average value of 25 cm. At age 10, the near point of the eye is typically approximately 18 cm. It increases to approximately 25 cm at age 20, to 50 cm at age 40, and to 500 cm or greater at age 60. The **far point** of the eye represents the greatest distance for which the lens of the relaxed eye can focus light on the retina. A person with normal vision can see very distant objects and therefore has a far point that can be approximated as infinity.

The retina is covered with two types of light-sensitive cells, called **rods** and **cones.** The rods are not sensitive to color but are more light sensitive than the cones. The rods are responsible for *scotopic vision,* or dark-adapted vision. Rods are spread throughout the retina and allow good peripheral vision for all light levels and motion detection in the dark. The cones are concentrated in the fovea. These cells are sensitive to different wavelengths of light. The three categories of these cells are called red, green, and blue cones because of the peaks of the color ranges to which they respond (Fig. 35.37). If the red and green cones are stimulated simultaneously (as would be the case if yellow light were shining on them), the brain interprets what is seen as yellow. If all three types of cones are stimulated by the separate colors red, blue, and green, white light is seen. If all three types of cones are stimulated by light that contains *all* colors, such as sunlight, again white light is seen.

Televisions and computer monitors take advantage of this visual illusion by having only red, green, and blue dots on the screen. With specific combinations of brightness in these three primary colors, our eyes can be made to see any color in the rainbow. Therefore, the yellow lemon you see in a television commercial is not actually yellow, it is red and green! The paper on which this page is printed is made of tiny, matted, translucent fibers that scatter light in all directions, and the resultant mixture of colors appears white to the eye. Snow, clouds, and white hair are not actually white. In fact, there is no such thing as a white pigment. The appearance of these things is a consequence of the scattering of light containing all colors, which we interpret as white.

When the eye suffers a mismatch between the focusing range of the lens–cornea system and the length of the eye, with the result that light rays from a near object reach the retina before they converge to form an image as shown in Figure 35.38a, the condition is known as **farsightedness** (or *hyperopia*). A farsighted person can

Figure 35.37 Approximate color sensitivity of the three types of cones in the retina.

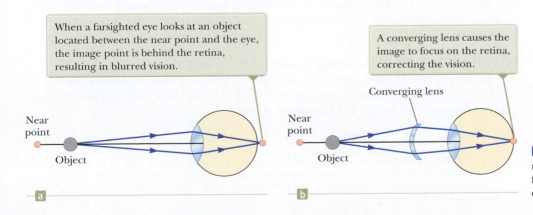

When a farsighted eye looks at an object located between the near point and the eye, the image point is behind the retina, resulting in blurred vision.

A converging lens causes the image to focus on the retina, correcting the vision.

Figure 35.38 (a) An uncorrected farsighted eye. (b) A farsighted eye corrected with a converging lens.

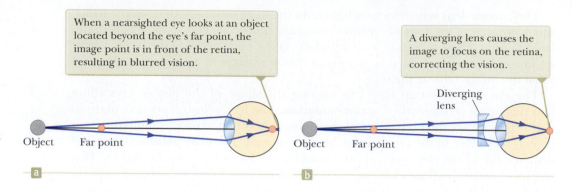

Figure 35.39 (a) An uncorrected nearsighted eye. (b) A nearsighted eye corrected with a diverging lens.

usually see faraway objects clearly but not nearby objects. Although the near point of a normal eye is approximately 25 cm, the near point of a farsighted person is much farther away. The refracting power in the cornea and lens is insufficient to focus the light from all but distant objects satisfactorily. The condition can be corrected by placing a converging lens in front of the eye as shown in Figure 35.38b. The lens refracts the incoming rays more toward the principal axis before entering the eye, allowing them to converge and focus on the retina.

A person with **nearsightedness** (or *myopia*), another mismatch condition, can focus on nearby objects but not on faraway objects. The far point of the nearsighted eye is not infinity and may be less than 1 m. The maximum focal length of the nearsighted eye is insufficient to produce a sharp image on the retina, and rays from a distant object converge to a focus in front of the retina. They then continue past that point, diverging before they finally reach the retina and causing blurred vision (Fig. 35.39a). Nearsightedness can be corrected with a diverging lens as shown in Figure 35.39b. The lens refracts the rays away from the principal axis before they enter the eye, allowing them to focus on the retina.

A number of people have difficulties with *color blindness*. Some individuals are *dichromats*, meaning that they only have functioning cones for two of the three colors in Figure 35.37. Another type of color blindness occurs in people who are *anomalous trichromats*. For these individuals, the range of sensitivity of, most often, red- and green-sensitive cones has shifted so that there is more overlap between the red and green curves in Figure 35.37. This makes it difficult to distinguish red and green.

A new type of glasses offers some relief for anomalous trichromats. The glasses are designed to filter out the wavelength regions in which the curves in Figure 35.37 are crossing, allowing the individual to see three distinct wavelength regions. Many people trying these new glasses report remarkable improvement in their perception of colors.

Optometrists and ophthalmologists usually prescribe lenses[1] measured in **diopters**: the **power** P of a lens in diopters equals the inverse of the focal length in meters: $P = 1/f$. For example, a converging lens of focal length +20 cm has a power of +5.0 diopters, and a diverging lens of focal length −40 cm has a power of −2.5 diopters.

QUICK QUIZ **35.7** Two campers wish to start a fire during the day. One camper is nearsighted, and one is farsighted. Whose glasses should be used to focus the Sun's rays onto some paper to start the fire? **(a)** either camper **(b)** the nearsighted camper **(c)** the farsighted camper

---

[1]The word *lens* comes from *lentil,* the name of an Italian legume. (You may have eaten lentil soup.) Early eyeglasses were called "glass lentils" because the biconvex shape of their lenses resembled the shape of a lentil. The first lenses for farsightedness and presbyopia appeared around 1280; concave eyeglasses for correcting nearsightedness did not appear until more than 100 years later.

The Simple Magnifier

The simple magnifier, or magnifying glass, consists of a single converging lens. This device increases the apparent size of an object.

Suppose an object is viewed at some distance p from the eye as illustrated in Figure 35.40. The size of the image formed at the retina depends on the angle θ subtended by the object at the eye. As the object moves closer to the eye, θ increases and a larger image is observed. An average normal human eye, however, cannot focus on an object closer than about 25 cm, the near point (Fig. 35.41a). Therefore, θ is maximum at the near point.

To further increase the apparent angular size of an object, a converging lens can be placed in front of the eye as in Figure 35.41b, with the object located at point O, immediately inside the focal point of the lens. At this location, the lens forms a virtual, upright, enlarged image. We define **angular magnification** m as the ratio of the angle subtended by an object with a lens in use (angle θ in Fig. 35.41b) to the angle subtended by the object placed at the near point with no lens in use (angle θ_0 in Fig. 35.41a):

$$m \equiv \frac{\theta}{\theta_0} \tag{35.22}$$

The angular magnification is a maximum when the image is at the near point of the eye, that is, when $q = -25$ cm. The object distance corresponding to this image distance can be calculated from the thin lens equation:

$$\frac{1}{p} + \frac{1}{-25 \text{ cm}} = \frac{1}{f} \rightarrow p = \frac{25f}{25 + f}$$

where f is the focal length of the magnifier in centimeters. If we make the small-angle approximations

$$\tan \theta_0 \approx \theta_0 \approx \frac{h}{25} \quad \text{and} \quad \tan \theta \approx \theta \approx \frac{h}{p} \tag{35.23}$$

Equation 35.22 becomes

$$m_{max} = \frac{\theta}{\theta_0} = \frac{h/p}{h/25} = \frac{25}{p} = \frac{25}{25f/(25 + f)}$$

$$m_{max} = 1 + \frac{25 \text{ cm}}{f} \tag{35.24}$$

Although the eye can focus on an image formed anywhere between the near point and infinity, it is most relaxed when the image is at infinity. For the image formed by the magnifying lens to appear at infinity, the object has to be at the focal point of the lens. In this case, Equations 35.23 become

$$\theta_0 \approx \frac{h}{25} \quad \text{and} \quad \theta \approx \frac{h}{f}$$

and the magnification is

$$m_{min} = \frac{\theta}{\theta_0} = \frac{25 \text{ cm}}{f} \tag{35.25}$$

With a single lens, such as that shown in Figure 35.42, it is possible to obtain angular magnifications up to about 4 without serious aberrations. Magnifications up to about 20 can be achieved by using one or two additional lenses to correct for aberrations.

The size of the image formed on the retina depends on the angle θ subtended at the eye.

Figure 35.40 An observer looks at an object at distance p.

Figure 35.41 (a) An object placed at the near point of the eye ($p = 25$ cm) subtends an angle $\theta_0 \approx h/25$ cm at the eye. (b) An object placed near the focal point of a converging lens produces a magnified image that subtends an angle $\theta \approx h'/25$ cm at the eye.

Figure 35.42 A simple magnifier, also called a magnifying glass, is used to view an enlarged image of a portion of a map.

The objective lens forms an image here.

The three-objective turret allows the user to choose from several powers of magnification.

The eyepiece lens forms an image here.

Figure 35.43 (a) Diagram of a compound microscope, which consists of an objective lens and an eyepiece lens. (b) A compound microscope.

mikeledray/Shutterstock

The Compound Microscope

A simple magnifier provides only limited assistance in inspecting minute details of an object. Greater magnification can be achieved by combining two lenses in a device called a **compound microscope** shown in Figure 35.43a. It consists of one lens, the *objective*, that has a very short focal length $f_o < 1$ cm and a second lens, the *eyepiece*, that has a focal length f_e of a few centimeters. The two lenses are separated by a distance L that is much greater than either f_o or f_e. The object, which is placed just outside the focal point of the objective, forms a real, inverted image at I_1, and this image is located at or close to the focal point of the eyepiece. The eyepiece, which serves as a simple magnifier, produces at I_2 a virtual, enlarged image of I_1. The lateral magnification M_1 of the first image is $-q_1/p_1$. Notice from Figure 35.43a that q_1 is approximately equal to L and that the object is very close to the focal point of the objective: $p_1 \approx f_o$. Therefore, the lateral magnification by the objective is

$$M_o \approx -\frac{L}{f_o}$$

The angular magnification by the eyepiece for an object (corresponding to the image at I_1) placed at the focal point of the eyepiece is, from Equation 35.25,

$$m_e = \frac{25 \text{ cm}}{f_e}$$

The overall magnification of the image formed by a compound microscope is defined as the product of the lateral and angular magnifications:

$$M = M_o m_e = -\frac{L}{f_o}\left(\frac{25 \text{ cm}}{f_e}\right) \tag{35.26}$$

The negative sign indicates that the image is inverted.

The microscope has extended human vision to the point where we can view previously unknown details of incredibly small objects. The capabilities of this instrument have steadily increased with improved techniques for precision grinding of lenses. A question often asked about microscopes is, "If one were extremely patient and careful, would it be possible to construct a microscope that would enable the human eye to see an atom?" The answer is no, as long as light is used to illuminate the object. For an object under an optical microscope (one that uses visible light) to be seen, the object must be at least as large as a wavelength of light. Because the diameter of any atom is many times smaller than the wavelengths of visible light, the mysteries of the atom must be probed using other types of "microscopes."

The objective lens forms an image here.

Eyepiece lens

Objective lens

θ_o

F_e F_o

θ_o

h'

I_1

θ

F_e'

f_e f_e

I_2

f_o

The eyepiece lens forms an image here.

a

b

doomu/Getty Images

Figure 35.44 (a) Lens arrangement in a refracting telescope, with the object at infinity. (b) A refracting telescope.

The Telescope

Two fundamentally different types of **telescopes** exist; both are designed to aid in viewing distant objects such as the planets in our solar system. The first type, the **refracting telescope,** uses a combination of lenses to form an image.

Like the compound microscope, the refracting telescope shown in Figure 35.44a has an objective and an eyepiece. The two lenses are arranged so that the objective forms a real, inverted image of a distant object very near the focal point of the eyepiece. Because the object is essentially at infinity, this point at which I_1 forms is the focal point of the objective. The eyepiece then forms, at I_2, an enlarged, inverted image of the image at I_1. To provide the largest possible magnification, the image distance for the eyepiece is infinite. Therefore, the image due to the objective lens, which acts as the object for the eyepiece lens, must be located at the focal point of the eyepiece. Hence, the two lenses are separated by a distance $f_o + f_e$, which corresponds to the length of the telescope tube.

The angular magnification of the telescope is given by θ/θ_o, where θ_o is the angle subtended by the object at the objective and θ is the angle subtended by the final image at the viewer's eye. Consider Figure 35.44a, in which the object is a very great distance to the left of the figure. The angle θ_o (to the *left* of the objective) subtended by the object at the objective is the same as the angle (to the *right* of the objective) subtended by the first image at the objective. Therefore,

$$\tan \theta_o \approx \theta_o \approx -\frac{h'}{f_o}$$

where the negative sign indicates that the image is inverted.

The angle θ subtended by the final image at the eye is the same as the angle that a ray coming from the tip of I_1 and traveling parallel to the principal axis makes with the principal axis after it passes through the lens. Therefore,

$$\tan \theta \approx \theta \approx \frac{h'}{f_e}$$

We have not used a negative sign in this equation because the final image is not inverted; the object creating this final image I_2 is I_1, and both it and I_2 point in

Figure 35.45 (a) A Newtonian-focus reflecting telescope. (b) A reflecting telescope. This type of telescope is shorter than that in Figure 35.44b.

the same direction. Therefore, the angular magnification of the telescope can be expressed as

$$m = \frac{\theta}{\theta_o} = \frac{h'/f_e}{-h'/f_o} = -\frac{f_o}{f_e} \tag{35.27}$$

This result shows that the angular magnification of a telescope equals the ratio of the objective focal length to the eyepiece focal length. The negative sign indicates that the image is inverted.

When you look through a telescope at such relatively nearby objects as the Moon and the planets, magnification is important. Individual stars in our galaxy, however, are so far away that they always appear as small points of light no matter how great the magnification. To gather as much light as possible, large research telescopes used to study very distant objects must have a large diameter. It is difficult and expensive to manufacture large lenses for refracting telescopes. Another difficulty with large lenses is that their weight leads to sagging, which is an additional source of aberration.

These problems associated with large lenses can be partially overcome by replacing the objective with a concave mirror, which results in the second type of telescope, the **reflecting telescope.** Because light is reflected from the mirror and does not pass through a lens, the mirror can have rigid supports on the back side. Such supports eliminate the problem of sagging.

Figure 35.45a shows the design for a typical reflecting telescope. The incoming light rays are reflected by a parabolic mirror at the base. These reflected rays converge toward point A in the figure, where an image would be formed. Before this image is formed, however, a small, flat mirror M reflects the light toward an opening in the tube's side and it passes into an eyepiece. This particular design is said to have a Newtonian focus because Newton developed it. Figure 35.45b shows such a telescope. Notice that the light never passes through glass (except through the small eyepiece) in the reflecting telescope. As a result, problems associated with chromatic aberration are virtually eliminated. The reflecting telescope can be made even shorter by orienting the flat mirror so that it reflects the light back toward the objective mirror and the light enters an eyepiece in a hole in the middle of the mirror.

The largest reflecting telescopes in the world are at the Gran Telescopio Canarias in the Canary Islands, Spain, and at the Keck Observatory on Mauna Kea, Hawaii. The Hawaii site includes two telescopes with diameters of 10 m, each containing 36 hexagonally shaped, computer-controlled mirrors that work together to form a large reflecting surface. In addition, the two telescopes can work together to provide a telescope with an effective diameter of 85 m. In contrast, the largest refracting telescope in the world, at the Yerkes Observatory in Williams Bay, Wisconsin, has a diameter of only 1 m.

Figure 35.46 shows a remarkable optical image from the Keck Observatory of a solar system around the star HR8799, located 129 light-years from the Earth. The planets labeled b, c, and d were seen in 2008 and the innermost planet, labeled e, was observed in December 2010. This photograph represents the first direct image of another solar system and was made possible by the adaptive optics technology used in the Keck Observatory.

Figure 35.46 A direct optical image of a solar system around the star HR8799, developed at the Keck Observatory in Hawaii.

Summary

▶ Definitions

The **lateral magnification** M of the image due to a mirror or lens is defined as the ratio of the image height h' to the object height h. It is equal to the negative of the ratio of the image distance q to the object distance p:

$$M \equiv \frac{\text{image height}}{\text{object height}} = \frac{h'}{h} = -\frac{q}{p} \qquad (35.1, 35.2, 35.19)$$

The **angular magnification** m is the ratio of the angle subtended by an object with a lens in use (angle θ in Fig. 35.41b) to the angle subtended by the object placed at the near point with no lens in use (angle θ_0 in Fig. 35.41a):

$$m \equiv \frac{\theta}{\theta_0} \qquad (35.22)$$

▶ Concepts and Principles

In the paraxial ray approximation, the object distance p and image distance q for a spherical mirror of radius R are related by the **mirror equation:**

$$\frac{1}{p} + \frac{1}{q} = \frac{2}{R} = \frac{1}{f} \qquad (35.4, 35.6)$$

where $f = R/2$ is the **focal length** of the mirror.

An image can be formed by refraction from a spherical surface of radius R. The object and image distances for refraction from such a surface are related by

$$\frac{n_1}{p} + \frac{n_2}{q} = \frac{n_2 - n_1}{R} \qquad (35.9)$$

where the light is incident in the medium for which the index of refraction is n_1 and is refracted in the medium for which the index of refraction is n_2.

The inverse of the **focal length** f of a thin lens surrounded by air is given by the **lens-makers' equation:**

$$\frac{1}{f} = (n-1)\left(\frac{1}{R_1} - \frac{1}{R_2}\right) \qquad (35.17)$$

Converging lenses have positive focal lengths, and **diverging lenses** have negative focal lengths.

For a thin lens, and in the paraxial ray approximation, the object and image distances are related by the **thin lens equation:**

$$\frac{1}{p} + \frac{1}{q} = \frac{1}{f} \qquad (35.18)$$

The maximum magnification of a single lens of focal length f used as a simple magnifier is

$$m_{max} = 1 + \frac{25\text{ cm}}{f} \qquad (35.24)$$

The overall magnification of the image formed by a compound microscope is

$$M = -\frac{L}{f_o}\left(\frac{25\text{ cm}}{f_e}\right) \qquad (35.26)$$

where f_o and f_e are the focal lengths of the objective and eyepiece lenses, respectively, and L is the distance between the lenses.

The angular magnification of a refracting telescope can be expressed as

$$m = -\frac{f_o}{f_e} \qquad (35.27)$$

where f_o and f_e are the focal lengths of the objective and eyepiece lenses, respectively. The angular magnification of a reflecting telescope is given by the same expression where f_o is the focal length of the objective mirror.

Think–Pair–Share

See the Preface for an explanation of the icons used in this problems set. For additional assessment items for this section, go to **WEBASSIGN** From Cengage

1. **ACTIVITY** In Figure 35.19, estimate the value of n_1 if air surrounds the material.

2. **BIO** Your group is training to become optician's assistants. The optician who is training you wants you to design a device that will create two images at the same position in space, one upright and the other inverted. When the patient looks into the device, he or she will see both images and remark on their relative color, shape, and brightness. This will allow the optician to reach some conclusions about differences in vision between the upper and lower parts of the retina. Figure TP35.2 (page 956) shows the optical system, along with the single object and two images. The optician has a lens with a focal length of $f_{lens} = 10.0$ cm. He wants

Figure TP35.2

the magnification of the two images to each have an absolute value of $M = 1.50$. (a) He asks you to determine the distance p_{lens} that the object must be placed from the lens. (b) He also

asks you to determine the focal length of the mirror that he needs, given that he has a housing in which he can mount the mirror at a distance of $d = 40.0$ cm from the lens. Work in your group to discuss and provide these quantities.

3. **ACTIVITY** With your group, perform the activity with the magnifying glass described in the opening storyline of the chapter. Carefully measure the two distances indicated in the description of the activity. From your understanding of the material in the chapter, construct an argument to explain why the two distances should be *similar*, but not *equal*. From the activity with the smartphone display, can you calculate some representative distance in the first activity that *should* be the same as the distance measured in the second activity?

Problems

See the Preface for an explanation of the icons used in this problems set. For additional assessment items for this section, go to :• **WEBASSIGN** From Cengage

SECTION 35.1 Images Formed by Flat Mirrors

1. (a) Does your bathroom mirror show you older or younger than you actually are? (b) Compute an order-of-magnitude estimate for the age difference based on data you specify.

2. Two flat mirrors have their reflecting surfaces facing each other, with the edge of one mirror in contact with an edge of the other, so that the angle between the mirrors is α. When an object is placed between the mirrors, a number of images are formed. In general, if the angle α is such that $n\alpha = 360°$, where n is an integer, the number of images formed is $n - 1$. Graphically, find all the image positions for the case $n = 6$ when a point object is between the mirrors (but not on the angle bisector).

3. A periscope (Fig. P35.3) is useful for viewing objects that cannot be seen directly. It can be used in submarines and when watching golf matches or parades from behind a crowd of people. Suppose the object is a distance p_1 from the upper mirror and the centers of the two flat mirrors are separated by a distance h. (a) What is the distance of the final image from the lower mirror? (b) Is the final image real or virtual? (c) Is it upright or inverted? (d) What is its magnification? (e) Does it appear to be left–right reversed?

Figure P35.3

4. Two plane mirrors stand facing each other, 3.00 m apart, **Q|C** and a woman stands between them. The woman looks at one of the mirrors from a distance of 1.00 m and holds her left arm out to the side of her body with the palm of her hand facing the closer mirror. (a) What is the apparent position of the closest image of her left hand, measured

perpendicularly from the surface of the mirror in front of her? (b) Does it show the palm of her hand or the back of her hand? (c) What is the position of the next closest image? (d) Does it show the palm of her hand or the back of her hand? (e) What is the position of the third closest image? (f) Does it show the palm of her hand or the back of her hand? (g) Which of the images are real and which are virtual?

SECTION 35.2 Images Formed by Spherical Mirrors

5. An object is placed 50.0 cm from a concave spherical mirror with focal length of magnitude 20.0 cm. (a) Find the location of the image. (b) What is the magnification of the image? (c) Is the image real or virtual? (d) Is the image upright or inverted?

6. An object is placed 20.0 cm from a concave spherical mirror having a focal length of magnitude 40.0 cm. (a) Use graph paper to construct an accurate ray diagram for this situation. (b) From your ray diagram, determine the location of the image. (c) What is the magnification of the image? (d) Check your answers to parts (b) and (c) using the mirror equation.

7. An object of height 2.00 cm is placed 30.0 cm from a convex spherical mirror of focal length of magnitude 10.0 cm. (a) Find the location of the image. (b) Indicate whether the image is upright or inverted. (c) Determine the height of the image.

8. *Why is the following situation impossible?* At a blind corner in an outdoor shopping mall, a convex mirror is mounted so pedestrians can see around the corner before arriving there and bumping into someone traveling in the perpendicular direction. The installers of the mirror failed to take into account the position of the Sun, and the mirror focuses the Sun's rays on a nearby bush and sets it on fire.

9. A large hall in a museum has a niche in one wall. On the floor plan, the niche appears as a semicircular indentation of radius 2.50 m. A tourist stands on the centerline of the niche, 2.00 m out from its deepest point, and whispers "Hello." Where is the sound concentrated after reflection from the niche?

10. A concave spherical mirror has a radius of curvature of magnitude 24.0 cm. (a) Determine the object position for which the resulting image is upright and larger than the object by a factor of 3.00. (b) Draw a ray diagram to determine the position of the image. (c) Is the image real or virtual?

11. An object 10.0 cm tall is placed at the zero mark of a meterstick. A spherical mirror located at some point on the meterstick creates an image of the object that is upright, 4.00 cm tall, and located at the 42.0-cm mark of the meterstick. (a) Is the mirror convex or concave? (b) Where is the mirror? (c) What is the mirror's focal length?

12. You are training to become an optician's assistant. One day, you are learning how to fit a contact lens to a patient's eye. You make a measurement with a *keratometer*, which is used to measure the curvature of the eye's front surface, the cornea. This instrument places an illuminated object of known size at a known distance p from the cornea. The cornea reflects some light from the object, forming an image of the object. The magnification M of the image is measured by using a small viewing telescope that allows comparison of the image formed by the cornea with a second calibrated image projected into the field of view by a prism arrangement. As part of your training, the optician has required that you do not use the automatic calculator associated with the machine, but must perform the calculations yourself. You must determine the radius of curvature R of the cornea for the measurements you make for the patient: $p = 30.0$ cm and $M = 0.013\ 0$.

13. A certain Christmas tree ornament is a silver sphere having a diameter of 8.50 cm. (a) If the size of an image created by reflection in the ornament is three-fourths the reflected object's actual size, determine the object's location. (b) Use a principal-ray diagram to determine whether the image is upright or inverted.

14. **Review.** A ball is dropped at $t = 0$ from rest 3.00 m directly above the vertex of a concave spherical mirror that has a radius of curvature of magnitude 1.00 m and lies in a horizontal plane. (a) Describe the motion of the ball's image in the mirror. (b) At what instant or instants do the ball and its image coincide?

15. You unconsciously estimate the distance to an object from the angle it subtends in your field of view. This angle θ in radians is related to the linear height of the object h and to the distance d by $\theta = h/d$. Assume you are driving a car and another car, 1.50 m high, is 24.0 m behind you. (a) Suppose your car has a flat passenger-side rearview mirror, 1.55 m from your eyes. How far from your eyes is the image of the car following you? (b) What angle does the image subtend in your field of view? (c) **What If?** Now suppose your car has a convex rearview mirror with a radius of curvature of magnitude 2.00 m (as suggested in Fig. 35.15). How far from your eyes is the image of the car behind you? (d) What angle does the image subtend at your eyes? (e) Based on its angular size, how far away does the following car appear to be?

16. A convex spherical mirror has a focal length of magnitude 8.00 cm. (a) What is the location of an object for which the magnitude of the image distance is one-third the magnitude of the object distance? (b) Find the magnification of the image and (c) state whether it is upright or inverted.

SECTION 35.3 Images Formed by Refraction

17. One end of a long glass rod ($n = 1.50$) is formed into a convex surface with a radius of curvature of magnitude 6.00 cm. An object is located in air along the axis of the rod. Find the image positions corresponding to object distances of (a) 20.0 cm, (b) 10.0 cm, and (c) 3.00 cm from the convex end of the rod.

18. The magnification of the image formed by a refracting surface is given by

$$M = -\frac{n_1 q}{n_2 p}$$

where n_1, n_2, p, and q are defined as they are for Figure 35.17 and Equation 35.9. A paperweight is made of a solid glass hemisphere with index of refraction 1.50. The radius of the circular cross section is 4.00 cm. The hemisphere is placed on its flat surface, with the center directly over a 2.50-mm-long line drawn on a sheet of paper. What is the length of this line as seen by someone looking vertically down on the hemisphere?

19. As shown in Figure P35.19, Ben and Jacob check out an aquarium that has a curved front made of plastic with uniform thickness and a radius of curvature of magnitude $R = 2.25$ m. (a) Locate the images of fish that are located (i) 5.00 cm and (ii) 25.0 cm from the front wall of the aquarium. (b) Find the magnification of images (i) and (ii) from the previous part. (See Problem 18 to find an expression for the magnification of an image formed by a refracting surface.) (c) Explain why you don't need to know the refractive index of the plastic to solve this problem. (d) If this aquarium were very long from front to back, could the image of a fish ever be farther from the front surface than the fish itself is? (e) If not, explain why not. If so, give an example and find the magnification.

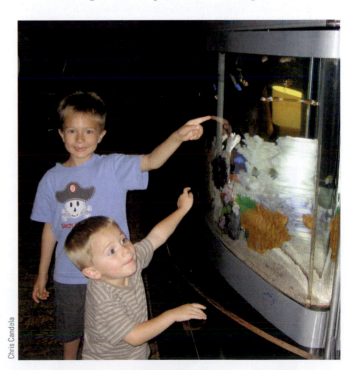

Chris Candela

Figure P35.19

20. Figure P35.20 (page 958) shows a curved surface separating a material with index of refraction n_1 from a material with index n_2. The surface forms an image I of object O. The ray shown in red passes through the surface along a radial line. Its angles of incidence and refraction are both zero, so its direction does not change at the surface. For the ray shown in blue, the direction changes according to Snell's law, $n_1 \sin \theta_1 = n_2 \sin \theta_2$. For paraxial rays, we assume θ_1 and θ_2 are small, so we may write $n_1 \tan \theta_1 = n_2 \tan \theta_2$. The magnification is defined as $M = h'/h$. Prove that the magnification is given by $M = -n_1q/n_2p$.

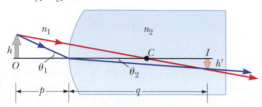

Figure P35.20

21. To dress up your dorm room, you have purchased a perfectly
CR spherical glass fishbowl to place on the windowsill. After placing the sand, decorations, and water in the bowl of diameter 40.0 cm, you transfer a single tropical fish from a plastic bag into the bowl. As you watch the fish, your roommate comes home. He watches the fish also and notices that the apparent size of the fish changes as it swims around in the bowl. (a) He is not taking a physics course, so he asks you to tell him the range of magnifications of the fish as it swims along a line from the back of the bowl along a line passing through the center of the bowl directly toward the observer. (b) Your roommate also asks you if the fish might be baked if it swims through a point at which the rays of the Sun focus at some point as they pass through the curved sides of the bowl. Should you worry about your fish being baked? Ignore the effect of the thin glass walls of the bowl; take only the water into consideration.

22. You are working for a solar energy company. Your supervisor
CR has asked you to investigate a new idea that has been proposed for a solar collector. A large sphere of glass focuses light on photocells, as shown in Figure P35.22. The photocells are moved by electronics along the curved track to the right of the sphere. Your supervisor would like to build a prototype of a material with index of refraction n, but needs for you to calculate the position at which the Sun's rays focus and, therefore, to find where to locate the curved track.

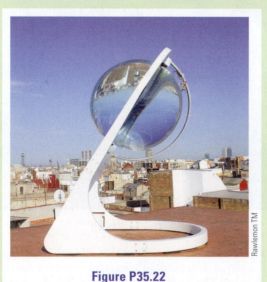

Rawlemon TM

Figure P35.22

SECTION 35.4 Images Formed by Thin Lenses

23. An object located 32.0 cm in front of a lens forms an image on a screen 8.00 cm behind the lens. (a) Find the focal

length of the lens. (b) Determine the magnification. (c) Is the lens converging or diverging?

24. An object's distance from a converging lens is 5.00 times the focal length. (a) Determine the location of the image. Express the answer as a fraction of the focal length. (b) Find the magnification of the image and indicate whether it is (c) upright or inverted and (d) real or virtual.

25. A contact lens is made of plastic with an index of refrac-
V tion of 1.50. The lens has an outer radius of curvature of +2.00 cm and an inner radius of curvature of +2.50 cm. What is the focal length of the lens?

26. A converging lens has a focal length of 10.0 cm. Con-
Q|C struct accurate ray diagrams for object distances of **(i)** 20.0 cm and **(ii)** 5.00 cm. (a) From your ray diagrams, determine the location of each image. (b) Is the image real or virtual? (c) Is the image upright or inverted? (d) What is the magnification of the image? (e) Compare your results with the values found algebraically. (f) Comment on difficulties in constructing the graph that could lead to differences between the graphical and algebraic answers.

27. A converging lens has a focal length of 10.0 cm. Locate the object if a real image is located at a distance from the lens of (a) 20.0 cm and (b) 50.0 cm. **What If?** Redo the calculations if the images are virtual and located at a distance from the lens of (c) 20.0 cm and (d) 50.0 cm.

28. Suppose an object has thickness dp so that it extends from
S object distance p to $p + dp$. (a) Prove that the thickness dq of its image is given by $(-q^2/p^2)dp$. (b) The longitudinal magnification of the object is $M_{\text{long}} = dq/dp$. How is the longitudinal magnification related to the lateral magnification M?

29. An object is placed 10.0 cm from a diverging lens of focal length -10.0 cm. (a) Find the location of the image. (b) Find the magnification of the image. (c) Comment on the difference between this situation and placing an object 10.0 cm from a converging lens of focal length 10.0 cm.

30. In Figure P35.30, a thin converging lens of focal length
Q|C 14.0 cm forms an image of the square $abcd$, which is $h_c = h_b = 10.0$ cm high and lies between distances of $p_d = 20.0$ cm and $p_a = 30.0$ cm from the lens. Let a', b', c', and d' represent the respective corners of the image. Let q_a represent the image distance for points a' and b', q_d represent the image distance for points c' and d', h_b' represent the distance from point b' to the axis, and h_c' represent the height of c'. (a) Find q_a, q_d, h_b', and h_c'. (b) Make a sketch of the image.

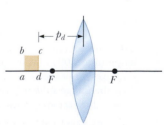

Figure P35.30

(c) The area of the object is 100 cm². By carrying out the following steps, you will evaluate the area of the image. Let q represent the image distance of any point between a' and d', for which the object distance is p. Let h' represent the distance from the axis to the point at the edge of the image between b' and c' at image distance q. Demonstrate that

$$|h'| = 10.0q\left(\frac{1}{14.0} - \frac{1}{q}\right)$$

where h' and q are in centimeters. (d) Explain why the geometric area of the image is given by

$$\int_{q_a}^{q_d} |h'| \, dq$$

(e) Carry out the integration to find the area of the image.

31. You are working for an electronics company that makes
CR devices for the home. Your supervisor has given you an assignment to help design the projection mechanism for a projection alarm clock. In this type of clock, a projection system is mounted on the body of the clock, as shown in Figure P35.31a, where the projection system is the silver cylinder, of radius $R = 3.25$ cm, mounted on the left side of the clock. A converging lens is mounted on the edge of the cylinder. Inside the cylinder, a small digital display of the time in red characters can be moved from the center of the cylinder outward radially toward the lens. The red light of the digital display can be seen in the lens in Figure P35.31a. As a result, an image of the time is projected in red onto the ceiling or wall of a darkened room (Fig. P35.31b). The range of distances for focused images of the digital display is from 0.500 m to 4.00 m, measured from the center of the cylinder. For the smallest value of the range, the digital display is at the center of the cylinder. You must determine for your supervisor the following parameters for the design of the projection system: (a) the focal length of the lens and (b) the distance of the digital display from the center of the cylinder for the largest value of the range.

a   **b**

Figure P35.31

32. *Why is the following situation impossible?* An illuminated object is placed a distance $d = 2.00$ m from a screen. By placing a converging lens of focal length $f = 60.0$ cm at two locations between the object and the screen, a sharp, real image of the object can be formed on the screen. In one location of the lens, the image is larger than the object, and in the other, the image is smaller.

SECTION 35.5 Lens Aberrations

33. Two rays traveling parallel to the principal axis strike a large plano-convex lens having a refractive index of 1.60 (Fig. P35.33). If the convex face is spherical, a ray near the edge does not pass through the focal point (spherical aberration occurs). Assume this face has a radius of curvature of $R = 20.0$ cm and the two rays are at distances $h_1 = 0.500$ cm and $h_2 = 12.0$ cm from the principal axis. Find the difference Δx in the positions where each crosses the principal axis.

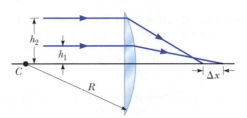

Figure P35.33

SECTION 35.6 Optical Instruments

34. Josh cannot see objects clearly beyond 25.0 cm (his far point). If he has no astigmatism and contact lenses are prescribed for him, what (a) power and (b) type of lens are required to correct his vision?

35. Figure 35.34 diagrams a cross section of a camera. It has a single lens of focal length 65.0 mm, which is to form an image on the CCD at the back of the camera. Suppose the position of the lens has been adjusted to focus the image of a distant object. How far and in what direction must the lens be moved to form a sharp image of an object that is 2.00 m away?

36. The refracting telescope at the Yerkes Observatory has
T a 1.00-m diameter objective lens of focal length 20.0 m. Assume it is used with an eyepiece of focal length 2.50 cm. (a) Determine the magnification of Mars as seen through this telescope. (b) Are the Martian polar caps right side up or upside down?

37. The distance between the eyepiece and the objective lens in a certain compound microscope is 23.0 cm. The focal length of the eyepiece is 2.50 cm and that of the objective is 0.400 cm. What is the overall magnification of the microscope?

38. What are (a) the maximum angular magnification that may be viewed clearly by the human eye with a magnifying glass having a focal length of 10 cm, and (b) the angular magnification of the image from this lens when the eye is relaxed?

39. A patient has a near point of 45.0 cm and far point of
BIO 85.0 cm. (a) Can a single pair of glasses correct the patient's
Q|C vision? Explain the patient's options. (b) Calculate the power lens needed to correct the near point so that the patient can see objects 25.0 cm away. Neglect the eye–lens distance. (c) Calculate the power lens needed to correct the patient's far point, again neglecting the eye–lens distance.

40. The intensity I of the light reaching the CCD in a camera is proportional to the area of the lens. Because this area is proportional to the square of the diameter D, it follows that I is also proportional to D^2. Because the area of the image is proportional to q^2 and $q \approx f$ (when $p \gg f$, so p can be approximated as infinite), we conclude that the intensity is also proportional to $1/f^2$ and therefore that $I \propto D^2/f^2$. The ratio f/D is called the *f-number* of a lens. Therefore, $I \propto 1/(f\text{-number})^2$. The *f*-number is often given as a description of the lens's "speed." The lower the *f*-number, the wider the aperture and the higher the rate at which energy from the light exposes the CCD; therefore, a lens with a low *f*-number is a "fast" lens. The conventional notation for an *f*-number is "*f*/" followed by the actual number. For example, "*f*/4" means an *f*-number of 4; it *does not* mean to divide *f* by 4! Suppose the lens of a digital camera has a focal length of

55 mm and a speed of $f/1.8$. The correct exposure time for this speed under certain conditions is known to be $\frac{1}{500}$ s. (a) Determine the diameter of the lens. (b) Calculate the correct exposure time if the f-number is changed to $f/4$ under the same lighting conditions.

41. A certain child's near point is 10.0 cm; her far point (with **BIO** eyes relaxed) is 125 cm. Each eye lens is 2.00 cm from the retina. (a) Between what limits, measured in diopters, does the power of this lens–cornea combination vary? (b) Calculate the power of the eyeglass lens the child should use for relaxed distance vision. Is the lens converging or diverging?

42. Astronomers often take photographs with the objective lens **Q|C** or mirror of a telescope alone, without an eyepiece. (a) Show that the image size h' for such a telescope is given by $h' = fh/(f - p)$, where f is the objective focal length, h is the object size, and p is the object distance. (b) **What If?** Simplify the expression in part (a) for the case in which the object distance is much greater than objective focal length. (c) The "wingspan" of the International Space Station is 108.6 m, the overall width of its solar panel configuration. When the station is orbiting at an altitude of 407 km, find the width of the image formed by a telescope objective of focal length 4.00 m.

43. A simple model of the human eye ignores its lens entirely. **BIO** Most of what the eye does to light happens at the outer surface of the transparent cornea. Assume that this surface has a radius of curvature of 6.00 mm and that the eyeball contains just one fluid with a refractive index of 1.40. Prove that a very distant object will be imaged on the retina, 21.0 mm behind the cornea. Describe the image.

ADDITIONAL PROBLEMS

44. A real object is located at the zero end of a meterstick. A large concave spherical mirror at the 100-cm end of the meterstick forms an image of the object at the 70.0-cm position. A small convex spherical mirror placed at the 20.0-cm position forms a final image at the 10.0-cm point. What is the radius of curvature of the convex mirror?

45. The distance between an object and its upright image is 20.0 cm. If the magnification is 0.500, what is the focal length of the lens being used to form the image?

46. The distance between an object and its upright image is d. If **S** the magnification is M, what is the focal length of the lens being used to form the image?

47. Andy decides to use an old pair of eyeglasses to make some **T** optical instruments. He knows that the near point in his left eye is 50.0 cm and the near point in his right eye is 100 cm. (a) What is the maximum angular magnification he can produce in a telescope? (b) If he places the lenses 10.0 cm apart, what is the maximum overall magnification he can produce in a microscope? *Hint:* Go back to basics and use the thin lens equation to solve part (b).

48. Two converging lenses having focal lengths of $f_1 = 10.0$ cm and $f_2 = 20.0$ cm are placed a distance $d = 50.0$ cm apart as shown in Figure P35.48. The image due to light passing through both lenses is to be located between the lenses at the position $x = 31.0$ cm indicated. (a) At what value of p should the object be positioned to the left of the first lens? (b) What is the magnification of the final image? (c) Is the final image upright or inverted? (d) Is the final image real or virtual?

Figure P35.48

49. Two lenses made of kinds of glass having different indices **S** of refraction n_1 and n_2 are cemented together to form an *optical doublet*. Optical doublets are often used to correct chromatic aberrations in optical devices. The first lens of a certain doublet has index of refraction n_1, one flat side, and one concave side with a radius of curvature of magnitude R. The second lens has index of refraction n_2 and two convex sides with radii of curvature also of magnitude R. Show that the doublet can be modeled as a single thin lens with a focal length described by

$$\frac{1}{f} = \frac{2n_2 - n_1 - 1}{R}$$

50. An object is originally at the $x_i = 0$ cm position of a meter- **Q|C** stick located on the x axis. A converging lens of focal length 26.0 cm is fixed at the position 32.0 cm. Then we gradually slide the object to the position $x_f = 12.0$ cm. (a) Find the location x' of the object's image as a function of the object position x. (b) Describe the pattern of the image's motion with reference to a graph or a table of values. (c) As the object moves 12.0 cm to the right, how far does the image move? (d) In what direction or directions?

51. An object is placed 12.0 cm to the left of a diverging lens **T** of focal length -6.00 cm. A converging lens of focal length 12.0 cm is placed a distance d to the right of the diverging lens. Find the distance d so that the final image is infinitely far away to the right.

52. An object is placed a distance p to the left of a diverging **S** lens of focal length f_1. A converging lens of focal length f_2 is placed a distance d to the right of the diverging lens. Find the distance d so that the final image is infinitely far away to the right.

53. In a darkened room, a burning candle is placed 1.50 m from **GP** a white wall. A lens is placed between the candle and the wall **T** at a location that causes a larger, inverted image to form on the wall. When the lens is in this position, the object distance is p_1. When the lens is moved 90.0 cm toward the wall, another image of the candle is formed on the wall. From this information, we wish to find p_1 and the focal length of the lens. (a) From the lens equation for the first position of the lens, write an equation relating the focal length f of the lens to the object distance p_1, with no other variables in the equation. (b) From the lens equation for the second position of the lens, write another equation relating the focal length f of the lens to the object distance p_1. (c) Solve the equations in parts (a) and (b) simultaneously to find p_1. (d) Use the value in part (c) to find the focal length f of the lens.

54. In many applications, it is necessary to expand or decrease
QC the diameter of a beam of parallel rays of light, which can be
accomplished by using a converging lens and a diverging lens
in combination. Suppose you have a converging lens of focal
length 21.0 cm and a diverging lens of focal length −12.0 cm.
(a) How can you arrange these lenses to increase the diameter of a beam of parallel rays? (b) By what factor will the diameter increase?

55. *Why is the following situation impossible?* Consider the lens–mirror
combination shown in Figure P35.55. The lens has a focal
length of $f_L = 0.200$ m, and the mirror has a focal length of
$f_M = 0.500$ m. The lens and mirror are placed a distance $d =$
1.30 m apart, and an object is placed at $p = 0.300$ m from the
lens. By moving a screen to various positions to the left of the
lens, a student finds two different positions of the screen that
produce a sharp image of the object. One of these positions
corresponds to light leaving the object and traveling to the
left through the lens. The other position corresponds to light
traveling to the right from the object, reflecting from the mirror and then passing through the lens.

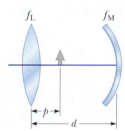

f_L f_M

←p→

←—— d ——→

Figure P35.55 Problems 55 and 57.

CHALLENGE PROBLEMS

56. A *zoom lens* system is a combination of lenses that produces
a variable magnification of a fixed object as it maintains a
fixed image position. The magnification is varied by moving one or more lenses along the axis. Multiple lenses are
used in practice, but the effect of zooming in on an object
can be demonstrated with a simple two-lens system. An
object, two converging lenses, and a screen are mounted on
an optical bench. Lens 1, which is to the right of the object,
has a focal length of $f_1 = 5.00$ cm, and lens 2, which is to
the right of the first lens, has a focal length of $f_2 = 10.0$ cm.
The screen is to the right of lens 2. Initially, an object is situated at a distance of 7.50 cm to the left of lens 1, and the
image formed on the screen has a magnification of +1.00.
(a) Find the distance between the object and the screen. (b)
Both lenses are now moved along their common axis while
the object and the screen maintain fixed positions until the
image formed on the screen has a magnification of +3.00.
Find the displacement of each lens from its initial position

in part (a). (c) Can the lenses be displaced in more than
one way?

57. Consider the lens–mirror arrangement shown in Fig-
S ure P35.55. There are two final image positions to the left of
the lens of focal length f_L. One image position is due to light
traveling from the object to the left and passing through
the lens. The other image position is due to light traveling
to the right from the object, reflecting from the mirror of
focal length f_M and then passing through the lens. For a
given object position p between the lens and the mirror and
measured with respect to the lens, there are two separation
distances d between the lens and mirror that will cause the
two images described above to be at the same location. Find
both positions.

58. A floating strawberry illusion is achieved with two par-
QC abolic mirrors, each having a focal length 7.50 cm, facing
each other as shown in Figure P35.58. If a strawberry is
placed on the lower mirror, an image of the strawberry is
formed at the small opening at the center of the top mirror,
7.50 cm above the lowest point of the bottom mirror. The
position of the eye in Figure P35.58a corresponds to the
view of the apparatus in Figure P35.58b. Consider the light
path marked A. Notice that this light path is blocked by the
upper mirror so that the strawberry itself is not directly
observable. The light path marked B corresponds to the eye
viewing the image of the strawberry that is formed at the
opening at the top of the apparatus. (a) Show that the final
image is formed at that location and describe its characteristics. (b) A very startling effect is to shine a flashlight beam
on this image. Even at a glancing angle, the incoming light
beam is seemingly reflected from the image! Explain.

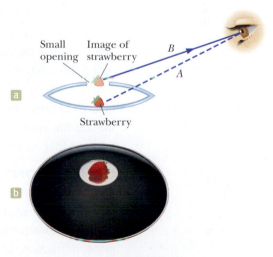

Small   Image of
opening   strawberry   B

a   A

Strawberry

b

Figure P35.58

36

Wave Optics

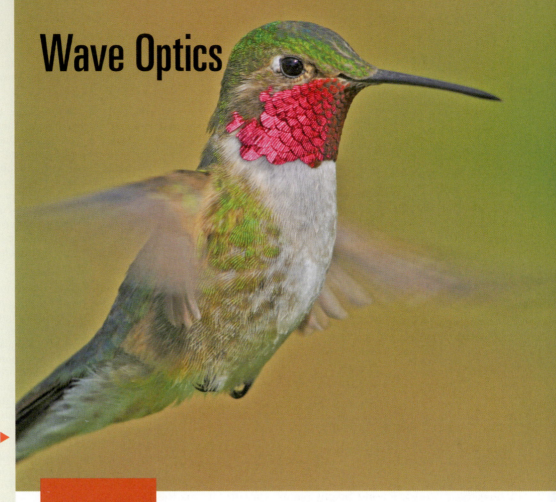

The colors in many of a hummingbird's feathers are not due to pigment. What do you think is the origin of these colors? (*Dec Hogan/Shutterstock*)

STORYLINE **Time to take a break from studying physics and just** chill out on your back patio! You are lying in your chaise lounge and enjoying the nice spring day. Suddenly, a hummingbird flies in, doesn't notice you, and lands just a few feet away. You hold still and watch quietly, amazed at the beautiful colors on the feathers of the bird, which seem to glisten. Then you notice, as the bird turns a bit, that the colors shift in their intensity and hue. You think, "Wait a minute! Why would that happen?" And then, in direct contradiction to your efforts to take a break from physics, you think, "Could there be some physics behind the appearance of the colors in this bird's feathers?" The startled bird flies away in fear as you reach for your smartphone and fire up the Internet.

CONNECTIONS In Chapter 35, we studied light rays passing through a lens or reflecting from a mirror to describe the formation of images. This discussion completed our study of *ray optics*. In this chapter and in Chapter 37, we are concerned with *wave optics*, sometimes called *physical optics*, the study of interference, diffraction, and polarization of light. We studied *interference* of sound waves in Chapter 17 and will look at the comparable effect for light in this chapter. We introduced the phenomenon of *diffraction* for light waves in Section 34.2. We did not discuss *polarization* in Chapter 17 because sound waves cannot be polarized. Light waves can be polarized, however, and we shall study that phenomenon in Chapter 37. These three phenomena cannot be adequately explained with the ray optics used in Chapters 34 and 35 because they depend on the fact that light is wavelike in nature. The discussion of interference leads to the historical development of the *Michelson interferometer*, one of the tools used to investigate relativity, leading in turn to the development of modern physics, which begins in Chapter 38.

36.1 Young's Double-Slit Experiment

In Chapter 17, we studied the waves in interference model and found that the superposition of two mechanical waves can be constructive or destructive. In constructive interference, the amplitude of the resultant wave is greater than that of either individual wave, whereas in destructive interference, the resultant amplitude is less than that of the larger wave. Light waves also interfere with one another. Fundamentally, all interference associated with light waves arises when the electromagnetic fields that constitute the individual waves combine.

Interference in light waves from two sources was first demonstrated by Thomas Young in 1801. A schematic diagram of the apparatus Young used is shown in Figure 36.1a. Plane light waves arrive at a barrier that contains two slits S_1 and S_2. The long dimension of the slits is perpendicular to the page in Figure 36.1a. The light rays from the two slits are in phase as they leave the slits. The light from S_1 and S_2 produces on a viewing screen a visible pattern of bright and dark parallel bands called **fringes** (Fig. 36.1b). When the light from S_1 and that from S_2 both arrive at a point on the screen such that constructive interference occurs at that location, a bright fringe appears. When the light from the two slits combines destructively at any location on the screen, a dark fringe results.

Figure 36.2 (page 964) shows some of the ways in which two waves can combine at the screen. In Figure 36.2a, the two waves strike the screen at the central point O. Because both waves travel the same distance, they arrive at O in phase. As a result, constructive interference occurs at this location and a bright fringe is observed. In Figure 36.2b, the two waves also start in phase, but here the lower wave has to travel one wavelength farther than the upper wave to reach point P. Because the lower wave falls behind the upper one by exactly one wavelength, they still arrive in phase at P and a second bright fringe appears at this location. At point R in Figure 36.2c, however, between points O and P, the lower wave has fallen half a wavelength behind the upper wave and a crest of the upper wave overlaps a trough

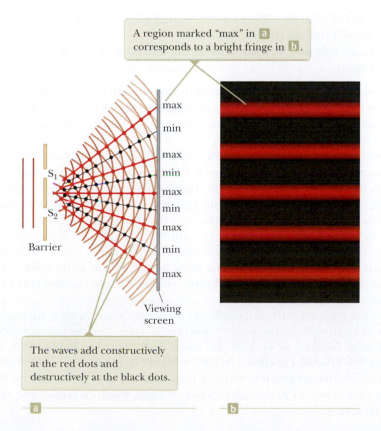

A region marked "max" in **a** corresponds to a bright fringe in **b**.

The waves add constructively at the red dots and destructively at the black dots.

S_1

S_2

Barrier

Viewing screen

max
min
max
min
max
min
max
min
max

a   **b**

Figure 36.1 (a) Schematic diagram of Young's double-slit experiment. Slits S_1 and S_2 behave as coherent sources of light waves that produce an interference pattern on the viewing screen (drawing not to scale). (b) A simulation of an enlargement of the center of a fringe pattern formed on the viewing screen.

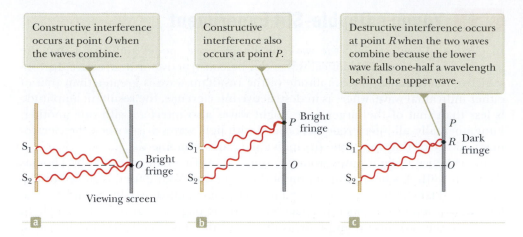

Constructive interference occurs at point O when the waves combine.

Constructive interference also occurs at point P.

Destructive interference occurs at point R when the two waves combine because the lower wave falls·one-half a wavelength behind the upper wave.

Viewing screen

Figure 36.2 Waves leave the slits and combine at various points on the viewing screen. (All figures not to scale.)

of the lower wave, giving rise to destructive interference at point R. A dark fringe is therefore observed at this location.

If two lightbulbs are placed side by side so that light from both bulbs combines, no interference effects are observed because the light waves from one bulb are emitted independently of those from the other bulb. The emissions from the two lightbulbs do not maintain a constant phase relationship with each other over time. Light waves from an ordinary source such as a lightbulb undergo random phase changes in time intervals of less than a nanosecond. Therefore, the conditions for constructive interference, destructive interference, or some intermediate state are maintained only for such short time intervals. Because the eye cannot follow such rapid changes, no interference effects are observed. Such light sources are said to be **incoherent.**

To observe interference of waves from two sources, the following conditions must be met:

- The sources must be **coherent;** that is, they must maintain a constant phase with respect to each other.
- The sources should be **monochromatic;** that is, they should be of a single wavelength.

As an example, single-frequency sound waves emitted by two side-by-side loud-speakers driven by a single amplifier can interfere with each other because the two speakers are coherent. In other words, they respond to the amplifier in the same way at the same time.

A common method for producing two coherent light sources is to use a monochromatic source to illuminate a barrier containing two small openings, usually in the shape of slits, as in the case of Young's experiment illustrated in Figure 36.1. The light emerging from the two slits is coherent because a single source produces the original light beam and the two slits serve only to separate the original beam into two parts (which, after all, is what is done to the sound signal from two side-by-side loudspeakers). Any random change in the light emitted by the source occurs in both beams at the same time. As a result, interference effects can be observed when the light from the two slits arrives at a viewing screen.

If the light traveled only in its original direction after passing through the slits as shown in Figure 36.3a, the waves would not overlap and no interference pattern would be seen. Instead, as we have discussed with regard to Figure 34.4, the waves spread out from the slits as shown in Figure 36.3b. In other words, the light deviates from a straight-line path and enters the region that would otherwise be shadowed. As noted in Section 34.2, this divergence of light from its initial line of travel is called **diffraction.**

Light passing through narrow slits does *not* behave this way.

Light passing through narrow slits *diffracts.*

Figure 36.3 (a) If light waves did not spread out after passing through the slits, no interference would occur. (b) The light waves from the two slits overlap as they spread out, filling what we expect to be shadowed regions with light and producing interference fringes on a screen placed to the right of the slits.

36.2 Analysis Model: Waves in Interference

We discussed the superposition principle for waves on strings in Section 17.1, leading to a one-dimensional version of the waves in interference analysis model. In Example 17.1 we briefly discussed a two-dimensional interference phenomenon for sound from two loudspeakers. In walking from point O to point P in Figure 17.5, the listener experienced a maximum in sound intensity at O and a minimum at P. This experience is exactly analogous to an observer looking at point O in Figure 36.2 and seeing a bright fringe and then sweeping his eyes upward to point R, where there is a minimum in light intensity.

Let's look in more detail at the two-dimensional nature of Young's experiment with the help of Figure 36.4. The viewing screen is located a perpendicular distance L from the barrier containing two slits, S_1 and S_2 (Fig. 36.4a). These slits are separated by a distance d, and the source is monochromatic. To reach any arbitrary point P in the upper half of the screen, a wave from the lower slit must travel farther than a wave from the upper slit. The extra distance traveled from the lower slit is the **path difference** δ (Greek letter delta). If we assume the rays labeled r_1 and r_2 are parallel (Fig. 36.5b), which is approximately true if L is much greater than d, then δ is given by

$$\delta = r_2 - r_1 = d \sin \theta \tag{36.1}$$

The value of δ determines whether the two waves are in phase when they arrive at point P. If δ is either zero or some integer multiple of the wavelength, the two waves are in phase at point P and constructive interference results. Therefore, the condition for bright fringes, or **constructive interference,** at point P is

$$d \sin \theta_{\text{bright}} = m\lambda \qquad m = 0, \pm 1, \pm 2, \ldots \tag{36.2}$$

◀ Condition for constructive interference

The number m is called the **order number.** For constructive interference, the order number is the same as the number of wavelengths that represents the path difference between the waves from the two slits. The central bright fringe at $\theta_{\text{bright}} = 0$ is called the *zeroth-order maximum.* The first maximum on either side, where $m = \pm 1$, is called the *first-order maximum,* and so forth.

When δ is an odd multiple of $\lambda/2$, the two waves arriving at point P are 180° out of phase and give rise to destructive interference. Therefore, the condition for dark fringes, or **destructive interference,** at point P is

$$d \sin \theta_{\text{dark}} = (m + \tfrac{1}{2})\lambda \qquad m = 0, \pm 1, \pm 2, \ldots \tag{36.3}$$

◀ Condition for destructive interference

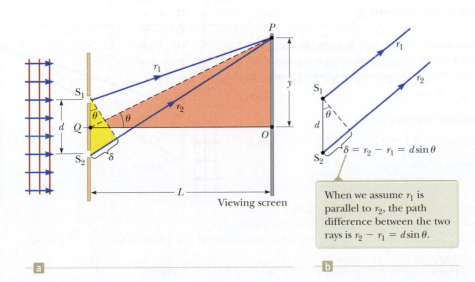

When we assume r_1 is parallel to r_2, the path difference between the two rays is $r_2 - r_1 = d \sin \theta$.

Viewing screen

Figure 36.4 (a) Geometric construction for describing Young's double-slit experiment (not to scale). (b) The slits are represented as sources, and the outgoing light rays are assumed to be parallel as they travel to P. To achieve that in practice, it is essential that $L \gg d$.

These equations provide the *angular* positions of the fringes. It is also useful to obtain expressions for the *linear* positions measured along the screen from O to P. From the triangle OPQ in Figure 36.4a, we see that

$$\tan \theta = \frac{y}{L} \tag{36.4}$$

Using this result, the linear positions of bright and dark fringes are given by

$$y_{\text{bright}} = L \tan \theta_{\text{bright}} \tag{36.5}$$

$$y_{\text{dark}} = L \tan \theta_{\text{dark}} \tag{36.6}$$

where θ_{bright} and θ_{dark} are given by Equations 36.2 and 36.3.

When the angles to the fringes are small, the positions of the fringes are linear near the center of the pattern. That can be verified by noting that for small angles, $\tan \theta \approx \sin \theta$, so Equation 36.5 gives the positions of the bright fringes as $y_{\text{bright}} = L \sin \theta_{\text{bright}}$. Incorporating Equation 36.2 gives

$$y_{\text{bright}} = L \frac{m\lambda}{d} \quad \text{(small angles)} \tag{36.7}$$

This result shows that y_{bright} is linear in the order number m, so the fringes are equally spaced for small angles. Similarly, for dark fringes,

$$y_{\text{dark}} = L \frac{(m + \frac{1}{2})\lambda}{d} \quad \text{(small angles)} \tag{36.8}$$

As demonstrated in Example 36.1, Young's double-slit experiment provides a method for measuring the wavelength of light. In fact, Young used this technique to do precisely that. In addition, his experiment gave the wave model of light a great deal of credibility. It was inconceivable that particles of light coming through the slits could cancel one another in a way that would explain the dark fringes.

The principles discussed in this section are the basis of the **waves in interference** analysis model. This model was applied to mechanical waves in one dimension in Chapter 17. Here we see the details of applying this model in three dimensions to light.

QUICK QUIZ 36.1 Which of the following causes the fringes in a two-slit interference pattern to move farther apart? **(a)** decreasing the wavelength of the light **(b)** decreasing the screen distance L **(c)** decreasing the slit spacing d **(d)** immersing the entire apparatus in water

---

ANALYSIS MODEL   **Waves in Interference**

Imagine a broad beam of light that illuminates a double slit in an otherwise opaque material. An interference pattern of bright and dark fringes is created on a distant screen. The condition for bright fringes **(constructive interference)** is

$$d \sin \theta_{\text{bright}} = m\lambda \quad m = 0, \pm 1, \pm 2, \dots \tag{36.2}$$

The condition for dark fringes **(destructive interference)** is

$$d \sin \theta_{\text{dark}} = (m + \tfrac{1}{2})\lambda \quad m = 0, \pm 1, \pm 2, \dots \tag{36.3}$$

The number m is called the **order number** of the fringe.

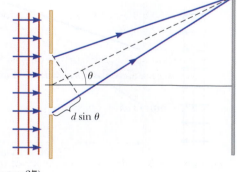

Examples:

- a thin film of oil on top of water shows swirls of color (Section 36.5)
- x-rays passing through a crystalline solid combine to form a Laue pattern (Chapter 37)
- a Michelson interferometer (Section 36.6) is used to search for the ether representing the medium through which light travels (Chapter 38)
- electrons exhibit interference just like light waves when they pass through a double slit (Chapter 39)

Example **36.1** **Measuring the Wavelength of a Light Source**

A viewing screen is separated from a double slit by 4.80 m. The distance between the two slits is 0.030 0 mm. Monochromatic light is directed toward the double slit and forms an interference pattern on the screen. The first dark fringe is 4.50 cm from the center line on the screen.

(A) Determine the wavelength of the light.

SOLUTION

Conceptualize Study Figure 36.4 to be sure you understand the phenomenon of interference of light waves. The distance of 4.50 cm is y in Figure 36.4. Because $L \gg y$, the angles for the fringes are small.

Categorize This problem is a simple application of the *waves in interference* model.

Analyze
Solve Equation 36.8 for the wavelength and substitute numerical values, taking $m = 0$ for the first dark fringe:

$$\lambda = \frac{y_{dark} d}{(m + \frac{1}{2})L} = \frac{(4.50 \times 10^{-2}\ \text{m})(3.00 \times 10^{-5}\ \text{m})}{(0 + \frac{1}{2})(4.80\ \text{m})}$$

$$= 5.62 \times 10^{-7}\ \text{m} = \boxed{562\ \text{nm}}$$

(B) Calculate the distance between adjacent bright fringes.

SOLUTION

Find the distance between adjacent bright fringes from Equation 36.7 and the results of part (A):

$$y_{m+1} - y_m = L\frac{(m+1)\lambda}{d} - L\frac{m\lambda}{d}$$

$$= L\frac{\lambda}{d} = 4.80\ \text{m}\left(\frac{5.62 \times 10^{-7}\ \text{m}}{3.00 \times 10^{-5}\ \text{m}}\right)$$

$$= 9.00 \times 10^{-2}\ \text{m} = \boxed{9.00\ \text{cm}}$$

Finalize For practice, find the wavelength of the sound in Example 17.1 using the procedure in part (A) of this example.

Example **36.2** **Separating Double-Slit Fringes of Two Wavelengths**

A light source emits visible light of two wavelengths: $\lambda = 430$ nm and $\lambda' = 510$ nm. The source is used in a double-slit interference experiment in which $L = 1.50$ m and $d = 0.025\ 0$ mm. Find the separation distance between the third-order bright fringes for the two wavelengths.

SOLUTION

Conceptualize In Figure 36.4a, imagine light of two wavelengths incident on the slits and forming two interference patterns on the screen. At some points, the fringes of the two colors might overlap, but at most points, they will not.

Categorize This problem is an application of the mathematical representation of the *waves in interference* analysis model.

Analyze
Use Equation 36.7 to find the fringe positions corresponding to these two wavelengths and subtract them:

$$\Delta y = y'_{bright} - y_{bright} = L\frac{m\lambda'}{d} - L\frac{m\lambda}{d} = \frac{Lm}{d}(\lambda' - \lambda)$$

Substitute numerical values:

$$\Delta y = \frac{(1.50\ \text{m})(3)}{0.025\ 0 \times 10^{-3}\ \text{m}}(510 \times 10^{-9}\ \text{m} - 430 \times 10^{-9}\ \text{m})$$

$$= 0.014\ 4\ \text{m} = \boxed{1.44\ \text{cm}}$$

Finalize Let's explore further details of the interference pattern in the following **What If?**

WHAT IF? What if we examine the entire interference pattern due to the two wavelengths and look for overlapping fringes? Are there any locations on the screen where the bright fringes from the two wavelengths overlap exactly?

continued

36.2 continued

Answer   Find such a location by setting the location of any bright fringe due to λ equal to one due to λ', using Equation 36.7:

$$L\frac{m\lambda}{d} = L\frac{m'\lambda'}{d} \quad\rightarrow\quad \frac{m'}{m} = \frac{\lambda}{\lambda'}$$

Substitute the wavelengths:

$$\frac{m'}{m} = \frac{430 \text{ nm}}{510 \text{ nm}} = \frac{43}{51}$$

Therefore, the 51st fringe of the 430-nm light overlaps with the 43rd fringe of the 510-nm light.

Use Equation 36.7 to find the value of y for these fringes:

$$y = (1.50 \text{ m})\left[\frac{51(430 \times 10^{-9} \text{ m})}{0.025\,0 \times 10^{-3} \text{ m}}\right] = 1.32 \text{ m}$$

This value of y is comparable to L, so the small-angle approximation used for Equation 36.7 is *not* valid. This conclusion suggests we should not expect Equation 36.7 to give us the correct result. If you use Equation 36.5, you can show that the bright fringes do indeed overlap when the same condition, $m'/m = \lambda/\lambda'$, is met (see Problem 30). Therefore, the 51st fringe of the 430-nm light does overlap with the 43rd fringe of the 510-nm light, but not at the location of 1.32 m. You are asked to find the correct location as part of Problem 30.

36.3 Intensity Distribution of the Double-Slit Interference Pattern

Notice that the edges of the bright fringes in Figure 36.1b are not sharp; rather, there is a gradual change from bright to dark. So far, we have discussed the locations of only the centers of the bright and dark fringes on a distant screen. Let's now direct our attention to the distribution of light intensity associated with the double-slit interference pattern.

Using an analysis of the electric fields of the light from the two slits, we can show (Problem 16) that the intensity of light on the screen in Figure 36.4 is given by

$$I = I_{max} \cos^2\left(\frac{\pi d \sin\theta}{\lambda}\right) \tag{36.9}$$

Alternatively, because $\sin\theta \approx y/L$ for small values of θ in Figure 36.4, we can write Equation 36.9 in the form

$$I = I_{max} \cos^2\left(\frac{\pi d}{\lambda L}y\right) \quad \text{(small angles)} \tag{36.10}$$

Constructive interference, which produces light intensity maxima, occurs when the quantity $\pi dy/\lambda L$ is an integral multiple of π, corresponding to $y = Lm\lambda/d$, where m is the order number. This result is consistent with Equation 36.7.

A plot of light intensity versus $d \sin\theta$ using Equation 36.9 is given in Figure 36.5 and compared to a photograph of the interference pattern. Figure 36.6 shows similar plots of light intensity versus $d \sin\theta$ for light passing through multiple slits. In this case, the pattern contains primary and secondary maxima. For three slits, the primary maxima are nine times more intense than the secondary maxima as measured by the height of the curve because the intensity varies as E^2 (see Eq. 33.27). For N slits, the intensity of the primary maxima is N^2 times greater than that for the secondary maxima. As the number of slits increases, the primary maxima increase in intensity and become narrower, while the secondary maxima decrease in intensity relative to the primary maxima. Figure 36.6 also shows that as the number of slits increases, the number of secondary maxima also increases. In fact, the number of secondary maxima is always $N - 2$, where N is the number of slits. In Section 37.4, we shall investigate the pattern for a very large number of slits in a device called a *diffraction grating*.

Figure 36.5 Light intensity versus $d \sin \theta$ for a double-slit interference pattern when the screen is far from the two slits ($L \gg d$).

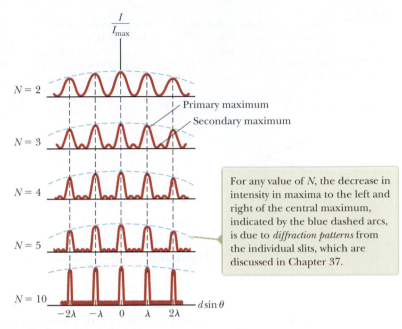

For any value of N, the decrease in intensity in maxima to the left and right of the central maximum, indicated by the blue dashed arcs, is due to *diffraction patterns* from the individual slits, which are discussed in Chapter 37.

Figure 36.6 Multiple-slit interference patterns. As N, the number of slits, is increased, the primary maxima (the tallest peaks in each graph) become narrower but remain fixed in position and the number of secondary maxima increases.

QUICK QUIZ 36.2 Using Figure 36.6 as a model, sketch the interference pattern from six slits.

36.4 Change of Phase Due to Reflection

Young's method for producing two coherent light sources involves illuminating a pair of slits with a single source. Another simple, yet ingenious, arrangement for producing an interference pattern with a single light source is known as *Lloyd's mirror*[1] (Fig. 36.7). A point light source S is placed close to a mirror, and a viewing screen is positioned some distance away and perpendicular to the mirror. Light waves can reach point P on the screen either directly from S to P (red) or by the path involving reflection from the mirror (blue). The reflected ray can be treated as a ray originating from a virtual source S′. As a result, we can think of this arrangement as a double-slit source where the distance d between sources S and S′ in Figure 36.7 is analogous to length d in Figure 36.4. Hence, at observation points far from the source ($L \gg d$), we expect waves from S and S′ to form an interference pattern exactly like the one formed by two real coherent sources. An interference pattern is indeed observed. The positions of the dark and bright fringes, however, are reversed relative to the pattern created by two real coherent sources (Young's experiment). Such a reversal can only occur if the coherent sources S and S′ differ in phase by 180°.

To illustrate further, consider point $P′$, the point where the mirror intersects the screen. This point is equidistant from sources S and S′. If path difference alone were responsible for the phase difference, we would see a bright fringe at $P′$ (because the path difference is zero for this point), corresponding to the central bright fringe of the two-slit interference pattern. Instead, a dark fringe is observed at $P′$. We therefore conclude that a 180° phase change must be produced by reflection from the

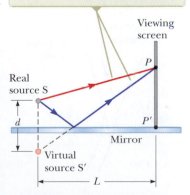

An interference pattern is produced on the screen as a result of the combination of the direct ray (red) and the reflected ray (blue).

Figure 36.7 Lloyd's mirror. The reflected ray undergoes a phase change of 180°.

---

[1]Developed in 1834 by Humphrey Lloyd (1800–1881), Professor of Natural and Experimental Philosophy, Trinity College, Dublin.

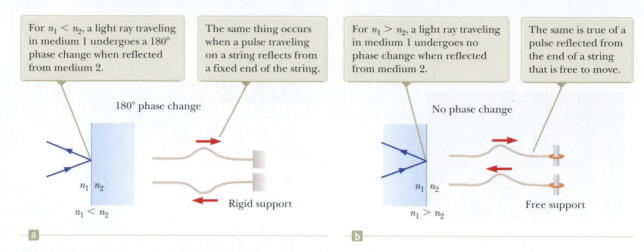

For $n_1 < n_2$, a light ray traveling in medium 1 undergoes a 180° phase change when reflected from medium 2.

The same thing occurs when a pulse traveling on a string reflects from a fixed end of the string.

For $n_1 > n_2$, a light ray traveling in medium 1 undergoes no phase change when reflected from medium 2.

The same is true of a pulse reflected from the end of a string that is free to move.

180° phase change

No phase change

$n_1 \mid n_2$

$n_1 < n_2$

Rigid support

$n_1 \mid n_2$

$n_1 > n_2$

Free support

a

b

Figure 36.8 Comparisons of reflections of light waves and waves on strings.

MarcelClemens/Shutterstock

Figure 36.9 Colors in a soap bubble due to interference.

Interference in light reflected from a thin film is due to a combination of rays 1 and 2 reflected from the upper and lower surfaces of the film.

180° phase change

No phase change

1   2

Air
$n = 1.00$

A

Film
n

t

B

Air
$n = 1.00$

3   4

Rays 3 and 4 lead to interference effects for light transmitted through the film.

Figure 36.10 Light paths through a thin film.

mirror. In general, **an electromagnetic wave undergoes a phase change of 180° upon reflection from a medium that has a higher index of refraction than the one in which the wave is traveling.**

It is useful to draw an analogy between reflected light waves and the reflections of a transverse pulse on a stretched string (Section 17.3). The reflected pulse on a string undergoes a phase change of 180° when reflected from the boundary of a denser string or a rigid support, but no phase change occurs when the pulse is reflected from the boundary of a less dense string or a freely supported end. Similarly, an electromagnetic wave undergoes a 180° phase change when reflected from a boundary leading to an optically denser medium (defined as a medium with a higher index of refraction), but no phase change occurs when the wave is reflected from a boundary leading to a less dense medium. These rules, summarized in Figure 36.8, can be deduced from Maxwell's equations, but the treatment is beyond the scope of this text.

36.5 Interference in Thin Films

Interference effects are commonly observed in thin films, such as thin layers of oil on water or the thin surface of a soap bubble as shown in Figure 36.9. The varied colors observed when white light is incident on such films result from the interference of waves reflected from the two surfaces of the film.

Consider a film of uniform thickness t and index of refraction n. The wavelength of light λ_n in the film (see Section 34.4) is

$$\lambda_n = \frac{\lambda}{n}$$

where λ is the wavelength of the light in free space and n is the index of refraction of the film material. Let's assume light rays traveling in air are nearly normal to the two surfaces of the film as shown in Figure 36.10.

Reflected ray 1, which is reflected from the upper surface (A) in Figure 36.10, undergoes a phase change of 180° with respect to the incident wave. Reflected ray 2, which is reflected from the lower film surface (B), undergoes no phase change because it is reflected from a medium (air) that has a lower index of refraction. Therefore, ray 1 is 180° out of phase with ray 2, which is equivalent to a path difference of $\lambda_n/2$. We must also consider, however, that ray 2 travels an extra distance $2t$ before the waves recombine in the air above surface A. (Remember that we are considering light rays that are close to normal to the surface. If the rays are not close to normal, the path difference is larger than $2t$.) If $2t = \lambda_n/2$, rays 1 and 2 recombine in phase and the result is constructive interference. In general, the condition for *constructive* interference in thin films is[2]

$$2t = (m + \tfrac{1}{2})\lambda_n \quad m = 0, 1, 2, \ldots \tag{36.11}$$

This condition takes into account two factors: (1) the difference in path length for the two rays (the term $m\lambda_n$) and (2) the 180° phase change upon reflection (the term $\tfrac{1}{2}\lambda_n$). Because $\lambda_n = \lambda/n$, we can write Equation 36.11 as

$$2nt = (m + \tfrac{1}{2})\lambda \quad m = 0, 1, 2, \ldots \tag{36.12}$$

If the extra distance $2t$ traveled by ray 2 corresponds to a multiple of λ_n, the two waves combine out of phase and the result is destructive interference. The general equation for *destructive* interference in thin films is

$$2nt = m\lambda \quad m = 0, 1, 2, \ldots \tag{36.13}$$

The foregoing conditions for constructive and destructive interference are valid when the medium above the top surface of the film is the same as the medium below the bottom surface or, if there are different media above and below the film, the index of refraction of both is less than n. If the film is placed between two different media, one with $n < n_{\text{film}}$ and the other with $n > n_{\text{film}}$, the conditions for constructive and destructive interference are reversed. In that case, either there is a phase change of 180° for both ray 1 reflecting from surface A and ray 2 reflecting from surface B or there is no phase change for either ray; hence, the net change in relative phase due to the reflections is zero. See Example 36.4 for a practical application of this effect.

Rays 3 and 4 in Figure 36.10 lead to interference effects in the light transmitted through the thin film. The analysis of these effects is similar to that of the reflected light. You are asked to explore the transmitted light in Problems 22 and 25.

QUICK QUIZ 36.3 One microscope slide is placed on top of another with their left edges in contact and a human hair under the right edge of the upper slide. As a result, a wedge of air exists between the slides. An interference pattern results when monochromatic light is incident on the wedge. What is at the left edges of the slides? **(a)** a dark fringe **(b)** a bright fringe **(c)** impossible to determine

Newton's Rings

Another method for observing interference in light waves is to place a plano-convex lens on top of a flat glass surface as shown in Figure 36.11a. With this arrangement, the air film between the glass surfaces varies in thickness from zero at the point of contact to some nonzero value at point P. If the radius of curvature R of the lens is

Courtesy of Bausch & Lomb

Figure 36.11 (a) The combination of rays reflected from the flat plate and the curved lens surface gives rise to an interference pattern known as Newton's rings. (b) Photograph of Newton's rings.

---

[2]The full interference effect in a thin film requires an analysis of an infinite number of reflections back and forth between the top and bottom surfaces of the film. We focus here only on a single reflection from the bottom of the film, which provides the largest contribution to the interference effect.

much greater than the distance r and the system is viewed from above, a pattern of light and dark rings is observed as shown in Figure 36.11b. These circular fringes, discovered by Newton, are called **Newton's rings.**

One important use of Newton's rings is in the testing of optical lenses. A circular pattern like that pictured in Figure 36.11b is obtained only when the lens is ground to a perfectly symmetric curvature. Variations from such symmetry produce a pattern with fringes that vary from a smooth, circular shape. These variations indicate how the lens must be reground and repolished to remove imperfections.

So what is going on with the colors from the hummingbird feathers in the opening storyline? The *iridescence* that makes the brilliant colors that often appear on the bird's throat and belly is due to an interference effect caused by light reflecting from microstructures in the feathers. The colors will vary with the viewing angle. Other organisms exhibiting iridescence include peacocks, Morpho butterflies, and some types of beetles and seashells.

PROBLEM-SOLVING STRATEGY **Thin-Film Interference**

The following features should be kept in mind when working thin-film interference problems.

1. Conceptualize. Think about what is going on physically in the problem. Identify the light source and the location of the observer.

2. Categorize. Confirm that you should use the techniques for thin-film interference by identifying the thin film causing the interference.

3. Analyze. The type of interference that occurs is determined by the phase relationship between the portion of the wave reflected at the upper surface of the film and the portion reflected at the lower surface. Phase differences between the two portions of the wave have two causes: differences in the distances traveled by the two portions and phase changes occurring on reflection. *Both* causes must be considered when determining which type of interference occurs. If the media above and below the film both have index of refraction larger than that of the film or if both indices are smaller, use Equation 36.12 for constructive interference and Equation 36.13 for destructive interference. If the film is located between two different media, one with $n < n_{\text{film}}$ and the other with $n > n_{\text{film}}$, reverse these two equations for constructive and destructive interference.

4. Finalize. Inspect your final results to see if they make sense physically and are of an appropriate size.

Example 36.3 **Interference in a Soap Film**

Calculate the minimum thickness of a soap-bubble film that results in constructive interference in the reflected light if the film is illuminated with light whose wavelength in free space is $\lambda = 600$ nm. The index of refraction of the soap film is 1.33.

SOLUTION

Conceptualize Imagine that the film in Figure 36.10 is soap, with air on both sides.

Categorize We determine the result using an equation from this section, so we categorize this example as a substitution problem.

The minimum film thickness for constructive interference in the reflected light corresponds to $m = 0$ in Equation 36.12. Solve this equation for t and substitute numerical values:

$$t = \frac{(0 + \frac{1}{2})\lambda}{2n} = \frac{\lambda}{4n} = \frac{(600 \text{ nm})}{4(1.33)} = 113 \text{ nm}$$

WHAT IF? What if the film is twice as thick? Does this situation produce constructive interference?

Answer Using Equation 36.12, we can solve for the thicknesses at which constructive interference occurs:

$$t = (m + \tfrac{1}{2})\frac{\lambda}{2n} = (2m + 1)\frac{\lambda}{4n} \quad m = 0, 1, 2, \ldots$$

The allowed values of m show that constructive interference occurs for *odd* multiples of the thickness corresponding to $m = 0$, $t = 113$ nm. Therefore, constructive interference does *not* occur for a film that is twice as thick.

Example **36.4** **Nonreflective Coatings for Solar Cells**

Solar cells—devices that generate electricity when exposed to sunlight—are often coated with a transparent, thin film of silicon monoxide (SiO, $n = 1.45$) to minimize reflective losses from the surface. Suppose a silicon solar cell ($n = 3.5$) is coated with a thin film of silicon monoxide for this purpose (Fig. 36.12a). Determine the minimum film thickness that produces the least reflection at a wavelength of 550 nm, near the center of the visible spectrum.

S O L U T I O N

Conceptualize Figure 36.12a helps us visualize the path of the rays in the SiO film that result in interference in the reflected light.

Categorize Based on the geometry of the SiO layer, we categorize this example as a thin-film interference problem.

Analyze The reflected light is a minimum when rays 1 and 2 in Figure 36.12a meet the condition of destructive interference. In this situation, *both* rays undergo a 180° phase change upon reflection: ray 1 from the upper SiO surface and ray 2 from the lower SiO surface. The net change in phase due to reflection is therefore zero, and the condition for a reflection minimum requires a path difference of $\lambda_n/2$, where λ_n is the wavelength of the light in SiO. Hence, $2nt = \lambda/2$, where λ is the wavelength in air and n is the index of refraction of SiO.

Figure 36.12 (Example 36.4) (a) Reflective losses from a silicon solar cell are minimized by coating the surface of the cell with a thin film of silicon monoxide. (b) The reflected light from a coated camera lens often has a reddish-violet appearance.

Solve the equation $2nt = \lambda/2$ for t and substitute numerical values:

$$t = \frac{\lambda}{4n} = \frac{550 \text{ nm}}{4(1.45)} = 94.8 \text{ nm}$$

Finalize A typical uncoated solar cell has reflective losses as high as 30%, but a coating of SiO can reduce this value to about 10%. This significant decrease in reflective losses increases the cell's efficiency because less reflection means that more sunlight enters the silicon to create charge carriers in the cell. No coating can ever be made perfectly nonreflecting because the required thickness is wavelength-dependent and the incident light covers a wide range of wavelengths.

Glass lenses used in cameras and other optical instruments are usually coated with a transparent thin film to reduce or eliminate unwanted reflection and to enhance the transmission of light through the lenses. The camera lens in Figure 36.12b has several coatings (of different thicknesses) to minimize reflection of light waves having wavelengths near the center of the visible spectrum. As a result, the small amount of light that is reflected by the lens has a greater proportion of the far ends of the spectrum and often appears reddish violet.

36.6 The Michelson Interferometer

The **interferometer,** invented by American physicist A. A. Michelson (1852–1931), splits a light beam into two parts and then recombines the parts to form an interference pattern. The device can be used to measure wavelengths or other lengths with great precision because a large and precisely measurable displacement of one of the mirrors is related to an exactly countable number of wavelengths of light.

A schematic diagram of the interferometer is shown in Figure 36.13 (page 974). A ray of light from a monochromatic source is split into two rays by mirror M_0, which is inclined at 45° to the incident light beam. Mirror M_0, called a *beam splitter,* transmits half the light incident on it and reflects the rest. One ray is reflected from M_0 to the right toward mirror M_1, and the second ray is transmitted vertically through M_0 toward mirror M_2. Hence, the two rays travel separate paths L_1 and L_2. After reflecting from M_1 and M_2, the two rays eventually recombine at M_0 to produce an interference pattern, which can be viewed through a telescope.

Figure 36.13 Diagram of the Michelson interferometer.

The interference condition for the two rays is determined by the difference in their path length. When the two mirrors are exactly perpendicular to each other, the interference pattern is a target pattern of bright and dark circular fringes. As M_1 is moved, the fringe pattern collapses or expands, depending on the direction in which M_1 is moved. For example, if a dark circle appears at the center of the target pattern (corresponding to destructive interference) and M_1 is then moved a distance $\lambda/4$ toward M_0, the path difference changes by $\lambda/2$. What was a dark circle at the center now becomes a bright circle. As M_1 is moved an additional distance $\lambda/4$ toward M_0, the bright circle becomes a dark circle again. Therefore, the fringe pattern shifts by one-half fringe each time M_1 is moved a distance $\lambda/4$. The wavelength of light is then measured by counting the number of fringe shifts for a given displacement of M_1. If the wavelength is accurately known, mirror displacements can be measured to within a fraction of the wavelength.

We will see an important historical use of the Michelson interferometer in our discussion of relativity in Chapter 38. Modern uses include the following two applications, Fourier transform infrared spectroscopy and the laser interferometer gravitational-wave observatory.

Fourier Transform Infrared Spectroscopy

Spectroscopy is the study of the wavelength distribution of radiation from a sample that can be used to identify the characteristics of atoms or molecules in the sample. Infrared spectroscopy is particularly important to organic chemists when analyzing organic molecules. Traditional spectroscopy involves the use of an optical element, such as a prism (Section 34.4) or a diffraction grating (Section 37.4), either of which spreads out various wavelengths in a complex optical signal from the sample into different angles. In this way, the various wavelengths of radiation and their intensities in the signal can be determined. These types of devices are limited in their resolution and effectiveness because they must be scanned through the various angular deviations of the radiation.

The technique of *Fourier transform infrared* (FTIR) *spectroscopy* is used to create a higher-resolution spectrum in a time interval of 1 second that may have required 30 minutes with a standard spectrometer. In this technique, the radiation from a sample enters a Michelson interferometer. The movable mirror is swept through the zero-path-difference condition, and the intensity of radiation at the viewing position is recorded. The result is a complex set of data relating light intensity as a function of mirror position, called an *interferogram*. Because there is a relationship between mirror position and light intensity for a given wavelength, the interferogram contains information about all wavelengths in the signal.

In Section 17.8, we discussed Fourier analysis of a waveform. The waveform is a function that contains information about all the individual frequency components that make up the waveform.[3] Equation 17.14 shows how the waveform is generated from the individual frequency components. Similarly, the interferogram can be analyzed by computer, in a process called a *Fourier transform*, to provide all the wavelength components. This information is the same as that generated by traditional spectroscopy, but the resolution of FTIR spectroscopy is much higher.

Laser Interferometer Gravitational-Wave Observatory

Einstein's general theory of relativity (Section 38.9) predicts the existence of *gravitational waves.* These waves propagate from the site of any gravitational disturbance, which could be periodic and predictable, such as the rotation of a double star around a center of mass, or unpredictable, such as the supernova explosion of a massive star.

In Einstein's theory, gravitation is equivalent to a distortion of space. Therefore, a gravitational disturbance causes an additional distortion that propagates through space in a manner similar to mechanical or electromagnetic waves. When gravitational waves from a disturbance pass by the Earth, they create a distortion of the local space. The laser interferometer gravitational-wave observatory (LIGO) apparatus is designed to detect this distortion. The apparatus employs a Michelson interferometer that uses laser beams with an effective path length of several kilometers. At the end of an arm of the interferometer, a mirror is mounted on a massive pendulum. When a gravitational wave passes by, the pendulum and the attached mirror move and the interference pattern due to the laser beams from the two arms changes.

Two sites for interferometers have been developed in the United States—in Richland, Washington, and in Livingston, Louisiana—to allow coincidence studies of gravitational waves. Figure 36.14 shows the Washington site. The two arms of the Michelson interferometer are evident in the photograph.

Despite the difficulties in detecting the very weak gravitational waves, the exciting announcement was made on 11 February 2016 that both the Washington and Louisiana sites had detected a gravitational wave on 14 September 2015. Analysis showed that the wave came from two massive black holes over 1 billion light-years away rotating around each other rapidly and then merging. Three solar masses of their combined mass was radiated away as gravitational waves. Estimates show that the peak power output of the event was about 50 times the power of the entire observable universe. Additional black-hole collisions were announced by LIGO in June 2016 and June 2017.

Courtesy Caltech/MIT/LIGO Laboratory

Figure 36.14 The Laser Interferometer Gravitational-Wave Observatory (LIGO) near Richland, Washington. Notice the two perpendicular arms of the Michelson interferometer.

---

[3]In acoustics, it is common to talk about the components of a complex signal in terms of frequency. In optics, it is more common to identify the components by wavelength.

Summary

> ## Concepts and Principles

Interference in light waves occurs whenever two or more waves overlap at a given point. An interference pattern is observed if (1) the sources are coherent and (2) the sources have identical wavelengths.

The **intensity** at a point in a double-slit interference pattern is

$$I = I_{max} \cos^2\left(\frac{\pi d \sin\theta}{\lambda}\right) \tag{36.9}$$

where I_{max} is the maximum intensity on the screen and the expression represents the time average.

A wave traveling from a medium of index of refraction n_1 toward a medium of index of refraction n_2 undergoes a 180° phase change upon reflection when $n_2 > n_1$ and undergoes no phase change when $n_2 < n_1$.

The condition for constructive interference in a film of thickness t and index of refraction n surrounded by air is

$$2nt = (m + \tfrac{1}{2})\lambda \quad m = 0, 1, 2, \ldots \tag{36.12}$$

where λ is the wavelength of the light in free space.
Similarly, the condition for destructive interference in a thin film surrounded by air is

$$2nt = m\lambda \quad m = 0, 1, 2, \ldots \tag{36.13}$$

> ## Analysis Models for Problem Solving

Waves in Interference. Young's double-slit experiment serves as a prototype for interference phenomena involving electromagnetic radiation. In this experiment, two slits separated by a distance d are illuminated by a single-wavelength light source. The condition for bright fringes (**constructive interference**) is

$$d \sin\theta_{bright} = m\lambda \quad m = 0, \pm1, \pm2, \ldots \tag{36.2}$$

The condition for dark fringes (**destructive interference**) is

$$d \sin\theta_{dark} = (m + \tfrac{1}{2})\lambda \quad m = 0, \pm1, \pm2, \ldots \tag{36.3}$$

The number m is called the **order number** of the fringe.

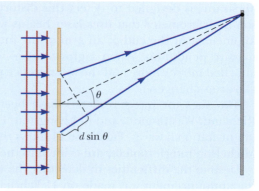

Think–Pair–Share

See the Preface for an explanation of the icons used in this problems set.
For additional assessment items for this section, go to 🔆 **WEBASSIGN** From Cengage

1. Your group is studying the *instrument landing system* used to guide aircraft to safe landings at some airports when the visibility is poor. It turns out that Young's double-slit experiment is used in this system. A pilot is trying to align her plane with a runway as suggested in Figure TP36.1. Two radio antennas (the black dots in the figure) are positioned adjacent to the runway, separated by a distance $d = 40.0$ m. The antennas broadcast unmodulated coherent radio waves at 30.0 MHz. The red lines in Figure TP36.1 represent paths along which maxima in the interference pattern of the radio waves exist. (a) Find the wavelength of these waves. The pilot "locks onto" the strong signal radiated along an interference maximum and steers the plane to keep the received signal strong. If she has found the central maximum, the plane will have precisely the correct heading to land when it reaches the runway, as exhibited by plane A

Figure TP36.1

in Figure TP36.1. (b) **What If?** Suppose the plane is flying along the first side maximum instead as is the case for plane B in the figure. How far to the side of the runway centerline will the plane be when it is 2.00 km from the antennas, measured along its direction of travel? (c) It is possible to tell the pilot that she is on the wrong maximum by sending out a second signal from each antenna and equipping the aircraft with a two-channel receiver. The ratio of the second frequency to that of the first must *not* be the ratio of small integers (such as $\frac{3}{4}$). Explain how this two-frequency system would work and why it would not necessarily work if the frequencies were related by an integer ratio.

2. Your group is working in an optoelectronics laboratory. **S** Your supervisor has given you the following technical task. A sheet of transparent plastic having an index of refraction n and thickness t is placed between the upper slit and the screen in an orientation such that the light passes through the plastic perpendicularly to its surfaces as shown in Figure TP36.2. When this is done, the central maximum of the interference pattern moves upward on the screen by a distance y'. Your supervisor asks your group to investigate this

situation and determine an expression for the distance y' in terms of d, L, n, and t.

3. **ACTIVITY** Your team is working for a chemical company. Your supervisor asks you to perform an experiment to measure the evaporation rate dV/dt of ethanol in mL/s. The physical setup is as follows. A sample of ethanol fills a shallow glass dish of radius $r = 5.00$ cm. A laser of wavelength 632.8 nm is placed above the dish, with the beam directed downward so that the laser beam strikes the surface of the ethanol at near-normal incidence. Next to the laser is a detector system that generates a graph of the intensity of the reflected light as a function of time, as shown in Figure TP36.3. The thin layer of ethanol acts as a thin film, so that the intensity of reflected light varies in time due to interference effects as the thickness of the layer decreases due to evaporation. The index of refraction of ethanol is 1.361.

Figure TP36.2

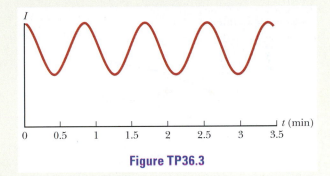

Figure TP36.3

Problems

See the Preface for an explanation of the icons used in this problems set. For additional assessment items for this section, go to WEBASSIGN From Cengage

SECTION 36.2 Analysis Model: Waves in Interference

Problems 1, 4, and 6 and online problems 17.2 and 17.4 in Chapter 17 can be assigned with this section.

1. Two slits are separated by 0.320 mm. A beam of 500-nm light strikes the slits, producing an interference pattern. Determine the number of maxima observed in the angular range $-30.0° < \theta < 30.0°$.

2. *Why is the following situation impossible?* Two narrow slits are separated by 8.00 mm in a piece of metal. A beam of microwaves strikes the metal perpendicularly, passes through the two slits, and then proceeds toward a wall some distance away. You know that the wavelength of the radiation is 1.00 cm ±5%, but you wish to measure it more precisely. Moving a microwave detector along the wall to study the interference pattern, you measure the position of the $m = 1$ bright fringe, which leads to a successful measurement of the wavelength of the radiation.

3. A laser beam is incident on two slits with a separation of 0.200 mm, and a screen is placed 5.00 m from the slits. An interference pattern appears on the screen. If the angle from the center fringe to the first bright fringe to the side is 0.181°, what is the wavelength of the laser light?

4. In a Young's double-slit experiment, two parallel slits with a slit separation of 0.100 mm are illuminated by light of wavelength 589 nm, and the interference pattern is observed on a screen located 4.00 m from the slits. (a) What is the difference in path lengths from each of the slits to the location of the center of a third-order bright fringe on the screen? (b) What is the difference in path lengths from the two slits to the location of the center of the third dark fringe away from the center of the pattern?

5. Light of wavelength 620 nm falls on a double slit, and the first bright fringe of the interference pattern is seen at an angle of 15.0° with the horizontal. Find the separation between the slits.

6. Light with wavelength 442 nm passes through a double-slit system that has a slit separation $d = 0.400$ mm. Determine how far away a screen must be placed so that dark fringes appear directly opposite both slits, with only one bright fringe between them.

7. A student holds a laser that emits light of wavelength 632.8 nm. The laser beam passes through a pair of slits separated by 0.300 mm, in a glass plate attached to the front of the laser. The beam then falls perpendicularly on a screen, creating an interference pattern on it. The student begins to walk directly toward the screen at 3.00 m/s. The central maximum on the screen is stationary. Find the speed of the 50th-order maxima on the screen.

8. A student holds a laser that emits light of wavelength λ. The laser beam passes through a pair of slits separated by a distance d, in a glass plate attached to the front of the laser. The beam then falls perpendicularly on a screen, creating an interference pattern on it. The student begins to walk directly toward the screen at speed v. The central maximum on the screen is stationary. Find the speed of the mth-order maxima on the screen, where m can be very large.

9. Coherent light rays of wavelength λ strike a pair of slits separated by distance d at an angle θ_1 with respect to the normal to the plane containing the slits as shown in Figure P36.9. The rays leaving the slits make an angle θ_2 with respect to the normal, and an interference maximum is formed by those rays on a screen that is a great distance from the slits. Show that the angle θ_2 is given by

$$\theta_2 = \sin^{-1}\left(\sin\theta_1 - \frac{m\lambda}{d}\right)$$

where m is an integer.

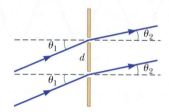

Figure P36.9

10. In Figure P36.10 (not to scale), let $L = 1.20$ m and $d = 0.120$ mm and assume the slit system is illuminated with monochromatic 500-nm light. Calculate the phase difference between the two wave fronts arriving at P when (a) $\theta = 0.500°$ and (b) $y = 5.00$ mm. (c) What is the value of θ for which the phase difference is 0.333 rad? (d) What is the value of θ for which the path difference is $\lambda/4$?

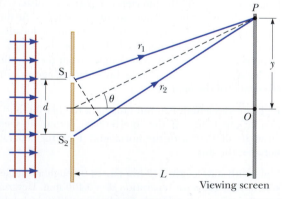

Figure P36.10

11. You are working in an optical research laboratory. One of your projects involves the use of a double slit through which you pass orange laser light of wavelength 590 nm. Unfortunately, because of budget cuts, there are a lot of researchers in the same room, with lots of equipment stuffed in the room, and, in particular, lots of laser beams flying around the room. One day, you find that a second laser beam of unknown origin and different color is entering your double slit along with your orange beam and you are seeing an interference pattern that is the sum of those due to the two beams. You notice that the combined pattern is pretty much a mess, but wait! The $m = 3$ maximum of your orange laser beam pattern is pure; there is absolutely no mixture of the other color at that point. From this fact, you determine the wavelength of the offending laser light so that you can figure out which other researcher to ask to modify the aiming of his laser.

12. You are operating a new radio telescope that has been installed on a tall cliff facing the ocean. You begin the testing of the telescope by facing the antenna toward the ocean, setting its receiving wavelength to 125 m, and sweeping its direction slowly from horizontal to straight up in the sky. Each sweep takes about an hour. When you review the data, you notice that the antenna received no signals when aimed at a certain angle above the horizontal. You continue to take data beginning at the same time each night and discover that the angle at which no signals are detected varies from night to night. Over a full month, the angle at which no signals are detected varies from 24.5° to 25.7°. You finally figure out that the loss of signal is due to destructive interference caused by the reflection of radio waves from the ocean surface, and the monthly variation is due to the changes caused by ocean tides. You inform the local oceanographic institute that you have a novel method of measuring tides. To verify your results, the institute asks for the variation in the heights of the tides during the previous month.

13. In the double-slit arrangement of Figure P36.13, $d = 0.150$ mm, $L = 140$ cm, $\lambda = 643$ nm, and $y = 1.80$ cm. (a) What is the path difference δ for the rays from the two slits arriving at P? (b) Express this path difference in terms of λ. (c) Does P correspond to a maximum, a minimum, or an intermediate condition? Give evidence for your answer.

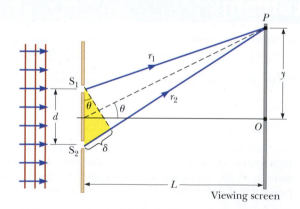

Figure P36.13

14. Monochromatic light of wavelength λ is incident on a pair of slits separated by 2.40×10^{-4} m and forms an interference pattern on a screen placed 1.80 m from the slits. The first-order bright fringe is at a position $y_{bright} = 4.52$ mm measured from the center of the central maximum. From this information, we wish to predict where the fringe for $n = 50$ would be located. (a) Assuming the fringes are laid out linearly along the screen, find the position of the $n = 50$ fringe by multiplying the position of the $n = 1$ fringe by 50.0. (b) Find the tangent of the angle the first-order bright fringe makes with respect to the line extending from the point midway between the slits to the center of the central maximum. (c) Using the result of part (b) and Equation 36.2,

calculate the wavelength of the light. (d) Compute the angle for the 50th-order bright fringe from Equation 36.2. (e) Find the position of the 50th-order bright fringe on the screen from Equation 36.5. (f) Comment on the agreement between the answers to parts (a) and (e).

SECTION 36.3 Intensity Distribution of the Double-Slit Interference Pattern

15. Show that the two waves with wave functions given by $E_1 = 6.00 \sin(100\pi t)$ and $E_2 = 8.00 \sin(100\pi t + \pi/2)$ add to give a wave with the wave function $E_R \sin(100\pi t + \phi)$. Find the required values for E_R and ϕ.

16. Show that the distribution of intensity in a double-slit pattern is given by Equation 36.9. Begin by assuming that the total magnitude of the electric field at point P on the screen in Figure 36.4 is the superposition of two waves, with electric field magnitudes

$$E_1 = E_0 \sin \omega t \qquad E_2 = E_0 \sin(\omega t + \phi)$$

The phase angle ϕ in E_2 is due to the extra path length traveled by the lower beam in Figure 36.4. Recall from Equation 33.27 that the intensity of light is proportional to the square of the amplitude of the electric field. In addition, the apparent intensity of the pattern is the time-averaged intensity of the electromagnetic wave. You will need to evaluate the integral of the square of the sine function over one period. Refer to Figure 32.5 for an easy way to perform this evaluation. You will also need the trigonometric identity

$$\sin A + \sin B = 2 \sin\left(\frac{A+B}{2}\right)\cos\left(\frac{A-B}{2}\right)$$

17. Green light ($\lambda = 546$ nm) illuminates a pair of narrow, parallel slits separated by 0.250 mm. Make a graph of I/I_{max} as a function of θ for the interference pattern observed on a screen 1.20 m away from the plane of the parallel slits. Let θ range over the interval from $-0.3°$ to $+0.3°$.

18. Monochromatic coherent light of amplitude E_0 and angular frequency ω passes through three parallel slits, each separated by a distance d from its neighbor. (a) Show that the time-averaged intensity as a function of the angle θ is

$$I(\theta) = I_{max}\left[1 + 2\cos\left(\frac{2\pi d \sin\theta}{\lambda}\right)\right]^2$$

(b) Explain how this expression describes both the primary and the secondary maxima. (c) Determine the ratio of the intensities of the primary and secondary maxima. *Hint:* See Problem 16.

SECTION 36.5 Interference in Thin Films

19. A material having an index of refraction of 1.30 is used as an antireflective coating on a piece of glass ($n = 1.50$). What should the minimum thickness of this film be to minimize reflection of 500-nm light?

20. A soap bubble ($n = 1.33$) floating in air has the shape of a spherical shell with a wall thickness of 120 nm. (a) What is the wavelength of the visible light that is most strongly reflected? (b) Explain how a bubble of different thickness could also strongly reflect light of this same wavelength. (c) Find the two smallest film thicknesses larger than 120 nm that can produce strongly reflected light of the same wavelength.

21. A film of MgF_2 ($n = 1.38$) having thickness 1.00×10^{-5} cm is used to coat a camera lens. (a) What are the three longest wavelengths that are intensified in the reflected light? (b) Are any of these wavelengths in the visible spectrum?

22. An oil film ($n = 1.45$) floating on water is illuminated by white light at normal incidence. The film is 280 nm thick. Find (a) the wavelength and color of the light in the visible spectrum most strongly reflected and (b) the wavelength and color of the light in the spectrum most strongly transmitted. Explain your reasoning.

23. When a liquid is introduced into the air space between the lens and the plate in a Newton's-rings apparatus, the diameter of the tenth ring changes from 1.50 to 1.31 cm. Find the index of refraction of the liquid.

24. You are working as an expert witness for an attorney who is suing a shipping company. The company operates ships that carry crude oil across the oceans. One ship suffered an oil spill, in which the spilled oil spreads out into a *slick*, forming a thin film that floats on the ocean surface. The legal issue is whether or not the ship spilled more or less than a volume of 10.0 m³ of oil into the ocean. You are reading documents that describe the oil slick on the ocean surface. In one document, you find out that reflection tests were performed on the oil slick. These tests showed that the ocean surface showed a maximum of interference for 500-nm light over a circular area of radius 4.25 km surrounding the location at which the spill occurred. At distances farther from the location, the ocean surface showed no constructive interference, indicating that no oil was present. The type of oil involved has an index of refraction of $n = 1.25$. Determine for the attorney the minimum amount of oil that was spilled.

25. Astronomers observe the chromosphere of the Sun with a filter that passes the red hydrogen spectral line of wavelength 656.3 nm, called the H_α line. The filter consists of a transparent dielectric of thickness d held between two partially aluminized glass plates. The filter is held at a constant temperature. (a) Find the minimum value of d that produces maximum transmission of perpendicular H_α light if the dielectric has an index of refraction of 1.378. (b) **What If?** If the temperature of the filter increases above the normal value, increasing its thickness, what happens to the transmitted wavelength? (c) The dielectric will also pass what near-visible wavelength? One of the glass plates is colored red to absorb this light.

26. A lens made of glass ($n_g = 1.52$) is coated with a thin film of MgF_2 ($n_s = 1.38$) of thickness t. Visible light is incident normally on the coated lens as in Figure P36.26. (a) For what

Figure P36.26

minimum value of t will the reflected light of wavelength 540 nm (in air) be missing? (b) Are there other values of t that will minimize the reflected light at this wavelength? Explain.

SECTION 36.6 The Michelson Interferometer

27. Mirror M_1 in Figure 36.13 is moved through a displacement **T** ΔL. During this displacement, 250 fringe reversals (formation of successive dark or bright bands) are counted. The light being used has a wavelength of 632.8 nm. Calculate the displacement ΔL.

ADDITIONAL PROBLEMS

28. Radio transmitter A operating at 60.0 MHz is 10.0 m from another similar transmitter B that is 180° out of phase with A. How far must an observer move from A toward B along the line connecting the two transmitters to reach the nearest point where the two beams are in phase?

29. In an experiment similar to that of Example 36.1, green light with wavelength 560 nm, sent through a pair of slits 30.0 μm apart, produces bright fringes 2.24 cm apart on a screen 1.20 m away. If the apparatus is now submerged in a tank containing a sugar solution with index of refraction 1.38, calculate the fringe separation for this same arrangement.

30. In the What If? section of Example 36.2, it was claimed that overlapping fringes in a two-slit interference pattern for two different wavelengths obey the following relationship even for large values of the angle θ:

$$\frac{m'}{m} = \frac{\lambda}{\lambda'}$$

(a) Prove this assertion. (b) Using the data in Example 36.2, find the nonzero value of y on the screen at which the fringes from the two wavelengths first coincide.

31. Two coherent waves, coming from sources at different locations, move along the x axis. Their wave functions are

$$E_1 = 860 \sin \left[\frac{2\pi x_1}{650} - 924\pi t + \frac{\pi}{6} \right]$$

and

$$E_2 = 860 \sin \left[\frac{2\pi x_2}{650} - 924\pi t + \frac{\pi}{8} \right]$$

where E_1 and E_2 are in volts per meter, x_1 and x_2 are in nanometers, and t is in picoseconds. When the two waves are superposed, determine the relationship between x_1 and x_2 that produces constructive interference.

32. Raise your hand and hold it flat. Think of the space between your index finger and your middle finger as one slit and think of the space between middle finger and ring finger as a second slit. (a) Consider the interference resulting from sending coherent visible light perpendicularly through this pair of openings. Compute an order-of-magnitude estimate for the angle between adjacent zones of constructive interference. (b) To make the angles in the interference pattern easy to measure with a plastic protractor, you should use an electromagnetic wave with frequency of what order of magnitude? (c) How is this wave classified on the electromagnetic spectrum?

33. In a Young's double-slit experiment using light of wave**S** length λ, a thin piece of Plexiglas having index of refraction n covers one of the slits. If the center point on the screen is a dark spot instead of a bright spot, what is the minimum thickness of the Plexiglas?

34. **Review.** A flat piece of glass is held stationary and horizon**AMT** tal above the highly polished, flat top end of a 10.0-cm-long vertical metal rod that has its lower end rigidly fixed. The thin film of air between the rod and glass is observed to be bright by reflected light when it is illuminated by light of wavelength 500 nm. As the temperature is slowly increased by 25.0°C, the film changes from bright to dark and back to bright 200 times. What is the coefficient of linear expansion of the metal?

35. Figure P36.35 shows a radio-wave transmitter and a receiver separated by a distance $d = 50.0$ m and both a distance $h = 35.0$ m above the ground. The receiver can receive signals both directly from the transmitter and indirectly from signals that reflect from the ground. Assume the ground is level between the transmitter and receiver and a 180° phase shift occurs upon reflection. Determine the longest wavelengths that interfere (a) constructively and (b) destructively.

Figure P36.35 Problems 35 and 36.

36. Figure P36.35 shows a radio-wave transmitter and a receiver **S** separated by a distance d and both a distance h above the ground. The receiver can receive signals both directly from the transmitter and indirectly from signals that reflect from the ground. Assume the ground is level between the transmitter and receiver and a 180° phase shift occurs upon reflection. Determine the longest wavelengths that interfere (a) constructively and (b) destructively.

37. In a Newton's-rings experiment, a plano-convex glass ($n = 1.52$) lens having radius $r = 5.00$ cm is placed on a flat plate as shown in Figure P36.37. When light of wavelength $\lambda = 650$ nm is incident normally, 55 bright rings are

Figure P36.37

observed, with the last one precisely on the edge of the lens. (a) What is the radius R of curvature of the convex surface of the lens? (b) What is the focal length of the lens?

38. Measurements are made of the intensity distribution within the central bright fringe in a Young's interference pattern (see Fig. 36.5). At a particular value of y, it is found that $I/I_{max} = 0.810$ when 600-nm light is used. What wavelength of light should be used to reduce the relative intensity at the same location to 64.0% of the maximum intensity?

39. A plano-concave lens having index of refraction 1.50 is placed on a flat glass plate as shown in Figure P36.39. Its curved surface, with radius of curvature 8.00 m, is on the bottom. The lens is illuminated from above with yellow sodium light of wavelength 589 nm, and a series of concentric bright and dark rings is observed by reflection. The interference pattern has a dark spot at the center that is surrounded by 50 dark rings, the largest of which is at the outer edge of the lens. (a) What is the thickness of the air layer at the center of the interference pattern? (b) Calculate the radius of the outermost dark ring. (c) Find the focal length of the lens.

Figure P36.39

40. *Why is the following situation impossible?* A piece of transparent material having an index of refraction $n = 1.50$ is cut into the shape of a wedge as shown in Figure P36.40. Both the top and bottom surfaces of the wedge are in contact with air. Monochromatic light of wavelength $\lambda = 632.8$ nm is normally incident from above, and the wedge is viewed from above. Let $h = 1.00$ mm represent the height of the wedge and $\ell = 0.500$ m its length. A thin-film interference pattern appears in the wedge due to reflection from the top and bottom surfaces. You have been given the task of counting the number of bright fringes that appear in the entire length ℓ of the wedge. You find this task tedious, and your concentration is broken by a noisy distraction after accurately counting 5 000 bright fringes.

Figure P36.40

41. Interference fringes are produced using Lloyd's mirror and a source S of wavelength $\lambda = 606$ nm as shown in Figure P36.41. Fringes separated by $\Delta y = 1.20$ mm are formed on a screen a distance $L = 2.00$ m from the source. Find the vertical distance h of the source above the reflecting surface.

Figure P36.41

42. A plano-convex lens has index of refraction n. The curved side of the lens has radius of curvature R and rests on a flat glass surface of the same index of refraction, with a film of index n_{film} between them, as shown in Figure P36.42. The lens is illuminated from above by light of wavelength λ. Show that the dark Newton's rings have radii given approximately by

$$r \approx \sqrt{\frac{m\lambda R}{n_{film}}}$$

where $r \ll R$ and m is an integer.

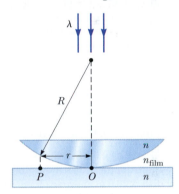

Figure P36.42

43. You are working for an electronics company that designs and manufactures digital drives and players. Your supervisor wishes to evaluate the feasibility of using shorter-wavelength lasers than those used in Blu-ray Discs (405 nm) to try to begin a new video revolution with *Ultraviolet-ray* discs. Figure P36.43 shows the general idea behind digital reading of CDs, DVDs, and Blu-ray Discs. The information is coded digitally in a plastic substrate of index of refraction 1.78 (green in Figure P36.43). Figure P36.43 shows areas called *flats,* which are undisturbed portions of the substrate, and *pits,* which are depressions in the substrate that represent the digital information. The surface of the substrate is

Figure P36.43

covered with a protective reflective coating (gray in Figure P36.43) to protect the surface from scratches. As the disc rotates, the laser beam, shown in violet coming in from below the disc, passes over a track of flats and pits. When there is a change from a flat to a pit or vice versa, part of the laser beam reflects from the flat and part from the pit, as shown in the figure. The depth of the pit is designed so that the reflections of the two halves of the laser beam from the flat and pit undergo destructive interference. Your supervisor wishes to use a laser with a vacuum wavelength of 200 nm. The optimal pit depth is as small as possible, but not less than the manufacturing limitation of 0.1 μm. He asks you to determine the minimum appropriate pit depth for an Ultraviolet-ray disc.

44. The quantity nt in Equations 36.12 and 36.13 is called the **QC** *optical path length* corresponding to the geometrical distance t and is analogous to the quantity δ in Equation 36.1, the path difference. The optical path length is proportional to n because a larger index of refraction shortens the wavelength, so more cycles of a wave fit into a particular geometrical distance. (a) Assume a mixture of corn syrup and water is prepared in a tank, with its index of refraction n increasing uniformly from 1.33 at $y = 20.0$ cm at the top to 1.90 at $y = 0$. Write the index of refraction $n(y)$ as a function of y. (b) Compute the optical path length corresponding to the 20.0-cm height of the tank by calculating

$$\int_0^{20\ cm} n(y)\,dy$$

(c) Suppose a narrow beam of light is directed into the mixture at a nonzero angle with respect to the normal to the surface of the mixture. Qualitatively describe its path.

45. Astronomers observe a 60.0-MHz radio source both directly **T** and by reflection from the sea as shown in Figure P36.45. If the receiving dish is 20.0 m above sea level, what is the angle of the radio source above the horizon at first maximum?

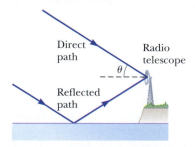

Figure P36.45

CHALLENGE PROBLEMS

46. A plano-convex lens having a radius of curvature of $r = 4.00$ m is placed on a concave glass surface whose radius of curvature is $R = 12.0$ m as shown in Figure P36.46.

Figure P36.46

Assuming 500-nm light is incident normal to the flat surface of the lens, determine the radius of the 100th bright ring.

47. Our discussion of the techniques for determining constructive and destructive interference by reflection from a thin film in air has been confined to rays striking the film at nearly normal incidence. **What If?** Assume a ray is incident at an angle of 30.0° (relative to the normal) on a film with index of refraction 1.38 surrounded by vacuum. Calculate the minimum thickness for constructive interference of sodium light with a wavelength of 590 nm.

48. The condition for constructive interference by reflection **S** from a thin film in air as developed in Section 36.5 assumes nearly normal incidence. **What If?** Suppose the light is incident on the film at a nonzero angle θ_1 (relative to the normal). The index of refraction of the film is n, and the film is surrounded by vacuum. Find the condition for constructive interference that relates the thickness t of the film, the index of refraction n of the film, the wavelength λ of the light, and the angle of incidence θ_1.

49. Both sides of a uniform film that has index of refraction n and thickness d are in contact with air. For normal incidence of light, an intensity minimum is observed in the reflected light at λ_2 and an intensity maximum is observed at λ_1, where $\lambda_1 > \lambda_2$. (a) Assuming no intensity minima are observed between λ_1 and λ_2, find an expression for the integer m in Equations 36.12 and 36.13 in terms of the wavelengths λ_1 and λ_2. (b) Assuming $n = 1.40$, $\lambda_1 = 500$ nm, and $\lambda_2 = 370$ nm, determine the best estimate for the thickness of the film.

50. Slit 1 of a double-slit is wider than slit 2 so that the light from slit 1 has an amplitude exactly 3 times that of the light from slit 2. Show that Equation 36.9 is replaced by the following equation for this situation:

$$I = I_{max}\left[1 + 3\cos^2\left(\frac{\pi d \sin\theta}{\lambda}\right)\right]$$

Begin by assuming that the total magnitude of the electric field at point P on the screen in Figure 36.4 is the superposition of two waves, with electric field magnitudes

$$E_1 = 3E_0 \sin\omega t \qquad E_2 = E_0 \sin(\omega t + \phi)$$

The phase angle ϕ in E_2 is due to the extra path length traveled by the lower beam in Figure 36.4. You will need to evaluate the integral of the square of the sine function over one period. Refer to Figure 32.5 for an easy way to perform this evaluation. You might find the following trigonometric identities helpful:

$$\sin A + \sin B = 2\sin\left(\frac{A+B}{2}\right)\cos\left(\frac{A-B}{2}\right)$$

$$\sin(A+B) = \sin A\cos B + \cos A\sin B$$

$$\cos A = 2\cos^2\left(\frac{A}{2}\right) - 1$$

Diffraction Patterns and Polarization

37

Listen carefully to the sound of the brass instruments next time you see a marching band. Compare how they sound when they are facing you to how they sound when facing away. (*Mark Herreid/Shutterstock*)

STORYLINE **You are in a football stadium, watching the marching** band doing their pre-game performance. As the band marches toward your position, you enjoy the rich, well-blended music. Then the band turns around and marches away from you, and you say, "Wait a minute!" You can still hear the clarinets and saxophones, but the trumpets and trombones have become very quiet. Why does that happen? You are determined to keep your mind on football today, so you set that phenomenon aside. You sit back in your seat and, while waiting for the game to start, you put on your polarized sunglasses. You try to ignore the nagging question in your mind about why sunglasses should be *polarized*. The fact that it must have something to do with physics creeps into your mind, and you pull out a pencil and a piece of paper. The football game begins, but you are drawing diagrams of waves.

CONNECTIONS In Chapter 34, we briefly introduced the notion of *diffraction*, a phenomenon that occurs when waves pass through an aperture or past an edge. In Chapter 36, we began our investigation into physical optics, in which we study the particular phenomena that occur due to the wave nature of light. Diffraction was important to the understanding of interference from double slits in that chapter. In this chapter, we will extend those discussions and expand our understanding of diffraction. We will also study the interesting phenomenon of *polarization*, which is important for light waves, but impossible for sound waves. In our discussion of quantum physics in Chapter 39, we will need the understanding of diffraction, because we find that *particles* will also diffract and interfere when they pass through apertures!

Figure 37.1 The diffraction pattern that appears on a screen when light passes through a narrow vertical slit. The pattern consists of a broad central fringe and a series of less intense and narrower side fringes.

37.1 Introduction to Diffraction Patterns

In Sections 34.2 and 36.1, we discussed that light of wavelength comparable to or larger than the width of a slit in a barrier spreads out in all directions beyond the barrier upon passing through the slit. This phenomenon is called *diffraction*. Other waves, such as sound waves and water waves, also have this property of spreading when passing through apertures or by sharp edges.

Based on this spreading of light as it passes through an opening, you might expect that the light passing through a small opening would simply result in a broad region of light on a screen. We find something more interesting, however. A **diffraction pattern** consisting of light and dark areas is observed, like that shown in Figure 37.1, and somewhat similar to the interference patterns discussed earlier. The pattern consists of a broad, intense central band (called the **central maximum**) flanked by a series of narrower, less intense additional bands (called **side maxima** or **secondary maxima**) and a series of intervening dark bands (or **minima**). Figure 37.2 shows a diffraction pattern associated with light passing by the edge of an object. Again we see bright and dark fringes, which is reminiscent of an interference pattern.

The wave–particle controversy about the nature of light, described at the beginning of Chapter 34, continued, even after Thomas Young's interference experiment in 1801. In 1818, a competition was established by the French Academy of Science to establish the true nature of light. One of the supporters of ray optics, Simeon Poisson, argued that if a new wave theory of light proposed by Augustin Fresnel for the competition were valid, a central bright spot should be observed in the shadow of a circular object illuminated by a point source of light. Light arriving at all points around the edge of the object would diffract inward into the shadow region (as well as outward to points outside the shadow). Because the center is equidistant from all points on the edge, the light from all points would interfere constructively there, causing a bright spot. Poisson considered this possibility to be absurd, because, in the particle theory, the particles of light would be blocked by the object. Furthermore, such a bright spot had never been observed. Dominique-François-Jean Arago, who was the head of the committee for the competition, performed the experiment suggested by Poisson and, much to Poisson's astonishment, observed the bright spot at the center of the shadow!

Figure 37.3 shows a modern version of this experiment using a penny and a laser. The bright spot is visible at the center of the shadow of the penny. In addition, we see several circular fringes extending outward from the edge of the shadow.

Viewing screen

Source

Opaque object

Figure 37.2 Light from a small source passes by the edge of an opaque object and continues on to a screen. A diffraction pattern consisting of bright and dark fringes appears on the screen in the region above the edge of the object.

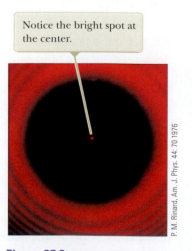

Notice the bright spot at the center.

Figure 37.3 Diffraction pattern created by the illumination of a penny with a laser, with the penny positioned midway between the screen and laser.

37.2 Diffraction Patterns from Narrow Slits

We understand an interference pattern produced by the interference of light from two separate slits. But how do we obtain a similar pattern of light and dark fringes from a *single* slit? Let's consider light passing through a narrow opening modeled as a slit and projected onto a screen that is far away. (This situation can also be achieved experimentally by using a converging lens to focus the parallel rays on a nearby screen.) In this model, the pattern on the screen is called a **Fraunhofer diffraction pattern.**[1]

Figure 37.4a shows light entering a single slit from the left and diffracting as it propagates toward a screen. Figure 37.4b shows the fringe structure of a Fraunhofer diffraction pattern. A bright fringe is observed along the axis at $\theta = 0$, with alternating dark and bright fringes on each side of the central bright fringe.

Until now, we have assumed slits are point sources of light. In this section, we abandon that assumption and see how the finite width of slits is the basis for understanding Fraunhofer diffraction. We can explain some important features of this phenomenon by examining waves coming from various portions of the slit as shown in Figure 37.5. According to Huygens's principle, each portion of the slit acts as a source of light waves. Hence, light from one portion of the slit can interfere with light from another portion, and the resultant light intensity on a viewing screen depends on the direction θ. Based on this analysis, we recognize that a diffraction pattern is actually an interference pattern in which the different sources of light are different portions of the single slit! Therefore, the diffraction patterns we discuss in this chapter are applications of the waves in interference analysis model.

To analyze the diffraction pattern, let's divide the slit into two halves as shown in Figure 37.5. Keeping in mind that all the waves are in phase as they leave the slit, consider rays 1 and 3. As these two rays travel toward a viewing screen far to the right of the figure, ray 1 travels farther than ray 3 by an amount equal to the path difference $(a/2) \sin \theta$, where a is the width of the slit. Similarly, the path

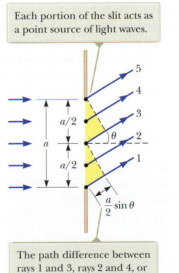

The pattern consists of a wide central bright fringe flanked by much weaker and narrower maxima alternating with dark fringes.

min
max
min
max
min
max
min

Slit

Incoming wave

Viewing screen

L

a   **b**

Each portion of the slit acts as a point source of light waves.

The path difference between rays 1 and 3, rays 2 and 4, or rays 3 and 5 is $(a/2) \sin \theta$.

Figure 37.4 (a) Geometry for analyzing the Fraunhofer diffraction pattern of a single slit. (Drawing not to scale.) (b) Simulation of a single-slit Fraunhofer diffraction pattern.

Figure 37.5 Paths of light rays that encounter a narrow slit of width *a* and diffract toward a screen in the direction described by angle θ (not to scale).

---

[1]If the screen is brought close to the slit (and no lens is used), the pattern is a *Fresnel* diffraction pattern. The Fresnel pattern is more difficult to analyze, so we shall restrict our discussion to Fraunhofer diffraction.

difference between rays 2 and 4 is also $(a/2) \sin \theta$, as is that between rays 3 and 5. If this path difference is exactly half a wavelength (corresponding to a phase difference of 180°), the pairs of waves cancel each other and destructive interference results. This cancellation occurs for any two rays that originate at points separated by half the slit width because the phase difference between two such points is 180°. Therefore, waves from the upper half of the slit interfere destructively with waves from the lower half when

$$\frac{a}{2} \sin \theta = \frac{\lambda}{2}$$

or, if we consider waves at angle θ both above the dashed line in Figure 37.5 and below,

$$\sin \theta = \pm \frac{\lambda}{a}$$

Dividing the slit into four equal parts and using similar reasoning, we find that the viewing screen is also dark when

$$\frac{a}{4} \sin \theta = \pm \frac{\lambda}{2} \quad \rightarrow \quad \sin \theta = \pm 2 \frac{\lambda}{a}$$

Likewise, dividing the slit into six equal parts shows that darkness occurs on the screen when

$$\frac{a}{6} \sin \theta = \pm \frac{\lambda}{2} \quad \rightarrow \quad \sin \theta = \pm 3 \frac{\lambda}{a}$$

Therefore, the general condition for destructive interference is

◀ Condition for destructive interference for a single slit

$$\sin \theta_{\text{dark}} = m \frac{\lambda}{a} \quad m = \pm 1, \pm 2, \pm 3, \dots \qquad (37.1)$$

This equation gives the values of θ_{dark} for which the diffraction pattern has zero light intensity, that is, when a dark fringe is formed. The general features of the intensity distribution are shown in Figure 37.4. A broad, central bright fringe is flanked by much weaker bright fringes alternating with dark fringes. The various dark fringes occur at the values of θ_{dark} that satisfy Equation 37.1. Notice that Equation 37.1 provides the location of the *dark* fringes in the single-slit pattern, unlike Equation 36.2, which provides locations for the *bright* fringes in a two-slit pattern. There is no equation for bright fringes in a single-slit pattern; each bright-fringe peak lies approximately halfway between its bordering dark-fringe minima. The central bright maximum is twice as wide as the secondary maxima. There is no central dark fringe, represented by the absence of $m = 0$ in Equation 37.1.

PITFALL PREVENTION 37.2

Similar Equation Warning! Equation 37.1 has exactly the same form as Equation 36.2, with d, the slit separation, used in Equation 36.2 and a, the slit width, used in Equation 37.1. Equation 37.2, however, describes the *bright* regions in a *two*-slit interference pattern, whereas Equation 37.1 describes the *dark* regions in a *single*-slit diffraction pattern.

QUICK QUIZ 37.1 Suppose the slit width in Figure 37.4 is made half as wide. Does the central bright fringe **(a)** become wider, **(b)** remain the same, or **(c)** become narrower?

Example 37.1   **Where Are the Dark Fringes?**

Light of wavelength 580 nm is incident on a slit having a width of 0.300 mm. The viewing screen is 2.00 m from the slit. Find the width of the central bright fringe.

SOLUTION

Conceptualize   Based on the problem statement, we imagine a single-slit diffraction pattern similar to that in Figure 37.4.

Categorize   We categorize this example as a straightforward application of our discussion of single-slit diffraction patterns, which comes from the *waves in interference* analysis model.

37.1 continued

Analyze Evaluate Equation 37.1 for the two dark fringes that flank the central bright fringe, which correspond to $m = \pm 1$:

$$\sin \theta_{\text{dark}} = \pm \frac{\lambda}{a}$$

Let y represent the vertical position along the viewing screen in Figure 37.4a, measured from the point on the screen directly behind the slit. Then, $\tan \theta_{\text{dark}} = y_1/L$, where the subscript 1 refers to the first dark fringe. Because θ_{dark} is very small, we can use the approximation $\sin \theta_{\text{dark}} \approx \tan \theta_{\text{dark}}$; therefore, $y_1 = L \sin \theta_{\text{dark}}$.

The width of the central bright fringe is twice the absolute value of y_1:

$$2|y_1| = 2|L \sin \theta_{\text{dark}}| = 2\left| \pm L \frac{\lambda}{a} \right| = 2L\frac{\lambda}{a} = 2(2.00 \text{ m})\frac{580 \times 10^{-9} \text{ m}}{0.300 \times 10^{-3} \text{ m}}$$

$$= 7.73 \times 10^{-3} \text{ m} = \boxed{7.73 \text{ mm}}$$

Finalize Notice that this value is much greater than the width of the slit. Let's explore below what happens if we change the slit width.

WHAT IF? What if the slit width is increased by an order of magnitude to 3.00 mm? What happens to the diffraction pattern?

Answer Based on Equation 37.1, we expect that the angles at which the dark bands appear will decrease as a increases. Therefore, the diffraction pattern narrows.

Repeat the calculation with the larger slit width:

$$2|y_1| = 2L\frac{\lambda}{a} = 2(2.00 \text{ m})\frac{580 \times 10^{-9} \text{ m}}{3.00 \times 10^{-3} \text{ m}} = 7.73 \times 10^{-4} \text{ m} = \boxed{0.773 \text{ mm}}$$

Notice that this result is *smaller* than the width of the slit. In general, for large values of a, the various maxima and minima are so closely spaced that only a large, central bright area resembling the geometric image of the slit is observed. This concept is very important in the performance of optical instruments such as telescopes.

Intensity of Single-Slit Diffraction Patterns

Analysis of the intensity variation in a diffraction pattern from a single slit of width a shows that the intensity is given by

$$I = I_{\text{max}} \left[\frac{\sin (\pi a \sin \theta/\lambda)}{\pi a \sin \theta/\lambda} \right]^2 \qquad (37.2)$$

◀ Intensity of a single-slit Fraunhofer diffraction pattern

where I_{max} is the intensity at $\theta = 0$ (the central maximum) and λ is the wavelength of light used to illuminate the slit.

Figure 37.6a represents a plot of the intensity in the single-slit pattern as given by Equation 37.2, and Figure 37.6b is a simulation of a single-slit Fraunhofer

A minimum in the curve in **a** corresponds to a dark fringe in **b**.

Figure 37.6 (a) A plot of light intensity I versus $(\pi/\lambda)a \sin \theta$ for the single-slit Fraunhofer diffraction pattern. (b) Simulation of a single-slit Fraunhofer diffraction pattern.

The diffraction pattern acts as an "envelope" (the blue dashed curve) that controls the intensity of the regularly spaced interference maxima.

Interference fringes

Diffraction minima

Figure 37.7 The combined effects of two-slit and single-slit interference. This pattern is produced when 650-nm light waves pass through two 3.0-μm slits that are 18 μm apart.

diffraction pattern. Notice that most of the light intensity is concentrated in the central bright fringe.

Intensity of Two-Slit Diffraction Patterns

When more than one slit is present, we must consider not only diffraction patterns due to the individual slits but also the interference patterns due to the waves coming from different slits. Notice the curved dashed lines in Figure 36.6 in Chapter 36, which indicate a decrease in intensity of the interference maxima as θ increases. This decrease is due to a diffraction pattern. The interference patterns in that figure are located entirely within the central bright fringe of the diffraction pattern, so the only hint of the diffraction pattern we see is the falloff in intensity toward the outside of the pattern. To determine the effects of both two-slit interference and a single-slit diffraction pattern from each slit from a wider viewpoint than that in Figure 36.6, we combine Equations 36.9 and 37.2:

$$I = I_{max} \cos^2 \left(\frac{\pi d \sin \theta}{\lambda} \right) \left[\frac{\sin (\pi a \sin \theta / \lambda)}{\pi a \sin \theta / \lambda} \right]^2 \tag{37.3}$$

Although this expression looks complicated, it merely represents the single-slit diffraction pattern (the factor in square brackets) acting as an "envelope" for a two-slit interference pattern (the cosine-squared factor) as shown in Figure 37.7. The dashed blue curve in Figure 37.7 represents the factor in square brackets in Equation 37.3. The cosine-squared factor by itself would give a series of peaks all with the same height as the highest peak of the red-brown curve in Figure 37.7. Because of the effect of the square-bracket factor, however, these peaks vary in height as shown.

37.3 Resolution of Single-Slit and Circular Apertures

The ability of optical systems such as telescopes to distinguish between closely spaced objects is limited because of the wave nature of light. To understand this limitation, consider Figure 37.8, which shows two light sources far from a narrow slit of width a. The sources can be two noncoherent point sources S_1 and S_2; for example, they could be two distant stars. If no interference occurred between light passing through different parts of the slit, two distinct bright spots (or images) would be observed on the viewing screen at the points where the straight blue lines in Figure 37.8 strike the screen. Because of interference, however, each source is imaged as a bright central region flanked by weaker bright and dark fringes, a diffraction pattern. What is observed on the screen is the sum of two diffraction patterns: one from S_1 and the other from S_2.

If the two sources are far enough apart to keep their central maxima from overlapping as in Figure 37.8a, their images can be distinguished and are said to be *resolved*. If the sources are close together as in Figure 37.8b, however, the two central maxima overlap and the images are not resolved. To determine whether two images are resolved, the following condition is often used:

> When the central maximum of one image falls on the first minimum of another image, the images are said to be just resolved. This limiting condition of resolution is known as **Rayleigh's criterion.**

From Rayleigh's criterion, we can determine the minimum angular separation θ_{min} subtended by the sources at the slit in Figure 37.8 for which the images are just resolved. Equation 37.1 indicates that the first minimum in a single-slit diffraction pattern occurs at the angle for which

$$\sin \theta = \frac{\lambda}{a} \tag{37.4}$$

where a is the width of the slit. According to Rayleigh's criterion, this expression gives the smallest angular separation for which the two images are resolved. Because $\lambda \ll a$ in most situations, $\sin \theta$ is small and we can use the approximation $\sin \theta \approx \theta$. Therefore, the limiting angle of resolution for a slit of width a is

$$\theta_{min} = \frac{\lambda}{a} \tag{37.5}$$

where θ_{min} is expressed in radians. Hence, the angle subtended by the two sources at the slit must be greater than λ/a if the images are to be resolved.

Many optical systems use circular apertures rather than slits. The diffraction pattern of a circular aperture as shown in the photographs of Figure 37.9 consists of a central circular bright disk surrounded by progressively fainter bright and dark rings. Figure 37.9 shows diffraction patterns for three situations in which light from

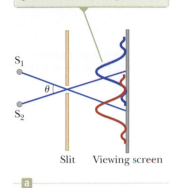

The angle subtended by the sources at the slit is large enough for the diffraction patterns to be distinguishable.

a

The angle subtended by the sources is so small that their diffraction patterns overlap, and the images are not well resolved.

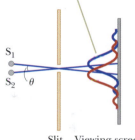

b

Figure 37.8 Two point sources far from a narrow slit each produce a diffraction pattern. (a) The sources are separated by a large angle. (b) The sources are separated by a small angle. (Notice that the angles are greatly exaggerated. The drawing is not to scale.)

The sources are far apart, and the patterns are well resolved.

The sources are closer together such that the angular separation satisfies Rayleigh's criterion, and the patterns are *just* resolved.

The sources are so close together that the patterns are not resolved.

Figure 37.9 Individual diffraction patterns of two point sources (solid curves) and the resultant patterns (dashed curves) for various angular separations of the sources as the light passes through a circular aperture. In each case, the dashed curve is the sum of the two solid curves.

two point sources passes through a circular aperture. When the sources are far apart, their images are well resolved (Fig. 37.9a). When the angular separation of the sources satisfies Rayleigh's criterion, the images are just resolved (Fig. 37.9b). Finally, when the sources are close together, the images are said to be unresolved (Fig. 37.9c) and the pattern looks like that of a single source.

Analysis shows that the limiting angle of resolution of the circular aperture is

▶ **Limiting angle of resolution for a circular aperture**

$$\theta_{min} = 1.22 \frac{\lambda}{D} \qquad (37.6)$$

where D is the diameter of the aperture. This expression is similar to Equation 37.5 except for the factor 1.22, which arises from a mathematical analysis of diffraction from the circular aperture.

Diffraction from circular apertures explains the question about the marching band in the opening storyline. While we have focused on light waves, sound waves also diffract when passing through apertures. In a brass instrument such as a trumpet or a trombone, the sound exits the flared *bell* at the end of the instrument. This is a relatively large opening, especially for high audio frequencies, so there is only a small amount of diffraction; much of the sound is directed forward in a concentrated beam directed in front of the marcher. On the other hand, the sound from woodwinds such as clarinets and saxophones exits the *tone holes* along the side of the instrument; almost nothing comes out the bell. The tone holes are small, so there is significant diffraction. What's more, the column of the instrument is generally held vertically, so that there is diffracted sound in all directions, including backward from the marcher. When the brass players turn around and march away from you, however, very little of their sound is directed behind them.

Ⓠ UICK QUIZ 37.2 Cat's eyes have pupils that can be modeled as vertical slits. At night, would cats be more successful in resolving **(a)** headlights on a distant car or **(b)** vertically separated lights on the mast of a distant boat?

Ⓠ UICK QUIZ 37.3 Suppose you are observing a binary star with a telescope and are having difficulty resolving the two stars. You decide to use a colored filter to maximize the resolution. (A filter of a given color transmits only that color of light.) What color filter should you choose? **(a)** blue **(b)** green **(c)** yellow **(d)** red

---

Example 37.2 **Resolution of the Eye**

Light of wavelength 500 nm, near the center of the visible spectrum, enters a human eye. Although pupil diameter varies from person to person, let's estimate a daytime diameter of 2 mm.

(A) Estimate the limiting angle of resolution for this eye, assuming its resolution is limited only by diffraction.

SOLUTION

Conceptualize Identify the pupil of the eye as the aperture through which the light travels. Light passing through this small aperture causes diffraction patterns to occur on the retina.

Categorize We determine the result using equations developed in this section, so we categorize this example as a substitution problem.

Use Equation 37.6, taking $\lambda = 500$ nm and $D = 2$ mm:

$$\theta_{min} = 1.22 \frac{\lambda}{D} = 1.22 \left(\frac{5.00 \times 10^{-7} \text{ m}}{2 \times 10^{-3} \text{ m}} \right)$$

$$= 3 \times 10^{-4} \text{ rad} \approx 1 \text{ min of arc}$$

where a minute of arc is 1/60 of a degree.

(B) Determine the minimum separation distance d between two point sources that the eye can distinguish if the point sources are a distance $L = 25$ cm from the observer (Fig. 37.10).

Figure 37.10 (Example 38.2) Two point sources separated by a distance d as observed by the eye.

37.2 continued

SOLUTION

Noting that θ_{min} is small, find d:

$$\sin \theta_{min} \approx \theta_{min} \approx \frac{d}{L} \rightarrow d = L\theta_{min}$$

Substitute numerical values:

$$d = (25 \text{ cm})(3 \times 10^{-4} \text{ rad}) = \boxed{8 \times 10^{-3} \text{ cm}}$$

This result is approximately equal to the thickness of a human hair.

WHAT IF?   What if you saw someone at the other end of the field at a football stadium? Could you recognize them?

Answer   A typical feature of someone's face might be of a size 3.0 cm, or even less. The face, if it is across the football field, is at least 120 yards ~ 120 m away. Therefore, the feature subtends an angle of 0.030 m/120 m = 2.5×10^{-4} rad. This is on the order of the limiting resolution of the eye found in part (A), so it is not likely that you would be able to recognize the person's face.

Example **37.3**   **Resolution of a Telescope**

Each of the two telescopes at the Keck Observatory on the dormant Mauna Kea volcano in Hawaii has an effective diameter of 10 m. What is its limiting angle of resolution for 600-nm light?

SOLUTION

Conceptualize   Identify the aperture through which the light travels as the opening of the telescope. Light passing through this aperture causes diffraction patterns to occur in the final image.

Categorize   We determine the result using equations developed in this section, so we categorize this example as a substitution problem.

Use Equation 37.6, taking $\lambda = 6.00 \times 10^{-7}$ m and $D = 10$ m:

$$\theta_{min} = 1.22 \frac{\lambda}{D} = 1.22 \left(\frac{6.00 \times 10^{-7} \text{ m}}{10 \text{ m}} \right)$$

$$= \boxed{7.3 \times 10^{-8} \text{ rad}} \approx \boxed{0.015 \text{ s of arc}}$$

where a second of arc is 1/60 of a minute = $(1/60)^2$ of a degree. Any two stars that subtend an angle greater than or equal to this value are resolved (if atmospheric conditions are ideal).

WHAT IF?   What if we consider radio telescopes? They are much larger in diameter than optical telescopes, but do they have better angular resolutions than optical telescopes? For example, the radio telescope at Arecibo, Puerto Rico, has a diameter of 305 m and is designed to detect radio waves of 0.75-m wavelength. How does its resolution compare with that of one of the Keck telescopes?

Answer   The increase in diameter might suggest that radio telescopes would have better resolution than a Keck telescope, but Equation 37.6 shows that θ_{min} depends on *both* diameter and wavelength. Calculating the minimum angle of resolution for the radio telescope, we find

$$\theta_{min} = 1.22 \frac{\lambda}{D} = 1.22 \left(\frac{0.75 \text{ m}}{305 \text{ m}} \right)$$

$$= 3.0 \times 10^{-3} \text{ rad} \approx 10 \text{ min of arc}$$

This limiting angle of resolution is measured in *minutes* of arc rather than the *seconds* of arc for the optical telescope. Therefore, the change in wavelength more than compensates for the increase in diameter. The limiting angle of resolution for the Arecibo radio telescope is more than 40 000 times larger (that is, *worse*) than the Keck minimum.

An Earth-based telescope such as the one discussed in Example 37.3 can never reach its diffraction limit because the limiting angle of resolution is always set by atmospheric blurring at optical wavelengths. This seeing limit is usually about 1 s of arc and is never smaller than about 0.1 s of arc. The atmospheric blurring is caused by variations in index of refraction with temperature variations in the air.

Figure 37.11 (a) The photograph on which Charon, the moon of Pluto, was discovered in 1978. From an Earth-based telescope, atmospheric blurring results in Charon appearing only as a subtle bump on the edge of Pluto. (b) A Hubble Space Telescope photo of Pluto and Charon, clearly resolving the two objects.

This blurring is one reason for the superiority of photographs from orbiting telescopes, which view celestial objects from a position above the atmosphere.

As an example of the effects of atmospheric blurring, consider telescopic images of Pluto and its moon, Charon. Figure 37.11a, an image taken in 1978, represents the discovery of Charon. In this photograph, taken from an Earth-based telescope, atmospheric turbulence causes the image of Charon to appear only as a bump on the edge of Pluto. In comparison, Figure 37.11b shows a photograph taken from the Hubble Space Telescope. Without the problems of atmospheric turbulence, Pluto and its moon are clearly resolved.

The distortion from atmospheric blurring can be reduced with the process of *adaptive optics*. This technique combines computer analysis with additional optical elements to improve the image. With adaptive optics, the resolution of the Keck telescope is improved from 1 second of arc to 30–60 milliseconds of arc, about a factor of 20 improvement. The image in Figure 35.46 is made possible by adaptive optics.

37.4 The Diffraction Grating

The **diffraction grating,** a useful device for analyzing light sources, consists of a large number of equally spaced parallel slits. A *transmission grating* can be made by cutting parallel grooves on a glass plate with a precision ruling machine. The spaces between the grooves are transparent to the light and hence act as separate slits. A *reflection grating* can be made by cutting parallel grooves on the surface of a reflective material. The reflection of light from the spaces between the grooves is specular, and the reflection from the grooves cut into the material is diffuse. Therefore, the spaces between the grooves act as parallel sources of reflected light like the slits in a transmission grating. Current technology can produce gratings that have very small slit spacings. Gratings are often labeled with the number of grooves per unit length, which is the inverse of the slit spacing d. For example, a typical grating ruled with 5 000 grooves/cm has a slit spacing $d = (1/5\,000)$ cm $= 2.00 \times 10^{-4}$ cm.

A section of a diffraction grating is illustrated in Figure 37.12. A plane wave is incident from the left, normal to the plane of the grating. The pattern observed on the screen far to the right of the grating is the result of the combined effects of interference and diffraction. Each slit produces diffraction, and the diffracted beams interfere with one another to produce the final pattern.

The waves from all slits are in phase as they leave the slits. For an arbitrary direction θ measured from the horizontal, however, the waves must travel different path lengths before reaching the screen. Notice in Figure 37.12 that the path difference δ between rays from any two adjacent slits is equal to $d \sin \theta$. If this path difference equals one wavelength or some integral multiple of a wavelength, waves from

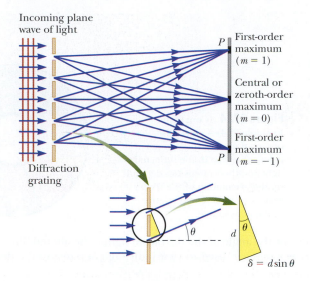

Incoming plane wave of light

P — First-order maximum ($m = 1$)

Central or zeroth-order maximum ($m = 0$)

First-order maximum ($m = -1$)
P

Diffraction grating

θ d θ

$\delta = d\sin\theta$

Figure 37.12 Side view of a diffraction grating. The slit separation is d, and the path difference between adjacent slits is $d\sin\theta$.

all slits are in phase at the screen and a bright fringe is observed. Therefore, the condition for *maxima* in the interference pattern at the angle θ_{bright} is identical to Equation 36.2:

$$d\sin\theta_{\text{bright}} = m\lambda \quad m = 0, \pm 1, \pm 2, \pm 3, \ldots \quad (37.7)$$

◄ Condition for interference maxima for a grating

We can use this expression to calculate the wavelength if we know the grating spacing d and the angle θ_{bright}. If the incident radiation contains several wavelengths, the mth-order maximum for each wavelength occurs at a specific angle. All wavelengths are seen at $\theta = 0$, corresponding to $m = 0$, the zeroth-order maximum. The first-order maximum ($m = 1$) is observed at an angle that satisfies the relationship $\sin\theta_{\text{bright}} = \lambda/d$, the second-order maximum ($m = 2$) is observed at a larger angle θ_{bright}, and so on. For the small values of d typical in a diffraction grating, the angles θ_{bright} are large, as we see in Example 37.5.

The intensity distribution for a diffraction grating obtained with the use of a monochromatic source is shown in Figure 37.13. Notice the sharpness of the principal maxima and the broadness of the dark areas compared with the broad bright fringes characteristic of the two-slit interference pattern (see Fig. 36.5). You should also review Figure 36.6, which shows that the width of the intensity maxima decreases as the number of slits increases. Because the principal maxima are so sharp, they are much brighter than two-slit interference maxima.

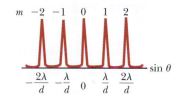

m -2 -1 0 1 2

$\sin\theta$

$-\dfrac{2\lambda}{d}$ $-\dfrac{\lambda}{d}$ 0 $\dfrac{\lambda}{d}$ $\dfrac{2\lambda}{d}$

Figure 37.13 Intensity versus $\sin\theta$ for a diffraction grating. The zeroth-, first-, and second-order maxima are shown.

QUICK QUIZ 37.4 Ultraviolet light of wavelength 350 nm is incident on a diffraction grating with slit spacing d and forms an interference pattern on a screen a distance L away. The angular positions θ_{bright} of the interference maxima are large. The locations of the bright fringes are marked on the screen. Now red light of wavelength 700 nm is used with a diffraction grating to form another diffraction pattern on the screen. Will the bright fringes of this pattern be located at the marks on the screen if **(a)** the screen is moved to a distance $2L$ from the grating, **(b)** the screen is moved to a distance $L/2$ from the grating, **(c)** the grating is replaced with one of slit spacing $2d$, **(d)** the grating is replaced with one of slit spacing $d/2$, or **(e)** nothing is changed?

| Conceptual Example **37.4** | **A DVD Is a Diffraction Grating** |

Light reflected from the surface of a video disc is multicolored as shown in Figure 37.14 (page 994). The colors and their intensities depend on the orientation of the DVD relative to the eye and relative to the light source. Explain how that works.

continued

37.4 continued

S O L U T I O N

The surface of a DVD has a spiral grooved track (with adjacent grooves having a separation on the order of 1 μm). Therefore, the surface acts as a reflection grating. The light reflecting from the regions between these closely spaced grooves interferes constructively only in certain directions that depend on the wavelength and the direction of the incident light. Any section of the DVD serves as a diffraction grating for white light, sending different colors in different directions. The different colors you see upon viewing one section change when the light source, the DVD, or you change position.

Figure 37.14 (Conceptual Example 37.4) A video disc observed under white light. The colors observed in the reflected light and their intensities depend on the orientation of the DVD relative to the eye and relative to the light source.

Carlos E. Santa Maria/Shutterstock

This change in position causes the angle of incidence or the angle of the diffracted light to be altered. Problem 23 describes an experiment that you can perform, in which a laser pointer is used to view individual maxima in the diffraction pattern from a DVD.

Example **37.5**   **The Orders of a Diffraction Grating**

Monochromatic light from a helium–neon laser ($\lambda = 632.8$ nm) is incident normally on a diffraction grating containing 6 000 grooves per centimeter. Find the angles at which the first- and second-order maxima are observed.

S O L U T I O N

Conceptualize   Study Figure 37.12 and imagine that the light coming from the left originates from the helium–neon laser. Let's evaluate the possible values of the angle θ for constructive interference.

Categorize   We determine results using equations developed in this section, so we categorize this example as a substitution problem.

Calculate the slit separation as the inverse of the number of grooves per centimeter:

$$d = \frac{1}{6\,000}\text{ cm} = 1.667 \times 10^{-4}\text{ cm} = 1\,667\text{ nm}$$

Solve Equation 37.7 for θ for an arbitrary value of m and substitute numerical values for the first-order maximum ($m = 1$) to find θ_1:

$$\theta_m = \sin^{-1}\left(\frac{m\lambda}{d}\right) = \sin^{-1}\left(\frac{632.8\text{ nm}}{1\,667\text{ nm}}m\right) = \sin^{-1}(0.379\,6m)$$

$$\theta_1 = \sin^{-1}[(0.379\,6)(1)] = \boxed{22.31°}$$

Repeat for the second-order maximum ($m = 2$):

$$\theta_2 = \sin^{-1}[(0.379\,6)(2)] = \boxed{49.39°}$$

W H A T   I F ?   What if you looked for the third-order maximum? Would you find it?

Answer   For $m = 3$, we find $\theta_3 = \sin^{-1}(1.139)$. Because $\sin\theta$ cannot exceed unity, this result does not represent a realistic solution. Hence, only zeroth-, first-, and second-order maxima can be observed for this situation.

Applications of Diffraction Gratings

A schematic drawing of a simple apparatus used to measure angles in a diffraction pattern is shown in Figure 37.15. This apparatus is a *diffraction grating spectrometer*. The light to be analyzed passes through a slit, and a collimated beam of light is incident on the grating. The diffracted light leaves the grating at angles that satisfy Equation 37.7, and a telescope is used to view the image of the slit. The wavelength can be determined by measuring the precise angles at which the images of the slit appear for the various orders.

The spectrometer is a useful tool in *atomic spectroscopy*, in which the light from an atom is analyzed to find the wavelength components. These wavelength

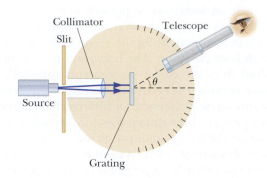

Photo by Ronald R. Erickson; hologram by Nicklaus Phillips

Figure 37.15 Diagram of a diffraction grating spectrometer.

components can be used to identify the atom. We shall investigate atomic spectra in Chapter 41 of the extended version of this text.

Another interesting application of diffraction gratings is **holography,** the production of three-dimensional images of objects. The physics of holography was developed by Dennis Gabor (1900–1979) in 1948 and resulted in the Nobel Prize in Physics for Gabor in 1971. The requirement of coherent light for holography delayed the realization of holographic images from Gabor's work until the development of lasers in the 1960s. Figure 37.16 shows the same hologram viewed from two different positions and the three-dimensional character of its image. Notice in particular the difference in the view through the magnifying glass in Figures 37.16a and 37.16b.

Figure 37.17 shows how a hologram is made. Light from the laser is split into two parts by a half-silvered mirror at B. One part of the beam reflects off the object to be photographed and strikes an ordinary photographic film. The other half of the beam is diverged by lens L_2, reflects from mirrors M_1 and M_2, and finally strikes the film. The two beams overlap to form an extremely complicated interference

Figure 37.16 In this hologram, a circuit board is shown from two different views. Notice the difference in the appearance of the measuring tape and the view through the magnifying lens in (a) and (b).

Figure 37.17 Experimental arrangement for producing a hologram.

pattern on the film. Such an interference pattern can be produced only if the phase relationship of the two waves is constant throughout the exposure of the film. This condition is met by illuminating the scene with light coming through a pinhole or with coherent laser radiation. The hologram records not only the intensity of the light scattered from the object (as in a conventional photograph), but also the phase difference between the reference beam and the beam scattered from the object. Because of this phase difference, an interference pattern is formed that produces an image in which all three-dimensional information available from the perspective of any point on the hologram is preserved.

In a normal photographic image, a lens is used to focus the image so that each point on the object corresponds to a single point on the photograph. Notice that there is no lens used in Figure 37.17 to focus the light onto the film. Therefore, light from each point on the object reaches *all* points on the film. As a result, each region of the photographic film on which the hologram is recorded contains information about all illuminated points on the object, which leads to a remarkable result: if a small section of the hologram is cut from the film, the complete image can be formed from the small piece! (The quality of the image is reduced, but the entire image is present.)

Holograms are finding a number of applications. You may have a hologram on your credit card. This special type of hologram is called a *rainbow hologram* and is designed to be viewed in reflected white light.

37.5 Diffraction of X-Rays by Crystals

In principle, the wavelength of any electromagnetic wave can be determined if a grating of the proper spacing (on the order of λ) is available. X-rays, discovered by Wilhelm Roentgen (1845–1923) in 1895, are electromagnetic waves of very short wavelength (on the order of 0.1 nm). It would be impossible to construct a grating having such a small spacing by the cutting process described at the beginning of Section 37.4. The atomic spacing in a solid is known to be about 0.1 nm, however. In 1913, Max von Laue (1879–1960) suggested that the regular array of atoms in a crystal could act as a three-dimensional diffraction grating for x-rays. Subsequent experiments confirmed this prediction. The diffraction patterns from crystals are complex because of the three-dimensional nature of the crystal structure. Nevertheless, x-ray diffraction has proved to be an invaluable technique for elucidating these structures and for understanding the structure of matter.

Figure 37.18 shows one experimental arrangement for observing x-ray diffraction from a crystal. A collimated beam of monochromatic x-rays is incident on a crystal. The diffracted beams are very intense in certain directions, corresponding to constructive interference from waves reflected from layers of atoms in the crystal. The diffracted beams, which can be detected by a photographic film, form an array of spots known as a *Laue pattern* as in Figure 37.19a. One can deduce the crystalline structure by analyzing the positions and intensities of the various spots

Figure 37.18 Schematic diagram of the technique used to observe the diffraction of x-rays by a crystal. The array of spots formed on the film is called a Laue pattern.

Used with permission of Eastman Kodak Company

SPL/Science Source

a

b

Figure 37.19 (a) A Laue pattern of a single crystal of the mineral beryl (beryllium aluminum silicate). Each dot represents a point of constructive interference. (b) A Laue pattern of the enzyme Rubisco, produced with a wide-band x-ray spectrum. This enzyme is present in plants and takes part in the process of photosynthesis. The Laue pattern is used to determine the crystal structure of Rubisco.

in the pattern. Figure 37.19b shows a Laue pattern from a crystalline enzyme, using a wide range of wavelengths so that a swirling pattern results.

The arrangement of atoms in a crystal of sodium chloride (NaCl) is shown in Figure 37.20. Each unit cell (the geometric solid that repeats throughout the crystal) is a cube having an edge length a. A careful examination of the NaCl structure shows that the ions lie in discrete planes (the shaded areas in Fig. 37.20). Now suppose an incident x-ray beam makes an angle θ with one of the planes as in Figure 37.21. The beam can be reflected from both the upper plane and the lower one, but the beam reflected from the lower plane travels farther than the beam reflected from the upper plane. The effective path difference is $2d \sin \theta$. The two beams reinforce each other (constructive interference) when this path difference equals some integer multiple of λ. The same is true for reflection from the entire family of parallel planes. Hence, the condition for *constructive* interference (maxima in the reflected beam) is

$$2d \sin \theta = m\lambda \quad m = 1, 2, 3, \ldots \quad (37.8)$$

This condition is known as **Bragg's law,** after W. L. Bragg (1890–1971), who first derived the relationship. If the wavelength and diffraction angle are measured, Equation 37.8 can be used to calculate the spacing between atomic planes.

PITFALL PREVENTION 37.4

Different Angles Notice in Figure 37.21 that the angle θ is measured from the reflecting surface rather than from the normal as in the case of the law of reflection in Chapter 34. With slits and diffraction gratings, we also measured the angle θ from the normal to the array of slits. Because of historical tradition, the angle is measured differently in Bragg diffraction, so interpret Equation 37.8 with care.

◀ Bragg's law

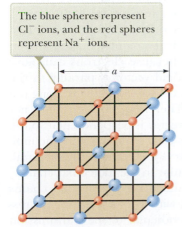

The blue spheres represent Cl⁻ ions, and the red spheres represent Na⁺ ions.

a

Figure 37.20 Crystalline structure of sodium chloride (NaCl). The length of the cube edge is $a = 0.562\ 737$ nm.

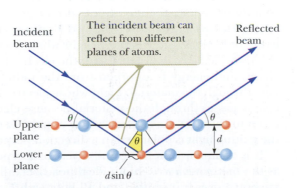

The incident beam can reflect from different planes of atoms.

Incident beam

Reflected beam

Upper plane

Lower plane

θ

θ

θ

d

$d \sin \theta$

Figure 37.21 A two-dimensional description of the reflection of an x-ray beam from two parallel crystalline planes separated by a distance d. The beam reflected from the lower plane travels farther than the beam reflected from the upper plane by a distance $2d \sin \theta$.

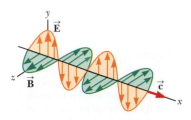

Figure 37.22 Schematic diagram of an electromagnetic wave propagating at velocity $\vec{c}$ in the x direction. The electric field vibrates in the xy plane, and the magnetic field vibrates in the xz plane.

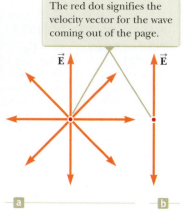

The red dot signifies the velocity vector for the wave coming out of the page.

Figure 37.23 (a) A representation of an unpolarized light beam viewed along the direction of propagation. The transverse electric field can vibrate in any direction in the plane of the page with equal probability. (b) A linearly polarized light beam with the electric field vibrating in the vertical direction.

37.6 Polarization of Light Waves

In Chapter 33, we described the transverse nature of light and all other electromagnetic waves. Polarization, discussed in this section, is firm evidence of this transverse nature.

An ordinary beam of light consists of a large number of individual waves emitted by the atoms of the light source. Each atom produces a wave having some particular orientation of the electric field vector $\vec{E}$, corresponding to the direction of atomic vibration. The *direction of polarization* of each individual wave is defined to be the direction in which the electric field is vibrating. For the individual wave shown in Figure 37.22, this direction happens to lie along the y axis. Now imagine *all* the individual waves that leave the light source in a beam of light directed along the x axis. Each individual electromagnetic wave will have an $\vec{E}$ vector parallel to the yz plane, but this vector could be at *any* possible angle with respect to the y axis. Because all directions of vibration from a wave source are possible, the resultant electromagnetic wave is a superposition of waves vibrating in many different directions. The result is an **unpolarized** light beam, represented in Figure 37.23a. The direction of wave propagation in this figure is perpendicular to the page. The arrows show a few possible directions of the electric field vectors for the individual waves making up the resultant beam. At any given point and at some instant of time, all these individual electric field vectors add to give one resultant electric field vector.

A beam of light is said to be **linearly polarized** if the resultant electric field $\vec{E}$ vibrates in the same direction *at all times* at a particular point as shown in Figure 37.23b. (Sometimes, such a wave is described as *plane-polarized*, or simply *polarized*.) The plane formed by $\vec{E}$ and the direction of propagation is called the *plane of polarization* of the wave. If the wave in Figure 37.22 represents the resultant of all individual waves, the plane of polarization is the xy plane.

A linearly polarized beam can be obtained from an unpolarized beam by removing all waves from the beam except those whose electric field vectors oscillate in a single plane. We now discuss four processes for producing polarized light from unpolarized light.

Polarization by Selective Absorption

The most common technique for producing polarized light is to use a material that transmits waves whose electric fields vibrate in a plane parallel to a certain direction and that absorbs waves whose electric fields vibrate in all other directions.

In 1938, E. H. Land (1909–1991) discovered a material, which he called *Polaroid*, that polarizes light through selective absorption. This material is fabricated in thin sheets of long-chain hydrocarbons. The sheets are stretched during manufacture so that the long-chain molecules align. After a sheet is dipped into a solution containing iodine, the molecules become good electrical conductors. Conduction takes place primarily along the hydrocarbon chains because electrons can move easily only along the chains. If light whose electric field vector is parallel to the chains is incident on the material, the electric field accelerates electrons along the chains and energy is absorbed from the radiation. Therefore, the light does not pass through the material. Light whose electric field vector is perpendicular to the chains passes through the material because electrons cannot move from one molecule to the next. As a result, when unpolarized light is incident on the material, the exiting light is polarized in a direction perpendicular to the molecular chains.

It is common to refer to the direction perpendicular to the molecular chains as the *transmission axis*. In an ideal polarizer, all light with $\vec{E}$ parallel to the transmission axis is transmitted and all light with $\vec{E}$ perpendicular to the transmission axis is absorbed.

Figure 37.24 represents an unpolarized light beam incident on a first polarizing sheet, called the *polarizer*. Because the transmission axis is oriented vertically in the figure, the light transmitted through this sheet is polarized vertically. A second

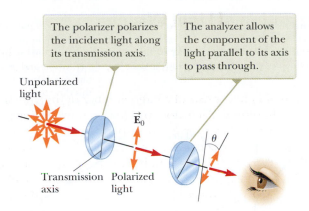

The polarizer polarizes the incident light along its transmission axis.

The analyzer allows the component of the light parallel to its axis to pass through.

Unpolarized light

$\vec{E}_0$

θ

Transmission axis   Polarized light

Figure 37.24 Two polarizing sheets whose transmission axes make an angle θ with each other. Only a fraction of the polarized light incident on the analyzer is transmitted through it.

polarizing sheet, called the *analyzer,* intercepts the beam. In Figure 37.24, the analyzer transmission axis is set at an angle θ to the polarizer axis. We call the electric field vector of the first transmitted beam $\vec{E}_0$. The component of $\vec{E}_0$ perpendicular to the analyzer axis is completely absorbed. The component of $\vec{E}_0$ parallel to the analyzer axis, which is transmitted through the analyzer, is $E_0 \cos \theta$. Because the intensity of the transmitted beam varies as the square of its magnitude, we conclude that the intensity I of the (polarized) beam transmitted through the analyzer varies as

$$I = I_{max} \cos^2 \theta \qquad (37.9)$$ ◀ Malus's law

where I_{max} is the intensity of the polarized beam incident on the analyzer. This expression, known as **Malus's law,**[2] applies to any two polarizing materials whose transmission axes are at an angle θ to each other. This expression shows that the intensity of the transmitted beam is maximum when the transmission axes are parallel ($\theta = 0$ or $180°$) and is zero (complete absorption by the analyzer) when the transmission axes are perpendicular to each other. This variation in transmitted intensity through a pair of polarizing sheets is illustrated in Figure 37.25. Because the average value of $\cos^2 \theta$ is $\frac{1}{2}$, the intensity of initially unpolarized light is reduced by a factor of one-half as the light passes through a single ideal polarizer.

Polarization by Reflection

When an unpolarized light beam is reflected from a surface, the polarization of the reflected light depends on the angle of incidence. If the angle of incidence is $0°$, the reflected beam is unpolarized. For other angles of incidence, the reflected light is polarized to some extent, and for one particular angle of incidence, the

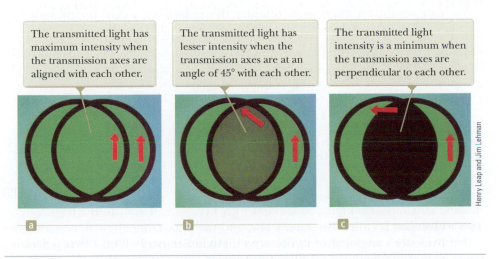

The transmitted light has maximum intensity when the transmission axes are aligned with each other.

The transmitted light has lesser intensity when the transmission axes are at an angle of 45° with each other.

The transmitted light intensity is a minimum when the transmission axes are perpendicular to each other.

a   b   c

Henry Leap and Jim Lehman

Figure 37.25 The intensity of light transmitted through two polarizers depends on the relative orientation of their transmission axes. The red arrows indicate the transmission axes of the polarizers.

[2]Named after its discoverer, E. L. Malus (1775–1812), a French mathematician and physicist. Malus discovered that reflected light was polarized by viewing it through a calcite ($CaCO_3$) crystal.

Figure 37.26 (a) When unpolarized light is incident on a reflecting surface, the reflected and refracted beams are partially polarized. (b) The reflected beam is completely polarized when the angle of incidence equals the polarizing angle θ_p, which satisfies Equation 37.10. At this incident angle, the reflected and refracted rays are perpendicular to each other.

reflected light is completely polarized. Let's now investigate reflection at that special angle.

Suppose an unpolarized light beam is incident on a surface as in Figure 37.26a. The electric field vector for each individual wave can be resolved into two components: one parallel to the surface (and perpendicular to the page in Fig. 37.26, represented by the dots) and the other (represented by the orange arrows) perpendicular both to the first component and to the direction of propagation. Therefore, the polarization of the entire beam can be described by two electric field components in these directions. It is found that the parallel component represented by the dots reflects more strongly than the other component represented by the arrows, resulting in a partially polarized reflected beam. Furthermore, the refracted beam is also partially polarized.

Now suppose the angle of incidence θ_1 is varied until the angle between the reflected and refracted beams is 90° as in Figure 37.26b. At this particular angle of incidence, the reflected beam is completely polarized (with its electric field vector parallel to the surface) and the refracted beam is still only partially polarized. The angle of incidence at which this polarization occurs is called the **polarizing angle θ_p**.

We can obtain an expression relating the polarizing angle to the index of refraction of the reflecting substance by using Figure 37.26b. From this figure, we see that $\theta_p + 90° + \theta_2 = 180°$; therefore, $\theta_2 = 90° - \theta_p$. Using Snell's law of refraction (Eq. 34.7) gives

$$\frac{n_2}{n_1} = \frac{\sin \theta_1}{\sin \theta_2} = \frac{\sin \theta_p}{\sin \theta_2}$$

Because $\sin \theta_2 = \sin (90° - \theta_p) = \cos \theta_p$, we can write this expression as $n_2/n_1 = \sin \theta_p/\cos \theta_p$, which means that

Brewster's law ▶

$$\tan \theta_p = \frac{n_2}{n_1} \tag{37.10}$$

This expression is called **Brewster's law,** and the polarizing angle θ_p is sometimes called **Brewster's angle,** after its discoverer, David Brewster (1781–1868), a Scottish physicist and mathematician. Because n varies with wavelength for a given substance, Brewster's angle is also a function of wavelength.

We can understand polarization by reflection by imagining that the electric field in the incident light sets electrons at the surface of the material in Figure 37.26b into oscillation. The component directions of oscillation are (1) parallel to the arrows shown on the refracted beam of light and therefore parallel to the reflected beam and (2) perpendicular to the page. The oscillating electrons act as dipole antennas radiating light with a polarization parallel to the direction of oscillation. Consult Figure 33.12, which shows the pattern of radiation from a dipole antenna. Notice that there is no radiation at an angle of $\theta = 0$, that is, along the oscillation direction of the antenna. Therefore, for the oscillations in direction 1, there is no radiation in the direction along the reflected ray. For oscillations in direction 2, the electrons radiate light with a polarization perpendicular to the page. Therefore, the light reflected from the surface at this angle is completely polarized parallel to the surface.

Polarization by reflection is a common phenomenon. Sunlight reflected from water, glass, and snow is partially polarized. If the surface is horizontal, the electric field vector of the reflected light has a strong horizontal component. Sunglasses made of polarizing material reduce the glare of reflected light. This is the answer to your nagging question in the storyline. The transmission axes of such lenses are oriented vertically so that they absorb the strong horizontal component of the reflected light. If you rotate sunglasses through 90°, they are not as effective at blocking the glare from shiny horizontal surfaces.

Polarization by Double Refraction

Solids can be classified on the basis of internal structure. Those in which the atoms are arranged in a specific order are called *crystalline;* the NaCl structure of Figure 37.20 is one example of a crystalline solid. Those solids in which the atoms are distributed randomly are called *amorphous.* When light travels through an amorphous material such as glass, it travels with a speed that is the same in all directions. That is, glass has a single index of refraction. In certain crystalline materials such as calcite and quartz, however, the speed of light is not the same in all directions. In these materials, the speed of light depends on the direction of propagation relative to the planes of the crystal structure *and* on the plane of polarization of the light. Such materials are characterized by two indices of refraction. Hence, they are often referred to as **double-refracting** or **birefringent** materials.

When unpolarized light enters a birefringent material, it may split into an **ordinary (O) ray** and an **extraordinary (E) ray.** These two rays have mutually perpendicular polarizations and travel at different speeds through the material. The two speeds correspond to two indices of refraction, n_O for the ordinary ray and n_E for the extraordinary ray.

There is one direction, however, called the **optic axis,** along which the ordinary and extraordinary rays have the same speed. If light enters a birefringent material from an arbitrary direction, the different indices of refraction will cause the two polarized rays to split and travel in different directions as shown in Figure 37.27.

The index of refraction n_O for the ordinary ray is the same in all directions. If one could place a point source of light inside the crystal as in Figure 37.28, the ordinary waves would spread out from the source as spheres. The index of refraction n_E varies with the direction of propagation. A point source sends out an extraordinary wave having wave fronts that are elliptical in cross section. The difference in speed for the two rays is a maximum in the direction perpendicular to the optic axis. For example, in calcite, $n_O = 1.658$ at a wavelength of 589.3 nm and n_E varies from 1.658 along the optic axis to 1.486 perpendicular to the optic axis. Values for n_O and the extreme value of n_E for various double-refracting crystals are given in Table 37.1 (page 1002).

If you place a calcite crystal on a sheet of paper and then look through the crystal at any writing on the paper, you would see two images as shown in Figure 37.29. As can be seen from Figure 37.27, these two images correspond to one formed by the ordinary ray and one formed by the extraordinary ray. If the two images are

These two rays are polarized in mutually perpendicular directions.

Figure 37.27 Unpolarized light incident at an arbitrary direction on a calcite crystal splits into an ordinary (O) ray and an extraordinary (E) ray (not to scale).

The E and O rays propagate with the same velocity along the optic axis.

Figure 37.28 A point source S inside a double-refracting crystal produces a spherical wave front corresponding to the ordinary (O) ray and an elliptical wave front corresponding to the extraordinary (E) ray.

Henry Leap and Jim Lehman

Figure 37.29 A calcite crystal produces a double image because it is a birefringent (double-refracting) material.

TABLE 37.1 Indices of Refraction for Some Double-Refracting Crystals at a Wavelength of 589.3 nm

Crystal	n_O	n_E	n_O/n_E
Calcite (CaCO$_3$)	1.658	1.486	1.116
Quartz (SiO$_2$)	1.544	1.553	0.994
Sodium nitrate (NaNO$_3$)	1.587	1.336	1.188
Sodium sulfite (NaSO$_3$)	1.565	1.515	1.033
Zinc chloride (ZnCl$_2$)	1.687	1.713	0.985
Zinc sulfide (ZnS)	2.356	2.378	0.991

viewed through a sheet of rotating polarizing glass, they alternately appear and disappear because the ordinary and extraordinary rays are plane-polarized along mutually perpendicular directions.

Polarization by Scattering

When light is incident on any material, the electrons in the material can absorb and reradiate part of the light. Such absorption and reradiation of light by electrons in the gas molecules that make up air is what causes sunlight reaching an observer on the Earth to be partially polarized. You can observe this effect—called **scattering**—by looking directly up at the sky through a pair of sunglasses whose lenses are made of polarizing material. Less light passes through at certain orientations of the lenses than at others.

Figure 37.30 illustrates how sunlight becomes polarized when it is scattered. The phenomenon is similar to that creating completely polarized light upon reflection from a surface at Brewster's angle. An unpolarized beam of sunlight traveling in the horizontal direction (parallel to the ground) strikes a molecule of one of the gases that make up air, setting the electrons of the molecule into vibration. These vibrating charges act like the vibrating charges in an antenna. The horizontal component of the electric field vector in the incident wave results in a horizontal component of the vibration of the charges, and the vertical component of the vector results in a vertical component of vibration. If the observer in Figure 37.30 is looking in a direction perpendicular to the original direction of propagation of the light, the vertical oscillations of the charges send no radiation toward the observer (See Fig. 33.12.). Therefore, the observer sees light that is completely polarized in the horizontal direction as indicated by the orange arrows. If the observer looks in other directions, the light is partially polarized in the horizontal direction.

Variations in the color of scattered light in the atmosphere can be understood as follows. When light of various wavelengths λ is incident on gas molecules of diameter d, where $d \ll \lambda$, the relative intensity of the scattered light varies as $1/\lambda^4$. The condition $d \ll \lambda$ is satisfied for scattering from oxygen (O$_2$) and nitrogen (N$_2$) molecules in the atmosphere, whose diameters are about 0.2 nm. Hence, short wavelengths (violet light) are scattered more efficiently than long wavelengths (red light). Therefore, when sunlight is scattered by gas molecules in the air, the short-wavelength radiation (violet) is scattered more intensely than the long-wavelength radiation (red).

When you look up into the sky in a direction that is not toward the Sun, you see the scattered light, which is predominantly violet. Your eyes, however, are not very sensitive to violet light. Light of the next color in the spectrum, blue, is scattered with less intensity than violet, but your eyes are far more sensitive to blue light than to violet light. Hence, you see a blue sky. If you look toward the west at sunset (or toward the east at sunrise), you are looking in a direction toward the Sun and are seeing light that has passed through a large distance of air. Most of the blue light has been scattered by the air between you and the Sun. The light that survives this trip through the air to you has had much of its blue component scattered and is therefore heavily weighted toward the red end of the spectrum; as a result, you see the red and orange colors of sunset (or sunrise).

The scattered light traveling perpendicular to the incident light is plane-polarized because the vertical vibrations of the charges in the air molecule send no light in this direction.

Unpolarized light

Air molecule

Figure 37.30 The scattering of unpolarized sunlight by air molecules.

QUICK QUIZ 37.5 A polarizer for microwaves can be made as a grid of parallel metal wires approximately 1 cm apart. Is the electric field vector for microwaves transmitted through this polarizer **(a)** parallel or **(b)** perpendicular to the metal wires?

QUICK QUIZ 37.6 You are walking down a long hallway that has many light fixtures in the ceiling and a very shiny, newly waxed floor. When looking at the floor, you see reflections of every light fixture. Now you put on sunglasses that are polarized. Some of the reflections of the light fixtures can no longer be seen. (Try it!) Are the reflections that disappear those **(a)** nearest to you, **(b)** farthest from you, or **(c)** at an intermediate distance from you?

Summary

> ## Concepts and Principles

Diffraction is the deviation of light from a straight-line path when the light passes through an aperture or around an obstacle. Diffraction is due to the wave nature of light.

The **Fraunhofer diffraction pattern** produced by a single slit of width a on a distant screen consists of a central bright fringe and alternating bright and dark fringes of much lower intensities. The angles θ_{dark} at which the diffraction pattern has zero intensity, corresponding to destructive interference, are given by

$$\sin \theta_{dark} = m\frac{\lambda}{a} \quad m = \pm 1, \pm 2, \pm 3, \ldots \quad (37.1)$$

Rayleigh's criterion, which is a limiting condition of resolution, states that two images formed by an aperture are just distinguishable if the central maximum of the diffraction pattern for one image falls on the first minimum of the diffraction pattern for the other image. The limiting angle of resolution for a slit of width a is $\theta_{min} = \lambda/a$, and the limiting angle of resolution for a circular aperture of diameter D is given by $\theta_{min} = 1.22\lambda/D$.

A **diffraction grating** consists of a large number of equally spaced, identical slits. The condition for intensity maxima in the interference pattern of a diffraction grating for normal incidence is

$$d \sin \theta_{bright} = m\lambda \quad m = 0, \pm 1, \pm 2, \pm 3, \ldots \quad (37.7)$$

where d is the spacing between adjacent slits and m is the order number of the intensity maximum.

When polarized light of intensity I_{max} is emitted by a polarizer and then is incident on an analyzer, the light transmitted through the analyzer has an intensity equal to $I_{max} \cos^2 \theta$, where θ is the angle between the polarizer and analyzer transmission axes.

In general, reflected light is partially polarized. Reflected light, however, is completely polarized when the angle of incidence is such that the angle between the reflected and refracted beams is 90°. This angle of incidence, called the **polarizing angle** θ_p, satisfies **Brewster's law:**

$$\tan \theta_p = \frac{n_2}{n_1} \quad (37.10)$$

where n_1 is the index of refraction of the medium in which the light initially travels and n_2 is the index of refraction of the reflecting medium.

Think–Pair–Share

See the Preface for an explanation of the icons used in this problems set. For additional assessment items for this section, go to ⟡ **WEBASSIGN** From Cengage

1. Your team is working in an optical research laboratory. Part of the apparatus you are working on involves polarized beams of light. Your supervisor wants to rotate the plane of polarization of one polarized beam by 45.0°. She can do that by inserting a single polarizer at 45.0° to the plane of polarization of the original beam, but, according to Malus's law, she then loses 50% of the intensity of the light. She gives you the task of designing a *stack* of polarizers, each with its axis at the same angle with the axis of the previous polarizer,

that will rotate the plane of polarization of the beam without losing more than 10.0% of the intensity of the original beam. She wants to know by quitting time tonight (a) how many polarizers she needs in the stack and (b) the angle between adjacent polarizers.

2. **ACTIVITY** A beam of unpolarized light is directed through a stack of three polarizers. The first polarizer is at an angle θ_1 with respect to a reference direction that is defined as 0°. The third polarizer is at an angle θ_3 with respect to the reference direction. The angle θ_2 of the second polarizer starts at the reference direction 0° and is rotated through 180°. The

graph in Figure TP37.2 shows the behavior of the intensity of the light passing through the stack of polarizers as a function of the direction of the second polarizer. (a) Discuss this situation in your group and find the angles that the first and third polarizers make with respect to the reference direction. (b) Can you determine which of the angles in part (a) corresponds to the first polarizer and which to the third?

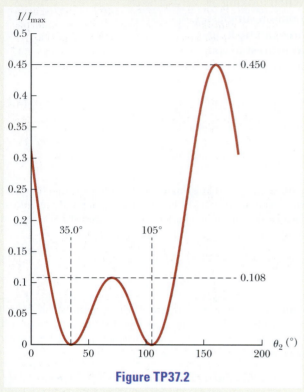

Figure TP37.2

3. [ACTIVITY] Your group is performing an experiment in which light of wavelength 632.8 nm illuminates a single slit, and a diffraction pattern is formed on a screen 1.00 m from the slit. You record relative intensity as a function of distance from the central maximum and generate the data in the following table. (a) Plot the data in the table. (b) From your plot, determine the width a of the single slit.

Position Relative to Central Maximum (mm)	Relative Intensity
0	1.00
0.8	0.95
1.6	0.80
3.2	0.39
4.8	0.079
6.5	0.003
8.1	0.036
9.7	0.043
11.3	0.013
12.9	0.000 3
14.5	0.012
16.1	0.015
17.7	0.004 4
19.3	0.000 3

Problems

See the Preface for an explanation of the icons used in this problems set. For additional assessment items for this section, go to ❖ **WEBASSIGN** From Cengage

SECTION 37.2 Diffraction Patterns from Narrow Slits

1. Helium–neon laser light ($\lambda = 632.8$ nm) is sent through a 0.300-mm-wide single slit. What is the width of the central maximum on a screen 1.00 m from the slit?

2. From Equation 37.2, find an expression for the sine of the angles at which the minimum intensity occurs in a single-slit diffraction pattern. Compare the result to Equation 37.1.

3. Light of wavelength 540 nm passes through a slit of width 0.200 mm. (a) The width of the central maximum on a screen is 8.10 mm. How far is the screen from the slit? (b) Determine the width of the first bright fringe to the side of the central maximum.

4. In Figure 37.7, show mathematically how many interference maxima are enclosed by the central diffraction maximum in the pattern. Notice that the diagram is generated by using 650-nm light to illuminate two 3.0-μm slits separated by 18 μm.

5. Assume light of wavelength 650 nm passes through two slits 3.00 μm wide, with their centers 9.00 μm apart.

Make a sketch of the combined diffraction and interference pattern in the form of a graph of intensity versus $\phi = (\pi a \sin \theta)/\lambda$. You may use Figure 37.7 as a starting point.

6. **What If?** Suppose light strikes a single slit of width a at an angle β from the perpendicular direction as shown in Figure P37.6. Show that Equation 37.1, the condition for destructive interference, must be modified to read

$$\sin \theta_{dark} = m\frac{\lambda}{a} - \sin \beta$$

$$m = \pm 1, \pm 2, \pm 3, \dots$$

Figure P37.6

7. A diffraction pattern is formed on a screen 120 cm away from a 0.400-mm-wide slit. Monochromatic 546.1-nm light is used. Calculate the fractional intensity I/I_{max} at a point on the screen 4.10 mm from the center of the principal maximum.

8. Coherent light of wavelength 501.5 nm is sent through two
GP parallel slits in an opaque material. Each slit is 0.700 μm
wide. Their centers are 2.80 μm apart. The light then falls on
a semicylindrical screen, with its axis at the midline between
the slits. We would like to describe the appearance of the
pattern of light visible on the screen. (a) Find the direction
for each two-slit interference maximum on the screen as
an angle away from the bisector of the line joining the slits.
(b) How many angles are there that represent two-slit interference maxima? (c) Find the direction for each single-slit
interference minimum on the screen as an angle away from
the bisector of the line joining the slits. (d) How many angles
are there that represent single-slit interference minima?
(e) How many of the angles in part (d) are identical to those
in part (a)? (f) How many bright fringes are visible on the
screen? (g) If the intensity of the central fringe is $I_{\max}$, what is
the intensity of the last fringe visible on the screen?

SECTION 37.3 Resolution of Single-Slit and Circular Apertures

Note: In text problems 11, 15, and 16, and online-only problem
37.9, you may use the Rayleigh criterion for the limiting angle
of resolution of an eye. The standard may be overly optimistic
for human vision.

9. The objective lens of a certain refracting telescope has a diameter of 58.0 cm. The telescope is mounted in a satellite that
orbits the Earth at an altitude of 270 km to view objects on the
Earth's surface. Assuming an average wavelength of 500 nm,
find the minimum distance between two objects on the
ground if their images are to be resolved by this lens.

10. Yellow light of wavelength 589 nm is used to view an object
Q|C under a microscope. The objective lens diameter is 9.00 mm.
(a) What is the limiting angle of resolution? (b) Suppose it
is possible to use visible light of any wavelength. What color
should you choose to give the smallest possible angle of resolution, and what is this angle? (c) Suppose water fills the
space between the object and the objective. What effect
does this change have on the resolving power when 589-nm
light is used?

11. What is the approximate size of the smallest object on the
Earth that astronauts can resolve by eye when they are orbiting 250 km above the Earth? Assume $\lambda = 500$ nm and a
pupil diameter of 5.00 mm.

12. A helium–neon laser emits light that has a wavelength of
T 632.8 nm. The circular aperture through which the beam
emerges has a diameter of 0.500 cm. Estimate the diameter
of the beam 10.0 km from the laser.

13. To increase the resolving power of a microscope, the object
T and the objective are immersed in oil ($n = 1.5$). If the limiting angle of resolution without the oil is 0.60 μrad, what is
the limiting angle of resolution with the oil? *Hint:* The oil
changes the wavelength of the light.

14. You are working for a new assistant professor in astronomy
CR who is interested in exoplanets. One day, a scientific rumor
begins circulating that there is a Jupiter-sized planet around
Alpha Centauri, 4.28 light-years away. Your professor has
access to viewing privileges for the Hubble Space Telescope (aperture diameter 2.4 m, 100 nm to 2 400 nm), the
Hale Telescope on Palomar Mountain in California (aperture diameter 5.08 m, visible light), the Keck Telescope on
Mauna Lea, Hawaii (aperture diameter 10.0 m, visible light),
and the Arecibo Radio Telescope in Puerto Rico (aperture
diameter 305 m, 75-cm radio waves). He asks you to advise
him as soon as possible as to which telescope he should
request time on in order to resolve an image of the planet.

15. Impressionist painter Georges Seurat created paintings
T with an enormous number of dots of pure pigment, each
of which was approximately 2.00 mm in diameter. The idea
was to have colors such as red and green next to each other
to form a scintillating canvas, such as in his masterpiece, *A
Sunday Afternoon on the Island of La Grande Jatte* (Fig. P37.15).
Assume $\lambda = 500$ nm and a pupil diameter of 5.00 mm.
Beyond what distance would a viewer be unable to discern
individual dots on the canvas?

SuperStock / SuperStock

Figure P37.15

16. Narrow, parallel, glowing gas-filled tubes in a variety of col-
Q|C ors form block letters to spell out the name of a nightclub.
Adjacent tubes are all 2.80 cm apart. The tubes forming
one letter are filled with neon and radiate predominantly
red light with a wavelength of 640 nm. For another letter,
the tubes emit predominantly blue light at 440 nm. The
pupil of a dark-adapted viewer's eye is 5.20 mm in diameter.
(a) Which color is easier to resolve? State how you decide.
(b) If she is in a certain range of distances away, the viewer
can resolve the separate tubes of one color but not the
other. The viewer's distance must be in what range for her
to resolve the tubes of only one of these two colors?

SECTION 37.4 The Diffraction Grating

Note: In the following problems, assume the light is incident
normally on the gratings.

17. Consider an array of parallel wires with uniform spacing of
1.30 cm between centers. In air at 20.0°C, ultrasound with a
frequency of 37.2 kHz from a distant source is incident perpendicular to the array. (a) Find the number of directions
on the other side of the array in which there is a maximum
of intensity. (b) Find the angle for each of these directions
relative to the direction of the incident beam.

18. Three discrete spectral lines occur at angles of 10.1°, 13.7°,
V and 14.8° in the first-order spectrum of a grating spectrometer. (a) If the grating has 3 660 slits/cm, what are the wavelengths of the light? (b) At what angles are these lines found
in the second-order spectrum?

19. A grating with 250 grooves/mm is used with an incandescent light source. Assume the visible spectrum to range in

wavelength from 400 nm to 700 nm. In how many orders can one see (a) the entire visible spectrum and (b) the short-wavelength region of the visible spectrum?

20. Show that whenever white light is passed through a diffraction grating of any spacing size, the violet end of the spectrum in the third order on a screen always overlaps the red end of the spectrum in the second order.

21. Light from an argon laser strikes a diffraction grating that **V** has 5 310 grooves per centimeter. The central and first-order principal maxima are separated by 0.488 m on a wall 1.72 m from the grating. Determine the wavelength of the laser light.

22. A wide beam of laser light with a wavelength of 632.8 nm is **QC** directed through several narrow parallel slits, separated by 1.20 mm, and falls on a sheet of photographic film 1.40 m away. The exposure time is chosen so that the film stays unexposed everywhere except at the central region of each bright fringe. (a) Find the distance between these interference maxima. The film is printed as a transparency; it is opaque everywhere except at the exposed lines. Next, the same beam of laser light is directed through the transparency and allowed to fall on a screen 1.40 m beyond. (b) Argue that several narrow, parallel, bright regions, separated by 1.20 mm, appear on the screen as real images of the original slits. (A similar train of thought, at a soccer game, led Dennis Gabor to invent holography.)

23. You are working as a demonstration assistant for a physics **CR** professor. For an upcoming lecture on diffraction gratings, he wishes to perform a demonstration where he shines a laser pointer at normal incidence onto the recorded surface of a DVD that is laying flat on the demonstration table. (a) He asks you to determine how many additional maxima beyond the normal reflection (which will be blocked by his hand holding the laser pointer) will be projected onto the ceiling or walls of the room if he uses a laser pointer with a wavelength of 632.8 nm. (b) He also asks you if he can show more maxima by using a laser pointer of another visible color. The tracks of pits on a DVD are separated by 0.800 μm.

SECTION 37.5 Diffraction of X-Rays by Crystals

24. Monochromatic x-rays ($\lambda = 0.166$ nm) from a nickel target are incident on a potassium chloride (KCl) crystal surface. The spacing between planes of atoms in KCl is 0.314 nm. At what angle (relative to the surface) should the beam be directed for a second-order maximum to be observed?

25. The first-order diffraction maximum is observed at 12.6° for a crystal having a spacing between planes of atoms of 0.250 nm. (a) What wavelength x-ray is used to observe this first-order pattern? (b) How many orders can be observed for this crystal at this wavelength?

26. You are performing research in an x-ray diffraction laboratory. **CR** In one of your experiments, you wish to study x-ray diffraction from a crystal of NaCl using x-rays of wavelength 0.136 nm. (a) For how many angles do you expect to detect a diffraction maximum from the crystal if your x-rays are reflecting from the shaded planes in Figure 37.20? (b) In another experiment, the crystal is rotated so that the reflections of x-rays arise from parallel planes of sodium and chlorine ions. Figure P37.26 shows portions of these planes containing atoms within the unit cell. Imagine extending these portions outward to form

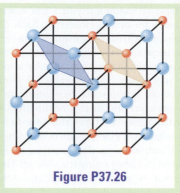

Figure P37.26

large planes, one with only sodium ions and the other with only chlorine ions. Considering these planes, for how many angles do you expect to detect a diffraction maximum from the crystal using the same x-rays?

SECTION 37.6 Polarization of Light Waves

Online-Only Problem 33.25 can be assigned with this section.

27. Two handheld radio transceivers with dipole antennas are separated by a large fixed distance. If the transmitting antenna is vertical, what fraction of the maximum received power will appear in the receiving antenna when it is inclined from the vertical by (a) 15.0°, (b) 45.0°, and (c) 90.0°?

28. *Why is the following situation impossible?* A technician is measuring the index of refraction of a solid material by observing the polarization of light reflected from its surface. She notices that when a light beam is projected from air onto the material surface, the reflected light is totally polarized parallel to the surface when the incident angle is 41.0°.

29. The critical angle for total internal reflection for sapphire surrounded by air is 34.4°. Calculate the polarizing angle for sapphire.

30. For a particular transparent medium surrounded by air, find **S** the polarizing angle θ_p in terms of the critical angle for total internal reflection θ_c.

31. You are working in a laser laboratory, assisting with an **CR** experiment involving gas lasers. Your supervisor explains that the ends of the glass tube containing the lasing gas are sealed with *Brewster windows*. Figure P37.31 shows such a window at one end of a glass laser tube. The laser light reflected from the first surface is shown as the dashed line in the figure, and is completely polarized parallel to the plane of the

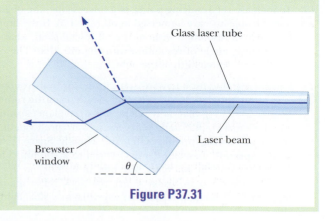

Figure P37.31

surface (perpendicular to the page). The light that transmits into the Brewster window and then out the opposite side is partially polarized parallel to the page. (a) Your supervisor asks you to determine the angle θ that a Brewster window made from ZnSe (n = 2.67) must make with the horizontal for these conditions to be satisfied. The index of refraction of the gas in the tube is 1.00. (b) After you report your angle to your supervisor, he says that he has seen a possible problem—what about the reflection at the *second* surface, as the beam leaves the Brewster window? He is afraid that some of the desired polarization of the beam (parallel to the page) will be lost at that surface. Convince him that he does not need to worry about that issue.

32. An unpolarized beam of light is incident on a stack of ideal polarizing filters. The axis of the first filter is perpendicular to the axis of the last filter in the stack. Find the fraction by which the transmitted beam's intensity is reduced in the three following cases. (a) Three filters are in the stack, each with its transmission axis at 45.0° relative to the preceding filter. (b) Four filters are in the stack, each with its transmission axis at 30.0° relative to the preceding filter. (c) Seven filters are in the stack, each with its transmission axis at 15.0° relative to the preceding filter. (d) Comment on comparing the answers to parts (a), (b), and (c).

ADDITIONAL PROBLEMS

33. In a single-slit diffraction pattern, assuming each side maximum is halfway between the adjacent minima, find the ratio of the intensity of (a) the first-order side maximum and (b) the second-order side maximum to the intensity of the central maximum.

34. Laser light with a wavelength of 632.8 nm is directed through one slit or two slits and allowed to fall on a screen 2.60 m beyond. Figure P37.34 shows the pattern on the screen, with a centimeter ruler below it. (a) Did the light pass through one slit or two slits? Explain how you can determine the answer. (b) If one slit, find its width. If two slits, find the distance between their centers.

Figure P37.34

35. In water of uniform depth, a wide pier is supported on pilings in several parallel rows 2.80 m apart. Ocean waves of uniform wavelength roll in, moving in a direction that makes an angle of 80.0° with the rows of pilings. Find the three longest wavelengths of waves that are strongly reflected by the pilings.

36. Two motorcycles separated laterally by 2.30 m are approaching an observer wearing night-vision goggles sensitive to infrared light of wavelength 885 nm. (a) Assume the light propagates through perfectly steady and uniform air. What aperture diameter is required if the motorcycles' headlights are to be resolved at a distance of 12.0 km? (b) Comment on how realistic the assumption in part (a) is.

37. The *Very Large Array* (VLA) is a set of 27 radio telescope dishes in Catron and Socorro counties, New Mexico (Fig. P37.37). The antennas can be moved apart on railroad tracks, and their combined signals give the resolving power of a synthetic aperture 36.0 km in diameter. (a) If the detectors are tuned to a frequency of 1.40 GHz, what is the angular resolution of the VLA? (b) Clouds of interstellar hydrogen radiate at the frequency used in part (a). What must be the separation distance of two clouds at the center of the galaxy, 26 000 light-years away, if they are to be resolved? (c) **What If?** As the telescope looks up, a circling hawk looks down. Assume the hawk is most sensitive to green light having a wavelength of 500 nm and has a pupil of diameter 12.0 mm. Find the angular resolution of the hawk's eye. (d) A mouse is on the ground 30.0 m below. By what distance must the mouse's whiskers be separated if the hawk can resolve them?

Figure P37.37

38. Two wavelengths λ and λ + Δλ (with Δλ << λ) are incident on a diffraction grating. Show that the angular separation between the spectral lines in the *m*th-order spectrum is

$$\Delta\theta = \frac{\Delta\lambda}{\sqrt{(d/m)^2 - \lambda^2}}$$

where *d* is the slit spacing and *m* is the order number.

39. **Review.** A beam of 541-nm light is incident on a diffraction grating that has 400 grooves/mm. (a) Determine the angle of the second-order ray. (b) **What If?** If the entire apparatus is immersed in water, what is the new second-order angle of diffraction? (c) Show that the two diffracted rays of parts (a) and (b) are related through the law of refraction.

40. *Why is the following situation impossible?* A technician is sending laser light of wavelength 632.8 nm through a pair of slits separated by 30.0 μm. Each slit is of width 2.00 μm. The screen on which he projects the pattern is not wide enough, so light from the m = 15 interference maximum misses the edge of the screen and passes into the next lab station, startling a coworker.

41. Light in air strikes a water surface at the polarizing angle. The part of the beam refracted into the water strikes a submerged slab of material with refractive index n = 1.62 as shown in Figure P37.41 (page 1008). The light reflected from the upper surface of the slab is completely polarized. Find the angle θ between the water surface and the surface of the slab.

Figure P37.41
Problems 41 and 42.

42. Light in air (assume $n = 1$) strikes the surface of a liquid of
S index of refraction n_ℓ at the polarizing angle. The part of the
beam refracted into the liquid strikes a submerged slab of
material with refractive index n as shown in Figure P37.41.
The light reflected from the upper surface of the slab is completely polarized. Find the angle θ between the water surface
and the surface of the slab as a function of n and n_ℓ.

43. A pinhole camera has a small circular aperture of diameter
D. Light from distant objects passes through the aperture
into an otherwise dark box, falling on a screen located a distance L away. If D is too large, the display on the screen will
be fuzzy because a bright point in the field of view will send
light onto a circle of diameter slightly larger than D. On the
other hand, if D is too small, diffraction will blur the display
on the screen. The screen shows a reasonably sharp image if
the diameter of the central disk of the diffraction pattern,
specified by Equation 37.6, is equal to D at the screen. (a)
Show that for monochromatic light with plane wave fronts
and $L \gg D$, the condition for a sharp view is fulfilled if
$D^2 = 2.44\lambda L$. (b) Find the optimum pinhole diameter for
500-nm light projected onto a screen 15.0 cm away.

44. Iridescent peacock feathers are shown in Figure P37.44a.
BIO The surface of one microscopic barbule is composed of
QC transparent keratin that supports rods of dark brown melanin in a regular lattice, represented in Figure P37.44b.
(Your fingernails are made of keratin, and melanin is the
dark pigment giving color to human skin.) In a portion of
the feather that can appear turquoise (blue-green), assume
the melanin rods are uniformly separated by 0.25 μm,
with air between them. (a) Explain how this structure can
appear turquoise when it contains no blue or green pigment. (b) Explain how it can also appear violet if light falls
on it in a different direction. (c) Explain how it can present
different colors to your two eyes simultaneously, which is a
characteristic of iridescence. (d) A compact disc can appear
to be any color of the rainbow. Explain why the portion of
the feather in Figure P37.44b cannot appear yellow or red.
(e) What could be different about the array of melanin rods
in a portion of the feather that does appear to be red?

Drop of Light/Shutterstock

a   b

Figure P37.44

45. The *scale* of a map is a number of kilometers per centimeter specifying the distance on the ground that any distance
on the map represents. The scale of a spectrum is its *dispersion*, a number of nanometers per centimeter, specifying the
change in wavelength that a distance across the spectrum
represents. You must know the dispersion if you want to compare one spectrum with another or make a measurement
of, for example, a Doppler shift. Let y represent the position
relative to the center of a diffraction pattern projected onto
a flat screen at distance L by a diffraction grating with slit
spacing d. The dispersion is $d\lambda/dy$. (a) Prove that the dispersion is given by

$$\frac{d\lambda}{dy} = \frac{L^2 d}{m(L^2 + y^2)^{3/2}}$$

(b) A light with a mean wavelength of 550 nm is analyzed
with a grating having 8 000 rulings/cm and projected onto
a screen 2.40 m away. Calculate the dispersion in first order.

46. (a) Light traveling in a medium of index of refraction n_1
S is incident at an angle θ on the surface of a medium of
index n_2. The angle between reflected and refracted rays is
β. Show that

$$\tan \theta = \frac{n_2 \sin \beta}{n_1 - n_2 \cos \beta}$$

(b) **What If?** Show that this expression for $\tan \theta$ reduces to
Brewster's law when $\beta = 90°$.

47. The intensity of light in a diffraction pattern of a single slit
is described by the equation

$$I = I_{\max} \frac{\sin^2 \phi}{\phi^2}$$

where $\phi = (\pi a \sin \theta)/\lambda$. The central maximum is at $\phi = 0$,
and the side maxima are *approximately* at $\phi = (m + \frac{1}{2})\pi$ for
$m = 1, 2, 3, \ldots$. Determine more precisely (a) the location
of the first side maximum, where $m = 1$, and (b) the location of the second side maximum. *Suggestion:* Observe in Figure 37.6a that the graph of intensity versus ϕ has a horizontal tangent at maxima and also at minima.

48. How much diffraction spreading does a light beam undergo?
One quantitative answer is the *full width at half maximum* of
the central maximum of the single-slit Fraunhofer diffraction pattern. You can evaluate this angle of spreading in
this problem. (a) In Equation 37.2, define $\phi = \pi a \sin \theta/\lambda$
and show that at the point where $I = 0.5I_{\max}$ we must have
$\phi = \sqrt{2} \sin \phi$. (b) Let $y_1 = \sin \phi$ and $y_2 = \phi/\sqrt{2}$. Plot y_1 and
y_2 on the same set of axes over a range from $\phi = 1$ rad to
$\phi = \pi/2$ rad. Determine ϕ from the point of intersection of
the two curves. (c) Then show that if the fraction λ/a is not
large, the angular full width at half maximum of the central
diffraction maximum is $\theta = 0.885\lambda/a$. (d) **What If?** Another
method to solve the transcendental equation $\phi = \sqrt{2} \sin \phi$
in part (a) is to guess a first value of ϕ, use a computer or
calculator to see how nearly it fits, and continue to update
your estimate until the equation balances. How many steps
(iterations) does this process take?

49. Two closely spaced wavelengths of light are incident on a diffraction grating. (a) Starting with Equation 37.7, show that
the angular dispersion of the grating is given by

$$\frac{d\theta}{d\lambda} = \frac{m}{d \cos \theta}$$

(b) A square grating 2.00 cm on each side containing 8 000 equally spaced slits is used to analyze the spectrum of mercury. Two closely spaced lines emitted by this element have wavelengths of 579.065 nm and 576.959 nm. What is the angular separation of these two wavelengths in the second-order spectrum?

CHALLENGE PROBLEMS

50. A spy satellite can consist of a large-diameter concave mirror forming an image on a digital-camera detector and sending the picture to a ground receiver by radio waves. In effect, it is an astronomical telescope in orbit, looking down instead of up. (a) Can a spy satellite read a license plate? (b) Can it read the date on a dime? Argue for your answers by making an order-of-magnitude calculation, specifying the data you estimate.

51. Figure P37.51a is a three-dimensional sketch of a birefringent crystal. The dotted lines illustrate how a thin, parallel-faced slab of material could be cut from the larger specimen with the crystal's optic axis parallel to the faces of the plate. A section cut from the crystal in this manner is known as a *retardation plate*. When a beam of light is incident on the plate perpendicular to the direction of the optic axis as shown in Figure P37.51b, the O ray and the E ray travel along a single straight line, but with different speeds. The figure shows the wave fronts for the two rays. (a) Let the thickness of the plate be d. Show that the phase difference between the O ray and the E ray after traveling the thickness of the plate is

$$\theta = \frac{2\pi d}{\lambda} |n_O - n_E|$$

where λ is the wavelength in air. (b) In a particular case, the incident light has a wavelength of 550 nm. Find the minimum value of d for a quartz plate for which $\theta = \pi/2$. Such a plate is called a *quarter-wave plate*. Use values of n_O and n_E from Table 37.1.

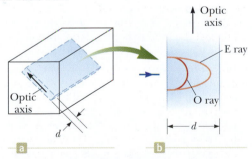

Figure P37.51

52. In Figure P37.52, suppose the transmission axes of the left and right polarizing disks are perpendicular to each other. Also, let the center disk be rotated on the common axis with an angular speed ω. Show that if unpolarized light is incident on the left disk with an intensity I_{max}, the intensity of the beam emerging from the right disk is

$$I = \tfrac{1}{16}I_{max}(1 - \cos 4\omega t)$$

This result means that the intensity of the emerging beam is modulated at a rate four times the rate of rotation of the center disk. *Suggestion:* Use the trigonometric identities $\cos^2 \theta = \tfrac{1}{2}(1 + \cos 2\theta)$ and $\sin^2 \theta = \tfrac{1}{2}(1 - \cos 2\theta)$.

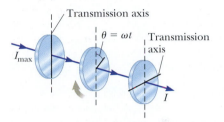

Figure P37.52

53. Consider a light wave passing through a slit and propagating toward a distant screen. Figure P37.53 shows the intensity variation for the pattern on the screen. Give a mathematical argument that more than 90% of the transmitted energy is in the central maximum of the diffraction pattern. *Suggestion:* You are not expected to calculate the precise percentage, but explain the steps of your reasoning. You may use the identification

$$\frac{1}{1^2} + \frac{1}{3^2} + \frac{1}{5^2} + \cdots = \frac{\pi^2}{8}$$

Figure P37.53

PART 6

Modern Physics

At the end of the 19th century, many scientists believed they had learned most of what there was to know about physics. Kinematics and dynamics, universal gravitation, principles of electricity and magnetism, the laws of thermodynamics and kinetic theory, and the principles of optics were highly successful in explaining a variety of phenomena. Thus far in this book, we have studied these ideas and have found that we can describe physical phenomena with separate sets of analysis models based on simplification models, many based on *particles* and others on *waves*.

At the turn of the 20th century, however, a major revolution shook the world of physics. In 1900, Max Planck provided the basic ideas that led to the formulation of the *quantum theory*, and in 1905, Albert Einstein formulated his *special theory of relativity*. Both theories were to have a profound effect on our understanding of nature.

Relativity tells us that concepts of kinematics and dynamics are not as we thought when we consider speeds close to that of light. One of the most startling results from quantum theory tells us that such entities as electrons (particles) and light (waves) have *both* particle-like and wave-like properties!

In Chapter 38, we shall introduce the special theory of relativity. Although the predictions of this theory often violate our common sense, the theory correctly describes the results of experiments involving speeds near the speed of light. The extended version of this textbook, *Physics for Scientists and Engineers with Modern Physics,* covers the basic concepts of quantum mechanics and their application to atomic and molecular physics, condensed matter physics, nuclear physics, particle physics, and cosmology.

Even though the physics that was developed during the 20th century has led to a multitude of important technological achievements, the story is still incomplete. Discoveries will continue to evolve during our lifetimes. ■

The Compact Muon Solenoid (CMS) Detector is part of the Large Hadron Collider at the European Laboratory for Particle Physics operated by CERN. It is one of several detectors that search for elementary particles. For a sense of scale, the green structure to the left of the detector and extending to the top is five stories high. (*CERN*)

1011

38 Relativity

In this chapter, we discuss the twin paradox, a standard example of the effects of relativity. Maybe these young twin sisters will be the first to test it out! It looks like they are already discussing it! (*eukukulka/Shutterstock*)

STORYLINE **You are excited to be beginning your study of modern** physics. While looking ahead in Chapter 38, you see a discussion of the *twin paradox*, where one twin stays on Earth and the other travels to a distant star and then back home at near the speed of light. When the traveling twin arrives back home, their ages are different! You dream about performing this experiment and testing the results. Of course, you want to be the traveling twin, so that you can visit another star! You would need to be paid for your services, so that you could support yourself once you return to Earth. This gets you thinking. To record your daily work shifts and document what your salary should be, should you take your timecard along with you and punch in for work daily, or should you leave your timecard at home and have your boss punch in for you each day?

CONNECTIONS Our everyday experiences and observations involve objects that move at speeds much less than the speed of light. Newtonian mechanics was formulated by observing and describing the motion of such objects, and this formalism is very successful in describing a wide range of phenomena that occur at low speeds. Nonetheless, it fails to describe properly the motion of objects whose speeds approach that of light. Although Albert Einstein made many other important contributions to science, the special theory of relativity alone represents one of the greatest intellectual achievements of all time. With this theory, experimental observations can be correctly predicted over the range of speeds from $v = 0$ to speeds approaching the speed of light. This chapter gives an introduction to the special theory of relativity, with emphasis on some of its predictions. In addition to its well-known and essential role in theoretical physics, the special theory of relativity has practical applications, including the design of nuclear power plants and modern global positioning system (GPS) units. These devices, and others we see in upcoming chapters, depend on relativistic principles for proper design and operation.

38.1 The Principle of Galilean Relativity

In Section 4.6, we studied observations made from different frames of reference. In Chapter 5, we discussed that an *inertial* frame of reference is one in which an object is observed to have no acceleration when no forces act on it. Furthermore, any frame moving with constant velocity with respect to an inertial frame must also be an inertial frame.

There is no absolute inertial reference frame. Therefore, the laws describing a phenomenon as determined by an observer in a vehicle moving with uniform velocity relative to an inertial reference frame must be identical to the laws determined by an observer in a vehicle that is stationary relative to the same frame. The formal statement of this result is called the **principle of Galilean relativity:**

> The laws of mechanics must be the same in all inertial frames of reference.

◀ Principle of Galilean relativity

It is important to point out that this statement says that the *laws* are the same, not the *results* of an experiment. Let's consider an observation that illustrates the equivalence of the laws of mechanics in different inertial frames. The pickup truck in Figure 38.1a moves with a constant velocity with respect to the ground. If a passenger in the truck throws a ball straight up and if air resistance is neglected, the passenger observes that the ball moves upward in a vertical path and then falls back into the observer's hand. The motion of the ball appears to be precisely the same as if the ball were thrown while the truck were at rest. The law of universal gravitation and the equations of motion under constant acceleration are obeyed whether the truck is at rest or in uniform motion.

Now consider an observer on the ground as in Figure 38.1b. Both observers agree on the laws of physics: the observer in the truck throws a ball straight up, and it rises and falls back into his hand according to the particle under constant acceleration model. Do the observers agree, however, on the path of the ball thrown by the observer in the truck? The observer on the ground sees the path of the ball as a parabola as illustrated in Figure 38.1b, whereas, as mentioned earlier, the observer in the truck sees the ball move in a vertical path. Furthermore, according to the observer on the ground, the ball has a horizontal component of velocity equal to the velocity of the truck, and the horizontal motion of the ball is described by the particle under constant velocity model. Although the two observers disagree on certain aspects of the situation, *they agree on the validity of Newton's laws* and on the

The observer in the moving truck sees the ball travel in a vertical path when thrown upward.

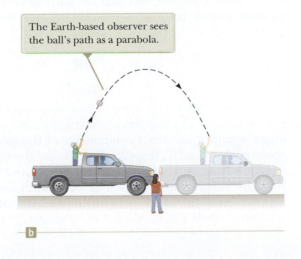

The Earth-based observer sees the ball's path as a parabola.

a

b

Figure 38.1 Two observers watch the path of a thrown ball and obtain different results.

results of applying appropriate analysis models that we have learned. This agreement implies that no mechanical experiment can detect any difference between the two inertial frames. The only thing that can be detected is the relative motion of one frame with respect to the other.

QUICK QUIZ **38.1** Which observer in Figure 38.1 sees the ball's *correct* path? **(a)** the observer in the truck **(b)** the observer on the ground **(c)** both observers

Suppose some physical phenomenon, which we call an *event,* occurs and is observed by an observer at rest in an inertial reference frame. The wording "in a frame" means that the observer is at rest with respect to the origin of that frame. The event's location and time of occurrence can be specified by the four coordinates (x, y, z, t). We would like to be able to transform these coordinates from those of an observer in one inertial frame to those of another observer in a frame moving with uniform relative velocity compared with the first frame.

Consider two inertial frames S and S′ (Fig. 38.2). The S frame is the one in which the first observer in the previous paragraph resides. The S′ frame, in which the second observer resides, moves with a constant velocity $\vec{v} = v\hat{i}$ along the common x and x' axes, where $\vec{v}$ is measured relative to S. We assume the origins of S and S′ coincide at $t = 0$ and an event occurs at point P in space at some instant of time. For simplicity, we show the observer O in the S frame and the observer O' in the S′ frame as blue dots at the origins of their coordinate frames in Figure 38.2, but that is not necessary: either observer could be at any fixed location in his or her frame. Observer O describes the event with space–time coordinates (x, y, z, t), whereas observer O' in S′ uses the coordinates (x', y', z', t') to describe the same event. Model the origin of S′ as a particle under constant velocity relative to the origin of S. As we see from the geometry in Figure 38.2 and the particle under constant velocity model, the relationships among these various coordinates can be written

$$x' = x - vt \qquad y' = y \qquad z' = z \qquad t' = t \tag{38.1}$$

◀ Galilean space–time transformation equations

These equations are the **Galilean space–time transformation equations.** Note that time is assumed to be the same in both inertial frames. That is, within the framework of classical mechanics, all clocks run at the same rate, regardless of their velocity, so the time at which an event occurs for an observer in S is the same as the time for the same event in S′. Consequently, the time interval between two successive events should be the same for both observers. Although this assumption may seem obvious, it turns out to be incorrect in situations where v is comparable to the speed of light, as we shall see.

Now suppose a particle moves through a displacement of magnitude dx along the x axis in Figure 38.2 in a time interval dt as measured by an observer in S. It follows from Equation 38.1 that the corresponding displacement dx' measured by an observer in S′ is $dx' = dx - v\,dt$. Because $dt = dt'$, we find that

$$\frac{dx'}{dt'} = \frac{dx}{dt} - v$$

or

$$u'_x = u_x - v$$

where u_x and u'_x are the x components of the velocity of the particle measured by observers in S and S′, respectively. (We use the symbol u rather than v for particle velocity because v is already used for the relative velocity of two reference frames.) We write the previous equation in vector form and solve it for the speed of the particle as seen by the observer in the unprimed frame:

$$\vec{u}_x = \vec{u}'_x + \vec{v} \tag{38.2}$$

◀ Galilean velocity transformation equation

Figure 38.2 An event occurs at a point P. The event is seen by two observers in inertial frames S and S′, where S′ moves with a velocity $\vec{v}$ relative to S.

PITFALL PREVENTION 38.1

The Relationship Between the S and S′ Frames Many of the mathematical representations in this chapter are true *only* for the specified relationship between the S and S′ frames. The x and x' axes coincide, except their origins are different. The y and y' axes (and the z and z' axes) are parallel, but they only coincide at one instant due to the time-varying position of the origin of S′ with respect to that of S. We choose the time $t = 0$ to be the instant at which the origins of the two coordinate systems coincide. If the S′ frame is moving in the positive x direction relative to S, then v is positive; otherwise, it is negative.

Equation 38.2 is the **Galilean velocity transformation equation** and is identical to Equation 4.30. It is consistent with our intuitive notion of time and space as well as with our discussions in Section 4.6. As we shall soon see, however, it leads to serious contradictions when applied to electromagnetic waves.

QUICK QUIZ **38.2** A baseball pitcher with a 90-mi/h fastball throws a ball while standing on a railroad flatcar moving at 110 mi/h. The ball is thrown in the same direction as that of the velocity of the train. If you apply the Galilean velocity transformation equation to this situation, is the speed of the ball relative to the Earth **(a)** 90 mi/h, **(b)** 110 mi/h, **(c)** 20 mi/h, **(d)** 200 mi/h, or **(e)** impossible to determine?

The Speed of Light

It is quite natural to ask whether the principle of Galilean relativity also applies to electricity, magnetism, and optics. Experiments indicate that the answer is no. Recall from Chapter 33 that Maxwell showed that the speed of light in free space is $c = 3.00 \times 10^8$ m/s. Physicists of the late 1800s thought light waves move through a medium filling the universe and called the *ether*. According to this model, the speed of light is measured to be c only in a special, absolute frame at rest with respect to the ether. The Galilean velocity transformation equation was expected to hold for observations of light made by an observer in any frame moving at speed v relative to the absolute ether frame.

Because the existence of a preferred, absolute ether frame would show that light is similar to other classical waves that require a medium and that Newtonian ideas of an absolute frame are true, considerable importance was attached to establishing the existence of the ether frame. Starting in about 1880, scientists decided to use the Earth as the moving frame in an attempt to improve their chances of detecting these small changes in the speed of light.

Observers fixed on the Earth can take the view that they are stationary and that the absolute ether frame containing the medium for light propagation moves past them with speed v. In Equation 38.2, the observed entity that is moving is light, so let $\vec{u}'_x = \vec{c}$, where the primed frame is attached to the ether. Then the speed of light as measured by an observer on Earth, the unprimed frame, is $\vec{u}_x = \vec{c} + \vec{v}$, where $\vec{v}$ is the velocity of the ether with respect to the Earth. Determining the speed of light under these circumstances is similar to determining the speed of an aircraft traveling in a moving air current, or wind; consequently, we speak of an "ether wind" blowing through our apparatus fixed to the Earth.

A direct method for detecting an ether wind would use an apparatus fixed to the Earth to measure the ether wind's influence on the speed of light. If v is the speed of the ether relative to the Earth, light should have its maximum speed $c + v$ when propagating downwind as in Figure 38.3a. Likewise, the speed of light should have its minimum value $c - v$ when the light is propagating upwind as in Figure 38.3b and an intermediate value $(c^2 - v^2)^{1/2}$ when the light is directed such that it travels perpendicular to the ether wind as in Figure 38.3c. In this latter case, the vector $\vec{c}$ must be aimed upstream so that the resultant velocity is perpendicular to the wind, like the boat in Figure 4.22b. If the Sun is assumed to be at rest in the ether, the velocity of the ether wind would be equal to the orbital velocity of the Earth around the Sun, which has a magnitude of approximately 30 km/s or 3×10^4 m/s. Because $c = 3 \times 10^8$ m/s, it is necessary to detect a change in speed of approximately 1 part in 10^4 for measurements in the upwind or downwind directions. Although such a change is experimentally measurable, all attempts to detect such changes and establish the existence of the ether wind (and hence the absolute frame) proved futile! We shall discuss the classic experimental search for the ether in Section 38.2.

The principle of Galilean relativity refers only to the laws of mechanics. If it is assumed the laws of electricity and magnetism are the same in all inertial frames, a paradox concerning the speed of light immediately arises. That can be understood

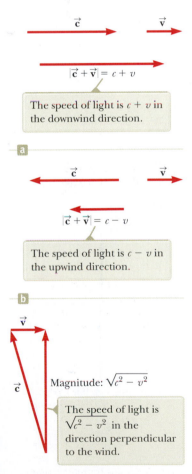

The speed of light is $c + v$ in the downwind direction.

a

The speed of light is $c - v$ in the upwind direction.

b

Magnitude: $\sqrt{c^2 - v^2}$

The speed of light is $\sqrt{c^2 - v^2}$ in the direction perpendicular to the wind.

c

Figure 38.3 If the velocity of the ether wind relative to the Earth is $\vec{v}$ and the velocity of light relative to the ether is $\vec{c}$, the speed of light relative to the Earth depends on the direction of the Earth's velocity.

by recognizing that Maxwell's equations imply that the speed of light always has the fixed value 3.00×10^8 m/s in all inertial frames, a result in direct contradiction to what is expected based on the Galilean velocity transformation equation. According to Galilean relativity, the speed of light should *not* be the same in all inertial frames.

To resolve this contradiction in theories, we must conclude that either (1) the laws of electricity and magnetism are not the same in all inertial frames or (2) the Galilean velocity transformation equation is incorrect. If we assume the first alternative, a preferred reference frame in which the speed of light has the value c must exist and the measured speed must be greater or less than this value in any other reference frame, in accordance with the Galilean velocity transformation equation. If we assume the second alternative, we must abandon the notions of absolute time and absolute length that form the basis of the Galilean space–time transformation equations.

38.2 The Michelson–Morley Experiment

The most famous experiment designed to detect small changes in the speed of light was first performed in 1881 by A. A. Michelson (see Section 36.6) and later repeated under various conditions by Michelson and Edward W. Morley (1838–1923). As we shall see, the outcome of the experiment contradicted the ether hypothesis.

The experiment was designed to determine the velocity of the Earth relative to that of the hypothetical ether. The experimental tool used was the Michelson interferometer, which was discussed in Section 36.6 and is shown again in Figure 38.4. Arm 2 is aligned along the direction of the Earth's motion through space. The Earth moving through the ether at speed v is equivalent to the ether flowing past the Earth in the opposite direction with speed v. This ether wind blowing in the direction opposite the direction of the Earth's motion should cause the speed of light measured in the Earth frame to be $c + v$ as the light approaches mirror M_2 and $c - v$ after reflection, where c is the speed of light in the ether frame.

The two light beams reflect from M_1 and M_2 and recombine, and an interference pattern is formed as discussed in Section 36.6. The interference pattern is then observed while the interferometer is rotated through an angle of $90°$. This rotation interchanges the speed of the ether wind between the arms of the interferometer. The rotation should cause the fringe pattern to shift slightly but measurably. Measurements failed, however, to show any change in the interference pattern! The Michelson–Morley experiment was repeated at different times of the year when the ether wind was expected to change direction and magnitude, but the results were always the same: no fringe shift of the magnitude required was *ever* observed.[1]

The negative results of the Michelson–Morley experiment not only contradicted the ether hypothesis, but also showed that it is impossible to measure the absolute velocity of the Earth with respect to the ether frame. Einstein, however, offered a postulate for his special theory of relativity that places quite a different interpretation on these null results. In later years, when more was known about the nature of light, the idea of an ether that permeates all of space was abandoned. Light is now understood to be an electromagnetic wave, which requires no medium for its propagation. As a result, the idea of an ether in which these waves travel became unnecessary.

Many efforts were made to explain the null results of the Michelson–Morley experiment and to save the ether frame concept and the Galilean velocity transformation equation for light. All proposals resulting from these efforts have been

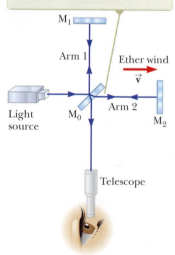

According to the ether wind theory, the speed of light should be $c + v$ as the beam approaches mirror M_2 and $c - v$ after reflection.

Figure 38.4 A Michelson interferometer is used in an attempt to detect the ether wind.

---

[1] From an Earth-based observer's point of view, changes in the Earth's speed and direction of motion in the course of a year are viewed as ether wind shifts. Even if the speed of the Earth with respect to the ether were zero at some time, six months later the speed of the Earth would be 60 km/s with respect to the ether and as a result a fringe shift should be noticed. No shift has ever been observed, however.

shown to be wrong. No experiment in the history of physics received such valiant efforts to explain the absence of an expected result as did the Michelson–Morley experiment. The stage was set for Einstein, who solved the problem in 1905 with his special theory of relativity.

Details of the Michelson–Morley Experiment

To understand the outcome of the Michelson–Morley experiment, let's assume the two arms of the interferometer in Figure 38.4 are of equal length L. We shall analyze the situation as if there were an ether wind because that is what Michelson and Morley expected to find. As noted above, the speed of the light beam along arm 2 should be $c + v$ as the beam approaches M_2 and $c - v$ after the beam is reflected. We model a pulse of light as a particle under constant speed. Therefore, the time interval for travel to the right for the pulse is $\Delta t = L/(c + v)$ and the time interval for travel to the left is $\Delta t = L/(c - v)$. The total time interval for the round trip along arm 2 is

$$\Delta t_{\text{arm 2}} = \frac{L}{c + v} + \frac{L}{c - v} = \frac{2Lc}{c^2 - v^2} = \frac{2L}{c}\left(1 - \frac{v^2}{c^2}\right)^{-1}$$

Now consider the light beam traveling along arm 1, perpendicular to the ether wind. Because the speed of the beam relative to the Earth is $(c^2 - v^2)^{1/2}$ in this case (see Fig. 38.3c), the time interval for travel for each half of the trip is $\Delta t = L/(c^2 - v^2)^{1/2}$ and the total time interval for the round trip is

$$\Delta t_{\text{arm 1}} = \frac{2L}{(c^2 - v^2)^{1/2}} = \frac{2L}{c}\left(1 - \frac{v^2}{c^2}\right)^{-1/2}$$

The time difference Δt between the horizontal round trip (arm 2) and the vertical round trip (arm 1) is

$$\Delta t = \Delta t_{\text{arm 2}} - \Delta t_{\text{arm 1}} = \frac{2L}{c}\left[\left(1 - \frac{v^2}{c^2}\right)^{-1} - \left(1 - \frac{v^2}{c^2}\right)^{-1/2}\right]$$

Because $v^2/c^2 \ll 1$, we can simplify this expression by using the following binomial expansion after dropping all terms higher than second order:

$$(1 - x)^n \approx 1 - nx \quad (\text{for } x \ll 1)$$

In our case, $x = v^2/c^2$, and we find that

$$\Delta t = \Delta t_{\text{arm 2}} - \Delta t_{\text{arm 1}} \approx \frac{Lv^2}{c^3} \tag{38.3}$$

This time difference between the two instants at which the reflected beams arrive at the viewing telescope gives rise to a phase difference between the beams, producing an interference pattern when they combine at the position of the telescope. A shift in the interference pattern should be detected when the interferometer is rotated through 90° in a horizontal plane so that the two beams exchange roles. This rotation results in a time difference twice that given by Equation 38.3. Therefore, the path difference that corresponds to this time difference is

$$\Delta d = c(2\,\Delta t) = \frac{2Lv^2}{c^2}$$

Because a change in path length of one wavelength corresponds to a shift of one fringe, the corresponding fringe shift is equal to this path difference divided by the wavelength of the light:

$$\text{Shift} = \frac{2Lv^2}{\lambda c^2} \tag{38.4}$$

In the experiments by Michelson and Morley, each light beam was reflected by mirrors many times to give an effective path length L of approximately 11 m. Using this value, taking v to be equal to 3.0×10^4 m/s (the speed of the Earth around the Sun), and using 500 nm for the wavelength of the light, we expect a fringe shift of

$$\text{Shift} = \frac{2(11 \text{ m})(3.0 \times 10^4 \text{ m/s})^2}{(5.0 \times 10^{-7} \text{ m})(3.0 \times 10^8 \text{ m/s})^2} = 0.44$$

The instrument used by Michelson and Morley could detect shifts as small as 0.01 fringe, but it detected no shift whatsoever in the fringe pattern! The experiment has been repeated many times since by different scientists under a wide variety of conditions, and no fringe shift has ever been detected. Therefore, it was concluded that the motion of the Earth with respect to the postulated ether cannot be detected.

38.3 Einstein's Principle of Relativity

In the previous section, we noted the impossibility of measuring the speed of the ether with respect to the Earth and the failure of the Galilean velocity transformation equation in the case of light. Einstein proposed a theory that boldly removed these difficulties and at the same time completely altered our notion of space and time.[2] He based his special theory of relativity on two postulates:

1. **The principle of relativity:** The laws of physics must be the same in all inertial reference frames.
2. **The constancy of the speed of light:** The speed of light in vacuum has the same value, $c = 3.00 \times 10^8$ m/s, in all inertial frames, regardless of the velocity of the observer or the velocity of the source emitting the light.

The first postulate asserts that *all* the laws of physics—those dealing with mechanics, electricity and magnetism, optics, thermodynamics, and so on—are the same in all reference frames moving with constant velocity relative to one another. This postulate is a generalization of the principle of Galilean relativity, which refers only to the laws of mechanics. From an experimental point of view, Einstein's principle of relativity means that any kind of experiment (measuring the speed of light, for example) performed in a laboratory at rest must give the same result when performed in a laboratory moving at a constant velocity with respect to the first one. Hence, no preferred inertial reference frame exists, and it is impossible to detect absolute motion.

Note that postulate 2 is required by postulate 1: if the speed of light were not the same in all inertial frames, measurements of different speeds would make it possible to distinguish between inertial frames. As a result, a preferred, absolute frame could be identified, in contradiction to postulate 1.

Although the Michelson–Morley experiment was performed before Einstein published his work on relativity, it is not clear whether or not Einstein was aware of the details of the experiment. Nonetheless, the null result of the experiment can be readily understood within the framework of Einstein's theory. According to his principle of relativity, the premises of the Michelson–Morley experiment were incorrect. In the process of trying to explain the expected results, we stated that when light traveled with the ether wind, its speed was $c + v$, in accordance with the Galilean velocity transformation equation. If the state of motion of the observer or of the source has no influence on the value found for the speed of light, however, one always measures the value to be c. Likewise, the light makes the return

Albert Einstein
German-American Physicist
(1879–1955)
Einstein, one of the greatest physicists of all time, was born in Ulm, Germany. In 1905, at age 26, he published four scientific papers that revolutionized physics. Two of these papers were concerned with what is now considered his most important contribution: the special theory of relativity.

In 1916, Einstein published his work on the general theory of relativity in *Annalen der Physik*. The most dramatic prediction of this theory is the degree to which light is deflected by a gravitational field. Measurements made by astronomers on bright stars in the vicinity of the eclipsed Sun in 1919 confirmed Einstein's prediction, and Einstein became a world celebrity as a result. Einstein was deeply disturbed by the development of quantum mechanics in the 1920s despite his own role as a scientific revolutionary. In particular, he could never accept the probabilistic view of events in nature that is a central feature of quantum theory. The last few decades of his life were devoted to an unsuccessful search for a unified theory that would combine gravitation and electromagnetism.

---

[2]A. Einstein, "On the Electrodynamics of Moving Bodies," *Ann. Physik* **17**:891, 1905. For an English translation of this article and other publications by Einstein, see the book by H. Lorentz, A. Einstein, H. Minkowski, and H. Weyl, *The Principle of Relativity* (New York: Dover, 1958).

trip after reflection from the mirror at speed c, not at speed $c - v$. Therefore, the motion of the Earth does not influence the interference pattern observed in the Michelson–Morley experiment, and a null result should be expected.

If we accept Einstein's theory of relativity, we must conclude that relative motion is unimportant when measuring the speed of light. At the same time, we must alter our commonsense notion of space and time and be prepared for some surprising consequences. As you read the pages ahead, keep in mind that our commonsense ideas are based on a lifetime of everyday experiences and not on observations of objects moving at hundreds of thousands of kilometers per second. Therefore, these results may seem strange, but that is only because we have no experience with them.

38.4 Consequences of the Special Theory of Relativity

As we examine some of the consequences of relativity in this section, we restrict our discussion to the concepts of simultaneity, time intervals, and lengths. All three of these are quite different in relativistic mechanics from what they are in Newtonian mechanics.

Simultaneity and the Relativity of Time

A basic premise of Newtonian mechanics is that a universal time scale exists that is the same for all observers. Newton and his followers took simultaneity for granted. In his special theory of relativity, Einstein abandoned this assumption.

Einstein devised the following thought experiment to illustrate this point. A box-car moves with uniform velocity, and two bolts of lightning strike its ends as illustrated in Figure 38.5a, leaving marks on the boxcar and on the ground. The marks on the boxcar are labeled A' and B', and those on the ground are labeled A and B. An observer O' moving with the boxcar is midway between A' and B', and a ground observer O is midway between A and B. The events recorded by the observers are the striking of the boxcar by the two lightning bolts.

The light signals emitted from A and B at the instant at which the two bolts strike later reach observer O at the same time as indicated in Figure 38.5b. This observer realizes that the signals traveled at the same speed over equal distances and so concludes that the events at A and B occurred simultaneously. Now consider the same events as viewed by observer O'. By the time the signals have reached observer O, observer O' has moved as indicated in Figure 38.5b. Therefore, the signal from B' has already swept past O', but the signal from A' has not yet reached O'. In other words, O' sees the signal from B' before seeing the signal from A'.

The events appear to be simultaneous to the stationary observer O who is standing midway between A and B.

The events do not appear to be simultaneous to observer O', who claims that the front of the car is struck before the rear.

Figure 38.5 (a) Two lightning bolts strike the ends of a moving boxcar. (b) At a later time, the leftward-traveling light signal has already passed O', but the rightward-traveling signal has not yet reached O'.

According to Einstein, *the two observers must find that light travels at the same speed.* Therefore, observer O' concludes that one lightning bolt strikes the front of the boxcar *before* the other one strikes the back.

This thought experiment clearly demonstrates that the two events that appear to be simultaneous to observer O do *not* appear to be simultaneous to observer O'. Simultaneity is not an absolute concept but rather one that depends on the state of motion of the observer. Einstein's thought experiment demonstrates that two observers can disagree on the simultaneity of two events. This disagreement, however, depends on the transit time of light to the observers and therefore does *not* demonstrate the deeper meaning of relativity. In relativistic analyses of high-speed situations, simultaneity is relative even when the transit time is subtracted out. In fact, in all the relativistic effects we discuss, we ignore differences caused by the transit time of light to the observers.

Time Dilation

To illustrate that observers in different inertial frames can measure different time intervals between a pair of events, consider a vehicle moving to the right with a speed v such as the boxcar shown in Figure 38.6a. A mirror is fixed to the ceiling of the vehicle, and observer O' at rest in the frame attached to the vehicle holds a flashlight a distance d below the mirror. At some instant, the flashlight emits a pulse of light directed toward the mirror (event 1), and at some later time after reflecting from the mirror, the pulse arrives back at the flashlight (event 2). Observer O' carries a clock and uses it to measure the time interval Δt_p between these two events. (The subscript p stands for *proper,* as we shall see in a moment.) We model the pulse of light as a particle under constant speed. Because the light pulse has a speed c, the time interval required for the pulse to travel from O' to the mirror and back is

$$\Delta t_p = \frac{\text{distance traveled}}{\text{speed}} = \frac{2d}{c} \tag{38.5}$$

Now consider the same pair of events as viewed by observer O in a second frame at rest with respect to the ground as shown in Figure 38.6b. According to this observer, the mirror and the flashlight are moving to the right with a speed v, and as a result, the sequence of events differs significantly. By the time the light from the flashlight

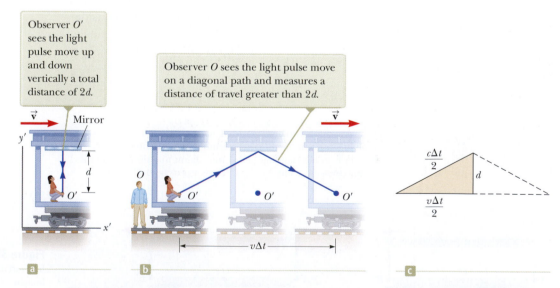

Figure 38.6 (a) A mirror is fixed to a moving vehicle, and a light pulse is sent out by observer O' at rest in the vehicle. (b) Relative to a stationary observer O standing alongside the vehicle, the mirror and O' move with a speed v and the light pulse follows a diagonal path. (c) The right triangle for calculating the relationship between Δt and Δt_p.

reaches the mirror, the mirror has moved to the right a distance $v\,\Delta t/2$, where Δt is the time interval required for the light to travel from O' to the mirror and back to O' as measured by O. Observer O concludes that because of the motion of the vehicle, if the light is to hit the mirror, it must leave the flashlight at an angle with respect to the vertical direction. Comparing Figure 38.6a with Figure 38.6b, we see that the light must travel farther in part (b) than in part (a). (Notice that neither observer "knows" that he or she is moving. Each is at rest in his or her own inertial frame.)

According to the second postulate of the special theory of relativity, both observers must measure c for the speed of light. Because the light travels farther according to O, the time interval Δt measured by O is longer than the time interval Δt_p measured by O'. To obtain a relationship between these two time intervals, let's use the right triangle shown in Figure 38.6c. The Pythagorean theorem gives

$$\left(\frac{c\,\Delta t}{2}\right)^2 = \left(\frac{v\,\Delta t}{2}\right)^2 + d^2$$

Solving for Δt gives

$$\Delta t = \frac{2d}{\sqrt{c^2 - v^2}} = \frac{2d}{c\sqrt{1 - \dfrac{v^2}{c^2}}} \qquad (38.6)$$

Because $\Delta t_p = 2d/c$, we can express this result as

$$\Delta t = \frac{\Delta t_p}{\sqrt{1 - \dfrac{v^2}{c^2}}} = \gamma\,\Delta t_p \qquad (38.7)$$

◀ Time dilation

where

$$\gamma = \frac{1}{\sqrt{1 - \dfrac{v^2}{c^2}}} \qquad (38.8)$$

Because γ is always greater than unity, Equation 38.7 shows that the time interval Δt measured by an observer moving with respect to a clock is longer than the time interval Δt_p measured by an observer at rest with respect to the clock. This effect is known as **time dilation.**

Time dilation is not observed in our everyday lives, which can be understood by considering the factor γ. This factor deviates significantly from a value of 1 only for very high speeds as shown in Figure 38.7 and Table 38.1. All of us, including astronauts, have spent our lives between the first two entries in Table 38.1 and at a point on the horizontal axis of the graph in Figure 38.7 that is, based on the scale of the figure, contained within the thickness of the line used to represent the vertical

TABLE 38.1 Approximate Values for γ at Various Speeds

v/c	γ
0	1
0.001 0	1.000 000 5
0.010	1.000 05
0.10	1.005
0.20	1.021
0.30	1.048
0.40	1.091
0.50	1.155
0.60	1.250
0.70	1.400
0.80	1.667
0.90	2.294
0.92	2.552
0.94	2.931
0.96	3.571
0.98	5.025
0.99	7.089
0.995	10.01
0.999	22.37

Figure 38.7 Graph of γ versus v. As the speed approaches that of light, γ increases rapidly.

axis. For a speed of 0.1*c*, which is far higher than any speed that humans have experienced, the value of γ is 1.005. Therefore, there is a time dilation of only 0.5% at one-tenth the speed of light.

The time interval Δt_p in Equations 38.5 and 38.7 is called the **proper time interval.** (Einstein used the German term *Eigenzeit*, which means "own-time.") In general, the proper time interval is the time interval between two events measured by an observer *who sees the events occur at the same point in space.*

If a clock is moving with respect to you, the time interval between ticks of the moving clock is observed by you to be longer than the time interval between ticks of an identical clock in your reference frame. Therefore, it is often said that a moving clock is measured to run more slowly than a clock in your reference frame by a factor γ. We can generalize this result by stating that all physical processes, including mechanical, chemical, and biological ones, are measured to slow down when those processes occur in a frame moving with respect to the observer. For example, the heartbeat of an astronaut moving through space keeps time with a clock inside the spacecraft. Both the astronaut's clock and heartbeat are measured to slow down relative to a clock back on the Earth (although the astronaut would have no sensation of life slowing down in the spacecraft).

QUICK QUIZ 38.3 Suppose the observer O' on the train in Figure 38.6 aims her flashlight at the far wall of the boxcar and turns it on and off, sending a pulse of light toward the far wall. Both O' and O measure the time interval between when the pulse leaves the flashlight and when it hits the far wall. Which observer measures the proper time interval between these two events? **(a)** O' **(b)** O **(c)** both observers **(d)** neither observer

QUICK QUIZ 38.4 A crew on a spacecraft watches a movie that is two hours long. The spacecraft is moving at high speed through space. Does an Earth-based observer watching the movie screen on the spacecraft through a powerful telescope measure the duration of the movie to be **(a)** longer than, **(b)** shorter than, or **(c)** equal to two hours?

An interesting example of time dilation involves the observation of *muons*, unstable elementary particles that have a charge equal to that of the electron and a mass 207 times that of the electron. Muons can be produced by the collision of cosmic radiation with atoms high in the atmosphere. Slow-moving muons in the laboratory have a lifetime that is measured to be the proper time interval $\Delta t_p = 2.2\ \mu$s. Muons created by cosmic radiation move at speeds very close to that of light. Let's choose a typical speed of 0.9997*c*. At this speed, we find that the distance the muon can travel during its 2.2-μs laboratory-measured lifetime is $(0.9997)(3.0 \times 10^8$ m/s$)(2.2 \times 10^{-6}$ s$) = 6.6 \times 10^2$ m before they decay (Fig. 38.8a). Hence, they are unlikely to reach the surface of the Earth from high in the atmosphere where they are produced. Experiments show, however, that a large number of muons *do* reach the surface. The phenomenon of time dilation explains this effect. As measured by an observer on the Earth, the muons have a dilated lifetime equal to $\gamma\,\Delta t_p$. For example, for $v = 0.9997c$, $\gamma \approx 41$, and $\gamma\,\Delta t_p \approx 90\ \mu$s. Hence, the average distance traveled by the muons in this time interval as measured by an observer on the Earth is approximately $(0.9997)(3.0 \times 10^8$ m/s$)(90 \times 10^{-6}$ s$) \approx 27 \times 10^3$ m as indicated in Figure 38.8b. This distance is larger than the typical height above the surface at which muons are produced, showing that they *can* reach the surface when time dilation is taken into account.

In 1976, at the laboratory of the European Council for Nuclear Research (CERN) in Geneva, muons injected into a large storage ring reached speeds of approximately 0.999 4*c*. Electrons produced by the decaying muons were detected by counters around the ring, enabling scientists to measure the decay rate and hence the muon lifetime. The lifetime of the moving muons was measured to be approximately 30 times as long as that of the stationary muon, in agreement with the prediction of relativity to within two parts in a thousand.

Without relativistic considerations, according to an observer on the Earth, muons created in the atmosphere and traveling downward with a speed close to c travel only about 6.6×10^2 m before decaying with an average lifetime of 2.2 μs. Therefore, very few muons would reach the surface of the Earth.

With relativistic considerations, the muon's lifetime is dilated according to an observer on the Earth. Hence, according to this observer, the muon can travel about 27×10^3 m before decaying. The result is many of them arriving at the surface.

Muon is created

$\approx 6.6 \times 10^2$ m

Muon decays

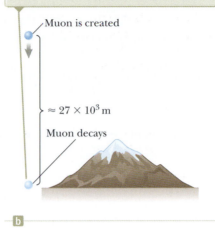

Muon is created

$\approx 27 \times 10^3$ m

Muon decays

a

b

Figure 38.8 Travel of muons according to an Earth-based observer.

Example 38.1 | What Is the Period of the Pendulum?

The period of a pendulum is measured to be 3.00 s in the reference frame of the pendulum. What is the period when measured by an observer moving at a speed of 0.960c relative to the pendulum?

SOLUTION

Conceptualize Let's change frames of reference. Instead of the observer moving at 0.960c, we can take the equivalent point of view that the observer is at rest and the pendulum is moving at 0.960c past the stationary observer. Hence, the pendulum is an example of a clock moving at high speed with respect to an observer. Because the pendulum is at rest in the reference frame of the clock that measured its period, the 3.00-s period is the proper time interval, and the observer who sees the pendulum moving will measure a dilated time interval.

Categorize Based on the Conceptualize step, we can categorize this example as a substitution problem involving relativistic time dilation.

The proper time interval, measured in the rest frame of the pendulum, is $\Delta t_p = 3.00$ s.

Use Equation 38.7 to find the dilated time interval:

$$\Delta t = \gamma \, \Delta t_p = \frac{1}{\sqrt{1 - \dfrac{(0.960c)^2}{c^2}}} \, \Delta t_p = \frac{1}{\sqrt{1 - 0.921\,6}} \, \Delta t_p$$

$$= 3.57(3.00 \text{ s}) = \boxed{10.7 \text{ s}}$$

This result shows that a moving pendulum is indeed measured to take longer to complete a period than a pendulum at rest does. The period increases by a factor of $\gamma = 3.57$.

WHAT IF? What if the speed of the observer increases by 4.00%? Does the dilated time interval increase by 4.00%?

Answer Based on the highly nonlinear behavior of γ as a function of v in Figure 38.7, we would guess that the increase in Δt would be different from 4.00%.

Find the new speed if it increases by 4.00%:

$$v_{\text{new}} = (1.040\,0)(0.960c) = 0.998\,4c$$

Perform the time dilation calculation again:

$$\Delta t = \gamma \, \Delta t_p = \frac{1}{\sqrt{1 - \dfrac{(0.998\,4c)^2}{c^2}}} \, \Delta t_p = \frac{1}{\sqrt{1 - 0.996\,8}} \, \Delta t_p$$

$$= 17.68(3.00 \text{ s}) = 53.1 \text{ s}$$

Therefore, the 4.00% increase in speed results in almost a 400% increase in the dilated time!

Example 38.2   **How Long Was Your Trip?**

Suppose you are driving your car on a business trip and are traveling at 30 m/s. Your boss, who is waiting at your destination, expects the trip to take 5.0 h. When you arrive late, your excuse is that the clock in your car registered the passage of 5.0 h but that you were driving fast and so your clock ran more slowly than the clock in your boss's office. If your car clock actually did indicate a 5.0-h trip, how much time passed on your boss's clock, which was at rest on the Earth?

SOLUTION

Conceptualize   The observer is your boss standing stationary on the Earth. The clock is in your car, moving at 30 m/s with respect to your boss.

Categorize   The low speed of 30 m/s suggests we might categorize this problem as one in which we use classical concepts and equations. Based on the problem statement that the moving clock runs more slowly than a stationary clock, however, we categorize this problem as one involving time dilation.

Analyze   The proper time interval, measured in the rest frame of the car, is $\Delta t_p = 5.0$ h.

Use Equation 38.8 to evaluate γ:

$$\gamma = \frac{1}{\sqrt{1 - \dfrac{v^2}{c^2}}} = \frac{1}{\sqrt{1 - \dfrac{(3.0 \times 10^1 \text{ m/s})^2}{(3.0 \times 10^8 \text{ m/s})^2}}} = \frac{1}{\sqrt{1 - 10^{-14}}}$$

If you try to determine this value on your calculator, you will probably obtain $\gamma = 1$. Instead, perform a binomial expansion:

$$\gamma = (1 - 10^{-14})^{-1/2} \approx 1 + \tfrac{1}{2}(10^{-14}) = 1 + 5.0 \times 10^{-15}$$

Use Equation 38.7 to find the dilated time interval measured by your boss:

$$\Delta t = \gamma \, \Delta t_p = (1 + 5.0 \times 10^{-15})(5.0 \text{ h})$$
$$= 5.0 \text{ h} + 2.5 \times 10^{-14} \text{ h} = \boxed{5.0 \text{ h} + 0.090 \text{ ns}}$$

Finalize   Your boss's clock would be only 0.090 ns ahead of your car clock. You might want to think of another excuse!

The Twin Paradox

An intriguing consequence of time dilation is the *twin paradox* (Fig. 38.9). Consider an experiment involving a set of twins named Speedo and Goslo. When they are 20 years old, Speedo, the more adventuresome of the two, sets out on an epic journey from the Earth to Planet X, located 20 light-years away. One light-year (ly) is the distance light travels through free space in 1 year. Furthermore, Speedo's spacecraft is capable of reaching a speed of $0.95c$ relative to the inertial frame of his twin brother back home on the Earth. After reaching Planet X, Speedo becomes

As Speedo (on the left) leaves his brother on Earth, both twins are the same age.

When Speedo returns from his journey, Goslo (on the right) is much older than Speedo.

Figure 38.9   The twin paradox. Speedo takes a journey to a star 20 light-years away and returns to the Earth.

homesick and immediately returns to the Earth at the same speed 0.95c. Upon his return, Speedo is shocked to discover that Goslo has aged 42 years and is now 62 years old. Speedo, on the other hand, has aged only 13 years.

The paradox is *not* that the twins have aged at different rates. Here is the apparent paradox. From Goslo's frame of reference, he was at rest while his brother traveled at a high speed away from him and then came back. According to Speedo, however, he himself remained stationary while Goslo and the Earth raced away from him and then headed back. Therefore, we might expect Speedo to claim that Goslo ages more slowly than himself. The situation appears to be symmetrical from either twin's point of view. Which twin *actually* ages more slowly?

The situation is actually not symmetrical. Consider a third observer moving at a constant speed relative to Goslo. According to the third observer, Goslo never changes inertial frames. Goslo's speed relative to the third observer is always the same. The third observer notes, however, that Speedo accelerates during his journey when he speeds up from zero to a speed of 0.95c relative to Earth and then again when he slows down and starts moving back toward the Earth, *changing reference frames in the process*. From the third observer's perspective, there is something very different about the motion of Goslo when compared to Speedo. Therefore, there is no paradox: only Goslo, who is always in a single inertial frame, can make correct predictions based on special relativity. Goslo finds that instead of aging 42 years, Speedo ages only $(1 - v^2/c^2)^{1/2}(42 \text{ years}) = [1 - (0.95)^2]^{1/2} (42 \text{ years}) = 13$ years. Of these 13 years, Speedo spends 6.5 years traveling to Planet X and 6.5 years returning.

So, in the opening storyline, who should you have punch your timecard if you are the traveling twin? You should definitely leave the timecard at home and have your boss punch it. More time will pass on Earth than for you, and you will end up much richer! You will have collected 42 years of pay, but you have only aged by 13 years, so you will have many years left to spend your treasure!

Length Contraction

The measured distance between two points in space also depends on the frame of reference of the observer. The **proper length** L_p of an object is the length measured by an observer *at rest relative to the object*. The length of an object measured by someone in a reference frame that is moving with respect to the object is always less than the proper length. This effect is known as **length contraction.**

To understand length contraction, consider a spacecraft traveling with a speed v from one star to another. There are two observers: one on the Earth and the other in the spacecraft. The observer at rest on the Earth (and also assumed to be at rest with respect to the two stars) measures the distance between the stars to be the proper length L_p. According to this observer, the time interval required for the spacecraft to complete the voyage is given by the particle under constant velocity model as $\Delta t = L_p/v$. The passages of the two stars by the spacecraft occur at the same position for the space traveler. Therefore, the space traveler measures the proper time interval Δt_p. Because of time dilation, the proper time interval is related to the Earth-measured time interval by $\Delta t_p = \Delta t/\gamma$. Because the space traveler reaches the second star in the time Δt_p, he or she concludes that the distance L between the stars is

$$L = v \, \Delta t_p = v \frac{\Delta t}{\gamma}$$

Because the proper length is $L_p = v \, \Delta t$, we see that

$$L = \frac{L_p}{\gamma} = L_p \sqrt{1 - \frac{v^2}{c^2}} \qquad (38.9)$$

◀ Length contraction

where $\sqrt{1 - v^2/c^2}$ is a factor less than unity. If an object has a proper length L_p when it is measured by an observer at rest with respect to the object, its length

PITFALL PREVENTION 38.4
The Proper Length As with the proper time interval, it is *very* important in relativistic calculations to correctly identify the observer who measures the proper length. The proper length between two points in space is always the length measured by an observer at rest with respect to the points. Often, the proper time interval and the proper length are *not* measured by the same observer.

A meterstick measured by an observer in a frame attached to the stick has its proper length L_p.

a

A meterstick measured by an observer in a frame in which the stick has a velocity relative to the frame is measured to be shorter than its proper length.

b

Figure 38.10 The length of a meterstick is measured by two observers.

L when it moves with speed v in a direction parallel to its length is measured to be shorter according to Equation 38.9.

For example, suppose a meterstick moves past a stationary Earth-based observer with speed v as in Figure 38.10. The length of the meterstick as measured by an observer in a frame attached to the stick is the proper length L_p shown in Figure 38.10a. The length of the stick L measured by the Earth observer is shorter than L_p by the factor $(1 - v^2/c^2)^{1/2}$ as suggested in Figure 38.10b. Notice that length contraction takes place only along the direction of motion.

Length contraction can also be applied to the twin paradox. Speedo measures the distance to Planet X as $L = L_p(1 - v^2/c^2)^{1/2} = (20\ \text{ly})[1 - (0.95)^2]^{1/2} = 6.2$ ly. At a speed of 0.95c, this trip requires a time interval of $(6.2\ \text{ly})/[(0.95)(1\ \text{ly/yr})] = 6.6$ yr. Multiplying by 2 to include both the outgoing and return trips gives 13 years, as we found using time dilation.

The proper length and the proper time interval are defined differently. The proper length is measured by an observer for whom the endpoints of the length remain fixed in space. The proper time interval is measured by someone for whom the two events take place at the same position in space. As an example of this point, let's return to the decaying muons moving at speeds close to the speed of light. An observer in the muon's reference frame measures the proper lifetime, whereas an Earth-based observer measures the proper length (the distance between the creation point and the decay point in Fig. 38.8b). In the muon's reference frame, there is no time dilation, but the distance of travel to the surface is shorter when measured in this frame. Likewise, in the Earth observer's reference frame, there is time dilation, but the distance of travel is measured to be the proper length. Therefore, when calculations on the muon are performed in both frames, the outcome of the experiment in one frame is the same as the outcome in the other frame: more muons reach the surface than would be predicted without relativistic effects.

Ⓠ UICK QUIZ 38.5 You are packing for a trip to another star. During the journey, you will be traveling at 0.99c. You are trying to decide whether you should buy smaller sizes of your clothing because you will be thinner on your trip due to length contraction. You also plan to save money by reserving a smaller cabin to sleep in because you will be shorter when you lie down. Should you **(a)** buy smaller sizes of clothing, **(b)** reserve a smaller cabin, **(c)** do neither of these things, or **(d)** do both of these things?

Ⓠ UICK QUIZ 38.6 You are observing a spacecraft moving away from you. You measure it to be shorter than when it was at rest on the ground next to you. You also see a clock through the spacecraft window, and you observe that the passage of time on the clock is measured to be slower than that of the watch on your wrist. Compared with when the spacecraft was on the ground, what do you measure if the spacecraft turns around and comes *toward* you at the same speed? **(a)** The spacecraft is measured to be longer, and the clock runs faster. **(b)** The spacecraft is measured to be longer, and the clock runs slower. **(c)** The spacecraft is measured to be shorter, and the clock runs faster. **(d)** The spacecraft is measured to be shorter, and the clock runs slower.

Space–Time Graphs

It is sometimes helpful to represent a physical situation with a **space–time graph,** in which ct is the ordinate and position x is the abscissa. The twin paradox is displayed in such a graph in Figure 38.11 from Goslo's point of view. A path through space–time is called a **world-line.** At the origin, the world-lines of Speedo (blue) and Goslo (green) coincide because the twins are in the same location at the same time. After Speedo leaves on his trip, his world-line diverges from that of his brother. Goslo's world-line is vertical because he remains fixed in location with respect to the Earth. At Goslo and Speedo's reunion, the two world-lines again come together. It would

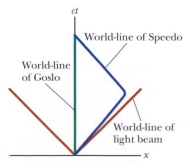

Figure 38.11 The twin paradox on a space–time graph. The twin who stays on the Earth has a world-line along the ct axis (green). The path of the traveling twin through space–time is represented by a world-line that changes direction (blue). The red-brown lines are world-lines for light beams traveling in the positive x direction (on the right) or the negative x direction (on the left).

be impossible for Speedo to have a world-line that crossed the path of a light beam that left the Earth when he did. To do so would require him to have a speed greater than c (which, as shown in Sections 38.6 and 38.7, is not possible).

World-lines for light beams are diagonal lines on space–time graphs, typically drawn at 45° to the right or left of vertical (assuming the x and ct axes have the same scales), depending on whether the light beam is traveling in the direction of increasing or decreasing x. All possible future events for Goslo and Speedo lie above the x axis and between the red-brown lines in Figure 38.11 because neither twin can travel faster than light. The only past events that Goslo and Speedo could have experienced occur between two similar 45° world-lines that approach the origin from below the x axis.

If Figure 38.11 is rotated about the ct axis, the red-brown lines sweep out a cone, called the *light cone*, which generalizes Figure 38.11 to two space dimensions. The y axis can be imagined coming out of the page. All future events for an observer at the origin must lie within the light cone. We can imagine another rotation that would generalize the light cone to three space dimensions to include z, but because of the requirement for four dimensions (three space dimensions and time), we cannot represent this situation in a two-dimensional drawing on paper or on a computer screen.

Example 38.3 **A Voyage to Sirius**

An astronaut takes a trip to Sirius, which is located a distance of 8 light-years from the Earth. The astronaut measures the time of the one-way journey to be 6 years. If the spaceship moves at a constant speed of 0.8c, how can the 8-ly distance be reconciled with the 6-year trip time measured by the astronaut?

SOLUTION

Conceptualize An observer on the Earth measures light to require 8 years to travel between Sirius and the Earth. The astronaut measures a shorter time interval for his travel of only 6 years. Is the astronaut traveling faster than light?

Categorize Because the astronaut is measuring a length of space between the Earth and Sirius that is in motion with respect to her, we categorize this example as a length contraction problem. We also model the astronaut as a *particle under constant velocity*.

Analyze The distance of 8 ly represents the proper length from the Earth to Sirius measured by an observer on the Earth seeing both objects nearly at rest.

Calculate the contracted length measured by the astronaut using Equation 38.9:

$$L = \frac{8 \text{ ly}}{\gamma} = (8 \text{ ly}) \sqrt{1 - \frac{v^2}{c^2}} = (8 \text{ ly}) \sqrt{1 - \frac{(0.8c)^2}{c^2}} = 5 \text{ ly}$$

Use the particle under constant velocity model to find the travel time measured on the astronaut's clock:

$$\Delta t = \frac{L}{v} = \frac{5 \text{ ly}}{0.8c} = \frac{5 \text{ ly}}{0.8(1 \text{ ly/yr})} = 6 \text{ yr}$$

Finalize Notice that we have used the value for the speed of light as $c = 1$ ly/yr. The trip takes a time interval shorter than 8 years for the astronaut because, to her, the distance between the Earth and Sirius is measured to be shorter.

WHAT IF? What if this trip is observed with a very powerful telescope by a technician in Mission Control on the Earth? At what time will this technician *see* that the astronaut has arrived at Sirius?

Answer The time interval the technician measures for the astronaut to arrive is

$$\Delta t = \frac{L_p}{v} = \frac{8 \text{ ly}}{0.8c} = 10 \text{ yr}$$

For the technician to *see* the arrival, the light from the scene of the arrival must travel back to the Earth and enter the telescope. This travel requires a time interval of

$$\Delta t = \frac{L_p}{v} = \frac{8 \text{ ly}}{c} = 8 \text{ yr}$$

Therefore, the technician sees the arrival after 10 yr + 8 yr = 18 yr. If the astronaut immediately turns around and comes back home, she arrives, according to the technician, 20 years after leaving, only 2 years *after the technician saw her arrive!* In addition, the astronaut would have aged by only 12 years.

Example **38.4**   The Pole-in-the-Barn Paradox

The twin paradox, discussed earlier, is a classic "paradox" in relativity. Another classic "paradox" is as follows. Suppose a runner moving at $0.75c$ carries a horizontal pole 15 m long toward a barn that is 10 m long. The barn has front and rear doors that are initially open. An observer on the ground can instantly and simultaneously close and open the two doors by remote control. When the runner and the pole are inside the barn, the ground observer closes and then immediately opens both doors so that the runner and pole are momentarily captured inside the barn and then proceed to exit the barn from the back doorway. Do both the runner and the ground observer agree that the runner makes it safely through the barn?

SOLUTION

Conceptualize   From your everyday experience, you would be surprised to see a 15-m pole fit inside a 10-m barn, but we are becoming used to surprising results in relativistic situations.

Categorize   The pole is in motion with respect to the ground observer so that the observer measures its length to be contracted, whereas the stationary barn has a proper length of 10 m. We categorize this example as a length contraction problem. The runner carrying the pole is modeled as a *particle under constant velocity*.

Analyze   Use Equation 38.9 to find the contracted length of the pole according to the ground observer:

$$L_{pole} = L_p \sqrt{1 - \frac{v^2}{c^2}} = (15 \text{ m})\sqrt{1 - (0.75)^2} = 9.9 \text{ m}$$

Therefore, the ground observer measures the pole to be slightly shorter than the barn and there is no problem with momentarily capturing the pole inside it. The "paradox" arises when we consider the runner's point of view.

Use Equation 38.9 to find the contracted length of the barn according to the running observer:

$$L_{barn} = L_p \sqrt{1 - \frac{v^2}{c^2}} = (10 \text{ m})\sqrt{1 - (0.75)^2} = 6.6 \text{ m}$$

Because the pole is in the rest frame of the runner, the runner measures it to have its proper length of 15 m. Now the situation looks even worse: How can a 15-m pole fit inside a *6.6-m* barn? Although this question is the classic one that is often asked, it is not the question we have asked because it is not the important one. We asked, "*Does the runner make it safely through the barn?*"

The resolution of the "paradox" lies in the relativity of simultaneity. The closing of the two doors is measured to be simultaneous by the ground observer. Because the doors are at different positions, however, they do not close simultaneously as measured by the runner. Both doors are open as the pole enters the barn. The rear door closes first and then opens, allowing the leading end of the pole to exit. The front door of the barn remains open in the meantime and does not close until the trailing end of the pole passes by.

We can analyze this "paradox" using a space–time graph. Figure 38.12a is a space–time graph from the ground observer's point of view. We choose $x = 0$ as the position of the front doorway of the barn and $t = 0$ as the instant at which the leading end of the pole is located at the front doorway of the barn. The world-lines for the two doorways of the barn are separated by 10 m and are vertical because the barn is not moving relative to this observer. For the pole, we follow two tilted world-lines, one for each end of the moving pole. These world-lines are 9.9 m apart horizontally, which is the contracted length seen by the ground observer. As seen in Figure 38.12a, the pole is entirely within the barn at some time.

Figure 38.12b shows the space–time graph according to the runner. Here, the world-lines for the pole are separated by 15 m and are vertical because the pole is at rest in the runner's frame of reference. The barn is hurtling *toward* the runner, so the world-lines for the front and rear doorways of the barn are tilted to the left. The world-lines for the barn are separated by 6.6 m, the contracted length as seen by the runner. The leading end of the pole leaves the

a

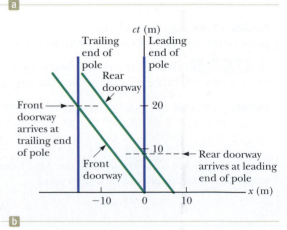

b

Figure 38.12   (Example 38.4) Space–time graphs for the pole-in-the-barn paradox (a) from the ground observer's point of view and (b) from the runner's point of view.

38.4 c o n t i n u e d

rear doorway of the barn long before the trailing end of the pole enters the barn. Therefore, the closing and opening of the rear door occur before the closing of the front door.

From the ground observer's point of view, use the particle under constant velocity model to find the time after $t = 0$ at which the trailing end of the pole enters the barn:

$$(1) \quad t = \frac{\Delta x}{v} = \frac{9.9 \text{ m}}{0.75c} = \frac{13.2 \text{ m}}{c}$$

From the runner's point of view, use the particle under constant velocity model to find the time at which the leading end of the pole leaves the barn:

$$(2) \quad t = \frac{\Delta x}{v} = \frac{6.6 \text{ m}}{0.75c} = \frac{8.8 \text{ m}}{c}$$

Find the time at which the trailing end of the pole enters the front door of the barn:

$$(3) \quad t = \frac{\Delta x}{v} = \frac{15 \text{ m}}{0.75c} = \frac{20 \text{ m}}{c}$$

Finalize   From Equation (1), the pole should be completely inside the barn at a time corresponding to $ct = 13.2$ m. So this is the time at which the doors quickly close and open. This situation is consistent with the point on the ct axis in Figure 38.12a where the pole is inside the barn. Now let's move to the runner's point of view. From Equation (2), the leading end of the pole leaves the barn at $ct = 8.8$ m. The closing and re-opening of the rear doors occur just before this time. This situation is consistent with the point on the ct axis in Figure 38.12b where the rear doorway of the barn arrives at the leading end of the pole. Equation (3) gives $ct = 20$ m, which agrees with the instant shown in Figure 38.12b at which the front doorway of the barn arrives at the trailing end of the pole. The front doors close just after this time.

The Relativistic Doppler Effect

Another important consequence of time dilation is the shift in frequency observed for light emitted by atoms in motion as opposed to light emitted by atoms at rest. This phenomenon, known as the Doppler effect, was introduced in Chapter 16 as it pertains to sound waves. In the case of sound, the velocity v_S of the source with respect to the medium of propagation (the air) can be distinguished from the velocity v_O of the observer with respect to the medium. Light waves must be analyzed differently, however, because *they require no medium of propagation,* and no method exists for distinguishing the velocity of a light source from the velocity of the observer. The only measurable velocity is the *relative velocity* v between the source and the observer.

If a light source and an observer approach each other with a relative speed v, the frequency f' measured by the observer is

$$f' = \frac{\sqrt{1 + v/c}}{\sqrt{1 - v/c}} f \qquad (38.10)$$

where f is the frequency of the source measured in its rest frame. This relativistic Doppler shift equation, unlike the Doppler shift equation for sound, depends only on the relative speed v of the source and observer and holds for relative speeds as great as c. As you might expect, the equation predicts that $f' > f$ when the source and observer approach each other (that is, when v is positive). Such a change to higher frequencies, or lower wavelengths, is called a *blueshift.* We obtain the expression for the case in which the source and observer recede from each other by substituting negative values for v in Equation 38.10. In this case, the shift is to lower frequencies, or longer wavelengths, and is described as a *redshift.*

The most spectacular and dramatic use of the relativistic Doppler effect is the measurement of shifts in the frequency of light emitted by a moving astronomical object such as a galaxy. Light emitted by atoms and normally found in the extreme violet region of the spectrum is shifted toward the red end of the spectrum for atoms in other galaxies, indicating that these galaxies are *receding* from us. American astronomer Edwin Hubble (1889–1953) performed extensive measurements of this redshift to confirm that most galaxies are moving away from us, indicating that the Universe is expanding (Chapter 44).

Figure 38.13 Events occur at points P and Q and are observed by an observer at rest in the S frame and another in the S′ frame, which is moving to the right with a speed v.

38.5 The Lorentz Transformation Equations

We looked at time intervals and lengths in Section 38.4. Let's now look at observations of specific positions in space and instants of time made by different observers, with an eye toward replacing the Galilean transformation equations with something more general.

Suppose two events occur at points P and Q and are reported by two observers, one at rest in a frame S and another in a frame S′ that is moving to the right with speed v as in Figure 38.13. The observer in S reports the events with space–time coordinates denoted as (x, y, z, t), and the observer in S′ reports the same events using the coordinates denoted as (x', y', z', t'). Equation 38.1 predicts that the distance between the two points P and Q in space at which the events occur does not depend on motion of the observer: $\Delta x = \Delta x'$. Because this prediction is contradictory to the notion of length contraction, the Galilean transformation is not valid when v approaches the speed of light. In this section, we present the correct transformation equations that apply for all speeds in the range $0 < v < c$.

The equations that are valid for all speeds and that enable us to transform coordinates from S to S′ are the **Lorentz transformation equations:**

◀ Lorentz transformation for S → S′

$$x' = \gamma(x - vt) \qquad y' = y \qquad z' = z \qquad t' = \gamma\left(t - \frac{v}{c^2}x\right) \qquad (38.11)$$

These transformation equations were developed by Hendrik A. Lorentz (1853–1928) in 1890 in connection with electromagnetism. It was Einstein, however, who recognized their physical significance and took the bold step of interpreting them within the framework of the special theory of relativity.

Notice the difference between the Galilean and Lorentz time equations. In the Galilean case, $t = t'$. In the Lorentz case, however, the value for t' assigned to an event by an observer O' in the S′ frame in Figure 38.13 depends both on the time t and on the coordinate x as measured by an observer O in the S frame, which is consistent with the notion that an event is characterized by four space–time coordinates (x, y, z, t). In other words, in relativity, space and time are *not* separate concepts but rather are closely interwoven with each other into something called *spacetime*.

If you wish to transform coordinates in the S′ frame to coordinates in the S frame, simply replace v by $-v$ and interchange the primed and unprimed coordinates in Equation 38.11:

◀ Inverse Lorentz transformation for S′ → S

$$x = \gamma(x' + vt') \qquad y = y' \qquad z = z' \qquad t = \gamma\left(t' + \frac{v}{c^2}x'\right) \qquad (38.12)$$

When $v \ll c$, the Lorentz transformation equations should reduce to the Galilean equations. As v approaches zero, $v/c \ll 1$; therefore, $\gamma \to 1$ and Equation 38.11 indeed reduces to the Galilean space–time transformation equations in Equation 38.1.

In many situations, we would like to know the difference in coordinates between two events or the time interval between two events as seen by observers O and O'. From Equations 38.11 and 38.12, we can express the differences between the four variables x, x', t, and t' in the form

$$\left.\begin{array}{l}\Delta x' = \gamma(\Delta x - v\,\Delta t) \\[2mm] \Delta t' = \gamma\left(\Delta t - \dfrac{v}{c^2}\Delta x\right)\end{array}\right\} \text{S} \rightarrow \text{S}' \qquad (38.13)$$

$$\left.\begin{array}{l}\Delta x = \gamma(\Delta x' + v\,\Delta t') \\[2mm] \Delta t = \gamma\left(\Delta t' + \dfrac{v}{c^2}\Delta x'\right)\end{array}\right\} \text{S}' \rightarrow \text{S} \qquad (38.14)$$

where $\Delta x' = x'_2 - x'_1$ and $\Delta t' = t'_2 - t'_1$ are the differences measured by observer O' and $\Delta x = x_2 - x_1$ and $\Delta t = t_2 - t_1$ are the differences measured by observer O. (We have not included the expressions for relating the y and z coordinates because they are unaffected by motion along the x direction.[3])

Example 38.5 Simultaneity and Time Dilation Revisited

(A) Use the Lorentz transformation equations in difference form to show that simultaneity is not an absolute concept.

SOLUTION

Conceptualize Imagine two events that are simultaneous and separated in space as measured in the S′ frame such that $\Delta t' = 0$ and $\Delta x' \neq 0$. These measurements are made by an observer O' who is moving with speed v relative to O.

Categorize The statement of the problem tells us to categorize this example as one involving the use of the Lorentz transformation.

Analyze From the expression for Δt given in Equation 38.14, find the time interval Δt measured by observer O:

$$\Delta t = \gamma\left(\Delta t' + \frac{v}{c^2}\Delta x'\right) = \gamma\left(0 + \frac{v}{c^2}\Delta x'\right) = \gamma\frac{v}{c^2}\Delta x'$$

Finalize The time interval for the same two events as measured by O is nonzero, so the events do not appear to be simultaneous to O.

(B) Use the Lorentz transformation equations in difference form to show that a moving clock is measured to run more slowly than a clock that is at rest with respect to an observer.

SOLUTION

Conceptualize Imagine that observer O' carries a clock that he uses to measure a time interval $\Delta t'$. He finds that two events occur at the same place in his reference frame ($\Delta x' = 0$) but at different times ($\Delta t' \neq 0$). Observer O' is moving with speed v relative to O.

Categorize The statement of the problem tells us to categorize this example as one involving the use of the Lorentz transformation.

Analyze From the expression for Δt given in Equation 38.14, find the time interval Δt measured by observer O:

$$\Delta t = \gamma\left(\Delta t' + \frac{v}{c^2}\Delta x'\right) = \gamma\left[\Delta t' + \frac{v}{c^2}(0)\right] = \gamma\,\Delta t'$$

Finalize This result is the equation for time dilation found earlier (Eq. 38.7), where $\Delta t' = \Delta t_p$ is the proper time interval measured by the clock carried by observer O'. Therefore, O measures the moving clock to run slow. Notice that the two events must occur *at the same location* in S′ in order to reproduce Equation 38.7.

38.6 The Lorentz Velocity Transformation Equations

Now that we have modified Equation 38.1 to be correct relativistically, let's see how to modify the Galilean velocity transformation in Equation 38.2. Suppose two observers in relative motion with respect to each other are both observing an object's motion. Previously, we defined an event as occurring at an instant of time. Now let's interpret the "event" as the object's motion. We know that the Galilean velocity transformation (Eq. 38.2) is valid for low speeds. How do the observers' measurements of the velocity of the object relate to each other if the speed of the object or the relative speed of the observers is close to that of light? Once again, S′

---

[3]Although relative motion of the two frames along the x axis does not change the y and z coordinates of an object, it does change the y and z *velocity* components of an object moving in either frame as noted in Section 38.6.

is our frame moving at a speed v relative to S. Suppose an object has a velocity component u_x' measured in the S' frame, where

$$u_x' = \frac{dx'}{dt'} \tag{38.15}$$

Using Equation 38.11, we have

$$dx' = \gamma(dx - v\,dt)$$

$$dt' = \gamma\left(dt - \frac{v}{c^2}\,dx\right)$$

Substituting these values into Equation 38.15 gives

$$u_x' = \frac{dx - v\,dt}{dt - \dfrac{v}{c^2}\,dx} = \frac{\dfrac{dx}{dt} - v}{1 - \dfrac{v}{c^2}\dfrac{dx}{dt}}$$

The term dx/dt, however, is simply the velocity component u_x of the object measured by an observer in S, so this expression becomes

◀ Lorentz velocity transformation for S → S′

$$u_x' = \frac{u_x - v}{1 - \dfrac{u_x v}{c^2}} \tag{38.16}$$

If the object has velocity components along the y and z axes, the components as measured by an observer in S′ are

$$u_y' = \frac{u_y}{\gamma\left(1 - \dfrac{u_x v}{c^2}\right)} \quad \text{and} \quad u_z' = \frac{u_z}{\gamma\left(1 - \dfrac{u_x v}{c^2}\right)} \tag{38.17}$$

Notice that u_y' and u_z' do not contain the parameter v in the numerator because the relative velocity is along the x axis.

When v is much smaller than c (the nonrelativistic case), the denominator of Equation 38.16 approaches unity and so $u_x' \approx u_x - v$, which is the Galilean velocity transformation equation. In another extreme, when $u_x = c$, Equation 38.16 becomes

$$u_x' = \frac{c - v}{1 - \dfrac{cv}{c^2}} = \frac{c\left(1 - \dfrac{v}{c}\right)}{1 - \dfrac{v}{c}} = c$$

This result shows that a speed measured as c by an observer in S is also measured as c by an observer in S′, independent of the relative motion of S and S′. This conclusion is consistent with Einstein's second postulate in Section 38.3: the speed of light must be c relative to all inertial reference frames. Furthermore, we find that the speed of an object can never be measured as larger than c. That is, *the speed of light is the ultimate speed*. We shall return to this point later.

To obtain u_x in terms of u_x', we replace v by $-v$ in Equation 38.16 and interchange the roles of u_x and u_x':

$$u_x = \frac{u_x' + v}{1 + \dfrac{u_x' v}{c^2}} \tag{38.18}$$

QUICK QUIZ **38.7** You are driving on a freeway at a relativistic speed. **(i)** Straight ahead of you, a technician standing on the ground turns on a searchlight and a beam of light moves exactly vertically upward as seen by the technician. As you observe the beam of light, do you measure the magnitude of the vertical component of its velocity as (a) equal to c, (b) greater than c, or (c) less than c? **(ii)** If the technician aims the searchlight directly at you instead of upward, do you measure the magnitude of the horizontal component of its velocity as (a) equal to c, (b) greater than c, or (c) less than c?

Example 38.6 Relative Velocity of Two Spacecraft

Two spacecraft A and B are moving in opposite directions as shown in Figure 38.14. An observer on the Earth measures the speed of spacecraft A to be $0.750c$ and the speed of spacecraft B to be $0.850c$. Find the velocity of spacecraft B as observed by the crew on spacecraft A.

Figure 38.14 (Example 38.6) Two spacecraft A and B move in opposite directions. The speed of spacecraft B relative to spacecraft A is *less* than c and is obtained from the relativistic velocity transformation equation.

SOLUTION

Conceptualize There are two observers, one (O) on the Earth and one (O') on spacecraft A. The event is the motion of spacecraft B.

Categorize Because the problem asks to find an observed velocity, we categorize this example as one requiring the Lorentz velocity transformation.

Analyze The Earth-based observer at rest in the S frame makes two measurements, one of each spacecraft. We want to find the velocity of spacecraft B as measured by the crew on spacecraft A. Therefore, $u_x = -0.850c$. The velocity of spacecraft A is also the velocity of the observer at rest in spacecraft A (the S′ frame) relative to the observer at rest on the Earth. Therefore, $v = 0.750c$.

Obtain the velocity u'_x of spacecraft B relative to spacecraft A using Equation 38.16:

$$u'_x = \frac{u_x - v}{1 - \frac{u_x v}{c^2}} = \frac{-0.850c - 0.750c}{1 - \frac{(-0.850c)(0.750c)}{c^2}} = \boxed{-0.977c}$$

Finalize The negative sign indicates that spacecraft B is moving in the negative x direction as observed by the crew on spacecraft A. Is that consistent with your expectation from Figure 38.14? Notice that the speed is less than c. That is, an object whose speed is less than c in one frame of reference must have a speed less than c in any other frame. (Had you used the Galilean velocity transformation equation in this example, you would have found that $u'_x = u_x - v = -0.850c - 0.750c = -1.60c$, which is impossible. The Galilean transformation equation does not work in relativistic situations.)

WHAT IF? What if the two spacecraft pass each other? What is their relative speed now?

Answer The calculation using Equation 38.16 involves only the velocities of the two spacecraft and does not depend on their locations. After they pass each other, they have the same velocities, so the velocity of spacecraft B as observed by the crew on spacecraft A is the same, $-0.977c$. The only difference after they pass is that spacecraft B is receding from spacecraft A, whereas it was approaching spacecraft A before it passed.

Example 38.7 Relativistic Leaders of the Pack

Two motorcycle pack leaders named Bethany and Keilah are racing at relativistic speeds along perpendicular paths as shown in Figure 38.15. How fast does Keilah recede as seen by Bethany over her right shoulder?

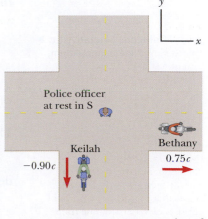

Figure 38.15 (Example 38.7) Bethany moves east with a speed $0.75c$ relative to the police officer, and Keilah travels south at a speed $0.90c$ relative to the officer.

SOLUTION

Conceptualize The two observers are Bethany and the police officer in Figure 38.15. The event is the motion of Keilah. Figure 38.15 represents the situation as seen by the police officer at rest in frame S. Frame S′ moves along with Bethany.

Categorize Because the problem asks to find an observed velocity, we categorize this problem as one requiring the Lorentz velocity transformation. The motion takes place in two dimensions.

continued

38.7 c o n t i n u e d

Analyze Identify the velocity components for Bethany and Keilah according to the police officer:

Bethany: $v_x = v = 0.75c$ $v_y = 0$
Keilah: $u_x = 0$ $u_y = -0.90c$

Using Equations 38.16 and 38.17, calculate u_x' and u_y' for Keilah as measured by Bethany:

$$u_x' = \frac{u_x - v}{1 - \frac{u_x v}{c^2}} = \frac{0 - 0.75c}{1 - \frac{(0)(0.75c)}{c^2}} = -0.75c$$

$$u_y' = \frac{u_y}{\gamma\left(1 - \frac{u_x v}{c^2}\right)} = \frac{\sqrt{1 - \frac{(0.75c)^2}{c^2}}(-0.90c)}{1 - \frac{(0)(0.75c)}{c^2}} = -0.60c$$

Using the Pythagorean theorem, find the speed of Keilah as measured by Bethany:

$$u' = \sqrt{(u_x')^2 + (u_y')^2} = \sqrt{(-0.75c)^2 + (-0.60c)^2} = \boxed{0.96c}$$

Finalize This speed is less than c, as required by the special theory of relativity.

38.7 Relativistic Linear Momentum

In the last few sections, we have investigated relativistic versions of kinematic variables: time, position, and velocity. Let us now move into the realm of dynamics and see what changes must be made to generalize the concepts of momentum and energy in relativity. We will find that we need new definitions of both of these quantities. These generalized definitions should reduce to the classical (nonrelativistic) definitions for $v \ll c$.

First, recall from the isolated system model that when two particles (or objects that can be modeled as particles) collide, the total momentum of the isolated system of the two particles remains constant. Suppose we observe this collision in a reference frame S and confirm that the momentum of the system is conserved. Now imagine that the momenta of the particles are measured by an observer in a second reference frame S' moving with velocity $\vec{v}$ relative to the first frame. Using the Lorentz velocity transformation equation and the classical definition of linear momentum, $\vec{p} = m\vec{u}$ (where $\vec{u}$ is the velocity of a particle), we find that linear momentum of the system is *not* measured to be conserved by the observer in S'. Because the laws of physics are the same in all inertial frames, however, linear momentum of the system *must* be conserved in all frames. We have a contradiction. In view of this contradiction and assuming the Lorentz velocity transformation equation is correct, we must modify the definition of linear momentum so that the momentum of an isolated system is conserved for all observers. For any particle, the correct relativistic equation for linear momentum that satisfies this condition is

Definition of relativistic ▶
linear momentum

$$\vec{p} \equiv \frac{m\vec{u}}{\sqrt{1 - \frac{u^2}{c^2}}} = \gamma m\vec{u} \tag{38.19}$$

where m is the mass of the particle and $\vec{u}$ is the velocity of the particle. When u is much less than c, $\gamma = (1 - u^2/c^2)^{-1/2}$ approaches unity and $\vec{p}$ approaches $m\vec{u}$. Therefore, the relativistic equation for $\vec{p}$ reduces to the classical expression when u is much smaller than c, as it should.

The relativistic force $\vec{F}$ acting on a particle whose linear momentum is $\vec{p}$ is defined as

$$\vec{F} \equiv \frac{d\vec{p}}{dt} \tag{38.20}$$

where $\vec{p}$ is given by Equation 38.19. This expression, which is the relativistic form of Newton's second law, is reasonable because it preserves classical mechanics in the limit of low velocities and is consistent with conservation of linear momentum for an isolated system ($\vec{F}_{ext} = 0$) both relativistically and classically.

It is left as an end-of-chapter problem (Problem 54) to show that under relativistic conditions, the acceleration $\vec{a}$ of a particle decreases under the action of a constant force, such that $a \propto (1 - u^2/c^2)^{3/2}$. This proportionality shows that as the particle's speed approaches c, the acceleration caused by any finite force approaches zero. Hence, it is impossible to accelerate a particle from rest to a speed $u \geq c$. This argument reinforces that the speed of light is the ultimate speed, the speed limit of the Universe. It is the maximum possible speed for a particle with mass as well as for information transfer.

Example 38.8 | **Linear Momentum of an Electron**

An electron, which has a mass of 9.11×10^{-31} kg, moves with a speed of $0.750c$. Find the magnitude of its relativistic momentum and compare this value with the momentum calculated from the classical expression.

SOLUTION

Conceptualize Imagine an electron moving with high speed. The electron carries momentum, but the magnitude of its momentum is not given by $p = mu$ because the speed is relativistic.

Categorize We categorize this example as a substitution problem involving a relativistic equation.

Use Equation 38.19 with $u = 0.750c$ to find the magnitude of the momentum:

$$p = \frac{m_e u}{\sqrt{1 - \dfrac{u^2}{c^2}}}$$

$$p = \frac{(9.11 \times 10^{-31}\ \text{kg})(0.750)(3.00 \times 10^8\ \text{m/s})}{\sqrt{1 - \dfrac{(0.750c)^2}{c^2}}}$$

$$= 3.10 \times 10^{-22}\ \text{kg} \cdot \text{m/s}$$

The classical expression (used incorrectly here) gives $p_{classical} = m_e u = 2.05 \times 10^{-22}$ kg · m/s. Hence, the correct relativistic result is 50% greater than the classical result!

38.8 Relativistic Energy

We have seen that the definition of linear momentum requires generalization to make it compatible with Einstein's postulates. This conclusion implies that the definition of kinetic energy must most likely be modified also.

To derive the relativistic form of the work–kinetic energy theorem, imagine a particle moving in one dimension along the x axis. A force in the x direction causes the momentum of the particle to change according to Equation 38.20. In what follows, we assume the particle is accelerated from rest to some final speed u. The work done by the force F on the particle is

$$W = \int_{x_1}^{x_2} F\, dx = \int_{x_1}^{x_2} \frac{dp}{dt}\, dx \qquad (38.21)$$

To perform this integration and find the work done on the particle and the relativistic kinetic energy as a function of u, we first evaluate dp/dt:

$$\frac{dp}{dt} = \frac{d}{dt}\frac{mu}{\sqrt{1 - \dfrac{u^2}{c^2}}} = \frac{m}{\left(1 - \dfrac{u^2}{c^2}\right)^{3/2}}\frac{du}{dt}$$

Substituting this expression for dp/dt and $dx = u\,dt$ into Equation 38.21 gives

$$W = \int_0^t \frac{m}{\left(1 - \dfrac{u^2}{c^2}\right)^{3/2}} \frac{du}{dt}(u\,dt) = m\int_0^u \frac{u}{\left(1 - \dfrac{u^2}{c^2}\right)^{3/2}}\,du$$

where we use the limits 0 and u in the integral because the integration variable has been changed from t to u. Evaluating the integral gives

$$W = \frac{mc^2}{\sqrt{1 - \dfrac{u^2}{c^2}}} - mc^2 \qquad (38.22)$$

Recall from Chapter 7 that the work done by a force acting on a system consisting of a single particle equals the change in kinetic energy of the particle: $W = \Delta K$. Because we assumed the initial speed of the particle is zero, its initial kinetic energy is zero, so $W = K - K_i = K - 0 = K$. Therefore, the work W in Equation 38.22 is equivalent to the relativistic kinetic energy K:

Relativistic kinetic energy ▶

$$K = \frac{mc^2}{\sqrt{1 - \dfrac{u^2}{c^2}}} - mc^2 = \gamma mc^2 - mc^2 = (\gamma - 1)mc^2 \qquad (38.23)$$

This equation is routinely confirmed by experiments using high-energy particle accelerators.

At low speeds, where $u/c \ll 1$, Equation 38.23 should reduce to the classical expression $K = \frac{1}{2}mu^2$. We can check that by using the binomial expansion $(1 - \beta^2)^{-1/2} \approx 1 + \frac{1}{2}\beta^2 + \cdots$ for $\beta \ll 1$, where the higher-order powers of β are neglected in the expansion. (In treatments of relativity, β is a common symbol used to represent u/c or v/c.) In our case, $\beta = u/c$, so

$$\gamma = \frac{1}{\sqrt{1 - \dfrac{u^2}{c^2}}} = \left(1 - \frac{u^2}{c^2}\right)^{-1/2} \approx 1 + \frac{1}{2}\frac{u^2}{c^2}$$

Substituting this result into Equation 38.23 gives

$$K \approx \left[\left(1 + \frac{1}{2}\frac{u^2}{c^2}\right) - 1\right]mc^2 = \frac{1}{2}mu^2 \quad \text{(for } u/c \ll 1\text{)}$$

which is the classical expression for kinetic energy. A graph comparing the relativistic and nonrelativistic expressions is given in Figure 38.16. In the relativistic case, the particle speed never exceeds c, regardless of the kinetic energy. The two curves are in good agreement when $u \ll c$.

The constant term mc^2 in Equation 38.23, which is independent of the speed of the particle, is called the **rest energy** E_R of the particle:

$$E_R = mc^2 \qquad (38.24)$$

Equation 38.24 shows that **mass is a form of energy,** where c^2 is simply a constant conversion factor. This expression also shows that a small mass corresponds to an enormous amount of energy, a concept fundamental to nuclear and elementary-particle physics.

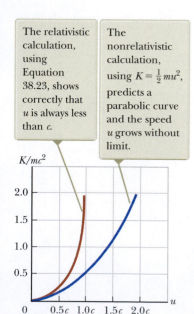

The relativistic calculation, using Equation 38.23, shows correctly that u is always less than c.

The nonrelativistic calculation, using $K = \frac{1}{2}mu^2$, predicts a parabolic curve and the speed u grows without limit.

K/mc^2

2.0

1.5

1.0

0.5

0   0.5c 1.0c 1.5c 2.0c u

Figure 38.16 A graph comparing relativistic and nonrelativistic kinetic energy of a moving particle. The energies are plotted as a function of particle speed u.

The term γmc^2 in Equation 38.23, which depends on the particle speed, is the sum of the kinetic and rest energies. It is called the **total energy** E:

Total energy = kinetic energy + rest energy

$$E = K + mc^2 \qquad (38.25)$$

or

$$E = \frac{mc^2}{\sqrt{1 - \dfrac{u^2}{c^2}}} = \gamma mc^2 \qquad (38.26)$$

◀ Total energy of a relativistic particle

In many situations, the linear momentum or energy of a particle rather than its speed is measured. It is therefore useful to have an expression relating the total energy E to the relativistic linear momentum p, which is accomplished by using the expressions $E = \gamma mc^2$ and $p = \gamma mu$. By squaring these equations and subtracting, we can eliminate u (Problem 36). The result, after some algebra, is[4]

$$E^2 = p^2c^2 + (mc^2)^2 \qquad (38.27)$$

◀ Energy–momentum relationship for a relativistic particle

When the particle is at rest, $p = 0$, so $E = E_R = mc^2$.

In Section 34.1, we discussed the fact that some early scientists believed in a particle nature of light. Since then, we have developed descriptions of the behavior of light using a wave theory. In Chapter 39, we will find that light indeed does have a particle nature! A particle of light has zero mass and is called a **photon.** For particles that have zero mass, such as photons, we set $m = 0$ in Equation 38.27 and find that

$$E = pc \qquad (38.28)$$

◀ Energy-momentum relationship for a photon

This equation is an exact expression relating total energy and linear momentum for photons, which always travel at the speed of light (in vacuum).

Finally, because the mass m of a particle is independent of its motion, m must have the same value in all reference frames. For this reason, m is often called the **invariant mass.** On the other hand, because the total energy and linear momentum of a particle both depend on velocity, these quantities depend on the reference frame in which they are measured.

When dealing with subatomic particles, it is convenient to express their energy in electron volts (Section 24.1) because the particles are usually given this energy by acceleration through a potential difference. The conversion factor, as you recall from Equation 24.5, is

$$1 \text{ eV} = 1.602 \times 10^{-19} \text{ J}$$

For example, the mass of an electron is 9.109×10^{-31} kg. Hence, the rest energy of the electron is

$$m_e c^2 = (9.109 \times 10^{-31} \text{ kg})(2.998 \times 10^8 \text{ m/s})^2 = 8.187 \times 10^{-14} \text{ J}$$

$$= (8.187 \times 10^{-14} \text{ J})(1 \text{ eV}/1.602 \times 10^{-19} \text{ J}) = 0.511 \text{ MeV}$$

Another way to represent this same idea is to express the mass in units of MeV/c^2 by dividing both sides of the previous equation by c^2:

$$m_e = 0.511 \frac{\text{MeV}}{c^2}$$

PITFALL PREVENTION 38.6
Watch Out for "Relativistic Mass"
Some older treatments of relativity maintained the conservation of momentum principle at high speeds by using a model in which a particle's mass increases with speed. You might still encounter this notion of "relativistic mass" in your outside reading, especially in older books. Be aware that this notion is no longer widely accepted; today, mass is considered as *invariant*, independent of speed. The mass of an object in all frames is considered to be the mass as measured by an observer at rest with respect to the object.

---

[4]One way to remember this relationship is to draw a right triangle having a hypotenuse of length E and legs of lengths pc and mc^2.

QUICK QUIZ 38.8 The following *pairs* of energies—particle 1: E, $2E$; particle 2: E, $3E$; particle 3: $2E$, $4E$—represent the rest energy and total energy of three different particles. Rank the particles from greatest to least according to their **(a)** mass, **(b)** kinetic energy, and **(c)** speed.

Example **38.9**   **The Energy of a Speedy Proton**

(A) Find the rest energy of a proton in units of electron volts.

SOLUTION

Conceptualize   Even if the proton is not moving, it has energy associated with its mass. If it moves, the proton possesses more energy, with the total energy being the sum of its rest energy and its kinetic energy.

Categorize   The phrase "rest energy" suggests we must take a relativistic rather than a classical approach to this problem.

Analyze   Use Equation 38.24 to find the rest energy:

$$E_R = m_p c^2 = (1.672\ 6 \times 10^{-27}\ \text{kg})(2.998 \times 10^8\ \text{m/s})^2$$

$$= (1.503 \times 10^{-10}\ \text{J})\left(\frac{1.00\ \text{MeV}}{1.602 \times 10^{-13}\ \text{J}}\right) = \boxed{938\ \text{MeV}}$$

(B) If the total energy of a proton is three times its rest energy, what is the speed of the proton?

SOLUTION

Use Equation 38.26 to relate the total energy of the proton to the rest energy:

$$E = 3m_p c^2 = \frac{m_p c^2}{\sqrt{1 - \dfrac{u^2}{c^2}}} \rightarrow 3 = \frac{1}{\sqrt{1 - \dfrac{u^2}{c^2}}}$$

Solve for u:

$$1 - \frac{u^2}{c^2} = \frac{1}{9} \rightarrow \frac{u^2}{c^2} = \frac{8}{9}$$

$$u = \frac{\sqrt{8}}{3}c = 0.943c = \boxed{2.83 \times 10^8\ \text{m/s}}$$

(C) Determine the kinetic energy of the proton in units of electron volts.

SOLUTION

Use Equation 38.25 to find the kinetic energy of the proton:

$$K = E - m_p c^2 = 3m_p c^2 - m_p c^2 = 2m_p c^2$$

$$= 2(938\ \text{MeV}) = \boxed{1.88 \times 10^3\ \text{MeV}}$$

(D) What is the proton's momentum?

SOLUTION

Use Equation 38.27 to calculate the momentum:

$$E^2 = p^2 c^2 + (m_p c^2)^2 = (3m_p c^2)^2$$

$$p^2 c^2 = 9(m_p c^2)^2 - (m_p c^2)^2 = 8(m_p c^2)^2$$

$$p = \sqrt{8}\ \frac{m_p c^2}{c} = \sqrt{8}\ \frac{938\ \text{MeV}}{c} = \boxed{2.65 \times 10^3\ \text{MeV/}c}$$

Finalize   The unit of momentum in part (D) is written MeV/c, which is a common unit in particle physics. For comparison, you might want to solve this example using classical equations.

WHAT IF?   In classical physics, if the momentum of a particle doubles, the kinetic energy increases by a factor of 4. What happens to the kinetic energy of the proton in this example if its momentum doubles?

Answer   Based on what we have seen so far in relativity, it is likely you would predict that its kinetic energy does not increase by a factor of 4.

Find the new doubled momentum:

$$p_{\text{new}} = 2\left(\sqrt{8}\ \frac{m_p c^2}{c}\right) = 4\sqrt{2}\ \frac{m_p c^2}{c}$$

38.9 continued

Use this result in Equation 38.27 to find the new total energy:

$$E_{new}^2 = p_{new}^2 c^2 + (m_p c^2)^2$$

$$E_{new}^2 = \left(4\sqrt{2}\,\frac{m_p c^2}{c}\right)^2 c^2 + (m_p c^2)^2 = 33(m_p c^2)^2$$

$$E_{new} = \sqrt{33}\,m_p c^2 = 5.7 m_p c^2$$

Use Equation 38.25 to find the new kinetic energy:

$$K_{new} = E_{new} - m_p c^2 = 5.7 m_p c^2 - m_p c^2 = 4.7 m_p c^2$$

This value is a little more than twice the kinetic energy found in part (C), not four times. In general, the factor by which the kinetic energy increases if the momentum doubles depends on the initial momentum, but it approaches 4 as the momentum approaches zero. In this latter situation, classical physics correctly describes the situation.

38.9 The General Theory of Relativity

Up to this point, we have sidestepped a curious puzzle. Mass has two seemingly different properties: a *gravitational attraction* for other masses and an *inertial* property that represents a resistance to acceleration. We first discussed these two attributes for mass in Section 5.5. To designate these two attributes, we add subscripts g and i to the masses and write modified versions of Equations 5.6 and 5.2:

Gravitational property (Eq. 5.6): $F_g = m_g g$

Inertial property (Eq. 5.2): $\sum F = m_i a$

The value for the gravitational constant G was chosen to make the magnitudes of m_g and m_i numerically equal. Regardless of how G is chosen, however, the strict proportionality of m_g and m_i has been established experimentally to an extremely high degree: a few parts in 10^{12}. Therefore, it appears that gravitational mass and inertial mass may indeed be exactly proportional.

Why, though? They seem to involve two entirely different concepts: a force of mutual gravitational attraction between two masses and the resistance of a single mass to being accelerated. This question, which puzzled Newton and many other physicists over the years, was answered by Einstein in 1916 when he published his theory of gravitation in *Annalen der Physik*, known as the *general theory of relativity*. Because it is a mathematically complex theory, we offer merely a hint of its elegance and insight.

In Einstein's view, the dual behavior of mass was evidence for a very intimate and basic connection between the two behaviors. He pointed out that no mechanical experiment (such as dropping an object) could distinguish between the two situations illustrated in Figures 38.17a and 38.17b (page 1040). In Figure 38.17a, a person standing in an elevator on the surface of a planet feels pressed into the floor due to the gravitational force. If he releases his briefcase, he observes it moving toward the floor with acceleration $\vec{\mathbf{g}} = -g\,\hat{\mathbf{j}}$. In Figure 38.17b, the person is in an elevator in empty space accelerating upward with $\vec{\mathbf{a}}_{el} = +g\,\hat{\mathbf{j}}$. The person feels pressed into the floor with the same force as in Figure 38.17a. If he releases his briefcase, he observes it moving toward the floor with acceleration g, exactly as in the previous situation. In Figure 38.17a, the person is at rest in an inertial frame in a gravitational field due to the planet. In Figure 38.17b, the person is in a noninertial frame accelerating in gravity-free space. Einstein's claim is that these two situations are completely equivalent.

Einstein carried this idea further and proposed that *no* experiment, mechanical or otherwise, could distinguish between the two situations. This extension to include all phenomena (not just mechanical ones) has interesting consequences. For example, suppose a light pulse is sent horizontally across the elevator as in Figure 38.17c, in which the elevator is accelerating upward in empty space. From the point of view of an observer in an inertial frame outside the elevator, the light

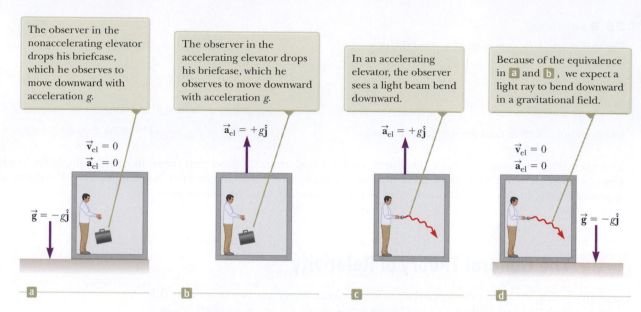

The observer in the nonaccelerating elevator drops his briefcase, which he observes to move downward with acceleration g.

The observer in the accelerating elevator drops his briefcase, which he observes to move downward with acceleration g.

In an accelerating elevator, the observer sees a light beam bend downward.

Because of the equivalence in **a** and **b**, we expect a light ray to bend downward in a gravitational field.

$\vec{\mathbf{v}}_{el} = 0$
$\vec{\mathbf{a}}_{el} = 0$

$\vec{\mathbf{a}}_{el} = +g\hat{\mathbf{j}}$

$\vec{\mathbf{a}}_{el} = +g\hat{\mathbf{j}}$

$\vec{\mathbf{v}}_{el} = 0$
$\vec{\mathbf{a}}_{el} = 0$

$\vec{\mathbf{g}} = -g\hat{\mathbf{j}}$

$\vec{\mathbf{g}} = -g\hat{\mathbf{j}}$

a   **b**   **c**   **d**

Figure 38.17 (a) The observer is at rest in an elevator in a uniform gravitational field $\vec{\mathbf{g}} = -g\,\hat{\mathbf{j}}$, directed downward. (b) The observer is in a region where gravity is negligible, but the elevator moves upward with an acceleration $\vec{\mathbf{a}}_{el} = +g\,\hat{\mathbf{j}}$. According to Einstein, the frames of reference in (a) and (b) are equivalent in every way. No local experiment can distinguish any difference between the two frames. (c) An observer watches a beam of light in an accelerating elevator. (d) Einstein's prediction of the behavior of a beam of light in a gravitational field.

travels in a straight line while the floor of the elevator accelerates upward. According to the observer on the elevator, however, the trajectory of the light pulse bends downward as the floor of the elevator (and the observer) accelerates upward. Therefore, based on the equality of parts (a) and (b) of the figure, Einstein proposed that a beam of light should also be bent downward by a gravitational field as in Figure 38.17d. Experiments have verified the effect, although the bending is small. A laser aimed at the horizon falls less than 1 cm after traveling 6 000 km in Earth's gravitational field. (No such bending is predicted in Newton's theory of gravitation.)

Einstein's **general theory of relativity** has two postulates:

- All the laws of nature have the same form for observers in any frame of reference, whether accelerated or not.
- In the vicinity of any point, a gravitational field is equivalent to an accelerated frame of reference in gravity-free space (the **principle of equivalence**).

One interesting effect predicted by the general theory is that time is altered by gravity. A clock in the presence of gravity runs slower than one located where gravity is negligible. Consequently, the frequencies of radiation emitted by atoms in the presence of a strong gravitational field are *redshifted* to lower frequencies when compared with the same emissions in the presence of a weak field. This gravitational redshift has been detected in spectral lines emitted by atoms in massive stars. It has also been verified on the Earth by comparing the frequencies of gamma rays emitted from nuclei separated vertically by about 20 m.

The second postulate suggests a gravitational field may be "transformed away" at any point if we choose an appropriate accelerated frame of reference, a freely falling one. Einstein developed an ingenious method of describing the acceleration necessary to make the gravitational field "disappear." He specified a concept, the *curvature of spacetime,* that describes the gravitational effect at every point. In Section 38.5, it was mentioned that space and time are not separate concepts, but are interwoven. The model of *spacetime* describes the Universe as having four *inseparable* dimensions, with three representing our classical notion of space and the fourth related to time. The curvature of spacetime completely replaces Newton's

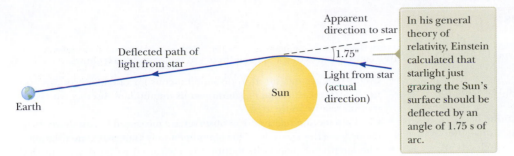

Apparent direction to star

Deflected path of light from star

1.75"

Light from star (actual direction)

Sun

Earth

In his general theory of relativity, Einstein calculated that starlight just grazing the Sun's surface should be deflected by an angle of 1.75 s of arc.

Figure 38.18 Deflection of starlight passing near the Sun. Because of this effect, the Sun or some other remote object can act as a *gravitational lens*.

gravitational theory. According to Einstein, there is no such thing as a gravitational force. Rather, the presence of a mass causes a curvature of spacetime in the vicinity of the mass, and this curvature dictates the spacetime path that all freely moving objects must follow. Therefore, the notion of gravitational mass is not needed. An object follows a path according to its inertial mass in curved spacetime.

As an example of the effects of curved spacetime, imagine two travelers moving on parallel paths a few meters apart on the surface of the Earth and maintaining an exact northward heading along two longitude lines. As they observe each other near the equator, they will claim that their paths are exactly parallel. As they approach the North Pole, however, they notice that they are moving closer together and will meet at the North Pole. Therefore, they claim that they moved along parallel paths, but moved toward each other, *as if there were an attractive force between them.* The travelers make this conclusion based on their everyday experience of moving on flat surfaces. From our mental representation, however, we realize they are walking on a curved surface, and it is the geometry of the curved surface, rather than an attractive force, that causes them to converge. In a similar way, general relativity replaces the notion of forces with the movement of objects through curved spacetime.

One prediction of the general theory of relativity is that a light ray passing near the Sun should be deflected in the curved spacetime created by the Sun's mass. This prediction was confirmed when astronomers detected the bending of starlight near the Sun during a total solar eclipse that occurred shortly after World War I (Fig. 38.18). When this discovery was announced, Einstein became an international celebrity.

If the concentration of mass becomes very great as is believed to occur when a large star exhausts its nuclear fuel and collapses to a very small volume, a **black hole** may form as discussed in Chapter 13. Here, the curvature of spacetime is so extreme that within a certain distance from the center of the black hole all matter and light become trapped as discussed in Section 13.6.

Summary

▶ **Definitions**

The relativistic expression for the **linear momentum** of a particle moving with a velocity $\vec{u}$ is

$$\vec{p} \equiv \frac{m\vec{u}}{\sqrt{1 - \dfrac{u^2}{c^2}}} = \gamma m \vec{u} \qquad (38.19)$$

The relativistic force $\vec{F}$ acting on a particle whose linear momentum is $\vec{p}$ is defined as

$$\vec{F} \equiv \frac{d\vec{p}}{dt} \qquad (38.20)$$

continued

❯ Concepts and Principles

The two basic postulates of the special theory of relativity are as follows:

- The laws of physics must be the same in all inertial reference frames.
- The speed of light in vacuum has the same value, $c = 3.00 \times 10^8$ m/s, in all inertial frames, regardless of the velocity of the observer or the velocity of the source emitting the light.

Three consequences of the special theory of relativity are as follows:

- Events that are measured to be simultaneous for one observer are not necessarily measured to be simultaneous for another observer who is in motion relative to the first.
- Clocks in motion relative to an observer are measured to run slower by a factor $\gamma = (1 - v^2/c^2)^{-1/2}$. This phenomenon is known as **time dilation.**
- The lengths of objects in motion are measured to be shorter in the direction of motion by a factor $1/\gamma = (1 - v^2/c^2)^{1/2}$. This phenomenon is known as **length contraction.**

To satisfy the postulates of special relativity, the Galilean transformation equations must be replaced by the **Lorentz transformation equations:**

$$x' = \gamma(x - vt) \quad y' = y \quad z' = z \quad t' = \gamma\left(t - \frac{v}{c^2}x\right) \quad \text{(38.11)}$$

where $\gamma = (1 - v^2/c^2)^{-1/2}$ and the S′ frame moves in the x direction at speed v relative to the S frame.

The relativistic form of the **Lorentz velocity transformation equation** is

$$u'_x = \frac{u_x - v}{1 - \frac{u_x v}{c^2}} \quad \text{(38.16)}$$

where u'_x is the x component of the velocity of an object as measured in the S′ frame and u_x is its component as measured in the S frame.

The relativistic expression for the **kinetic energy** of a particle is

$$K = \frac{mc^2}{\sqrt{1 - \frac{u^2}{c^2}}} - mc^2 = (\gamma - 1)mc^2 \quad \text{(38.23)}$$

The constant term mc^2 in Equation 38.23 is called the **rest energy** E_R of the particle:

$$E_R = mc^2 \quad \text{(38.24)}$$

The **total energy** E of a particle is given by

$$E = \frac{mc^2}{\sqrt{1 - \frac{u^2}{c^2}}} = \gamma mc^2 \quad \text{(38.26)}$$

The relativistic linear momentum of a particle is related to its total energy through the equation

$$E^2 = p^2c^2 + (mc^2)^2 \quad \text{(38.27)}$$

Think–Pair–Share

See the Preface for an explanation of the icons used in this problems set. For additional assessment items for this section, go to ⟆ WEB**ASSIGN** From Cengage

1. Your team is working on the detection of gravitational waves. The first detection of gravitational waves occurred in September 2015 and was announced in February 2016. Data from the detection indicated that two black holes of 29 and 36 solar masses had combined to form a black hole of 62 solar masses. The final collapse occurred in a time interval of 0.2 s, leading your team to claim that the power output of the collapse during that time interval was 50 times that of all the stars in the observable Universe. In June 2016, detection of a second combination of black holes was announced. In this case, black holes of 14.2 and 7.5 solar masses combined into a black hole of 20.8 solar masses in 1.0 s. Work again with your team to determine the following: How many times greater is the power released in this second event compared to that of all the stars in the observable Universe?

2. **ACTIVITY** The neutrino was initially considered to have zero mass. If the neutrino were massless, then it would travel at the speed of light, regardless of energy. Later experiments, however, indicate that the neutrino has a very small rest energy and therefore a small but finite mass. One piece of evidence for neutrino mass comes from arrivals of neutrinos from a supernova explosion. Consider the following data on arrival times at an Earth-based facility of two neutrinos from a supernova explosion. The supernova occurred in a star located 1.64×10^5 light-years from Earth.

Time of Arrival	Neutrino Energy (MeV)
7:35:41.37	38
7:35:46.96	24

Discuss these data in your group and determine an estimate for the mass of the neutrino.

Problems

See the Preface for an explanation of the icons used in this problems set. For additional assessment items for this section, go to ⚡ **WEB**ASSIGN From Cengage

SECTION 38.1 The Principle of Galilean Relativity

> Problems 28–31, 33, and 45 in Chapter 4, and Online-Only Problems 4.23 and 4.24 can be assigned with this section.

1. In a laboratory frame of reference, an observer notes that Newton's second law is valid. Assume forces and masses are measured to be the same in any reference frame for speeds small compared with the speed of light. (a) Show that Newton's second law is also valid for an observer moving at a constant speed, small compared with the speed of light, relative to the laboratory frame. (b) Show that Newton's second law is *not* valid in a reference frame moving past the laboratory frame with a constant acceleration.

2. **AMT** A car of mass 2 000 kg moving with a speed of 20.0 m/s collides and locks together with a 1 500-kg car at rest at a stop sign. Show that momentum is conserved in a reference frame moving at 10.0 m/s in the direction of the moving car.

SECTION 38.4 Consequences of the Special Theory of Relativity

> Problem 51 in Chapter 4 can be assigned with this section.

3. **Q|C** A meterstick moving at $0.900c$ relative to the Earth's surface approaches an observer at rest with respect to the Earth's surface. (a) What is the meterstick's length as measured by the observer? (b) Qualitatively, how would the answer to part (a) change if the observer started running toward the meterstick?

4. A muon formed high in the Earth's atmosphere is measured by an observer on the Earth's surface to travel at speed $v = 0.990c$ for a distance of 4.60 km before it decays into an electron, a neutrino, and an antineutrino ($\mu^- \rightarrow e^- + \nu + \bar{\nu}$). (a) For what time interval does the muon live as measured in its reference frame? (b) How far does the Earth travel as measured in the frame of the muon?

5. A deep-space vehicle moves away from the Earth with a speed of $0.800c$. An astronaut on the vehicle measures a time interval of 3.00 s to rotate her body through 1.00 rev as she floats in the vehicle. What time interval is required for this rotation according to an observer on the Earth?

6. **BIO** An astronaut is traveling in a space vehicle moving at $0.500c$ relative to the Earth. The astronaut measures her pulse rate at 75.0 beats per minute. Signals generated by the astronaut's pulse are radioed to the Earth when the vehicle is moving in a direction perpendicular to the line that connects the vehicle with an observer on the Earth. (a) What pulse rate does the Earth-based observer measure? (b) **What If?** What would be the pulse rate if the speed of the space vehicle were increased to $0.990c$?

7. For what value of v does $\gamma = 1.0100$? Observe that for speeds lower than this value, time dilation and length contraction are effects amounting to less than 1%.

8. **CR** You have been hired as an expert witness for an attorney who is representing a speeding driver. The driver of the car was given a ticket for running a red light at an intersection. According to the driver, who has taken some courses in physics, when he was looking at the red light as he approached the intersection, the Doppler shift made the light of wavelength 650 nm appear to be green light of wavelength 520 nm. Therefore, according to the driver, he should not be charged with running a red light because it appeared green to him. What advice do you give the attorney?

9. **AMT** **T** A spacecraft with a proper length of 300 m passes by an observer on the Earth. According to this observer, it takes 0.750 μs for the spacecraft to pass a fixed point. Determine the speed of the spacecraft as measured by the Earth-based observer.

10. **S** A spacecraft with a proper length of L_p passes by an observer on the Earth. According to this observer, it takes a time interval Δt for the spacecraft to pass a fixed point. Determine the speed of the object as measured by the Earth-based observer.

11. A light source recedes from an observer with a speed v_S that is small compared with c. (a) Show that the fractional shift in the measured wavelength is given by the approximate expression

$$\frac{\Delta\lambda}{\lambda} \approx \frac{v_S}{c}$$

This phenomenon is known as the *redshift* because the visible light is shifted toward the red. (b) Spectroscopic measurements of light at $\lambda = 397$ nm coming from a galaxy in Ursa Major reveal a redshift of 20.0 nm. What is the recessional speed of the galaxy?

12. A cube of steel has a volume of 1.00 cm³ and mass 8.00 g when at rest on the Earth. If this cube is now given a speed $u = 0.900c$, what is its density as measured by a stationary observer? Note that relativistic density is defined as E_R/c^2V.

13. **Q|C** **Review.** In 1963, astronaut Gordon Cooper orbited the Earth 22 times. The press stated that for each orbit, he aged two-millionths of a second less than he would have had he remained on the Earth. (a) Assuming Cooper was 160 km above the Earth in a circular orbit, determine the difference in elapsed time between someone on the Earth and the orbiting astronaut for the 22 orbits. You may use the approximation

$$\frac{1}{\sqrt{1-x}} \approx 1 + \frac{x}{2}$$

for small x. (b) Did the press report accurate information? Explain.

14. **CR** You have an assistantship with a math professor in a future world where space travel is common and spacecraft regularly achieve near-light speeds. A spacecraft has taken off recently to carry individuals to colonize an Earth-like planet around a nearby star. Your professor, who remains on Earth, is teaching the students on the spacecraft via the future version of distance learning. It is time for the students on the spacecraft to take a math exam. The professor wishes the students to have a time interval $\Delta t_p = 2.00$ h to complete the exam, so just as the spacecraft passes Earth on its last trip around the Sun at its constant cruising speed of $0.960c$, she

sends a signal to the proctor to have the students begin the exam. Knowing of your experience in physics courses, the professor asks you to determine the time interval through which she should wait before sending a radio signal to the departing spacecraft to tell the proctor to have the students stop working on the exam.

15. Police radar detects the speed of a car (Fig. P38.15) as follows. Microwaves of a precisely known frequency are broadcast toward the car. The moving car reflects the microwaves with a Doppler shift. The reflected waves are received and combined with an attenuated version of the transmitted wave. Beats occur between the two microwave signals. The beat frequency is measured. (a) For an electromagnetic wave reflected back to its source from a mirror approaching at speed v, show that the reflected wave has frequency

$$f' = \frac{c + v}{c - v} f$$

where f is the source frequency. (b) Noting that v is much less than c, show that the beat frequency can be written as $f_{beat} = 2v/\lambda$. (c) What beat frequency is measured for a car speed of 30.0 m/s if the microwaves have frequency 10.0 GHz? (d) If the beat frequency measurement in part (c) is accurate to ± 5.0 Hz, how accurate is the speed measurement?

Figure P38.15

SECTION 38.5 The Lorentz Transformation Equations

16. Shannon observes two light pulses to be emitted from the same location, but separated in time by 3.00 μs. Kimmie observes the emission of the same two pulses to be separated in time by 9.00 μs. (a) How fast is Kimmie moving relative to Shannon? (b) According to Kimmie, what is the separation in space of the two pulses?

17. A moving rod is observed to have a length of $\ell = 2.00$ m and to be oriented at an angle of $\theta = 30.0°$ with respect to the direction of motion as shown in Figure P38.17. The rod has a speed of 0.995c. (a) What is the proper length of the rod? (b) What is the orientation angle in the proper frame?

Figure P38.17

18. A rod moving with a speed v along the horizontal direction is observed to have length ℓ and to make an angle θ with respect to the horizontal as shown in Figure P38.17. (a) Show that the length of the rod as measured by an observer at rest with respect to the rod is $\ell_p = \ell[1 - (v^2/c^2)\cos^2\theta]^{1/2}$. (b) Show that the angle θ_p that the rod makes with the x axis according to an observer at rest with respect to the rod can be found from $\tan \theta_p = \gamma \tan \theta$. These results show that the rod is observed to be both contracted and rotated. (Take the lower end of the rod to be at the origin of the coordinate system in which the rod is at rest.)

19. A red light flashes at position $x_R = 3.00$ m and time $t_R = 1.00 \times 10^{-9}$ s, and a blue light flashes at $x_B = 5.00$ m and $t_B = 9.00 \times 10^{-9}$ s, all measured in the S reference frame. Reference frame S′ moves uniformly to the right and has its origin at the same point as S at $t = t' = 0$. Both flashes are observed to occur at the same place in S′. (a) Find the relative speed between S and S′. (b) Find the location of the two flashes in frame S′. (c) At what time does the red flash occur in the S′ frame?

SECTION 38.6 The Lorentz Velocity Transformation Equations

20. You have been hired as an expert witness in the future by an attorney representing the driver of a spacecraft. The driver is accused of exceeding the galactic speed limit of 0.700c relative to the Earth while being chased by a galactic police spacecraft. The driver claims he is innocent, that his speed was well below that limit. You have been provided with the following data: the police spacecraft was traveling at 0.600c while chasing the driver and a technician on the police spacecraft measured the suspected spacecraft as traveling at 0.300c relative to the police spacecraft. What advice should you give the attorney?

21. Figure P38.21 shows a jet of material (at the upper right) being ejected by galaxy M87 (at the lower left). Such jets are believed to be evidence of supermassive black holes at the center of a galaxy. Suppose two jets of material from the center of a galaxy are ejected in opposite directions. Both jets move at 0.750c relative to the galaxy center. Determine the speed of one jet relative to the other.

Figure P38.21

22. A spacecraft is launched from the surface of the Earth with a velocity of 0.600c at an angle of 50.0° above the horizontal positive x axis. Another spacecraft is moving past with a velocity of 0.700c in the negative x direction. Determine the magnitude and direction of the velocity of the first spacecraft as measured by the pilot of the second spacecraft.

SECTION 38.7 Relativistic Linear Momentum

23. Calculate the momentum of an electron moving with a
[V] speed of (a) 0.010 0*c*, (b) 0.500*c*, and (c) 0.900*c*.

24. Show that the speed of an object having momentum of mag-
[S] nitude *p* and mass *m* is

$$u = \frac{c}{\sqrt{1 + (mc/p)^2}}$$

25. (a) Calculate the classical momentum of a proton traveling
[QC] at 0.990*c*, neglecting relativistic effects. (b) Repeat the calculation while including relativistic effects. (c) Does it make
sense to neglect relativity at such speeds?

26. The speed limit on a certain roadway is 90.0 km/h. Suppose speeding fines are made proportional to the amount
by which a vehicle's momentum exceeds the momentum
it would have when traveling at the speed limit. The fine
for driving at 190 km/h (that is, 100 km/h over the speed
limit) is $80.0. What, then, is the fine for traveling (a) at
1 090 km/h? (b) At 1 000 000 090 km/h?

27. An unstable particle at rest spontaneously breaks into two
[T] fragments of unequal mass. The mass of the first fragment
is 2.50×10^{-28} kg, and that of the other is 1.67×10^{-27} kg. If
the lighter fragment has a speed of 0.893*c* after the breakup,
what is the speed of the heavier fragment?

SECTION 38.8 Relativistic Energy

28. (a) Find the kinetic energy of a 78.0-kg spacecraft launched
[QC] out of the solar system with speed 106 km/s by using the
classical equation $K = \frac{1}{2}mu^2$. (b) **What If?** Calculate its
kinetic energy using the relativistic equation. (c) Explain
the result of comparing the answers of parts (a) and (b).

29. Determine the energy required to accelerate an electron
from (a) 0.500*c* to 0.900*c* and (b) 0.900*c* to 0.990*c*.

30. Show that for any object moving at less than one-tenth
the speed of light, the relativistic kinetic energy agrees
with the result of the classical equation $K = \frac{1}{2}mu^2$ to within
less than 1%. Therefore, for most purposes, the classical
equation is sufficient to describe these objects.

31. Protons in an accelerator at the Fermi National Laboratory
near Chicago are accelerated to a total energy that is 400
times their rest energy. (a) What is the speed of these protons in terms of *c*? (b) What is their kinetic energy in MeV?

32. You are working for an alternative energy company. Your
[CR] supervisor has an idea for a new energy source. He wants to
build a matter-antimatter reactor that will convert the entire
mass of the matter and antimatter into recoverable energy,
with *no* waste. He has lofty ideas; he wants his reactor to
provide energy to the *entire* world, replacing coal, fossil fuel,
hydroelectric, wind, thermal, and nuclear energy sources in
all countries. (a) He asks you to determine the masses of the
supply of matter and antimatter that will need to be combined to provide the world's needs for one year. (b) He also
asks you to determine how large the storage containers must
be to hold a 5.0-yr supply of the matter and antimatter while
it is waiting to be used in the reactor. The current energy
consumption worldwide is about 4.0×10^{20} J per year, and
the matter and antimatter will have approximately the density of aluminum, 2.70 g/cm^3.

33. The total energy of a proton is twice its rest energy. Find the
momentum of the proton in MeV/*c* units.

34. When 1.00 g of hydrogen combines with 8.00 g of oxygen,
[QC] 9.00 g of water is formed. During this chemical reaction,
[V] 2.86×10^5 J of energy is released. (a) Is the mass of the water
larger or smaller than the mass of the reactants? (b) What
is the difference in mass? (c) Explain whether the change in
mass is likely to be detectable.

35. The rest energy of an electron is 0.511 MeV. The rest energy
of a proton is 938 MeV. Assume both particles have kinetic
energies of 2.00 MeV. Find the speed of (a) the electron
and (b) the proton. (c) By what factor does the speed of the
electron exceed that of the proton? (d) Repeat the calculations in parts (a) through (c) assuming both particles have
kinetic energies of 2 000 MeV.

36. Show that the energy–momentum relationship in Equation
[S] 38.27, $E^2 = p^2c^2 + (mc^2)^2$, follows from the expressions $E = \gamma mc^2$ and $p = \gamma mu$.

37. Massive stars ending their lives in supernova explosions
produce the nuclei of all the atoms in the bottom half of
the periodic table by fusion of smaller nuclei. This problem
roughly models that process. A particle of mass $m = 1.99 \times 10^{-26}$ kg moving with a velocity $\vec{u} = 0.500c\hat{i}$ collides head-on
and sticks to a particle of mass $m' = m/3$ moving with the
velocity $\vec{u} = -0.500c\hat{i}$. What is the mass of the resulting
particle?

38. Massive stars ending their lives in supernova explosions
[QC] produce the nuclei of all the atoms in the bottom half of
[S] the periodic table by fusion of smaller nuclei. This problem roughly models that process. A particle of mass *m* moving along the *x* axis with a velocity component +*u* collides
head-on and sticks to a particle of mass *m*/3 moving along
the *x* axis with the velocity component −*u*. (a) What is the
mass *M* of the resulting particle? (b) Evaluate the expression from part (a) in the limit $u \rightarrow 0$. (c) Explain whether
the result agrees with what you should expect from nonrelativistic physics.

39. Consider a car moving at highway speed *u*. Is its actual
kinetic energy larger or smaller than $\frac{1}{2}mu^2$? Make an
order-of-magnitude estimate of the amount by which its
actual kinetic energy differs from $\frac{1}{2}mu^2$. In your solution,
state the quantities you take as data and the values you
measure or estimate for them. You may find Appendix B.5
useful.

40. An unstable particle with mass $m = 3.34 \times 10^{-27}$ kg is
[GP] initially at rest. The particle decays into two fragments
that fly off along the *x* axis with velocity components
$u_1 = 0.987c$ and $u_2 = -0.868c$. From this information, we
wish to determine the masses of fragments 1 and 2. (a) Is
the initial system of the unstable particle, which becomes
the system of the two fragments, isolated or nonisolated?
(b) Based on your answer to part (a), what two analysis
models are appropriate for this situation? (c) Find the values of γ for the two fragments after the decay. (d) Using
one of the analysis models in part (b), find a relationship between the masses m_1 and m_2 of the fragments. (e)
Using the second analysis model in part (b), find a second
relationship between the masses m_1 and m_2. (f) Solve the
relationships in parts (d) and (e) simultaneously for the
masses m_1 and m_2.

SECTION 38.9 The General Theory of Relativity

41. **Review.** A global positioning system (GPS) satellite moves in a circular orbit with period 11 h 58 min. (a) Determine the radius of its orbit. (b) Determine its speed. (c) The nonmilitary GPS signal is broadcast at a frequency of 1 575.42 MHz in the reference frame of the satellite. When it is received on the Earth's surface by a GPS receiver (Fig. P38.41), what is the fractional change in this frequency due to time dilation as described by special relativity? (d) The gravitational "blueshift" of the frequency according to general relativity is a separate effect. It is called a blueshift to indicate a change to a higher frequency. The magnitude of that fractional change is given by

$$\frac{\Delta f}{f} = \frac{\Delta U_g}{mc^2}$$

where U_g is the change in gravitational potential energy of an object–Earth system when the object of mass m is moved between the two points where the signal is observed. Calculate this fractional change in frequency due to the change in position of the satellite from the Earth's surface to its orbital position. (e) What is the overall fractional change in frequency due to both time dilation and gravitational blueshift?

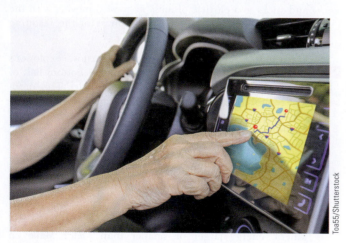

Toa55/Shutterstock

Figure P38.41

ADDITIONAL PROBLEMS

42. *Why is the following situation impossible?* On their 40th birthday, twins Speedo and Goslo say good-bye as Speedo takes off for a planet that is 50 ly away. He travels at a constant speed of $0.85c$ and immediately turns around and comes back to the Earth after arriving at the planet. Upon arriving back at the Earth, Speedo has a joyous reunion with Goslo.

43. An astronaut wishes to visit the Andromeda galaxy, making a one-way trip that will take 30.0 years in the spaceship's frame of reference. Assume the galaxy is 2.00 million light-years away and his speed is constant. (a) How fast must he travel relative to Earth? (b) What will be the kinetic energy of his spacecraft, which has mass of 1.00×10^6 kg? (c) What is the cost of this energy if it is purchased at a typical consumer price for electric energy, 13.0¢ per kWh? The following approximation will prove useful:

$$\frac{1}{\sqrt{1+x}} \approx 1 - \frac{x}{2} \text{ for } x \ll 1$$

44. The equation

$$K = \left(\frac{1}{\sqrt{1 - u^2/c^2}} - 1\right)mc^2$$

gives the kinetic energy of a particle moving at speed u. (a) Solve the equation for u. (b) From the equation for u, identify the minimum possible value of speed and the corresponding kinetic energy. (c) Identify the maximum possible speed and the corresponding kinetic energy. (d) Differentiate the equation for u with respect to time to obtain an equation describing the acceleration of a particle as a function of its kinetic energy and the power input to the particle. (e) Observe that for a nonrelativistic particle we have $u = (2K/m)^{1/2}$ and that differentiating this equation with respect to time gives $a = P/(2mK)^{1/2}$. State the limiting form of the expression in part (d) at low energy. State how it compares with the nonrelativistic expression. (f) State the limiting form of the expression in part (d) at high energy. (g) Consider a particle with constant input power. Explain how the answer to part (f) helps account for the answer to part (c).

45. Consider the astronaut planning the trip to Andromeda in Problem 43. (a) To three significant figures, what is the value for γ for the speed found in part (a) of Problem 43? (b) Just as the astronaut leaves on his constant-speed trip, a light beam is also sent in the direction of Andromeda. According to the Earth observer, how much later does the astronaut arrive at Andromeda after the arrival of the light beam?

46. The motion of a transparent medium influences the speed of light. This effect was first observed by Fizeau in 1851. Consider a light beam in water. The water moves with speed v in a horizontal pipe. Assume the light travels in the same direction as the water moves. The speed of light with respect to the water is c/n, where $n = 1.33$ is the index of refraction of water. (a) Use the velocity transformation equation to show that the speed of the light measured in the laboratory frame is

$$u = \frac{c}{n}\left(\frac{1 + nv/c}{1 + v/nc}\right)$$

(b) Show that for $v \ll c$, the expression from part (a) becomes, to a good approximation,

$$u \approx \frac{c}{n} + v - \frac{v}{n^2}$$

(c) Argue for or against the view that we should expect the result to be $u = (c/n) + v$ according to the Galilean transformation and that the presence of the term $-v/n^2$ represents a relativistic effect appearing even at "nonrelativistic" speeds. (d) Evaluate u in the limit as the speed of the water approaches c.

47. An object disintegrates into two fragments. One fragment has mass 1.00 MeV/c^2 and momentum 1.75 MeV/c in the positive x direction, and the other has mass 1.50 MeV/c^2 and momentum 2.00 MeV/c in the positive y direction. Find (a) the mass and (b) the speed of the original object.

48. *Why is the following situation impossible?* An experimenter is accelerating electrons for use in probing a material. She finds that when she accelerates them through a potential difference of 84.0 kV, the electrons have half the speed she wishes. She quadruples the potential difference to 336 kV, and the electrons accelerated through this potential difference have her desired speed.

49. **Review.** Around the core of a nuclear reactor shielded by a large pool of water, Cerenkov radiation appears as a blue glow. (See Fig. P16.39 on page 448.) Cerenkov radiation occurs when a particle travels faster through a medium than the speed of light in that medium. It is the electromagnetic equivalent of a bow wave or a sonic boom. An electron is traveling through water at a speed 10.0% faster than the speed of light in water. Determine the electron's (a) total energy, (b) kinetic energy, and (c) momentum. (d) Find the angle between the shock wave and the electron's direction of motion.

50. (a) Prepare a graph of the relativistic kinetic energy and the classical kinetic energy, both as a function of speed, for an object with a mass of your choice. (b) At what speed does the classical kinetic energy underestimate the experimental value by 1%? (c) By 5%? (d) By 50%?

51. Imagine that the entire Sun, of mass M_S, collapses to a sphere of radius R_g such that the work required to remove a small mass m from the surface would be equal to its rest energy mc^2. This radius is called the *gravitational radius* for the Sun. (a) Use this approach to show that $R_g = GM_S/c^2$. (b) Find a numerical value for R_g.

52. A ^{57}Fe nucleus at rest emits a 14.0-keV photon. Use conservation of energy and momentum to find the kinetic energy of the recoiling nucleus in electron volts. Use $Mc^2 = 8.60 \times 10^{-9}$ J for the final state of the ^{57}Fe nucleus.

CHALLENGE PROBLEMS

53. The creation and study of new and very massive elementary particles is an important part of contemporary physics. To create a particle of mass M requires an energy Mc^2. With enough energy, an exotic particle can be created by allowing a fast-moving proton to collide with a similar target particle. Consider a perfectly inelastic collision between two protons: an incident proton with mass m_p, kinetic energy K, and momentum magnitude p joins with an originally stationary target proton to form a single product particle of mass M. Not all the kinetic energy of the incoming proton is available to create the product particle because conservation of momentum requires that the system as a whole still must have some kinetic energy after the collision. Therefore, only a fraction of the energy of the incident particle is available to create a new particle. (a) Show

that the energy available to create a product particle is given by

$$Mc^2 = 2m_p c^2 \sqrt{1 + \frac{K}{2m_p c^2}}$$

This result shows that when the kinetic energy K of the incident proton is large compared with its rest energy $m_p c^2$, then M approaches $(2m_p K)^{1/2}/c$. Therefore, if the energy of the incoming proton is increased by a factor of 9, the mass you can create increases only by a factor of 3, not by a factor of 9 as would be expected. (b) This problem can be alleviated by using *colliding beams* as is the case in most modern accelerators. Here the total momentum of a pair of interacting particles can be zero. The center of mass can be at rest after the collision, so, in principle, all the initial kinetic energy can be used for particle creation. Show that

$$Mc^2 = 2mc^2 \left(1 + \frac{K}{mc^2}\right)$$

where K is the kinetic energy of each of the two identical colliding particles. Here, if $K \gg mc^2$, we have M directly proportional to K as we would desire.

54. A particle with electric charge q moves along a straight line in a uniform electric field $\vec{E}$ with speed u. The electric force exerted on the charge is $q\vec{E}$. The velocity of the particle and the electric field are both in the x direction. (a) Show that the acceleration of the particle in the x direction is given by

$$a = \frac{du}{dt} = \frac{qE}{m}\left(1 - \frac{u^2}{c^2}\right)^{3/2}$$

(b) Discuss the significance of the dependence of the acceleration on the speed. (c) **What If?** If the particle starts from rest at $x = 0$ at $t = 0$, how would you proceed to find the speed of the particle and its position at time t?

55. Suppose our Sun is about to explode. In an effort to escape, we depart in a spacecraft at $v = 0.800c$ and head toward the star Tau Ceti, 12.0 ly away. When we reach the midpoint of our journey from the Earth, we see our Sun explode, and, unfortunately, at the same instant, we see Tau Ceti explode as well. (a) In the spacecraft's frame of reference, should we conclude that the two explosions occurred simultaneously? If not, which occurred first? (b) **What If?** In a frame of reference in which the Sun and Tau Ceti are at rest, did they explode simultaneously? If not, which exploded first?

Appendix A   Tables

TABLE A.1   Conversion Factors

Length

	m	cm	km	in.	ft	mi
1 meter	1	10^2	10^{-3}	39.37	3.281	6.214×10^{-4}
1 centimeter	10^{-2}	1	10^{-5}	0.393 7	3.281×10^{-2}	6.214×10^{-6}
1 kilometer	10^3	10^5	1	3.937×10^4	3.281×10^3	0.621 4
1 inch	2.540×10^{-2}	2.540	2.540×10^{-5}	1	8.333×10^{-2}	1.578×10^{-5}
1 foot	0.304 8	30.48	3.048×10^{-4}	12	1	1.894×10^{-4}
1 mile	1 609	1.609×10^5	1.609	6.336×10^4	5 280	1

Mass

	kg	g	slug	u
1 kilogram	1	10^3	6.852×10^{-2}	6.024×10^{26}
1 gram	10^{-3}	1	6.852×10^{-5}	6.024×10^{23}
1 slug	14.59	1.459×10^4	1	8.789×10^{27}
1 atomic mass unit	1.660×10^{-27}	1.660×10^{-24}	1.137×10^{-28}	1

Note: 1 metric ton = 1 000 kg.

Time

	s	min	h	day	yr
1 second	1	1.667×10^{-2}	2.778×10^{-4}	1.157×10^{-5}	3.169×10^{-8}
1 minute	60	1	1.667×10^{-2}	6.994×10^{-4}	1.901×10^{-6}
1 hour	3 600	60	1	4.167×10^{-2}	1.141×10^{-4}
1 day	8.640×10^4	1 440	24	1	2.738×10^{-5}
1 year	3.156×10^7	5.259×10^5	8.766×10^3	365.2	1

Speed

	m/s	cm/s	ft/s	mi/h
1 meter per second	1	10^2	3.281	2.237
1 centimeter per second	10^{-2}	1	3.281×10^{-2}	2.237×10^{-2}
1 foot per second	0.304 8	30.48	1	0.681 8
1 mile per hour	0.447 0	44.70	1.467	1

Note: 1 mi/min = 60 mi/h = 88 ft/s.

Force

	N	lb
1 newton	1	0.224 8
1 pound	4.448	1

(Continued)

TABLE A.1   Conversion Factors (*continued*)

Energy, Energy Transfer

	J	ft · lb	eV
1 joule	1	0.737 6	6.242×10^{18}
1 foot-pound	1.356	1	8.464×10^{18}
1 electron volt	1.602×10^{-19}	1.182×10^{-19}	1
1 calorie	4.186	3.087	2.613×10^{19}
1 British thermal unit	1.055×10^3	7.779×10^2	6.585×10^{21}
1 kilowatt-hour	3.600×10^6	2.655×10^6	2.247×10^{25}

	cal	Btu	kWh
1 joule	0.238 9	9.481×10^{-4}	2.778×10^{-7}
1 foot-pound	0.323 9	1.285×10^{-3}	3.766×10^{-7}
1 electron volt	3.827×10^{-20}	1.519×10^{-22}	4.450×10^{-26}
1 calorie	1	3.968×10^{-3}	1.163×10^{-6}
1 British thermal unit	2.520×10^2	1	2.930×10^{-4}
1 kilowatt-hour	8.601×10^5	3.413×10^2	1

Pressure

	Pa	atm
1 pascal	1	9.869×10^{-6}
1 atmosphere	1.013×10^5	1
1 centimeter mercury[a]	1.333×10^3	1.316×10^{-2}
1 pound per square inch	6.895×10^3	6.805×10^{-2}
1 pound per square foot	47.88	4.725×10^{-4}

	cm Hg	lb/in.²	lb/ft²
1 pascal	7.501×10^{-4}	1.450×10^{-4}	2.089×10^{-2}
1 atmosphere	76	14.70	2.116×10^3
1 centimeter mercury[a]	1	0.194 3	27.85
1 pound per square inch	5.171	1	144
1 pound per square foot	3.591×10^{-2}	6.944×10^{-3}	1

[a]At 0°C and at a location where the free-fall acceleration has its "standard" value, 9.806 65 m/s².

TABLE A.2   Symbols, Dimensions, and Units of Physical Quantities

Quantity	Common Symbol	Unit[a]	Dimensions[b]	Unit in Terms of Base SI Units
Acceleration	$\vec{a}$	m/s²	L/T^2	m/s²
Amount of substance	n	MOLE		mol
Angle	θ, ϕ	radian (rad)		
Angular acceleration	$\vec{\alpha}$	rad/s²	T^{-2}	s⁻²
Angular frequency	ω	rad/s	T^{-1}	s⁻¹
Angular momentum	$\vec{L}$	kg · m²/s	ML^2/T	kg · m²/s
Angular velocity	$\vec{\omega}$	rad/s	T^{-1}	s⁻¹
Area	A	m²	L^2	m²
Atomic number	Z			
Capacitance	C	farad (F)	Q^2T^2/ML^2	$A^2 \cdot s^4/kg \cdot m^2$
Charge	q, Q, e	coulomb (C)	Q	A · s

TABLE A.2 Symbols, Dimensions, and Units of Physical Quantities (*continued*)

Quantity	Common Symbol	Unit[a]	Dimensions[b]	Unit in Terms of Base SI Units
Charge density				
Line	λ	C/m	Q/L	$A \cdot s/m$
Surface	σ	C/m^2	Q/L^2	$A \cdot s/m^2$
Volume	ρ	C/m^3	Q/L^3	$A \cdot s/m^3$
Conductivity	σ	$1/\Omega \cdot m$	Q^2T/ML^3	$A^2 \cdot s^3/kg \cdot m^3$
Current	I	AMPERE	Q/T	A
Current density	J	A/m^2	Q/TL^2	A/m^2
Density	ρ	kg/m^3	M/L^3	kg/m^3
Dielectric constant	κ			
Electric dipole moment	$\vec{\mathbf{p}}$	$C \cdot m$	QL	$A \cdot s \cdot m$
Electric field	$\vec{\mathbf{E}}$	V/m	ML/QT^2	$kg \cdot m/A \cdot s^3$
Electric flux	Φ_E	$V \cdot m$	ML^3/QT^2	$kg \cdot m^3/A \cdot s^3$
Electromotive force	ε	volt (V)	ML^2/QT^2	$kg \cdot m^2/A \cdot s^3$
Energy, energy transfer	E, U, K, T	joule (J)	ML^2/T^2	$kg \cdot m^2/s^2$
Entropy	S	J/K	ML^2/T^2K	$kg \cdot m^2/s^2 \cdot K$
Force	$\vec{\mathbf{F}}$	newton (N)	ML/T^2	$kg \cdot m/s^2$
Frequency	f	hertz (Hz)	T^{-1}	s^{-1}
Heat	Q	joule (J)	ML^2/T^2	$kg \cdot m^2/s^2$
Inductance	L	henry (H)	ML^2/Q^2	$kg \cdot m^2/A^2 \cdot s^2$
Length	ℓ, L	METER	L	m
Displacement	$\Delta x, \Delta\vec{\mathbf{r}}$			
Distance	d, h			
Position	$x, y, z, \vec{\mathbf{r}}$			
Width, height, radius	w, h, r, R, a, b			
Magnetic dipole moment	$\vec{\boldsymbol{\mu}}$	$N \cdot m/T$	QL^2/T	$A \cdot m^2$
Magnetic field	$\vec{\mathbf{B}}$	tesla (T) $(= Wb/m^2)$	M/QT	$kg/A \cdot s^2$
Magnetic flux	Φ_B	weber (Wb)	ML^2/QT	$kg \cdot m^2/A \cdot s^2$
Mass	m, M	KILOGRAM	M	kg
Moment of inertia	I	$kg \cdot m^2$	ML^2	$kg \cdot m^2$
Momentum	$\vec{\mathbf{p}}$	$kg \cdot m/s$	ML/T	$kg \cdot m/s$
Period	T	s	T	s
Permeability of free space	μ_0	N/A^2 $(= H/m)$	ML/Q^2	$kg \cdot m/A^2 \cdot s^2$
Permittivity of free space	ϵ_0	$C^2/N \cdot m^2$ $(= F/m)$	Q^2T^2/ML^3	$A^2 \cdot s^4/kg \cdot m^3$
Potential	V	volt (V) $(= J/C)$	ML^2/QT^2	$kg \cdot m^2/A \cdot s^3$
Power	P	watt (W) $(= J/s)$	ML^2/T^3	$kg \cdot m^2/s^3$
Pressure	P	pascal (Pa) $(= N/m^2)$	M/LT^2	$kg/m \cdot s^2$
Resistance	R	ohm $(\Omega)(= V/A)$	ML^2/Q^2T	$kg \cdot m^2/A^2 \cdot s^3$
Specific heat	c	$J/kg \cdot K$	L^2/T^2K	$m^2/s^2 \cdot K$
Speed	υ	m/s	L/T	m/s
Temperature	T	KELVIN	K	K
Time	t	SECOND	T	s
Torque	$\vec{\boldsymbol{\tau}}$	$N \cdot m$	ML^2/T^2	$kg \cdot m^2/s^2$
Velocity	$\vec{\mathbf{v}}$	m/s	L/T	m/s
Volume	V	m^3	L^3	m^3
Wavelength	λ	m	L	m
Work	W	joule (J) $(= N \cdot m)$	ML^2/T^2	$kg \cdot m^2/s^2$

[a]The base SI units are given in uppercase letters.

[b]The symbols M, L, T, K, and Q denote mass, length, time, temperature, and charge, respectively.

Appendix B Mathematics Review

This appendix in mathematics is intended as a brief review of operations and methods. Early in this course, you should be totally familiar with basic algebraic techniques, analytic geometry, and trigonometry. The sections on differential and integral calculus are more detailed and are intended for students who have difficulty applying calculus concepts to physical situations.

B.1 Scientific Notation

Many quantities used by scientists often have very large or very small values. The speed of light, for example, is about 300 000 000 m/s, and the ink required to make the dot over an i in this textbook has a mass of about 0.000 000 001 kg. Obviously, it is very cumbersome to read, write, and keep track of the numbers of zeros in such quantities. We avoid this problem by using a method incorporating powers of the number 10:

$$10^0 = 1$$

$$10^1 = 10$$

$$10^2 = 10 \times 10 = 100$$

$$10^3 = 10 \times 10 \times 10 = 1\ 000$$

$$10^4 = 10 \times 10 \times 10 \times 10 = 10\ 000$$

$$10^5 = 10 \times 10 \times 10 \times 10 \times 10 = 100\ 000$$

and so on. The number of zeros corresponds to the power to which ten is raised, called the **exponent** of ten. For example, the speed of light, 300 000 000 m/s, can be expressed as 3.00×10^8 m/s.

In this method, some representative numbers smaller than unity are the following:

$$10^{-1} = \frac{1}{10} = 0.1$$

$$10^{-2} = \frac{1}{10 \times 10} = 0.01$$

$$10^{-3} = \frac{1}{10 \times 10 \times 10} = 0.001$$

$$10^{-4} = \frac{1}{10 \times 10 \times 10 \times 10} = 0.000\ 1$$

$$10^{-5} = \frac{1}{10 \times 10 \times 10 \times 10 \times 10} = 0.000\ 01$$

In these cases, the number of places the decimal point is to the left of the digit 1 equals the value of the (negative) exponent. Numbers expressed as some power of ten multiplied by another number between one and ten are said to be in **scientific notation.** For example, the scientific notation for 5 943 000 000 is 5.943×10^9 and that for 0.000 083 2 is 8.32×10^{-5}.

When numbers expressed in scientific notation are being multiplied, the following general rule is very useful:

$$10^n \times 10^m = 10^{n+m} \tag{B.1}$$

where n and m can be *any* numbers (not necessarily integers). For example, $10^2 \times 10^5 = 10^7$. The rule also applies if one of the exponents is negative: $10^3 \times 10^{-8} = 10^{-5}$.

When dividing numbers expressed in scientific notation, note that

$$\frac{10^n}{10^m} = 10^n \times 10^{-m} = 10^{n-m} \tag{B.2}$$

Exercises

With help from the preceding rules, verify the answers to the following equations:

1. $86\ 400 = 8.64 \times 10^4$
2. $9\ 816\ 762.5 = 9.816\ 762\ 5 \times 10^6$
3. $0.000\ 000\ 039\ 8 = 3.98 \times 10^{-8}$
4. $(4.0 \times 10^8)(9.0 \times 10^9) = 3.6 \times 10^{18}$
5. $(3.0 \times 10^7)(6.0 \times 10^{-12}) = 1.8 \times 10^{-4}$
6. $\dfrac{75 \times 10^{-11}}{5.0 \times 10^{-3}} = 1.5 \times 10^{-7}$
7. $\dfrac{(3 \times 10^6)(8 \times 10^{-2})}{(2 \times 10^{17})(6 \times 10^5)} = 2 \times 10^{-18}$

B.2   Algebra

Some Basic Rules

When algebraic operations are performed, the laws of arithmetic apply. Symbols such as x, y, and z are usually used to represent unspecified quantities, called the **unknowns.**

First, consider the equation

$$8x = 32$$

If we wish to solve for x, we can divide (or multiply) each side of the equation by the same factor without destroying the equality. In this case, if we divide both sides by 8, we have

$$\frac{8x}{8} = \frac{32}{8}$$

$$x = 4$$

Next, consider the equation

$$x + 2 = 8$$

In this type of expression, we can add or subtract the same quantity from each side. If we subtract 2 from each side, we have

$$x + 2 - 2 = 8 - 2$$

$$x = 6$$

In general, if $x + a = b$, then $x = b - a$.

Now consider the equation

$$\frac{x}{5} = 9$$

If we multiply each side by 5, we are left with x on the left by itself and 45 on the right:

$$\left(\frac{x}{5}\right)(5) = 9 \times 5$$

$$x = 45$$

In all cases, *whatever operation is performed on the left side of the equality must also be performed on the right side.*

The following rules for multiplying, dividing, adding, and subtracting fractions should be recalled, where a, b, c, and d are four numbers:

	Rule	Example
Multiplying	$\left(\dfrac{a}{b}\right)\left(\dfrac{c}{d}\right) = \dfrac{ac}{bd}$	$\left(\dfrac{2}{3}\right)\left(\dfrac{4}{5}\right) = \dfrac{8}{15}$
Dividing	$\dfrac{(a/b)}{(c/d)} = \dfrac{ad}{bc}$	$\dfrac{2/3}{4/5} = \dfrac{(2)(5)}{(4)(3)} = \dfrac{10}{12}$
Adding	$\dfrac{a}{b} \pm \dfrac{c}{d} = \dfrac{ad \pm bc}{bd}$	$\dfrac{2}{3} - \dfrac{4}{5} = \dfrac{(2)(5) - (4)(3)}{(3)(5)} = -\dfrac{2}{15}$

Exercises

In the following exercises, solve for x:

Answers

1. $a = \dfrac{1}{1 + x}$ $\qquad\qquad$ $x = \dfrac{1 - a}{a}$

2. $3x - 5 = 13$ $\qquad\qquad$ $x = 6$

3. $ax - 5 = bx + 2$ $\qquad\qquad$ $x = \dfrac{7}{a - b}$

4. $\dfrac{5}{2x + 6} = \dfrac{3}{4x + 8}$ $\qquad\qquad$ $x = -\dfrac{11}{7}$

Powers

When powers of a given quantity x are multiplied, the following rule applies:

$$x^n x^m = x^{n+m} \tag{B.3}$$

For example, $x^2 x^4 = x^{2+4} = x^6$.

When dividing the powers of a given quantity, the rule is

$$\frac{x^n}{x^m} = x^{n-m} \tag{B.4}$$

For example, $x^8/x^2 = x^{8-2} = x^6$.

A power that is a fraction, such as $\frac{1}{3}$, corresponds to a root as follows:

$$x^{1/n} = \sqrt[n]{x} \tag{B.5}$$

For example, $4^{1/3} = \sqrt[3]{4} = 1.587\,4$. (A scientific calculator is useful for such calculations.)

Finally, any quantity x^n raised to the mth power is

$$(x^n)^m = x^{nm} \tag{B.6}$$

Table B.1 summarizes the rules of exponents.

TABLE B.1   Rules of Exponents

$x^0 = 1$
$x^1 = x$
$x^n x^m = x^{n+m}$
$x^n/x^m = x^{n-m}$
$x^{1/n} = \sqrt[n]{x}$
$(x^n)^m = x^{nm}$

Exercises

Verify the following equations:

1. $3^2 \times 3^3 = 243$
2. $x^5 x^{-8} = x^{-3}$

3. $x^{10}/x^{-5} = x^{15}$
4. $5^{1/3} = 1.709\ 976$ (Use your calculator.)
5. $60^{1/4} = 2.783\ 158$ (Use your calculator.)
6. $(x^4)^3 = x^{12}$

Factoring

Some useful formulas for factoring an equation are the following:

$ax + ay + az = a(x + y + z)$ common factor

$a^2 + 2ab + b^2 = (a + b)^2$ perfect square

$a^2 - b^2 = (a + b)(a - b)$ differences of squares

Quadratic Equations

The general form of a quadratic equation is

$$ax^2 + bx + c = 0 \qquad \text{(B.7)}$$

where x is the unknown quantity and a, b, and c are numerical factors referred to as **coefficients** of the equation. This equation has two roots, given by

$$x = \frac{-b \pm \sqrt{b^2 - 4ac}}{2a} \qquad \text{(B.8)}$$

If $b^2 \geq 4ac$, the roots are real.

Example B.1

Find the roots of the equation $x^2 + 5x + 4 = 0$.

SOLUTION

Use Equation B.8 to find the roots:

$$x = \frac{-5 \pm \sqrt{5^2 - (4)(1)(4)}}{2(1)} = \frac{-5 \pm \sqrt{9}}{2} = \frac{-5 \pm 3}{2}$$

Evaluate the root for each of the two possibilities of the sign:

$$x_+ = \frac{-5 + 3}{2} = -1 \qquad x_- = \frac{-5 - 3}{2} = -4$$

where x_+ refers to the root corresponding to the positive sign and x_- refers to the root corresponding to the negative sign.

Exercises

Solve the following quadratic equations:

		Answers	
1. $x^2 + 2x - 3 = 0$	$x_+ = 1$		$x_- = -3$
2. $2x^2 - 5x + 2 = 0$	$x_+ = 2$		$x_- = \frac{1}{2}$
3. $2x^2 - 4x - 9 = 0$	$x_+ = 1 + \sqrt{22}/2$		$x_- = 1 - \sqrt{22}/2$

Linear Equations

A linear equation has the general form

$$y = mx + b \qquad \text{(B.9)}$$

where m and b are constants. This equation is referred to as linear because the graph of y versus x is a straight line as shown in Figure B.1. The constant b, called the **y-intercept,** represents the value of y at which the straight line intersects

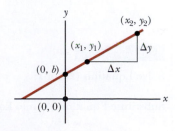

Figure B.1 A straight line graphed on an xy coordinate system. The slope of the line is the ratio of Δy to Δx.

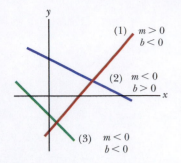

Figure B.2 The brown line has a positive slope and a negative y-intercept. The blue line has a negative slope and a positive y-intercept. The green line has a negative slope and a negative y-intercept.

the y axis. The constant m is equal to the **slope** of the straight line. If any two points on the straight line are specified by the coordinates (x_1, y_1) and (x_2, y_2) as in Figure B.1, the slope of the straight line can be expressed as

$$\text{Slope} = \frac{y_2 - y_1}{x_2 - x_1} = \frac{\Delta y}{\Delta x} \tag{B.10}$$

Note that m and b can have either positive or negative values. If $m > 0$, the straight line has a *positive* slope as in Figure B.1. If $m < 0$, the straight line has a *negative* slope. In Figure B.1, both m and b are positive. Three other possible situations are shown in Figure B.2.

Exercises

1. Draw graphs of the following straight lines: (a) $y = 5x + 3$ (b) $y = -2x + 4$ (c) $y = -3x - 6$
2. Find the slopes of the straight lines described in Exercise 1.

Answers (a) 5 (b) -2 (c) -3

3. Find the slopes of the straight lines that pass through the following sets of points: (a) $(0, -4)$ and $(4, 2)$ (b) $(0, 0)$ and $(2, -5)$ (c) $(-5, 2)$ and $(4, -2)$

Answers (a) $\frac{3}{2}$ (b) $-\frac{5}{2}$ (c) $-\frac{4}{9}$

Solving Simultaneous Linear Equations

Consider the equation $3x + 5y = 15$, which has two unknowns, x and y. Such an equation does not have a unique solution. For example, $(x = 0, y = 3)$, $(x = 5, y = 0)$, and $(x = 2, y = \frac{9}{5})$ are all solutions to this equation.

If a problem has two unknowns, a unique solution is possible only if we have *two* pieces of information. In most common cases, those two pieces of information are equations. In general, if a problem has n unknowns, its solution requires n equations. To solve two simultaneous equations involving two unknowns, x and y, we solve one of the equations for x in terms of y and substitute this expression into the other equation.

In some cases, the two pieces of information may be (1) one equation and (2) a condition on the solutions. For example, suppose we have (1) the equation $m = 3n$ and (2) the condition that m and n must be the smallest positive nonzero integers possible. Then, the single equation does not allow a unique solution, but the addition of the condition gives us that $n = 1$ and $m = 3$.

Example **B.2**

Solve the two simultaneous equations

$$(1) \quad 5x + y = -8$$

$$(2) \quad 2x - 2y = 4$$

SOLUTION

Solve Equation (2) for x:

$$(3) \qquad x = y + 2$$

Substitute Equation (3) into Equation (1):

$$5(y + 2) + y = -8$$

$$6y = -18$$

$$y = \boxed{-3}$$

Use Equation (3) to find x:

$$x = y + 2 = \boxed{-1}$$

B.2 continued

Alternative Solution

Multiply each term in Equation (1) by 2: $10x + 2y = -16$
Add Equation (2): $2x - 2y = 4$

$$12x \qquad = -12$$

Solve for x: $x = -1$
Use Equation (3) to find y: $y = x - 2 = -3$

Two linear equations containing two unknowns can also be solved by a graphical method. If the straight lines corresponding to the two equations are plotted in a conventional coordinate system, the intersection of the two lines represents the solution. For example, consider the two equations

$$x - y = 2$$
$$x - 2y = -1$$

These equations are plotted in Figure B.3. The intersection of the two lines has the coordinates $x = 5$ and $y = 3$, which represents the solution to the equations. You should check this solution by the analytical technique discussed earlier.

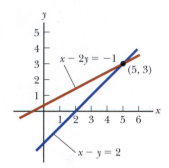

Figure B.3 A graphical solution for two linear equations.

Exercises

Solve the following pairs of simultaneous equations involving two unknowns:

Answers

1. $x + y = 8$ $x = 5, y = 3$
   $x - y = 2$

2. $98 - T = 10a$ $T = 65, a = 3.27$
   $T - 49 = 5a$

3. $6x + 2y = 6$ $x = 2, y = -3$
   $8x - 4y = 28$

Logarithms

Suppose a quantity x is expressed as a power of some quantity a:

$$x = a^y \tag{B.11}$$

The number a is called the **base** number. The **logarithm** of x with respect to the base a is equal to the exponent to which the base must be raised to satisfy the expression $x = a^y$:

$$y = \log_a x \tag{B.12}$$

Conversely, the **antilogarithm** of y is the number x:

$$x = \text{antilog}_a y \tag{B.13}$$

In practice, the two bases most often used are base 10, called the *common* logarithm base, and base $e = 2.718\,282$, called Euler's constant or the *natural* logarithm base. When common logarithms are used,

$$y = \log_{10} x \quad (\text{or } x = 10^y) \tag{B.14}$$

When natural logarithms are used,

$$y = \ln x \quad (\text{or } x = e^y) \tag{B.15}$$

For example, $\log_{10} 52 = 1.716$, so $\text{antilog}_{10} 1.716 = 10^{1.716} = 52$. Likewise, $\ln 52 = 3.951$, so $\text{antiln } 3.951 = e^{3.951} = 52$.

In general, note you can convert between base 10 and base e with the equality

$$\ln x = (2.302\ 585)\log_{10} x \qquad \textbf{(B.16)}$$

Finally, some useful properties of logarithms are the following:

$$\left.\begin{array}{l} \log(ab) = \log a + \log b \\[4pt] \log(a/b) = \log a - \log b \\[4pt] \log(a^n) = n\log a \end{array}\right\} \text{any base}$$

$$\ln e = 1$$

$$\ln e^a = a$$

$$\ln\left(\frac{1}{a}\right) = -\ln a$$

B.3   Geometry

The **distance** d between two points having coordinates (x_1, y_1) and (x_2, y_2) is

$$d = \sqrt{(x_2 - x_1)^2 + (y_2 - y_1)^2} \qquad \textbf{(B.17)}$$

Two angles are equal if their sides are perpendicular, right side to right side and left side to left side. For example, the two angles marked θ in Figure B.4 are the same because of the perpendicularity of the sides of the angles. To distinguish the left and right sides of an angle, imagine standing at the angle's apex and facing into the angle.

Radian measure: The arc length s of a circular arc (Fig. B.5) is proportional to the radius r for a fixed value of θ (in radians):

$$s = r\theta$$

$$\theta = \frac{s}{r} \qquad \textbf{(B.18)}$$

Table B.2 gives the **areas** and **volumes** for several geometric shapes used throughout this text.

Figure B.4 The angles are equal because their sides are perpendicular.

Figure B.5 The angle θ in radians is the ratio of the arc length s to the radius r of the circle.

TABLE B.2   Useful Information for Geometry

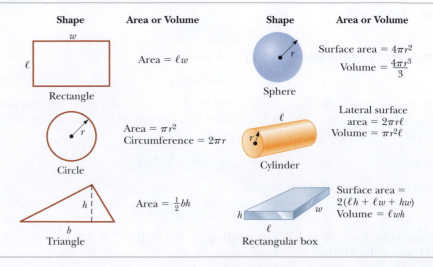

Shape	Area or Volume	Shape	Area or Volume
Rectangle	Area $= \ell w$	Sphere	Surface area $= 4\pi r^2$ Volume $= \dfrac{4\pi r^3}{3}$
Circle	Area $= \pi r^2$ Circumference $= 2\pi r$	Cylinder	Lateral surface area $= 2\pi r\ell$ Volume $= \pi r^2 \ell$
Triangle	Area $= \frac{1}{2}bh$	Rectangular box	Surface area $= 2(\ell h + \ell w + hw)$ Volume $= \ell wh$

The equation of a **straight line** (Fig. B.6) is

$$y = mx + b \qquad \text{(B.19)}$$

where b is the y-intercept and m is the slope of the line.

The equation of a **circle** of radius R centered at the origin is

$$x^2 + y^2 = R^2 \qquad \text{(B.20)}$$

The equation of an **ellipse** having the origin at its center (Fig. B.7) is

$$\frac{x^2}{a^2} + \frac{y^2}{b^2} = 1 \qquad \text{(B.21)}$$

where a is the length of the semimajor axis (the longer one) and b is the length of the semiminor axis (the shorter one).

The equation of a **parabola** the vertex of which is at $y = b$ (Fig. B.8) is

$$y = ax^2 + b \qquad \text{(B.22)}$$

The equation of a **rectangular hyperbola** (Fig. B.9) is

$$xy = \text{constant} \qquad \text{(B.23)}$$

Figure B.6 A straight line with a slope of m and a y-intercept of b.

Figure B.7 An ellipse with semimajor axis a and semiminor axis b.

B.4 Trigonometry

That portion of mathematics based on the special properties of the right triangle is called trigonometry. By definition, a right triangle is a triangle containing a $90°$ angle. Consider the right triangle shown in Figure B.10, where side a is opposite the angle θ, side b is adjacent to the angle θ, and side c is the hypotenuse of the triangle. The three basic trigonometric functions defined by such a triangle are the sine (sin), cosine (cos), and tangent (tan). In terms of the angle θ, these functions are defined as follows:

$$\sin \theta = \frac{\text{side opposite } \theta}{\text{hypotenuse}} = \frac{a}{c} \qquad \text{(B.24)}$$

$$\cos \theta = \frac{\text{side adjacent to } \theta}{\text{hypotenuse}} = \frac{b}{c} \qquad \text{(B.25)}$$

$$\tan \theta = \frac{\text{side opposite } \theta}{\text{side adjacent to } \theta} = \frac{a}{b} \qquad \text{(B.26)}$$

The Pythagorean theorem provides the following relationship among the sides of a right triangle:

$$c^2 = a^2 + b^2 \qquad \text{(B.27)}$$

From the preceding definitions and the Pythagorean theorem, it follows that

$$\sin^2 \theta + \cos^2 \theta = 1$$

$$\tan \theta = \frac{\sin \theta}{\cos \theta}$$

The cosecant, secant, and cotangent functions are defined by

$$\csc \theta = \frac{1}{\sin \theta} \qquad \sec \theta = \frac{1}{\cos \theta} \qquad \cot \theta = \frac{1}{\tan \theta}$$

Figure B.8 A parabola with its vertex at $y = b$.

Figure B.9 A hyperbola.

a = opposite side
b = adjacent side
c = hypotenuse

Figure B.10 A right triangle, used to define the basic functions of trigonometry.

TABLE B.3 Some Trigonometric Identities

$\sin^2 \theta + \cos^2 \theta = 1$	$\csc^2 \theta = 1 + \cot^2 \theta$
$\sec^2 \theta = 1 + \tan^2 \theta$	$\sin^2 \dfrac{\theta}{2} = \tfrac{1}{2}(1 - \cos \theta)$
$\sin 2\theta = 2 \sin \theta \cos \theta$	$\cos^2 \dfrac{\theta}{2} = \tfrac{1}{2}(1 + \cos \theta)$
$\cos 2\theta = \cos^2 \theta - \sin^2 \theta$	$1 - \cos \theta = 2 \sin^2 \dfrac{\theta}{2}$
$\tan 2\theta = \dfrac{2 \tan \theta}{1 - \tan^2 \theta}$	$\tan \dfrac{\theta}{2} = \sqrt{\dfrac{1 - \cos \theta}{1 + \cos \theta}}$

$$\sin (A \pm B) = \sin A \cos B \pm \cos A \sin B$$
$$\cos (A \pm B) = \cos A \cos B \mp \sin A \sin B$$
$$\sin A \pm \sin B = 2 \sin \left[\tfrac{1}{2}(A \pm B)\right] \cos \left[\tfrac{1}{2}(A \mp B)\right]$$
$$\cos A + \cos B = 2 \cos \left[\tfrac{1}{2}(A + B)\right] \cos \left[\tfrac{1}{2}(A - B)\right]$$
$$\cos A - \cos B = 2 \sin \left[\tfrac{1}{2}(A + B)\right] \sin \left[\tfrac{1}{2}(B - A)\right]$$

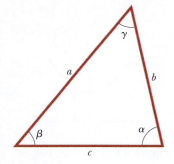

Figure B.11 An arbitrary, non-right triangle.

The following relationships are derived directly from the right triangle shown in Figure B.10:

$$\sin \theta = \cos (90° - \theta)$$
$$\cos \theta = \sin (90° - \theta)$$
$$\cot \theta = \tan (90° - \theta)$$

Some properties of trigonometric functions are the following:

$$\sin (-\theta) = -\sin \theta$$
$$\cos (-\theta) = \cos \theta$$
$$\tan (-\theta) = -\tan \theta$$

The following relationships apply to *any* triangle as shown in Figure B.11:

$$\alpha + \beta + \gamma = 180°$$

$$\text{Law of cosines} \begin{cases} a^2 = b^2 + c^2 - 2bc \cos \alpha \\ b^2 = a^2 + c^2 - 2ac \cos \beta \\ c^2 = a^2 + b^2 - 2ab \cos \gamma \end{cases}$$

$$\text{Law of sines} \qquad \frac{a}{\sin \alpha} = \frac{b}{\sin \beta} = \frac{c}{\sin \gamma}$$

Table B.3 lists a number of useful trigonometric identities.

Example **B.3**

Consider the right triangle in Figure B.12 in which $a = 2.00$, $b = 5.00$, and c is unknown. **(A)** Find c.

Figure B.12 (Example B.3)

SOLUTION

Use the Pythagorean theorem: $c^2 = a^2 + b^2 = 2.00^2 + 5.00^2 = 4.00 + 25.0 = 29.0$

$$c = \sqrt{29.0} = 5.39$$

(B) Find the angle θ.

Use the tangent function: $\tan \theta = \dfrac{a}{b} = \dfrac{2.00}{5.00} = 0.400$

B.3 continued

Use your calculator to find the angle: $\theta = \tan^{-1}(0.400) = \boxed{21.8°}$

where $\tan^{-1}(0.400)$ is the notation for "angle whose tangent is 0.400," sometimes written as arctan (0.400).

Exercises

1. In Figure B.13, identify (a) the side opposite θ (b) the side adjacent to ϕ and then find (c) $\cos \theta$, (d) $\sin \phi$, and (e) $\tan \phi$.

Answers   (a) 3   (b) 3   (c) $\frac{4}{5}$ (d) $\frac{4}{5}$ (e) $\frac{4}{3}$

2. In a certain right triangle, the two sides that are perpendicular to each other are 5.00 m and 7.00 m long. What is the length of the third side?

Answer   8.60 m

3. A right triangle has a hypotenuse of length 3.0 m, and one of its angles is 30°. (a) What is the length of the side opposite the 30° angle? (b) What is the side adjacent to the 30° angle?

Answers   (a) 1.5 m   (b) 2.6 m

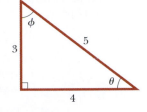

Figure B.13 (Exercise 1)

B.5   Series Expansions

$$(a + b)^n = a^n + \frac{n}{1!} a^{n-1} b + \frac{n(n-1)}{2!} a^{n-2} b^2 + \cdots$$

$$(1 + x)^n = 1 + nx + \frac{n(n-1)}{2!} x^2 + \cdots$$

$$e^x = 1 + x + \frac{x^2}{2!} + \frac{x^3}{3!} + \cdots$$

$$\ln (1 \pm x) = \pm x - \tfrac{1}{2} x^2 \pm \tfrac{1}{3} x^3 - \cdots$$

$$\left.\begin{array}{l} \sin x = x - \dfrac{x^3}{3!} + \dfrac{x^5}{5!} - \cdots \\[2mm] \cos x = 1 - \dfrac{x^2}{2!} + \dfrac{x^4}{4!} - \cdots \\[2mm] \tan x = x + \dfrac{x^3}{3} + \dfrac{2x^5}{15} + \cdots \quad |x| < \dfrac{\pi}{2} \end{array}\right\} x \text{ in radians}$$

The following approximations can be used:

For $x \ll 1$: $(1 + x)^n \approx 1 + nx$ For $x \leq 0.1$ rad: $\sin x \approx x$

$e^x \approx 1 + x$ $\cos x \approx 1$

$\ln (1 \pm x) \approx \pm x$ $\tan x \approx x$

B.6   Differential Calculus

In various branches of science, it is sometimes necessary to use the basic tools of calculus, invented by Newton, to describe physical phenomena. The use of calculus is fundamental in the treatment of various problems in Newtonian mechanics, electricity, and magnetism. In this section, we simply state some basic properties and "rules of thumb" that should be a useful review to the student.

Figure B.14 The lengths Δx and Δy are used to define the derivative of this function at a point.

First, a **function** must be specified that relates one variable to another (e.g., position as a function of time). Suppose one of the variables is called y (the dependent variable), and the other x (the independent variable). We might have a function relationship such as

$$y(x) = ax^3 + bx^2 + cx + d$$

If a, b, c, and d are specified constants, y can be calculated for any value of x. We usually deal with continuous functions, that is, those for which y varies "smoothly" with x.

The **derivative** of y with respect to x is defined as the limit as Δx approaches zero of the slopes of chords drawn between two points on the y versus x curve. Mathematically, we write this definition as

$$\frac{dy}{dx} = \lim_{\Delta x \to 0} \frac{\Delta y}{\Delta x} = \lim_{\Delta x \to 0} \frac{y(x + \Delta x) - y(x)}{\Delta x} \qquad \textbf{(B.28)}$$

where Δy and Δx are defined as $\Delta x = x_2 - x_1$ and $\Delta y = y_2 - y_1$ (Fig. B.14). Note that dy/dx *does not* mean dy divided by dx, but rather is simply a notation of the limiting process of the derivative as defined by Equation B.28.

A useful expression to remember when $y(x) = ax^n$, where a is a *constant* and n is *any* positive or negative number (integer or fraction), is

$$\frac{dy}{dx} = nax^{n-1} \qquad \textbf{(B.29)}$$

If $y(x)$ is a polynomial or algebraic function of x, we apply Equation B.29 to *each* term in the polynomial and take $d[\text{constant}]/dx = 0$. In Examples B.4 through B.7, we evaluate the derivatives of several functions.

Special Properties of the Derivative

A. Derivative of the product of two functions If a function $f(x)$ is given by the product of two functions—say, $g(x)$ and $h(x)$—the derivative of $f(x)$ is defined as

$$\frac{d}{dx} f(x) = \frac{d}{dx} [g(x)h(x)] = g\frac{dh}{dx} + h\frac{dg}{dx} \qquad \textbf{(B.30)}$$

B. Derivative of the sum of two functions If a function $f(x)$ is equal to the sum of two functions, the derivative of the sum is equal to the sum of the derivatives:

$$\frac{d}{dx} f(x) = \frac{d}{dx} [g(x) + h(x)] = \frac{dg}{dx} + \frac{dh}{dx} \qquad \textbf{(B.31)}$$

C. Chain rule of differential calculus If $y = f(x)$ and $x = g(z)$, then dy/dz can be written as the product of two derivatives:

$$\frac{dy}{dz} = \frac{dy}{dx}\frac{dx}{dz} \qquad \textbf{(B.32)}$$

D. The second derivative The second derivative of y with respect to x is defined as the derivative of the function dy/dx (the derivative of the derivative). It is usually written as

$$\frac{d^2 y}{dx^2} = \frac{d}{dx}\left(\frac{dy}{dx}\right) \qquad \textbf{(B.33)}$$

Some of the more commonly used derivatives of functions are listed in Table B.4.

TABLE B.4 Derivative for Several Functions

$\dfrac{d}{dx}(a) = 0$

$\dfrac{d}{dx}(ax^n) = nax^{n-1}$

$\dfrac{d}{dx}(e^{ax}) = ae^{ax}$

$\dfrac{d}{dx}(\sin ax) = a\cos ax$

$\dfrac{d}{dx}(\cos ax) = -a\sin ax$

$\dfrac{d}{dx}(\tan ax) = a\sec^2 ax$

$\dfrac{d}{dx}(\cot ax) = -a\csc^2 ax$

$\dfrac{d}{dx}(\sec x) = \tan x \sec x$

$\dfrac{d}{dx}(\csc x) = -\cot x \csc x$

$\dfrac{d}{dx}(\ln ax) = \dfrac{1}{x}$

$\dfrac{d}{dx}(\sin^{-1} ax) = \dfrac{a}{\sqrt{1 - a^2 x^2}}$

$\dfrac{d}{dx}(\cos^{-1} ax) = \dfrac{-a}{\sqrt{1 - a^2 x^2}}$

$\dfrac{d}{dx}(\tan^{-1} ax) = \dfrac{a}{1 + a^2 x^2}$

Note: The symbols a and n represent constants.

Example **B.4**

Use Equation B.28 to find the derivative of the following function: $y(x) = ax^3 + bx + c$, where a, b, and c are constants.

SOLUTION

Evaluate the function at $x + \Delta x$:

$$y(x + \Delta x) = a(x + \Delta x)^3 + b(x + \Delta x) + c$$

$$= a(x^3 + 3x^2\,\Delta x + 3x\,\Delta x^2 + \Delta x^3) + b(x + \Delta x) + c$$

Evaluate the numerator of Equation B.28:

$$\Delta y = y(x + \Delta x) - y(x) = a(3x^2\,\Delta x + 3x\,\Delta x^2 + \Delta x^3) + b\Delta x$$

Substitute into Equation B.28 and take the limit:

$$\frac{dy}{dx} = \lim_{\Delta x \to 0}\frac{\Delta y}{\Delta x} = \lim_{\Delta x \to 0}\left[a(3x^2 + 3x\,\Delta x + \Delta x^2)\right] + b$$

$$\frac{dy}{dx} = 3ax^2 + b$$

Example **B.5**

Find the derivative of

$$y(x) = 8x^5 + 4x^3 + 2x + 7$$

SOLUTION

Apply Equation B.29 to each term separately and remember that the derivative of a constant is zero:

$$\frac{dy}{dx} = 8(5)x^4 + 4(3)x^2 + 2(1)x^0 + 0$$

$$\frac{dy}{dx} = 40x^4 + 12x^2 + 2$$

Example **B.6**

Find the derivative of $y(x) = x^3/(x + 1)^2$ with respect to x.

SOLUTION

Rewrite the function as a product:

$$y(x) = x^3(x + 1)^{-2}$$

Use Equation B.30 to find the derivative:

$$\frac{dy}{dx} = (x + 1)^{-2}\frac{d}{dx}(x^3) + x^3\frac{d}{dx}(x + 1)^{-2}$$

$$= (x + 1)^{-2}\,3x^2 + x^3\,(-2)(x + 1)^{-3}$$

$$\frac{dy}{dx} = \frac{3x^2}{(x + 1)^2} - \frac{2x^3}{(x + 1)^3} = \frac{x^2(x + 3)}{(x + 1)^3}$$

Example **B.7**

A useful formula that follows from Equation B.30 is the derivative of the quotient of two functions. Show that

$$\frac{d}{dx}\left[\frac{g(x)}{h(x)}\right] = \frac{h\dfrac{dg}{dx} - g\dfrac{dh}{dx}}{h^2}$$

continued

B.6 continued

SOLUTION

Write the quotient as gh^{-1} and use Equations B.29 and B.30:

$$\frac{d}{dx}\left(\frac{g}{h}\right) = \frac{d}{dx}(gh^{-1}) = g\frac{d}{dx}(h^{-1}) + h^{-1}\frac{d}{dx}(g)$$

$$= -gh^{-2}\frac{dh}{dx} + h^{-1}\frac{dg}{dx}$$

$$= \frac{h\dfrac{dg}{dx} - g\dfrac{dh}{dx}}{h^2}$$

B.7 Integral Calculus

We think of integration as the inverse of differentiation. As an example, consider the expression

$$f(x) = \frac{dy}{dx} = 3ax^2 + b \tag{B.34}$$

which was the result of differentiating the function

$$y(x) = ax^3 + bx + c$$

in Example B.4. We can write Equation B.34 as $dy = f(x)\,dx = (3ax^2 + b)\,dx$ and obtain $y(x)$ by "summing" over all values of x. Mathematically, we write this inverse operation as

$$y(x) = \int f(x)\,dx$$

For the function $f(x)$ given by Equation B.34, we have

$$y(x) = \int (3ax^2 + b)\,dx = ax^3 + bx + c$$

where c is a constant of the integration. This type of integral is called an *indefinite integral* because its value depends on the choice of c.

A general **indefinite integral** $I(x)$ is defined as

$$I(x) = \int f(x)\,dx \tag{B.35}$$

where $f(x)$ is called the *integrand* and $f(x) = dI(x)/dx$.

For a *general continuous* function $f(x)$, the integral can be interpreted geometrically as the area under the curve bounded by $f(x)$ and the x axis, between two specified values of x, say, x_1 and x_2, as in Figure B.15.

The area of the blue element in Figure B.15 is approximately $f(x_i)\,\Delta x_i$. If we sum all these area elements between x_1 and x_2 and take the limit of this sum as $\Delta x_i \rightarrow 0$, we obtain the *true* area under the curve bounded by $f(x)$ and the x axis, between the limits x_1 and x_2:

$$\text{Area} = \lim_{\Delta x_i \rightarrow 0} \sum_i f(x_i)\Delta x_i = \int_{x_1}^{x_2} f(x)\,dx \tag{B.36}$$

Integrals of the type defined by Equation B.36 are called **definite integrals.**

Figure B.15 The definite integral of a function is the area under the curve of the function between the limits x_1 and x_2.

One common integral that arises in practical situations has the form

$$\int x^n \, dx = \frac{x^{n+1}}{n+1} + c \quad (n \neq -1) \tag{B.37}$$

This result is obvious, being that differentiation of the right-hand side with respect to x gives $f(x) = x^n$ directly. If the limits of the integration are known, this integral becomes a *definite integral* and is written

$$\int_{x_1}^{x_2} x^n \, dx = \frac{x^{n+1}}{n+1} \Bigg|_{x_1}^{x_2} = \frac{x_2^{n+1} - x_1^{n+1}}{n+1} \quad (n \neq -1) \tag{B.38}$$

Exercises

In the following exercises, evaluate the integral:

	Answer		**Answer**
1. $\displaystyle\int_0^a x^2 \, dx$	$\dfrac{a^3}{3}$	3. $\displaystyle\int_3^5 x \, dx$	8
2. $\displaystyle\int_0^b x^{3/2} \, dx$	$\dfrac{2}{5} b^{5/2}$		

Partial Integration

Sometimes it is useful to apply the method of *partial integration* (also called "integrating by parts") to evaluate certain integrals. This method uses the property

$$\int u \, dv = uv - \int v \, du \tag{B.39}$$

where u and v are *carefully* chosen so as to reduce a complex integral to a simpler one. In many cases, several reductions have to be made. Consider the function

$$I(x) = \int x^2 \, e^x \, dx$$

which can be evaluated by integrating by parts twice. First, if we choose $u = x^2$, $v = e^x$, we obtain

$$\int x^2 \, e^x \, dx = \int x^2 \, d(e^x) = x^2 \, e^x - 2\int e^x \, x \, dx + c_1$$

Now, in the second term, choose $u = x$, $v = e^x$, which gives

$$\int x^2 \, e^x \, dx = x^2 \, e^x - 2x \, e^x + 2\int e^x \, dx + c_1$$

or

$$\int x^2 \, e^x \, dx = x^2 \, e^x - 2xe^x + 2e^x + c_2$$

The Perfect Differential

Another useful method to remember is that of the *perfect differential*, in which we look for a change of variable such that the differential of the function is the differential of the independent variable appearing in the integrand. For example, consider the integral

$$I(x) = \int \cos^2 x \sin x \, dx$$

This integral becomes easy to evaluate if we rewrite the differential as $d(\cos x) = -\sin x \, dx$. The integral then becomes

$$\int \cos^2 x \sin x \, dx = -\int \cos^2 x \, d(\cos x)$$

If we now change variables, letting $y = \cos x$, we obtain

$$\int \cos^2 x \sin x \, dx = -\int y^2 \, dy = -\frac{y^3}{3} + c = -\frac{\cos^3 x}{3} + c$$

Table B.5 lists some useful indefinite integrals. Table B.6 gives Gauss's probability integral and other definite integrals. A more complete list can be found in various handbooks, such as *The Handbook of Chemistry and Physics* (Boca Raton, FL: CRC Press, published annually).

TABLE B.5   Some Indefinite Integrals (An arbitrary constant should be added to each of these integrals.)

$\displaystyle\int x^n \, dx = \frac{x^{n+1}}{n+1}$ (provided $n \neq 1$) $\qquad\qquad \displaystyle\int \ln \, ax \, dx = (x \ln \, ax) - x$

$\displaystyle\int \frac{dx}{x} = \int x^{-1} \, dx = \ln x$ $\qquad\qquad \displaystyle\int xe^{ax} \, dx = \frac{e^{ax}}{a^2}(ax - 1)$

$\displaystyle\int \frac{dx}{a + bx} = \frac{1}{b} \ln \, (a + bx)$ $\qquad\qquad \displaystyle\int \frac{dx}{a + be^{cx}} = \frac{x}{a} - \frac{1}{ac} \ln \, (a + be^{cx})$

$\displaystyle\int \frac{x \, dx}{a + bx} = \frac{x}{b} - \frac{a}{b^2} \ln \, (a + bx)$ $\qquad\qquad \displaystyle\int \sin \, ax \, dx = -\frac{1}{a} \cos \, ax$

$\displaystyle\int \frac{dx}{x(x + a)} = -\frac{1}{a} \ln \frac{x + a}{x}$ $\qquad\qquad \displaystyle\int \cos \, ax \, dx = \frac{1}{a} \sin \, ax$

$\displaystyle\int \frac{dx}{(a + bx)^2} = -\frac{1}{b(a + bx)}$ $\qquad\qquad \displaystyle\int \tan \, ax \, dx = -\frac{1}{a} \ln \, (\cos \, ax) = \frac{1}{a} \ln \, (\sec \, ax)$

$\displaystyle\int \frac{dx}{a^2 + x^2} = \frac{1}{a} \tan^{-1} \frac{x}{a}$ $\qquad\qquad \displaystyle\int \cot \, ax \, dx = \frac{1}{a} \ln \, (\sin \, ax)$

$\displaystyle\int \frac{dx}{a^2 - x^2} = \frac{1}{2a} \ln \frac{a + x}{a - x} \, (a^2 - x^2 > 0)$ $\qquad\qquad \displaystyle\int \sec \, ax \, dx = \frac{1}{a} \ln \, (\sec \, ax + \tan \, ax) = \frac{1}{a} \ln \left[\tan \left(\frac{ax}{2} + \frac{\pi}{4} \right) \right]$

$\displaystyle\int \frac{dx}{x^2 - a^2} = \frac{1}{2a} \ln \frac{x - a}{x + a} \, (x^2 - a^2 > 0)$ $\qquad\qquad \displaystyle\int \csc \, ax \, dx = \frac{1}{a} \ln \, (\csc \, ax - \cot \, ax) = \frac{1}{a} \ln \left(\tan \frac{ax}{2} \right)$

continued

TABLE B.5   Some Indefinite Integrals *(continued)*

$$\int \frac{x\,dx}{a^2 \pm x^2} = \pm\frac{1}{2}\ln(a^2 \pm x^2)$$

$$\int \frac{dx}{\sqrt{a^2 - x^2}} = \sin^{-1}\frac{x}{a} = -\cos^{-1}\frac{x}{a}\ (a^2 - x^2 > 0)$$

$$\int \frac{dx}{\sqrt{x^2 \pm a^2}} = \ln(x + \sqrt{x^2 \pm a^2})$$

$$\int \frac{x\,dx}{\sqrt{a^2 - x^2}} = -\sqrt{a^2 - x^2}$$

$$\int \frac{x\,dx}{\sqrt{x^2 \pm a^2}} = \sqrt{x^2 \pm a^2}$$

$$\int \sqrt{a^2 - x^2}\,dx = \frac{1}{2}\left(x\sqrt{a^2 - x^2} + a^2 \sin^{-1}\frac{x}{|a|}\right)$$

$$\int x\sqrt{a^2 - x^2}\,dx = -\frac{1}{3}(a^2 - x^2)^{3/2}$$

$$\int \sqrt{x^2 \pm a^2}\,dx = \frac{1}{2}x\sqrt{x^2 \pm a^2} \pm a^2 \ln(x + \sqrt{x^2 \pm a^2})$$

$$\int x(\sqrt{x^2 \pm a^2})\,dx = \frac{1}{3}(x^2 \pm a^2)^{3/2}$$

$$\int e^{ax}\,dx = \frac{1}{a}e^{ax}$$

$$\int \sin^2 ax\,dx = \frac{x}{2} - \frac{\sin 2ax}{4a}$$

$$\int \cos^2 ax\,dx = \frac{x}{2} + \frac{\sin 2ax}{4a}$$

$$\int \frac{dx}{\sin^2 ax} = -\frac{1}{a}\cot ax$$

$$\int \frac{dx}{\cos^2 ax} = \frac{1}{a}\tan ax$$

$$\int \tan^2 ax\,dx = \frac{1}{a}(\tan ax) - x$$

$$\int \cot^2 ax\,dx = -\frac{1}{a}(\cot ax) - x$$

$$\int \sin^{-1} ax\,dx = x(\sin^{-1} ax) + \frac{\sqrt{1 - a^2 x^2}}{a}$$

$$\int \cos^{-1} ax\,dx = x(\cos^{-1} ax) - \frac{\sqrt{1 - a^2 x^2}}{a}$$

$$\int \frac{dx}{(x^2 + a^2)^{3/2}} = \frac{x}{a^2\sqrt{x^2 + a^2}}$$

$$\int \frac{x\,dx}{(x^2 + a^2)^{3/2}} = -\frac{1}{\sqrt{x^2 + a^2}}$$

TABLE B.6   Gauss's Probability Integral and Other Definite Integrals

$$\int_0^\infty x^n e^{-ax}\,dx = \frac{n!}{a^{n+1}}$$

$$I_0 = \int_0^\infty e^{-ax^2}\,dx = \frac{1}{2}\sqrt{\frac{\pi}{a}}\quad\text{(Gauss's probability integral)}$$

$$I_1 = \int_0^\infty xe^{-ax^2}\,dx = \frac{1}{2a}$$

$$I_2 = \int_0^\infty x^2 e^{-ax^2}\,dx = -\frac{dI_0}{da} = \frac{1}{4}\sqrt{\frac{\pi}{a^3}}$$

$$I_3 = \int_0^\infty x^3 e^{-ax^2}\,dx = -\frac{dI_1}{da} = \frac{1}{2a^2}$$

$$I_4 = \int_0^\infty x^4 e^{-ax^2}\,dx = \frac{d^2 I_0}{da^2} = \frac{3}{8}\sqrt{\frac{\pi}{a^5}}$$

$$I_5 = \int_0^\infty x^5 e^{-ax^2}\,dx = \frac{d^2 I_1}{da^2} = \frac{1}{a^3}$$

$$\vdots$$

$$I_{2n} = (-1)^n \frac{d^n}{da^n} I_0$$

$$I_{2n+1} = (-1)^n \frac{d^n}{da^n} I_1$$

B.8   Propagation of Uncertainty

In laboratory experiments, a common activity is to take measurements that act as raw data. These measurements are of several types—length, time interval, temperature, voltage, and so on—and are taken by a variety of instruments. Regardless of the measurement and the quality of the instrumentation, **there is always uncertainty associated with a physical measurement.** This uncertainty is a combination of that associated with the instrument and that related to the system being measured. An example of the former is the inability to exactly determine the position of a length measurement between the lines on a meterstick. An example of uncertainty related to the system being measured is the variation of temperature within a sample of water so that a single temperature for the sample is difficult to determine.

Uncertainties can be expressed in two ways. **Absolute uncertainty** refers to an uncertainty expressed in the same units as the measurement. Therefore, the length of a computer disk label might be expressed as (5.5 ± 0.1) cm. The uncertainty of ± 0.1 cm by itself is not descriptive enough for some purposes, however. This uncertainty is large if the measurement is 1.0 cm, but it is small if the measurement is 100 m. To give a more descriptive account of the uncertainty, **fractional uncertainty** or **percent uncertainty** is used. In this type of description, the uncertainty is divided by the actual measurement. Therefore, the length of the computer disk label could be expressed as

$$\ell = 5.5 \text{ cm} \pm \frac{0.1 \text{ cm}}{5.5 \text{ cm}} = 5.5 \text{ cm} \pm 0.018 \quad \text{(fractional uncertainty)}$$

or as

$$\ell = 5.5 \text{ cm} \pm 1.8\% \quad \text{(percent uncertainty)}$$

When combining measurements in a calculation, the percent uncertainty in the final result is generally larger than the uncertainty in the individual measurements. This is called **propagation of uncertainty** and is one of the challenges of experimental physics.

Some simple rules can provide a reasonable estimate of the uncertainty in a calculated result:

Multiplication and division: When measurements with uncertainties are multiplied or divided, add the *percent uncertainties* to obtain the percent uncertainty in the result.

Example **B.8**

Find the area, with associated uncertainty, of a rectangular plate of dimensions 5.5 cm $\pm$ 1.8% by 6.4 cm $\pm$ 1.6%.

SOLUTION

Because the result is a multiplication, add the percent uncertainties:

$$A = \ell w = (5.5 \text{ cm} \pm 1.8\%)(6.4 \text{ cm} \pm 1.6\%)$$
$$= 35 \text{ cm}^2 \pm 3.4\% = (35 \pm 1) \text{ cm}^2$$

Addition and subtraction: When measurements with uncertainties are added or subtracted, add the *absolute uncertainties* to obtain the absolut uncertainty in the result.

Example B.9

Find the change in temperature, with associated uncertainty, when the temperature increases from $(27.6 \pm 1.5)°C$ to $(99.2 \pm 1.5)°C$

SOLUTION

Because the result is a subtraction, add the absolute uncertainties:

$$\Delta T = T_2 - T_1 = (99.2 \pm 1.5)°C - (27.6 \pm 1.5)°C$$
$$= (71.6 \pm 3.0)°C = \boxed{1.6°C \pm 4.2\%}$$

Powers: If a measurement is taken to a power, the percent uncertainty is multiplied by that power to obtain the percent uncertainty in the result.

Example B.10

Find the volume of a sphere of radius 6.20 cm $\pm$ 2.0%.

SOLUTION

Because the result is determined by raising a quantity to a power, multiply the power by the percent uncertainty:

$$V = \tfrac{4}{3}\pi r^3 = \tfrac{4}{3}\pi (6.20 \text{ cm} \pm 2.0\%)^3$$
$$= 998 \text{ cm}^3 \pm 6.0\% = \boxed{(998 \pm 60) \text{ cm}^3}$$

For complicated calculations, many uncertainties are added together, which can cause the uncertainty in the final result to be undesirably large. Experiments should be designed such that calculations are as simple as possible.

Notice that uncertainties in a calculation always add. As a result, an experiment involving a subtraction should be avoided if possible, especially if the measurements being subtracted are close together. The result of such a calculation is a small difference in the measurements and uncertainties that add together. It is possible that the uncertainty in the result could be larger than the result itself!

Appendix C Periodic Table of the Elements

Transition elements

Legend example:
Symbol — **Ca** 20 — Atomic number
Atomic mass† — 40.078
$4s^2$ — Electron configuration

Group I	Group II								
H 1 1.007 9 $1s$									
Li 3 6.941 $2s^1$	**Be** 4 9.0122 $2s^2$								
Na 11 22.990 $3s^1$	**Mg** 12 24.305 $3s^2$								
K 19 39.098 $4s^1$	**Ca** 20 40.078 $4s^2$	**Sc** 21 44.956 $3d^14s^2$	**Ti** 22 47.867 $3d^24s^2$	**V** 23 50.942 $3d^34s^2$	**Cr** 24 51.996 $3d^54s^1$	**Mn** 25 54.938 $3d^54s^2$	**Fe** 26 55.845 $3d^64s^2$	**Co** 27 58.933 $3d^74s^2$	
Rb 37 85.468 $5s^1$	**Sr** 38 87.62 $5s^2$	**Y** 39 88.906 $4d^15s^2$	**Zr** 40 91.224 $4d^25s^2$	**Nb** 41 92.906 $4d^45s^1$	**Mo** 42 95.96 $4d^55s^1$	**Tc** 43 (98) $4d^55s^2$	**Ru** 44 101.07 $4d^75s^1$	**Rh** 45 102.91 $4d^85s^1$	
Cs 55 132.91 $6s^1$	**Ba** 56 137.33 $6s^2$	57–71*	**Hf** 72 178.49 $5d^26s^2$	**Ta** 73 180.95 $5d^36s^2$	**W** 74 183.84 $5d^46s^2$	**Re** 75 186.21 $5d^56s^2$	**Os** 76 190.23 $5d^66s^2$	**Ir** 77 192.2 $5d^76s^2$	
Fr 87 (223) $7s^1$	**Ra** 88 (226) $7s^2$	89–103**	**Rf** 104 (267) $6d^27s^2$	**Db** 105 (268) $6d^37s^2$	**Sg** 106 (269) $6d^47s^2$	**Bh** 107 (270) $6d^57s^2$	**Hs** 108 (277) $6d^67s^2$	**Mt**†† 109 (278) $6d^77s^2$	

*Lanthanide series

La 57 138.91 $5d^16s^2$	Ce 58 140.12 $5d^14f^16s^2$	Pr 59 140.91 $4f^36s^2$	Nd 60 144.24 $4f^46s^2$	Pm 61 (145) $4f^56s^2$	Sm 62 150.36 $4f^66s^2$

**Actinide series

Ac 89 (227) $6d^17s^2$	Th 90 232.04 $6d^27s^2$	Pa 91 231.04 $5f^26d^17s^2$	U 92 238.03 $5f^36d^17s^2$	Np 93 (237) $5f^46d^17s^2$	Pu 94 (244) $5f^67s^2$

Note: Atomic mass values given are averaged over isotopes in the percentages in which they exist in nature.

†For an unstable element, mass number of the most stable known isotope is given in parentheses.

††For elements 109 and higher, electron configurations are theoretically predicted.

	Group III	Group IV	Group V	Group VI	Group VII	Group 0
					H 1 1.007 9 $1s^1$	He 2 4.002 6 $1s^2$
	B 5 10.811 $2p^1$	C 6 12.011 $2p^2$	N 7 14.007 $2p^3$	O 8 15.999 $2p^4$	F 9 18.998 $2p^5$	Ne 10 20.180 $2p^6$
	Al 13 26.982 $3p^1$	Si 14 28.086 $3p^2$	P 15 30.974 $3p^3$	S 16 32.066 $3p^4$	Cl 17 35.453 $3p^5$	Ar 18 39.948 $3p^6$

Ni 28 58.693 $3d^84s^2$	Cu 29 63.546 $3d^{10}4s^1$	Zn 30 65.39 $3d^{10}4s^2$	Ga 31 69.723 $4p^1$	Ge 32 72.64 $4p^2$	As 33 74.922 $4p^3$	Se 34 78.96 $4p^4$	Br 35 79.904 $4p^5$	Kr 36 83.80 $4p^6$
Pd 46 106.42 $4d^{10}$	Ag 47 107.87 $4d^{10}5s^1$	Cd 48 112.41 $4d^{10}5s^2$	In 49 114.82 $5p^1$	Sn 50 118.71 $5p^2$	Sb 51 121.76 $5p^3$	Te 52 127.60 $5p^4$	I 53 126.90 $5p^5$	Xe 54 131.29 $5p^6$
Pt 78 195.08 $5d^96s^1$	Au 79 196.97 $5d^{10}6s^1$	Hg 80 200.59 $5d^{10}6s^2$	Tl 81 204.38 $6p^1$	Pb 82 207.2 $6p^2$	Bi 83 208.98 $6p^3$	Po 84 (209) $6p^4$	At 85 (210) $6p^5$	Rn 86 (222) $6p^6$
Ds 110 (281) $6d^87s^2$	Rg 111 (282) $6d^97s^2$	Cn 112 (285) $6d^{10}7s^2$	Nh 113 (286) $7p^1$	Fl 114 (289) $7p^2$	Mc 115 (289) $7p^3$	Lv 116 (293) $7p^4$	Ts 117 (294) $7p^5$	Og 118 (294) $7p^6$

Eu 63 151.96 $4f^76s^2$	Gd 64 157.25 $4f^75d^16s^2$	Tb 65 158.93 $4f^85d^16s^2$	Dy 66 162.50 $4f^{10}6s^2$	Ho 67 164.93 $4f^{11}6s^2$	Er 68 167.26 $4f^{12}6s^2$	Tm 69 168.93 $4f^{13}6s^2$	Yb 70 173.04 $4f^{14}6s^2$	Lu 71 174.97 $4f^{14}5d^16s^2$
Am 95 (243) $5f^77s^2$	Cm 96 (247) $5f^76d^17s^2$	Bk 97 (247) $5f^86d^17s^2$	Cf 98 (251) $5f^{10}7s^2$	Es 99 (252) $5f^{11}7s^2$	Fm 100 (257) $5f^{12}7s^2$	Md 101 (258) $5f^{13}7s^2$	No 102 (259) $5f^{14}7s^2$	Lr 103 (262) $5f^{14}6d^17s^2$

Note: For a description of the atomic data, visit *physics.nist.gov/PhysRefData/Elements/per_text.html*.

Appendix D SI Units

TABLE D.1 SI Units

Base Quantity	SI Base Unit	
	Name	Symbol
Length	meter	m
Mass	kilogram	kg
Time	second	s
Electric current	ampere	A
Temperature	kelvin	K
Amount of substance	mole	mol
Luminous intensity	candela	cd

TABLE D.2 Some Derived SI Units

Other Quantity	Name	Symbol	Expression in Terms of Base Units	Expression in Terms of SI Units
Plane angle	radian	rad	m/m	
Frequency	hertz	Hz	s^{-1}	
Force	newton	N	$kg \cdot m/s^2$	J/m
Pressure	pascal	Pa	$kg/m \cdot s^2$	N/m^2
Energy	joule	J	$kg \cdot m^2/s^2$	$N \cdot m$
Power	watt	W	$kg \cdot m^2/s^3$	J/s
Electric charge	coulomb	C	$A \cdot s$	
Electric potential	volt	V	$kg \cdot m^2/A \cdot s^3$	W/A
Capacitance	farad	F	$A^2 \cdot s^4/kg \cdot m^2$	C/V
Electric resistance	ohm	Ω	$kg \cdot m^2/A^2 \cdot s^3$	V/A
Magnetic flux	weber	Wb	$kg \cdot m^2/A \cdot s^2$	$V \cdot s$
Magnetic field	tesla	T	$kg/A \cdot s^2$	
Inductance	henry	H	$kg \cdot m^2/A^2 \cdot s^2$	$T \cdot m^2/A$

Answers to Quick Quizzes and Odd-Numbered Problems

Chapter 1

Answers to Quick Quizzes

1. (a)
2. False
3. (b)

Answers to Odd-Numbered Problems

1. (a) 5.52×10^3 kg/m^3 (b) It is between the density of aluminum and that of iron and is greater than the densities of typical surface rocks.
3. 7.69 cm
5. The angle subtended by the Great Wall is less than the visual acuity of the eye.
7. 0.141 nm
9. (b) only
11. 11.4×10^3 kg/m^3
13. 2.86 cm
15. 151 μm
17. (a) $\sim 10^2$ kg (b) $\sim 10^3$ kg
19. The average distance between asteroids in the asteroid belt is about 400 000 km.
21. 31 556 926.0 s
23. 19
25. 63
27. ± 3.46
29. 316 m
31. 10^{11} stars
33. Answers may vary. (a) $\sim 10^{29}$ prokaryotes (b) $\sim 10^{14}$ kg
35. (a) 478 cm^3/s (b) 0.225 cm/s (c) When the balloon radius is twice as large, its surface area is four times larger. The new volume added in one second in the inflation process is equal to this larger area times an extra radial thickness that is one-fourth as large as it was when the balloon was smaller.
37. $V = 0.579t + (1.19 \times 10^{-9})t^2$, where V is in cubic feet and t is in seconds
39. $\dfrac{d \tan\phi \tan\theta}{\tan\phi - \tan\theta}$

Chapter 2

Answers to Quick Quizzes

1. (b)
2. (c)
3. (b)
4. False. Your graph should look something like the one shown in the next column. This v_x–t graph shows that the maximum speed is about 5.0 m/s, which is 18 km/h (= 11 mi/h), so the driver was not speeding.

5. (b)
6. (c)
7. (a)–(e), (b)–(d), (c)–(f)
8. (i) (e) (ii) (d)

Answers to Odd-Numbered Problems

1. 0.02 s
3. (a) 2.30 m/s (b) 16.1 m/s (c) 11.5 m/s
5. (a) -2.4 m/s (b) -3.8 m/s (c) 4.0 s
7. (a) 2.80 h (b) 218 km
9. (a) 1.3 m/s^2 (b) $t = 3$ s, $a = 2$ m/s^2 (c) $t = 6$ s, $t > 10$ s (d) $a = -1.5$ m/s^2, $t = 8$ s
11. (a) 20 m/s, 5 m/s (b) 263 m
13.

15. (a) 9.00 m/s (b) −3.00 m/s (c) 17.0 m/s (d) The graph of velocity versus time is a straight line passing through 13 m/s at 10:05 a.m. and sloping downward, decreasing by 4 m/s for each second thereafter. (e) If and only if we know the object's velocity at one instant of time, knowing its acceleration tells us its velocity at every other moment as long as the acceleration is constant.

17. −16.0 cm/s^2

19. (a) The idea is false unless the acceleration is zero. We define constant acceleration to mean that the velocity is changing steadily in time. So, the velocity cannot be changing steadily in space. (b) This idea is true. Because the velocity is changing steadily in time, the velocity halfway through an interval is equal to the average of its initial and final values.

21. (a) 19.7 cm/s (b) 4.70 cm/s^2 (c) The length of the glider is used to find the average velocity during a known time interval.

23. (a) 3.75 s (b) 5.50 cm/s (c) 0.604 s (d) 13.3 cm, 47.9 cm (e) The cars are initially moving toward each other, so they soon arrive at the same position x when their speeds are quite different, giving one answer to (c) that is not an answer to (a). The first car slows down in its motion to the left, turns around, and starts to move toward the right, slowly at first and gaining speed steadily. At a particular moment its speed will be equal to the constant rightward speed of the second car, but at this time the accelerating car is far behind the steadily moving car; thus, the answer to (a) is not an answer to (c). Eventually the accelerating car will catch up to the steadily coasting car, but passing it at higher speed, and giving another answer to (c) that is not an answer to (a).

25. David will be unsuccessful. The average human reaction time is about 0.2 s (research on the Internet) and a dollar bill is about 15.5 cm long, so David's fingers are about 8 cm from the end of the bill before it is dropped. The bill will fall about 20 cm before he can close his fingers.

27. 7.96 s

29. (a) 10.0 m/s up (b) 4.68 m/s down

31. (a) The box could reach the window according to the data provided. (b) Answers will vary.

33. (a) $a_x(t) = a_{xi} + Jt$; $v_x(t) = v_{xi} + a_{xi}t + \frac{1}{2}Jt^2$; $x(t) = x_i + v_{xi}t + \frac{1}{2}a_{xi}t^2 + \frac{1}{6}Jt^3$

35. (a) 4.00 m/s (b) 1.00 ms (c) 0.816 m

37. (a) Here, v_1 must be greater than v_2 and the distance between the leading athlete and the finish line must be great enough so that the trailing athlete has time to catch up.

(b) $t = \dfrac{d_1}{v_1 - v_2}$ (c) $d_2 = \dfrac{v_2 d_1}{v_1 - v_2}$

39. (a) 5.46 s (b) 73.0 m (c) $v_{\text{Stan}} = 22.6$ m/s, $v_{\text{Kathy}} = 26.7$ m/s

41. 1.60 m/s^2

43. (a) 5.32 m/s^2 for Laura and 3.75 m/s^2 for Healan (b) 10.6 m/s for Laura and 11.2 m/s for Healan (c) Laura, by 2.63 m (d) 4.47 m at $t = 2.84$ s

Chapter 3

Answers to Quick Quizzes

1. vectors: (b), (c); scalars: (a), (d), (e)
2. (c)
3. (b) and (c)
4. (b)
5. (c)

Answers to Odd-Numbered Problems

1. (a) 8.60 m (b) 4.47 m, −63.4°; 4.24 m, 135°
3. (a) (−3.56 cm, −2.40 cm) (b) ($r = 4.30$ cm, $\theta = 326°$) (c) ($r = 8.60$ cm, $\theta = 34.0°$) (d) ($r = 12.9$ cm, $\theta = 146°$)
5. This situation can *never* be true because the distance is the length of an arc of a circle between two points, whereas the magnitude of the displacement vector is a straight-line chord of the circle between the same points.
7. 9.5 N, 57° above the x axis
9. (a) 5.2 m at 60° (b) 3.0 m at 330° (c) 3.0 m at 150° (d) 5.2 m at 300°
11. (a) yes (b) The speed of the camper should be 28.3 m/s or more to satisfy this requirement.
13. 9.48 m at 166°
15. (a) 185 N at 77.8° from the positive x axis (b) $(-39.3\hat{\mathbf{i}} - 181\hat{\mathbf{j}})$ N
17. (a) 2.83 m at $\theta = 315°$ (b) 13.4 m at $\theta = 117°$
19. (a) $8.00\hat{\mathbf{i}} + 12.0\hat{\mathbf{j}} - 4.00\hat{\mathbf{k}}$ (b) $2.00\hat{\mathbf{i}} + 3.00\hat{\mathbf{j}} - 1.00\hat{\mathbf{k}}$ (c) $-24.0\hat{\mathbf{i}} - 36.0\hat{\mathbf{j}} + 12.0\hat{\mathbf{k}}$
21. (a) $-3.00\hat{\mathbf{i}} + 2.00\hat{\mathbf{j}}$ (b) 3.61 at 146° (c) $3.00\hat{\mathbf{i}} - 6.00\hat{\mathbf{j}}$
23. (a) $a = 5.00$ and $b = 7.00$ (b) For vectors to be equal, all their components must be equal. A vector equation contains more information than a scalar equation.
25. $(2.60\hat{\mathbf{i}} + 4.50\hat{\mathbf{j}})$m
27. 196 cm at 345°
29. (a) $(-20.5\hat{\mathbf{i}} + 35.5\hat{\mathbf{j}})$ m/s (b) $25.0\hat{\mathbf{j}}$ m/s (c) $(-61.5\hat{\mathbf{i}} + 107\hat{\mathbf{j}})$ m (d) $37.5\hat{\mathbf{j}}$ m (e) 157 km
31. 1.43×10^4 at 32.2° above the horizontal
33. (a) $(5 + 11f)\hat{\mathbf{i}} + (3 + 9f)\hat{\mathbf{j}}$ meters (b) $(5 + 0)\hat{\mathbf{i}} + (3 + 0)\hat{\mathbf{j}}$ meters (c) This is reasonable because it is the location of the starting point, $5\hat{\mathbf{i}} + 3\hat{\mathbf{j}}$ meters. (d) $16\hat{\mathbf{i}} + 12\hat{\mathbf{j}}$ meters (e) This is reasonable because we have completed the trip, and this is the position vector of the endpoint.
35. 240 m at 237°
37. 1.15°
39. (a) 25.4 s (b) 15.0 km/h
41. (a) The x, y, and z components are, respectively, 2.00, 1.00, and 3.00. (b) 3.74 (c) $\theta_x = 57.7°$, $\theta_y = 74.5°$, $\theta_z = 36.7°$
43. (a) $-2.00\hat{\mathbf{k}}$ m/s (b) its velocity vector
45. (a) $\vec{\mathbf{R}}_1 = a\hat{\mathbf{i}} + b\hat{\mathbf{j}}$ (b) $R_1 = (a^2 + b^2)^{1/2}$ (c) $\vec{\mathbf{R}}_2 = a\hat{\mathbf{i}} + b\hat{\mathbf{j}} + c\hat{\mathbf{k}}$

Chapter 4

Answers to Quick Quizzes

1. (a)
2. (i) (b) (ii) (a)
3. 15°, 30°, 45°, 60°, 75°

4. (i) (d) **(ii)** (b)

5. (i) (b) **(ii)** (d)

Answers to Odd-Numbered Problems

1. (a) $(1.00\hat{\mathbf{i}} + 0.750\hat{\mathbf{j}})$ m/s   (b) $(1.00\hat{\mathbf{i}} + 0.500\hat{\mathbf{j}})$ m/s, 1.12 m/s

3. (a) $\vec{\mathbf{v}} = -12.0t\,\hat{\mathbf{j}}$, where $\vec{\mathbf{v}}$ is in meters per second and t is in seconds (b) $\vec{\mathbf{a}} = -12.0\hat{\mathbf{j}}$ m/s^2 (c) $\vec{\mathbf{r}} = (3.00\hat{\mathbf{i}} - 6.00\hat{\mathbf{j}})$ m; $\vec{\mathbf{v}} = -12.0\hat{\mathbf{j}}$ m/s

5. (a) $\vec{\mathbf{v}}_f = (3.45 - 1.79t)\hat{\mathbf{i}} + (2.89 - 0.650t)\hat{\mathbf{j}}$

(b) $\vec{\mathbf{r}}_f = (-25.3 + 3.45t - 0.893t^2)\hat{\mathbf{i}} + (28.9 + 2.89t - 0.325t^2)\hat{\mathbf{j}}$

7. 12.0 m/s

9. 67.8°

11. $d\tan\theta_i - \dfrac{gd^2}{2v_i^2\cos^2\theta_i}$

13. (a) $(0, 50.0$ m$)$ (b) $v_{xi} = 18.0$ m/s; $v_{yi} = 0$ (c) Particle under constant acceleration (d) Particle under constant velocity (e) $v_{xf} = v_{xi}$; $v_{yf} = -gt$ (f) $x_f = v_{xi}t$; $y_f = y_i - \frac{1}{2}gt^2$ (g) 3.19 s (h) 36.1 m/s, −60.1°

15. (a) 41.7 m/s (b) 3.81 s (c) $v_x = 34.1$ m/s, $v_y = -13.4$ m/s, $v = 36.7$ m/s

17. 1.92 s

19. 7.58×10^3 m/s, 5.80×10^3 s

21. 377 m/s^2

23. (a) Yes. The particle can be either speeding up or slowing down, with a tangential component of acceleration of magnitude $\sqrt{6^2 - 4.5^2} = 3.97$ m/s^2. (b) No. The magnitude of the acceleration cannot be less than $v^2/r = 4.5$ m/s^2.

25. (a) 9.80 m/s^2 down and 2.50 m/s^2 south (b) 9.80 m/s^2 down (c) The bolt moves on a parabola with its axis downward and tilting to the south. It lands south of the point directly below its starting point. (d) The bolt moves on a parabola with a vertical axis.

27. 18.2°

29. 15.3 m

31. (a) $\dfrac{2d/c}{1 - v^2/c^2}$ (b) $\dfrac{2d}{c}$

(c) The trip in flowing water takes a longer time interval. The swimmer travels at the low upstream speed for a longer time interval, so his average speed is reduced below c. Mathematically, $1/(1 - v^2/c^2)$ is always greater than 1. In the extreme, as $v \rightarrow c$, the time interval becomes infinite. In that case, the student can never return to the starting point because he cannot swim fast enough to overcome the river current.

33. (a) straight up, at 0° to the vertical (b) 8.25 m/s (c) a straight up and down line (d) a symmetric parabola opening downward (e) 12.6 m/s north at $\tan^{-1}(8.25/9.5) = 41.0°$ above the horizontal

35. The relationship between the height h and the walking speed is $h = (4.16 \times 10^{-3})v_x^2$, where h is in meters and v_x is in meters per second. At a typical walking speed of 4 to 5 km/h, the ball would have to be dropped from a height of about 1 cm, clearly much too low for a person's hand. Even at Olympic-record speed for the 100-m run (confirm on the Internet), this situation would only occur if the ball is dropped from about 0.4 m, which is also below the hand of a normally proportioned person.

37. (a) 26.9 m/s (b) 67.3 m (c) $(2.00\hat{\mathbf{i}} - 5.00\hat{\mathbf{j}})$ m/s^2

39. The initial height of the ball when struck is 3.94 m, which is too high for the batter to hit the ball.

41. (a) 1.69 km/s (b) 1.80 h

43. (a) $x = v_i(0.164\ 3 + 0.002\ 299v_i^2)^{1/2} + 0.047\ 94v_i^2$, where x is in meters and v_i is in meters per second (b) 0.041 0 m (c) 961 m (d) $x \approx 0.405v_i$ (e) $x \approx 0.095\ 9v_i^2$ (f) The graph of x versus v_i starts from the origin as a straight line with slope 0.405 s. Then it curves upward above this tangent line, becoming closer and closer to the parabola $x = 0.095\ 9v_i^2$, where x is in meters and v_i is in meters per second.

45. (a) 4.00 km/h (b) 4.00 km/h

47. $\sim 10^2$ m/s^2

49. (a) 43.2 m (b) $(9.66\hat{\mathbf{i}} - 25.6\,\hat{\mathbf{j}})$ m/s (c) Air resistance would ordinarily make the jump distance smaller and the final horizontal and vertical velocity components both somewhat smaller. If a skilled jumper shapes her body into an airfoil, however, she can deflect downward the air through which she passes so that it deflects her upward, giving her more time in the air and a longer jump.

51. (a) $\Delta t_1 = \dfrac{L}{c + v} + \dfrac{L}{c - v} = \dfrac{2L/c}{1 - v^2/c^2}$

(b) $\Delta t_2 = \dfrac{2L}{\sqrt{c^2 - v^2}} = \dfrac{2L/c}{\sqrt{1 - v^2/c^2}}$

(c) Sarah, who swims cross-stream, returns first.

53. $\tan^{-1}\left(\dfrac{\sqrt{2gh}}{v}\right)$

Chapter 5

Answers to Quick Quizzes

1. (d)

2. (a)

3. (d)

4. (b)

5. (i) (c) **(ii)** (a)

6. (b)

7. (b) Pulling up on the rope decreases the normal force, which, in turn, decreases the force of kinetic friction.

Answers to Odd-Numbered Problems

1. 8.71 N

3. (a) $(6.00\hat{\mathbf{i}} + 15.0\hat{\mathbf{j}})$ N (b) 16.2 N

5. (a) $(-45.0\hat{\mathbf{i}} + 15.0\hat{\mathbf{j}})$ m/s (b) 162° from the $+x$ axis (c) $(-225\hat{\mathbf{i}} + 75.0\,\hat{\mathbf{j}})$ m (d) $(-227\hat{\mathbf{i}} + 79.0\,\hat{\mathbf{j}})$ m

7. (a) $\hat{\mathbf{a}}$ is at 181° (b) 11.2 kg (c) 37.5 m/s (d) $(-37.5\hat{\mathbf{i}} - 0.893\hat{\mathbf{j}})$ m/s

9. (a) 1.53 m (b) 24.0 N forward and upward at 5.29° with the horizontal

11. (a) 3.64×10^{-18} N (b) 8.93×10^{-30} N is 408 billion times smaller

13. (a) $\sim 10^{-22}$ m/s^2 (b) $d \sim 10^{-23}$ m

15. (a) 3.43 kN (b) 0.967 m/s horizontally forward

17. (a)

(b) -2.54 m/s^2 (c) 3.19 m/s

19. (a)

(b) 613 N

21. (a) $a = g \tan \theta$ (b) 4.16 m/s^2
23. (a) $F_x > 19.6$ N   (b) $F_x \leq -78.4$ N
(c)

25. (a) $a_2 = 2a_1$ (b) $T_2 = \dfrac{m_1 m_2}{2m_2 + \frac{1}{2}m_1}g$ and $T_2 = \dfrac{m_1 m_2}{m_2 + \frac{1}{4}m_1}g$

(c) $\dfrac{m_1 g}{2m_2 + \frac{1}{2}m_1}$ and $\dfrac{m_1 g}{4m_2 + m_1}$

27. (a) 14.7 m   (b) neither mass is necessary
29. 37.8 N
31. (a)

(b) 2.31 m/s^2, down for m_1, left for m_2, and up for m_3
(c) $T_{12} = 30.0$ N and $T_{23} = 24.2$ N

(d) T_{12} decreases and T_{23} increases
33. Driver was traveling at 67.1 mi/h
35. 834 N
37. (a) 3.43 m/s^2 toward the scrap iron   (b) 3.43 m/s^2 toward the scrap iron; -6.86 m/s^2 toward the magnet
39. (a)

(b) F (c) $F - P$ (d) P (e) m_1: $F - P = m_1 a$; m_2: $P = m_2 a$

(f) $a = \dfrac{F - \mu_1 m_1 g - \mu_2 m_2 g}{m_1 + m_2}$

(g) $P = \dfrac{m_2}{m_1 + m_2}[F + m_1(\mu_2 - \mu_1)g]$

41. (a)

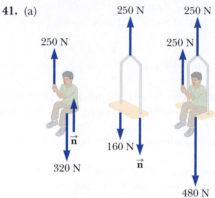

(b) 0.408 m/s^2 (c) 83.3 N
43. (a)

(b) 2.00 m/s^2 to the right   (c) 4.00 N on m_1, 6.00 N right on m_2, 8.00 N right on m_3 (d) 14.0 N between m_1 and m_2, 8.00 N between m_2 and m_3 (e) The m_2 block models the heavy block of wood. The contact force on your back is modeled by the force between the m_2 and the m_3 blocks, which is much less than the force F. The difference between F and this contact force is the net force causing the acceleration of the 5-kg pair of objects. The acceleration is real and nonzero, but it lasts for so short a time that it is never associated with a large velocity. The frame of the building and your legs exert forces, small in magnitude relative to the hammer blow, to bring the partition, block, and you to rest again over a time interval large relative to the hammer blow.
45. (b) If θ is greater than $\tan^{-1}(1/\mu_s)$, motion is impossible.
47. Ship requires 1.5 km to come to rest.
49. $(M + m_1 + m_2)(m_1 g/m_2)$
51. (a) 0.931 m/s^2 (b) From a value of 0.625 m/s^2 for large x, the acceleration gradually increases, passes through a

maximum, and then drops more rapidly, becoming negative and reaching -2.10 m/s^2 at $x = 0$.   (c) 0.976 m/s^2 at $x = 25.0$ cm   (d) 6.10 cm

53. (a) $m_2 g \left[\dfrac{m_1 M}{m_2 M + m_1(m_2 + M)} \right]$ (b) $\left[\dfrac{g m_1(m_2 + M)}{m_2 M + m_1(m_2 + M)} \right]$

(c) $\left[\dfrac{m_1 m_2 g}{m_2 M + m_1(m_2 + M)} \right]$ (d) $\left[\dfrac{m_1 M g}{m_2 M + m_1(m_2 + M)} \right]$

55. $\vec{\mathbf{R}} = [m\cos\theta \sin\theta \hat{\mathbf{i}} + (M + m\cos^2\theta)\hat{\mathbf{j}}]g$, where the x axis is horizontal and the y axis is vertical in Figure P5.55.

Chapter 6

Answers to Quick Quizzes

1. (i) (a)   (ii) (b)

2. (i) Because the speed is constant, the only direction the force can have is that of the centripetal acceleration. The force is larger at Ⓒ than at Ⓐ because the radius at Ⓒ is smaller. There is no force at Ⓑ because the wire is straight.   (ii) In addition to the forces in the centripetal direction in part (a), there are now tangential forces to provide the tangential acceleration. The tangential force is the same at all three points because the tangential acceleration is constant.

3. (c)

4. (a)

Answers to Odd-Numbered Problems

1. (a) 8.33×10^{-8} N toward the nucleus
(b) 9.15×10^{22} m/s^2 inward

3. (a) $(-0.233\hat{\mathbf{i}} + 0.163\hat{\mathbf{j}})$ m/s^2
(b) 6.53 m/s, $(-0.181\hat{\mathbf{i}} + 0.181\hat{\mathbf{j}})$ m/s^2

5. 6.22×10^{-12} N

7. (a) no   (b) yes

9. (a) 1.33 m/s^2 (b) 1.79 m/s^2 at 48.0° inward from the direction of the velocity

11. (a) $v = \sqrt{R\left(\dfrac{2T}{m} - g\right)}$ (b) $2T$ up

13. (a) 8.62 m   (b) Mg, downward   (c) 8.45 m/s^2 (d) Calculation of the normal force shows it to be negative, which is impossible. We interpret it to mean that the normal force goes to zero at some point and the passengers will fall out of their seats near the top of the ride if they are not restrained in some way. We could arrive at this same result without calculating the normal force by noting that the acceleration in part (c) is smaller than that due to gravity. The teardrop shape has the advantage of a larger acceleration of the riders at the top of the arc for a path having the same height as the circular path, so the passengers stay in the cars.

15. (a) 491 N   (b) 50.1 kg   (c) 2.00 m/s^2

17. 0.527°

19. (a) 2.03 N down   (b) 3.18 m/s^2 down   (c) 0.205 m/s down

21. (a) 1.47 N · s/m   (b) 2.04×10^{-3} s   (c) 2.94×10^{-2} N

23. 10^1 N

25. 781 N

27. (a) $mg - \dfrac{mv^2}{R}$ (b) $\sqrt{gR}$

29. (a) $v = v_i\, e^{-bt/m}$ (b)

(c) In this model, the object keeps moving forever.
(d) It travels a finite distance in an infinite time interval.

31. (a) the downward gravitational force and the tension force in the string, always directed toward the center of the path

(b)   (c) 6.05 N   (d) 7.82 m/s

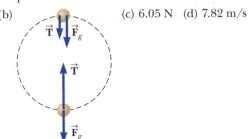

33. (a) 1 975 lb, directed upward   (b) 647 lb, directed downward   (c) When $F'_g = 0$, then $mg = \dfrac{mv^2}{R}$.

35. (a) The only horizontal force on the car is the force of friction, with a maximum value determined by the surface roughness (described by the coefficient of static friction) and the normal force (here equal to the gravitational force on the car).   (b) 34.3 m   (c) 68.6 m   (d) Braking is better. You should not turn the wheel. If you used any of the available friction force to change the direction of the car, it would be unavailable to slow the car and the stopping distance would be greater.   (e) The conclusion is true in general. The radius of the curve you can barely make is twice your minimum stopping distance.

37. (a) 735 N   (b) 732 N   (c) The gravitational force is larger. The normal force is smaller, just like it is when going over the top of a Ferris wheel.

39. (a) $\sum \vec{\mathbf{F}} = km\vec{\mathbf{v}}$ (b) In general, the possibility of k positive is unrealistic in nature. You might be able to imagine some device with a feedback mechanism that could be used to apply a force to cause the velocity to increase in magnitude. In this case the speed would increase exponentially, so such a situation could only exist temporarily.   (c) Think of a duck landing on a lake, where the water exerts a resistive force on the duck proportional to its speed.

41. (a) $v_{\min} = \sqrt{\dfrac{Rg(\tan\theta - \mu_s)}{1 + \mu_s \tan\theta}}$, $v_{\max} = \sqrt{\dfrac{Rg(\tan\theta + \mu_s)}{1 - \mu_s \tan\theta}}$

(b) $\mu_s = \tan\theta$

43. (a) Particle under constant acceleration   (b) $\Delta t = \dfrac{2v}{g}$

(c) Particle in uniform circular motion   (d) $T = \dfrac{2\pi R}{v}$
(e) $v = \sqrt{\pi Rg}$ (f) $F = \pi mg$
45. (a) 78.3 m/s   (b) 11.1 s   (c) 121 m
47. (a) 8.04 s   (b) 379 m/s   (c) 1.19×10^{-2} m/s   (d) 9.55 cm
49. 0.092 8°

Chapter 7

Answers to Quick Quizzes

1. (a)
2. (c), (a), (d), (b)
3. (d)
4. (a)
5. (b)
6. (c)
7. (i) (c)   (ii) (a)
8. (d)

Answers to Odd-Numbered Problems

1. (a) 1.59×10^{3} J   (b) smaller   (c) the same
3. (a) 472 J   (b) 2.76 kN
7. 5.33 J
9. (a) 7.50 J   (b) 15.0 J   (c) 7.50 J   (d) 30.0 J
11. (a) 0.938 cm   (b) 1.25 J
13. Each spring should have a spring constant of 316 N/m.
15. (b) mgR
17. (a)

F (N)

(b) The slope of the line is 116 N/m.   (c) We use all the points listed and also the origin. There is no visible evidence for a bend in the graph or nonlinearity near either end.   (d) 116 N/m   (e) 12.7 N
19. (a) 50.0 J   (b) 87.5 J; path independent.
21. (a) 1.20 J   (b) 5.00 m/s   (c) 6.30 J
23. 878 kN up
25. (a) 4.56 kJ   (b) 4.56 kJ   (c) 6.34 kN
(d) 422 km/s²   (e) 6.34 kN   (f) The two theories agree.
27. (a) 97.8 J   (b) $(-4.31\hat{\mathbf{i}} + 31.6\,\hat{\mathbf{j}})$ N   (c) 8.73 m/s
29. (a) 2.5 J   (b) −9.8 J   (c) −12 J
31. (a) −196 J   (b) −196 J   (c) −196 J   (d) The gravitational force is conservative.
33. (a) 125 J   (b) 50.0 J   (c) 66.7 J   (d) nonconservative
(e) The work done on the particle depends on the path followed by the particle.
35. (a) 40.0 J   (b) −40.0 J   (c) 62.5 J
37. A/r^{2} away from the other particle
39.

Stable   Unstable   Neutral

41. 0.559 m/s
43. 0.799 N · m
45. (a) $\vec{\mathbf{F}}_1 = (20.5\hat{\mathbf{i}} + 14.3\hat{\mathbf{j}})$ N, $\vec{\mathbf{F}}_2 = (-36.4\hat{\mathbf{i}} + 21.0\hat{\mathbf{j}})$ N
(b) $\sum \vec{\mathbf{F}} = (-15.9\hat{\mathbf{i}} + 35.3\hat{\mathbf{j}})$ N
(c) $\vec{\mathbf{a}} = (-3.18\hat{\mathbf{i}} + 7.07\hat{\mathbf{j}})$ m/s²
(d) $\vec{\mathbf{v}} = (-5.54\hat{\mathbf{i}} + 23.7\hat{\mathbf{j}})$ m/s
(e) $\vec{\mathbf{r}} = (-2.30\hat{\mathbf{i}} + 39.3\hat{\mathbf{j}})$ m   (f) 1.48 kJ   (g) 1.48 kJ
(h) The work–kinetic energy theorem is consistent with Newton's second law.
47. 0.131 m
49. (a) 19.3°   (b) 1.39×10^{4} J

Chapter 8

Answers to Quick Quizzes

1. (i) (b)   (ii) (b)   (iii) (a)
2. (a)
3. $v_1 = v_2 = v_3$
4. (c)

Answers to Odd-Numbered Problems

1. (a) $\Delta K + \Delta U = 0$, $v = \sqrt{2gh}$ (b) $\Delta K = W$, $v = \sqrt{2gh}$
3. (a) 5.94 m/s, 7.67 m/s   (b) 147 J
5. 5.49 m/s
7. (a) −168 J   (b) 184 J   (c) 500 J   (d) 148 J   (e) 5.65 m/s
9. (a) 5.60 J   (b) 2.29 rev
11. (a) 22.0 J, 40.0 J   (b) Yes   (c) The total mechanical energy has decreased, so a nonconservative force must have acted.
13. (a) Isolated. The only external influence on the system is the normal force from the slide, but this force is always perpendicular to its displacement so it performs no work on the system.   (b) No, the slide is frictionless.
(c) $E_{\text{system}} = mgh$ (d) $E_{\text{system}} = \frac{1}{5}mgh + \frac{1}{2}mv_i^2$
(e) $E_{\text{system}} = mgy_{\text{max}} + \frac{1}{2}mv_{xi}^2$
(f) $v_i = \sqrt{\dfrac{8gh}{5}}$ (g) $y_{\text{max}} = h(1 - \frac{4}{5}\cos^2\theta)$ (h) If friction is
present, mechanical energy of the system would *not* be conserved, so the child's kinetic energy at all points after leaving the top of the waterslide would be reduced when compared with the frictionless case. Consequently, her launch speed and maximum height would be reduced as well.
15. Both trails result in the same speed.
17. \$145
19. ~ 10^4 W
21. (a) 423 mi/gal   (b) 776 mi/gal
23. (a) 0.225 J   (b) −0.363 J   (c) no   (d) It is possible to find an effective coefficient of friction but not the actual value of μ since n and f vary with position.
25. (a) 1.29×10^4 N   (b) 45.4 m/s   (c) 3.72×10^4 N ; 46.1 m/s
(d) 45 m   (e) No
27. (a) $x = -4.0$ mm   (b) −1.0 cm
29. (a) -6.08×10^3 J   (b) -4.59×10^3 J   (c) 4.59×10^3 J
31. (a) 1.38×10^4 J   (b) 5.51×10^3 W
(c) The value in part (b) represents only energy that leaves the engine and is transformed to kinetic energy of the car. Additional energy leaves the engine by sound and heat. More energy from the engine is transformed to internal energy by friction forces and air resistance.

33. (a) 0.403 m or −0.357 m   (b) From a perch at a height of 2.80 m above the top of a pile of mattresses, a 46.0-kg child jumps upward at 2.40 m/s. The mattresses behave as a linear spring with force constant 19.4 kN/m. Find the maximum amount by which they are compressed when the child lands on them.   (c) 0.023 2 m   (d) This result is the distance by which the mattresses compress if the child just stands on them. It is the location of the equilibrium position of the oscillator.

35. (a) 1.53 J at $x = 6.00$ cm, 0 J at $x = 0$ (b) 1.75 m/s   (c) 1.51 m/s   (d) The answer to part (c) is not half the answer to part (b), because the equation for the speed of an oscillator is not linear in position

37. 48.2°

39. (a) No, mechanical energy is not conserved in this case.   (b) 77.0 m/s

43. (b) 0.342

45. (a) $-\mu_k gx/L$ (b) $(\mu_k gL)^{1/2}$

47. Less dangerous

Chapter 9

Answers to Quick Quizzes

1. (d)
2. (b), (c), (a)
3. (i) (c), (e)   (ii) (b), (d)
4. (a) All three are the same. (b) dashboard, seat belt, air bag
5. (a)
6. (b)
7. (b)
8. (i) (a)   (ii) (b)

Answers to Odd-Numbered Problems

1. (b) $p = \sqrt{2mK}$

3. $\vec{\mathbf{F}}_{\text{on bat}} = (3.26\hat{\mathbf{i}} - 3.99\hat{\mathbf{j}})$ kN, where positive x is from the pitcher toward home plate and positive y is upward.

5. (a) $-6.00\hat{\mathbf{i}}$ m/s   (b) 8.40 J   (c) The original energy is in the spring.   (d) A force had to be exerted over a displacement to compress the spring, transferring energy into it by work. The cord exerts force, but over no displacement.   (e) System momentum is conserved with the value zero.   (f) The forces on the two blocks are internal forces, which cannot change the momentum of the system; the system is isolated.   (g) Even though there is motion afterward, the final momenta are of equal magnitude in opposite directions, so the final momentum of the system is still zero.

7. (c) no difference

9. (a) 9.60×10^{-2} s   (b) 3.65×10^5 N   (c) $26.6g$

11. 16.5 N

13. (a) 2.50 m/s   (b) 37.5 kJ

15. (a) $v_f = \dfrac{1}{3}(v_1 + 2v_2)$ (b) $\Delta K = -\dfrac{m}{3}(v_1^2 + v_2^2 - 2v_1 v_2)$

17. (a) 4.85 m/s   (b) 8.41 m

19. The defendant was traveling at 41.5 mi/h.

21. $v_O = v_i \cos\theta$, $v_Y = v_i \sin\theta$

23. $v = \dfrac{v_i}{\sqrt{2}}$, 45.0°, −45.0°

25. 3.57×10^8 J

27. (a)

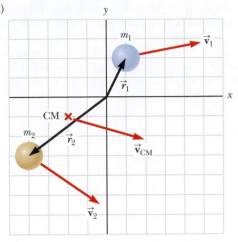

(b) $(-2.00\hat{\mathbf{i}} - 1.00\hat{\mathbf{j}})$ m   (c) $(3.00\hat{\mathbf{i}} - 1.00\hat{\mathbf{j}})$ m/s
(d) $(15.0\hat{\mathbf{i}} - 5.00\hat{\mathbf{j}})$ kg · m/s

29. The drone was struck by a meteorite.

31. (a) yes   (b) no   (c) 103 kg · m/s, up   (d) yes   (e) 88.2 J   (f) No, the energy came from potential energy stored in the person from previous meals.

33. (a) 442 metric tons   (b) 19.2 metric tons   (c) It is much less than the suggested value of 442/2.50. Mathematically, the logarithm in the rocket propulsion equation is not a linear function. Physically, a higher exhaust speed has an extra-large cumulative effect on the rocket body's final speed by counting again and again in the speed the body attains second after second during its burn.

35. (a) She moves, just like the archer in Example 9.1.
(b) $-\left(\dfrac{m}{M - m}\right)\vec{\mathbf{v}}_{\text{gloves}}$ (c) As she throws the gloves and exerts a force on them, the gloves exert an equal and opposite force on her that causes her to accelerate from rest to reach the velocity $\vec{\mathbf{v}}_{\text{girl}}$.

37. (a) $1.33\hat{\mathbf{i}}$ m/s   (b) $-235\hat{\mathbf{i}}$ N   (c) 0.680 s   (d) $-160\hat{\mathbf{i}}$ N · s and $+160\hat{\mathbf{i}}$ N · s   (e) 1.81 m   (f) 0.454 m   (g) −427 J   (h) +107 J   (i) Let's imagine an ideal situation in which the person and the cart have a perfect thermal insulator between them, so that no energy can transfer by heat Q between the person and the cart. Then, the change in kinetic energy of one member of the system, according to Equation 8.2, will be equal to the negative of the change in internal energy for that member: $\Delta K = -\Delta E_{\text{int}}$. The change in internal energy, in turn, is the product of the friction force and the distance through which the member moves while experiencing that force. Equal-magnitude friction forces act on the person and the cart, but the person and the cart move through different distances, as we see in parts (e) and (f). Therefore, there are different changes in internal energy for the person and the cart and, in turn, different changes in kinetic energy. The person and the cart will experience different changes in internal energy and, therefore, in temperature, which, in the real situation without the thermal insulator, will equalize after the event by means of the transfer of energy by heat Q between the person and the cart. The total change in kinetic energy of the system, −320 J, becomes +320 J of extra internal energy in the entire system in this perfectly inelastic collision.

39. (a) Momentum of the bullet–block system is conserved in the collision, so you can relate the speed of the block and bullet immediately after the collision to the initial speed of the bullet. Then, you can use conservation of mechanical energy for the bullet–block–Earth system to relate the speed after the collision to the maximum height. (b) 521 m/s upward

41. (a) $\dfrac{m_1 v_1 + m_2 v_2}{m_1 + m_2}$ (b) $(v_1 - v_2)\sqrt{\dfrac{m_1 m_2}{k(m_1 + m_2)}}$

(c) $v_{1f} = \dfrac{(m_1 - m_2)v_1 + 2m_2 v_2}{m_1 + m_2}$,

$v_{2f} = \dfrac{2m_1 v_1 + (m_2 - m_1)v_2}{m_1 + m_2}$

43. (a) 6.29 m/s (b) 6.16 m/s (c) Most of the 2% difference between the values for speed could be accounted for by air resistance.

45. 143 m/s
47. (a) 0; inelastic (b) $(-0.250\,\hat{\mathbf{i}} + 0.75\,\hat{\mathbf{j}} - 2.00\,\hat{\mathbf{k}})$ m/s; perfectly inelastic (c) either $a = -6.74$ with $\vec{\mathbf{v}} = -0.419\,\hat{\mathbf{k}}$ m/s or $a = 2.74$ with $\vec{\mathbf{v}} = -3.58\,\hat{\mathbf{k}}$ m/s

49. (a) $-0.256\,\hat{\mathbf{i}}$ m/s and $0.128\,\hat{\mathbf{i}}$ m/s
(b) $-0.064\,2\,\hat{\mathbf{i}}$ m/s and 0 (c) 0 and 0
51. (a) $(20.0\,\hat{\mathbf{i}} + 7.00\,\hat{\mathbf{j}})$ m/s (b) $4.00\,\hat{\mathbf{i}}$ m/s^2 (c) $4.00\,\hat{\mathbf{i}}$ m/s^2
(d) $(50.0\,\hat{\mathbf{i}} + 35.0\,\hat{\mathbf{j}})$ m (e) 600 J (f) 674 J (g) 674 J
(h) The accelerations computed in different ways agree. The kinetic energies computed in different ways agree. The three theories are consistent.
53. (a) particle of mass m: $\sqrt{2}v_i$; particle of mass $3m$: $\sqrt{\tfrac{2}{3}}v_i$
(b) 35.3°

Chapter 10

Answers to Quick Quizzes

1. (i) (c) (ii) (b)
2. (b)
3. (i) (b) (ii) (a)
4. (b)
5. (b)
6. (a)
7. (b)

Answers to Odd-Numbered Problems

1. (a) 7.27×10^{-5} rad/s (b) Because of its angular speed, the Earth bulges at the equator.
3. (a) 4.00 rad/s^2 (b) 18.0 rad
5. (a) 8.21×10^2 rad/s^2 (b) 4.21×10^3 rad

7. (a) $\omega h^{3/2}\left(\dfrac{2}{g}\right)^{1/2}$ (b) 1.16 cm (c) The deflection is only 0.02% of the original height, so it is negligible in many practical cases. (d) decrease

9. (a) 25.0 rad/s (b) 39.8 rad/s^2 (c) 0.628 s
11. (a) 54.3 rev (b) 12.1 rev/s
13. (a) 3.47 rad/s (b) 1.74 m/s (c) 2.78 s (d) 1.02 rotations
15. (a) 1.03 s (b) 10.3 rev
17. (a) 24.0 N · m (b) 0.035 6 rad/s^2 (c) 1.07 m/s^2
19. (a) 0.312 (b) 117 N

21. $\tau_f = -0.039\,8$ N · m

23. $I_{y'} = \displaystyle\int_{\text{all mass}} r^2\,dm = \int_0^L x^2 \dfrac{M}{L}\,dx = \dfrac{M}{L}\dfrac{x^3}{3}\Big|_0^L = \tfrac{1}{3}ML^2$

25. (a) 92.0 kg · m^2 (b) 184 J (c) 6.00 m/s, 4.00 m/s, 8.00 m/s (d) 184 J (e) The kinetic energies computed in parts (b) and (d) are the same.
27. 1.03×10^{-3} J
29. (a) 11.4 N (b) 7.57 m/s^2 (c) 9.53 m/s (d) 9.53 m/s
31. (a) $2(Rg/3)^{1/2}$ (b) $4(Rg/3)^{1/2}$ (c) $(Rg)^{1/2}$
33. (a) 2.38 m/s (b) The centripetal acceleration at the top is $\dfrac{v_2^2}{r} = \dfrac{(2.38\ \text{m/s})^2}{0.450\ \text{m}} = 12.6\ \text{m/s}^2 > g$. Therefore, the ball must be in contact with the track, with the track pushing downward on it. (c) 4.31 m/s (d) The speed of the ball turns out to be imaginary. (e) When the ball is projected with the same speed as before, but with only translational kinetic energy, there is insufficient kinetic energy for the ball to arrive at the top of the track.
35. (a) 1.21×10^{-4} kg · m^2 (b) Knowing the height of the can is unnecessary. (c) The mass is not uniformly distributed; the density of the metal can is larger than that of the soup.
37. (a) 12.5 rad/s (b) 128 rad

39. (a) $d = (1\,890 + 80n)\left(\dfrac{0.459\ \text{m}}{80n - 150}\right)$ (b) 94.1 m (c) 1.62 m
(d) -5.79 m (e) The rising car will coast to a stop only for $n \geq 2$. (f) For $n = 0$ or $n = 1$, the mass of the elevator is less than the counterweight, so the car would accelerate upward if released. (g) 0.459 m
43. 54.0°
45. (b) to the left

Chapter 11

Answers to Quick Quizzes

1. (d)
2. (i) (a) (ii) (c)
3. (b)
4. (a)

Answers to Odd-Numbered Problems

1. $\hat{\mathbf{i}} + 8.00\,\hat{\mathbf{j}} + 22.0\,\hat{\mathbf{k}}$
3. 45.0°
5. (a) $F_3 = F_1 + F_2$ (b) no
7. (a) $(-10.0\ \text{N} \cdot \text{m})\hat{\mathbf{k}}$ (b) yes (c) yes (d) yes (e) no
(f) $5.00\,\hat{\mathbf{j}}$ m
9. $m(xv_y - yv_x)\hat{\mathbf{k}}$
11. (a) zero (b) $(-mv_i^3 \sin^2\theta \cos\theta/2g)\hat{\mathbf{k}}$
(c) $(-2mv_i^3 \sin^2\theta \cos\theta/g)\hat{\mathbf{k}}$
(d) The downward gravitational force exerts a torque on the projectile in the negative z direction.
13. $mvR[\cos(vt/R) + 1]\hat{\mathbf{k}}$
15. (a) $-m\ell gt \cos\theta\,\hat{\mathbf{k}}$ (b) The Earth exerts a gravitational torque on the ball. (c) $-mg\ell \cos\theta\,\hat{\mathbf{k}}$
17. (a) 0.360 kg · m^2/s (b) 0.540 kg · m^2/s
19. 1.20 kg · m^2/s
21. 8.63 m/s^2
23. (a) The mechanical energy of the system is not constant. Some potential energy in the woman's body from

previous meals is converted into mechanical energy. (b) The momentum of the system is not constant. The turntable bearing exerts an external northward force on the axle.   (c) The angular momentum of the system is constant.   (d) 0.360 rad/s counterclockwise   (e) 99.9 J

25. (a) 11.1 rad/s counterclockwise   (b) No; 507 J is transformed into internal energy in the system.   (c) No; the turntable bearing promptly imparts impulse 44.9 kg · m/s north into the turntable–clay system and thereafter keeps changing the system momentum as the velocity vector of the clay continuously changes direction.

27. (a) $mv\ell$ down   (b) $M/(M + m)$

29. (a) $\omega = 2mv_i d/[M + 2m]R^2$ (b) No; some mechanical energy of the system changes into internal energy.   (c) The momentum of the system is not constant. The axle exerts a backward force on the cylinder when the clay strikes.

31. 5.46×10^{22} N · m

33. (a) 2.35 rad/s   (b) 0.498 rad/s   (c) 5.58°

35. 7.50×10^{-11} s

37. (a) $7md^2/3$ (b) $mgd\hat{\mathbf{k}}$ (c) $3g/7d$ counterclockwise   (d) $2g/7$ upward   (e) mgd (f) $\sqrt{6g/7d}$ (g) $m\sqrt{14gd^3/3}$ (h) $\sqrt{2gd/21}$

39. (a) 3 750 kg · m²/s   (b) 1.88 kJ   (c) 3 750 kg · m²/s   (d) 10.0 m/s   (e) 7.50 kJ   (f) 5.62 kJ

41. (a) $2mv_0$ (b) $2v_0/3$ (c) $4m\ell v_0/3$ (d) $4v_0/9\ell$ (e) mv_0^2 (f) $26mv_0^2/27$ (g) No horizontal forces act on the bola from outside after release, so the horizontal momentum stays constant. Its center of mass moves steadily with the horizontal velocity it had at release. No torques about its axis of rotation act on the bola, so the angular momentum stays constant. Internal forces cannot affect momentum conservation and angular momentum conservation, but they can affect mechanical energy.

43. 10.7 m/s

45. an increase of 6.368×10^{-4} % or 0.550 s, which is not significant

47. 14.0 s

49. (a) 2.0 m/s   (b) 1.0 rad/s

Chapter 12

Answers to Quick Quizzes

1. (a)
2. (b)
3. (b)
4. **(i)** (b)   **(ii)** (a)   **(iii)** (c)

Answers to Odd-Numbered Problems

1. Safe arrangements: 2-3-1, 3-1-2, 3-2-1; dangerous arrangements: 1-2-3, 1-3-2, 2-1-3
3. (3.85 cm, 6.85 cm)
5. $x = 0.750$ m
7. 177 kg
9. (a) 29.9 N   (b) 22.2 N
11. (a) 27.7 kN   (b) 11.5 kN   (c) 4.19 kN
13. (a) 1.04 kN at 60.0° upward and to the right   (b) $(370\hat{\mathbf{i}} + 910\hat{\mathbf{j}})$ N
15. (a) 859 N   (b) 1.04 kN at 36.9° to the left and upward
17. (a) $-0.053\,8$ m³   (b) 1.09×10^3 kg/m³   (c) With only a 5% change in volume in this extreme case, liquid water can be modeled as incompressible in biological and student laboratory situations.

19. 23.8 μm

21. (a) 3.14×10^4 N   (b) 6.28×10^4 N

23. 9.85×10^{-5}

25. $n_A = 5.98 \times 10^5$ N, $n_B = 4.80 \times 10^5$ N

27. (a) 0.400 mm   (b) 40.0 kN   (c) 2.00 mm   (d) 2.40 mm   (e) 48.0 kN

29. (a)

(b) $T = 343$ N, $R_x = 171$ N to the right, $R_y = 683$ N up   (c) 5.14 m

31. (a) $T = F_g(L + d)/[\sin\theta\,(2L + d)]$ (b) $R_x = F_g(L + d)\cot\theta/(2L + d)$; $R_y = F_g L/(2L + d)$

33. (a) 5.08 kN   (b) 4.77 kN   (c) 8.26 kN

35. (a) $\frac{1}{2}m\left(\dfrac{2\mu_s\sin\theta - \cos\theta}{\cos\theta - \mu_s\sin\theta}\right)$ (b) $(m + M)g\sqrt{1 + \mu_s^2}$ (c) $g\sqrt{M^2 + \mu_s^2(m + M)^2}$

37. (a) 9.28 kN   (b) The moment arm of the force $\vec{\mathbf{F}}_h$ is no longer 70 cm from the shoulder joint but only 49.5 cm, therefore reducing $\vec{\mathbf{F}}_m$ to 6.56 kN.

39. (a) 66.7 N   (b) increasing at 0.125 N/s

41. (a) $\dfrac{1}{\sqrt{15}}\dfrac{mgd}{\ell}$ (b) $n_A = mg\left(\dfrac{2\ell - d}{2\ell}\right)$, $n_B = \dfrac{mgd}{2\ell}$ (c) $R_x = \dfrac{1}{\sqrt{15}}\dfrac{mgd}{\ell}$ to the right, $R_y = \dfrac{mgd}{2\ell}$ downward

43. (a) $P_1 = P_3 = 1.67$ N, $P_2 = 3.33$ N   (b) 2.36 N

45. 5.73 rad/s

47. (a) 443 N   (b) 221 N (to the right), 217 N (upward)

49. (b) 60.0°   (c) unstable

Chapter 13

Answers to Quick Quizzes

1. (e)
2. (c)
3. (a)
4. (a) perihelion   (b) aphelion   (c) perihelion   (d) all points

Answers to Odd-Numbered Problems

1. 7.41×10^{-10} N
3. $\sim 10^{-7}$ N
5. (a) 7.61 cm/s²   (b) 363 s   (c) 3.08 km   (d) 28.9 m/s at 72.9° below the horizontal
7. (a) 1.31×10^{17} N   (b) 2.62×10^{12} N/kg
9. (a) 0.708 yr   (b) 0.399 yr
11. 4.99 days
13. (a) yes   (b) 3.93 yr
15. 4.17×10^{10} J
17. (a) -1.67×10^{-14} J   (b) The particles collide at the center of the triangle.
19. 1.58×10^{10} J
21. 1.78×10^3 m
23. (a) same size force   (b) 15.6 km/s

25. 492 m/s

27. 1.30×10^3 m/s

29. 2.25×10^{-7}

31. (a) 1.00×10^7 m (b) 1.00×10^4 m/s

33. (a) 15.3 km (b) 1.66×10^{16} kg (c) 1.13×10^4 s (d) No; its mass is so large compared with yours that you would have a negligible effect on its rotation.

35. (c) 1.85×10^{-5} m/s^2

37. (a) 2×10^8 yr (b) $\sim 10^{41}$ kg (c) 10^{11}

39. (a) 2.93×10^4 m/s (b) $K = 2.74 \times 10^{33}$ J, $U = -5.39 \times 10^{33}$ J (c) $K = 2.56 \times 10^{33}$ J, $U = -5.21 \times 10^{33}$ J (d) Yes; $E = -2.65 \times 10^{33}$ J at both aphelion and perihelion.

41. (a) $(2.77 \text{ m/s}^2)\left(1 + \dfrac{m}{5.98 \times 10^{24} \text{ kg}}\right)$ (b and c) 2.77 m/s^2 (d) 3.70 m/s^2 (e) Any object with mass small compared to the Earth starts to fall with acceleration 2.77 m/s^2. As m increases to become comparable to the mass of the Earth, the acceleration increases and can become arbitrarily large. It approaches a direct proportionality to m.

43. 18.2 ms

Chapter 14

Answers to Quick Quizzes

1. (a)

2. (a)

3. (c)

4. (b) or (c)

5. (a)

Answers to Odd-Numbered Problems

1. 2.96×10^6 Pa

3. 5.27×10^{18} kg

5. 7.74×10^{-3} m^2

7. 0.072 1 mm

9. (a) 10.5 m (b) No. The vacuum is not as good because some alcohol and water in the wine will evaporate. The equilibrium vapor pressures of alcohol and water are higher than the vapor pressure of mercury.

11. 3.33×10^3 kg/m^3

13. (a) 1 250 kg/m^3 (b) 500 kg/m^3

15. (a) 408 kg/m^3 (b) When m is less than 0.310 kg, the wooden block will be only partially submerged in the water. (c) When m is greater than 0.310 kg, the wooden block and steel object will sink.

17. (a) 11.6 cm (b) 0.963 g/cm^3 (c) No; the density ρ is not linear in h.

19. 20.0 g

21. (b) 616 MW

23. (a) 15.1 MPa (b) 2.95 m/s

25. (a) 28.0 m/s (b) 28.0 m/s (c) The answers agree precisely. The models are consistent with each other. (d) 2.11 MPa

27. 0.120 N

29. 0.200 mm

31. (a) 4.43 m/s (b) 10.1 m

33. (a) particle in equilibrium
(b) $\sum F_y = B - F_b - F_{\text{He}} - F_s = 0$

(c) $m_s = \frac{4}{3}(\rho_{\text{air}} - \rho_{\text{He}})\pi r^3 - m_b$

(d) 0.023 7 kg (e) 0.948 m

35. $\sim 10^4$

37. (a) 8.01 km (b) yes

39. 91.64%

41. (a) 3.307 g (b) 3.271 g (c) 3.48×10^{-4} N

43. 18.1 N

45. 758 Pa

47. 4.43 m/s

Chapter 15

Answers to Quick Quizzes

1. (d)

2. (f)

3. (a)

4. (b)

5. (c)

6. (i) (a) **(ii)** (a)

Answers to Odd-Numbered Problems

1. (a) 17 N to the left (b) 28 m/s^2 to the left

3. (a) 1.50 Hz (b) 0.667 s (c) 4.00 m (d) π rad (e) 2.83 m

5. (a) -2.34 m (b) -1.30 m/s (c) $-0.076\,3$ m (d) 0.315 m/s

7. (a) $x = 2.00 \cos (3.00\pi t - 90°)$ or $x = 2.00 \sin (3.00\pi t)$ where x is in centimeters and t is in seconds (b) 18.8 cm/s (c) 0.333 s (d) 178 cm/s^2 (e) 0.500 s (f) 12.0 cm

9. (a) yes (b) The value of k in Equation 15.13 is proportional to the mass m, so the mass cancels in the equation, leaving only the extension of the spring and the acceleration due to gravity in the equation: $T = 0.859$ s.

11. 2.60 cm or -2.60 cm

13. (a) $\frac{8}{9}E$ (b) $\frac{1}{9}E$ (c) $x = \pm\sqrt{\frac{2}{3}}A$
(d) No; the maximum potential energy is equal to the total energy of the system. Because the total energy must remain constant, the kinetic energy can never be greater than the maximum potential energy.

15. (a) 4.58 N (b) 0.125 J (c) 18.3 m/s^2 (d) 1.00 m/s (e) smaller (f) the coefficient of kinetic friction between the block and surface (g) 0.934

17. (a) 1.50 s (b) 0.559 m

19. 0.944 kg · m^2

21. (a) 0.820 m/s (b) 2.57 rad/s^2 (c) 0.641 N (d) $v_{\text{max}} = 0.817$ m/s, $\alpha_{\text{max}} = 2.54$ rad/s^2, $F_{\text{max}} = 0.634$ N (e) The answers are close but not exactly the same. The angular amplitude of 15° is not a small angle, so the simple harmonic oscillation model is not accurate. The answers computed from conservation of energy and from Newton's second law are more accurate.

23. (a) 5.00×10^{-7} kg · m^2 (b) 3.16×10^{-4} N · m/rad

27. (a) 3.16 s^{-1} (b) 6.28 s^{-1} (c) 5.09 cm

29. (a) 0.349 kg · m^2 (b) too low

31. (a) 2.09 s (b) 0.477 Hz (c) 36.0 cm/s (d) $E = 0.064\,8m$, where E is in joules and m is in kilograms (e) $k = 9.00m$, where k is in newtons/meter and m is in kilograms (f) Period, frequency, and maximum speed are all independent of mass in this situation. The energy and the force constant are directly proportional to mass.

33. (a) 2.00 cm (b) 4.00 s (c) $\dfrac{\pi}{2}$ rad/s (d) π cm/s (e) 4.93 cm/s^2 (f) $x = 2.00 \sin\left(\dfrac{\pi}{2}t\right)$, where x is in centimeters and t is in seconds

35. $\dfrac{1}{2\pi L}\sqrt{gL+\dfrac{kh^2}{M}}$

37. (a) 1.26 m   (b) 1.58   (c) The energy decreases by 120 J.
(d) Mechanical energy is transformed into internal energy in the perfectly inelastic collision.

41. (b) $T=\dfrac{2}{r}\sqrt{\dfrac{\pi M}{\rho g}}$

43. 13.0 s

47. (a) $x=2\cos\left(10t+\dfrac{\pi}{2}\right)$ (b) ± 1.73 m   (c) 0.105 s = 105 ms
(d) 0.098 0 m

49. (a) $y_f=-0.110$ m   (b) greater

51. (a) $\dfrac{2\pi}{\sqrt{g}}\sqrt{L_i+\dfrac{1}{2\rho a^2}\left(\dfrac{dM}{dt}\right)t}$ (b) $2\pi\sqrt{\dfrac{L_i}{g}}$

Chapter 16

Answers to Quick Quizzes

1. **(i)** (b)   **(ii)** (a)
2. **(i)** (c)   **(ii)** (b)   **(iii)** (d)
3. (c)
4. (f) and (h)
5. (d)
6. (c)
7. (b)
8. (b)
9. (e)
10. (e)
11. (b)

Answers to Odd-Numbered Problems

1. 184 km
3. (a) $L=(380 \text{ m/s})\Delta t$ (b) 48.2 m   (c) 48 cm
5. 2.40 m/s
7. ± 6.67 cm
9. (a) $y=0.080\ 0\sin(2.5\pi x+6\pi t)$
   (b) $y=0.080\ 0\sin(2.5\pi x+6\pi t-0.25\pi)$
11. 13.5 N
13. (a) 0.051 0 kg/m   (b) 19.6 m/s
15. (a) 1   (b) 1   (c) 1   (d) increased by a factor of 4
17. (a) $y=0.075\sin(4.19x-314t)$, where x and y are in meters and t is in seconds   (b) 625 W
19. $\sqrt{2}P_0$
23. (a) 2.00 μm   (b) 40.0 cm   (c) 54.6 m/s   (d) -0.433 μm
    (e) 1.72 mm/s
25. 5.81 m
27. 335 m/s
29. (a) 27.2 s   (b) 25.7 s; the time interval in part (a) is longer.
31. (a) 3.75 W/m^2 (b) 0.600 W/m^2
33. (a) 0.691 m   (b) 691 km
35. 4.28 m
39. 2.82×10^8 m/s
41. (a) 441 Hz   (b) 439 Hz   (c) 54.0 dB
43. 14.7 kg
45. 0.883 cm
47. (a) 375 m/s^2 (b) 0.045 0 N   (c) The maximum transverse force is very small compared to the tension of 46.9 N in the string, more than a thousand times smaller.

49. (a) $v=\sqrt{\dfrac{T}{\rho(1.00\times 10^{-5}\,x+1.00\times 10^{-6})}}$, where v is in meters per second, T is in newtons, ρ is in kilograms per meter cubed, and x is in meters   (b) $v(0)=94.3$ m/s, $v(10.0 \text{ m})=9.38$ m/s

51. (a) $\dfrac{\mu\omega^3}{2k}A_0^2 e^{-2bx}$ (b) $\dfrac{\mu\omega^3}{2k}A_0^2$ (c) e^{-2bx}

53. It is unreasonable, implying a sound level of 123 dB. Nearly all the decrease in mechanical energy becomes internal energy in the latch.

55. 1.34×10^4 N

57. (a) $\mu(x)=\dfrac{(\mu_L-\mu_0)x}{L}+\mu_0$

(b) $\Delta t=\dfrac{2L}{3\sqrt{T}\,(\mu_L-\mu_0)}(\mu_L^{3/2}-\mu_0^{3/2})$

Chapter 17

Answers to Quick Quizzes

1. (c)
2. **(i)** (a)   **(ii)** (d)
3. (d)
4. (b)
5. (c)

Answers to Odd-Numbered Problems

1. (a) -1.65 cm   (b) -6.02 cm   (c) 1.15 cm
3.

5. (a) y_1: positive x direction; y_2: negative x direction
    (b) 0.750 s   (c) 1.00 m
7. (a) 2.72 rad = 156°   (b) 0.058 4 cm
9. (a) The separation of adjacent nodes is $\Delta x = \dfrac{\pi}{k} = \dfrac{\lambda}{2}$. The
    nodes are still separated by half a wavelength   (b) Yes.
    The nodes are located at $kx + \dfrac{\phi}{2} = n\pi$, so that
    $x = \dfrac{n\pi}{k} - \dfrac{\phi}{2k}$, which means that each node is shifted $\dfrac{\phi}{2k}$
    to the left by the phase difference between the traveling
    waves in comparison to the case in which $\phi = 0$.
11. (a) 0.600 m   (b) 30.0 Hz
13. (a) 78.6 Hz   (b) 157 Hz, 236 Hz, 314 Hz
15. 1.86 g
17. (a) 3.8 cm   (b) 3.85%
19. The resonance frequency of the bay calculated from the
    data provided is 12 h, 24 min. The natural frequency of
    the water sloshing in the bay agrees precisely with that
    of lunar excitation, so we identify the extra-high tides as
    amplified by resonance.
21. (a) 0.656 m   (b) 1.64 m
23. (a) 349 m/s   (b) 1.14 m
25. $n(0.252 \text{ m})$ with $n = 1, 2, 3, \ldots$
27. 158 s
29. $-10.0°$C
31. (a) 1.99 beats/s   (b) 3.38 m/s
33. The coefficients beyond $n = 1$ are approximate: $A_1 = 100$,
    $A_2 = 156$, $A_3 = 62$, $A_4 = 104$, $A_5 = 52$, $A_6 = 29$, $A_7 = 25$.

35. 800 m
37. (a) larger   (b) 2.43
39. (a) $r = 0.078\ 2\left(1 - \dfrac{4}{n^2}\right)^{1/3}$ (b) 3   (c) 0.078 2 m
    (d) The sphere floats on the water.
41. (a) 3.99 beats/s   (b) 3.99 beats/s
43. (a) Frequency should be halved.   (b) $\left[\dfrac{n}{n+1}\right]^2 T$
    (c) $\dfrac{T'}{T} = \dfrac{9}{16}$
45. 283 Hz
47. (a) 78.9 N   (b) 211 Hz
49. (b) $A = 11.2$ m, $\phi = 1.11$ rad $= 63.4°$

Chapter 18

Answers to Quick Quizzes

1. (c)
2. (c)
3. (c)
4. (c)
5. (a)
6. (b)

Answers to Odd-Numbered Problems

1. (a) $-738°$N   (b) $-105°$N   (c) 270°N   (d) 153°N
3. (a) $-109°$F, 195 K   (b) 98.6°F, 310 K
5. (a) 56.7°C and $-62.1°$C   (b) 330 K and 211 K
7. 3.27 cm
9. 1.54 km. The pipeline can be supported on rollers. In
    addition, Ω-shaped loops can be built between straight
    sections; these loops bend as the steel changes length.
11. 2.74 m
13. (a) 437°C   (b) $2.1 \times 10^3\ °$C   (c) No; aluminum melts at
    660°C (Table 19.2). Also, although it is not in Table 19.2,
    Internet research shows that brass (an alloy of copper
    and zinc) melts at about 900°C.
15. (a) 99.8 mL   (b) It lies below the mark. The acetone
    has reduced in volume, and the flask has increased in
    volume.
17. (a) 396 N   (b) $-101°$C   (c) The original length divides
    out of the equations in the calculation, so the answers
    would not change.
19. 1.50×10^{29} molecules
21. (a) 41.6 mol   (b) 1.20 kg   (c) This value is in agreement
    with the tabulated density.
23. 2.42×10^{11} molecules
25. 473 K
27. $\sim 10^2$ kg
31. (a) $\theta = 2\sin^{-1}\left(\dfrac{1 + \alpha_{\text{Al}}T_\text{C}}{2}\right)$ (b) yes   (c) yes
    (d) $\theta = 2\sin^{-1}\left(\dfrac{1 + \alpha_{\text{Al}}T_\text{C}}{2(1 + \alpha_{\text{invar}}T_\text{C})}\right)$ (e) 61.0°   (f) 59.6°
33. (a) 94.97 cm   (b) 95.03 cm
35. (b) As the temperature increases, the density decreases
    (assuming β is positive).   (c) $5 \times 10^{-5}\ (°\text{C})^{-1}$
    (d) $-2.5 \times 10^{-5}\ (°\text{C})^{-1}$
37. (b) It assumes $\alpha\ \Delta T$ is much less than 1.
39. (a) yes, as long as the coefficients of expansion remain
    constant   (b) The lengths L_{Cu} and L_{St} at 0°C need to
    satisfy $17L_{\text{Cu}} = 11L_{\text{St}}$. Then the steel rod must be longer.
    With $L_{\text{St}} - L_{\text{Cu}} = 5.00$ cm, the only possibility is $L_{\text{St}} =$
    14.2 cm and $L_{\text{Cu}} = 9.17$ cm.
41. (a) 0.34%   (b) 0.48%   (c) All the moments of inertia
    have the same mathematical form: the product of a con-
    stant, the mass, and a length squared.
45. 4.54 m

Chapter 19

Answers to Quick Quizzes

1. (i) iron, glass, water   (ii) water, glass, iron
2. The figure on the next page shows a graphical represen-
    tation of the internal energy of the system as a function
    of energy added. Notice that this graph looks quite dif-
    ferent from Figure 19.3 in that it doesn't have the flat
    portions during the phase changes. Regardless of how
    the temperature is varying in Figure 19.3, the internal
    energy of the system simply increases linearly with energy
    input; the line in the graph below has a slope of 1.

3.

Situation	System	Q	W	ΔE_{int}
(a) Rapidly pumping up a bicycle tire	Air in the pump	0	+	+
(b) Pan of room-temperature water sitting on a hot stove	Water in the pan	+	0	+
(c) Air quickly leaking out of a balloon	Air originally in the balloon	0	−	−

4. Path A is isovolumetric, path B is adiabatic, path C is isothermal, and path D is isobaric.

5. (b)

Answers to Odd-Numbered Problems

1. (a) 2.26×10^6 J   (b) 2.80×10^4 steps   (c) 6.99×10^3 steps
3. 23.6°C
5. 0.918 kg
7. (a) 1 822 J/kg · °C   (b) We cannot make a definite identification. It might be beryllium.   (c) The material might be an unknown alloy or a material not listed in the table.
9. (a) 25.8°C   (b) The symbolic result from part (a) shows no dependence on mass. Both the change in gravitational potential energy and the change in internal energy of the system depend on the mass, so the mass cancels.
11. 2.27 km
13. (a) 0°C   (b) 114 g
15. (a) $-4P_i V_i$ (b) According to $T = (P_i/nRV_i)V^2$, it is proportional to the square of the volume.
17. 720 J
19. (a) 0.041 0 m³   (b) +5.48 kJ   (c) −5.48 kJ
21. (a) −0.048 6 J   (b) 16.2 kJ   (c) 16.2 kJ
23. 74.8 kJ
25. (a) 1.19   (b) 1.19
27. (a) 1.85 ft² · °F · h/Btu   (b) 1.78
29. (a) -6.08×10^5 J   (b) 4.56×10^5 J
31. 888 K
33. 1.90×10^3 J/kg · °C
35. (a) 9.31×10^{10} J   (b) -8.47×10^{12} J   (c) 8.38×10^{12} J
37. (a) First, energy must be removed from the liquid water to cool it to 0° C. Next, energy must be removed from the water at 0° C to freeze it, which corresponds to a liquid-to-solid phase transition. Finally, once all the water has frozen, additional energy must be removed from the ice to cool it from 0° to −8.00°C   (b) 32.5 kJ
39. (a) 2 000 W   (b) 4.46°C
41. (a) 3.16×10^{22} W   (b) 3.17×10^{22} W   (c) It is 0.408% larger.   (d) 5.78×10^3 K
43. 1.44 kg
45. (b) 9.32 kW
47. 3.66×10^4 s = 10.2 h

Chapter 20

Answers to Quick Quizzes

1. (i) (b)   (ii) (a)
2. (i) (a)   (ii) (c)
3. (d)
4. (c)

Answers to Odd-Numbered Problems

1. 3.32 mol
3. 5.05×10^{-21} J
5. (a) 2.28 kJ   (b) 6.21×10^{-21} J
7. 17.4 kPa
9. 74.8 J
11. (a) $W = 0$ (b) $\Delta E_{int} = 209$ J   (c) 317 K
13. between 10^{-3} °C and 10^{-2} °C
15. (a) 1.08   (b) no
17. 5.74×10^6 Pa
19. 227 K
21. (a) 2.45×10^{-4} m³   (b) 9.97×10^{-3} mol   (c) 9.01×10^5 Pa   (d) 5.15×10^{-5} m³   (e) 560 K   (f) 53.9 J   (g) 6.79×10^{-6} m³   (h) 53.3 g   (i) 2.24 K
23. (a) 2.37×10^4 K   (b) 1.06×10^3 K
25. (b) 0.278
27. (a) 3.90 km/s   (b) 4.18 km/s
29. (a) 7.89×10^{26} molecules   (b) 37.9 kg   (c) 6.07×10^{-21} J   (d) 503 m/s   (e) 0   (f) When the furnace operates, air expands and some of it leaves the room. The smaller mass of warmer air left in the room contains the same internal energy as the cooler air initially in the room.
31. (a) 367 K   (b) The rms speed of nitrogen would be higher because the molar mass of nitrogen is less than that of oxygen.   (c) 572 m/s
33. Sulfur dioxide is the gas with the greatest molecular mass of those listed. If the effective spring constants for various chemical bonds are comparable, SO_2 can then be expected to have low frequencies of atomic vibration. Vibration can be excited at lower temperature than for other gases. Some vibration may be going on at 300 K. With more degrees of freedom for molecular motion, the material has higher specific heat.
35. (a) 300 K   (b) 1.00 atm
37. (a) 7.27×10^{-20} J   (b) 2.20 km/s   (c) 3.51×10^3 K   (d) The evaporating particles emerge with much less kinetic energy, as negative work is performed on them by restraining forces as they leave the liquid. Much of the initial kinetic energy is used up in overcoming the latent heat of vaporization. There are also very few of these escaping at any moment in time.

39. (a) 1.09×10^{-3} (b) 2.69×10^{-2} (c) 0.529 (d) 1.00
    (e) 0.199 (f) 1.01×10^{-41} (g) $1.25 \times 10^{-1\,082}$
43. (a) 0.510 m/s   (b) 20 ms
45. (c) $2 - 2a + a^2$

The graph above shows the behavior of the factor in parentheses in part (b) between the possible limits of $a = 0$ and $a = 2$. Except at the value of $a = 1$, the factor is always greater than 1. Therefore, the equation shows that, in general (except for the special case of $a = 1$), $v_{\text{rms}} > v_{\text{avg}}$. (d) $a = 1$

Chapter 21

Answers to Quick Quizzes

1. **(i)** (c)   **(ii)** (b)
2. (d)
3. C, B, A
4. (a) one   (b) six
5. (a)
6. false (The adiabatic process must be *reversible* for the entropy change to be equal to zero.)

Answers to Odd-Numbered Problems

1. (a) 10.7 kJ   (b) 0.533 s
3. 55.4%
5. (a) 4.51×10^6 J   (b) 2.84×10^7 J   (c) 68.1 kg
7. (a) 67.2% (b) 58.8 kW
9. 1.86
11. (a) $564°$C   (b) No; a real engine will always have an efficiency *less* than the Carnot efficiency because it operates in an irreversible manner.
13. (a) 5.12% (b) 5.27×10^{12} J/h   (c) 5.68×10^4
    (d) 4.50×10^6 m^2 (e) yes   (f) numerically, yes; feasibly, probably not
15. (a) $\dfrac{Q_c}{\Delta t} = 1.40 \left(\dfrac{0.5\,T_h + 383}{T_h - 383} \right)$, where $Q_c/\Delta t$ is in mega-

watts and T_h is in kelvins   (b) The exhaust power decreases as the firebox temperature increases.   (c) 1.87 MW (d) 3.84×10^3 K   (e) No answer exists. The energy exhaust cannot be that small.
17. 1.17
21. (a)

Macrostate	Microstates	Number of ways to draw
All R	RRR	1
2 R, 1 G	GRR, RGR, RRG	3
1 R, 2 G	GGR, GRG, RGG	3
All G	GGG	1

(b)

Macrostate	Microstates	Number of ways to draw
All R	RRRR	1
4R, 1G	GRRRR, RGRRR, RRGRR, RRRGR, RRRRG	5
3R, 2G	GGRRR, GRGRR, GRRGR, GRRRG, RGGRR, RGRGR, RGRRG, RRGGR, RRGRG, RRRGG	10
2R, 3G	RRGGG, RGRGG, RGGRG, RGGGR, GRRGG, GRGRG, GRGGR, GGRRG, GGRGR, GGGRR	10
1R, 4G	RGGGG, GRGGG, GGRGG, GGGRG, GGGGR	5
All G	GGGGG	1

23. 1.02 kJ/K
25. 195 J/K
27. (a) -3.45 J/K   (b) $+8.06$ J/K   (c) $+4.62$ J/K
29. 1 W/K
31. (a) $\frac{1}{3}$ (b) $\frac{2}{3}$
33. (a) $3nRT_i$ (b) $3nRT_i \ln 2$ (c) $-3nRT_i$ (d) $-nRT_i \ln 2$
    (e) $3nRT_i(1 + \ln 2)$ (f) $2nRT_i \ln 2$ (g) 0.273
35. (a) 39.4 J   (b) 65.4 rad/s $= 625$ rev/min
    (c) 293 rad/s $= 2.79 \times 10^3$ rev/min
37. (a) 4.10×10^3 J   (b) 1.42×10^4 J   (c) 1.01×10^4 J
    (d) 28.8% (e) Because $e_C = 80.0\%$, the efficiency of the cycle is much lower than that of a Carnot engine operating between the same temperature extremes.
39. (a) 0.476 J/K   (b) 417 J
41. (a) 5.97 K   (b) higher   (c) 22.6 K
43. (a) 13.4 J/K   (b) 310 K   (c) 13.3 J/K   (d) smaller by less than 1%
45. (b) yes   (c) No; the second law refers to an engine operating in a cycle, whereas this problem involves only a single process.
47. (a)

	T (K)	P (kPa)	V (cm³)
A	293	100	500
B	673	1.84×10^3	62.5
C	1 023	2.79×10^3	62.5
D	445	152	500

(b)

	Q	W_{eng}	ΔE_{int}
A→B	0	−162	162
B→C	149	0	149
C→D	0	246	−246
D→A	−65.0	0	−65.0
ABCD	84.3	84.3	0

(c) 149 J   (d) 65.0 J   (e) 84.3 J   (f) 0.565
(g) 1.42×10^3 rev/min

Chapter 22

Answers to Quick Quizzes

1. (a), (c), (e)
2. (e)
3. (b)
4. (a)
5. A, B, C

Answers to Odd-Numbered Problems

1. (a) $+1.60 \times 10^{-19}$ C, 1.67×10^{-27} kg
   (b) $+1.60 \times 10^{-19}$ C, 3.82×10^{-26} kg
   (c) -1.60×10^{-19} C, 5.89×10^{-26} kg
   (d) $+3.20 \times 10^{-19}$ C, 6.65×10^{-26} kg
   (e) -4.80×10^{-19} C, 2.33×10^{-26} kg
   (f) $+6.40 \times 10^{-19}$ C, 2.33×10^{-26} kg
   (g) $+1.12 \times 10^{-18}$ C, 2.33×10^{-26} kg
   (h) -1.60×10^{-19} C, 2.99×10^{-26} kg
3. 3.60×10^6 N downward
5. (a) 8.74×10^{-8} N   (b) repulsive
7. (a) 0.951 m   (b) yes, if the third bead has positive charge
9. (a) 8.24×10^{-8} N   (b) 2.19×10^6 m/s
11. $k_e \dfrac{Q^2}{d^2} \left[\dfrac{1}{2\sqrt{2}} \hat{\mathbf{i}} + \left(2 - \dfrac{1}{2\sqrt{2}} \right) \hat{\mathbf{j}} \right]$
13. (b) $\dfrac{\pi}{2} \sqrt{\dfrac{md^3}{k_e qQ}}$ (c) $4a \sqrt{\dfrac{k_e qQ}{md^3}}$
15. (a) $-(5.58 \times 10^{-11}$ N/C$)\hat{\mathbf{j}}$ (b) $(1.02 \times 10^{-7}$ N/C$)\hat{\mathbf{j}}$
17. (a) $k_e \dfrac{Q}{d^2} [(1 - \sqrt{2})\,\hat{\mathbf{i}} + \sqrt{2}\,\hat{\mathbf{j}}]$

   (b) $-k_e \dfrac{Q}{4d^2} [(1 + 4\sqrt{2})\,\hat{\mathbf{i}} + 4\sqrt{2}\,\hat{\mathbf{j}}]$
19. (a) 1.80×10^4 N/C to the right   (b) 8.98×10^{-5} N to the left
21. (a) $(-0.599\hat{\mathbf{i}} - 2.70\hat{\mathbf{j}})$ kN/C   (b) $(-3.00\hat{\mathbf{i}} - 13.5\hat{\mathbf{j}})$ μN
23. (a)

   (b) at the center   (c) $1.7k_e \dfrac{q}{a^2}$
   (d) upward in the plane of the page
25. (a) 111 ns   (b) 5.68 mm   (c) $(450\hat{\mathbf{i}} + 102\hat{\mathbf{j}})$ km/s
27. 4.52×10^{-14} C
29. $-\dfrac{\pi^2 k_e q}{6a^2} \hat{\mathbf{i}}$
31. (a) $\dfrac{mg}{|Q|} \sin \theta$ (b) 3.19×10^3 N/C down the incline
33. (a) 1.09×10^{-8} C   (b) 5.44×10^{-3} N
35. (a) $24.2\hat{\mathbf{i}}$ N/C   (b) $(-4.21\hat{\mathbf{i}} + 8.42\hat{\mathbf{j}})$ N/C

37. 25.9 cm
39. 1.67×10^{-5} C
41. 1.98 μC
43. 1.14×10^{-7} C on one sphere and 5.69×10^{-8} C on the other
45. (a) $\theta_1 = \theta_2$
47. (a) 0.307 s   (b) Yes; the downward gravitational force is not negligible in this situation, so the tension in the string depends on both the gravitational force and the electric force.
49. (a) $\vec{\mathbf{E}} = \dfrac{935x}{(0.062\,5 + x^2)^{3/2}} \hat{\mathbf{i}}$ where $\vec{\mathbf{E}}$ is in newtons per coulomb and x is in meters   (b) $4.00\hat{\mathbf{i}}$ kN/C
   (c) $x = 0.016\,8$ m and $x = 0.916$ m
   (d) nowhere is the field as large as 16 000 N/C

Chapter 23

Answers to Quick Quizzes

1. (e)
2. (b) and (d)

Answers to Odd-Numbered Problems

1.

3. (a) 6.64×10^6 N/C away from the center of the ring
   (b) 2.41×10^7 N/C away from the center of the ring
   (c) 6.39×10^6 N/C away from the center of the ring
   (d) 6.64×10^5 N/C away from the center of the ring
5. (a) 9.35×10^7 N/C away from the center of the disk
   (b) 1.04×10^8 N/C away from the center of the disk (about 11% higher)
   (c) 5.15×10^5 N/C away from the center of the disk
   (d) 5.19×10^5 N/C away from the particle (about 0.7% higher)
7. (a) $k_e \dfrac{\lambda_0}{x_0}$ (b) to the left
9. (a) $\dfrac{k_e Q \hat{\mathbf{i}}}{h} \left[\dfrac{1}{(d^2 + R^2)^{1/2}} - \dfrac{1}{\left((d + h)^2 + R^2\right)^{1/2}} \right]$

   (b) $\dfrac{2k_e Q \hat{\mathbf{i}}}{R^2 h} \left[h + (d^2 + R^2)^{1/2} - \left((d + h)^2 + R^2\right)^{1/2} \right]$
11. (a) 1.98×10^6 N $\cdot$ m^2/C   (b) 0
13. 28.2 N $\cdot$ m^2/C
15. $-Q/\epsilon_0$ for S_1; 0 for S_2; $-2Q/\epsilon_0$ for S_3; 0 for S_4
17. (a) 339 N $\cdot$ m^2/C   (b) No. The electric field is not uniform on this surface, so the integral in Equation 23.7 cannot be evaluated.
19. -18.8 kN $\cdot$ m^2/C
21. (a) $\dfrac{q}{2\epsilon_0}$ (b) $\dfrac{q}{2\epsilon_0}$ (c) The fluxes are the same. The plane and the square look the same to the charge.
23. (a) $EA \cos \theta$ (b) $-EA \sin \theta$ (c) $-EA \cos \theta$ (d) $EA \sin \theta$
   (e) 0 for both faces   (f) 0   (g) 0
25. 3.50 kN

27. 508 kN/C up

29. (a) 51.4 kN/C outward   (b) 645 N · m²/C

31. $\vec{E} = \rho r / 2\epsilon_0 = 2\pi k_e \rho r$ away from the axis

33. (a) 0   (b) 3.65×10^5 N/C   (c) 1.46×10^6 N/C
(d) 6.49×10^5 N/C

35. (a) $r = a\left(\dfrac{-q}{4Q}\right)^{1/3}$ (b) Yes, it is possible for *any* value of $r > a$.

37. 0.438 N · m²/C

39. $-\dfrac{k_e \lambda_0}{2x_0}\,\hat{\mathbf{i}}$

41. $-0.706\,\hat{\mathbf{i}}$ N

43. 8.27×10^5 N · m²/C

45. (a) $E = \dfrac{Cd^3}{24\epsilon_0}$ to the right for $x > d/2$ and to the left for
$x < -d/2$ (b) $\vec{E} = \dfrac{Cx^3}{3\epsilon_0}\,\hat{\mathbf{i}}$

47. (a) $\dfrac{\rho_0 r}{2\epsilon_0}\left(a - \dfrac{2r}{3b}\right)$ (b) $\dfrac{\rho_0 R^2}{2\epsilon_0 r}\left(a - \dfrac{2R}{3b}\right)$

Chapter 24

Answers to Quick Quizzes

1. (i) (b)   (ii) (a)
2. Ⓑ to Ⓒ, Ⓒ to Ⓓ, Ⓐ to Ⓑ, Ⓓ to Ⓔ
3. (i) (c)   (ii) (a)
4. (i) (a)   (ii) (a)

Answers to Odd-Numbered Problems

1. 1.35 MJ

3. (a) 1.13×10^5 N/C   (b) 1.80×10^{-14} N   (c) 4.37×10^{-17} J

5. (a) 0.400 m/s   (b) It is the same. Because the electric
field is uniform, each bit of the rod feels a force of the
same size as before.

7. $6.93 k_e \dfrac{Q}{d}$

9. $-k_e \dfrac{Q}{R}$

11. (a) $4\sqrt{2}k_e \dfrac{Q}{a}$ (b) $4\sqrt{2}k_e \dfrac{qQ}{a}$

15. (a) no point   (b) $\dfrac{2k_e q}{a}$

17. (a) 10.8 m/s and 1.55 m/s   (b) They would be greater.
The conducting spheres will polarize each other, with
most of the positive charge of one and the negative
charge of the other on their inside faces. Immediately
before the spheres collide, their centers of charge will be
closer than their geometric centers, so they will have less
electric potential energy and more kinetic energy.

19. $22.8 k_e \dfrac{q^2}{s}$

21. $E_y = \dfrac{k_e Q}{y\sqrt{\ell^2 + y^2}}$

23. (a) $E_A > E_B$ The electric field can be interpreted as the rate
of change of electric potential in space. The equipotential
surfaces are closer together at A than at B, so the potential

is changing more rapidly in space at A.   (b) The magnitude of the electric field at B is approximately 200 V/m
based on the rate of change of the electric potential in
space and Equation 24.16.

(c)

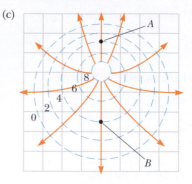

25. (a) C/m²   (b) $k_e \alpha\left[L - d\,\ln\left(1 + \dfrac{L}{d}\right)\right]$

27. $k_e \lambda(\pi + 2\ln 3)$

29. No. A conductor of any shape forms an equipotential surface. However, if the surface varies in shape, there is no
clear way to relate electric field at a point on the surface
to the potential of the surface.

31. $\dfrac{\sigma}{\epsilon_0}$

33. $E_{\text{glass}} = E_{\text{Al}}$

35. (a) 0, 1.67 MV   (b) 5.84 MN/C away, 1.17 MV
(c) 11.9 MN/C away, 1.67 MV

37. Using Equation 24.13 for the potential energy of the
atom and using the numerical values provided, n does
not turn out to be an integer. Therefore, the problem
does *not* describe an allowed state of the atom.

39. (a) $-\dfrac{k_e q}{4a}$ (b) The approximate expression $-2k_e qa/x^2$
gives $-k_e q/4.5$, which is different by only 11.1%.

41. $k_e \lambda \ln\left[\dfrac{a + L + \sqrt{(a+L)^2 + b^2}}{a + \sqrt{a^2 + b^2}}\right]$

43. (a) 4.07 kV/m   (b) 488 V   (c) 7.82×10^{-17} J   (d) 306 km/s
(e) 3.89×10^{11} m/s² toward the negative plate
(f) 6.51×10^{-16} N toward the negative plate
(g) 4.07 kV/m   (h) They are the same.

45. (a) $Q\left(\dfrac{r}{a}\right)^3$ (b) $k_e \dfrac{Qr}{a^3}$ (c) Q (d) $k_e \dfrac{Q}{r^2}$ (e) $E = 0$
(f) $-Q$ (g) $+Q$ (h) inner surface of radius b

47. (a) -4.01 nC   (b) $+9.57$ nC   (c) $+4.01$ nC   (d) $+5.56$ nC

49. $\pi k_e C\left[R\sqrt{R^2 + x^2} + x^2\,\ln\left(\dfrac{x}{R + \sqrt{R^2 + x^2}}\right)\right]$

51. (a) $\dfrac{k_e Q}{h}\,\ln\left[\dfrac{d + h + \sqrt{(d+h)^2 + R^2}}{d + \sqrt{d^2 + R^2}}\right]$

(b) $\dfrac{k_e Q}{R^2 h}\Bigg[(d+h)\sqrt{(d+h)^2 + R^2} - d\sqrt{d^2 + R^2}$

$-2dh - h^2 + R^2\,\ln\left(\dfrac{d + h + \sqrt{(d+h)^2}}{d + \sqrt{d^2 + R^2}}\right)\Bigg]$

Chapter 25

Answers to Quick Quizzes

1. (d)
2. (a)
3. (a)
4. (b)
5. (a)

Answers to Odd-Numbered Problems

1. (a) 9.00 V (b) 12.0 V
3. 4.43 μm
5. $\dfrac{(2N-1)\epsilon_0(\pi-\theta)R^2}{d}$
7. (a) 2.81 μF (b) 12.7 μF
9. ten
11. (a) 5.96 μF (b) 89.5 μC on 20 μF, 63.2 μC on 6 μF, and 26.3 μC on 15 μF and 3 μF
13. 12.9 μF
15. 6.00 pF and 3.00 pF
17. (a) 216 μJ (b) 54.0 μJ
19. (a) 2.50×10^{-2} J (b) 66.7 V (c) 3.33×10^{-2} J (d) Positive work is done by the agent pulling the plates apart.
21. (a)

(b) 0.150 J (c) 268 V
(d)

$$25.0\ \mu\text{F}$$
$$268\text{ V}$$
$$5.00\ \mu\text{F}$$

23. (b) $\dfrac{k_e q_1^2}{2R_1} + \dfrac{K_e(Q-q_1)^2}{2R_2}$ (c) $\dfrac{R_1 Q}{R_1+R_2}$ (d) $\dfrac{R_2 Q}{R_1+R_2}$

(e) $V_1 = \dfrac{k_e Q}{R_1+R_2}$ and $V_2 = \dfrac{k_e Q}{R_1+R_2}$ (f) 0

25. (a) 81.3 pF (b) 2.40 kV
27. 1.04 m
29. (a) 40.0 μJ (b) 500 V
33. (a) 100 pF (b) 0.22 μC (c) 2.2 kV
35. 2.51×10^{-3} m^3 = 2.51 L
37. (a) 25.0 μF$(1-0.846f)^{-1}$ (b) 25.0 μF, the general expression agrees (c) 162 μF; the general expression agrees.
39. (a) $\dfrac{Q_0^2 d(\ell-x)}{2\epsilon_0 \ell^3}$ (b) $\dfrac{Q_0^2 d}{2\epsilon_0 \ell^3}$ to the right (c) $\dfrac{Q_0^2}{2\epsilon_0 \ell^4}$

(d) $\dfrac{Q_0^2}{2\epsilon_0 \ell^4}$ (e) They are precisely the same.

41. 579 V
43. (a) One capacitor cannot be used by itself—it would burn out. The technician can use two capacitors in series, connected in parallel to another two capacitors in series. Another possibility is two capacitors in parallel, connected in series to another two capacitors in parallel. In either case, one capacitor will be left over in the box (b) Each of the four capacitors will be exposed to a maximum voltage of 45 V.
45. $k = \dfrac{C(\Delta V)^2}{f^2(1-f)d^2}$
47. $\dfrac{C_0}{2}(\sqrt{3}-1)$
49. (a) $\dfrac{\epsilon_0 \ell}{d}[\ell + x(\kappa-1)]$ (b) $\dfrac{Q^2 d}{2\epsilon_0 \ell[\ell + x(\kappa-1)]}$

(c) $\dfrac{Q^2 d(\kappa-1)}{2\epsilon_0 \ell[\ell + x(\kappa-1)]^2}\hat{\mathbf{i}}$ (d) $205\hat{\mathbf{i}}$ μN

Chapter 26

Answers to Quick Quizzes

1. (a) > (b) = (c) > (d)
2. (b)
3. (b)
4. (a)

Answers to Odd-Numbered Problems

1. 27.0 yr
3. 1.05 mA
5. (a) $0.632 I_0 \tau$ (b) $0.999\,95 I_0 \tau$ (c) $I_0 \tau$
7. (a) 17.0 A (b) 85.0 kA/m^2
9. 0.256 C
11. 8.89 Ω
13. (a) 1.82 m (b) 280 μm
15. 6.00×10^{-15} $(\Omega \cdot \text{m})^{-1}$
17. 0.12
19. (a) 31.5 n$\Omega \cdot$ m (b) 6.35 MA/m^2 (c) 49.9 mA
(d) 658 μm/s (e) 0.400 V
21. 227°C
23. (a) 3.00×10^8 W (b) 1.75×10^{17} W
25. 36.1%
27. (a) $1.48 (b) $0.005 34 (c) $0.381
29. $0.494/day
31. (a) 4.75 m (b) 340 W
33. ~ $10
35. 50.0 MW
37. (a) $\dfrac{Q}{4C}$ (b) $\dfrac{Q}{4}$ on C, $\dfrac{3Q}{4}$ on $3C$

(c) $\dfrac{Q^2}{32C}$ in C, $\dfrac{3Q^2}{32C}$ in $3C$ (d) $\dfrac{3Q^2}{8C}$

39. (a) 8.00 V/m in the positive x direction (b) 0.637 Ω
(c) 6.28 A in the positive x direction (d) 200 MA/m^2
41. (a) Any diameter d and length ℓ related by $d^2 = (4.77 \times 10^{-8})\ell$, where d and ℓ are in meters (b) Yes; for $V = 0.500$ cm^3 of Nichrome, $\ell = 3.65$ m and $d = 0.418$ mm.
43. (b) Charges flow in the direction of decreasing voltage. Energy flows by heat in the direction of decreasing temperature.
45. (a) $\dfrac{\epsilon_0 \ell}{2d}(\ell + 2x + \kappa\ell - 2\kappa x)$ (b) $\dfrac{\epsilon_0 \ell v \Delta V}{d}(\kappa-1)$ clockwise

47. The value of 11.4 A is what results from substituting the given voltage and resistance into Equation 26.7. However, the resistance measured for a lightbulb with an ohmmeter is not the resistance at which it operates, because of the change in resistivity with temperature. The higher resistance of the filament at the operating temperature brings the current down significantly.

Chapter 27

Answers to Quick Quizzes

1. (a)
2. (b)
3. (a)
4. (i) (b)   (ii) (a)   (iii) (a)   (iv) (b)
5. (i) (c)   (ii) (d)

Answers to Odd-Numbered Problems

1. (a) $4.59 \ \Omega$ (b) 8.16%
3. (a) 75 W   (b) 100 W   (c) 175 W   (d) Two: switch positions 3 and 4. In both cases, the power is 100 W.
5. (a) $I_A = \mathcal{E}/R$, $I_B = I_C = \mathcal{E}/2R$ (b) B and C have the same brightness because they carry the same current.   (c) A is brighter than B or C because it carries twice as much current.
7. $0.6 \ \Omega < R_{extra} < 1.6 \ \Omega$ and $0.672 \ k\Omega < R_{extra} < 1.74 \ k\Omega$
9. (a) $1.00 \ k\Omega$ (b) $2.00 \ k\Omega$ (c) $3.00 \ k\Omega$
11. (a) The single hot dog and the two in parallel will all cook first. (b) single hot dog and the two in parallel: 57.3 s; two hot dogs in series: 229 s
13. 14.2 W to $2.00 \ \Omega$, 28.4 W to $4.00 \ \Omega$, 1.33 W to $3.00 \ \Omega$, 4.00 W to $1.00 \ \Omega$
15. (a) $\Delta V_1 = \dfrac{\mathcal{E}}{3}$, $\Delta V_2 = \dfrac{2\mathcal{E}}{9}$, $\Delta V_3 = \dfrac{4\mathcal{E}}{9}$, $\Delta V_4 = \dfrac{2\mathcal{E}}{3}$

    (b) $I_1 = I$, $I_2 = I_3 = \dfrac{I}{3}$, $I_4 = \dfrac{2I}{3}$ (c) I_4 increases and I_1, I_2,

    and I_3 decrease   (d) $I_1 = \dfrac{3I}{4}$, $I_2 = I_3 = 0$, $I_4 = \dfrac{3I}{4}$
17. (a) 0.846 A down in the 8.00-Ω resistor, 0.462 A down in the middle branch, 1.31 A up in the right-hand branch   (b) -222 J by the 4.00-V battery, 1.88 kJ by the 12.0-V battery   (c) 687 J to $8.00 \ \Omega$, 128 J to $5.00 \ \Omega$, 25.6 J to the 1.00-Ω resistor in the center branch, 616 J to $3.00 \ \Omega$, 205 J to the 1.00-Ω resistor in the right branch   (d) Chemical potential energy in the 12.0-V battery is transformed into internal energy in the resistors. The 4.00-V battery is being charged, so its chemical potential energy is increasing at the expense of some of the chemical potential energy in the 12.0-V battery.   (e) 1.66 kJ
19. 50.0 mA from a to e
21. (a) No. The circuit cannot be simplified further, and Kirchhoff's rules must be used to analyze it.   (b) $I_1 = 3.50$ A   (c) $I_2 = 2.50$ A   (d) $I_3 = 1.00$ A
23. (a) 2.00 ms   (b) 1.80×10^{-4} C   (c) 1.14×10^{-4} C
25. (a) 1.50 s   (b) 1.00 s   (c) $i = 200 + 100e^{-t}$, where i is in microamperes and t is in seconds
27. $587 \ k\Omega$
29. No.
31. (a) $\sim 10^{-14}$ (b) $\sim 10^{-10}$ V
33. $7.49 \ \Omega$

35. (a) 1.02 A down   (b) 0.364 A down   (c) 1.38 A up   (d) 0   (e) $66.0 \ \mu C$
37. (a) 4.00 V   (b) Point a is at the higher potential.
39. $6.00 \ \Omega$, $3.00 \ \Omega$
41. (a) $q = 240(1 - e^{-t/6})$ (b) $q = 360(1 - e^{-t/6})$, where in both answers, q is in microcoulombs and t is in milliseconds
43. (a) $4.40 \ \Omega$ (b) 32.0 W   (c) 9.60 W   (d) 70.4 W   (e) 48.0 W
45. (a) 9.30 V   (b) $2.51 \ \Omega$ (c) 18.6 V   (d) 3.70 A   (e) 1.09 A   (f) 14.3 W   (g) 8.54 W   (h) Because of the internal resistance of the batteries, the terminal voltage of the pair of batteries is not the same in both cases.
47. (a) 0 in 3 kΩ, 333 μA in 12 kΩ and 15 kΩ   (b) $50.0 \ \mu C$ (c) $i(t) = 278 \ e^{-t/0.180}$, where i is in microamperes and t is in seconds   (d) 290 ms
49. (a) $R_x = R_2 - \frac{1}{4}R_1$ (b) No; $R_x = 2.75 \ \Omega$, so the station is inadequately grounded.
51. $(R_1 + 2R_2) C \ln 2$

Chapter 28

Answers to Quick Quizzes

1. (e)
2. (i) (b)   (ii) (a)
3. (c)
4. (i) (c), (b), (a)   (ii) (a) = (b) = (c)

Answers to Odd-Numbered Problems

1. Gravitational force: 8.93×10^{-30} N down, electric force: 1.60×10^{-17} N up, and magnetic force: 4.80×10^{-17} N down.
3. (a) into the page   (b) toward the right   (c) toward the bottom of the page
5. (a) 1.25×10^{-13} N   (b) 7.50×10^{13} m/s^2
7. $-20.9 \ \hat{\mathbf{j}}$ mT
9. (a) $\sqrt{2}r_p$ (b) $\sqrt{2}r_p$
11. 115 keV
13. (a) 5.00 cm   (b) 8.79×10^6 m/s
15. 1.56×10^5
17. (a) 7.66×10^7 s^{-1} (b) 2.68×10^7 m/s   (c) 3.75 MeV   (d) 3.13×10^3 revolutions   (e) 2.57×10^{-4} s
19. (a) Yes. The constituent of the beam is present in all kinds of atoms   (b) Yes. Everything in the beam has a single charge-to-mass ratio   (c) In a charged macroscopic object most of the atoms are uncharged. Therefore, its charge-to-mass ratio is tiny, on the order of 10^{-6} C/kg. A molecule never has all of its atoms ionized. Any atoms other than hydrogen contain neutrons and so has more mass per charge if it is ionized than hydrogen does. Therefore, the greatest charge-to-mass ratio Thomson could expect was for ionized hydrogen, 1.6×10^{-19} C/1.67×10^{-27} kg $\sim 10^8$ C/kg, smaller than the value e/m he measured, 1.6×10^{-19} C/9.11×10^{-31} kg $\sim 10^{11}$ C/kg, by a factor of 1836. The particles in his beam could not be whole atoms but rather must be much smaller in mass   (d) No. The particles move with speed on the order of ten million meters per second, so they fall by an immeasurably small amount over a distance of less than 1 m.
21. $-2.88 \ \hat{\mathbf{j}}$ N
23. 1.07 m/s
25. (a) east   (b) 0.245 T

27. (a) $2\pi r I B \sin\theta$ (b) up, away from magnet
29. (a) north at 48.0° below the horizontal
    (b) south at 48.0° above the horizontal (c) 1.07 μJ
31. (a) 0.713 A (b) Current is independent of angle.
33. (a) 9.98 N · m (b) clockwise as seen looking down from a position on the positive y axis
35. (a) 118 μN · m (b) $-118\,\mu\text{J} \le U_B \le +118\,\mu\text{J}$
37. 2.75 Mrad/s
39. (a) 12.5 km (b) It will not arrive at the center. Because the radius of curvature of the proton's path is much smaller than the radius of the cylinder, the proton enters the magnetic field only for a short distance before turning around and exiting the field.
41. $3R/4$
43. (a) the positive z direction (b) 0.696 m (c) 1.09 m (d) 54.7 ns
45. (a) $B \sim 10^{-1}$ T (b) $\tau \sim 10^{-1}$ N · m (c) $I \sim 1$ A $= 10^0$ A (d) $A \sim 10^{-3}$ m^2 (e) $N \sim 10^3$
47. (a) 1.33 m/s (b) Positive ions carried by the blood flow experience an upward force resulting in the upper wall of the blood vessel at electrode A becoming positively charged and the lower wall of the blood vessel at electrode B becoming negatively charged. (c) No. Negative ions moving in the direction of v would be deflected toward point B, giving A a higher potential than B. Positive ions moving in the direction of v would be deflected toward A, again giving A a higher potential than B. Therefore, the sign of the potential difference does not depend on whether the ions in the blood are positively or negatively charged.
49. 3.71×10^{-24} N · m
51. (a) 0.128 T (b) 78.7° below the horizontal

Chapter 29

Answers to Quick Quizzes

1. $B > C > A$
2. (a)
3. $c > a > d > b$
4. $a = c = d > b = 0$
5. (c)

Answers to Odd-Numbered Problems

1. 1.60×10^{-6} T
3. 12.5 T
5. $\dfrac{\mu_0 I}{2r}\left(\dfrac{1}{\pi} + \dfrac{1}{4}\right)$
7. (a) 53.3 μT toward the bottom of the page (b) 20.0 μT toward the bottom of the page (c) zero
9. $\dfrac{\mu_0 I}{2\pi a d}(\sqrt{d^2 + a^2} - d)$ into the page
11. (a) 4.00 μT toward the bottom of the page (b) 6.67 μT at 167.0° from the positive x axis
13. (a) 3.00×10^{-5} N/m (b) attractive
15. $k = \dfrac{\mu_0 I^2 L}{4\pi d(d + \ell)}$
17. (a) opposite directions (b) 67.8 A (c) It would be smaller. A smaller gravitational force would be pulling

down on the wires, requiring less magnetic force to raise the wires to the same angle and therefore less current.
19. (a) 3.60 T (b) 1.94 T
21. (a) 4.00 m (b) 7.50 nT (c) 1.26 m (d) zero
23. 31.8 mA
25. 5.96×10^{-2} T
27. (a) $-\pi B R^2 \cos\theta$ (b) $\pi B R^2 \cos\theta$
29. (a) 7.40 μWb (b) 2.27 μWb
31. 3.18 A
33. (a) $\sim 10^{-5}$ T
    (b) It is $\sim 10^{-1}$ as large as the Earth's magnetic field.
35. 143 pT
37. (a) $\mu_0 \sigma v$ into the page (b) zero (c) $\frac{1}{2}\mu_0 \sigma^2 v^2$ up toward the top of the page (d) $\dfrac{1}{\sqrt{\mu_0 \epsilon_0}}$; we will find out in Chapter 33 that this speed is the speed of light. We will also find out in Chapter 38 that this speed is not possible for the capacitor plates.
39. 1.80 mT
41. (b) 3.20×10^{-13} T (c) 1.03×10^{-24} N (d) 2.31×10^{-22} N
43. $B = 4.36 \times 10^{-4}\,I$, where B is in teslas and I is in amperes
45. (a) $\dfrac{\mu_0 I N}{2\ell}\left[\dfrac{\ell - x}{\sqrt{(\ell - x)^2 + a^2}} + \dfrac{x}{\sqrt{x^2 + a^2}}\right]$
47. (b) $\dfrac{\mu_0 I}{4\pi}(1 - e^{-2\pi})$ out of the page
49. (a) $\dfrac{\mu_0 I(2r^2 - a^2)}{\pi r(4r^2 - a^2)}$ to the left (b) $\dfrac{\mu_0 I(2r^2 + a^2)}{\pi r(4r^2 + a^2)}$ toward the top of the page
51. (b) 5.92×10^{-8} N

Chapter 30

Answers to Quick Quizzes

1. (c)
2. (c)
3. (b)
4. (a)

Answers to Odd-Numbered Problems

1. 2.26 mV
3. 1.89×10^{-11} V
5. (a) 1.60 A counterclockwise when viewed from the left of the figure (b) 20.1 μT (c) left
7. 272 m
9. $\mathcal{E} = 0.422 \cos 120\pi t$, where $\mathcal{E}$ is in volts and t is in seconds
11. 2.83 mV
13. $\dfrac{Rmv}{B^2 \ell^2}$
15. (a) 0.729 m/s (b) counterclockwise (c) 0.650 mW (d) Work must be done by an external force if the bar is to move with constant speed. This input of energy by work appears as internal energy in the resistor.
17. 3.32×10^3 rev/min
19. 1.00 T
21. (a) 8.01×10^{-21} N (b) tangent to a circle of radius r, in a clockwise direction (c) $t = 0$ or $t = 1.33$ s
23. 13.3 V

25. (a) $\Phi_B = 8.00 \times 10^{-3} \cos 120\pi t$, where Φ_B is in T · m² and t is in seconds (b) $\mathcal{E} = 3.02 \sin 120\pi t$, where $\mathcal{E}$ is in volts and t is in seconds (c) $I = 3.02 \sin 120\pi t$, where I is in amperes and t is in seconds (d) $P = 9.10 \sin^2 120\pi t$, where P is in watts and t is in seconds
 (e) $\tau = 0.024\ 1 \sin^2 120\pi t$, where τ is in newton meters and t is in seconds

29. 3.79 mV

31. 8.80 A

33. $\mathcal{E} = -7.22 \cos 1\ 046\pi t$, where $\mathcal{E}$ is in millivolts and t is in seconds

35. (a) 3.50 A up in 2.00 Ω and 1.40 A up in 5.00 Ω (b) 34.3 W (c) 4.29 N

37. 2.29 μC

39. (a) 0.125 V clockwise (b) 0.020 0 A clockwise

41. (a) We would need to know if the field is increasing or decreasing (b) 248 μΩ (c) Higher resistance would reduce the power delivered.

43. (a) $NB\ell v$ (b) $\dfrac{NB\ell v}{R}$ (c) $\dfrac{N^2 B^2 \ell^2 v^2}{R}$ (d) $\dfrac{N^2 B^2 \ell^2 v}{R}$
 (e) clockwise (f) directed to the left

45. $\mathcal{E} = -87.1 \cos (200\pi t + \phi)$, where $\mathcal{E}$ is in millivolts and t is in seconds

47. (a) $\dfrac{(1.18 \times 10^{-4})t}{0.800 - 4.90t^2}$ (b) zero (c) infinity (d) 98.3 μV

51. $\dfrac{MgR}{B^2 \ell^2}\left[1 - e^{-B^2 \ell^2 t/R(M + m)}\right]$

Chapter 31

Answers to Quick Quizzes

1. (c), (f)
2. (i) (b) (ii) (a)
3. (a), (d)
4. (a)
5. (i) (b) (ii) (c)

Answers to Odd-Numbered Problems

1. 100 V
3. 19.2 μT · m²
5. $\dfrac{\mathcal{E}_0}{Lk^2}$
7. $\mathcal{E} = -18.8 \cos 120\pi t$, where $\mathcal{E}$ is in volts and t is in seconds
9. (a) 5.90 mH (b) 23.6 mV
11. (a) 1.00 kΩ (b) 3.00 ms
13. (a) 20.0% (b) 4.00%
15. (a) $i_L = 0.500(1 - e^{-10.0t})$, where i_L is in amperes and t is in seconds (b) $i_S = 1.50 - 0.250e^{-10.0t}$, where i_S is in amperes and t is in seconds
17. (a) 6.67 A/s (b) 0.332 A/s
19. For $t \leq 0$, the current in the inductor is zero; for $0 \leq t \leq 200$ μs, $i_L = 10.0(1 - e^{-10\ 000t})$, where i_L is in amperes and t is in seconds; for $t \geq 200$ μs, $i_L = 63.9e^{-10\ 000t}$, where i_L is in amperes and t is in seconds
21. 2.44 μJ
23. (a) 18.0 J (b) 7.20 J
25. 80.0 mH
27. (a) $M_{12} = \mu_0 \pi R_2^2 N_1 N_2 / \ell$ (b) $M_{21} = \mu_0 \pi R_2^2 N_1 N_2 / \ell$ (c) They are the same.

29. 281 mH

31. 20.0 V

33. (a) 2.51 kHz (b) 69.9 Ω

35. (a) $0.693 \left(\dfrac{2L}{R}\right)$ (b) $0.347 \left(\dfrac{2L}{R}\right)$

37. $\dfrac{Q}{2N} \sqrt{\dfrac{3L}{C}}$

39. (a) $\frac{1}{2}\mu_0 \pi N^2 R$ (b) $\sim 10^{-7}$ H (c) $\sim 10^{-9}$ s

41. 1.20

43. 3.67×10^{-5} C

45. Δv_{ab} (mV)

47. (a) 50.0 mT (b) 20.0 mT (c) 2.29 MJ (d) 318 Pa

51. (a) $\dfrac{2\pi B_0^2 R^3}{\mu_0}$ (b) 2.70×10^{18} J

53. $\dfrac{L_1 L_2 - M^2}{L_1 + L_2 - 2M}$

Chapter 32

Answers to Quick Quizzes

1. (i) (c) (ii) (b)
2. (b)
3. (a)
4. (b)
5. (a) $X_L < X_C$ (b) $X_L = X_C$ (c) $X_L > X_C$
6. (c)
7. (c)

Answers to Odd-Numbered Problems

1. (a) 193 Ω (b) 144 Ω
3. 14.6 Hz
5. (a) 25.3 rad/s (b) 0.114 s
7. 5.60 A
9. (a) 12.6 Ω (b) 6.21 A (c) 8.78 A
11. 32.0 A
13. (a) 141 mA (b) 235 mA
15.

$X_L = 200\ \Omega$

$X_L - X_C = 109\ \Omega$ $\left\{ \vphantom{\begin{array}{c}a\\b\end{array}} \right.$

$Z = 319\ \Omega$

$\phi = 20.0°$

$R = 300\ \Omega$

$X_C = 90.9\ \Omega$

17. 11.1 A
19. (a) 17.4° (b) the voltage
21. 353 W

23. 88.0 W
25. (a) 156 pH   (b) 8.84 Ω
27. 1.41×10^5 rad/s
29. $\dfrac{4\pi RC\sqrt{LC}(\Delta V_{rms})^2}{4R^2C + 9L}$
31. 1.88 V
33. The resonance frequency for this circuit is not in the North American AM frequency range.
35. 2.6 cm
37. (b) 31.6
39. (a) 19.7 cm at 35.0°   (b) 19.7 cm at 35.0°
   (c) The answers are identical.   (d) 9.36 cm at 169°
41. (a) Tension T and separation d must be related by $T = 274d^2$, where T is in newtons and d is in meters.
   (b) One possibility is $T = 10.9$ N and $d = 0.200$ m.
43. (a) 78.5 Ω   (b) 1.59 kΩ   (c) 1.52 kΩ   (d) 138 mA
   (e) −84.3°   (f) 0.098 7   (g) 1.43 W
45. (a) capacitor   (b) resistor   (c) $\dfrac{R}{\sqrt{R^2 + \left(\dfrac{1}{\omega C}\right)^2}}$
   (d) $\dfrac{1}{\sqrt{R^2C^2\omega^2 + 1}}$
47. 1.00×10^4 rad/s
49. (a) 580 μH   (b) 54.6 μF   (c) 1.00   (d) 894 Hz   (e) At 200 Hz, $\phi = -60.0°$ (Δv_{out} leads Δv_{in}); at f_0, $\phi = 0$ (Δv_{out} is in phase with Δv_{in}); and at 4.00×10^3 Hz, $\phi = +60.0°$ (Δv_{out} lags Δv_{in}).   (f) At 200 Hz and at 4.00×10^3 Hz, $P = 1.56$ W; and at f_0, $P = 6.25$ W.   (g) 0.408
51. 58.7 Hz or 35.9 Hz. The circuit can be either above or below resonance.

Chapter 33

Answers to Quick Quizzes

1. (i) (b)   (ii) (c)
2. (c)
3. (c)
4. (b)
5. (a)
6. (c)
7. (a)

Answers to Odd-Numbered Problems

1. (a) 7.19×10^{11} V/m · s   (b) 2.00×10^{-7} T
3. $(-2.87\hat{\mathbf{j}} + 5.75\hat{\mathbf{k}}) \times 10^9$ m/s²
5. (a) 681 yr   (b) 8.32 min   (c) 2.56 s
7. 2.25×10^8 m/s
9. 2.9×10^8 m/s ±5%
11. The ratio of ω to k is higher than the speed of light in a vacuum, so the wave as described is impossible.
13. 3.34 μJ/m³
15. (a) 2.33 mT   (b) 650 MW/m²   (c) 511 W
17. $\sim 1 \times 10^4$ m²
19. 5.16 m
21. 5.31×10^{-5} N/m²
23. (a) 1.90 kN/C   (b) 50.0 pJ   (c) 1.67×10^{-19} kg · m/s
25. (a) $1.60 \times 10^{-10}\hat{\mathbf{i}}$ kg · m/s each second   (b) $1.60 \times 10^{-10}\hat{\mathbf{i}}$ N
   (c) The answers are the same. Force is the time rate of momentum transfer (Eq. 9.3).
27. (a) 1.00×10^3 km or 621 mi   (b) While the project may be theoretically possible, it is not very practical, due to the required size of the antenna.

29. 56.2 m
31. (a) $\sim 10^8$ Hz radio wave   (b) $\sim 10^{13}$ Hz infrared
33. (a) 3.85×10^{26} W   (b) 1.02 kV/m and 3.39 μT
35. 5.50×10^{-7} m
37. 75.0 MHz
39. $\sim 10^6$ J
41. 378 nm
43. (a) 625 kW/m²   (b) 21.7 kV/m   (c) 72.4 μT   (d) 17.8 min
45. (a) 388 K   (b) 363 K
47. -1.25×10^{-7} rad/s
49. (a) 0.161 m   (b) 0.163 m²   (c) 76.8 W   (d) 470 W/m²
   (e) 595 V/m   (f) 1.98 μT   (g) 119 W
51. (a) 3.33 m   (b) 11.1 ns   (c) 6.67 pT
   (d) $\vec{\mathbf{E}} = (2.00 \times 10^{-3})\cos 2\pi\left(\dfrac{x}{3.33} - 90.0 \times 10^6 t\right)\hat{\mathbf{j}}$ and
   $\vec{\mathbf{B}} = (6.67 \times 10^{-12})\cos 2\pi\left(\dfrac{x}{3.33} - 90.0 \times 10^6 t\right)\hat{\mathbf{k}}$
   (e) 5.31×10^{-9} W/m²   (f) 1.77×10^{-17} J/m²
   (g) 3.54×10^{-17} Pa

Chapter 34

Answers to Quick Quizzes

1. (d)
2. Beams ② and ④ are reflected; beams ③ and ⑤ are refracted.
3. (c)
4. (c)
5. (i) (b)   (ii) (b)

Answers to Odd-Numbered Problems

1. 114 rad/s
3. 2.27×10^8 m/s
5. $\beta = 2\delta$
7. (a) 1.94 m   (b) 50.0° above the horizontal
9. (a) 1.81×10^8 m/s   (b) 2.25×10^8 m/s
   (c) 1.36×10^8 m/s
11. (a) 29.0°   (b) 25.8°   (c) 32.0°
13. (a) 1.52   (b) 417 nm   (c) 4.74×10^{14} Hz   (d) 198 Mm/s
15. $\sim 10^{-11}$ s, $\sim 10^3$ wavelengths
17. $n = 1.55$
19. (a) 1.67 m   (b) yes
21. The index of refraction of the atmosphere decreases with increasing altitude because of the decrease in density of the atmosphere with increasing altitude, just like the index of refraction of the slabs as you move upward from the bottom in Figure P34.21. Imagine that the Sun is the source of light at the upper left of the diagram. Imagine yourself to be at the point where the light strikes the lower surface of the bottom slab. The direction from which the refracted light from the Sun comes to you is higher in angle relative to the horizontal than the actual geometric position of the Sun.
23. $\tan^{-1}(n_g)$
25. $\sin^{-1}\left\{n_V \sin\left[\Phi - \sin^{-1}\left(\dfrac{\sin\theta}{n_V}\right)\right]\right\} -$
   $\sin^{-1}\left\{n_R \sin\left[\Phi - \sin^{-1}\left(\dfrac{\sin\theta}{n_R}\right)\right]\right\}$

27. (a) 27.0° (b) 37.1° (c) 49.8°
29. (a) 10.7° (b) air (c) Looking at Table 16.1, we see that the speeds of sound for solids are an order of magnitude larger than the speed of sound in air. Therefore, we can estimate the critical angle for the air-concrete interface by using Equation 34.9 and letting the ratio of indices of refraction be ~0.1. This gives a critical angle of about 6°. Therefore, all sound striking the wall at angles greater than 6° is completely reflected.
31. (a) $\dfrac{nd}{n-1}$ (b) $R_{min} \to 0$. Yes; for very small d, the light strikes the interface at very large angles of incidence. (c) R_{min} decreases. Yes; as n increases, the critical angle becomes smaller. (d) $R_{min} \to \infty$. Yes; as $n \to 1$, the critical angle becomes close to 90° and any bend will allow the light to escape. (e) 350 μm
33. five times from the right-hand mirror and six times from the left
35. The angle of 38.0° above the horizontal is equivalent to 52.0° with respect to the normal at the water surface. As found in the What If? of Example 34.6, all the light from above the water is seen by the scuba diver in a circle corresponding to an angle of 48.8° with respect to the normal. Therefore, the Sun would be seen within this circle. At 52.0° with respect to the normal at the water surface, or 38.0° above the horizontal, the diver would see a reflection of the bottom of the lake.
37. (a) 0.042 6 or 4.26% (b) no difference
39. (a) 334 μs (b) 0.014 6%
41. (a) Total internal reflection occurs for all values of θ, or the maximum angle is 90°. (b) 30.3° (c) Total internal reflection never occurs as the light moves from lower-index polystyrene to higher-index carbon disulfide.
43. $\sin^{-1}\left[\dfrac{L}{R^2}\left(\sqrt{n^2R^2 - L^2} - \sqrt{R^2 - L^2}\right)\right]$ or

$\sin^{-1}\left[n\sin\left(\sin^{-1}\dfrac{L}{R} - \sin^{-1}\dfrac{L}{nR}\right)\right]$

45. (a) 53.1° (b) $\theta_1 \geq 38.7°$
47. 36.5°
49. (a) $n = \sqrt{1 + \left(\dfrac{4t}{d}\right)^2}$ (b) 2.10 cm (c) violet
51. (a) 0.172 mm/s (b) 0.345 mm/s (c) and (d) northward and downward at 50.0° below the horizontal.
53. (a) $\left(\dfrac{4x^2 + L^2}{L}\right)\omega$ (b) 0 (c) $L\omega$ (d) $2L\omega$ (e) $\dfrac{\pi}{8\omega}$
57. 70.6%

Chapter 35

Answers to Quick Quizzes

1. false
2. (b)
3. (b)
4. (d)
5. (a)
6. (b)
7. (c)

Answers to Odd-Numbered Problems

1. (a) younger (b) ~ 10^{-9} s younger
3. (a) $p_1 + h$, behind the lower mirror (b) virtual (c) upright (d) 1.00 (e) no
5. (a) 33.3 cm in front of the mirror (b) −0.666 (c) real (d) inverted
7. (a) 7.50 cm behind the mirror (b) upright (c) 0.500 cm
9. 3.33 m out from the deepest point in the niche
11. (a) convex (b) at the 30.0-cm mark (c) −20.0 cm
13. (a) 0.708 m in front of the sphere (b) upright
15. (a) 25.6 m (b) 0.058 7 rad (c) 2.51 m (d) 0.023 9 rad (e) 62.8 m
17. (a) 45.1 cm (b) −89.6 cm (c) −6.00 cm
19. (a) (i) 3.77 cm from the front of the wall, in the water, (ii) 19.3 cm from the front wall, in the water (b) (i) +1.01, (ii) +1.03 (c) The plastic has uniform thickness, so the surfaces of entry and exit for any particular ray are very nearly parallel. The ray is slightly displaced, but it would not be changed in direction by going through the plastic wall with air on both sides. Only the difference between the air and water is responsible for the refraction of the light (d) yes (e) If $p = |R|$, then $q = -p = -|R|$; if $p > |R|$, then $|q| > |R|$. For example, if $p = 2|R|$, then $q = -3.00|R|$ and $M = +2.00$.
21. (a) $1.00 < M < 1.99$ (b) No; the light from the Sun does not focus within the bowl.
23. (a) 6.40 cm (b) −0.250 (c) converging
25. 20.0 cm
27. (a) 20.0 cm from the lens on the front side
    (b) 12.5 cm from the lens on the front side
    (c) 6.67 cm from the lens on the front side
    (d) 8.33 cm from the lens on the front side
29. (a) −5.00 cm (b) +0.500 (c) The image from a converging lens of an object placed at the focal point is infinitely far away and of infinite magnification.
31. (a) 3.05 cm (b) 0.17 cm
33. 21.3 cm
35. 2.18 mm away from the CCD
37. −575
39. (a) Yes, if the lenses are bifocal.
    (b) +1.78 diopters (c) −1.18 diopters
41. (a) + 50.8 diopters $\leq P \leq$ 60.0 diopters
    (b) −0.800 diopters, diverging
43. The image is inverted, real, and diminished in size.
45. −40.0 cm
47. (a) 1.50 (b) 1.90
51. 8.00 cm
53. (a) $\dfrac{1}{f} = \dfrac{1}{p_1} + \dfrac{1}{1.50 - p_1}$ (b) $\dfrac{1}{f} = \dfrac{1}{p_1 + 0.900} + \dfrac{1}{0.600 - p_1}$
    (c) 0.300 m (d) 0.240 m
55. Both images form at the *same* position, and there are not two locations at which the student can hold a screen to see images formed by this system.
57. $d = p$ and $d = p + 2f_M$

Chapter 36

Answers to Quick Quizzes

1. (c)
2. The graph is shown on the next page. The width of the primary maxima is slightly narrower than the $N = 5$

primary width but wider than the $N = 10$ primary width. Because $N = 6$, the secondary maxima are $\frac{1}{36}$ as intense as the primary maxima.

3. (a)

Answers to Odd-Numbered Problems

1. 641
3. 632 nm
5. 2.40 μm
7. 0.318 m/s
11. 506 nm
13. (a) 1.93 μm   (b) 3.00λ (c) It corresponds to a maximum. The path difference is an integer multiple of the wavelength.
15. $E_R = 10.0$ and $\phi = 53.1°$
17.

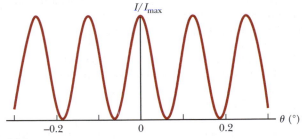

19. 96.2 nm
21. (a) 276 nm, 138 nm, 92.0 nm   (b) No visible wavelengths are intensified.
23. 1.31
25. (a) 238 nm   (b) The wavelength of the transmitted light increases.   (c) 328 nm
27. 39.6 μm
29. 1.62 cm
31. $x_1 - x_2 = (m - \frac{1}{48})650$, where x_1 and x_2 are in nanometers and $m = 0, \pm 1, \pm 2, \pm 3, \ldots$
33. $\dfrac{\lambda}{2(n-1)}$
35. (a) 72.0 m   (b) 36.0 m
37. (a) 70.6 m   (b) 136 m
39. (a) 14.7 μm   (b) 1.53 cm   (c) −16.0 m
41. 0.505 mm
43. 140 nm
45. 3.58°
47. 115 nm
49. (a) $m = \dfrac{\lambda_1}{2(\lambda_1 - \lambda_2)}$ (b) 266 nm

Chapter 37

Answers to Quick Quizzes

1. (a)
2. (b)
3. (a)

4. (c)
5. (b)
6. (c)

Answers to Odd-Numbered Problems

1. 4.22 mm
3. (a) 1.50 m   (b) 4.05 mm
5.

7. 1.62×10^{-2}
9. 0.284 m
11. 30.5 m
13. 0.40 μrad
15. 16.4 m
17. (a) three   (b) 0°, +45.2°, −45.2°
19. (a) five   (b) ten
21. 514 nm
23. (a) two, at ±52.3°   (b) no
25. (a) 0.109 nm   (b) four
27. (a) 93.3%   (b) 50.0%   (c) 0.00%
29. 60.5°
31. (a) 20.5°   (b) The refracted beam arrives at the second surface at Brewster's angle.
33. (a) 0.045 0   (b) 0.016 2
35. 5.51 m, 2.76 m, 1.84 m
37. (a) 7.26 μrad = 1.50 arc seconds   (b) 0.189 ly   (c) 50.8 μrad   (d) 1.52 mm
39. (a) 25.6°   (b) 18.9°
41. 13.7°
43. (b) 428 μm
45. (b) 3.77 nm/cm
47. (a) $\phi = 4.49$ rad compared with the prediction from the approximation of $1.5\pi = 4.71$ rad   (b) $\phi = 7.73$ rad compared with the prediction from the approximation of $2.5\pi = 7.85$ rad
49. (b) 0.001 90 rad = 0.109°
51. (b) 15.3 μm

Chapter 38

Answers to Quick Quizzes

1. (c)
2. (d)
3. (d)
4. (a)
5. (c)
6. (d)
7. (i) (c)   (ii) (a)
8. (a) $m_3 > m_2 = m_1$ (b) $K_3 = K_2 > K_1$ (c) $u_2 > u_3 = u_1$

Answers to Odd-Numbered Problems

3. (a) 0.436 m   (b) less than 0.436 m
5. 5.00 s

7. $0.140c$

9. $0.800c$

11. (b) $0.050\ 4c$

13. (a) $39.2\ \mu s$ (b) accurate to one digit

15. (c) $2.00\ kHz$ (d) $0.075\ m/s = 0.168\ mi/h\ (0.250\%)$

17. (a) $17.4\ m$ (b) $3.30°$

19. (a) $2.50 \times 10^8\ m/s = 0.834c$ (b) $4.98\ m$ (c) $-1.33 \times 10^{-8}\ s$

21. $0.960c$

23. (a) $2.73 \times 10^{-24}\ kg \cdot m/s$ (b) $1.58 \times 10^{-22}\ kg \cdot m/s$
(c) $5.64 \times 10^{-22}\ kg \cdot m/s$

25. (a) $929\ MeV/c$ (b) $6.58 \times 10^3\ MeV/c$ (c) No

27. $0.285c$

29. (a) $0.582\ MeV$ (b) $2.45\ MeV$

31. (a) $0.999\ 997c$ (b) $3.74 \times 10^5\ MeV$

33. $1.63 \times 10^3\ MeV/c$

35. (a) $0.979c$ (b) $0.065\ 2c$ (c) 15.0
(d) $0.999\ 999\ 97c$; $0.948c$; 1.06

37. $2.97 \times 10^{-26}\ kg$

39. larger; $\sim 10^{-9}\ J$

41. (a) $2.66 \times 10^7\ m$ (b) $3.87\ km/s$ (c) -8.35×10^{-11}
(d) 5.29×10^{-10} (e) $+4.46 \times 10^{-10}$

43. (a) $v/c = 1 - 1.12 \times 10^{-10}$ (b) $6.00 \times 10^{27}\ J$
(c) $\$2.17 \times 10^{20}$

45. (a) 6.67×10^4 (b) $1.97\ h$

47. (a) $3.65\ MeV/c^2$ (b) $0.589c$

49. (a) $0.905\ MeV$ (b) $0.394\ MeV$
(c) $0.747\ MeV/c = 3.99 \times 10^{-22}\ kg \cdot m/s$ (d) $65.4°$

51. (b) $1.48\ km$

55. (a) Tau Ceti exploded 16.0 years before the Sun. (b) The two stars blew up simultaneously.

Index

Solar System Data

Body	Mass (kg)	Mean Radius (m)	Period (s)	Mean Distance from the Sun (m)
Mercury	3.30×10^{23}	2.44×10^{6}	7.60×10^{6}	5.79×10^{10}
Venus	4.87×10^{24}	6.05×10^{6}	1.94×10^{7}	1.08×10^{11}
Earth	5.97×10^{24}	6.37×10^{6}	3.156×10^{7}	1.496×10^{11}
Mars	6.42×10^{23}	3.39×10^{6}	5.94×10^{7}	2.28×10^{11}
Jupiter	1.90×10^{27}	6.99×10^{7}	3.74×10^{8}	7.78×10^{11}
Saturn	5.68×10^{26}	5.82×10^{7}	9.29×10^{8}	1.43×10^{12}
Uranus	8.68×10^{25}	2.54×10^{7}	2.65×10^{9}	2.87×10^{12}
Neptune	1.02×10^{26}	2.46×10^{7}	5.18×10^{9}	4.50×10^{12}
Pluto[a]	1.25×10^{22}	1.20×10^{6}	7.82×10^{9}	5.91×10^{12}
Moon	7.35×10^{22}	1.74×10^{6}	—	—
Sun	1.989×10^{30}	6.96×10^{8}	—	—

[a]In August 2006, the International Astronomical Union adopted a definition of a planet that separates Pluto from the other eight planets. Pluto is now defined as a "dwarf planet" (like the asteroid Ceres).

Physical Data Often Used

Average Earth–Moon distance	3.84×10^{8} m
Average Earth–Sun distance	1.496×10^{11} m
Average radius of the Earth	6.37×10^{6} m
Density of air (20°C and 1 atm)	1.20 kg/m^3
Density of air (0°C and 1 atm)	1.29 kg/m^3
Density of water (20°C and 1 atm)	1.00×10^{3} kg/m^3
Free-fall acceleration on the Earth	9.80 m/s^2
Mass of the Earth	5.97×10^{24} kg
Mass of the Moon	7.35×10^{22} kg
Mass of the Sun	1.99×10^{30} kg
Standard atmospheric pressure on the Earth	1.013×10^{5} Pa

Note: These values are the ones used in the text.

Some Prefixes for Powers of Ten

Power	Prefix	Abbreviation	Power	Prefix	Abbreviation
10^{-24}	yocto	y	10^{1}	deka	da
10^{-21}	zepto	z	10^{2}	hecto	h
10^{-18}	atto	a	10^{3}	kilo	k
10^{-15}	femto	f	10^{6}	mega	M
10^{-12}	pico	p	10^{9}	giga	G
10^{-9}	nano	n	10^{12}	tera	T
10^{-6}	micro	μ	10^{15}	peta	P
10^{-3}	milli	m	10^{18}	exa	E
10^{-2}	centi	c	10^{21}	zetta	Z
10^{-1}	deci	d	10^{24}	yotta	Y

Standard Abbreviations and Symbols for Units

Symbol	Unit	Symbol	Unit
A	ampere	K	kelvin
u	atomic mass unit	kg	kilogram
atm	atmosphere	kmol	kilomole
Btu	British thermal unit	L	liter
C	coulomb	lb	pound
°C	degree Celsius	ly	light-year
cal	calorie	m	meter
d	day	min	minute
eV	electron volt	mol	mole
°F	degree Fahrenheit	N	newton
F	farad	Pa	pascal
ft	foot	rad	radian
G	gauss	rev	revolution
g	gram	s	second
H	henry	T	tesla
h	hour	V	volt
hp	horsepower	W	watt
Hz	hertz	Wb	weber
in.	inch	yr	year
J	joule	Ω	ohm

Mathematical Symbols Used in the Text and Their Meaning

Symbol	Meaning
$=$	is equal to
$\equiv$	is defined as
$\neq$	is not equal to
$\propto$	is proportional to
$\sim$	is on the order of
$>$	is greater than
$<$	is less than
$\gg (\ll)$	is much greater (less) than
$\approx$	is approximately equal to
Δx	the change in x
$\displaystyle\sum_{i=1}^{N} x_i$	the sum of all quantities x_i from $i = 1$ to $i = N$
$\lvert x \rvert$	the absolute value of x (always a nonnegative quantity)
$\Delta x \rightarrow 0$	Δx approaches zero
$\dfrac{dx}{dt}$	the derivative of x with respect to t
$\dfrac{\partial x}{\partial t}$	the partial derivative of x with respect to t
$\displaystyle\int$	integral

Conversions

Length

1 in. = 2.54 cm (exact)
1 m = 39.37 in. = 3.281 ft
1 ft = 0.304 8 m
12 in. = 1 ft
3 ft = 1 yd
1 yd = 0.914 4 m
1 km = 0.621 mi
1 mi = 1.609 km
1 mi = 5 280 ft
1 μm = 10^{-6} m = 10^3 nm
1 ly (light-year) = 9.461×10^{15} m
1 pc (parsec) = 3.26 ly = 3.09×10^{16} m

Area

1 m^2 = 10^4 cm^2 = 10.76 ft^2
1 ft^2 = 0.092 9 m^2 = 144 in.^{2}
1 in.^{2} = 6.452 cm^2
1 ha (hectare) = 1.00×10^4 m^2

Volume

1 m^3 = 10^6 cm^3 = 6.102×10^4 in.^{3}
1 ft^3 = 1 728 in.^{3} = 2.83×10^{-2} m^3
1 L = 1 000 cm^3 = 1.057 6 qt = 0.035 3 ft^3
1 ft^3 = 7.481 gal = 28.32 L = 2.832×10^{-2} m^3
1 gal = 3.786 L = 231 in.^{3}

Mass

1 000 kg = 1 t (metric ton)
1 slug = 14.59 kg
1 u = 1.66×10^{-27} kg = 931.5 MeV/c^2

Force

1 N = 0.224 8 lb
1 lb = 4.448 N

Velocity

1 mi/h = 1.47 ft/s = 0.447 m/s = 1.61 km/h
1 m/s = 100 cm/s = 3.281 ft/s
1 mi/min = 60 mi/h = 88 ft/s

Acceleration

1 m/s^2 = 3.28 ft/s^2 = 100 cm/s^2
1 ft/s^2 = 0.304 8 m/s^2 = 30.48 cm/s^2

Pressure

1 bar = 10^5 N/m^2 = 14.50 lb/in.^{2}
1 atm = 760 mm Hg = 76.0 cm Hg
1 atm = 14.7 lb/in.^{2} = 1.013×10^5 N/m^2
1 Pa = 1 N/m^2 = 1.45×10^{-4} lb/in.^{2}

Time

1 yr = 365 days = 3.16×10^7 s
1 day = 24 h = 1.44×10^3 min = 8.64×10^4 s

Energy

1 J = 0.738 ft·lb
1 cal = 4.186 J
1 Btu = 252 cal = 1.054×10^3 J
1 eV = 1.602×10^{-19} J
1 kWh = 3.60×10^6 J

Power

1 hp = 550 ft·lb/s = 0.746 kW
1 W = 1 J/s = 0.738 ft·lb/s
1 Btu/h = 0.293 W

Some Approximations Useful for Estimation Problems

1 m ≈ 1 yd
1 kg ≈ 2 lb
1 N ≈ $\frac{1}{4}$ lb
1 L ≈ $\frac{1}{4}$ gal

1 m/s ≈ 2 mi/h
1 yr ≈ $\pi \times 10^7$ s
60 mi/h ≈ 100 ft/s
1 km ≈ $\frac{1}{2}$ mi

Note: See Table A.1 of Appendix A for a more complete list.

The Greek Alphabet

Alpha	A	α	Iota	I	ι	Rho	P	ρ
Beta	B	β	Kappa	K	κ	Sigma	Σ	σ
Gamma	Γ	γ	Lambda	Λ	λ	Tau	T	τ
Delta	Δ	δ	Mu	M	μ	Upsilon	Υ	υ
Epsilon	E	ϵ	Nu	N	ν	Phi	Φ	ϕ
Zeta	Z	ζ	Xi	Ξ	ξ	Chi	X	χ
Eta	H	η	Omicron	O	o	Psi	Ψ	ψ
Theta	Θ	θ	Pi	Π	π	Omega	Ω	ω

Pedagogical Color Chart

Mechanics and Thermodynamics

Displacement and
  position vectors

Displacement and position
  component vectors

Linear ($\vec{\mathbf{v}}$) and angular ($\vec{\boldsymbol{\omega}}$)
  velocity vectors

Velocity component vectors

Force vectors ($\vec{\mathbf{F}}$)

Force component vectors

Acceleration vectors ($\vec{\mathbf{a}}$)

Acceleration component vectors

Energy transfer arrows

W_{eng}

Q_c

Q_h

Process arrow

Linear ($\vec{\mathbf{p}}$) and
  angular ($\vec{\mathbf{L}}$)
  momentum vectors

Linear and
  angular momentum
  component vectors

Torque vectors ($\vec{\boldsymbol{\tau}}$)

Torque component
  vectors

Schematic linear or
  rotational motion
  directions

Dimensional rotational
  arrow

Enlargement arrow

Springs

Pulleys

Electricity and Magnetism

Electric field lines

Electric field vectors

Electric field component vectors

Magnetic field lines

Magnetic field vectors

Magnetic field
  component vectors

Positive charges

Negative charges

Resistors

Batteries and other
  DC power supplies

Switches

Capacitors

Inductors (coils)

Voltmeters

Ammeters

AC Sources

Lightbulbs

Ground symbol

Current

Light and Optics

Light rays

Extension of light ray

Converging lens

Diverging lens

Flat mirror

Curved mirror

Objects

Images

e Physical Constants

ntity	Symbol	Value[a]
nic mass unit	u	$1.660\ 539\ 040\ (20) \times 10^{-27}$ kg
		$931.494\ 095\ 4\ (57)$ MeV/c^2
gadro's number	N_A	$6.022\ 140\ 857\ (74) \times 10^{23}$ particles/mol
r magneton	$\mu_B = \dfrac{e\hbar}{2m_e}$	$9.274\ 009\ 994\ (57) \times 10^{-24}$ J/T
r radius	$a_0 = \dfrac{\hbar^2}{m_e e^2 k_e}$	$5.291\ 772\ 106\ 7\ (12) \times 10^{-11}$ m
zmann's constant	$k_B = \dfrac{R}{N_A}$	$1.380\ 648\ 52\ (79) \times 10^{-23}$ J/K
apton wavelength	$\lambda_C = \dfrac{h}{m_e c}$	$2.426\ 310\ 236\ 7\ (11) \times 10^{-12}$ m
lomb constant	$k_e = \dfrac{1}{4\pi\epsilon_0}$	$8.987\ 551\ 788\ldots \times 10^9$ N·m²/C² (exact)
teron mass	m_d	$3.343\ 583\ 719\ (41) \times 10^{-27}$ kg
		$2.013\ 553\ 212\ 745\ (40)$ u
tron mass	m_e	$9.109\ 383\ 56\ (11) \times 10^{-31}$ kg
		$5.485\ 799\ 090\ 70\ (16) \times 10^{-4}$ u
		$0.510\ 998\ 946\ 1\ (31)$ MeV/c^2
tron volt	eV	$1.602\ 176\ 620\ 8\ (98) \times 10^{-19}$ J
nentary charge	e	$1.602\ 176\ 620\ 8\ (98) \times 10^{-19}$ C
constant	R	$8.314\ 459\ 8\ (48)$ J/mol·K
vitational constant	G	$6.674\ 08\ (31) \times 10^{-11}$ N·m²/kg²
tron mass	m_n	$1.674\ 927\ 471\ (21) \times 10^{-27}$ kg
		$1.008\ 664\ 915\ 88\ (49)$ u
		$939.565\ 413\ 3\ (58)$ MeV/c^2
lear magneton	$\mu_n = \dfrac{e\hbar}{2m_p}$	$5.050\ 783\ 699\ (31) \times 10^{-27}$ J/T
neability of free space	μ_0	$4\pi \times 10^{-7}$ T·m/A (exact)
mittivity of free space	$\epsilon_0 = \dfrac{1}{\mu_0 c^2}$	$8.854\ 187\ 817\ldots \times 10^{-12}$ C²/N·m² (exact)
ick's constant	h	$6.626\ 070\ 040\ (81) \times 10^{-34}$ J·s
	$\hbar = \dfrac{h}{2\pi}$	$1.054\ 571\ 800\ (13) \times 10^{-34}$ J·s
ton mass	m_p	$1.672\ 621\ 898\ (21) \times 10^{-27}$ kg
		$1.007\ 276\ 466\ 879\ (91)$ u
		$938.272\ 081\ 3\ (58)$ MeV/c^2
berg constant	R_H	$1.097\ 373\ 156\ 850\ 8\ (65) \times 10^7$ m^{-1}
ed of light in vacuum	c	$2.997\ 924\ 58 \times 10^8$ m/s (exact)

These constants are the values recommended in 2014 by CODATA, based on a least-squares adjustment of data from different measurements. For a more plete list, see P. J. Mohr, B. N. Taylor, and D. B. Newell, "CODATA Recommended Values of the Fundamental Physical Constants: 2014." *Rev. Mod. Phys.* **88**:3, 09, 2016.

numbers in parentheses for the values represent the uncertainties of the last two digits.